Impact of Microbiology on Society

YEAR AWARDED	SCIENTIST	PRIZE	EVENT OR DISCOVERY
1968	Robert W. Holley, Har Gobind Khorana, Marshall W. Nirenberg	Nobel Prize Physiology or Medicine	Interpretation of genetic code
1969	Max Delbrück, Alfred D. Hershey, Salvador E. Luria	Nobel Prize Physiology or Medicine	DNA replication and bacteriophage genetics
1970	Cornelis Bernardus van Niel	Leeuwenhoek Medal	Demonstration of bacterial photosynthesis as a light-dependent redox reaction
1975	David Baltimore, Renato Dulbecco, Howard Martin Temin	Nobel Prize Physiology or Medicine	Transformation by tumor viruses and discovery of reverse transcriptase
1976	Baruch S. Blumberg, D. Carleton Gajdusek	Nobel Prize Physiology or Medicine	Studies of kuru infection
1978	Peter D. Mitchell	Nobel Prize Chemistry	Formulation of the chemiosmotic theory
1978	Werner Arber, Daniel Nathans, Hamilton O. Smith	Nobel Prize Physiology or Medicine	Discovery of restriction enzymes
1980	Baruj Benacerraf, Jean Dausset, George D. Snell	Nobel Prize Physiology or Medicine	Discovery of the major histocompatibility complex
1980	Paul Berg, Walter Gilbert, Frederick Sanger	Nobel Prize Chemistry	Recombinant DNA and DNA sequencing
1981	Roger Yate Stanier	Leeuwenhoek Medal	Classification of the cyanobacteria
1984	Niels K. Jerne, Georges J. F. Köhler, César Milstein	Nobel Prize Physiology or Medicine	Development of technique for production of monoclonal antibodies
1986	Ernst Ruska	Nobel Prize Physics	Design of the first electron microscope
1987	Susumu Tonegawa	Nobel Prize Physiology or Medicine	Discovery of genetic mechanism for generation of antibody diversity
1989	J. Michael Bishop, Harold E. Varmus	Nobel Prize Physiology or Medicine	Discovery of oncogenes
1989	Sidney Altman, Thomas R. Cech	Nobel Prize Chemistry	Discovery of catalytic properties of RNA
1990	Allen Kerr, Eugene Nester, Jeff Schell	Australia Prize	Biological control of crown gall, and use of *Agrobacterium* to engineer plants
1993	Kary B. Mullis, Michael Smith	Nobel Prize Chemistry	Development of PCR and site-directed mutagenesis techniques
1996	Peter C. Doherty, Rolf M. Zinkernagel	Nobel Prize Physiology or Medicine	Discovery of T-cell recognition of target antigens
1997	Stanley B. Prusiner	Nobel Prize Physiology or Medicine	Discovery of prions
2003	Carl R. Woese	Crafoord Prize	Discovery of the third domain of life
2003	Karl Stetter	Leeuwenhoek Medal	Isolation and characterization of archaeons
2005	Barry J. Marshall, J. Robin Warren	Nobel Prize Physiology or Medicine	Discovery of the bacterium *Helicobacter pylori* and its role in gastritis and peptic ulcer disease
2008	Harald zur Hausen, Françoise Barré-Sinoussi, Luc Montagnier	Nobel Prize Physiology or Medicine	Discovery of human papillomaviruses and HIV
2008	J. Craig Venter	Kistler Prize	Advances in genomics
2009	Lucy Shapiro, Richard Losick	Gairdner International Award	Discovery of mechanisms defining cell polarity and asymmetry
2009	Elizabeth H. Blackburn, Carol W. Greider, Jack W. Szostak	Nobel Prize Physiology or Medicine	Discovery of telomerase
2010	Leroy Hood	Kistler Prize	Invention of proteomics and genomics instruments
2011	Bruce A. Beutler, Jules A. Hoffman, Ralph M. Steinman	Nobel Prize Physiology or Medicine	Discovery of activation of innate immunity and role of dendritic cell in adaptive immunity

The Three Pillars of Microbiology

Genetics
Genes possessed
and expressed

Physiology
Metabolism
and structure

Ecology
Relationship with
the environment
and other
organisms

WileyPLUS

WileyPLUS is a research-based online environment for effective teaching and learning.

WileyPLUS builds students' confidence because it takes the guesswork out of studying by providing students with a clear roadmap:

- **what to do**
- **how to do it**
- **if they did it right**

It offers interactive resources along with a complete digital textbook that help students learn more. With *WileyPLUS*, students take more initiative so you'll have greater impact on their achievement in the classroom and beyond.

or more information, visit www.wileyplus.com

WileyPLUS

ALL THE HELP, RESOURCES, AND PERSONAL SUPPORT YOU AND YOUR STUDENTS NEED!
www.wileyplus.com/resources

1st DAY OF CLASS ... AND BEYOND!

2-Minute Tutorials and all of the resources you and your students need to get started

WileyPLUS Student Partner Program

Student support from an experienced student user

Wiley Faculty Network

Collaborate with your colleagues, find a mentor, attend virtual and live events, and view resources
www.WhereFacultyConnect.com

WileyPLUS Quick Start

Pre-loaded, ready-to-use assignments and presentations created by subject matter experts

Technical Support 24/7 FAQs, online chat, and phone support
www.wileyplus.com/support

© Courtney Keating/ iStockphoto

Your *WileyPLUS* Account Manager, providing personal training and support

MICROBIOLOGY

MICROBIOLOGY

DAVID R. WESSNER
Davidson College

CHRISTINE DUPONT
University of Waterloo

TREVOR C. CHARLES
University of Waterloo

WILEY

VICE PRESIDENT AND EXECUTIVE PUBLISHER	Kaye Pace	ILLUSTRATION EDITOR	Kathy Naylor
DIRECTOR OF PRODUCT AND CONTENT DEVELOPMENT	Barbara Heaney	SENIOR PRODUCT DESIGNER	Bonnie Roth
		MEDIA ASSOCIATE EDITOR	Lauren Morris
SENIOR DEVELOPMENTAL EDITOR	Mary O'Sullivan	MEDIA SPECIALIST	Margarita Valdez
EXECUTIVE EDITOR	Kevin Witt	PHOTO MANAGER	Hilary Newman
MARKETING MANAGER	Carrie Ayers	PHOTO RESEARCHERS	Sara Wight, Teri Stratford
PRODUCTION EDITOR	Barbara Russiello	DESIGN DIRECTOR	Harry Nolan
ASSISTANT EDITOR	Lauren Samuelson	SENIOR DESIGNER AND COVER DESIGNER	Maureen Eide
TEXT DEVELOPER	Deborah Allen	COVER ILLUSTRATION	Janet Iwasa

REPEATED DESIGN ELEMENT PHOTO CREDITS:

OPENER BORDER AND ICON TAB — Henrik Jonsson/iStockphoto
TOOLBOX, MICROBES IN FOCUS, PERSPECTIVE, MINI-PAPER — Sergey Panteleev/iStockphoto
MICROBES IN FOCUS MAGNIFYING GLASS PHOTO — mustafa deliormanli/iStockphoto

This book was set in 10/12 BT Baskerville by cMPreparé. Book and cover are printed and bound by Quad Graphics/Versailles. This book is printed on acid-free paper. ∞

Founded in 1807, John Wiley & Sons, Inc. has been a valued source of knowledge and understanding for more than 200 years, helping people around the world meet their needs and fulfill their aspirations. Our company is built on a foundation of principles that include responsibility to the communities we serve and where we live and work. In 2008, we launched a Corporate Citizenship Initiative, a global effort to address the environmental, social, economic, and ethical challenges we face in our business. Among the issues we are addressing are carbon impact, paper specifications and procurement, ethical conduct within our business and among our vendors, and community and charitable support. For more information, please visit our website: www.wiley.com/go/citizenship.

Library of Congress Cataloging-in-Publication Data

Wessner, David R., 1963-
    Microbiology / David R. Wessner, Davidson College, Christine Dupont,
University of Waterloo, Trevor Charles, University of Waterloo.
        pages cm
    Includes bibliographical references and index.
    ISBN 978-0-471-69434-2 (cloth)
    ISBN 978-1-118-12924-1 (binder ready version)
    1. Microbiology. I. Dupont, Christine, 1961- II. Charles, Trevor, 1963-
III. Title.
    QR41.2.W47 2013
    579—dc23

                        2012038562

Printed in the United States of America

10 9 8 7 6 5 4 3 2 1

DEDICATIONS

DAVID R. WESSNER

Many thanks to Dick Fluck, Carl Pike, and Bernie Fields, all of whom taught me to think like a biologist. Thanks also to my great colleagues and the wonderful students at Davidson College for their support, guidance, and help. Finally, and most importantly, thanks to Connie and Ian.

CHRISTINE DUPONT

To my family and friends who did without me for so long during this project. I hope I can make the time up.

TREVOR C. CHARLES

I am fortunate to have had exceptional mentors, Turlough and Gene, who recognized the true value of good science. The members of my research lab have kept me in touch with the excitement of discovery. Most importantly, I thank my family, Cheryl, Claire, and Tarin, simply for being there for me, and my parents, Carlos and Norma, for nurturing my interests and leading by example.

About the Authors

Courtesy of Connie Wessner

DAVID R. WESSNER

Professor of Biology and Associate Director of the Center for Interdisciplinary Studies at Davidson College, **David R. Wessner** teaches introductory biology and courses on microbiology, genetics, and HIV/AIDS. His research focuses on viral pathogenesis. He is a member of the Charlotte Teachers Institute University Advisory Committee and the American Society for Microbiology Committee for K–12 Education. He also is a coauthor of Vision and Change in Undergraduate Education: A Call to Action. Prior to joining the faculty at Davidson, David conducted research at the Navy Medical Center. He earned his Ph.D. in Microbiology and Molecular Genetics from Harvard University and his B.A. in Biology from Franklin and Marshall College.

Courtesy of John Lumsden

CHRISTINE DUPONT

Lecturer in the Department of Biology at the University of Waterloo in southern Ontario, Canada, **Christine Dupont** teaches undergraduate courses in genetics, biotechnology, virology, and bacterial pathogenesis. She earned her Ph.D. from Massey University, faculty of Veterinary Science, New Zealand; B.Ed. from the University of Windsor, Ontario; and M.Sc. and B.Sc. in Microbiology from the University of Guelph, Ontario. Prior to her Ph.D. studies, Christine taught high school science for several years, developing a passion for teaching and working with students.

Courtesy of Trevor Charles

TREVOR C. CHARLES

A professor in the Department of Biology, University of Waterloo, Trevor Charles teaches undergraduate courses in microbiology and synthetic biology, and runs a research program that focuses on plant-microbe interactions and functional metagenomics. Prior to joining the faculty at Waterloo, he held a faculty position at McGill University and did postdoctoral research at the University of Washington. He earned his Ph.D. from McMaster University, and his B.Sc. in Microbiology from the University of British Columbia.

CONTRIBUTING AUTHORS

Courtesy of Janet Iwasa

Janet Iwasa (contributor) received her Ph.D. from the University of California, San Francisco. She joined the faculty at Harvard Medical School in 2008 and will be joining the faculty at the University of Utah in 2013. She started working with visualizations when she saw her first animated molecule five years ago. "Just listening to scientists describe how the molecule moved in words wasn't enough for me," she said. "What brought it to life was really seeing it in motion." In 2006, with a grant from the National Science Foundation, she spent three months at the Gnomon School of Visual Effects, an animation school in Hollywood, CA.

Courtesy of Kelli A. Prior

Kelli Prior (contributor) is a Professor of Biology, and has been with Finger Lakes Community College since 2002. Prior served as Coordinator of FLCC's Gladys M. Snyder Center for Teaching and Learning from 2005 to 2009. She received the State University of New York Chancellor's Award for Excellence in Teaching in 2009. In addition to her academic specialty in microbiology, she has conducted research and presented at conferences and workshops with an emphasis on teaching and pedagogy. She is also a reviewer and contributing author for Wiley and Prentice Hall. She earned a master's degree and a Ph.D. in Microbiology and Immunology from the University of Rochester, as well as a bachelor's degree in Biology from Nazareth College.

Why We Wrote This Book

We wrote this book for one simple reason—we have a passion for microbiology. Moreover, we want to share this passion with as many students as possible. First, we want to show students that microbiology is a dynamic discipline. We were introduced to the field when DNA sequencing was not widespread and personal computers were rare. Today, hundreds of complete genome sequences have been determined and the personal computer is one of the microbiologist's most important tools. The field has changed dramatically over the past 25 years, and we can only guess what remarkable changes the future will bring. Second, we want to show students that experimentation is at the very heart of microbiology. Since the development of the microscope over 300 years ago, microbiologists have asked probing questions, developed elegant experiments, and formulated testable hypotheses. This scientific exploration continues today. To achieve these two major goals, we have written this textbook as an engaging narrative that brings the story of microbiology to life. Our textbook not only provides students with a historical understanding of the field but also gives them an appreciation for the dynamic, exciting nature of today's microbiology. Indeed, we hope that our textbook will make students as passionate as we are about the science of microbiology.

Exploration and Experimentation

To achieve our goals, we present material within the context of exploration and experimentation. Knowing myriad facts about microbiology is not sufficient. To have a true understanding of this subject, students need to understand experimentation and be able to critically analyze information. Several features within the book will help our students understand and appreciate the science behind our knowledge.

First, each chapter begins with an opening vignette—a short story that frames a basic question within the context of both contemporary and historical issues—visually supported by a dynamic illustration. As the chapter unfolds, references back to the opening vignette are made repeatedly. At the conclusion of each chapter, an additional feature, The Rest of the Story, again refers back to this opening vignette and art. This also ties into an active learning feature, Image in Action, that includes several critical-thinking questions.

Second, each chapter contains a Mini-Paper, a synopsis of a scientific journal article that includes original data and Questions for Discussion. Through the use of this feature, students will see how microbiologists ask intelligent questions, rationally design experiments, and evaluate data. Again, this feature will improve students' critical-thinking skills and show them that our knowledge is evidence-based.

Third, several chapters contain scientific methodology figures, diagrams that provide a brief overview of a specific experiment, including the original observation, hypothesis, experiment, results, and conclusions. Again, this feature will emphasize to our students the science behind our knowledge.

Three Pillars: Genetics, Physiology, and Ecology

Throughout the textbook, we frame information around the three pillars of genetics, physiology, and ecology. The pillars help weave concepts in evolution, structure and function, and the interactions between individual microbes, between different types of microbes, between microbes and other organisms, and between microbes and the environment. Often, the importance of these key interactions are presented from different perspectives at different places within the textbook: from that of the molecular biologist, of the ecologist, of the physician, of the engineer, or even of the microbe itself.

An Emphasis on Connections

Finally, we make explicit the interconnectedness of topics, both within a chapter and between chapters. This goal is achieved in several ways. Important concepts are reintroduced throughout appropriate chapters with Connection notes. Through such layering, students gradually deepen their understanding of complex topics as the semester progresses. Gene regulation, for example, is introduced during the discussion of transcription. Aspects of gene regulation also are discussed in subsequent sections on biofilms and virulence. Thus, students gain an appreciation of gene regulation by seeing how it affects different processes.

Ultimately, of course, we want our students to gain a more complete understanding of and appreciation for microbiology. By presenting information within the dual contexts of experimentation and the interconnectedness of topics, we encourage students to learn important material not as discrete bits of information, but as pieces of a much larger body of knowledge. Through this approach, students will better understand and connect the basic concepts of this exciting, dynamic field.

Thank you for taking the time to read these opening remarks and learn more about our book. We hope our approach fosters a passion in your students for microbiology and an appreciation for research and critical thinking. Hopefully, our book will help you share your fascination with microbiology with your students. Who knows? Maybe an undergraduate student in today's microbiology class will make tomorrow's big discovery!

Best,

DAVE WESSNER

CHRISTINE DUPONT

TREVOR CHARLES

Special Features

Mini-Papers: In every chapter, primary research articles are summarized and interpreted, helping students sharpen critical thinking skills and deepen their understanding of what microbiologists do.

Mini-Paper: A Focus on the Research
THE THREE DOMAINS OF LIFE

C. R. Woese, O. Kandler, and M. L. Wheelis. 1990. Towards a natural system of organisms: Proposal for the domains Archaea, Bacteria, and Eucarya. Proc Natl Acad Sci USA 87: 4576–4579.

Context

The ideal biological taxonomy should accurately reflect phylogenetic relationships. This union between taxonomy and phylogeny, however, has been a major challenge for microbiologists. Aristotle's categorization of life into just two fundamental groups, animals and plants, persisted until the dawn of microbiology as a science. In 1868, two centuries after van Leeuwenhoek's discovery of microbial life, German biologist Ernst Haeckel proposed a third kingdom, Protista, for microscopic life forms. In 1938, Herbert Copeland suggested that microbes should actually be divided into two kingdoms, Protista and Bacteria, thereby recognizing the fundamental difference between eukaryotic and prokaryotic cells. Twenty years later, Robert Whittaker advocated further separation of eukaryotic microbes into kingdoms of Fungi and Protista, but kept prokaryotic cells in a single kingdom called Monera. This five kingdom taxonomic system—Animalia, Plantae, Fungi, Protista, and Monera—became the accepted standard for the next few decades until DNA and protein sequences became widely accessible.

Carl Woese, a microbiologist at the University of Illinois, was fasci-

The sequence of this molecule changes very slowly because of the functional constraints on the molecule. Random mutations that occur within the gene encoding the small subunit rRNA often have serious negative consequences, so relatively few changes are passed on to subsequent generations. Nevertheless, there are enough differences in the roughly 1,600 nucleotide sequence to differentiate between species to map patterns of similarity. If one assumes that overall mutation rates are similar between species (which seems to be true with respect to rRNA genes), then one can quantify sequence differences between SSU rRNA genes in multiple species to infer relationships. Ultimately, Woese discovered that the methane-producing microorganisms were no more closely related to other bacteria than they were to the eukaryotes. Let's examine the scientific work that led to this conclusion.

Experiments

The 1990 Woese et al. paper actually presented no new experimental data. Its importance was in articulating a new view of the phylogeny of life. To understand the genesis of this idea, we should step back and examine the data. The biggest challenge Woese faced in the 1970s in developing ribosomal RNA sequences as a tool for phylogenetic analysis was the difficulty in determining such sequences. Woese and colleagues developed a laborious method to infer the sequence of 16S rRNA mol... which they described in a 1977 article, "Comparative cataloging ...omal ribonucleic acid: A molecular approach to procaryotic

Toolbox 5.1
CELL CULTURE TECHNIQUES

Because viruses only can replicate within living host cells, virologists are presented with an interesting problem: before researchers can effectively study viruses in the laboratory, they must be able to grow appropriate host cells. Prior to the development of mammalian cell culture methodologies, the propagation of animal viruses occurred via serial passage of the virus in animals. A susceptible animal would be inoculated with a small sample of the virus of interest. At a later time, perhaps when the infected animal became sick, virus would be isolated from this animal and a second susceptible animal would be inoculated, thereby allowing the virus to continue replicating. Clearly, this method of viral propagation was not terribly practical; it required the researcher to maintain appropriate animal hosts. As you can imagine, this requirement greatly reduced the study of viruses that infect humans!

During the 1940s and 1950s, this difficulty was overcome with the development of cell culture techniques. In 1949, John Enders, Thomas Weller, and Frederick Robbins demonstrated that poliovirus could replicate in various embryonic tissues. This discovery directly led to the development of the poliovirus vaccine a few years later and earned the three researchers the Nobel Prize in Physiology or Medicine in 1954. George Gey and colleagues cultured the first human cell line in 1952. This cell line, derived from a human cervical carcinoma, was named HeLa, in recognition of Henrietta Lacks, the young woman from whom the cells were isolated. She died of cervical carcinoma in 1951. The HeLa cell line, though, still is widely used by researchers throughout the world. A final important contribution to the budding field of virology came in 1955 when Harry Eagle and colleagues developed a well-characterized nutrient medium that could be used for the maintenance of cells in the laboratory. With the availability of defined growth media and several different cell lines, researchers easily could propagate animal viruses and conduct controlled experiments. Thus, the history of animal virol-

Courtesy David Wessner

Figure B5.1. Propagation of mammalian cells In the laboratory, mammalian cells typically are grown at 37°C in an atmosphere containing 5 percent CO_2. Here, a researcher is dislodging cells from a flask and transferring them to another flask. She is working in a laminar flow hood to prevent contamination of the cells.

Toolboxes: Every chapter contains Toolboxes— in-depth descriptions of experimental techniques relevant to the chapter.

Microbes in Focus 20.1
SLIPPERY *HAEMOPHILUS INFLUENZAE*

Habitat: Strictly human inhabitant carried asymptomatically in the nasopharynx by approximately 40 percent of adults

Description: Gram-negative coccobacillus, 0.3 to 0.8 μm in length

Key Features: Pathogenic strains all possess one of six different types of capsule carbohydrate. Strains carrying type b capsule cause 95 percent of *H. influenzae* disease in children, predominantly meningitis and pneumonia. The capsule interferes with phagocytosis by preventing C3b opsonization of the bacterial cells (see Sections 19.4, 19.5, and 21.1). Antibody produced against the polysaccharide capsule successfully opsonizes the bacteria and provides protection.

SEM

© Eye of Science/Photo Researchers, Inc.

Microbes in Focus Examples: These examples provide students with more detailed information about the habitats and key features of microbes mentioned in the chapter.

Perspective Examples: In these examples, students see how topics may be viewed or used by different groups.

Perspective 17.2
COWS CONTRIBUTE TO CLIMATE CHANGE

In reading the previous section, you may have recognized that the major gases produced by cows—carbon dioxide and methane—are greenhouse gases. When present in the atmosphere, these gasses trap radiant energy, preventing its dissipation to space. Thus, cows pose a concern for climate change as methane is a potent greenhouse gas. Cow populations have increased greatly over the past century as a result of human demand for meat and dairy products. Currently, there are an estimated 1.3 billion cattle on the planet—a source of approximately 15 percent of the total methane released into the atmosphere. When you consider the biochemistry involved in methane production in cows, it is their microbes that present the problem, although the two can hardly be separated. Researchers have been trying to find ways to reduce methane production in cows. They have found that high quality feeds, like alfalfa pasture, results in significantly less methane than cattle kept on grass pasture. There is a cost in this, however. High protein diets are more costly to supply and increase the danger of bloat.

Courtesy Christine Dupont

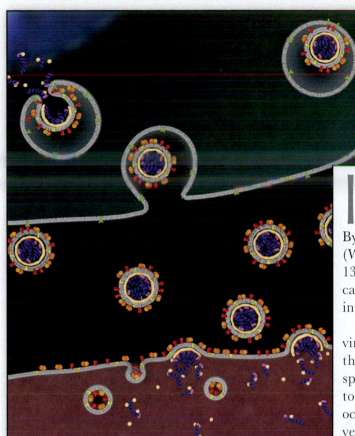

In the spring of 2009, outbreaks of a novel influenza virus infection in Mexico were reported. By April 26, 2009, 38 cases (20 in the United States and 18 in Mexico) had been confirmed. By the beginning of May 2009, the World Health Organization (WHO) reported that this new virus had infected 367 people in 13 countries. On June 11, 2009, the WHO reported over 28,000 cases in 74 countries. Also on this date, the WHO declared an influenza virus pandemic.

Initially referred to as swine flu, this strain of influenza virus—officially known as influenza A(H1N1)—quickly spread throughout the world. With its spread, fear and uncertainty also spread. During the spring of 2009, many schools were closed to avoid transmission of the virus. In Mexico, soccer matches occurred in empty stadiums when the Mexican government prevented fans from attending. Several countries urged caution to citizens traveling to the United States and Mexico. Common questions were hotly debated: Could this virus cause widespread

The Rest of the Story

As we noted in the opening of this chapter, Tamiflu reduces the severity and length of the influenza disease. But how does it work? Tamiflu (oseltamivir) is a neuraminidase inhibi The neuraminidase protein (NA) is located in the influ envelope, along with the hemagglutinin protein (HA virion first encounters a host cell, HA binds to sialic acid the virus to enter the cell. During budding of newly viral particles, NA cleaves sialic acid residues present fected cell's plasma membrane. So? The cleavage of residues by NA prevents the newly formed viruses from stuck to the infected cell. In other words, NA increases hood that new virus particles will move to other cells **(Fi**

By inhibiting neuraminidase, Tamiflu greatly reduc to-cell spread of influenza virus. Because the spread of t limited, the number of infected cells is limited, and the

Image in Action

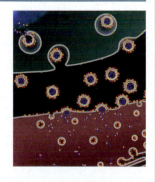

WileyPLUS This image depicts the replication cycle of influenza viruses leaving one host cell (*bottom*) and infecting another (*top*).

1. Describe the molecular steps that occur when a virus initially interacts with a host cell. Identify and include a description of the specific roles of the green- and red-colored components.

2. Imagine that Tamiflu was administered and present now in this situation. Describe the antiviral effects of Tamiflu and outline how this image would look different if this drug were present.

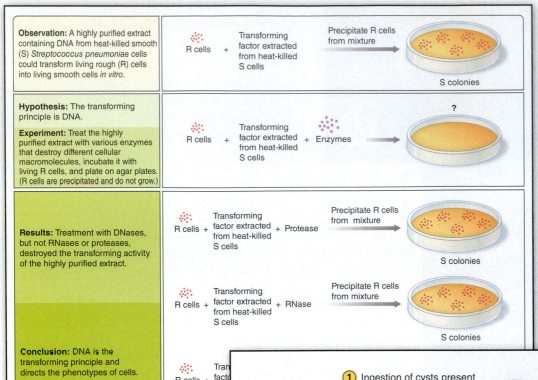

Observation: A highly purified extract containing DNA from heat-killed smooth (S) *Streptococcus pneumoniae* cells could transform living rough (R) cells into living smooth cells *in vitro*.

R cells + Transforming factor extracted from heat-killed S cells → Precipitate R cells from mixture → S colonies

Hypothesis: The transforming principle is DNA.

Experiment: Treat the highly purified extract with various enzymes that destroy different cellular macromolecules, incubate it with living R cells, and plate on agar plates. (R cells are precipitated and do not grow.)

R cells + Transforming factor extracted from heat-killed S cells + Enzymes → ?

Results: Treatment with DNases, but not RNases or proteases, destroyed the transforming activity of the highly purified extract.

R cells + Transforming factor extracted from heat-killed S cells + Protease → Precipitate R cells from mixture → S colonies

R cells + Transforming factor extracted from heat-killed S cells + RNase → Precipitate R cells from mixture → S colonies

Conclusion: DNA is the transforming principle and directs the phenotypes of cells.

R cells + Transforming factor from heat-killed S cells

Experimental Figures: These figures allow students to better understand how an experiment proceeds, from Observation to Hypothesis to Results to Conclusions.

Life-cycle Figures: Often combining art and photographs, these figures provide a clear overview of how microbes replicate.

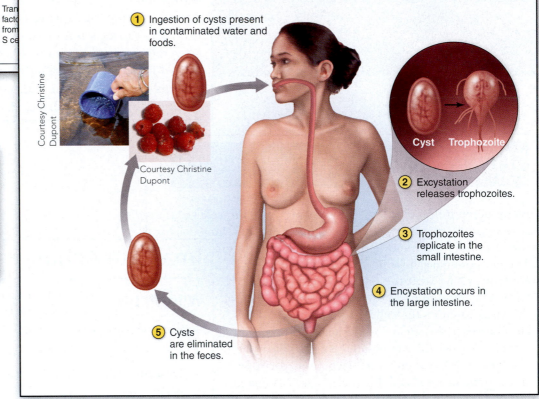

Courtesy Christine Dupont

① Ingestion of cysts present in contaminated water and foods.

Courtesy Christine Dupont

Cyst   Trophozoite

② Excystation releases trophozoites.

③ Trophozoites replicate in the small intestine.

④ Encystation occurs in the large intestine.

⑤ Cysts are eliminated in the feces.

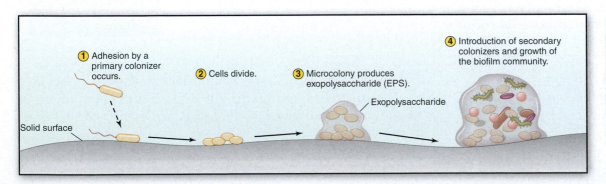

① Adhesion by a primary colonizer occurs.

② Cells divide.

③ Microcolony produces exopolysaccharide (EPS).

④ Introduction of secondary colonizers and growth of the biofilm community.

Exopolysaccharide

Solid surface

Process Diagrams: These figures lend clarity to the flow of complex processes.

CHAPTER NAVIGATOR

As you study the key topics, make sure you review the following elements:

Microbes used for biotechnology may be obtained from existing collections or isolated from nature.

- Figure 12.2: Microbial culture collections
- Perspective 12.1: Bioprospecting: Who owns the microbes?
- Figure 12.4: Types of bioreactors

To improve their usefulness, microbes can be genetically modified in numerous ways.

- 3D Animation: Using molecular biology tools to improve microbial strains
- Microbes in Focus 12.2: *Penicillium chrysogenum*: The mold that started the antibiotic revolution
- Toolbox 12.1: Site-directed mutagenesis
- Figure 12.8: Process Diagram: Directed enzyme evolution
- Figure 12.13: Process Diagram: Design and construction of synthetic organism
- Mini-Paper: Making a synthetic genome

Red biotech involves the use of biotechnology in the pharmaceutical sector.

- Figure 12.15: Production of recombinant insulin

White biotech involves the use of biotechnology in the industrial sector.

- Figure 12.18: Commercial ethanol production using different feedstocks
- Perspective 12.3: Biofuels: Biodiesel and algae
- Figure 12.22: Structures of bacterial polyhydroxyalkanoates (PHAs)

Green biotech involves the use of biotechnology in the agricultural sector.

- 3D Animation: *Agrobacterium* in agricultural biotechnology
- Toolbox 12.3: Plant transformation using bacteria
- Figure 12.33: Process Diagram: *Bacillus thuringiensis* crystal and mode of action

CONNECTIONS for this chapter:

Using metagenomics to find potential genes for biotechnolo applications in unculturable microbes (**Section 10.4**)

Use of GFP fusion proteins (**Section 3.1**)

Effects of endotoxin on humans (**Section 19.3**)

Interactions between plants and soil microbes (**Section 17.2**)

Chapter Navigator: This feature organizes the chapter by key concepts and directs students to the most important elements.

CONNECTIONS In **Section 17.2**, we will investigate the intimate associations that many plants have formed with soil microbes that live as endosymbionts inside the plant. Some of these bacteria can fix nitrogen, and the cultivation of several important crops takes advantage of this bacterial capability. Because of these bacteria, the application of nitrogen fertilizer is not required for optimal crop yield. The agricultural inoculant industry provides cultures of endosymbiotic nitrogen-fixing root nodule bacteria, as well as free-living bacteria and fungi with plant growth-promoting properties such as phosphate solubilization.

Connections: Throughout each chapter, we highlight the interconnectedness of topics, both within and between chapters.

Fact Check: Each section ends with a few questions to test students' understanding of the content.

12.1 Fact Check

1. How are culture collections used in biotechnology?
2. What is bioprospecting?
3. Explain how bioreactors are used for biotechnology fermentation processes.
4. Distinguish between fed-batch bioreactors and chemostats.
5. Distinguish between a primary and a secondary metabolite.

Summary

23.1: How do eukaryal microbes cause disease?

Eukaryal microbes can be transmitted between hosts in various ways.

- For eukaryal pathogens like *Ophiostoma novo-ulmi*, insect vectors passively transport **spores** between hosts.
- Some eukaryal pathogens have a **complex life cycle**, in which they undergo sexual reproduction in the **definitive host** and asexual reproduction and differentiation in the **intermediate host**.
- The eukaryal pathogen *Plasmodium falciparum* must replicate and develop within the mosquito to be transmitted.
- The eukaryal pathogen *Giardia lamblia* undergoes a **simple life cycle**. It is transmitted via the ingestion of **cysts**, which then give rise to **trophozoites**.

Upon entering an appropriate host, eukaryal microbes must evade the host defenses.

- Some opportunistic pathogens, like *Pneumocystis jirovecii*, only cause disease in immunocompromised individuals, such as people with HIV disease or people on **immunosuppressive drugs**.
- Other opportunistic pathogens, like *Candida albicans*, cause disease when the normal microbial inhabitants of the host change.
- Some pathogens, like *Trypanosoma brucei* spp., actively subvert the host defenses through **antigenic variation**. These

- Sporozoites can be transmitted to a human when an infected mosquito bites.
- In the human, these sporozoites initially infect liver cells, where they replicate, releasing diploid **merozoites**, which then infect erythrocytes.
  In an infected human, *P. falciparum* replication leads to malaria.
- To facilitate attachment to red blood cells, the merozoites express a series of merozoite surface proteins (MSPs).
- Once inside an erythrocyte, the merozoites obtain hemoglobin from the host cell through **cytostomes**. Digestion of hemoglobin by the pathogen releases **heme**, which *P. falciparum* then converts to **hemozoin**.
- Replication of merozoites within erythrocytes leads to the destruction of these cells, resulting in **anemia**.

While malaria remains a huge global problem, methods of preventing malaria and treating it do exist.

- Insecticide-treated mosquito sleeping nets can prevent the transmission of *P. falciparum*.
- The antimalarial drug **chloroquine** blocks the formation of hemozoin.

Summary: Organized by section, the Summary lists the introductory questions, along with short descriptions of the key points. Key terms also are incorporated and highlighted.

Application Questions: Each chapter ends with a series of thought-provoking questions that test the students' factual understanding of the chapter and their ability to apply their knowledge.

Application Questions

1. Explain, from a virus's perspective, the potential advantages of acute and persistent infections.
2. Why might stress cause the reactivation of a latent human virus?
3. Explain why apoptosis of infected cells may be beneficial to an organism.
4. We mentioned that cancer may be an unintended outcome of viruses attempting to obtain resources. Explain.
5. Define antigenic drift and antigenic shift. How are these two events similar? How are they different?
6. Explain how bacteriophage lysogeny occurs.
7. What is an endogenous retrovirus?
8. An HIV-infected cell can fuse with adjoining, non-infected cells, forming a syncytium. Explain how this process might occur.
9. Some researchers are concerned that the current H1N1 human influenza virus and the H5N1 avian influenza virus may give rise to a more deadly human virus. How could this occur?
10. For a virus, which is more advantageous: evolving to become highly virulent or evolving to become less virulent? Defend your answer. Provide examples.
11. For a respiratory virus like rhinovirus, transmission often occurs through the air—viral particles exit an infected individual when the host coughs or sneezes and then are inhaled by another susceptible host. The viral particles, then, must remain infectious during this exposure to the outside environment. Devise an experiment that would allow you to measure infectivity of a virus after exposure to various environmental insults.

Key Features

- **CHAPTER VIGNETTE WITH ORIGINAL ART** Every chapter begins with an opening vignette that frames a basic question within the context of both contemporary and historical issues. Each vignette is visually supported by original art created by Janet Iwasa, Harvard Medical School.

- **CHAPTER NAVIGATOR** Organizes the chapter by key concepts and directs students to the most important elements (such as examples, animations, tables) within the chapter to support each concept.

- **MINI-PAPERS** Each chapter includes a Mini-Paper example in which key research papers are summarized and original data is presented and interpreted. Critical-thinking questions, Questions for Discussion, are offered at the end of each Mini-Paper to help students learn to ask intelligent questions and rationally design and evaluate experiments.

- **TOOLBOX** Toolbox examples present and explain important techniques and/or experiments and are accompanied by Test Your Understanding questions.

- **PERSPECTIVE** Perspective examples examine how someone, such as an ecologist or geneticist, would study or use the information presented in the chapter.

- **MICROBES IN FOCUS** Microbes in Focus examples provide a more detailed description of the habitat and key features of specific microbes that are mentioned in the narrative of the chapter.

- **FACT CHECKS** Each section of the chapter ends with 3 to 5 questions to test students' understanding of the content covered in that section.

- **CONNECTIONS** Connections emphasize the interconnectedness of topics, both within a chapter and between chapters. These notes serve as reminders to the students that related material was covered previously or that related material will be introduced again in a subsequent chapter.

- **PROCESS DIAGRAMS** Each chapter includes selected, numbered Process Diagrams to lend clarity to the flow of complex processes.

- **ONLINE RESOURCES** Icons throughout the chapters indicate where a concept has further online resources, such as a 3D animation or extra resources found in WileyPLUS. These resources augment the information presented in the text and make the learning process more interactive.

- **THE REST OF THE STORY** This offers a conclusion to the opening vignette and links it more explicitly to the chapter content.

- **IMAGE IN ACTION** Here we revisit the chapter opener art and ask students questions about how the art and the chapter content are connected.

- **SUMMARY** Organized by section, introductory questions are included along with short narrative descriptions of the key points, allowing students to gauge their understanding of the topics. Key terms are also incorporated as appropriate throughout the Summary.

- **APPLICATION QUESTIONS** Each chapter ends with 10 to 15 thought-provoking questions that not only test students' factual understanding of the chapter but also test their ability to apply their knowledge.

- **SUGGESTED READING** A rich resource of additional materials to enhance understanding is offered for the students.

Additional Resources for Instructors and Students

Key Student Resources Available in WileyPLUS:

ANIMATIONS Over 125 animations, many of which have been developed using 3D animation techniques by Janet Iwasa, Harvard Medical School. Following the unique style of the chapter opener illustrations and developed to support learning objectives in the text, these animations help students visualize and master the toughest topics in microbiology. A series of system-gradable assessments specific to the animations can be assigned by the instructor in WileyPLUS.

For a look at a sample 3D animation and a list of topics to be covered in 3D, please visit: http://www.wiley.com/college/sc/wessner

MINI-PAPER PROJECT ACTIVITIES All chapters contain one Mini-Paper in which key research articles are summarized and interpreted. Through the use of this feature, students will learn to ask intelligent questions and rationally design and evaluate experiments, thereby greatly enhancing their critical-thinking skills. Students can then view a related paper, animation, or video within WileyPlus.

READING AND UNDERSTANDING THE PRIMARY LITERATURE As students prepare lab reports or science research papers, it is important to know what types of information are relevant, appropriate, and available to assist in developing a meaningful connection with the material through the incorporation of primary literature. The goal of this resource is to assist students in navigating this wealth of information.

ILLUSTRATION PODCASTS A set of downloadable audio podcasts walk students through the detail in many of the chapter opener illustrations throughout the text.

Other student features included in WileyPLUS are the Biology Newsfinder, a set of Flashcards with audio, answers to the Fact Check questions, and Practice Quizzes.

Instructor Resources Available in WileyPLUS:

ASSESSMENT: An expansive selection of system-gradable Assessment content with prebuilt Assignments are available to instructors using WileyPLUS, including the following banks of questions for each chapter:

- **Image in Action Questions** based on the chapter opener art for every chapter
- **Animation Questions** based on many of the animations available
- **Prelecture Quiz and Postlecture Quiz** by James Bader, Case Western Reserve University
- **Practice Quiz** by Rebecca Sparks-Thissen, Wabash College
- **Application Questions** by John Steiert, Missouri State University
- **Research Questions** Each chapter also contains key questions with a research focus designed to test the students' ability to more fully analyze and evaluate the material.
- **Fact Check Questions** by William Navarre, University of Toronto
- **Testbank** (also available in Respondus and RTF formats) by John Steiert, Missouri State University

OTHER TEACHING RESOURCES

- **Instructor's Manual** by Evelyn Biluk, Chippewa Valley Technical College
- **PowerPoint slides** for lectures, as well as full slide decks containing all the text images and animations by Benjamin Rowley, University of Central Arkansas
- **Clicker Questions**
- **Life Sciences Visual Library** gives instructors access to a database of searchable and downloadable images from a variety of Wiley texts to use in lectures and presentations.

Mini-Papers: A Focus on the Research

Mini-Paper examples are brief, edited versions of actual scientific papers designed to expose students to highly relevant and current research. The following is a list of each Mini-Paper per chapter. Links to additional topics can be found at www.wiley.com/college/wessner or through WileyPLUS.

Toolbox, Perspective, and Microbes in Focus Examples

Toolboxes in Microbiology

WILEY

Wiley's Enhanced E-Textbooks

Kno delivers the same e-textbook capabilities that
professors and students have come to rely on, only better.
With over 70 free, interactive features, Kno makes
learning more engaging, "hands-on," and efficient.

Portability + Functionality + Lower Prices = Better Learning Experience
With e-textbooks, students not only benefit from being able to access course
materials and content anytime and anywhere but they also can improve
organization and study smarter. Students say the following capabilities
help them study more effectively:

• Content searchability

• Note-taking ability

• Ability to highlight key materials

• One-stop location for all their work allows more efficient studying

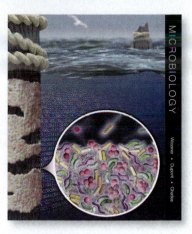

Wessner, Dupont, and Charles
Microbiology
978-0-471-69434-2

Some reasons why studying just got better with Kno:

Tools for Instructors

- **Social Sharing:** A great way to share the most important and relevant information with people
  who can benefit from this content. Professors can share their highlights, sticky notes, and bookmarks directly with their students, colleagues, or TAs. It's easy, simple, and done instantaneously.

- **Smart Links:** In-context instructional video, images, and photos to help explain key concepts.
  Smart Links is a great tool for professors to guide students through difficult concepts.

- **Quiz Me:** Double tap on any diagram in the book and create an instant multiple-choice quiz. As
  a teacher, you can use interactive quizzes to see how much the class has retained in a fun and
  exciting way.

Tools for Students

- **Journal:** Automatically create your own study guide from your highlights, sticky notes,
  bookmarks, and notes.

- **Advanced Search:** Get to the answer quickly with a more powerful search
  engine. Search through all your Kno books, courses, terms, highlights, or notes, with advanced
  search doing even more.

- **Flashcards:** Flashcards are automatically created from glossary terms throughout the book.
  It's the fastest way to retain key concepts without the added work.

Available on multiple devices; visit kno.com for more details.

Learn more about your options at **www.wiley.com/college/wileyflex**

Acknowledgments

We acknowledge the invaluable contributions of our colleagues: Barbara Butler, Brian Driscoll, Heidi Elmendorf, Craig Stephens, and Dave Westenberg. We also thank the editorial and production team at Wiley: Executive Editor Kevin Witt, Senior Developmental Editor Mary O'Sullivan, Freelance Developmental Editors Kathy Naylor and Deborah Allen, Production Editor Barbara Russiello, Product Designer Bonnie Roth, Marketing Manager Carrie Ayers, Project Editors Lauren Morris and Lauren Stauber, and Photo Research Manager Hilary Newman for their wonderful guidance and support. Many thanks as well to Adriane Ruggiero, copyeditor, for making sense of our manuscript, and to Lilian Brady, proofreader, who made sure the content still made sense in pages. Thanks to our indexer, WordCo, for giving us a map to this first edition. We wouldn't have a book without you! We would like to thank Sarah J. VanVickle-Chavez, Ph.D., of Washington University in St. Louis who wrote Appendix A, Reading and Understanding the Primary Literature.

A very special thank you to the members of our advisory board and to all the instructors who helped along the way as we wrote our book.

Microbiology 1e Advisory Board

DWAYNE BOUCAUD	Quinnipiac University
JOANNA BROOKE	DePaul University
ANN BUCHMANN	Chadron State University
SILVIA CARDONA	University of Manitoba
WENDY DUSTMAN	University of Georgia
KATHLEEN FELDMAN	University of Connecticut, Storrs
SANDRA GIBBONS	University of Illinois, Chicago
JANICE HAGGART	North Dakota State University
MICHAEL IBBA	Ohio State University
ROSS JOHNSON	Chicago State University
WILLIAM NAVARRE	University of Toronto
REBECCA SPARKS-THISSEN	Wabash College
JOHN STEIERT	Missouri State University

Reviewers, Class Testers, Focus Group Participants

TAMARAH ADAIR	Baylor University
ERIC ALLEN	University of California, San Diego
EMMA ALLEN-VERCOE	University of Guelph
SHIVANTHI ANANDAN	Drexel University
JAMES BARBAREE	Auburn University
JOHN BASSO	University of Ottawa
JEFFREY BECKER	University of Tennessee, Knoxville
LORI BERGERON	New England College
BENJIE BLAIR	Jacksonville State University
PAUL BLUM	University of Nebraska, Lincoln
ADAM BOGDANOVE	Iowa State University
NANCY BOURY	Iowa State University
DERRICK BRAZILL	Hunter College–CUNY
GRACIELA BRELLES-MARINO	California State Polytechnic University, Pomona
AMY BRIGGS	Beloit College
GINGER BRININSTOOL	Louisiana State University A&M
STACIE BROWN	Texas State University, San Marcos
CARROLL BOTTOMS	Collin College
ALISON BUCHAN	University of Tennessee, Knoxville
KELLY BURKE	College of the Canyons
MARTHA SMITH CALDAS	Kansas State University
GARY CHILDERS	Southeastern Louisiana University
THOMAS CHRZANOWSKI	University of Texas at Arlington
TODD CICHE	Michigan State University
PAUL COBINE	Auburn University
JIM COLLINS	University of Arizona
CHESTER COOPER	Youngstown State University
VAUGHAN COOPER	University of New Hampshire
SIDNEY CROW	Georgia State University
JOAN CUNNINGHAM	Ohio University
CHARLES DANIELS	Ohio State University, Columbus
WENDY DIXON	California State Polytechnic University, Pomona
JANET DONALDSON	Mississippi State University
DIANA DOWNS	University of Wisconsin, Madison
PAUL DUNLAP	University of Michigan, Ann Arbor
LEHMAN ELLIS	Our Lady of Holy Cross College
CLIFTON FRANKLUND	Ferris State University
BERNARD FRYE	University of Texas at Arlington
DANIEL GAGE	University of Connecticut, Storrs
CHRISTINA GAN	Highline Community College
DONALD GLASSMAN	Des Moines Area Community College
MARYANN GLOGOWSKI	Loyola University, Chicago
JIM GOLDEN	University of California, San Diego
STJEPKO GOLUBIC	Boston University
ENID GONZALEZ	California State University, Sacramento
DOUG GRAHAM	Grand Valley State University
HAIDONG GU	Wayne State University
NICK HACKETT	Moraine Valley Community College
ANTHONY HAYS	Cornell University
IVAN HIRSHFIELD	St. John's University
RANDY HUBBARD	Liberty University
SHEELA HUDDLES	Harrisburg Area Community College
WAYNE HYNES	Old Dominion University
EDWARD ISHIGURO	University of Victoria
DIANA IVANKOVIC	Anderson University
NARVEEN JANDU	Harvard Medical School
CARL JOHNSTON	Youngstown State University
CHARLES KASPAR	University of Wisconsin, Madison
SUE KATZ	Rogers State University
DANIEL KEARNS	Indiana University, Bloomington
STEVEN KEATING	Penn State University, University Park
BESSIE KEBAARA	Baylor University
WENDY KEENLEYSIDE	University of Guelph
JOHN KELLY	Loyola University, Chicago
JUDY KIPE-NOLT	Bloomsburg University of Pennsylvania
CHRISTINE KIRVAN	California State University, Sacramento
RENU KUMAR	Minneapolis Community & Technical College
JOHN M. LAMMERT	Gustavus Adolphus College
WEI-JEN LIN	California State Polytechnic University, Pomona
ANDREW LLOYD	University of Delaware
GERALDYNE LOPEZ DE VICTORIA	Midlands Technical College
JOHN MAKEMSON	Florida International University
CAROLYN MATHUR	York College of Pennsylvania
ANN MATTHYSSE	University of North Carolina, Chapel Hill
BARBARA MAY	St. John's University
RENEE McFARLANE	Clayton State University
DONALD McGAREY	Kennesaw State University

CATHERINE MCVAY	Auburn University	DEBORAH SIEGELE	Texas A & M University
SCOTT MINNICH	University of Idaho	LYLE SIMMONS	University of Michigan, Ann Arbor
ROBERT THOMAS MORRIS	Wayne State University	OLANREWAJ SODEINDE	Penn State University
XIAOZHEN MOU	Kent State University	JOE SORG	Texas A & M University
CRAIG MOYER	Western Washington University	SHELDON STEINER	University of Kentucky
SCOTT MULROONEY	Michigan State University	ANN STEVENS	Virginia Tech
LALITHA RAMAMOORTHY	Marquette University	SHERRY STEWART	Navarro College
DAVID NAGLE	University of Oklahoma, Norman	VALERIE STOUT	Arizona State University
ANTHONY NIEUWKOOP	Grand Valley State University	ERICA SUCHMAN	Colorado State University
JANE NOBLE-HARVEY	University of Delaware	KAREN SUE SULLIVAN	Louisiana State University
TANYA NOEL	York University		and A & M College
VALERIE OKE	University of Pittsburgh	DONALD TAKEDA	College of the Canyons
EDITH PORTER	California State University, Los Angeles	MONICA TISCHLER	Benedictine University
		DEREK THOMAS	Grand Valley State University
RONALD PORTER	Penn State University	CLAIRE VIEILLE	Michigan State University
MAMTA RAWAT	California State University, Fresno	JAMES WALKER	University of Texas at Austin
SABINE RECH	San Jose State University	RACHEL WATSON	University of Wyoming
JORGE RODRIGUES	University of Texas at Arlington	BRIAN WEAVER	Missouri State University
BENJAMIN ROWLEY	University of Central Arkansas	KARRIE WEBER	University of Nebraska
KRISTA RUDOLPH	Clemson University	ROSEANN WHITE	University of Central Florida
KATHLEEN RYAN	University of California, Berkeley	BETSY WILSON	University of North Carolina, Asheville
PRATIBHA SAXENA	University of Texas at Austin		
MATTHEW SCHMIDT	Stony Brook University, SUNY	HELEN WING	University of Nevada, Las Vegas
MATT SCHRENK	East Carolina University	PETER WONG	Kansas State University
JULIE SHAFFER	University of Nebraska	ALICE WRIGHT	California State University, Fresno
PETER SHERIDAN	Idaho State University	JIANMIN ZHONG	Humboldt State University
LOUIS SHERMAN	Purdue University	RACHEL ZUFFEREY	St. John's University

Brief Contents

Contents

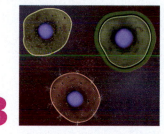

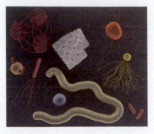

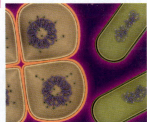

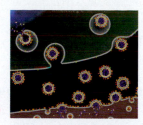
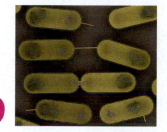
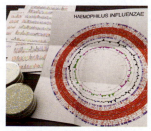
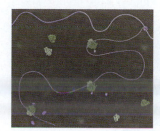

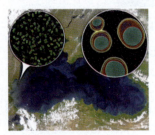

17

Microbial Symbionts   560

PART IV MICROBES AND DISEASE

18

Introduction to Infectious Diseases   600

19

Innate Host Defenses Against Microbial Invasion   636

24

Control of Infectious Diseases 822

MICROBIOLOGY

The Microbial World

Anton van Leeuwenhoek was a successful textile merchant in the city of Delft, in the Netherlands, in the late seventeenth century. He used magnifying lenses in his trade to examine cloth, but in 1665, after reading Robert Hooke's book *Micrographia*, van Leeuwenhoek became fascinated with using microscopes to explore the natural world. Hooke, an Englishman of about the same age as van Leeuwenhoek, had laboriously constructed microscopes that magnified objects roughly 30 times, and used them to examine the fine structure of materials both living and dead. His greatest contribution to biology was the discovery of *cells* (which he first observed in cork slices) as the units from which living organisms are assembled. Hooke's writings inspired van Leeuwenhoek, who enjoyed blowing glass and grinding tiny lenses, to fabricate simple but remarkably powerful microscopes. Some of the 400 or so microscopes that van Leeuwenhoek built magnified images almost 300-fold, and could be used to observe objects one-tenth the size that Hooke had seen. If we consider that the best modern light microscopes of today are limited to around 1000-fold magnification, van Leeuwenhoek's accomplishments are even more astounding!

With his extraordinary lenses, van Leeuwenhoek pushed the frontiers of human knowledge to ever-smaller dimensions. No one had imagined living creatures so small they could not be seen by the human eye, yet van Leeuwenhoek saw them all around us, on us, even *inside us*. In a letter to the Royal Society of London in 1684, he related that:

> *The number of these animals in the scurf of a mans Teeth, are so many that I believe they exceed the number of Men in a kingdom. For upon the examination of a small parcel of it, no thicker than a Horse-hair, I found too many living Animals therein, that I guess there might have been 1000 in a quantity of matter no bigger than the 1/100 part of a sand.*

In another letter, he confided with amazement that:

> *Some of these are so exceedingly small that millions of millions might be contained in a single drop of water. I was much surprized at this wonderful spectacle, having never seen any living creature comparable to those for smallness; nor could I indeed imagine that nature had afforded instances of so exceedingly minute animal proportions.*

Thus, this modest Dutch merchant revealed a whole new *microscopic* world to humanity. Van Leeuwenhoek discovered microorganisms.

Introduction

With wonder in his voice, Anton van Leeuwenhoek shared his observations of microbial life with a skeptical public. In the three centuries since van Leeuwenhoek first viewed these "animalcules," the scientific community and the general public have become much more appreciative of the importance of microbes. We now know that microscopic life on Earth is enormously abundant and diverse, that microbes appeared billions of years before humans, and that the health of the entire biosphere depends on its tiniest inhabitants, the microbes. We also know that microbes interact with each other and with multicellular organisms, including humans, in many ways. Because of our increased understanding of microbes, we now can use them to help us in many agricultural and industrial settings. Because of our increased understanding of microbes, we now better understand how our own bodies work. We also have learned to fear microbes; some of them cause diseases that have resulted in the suffering and death of untold millions of people through the ages.

Throughout this book, we will explore all of these aspects of microbiology. While we will use specific examples to illustrate our points, we will focus on the general principles. We will emphasize the evolutionary relationships between microbes and the evolutionary history of biological processes. We also will learn how the study of microbes relates to various other disciplines,

like genetics, chemistry, and environmental science. Finally, we will see that microbiology itself is a dynamic, evolving science. Our knowledge of microbiology is predicated on thoughtful, interesting, exciting experiments. Much more still awaits our discovery. As we will note throughout this book, we do not know all the answers. We probably do not even know all the questions! The field of microbiology is ever changing. Today's basic research will lead to tomorrow's revelations.

So, let's start our exploration of this dynamic field. In this chapter and in the book as a whole, we first will learn about the microbes. Then, we will examine the genetics of microbes. Next, we will look at the metabolism of microorganisms and how microbes interact with their environment. Finally, we will explore the role of microbes in disease. We can frame our initial discussion, then, around these questions:

What is microbiology? (1.1)

What do we know about the evolution of life and the genetics of microbes? (1.2)

How do microbes get energy and interact with the world around them? (1.3)

How are microbes associated with disease? (1.4)

1.1 The microbes

What is microbiology?

Microorganisms are microscopic forms of life—organisms that are too small to see with the unaided eye. They usually consist of a single cell and include bacteria, archaeons, fungi, protozoa, and algae. We will include viruses in many of our discussions as well. Viruses are not living, but they are microscopic; they utilize biological molecules and cellular machinery (borrowed from their host) to replicate, and they can cause infectious diseases like some microorganisms. While viruses are not microorganisms, we can refer to them as **microbes**, a more general term that includes microorganisms and viruses. **Microbiology**, then, is the study of microbes.

Our relationship with the microbial world is complex and dynamic. On one hand, harmful bacteria, viruses, fungi, and protozoa kill millions of people each year, and sicken billions. On the other hand, beneficial microbes associated with our bodies help us digest food, and protect us from potentially harmful microbial invaders **(Figure 1.1)**. Some microbes cause crops to fail, while others provide essential nitrogen to plant roots through symbiotic relationships. Some microbes cause food to rot, but others carry out fermentations that produce yogurt, wine, beer, and other foods and beverages **(Figure 1.2)**.

In the past few decades, we have learned so much about the molecular machinery of life through the study of microbes such

as the bacterium *Escherichia coli* that scientists now routinely alter microbial cells to produce high-value, lifesaving medical products **(Figure 1.3)**. Whether helpful or harmful, the microbial world is deeply intertwined with our lives, and with the very fabric of life on Earth. Let's begin our exploration of microbiology, then, by asking a very fundamental question. What is life?

The basis of life

So, what is life? This question has fascinated humans for millennia—perhaps since our ancestors first developed conscious, introspective thought. As biologists, we will focus on a practical definition of "life" that distinguishes living organisms from nonliving objects.

First, living organisms are composed of **cells**, the smallest units of life as we know it. Second, living organisms are capable of:

- **Metabolism**: a controlled set of chemical reactions that extract energy and nutrients from the environment, and transform them into new biological materials.

- **Growth**: an increase in the mass of biological material.

- **Reproduction**: the production of new copies of the organism.

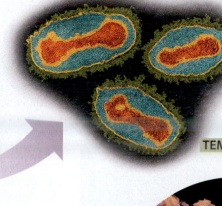

TEM

© Science Source/Photo Researchers, Inc.

A. Microbes and disease

© Eye of Science/Photo Researchers, Inc.

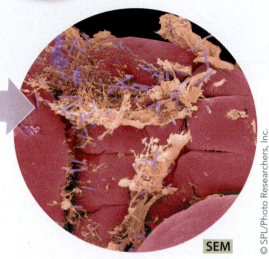

SEM

© SPL/Photo Researchers, Inc.

© Jose Luis Pelaez/Getty Images, Inc.

B. Microbes and digestion

Figure 1.1. Microbes and humans A. Some microbes cause horrific infectious diseases, like smallpox. Man with smallpox *(left);* color enhanced smallpox viruses *(right).* **B.** Other microbes, particularly those that reside in our gut, do not usually cause disease and help us digest the food that we eat *(left).* Food debris (yellow) and bacteria (purple) in the small intestine *(right).*

© Nigel Cattlin/Alamy

A. Some microbes infect important agricultural plants.

© Inga Spence/Photo Researchers, Inc.

B. Other microbes provide nutrients to plants.

© Erika Craddock/Photo Researchers, Inc.

C. Many microbes cause food to spoil.

© Julien Bastide/iStockphoto

D. Other microbes aid in food and beverage preparation.

Figure 1.2. Microbes and food A. Soybean rust, a disease caused by a fungus, causes significant crop losses every year. **B.** Nitrogen-fixing bacteria interact with the roots of certain plants, forming nodules. The bacteria provide essential nutrients to plants, thereby aiding in their growth. **C.** These rotting tomatoes show growth of fungi. **D.** For centuries, microbes have been used by humans to help us produce cheese, yogurt, wine, and beer.

Figure 1.3. Microbes and medicine Using recombinant DNA techniques, researchers can alter genes of microbes such that the microbes produce large quantities of medically important compounds. As we will see later in this chapter, human insulin today is produced by the bacterium *Escherichia coli*. Here, researchers monitor conditions in a large-scale bioreactor used to grow recombinant bacteria.

To accomplish these tasks, organisms contain a biological instruction set to guide their actions. These instructions need to be reproduced as the organism itself reproduces. Other features that living organisms share include:

- Genetic variation, allowing the possibility of **evolution**, or inherited change within a population, through natural selection over the course of multiple generations.

- Response to external stimuli and adaptation to the local environment (within genetic and physiological constraints).

- **Homeostasis**: active regulation of their internal environment to maintain relative constancy.

Does this list represent a complete description of what it means to be alive? Probably not. It's easy to come up with situations that challenge these criteria. Consider the curious case of bacterial endospores—specialized, metabolically inert cells produced by some bacterial species under highly stressful conditions. After shutting down metabolism, growth, and reproduction, the endospores can remain dormant for long periods of time, even thousands of years, awaiting a favorable environment to germinate. Is an endospore "alive" during this state of suspended animation? These spores have all the components of living cells and, when conditions are appropriate, they will again develop into cells that meet the criteria listed above. As we will see later in this section, viruses—subcellular microbes—represent an even more interesting anomaly to the standard definition of life. So, our definition of life should be applied holistically; an organism may not exhibit all of these traits at all times.

Most microorganisms live and function as single, autonomous cells. A free-living unicellular, or single-celled, organism can carry out all the necessary functions of metabolism, growth, and reproduction without physical connection to any other cells. Multicellular organisms, by contrast, are comprised of many physically connected and genetically identical cells. The constituent cells that contribute to a multicellular organism can have distinct, specialized functions. A complex organism like a human can have hundreds of cell types, organized into tissues and organs. While the distinction between unicellular and multicellular organisms seems obvious, you might rethink this issue later, as we learn more about the microbial world. Some unicellular organisms, for instance, only can survive in close association with cells of another species. Other unicellular microorganisms can communicate, behave socially, form three-dimensional structures containing millions of cells with different functions, and enter into dependent relationships with other cells—behaviors that blur the boundaries between unicellular and multicellular lifestyles. The slime mold *Dictyostelium discoideum*, for instance, exists as a rather typical unicellular organism when food is readily available. During periods of nutrient depletion, however, individual cells aggregate and form a complex structure, with cells differentiating to assume specialized tasks **(Figure 1.4)**. Before we investigate these more unusual arrangements, let's learn more about the chemical make-up of cells.

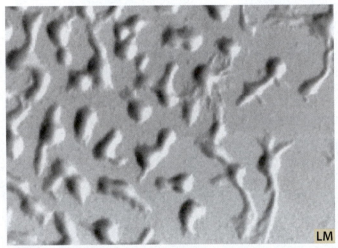

A. Unicellular form

B. Aggregation and differentiation

Figure 1.4. Developmental stages of *Dictyostelium discoideum* A. The slime mold *Dictyostelium discoideum* exists as a unicellular organism when food is plentiful. **B.** When its food supply becomes limited, cells aggregate in response to a cellular signal forming a multicellular slug. Cells then begin to differentiate, eventually forming a stalk and fruiting body.

TABLE 1.1 **Macromolecules in microbial cells**

Macromolecule	Subunits	Functions	Dry weight of cell (%)
Polypeptides	Amino acids	Enzymes catalyze the vast majority of biochemical reactions in the cell. Other proteins are structural components of cells.	50–55
Nucleic acids	Deoxyribonucleotides	Informational: DNA provides the instructions for assembly and reproduction of the cell.	2–5
	Ribonucleotides	Many functions, most of which are involved in the production of polypeptides. Some serve structural or catalytic functions.	15–20
Lipids	Diverse structures	Structural: make up cellular membranes that form physical boundary between the inside of cell and surroundings and membranes of internal organelles.	10
Polysaccharides	Sugars	Structural (such as cellulose and chitin) and energy storage (such as glycogen and starch).	6–7

Chemical make-up of cells

As we shall see in this section, all cells share some basic features. Notably, all cells are built from **macromolecules**—large, complex molecules composed of simpler subunits (**Table 1.1**). Macromolecules, in fact, make up over 90 percent of the dry weight, or weight after the removal of all water, of most cells. In this section, we will explore the four major types of macromolecules found in cells: polypeptides, nucleic acids, lipids, and polysaccharides. For each, we will look briefly at their structure and functions.

Polypeptides, polymers of amino acids, constitute the most abundant class of macromolecules. Polypeptides, also often referred to as proteins, fold into elaborate structures and can execute a vast array of important jobs. Some proteins function as **enzymes**, macromolecules that catalyze chemical reactions within the cell (**Figure 1.5**). Other proteins may facilitate the movement of material into or out of the cell. Still other proteins comprise critical structures such as microfilaments, that allow cell movement (**Table 1.2**).

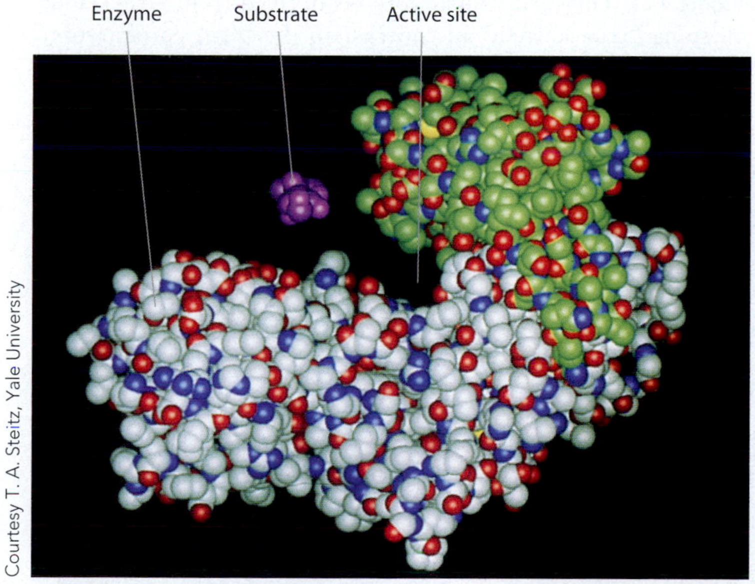

Nucleic acids, polymers of nucleotides, make up most of the remainder of the macromolecules within a cell. This category includes deoxyribonucleic acid (DNA), a polymer of deoxyribonucleotides, and ribonucleic acid (RNA), a polymer of ribonucleotides. Individual nucleotides, in turn, are composed of a sugar molecule (deoxyribose in DNA, ribose in RNA), a phosphate moiety, and one of four nitrogen-containing bases (abbreviated A, T, C, and

Figure 1.5. Structure and function of enzymes In cells, many polypeptides function as enzymes—macromolecules that can catalyze chemical reactions. The function of the enzyme depends on its structure. The three-dimensional shape creates an active site, with which the substrate interacts.

Courtesy T. A. Steitz, Yale University

Enzyme   Substrate   Active site

TABLE 1.2 **Selected functions of polypeptides**

Polypeptide	Location	Function
RNA polymerase	Cytoplasm of bacteria and archaeons, nucleus of eukarya	Produces RNA molecules from DNA template
Glycogen phosphorylase	Cytoplasm	Conversion of glycogen into glucose monomers
K^+ channel	Plasma membrane	Passive transport of K^+ across the membrane, from an area of high concentration to an area of low concentration
Na^+/K^+ ATPase	Plasma membrane	Active transport of Na^+ and K^+ across the membrane, from areas of low concentration to areas of high concentration
Flagellin	Bacterial flagellum	Monomers polymerize to form flagellum, which aids in bacterial motility
FtsZ	Associated with plasma membrane of bacteria	Key component of cell division machinery

Figure 1.6. Plasma membrane
All cells are enclosed within a lipid-based membrane that compartmentalizes the cell, allowing the composition of the inside and outside to differ. In most organisms, the membrane consists of a lipid bilayer. This membrane, however, is not impervious. Various polypeptides and polysaccharides are associated with the membrane. These macromolecules help the cell control the movement of materials into and out of the cell.

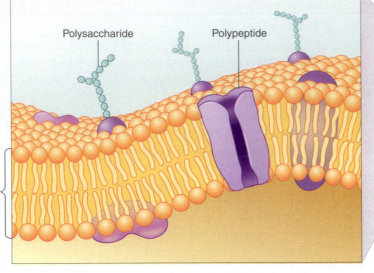

Polysaccharide

Polypeptide

Plasma membrane (lipid bilayer)

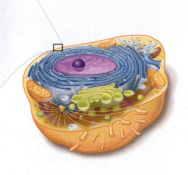

G in DNA; A, U, C, and G in RNA). In all cells, DNA constitutes the main informational molecule, containing instructions for the production of RNA molecules. These RNA molecules fulfill numerous functions within the cell, most of which are associated with protein production.

Lipids, hydrophobic hydrocarbon molecules, represent another important class of macromolecules. The primary role of lipids in most cells is to form the foundation of the plasma membrane, a barrier surrounding the cell that, quite simply, separates inside from outside. This membrane restricts the movement of materials into and out of the cell, thereby allowing the cell to capture and concentrate nutrients for metabolism and growth, and prevent the products of metabolism from escaping **(Figure 1.6)**.

Polysaccharides are polymers of monosaccharides, or sugars. These molecules are comprised entirely of carbon, hydrogen, and oxygen, with the general formula of $C_m(H_2O)_n$. Some polysaccharides serve as energy storage molecules. Starch and glycogen, for instance, both are polymers of the mono-

saccharide glucose ($C_6H_{12}O_6$). Other polysaccharides serve as structural molecules. Cellulose, the primary structural component of plant cell walls, also is a polymer of glucose monomers. Chitin, the primary structural component of fungal cell walls, consists of a derivative of glucose: *N*-acetylglucosamine. Many bacterial and archaeal cells use other polysaccharides for their cell walls.

The domains of life

While polypeptides, nucleic acids, lipids, and polysaccharides exist in all living organisms, major groups of organisms also differ in significant ways. Today, we categorize all living organisms, and, by extension, their cells, into three domains: Bacteria, Archaea, and Eukarya. Until the late 1900s, however, biologists divided cells into two types, **prokaryotes** and **eukaryotes** **(Figure 1.7)**. The term "eukaryote" is derived from Greek roots meaning "true kernel," in contrast to the term "prokaryote," which translates as "before kernel." The "kernel" refers to the

Figure 1.7. Prokaryotic and eukaryotic cells A. Prokaryotic cells as seen in this colorized micrograph of *Escherichia coli* lack a membrane-bound nucleus. They include organisms in the domains Bacteria and Archaea. **B.** Eukaryotic cells, conversely, contain a membrane-bound nucleus, seen in this artist's rendition of a plant cell in purple. Until the latter decades of the twentieth century, biologists divided all cells into these two main types: prokaryotes and eukaryotes. Today, we recognize that all living organisms really should be divided into three categories, or domains: Bacteria, Archaea, and Eukarya. Eukarya contain nuclei. Bacteria and archaeons do not.

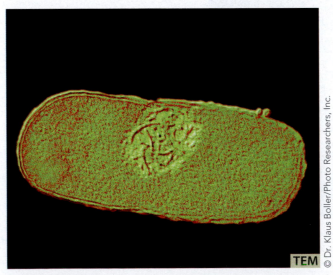

TEM

© Dr. Klaus Boller/Photo Researchers, Inc.

A. Prokaryotic cell

© Russell Kightley/Photo Researchers, Inc.

B. Eukaryotic cell

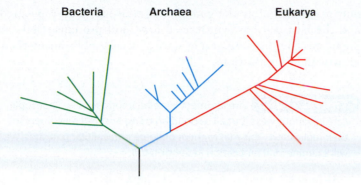

Bacteria   **Archaea**   **Eukarya**

Figure 1.8. Phylogenetic tree of life By comparing the sequences of small subunit (SSU) ribosomal RNA gene sequences, researchers now classify all living organisms into one of three domains—Bacteria, Archaea, or Eukarya. In this phylogenetic tree, bacteria are shown in green, archaeons are shown in blue, and eukarya are shown in red. The linear distance between the endpoints of any two lines is proportional to the sequence similarity of the SSU rRNA gene sequences from the organisms corresponding to the endpoints. Sequence similarity reflects evolutionary distance.

and polypeptides—the most ancient, important, and conserved processes in cells. In the 1970s, microbiologists studying some prokaryotes noted that their molecular machinery resembled that of eukaryotes more than it did other prokaryotes. Leading the way in these studies was Dr. Carl Woese of the University of Illinois. Woese focused his attention on the structure and sequence of one of the RNA molecules that serves as a scaffold for assembly of the ribosome—the small subunit (SSU) ribosomal RNA. This molecule is a critical component of the ribosome in all living organisms and interacts with the messenger RNA during translation (see Section 7.4). His work paved the way for a revolution in thinking about the **phylogeny**, or evolutionary history, of organisms **(Figure 1.8)**. His studies also led to a major revision in the taxonomy, or the classification, of living organisms. Because of Woese's work, we now categorize all living organisms into one of three domains: Bacteria, Archaea, or Eukarya **(Mini-Paper)**.

@ **Classification systems** ANIMATION

Thanks largely to the development of the **polymerase chain reaction (PCR)**, a technique that allows researchers to quickly amplify specific pieces of DNA **(Toolbox 1.1)**, we now have a richer, more accurate phylogenetic tree. This tree is consistent with the idea that the archaeal and eukaryal domains shared a common ancestor after they split from the bacterial domain. It probably is impossible, though, to determine when the divergence of these lineages actually occurred. Microorganisms don't fossilize well, but fossilized stromatolites, mineralized mats built up by layer upon layer of photosynthetic bacteria and other microbes in shallow marine habitats, have been observed in rock formations nearly 3.5 billion years old (see Figure 1.13). If such elaborate microbial communities, including bacteria capable of photosynthesis, existed 3.5 billion years ago, then the split between Bacteria and the Archaea/Eukarya domains probably occurred significantly earlier.

membrane-enclosed nucleus of eukaryal cells. The nucleus contains the genetic material of the eukaryal cell during most of the cell cycle and was clearly visible to microscopists in the 1800s. Its function as the organizer of the hereditary material was not understood until well into the 1900s.

Biologists noted other differences between these cell types. Additional membrane-enclosed organelles exist within eukaryal cells, with each organelle serving a unique and important function. Prokaryotes and eukaryotes also differ strikingly in the organization of their genetic material. Prokaryotes usually contain a single, circular chromosomal DNA molecule. Eukaryotes, in contrast, usually contain multiple, linear DNA molecules. At some point in their life cycle, most eukaryal organisms have two copies, or a $2n$ complement, of their genetic material. Most prokaryotes, in contrast, possess a single copy of their genetic material.

Conventional wisdom through the better part of the twentieth century stated that prokaryotes represented a fairly uniform group, until scientists started looking in more detail at the molecular machinery for the synthesis of DNA, RNA,

Although all cells share many features, studies have clearly demonstrated that bacteria, archaeons, and eukarya are evolutionarily distinct. Some of their differences are listed in **Table 1.3**. We will discuss each of these types of cells in much more detail in Chapters 2–4.

TABLE 1.3   **Selected characteristics of the three domains**

	Bacteria	Archaea	Eukarya
Nuclear membrane	No	No	Yes
Membrane-bound organelles	Rare, a few types found in a few species	Rare, a few types found in a few species	Multiple distinct types, found in all species
Plasma membrane	Similar to Eukarya	Different from Bacteria and Eukarya	Similar to Bacteria
Cell wall	Found in nearly all species, constructed of peptidoglycan	Found in nearly all species, constructed of various materials	Found in some species, constructed of various materials
RNA polymerases	Single polymerase	Single polymerase, Eukaryal-like RNA pol II	Three main polymerases (RNA pol I, II, and III)
Histones	Histone-like proteins	Yes	Yes

Viruses

Are viruses alive? They are not cellular, but they certainly replicate and evolve. Viruses, however, require host cells for replication. Outside of a host cell, virus particles are essentially inert. An isolated virus has no metabolism—it takes up no nutrients and extracts no energy from its environment. Viruses also lack most of the basic machinery needed for the synthesis of macromolecules. Viruses do not respond to stimuli, except perhaps when they bind to receptors on a new host cell, and they do not maintain internal homeostasis. When a virus enters a host cell, it does not grow and reproduce in the same sense that cellular organisms do. Virus particles are more or less completely disassembled in the host cell, and new virus particles are only assembled after the genetic material has been replicated and the host cell has synthesized new viral proteins. Cellular organisms have no comparable state of disassembly during their growth and reproduction.

CONNECTIONS Viruses infect all cellular forms of life. They replicate in various ways, but all depend on using host cell machinery for their replication. This makes them obligate intracellular parasites. We will examine viral replication in **Section 8.3**.

Although viruses are not cellular, they are still very important biological entities to study **(Figure 1.9)**. Viruses are molecular parasites that probably have been around since

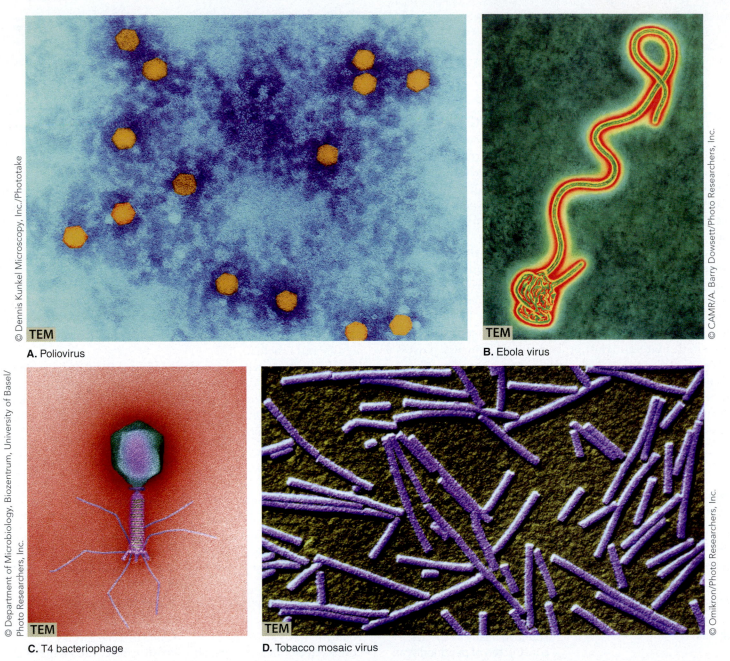

© Dennis Kunkel Microscopy, Inc./Phototake

TEM

A. Poliovirus

© CAMR/A. Barry Dowsett/Photo Researchers, Inc.

TEM

B. Ebola virus

© Department of Microbiology, Biozentrum, University of Basel/ Photo Researchers, Inc.

TEM

C. T4 bacteriophage

© Omikron/Photo Researchers, Inc.

TEM

D. Tobacco mosaic virus

Figure 1.9. Viruses Different viruses have different shapes and some can cause horrific infectious diseases. All images are artificially colored to enhance their appearance. **A.** Poliovirus, the cause of paralytic polio. **B.** Ebola virus, the cause of hemorrhagic fever, a rapidly progressing, highly fatal disease. **C.** T4 bacteriophage, a virus that infects bacteria and has been extensively used in research. **D.** Tobacco mosaic virus, a virus that infects plants. It was the first virus to be discovered.

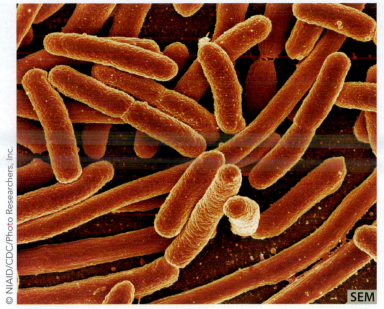

A. *Escherichia coli*

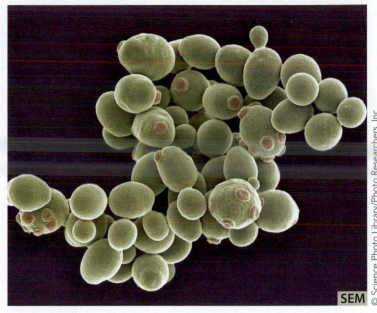

B. *Saccharomyces cerevisiae*

Figure 1.10. Microbes as model organisms Microbes have been used extensively in research. Because they replicate quickly, are cheap to grow, and have relatively simple structures, they have been used extensively to study basic cellular processes like DNA replication, transcription, and translation. Two of the most studied microbial model organisms are **A.** the bacterium *Escherichia coli* and **B.** the eukarya *Saccharomyces cerevisiae*, a yeast.

shortly after the first cells evolved. Microbiologists are interested in viruses not only because they cause many important infectious diseases in humans, crop plants, and livestock, but also because they are fascinating biological systems in their own right. Viruses have taught us a great deal about how cellular organisms function. As parasites, viruses must adapt to their host organism. To be taken up by host cells, most viruses have evolved to bind to host cell surface molecules, and often enter cells by hijacking host systems ordinarily used for taking up non-viral molecules. Many viruses rely on the host enzymes for the production of mRNA, and all viruses use host cell ribosomes for the production of proteins. By studying how viruses use the machinery of their host cells, scientists have gained insight into many critical processes in eukaryal, bacterial, and archaeal cells.

Microbes as research models

Basic research on the structure and function of microbes has laid a solid foundation for understanding the biology of all cells, including our own. Unicellular microorganisms generally possess the same genetic code and many of the same biochemical pathways as multicellular organisms. Additionally, microbes have many advantages for use in research:

- Many are easily cultivated in the lab; they grow rapidly to high cell density on cheap nutrient sources, using inexpensive equipment.

- They facilitate the production of enzymes, other proteins, and various biomolecules for industrial and medical uses.

- Most have relatively small numbers of genes to analyze. Even the largest bacterial and archaeal genomes are smaller than the smallest eukaryal genomes, and eukaryal

microbes have significantly fewer genes than complex multicellular eukarya.

- Many can be genetically manipulated much more easily than complex eukarya.

Popular microbial model systems for research include the intestinal bacterium *Escherichia coli* and the eukaryal yeast *Saccharomyces cerevisiae*, which is also known as "baker's yeast" or "brewer's yeast," because of its long historical use in food and beverage production (**Figure 1.10**). These model microbes have been subjected to the vast experimental armaments of the fields of biochemistry, genetics, molecular biology, and cell biology. Our current understanding of the complexities of biochemical pathways, DNA replication and cell division, the nature of genes, control of gene expression, and protein synthesis, folding, and function has arisen largely from studies of these microorganisms.

Research on the biology of microbial cells has virtually unlimited practical applications. For example, to understand how some antimicrobial drugs work against their microbial targets, while sparing host cells, we need to understand differences in structure between bacterial and eukaryal cells, or perhaps between fungal and human cells. Paul Ehrlich, a towering figure in the history of medicine and immunology, was among the first to recognize that such differences had medical implications. From his experience in the field of histology, Ehrlich was familiar with dyes that differentially stained bacterial and human cells. Based on this observation, he speculated that molecular "magic bullets" that specifically target microbial invaders were feasible. He had little knowledge of the actual structures present on or in cells of any kind, but this concept that certain drugs may adversely affect specific types of cells, while sparing other types of cells, remains at the heart of our drug development initiatives today.

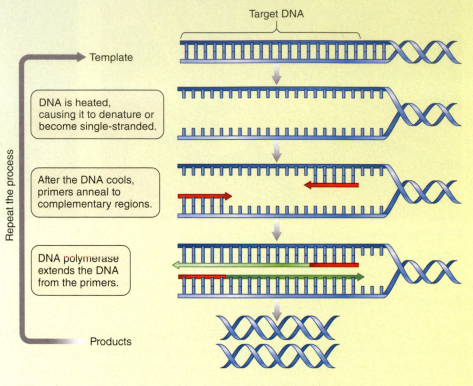

Figure B1.1. The polymerase chain reaction (PCR) After heating the DNA to denature, or separate, the two strands, the mixture is cooled, allowing the short, single-stranded DNA primers to anneal to their complementary regions. The DNA polymerase then extends these primers, using the opposite strand as a template. This process is repeated multiple times, resulting in the generation of many copies of the segment of DNA bounded by the primers. In this fashion, a small amount of input DNA can be amplified sufficiently to perform routine chemical analyses.

The method of ribosomal RNA sequencing developed by Carl Woese was extremely laborious. Fortunately, techniques just being developed in the 1970s and 1980s, as Woese and colleagues were initially developing the universal phylogenetic tree, made nucleic acid sequencing much simpler. The most important of these techniques was the 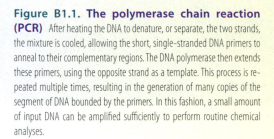 polymerase chain reaction (PCR). With this technique, researchers can create millions of copies of a specific piece of DNA.

Kary Mullis, then a scientist at Cetus Corporation, a biotechnology company in Emeryville, California, invented PCR in 1983 and was awarded the Nobel Prize in Chemistry in 1993 for this discovery. The technique basically mirrors the process of DNA replication utilized by all cells. Rather than replicating an entire DNA molecule, however, PCR results in the repeated replication, or amplification, of a small, defined segment of a larger DNA molecule. The reaction requires only a few basic reagents:

- DNA containing the sequence to be amplified
- Deoxyribonucleotides (dATP, dCTP, dTTP, and dGTP)
- DNA polymerase
- Oligonucleotide primers

The process begins with the denaturation of double-stranded DNA, making it single-stranded. This step is achieved by heating the DNA to around 95°C for a short period of time. The primers, 15–30 nucleotide-long pieces of single-stranded DNA synthesized in the laboratory, then bind to complementary regions on this newly denatured DNA. The primers are designed such that one primer binds to one strand of the denatured DNA, while the other primer binds to the other strand of the DNA. Additionally, the two primers bind to regions of the DNA flanking the sequence to be amplified **(Figure B1.1)**.

After the primers bind, then the DNA polymerase begins generating new DNA, using the denatured DNA as a template. We will see in Section 7.2 that DNA polymerases generate new DNA by attaching

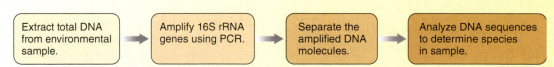

Figure B1.2. Use of PCR to identify microorganisms With PCR, microbial species can be identified simply by isolating a little of their DNA. As shown in this schematic, DNA can be isolated from an environmental source without isolating and growing pure cultures of specific bacterial species. By doing PCR with primers specific for the 16S rRNA gene, this region of the genome can be amplified and subsequently sequenced, thereby providing the investigator with enough information to determine which species are present.

deoxyribonucleotides to primers bound to a template strand. Because PCR usually employs only two primers that bind on either side of a specific region of DNA, these primers delineate which segment of the original DNA molecule will be replicated. The whole process, then, involves three steps:

- Denaturation, or melting, of the DNA
- Attachment, or annealing, of the primers
- Generation of new DNA by DNA polymerase

These steps are repeated multiple times, resulting in the amplification of the region bounded by the two primers. After 10 cycles, the number of these amplified DNA molecules increases over 1000-fold. After 20 cycles, the increase is over 1 million-fold, and 30 cycles will generate over 1 billion new copies!

The crucial development that made PCR widely usable was the discovery of thermostable DNA polymerases that could withstand the high temperature (> 90°C) used to separate DNA strands. The first thermostable polymerase used for PCR was "Taq" DNA polymerase, which came originally from the bacterium *Thermus aquaticus*, isolated from hot springs in Yellowstone National Park in the United States. Today, a variety of thermostable DNA polymerases are commercially available, several of which have been isolated from heat-loving archaeons. We will discuss these fascinating microorganisms more throughout Chapter 4.

PCR revolutionized SSU rRNA gene sequence analysis by making it possible to easily generate many copies of the gene encoding the 16S (or 18S) rRNA, or at least a segment of the gene. Just a bit of chromosomal DNA needs to be extracted from a microorganism to use as a template for PCR **(Figure B1.2)**. The PCR-amplified DNA then can be sequenced. The availability of rRNA gene sequences up to 1,600 bases long allows researchers to make much more sensitive and accurate comparisons between species, a big improvement over Woese's method of sequencing fragments of rRNA molecules.

The uses of PCR in microbiology extend much further. PCR-based tests have been developed to detect the human pathogen *Chlamydophila pneumoniae*, a bacterium that typically is difficult to identify. As we will see in Perspective 5.1, a form of PCR routinely is used to monitor the viral load, or amount of virus present, in people with HIV disease. PCR also allows us to learn more about microbes that currently cannot be grown in the laboratory, a topic we will explore in Section 6.3. Virtually all areas of microbiology have been affected by the conceptually simple polymerase chain reaction.

● Test Your Understanding ·················

Explain how a PCR reaction would differ if a standard DNA polymerase, instead of a thermostable DNA polymerase such as Taq, were used. What would be the result and why?

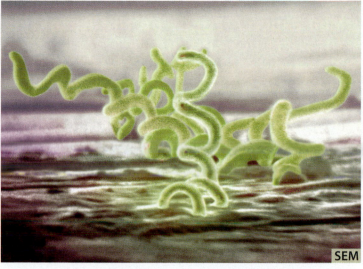

SEM

Figure 1.11. The development of antimicrobial drugs The bacterium *Treponema pallidum* causes syphilis, a sexually transmitted disease. Salvarsan, the first commercially available drug to combat syphilis, was developed in 1910. While Salvarsan prevented the replication of this bacterium, it also was toxic to human cells, and its use was curtailed after the development of penicillin.

Members of Ehrlich's research group discovered an organic arsenic-containing compound, arsphenamine, which in 1910 became the first effective commercial drug for the treatment of *Treponema pallidum*, the bacterium that causes the sexually transmitted disease syphilis **(Figure 1.11)**. Because it also exhibited toxicity to host cells, arsphenamine, known by its trade name Salvarsan, was abandoned in the 1940s in favor of penicillin, the first widely used antibiotic capable of killing many different kinds of bacteria.

Salvarsan's historical importance was in establishing that lethal agents specifically targeted at microbial cells are indeed possible. In the century since Salvarsan came on the market, an enormous amount has been learned about the molecular differences between bacterial and eukaryal cells. Hundreds of new antimicrobial and antiviral drugs have been discovered, and hopefully there will be more to come.

CONNECTIONS Basic research into the structure and replication of microbes has led to the development of numerous antimicrobial and antiviral drugs. Many of the currently approved drugs for the treatment of HIV, for instance, interfere with specific viral enzymes needed for the production of new virus particles. We will discuss how a particular class of these drugs—nucleoside analogs—works in **Section 24.2**.

1.1 Fact Check

1. What are the key features of living organisms?
2. Describe the macromolecules found in cells.
3. What are the three domains of living organisms?
4. Explain why microbes are useful model systems in research and provide examples of microbial model systems.

Mini-Paper: A Focus on the Research
THE THREE DOMAINS OF LIFE

C. R. Woese, O. Kandler, and M. L. Wheelis. 1990. Towards a natural system of organisms: Proposal for the domains Archaea, Bacteria, and Eucarya. Proc Natl Acad Sci USA 87: 4576–4579.

Context

The ideal biological taxonomy should accurately reflect phylogenetic relationships. This union between taxonomy and phylogeny, however, has been a major challenge for microbiologists. Aristotle's categorization of life into just two fundamental groups, animals and plants, persisted until the dawn of microbiology as a science. In 1868, two centuries after van Leeuwenhoek's discovery of microbial life, German biologist Ernst Haeckel proposed a third kingdom, Protista, for microscopic life forms. In 1938, Herbert Copeland suggested that microbes should actually be divided into two kingdoms, Protista and Bacteria, thereby recognizing the fundamental difference between eukaryotic and prokaryotic cells. Twenty years later, Robert Whittaker advocated further separation of eukaryotic microbes into kingdoms of Fungi and Protista, but kept prokaryotic cells in a single kingdom called Monera. This five kingdom taxonomic system—Animalia, Plantae, Fungi, Protista, and Monera—became the accepted standard for the next few decades until DNA and protein sequences became widely accessible.

Carl Woese, a microbiologist at the University of Illinois, was fascinated by a group of prokaryotes known at the time as "archaeabacteria." Initially, these strange microbes were found primarily in marginal environments—anaerobic sediments, hypersaline ponds, and hot springs, for example. Other than their curious ability to colonize extreme habitats, the feature of archaebacteria that generated the most interest among the wider biological community was the ability of some of these organisms to produce methane. These microbes remain the only organisms known to produce this gas.

To understand the phylogeny of these methane-producing microbes, Woese followed the lead of Zuckerkandl and Pauling, who had first demonstrated in the 1960s that comparisons of protein sequences could reveal evolutionary relationships. If two organisms were closely related, these researchers reasoned, then the amino acid sequence of a common protein in the organisms should be very similar. If two organisms were distantly related, conversely, then the amino acid sequence of a common protein should be more divergent. Woese and colleagues began to focus not on protein sequences, but on RNA sequences. Woese reasoned that the ribosomal RNAs (rRNA), because of their universal presence in all cells, could be excellent molecules to compare. In bacteria, the 16S rRNA molecule is part of the small subunit of the ribosome. In eukaryotes, the equivalent ribosomal RNA is the 18S rRNA. These molecules, collectively referred to as "small subunit (SSU) rRNA" molecules, are critical in the ribosome, helping to bring together the ribosomal structure, and interacting with messenger RNA.

The SSU rRNA gene sequence has been referred to as a "molecular chronometer," a slowly ticking clock that measures evolutionary time.

The sequence of this molecule changes very slowly because of the functional constraints on the molecule. Random mutations that occur within the gene encoding the small subunit rRNA often have serious negative consequences, so relatively few changes are passed on to subsequent generations. Nevertheless, there are enough differences in the roughly 1,600 nucleotide sequence to differentiate between species to map patterns of similarity. If one assumes that overall mutation rates are similar between species (which seems to be true with respect to rRNA genes), then one can quantify sequence differences between SSU rRNA genes in multiple species to infer relationships. Ultimately, Woese discovered that the methane-producing microorganisms were no more closely related to other bacteria than they were to the eukaryotes. Let's examine the scientific work that led to this conclusion.

Experiments

The 1990 Woese et al. paper actually presented no new experimental data. Its importance was in articulating a new view of the phylogeny of life. To understand the genesis of this idea, we should step back and examine the data. The biggest challenge Woese faced in the 1970s in developing ribosomal RNA sequences as a tool for phylogenetic analysis was the difficulty in determining such sequences. Woese and colleagues developed a laborious method to infer the sequence of 16S rRNA molecules, which they described in a 1977 article, "Comparative cataloging of 16S ribosomal ribonucleic acid: A molecular approach to procaryotic systematics," Journal of Bacteriology, vol. 27, pp. 44–57. First, RNA was extracted from cells. The isolated rRNA then was cut into small chunks using a ribonuclease enzyme that yields short fragments of nucleic acid usually 5–20 bases long. The sequence of each oligonucleotide was determined by further chemical and enzymatic analysis. In its original incarnation, this method did not actually yield a complete rRNA sequence, but rather a catalog of short oligonucleotide sequences present in the rRNA. Catalogs from different species then were compared.

The underlying assumption of molecular sequence comparisons, at least with respect to phylogeny and taxonomy, is that the number of nucleotide differences between two sequences is proportional to the time since the two species diverged from a common ancestor. Species that shared a more recent common ancestor will have fewer differences than species that have been separated for longer periods of time. Exactly how long ago two species separated depends on the rate at which mutations accumulate, which can be very difficult, if not impossible, to know. Fortunately, to determine phylogenetic relationships, we do not need to know *exact* times of divergence. We are just interested in *relative* times: if organisms A and B shared a common ancestor *after* they shared an ancestor with organism C, then A and B would have fewer sequence differences with each other than either would have with C. The Woese method essentially took each 16S rRNA sequence and compared it against all of the other species. A method for quantifying the similarity between sequences was developed to yield an "association coefficient" between 0 and 1. A perfect match between sequences would result in an association coefficient of 1, whereas no matches would give a score of 0. From these scores, a computer algorithm was used to plot the most likely phylogenetic tree.

Starting with papers in 1977 and continuing through the 1980s, Woese and colleagues built a "universal phylogenetic tree" through comparison of SSU rRNA sequences from diverse organisms, including members of each of the five kingdoms of life, as defined at that time. The great strength of this universal tree is that it compares all organisms using a common standard, a molecule they all possess. Woese noted that this universal tree supports *three* primary branches of life, not the five kingdoms previously accepted. In the 1990 paper, Woese and his coauthors propose that these very ancient branches— Bacteria, Eukarya (or Eucarya, as it is spelled in the 1990 paper), and Archaea—be referred to as domains (see Figure 1.8).

The domain Archaea is composed of species previously known as archaebacteria. Though archaeons lack a nucleus, they turned out to be no more similar to bacteria than they are to eukaryotic organisms, or eukarya. In fact, Woese's tree suggests that Archaea and Eukarya share a more recent common ancestor than either does with Bacteria. As we will see in Chapter 4, archaeons are similar in size and shape to bacteria. Many of the enzymes used by archaeons for DNA replication, transcription, and translation, however, more closely resemble the corresponding enzymes found in eukarya. Perhaps most interestingly, the plasma membrane of archaeons differs chemically from the plasma membranes of bacteria or eukarya. The exact evolutionary history of these organisms is yet to be determined.

In this three-domain phylogenetic tree, Monera and Protista disappear as kingdoms. In fact, if kingdoms are to be defined by equivalent depth of branching—which implies roughly equivalent evolutionary times since divergence—rRNA gene sequence comparisons support many more kingdoms than were previously known, most of which are populated by microorganisms. As we will see in Chapter 4, Woese and coauthors proposed two kingdoms within the Archaeal domain: the Crenarchaeota and the Euryarchaeota. The authors also noted the presence of heat-loving microbes in several other branches of life. It is quite possible, then, that the organism at the root of the tree (the last common ancestor of all life on Earth) was thermophilic, or heat-loving. We will return to this point in Section 1.2.

Impact

For the first time, biologists could create a natural taxonomic system in which all organisms are compared by the same criteria. The realization that the prokaryotes could not be united as a phylogenetic group raised many questions regarding the validity of the five kingdom system. Not surprisingly, there was initial resistance. Many scientists challenged the validity of this approach and the computational methods on which it was based. Since Woese first conducted his experiments, there have been methodological improvements, most significantly the ability to amplify entire rRNA genes using the polymerase chain reaction (PCR; see Toolbox 1.1), followed by rapid and straightforward DNA sequencing. The basic concept of using rRNA gene sequence comparisons to derive phylogenetic relationships is now accepted as an essential method of phylogenetic analysis.

Sequence-based phylogenies have had more impact on microbiology than any other branch of science. Since this paper was published in 1990, databases containing ribosomal RNA gene sequences have grown explosively. The universal phylogenetic tree is now much richer and more complex, but the three-domain organization remains unchallenged. Using PCR, microbiologists can characterize organisms using rRNA gene sequences even if the microorganisms cannot be grown in culture. Since the majority of microbes apparently will not grow in laboratory culture (see Section 6.3), this approach is enormously important for understanding the true diversity of life on Earth, and its evolutionary history. Proposals for new bacterial and archaeal kingdoms, based on rRNA gene analysis of uncultured organisms, have appeared regularly since 1990. Classification of eukaryal microorganisms also is affected by rRNA-based phylogenetics.

Ribosomal RNA sequences, however, do not tell the entire evolutionary story of an organism. In the last decade, the DNA sequences of hundreds of entire genomes have been determined, most of them from microorganisms. It is clear that microbes are rampant sharers of genes, which we will discuss more in Chapters 9, 10, and 21. While it is likely that rRNA genes are rarely shared and do accurately reflect the evolutionary history of the "core" genome, large fractions of genetic material in many organisms may have distinct histories—a finding that Carl Woese could scarcely have imagined when he began his revolutionary efforts to clarify microbial taxonomy.

● Questions for Discussion ....

1. What features make SSU rRNA gene sequences ideal for phylogenetic studies?

2. What drawbacks do you see with the use of rRNA for these studies?

3. If we discover forms of life on another planet, would studies of rRNA gene sequences be useful for categorizing these life forms?

1.2 Microbial genetics

What do we know about the evolution of life and the genetics of microbes?

While groups of microbes may be different from each other, they all share common information processes. Indeed, one of the most remarkable aspects of life as we know it is the constancy of these information processes. In all cells, the main informational molecule is double-stranded DNA. In all cells, a specific type of RNA, messenger RNA, or mRNA, serves as the conduit between the information in DNA and the actual production of proteins. In all cells, the code used to convert the information present in DNA to RNA to protein is the same. This conserved genetic code probably represents the most compelling evidence for evolution. All living organisms share a common informational pathway, suggesting that all living organisms share a common ancestor.

Because all living organisms, and the genetic processes of these organisms, are evolutionarily related, we will begin our exploration of microbial genetics by examining the origins of life. We then will look at how genetic processes occur in microbes and how microbiologists study these processes. We will end this section with a brief overview of how researchers today use these processes to learn more about living organisms.

The evolution of life on Earth

Earth is home to a huge variety of microbes. To understand how this incredible diversity evolved, we need to consider the history of Earth and the origins of life itself. The geochemical changes that have occurred on Earth in the past 4 billion years—in the oceans, on land, and in the atmosphere—have been dramatic. These changes have profoundly affected, and were profoundly affected by, microorganisms. The vast majority of the living organisms we see today, at least without microscopes, are large, multicellular eukarya that arose within the last few hundred million years, the last 10 percent of Earth's history. But most of the major evolutionary events that moved life toward today's world occurred in the distant past, when microbes alone ruled the planet. To get some perspective on this point, let's take a brief walk through the history of life.

Prebiotic Evolution

When Earth formed approximately 4.5 billion years ago (abbreviated ybp, for years before present), it was a hot, sterile place. Oceans of liquid water formed around 4 billion ybp, once the crust and atmosphere had cooled sufficiently for liquid water to condense (**Figure 1.12**). These oceans may have been partially or completely converted to steam on multiple occasions by the energy of asteroid impacts, which were far more common in the early solar system. Depending on when the first life forms evolved, such impacts could have resulted in mass extinctions, coupled with selection for life forms that could live in this extreme environment. By 3.8 billion ybp, life clearly had gained a permanent foothold. The first microorganisms appeared as life transformed from a semi-organized set of chemicals and

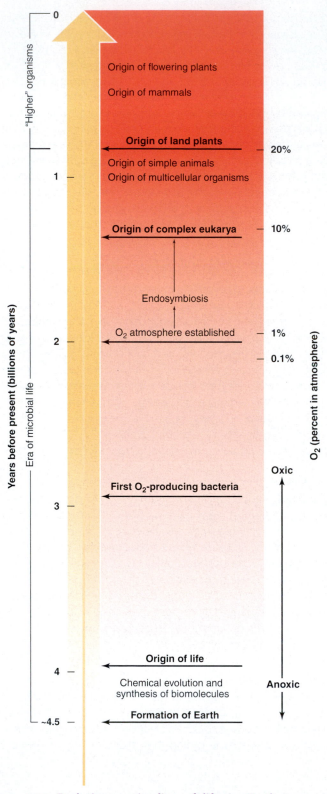

Figure 1.12. Evolutionary timeline of life on Earth Current evidence suggests that Earth formed roughly 4.5 billion years ago. Life appeared around 3.8 billion ybp. Some early eukarya first appeared around 1.5 billion ybp. Mammals first appeared about 200 million ybp. Oxygen did not become a major constituent of the atmosphere until the advent of oxygen-producing photosynthesis.

Figure 1.13. Ancient fossils A. This light micrograph shows fossils of cyanobacteria. The fossils were found in Wyoming and date back about 50 million years. **B.** Fossilized stromatolites 3.5 billion years old, such as these from western Australia, have been identified, indicating that photosynthetic bacteria existed on Earth at least this long ago.

A. Fossils of cyanobacteria

B. Fossilized stromatolites

reactions to a true cellular form. By 3.5 billion ybp, microbial cells were abundant on Earth, as is evident from fossilized stromatolites containing cyanobacteria-like structures **(Figure 1.13)**.

Cyanobacteria, we should note, are photosynthetic bacteria. The evolution of these organisms, and their oxygen-releasing photosynthetic capabilities, led to the eventual oxygenating of Earth's atmosphere. Given that multicellular algae and marine invertebrates are not evident in the fossil record until 0.5 billion ybp, it appears that microbial life ruled Earth for over 3 billion years. Only during the last 500 million years has Earth seen the rise of plants and animals! Our planet has changed drastically since its violent birth, but with the exception of dramatic events like asteroid impacts and volcanic eruptions, changes have occurred gradually. Microbes had plenty of time to evolve an incredible array of talents, allowing them to exploit every possible niche. Given the eons that have gone by, we can only imagine the diversity of microbial life that has existed since Earth's origins; we still do not fully comprehend the richness of microbial life on present-day Earth.

When life first appeared, Earth was a harsh place. The average temperature was quite hot, probably over 50°C. The composition of the atmosphere is not known for sure, but researchers hypothesize that it had a high concentration of CO_2, perhaps up to 30 percent. Other atmospheric gases may have included nitrogen (N_2) and hydrogen (H_2). Whether gases such as ammonia (NH_3), methane (CH_4), cyanide (HCN), and hydrogen sulfide (H_2S) were present in substantial concentrations is not known with certainty. It is

clear, though, that there was little or no molecular oxygen (O_2). The oceans probably were fairly acidic due to the high concentration of dissolved CO_2. By comparison, today's atmosphere consists of about 0.03 percent CO_2 and 21 percent O_2, with a moderate average temperature of 13°C. What changed the O_2 and CO_2 concentrations so dramatically since life began? Microbial activities over the past 4 billion years are part of the answer.

The First Microbial Life

It is generally assumed that life forms present on early Earth have not survived unaltered to modern times. Conditions on our planet have changed radically over the past four billion years, and it is reasonable to assume that evolutionary innovations incorporated into living systems during that time out-competed and displaced primitive cells long ago. Nevertheless, the biochemical origins of life are of great interest. As we look outward for life elsewhere in our solar system and beyond, simple living systems—microorganisms—are far more likely to be discovered than are advanced civilizations in flying saucers. The better we understand the evolution of life on Earth, the better idea we have of what to look for elsewhere.

Many hypotheses have addressed the origin of life. The Miller–Urey experiment envisioned an early Earth where organic molecules accumulated in the oceans, creating a rich "prebiotic soup" from which organized cellular life eventually emerged **(Perspective 1.1)**. Perhaps the organic molecules would have required a surface on which to accumulate, rather than simply floating in the open ocean. Günter Wächtershäuser

We cannot go back in time to the prebiotic world to watch how life actually evolved. We can imagine scenarios, formulate hypotheses, and test them in the laboratory to gain insight into potential pathways to life. In 1953, Stanley Miller described the first laboratory investigation intended to simulate prebiotic Earth. Miller, then a graduate student at the University of Chicago, designed a reactor with his mentor, Harold Urey, to test for abiotic production of biologically relevant molecules. The Miller–Urey experiment revolutionized the thinking of many scientists, and a fair number of non-scientists as well, about the origin of life on Earth.

The Miller–Urey experiment started with a water-filled flask, heated by a burner. The atmosphere in the apparatus was intended to simulate that of primitive Earth, and consisted of a mixture of ammonia (NH_3), methane (CH_4), and hydrogen (H_2). The water was heated to boiling, at which time water vapor mixed with the various gases in the atmosphere and circulated through the apparatus. A separate chamber discharged electrical sparks through the atmosphere, after which the tubing was cooled so that water vapor would condense and drip back into the original flask **(Figure B1.3)**.

After a few days of continuous operation, Miller observed that the water in the flask was changing color. Within a week, the solution in the flask had turned a deep red, and had become turbid. Miller examined the solution after a week and found that the solution contained organic molecules, or molecules containing carbon-hydrogen bonds. Most notably, the simple amino acids glycine and alanine were readily detectable, along with aspartic acid. These amino acids are among the 20 primary amino acids used by life on Earth for protein synthesis, and presumably were abundant as life evolved. Miller's flask provided the first evidence that such molecules could be synthesized from inorganic precursors under conditions believed to mimic primordial Earth, where energy inputs could have included electrical discharge from lightning strikes as well as heat from both the sun and Earth's crust.

The Miller–Urey experiment stimulated other scientists to design experiments to simulate hypothetical atmospheric, oceanic, and geological conditions on prebiotic Earth. Experiments done in hydrogen-rich reducing atmospheres similar to Miller's showed that many amino acids, nitrogenous bases, and other organic compounds could be produced abiotically. Questions about these experiments arose, however, as the consensus of scientists shifted in the 1970s to favor an atmosphere on early Earth that was much richer in CO_2 and N_2. In a Miller-type apparatus, these atmospheric conditions yield few or no organic molecules such as amino acids. Models of early Earth conditions continued to evolve. By 2005, scenarios with much higher levels of H_2 in early atmospheres, coexisting with CO_2, were tested. Jeffrey Bada, a professor at Scripps Institution of Oceanography and a former graduate student of Miller's, further altered the composition and atmosphere in a Miller-type reactor and was able to reproduce Miller's yield of organic material.

Controversies continue as we do not know the exact chemical and physical conditions on early Earth. Predicting the pathway by which life arose depends greatly on what conditions are assumed to have existed. To further complicate the situation, some evidence suggests that substantial amounts of organic molecules could have been delivered to Earth by comets and asteroids crashing into the planet during the tumultuous early days of the solar system. Hydrocarbons and complex nitrogen-containing organic molecules, including amino acids, have been detected on comets in space, and in meteorites on Earth. The magnitude of the contribution of these impacts remains unclear.

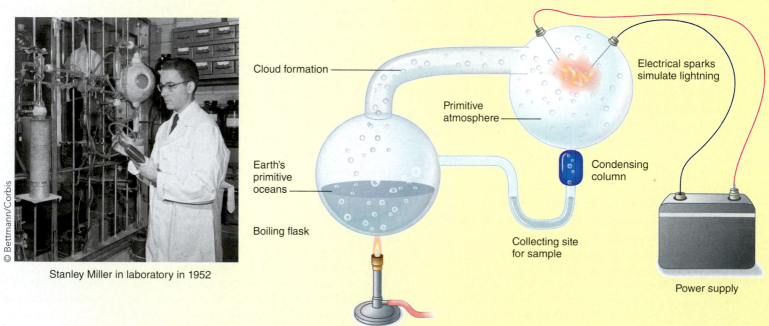

© Bettmann/Corbis

Stanley Miller in laboratory in 1952

Cloud formation

Electrical sparks simulate lightning

Primitive atmosphere

Earth's primitive oceans

Condensing column

Boiling flask

Collecting site for sample

Power supply

Figure B1.3. The Miller–Urey experiment To explore the origins of life, Miller (pictured here) and Urey created an early Earth atmosphere within a closed apparatus. After supplying heat and electricity to the apparatus (simulating energy sources on ancient Earth), they extracted samples for analysis. Their chemical analysis showed that organic molecules had been formed, including amino acids. This experiment suggested that the building blocks for living systems could have formed on Earth from small inorganic precursors.

has theorized that life evolved on iron-containing surfaces, such as iron pyrite (FeS_2), an insoluble, positively charged surface with affinity for organic compounds. Metabolic processes that occur in modern cells could have evolved from reactions among surface-bound organic compounds. Many modern proteins, such as cytochromes and hemes, bind iron and other metallic atoms; many enzymes, including DNA polymerases, require bound metals for activity. These interactions with metals may reflect a very ancient relationship.

As we try to envision the origins of life, the formation of chains of ribonucleotides (RNA) and amino acids (polypeptides) clearly represents a necessary event. These polymers, as

we noted, contain information and carry out cellular functions. Some researchers have suggested that information was initially stored in RNA molecules, rather than DNA. We now know that certain RNA molecules, **ribozymes**, can catalyze chemical reactions **(Perspective 1.2)** Early in the history of life, then, this macromolecule may have had dual functions.

We noted earlier that all cells contain a plasma membrane. As we think about the origins of life, let's ask another question. How did the first membranes form? We know that hydrocarbons coupled to charged groups, such as phosphates, form polar lipids, which can spontaneously organize into micelles and even bilayer membranes that close back upon themselves to form a sealed

Perspective 1.2
RIBOZYMES: EVIDENCE FOR AN RNA-BASED WORLD

From the 1920s, when the enzyme urease was first purified and crystallized, to the 1980s, enzymes were thought to be solely the domain of proteins. Proteins can be formed from 20 different amino acid monomers and can fold into complex three-dimensional shapes capable of forming a seemingly infinite array of active sites for catalytic activities. RNA has only four possible subunits—ribonucleotides containing adenine, guanine, cytosine, or uracil—and thus lacks the chemical diversity of proteins. Still, RNA is not locked into a double helical structure like double-stranded DNA, and can fold into more elaborate shapes. When researchers realized in the late 1960s that single-stranded RNA could assume complex shapes, some molecular biologists began to speculate that RNA might be capable of enzymatic activity, and thus represent an evolutionary link between genetic information and catalytic function.

In the 1980s, Tom Cech and colleagues at the University of Colorado showed that a purified ribosomal RNA molecule from the ciliated protozoan *Tetrahymena thermophila* can undergo a self-cleavage event, removing a segment of itself without the help of any proteins **(Figure B1.4)**. Simultaneously, Sidney Altman and colleagues at Yale were studying the enzyme RNase P that cuts transfer RNA, or tRNA, precursors, which are initially transcribed as part of a larger RNA molecule, into active tRNAs. RNase P is composed of protein and RNA components, but Altman's group showed that the RNA portion of the *E. coli* enzyme can carry out cleavage on its own. Cech and Altman won the Nobel Prize in 1989 for their demonstration of RNA catalysis.

Since the last decade of the twentieth century, molecular biologists have discovered more catalytic roles for RNA. Non-coding sections of messenger RNA that must be removed by splicing prior to translation, are ubiquitous in eukaryal genes. These regions are removed by the spliceosome, a protein-RNA complex in which RNA components are critical for catalysis. Peptide bond formation during translation also relies on RNA catalysis. Scientists and biotechnology companies have jumped on the catalytic RNA bandwagon, because the activity of synthetic ribozymes can be targeted to specific nu-

cleic acid sequences. One could envision creating an RNA molecule that cleaves a specific target. Potential targets for cleavage include the RNA of viruses such as HIV, and messenger RNAs from genes associated with cancer. None of these ideas have reached clinical application yet, but research in this exciting area continues.

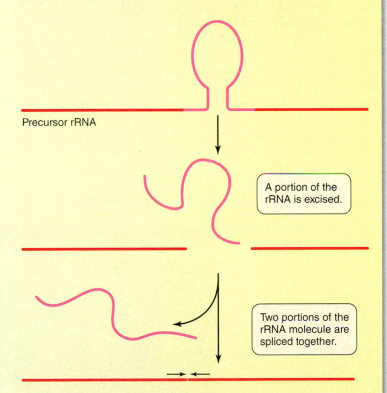

Precursor rRNA

A portion of the rRNA is excised.

Two portions of the rRNA molecule are spliced together.

Figure B1.4. Self-splicing RNA molecules As demonstrated by Tom Cech, a ribosomal RNA molecule from the protozoan *Tetrahymena thermophila* undergoes cleavage without the aid of an enzymatic protein. The RNA molecule itself is catalytic. Several different forms of self-splicing have since been observed in various species. As shown in this schematic, two portions of a precursor RNA molecule are spliced together. The intervening sequence is excised and ultimately degraded.

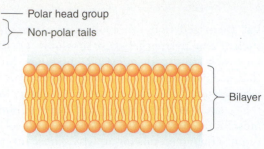

Figure 1.14. Micelles and bilayer membranes
Phospholipids, consisting of non-polar fatty acid tails and polar phosphate heads, can spontaneously form micelles, spherical units with polar surfaces and non-polar cores. These molecules also can form bilayers, in which the non-polar regions exist between two polar surfaces.

compartment **(Figure 1.14)**. Perhaps such primitive membranes initially formed around fragments of minerals, such as FeS_2, that also happened to serve as surfaces for the aggregation of organic molecules, as discussed above. If a membrane encapsulated an informational molecule and a catalytic molecule, like a ribozyme, then something like a cell would have been formed.

The Appearance of Eukarya

How did complex cells arise? Most biologists now agree that mitochondria and chloroplasts, two of the most distinctive organelles in eukaryal cells **(Figure 1.15)**, are derived from bacterial cells, through a process known as *endosymbiosis*. These organelles have their own circular genomes, though these genomes are far smaller than the genomes of contemporary bacteria. The genes present in the chloroplast and mitochondrial genomes include a gene encoding 16S rRNA, the SSU rRNA molecule present in bacterial ribosomes. Sequence analysis of the mitochondrial and chloroplast 16S rRNA gene indicates that mitochondria and chloroplasts are related to specific groups of bacteria.

@ Endosymbiosis ANIMATION

Lynn Margulis, a biologist at the University of Massachusetts–Amherst, has long championed this idea to explain the origins of mitochondria and chloroplasts. Indeed, substantial molecular evidence indicates that these organelles became part of eukaryal cells through endosymbiotic relationships that became permanent. Mitochondria probably were added first to a developing eukaryal cell; they are present in most, but not all, modern eukarya. We can imagine that the endosymbiont provided the host with extra ATP. In return, the host provided the endosymbiont with nutrients and a safe place to live. The extra ATP ultimately allowed for an increase in cell size, and for multicellular arrangements.

Chloroplasts probably were added later, leading to the evolution of algae and plants. A bacterium capable of photosynthesis, the ability to harvest the energy present in sunlight to produce organic molecules, would provide a host with a new energy source and allow the host to expand into new niches. During the evolution of eukarya, multiple endosymbiotic events leading to chloroplasts may have occurred. We now know that some differences exist between the chloroplasts present in different algal and plant groups. Regardless, photosynthesis brought solar power to eukaryal cells. Algae and plants, as a result, have been phenomenally successful, and now account for a major fraction of the biomass on land (plants) and sea (algae). Once these mitochondria and chloroplasts became permanently established in eukaryal hosts, their own genomes degenerated until only a few essential genes were left.

CONNECTIONS We will look more closely at the origins of eukaryal cells in **Section 3.4** and explore the evidence supporting the endosymbiotic theory. In **Chapter 17**, we'll see that endosymbiotic relationships between microbes and host cells are actually very common in nature. In most cases, the host cannot survive without the essential functions supplied by the endosymbiont.

In the earliest cells, RNA may have been the major informational and catalytic molecule. Eventually, double-stranded DNA supplanted RNA as the major informational molecule and proteins arose as the major catalytic molecules, giving rise to the first living organism. Carl Woese refers to this first living organism as a **progenote**, a cell hypothesized to store information in genes not yet linked together on chromosomes **(Figure 1.16)**.

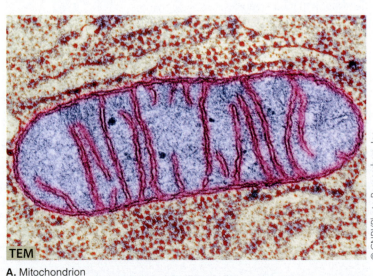

Figure 1.15. Mitochondria and chloroplasts These two distinctive eukaryal organelles probably originated via endosymbiosis. **A.** Mitochondrion, which may have originated when a developing eukaryal cell engulfed another bacterial cell capable of undergoing efficient aerobic respiration. **B.** Chloroplast, which may have originated in a similar fashion when an early eukaryal cell engulfed a cyanobacterial-like bacterium capable of photosynthesis.

A. Mitochondrion

© CNRI/Photo Researchers, Inc.

B. Chloroplast

© Biophoto Associates/Photo Researchers, Inc.

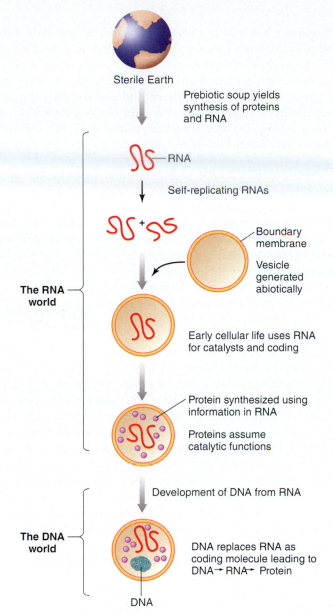

Figure 1.16. The origins of life We probably will never know for sure how life began, but we can make a reasonable hypothesis. Shortly after Earth was formed, natural processes yielded a variety of substances including RNA. Self-replicating, catalytic RNA molecules became enclosed within a boundary membrane, leading to some early version of cellular life. Eventually, proteins became the major catalytic molecules and DNA supplanted RNA as the major informational molecule.

When a progenote replicated itself, each of the progeny may have had a different subset of the parental genes, and thus a number of variations could have arisen in one generation. Genetic variation and mutation were probably frequent, given that these primitive cells would have lacked the sophisticated genetic repair mechanisms seen in modern cells. Though they might not have looked much like modern life, progenotes would have been just as subject to Darwinian evolution. Darwinian evolution, we should note, depends on (1) genetic variation in a population, (2) the environment exerting selective pressure(s), and (3) differential reproductive success among genetic variants as a result of that selective pressure. Primitive cells that not only survived, but reproduced more efficiently and passed gene sets to their progeny, would have spread most rapidly. Assuming that genetic variation was common in progenotes, the rate of evolutionary change could have been quite high.

By 500 million years ago, multicellular eukarya had begun to dominate the macroscopic landscape of life on Earth. Microbes had set the stage for them. As we will see in the next section, the evolution of oxygenic photosynthesis in cyanobacteria led to the abundance of molecular oxygen we see in the atmosphere today. Highly efficient aerobic respiration became possible. Aerobically respiring bacteria then entered into the symbioses that led to mitochondria in eukaryal cells. Without mitochondria to efficiently power cellular metabolism, large multicellular eukaryal organisms would never have appeared.

DNA and RNA: The genetic molecules

Today, in all organisms, genes are linked together in long molecules of double-stranded DNA. DNA is an elegant, if enormously long, molecule whose structure is beautifully suited for information storage and replication. Because each strand of a double-stranded DNA molecule is complementary to the other strand, one can easily imagine a mechanism of faithfully replicating the molecule. If the two strands denature, or come apart, then each strand contains the information necessary to re-form the other strand **(Figure 1.17)**. When a cell divides, identical copies of DNA can be produced so that each progeny cell will contain the exact same informational molecule.

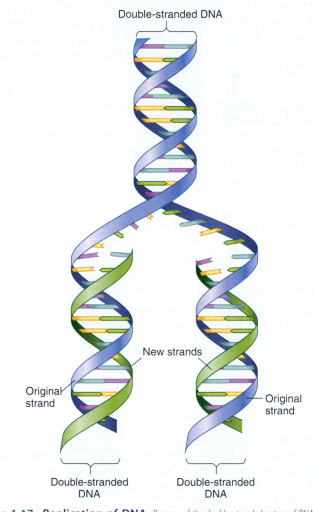

Figure 1.17. Replication of DNA Because of the double-stranded nature of DNA, and the fact that each strand within the double-stranded molecule is complementary to the other strand, a method of replication seems quite obvious. If the two strands are separated, then each strand can serve as a template for the formation of another new strand, resulting in the production of two identical double-stranded molecules.

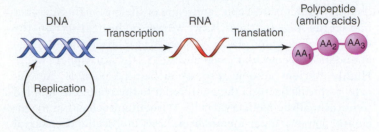

DNA —Transcription→ RNA —Translation→ Polypeptide (amino acids)

Replication

Figure 1.18. The flow of information in cells The main informational molecule, DNA, generally is replicated prior to cell division, ensuring that each daughter cell possesses the same genetic material. Information within DNA is copied into RNA during transcription. The information in RNA specifies a corresponding sequence of amino acids during translation of the RNA into a polypeptide.

The companion nucleic acid, RNA, also plays a critical role in the information flow in cells (**Figure 1.18**). Messenger RNA (mRNA) is synthesized from a DNA template during the process of transcription, and delivers instructions to the ribosome for the production of a specific chain of amino acids. Ribosomes are the protein synthesis factories of the cell. During translation (polypeptide synthesis), transfer RNAs (tRNAs) deliver amino acids to the ribosome. The ribosome itself is composed of proteins and ribosomal RNA (rRNA) molecules, which provide a structural framework. One of the rRNAs contributes catalytically to peptide bond formation. As we noted earlier, the functional versatility of RNA has suggested to scientists that it could have played a central role in the origin of living systems.

CONNECTIONS While most living organisms share a common genetic code, significant differences in some of the details of the code and various genetic processes exist between members of the domains Eukarya, Archaea, and Bacteria. We will examine translation more in **Section 7.4** and explore some of these domain and organism-specific differences.

Genetic analysis

The ability to faithfully replicate the genome and convert the information contained within the genome into functional proteins only represents part of the story. As we noted earlier, living organisms must contain some genetic variation in order for evolution to occur. The ultimate source of this variation is **mutation**, a heritable change in the genome. As we will first see

in Section 7.5, mutations can result from errors made during replication or from various physical or chemical insults to the DNA. Regardless of their source, these changes in the genotype, or genetic composition of an organism, can be passed on when the genome replicates and the cell divides (**Figure 1.19**). More importantly, these mutations may alter the genetic information in such a way that the proteins produced by the cell differ (**Figure 1.20**). These changes in the proteins, in turn, can alter the phenotype, or observable characteristics of the cell.

While bacteria do not undergo sexual reproduction by the formation and uniting of gametes, they do exchange genes between cells, a process referred to as horizontal gene transfer. As we will see in Section 9.5, genetic material can be transferred between bacteria in several ways. Moreover, more and more evidence suggests that genetic material can be transferred not only between different bacterial types, but also between organisms in different domains. This exchange of genetic material muddies somewhat our ability to construct a perfect phylogenetic tree of all living organisms (**Figure 1.21**).

@ Tree of Life ANIMATION

In recent years, improved DNA sequencing techniques have allowed researchers to determine the exact DNA sequences of entire genomes of organisms. Rather than just studying specific genes, we now can study entire genomes, a field of inquiry known as genomics. Along with these improved sequencing techniques, we also now have improved computing power. The computer tools available today allow us to analyze and compare genomes, a burgeoning field referred to as bioinformatics. With these new tools at our disposal, we are learning more about the evolutionary history of life, the diversity of organisms, and the functioning of our cells. The flood of microbial genome sequences in recent years has clearly shown that horizontal gene transfer is ubiquitous in microorganisms. In reality, most microbial genomes are composites of DNA fragments with distinct evolutionary histories. Microbial genomes, then, are quite plastic. It's probably more accurate to say that all members of a microbial type share a common pool of genes.

CONNECTIONS Genomics has revealed the profound impacts of gene transfer in regard to pathogen evolution. In **Section 21.3**, we'll explore several examples of how transfer events have led to the development of a pathogen.

Figure 1.19. Heritable changes in the DNA During DNA replication, errors may occur, such as in this example where G pairs with T, instead of with C. When this molecule of DNA undergoes a second round of replication, then one of the new molecules will contain the original CG base pair, while the other newly formed molecule will contain a mutant TA base pair at this location. The mutation, at this point, has become incorporated into this molecule of DNA and will appear in all progeny DNA.

Snustad, Simmons: *Principles of Genetics*, copyright 2009, John Wiley & Sons, Inc. This material is reproduced with permission of John Wiley & Sons, Inc.

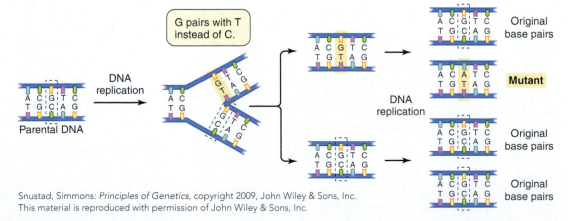

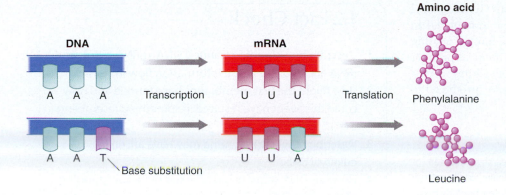

Figure 1.20. Effects of mutations Alterations in the DNA sequence can lead to changes in the mRNA and amino acids that constitute the resulting polypeptides. Here, a change in the DNA sequence results in a change in the amino acid present in the resulting polypeptide. Amino acid changes could alter the shape of the polypeptide, which, in turn, may affect the phenotype of the organism.

Biotechnology and industrial microbiology

Scientists studying genetics and molecular biology revolutionized all of biology in the late 1970s with the development of methods for producing **recombinant DNA** molecules—DNA sequences linked together to form a single molecule that never existed previously in the natural world. This revolutionary technology has allowed genes from humans and many other organisms to be inserted into bacteria and other microbes. This technique, in turn, has allowed microbes to be used as low-cost manufacturing plants for the production of valuable proteins such as human insulin and human growth hormone. The incredible range of applications for recombinant DNA methods could not have been realized without the availability of well-studied microorganisms such as *E. coli*.

The process of creating recombinant DNA is surprisingly simple. From bacteria, researchers first can isolate plasmids, small circular DNA molecules that replicate independently of the chromosome. Using restriction endonucleases, bacterial enzymes that cleave DNA at specific locations, researchers can insert foreign DNA into a plasmid, creating a new molecule.

Finally, this recombinant DNA molecule can be added back to bacteria. As the bacteria replicate, the recombinant plasmid also replicates. More importantly, foreign genes inserted into the plasmid will be expressed **(Figure 1.22)**.

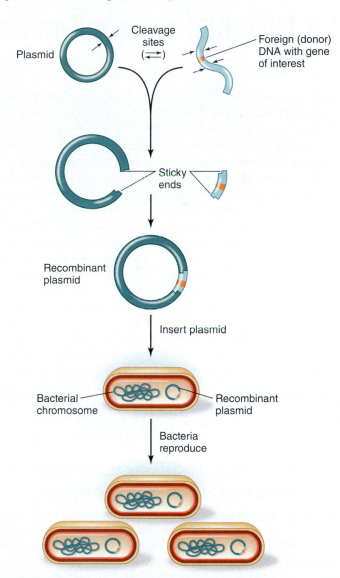

Figure 1.22. Recombinant DNA techniques Plasmid DNA and donor DNA can be cleaved with a restriction endonuclease, or restriction enzyme. When mixed together, the foreign DNA can become incorporated into the linearized plasmid, resulting in the formation of a recombinant plasmid. This plasmid can be introduced into bacteria. As the bacteria reproduce, the plasmid will be replicated, resulting in the replication of the foreign DNA. If the donor DNA contains a gene, then the bacteria can express that gene, resulting in the production of a foreign protein, coded by the donor (foreign) DNA.

Figure 1.21. Genetic exchange occurs between species and between domains Gene transfer usually takes place between members of the same species by sexual reproduction or other transfer of DNA. However, there now is evidence that gene transfer also can occur between bacteria, archaeons, and eukarya (indicated by thin lines). While this genetic exchange does not disrupt the phylogenetic tree, it does mean that individual genomes may contain segments of DNA obtained recently from distantly related organisms.

Following a related process, researchers at Genentech, a biotechnology company based in San Francisco, announced in 1978 that they had engineered *E. coli* to produce human insulin. This achievement marked the first time that a medical product was produced via recombinant DNA technology. Today, a vast majority of insulin used by people with diabetes is produced in an analogous fashion.

CONNECTIONS As we will see in **Section 12.2**, researchers now can use techniques like site-directed mutagenesis to intentionally modify natural gene products. In **Section 12.4**, we will investigate how biotechnology may help to alleviate our dependence on fossil fuels.

1.2 Fact Check

1. Describe how double-stranded DNA, mRNA, tRNA, and rRNA are involved in the information flow in cells.
2. How might the first membranes have formed and how could this event have resulted in something resembling a cell?
3. Describe how mitochondria may have arisen in eukaryal cells.
4. How has DNA sequencing affected our understanding of microbes?

1.3 Microbial metabolism and ecology

How do microbes get energy and interact with the world around them?

So far we have seen that microbes share many structural and functional features, but also exhibit a great deal of diversity. We also have seen, at least briefly, how life may have evolved and how genetic information is transmitted and interpreted within microbes. As we continue our introduction to microbiology, let's ask two more important questions. First, how do microorganisms get the energy needed to support biosynthesis and growth? Second, how do microorganisms interact with their environment? As we will see, these two questions are intricately linked.

Bacteria and archaeons have greater metabolic diversity and inhabit more diverse environments than eukarya, particularly multicellular eukarya. Environments inhabited by bacteria and archaeons range from deep sea thermal vents with temperatures of greater than 110°C, to Antarctic ice sheets, to deserts that rarely if ever see a drop of rain, to porous rocks a kilometer or more beneath Earth's surface. Microbes thrive in acidic hot springs with a pH of 1 and alkaline lakes with a pH of greater than 11. They live in distilled water taps, and saturated brine solutions in salt evaporating ponds. Their ability to live in these varied habitats reflects their ability to acquire energy from these environments. Their metabolic capabilities, in other words, dictate the habitats in which they can live.

Photosynthesis, respiration, and the appearance of atmospheric oxygen

All microorganisms, indeed, all living organisms, must acquire energy and produce macromolecules. Most macroscopic eukarya exhibit relatively limited types of metabolism. Microbes, as we will see briefly in this section and more fully in Chapter 13, exhibit more diverse types of metabolism. This metabolic diversity allows microbes to inhabit a wide range of habitats. Because different microbes can utilize various nutrients, they can exist in environments that may be uninhabitable by other organisms.

In a very basic sense, all living organisms need organic molecules. **Heterotrophs**, or "other" feeders, ingest organic molecules. These pre-formed molecules can be used for the biosynthesis of other macromolecules or as an energy source. **Autotrophs**, or "self" feeders, can produce their own organic molecules from an inorganic carbon source. We typically think of plants when we think of autotrophs. These photosynthetic organisms harvest the energy present in sunlight to convert the carbon in CO_2 into organic molecules.

In cyanobacteria, the presumed descendants of the organisms that gave rise to chloroplasts, membrane-bound pigment molecules absorb the energy present in sunlight. This absorbed energy results in the transfer of an electron from the pigment molecule to a series of membrane-bound proteins. Ultimately, the movement of this electron powers the formation of ATP, the energy currency of the cell (**Figure 1.23**). The cells can use this ATP to incorporate inorganic carbon from CO_2 into an organic molecule. To complete the process, the pigment molecule must regain an electron to replace the one that was lost. In cyanobacteria and plants, the pigment molecule regains this electron by removing electrons from water, resulting in the liberation of O_2:

$$2H_2O \longrightarrow 4H^+ + 4e^- + O_2$$

This liberation of oxygen via photosynthesis ultimately led to the increased atmospheric O_2 concentration that we discussed previously.

CONNECTIONS Not all photosynthetic organisms produce O_2 as a by-product. Some photosynthetic organisms gain electrons from hydrogen sulfide (H_2S), instead of water. For these organisms, elemental sulfur (S^0), not O_2, is liberated. This anoxygenic photosynthesis will be discussed in more detail in **Section 13.6**.

Regardless of how organisms acquire organic molecules, all living organisms also need a mechanism of oxidizing those molecules to generate ATP. One of the simplest means of acquiring

Energy and Reactants	Products

🔴 Sunlight +

$$H_2O + CO_2 + Nutrients \longrightarrow \text{“CH}_2\text{O”} + O_2$$

Water	Carbon dioxide	Nitrate (NO_3) Phosphate (PO_4^-) Iron Silica etc.	“Organic matter”	Oxygen

Figure 1.23. Overview of oxygenic photosynthesis Energy from sunlight is harvested by a membrane-bound pigment molecule, resulting in the loss of an electron. The harvested energy is used to form organic molecules from CO_2. The electron lost by the pigment molecule is replaced by the breakdown of water, resulting in the generation of O_2.

energy from organic molecules is glycolysis, the reaction in which glucose is converted to pyruvate, with the subsequent generation of two ATP molecules:

$$Glucose + 2\ ADP + 2\ P_i + 2\ NAD^+ \longrightarrow$$
$$2\ pyruvate + 2\ ATP + 2\ NADH + 2\ H^+$$

In some cases, glycolysis is coupled with fermentation, a process in which the NADH produced by glycolysis is converted back to NAD^+ and the pyruvate molecules are converted to a waste product, such as ethanol or lactate. While this system of obtaining energy from glucose is fairly simple, it is not particularly efficient; much of the potential energy present in the original glucose molecule is not converted to ATP **(Figure 1.24)**.

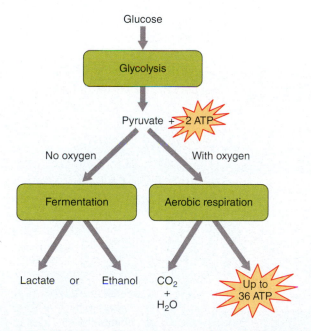

Figure 1.24. Glycolysis, fermentation, and aerobic respiration In almost all organisms, glucose can be converted to pyruvate, resulting in the production of some ATP. The process of aerobic respiration allows cells to generate much more ATP from pyruvate. In plants and animals, respiration only can occur in the presence of oxygen (aerobic respiration) and results in the generation of CO_2 and H_2O. If aerobic respiration cannot occur, some organisms undergo fermentation reactions. Two well-studied types of fermentation result in the production of lactate or ethanol. These products of fermentation generally are toxic to the microbe, but may be helpful to humans.

A more effective means of obtaining ATP from glucose involves respiration, a set of metabolic processes whereby the glucose is more completely utilized, resulting in the production of larger quantities of ATP. Plants, animals, and many bacteria undergo aerobic respiration, a form of respiration that involves the addition of electrons to O_2, resulting in the formation of H_2O. Some bacteria perform respiration in which a terminal electron acceptor other than oxygen is used. This type of respiration generally results in the formation of less ATP than respiration involving oxygen.

The presence of O_2 in the atmosphere had profound consequences for life on Earth. Respiration utilizing oxygen allows organisms to harvest a significant amount of energy from organic molecules. During this process, though, oxygen is converted to a series of toxic by-products. These by-products can damage cells through their ability to oxidize other molecules and their ability to generate even more potent toxins through interactions with light. Until oxygenic photosynthesis evolved, *all* life, by necessity, was anaerobic. When O_2 began accumulating in the atmosphere, organisms either had to develop strategies for defending themselves against these dangerous by-products, or retreat to niches where O_2 remained less abundant or absent. This remains true today. Anaerobic microbes are still with us, living in specialized habitats free from molecular oxygen.

CONNECTIONS During aerobic respiration, O_2 can be converted to hydrogen peroxide (H_2O_2). This reactive molecule can damage cells. As a result, most organisms that can survive in the presence of O_2 produce catalase, an enzyme that converts H_2O_2 to H_2O and O_2. We will learn more about toxic oxygen species in **Section 6.3**.

The increase of O_2 in the atmosphere also created an ozone layer in the stratosphere, approximately 10–25 miles above Earth's surface. Ozone (O_3) is generated naturally from O_2 and strongly absorbs short wavelength ultraviolet (UV) light. UV light is dangerous to cells because it causes damaging chemical reactions in DNA. Because UV light does not penetrate water very well, aquatic life is fairly well protected from its harmful effects. Terrestrial organisms, however, are not so lucky. The accumulation of ozone in the atmosphere reduced the amount of UV radiation reaching Earth's surface. The rise of O_2 in the atmosphere, therefore, indirectly facilitated the colonization of land by microorganisms.

Microorganisms and biogeochemical cycling

Microbes have been intimately involved in modulating conditions within the **biosphere**, those regions of Earth that can support life. Not only did microbes, through photosynthesis, create the oxygen-rich atmosphere on which most life on Earth relies, but they are also involved in **biogeochemical cycling**, the transitioning of various chemicals between organic and inorganic forms. The amount of carbon contained within living bacteria on Earth, for instance, is estimated to be nearly as great as the amount of carbon in all the multicellular organisms combined. Photosynthetic cyanobacteria convert CO_2 from the atmosphere into organic molecules. Microbial metabolism ultimately converts much of this organic carbon back into CO_2.

As we saw in Section 1.1, nucleic acids and polypeptides, two important categories of cellular macromolecules, contain nitrogen. Only certain types of bacteria can convert nitrogen gas (N_2) from the atmosphere into forms that can be readily used by other organisms to form these molecules. This nitrogen fixation is accomplished both by free-living bacteria, and by bacteria living in symbiotic associations with plants **(Figure 1.25)**. Bacteria also convert nitrogen present in organic material back to N_2 gas, through both denitrification and ammonia oxidation. Denitrification is primarily a terrestrial process that is of special concern to farmers, as it decreases the fertility of agricultural soils. Ammonia oxidation is largely a marine phenomenon, which limits ocean productivity.

Microbial interactions

Before we leave this section, we should note that microbes do not exist in isolation. As we will investigate more thoroughly in Chapter 15, microbes exist in diverse communities of organisms that interact with each other and the environment. As we will see in Chapter 3, vast numbers of eukaryal microbes exist within the Rio Tinto, a river in Spain with a pH of approximately 2. As we will see in Chapter 4, communities of organisms live in the waters surrounding deep sea thermal vents, dependent, in large part, on archaeons that thrive in the superheated water near these thermal vents. As we will see in Chapter 17, many microbes live with, in, or on various plants, invertebrates, and vertebrates.

Microbes, of course, also live in and on humans. Van Leeuwenhoek, remember, described microorganisms living on a person's teeth over 300 years ago. In fact, we can view the human body as a complex ecosystem. Microbes live throughout our bodies, often in a symbiotic relationship with us. Microbes within our body produce vitamins for us and help us digest our food. Some microbes, as we will discuss in Chapter 23, even help us control the replication of other, unwanted microbes, thereby aiding in the prevention of disease.

1.3 Fact Check

1. Differentiate between the terms heterotroph and autotroph.
2. Describe the roles of glycolysis, fermentation, and respiration in energy production in different environments.
3. Explain how bacteria are involved in nitrogen cycling.

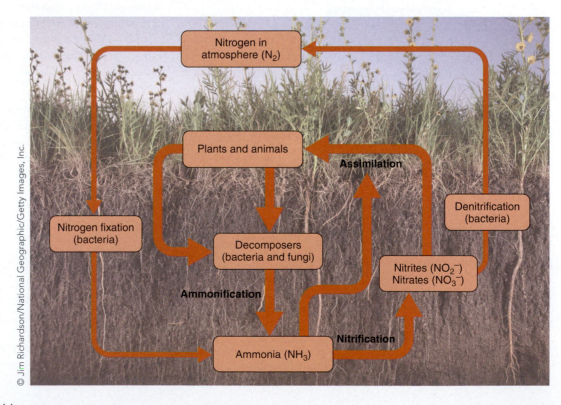

Figure 1.25. Role of microbes in the global nitrogen cycle Only certain bacteria can convert gaseous nitrogen (N_2) into forms that can be utilized by other living organisms. Nitrogen incorporated into ammonia (NH_3), nitrites (NO_2^-), and nitrates (NO_3^-) can be used to synthesize amino acids. These molecules also can be used in the synthesis of nucleic acids. Other microbes serve as decomposers, converting various forms of nitrogen back into N_2, thereby completing the nitrogen cycle.

© Jim Richardson/National Geographic/Getty Images, Inc.

1.4 Microbes and disease

How are microbes associated with disease?

Today, we know that microbes can cause diseases. Throughout most of human history, however, people thought that diseases had various causes from angry gods to bad air. These views prevailed mostly because people could not see the bacteria associated with bacterial infections. Even after the development of the microscope, people still did not understand that microbes could be transmitted from person to person. Rather, many assumed that microbes arose from inanimate materials, a process known as spontaneous generation. This belief negated a need to even consider the transmission or prevention of microbial diseases. If microbes arose spontaneously, then there was no reason to investigate how a person became infected.

While many scientists worked to disprove the widely held belief in spontaneous generation, Louis Pasteur's experiments in the mid-1800s provided the most compelling evidence refuting this idea. As with so many classic experiments, his experiment was both simple and elegant. Pasteur added nutrient broths to swan-necked flasks and then boiled the broths to kill any contaminating microorganisms. He then observed the broths for signs of microbial growth **(Figure 1.26)**.

With this approach, Pasteur reasoned, outside air could enter the flask. Bacteria present in the outside air, though, would become trapped in the neck of the flask, never coming in contact with the sterile broth. No bacteria, he hypothesized, would grow. If the flask were tilted such that the broth traveled to the neck of the flask, however, then bacteria trapped there would gain access to the nutrient broth and growth would occur. Microbial life in the broth, in other words, would only result from microbial life present in the neck of the flask. It would not arise spontaneously.

> **CONNECTIONS** What signs of microbial growth did Pasteur observe? In **Section 6.2**, we will examine how bacteria grow under standard laboratory conditions. In **Section 6.5**, we will explore how we can block this growth.

Microbes in Focus 1.1
BACILLUS ANTHRACIS

Habitat: Infects various mammals, including cattle, sheep, and horses. Also can infect humans. Spores can be found in soil.

Description: Gram-positive rod-shaped bacterium, measuring approximately 1 μm in width and 3 μm in length.

Key Features: When nutrients are lacking, *B. anthracis* can form endospores, metabolically inert structures that are largely resistant to harsh environmental conditions. The endospores can begin replicating again when conditions improve, even after years in the endospore state. While naturally occurring human infections with *B. anthracis* are rare, this bacterium still remains actively studied by researchers, mainly because of its potential use as a bioweapon. We will explore this topic more in Chapter 6.

LM

Courtesy Center for Disease Control

The identification of infectious agents

About 200 years after van Leeuwenhoek's first observations of microbes, and just 15 years after Pasteur showed that microorganisms do not arise by spontaneous generation, Robert Koch provided the first clear proof that a bacterium, *Bacillus anthracis* **(Microbes in Focus 1.1)**, was the cause of a specific disease, anthrax, in livestock.

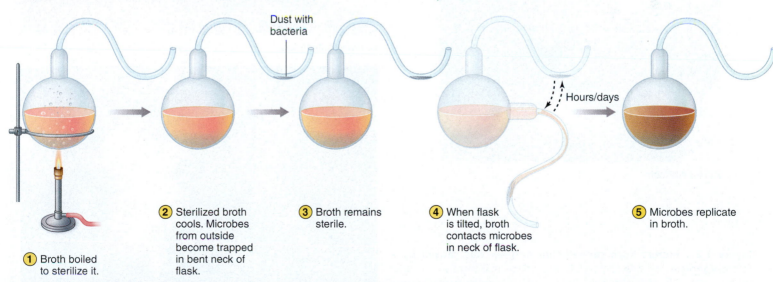

Dust with bacteria

Hours/days

② Sterilized broth cools. Microbes from outside become trapped in bent neck of flask.

③ Broth remains sterile.

④ When flask is tilted, broth contacts microbes in neck of flask.

⑤ Microbes replicate in broth.

① Broth boiled to sterilize it.

Figure 1.26. Process Diagram: Pasteur refutes spontaneous generation By showing that bacterial growth did not occur in sterile broth within swan-necked flasks, but did occur when the sterile broth contacted outside contaminants, Pasteur demonstrated that microorganisms did not arise spontaneously.

Robert Koch was a German physician and scientist, and a true pioneer in the field of microbiology. Aside from his anthrax work **(Figure 1.27)**, he was responsible not only for developing important laboratory techniques—for example, the use of media solidified with agar for the isolation of bacteria in the laboratory—but also for developing clear criteria linking particular microorganisms to specific diseases. We will explore the details of these criteria, referred to as Koch's postulates, in Section 18.4.

Koch used these criteria to establish the cause of other important diseases, including tuberculosis, which is caused by the bacterium *Mycobacterium tuberculosis*. The identification of other infectious microbes quickly followed. The late 1800s, in fact, were a golden age of sorts, in which medical science finally began to achieve real insight into many infectious diseases. Many disease agents, though, continued to be elusive, and even Koch realized his criteria had limitations. For example, he knew that many microorganisms suspected to cause disease, such as the agent of cholera, *Vibrio cholerae*, could be isolated from both sick and healthy people, invalidating his first postulate.

CONNECTIONS As our understanding of infectious diseases, and our experimental techniques, have changed, so too have our interpretations of Koch's postulates. Molecular correlates of his basic postulates are explored in **Section 18.4**.

The effects of infectious diseases

As van Leeuwenhoek observed, many microbes exist in and on our bodies. Most of the time, these microbial partners are just that—partners. When the relationship between microbes and host becomes unbalanced, then disease can arise. Microbial diseases, as we all know, have profoundly affected human life. As we will see in the final section of this book, our understanding of how microbes cause disease—and how we can prevent these diseases or alleviate their effects—is constantly changing.

We all feel the effects of microbial infections through the course of our lives. A runny nose and a cough often are signs of "a cold," which may be due to a rhinovirus infection of the upper respiratory tract. The onset of winter portends the flu

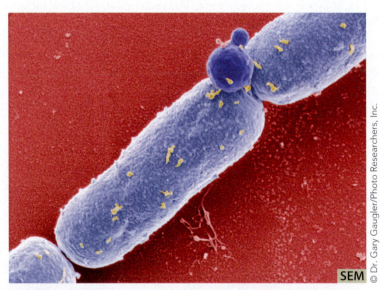

B. *Bacillus anthracis*

© Dr. Gary Gaugler/Photo Researchers, Inc.

© Mary Evans Picture Library/Alamy

A. Robert Koch

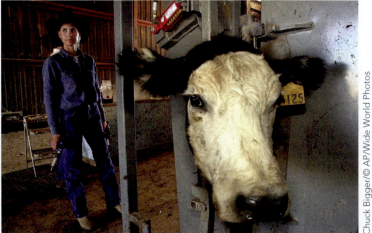

C. Vaccination

Chuck Bigger/© AP/Wide World Photos

Figure 1.27. Robert Koch proved that anthrax was caused by a microorganism Koch showed that a specific microbe, the bacterium *Bacillus anthracis*, caused anthrax in livestock. Koch also developed Koch's postulates, a series of criteria that needed to be met to prove that a specific microorganism caused a specific disease. **A.** Robert Koch. **B.** *B. anthracis* (color enhanced). **C.** Cow about to receive anthrax vaccination.

season, when hundreds of millions of people each year suffer from fevers, headaches, coughs, and muscle aches caused by the influenza virus. The average person in the United States experiences several bouts of diarrhea per year, which could have any number of different causes, including viruses, bacteria, or protozoa, all of which can be acquired through food or drink. If you are studying this book in the United States, Canada, or Europe, it's unlikely (but not unheard of) that you will have to deal with the constant threat of infections with serious, or even life-threatening consequences. If you live in one of the many regions of the world where affordable effective public health measures are lacking, infectious diseases may have a more prominent role in your life. AIDS, malaria, tuberculosis, and other diseases of microbial origin take millions of lives per year in these regions; we are far from a solution to this enormous problem.

History is full of examples of the powerful impact of infectious diseases. Modern medicine has greatly improved our ability to deal with these diseases. We now understand how infectious diseases spread, and the widespread availability of vaccines and antimicrobial drugs facilitates prevention and treatment. But

we have not eliminated the threat of new infectious diseases. The human immunodeficiency virus (HIV), which was only discovered in the 1980s, has caused over 40 million deaths since then, and there is still no vaccine to prevent HIV infection. The influenza virus has caused devastating pandemics, or worldwide outbreaks, in centuries past, and now there is great concern that the avian influenza serotype H5N1 or the recently identified H1N1 strain will mutate to become more deadly, with potentially catastrophic consequences.

CONNECTIONS As we saw during the summer and fall of 2009, the H1N1 strain of influenza virus quickly spread throughout the world. What factors led to the rapid dissemination of this microbe? We will discuss epidemiology, or the study of how diseases spread in populations, in **Section 18.3**.

An exceptionally devastating pandemic occurred when plague, popularly referred to as "The Black Death," killed a third or more of the population of Europe, Asia, and North Africa in a 60-year span from 1340 to 1400 (**Figure 1.28**). Plague

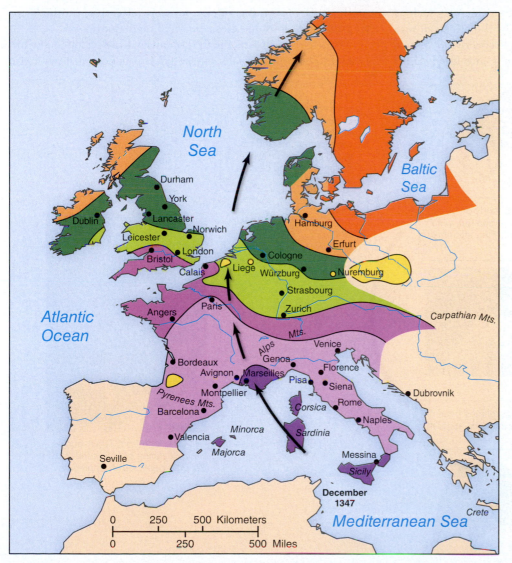

Figure 1.28. The Great Plague Also referred to as The Black Death, the plague spread rapidly throughout Europe in the 1300s. In a 60-year span, roughly a third of the European population died from this infectious disease. The dates on the map show how rapidly the disease moved through Europe in an era without the modern means of rapid mass transport. Poor sanitation and a lack of knowledge about the causes of infectious diseases contributed to its spread.

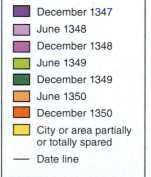

	December 1347
	June 1348
	December 1348
	June 1349
	December 1349
	June 1350
	December 1350
	City or area partially or totally spared
—	Date line

is caused by the bacterium *Yersinia pestis*, which infects rodents and humans, and is transmitted between them by fleas (**Figure 1.29**). This bacterium is commonly present in rodents such as mice, rats, and squirrels in many parts of the world. The plague pandemic of the 1300s is thought to have originated in central Asia or China, and probably moved westward with trading caravans and/or Mongol armies. Historical records indicate that plague was absolutely devastating. Up to 60 million people, perhaps half of its population, may have died from plague in China in the 1300s. In Europe, 25–50 million people perished, largely between 1347 and 1351. Several million more people were killed by plague in the Middle East and northern Africa.

To put the global death toll due to the plague pandemic of the 1300s into perspective, imagine a modern pandemic that resulted in the death of *over 1 billion people* within a few years! Even the modern AIDS and influenza pandemics pale by comparison. In the Middle Ages, people did not understand the underlying cause of this horrifying disease. Supernatural explanations of disease were universally accepted; diseases were punishments from God or the results of curses, witchcraft, or "bad air." The true cause of plague, in fact, was not discovered until 1893, when Alexandre Yersin isolated the bacterium *Yersinia pestis* during a plague outbreak in Hong Kong. Today, plague is rare and easily treatable with antibiotics.

The massive social disruptions caused by epidemics such as plague in the Middle Ages are rivaled only by the effects of war, famine, and natural disasters. From the perspective of public health, we should note, these events are not unrelated. Until very recently, most people who died during wartime, both military and civilian, were victims of microbes. Before the advent of antibiotics during World War II, battlefield wounds were highly likely to become infected, the end result of which was often death. Troops in the field for prolonged periods often were affected by disease. Cholera, typhus, and dysentery were common, due to the poor sanitation conditions. Diseases spread by carriers such as lice and fleas, which thrive in crowded conditions, also are a great risk to soldiers and displaced civilians (refugees) during wartime. Malnutrition resulting from war or famine (or both) blunts the effectiveness of the immune system, leaving soldiers and civilians more susceptible to disease.

Control of infectious diseases

As researchers learned more about the causes of infectious diseases in the late 1800s and early 1900s, the medical treatment of these diseases gained a scientific basis. Even before Koch linked specific microbes to specific diseases, the British physician Joseph Lister had discovered the value of cleanliness and disinfection measures in reducing mortality from post-surgical and post-childbirth infections. Doctors' offices and hospitals strove for the universal application of such disinfection measures to decrease the incidence of disease. While these methods of preventing infections were effective, methods for treating infectious diseases remained inadequate.

While Salvarsan was the first scientifically developed antimicrobial to be commercially marketed, it was not until penicillin

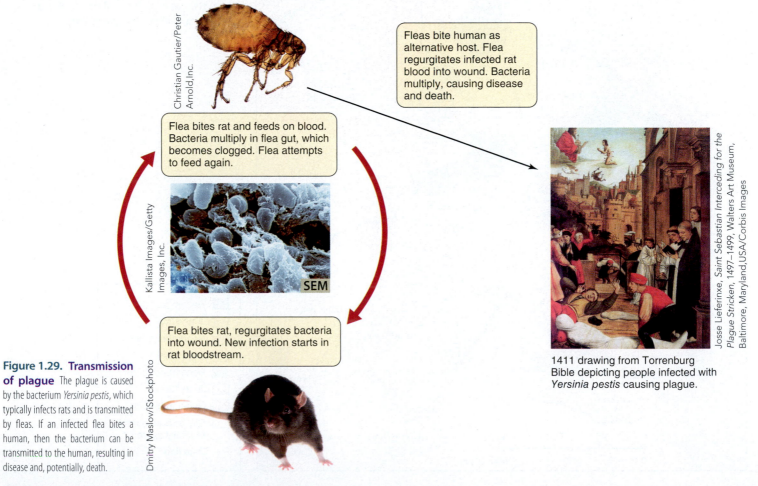

Christian Gautier/Peter Arnold,Inc.

Fleas bite human as alternative host. Flea regurgitates infected rat blood into wound. Bacteria multiply, causing disease and death.

Flea bites rat and feeds on blood. Bacteria multiply in flea gut, which becomes clogged. Flea attempts to feed again.

Kallista Images/Getty Images, Inc.

SEM

Flea bites rat, regurgitates bacteria into wound. New infection starts in rat bloodstream.

Figure 1.29. Transmission of plague The plague is caused by the bacterium *Yersinia pestis*, which typically infects rats and is transmitted by fleas. If an infected flea bites a human, then the bacterium can be transmitted to the human, resulting in disease and, potentially, death.

Dmitry Maslov/iStockphoto

Josse Lieferinxe, Saint Sebastian Interceding for the Plague Stricken, 1497–1499, Walters Art Museum, Baltimore, Maryland,USA/Corbis Images

1411 drawing from Torrenburg Bible depicting people infected with *Yersinia pestis* causing plague.

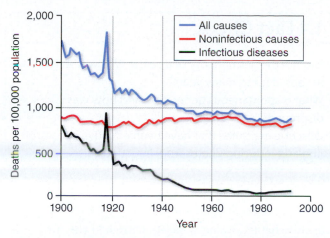

Figure 1.30. **Infectious disease deaths in the United States during the twentieth century** Deaths associated with infectious diseases decreased dramatically in the United States over the past century. The spike in deaths due to infectious diseases around 1920 represents the increased deaths caused by the 1918 influenza pandemic.

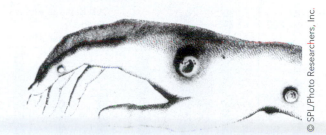

A. Skin blisters caused by smallpox infection often left permanent scars.

and sulfa drugs came into use in the 1940s that antimicrobial drugs had a major impact on the treatment of infectious diseases. Indeed, people in developed countries saw a dramatic decrease in deaths due to infectious disease during the twentieth century, because of improved sanitation and the development of antimicrobial drugs and vaccines (**Figure 1.30**). A new problem, however, now faces us. Increasingly, infections result from antibiotic-resistant bacteria. Our existing antibiotics are not effective against these microorganisms, and our ability to treat some infectious diseases is declining.

CONNECTIONS Recall that bacteria are masters at sharing genes through horizontal gene transfer. This facilitates the passing of genes for antimicrobial resistance. History has shown that the more effective an antimicrobial drug is, the more it is prescribed, and the faster resistance spreads, rendering the drug ineffective. See **Section 24.3** for more on the acquisition of drug resistance.

Vaccines also have had an enormous impact in reducing the sickness and death associated with infectious diseases. Vaccination involves exposing a person to an inactivated or weakened version of a microbe, or even just a part of the microbe, to create immunity to a disease. The past century has seen vaccines developed for many deadly diseases, including polio, diphtheria, rabies, and many others. Historically, vaccination began roughly 2,000 years ago in China and India as a defense against smallpox. Smallpox, a viral disease, was common in Europe and Asia, in some areas being responsible for 20 percent of all deaths. It was fatal to 1 of every 4 people infected. Those who survived carried characteristic smallpox scars for the rest of their lives, but they also carried a lifelong protection from the disease.

Vaccination against smallpox was occasionally practiced in eighteenth century Europe, but it was the famous experiment of English physician Edward Jenner in 1796 that popularized the procedure. Jenner used material from a cowpox infection of a milkmaid to inoculate a boy, who was later shown to be immune to smallpox (**Figure 1.31**). We now know

B. Painting depicting Jenner inoculating a boy against smallpox

C. The virus that causes smallpox

Figure 1.31. **Edward Jenner and the development of vaccination** While the Chinese practiced forms of vaccination against smallpox for hundreds of years, vaccination was not practiced in Europe until the end of the eighteenth century. In 1796, Edward Jenner showed that he could inoculate a boy with material obtained from the pox marks on a woman infected with cowpox and provide the boy with immunity to smallpox. Global vaccinations have eliminated smallpox. **A.** 1798 illustration from Jenner of cowpox lesions on a milkmaid's hand. **B.** Painting of Jenner vaccinating a person against smallpox. **C.** Vaccinia virus, the causative agent of smallpox.

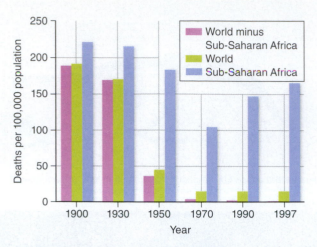

Figure 1.32. Impact of malaria in sub-Saharan Africa Worldwide, deaths associated with malaria, an infectious disease caused by the protozoan *Plasmodium falciparum*, have decreased dramatically. This decrease, however, has not been evident in sub-Saharan Africa, where malaria remains a major problem. Today, infectious diseases disproportionately affect people in developing countries and people without clean water or access to adequate health care.

that cowpox virus is closely related to the smallpox virus, explaining why exposure to it generated immunity to smallpox. Thanks to the smallpox vaccine, in fact, no naturally occurring cases of smallpox have occurred since 1977. In one of the greatest achievements in medical history, we have eliminated this scourge from the face of the earth.

People in developing regions of the world, particularly in sub-Saharan Africa, continue to suffer from infectious diseases, particularly AIDS, tuberculosis, and malaria, which combined take nearly 5 million lives per year **(Figure 1.32)**. Vaccines are not yet available that can prevent HIV or malaria infection, and the current tuberculosis vaccine is far from ideal. Drugs to treat these three diseases are available, but they are expensive, which means that many people do not have adequate access to these therapies. Additionally, it is not uncommon to find people in developing countries who are simultaneously affected by at least two of these diseases, lowering their life expectancy even further.

Other major historical factors that have reduced death from infectious diseases include improvements in personal hygiene, public sanitation, and food and water safety. Indoor plumbing, water treatment measures, and large-scale sewage disposal systems have led to a decrease in water-borne infectious diseases. **Pasteurization**, the process in which milk is heated briefly to kill most microorganisms, freezers, and refrigerators all have contributed to the increased safety of food. Many of us take these advances for granted. Unfortunately, these simple measures are not universally available. As a result, we now are faced with what Dr. Paul Farmer refers to as the "great epi (epidemiology) divide." People in developing countries and people in developed countries without access to adequate health care suffer a disproportionate infectious disease burden.

1.4 Fact Check

1. Discuss the germ theory of disease. Why was it not accepted for so many years?
2. What is spontaneous generation and how was it disproved?
3. Describe measures that have been effective in controlling and reducing deaths from infectious diseases.

In this chapter, we have barely scratched the surface of microbiology. We have been introduced to the microbes. We have learned about the evolution of life and the genetics of microbes. We have begun to explore how microorganisms acquire energy. Finally, we have seen how microbes can cause diseases. In later chapters, we will further explore these topics. We will see how important microbes are in our world. As we investigate these topics, we will see that extensive experimental evidence supports our current understanding of the microbial world. We also will see that many questions and ambiguities still remain.

Throughout this book, we will see that microbiology is a dynamic science. Microbiologists continue to make new discoveries using an expanding selection of laboratory tools. They also continue to refine, modify, and, occasionally, reject existing hypotheses. As we saw in the Mini-Paper, inquisitive minds, continuous experimentation, and the development of new tools can combine to result in completely new ways of seeing the world around us. Such dramatic advances have occurred throughout the history of microbiology and certainly will continue to occur **(Table 1.4)**.

TABLE 1.4	**Selected advances in microbiology**	
Year	**Scientist**	**Advance**
Late 1600s	Anton van Leeuwenhoek	Uses microscope to see microorganisms
1860s	Louis Pasteur	Disproves idea of spontaneous generation
1860s	Joseph Lister	Practices infection control
1876	Robert Koch	Identifies *Bacillus anthracis* as cause of anthrax
1928	Alexander Fleming	Discovers penicillin
1950s	Jonas Salk and Albert Sabin	Develop poliovirus vaccines
1966	Lynn Margulis	Proposes endosymbiotic theory
1983	Kary Mullis	Invents PCR
1990	Carl Woese	Proposes three-domain classification of living organisms
1995	Craig Venter	Publishes first complete bacterial genome sequence

During our exploration, we will make connections between topics. We will see how intimately related microbes are to each other, to their environment, and to us. We will see how the genetics, physiology, and habitat of microbes are intertwined. We also will see how, as the evolutionary biologist Theodosius Dobzhansky wrote in 1973, "Nothing in biology makes sense except in the light of evolution." Most importantly, we will see that the microbial world presents as many surprises to us today as it did to van Leeuwenhoek over 300 years ago.

human microbiota). As part of this project, researchers amplified and sequenced bacterial small subunit ribosomal RNA (SSU rRNA) genes, as described in Section 1.1, from various body sites on nearly 250 people. The initial results, published in June of 2012, were astounding. Every one of us is home to hundreds of different bacterial species and the inhabitants of one person often differ from the inhabitants of another person. This "non-cultivation-based" approach ultimately will facilitate the genetic sequencing and more complete understanding of the microbes that have a major influence on human health and disease.

The Rest of the Story

Van Leeuwenhoek noted that the number of microbes on a person's teeth may "exceed the number of Men in a kingdom." It most certainly does. Today, many researchers estimate that the human body consists of 10^{13} human cells. On us and in us, though, there may be 10^{14} microbial cells. As Dr. David Relman from Stanford University noted, "we are 10 parts microbe, and one part human." The study of these microbes has moved well beyond the use of the microscope. In this chapter, we reviewed how the microbes present today have evolved to occupy a plethora of environments and niches using a vast array of metabolic processes, even within the human body. Growth conditions for many of these microbes are well understood, and they have been isolated and grown in the laboratory. However, a large subset still remains a mystery because their specific growth microenvironments have not been determined, or cannot be reproduced experimentally.

In 2007, the Human Microbiome Project was launched with an aim to identify, analyze, and catalog the hundreds of microbial species residing in or on the human body (the

Image in Action

WileyPLUS Anton van Leeuwenhoek is examining a sample of pond water using the microscope he built by hand. Today, microbiologists also can analyze DNA sequences (represented in the upper left) to study environmental samples.

1. What types of microbes (and from what domains of life) did Anton van Leeuwenhoek most likely observe? What microbes was van Leeuwenhoek unable to observe with his simple microscope? Explain.

2. Imagine that you were given access to the same drop of pond water observed by van Leeuwenhoek. How could you use PCR to identify or characterize some of the microbes he observed or even those he didn't observe?

Summary

Section 1.1: What is microbiology?
Microbiology is the study of **microorganisms**, including bacteria, archaeons, and eukaryal microbes, and viruses. Collectively, we can refer to all of them as **microbes**.

- All living organisms share certain features, including **metabolism**, **growth**, **reproduction**, genetic variation resulting in **evolution**, response to outside stimuli, and internal **homeostasis**.
- A **cell** is the simplest structure capable of carrying out all the processes of life.
- All cells contain various **macromolecules**, including **polypeptides**, **nucleic acids**, **lipids**, and **polysaccharides**. Many polypeptides function as **enzymes**.

  Historically, all living organisms were classified as **eukaryotes** or **prokaryotes**, depending on whether they did, or did not, have a nucleus. Today, the taxonomy of living organisms consists of three domains, Bacteria, Archaea, and Eukarya.

- This classification scheme reflects the **phylogeny** of all living organisms.

- Our ability to classify microorganisms has been aided greatly by the **polymerase chain reaction (PCR)**.
- Viruses are sub-cellular and can be classified as microbes.

  Because of their relatively simple structures, microbes have been useful research models.

Section 1.2: What do we know about the evolution of life and the genetics of microbes?
All living organisms possess remarkably similar informational molecules and processes for converting this genetic information into functional molecules. This conservation of genetic processes provides compelling evidence that all living organisms are evolutionarily related.

- Life probably evolved on Earth around 3.8 billion years ago. Simple organic molecules, possibly associated with iron-containing surfaces, became enclosed within a lipid membrane.
- The identification of **ribozymes** lends support to the idea that the precursors of life, the so-called **progenote**, may have used RNA as the major informational molecule.

- Eukarya arose through endosymbiosis, in which free-living bacteria became engulfed within a developing eukaryal cell, providing the host with the ability to harvest sunlight or undergo aerobic respiration.

- In all living organisms, the main informational molecule is double-stranded DNA, a molecule ideally suited for containing information and for faithful replication. This DNA-based information is converted into functional molecules through the processes of transcription and translation.

- **Mutation** in the DNA allows for genetic variation, a prerequisite for evolution.

Our understanding of the basic genetic processes has led to the development of **recombinant DNA** technologies. We now can alter DNA molecules and produce human proteins in microbial hosts.

- Recent advances in DNA sequencing technologies and increased computing power have led to the emerging fields of genomics and bioinformatics.

Section 1.3: How do microbes get energy and interact with the world around them?

All living organisms must obtain organic molecules. **Heterotrophs** ingest them. **Autotrophs** produce their own organic molecules.

- Most autotrophs generate organic molecules through photosynthesis.

- Organic molecules generally are converted into ATP through the processes of glycolysis, fermentation, and respiration.

- Microbial metabolism has affected the **biosphere**, and microbes are intimately involved in the **biogeochemical cycling** of many chemicals, including nitrogen, phosphorus, and sulfur, in the biosphere.

- Microbes interact with each other and other organisms in many complex ways.

Section 1.4: How are microbes associated with disease?

The work of a number of microbiologists, including Louis Pasteur, led to the development and acceptance of the germ theory of disease.

- Koch's postulates, developed in the 1800s, provide a means of demonstrating that a particular microorganism causes a particular disease.

- Infectious diseases have had, and continue to have, a profound impact on humans.

- Today, we have an assortment of antibiotics, antivirals, and vaccines that treat or prevent many infectious diseases. The development of these therapies has depended, in large part, on our understanding of the structure of these microbes and their replication strategies. Other techniques, like **pasteurization**, also have led to a decrease in the incidence of certain infectious diseases.

- Unfortunately, infectious diseases remain horrific threats to people throughout the world.

● Application Questions ··

1. A researcher is studying a newly discovered microbe and must first determine whether this microbe should be considered alive. What characteristics will she examine to make this determination?

2. Given what we learned about proteins in Section 1.1, what might be the effects of an antimicrobial drug that halts protein synthesis in cells?

3. Imagine a researcher is examining several cell samples in the laboratory and needs to categorize each sample as either bacterial or eukaryal. What features should the researcher examine and why?

4. Inspired by the Miller–Urey experiment described in Perspective 1.1, a researcher continues to explore the origins of life on Earth. The first task is to design a simulation of prebiotic Earth. What conditions should be considered when creating this simulation?

5. According to the endosymbiotic theory, mitochondria and chloroplasts are derived from bacterial cells. What evidence exists to support this theory?

6. If a species of microbe were discovered that did not appear to mutate, scientists probably would hypothesize that its lack of mutation would be detrimental to its evolution. Explain why the scientists would make this prediction.

7. In Section 1.3, oxygenic photosynthesis and aerobic respiration were described. Based on our discussion of these two processes, explain why oxygenic photosynthesis must have evolved on Earth before aerobic respiration.

8. Oxygen is not always a "good thing." In fact, O_2 can be considered dangerous. Explain why oxygen can be harmful and what microbes must do to deal with the dangers of an oxygenated world.

9. Section 1.4 discusses the great epi (epidemiology) divide between developing and developed countries. List several policy and funding changes that could be implemented in developing countries to decrease the number of infectious disease cases. Explain each of these recommendations.

10. A research group has received funding to study the diversity and evolutionary relatedness of microbes in a marine ecosystem. The head of the laboratory has decided to sequence and analyze the SSU rRNA genes of the microbes in this ecosystem.
   a. What are SSU rRNA genes?
   b. Why are these gene sequences ideal for studying the evolutionary relationships of the microbes found there?
   c. How is this process conducted?
   d. After sequencing and analyzing three gene samples, the researchers conclude that two of the samples are from closely related microbes, while the third sample is very distantly related. Describe the sequencing results that would lead to this conclusion.

11. In Section 1.1, we discussed Paul Ehrlich's foundational research in which he noted differences between bacterial and human cells. Those observed differences became the basis for the development of antimicrobial drugs that "spare" eukaryal human cells. Based on these observed differences between bacterial and eukaryal cells:
   a. Identify an aspect of bacteria that would NOT be a good target for a drug because it would not "spare" the host. Explain why this bacterial component is not a good choice for an antimicrobial drug.
   b. Identify an aspect of bacteria that might be worth further exploration as a potential target for an antimicrobial drug that would not harm human cells. Explain.

Suggested Reading

Fleischmann, R. D., M. D. Adams, O. White, R. A. Clayton, E. F. Kirkness, A. R. Kerlavage, C. J. Bult, J. F. Tomb, B. A. Dougherty, J. M. Merrick, et al. 1995. Whole-genome random sequencing and assembly of *Haemophilus influenzae* Rd. Science 269:496–512.

Kruger, K., P. J. Grabowski, A. J. Zaug, J. Sands, D. E. Gottschling, and T. R. Cech. 1982. Self-splicing RNA: Autoexcision and autocyclization of the ribosomal RNA intervening sequence of *Tetrahymena*. Cell 31:147–157.

Miller, S. L. 1953. A production of amino acids under possible primitive Earth conditions. Science 117:528–529.

Miller, S. L., and H. C. Urey. 1959. Organic compound synthesis on the primitive earth. Science 130:245–251.

Turnbaugh, P. J., R. E. Ley, M. Hamady, C. M. Fraser-Liggett, R. Knight, and J. I. Gordon. 2007. The human microbiome project. Nature 449:804–810.

Woese, C. R., O. Kandler, and M. L. Wheelis. 1990. Towards a natural system of organisms: Proposal for the domains Archaea, Bacteria, and Eucarya. Proc Natl Acad Sci USA 87:4576–4579.

Bacteria

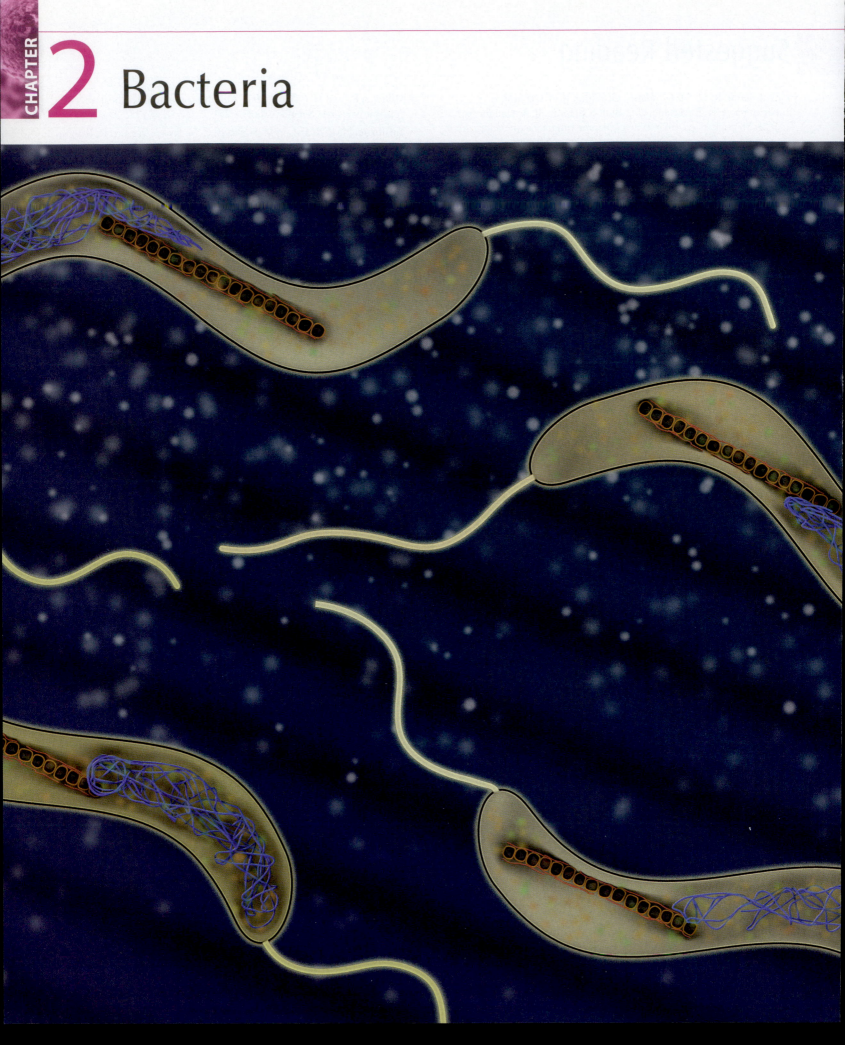

In 1974, Richard Blakemore, a graduate student at the University of Massachusetts–Amherst, was examining sediment samples with his microscope. He noticed something curious; within this muddy microbial menagerie, some bacteria consistently accumulated on one side of the slide. More interestingly, a magnet placed near the microscope stage altered the movement of these bacteria; placing one pole of the magnet directly next to the stage would cause them to swim toward or away from it, depending on which pole of the magnet was closest to the stage (Blakemore's bacteria turned out to be north-seeking). It was not simply magnetic attraction pushing or pulling the bacteria (dead bacteria aligned along their long axis in a magnetic field, like tiny compass needles, but otherwise were not active). The live bacteria used flagella—tiny propeller structures common in bacteria—to power their movement, and somehow used a magnetic field for steering. Blakemore called this phenomenon "magnetotaxis."

Magnetotaxis provided a simple way for Blakemore to capture these bacteria. He put mud in a container with a magnet next to one side, allowed the magnetotactic bacteria to accumulate, collected the cells, and then repeated this process until he had a nearly pure population of magnetic bacteria. With this basic strategy, plus some trial and error to figure out how to grow the bacteria, he eventually isolated pure cultures of these magnetotactic bacteria. Closer analysis of the cells found them to be rich in iron, due to chains of membrane-enclosed magnetite particles (dubbed "magnetosomes") lined up parallel to the long axis of the cell. The force exerted by the magnetosome chain, as it aligns with Earth's magnetic field, orients the bacterium with respect to the magnetic field.

Thirty years after their discovery, magnetotactic bacteria still are being studied by microbiologists. We continue our attempt to understand how these small and supposedly "simple" cells build such a fascinating structure. We will learn more about the magnetosome in this chapter, as we explore the structure and function of bacterial cells.

Introduction

In Chapter 1, we briefly explored the incredible world of microbes. In this chapter, we will look specifically at bacteria, followed in Chapters 3, 4, and 5 by more thorough examinations of eukarya, archaea, and viruses. Although bacterial cells appear to be less complex than eukaryal cells, they certainly are not "simple." Our tour of the bacterial cell will start with the nucleoid and cytoplasm, then progress outward through the plasma membrane and cell wall, and ultimately reach the cell surface. Along the way we will address several key questions:

What do bacteria look like? (2.1)
What is in the cytoplasm of bacterial cells? (2.2)
What kinds of internal structures help to organize bacterial cells? (2.3)
What are the critical structural and functional properties of the bacterial cell envelope? (2.4)
How do structures on the surface of bacterial cells allow for complex interactions with the environment? (2.5)
How are bacteria categorized and named? (2.6)

2.1 Morphology of bacterial cells

What do bacteria look like?

Bacteria exhibit several distinct shapes, or "morphologies" (**Figure 2.1**). The most common shapes, and the terms they are given by microbiologists, are:

- Spherical = **coccus** (plural: cocci, see Figure 2.1A)
- Rod = **bacillus** (plural: bacilli, see Figure 2.1B)
- Curved rod = **vibrio** (plural: vibrios, see Figure 2.1C)
- Spiral = **spirillum** (plural: spirilla, see Figure 2.1D)

The shape of bacterial cells is determined by the organization of the cell wall, the semi-rigid structure surrounding the cell. Morphology is a fairly reliable feature of most bacterial species.

© Science Source/Photo Researchers, Inc.

© Dr. Gary Gaugler/Photo Researchers, Inc.

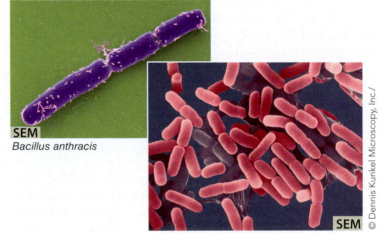

SEM
Staphylococcus aureus

SEM
Streptococcus pyogenes

© Eye of Science/Photo Researchers, Inc.

A. Cocci are spherical

SEM
Bacillus anthracis

SEM
Escherichia coli

© Dennis Kunkel Microscopy, Inc./Phototake

B. Bacilli are rod shaped

© Dennis Kunkel Microscopy, Inc./Phototake

SEM
Vibrio cholerae

C. Vibrio are slightly curved rods

© James Cavallini/Photo Researchers, Inc.

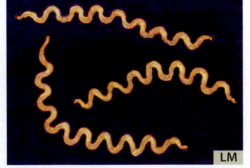

LM
Treponema pallidum

D. Spirilla are spiral shaped

Figure 2.1. Common shapes of bacteria
A. Normally found on the skin, *Staphylococcus aureus* cells are spherical, as are *Streptococcus pyogenes*, which may cause mild or serious infections, including scarlet fever. **B.** The rod-shaped bacterium *Bacillus anthracis* causes anthrax, while *Escherichia coli* typically is a non-disease causing inhabitant of the human intestine. **C.** *Vibrio cholerae* are curved and cause cholera in humans. **D.** Spiral-shaped bacteria include spirochetes like *Treponema pallidum*, which causes syphilis in humans.

For instance, *Escherichia coli* cells generally are straight rods, *Vibrio cholerae* cells are curved, *Staphylococcus aureus* cells are spherical, and *Treponema pallidum* cells are long, thin spirals. However, because many bacterial species have similar morphologies and because environmental conditions and stresses can sometimes cause changes in bacterial morphology, physical appearance is seldom conclusive for identifying bacterial species.

For many bacterial species, like *E. coli*, individual cells typically remain separate from each other. The cells of other bacteria stay physically connected after they divide. For example, the rod-shaped cells of *Bacillus anthracis*, the cause of anthrax, and the spherical cells of *Streptococcus pyogenes*, the cause of strep throat, often are seen in long chains. In contrast, *Staphylococci* cells tend to form irregular clusters rather than chains (see Figure 2.1).

Some bacteria do not exhibit regular shapes, but may exhibit highly variable shapes. These bacteria are referred to as **pleiomorphic**. Examples of pleiomorphic bacteria include members of the genus *Mycoplasma*, which do not make a cell wall and, as a result, do not have a regular shape **(Figure 2.2)**.

CONNECTIONS As we will discuss in **Section 6.5**, most bacteria can be removed from solutions by passing the solutions through fine filters. Because of their small size and lack of a cell wall, mycoplasma cells often are not retained by these fine filters.

Some bacteria grow in more complex multicellular arrangements **(Figure 2.3)**. Soil bacteria of the Actinomycete group grow as irregularly branching filaments called **hyphae** that are composed of chains of cells. Hyphae can form three-dimensional networks called **mycelia** that can rise above the substrate, penetrate down into soil, or both. Many fungi,

Figure 2.2. Pleiomorphic *Mycoplasma* These bacteria are very small (about one-tenth the size of *E. coli* cells). Lacking a cell wall, the cells may have an irregular shape—some may appear sphere-like, others more rod shaped. Although cells may stick together, *Mycoplasma* are not multicellular.

eukaryal organisms, form hyphae and mycelia superficially similar to the hyphae and mycelia formed by these bacterial species. A distinctive multicellular arrangement found in cyanobacteria is the formation of smooth, unbranched chains of cells called **trichomes** that may have a polysaccharide sheath coating the entire filament. In both trichomes and hyphae, the partitions between cells contain channels for intercellular passage of materials such as nutrients and signaling molecules.

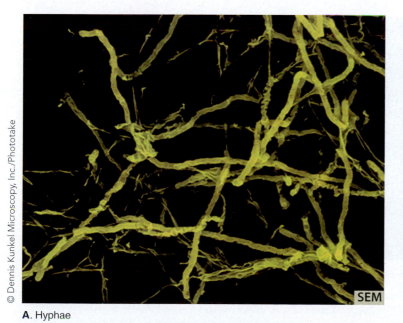

A. Hyphae

B. Trichomes

Figure 2.3. Multicellular arrangements of bacteria Some bacteria show more complex organization and specialization.
A. *Streptomyces* species grow into irregularly branching hyphae that can form a three-dimensional network, or mycelium, in their soil habitat.
B. Cells of the cyanobacterium *Spirulina* grow into a long, smooth, unbranched structure, or trichome, surrounded by a sheath.

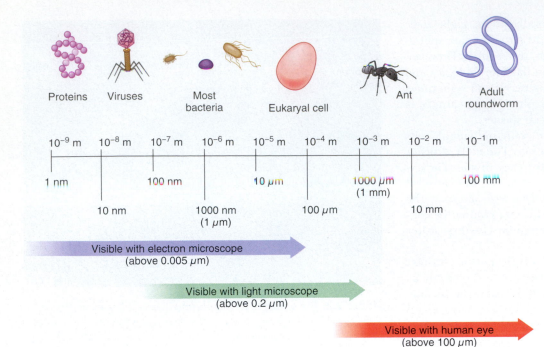

Figure 2.4. Relative sizes of microbes
Bacteria range in size from as small as 200 nm (0.2 μm) to several hundred μm, but the vast majority of individual bacterial cells are between 0.5 and 5 μm. Because of this small size, most bacteria only can be seen with a microscope. Objects need to be about 100 μm in diameter or larger to be seen by the unaided human eye. *NOTE*: m = meter, mm = millimeter, μm = micrometer, nm = nanometer.

Just as they come in a range of shapes, bacteria also come in a range of sizes **(Figure 2.4)**, with cells of most bacterial species being somewhere between 0.5 μm and 5 μm in length. Bacteria are usually smaller than eukaryal cells; even small eukaryal microbes such as yeast are typically at least 5 μm in diameter.

Most bacteria cannot be seen by the unaided human eye. Microscopy, therefore, is an integral tool of the microbiologist. As shown in Figure 2.4, different types of microscopes, like electron microscopes and light microscopes, allow us to see objects of different sizes. In Appendix B: Microscopy, we will explore these various types of microscopes. We will investigate how the microscopes work and also the uses of these microscopes. We already have seen images, or micrographs, taken with cameras linked to microscopes. Throughout this text, these micrographs will be marked with a microscopy icon that provides us with some information about the image. An "LM" label, for instance, indicates that the image was generated with a light microscope. Thanks to continuing advances in microscopy and photography, we can see an increasingly detailed view of the microbial world.

While most bacteria are less than 5 μm in diameter, some amazingly large bacteria have been discovered in recent years **(Figure 2.5)**. The spherical bacterium *Thiomargarita namibiensis*

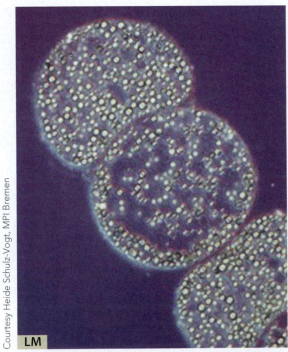

Courtesy Heide Schulz-Vogt, MPI Bremen

LM

A. *Thiomargarita namibiensis*

© Esther R. Angert, Ph.D./Phototake

LM

B. *Epulopiscium fishelsoni*

Figure 2.5. Giants of the bacterial world A. *Thiomargarita namibiensis* cells can be up to 700 μm in diameter. The nearly spherical cells, often filled with sulfur granules, are almost visible to the unaided human eye. **B.** Cigar-shaped *Epulopiscium fishelsoni* is so large (200–700 μm $\times$ 80 μm) that when it was discovered by Avi Fishelsoni in 1985 it was believed to be a eukaryal organism. This giant bacterium lives symbiotically in the guts of surgeonfish. It produces offspring internally, a unique mechanism of reproduction for a bacterium.

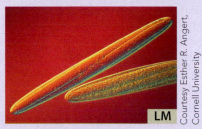
Microbes in Focus 2.1
EPULOPISCIUM FISHELSONI: REMARKABLE FOR MORE THAN ITS SIZE

Habitat: Only found within the gut of surgeonfish

Description: Gram-positive bacterium, 200–700 μm in length and up to 80 μm in diameter

Key Features: Aside from its size, this bacterium also exhibits a remarkable genome and replication cycle. Each cell appears to contain hundreds to thousands of copies of its genome, organized into many separate nucleoids located on the periphery of the cytoplasm. Stranger yet, unlike most bacteria, in which one cell divides into two progeny cells, *Epulopiscium* offspring arise within the parent cell, as seen in this image. Some types of *Epulopiscium* have been seen to have as many as 12 smaller cells inside the parent! How genomes and nucleoids are distributed into the offspring is unknown. The new cells appear to break out all at once by perforating the membrane of the parental cell, destroying it in the process.

LM

Courtesy Esther R. Angert, Cornell University

can be up to 700 μm in diameter, and the cigar-shaped *Epulopiscium fishelsoni* **(Microbes in Focus 2.1)** can be over 600 μm long. At the other extreme, some mycoplasmas are only 0.2 μm in diameter. Some scientists have speculated that even smaller nanobacteria exist, but 0.2 μm is probably near the lower size limit for cellular life as we know it. After all, even a single ribosome, the machinery for protein synthesis, is roughly 50 nm (0.05 μm) in diameter, and even the smallest cell would have to contain multiple ribosomes, a DNA genome, messenger RNAs, and various key enzymes in the cytoplasm.

As we have seen in this section, bacterial cells exhibit great variation in their shapes and sizes. All bacterial cells, however, also share many common properties. In the next section, we will begin to explore the various structures found within a typical bacterial cell, focusing on the functions of these structures and, in some cases, their evolutionary history.

2.1 Fact Check

1. What are the most common bacterial shapes?
2. Describe the various multicellular arrangements formed by some bacteria as a result of their cellular division.
3. What are the general size ranges of bacteria?

2.2 The cytoplasm

3D ANIMATION

What is in the cytoplasm of bacterial cells?

The **cytoplasm** of a bacterial cell, the aqueous environment within the plasma membrane, contains a diverse array of components **(Table 2.1)**. The largest single entity in the cytoplasm is the **nucleoid**, a convoluted mass of DNA (usually a single, circular chromosome) coated with proteins and RNA molecules still in the process of being synthesized. In contrast to the nucleus of

TABLE 2.1 Features of the bacterial cytoplasm

Organelle or molecules	Composition	Function
DNA nucleoid	DNA, RNA, protein	Genetic information storage and gene expression
Chromosome-packaging proteins	Protein	Protection and compaction of genomic DNA
Enzymes involved in synthesis of DNA, RNA	Protein	Replication of the genome, transcription
Regulatory factors	Protein, RNA	Control of replication, transcription, and translation
Ribosomes	RNA, protein	Translation (protein synthesis)
Plasmid(s)	DNA	Variable, encode non-chromosomal genes for a variety of functions
Enzymes involved in breaking down substrates	Protein	Energy production, providing anabolic precursors
Inclusion bodies	Various polymers	Storage of carbon, phosphate, nitrogen, sulfur
Gas vesicles	Protein	Buoyancy
Magnetosomes	Protein, lipid, iron	Orienting cell during movement
Cytoskeletal structures	Protein	Guiding cell wall synthesis, cell division, and possibly partitioning of chromosomes during replication

eukaryal cells, a membrane does not surround the nucleoid of bacterial cells (**Figure 2.6**).

If an average-size bacterial chromosome were stretched out in linear form, it would be hundreds of times longer than the bacterial cell in which it resides. To pack the DNA into a manageable form, bacteria use several strategies. Cations, such as Mg^{2+}, K^+, or Na^+, shield the negative charge on the sugar-phosphate backbone of each strand of the DNA helix, allowing the DNA molecule to pack more closely. Also, molecules of small, positively charged proteins bind to the chromosome to help maintain the condensed structure of the nucleoid. Finally, the topology of the DNA molecule is adjusted by **topoisomerases**, enzymes that encourage the chromosome to coil upon itself (**supercoiling**) in order to collapse it into a more compact mass.

CONNECTIONS In eukaryal cells, the DNA is compacted by histones (**Figure 7.5**). In **Section 4.2**, we will see that histones also exist in archaeal cells.

Many other proteins are located around the nucleoid, including enzymes involved in replicating and transcribing the DNA (DNA and RNA polymerases, respectively), and proteins that control gene expression. As RNA polymerases produce messenger RNA, ribosomes assemble on the mRNA and begin producing polypeptide chains. Ribosomes are thus indirectly associated with the nucleoid in living bacterial cells, although they fall off for the most part when the nucleoid is extracted from cells in the laboratory. The nucleoid isolated from *E. coli* after gentle cell lysis is roughly 60 percent DNA, 30 percent RNA, and 10 percent protein by weight.

Aside from the nucleoid, the cytoplasm is a complex water-based stew of macromolecules, with hundreds to thousands of proteins, transfer RNAs, messenger RNAs, and ribosomes. Most enzyme-catalyzed metabolic reactions occur in the cytoplasm. Some enzymes involved in the synthesis of membrane lipids and cell wall precursors are found in the cytoplasm, as are several proteins organized into filamentous structures that appear to have a guiding role in cell wall synthesis and cell division. We will talk more about these cytoskeletal proteins in the next section. In addition to macromolecules, a variety of smaller organic molecules (amino acids, nucleotides, and metabolic intermediates) and inorganic ions spice up the cytoplasmic soup.

Depending on the environment and growth conditions, bacterial cells sometimes store extra carbon, nitrogen, or phosphorus in **inclusion bodies** large enough to be seen by microscopy as granules within the cytoplasm. A separate membrane does not surround these inclusion bodies. As an example, polyhydroxybutyrate, or PHB, is a lipid polymer used for carbon storage. PHB granules can form over 50 percent of

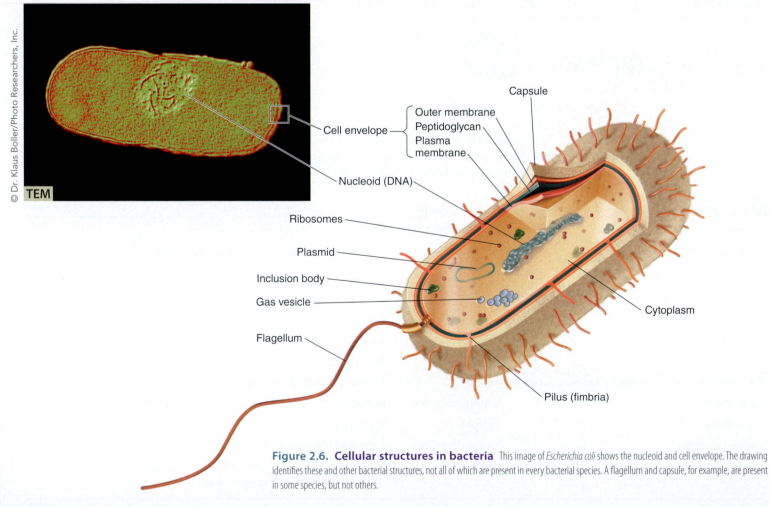

TEM

Cell envelope

Capsule

Outer membrane
Peptidoglycan
Plasma membrane

Nucleoid (DNA)

Ribosomes

Plasmid

Inclusion body

Gas vesicle

Flagellum

Cytoplasm

Pilus (fimbria)

Figure 2.6. Cellular structures in bacteria This image of *Escherichia coli* shows the nucleoid and cell envelope. The drawing identifies these and other bacterial structures, not all of which are present in every bacterial species. A flagellum and capsule, for example, are present in some species, but not others.

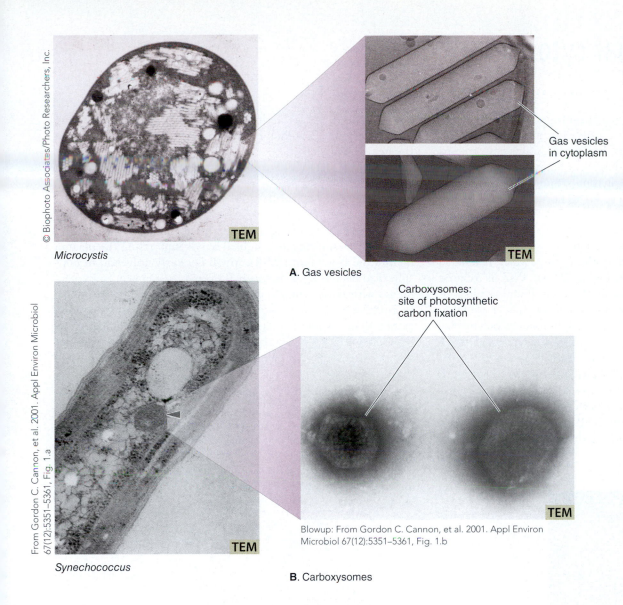

Figure 2.7. Specialized cytoplasmic organelles in bacteria

A. In this cross section of a *Microcystis* cell, gas vesicles, hollow microcompartments that permit gas to move in and out but prevent the entry of water, are visible. By regulating gas content in these vesicles, this cyanobacterium can regulate its position in a column of water for best access to light or nutrients. (blowup top: © N. Grigorieff, sample by A. E. Walsby; blowup bottom: © Linda Melanson, David J. Derosier, Marian Belenky, and Judith Herzfeld; reprinted from Belenky, et al., 2004, with permission of the Biophysical Journal)

B. Some cyanobacteria, like *Synechococcus*, contain carboxysomes (indicated by arrow), protein-encased structures containing enzymes involved in the photosynthetic transformation of CO_2 into organic material.

Microcystis

Gas vesicles in cytoplasm

A. Gas vesicles

Carboxysomes: site of photosynthetic carbon fixation

Blowup: From Gordon C. Cannon, et al. 2001. Appl Environ Microbiol 67(12):5351–5361, Fig. 1.b

Synechococcus

B. Carboxysomes

the cell's dry weight. PHB is an example of a broader group of polymers called polyhydroxyalkanoates that have been the subject of much industrial interest in recent years due to their ability to substitute for plastics that are derived from petrochemicals. **Sulfur globules**, composed of elemental sulfur (S^0), represent another form of intracellular storage. Cells of the giant bacterium *Thiomargarita namibiensis* (see Figure 2.5A) often contain large numbers of sulfur globules as a by-product of massive oxidation of sulfide (S^{2-}); the sulfur in these globules can be further oxidized to sulfate (SO_4^{2-}) to yield energy when sulfide is limiting.

Some bacteria produce protein-encased compartments within the cytoplasm that carry out specialized functions (Figure 2.7). Some aquatic photosynthetic bacteria produce air-filled **gas vesicles** that provide buoyancy to the cells. These gas vesicles can regulate the cell's position in a water column in response to light or nutrient levels. Photosynthetic cyanobacteria also can produce **carboxysomes** that contain the key enzymes involved in the conversion of inorganic carbon into organic matter during photosynthesis. **Magnetosomes**, the

magnetite-containing particles we introduced in the beginning of this chapter, are a rare example of a membrane-enclosed cytoplasmic organelle in a bacterial cell. Some researchers studying magnetosome development believe they may provide clues to the early evolution of membrane-enclosed organelles in ancestral cellular life-forms.

2.2 Fact Check

1. What is the nucleoid and how is it different from the eukaryal nucleus?
2. Describe the strategies used by bacterial cells to package the large chromosome into a "manageable form."
3. What additional components are contained within the cytoplasm of bacterial cells?
4. What are inclusion bodies and what is their role in the bacterial cell?

2.3 The bacterial cytoskeleton

What kinds of internal structures help to organize bacterial cells?

Just as our skeleton provides a structural framework for organizing our organs and tissues, the **cytoskeleton** is important for the internal organization of cells. In bacteria, a few key proteins recently have been discovered that form filamentous structures that organize cell growth and division (**Figure 2.8**). The major cytoskeletal proteins discovered so far in bacteria are located in the cytoplasm, but the crucial jobs of these proteins involve interacting with the plasma membrane and the cell wall, structures that will be discussed in the next section.

The FtsZ protein forms a ring, the **Z-ring**, which is needed for bacterial cell division (see Figure 2.8A). FtsZ is related to tubulin, a protein that serves as the main building block of eukaryal microtubules, a major component of the cytoskeleton of eukaryal cells (see Section 3.1). FtsZ monomers polymerize into filaments that bundle together to form the Z-ring. The formation of the Z-ring on the inner face of the plasma membrane is a critical step in cell division, because this ring interacts with membrane proteins that ultimately direct synthesis of the bacterial cell wall. As the Z-ring contracts through controlled release of subunits from FtsZ polymers by GTP hydrolysis, the cell envelope is forced inward at the eventual division site by reoriented cell wall synthesis. By the time cell division is finished, the Z-ring has disappeared. It is then rebuilt from cytoplasmic FtsZ during the next cell cycle.

The MreB protein is another important cytoskeletal component in bacteria. MreB is evolutionarily related to actin, a eukaryal cytoskeletal protein. MreB polymerizes into filaments that look strikingly similar to actin filaments, the heart of cytoskeletal structures called microfilaments in eukaryal cells. In bacteria, MreB forms long helical bands underlying the plasma membrane (see Figure 2.8B). MreB is nearly universal in nonspherical bacteria, but is rarely if ever present in cocci. Experiments with *E. coli* and *Bacillus subtilis* suggest that MreB helps to guide cell wall formation to produce an elongated cylinder, rather than a spherical shape. Scientists are still figuring out how bacterial cytoskeletal proteins help to guide the synthesis and organization of the cell wall, and there is considerable interest in how such processes might be modified in bacteria that form more elaborate shapes, such as vibrios, spirilla (e.g., *Magnetospirillum*), and hyphae.

Cytoskeletal structures may aid in the distribution of cytoplasmic components during bacterial cell division. We will examine

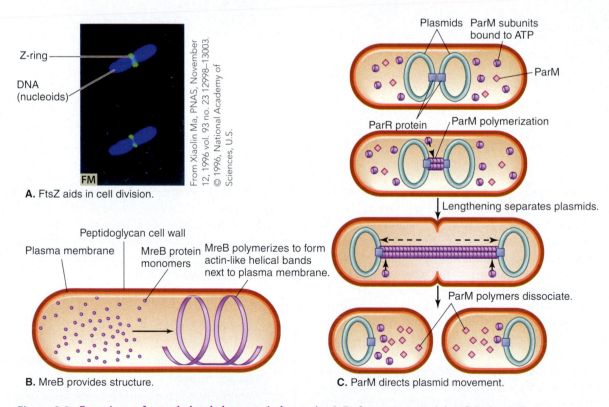

From Xiaolin Ma, PNAS, November 12, 1996 vol. 93 no. 23 12998–13003. © 1996, National Academy of Sciences, U.S.

A. FtsZ aids in cell division.

B. MreB provides structure.

C. ParM directs plasmid movement.

Figure 2.8. Functions of cytoskeletal elements in bacteria **A.** This fluorescence micrograph shows *Escherichia coli* cells undergoing cell division. The FtsZ "Z-ring" (green) is seen at the middle of each cell. Nucleoids that have already divided can be seen in blue on either side. **B.** MreB forms filaments in helical patterns on the inner face of the plasma membrane, aiding in providing shape to the cell. **C.** ParM forms filaments that direct plasmid movement to either side of the cell, ensuring plasmid segregation. ParM subunits bound to ATP (blue cylinder) polymerize into a rod, propelling plasmid molecules toward the poles as the ParM grows outward. Blue squares are the ParR protein, which recognizes the plasmid DNA and connects with the ParM filament. After division, ParM dissociates from the filament (pink diamonds).

more closely the role of actin-like filaments in organizing magnetosomes **(Perspective 2.1)**, but the distribution of genetic information during cell division is a more critical function. Like MreB, the ParM protein forms actin-like filaments. The *parM* gene is found on certain plasmids, small DNA molecules that replicate independently from the chromosome. Experiments have shown that ParM filaments aligned with the long axis of the *E. coli* cell are responsible for moving copies of these plasmids to opposite sides of the cell. This separation ensures that copies of the plasmid will be found on either side of the division site, so they will be passed along to each of the new cells (see Figure 2.8C). There is growing evidence that, as with plasmids, bacterial chromosomes are carefully positioned in the cell prior to cell division, and newly replicated origins of replication, the site at which DNA synthesis starts, move rapidly and specifically to opposite sides of the cell. These observations suggest that cytoskeletal filaments could actively participate in the positioning and segregation of bacterial chromosomes during cell division.

CONNECTIONS As we have seen briefly in this section, the distribution of genetic material within a dividing cell is well regulated. The initiation of DNA replication represents another well-regulated event. We will examine the factors controlling where bacterial DNA replication begins and how DNA replication is initiated in **Section 7.2**.

2.3 Fact Check

1. How are cytoskeletal-like proteins involved in magnetotaxis?
2. Describe the functions of the FtsZ protein and the Z-ring in bacterial cells.
3. What is the purpose of the MreB protein in bacteria?
4. What role do ParM proteins play in bacterial cell division?

2.4 The cell envelope

3D ANIMATION

What are the critical structural and functional properties of the bacterial cell envelope?

All cells are spatially defined by at least one membrane, the plasma membrane **(Figure 2.9)**. Most bacterial cells also contain a semi-rigid cell wall made of peptidoglycan and some bacteria contain a second membrane, the outer membrane. These layers in total are referred to as the **cell envelope**. In this section, we will examine the structure and function of these layers. In Section 2.5, we will examine other structures that may surround bacterial cells.

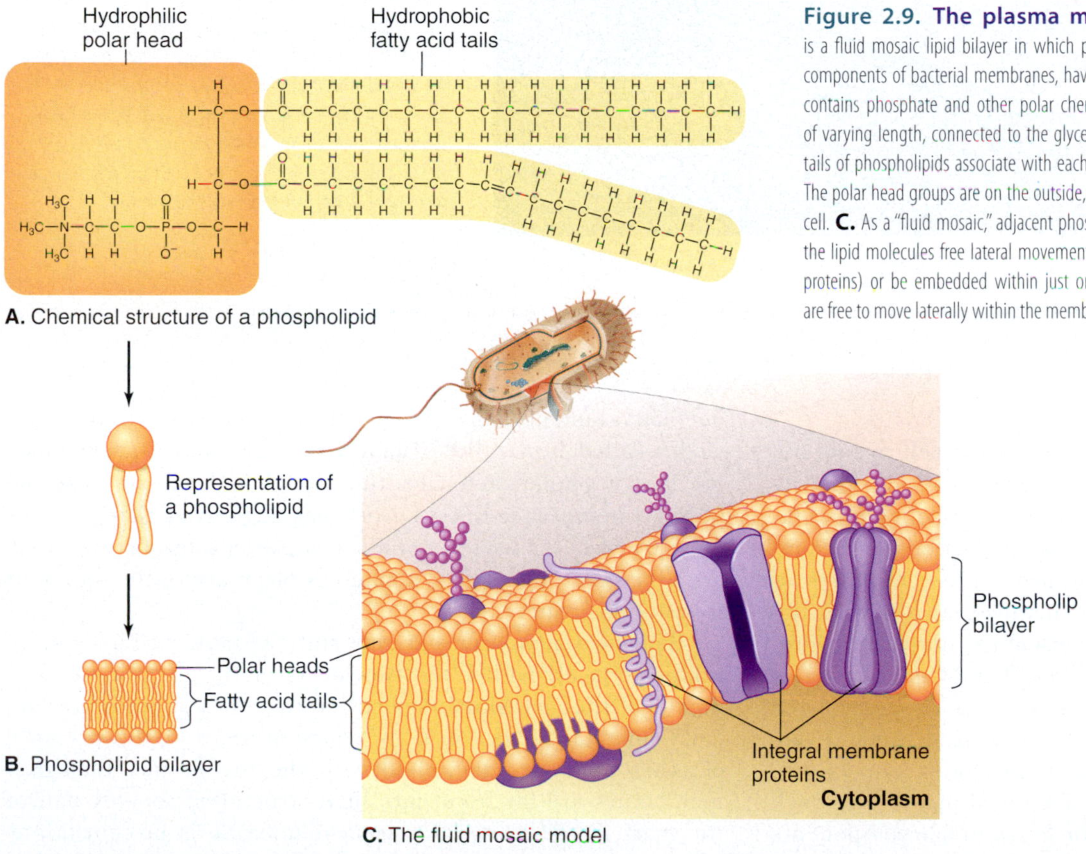

Figure 2.9. The plasma membrane of bacteria The plasma membrane is a fluid mosaic lipid bilayer in which proteins are embedded. **A.** Phospholipids, the major lipid components of bacterial membranes, have both polar and non-polar regions. The hydrophilic "head" contains phosphate and other polar chemical groups. The hydrophobic "tails" are fatty acid chains of varying length, connected to the glycerol backbone. **B.** In an aqueous solution, the hydrophobic tails of phospholipids associate with each other to exclude water, creating the interior of the bilayer. The polar head groups are on the outside, interacting with the aqueous cytoplasm and exterior of the cell. **C.** As a "fluid mosaic," adjacent phospholipids are not covalently linked to each other, allowing the lipid molecules free lateral movement. Proteins may span the entire bilayer (integral membrane proteins) or be embedded within just one leaf of the bilayer (peripheral proteins). These proteins are free to move laterally within the membrane.

Hydrophilic polar head

Hydrophobic fatty acid tails

A. Chemical structure of a phospholipid

Representation of a phospholipid

Polar heads
Fatty acid tails

B. Phospholipid bilayer

Phospholipid bilayer

Integral membrane proteins

Cytoplasm

C. The fluid mosaic model

MARVELOUS MAGNETOSOMES!

Bacteria have various systems to direct their movement, but magnetotaxis is undoubtedly one of the most unusual. As we described in the opening story, magnetotaxis involves orienting movement in response to Earth's magnetic field. The bacteria achieve this orientation through the use of organelles called magnetosomes, tiny magnets made of magnetite (Fe_3O_4) crystals that align in chains in the cytoplasm of magnetotactic bacteria. Magnetosomes have a membrane around them **(Figure B2.1)**, which high-resolution imaging experiments have shown to be formed from the plasma membrane. The magnetosome chain acts like a compass needle powerful enough to physically align the long axis of the cell with Earth's geomagnetic field. To generate enough force to orient the cell body, magnetosomes must be organized into a linear chain, in which the total dipole moment is summed across the individual magnetosomes. In *Magnetospirillum*, typical cells have around 20 magnetosomes in a chain. These chains probably do not form spontaneously. Indeed, protein filaments can be found alongside the magnetosome chains and may play a role in organizing them.

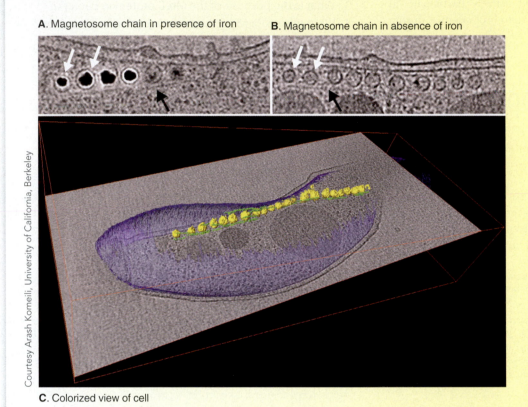

A. Magnetosome chain in presence of iron

B. Magnetosome chain in absence of iron

C. Colorized view of cell

Courtesy Arash Komeili, University of California, Berkeley

Figure B2.1. Magnetosome organization
A. *Magnetospirillum magneticum* cell showing a chain of magnetosomes (white arrows). Each magnetosome is enclosed by a membrane vesicle. One or more cytoskeletal filaments (black arrow) flank the chain. **B.** Similar view of a magnetosome chain from a cell grown in the absence of iron, which prevents the formation of magnetite crystals. Empty vesicles and filaments are still present. **C.** Three-dimensional reconstruction of a *Magnetospirillum* cell (blue), showing magnetosomes (yellow) and their associated filaments (green). A form of electron microscopy, cryo-EM, was used to generate these images.

The plasma membrane

The **plasma membrane**, also referred to as the cell membrane or the cytoplasmic membrane, is a bilayer composed primarily of phospholipids. The structure of these phospholipids is amphipathic, meaning that they have a polar portion and a non-polar portion. Most cellular membranes contain mixtures of phospholipids that can vary in the chemical structure of the polar head group, as well as in the length of the non-polar hydrocarbon tail and the frequency and position of double bonds within the hydrocarbon chain (see Figure 2.9). The polar head is hydrophilic, and thus interacts with water inside and outside the cell, while the non-polar hydrocarbon chains associate to form the interior of the membrane. Bacterial membranes lack sterol lipids, such as cholesterol, which are major components of eukaryal membranes. Some bacteria produce sterol-like molecules called **hopanoids** (Figure 2.10). Like sterols, hopanoids are largely planar molecules that are thought to stabilize the plasma membrane. Recent work has suggested that perhaps only 10 percent of bacteria produce hopanoids, but because they are very stable molecules, hopanoids often are quite abundant in soils and sediments.

Biological membranes are not static structures. Lipids move relatively freely within the two layers of the membrane. The fluidity of the membrane depends on the types of lipids present, environmental factors such as temperature, and the presence of other molecules associated with the membrane. Biological membranes are far from pure lipid structures; roughly half of the mass of the bacterial plasma membrane is protein. Many

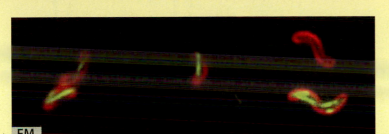

A. *Magnetospirillum* cells expressing green fluorescent MamK

From Komeli et al. 2006. Science 311:242–245. Reprinted with permission from AAAS.

FM

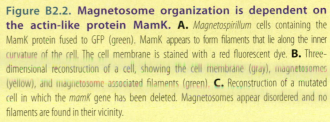

Figure B2.2. Magnetosome organization is dependent on the actin-like protein MamK. A. *Magnetospirillum* cells containing the MamK protein fused to GFP (green). MamK appears to form filaments that lie along the inner curvature of the cell. The cell membrane is stained with a red fluorescent dye. **B.** Three-dimensional reconstruction of a cell, showing the cell membrane (gray), magnetosomes (yellow), and magnetosome associated filaments (green). **C.** Reconstruction of a mutated cell in which the *mamK* gene has been deleted. Magnetosomes appear disordered and no filaments are found in their vicinity.

B. Cell expressing MamK

C. Cell with disrupted *mamK* gene

The amino acid sequence of the *Magnetospirillum* MamK protein is similar to actin and the bacterial cytoskeletal protein MreB, and the filaments observed alongside the magnetosome chains are approximately the same dimension as actin filaments. Convincing evidence that MamK assembles into filaments comes from experiments using GFP fusion proteins. This method, which has been tremendously useful to cell biologists, involves first creating a gene fusion in which the coding sequence of a target protein is linked, or fused, to the coding sequence of green fluorescent protein (GFP), a jellyfish protein that naturally fluoresces when illuminated with certain wavelengths of light. When inserted into a cell, the fused gene creates a hybrid protein that fluoresces, thereby allowing researchers to determine the cellular location of the protein of interest by viewing the cells with high-resolution fluorescence microscopy (see Toolbox 3.1). When scientists examined *Magnetospirillum* cells containing

MamK-GFP fusions (**Figure B2.2**), they observed beautiful green filaments coinciding with magnetosome chains. Furthermore, when the *mamK* gene is removed from the *Magnetospirillum* genome, these magnetosome-associated filaments disappear, and magnetosomes disperse throughout the cell.

The story does not end there. Another group of scientists studying MamJ, a highly acidic and repetitive protein thought to possibly be involved in magnetite crystal formation, found that when the *mamJ* gene is removed, magnetosomes still form, but the magnetosomes do not arrange in a line. In contrast to the MamK observations, the organizing filament is still present in the *mamJ* deletion mutant. In fact, the MamJ-GFP fusion is localized to this filament in normal cells. It appears that MamJ may facilitate the interaction between the magnetosomes and the filament. How might MamJ carry out this function? How could you test your hypothesis?

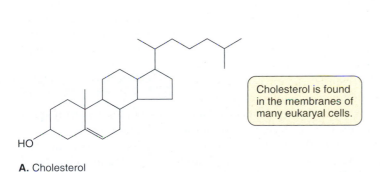

Cholesterol is found in the membranes of many eukaryal cells.

A. Cholesterol

Hopanoids are found in the membranes of some, but not all, bacteria.

B. Bacteriohopanetetrol

Figure 2.10. Bacterial membranes lack cholesterol, but may contain hopanoids. A. Cholesterol is a common example of sterol lipids found in many eukaryal membranes. **B.** Hopanoids, such as this bacteriohopanetetrol, are present in the membranes of some bacteria. These molecules are structurally similar to cholesterol and provide some of the same functions, strengthening the membrane.

proteins either cross or are integrated into the membrane bilayer (see Figure 2.9). These proteins have hydrophobic surfaces that interact with the interior of the membrane, and hydrophilic domains exposed to the cytoplasmic or outside environments. Key functions of the plasma membrane proteins include:

- controlling access of materials to the cytoplasm through differential permeability
- capturing and/or storing energy through photosystems, oxidative electron transport, and maintenance of chemical and electrical gradients
- environmental sensing and signal transduction

Permeability of the Plasma Membrane to Water

A fundamental property of the plasma membrane is its differential or selective permeability. Small, uncharged molecules such as O_2 and CO_2 can diffuse relatively freely across a phospholipid bilayer. Larger compounds, or molecules that are more polar or charged, cross the membrane less easily. Although they are polar, water molecules are small enough to cross phospholipid bilayers. Their transit may be facilitated, and perhaps regulated, by protein channels called **aquaporins**. Water concentration differences between the interior and exterior of the cell create an osmotic gradient. The cell is in a hypotonic solution if the cytoplasm has a higher solute concentration than the external environment, causing water to move into the cell. Because phospholipid bilayers are quite flexible, the cell expands, like a balloon being inflated. Without a structure to provide stability, cells would risk explosive destruction. The cell wall, however, largely prevents this outcome. Conversely, the cell is in a hypertonic solution when lower dissolved ion concentrations in the cytoplasm relative to the external environment lead to a net loss of water from the cell. Like a deflating balloon, the cell could experience structural collapse—a crisis that, again, a protective cell wall helps to avert.

 Osmosis ANIMATION

CONNECTIONS Salt-loving archaeons are found in environments that would draw water from most cells. **Figures 4.25** and **4.26** show the effects of tonicity on living cells, and how archaeons deal with the osmotic challenge.

Nutrient Transport

Cellular metabolism depends on the cell's ability to obtain nutrients from the surrounding environment, and to a lesser extent, on its ability to export waste products. Because movement of charged or polar molecules, which include many potential nutrients, is greatly restricted by the limited permeability of the phospholipid bilayer, transport proteins embedded in the membrane are needed to provide routes of entry into the cell for such compounds. Transport systems are often very specific in the molecules they admit, so that the cell takes in useful compounds, but ignores molecules that are not useful, or are harmful. Because of this specificity, microbial cells generally use many different transporters to bring in the mix of nutrients they need **(Figure 2.11)**.

Some transport proteins simply facilitate diffusion, allowing specific molecules to pass through the membrane from higher

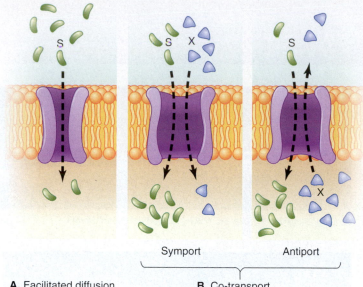

Symport   Antiport

A. Facilitated diffusion   **B.** Co-transport

Figure 2.11. Movement across a membrane by facilitated diffusion and co-transport A. In facilitated diffusion, one kind of molecule (labeled S) moves across the membrane. Diffusion allows a molecule to move across the membrane from the side of higher concentration to the side of lower concentration. **B.** Co-transporters can move a substance (S) against its concentration gradient, by coupling that movement to the diffusion of another substance (X) down its concentration gradient. Symport proteins facilitate the movement of both substances in the same direction. Antiport proteins facilitate the movement of both substances in opposite directions.

to lower concentration. These transport proteins require no investment of energy, but their role in nutrient uptake is limited to molecules that are abundant in the environment. If there is not a favorable concentration gradient to drive a particular nutrient into the cell at the rate an organism needs, another mechanism must be used. **Active transport systems** in the plasma membrane are driven by the expenditure of energy to drive the movement of solutes against a concentration gradient. One way this transport can be done is to bring the desired molecule into the cell along with another molecule that is traveling down a favorable concentration gradient. As the one molecule travels down its concentration gradient, energy is released that, in turn, can be used to move the primary molecule against its gradient. This hitchhiking strategy is called **symport**. The most common type of bacterial symport system couples the import of a substrate, such as a sugar, to the uptake of one or more protons. This strategy takes advantage of the favorable proton concentration gradient across the membrane generated by respiratory electron transport (see Chapter 13 for description of the electron transport system). With **antiporters**, the energy-requiring uptake of one molecule is driven by the energetically favorable ejection of another. Both symport and antiport mechanisms are examples of co-transport, in which two different molecules are being moved.

 Molecular movement ANIMATION

One of the most common types of active transport involves the **ABC transporters**. ABC stands for "ATP-binding cassette," reflecting the fact that these proteins include a nucleotide binding domain by which ATP is hydrolyzed to provide the energy for transport. Each ABC transporter is specific for a given compound

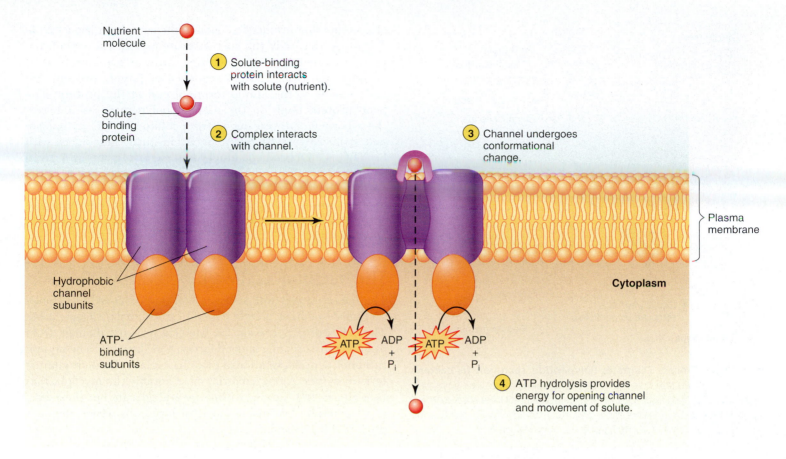

Figure 2.12. Process Diagram: Active transport systems require an input of cellular energy, often as ATP. ABC (ATP-binding cassette) transport systems carry out active transport using ATP hydrolysis to open a specific transmembrane channel. The solute to be transported, perhaps a nutrient molecule, first is bound by a high-affinity solute-binding protein outside the membrane. The solute binding protein delivers the molecule to an integral protein, which then changes conformation to open a channel allowing the molecule to enter the cell. ATP hydrolysis is necessary for the cycle of opening and closing.

or a range of compounds, and consists of four subunits (**Figure 2.12**). The membrane channel is formed by two hydrophobic subunits, while two hydrophilic subunits are located on the cytoplasmic surface of the membrane and contain the ATP-binding domain. An associated subunit is a high-affinity solute binding protein that binds to substrates outside the plasma membrane and delivers them to the membrane-bound components, effectively increasing the affinity of the transport system for substrates.

Energy Capture

The plasma membrane hosts critical parts of the energy-capturing machinery of many bacterial cells. These include cytochromes and other membrane-soluble electron carriers that form the electron transport system of microbes that utilize respiration, and the light-capturing machinery of photosynthetic microbes. Both respiratory electron transport and photosynthetic light capture result in the ejection of protons from the cytoplasm, across the membrane, to the external environment. As they accumulate outside the cell, these protons create concentration and charge gradients that combine to create a proton motive force that we will discuss more in Chapters 6 and 13. The proton gradient across the membrane

is a very useful energy source. We mentioned in the previous section that it can be used to drive active transport of nutrients across the membrane. The proton gradient also powers rotation of the flagellum in most bacteria. Perhaps most importantly, protons also can be pushed back into the cell through a membrane protein complex called ATP synthase, which couples proton re-entry to the production of ATP, the main energy source for biochemical reactions inside the cell.

Sensory Systems

The plasma membrane senses the environment around bacterial cells with sensor proteins that transduce signals to cytoplasmic response systems. These signals often result in changes in gene expression that ensure production of an appropriate set of proteins. A particular transporter, for instance, might be made only when its substrate is present. Not all of these signaling processes, we should note, necessarily happen at or in the membrane. Bacteria also utilize many cytoplasmic sensors.

Protein Secretion

The plasma membrane contains proteins for the **general secretory pathway**, allowing the transmembrane movement

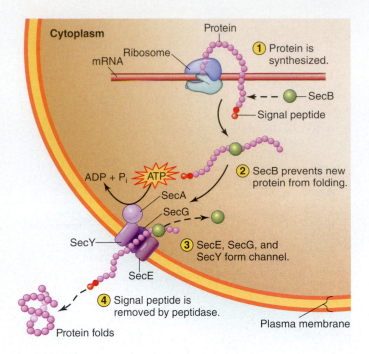

Figure 2.13. Process Diagram: Polypeptide secretion in bacteria
SecB binds to polypeptides to be secreted as they leave the ribosome and delivers them to SecA, which associates with SecYEG. Using energy derived from the hydrolysis of ATP, SecA facilitates movement of the polypeptide through the SecYEG channel. After the polypeptide enters the periplasm, a signal peptidase removes the signal peptide and the polypeptide assumes its functional conformation.

of proteins that are needed outside the cytoplasm (**Figure 2.13**). To ensure that only the correct proteins exit the cytoplasm, proteins targeted to this secretory pathway are identified by a **signal peptide**, a short sequence of largely hydrophobic amino acids at the amino terminal end of the protein. The SecB protein binds to the nascent polypeptide as it leaves the ribosome, preventing it from folding in the cytoplasm and delivering it to SecA. SecA, in turn, associates with SecYEG, a membrane channel complex. Using energy derived from the hydrolysis of ATP, SecA facilitates movement of the polypeptide through the SecYEG channel. Once the polypeptide enters the periplasm, a signal peptidase removes the signal peptide and the protein assumes its functional conformation.

The bacterial cell wall

The bacterial cell wall consists of a highly crosslinked polysaccharide-peptide matrix called **peptidoglycan** (**Figure 2.14**). The cell wall is necessary for bacteria to resist damage from osmotic pressure, mechanical forces, and shearing. The organization of peptidoglycan also gives bacterial cells their characteristic shapes. A few bacteria, most notably the mycoplasmas (see Figure 2.2), survive without peptidoglycan, but these bacteria generally live inside eukaryal host cells where they are protected from osmotic stress.

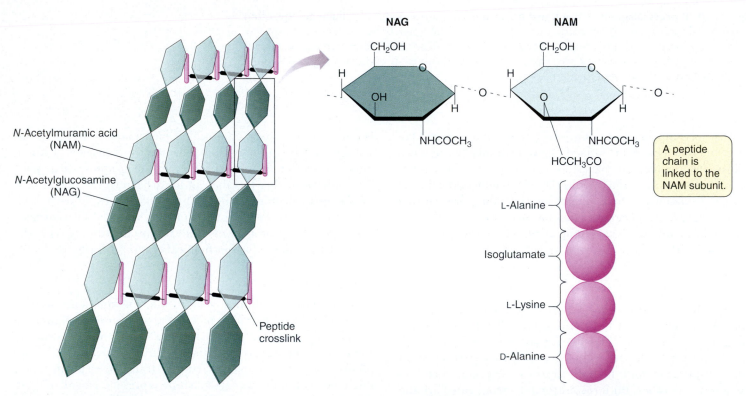

A. Crosslinked peptidoglycan

B. Disaccharide backbone with peptide chain

Figure 2.14. The structure of the peptidoglycan cell wall A. Layers of polysaccharide (*N*-acetylglucosamine and *N*-acetylmuramic acid) are linked together by short peptide chains (indicated by solid black lines). The chemical basis of this linkage differs between species. **B.** The basic peptidoglycan subunit consists of *N*-acetylmuramic acid (NAM) and *N*-acetylglucosamine (NAG) monomers connected by β-1,4-glycosidic bonds. The NAM group is linked to a short peptide chain (indicated by purple spheres), which crosslinks to another NAM molecule. This peptide chain consists of different amino acids in different species.

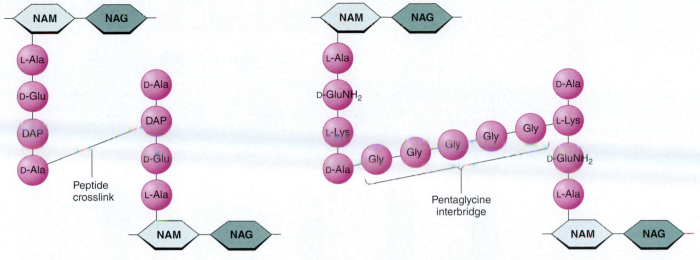

A. Peptidoglycan crosslinking in *E. coli*

B. Peptidoglycan crosslinking in *S. aureus*

Figure 2.15. Crosslinking of peptide chains provides strength to the peptidoglycan. To increase the strength of the peptidoglycan, crosslinking between the peptide chains occurs. **A.** In *E. coli*, peptide chains are directly linked by a covalent bond between a D-alanine (D-Ala) amino acid and a *meso*-diaminopimelic (x2) (DAP) amino acid. **B.** In *S. aureus*, a pentaglycine interbridge connects the D-Ala amino acid on one chain to the L-lysine (L-Lys) on another chain.

Peptidoglycan forms a net-like structure composed of a glycan backbone made up of alternating molecules of *N*-acetylglucosamine (NAG) and *N*-acetylmuramic acid (NAM), connected by β-1,4-glycosidic bonds. The *N*-acetylmuramic acid carries a short peptide chain that is used to crosslink peptidoglycan strands, creating the protective network surrounding the cell. The sequence of amino acids in the peptide chain can vary somewhat between species; *Escherichia coli* has a slightly different sequence than *Staphylococcus aureus*.

Crosslinking is very important for the strength of the peptidoglycan network, but bacterial species can achieve this crosslinking through different mechanisms. For example, in *Escherichia coli*, the D-alanine (D-Ala) amino acid attached to one NAM residue is linked directly to the *meso*-diaminopimelic (DAP) amino acid

present on another NAM residue. In *Staphylococcus aureus*, tetrapeptides are crosslinked through a peptide interbridge between D-alanine on one strand and L-lysine (L-Lys) on the other **(Figure 2.15)**. As shown in this figure, *Staphylococcus aureus* uses a pentaglycine interbridge, but the interbridge amino acids vary between species.

We should note that several amino acids found in peptidoglycan are rarely found in proteins, including *meso*-diaminopimelic acid, and the D-stereoisomer of several amino acids. Stereoisomers are molecules with the same chemical formula and bond structure, but different arrangements of atoms **(Figure 2.16)**. The amino acids used by ribosomes for polypeptide synthesis are almost always the L-stereoisomer.

The production of peptidoglycan starts in the cytoplasm, where enzymes first link *N*-acetylmuramic acid (NAM) and uridine

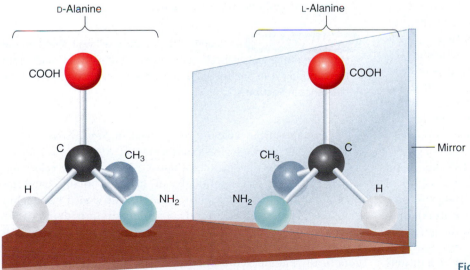

Karp: *Cell and Molecular Biology: Concepts and Experiments*, copyright 2010, John Wiley & Sons, Inc. This material is reproduced with permission of John Wiley & Sons, Inc.

Figure 2.16. Stereoisomers are mirror images of each other. D-Alanine and L-alanine have the same chemical formula. The arrangement of the bonds on the central carbon, though, differs.

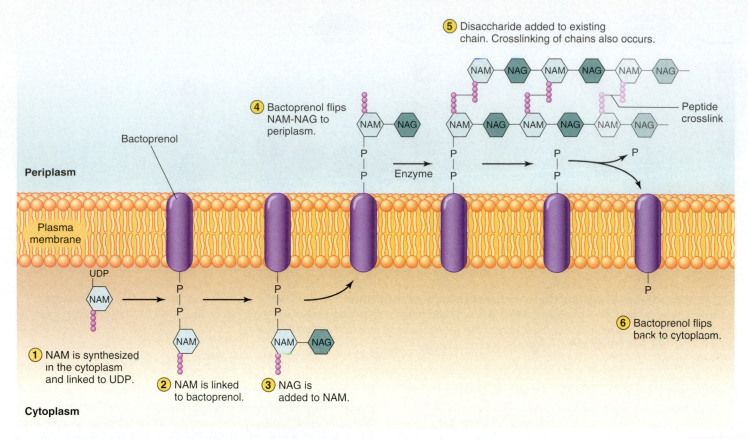

5 Disaccharide added to existing chain. Crosslinking of chains also occurs.

4 Bactoprenol flips NAM-NAG to periplasm.

Peptide crosslink

Bactoprenol

Periplasm

Plasma membrane

Enzyme

6 Bactoprenol flips back to cytoplasm.

UDP

1 NAM is synthesized in the cytoplasm and linked to UDP.

2 NAM is linked to bactoprenol.

3 NAG is added to NAM.

Cytoplasm

Figure 2.17. Process Diagram: Synthesis of peptidoglycan Synthesis begins in the cytoplasm, where NAM is linked to UDP, and then coupled to a short peptide chain. This complex associates with the integral membrane protein bactoprenol. NAG then is linked to the NAM, forming the NAM-NAG disaccharide. This disaccharide flips across the membrane, transporting the NAM-NAG complex to the periplasm. The NAM-NAG disaccharide is linked to the growing polysaccharide chain. The peptide chain of the newly added NAM-NAG subunit is crosslinked by a transpeptidase enzyme to another strand of peptidoglycan to form the final structure.

diphosphate (UDP). This molecule then is linked to a pentapeptide (Figure 2.17). The UDP-NAM-peptide complex is coupled to a lipid carrier called bactoprenol, which holds it on the cytoplasmic face of the membrane. A second sugar, *N*-acetylglucosamine, then is added to form the basic disaccharide-peptide unit. This complex then pulls a remarkable trick, still not well understood, in which the bactoprenol carrier flips the disaccharide-peptide complex to the other side of the membrane. Only there can the new subunit be attached to the terminal sugar of a growing peptidoglycan chain. Other enzymes, most notably transpeptidase, then crosslink the pentapeptide precursor to another strand, a transpeptidation reaction in which the terminal D-alanine of the pentapeptide precursor is discarded, leaving a tetrapeptide.

Many potential enemies of bacteria, including bacteriophages and animals, produce the enzyme **lysozyme** that degrades peptidoglycan. Lysozyme hydrolyzes the β-1,4-glycosidic bond between NAG and NAM, profoundly weakening the cell wall (Figure 2.18). If a bacterium encounters lysozyme under isotonic conditions (water leaves the cell at the same rate it enters), then it may survive as a "protoplast," the term given to the generally fragile form of the cell with its protective cell wall removed. If a bacterium encounters lysozyme under hypotonic conditions (solute concentrations are lower outside the cell than in the

cytoplasm), then water may enter the protoplast, swelling it until it bursts. Lysozyme in human tears and mucus secretions helps to control bacterial populations in and on our body. We will see later in this chapter that some bacteria (Gram-negative bacteria) cover their peptidoglycan with another membrane, effectively shielding the cell wall from lysozyme.

CONNECTIONS Archaeons also possess a cell wall. The chemical structure of the archaeal cell wall, however, differs significantly from the peptidoglycan present in bacteria. We will explore this structural difference in **Section 4.2**.

Lysostaphin, an enzyme discovered in *Staphylococcus simulans*, cuts the pentaglycine crossbridge of *Staphylococcus aureus* and some related species (see Figure 2.18B). Unlike lysozyme, which degrades the peptidoglycan layer of many bacteria, lysostaphin is active only against *S. aureus*. The cloned lysostaphin gene has been experimentally engineered into dairy cattle to generate cows that suffer much less frequently from mastitis, a painful udder infection usually caused by *S. aureus*. Such genetically engineered animals, resistant to various infectious organisms, may become common in our future.

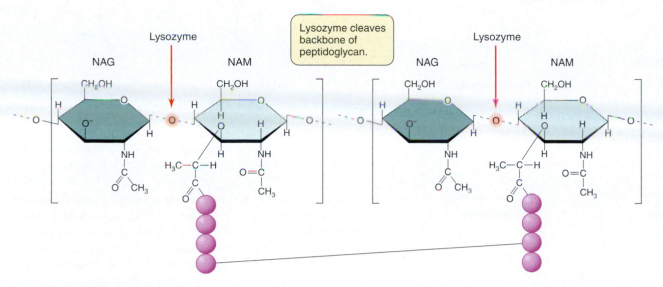

A. Chemical action of lysozyme

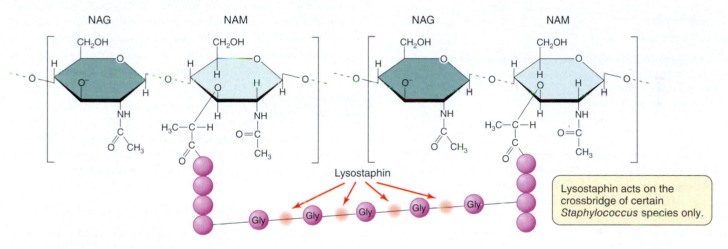

B. Chemical action of lysostaphin

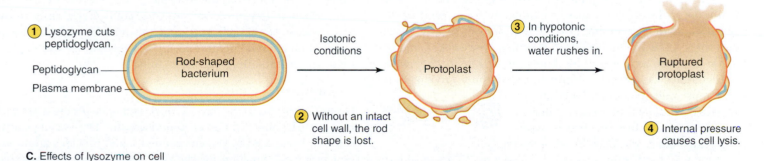

C. Effects of lysozyme on cell

Figure 2.18. Process Diagram: Bacterial wall disruption and its consequences A. Lysozyme enzymatically cleaves the β-1,4-glycosidic bonds linking NAG and NAM residues (highlighted), thereby disrupting the polysaccharide backbone of peptidoglycan. **B.** Lysostaphin cleaves the glycine-glycine linkages (highlighted) of the pentaglycine interbridge specifically found in some *Staphylococci* species. **C.** As the peptidoglycan network surrounding the cell is degraded, the cell wall loses its mechanical strength. Non-spherical cells will become round protoplasts. If the external medium is hypotonic (lower solute concentration) relative to the cytoplasm of the protoplast, the pressure created by the osmotic influx of water will likely cause the cell to rupture.

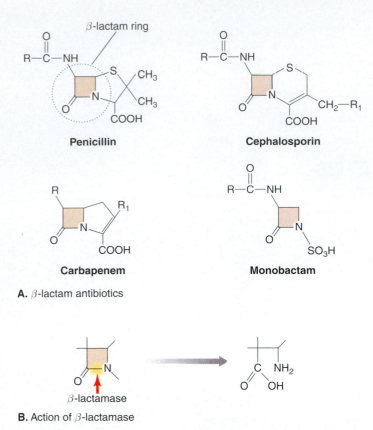

Penicillin

Cephalosporin

Carbapenem

Monobactam

A. β-lactam antibiotics

B. Action of β-lactamase

Figure 2.19. Beta-lactam antibiotics inhibit cell wall formation.
A. All antibiotics within the beta-lactam family contain a β-lactam ring, which is shaded in these drawings.
B. In some bacteria, β-lactamases hydrolyze the C—N bond of the β-lactam ring and destroy the activity of these antibiotics. *NOTE*: R represents a variable chemical group.

Several important families of antibiotics inactivate transpeptidase enzymes that crosslink peptidoglycan. The **β-lactam** family, for example, is represented in **Figure 2.19**. The three-dimensional structure of the β-lactam ring mimics the terminal D-alanine of peptidoglycan precursors well enough to fool peptidoglycan crosslinking enzymes. These enzymes bind to β-lactam antibiotics and attempt to catalyze the crosslinking reaction, but instead attach the antibiotic molecule covalently to their active site, permanently destroying the enzyme. This inhibition of the enzyme causes big problems for growing bacterial cells, as they try to integrate new strands of peptidoglycan into their cell wall. The cell wall becomes ever weaker, and osmotic pressure can eventually burst the cell, as it does when lysozyme attacks. A key difference is that only growing cells that must add new peptidoglycan are sensitive to β-lactam drugs, whereas lysozyme, which directly cleaves peptidoglycan chains, also attacks non-growing cells.

@ Peptidoglycan ANIMATION

One way for bacteria to become resistant to an antibiotic is to produce enzymes that modify or destroy the antibiotic. **β-lactamases** hydrolyze the C—N bond in the β-lactam ring of antibiotics, such as penicillin and cephalosporin (see Figure 2.19A). This ring structure is necessary for these antibiotics to bind to their target. Cleavage of the ring inactivates the antibiotics. Soil fungi have produced these antibiotics for millions of years. In response, some bacteria long ago evolved β-lactamases to combat these fungal weapons.

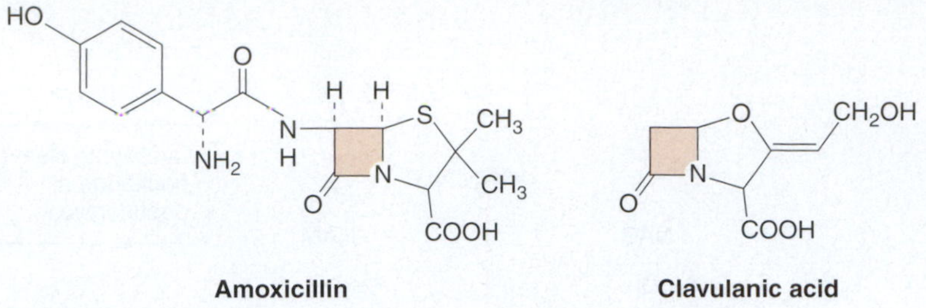

Amoxicillin

Clavulanic acid

Figure 2.20. Combination antibiotics can be manufactured. Augmentin, a combination antibiotic, is designed to overcome the resistance of bacteria that produce β-lactamases. This drug contains amoxicillin (*left*) and clavulanic acid (*right*). Both amoxicillin and clavulanic acid contain the β-lactam ring. Clavulanic acid has little or no antibiotic activity by itself but binds to and inhibits the activity of many β-lactamases, allowing the accompanying amoxicillin to attack peptidoglycan synthesis.

A few short years after the introduction of β-lactam antibiotics into medical practice, resistant bacteria that produce β-lactamases became fairly common. Modified β-lactam antibiotics that are more resistant to β-lactamase attack have been developed over the past few decades, but bacteria have responded with altered β-lactamases that also inactivate these newer drugs. An alternative strategy for subverting bacterial defenses was spurred by the discovery of β-lactam variants such as clavulanic acid (**Figure 2.20**) that bind to and strongly inhibit β-lactamases, even though they do not function as antibiotics by themselves. Clavulanic acid is now combined with β-lactam antibiotics such as amoxicillin (in the popular combination marketed as Augmentin) or ticarcillin (in Timentin) to protect these drugs from inactivation.

CONNECTIONS Chemists have designed semi-synthetic versions of penicillin that can more readily pass through the outer membrane of Gram-negative organisms, thereby making these drugs more effective against a wider range of bacterial pathogens. These advances in drug design will be discussed more fully in **Section 24.2**.

Variation in the bacterial cell envelope

The **Gram stain**, a technique for staining bacterial cells developed by Hans Christian Gram in 1884, allows us to differentiate two types of bacteria: **Gram-positive** and **Gram-negative** (**Toolbox 2.1**). Electron microscopic examinations of bacteria have revealed that these two categories of bacteria differ in the organization of the cell wall, and the presence or absence of an additional outer membrane. In this section, we will explore these structural differences more fully.

Gram-positive Cells

Gram-positive bacteria have a thick cell wall (**Figure 2.21**) with many overlapping strands of peptidoglycan. The Gram-positive cell wall is exposed to the environment, and provides an important protective function for the cell, but it is not impermeable. After all, nutrients must be able to get to the plasma membrane, where they can be taken into the cell. Peptidoglycan is the most abundant polymer in the Gram-positive cell wall, but a mixture of other polymers can constitute up to half of the dry weight of the cell wall, depending on the species. A charged polymer called **teichoic acid** (not found in Gram-negative bacteria) is intermingled with the peptidoglycan. Teichoic acids are polymers of either ribitol phosphate or glycerol phosphate (see Figure 2.21).

Hans Christian Gram developed the Gram-stain technique in 1884 for detecting bacteria in lung tissue samples of patients who had died of pneumonia. Over a century after it was first developed, the Gram stain remains a key step in bacterial identification strategies in clinical microbiology laboratories. The Gram stain is an excellent example of a differential stain; bacterial species differ in their response to the staining procedure, depending on the structure of the cell envelope. This staining method classifies bacteria into two major groups, Gram-positive and Gram-negative. The structural basis for these staining patterns is described in the text. The Gram-stain procedure consists of just a few simple steps (**Figure B2.3**).

Staining bacteria ANIMATION

First, the specimen is placed on a microscope slide. Then, the specimen is fixed, or attached, to the slide by heating the underside of the slide. The cells first are stained with crystal violet, the primary stain—virtually all bacteria are stained by crystal violet. The cells then are stained with iodine, which interacts with the crystal violet, forming a larger, more stable complex. The slide then is immersed in an alcohol, or an alcohol/acetone mixture, often referred to as a decolorizer. Exposure to this reagent results in the loss of the crystal violet/iodine complex from Gram-negative cells. In Gram-positive cells, conversely, the crystal violet/iodine complex remains. The slide then is dipped into a counterstain, safranin. This stain provides a light red color to cells, thereby allowing the viewer to see cells from which the primary stain was removed. The stained cells then can be viewed with a light microscope (**Figure B2.4**).

Interpretation

Gram-positive = Purple (retains the crystal violet primary stain)

Gram-negative = Pink or light red (loses the crystal violet but is stained with safranin)

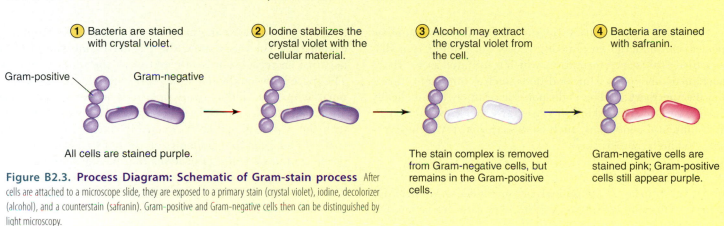

① Bacteria are stained with crystal violet.

② Iodine stabilizes the crystal violet with the cellular material.

③ Alcohol may extract the crystal violet from the cell.

④ Bacteria are stained with safranin.

Gram-positive   Gram-negative

All cells are stained purple.

The stain complex is removed from Gram-negative cells, but remains in the Gram-positive cells.

Gram-negative cells are stained pink; Gram-positive cells still appear purple.

Figure B2.3. Process Diagram: Schematic of Gram-stain process After cells are attached to a microscope slide, they are exposed to a primary stain (crystal violet), iodine, decolorizer (alcohol), and a counterstain (safranin). Gram-positive and Gram-negative cells then can be distinguished by light microscopy.

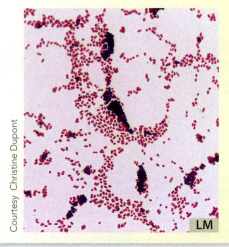

Courtesy Christine Dupont

LM

Figure B2.4. Gram-stained mixture of bacterial cells In this image, a mixture of Gram-negative (pink) and Gram-positive (purple) bacteria can be seen. Developed over 125 years ago, the Gram staining technique remains an easy and useful means of classifying, and, potentially, identifying bacteria.

● Test Your Understanding ....

A student carried out the Gram stain on a culture of Gram-negative cells but forgot to add the safranin in the last step. What would the student see when she examines the stained cells under the microscope? Explain.

Many of the teichoic acid molecules are firmly anchored in the cell wall by covalent attachment to peptidoglycan chains. Alternatively, some teichoic acids are connected to the plasma membrane, through the covalent addition of a lipid tail. This subset of teichoic acid molecules is called **lipoteichoic acid (LTA)**.

CONNECTIONS During an infection, the lipoteichoic acid of Gram-positive bacteria can induce a strong inflammatory response in the host. The mechanism of this process, and its potential dangers, are addressed in **Section 21.1**.

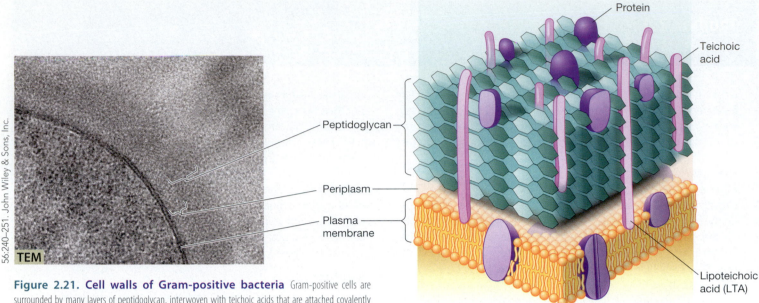

From V. R. F. Matias et al. 2005. Mol Microbial 56:240–251. John Wiley & Sons, Inc.

Protein

Teichoic acid

Peptidoglycan

Periplasm

Plasma membrane

Lipoteichoic acid (LTA)

TEM

Figure 2.21. Cell walls of Gram-positive bacteria Gram-positive cells are surrounded by many layers of peptidoglycan, interwoven with teichoic acids that are attached covalently to the peptidoglycan. Lipoteichoic acids (LTAs) have a lipid tail that is inserted into the plasma membrane. Various proteins also are inserted into the Gram-positive cell wall.

Some Gram-positive bacteria, such as *Bacillus* and *Clostridium* species, can undergo an amazing process of cellular remodeling called **endospore** formation, which is initiated as a survival mechanism under stressful conditions, such as imminent starvation **(Perspective 2.2)**. The endospores are largely metabolically inert structures that exhibit increased resistance to many harsh environmental conditions, such as desiccation, UV light exposure, and high temperatures.

Gram-negative Cells

A relatively thin layer of peptidoglycan surrounds Gram-negative bacteria. This slim coating may seem like scant protection, but it is supplemented by another structure, the **outer membrane (Figure 2.22)**. For this reason, the plasma membrane of Gram-negative bacteria often is referred to as the inner membrane. The space between the inner and outer membranes is called the **periplasm**. The width of the periplasm is not clear, and probably fluctuates; the *E. coli* periplasm is estimated to be 13–25 nm wide, with the peptidoglycan layer occupying roughly a third of that space. In addition to peptidoglycan assembly enzymes, the periplasm contains a roster of proteins that includes proteins to aid in nutrient uptake and components of protein export systems. The periplasm also can contain oligosaccharides that help the bacteria adjust to changes in the osmolarity of the medium.

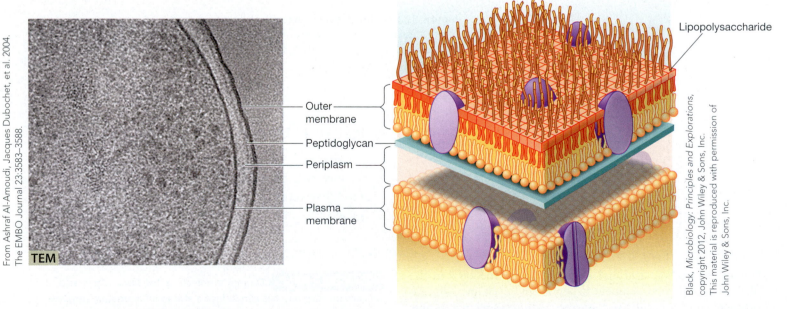

From Ashraf Al-Amoudi, Jacques Dubochet, et al. 2004. The EMBO Journal 23:3583–3588.

Lipopolysaccharide

Outer membrane

Peptidoglycan

Periplasm

Plasma membrane

TEM

Figure 2.22. Gram-negative bacteria have an outer membrane In Gram-negative bacteria, a thin peptidoglycan layer is surrounded by an outer membrane. The presence of this second membrane creates the periplasm as a separate compartment. The outer membrane contains phospholipids on its inner surface and lipopolysaccharide (LPS) on its outer surface. Like the plasma membrane, it also contains proteins.

Endospores have dramatically thickened cell envelopes (**Figure B2.5**), with additional layers of protein outside of the peptidoglycan. This thick protective coat makes them resistant to desiccation and chemical attack. Endospores shut down their metabolism completely, and compact chromosomal DNA tightly with protective proteins. As a consequence, they are incredibly durable and can remain viable for many years. Exactly how long endospores can survive is not known; thousands of years seems likely, but in recent years work has been published suggesting that endospore survival for millions of years may be possible. To learn more about this debate, see the Suggested Reading resources at the end of this chapter.

Regardless of the precise longevity of endospores, the formation of these specialized cells has fascinated many microbiologists as a model for cellular remodeling and development. The cellular remodeling involved in endospore formation requires major changes in gene expression. Studies of endospore formation, as a result, have provided important insight into basic properties of gene regulation. Endospores also have historic and practical relevance in the field of microbiology. Robert Koch, one of the founding fathers of microbiology, first observed endospore formation while studying *Bacillus anthracis*, the bacterium that causes anthrax.

The formation of endospores by *B. anthracis* has significant implications for the natural transmission of this pathogen. Usually, a person becomes infected by inhaling or ingesting endospores. These endospores can be present in the soil or associated with animal products, such as wool. Terrorists, intent on spreading this disease, also can use these endospores. During the fall of 2001, letters containing *B. anthracis* endospores were mailed to several people in the United States by a terrorist or terrorist group. Upon handling or opening these letters, several people became infected, five of whom died.

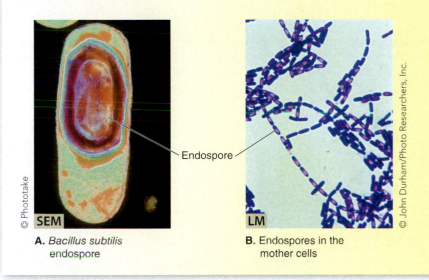

© Phototake

SEM

A. *Bacillus subtilis* endospore

— Endospore

© John Durham/Photo Researchers, Inc.

LM

B. Endospores in the mother cells

Figure B2.5. Endospores A. This colorized image of a *Bacillus subtilis* endospore shows the dramatically modified cell structure. The cytoplasm is shrunken and almost completely dehydrated, with densely packed chromosomal DNA and ribosomes. A thick protective coat surrounds the endospore. **B.** In this micrograph of *Bacillus subtilis* cells undergoing sporulation, the endospores (which appear as light purple ovals within the darkly stained cells) are visible within the mother cell.

The outer membrane of Gram-negative bacteria is a bilayer, but it only contains phospholipids in the inner leaf of the bilayer. The outer leaf is composed of a molecule called **lipopolysaccharide (LPS)** (**Figure 2.23**). LPS molecules have three distinct parts: lipid A, a core polysaccharide, and the O side chain. The hydrophobic portion of LPS constitutes the outer layer of the membrane bilayer, while the polysaccharide portion is exposed on the cell surface. The lipid A component, which is very similar in all Gram-negative bacteria, includes the hydrophobic hydrocarbon chains. The organization of the core polysaccharide also is conserved, although the identities and sequence of the sugar building blocks can vary somewhat between species. In contrast, the sugars in the O (outer) side chain can vary dramatically, even between different variants of bacteria that clearly are members of the same species. Like the LTA of Gram-positive bacteria, LPS from invading Gram-negative bacteria triggers a powerful inflammatory response in the human body that is responsible for many of the symptoms of infections.

The outer membrane of Gram-negative bacteria provides protection from at least some environmental threats. Because large polar molecules generally cannot cross this membrane, it serves to retain periplasmic proteins, and can keep proteins from the outside world out of the periplasm and away from the plasma membrane. We noted earlier that lysozyme is one of the human body's defenses against bacteria, but the outer membrane protects Gram-negative bacteria from its attack.

The outer membrane is attached to the cell wall by numerous proteins. The most abundant of these proteins is murein lipoprotein, which is covalently linked to the peptidoglycan through its carboxyl terminal amino acid (a lysine), and is covalently linked to lipid chains embedded in the outer membrane through its amino-terminal amino acid (a cysteine). Both murein lipoprotein and another attachment protein, peptidoglycan-associated lipoprotein, have been shown to be critically important for the stability of the outer membrane in Gram-negative bacteria.

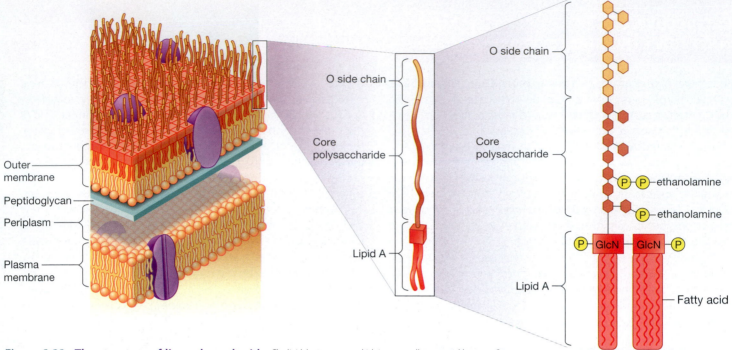

Figure 2.23. The structure of lipopolysaccharide The lipid A component, which is structurally conserved between Gram-negative species, contains the hydrocarbon chains comprising the hydrophobic interior of the membrane. The sequence of the sugar building blocks in the core polysaccharide is less conserved between species, while the sugars in the outer O side chain can vary even more dramatically in sequence, even between variants of a species. Representative sugars are shown here.

Nutrient Transport Through the Outer Membrane The outer membrane of Gram-negative bacteria, of course, cannot be completely impervious. Various nutrients, for instance, must cross this barrier and then the plasma membrane to be used by the cell. To allow the entry of these essential molecules, the outer membrane of Gram-negative bacteria contains numerous channels. The most common of these channels are the **porins**. Porin proteins typically form trimeric, or three subunit, pores through the outer membrane. These pores allow diffusion of small polar molecules, including nutrients, across the outer membrane from the external environment into the periplasm, where they are available to plasma membrane transport systems **(Figure 2.24)**. In *E. coli*, glucose, for instance, enters the periplasm by passing through a porin channel. Once inside the periplasm, glucose then is transported into the periplasm via an active transport system.

The largest porin channel will allow entry of molecules up to a molecular weight of approximately 600 daltons. For reference, the disaccharide sucrose has a molecular weight of 342 Da. This molecular weight restriction helps to explain why Gram-negative bacteria are naturally resistant to large peptidoglycan-targeting antibiotics like vancomycin (MW ~1.5 kDa), as well as proteins such as lysozyme (MW ~14.3 kDa). Gram-positive bacteria, which have no protective outer membrane, are generally susceptible to vancomycin. Some antibiotics are small enough to pass through outer membrane pores, but under the selective pressure of antibiotic exposure, bacteria containing mutant porins have been isolated. In these cases, the altered porins are apparently even more exclusive, keeping the antibiotics out that normally would have entered the periplasm.

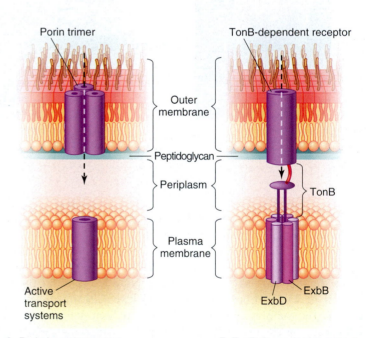

A. Porin-based transport   **B.** TonB-dependent transport

Figure 2.24. Porins and TonB-dependent receptors facilitate movement across the outer membrane. A. Porins create trimeric (three subunit) protein channels for the facilitated diffusion of hydrophilic molecules across the outer membrane. Once in the periplasm, these molecules then may cross the plasma membrane via an active transport system. **B.** The TonB-dependent transporter system allows high affinity active transport of substrates through the outer membrane. The TonB receptor binds to its ligand. Energy from the proton motive force allows TonB, complexed with ExbB and ExbD, to interact with this receptor and triggers the movement of the substance across the outer membrane. Again, a plasma membrane active transport system then facilitates the movement of the substance into the cytoplasm.

When nutrients are present in low concentrations in the environment, passive nutrient uptake systems such as porins that depend on favorable concentration gradients may be inadequate to supply the cell's needs. Bacteria then rely on high-affinity transporters. In addition to porins, the outer membrane contains proteins called TonB-dependent receptors that bind scarce nutrients (such as iron and vitamin B_{12}) with high affinity, and deliver them into the periplasm by active transport. In addition to the TonB-dependent receptor in the outer membrane, this active transport system consists of three other proteins. The ExbB and ExbD proteins are embedded in the plasma membrane and seem to form a complex with TonB (see Figure 2.24B). Active transport across the outer membrane requires a source of energy, which the TonB system obtains from the proton motive force across the plasma membrane. The energy in this proton gradient allows TonB to interact with the TonB receptor, triggering it to change its conformation and release its substrate into the periplasm. How energy actually is used in this process is not well understood. From the periplasm, the substrate then can be delivered into the cytoplasm by another plasma membrane transport system, such as the ABC transporter used for B_{12} in *E. coli*. Gram-negative bacteria that live in mammalian guts have just a few kinds of TonB-dependent receptors, but free-living bacteria from less nutrient-rich habitats often have dozens of different kinds of these receptors.

Protein Export and the Outer Membrane For Gram-negative bacteria, secreted proteins require special export systems to cross the outer membrane. Research has revealed that several distinct mechanisms seem to be involved in secretion. Often the general secretory pathway (see Figure 2.13) is first used to transport the protein into the periplasm. Some proteins, called autotransporters, then catalyze their own transit across the outer membrane, but most proteins seem to need the assistance of export systems.

Some exported proteins avoid the general secretory pathway, as they lack a hydrophobic signal peptide at the amino-terminus of the polypeptide chain. These proteins cross the plasma membrane and outer membrane in a single step, with no periplasmic intermediate. One of these single-step export systems, called the type III secretion pathway, has a particularly unusual evolutionary history. The secretory apparatus resembles a hypodermic syringe projecting from the surface of the bacterium **(Figure 2.25).** This form is appropriate as it literally injects proteins such as toxins into host cells. Several of the components of the type III secretion system are evolutionarily related to proteins involved in making flagella. Flagellar proteins and proteins exported through the type III secretion system appear to be secreted by a similar mechanism, entering a pore at the base of the flagellum or type III needle, then moving through a channel in the core of the flagellar filament or secretory syringe to emerge at the other end. Flagellins assemble at the tip of the growing filament, while type III substrates are released outside the cell, or directly into a host cell penetrated by the secretory needle. Flagella are common in both Gram-positive and Gram-negative bacteria, so the flagellum is probably an ancient invention. Only Gram-negative bacteria, however, possess the type III secretion system, so it may have evolved by a duplication of genes encoding flagellar secretory components in a Gram-negative ancestor.

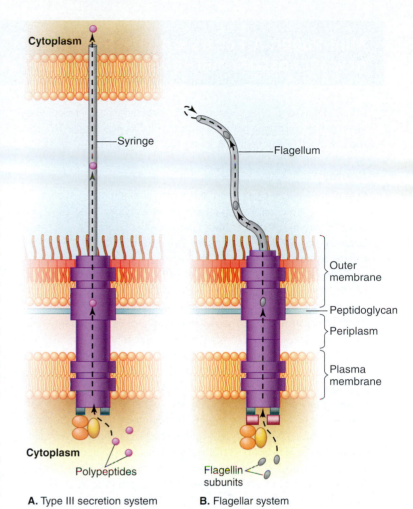

Figure 2.25. Type III secretion pathway and flagellar synthesis
A. The type III secretion apparatus consists of several cytoplasmic and integral membrane proteins, as well as a hollow, syringe-like extension. Newly synthesized polypeptides in the cytoplasm are delivered to the secretory channel in the plasma membrane. They pass through a gated channel created by integral membrane proteins, then continue out through the hollow "syringe." The syringe is capable of penetrating the surface of host cells and injecting various cytotoxins. **B.** Several proteins involved in flagellar assembly are evolutionarily related to the type III system, particularly the plasma membrane components that contribute to the secretory channel. Flagellin subunits, the primary substrates for this secretory system, move through the hollow filament and are added to the tip of a growing flagellum.

Many questions remain about protein export in Gram-negative bacteria. Why are there so many routes to get out of the cell? How are the substrates for these pathways identified? What energizes protein export across the outer membrane, given that there is no ATP in the periplasm, and no proton motive force across the outer membrane? How are export systems configured in the outer membrane? Finally, can we determine a way to block protein export, to possibly disarm pathogens that rely on it?

CONNECTIONS Many bacterial pathogens cause disease by secreting proteases, toxins, and other factors that damage host cells. As we will see in **Section 21.1**, these molecules often benefit the pathogen by allowing it to obtain resources from the host or evade host defense mechanisms.

V. R. F. Matias and T. J. Beveridge. 2005. Cryo-electron microscopy reveals native polymeric cell wall structure in *Bacillus subtilis* 168 and the existence of a periplasmic space. Mol Microbiol 56:240–251.

Context

One of the earliest recognized differential characteristics of bacteria was the structure of the cell envelope. As we saw earlier, the Gram stain allows us to differentiate Gram-positive and Gram-negative bacteria. The thick peptidoglycan cell wall of Gram-positive cells retains crystal violet stain following treatment with iodine and washing with ethanol, while the thinner cell wall and outer membrane of the Gram-negative cells does not retain this stain. Transmission electron microscopy (TEM) has been used to investigate more thoroughly the structure of the cell envelopes of Gram-negative and Gram-positive cells. From these studies, it is clear that Gram-negative cells have a space between the plasma membrane and the outer membrane called the periplasm, which is partially filled by the thin peptidoglycan cell wall. In contrast, Gram-positive cells have no outer membrane, and their cell wall consists of peptidoglycan, polymers such as teichoic acids, and proteins. There is no periplasm detectable in these cells by TEM. Recognizing that the preparation of cells for visualization by TEM is an exacting process involving harsh conditions, researchers worried that the preparation process might result in a flawed visualization of the structure of the Gram-positive cell envelope. To address this concern, Professor Terry Beveridge and graduate student Valério Matias at the University of Guelph in Canada used a powerful technique of sample preparation in which cells are frozen and then visualized in the TEM in this frozen state. The resulting images provide a clearer view of the structure of the Gram-positive cell envelope, and reveal that these cells do indeed have a periplasm!

Experiment

The work presented in this paper consisted of the application of an ultralow temperature method of sample preparation and examination called cryo-electron microscopy (CEM). This method involves rapid freezing of the samples to temperatures of −140 to −180°C in the presence of glycerol to prevent ice crystal formation, and then cutting of thin sections immediately before viewing in the electron microscope. It is a challenging method, since it is not possible to stain the sections before viewing, and the use of a dry knife at very low temperatures to prepare the thin sections tends to introduce knife marks and sample compressions, as apparent in **Figure B2.6**.

The sample preparation began with broth cultures of *Bacillus subtilis*, from which cell pellets were recovered by centrifugation. The pellets then were injected into a thin copper cooling tube and rapidly frozen under high pressure. Cross sections, approximately 50 nm thick, were produced by cutting with a diamond knife, and then viewed in the electron microscope at −170°C.

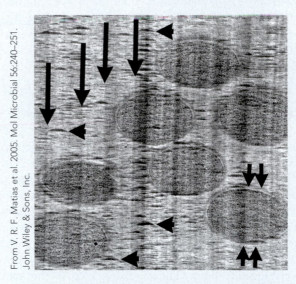

From V. R. F. Matias et al. 2005. Mol Microbial 56:240–251. John Wiley & Sons, Inc.

Figure B2.6. Artifacts associated with cryo-electron microscopy The cryo-electron microscopy technique results in many artifacts, including knife marks, crevasses (short arrows), and compressions (double arrows). The pictures are not always pretty, but they can be very informative.

Summary: Gram-negative Versus Gram-positive Cell Envelopes

The cell envelopes of Gram-negative and Gram-positive bacteria differ significantly. As we have seen, Gram-positive bacteria, with their multiple layers of peptidoglycan, have a much thicker cell wall than do Gram-negative bacteria. Gram-negative bacteria, on the other hand, have an additional outer membrane layer not found in Gram-positive species. Clearly the surfaces these cells expose to the outside world are distinctive—peptidoglycan, teichoic acids, and proteins in the case of Gram-positives, versus lipopolysaccharides and proteins for Gram-negative bacteria. Although these bacterial lineages diverged from each other billions of years ago, both types of cell envelope have been successful in evolutionary terms. Both Gram-negative and Gram-positive bacteria are abundant on Earth today, and most contemporary microbial communities include at least some species of both types of bacteria. Furthermore, our understanding of these basic structures continues to advance (**Mini-Paper**).

Now that we have discussed the structural differences between these kinds of cells, you might ask why the Gram-positive

The images showed something that had never been seen before. The cell walls, rather than being uniformly dense, had two well-defined layers. The less dense inner layer, about 22 nm wide, was called the inner wall zone (IWZ), while the denser outer zone, about 33 nm wide, was called the outer wall zone (OWZ) (**Figure B2.7**). Based on these images, the researchers concluded that the OWZ is the true peptidoglycan cell wall, while the IWZ is a periplasm. Although the IWZ appears to be very low density, it is likely filled with important cellular components such as enzymes, secreted proteins, and proteins involved in building the peptidoglycan cell wall.

Impact

The real breakthrough of this study was using a method that allowed the direct comparison of material density within the cell envelope. Commonly used methods of sample preparation for TEM involve embedding samples into plastic followed by cutting of sections and the use of heavy metal stains to bind charged molecules to detect beyond the background of the plastic. The resulting images are therefore not representative of the actual differences in material density, but rather indicate where the staining occurred. In comparison, the direct visualization of frozen sections that had not been embedded in plastic allows a truer view of the actual density.

Since this study was published, similar studies have been done in other Gram-positive bacteria such as *Staphylococcus aureus*, and the results have been similar. Perhaps, a periplasm is a common feature of Gram-positive organisms. Given the diversity of microbes that are known, it will be interesting to use this type of microscopy to provide a more accurate view of the cell envelope ultrastructure. Interestingly, other work in which this technique has been applied to visualization of the Gram-negative cell envelope has suggested that the Gram-negative periplasm is very low density, rather than the higher density gel that was previously thought to be the case.

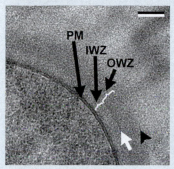

From V. R. F. Matias et al. 2005. Mol Microbial 56:240–251. John Wiley & Sons, Inc.

A. Ribosomes

B. Plasma membrane

Figure B2.7. Cryo-electron microscopy cross sections of *B. subtilis* cells A. Ribosomes are visible in the cytoplasm. **B.** Three distinct layers can be seen in the cell envelope: the plasma membrane (PM), the inner wall zone (IWZ), and the outer wall zone (OWZ).

Questions for Discussion

1. What molecular activities do you think are carried out in the Gram-positive periplasm?

2. Do you think that a periplasm might be a requirement for all cells that have a cell wall? Why or why not?

cell envelope allows bacteria to retain the crystal violet, even after being washed with ethanol, while the Gram-negative cell envelope does not. The alcohol probably acts as a solvent to dissolve the outer membrane of Gram-negative bacteria, helping the crystal violet to leach out of the thin peptidoglycan layer. Although there is no outer membrane to protect the Gram-positive cell, the cell wall is much thicker, and the alcohol may toughen it further by dehydrating the structure, helping to trap enough crystal violet that the cells retain the intense purple color. Bacteria that lack a peptidoglycan layer, like the mycoplasmas that we discussed earlier, cannot be Gram-stained and are not classified as Gram-negative or Gram-positive.

2.4 Fact Check

1. What are the key components of the bacterial plasma membrane and what are its functions?
2. Describe the structure and function of the bacterial cell wall.
3. Differentiate between Gram-positive and Gram-negative bacterial cell envelopes.
4. Describe the role of porins and TonB-dependent receptors in Gram-negative bacteria.
5. What is the type III secretion pathway?

2.5 The bacterial cell surface

How do structures on the surface of bacterial cells allow for complex interactions with the environment?

Cells directly interact with the outside world at the cell surface. Molecules present on the cell surface perform many jobs: to allow movement, stick to surfaces, sense the environment, and acquire nutrients **(Table 2.2)**. In this section, we will discuss the specialized surface structures and organelles of bacterial cells that mediate these activities.

Motility from flagella

For active movement, bacteria most commonly use **flagella**, spiral filaments that stick out from the surface of the cell and rotate to propel the cell **(Figure 2.26)**. Some bacteria have flagella only at the ends of the cell, in which case they are called polar flagella. **Monotrichous** bacteria have only a single flagellum (*trich* being Latin for "hair"). **Lophotrichous** bacteria have more than one flagellum at one or both ends of the cell, and **peritrichous** bacteria have multiple flagella spread all over the surface of the cell.

The flagellar filament is built from many copies of the protein flagellin. Often 5 to 10 μm in length, the flagellar filament is usually longer than the cell itself. The flagellum is anchored in the cell envelope by the basal body, a disc-like structure that serves as the structural foundation of the flagellum and interfaces with the motor that drives rotation of the flagellar filament. The central rod emerging from the basal body transitions to a curved hook, to which the filament is attached. The helical filament acts as a screw propeller, driving or pulling the bacterium through the surrounding fluid like a submarine. The flagellar motor is located in the plasma membrane, where it converts energy from the proton motive force to drive rotation of the filament. Bacterial flagella are complex structures that involve more than 40 distinct proteins, assembled in modular fashion from the plasma membrane outward. As we noted earlier in this chapter, most of the protein subunits that contribute to the flagellum structure exit the cytoplasm through a pore in the center of the basal body, before assembling at the tip. The expression, secretion, and assembly of flagellar subunits in bacteria comprise a highly ordered process that has served as a model system for understanding developmental regulation in microbes.

TABLE 2.2 Molecules and structures on bacterial cell surfaces

Organelle or molecule	Composition	Location	Function
Lipopolysaccharide (LPS)	Lipid, polysaccharide	Outer layer of Gram-negative outer membrane; lipid portion embedded in membrane; polysaccharide exposed on surface	Stabilizes membrane; elicits an inflammatory response in the human body
Lipoteichoic acid (LTA)	Lipid, polysaccharide	Found in peptidoglycan layer of Gram-positive bacteria	Unknown; elicits an inflammatory response in the human body
Peptidoglycan	Polysaccharide backbone crosslinked with peptides	In Gram-positive bacteria, usually exposed to environment In Gram-negative bacteria, covered by the outer membrane	Maintains shape and provides structural integrity to cell
Porins	Proteins	Embedded in Gram-negative outer membrane	Form pores that allow diffusion of nutrients and water through outer membrane
TonB-dependent receptors	Proteins	Embedded in Gram-negative outer membrane	Catalyze high-affinity active transport of molecules across outer membrane
Flagella	Protein subunits	Extend outward from surface, except in spirochetes, where periplasmic flagella wrap around cell	Provide motility
Pili	Protein subunits	Extend outward from cell	Allow attachment; tip often binds to specific molecules. In some bacteria, pili are retractable and allow "twitching motility."
Capsule	Usually loose network of polysaccharides	Covers surface of cell	Protects from phagocytes; contributes to biofilm formation
Surface array (S-layer)	Protein	Covers surface of cell	May protect from bacteriophage

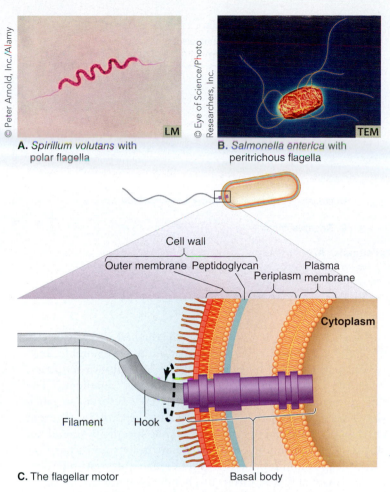

A. *Spirillum volutans* with polar flagella

B. *Salmonella enterica* with peritrichous flagella

Cell wall

Outer membrane Peptidoglycan Plasma membrane
                 Periplasm

Cytoplasm

Filament Hook

C. The flagellar motor

Basal body

Figure 2.26. Bacterial flagella all have a common molecular structure. **A.** Bacteria with polar flagella, such as the *Spirillum volutans* cell, only have flagella emerging from the ends of the cell. **B.** Peritrichous flagella, as seen on this *Salmonella enterica* cell, are distributed across the cell surface. **C.** The flagellum is embedded in the cell wall and plasma membrane through the basal body. The flagellar motor, a protein complex embedded in the plasma membrane, rotates the flagellum.

Structures of bacterial, archaeal, and eukaryal flagella are distinct. Archaeal flagellar filaments look superficially similar to bacterial flagella, and likewise function as rotating propellers. Archaeal flagella seem to be constructed differently, though, growing from the base rather than the tip, and lacking rings in the basal body. The known protein components of archaeal flagella are generally not related to bacterial flagellar subunits, so similarities in structure and function between bacterial and archaeal flagella probably resulted from convergent evolution. In other words, the structures arose independently in both groups of organisms to serve the same function. Eukaryal flagella are structurally and functionally distinct from both bacterial and archaeal structures; they are surrounded by the plasma membrane and do not actually rotate. Internally, eukaryal flagella are built from a bundle of microtubules. We will talk about eukaryal flagella in more detail in Section 3.1.

Bacteria with a single polar flagellum will be pushed or pulled by the flagellum, depending on which direction it rotates. It is a bit more complex for bacteria such as *E. coli* and *Salmonella* Typhimurium, which have peritrichous flagella. In this case, rotation of the flagellar motor in one direction (counterclockwise for *E. coli* and related Gram-negative bacteria) causes the flagella to bundle together and push the cell forward. Rotation in the other direction causes the flagella to fly apart and point in different directions, which causes the cell to tumble in place (**Figure 2.27**). This tumbling randomly reorients the cell, so that when flagellar rotation is reversed once again, the cell sets off in a new direction.

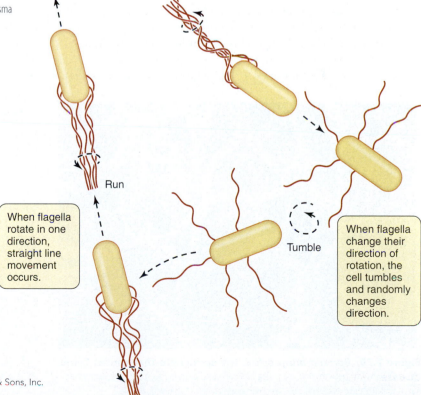

Run

When flagella rotate in one direction, straight line movement occurs.

Tumble

When flagella change their direction of rotation, the cell tumbles and randomly changes direction.

Figure 2.27. Bacterial movement by straight line "runs" and random "tumbles" Peritrichous flagella rotating in one direction will bundle together and cause the bacterium to move in a straight direction. When the direction of rotation of the flagella is changed, the bacterium tumbles and randomly reorients. When flagellar rotation switches again, the cell takes off in a new direction. Bacterial cells with a single polar flagellum are pushed or pulled depending on the direction of flagellar rotation.

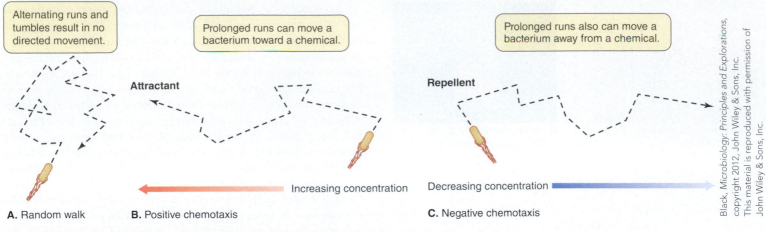

Alternating runs and tumbles result in no directed movement.

Prolonged runs can move a bacterium toward a chemical.

Prolonged runs also can move a bacterium away from a chemical.

Attractant

Repellent

Increasing concentration

Decreasing concentration

A. Random walk   **B.** Positive chemotaxis   **C.** Negative chemotaxis

Black, *Microbiology: Principles and Explorations,* copyright 2012, John Wiley & Sons, Inc. This material is reproduced with permission of John Wiley & Sons, Inc.

Figure 2.28. Directed motility lets bacteria respond to outside chemical signals. A. A peritrichously flagellated bacterium such as *E. coli*, detecting no chemical signals, will undergo alternating runs and tumbles, resulting in no directed movement. This type of non-directed movement often is referred to as a "random walk." **B.** Prolonging the swimming mode ("running") by delaying tumbles will move the bacterium toward higher concentrations of a desirable compound, resulting in positive chemotaxis. **C.** Similarly, bacteria can move away from undesirable compounds, resulting in negative chemotaxis.

Bacterial cells sense many chemicals in their environment through receptors in the plasma membrane that communicate indirectly with the flagellar motor through intermediary proteins in the cytoplasm. This process of using chemical signals from the environment to direct motility is called **chemotaxis (Figure 2.28).** The cytoplasmic control system ensures that all of the motors in cells with multiple flagella tend to rotate in the same direction. If the chemoreceptors detect a higher attractant concentration as a bacterial cell moves, then the flagellar motors tend to continue rotating in the direction causing forward motion. The bacterium, in other words, moves up the gradient of the desired nutrient. This type of movement is referred to as a run. If the concentration of an attractant is not increasing, or if the concentration of a repellent, a chemical the cell wants to avoid, is increasing, then the flagellar motor switches more often to the direction that results in a tumble. The tumbling state is transient, and the cell sets off on a run in a new direction.

Chemotactic attractants are often molecules that can be productively metabolized (e.g., sugars and amino acids), whereas chemicals that could damage the cell (e.g., acids and alcohols) are more likely to be interpreted as repellents. Bacterial species often have many different **chemoreceptors,** receptors that detect these attractants and repellents. The different chemoreceptors, in turn, each have different specificities. They interact with, and thus signal a response to, different chemical signals. This degree of receptor specificity should not surprise us. Depending on where bacteria live, and what they can transport and productively metabolize, different microbes might reasonably find different chemicals attractive or repellent. Several bacterial pathogens rely on motility and chemotaxis to establish infections, including some variants of *E. coli* that cause diarrhea. These **strains,** or genetically and often phenotypically distinct subtypes, of *E. coli* attach to the lining of the intestine. Chemotaxis may help them to reach this surface before they are flushed through the GI tract.

Spirochetes are long, thin, corkscrew-like cells **(Figure 2.29).** These unusually shaped cells do not have surface flagella. Surprisingly, they do depend on flagella for motility, but they

The periplasmic flagella (arrow) occupy the periplasm and extend the length of the bacterium.

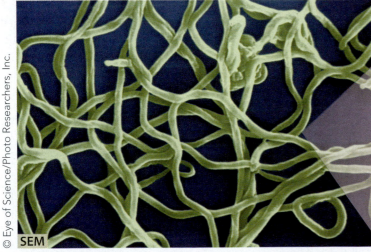

SEM

Figure 2.29. *Borrelia burgdorferi,* **the spirochete that causes Lyme disease** As seen in this colorized image, flagella are not visible on the surface of *Borrelia burgdorferi* cells. A cross section of these cells (inset), however, shows a large number of periplasmic flagella.

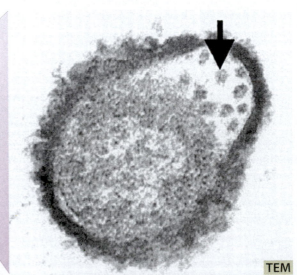

TEM

From Nyles W. Charon, PNAS, September 26, 2000, vol. 97, no. 20, 10899–10904. © 2000 National Academy of Sciences, U.S.A.

keep the flagellar filament shielded inside the periplasm. The filament is basically around the cell body. When the flagellum rotates, it causes the entire cell to spin like a tiny screw propeller through the medium. Spirochetes move best in viscous environments, such as in mucus that lines the intestinal tract. Some spirochetes, such as *Borrelia burgdorferi*, the cause of Lyme disease, and *Treponema pallidum*, which causes syphilis, are important pathogens of humans. This unique form of motility probably affects their ability to invade host tissues.

Flagellar motility works best in liquid environments, but some bacteria, like the urinary tract pathogen *Proteus mirabilis*, adapt to contact with surfaces by becoming long, hyper-flagellated swarmer cells **(Figure 2.30)**. In the laboratory, *P. mirabilis* swarming can be easily observed. Colonies growing on agar plates rapidly spread over the entire surface of the agar in the course of a few hours. *Proteus mirabilis* **(Microbes in Focus 2.2)** coordinates swarming with the expression of other genes needed for infection, suggesting that this mode of motility is useful during infection, perhaps because it facilitates movement in host tissues.

Non-flagellar motility

Some forms of bacterial motility do not depend on flagella. During **gliding motility**, some non-flagellated bacteria, such as myxobacteria and cyanobacteria, slide smoothly over surfaces. The mechanism of gliding motility is not well understood, and there may be more than one mechanism. Other bacteria that exhibit surface-dependent motility rely on fibers called pili that help the bacteria move along surfaces by a slow, jerky process called twitching motility. The pili bind to the surface and appear to be capable of rapid retraction to pull the cell body along. *Neisseria meningitidis* and *Pseudomonas aeruginosa* can use this type of motility, although *P. aeruginosa* usually relies on flagella for motility. Only some pili-forming bacteria are capable of twitching motility, suggesting that this action is not an intrinsic property of all pili.

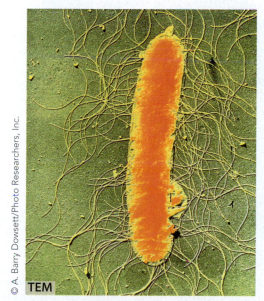

TEM

Figure 2.30. Hyper-flagellated *Proteus mirabilis* *Proteus mirabilis* swarmer cells become elongated and can produce hundreds of flagella per cell. In this form, the cells are able to move effectively on solid surfaces and rapidly establish many colonies.

Microbes in Focus 2.2
PROTEUS MIRABILIS: A SWARMING BACTERIUM

Habitat: Exists in soil and water. Also commonly found in the GI tract of humans. Can cause urinary tract infections.

Description: Gram-negative, rod-shaped, motile bacterium, approximately 0.5 μm wide and 2.0 μm long

Key Features: In the laboratory, *P. mirabilis* can be identified by its swarming abilities. In an aqueous environment, *P. mirabilis* cells typically exist as swimmers—relatively short cells with about 10 flagella. Upon contact with a solid surface, some cells differentiate into swarmer cells that are much longer and contain numerous flagella. These swarmers cause colonies of *P. mirabilis* on agar plates to spread out over the plate.

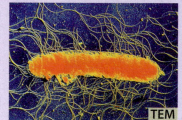

TEM

Another fascinating mechanism of motility has been observed in some bacteria that invade eukaryal cells. *Shigella dysenteriae*, a Gram-negative, rod-shaped bacterium that causes bloody diarrhea (dysentery), and *Listeria monocytogenes*, a Gram-positive bacterium that can cause serious food poisoning, have evolved a strategy for hijacking the mammalian actin cytoskeleton. When these bacteria invade cells, they induce actin polymers to form at one end of the bacterium. Polymerization of actin propels the bacterium through the cytoplasm of the host cell **(Figure 2.31)**, and the force generated can actually shoot the bacterial cell through the host plasma membrane into an adjacent cell, which may be the real purpose of this actin-based motility. When they take up residence within intestinal cells, these bacteria are protected from antibodies produced by the immune system. By spreading directly from one cell to another, they effectively evade the host immune response, thereby taking full advantage of this sheltered niche.

Adherence

Many bacteria have fibers called **pili** sticking out from the cell surface. These fibers are constructed from a single type of protein subunit, called pilin. As with flagella, some bacteria display pili only at the cell poles, while others have them spread across the entire cell surface. Although we mentioned pili in the context of motility above, the more common purpose of pili is to allow bacteria to attach to surfaces, including other cells. The tip of the pilus usually contains distinct proteins that serve as adhesins designed to bind specific molecules on target surfaces. Pili can be very important for pathogenic microbes, because adherence to target cells in the host is often an early step in the infection process. Even in non-pathogens, pili that mediate surface attachment can be important for persistence in any environment where there is fluid flow.

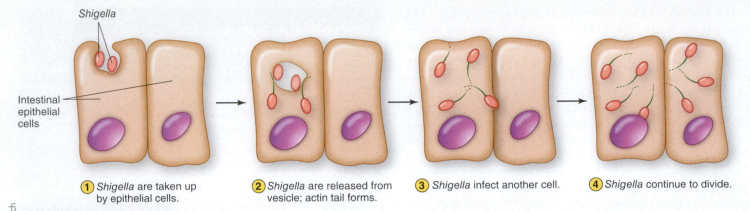

Shigella

Intestinal epithelial cells

① *Shigella* are taken up by epithelial cells.

② *Shigella* are released from vesicle; actin tail forms.

③ *Shigella* infect another cell.

④ *Shigella* continue to divide.

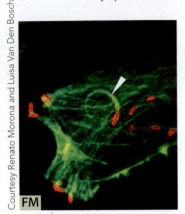

Figure 2.31. Process Diagram: *Shigella* move within and between infected cells. *Shigella flexneri* cell (red) within a HeLa cell develops an actin tail (green) to propel the bacterium. Initially, *Shigella* cells are taken up in a vesicle by an epithelial cell. The *Shigella* are released from the vesicle and recruit actin from the cytoskeleton to build an actin "tail" on one end of the bacterial cell. The growing tail (driven by actin polymerization) pushes the bacterium through the cytoplasm of the human cell and can propel it into adjacent cells as well, where it continues to divide.

Courtesy Renato Morona and Luisa Van Den Bosch

A special type of pilus, the **sex pilus**, or conjugal pilus, is used to connect bacterial cells for the transfer of plasmid DNA. Genes for constructing sex pili are found on some large plasmids that have evolved to move between bacteria. Both adhesive and conjugal pili may be found on the same *E. coli* cell. Some microbiologists prefer to use the term "pilus" exclusively for the structures involved in conjugation, and assign the alternative term **fimbria** (plural: fimbriae) to cell surface fibers used for adhesion, and lacking the associated intracellular machinery for conjugal plasmid transfer **(Figure 2.32)**.

Gram-negative bacteria such as *Caulobacter* and *Hyphomonas* attach to surfaces using a **stalk**. The stalk is not simply a protein assembly (like flagella and pili), but is a tubular extension of the entire cell envelope **(Figure 2.33)**. The stalk is often tipped

by an adhesive structure, called a holdfast, made from secreted polysaccharides. Because the stalk extends the cell envelope and contains nutrient transport systems, it effectively increases the surface-to-volume ratio of the cell and boosts nutrient uptake capacity. Stalked bacteria are common in aquatic habitats, particularly where nutrient concentrations are very low. The stalks, researchers hypothesize, serve as extensions of the cell surface that can aid in nutrient acquisition.

Capsules

Colonies of some pathogenic bacteria, such as *Streptococcus pneumoniae*, a cause of bacterial pneumonia, and *Neisseria meningitidis*, a major cause of bacterial meningitis, have a smooth, glistening appearance. A thick layer of polysaccharides, called a **capsule**, surrounds these cells **(Figure 2.34)**. Many pathogens use capsules to shield themselves from host defense systems, particularly host cells that are capable of engulfing and destroying bacteria. Phagocytic cells often cannot recognize encapsulated microbes as foreign invaders

@ Bacterial cell structure ANIMATION

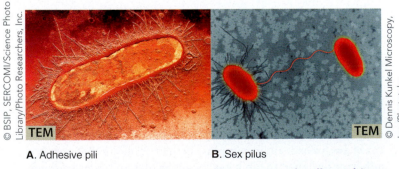

A. Adhesive pili

B. Sex pilus

© BSIP, SERCOMI/Science Photo Library/Photo Researchers, Inc.

© Dennis Kunkel Microscopy, Inc./Phototake

Figure 2.32. Bacterial cell surface fibers can attach cells to objects or other cells. A. Pili resemble hairs on the cell surface in this image of an *E. coli* cell. Pili are used for adhesion. **B.** The *E. coli* cell on the left is expressing adhesive pili but is also producing a thicker conjugal (sex) pilus, which has attached to the cell on the right. The conjugal pilus is involved in delivering genetic material (usually plasmids) from one cell to another. The pili in both micrographs have been shadowed, making them appear thicker than they are in reality.

Stalk

© Martin Oeggerli/Photo Researchers, Inc.

Figure 2.33. *Caulobacter crescentus* cells alternate between free swimming and attached forms. Stalks can be seen emerging from the lower poles of cells. Caulobacters have a unique life cycle in which a flagellated swarmer cell is produced each cell cycle from a non-motile stalked cell at division. The swarmer cells, with a single polar flagellum, can move in search of nutrients and differentiate into stalked cells.

worthy of destruction. Capsules also may serve other purposes. Because most capsules are made of hydrophilic material that retains water, capsules may be useful for surviving desiccation.

CONNECTIONS In **Section 7.1**, we will explore one of the classic experiments of modern biology—Griffith's demonstration of natural transformation, or the transfer of genetic material from one cell to another. Although he did not know it at the time, Griffith observed the transfer of genetic material needed for the production of a capsule from one strain of *S. pneumoniae* to another.

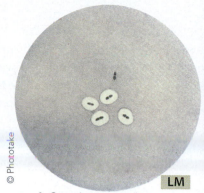
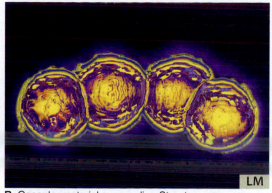

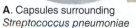

A. Capsules surrounding *Streptococcus pneumoniae*

B. Capsular material surrounding *Streptococcus zooepidemicus*

Figure 2.34. Capsules provide protection to cells. A. Photomicrograph of *Streptococcus pneumoniae* cells with capsules. The clear area around each cell shows the extent of the capsule. **B.** A colorized light micrograph showing amorphous capsular material on the surface of the *Streptococcus zooepidemicus* cells.

Polysaccharide coatings on microbes can aid in adherence of the microbes to surfaces. In wet habitats, microbes can form attached communities called **biofilms (Figure 2.35)**. Biofilms often contain more than one kind of microorganism. A biofilm may be initiated by a bacterium sticking to a surface using pili or some other attachment mechanism. The number of adherent cells grows as a result of division, or aggregation from the surrounding environment. As the cells pile up, extracellular polysaccharides form a matrix holding the biofilm together. Over time, the biofilm may get deeper and spread outward, developing into a complex structure with distinct microenvironments.

CONNECTIONS Biofilms represent complex communities in which various members of the community benefit from the presence of the other members. We will examine in **Section 15.1** how individual members of a biofilm community contribute to and benefit from the association with other members.

Surface arrays ("S-layers")

A regular, crystalline-like layer of protein referred to as a **surface array** or **S-layer (Figure 2.36)** has been observed in many bacterial cells, both Gram-positive and Gram-negative. As you might imagine, making this protein coat represents a tremendous investment of resources by the cell. Surface arrays are generally thought to act as a suit of armor, providing protective functions such as:

- preventing infection by bacteriophages (viruses that attack bacteria)
- blocking penetration by predatory bacteria such as *Bdellovibrio*
- shielding the cell from attack by a host's immune system

As we can see from this brief list, the presence of the S-layer can be quite useful. The expenditure of resources, then, has its benefits.

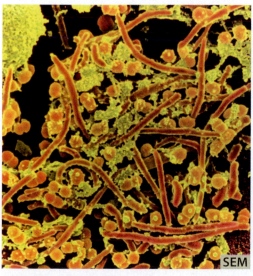
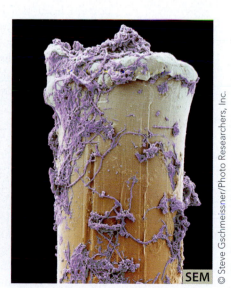
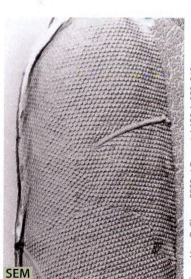

A. Biofilm on human tooth

B. Biofilm on a toothbrush bristle

Figure 2.35. Biofilms These colorized images show microbial biofilms **A.** on the surface of a human tooth and **B.** on a toothbrush bristle. Note the variety of different cell sizes and morphologies. Inside the mouth, the plaque would be covered with a film of polysaccharide and other material, but the matrix becomes dehydrated during preparation for microscopy and is not visible here.

Figure 2.36. The bacterial surface array, or S-layer The regular organization of the S-layer proteins forms a crystalline matrix covering the tip of the bacterial cell shown here.

2.6 Bacterial taxonomy

How are bacteria categorized and named?

In Chapter 1, we briefly explored the three-domain classification of life on Earth. Bacteria make up the largest and most ubiquitous domain. These organisms live in virtually all habitats, and we know far more about them than we know about the Archaea and Eukarya. Despite this wealth of information, the majority of the members of the Bacteria domain remain to be discovered and described. Because bacteria have evolved for over four billion years on a planet that has undergone continual change, bacteria exhibit incredible diversity. This diversity can be considered at several different levels, including physical, biochemical, physiological, and genetic. All of this diversity must be taken into account when describing the different types of organisms, and giving them names. In this section, we will examine how bacteria are classified and named. We will consider only those bacteria for which pure cultures have been obtained. We should note, though, that the majority of bacteria have not yet been cultivated. As microbiologists succeed in isolating more organisms in pure culture and sequencing the genes of uncultivated bacteria, more organisms will be added to the taxonomic categories.

> **CONNECTIONS** See **Perspective 15.1** to learn how scientists can communicate their findings about uncultivated bacteria using the designation *Candidatus.*

Classification and nomenclature

Why classify bacteria? The simplest answer to this question is that microbiologists, health care professionals, biotechnology workers, and other individuals working with bacteria need a mechanism to describe and keep track of the organisms. To aid in this record keeping, and to minimize confusion, a system of nomenclature has been instituted, and it is regulated by the International Code of Nomenclature of Bacteria. Sometimes referred to as the Bacteriological Code, this regulatory system is overseen by the International Committee on Systematics of Prokaryotes, and the list of valid names is updated on a continual basis. For new names to be officially registered, the name and the corresponding organism description must be published in the *International Journal of Systematic and Evolutionary Microbiology*. If the information has been published elsewhere, this publication must be reported in the journal.

Bacteria are classified using a hierarchical taxonomic system following the same general outline that is used for plants and animals. Thus, they conform to the binomial naming system of Linnaeus; each organism is given a Latinized scientific name, which is always italicized. The basic taxonomic group of **species** can be regarded as a grouping of strains that share common features, while differing considerably from other strains. A genus consists of closely related species. Organizational categories above the Genus level are Family, Order, Class, Phylum, and finally Domain **(Table 2.3)**, and each of these groups is named as well. Each named grouping is considered a **taxon** (plural: taxa).

Bacterial classification and the corresponding nomenclature initially depended on comparison of many of the physical characteristics of cells that we have discussed in the earlier sections of this chapter. Scientists grouped organisms using as much information as was available, including cell size and shape, Gram stain properties, colony morphology, and presence and type of structures such as flagella or endospores. Of course, much of the diversity of bacteria is physiological in nature, and comparison of metabolic characteristics provided another important level of differentiation. We will examine some of this physiological diversity in Chapter 13. The availability of DNA sequence data has affected classification tremendously, in many cases helping to resolve ambiguous relationships by comparing the sequences of shared genes, and greatly improving the reliability of identification methods.

Type strains and culture collections

For each new taxon that is named, a corresponding culture must be deposited in at least two different culture collections. The World Federation for Culture Collections maintains a database of more than 500 collections from over 60 countries. These cultures represent specimens that are available so that other scientists can make direct comparisons, without relying solely on the published descriptions. The availability of culture collections ensures that as scientists develop new analysis techniques, those techniques can be applied to the type strains that define the taxon. For example, taxa that initially were defined by physical and metabolic analyses can be further defined by DNA sequence analysis, and even complete genome sequencing (see Chapter 10).

TABLE 2.3 Classification hierarchy within the Domain Bacteria, using *Brucella melitensis* as an example

Taxonomic level	Example
Phylum	Proteobacteria
Class	Alphaproteobacteria
Order	Rhizobiales
Family	Brucellaceae
Genus	*Brucella*
Species	*melitensis*

Names should be static labels for organisms. Once a name has been given to an organism and validated, that name should not change, even if further research indicates that the name is no longer descriptive. The application of new analysis methods sometimes provides evidence that taxa that were previously thought to be distinct should in fact be considered to be the same. So what happens if researchers determine that independently named organisms actually fall within the same taxon? Which name should be used? When this situation occurs, taxonomists invoke the principle of priority; the first validly described name takes precedence. As a result, some interesting name changes can occur. For example, the use of newer techniques and better analyses have caused the species *Alcaligenes eutrophus* to be renamed *Ralstonia eutropha*, then *Wautersia eutropha*, and then *Cupriavidus necator*. Bacterial taxonomy remains fluid. Hundreds of new taxa are described each year, and new information results in the recognition of new relationships between organisms.

2.6 Fact Check

1. Identify the taxonomic groups used to classify bacteria.
2. Each bacterium's scientific name is binomial and based on which taxonomic categories?
3. How is a species defined in the bacterial classification system?
4. What characteristics are used to classify bacteria?

The Rest of the Story

Since their discovery in 1974, magnetotactic bacteria have been found in freshwater and marine sediments worldwide. As we discussed in Perspective 2.1, scientists are beginning to understand how the magnetosomes that control this movement are formed. Another question still remains. How could this trait be useful (and thus evolutionarily advantageous) to a microbe?

The magnetotactic bacteria isolated so far prefer habitats with no O_2 (anoxic) or very low O_2 concentrations. Most microbiologists now hypothesize that magnetotaxis tells the bacteria which direction is "down." In the Northern Hemisphere, following Earth's geomagnetic field toward the North Pole would result in a downward trajectory. In a water column, this downward trajectory should direct a bacterium toward the sediment and more anoxic conditions. The bacteria, then, utilize a process of magneto-aerotaxis; they move along a magnetic field toward their preferred concentrations of oxygen.

Research now focuses on understanding magneto-aerotaxis and the role magnetosomes play in this process. In the laboratory, researchers have compared naturally occurring magnetotactic bacteria with an engineered magnetosome-free strain. The magnetosome-free strain still responded to an oxygen gradient in the absence of a magnetic field. However, without the magnetic advantage, aerotaxis took twice as long in the experiments, compared to the aerotaxis of the bacteria containing magnetosomes. While research continues to unravel the details of this magneto-aerotaxis, these results indicate that magnetosomes provide bacteria with the selective advantage of reaching their ideal environment more quickly. A greater understanding of these fascinating microbial structures will not only provide insight into this process, but also set the stage for industrial and commercial applications in the future.

Image in Action

WileyPLUS Within the Northern Hemisphere, Earth's geomagnetic field exerts a downward force, as indicated by the black lines in this illustration. Magnetotactic bacteria align themselves with this force. Within these bacteria, the magnetosome organelle, composed of MamK protein and vesicles containing magnetite (dark gray with orange), facilitates this behavior. 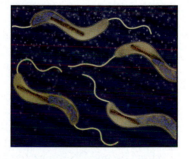 As the bacteria move in this direction, they not only move lower in the water column, but also move toward areas of lower oxygen content (indicated by white circles).

1. The microbes in this diagram are classified as bacteria but contain a cytoplasmic organelle. Why might this observation be considered "confusing" to a student studying them?

2. Imagine you are studying the MamK protein shown in orange in this image. When using bioinformatics tools to compare it to other known proteins, what would you likely determine?

3. Imagine the cell at the top was missing the filamentous MamK protein. Describe how that cell would appear different in this diagram compared to the other cells.

Summary

Section 2.1: What do bacterial cells look like?

Generally, the morphology of a given species is fairly consistent.

- Bacterial cells can have several distinct morphologies: spheres (**cocci**), rods (**bacilli**), curved rods (**vibrios**), and spirals (**spirilla**).
- Some bacterial species are **pleiomorphic**; they exhibit highly variable shapes.
    In many bacterial species, cells can stay attached after cell division.
- Within a given species, cells may form clusters, chains, and branching filaments.
- Some bacteria grow more complex multicellular arrangements, forming **hyphae**, **mycelia**, or **trichomes**.
    Bacterial cells range in size from 0.2 μm to 700 μm in diameter, but most are in the range of 0.5–5 μm.

Section 2.2: What is in the cytoplasm of bacterial cells?

The **nucleoid**, which is comprised primarily of chromosomal DNA and associated proteins, is the most massive component of the **cytoplasm**.

- The nucleoid is not partitioned from the cytoplasm by a membrane.
- Chromosomal DNA is compacted significantly, through the action of **topoisomerases**, which cause **supercoiling** of the DNA, and interaction with packaging proteins.
    The cytoplasm contains many organelles, along with proteins and RNA molecules that carry out the work of the cell.
- Ribosomes produce proteins within the cytoplasm.
- The cytoplasm contains many different soluble organic metabolites and inorganic ions and may contain large **inclusion bodies** storing various nutrients.
- Other structures, such as **sulfur globules**, **gas vesicles**, **carboxysomes**, and **magnetosomes**, also may exist in the cytoplasm.

Section 2.3: What kinds of internal structures help to organize bacterial cells?

Bacterial cells contain **cytoskeleton**-like structures that are important for cell shape and division.

- The dynamic **Z-ring**, formed from the tubulin-like FtsZ protein, guides cell division. It is associated with the inner face of the plasma membrane and, as it undergoes depolymerization, causes the membrane to constrict.
- Actin-like proteins, such as MreB, form filaments that control cell shape in many bacteria. A related protein, ParM, ensures that plasmids are evenly distributed during cell division.
- Other cytoskeletal proteins may have roles in distribution of chromosomal DNA in bacteria.

Section 2.4: What are the critical structural and functional properties of the bacterial cell envelope?

The **plasma membrane** is a phospholipid bilayer in which many proteins and, in some species, **hopanoids** are embedded. This membrane mediates many critical cellular processes, including nutrient transport, energy metabolism, environmental sensing, and protein secretion.

- Protein channels called **aquaporins** regulate the transit of water across the plasma membrane.
- **Active transport systems**, like the **ABC transporters**, aid in transporting nutrients across the membrane. These molecules may cross the membrane through **symport** or **antiport** mechanisms.
- Protein secretion involves a **general secretory pathway** and the presence of **signal peptides** on proteins targeted for secretion.
    The bacterial cell wall, which lies outside the plasma membrane, is comprised of a crosslinked network of **peptidoglycan** that determines shape and provides mechanical strength and protection for bacterial cells.

- **Lysozyme**, an enzyme found in human tears and mucus, hydrolyzes specific bonds within the peptidoglycan.
- **Lysostaphin**, an enzyme produced by *Staphylococcus simulans*, affects the peptidoglycan of *Staphylococcus aureus*.
- Antibiotics in the β-**lactam** family also attack peptidoglycan. β-**lactamases** destroy these antibiotics, providing bacteria with resistance to them.
    There are two major types of **cell envelopes** in bacteria, which can be distinguished by the **Gram stain**.
- **Gram-positive bacteria** have a thick cell wall composed of multiple layers of peptidoglycan, along with **teichoic** and **lipoteichoic acids**. Some Gram-positive bacteria form structurally modified, metabolically inactive cells called **endospores** under stressful conditions.
- **Gram-negative bacteria** have an additional **outer membrane** outside a relatively thin layer of peptidoglycan. The space between the plasma and outer membranes of Gram-negative cells is called the **periplasm**.
- The outer membrane of Gram-negative cells contains **lipopolysaccharide (LPS)** and transport systems such as **porins** and TonB-dependent transporters for nutrient passage.

Section 2.5: How do structures on the surface of bacterial cells allow for complex interactions with the environment?

Flagella are cell surface structures that propel bacterial cells through liquid environments.

- Bacteria with flagella may be **monotrichous**, **lophotrichous**, or **peritrichous**.
- Bacterial, archaeal, and eukaryal flagella are structurally and evolutionarily distinct.
    Chemotaxis is used to direct bacterial motility in response to concentration gradients of attractants and/or repellants and involves **chemoreceptors**. Other forms of taxis also exist.
    Some bacteria utilize non-flagellar-based motility, including pilus-mediated twitching, **gliding motility** across solid surfaces, and actin-based motility inside mammalian host cells.
    Adherence can be mediated by cell surface proteins, **pili**, **stalks**, **fimbria**, **sex pilus**, or **capsules**.
- **Strains** may differ in their ability to adhere.
- Surface adhesion can be the first step in the creation of **biofilms**, adherent communities of microorganisms.
    Crystalline-like **surface arrays** or **S-layers** surround some bacterial cells.
- These coats may provide protection from bacteriophages.
- These coats may also shield a bacterial pathogen from the attacks of the host immune response.

Section 2.6: How are bacteria categorized and named?

The evolution of bacteria over four billion years has resulted in great physical, biochemical, physiological, and genetic diversity.

- A regulated system of nomenclature ensures that bacteria are effectively described and named.
- The classification system for bacteria is a hierarchical taxonomic system, in which the basic taxonomic level of **species** refers to groups of strains that share common physical, metabolic, and genetic features.
- Culture collections contain reference specimens called type strains that are representative of each **taxon**.
- The principle of priority ensures that the first validly named described name takes precedence.
- Bacterial taxonomy is not static, as new taxa are continually described, and new relationships are discovered.

Application Questions

1. Under the microscope, a student observes a mixed culture consisting of *E. coli* and *B. anthracis*. Both are bacillus shaped, but one type has formed a long chain of cells while the other type of bacteria is scattered individually across the slide. Explain what might cause these different arrangements.

2. Mammalian cell cultures often are used for experimentation. The cell culture medium typically is supplemented with penicillin to prevent bacterial contamination of the cell line. Researchers know that this antibiotic will not prevent *Mycoplasma* contamination, however. Explain why penicillin does not protect the culture from *Mycoplasma* infection.

3. The nucleoid of *E. coli* contains roughly 60 percent DNA, 30 percent RNA, and 10 percent proteins, by weight. Identify the proteins that would be nucleoid associated and explain their functions.

4. In the laboratory, a researcher identified mutant bacteria deficient in the FtsZ protein. Their growth appears filamentous, displaying an incomplete cell division. Explain the role of the FtsZ protein and how a deficiency would account for this altered growth.

5. Researchers analyze an engineered strain of *E. coli* with a nonfunctional *parM* gene and an *E. coli* strain containing a functional *parM* gene. How might the growth of these two strains differ? Why?

6. Lysozyme is added to certain cheeses as an antibacterial agent. Explain how this enzyme works to destroy certain bacterial contaminants in the cheese and why it does not harm humans.

7. When carrying out the Gram stain procedure, adding too little or too much alcohol to the sample is a common mistake. Suppose a student added too little alcohol to a sample of Gram-negative bacteria. What would she observe? Why?

8. Some antibacterial drugs interrupt lipid A biosynthesis. What types of bacteria would these drugs affect? Why?

9. Some Gram-negative bacterial cells contain porins and TonB-dependent transporters in their outer membranes. Based on your understanding of each of these structures, explain how their mechanisms of transport differ and why both might be needed in the cell.

10. Imagine a company develops a chemical that inhibits bacterial capsule formation and is non-toxic in humans. Based on your understanding of capsules, explain some medical or commercial applications of this chemical.

11. Suppose researchers want to design an antibacterial compound that will target the bacterial cell surface.
    a. Select a structure on the bacterial cell surface and describe its normal function.
    b. Imagine a compound is engineered that interferes with this surface structure. Describe how the compound would alter the normal function of the structure.
    c. Describe how researchers could test the effectiveness of this compound in the laboratory.

12. *Borrelia burgdorferi* is a motile bacterium that causes Lyme disease. To study the chemotaxis of *B. burgdorferi*, researchers added an attractant to one end of a capillary tube, and bacteria to the opposite end of the tube (capillary tube assay). The concentration of bacteria that entered the capillary tube toward the attractant was then measured.
    a. What is an attractant and what effect does it have on the flagella and movement of bacteria?
    b. Several mutant *B. burgdorferi* strains were generated that were deficient in different chemotaxis-regulating genes. How do you think the chemotaxis responses of these mutants compared to the chemotaxis response of the wild-type strain?
    c. The LC-A2 strain is deficient in a certain chemotaxis protein known as CheA2. Based on these results, what conclusion can be drawn about this chemotaxis protein?

Suggested Reading

Angert, E., K. Clements, and N. Pace. 1993. The largest bacterium. Nature 362:239–241.

Cano, R. J., and M. K. Borucki. 1995. Revival and identification of bacterial spores in 25- to 40-million-year-old Dominican amber. Science 268:1060–1064.

Euzéby, J. P. 1997. List of bacterial names with standing in nomenclature: A folder available on the Internet. Int J Syst Bacteriol 47:590–592.

Flint, J., D. Drzymalski, W. Montgomery, G. Southam, and E. Angert. 2005. Nocturnal production of endospores in natural populations of epulopiscium-like surgeonfish symbionts. Journal of Bacteriology 187:7460–7470.

Hazen, R., and E. Roedder. 2001. Biogeology: How old are bacteria from the Permian age? Nature 411:155.

Johnson, S., et al. 2007. Ancient bacteria show evidence of DNA repair. Proc Natl Acad Sci USA 104:14401–14405.

Komeili, A., Z. Li, D. Newman, and G. Jensen. 2006. Magnetosomes are cell membrane invaginations organized by the actin-like protein MamK. Science 311:242–245.

Konstantinidis, K. T., and J. M. Tiedje. 2007. Prokaryotic taxonomy and phylogeny in the genomic era: Advancements and challenges ahead. Curr Opin Microbiol 10:504–509.

Scheffel, A., M. Gruska, D. Faivre, A. Linaroudis, J. Plitzko, and D. Schuler. 2006. An acidic protein aligns magnetosomes along a filamentous structure in magnetotactic bacteria. Nature 440:110–114.

Simmons, S., D. Bazylinski, and K. Edwards. 2006. South-seeking magnetotactic bacteria in the Northern Hemisphere. Science 311:371–374.

Tindall, B. J., and G. M. Garrity. 2008. Proposals to clarify how type strains are deposited and made available to the scientific community for the purpose of systematic research. Int J Syst Evol Microbiol 58:1987–1990.

Vandamme, P., and T. Coenye. 2004. Taxonomy of the genus *Cupriavidus*: A tale of lost and found. Int J Syst Evol Microbiol 54:2285–2289.

Vreeland, R. H., W. D. Rosenzweig, and D. W. Powers. 2000. Isolation of a 250 million-year-old halotolerant bacterium from a primary salt crystal. Nature 407(6806):897–900.

Eukaryal Microbes

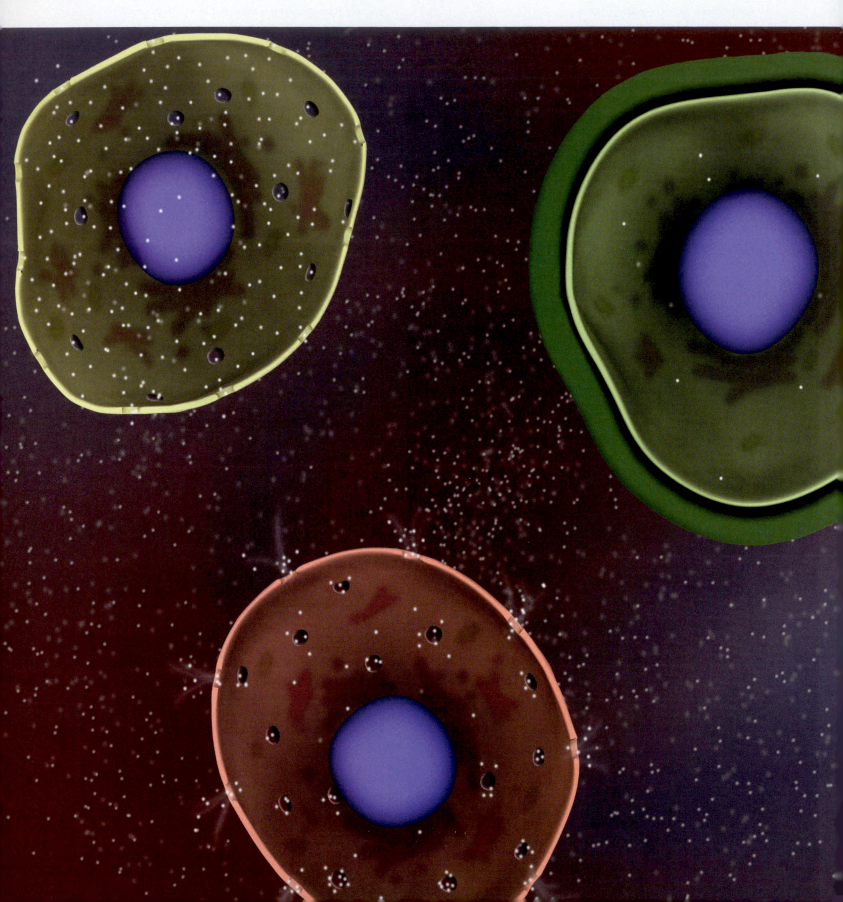

The Rio Tinto ("Red River") in Spain hovers at a pH of 1.7 to 2.5 because of an abundance of acidic iron compounds and other heavy metals that are typically toxic to cells. So, it might seem safe to conclude that this harsh environment would be virtually devoid of life. Or, if you remember the discussion in Chapter 2 about the relatively durable cell wall of bacteria, you might think that any life-forms found would be bacterial cells. What scientists discovered in the Rio Tinto, however, was a wide variety of eukaryal microbes, including protozoa, fungi, and algae. This diversity was observed using a combination of molecular, microscopic, and culture techniques. From water and biofilm samples, they isolated and sequenced DNA, observed the samples under the microscope, and cultured samples in the laboratory.

These findings raise many questions that we can start to address in this chapter. What kinds of eukaryal microorganisms exist in the world? How do they vary in cell structure and function? How are they related to each other? Considering the diversity observed in the Rio Tinto, can we make predictions about how that microbial ecosystem operates? Algae, such as the *Chlamydomonas* found in the Rio Tinto, typically are the primary producers of a community. They contain chloroplasts that allow them to bring energy into the system via photosynthesis and to bring organic carbon into the system via carbon fixation. Fungi and protozoa typically are heterotrophic. In other words, they act as consumers or decomposers. Fungi, for instance, often decompose organic matter and recycle it back into the community. In the case of the Rio Tinto, the fungi serve another notable role of sequestering the toxic heavy metals, thereby protecting more sensitive cell types.

How do the microbes of the Rio Tinto withstand the acidic conditions? Three possibilities seem likely based on research in other systems. First, the cells could let the hydrogen ions flow in and, as a result, have a reduced intracellular pH (most cells maintain a cytosolic, or intracellular, pH close to 7). Second, they could erect a barrier against hydrogen ions. Third, they could actively defend themselves against the influx of hydrogen ions. Resolution of this question would require further research.

CHAPTER NAVIGATOR

Introduction

This chapter introduces another domain of life: Eukarya. Historically, eukaryal microbes (single-celled, nucleated microorganisms) have been underappreciated because of our limited understanding of their remarkable diversity and their evolutionary relationships. Today, however, scientists are particularly interested in defining the spectrum of their diversity: How many different types of eukaryal microbes are there? What unique structural and functional properties define them? How did such diversity evolve? Where are they found? What roles do they play in our world? The opening story about the discovery of eukaryal microbes in the Rio Tinto exemplifies this area of research by raising questions not only about eukaryal microbial diversity,

but also the relationship between cell structure and function. In this chapter, we will provide an overview of this exciting and important group of microorganisms by describing some of its members, their activities, and their relationships to each other. Along the way we will focus on several questions:

What do eukaryal cells look like? (3.1)
What are the different types of eukaryal microbes? (3.2)
How do eukaryal microbes replicate? (3.3)
How did eukaryal microbes originate? (3.4)
What harmful and beneficial roles do eukaryal microbes play? (3.5)

3.1 The morphology of typical eukaryal cells

What do eukaryal cells look like?

Before examining the morphology of eukaryal microbes, let's address a more fundamental question: What are eukaryal cells? Members of the domain **Eukarya** are defined by the presence of a membrane-bound nucleus. The term "eukaryote," in fact, means "true nucleus." Thus, while the chromosome of bacteria and archaeons exists within a nucleoid region, only eukarya possess true nuclei in which chromosomes are packaged in a membrane-bound compartment that forces a spatial and temporal separation of transcription and translation.

Eukaryal cells typically are larger than bacterial and archaeal cells and contain other intracellular compartments, or **organelles (Figure 3.1)**.

Within these organelles, eukarya compartmentalize different biochemical activities that in turn have important implications for cellular processes **(Table 3.1)**. The entire cell is surrounded by a plasma membrane and, in some cases, a cell wall. Finally, a complex cytoskeleton provides both structure and flexibility to eukaryal cells. In this section, we will examine these various structures. More importantly, we will investigate specific functions that eukaryal cells perform and examine which cell structures accomplish those functions and in what ways structure and function are linked.

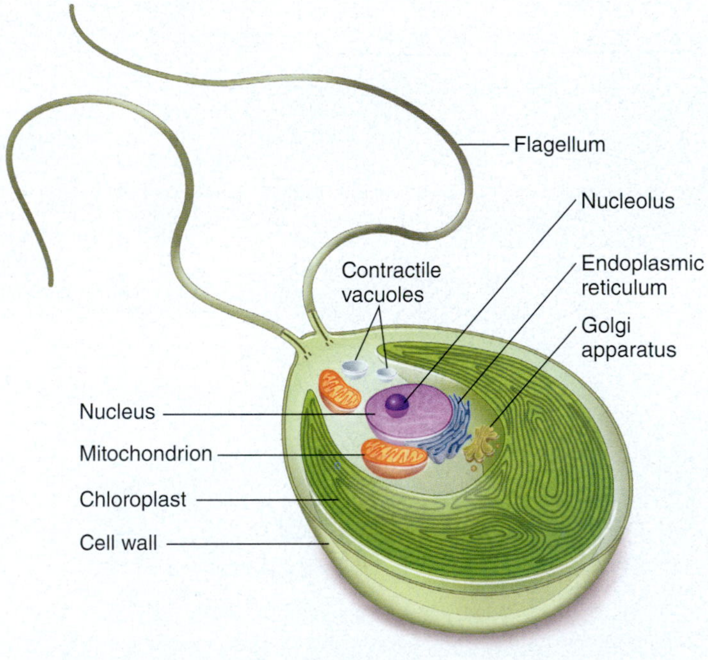

Figure 3.1. Organelles The algae *Chlamydomonas* displays general eukaryal cell structures, including a nucleus, Golgi apparatus, and mitochondria. You might conclude that *Chlamydomonas* is a photosynthetic cell because it has a chloroplast, and that it is motile because it has two flagella.

- Flagellum
- Nucleolus
- Contractile vacuoles
- Endoplasmic reticulum
- Golgi apparatus
- Nucleus
- Mitochondrion
- Chloroplast
- Cell wall

The nucleus: A role in the storage and expression of information

The **nucleus** is the defining organelle of eukaryal cells **(Figure 3.2A)**. Surrounded by a double phospholipid bilayer (the nuclear membrane), it contains the genomic content of the cells with DNA packaged into linear chromosomes. Additionally, the nucleolus, a non-membrane bound structure, exists within the nucleus. Ribosome synthesis begins here. In bacterial and archaeal cells, the genomic DNA sits within the cytoplasm. Thus, the processes of transcription and translation (the production of mRNA and proteins, respectively) can occur at the same time. In other words, as the information from a gene is being transcribed into mRNA, that same strand of mRNA can be translated into protein by ribosomes. In contrast, in eukaryal cells the genomic DNA is within the nucleus, and thus the process of transcription occurs first within the nucleus, and the completed mRNA strand must be exported through nuclear pores into the cytoplasm before translation can occur **(Figure 3.2B)**. Once mRNA molecules enter the cytoplasm, translation is initiated.

TABLE 3.1 Selected internal organelles of Eukarya

Organelle	Main function	Interesting features
Nucleus	Contains most of cell's DNA, site of transcription	Double membrane containing pores Outer membrane continuous with endoplasmic reticulum
Mitochondrion	Energy production	Double membrane Contains DNA Independent replication Not present in amitochondriates
Chloroplast	Photosynthesis	Double membrane Contains DNA Independent replication Unique to photosynthetic organisms
Rough endoplasmic reticulum (ER)	Site of translation and protein folding	Rough ER has protein-synthesizing ribosomes attached to it
Golgi apparatus	Modifies, sorts, and transports proteins	Connected to the ER through a series of vesicles
Vacuole	Storage and structure	Food vacuoles serve as sites of digestion Contractile vacuoles help maintain water balance
Lysosome	Digestion of macromolecules	Contains digestive enzymes
Peroxisome	Breakdown of fatty acids	Contains various oxidative enzymes, like catalase and oxidase
Hydrogenosome	Production of H_2 and ATP	Double membrane Found in some amitochondriates May be remnant of mitochondrion

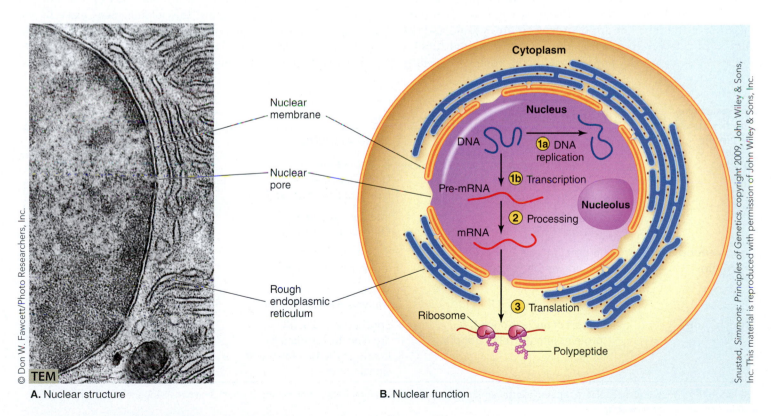

A. Nuclear structure

B. Nuclear function

Figure 3.2. Process Diagram: The nucleus A. Note the nuclear membrane, the nuclear pores, and the surrounding rough endoplasmic reticulum. **B.** DNA replication and transcription occur within the nucleus. Following transcription, mRNA leaves the nucleus through the nuclear pores. Translation occurs in the cytoplasm. Ribosome synthesis begins in the nucleolus.

In eukarya, each mRNA generally results in the production of one polypeptide. In bacteria, however, several different polypeptides can be translated from a single mRNA. We will explore this fundamental difference in **Section 7.4**.

CONNECTIONS In eukarya, each mRNA generally results in the production of one polypeptide. In bacteria, however, several different polypeptides can be translated from a single mRNA. We will explore this fundamental difference in **Section 7.4**.

The secretory pathway: A role in protein trafficking

Newly made proteins in eukarya have a wide array of possible destinations: the cytoplasm, various organelles, or destinations outside of the cell. As with bacterial proteins, destination is dictated by signal sequences contained in the amino acid code within individual proteins. One key difference between the trafficking of proteins in eukarya and bacteria is that with so many different destinations, a much larger set of signal sequences is needed (Table 3.2). Many times, multiple signals will distinguish a precise destination within an organelle. Some proteins, for instance, may be targeted to the lumen of the mitochondrion, while other proteins may be targeted specifically to the inner membrane, the intermembrane space, or the outer membrane of the mitochondrion. Typically, signal sequences are removed by specific cellular proteases once the protein has reached its destination. Removal of signal sequences can be detected in the laboratory by observing size differences between pre-proteins (the larger version of the proteins containing the signal sequences) and mature proteins (the smaller version of the proteins from which signal sequences have been removed).

To examine this trafficking process in more detail, let's follow one important subset of proteins: those proteins that will enter into the **secretory pathway**, a series of organelles consisting of the rough **endoplasmic reticulum (ER)** and the **Golgi apparatus** that are connected via vesicles and tubules (**Figure 3.3**). The "rough" appearance of the rough ER results from ribosomes that stud the surface of this structure. As translation occurs, proteins leave these ribosomes and enter the rough ER. These proteins eventually may be secreted from the cell at the plasma membrane (see Step 5, Figure 3.3B) or become embedded in the plasma membrane or membranes of organelles. One protein signal—a stretch of approximately 20 hydrophobic amino acids at the N-terminus of the protein—signals entry into the secretory pathway.

This first signal helps ribosomes translating proteins of the secretory pathway to dock at the surface of the rough ER and then further directs the protein to be transported into the lumen (inside) of the rough ER. At this stage, the first signaling peptide usually is removed from the protein. Additional signals then determine whether proteins should remain within one of the organelles of the secretory system. At this stage, proteins often proceed to the Golgi apparatus. Here, the proteins may be modified or packaged before being shipped to their final destination. Proteins transported to the plasma membrane may be secreted from the cell, a process known as exocytosis, or remain associated with the membrane. As shown in Figure 3.3B, eukaryal cells also contain another structure, the smooth endoplasmic reticulum. The function of the smooth ER differs in various cell types, but most typically is involved in lipid and carbohydrate synthesis. As we showed with the rough ER, vesicles can move between the smooth ER and the Golgi apparatus. Unlike the rough ER, however, the surface of the smooth ER is not studded with ribosomes.

One characteristic of protein secretion within eukaryal cells is the extensive modification that proteins undergo post-translationally. Following translation, a protein folds into a complex structure. For secreted proteins, this folding takes place in the lumen of the endoplasmic reticulum and involves the assistance of chaperone proteins. These molecular chaperones transiently interact with growing proteins and assist in the folding process.

TABLE 3.2 Protein trafficking and signal sequences

Organelle	Purpose of signal	Signal properties	Sample signal sequence[a]
Nucleus	Import into nucleus (Nuclear Localization Signal or NLS)	Internal sequence One or two short series of lysine (K) and arginine (R)	KKKRK
	Export from nucleus (Nuclear Export Signal or NES)	Internal sequence Enriched in leucine (L)	$LX_{1-3} LX_{2-3} LXL$[b]
Rough endoplasmic reticulum	Import into rough ER Binding by Signal Recognition Particle or SRP	5–10 hydrophobic amino acids at N-terminus	H_2N-MSFVFLLLVGILFWAGA
	Retention in rough ER	4 amino acids at C-terminus	KDEL
Mitochondria	Import of nuclear-encoded proteins into the mitochondria	Enriched in lysine (K), arginine (R), histidine (H) Cleavage site for removal of signal sequence	RSIYHSHHPT RKLKFSPIKY RX(F/L/I)XX(G/S/T)XXXX

[a]Letters in the signal sequence refer to the single letter amino acid codes. There are typically other variations not shown here. Sequences can vary both within a species and between species.
[b]The letter X in a sequence refers to any amino acid. H_2N: N-terminus of protein. COOH: C-terminus of protein.

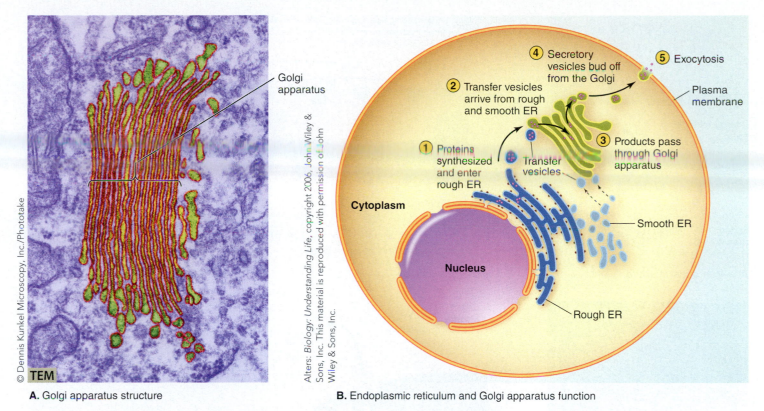

A. Golgi apparatus structure

© Dennis Kunkel Microscopy, Inc./Phototake

TEM

Alters: *Biology: Understanding Life*, copyright 2006, John Wiley & Sons, Inc. This material is reproduced with permission of John Wiley & Sons, Inc.

B. Endoplasmic reticulum and Golgi apparatus function

In the diagram:
- ① Proteins synthesized and enter rough ER
- ② Transfer vesicles arrive from rough and smooth ER
- ③ Products pass through Golgi apparatus
- ④ Secretory vesicles bud off from the Golgi
- ⑤ Exocytosis
- Plasma membrane
- Cytoplasm
- Transfer vesicles
- Smooth ER
- Nucleus
- Rough ER

Figure 3.3. Process Diagram: The protein secretory pathway A. The endoplasmic reticulum and Golgi apparatus are a series of distinct membrane-bound compartments. Products can be shuttled between these compartments via vesicles and tubules. **B.** Proteins destined to be secreted from the cell pass through the rough ER and Golgi apparatus via transfer vesicles. Secretory vesicles filled with these proteins bud off of the Golgi apparatus and fuse with the plasma membrane, secreting their contents into the extracellular space, a process known as exocytosis. Products from the smooth ER also can be transferred to the Golgi apparatus via transfer vesicles.

The folding involves the formation of hydrogen bonds and, in some cases, disulfide bridges. The proteins also are modified by the addition of sugars, lipids, and/or other functional groups (such as acetyl or phosphate groups) as the proteins pass through the rough ER and Golgi. These functional groups affect the overall activity of the protein; most fully mature proteins are abundantly decorated with a thick and diverse array of these chemical additions.

CONNECTIONS Molecular chaperones are present in all domains of life. High temperatures typically cause proteins to unfold. In **Section 4.3**, we will explore how these molecular chaperones may be utilized by archaeons that live at high temperatures to maintain the integrity of their proteins.

The mitochondria and chloroplasts: A role in cell metabolism

Eukarya compartmentalize energetic processes within the **mitochondria** and **chloroplasts** (**Figure 3.4**). The key to ATP synthesis in both cellular respiration and photosynthesis is the passage of electrons along an electron transport chain and the use of energy released during this passage to generate a proton gradient that, in turn, drives the synthesis of ATP. This process is known as **chemiosmosis** and requires a membrane barrier

for the creation of the proton gradient. Thus, mitochondria and chloroplasts contain inner membranes that maximize surface area and facilitate energy production.

Mitochondria maximize the amount of energy that cells can derive from the breakdown of organic molecules. This process of **cellular respiration** is the means by which heterotrophs acquire both energy and building blocks for cellular activities. To generate a proton gradient, electrons move along a series of electron transport proteins that are embedded within the mitochondrial inner membrane. As the electrons move from protein to protein, protons (H^+) are moved across this membrane, into the intermembrane space. The proton gradient then forms in the space between the inner and outer membranes. As the protons flow down the resulting concentration gradient, through a protein known as *ATP synthase*, inorganic phosphate is added to ADP, forming ATP (see Figure 3.4B).

Chloroplasts convert solar energy into cellular energy and reducing power that can drive the fixation of carbon dioxide into organic molecules by the process known as photosynthesis. In chloroplasts, the proton gradient forms within the thylakoids, internal membrane-bound structures. Electron transport chain proteins are embedded within the thylakoid membrane and protons are transported into the interior, or lumen, of the thylakoid. As protons flow down their concentration gradient, again through ATP synthase, ATP is generated (see Figure 3.4D).

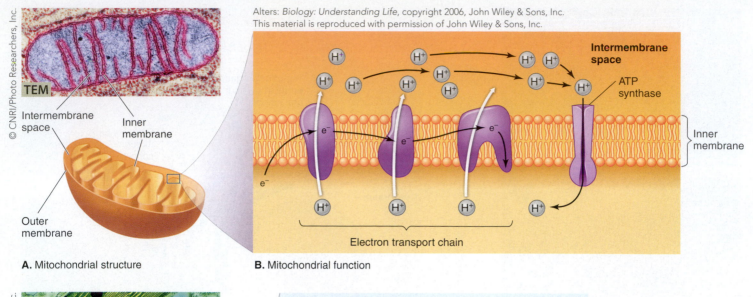

Alters: *Biology: Understanding Life*, copyright 2006, John Wiley & Sons, Inc. This material is reproduced with permission of John Wiley & Sons, Inc.

A. Mitochondrial structure

B. Mitochondrial function

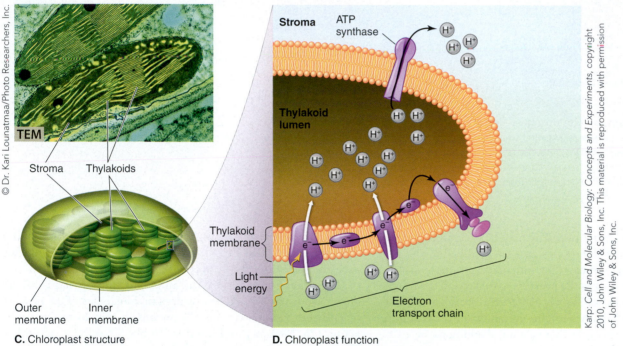

Karp: *Cell and Molecular Biology: Concepts and Experiments*, copyright 2010, John Wiley & Sons, Inc. This material is reproduced with permission of John Wiley & Sons, Inc.

C. Chloroplast structure

D. Chloroplast function

Figure 3.4. Mitochondria and chloroplasts A. The extensive, deeply involuted, inner membrane increases surface area. **B.** Electrons move along the electron transport chain proteins embedded in the inner membrane and help to transport protons into the intermembrane space. As these protons flow back through the inner membrane, through ATP synthase, ATP is generated. **C.** Chloroplasts contain outer and inner membranes and a series of internal membrane-bound structures known as thylakoids. **D.** Harnessing energy from sunlight, electrons move along electron transport chain proteins embedded in the thylakoid membranes and help to transport protons into the interior, or lumen, of the thylakoids. ATP is generated in the stroma, as protons flow out of the thylakoids, through ATP synthase.

Please remember that although bacteria and archaeons do not possess mitochondria or chloroplasts, they still perform almost identical metabolic processes by simply substituting the plasma membrane for the organelle membrane.

The early steps of cellular respiration, glycolysis and fermentation, occur in the cytoplasm of eukaryal cells. The later, more energetically favorable, steps occur in the mitochondria. The tricarboxylic acid (TCA) cycle takes place within the matrix (the innermost area) of the mitochondria, while the conversion of the reducing power stored in the form of NADH and $FADH_2$

is converted to ATP through the process of chemiosmosis across the inner membrane. Cells have different numbers of mitochondria and their abundance is typically linked to the energetic needs of the cell. Some intracellular pathogens will gather the mitochondria of the host cell around themselves to gain first access to the cell's ATP production.

Some eukaryal microbes do not possess mitochondria. These organisms, termed **amitochondriates**, are discussed in more detail when we discuss the evolution of eukaryal cells. Some of these cells appear to have simply lost mitochondria they once possessed (perhaps through disuse in an anaerobic lifestyle) and survive on glycolysis, while others possess novel organelles such as the glycosome and hydrogenosome. These organelles share a common purpose in helping these anaerobic cells derive the most energy possible without chemiosmosis.

Mitochondria and chloroplasts are semi-autonomous organelles. They contain some DNA that codes for certain organelle proteins and have transcription and translation machinery for these genes. They also can replicate independently of the host cell. This feature enables the cell to rapidly regulate the number of mitochondria in a cell in response to cell needs. However, most proteins required by mitochondria and chloroplasts are coded for in the nucleus, synthesized on ribosomes free in the cytoplasm, targeted to the organelles by specific signal sequences, and then imported after translation.

The plasma membrane: A role in homeostasis

A key premise of cell function is homeostasis—the ability of cells to maintain an internal equilibrium despite a changing external environment. The **plasma membrane** contributes to homeostasis by separating the environment within compartments from that on the outside and also regulating the passage of molecules. For the organisms living in the Rio Tinto, these basic functions must occur at a very low pH.

CONNECTIONS In **Section 2.4**, we examined the structure of the bacterial plasma membrane. As we noted in that section, bacteria possess several distinct mechanisms for transporting molecules across this membrane.

Proteins can be embedded in (integral membrane proteins) or affixed to the surface (peripheral membrane proteins) of the plasma membrane (see Figure 2.9). You can think of the proteins floating in a sea of phospholipids—an analogy that captures the essence of the official language used to refer to membranes as a **fluid mosaic model**. Because the hydrophobic core of the phos-

pholipid bilayer does not allow large polar or charged molecules to cross it, transmembrane proteins are essential to the transport of these molecules into and out of cells. These transmembrane proteins typically function with great specificity, generally transporting only a single class of molecules. They can act as simple tunnels through which molecules move down a chemical gradient (solutes diffuse from areas of higher to lower concentration) by a process known as *facilitated diffusion*. Alternatively, the proteins may expend energy to move molecules against a gradient (to accumulate a higher concentration of a solute on one side of the membrane than the other) through a process known as *active transport*.

Historically, cell biologists assumed that the distribution of various types of phospholipids within the cell membrane was relatively random. If these molecules could move freely within the membrane, then their distribution should not be organized. It now seems clear, however, that localized pockets of specific lipids—**lipid rafts**—exist within the membrane. Moreover, specific proteins appear to localize to these lipid rafts. Recent evidence suggests that these lipid microdomains play important roles in protein movement, cell signaling, and the entry and exit of some viruses from cells **(Mini-Paper)**.

As we saw in Chapter 2, bacteria also contain membranes that separate inside from outside. Important differences in the structures of plasma membranes found in bacteria, archaeons (as we will see in Chapter 4), and eukarya exist, however, and these structural differences affect the strength and flexibility of the membranes **(Table 3.3)**. Comparisons of membrane characteristics in various organisms have led researchers to hypothesize that membranes may have a complicated evolutionary history. Eukarya and bacteria, for instance, both have *phospholipids* as the dominant membrane lipid with *ester* linkages between the glycerol molecule in the head group and *straight* chain fatty acid tails. As we will see in Chapter 4, archaeons, in contrast, have a wide array of different membrane lipids. Additionally, eukaryal membranes contain sterols (e.g., cholesterol) and have a low protein content. The membranes of bacteria and archaeons, on the other hand, lack sterols and have a high protein content. The geometry of the fatty acid chains and the abundance of sterols and proteins all

TABLE 3.3 Comparison of plasma membranes in Bacteria, Archaeons, and Eukarya

Characteristic	Function	Bacteria	Archaeons	Eukarya
Membrane structure	Membrane assembly Hydrophilic surface Hydrophobic core	Phospholipid bilayer	Bilayer or monolayer Diverse lipid composition (sulfo-, glyco-, isoprenoid-)	Phospholipid bilayer
Lipid structure	Membrane fluidity	Ester linkage Straight fatty acid chains[a]	Ether linkage Branched isoprenoid chains[a]	Ester linkage Straight fatty acid chains[a]
Sterols	Membrane stability	No[b]	No	Yes
Proteins	Structural	High abundance	High abundance	Low abundance

[a]This designation refers to whether the main backbone of the fatty acid chain is straight or branched. It does not specify whether the fatty acids are saturated or unsaturated. Both branched and unbranched fatty acids can be saturated or unsaturated.
[b]A few bacteria have sterol-like compounds for membrane stability: Cyanobacteria have hopanoids and Mycobacteria (a wall-less bacterium) integrate sterols from host cell membranes.

P. A. Scheiffele, A. Rietveld, T. Wilk, and K. Simons. 1999. Influenza viruses select ordered lipid domains during budding from the plasma membrane. J. Biol. Chem. 274:2038–2044.

Context

With the initial description of the fluid mosaic model of the cell membrane by Singer and Nicolson in 1972, it became easy to envision the lipid component of the cell membrane as a relatively uniform "sea" in which proteins were floating. The lipids, one could imagine, served only to house the proteins. Experiments showed that there are regions of the membrane that, unlike the rest of the membrane, are resistant to detergent extraction, so they are not soluble in detergent. These regions are rich in cholesterols and sphingolipids, and harbor high concentrations of membrane proteins. As we have seen, these regions were called "rafts" and appeared to be involved in the functions of the membrane proteins such as detection and transduction of signals, and determination of location in the cell. Other studies had demonstrated that the lipid composition of some enveloped viruses that bud through the cell membrane is different than the overall lipid composition of those membranes. This finding suggests that the budding might occur at sites with different lipid composition such as the lipid rafts. This paper presents a series of experiments in which three different kinds of viruses were compared to see whether they incorporated lipid rafts in their membranes.

Experiments

The three types of viruses used in this study were an influenza virus called fowl plague virus (FPV), vesicular stomatitis virus (VSV) of the Rabies family, and Semliki Forest virus (SFV). All three viruses were propagated in baby hamster kidney (BHK) cells. The envelope of each of these viruses is derived from the host cell plasma membrane as the virus leaves the cell by budding, and also contains the virus glycoproteins. In the first set of experiments, researchers attempted to determine if lipid raft-derived material is found in the envelope of these viruses. Viruses produced after infection of the BHK cells were isolated, and proteins were extracted from the envelope using detergent. While the VSV and SFV glycoproteins were solubilized easily by the detergent, the FPV glycoproteins were not soluble. The presence of the FPV glycoproteins in these detergent insoluble complexes (called detergent-insoluble glycolipid-enriched complexes, or DIGs, in the paper) suggested that the FPV glycoproteins might be inserted in the envelope that was derived from detergent insoluble lipid rafts, while the glycoproteins of the two other viruses were inserted in the envelope that had much less lipid raft-derived material.

Next, the solubility of cholesterol, which is known to be enriched in detergent insoluble complexes of cellular membranes, was measured in the different viral envelopes. This experiment required that the BHK cells first were labeled with radioactive cholesterol, thereby making tracking easier. Viruses were produced using those labeled cells, and the solubility of the cholesterol in the resulting viruses was determined by tracking the radioactivity. The results (Figure B3.1) clearly show that the FPV cholesterol was much more insoluble in detergent than was the VSV and SFV cholesterol.

To examine the lipids present in the envelopes, the BHK host cells were incubated with radioactive phosphate before the virus was produced, resulting in labeling of the membrane lipids which could be separated by thin layer chromatography. As shown in Figure B3.2, the sphingomyelin lipid (SM) that is known to be associated with lipid rafts was found to be mostly in the detergent insoluble fraction of the FPV (in pellet, P), while mostly in the detergent soluble fraction (in supernatant, S) of the VSV and SFV viral particles.

Other lipids that were measured, phosphatidylethanolamine (PE) and phosphatidylcholine (PC), were distributed between the soluble and insoluble fractions in a manner roughly similar to their distribution in the membranes of the BHK host cells. It was further shown that removal of cholesterol from the membranes using cyclodextrin resulted in improved solubility of the SM in the FPV envelope. The overall conclusion was that cholesterol is involved in the formation of the detergent

---

help to dictate the fluidity of the membrane by affecting packing of molecules. Thus, branched and unsaturated fatty acid chains make membranes more fluid, while sterols and proteins typically make membranes less fluid. The membrane composition often changes in response to changing environmental conditions. At low temperatures, some cells alter the fatty acid profile to increase unsaturated fatty acids and maintain fluidity. Other molecules, such as sterols, behave differently at different temperatures and can increase membrane fluidity as the temperature drops.

One final point worth noting is that the swimming appendages of eukaryal cells, the **cilia** and **flagella**, are encased within the phospholipid bilayer of the plasma membrane. In bacteria and archaeons, the flagella protrude through the plasma membrane. As we will see later in this chapter, the swimming appendages of eukaryal microbes differ in several ways from their archaeal and bacterial counterparts.

The cell wall: A role in cell support

Cellular support can be provided either by an external casing (cell wall) or by an internal scaffold (cytoskeleton), and the choice of which a cell uses has a large impact on the cell's life cycle. Thus, eukarya fall into two broad categories: those with cell walls and those without cell walls. As always, these simplified categories obscure important details. The term **cell wall** refers to a semi-rigid structure at the cell periphery that holds the cell in a certain shape, limits cellular access to chemical compounds, and protects it from environmental pressures (osmotic stress as well as physical stresses). A wide array of chemical structures can form a wall and a fascinating diversity exists within eukarya (Table 3.4). As illustrated in Toolbox 3.1, microscopy and molecular biology techniques can be combined to give us a richer understanding of these structures and their roles in the dynamic life of a cell.

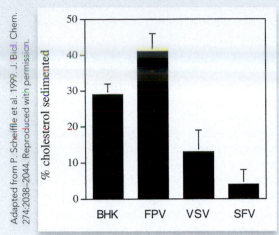

Adapted from P. Scheiffle et al. 1999. J. Biol. Chem. 274:2038–2044. Reproduced with permission.

Figure B3.1. Cholesterol in membranes of viruses and host cell This graph compares amount of cholesterol (percentage of cholesterol sedimented) present in different virus envelopes (FPV, VSV, and SFV) compared to the host BHK cell membrane.

insoluble lipid rafts and the association of the proteins with them. Influenza viruses such as FPV select the lipid rafts during budding.

Impact

This study was a very good demonstration of the use of viruses to examine membrane structure. Clearly, viruses differ in the extent of their association with lipid rafts during budding. These differences were used to emphasize that such lipid rafts, as determined by detergent insolubility, are genuine features of cell membranes. The lipid rafts depend on cholesterol for their insolubility. It is now recognized that many intracellular bacterial and viral pathogens invade host cells via lipid rafts, making them potentially important host virulence factors.

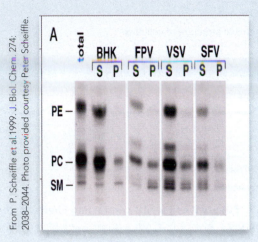

From P. Scheiffle et al.1999. J. Biol. Chem. 274: 2038–2044. Photo provided courtesy Peter Scheiffle.

Figure B3.2. Distribution of specific lipids in viral envelopes and host cell membrane Cells were incubated with radioactive phosphate, which would result in labeling of the membrane lipids. Viruses were produced in these labeled cells and isolated. Lipids from the viruses and host cells then were separated into detergent insoluble (in pellet, P) and detergent soluble (in supernatant, S) fractions and analyzed. Sphingomyelin lipid (SM) is found primarily in the supernatant fraction of BHK cells, VSV, and SFV viruses, but in the pellet fraction of FPV.

● Questions for Discussion ....

1. If lipid rafts make host cells susceptible to infection by intracellular pathogens, why do you think they are present in these cells?

2. Different viruses appear to have different affinities for lipid rafts during budding. What might account for these differences?

TABLE 3.4 Comparison of cell walls in Bacteria, Archaeons, and Eukarya

| | Bacteria | | Archaea | Eukarya | | |
	Gram-positive	Gram-negative		Fungi	Algae	Protozoa
Basic structural components	Thick peptidoglycan[a] layer (~40 nm) comprises 40–80% of cell wall dry weight Surface of peptidoglycan layer decorated by teichoic acid	Thin peptidoglycan[a] layer (~2 nm) comprises 5% of cell wall dry weight Outer membrane has phospholipid inner leaflet and lipopolysaccharide outer leaflet	Varied Methanogens: glycopeptides or pseudopeptidoglycan Halogens and Hyperthermophilic: glycopeptides	Chitin	Cellulose[b]	None[c]

[a]The precise chemical structure of peptidoglycan varies between species.
[b]Diatoms have cell walls (termed frustules) made of silicon dioxide, proteins, and polysaccharides.
[c]Glycoproteins may be assembled into a cell wall during particular developmental stages to produce cysts or spores.

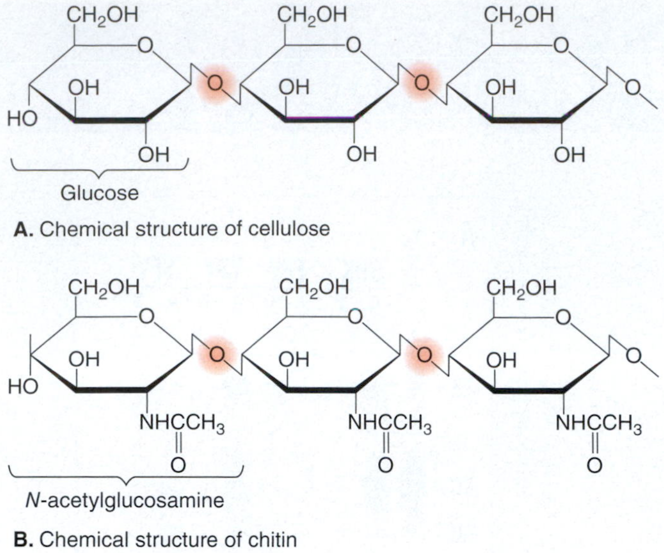

A. Chemical structure of cellulose

B. Chemical structure of chitin

Figure 3.5. Eukaryal cell walls: contrasting chemical compositions and structures A. In cellulose, glucose monomers are linked by β-1,4-glycosidic bonds. **B.** In chitin, N-acetylglucosamine monomers are linked by β-1,4-glycosidic bonds. In both molecules, the β-1,4-glycosidic bonds are highlighted.

All algae have cell walls, which are composed predominantly of **cellulose** like the cell walls of their plant relatives. Cellulose is a linear carbohydrate polymer (polysaccharide) in which glucose molecules are linked via β-1,4-glycosidic bonds **(Figure 3.5)**. Individual cellulose chains then are crosslinked by hydrogen bonds to increase their tensile strength. One exception to this general rule is the diatoms whose cell wall is made of a silicon compound. Researchers hypothesize that this silicon compound may be an energy-efficient alternative to cellulose, because silica bonds are weaker and require less energy to assemble into polymers.

Most fungi also have cell walls, but their cell walls are usually made of chitin (see Figure 3.5). Chitin is the same compound found in the exoskeleton of insects. Like cellulose, chitin is a polysaccharide, with sugar monomers again linked via β-1, 4-glycosidic bonds. In chitin, however, the monomer building blocks are N-acetylglucosamine. This is a modified glucose molecule in which an acetylamino group ($-NH-CO-CH_3$) replaces a hydroxyl group ($-OH$). Again, linear chains of chitin are crosslinked to one another with hydrogen bonds, but the acetylamino groups permit more hydrogen bonds. Thus, chitin is a stronger polymer than cellulose. Glucosamine may be familiar to you in a different guise: it is a compound your body makes to stimulate the formation of cartilage and is commonly taken as a dietary supplement by people with arthritis.

Although protozoa are generally considered to lack cell walls (a feature commonly used to distinguish members of this group) some protozoa have cell walls at certain points in their cell cycle. *Giardia*, *Cryptosporidium*, and *Entamoeba* are three pathogenic protozoa that have two different life-cycle stages: the trophozoite (*Giardia* and *Cryptosporidium*) and amoeba (*Entamoeba*) stages in which the parasites cause disease within the intestinal tract of hosts, and the cyst stage required for transmission between hosts. During the trophozoite and amoeba stages, these organisms lack cell walls. During the cyst stages, however, these

Toolbox 3.1
USING MICROSCOPY TO EXAMINE CELL STRUCTURE

What structures and processes might we want to visualize when studying cells? We may want to see how the cytoskeleton is configured within cells. We may want to study the fine structure of the components of the cytoskeleton. We may want to identify which proteins comprise the cytoskeleton. We also may want to observe the dynamic and flexible properties of the cytoskeleton that give cells structure, shape, and movement. Microscopy can allow us to address all of these wants.

One essential tool in cell biology is immunofluorescence microscopy. In this technique, cells are killed and their structures fixed in space and time using chemicals (usually formaldehyde or gluteraldehyde) to crosslink molecules with a network of molecular spot welds. The permeability of the cells is then increased (often by treatment with detergents), thereby allowing reagents to pass through the cell membrane and enter the interior of the cell. Now, specific proteins can be detected using antibodies. Antibodies are proteins made by the immune system designed to bind to a specific region of another molecule. The immune system then triggers the attack and destruction of the bound molecule. In this research technique, the antibodies bind to the protein the scientist is interested in locating. In direct immunofluorescence, the antibody that binds to the protein of interest can be tagged with a fluorescent molecule. In a related technique called "indirect immunofluorescence," the antibody that binds to the protein of interest may not be tagged but is the target of a second antibody that is tagged with the fluorescent molecule. In either case, when a light of the proper wavelength is shone on the sample, the fluorescent molecule will glow and the location will literally light up like a beacon. Microtubules and microfilaments in a human cell can be seen in **Figure B3.3**.

A newer approach allows scientists to study proteins in living cells. With this technique, researchers can learn not only about protein localization, but also protein dynamics. In this technique, scientists make use of a naturally fluorescent protein (green fluorescent protein or GFP, and its many colorful relatives). Using various molecular biology techniques, scientists modify a cell so that a protein of interest is produced that is attached to GFP. When the cell is illuminated by light of the proper wavelength, the GFP-tagged protein will glow **(Figure B3.4)**. The dynamics of the fusion protein can now be watched over time.

Studies of various GFP fusion proteins have been instrumental in furthering our understanding of cellular processes. By inserting into cells recombinant genes that produce GFP-tubulin fusion proteins, researchers can see the assembly and disassembly, in real time, of the spindle fibers involved in chromosome separation during mitosis, a process shown in Figure 3.13. Using similar approaches, researchers have been able to better understand how viruses move within cells. After entering an appropriate host cell, some portion of the virion must travel within the cell to its site of replication. As we will see in Perspective 3.1, herpes viruses travel along host cell microtubules from the plasma membrane to the cell's nucleus. GFP-tagged proteins have more clearly revealed the dynamics of this remarkable process.

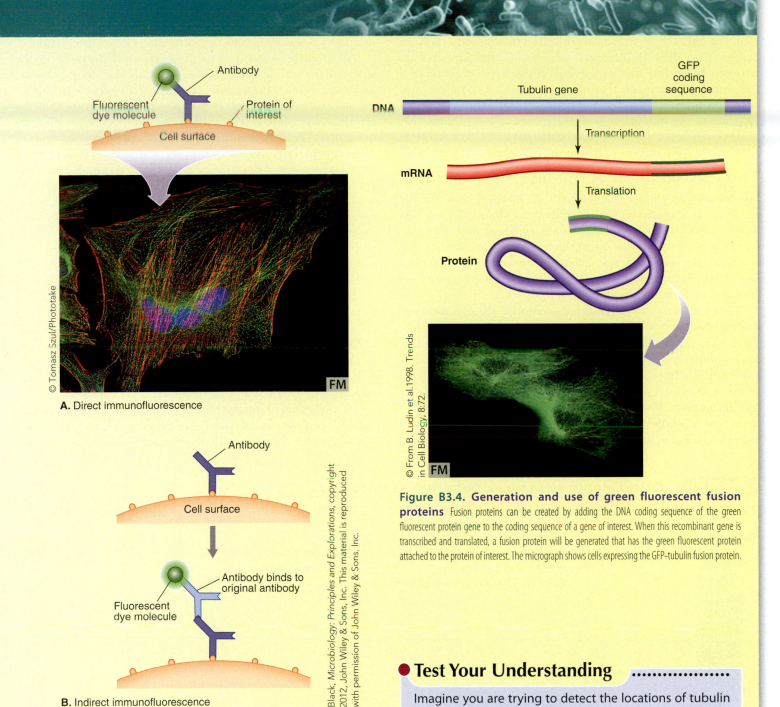

A. Direct immunofluorescence

© Tomasz Szul/Phototake

B. Indirect immunofluorescence

Black, *Microbiology: Principles and Explorations,* copyright 2012, John Wiley & Sons, Inc. This material is reproduced with permission of John Wiley & Sons, Inc.

Figure B3.3. Direct and indirect immunofluorescence A. With direct immunofluorescence, an antibody that binds to the protein of interest is labeled with a fluorescent marker. In this photo of a multinucleated human HeLa cell, microtubules were stained with an antibody linked to a green fluorescent molecule, while microfilaments were stained with an antibody linked to a red fluorescent molecule. A dye stained the nuclei blue. **B.** With indirect immunofluorescence, a secondary antibody that binds to the initial, or primary, antibody is labeled with a fluorescent marker.

© From B. Ludin et al. 1998. Trends in Cell Biology, 8:72.

Figure B3.4. Generation and use of green fluorescent fusion proteins Fusion proteins can be created by adding the DNA coding sequence of the green fluorescent protein gene to the coding sequence of a gene of interest. When this recombinant gene is transcribed and translated, a fusion protein will be generated that has the green fluorescent protein attached to the protein of interest. The micrograph shows cells expressing the GFP-tubulin fusion protein.

● Test Your Understanding ···················

Imagine you are trying to detect the locations of tubulin protein in a cell, but no tagged antibody is commercially available that is specific for tubulin. You do have an untagged tubulin antibody, however. What would you have to do to detect the tubulin using this untagged antibody if you wanted to utilize immunofluorescence?

organisms possess cell walls made of carbohydrate-modified proteins. The absence of a cell wall in trophozoites and amoebas permits parasite motility and enhances the interaction with the host intestine. The cell walls in cysts are thought to be important not only for survival in fresh water that would otherwise destroy the cells with osmotic pressure when the parasites are between hosts, but also for survival in the harsh conditions of the stomach after a new host ingests the parasites. Indeed, this trio of pathogenic protozoa appears to actually use the acids and enzymes of the stomach to start the process of excystation that destroys the cyst walls and allows the parasites to emerge in time to establish infections within the intestinal tract.

The cytoskeleton: A role in cell structure

Historically, a defining characteristic of eukaryal cells was the presence of a **cytoskeleton**, literally a "cellular skeleton." As we saw in Chapter 2, though, recent evidence indicates that bacteria may also possess cytoskeletons. In this section, we will examine the eukaryal cytoskeleton structure and functions more fully. Before addressing these points, let's ask a more basic question: What is the cytoskeleton?

Three major types of structures comprise the cytoskeleton: **microtubules**, **microfilaments**, and **intermediate filaments**. These structures differ in composition and function **(Table 3.5)**. Microtubules consist of polymers of α- and β-tubulin dimers that assemble into fibers about 24 nm in diameter. They typically act as an intracellular organization network onto which organelles and chromosomes are attached. Microfilaments are polymers of actin that assemble in fibers about 7 nm in diameter. They typically can be found at the periphery of cells providing cell shape.

Finally, intermediate filaments are polymers of a diverse array of proteins rich in alpha-helices (regions that display a spiraled structure) that form fibers 9–11 nm in diameter. Intermediate filaments can exist in the nucleus to give it shape or at the cell periphery in some animals to coordinate cell-cell interactions.

All of these structures, to some extent, contribute to the shape of the cell. In cells without cell walls (i.e., most protozoa and animal cells), the cytoskeleton serves a vital role in providing structure and support. It is important to recognize that the cytoskeleton cannot protect cells to the same extent as cell walls can protect them. Without cell walls, cells are more vulnerable to environmental stresses. One dramatic consequence is manifested by cell behavior in water. When cells are placed in pure water, water enters the cell by osmosis because the cell has a higher solute concentration. If the cell is a wall-less eukaryal cell, the water entering the cell will cause the cell to burst; the water pressure entering the cell exceeds the strength of the membrane. In contrast, if cells with walls are placed in pure water, the entering water pressure is less than the strength of the cell wall; the cell will swell within the wall but then stop filling with water.

Components of the cytoskeleton participate in a variety of intracellular dynamics, such as the assortment of organelles and chromosomes during cell division and protein trafficking in eukaryal cells. Generally, this movement occurs through one of two mechanisms: cycles of polymerization and depolymerization (a process referred to as "turn-over") or the action of molecular motors. The classic example of movement achieved by turnover is the action of spindle fibers (a type of microtubule) during mitosis. During metaphase, these fibers are attached at one end to specific chromosomal structures, the *centromeres*, and are

---

TABLE 3.5 Cytoskeleton structures and proteins in eukarya

Fiber	Functions	Structure	Building Blocks	Motors
Microtubules	Intracellular transport Separation of chromosomes in mitosis and meiosis Cell movement (e.g., cilia and flagella)	25 nm 13 protofilaments that form a hollow tube	α- and β-tubulin	Dyneins Kinesins
Microfilaments	Maintain cell shape Create division furrow in cytokinesis Cell movement (e.g., pseudopods)	7-nm diameter 2 protofilaments twisted around each other in a helix	Actin	Myosins
Intermediate filaments	Nuclear structure Cell-cell interactions	8–11 nm diameter	Varied: lamin, keratin, vimentin	None

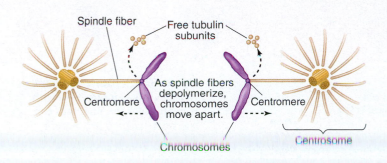

Figure 3.6. Disassembly of spindle fibers During anaphase, tubulin monomers are removed from the spindle fibers through a process of depolymerization. As the spindle fibers shorten, the chromosomes move toward the centrosomes.

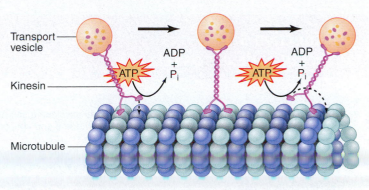

Figure 3.7. Molecular motors convert chemical energy into motion. Intracellularly, dynein and kinesin transport vesicles along microtubules, using energy derived from the conversion of ATP to ADP. As shown here for kinesin, a transport vesicle, or cargo, attaches to one end of the molecule while the other end of the molecule allows the complex to "walk" along the microtubule.

associated at their other ends with *centrosomes*. In animal cells, the centrosomes contain two *centrioles*, tube-like structures that serve as the organizing centers of microtubules. During anaphase, tubulin monomers are removed one by one from the chromosomal ends of the spindle fiber microtubules. This disassembly causes the spindle fibers to shorten, and the copies of each chromosome are pulled toward the centrosomes at opposite poles of the cell **(Figure 3.6)**. This action results in the accurate assortment of chromosomes into daughter cells during cell division. Finally, we should note that bacterial and viral pathogens also make use of microtubules present in their host cells **(Perspective 3.1)**.

Motors provide a diverse range of cell movements. Two different motors, dynein and kinesin, interact with and move along microtubules **(Figure 3.7)**. These motors are responsible for moving vesicles along microtubules during intracellular protein trafficking. Generally, dyneins move along microtubules toward the center of the cell. Kinesins, conversely, tend to move from the center of the cell toward the periphery. Dyneins also are involved in the movement of cilia and flagella, in a process quite different from the flagellar motion in bacteria that we explored in Chapter 2. Eukaryal cilia and flagella emerge from basal bodies (the flagella microtubule organizing centers) within the cell body and consist of an array of

microtubules in which nine microtubule doublets form a tube around a core pair of microtubules This array is referred to as the axoneme. The motion of axonemes resembles that of a cracked whip with undulating motion created by the sliding of microtubules past one another in the axoneme. Dynein and associated proteins perform this sliding action **(Table 3.6)**.

3.1 Fact Check

1. What are the key features of members of the domain Eukarya?
2. What is the secretory pathway and what post-translational modifications are made to proteins being secreted via this pathway?
3. Describe the process of chemiosmosis.
4. Describe the components of a eukaryal plasma membrane and each of their roles in cell function.
5. Which eukarya have cell walls? How do their cell walls differ from those of bacteria?

TABLE 3.6 **Comparison of flagella in Bacteria and Eukarya**

Feature	Bacteria	Eukarya
Structure	Nonflexible hollow helical filament Extends outside of cell membrane and cell wall Composed of flagellin subunits	Flexible Covered by the cell membrane Composed of a 9+2 array of microtubules termed the "axoneme"
Assembly	At distal tip Flagellin monomers are transported out inside of hollow flagella	At distal tip Microtubule segments are transported out along the outside of the axoneme via an intraflagellar transport system
Arrangement	Polar (one or multiple) Peritrichous (multiple)	Varies by cell type
Motion	Motor in membrane turns flagella in a screw-like motion Energy from a proton motive force	Dynein spokes slide microtubules along each other within the axoneme. This bends the axoneme and gives a whip-like motion to the flagella Energy from hydrolysis of ATP

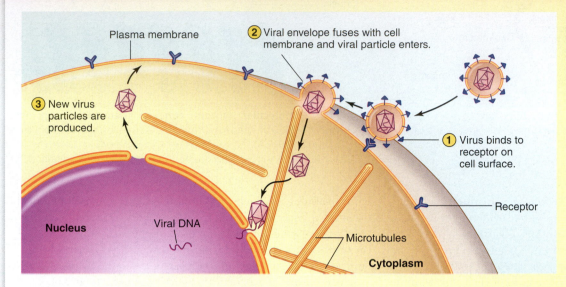

Plasma membrane

② Viral envelope fuses with cell membrane and viral particle enters.

③ New virus particles are produced.

Nucleus

Viral DNA

① Virus binds to receptor on cell surface.

Receptor

Microtubules

Cytoplasm

Figure B3.5. Process Diagram: Herpesvirus interacts with microtubules in infected cell. Upon infecting a cell, herpesvirus must travel to the nucleus, where the viral genome replicates. Evidence suggests that the viral particle interacts with dynein, which facilitates the rapid movement of the virus along microtubules to the nucleus.

Humans aren't the only ones who have noticed that the cytoskeleton guides dynamic processes in eukaryal cells. A number of bacterial and viral intracellular pathogens use the cytoskeleton to facilitate their movement during infections of eukaryal cells. Many viruses, including herpes simplex virus, poliovirus, vaccinia virus, and adenovirus, all hitch rides on microtubules to speed and direct their intracellular movement **(Figure B3.5)**. Think about poliovirus traveling along the length of a neuronal axon from a peripheral site to the central nervous system, and you can quickly understand the benefits of efficient and directed movement.

The intracellular bacterium *Listeria monocytogenes* employs another fascinating means of locomotion. *Listeria* causes a dangerous foodborne infection. Once inside a host cell, *Listeria* depletes cell resources as it multiplies, but then must spread its progeny to new host cells to expand the infection. To accomplish this movement, *Listeria* actually hijacks actin proteins from the host cell, assembling them rapidly into a "tail" at the base of the bacterium **(Figure B3.6)**. The growth of the tail pushes the bacterium, attaining such a force and velocity that it can actually inject the bacterium from one cell into a neighboring cell.

Another bacterium, *Shigella flexneri*, also utilizes the host cell's actin to form a "tail" and spread from one cell to another. The bacterial proteins necessary for actin polymerization in these two species, however, share no sequence similarity, even though they perform basically the same function. This finding has suggested to researchers that *Listeria* and *Shigella* evolved this mechanism of motility independently.

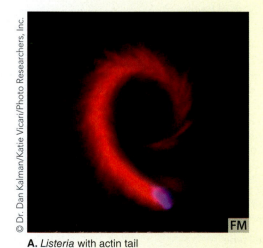

© Dr. Dan Kalman/Katie Vicari/Photo Researchers, Inc.

FM

A. *Listeria* with actin tail

B. Actin tail propelling *Listeria*

TEM

© The Rockefeller University Press. 2004. The Journal of Cell Biology 165:233–242, Fig. 2

Figure B3.6. Interactions between *Listeria* and actin A. Immunofluorescence image showing *Listeria* bacterium (stained blue) and the red actin tail that it assembles. The actin in host cell microfilaments is also stained in red. **B.** Electron micrograph of the actin tail pushing a *Listeria* bacterium outward from the periphery of a cell. The action of the actin tail may be sufficient to propel the bacterium into a neighboring cell, spreading the infection.

3.2 Diversity of eukaryal microbes

What are the different types of eukaryal microbes?

Eukarya encompasses a remarkable diversity of organisms, from unicellular microbes to multicellular animals, plants, and fungi that comprise the visible biological world. Indeed, Eukarya is the only domain of life that includes multicellular organisms. In this section, we will examine the phylogeny of the eukaryal microbes and explore in depth several model organisms within this group. To start our examination of the diversity of eukaryal microbes, let's first look at the phylogeny of these organisms.

The phylogeny of eukaryal microbes

As we saw in Chapter 1, we currently classify all living organisms into three domains: Bacteria, Archaea, and Eukarya (see Figure 1.8). Classifications within the domain Eukarya traditionally have been based on overt similarities; animals are multicellular, heterotrophic, and capable of voluntary motion, while plants are multicellular, photosynthetic, and not capable of voluntary motion. What can SSU rRNA gene sequence comparisons tell us about phylogeny within the domain Eukarya? When we examine the 18S rRNA gene sequences of various eukarya, we see that several traditional groups, most notably the plants, animals, and fungi, do indeed represent monophyletic groups; all species within these groups share a common ancestor. Other traditional groups, most notably the algae, seem to appear at various places within the phylogenetic tree **(Figure 3.8)**. Green algae are closely related to land plants. Red algae and brown algae, however, seem to be more distantly related. Thus, the algae, a traditional classification within the eukarya, do not represent a monophyletic group.

Another interesting group of organisms, the amitochondriates, seem to be very distantly related to the other eukarya.

As we will see later in this chapter, the amitochondriates lack mitochondria. Compelling evidence indicates that these organisms may once have contained mitochondria. Does that make them the earliest eukarya? The answer to that question also remains to be determined.

CONNECTIONS Analysis of SSU rRNA gene sequences demonstrated that archaeons are genetically distinct from bacteria. This realization led to the development of the three-domain classification scheme. Details about this major shift in the classification of organisms are described in **Section 1.1**.

Despite the important information about phylogeny that we can glean from SSU rRNA gene sequence comparisons, these studies may not provide us with the perfect answer. To determine more accurately the phylogeny of eukarya, researchers have compared the sequences of genes for several other common proteins. Most notably, researchers have constructed phylogenetic trees based on the sequences of several proteins, including tubulins, elongation factor 1-alpha and, most recently, heat shock proteins. While the major groupings remain in these alternate trees, these trees often differ from the SSU rRNA tree, and from each other. These alternate trees raise some interesting issues. Phylogenetic trees based on these alternative markers show all of the eukarya radiating from a single point. In other words, it does not appear that plants, animals, and fungi are advanced eukarya and amitochondriates are more primitive eukarya. Some researchers have used information like this to postulate that all major groups of eukarya arose relatively quickly after the appearance of the mitochondrion, an event referred to as an evolutionary Big Bang.

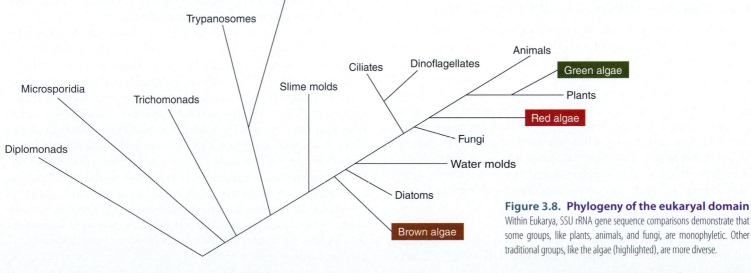

Figure 3.8. Phylogeny of the eukaryal domain
Within Eukarya, SSU rRNA gene sequence comparisons demonstrate that some groups, like plants, animals, and fungi, are monophyletic. Other traditional groups, like the algae (highlighted), are more diverse.

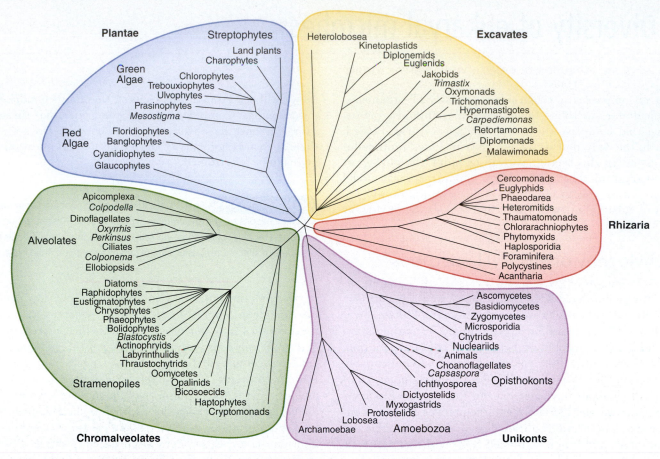

Figure 3.9. Alternative classification of Eukarya By combining sequence data from several molecular markers with morphological data, researchers have developed an alternative eukaryal tree. In this model, all Eukarya are grouped into five major super-groups. While strong data support this model, uncertainties still remain.

Because of these new analyses, researchers are beginning to classify eukaryal organisms differently. **Figure 3.9** shows one of these new classification schemes. To construct this phylogenetic tree, researchers combined the information gleaned from several molecular marker studies and morphological and structural data. The resulting tree divides Eukarya into five major groups, the Plantae, Excavates, Rhizaria, Unikonts, and Chromalveolates. While there is evidence supporting this classification, uncertainty still exists. Our understanding of the phylogeny of Eukarya remains in flux.

Eukaryal microbes: Model organisms

As we have seen in the previous section, questions still remain regarding the phylogeny of eukarya. In this section, we'll look at four different eukaryal microbes representing the traditional groups. We will focus on a fungus, *Saccharomyces cerevisiae*; a protozoan, *Giardia lamblia*; a slime mold, *Dictyostelium discoideum*; and an alga, *Chlamydomonas*. These four organisms are chosen because together they represent traditional groups within the eukaryal microbial world while having unique features of interest **(Table 3.7)**. They are called **model organisms** because they have been so thoroughly studied that their traits define much of our scientific understanding about eukaryal cells.

Fungi: *Saccharomyces cerevisiae*

Fungi are a diverse group of eukarya, not all of which are single-celled microbes **(Figure 3.10)**. Like all other eukarya,

fungi compartmentalize cell functions into membrane-bound compartments termed organelles, such as the nucleus, endoplasmic reticulum, mitochondria, and Golgi. Also, like all other eukarya, fungi control cell shape and dynamics via proteins of the cytoskeleton. Knowing that a cell is a fungus immediately helps to further define both structural and functional features of the cell. A distinguishing feature of all fungi is their cell wall, which generally is a chitin-based structure quite distinct from either the peptidoglycan-based cell wall of bacteria or the cellulose-based structure of algae and plants. Also, all fungi are heterotrophic, meaning that they derive energy and gain building blocks by ingesting organic compounds.

Saccharomyces cerevisiae is a natural choice as an exemplar eukaryal microbe (see Figure 1.10). This staple of applied microbiology is sometimes called brewer's yeast because of its role in the fermentation of alcoholic beverages, or baker's yeast because it makes bread rise. It is one of the best studied model organisms. You name the problem, and some researcher somewhere probably uses *S. cerevisiae* to study it! Perhaps most notable is the work done with yeast to develop sophisticated molecular tools that can aid researchers studying various questions in different organisms. Because it is inexpensive to study and easy to manipulate, scientists use *Saccharomyces* to address fundamental questions about genes, gene expression, and protein interactions. Information from these studies has provided important insight into the molecular workings of other eukaryal cells, including human cells.

TABLE 3.7

TABLE 3.7 Traditional categories of eukaryal microbes and their characteristics

	Fungi	Protozoa	Slime molds	Algae
Cell functions				
Metabolism	Heterotrophic	Heterotrophic	Heterotrophic	Phototrophic
Motility	Typically non-motile	Varied: Swimming: cilia/flagella Amoeboid: pseudopods	Amoeboid: pseudopods	Non-motile or Swimming: flagella
Cell structures				
Cell wall	Typically Yes Chitin	Typically No[a]	Typically No[a]	Typically Yes Cellulose
Nucleus	Yes	Yes	Yes	Yes
Mitochondria	Yes	Typically Yes	Yes	Yes
Chloroplast	No	No	No	Yes

[a]Spore and cyst stages of development for these cells often have cell walls. These are typically glycoprotein-based for protozoan and cellulose and protein-based for slime molds.

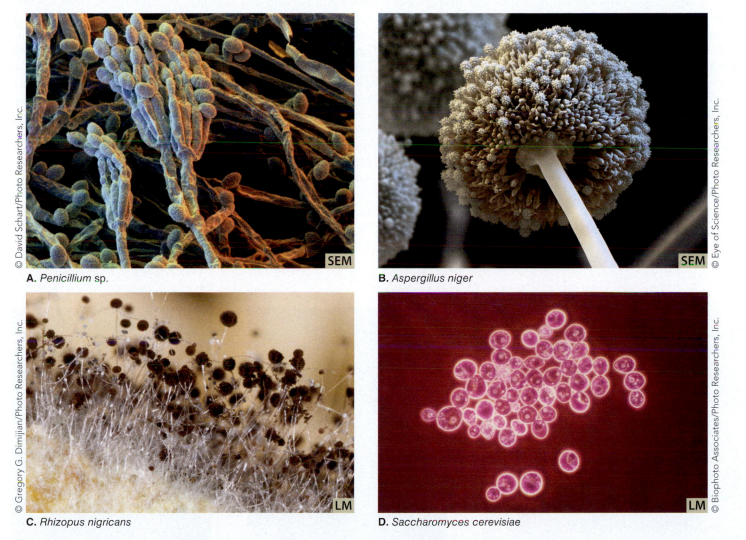

A. *Penicillium* sp.   SEM   © David Schart/Photo Researchers, Inc.

B. *Aspergillus niger*   SEM   © Eye of Science/Photo Researchers, Inc.

C. *Rhizopus nigricans*   LM   © Gregory G. Dimijian/Photo Researchers, Inc.

D. *Saccharomyces cerevisiae*   LM   © Biophoto Associates/Photo Researchers, Inc.

Figure 3.10. Diversity of microbial fungi **A.** Species of the genus *Penicillium* produce the antibiotic penicillin. Other species are present in and contribute to the tastes of Camembert and Roquefort cheeses. **B.** Several species of the genus *Aspergillus* cause disease in humans. *A. flavus* produces aflatoxin, a toxin occasionally found in various agricultural products. **C.** Members of the genus *Rhizopus* can be found in many environments. Some species are associated with human diseases. Other species are used in the production of cheeses. **D.** The most famous member of this genus, *Saccharomyces cerevisiae*, is used in baking. It also has been used extensively by biologists as a model eukaryal organism.

CONNECTIONS Studies of ribosomes in the eukaryal microbe *Saccharomyces cerevisiae*, the archaeon *Sulfolobus solfataricus*, and the bacterium *Escherichia coli* have provided evidence that translation in archaeons may be more similar to translation in eukarya than in bacteria. This experiment is discussed in **Section 7.4**.

will see later in this chapter, *Giardia* even may provide clues to the origins of eukaryal cells; unlike most eukaryal organisms, *Giardia* does not possess mitochondria. Finally, *Giardia* causes significant health problems in humans worldwide. Indeed, the World Health Organization estimates that this microorganism causes intestinal problems in up to 1 billion individuals annually.

Protozoa: *Giardia lamblia*

The protozoan group is perhaps the broadest category of eukaryal microbes and one that is losing meaning as we come to better understand the evolutionary diversity of the group. As you may guess from examination of the representative protozoa in **Figure 3.11**, there is little we can designate to define the group. Protozoa can be either heterotrophic (gaining energy and cellular building blocks by consuming organic food) or photosynthetic (gaining energy from sunlight and cellular building blocks by fixing carbon dioxide from the atmosphere). While most protozoa never have cell walls, some protozoa have cell walls throughout their life cycle, and others develop cell walls only at certain stages. Protozoa also have different strategies of motility. *Giardia lamblia*, a commonly studied model organism, is a flagellate (a swimmer like other ciliates and flagellates), but other cells are amoeboid and move through extension and contraction of pseudopods (literally "false feet"). Others are not self-propelled but rely on external physical forces to move them. The means of reproduction within the protozoa are quite varied. In fact, some ciliated protozoa even have two nuclei, one that is primarily used for gene expression and one that serves as the basis for genetic inheritance between generations.

Giardia is fascinating because of its evolutionary position. Estimates from rRNA and protein phylogenies suggest that *Giardia* diverged from the main branch of the eukaryal domain over 1 billion years ago, making it one of the earliest-diverging eukarya that we can readily culture in the laboratory. Studies of its biology may illuminate many aspects of eukaryal evolution. As we

Slime Molds: *Dictyostelium discoideum*

Unlike other protozoa, slime molds come in two basic types: the cellular forms, represented by the model organism *Dictyostelium discoideum*, and the acellular forms, represented by species in the genus *Physarum*. Slime molds generally spend most of their lives as individual cells. At times, though, these cells form a larger structure that facilitates eating and reproduction. For the cellular slime molds, individual cells aggregate to form a multicellular mass. In the acellular slime molds, in contrast, the individual cells fuse, forming a large, multinucleated aggregate called a plasmodium **(Figure 3.12)**.

Dictyostelium has been a model organism for studying the basis of amoeboid cell motility. Additionally, its complex developmental cycle has provided key insights into intercellular communication. During a developmental cycle, individual cells aggregate, thus allowing the organism to hover on the brink between single-celled and multicellular status (see Figure 1.4). This aggregation of *Dictyostelium* occurs via a cyclic adenosine monophosphate (cAMP) signaling pathway, a pathway also common in humans. The study of *Dictyostelium* organization, therefore, may provide answers to questions about the evolution of multicellularity and cell signaling. Finally, the role that *Dictyostelium* plays in soil ecology is a key area of research. Because the slime molds feed primarily on other microorganisms present on decaying vegetation, they may be a key player in nutrient recycling. As we seek more natural solutions to maintaining soil health for crop growth, a better understanding of natural soil ecology is imperative.

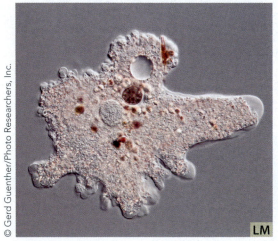

A. *Amoeba proteus*

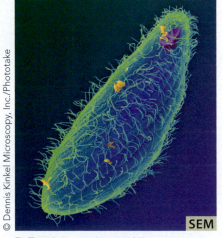

B. *Tetrahymena thermophila*

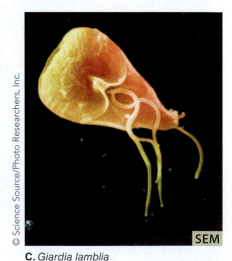

C. *Giardia lamblia*

Figure 3.11. Diversity of protozoa Protozoa exhibit a wide range of morphologies. These groups also show distinct methods of motility. **A.** Amoeba exhibit an amorphous shape. Motility seems to rely on the extension and contraction of pseudopods and subsequent movement of cytoplasm. **B.** The ciliated *Tetrahymena* have a more defined, teardrop shape. Beating cilia provide motility. **C.** *Giardia* also has a more well-defined shape. Flagella power the movement of these organisms.

A. *Dictyostelium discoideum*

B. *Physarum* sp.

Figure 3.12. Diversity of slime molds **A.** In cellular slime molds, like *Dictyostelium discoideum*, individual cells come together, forming a multicellular aggregate. Shown here is a developing fruiting body. These fruiting bodies can be 1 to 2 mm tall. **B.** In acellular slime molds, like *Physarum polycephalum*, individual cells fuse, forming a multinucleated giant cell. In the structure shown here, the cytoplasm is continuous.

Algae: *Chlamydomonas*

Algae represent a diverse category of photosynthetic eukarya, not all of which are single-celled microbes. *Chlamydomonas*, for instance, is a single-celled alga **(Microbes in Focus 3.1)**. Other algae, in contrast, form multicellular colonies. Different algae also may vary greatly in size, shape, and motility. Like all algae, *Chlamydomonas* contains chloroplasts and converts solar energy to cellular energy via the light reaction of photosynthesis. During the corresponding dark reaction, it converts carbon dioxide into organic forms of carbon. Thus, algae are autotrophic and, unlike heterotrophic organisms, can survive in relatively nutrient-poor water. Of course, there are limits to their self-sufficiency. *Chlamydomonas* and other algae must scavenge from the environment nitrogen, sulfur, phosphorus, and other elements required to build organic molecules.

CONNECTIONS *Chlamydomonas*, like all algae and plants, undergoes oxygenic photosynthesis: oxygen is generated as a by-product. As we will see in **Section 13.6**, some bacteria undergo anoxygenic photosynthesis. In these organisms, oxygen is not released.

Microbes in Focus 3.1
CHLAMYDOMONAS REINHARDTII
AND THE STUDY OF CELL MOTILITY

Habitat: Soil and freshwater throughout the world

Description: Single-celled alga, approximately 10 μm in diameter, with two flagella and an "eyespot" that allows it to move toward light sources

Key Features: *Chlamydomonas reinhardtii* can be induced to shed their flagella by exposing the cells to a low pH medium. In a short time, the cells regrow the flagella. Studies of this flagellar regeneration have provided researchers with important insight into the assembly of microtubules in eukaryal organisms. Additionally, studies with laser-based "optical tweezers" have allowed researchers to measure the swimming force of these flagella, thereby providing insight into the mechanism of cellular motility.

Like all other algae, *Chlamydomonas* has a cell wall made of cellulose that protects the cell from toxic molecules, osmotic stresses, and physical damage. *Chlamydomonas* has two flagella and forms zoospores—motile, flagellated, reproductive structures that can withstand harsh environmental conditions and remain dormant until the conditions become more favorable.

Like *Saccharomyces*, *Chlamydomonas* has contributed to our understanding of the biology of eukaryal cells. We know a great deal about it and can ask very detailed questions about cellular function that can then be extrapolated to algae and other eukaryal cells. Researchers are investigating the genetic content of *Chlamydomonas*, the gene expression patterns of this organism, and the protein composition of its flagella and associated structures. *Chlamydomonas* is the definitive model organism for understanding the biogenesis and function of eukaryal flagella. *Chlamydomonas* also has some unique properties that earned it the name of "cockroach of the algae world." This title easily could be interpreted as a stinging critique of the cell, but rather it is a tribute to the properties that make it an extraordinarily durable cell.

3.2 Fact Check

1. What eukaryal microbes are studied as model organisms?
2. Describe the benefits and drawbacks of studying model organisms.
3. Algae do not seem to form a monophyletic group. How could you explain this?
4. Why are slime molds logical organisms in which to study cell aggregation?

3.3 Replication of eukaryal microbes

How do eukaryal microbes replicate?

The genetic mechanisms of eukaryal cells differ significantly from those of their bacterial and archaeal counterparts. Unlike bacteria and archaeons, eukaryal microbes may be **haploid** (have a single copy of their chromosomes) at some stages of their life cycle and **diploid** (have two copies of their chromosomes) at other stages. They can replicate asexually through the process of mitosis and also can reproduce sexually through the process of meiosis (true sexual recombination). This life cycle is in contrast to bacteria and archaeons that replicate asexually by binary fission and for which meiosis is impossible. In this section, we will examine the processes of mitosis and meiosis. We also will examine the life cycles of several eukaryal microbes, focusing on their transitions between haploid and diploid genomes.

Mitosis

Mitosis is a type of nuclear division that produces two nuclei with identical chromosomes, typically followed by division of the cytoplasm forming two genetically identical daughter cells, a process known as **cytokinesis**. In animals and plants, growth occurs by mitosis. In single-celled eukaryal microbes, mitosis results in asexual replication leading to population growth. In either case, the process begins with DNA replication, resulting in the production of two identical copies of the cell's genetic material **(Figure 3.13)**.

CONNECTIONS Cells possess several mechanisms to limit the number of mutations, or mistakes, that occur during DNA replication. In **Section 7.5**, we will explore some of these DNA repair pathways.

Although we often focus exclusively on the "dance of the chromosomes" in mitosis, it is important to remember that eukaryal cells also must ensure that representatives of each type of organelle are distributed to the daughter cells. Typically, the organelles are not counted nearly as carefully as chromosomes. Because mitochondria and chloroplasts can replicate autonomously, as long as a single one makes it into a daughter cell, it can divide by fission as many times as needed to generate a sufficient number. Other membrane-bound organelles, like the endoplasmic reticulum and Golgi apparatus, break down into small vesicles prior to mitosis, and it is left to chance that this pool of vesicles will be adequately distributed. The organelles then reassemble in the daughter cells.

Meiosis

Meiosis is a process that results in a reduction of the genome from diploid to haploid. This process results in the rearrangement of the genetic material and the production of cells that may be involved in sexual reproduction. These two goals can be achieved in meiosis because two rounds of cell division follow a single round of DNA replication.

Consequently, the resulting cells are haploid ($1n$); they contain only a single copy of each chromosome **(Figure 3.14)**. Moreover, the genetic material can be rearranged in meiosis, as shown in Figure 3.14. This rearrangement occurs in two ways. First the *independent assortment* of paternal and maternal chromosomes

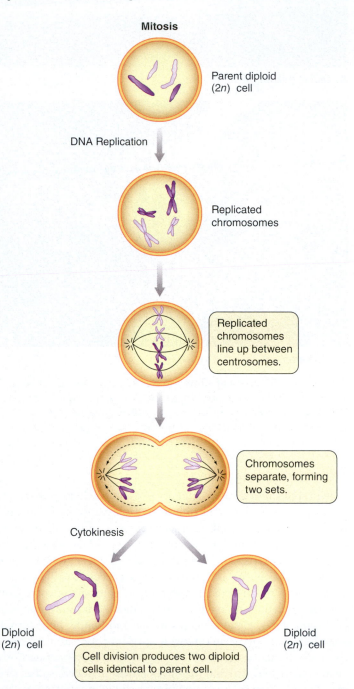

Figure 3.13. Mitosis yields genetically identical cells. Following replication of the DNA, replicated chromosomes align in the middle of the cell, between the centrosomes and then are separated. Following cytokinesis, or cell division, two cells with genetic complements identical to the parent cell are created. The newly produced cells, in other words, have the same amount and kinds of chromosomes as the original cell. In this example, the initial cell is diploid, containing two copies (light purple and dark purple) of the large chromosome and two copies (light purple and dark purple) of the short chromosome.

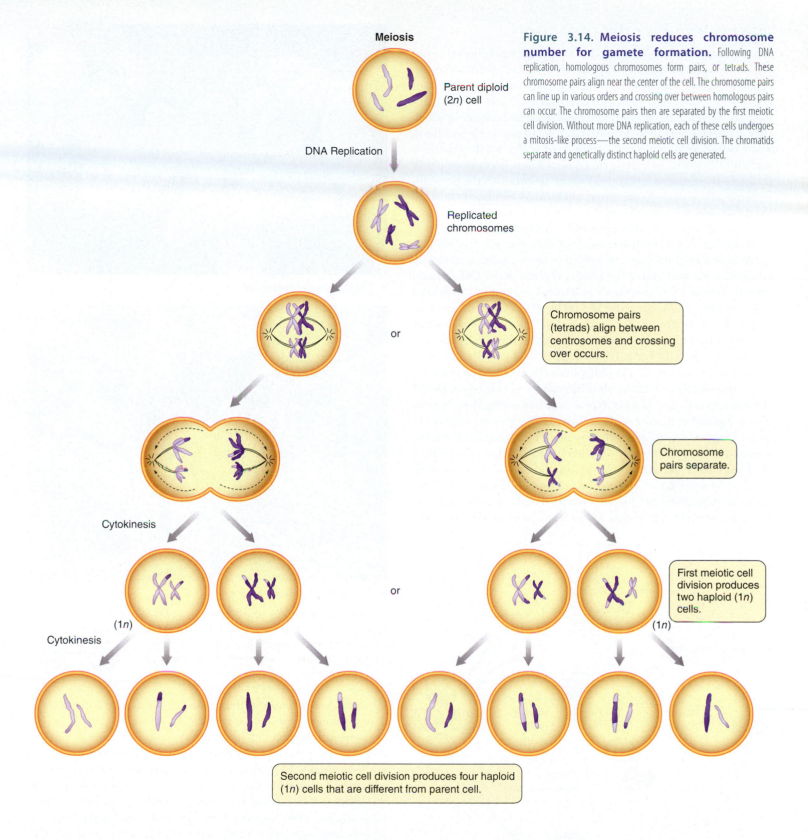

Figure 3.14. Meiosis reduces chromosome number for gamete formation. Following DNA replication, homologous chromosomes form pairs, or tetrads. These chromosome pairs align near the center of the cell. The chromosome pairs can line up in various orders and crossing over between homologous pairs can occur. The chromosome pairs then are separated by the first meiotic cell division. Without more DNA replication, each of these cells undergoes a mitosis-like process—the second meiotic cell division. The chromatids separate and genetically distinct haploid cells are generated.

Meiosis

Parent diploid (2n) cell

DNA Replication

Replicated chromosomes

or

Chromosome pairs (tetrads) align between centrosomes and crossing over occurs.

Cytokinesis

Chromosome pairs separate.

or

First meiotic cell division produces two haploid (1n) cells.

(1n)

(1n)

Cytokinesis

Second meiotic cell division produces four haploid (1n) cells that are different from parent cell.

within a cell (indicated by the light purple and dark purple copies of each chromosome) allows for the production of haploid cells that contain various combinations of chromosomes. Furthermore, *crossing over*, a form of genetic recombination, also occurs in meiosis, as indicated by the exchange of light purple and dark purple material in Figure 3.14. The resulting haploid cells, then, differ in their genetic content.

In sexual reproduction, one haploid cell, or gamete, combines with another haploid cell to form a new diploid offspring. During this event, new pairings of gene variants result. Remember that the new combinations are not necessarily good; the results can be either advantageous or detrimental. The rearrangement of genetic material, however, is one of the key sources of genetic variation upon which natural selection acts in evolution.

Life cycles of model organisms

Alternating cycles of mitosis and meiosis are closely linked to developmental cycles in eukaryal microbes. These cycles serve two functions. First, they generate the haploid and diploid stages of the cells that are involved in asexual replication and sexual reproduction. Second, they permit the cells to enter a phase in which they are better able to withstand harsh environmental conditions. Some cells develop protective spore coats under harsh conditions and enter a quiescent metabolic phase. Others become motile to facilitate their escape. Still others enter meiosis with the possibility that a new genetic combination will prove to be more beneficial. While all eukaryal microbes undergo mitosis and meiosis, the specific life cycles of these organisms differ quite dramatically.

Fungi vary widely in their mode of reproduction. Some fungi spend much of their lives as diploid cells and only occasionally enter into haploid states for sexual reproduction. Other fungi spend virtually all of their lives as haploid cells and form diploid cells only briefly. What is the life cycle of *Saccharomyces cerevisiae* **(Figure 3.15)**? Diploid cells can reproduce asexually via mitosis. Usually during times of environmental stress, these cells undergo meiosis with each diploid cell forming four haploid cells, arranged in a spore case, or **ascus**. These haploid cells are distinguished not by gender—male or female—but instead by mating types; half of the haploid cells are mating type *a* and half are mating type α. These haploid cells can replicate asexually through mitosis, or fuse to form a new diploid cell. Because of structural differences between these cells, type *a* cells can only fuse with type α cells.

Fungi show variation even in simple mitosis. *Saccharomyces cerevisiae* is sometimes referred to as the "budding yeast" because it reproduces by forming small daughter cells that bud from the mother cell. *Schizosaccharomyces pombe* is known as the

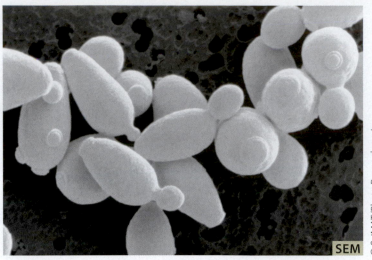

A. *Saccharomyces cerevisiae*

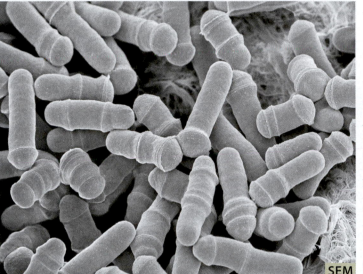

B. *Schizosaccharomyces pombe*

Figure 3.16. Budding and fission in yeast A. In *Saccharomyces cerevisiae*, or budding yeast, cytokinesis (or cell division), occurs asymmetrically, resulting in what appears to be a smaller cell budding off of the original cell. **B.** In *Schizosaccharomyces pombe*, or fission yeast, cytokinesis occurs symmetrically, resulting in two daughter cells of equivalent sizes.

@ Budding ANIMATION

"fission yeast" because cell division results in two daughter cells of approximately the same size **(Figure 3.16)**.

During favorable environmental conditions, motile haploid cells of *Chlamydomonas* replicate asexually, as shown in the left circle of **Figure 3.17**. When growth conditions deteriorate (as indicated in the right circle of Figure 3.17), these haploid cells differentiate, becoming reproductive cells, or gametes. These haploid cells fuse, forming a diploid cell that quickly loses its flagella and develops a thick cell wall, becoming a spore. This spore can exist, basically inert, for extended periods. When environmental conditions improve, the spore undergoes meiosis, producing four motile, haploid cells.

Among the eukaryal microbes, slime molds like *Dictyostelium* may have the most interesting life cycles (see Figure 1.4). Haploid, amoeboid forms of *Dictyostelium* typically feed on bacteria and divide by mitosis. When they begin depleting the existing food supplies, these cells aggregate, forming a multicellular slug. As we

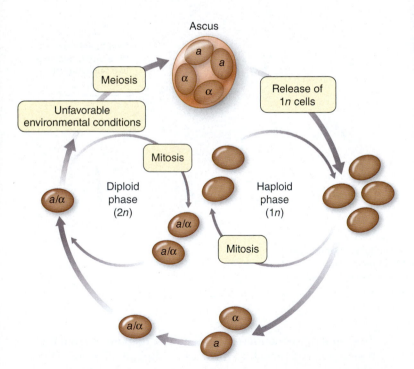

Figure 3.15. Life cycle of *Saccharomyces* Haploid cells, which can be *a* or α mating types, replicate via mitosis. Two haploid cells of different mating types can fuse, producing a diploid cell. This diploid cell also can replicate via mitosis. Under times of stress, diploid cells can undergo meiosis, resulting in the formation of four haploid cells, which remain within a protective spore case, or ascus. When conditions are favorable, the four haploid cells are released from the ascus.

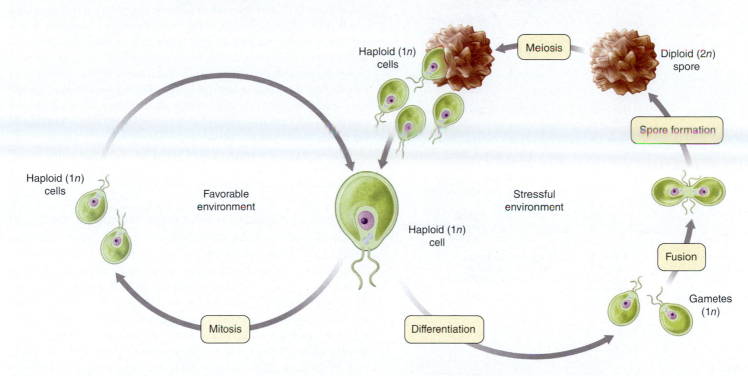

Figure 3.17. Life cycle of *Chlamydomonas* Haploid cells replicate via mitosis, as shown in the left circle. During times of stress, these haploid cells differentiate, becoming gametes that can fuse, resulting in a diploid cell. The diploid cell develops into an environmentally stable spore. When conditions improve, the diploid cell undergoes meiosis, forming four haploid cells, which are released and can replicate again via mitosis.

mentioned earlier, this aggregation results from a complex and quite fascinating cAMP-mediated cell signaling mechanism. Once formed, the slug can move short distances. During this period of the life cycle, cell differentiation also occurs. Some cells form a stalk and others form a fruiting body. Spores form within the fruiting body and then are released. The released cells are dispersed and resume the life cycle as amoeboid cells. Two of these haploid amoeboid cells can fuse, engulf neighboring cells, and form a diploid macrocyst. Meiosis occurs within the macrocyst, resulting in the formation of new haploid cells.

3.3 Fact Check

1. What are the results of mitosis and meiosis? For what purposes are they used?

2. How are the processes of meiosis and mitosis involved in the life cycle of *Saccharomyces*?

3. Describe the life cycle of *Dictyostelium*.

3.4 The origin of eukaryal cells

How did eukaryal microbes originate?

It is generally thought that Eukarya emerged as a new domain of life between 2.1 and 1.6 billion years ago. Thus, the world was exclusively prokaryotic for the first 2–3 billion years of life. The emergence of eukarya was a new wrinkle in the biological landscape, with the complexity of internal structures making possible an extraordinary diversity of cell morphologies and, eventually, multicellular organisms. It is equally important to remember that eukarya do not necessarily represent an evolutionary step forward. We often are tempted to think about complexity as a sign of progress (reflections of our own technological focus as humans, perhaps), but in life, evolution selects among varied options

through the relative reproductive success of the best adapted organisms. Those organisms better suited to the time and place will have more offspring. Numerically, bacteria and archaeons are much more prevalent than eukarya and are therefore at least as successful, if not more so, than their nucleated cousins.

The endosymbiotic theory

Because we traditionally define eukarya by the presence of a nucleus and other organelles, when we consider the origin of eukaryal cells, we are primarily concerned with understanding

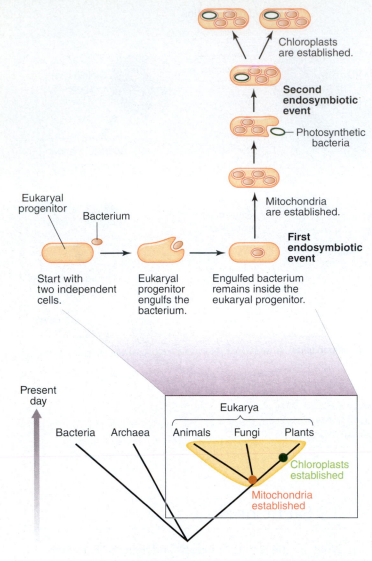

Figure 3.18. Endosymbiosis and the evolution of eukarya Mitochondria appeared in developing eukaryal cells before animals, fungi, and plants differentiated from each other. The progenitor eukaryal cell engulfed a bacterial cell. Rather than being destroyed, the engulfed cell was maintained in a symbiotic relationship, eventually becoming the mitochondrion. In a subsequent event, a photosynthetic bacterium entered a eukaryal cell, and eventually became the chloroplast. Chloroplasts appeared in the plant lineage after plants differentiated from the other eukaryals that had already acquired mitochondria.

how organelles evolved. Most researchers agree that eukaryal microbes arose by a process known as **endosymbiosis (Figure 3.18)**. The basic theory is that a single cell engulfed another cell, likely for the purpose of eating it for energy and building blocks. The engulfed cell, however, escaped destruction and set up housekeeping within its new host cell. Each member brought new capabilities to this partnership that enhanced the survival of the new combined cell, giving it evolutionary advantages and thus ensuring its survival. "Endosymbiosis" thus indicates that the two original cells have an intimate relationship with each other ("-symbiosis"), and one cell is fully inside of the other ("endo"). Endosymbiosis is the best theory for the origins of the mitochondrion and chloroplast within eukaryal cells, and perhaps other organelles as well.

Because virtually all eukarya contain mitochondria, but only a subset of eukarya contain chloroplasts, it is believed that the first endosymbiotic event was the engulfment of a bacterium

that led to the presence of the mitochondrion (see Figure 3.18A). Subsequently, a second endosymbiotic event occurred in which a cell that already contained the mitochondrion engulfed a second, autotrophic bacterium that gave rise to the chloroplast. In both cases, the benefit to the host cell involved the expansion of energetic capabilities (the mitochondrion permits aerobic respiration and the chloroplast permits photosynthesis) that would have made the host cell more competitive in an environment with scarce energy resources. In parallel, the engulfed bacterium would have benefited from the protection offered by the larger host cell. Note that over time the engulfed cells, today's mitochondria and chloroplasts, have lost so much of their genetic material that they could no longer function as independent cells.

Evidence in Support of the Endosymbiotic Theory

Where can we find evidence to support the theory of endosymbiosis? How do we find evidence for events that are proposed to have taken place over a billion years ago? The key is finding evidence of ancient cellular processes and life forms that allow us to retrace the paths of evolution. Historically, such evidence has been physical, usually involving fossils of life forms from microbes to dinosaurs. Today, however, scientists look for molecular and cellular evidence. By examining protein structures and DNA sequences, we can identify fingerprints, or leftover remnants, of the predecessors to current cells.

The hypothesis that endosymbiosis is responsible for the origins of the chloroplast and mitochondrion was first proposed, respectively, by the Russian scientist Konstantin Mereschkowsky and the American scientist Ivan Wallin in the early 1900s. However, Lynn Margulis is most closely associated today with the term, and it has been her work for the past 40 years that has provided some of the key evidence to move endosymbiosis from a fringe hypothesis into an accepted theory. Her first book on the topic, *The Origin of Eukaryal Cells*, was published in 1970 and became a seminal work in shaping our understanding of the evolution of cells.

Three main lines of evidence from Margulis and hundreds of other researchers support the theory of endosymbiosis. First, mitochondria and chloroplasts resemble bacteria both in shape and size. Both organelles are rod-shaped and usually measure between 0.5 μm and 5 μm in length. Second, the arrangement of membranes around these organelles is consistent with the endosymbiotic theory. Transmission electron micrographs clearly show that two membranes surround mitochondria and chloroplasts, while a single membrane surrounds other cell organelles. This observation correlates nicely with a model in which a parent cell engulfed a bacterium during the original endosymbiotic event and the resulting mitochondrion (or chloroplast) retained both the original bacterial membrane as well as the piece of the host cell plasma membrane that surrounded the bacterium when it was engulfed. Moreover, the lipid composition of the membranes of these organelles is bacteria-like, in contrast to the eukaryal-like lipid composition of other organelle membranes.

The third critical line of evidence supporting endosymbiosis came with the discovery that mitochondria and chloroplasts contain their own DNA (mitochondrial DNA is represented as mtDNA, and chloroplast DNA is known as ctDNA). This DNA is essential for the activities and reproduction of the mitochondria and chloroplasts, suggesting a degree of autonomy otherwise unique among cell organelles. Furthermore, the gene sequences and organization of the mitochondrial and chloroplast genomes are much more analogous to those of bacteria than to those of eukaryal cells. Indeed, complete sequencing of mitochondrial genomes and numerous bacterial genomes has indicated that all mitochondria are derived from the endosymbiosis of a particular type of bacteria: an alpha-proteobacterium. Likewise, all chloroplasts are derived from the endosymbiosis of a cyanobacterium. Finally, mitochondria and chloroplasts divide by fission, much like bacteria. Mitochondria and chloroplasts control their own division, dividing on their own timetable, independent of the cell cycle.

Mitochondria and chloroplasts contain only a small subset of genes found in alpha-proteobacteria and cyanobacteria. What happened to the missing genes? The answer is twofold. One likely reason for the missing genes is that the alpha-proteobacteria or cyanobacteria live their lives exclusively within a host cell and thus have diminished needs for independent function since the host cell will be providing them with some materials for life. Once a gene product is no longer needed, the extra DNA content becomes a burden during replication and can be lost due to the selective pressure of evolution. In other words, genes that became redundant or obsolete were lost. With a smaller genome the mitochondria can replicate faster. The second, and more important, reason is that the majority of genes (well over 1,000) that originated in the alpha-proteobacterium or cyanobacterium moved during the course of evolution from the mitochondrion and chloroplast into the nucleus. These genes are transcribed in the nucleus, translated on ribosomes in the cytoplasm, and imported post-translationally into the mitochondria.

Eukaryal microbes termed "amitochondriates" lack mitochondria but are not otherwise closely related. Being non-photosynthetic, they also lack chloroplasts. Several medically important eukaryal microbes, including *Giardia*, *Trichomonas*, and *Entamoeba*, are amitochondriates. It once was believed that these organisms evolved prior to that first endosymbiotic event that gave rise to the mitochondrion. Two lines of recent evidence, however, cast doubt on this hypothesis. First, the amitochondriates have been shown to contain remnant organelles. Second, genes within the amitochondriates' nuclear genome appear to have been derived from the alpha-proteobacterium genome. Based on these pieces of evidence, one could hypothesize that the amitochondriates once had mitochondria, but that they have been lost through evolution.

Endosymbiosis in Modern Cells

More experimental evidence supporting the endosymbiotic model is provided by the work of Kwang Jeon at the University of Tennessee. In the mid-1960s, Jeon began investigating a group of amoebas that had become infected with bacteria he termed "x-bacteria." Most of the amoebas died from the infection, but a few survived. Jeon's continued study of the surviving amoebas led to a pair of surprising discoveries. First, he found that the x-bacteria were still living inside the surviving amoebas. Second, when he then treated the amoebas with antibiotics that specifically targeted the bacteria to cure the amoebas of the infection, both the x-bacteria and the amoebas died (**Figure 3.19**).

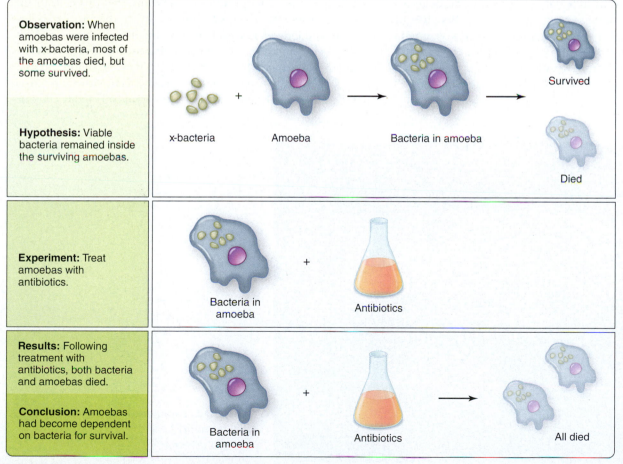

Observation: When amoebas were infected with x-bacteria, most of the amoebas died, but some survived.

Hypothesis: Viable bacteria remained inside the surviving amoebas.

x-bacteria   +   Amoeba   Bacteria in amoeba   Survived   Died

Experiment: Treat amoebas with antibiotics.

Bacteria in amoeba   +   Antibiotics

Results: Following treatment with antibiotics, both bacteria and amoebas died.

Conclusion: Amoebas had become dependent on bacteria for survival.

Bacteria in amoeba   +   Antibiotics   All died

Figure 3.19.
Experimental evidence for endosymbiosis: Amoeba and x-bacteria
Following treatment with antibiotics targeting the x-bacteria, both the bacteria and the amoebas died. What killed the amoebas? Jeon concluded that the amoeba had become dependent upon the bacteria. The loss of the bacteria deprived the amoeba of some needed substance.

These results indicated that not only had the amoebas adapted to live with the x-bacteria inside of them, but the x-bacteria actually became essential for the survival of the amoebas. Jeon's further research has indicated that the x-bacteria alter patterns of gene expression in the amoebas to cause this endosymbiotic interdependence. This example, and its ability to be reproduced and studied in real-time in the laboratory, provides evidence that endosymbiosis can give rise to new organisms, though the amoeba-bacteria hybrid falls far short of the extremely complex cell-organelle relationship found in eukaryal cells.

An additional example of a naturally occurring transient endosymbiotic relationship can be found in a common protozoan, *Paramecium bursaria*. This paramecium can derive energy and building blocks from either the ingestion of other cells or organic particles (heterotrophic) or through the use of sunlight (phototrophic). Although it has mitochondria, it doesn't have chloroplasts, so how can it perform photosynthesis? The answer is that, in the presence of sunlight, *Paramecium bursaria* ingests a type of algae known as *Zoochlorella* and maintains the algae within its cytoplasm, a phenomenon sometimes referred to as "algal farming." The algae perform photosynthesis and provide the *Paramecium* with energy and sugars, while the *Paramecium* provides the algae with transportation and safety **(Figure 3.20)**. *Paramecium bursaria* is fickle, however, and once the sunlight is gone, it becomes heterotrophic and digests the algae for energy and food. In **Perspective 3.2**, we will investigate yet another type of endosymbiosis—secondary endosymbiosis.

An important medical consequence of secondary endosymbiosis lies in the potential for development of new classes of chemotherapeutics (drugs) to treat malaria. Because of the presence of an algal-derived organelle, the malaria parasite has structures and biochemical processes in common with both bacteria and plants. Research is underway to evaluate the effectiveness of certain antibiotics and herbicides as treatments for malaria.

Lingering Questions about the Endosymbiotic Theory

The theory of endosymbiosis has good supporting evidence and seems straightforward. Like most ideas in science, however, the deeper you look into the question, the more interesting and complex it becomes. So, while endosymbiosis is a solid theory to explain the origin of eukaryal cells, some surprises and details remain unresolved, awaiting further investigation. Sequence analysis of mitochondrial and chloroplast genomes indicates that only two evolutionarily successful endosymbiotic events occurred: one that gave rise to mitochondria and one that gave rise to chloroplasts. Yet experiments such as those involving Jeon's amoeba and the *Paramecium bursaria* indicate that endosymbiosis may occur readily. Why, then, have stable endosymbiotic events been so rare in the course of evolution?

Another important question remaining is the identity of the original host cell that engulfed the alpha-proteobacterium. Scientists debate whether it was a bacterium or an archaeon, and look to evidence of gene sequences and biochemical processes for a better understanding.

A dilemma comes from our ready use of the term "engulf" when referring to the process of endosymbiosis. We typically envision something like phagocytosis, the process by which cells of our immune system eat infectious bacteria. Phagocytosis requires cells to have both a flexible periphery and a cytoskeleton capable of forming a vacuole; neither of these properties typically exists in bacteria or archaeons that are surrounded by a rigid cell wall. It is possible, of course, to imagine that the engulfing cell lacked a wall. Possibly, this host cell was not the initiator of the endosymbiosis. Maybe it didn't eat the alpha-proteobacterium, but instead the alpha-proteobacterium invaded this cell. In this case, the need for a cytoskeleton to direct the process is less critical, and indeed, researchers have identified a number of disease-causing microbes that force their way inside host cells in a manner that does not involve the host cell cytoskeleton. Another possible scenario is a progenitor eukaryal cell with many attributes we commonly associate with eukaryal cells today (flexible

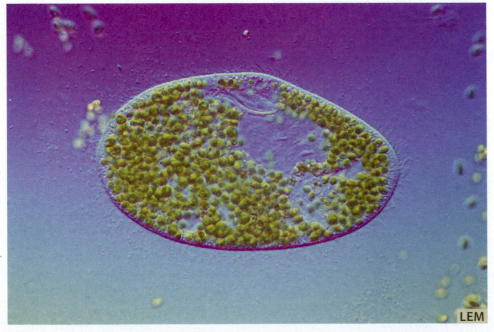

Figure 3.20. Paramecium and endosymbiosis: Algal farming This image shows a *Paramecium bursaria* cell containing numerous *Zoochlorella* algae. The algae perform photosynthesis for the benefit of both cell types.

LEM

A wonderfully rich example of the correlation between double membranes and phagocytic events was recently discovered in an organelle of *Plasmodium falciparum*, the protozoan that causes malaria. This organelle (termed the "apicoplast") was characterized by two surprising discoveries. First, its genome contains genes most like those of chloroplasts in sequence and organization. Second, it is surrounded by four membranes! How do you end up with four membranes around an organelle? Scientists now believe that the ancestor of *Plasmodium* ate an algal cell that already contained a chloroplast. Thus, the membranes of the apicoplast consist of (from the inside out): the membrane of the cyanobacterium, the membrane from the vacuole surrounding the cyanobacterium after the first endosymbiosis, the membrane of the alga, and the membrane from the vacuole surrounding the alga after the second endosymbiosis (**Figure B3.7**). This event, logically termed secondary endosymbiosis, was first proposed by Sarah Gibbs more than 25 years ago in her work on *Euglena*, although it took genome sequencing to bring the idea into mainstream biology. Similar evidence for secondary endosymbiosis can be found in diatoms, in which four membranes surround a chloroplast that was originally a cyanobacterium, eaten by a red alga, which in turn was eaten by the ancestor of the diatom. Again, the presence of genes homologous to genes found in both the cyanobacterium and red algae can be found within the chloroplast genome.

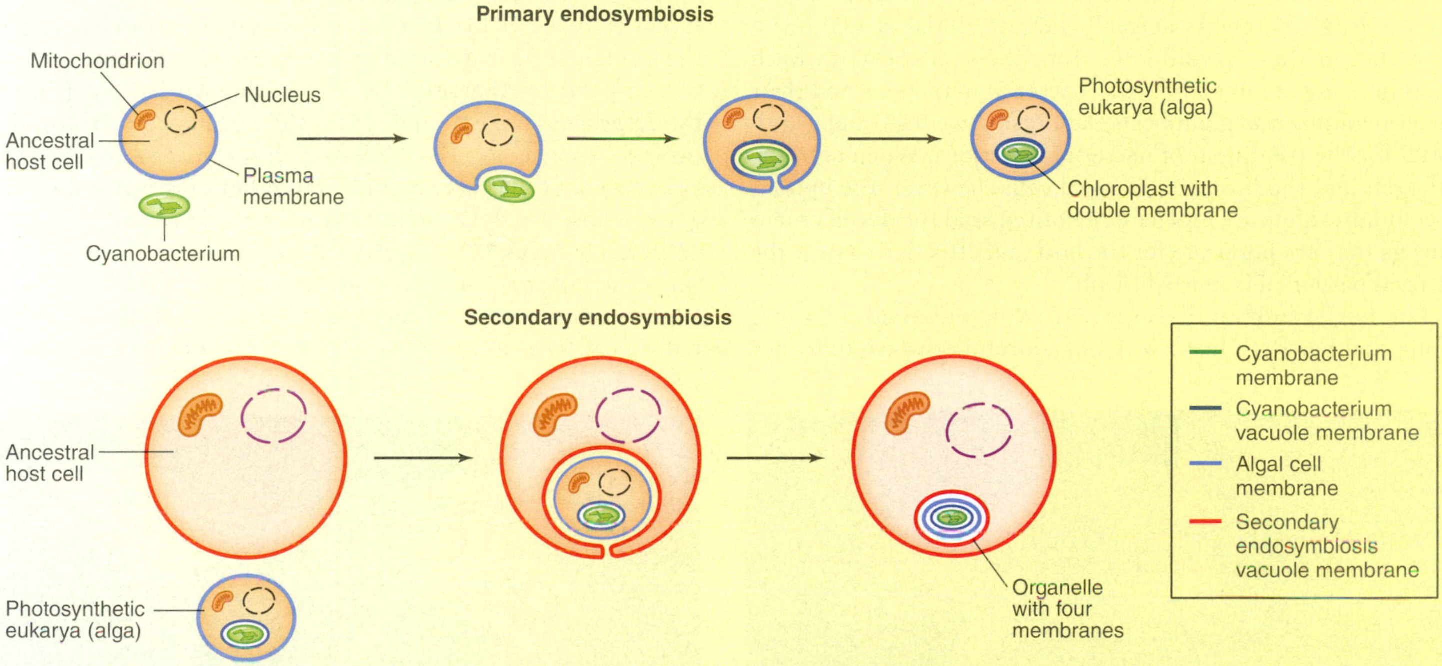

Primary endosymbiosis

Secondary endosymbiosis

Figure B3.7. Secondary endosymbiosis: Generating an organelle with four membranes The process by which the product of a primary endosymbiosis leads to secondary endosymbiosis and results in the formation of a new organelle with four peripheral membranes.

membrane, complex cytoskeleton) and was simply missing the full complement of internal organelles.

Might other organelles also be derived from endosymbiotic events? Is it possible that the nucleus itself might be the result of an endosymbiotic event? Remember that it too is surrounded by a double membrane and contains DNA. The origins of other organelles are less clear. Because other organelles lack DNA—the key piece of evidence of a former life as an independent cell—some think that they may be the result of invagination of membranes and subsequent localization of specific cell functions within those isolated compartments. Lynn Margulis and other scientists believe that endosymbiosis may have been more widespread and are actively seeking evidence that other organelles also might be endosymbiotic legacies.

CONNECTIONS Several lines of evidence support the hypothesis that the eukaryal nucleus may have resulted from endosymbiosis of an archaeon. In **Section 7.2**, we will see that many of the enzymes involved in archaeal DNA replication show a similarity to enzymes utilized by eukarya.

3.4 Fact Check

1. What is the endosymbiotic theory, and how is it related to the origins of mitochondria and chloroplasts?
2. What evidence supports the endosymbiotic theory?

3.5 Interactions between eukaryal microbes and animals, plants, and the environment

What harmful and beneficial roles do eukaryal microbes play?

Given the vast number and great diversity of eukaryal microbes, it should come as no surprise that these organisms play vital roles in our world. In fact, it's hard to imagine what life on Earth would be like without these organisms. Certainly, the ecosystem of the Rio Tinto would be very different! In this last section, we will examine some of the roles eukaryal microbes play, focusing first on diseases caused by these organisms and second on the roles they play in the environment.

Diseases caused by eukaryal microbes

Eukaryal microbes cause a wide array of diseases in both animals and plants. Their role as pathogens raises particular problems because drug treatments normally target cellular structures in the pathogen that are different from those of the host. Such differences are common between bacterial pathogens and their eukaryal animal and plant hosts. As a result, we have many antibiotics for the treatment of bacterial infections. When both the pathogen and the host are eukaryal cells, however, the options for cellular targets are much more limited, and the development of drugs that are both safe for the host and effective against the eukaryal pathogen is more difficult.

The list of human diseases caused by eukaryal microbes is long and has both historical and contemporary significance

(Figure 3.21). One of the top three infectious disease killers in the world today is cerebral **malaria**, a disease caused by the parasite *Plasmodium falciparum* that is transmitted between humans by the *Anopheles* mosquito. Malaria kills around 1 million people per year. In countries with high rates of infection, largely in sub-Saharan Africa, up to 50 percent of public health costs and hospital visits are attributable to malaria. The parasite has developed resistance to current drug treatments in many regions and no effective vaccine against the disease exists.

Eukaryal microbes cause many other insect-borne diseases. Trypanosomatids cause several diseases: African sleeping sickness (caused by *Trypanosoma brucei* and transmitted by the tsetse fly), Leishmaniasis (caused by *Leishmania* and transmitted by the sand fly), and Chagas disease (caused by *Trypanosoma cruzi* and transmitted by the triatomine insects known as the "kissing bugs"). Disease-causing eukaryal microbes also include a number of intestinal pathogens transmitted in food and water: *Entamoeba histolytica* **(Microbes in Focus 3.2)**, *Cryptosporidium parvum*, and *Giardia lamblia*. One eukaryal microbe, *Toxoplasma gondii*, which is related to *Plasmodium*, causes a common infection typically transmitted via cat feces or raw meat. The infection is most commonly

© Science Picture Co./Getty Images, Inc.

SEM

A. *Plasmodium falciparum*

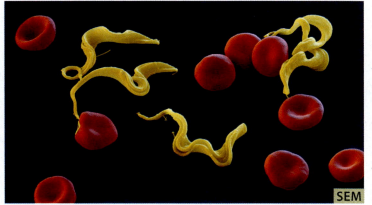

© Eye of Science/Photo Researchers, Inc.

SEM

B. *Trypanosoma brucei*

© 3DMedical/Phototake

SEM

C. *Giardia lamblia*

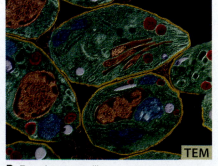

© Eye of Science/Photo Researchers, Inc.

TEM

D. *Toxoplasma gondii*

Figure 3.21. Parasitic protozoa cause human diseases. A. *Plasmodium falciparum*, the cause of malaria, emerging from infected red blood cells. **B.** *Trypanosoma brucei*, the cause of African sleeping sickness, swimming within human blood vessels. **C.** *Giardia lamblia*, a cause of diarrheal disease, displaying the disk on their ventral surface. **D.** *Toxoplasma gondii*, seen here with membranous structures highlighted, can cause severe disease in immunocompromised people, such as those with HIV/AIDS.

Microbes in Focus 3.2

ENTAMOEBA HISTOLYTICA: THE CAUSE OF AMOEBIC DYSENTERY

Habitat: Infects humans and other primates. Cysts can exist in water, soil, and on foods. Infection occurs by ingestion of cysts. Found worldwide, but infections are more common in developing countries.

Description: Parasitic unicellular microbe; lacks mitochondria

Key Features: It was once thought that pathogenic (disease-causing) and non-pathogenic strains of *E. histolytica* existed. Now, researchers recognize that these strains are actually two distinct species. *E. histolytica* is pathogenic, *E. dispar* is not.

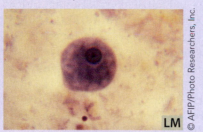

© AFIP/Photo Researchers, Inc.

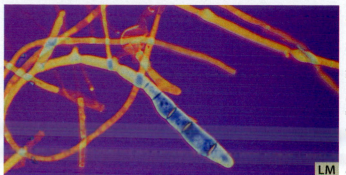

A. *Epidermophyton floccosum*

© Joaquin Carrillo-Fargo/Photo Researchers, Inc.

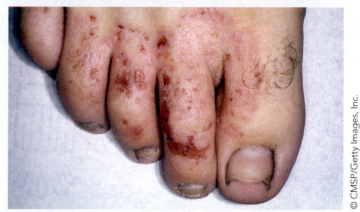

B. Athlete's foot

© CMSP/Getty Images, Inc.

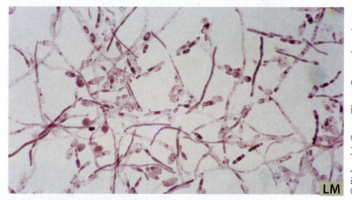

C. *Candida albicans*

© Richard J. Green/Photo Researchers, Inc.

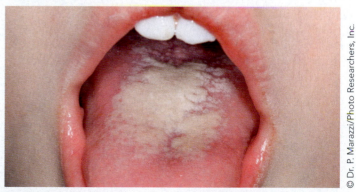

D. Oral thrush

© Dr. P. Marazzi/Photo Researchers, Inc.

asymptomatic, but if a woman becomes infected for the first time at the beginning of a pregnancy, it can cause significant neurological problems in a small percentage of fetuses. Additionally, the parasite remains within the host and can re-emerge to cause neurological damage if the host immune system is suppressed, for example in people with HIV/AIDS or those who have received organ transplants.

The list of fungi that cause diseases in healthy people is quite limited. Fungi do, however, cause some superficial skin infections **(Figure 3.22)**. Fungal infections can cause athlete's foot, ringworm, or jock itch, depending on the species of fungus. *Candida* infections can also cause different diseases, ranging from an oral disease termed "thrush" in infants to vaginal yeast infections in adult women. Many groups of fungi, however, can cause disease in people with suppressed immune systems. Indeed, one of the features of HIV/AIDS that first attracted medical interest was the appearance of fungal infections in these patients.

The list of plant diseases caused by eukaryal microbes is long and has both historical and modern significance. While humans more often suffer infections from protozoa, plants are more often impacted by fungi. Fungi are particularly well-suited to causing plant infections because the durable nature of their spores readily permits them to survive in soil between crops and during inclement weather conditions. These organisms cause a number of plant diseases, descriptively named rusts, rots, mildews, and smuts, that infect many grains and other crops, exacting a devastating toll on both crop yields and economies. Dutch elm disease, for instance, is a fungal disease of plants that has dramatically altered the landscape of American cities and towns (see Section 23.1). The most famous example of a plant disease is the infection of potato plants with *Phytophthora infestans* **(Microbes in Focus 3.3)**, a

Figure 3.22. Parasitic fungi cause human diseases. A. The fungus *Epidermophyton floccosum* showing reproductive spore structures. This fungus can cause athlete's foot. Related fungi also can cause athlete's foot, ringworm, and jock itch. **B.** Athlete's foot resulting from a fungal infection. **C.** The fungus *Candida* showing reproductive spore structures. *Candida* can cause a variety of diseases, ranging from oral thrush to vaginal yeast infections. **D.** Oral thrush resulting from a *Candida* infection.

water mold (a protozoa most closely related to brown algae and diatoms) responsible for potato blight **(Figure 3.23)**. Infections with *Phytophthora infestans* caused widespread famine in Ireland in the mid-1800s, resulting in 0.5–1 million deaths and at least twice as many emigrants, in turn shaping the demographics of the United States. Today, in the United States alone, there are more than 50,000 identified infectious diseases of plants with an annual cost of tens of billions of dollars spent on prevention and loss to crop damage.

CONNECTIONS *Ophiostoma novo-ulmi*, the causative agent of Dutch elm disease, produces several enzymes that help it degrade the plant cell wall. We will explore the roles of these enzymes in **Section 23.1**.

Beneficial roles of eukaryal microbes

As mentioned earlier, not all microbes are associated with disease. Eukaryal microbes also play essential beneficial roles. In most ecosystems, eukaryal microbes, bacteria, and archaeons serve as primary producers and biodegraders, keeping both energy and nutrient cycles going. Indeed, any cupful of water from a pond will reveal a world teeming with an enormous diversity of protozoa, algae, and fungi—a complete and self-sustaining microbial ecosystem. The description of life in the Rio Tinto presents this fundamental point. Photosynthetic algae harvest the energy of sunlight. Protozoa may survive by ingesting these algae or bacteria. Fungi decompose wastes, releasing nutrients. A complete and self-sustaining ecosystem thrives in this apparently hostile environment.

Our mental image of photosynthesis is usually one of large green trees. In reality, however, it is the open oceans that account for the majority of photosynthesis, followed closely by tropical rainforests. Algae and photosynthetic bacteria play key roles in photosynthesis in the ocean. Thus, photosynthetic microbes are essential for maintaining an aerobic atmosphere and for providing energy from sunlight and organic carbon from carbon dioxide.

Eukaryal microbes also serve a critical function in biodegradation of organic matter. They are particularly important in the breakdown of cellulose, an abundant structural polysaccharide in plants that is poorly digested by animals. By degrading cellulose, eukaryal microbes are vitally important in the decomposition of plant matter and provide a critical

© Science Photo Library/Photo Researchers, Inc.

SEM

A. *Phytophthora infestans*

© Science Photo Library/Photo Researchers, Inc.

B. Leaves affected by potato blight

Figure 3.23. Potato blight A. The spore cases of *Phytophthora infestans* emerging from leaf tissue. **B.** Resulting damage to a potato plant from infection with *Phytophthora infestans*.

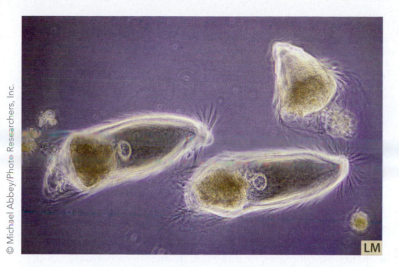

© Michael Abbey/Photo Researchers, Inc.

LM

Figure 3.24. Termite gut contains a protist that digests cellulose. The flagellated protist *Trichonympha* in the gut of the termite is responsible for digestion of cellulose. Despite being a danger to wooden structures, the protist's breakdown of cellulose is ecologically important and difficult chemically. The β-1,4 linkage between the glucose units is more stable to chemical breakdown than is the α-1,4 linkage found in starch. Enzymes capable of degrading cellulose are found in some microbes, such as *Trichonympha*.

step in rendering this plant matter usable for energy and nutrients by other organisms. Eukaryal microbes also can be found in great numbers in ruminant and termite digestive tracts, where, again, they facilitate the degradation of cellulose **(Figure 3.24)**.

3.5 Fact Check

1. Why is it difficult to treat infections by eukaryal microbes?
2. What are examples of human infections caused by fungal pathogens?
3. List common fungal infections in plants.
4. What are some beneficial environment roles of eukaryal microbes?

Image in Action

WileyPLUS Eukaryal organisms could use various mechanisms to withstand acidic environments such as in the Rio Tinto. These mechanisms may include allowing for the flow of protons (white spheres) freely into the cell (yellow, *top left*) and coping with the cytoplasmic acidity,

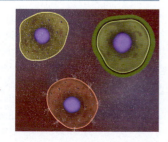

maintaining a thick cell wall to prevent the influx of protons (green, *top right*), or maintaining porous membranes to allow protons to enter the cell, and then pumping them back out (red, *bottom center*).

1. You have been asked to add additional detail to this image to convey to the viewer that these cells are eukaryal cells. What features are already present and what detail or labeling would you add to convey this point? What features might be specifically used to point out that these cells are a type of algae?

2. If a toxin that blocked H⁺ pumps in the cell membrane were dumped into the Rio Tinto, predict how this event would affect the metabolism and survival of the cells in this image and, by extension, the overall Rio Tinto ecosystem. Explain your reasoning.

The Rest of the Story

When scientists first discovered the rich diversity of the Rio Tinto ecosystem described in the opening story, they were struck by the dominant presence of eukaryal microbes and wondered at their ability to survive in such harsh conditions. In particular, they questioned how the eukaryal cells could survive in very low pH conditions and proposed the three possibilities mentioned earlier. Our understanding of the role of membranes in maintaining homeostasis will illuminate the answer to this question.

Some researchers hypothesized that the eukaryal microbes within the Rio Tinto simply let hydrogen ions flow in and, as a result, had a reduced intracellular pH. The intracellular pH of one particular species (a new species in the *Chlamydomonas* genus), though, is close to pH 6.6. Remember that the pH scale is logarithmic, so the hydrogen ion concentration outside the cells, where the pH is around 2.2, is close to 40,000 times higher than inside of the cells! This finding certainly does not support the first option.

It seems, then, that the cells must erect barriers against hydrogen ions or actively defend themselves against the influx of hydrogen ions. To distinguish between these options, scientists turned their attention to a novel set of hydrogen ion ATPases. They found that *Chlamydomonas* grows better at higher extracellular pH than at lower extracellular pH, and they measured the cost at 7 percent more ATP consumed per second by the cell under acidic conditions. Based on these data, it seems that the cells do not simply have a barrier that protects them from the hydrogen ions. Rather, these data provide good evidence that the hydrogen ions are actively transported against a chemical gradient out of the *Chlamydomonas* cytoplasm by using ATP for energy. When the extracellular pH is lower, this chemical gradient is greater and the cell must use more energy to pump the hydrogen ions out of the cell. Thus, the cells maintain homeostasis, but at an energy cost. Possibly, the protons are pumped from the cytoplasm into an internal organelle. In this case, the H⁺ pumps would not be exposed to the acidic extracellular environment and could function just like pumps found in *Chlamydomonas* species that live under less acidic conditions. As we can see from this one example, our understanding of cellular structure helps us to formulate hypotheses about cellular function.

Summary

Section 3.1: What do eukaryal cells look like?

Eukarya possess a membrane-bound nucleus. Additionally, eukaryal cells also contain other internal **organelles** and a complex cytoskeleton.

- The **nucleus** contains the genetic material. Within this double membrane-bound organelle, DNA replication and transcription occur.

- The **secretory pathway**, which includes the **endoplasmic reticulum (ER)** and the **Golgi apparatus**, is involved in protein trafficking.

- **Mitochondria** and **chloroplasts** are responsible for cellular energetics. **Cellular respiration** occurs within the mitochondria. Conversion of solar energy to chemical energy and carbon fixation occurs within the chloroplast. Both organelles generate ATP via **chemiosmosis**. The **amitochondriates** are eukarya that lack mitochondria.

- The **plasma membrane** effectively separates *inside* from *outside*. While bacteria and archaeons also possess plasma membranes, structural differences in the membranes of these domains do exist. Proteins exist within the plasma membrane and can move within the membrane, which has led to the **fluid mosaic model** of the plasma membrane. Localized pockets of specific lipids, **lipid rafts**, exist within cellular membranes.

- **Cilia** and **flagella** of eukaryal cells are encased within the plasma membrane.

- The **cell wall** plays a role in cell support. Not all eukarya possess cell walls. Algae have cell walls composed primarily of **cellulose**. Fungi generally have cell walls composed of chitin.

- The **cytoskeleton** provides shape and support to eukaryal cells and is comprised of three major types of structures: **microtubules**, **microfilaments**, and **intermediate filaments**. The cytoskeleton directs the movement of material within the cell and the movement of the cell itself.

Section 3.2: What are the different types of eukaryal microbes?

Eukaryal microbes are very diverse and their classification has been problematic.

- Traditional groupings have been based mainly on morphological features.

- Currently, phylogeny is based on SSU rRNA sequence comparisons. These comparisons demonstrate that some traditional groups, like the fungi, are, in fact, monophyletic. Other traditional groups, like the algae, are much more diverse.

- Phylogenetic trees based on other molecular markers, like Hsp70, reveal somewhat different evolutionary relationships among the eukaryal microbes.

  Much of what we know about eukaryal microbes is based on studies of **model organisms**.

- The fungi, like *Saccharomyces cerevisiae*, are distinguished by their cell wall, a chitin-based structure. Fungi are all heterotrophs.

- The protozoa, like *Giardia lamblia*, represent a broad and poorly defined group of eukaryal microbes. While some are heterotrophic, others are photosynthetic. Some contain a cell wall, while others do not.

- Slime molds, like *Dictyostelium discoideum*, often are classified as protozoa. The slime molds, though, always are heterotrophic and usually exist in multicellular aggregates.

- Algae, like *Chlamydomonas reinhardtii*, contain chloroplasts and are autotrophic.

Section 3.3: How do eukaryal microbes replicate?

Unlike bacteria and archaeons, eukaryal microbes have **haploid** (1n) and **diploid** (2n) genetic complements at different stages of their life cycles.

- In **mitosis**, genetically identical copies of the parent cell are produced. **Cytokinesis** generally follows mitosis.

- In **meiosis**, a reduction in the genetic complement, from 2n to 1n, occurs, along with a recombination of genetic material.

- Diploid cells of *Saccharomyces cerevisiae* reproduce asexually via mitosis. During periods of stress, the diploid cells undergo meiosis, producing haploid ascospores arranged in an **ascus**. These haploid cells can mate, resulting in the formation of a new diploid cell.

- Haploid cells of *Chlamydomonas* reproduce asexually via mitosis. During periods of stress, these haploid cells undergo morphological changes and become gametes. These gametes can fuse, producing a diploid spore. When conditions improve, this spore can undergo meiosis, resulting in four new haploid cells.

- The slime mold *Dictyostelium* has a very interesting and complex life cycle. Haploid amoeboid cells aggregate when food sources are depleted. The aggregated cells then undergo a type of differentiation, eventually forming spores that are released in the environment.

Section 3.4: How did eukaryal microbes originate?

Eukaryal cells originated when one cell engulfed another. The engulfed cell escaped destruction and became a necessary component of this developing eukaryal cell, a process known as **endosymbiosis**. Three main pieces of evidence support the endosymbiotic theory.

- Mitochondria and chloroplasts resemble bacteria both in size and shape and these organelles replicate independently of the cell cycle via fission.

- A double membrane surrounds mitochondria and chloroplasts, which is consistent with phagocytosis.

- Mitochondria and chloroplasts contain their own DNA and this DNA is more analogous to bacterial DNA than it is to the DNA of eukaryal cells.

  Questions about the origins of eukaryal cells, including how organelles other than mitochondria and chloroplasts originated, still remain.

Section 3.5: What harmful and beneficial roles do eukaryal microbes play?

Eukaryal microbes cause a wide array of diseases in both plants and animals, but also serve many beneficial roles.

- One of the major infectious diseases in humans is **malaria**, which is caused by the parasite *Plasmodium falciparum*.

- Photosynthetic eukaryal algae account for a significant amount of the photosynthesis that occurs in the oceans.

- Eukaryal microbes serve a critical function in the degradation of organic matter.

● Application Questions

1. Imagine you identify a cell and must characterize it as a eukaryal microbe, a bacterium, or an archaeal microbe by examining it under the microscope. Explain how you would make this determination.

2. What genes might you look for in the genome of a newly identified cell to identify it as a eukaryal microbe rather than a bacterial or archaeal microbe? Explain.

3. In the process of studying a given eukaryal microbe, you have disabled its Golgi apparatus. What would be the functional consequences for the eukaryal cell in this experiment?

4. While examining cellulose synthesis in the green algae *Chlamydomonas reinhardtii*, a researcher has deleted the genes involved in the process. What would be the structural and functional consequences of deleting cellulose synthesis genes in these algae? Explain.

5. The fungus *Candida albicans* causes vaginal yeast infections in women. This microbe does have a cell wall, but penicillin is inactive against it. Explain why.

6. The bacterium *Escherichia coli* and the parasitic protozoan *Giardia lamblia* both have flagella that help them swim. Although they have the same name, the structure and mechanism of action of the two flagella are quite different. Describe these differences.

7. Kwang Jeon found that when he treated the x-bacteria-infected amoebas with antibiotics, both the bacteria and amoebas died. What is the significance of this finding and how does it support the theory explaining endosymbiotic events that happened billions of years ago?

8. You read about a drug that disrupts the enzyme chitin synthase (involved in synthesizing chitin in cells). What type of microbes would this drug combat? Explain.

9. A scientist is writing a research grant to obtain funds to study the metabolism and activity of the slime mold *Dictyostelium*. He notes that his research could have many applications to understanding both human health and soil health. Explain how studying this one organism could provide information in such different scientific fields.

10. Many cases of disease outbreaks caused by the protozoan *Cryptosporidium parvum* in public pools have been documented, even though the pools were chlorinated. Why are these organisms not destroyed by the osmotic pressure or the chlorine in the pools?

11. You accompany scientists on their next trip to the Rio Tinto and carefully scrape a crusty sample from a rock in shallow water at the edge of the river. Upon returning to the laboratory, you are rewarded by the sight of many large oblong cells.
    a. First, describe what cell features you would look for to determine whether these cells were bacterial or eukaryal microbes.
    b. Second, assume that you have identified them as eukaryal microbes. Now, describe what structures/functions you would examine to determine whether the cells are fungi, algae, slime molds, or protozoa.
    c. Describe how you might use immunofluorescence techniques to determine your answer in part b.

Suggested Reading

Amaral Zettler, L. A., F. Gomez, E. Zettler, B. G. Keenan, R. Amils, and M. L. Sogin. 2002. Microbiology: Eukaryotic diversity in Spain's River of Fire. Nature 417:137.

Jeon, K. W. 1995. The large, free-living amoebae: Wonderful cells for biological studies. J Eukaryot Microbiol 42:1–7.

Jeon. K. W. 2004. Genetic and physiological interactions in the amoeba-bacteria symbiosis. J Eukaryot Microbiol 51:502–508.

Kohler, S., C. F. Delwiche, P. W. Denny, L. G. Tilney, P. Webster, R. J. Wilson, J. D. Palmer, and D. S. Roos. 1997. A plastid of probable green algal origin in apicomplexan parasites. Science 275:1485–1489.

Messerli, M. A., L. A. Amaral-Zettler, E. Zettler, S. K. Jung, P. J. Smith, and M. L. Sogin. 2005. Life at acidic pH imposes an increased energetic cost for a eukaryal acidophile. J Exp Biol 208:2569–2579.

Waller, R. F., and G. I. McFadden. 2005. The apicoplast: A review of the derived plastid of apicomplexan parasites. Curr Issues Mol Biol 7:57–79.

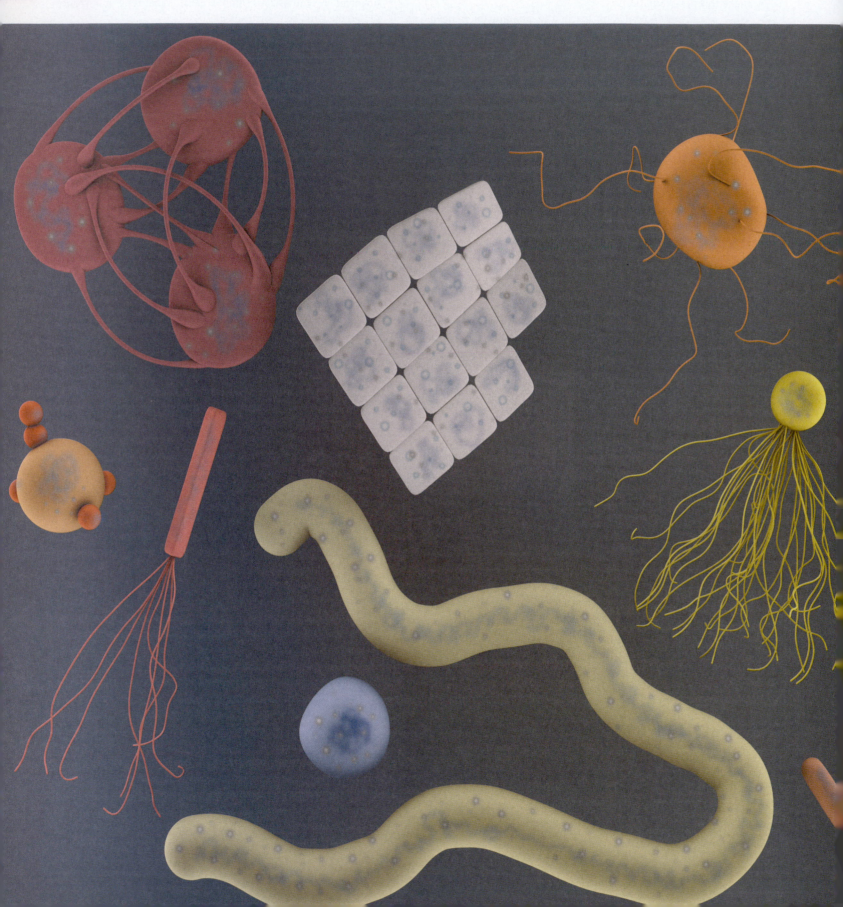

During explorations of hydrothermal vents off the coast of Iceland, microbiologists isolated a new microbial species, *Ignicoccus hospitalis*. They discovered that this single-celled, coccus-shaped microorganism lacks a nucleus, like the bacteria that we explored in Chapter 2. *I. hospitalis*, though, is not a bacterium. Rather, it is a member of another domain of life, Archaea.

In the laboratory, researchers could cultivate this microbe, but only under some fairly atypical conditions. Like the only other members of the *Ignicoccus* genus currently categorized, *I. islandicus* and *I. pacificus*, this microorganism grows optimally at 90°C. Additionally, these three species all grow best in slightly acidic and somewhat salty environments. They also gain energy by reducing elemental sulfur and, in the process, forming hydrogen sulfide. Structurally, *I. hospitalis*, *I. islandicus*, and *I. pacificus* differ from all other known archaeons. For these species, the cell envelope consists of a plasma membrane, a periplasm, and an outer membrane, similar in many respects to the cell envelope of the Gram-negative bacteria that we studied in Chapter 2.

While *I. hospitalis* exhibited unusual growth requirements, another aspect of this microbe interested the researchers even more. They noticed that very tiny, coccus-shaped cells were attached to many of the *Ignicoccus* cells. This second microbial species, named *Nanoarchaeum equitans*, also is an archaeon and appears to grow only in association with *I. hospitalis*. The exact nature of the relationship between these microbes remains a topic of great interest. The cells of *N. equitans* are only 0.4 μm in diameter and its genome is less than 500,000 base pairs long, making it one of the smallest known living organisms. Studies of this unusual organism may provide us with new insight into life under extreme conditions, interspecies relationships, and the minimal requirements for life. Oh, and one other thing—some researchers have proposed that *N. equitans* represents the sole member of an entirely new phylum within the archaeal domain.

Introduction

In this chapter, we will explore another domain of life—Archaea. Like the bacteria that we examined in Chapter 2, members of this domain, archaeons, lack a true nucleus. These organisms, though, differ in very fundamental ways from both bacteria and eukarya. While we still know relatively little about archaeons, our studies of these organisms already have shed light on important aspects of how proteins and cells function. Future studies of archaeons may shed light on the evolution of all living organisms. The story of the domain Archaea is both exciting and exotic. It begins in some of the most inhospitable environments on Earth.

Since the discovery of the first organisms that grow optimally in environments with extreme chemical and/or physical properties, microbiologists have searched for signs of life in almost every imaginable environment on Earth. Perhaps surprisingly, or perhaps not, we have found living organisms that thrive in every environment that has been explored. From the boiling waters of hot springs to the frigid environments of Antarctica, and from the low pH, sulfur-rich waters often associated with volcanoes to the high pH waters of lakes containing sodium carbonate, life, it seems, exists everywhere.

Underwater, hydrothermal vents are no exception. Often referred to as black smokers, these vents actually represent fissures in Earth's crust through which geothermally super-heated water escapes. Often topping 400°C initially, this mineral-laden water quickly cools, causing the precipitation of its mineral contents, resulting in the formation of the characteristic rock chimneys. Studies of black smokers in oceans throughout the world have demonstrated that a wonderful array of life calls these very hot environments home (**Figure 4.1**), including many archaeons.

Because archaeons superficially resemble bacteria, Archaea was not recognized as a domain of life until quite recently. In 1977, Carl Woese and colleagues proposed that these organisms were distinct from bacteria and, in 1990, argued that Archaea represents a unique domain of life, separate from Bacteria and

Courtesy Verena Tunnicliffe, Department of Biology, University of Victoria, B.C.

Figure 4.1. Black smoker in the Pacific Ocean Arising from the sea floor over a mile beneath the surface, black smokers may seem inhospitable. The water emerging from them may be over 400°C and laden with various sulfur compounds. Nonetheless, a rich community of life, including numerous tube worms, grows in the super-heated waters of these 10- to 15-foot-tall formations. Some of the most abundant inhabitants of these environments are archaeons that have evolved to survive in these hot waters.

Eukarya. In subsequent chapters, we will explore how archaeons differ from bacteria and eukarya. In this chapter, we will address this issue briefly. We also will tour an archaeal cell, much like we toured the bacterial cell in Chapter 2. Finally, we will begin to investigate the diverse groups of Archaea. So let's begin by asking a few questions:

How do we know that Archaea is a distinct domain of life? (4.1)
What do archaeal cells look like? (4.2)
What are the major groups of archaeons? (4.3)

4.1 Distinctive properties of Archaea

How do we know that Archaea is a distinct domain of life?

It's hard to imagine that a whole domain of life, one of the three domains into which all living organisms are categorized, was not recognized by most biologists until the last decades of the twentieth century. The entire story of the domain Archaea, though, is somewhat hard to imagine. For many years, members of this domain were lumped together with bacteria into a single kingdom. Indeed, **archaeons** bear striking similarities to bacteria. Most notably, archaeons and bacteria are very similar in size and have chromosomes of similar size and organization. Members of both groups also are prokaryotic—they lack a membrane-bound nucleus. Genetic studies have shown, though, that prokaryotic organisms form two evolutionarily distinct groups. Additional studies have confirmed

that bacteria and archaeons are distinct; archaeons differ significantly from both bacteria and eukarya. In Section 4.2, we will examine the structure of archaeal cells, focusing on how common structures in archaeons, bacteria, and eukarya differ. In this section, we will focus on a more fundamental question. How do we know that Archaea is a distinct domain of life?

Phylogeny

The question posed above can best be answered by phylogenetic studies, or studies of the evolutionary relatedness of organisms. The question, then, becomes: how can we determine the evolutionary relatedness of all living organisms? In 1977, Carl Woese and George Fox, then researchers at the University of Illinois, Urbana, answered this question in a most elegant way. In their landmark paper (PNAS 74:5088), the authors noted that:

> Phylogenetic relationships cannot be reliably established in terms of noncomparable properties....To determine relationships covering the entire spectrum of extant living systems, one optimally needs a molecule of appropriately broad distribution. None of the readily characterized proteins fits this requirement. However, ribosomal RNA does.

In other words, by comparing **ribosomal RNA (rRNA)** nucleotide sequences, we can determine the evolutionary relatedness of all living organisms (see the Mini-Paper in Chapter 1 for more details about this important study). Organisms with closely related rRNA sequences must be closely related evolutionarily. More specifically, Woese and Fox examined the sequence of the 16S or 18S ribosomal RNA molecules, the small subunit rRNA, or SSU rRNA. This molecule, they reasoned, is present in all known living organisms and fulfills basically the same biological function in all living organisms. Woese and Fox reasoned that because rRNA fulfills a common, universal function, it will show little variation among different organisms (see Mini-Paper, Chapter 1). Minor sequence variations, resulting from random mutations, will exist and the number of differences that exist between the rRNA sequences of two species should reflect the evolutionary distance between those species. Ribosomal RNA sequences, in other words, can be used to compare all living organisms.

Their findings, quite literally, resulted in the re-writing of biology textbooks. Ribosomal RNA sequence analyses showed that all eukaryal organisms, as Woese and Fox expected, were related. Prokaryotic organisms, however, formed two distinct groups. All of the well-characterized bacteria, like *Escherichia coli* and *Staphylococcus aureus*, formed a distinct group. Not all prokaryotic organisms, however, were related to these well-characterized bacteria. The rRNA sequence analysis of a then poorly characterized group of organisms capable of producing methane showed them to be evolutionarily distinct from bacteria **(Figure 4.2)**. As Woese and Fox noted (PNAS 74:5088), "These 'bacteria' appear to be no more related to typical bacteria than they are to eukaryal cytoplasms." These methane-producing organisms, Woese concluded, constituted a separate domain of life—Archaea.

CONNECTIONS Methane-producing archaeons remain an interesting—and well-studied—group of organisms. Methane-producing archaeons often live in the gastrointestinal tract of mammals, aiding in the digestion of food. As we will see in **Section 17.4**, cows produce a significant amount of methane, thanks to their archaeal partners.

We now know that the domain Archaea includes more than just these methane-producing organisms. Many archaeons have been identified in environments that seem extreme to us **(Table 4.1)**. In fact, for quite some time, many scientists assumed that all archaeons lived in atypical environments and referred

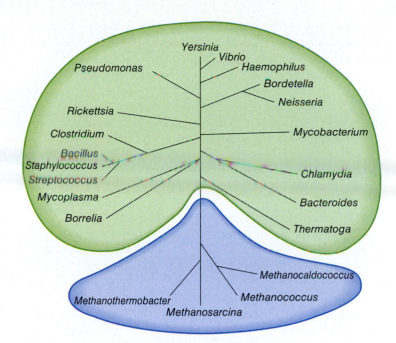

Figure 4.2. Methanogens are genetically distinct from bacteria. In 1977, Woese and Fox proposed that methanogens were distinct from all commonly known bacteria. Subsequent SSU rRNA gene sequence analysis has confirmed this hypothesis. As shown in this schematic, when the 16S rRNA sequences of various bacteria and the four methane-producing microbes are compared, the methanogenic microbes form a distinct cluster.

to them as extremophiles. We should note that "extreme" lies in the eyes of the beholder; what may seem extreme to us is quite normal for other organisms. More recently, archaeons have been identified in environments that humans consider less extreme. All of these organisms are phylogenetically related. All of them exhibit significant 16S rRNA gene sequence similarity.

Structure

The 16S rRNA sequence analysis shows us quite clearly that archaeons are distinct from bacteria and eukarya. In some respects, archaeons almost seem to be an amalgam of bacteria and eukarya. Certainly, archaeons look like bacteria. The typical archaeal cell is similar in size (between 0.5 μm and 5 μm in diameter) to the typical bacterial cell. Cells of species in both

TABLE 4.1	Selected archaeons and their preferred growth requirements	
Organism	**Notable growth requirement(s)**	**Natural habitat**
Halobacterium salinarium	3.0–5.0 M NaCl	Dead Sea, salted foods
Pyrococcus furiosus	100°C	Hydrothermal vents
Picrophilus oshimae	0.7 pH	Sulfur-rich volcanic regions
Methanogenium frigidum	15°C	Ace Lake, Antarctica

domains also exhibit similar shapes. Both archaeons and bacteria generally possess single, circular chromosomes. Archaeons and bacteria also lack a membrane-bound nucleus.

Other features of archaeons, however, more closely resemble eukarya. Specifically, the DNA replication, transcription, and translation processes in archaeons seem to be eukarya-like. Most obviously, as we will discuss in the next section, archaeal DNA is complexed with **histones**, DNA-binding proteins traditionally considered to be a defining characteristic of eukarya. Additionally, many of the archaeal enzymes involved in DNA replication, like DNA polymerase, primase, and helicase, show similarity to their eukaryal counterparts. We will investigate the replication of DNA in archaeons more fully in Section 7.2.

While archaeons resemble both bacteria and eukarya in some respects, they also differ from both bacteria and eukarya in at least one fundamental way. The plasma membrane of all archaeons differs from the plasma membrane found in all bacteria and eukarya. As we saw in Chapters 2 and 3, a phospholipid bilayer encloses all bacterial and eukaryal cells. A lipid-based membrane also encloses all archaeal cells. As we noted in the opening story, both a plasma membrane *and* an outer membrane surround cells of organisms in the *Ignicoccus* genus. The chemistry of archaeal membranes, however, differs fundamentally from bacterial and eukaryal membranes. We'll investigate this difference more fully in the next section.

Evolution

What do we know about the evolutionary history of archaeons, and what can these organisms tell us about the evolutionary history of life? As with so many aspects of the biology of archaeons, we still are developing answers to these critical questions. Initially, many people assumed that archaeons were ancient organisms, perhaps even the descendants of the first life forms. After all, the first archaeons identified were extremophiles, and conditions on Earth when life arose were quite extreme.

SSU rRNA sequence analyses, however, paint a different picture. These analyses clearly show that Bacteria, Archaea, and Eukarya form three distinct clusters. These analyses also suggest that Archaea and Eukarya may have branched off of Bacteria (see Figure 1.8). Bacteria, then, may be the closest relatives of the original life-form. Some researchers have hypothesized that the appearance of histones in a bacterial precursor led to the evolution of archaeons and eukarya. Histones certainly can protect DNA in high temperature environments and can allow larger amounts of DNA to be packaged into cells. The appearance of histones could have led to a rapid evolutionary burst. It is unclear, though, what event may have resulted in the switch from a bacterial to an archaeal/eukaryal DNA replication machinery.

Any discussion of the evolution of archaeons faces an even more perplexing issue—the altered plasma membrane found in members of this domain. We could surmise that the archaeal membrane provides increased thermal stability. Certainly, that would benefit species that live in high temperature environments. Not all archaeons, though, can be found in high temperature environments. In fact, more and more evidence suggests that archaeons exist in a wide range of environments. Furthermore, many bacteria live at high temperatures. The archaeal membrane structure, then, is not absolutely necessary for survival at elevated temperatures. This fundamental question about the evolutionary history of Archaea remains unanswered.

4.1 Fact Check

1. How did Woese and Fox determine evolutionary relatedness, and why did they consider their approach best?
2. How are archaeons similar to and different from bacteria?
3. How are archaeons similar to and different from eukaryal cells?
4. What do SSU rRNA sequence analyses show about the timing of archaeon evolution? Explain.

4.2 Archaeal cell structure

What do archaeal cells look like?

In Chapters 2 and 3, we investigated the structures of bacterial and eukaryal cells, paying particular attention to the functions performed by various structures and the evolutionary relationships of these structures. In this section, we will investigate the structure of archaeal cells. Again, we will investigate the biological functions performed by these structures and begin to explore the interesting evolutionary history of these organisms. We will begin by describing the archaeal cell. Then, we will look at specific parts of these cells, beginning with the cytoplasm and progressing to the plasma membrane, the cell wall, and, finally, the cell surface.

Morphology of archaeal cells

Like bacteria, most archaeons range in size from 0.5 micrometers (μm) to 5 μm, about a tenth of the size of typical eukaryal cells. As we saw with bacteria in Chapter 2, though, the sizes of different species of archaeons can vary widely. *Nanoarchaeum*

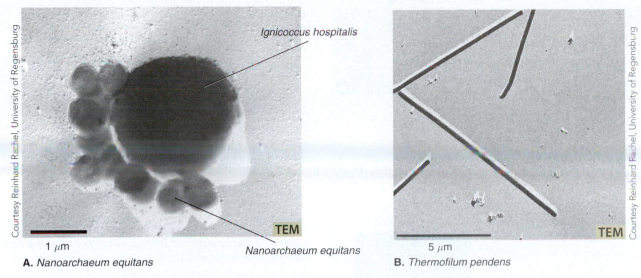

Ignicoccus hospitalis

TEM

1 μm

A. *Nanoarchaeum equitans*

Nanoarchaeum equitans

TEM

5 μm

B. *Thermofilum pendens*

Courtesy Reinhard Rachel, University of Regensburg

Courtesy Reinhard Rachel, University of Regensburg

Figure 4.3. Size range of archaeal species Like bacteria, archaeons may differ dramatically in size. *Nanoarchaeum equitans* cells are only 0.4 μm long. Cells of species in the *Thermofilum* genus, conversely, may be 100 μm long. For comparison, cells of the bacterium *Escherichia coli* are approximately 2 μm in length. **A.** *N. equitans* (small cells) associated with an *Ignicoccus* cell (scale bar = 1 μm). Both of these organisms have been isolated from marine hydrothermal vents. **B.** *Thermofilum pendens* initially isolated from slightly acidic hot springs associated with volcanoes (scale bar = 5 μm).

equitans, the microbe we mentioned in the opening story, is only about 0.4 μm in diameter and its internal volume is only about 1 percent of the volume of an *Escherichia coli* cell. *N. equitans*, then, is among the smallest living organisms. Species in the genera *Thermoproteus* and *Thermofilum*, conversely, may have cells up to 100 μm long **(Figure 4.3)**.

Also like bacteria, archaeons display various distinct shapes, or morphologies. Many characterized archaeons exist as rods or spheres. Additionally, some archaeal species display spiral-shaped cells. Other archaeal species exhibit more unusual shapes. Members of the *Sulfolobus* genus often have an irregular shape, while members of the *Thermoproteus* and *Pyrobaculum* genera appear to be rectangular in shape **(Figure 4.4)**. Researchers even have identified an archaeal species, *Haloquadratum walsbyi*, in which the cells grow as very thin, almost flat, squares **(Microbes in Focus 4.1)**. While *H. walsbyi* first was observed in 1980, we still have no direct explanation for how the cells maintain their unusual shape. As our studies of archaeons continue, we may discover other unusually shaped species and, eventually, better understand the basic properties affecting cell morphology.

Figure 4.4. Shapes of archaeal cells While most archaeons are rod- or coccus-shaped, some exhibit more unusual shapes. **A.** *Sulfolobus acidocaldarius* cells often exhibit irregular shapes. This organism grows optimally in high temperature, low pH environments. **B.** *Thermoproteus tenax* cells have a rectangular shape. Like *S. acidocaldarius*, this organism grows optimally at high temperatures. Additionally, species in the *Thermoproteus* genus can utilize sulfur compounds for energy.

TEM

A. *Sulfolobus acidocaldarius*

Courtesy Dennis Grogan, University of Cincinnati

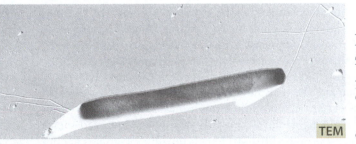

B. *Thermoproteus tenax*

TEM

Courtesy Reinhard Rachel, University of Regensburg

Microbes in Focus 4.1
HALOQUADRATUM WALSBYI: THE SQUARE MICROBE

Habitat: Originally identified in 1980 in a salt pool near the Red Sea; subsequently found in salt pools throughout the world.

Description: Halophilic archaeon in the Euryarchaeota phylum. Distinguished by its flat (only 0.1 to 0.5 μm thick), square (2–5 μm per side) shape. Also contains a large number of gas vesicles.

Key Features: Growth in the laboratory occurs in medium containing 3.3 M NaCl and 2.2 M KCl. Researchers are interested in determining how this microbe grows in such a high salt environment and what factors determine its unusual shape. Although the 3.1 Mbp genome was sequenced in 2006, researchers still cannot adequately explain how these cells maintain their unusual shape.

Courtesy Mike Dyall-Smith, University of Melbourne, Australia

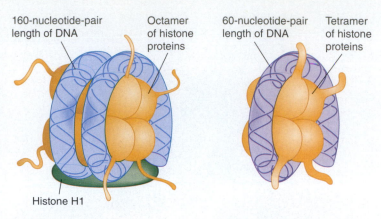

A. Eukaryal nucleosome   **B.** Archaeal nucleosome

Figure 4.5. Compaction of DNA in eukarya and some archaeons by histones Positively charged histone polypeptides interact with the negatively charged DNA, compacting the DNA. **A.** In eukarya, a histone octamer associates with approximately 160 bp of DNA. The H1 histone interacts with this octamer and the DNA. **B.** In archaeons, a histone tetramer associates with approximately 60 bp of DNA. The presence of histones in archaeaons and eukarya, but not bacteria, suggests that histones evolved after the split between bacteria and archaeons, but before eukarya evolved.

The cytoplasm

The cytoplasm of an archaeal cell, like the cytoplasm of a bacterial cell, contains a diverse mixture of molecules. As in bacterial cells, the largest, and arguably the most important entity within an archaeal cell is the nucleoid. Like the DNA of bacteria, the DNA of archaeons generally exists as a single, circular molecule. Also, as we saw in bacteria, no membrane surrounds the DNA. Unlike bacteria, though, at least some archaeons produce histones, basic proteins that serve to condense, or compact, the DNA, thereby allowing more DNA to be packaged within a smaller area.

Histones exist in virtually all eukaryal species. Typically, two copies each of four different histone polypeptides (H2A, H2B, H3, and H4) form an octameric complex. Approximately 160 base pairs of DNA wraps each histone octamer, thereby compacting the DNA. A separate histone protein, the H1 histone, interacts with this octamer and the DNA. In some archaeons, histones similar to the eukaryal H3 and H4 histones have been identified. Current research indicates that, in some archaeal species, approximately 60 base pairs of DNA wrap around a tetramer of these polypeptides **(Figure 4.5)**. Until 2005, histones only had been identified in one phylum of Archaea—Euryarchaeota. Researchers now have identified histones in members of the archaeal phylum Crenarchaeota and in *Nanoarchaeum equitans*, which is the only currently identified member of a proposed third archaeal phylum—Nanoarchaeota (see Section 4.3).

The existence of histones in species from all known phyla of Archaea has led some researchers to propose that histones (and consequently a method for efficiently packaging DNA) evolved early in the history of life. According to this model, histones evolved after the divergence of bacteria and archaeons but before the emergence of eukarya. Furthermore, the existence of histones, and the ability to package DNA, may have been a necessary prerequisite to the increased genomic size that we see today in many eukarya. Macroscopic life, in other words, may exist because of these DNA-binding proteins.

In addition to the nucleoid, the cytoplasm of archaeons contains many other molecules that we see in the cytoplasm of bacteria. DNA and RNA polymerases, the enzymes needed for DNA replication and transcription, respectively, exist in the cytoplasm. Ribosomes, the complex structures involved in translation, or protein synthesis, also exist here. Other enzymes needed for various cellular functions reside within the cytoplasm, as do other important molecules, like transfer RNA, ions, and sugars. More complex structures, like gas vesicles, also have been identified in archaeons. While the biochemical nature of some of these components may differ between bacteria and archaeons, the types of molecules and their basic functions are remarkably similar.

The archaeal cytoskeleton

Historically, the cytoskeleton, an intracellular framework of proteins, was considered to be a defining characteristic of eukarya. As we discussed in Chapter 2, though, cytoskeleton-like structures exist in bacteria. The bacterial FtsZ, MreB, and ParM proteins show similarity to eukaryal tubulin (FtsZ) and actin (MreB and ParM). Researchers also have identified a cytoskeleton-like element in archaeons. An actin homolog, Ta0583, was identified in *Thermoplasma acidophilum* **(Microbes in Focus 4.2)**, a species in the Euryarchaeota phylum. Genetic and biochemical studies have shown that Ta0583

Journal of Bacteriology 189:2039–2045, 2007. Reproduced/amended with permission from American Society for Microbiology.

Microbes in Focus 4.2
LIVING IN "HOT VINEGAR":
THERMOPLASMA ACIDOPHILUM

Habitat: Originally isolated from a coal refuse pile. Has been isolated from a variety of high temperature, low pH environments since.

Description: Archaeon approximately 1 μm in diameter in the Euryarchaeota phylum. Lacks a cell wall and, probably as a result, displays a variety of shapes. Grows optimally at 56°C, pH 1.8.

Key Features: Some researchers have described its preferred habitat as "hot vinegar." The cell membrane of this thermoacidophilic microbe consists of a tetra-ether lipid monolayer, rather than a more typical phospholipid bilayer. Presumably, this monolayer arrangement provides increased stability in the presence of elevated temperatures and a low pH.

TEM

1 μm

Courtesy Reinhard Rachel, University of Regensburg

may more closely resemble eukaryal actin than bacterial MreB or ParM proteins **(Figure 4.6)**. Cytoskeletal-like proteins isolated from the archaeal species *Methanobacterium thermoautotrophicum* and *Methanopyrus kandleri*, conversely, show greater similarity to the bacterial MreB protein. From these results, it should be obvious that much still needs to be learned about the evolution, distribution, and function of cytoskeletal proteins.

The cell envelope

All archaeons, like all bacteria and eukarya, contain a plasma membrane, a semi-permeable barrier surrounding the cytoplasm. Most archaeons, like most bacteria and some eukarya, also contain a cell wall, a more rigid structure surrounding the plasma membrane that helps provide shape and stability to the cells. The chemical compositions of the archaeal plasma membrane and cell wall, however, differ significantly from their bacterial and eukaryal counterparts. These structures, then, provide defining characteristics of archaeons.

The Plasma Membrane

As we have seen in bacterial and eukaryal cells, a membrane surrounds the cytoplasm of archaeal cells. Functionally, the membranes in all of these cells are equivalent. In all cases, the plasma membrane forms a differentially, or selectively, permeable barrier that separates inside from outside. Proteins embedded within the membrane allow both active and passive transport of important molecules. Other integral membrane proteins serve in cell-signaling capacities. Still other proteins control secretory events, facilitating endocytosis and exocytosis events.

While the functions of the plasma membranes in bacteria, eukarya, and archaeons are quite similar, the structure of the archaeal plasma membrane differs quite significantly from the plasma membranes of bacteria and eukarya. The plasma membranes in bacteria and eukarya consist of a phospholipid bilayer. In all cases, this membrane consists of fatty acids linked to glycerol 3-phosphate (G3P) molecules. In archaeons, though, the membrane consists of **isoprenoids**, hydrocarbon molecules built from 5 carbon isoprene subunits, attached to glycerol 1-phosphate (G1P), a stereoisomer, or mirror image, of the molecule found in bacteria and eukarya. Moreover, these hydrophobic molecules are linked via ether linkages to the glycerol, as opposed to the ester linkages seen in bacterial and eukaryal phospholipid bilayers **(Figure 4.7)**.

In many archaeons, the most commonly used isoprene polymer is **phytanyl**, a 20-carbon hydrocarbon. Alternatively, some archaeons contain **biphytanyl**, a 40-carbon hydrocarbon. Among species that contain biphytanyl, a phosphoglycerol molecule often is linked to both ends of the isoprene polymer. The membrane, then, is a phospholipid monolayer, rather than a bilayer (see Figure 4.7). This formation is seen most commonly among archaeal species that live at very high temperatures. The monolayer formation, presumably, is more stable at high temperatures than its bilayer counterpart.

As we noted in the opening story, species in the *Ignicoccus* genus possess both a plasma membrane and an outer membrane, a cell envelope organization unique among archaeons thus far categorized. The periplasm appears to be quite large, measuring

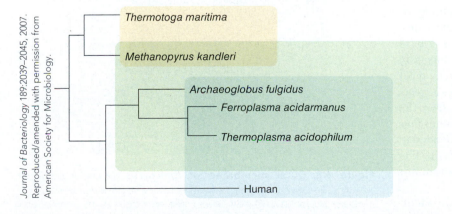

Figure 4.6. Protein sequence comparison of eukaryal actin, bacterial MreB, and archaeal homologues In this comparative tree, the amino acid sequences of archaeal homologues to MreB were compared to bacterial MreB and eukaryal actin amino acid sequences. The linear distance between two samples reflects the likely evolutionary distance between those samples. The homologue found in the methanogenic archaeon *Methanopyrus kandleri* shows more similarity to a bacterial MreB isolate from *Thermotoga maritima* than to eukaryal actin. Homologues in the thermophilic archaeons *Archaeoglobus fulgidus*, *Ferroplasma acidarmanus*, and *Thermoplasma acidophilum* are more similar to human actin than to bacterial MreB.

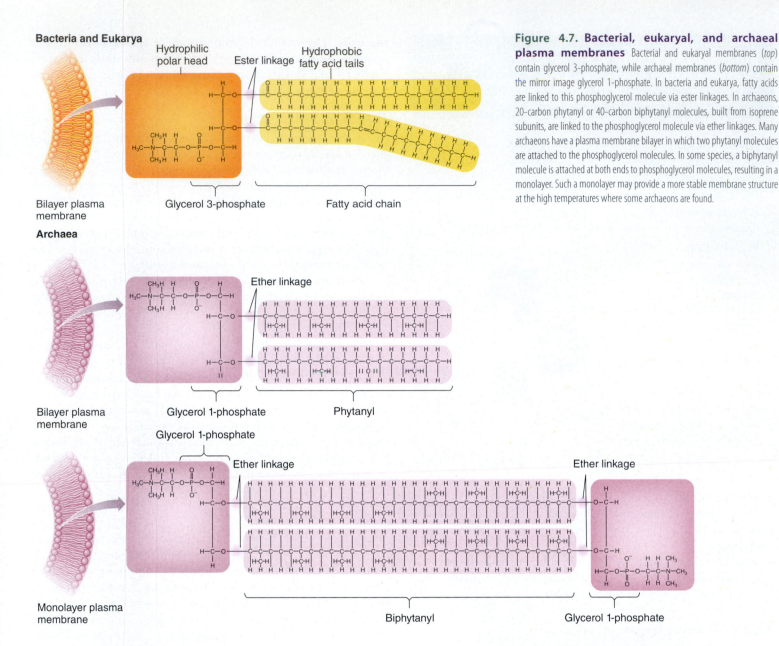

Bacteria and Eukarya

Hydrophilic polar head
Ester linkage
Hydrophobic fatty acid tails

Bilayer plasma membrane

Glycerol 3-phosphate
Fatty acid chain

Archaea

Ether linkage

Bilayer plasma membrane

Glycerol 1-phosphate
Phytanyl

Glycerol 1-phosphate

Ether linkage
Ether linkage

Monolayer plasma membrane

Biphytanyl
Glycerol 1-phosphate

Figure 4.7. Bacterial, eukaryal, and archaeal plasma membranes Bacterial and eukaryal membranes (*top*) contain glycerol 3-phosphate, while archaeal membranes (*bottom*) contain the mirror image glycerol 1-phosphate. In bacteria and eukarya, fatty acids are linked to this phosphoglycerol molecule via ester linkages. In archaeons, 20-carbon phytanyl or 40-carbon biphytanyl molecules, built from isoprene subunits, are linked to the phosphoglycerol molecule via ether linkages. Many archaeons have a plasma membrane bilayer in which two phytanyl molecules are attached to the phosphoglycerol molecules. In some species, a biphytanyl molecule is attached at both ends to phosphoglycerol molecules, resulting in a monolayer. Such a monolayer may provide a more stable membrane structure at the high temperatures where some archaeons are found.

between 20 and 400 nm **(Figure 4.8)**. In contrast, the periplasms of Gram-negative bacteria like *Escherichia coli* and *Pseudomonas aeruginosa* appear to be only 20–25 nm wide.

The functional role of this outer membrane may be more unusual than its mere existence in these organisms. It appears that *Ignicoccus* species make energy across this outer membrane. In Section 2.4, we briefly examined how bacteria create energy by transporting protons across the plasma membrane. As these protons then travel back into the cytoplasm through ATP synthase molecules embedded in the plasma membrane, ATP forms in the cytoplasm. This same process appears to be utilized by most archaeons. Among *Ignicoccus* species, however, evidence suggests that protons are transported across the outer membrane, from the periplasm to the extracellular environment. As these protons flow back into the cell, through ATP synthase molecules embedded in the outer membrane, ATP is generated within the periplasm. The evolutionary significance of this atypical arrangement remains a topic of active investigation.

While it is interesting to note these differences in membrane composition between bacteria and archaeons, let's ask a

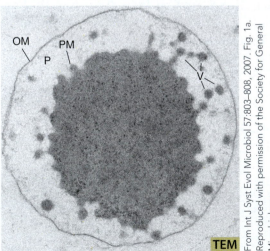

From Int J Syst Evol Microbiol 57:803–808, 2007. Fig. 1a. Reproduced with permission of the Society for General Microbiology

Figure 4.8. Outer membrane of *Ignicoccus* Like Gram-negative bacteria, archaeons in the genus *Ignicoccus* possess a plasma membrane (PM) and an outer membrane (OM). A relatively large periplasm (P) exists between these two membranes and membrane-bound vesicles (V) appear to exist within the periplasm.

more fundamental question: What is the significance of these differences? As we just noted, a lipid monolayer, presumably, would be more stable at elevated temperatures than a lipid bilayer. Indeed, researchers have demonstrated that monolayer **liposomes**, tiny vesicles made from cell membrane material, are more stable at high temperatures than are bilayer liposomes. Additionally, ether linkages are more resistant than ester linkages to high temperatures, oxidation, and alkaline degradation. The unusual chemical nature of the archaeal plasma membrane, then, may be an adaptation to life in extreme environments. Additionally, researchers are exploring the potential medical uses of archaeal liposomes (**Toolbox 4.1**).

Toolbox 4.1
VACCINE DELIVERY STRATEGIES

In Perspective 4.1, we will examine the many industrial uses for enzymes isolated from archaeons. A less obvious component of these organisms has generated some medical interest—their lipids. As we noted in this section, the plasma membranes of archaeons differ fundamentally from the plasma membranes of bacteria and eukarya. These chemical differences, most notably the ether linkages present in archaeal membranes, confer the archaeal lipids with increased stability. Recently, researchers have explored the use of archaeosomes, liposomes composed of archaeal lipids (**Figure B4.1**), as vaccine additives. These archaeosomes are quite stable and serve as strong adjuvants, agents that augment the immune response to a vaccine (see Section 20.4).

To explore the efficacy of archaeosomes as vaccine adjuvants and delivery modules, scientists at the National Research Council of Canada created lipid vesicles from archaeal membrane lipids and lipid vesicles from bacterial membrane lipids. Specifically, they used lipids extracted from the archaeon *Methanobrevibacter smithii* (see Microbes in Focus 4.4). Additionally, these vesicles contained bovine serum albumin (BSA), a protein commonly available in laboratories.

Mice injected with the archaeosomes produced a stronger immune response to BSA than did mice injected with liposomes made from bacterial lipids. Thus, these studies showed that archaeosomes can serve as effective vaccine delivery modules. Because of the increased stability of archaeal lipids, archaeosomes may be particularly useful in resource-limited areas. Their increased stability may mean that vaccines can be stored longer or stored under less than ideal conditions.

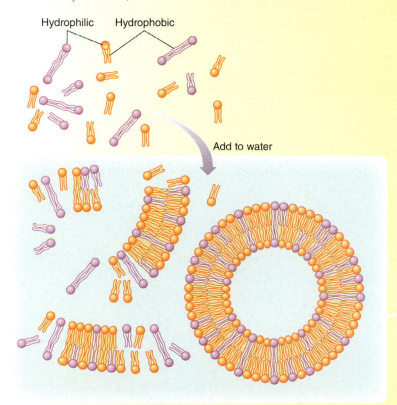

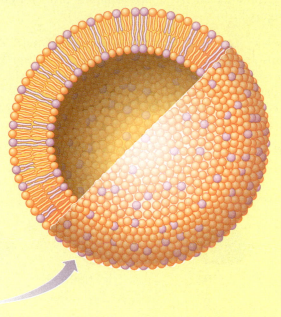

Figure B4.1. Archaeosomes: Liposomes made from archaeal lipids
Researchers are exploring the usefulness of archaeosomes—lipid vesicles constructed from archaeal lipids. When bacterial and archaeal phospholipids are combined in water, vesicles form. These lipid vesicles can be used for many purposes. They have proven to be very useful for studying basic properties of cell membranes. Additionally, they are being explored as drug delivery vesicles. Therapeutic agents can be enclosed within the liposome or embedded within the liposome bilayer. Archaeosomes are more stable than liposomes constructed from bacterial lipids. Also, archaeosomes used as vaccine delivery vesicles function as effective adjuvants.

● Test Your Understanding   ....

You are asked to develop novel archaeosomes that could be used as vaccine delivery vesicles in tropical countries. From what species might you obtain archaeal lipids? Why?

The Cell Wall

Many archaeons also possess a cell wall. Like the cell walls of bacteria, plants, and fungi, the archaeal cell wall gives shape to the cells and provides them with protection from, among other things, osmotic pressure and mechanical stress. While all cell walls perform these similar functions, the chemical composition of cell walls may differ significantly. In bacteria, the cell wall consists largely of peptidoglycan (see Section 2.4), while plants and fungi have cell walls of cellulose and chitin, respectively (see Section 3.1).

Archaeons may contain one of several different types of cell walls. In some archaeons, most notably species in the *Methanobacterium* genus, the cell wall consists of **pseudomurein** (also referred to as pseudopeptidoglycan), which is chemically similar to the peptidoglycan seen in bacteria. Both peptidoglycan and pseudomurein consist of a polymer of alternating sugar derivatives to which are attached crosslinked short peptide chains. Despite these similarities, the cell walls of archaeons and bacteria differ in several ways **(Figure 4.9)**. The cell walls of archaeons have the following characteristics:

- *N*-acetyltalosaminuronic acid replaces *N*-acetylmuramic acid.

- β-1,3-glycosidic bonds, instead of β-1,4-glycosidic bonds, link the sugar derivatives in pseudomurein.

- All amino acids in pseudomurein are L-stereoisomers, while peptidoglycan frequently contains D-stereoisomers (see Figure 2.16).

It is unclear if these chemically distinct, yet closely related structures arose through divergent evolution or represent an example of convergent evolution. In other words, the cell walls in archaeons and bacteria may represent structures that have evolved, and diverged, from a structure that existed in a common ancestral organism. Alternatively, the cell walls in archaeons and bacteria may have evolved independently.

CONNECTIONS As we will see in **Section 24.2**, penicillin and related β-lactam antibiotics interfere with cell wall synthesis in bacteria. Because of the chemical differences between peptidoglycan and pseudomurein, these agents are not effective against archaeons.

Finally, a few archaeons, like the *Thermoplasma acidophilum* that we mentioned earlier, contain no cell wall. In fact, the absence of a cell wall in these archaeons, coupled with their non-spherical, irregular shape, led researchers to postulate that *T. acidophilum* might contain a cytoskeleton. In the absence of a cell wall, they reasoned, a cytoskeleton would be needed to help the cells maintain a non-spherical appearance.

CONNECTIONS As we discussed in **Section 2.4**, bacteria in the genus *Mycoplasma* also lack a cell wall. Again, this lack of a cell wall led researchers to hypothesize that *Mycoplasma* species must contain a cytoskeleton.

The cell surface

As we saw for bacteria and eukarya, archaeons interact with each other and with the outside environment in many ways. Archaeal cells must, among other tasks, move around, stick to surfaces, sense the environment, and acquire nutrients. To a large extent, these activities are mediated by structures that exist on the cell surface. Many of these structures, like receptors and channels, may be quite familiar to us. Some archaeons, like many bacteria, contain a layer of protein, glycoprotein, or, in some cases, polysaccharide, often referred to collectively as the S-layer. Generally, the S-layer is composed of multiple copies of a single molecule

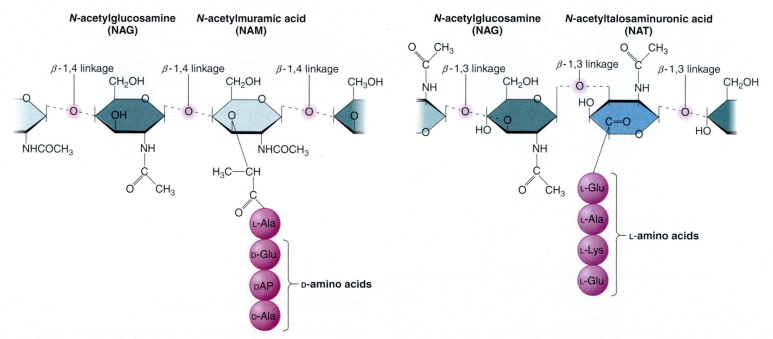

A. Peptidoglycan in bacterial cell walls

B. Pseudomurein in archaeal cell walls

Figure 4.9. Cell walls of bacteria and archaeons **A.** The peptidoglycan in bacterial cell walls consists of *N*-acetylglucosamine (NAG) and *N*-acetylmuramic (NAM) acid sugars, linked by β-1,4-glycosidic bonds. **B.** The pseudomurein in archaeal cell walls consists of *N*-acetylglucosamine (NAG) and *N*-acetyltalosaminuronic (NAT) acid sugars, linked by β-1,3-glycosidic bonds.

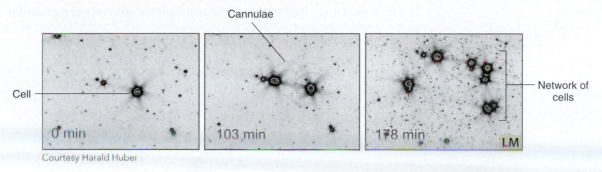

Courtesy Harald Huber

Figure 4.10. Hollow tubes connect _Pyrodictium abyssi._ As these cells grow, hollow, glycoprotein tubes, the cannulae, are produced that connect the individual cells.

arranged in a crystalline fashion. The exact role of this structure is unclear. This outer shield may protect the cells from viruses, provide environmental stability, or aid in adhesion.

Other archaeal surface structures may seem more exotic. Among the _Pyrodictium_, for example, **cannulae**, hollow glycoprotein tubes, connect individual cells together, forming a complex network **(Figure 4.10)**. The function of these tubes, and the resulting network of cells, has not been determined. The network may permit the cells to exchange nutrients. Some evidence indicates that the tubes may extend into the periplasm, but not the cytoplasm, of connected cells.

As we conclude our introduction to archaeal cell structure, we will focus on one particular cell surface structure—the flagella. Like some bacteria, some archaeons contain flagella. Like bacteria, some archaeons contain a single flagellum, some contain multiple flagella, and some lack flagella entirely **(Figure 4.11)**. Also, as we saw in bacteria, the flagella of archaeons rotate and are used primarily for movement. Despite these outward similarities in appearance and function, though, the bacterial and archaeal flagella display significant structural differences.

Archaeal flagella tend to be thinner than their bacterial counterparts (10 to 14 nm versus 20 to 24 nm). Additionally, archaeal flagella often are composed of multiple copies of two or more different versions of **flagellin**, the polypeptide that makes up the shaft of the flagellum. In bacteria, conversely, multiple copies of a single flagellin species typically comprise the flagella. Some archaeal flagellins exhibit _N_-linked glycosylation; only _O_-linked glycosylation has been observed among bacterial flagellins. Finally, the assembly of archaeal and bacterial flagella appears to differ. As we noted in Chapter 2, bacterial flagella are assembled via the Type III secretory pathway. Flagellin monomers travel through the developing flagellar filament and assembly occurs at the tip of the growing filament. Current evidence suggests that in archaeons, conversely, flagellin monomers are added to the base of the growing structure **(Figure 4.12)**.

Interestingly, this mode of archaeal flagellar construction mirrors the assembly process exhibited by some types of bacterial pili. Certain bacterial pili also grow from the base rather than the tip. Indeed, genetic and biochemical evidence shows a link between archaeal flagella and bacterial pili. First, the archaeal flagellins and bacterial pili exhibit some amino acid sequence similarity. Second, homologous assembly proteins have been identified in both systems. Conversely, a number of bacterial genes necessary for the assembly of flagella and pili appear to have no homologues in archaeons, and several genes

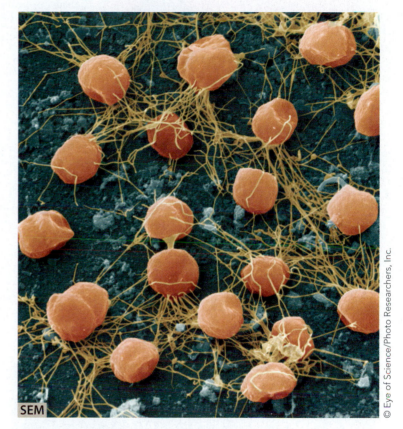

© Eye of Science/Photo Researchers, Inc.

Figure 4.11. Flagella of _Pyrococcus furiosus_ Cells of this species may have as many as 70 flagella per cell. These structures may be used for attachment, in addition to motility.

essential for flagellar assembly in archaeons have been identified that have no homologues in bacteria. The exact evolutionary relationship among bacterial flagella, bacterial pili, and archaeal flagella, therefore, remains unclear.

While the flagella of bacteria and archaeons seem to be evolutionarily distinct, it is interesting to note that they may utilize related chemotaxis pathways. Bacterial chemotaxis, the process of directed motility in response to a chemical signal, depends ultimately on switching the rotational direction of the flagella (see Figure 2.28). The cell-signaling pathway in bacteria involves several proteins, most notably CheA, CheB, and CheY. The binding of an extracellular signal to a receptor protein on the cell surface can initiate a cell-signaling cascade involving these proteins. Changes in these proteins, in turn, affect the rotational direction of the flagella (see Figure 11.29). Studies

Figure 4.12. Bacterial and archaeal flagellar structure and assembly Functionally, archaeal flagella closely resemble their bacterial counterparts. Structurally, though, these appendages differ. **A.** Bacterial flagella are composed of multiple copies of a single flagellin subunit. During assembly, flagellin monomers travel through the flagella and attach to the tip of the growing flagellum. **B.** Archaeal flagella, in contrast, consist of several different subunits. During assembly, new subunits are attached to the base of the growing flagellum. As shown in this schematic, the yellow and blue subunits first form the flagellum. The red subunits then form the hook.

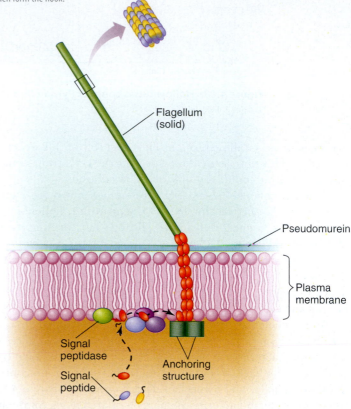

Flagellum (hollow)

Outer membrane

Peptidoglycan

Periplasm

Plasma membrane

Flagellin subunits

A. Bacterial flagellum

Flagellum (solid)

Pseudomurein

Plasma membrane

Signal peptidase

Signal peptide

Anchoring structure

B. Archaeal flagellum

with *Halobacterium salinarum* (**Microbes in Focus 4.3**), an archaeon that exhibits chemotaxis, have shown that this species contains similar genes. Furthermore, when researchers mutated these genes in *H. salinarum*, chemotaxis was disrupted.

CONNECTIONS This cell-signaling cascade involves the phosphorylation or methylation of a series of proteins. In **Section 11.7**, we will see how the flagellar rotation is affected by these molecular changes.

4.2 Fact Check

1. In some archaeons, DNA is complexed with histones. What may the presence of histones indicate about the evolutionary history of archaeons?
2. How does the archaeal plasma membrane differ from the plasma membrane in bacteria and eukarya?
3. Would the antibiotic penicillin be effective against archaeons? Defend your answer.
4. Discuss the similarities between archaeal and bacterial flagella.

4.3 Diversity of Archaea

What are the major groups of archaeons?

As we have seen throughout this chapter, we still have a lot to learn about archaeons. Our understanding of the phylogeny of these microbes is no exception. Based on 16S rRNA sequence analyses, researchers have grouped most well-characterized archaeons into two phyla: Euryarchaeota and Crenarchaeota. Analyses of 16S rRNA sequences from non-cultured organisms have led a growing number of microbiologists to propose another archaeal phylum: Korarchaeota. A fourth phylum—Nanoarchaeota—also has been proposed. This proposed phylum has only a single potential member—*Nanoarchaeum equitans*, the organism we mentioned in the opening story of this chapter. As we learn more about these microbes, and develop better methods of culturing more of them, the current classification scheme of archaeons almost certainly will change. For now, though, we will briefly examine these four phyla, emphasizing some unique features of each.

Crenarchaeota

Members of the Crenarchaeota are phylogenetically related, but researchers have identified few, if any, structural or biochemical features unique to all members of this phylum. For some time, it was assumed that all species in this phylum were **thermophiles** or **hyperthermophiles**, organisms whose optimal growth temperatures are greater than 55°C or 80°C, respectively **(Table 4.2)**. Indeed, most of the cultured crenarchaeotes fall into one of these two categories. Since the early 1990s, though, researchers have reported the isolation of crenarchaeotal DNA from a variety of marine and terrestrial environments, indicating that members of this phylum may be widespread in nature. In 2005, researchers first reported the cultivation of a mesophilic crenarchaeote—*Nitrosopumilus maritimus*. The initial assumption by researchers that all crenarchaeotes thrive at elevated temperatures appears not to be correct. In this section, we will briefly examine the major groups of crenarchaeotes.

The Thermophiles and Hyperthermophiles

Most thermophilic and hyperthermophilic crenarchaeotes have been isolated from thermal springs and geysers, such as those in Yellowstone National Park, and volcanic regions **(Figure 4.13)**. Often, these areas contain large amounts of sulfur, and are quite acidic. Many of these hyperthermophiles also are **acidophiles**, growing in low pH environments. The crenarchaeote *Sulfolobus solfataricus* represents a prime example of a hyperthermophilic acidophile. Isolated from a volcanic crater near Naples, Italy, this microbe has an optimal growth temperature of approximately 80°C. Additionally, it grows optimally at pH 3.0. Other members of this phylum have been isolated from deep-sea hydrothermal vents. These species grow not only at elevated temperatures, but also at high pressure and are referred to as **barophiles**.

Life at high temperatures requires certain adaptations. For most organisms, elevated temperatures lead to the disruption of the plasma membrane. Plasma membranes containing tetraether lipids or lipid monolayers, as we discussed earlier, may provide extra stability at high temperatures and many crenarchaeotes possess this feature. Additionally, elevated temperatures generally cause the denaturation of proteins. Hyperthermophiles, as a result, must possess proteins that remain functional in extreme environments. It appears that hyperthermophiles may employ two distinct mechanisms to maintain protein integrity.

The proteins in thermophiles and hyperthermophiles differ slightly, but significantly, from their mesophilic counterparts. Researchers at the National Cancer Institute in the United States and Tel Aviv University in Israel recently compared the biochemical properties of proteins isolated from thermophilic organisms with the homologous proteins from mesophilic organisms. They observed a few interesting differences. First, the proteins from thermophiles have a larger percentage of α-helical regions. Second, these proteins contain more salt bridges and side chain interactions. Third, the amino acids arginine and tyrosine are more common in thermophilic proteins than in their mesophilic counterparts, while the amino acids

Figure 4.13. Some crenarchaeotes thrive in hot springs above 55°C. Located over a large volcanic region spread over three states, Yellowstone National Park is home to a great variety of geysers, hot springs, and mudpots. Numerous thermophilic archaeons have been isolated from these locations.

© Interfoto/Alamy

TABLE 4.2 Representative thermophilic and hyperthermophilic crenarchaeotes

Species	Temperature range (°C)	Temperature optimum (°C)
Ignicoccus hospitalis	75–98	90
Nanoarchaeum equitans	75–98	85–90
Sulfolobus solfataricus	50–87	80
Pyrolobus fumarii	90–113	106
Archaeoglobus fulgidus	60–95	83

TABLE 4.3 Usage (%) of selected amino acids in mesophiles and thermophiles

Amino acids	Mesophiles (%)	Thermophiles (%)
Alanine	9.2	8.9
Proline	4.2	4.2
Valine	8.2	8.2
Cysteine[a]	1.0	0.6
Serine[a]	5.5	4.0
Arginine[a]	3.6	4.6
Tyrosine[a]	3.7	4.5

[a]Statistically significant difference

Source: S. Kumar, C.-J. Tsai, and R. Nussinov. 2000. Factors enhancing protein thermostabililty. Prot. Engin. 13:179–191.

cysteine and serine are less common (Table 4.3). The researchers postulate that these differences contribute to the increased thermostability of the proteins isolated from thermophiles. The increased α-helical content and increased proportion of arginine and tyrosine, they note, may lead to strengthened interactions between amino acids. These strengthened interactions allow the protein to maintain its shape at elevated temperatures. Cysteine, on the other hand, is a thermolabile amino acid. Its decreased usage may diminish the potentially damaging effects of high temperatures.

The hyperthermophiles also rely on a series of **molecular chaperones**, proteins that help fold proteins or refold denatured proteins, to maintain the functionality of their proteins. While molecular chaperones have been identified in species from all three domains, the archaeal chaperonins more closely resemble eukaryal chaperonins than their bacterial counterparts. Often referred to as the **thermosome** in hyperthermophilic archaeons, this protein complex appears to be quite abundant and instrumental in these organisms. In *Ignicoccus hospitalis*, the host of *Nanoarchaeum equitans*, thermosome proteins are some of the most abundant proteins in the cytoplasm. The study of these thermostable proteins, we should note, could have profound effects on our everyday life; enzymes that function under extreme conditions could have many industrial uses (Perspective 4.1).

CONNECTIONS Humans have utilized microbes, intentionally or unintentionally, for thousands of years in the production of various foods and drinks. We will learn about current industrial uses of microbes in **Section 12.4**. Specifically, we will investigate how microbes could be utilized to produce renewable biofuels.

In addition to maintaining the integrity of their proteins, hyperthermophiles also must maintain the integrity of their DNA. Some species possess thermostable DNA-binding proteins that increase the melting temperature of double-stranded DNA. More importantly, all known hyperthermophiles possess **reverse DNA gyrase**, an enzyme that increases the supercoiling of the DNA. By increasing the degree of supercoiling, these enzymes greatly increase the temperature at which the DNA unwinds and denatures.

The Mesophiles and Psychrophiles

While the crenarchaeotes traditionally have been considered to be thermophiles or hyperthermophiles, evidence of crenarchaeotes that are **mesophiles** and **psychrophiles** abounds. These organisms, literally "intermediate loving" and "cold loving," grow optimally between 15°C and 40°C or at temperatures less than 15°C, respectively. Since the early 1990s, researchers have detected crenarchaeote rRNA sequences in various temperate and cold marine environments.

Given the apparent ubiquitous nature of the organisms, they may be major contributors to the biogeochemical cycles in the oceans. Several pieces of evidence suggest that these organisms are major contributors to carbon cycling in the oceans. Analysis of *Nitrosopumilus maritimus*, a marine crenarchaeote first isolated in 2005, indicates that this organism also may contribute to nitrogen cycling. *N. maritimus* is an ammonia oxidizer, obtaining energy by oxidizing ammonia to nitrite. Genome sequencing studies have shown that it contains an ammonia monooxygenase gene (*amoA*), the product of which converts ammonia to hydroxylamine, the first step in the nitrification process.

By examining SSU rRNA gene sequences, most researchers have concluded that the mesophilic and psychrophilic archaeons belong in the Crenarchaeota phylum. Recently, though, some researchers have proposed that the mesophilic and psychrophilic crenarchaeotes should be placed within a new archaeal phylum, Thaumarchaeota. In 2006, the first complete genome sequence of a psychrophilic crenarchaeote, *Cenarchaeum symbiosum*, was reported. This species, which still has not been grown in pure culture, was identified in 1996 and resides within a marine sponge. By analyzing this genome, researchers observed that *C. symbiosum* contains some genes previously identified only in crenarchaeotes and some genes previously identified only in euryarchaeotes. Based on these findings, the researchers proposed that *C. symbiosum* represents a member of a new phylum, distinct from either of the two recognized phyla. Clearly, the data supporting the creation of this new phylum are limited. As we learn more about these organisms, our view of archaeal taxonomy almost certainly will change.

Euryarchaeota

Like members of the phylum Crenarchaeota, members of the phylum Euryarchaeota share significant 16S rRNA gene sequence homology, but also exhibit quite diverse morphologies and biochemical properties. As with the Crenarchaeota, one would struggle to list even a single defining characteristic for this phylum, aside from their phylogenetic relatedness. In this section, we will investigate two major groups of euryarchaeotes, the methanogens and the halophiles.

Perspective 4.1
EXTREMOPHILES AND BIOTECHNOLOGY

Since the discovery of the first thermophilic microbes, researchers have been very interested in determining how organisms thrive in extreme environments. These environments, whether they are characterized by extreme temperatures, pH, or salinity, present certain challenges to living organisms. The proteins of thermophiles and hyperthermophiles, for example, must remain active at elevated temperatures. Studies of these organisms may provide insight into many aspects of biology, including protein folding and enzymatic activity. Additionally, enzymes isolated from these organisms may have significant industrial applications.

You should be familiar with one very important enzyme isolated from a thermophilic organism: Taq polymerase. As we learned in Section 1.1, this enzyme, isolated from the thermophilic bacterium *Thermus aquaticus*, often is used by researchers for the polymerase chain reaction (PCR). While Taq was the first thermostable DNA polymerase to be isolated and used for PCR, other choices now are available. Pfu and Vent, DNA polymerases isolated from the hyperthermophilic archaeons *Pyrococcus furiosus* and *Thermococcus litoralis*, respectively, are widely used **(Table B4.1)**.

Other enzymes isolated from hyperthermophiles also may be commercially useful. Thermostable proteases, enzymes that degrade proteins, and lipases, enzymes that degrade lipids, are being investigated as detergent additives. The stability of these enzymes at elevated temperatures would allow them to remain active during washing in hot water. Proteases and lipases from alkaliphiles, organisms that grow optimally in high pH environments, also may prove to be useful detergent additives. Because laundry detergent is basic, the activity of these enzymes in high pH environments would be ideal.

TABLE B4.1 Thermostable enzymes

Enzyme	Source (species/domain)	Commercial name	95°C half-life (min)
Taq	*Thermus aquaticus* Bacteria	Taq AmpliTaq	40
Pfu	*Pyrococcus furiosus* Archaea	Pfu Pfu Turbo	120
Tli	*Thermococcus littoralis* Archaea	Vent	400

The Methanogens

The **methanogens**, organisms that produce methane (CH_4), represent one of the more interesting groups of organisms in the entire living world. Despite the abundant metabolic diversity that exists within biology, very few organisms and only a few members of the phylum Euryarchaeota can produce methane. These organisms produce methane from a variety of substrates, such as carbon dioxide (CO_2) or various simple organic molecules like methanol (CH_3OH) and formate ($CHOO^-$). Most commonly, the production of methane involves the reduction of these substrates with H_2 serving as the electron donor.

$$CO_2 + 4H_2 \rightarrow CH_4 + 2H_2O$$

During the reduction of CO_2 in this reaction, enough free energy is released to drive the fixation of carbon. These organisms, then, are autotrophs, incorporating inorganic carbon into organic molecules.

CONNECTIONS All living organisms require organic molecules. Heterotrophs ingest organic molecules, while autotrophs manufacture their own organic molecules. A more detailed description of photosynthesis, the process by which the energy present in sunlight is captured by organisms and converted into chemical energy, can be found in **Section 13.6**.

Where can we find methanogens? Given the metabolic properties of these organisms, we can assume that methanogens can be found where their necessary substrates (carbon dioxide and hydrogen, for instance) are found. Moreover, all of the identified methanogens are strict anaerobes, organisms that can survive only in **anoxic**, or oxygen-free, environments. While we tend to think of the world as replete with oxygen, anoxic environments abound. From lake and ocean sediments, to hot springs, to the GI tracts of animals, environments lacking oxygen are quite common **(Figure 4.14)**. Methanogens probably can be found in all of these locations.

From Nature Reviews Microbiology 5(8), 572–573, 2007. Photo courtesy Jeffery Gordon, Center for Genome Sciences and Systems Biology, St. Louis, MO. Reprinted by permission from Macmillan Publishers, Ltd.

TEM

© Aimin Tang/iStockphoto

TEM

1 μm

Courtesy Bonnie Chaban, University of Saskatchewan

A. *Methanobrevibacter smithii* lives in the human gut.

B. *Methanococcus maripaludis* lives in salt marshes.

Figure 4.14. Anoxic environments and methanogens Because we are aerobic organisms, and spend virtually all of our time in an aerobic environment, we may think that anoxic environments are quite rare. This thought couldn't be further from the truth. In fact, anoxic environments are quite common. **A.** The human gut is largely anoxic and houses many anaerobic microbes, like the methanogen *Methanobrevibacter smithii* **B.** The sediment in ponds, swamps, and salt marshes also may be anoxic and house methanogens like *Methanococcus maripaludis*.

One of the first well-documented discoveries of methanogenesis occurred over 200 years ago, well before Archaea was recognized as a domain of life. The Italian physicist Alessandro Volta observed that gases stirred up from the bottom of a marsh were capable of being burned **(Figure 4.15)**. The gases, he proposed, were "combustible air." That gas was methane. The source of that methane was methanogens living in the anoxic sediment of the marsh.

The human gut represents another environment where methanogens can be found. The large intestine contains a great diversity of microbes, including the methanogen *Methanobrevibacter smithii* **(Microbes in Focus 4.4)**. In this location, *M. smithii* utilizes formate and H_2 produced as fermentation by-products by bacteria in our intestines. *M. smithii*, in turn, produces its own noticeable by-product: gas. Our gastrointestinal tract is a complex, fascinating ecosystem. As we ingest and begin digesting food, certain bacteria further process the food, releasing the nutrients needed by this archaeon. The roles of these various bacteria and this particular archaeon on our health are just now being investigated **(Mini-Paper)**.

Courtesy John Leigh, University of Washington, Seattle

Figure 4.15. Methanogenesis In the eighteenth century, Alessandro Volta observed that gas collected from a swamp burned. We now know that this "combustible air" is methane, produced in the anoxic environment of the swamp by archaeal methanogens. It is possible to repeat Volta's classic observation today. By placing a large, plugged funnel in a wetland and stirring up the underlying decaying material, methane can be trapped in the funnel. When the plug is removed, the methane can be ignited.

Microbes in Focus 4.4
METHANOBREVIBACTER SMITHII

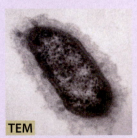

Habitat: Anoxic environments, including the GI tract of humans and other animals

Description: Rod-shaped archaeon within the Euryarchaeota phylum; mesophilic anaerobe that grows in pairs or short chains

Key Features: *Methanobrevibacter smithii*, a methane producer, is the most abundant archaeon in the human digestive tract. It utilizes formate and H_2 produced as by-products of fermentation by bacteria. In a 2009 study, researchers observed higher levels of *M. smithii* in the feces of people with anorexia than in the feces of obese or lean individuals. The authors speculated that the increased levels of this methanogenic microbe might be a response to the extremely low calorie diet.

TEM

From Nature Reviews Microbiology 5(8), 572–573, 2007. Photo courtesy Jeffery Gordon, Center for Genome Sciences and Systems Biology, St. Louis, MO. Reprinted by permission from Macmillan Publishers, Ltd.

The anoxic habitats in which methanogens exist can be fairly diverse. Perhaps it should not be surprising, then, that the shapes of various methanogens and their optimal growth temperatures also differ. Some methanogen species, like *Methanothermus fervidus*, are rods, while other species, like *Methanculleus olentangii*, are cocci. Some, like *Methanopyrus kandleri*, are hyperthermophilic, while other species are mesophilic or psychrophilic. While all of the methanogens share a unique metabolic property, they also exhibit remarkable diversity.

The Halophiles

While the methanogens share a common metabolic property, the salt-loving **halophiles** share a common environmental requirement. These organisms only grow in environments containing NaCl concentrations of at least 1.5 M. Many of the identified halophilic archaeons require even higher salt concentrations. As a brief introduction to these organisms, we will investigate where we can find these organisms and what adaptations they may have that allow them to survive in these unusual environments.

While environments favored by methanogens may be relatively common, high salt environments are more limited. Probably the two most well known high salt environments are the Great Salt Lake in Utah and the Dead Sea, located between Israel and Jordan **(Figure 4.16)**. Because these lakes have no outlets, water only leaves through evaporation. As a result, minerals brought in by the tributaries accumulate to high levels. The specific salinity in both lakes depends on the location within the lake, the season, and the recent weather. On average, the salinity in the Great Salt Lake varies from 5 to 25 percent, while the salinity of the Dead Sea may be as high as 34 percent. The salinity of the ocean, in contrast, is around 3.5 percent, or approximately 0.6 M. Because of these high salt concentrations, both the Great Salt Lake and the Dead Sea are largely devoid of macroscopic life, sustaining brine shrimp and some plankton, but no fish.

Other high salt environments do exist. A number of hypersaline lakes, like Deep Lake, have been characterized in Antarctica. As with the Great Salt Lake and the Dead Sea, the actual salt

Figure 4.16. The Dead Sea Halophilic archaeons are nearly the only life-forms in hypersalinic bodies of water. Located between Israel and Jordan, the Dead Sea is over 1,300 feet below sea level and is one of the world's saltiest environments. It is largely devoid of macroscopic life, but harbors various archaeal species.

concentration varies spatially and temporally within these lakes. Additionally, the temperatures are very low. Halophiles isolated from these locations, then, must be psychrophiles, or psychrotolerant. Other locations, like Lake Magadi in Kenya, contain high concentrations of sodium carbonate (Na_2CO_3). The carbonate ions result in an alkaline environment. The halophiles isolated from these locations, then, grow optimally under high salt and high pH conditions.

Life in a high salt environment presents some interesting challenges. Most notably, these cells must avoid osmotic shock. In a hypertonic environment, such as the Dead Sea, ion concentrations outside of the cell are much higher than ion concentrations inside a typical cell. As a result, water, will flow out of the cell, causing it to shrink. In hypotonic environments, conversely, ion concentrations outside the cell are lower than inside the cell, causing water to flow into the cell, making it swell **(Figure 4.17)**. To combat this problem,

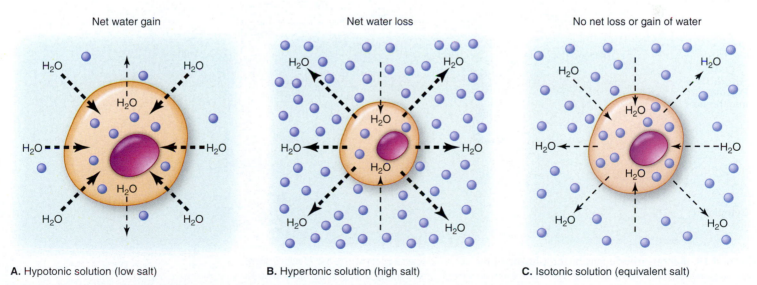

Figure 4.17. Osmotic balance in cells **A.** When cells are placed in hypotonic, or lower salt, solutions, water molecules will move into the cell, causing the cell to swell. In extreme cases, cells lacking a cell wall will lyse. **B.** When cells are placed in a hypertonic, or higher salt, solution, water molecules will move out of the cell, causing the cell to shrink. **C.** When the extracellular and intracellular salinities (indicated by blue spheres) are similar, there is no net movement of water across the membrane.

Ions balanced; no net gain or loss of water

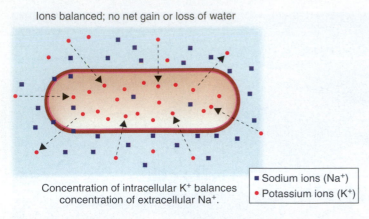

Concentration of intracellular K⁺ balances concentration of extracellular Na⁺.

■ Sodium ions (Na⁺)
● Potassium ions (K⁺)

Figure 4.18. Maintaining osmotic balance in halophiles To combat the movement of water out of their cells, some halophilic archaea, like *Halobacterium salinarum*, accumulate high intracellular potassium (K⁺) concentrations. The high levels of intracellular K⁺ balance the high extracellular Na⁺ concentrations, thereby preventing the efflux of water from the cell. Such maintenance of water balance and cell shape permits normal functioning.

Halobacterium salinarum, a well-studied halophile, maintains a high intracellular concentration of potassium ions (K⁺). The high intracellular potassium levels provide the cell with a type of osmotic balance, thereby preventing the efflux of water **(Figure 4.18)**.

High intracellular potassium concentrations, though, present other problems. First, high salt concentrations denature DNA. The increased potassium levels would disrupt the hydrogen bonds that hold together the two strands of a double-stranded DNA molecule. Second, the increased intracellular potassium levels also would adversely affect cytoplasmic proteins. Again, proteins tend to denature when placed in high salt environments. So how do halophiles avoid these problems?

Studies of *Halobacterium* sp. NRC-1 have provided some answers to this question. This archaeal species grows optimally in medium containing 4.5 M NaCl—slightly less salty than the Dead Sea but over seven times more salty than typical ocean water. In 2000, its complete genome was sequenced, making it the first halophile to be sequenced. Computational analyses of its genome have provided interesting insight into life in high salt environments. First, the genome has a high GC content. Within the chromosome, approximately 68 percent of the 2,000,000 base pairs (2 megabase pairs) are GC pairs. Because GC pairs are held together by three hydrogen bonds, as opposed to the two hydrogen bonds that connect AT pairs, a chromosome with a high GC content will be less likely to be denatured by the high intracellular potassium levels. Second, the proteins are highly acidic, containing a large number of aspartic acid and glutamic acid residues. Acidic proteins, biochemists have demonstrated, tend to remain more stable in high salt environments than do neutral or basic proteins. From these studies of *Halobacterium* sp. NRC-1, then, it appears that this species possesses several unique properties allowing it to flourish in a very atypical environment.

Perhaps the most interesting feature of *Halobacterium* species is not directly related to their adaptations to high salt environments. Members of this genus possess a novel means of obtaining energy. Species such as *H. salinarum* obtain energy via **phototrophy**, the acquisition of energy from sunlight. These organisms, however, do not use a chlorophyll-based form of phototrophy. Rather, the energy present in sunlight is harvested by another molecule—**bacteriorhodopsin**. This integral membrane protein, like the rhodopsin protein found in the human eye, surrounds a molecule of retinal. Absorption of light by retinal results in the transfer of an electron by bacteriorhodopsin directly into the periplasm. A proton motive force is generated without an electron transport system. These protons then flow back into the cell through a typical ATP synthase H⁺ channel, generating ATP **(Figure 4.19)**. The bacteriorhodopsin molecules also cause a characteristic

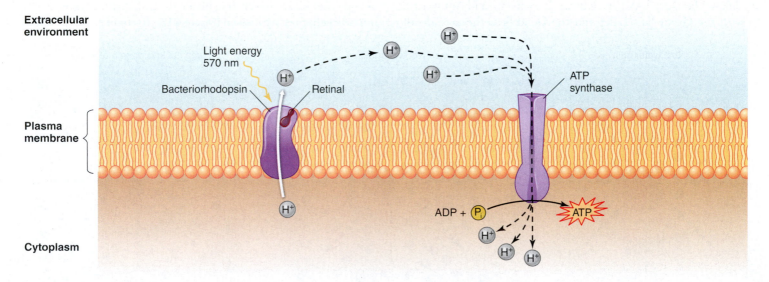

Figure 4.19. Bacteriorhodopsin-based phototrophy Many halophilic archaeons can utilize sunlight as an energy source. Unlike plants and most photosynthetic bacteria, these organisms do not contain chlorophyll. Rather, a molecule of retinal, located within the integral membrane protein bacteriorhodopsin, absorbs the energy present in sunlight. Bacteriorhodopsin functions directly as a proton pump, allowing the cell to create a proton motive force (PMF). The protons then pass through ATP synthase, resulting in the production of ATP. This process was demonstrated clearly when researchers inserted bacteriorhodopsin and ATP synthase into an artificial lipid vesicle. When this structure was exposed to light, protons accumulated in the vesicle and ATP was produced.

Figure 4.20. Red coloration by *Halobacterium* Because of water diversion for the city of Los Angeles, Owens Lake in eastern California is nearly dry and exhibits a high salt concentration. The large numbers of halophilic archaeons that thrive in this hypersaline environment give the lakebed a dramatic red color.

reddish hue in locations containing significant concentrations of *Halobacterium* species **(Figure 4.20)**.

CONNECTIONS In most living organisms, a proton motive force provides the potential energy needed for the generation of ATP. As the protons flow down their concentration gradient, through ATP synthase, ADP and inorganic phosphate are combined to form ATP. This process is described in more detail in **Section 13.4**.

The Other Euryarchaeota

While the methanogens and halophiles represent the most closely studied groups of Euryarchaeota, other members of this phylum do exist. Most of these species are thermophiles or hyperthermophiles, and many of them also are acidophiles, growing optimally in very acidic environments. *Pycrophilus* species, for example, have been isolated from volcanic areas replete with sulfuric acid in northern Japan and grow optimally at 60°C and a pH of 0.7. Certainly, as we have seen for the halophiles, these species must be uniquely adapted to their surroundings. Preliminary studies, for instance, have shown that the cell membranes of these species actually become

destabilized when the pH increases. More detailed biochemical studies of these species will provide us with interesting information about life under extreme conditions.

Emerging phyla

As we have noted throughout this chapter, our understanding of the Archaea is rapidly changing. The taxonomy of this domain is no exception. As more species are identified and cultivated, our classification of these organisms almost certainly will change. We noted earlier that some researchers have postulated that species currently classified as mesophilic crenarchaeotes represent members of a separate phylum: Thaumarchaeota. Other researchers have suggested that two other phyla, Korarchaeota and Nanoarchaeota, should be recognized **(Figure 4.21)**.

Korarchaeota

No species from this proposed phylum have been cultivated. Based on a series of 16S rRNA gene sequences obtained from hydrothermal environments, though, researchers have proposed this new archaeal phylum. Analysis of these sequences indicates that the organisms may be evolutionarily distinct from the euryarchaeotes and the crenarchaeotes.

Nanoarchaeota

Nanoarchaeum equitans, the organism we mentioned in the opening story, currently is the only member of this proposed phylum. As we saw with the proposed korarchaeotes, the 16S rRNA gene sequence of *N. equitans* indicates that this organism is distinct from the crenarchaeotes and the euryarchaeotes. This species also appears to be among the smallest living organisms and exhibits a potentially interesting symbiosis with a host organism. Future studies of this species may tell us more about the taxonomy of the archaeons and provide insight into the minimal requirements of life and how different species interact.

CONNECTIONS Some researchers have been studying the bacterium *Mycoplasma genitalium* (see **Microbes in Focus 10.1**) in attempts to define the minimal requirements of life. Their experimental approaches are described in **Section 10.3**.

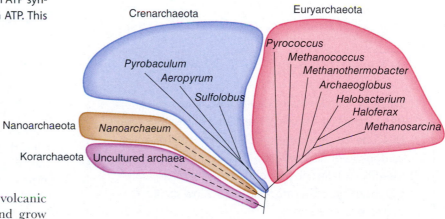

Figure 4.21. Phyla of Archaea Some researchers now postulate that Archaea contains four phyla. In this schematic, two proposed phyla are indicated by dashed lines. As more archaeal species are identified and their genomes sequenced, the evolutionary relatedness of members of this domain will become clearer and this phylogenetic tree undoubtedly will change.

B. S. Samuel and J. I. Gordon. 2006. A humanized gnotobiotic mouse model of host-archaeal-bacterial mutualism. Proc Natl Acad Sci USA 103:10011–10016.

Context

Our intestines contain trillions of microbes. Many of the bacteria residing in our large intestine feed off of indigestible material in our diet, often fermenting fiber, releasing various small organic molecules, H_2, and CO_2. The by-products of these fermentation reactions, in turn, can be absorbed by our cells, contributing to the amount of calories we ultimately derive from the food we eat. Studies with gnotobiotic or germfree mice, mice raised in sterile environments and lacking microbes within their GI tract, have demonstrated the role microbes play in this process. When the GI tracts of gnotobiotic mice became colonized with microbes from another animal, the previously germfree animals exhibited increased fat deposits, even though their food intake did not change. The microbes, then, allowed the host animals to gain more from the food they ate (see Toolbox 17.1).

The archaeon *Methanobrevibacter smithii* may constitute 10 percent of all anaerobic microorganisms present in the GI tract. Presumably, this microbe survives on the H_2 and formate produced by other fermenting organisms. Given this information, the researchers posed an interesting question. Does this archaeon affect how much energy we get from the food we eat? This question will be addressed again in the Chapter 17 Mini-Paper.

Experiments

To address the impact of *M. smithii* on digestion, Samuel and Gordon examined gnotobiotic mice. These mice were colonized with either *Bacteroides thetaiotaomicron*, a common bacterial inhabitant of the human GI tract known to scavenge various polysaccharides that we consume, or *M. smithii*, or both organisms together. At de-

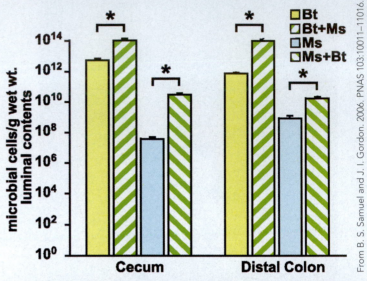

Figure B4.2. Mutually beneficial relationship between *Methanobrevibacter smithii* and *Bacteroides thetaiotaomicron* in the GI tract Following colonization or co-colonization of germfree mice, researchers examined the amount of *B. thetaiotaomicron* (Bt and Bt+Ms bars) or *M. smithii* (Ms and Ms+Bt bars) in the organisms. The colonization of each microbe was enhanced by the presence of the other microbe.

fined times after colonization, the researchers measured the amounts of these microbes present in the cecum and colon, two portions of the large intestine, of the previously germfree mice **(Figure B4.2)**. In mice that were co-colonized (Bt+Ms or Ms+Bt), the number of microbes present was significantly greater than in mice that were colonized with only one species (Bt or Ms), leading the researchers to conclude that the two species forged a "mutually beneficial relationship." Presumably, *M. smithii* allows *B. thetaiotaomicron* to ferment polysaccharides more

---

4.3 Fact Check

1. What are the two major phyla of Archaea?
2. What is the key feature of most known crenarchaeotes?
3. Describe the terms thermophile, hyperthermophile, acidophile, and basophile.
4. Identify and describe several of the chemical adaptations of hyperthermophiles that allow for life at extreme temperatures.
5. What are methanogens, and where are they found?
6. What is bacteriorhodopsin? Describe its role in the generation of energy.

Image in Action

WileyPLUS In this illustration, we see the diversity in morphologies present within the domain Archaea. The tiny *Nanoarchaeum equitans* lives with its host, *Ignicoccus hospitalis*. *Pyrodictium abyssi* produces a network of long hollow tubules called cannulae, and *Haloquadratum walsbyi*

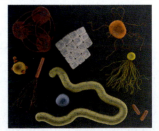

forms sheets of flat squares. Some archaeons are rod-shaped, while others are spherical. Some contain many flagella, while others contain few or no flagella.

1. If you were able to examine the archaeons in this image more closely, what key features might you expect to find?
2. Several of the archaeons in this image may be thermophilic. What adaptations would you expect to find in these organisms?

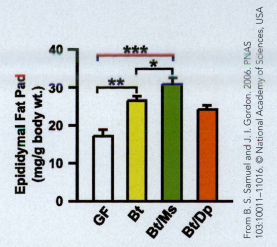

From B. S. Samuel and J. I. Gordon. 2006. PNAS 103:10011–11016. © National Academy of Sciences, USA

Figure B4.3. Presence of *M. smithii* and increased fat deposition in co-colonized mice Energy utilization, as measured by fat deposits, was assayed in germfree (GF) mice and mice colonized with *B. thetaiotaomicron* alone (Bt) or co-colonized with *B. thetaiotaomicron* and *M. smithii* (Bt/Ms) or *B. thetaiotaomicron* and *Desulfovibrio piger* (Bt/Dp). Co-colonization with *M. smithii*, but not *D. piger*, significantly increased the energy utilization of the animals.

efficiently. *M. smithii*, in turn, utilizes the formate produced by its partner for methanogenesis.

To explore the effects of this microbial relationship on the host, the researchers then looked at the amount of fat stored in these animals (**Figure B4.3**). As shown previously, the amount of energy the host derived from its diet, as evidenced by its fat pad weight, increased significantly when germfree mice were colonized with *B. thetaiotaomicron*. An even greater increase in fat pad weight was observed when mice were co-colonized with *B. thetaiotaomicron* and *M. smithii*. This co-colonization-associated increase, however, was not observed when germfree mice were co-colonized with *B. thetaiotaomicron* and *Desulfovibrio*

piger, another common bacterial inhabitant of the human GI tract. The increased fat pad weight, in other words, was not simply a factor of having two microbial species present.

Impact

These results indicate that a complex relationship exists between the various inhabitants of the GI tract. Not only do the microbes interact with each other, but they also interact with the host. From this specific set of experiments, it is clear that *B. thetaiotaomicron* and *M. smithii* have a relationship with each other and with their host that is mutually beneficial. The authors note that these results could provide insight into preventing obesity or, conversely, allowing underfed people to obtain more calories from the food they ingest. Altering our gut microbiota could affect the nutritional value of the food we eat.

● Questions for Discussion ···············

1. How valid is the germfree mouse model used by these researchers?

2. If specific microbes affect our ability to extract calories from food, then how could this information be used therapeutically?

3. Our GI tract, as we noted, contains trillions of microbes, representing a large number of species. How accurately can a two-species model, as described in this paper, shed light on our digestive processes?

The Rest of the Story ●●●●●○

Since the discovery of *Nanoarchaeum equitans* in 2002, we have learned much more about this unusual microbe. As we noted in Section 4.3, some researchers have proposed that *N. equitans* represents a member of a new archaeal phylum, Nanoarchaeota. Indeed, the SSU rRNA genes of this organism could not be detected by PCR using universal primers, primers thought to recognize all archaeal species. The SSU rRNA of *N. equitans* differs significantly from the SSU rRNA of other known members of this domain.

The relationship between *N. equitans* and its host, *Ignicoccus hospitalis*, has generated a great deal of interest. All attempts to

grow this microbe in the absence of a host have failed. Indeed, the relationship between *N. equitans* and *I. hospitalis* appears to be quite specific; attempts to grow *N. equitans* in the presence of other *Ignicoccus* species also have failed. It also appears that *N. equitans* only will grow when physically attached to *I. hospitalis*. Initially, researchers concluded that the relationship between these organisms was not parasitic. They observed that the growth of *I. hospitalis* was not altered in the presence of *N. equitans*. More recent studies, however, suggest that the relationship may be mildly parasitic. In other words, it may harm *I. hospitalis* in some way.

The *N. equitans* genome is approximately 490,000 bp, making it among the smallest genomes of any living organism. It appears to lack necessary genes for amino acid, nucleotide, and lipid synthesis. Most likely, it acquires these materials from its host. Indeed, analyses of the lipid composition of membranes from both species indicate that the *N. equitans* plasma membrane may be obtained from its host. Other studies have demonstrated that amino acids are transported from *I. hospitalis* to *N. equitans*.

Recent findings indicate that *N. equitans* may have other close relatives. SSU rRNA sequences closely related to *N. equitans* have been detected in Yellowstone National Park, United States, and Kamchatka, Russia, indicating that related species may be widely distributed. Further studies, in all likelihood, will help us better understand the relationships between these microbes and their hosts and help us to further elucidate that evolutionary history.

Summary

Section 4.1: How do we know that Archaea is a distinct domain of life?

Historically, all living organisms were divided into two broad groups—the eukaryotes and the prokaryotes—based on the presence or absence of a membrane-bound nucleus. **Archaeons**, however, now are recognized as a distinct domain of life.

- To determine the phylogeny of all living organisms, researchers often compare **ribosomal RNA(rRNA)** sequences.

- Small subunit (SSU) rRNA sequences comparison studies demonstrated that living organisms form three phylogenetically distinct groups, or domains: Eukarya, Bacteria, and Archaea.

- Members of the archaeal domain resemble bacteria in terms of size and shape.

- Many archaeal DNA replication, transcription, and translation proteins resemble their eukaryal counterparts.

- Current studies indicate that archaeons and eukarya may have evolved from a bacteria precursor.

Section 4.2: What do archaeal cells look like?

Archaeons, like bacteria, exhibit a multitude of shapes, although most appear to be spherical or rod-shaped.

- Most archaeons, like most bacteria, are between 0.5 μm and 5 μm in diameter.

- The cytoplasm of archaeons contains a mixture of molecules, most notably the DNA, which usually exists as a single, circular chromosome.

- Unlike bacteria, archaeons possess **histones**, DNA-binding proteins that help to package the DNA and may make it more thermostable.

- Some archaeons contain cytoskeletons that may help them maintain their shape. The identified cytoskeleton proteins resemble identified eukarya cytoskeleton proteins.

- The archaeal plasma membrane consists of glycerol 1-phosphate, a stereoisomer of the molecule present in bacterial and eukaryal membranes. Attached to the glycerol molecule are **isoprenoids**, usually **phytanyl** or **biphytanyl**, via ether linkages. Some archaeons possess lipid monolayers, rather than bilayers, which can form stable **liposomes**.

- Most archaeons possess a cell wall composed of **pseudomurein**, which is similar to, but chemically distinct from, the peptidoglycan seen in bacterial cell walls.

- Some archaeons possess **flagellin**. Again, the flagellin proteins in archaeons are distinct from the flagellin proteins used by bacteria.

- Some archaeons form networks, in which cells are connected by **cannulae**.

Section 4.3: What are the major groups of archaeons?

The domain Archaea is divided into two major phyla, Crenarchaeota and Euryarchaeota. Some researchers have proposed up to three additional phyla, Thaumarchaeota, Korarchaeota, and Nanoarchaeota.

- Most cultivated crenarchaeotes are **thermophiles** or **hyperthermophiles**. Many of these species also are **acidophiles** or **barophiles**.

- Hyperthermophiles possess several adaptations to life at high temperatures, including the presence of **molecular chaperones**, which form a complex referred to as the **thermosome** that facilitates protein folding and refolding, and **reverse DNA gyrase**, which increases the melting temperature of DNA.

- Recent molecular studies have revealed evidence of crenarchaeotes in temperate and arctic environments, suggesting that some crenarchaeotes are **mesophiles** or **psychrophiles**.

- Based on the SSU rRNA sequences isolated from these temperate and arctic regions, some researchers propose that the mesophilic and psychrophilic crenarchaeotes may constitute a new archaeal phylum, Thaumarchaeota.

- A major group of euryarchaeotes is the **methanogens**, organisms that produce methane.

- These organisms are found in **anoxic** environments throughout the world, including the human GI tract.

- Another major group of euryarchaeotes is the **halophiles**, organisms that grow in high salt environments. Halophiles possess several adaptations to life in high salt environments.

- Some halophiles undergo **phototrophy**, using **bacteriorhodopsin** to harvest the energy of sunlight.

- Based on SSU rRNA studies, two additional phyla, Korarchaeota and Nanoarchaeota, have been proposed. *Nanoarchaeum equitans*, one of the smallest known living organisms, would be a member of the latter phylum.

● Application Questions ....

1. Many hyperthermophilic organisms will die if exposed to lower temperatures. Why might these lower temperatures be too extreme for these organisms?

2. Thermophilic organisms must maintain the integrity of their proteins at elevated temperatures. Explain what would typically happen to bacterial proteins at elevated temperatures and how thermophilic bacteria are able to avoid this.

3. Imagine you are attempting to isolate and cultivate various extremophiles. What problems might you encounter in trying to accomplish this?

4. You are reading about archaeons and their importance in many industrial applications. Explain why enzymes isolated from some archaeons can be useful commercially. Provide an example.

5. Why might a cell contain a lipid monolayer as opposed to a lipid bilayer? Explain.

6. A scientist is studying the optimal growth conditions of a unique microbial cell. After testing many conditions, she finds that the cell grows best at 80°C and a pH of 3.0. How would this cell be categorized? Explain.

7. A microbe's cellular proteins are being analyzed. Nearly all of its proteins have a high percentage of α-helical regions, low levels of the amino acid cysteine, and increased levels of arginine. Based on these unique characteristics, what type of microbe might this be? Explain.

8. The search for life on Mars has focused on the chemistry of the subsurface as a potential environment for methanogens. Describe the conditions in this environment that might support the metabolic activities of methanogens.

9. You are examining a type of microbe that contains a membrane protein with very high similarity to the protein bacteriorhodopsin. Based on this similarity, what would you hypothesize is the function of this protein? Explain how it would function in the cell.

10. A microbe is observed with a gene that codes for reverse DNA gyrase. What would this enzyme do, and what selective advantage would it provide the microbe?

11. You detect what appears to be a new type of archaeon, although it does not live in an extreme environment.
    a. What structural characteristics would you need to observe to officially consider this microbe an archaeon?
    b. Explain how you would go about determining whether this organism is phylogenetically related to archaeons.

Suggested Reading

Brochier-Armanet, C., B. Boussau, S. Gribaldo, and P. Forterre. 2008. Mesophilic Crenarchaeota: Proposal for a third archaeal phylum, the Thaumarchaeota. Nat Rev Microbiol 6:245–252.

Huber, H., M. J. Hohn, R. Rachel, T. Fuchs, V. C. Wimmer, and K. O. Stetter. 2002. A new phylum of Archaea represented by a nanosized hyperthermophilic symbiont. Nature 417:63–67.

Ľubomíra, Č., K. Sandman, S. J. Hallam, E. F. DeLong, and J. N. Reeve. 2005. Histones in Crenarchaea. J Bacteriol 187:5482–5485.

Woese, C. R., and G. E. Fox. 1977. Phylogenetic structure of the prokaryotic domain: The primary kingdoms. Proc Natl Acad Sci USA 74:5088–5090.

Viruses

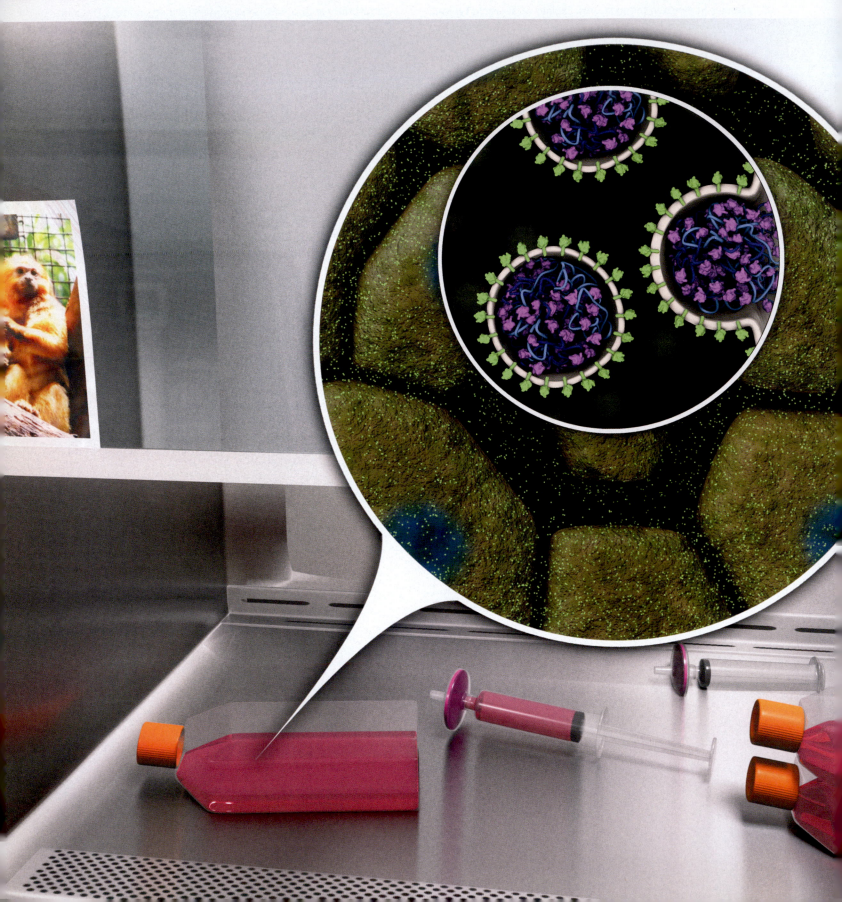

In the spring of 1991, Dr. Dick Montali, the chief veterinary pathologist at the National Zoo in Washington, DC, learned about a group of deaths among golden lion tamarins at a zoo in the United States. Similar outbreaks of a fatal hepatitis among these small primates had occurred at various U.S. zoos during the preceding several years. Certainly, the unexplained deaths of any animals at a zoo troubled Dr. Montali, but these deaths, perhaps, worried him more than most. The golden lion tamarins, or GLTs, you see, are endangered.

The brightly colored golden lion tamarins live in the Atlantic coastal forest of Brazil, where they eat various fruits, insects, and small reptiles and amphibians. Because of increased human activity in this area, their habitat has shrunk and their populations have plummeted. Today, researchers estimate that only about 1,500 GLTs remain in the wild. Another 500 live in zoos throughout the world.

To help stabilize the GLT populations, the National Zoo coordinates a GLT breeding and reintroduction to the wild program. At various zoos throughout the United States, GLTs are bred in captivity. Upon maturation, animals destined for reintroduction first are sent to the National Zoo. There, they become acclimated to life in the wild. The GLTs at the National Zoo live in a non-enclosed wooded area in the center of the zoo, where they are provided with food and nesting boxes. After a short period of acclimation, they are then transported to the Reserva Biologica de Poço das Antas, a protected reserve near Rio de Janeiro. Again, food is initially provided, but over time the released animals learn to forage for themselves. Since 1984, about 400 GLTs have been reintroduced into Brazil.

Obviously, the deaths of GLTs could adversely affect the reintroduction program, but veterinarians had a bigger concern: What if an infectious agent were killing these animals? Because zookeepers routinely transported GLTs between various zoos for breeding purposes, an infectious agent could easily be spread to zoos throughout the country. Even worse, if conservation biologists inadvertently reintroduced an infected animal into the Brazilian rain forest, then wild populations could become infected. Rather than saving a species from extinction, the efforts of concerned biologists could actually have deleterious effects. Finally, what if an infectious agent affecting the golden lion tamarins could be transmitted to humans?

CHAPTER NAVIGATOR

Introduction

"Virus." Many of us associate this word with disease or sickness. Viruses, after all, cause common colds. Viruses such as influenza, smallpox, polio, and HIV have caused an immeasurable amount of human suffering and death. Only a very small percentage of all viruses, though, are human pathogens, or disease-causing agents. Probably, the air we breathe, the water we drink, and the foods we eat contain millions of viral particles. Obviously, all of these viruses do not cause us to become ill. In fact, most of these viruses do not even infect us. Rather, they may infect other mammals, insects, plants, or even bacteria. Most likely, there are viruses that infect every type of living organism. In this chapter, we will discuss the history of virology and basic properties of viruses, explore theories about the origin of viruses, and learn about common techniques for cultivating, purifying, and quantifying viruses. Other topics, including viral classification and identification also will be addressed. We will investigate viral replication more thoroughly in Chapter 8. For now, let's focus on these questions:

What is a virus? (5.1)
From what did viruses arise? (5.2)
How can we "grow" and quantify viruses? (5.3)
What are the different types of viruses? (5.4)
Are viruses the simplest pathogens? (5.5)
What's next for virology? (5.6)

5.1 A basic overview of viruses

What is a virus?

Quite simply, **viruses** are small, acellular particles that only can replicate within living host cells. Viruses, then, are **obligate intracellular parasites**; they cannot replicate independently. They consist of an RNA or DNA genome enclosed within a protective protein shell, or **capsid**. Together, the genome and capsid comprise the **nucleocapsid**. Some viruses also contain a lipid bilayer, or envelope, that surrounds the capsid. During the production of new virus particles, all viruses require host cell enzymes for translation. Many viruses also require host cell enzymes for various aspects of transcription and/or genomic replication. Because of this dependence on the host cell, all viruses must possess a mechanism for entering a host cell, and newly formed viruses must possess a mechanism for then exiting the cell. As we will see later in this chapter and in Chapter 8, different viruses carry out these basic activities in very different ways. To begin our investigation of viruses, we will explore the history of virology.

History of virology

In some respects, the history of virology is nearly as old as the history of life itself. We know that different types of viruses exist that can infect all forms of life—plants and animals, fungi, bacteria, and archaeons. In a recent article, researchers even postulated that all viruses, like all living organisms, may have evolved from a common ancestor—what one might call the primordial virus. Although this hypothesis still needs more supportive data, it does raise an interesting question. Are all viruses related? We'll examine the question of viral origins in more detail later in this chapter.

Most likely, viruses that infect humans have existed since humans first evolved and have dramatically affected human history. Archeological and anthropological evidence suggests that pathogenic viruses, particularly smallpox virus, have infected humans throughout our history. Smallpox epidemics plagued ancient Egypt and may have caused the death of King Ramses V in the twelfth century B.C. **(Figure 5.1)**. Many historians today agree that the decimation of Native American populations after the invasion of the Europeans was more due to smallpox than warfare. Some historians argue that the outcome of World War I may, at least in part, have been determined by the 1918 influenza outbreak.

Virology as a science, on the other hand, is only a little over one hundred years old. Dimitri Ivanovski, in 1892, demonstrated that a disease of tobacco plants (see Figure 1.2) could be transmitted by extracts of a diseased plant that had been passed through a filter small enough to exclude the smallest known bacteria. A few years later, in 1898, Martinus Beijerinck

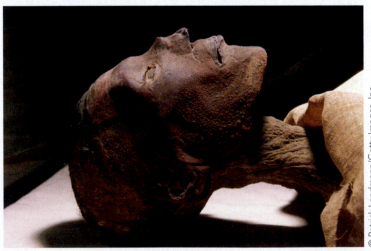

Figure 5.1. Pathogenic viruses and human history This photograph of the mummy of Ramses V, king of ancient Egypt in the twelfth century B.C., shows pustule-like markings on his face, suggesting that a smallpox infection probably led to his death. Smallpox plagued human populations for thousands of years, but the last recorded natural case occurred in Somalia in 1977.

described this agent (now known to be tobacco mosaic virus) as *contagium vivum fluidum*, a soluble living germ. Most virologists see this event as the official beginning of the field of virology

The advances in this new field have been amazingly rapid. By 1901, Walter Reed and colleagues demonstrated that a human disease—yellow fever—was caused by a virus and could be transmitted by mosquitoes **(Figure 5.2)**. During the early part of the twentieth century, Frederick Twort and Felix d'Herelle described viruses that infect bacteria, termed **bacteriophages** ("bacteria eaters") by d'Herelle. In the ensuing decades, researchers isolated a number of plant and animal viruses and conducted groundbreaking genetic and biochemical experiments with bacteriophages (or phages). Today, viruses that infect all forms of life continue to be identified. Molecular studies continue to provide us with a more detailed understanding of how viruses replicate and the effects that they have on their hosts.

CONNECTIONS If bacteriophages "eat" bacteria, then could they be used clinically as antibacterial agents? Maybe. This idea has been explored for many years. With the increased incidence of antibiotic-resistant bacteria, some researchers are intrigued by the therapeutic potential of bacteriophages. We will explore this topic in **Perspective 24.3**.

Structure of viruses

While all viruses tend to share several properties, perhaps the most notable characteristic of almost all viruses is their small size. Typically, the diameter of a viral particle is between 10 and 100 nanometers (1 nanometer = 10^{-9} meters or 1 billionth of a meter!). Poliovirus, for example, has a diameter of approximately 30 nm, while the oval-shaped poxviruses like smallpox may be over 200 nm long. Bacteria and archaeons, in contrast, typically have diameters from 1 to 10 micrometers (1 micrometer = 1,000 nanometers), and eukaryal cells generally have diameters from 10 to 100 micrometers **(Figure 5.3)**.

Not surprisingly, all viruses possess a genome, their genetic material. While every known bacterium, archaeon, and eukaryal organism possesses a double-stranded DNA genome, the genomic material of viruses may be DNA or RNA. Additionally, some viruses possess double-stranded genomes, while other viruses possess single-stranded genomes. Some viruses possess circular genomes, while other viruses possess linear genomes. Most viruses possess a single genetic molecule, but a few viruses contain segmented genomes. Influenza virus A, for example, contains eight segments of single-stranded RNA that code for a total of 11 different proteins. Despite these structural differences, almost all viral genomes are small, ranging from a few thousand nucleotides or base pairs in length for the smallest viruses to

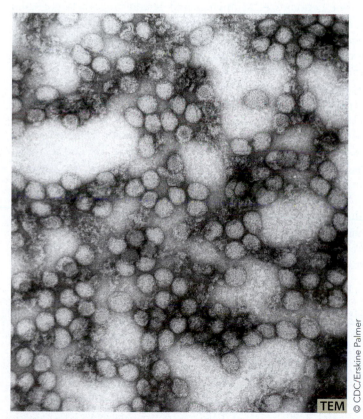

A. Yellow fever virus particles

© CDC/Erskine Palmer

B. Dr. Walter Reed

Bettmann/© Corbis

Figure 5.2. Yellow fever: First human disease linked to a virus In 1901, a team of researchers led by Walter Reed demonstrated that yellow fever was caused by a filterable agent, yellow fever virus, that was transmitted by mosquitoes. The last major outbreak in the United States occurred in 1905. Despite various public health efforts and an effective vaccine, yellow fever remains a significant problem in South America and Africa. **A.** Yellow fever virus is the causative agent of yellow fever. **B.** Walter Reed was head of the team that showed mosquitoes transmit yellow fever.

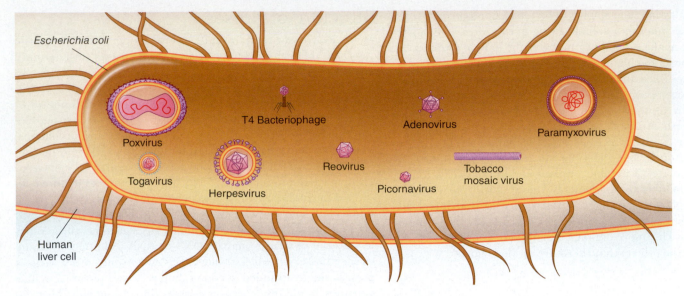

Figure 5.3. Sizes and shapes of selected viruses Specific viruses differ in size. Picornaviruses, like poliovirus, are quite small. Poxviruses, like smallpox virus, are relatively large. Virus structures also differ significantly. The models depicted here represent average sizes and shapes.

about 200,000 base pairs for the largest viruses (**Table 5.1**). Contrast this with the genome sizes of the bacterium *Escherichia coli* (4.6 million base pairs) and humans (3 billion base pairs).

CONNECTIONS When two genetically distinct strains of influenza virus infect the same cell, progeny viruses may be produced that contain some gene segments from one of the infecting viruses and some gene segments from the other infecting virus. As we will see in **Section 22.4**, this genetic reassortment can lead to influenza pandemics.

Of course, there are exceptions to every rule, and the general "rules" about viruses are no different. Recently, researchers have isolated and described several viruses whose physical properties certainly stretch the general rule about the size of viruses. *Cafeteria roenbergensis* virus, or CroV, infects certain marine single-celled eukaryal organisms and has a genome of approximately 730,000 base pairs. Mimivirus, a double-stranded DNA virus that infects amoeba, has a capsid approximately 400 nm in diameter and a genome of 1.2 megabase pairs (1.2×10^6 base pairs) that encodes 979 proteins. To put the size of Mimivirus in some perspective, consider this: at 400 nm in diameter, Mimivirus is

roughly the same size as the bacterium *Ureaplasma urealyticum* (**Figure 5.4**). Mimivirus, though, does not appear to be the world's largest virus. In October 2011, researchers reported the isolation of an even larger virus that also infects amoeba. Isolated off the coast of Chile, *Megavirus chilensis* has a genome of over 1.2 megabase pairs that researchers think encodes over 1,200 proteins.

The genomes of Mimivirus and Megavirus are larger than the genomes of numerous bacteria, including *Mycoplasma genitalium*, *Treponema pallidum*, and *Chlamydia trachomatis*. Perhaps more astonishingly, the Mimivirus genome contains nearly 2.5 times as many genes as are present in the *M. genitalium* genome! Clearly, the size of these viruses does not conform to the conventional descriptions of viruses. More interestingly, the genomes of these large viruses include genes needed for the synthesis of nucleotides and amino acids. No other known virus possesses these types of genes. Because of their unusual size and the presence of these genes, these large viruses may force virologists to re-examine the definition of "virus."

The structures of viruses also exhibit great variation, as we will see in the rest of this section. Some viruses have an almost spherical shape, while others are rod-like. Still other viruses, like the bacteriophages, exhibit a combination of structures. Poxviruses, conversely, exhibit more complex structures.

TABLE 5.1 Sizes and structural features of selected viruses

Virus	Host	Structure	Size (nm)	Genome size (bp)	Genetic material[a]
Poliovirus (Picornavirus)	Humans	Non-enveloped, icosahedral	30 (diameter)	7,700	ssRNA
Tobacco mosaic virus (TMV)	Tobacco and related plants	Non-enveloped, helical	300 × 18	6,400	ssRNA
T4 (bacteriophage)	*E. coli*	Non-enveloped, complex	200 × 90	170,000	dsDNA
Smallpox virus (poxvirus)	Humans	Enveloped, complex	300 × 250	186,000	dsDNA
Mimivirus	Amoeba	Enveloped, complex	400 (diameter)	1,200,000	dsDNA

[a]ssRNA: single-stranded RNA; dsDNA: double-stranded DNA

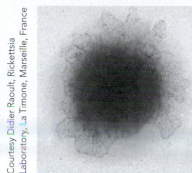 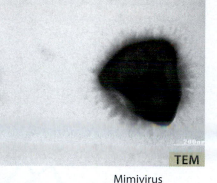

Ureaplasma urealyticum   Mimivirus

Figure 5.4. Mimivirus—An unusually large and complex virus While viruses generally are much smaller than even the smallest bacterial cells, Mimivirus is an exception. Mimivirus (*right*) is approximately 400 nm in diameter, about the same size as the bacterium *Ureaplasma urealyticum* (*left*), a mycoplasma that is part of the normal genital microbiota of about 70 percent of sexually active males and females. Mimivirus has not been shown to be pathogenic, but has been associated with cases of pneumonia. The genome of Mimivirus includes genes coding for the synthesis of nucleotides and amino acids. Such genes are unknown in other viral genomes.

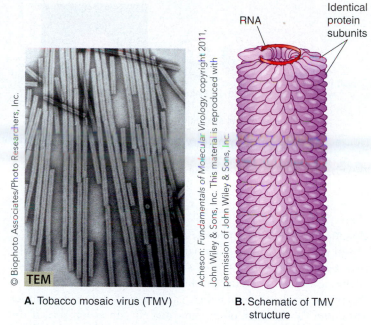

A. Tobacco mosaic virus (TMV)   **B.** Schematic of TMV structure

Figure 5.5. Helical symmetry of tobacco mosaic virus A viral capsid exhibits helical symmetry if it can be rotated about its long axis and shows no distinguishing features. **A.** Electron micrograph of TMV, showing the filamentous, rod-like shape of the virus. **B.** Schematic of TMV showing the presence of the RNA genome within the helical protein shell. The helical capsid is composed of multiple copies of a single polypeptide that interact with each other and the RNA genome.

Symmetry

In all viruses, one or more viral proteins surround the genome, forming the nucleocapsid or, more simply, the capsid. Biochemical and genetic studies have shown that the capsid of all viruses consists of many symmetrically arranged subunits, or **capsomeres**. Each capsomere, in turn, consists of one or more polypeptides. Electron microscopy and X-ray crystallography studies have demonstrated that the capsids of most viruses exist in one of two forms.

The capsids of some viruses, like tobacco mosaic virus (see Figure 5.3) exhibit a **helical morphology**, in which the capsomeres form a helix and the capsid resembles a hollow tube. The capsids of other viruses exhibit an **icosahedral morphology**, in which the capsomeres form an icosahedron, or 20-sided polygon, with each capsomere making up a face of the icosahedron. Many bacteriophages, like T4 (see Figure 5.3), exhibit both helical and icosahedral morphologies, containing an icosahedral "head" and a helical "tail." A few animal viruses, most notably the poxviruses, have a less well-defined capsid morphology.

Helical Symmetry The simplest structural arrangement found in virus capsids is that of a single polypeptide that directly coats the genome in a repetitive manner, as occurs for tobacco mosaic virus (TMV). The individual capsid proteins interact at regular intervals with the genome and also associate with each other, causing the structure to twist and take on a tube-like coiled form, much like a spiral staircase (**Figure 5.5**). A similar arrangement occurs in influenza viruses. Multiple copies of a single viral protein, NP or nucleoprotein, and a viral polymerase complex associate with each segment of RNA, resulting in eight separate nucleocapsids within each virus particle.

This helical symmetry gives viruses like TMV a filamentous, fiber-like, or rod-shaped morphology. In all viruses with helical capsid symmetry, there is an intimate association between the genome and the capsid proteins. Many phages and plant viruses consist of just helical nucleocapsids. In these viruses, the helical nucleocapsid *is* the **virion**, or complete viral particle. However, all known animal viruses with helical nucleocapsid symmetry, like influenza virus, have an additional membrane, or envelope, surrounding the nucleocapsid (**Figure 5.6**).

The vast majority of viruses with helical capsid symmetry contain single-stranded RNA genomes. There probably are molecular reasons for this pairing. Single-stranded RNA is quite unstable and subject to rapid degradation; directly coating the

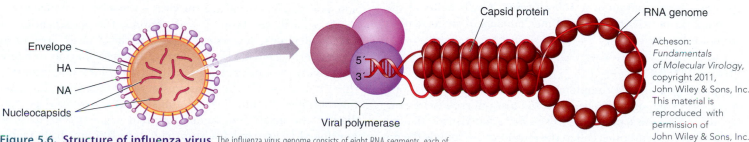

Figure 5.6. Structure of influenza virus The influenza virus genome consists of eight RNA segments, each of which is enclosed with capsid protein. The viral polymerase is bound to the end of each nucleocapsid. These nucleocapsids are surrounded by an envelope. Embedded in the envelope are two viral proteins, hemagglutinin (HA) and neuraminidase (NA).

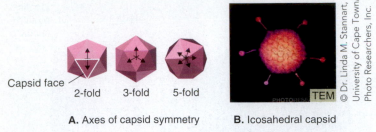

A. Axes of capsid symmetry

Capsid face | 2-fold | 3-fold | 5-fold

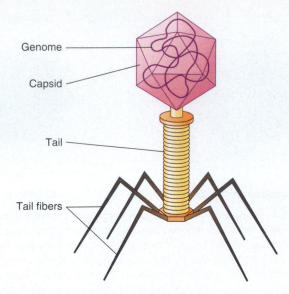

© Dr. Linda M. Stannart, University of Cape Town/ Photo Researchers, Inc.

B. Icosahedral capsid

Figure 5.7. Icosahedral symmetry in viruses An icosahedron has 20 faces and 12 vertices. **A.** As seen in this drawing, each face is an equilateral triangle. Icosahedrons exhibit twofold, threefold, and fivefold axes of symmetry. **B.** Adenovirus, depicting icosahedral symmetry, with attachment proteins protruding from the vertices.

genome in protein helps to protect it. In many viruses with helical capsid symmetry, the genome length determines the size of the virus because the capsid proteins faithfully coat the entire length of the new genomes being produced in the host cell. Because of this property, viruses with helical capsids, such as the filamentous TMV and rabies virus, often show diversity in virion size. If the length of the genome varies, then the size of the capsid also will vary. Conversely, viruses with icosahedral capsids have very strict limits on their genome size because of the packaging constraints imposed by the volume of the enclosing capsid. A defined amount of space is available within the capsid.

Icosahedral symmetry

A geometrically perfect icosahedron has 20 faces and 12 vertices forming a closed sphere and each face of the icosahedron is an equilateral triangle **(Figure 5.7)**. Proteins, however, are globular in shape, not triangular. The only way to make a triangle, then, is to use a minimum of three polypeptides per face. So, even the smallest icosahedral viruses must use 60 protein units to construct a capsid. In the simplest icosahedral viruses, such as parvoviruses, each face is formed from three identical proteins. Rhinovirus, which causes the common cold, uses three different polypeptides to construct its capsid.

Complex Symmetry

While some viruses of bacteria exhibit simple helical or icosahedral capsid symmetry, most of the known bacteriophages exhibit a more complex structure. These viruses possess an icosahedral head, which contains the genome, and a helical tail **(Figure 5.8)**. This structural arrangement has not been observed among any viruses of plants or animals. As we will see in the next section, this structure is intimately tied to the way in which phages infect their host cells.

A few large viruses of animals, most notably the poxviruses, have a very irregular shape. The poxviruses generally are enveloped and often contain over 100 proteins. The benefits of this complexity, and the biochemical mechanisms of its formation, are not fully understood.

Viral Envelopes

For many viruses, the complete infectious virions consist only of a capsid surrounding a genome and are referred to as **non-enveloped viruses**, or naked viruses. For other viruses, though, a cell-derived membrane, called the envelope, surrounds the nucleocapsid. Examples of **enveloped viruses** include influenza virus and HIV **(Figure 5.9)**.

Genome

Capsid

Tail

Tail fibers

Figure 5.8. Complex structure of bacteriophages Viruses of bacteria typically possess an icosahedral head, which houses the genome, and a helical tail. Attached to the tail are attachment fibers that facilitate binding to host cells.

The envelope is added as progeny virus particles bud out of cellular membranes on their transit out of the host cell. In so doing, they take part of the cellular membrane with them. Budding begins where specialized viral envelope proteins have been inserted into the membrane (see Figure 8.22). For some viruses, like rabies virus, other proteins associate in a layer under the membrane where they interact with envelope proteins and together help target the newly synthesized viral nucleocapsids to the membrane for final packaging and exit. While many animal viruses are enveloped, very few enveloped viruses of plants or bacteria have been identified.

For enveloped viruses, the viral envelope is usually required for successful infection of a new host. Thus, any viral proteins that are needed for attachment and entry into the new host cell must be present on the outer surface of the envelope. These viral attachment and entry proteins are transcribed and translated from the viral genome in the infected host cell, and are inserted into the host cell membrane that is destined to become the viral envelope. In acquiring an envelope, the virus also may take with it some host cell-derived membrane proteins. These proteins may help disguise the virus from immune attack by making it appear similar to a host cell, at least on the outside. While this strategy has an obvious advantage to the virus, the drawback is that the envelope, being a biological membrane, is usually easily destroyed in the external environment through desiccation and exposure to chemicals. Without an intact envelope, the proteins used for viral attachment are absent or non-functional and the virus is rendered non-infectious.

As we will see later in this chapter, the classification of viruses depends, to some extent, on these basic structural properties. Virologists used this type of information to begin identifying the infectious agent affecting golden lion tamarins that we discussed in the opening story of this chapter. Electron micrographs of thin liver sections isolated from GLTs that died of the mysterious agent revealed what appeared to be enveloped virus particles approximately 100 nm in diameter exiting the GLT hepatocytes, or liver cells. Furthermore, the envelopes contained very distinct spikes. Based on this simple

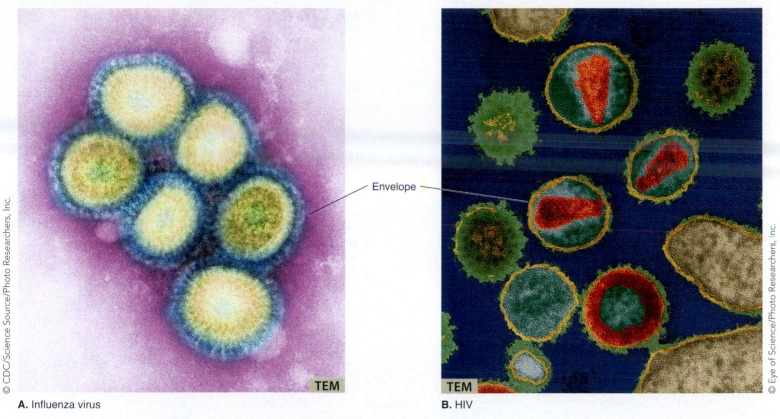

A. Influenza virus

B. HIV

Figure 5.9. Electron micrographic identification of the viral envelope The envelope surrounds the nucleocapsids and contains viral proteins that extend from the envelope. Shown here are **A.** influenza virus and **B.** HIV.

piece of evidence, virologists began their quest to identify the infectious agent which appeared to be a virus in the *Coronaviridae* (see Figure 5.25) or *Arenaviridae* family of viruses.

Replication cycle

To enter an appropriate host cell, most animal viruses bind to the cell. Usually, this binding occurs through a specific interaction between the viral attachment protein, a protein or proteins present on the surface of the virus, and the receptor, a protein present on the surface of the cell. For enveloped viruses, the attachment protein generally resembles a spike embedded within the lipid bilayer envelope of the virion. For some non-enveloped viruses, like the adenoviruses, the attachment protein exists as a spike protruding from the vertices of the icosahedron (see Figure 5.7B). For other non-enveloped viruses, like poliovirus, the cellular receptor interacts with amino acid residues that exist in an indentation, or canyon, within the capsid.

CONNECTIONS If the virus binds to a host cell through a specific protein–protein interaction, then shouldn't changes in the host cell protein affect the cell's susceptibility to certain viruses? Yes. We will see how in **Section 8.1**.

Once a viral particle has attached to a host cell, some form of the virion, capsid, or viral genome penetrates the cellular membrane and enters the cytoplasm of the cell **(Figure 5.10)**. For non-enveloped animal viruses like rhinovirus, the entire viral particle enters the cell through a type of endocytosis referred

to as receptor-mediated endocytosis (see Figure 5.10A). Following endocytosis, conformational changes in the capsid allow the viral genome to enter the cytoplasm.

For most enveloped animal viruses, entry into the cell is mediated by the fusion of the viral and cellular membranes. This event can occur at the plasma membrane or within an intracellular vesicle. HIV, for example, fuses at the plasma membrane (see Figure 5.10B). The virus initially attaches to its primary receptor—the CD4 molecule. Subsequently, the virus interacts with a secondary receptor, or co-receptor, typically CCR5 or CXCR4, integral membrane proteins found primarily in certain cells of the human immune system. The interaction with the co-receptor triggers conformational changes in the viral proteins embedded within the viral envelope. Following these conformational changes, the viral envelope fuses with the plasma membrane. In Figure 8.6, we describe this process in more detail.

The envelope of influenza virus, on the other hand, fuses with an intracellular membrane (see Figure 5.10C). Again, the process begins with an interaction between the virus and a cell surface receptor. After binding, the virus is endocytosed. The acidic pH of this endosomal vesicle causes conformational changes in the viral proteins. Fusion between the viral envelope and the vesicle membrane then occurs. After entry, the viral capsid usually undergoes a specific set of disassembly, or uncoating, events in the cytoplasm that result in the release of the genome. More detail of this process is shown in Figure 8.5.

Plant viruses often enter cells through disruptions in the cuticle covering of the plant tissue. These disruptions can be caused, for instance, by farm machinery, animals grazing on the

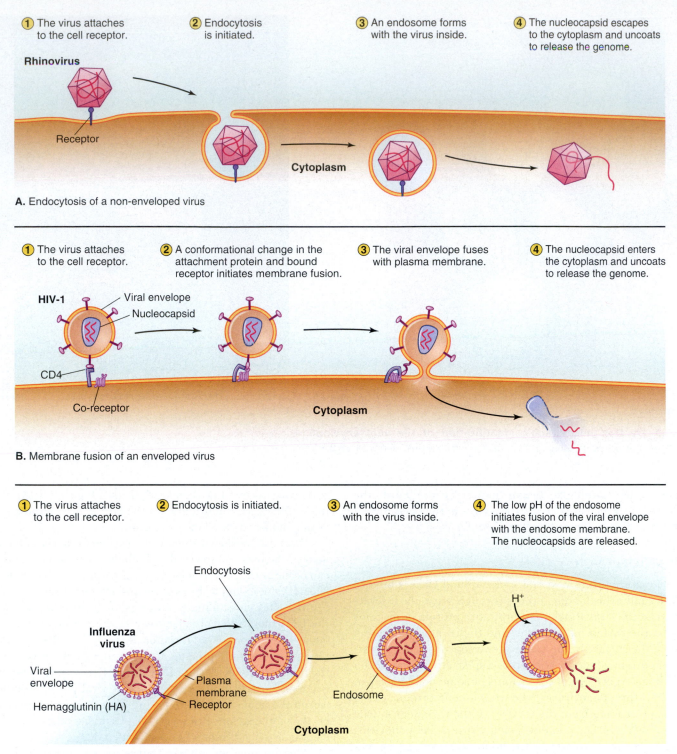

① The virus attaches to the cell receptor.　**② Endocytosis is initiated.**　**③ An endosome forms with the virus inside.**　**④ The nucleocapsid escapes to the cytoplasm and uncoats to release the genome.**

Rhinovirus

Receptor

Cytoplasm

A. Endocytosis of a non-enveloped virus

① The virus attaches to the cell receptor.　**② A conformational change in the attachment protein and bound receptor initiates membrane fusion.**　**③ The viral envelope fuses with plasma membrane.**　**④ The nucleocapsid enters the cytoplasm and uncoats to release the genome.**

HIV-1　Viral envelope
Nucleocapsid

CD4

Co-receptor　　Cytoplasm

B. Membrane fusion of an enveloped virus

① The virus attaches to the cell receptor.　**② Endocytosis is initiated.**　**③ An endosome forms with the virus inside.**　**④ The low pH of the endosome initiates fusion of the viral envelope with the endosome membrane. The nucleocapsids are released.**

Endocytosis

Influenza virus

H⁺

Viral envelope

Plasma membrane
Receptor

Endosome

Hemagglutinin (HA)

Cytoplasm

C. Endocytosis of an enveloped virus

Figure 5.10. Process Diagram: Entry mechanisms of animal viruses Following attachment to a host cell, some part of the virion or viral genome must gain access to the cell. **A.** Non-enveloped viruses like rhinovirus often enter through endocytosis. **B.** Enveloped viruses like HIV undergo a membrane fusion event at the cell surface. **C.** Enveloped viruses like influenza first enter the cell via endocytosis. A fusion event between the viral envelope and vesicle membrane then occurs.

plants, or insects feeding on the plants. Once inside a cell, the viral capsid undergoes a disassembly process, thereby releasing its genome. The genome and newly formed virions can spread to other cells within the plant through the plasmodesmata, or small channels, that connect plant cells and serve as conduits for the transport of necessary materials. Unlike animal viruses, plant viruses do not recognize specific cellular receptors on their

host cells. We'll investigate some of the implications of this in later chapters.

Like animal viruses, bacteriophages interact with various components of the bacterial surface, including specific polysaccharide residues. For the phages T2 **(Microbes in Focus 5.1)** and T4, this initial contact occurs between the tail fibers and, most likely, specific sugar residues on the bacterial cell surface. After relatively stable

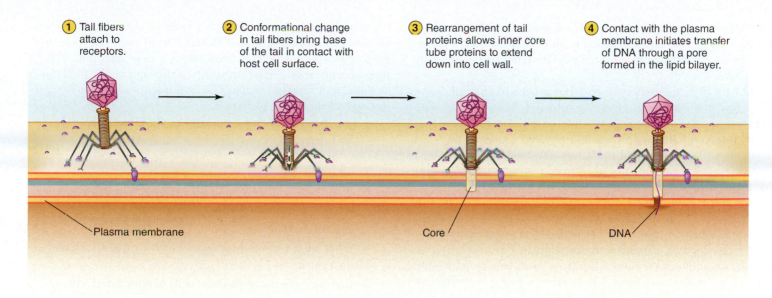

① Tail fibers attach to receptors.

② Conformational change in tail fibers bring base of the tail in contact with host cell surface.

③ Rearrangement of tail proteins allows inner core tube proteins to extend down into cell wall.

④ Contact with the plasma membrane initiates transfer of DNA through a pore formed in the lipid bilayer.

Plasma membrane

Core

DNA

Figure 5.11. Process Diagram: Entry mechanisms of bacteriophages Following attachment to a host cell, a series of conformational changes occur in the protein units of the tail, allowing the DNA to move from the capsid head directly into the cytoplasm of the host cell. The virus particle does not actually enter the host cell.

binding has occurred, a conformational change results in the contraction of the bacteriophage tail. With the action of specific enzymes, the phage core proteins penetrate the cell membrane and the bacteriophage injects its genetic material into the host cell, much like a syringe injects its contents. Thus, the viral proteins remain outside the cell, never crossing the cell membrane **(Figure 5.11)**.

For all viruses, replication of the viral genome and production of viral proteins follows attachment and entry. These newly formed components then assemble to form new virus particles, which exit the infected cell **(Figure 5.12)**. This assembly process often is referred to as a sequential and irreversible process. The

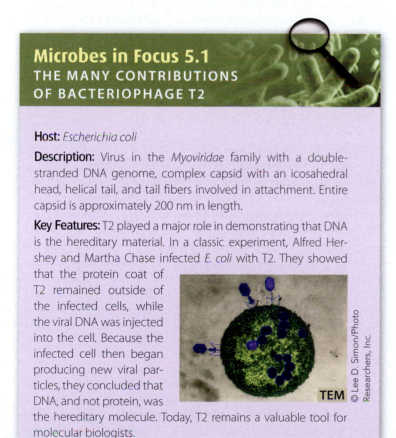

Microbes in Focus 5.1
THE MANY CONTRIBUTIONS OF BACTERIOPHAGE T2

Host: *Escherichia coli*

Description: Virus in the *Myoviridae* family with a double-stranded DNA genome, complex capsid with an icosahedral head, helical tail, and tail fibers involved in attachment. Entire capsid is approximately 200 nm in length.

Key Features: T2 played a major role in demonstrating that DNA is the hereditary material. In a classic experiment, Alfred Hershey and Martha Chase infected *E. coli* with T2. They showed that the protein coat of T2 remained outside of the infected cells, while the viral DNA was injected into the cell. Because the infected cell then began producing new viral particles, they concluded that DNA, and not protein, was the hereditary molecule. Today, T2 remains a valuable tool for molecular biologists.

TEM

© Lee D. Simon/Photo Researchers, Inc.

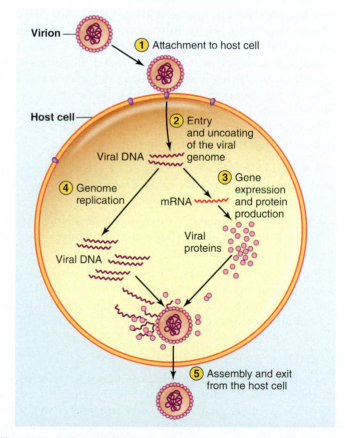

Virion

① Attachment to host cell

Host cell

② Entry and uncoating of the viral genome

Viral DNA

③ Gene expression and protein production

④ Genome replication

mRNA

Viral proteins

Viral DNA

⑤ Assembly and exit from the host cell

Figure 5.12. Process Diagram: General steps in a viral replication cycle Despite the many differences that exist in viral replication strategies, the underlying process is the same for all viruses. This schematic depicts the replication of a virus with a DNA genome.

assembly steps occur in an ordered fashion and, once completed, a step cannot be reversed. Recent studies with bacteriophage T4, however, challenge this simple explanation of viral assembly **(Mini-Paper)**.

Once new virus particles form, they must exit the cell. Some viruses exit by budding off of the infected cell. Other viruses exit by lysing, or rupturing, the infected cell. In this manner, a single infected cell may produce up to 10,000 new viral particles. These newly released viral particles may bind to another host cell within the organism or be shed from the organism and enter the environment.

Not surprisingly, the site of entry, the location of appropriate host cells, and the site of exit for different viruses differ tremendously. The types of disease associated with different pathogenic viruses also differ tremendously. Respiratory viruses, such as influenza, cause respiratory problems by damaging cells of the respiratory tract. These viruses exit a host through coughing and sneezing and, if inhaled by another person, may begin a new infection. Enteric viruses, such as rotavirus, cause

intestinal problems by damaging cells of the GI tract. These viruses generally exit a host through fecal material and, if ingested by another person, may begin a new infection. HIV, a blood-borne virus, can be transmitted through bodily fluids such as blood and semen. It causes an immunodeficiency by destroying cells of the immune system.

5.1 Fact Check

1. What are some key properties of viruses?
2. Identify the different types of viral genomes.
3. Contrast the capsid structures of helical and icosahedral viruses.
4. What are the different ways in which viruses attach to host cells?
5. Compare the entry into host cells of enveloped and non-enveloped viruses.

5.2 Origins of viruses

From what did viruses arise?

How did viruses originate? The answer to this basic question remains unresolved. Our understanding of the structure of viruses and how viruses replicate, however, allows us to make educated guesses about their origin. In Chapter 8, we will examine in more detail the replication strategies of specific viruses. For now, though, it is sufficient to remember that all viruses only can replicate within living host cells, using various resources of their host. Thus, the origin of viruses must be tightly entwined with the host cells.

Given even this limited information about viruses, one still can formulate hypotheses about their origin.

- **Coevolution hypothesis**: Viruses may have originated prior to or at the same time as the primordial cell and have continued to coevolve with these hosts.

- **Regressive hypothesis**: Viruses may represent a form of "life" that has lost some of its essential features and has become dependent on a host.

- **Progressive hypothesis**: Viruses may have originated when genetic material in a cell gained functions that allowed the DNA or RNA to replicate and be transmitted in a semi-autonomous fashion.

Coevolution hypothesis

Perhaps viruses evolved along with their host cells or arose before the first cells. Most biologists agree that life arose billions of years ago when a self-replicating RNA molecule first developed in the primordial soup present on the surface of Earth. At some point, this RNA molecule gave way to a DNA genome capable of directing the synthesis of proteins via an RNA intermediate. With the encapsulation

of this DNA:RNA:protein system within a lipid envelope, the ancestor of the first cell arose. One can imagine that the first virus evolved along with the evolution of the first cell. A self-replicating RNA molecule may have evolved to make use of the protein synthesizing machinery of the nascent cell.

CONNECTIONS As we saw in **Section 1.2**, certain RNA molecules, ribozymes, can catalyze chemical reactions. This RNA-based enzymatic activity has led some researchers to hypothesize that life on Earth began with a self-replicating RNA molecule.

Other researchers have speculated that primordial viruses gave rise to the eukaryal nucleus. According to proponents of this hypothesis, a large, complex DNA virus entered a bacterial or archaeal cell. Rather than replicating within the host, the virus became permanently established in a type of endosymbiotic event. The existence of a specific group of viruses, the nucleocytoplasmic large DNA viruses (NCLDVs) lends some support to this hypothesis. These viruses, which include herpesviruses, poxviruses, and the recently discovered Mimivirus (see Figure 5.4), exhibit a complex DNA genome and some genes associated with translation. Perhaps, the modern nucleus arose from NCLDV ancestors.

Regressive hypothesis

As an alternative to the coevolution hypothesis, one can hypothesize that viruses originated from fully functional cells that became parasites of other cells. Over time, these parasites gradually lost some of their replicating and metabolic capabilities, eventually becoming dependent on the host cells for their complete replication. Several examples of this type of regressive evolution exist within the biological world. As we discussed in Section 3.4, the endosymbiotic theory states that mitochondria and chloroplasts present in eukaryal cells originated when fully functional bacteria were engulfed by a developing eukarya. Rather than being destroyed by the engulfing cell, though, the engulfed cells became permanent residents. Initially, these cells may have remained fully autonomous within their host. Eventually, these engulfed cells lost some of their functions, thus becoming dependent on the host.

Chlamydia represent another interesting example of potential regressive evolution. These bacteria are very small, about the size of some large animal viruses. Like viruses, they are obligate intracellular parasites. While chlamydia possess DNA, RNA, translational machinery, and some metabolic pathways, they can reproduce only within a host cell. It appears that members of this phylum lack key components of the electron transport chain and, as a result, cannot synthesize ATP. Rather, they depend on the host for energy. Most likely, these organisms evolved from a fully functional organism that, over time, lost some essential features. In some ways, chlamydia could be viewed as intermediates between fully independent cells and viruses. As with mitochondria and chloroplasts, though, the genomes of chlamydia are most definitely bacterial.

The existence of organelles like mitochondria and chloroplasts, and bacteria like chlamydia provide some support for regressive evolution. Viruses, perhaps, are evolutionary descendents of life-forms that have lost most of their genetic code and now are non-living parasites, completely dependent on a host cell. While some evidence supports this hypothesis, some discrepancies also exist. Most importantly, sequence analysis of viral genomes reveals that the genomes of animal viruses show great similarity to regions of their host genomes, and almost no similarity to bacterial genomes. This finding stands in stark contrast to the evidence supporting the regressive evolution of mitochondria, chloroplasts, and chlamydia. The regressive evolution from a fully functional bacterium to a chlamydia-like bacterium to an animal virus thus seems unlikely.

CONNECTIONS To determine the relationships between bacteria and viruses that infect mammals, we need to compare their genomes. This burgeoning field of comparative genomics is explored in **Section 10.3**.

The NCLDVs that we mentioned earlier, though, may have arisen via a regressive pathway. These viruses have a more complex genome than other viruses and, as we just noted, possess some genes associated with translation. Mimivirus, in fact, contains genes that may encode aminoacyl-tRNA synthetases and translation initiation factors, two key components of the translational machinery. Additionally, the genome of this giant virus contains other genes that show homology to eukaryal genes associated with DNA repair and protein folding. It seems plausible that this virus could have evolved from a previously free-living ancestor.

Progressive hypothesis

According to the progressive hypothesis of viral origins, viruses evolved from existing genetic elements that, over time, gained

the functions necessary to allow them to be transmitted between organisms and undergo some aspects of replication. The bacteriophages, for example, may have evolved from preexisting, autonomously replicating plasmids. If these plasmids acquired a gene coding for the production of a capsid protein, then a virus-like particle might form.

In a similar manner, retroviruses may have evolved from eukaryal **retrotransposons**, pieces of DNA capable of moving to new locations within a genome. For these mobile genetic elements to move within a genome, their DNA first is converted to RNA. This free-floating RNA then is converted back into DNA by a reverse transcriptase enzyme before being reinserted into the genome (**Figure 5.13**). Retroviruses also undergo an RNA to DNA conversion, and the genetic organization of retroviruses resembles the genetic organization of retrotransposons.

CONNECTIONS As we will see in **Section 8.3**, the replication cycle of retroviruses is remarkably similar to the steps involved in the movement of retrotransposons. A key component of both processes—reverse transcriptase—is the target of several antiretroviral drugs.

What about DNA viruses? In a 2011 report, researchers from the University of British Columbia presented compelling evidence that some viruses with DNA genomes may be evolutionarily related to another class of mobile genetic elements. The genome of Mavirus, a small virus associate with CroV, the giant virus mentioned in Section 5.1, exhibits significant similarity to

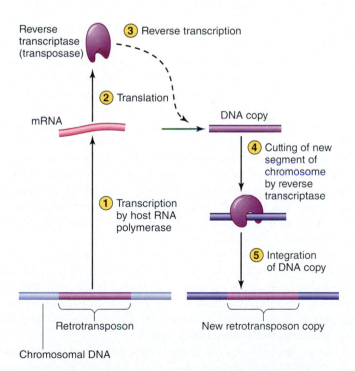

Figure 5.13. Process Diagram: Replication of retrotransposons While the origin of viruses still is open for debate, the similarities between the replication of retroviruses and the replication of retrotransposons seem compelling. Retrotransposons copy and insert themselves within the genome via an RNA intermediate that is converted into DNA by reverse transcriptase—a mechanism very similar to that used by retroviruses like HIV whose RNA genome is copied into DNA before replicating their RNA genomes.

the Maverick/Polinton class of transposons. Perhaps, this virus evolved from a transposon. Alternatively, the authors note, these transposons may have evolved from ancestors of this virus.

Compelling evidence supports the coevolution, progressive, and regressive hypotheses of viral origins. No single hypothesis, though, seems completely adequate. Perhaps viruses have originated repeatedly over the 4-billion-year history of life on Earth. All three hypotheses about the origin of viruses, then, may be correct. Maybe, other, more compelling hypotheses about the origin of viruses are yet to be formulated.

5.2 Fact Check

1. Discuss the various hypotheses regarding the evolutionary origins of viruses.
2. What evidence does *not* support the regressive evolution of viruses?
3. Under the progressive hypothesis, describe how bacteriophages and animal viruses may have evolved.
4. Differentiate between transposons and retrotransposons.

5.3 Cultivation, purification, and quantification of viruses

How can we "grow" and quantify viruses?

Several hallmark features of viruses make working with these particles somewhat difficult. Most notably, as we discussed previously, viruses only replicate within appropriate host cells. The cultivation of viruses, therefore, requires the availability of these cells. Additionally, the small size of viruses affects our ability to purify and count them. In this section, we will examine some of the techniques used by virologists to cultivate, purify, and quantify viruses. We also will examine the development of cell culture techniques—a major advancement in the field of virology—and the clinical importance of viral quantification.

Viral cultivation

As mentioned previously, viruses only can replicate within appropriate host cells. To cultivate viruses, then, a researcher must inoculate appropriate cells with the virus and then harvest the resulting progeny viruses. For bacteriophages (or phages), cultivation can be achieved by inoculating a liquid culture of actively growing bacteria with a small amount of phage. Upon continued incubation, the phage will replicate in the host cells. With virulent, or **lytic bacteriophages**, the viruses replicate within the bacteria and eventually lyse, or destroy, the infected cells. When the host cells lyse, the newly produced phage particles are released, and then can infect other bacterial cells. As more and more bacterial cells in the culture are destroyed, the liquid in the flask containing the bacteria and phage turns from turbid to clear, indicating a decrease in the number of intact bacterial cells (**Figure 5.14**). After a period of incubation, the solution can be filtered to remove any remaining intact bacteria and relatively large cellular debris. The resulting cell-free filtrate will contain the amplified phage.

Temperate, or **lysogenic bacteriophages**, generally can replicate lytically, resulting in the lysis of the host cell. Alternatively, they can exist in a non-replicative, latent state within the host cell as the phage genome usually becomes integrated into the host cell's genome. The phage genome is replicated along with the host cell's DNA, but transcription of phage genes is repressed. In this state, the virus typically is referred to as a **prophage** and

the harboring cell is referred to as a **lysogen**. Under certain conditions, lysogeny is terminated and the prophage can begin a lytic replication cycle. As we will see in Section 8.3, a complex and fascinating set of events often controls the switch between the lytic and lysogenic states.

Since the advent of mammalian cell culture techniques **(Toolbox 5.1)**, the propagation of animal viruses has become almost as simple as the propagation of lytic bacteriophages. Typically,

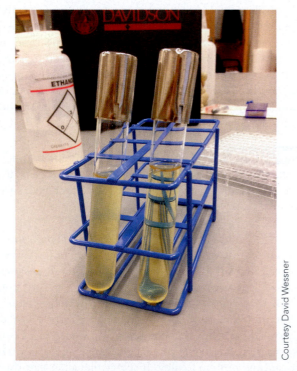

Figure 5.14. Cultivation of bacteriophage A small sample of phage can be used to inoculate a tube of actively growing bacteria. As the phage replicates and lyses the bacterial cells, the growth medium may turn from turbid to clear, indicating that intact bacterial cells are being disrupted and scattering less light. After incubation, the medium is centrifuged and the resulting supernatant is filtered through a 0.2-micrometer filter. This will prevent the passage of bacteria, but allow the passage of bacteriophage. The resulting filtrate is an amplified, cell-free sample of phage.

Courtesy David Wessner

Toolbox 5.1
CELL CULTURE TECHNIQUES

Because viruses only can replicate within living host cells, virologists are presented with an interesting problem: before researchers can effectively study viruses in the laboratory, they must be able to grow appropriate host cells. Prior to the development of mammalian cell culture methodologies, the propagation of animal viruses occurred via serial passage of the virus in animals. A susceptible animal would be inoculated with a small sample of the virus of interest. At a later time, perhaps when the infected animal became sick, virus would be isolated from this animal and a second susceptible animal would be inoculated, thereby allowing the virus to continue replicating. Clearly, this method of viral propagation was not terribly practical; it required the researcher to maintain appropriate animal hosts. As you can imagine, this requirement greatly reduced the study of viruses that infect humans!

During the 1940s and 1950s, this difficulty was overcome with the development of cell culture techniques. In 1949, John Enders, Thomas Weller, and Frederick Robbins demonstrated that poliovirus could replicate in various embryonic tissues. This discovery directly led to the development of the poliovirus vaccine a few years later and earned the three researchers the Nobel Prize in Physiology or Medicine in 1954. George Gey and colleagues cultured the first human cell line in 1952. This cell line, derived from a human cervical carcinoma, was named HeLa, in recognition of Henrietta Lacks, the young woman from whom the cells were isolated. She died of cervical carcinoma in 1951. The HeLa cell line, though, still is widely used by researchers throughout the world. A final important contribution to the budding field of virology came in 1955 when Harry Eagle and colleagues developed a well-characterized nutrient medium that could be used for the maintenance of cells in the laboratory. With the availability of defined growth media and several different cell lines, researchers easily could propagate animal viruses and conduct controlled experiments. Thus, the history of animal virology is inextricably linked to the development of *in vitro* cell culture methodologies.

Today, researchers can use many different types of cultured cells. Mammalian cells typically are incubated at 37°C (normal human body temperature) in an atmosphere containing 5 percent CO_2 (to mimic blood gas levels and maintain an appropriate pH). Once attached to a substrate, the cells will begin to divide. These cells will continue to divide until they completely cover the substrate, or become confluent, at which point most animal cells cease to divide (see Figure 5.15A). For the cells to continue growing, their density needs to be decreased. The growth medium is removed and trypsin, a protease, usually is added to the cells to dislodge them from the surface. These dislodged cells are diluted in fresh medium and plated, at a lower density, in new cell culture dishes, thereby allowing the cells to continue replicating (**Figure B5.1**).

Most animal cells can be grown in this fashion, but only for a limited number of rounds of replication, at which point the cells will begin to die. In some cases, though, a few cells within the culture will become immortalized. These cells have undergone genetic

Courtesy David Wessner

Figure B5.1. Propagation of mammalian cells In the laboratory, mammalian cells typically are grown at 37°C in an atmosphere containing 5 percent CO_2. Here, a researcher is dislodging cells from a flask and transferring them to another flask. She is working in a laminar flow hood to prevent contamination of the cells.

changes that allow them to grow indefinitely. At this point, the cells are referred to as a cell line or a continuous cell line. While these cells can be maintained in the laboratory indefinitely, they do differ significantly from "normal" cells. Immortalized cells often are aneuploid, that is, they possess an incorrect chromosomal complement. HeLa cells, for example, usually possess 82 chromosomes, rather than the 46 chromosomes normally found in human cells. These cells also often exhibit other aberrant properties. Thus, while cell lines allow researchers to conduct exquisitely detailed biochemical and genetic experiments, the results obtained with cell lines, one could argue, may differ from that which occurs *in vivo*.

● Test Your Understanding ....

Imagine you are culturing a line of mammalian cells and mistakenly added the enzyme trypsin to the growth medium. What would you observe in your culture flask. Why?

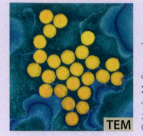
infected cells or the fusion of individual, infected cells into a large, multinucleated mass known as a **syncytium** (Figure 5.15). Researchers often use this CPE as a marker of the extent of viral replication. After a sufficient period of incubation, the medium, containing the released virions, can be harvested.

Techniques similar to those used to propagate viruses can be used to generate clonal populations of viruses. For bacteriophages, the phage is mixed with molten agar and this mixture then is added to host bacteria growing on a nutrient agar plate. For animal viruses, the virus is added to appropriate host cells growing in a small dish and then a nutrient agar is layered on top of the infected cells. In both cases, the viruses replicate within the host cells. As progeny viruses exit an infected cell, however, they are spatially constrained by the agar and only can infect adjoining cells. As a result, a localized area of cell death, a **plaque**, results (Figure 5.16).

Each plaque potentially contains thousands of viral particles. Because only a small number of virus particles initially were used to infect the cells, though, one can assume that each plaque arose from a single virion. Thus, each plaque contains a clonal population of virus. These clonal viruses can be easily isolated. A researcher can extract the plug of agar over a plaque, which contains many viral particles, with a pipette. The particles then can be isolated, providing the researcher with a clonal population of virions.

Viral purification

For many experimental protocols, it may be necessary to obtain purified preparations of a virus. Of course, different levels of purification can be achieved, depending on the experimental needs. Most purification strategies make use of the small size of viruses. For simple, relatively crude purifications, a viral suspension can be centrifuged and then filtered through a very small pore filter.

Virologists working with the National Zoo initially used this very simple technique to determine that a virus caused

virologists add a small amount of virus to appropriate host cells growing in a flask. As each virion infects a cell and replicates, new virions are released from the infected cell into the medium and infect other cells. The replication of many different animal viruses, like poliovirus (Microbes in Focus 5.2), can cause **cytopathic effects (CPE)**, visible changes in cellular morphology often associated with cell damage or death. These morphological changes may involve the rounding and detachment of the

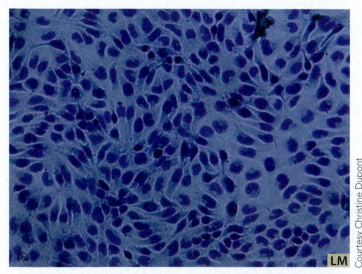

Courtesy Christine Dupont

LM

A. Uninfected fish fibroblast cells

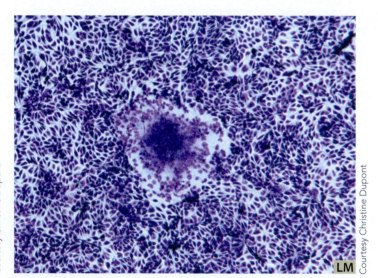

Courtesy Christine Dupont

LM

B. Fish fibroblast cells infected with a reovirus

Figure 5.15. Cytopathic effects (CPE) in virally infected cells When removed from a living tissue and placed in a flask with well-defined nutrient medium, animal cells can grow rapidly and may form a layer attached to the substrate. **A.** A layer of healthy uninfected cells grown in cell culture showing a uniform, elongated form. The nuclei are stained blue. **B.** Cells infected with virus show fused cells called syncytia (*middle*) containing numerous nuclei.

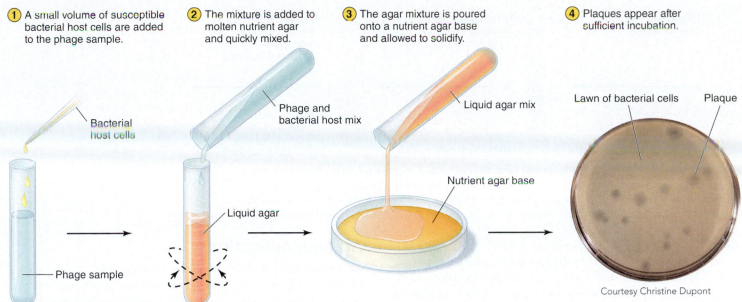

① A small volume of susceptible bacterial host cells are added to the phage sample.

② The mixture is added to molten nutrient agar and quickly mixed.

③ The agar mixture is poured onto a nutrient agar base and allowed to solidify.

④ Plaques appear after sufficient incubation.

Bacterial host cells

Phage and bacterial host mix

Liquid agar mix

Lawn of bacterial cells   Plaque

Liquid agar

Nutrient agar base

Phage sample

Courtesy Christine Dupont

Figure 5.16. Process Diagram: Producing a viral clone Susceptible bacterial cells are mixed with a phage sample. After phages have attached to the cells, a molten nutrient agar is added to the cells. This mixture then is poured onto a plate containing a nutrient agar base and allowed to harden. The plates are incubated, permitting viruses to replicate and infect neighboring cells. The layer of agar inhibits the free movement of released progeny viruses during incubation. Each virus that infected a cell produces a distinctive visible plaque of dead cells. Each discrete plaque contains the progeny of a single virus and thus is a clone. Such clones are essential to doing viral genetics.

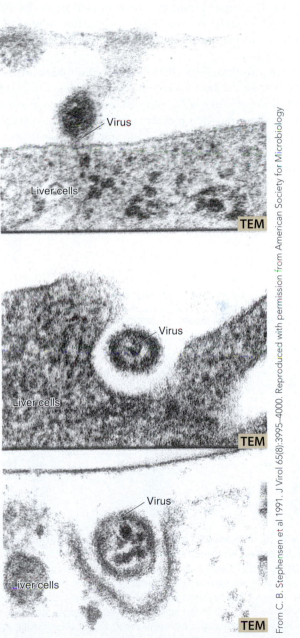

the acute hepatitis affecting the golden lion tamarins (GLTs). They disrupted liver sections from affected GLTs and filtered this cell suspension through a 0.2-micrometer filter. They then inoculated common marmosets, a species closely related to GLTs, and various cell lines with this bacteria-free filtrate. The marmosets developed an acute hepatitis. Additionally, marmoset and monkey cell lines produced noticeable virus particles when inoculated with this filtrate **(Figure 5.17)**. Thus, the researchers concluded that the infectious agent was a virus.

Filtration, as you can see, may separate the virions from eukaryal cells, large cellular debris, and bacteria, but does not concentrate the virions. The concentration of viruses generally requires some form of ultracentrifugation. In ultracentrifugation, centrifugal forces of up to one million times the force of gravity can be generated. These extreme forces cause particles even as small as viruses to move through the liquid, toward the bottom of the centrifuge tube. In a process known as **differential centrifugation**, virologists subject medium containing virions and cells to centrifugation at a relatively low speed,

Figure 5.17. Virus-like particles produced in cells inoculated with a bacteria-free liver extract Liver from an infected common marmoset, a species closely related to the golden lion tamarin, was homogenized and filtered through a 0.2-micrometer filter. Cells then were inoculated with this bacteria-free filtrate. Electron micrographs show virus particles associated with inoculated marmoset hepatocytes.

From C. B. Stephensen et al 1991. J Virol 65(8):3995–4000. Reproduced with permission from American Society for Microbiology

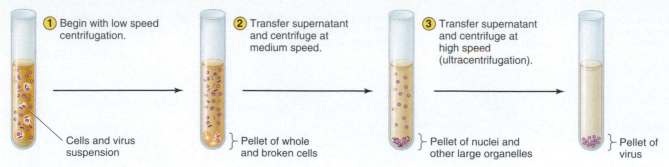

① Begin with low speed centrifugation.

Cells and virus suspension

Pellet of whole and broken cells

② Transfer supernatant and centrifuge at medium speed.

Pellet of nuclei and other large organelles

③ Transfer supernatant and centrifuge at high speed (ultracentrifugation).

Pellet of virus

Figure 5.18. Process Diagram: Differential centrifugation to separate particles based on mass
A low speed centrifugation results in the pelleting of large objects, including whole and damaged cells. The supernatant above the pellet can be re-centrifuged at a higher spin, resulting in the pelleting of smaller objects, such as cell organelles. With even higher centrifugation rates, researchers can pellet objects as small as viruses. Separated from other cellular components, this purified preparation of a virus can be used for experiments.

causing intact cells and large cellular debris to pellet, or collect at the bottom of the tube **(Figure 5.18)**. The supernatant can be harvested and subjected to ultracentrifugation, causing the virions to pellet. Once virions have formed a pellet in the tube, the supernatant can be removed from the tube and the particles can be resuspended in a smaller volume of liquid, thereby resulting in a purified and much more concentrated viral solution.

Viruses also can be purified based on their size and density. In **gradient centrifugation**, medium containing virus particles is added to a centrifuge tube containing a linear gradient of a salt or sugar solution (often sucrose) and subjected to ultracentrifugation. Particles present in the sample will travel through the gradient until they reach a point where the density of the gradient solvent impedes their downward movement. Even particles of very similar sizes and densities can be effectively separated with this technique. After a period of centrifugation, the tubes can be examined and the band of virions identified within the gradient **(Figure 5.19)**.

① The tube is successively filled with layers of decreasing concentrations of sucrose.

② The suspension containing virus is layered on top.

③ The preparation is centrifuged.

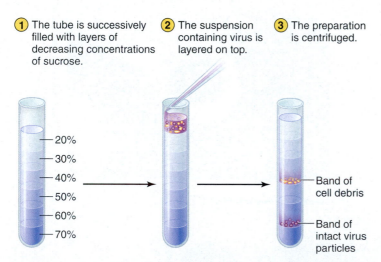

20%
30%
40%
50%
60%
70%

Band of cell debris

Band of intact virus particles

Figure 5.19. Process Diagram: Gradient centrifugation to separate particles based on density A density gradient consisting of sucrose concentrations varying from 20 to 70 percent is formed in a centrifuge tube. The sample is layered on top of this gradient. During centrifugation, the gradient is maintained, but suspended particles will move down the gradient until they encounter a sucrose concentration (density) that is equivalent to their particle density. Particles of different densities form distinct bands at different locations in the tube. The band of particles then can be removed from the centrifugation tube. In this fashion a purified preparation of particles is obtained for further investigations.

Viral quantification

Not surprisingly, it often is useful to know how much virus a specific sample contains. It is necessary for a researcher to know the viral **titer**, or concentration of a virus preparation, in order to accurately plan experiments. If a researcher wishes to repeat a given experiment, then he or she must know how much virus was used initially and be able to accurately measure the same amount of virus when repeating the protocol. For the clinician, knowing the amount of virus present in an infected person may affect the treatment options. When treating people with HIV, for example, clinicians routinely monitor the viral load, or amount of virus in the blood of the infected person **(Perspective 5.1)**. The viral load, it turns out, presents a fairly good indicator of the overall health of the individual.

At first glance, one might think that viral quantification should be very straightforward; simply count the number of viral particles in a given solution. But, of course, nothing ever is as simple as it first seems. Let's start by examining a few basic questions. First, how can we "see" viruses to count them? Second, are all viral particles infectious? In other words, if some viral particles do not infect anything, then should they be included in the count? Third, what do we mean by "infectious"? Should we only count viral particles capable of causing an infection in a human being? How could we possibly conduct such a count? As you can see, viral quantification is not that simple. In this section, we will explore some of the various techniques used to determine viral titers.

Direct Count

To determine the absolute number of total viral particles in a sample, a direct counting method can be utilized. In this procedure, a known quantity of microscopic markers, such as small latex beads, first is added to a sample of the purified virus being tested. This mixture then is placed onto a grid and viewed with an electron microscope (EM). When the sample is viewed, the viral particles and beads are counted.

By determining the number of beads on the grid, one can determine the exact volume of the sample applied to the grid. If 1,000 beads were initially added to 1,000 μL of liquid and 10 beads were present on the grid, then we can conclude that 10 μL of liquid was applied to the grid. And if a total of 37 viral particles also were present on the grid, then we can conclude that the viral concentration is 37 particles per 10 μL, or 3,700 particles/mL.

Determining the amount of virus in a patient, or the viral load, can be very important to the clinician. Today, infectious disease specialists routinely monitor the viral load of people infected with HIV. For the clinician, this information can be vital in determining the best course of treatment.

The human immunodeficiency virus, or HIV, is a retrovirus. Viruses in this family possess several defining characteristics, and their replication will be discussed in more detail in Section 8.3. For now, it is important to know that HIV contains an RNA genome and can be found in the blood of an infected person. To determine the amount of virus present in the bloodstream, researchers use a technique referred to as reverse transcriptase polymerase chain reaction, or RT-PCR (see Toolbox 5.2). Basically, a sample, usually plasma from an infected person, is obtained and treated to disrupt any viral particles, thereby releasing the viral RNA. The enzyme reverse transcriptase is used to convert the HIV RNA into DNA. The viral DNA then is amplified and the amount of the resulting PCR product is quantified. Usually, the results are presented as copies of HIV RNA per milliliter of blood. Currently, available tests can detect fewer than 50 copies per

milliliter (cpm). It follows, then, that these tests do have their limits. A person infected with HIV may have undetectable levels of the virus in his or her blood. This result does not mean that the person contains no HIV. Rather, it means that the person contains lower levels of the virus than can be detected.

Numerous studies have shown that the viral load of a person with HIV represents a good indicator of disease progression; higher viral loads are associated with a more rapid progression to AIDS. A report published in 1996 by John Mellors and colleagues at the University of Pittsburgh, for example, showed that people with a plasma viral load of 4,500 cpm have a typical progression time to AIDS of 10 years. For people with a viral load of over 36,300 cpm, though, the typical time until the development of AIDS is 3.5 years. Similarly, viral load correlates with transmission; higher viral loads correlate with a greater risk of transmission. This finding has been especially useful in decreasing the incidence of mother-to-child transmission of HIV. By providing pregnant women who are HIV+ with antiretroviral drugs, clinicians can decrease their viral load prior to delivery and decrease the likelihood of the infant becoming HIV+ during birth.

While this procedure seems very straightforward, it has two major drawbacks. First, it requires the use of a very expensive microscope. Second, it does not allow the researcher to differentiate between infectious and non-infectious viral particles—all particles get counted. After a direct count is made initially, though, it may be possible to estimate the number of viral particles in similar samples using simpler biochemical or spectrophotometric assays. For instance, after doing a direct count of a sample of purified virus, a researcher could measure the absorbance at a particular wavelength of the same sample. Potentially, one could calculate a conversion factor that converts the absorbance reading to a

direct count value. In the future, then, one simply could measure the absorbance of samples prepared in the same manner and use this information to calculate the virion concentration.

Hemagglutination Assay

Many viruses bind to the surface of erythrocytes, or red blood cells (RBCs) and cause the RBCs to clump together, a process referred to as **hemagglutination**. When a sufficient number of viral particles combine with a specific amount of RBCs, the virus:RBC interactions result in the agglutination, or binding together, of the red blood cells. This hemagglutination event easily can be seen with the naked eye. If too few viral particles are added to the RBCs, though, then hemagglutination does not occur. Thus, by adding serial dilutions of a viral preparation to constant numbers of RBCs, a researcher can determine the hemagglutination titer of the viral preparation—the maximum dilution of the viral preparation that still results in complete hemagglutination (**Figure 5.20**).

Again, this technique generally does not allow one to differentiate infectious and non-infectious particles. Additionally, a hemagglutination titer does not accurately reflect the total number of viral particles present in a sample; many viral particles usually need to be attached to each RBC to result in agglutination. Finally, for some viruses, hemagglutination may occur in the absence of intact viral particles; specific viral proteins alone may foster this clumping of red blood cells. Despite these drawbacks, though, it is a very simple and cheap assay that can provide useful information about virus concentration.

Plaque Assay

In most instances, it is more useful to determine the titer of infectious viral particles, rather than the total number of infectious

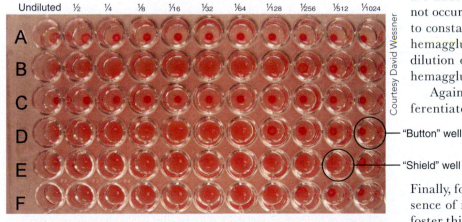

Dilution of virus sample

Undiluted ½ ¼ ⅛ ¹⁄₁₆ ¹⁄₃₂ ¹⁄₆₄ ¹⁄₁₂₈ ¹⁄₂₅₆ ¹⁄₅₁₂ ¹⁄₁₀₂₄

A B C D E F

Virus isolates

Courtesy David Wessner

"Button" well

"Shield" well

Figure 5.20. Using viral hemagglutination to determine a viral titer
Serial dilutions of the virus are mixed with a constant amount of red blood cells (RBCs) and added to wells of an assay plate. When high numbers of viral particles are present, the virions bind to the RBCs and form a diffuse mesh, or "shield." When fewer viral particles are present, the RBCs settle to the very bottom of the well, forming a "button." In this figure, each row contains serial twofold dilutions of reovirus isolates, added to wells containing bovine RBCs. The isolate in row B has an HA titer of 128 (greatest dilution still capable of forming a shield). A titer cannot be determined for the isolate in row C.

Virus dilutions: Undiluted $\frac{1}{10}$ $\frac{1}{100}$

Courtesy Christine Dupont

Figure 5.21. Plaque assay to determine viral titer By plating serial dilutions of a virus on susceptible host cells, one can determine the number of plaque-forming units (PFUs) in the original stock. In this assay, tenfold serial dilutions of a virus stock were added to each plate. As you can see, approximately tenfold fewer plaques are observed when the virus is diluted tenfold.

and non-infectious particles. The plaque assay commonly is used to determine the infectious titer. As discussed previously, a virologist can inoculate a series of dishes containing appropriate host cells with serial dilutions of the virus and then add a nutrient agar, or gel-like medium, to the cells to constrain the movement of progeny viruses. As a result, newly formed viruses only can infect neighboring cells and discrete areas of dead cells, or plaques, form. By counting the plaques, then, the virologist can determine the plaque-forming unit (PFU) titer of the original viral suspension **(Figure 5.21)**. This simple technique provides a good measurement of the infectious viruses. Of course, it does depend on the ability of the virus being studied to lyse available host cells.

A similar type of assay can be used to determine the infectious titer of plant viruses. Briefly, serial dilutions of a viral

© Nigel Cattlin/Photo Researchers, Inc.

Discrete areas of cell death

Figure 5.22. Using viral lesions on a leaf to determine titer of plant virus The plant leaf shows distinct lesions caused by ringspot virus. If a plant leaf is experimentally inoculated with virus, then these areas of localized cell death, analogous to plaques that form in a standard plaque assay, can be counted, thereby allowing one to determine the infectious titer of the virus.

suspension are applied to previously scarified, or scratched, leaves. The scarification serves to break the cell walls, thereby allowing the viruses to enter the leaf cells. After a sufficient period of incubation, the numbers of lesions present on the leaves are counted. If one assumes that each lesion arose initially from the infection of a single cell by a single virion, then the number of lesions can be used to determine the concentration of the infectious viral particles **(Figure 5.22)**.

A plaque assay of a particular viral sample may provide a titer that differs dramatically from a direct count of the same sample. For some viruses, the direct count may yield a value for the number of viral particles that is 100 or 1,000 times greater than the PFU titer. How can we interpret this apparent discrepancy? Let's look at what each assay actually measures. In a direct count, every particle that looks like a virus is counted. In a plaque assay, every virus that results in the death of host cells is counted. In other words, the direct count measures the total number of viral particles, while the plaque assay measures the number of *infectious* particles. So, let's reevaluate our apparent contradiction: if the direct count exceeds the PFU titer by a factor of 100, then we can assume that only one out of every 100 viral particles actually is infectious. For some reason, the other 99 particles are unable to infect cells. These defective viral particles may have been assembled incorrectly, are missing part of their genome, or contain mutations that preclude further replication. As we will see in Chapter 8, viruses, especially RNA viruses, have very high mutation rates, allowing them to evolve at a quick pace. A large number of defective particles may be the price that viruses pay for this genomic plasticity.

Endpoint Assays

In some instances, viruses do not form plaques in cultured cells and do not agglutinate red blood cells, but do have some observable effect on inoculated cells or animals. These viruses may be

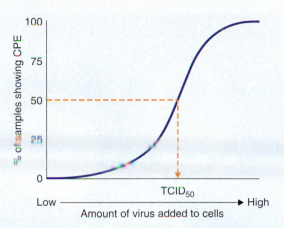

Figure 5.23. Endpoint assays determine tissue culture infectious dose 50 (TCID$_{50}$). In an endpoint assay, the effects of various dilutions of the virus are assayed for a specific outcome, such as cytopathic effects (CPE). To determine the tissue culture infectious dose 50 (TCID$_{50}$), for example, dilutions of virus are used to infect cells growing in culture. The cells then are monitored for CPE. By analyzing the data, one can determine how much virus is needed to cause CPE 50 percent of the time.

quantified by an endpoint assay. Serial dilutions of the virus are examined for their ability to cause the observable effect. A statistical analysis of the data then is used to estimate the amount of virus needed to cause the desired effect 50 percent of the time.

For instance, if a virus can cause the death of a certain type of animal, then it would be possible to determine the amount of virus that kills 50 percent of infected animals: the **lethal dose 50 (LD$_{50}$)** of that virus. Again, serial dilutions of the virus are

made. Susceptible animals are inoculated with each of the viral dilutions, the animals are monitored, and the number of animals that died after receiving each dose of virus is recorded. High amounts of virus, presumably, would result in high numbers of deaths, while low amounts of virus would result in few deaths. While this measurement may provide very valuable information about the virus being studied, it also requires the use of a large number of appropriate host animals. Because of concerns about the use of animals in the laboratory, this assay is not typically used. Alternatively, one can determine the **tissue culture infectious dose 50 (TCID$_{50}$)** of a virus. Serial dilutions of the virus would be used to inoculate cells and these cells then would be observed for cytopathic effects. Statistical analysis would be used to determine the amount of virus typically required to produce an infection 50 percent of the time **(Figure 5.23)**.

5.3 Fact Check

1. Describe how bacteriophages are cultivated in the lab.
2. How are animal viruses cultivated and harvested?
3. What is meant by the term "cytopathic effect"?
4. Explain how filtration and centrifugation can be used to purify viruses.
5. What techniques can be used to quantify viruses?
6. Differentiate between LD$_{50}$ and TCID$_{50}$.

5.4 Diversity of viruses

What are the different types of viruses?

Every student of biology is at least marginally familiar with the Linnaean classification scheme. We all know that birds and mosquitoes and humans can be categorized into specific phyla, classes, orders, families, genera, and species. Bacteria and archaeons are classified in a similar way. The classification of viruses, though, is more problematic for two main reasons. First, the evolutionary history and relatedness of viruses still is poorly understood. Second, no single genetic yardstick, comparable to the ribosomal RNA sequences of bacteria, archaeons, and eukarya, exists for establishing relationships between mammalian viruses or between mammalian viruses and the viruses of plants, insects, eukaryal microbes, bacteria, or archaeons. No single gene exists in all viruses.

Despite these obstacles, it still is useful, perhaps even necessary, for virologists to categorize viruses in some meaningful way. To explore this important topic of viral classification more thoroughly, we will examine how viruses are named and how viruses are classified.

Virus names

Historically, viruses have been named in a variety of ways. Many bacteriophages, such as T2 and λ, have been named with a combination of Roman or Greek letters and numbers.

Some mammalian viruses have been named after the location where they were first isolated. The Ebola virus was identified after an outbreak near the Ebola River in Zaire **(Figure 5.24)**. Other mammalian viruses, such as hepatitis A virus, have been named after the disease they cause. Most plant viruses, such

Figure 5.24. Ebola River, Zaire The first documented cases of Ebola infection occurred near the Ebola River. The virus, as a result, was named Ebola virus.

as tobacco mosaic virus (see Figure 1.9), have been named after the appearance of the infected plant. The names of other viruses are based on the shape of the virus. Coronaviruses, for instance, have characteristic crown-like or corona projections on their envelope (**Figure 5.25**). In some cases, the names convey useful information about the virus. In other cases, the names reveal an interesting, though not particularly important, factoid. In still other cases, the names are just names (**Table 5.2**).

For biologists, though, names ideally should reflect the relatedness of organisms. We know that all organisms within the genus *Drosophila*, for example, are fruit flies. Similarly, the virus that causes herpes cold sores has the species name *Human herpesvirus* 1 and is a member of the genus *Simplexvirus* and the family *Herpesviridae*. In this chapter, for simplicity, we will use primarily the family name and the species name when referring to viruses.

Virus classification

While a name may be nothing more than a name, a systematic means of classifying entities is very important to biologists. Again, viruses historically have been classified in a variety of ways. In some cases, the mode of spread of a virus (by insect, for example) has played a role in classification. In other cases, the disease caused by a virus (hepatitis, for example) has played a role in classification. These types of classification strategies may have some merit. For the clinician, it may be very logical to group together all the viruses that cause hepatitis, or damage to the liver. Ultimately, though, such a classification strategy does not have much scientific usefulness. Hepatitis A virus, hepatitis B virus, and hepatitis C virus, for example, share very few attributes other than causing hepatitis.

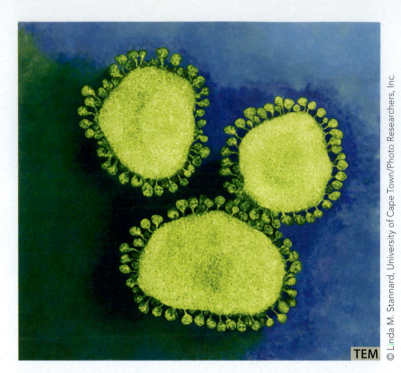

TEM

Figure 5.25. Crown-like or "corona" appearance of coronaviruses
The viral attachment proteins, or spikes, extending from the surface of coronaviruses give them a crown-like appearance.

The ICTV Classification Scheme

In 1971, the International Committee on Taxonomy of Viruses (ICTV), a committee of the International Union of Microbiological Societies, developed an official, comprehensive classification system for viruses. This system classifies viruses into the following groups: Order, Family, Sub-Family, Genus, and Species. By convention, the order name ends in "-virales"

TABLE 5.2	Many different strategies are used to name viruses	
Strategy	**Example**	**Notes**
Location	Ebola virus	The first recognized outbreak of Ebola occurred in Zaire, near the Ebola River. Most subsequent outbreaks have occurred in central Africa.
	West Nile virus	Initially isolated from a person living near the Nile River in Uganda, this virus first appeared in the United States in 1999 and now is firmly established in North America.
Disease	Tobacco mosaic virus	Plants infected with TMV show a distinct discoloration, or mosaic pattern. The disease was first described in the 1880s and still causes significant crop losses today.
	Hepatitis A virus	A number of viruses, including hepatitis A virus, hepatitis B virus, and hepatitis C virus, have been identified. These viruses all cause damage to cells of the liver but differ significantly from each other.
Physical Characteristic	*Coronaviridae*	Viruses in this family have projections on their surface that give them a crown-like or corona appearance.
	Picornaviridae	These viruses all have very small, or pico, RNA genomes.

and the family name ends in "-viridae" (*Picornaviridae* and *Herpesviridae*, for example). The genus and species names both end in "-virus." It should be noted, though, that the concept of species is poorly defined for viruses. Certainly, the standard definition of a species as a group of interbreeding individuals does not apply to viruses (or, for that matter, bacteria or archaeons). In fact, individual virus particles within a species may differ from each other. Many virologists now talk about viral swarms or quasi-species. All viruses within a quasi-species have highly similar sequences, but the sequences of individual genomes differ slightly from each other.

Because viruses most likely do not share a common ancestor, this viral taxonomy does not necessarily reflect evolutionary relatedness. Rather, each group represents a collection of viruses that possess a number of shared attributes. More specifically, some of the attributes considered by the ICTV are:

- Virion morphology
  - Size
  - Capsid symmetry
  - Presence or absence of an envelope
- Genome structure
  - Double-stranded DNA
  - Single-stranded DNA
  - Double-stranded RNA
  - Single-stranded RNA
- Biological features
  - Replication strategy
  - Host range
  - Pathogenicity

In its 2008 report, the ICTV recognized five orders, containing a total of 20 families. Another 64 families have not been assigned to a specific order **(Table 5.3)**.

TABLE 5.3 Orders and selected examples of viruses in ICTV taxonomy[a]

Order	Selected families	Selected species
Caudovirales	*Myoviridae*	*Enterobacteria phage T4*
	Siphoviridae	*Enterobacteria phage* λ
Herpesvirales	*Herpesviridae*	*Human herpesvirus 1*
	Alloherpesviridae	*Ictalurid herpesvirus 1*
Mononegavirales	*Filoviridae*	*Ebola virus*
	Paramyxoviridae	*Measles virus*
Nidovirales	*Coronaviridae*	*Murine hepatitis virus*
	Arteriviridae	*Equine arteritis virus*
Picornavirales	*Picornaviridae*	*Human rhinovirus A*
	Comoviridae	*Tomato ringspot virus*
Unclassified	*Caulimoviridae*	*Cauliflower mosaic virus*
	Bunyaviridae	*Hantaan virus*

[a]The complete ICTV taxonomy can be accessed at
http://www.ICTVonline.org/virusTaxonomy.asp?version=2008

TABLE 5.4 Baltimore classification scheme

Class	Description	Example
I	dsDNA genome	Human herpesvirus
II	ssDNA genome	Parvoviruses
III	dsRNA genome	Reoviruses
IV	ssRNA genome, positive sense	Poliovirus
V	ssRNA genome, negative sense	Influenza viruses
VI	ssRNA genome, DNA intermediate	HIV
VII	dsDNA genome, RNA intermediate	Hepatitis B virus

The Baltimore Classification Scheme

While the ICTV system represents the official viral taxonomy, many virologists categorize viruses based on a classification scheme developed primarily by Nobel laureate David Baltimore. As we noted earlier in this chapter, all viruses rely on the host cell machinery for translation. All viruses, then, must generate mRNA that can be recognized by the ribosomes of their host cells. Whether the virus has a DNA or RNA genome, or infects bacteria, archaeons, or eukarya, the virus must generate usable mRNA molecules. Thus, the generation of mRNA can be thought of as a common feature by which we can compare all viruses. Using this single defining feature, Baltimore has classified all viruses into seven groups **(Table 5.4)**. We will explore this method of categorizing viruses more extensively in Section 8.3, when we examine methods of viral replication.

Virus identification

Today, researchers employ a number of biochemical and molecular tests to more clearly identify viral isolates. In this section, we will examine two specific techniques: electron microscopy and nucleic acid analyses. Certainly, nucleic acid analyses provide important information about the identity of a virus, but we will start with the more straightforward approach: What does the virus look like?

Electron Microscopy

As we discussed earlier in this chapter, viruses differ a great deal physically. Poliovirus particles, on the one hand, are non-enveloped icosahedrons approximately 30 nm in diameter. In contrast, variola viruses, which cause smallpox, are enveloped, brick-shaped, and over 200 nm long. Clearly, these two viruses look different. By viewing virus preparations with an electron microscope, we can see these physical differences **(Figure 5.26)**. It shouldn't come as any surprise, then, that poliovirus virions and smallpox virus virions easily can be differentiated by electron microscopy.

Electron microscopy is not just useful for differentiating vastly different viruses. Many viral groups have distinctive shapes. As a result, electron microscopy can be a fairly useful tool in identifying viruses. Not surprisingly, though, electron microscopy also can be fairly imprecise. Many viruses "look" similar, at least

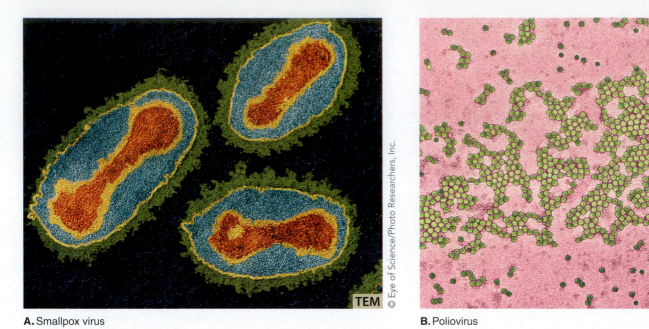

A. Smallpox virus

© Eye of Science/Photo Researchers, Inc.

B. Poliovirus

© NIBSC/Photo Researchers, Inc.

Figure 5.26. Using electron microscopy to differentiate viruses While identification of a virus by EM is not always possible, researchers certainly can use EM to differentiate a large, enveloped virus from a small, non-enveloped virus. **A.** Smallpox virus, an enveloped virus nearly 220 nm in diameter. **B.** Poliovirus, a non-enveloped virus approximately 30 nm in diameter.

superficially or to the untrained eye. Viruses in both the *Paramyxoviridae* and *Orthomyxoviridae* families, for example, are enveloped, spherical, and about 100 to 150 nm in diameter **(Figure 5.27)**. Paramyxoviruses, though, enter the cell by fusing with the cell membrane and contain a linear single-stranded RNA genome of 16,000 to 20,000 nucleotides. Orthomyxoviruses enter the cell by endocytosis, followed by fusion between the viral envelope and the endosomal membrane. Additionally, orthomyxoviruses possess a segmented genome. These viruses contain eight segments of single-stranded RNA, rather than a single RNA molecule. Clearly, the electron microscope cannot be used to observe these important biological features of the viruses.

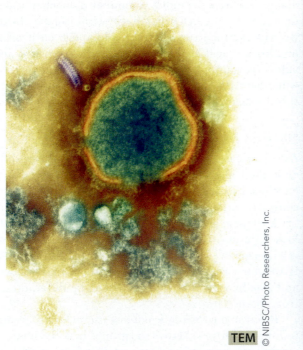

A. Measles virus

© NIBSC/Photo Researchers, Inc.

B. Human influenza virus

© NIBSC/Photo Researchers, Inc.

Figure 5.27. Using electron microscopy to identify viruses Similarly shaped viruses may be difficult to distinguish by electron microscopy. **A.** Measles virus is an enveloped virus approximately 100 to 150 nm in diameter in the *Paramyxoviridae* family. **B.** Human influenza virus is an enveloped virus approximately 100 to 150 nm in diameter in the *Orthomyxoviridae* family.

The severe acute respiratory syndrome, or SARS, outbreak provides an excellent case study in viral identification and demonstrates the advantages and disadvantages of electron microscopy. In the winter and spring of 2003, the world quickly became aware of a threatening new infectious disease. In February 2003, people in Guangdong province of China showed signs of a severe respiratory disease. By the end of March, people with similar symptoms were identified in countries throughout the world, including Vietnam, Hong Kong, Thailand, and Canada. All of these people had SARS.

CONNECTIONS We now know that the SARS virus is a coronavirus. As we will see in **Section 22.4**, researchers hypothesize that strains of this virus that infect bats recombined, giving rise to the human pathogen.

To identify the causative agent of SARS, clinicians isolated samples from the lungs of people with the syndrome and sent them to electron microscopists. The electron micrographs, images obtained by the electron microscopists, clearly revealed virus-like particles in bronchial washes. These particles appeared to be enveloped virions approximately 100 nm in diameter. Based on these initial electron micrographs, some virologists hypothesized that the causative agent of SARS was a paramyxovirus. Further studies, however, showed that this supposition was incorrect—a coronavirus causes SARS. The initial electron micrographs were not specific enough to allow researchers to distinguish between these similar viruses. More precise tests were needed to confirm the identity of the SARS virus.

CONNECTIONS As we have seen in this section, electron microscopy can be a very useful tool for visualizing viruses. More information about electron microscopy, and other types of microscopy, can be found in **Appendix B: Microscopy**.

Nucleic Acid Analysis

As we have seen previously, sequence analyses can provide us with a wealth of information about bacteria, archaeons, and eukaryal microbes. The same holds true for viruses. Physical appearances can be misleading because many viruses have similar shapes. Indeed, various assays may not differentiate closely related viruses within the same virus family. Nucleic acid analyses, though, generally are very specific and provide more exact information.

The polymerase chain reaction, or PCR, is one type of nucleic acid analysis that is very useful in identifying viruses (see Section 1.1 for a review of PCR). By using primers specific for a given virus, researchers can determine if that virus's nucleic acid is present in a tissue sample. Let's return to the SARS outbreak. In May 2003, researchers reported in the *New England Journal of Medicine* that the SARS virus was a novel coronavirus. As part of their analysis, they isolated RNA from cells infected with the SARS virus. Using primers designed to hybridize to sequences conserved among all known coronaviruses, they used **reverse transcriptase (RT-PCR) (Toolbox 5.2)** to try to amplify

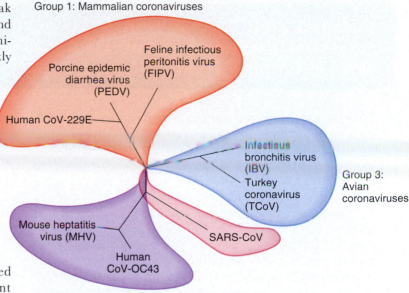

Figure 5.28. Genome sequence analysis of SARS coronavirus As shown in this relatedness diagram based on sequence comparison of the gene encoding the replicase enzyme, the genetic sequence of the SARS genome (SARS CoV) is similar to, but distinct from, the other known groups of coronaviruses.

coronavirus-specific RNA sequences. They succeeded, thereby demonstrating that the SARS virus was a coronavirus.

CONNECTIONS Researchers commonly use a reverse transcriptase enzyme to make DNA complementary to a molecule of RNA. As we will see in **Section 10.2**, the use of reverse transcriptase often is used in microarray studies.

By using RT-PCR, researchers determined conclusively that a coronavirus caused SARS. A more pressing question still remained: what coronavirus was it? To address this question, researchers sequenced the amplified viral cDNA and used sequence comparison software programs to compare the sequence of this unidentified virus to the sequences of other, known coronaviruses. Interestingly, the SARS coronavirus was related to, but definitely distinct from, all other known coronaviruses **(Figure 5.28)**. These sequence comparisons led researchers to conclude that SARS was caused by a newly identified coronavirus.

5.4 Fact Check

1. Describe the strategies used to name viruses.
2. Identify the categories used by the ICTV classification scheme.
3. What is a viral swarm or quasi-species?
4. What viral attributes are considered by the ICTV when classifying viruses?
5. How does the Baltimore classification scheme categorize viruses?
6. Describe how electron microscopy and nucleic acid analysis are each used to identify viruses.

As we discussed in Chapter 1, the polymerase chain reaction, or PCR, is a powerful tool for the molecular biologist. By combining a small amount of DNA with thermostable DNA polymerase, nucleotides, and specific primers, we can quickly amplify a specific region of the DNA. Since its development in 1983, PCR has become one of the most widely used molecular biology techniques.

Often, though, a researcher may need to amplify a piece of RNA, rather than DNA. Generally, it is desirable to convert the RNA of interest into DNA. A DNA copy of the RNA is advantageous for two main reasons. First, DNA can be used in further applications such as sequencing or cloning. Second, DNA is more stable than RNA. Obviously, if we can convert our RNA of interest into DNA, then PCR can be used to amplify this newly generated DNA. One important problem remains to be addressed. How can we convert the RNA into DNA?

As happens so often in research, nature already has solved this problem. In 1970, David Baltimore and Howard Temin independently reported the identification and characterization of reverse transcriptase, an enzyme that converts single-stranded RNA into double-stranded DNA (see Mini-Paper in Chapter 8). In retroviruses like HIV, this enzyme converts the single-stranded RNA genome of the virus into double-stranded DNA, which then becomes integrated into the host cell's genome. Researchers can use this enzyme to convert any RNA molecule into complementary DNA, or cDNA. Once cDNA has been generated, then a normal PCR procedure can be used to amplify it.

Like other DNA polymerases, reverse transcriptase cannot simply add nucleotides onto a piece of single-stranded nucleic acid. Rather, it requires a short oligonucleotide primer to which it can add new nucleotides in a 5' to 3' fashion. If a specific region of RNA needs to be amplified, then a specific primer, complementary to a short region of the RNA, can be used for the reverse transcription. Alternatively, researchers often use a series of random hexamer primers—six nucleotide-long primers of random sequences. These random primers will anneal to the RNA at various locations, thereby allowing the reverse transcriptase to convert all of the RNA into cDNA **(Figure B5.2)**.

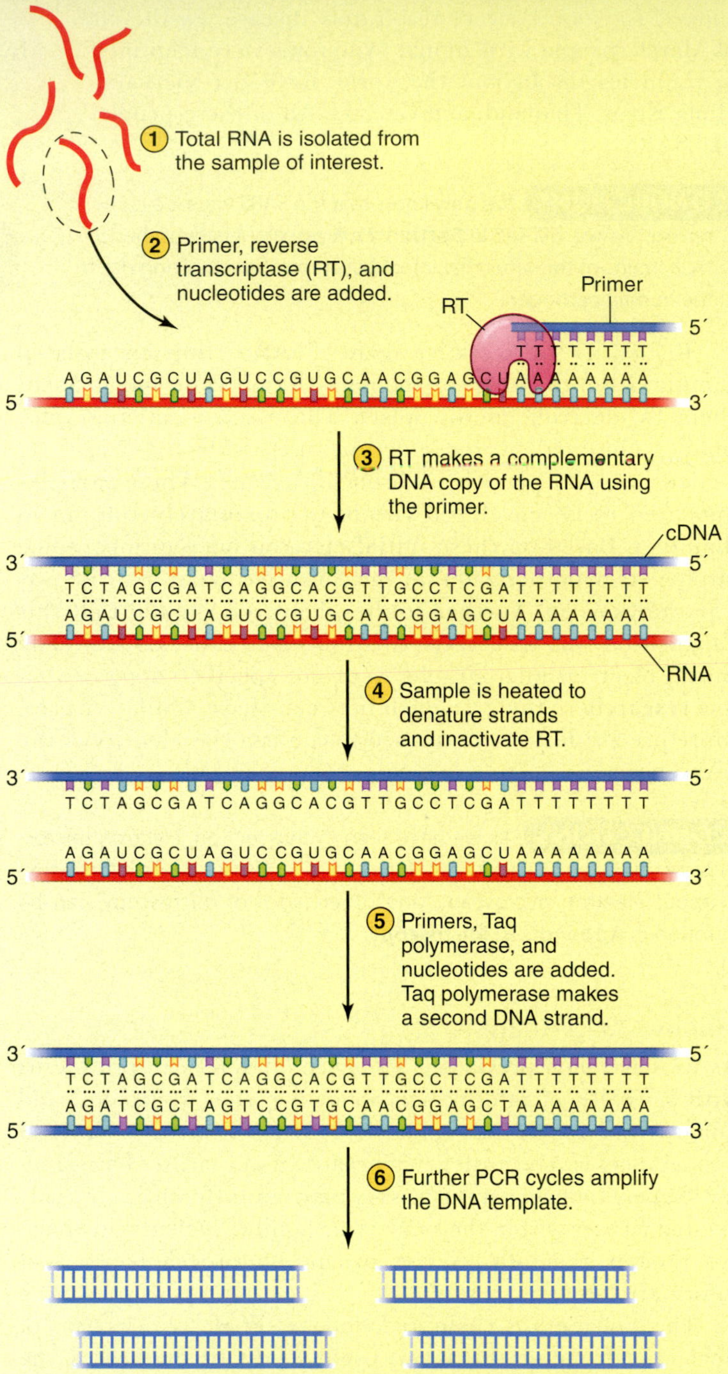

● **Test Your Understanding** ....

You are troubleshooting an RT-PCR reaction that did not work and did not produce a DNA product. You realize that you forgot to add one of the reagents: the oligonucleotide primer. Why would this mistake result in a lack of DNA product?

① Total RNA is isolated from the sample of interest.

② Primer, reverse transcriptase (RT), and nucleotides are added.

③ RT makes a complementary DNA copy of the RNA using the primer.

④ Sample is heated to denature strands and inactivate RT.

⑤ Primers, Taq polymerase, and nucleotides are added. Taq polymerase makes a second DNA strand.

⑥ Further PCR cycles amplify the DNA template.

Figure B5.2. Process Diagram: Reverse transcriptase PCR Using RT-PCR, DNA can be formed from RNA and amplified. First, reverse transcriptase is used to generate complementary DNA (cDNA) from the RNA. Random primers can be used to form cDNA from all RNA or a specific primer can be used to generate cDNA from a particular RNA species. Following the denaturation of the RNA:DNA molecule and inactivation of the RT, a thermostable DNA polymerase like Taq and specific primers can be used to amplify the DNA fragment of interest.

5.5 Virus-like particles

Are viruses the simplest pathogens?

Viruses represent a very diverse set of sub-cellular particles. While many different types of viruses exist, they all share some common features. A number of other sub-cellular infectious agents, which do not share all of these characteristics, also have been identified. Quite possibly, more will be discovered.

Viroids

In 1967, researchers isolated an infectious agent that causes potatoes to grow abnormally (**Figure 5.29**). Biochemical tests revealed that this sub-cellular agent, now known as potato spindle tuber viroid (PSTVd), differed significantly from all known viruses. Today, a number of **viroids** that infect plants have been isolated and characterized. All of them share several characteristics and differ from viruses in significant ways. Viroids:

- Consist only of naked RNA
- Are extremely small (less than 400 nucleotides)
- Exhibit a great deal of internal complementarity
- Exhibit increased resistance to ribonucleases, enzymes that degrade RNA

In other words, certain regions of the genome could form base-pair interactions with other regions of the genome. A string of Gs in the single-stranded molecule, for example, could interact with a string of Cs in the same molecule, resulting in a short region of double-stranded RNA. As a result of this intramolecular base pairing, the molecule exhibits a complex structure (**Figure 5.30**). This extensive secondary structure probably accounts for the resistance to ribonucleases.

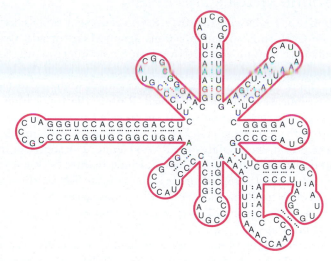

Figure 5.30. Structure of viroid RNA The RNA genomes of viroids are circular and single-stranded, but intramolecular base-pairing results in double-stranded regions.

The replication cycle, pathogenesis, and transmission of viroids have been areas of intense research. It appears that viroid genomes do not contain any genes and, consequently, viroids do not produce any proteins. So, what enzymes are involved in the replication of the viroid genome? Several studies, using inhibitors of various cellular polymerases, indicate that one or more of the host cell RNA polymerases may be involved. The pathology associated with these agents may result from this utilization of the host RNA polymerases. It is postulated that replication of the viroids may divert essential resources away from cellular transcription, thereby adversely affecting the health of the plant.

While this logical, though rather mundane, explanation of viroid pathogenesis may, indeed, be correct, another interesting explanation has been proposed. The genomes of several viroids, including PSTVd, contain a region of nucleotides that are complementary to a sequence found in signal recognition particle (SRP) RNA. The eukaryal SRP consists of several polypeptides and an RNA molecule, referred to as the 7S RNA. This particle directs nascent polypeptides in the cytoplasm into the endoplasmic reticulum. Possibly, then, the viroid genome can bind to the 7S RNA, interfere with the formation of the SRP, and disrupt the SRP-directed localization of proteins in the infected cell.

CONNECTIONS Following translation, polypeptides may be transported to various areas within the cell, inserted into the plasma membrane, or secreted from the cell. In **Section 3.1**, we investigated the role of signaling sequences in this cell-sorting process.

Finally, transmission of viroids from one plant to another often occurs through insects and may be facilitated by human activities, such as farming. When an infected plant is pruned, for example, the pruning shears may become contaminated

© Nigel Cattlin/Photo Researchers, Inc.

Figure 5.29. Infectious, naked RNA and potato disease Potato spindle tuber viroid (PSTV) contains an RNA genome of less than 400 nucleotides. The genome is not associated with any protein and appears not to code for any protein, but this sub-viral infectious agent can cause severe pathology. PSTV has caused this potato to be cracked and misshapen.

with viroids. When those same shears are used to prune another plant, the integrity of the second plant will be damaged, thereby providing a gateway for the viroids.

Satellite viruses and RNAs

Satellite viruses and **satellite RNAs** represent other classes of non-viral, sub-cellular infectious agents. Like viroids, these agents both contain small RNA genomes and infect plants. Satellite viruses and RNAs, though, differ from viroids in two basic ways. First, satellite viruses and RNAs cannot undergo autonomous replication within a host cell. Rather, the replication of these agents depends on functions provided by a helper virus that coinfects the host cell. Second, these agents contain a protein coat that encapsulates the RNA genome. The source of this protein provides the major distinguishing feature between satellite RNAs and satellite viruses. For satellite RNAs, this coat consists of a protein encoded by the helper virus. Satellite viruses, on the other hand, possess a gene that encodes their own capsid protein.

A human infectious agent very similar to satellite viruses has been identified. Hepatitis delta virus (HDV) contains a single-stranded RNA genome approximately 1,700 nucleotides in length that, like viroids, contains many regions of intrastrand complementarity. HDV only can replicate and form infectious progeny in a cell that is coinfected with hepatitis B virus (HBV). Infectious HDV particles are enveloped and the envelope contains proteins derived from hepatitis B virus. The HDV genome, though, is complexed with the large form of hepatitis delta antigen, a protein encoded by a gene in the HDV genome. Clinical studies suggest that coinfection with HBV and HDV results in more severe disease than is seen with HBV infection alone **(Figure 5.31)**.

Prions

While satellite viruses and satellite RNAs may seem a bit unusual, **prions** probably represent the most unusual infectious agents identified to date. Prions, or *proteinaceous infectious particles*, appear to contain no DNA or RNA! Despite the absence of a genome, these agents can replicate, in a manner of speaking, in the host, causing disease. In fact, researchers now recognize prions as the causative agents of several diseases, including kuru and Creutzfeldt-Jakob disease (CJD) in humans, scrapie in sheep, and bovine spongiform encephalopathy (also referred to as mad cow disease) in cattle **(Figure 5.32)**. Collectively, these diseases are referred to as **transmissible spongiform encephalopathies (TSEs)**, progressive neurological diseases characterized by impaired mental functions and sponge-like holes in the brain. In humans, an inherited form of CJD has been observed. An infectious form, vCJD, also has been observed.

During the initial studies of these diseases, most researchers assumed that the causative agents were viruses. Indeed, these diseases *seemed* to be caused by viruses. Studies clearly showed that scrapie could be transmitted from animal to animal, indicating that it was caused by an infectious agent. Moreover, brain homogenates passed through a 0.2-micrometer filter still were infectious, indicating that the responsible agent probably was not a bacterium. Experiments by Stanley Prusiner at the University of California at San Francisco and others demonstrated that treatments that destroy nucleic acids did not alter the infectivity of the agent. Treatments with proteases, like trypsin, or protein-denaturing chemicals, like phenol, however, did inactivate the agent. Prusiner concluded that the agent must contain protein but not nucleic acid. In 1997, he received the Nobel Prize for the characterization of this truly unusual pathogen.

But how can an agent that does not contain DNA or RNA replicate? It appears that prions are misshapen forms of naturally

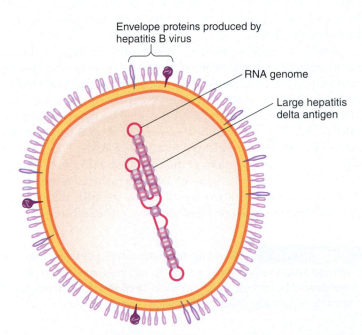

Figure 5.31. Hepatitis delta virus Hepatitis D virus (HDV) is an enveloped virus-like agent, with a circular, single-stranded RNA genome with extensive base-pairing. The genome is associated with the large form of the hepatitis delta antigen, encoded by HDV. HDV only can replicate in cells that are also infected with hepatitis B virus (HBV). HBV provides the proteins necessary for the HDV envelope.

Envelope proteins produced by hepatitis B virus

RNA genome

Large hepatitis delta antigen

© C.E.V./Photo Researchers, Inc.

Figure 5.32. Proteinaceous infectious particles (prions) and transmissible spongiform encephalopathy (TSE) Signs of mad cow disease include various neurological problems, aggressive behavior, and death. It is related to scrapies in sheep and variant Creutzfeld-Jakob disease (vCJD) in humans. Evidence suggests that vCJD can occur as a result of eating beef from affected cattle.

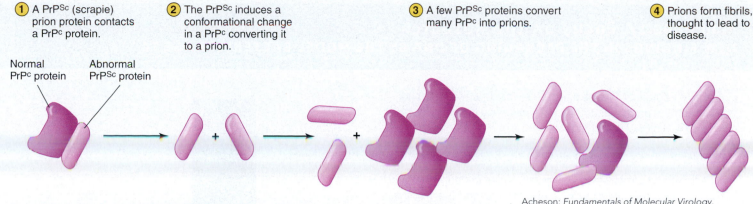

① A PrP^Sc (scrapie) prion protein contacts a PrP^c protein.

② The PrP^Sc induces a conformational change in a PrP^c converting it to a prion.

③ A few PrP^Sc proteins convert many PrP^c into prions.

④ Prions form fibrils, thought to lead to disease.

Normal PrP^c protein   Abnormal PrP^Sc protein

Figure 5.33. Process Diagram: Model of prion replication When pathogenic PrP^Sc (scrapie) protein molecules come in contact with PrP^c (normal) molecules, they cause the PrP^c molecules to change shape, becoming PrP^Sc molecules. In this fashion, the prion protein, devoid of any nucleic acid, can replicate and function as an infectious agent.

Acheson: *Fundamentals of Molecular Virology*, copyright 2011, John Wiley & Sons, Inc. This material is reproduced with permission of John Wiley & Sons, Inc.

occurring proteins. This protein normally exists on the surface of neurons and can exist in one of two different conformations, the more common PrP^c (cellular) form and the pathogenic PrP^Sc (scrapie) form. A molecule of PrP^Sc can facilitate the conversion of PrP^c molecules to PrP^Sc. Thus, PrP^Sc replicates (**Figure 5.33**).

Obviously, too, PrP^Sc can be classified as an infectious agent. If PrP^Sc is transferred to a susceptible individual, then PrP^c molecules within that individual will be converted to PrP^Sc. When enough PrP^c molecules have assumed the PrP^Sc conformation, the encephalopathy associated with prion diseases occurs. Certain mutations in the *PrP* gene make the protein more likely to assume the PrP^Sc conformation. Thus, some people develop Creutzfeldt-Jakob disease, for instance, not because they were infected by PrP^Sc molecules, but because they

inherited a mutant allele of the *PrP* gene. As a result, prion-associated diseases can be both infectious and hereditary!

5.5 Fact Check

1. How are viroids different from viruses?
2. What are the characteristics of satellite viruses and satellite RNAs?
3. What are the key features of prions?
4. What diseases are known to be caused by prions? Describe.

5.6 Virology today

What's next for virology?

The field of virology has progressed greatly during its relatively brief history. The impact of this explosion of knowledge has affected many disparate areas, including genetics, molecular biology, and medicine. These advances continue to occur at a dizzying pace. Today, virology remains an extremely exciting and dynamic field.

Because viruses are obligate intracellular parasites and, as a result, use various components of the host cell's transcriptional and translational machinery, they represent wonderful tools for studying the inner workings of our own cells. Indeed, since the advent of molecular biology in the 1950s, viruses have been used extensively to further our understanding of DNA replication and gene regulation. Some of our first detailed glimpses into gene regulation, in fact, came from studies of simian virus 40 (SV40) and the attendant discovery of enhancers, genetic elements that increase the expression of genes. Additional studies with SV40 and other viruses led to a better understanding of promoters, transcription factors, RNA splicing, and translation.

Indeed, studies with bacteriophages showed that DNA, not proteins, was the genetic material. Clearly, our understanding of genetics owes a lot to studies of these simple particles.

While molecular studies with viruses have contributed greatly to our understanding of how our cells work, viruses also have proved essential in developing the molecular techniques used to address fundamental questions about cell biology. Indeed, many of the most influential molecular biology experiments initially were conducted with viruses. Studies of viral replication, for instance, led to the discovery by David Baltimore and Howard Temin of reverse transcriptase, the enzyme utilized by retroviruses to convert RNA into DNA (see Mini-Paper in Chapter 8). Today, this enzyme is widely used in molecular biology, most notably for converting messenger RNA into complementary DNA (cDNA). Because mRNA is unstable and easily degraded, its conversion to cDNA makes experimentation with mRNA sequences more feasible.

Studies with viruses also have led to profound advances in various medical fields. Our understanding of cancer owes

Mini-Paper: A Focus on the Research
NEW FINDINGS IN THE PACKAGING OF DNA BY THE MODEL BACTERIOPHAGE T4

Z. Zhang, V. I. Kottadiel, R. Vafabakhsh, L. Dai, Y. R. Chemla, T. Ha, and V. B. Rao. 2011. A promiscuous DNA-packaging machine from bacteriophage T4. PLoS Biol 9(2):e1000592.

Context

The T4 bacteriophage is widely distributed and abundant in nature. It also has been the subject of extensive genetic and biochemical studies. Indeed, much of our understanding of bacteriophage capsid assembly results from studies of T4. The capsid of T4 includes a head that packages the 170-kb dsDNA genome, and a contractile neck and tail that is used to inject the DNA into the bacterial host cell.

A key event in T4 assembly involves the packaging of the DNA into the head, a process carried out by a molecular "motor" called the terminase. This motor consists of two proteins, gp17 and gp16, and its activity is fueled by the hydrolysis of ATP. When T4 DNA replication occurs, a concatamer, or string of complete genome copies arranged end to end, is produced. This long molecule is stuffed into the bacteriophage head until it becomes full, at which time the terminase cleaves the DNA and then falls off the developing structure. The neck and tail proteins then attach to the filled head to complete the assembly of the infectious virion in a highly ordered process. The gp17 protein is responsible for the ATPase and nuclease activities of the terminase, while gp16 regulates the activity of gp17. Because the sequence of events in virion assembly appears to be strictly controlled, researchers thought that the terminase only would introduce DNA into empty heads. In this study, they tested this hypothesis.

Experiments

To determine how DNA enters bacteriophage heads, the researchers examined T4 mutants that could not assemble the neck and tail. Cells infected by these mutants produced bacteriophage heads containing DNA. Researchers observed that two types of heads could be isolated by differential centrifugation—heads that contained genome-length DNA (termed "full" heads by the investigators) and heads that contained approximately 8 kbp of DNA (termed "partial" heads) (Figure B5.3).

If these heads have already packaged DNA in the past, then would it be possible for them to package DNA again? To answer this question, the investigators used both the full and partial heads in an

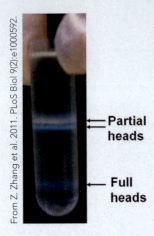

From Z. Zhang et al. 2011. PLoS Biol 9(2):e1000592.

Figure B5.3. Identification of full and partial T4 heads Following infection with T4 mutants unable to produce necks and tails, *E. coli* cells were lysed to release bacteriophage heads. CsCl density gradient centrifugation was used to isolate these structures. Low density "partial" heads and high density "full" heads were observed.

in vitro DNA packaging assay. Empty heads that had never packaged DNA were used as a control. Surprisingly, both the full and the partial heads packaged short DNA fragments of 50 to 766 bp in length, as shown by gel electrophoresis analysis (Figure B5.4).

To ensure that they were detecting DNA that had been packaged, the researchers treated the T4 heads first with DNase I and then with proteinase K. Why? DNase I digests exposed DNA. Treatment with this enzyme, then, would degrade any non-packaged DNA. Proteinase K digests proteins. Treatment with this enzyme, then, would degrade the previously added DNase I and degrade the proteinaceous bacteriophage head, freeing any encapsulated DNA. This combination of enzymes, in other words, ensured the researchers that the detected bands represented DNA that had been packaged.

When a similar packaging assay was done with 48.5-kb phage λ DNA, the partial heads successfully incorporated this DNA, but the full heads did not (Figure B5.5). Most likely, the full heads contained limited free space and could not accommodate an additional DNA molecule of this size.

---

a great deal to virology. Studies by Harold Varmus and J. Michael Bishop in the 1970s showed that Rous sarcoma virus, a retrovirus, caused cancer in experimental animals because the virus contained a mutated form of an animal gene. This discovery led to their description of **proto-oncogenes**, genes involved in the normal regulation of the cell cycle, and **oncogenes**, altered forms of proto-oncogenes that can lead to uncontrolled cell growth and the development of tumors. Today, a host of proto-oncogenes and oncogenes have been characterized, many through the study of retroviruses. The

characterization of these genes is revolutionizing our approaches to the treatment of cancer.

CONNECTIONS Viruses cause cancer? In some cases, yes. Human papillomavirus infection represents the major cause of cervical cancer. The mechanism of this process is explored more fully in **Section 22.3**.

Not only have viruses been utilized to further our understanding of cancer, but also they are being investigated as anti-cancer agents. **Oncolytic viruses** infect and kill cancerous

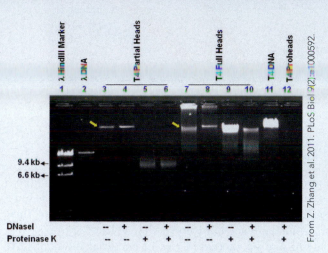

Figure B5.4. Full and partial heads package additional DNA. Following *in vitro* DNA packaging reactions, T4 heads were treated first with DNase I and then with proteinase K to release the encapsulated DNA. Agarose gel electrophoresis analysis demonstrated that both full and partial heads efficiently package a mixture of short fragments of DNA.

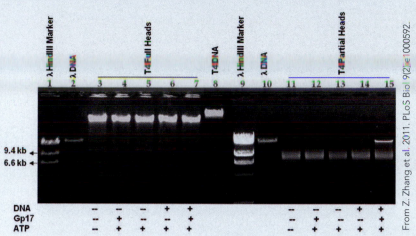

Figure B5.5. Packaging of λ DNA In a separate *in vitro* DNA packaging assay, researchers exposed full and partial T4 heads to 48.5 kbp λ DNA. Gel electrophoresis analysis demonstrates that partial heads can package these larger DNA fragments, but the full heads cannot.

Impact

The results of these experiments challenge the central idea that virus assembly occurs sequentially and irreversibly. Virologists long have assumed that virion assembly occurs in a highly ordered fashion; each step depends on the successful completion of the immediately preceding step, and each step is irreversible. After all, the entire assembly of the virion is a very intricate process that occurs rapidly. It stands to reason that each step must be completed before the next step begins. Virologists had assumed, then, that once a bacteriophage T4 head had been filled with DNA and the terminase was released, then the developing capsid would not be able to reassociate with terminase and receive more DNA.

The results reported in this article demonstrate that the interaction of a full head with the terminase is not inhibited. Because the terminase appears not to discriminate between empty and full heads, the authors refer to this type of interaction as "promiscuous." We now need to re-examine certain assumptions about the assembly of virions. We also need to investigate the possibility of using full or partial heads for *in vitro* packaging reactions. These bacteriophage components could be used to incorporate various types of foreign DNA of potential use for different biomedical applications.

● Questions for Discussion ....

1. This study used differential centrifugation to separate full bacteriophage heads from partial bacteriophage heads. How does this separation method occur?

2. The small DNA fragments that were used in the *in vitro* packaging assay did not have any special qualities. Based on this point, do you think there are any specific DNA sequences that are required for the DNA to be packaged?

cells without harming normal cells. Recent studies with reovirus have been especially noteworthy. These non-enveloped, icosahedral, double-stranded RNA viruses preferentially infect cells in which the Ras signaling pathway has been activated. Intracellular Ras proteins often become activated in response to an extracellular signal, such as a growth hormone binding to a membrane-bound receptor. Once activated, Ras proteins normally interact with a series of intracellular proteins, eventually leading to increased gene expression and, often, increased cell proliferation. Abnormal activation of the Ras pathway has been observed in a large percentage of human cancers. Because reovirus preferentially infects cells with an activated Ras pathway, cancer cells should be susceptible to reovirus infection. Several clinical trials currently are underway investigating the ability of reovirus to combat different cancers.

Viruses also are central players in the burgeoning field of **gene therapy**. With gene therapy, researchers and clinicians hope to someday actually correct genetic defects in humans. The idea is simple: if a genetic defect is identified, then one could insert a "good" copy of the gene into a person with the

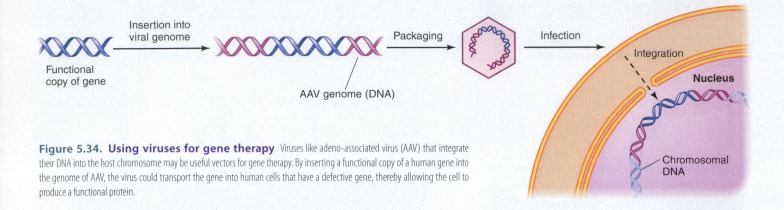

Figure 5.34. Using viruses for gene therapy Viruses like adeno-associated virus (AAV) that integrate their DNA into the host chromosome may be useful vectors for gene therapy. By inserting a functional copy of a human gene into the genome of AAV, the virus could transport the gene into human cells that have a defective gene, thereby allowing the cell to produce a functional protein.

defect, thereby eliminating the problem. Of course, the process is not that simple in reality. One big problem, obviously, involves inserting the correct gene into the defective cells. Again, because viruses naturally enter cells, they are being investigated as potential shuttles of genetic information. In one sense, viruses can be thought of as microscopic containers. The capsid, or container, carries the infectious viral genome—its cargo. With biochemical and genetic manipulations, though, we can alter this cargo. For example, we could remove part of a viral genome and replace it with a copy of the human CFTR gene,

which encodes the cystic fibrosis transmembrane conductance regulator protein. In people with cystic fibrosis, this gene is mutated, producing a non-functional protein. As a result of this mutation, a thick mucus builds in the lungs, leading to respiratory problems and an increased risk of bacterial infections. A virus containing a wild-type copy of the CFTR gene, though, would house a very beneficial cargo. If this altered virus infects lung epithelial cells in a person with CF, then, at least in theory, the virus could deliver a new, correct copy of the defective gene, thereby alleviating the underlying genetic problem **(Figure 5.34)**.

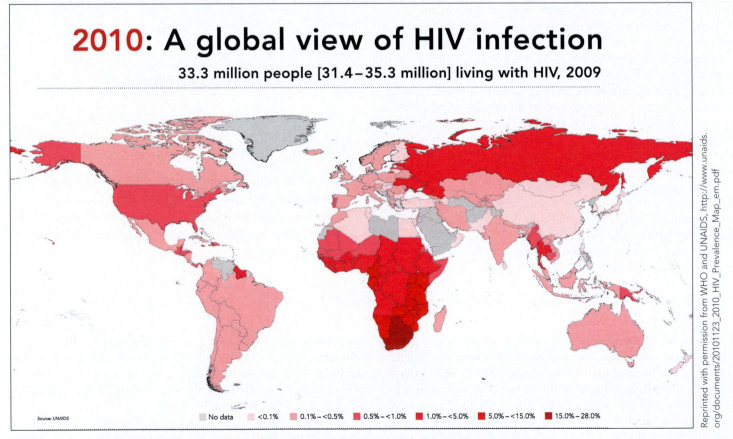

Reprinted with permission from WHO and UNAIDS, http://www.unaids.org/documents/20101123_2010_HIV_Prevalence_Map_em.pdf

Figure 5.35. Numbers of people infected with HIV worldwide Despite the progress we have made in HIV/AIDS research, there still is not a vaccine or a cure. Worldwide, approximately 33 million people are infected with HIV. Most of the people living with HIV are in sub-Saharan Africa.

As microscopic containers, viruses also are being used in the expanding field of **nanotechnology**. Briefly, nanotechnology involves the development and use of nano-devices—machines that are nanometers in size. So how do viruses fit into this field? Given what we have discussed about viruses, you probably can imagine the viral capsid serving as a container for a desirable cargo. In one recent set of experiments, cowpea chlorotic mottle virus (CCMV), a plant virus, has been used to facilitate the formation of complex inorganic molecules. By slightly altering the CCMV capsid, researchers were able to use the capsid as a nano-staging area for the desired chemical reaction. The capsid not only spatially restricted the precursors but also sequestered them from the host cell. Because of this natural function of the viral capsid as an ultra-small vesicle, viruses certainly will play an ever-increasing role in the exciting field of nanotechnology.

Finally, sadly, viruses remain a public health threat throughout the world. Currently, HIV infects approximately 33 million people worldwide **(Figure 5.35)**. Millions of people die every year from virally caused diarrheal diseases. The SARS outbreak of 2003 and the H1N1 influenza pandemic of 2009 showed everyone that we do, indeed, live in a global village and viruses that infect humans *any*where can infect humans *every*where. But we can be hopeful. Our increased understanding of viruses is leading to the development of new antiviral drugs. New vaccines against pathogenic viruses also are being developed and tested. The research of today will lead to more treatment and prevention options tomorrow.

CONNECTIONS Our understanding of virus structures and replication strategies has led to the development of some antiviral drugs. In **Section 8.5**, we will explore how antiviral drugs like AZT function to inhibit the replication of HIV.

5.6 Fact Check

1. How can viruses be used as molecular biology tools? Provide examples.
2. How are the terms "proto-oncogenes" and "oncogenes" related? How were viruses used to characterize these genes?
3. What are oncolytic viruses and how could they be used as anticancer agents?
4. How can viruses be used for gene therapy? Give examples.

Image in Action

WileyPLUS A cell culture hood is being used to study the lymphocytic choriomeningitis virus (LCMV) outbreak in golden lion tamarins. The flasks contain primate cells. The syringes attached to filters help the researchers isolate the viruses. A zoom-in window on the flask reveals cells that are at late stages of LCMV infection. In the second zoom-in window on the cells, a new viral particle is shown budding off a cell (*right*).

1. To isolate viruses from samples obtained from infected animals, a filter is being used. Explain how this technique would aid in the isolation of viruses.

2. The researchers needed to determine which cells in the flask were appropriate to use. What cell characteristics did they need to consider, and what would they look for to determine whether their viral culture attempt was successful?

The Rest of the Story

With the help of several microbiologists using a variety of tools, Dr. Montali at the National Zoo investigated the golden lion tamarin (GLT) deaths. Electron micrographs depicted viral particles within the tissues of the diseased GLTs. Viruses could be purified from these tissues and replicated in the laboratory. All of these findings confirmed Dr. Montali's initial fears—an infectious agent was killing the GLTs.

The research team replicated these viruses in a monkey epithelial cell line. Using antibody binding and nucleic acid hybridization assays, virologists determined that the golden lion tamarins were infected by a strain of lymphocytic choriomeningitis virus (LCMV). LCMV, a large, enveloped virus with a single-stranded RNA genome in the *Arenaviridae* family of viruses, typically causes a mild or no apparent disease in rodent species. When transmitted from rodents to humans, though, these viruses can cause serious disease. Lassa fever and Bolivian hemorrhagic fever, for example, both are caused by arenaviruses. LCMV itself has been associated with serious disease in humans.

Researchers considered how rodents were involved in these GLT infections. It turned out that animal keepers often provided GLTs with mice along with their food as a supplemental protein source. Occasionally, these feed mice were infected with LCMV. When the tamarins ingested these mice, they, too, became infected, with dire consequences. The solution to the problem proved to be obvious and very simple—stop providing mice to the GLTs. Quickly, zoo personnel throughout the country were apprised of the situation. Just as quickly, zoo personnel stopped providing mice to their tamarins. The outbreaks ended.

Summary

Section 5.1: What is a virus?

Viruses probably have existed since life itself first evolved. We have isolated and characterized viruses that infect every known type of life, from bacteria to plants to humans.

- **Viruses** are small **obligate intracellular parasites**, with diameters typically between 10 and 100 nanometers and genomes typically between a few thousand and about 200,00 nucleotides in length.

- Viruses can contain single-stranded or double-stranded DNA or RNA genomes. They have **capsids** surrounding the genome, forming **nucleocapsids**, which generally exhibit **icosahedral** or **helical morphology**. The capsids are constructed from symmetrically arranged **capsomeres**. **Bacteriophages** often exhibit a complex morphology, with **virions** containing an icosahedral head and a helical tail. Some large animal viruses have more irregular shapes.

- In **enveloped viruses**, a lipid bilayer surrounds the capsid. **Non-enveloped viruses** lack this feature. These various characteristics are used to classify viruses.

- Viral replication involves recognition of and attachment to an appropriate host cell, penetration of the host cell membrane, disassembly of the capsid (for viruses of eukaryal cells), replication of the viral genome and production of viral proteins, assembly of new virus particles, and exit from the cell of these new particles.

Section 5.2: From what did viruses arise?

While the evolutionary history of viruses remains unclear, several hypotheses have emerged.

- Viruses may have evolved along with their appropriate host cells. While this **coevolution hypothesis** is very simple and could explain the origins of the many RNA viruses, there is very little data supporting it.

- Viruses may represent cells that, over time, lost some of their replicative and metabolic capabilities, eventually becoming dependent on host cells. While examples supporting this **regressive hypothesis** exist in the biological world, this hypothesis does not offer an explanation for the many RNA viruses.

- Most virologists currently think that viruses arose from existing genetic elements that eventually gained the ability to move from cell to cell and organism to organism. The similarities between transposons and certain DNA viruses and **retrotransposons** and retroviruses perhaps provide the best evidence for this **progressive hypothesis**.

Section 5.3: How can we "grow" and quantify viruses?

Because viruses are very small and only replicate within appropriate host cells, various strategies are used to amplify and quantify viruses.

- Viral cultivation only can occur within appropriate host cells. For **lytic bacteriophages**, a culture of bacteria can be inoculated with a sample of phage. After a period of incubation, the phage will replicate within the bacteria.

- **Lysogenic bacteriophages** can integrate their genome into the genome of their host, forming **prophages** in **lysogens**.

- For animal viruses, similarly, a sample of virus can be added to cells of an appropriate cell line. Many animal virus cause characteristic

cytopathic effects (CPE) in infected host cells, often resulting in the formation of **plaques** or a **syncytium**.

- Viruses can be purified in several different ways. Filtration can be used to separate viruses from eukaryal, bacterial, and archaeal cells. **Differential centrifugation** or **gradient centrifugation** then can be used to purify virus particles.

- Virus **titers** can be quantified in many different ways. Electron microscopy can be used to count the actual number of particles. Various assays, like the **hemagglutination** assay, **tissue culture infectious dose 50 (TCID$_{50}$)** or **lethal dose 50 (LD$_{50}$)** assay can be used to measure some specific property displayed by the virus particles. Plaque assays allow a researcher to measure the number of infectious virus particles.

Section 5.4: What are the different types of viruses?

Viruses have been classified in many ways. The most useful classification schemes are based on structure and replication strategies.

- Viruses have been named in a variety of ways, ranging from descriptions of the disease caused by the virus, to the location where the virus was first isolated, to simple numbers or letters.

- The ICTV has developed a systematic taxonomy for viruses in which viruses are classified based on a series of shared attributes, including virion structure, genome composition, and host range.

- The Baltimore classification system organizes all viruses based on the process by which they generate mRNA.

- Researchers can begin to identify viruses by examining their structure.

- Techniques like **reverse transcriptase PCR (RT-PCR)** and nucleic acid sequence comparisons provide a more definitive means of identifying viruses.

Section 5.5: Are viruses the simplest pathogens?

A variety of subviral entities have been discovered and characterized.

- **Viroids** consist only of naked RNA; they lack a protein coat. The RNA genome typically is less than 400 nucleotides long and does not encode any proteins. Host cell polymerases probably replicate the genome.

- **Satellite viruses** and **satellite RNAs** are dependent on a helper virus for replication. Also, these agents contain a protein coat. For satellite viruses, the agent's genome encodes the coat protein. For satellite RNAs, the coat protein is encoded by the helper virus.

- **Prions** do not contain a genome. These infectious agents, responsible for **transmissible spongiform encephalopathies (TSEs)**, such as mad cow disease, consist only of protein.

Section 5.6: What's next for virology?

The future of virology certainly will include many new advances and, hopefully, an increasing array of treatments to prevent viral diseases.

- Studies of viruses will continue to help us learn more about the normal workings of eukaryal cells and about problems associated with these cells. These studies will continue to lead to the development of new molecular biology techniques.

- Viral research has increased our understanding of **proto-oncogenes** and **oncogenes**.
- Studies of **oncolytic viruses** may lead to new cancer treatments.
- Virologists are becoming actively involved in the burgeoning fields of **gene therapy** and **nanotechnology**. Because of their small size and ability to package a genome, viruses represent ideal tools for these two exciting new areas of research.
- Viruses continue to contribute to human mortality and suffering. Studies of viral pathogenesis remain critically important.

● Application Questions

1. Many microbiologists consider viruses to be non-living particles. Explain their rationale for this.

2. Enveloped and non-enveloped viruses enter host cells differently. Why is this, and what are their different approaches for entry?

3. Viral attachment obviously is a key step in a viral infection. What strategies could be employed by an antiviral drug to prevent this step?

4. You are working in the lab and must quantify the viruses in your sample. Which would you consider a better approach to use: a direct count of viral particles or a count of infectious viral particles? Explain your answer.

5. Currently, many different cell lines have been characterized and are available for use by researchers. Why might virologists need many different cell lines to do their work?

6. A connection has been made between helical morphology of the viral capsid and the presence of an RNA genome in a virus. Explain why this might be.

7. When examining a new virus, what characteristics would you need to observe in order to determine the type or classification of that virus?

8. Many chemical biocides (disinfectants) target the lipid envelopes of viruses. Why would the chemical destruction of the envelope render the virus "non-infectious"?

9. Some viruses cause cancer. Given what you know about viral replication, would the induction of cancer be beneficial to the virus?

10. Phage therapy involves the therapeutic use of bacteriophages to treat bacterial infections in animals. Based on your understanding of bacteriophages, would the phage used for this therapy be able to destroy or replicate inside the animal cells?

11. Examinations of golden lion tamarins (GLTs) that died at zoos in the United States revealed a severe hepatitis, or inflammation of the liver. Electron microscopy of the liver tissue revealed significant tissue damage and the presence of virions.
    a. Initially, how could you use filtration to determine if the liver was infected with viruses or bacteria?
    b. How would you go about isolating and culturing the virus in the lab? What type of cell line would you need to use?
    c. What technique would you use to determine the viral load in the GLTs? Explain.

Suggested Reading

Henig, Robin Marantz. A Dancing Matrix: Voyages Along the Viral Frontier. 1993. Alfred A. Knopf, Inc., New York.

Montali, R. J., E. C. Ramsay, C. B. Stephensen, M. Worley, J. A. Davis, and K. V. Holmes. 1989. A new transmissible viral hepatitis of marmosets and tamarins. J Infect Dis 160:759–765.

Montali , R. J., C. A. Scanga, D. Pernikoff, D. R. Wessner, R. Ward, and K. V. Holmes. 1993. A common-source outbreak of callitrichid hepatitis in captive tamarins and marmosets. J Infect Dis 167:946–950.

Stephensen, C. B., J. R. Jacob, R. J. Montali, K. V. Holmes, E. Muchmore, R. W. Compans, E. D. Arms, M. J. Buchmeier, and R. E. Lanford. 1991. Isolation of an arenavirus from a marmoset with callitrichid hepatitis and its serologic association with disease. J Virol 65:3995–4000.

6 Cultivating Microorganisms

"Anthrax Tower," otherwise known as Building 470 or the "Pilot Plant," was a seven-story brick building at Fort Detrick, Maryland. Built in 1953 by the Department of Defense, it served as a pilot plant for the production of biological weapons, initiated in response to the perceived threat of biological warfare during the Cold War. Between 1953 and 1965, researchers cultivated *Bacillus anthracis*, the cause of anthrax disease, in two 2,500-gallon, three-story-high fermenters housed within the tower, similar to brewery fermentation vats. To prevent the unintentional escape of pathogens to the outside world, a powerful ventilation system maintained negative pressure in the building. If a door opened, or a crack occurred in the structure, air would rush into the building, not out.

Bacillus anthracis can be cultivated easily under aerobic conditions in a rich nutrient medium at 35–37°C. When nutrients in the medium become depleted, however, the cells undergo a morphological change, resulting in the formation of endospores. It is the endospores that are coveted for use as a biological weapon. *B. anthracis* spores are particularly resilient to extreme temperatures, desiccation, harsh chemicals, and sunlight. They can survive for decades in soils. In 1943, the British government tested the viability of anthrax spores on an uninhabited, remote island off the coast of Scotland. In 1980, nearly 40 years after the initial experiments, sheep were moved onto the island and promptly died of anthrax. The endospores had remained viable.

Bacillus anthracis spores germinate rapidly in response to specific environmental conditions found in mucosal surfaces and wounds of mammalian hosts. The infection process can begin with inhalation, ingestion, or contamination of a cut with just a few spores. It seemed to be the perfect biological weapon but thankfully never was deployed. Eventually the Cold War ended. In 1965, researchers stopped production of biological agents at Fort Detrick and Building 470 was sealed. But what to do with a building full of anthrax spores? The very characteristics of the spores that made them once so desirable now became a detriment.

CHAPTER NAVIGATOR

As you study the key topics, make sure you review the following elements:

Microbes obtain nutrients and energy in diverse ways.

- Figure 6.1: Energy, electron, and carbon sources for metabolism
- Figure 6.2: Incorporation of ammonia into amino acids

Nutritional requirements and environmental factors affect microbial growth.

- Toolbox 6.1: Phenotype MicroArrays for examining microbial growth
- Microbes in Focus 6.1: The metabolic diversity of *Pseudomonas aeruginosa*
- Table 6.1: Toxic oxygen species
- Figure 6.7: Acceptable temperature ranges of organisms in the three domains

Using various growth conditions, researchers can grow some microorganisms in the laboratory.

- 3D Animation: How microorganisms grow in the laboratory
- 3D Animation: Other roles of bacteriological media
- 3D Animation: Obtaining a pure culture of a microorganism
- Figure 6.10: Growth on selective and differential media
- Figure 6.12: Streaking plates to isolate cells and separate colonies
- Perspective 6.1: The discovery of *Helicobacter pylori*
- Toolbox 6.2: FISHing for uncultivated microbes
- Mini-Paper: Bringing to life the previously unculturable using the soil substrate membrane system (SSMS)

In the laboratory, researchers can monitor the growth of microbes.

- 3D Animation: Measuring microbial populations
- Table 6.4: Measuring viable cell density in a bacterial culture by spread plating
- Perspective 6.3: The human intestine—a continuous culture

Various strategies prevent or limit microbe growth.

- Figure 6.23: Ultraviolet radiation for disinfection or sterilization
- Table 6.5: Commonly used disinfectants
- Figure 6.26: Decimal reduction time (D value)

CONNECTIONS for this chapter:

Role of iron-binding proteins in pathogens (Section 21.1)
Nucleoside acquisition by *Plasmodium falciparum* (Section 23.1)
Archaeal adaptations to high salt conditions (Section 4.3)
Metagenomic analysis of non-cultivated microbes (Section 10.4)
Process of pasteurization (Section 16.2)

Introduction

Microorganisms are capable of reproducing faster than any of the larger forms of life on Earth, and microbial growth has enormous consequences for the biosphere and human health. As we saw in Section 4.3, the archaeon *Methanobrevibacter smithii* may play an integral role in our digestive system. As we will learn in Sections 14.2 and 14.3, microbes play integral roles in biogeochemical cycling, transferring important elements like carbon and nitrogen from one molecular state to another. Other microbes, like the bacterium *Streptococcus pyogenes*, can cause various diseases in humans, as we will see in Section 21.2. For us to better understand the actions of these microbes, we need to work with them in the laboratory. To facilitate this research, we need to understand the growth requirements of these microbes.

Understanding how microbes grow can be surprisingly challenging. Like animals in a zoo, some microbes replicate well in the laboratory, while other microbes do not. In this chapter, we will explore how nutritional and environmental factors affect population growth. We also will explore methods for studying these effects. We will learn about methods for isolating and cultivating microbes in the laboratory. We also will learn about methods for studying microbes that we cannot yet grow. Finally, we will examine methods for preventing the growth of microbes, and destroying undesirable microbes. Along the way, we will address the following questions:

What do microorganisms generally need to grow? (6.1)

How do nutritional and environmental factors affect microbial growth? (6.2)

How can different types of microorganisms be grown in the laboratory? (6.3)

How can we measure and monitor microbial populations? (6.4)

How can we eliminate microbes or inhibit their growth? (6.5)

6.1 Nutritional requirements of microorganisms

What do microorganisms generally need to grow?

As we saw in Section 1.1, all living organisms contain certain macromolecules, primarily proteins, nucleic acids, lipids, and carbohydrates (see Table 1.1). These macromolecules, in turn, are composed primarily of carbon (C), nitrogen (N), phosphorus (P), sulfur (S), oxygen (O), and hydrogen (H). Cells also require several elements that are not major components of macromolecules. Potassium, sodium, magnesium, calcium, and iron (K$^+$, Na$^+$, Mg^{2+}, Ca^{2+}, and Fe^{2+} or Fe^{3+}) are the most abundant cations in microbial cells. Chloride (Cl$^-$) is the major anion. K$^+$, Na$^+$, and Cl$^-$ have important functions in controlling osmotic balance, and in charge-dependent interactions with macromolecules. Mg^{2+} associates with ATP, and is a key component of many enzymes and protein-nucleic acid complexes, such as the ribosome. DNA polymerases require Mg^{2+} or Mn^{2+} as a cofactor for their activity.

Several enzymes and cytochromes, the membrane-bound electron transfer proteins that have critical roles in energy metabolism, require iron for their activity. Despite its importance, iron frequently is limiting for growth in natural habitats, where soluble iron can be in short supply. The mammalian body goes to great lengths to sequester iron in tightly bound protein complexes, both to minimize iron toxicity and to keep it away from microbial pathogens.

CONNECTIONS Because free iron typically is in short supply, many pathogens possess siderophores—iron-binding proteins. Additionally, many pathogens possess specific mechanisms for getting iron from their hosts. We'll explore some of these processes in **Section 21.1**.

Finally, various **micronutrients**, such as zinc, cobalt, molybdenum, copper, and manganese (Zn^{2+}, Co^{2+}, Mo$^+$, Cu^{2+}, and Mn^{2+}), also are needed for microbial growth. Certain enzymes and electron transfer protein complexes require small amounts of these metals. Most microbes need so little of these metals that they often are not intentionally added to culture media. The microbe's needs are supplied as these elements contaminate other chemicals used to make media, or leach from glass culture vessels.

To grow microorganisms in the laboratory, we must make the appropriate nutrients available in the appropriate forms. In Section 6.2, we will examine various ways of cultivating bacteria. In this section, though, we first will examine how microbes naturally acquire these various nutrients. Then, we will explore how microbes acquire energy. Finally, we will investigate the electron sources utilized by various microbes.

Acquisition of nutrients

Because carbon is the most abundant element in biological macromolecules, acquiring carbon and incorporating it into macromolecules is critical for a growing cell. The general process by which cells import a molecule and incorporate it into cellular constituents is called **assimilation**. **Autotrophs** assimilate carbon from inorganic sources. They convert the carbon present in molecules such as carbon dioxide (CO$_2$) into organic molecules **(Figure 6.1)**, a process known as **carbon fixation**. Autotrophic organisms generate most of Earth's biomass. Additionally, as we saw in Section 1.3, these autotrophic organisms generally are responsible for the oxygen-rich atmosphere that we have today.

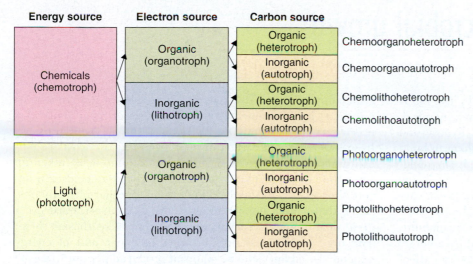

Figure 6.1. Energy, electron, and carbon sources for metabolism Metabolic types of organisms are categorized on the basis of how they obtain their energy, electrons, and carbon.

Acquisition of energy

In addition to acquiring nutrients such as carbon and nitrogen for building new biomolecules, cells need energy to drive their metabolism. Microorganisms acquire energy in various ways. As we will see in Section 13.6, **phototrophs** (see Figure 6.1) capture light energy, or photons, through the process of photosynthesis, to generate chemical energy, such as ATP. This ATP, in turn, is used for metabolic processes. Many phototrophs, including certain bacteria, eukaryal algae, and green plants, use the energy generated from photosynthesis for carbon fixation, incorporating CO_2 into biological molecules, such as carbohydrates. For these organisms, then, an external source of organic carbon is not needed. They simply need access to adequate light and CO_2 to acquire both carbon and energy.

In contrast to phototrophs, **chemotrophs** acquire energy through the oxidation of reduced organic or inorganic compounds they have acquired from the environment. In many cases where organic molecules are used as a source of energy, the same molecule serves as both the carbon and energy source. A good example is glucose, which heterotrophic microbes break down through glycolysis and the tricarboxylic acid cycle to provide essential precursors for the synthesis of macromolecules and for powering the production of ATP. Glucose is a common constituent in synthetic media meant for the culture of a wide variety of heterotrophic microbes.

Heterotrophs cannot assimilate carbon from inorganic carbon molecules. They must obtain it in a preexisting organic form (see Figure 6.1). Heterotrophic microbes have evolved to acquire carbon from many kinds of molecules—carbohydrates, amino acids, lipids, organic acids, alcohols, and more. As a result, we see microorganisms growing in every conceivable niche on Earth.

CONNECTIONS Microorganisms can be categorized on the basis of how they obtain carbon, energy, and electrons. These metabolic categories are described in **Section 13.2** and summarized in **Table 13.1**.

Nitrogen also is abundant in biological macromolecules, and actively growing microbes must acquire it in a usable form. As we will see in Section 13.7, certain microbes can convert the dinitrogen gas (N_2) present in the atmosphere into ammonia (NH_3), through a process known as nitrogen fixation. Other organisms assimilate nitrogen in the form of ammonia (NH_3). Still other microbes assimilate nitrate (NO_3^-) and nitrite (NO_2^-), which they then convert to ammonia. Once a cell obtains NH_3, through any of these avenues, it can incorporate the nitrogen into the amino acids glutamate and glutamine **(Figure 6.2)**. These molecules can serve as the precursors for the generation of other amino acids and additional molecules that contain nitrogen.

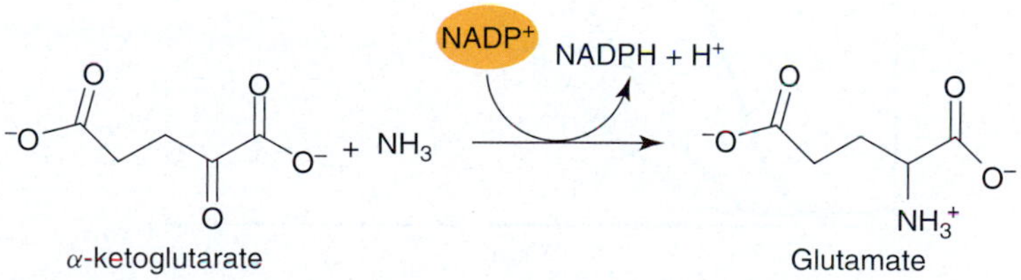

Electron sources

In addition to nutrients and a source of energy, all microorganisms require an appropriate source of electrons. Electron donors, or reducing agents, are needed for biochemical oxidation and reduction reactions, which will be covered in Chapter 13. **Organotrophs** acquire electrons from organic molecules, such as glucose (see Figure 6.1). **Lithotrophs** (literally, rock eaters) remove electrons from inorganic reduced molecules, such as ferrous iron (Fe^{2+}), elemental sulfur(S^0), hydrogen gas (H_2), hydrogen sulfide (H_2S), ammonium (NH_4^+), and nitrite (NO^{2-}). Lithotrophic bacteria and archaeons contribute significantly to biogeochemical cycling of sulfur, nitrogen, and other elements, as we will see in Chapter 14.

Figure 6.2. Incorporation of ammonia into amino acids Nitrogen can be incorporated into organic molecules through the combination of ammonia (NH_3) with α-ketoglutarate, resulting in the formation of the amino acid glutamate. This amino acid can serve as a precursor for the biosynthesis of other amino acids.

6.1 Fact Check

1. What roles do ions play in cells?
2. Contrast autotrophic and heterotrophic organisms.
3. From what sources can organisms acquire electrons?
4. How might understanding the natural growth requirements of a microorganism help a researcher?

6.2 Factors affecting microbial growth

How do nutritional and environmental factors affect microbial growth?

Microbial growth rate and yield are determined by the interaction of environmental factors, both physical and chemical, and the genetically encoded traits of an organism. To grow and divide, microbes must acquire nutrients and use them to build biomass, while extracting useable energy from their surroundings to drive chemical reactions. The growth of a microbe in any environment depends on whether, and how quickly, it can capture the necessary nutrients and energy sources. The chemical composition of the environment determines nutrient availability, and can affect cell structure and function. Physical parameters such as temperature, pressure, and illumination also can affect cell structure and function, and thereby influence growth rates. In this section, first we will explore how nutritional factors affect microbial growth. Then we will investigate the effects of chemical and physical factors, such as oxygen availability and pH, on microbial growth.

Effects of nutritional factors on microbial growth

When a molecule enters the cytoplasm of a cell, it may contribute to growth if it is used to synthesize cellular components, or it may contribute to ATP production. It may provide another useful function, such as serving as an enzyme cofactor or enhancing the stability of the cell membrane. Ultimately, though, the rate at which a cell can acquire critical nutrients, process them into macromolecular building blocks, and then assemble essential macromolecules determine the cell's growth rate. Before it can divide, a cell must (1) fully duplicate its genome, and (2) approximately double its mass, which is largely composed of proteins, nucleic acids, lipids, and cell wall. If these tasks are not executed fully and faithfully, then progeny cells will not receive the complete repertoire of genetic information or cellular machinery, or cell size will steadily decrease every generation. Neither result would be sustainable. As a result, cells have regulatory mechanisms linking replication and cell growth to cell division, thereby making sure these critical processes work together.

Nutrient Type

The types of nutrients a microbe can access profoundly affect its growth rate and yield. **Prototrophs** can synthesize all necessary cellular constituents from a single organic carbon source and inorganic precursors. These organisms, in other words, possess biosynthetic pathways to build all the macromolecular precursors they need. An **auxotroph**, in contrast, cannot build all the precursors that it needs. Rather, certain organic precursors must be acquired from the environment. The growth of these organisms, as a result, is limited to niches in which the appropriate organic molecules can be found. Determining what nutrients an organism requires **(Toolbox 6.1)**, then, can improve greatly our understanding of the natural ecology of a microorganism.

While prototrophic organisms can grow in environmental niches with low concentrations of organic nutrients, growth in

such habitats is typically slow, or at least slower than in a more organic-rich environment. Why? Synthesis of macromolecular precursors such as amino acids "from scratch," using nitrogen from ammonia and carbon derived from glucose, costs significantly more ATP than using preformed amino acids taken up from the environment. Cells that import amino acids can devote their energy to simply linking them together into proteins. A cell that cannot import amino acids, in contrast, must use a great deal of energy, carbon, and nitrogen to synthesize them. The net result is less protein assembled per unit of energy expended, and less cellular material produced per unit time.

CONNECTIONS *Plasmodium falciparum*, the eukaryal microbe that causes malaria, cannot manufacture certain nucleosides, important precursor components of DNA. Instead, it possesses a well-characterized mechanism for importing nucleosides from host cells. We will discuss this process in **Section 23.1**.

Nutrient Concentration

Within certain limits, growth rate is sensitive to the concentrations of key nutrients in the environment **(Figure 6.3)**. If the concentration of a critical nutrient is low, then one or more biochemical pathways may operate at less than their maximal potential. The growth rate of the microbe, as a result, will be less than maximal. To illustrate the relationship between growth rate and nutrient concentration, consider an experiment in which a bacterium such as *Escherichia coli* is inoculated into a medium lacking a carbon source. The initial growth rate would be 0. Glucose then is added to the medium at progressively higher concentrations, and the growth rate is measured to generate the curves in Figure 6.3. The growth rate initially increases proportionally to glucose concentration, but this relationship breaks

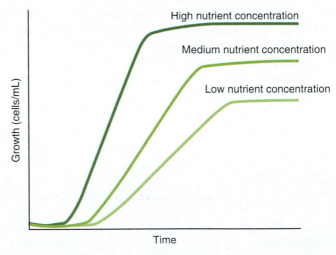

Figure 6.3. Effect of nutrient concentration on growth Increases in nutrient concentrations lead to increases in the growth rate of microbial cells. At high nutrient levels, growth rate reaches a maximum determined by the rate at which cells can acquire nutrients, the efficiency with which they can use those nutrients, and the space available for cells to occupy.

Toolbox 6.1
PHENOTYPE MICROARRAYS FOR EXAMINING MICROBIAL GROWTH

When investigating a particular microorganism, a microbiologist may want to explore its physiology, such as its ability to metabolize various nutrient sources. Testing potential growth substrates can be very tedious, but the Phenotype MicroArrays (Biolog, Hayward, CA), shown in **Figure B6.1**, allow microbiologists to easily and systematically determine whether a microbe can utilize hundreds of potential carbon, nitrogen, phosphorus, and sulfur sources. Phenotype MicroArrays (PMs) also can be used to examine the ability of a microbe to tolerate a wide range of chemically distinct environments.

How do Phenotype MicroArrays work? The system starts with a solution containing a standard set of nutrients that support the growth of most commonly cultivated bacteria and fungi. The medium is supplied in a dehydrated format in 96 well plates. To test the ability of a heterotrophic microbe to utilize various carbon sources, we use a set of plates in which each well contains a different carbon source, along with the basal medium supplying N, P, S, and micronutrients. One of the wells on each plate has no carbon source added, and serves as a negative control. Two different carbon source plates allow the researcher to examine the ability of a microbe to use 190 different organic compounds. Another plate tests nitrogen sources, and yet another tests phosphorus and sulfur utilization. Additional plates can be used to examine the effects of varying pH, solute concentration, and the presence of potential growth inhibitors in the medium.

To use these plates, a sample of the microbe being tested is suspended in inoculating fluid and added to each well. The plates are covered and incubated at the organism's preferred temperature. The standard inoculating fluid contains a tetrazolium dye that can be reduced by the electron transport systems of many microbes. During active metabolism, as electrons flow through the membrane-bound electron transport system, some of the electrons are diverted to dye molecules. Reduction of the dye generates an intense color that can be monitored in a spectrophotometer. By using an electron-sensitive dye like tetrazolium, the investigator can determine if active metabolism is occurring in the wells, even if growth is quite slow.

Phenotype MicroArrays have many uses. They allow a broad assessment of the metabolic capabilities and environmental tolerances of an organism. Is a microbe versatile in its nutritional requirements, or finicky? Does it prefer sugars, amino acids, lipids, or other carbon compounds? To what sorts of chemical and environmental conditions can it adapt? Such information can provide insight into environmental niches the microbe might occupy, or may be used to improve laboratory media upon which it can grow. It also can provide support for metabolic predictions based on annotated genes in a sequenced genome, a topic that will be discussed further in Chapter 10. Finally, for microbes that can be genetically manipulated, Phenotype MicroArrays can be valuable for broadly assessing the effects of mutations on physiology and stress responses.

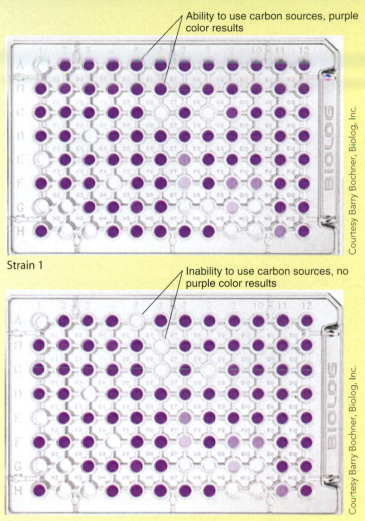

Figure B6.1. Analyzing metabolic properties with the Biolog Phenotype MicroArray Two strains of a bacterium were added to wells of Biolog Phenotype MicroArray plates to examine their metabolic properties. Cellular activity in a well leads to a purple color change. In the absence of cellular metabolism, wells remain clear. As seen in this example, these strains differed in their ability to grow under two different conditions. Strain 1 grew in the two highlighted wells, while Strain 2 did not.

● Test Your Understanding ....

You isolate two new species of bacteria. It appears that one of these newly discovered microorganisms thrives in many different environments. The other one, in contrast, seems to grow only in a limited number of environments. You test both with a Phenotype MicroArray. Predict your results.

Microbes in Focus 6.1
THE METABOLIC DIVERSITY OF *PSEUDOMONAS AERUGINOSA*

Habitat: Normal inhabitant of soil and water, occasionally found on the skin of various animals. A noted opportunistic pathogen of humans, often causing respiratory disease in people living with cystic fibrosis.

Description: Gram-negative, motile rod-shaped bacterium, approximately 1.5 to 3.0 μm in length

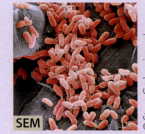
SEM
© Steve Gschmeissner/Photo Researchers, Inc.

Key Features: Members of the *Pseudomonas* genus exhibit a great deal of metabolic diversity. While they all undergo respiration, they can utilize many different organic molecules as carbon sources. Because of this metabolic diversity, members of this genus are being investigated as possible bioremediation agents. Strains of pseudomonads that can consume toxic waste materials may be identified.

---

obligate aerobes. Obligate implies an absolute requirement. Common bacterial examples of obligate aerobes are found in the genus *Pseudomonas*, including the occasional pathogen *Pseudomonas aeruginosa* (Microbes in Focus 6.1). Growth rates for most obligate aerobes are maximal at or near atmospheric O_2 concentrations, and decline as the oxygen concentration drops.

Anaerobic growth occurs without the use of oxygen. **Aerotolerant anaerobes** do not use oxygen for respiration, but are not harmed by it. **Obligate anaerobes**, in contrast, cannot grow in the presence of oxygen. **Facultative anaerobes** can use oxygen for respiration when it is present, but also can grow in the absence of oxygen. What accounts for these differences? Molecular oxygen can generate by-products that act as powerful oxidants capable of attacking proteins, nucleic acids, and lipids. These reactive oxygen species (Table 6.1) include superoxide, hydroxyl radical, hydrogen peroxide, and singlet oxygen, a transient high-energy form of O_2. Aerobic microbes depend on multiple defenses against oxidative damage, since reduction of O_2 during respiration generates some of these potentially dangerous by-products. To live in aerobic habitats, even microbes that do not undergo aerobic respiration need defenses against toxic oxygen species generated by various biochemical and abiotic reactions. Obligate anaerobes lack adequate defenses.

Molecular oxygen is not ubiquitous in the biosphere; many habitats have reduced O_2 levels, or are completely anaerobic. For example, the upper portion of the human digestive tract is aerobic, largely due to swallowed air, but quickly becomes anaerobic in the intestines. It is impossible for obligate aerobes to grow in the colon, so the colonic microbiota must consist of aerotolerant, facultative, and obligate anaerobes. There are several ways to grow anaerobic microbes in the laboratory. One approach is to evacuate the normal O_2-containing atmosphere and replace it with an artificial atmosphere lacking O_2. An alternative small-scale method employs the anaerobic jar used in many diagnostic labs (Figure 6.4). After the agar plates are placed in the bottom of the jar, the gas generator envelope is opened, and water is added to it. A chemical reaction in the envelope generates hydrogen (H_2) and carbon dioxide (CO_2) gases, which escape into the atmosphere of the jar. At this point, the jar is sealed; an airtight gasket prevents atmospheric O_2 from entering, while the O_2 in the jar is consumed by reaction with the H_2 gas.

---

down as the growth rate approaches a maximum. Once the bacteria are dividing as fast as they possibly can, further increases in the nutrient concentration have no effect on growth rate.

Effects of oxygen on microbial growth

Aerobes require oxygen for growth. Molecular oxygen (O_2) is the only electron acceptor that can be used by their respiratory electron transport system. As we will discuss more in Chapter 13, the favorable redox properties of O_2 allow aerobic respiration to extract more energy from oxidative metabolism than any other terminal electron acceptor. In habitats where O_2 is present at substantial levels, the dominant organisms almost always will be aerobes. Organisms that absolutely depend on oxygen are called

---

TABLE 6.1	Toxic oxygen species	
Toxic species	**Sources**	**Cellular defenses**
Singlet oxygen: $^1O_2\cdot$	Photochemical reaction; product of peroxidase enzymes	Antioxidants such as carotenoid pigments
Superoxide anion: O_2^-	By-products of reduction of O_2 during respiration and other biochemical redox reactions	Superoxide dismutase, superoxide reductase enzymes
Hydroxyl radical: $OH\cdot$		Antioxidants such as glutathione
Hydrogen peroxide: H_2O_2		Catalase and peroxidase enzymes

Figure 6.4. Anaerobic culture system A simple, sealed container, an anaerobic jar, can be used for cultivating anaerobic microorganisms on agar plates. Chemical reactions within the jar consume virtually all of the O_2, creating an anoxic environment.

© Scott Coutts/Alamy

Microaerophiles show maximal growth rates at O_2 concentrations lower than are typically present in the atmosphere. These microbes are adapted to relatively stable zones where O_2 is depleted by microbial growth, and diffusion of atmospheric O_2 is restricted. Microaerobic habitats can be found in dense soils, aquatic sediments, and relatively deep marine or freshwater zones.

Effects of pH on microbial growth

Basically, the term "pH" expresses the hydrogen ion activity, or concentration of H^+ ions of a solution. The pH scale is an inverse logarithmic function ($pH = -\log[H^+]$), ranging from 0 to 14. Pure water at 25°C has a pH of 7, which is defined as neutral pH. Solutions with pH values lower than 7 are acidic, solutions with higher pH than 7 are alkaline, or basic. Because pH is a log function, a solution with $pH = 6$ has a tenfold higher H^+ concentration than a solution with $pH = 7$. Microbial growth rates can be affected significantly by pH. Natural habitats span a wide range of pH **(Figure 6.5)**. In addition to affecting macromolecular structures, pH affects the transmembrane electrochemical gradients that are necessary for ATP synthesis and many transport processes.

Neutrophiles grow optimally in environments with pH close to neutrality (5.5 to 8.5). **Acidophiles** prefer significantly acidic habitats (pH < 5.5). **Alkalophiles** prefer alkaline habitats (pH > 8.5) (see Figure 6.5). The most thoroughly studied acidophilic microbes are the lactic acid bacteria, composed of Gram-positive bacteria of the genus *Streptococcus* and related species. These aerotolerant anaerobes generate energy through fermentation, often producing lactic acid as a primary fermentation product. These bacteria have many nutritional requirements. Because their fermentative metabolism is so inefficient, they grow best in environments with high sugar concentrations. Under these conditions, these bacteria produce abundant organic acids that can acidify the medium to pH 4–5. These bacteria are often found in anaerobic habitats rich in sugars and other organic materials, such as decaying fruits and vegetables.

CONNECTIONS While some lactic acid bacteria may be associated with rotting fruit, other lactic acid bacteria are used in the production of foods. As we will see in **Section 16.3**, various lactic acid bacteria are used in the production of cheeses, sour cream, and yogurt.

Although geochemical processes determine the pH of many habitats, microbes also can play an active role in altering the pH of their environment. In Chapter 3, the extremely acidic environment (pH 1.5 to 3) of the Rio Tinto results from the metabolism of lithotrophic bacteria such as *Acidithiobacillus ferrooxidans* that oxidize pyrite (FeS_2) and ferrous iron (Fe^{2+}), and in the process release sulfuric acid (H_2SO_4), a strong acid. This process contributes greatly to acid drainage from thousands of abandoned, flooded mines around the world, including many in the United States.

CONNECTIONS As we have seen in this section, pH can affect the growth of microorganisms. Changes in the pH of an environment, then, can affect the microbial inhabitants of a location. In **Section 14.3**, we will examine in more depth the effects of pH changes caused by acid mine drainage.

Effects of osmotic pressure and water availability on microbial growth

Microbes inhabit environments with a wide range of solute concentrations. Because cells have semi-permeable membranes, a difference in solute concentration between the cytoplasm and the surrounding environment causes water to move in or out of the cell (see Figure 4.21). If solute concentrations are higher in the cytoplasm as, for example, for microbes living in freshwater habitats with low levels of dissolved salts and organics, net water flow will be into the cell, resulting in pressure outward against the cell envelope. It is common for microbes to maintain higher solute concentrations in their cytoplasm than in the surrounding environment, so some osmotic pressure is normal. Membranes have some flexibility for expansion, but ultimately a rigid cell wall, whether peptidoglycan in bacteria or chitin or cellulose in eukaryal microbes, is critical for withstanding osmotic pressure.

CONNECTIONS The cytoplasm of the halophilic archaeon *Halobacterium salinarium* contains high levels of potassium, which provides the cell with osmotic balance against high salt concentrations. Additional adaptations to life in salty environments are discussed in **Section 4.3**.

Osmotic pressure is closely affiliated with water availability. For microorganisms to grow, water must not only be present, it also must be accessible. To determine the availability of water, researchers may calculate the **water activity (a_w)** of an environment. This term refers to the vapor pressure of a liquid relative to the vapor pressure of pure water, under identical conditions. Pure water, then, has a water activity of 1.0. When solutes are added to water, though, they often associate with the water in some way, decreasing the

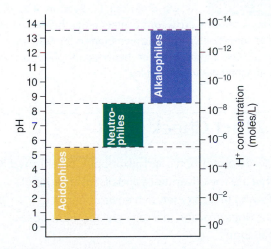

Figure 6.5. Effects of pH on microbial growth The pH scale ranges from extreme acidity (*bottom*) to extreme alkalinity (*top*). Microorganisms can be categorized as acidophiles, neutrophiles, or alkalophiles, based on their pH preferences.

vapor pressure of the liquid. As a result, the a_w decreases. The a_w of seawater is approximately 0.98. The a_w of honey is approximately 0.6. Most bacteria require an a_w of greater than 0.9. The relationship between water activity and food spoilage is discussed in Section 16.1.

Effects of temperature on microbial growth

As we discovered in Section 4.3, microbes inhabit environments with a wide range of temperatures. Mesophiles grow at temperatures that we consider "normal"—between 10°C and 40°C. Psychrophiles grow at temperatures below 10°C (Figure 6.6), while thermophiles and hyperthermophiles grow optimally at temperatures above 55°C and 80°C, respectively. What cellular or molecular factors determine temperature tolerance? First, remember that temperature affects the rate of chemical reactions. Increasing temperature raises the thermal energy of molecules, making them more likely to react. Decreasing temperature lowers the available energy. So in a broad sense, we could expect growth to slow as temperature declines below a microbe's optimal growth temperature, as biochemical reactions slow.

Of course, biochemical reactions in cells are not simply reactions between free molecules. Enzymes, complex molecules whose conformations are affected by temperature, catalyze the reactions. Protein folding and activity are important determinants of the upper and lower temperature boundaries of an organism. At lower temperatures, proteins fold more slowly, and dynamic changes in structure that may be necessary for proper function are restricted. Conversely, at higher temperatures, the effects of thermal energy may overcome some of the weaker chemical interactions that help to stabilize protein structure. The cumulative effects of such structural disruptions can change a finely tuned enzyme structure into a non-functional polypeptide chain. Between

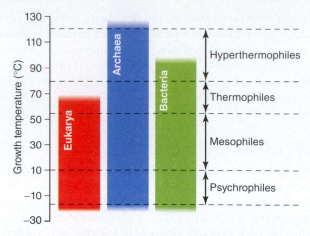

Figure 6.7. Acceptable temperature ranges of organisms in the three domains Growth at elevated temperatures is quite rare among eukaryal organisms. It is more common among bacteria and archaeons.

the upper and lower temperature boundaries lies the temperature range at which an organism's enzyme set has evolved to function at or near maximal efficiency.

Fluidity of the membrane bilayer also is affected by temperature. Fluidity decreases as the temperature drops. The progressive "freezing" of a membrane can negatively affect critical transmembrane processes, such as nutrient transport, electron transport, and maintenance of electrochemical gradients. A bilayer richer in unsaturated lipids will maintain a higher degree of fluidity at low temperatures than a membrane that is richer in saturated lipids, which pack more tightly. The organization and structural integrity of a membrane bilayer, including its constituent proteins, also can break down at high temperatures. Because the effects of both low and high temperatures on membranes are somewhat dependent on lipid composition, many microbes actively adjust lipid composition in response to temperature. Differences in the temperature extremes to which bacteria, archaeons, and eukarya can adapt (Figure 6.7) may relate to differences in membrane lipid composition.

CONNECTIONS As we discussed in **Section 4.2**, the plasma membranes of some archaeons contain tetraether-linked glycophospholipids. These lipids generate a membrane monolayer with enhanced thermostability. Membranes of hyperthermophilic archaeons are greatly enriched in such molecules.

Figure 6.6. The psychrophilic alga *Chlamydomonas nivalis* The bright red pigment astaxanthin is concentrated in the spores of this alga and produces a pinkish tinge to snow, known as "watermelon snow." The alga thrives in cold environments, such as glaciers and snow-covered mountain peaks. Astaxanthin is produced under conditions of low nitrogen and high UV light. The pigment aids in resistance to UV light damage.

> ## 6.2 Fact Check
>
> 1. How do prototrophic and auxotrophic organisms differ?
> 2. What are the advantages and disadvantages of growing in an oxygen-rich environment?
> 3. Why does growing in a salty environment pose a challenge?
> 4. Name a couple of cellular components or processes that may be affected by changes in temperature.

6.3 Growing microorganisms in the laboratory

As we have seen in the preceding sections, different microbes have different physicochemical requirements for growth. Different microbes grow at different temperatures and utilize different carbon sources, to provide just two examples. To grow a specific microbe in a laboratory setting, then, we must understand the growth requirements of that species and also adequately recreate those requirements in the laboratory. In this section, we will examine how we can grow microbes. We will focus our exploration of this topic on four main areas. First, we will discuss properties of general microbial media. Second, we will examine the interesting features and uses of some specialized media. Third, we will look at techniques used for isolating pure cultures. Finally, we will investigate how we can characterize microbes that we currently cannot grow in the laboratory.

Media for microbial growth

In the laboratory, both liquid and solid forms of growth media, combinations of nutrients designed to allow the growth of microorganisms, may be used. Liquid media, sometimes referred to as broths, contain dissolved nutrients. A solid medium usually contains the same nutrients, along with a solidifying agent. Today, **agar**, a polysaccharide purified from red algae, is the preferred solidifying agent **(Figure 6.8)**.

Purified dehydrated agar, typically at a final concentration of 1.5 percent, is mixed with water and other media components. When this solution then is heated to a high temperature, the dehydrated agar melts. The high temperature also destroys most, if not all microbes present in the solution, thereby sterilizing it. When this molten agar begins to cool, it typically is poured into Petri dishes. The molten agar will transition to a solid gel when the medium cools to around 40°C. This solidified, hydrated agar provides a smooth, firm, moist surface on which microbes can grow. If the medium contains the nutrients a microbe needs, and other environmental conditions are acceptable, a cell deposited

on the agar surface will divide repeatedly to eventually produce a **colony**—a mound of clonal cells that grows large enough to be seen without a microscope **(Figure 6.9)**.

Pathogenic microorganisms generally are adapted to the complex mix of nutrients available in or on a host organism. In the laboratory, pathogens often prefer, or even require, a rich medium with abundant sugars, amino acids, and vitamins. Media used in clinical laboratories to grow bacteria such as *Neisseria meningitidis* and *Streptococcus pneumoniae*, two species that can cause meningitis, contain blood in part because hemoglobin is a good source of amino acids and iron. A medium containing blood is considered a **complex medium**, because the chemical components of the medium are not precisely defined in identity or concentration. Other common components of complex media include yeast extract composed of hydrolyzed yeast cells, and beef extract composed of beef muscle tissue that has been soaked and boiled and the soluble components extracted. Both these preparations are rich in amino acids, nucleosides, and various mineral-containing salts as they are extracted from cells. Peptone is prepared from acid hydrolysis of casein, the major protein in milk, and tryptone is prepared from trypsin treatment of casein. Both are rich in amino acids and peptides. By contrast, in a **defined medium**, the chemical identity and concentration of all of the constituents is known. **Table 6.2** compares the makeup of a complex medium (lysogeny broth, or LB) and a defined medium (M9), both of which can be used for the growth of *Escherichia coli*.

Specialized media

In some cases, the general media that we have just described may not be sufficient. A researcher, for example, may want to cultivate bacteria under conditions in which some types of bacteria will grow, but not others. Alternatively, a researcher may want several types of bacteria to grow, but be able to distinguish one type of bacterium from another. Perhaps the researcher wants to grow a

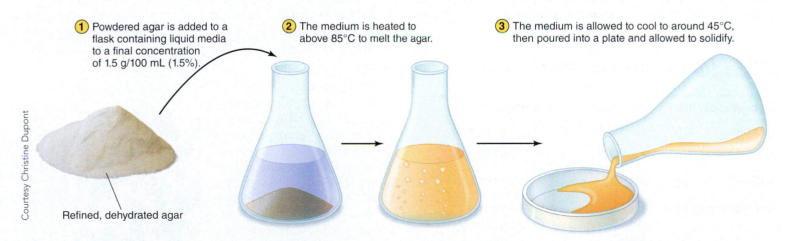

① Powdered agar is added to a flask containing liquid media to a final concentration of 1.5 g/100 mL (1.5%).

② The medium is heated to above 85°C to melt the agar.

③ The medium is allowed to cool to around 45°C, then poured into a plate and allowed to solidify.

Courtesy Christine Dupont

Refined, dehydrated agar

Figure 6.8. Process Diagram: Making agar plates to support microbial growth Agar is a purified polysaccharide used as a solidifying agent for solid media. Dehydrated agar powder is added to microbiological nutrient growth media, typically at a concentration of 1.5 percent. Following melting, the agar cools to form a stable gel.

Courtesy Christine Dupont

A. Compact circular colonies

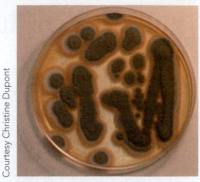

Courtesy Christine Dupont

B. Filamentous colonies

Figure 6.9. Colony growth on agar media When plated on an appropriate nutrient agar, microorganisms will grow, eventually forming a colony that is visible to the naked eye. Each colony is derived from a single cell deposited on the agar surface. **A.** Many microbes, including *E. coli* shown here, form compact circular colonies on the surface of agar plates. **B.** Some microbes, particularly fungi shown here, and soil bacteria such as *Actinomycetes*, form colonies composed of filaments that spread across the surface of the agar. **C.** Colonies can have a variety of morphologies and colors. Colony morphology, however, usually is not a reliable characteristic for identification.

Courtesy Christine Dupont

C. Variations in form, texture, elevation, and color

mixture of bacteria under conditions that allow one specific type to grow more rapidly than other types. In all of these scenarios, to borrow a phrase, there's a medium for that **(Table 6.3)**.

Selective Media

Clinical samples, like samples from any natural habitat, often contain many different species of microorganisms, with different growth needs and traits. A scientist interested in one particular type of microbe could use a **selective medium** designed to allow the growth of only certain target organisms. Selective media could have a limited set of nutrients, an unusually acidic or alkaline pH, unusual solute concentrations, or other chemical components that limit the growth of non-target species. A good

example is a medium that contains one or more antibiotics, so that only resistant microbes can grow.

Selective media are quite useful in the cultivation of several human pathogens, like *Bordetella pertussis* **(Microbes in Focus 6.2)**,

TABLE 6.2	Complex and defined media for growing *Escherichia coli*	
LB broth (per liter)	**M9 minimal salts broth (per liter)**	
Peptone	10 g	Glucose ($C_6H_{12}O_6$)	2 g (11 mM)
Yeast extract	10 g	Na_2HPO_4	6 g (42 mM)
NaCl	5 g	KH_2PO_4	3 g (22 mM)
(pH adjusted to 7.0)		NaCl	0.5 g (9 mM)
		NH_4Cl	5 g (93 mM)
		$MgSO_4$	1 mM
		$CaCl_2$	0.1 mM
		(pH adjusted to 7.0)	

Microbes in Focus 6.2
A PICKY EATER— *BORDETELLA PERTUSSIS*

Habitat: Respiratory tract of humans

Description: Gram-negative, coccus-shaped bacterium, approximately 0.4 to 0.8 μm in diameter. Expresses several virulence factors.

Key Features: This bacterium causes whooping cough (pertussis), a potentially severe respiratory disease. The availability of an effective vaccine has decreased dramatically the number of cases in developed countries, but it remains a major problem worldwide. It is a fastidious organism and typically is cultivated in the laboratory on solid medium supplemented with horse or sheep blood.

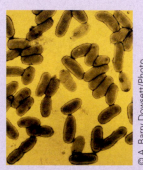

© A. Barry Dowsett/Photo Researchers, Inc.

TABLE 6.3 Common specialized media

Medium	Organism(s) identified	Selectivity and/or differentiation achieved
Brilliant green agar	*Salmonella*	**Selective** Brilliant green dye inhibits Gram-positive bacteria and thus selects Gram-negative ones. **Differential** Differentiates *Shigella* colonies (which do not ferment lactose or sucrose and are red to white) from other organisms that do ferment one of those sugars and are yellow to green.
Eosin methylene blue agar (EMB)	Gram-negative enterics (*Enterobacteriaceae*)	**Selective** Medium partially inhibits Gram-positive bacteria. **Differential** Eosin and methylene blue differentiate among organisms: *Escherichia coli* colonies are purple and typically have a metallic green sheen; *Enterobacter aerogenes* colonies are pink, indicating that they ferment lactose; and colonies of other organisms are colorless, indicating they do not ferment lactose.
MacConkey agar	Gram-negative enterics	**Selective** Crystal violet and bile salts inhibit Gram-positive bacteria. **Differential** Lactose and the pH indicator neutral red (red when acidic) identify lactose fermenters as red colonies and non-fermenters as white or tan. Most intestinal pathogens are non-fermenters and hence do not produce acid.
Triple sugar-iron agar (TSI)	Gram-negative enterics	**Not Selective** **Differential** Used in agar slants (tubes cooled in slanted position), where differentiation is based on both aerobic surface growth (slant) and anaerobic growth in agar in base of tube (butt). Medium contains specific amounts of glucose, sucrose, and lactose, sulfur-containing amino acids, iron, and a pH indicator, so relative use of each sugar and H_2S formation can be detected. **1** Uninoculated tube of TSI. **2** Inoculated: red slant and red butt = no change; no sugar fermented. **3** Yellow slant and yellow butt = lactose and glucose fermented to acid; trapped bubbles in butt indicate fermentation to acid and gas. **4** Red slant (lactose not fermented) and yellow butt (glucose fermented to acid); black precipitate = H_2S produced; sometimes obscures yellow butt. Almost all enteric pathogens produce red slant and yellow butt, with or without H_2S and/or gas.

Photo credits (vertical): © Fancy/Alamy Limited; Carolina Biological Supply Company/Phototake; © Science PR/Getty Images, Inc.; © LeBeau/Custom Medical Stock Photo; © CDC; © Elizabeth Fitzgerald

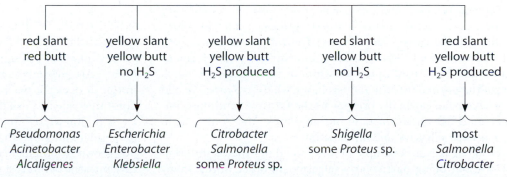

Differentiation of intestinal bacilli based on TSI

red slant red butt	yellow slant yellow butt no H_2S	yellow slant yellow butt H_2S produced	red slant yellow butt no H_2S	red slant yellow butt H_2S produced
Pseudomonas *Acinetobacter* *Alcaligenes*	*Escherichia* *Enterobacter* *Klebsiella*	*Citrobacter* *Salmonella* some *Proteus* sp.	*Shigella* some *Proteus* sp.	most *Salmonella* *Citrobacter*

the cause of whooping cough, and *Neisseria meningitidis*, a potential cause of meningitis. Both of these species require very nutrient-rich media, but grow so slowly that other bacteria and fungi easily overwhelm them. To avoid this conundrum, clinicians often streak *N. meningitidis* on a medium, often referred to as Thayer Martin medium, that contains four antibiotics: nystatin, vancomycin, colistin, and trimethoprim. Nystatin kills most fungi, and vancomycin kills most Gram-positive bacteria. Colistin and trimethoprim are effective against many Gram-negative bacteria, except for *Neisseria*. Trimethoprim is added especially to kill *Proteus* bacteria, Gram-negative bacteria that also happen to be resistant to colistin. In the presence of these antibiotics, *Neisseria* have a chance to grow and be detected. This specialized medium, in other words, allows researchers to selectively grow *Neisseria* species.

Another important application of selective media in the clinical laboratory involves the isolation of possible pathogens from fecal specimens that can contain hundreds of species of bacteria. A clinician interested in determining whether a patient harbors diarrheal pathogens such as bacteria of the genera *Salmonella* or *Shigella* might streak a fecal sample on a selective medium that allows the growth of these species, but inhibits the growth of many other genera. MacConkey agar, a complex medium, is selective because it contains bile salts, steroid-derived compounds that mammals secrete in the small intestine to emulsify (break up) fats and promote digestion. Bile salts also disrupt membranes and are inhibitory to most microbes, but many intestinal bacteria and Gram-negative pathogens have developed resistance to these detergents. This medium, then, allows for the selective growth of *Salmonella* and *Shigella* species.

Uninoculated MacConkey agar

Lactose-positive reaction
(*E. coli*)

Lactose-negative reaction
(*Salmonella enterica* Typhimurium)

Inhibited growth
(*Staphylococcus aureus*)

Courtesy Christine Dupont

Figure 6.10. Growth on selective and differential media MacConkey agar is both selective and differential. Bile salts and crystal violet dye inhibit the growth of most Gram-positive bacteria, such as *Staphylococcus aureus*. This medium, in other words, selects for Gram-negative bacteria. Because of the presence of lactose and the pH-sensitive dye neutral red in this medium, it also can be used to differentiate between lactose-fermenting and non-fermenting bacteria. Bacteria such as *Escherichia coli* ferment lactose and generate acids that lower the pH that converts the dye to its red form. Colonies of lactose-fermenting bacteria therefore appear dark red. Colonies of bacteria that cannot ferment lactose, like *Salmonella enterica* Typhimurium, appear white, gray, or tan.

Differential Media

While selective media allow researchers to selectively cultivate one or more types of bacteria, a **differential medium** allows a researcher to distinguish two, often related, types of bacteria. We already have been introduced to a differential medium. MacConkey agar, the selective agar that we just discussed, also allows some important species to be differentiated by colony color **(Figure 6.10)**. On MacConkey agar, *Escherichia coli* colonies are red because *E. coli* ferments lactose to produce acid, which lowers the pH and causes an indicator dye in the medium to turn red. *Salmonella* and *Shigella* species, in contrast, do not ferment lactose and, as a result, produce white or tan colonies. *E. coli* is part of the normal intestinal microbiota, whereas *Salmonella* and *Shigella* are not normally present in the GI tract, and the clinician would like them to stand out on plates. MacConkey agar, then, is both selective and differential.

Blood agar is an example of a rich medium that is differential, without being particularly selective. In other words, many

organisms can grow on the medium, but some can be specifically recognized. The differential property derives from the ability of some microbes to release enzymes called "hemolysins" that attack the cytoplasmic membrane of the red blood cells suspended in the medium, causing the cells to lyse. This hemolysis creates either a clear or a greenish zone around the bacterial colony (see Figure 21.21). Hemolytic activity on blood agar is particularly useful for differentiating species of the genus *Streptococcus* **(Figure 6.11)**. Hemolytic activity is a reliable characteristic of *Streptococcus pyogenes*, a common cause of throat infections ("strep throat"), as well as a few other types of infections in humans. In contrast, other streptococci and related Gram-positive cocci cause partial or no hemolysis on blood agar. Streaking a throat swab on a blood agar plate is a useful initial diagnostic for patients suspected to have strep throat.

Enrichment Media

An **enrichment medium** is one that contains nutrients or other components designed to favor the growth of particular microbes. An enrichment medium differs from a selective medium in that it does not necessarily prevent the growth or survival of non-target species, usually because there is no good way to selectively inhibit these other species. Enrichment media are used to help isolate microbes that are rare in a population, and might be overgrown by other microbes in a non-enrichment culture. If a rare microbe of interest grows better on a particular enrichment

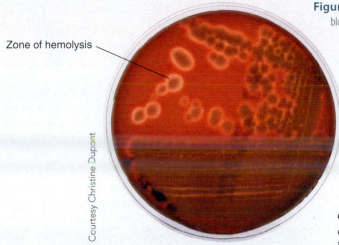

Zone of hemolysis

Courtesy Christine Dupont

Figure 6.11. Hemolytic activity Many types of bacteria can cause lysis of red blood cells, indicated by a distinctive clear zone around the colonies when cultured on blood agar. Shown here is *Bacillus cereus* on sheep blood agar.

medium than other, initially more numerous microbes, then the concentration of the initially rare microorganism will increase over time.

CONNECTIONS Enrichment media can be used to isolate bacteria that metabolize a specific carbon source, such as oil. These bacteria, in turn, could be useful in cleaning up oil spills. The use of bacteria in this way will be discussed in **Section 15.4**.

When trying to cultivate microbes from environmental samples, microbiologists must recognize that every medium and incubation method is selective and/or enriching in some way. A rich medium does not make every microbe equally likely to grow. Rich media enrich for microbes that are nutritionally versatile, have the fastest and/or highest capacity transport systems, and are adapted to high nutrient fluxes. But chemicals that are used as nutrients by some microbes inhibit the growth of others. If we want to grow microbes that are adapted to very low concentrations of organic carbon or nitrogen, for example, then we can use a medium with appropriately low concentrations of these materials—as long as we recognize that these microbes tend to grow very slowly, and plan our experiment accordingly.

Aside from nutrient composition, environmental conditions also exert powerful selective forces on microbial growth. Incubating plates aerobically in a standard atmosphere of 20 percent O_2 selects against anaerobes and microaerophiles, while incubating anaerobically does just the opposite. As described in **Perspective 6.1**, lowering the oxygen concentration helped select the pathogen *Helicobacter pylori* from biopsies of gastric ulcers. Any given temperature will enrich for microbes evolutionarily optimized for that temperature, and disfavor species optimized for higher or lower temperatures. Incubating in the dark selects against obligate phototrophs, while incubating in the light can enrich the growth of phototrophs and inhibit growth of organisms with inadequate systems to repair cellular damage caused by ultraviolet light.

Obtaining a pure culture

To isolate a microbe on a solid medium, microbiologists can produce a **streak plate** (**Figure 6.12**). This process involves spreading microbial cells across the surface of the plate, so that individual bacterial cells become spatially separated. If these cells can use the nutrients in the medium to grow and divide, then each cell eventually produces a colony (see Figure 6.9). The advantage of a solid medium is that cells are held in place. Isolated colonies should contain a single type of microorganism; all cells within the colony should be descended from a single cell. It takes millions of microbial cells piled together to form a colony that we can see without a microscope. Some microbes, such as *Escherichia coli* growing on a rich medium, grow fast enough to form visible colonies overnight. Other species, such as *Mycobacterium tuberculosis*, may take weeks to form visible colonies.

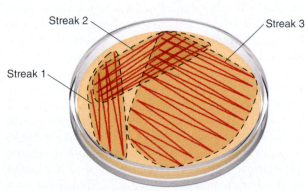

Streak 2

Streak 3

Streak 1

Black, *Microbiology: Principles and Explorations*, copyright 2012, John Wiley & Sons, Inc. This material is reproduced with permission of John Wiley & Sons, Inc.

A. Streak plate method for isolation of colonies

Courtesy Christine Dupont

B. Isolated colonies after incubation

Figure 6.12. Streaking plates to isolate cells and separate colonies A. Petri plates containing an agar medium are divided into sectors. A loop of a culture is inoculated near the edge of the plate (Streak 1). The loop then is sterilized by flaming to kill any microorganisms remaining on it. The plate is rotated, and the sterile loop is streaked across a second sector that crosses the edge of the previous streak to transfer some cells (Streak 2). This procedure is repeated one (Streak 3) or two more times. The loop leaves progressively fewer cells in each sector. **B.** After the plate has been inoculated, it is incubated. Isolated colonies of bacteria should appear in the latter sectors.

In the early 1980s, Barry Marshall was a young physician working at a hospital in Western Australia with Robin Warren, a pathologist. The two men spent endless hours peering into microscopes examining biopsy samples from patients with gastritis, an inflammation of the lining of the stomach. Marshall and Warren frequently saw curved and spiral bacteria **(Figure B6.2)** associated with the gastric mucosa, the layer of mucus-secreting cells lining the stomach. This observation puzzled the researchers; bacteria rarely had been cultured from this site. The pH 2 acidity of the stomach creates a very harsh environment. Exposure to such acidity kills most microorganisms as food passes through the stomach. Could bacteria really colonize this niche, and could they be a cause of gastric pathologies?

Marshall attempted to culture bacteria from these biopsy specimens on rich media designed for *Campylobacter*, a genus of curved, polarly flagellated bacteria sometimes found in the human intestinal tract, which these bacteria resembled. He also reduced the oxygen concentration in the incubator to the low levels that microaerophilic *Campylobacter* prefers, but still no colonies appeared. A few days later, Marshall had an unexpected surprise when he returned from a long Easter weekend. The plates he had left in the incubator for five days now had colonies on them! Marshall and Warren initially named these slowly growing bacteria *Campylobacter pylori*. The name was later changed to *Helicobacter pylori*, as molecular evidence made it clear that, despite the morphological similarity, these bacteria are not phylogenetically aligned with the Campylobacters.

Marshall and Warren's discovery turned out to be an enormous breakthrough. Evidence has accumulated that *Helicobacter pylori* is the underlying cause of most gastric (stomach) and duodenal (intestinal) ulcers in humans. This bacterium has been linked to gastric cancers as well. Antibiotic treatment targeted at *H. pylori* is now a routine part of ulcer treatment, and is much more effective than simply using drugs that block stomach acid production, as was done previously. Millions of people have benefited from this discovery, for which Marshall and Warren were awarded the Nobel Prize in Physiology or Medicine in 2005.

If Barry Marshall had not been persistent in his attempts to cultivate the bacteria he observed in the microscope, then where would we be now? Vast arrays of microbial species on Earth are poorly understood because they have not been cultivated in the laboratory, although molecular tools are changing that. The *Helicobacter* story teaches us not to make too many assumptions about where microbes can or cannot grow—and that we must be persistent, patient, and clever in our efforts to study such microbes.

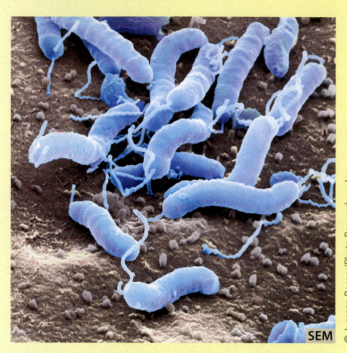

© Juergen Berger/Photo Researchers, Inc.

Figure B6.2. *Helicobacter pylori* **and ulcers** These spiral-shaped bacteria were observed in biopsies from patients with gastritis. These acidophilic bacteria survive the highly acidic (pH 2) environment of the stomach on their transit to colonize the more neutral tissues that line the stomach where they contribute to the development of ulcers.

Isolated colonies also can be generated by the spread or pour plate procedures **(Figure 6.13)**. First, dilutions of a bacterial culture are made. In the spread plate procedure, aliquots of each dilution are added to plates containing solidified nutrient agar. The aliquots then are spread evenly over the surface of the plate. In the pour plate procedure, aliquots of dilutions are added to tubes of melted nutrient agar. The tubes then are mixed and added to empty Petri plates. In both cases, following incubation, colonies should be evenly distributed across the plates. As we will see in the next section, this method can be used to quantify the number of bacteria in a sample.

How do these methods compare? Spreading a sample on a plate is easy and does not require molten agar at the right temperature. A defined volume of the sample simply is added to the plate and dispersed over the surface. The disadvantage of this procedure is that colonies may overlap on the surface of the agar, which makes counting difficult and error-prone. The pour plate method embeds colonies in agar, which keeps them smaller and reduces overlap, so that more discrete colonies can be counted. Microbes that are particularly sensitive to heat, though, might suffer some mortality after even brief exposure to molten agar.

When attempting to initially isolate and cultivate microbes in the laboratory, or measure microbial populations, researchers may utilize relatively rich media. After all, if the goal is rapid growth for detection, isolation, or counting, then it seems reasonable to offer microbes lots of food, with perhaps a buffet of different menu items. This approach often works well, but there are situations in which an excessively rich medium can suppress growth, or even be lethal. This apparent contradiction is particularly true for microbes adapted to life in nutrient-poor environments. The bottom line is that the medium should be well suited to the microbe you want to isolate. A problem, though, still remains. We don't know what growth conditions all bacteria prefer.

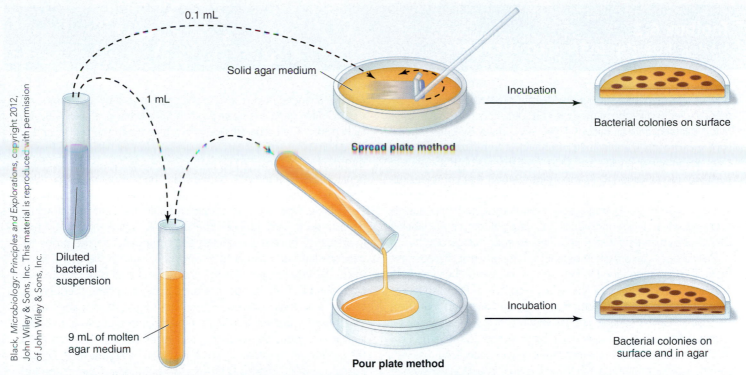

0.1 mL

Solid agar medium

Incubation

Bacterial colonies on surface

Spread plate method

1 mL

Diluted bacterial suspension

9 mL of molten agar medium

Incubation

Bacterial colonies on surface and in agar

Pour plate method

Figure 6.13. Spread plate and pour plate protocols To generate evenly dispersed colonies, aliquots of dilution of a bacterial culture can be grown by the spread plate method (*top*), or the pour plate method (*bottom*). After incubation, individual colonies should be visible on plates inoculated with certain dilutions of the original culture. The number of viable microorganisms in the original culture can be calculated from the colony counts.

Unculturable bacteria

When examining microbial populations in environmental samples, microbiologists often note that, based on microscopic observations, there appear to be far more cells present in the sample than can be cultivated as colonies on plates. Often the difference is extreme, with the apparent density of cells in the sample being 10 to 100-fold greater than the density calculated based on the colonies that grow when the sample is plated on laboratory media. This stark discrepancy has become known as the "great plate count anomaly," a phrase coined by microbiologists James Staley and Allan Konopka in describing their work on aquatic and soil samples. This anomaly applies even to the microbes of our own bodies. Only about 30 percent of the microbial inhabitants of the human colon form colonies in the laboratory under anaerobic conditions at 37°C, using a variety of media. Under aerobic conditions, less than 10 percent form colonies.

What accounts for this anomaly? We've discussed many factors that affect microbial growth, from nutrient composition to environmental factors such as pH, osmosis, oxygen, temperature, and light. If critical nutritional needs for an organism are not satisfied, or if one or more environmental parameters are not within acceptable ranges, then growth may not happen at an appreciable rate. For a macroscopically visible colony to arise from a single cell, the starting cell must grow and divide for 20 generations or more. Colony formation seems so simple, because it happens readily for the few thousand species of familiar microorganisms. And yet, the evidence is undeniable that even more species of microbes elude cultivation because we have not found the right combination of nutritional and environmental conditions.

To compensate for the limitations of laboratory cultivation, **cultivation-independent methods** of analyzing microbes have been developed. It has been estimated that 0.1 percent or less of the bacteria in the open ocean form colonies by standard laboratory plating techniques. In the 1990s, Steven Giovannoni and colleagues identified a type of bacterium, which they designated SAR11, which had not previously been isolated or cultivated. SAR11 was identified initially by using the polymerase chain reaction (PCR) to amplify ribosomal RNA genes from DNA isolated from microorganisms collected on filters from ocean water samples. Nucleotide sequences were determined for a large number of the amplified 16S rRNA genes. Amplification, cloning, and sequencing of ribosomal RNA genes is a common method for assessing microbial diversity in an environmental sample, because it does not require microorganisms actually to grow in the laboratory (see Figure B1.4).

As we discussed in Chapter 1, ribosomal RNA sequences can be compared to place organisms within the overall phylogeny of life. The number of different rRNA genes isolated, and their relative abundance in the collection, can give rough estimates of diversity and population structure. Fluorescent staining methods, like the one shown in **Toolbox 6.2** using labeled DNA targeted to specific microbes can then be used to further investigate population structure. Such techniques indicate that SAR11 may constitute a quarter or more of the bacterial cells in marine environments worldwide. In the last decade, similar studies have been undertaken in hundreds of other habitats, including the human intestine (see Section 7.4). These surveys reveal the types of microbes present, regardless of which species can be cultivated in the laboratory.

How can we learn more about the identity, physiology, and environmental roles of microbes if they can't be cultivated in the laboratory? Many bacteria and archaeons look very similar by conventional light microscopy. Identification of uncultivated microbes therefore takes advantage of nucleic acid sequences that are both unique and useful for identification. One important method that combines microscopy and detection of specific nucleic acid sequences is FISH (Fluorescence *In Situ* Hybridization). FISH detects particular nucleic acid sequences inside individual cells using fluorescently labeled DNA probes that are detectable by fluorescence microscopy.

FISH probes can assign an organism to a particular phylogenetic group if they are designed to be complementary to ribosomal RNA sequences characteristic of that group. Ribosomal RNA sequences are highly conserved over broad phylogenetic distances, but there are some relatively variable regions within which sequences can be targeted that are only shared by closely related species. The FISH method starts with a sample fixed onto a glass slide, such that cells are physically immobilized. The sample is then chemically treated to make the cell membranes permeable, and to denature their DNA.

A single-stranded nucleic acid probe complementary to a portion of the 16S rRNA gene is prepared with a fluorescent dye covalently attached to it. The prepared slide is incubated with the probe, under conditions where the probe DNA will "hybridize" with its target. That is, the probe will bind specifically to the complementary sequence of the denatured DNA in the cells on the slide. The non-bound probe then is washed away, and the slide is viewed by fluorescence microscopy.

Multiple types of microorganisms can be detected simultaneously using this method. The ultraviolet light emitted by the fluorescence microscope excites the compound on the probe to generate light of a particular wavelength, causing the cell to fluoresce. The wavelength, or color, that is produced is unique to the compound. Different probes designed to detect different sequences can be individually labeled with different fluorescent compounds. When used together in a single hybridization procedure, the collection of probes can detect a variety of different organisms in the sample, an example of which is shown in **Figure B6.3**. Digital imaging systems then can be used to photograph the image for subsequent analysis of community composition and organization.

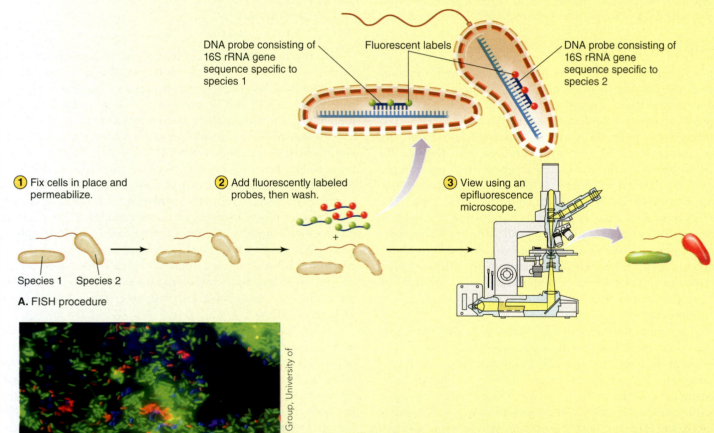

DNA probe consisting of 16S rRNA gene sequence specific to species 1

Fluorescent labels

DNA probe consisting of 16S rRNA gene sequence specific to species 2

① Fix cells in place and permeabilize.

② Add fluorescently labeled probes, then wash.

③ View using an epifluorescence microscope.

Species 1   Species 2

A. FISH procedure

B. FISH image of mixed bacterial sample

Courtesy Wei Huang Group, University of Sheffield

Figure B6.3. Process Diagram: Fluorescence *in situ* hybridization (FISH) To distinguish species of microorganisms in a mixture, fluorescently labeled nucleic acid probes can be used. **A.** Cells first are fixed and permeabilized on a solid support. Fluorescently labeled nucleic acid probes, such as those designed to detect 16S rRNA genes specific for particular species, are added to the cells. Non-bound probe is washed away and the cells are viewed with a fluorescent microscope. If probes specific for different species are labeled with different fluorescent markers (green and red in this example), then cells from these species can be distinguished. **B.** Image of microbial mixture in which three different nucleic acid probes (labeled green, red, and blue) were used.

As described in **Toolbox 6.2**, fluorescently labeled DNA probes can be used to identify cells. In **Toolbox 15.3**, we will examine another powerful technique involving fluorescence—flow cytometry. With this technique, researchers can identify and then isolate specific cells for further experimentation.

Metagenomics circumvents the need for individual culturing of microbes by using direct genome analysis from whole communities of microbes to search for genes. To identify different genera and species in microbial populations, often ribosomal RNA genes are isolated and sequenced as will be discussed in **Section 10.4**.

The use of rRNA sequences for phylogenetic identification of uncultivated microbes has practical applications. For example, it can suggest what kinds of genes, metabolic pathways, and structures a microbe may possess. This information, in turn, could suggest media composition and environmental conditions for cultivation efforts, based on what works for related microbes that can be cultured.

In the past decade, scientists have taken an even bigger step toward gaining information from uncultivated microbes. In **metagenomics**, DNA isolated from an environmental sample is randomly cloned, and sequences are determined from the cloned DNA fragments. The gene content of microorganisms inhabiting a particular site can be examined, regardless of whether they can be cultivated. Our accumulated knowledge of metabolic genes, nutrient transport, regulation, and cell structure is used as a framework for interpreting the metagenomic data set. This approach gives greater insight into the lifestyles of resident microorganisms than ribosomal RNA profiles alone, although metagenomic data sets tend to be biased toward the numerically dominant species in the habitat.

We should note, though, that metagenomics does have limitations. While metagenomics provides information about the genes present in the uncultivated fraction of microbial communities, and the functions of these genes can be studied in surrogate host organisms, a more complete understanding of the biology of these organisms requires that they be isolated in pure culture. To achieve this goal, it is necessary to devise new methods for cultivation **(Mini-Paper)**.

Before leaving our discussion of unculturable microbes, let's address one more factor affecting our ability to grow microbes in the laboratory. Microbes commonly live in communities containing many species. Often, these communities take the form of a biofilm, a complex community of microorganisms adhered to a solid surface. In situations where communities are stable over long periods, microorganisms may evolve dependent relationships. For example, the by-products of one species' metabolism could provide a nutrient source for another species. Alternatively, the activities of some organisms could affect the pH of a local environment, thereby making the area more hospitable to other organisms. A microbial community in which the growth or survival of members is dependent on the community is referred to as a **consortium**.

Microbial consortia are often present in stable sediments. **Figure 6.14** shows a shallow hypersaline pond in Baja California Sur (Mexico) where microbial consortia have been studied. The high salt concentration of the pond reduces grazing by multicellular organisms, allowing thick microbial mats to develop on the bottom. Such mats often have photosynthetic cyanobacteria on top if the water column is shallow enough for light penetration to the sediment surface. The light-harvesting bacteriochlorophyll from these cyanobacteria tints the top layer of the sediment green. Some of the O_2 generated by oxygenic photosynthesis diffuses downward, where aerobic heterotrophic microbes quickly consume it. Typically, O_2 levels are depleted significantly within a few millimeters of the sediment surface in aquatic habitats. If the sediment also contains significant sulfate (SO_4^-) or sulfur

A. Saline pond

B. Mat surface

Figure 6.14. Microbial consortia show dependence of species on the community. Microbial consortia are often observed in stable, layered sediments, such as in hypersaline ponds in the Exportadora de Sal saltern system, Guerrero Negro, Baja California Sur, Mexico. Sampling the mats reveals a variety of interdependent microbial communities. **A.** Thick microbial mats line the bottom of this entire pond. **B.** The surface of a microbial mat from the pond. **C.** Many layers are present in the mat. The smallest graduations in the ruler (*left*) are millimeters. **D.** Organisms of the mat include filamentous cells of the cyanobacterium *Microcoleus chthonoplastes* and small dark spheres that are probably purple sulfur bacterium.

C. Cross section of mat

D. Micrograph of a sample of the mat

Mini-Paper: A Focus on the Research
BRINGING TO LIFE THE PREVIOUSLY UNCULTURABLE USING THE SOIL SUBSTRATE MEMBRANE SYSTEM (SSMS)

B. C. Ferrari, S. J. Binnerup, and M. Gillings. 2005. Microcolony cultivation on a soil substrate membrane system selects for previously uncultured soil bacteria. Appl Environ Microbiol 71(12):8714–8720.

Context

Although advances in the science of microbiology depended on the cultivation of microbes, a large fraction of the microbes in any given habitat have not been cultured. How can we learn about the functions and roles of the diverse uncultured members of microbial communities and why is it so difficult to cultivate many members of microbial communities? First, traditional culture media favor organisms that grow fast and tolerate high levels of nutrients. Many of the uncultured microbes do not grow quickly. Second, many of these organisms also may be adapted to growth under very low nutrient conditions; they may be unable to tolerate high nutrient levels. For these reasons, attempts have been made to use low nutrient agar to successfully grow microbes that have been difficult to grow, but often these organisms form microcolonies that are only visible under the microscope, even after several weeks of incubation.

The study presented in this paper describes a method for cultivation of soil bacteria in which non-sterile soil, and the associated natural microbial community, provides nutrients for the growing microcolonies.

The growing colonies then could be visualized by microscopy, but were not visible to the unaided eye. This method, called the soil substrate membrane system (SSMS), mimics the natural soil environment, allowing the cultivation of microbes that have not been successfully cultivated using other methods.

Experiment

These researchers made use of an apparatus that was originally developed for the cultivation of mammalian cells whose growth is dependent on binding to a solid surface to supply growing cells with nutrients that closely reflect the natural habitat of soil microbes. This apparatus, called a tissue culture insert, consists of a vessel that is sealed with a filter. Garden soil was diluted 1:200 in water and mixed vigorously to prepare the inoculum, which then was collected by filtration onto a membrane. This membrane, containing cells of soil microbes on its surface, then was placed face up on top of another membrane on a tissue culture insert apparatus that contained soil. Nutrients were able to diffuse through the membranes, feeding the inoculum that was on the membrane. Incubation was carried out at room temperature, in the dark, for up to 10 days.

Following incubation, the membrane containing cells was removed from the apparatus, stained with SYBR Green, which stains

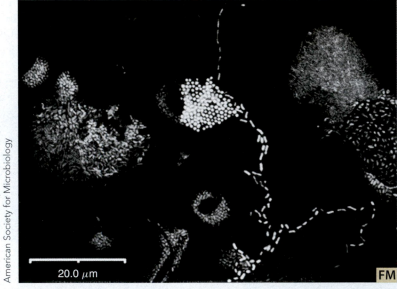

A. Diverse cell morphologies within several colonies

B. Uniform morphology within a single colony

Figure B6.4. Microscopic appearance of microcolonies Following incubation, sections of the polycarbonate filter were stained and viewed with a confocal microscope. **A.** Several different cell morphologies were present. **B.** Clusters of cells exhibiting a common morphology also were evident.

DNA within cells, and visualized by confocal laser scanning microscopy (see Appendix B: Microscopy). Microcolonies, which were defined as three or more cells that were closely associated, were clearly observed (**Figure B6.4**), and these microcolonies contained up to 200 cells. Although many different cell types were observed, in most cases each microcolony contained a single cell type. Larger colonies, visible to the naked eye, were not observed on any of the membranes.

To determine the identity of the bacteria in the microcolonies, cells were washed from the membranes, DNA was extracted and PCR was amplified using primers for 16S rRNA genes. Subsequent DNA sequence analysis demonstrated that the majority of the sequences were from uncultivated members of the Betaproteobacteria, with some Gammaproteobacteria, Actinobacteria, and TM7. Of particular interest was TM7, a lineage of bacteria that has never been cultivated but has been found in many different habitats, including soil, dental plaque, and wastewater treatment systems. FISH (see Toolbox 6.2) with TM7-specific probes was used to identify TM7 microcolonies on 7-day-old membranes (**Figure B6.5**). Two TM7 microcolony types were detected, one containing short rods, and another containing long, filamentous rods.

Is it possible to cultivate the isolated microcolonies in pure culture? Eight portions of membrane containing distinct microcolonies were cut out, the cells suspended in sterile buffer, and then each transferred to a new SSMS apparatus and incubated. Only one of these subcultures resulted in growth, and this isolate was determined by 16S rRNA sequence analysis to be a member of the Betaproteobacteria. It was not capable of growth on rich medium.

Impact

The microcolony-forming bacteria detected in these studies reflect an interesting growth strategy that is specialized for very low nutrient conditions. The use of natural soil as nutrient ensures that not only are appropriate sources of carbon and energy provided, but also signaling molecules that might influence growth. Further attempts in subsequent studies have resulted in the successful cultivation of many more of the isolated microcolonies on low nutrient agar media. While this strategy was demonstrated using garden soil, it could be fairly easily adapted to the cultivation of microbes from other habitats. It will be particularly interesting to compare the microbes that are cultivated using this approach with the sequences that are discovered through metagenomics.

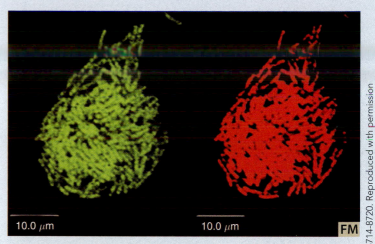

A. Short rods of TM7 bacteria

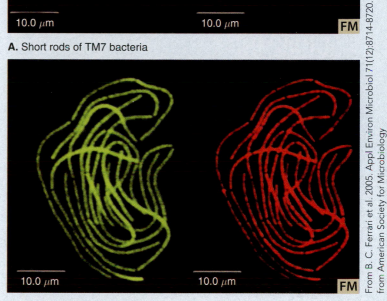

B. Filamentous TM7 bacteria

Figure B6.5. FISH analysis of microcolonies FISH was used to tentatively identify microorganisms in microcolonies. A DNA probe specific for TM7 bacteria (red) and a more general DNA probe for most bacteria (green) were used. **A.** Some colonies consisted of short, rod-shaped cells. **B.** Other colonies consisted of long rods in looping chains.

● Questions for Discussion ....

1. Why do you think non-sterile soil slurry was used as the nutrient source rather than sterilized soil?

2. The isolated microcolonies were not able to form colonies on rich media. What is the significance of this finding?

(S^0) along with organic carbon, the anoxic zone may be dominated by heterotrophs that consume organic material and use sulfur compounds as terminal electron acceptors for anaerobic respiration. This type of respiration generates H_2S, which can be toxic, but may be consumed in turn by anaerobic phototrophs that capture residual light penetrating into the anoxic zone. These microbes strip electrons from H_2S to fuel photosynthetic reduction of CO_2 for autotrophic growth. It is clear that fascinating and complex biochemical transactions occur in these microbial communities. Indeed, researchers working with NASA are studying these microbial mats because similar habitats and microbial communities may have been present on ancient Earth.

Microbes adapted to living in consortia may be difficult or impossible to cultivate as isolated species. If we know how the various species interact with each other, though, then clever design of the medium may allow some of the dependencies to be eliminated or chemically substituted. Some members of the community may be eliminated if we can artificially add their contribution to the community. Given that we often do not understand the complex interrelationships in microbial communities, it remains difficult or impossible to grow many members of natural communities and consortia in pure culture.

6.3 Fact Check

1. How do complex and defined media differ?
2. Give examples of selective and differential media.
3. How can a pure culture be obtained?
4. How can we characterize unculturable microorganisms?

6.4 Measuring microbial population growth

How can we measure and monitor microbial populations?

As we have seen in the previous section, various media have been developed to facilitate the growth of microbes in laboratory settings. Likewise, several techniques exist for the measurement of microbial growth. First, we will examine techniques that allow us to determine the number of microbes present in a specific location. Second, we will explore techniques that allow us to measure the rate at which microbial populations grow. Third, we will examine the mathematical analysis of microbial growth. Finally, we will investigate a specific growth scenario in which nutrients are supplied continuously. Let's begin by examining how we can determine how many microbes are present.

Direct counts

There are several ways to measure microbial populations. Most simply, we can count the number of cells that are present. The most direct method is to count cells using a microscope. To convert such data to population density, researchers must count the number of cells in a defined volume. This type of measurement can be done using a **Petroff–Hausser counting chamber (Figure 6.15)**, a specially constructed microscope slide with a chamber of defined depth, overlaid by a cover slip with a grid marking off squares of defined area. When the user counts the

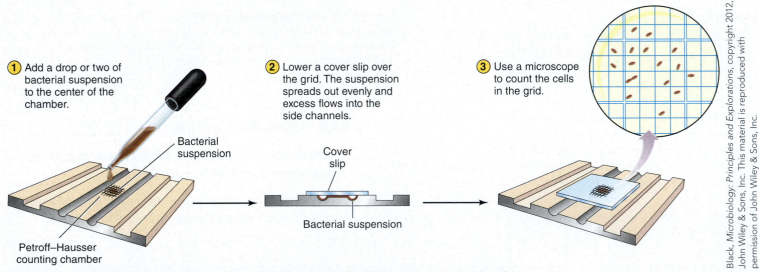

① Add a drop or two of bacterial suspension to the center of the chamber.

Bacterial suspension

Petroff–Hausser counting chamber

② Lower a cover slip over the grid. The suspension spreads out evenly and excess flows into the side channels.

Cover slip

Bacterial suspension

③ Use a microscope to count the cells in the grid.

Black, *Microbiology: Principles and Explorations*, copyright 2012, John Wiley & Sons, Inc. This material is reproduced with permission of John Wiley & Sons, Inc.

Figure 6.15. Process Diagram: Counting bacterial cells with a Petroff–Hausser counting chamber
This specialized microscope slide provides a defined volume within which to count cells. The smallest grid squares are typically 50 μm per side, and the chamber is 20 μm deep. Knowing the dimensions of the chamber allows the microbiologist to convert the number of microbes observed per grid into cell density (cells/mL).

number of cells in several grid squares, the average cell density in the chamber can be calculated.

This approach, though, presents several problems. As you might imagine, it can be quite labor intensive. Additionally, the cells counted may not be alive. Generally, a researcher will be more interested in the number of live cells rather than the number of total cells present in a sample. So, how can we identify live cells? A common way of counting living cells is to add dilutions of a culture sample to a solid medium that supports microbial growth **(Figure 6.16)**. Following a period of incubation, each viable cell will give rise to a visible colony. By counting the number of colonies, then, we can determine the number of viable cells added to the plate.

Plate count data are used to calculate microbial population density as **colony-forming units (CFUs)** per milliliter of culture. To illustrate this point, let's say a microbiologist wants to examine the growth of *Escherichia coli* in M9 medium with glucose as the carbon and energy source. At several intervals after the initial inoculation of *E. coli* into a flask of M9 medium, she takes a sample and serially dilutes it in tubes of sterile saline solution, as shown in **Table 6.4**. She then spreads 0.1 mL of each dilution on lysogeny broth (LB) agar plates, and incubates the plates overnight at 37°C. The next day, she observes colonies on all of the plates and attempts to count them. The concentration of cells in the M9 culture, in colony-forming units per milliliter, is extrapolated for each time point using the equation:

$$\text{Culture density (CFU/mL)} = \frac{\text{number of colonies}}{(\text{dilution}) \times (\text{volume plated, in mL})}$$

Samples were taken at the indicated times after initial inoculation. Dilutions were done by initially mixing 1 mL of culture with 9 mL of sterile saline, to give a final volume of 10 mL. This is referred to as a 1:10 dilution, or in mathematical terms a 1×10^{-1} dilution. For serial dilutions, 1 mL of the

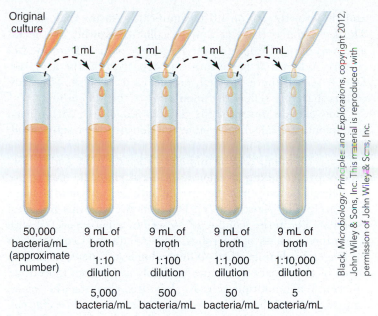

50,000 bacteria/mL (approximate number)

9 mL of broth
1:10 dilution
5,000 bacteria/mL

9 mL of broth
1:100 dilution
500 bacteria/mL

9 mL of broth
1:1,000 dilution
50 bacteria/mL

9 mL of broth
1:10,000 dilution
5 bacteria/mL

Figure 6.16. Serial dilution of cultures to isolate colonies and determine population density A series of 1:10 dilutions is made from an original bacterial culture. Each sample, then, should have tenfold fewer bacteria than the previous dilution. If a known volume of each dilution is plated, then the number of viable cells per milliliter in the original sample can be calculated.

1×10^{-1} dilution was added to another 9 mL of sterile saline to give a 1:100, or 1×10^{-2} dilution, and so forth. TNTC is shorthand for "too numerous to count." The asterisk * indicates that the cell density could not be calculated accurately because the plates either had too many colonies to be counted (TNTC), too many to be counted accurately (>300), or too few colonies to be meaningful (<10).

Why did the microbiologist use LB agar for plating, when her experiment was intended to monitor growth in M9 medium? Agar plates made with M9 medium could have been used for plating. LB agar was chosen simply because it is a rich medium,

TABLE 6.4	Measuring viable cell density in a bacterial culture by spread plating				
Time	**Dilution (before plating)**	**Volume spread on plate**	**Number of colonies observed (replicate plates)**		**Calculated CFU/mL**
1 hr	none	0.1 mL	TNTC	TNTC	*
	1:10 (1×10^{-1})	0.1 mL	499	518	*
	1:100 (1×10^{-2})	0.1 mL	83	87	8.5×10^4
	1:1,000 (1×10^{-3})	0.1 mL	10	8	*
2 hr	1×10^{-1}	0.1 mL	TNTC	TNTC	*
	1×10^{-2}	0.1 mL	190	179	1.85×10^5
	1×10^{-3}	0.1 mL	25	26	2.55×10^5
	1×10^{-4}	0.1 mL	4	2	*
3 hr	1×10^{-2}	0.1 mL	408	435	*
	1×10^{-3}	0.1 mL	66	60	6.3×10^5
	1×10^{-4}	0.1 mL	8	10	*
	1×10^{-5}	0.1 mL	1	0	*

and the growth rate on LB is significantly faster than on M9. The key point is that each of the colonies growing on the LB agar was derived from a single *Escherichia coli* cell from the M9 culture. When that bacterium was deposited on the agar surface, it stayed there, divided, and formed a colony. The plate counts therefore reflect the population of the M9 culture.

How can we interpret the data in Table 6.4? Within each time point, the calculated viable cell densities vary substantially. These variations may be caused by several factors, including inaccuracies in pipeting or mixing of each dilution, and the degree of overcrowding of colonies on each plate. If colonies are too densely packed, then they are difficult or impossible to count accurately. The shorthand TNTC acknowledges this potential problem. The term "confluent" describes a plate covered with a continuous mass of microbes, where individual colonies can no longer be distinguished. When colonies are densely packed, many of the apparent colonies probably arose from more than one cell. When multiple cells are deposited close together on an agar surface and their progeny merge to form a single visible colony, subsequent colony counts underestimate the true number of cells plated. But how do you know when colonies are "too dense"? The answer is somewhat arbitrary, but most microbiologists consider more than 300 or so colonies on a standard 100-mm Petri dish to be too many to give reliable data. Plates with less than 10 colonies are often excluded in calculations, unless several plates are counted at that dilution. Why? In statistical terms, small numbers lead to high variability and, arguably, unreliable data, unless the sample size (the number of plates counted) is large.

With respect to data interpretation, let's first apply the data standards discussed above. Plates containing more than 300 colonies or fewer than 10 colonies, will be excluded. From the remaining data, the means of the replicate samples for each dilution and time point are determined. To calculate CFU/mL, only one usable dilution exists for the 1- and 3-hour time points, giving us CFU/mL values of 8.5×10^4 (at 1 hour) and 6.3×10^5 (at 3 hours). For the 2-hour time point, both dilutions have valid data, so the mean is taken of the calculated CFU/mL values for both dilutions, yielding 2.2×10^5 CFU/mL. As we can see from these data, the number of viable cells in the culture increased over time.

Plate counts of this sort do not work for all microbes. If cells tend to grow in filaments or chains, or aggregate into clumps that cannot be separated easily into individual cells, then plate counts will underestimate cell density in the parent culture because a colony could have been derived from multiple cells deposited in a single site on the plate. The fate of the cells after plating also can be an issue. For discrete colonies to form, cells need to remain fixed in place after deposition on the agar. Many microbes, though, can move through a film of water remaining after plating, or can actively move on solid surfaces. If cells are still moving as they are growing and dividing, then we are more likely to observe either large and/or confluent colonies, or more colonies than should be present given the number of cells initially deposited on the plate. If such problems arise, using the pour plate method may work better than spread plates, since pour plates trap cells and the resultant colonies inside the agar, where they cannot move or spread as readily.

Environmental samples, such as water samples from a pond, may have low microbial population densities, or low densities of specific microbes of interest. If the population density is too low to be practical for plate counts or microscopic analysis, then cells can be concentrated in the sample before plating. Membrane filtration is a convenient way to concentrate bacterial cells in liquid samples. A membrane with pores smaller than the diameter of most bacterial cells (0.2 μm or 0.45 μm) will trap the bacteria in the sample as the fluid passes through the filter. The filter then can be placed directly on an agar plate. The colonies that form can be counted (**Figure 6.17**).

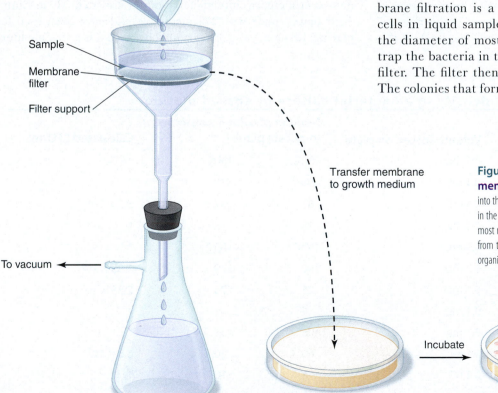

Sample

Membrane filter

Filter support

To vacuum

Transfer membrane to growth medium

Incubate

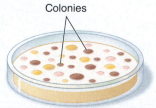

Colonies

Figure 6.17. Concentrating microorganisms with membrane filtration A known volume of a liquid sample is poured into the filter apparatus, and forced by suction through the filter. The pore size in the filters is generally either 0.2 or 0.45 μm, smaller than the diameter of most microbial cells, so cells in the sample are trapped. The filter is removed from the apparatus and placed on growth medium, so that trapped microorganisms, such as bacteria, grow into colonies.

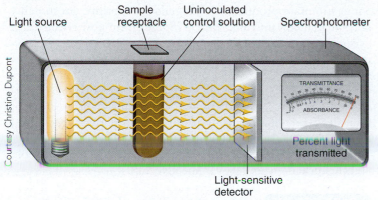

A. Light transmission for uninoculated culture media

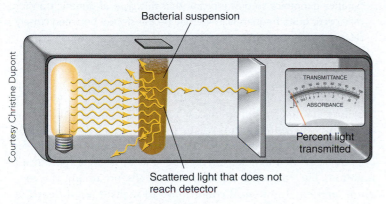

B. Light transmission for a bacterial culture

Figure 6.18. Measuring culture turbidity A spectrophotometer can be used to quantify the amount of light passing through a turbid liquid bacterial culture. The light transmitted at various cell densities can be used to measure the number of cells present. **A.** The culture tube is sterile and clear. Almost all the incident light is transmitted. **B.** After bacterial growth, the solution becomes turbid (cloudy) because the microbial cells scatter the incident light and less light is transmitted. Usually, turbidity is reported as absorbance or optical density, which is inversely related to transmittance.

Turbidity

Direct counts, as we just saw, provide us with an absolute number of viable cells within a sample. Microbial populations also can be monitored by measuring the **turbidity** of a liquid culture using a spectrophotometer set at a wavelength of 550 to 600 nm **(Figure 6.18)**. As the light beam of the spectrophotometer passes through the liquid, some photons strike cells and are absorbed or scattered, so that they do not reach the photodetector. The greater the density of cells, the less light gets through, and the higher the **absorbance** or **optical density** reading.

Figure 6.19. Microbial growth curve in a batch culture This graph illustrates growth of a hypothetical microbial population over time in a culture vessel with a fixed supply of nutrients. The x-axis shows time on a linear scale. The left y-axis shows the population or population density, plotted on a logarithmic scale. The right y-axis gives the relative optical density, a measure of the turbidity related to population density. The actual length of each of the four phases of the growth curve (lag, log, stationary, and death) depends on many factors—the organism, the nutrient composition of the medium, and various environmental factors.

Within a certain range, the optical density of the culture is roughly proportional to the population density. Optical density is easy to measure in a spectrophotometer, so this technique represents an easy way to follow population growth in a culture. However, it is only applicable over a relatively narrow range of cell concentrations. For most bacteria, the culture must have at least ten million (1×10^7) cells per milliliter to see detectable turbidity. Below this concentration, the optical density is so low that an accurate reading is difficult or impossible. When the culture density exceeds roughly a billion (1×10^9) cells per milliliter, the optical density likely will be greater than 1.0. Very little light passes through the fluid, and the linear relationship of optical density to culture density breaks down.

The microbial growth curve

In addition to determining the amount of bacteria in a sample at one point in time, we may, instead, want to determine how the amount of bacteria in a sample changes over time. We may, in other words, want to measure the growth of a population. To track population density over time, we can analyze turbidity data obtained with a spectrophotometer. Inoculating cells into a vessel containing liquid medium initiates growth. Nutrients are consumed and the microbial population expands, resulting in an increase in turbidity over time. Viable cell density is more difficult to track in real time, but we can take samples periodically and carry out serial dilutions and plating (see Figures 6.13 and 6.16). Once colonies have a chance to grow, the viable cell density can be retrospectively determined. The data from such an experiment generate a **growth curve** that reflects the behavior of the population in a closed system, or **batch culture (Figure 6.19)**. Typically, these growth curves display a few common features: a lag phase, a log phase, a stationary phase, and a death phase.

Lag Phase

Microbial cultures often grow very slowly immediately after inoculation, particularly if the starting cells had previously slowed or stopped growing, or were in a different medium. During this initial **lag phase** (see Figure 6.19), the cells are adjusting to

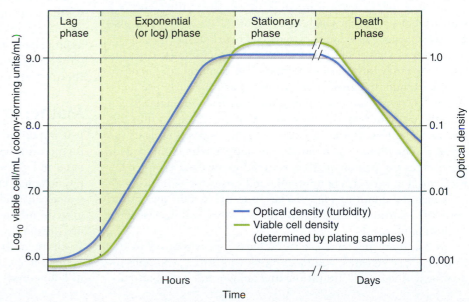

their new surroundings, which may require altering gene expression, protein content, and membrane composition. New transport systems and metabolic pathways may be needed. Depending on the organism, the medium, and environmental conditions, this lag phase may be virtually absent, or it could last minutes, hours, or longer.

CONNECTIONS Changes in gene expression in response to environmental changes isn't just a laboratory phenomenon. As we will discuss in **Section 7.3**, bacteria can alter which classes of genes are expressed by changing which sigma factors the cells produce. Changes in sigma-factor production often reflect changes in environmental conditions.

Log Phase

As cells adjust to their new surroundings, intervals between cell divisions become progressively shorter until the population reaches its maximal growth rate. At this point, the average time interval from one division to the next, known as the **generation time**, is constant. The rate at which cells divide is determined by the physiological capabilities of the organism, the nature of the medium, and environmental conditions. When microbes are dividing by binary fission—one cell producing two cells by each division—the rate of increase of the population is exponential. This stage of the culture is referred to as the **exponential phase**, or **log phase** (see Figure 6.19).

Stationary Phase

Following the log phase, the growth rate slows, until net population growth stops completely, and the culture has reached **stationary phase**. In a batch culture, the concentration of available nutrients in the medium is continually declining as the number and total mass of cells increases. At some point, the concentration of one or more critical nutrients will drop below the level necessary to sustain the maximal growth rate the culture experiences in log phase. Basically, there will not be enough of some critical nutrient for the cells to keep operating macromolecular synthesis fast enough to sustain rapid growth. As the level of the critical nutrient drops, the culture enters a transitional period in which growth rate declines, due to progressively longer intervals between cell divisions.

Besides exhaustion of nutrients, other factors also can contribute to entry into stationary phase. For example, metabolism during exponential growth can produce harmful waste products. When *Escherichia coli* consumes sugars without enough oxygen for fully aerobic respiration, some carbon is directed into fermentative pathways that generate organic acids, lowering the pH of the medium. The opposite can happen when amino acids are used as the primary carbon source, because excess nitrogen is released in the form of ammonia and raises the pH. Although microbes can adapt to some variation in pH, a shift away from optimal conditions reduces growth rate, as we will discuss later in this chapter.

Microorganisms have evolved a variety of responses to stationary phase stresses, and most employ multiple strategies for dealing with resource limitations and unfavorable environments.

Some stimulate production of flagella for motility, presumably to try to find new nutrient sources. Some Gram-positive bacteria become competent to take up extracellular DNA from other dying cells, which can be consumed for nutrients. Other microbes synthesize antimicrobial compounds that kill competitors which, again, frees up nutrients. Some bacteria, like *Bacillus anthracis* that we discussed in the opening story of this chapter, radically remodel their cells to form endospores—specialized cells that are metabolically inactive and highly resistant to environmental insults, including high temperatures, dehydration, and chemical damage (see Perspective 2.2).

CONNECTIONS The transfer of DNA between organisms is referred to as horizontal gene transfer. As we will see in **Section 10.3**, comparative genomics studies indicate that this type of genetic transfer may occur quite frequently. DNA may be transferred between closely related organisms and between more distantly related organisms.

Death Phase

If a culture stays in stationary phase long enough, then cells may lose viability. When the concentration of viable cells starts to decline, the culture enters **death phase** (see Figure 6.19). The rate of decline in a cell population during death phase, like the rate of increase during log phase, is an exponential function. The rate of death depends on the organism, the medium, and environmental conditions. To measure cell death, a method such as plate counting must be used that actually tests for viability. Direct counts by microscopy, or culture turbidity measurements, may be misleading if non-viable cells remain intact, at least initially.

Why do cells eventually die in a batch culture? To a certain extent, it depends on why they stopped growing in the first place. If a culture is depleted for an energy source, then several metabolic crises will develop. Transmembrane ion and electrical gradients diminish, and the cell depletes its supply of ATP. Metabolic systems are unable to continue to function normally and synthesis of macromolecules such as proteins, RNA, DNA, and peptidoglycan (in the case of bacteria) slows drastically. The final, fatal blow may come when the cell can no longer repair routine damage to the genome, cell wall, or membrane, either because of insufficient energy or raw materials, or because the appropriate enzymes can no longer be produced. The various stress responses induced in stationary phase attempt to stave off these events for as long as possible.

Mathematical analysis of growth

It is important for microbiologists to understand basic methods for quantitative analysis of microbial growth. Comparing **growth rate** and **growth yield** can, for example, provide insight into physiological processes, genetic differences, and interactions between cells and their environments. Let's begin our discussion of growth rate by considering a bacterial population in which cells are dividing with a generation time of 1 hour. If we start with 1 cell, then after 1 hour we would expect 2 cells. After another hour (2 hours total), there would be 4 cells; after another hour (3 hours total), 8 cells, and so on. In a real culture, though, cells do not all divide at precise intervals. The generation time we calculate for a culture is an "average" for the entire population of cells over a defined time period. Furthermore, the generation time for the culture is only constant if the culture is in log phase throughout the time period included in the calculation.

To see how we can determine generation time (g), we will introduce a simple equation that describes the growth of a population in log phase:

$$N_t = N_0 \times 2^n$$

where N_t = population at time t

N_0 = initial population at time "0"

n = number of generations that have elapsed between time "0" and time t

What if we do not know g or n? Generation time is not a universal constant—it varies between microbial species and strains, and is dependent on medium composition and environmental conditions. We can calculate n and g for a culture from experimental data. The first step is to solve for n by transforming the equation $N_t = N_0 \times 2^n$. Taking the logarithm of both sides:

$$\log N_t = \log N_0 + (n)(\log 2)$$

$$(\log 2)(\log N_t) = (\log 2)(\log N_0) + n$$

$$n = \frac{\log N_t - \log N_0}{\log 2} = \frac{\log N_t - \log N_0}{0.301} = 3.32(\log N_t - \log N_0)$$

To calculate g, we can substitute for n in the equation $g = t/n$ as follows:

$$g = t/[3.32(\log N_t - \log N_0)]$$

The value of g also can be approximated by inspection of a semi-logarithmic data plot. From a convenient starting point during log phase, measure the time for the population, or population density, to increase by a factor of 2. To determine g during logarithmic growth, both points must be during the period when the culture shows a straight line on the semi-log plot.

In some circumstances, it is useful to express growth rate as the number of generations per unit time. This term is referred to as the "mean growth rate, k," usually expressed as generations per hour.

$$k = n/t, \text{ where } t \text{ is measured in hours}$$

or

$$k = 3.32(\log N_t - \log N_0)/t$$

Thus, generation time is the inverse of the mean growth rate. In mathematical terms, $g = 1/k$. If g is less than 1 hour, it can be expressed in units of minutes. For example, for *Escherichia coli* growing in a rich, well-aerated medium at 37°C, g can be described as 20 minutes, rather than 0.33 hour. On the other hand, it is more practical to express k as 3 generations/hour, rather than 0.05 generation/minute.

In the laboratory, it often is faster and easier to measure the turbidity of a culture, rather than the actual number of cells. If all the cells in a population are of similar size and composition over a specified time interval, then they have approximately the same ability to absorb or scatter light, and cell density will have a constant relationship to turbidity. For cultures in log phase, this assumption generally is true. Under these circumstances, turbidity measurements can be used to calculate g and k. To calculate generation time, "N_0" is replaced with OD_0, the optical density at a time point early in log phase, and "N_t" is replaced with OD_t, the optical density at a time point later in log phase:

$$n = \frac{\log(OD_t) - \log(OD_0)}{\log 2} = 3.32[\log(OD_t) - \log(OD_0)]$$

Thus, $g = t/[3.32(\log OD_t - \log OD_0)]$

Rates of growth of microorganisms vary widely, both in the laboratory and in natural environments. The most rapidly reproducing organism known so far is the marine bacterium *Vibrio natriegens*, which can divide in as little as 10 minutes. At the other extreme, some bacteria take days to divide even under "optimal" conditions in the lab. One of the slowest growing bacteria known is *Mycobacterium leprae*, which takes at least two weeks to divide (Perspective 6.2). *Bacillus anthracis*, by comparison, divides in roughly 20 minutes. Archaeons tend to grow more slowly than bacteria; one of the most rapidly growing archaeons known is *Pyrococcus furiosus*, which has a generation time of just under 40 minutes in ideal conditions. Eukaryal microbes also tend to grow more slowly than bacteria. Among the fastest growing eukaryal microbes are yeasts such as *Saccharomyces cerevisiae*, whose generation time under optimal laboratory conditions is around two hours.

Why might a microbiologist be interested in determining growth rate in the laboratory? Let's consider some examples. We could compare the relative ability of a microbe to utilize various nutrients—which sugars will *Escherichia coli* use as carbon and energy source, and what is its growth rate on each? Which fatty acids support the fastest growth by *Staphylococcus aureus*? What nitrogen-containing compounds support growth of a particular soil bacterium? Experimental answers to such questions could provide insight into what nutrients are available in the environmental niches these organisms inhabit, and/or the transport or metabolic pathways utilized. Examining the effect of temperature, pH, or other environmental parameters on growth can generate insight into conditions to which the organism is evolutionarily and physiologically adapted. One could compare the growth of genetically distinct isolates of a species, even genetically engineered strains, to see how particular genes affect growth under various conditions.

Once we understand the mathematics of population growth, the remarkable power of exponential growth becomes clear. Let's look at an example. Assuming unlimited nutrients, and an *Escherichia coli* culture that sustained logarithmic growth indefinitely with a generation time of 20 minutes, how long would it take to go from a single bacterial cell to a mass of bacteria weighing as much as Earth?

MYCOBACTERIUM LEPRAE, AN EXTRAORDINARILY SLOW-GROWING PATHOGEN

Mycobacteria are unusual bacteria, evolutionarily related to Gram-positive bacteria, but lacking the thick peptidoglycan layer characteristic of this group. Instead, Mycobacteria have a distinctive thick, waxy, lipid-like layer protecting their cells. Some Mycobacteria are well known for the diseases they cause, but these bacteria are exasperating to work with in the lab. Cells of *Mycobacterium tuberculosis*, the cause of tuberculosis (discussed in Chapter 21), take a full day to divide in the laboratory. This slow generation time pales in comparison to *Mycobacterium leprae*, the cause of Hansen's disease, also known as leprosy, which takes roughly *two weeks* to divide! Furthermore, unlike *M. tuberculosis*, which forms colonies on specialized laboratory media, *M. leprae* has so far only been cultivated in live armadillos and mice.

Why is *Mycobacterium leprae* so difficult to cultivate in the laboratory? This bacterium normally invades host cells and lives inside them as an *intracellular pathogen*, slowly growing and dividing where it is largely protected from the immune system. It infects Schwann cells that cover the nerves, but because the bacteria replicate so slowly, the symptoms of Hansen's disease are gradual, progressive nerve damage. The bacteria acquire nutrients from the cytoplasm of the host cell, which effectively acts as a rich, complex medium. Adaptation to an intracellular life where it can rely on the host cell for nutrients has allowed the *M. leprae* genome to itself suffer progressive decay by mutation, to the point that it now has less than half the number of functional genes as *M. tuberculosis*. The remaining *M. leprae* genome encodes a limited repertoire of metabolic and energy-generation pathways, severely compromising its growth rate and fitness outside of host cells.

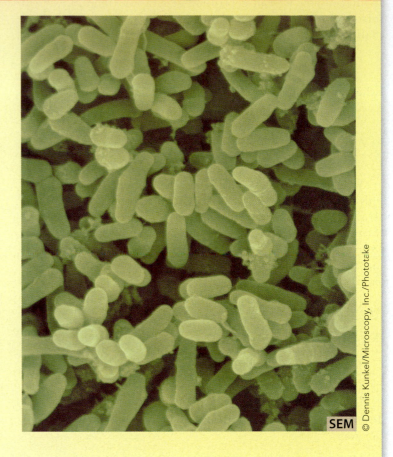

© Dennis Kunkel/Microscopy, Inc./Phototake

SEM

To answer this question, you need to know that an *Escherichia coli* cell has a mass of approximately 1 pg (1×10^{-12} g), and that Earth has a mass of approximately 5.972×10^{27} g. How many bacteria are needed to equal the mass of Earth?

$$(5.972 \times 10^{27})/(1 \times 10^{-12}) = 5.972 \times 10^{39} \text{ bacteria}$$

How long will it take to generate that many bacteria? Using the following equation

$$n = \frac{\log N_t - \log N_0}{0.301}$$

we have

$$n = \frac{\log(5.972 \times 10^{39}) - \log(1)}{0.301}$$

$$= \frac{39.776}{0.301} = 132 \text{ generations}$$

Thus, 132 generations $\times$ 20 min = 2,640 min = 44 hr

So, in less than two days, this theoretical population doubling every 20 minutes, which is roughly the maximum growth rate of *Escherichia coli* in rich medium, would have accumulated enough cells to equal the mass of Earth! Clearly, this type of growth does not happen in the real world. Although the mass of microbial cells on Earth is enormous, it is basically stable. No natural population of microorganisms could grow at anywhere near this rate for more than a few hours. Nutrients are virtually always limiting. There may be short periods when a microbial population experiences explosive growth, but these ideal conditions never last long!

When characterizing the behavior of a microbial culture, we also may be interested in growth yield, an expression of the maximum population density achieved, and/or the amount of cellular material generated by the culture. This measurement would be particularly important if, for example, we were using a microorganism as an industrial production system for a valuable enzyme or other biological product, and wanted to maximize the yield. At the cellular level, growth yield also can indicate the pathways and efficiency with which cells are converting nutrients into macromolecules.

Growth yield can be measured and expressed in several ways, including (1) the maximum optical density of the culture at a specified wavelength; (2) maximum viable cell density, as determined by plate counts; or (3) the dry weight of cells collected from a defined volume of culture. In the latter approach, cells are harvested by centrifugation or filtration and water is removed by evaporation, in order to obtain the dry weight of the cellular material.

Continuous culture

So far, our discussion of microbial growth has focused on batch cultures. In such closed systems, there are finite resources, and exponential growth stops when these resources are depleted or when waste products accumulate. For microbes in a batch culture, life is boom and bust—rapid logarithmic growth when resources are abundant, followed by stasis and perhaps death when resources are exhausted. This pattern is common in dynamic habitats. For example, when an animal or plant dies, microbes take advantage of the abundant nutrients in the dead material and proliferate madly as they consume it. The microbial population expands dramatically as the dead material decays, but as more and more microbes compete for the declining pool of nutrients, the growth rate of the population as a whole inevitably slows down, and the population eventually crashes.

Microbes living in more stable environments may rarely experience boom and bust cycles. Instead, they face constant slow growth under nutrient-limited conditions. This situation can be re-created in the laboratory with a **continuous culture**. A continuous culture is an open system in which nutrients (sterile growth medium) are fed into a culture vessel at a constant rate. To keep the culture volume constant, excess medium is drained at the same rate, removing cells and waste products from the culture. Under such conditions, the microbial population initially will grow exponentially at the maximum rate the medium and environmental conditions will support, until it has nearly exhausted one or more critical nutrients, just as it would in a batch culture. At this point, the growth rate becomes limited by the rate at which new nutrients are added, in the form of fresh medium. If the culture is rapidly mixed by vigorous stirring so that input nutrients are quickly distributed throughout the culture vessel, then the culture achieves a state of equilibrium in which nutrient and cell concentrations are stable.

A **chemostat (Figure 6.20)** is a practical continuous culture system in which growth rate and cell density can be manipulated through the nutrient composition of the medium and the flow rate of fresh medium into the culture vessel. The higher the concentration of nutrients in the medium, the greater the steady-state cell density. Once maximal population density is achieved, growth rate can be manipulated by adjusting the flow of fresh media into the vessel. The mathematical term describing this state is "dilution rate," the time it takes to replace the entire volume of the culture with fresh medium. For example, if fresh medium is added at a rate of 100 mL per hour (accompanied by the same rate of drainage of effluent) into a 1-liter culture, the dilution rate is (100 mL/hr)/(1,000 mL) = 0.1/hr. If population density remains constant, with new growth balanced by cells lost in the effluent, the average growth rate of cells in this culture is equivalent to the dilution rate. To achieve

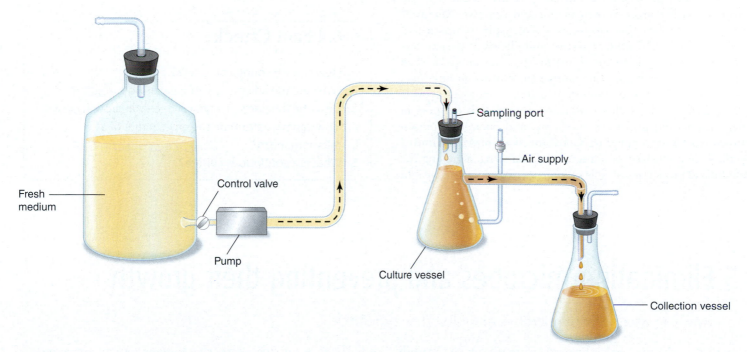

Figure 6.20. A continuous culture system Chemostats provide a means of maintaining exponentially growing cells under constant conditions for the study of microbial physiology. The most basic form consists of a culture vessel with an inlet port for addition of sterile medium, and an outlet port for draining culture from the vessel at the same rate as fresh media is added, to keep the culture at a constant volume. The apparatus shown here also has a port for sterile addition of air, a necessity when growing organisms requiring aerobic metabolism.

Every day, a human eliminates billions of bacteria in feces. In fact, roughly half of the weight of human fecal material is microbial. Every day, the microbes that are lost are replaced through new microbial growth. The colon of a healthy, well-fed person approximates a continuous culture. Nutrients enter in the form of digested food remnants. Microorganisms grow and divide, and waste products and some microbial cells are eliminated. The input of digested food and elimination of feces roughly balance, so the microbial population remains fairly constant.

What is the growth rate of microbes in this continuous culture? The rate of transit of materials through the entire gastrointestinal tract is typically 1.5 to 2 days, most of which is spent in the colon. Let's assume that material spends a day moving through the colon before being eliminated as feces, the "effluent" of this continuous culture. The "dilution rate" of the microbial culture contained in the

colon is therefore once per day. So on average, the stable members of the colonic microbiota have generation times of approximately 24 hours when growing in the colon. For a bacterium like *Escherichia coli*, this generation time is very slow; this organism has the ability to divide every 20 minutes in rich laboratory media. What limits the growth rate of microbes in the colon? First, nutrients in the biomass passing through the digestive system are continually being absorbed by our intestinal cells or competing microbes, so the *E. coli* in the colon are unlikely to have access to optimal nutrient concentrations. Also, by the time ingested material reaches the colon, the molecular oxygen available for respiration has been virtually eliminated, so bacteria must rely on anaerobic metabolism that yields less energy, and utilizes nutrients less efficiently. These two factors greatly limit the generation time of *E. coli* in our colon.

---

stability, the dilution rate must be *less* than the maximal growth rate of the microbe for the medium and environmental conditions. If a microbe cannot divide fast enough to keep up with the dilution rate, then the chemostat will experience washout, in which cells are flushed out of the vessel faster than the culture can replenish them.

Why would a microbiologist use a continuous culture? Chemostats can be set up so that cultures are monitored and controlled, with pumps to feed media and remove effluent at steady rates. Probes that continuously measure temperature, pH, oxygenation, and other physical and chemical parameters can be coupled to systems for adjusting these parameters when necessary. Ports can be used to sample the culture at any time. For studying microbial physiology, continuous cultures provide more consistent, controlled conditions than batch cultures in which exponential growth is transient, and chemical concentrations are always changing. Constant slow, nutrient-limited growth is characteristic of many natural habitats. Studying the physiological properties of cells growing this way gives us insight

into their lifestyle. In industrial microbiology, continuous cultures can be more cost-effective than batch cultures. Products from the culture can be harvested continuously without the downtime or time lag associated with repeatedly starting new batch cultures. **Perspective 6.3** discusses a natural continuous culture, that of the human intestine.

6.4 Fact Check

1. How can we count viable cells?
2. What are the advantages of measuring turbidity?
3. Describe the parts of a typical bacterial growth curve.
4. How can we determine the growth rate of a microorganism?
5. What is a continuous culture?

6.5 Eliminating microbes and preventing their growth

How can we eliminate microbes or inhibit their growth?

So far in this chapter, we've focused primarily on the growth of microorganisms. In many situations, though, we may want to prevent the growth of microbes, kill them, or physically remove them. In the laboratory, we want to be certain that unwanted bacteria do not contaminate glassware and reagents. In

a hospital or nursing home, where many patients have limited resistance to microbial infections, it is very important to minimize the exposure of patients to potentially infectious agents. Similarly, it is important to minimize the numbers of microbes in some public places, like swimming pools.

In this section, we will examine methods for removing microbes or limiting their growth. We will focus primarily on removing microbes from inanimate objects, such as liquids, surfaces, and foods. Various ways of preventing infections of our bodies and eliminating microbes from our bodies after an infection will be covered in Chapter 24. For now, we will explore how we can use filtration, heat, electromagnetic radiation, and various chemicals to remove or inactivate microbes from various substances. We also will examine a few practical issues associated with eliminating microbes.

Physical removal of microbes by filtration

It is not practical to physically remove microorganisms from solid materials. Most bacteria, archaeons, and eukarya, however, can be removed from liquids by filtering the liquid through a material with pores (holes) that are too small for microbial cells to pass through. You might recall from Chapter 5 that one of the first indications that viruses were distinct from bacteria is that they could pass through porcelain filters known to capture bacteria and larger microbes. "Low tech" water filtration systems made from sand, diatomaceous earth, charcoal, or cloth (or some combination of the above) have been used for centuries to remove sediment and improve the smell, color, and taste of drinking water. Only in the last century have we appreciated that proper filtration could also remove infectious microbes and thus make water safer for human consumption. Modern membrane filters used for microbiological applications are made of tough synthetic materials such as nylon, polyvinylidene difluoride (PVDF), and polytetrafluoroethylene (PTFE, also known as Teflon). Cheaper, less durable filters are made of biologically based materials such as cellulose nitrate or cellulose acetate.

Filters used for sterilization of fluids in the laboratory have an average pore size of 0.2 or 0.45 μm, which is sufficient to trap all eukaryal microbes and most bacteria. A 0.2-μm pore still allows many viruses and some very small bacteria, such as *Mycoplasma* and *Chlamydia*, to pass through. Ultrafiltration methods capable of virus removal are used in the pharmaceutical industry. Ultrafiltration pore sizes are reduced to 10 to 100 nm, the size range of macromolecules. These tiny pores result in very slow flow rates, and high pressure is used to force liquids through the filters. Liquids that are significantly more viscous than water are very difficult to filter. Solutions that contain large particles are also a problem for filtration, because they can easily clog filters. Using a pre-filter screen to remove these large particles can solve this problem.

Temperature manipulation

Heating is a simple and effective method for destroying microbes. Elevated temperatures cause proteins to denature—to lose their native structure and become non-functional. Other macromolecules are also sensitive to heat-induced denaturation. Nucleic acids rely extensively on hydrogen bonding to maintain functional structure; double-stranded chromosomal DNA and folded RNA molecules such as ribosomal and transfer RNAs can be structurally disrupted by heat.

Heating to 100°C, the boiling point of water, is sufficient to kill most microbes within minutes. Boiling is not sufficient for

Figure 6.21. An autoclave for heat sterilization Autoclaves are used in nearly all microbiology laboratories, as well as clinics and hospitals, for heat sterilization of instruments, media, and wastes.

complete sterilization, as some microbial cells can survive such conditions. Some hyperthermophilic microbes, as we discussed in Chapter 4, withstand temperatures above 100°C, but hyperthermophiles are not common outside of specialized niches like hot springs and thermal vents. Of greater concern are bacterial endospores, the specialized and highly durable cells produced by several common Gram-positive bacteria under stress conditions. Endospores are resistant to heat and other environmental stresses, and some can survive boiling for hours! In order to inactivate endospores, an **autoclave** is used in the laboratory to apply heat to media, instruments, and contaminated waste **(Figure 6.21)**.

An autoclave is essentially a pressure cooker—the chamber fills with superheated steam to reach a temperature of 121°C at a pressure of 15 pounds per square inch. These conditions are well above the boiling point of water at atmospheric pressure, but the elevated pressure in the chamber keeps liquids from boiling away. Like liquid water, steam transfers heat very efficiently. Vegetative cells and endospores rapidly lose viability under these conditions, and 15 minutes of exposure is sufficient for sterilization of small samples or volumes. When autoclaving large volumes of liquids or large masses of solids, longer exposures are used to ensure that the entire mass/volume reaches 121°C long enough for sterilization.

Some materials cannot withstand the high temperatures and pressure used in autoclaving, so alternative heating methods, such as pasteurization, have been developed to reduce microbial content. Today, pasteurization is most familiar to the general public as it relates to milk and dairy products. In this process, liquids are heated for a short period of time. This treatment greatly reduces the number of viable bacteria in milk. We should note, though, that mammalian milk is rich in nutrients, and microbes that survive pasteurization, or are introduced when a container is opened, will grow if given a chance. Milk turns sour because of a mixture of acidic by-products from bacterial metabolism. If we keep the temperature of milk low, as in a refrigerator set at 4°C, microbial metabolism and growth rates are very slow. Refrigeration inhibits the growth of most microbes, and thus is valuable for storage of perishable products subject to microbial degradation, including non-food products such as pharmaceuticals and medical supplies. Refrigeration is *not*, however, a method for disinfection or sterilization.

> **CONNECTIONS** Several forms of pasteurization, including high-temperature, short-time (HTST) and ultra-high temperature (UHT), may be utilized. We will learn more about pasteurization in **Section 16.2**.

Freezing can completely stop the growth of microbes. It converts liquid water, the aqueous environment in which most biochemical reactions happen, to a crystalline form and dramatically slows the rate of enzymatic reactions. Storing food at temperatures of −20°C or lower is therefore an excellent method for long-term preservation without spoilage. Freezing can be lethal to cells, as ice crystals can cause severe damage to cellular structures, particularly in cells without strong walls. Microbial populations exposed to freezing conditions generally suffer significant reductions in population. It is, however, possible for cells to undergo adaptive changes in cell composition that reduce the formation of ice crystals. Psychrophilic microbes in particular express adaptations that increase the likelihood that cells will survive freezing, including synthesis or accumulation of cryoprotectants such as sugars or glycerol that prevent formation of ice crystals. Microbiology laboratories freeze samples of microbes for long-term storage after adding cryoprotectants such as dimethylsulfoxide (DMSO) or glycerol. Thus, freezing can be fatal to cells, or it can be a preservative. It's a useful way to shut down microbial growth but like refrigeration should not be thought of as a disinfection or sterilization method **(Figure 6.22)**.

Before refrigeration was widely available, people learned to ferment dairy products, juices, and grains—not just because the alcohol produced in some fermentations made them feel good, but because alcohols and acids limited further spoilage! Drying foods eliminates water, thereby inhibiting microbial growth. Adding salt to meat extracts water and increases salinity, both of which reduce microbial growth and slow spoilage. These food preservation strategies were developed through observation and trial and error during millennia of human cultural evolution, long before the microbiology of food spoilage was understood. We'll learn more about food microbiology in Chapter 16.

Using electromagnetic radiation to control microbes

Exposure to various forms of electromagnetic radiation can damage or kill microorganisms. Ultraviolet (UV) radiation is effective for disinfection and sterilization. Photons in the ultraviolet range have wavelengths of 200 to 400 nm. Ultraviolet radiation at 260 to 280 nm is strongly absorbed by nucleic acids, and induces chemical reactions. The most common form of UV-induced damage in DNA is formation of thymine dimers, in which adjacent thymine bases become chemically linked.

Courtesy Christine Dupont

Figure 6.22. Cryopreservation of cells For long-term storage of cells, researchers often store vials of cells in liquid nitrogen. To avoid the cellular disruption typically associated with freezing, researchers add cryoprotectants, such as DMSO or glycerol to the cells prior to freezing them.

These dimers disrupt the normal structure of DNA, and prevent its replication, which in turn prevents cell division. Cells have mechanisms for repairing such damage on a small scale, but massive damage to the genome is generally lethal. In the laboratory, UV light, which has little penetrating power in solids, can be used to disinfect surfaces, such as the working areas of specialized laminar flow hoods used for microbiological or cell culture work **(Figure 6.23)**.

> **CONNECTIONS** The bacterium *Deinococcus radiodurans* can withstand extremely high levels of radiation. This radiation resistance may come from its ability to effectively repair double-strand breaks in its DNA. We'll examine this remarkable organism in **Section 7.5**.

Sunlight contains a significant UV component, much of which is absorbed by Earth's atmosphere but some of which penetrates. Small-scale "solar disinfection" of drinking water by sunlight is remarkably effective and useful in rural areas in developing countries, which often lack public water treatment and distribution systems. For solar disinfection, water that is

Figure 6.23. Ultraviolet radiation for disinfection or sterilization Hoods used for cell culture research in the laboratory typically have UV bulbs installed for rapid disinfection of surfaces. These hoods also have laminar air flow systems blowing filter-sterilized air over the working surface.

relatively clear of sediment, which blocks UV penetration, is put in clear plastic bottles, shaken for aeration, and exposed to full sunlight for several hours (**Figure 6.24**). The hotter the water gets, the faster the disinfection. The results are stunning—bacterial counts can be reduced by a million-fold in just a few hours in a warm, sunny area.

Direct UV damage to the microbial cell is not the sole cause of death. UV radiation between 300 and 400 nm interacts with dissolved oxygen in water to generate some of the harmful reactive oxygen species we mentioned earlier in Table 6.1. These oxygen species attack microbes as well, magnifying the killing effect of UV light. In recent years, UV exposure has been increasingly used as a component of large-scale treatment of drinking water as well, in part because it contributes no lasting residual chemicals that alter the taste of the water.

Figure 6.24. Solar water disinfection (SODIS) In resource-poor areas, sunlight can be used to partially purify water. Water simply is placed in a clear plastic bottle and exposed to sunlight. Over time, the UV irradiation and increased temperature destroy a significant proportion of microorganisms originally in the water.

Many types of perishable products, including foods and various medical supplies, cannot withstand high temperatures, and cannot be penetrated by UV light. For these products, ionizing radiation has excellent penetrating and sterilizing capabilities. Ionizing radiation, which includes X-rays and gamma rays, are highly energetic photons or other particles that can remove electrons from molecules. Gamma rays are photons with wavelengths from 0.01 to 0.1 nm, much shorter in wavelength and higher in energy than visible and UV light, and thus capable of much greater penetration of solid materials.

The high energy of gamma rays causes breakage of DNA and RNA strands, which leads rapidly to cell death. The overall chemical composition of gamma-irradiated products is basically unchanged. Gamma irradiation is widely used for sterilization in the pharmaceutical and medical supply industries, but has not gained traction in the food industry. Irradiation of food is a safe and effective way of improving the microbiological safety of food and prolonging shelf life. The gamma ray source (Cobalt-60) never actually comes into contact with the food, and there is no residual radioactivity associated with the product. Nevertheless, the general public does not understand the process well. While the irradiated food certainly does not become radioactive, a general fear among the public of all things associated with "radioactivity" has limited the ability of the food industry to employ this technique.

Chemical methods of controlling microbes

Many chemicals can kill microbes, or prevent their growth. We will focus on only a few agents commonly used at home, in the microbiology laboratory, or in health-care settings. A **disinfectant** is a chemical that is applied to non-living objects to kill potentially infectious microorganisms. Sterilization (the complete destruction or inactivation of living organisms and viruses) accomplishes disinfection, but most chemical disinfectants used for non-medical applications are unlikely to accomplish sterilization of a surface. **Antiseptics** are chemicals that kill infectious microbes, and can be used on living tissue, or at least on surfaces such as skin. Most antiseptics are not used internally, because they would be toxic to tissues when ingested.

CONNECTIONS To kill microbes within the human body, we use agents that are selectively toxic to microbial targets, but spare our own cells. In **Section 24.2**, we will investigate several drugs that are used to combat infectious agents.

What makes a good chemical disinfectant? First, it has to kill microbes. It should be able to destroy a wide range of infectious agents: bacteria (Gram-negative, Gram-positive, and endospores), fungi (including spores), other eukaryal microbes, and diverse viruses. Of course, there are many harsh chemicals that are toxic enough to quickly kill any cell, and destroy viruses as well. So we need additional criteria. For example, we might not want chemicals that are highly corrosive, as they would damage too many of the materials we might need to treat. These chemicals shouldn't leave residues behind, or they should be easily washable with water. They should not smell too bad, or emit toxic fumes, or be overly toxic to handle, unless they are going to be used in highly restricted, controlled settings. An antiseptic must meet an even higher standard of not being toxic, corrosive, or excessively irritating when applied to skin. In this section, we will look at the properties of some commonly used chemical disinfectants (Table 6.5).

Alcohols

Ethanol and isopropanol (rubbing alcohol) are commonly used as antiseptics. Alcohols dissolve and disrupt cell membranes, and are effective against many bacteria and viruses, but are less active against fungi and bacterial spores. Although much more potent disinfectants are available, these alcohols have the advantage of being inexpensive, acting quickly but evaporating rapidly, and leaving no residue behind. Ethanol and isopropanol are the active ingredients in most gel or foam-type hand sanitizers marketed to the general public. In the United States, the Food and Drug Administration (FDA) and the Centers for Disease Control and Prevention (CDC) both recommend that hand sanitizers contain at least 60 percent alcohol for maximum efficacy. Although these hand sanitizers do not sterilize skin surfaces, they can reduce bacterial load on the skin surface by 90 percent or more when used properly, and they can inactivate many viruses. Routine use of effective alcohol-based hand sanitizers has been shown to reduce the spread of infectious diseases in schools and daycare centers.

Phenolic Compounds

Phenol, known at the time as carbolic acid, was one of the first antiseptics used in medical practice. In the mid-1800s, the surgeon Joseph Lister pioneered the use of the antiseptic and dramatically reduced post-surgical infections as a result. The antimicrobial activity of phenol and derivatives is associated with their aromatic ring (see Table 6.5), which is non-polar and capable of inserting into the hydrophobic interiors of both lipid bilayers and proteins, and disrupting their function.

Phenol is prepared as a dilute aqueous solution for use as an antiseptic, since it is quite toxic and corrosive in concentrated form. Because of phenol's historic importance as a disinfectant, it has been used as a standard by which to compare the effectiveness of other disinfectants. The term "phenol coefficient" refers to the effectiveness of a disinfectant at killing bacteria, relative to a standard phenol solution. Less corrosive derivatives of phenol include chloroxylenol (an ingredient of antibacterial soaps) and O-phenylphenol (used in some household disinfectants, and as an agricultural fungicide).

Triclosan, or 5-chloro-2-(2,4-dichlorophenoxy) phenol, is a structurally complex phenol derivative that is widely used in consumer products because of its stability, and its relatively low toxicity to animals and humans. Triclosan can kill bacteria and fungi by the same mechanism as other phenolic compounds, attacking cell membranes. Unlike most other phenolic compounds, triclosan also binds to and reduces the activity of a

TABLE 6.5 Commonly used disinfectants

Class	Example	Structure	Notes
Alcohols	Ethanol		Routinely used in laboratory settings; also present in most hand sanitizers
Phenolic compounds	Triclosan		Added to numerous products, including some soaps, deodorants, and cosmetics
Oxidizing agents	Sodium hypochlorite		Commonly added to swimming pools and hot tubs to inhibit microbial growth
Others	Benzalkonium chloride	$R = C_8H_{17} - C_{18}H_{37}$	Major ingredient in Lysol®
	Glutaraldehyde		Often used to prepare biological specimens

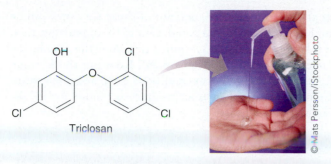

Figure 6.25. Increased use of triclosan in consumer products Since its development in 1972, the phenolic compound triclosan has been used in an expanding range of consumer products, such as liquid hand sanitizer, shown here.

specific enzyme, enoyl-ACP reductase, which is needed for fatty acid synthesis. As a consequence, triclosan can inhibit microbial growth even at very low concentrations. Mutations in the gene encoding enoyl-ACP reductase, however, can result in bacteria that are resistant to the effects of triclosan. As triclosan has been incorporated into more and more consumer products, some scientists have expressed concern that its excessive use will stimulate the proliferation of triclosan-resistant microbes **(Figure 6.25)**.

Oxidizing Agents

Oxidizing agents strip electrons from cellular molecules, including lipids and proteins. This action leads to their disinfectant properties. A diverse range of chemicals falls into this category. For example, hydrogen peroxide (H_2O_2) is used as a skin or wound disinfectant as a 3 percent solution. Another oxidizer used for disinfection of minor skin wounds is iodine, dissolved in a water-alcohol solution ("tincture of iodine"). Iodophores attach iodine to larger carrier molecules such as polyvinylpyrrolidone, which are water soluble and eliminate the need for alcohol. An example of this type of agent is Betadine®, a topical disinfectant. Iodine is also used to disinfect drinking water, though usually only on a small scale and in emergencies, since it leaves a rather unpleasant taste.

Chlorine-based compounds that act as oxidizing agents also are used for disinfection. Sodium hypochlorite is the active ingredient in household bleach (such as Clorox®), which in diluted form is an effective disinfectant for household surfaces. Sodium hypochlorite and calcium hypochlorite are added to swimming pools, hot tubs, and fountains to control microbial growth. For larger-scale treatment of drinking water, chlorine gas, chloramines, and chlorine dioxide are all used, depending on the scale, location, and context of the system. Sanitation of water by chlorine treatment, called "chlorination," is fast, effective, and relatively inexpensive, but must be carefully controlled. Chlorine ions are consumed when they react with organic molecules, including proteins in microbial cells, so chlorination is used at a late stage in water treatment, after most organic material has already been removed. Municipal water treatment systems add enough chlorine to ensure that low levels are still present in the water as it enters the distribution system, to provide residual

protection from microbial contamination as the water is in transit to the end user.

Ozone (O_3) is another strong oxidizing agent and often is used in wastewater treatment. Ozone treatment is very effective, as it damages cell walls, proteins, and nucleic acids. Ozone is highly reactive and unstable, decomposing to dioxygen (O_2) fairly quickly. Because it is unstable, ozone is generated at the site of application, typically by high voltage electrical discharge into an O_2-containing atmosphere. The instability of ozone means that it cannot provide the residual protection of water that chlorination does. On the other hand, the fact that it converts into O_2 is an advantage for applications where the user does not want residual chemicals in the water.

Other Chemical Disinfectants

Quaternary ammonium compounds can act as both detergents and disinfectants. The most common member of this class is benzalkonium chloride, or alkyl dimethyl benzyl ammonium chloride. Benzalkonium chloride is not actually a single molecular species, as the R alkyl group can be from 8 to 18 carbons long (see Table 6.5). Commercial formulations are a mixture of molecules with different R-group lengths. Benzalkonium chloride is the most potent active ingredient in Lysol®, a widely used household disinfectant and cleaning solution in the United States. Lysol includes other active ingredients as well, including alcohols (ethanol and isopropanol) and a phenol derivative (p-Chloro-o-benzylphenol), to enhance its spectrum of disinfectant activity.

Chlorhexidine is the active ingredient of several disinfectants used for skin and oral applications, including toothpastes and mouthwashes. Chlorhexidine has particularly strong antibacterial activity, and is thought to disrupt membranes.

Glutaraldehyde is used as a disinfectant on medical equipment. Glutaraldehyde and the related compound formaldehyde react with both proteins and nucleic acids, generating cross-linking chemical bonds that effectively tie macromolecules together. This method of killing cells preserves their structure, which is why glutaraldehyde and formaldehyde are used as fixatives for biological specimens. These chemicals are also used in embalming (along with alcohols), because their toxicity to microbes prevents, or at least slows, the ordinary process of tissue degradation by microbes. Glutaraldehyde and formaldehyde are too toxic for household use.

Practical issues for destroying microbes or preventing their growth

As we have investigated these various methods of eliminating microbes or inhibiting their growth, you hopefully have realized that many practical issues must be considered before deciding which method to use. Some important questions to consider include:

- What microbes need to be controlled? Are bacteria, fungi, protozoa, or viruses present? Are all of these microbes present? Do you need to be concerned about especially durable forms such as endospores?

- How many microbes are present? Do you need to kill them all? If the requirement is for sterilization, the choice of methods will be much different than if a less extreme reduction in microbial load is acceptable.

- What kind of objects, surfaces, and materials are going to be treated? How durable are they to various treatment options? How much collateral damage can be tolerated? Are the target microbes mixed with, or protected inside, organic or inorganic materials that might interfere with treatment?

- If a physical method is applied, how long and how intense must it be? If a chemical approach is considered, how powerful should the active agent be, and how long must it be present?

- Is toxicity to humans or other non-microbial targets an issue?

The answers to such questions are critical for deciding how to apply microbial control methods to particular situations.

In some cases, specific measurements of microbial control effectiveness have been defined. Intense consumer reaction to foodborne botulism outbreaks led to research by commercial canners in the 1920s to establish protocols designed to vanquish endospore survival in canned products. The concept of **decimal reduction time (D value)** arose from this research. The D value is the time required to kill 90 percent of the target organisms under specific conditions. The decimal reduction time is the time needed to reduce the number of microorganisms present by one order of magnitude, that is, by a factor of 10, or one decimal place. For example, if the D value for an organism under specified conditions is 3 minutes, then 90 percent of the organisms will be killed in 3 minutes; 10 percent remains.

In the next three minutes, 90 percent of those survivors will be killed, and so on. If the initial population is 10^5 organisms, then it would take 5 D values or 15 minutes to reduce the population to only one (10^0) organism **(Figure 6.26)**. Knowing the D value for a target organism under set conditions, such as temperature or pH, allows treatment of that product to render it safe for consumption.

Image in Action

WileyPLUS On this laboratory bench, we see different ways of growing bacterial cultures. On the left are two large flasks containing liquid growth media. The flask in the foreground contains a culture in the early stages of exponential growth, while the flask in the background contains a much older culture, with the cells forming endospores. On the right, a nutrient agar plate streaked with bacteria is displayed, in front of two stacks of uninoculated plates. The plates on the left (yellow-green) and on the right (pink) contain different kinds of media.

1. The cells in the flask near the back of the bench are forming endospores. Explain what conditions would cause this change to occur. What other changes might cause the formation of endospores?

2. The yellow nutrient agar may be a general medium, while the pink agar may be a selective and differential medium. Compare and contrast these two types of media and explain how the growth of common bacteria such as Gram-negative *Salmonella* species and Gram-positive *Staphylococcus* species might differ on these plates.

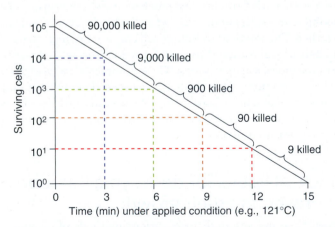

Figure 6.26. Decimal reduction time (D value) The decimal reduction time is the time required to kill 90 percent of the microbial cells under specified conditions. In this example, the population is reduced by 90 percent every three minutes (D value = 3 minutes), requiring 15 minutes to reduce the surviving population to one cell.

Following the quarantine of Anthrax Tower, investigators tried three times during the late 1960s and early 1970s to decontaminate the building. The sealed building was flooded with paraformaldehyde gas and then the army conducted thorough culture testing at about 1,500 locations in the building. In 1971, after finding no evidence of *Bacillus anthracis*, they gave the all-clear for re-occupation, but not destruction; they could not be absolutely sure that spores were not lurking in the cracks and crevices that were not sampled. The building sat unused, a reminder to all of the peril faced in the pursuit of biological warfare. Years went by and the building fell into such disrepair that demolition was necessary. In 2000, experts used culture and PCR to test several hundred samples and confirmed there was no evidence of *B. anthracis*. Anthrax Tower was finally demolished in 2003—but that was not the last of anthrax.

"Amerithrax," as it was called by the FBI, hit the United States in 2001, in the wake of the Al-Qaeda terrorist attacks. Letters containing anthrax spores were mailed to two U.S. senators and several news media offices, killing five people and infecting 17 others. Millions of citizens feared that a new kind of attack was underway, but the letters ended and the cases of anthrax stopped. The cleanup began. Chlorine dioxide (ClO_2), a powerful oxidizing agent, was chosen for the job.

As a result of the letters containing anthrax spores, dozens of buildings needed decontamination. Postal workers at facilities who handled the letters during transit were some of the victims. Spores were found on keyboards in the offices where some of the letters were opened. The cleanup of the Brentwood postal facility in Washington took 26 months and cost $130 million. Decontaminating the Hamilton, New Jersey, postal facility cost $65 million. Decontaminating the Senate office buildings cost $27 million. The affected buildings had to be sealed and chlorine dioxide gas was generated by mixing five separate chemicals: sodium chlorite, sodium hypochlorite, hydrochloric acid, sodium hydroxide, and sodium bisulfite. Exhaustive PCR and culture tests were conducted. It took more than two years to declare some of the buildings safe for occupation.

To preempt the costly cleanup of such future hazards, researchers have begun to develop rapid, cheap, and effective methods for decontamination. A promising mobile sterilization system produces X-rays and ultraviolet-C light. As we have seen in Section 6.5, irradiation is already widely used to sterilize medical equipment and food. Both types of radiation result in irreversible DNA damage. The X-rays can penetrate crevices and many soft materials, such as upholstery. The ultraviolet-C light sterilizes surfaces. Both can penetrate spores, and research has shown the system can kill *Bacillus anthracis* spores in two to three hours. Unlike chlorine dioxide, it leaves no toxic fumes or residues.

What do we know about the perpetrator(s) of these crimes? Sequence analysis of the *B. anthracis* strain delivered in the letters showed it was extremely similar to a strain cultured in Anthrax Tower, called the Ames strain, which now resides in several government facilities. With this information, investigators were able to confirm the strain originated from a government research lab in the United States. The FBI focused their suspicions on a few scientists who worked at the biodefense labs at Fort Detrick, but a conviction was never made.

Summary

Section 6.1: What do microorganisms generally need to grow?

All living organisms need to acquire basic nutrients and energy. Microbes have evolved several different strategies for obtaining both of these necessary ingredients.

- All cells need access to carbon, nitrogen, phosphorus, sulfur, and oxygen to build macromolecules.
- Microbes also require various **micronutrients**, including several metal ions that have roles in protein structure or enzyme activity.
- Cells must undergo the process of **assimilation** to incorporate nutrients from their surroundings. **Autotrophs** assimilate carbon from inorganic sources, through a process known as **carbon fixation**, while **heterotrophs** obtain carbon in organic forms.
- **Phototrophs** capture light energy through the process of photosynthesis. **Chemotrophs** acquire energy through the oxidation of organic or inorganic compounds. **Organotrophs** acquire electrons from organic molecules. **Lithotrophs** acquire electrons from inorganic molecules.

Section 6.2: How do nutritional and environmental factors affect microbial growth?

Microbes live in disparate niches throughout Earth. Their preferred growth conditions depend on how they obtain nutrients and on environmental factors.

- **Prototrophs** can synthesize all the needed macromolecular precursors from a single carbon source and inorganic molecules.
- **Auxotrophs** cannot synthesize all the needed precursors from a single carbon source.
- **Aerobes** grow in the presence of oxygen. **Obligate aerobes** absolutely depend on oxygen. **Microaerophiles** grow optimally at oxygen concentrations lower than atmospheric oxygen.
- **Anaerobic growth** occurs without the use of oxygen. **Aerotolerant anaerobes** do not use oxygen, but are not harmed by it. **Obligate anaerobes** cannot grow in the presence of oxygen. **Facultative anaerobes** can utilize oxygen, but also can grow in the absence of oxygen.

- Different microorganisms grow preferentially at different pHs. **Neutrophiles** grow optimally when the pH is close to neutral (7.0). **Alkalophiles** grow optimally in basic environments (pH > 8.5), while **acidophiles** grow optimally in acidic environments (pH < 5.5).
- Osmotic pressure, **water activity (a_w)**, and temperature also affect microbial growth.

Section 6.3: How can different types of microorganisms be grown in the laboratory?

Various strategies have been developed for growing microorganisms under controlled conditions in the laboratory.

- Microorganisms can be grown in liquid media or solid media containing **agar**. **Complex media** or **defined media** can be used. Under appropriate conditions, a viable cell deposited on solid medium will give rise to a **colony**.
- A **selective medium** allows for isolation of microorganisms with specific properties. A **differential medium** allows certain microbes to be recognized based on a visual reaction in the medium. **Enrichment media** are used to increase the population of microbes with specific properties.
- The **streak plate** is a simple method for isolating microorganisms in the laboratory.
- **Cultivation-independent methods** based on nucleic acid detection and **metagenomics** can allow identification and characterization of uncultivated microbes.
- Members of microbial **consortia** can evolve obligate dependencies with other members of the community that preclude their isolation as individual species.

Section 6.4: How can we measure and monitor microbial populations?

In the laboratory, microbial growth can be controlled and monitored. Such monitoring is instrumental to experimental analysis of microbes.

- The actual number of cells in a sample can be determined by a direct count, usually done with a **Petroff–Hausser counting chamber**.

- The number of viable cells can be determined by adding bacteria to an agar medium, either via the spread plate or pour plate method. **Colony-forming units (CFUs)** then can be calculated.
- **Turbidity** can be used to measure the density of a microbial population growing in a liquid medium. As the population density increases, the **absorbance**, or **optical density**, increases.
- Microbial populations growing in a **batch culture** tend to follow a **growth curve** that includes four phases: **lag**, **log** or **exponential**, **stationary**, and **death**. By analyzing the growth kinetics, the **generation time**, **growth rate**, and **growth yield** of the microorganism can be calculated.
- **Continuous cultures** operate as open systems. **Chemostats** are practical continuous culture vessels for physiological studies, and are sometimes used for industrial production.

Section 6.5: How can we eliminate microbes or inhibit their growth?

While much research depends on the rapid growth of microorganisms, there also is a need to limit their growth or remove them from certain locations.

- Many microorganisms can be removed from liquids by filtration.
- Microorganisms can be killed by heating. In the laboratory, this destruction occurs in an **autoclave**. Pasteurization uses brief exposure to sub-boiling temperatures to destroy most cells.
- Ultraviolet (UV) radiation, ionizing radiation, and gamma rays profoundly damage DNA and are useful for sterilization.
- **Disinfectants** are chemicals that can be applied to non-living objects to kill microorganisms. **Antiseptics** can be used on skin.
- Antimicrobial compounds can be classified by structure or mechanism of action. Important classes include alcohols, phenolic compounds, and oxidizing agents.
- The **decimal reduction time (D value)** provides a specific measurement of the effectiveness of these agents.

● Application Questions

1. The most common autotrophic organisms on Earth are photoautotrophs such as plants, eukaryotic algae, and cyanobacteria. Photoautotrophs, by definition, capture light energy and use it to drive CO_2 fixation to carbohydrates. Is it possible for a non-phototrophic microbe to be an autotroph?

2. A microbiologist attempting to cultivate bacteria from human feces chooses to use LB agar, rationalizing that it is a rich medium. When he compares microscopic counts of microbial population density to the counts he obtains from cultivation of colonies on agar plates, he realizes he has cultured only a small fraction of the colonic bacteria. What was wrong with his approach? What incubation conditions would you suggest he use to maximize recovery of the colonic microbiota?

3. A scientist takes a water sample and dilutes it 1:1,000. Then 0.1 mL of the diluted sample is plated on an agar plate and incubated, yielding 50 colonies. What was the concentration of bacteria that grew on this medium in the original mixture? How many colonies would you expect to see on a plate on which 0.1 mL from a 1:10,000 dilution was plated?

4. A scientist analyzing the growth of *Escherichia coli* in the laboratory takes a 1-mL culture sample and dilutes it into 99 mL of sterile saline. She then takes a 1-mL sample of the first dilution and mixes it with another 99 mL of saline. She then takes a 1-mL sample of the second dilution and mixes it with another 99 mL of saline. She then spreads 200 μL of the final dilution on a nutrient agar plate. The next day, she counts 32 colonies on the plate. What was the cell density (CFU/mL) in the culture at the time of sampling?

5. What characteristics of microbes could make it difficult to get an accurate count using a Petroff–Hausser counting chamber? What steps could be taken to make such counting easier?

6. A microbiologist wants to study microorganisms in groundwater. Her first goal is to culture strains from groundwater samples. What are some significant chemical or environmental parameters she should investigate first, in order to develop media and

culture conditions likely to give the largest number of viable microbes?

7. Laboratory microbiologists almost always prefer working with pure cultures. Scientists working on processes such as bioremediation (removal or immobilization of an undesirable compound by microorganisms) have sometimes found that degradation of target compounds by pure cultures in the laboratory, even under optimal conditions, is considerably slower and less efficient than degradation by microbial communities in the field. How might this be explained?

8. When preparing microbiological media in the laboratory, why isn't it sufficient to simply boil the medium at 100°C for 15 minutes to sterilize it? Is an autoclave really necessary?

9. Laboratory hoods used for cell manipulations have UV lights built into them. What is the purpose of these lights?

10. Microbiologists studying physiological processes often prefer to study continuous cultures instead of batch cultures. What experimental advantages can continuous cultures offer?

11. Cultures of two very similar bacterial strains were inoculated into fresh medium at a 1:1 ratio. Strain A has a generation time of 20 minutes in this medium, while Strain B has a generation time of 21 minutes under the same conditions. Assuming that both strains skip lag phase and immediately start growing and dividing at their maximal rate, and assuming that they are able to maintain constant exponential growth rate over the course of the experiment, what fraction of the total population would Strain A represent after 5 hours? 10 hours? How long would it take before Strain A constituted 99 percent of the population?

Suggested Reading

Ben-Amor, K., H. Heilig, H. Smidt, E. E. Vaughan, T. Abee, and W. M. de Vos. 2005. Genetic diversity of viable, injured, and dead fecal bacteria assessed by fluorescence-activated cell sorting and 16S rRNA gene analysis. Appl Environ Microbiol 71:4679–4689.

Cole, S.T., K. Eiglmeier, J. Parkhill, K. D. James, N. R. Thomson, P. R. Wheeler, N. Honoré, T. Garnier, C. Churcher, D. Harris, K. Mungall, D. Basham, D. Brown, T. Chillingworth, R. Connor, R. M. Davies, K. Devlin, S. Duthoy, T. Feltwell, A. Fraser, N. Hamlin, S. Holroyd, T. Hornsby, K. Jagels, C. Lacroix, J. Maclean, S. Moule, L. Murphy, K. Oliver, M. A. Quail, M.-A. Rajandream, K. M. Rutherford, S. Rutter, K. Seeger, S. Simon, M. Simmonds, J. Skelton, R. Squares, S. Squares, K. Stevens, K. Taylor, S. Whitehead, J. R. Woodward, and B. G. Barrell. 2001. Massive gene decay in the leprosy bacillus. Nature 409:1007–1011.

Eckburg, P. B., E. M. Bik, C. N. Bernstein, E. Purdom, L. Dethlefsen, M. Sargent, S. R. Gill, K. E. Nelson, and D. A. Relman 2005. Diversity of the human intestinal microbial flora. Science 308:1635–1638.

Gill, S. R., M. Pop, R. T. Deboy, P. B. Eckburg, P. J. Turnbaugh, B. S. Samuel, J. I. Gordon, D. A. Relman, C. M. Fraser-Liggett, and K. E. Nelson. 2006. Metagenomic analysis of the human distal gut microbiome. Science 312:1355:1359.

Hickey, D., and G. A. C. Singer. 2004. Genomic and proteomic adaptations to growth at high temperature. Genome Biol 5:117.

Morris, R. M., M. S. Rappé, S. A. Connon, K. L. Vergin, W. A. Siebold, C. A. Carlson, and S. J. Giovannoni. 2002. SAR11 clade dominates ocean surface bacterioplankton communities. Nature 420:806–810.

Schleper, C., G. Jurgens, and M. Jonuscheit. 2005. Genomic studies of uncultivated archaea. Nature Rev Microbiol 3:479–488.

Staley, J. T., and A. Konopka. 1985. Measurement of in situ activities of nonphotosynthetic microorganisms in aquatic and terrestrial habitats. Annu Rev Microbiol 39:321–346.

Warren, J., and B. Marshall. 1983. Unidentified curved bacilli on gastric epithelium in active chronic gastritis. The Lancet 321:1273–1275.

DNA Replication and Gene Expression

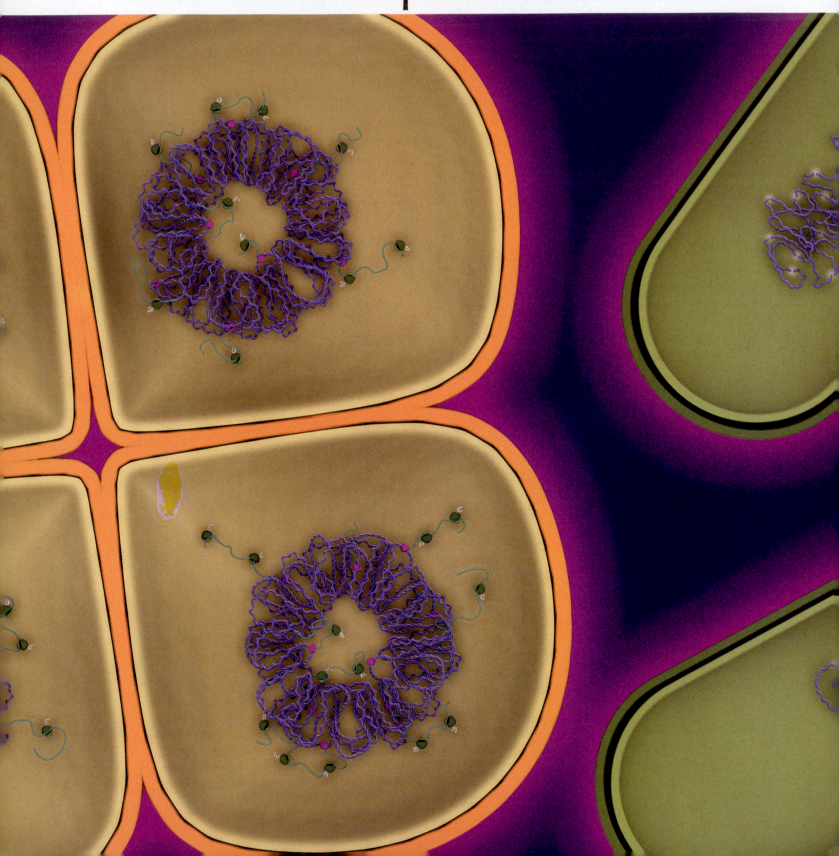

What would it take for a microorganism to make it into the Guinness World Records? How about surviving nearly 1,000 times the amount of radiation that would kill a person and 100 times the amount of radiation that would kill *E. coli*? Nicknamed Conan the Bacterium, *Deinococcus radiodurans* can do just that. Oh, and it also can withstand exposure to UV light, hydrogen peroxide, and desiccation.

Researchers first isolated *Deinococcus radiodurans* in 1956. Like a surprising number of events in science, the isolation happened quite by accident. Researchers were investigating the use of gamma radiation to sterilize food. A can of meat was exposed to a supposedly lethal dose of radiation. To the researchers' surprise, something survived—a bacterium later classified as *Deinococcus radiodurans*. This bacterium appears to be Gram-positive when stained, although it possesses an outer cell membrane typically associated with Gram-negative bacteria. Cells of *D. radiodurans* often exist in groups of four. The species can be isolated from a variety of environments, including soil, raw meat, and sewage.

High doses of radiation kill organisms primarily by causing extensive damage to DNA. Specifically, ionizing radiation causes double-stranded breaks in the DNA. When the DNA of a cell is extensively disrupted in this way, the cell cannot repair the damage to its genetic material and, ultimately, the cell dies. So how does *D. radiodurans* survive such horrific insults? Perhaps the cells, somehow, are impervious to radiation. Alternatively, the DNA of this species may be resistant to the effects of radiation. Maybe *D. radiodurans* has developed a novel means of repairing damage to its DNA. Not surprisingly, researchers are investigating the DNA of this species in an attempt to solve this intriguing mystery.

CHAPTER NAVIGATOR

As you study the key topics, make sure you review the following elements:

DNA is the hereditary molecule in all living organisms and exhibits a similar composition and structure in all organisms.

- Figure 7.2: Biochemical evidence that Griffith's transforming factor is DNA
- Microbes in Focus 7.1: *Streptococcus pneumoniae* in the laboratory and the hospital
- Figure 7.4: The composition and structure of DNA
- Table 7.1: Characteristics of DNA molecules

DNA replication is semiconservative and involves precise initiation, elongation, and termination events.

- 3D Animation: DNA Replication
- Figure 7.7: Semiconservative replication of DNA
- Figure 7.11: Bidirectional replication of DNA
- Table 7.2: Major DNA replication enzymes

Transcription involves the production of RNA, using DNA as a template.

- 3D Animation: Transcription
- Table 7.3: RNA molecules
- Toolbox 7.1: Using a gel shift assay to identify DNA-binding proteins
- Figure 7.21: Process Diagram: Post-transcriptional processing of eukaryal RNA

Translation involves the production of protein, using RNA as a template.

- 3D Animation: Translation
- Figure 7.25: Role of the Shine-Dalgarno sequence in bacteria
- Microbes in Focus 7.2: The extreme life of *Sulfolobus solfataricus*
- Figure 7.29: Spatial separation of transcription and translation in eukarya, but not bacteria

Changes in the DNA sequence can change the phenotype of a cell and make evolutionary change possible.

- Figure 7.30: Possible effects of point mutations
- Mini-Paper: Telomeres with promoter activity
- Perspective 7.1: Using mutations to control viral infections

CONNECTIONS for this chapter:

Artificial transformation and molecular cloning (Section 9.4)
Role of regulatory RNA molecules in gene expression (Section 11.4)
Protein folding and disease (Section 23.4)
Chaperonins and protein folding in hyperthermophilic archaeons (Section 4.3)
Use of ionizing radiation in sterilizing food (Section 16.2)

Introduction

We truly live in a golden era of genetics. Whole genomes can be sequenced and compared. New species can be identified solely on the basis of their sequences. We can use bacterial mutants to explore the inner workings of cells. We can manipulate the DNA of bacteria and have them produce useful proteins. In the very near future, we may be able to engineer bacteria to degrade toxic waste or generate a clean form of energy. We may also be able to treat deadly diseases by actually altering the genetic make-up of a person. All of these amazing possibilities have resulted from our understanding of how DNA works. Most of our understanding of how DNA works has resulted from studies done with bacteria and viruses.

In this chapter, we will investigate several key genetic molecules and examine the major steps in the flow and processing of genetic information. First, we will investigate the genetic material itself. For all bacteria, archaeons, and eukarya, the major information-carrying molecule is DNA, a double-stranded macromolecule. As we will see, each strand of DNA is a polymer of nucleotides linked together by covalent bonds. Each nucleotide, in turn, consists of a sugar molecule, a phosphate group, and nitrogen-containing base. Four different nucleotides, containing four different bases, exist in DNA. The two strands of a double-stranded DNA molecule are held together by hydrogen bonds that form between the nitrogenous bases. When a cell replicates, the genetic information needs to be replicated as well. In DNA replication, the double-stranded molecule transiently becomes single-stranded, and each strand serves as a template, or model, for the formation of a new, identical double-stranded molecule.

Not only does the genetic material—DNA—need to be replicated, but it also needs to be interpreted; functional molecules need to be produced from this informational molecule. This process involves two steps. First, specific regions of the DNA, genes, serve as templates for the formation of another genetic molecule—RNA. This process of producing RNA from DNA is known as transcription. Like DNA, RNA is a polymer of nucleotides. The nucleotides that comprise RNA, however, contain a slightly different sugar molecule. One of the nitrogenous bases in RNA also differs from its counterpart in DNA. Additionally, RNA is usually single-stranded.

Some RNA molecules are themselves, functional. Other RNA molecules, messenger RNA (mRNA), serve as templates for the production of proteins, a process known as translation. Each group of three bases, or a codon, in these RNA molecules corresponds to a specific amino acid. Thus, the string of codons that composes an RNA molecule determine which amino acids will be used to construct a polypeptide, or amino acid polymer. Through the processes of transcription and translation, the information present in the DNA molecule is used to construct the functional proteins required by the cell (see Figure 1.18).

Not surprisingly, DNA replication, transcription, and translation are quite a bit more complicated than we have presented them in this short introduction. So, let's investigate these processes in more detail. We'll frame our introduction to genome replication and gene expression around these questions:

How do we know DNA is the hereditary molecule? (7.1)
How does DNA replicate? (7.2)
How are genes transcribed? (7.3)
How are proteins made? (7.4)
What effects do changes in the DNA sequence have? (7.5)

7.1 The role of DNA

How do we know DNA is the hereditary molecule?

Today, we all know that DNA is the hereditary molecule. Our understanding of DNA and its role as the genetic material, however, was not always so clear. During the first half of the twentieth century, many biologists thought that the DNA molecule was too simple to serve as the proverbial blueprint of life. As we mentioned, DNA is a polymer of nucleotides, each of which contains one of only four different nitrogenous bases. Proteins, conversely, are composed of at least 20 different amino acids that can be arranged in various orders to produce molecules of various shapes and sizes.

Several landmark experiments demonstrated that DNA affected the phenotype of an organism and could direct the synthesis of more complex structures. Not surprisingly, researchers made these important discoveries using microorganisms. In this section, we will examine three experiments providing evidence that DNA is the genetic molecule. We also will review the structure of DNA.

The Griffith experiment

The identification of DNA as the genetic material arose from a seemingly odd observation. In 1928, Fred Griffith published a report in which he examined the virulence, or disease-causing capability, of two strains of the bacterium *Streptococcus pneumoniae*. The strains were referred to as smooth, or S, and rough, or R, based on their colony appearance when grown on agar plates. When live S cells were injected into mice, the mice died. When live R cells were injected into mice, the mice lived. Not surprisingly, when S cells were heat-killed prior to being injected into mice, the mice did not die. From these observations, one could

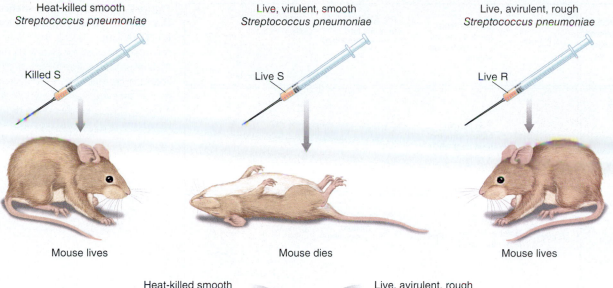

Heat-killed smooth
Streptococcus pneumoniae

Killed S

Mouse lives

Live, virulent, smooth
Streptococcus pneumoniae

Live S

Mouse dies

Live, avirulent, rough
Streptococcus pneumoniae

Live R

Mouse lives

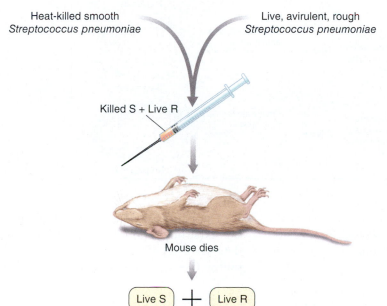

Heat-killed smooth
Streptococcus pneumoniae

Live, avirulent, rough
Streptococcus pneumoniae

Killed S + Live R

Mouse dies

Live S   +   Live R

Live, smooth *Streptococcus pneumoniae* plus live, rough *Streptococcus pneumoniae* isolated from dead mouse

Figure 7.1. Griffith's demonstration of genetic transfer When injected into mice, *Streptococcus pneumoniae* cells from smooth colonies were virulent. Cells from rough colonies were avirulent. When smooth, virulent *S. pneumoniae* cells were heat-killed and mixed with live, rough, avirulent cells, a factor from the smooth cells altered (or transformed) the phenotype of the rough cells, producing some smooth cells.

easily conclude that the S strain was pathogenic, or virulent, and the R strain was non-pathogenic, or avirulent. Griffith did one more experiment. He injected into mice both heat-killed S bacteria and live R bacteria. The result was quite interesting. First, the mice died. Second, Griffith could isolate from the mice live S bacteria **(Figure 7.1)**.

How can we interpret this result? Griffith assumed (understandably and quite correctly!) that the S cells did not come back to life. Rather, he postulated that something released from the killed S cells altered the phenotype of the   R cells, converting, or transforming, them from non-pathogenic R cells into pathogenic S cells. He termed this process **transformation**.

CONNECTIONS Griffith observed natural transformation. Today, we can intentionally insert DNA from one organism into a bacterial cell. This process of artificial transformation is presented in more detail in **Section 9.4**.

The Avery, MacLeod, and McCarty experiment

Griffith's experiments with *Streptococcus pneumoniae* clearly showed that some cellular factor could be transferred to another cell and affect the characteristics of the recipient cell. Of course, Griffith left unanswered a major question: What is this transforming agent? This part of the puzzle was answered in another classic experiment. During the early 1940s, Oswald Avery, Colin MacLeod, and Maclyn McCarty, researchers at the Rockefeller Institute (now known as Rockefeller University) and New York University, investigated this very question. Using various biochemical techniques, they purified the transforming factor from S cells and determined that it was mostly DNA. The purified transforming factor, though, may have contained some protein or RNA. To show that DNA, and DNA alone, was the transforming factor, they demonstrated that its ability to alter the phenotype of the recipient cell was not affected by treatment with proteases (enzymes that degrade proteins) or RNases (enzymes that degrade RNA).

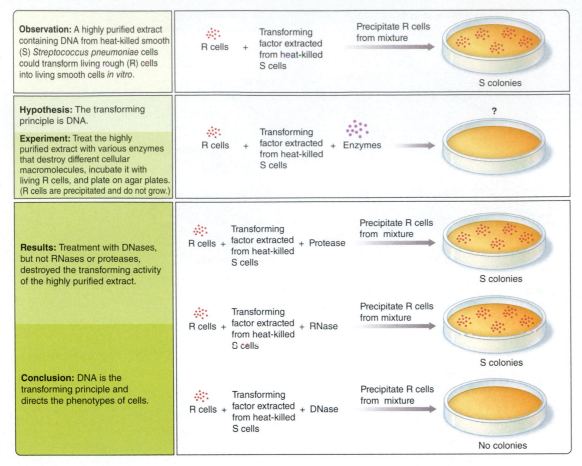

Observation: A highly purified extract containing DNA from heat-killed smooth (S) *Streptococcus pneumoniae* cells could transform living rough (R) cells into living smooth cells *in vitro*.

R cells + Transforming factor extracted from heat-killed S cells → Precipitate R cells from mixture → S colonies

Hypothesis: The transforming principle is DNA.

Experiment: Treat the highly purified extract with various enzymes that destroy different cellular macromolecules, incubate it with living R cells, and plate on agar plates. (R cells are precipitated and do not grow.)

R cells + Transforming factor extracted from heat-killed S cells + Enzymes → ?

Results: Treatment with DNases, but not RNases or proteases, destroyed the transforming activity of the highly purified extract.

R cells + Transforming factor extracted from heat-killed S cells + Protease → Precipitate R cells from mixture → S colonies

R cells + Transforming factor extracted from heat-killed S cells + RNase → Precipitate R cells from mixture → S colonies

Conclusion: DNA is the transforming principle and directs the phenotypes of cells.

R cells + Transforming factor extracted from heat-killed S cells + DNase → Precipitate R cells from mixture → No colonies

Figure 7.2. Biochemical evidence that Griffith's transforming factor is DNA Avery, MacLeod, and McCarty demonstrate that the transforming factor is DNA. Live R cells were incubated with the transforming factor isolated from heat-killed S cells. Serum that would precipitate and remove R cells was added and the mixture was plated on agar plates. DNases, but not RNases or proteases, eliminated transformation.

However, DNases (enzymes that degrade DNA) did destroy the transforming ability of their isolated factor (**Figure 7.2**).

Avery described the impact of this finding in a typically understated way:

> If we are right, and of course that is not yet proven, then it means that nucleic acids are not merely structurally important but functionally active substances in determining the biochemical activities and specific characteristics of cells and that by means of a known chemical substance it is possible to induce *predictable and hereditary* changes in cells. This is something that has long been the dreams of geneticists. (Oswald Avery, 1943)

The researchers published their results in the *Journal of Experimental Medicine* in the spring of 1944.

Today, we have a much better understanding of both this experiment and the Griffith experiment. The pathogenic S strain of *Streptococcus pneumoniae* (**Microbes in Focus 7.1**) contains a gene that results in the production of a polysaccharide capsule (see Figure 2.34). This capsule makes the bacterial colonies appear smooth. It also helps the bacteria evade the host immune response, thus making this strain pathogenic. The R strain, conversely, lacks this gene. As a result, the cells do not make a polysaccharide capsule, colonies appear rough, and the strain is non-pathogenic. When Griffith combined heat-killed S strain cells with live R strain cells, random fragments of S strain DNA, released from the killed cells, were incorporated by the live R cells. In a few cases, cells incorporated a piece of DNA that included the gene necessary for capsule production. These cells, then, were transformed from rough, non-pathogenic cells to smooth, pathogenic cells.

The implications of Avery's findings are staggering. First, this experiment provided the most conclusive evidence to date that DNA was the hereditary molecule. Perhaps more importantly, this experiment opened the door for the molecular

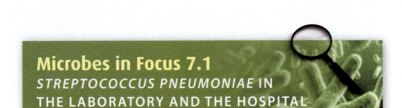

Microbes in Focus 7.1
STREPTOCOCCUS PNEUMONIAE IN THE LABORATORY AND THE HOSPITAL

Habitat: Commonly found in the nasopharynx of humans

Description: Gram-positive coccus, approximately 0.8 μm in diameter, that often appears in short chains

Key Features: *Streptococcus pneumoniae* played an integral role in defining DNA as the genetic material. *S. pneumoniae* also is a significant human pathogen, causing not only pneumonia, but also sinus and ear infections and meningitis. Its oligosaccharide capsule interferes with phagocytosis of the bacteria by immune cells, thereby enhancing its virulence.

© Luis M.de la Maza/ Phototake

biology revolution. As Avery noted, transformation makes it "possible to induce *predictable and hereditary* changes in cells." If we can identify pieces of DNA associated with specific phenotypes, then we can insert that DNA into recipient cells and alter the characteristics of those cells.

The Hershey–Chase experiment

While the conclusions drawn by Avery and his colleagues may seem quite evident, some biologists still questioned the preeminent role of DNA as the hereditary molecule. All lingering doubts were put to rest by a third classic experiment, the Hershey–Chase experiment. Shortly after Avery, MacLeod, and McCarty published their paper, Alfred Hershey and Martha Chase, researchers at the Cold Spring Harbor Laboratory, used the bacteriophage T2 to investigate the role of DNA in heredity. Their approach was elegant in its simplicity.

CONNECTIONS Replication of bacteriophage T2 begins with a specific binding event between a viral attachment protein and a cellular receptor. Research investigating the molecular determinants of this process is presented in **Section 8.1**.

Hershey and Chase knew that the genetic material of viruses entered cells and directed the infected cells to produce more virus particles. They also knew that T2 was composed of just two components—protein and DNA. During the infection process, the protein capsid of bacteriophage T2 attached to, but did not enter, the infected cell. Finally, biochemical studies had shown that nucleic acids contained phosphorus (P), but not sulfur (S). Proteins, conversely, contained lots of sulfur but little phosphorus. Using these pieces of information, they developed their experiment. Some phages were labeled with ^{35}S, radioactive sulfur. Because S is present in proteins, these phages contained radiolabeled protein. Other phages were labeled with ^{32}P, radioactive phosphorus. Because P is present in nucleic acid, these phages contained radiolabeled DNA. In separate experiments, these phages were incubated with *Escherichia coli*, their host cells, for a few minutes and then the mixture was placed in a blender and agitated to remove the capsids from the surface of the bacteria. The mixture then was centrifuged, causing the bacterial cells to pellet, and the amount of radioactivity present in the resulting pellet and supernatant was determined. They found that the ^{35}S was present in the supernatant and the ^{32}P was present in the pellet **(Figure 7.3)**.

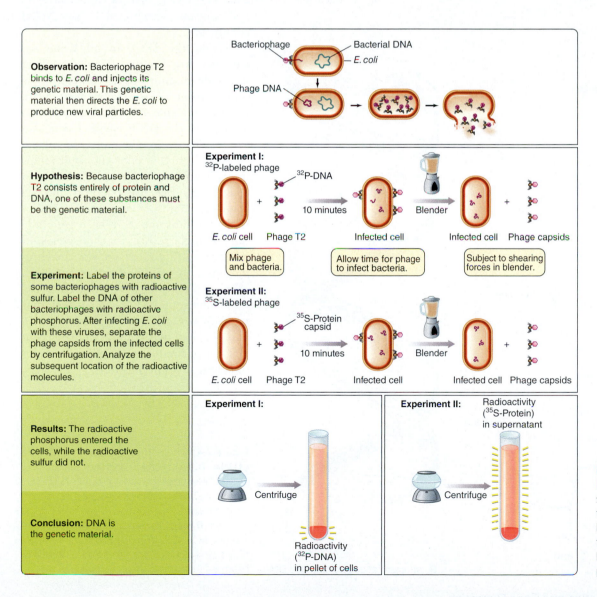

Figure 7.3. Confirmation that DNA is the genetic molecule By separately labeling with radioactive isotopes the proteins of bacteriophage T2 or the DNA of bacteriophage T2, Hershey and Chase demonstrated that DNA, and not protein, enters the host bacterium, *E. coli*, and directs the synthesis of new phage particles.

In other words, the phage capsids, which contained protein, did not enter the cells. The DNA, conversely, was injected into the host cells. The injected DNA, then, directed the production of new phage particles within the cells. The hereditary material of bacteriophage T2, Hershey and Chase concluded, was DNA.

Structure of DNA

Around the same time that Avery, MacLeod, and McCarty and Hershey and Chase were demonstrating that DNA was the hereditary material of cells, many researchers were investigating the structure of this molecule. In 1953, James Watson and Francis Crick solved the puzzle and published their history-making paper. They noted that DNA consists of two chains. Each chain is a polymer of nucleotides. Each nucleotide, in turn, consists of the following:

- A five-carbon sugar, **2-deoxyribose**

- A **phosphate** group that covalently links the 3′ carbon of one 2-deoxyribose to the 5′ carbon of the next sugar

- One of four nitrogen-containing bases, which are attached to the 1′ carbon of the sugar

A nucleoside, in contrast, consists only of a nitrogenous base connected to a sugar. The nitrogenous bases are classified as **purines** or **pyrimidines**. The purines, **adenine (A)** and **guanine (G)** contain a double-ring structure. The pyrimidines, **cytosine (C)** and **thymine (T)** contain a single-ring structure (**Figure 7.4**). Note the conventional numbering of carbon atoms within these molecules indicated in the figure.

Each linear chain of DNA has a 5′ end and a 3′ end. The two chains of a double-stranded molecule align in an **antiparallel** fashion, with the 5′ end of one chain aligned with the 3′ end of the other chain. More importantly, the nucleotides pair together in a very precise way: A always aligns with T and C always aligns with G. The two strands are **complementary**. If we know the sequence of nitrogenous bases in one strand, then we can determine the sequence of bases in the other strand. Hydrogen bonds hold together the two chains, with two H bonds linking A to T and three H bonds linking C to G. The entire double-stranded molecule forms a very precise spiral, or helix. Because of their discovery of the double-stranded structure of DNA, Watson and Crick, along with their collaborator Maurice Wilkins, received the Nobel Prize in Physiology or Medicine in 1962. Sadly, Rosalind Franklin, a coworker of Wilkins who generated most of the data used by Watson and Crick, died of cancer in 1958 and was not recognized by the Nobel committee.

While the DNA in all living organisms is a double-stranded, helical molecule, significant differences in the organization of the DNA have been identified (**Table 7.1**). In bacteria and archaeons, the DNA usually exists as a single, closed circular molecule. Supercoiling of this circular molecule helps to compact its size, at least in bacteria (**Figure 7.5**). Because bacteria and archaeons usually have a single copy of their chromosome, they also generally have a single copy of all their genes. In addition to this single copy of the chromosome, bacteria and archaeons often contain **plasmids**—small, circular extrachromosomal DNA molecules that typically contain non-essential genes. A bacterial cell may contain multiple copies of a particular plasmid and several different types of plasmids.

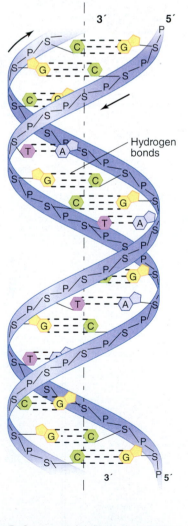

A. Structure of nucleotide polymer

B. Structure of double-stranded DNA

Figure 7.4. The composition and structure of DNA A. Each nucleotide consists of a five-carbon deoxyribose molecule, a phosphate group, and a nitrogenous base. The phosphate group links the 3′ carbon of one deoxyribose sugar to the 5′ carbon of the next deoxyribose. Nitrogenous bases are linked to the 1′ carbons of the sugars. **B.** Two strands align in an antiparallel fashion forming a double helix, in which adenine (A) always pairs with thymine (T), and cytosine (C) always pairs with guanine (G). Dashed lines between the nitrogenous bases represent H bonds.

TABLE 7.1 Characteristics of DNA molecules

	Organization	Plasmids	Histones	Representative sizes (kbp)	Notable features
Bacteria	Single, circular molecule	Common	No	580 (*Mycoplasma genitalium*) 4,600 (*Escherichia coli*)	Some bacteria, like *Borrelia burgdorferi*, have linear chromosomes.
Archaea	Single, circular molecule	Common	Yes	1,740 (*Pyrococcus horikoshii*) 3,000 (*Sulfolobus solfataricus*)	Hyperthermophilic archaeons possess reverse gyrase, an enzyme that increases the thermostability of the chromosome.
Eukarya	Linear molecules, 1*n* and 2*n* during replication cycle	Rare	Yes	12,495 (*Saccharomyces cerevisiae*) 3,000,000 (*Homo sapiens*)	Plasmids have been identified in fungi and plants.

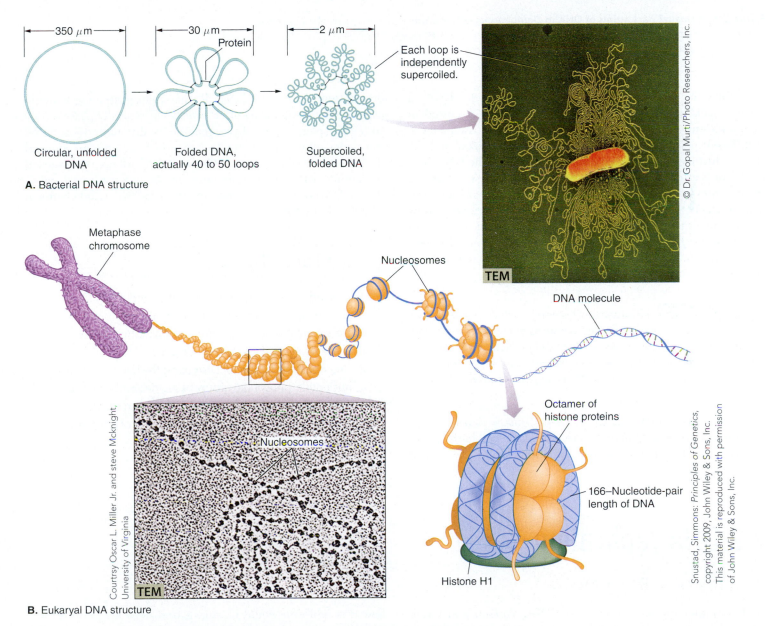

A. Bacterial DNA structure

B. Eukaryal DNA structure

Figure 7.5. The structural organization of DNA in bacteria and eukarya A. The circular chromosome is supercoiled, causing the molecule to form a tightly compacted structure. Diameter measurements attest to the compaction accomplished by folding and supercoiling the DNA. **B.** The DNA is organized around a core of eight histone proteins that interact with the DNA in a very specific fashion, forming nucleosomes. An additional histone, H1, stabilizes the nucleosome structure.

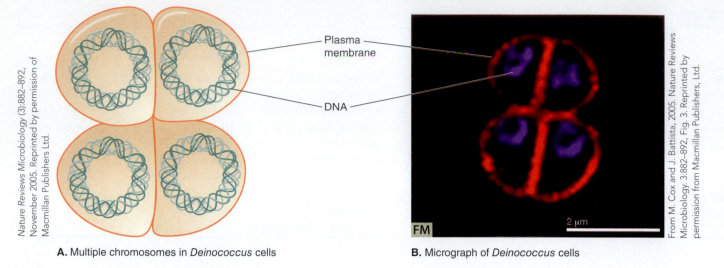

Nature Reviews Microbiology (3):882–892, November 2005. Reprinted by permission of Macmillan Publishers Ltd.

Plasma membrane

DNA

From M. Cox and J. Battista, 2005. Nature Reviews Microbiology. 3:882–892, Fig. 3. Reprinted by permission from Macmillan Publishers, Ltd.

FM

2 μm

A. Multiple chromosomes in *Deinococcus* cells

B. Micrograph of *Deinococcus* cells

Figure 7.6. Unusual packaging of DNA in *Deinococcus radiodurans* Actively growing cells often possess four to ten copies of the genome. **A.** The DNA in *Deinococcus* cells adopts a doughnut-like conformation. **B.** The DNA is stained with DAPI, causing it to appear blue in this epifluorescence image of *Deinococcus* cells. The plasma membrane is stained red.

Unlike the DNA in bacteria and archaeons, the DNA in eukarya usually exists as linear molecules. As we discussed in Chapter 3, eukarya typically have a 1n and 2n genetic complement at different points in their replication cycle. The eukaryal DNA binds to histones, basic proteins that interact with the DNA. The DNA wraps around these histones in a very precise way, forming nucleosomes (see Figure 7.5B). Again, this level of organization compacts the DNA and also protects it from degradation by various nucleases, enzymes that destroy nucleic acids. In archaeons, the DNA also appears to be organized around histones. Recent sequence analysis indicates that the archaeal and eukaryal histones share an evolutionary history.

CONNECTIONS The existence of histones in archaeons and eukarya, but not bacteria, raises interesting questions about the evolutionary history of these organisms. As we discussed in **Section 4.2**, some researchers hypothesize that histones evolved after the divergence of bacteria and archaeons, but before the emergence of eukarya.

It should be noted that the DNA of *Deinococcus radiodurans*, the radiation-resistant bacterium that we discussed in the opening story of this chapter, differs somewhat from the typical description of bacterial DNA. Rather than possessing a single copy of its genome, cells of *D. radiodurans* typically possess four to ten complete copies of the genome. Additionally, these DNA molecules appear to be stacked together, forming an unusual doughnut-like conformation **(Figure 7.6)**. Perhaps this stacked packaging of the DNA contributes to the extraordinary stability the DNA of this bacterium shows to radiation and chemical assaults that would severely damage other microbes.

7.1 Fact Check

1. Describe the experiments of Griffith and Avery that showed DNA was the hereditary material.
2. How were radioactive sulfur and phosphorus used in the Hershey and Chase experiment?
3. Describe the structure of DNA using the following terms: nucleotide, phosphate group, bases, deoxyribose sugar, antiparallel, complementary, and hydrogen bonds.
4. Compare and contrast the organization of DNA in archaeons, bacteria, and eukarya.

7.2 DNA replication

How does DNA replicate?

In their paper on the structure of DNA, Watson and Crick rather slyly noted that their proposed structure, "immediately suggests a possible copying mechanism for the genetic material." Indeed, by looking at the structure, one can imagine how this molecule could be duplicated. If A always interacts with T and if C always interacts with G, then each chain of a double-stranded molecule could serve as a template, or pattern, for the production of a new, complementary second chain. By

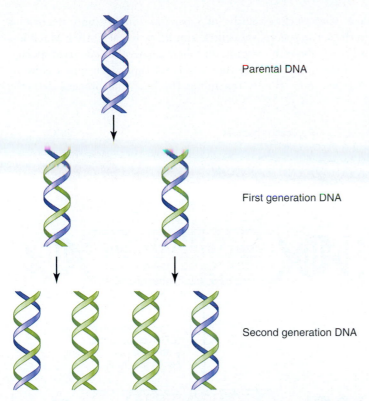

Parental DNA

First generation DNA

Second generation DNA

Figure 7.7. Semiconservative replication of DNA As the double-stranded DNA molecule unwinds, each strand serves as a template for the formation of a new, complementary strand. As a result, two new double-stranded molecules are formed, each identical to the original molecule. In each new double-stranded molecule, one strand is derived from the original molecule (blue), while the other strand is newly synthesized (green).

separating the two chains of a double-stranded molecule, two new double-stranded molecules could be generated, each identical to the original DNA. This mechanism of DNA replication is referred to as **semiconservative replication** because one strand of the original DNA is conserved in the newly formed molecule **(Figure 7.7)**. It is worth noting, too, that this mechanism of replication preserves the sequence of the bases. As we will see later in this chapter, maintaining the fidelity of the genetic molecule is critically important.

Of course, the actual process isn't quite as simple as Watson and Crick made it seem. In this section, we will briefly examine the process of DNA replication. While the basics of DNA replication are the same in all organisms, important and fundamental differences do exist among different groups of organisms. We will look at some of the more notable differences that exist between bacteria, eukarya, and archaeons. The replication of viruses will be addressed in Chapter 8.

DNA replication: Origins of replication

Ten years after Watson and Crick described the structure of DNA, Jacob, Brenner, and Cuzin proposed an elegant model for the initiation of DNA replication, which they termed the replicon model. They proposed that the process would require two main components—a replicator and an initiator. The replicator, they hypothesized, would be the site on the DNA molecule where replication started. The initiator would be a *trans*-acting

factor—a factor that could act upon another molecule. Most likely, they postulated, this factor was a protein that could interact with the replicator, and begin the process.

Subsequent studies with *E. coli* have confirmed the basics of this model. In bacteria, DNA replication begins at a specific site on the chromosome, the origin of replication, or ***oriC***. This sequence, the replicator proposed by Jacob and colleagues, has been very well characterized. In *E. coli*, it consists of roughly 245 base pairs (bp) and contains four repeats of a 9-bp sequence and three repeats of a 13-bp, AT-rich, sequence **(Figure 7.8)**. In other species of bacteria, these same recurring elements, or motifs, also are present, although the number of repeats and their exact positioning within *oriC* may differ. The functional importance of *oriC* can be demonstrated in a simple way. When *oriC* is added to an artificially constructed closed circular DNA molecule, then that molecule can be replicated within an *E. coli* cell. Finally, it appears that every bacterial chromosome contains a single *oriC*. In other words, replication of the bacterial chromosome begins at a single location.

The 9-bp repeats in *oriC*, often referred to as DnaA boxes, serve as binding sites for **DnaA**, a DNA-binding protein produced by the *dnaA* gene. Multiple copies of DnaA, the initiator proposed by Jacob, bind to the DnaA boxes and then interact with the 13-bp repeat regions, resulting in strand separation and the formation of a single-stranded region. Several studies have shown that DnaA is an essential protein; without a functional DnaA, DNA replication does not begin. After DnaA begins denaturing, or "melting," the DNA, a series of other proteins, including DnaB (often referred to as **helicase**) and DnaC (often referred to as a helicase loader) are recruited to this site to aid in the unwinding. **DnaG**, or **primase**, synthesizes short segments of RNA needed to prime DNA replication. **Single-stranded DNA-binding proteins (SSB)** attach to the newly formed single-stranded DNA to keep the strands from reannealing and a single-stranded **replication bubble** forms

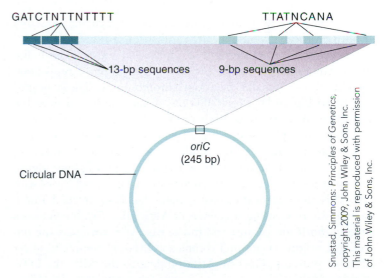

GATCTNTTNTTTT   TTATNCANA

13-bp sequences   9-bp sequences

oriC
(245 bp)

Circular DNA

Snustad, Simmons: *Principles of Genetics*, copyright 2009, John Wiley & Sons, Inc. This material is reproduced with permission of John Wiley & Sons, Inc.

Figure 7.8. Origin of replication in *E. coli* This AT-rich region contains four 9-base-pair (bp) and three 13-bp repeated sequences. Other bacteria contain similar motifs, although the exact numbers and their positioning may differ. In order to replicate in a cell, every circular DNA needs an origin of replication. *NOTE:* "N" indicates that any base may be at that location.

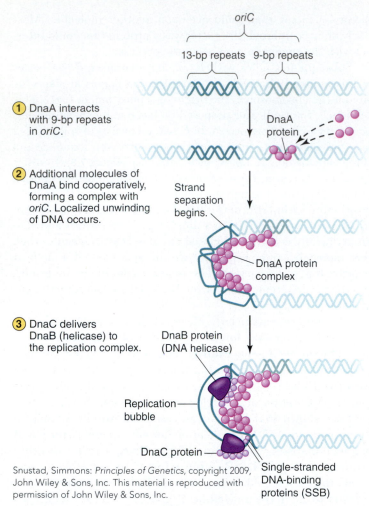

① DnaA interacts with 9-bp repeats in *oriC*.

② Additional molecules of DnaA bind cooperatively, forming a complex with *oriC*. Localized unwinding of DNA occurs.

③ DnaC delivers DnaB (helicase) to the replication complex.

Snustad, Simmons: *Principles of Genetics*, copyright 2009, John Wiley & Sons, Inc. This material is reproduced with permission of John Wiley & Sons, Inc.

Figure 7.9. Process Diagram: Initiation of DNA replication in bacteria A series of proteins, including DnaA, DnaB, and DnaC interact with *oriC* to help form a replication bubble. DnaA binds to the 9-base-pair (bp) repeats. Additional molecules of DnaA bind cooperatively, forming a complex with *oriC*, and begin denaturing the DNA. DnaC brings DnaB (also referred to as DNA helicase) to the complex and further unwinding of the double-stranded DNA occurs, forming a replication bubble.

(Figure 7.9). DNA polymerases then begin replicating the DNA at both ends of this bubble, the replication forks.

In eukarya, many origins of replication exist in each linear chromosome, in contrast to the single origin of replication present in the circular bacterial chromosomes. These origins may be within 10 kbp, or 10,000 bp, of one another or over 100 kbp apart. Most likely, multiple origins of replication are needed to effectively replicate the larger genomes of eukaryal cells in a reasonable amount of time. Studies with *Saccharomyces cerevisiae*, a model eukaryal organism introduced in Chapter 3, have identified a eukaryal replicator that shares functional similarities with *oriC*. This element in the yeast genome has been termed an autonomously replicating sequence, or ARS. Like *oriC*, it contains a binding motif for an initiator. In the case of *S. cerevisiae*, the initiator, or origin recognition complex (ORC), is composed of six different proteins (Orc1-6). After binding to the ARS, the ORC recruits additional proteins necessary for replication, including Cdc6, Cdt1, and the minichromosomal maintenance (MCM) complex, a group of six proteins that functions as the helicase **(Figure 7.10)**. These proteins then begin the important task of unwinding the DNA. An origin of replication also has been identified

in *S. pombe*, a close relative of *S. cerevisiae*. Surprisingly, the origins in these two species show little similarity to each other. Much less is known about the origins of replication in other eukaryal species.

One might assume that the DNA replication process in archaeons more closely resembles the bacterial process than the

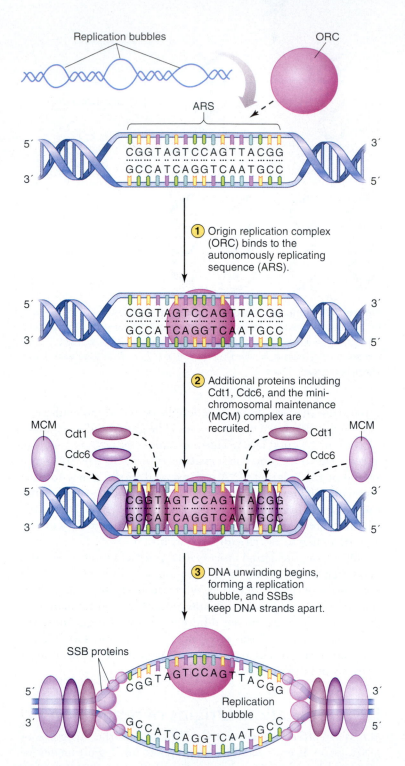

① Origin replication complex (ORC) binds to the autonomously replicating sequence (ARS).

② Additional proteins including Cdt1, Cdc6, and the mini-chromosomal maintenance (MCM) complex are recruited.

③ DNA unwinding begins, forming a replication bubble, and SSBs keep DNA strands apart.

Figure 7.10. Process Diagram: Initiation of DNA replication in eukarya In eukarya, DNA replication initiates at several spots along the linear chromosome. In *Saccharomyces cerevisiae*, the origin recognition complex (ORC) binds to the origin, or ARS. Additional proteins, including Cdc6 and Cdt1, are recruited. The minichromosomal maintenance (MCM) complex also binds and functions as the helicase. Additional proteins then locate to the replication bubble and begin synthesizing new DNA.

eukaryal process. After all, archaeons, like bacteria, generally possess a single, small, circular chromosome. Recent evidence, however, indicates that the replication machinery of archaeons may more closely resemble the replication machinery of eukarya. In *Pyrococcus horikoshii*, a hyperthermophilic archaeon, a single origin of replication has been identified, much as we see in bacteria. The initiator protein, though, appears to be closely related to Orc1 and Cdc6, two of the eukaryal replication complex proteins. *Sulfolobus solfataricus*, another hyperthermophilic archaeon, presents an even more interesting case. The *S. solfataricus* chromosome appears to possess two functional origins of replication, *oriC1* and *oriC2*. Additionally, three *cdc6*-like genes (*cdc6-1*, *cdc6-2*, and *cdc6-3*) have been identified in the genome of this species. It has been hypothesized that the Cdc6-1 and Cdc6-3 proteins may be positive regulators of replication, while Cdc6-2 serves as a negative regulator, or inhibitor, of replication. Thus, DNA replication in archaeons and bacteria exhibit some fundamental differences. This finding provides more conclusive evidence in support of the three domain tree of life hypothesis. This finding also further supports the idea that archaeons may be more closely related to eukarya than to bacteria. The *S. solfataricus* data suggest that archaeons may ultimately prove to be useful model organisms for the study of complex eukaryal events like DNA replication.

DNA replication: Initiation and elongation

Following the attachment of the initiator and helicase proteins to the origin of replication, the production of new DNA strands, complementary to the old strands, can occur. A series of enzymes, known as **DNA polymerases**, catalyze this task. DNA polymerases, however, cannot simply add new nucleotides to a piece of single-stranded DNA. Rather, these enzymes only can add new nucleotides onto an existing piece of DNA or RNA, using another strand as the template. Furthermore, all DNA polymerases only add new nucleotide triphosphates onto the free 3′—OH group of a nucleotide. So how does this process ultimately begin? Again, the basic mechanism is remarkably similar in bacteria, eukarya, and archaeons.

To initiate DNA elongation, primase, a special RNA polymerase introduced earlier, synthesizes a short piece of single-stranded RNA that attaches to the exposed single-stranded DNA present at the replication bubble. Once this **primer** has been formed, DNA polymerase then can add new bases onto this fragment, in a 5′ to 3′ fashion, using the existing DNA molecule as a template. This elongation occurs bidirectionally, and the helicase proteins described in the previous section continue to unwind the existing double-stranded DNA as the elongation events proceed. Because DNA polymerase only can elongate DNA in a 5′ to 3′ direction, only one strand of single-stranded DNA at each replication fork can be extended continuously **(Figure 7.11)**. These strands are referred to as the **leading strands**. On the opposite strands, new primers must be added as the helicase unwinds more of the DNA. These strands, then, exhibit discontinuous elongation and are referred to as **lagging strands**. These individual pieces of newly synthesized DNA and their corresponding RNA primers often are called **Okazaki fragments** after their discoverer, Reiji Okazaki.

In bacteria, DnaG, the product of the *dnaG* gene, serves as the primase, synthesizing short segments of RNA that bind to

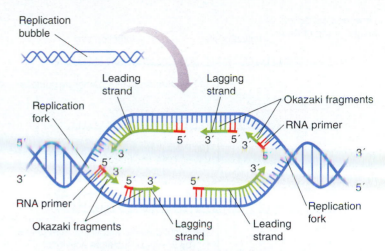

Figure 7.11. Bidirectional replication of DNA As the DNA unwinds, DNA synthesis occurs in a 5′ to 3′ direction, extending from short RNA primers produced by primase. At each replication fork, one strand can elongate continuously (leading strand), while the other strand elongates discontinuously (lagging strand).

the single-stranded DNA. The major DNA polymerase involved in replication is the DNA polymerase III (DNA pol III) holoenzyme, a large, multisubunit complex. This enzyme adds a deoxynucleotide triphosphate to the free 3′—OH present on an existing nucleotide, removing two phosphates in the process. In this manner, the new DNA strand elongates in a 5′ to 3′ direction. Elongation occurs at an impressive rate. The *E. coli* DNA polymerase III can synthesize DNA at a rate of 1,000 bases per second! In the lagging strand, this activity continues until the DNA polymerase and its newly made strand of DNA run into an RNA primer. At this point, a different DNA polymerase, DNA polymerase I (DNA pol I), removes the RNA primer, using a 5′ exonuclease activity. DNA pol I also fills in the resulting short gap. One last step remains. DNA **ligase** links, or ligates, the final 5′—phosphate and 3′—OH ends. The new DNA strand now is complete **(Figure 7.12)**.

Mechanistically, the process of DNA replication in eukarya is very similar to DNA replication in bacteria. Again, a primase adds short stretches of RNA to the single-stranded DNA present at the origin of replication. A DNA polymerase then begins adding nucleotides to the free 3′—OH group on the terminal nucleotide of the primer. Elongation occurs in a 5′ to 3′ direction and, as we described for bacteria, it occurs bidirectionally, with a leading strand and a lagging strand present on each arm of the replication fork. Initially, the DNA polymerase α/primase complex synthesizes the RNA primer (see Table 7.2). Quickly, though, this polymerase is replaced by two other polymerases, DNA pol ε and DNA pol δ. DNA pol ε specifically synthesizes the leading strand, while DNA pol δ synthesizes the lagging strand. As in bacteria, DNA ligase joins together adjacent lagging strands.

The general process of DNA replication in archaeons does not vary much from this standard paradigm. The specific proteins involved, however, more closely resemble the proteins used by eukarya than bacteria. As in bacteria and eukarya, a primase first adds an RNA primer to the single-stranded DNA at the origin of replication. This enzyme bears significant sequence homology to the eukaryal protein p48, the component of the DNA pol α/primase complex responsible for synthesizing the RNA primers. The major replicative enzyme in archaeons

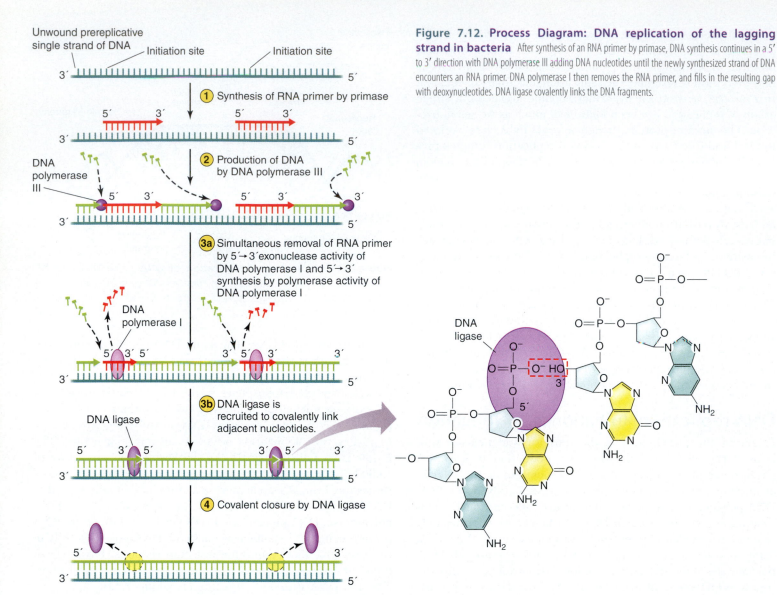

Figure 7.12. Process Diagram: DNA replication of the lagging strand in bacteria After synthesis of an RNA primer by primase, DNA synthesis continues in a 5′ to 3′ direction with DNA polymerase III adding DNA nucleotides until the newly synthesized strand of DNA encounters an RNA primer. DNA polymerase I then removes the RNA primer, and fills in the resulting gap with deoxynucleotides. DNA ligase covalently links the DNA fragments.

Labels within the figure:

- Unwound prereplicative single strand of DNA
- Initiation site
- Initiation site
- ① Synthesis of RNA primer by primase
- ② Production of DNA by DNA polymerase III
- DNA polymerase III
- ③a Simultaneous removal of RNA primer by 5′→3′ exonuclease activity of DNA polymerase I and 5′→3′ synthesis by polymerase activity of DNA polymerase I
- DNA polymerase I
- ③b DNA ligase is recruited to covalently link adjacent nucleotides.
- DNA ligase
- ④ Covalent closure by DNA ligase
- DNA ligase

also shows significant sequence homology to the replicative enzymes found in eukarya, suggesting an evolutionary link between the replication machinery of archaeons and eukarya. A summary of the major enzymes involved in DNA replication is provided in **Table 7.2**.

DNA replication: Termination

As the two arms of the replication fork proceed around the circular bacterial chromosome, they eventually will meet on the opposite side of the circle. At this point, DNA elongation stops. In *E. coli*, this termination event is not random. Rather, a

TABLE 7.2 Major DNA replication enzymes

	Bacteria	Eukarya	Archaea	Function
Origin recognition	DnaA	Origin recognition complex (ORC) proteins 1–6	ORC-like proteins	Site at which replication begins. Typically AT-rich
Primase	DnaG	DNA pol α	DNA pol α-like enzyme	Synthesizes short RNA oligonucleotides to which DNA polymerase adds new nucleotides
Helicase	DnaB	Dna2	Dna2-like enzyme	Aids in unwinding double-stranded DNA
Replication enzyme	DNA pol III	DNA pol ε (leading strand) DNA pol ε (lagging strand)	Eukaryal-like Family B DNA polymerase	Major enzymes involved in synthesizing new DNA
Removal of primers	DNA pol I	FEN1/Rad2	FEN1/Rad2	Excise short pieces of RNA used as primers
Ligation	DNA ligase	DNA ligase I	DNA ligase	Covalently links adjacent lagging strands

Source: Adapted from *Cell* 89:995–998.

series of termination sequences, or *ter* sites, exist on the chromosome roughly opposite of *oriC*. A Tus protein binds to one of the *ter* sites and stops the forward progress of one of the replication forks. When the other replication fork arrives at this termination locus, the two halves of the newly synthesized DNA are joined, resulting in two identical circular double-stranded DNA molecules. Because these two circular chromosomes are linked together, one more enzyme is needed—topoisomerase. Topoisomerase II forms a transient double-stranded break in the DNA molecules, thereby allowing the two circles to become disentangled **(Figure 7.13)**.

The termination of DNA elongation in eukarya presents a more challenging problem. Think about the scenario described in Figure 7.10. As bubbles of replication along the linear chromosome extend, the replication forks eventually will merge. Through the actions of exonucleases and DNA ligase, the RNA primers needed to initiate the elongation process will be removed and replaced with DNA. In this manner, the leading strand can extend to the very 5′ end of its complementary template strand. Completion of the lagging strand, however, is not so simple. An RNA primer will remain at the 3′ end of the newly synthesized strand of DNA. The RNA in this RNA:DNA heteroduplex will be degraded by exonucleases, resulting in a short stretch of single-stranded DNA at the end of the chromosome **(Figure 7.14)**. If this stretch of DNA were to remain single-stranded, then it would be degraded, resulting in a shortening of the chromosome. Filling in this stretch of single-stranded DNA, however, is problematic; DNA polymerase only can add new nucleotides onto the 3′ end of an existing segment of DNA. As shown in Figure 7.14, the single-stranded region at the terminus of a linear chromosome has a 5′ end.

To solve this problem, linear chromosomes of eukarya end in a special sequence known as a **telomere**. Basically, the telomere consists of a short sequence of bases repeated many times. During replication, a unique DNA polymerase, **telomerase**, interacts with the telomere. Telomerase consists of several proteins and a short piece of RNA. The RNA, it turns out, is complementary to the single-stranded DNA that forms at the end of a linear chromosome. Through complementary base-pairing, telomerase attaches to the telomere. It attaches in such a way that part of the telomerase RNA (TER) extends beyond the 3′ end of the telomere. Telomerase extends this 3′ end, using TER as a template. In other words, telomerase is an RNA-dependent DNA polymerase. After the 3′ end of the chromosome has been extended, the enzyme moves down the newly synthesized molecule, thereby extending the single-stranded DNA of the telomere and lengthening the chromosome. Eventually, primase can add an RNA primer to this newly created DNA, and DNA polymerases can generate a complementary strand of DNA. The final RNA primer eventually will be removed, generating the single-stranded region necessary for telomerase binding during the next round of DNA replication.

While the process by which telomerase extends the end of a linear chromosome has been well established, the exact role of telomerase in the life of an individual remains an interesting topic of research. In studies involving various vertebrate species, several groups of researchers have shown that telomerase is active in cells early in development, but not in the somatic cells of adult individuals. In contrast, telomerase is active in tumor cells

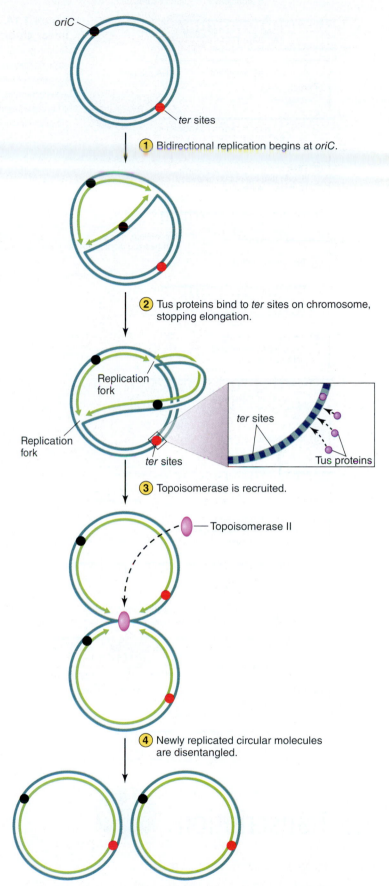

Figure 7.13. Process Diagram: The termination of replication for circular chromosomes In *E. coli*, a series of *ter*, or termination, sites exist roughly opposite of the origin of replication (*oriC*). After bidirectional replication begins at *oriC*, the Tus protein binds to these *ter* sites and stops the progress of the replication forks. Topoisomerase II then forms a transient break in the DNA, allowing the two circles to become disentangled.

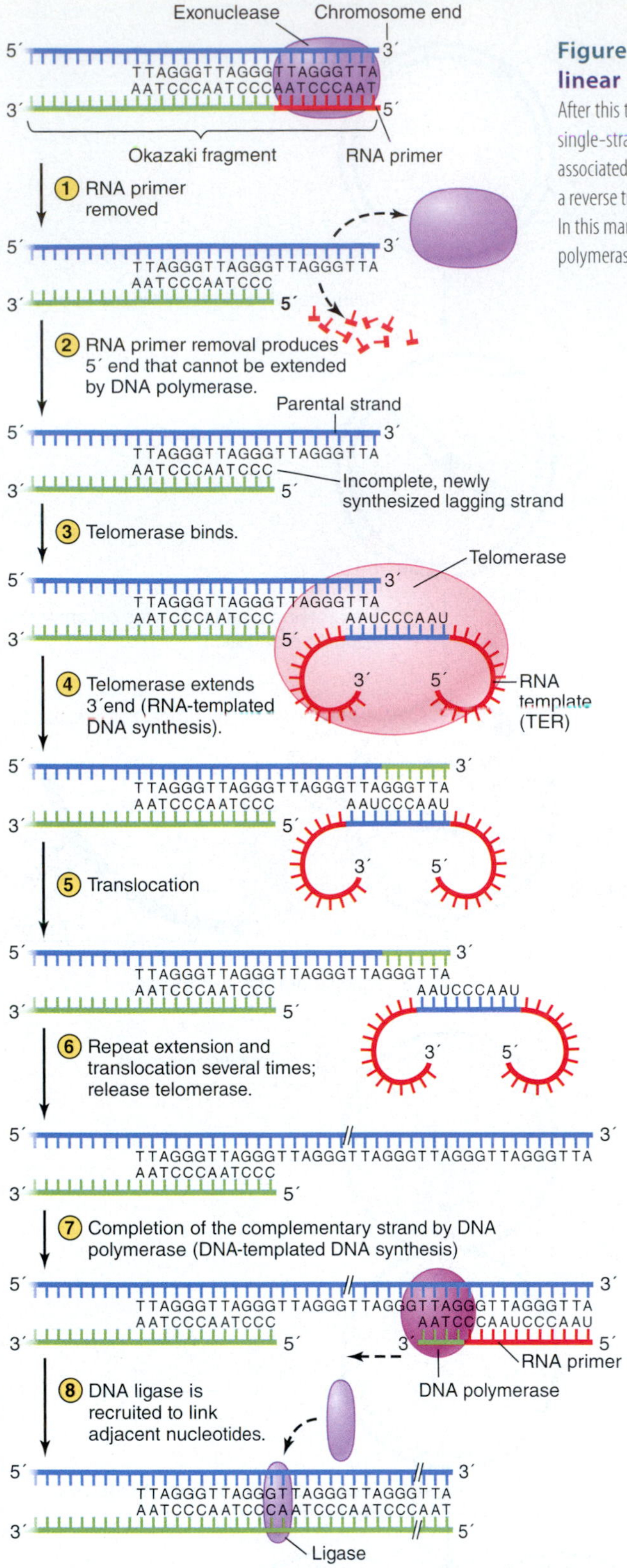

Exonuclease
Chromosome end
5′
3′
TTAGGGTTAGGGTTAGGGTTA
AATCCCAATCCCAATCCCAAT
3′
5′
Okazaki fragment
RNA primer
1 RNA primer removed
5′
3′
TTAGGGTTAGGGTTAGGGTTA
AATCCCAATCCC
3′
5′
2 RNA primer removal produces 5′ end that cannot be extended by DNA polymerase.
Parental strand
5′
3′
TTAGGGTTAGGGTTAGGGTTA
AATCCCAATCCC
3′
5′
Incomplete, newly synthesized lagging strand
3 Telomerase binds.
Telomerase
5′
3′
TTAGGGTTAGGGTTAGGGTTA
AATCCCAATCCC
AAUCCCAAU
3′
5′
4 Telomerase extends 3′end (RNA-templated DNA synthesis).
RNA template (TER)
5′
3′
TTAGGGTTAGGGTTAGGGTTAGGGTTA
AATCCCAATCCC
AAUCCCAAU
3′
5′
5 Translocation
5′
3′
TTAGGGTTAGGGTTAGGGTTAGGGTTA
AATCCCAATCCC
AAUCCCAAU
3′
5′
6 Repeat extension and translocation several times; release telomerase.
5′
3′
TTAGGGTTAGGGTTAGGGTTAGGGTTAGGGTTAGGGTTA
AATCCCAATCCC
3′
5′
7 Completion of the complementary strand by DNA polymerase (DNA-templated DNA synthesis)
5′
3′
TTAGGGTTAGGGTTAGGGTTAGGGTTAGGGTTAGGGTTA
AATCCCAATCCC
AATCCCAAUCCCAAU
3′
5′
RNA primer
DNA polymerase
8 DNA ligase is recruited to link adjacent nucleotides.
5′
3′
TTAGGGTTAGGGTTAGGGTTAGGGTTA
AATCCCAATCCCAATCCCAATCCCAAT
3′
5′
Ligase

Figure 7.14. Process Diagram: Role of telomerase in replication of the ends of linear chromosomes On the lagging strand, a terminal RNA primer reaches the very end of the chromosome. After this terminal RNA primer is removed, a short segment of single-stranded DNA will remain. To prevent the degradation of this single-stranded region, the enzyme telomerase interacts with the telomere sequence of the single-stranded DNA. A piece of RNA associated with the telomerase enzyme binds to the single-stranded DNA of the telomere. Telomerase then extends the DNA, using a reverse transcriptase-like activity. The enzyme subsequently moves to the newly synthesized piece of DNA and repeats the process. In this manner, telomerase extends the length of the telomere. An RNA primer attaches to the newly formed piece of DNA, and DNA polymerase synthesizes new DNA, filling in the single-stranded gap.

and immortal cell lines. Based on these pieces of evidence, many researchers now hypothesize that chromosomal shortening that occurs in the absence of telomerase activity may play a role in the aging process.

Let's address one final question before ending our discussion of the termination of DNA replication. What about bacteria with linear chromosomes? While a vast majority of known bacteria and archaeons possess circular chromosomes, several bacterial species possess linear chromosomes. Presumably, replication of these linear chromosomes presents the same challenge that we just explored with linear chromosomes in eukarya. This issue has been studied most extensively in *Streptomyces* species. *Streptomyces* are Gram-positive bacteria commonly found in soil. Species within this genus possess linear plasmids and a linear chromosome. These linear DNA molecules end in inverted repeat sequences of differing lengths in different species. Conserved terminal proteins bind to these sequences. While this arrangement may sound similar to the telomere/telomerase arrangement we described for eukarya, the two systems do not appear to be homologous. Recent evidence, however, has shed some light on how this system functions **(Mini-Paper)**.

7.2 Fact Check

1. Why is DNA replication described as semi-conservative?
2. Describe the replicon model of DNA replication.
3. Explain how *oriC*, DnaA, helicase, and DnaC play a role in DNA replication.
4. What is the role of the enzymes primase and DNA polymerase?
5. Explain the role of topoisomerase.
6. What is the relationship between telomeres and telomerase?

7.3 Transcription

3D ANIMATION

How are genes transcribed?

As we have seen in the previous section, the structure of DNA clearly determines the mechanism of replication and allows for this replication to produce two new molecules identical to the original molecule. In this section, we will examine **transcription**, the process through which segments of DNA serve as templates for the production of complementary strands of single-stranded RNA. As we will see in the next section, these RNA molecules subsequently perform a variety of functions, most notably serving

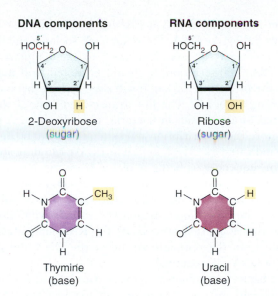

Thymine
(base)

Uracil
(base)

Figure 7.15. Chemical composition differences between DNA and RNA These molecules differ in two fundamental ways. First, the sugar in RNA is ribose, not 2-deoxyribose as seen in DNA. Second, the nitrogenous base thymine in DNA is replaced with uracil in RNA.

as templates for the production of proteins. For now, we will focus on the basics of transcription and the differences that exist in transcription among bacteria, eukarya, and archaeons. Let's start with a more fundamental question: What parts of the DNA get transcribed?

What is a gene?

A **gene**, quite simply, is any segment of DNA that gets transcribed, or copied into a piece of single-stranded RNA, as well as the associated DNA elements that direct its transcription. RNA is a polynucleotide, just like DNA. Unlike DNA, though, the sugar in RNA is **ribose**, not 2-deoxyribose. Additionally, RNA includes the nitrogenous base **uracil** (U) in place of thymine (T) **(Figure 7.15)**. Like thymine, uracil interacts with adenine.

In bacteria, most of the genome gets transcribed. In eukarya, conversely, it appears that long stretches of DNA do not get transcribed. In all organisms, different types of genes give rise to different classes of RNA molecules, which serve many different functions **(Table 7.3)**. Many genes give rise to **messenger**

RNA, or **mRNA**, molecules. As we will see in the next section, mRNA subsequently is translated, producing a polypeptide. Transcription of other genes results in RNA molecules that do not get translated. Two major classes of non-translated RNAs are **transfer RNA (tRNA)** and **ribosomal RNA (rRNA)** molecules, critical components of the translation process. Regulatory RNA molecules constitute yet another category of RNA. These molecules include **micro RNAs (miRNAs)**, small RNA molecules that regulate the expression of genes. Regardless of the ultimate role of the RNA molecule, the process of RNA production remains basically the same.

CONNECTIONS Regulatory RNA molecules, like miRNAs, regulate gene expression by interacting with mRNA molecules. In **Section 11.4**, we will see how a regulatory RNA molecule affects the level of RpoS, a protein in *E. coli* that affects gene expression.

Transcription: Initiation and elongation

For every gene, the transcription machinery must identify the beginning and the end of the segment to be transcribed. Once the initiation site has been identified, transcription, much like DNA replication, begins with the transient unwinding, or melting of the DNA. As we saw with DNA replication, this unwinding involves specialized proteins and occurs at a very specific location. In all organisms, the enzyme that transcribes DNA is an **RNA polymerase**, or, more appropriately, a DNA-dependent RNA polymerase. The transcription process begins at regions of DNA known as **promoters**. Once the RNA polymerase binds to the promoter, it unwinds the double-stranded DNA, much like helicases unwind DNA during DNA replication. The RNA polymerase then makes a strand of single-stranded RNA that is complementary (and antiparallel) to one of the strands of DNA, the template strand. The other strand of DNA is called the coding strand. As shown in **Figure 7.16**, the sequence of the transcribed RNA is the same as the sequence of the coding strand, except for the substitution of U for T. Like DNA polymerase, RNA

TABLE 7.3	RNA molecules	
Name	**Approximate size (nucleotides)**	**Function(s)**
Messenger RNA (mRNA)	500–10,000	Coding molecules Translated into proteins
Transfer RNA (tRNA)	75–100	Involved in translation Charged with amino acids
Ribosomal RNA (rRNA)	1,500–1,900 (small subunit) 2,900–4,700 (large subunit)	Structural components of ribosomes
Micro RNA (miRNA)	< 100	Various regulatory functions

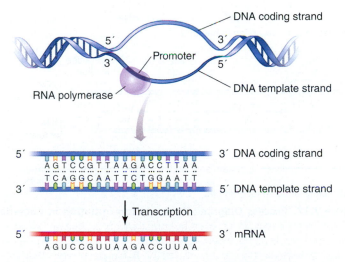

Figure 7.16. Comparison of DNA and transcribed RNA sequences RNA polymerase binds to the promoter region of the DNA and uses the template strand of the DNA to make an RNA molecule identical to the coding strand of the DNA, except that each T in the coding strand is replaced by U. The RNA is elongated in a 5′ to 3′ direction.

polymerase elongates the growing RNA in a 5′ to 3′ fashion, adding a new ribonucleotide to the 3′—OH of the previous ribose. Unlike DNA polymerase, however, RNA polymerase does not need a primer. The enzyme can add the first ribonucleotide directly to the template strand of DNA.

@ **Protein synthesis** ANIMATION

The bacterial RNA polymerase consists of five different polypeptides, α, β, β′, ω, and σ. The core enzyme is a complex of β, β′, ω, and two copies of α. The **sigma factor (σ)** combines more weakly with this core enzyme, forming the RNA polymerase holoenzyme, and recognizes promoter sequences on the DNA (**Figure 7.17**). Once the sigma factor has recognized a promoter and the RNA polymerase has bound, the enzyme will begin unwinding the DNA. The σ factor dissociates and the core enzyme creates RNA complementary to the template strand of the DNA. Transcription, it should be noted, will not begin right at the promoter. Rather, it will start a short distance from this binding sequence. The first transcribed base of the gene is referred to as +1 and, in bacteria, typically is an A or a G. Bases before the transcription start site are said to be upstream and are indicated as −1, −2, and so forth. We can describe these upstream sequences as being 5′ of the transcription start site. Downstream sequences, then, are in a 3′ direction. Shortly after the elongation process has begun, the sigma factor will dissociate from the other subunits and the core enzyme alone will continue the elongation process. As the RNA elongates, it will dissociate from the template strand of the DNA and the two strands of DNA will reanneal.

Most bacterial promoters possess two defined regions, one at position −10 and one at position −35, to which the RNA polymerase binds (see Figure 7.17). This binding can be examined in several ways, such as a **gel shift assay** (**Toolbox 7.1**). While the specific sequences of these genetic elements may differ between bacterial species and even between genes within a species, the sequences are fairly well conserved and we can determine the "typical" or consensus promoter sequence. The consensus sequence for the −10 element, or Pribnow box, in the coding strand of most commonly expressed genes is TATAAT. Because of its sequence, it often is referred to as a TATA box. The consensus sequence in the coding strand for the −35 element of commonly expressed genes is TTGACA. More recently, additional promoter elements have been identified. The UP-element is an AT-rich region located upstream of the −35 element. The discriminator is a GC-rich motif located between the −10 element and the transcription start site.

As we mentioned previously, the sequences of promoter elements can vary. Differences in the sequences can alter the affinity of the sigma factor for the promoter. Differences in the sequences of these elements, then, can alter the level of transcription. If RNA polymerase cannot bind well to a promoter, then it is less likely to initiate transcription of that gene. So what is the practical importance of this information? Some promoters will interact strongly with sigma factors and undergo relatively high levels of transcription. Other promoters will interact less well with sigma factors and undergo lower levels of transcription. The amount of RNA produced from a gene, in other words, depends to some extent on the sequence of a gene's promoter. The story, however, doesn't stop there. Most bacteria can produce several different sigma factors. *Bacillus subtilis*, for example, produces at least 18 different sigma factors under different conditions. These different sigma factors, it turns out, preferentially interact with different classes of promoters. So, by simply regulating which sigma factor it is producing, a cell can regulate which group of genes it will transcribe.

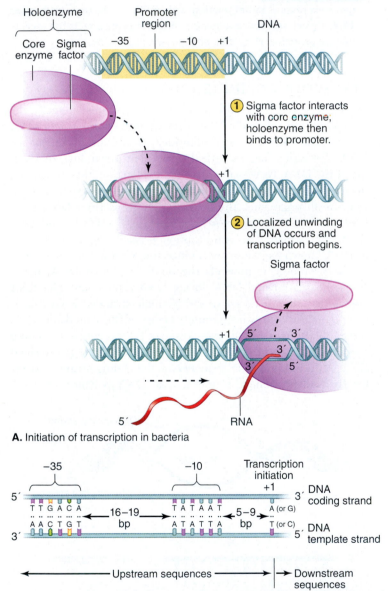

A. Initiation of transcription in bacteria

B. Structure of bacterial promoter

Figure 7.17. Process Diagram: Transcription initiation in bacteria
A. The core enzyme interacts with the sigma factor, forming the functional holoenzyme. This complete enzyme interacts with the promoter, facilitates localized unwinding of the DNA, and creates a molecule of RNA that is complementary to the template strand of the DNA. **B.** To ensure that transcription begins at the correct location, the holoenzyme recognizes a specific region of DNA—the promoter. In bacteria, the promoter consists of several elements, most notably the −10 and −35 elements, which are located 10 base pairs (bp) and 35 bp upstream of transcription initiation. The site of transcription initiation is referred to as the +1 position and, in bacteria, typically is an A or a G.

CONNECTIONS When environmental conditions change, bacteria can change the expression of sigma factors. By changing the expression of their sigma factors, bacteria can alter which genes are expressed. More information about how bacteria can regulate gene expression is presented in **Section 11.3**.

USING A GEL SHIFT ASSAY TO IDENTIFY DNA-BINDING PROTEINS

We have seen that numerous proteins bind to DNA. Histones bind to DNA in eukarya to facilitate condensing the molecule. DnaA in bacteria and the Orc proteins in eukarya bind to specific sites on the DNA molecule to initiate replication. Sigma factors initiate transcription by recognizing and binding to promoter regions. How do we know that the σ factor binds to the promoter region of a gene? Or, more generally, how can we identify any DNA-binding proteins? One assay commonly used to address this question is the gel shift assay.

The gel shift, or electrophoretic mobility shift assay (EMSA) often is used to identify proteins that bind to DNA. First, a piece of DNA suspected of containing a protein binding domain (a promoter, for instance) is radiolabeled. Second, two aliquots of this labeled DNA are prepared, one that is preincubated with the potential DNA-binding protein (a sigma factor, for instance) and one that is not exposed to the protein. These DNA aliquots then are run on a non-denaturing electrophoresis gel. Gel electrophoresis allows researchers to separate molecules based on their sizes; small molecules travel farther in a given period of time than do large molecules. Non-denaturing conditions allow weak molecular interactions to remain intact. In other words, protein:DNA-binding interactions would not be disrupted under these conditions. After the DNA samples are run on the gel, their locations on the gel are determined. A piece of DNA with protein bound to it will not migrate as far on the gel as the same piece of DNA without protein bound to it. A piece of DNA with protein bound to it, then, will migrate more slowly than a similar piece of DNA without protein bound to it. In other words, protein binding will result in a gel shift (**Figure B7.1**).

This type of assay can be used to identify proteins that bind to a specific piece of DNA. Similarly, this assay can be used to determine the specific sequence of DNA necessary for protein binding. If a DNA-binding protein and its recognition sequence have been identified, then researchers can experimentally modify the DNA sequence, run a gel shift assay with the altered DNA sequences, and determine what changes in the DNA sequence decrease or eliminate protein binding.

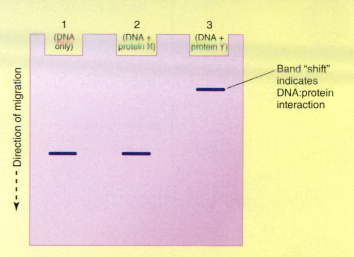

Figure B7.1. Gel shift assay to detect DNA : protein interactions
In this assay, DNA or DNA incubated with protein are electrophoresed on a non-denaturing gel. DNA complexed with protein will migrate more slowly than the same fragment of DNA not complexed with protein. In this example, lane 1 contains DNA only, lane 2 contains DNA incubated with protein X, and lane 3 contains DNA incubated with protein Y. In all three cases, the DNA was labeled so that it could be detected. In lane 3, the DNA band did not migrate as far, indicating that it was complexed with protein Y.

● Test Your Understanding ·················

You are trying to determine key bases in a bacterial promoter that are needed for attachment of a sigma factor. You test two different mutant promoters, each with a different single base change, for their ability to interact with the sigma factor using a gel shift assay. What specific results would you see on the gel if mutant A did not allow sigma factor binding, but mutant B did allow binding? Be specific and explain your answer.

In eukarya, the basic process of transcription mirrors the bacterial process. Instead of relying on a single RNA polymerase, though, eukarya possess three different RNA polymerases, RNA pol I, II, and III. RNA pol II is the best studied of these enzymes and transcribes all protein-coding genes. RNA pol I transcribes the ribosomal RNA genes (with the exception of the 5S rRNA), while RNA pol III transcribes tRNA genes, the 5S rRNA gene, and the genes encoding various other small, non-coding RNA molecules. In eukarya, RNA polymerase II does not bind directly with the promoter. Rather, a series of **transcription factors**, DNA-binding proteins, first must bind to the promoter (**Figure 7.18**). TATA-binding protein (TBP) binds to the TATA box found in most eukaryal promoters. Other transcription factors, such as TFIIB, also bind, forming a transcription initiation complex. These bound transcription factors then recruit the RNA polymerase

and facilitate its interaction with the DNA. As we saw in bacteria, promoters, located upstream of the transcription start site, mark the locations of genes. Eukaryal promoters contain more elements than bacterial promoters and exhibit more variability than bacterial promoters. Most structural genes contain a TATA box approximately 25 bp upstream of the transcription initiation site. The presence, numbers, and locations of the other promoter elements may vary.

While the TATA box usually is around 25 bp upstream of the transcription initiation site, the relative positions of the other elements may vary in different genes. The sequences shown are consensus, or most typical, sequences. The exact sequences differ somewhat in different genes and different organisms.

Archaeons, like bacteria, possess a single RNA polymerase. This archaeal RNA polymerase, however, looks more like RNA

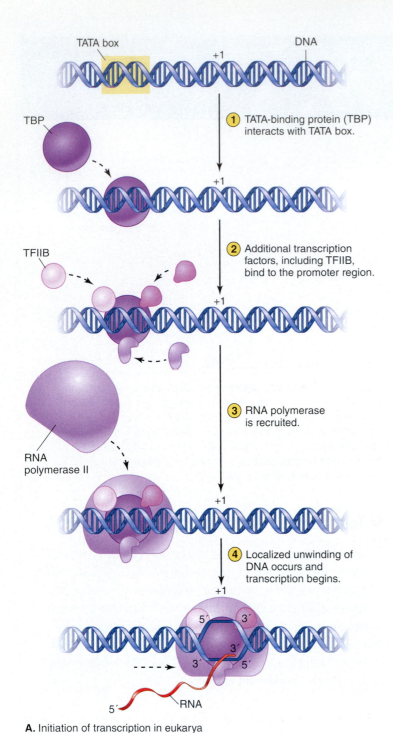

① TATA-binding protein (TBP) interacts with TATA box.

② Additional transcription factors, including TFIIB, bind to the promoter region.

③ RNA polymerase is recruited.

④ Localized unwinding of DNA occurs and transcription begins.

A. Initiation of transcription in eukarya

pol II of eukarya than the RNA polymerase of bacteria. While the bacterial RNA polymerase core enzyme consists of four different polypeptides, RNA polymerases characterized from archaeons possess eight or more subunits. Additionally, the archaeal RNA polymerases do not bind directly to DNA. Rather, the enzyme is recruited to the promoter by various transcription factors, including TBP and TFB **(Figure 7.19)**. These transcription factors resemble the eukaryal transcription factors TBP and TFIIB. The archaeal promoter itself is more eukarya-like. One final note about the RNA polymerase of archaeons; it is not inhibited by rifampicin, an antibiotic that effectively inhibits the RNA polymerase of bacteria.

Transcription: Termination

Not surprisingly, specific mechanisms control the termination of transcription. In bacteria, two main mechanisms of termination have been identified, rho-independent and rho-dependent. Genes that undergo rho-independent, or intrinsic termination, end with a short segment of inverted GC-rich repeats, followed by a string of adenines (A). As RNA polymerase is transcribing the gene, the bases in the inverted repeats of the growing RNA molecule can hydrogen bond to each other, allowing the nascent RNA molecule to adopt a hairpin structure. It is thought that this secondary structure in the RNA molecule causes the RNA polymerase to stall and fall off of the DNA. The bases A and U bind weakly to each other. Most likely, then, the A:U interactions at the end of the RNA molecule are too weak to keep the newly made RNA attached to the DNA. As a result, the RNA molecule dissociates from the DNA **(Figure 7.20)**.

Other bacterial genes undergo rho-dependent termination. In these genes, the transcript does not end with a string of Us and often does not contain the inverted repeats seen in rho-independent termination. A separate protein, the rho factor, mediates transcription termination. This protein binds to the nascent RNA molecule and travels along the molecule. The RNA polymerase will slow down or pause when it reaches a specific termination site. The rho factor then will displace the RNA polymerase, causing transcription to end.

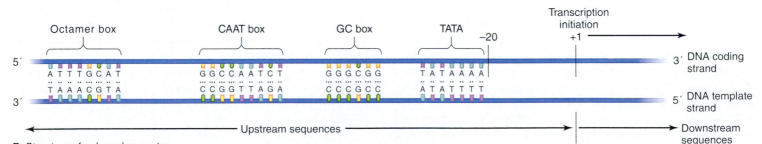

B. Structure of eukaryal promoter

Figure 7.18. Process Diagram: Transcription initiation in eukarya A. RNA polymerase II does not bind directly to DNA. Rather, its binding is facilitated by various transcription factors, including TATA-binding protein (TBP) and transcription factor IIB (TFIIB). **B.** The promoters in eukaryal structural genes are complex, consisting of several motifs, including TATA, GC, CAAT, and octamer (eight nucleotide) boxes. Different eukaryal genes contain various numbers of these elements, arranged in various orders. This schematic representation shows the major elements of a eukaryal promoter.

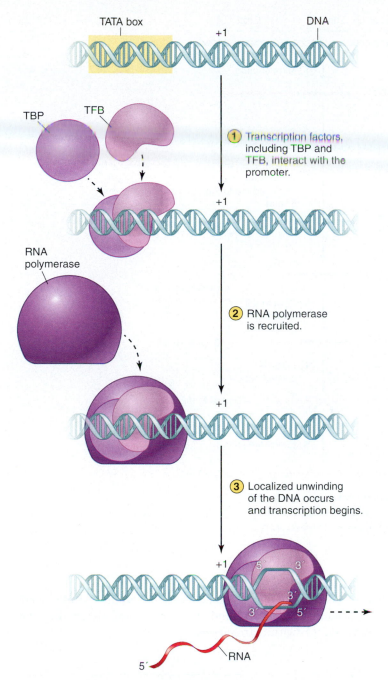

① Transcription factors, including TBP and TFB, interact with the promoter.

② RNA polymerase is recruited.

③ Localized unwinding of the DNA occurs and transcription begins.

Figure 7.19. Process Diagram: Transcription initiation in archaeons As in eukarya, the RNA polymerase of archaeons does not bind directly to the DNA. The transcription factors TBP and TFB, similar to the eukaryal TATA-binding protein and transcription factor IIB, respectively, interact with the DNA and recruit the RNA polymerase.

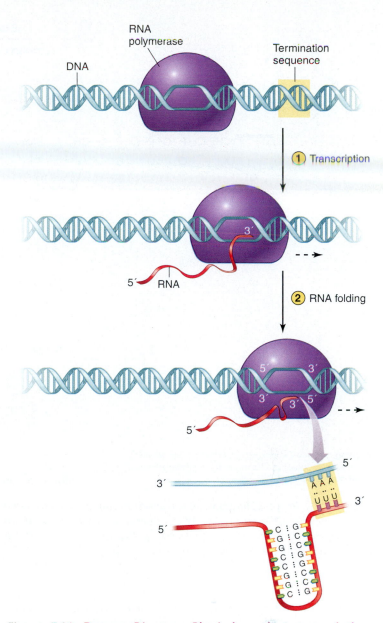

① Transcription

② RNA folding

Figure 7.20. Process Diagram: Rho-independent transcription termination in bacteria As RNA polymerase nears the end of the transcriptional unit, GC-rich inverted repeats in the RNA molecule form a hairpin loop. Formation of this loop causes the RNA polymerase to pause and, eventually, fall off of the DNA, thereby terminating transcription. The weak A:U interactions at the end of the transcript (highlighted in yellow) facilitate dissociation of the RNA molecule from the DNA.

Transcription termination in eukarya is more complex. Termination of transcription by RNA pol I resembles the bacterial rho-dependent termination, while termination of transcription by RNA pol III more closely resembles rho-independent termination. For RNA pol II, the elongating RNA molecule is cleaved by a specific endonuclease, thereby generating the 3′ of the RNA molecule.

Within the nucleus, the RNA pol II products, or pre-mRNA molecules, are modified in three major ways before becoming functional messenger RNA (mRNA) molecules. First, a special modified nucleotide, 7-methylguanosine, is added to the

5′ end of the transcript, giving the RNA molecule a **5′ cap**. Not only is this terminal base modified, but also it is connected to the end of the RNA molecule via an atypical 5′ to 5′ triphosphate linkage. Usually, this modification of the RNA occurs soon after elongation has begun. Second, poly(A) polymerase adds a string of adenosine monophosphates to the 3′ end of the RNA molecule. Often, this **3′ poly(A) tail** can be around 200 nucleotides long. Both the 5′ cap and the 3′ poly(A) tail protect the RNA from degradation. As we will see in the next section, the 5′ cap also serves as a necessary signal for the start of translation. Third, eukaryal RNA molecules often contain

@ **Eukaryal genes contain introns** ANIMATION

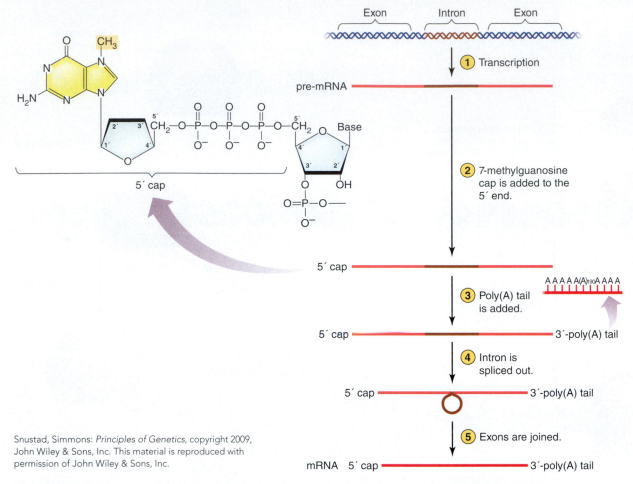

Figure 7.21. Process Diagram: Post-transcriptional processing of eukaryal RNA In eukarya, primary RNA pol II transcripts, or pre-mRNA molecules, undergo three major forms of processing before leaving the nucleus. A 7-methylguanosine cap is added to the 5′ end via a 5′ to 5′ triphosphate linkage (shown in enlarged view). Also, a poly(A) tail is added to the 3′ end. Finally, intervening sequences, or introns, are spliced out, bringing together the exons, or coding sequences.

non-protein coding regions, referred to as **introns** (intervening sequences), which are removed. The coding regions, or **exons**, are spliced together, producing a functional mRNA that contains only protein coding information **(Figure 7.21)**. Because these processing events occur after transcription, they are referred to as **post-transcriptional modifications**. Only after these modifications occur do the RNA molecules leave the nucleus.

Relatively little is known about transcription termination in archaeons. Several studies, however, indicated that a string of Ts in a transcriptional unit, a region of DNA to be transcribed, may represent a termination signal. To test the importance of these oligo(T) stretches, researchers from Ohio State University genetically modified strains of the archaeon *Thermococcus kodakarensis*. They inserted oligo(T) units between a strong promoter and a reporter gene, a gene whose protein product easily can be detected and monitored. If the oligo(T) units served as transcription termination signals, they reasoned, then expression of the reporter gene would decrease when these oligo(T) units were present in the DNA. Indeed, a 5′-TTTTTTTT-3′ (T_8) insertion almost completely inhibited

the reporter gene expression. Additional studies will be needed to more fully understand this process and determine how universal it is within archaeons.

7.3 Fact Check

1. Contrast the structures of DNA and RNA.
2. Distinguish between the different types of RNA molecules.
3. Describe the role of the RNA polymerase in the process of transcription.
4. What is a promoter, and what are the key regions of most promoters?
5. Differentiate between rho-dependent and rho-independent termination of transcription.
6. What is involved in post-transcriptional modification of eukaryal mRNA?

7.4 Translation

3D ANIMATION

How are proteins made?

Translation, the process of protein synthesis, involves using the mRNA sequence as a template to produce a specific amino acid sequence. Again, the basic process is remarkably similar in bacteria, eukarya, and archaeons. Briefly, **ribosomes** interact with the mRNA and "read" the mRNA sequence, three nucleotides at a time. These groups of three nucleotides are called **codons**. For every codon, a corresponding transfer RNA (tRNA) exists in the cell that possesses an **anticodon**, three nucleotides complementary to a given codon. A specific **amino acid** is attached to each tRNA. Each tRNA, as a result, has attached to it an amino acid defined by the tRNA anticodon. So, as the ribosome "reads" the mRNA sequence, it recruits a tRNA for each codon. In this fashion, specific amino acids are brought together. Peptide bonds form between these amino acids, and a polypeptide forms **(Figure 7.22)**.

Protein synthesis ANIMATION

With four bases in mRNA, 64 different three-base codons can be formed. Sixty-one of these codons code for the twenty amino acids commonly found in proteins. As we will see later, one of these codons, AUG, serves as the start codon, allowing translation to begin at the correct location. The remaining three codons, as we also will see later, do not code for amino acids; rather, they serve as signals for the termination of translation. The genetic code, therefore, is degenerate; most amino acids are specified by several different codons **(Figure 7.23)**. The amino acid arginine, for example, is coded for by CGU, CGC, CGA, CGG, AGA, and AGG (codons always are written in a 5′ to 3′ orientation). In fact, only two amino acids, methionine and tryptophan, have only one codon each. Additionally, all living organisms share the same genetic code. This finding alone provides compelling evidence for the evolutionary relatedness of all living organisms. We should note, however, that interesting exceptions have been identified. In the mitochondria of animals, for example, the traditional stop codon UGA codes for the amino acid tryptophan. The evolutionary implications of this and other exceptions remain to be determined.

As we saw for DNA replication and transcription, specific initiation and termination sites for translation must exist. In this section, we will briefly examine the translation initiation and termination events, paying special attention to how these events differ in bacteria, eukarya, and archaeons. Finally, we will examine how the presence of the membrane-bound nucleus in eukarya affects the processes of transcription and translation.

Translation: Initiation and elongation

Before we investigate how translation begins, let's first examine two major components of the process, tRNA and ribosomes. Transfer RNA molecules are about 70 nucleotides long. Because of their sequences, these molecules can undergo intramolecular base-pairing, resulting in a molecule with a complex shape often referred to as a cloverleaf. At one end of this molecule, several bases remain unpaired. These bases are the anticodon.

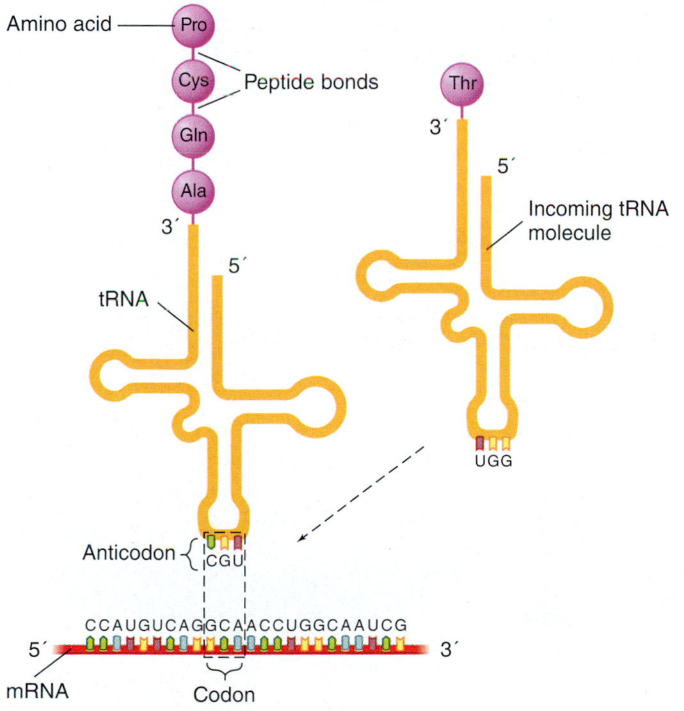

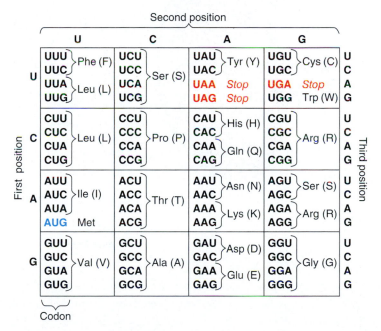

Figure 7.22. Translation of mRNA Ribosomes interact with the mature mRNA and "read" groups of three nucleotides, or codons. Transfer RNA molecules containing complementary anticodons bring in the appropriate amino acids. As shown here, the mRNA codons are read in a 5′ to 3′ direction.

First position	Second position: U	C	A	G	Third position
U	UUU Phe (F)	UCU Ser (S)	UAU Tyr (Y)	UGU Cys (C)	U
	UUC Phe (F)	UCC Ser (S)	UAC Tyr (Y)	UGC Cys (C)	C
	UUA Leu (L)	UCA Ser (S)	UAA Stop	UGA Stop	A
	UUG Leu (L)	UCG Ser (S)	UAG Stop	UGG Trp (W)	G
C	CUU Leu (L)	CCU Pro (P)	CAU His (H)	CGU Arg (R)	U
	CUC Leu (L)	CCC Pro (P)	CAC His (H)	CGC Arg (R)	C
	CUA Leu (L)	CCA Pro (P)	CAA Gln (Q)	CGA Arg (R)	A
	CUG Leu (L)	CCG Pro (P)	CAG Gln (Q)	CGG Arg (R)	G
A	AUU Ile (I)	ACU Thr (T)	AAU Asn (N)	AGU Ser (S)	U
	AUC Ile (I)	ACC Thr (T)	AAC Asn (N)	AGC Ser (S)	C
	AUA Ile (I)	ACA Thr (T)	AAA Lys (K)	AGA Arg (R)	A
	AUG Met	ACG Thr (T)	AAG Lys (K)	AGG Arg (R)	G
G	GUU Val (V)	GCU Ala (A)	GAU Asp (D)	GGU Gly (G)	U
	GUC Val (V)	GCC Ala (A)	GAC Asp (D)	GGC Gly (G)	C
	GUA Val (V)	GCA Ala (A)	GAA Glu (E)	GGA Gly (G)	A
	GUG Val (V)	GCG Ala (A)	GAG Glu (E)	GGG Gly (G)	G

Codon

Figure 7.23. The genetic code Each group of three nucleotides, or codon, in an mRNA molecule codes for a specific amino acid. Three codons (UAA, UAG, and UGA) are stop codons that do not code for any amino acid. Rather, they signal where translation should end. In this table, the first letter of each codon is on the left side of the table, the second letter is on the top, and the third letter is on the right side. The three-letter abbreviation and one-letter code for the amino acids are indicated within the table.

A specific enzyme, aminoacyl-tRNA-synthetase, adds the amino acid appropriate for the anticodon on that molecule to the other end of the tRNA (Figure 7.24). The specificities of tRNA molecules are indicated by a superscript—tRNA^{ala}. This example indicates the tRNA specific for the amino acid alanine. When the amino acid becomes attached to the tRNA, the molecule is referred to as Ala-tRNA^{ala}.

In all organisms, ribosomes consist of two parts, the small and large subunits. Each subunit, in turn, consists of many polypeptides and one or more rRNA molecules. The exact components of the ribosomes differ in organisms from different domains (Table 7.4). The ribosomes in archaeons are similar in size to the bacterial ribosomes. The ribosomal polypeptides and rRNA molecules, though, seem to more closely resemble the constituents of eukaryal ribosomes (see Figure 4.2).

The initiation of translation, not surprisingly, depends on the recognition of mRNA by ribosomes. In bacteria, this initial recognition event involves interactions between a specific sequence in the mRNA, the **Shine–Dalgarno (SD) sequence**, or ribosome binding site (5′...AGGAGG...3′) and a complementary sequence in the 16S rRNA, which is part of the 30S small subunit of the ribosome. Translation in bacteria then begins at

TABLE 7.4 Ribosome components of Eukarya and Bacteria

Eukarya	Bacteria
Small (40S[a]) subunit	Small (30S) subunit
33 polypeptides	21 polypeptides
18S rRNA	16S rRNA
Large (60S) subunit	Large (50S) subunit
49 polypeptides	31 polypeptides
28S rRNA	23S rRNA
5.8S rRNA	5S rRNA
5S rRNA	

[a]S refers to Svedberg units, a unit of measurement based on a particle's sedimentation rate. Larger particles have greater Svedberg values.

the first AUG after the Shine–Dalgarno (SD) sequence. The 16S rRNA contains a sequence (3′...UCCUCC...5′) that is complementary to the SD sequence. It seems logical then, to infer that the 16S rRNA interacts with the SD sequence, thereby aligning the ribosome with the mRNA (Figure 7.25).

Three pieces of evidence support this hypothesis. First, mutations within the Shine–Dalgarno sequence eliminate or greatly reduce the translation of mRNA. Second, mutations within the complementary 16S rRNA sequence can affect overall protein synthesis within a cell. Third, a specific mutation within the Shine–Dalgarno sequence can be "corrected" by a complementary mutation in the 16S rRNA sequence. In other words, if the Shine–Dalgarno sequence is altered from 5′...AGGAGG...3′ to 5′...AGCAGG...3′, then a change in the 16S rRNA sequence from 3′...UCCUCC...5′ to 3′...UCGUCC...5′ will restore translation levels of that mRNA to normal.

A special tRNA, often referred to as the initiator tRNA, interacts with the first codon. The codon AUG, as shown in Figure 7.23, codes for the amino acid methionine. The first AUG after the Shine–Dalgarno sequence, however, interacts with fMet-tRNA^{fMet}. The amino acid attached to this tRNA is a

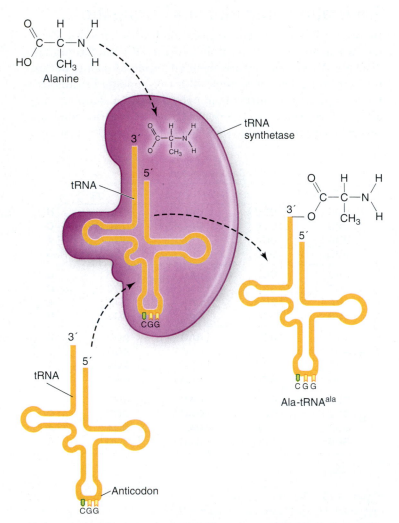

Figure 7.24. Role of transfer RNA Because of intramolecular complementarity, the tRNA molecule assumes a cloverleaf conformation. Three unpaired nucleotides form the anticodon and allow each tRNA to bind to a specific codon on an mRNA molecule. In this case, an aminoacyl-tRNA synthetase specific for both the amino acid alanine and the tRNA with the anticodon CGG attaches alanine to its tRNA.

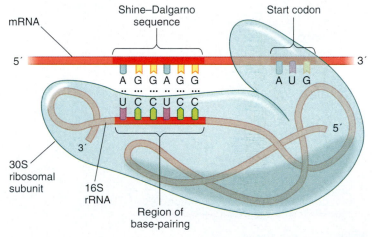

Figure 7.25. Role of the Shine–Dalgarno sequence in bacteria In bacteria, the 30S subunit of the ribosome attaches to the mRNA through base-pair interactions between the 16S rRNA and the Shine–Dalgarno sequence. The 50S ribosome subunit then binds, and translation begins at the first AUG (translation initiation site) after the Shine–Dalgarno sequence. The Shine–Dalgarno sequence and the translation initiation site typically are 6 to 10 nucleotides apart.

modified version of methionine, N-formylmethionine. All other AUG codons in the mRNA interact with tRNA attached to a non-modified methionine. Initiation factor 2 (IF-2), one of three initiation factors involved in bacterial translation, mediates the interaction between the start codon and fMet-tRNA^{fMet}. After the initiator tRNA has interacted with the start codon, the ribosome "reads" the next codon on the mRNA. Another tRNA is recruited and a peptide bond forms between the two amino acids, linking the amino, or N, terminus of the second amino acid to the carboxy, or C, terminus of the first amino acid. As a result, every polypeptide, has an N-terminus and a C-terminus. The ribosome continues to move along the mRNA, recruiting the appropriate tRNA molecules, and the polypeptide chain continues to elongate (Figure 7.26).

The process of translation in eukarya, in most respects, is similar to the process in bacteria. Three major differences, however, do exist. First, the initial amino acid in eukaryal polypeptides is not f-methionine. Rather, a non-modified methionine is used. A unique tRNA, often referred to as the initiator tRNA, or tRNA$_i^{Met}$, recognizes the AUG start codon. Second, eukaryal mRNAs do not contain a Shine–Dalgarno-like sequence. Rather, the eukaryal ribosome generally initiates translation at the first AUG on the 5′ end of the molecule. Finally, eukaryal cells use many more initiation factors. Recognition of the mRNA by the ribosome requires several different factors. A cap-binding protein, or CBP, first binds to the 5′ cap. Additional polypeptides then bind before the 40S subunit of the ribosome can attach.

Let's think about this process of translation initiation a little more. If translation of eukaryal mRNA begins at the first AUG after the 5′ cap, then how many translation initiation sites can a molecule of eukaryal mRNA contain? One, of course. There only can be one first AUG after the 5′ cap. The situation differs in bacteria. A molecule of mRNA could contain multiple Shine–Dalgarno sequences. Each SD sequence could be followed by an AUG. In bacteria, then, a molecule of mRNA could contain several translation units. In other words, a molecule of mRNA in bacteria, could produce several different polypeptides, or be polycistronic (Figure 7.27). This process allows the cell to produce two or more different polypeptides from a single mRNA.

Eukaryal mRNA molecules, on the other hand, generally only produce a single polypeptide and are said to be **monocistronic**. Recently, though, it has been shown that a second form of translation initiation can occur in some eukaryal mRNAs. In some cases, the ribosome binds to the mRNA in a 5′ cap-independent fashion, utilizing an internal ribosome entry site, or IRES. These internal ribosome entry sites have been identified in the mRNAs of several viruses of mammals, as well as several mammalian mRNAs. In viruses, the presence of an IRES may facilitate production of viral proteins even when traditional 5′ cap-dependent translation initiation is inhibited. In mammalian mRNAs, the presence of an IRES may allow the production of two distinct polypeptides from a single mRNA. The exact mechanism of ribosome binding and the required IRES sequence have not been fully elucidated.

CONNECTIONS With polycistronic mRNA, bacteria can coordinately regulate the production of several polypeptides. We will examine this mechanism of regulation in **Section 11.2**.

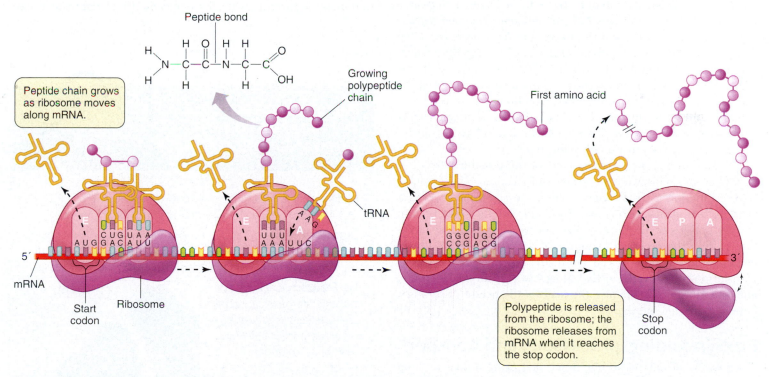

Peptide bond

Peptide chain grows as ribosome moves along mRNA.

Growing polypeptide chain

First amino acid

tRNA

E P A

mRNA

Start codon

Ribosome

Polypeptide is released from the ribosome; the ribosome releases from mRNA when it reaches the stop codon.

Stop codon

Figure 7.26. Protein synthesis Following the initiation of translation, the ribosome moves along the mRNA, "reading" three base units, or codons. For each successive codon, a tRNA with a complementary anticodon, and charged with the appropriate amino acid, is recruited and enters the A (acceptor) site on the ribosome. Peptide bonds form between adjacent amino acids, resulting in polypeptide formation as the tRNA enters the P (peptide) site. After the peptide bonds have formed, the amino acid dissociates from its tRNA and the tRNA leaves the ribosome from the E (exit) site. When the ribosome encounters a stop codon, translation ceases and the polypeptide leaves the ribosome.

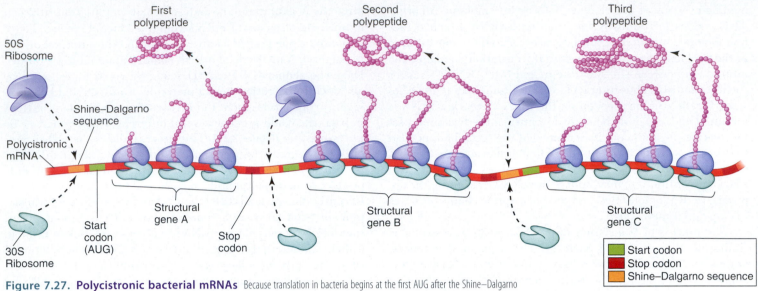

Figure 7.27. Polycistronic bacterial mRNAs Because translation in bacteria begins at the first AUG after the Shine–Dalgarno sequence, more than one different polypeptide can be generated from a single mRNA. If multiple Shine–Dalgarno sequences exist on a molecule of mRNA, then ribosomes will bind at multiple locations on a single RNA molecule, and begin translation at multiple sites.

While we know less about translation in archaeons, it appears that the process of translation in archaeons mirrors more closely the process of translation in eukarya. Most of the ribosomal proteins in archaeons are closely related to eukaryal ribosome proteins, suggesting that the two systems share an evolutionary link. Additionally, the initiator amino acid in archaeons is methionine, not formylmethionine. One particular set of experiments provides compelling evidence for the evolutionary relatedness of the translation machinery in archaeons and eukarya. Researchers purified ribosomal subunits from the archaeon *Sulfolobus solfataricus* **(Microbes in Focus 7.2)**, the eukaryal microbe *Saccharomyces cerevisiae* (yeast), and the bacterium *Escherichia coli*. They demonstrated that combinations of the *S. solfataricus* 30S subunit and *S. cerevisiae* 60S subunit or the *S. solfataricus* 50S subunit and the *S. cerevisiae* 40S subunit resulted in the formation of polypeptides in an *in vitro* translation assay. Combinations of the yeast subunits and *E. coli* subunits, however, were non-functional. This study suggests that parts of the archaeal and eukaryal translation machinery are functionally equivalent.

We mentioned previously that 61 of the possible 64 codons specify amino acids, while the other three codons—UAA, UGA, UAG—function as stop codons. When the ribosome reaches a **stop codon**, translation ceases. In *E. coli*, two different release factors, RF-1 and RF-2, recognize the stop codons and facilitate ending the process. A single release factor, eRF, has been identified in eukarya. At this point, the polypeptide leaves the translation complex, the mRNA moves away from the ribosome, and the ribosomal subunits themselves dissociate. The story, however, doesn't end there.

Protein folding, processing, and transport

The process of making a fully functional protein does not end with the termination of translation. The polypeptide chain produced during translation represents, in a way, merely the basic architecture of a functional protein. To complete the process, the string of linked amino acids must be properly folded into a complex three-dimensional shape. Additionally, the polypeptide often is modified in one or more ways. Finally, the protein must be transported to its intended location. We'll briefly examine these events in this section.

Protein Folding

A linear string of amino acids produced during translation hardly resembles a protein. Proteins, as we have noted throughout the preceding chapters, have complex three-dimensional shapes that, in large part, allow them to carry out their many functions.

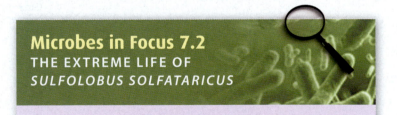

Microbes in Focus 7.2
THE EXTREME LIFE OF
SULFOLOBUS SOLFATARICUS

Habitat: Found throughout the world, usually in areas of volcanic activity with a temperature of ~80°C, a pH of ~3, and sulfur.

Description: Archaeon about 1-3 μm in diameter with a single circular chromosome approximately 3 million base pairs in length; cell membrane contains tetraether lipids, making it a monolayer.

Key Features: One of the first extreme thermophiles to be characterized, *S. solfataricus* grows optimally at 80°C and a pH of 3. *S. solfataricus* and related species have been isolated from volcanic hot springs throughout the world.

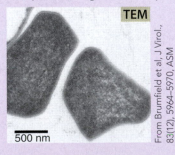

TEM

500 nm

From Brumfield et al, J Virol., 83(12), 5964–5970, ASM

Chemical interactions between individual amino acids within a polypeptide chain cause the linear chain to assume a specific globular shape. Many of these interactions are non-covalent and include hydrogen bonding and ionic interactions. In some cases, covalent disulfide bonds form between cysteine amino acids (**Figure 7.28**). Additionally, two or more polypeptides may interact with each other to form a mature, functioning protein.

CONNECTIONS When proteins do not fold correctly, disease can result. We saw an example of this in **Section 5.5** when we discussed prions. We'll see another example in **Section 23.4** during our discussion of malaria and sickle cell disease.

While these folding events had been thought to occur spontaneously, it now appears that the folding process may be more directed. Molecular chaperones, or chaperonins, are proteins identified in species from all three domains of life that aid in the correct folding of polypeptides. Originally referred to as **heat shock proteins** because their levels become elevated when cells are transiently exposed to high temperatures, the chaperonins appear to interact with nascent polypeptides. Most likely, they bind to stretches of hydrophobic amino acids that are exposed only before the polypeptide folds. After binding, the chaperonins then aid in the correct folding of the polypeptide.

CONNECTIONS For microbes that survive in harsh environmental conditions, keeping proteins properly folded may be a challenge. As we saw in **Section 4.3**, chaperonins are quite abundant in hyperthermophilic archaeons.

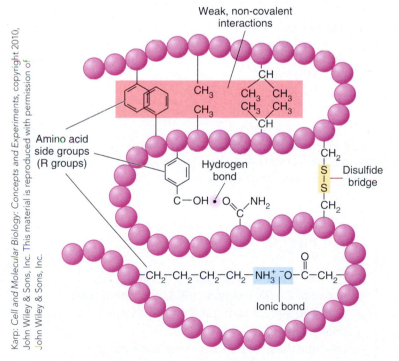

Figure 7.28. Non-covalent and covalent interactions in the folding of polypeptides Non-covalent interactions between the amino acids, such as H bonds, ionic bonds, and other weak interactions, affect the shape and functional properties of the polypeptides. Covalent disulfide bridges can form between two cysteine amino acids, also affecting the shape and function of the polypeptide.

Protein Processing

In addition to folding correctly, a polypeptide may undergo other **post-translational modifications**. Many proteins undergo **phosphorylation**, a process in which a phosphate group ($—PO_4$) is covalently added to the —OH group present on the amino acids threonine, serine, and tyrosine. Addition or removal of phosphate groups can alter the structure and, as a result, function of a protein. Other amino acid modifications, including the addition of methyl ($—CH_3$) and acetyl ($—COCH_3$) groups also occurs. **Glycosylation**, the addition of complex sugar residues to polypeptides, appears to be quite common in eukarya. Again, these modifications alter the function of the non-modified polypeptide.

Protein Transport

Finally, proteins must be correctly located. Many proteins function in the cytoplasm. Other proteins need to be embedded within the cytoplasmic membrane. In bacterial cells, proteins may function within the periplasm, while in eukaryal cells, certain proteins may need to get inside specific organelles or become embedded within the membranes of these organelles. For all organisms, some proteins need to be secreted out of the cell entirely.

How do proteins get to their desired locations? While the details of this transport process differ between bacteria, archaeons, and eukarya, the basic steps are fairly similar. Proteins that need to be embedded within or cross a membrane contain a short stretch of hydrophobic amino acids, known as the signal peptide, at their N-terminus. The signal recognition particle (SRP), a complex ribonucleoprotein, recognizes and binds to the signal peptide as the polypeptide is being translated. The SRP then transports the polypeptide to its appropriate location, based on the specific amino acids present in the signal peptide. The signal peptide, in other words, acts a type of zip code for the protein. Once the polypeptide has arrived at the correct location, a signal peptidase cleaves the signal peptide from the polypeptide.

The role of the nuclear membrane

In the preceding sections, we have examined the mechanics of transcription and translation in bacteria, eukarya, and archaeons. While the details of these events may differ, the basic processes show a great deal of similarity. Before leaving this topic, however, let's consider another question: What role does the nuclear membrane play in these events?

The nuclear membrane itself may not be actively involved in transcription or translation, but the presence of a nuclear membrane in eukarya has a profound effect on these two processes. Transcription, of course, must occur where the DNA is located. For eukarya, the DNA resides within the nucleus. The ribosomes, however, reside outside of the nucleus, in the cytoplasm. The nuclear membrane, then, separates the locations of transcription and translation. This physical separation also means that transcription and translation must be temporally separated; transcription must end (and the RNA molecule must travel out of the nucleus) before translation can occur

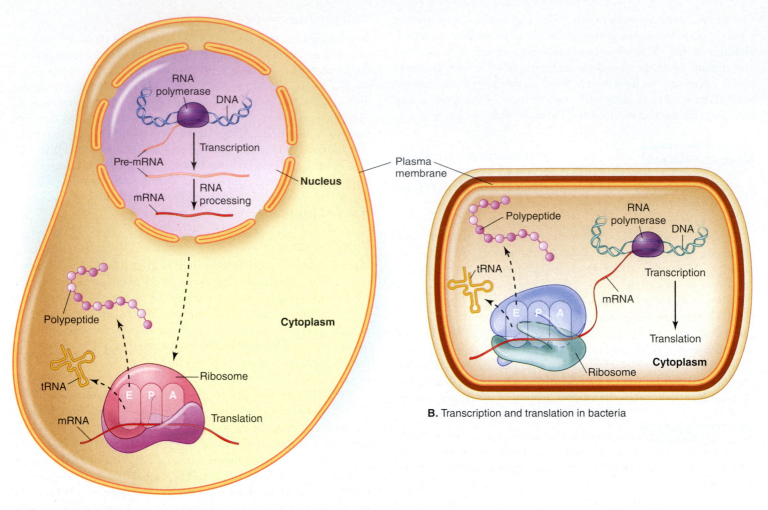

A. Transcription and translation in eukarya

Figure 7.29. Spatial separation of transcription and translation in eukarya, but not bacteria
A. Transcription occurs in the nucleus. After the pre-mRNA has been modified, or processed, the mature mRNA is transported into the cytoplasm where translation occurs. The two events, then, are separated temporally and spatially, producing a time lag between them. **B.** No physical separation exists between transcription and translation. As a result, translation can begin before transcription ends. This speeds the bacterial capacity to adapt to environmental changes by altering changes in protein production.

(**Figure 7.29**). As we saw during our discussion of transcription, eukaryal RNA undergoes several forms of post-transcriptional modification prior to translation. A 5′ cap is added, a 3′ poly(A) tail is added, and the introns are spliced out. All of these modifications occur within the nucleus. Only after these events have occurred will the mature mRNA leave the nucleus, exiting through nuclear pores.

In bacteria, conversely, transcription and translation are not spatially separated by a nuclear membrane and, as a result, are not temporally separated. The ribosomes exist in close proximity to the DNA. The ribosomes, then, can bind to mRNA even before the transcription process has been completed. In other words, transcription and translation are tightly coupled and can occur almost simultaneously. What does this mean for the organism? In bacteria, the time frame from DNA to mRNA to protein is shorter. Even if all the transcription and translation enzymes in bacteria and eukarya work at equivalent rates, a bacterium will be able to produce a protein from a gene more quickly than its eukaryal counterpart because it can begin translation before

completing transcription. This rapid protein production strategy probably benefits bacteria that may need to respond very quickly to environmental changes.

7.4 Fact Check

1. How are the following involved in translation: ribosomes, mRNA, codons, tRNA, and amino acids?
2. Compare and contrast translation initiation in bacteria and eukarya.
3. Describe the post-translational modifications made to polypeptides in eukaryal organisms.
4. Why are transcription and translation spatially and temporally separated in eukaryal cells but not in bacteria or archaeons?

7.5 The effects of mutations

What effects do changes in the DNA sequence have?

When we discussed the replication of DNA earlier in this chapter, we mentioned that the replication process ensures that each newly produced double-stranded DNA molecule contains the exact same nucleotide sequence as the original molecule. Because A always binds to T and C always binds to G, each strand of a double-stranded molecule can serve as the template for the synthesis of a complementary strand. The importance of this fidelity now should be obvious. If the DNA sequence of the coding region of a gene changes, then the mRNA sequence will change and the amino acid sequence of the resulting protein may change. A change in the amino acid sequence of a protein may alter the function of that protein. Thus, changes in the DNA sequence may alter the cell's ability to function.

Changes in the DNA sequence, however, do occur. These changes, or mutations, can arise during DNA replication. The DNA polymerase simply may incorporate the wrong nucleotide. Such spontaneous mutations happen all the time, although at a low frequency. Mutations also may be caused by outside agents, or mutagens, that alter the DNA sequence. As we will see in Chapter 22, viruses sometimes can alter the DNA sequence of the cells they infect. In this section, we will briefly examine different types of mutations, some common causes of these mutations, and ways in which bacteria can correct these errors. Before we address these points, though, let's not forget one very important point—in the grand scheme of things, mutations are not bad. Mutations provide the raw material for evolution. The great diversity of life that we see all around us exists thanks to changes that have occurred in DNA and were acted upon by the natural forces of evolution. Think about life over the 5-billion-year history of Earth if absolutely no mutations ever occurred in the original molecule of nucleic acid. It's a pretty bleak picture, isn't it?

Types of mutations

Probably the simplest type of mutation is a **base substitution**, a mutation in which a single base in the wild-type, or "normal" sequence, changes. Transitions involve purine to purine or pyrimidine to pyrimidine changes, while transversions involve purine to pyrimidine or pyrimidine to purine changes. The effects of these changes can differ dramatically, depending on the bases involved and their location. Changes that alter codons could have an obvious effect on a protein. If a mutation alters the specificity of a codon, then the amino acid sequence of the protein will be altered. Such changes are referred to as **missense mutations**. We should note, however, that even if the amino acid sequence of a protein is altered, the overall shape of the protein and consequently, its function, may not change. Because of the redundancy of the genetic code, some changes in codons will not alter the amino acid sequence. These changes are referred to as **silent mutations**. Finally, point mutations can lead to the conversion of an amino acid-coding codon into a stop codon. These changes are referred to as **nonsense mutations** (Figure 7.30).

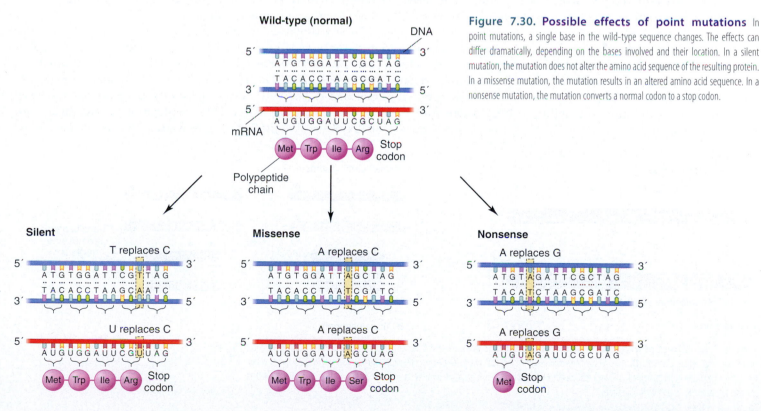

Figure 7.30. Possible effects of point mutations In point mutations, a single base in the wild-type sequence changes. The effects can differ dramatically, depending on the bases involved and their location. In a silent mutation, the mutation does not alter the amino acid sequence of the resulting protein. In a missense mutation, the mutation results in an altered amino acid sequence. In a nonsense mutation, the mutation converts a normal codon to a stop codon.

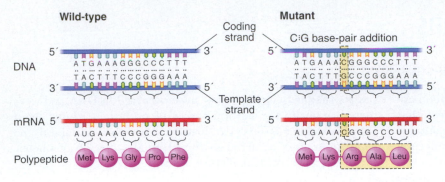

Figure 7.31. Effects of insertions and deletions The insertion or deletion of one or two bases from the coding region of a gene will alter the reading frame of the subsequent bases. As shown here, all codons after the insertion are altered.

Substitutions in the non-transcribed regulatory regions of genes also can have profound effects. As we noted during our discussion of transcription initiation, RNA polymerases recognize specific promoter sequences. Mutations in the promoter region, then, could dramatically decrease, or increase, the affinity of the RNA polymerase for that promoter. Expression of the gene, as a result, would be altered. So, while the structure of the resulting protein would not be altered, the amount produced would change. Again, this change could have dramatic consequences for the cell.

Instead of substitutions, **insertions** and **deletions** also may occur. With these mutations, one or more bases are added to, or removed from, the original DNA sequence. As with substitutions, the effects of these mutations may be most obvious if they occur within the coding region of a gene. Insertions or deletions within the coding region of a gene will alter the reading frame of the bases. In other words, all the codons after an insertion or deletion will be different. Subsequently, most amino acids after the insertion or deletion probably will be altered. These mutations are **frameshift mutations (Figure 7.31)**.

As we saw with substitutions, insertions and deletions also can cause problems in regulatory regions. The −10 and −35 elements of the bacterial promoter, for example, need to be about 15 base pairs apart. A lengthy insertion or deletion between these elements could alter the ability of RNA polymerase to bind to the promoter.

Finally, more extensive alterations in the DNA sequence can occur. In **inversions**, sections of DNA literally are inverted and reinserted into the genome. In **translocations**, a segment of a chromosome breaks off and reattaches to a different chromosome or a different location within the same chromosome **(Figure 7.32)**. The effects of inversions and translocations depend on the sequences involved and, in the case of translocations, the new location of the translocated DNA. For example, a gene could be moved closer to a regulatory element, resulting in increased transcription of that gene.

@ **Mutations** ANIMATION

Causes of mutations

Because of the importance associated with maintaining the correct DNA sequence, DNA polymerases rarely make mistakes. The DNA polymerase in *E. coli*, as we mentioned before, produces DNA at a rate of 1,000 nucleotides per second. Yet it makes a mistake—incorporating the incorrect base into the growing strand of DNA—only about 1 in every 10^9 bases (10^9 = 1 billion). Imagine doing that well on your tests! This amazing accuracy occurs due to several processes. First, most DNA polymerases possess a **proofreading activity**. As the enzymes elongate DNA, they also can check their work. Using a 3′ to 5′ exonuclease activity, the polymerase can identify and remove an incorrect base that has been incorporated into the growing chain. Second, bacteria often use **mismatch repair** to correct errors that occurred during replication. In *E. coli*, a series of proteins, including MutS, MutL, and MutH, will bind to the mismatched DNA and facilitate the excision of several bases around the mismatched base by exonucleases. DNA polymerase I then fills in the resulting gap and DNA ligase links together the adjoining fragments.

Let's think about this repair process a little more. How can these repair proteins determine which strand of DNA contained the correct base and which strand contained the incorrect base? The correct "decision" could be vitally important to the cell. Not surprisingly, bacteria have evolved a clever solution. After DNA replication occurs in *E. coli*, deoxyadenosine methylase methylates, that is adds a methyl

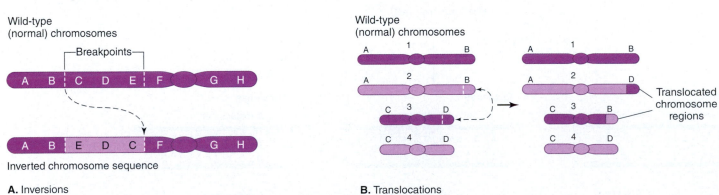

A. Inversions

B. Translocations

Figure 7.32. Effect of inversions and translocations A. A section of DNA becomes inverted, resulting in a different order of genes and other genetic elements. **B.** A segment of one chromosome breaks off and reattaches to a different chromosome or a different location within the same chromosome. This schematic shows a reciprocal translocation, in which pieces of chromosomes 2 and 3 are switched. Note that the letters represent different genes on the chromosomes.

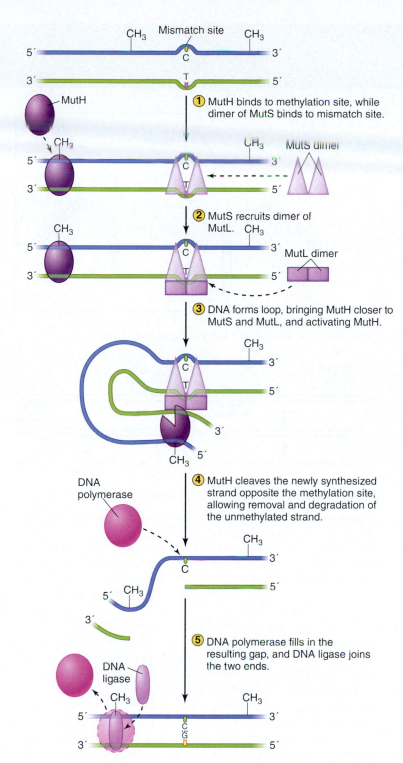

① MutH binds to methylation site, while dimer of MutS binds to mismatch site.

② MutS recruits dimer of MutL.

③ DNA forms loop, bringing MutH closer to MutS and MutL, and activating MutH.

④ MutH cleaves the newly synthesized strand opposite the methylation site, allowing removal and degradation of the unmethylated strand.

⑤ DNA polymerase fills in the resulting gap, and DNA ligase joins the two ends.

Figure 7.33. Process Diagram: Mismatch repair in *E. coli* When an error occurs during DNA replication, a mismatched base pair results. In this schematic, a TC mismatch is shown as an example. To correct this mistake, the cell must distinguish the correct (original) strand of DNA from the mutated (newly synthesized) strand of DNA. The MutH/MutS/MutL system accomplishes this task by detecting the methylation pattern of the DNA. For a brief period following DNA replication, the old strand is methylated but the new strand is not. MutH binds to these sites of methylation. A dimer of MutS binds to a mismatched base pair and recruits a dimer of MutL. MutH interacts with the MutS and MutL dimers, which activates MutH, allowing it to cleave the unmethylated strand of DNA opposite the methylation site. The DNA transiently unwinds, and the newly synthesized piece of DNA is degraded from this cleavage site to the mismatch site. DNA polymerase then fills in the resulting gap, and DNA ligase joins the two ends.

strand of DNA at the methylation site, and exonucleases remove a stretch of nucleotides, including the mismatched base. DNA polymerase fills in the resulting gap (**Figure 7.33**).

While DNA polymerases occasionally make mistakes, other mutations result from external agents that alter the DNA—**mutagens**. Nitrous acid, for example, chemically alters bases, removing amino groups from them. Deamination of cytosine changes it to uracil, while deamination of adenine converts it to hypoxanthine and deamination of guanine converts it to xanthine (**Figure 7.34**). During subsequent DNA replication, DNA polymerase will pair an adenine with this newly created uracil and a cytosine with the hypoxanthine and xanthine, ultimately causing changes in the DNA. Similarly, chemicals like methyl-nitrosoguanidine add methyl (—CH₃) groups to bases. When guanine is methylated in this fashion, it may pair with thymine, instead of cytosine, ultimately resulting in a GC to AT change.

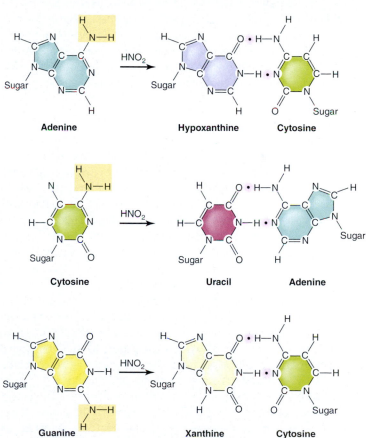

Figure 7.34. Effects of nitrous acid on DNA Nitrous acid removes amino groups (highlighted) from bases. Deamination of adenine converts it to hypoxanthine, while the deamination of cytosine converts it to uracil, and deamination of guanine converts it to xanthine.

group, to A nucleotides found in 5′-GATC-3′ regions. This methylation, though, does not occur immediately after replication; a short period of time elapses between replication of the DNA and addition of the methyl groups. As a result, the newly formed double-stranded DNA is hemimethylated, or methylated on only one strand, the original parental strand, for a short period of time. MutH binds to the DNA at this hemimethylated site. If a mismatch occurs, then a dimer of MutS binds to the mismatched base pair and recruits a dimer of MutL. The DNA then forms a loop, resulting in a MutH/MutS/MutL complex. MutH cleaves the unmethylated, or new,

Y-R. Lin, M-Y. Hahn, J-H. Roe, T-W. Huang, H-H. Tsai, Y-F. Lin, T-S. Su, Y-J. Chan, and C.W. Chen. 2009. *Streptomyces* telomeres contain a promoter. J. Bacteriol. 191:773–781.

Context

Unlike most bacteria, the *Streptomyces* are well known for having linear, rather than circular, chromosomes. Attached to the 5' ends of these linear chromosomes are well-conserved 185 amino acid proteins, called terminal proteins (referred to as TP and Tpg in this paper). These *Streptomyces* linear chromosomes end in telomeres, but telomeres with a different structure than the better-known eukaryal telomeres. The *Streptomyces* telomere sequences are highly conserved, and tend to form complex secondary structures. Each *Streptomyces* linear chromosome contains an internal origin of replication. DNA replication from the origin in both directions toward the telomeres results in single-stranded gaps about 300 bp from the end, leaving 3' overhangs. TP is responsible for patching these gaps. The mechanism of this end patching is not understood; it is not even known which DNA polymerase is used.

In addition to TP and Tpg, another protein, called Tap, also binds to the telomeres. More specifically, Tap binds to the secondary structure of the 3' overhangs, and is thought to help bring TP to the telomere so that it can patch the single-stranded gap. This study started with the goal of determining what other proteins bind to *Streptomyces* telomeres. Along the way, though, the researchers made another discovery—genes exist within the telomeres. The discovery that the telomere sequences have genes arose naturally from the unexpected finding that RNA polymerase, the enzyme responsible for making mRNA from genes, was one of the proteins that bound telomeres.

Experiments

To isolate telomere binding proteins, the researchers first used PCR amplification to prepare double-stranded and single-stranded DNA molecules corresponding to the terminal 167 bp of the *Streptomyces lividans* chromosome. Biotin, also known as vitamin H or B$_7$, was covalently attached to one of the primers used in PCR, resulting in the biotin-labeling of the resulting PCR products **(Figure B7.2)**.

The biotin-labeled DNA fragments were incubated with crude extract from *S. lividans* cells. During this incubation, proteins that interact with the DNA should bind to it. Streptavidin-coated beads then were added to the mixture. Because biotin and streptavidin bind to each other with high affinity, any DNA:protein complexes would bind to these beads. The beads were washed with buffer, and bound proteins were removed from the beads with a salt solution. The liberated proteins then were separated by SDS-PAGE **(Figure B7.3)**. Three proteins attached to the labeled DNA fragments. Note that the bands marked by the * were shown to be due to non-specific binding.

The researchers then tried to identify these three telomere binding proteins. The proteins, two of 150 kDa, and one of 40 kDa, match the expected sizes of the proteins that make up the core RNA polymerase subunits β, β', and α_2. Consistent with this finding, the 40-kDa band reacted with an anti-α antibody. Another antibody against a major component

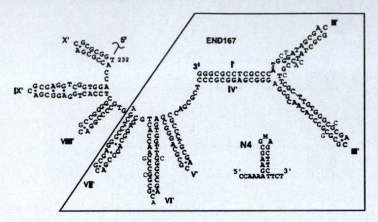

A. Secondary structure in *S. lividans* chromosome

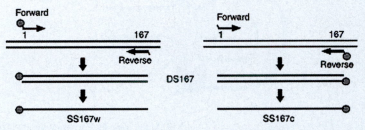

B. PCR amplification with biotin-labeled primers

Figure B7.2. Preparation of biotin-labeled telomere probes

A. Computer modeling indicates that extensive secondary structures occur at the 3' ends of *S. lividans* chromosome. The shaded region indicates the terminal 167 nucleotides of the chromosome. **B.** PCR, using biotin-labeled oligonucleotide primers (indicated by circles attached to forward and reverse primers), was used to amplify this region of the chromosome. Biotin labeling was used because biotin binds very tightly to Streptavidin protein-coated magnetic beads, enabling easy separation of the labeled strands.

From Yuh-ru Lin, 2009. J Bacteriol. 191:773–781. Fig. 1. Image provided courtesy of Canton W. Chen

of the bacterial RNA polymerase also detected bound proteins that were not visible by SDS-PAGE. In addition to the proteins that bound to all DNA fragments, a 100-kDa protein was found to bind to the double-stranded fragment only, and later was identified as DNA polymerase I.

Taken together, these results suggest that RNA polymerase binds to the telomere of *S. lividans*. If this enzyme binds to these sequences, then the likelihood is strong that it is carrying out transcription, and the telomere region, by extension, contains genes. The researchers tested this hypothesis with an *in vitro* transcription assay using the double-stranded 167-bp telomere fragment as a template and a purified RNA polymerase. In this assay, mRNA transcripts of 100 and 130 nucleotides in length were produced. The telomere sequences, in other words, contain genes. Further *in vivo* experiments were conducted, in which the researchers inserted the telomere DNA sequences in front of genes that lacked promoters, the DNA sequences typically required for the initiation of transcription. These telomere sequences led to activity in the bacteria *Streptomyces* and *E. coli*, and also activated transcription in the eukarya, *Saccharomyces cerevisiae*.

If you look at the telomere DNA sequences, you won't find sequences typically associated with *Streptomyces* promoters. What,

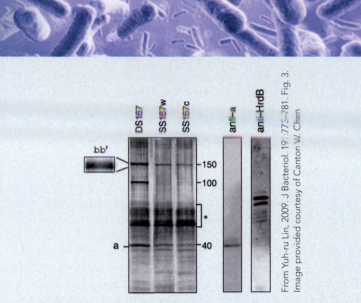

From Yuh-ru Lin, 2009. J Bacteriol. 19 :777—781. Fig. 3. Image provided courtesy of Canton W. Chen

Figure B7.3. Isolation of telomere binding proteins The biotin-labeled DNA fragments were incubated with proteins from *S. lividans* cells. DNA:protein complexes were isolated with streptavidin-coated beads. Bound proteins were separated by SDS-PAGE.

then, is responsible for this promoter activity? Could the secondary structure of the telomere lead to gene expression? Is the promoter activity biologically meaningful? These and other important questions still need to be answered.

Impact

The authors of this paper proposed a few ideas about the possible functions of this promoter activity. First, it may be involved in conjugation, a process whereby bacteria can transfer DNA between cells. The F plasmid promoter is involved in conjugation, and its activity depends on its secondary structure. Second, the telomere promoter might be involved in the synthesis of the primer of the terminal Okazaki DNA fragment. Third, it might be involved in the transcription of a specific gene. Indeed, a gene called *ttrA* that is involved in the conjugal transfer of DNA is located in this region of the telomeres. It will be very interesting to see whether the activity of the promoter is subject to regulation, and whether the promoter activity is affected by whether the telomeres are single- or double-stranded.

● Questions for Discussion ....

1. While the telomere sequences function as a promoter in bacteria, they do not work as promoters in yeast. They do, however, augment gene activity in these eukarya. Are you surprised by this difference? Why or why not?

2. If promoter activity was demonstrated by *in vitro* transcription assays, why was it also necessary to show that it occurred *in vivo*?

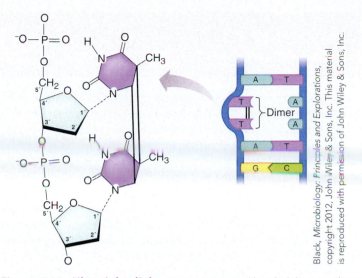

Figure 7.35. Ultraviolet light as a mutagen Ultraviolet light causes adjacent thymines to form covalent bonds, generating thymine dimers, which distort the shape of the DNA. If not repaired, these dimers result in DNA damage during the next round of DNA replication.

Physical agents also can damage DNA. For example, ultraviolet (UV) light causes adjacent thymine molecules to bond to each other, forming a T—T dimer **(Figure 7.35)**. These dimers cause a deformation in the double-stranded DNA and can lead to subsequent errors during the next round of DNA replication.

Ionizing radiation, on the other hand, causes double-stranded breaks (DSBs) in the phosphate-sugar backbone of the DNA molecule. In some respects, DSBs may be the most dangerous type of DNA damage; unless the DSBs are repaired, the entire molecule will be non-functional. As we mentioned in the opening story, *Deinococcus radiodurans* exhibits extreme resistance to this type of DNA damage, withstanding the generation of approximately 100 DSBs per genome. *Shewanella oneidensis*, conversely, only can survive levels of ionizing radiation that result in about 1 DSB per genome.

CONNECTIONS The effects of ionizing radiation, and its uses in sterilizing food, are discussed in **Section 16.2**. In **Figure 16.9**, we review the electromagnetic spectrum.

As we have seen in this section, various mutagens can cause mutations in a cell's DNA. Researchers intentionally use mutagens to introduce mutations into model organisms. Food safety experts intentionally use mutagens to sterilize foods. Our own cells even use mutagens. Intracellular mutagens may help control viral infections **(Perspective 7.1)**.

Repair of mutations

While evolution depends on mutations, individual organisms probably would prefer not to have their genomes altered. Not surprisingly, then, all living organisms possess elaborate mechanisms to repair their DNA. As we mentioned previously, the 3′ to 5′ exonuclease activity of DNA polymerases and the methylation-directed mismatch repair mechanism allow cells to correct most of the errors that occur during DNA replication. Other systems exist to repair specific types of DNA damage. The photoreactivation pathway, for example, specifically repairs thymine dimers caused by UV light. **Photolyase** cleaves the covalent bonds linking

Mutations are necessary for evolution, but can be detrimental to an organism. Certainly, a very high level of mutations would alter many genes, change many proteins, and have a severe effect on the organism. So, could we control pathogens by intentionally causing mutations in an infectious agent? It turns out that our cells do just that as a way of controlling some viral infections. Perhaps we can make use of this strategy to develop new antiviral agents.

This story has rather humble beginnings. Quite some time ago, researchers observed that HIV-1, the virus that causes AIDS, does not replicate in all cell lines. These cells were classified as *non-permissive*. A series of experiments demonstrated that these non-permissive cells produced an antiviral factor, later identified as APOBEC3G. In infected cells, this host cell protein can become incorporated in newly formed virions **(Figure B7.4)**.

When these APOBEC3G-containing virions infect other cells, APOBEC3G goes to work. It functions as a cytidine deaminase, removing the amino group from the cytosine base on cytidine nucleotides. This chemical modification affects replicating viral genome. So why would a cytidine deaminase block viral replication? As we discussed earlier in this section, deamination of cytosine converts it into uracil (see Figure 7.34). When a C is converted into a U on one strand of DNA, the other strand will change from a G to an A. In other words, a highly active cytidine deaminase would cause a large number of CG to TA mutations. Indeed, in cells expressing APOBEC3G, 1 percent–2 percent of all Cs in the viral genome may be changed to Us. This extraordinarily high level of mutations leads to the production of non-functional viruses. Our cells, then, induce mutations in the virus as a way of protecting themselves.

The story doesn't stop there. HIV-1, it turns out, produces a protein of its own designed to counteract APOBEC3G. The viral protein Vif, or virus infectivity factor, prevents APOBEC3G from becoming incorporated into the virion, thereby inhibiting the activity of APOBEC3G **(Figure B7.5)**.

It appears that Vif from HIV-1 only inhibits the activity of human APOBEC3G and APOBEC3G from very closely related species. HIV-1 Vif does not inhibit APOBEC3G from more distantly related species. It seems, then, that the *vif* gene in HIV-1 has evolved specifically to interact with the form of APOBEC3G in its host organism.

By analyzing the activity of APOBEC3G, we may learn more about the origins and evolution of HIV. More importantly, we may be able to use this information to design new antimicrobial agents that function by mutating selectively the genomes of specific pathogens.

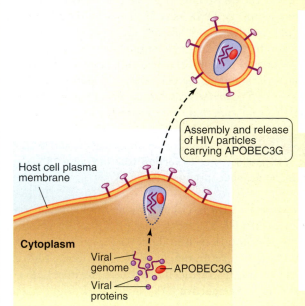

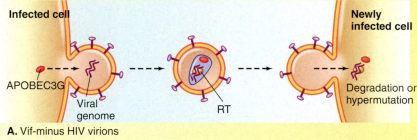

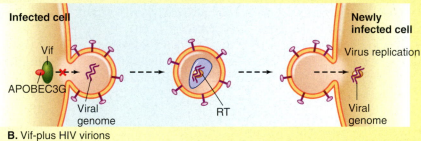

Figure B7.4. Incorporation of the protein APOBEC3G into newly produced virions When a host cell produces the protein APOBEC3G, this protein gets incorporated into newly formed HIV particles, along with the viral genome and key viral proteins. These newly formed virions contain this host protein and transfer it, along with their genome, into newly infected cells.

Figure B7.5. The HIV-1 Vif protein counters the action of the APOBEC3G protein **A.** When an HIV-1 virion lacking the *vif* gene replicates within a host cell, the cellular protein APOBEC3G becomes incorporated into newly produced virions. When these virions then infect another cell, the APOBEC3G enzyme deaminates Cs in the replicating viral genome, leading to hypermutations in or degradation of the viral genome, thereby destroying its capacity to effectively replicate. **B.** When an HIV-1 virion containing the *vif* gene replicates within a host cell, the viral Vif protein prevents the incorporation of APOBEC3G into the newly produced virions, thereby preventing the damaging hypermutations or degradation of the viral genome in cells that subsequently become infected by the virus.

the adjacent thymines, thereby correcting the error (**Figure 7.36**). Various **alkyltransferases** remove methyl groups added to bases by DNA modifying agents, thereby returning the bases to their normal structure. Researchers have identified over 100 different alkyltransferases in bacteria, archaeons, and eukarya, suggesting that this repair mechanism is quite important to all living organisms. Finally, a series of proteins recognizes and corrects double-stranded breaks, the type of damage caused by ionizing radiation. As a result of these and other repair pathways, the sequence of the genetic material remains fairly constant, yet mutates enough to provide the variation needed for evolutionary change.

> **@WWW** **Thymine dimer repair** ANIMATION

7.5 Fact Check

1. Differentiate between the effects of these types of substitution mutations: missense mutations, silent mutations, and nonsense mutations.
2. How would the result of an insertion of two nucleotides differ from that of the insertion of three nucleotides?
3. How can spontaneous mutations that occur during DNA replication be repaired?
4. Provide several examples of mutagens and their effects.

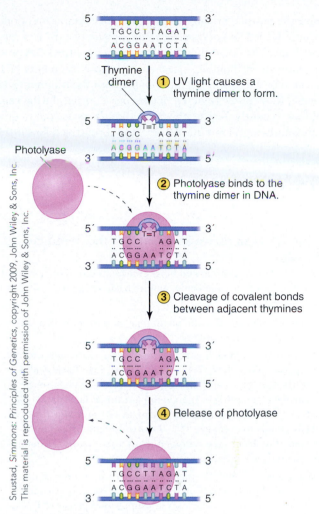

Snustad, Simmons: Principles of Genetics, copyright 2009, John Wiley & Sons, Inc. This material is reproduced with permission of John Wiley & Sons, Inc.

① UV light causes a thymine dimer to form.

② Photolyase binds to the thymine dimer in DNA.

③ Cleavage of covalent bonds between adjacent thymines

④ Release of photolyase

Figure 7.36. Process Diagram: Photolyase can correct thymine dimers. Living organisms have evolved a complex set of pathways to correct different types of DNA damage. Photolyase cleaves the covalent bonds in thymine dimers, correcting the errors associated with exposure to UV light.

Image in Action

WileyPLUS The *Deinococcus radiodurans* cells shown here are exposed to ionizing radiation (pink halo) and still are able to undergo normal cell processes, including transcription and translation. In contrast, the irradiation has caused double-stranded breaks in the genome of *E. coli* cells on the right that cannot be properly repaired.

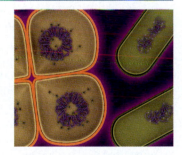

1. Researchers recognize that the DNA of *Deinococcus radiodurans* also is broken when exposed to irradiation. However, the arrangement or packaging of the DNA may play a key protective role. Describe how the DNA is packaged in this organism and how it differs from the *E. coli* DNA shown on the right.

2. Because the DNA of *E. coli* is not packaged the same as *D. radiodurans*, the breakages may result in mutations that can have significant impacts on the cells. Describe some of the mutations that now may occur in the damaged *E. coli* cells in the image.

The Rest of the Story

So, how can *Deinococcus radiodurans* survive exposure to extreme levels of ionizing radiation? It seemed very unlikely that *D. radiodurans* was somehow impervious to the effects of radiation. Its structure does not differ dramatically from other bacteria and the chemical make-up of its DNA does not differ at all from the DNA of any other organism. The DNA of *D. radiodurans* appears to undergo numerous double-stranded breaks when exposed to ionizing radiation, just like the DNA of any other organism. Unlike other organisms, though, scientists determined that *D. radiodurans* must possess a mechanism for repairing this damage.

Sequence analyses of the *D. radiodurans* genome indicated that it contains a normal set of DNA repair genes. A series of experiments demonstrated that DNA repair genes from radiation-sensitive bacteria could restore radiation resistance to several radiation-sensitive mutants of *D. radiodurans*. So how did this bacterium, with standard DNA repair genes survive exposure to high levels of ionizing radiation? The answer seemed to first involve the unusual arrangement of its DNA. As was mentioned in Section 7.1, the DNA of *D. radiodurans* appears to exist in a very unusual shape. The doughnut-like arrangement of the DNA provides a mechanism for holding together the many fragments of DNA generated by exposure to radiation. Because the fragments do not disperse, they can be rejoined in the correct order during repair.

More recently, it was observed that *D. radiodurans* contains an unusually high amount of Mn^{2+} and a relatively low intracellular iron concentration. Could the high Mn/Fe ratio account for the radiation resistance? Professor Michael Daly and colleagues at the Uniformed Services University of the Health Sciences hypothesized that the Mn^{2+} may protect critical DNA repair enzymes from damage caused by radiation. The DNA repair enzymes in *D. radiodurans*, then, can continue to repair the DNA fragments being held together, while homologous enzymes in other species simply become inactivated.

Summary

Section 7.1: How do we know DNA is the hereditary molecule?

A series of elegant genetic and biochemical studies demonstrated that DNA is the hereditary molecule, and its basic structure is similar in all living organisms.

- Foreign DNA can alter the phenotype of another cell, a process known as **transformation**.

- DNA exists as a double-stranded molecule, with each strand composed of a **2-deoxyribose** and **phosphate** backbone. A **purine** or **pyrimidine** nitrogenous base is attached to the 1′ carbon of each sugar. In a double-stranded DNA molecule, **adenine (A)** always binds to **thymine (T)** and **guanine (G)** always binds to **cytosine (C)**. The two strands of a double-stranded molecule are arranged in an **antiparallel** fashion, and the strands are **complementary** to each other.

- Bacteria usually have a single, circular chromosome that is supercoiled for compaction. Bacteria also commonly contain **plasmids**, small, circular extrachromosomal pieces of DNA.

- Eukarya usually have linear DNA chromosomes, in which the DNA is complexed with histones, forming nucleosomes.

- Archaeons usually have a single, circular chromosome, but it often is complexed with histone-like proteins.

Section 7.2: How does DNA replicate?

The replication of DNA is precisely regulated within the cell and occurs with great fidelity.

- DNA undergoes **semiconservative replication**. Each strand of a double-stranded molecule serves as a template for the formation of a new, complementary second strand.

- In *E. coli*, replication begins at the origin of replication, or *oriC*. A series of proteins, including **DnaA** and **helicase**, interact with *oriC* and initiate replication.

- After a **replication bubble** has formed at the origin of replication, **single-stranded DNA-binding proteins (SSB)** attach to the DNA, and a **DnaG**, or **primase**, adds a short RNA **primer** to the newly formed single-stranded DNA. **DNA polymerase** then adds nucleotides to the primer, replicating the DNA bidirectionally. One new strand, the **leading strand**, elongates continuously. The **lagging strand** elongates discontinuously, forming **Okazaki fragments**.

- The RNA primers are removed by an exonuclease, and a **ligase** links the new fragments into a complete molecule.

- In circular DNA molecules, replication continues until the two replication forks meet. In linear molecules, specialized features, **telomeres**, exist on the ends of the molecules, and a specialized enzyme, **telomerase**, completes the replication process.

Section 7.3: How are genes transcribed?

Like DNA replication, the process of transcription is a closely regulated process. Specific initiation, elongation, and termination processes occur.

- **Transcription** is the process by which **genes** are converted into complementary strands of single-stranded RNA by **RNA polymerases**. These RNA molecules include **messenger RNA (mRNA)**, **transfer RNA (tRNA)**, **ribosomal RNA (rRNA)**, and various regulatory **micro RNAs (miRNAs)**. The RNA contains the sugar **ribose** and the base **uracil**.

- The transcription process begins at a **promoter**, a specific sequence recognized by the RNA polymerase. In bacteria, **sigma (σ) factor**, a transient component of RNA polymerase, binds to the promoter. In eukarya, **transcription factors** first bind to the promoter and then recruit RNA polymerase. These binding interactions can be observed with a **gel shift assay**.

- Transcription ends at specific sequences. Eukaryal messenger RNAs undergo **post-transcriptional modifications**. A **5′ cap** and **3′ poly(A) tail** are added, **introns** are spliced out, and **exons** are joined together.

Section 7.4: How are proteins made?

In translation, the information within the mRNA molecules is used to make proteins. Again, initiation, elongation, and termination are closely regulated.

- **Translation** is the process of protein synthesis. **Ribosomes** interact with the mRNA, "reading" sequential **codons** and recruiting tRNA with complementary **anticodons**. Each tRNA is associated with a specific **amino acid**.

- In bacteria, translation begins at the first AUG codon after the **Shine–Dalgarno (SD) sequence**, a short sequence in the mRNA recognized by the ribosome. In eukarya, translation begins at the first AUG after the 5′ cap. As a result, bacterial genes may be **polycistronic**, while eukaryal genes usually are **monocistronic**.

- Translation stops when the ribosome reaches a **stop codon**—UAA, UAG, or UGA.

- Following translation, nascent polypeptides must be correctly folded and, in many cases, are modified and transported.

- Chaperonins, like **heat shock proteins**, aid in the correct folding of polypeptides.

- **Post-translational modifications**, such as **phosphorylation** and **glycosylation**, may occur during protein processing.

- In bacteria, transcription and translation occur almost simultaneously. In eukarya, the two events are spatially and, as a result, temporally separated.

Section 7.5: What effects do changes in the DNA sequence have?

While the DNA replication occurs with great fidelity, mistakes do occur. These mistakes provide the genetic diversity needed for evolutionary change.

- Mutations are changes in the DNA sequence. These changes can alter the coding portion of a gene, resulting in an aberrantly functioning protein.

- Common types of mutations include **base substitutions**, such as **missense**, **silent**, and **nonsense mutations**, **insertions**, and **deletions**. Insertions and deletions within the coding region of a gene will alter the codon reading frame and are referred to as **frameshift mutations**.

- **Inversions**, in which segments of DNA are inverted and reinserted into the genome, and **translocations**, in which pieces of DNA break off and reattach at another location, also can occur, and can have varying effects, depending on the sequences involved.

- Some mutations result from DNA polymerase errors during DNA replication. External agents, or **mutagens**, can cause other mutations.

- Most organisms have evolved complex mechanisms for repairing DNA damage. DNA polymerase, for example, possesses a 3' to 5' exonuclease **proofreading activity** that allows it to scan newly synthesized DNA and correct most errors. Other mechanisms, including **mismatch repair**, **photolyase**, and **alkyltransferase**, exist to correct various types of DNA damage.

● Application Questions

1. The experiments of Griffith and Avery, MacLeod, and McCarty showed that DNA from one organism could alter, or transform, the characteristics of another organism. Using what you have learned in this chapter, explain how this "transformation" actually occurs.

2. You are given macromolecules isolated from a culture of cells. You are asked to determine whether the molecules in the sample are DNA, RNA, or proteins. Using what you've learned in this chapter, how could you go about determining this?

3. Replication of linear DNA encounters problems that are not an issue for circular bacterial chromosomes. Describe these problems and the solution the cell uses to deal with this issue.

4. Do mutations always alter protein function? Explain.

5. A microbe is observed that appears to mutate at twice the normal rate of other similar microbes. Explain how this characteristic could be considered good and bad for this organism.

6. The antibiotic Ciprofloxacin (Cipro) inhibits DNA gyrase, a bacterial enzyme involved in DNA replication. What direct effects would this have on a bacterial cell exposed to this drug?

7. As described in Perspective 7.1, APOBEC3G functions as a cytidine deaminase. What effect is implied by its name? How can it be used to render viruses nonfunctional?

8. Imagine you are attempting to create a collection of mutants that will be used to study and specifically compare to non-mutated strains. How might you use a chemical to generate this collection of mutants? Explain.

9. Researchers isolated *Deinococcus radiodurans* after subjecting canned meat to a supposedly lethal dose of ionizing radiation.
   a. Describe how you might go about isolating from a soil sample a mutant bacterium that exhibits increased resistance to UV light.
   b. How would you quantify this microbe's level of UV resistance?
   c. After isolating this organism, how could you determine if this microorganism exhibits increased resistance to other chemical or physical agents? (Use this chapter's information on *Deinococcus radiodurans* as a guide.)

Suggested Reading

Avery, O. T., C. M. MacLeod, and M. McCarty. 1944. Studies of the chemical nature of the substance inducing transformation of pneumococcal types. J Exp Med 79:137–158.

Cox, M. M., and J. R. Battista. 2005. *Deinococcus radiodurans*—the consummate survivor. Nat Rev Microbiol 3:882–892.

Hershey, A. D., and M. Chase. 1952. Independent functions of viral protein and nucleic acid in growth of bacteriophage. J Gen Physiol 36:39–56.

Lin, Y.-R., M-Y. Hahn, J-H. Roe, T-W. Huang, H-H. Tsai, Y-F. Lin, T-S. Su, Y-J. Chan, and C.W. Chen. 2009. *Streptomyces* telomeres contain a promoter. J Bacteriol 191:773–781.

Santangelo, T. J., L. Cubonova, K. M. Skinner, and J. N. Reeve. 2009. Archaeal intrinsic transcription termination *in vivo*. J Bacteriol 191:7102–7108.

Sheehy, A. M., N. C. Gaddis, J. D. Choi, and M. H. Malim. 2002. Isolation of a human gene that inhibits HIV-1 infection and is suppressed by the viral Vif protein. Nature 418:646–650.

Watson, J., and F. Crick. 1953. A structure for deoxyribose nucleic acid. Nature 171:737–738.

8 Viral Replication Strategies

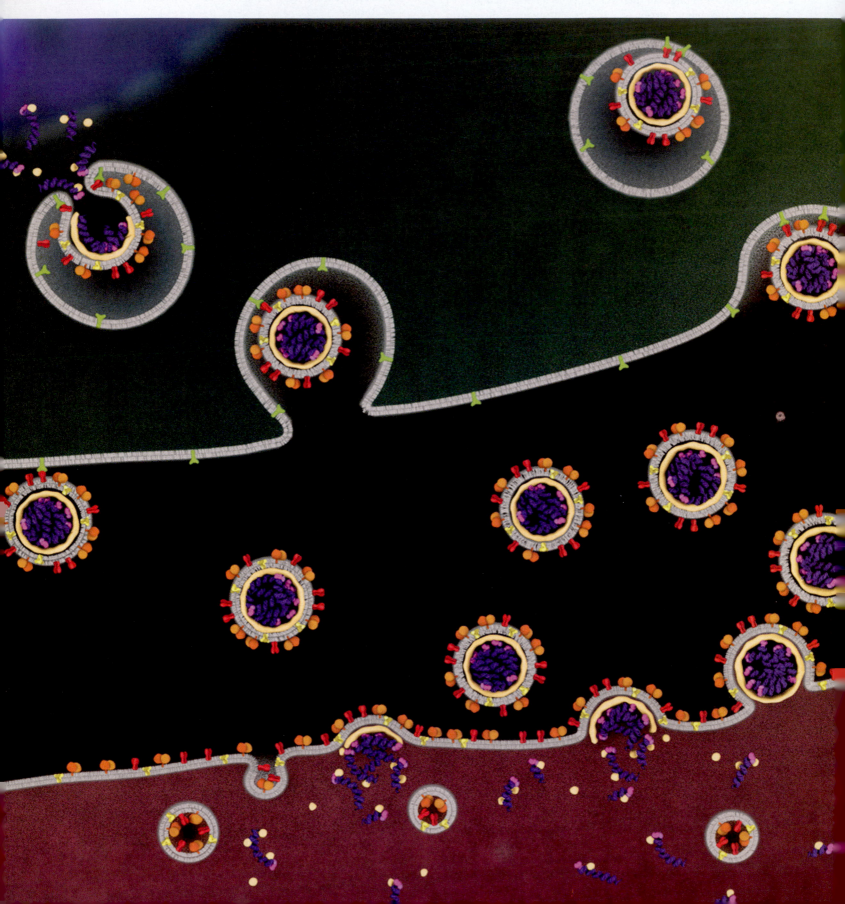

In the spring of 2009, outbreaks of a novel influenza virus infection in Mexico were reported. By April 26, 2009, 38 cases (20 in the United States and 18 in Mexico) had been confirmed. By the beginning of May 2009, the World Health Organization (WHO) reported that this new virus had infected 367 people in 13 countries. On June 11, 2009, the WHO reported over 28,000 cases in 74 countries. Also on this date, the WHO declared an influenza virus pandemic.

Initially referred to as swine flu, this strain of influenza virus—officially known as influenza A(H1N1)—quickly spread throughout the world. With its spread, fear and uncertainty also spread. During the spring of 2009, many schools were closed to avoid transmission of the virus. In Mexico, soccer matches occurred in empty stadiums when the Mexican government prevented fans from attending. Several countries urged caution to citizens traveling to the United States and Mexico. Common questions were hotly debated: Could this virus cause widespread death throughout the world? Should schools, sporting events, and conferences be cancelled? How can we protect ourselves?

Approximately a year later, on August 10, 2010, the WHO declared an end to the 2009 influenza pandemic. This declaration did not indicate that the H1N1 strain of influenza had disappeared. Rather, the announcement that the pandemic had ended simply reflected that the number of new influenza infections had receded to normal levels. Indeed, the threat of influenza remains. When the 64th World Health Assembly, the annual meeting of the WHO, opened in May of 2011, global influenza preparedness remained a primary topic.

So, how can we protect ourselves? During the height of the 2009 pandemic, many individuals tried to protect themselves from infection by purchasing hand sanitizers and protective masks. Governments made plans for dealing with increased demands on the health care industry. The WHO began distributing the antiviral drug oseltamivir, also known by the trade name Tamiflu, to various developing countries.

Numerous studies have shown that Tamiflu can reduce the severity of influenza disease and decrease the duration of the symptoms. The Food and Drug Administration (FDA) also has approved it for use as a preventive drug. But what does it do? To answer this question, we need an understanding of influenza virus and how it replicates.

CHAPTER NAVIGATOR

As you study the key topics, make sure you review the following elements:

Viruses generally recognize host cells through specific protein interactions.

- 3D Animation: How do viruses recognize and attach to host cells?
- Table 8.1: Cellular receptors for selected viruses
- Figure 8.1: Strain specificity of bacteriophage T2 dependent on its attachment protein
- Figure 8.3: Strain specificity of mouse hepatitis virus dependent on the cellular receptor
- Toolbox 8.1: The Western blot

Viruses enter cells through different mechanisms.

- 3D Animation: How do viruses enter host cells?
- Figure 8.5: Binding and entry of influenza virus
- Figure 8.6: HIV entry
- Figure 8.7: Mechanism of action of Fuzeon
- Figure 8.9: Entry of bacteriophages

Viruses replicate in various ways.

- 3D Animation: How do viruses replicate their genome?
- Figure 8.12: Classification of viruses
- Microbes in Focus 8.2: Human papillomaviruses (HPVs) and cervical cancer
- Figure 8.19: Replication of retroviruses
- Mini-Paper: The discovery of reverse transcriptase
- Figure 8.21: Phage lambda: Lysogenic or lytic phase of replication, depending on the ratio of two repressor proteins.

Newly formed viruses must exit the infected cell.

- 3D Animation: How do replicated viruses exit their host cells?
- Animation: Enveloped virus replication

Antiviral agents often disrupt specific aspects of viral replication.

- Perspective 8.1: DNA microarrays and the SARS virus
- Figure 8.23: Antiviral drugs as nucleoside analogs

CONNECTIONS for this chapter:

Segmented genome of influenza virus (Section 22.4)

Evolutionary relationship between bacteriophage tail proteins and bacterial secretion apparatus proteins (Section 2.4)

Viral replication and the development of cancer (Section 20.2)

Role of antiviral drugs and vaccines (Section 24.5)

Introduction

Most of us have a pretty good understanding of how plants and animals reproduce. Yes, there are exceptions to the general rules, but most macroscopic organisms reproduce sexually and sexual replication is fairly standard (at least microbiologists think so!). Bacterial and archaeal replication also is fairly standard. Bacteria and archaeons generally replicate via binary fission. The replication of viruses, however, is not uniform.

As we saw in Chapter 5, viruses exhibit a great deal of diversity. Not only do viruses differ in size and shape, but also they differ in the type of genetic material they contain. Some viruses have DNA genomes, some have RNA genomes, some have double-stranded genomes, and some have single-stranded genomes. This genetic diversity means that the replication strategies employed by viruses are quite diverse.

Perhaps surprisingly, our understanding of how viruses replicate is relatively limited. Despite the fact that viruses outnumber all other biological entities on Earth, most of our knowledge about virus replication is based on studies of only a few species. It seems clear, though, that all viruses must carry out several other basic functions in order to replicate. All viruses must:

- Recognize appropriate host cells
- Make their genomes accessible to the host cell machinery
- Replicate their genomes and make multiple copies of necessary proteins
- Assemble new viral particles and exit the infected cell

In this chapter, we will begin to explore how viruses of bacteria (bacteriophages) and viruses of eukarya complete these processes, focusing on the strategies utilized by several well-studied viruses. Furthermore, we will examine how this information can be used in a clinical setting. Let's start by asking these questions:

How do viruses recognize appropriate cells? (8.1)
How do viruses enter cells and uncoat? (8.2)
How do viruses replicate? (8.3)
How do virus particles form and exit cells? (8.4)
How do antiviral drugs work? (8.5)

8.1 Recognition of host cells

How do viruses recognize appropriate cells?

For most viruses, replication only occurs within very specific cells. A plant virus, for instance, generally can infect certain plant, but not human, cells. Human hepatitis A virus can infect human liver cells but not lung cells. To a large extent, this cell specificity is determined by the virus's ability to recognize and bind to the surface of an appropriate cell. To facilitate this binding event, the virus contains a protein, often referred to as the **viral attachment protein**, which allows it to attach to the host cell. Generally, this viral attachment protein interacts with a molecule on the surface of the host cell, the **receptor** (Table 8.1). These receptors normally are used by the host cell for communication and signaling purposes but essentially are hijacked by the virus in order to facilitate attachment.

For most bacteriophages, portions of the tail fibers serve as the attachment proteins. The virus, most likely, randomly "bumps" into a potential host cell. When one tail fiber recognizes an appropriate receptor, the virus becomes weakly anchored to the cell. The remaining tail fibers then can bind to additional receptor molecules, thus firmly attaching the virus to the surface

TABLE 8.1 Cellular receptors for selected viruses

Receptor	Cellular function	Virus (family)
CD4	Immune cell interactions	HIV (*Retroviridae*)
ICAM-1	Cell adhesion	Rhinovirus (*Picornaviridae*)
Bgp1[a]	Cell adhesion	MHV-A59 (*Coronaviridae*)
CR2	B cell activation	Epstein-Barr virus (*Herpesviridae*)
Sialic acid	Various functions	Influenza virus (*Orthomyxoviridae*)
OmpF	Transmembrane channel	T2 (*Myoviridae*)

Microbes in Focus 8.1
MOUSE HEPATITIS VIRUS

Habitat: Mice

Description: Single-stranded RNA, enveloped virus roughly 100 nm in diameter in the *Coronaviridae* family with surface glycoproteins that provide a characteristic crown-like or corona appearance

Key Features: Several genetically and phenotypically distinct strains of MHV have been identified. Many of these strains are highly transmissible and can easily be spread within laboratory colonies of mice. Recent studies have shown that the causative agent of SARS is a coronavirus closely related to MHV.

TEM

Courtesy Frederick A. Murphy

of the cell. The bacteriophage T2, which we discussed in Chapter 5, recognizes two molecules on the host cell, lipopolysaccharide and the outer membrane protein OmpF of *E. coli* K12 and related strains. T2 cannot, however, infect *E. coli* O157:H7, a strain of *E. coli* associated with severe intestinal disorders in humans. We could hypothesize that T2 does not infect this strain because of strain-specific differences in the receptors. To test this hypothesis, researchers, using standard genetic engineering techniques, replaced gp37 and gp38, two proteins present at the tips of the T2 tail fibers, with the corresponding proteins from a phage that does infect *E. coli* O157:H7. The altered T2 phage (referred to as T2ppD1) gained the ability to infect O157:H7 and lost the ability to infect K12, its original host **(Figure 8.1)**. Thus, one may conclude that the host specificity, or **host range**, of this phage is determined by the interactions between the viral attachment proteins and the host cell receptors.

Most enveloped mammalian viruses contain a specific protein embedded within the envelope that serves as the attachment protein. Again, this spike interacts with an appropriate host cell receptor, thereby anchoring the virus to the cell. For influenza viruses, like the H1N1 influenza virus that we discussed in the opening of this chapter, the HA, or hemagglutinin, protein serves as this attachment protein. Non-enveloped mammalian viruses bind to cells in a similar fashion. Some non-enveloped viruses, like adenoviruses, have spikes, or attachment fiber proteins, that extend from the viral surface. For non-enveloped viruses like poliovirus, the viral capsid interacts with the receptor **(Figure 8.2)**.

As we saw with the bacteriophage T2, the viral attachment protein:cellular receptor interaction also is very specific for mammalian viruses. To understand this specificity more fully, let's examine a particular virus: mouse hepatitis virus, or MHV **(Microbes in Focus 8.1)**. As the name suggests, this coronavirus

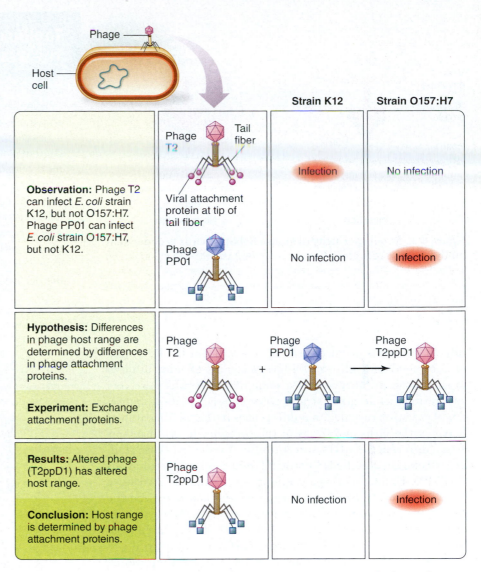

Figure 8.1. Strain specificity of bacteriophage T2 is dependent on its attachment protein When proteins present at the tips of the T2 tail fibers (●) were replaced by the PP01 versions of these proteins (■), the altered T2 phage (T2ppD1) gained the ability to infect strain O157:H7 but lost its ability to infect strain K12.

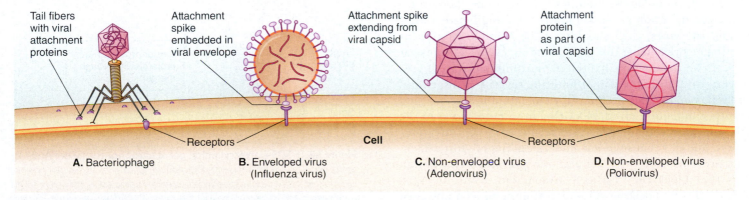

Figure 8.2. Attachment of viruses A. Bacteriophages attach to host cell receptors by viral attachment proteins on the tail fibers. **B.** Enveloped viruses attach by a specific viral protein embedded within the envelope. **C.** Some non-enveloped viruses attach by a protein extending from the capsid. **D.** Other non-enveloped viruses attach by a protein that forms part of the capsid itself.

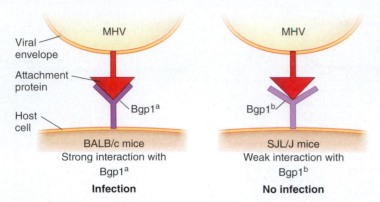

Figure 8.3. Strain specificity of mouse hepatitis virus is dependent on the cellular receptor Mouse hepatitis virus (MHV) can infect BALB/c mice, but not SJL/J mice. The receptor in BALB/c mice is the cell membrane protein Bgp1[a]. An allelic variant, Bgp1[b], is found in SJL/J mice. Amino acid differences between these two molecules cause MHV to interact strongly with Bgp1[a], but only weakly with Bgp1[b].

infects liver cells of mice, causing severe hepatitis. Researchers observed, though, that different strains of mice differed in their susceptibility to MHV; some mouse strains were very susceptible, while other strains were largely resistant. Subsequent research revealed that differences in the binding affinity between the virus and forms of the receptor determined this strain difference in MHV susceptibility. To conduct these studies, virologists used SDS-polyacrylamide gel electrophoresis (SDS-PAGE) and another technique very similar to a Western blot **(Toolbox 8.1)**. Briefly, they isolated and characterized the MHV receptor from BALB/c mice, a strain of mice susceptible to MHV. A structurally similar protein was isolated from SJL/J mice, a genetically distinct, MHV-resistant strain of mice. While these two proteins "looked" similar, their amino acid sequences differed somewhat. More importantly, these proteins differed dramatically in their ability to bind to MHV, presumably because the amino acid differences resulted in slight conformational differences. These observed differences in binding affinities led researchers to conclude that the viral attachment protein:cellular receptor interaction probably determined the susceptibility of these mouse strains to MHV **(Figure 8.3)**.

Before leaving this section, let's address one more question. If the infection of a cell requires binding between that cell and the virus, then shouldn't it be possible to prevent infection by preventing this initial binding event? The answer is yes. As we will see in Chapter 20, our immune system uses this very strategy to prevent or limit the severity of infections. Antibodies, molecules produced by our immune system, bind to viruses and, quite often, block the interaction between the virus and host cell, thereby interfering with the infectious process. Today, because of our understanding of viral attachment proteins and their cellular receptors, we can design drugs that block the virus:cell attachment event. In April 2007, the Food and Drug Administration approved one such drug. That drug was maraviroc.

Toolbox 8.1
THE WESTERN BLOT

To investigate the interactions between mouse hepatitis virus (MHV) and the presumptive receptor molecules, the researchers used a modification of a standard molecular biology technique—the Western blot. In this technique, protein samples first are mixed with the detergent SDS, which disrupts their three-dimensional shape and coats them with a negative charge. The protein/SDS mixture generally also is boiled, further disrupting any intramolecular interactions and fully denaturing the proteins. The samples then are analyzed by SDS-polyacrylamide gel electrophoresis, or SDS-PAGE. Briefly, the protein samples are loaded onto a polyacrylamide gel and an electric current is applied to the gel. Because the proteins have been coated with the negatively charged SDS, they will migrate toward the positive

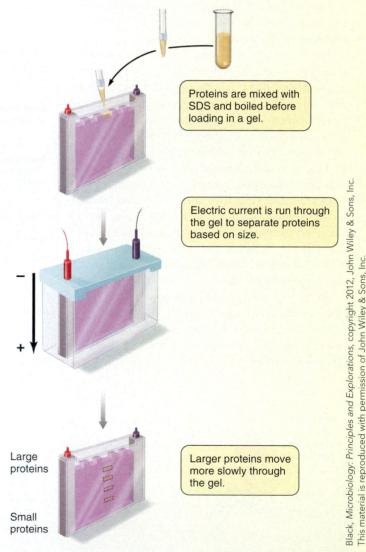

Proteins are mixed with SDS and boiled before loading in a gel.

Electric current is run through the gel to separate proteins based on size.

Larger proteins move more slowly through the gel.

Large proteins

Small proteins

Black, *Microbiology: Principles and Explorations*, copyright 2012, John Wiley & Sons, Inc. This material is reproduced with permission of John Wiley & Sons, Inc.

Figure B8.1. SDS-polyacrylamide gel electrophoresis (SDS-PAGE) To separate proteins based on size, the proteins first are mixed with SDS and boiled. This treatment denatures the proteins and coats them with a negative charge. The proteins then are added to a polyacrylamide gel. An electric current is applied to the gel and individual proteins migrate toward the positive electrode. Their rate of migration is proportional to their size and separation is achieved.

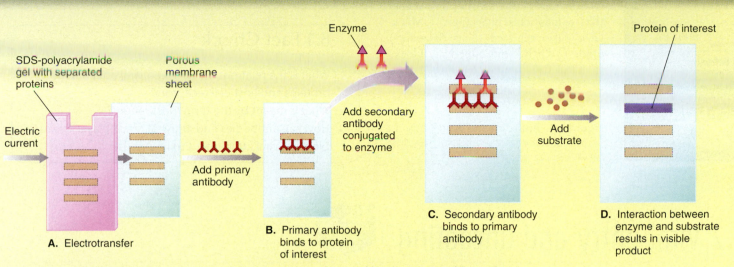

Enzyme

SDS-polyacrylamide gel with separated proteins

Porous membrane sheet

Protein of interest

Electric current

Add primary antibody

Add secondary antibody conjugated to enzyme

Add substrate

A. Electrotransfer

B. Primary antibody binds to protein of interest

C. Secondary antibody binds to primary antibody

D. Interaction between enzyme and substrate results in visible product

Figure B8.2. After electrophoretic separation, proteins are detected by the Western blot technique **A.** Following electrophoresis, the proteins are transferred to a porous membrane, typically composed of nitrocellulose or polyvinylidene fluoride (PVDF). **B.** A primary antibody that specifically binds to the protein of interest then is added to the membrane. **C.** A secondary antibody that binds to the primary antibody and contains a detectable marker, such as an enzyme, then is added to the membrane. **D.** A substrate is added next. The enzyme catalyzes the conversion of the substrate to a visible product, thereby permitting localization of the protein of interest.

electrode. The mesh-like consistency of the gel impedes the migration of the proteins in a size-dependent fashion; the migration of large proteins is hindered more than small proteins. So small proteins migrate through the gel more quickly than large proteins. In other words, the proteins separate based on their size **(Figure B8.1)**.

Following this electrophoretic size separation, the proteins are transferred from the gel to a nitrocellulose or polyvinylidene fluoride (PVDF) membrane for the next step, which is the actual Western blotting step. Following this step, the proteins are positioned on top of the membrane, rather than being embedded within the gel. This transfer of the proteins to a membrane makes them more accessible to reagents used in subsequent steps.

The Western blot allows the researcher to detect a specific protein. Antibodies are generally used for this purpose because they bind strongly and specifically to other molecules (see Section 20.6). If a specific antibody is added to the membrane, it will bind to a specific protein—the protein the researcher is interested in detecting. By itself, this primary antibody is not visible to the researcher, so a secondary antibody that attaches to the primary antibody usually is employed. This secondary antibody typically is conjugated to an enzyme so that it can be detected **(Figure B8.2)**.

To study the MHV receptor, the researchers used a slightly different and creative approach to Western blotting. Rather than detecting the receptor protein with an antibody, they detected the receptor protein with the virus (MHV). Cellular proteins from BALB/c mice and cellular proteins from SJL/J mice were separated by SDS-PAGE and then transferred to a membrane. The researchers then added MHV to the membrane, using the virus to identify its own receptor protein. Viral strain MHV-A59 bound very strongly to two differently sized forms of Bgp1[a], a protein present in BALB/c mice, but only weakly to Bgp1[b], a protein present in SJL/J mice **(Figure B8.3)**.

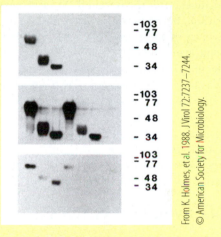

From K. Holmes, et al. 1988. J Virol 72:7237–7244. © American Society for Microbiology.

Figure B8.3. Analysis of MHV receptors Three size variants of Bgp1[a] (*left*, lanes 1–3) and Bgp1[b] (*right*, lanes 4–6) were separated based on size by SDS-PAGE. The proteins then were transferred to a membrane and probed with an antibody that recognizes Bgp1[a] (*top*) an antibody that recognizes Bgp1[a] and Bgp1[b] (*middle*) or viral strain MHV-A59 (*bottom*). As seen in the bottom panel, the virus strongly bound to the large and small forms of Bpg1[a] (lanes 1 and 3) but only weakly bound to the large form of Bgp1[b] (lane 4). *Source:* B. D. Zelus et al. 1998. J. Virol. 72(9):1737–7244.

● Test Your Understanding ••••••••••••••••••

Based on the result of this experiment, which of these mice would the researchers have concluded to be susceptible to MHV and which of these mice would they conclude to be resistant to MHV? Explain how the data from this technique provided that conclusion.

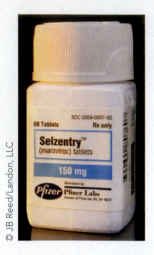

Maraviroc is used in the treatment of HIV disease. It blocks the interaction between the retrovirus human immunodeficiency virus (HIV) and CCR5, one of the virus's co-receptors that we mentioned in Section 5.1. Maraviroc (also referred to as Selzentry) binds to CCR5 and effectively blocks the virus from binding **(Figure 8.4)**.

Figure 8.4. Antiretroviral drug Maraviroc (or Selzentry) blocks the interaction of some strains of HIV with the host cell slowing the spread of infection.

Currently, maraviroc is used only to slow the progression of HIV disease in people already infected with the virus and not as a preventive measure. Additionally, this drug is not effective against strains of HIV-1 that utilize CXCR4, another co-receptor.

8.1 Fact Check

1. What is the relationship between viral attachment proteins and host cell receptors?
2. What is meant by the term "host range"?
3. How can antibodies and antiviral drugs play a role in blocking viral attachment to host cells?

8.2 Viral entry and uncoating

How do viruses enter cells and uncoat?

Binding to an appropriate host cell represents only the first step in viral replication. After binding, the virus, or, at the very least, its genome, must enter the host cell. As we discussed in Chapters 2, 3, and 4, though, the plasma membranes of bacteria, eukarya, and archaeons provide these cells with a fairly tight casing, effectively separating the outside of the cell from the inside. Even small molecules cannot freely pass through the membranes; something as large as a virus or a viral genome certainly will not cross this barrier. So, how do viruses or their genomes get inside cells? Actually, several different mechanisms for entry have been discovered. As might be expected, non-enveloped viruses and enveloped viruses display very different strategies. More surprisingly, bacteriophages do not enter their host cells at all. We'll investigate this apparent contradiction shortly.

Entry of enveloped viruses

Consider the entry of an enveloped virus into a host cell. Ultimately, the viral genome must become accessible to assorted host cell molecules. Even after the virus binds to the surface of the cell, though, two lipid bilayers—the plasma membrane and the viral envelope—separate the viral genome from the interior of the cell. For the genome to enter the cell, then, these membranes must fuse. Some mammalian viruses, like the influenza viruses, undergo **endocytosis**. During the process, the viral envelope fuses with the endocytic vesicle membrane. Other mammalian viruses, like the retrovirus HIV, fuse at the plasma membrane. Almost always, a viral **fusion peptide**, a short string of hydrophobic amino acids, mediates the actual fusion event.

With influenza virus, the initial interaction between the viral attachment protein, or hemagglutinin (HA), and the cellular receptor (sialic acid-containing glycoproteins) triggers receptor-mediated endocytosis, resulting in the formation of an endocytic vesicle, or endosome, containing the virus. Normal cellular processes result in the acidification of the vesicle. The resulting low pH of the endosome triggers a conformational change in HA, which exposes a fusion peptide, previously structurally hidden within the HA molecule. This fusion peptide then facilitates the fusion of the viral envelope and the endosomal membrane, thereby allowing the nucleocapsid to enter the cytoplasm **(Figure 8.5)**. Because of its dependence on the low pH of the endosome, influenza entry often is referred to as being pH dependent.

CONNECTION As we discussed in **Section 5.1**, the influenza virus genome consists of eight segments of single-stranded RNA, each of which is enclosed within its own helical nucleocapsid. In **Section 22.4**, we will explore how the segmented nature of the influenza virus genome contributes to its virulence.

When HIV binds to a host cell, through the specific interaction between gp120, the viral attachment protein, and CD4, the cellular receptor, a conformational change occurs in gp120 that allows it to bind to a co-receptor **(Figure 8.6)**. This binding event causes additional conformational changes, which ultimately expose the fusion peptide on another surface protein, gp41. This exposed fusion domain then facilitates fusion between the viral envelope and the plasma membrane, thereby allowing the viral genome to enter the cell.

As we did in the previous section, let's end this section by asking an obvious question. If the infection of cells by most enveloped viruses generally requires a membrane fusion event, then shouldn't it be possible to prevent infection by preventing this fusion event? Again, the answer is yes. Again, we can look to HIV for an example. Approved by the FDA in 2003, Fuzeon was the first antiretroviral drug designed to block fusion. It functions by binding to gp41 and preventing the folding of this molecule, thereby effectively blocking fusion **(Figure 8.7)**. Like maraviroc, it is currently used to slow the progression of HIV disease in people infected with the virus.

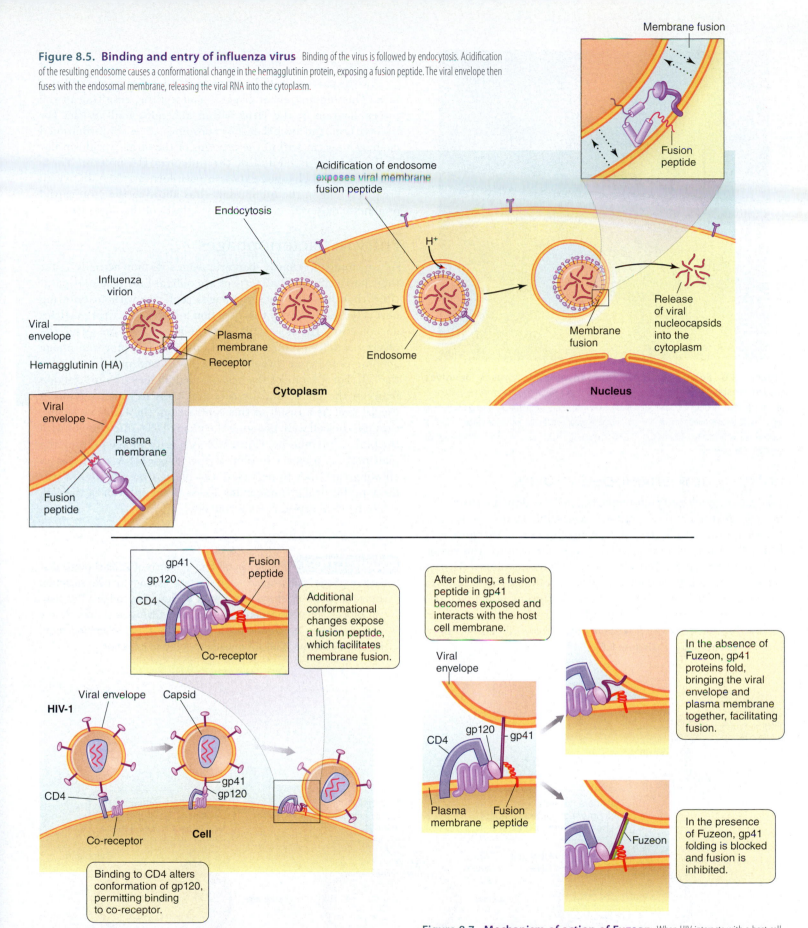

Figure 8.5. Binding and entry of influenza virus Binding of the virus is followed by endocytosis. Acidification of the resulting endosome causes a conformational change in the hemagglutinin protein, exposing a fusion peptide. The viral envelope then fuses with the endosomal membrane, releasing the viral RNA into the cytoplasm.

Membrane fusion

Fusion peptide

Acidification of endosome exposes viral membrane fusion peptide

Endocytosis

H+

Influenza virion

Viral envelope

Plasma membrane

Receptor

Hemagglutinin (HA)

Endosome

Membrane fusion

Release of viral nucleocapsids into the cytoplasm

Cytoplasm

Nucleus

Viral envelope

Plasma membrane

Fusion peptide

gp41
gp120
CD4
Fusion peptide

Co-receptor

Additional conformational changes expose a fusion peptide, which facilitates membrane fusion.

After binding, a fusion peptide in gp41 becomes exposed and interacts with the host cell membrane.

Viral envelope

In the absence of Fuzeon, gp41 proteins fold, bringing the viral envelope and plasma membrane together, facilitating fusion.

HIV-1

Viral envelope

Capsid

CD4

gp41
gp120

gp120
gp41

CD4

Co-receptor

Cell

Plasma membrane

Fusion peptide

Binding to CD4 alters conformation of gp120, permitting binding to co-receptor.

In the presence of Fuzeon, gp41 folding is blocked and fusion is inhibited.

Fuzeon

Figure 8.6. HIV entry Entry of HIV requires the binding of HIV to two receptor molecules on the host cell: CD4 and a co-receptor. The initial interaction between the viral polypeptide gp120 and the primary receptor CD4 facilitates a subsequent interaction between the virus and a co-receptor. This interaction then facilitates membrane fusion, which is mediated by the viral protein gp41.

Figure 8.7. Mechanism of action of Fuzeon When HIV interacts with a host cell, a conformational change occurs in the viral gp120/gp41 complex. A fusion peptide in gp41 becomes exposed and interacts with the host cell plasma membrane. Folding of gp41 then brings together the viral envelope and plasma membrane, facilitating fusion. Fuzeon binds to gp41, preventing the conformational change and fusion.

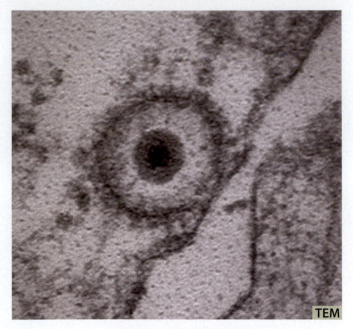

Figure 8.8. Entry of most non-enveloped viruses involves endocytosis Viral binding to a receptor on the cell surface triggers receptor-mediated endocytosis as with this reovirus. Generally, acidification of the endosome causes conformational changes in the viral capsid that facilitates transfer of the viral capsid or viral genome from the endosome to the cytoplasm. *Source*: From M. Maginnis, T. Dermody, et al. 2008. J. Virol. 82(7):3181–3191, Fig. 6b. © 2008, American Society for Microbiology.

Entry of non-enveloped viruses

Like their enveloped counterparts, non-enveloped viruses ultimately must make their genomes accessible to the replication, transcription, and/or translation machinery of the host cell. Entry for these viruses, though, does not require the fusion of two membranes. Rather, entry of non-enveloped viruses requires the direct penetration of a cellular membrane by the virus particle or genome. Again, different non-enveloped viruses achieve this step through a variety of mechanisms. Generally, non-enveloped viruses enter host cells via receptor-mediated endocytosis **(Figure 8.8)**.

A virus particle initially binds to its receptor on the cell surface. This binding event triggers endocytosis, resulting in the internalization of the virus within an endosomal vesicle. For most viruses, the low pH of the endosome causes conformational changes to occur within the proteins of the viral capsid, similar to the changes exhibited by the influenza HA protein. These conformational changes expose a pore-forming domain, which creates a hole in the membrane, thus allowing the viral capsid or genetic material to enter the cytoplasm of the host cell.

Entry of bacteriophages

As we noted previously, bacteriophages do not actually enter their host cells. So, if all viruses are obligate intracellular parasites, then how can bacteriophages replicate without entering host cells? Let's clarify this question a little. The viral particles do not enter the host cells, but their genomes do. After the phage particle interacts with its host cell receptor, the phage genome is injected through the cell wall and membrane, into the cell. For T4, one of the best-studied phages, the interaction between the phage and its receptor leads to a contraction of the T4 tail. As a result of this contraction, the tail core pushes through the cell wall, bringing the phage DNA into contact with the plasma membrane **(Figure 8.9)**. A specific phage protein, the pilot protein, appears to bind to the phage DNA and assist in allowing the DNA to penetrate the plasma membrane. In this fashion, the viral genome gains access to the interior of the cell, while the viral capsid remains outside.

CONNECTION A study published in March of 2009 showed that bacteriophage tail proteins and certain Gram-negative secretion apparatus proteins share a common evolutionary origin. The general process of translocating proteins and DNA across the plasma membrane may be conserved. In **Section 2.4**, we examined more thoroughly the mechanisms of bacterial protein secretion.

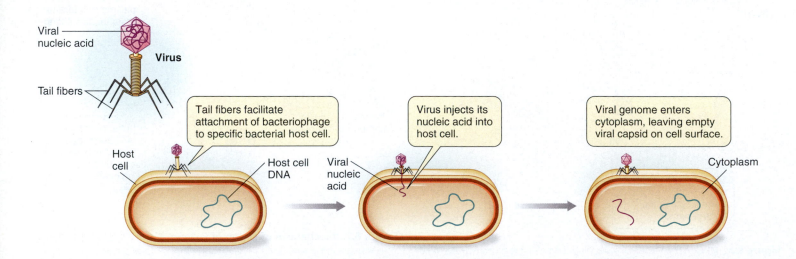

Figure 8.9. Entry of bacteriophages Bacteriophages inject the viral genome into a host cell, but capsids do not enter cells.

Entry of plant viruses

Before leaving the topic of viral entry, let's briefly consider plant viruses. Like bacterial cells, plant cells generally contain a sturdy cell wall. Additionally, a cuticle, or thick waxy covering, usually protects plant leaves and new shoots. Like bacteriophages, then, plant viruses are faced with a major obstacle when entering a host cell. Plant viruses, though, do not inject their genomes into cells as do the bacteriophages. Rather, plant viruses often rely on the help of an outside force. When insects feed on a plant, for instance, they typically damage the cuticle and plant cell wall **(Figure 8.10)**.

When feeding on an infected plant, insects may acquire viruses. When the insect feeds on another plant, it can transmit these particles, thereby infecting a new plant. Once some cells within a plant are infected in this manner, the virus can spread to other cells within that plant through cytoplasmic connections, or plasmodesmata, between the cells.

© Nigel Cattlin/Alamy

Figure 8.10. Entry of plant viruses Damage to plant surface cuticle and cellulose wall is required for viruses to enter plant cell cytoplasm. Whiteflies shown here transmit several different plant viruses, including tomato yellow leaf curl virus.

Viral uncoating

Before viral replication can begin, the virus not only must enter the cell, but also must make its genetic material accessible to at least some components of the host cell. For bacteriophages, this actually represents a false problem. Because most bacteriophages inject their genome directly into the host cell, the viral genome is immediately accessible. Replication and/or transcription can begin as soon as the genetic material enters.

For most mammalian viruses, some form of **uncoating**, or viral capsid disassembly, must occur to allow the viral genome to interact with the required host cell machinery. This disassembly often involves conformational changes in or proteolytic processing of capsid proteins. The capsid of poliovirus, for instance, undergoes an irreversible change after binding to its receptor, the poliovirus receptor (PVR). The native virion, or 160S particle, forms an altered, or 135S, particle soon after binding to the receptor. As part of this conversion, lipophilic portions of VP4, an internal capsid protein, and VP1, another capsid protein, become exposed. These domains probably interact with the host cell membrane, forming a channel in the membrane through which the viral RNA is extruded **(Figure 8.11)**. In 2007, researchers provided evidence that this series of events occurs very near the surface of the cell, either in discrete vesicles or plasma membrane invaginations.

Several lines of experimental evidence support this model for poliovirus uncoating. Most notably, it has been demonstrated that altered, or A, particles can induce transient pore formation in artificial lipid bilayers. This result supports the hypothesis that the altered particles may form pores in a native lipid bilayer, through which the viral RNA then can enter the host cell cytoplasm. It still is unclear, though, if the viral RNA enters the cell at the plasma membrane, or if the viral particle enters the cell via endocytosis and the RNA enters the cytoplasm by passing through the endosomal membrane. For many

viruses that enter cells via receptor-mediated endocytosis, drugs that prevent the acidification of the endosomes also prevent viral replication. Such drugs, however, do not inhibit poliovirus replication. Additionally, blocking endocytosis itself does not inhibit poliovirus replication. These findings suggest that poliovirus RNA may enter the cytoplasm through the plasma membrane or, possibly, through either the plasma membrane or an endosomal membrane.

8.2 Fact Check

1. What is membrane fusion and how is it involved in enveloped virus entry?
2. What is receptor-mediated endocytosis and how is it involved in virus entry?
3. Explain why viruses of mammals generally undergo some form of uncoating, but viruses of bacteria usually do not.

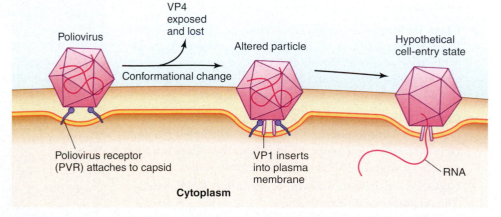

Figure 8.11. Disassembly of poliovirus is necessary for subsequent replication After binding to the cellular receptor, poliovirus particles undergo a conformational change resulting in the loss of VP4, a capsid protein. This conformational change allows another capsid protein, VP1, to interact with the host cell's plasma membrane, presumably creating a pore through the membrane. The viral RNA leaves the partially disassembled capsid and enters the cytoplasm through the VP1-induced channel.

How do viruses replicate?

As we have discussed previously, some viruses are enveloped and some are non-enveloped. Some viruses exhibit icosahedral symmetry and some exhibit helical symmetry. Some viruses contain DNA genomes and some contain RNA genomes. These genomes may be circular or linear. Furthermore, the nucleic acid itself may be single-stranded or double-stranded. A few known viruses, most notably hepatitis B virus, even have genomes that are partly double-stranded and partly single-stranded. These fundamental genetic differences necessitate fundamentally different replication strategies.

Before we look at specific aspects of these replication strategies, let's ask a fairly simple question: For a virus to replicate, what needs to occur? We can provide a simple, though not overly useful, answer to this question: More copies of the genome need to be made, and multiple copies of the viral proteins need to be made and assembled. So let's ask another, more complex question: How does this occur? Several events are necessary. Genome replication must occur. The specific details of this process will differ depending on the type of genome possessed by a given virus. Additionally, mRNA that can be recognized and translated by the host cell must be produced. Again, how this mRNA is produced will depend on the type of genome possessed by the virus. These processes together lead to the production of new virus particles, a process collectively referred to as viral replication.

As we mentioned in Chapter 5, viruses can be classified based on how they replicate or, more precisely, how they generate mRNA. In the Baltimore classification scheme presented here, all viruses are categorized into seven distinct classes:

- **Class I: Double-stranded DNA viruses.** DNA serves as a template for synthesis of mRNA. Genome replication usually occurs in the nucleus of the host cell and uses the host DNA-dependent DNA polymerase. Some viruses in this class, like the poxviruses, replicate in the cytoplasm, using viral enzymes.

- **Class II: Single-stranded DNA viruses.** Messenger RNA forms from a double-stranded DNA intermediate. Genome replication occurs as the single-stranded genome becomes converted to double-stranded DNA, which then serves as a template for the production of more genomes.

- **Class III: Double-stranded RNA viruses.** Messenger RNA is generated from one strand of the double-stranded genome. Genome replication generally occurs in the cytoplasm and requires a viral RNA-dependent RNA polymerase.

- **Class IV: Positive-sense single-stranded RNA viruses.** The genome can be recognized by the host ribosomes and thus functions directly as mRNA; it is referred to as positive-sense (same as mRNA). Genome replication usually occurs in the cytoplasm and requires a viral RNA-dependent RNA polymerase.

- **Class V: Negative-sense single-stranded RNA viruses.** The genome serves as the template for formation of mRNA and is referred to as negative-sense (complementary to mRNA). A viral RNA-dependent RNA polymerase produces the mRNA. This enzyme also is required for genome replication, which usually occurs in the cytoplasm.

- **Class VI: Single-stranded RNA viruses that utilize reverse transcriptase.** The viral enzyme reverse transcriptase, an RNA-dependent DNA polymerase, converts the ssRNA genome into dsDNA, which then becomes integrated into the host genome. Messenger RNA and new genomic RNA is generated from this integrated DNA.

- **Class VII: Double-stranded DNA viruses that utilize reverse transcriptase.** The genome serves as a template for the generation of mRNA and full-length RNA molecules. A viral reverse transcriptase generates genomic DNA from this RNA intermediate.

Figure 8.12 provides a basic overview of the replication strategies employed by these seven classes of viruses.

Microbes in Focus 8.2
HUMAN PAPILLOMAVIRUSES (HPVs) AND CERVICAL CANCER

Habitat: Humans

Description: Double-stranded DNA, non-enveloped virus about 50–55 nm in diameter with icosahedral symmetry in the *Papillomaviridae* family. Approximately 130 different strains have been identified.

Key Features: Over 40 strains of HPV are sexually transmitted and infect the genital areas of men and women. In the United States, roughly half of all sexually active men and women will become infected with HPV at some point in their life. Several strains of HPV lead to the development of genital warts, and a few strains have been associated with the development of cervical, vaginal, penile, anal, throat, and mouth cancers. In 2006, the FDA approved for use in girls and women Gardasil, a vaccine effective against the most common strains of HPV associated with cervical cancer. This vaccine was approved for use in boys and men in 2009. A second vaccine effective against HPV, Cervarix, received FDA approval for use in girls and women in 2009.

TEM

© Kwangshin Kim/Photo Researchers, Inc.

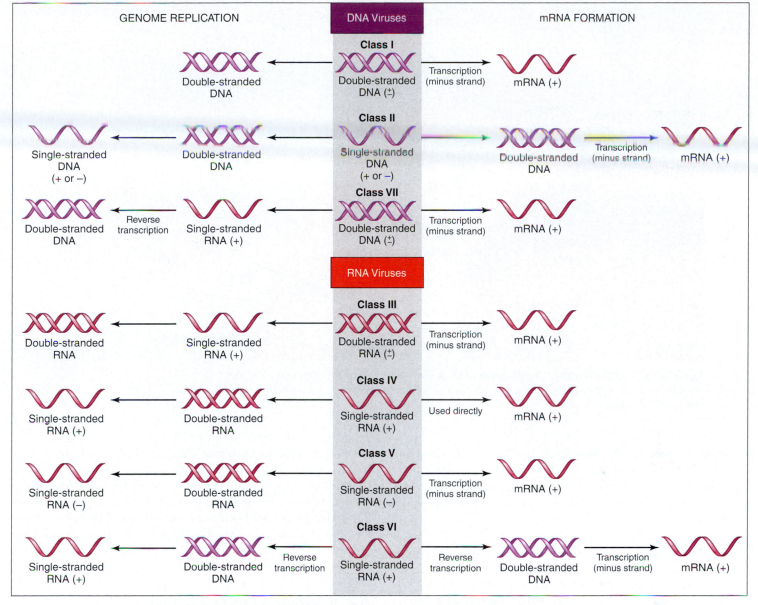

Figure 8.12. Classification of viruses All viruses must both replicate their genome (shown on the *left*) and produce mRNA that functions in the host cell (shown on the *right*). The Baltimore classification scheme is based primarily on how the mRNA of different viruses is generated.

In this section, we will examine more closely the replication of a few selected mammalian DNA and RNA viruses. We also will examine some of the interesting features of bacteriophage replication.

Replication of eukaryal DNA viruses

The genetic material of DNA viruses resembles the genetic material of their host cells. As a result, replication of the viral genome and transcription of viral genes occur much like they do in the host cell. For most DNA viruses, some portion of the viral particle migrates to the nucleus. Here, viral genes are transcribed and the viral genome is replicated **(Figure 8.13**, see next page).

For SV40, a well-studied polyomavirus with a circular DNA genome, the viral protein large T antigen unwinds the viral DNA, while host primases and the host DNA polymerase replicate

DNA virus replication ANIMATION

the viral genome in a bidirectional manner (refer back to Chapter 7 for a refresher on DNA replication). Numerous studies have shown that the large T antigen also interacts with cell cycle regulatory proteins and drives the cell toward DNA replication. By priming the cell for DNA replication, the large T antigen also facilitates viral replication. The cell begins producing materials needed for cellular DNA replication—DNA polymerase molecules and nucleotides, for instance. The virus needs these same materials for viral replication. Thus, by causing the cell to begin DNA replication, the virus ensures that adequate supplies for viral replication will be present. In some cases, this viral "trick" can result in the immortalization of cells in culture. Human papillomaviruses (HPVs) also contain specific proteins that drive the cell toward DNA replication. Certain strains of these viruses have been strongly associated with cervical cancers **(Microbes in Focus 8.2)**. For polyomaviruses

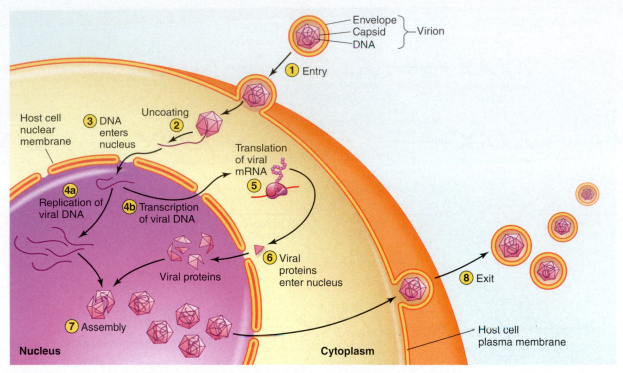

Figure 8.13. Process Diagram: Replication of DNA viruses Generally, replication of DNA viruses occurs in the nucleus. Usually, the host cell DNA polymerase is used to produce new copies of the viral genome and the host cell RNA polymerase is used to produce viral mRNA molecules.

and papillomaviruses, the host RNA polymerase II is responsible for transcription of the viral genes. It appears that high molecular weight precursor RNAs are transcribed and then spliced to form individual mRNA molecules that then are translated.

CONNECTION A number of viruses have been linked to various types of cancer. In most cases, the viruses alter normal cell cycle controls, thereby leading to altered cell replication. More about the link between viruses and cancer will be discussed in **Section 22.3**.

While most eukaryal DNA viruses replicate in the nucleus, using the host cell DNA polymerase, there are, of course, exceptions to every rule. Poxviruses represent a very interesting exception to this general rule. Poxviruses have a double-stranded DNA genome, but their DNA replication and transcription occur entirely within the cytoplasm of the host cell. Thus, the poxviruses cannot rely on any host proteins normally found in the nucleus. Let's look at this problem another way. If the poxviruses cannot rely on any host proteins normally found in the nucleus, then the viral genome must encode all the proteins needed for genomic replication and transcription. In fact, the relatively large poxvirus genome does encode all of these necessary enzymes and associated factors. Merely possessing genes that encode all the proteins needed for replication and transcription, however, is not enough. How would the necessary viral polymerases get produced initially? To solve this apparent problem, the virion contains an RNA polymerase and assorted transcription factors. This

polymerase produces eukaryal-like mRNAs, which give rise to the necessary viral proteins **(Figure 8.14)**.

Initially, "early" genes are transcribed and translated, producing the DNA polymerase and other proteins necessary for replication. "Late" genes encode structural proteins needed for capsid formation. As we'll see in the next section, many RNA viruses follow a similar strategy and bring essential enzymes into the host cell with the genome.

Replication of RNA viruses

For all RNA viruses, the replication of an RNA genome presents an interesting problem. Through some mechanism, exact RNA copies of an RNA template must be made. All cells can make DNA copies of DNA templates. No known cells routinely make RNA copies of RNA. So, all RNA viruses must employ a virally encoded RNA polymerase. For most RNA viruses, this polymerase is an **RNA-dependent RNA polymerase**. In other words, it produces RNA from an RNA template. For many RNA viruses, this RNA-dependent RNA polymerase manufactures not only genomic RNA, but also messenger RNA that can be recognized and translated by the host cell's ribosomes. Thus, replication of RNA viruses that infect eukarya occurs independent of host proteins found in the nucleus. It follows, then, that replication of these viruses can (and, in fact, usually does) occur in the cytoplasm. To understand this process a bit more, let's examine three distinct types of RNA viruses: positive-sense single-stranded RNA viruses, negative-sense single-stranded RNA viruses, and retroviruses.

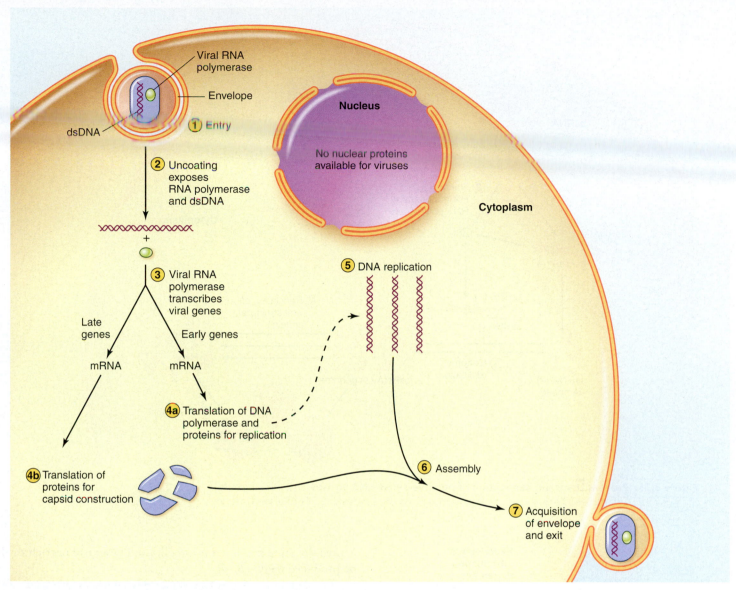

Figure 8.14. Process Diagram: Replication of poxviruses Unlike most DNA viruses, poxviruses replicate entirely within the cytoplasm of host cells. To facilitate this process, the virion brings a viral RNA polymerase and several transcription factors with it into the cell. Viral mRNAs are generated in the cytoplasm and translated by host cell ribosomes. One of the resulting viral proteins is a viral DNA polymerase that then replicates the viral genome.

Positive-Sense Single-stranded RNA Viruses

For some single-stranded RNA viruses, like poliovirus, the genomic RNA "looks" like messenger RNA. In other words, the genomic RNA can be translated directly by the host cell translation machinery. Thus, for these single-stranded **positive-sense RNA**, or **ssRNA (+)**, viruses, the initial production of new viral proteins can occur in the absence of any existing viral proteins within the host cell. As we've seen earlier, production of new viral genomes in most RNA viruses (retroviruses represent an interesting exception) requires a virally encoded RNA-dependent RNA polymerase. This enzyme, though, can be produced

Positive-sense RNA virus replication
ANIMATION

from the genome of an ssRNA (+) virus after the virus has entered its host cell. The virus does not need to bring this enzyme with it into the cell. Once the RNA-dependent RNA polymerase is produced, then it can produce new copies of the viral genome, using the original genome as a template (**Figure 8.15**, see next page).

For researchers, a positive-sense RNA genome presents many exciting possibilities. Because the viral genome "looks" like mRNA, it can be translated directly by the host cell, resulting in the production of viral proteins. These resulting viral proteins, then, can replicate the viral genome and new viral particles can be assembled. Let's examine the implications of this process. If a researcher isolates the RNA from poliovirus and

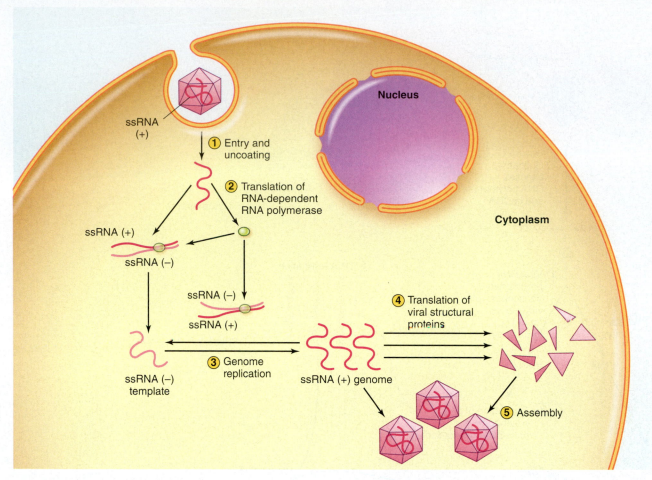

Figure 8.15. Process Diagram: Replication of positive-sense RNA viruses Because the viral genome is positive-sense, it can be translated directly by the host cell. The necessary viral polymerase is not brought into the cell by the virion but is synthesized inside the cell.

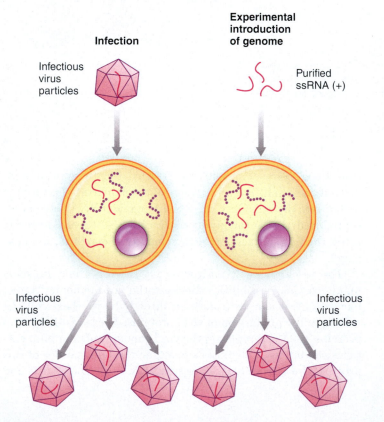

artificially injects it into a host cell, then new viral particles will be produced (**Figure 8.16**).

The genomes of positive-sense RNA viruses often are referred to as **infectious RNA**; the RNA itself can lead to the production of new viral particles. More importantly, researchers observed that if the genomic RNA is converted into double-stranded DNA and cloned into a plasmid, then the plasmid, too, is infectious. Using standard molecular biology techniques, researchers easily can mutate specific bases within such a plasmid, thus effectively mutating the viral genome. By introducing the mutated plasmid into host cells, virologists can determine the effects of the specific mutations on the virus. These types of studies have provided many insights into viral replication and pathogenesis.

CONNECTION As we will see in **Section 9.2**, plasmids are common components of the bacterial genome. Additionally, as we will see in Chapter 12, plasmids are used extensively in biotechnology.

Figure 8.16. Infectious RNA When the RNA genome from a positive-sense RNA virus is isolated and then introduced into a permissive host cell, the host cell machinery can translate the viral RNA. Viral proteins, including the viral RNA-dependent RNA polymerase, are produced. This enzyme can produce new copies of the viral genome and new infectious viral particles then assemble.

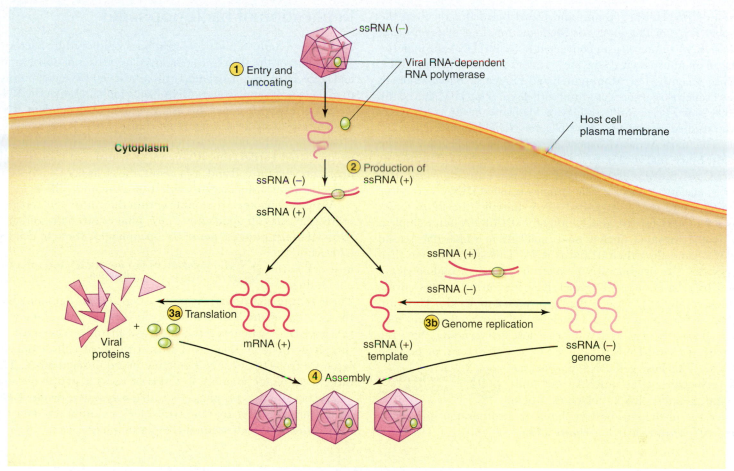

Figure 8.17. Process Diagram: Replication of negative-sense RNA viruses The basic replication strategy employed by negative-sense RNA viruses is very similar to that seen with positive-sense RNA viruses. The major difference, though, is that the negative-sense RNA genome cannot be translated directly by the host cell. A complementary copy first must be made. Because cellular enzymes cannot make an RNA copy of RNA, the virus must bring an RNA-dependent RNA polymerase with it into the host cell.

Negative-Sense Single-stranded RNA Viruses

In contrast to the genomes of positive-sense single-stranded RNA viruses, the genomes of single-stranded **negative-sense RNA**, or **ssRNA (−)**, viruses are complementary to the viral messenger RNA. In other words, complementary copies of the viral genes must be produced that can be recognized by the host cell translational machinery. Because the viral genome is RNA, this genomic RNA must serve as a template for the production of messenger RNA. As we've seen earlier, production of RNA from RNA requires a viral RNA-dependent RNA polymerase. For ssRNA (−) viruses, then, this polymerase must enter the cell with the viral genome **(Figure 8.17)**.

Retroviruses

Today, the word "retrovirus" has become nearly synonymous with HIV. Indeed, since HIV was determined to be the causative agent of AIDS in 1983, HIV has become the most famous retrovirus. It's important to note, however, that the study of retroviruses has a much longer, and quite interesting, history. Although he didn't know it at the time, Peyton Rous was studying a retrovirus back in 1911 when he described the ability of a virus to cause sarcomas (connective tissue tumors) in chickens. Rous received the Nobel Prize in Physiology or Medicine in 1966 for this discovery of an oncogenic virus, a virus that causes cancer. Until the AIDS pandemic was recognized in the early 1980s, retroviruses

were studied primarily because of their ability to form tumors. In 1976, Harold Varmus and J. Michael Bishop, working together at UCSF, published a series of papers detailing their concept of how retroviruses can convert proto-oncogenes, normal genes typically encoding proteins involved in cell division, into oncogenes, mutated proto-oncogenes associated with the development of cancer. They received the Nobel Prize in Physiology or Medicine in 1989 **(Figure 8.18)**.

Figure 8.18. Nobel Prize winners J. Michael Bishop (*left*) and Harold Varmus (*right*) received the Nobel Prize in Physiology or Medicine in 1989 for their work with Rous sarcoma virus, an oncogenic retrovirus.

In 1975, Howard Temin and David Baltimore received the Nobel Prize in Physiology or Medicine for their discovery that these RNA tumor viruses converted RNA into DNA through the use of an unusual viral enzyme—**reverse transcriptase (RT)**. More recently, Luc Montagnier and Françoise Barré-Sinoussi, the French researchers credited with discovering HIV, shared the 2008 Nobel Prize in Physiology or Medicine. Seven researchers who studied retroviruses have become Nobel laureates.

The most unusual aspect of retrovirus replication is the multi-functional viral reverse transcriptase, an **RNA-dependent DNA polymerase** (see **Mini-Paper**). Upon entering a cell, this viral enzyme converts the single-stranded RNA retrovirus genome into double-stranded DNA. Initially, the RNA serves as a template for the generation of a DNA:RNA heteroduplex, a double-stranded nucleic acid molecule in which one strand is DNA and the other strand is RNA. This same enzyme then degrades the original RNA molecule, using an endonuclease activity referred to as RNase H. RT next converts the resulting single-stranded DNA to a double-stranded molecule. This viral DNA is transported to the nucleus and another viral enzyme, **integrase**, integrates it into the host cell genome. Now the integrated viral DNA is referred to as a **provirus**, or proviral DNA. The host cell RNA polymerase II generates viral mRNA and full-length genomic RNA molecules, which

HIV replication ANIMATION

leave the nucleus to be translated or incorporated into newly assembled viral particles **(Figure 8.19)**.

Replication of bacteriophages

To a large extent, replication strategies employed by bacteriophages mirror the strategies employed by viruses of eukarya. By and large, DNA bacteriophages utilize host cell DNA polymerase to make new genome copies. RNA bacteriophages, like the RNA viruses of eukarya, generally use a viral RNA-dependent RNA polymerase to make new genome copies. Most RNA bacteriophages currently known have a positive-sense RNA genome. As we discussed earlier, this RNA can be translated immediately upon entering the cell and, as a result, the virus does not need to bring the replication enzyme with it into the host cell.

Viruses of bacteria assemble within their host cell and generally lyse this cell upon exit. These phages often are referred to as **virulent phages**. Some bacteriophages, though, can exist within the host cell without lysing it.

Replication of temperate bacteriophages ANIMATION

These **temperate phages** usually integrate their genome into the genome of the host cell, where it is replicated along with the host DNA whenever the host cell divides. Alternatively, the phage genome may exist within the host cell cytoplasm separate from the host cell genome in a quiescent state. The integrated phage genome is referred to as a **prophage** and a cell containing a prophage is called a **lysogen** (Figure 8.20). To explore this process more fully, let's examine lambda (λ) phage, the most well studied temperate phage.

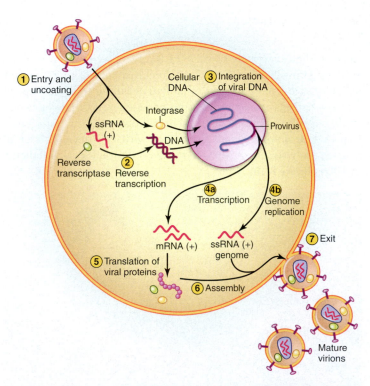

Figure 8.19. Process Diagram: Replication of retroviruses The single-stranded RNA genome is converted to double-stranded DNA by the viral reverse transcriptase. This viral DNA genome then moves to the nucleus of the host cell, where the viral integrase inserts it into the cellular DNA. The host cell RNA polymerase transcribes the integrated viral genome, producing mRNA molecules that are translated to produce viral proteins and new viral RNA genomes.

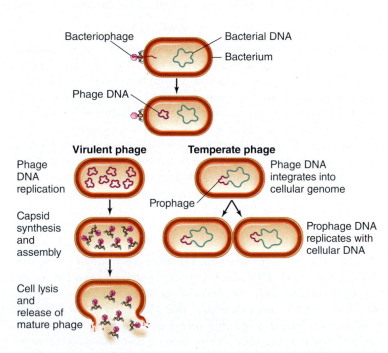

Figure 8.20. Virulent and temperate phages Virulent, or lytic, phages replicate within the host cell, eventually lysing it. Most temperate phages insert their genomes into the host cell DNA, forming prophages.

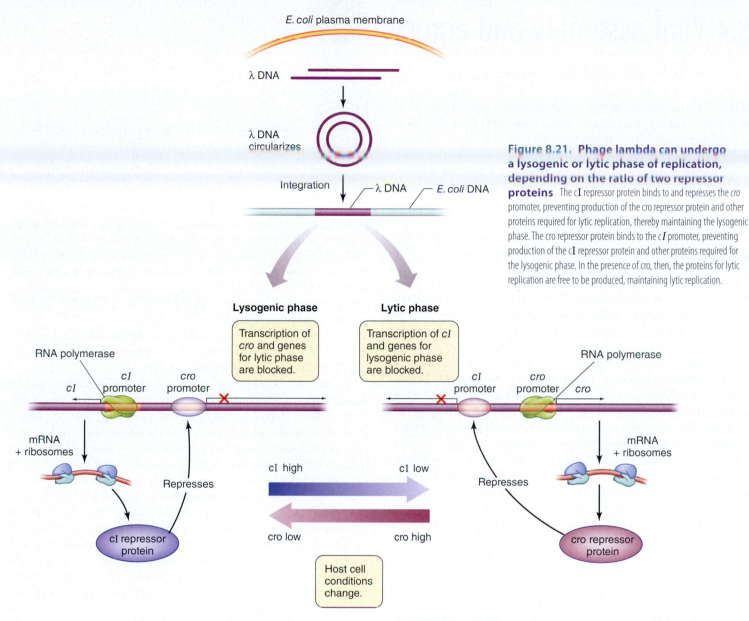

E. coli plasma membrane

λ DNA

λ DNA circularizes

Integration — λ DNA — E. coli DNA

Figure 8.21. Phage lambda can undergo a lysogenic or lytic phase of replication, depending on the ratio of two repressor proteins The cI repressor protein binds to and represses the *cro* promoter, preventing production of the cro repressor protein and other proteins required for lytic replication, thereby maintaining the lysogenic phase. The cro repressor protein binds to the *cI* promoter, preventing production of the cI repressor protein and other proteins required for the lysogenic phase. In the presence of cro, then, the proteins for lytic replication are free to be produced, maintaining lytic replication.

Lysogenic phase

Transcription of *cro* and genes for lytic phase are blocked.

RNA polymerase

cI   *cI* promoter   *cro* promoter

mRNA + ribosomes

Represses

cI repressor protein

Lytic phase

Transcription of *cI* and genes for lysogenic phase are blocked.

RNA polymerase

cI promoter   *cro* promoter   cro

mRNA + ribosomes

Represses

cro repressor protein

cI high — cI low

cro low — cro high

Host cell conditions change.

When the linear λ DNA enters a host cell, it almost immediately circularizes and transcription of viral genes begins. A specific viral enzyme, integrase, then facilitates the integration of the viral genome into the host genome through a recombination process. A virally encoded repressor protein, cI, binds to promoters of certain genes, including *cro*, required for lytic replication and prevents their transcription, thereby maintaining the lysogenic state. Under certain circumstances, the cI levels will decline. Usually, the cI levels drop in response to damage to the host cell, perhaps caused by UV light or chemical mutagens. As the levels of cI drop, another virally encoded repressor protein, cro, binds to and blocks the *cI* promoter, causing an even greater decline in the concentration of cI. Freed from cI repression, other phage genes such as *cro* now can be expressed, leading to lytic replication of the phage **(Figure 8.21)**. To summarize, when levels of the cI repressor protein are high, genes leading to the lytic phase are blocked. When levels of the cro repressor protein are high, genes leading to the lysogenic phase are blocked. Thus, temperate phages like λ can toggle between lytic and lysogenic replication in response to the overall health of their host cell.

CONNECTION As we have just discussed, some bacteriophages can alter their replication strategies within a host cell. Bacteria also have their own means of regulating, or completely blocking, bacteriophage replication. These bacterial defense systems are discussed in **Section 18.2**.

8.3 Fact Check

1. What are the seven classes of viruses?
2. How and where do many DNA viruses such as SV40 replicate in the host cell?
3. Contrast the replication strategy of positive-sense RNA viruses with that of negative-sense RNA viruses.
4. How are the enzymes reverse transcriptase and integrase involved in retrovirus replication in host cells?
5. Describe the lytic and lysogenic replication cycles of bacteriophages.

8.4 Viral assembly and egress

How do virus particles form and exit cells?

Making many copies of the viral genome and all the viral proteins is not the end of the story. New viral particles must be assembled and then leave the infected cell. How do these processes occur? Again, we can provide a very cursory answer: newly formed viral proteins aggregate around a newly produced viral genome. This newly formed viral particle then leaves the cell. Of course, the actual processes employed by viruses are a bit more complex.

Assembly

For all viruses, assembly of new viral particles involves a well-orchestrated set of events. Most often, the individual capsid proteins undergo a self-assembly process. Specific chemical interactions cause individual polypeptides to bind to each other, forming more complex structures like the capsomeres, or faces of the capsid (in icosahedral viruses), and, eventually, the capsid. Alternatively, assembly may require the function of specific viral proteins that then do not reside within the mature viral particle. In some cases, a non-infectious, precursor form of the viral particle may assemble initially. Certain proteins of this precursor virion then undergo proteolytic processing. Cellular or viral enzymes cleave specific proteins in the precursor virion, causing the virion to undergo some conformational changes. These changes in shape result in the formation of the mature, infectious virion. Not surprisingly, disruption of this proteolytic processing disrupts the formation of infectious viruses. The antiretroviral **protease inhibitor** drugs, agents that block the activity of a proteolytic processing enzyme required for the formation of infectious HIV particles, decrease HIV replication in exactly this manner.

Viral genomes seem to get into the viral particles in one of two ways. In some cases, the proteins assemble around the viral nucleic acid. For these viruses, the genomic RNA or DNA most likely possesses a sequence-specific assembly motif, or **packaging sequence**. The viral proteins interact with this sequence, thereby causing the capsid to coalesce around the nucleic acid. In other cases, "empty" capsids are partially assembled first and then the viral genome is inserted into these capsids. Detailed studies of tobacco mosaic virus (TMV), a virus exhibiting helical symmetry (see Figure 5.4), have shown that this virus undergoes the former process. Poliovirus, a virus exhibiting icosahedral symmetry, undergoes the latter process. Initially, the major poliovirus structural proteins VP1, VP3, and VP0 assemble into capsomeres, the basic structural unit of the capsid. Groups of five capsomeres then assemble into pentamers, the faces of the capsid. It is thought that these pentamers then coalesce around viral genomic RNA, forming an immature viral particle. Subsequent cleavage of VP0 into VP2 and VP4 results in the formation of a mature, infectious virion.

Egress

Finally, the newly formed viral particles must leave the cell. For enveloped viruses, this process of egress necessarily involves several steps. First, proteins destined for the viral envelope must be inserted into the cellular membrane. Like the integral membrane proteins of the host cell, these proteins contain a signal sequence that directs them to the appropriate membrane location. Second, the nucleocapsid, formed within the cell, must migrate to the region of the cellular membrane containing the viral envelope proteins. Usually, certain amino acids within the tail region of these transmembrane proteins interact with regions of the nucleocapsid, thereby bringing the nucleocapsid into proximity

with the membrane. Third, budding of the nucleocapsid occurs, thus providing the virus with a lipid bilayer that contains the required envelope proteins **(Figure 8.22)**.

In almost all cases, egress of non-enveloped viruses occurs through lysis of the host cell. This lysis event differs somewhat between bacteriophages and mammalian viruses. Bacteriophages generally produce specific enzymes that actively destroy the host cell's plasma membrane and cell wall. T4, for instance, produces lysozyme, an enzyme that degrades the peptidoglycan present in the cell wall. Non-enveloped mammalian viruses, conversely, rarely produce specific proteins necessary for egress. Rather, the host cell becomes progressively unstable because of the replication of the virus, leading to the disintegration of the cell and release of the new virions. Viruses of plants, as we alluded to earlier, do not lyse the host cell. These viruses travel from one cell to another within a plant through the cytoplasmic connections between cells. Spread to a new plant requires the disruption of the cell wall and cuticle by outside forces, such as insects.

CONNECTION The insertion of viral envelope proteins into the host cell plasma membrane exposes these viral proteins on the surface of infected host cells. These foreign proteins can be recognized by the immune system. Cytotoxic T cells (**Section 20.2**) are immune cells designed to destroy infected cells.

8.4 Fact Check

1. How can protease inhibitor drugs specifically disrupt viral assembly?
2. Describe two ways that viral genomes can become enclosed within capsids during assembly.
3. How is the egress of new virus particles different for an enveloped virus versus a non-enveloped virus? Explain each process.

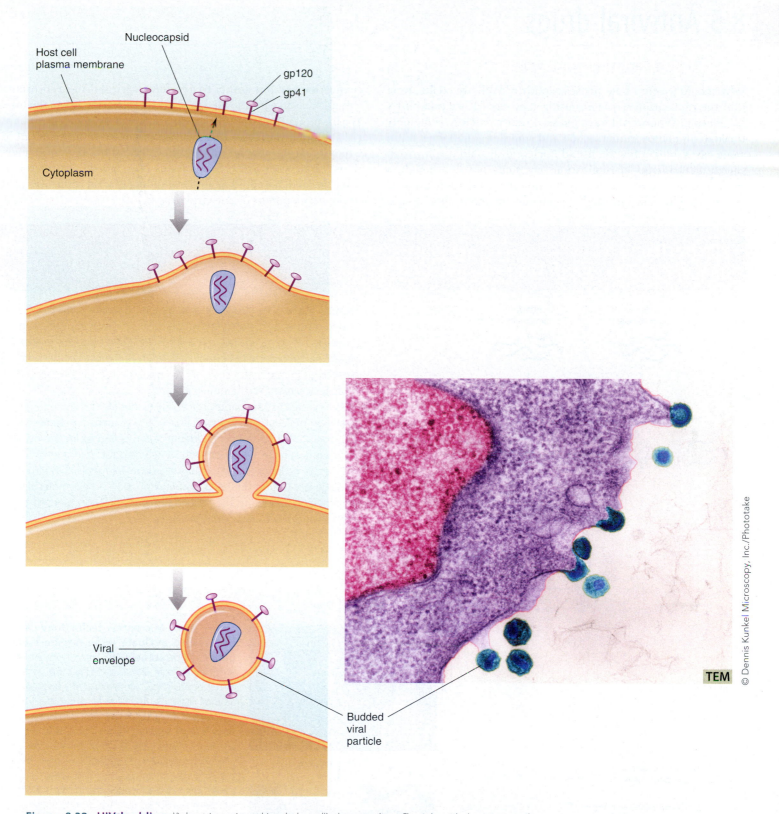

Figure 8.22. HIV budding Viral proteins are inserted into the host cell's plasma membrane. The viral particle then migrates to the membrane and buds off of the cell, thereby acquiring an envelope that contains the viral proteins gp120 and gp41. The electron micrograph shows HIV particles budding from lymphoid tissue.

Host cell plasma membrane

Nucleocapsid

gp120

gp41

Cytoplasm

Viral envelope

Budded viral particle

TEM

© Dennis Kunkel Microscopy, Inc./Phototake

8.5 Antiviral drugs

How do antiviral drugs work?

Why should we care how viruses replicate? Well, we could argue that the replication strategies utilized by viruses are inherently fascinating. Who would have guessed that particles visible only through an electron microscope and containing genomes of, in some cases, only a few thousand nucleotides could display such varied mechanisms of replication? For some people, though, a more utilitarian reason for understanding viral replication is necessary. The relationship between our understanding of viral replication and our ability to develop antiviral drugs represents a major reason why knowledge of viral replication is worthwhile. We have mentioned a few specific antiviral drugs in this chapter and other antiviral drugs will be discussed in more detail in Chapter 24. For now, let's address a couple of fundamental questions. First, what are the attributes of a "good" antimicrobial drug? Second, how does an understanding of viral replication aid in the development of antiviral drugs?

Perspective 8.1
DNA MICROARRAYS AND THE SARS VIRUS

Because of the vast structural and genetic diversity exhibited by viruses, most antivirals are very virus-specific. A given antiviral drug generally is effective against only a limited number of closely related viruses. As our repertoire of antivirals expands, then, it will become increasingly important for clinicians to rapidly and accurately identify the specific viruses infecting their patients.

As indicated in this chapter, our ability to identify viruses depends greatly on molecular techniques. From DNA sequencing studies to PCR to *in situ* hybridization, the identification of viruses increasingly has become a molecular endeavor. Some virologists might even argue that we no longer need to study viruses in animals or cell culture for diagnosis. During the SARS outbreak of 2003, a new molecular technique was added to the list of techniques that can be used to identify viruses: DNA microarray analysis.

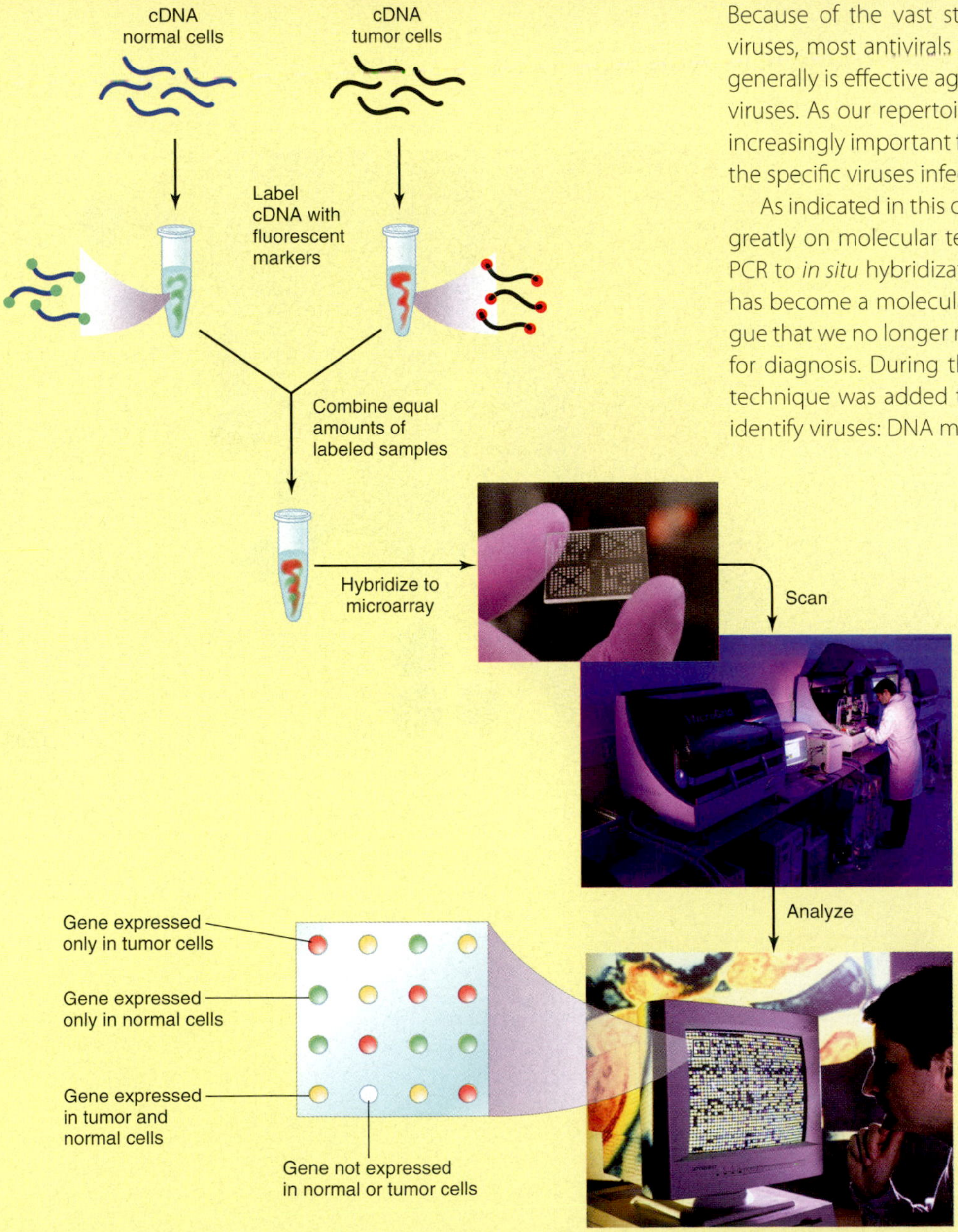

Figure B8.4. Microarray technology can be used to evaluate large numbers of genes simultaneously First, oligonucleotides representing the target DNA are attached to a slide. In this example, oligonucleotides corresponding to every human gene are placed on the slide. Next, the nucleic acids of interest are fluorescently labeled. In this example, cDNA (corresponding to expressed genes), derived from normal cells, is labeled with a green fluorescent marker. cDNA derived from tumor cells is labeled with a red fluorescent marker. These tagged cDNAs then are added to the microarray and the microarray slide is scanned. Using software, researchers can analyze the fluorescence associated with each spot on the slide. If both probes bind (yellow spots), then we know that the gene is expressed in normal and tumor cells. Red spots indicate genes that are only expressed in tumor cells. Green spots indicate genes expressed only in normal cells. (*Top*: © BSIP/Phototake; *middle* and *bottom*: © Wellcome Trust Library/ Custom Medical Stock Photo, Inc.)

Obviously, all antivirals and, in fact, all drugs should meet certain basic criteria. Ideally, they should:

- Be easily administered
- Be easily produced
- Be relatively affordable
- Have few side effects
- Decrease the severity of the disease

For an antimicrobial drug to be effective and have few side effects, it should target some unique attribute of the pathogen. In other words, the drug should inhibit or interfere with some property of the disease-causing agent that does not exist in or is not essential to the host organism. For viruses, developing a drug that targets a unique attribute of the pathogen is a bit difficult. As we have seen in this chapter, viruses utilize many host cell enzymes during the production of new viral particles. Certainly, all viruses utilize the host cell translational machinery to make viral proteins. Antiviral drugs, therefore, cannot target this process. Many viruses also utilize host cell proteins for DNA replication and transcription. For these viruses, antiviral drugs ideally should not target these processes. So what processes can an antiviral drug target? To answer this question, we must be able to identify the virus of interest (**Perspective 8.1**).

The basic concept behind DNA microarray technology is almost deceptively simple. First, a series of specific oligonucleotides, or short segments of single-stranded DNA, are attached to a solid support, forming an array. Then, single-stranded DNA of interest is added to this array. If the DNA of interest is complementary to any of the oligonucleotides, then it will bind to that specific piece of DNA. By determining to which oligonucleotide the DNA binds, a researcher could gain important information about the DNA of interest (**Figure B8.4**).

In 2003, researchers at the University of California at San Francisco, Washington University, and the Centers for Disease Control and Prevention demonstrated that this same technique could be used to rapidly identify an unknown virus. To identify an unknown virus, the researchers developed a "pan-viral microarray." This microarray consisted of approximately ten 70-nucleotide-long synthetic DNA oligonucleotides for every viral genome currently referenced in GenBank, the national repository of DNA sequence data. In total, then, the microarray consisted of 10,000 oligomers representing 1,000 different viruses (**Figure B8.5**).

Total nucleic acid was isolated from uninfected cells and cells infected with the unknown virus. These nucleic acid samples then were PCR amplified using random primers and labeled with a fluorescent dye. The outcome of this procedure was that a collection of short pieces of amplified DNA, corresponding to all the DNA and RNA sequences present in the cells, was generated for both the uninfected and infected cells. These sequences were added in parallel to the pan-viral microarrays and allowed to bind, or hybridize, to any complementary oligonucleotides.

Following this hybridization step, the fluorescent intensity associated with each oligonucleotide, or spot, on both microarrays was analyzed. The researchers were most interested in spots for which the infected cell nucleic acid exhibited a much stronger fluorescence than did the uninfected cell nucleic acid. Such a result would indicate that those spots contained oligomers corresponding to the unknown virus. The results were very straightforward; the oligomers that hybridized with infected cell DNA all represented coronaviruses. The unknown virus was a member of the coronavirus family.

To confirm and extend this finding, researchers then isolated the hybridized nucleic acid by physically removing it from the microarray

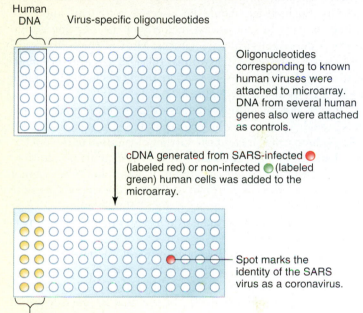

Figure B8.5. Identification of SARS virus with an innovative tool: A pan-viral microarray Synthetic single-stranded DNA oligonucleotides corresponding to all known human viruses were attached to the microarray. Nucleic acid from SARS-infected and uninfected cells was isolated and differentially labeled. The labeled nucleic acid then was added to the microarray. Because the labeled nucleic acid from the SARS-infected cells hybridized to the microarray spot containing coronavirus DNA, the researchers could conclude that the cells had been infected with a coronavirus.

and sequenced it. When they compared the sequence to the existing nucleic acid sequence database, the new sequence was very similar to, but different from, mouse hepatitis virus, a well-characterized coronavirus. The sequence data corroborated their previous finding—the SARS virus was a novel coronavirus.

Using this DNA microarray methodology, these researchers were able to demonstrate the power of this newly emerging technology. Because of recent advances in areas such as bioinformatics, robotics, and fluorescent detection technologies, we now can usually identify a virus in a matter of days.

As we will see in this section, antiviral drugs may decrease the severity or length of a disease. These drugs generally do not, however, prevent infections. Vaccines, conversely, can prevent infections. Types of vaccines will be explored in **Section 24.5**.

As we have discovered in this chapter, several classes of viruses require virus-specific DNA or RNA polymerases for their genomic replication. The single-stranded RNA viruses use a viral RNA-dependent RNA polymerase. The retroviruses use a viral RNA-dependent DNA polymerase (see Mini-Paper). Even the herpesviruses and poxviruses, which contain double-stranded DNA genomes, use their own DNA-dependent DNA polymerases. These viral polymerases tend to be less specific than our cells' DNA polymerase. As a result, the viral polymerases are more likely to incorporate chemically incorrect bases when replicating the viral genome.

Because these viruses utilize their own DNA or RNA polymerases, it should be possible to design drugs that target these enzymes. Such drugs should be effective antivirals and have limited toxicity. In fact, a vast majority of the existing antiviral drugs do target viral polymerases. Most of these drugs are **nucleoside analogs**. Structurally, these drugs are very similar to normal nucleosides, precursors to the nucleotide building blocks of DNA and RNA **(Figure 8.23)**.

Most of the nucleoside analogs differ from normal nucleosides in such a way that, when incorporated into a growing DNA or RNA chain, they prevent the further elongation of the nucleic acid chain. Because of the structures of these analogs, additional bases cannot be added onto them. If additional bases cannot be added onto the growing DNA or RNA chain, then a complete DNA or RNA molecule cannot be made. If a complete DNA or RNA molecule cannot be made, then new, infectious viral particles cannot be made.

Probably the most well-known antiviral currently available is azidothymidine or AZT (also called zidovudine), the first drug approved for use in people infected with HIV. This drug is a thymidine analog (see Figure 8.23). Rather than having a hydroxyl group ($-OH$) attached to the 3′ carbon of the ribose sugar, AZT contains an azide group ($-N_3$) in this position. Because a free 3′$-OH$ group is necessary for DNA elongation (see Section 7.2 for a review of DNA replication), no additional nucleotides can be added to AZT if it becomes incorporated into a growing DNA chain. Thus, this drug serves as a chain terminator. Furthermore, the HIV reverse transcriptase has a roughly 100-fold greater affinity for AZT than does our DNA polymerase. In other words, we could say that our DNA polymerase is more selective than HIV's RT; our DNA polymerase usually can distinguish AZT from thymine. The selectivity of our DNA polymerase, though, is not perfect. If a person infected with HIV is taking AZT, then the drug occasionally will become incorporated into replicating cellular DNA, thereby preventing the complete replication of the DNA and severely damaging the cell. As a result, serious side effects are associated with AZT. Additionally, viral mutants resistant to the drug routinely develop. Thus, AZT is far from a perfect drug. Today, the ideal treatment regimen for a person infected with HIV involves three different drugs: a triple drug cocktail. This mixture of drugs more effectively inhibits the replication of the virus. To some degree, this combination therapy also decreases the frequency with which drug-resistant viral mutants develop.

Antiviral drugs can target other aspects of viral replication. The initial binding of the virus to a host cell can be targeted. Penetration of the plasma membrane could be targeted. For viruses that replicate in the host cell nucleus, movement through the cytoplasm to the nucleus could be targeted. Even aspects of assembly and exit from the cell could be targeted. The options, however, are somewhat limited. Because of these limitations, the rational design and development of any antiviral drug only is possible if one has a detailed understanding of how viruses replicate.

8.5 Fact Check

1. What are nucleoside analogs and how do they function as antiviral drugs?
2. The drug AZT, used to treat HIV, inhibits viral DNA production. How does it do this?
3. In addition to the replication of the viral genome, what are some additional steps in the viral replication cycle that could be targeted by antiviral drugs?

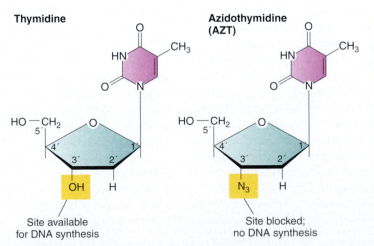

Thymidine   Azidothymidine (AZT)

Site available for DNA synthesis   Site blocked; no DNA synthesis

Figure 8.23. Many antiviral drugs are nucleoside analogs Structurally, azidothymidine (AZT) resembles the nucleoside thymidine. When the triphosphorylated form of AZT is added to a growing DNA strand, however, it acts as a chain terminator. Additional nucleotides cannot be added to AZT because it lacks an available 3′ — OH group.

Image in Action

WileyPLUS This image depicts the replication cycle of influenza viruses leaving one host cell (*bottom*) and infecting another (*top*).

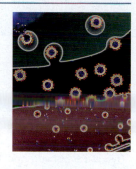

1. Describe the molecular steps that occur when a virus initially interacts with a host cell. Identify and include a description of the specific roles of the green- and red-colored components.

2. Imagine that Tamiflu was administered and present now in this situation. Describe the antiviral effects of Tamiflu and outline how this image would look different if this drug were present.

As we noted in the opening of this chapter, Tamiflu reduces the severity and length of the influenza disease. But how does it work? Tamiflu (oseltamivir) is a neuraminidase inhibitor (NAI). The neuraminidase protein (NA) is located in the influenza virus envelope, along with the hemagglutinin protein (HA). When a virion first encounters a host cell, HA binds to sialic acid, allowing the virus to enter the cell. During budding of newly replicated viral particles, NA cleaves sialic acid residues present on the infected cell's plasma membrane. So? The cleavage of sialic acid residues by NA prevents the newly formed viruses from getting stuck to the infected cell. In other words, NA increases the likelihood that new virus particles will move to other cells **(Figure 8.24)**.

By inhibiting neuraminidase, Tamiflu greatly reduces the cell-to-cell spread of influenza virus. Because the spread of the virus is limited, the number of infected cells is limited, and the severity of

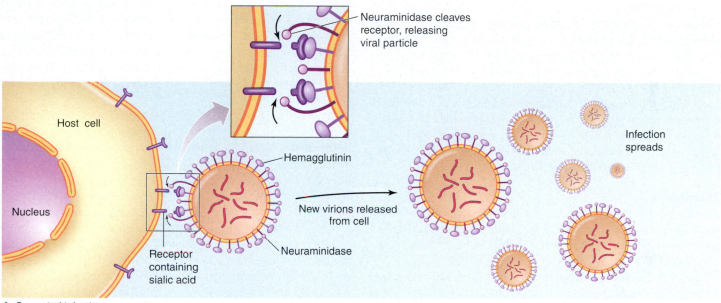

A. Spread of infection

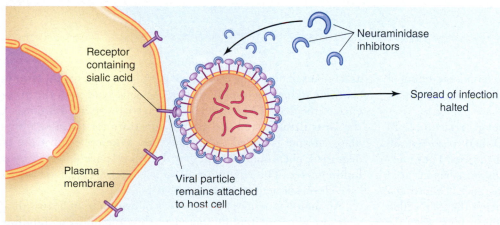

B. Infection halted

Figure 8.24. Action of Tamiflu A. Neuraminidase cleaves sialic acid allowing newly formed virions to disperse from the infected cell.
B. Tamiflu functions as a neuraminidase inhibitor, preventing the release of virions from infected cells and limiting the spread of the virus to other cells.

D. Baltimore. 1970. RNA-dependent DNA polymerase in virions of RNA tumour viruses. Nature 226:1209–1211.

H. Temin and S. Mizutani. 1970. RNA-dependent DNA polymerase in virions of Rous sarcoma virus. Nature 226:1211–1213.

Context

By the 1960s, several lines of evidence indicated that the replication of RNA tumor viruses, like the tumorigenic virus first described by Peyton Rous in 1911, included a DNA intermediate. First, researchers observed that DNA synthesis inhibitors block replication of these viruses, but only if these inhibitors are added to cells early after their infection. Second, researchers observed that actinomycin D, a drug that inhibits the DNA-dependent RNA polymerase responsible for transcription in mammalian cells, also prevents the formation of new virus particles. Third, several experiments showed that cells infected by RNA tumor viruses contain DNA that hybridizes with, or binds to, viral RNA.

Based on these observations, Baltimore and Temin independently postulated that the replication of RNA tumor viruses proceeds through a DNA intermediate. In other words, the RNA genome is converted into DNA, which later is converted back into RNA genomes. This model would explain the observations previously described. This model also would require the existence of a RNA-dependent DNA polymerase, or an enzyme that could convert RNA into DNA. Such an enzyme had never been observed in any organism. Furthermore, such an enzyme clearly would violate the central dogma of biology: that DNA is converted to RNA, which is converted to protein.

Experiments

To determine if RNA tumor viruses contained RNA-dependent DNA polymerases, the researchers asked two fairly straightforward questions. First, do the viruses produce DNA? Second, what is the template from which this DNA is derived?

To address the first question, both groups started with purified preparations of two RNA tumor viruses: Rauscher mouse leukemia virus (R-MLV) or Rous sarcoma virus (RSV). Refer to Chapter 5 to review how viruses can be isolated, purified, and quantified. Next, they added these viruses to a standard DNA polymerase assay. The ingredients of this assay included the four deoxyribonucleotide triphosphates (dATP, dTTP, dCTP, and dGTP), the precursors of DNA. Additionally, one of the nucleotides, thymidine, was labeled with ^{3}H, a radioactive form of hydrogen. Following an incubation, trichloroacetic acid was added to the reaction and the entire reaction then was filtered through a fine filter. Previous research had shown that DNA was acid-insoluble. In other words, any DNA produced in the reaction would precipitate out of solution when the acid was added and would be retained on the filter. By simply measuring the amount of radioactivity retained on the filter, the researchers could determine if any DNA had been produced.

The reaction did, indeed, result in the formation of an acid-insoluble product. To confirm that this product was DNA, Baltimore then treated the completed reaction with deoxyribonuclease, an enzyme that destroys DNA, ribonuclease, an enzyme that destroys RNA, and micrococcal nuclease, a relatively non-specific nuclease that destroys both RNA and DNA. Deoxyribonuclease and micrococcal nuclease, but not ribonuclease, digested the radioactive product. These two results support the conclusion that purified preparations of R-MLV could produce DNA.

To address the second question, Baltimore and Temin both investigated the sensitivity of the virus to ribonuclease (**Figure B8.6**). In the absence of ribonuclease, the amount of DNA produced increased over time (*line 1*). A similar increase was observed when ribonuclease was added after the virus and deoxynucleotides were allowed to incubate together for several minutes (*line 2*). When ribonuclease was included initially in the reaction mixture, however, the amount of DNA produced decreased markedly (*line 3*). When the virus was pre-incubated with ribonuclease before being added to the polymerase assay reaction, the amount of DNA produced decreased even more dramatically (line 4). These results support the conclusion that RNA is the template for DNA production.

---

the disease also is limited. Thus, a very detailed understanding of how influenza virus replicates was essential for the development of this effective antiviral medication.

Influenza viruses, however, may develop resistance to this drug. By October 2009, the WHO had reported 39 oseltamivir-resistant isolates of H1N1 influenza virus. The emergence of these drug-resistant strains of influenza has created concern and a need for rapid detection of resistance. Following a significant increase in Tamiflu resistance during the 2007–2008 flu season, researchers have been closely monitoring resistance rates. The method they use to screen for increased oseltamivir resistance is called the neuraminidase inhibition assay. This assay allows scientists to quantitate viral neuraminidase activity as a means of detecting resistance to the drug. Resistant strains continue to have neuraminidase (NA) activity even in the presence of the drug. Oseltamivir is mixed with varying dilutions of the virus being tested and then a substrate for the neuraminidase enzyme is added to the mixture. When the substrate is degraded, chemiluminescence or light levels are detected by a luminometer. This chemiluminescence, therefore, allows you to "see" levels of neuraminidase activity.

An understanding of the process of evolution indicates that widespread use of Tamiflu increases the probability of Tamiflu-resistant influenza, as those strains will be selected for in the population. Surveillance of Tamiflu resistance will be a key strategy in monitoring and limiting resistance.

Impact

The identification of this enzyme that converts RNA to DNA, now referred to as reverse transcriptase (RT), had an immediate and profound effect on the fields of virology and, more generally, biology itself. Indeed, as the editors of *Nature* noted in a comment preceding these two articles, "This discovery, if upheld, will have important implications . . . for the general understanding of genetic transcription." The discovery has been upheld. Today, two classes of viruses, the retroviruses and the Class VII double-stranded DNA viruses, have been shown to utilize RT. Eukaryal cells, too, utilize forms of this enzyme both for the maintenance of the ends of linear chromosomes (see Section 7.2) and the movement of certain transposable elements.

Reverse transcriptase also has become an indispensable tool for the molecular biologist. As we saw in Chapter 5, RT is used to convert the RNA of HIV into DNA as part of a standard viral load test. In addition, it is used routinely in the laboratory to convert messenger RNA into DNA. Not only did the discovery of reverse transcriptase change the way biologists viewed the flow of information within cells, but it also changed the way we can do experiments within the laboratory.

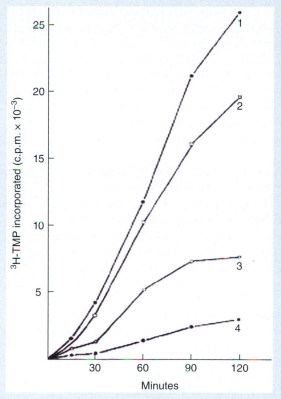

Figure B8.6. Incorporation of radiolabeled dTTP by purified virions Virus particles were incubated with deoxynucleotides, one of which was radiolabeled. At specified times, the amount of acid-insoluble radioactivity was measured. As seen in line 1, the amount of incorporated radioactivity increased over time. The amount of incorporated radioactivity was greatly diminished if the virus was pre-incubated with ribonuclease (line 4), indicating that the template for DNA production was the viral RNA. (From D. Baltimore. Nature 226:1209–1211, 1970. Figure 1. Reproduced with permission.)

● Questions for Discussion ············

1. Given the information presented here, would you hypothesize that retroviruses bring RT into the cell or produce it after infecting a cell?

2. Do you think that RT would be a viable target for antiretroviral drugs? Give reasons why or why not.

3. We mentioned that earlier experiments showed that cells infected by RNA tumor viruses contained DNA that hybridizes with viral RNA. One explanation for this finding is that the viral RNA is converted to DNA within the infected cell. Provide another possible explanation for this initial finding.

Summary

Section 8.1. How do viruses recognize appropriate cells?

Most viruses interact with appropriate host cells through a specific interaction between the **viral attachment protein** and a cellular **receptor**.

- For enveloped viruses, this attachment protein generally exists as a spike embedded in the envelope.
- For non-enveloped viruses, the capsid may contain a spike protein or, alternatively, a less defined component of the capsid may interact with the host cell.
- For bacteriophages, the tail fibers generally serve as the attachment proteins.

This interaction, in many cases, determines the **host range** of a virus and which tissues within a host a virus may infect.

- Studies with the bacteriophage T2 have demonstrated that altering the tail fibers of the phage can alter the host specificity of the phage.
- Studies with the mammalian virus mouse hepatitis virus have demonstrated that strains of mice are more or less susceptible to this virus depending on which form of a surface protein they express.
- Information about viral attachment proteins and receptors can be gleaned from two related techniques: SDS-PAGE and Western blotting.

- Drugs that block this initial interaction between the virus and its host cell, like the antiretroviral drug Maraviroc, can prevent or slow viral infections.

Section 8.2. How do viruses enter cells and uncoat?

For enveloped viruses, the viral envelope must fuse with a cellular membrane in order for the virus to gain access to the cell.

- Membrane fusion can occur at the cell surface.
- Alternatively, the virus may undergo **endocytosis** and fusion then occurs between the viral envelope and the endosomal membrane.
- In both cases, a viral **fusion peptide** facilitates the membrane fusion event.

Non-enveloped mammalian viruses usually enter a cell through endocytosis.

- Within the endocytic vesicle, conformational changes in the viral capsid generally expose some type of fusion peptide.
- Again, the fusion peptide allows part of the virus or its genome to penetrate the endosomal membrane.

Bacteriophages do not technically enter their host cells. Rather, they inject their genome through the cell wall.

Plant viruses often rely on an outside force—a feeding insect, for instance—to damage the cell cuticle and wall, thereby giving the virus access to the cytoplasm.

Following entry, some **uncoating** of the viral particle occurs to make the genome accessible to necessary cellular components.

Section 8.3. How do viruses replicate?

DNA viruses usually replicate within the nucleus, using the DNA and RNA polymerases of the host cell.

Most RNA viruses require a virally encoded **RNA-dependent RNA polymerase** to replicate their genome.

The genomes of **positive-sense RNA** viruses can be translated directly by the cell and are said to be **infectious RNA**.

The genomes of **negative-sense RNA** viruses cannot be translated directly. These viruses bring a polymerase with them into the cell during infection. The polymerase then makes RNA that can be read by the cellular translational machinery.

Retroviruses convert their RNA genome into DNA, using an **RNA-dependent DNA polymerase** known as **reverse transcriptase (RT)**. The viral DNA then gets integrated into the host cell's genome, through the action of the viral enzyme **integrase** and is referred to as a **provirus**. This main point starts with "Virulent phages..."

Virulent phages replicate in and lyse the host cell. Most **temperate phages** can insert their genome into the host cell's genome. The integrated phage DNA is referred to as a **prophage** and a cell containing a prophage is referred to as a **lysogen**.

Section 8.4. How do virus particles form and exit cells?

Newly made viral proteins generally self-assemble to form new viral capsids within an infected cell.

- In some cases, proteolytic processing of viral polypeptides is necessary for assembly to occur. The antiretroviral **protease inhibitor** drugs block this step of the assembly process.
- The proteins may assemble around the viral RNA or DNA, attaching to a **packaging sequence**.

Non-enveloped viruses usually exit the cell through lysis of the host cell.

- Bacteriophages like T4 release lysozyme, an enzyme that degrades the cell wall, thereby aiding in cell lysis.
- Many non-enveloped mammalian viruses disrupt the infected host cell without the use of specialized exit proteins.
- Plant viruses often do not lyse the infected cell, but travel to adjoining cells through cytoplasmic connections between cells.

Enveloped viruses exit the cell by budding off of the cell, thereby acquiring their envelope from the host cell's plasma membrane.

Section 8.5. How do antiviral drugs work?

Many antiviral drugs work by interfering with the viral polymerase. Such drugs usually are **nucleoside analogs** that inhibit viral replication.

Other antiviral drugs interfere with other steps in the replication process, including receptor binding, disassembly, or budding.

● Application Questions

1. Obviously, host cells do not "want" to be infected by viruses, yet they contain receptors that allow viral entry. Why do these receptors exist on the host cells and why can't host cells eliminate the receptors to prevent viral attachment?

2. Rabies virus is somewhat unusual in that it can infect a wide range of mammalian species. Given this information, what conclusions might you draw about the host cell receptor used by this virus? Explain.

3. In the lab, you expose a culture of host cells to ammonium chloride and note that the pH of endosomes within the cells increases. Based on what you learned in Section 8.2, how would this pH change affect the replication of an influenza virus in the cell?

4. A television show highlights the HIV drug Fuzeon, described in Section 8.2. The show mentions that Fuzeon is not a cure for HIV/AIDS, but that it simply "slows down the disease process" by not allowing the viruses to fuse with host cell plasma membranes. Use what you have learned to clarify how this drug will slow down the disease by explaining its effect on the overall viral replication process.

5. You are attending a presentation at your local gardening supply store. The speaker warns that insects such as aphids that infest your garden may cause viral plant infections. Explain the link between insects and plant viruses.

6. After re-reading this chapter you notice that some types of viruses must bring along their own RNA polymerase or transcription factors into a host cell, while others do not. Explain why.

7. Some viruses manipulate the host cell machinery so that the cell continually replicates DNA and makes additional nucleotides. Explain how this viral manipulation may lead to perpetuations of the normal cell cycle.

8. You are assigned the job of determining whether DNA replication and transcription in a newly discovered virus occur in the cytoplasm or in the nucleus of its host cell. Since you can't see where the processes are taking place, what could you do or look for to determine your answer? Explain.

9. For many viral diseases, diagnosis can be made in the lab by growing host cells in culture and adding a viral sample from a patient. If the virus in question is present in the patient's sample, it will

cause visible changes in the cultured host cells. Based on what you learned in Section 8.4, how would the cultured host cells look different following this process if the patient's sample contained a non-enveloped virus? Explain.

10. Developing drugs against microbes involves finding unique targets that involve only the microbe and are not part of the host organism. Explain why an antiviral drug targeting the translational process of producing new viral proteins is not an ideal target.

11. You are studying a new strain of virus and are in the process of identifying the host receptor protein utilized by that virus. You are told to use the Western blot technique, as described in Toolbox 8.1, to demonstrate that the protein in question is the receptor.

a. How will the host protein receptor be prepared and analyzed for this Western blot procedure?
b. What will be the role of the virus in this procedure?
c. What result will you observe if the protein is in fact the correct receptor?

12. You are assigned the task of determining whether the genome of a newly discovered virus contains infectious RNA. You are given some host cells and a sample of the virus.

a. First, what is meant by the term infectious RNA?
b. What type of virus contains a genome of infectious RNA?
c. What could you do in the lab with the samples provided to test whether the RNA in this virus is infectious?
d. What would be the observed result if it is infectious RNA?

Suggested Reading

Brandenburg, B., L. Y. Lee, M. Lakadamyali, M. J. Rust, X. Zhuang, and J. M. Hogle. 2007. Imaging poliovirus entry in live cells. PLoS Biol 5:e183.

Drosten, C., S. Gunther, W. Preiser, S. van der Werf, H. R. Brodt, S. Becker, H. Rabenau, M. Panning, L. Kolesnikova, R. A. Fouchier, A. Berger, A. M. Burguiere, J. Cinatl, M. Eickmann, N. Escriou, K. Grywna, S. Kramme, J. C. Manuguerra, S. Muller, V. Rickerts, M. Sturmer, S. Vieth, H. D. Klenk, A. D. Osterhaus, H. Schmitz, and H. W. Doerr. 2003. Identification of a novel coronavirus in patients with severe acute respiratory syndrome. N Engl J Med 348:1967–1976.

Ksiazek, T. G., D. Erdman, C. S. Goldsmith, S. R. Zaki, T. Peret, S. Emery, S. Tong, C. Urbani, J. A. Comer, W. Lim, P. E. Rollin, S. F. Dowell, A. E. Ling, C. D. Humphrey, W. J. Shieh, J. Guarner, C. D. Paddock, P. Rota, B. Fields, J. DeRisi, J. Y. Yang, N. Cox, J. M. Hughes, J. W. LeDuc, W. J. Bellini, and L. J. Anderson. 2003. A novel coronavirus associated with severe acute respiratory syndrome. N Engl J Med 348:1953–1966.

Leiman, P. G., M. Basler, U. A. Ramagopal, J. B. Bonanno, J. M. Sauder, S. Pukatzki, S. K. Burley, S. C. Almo, and J. J. Mekalanos. 2009. Type VI secretion apparatus and phage tail-associated protein complexes share a common evolutionary origin. Proc Natl Acad Sci USA 106:4154–4159.

Wang, D., L. Coscoy, M. Zylberberg, P. C. Avila, H. A. Boushey, D. Ganem, and J. L. DeRisi. 2002. Microarray-based detection and genotyping of viral pathogens. Proc Natl Acad Sci USA 99:15687–15692.

Wang, D., A. Urisman, Y. T. Liu, M. Springer, T. G. Ksiazek, D. D. Erdman, E. R. Mardis, M. Hickenbotham, V. Magrini, J. Eldred, J. P. Latreille, R. K. Wilson, D. Ganem, and J. L. DeRisi. 2003. Viral discovery and sequence recovery using DNA microarrays. PLoS Biol 1:e2.

Yoichi, M., M. Abe, K. Miyanaga, H. Unno, and Y. Tanji. 2005. Alteration of tail fiber protein gp38 enables T2 phage to infect Escherichia coli O157:H7. J Biotechnol 115:101–107.

Zelus, B. D., D. R. Wessner, R. K. Williams, M. N. Pensiero, F. T. Phibbs, M. deSouza, G. S. Dveksler, and K. V. Holmes. 1998. Purified, soluble recombinant mouse hepatitis virus receptor, Bgp1(b), and Bgp2 murine coronavirus receptors differ in mouse hepatitis virus binding and neutralizing activities. J Virol 72:7237–7244.

CHAPTER 9

Bacterial Genetic Analysis

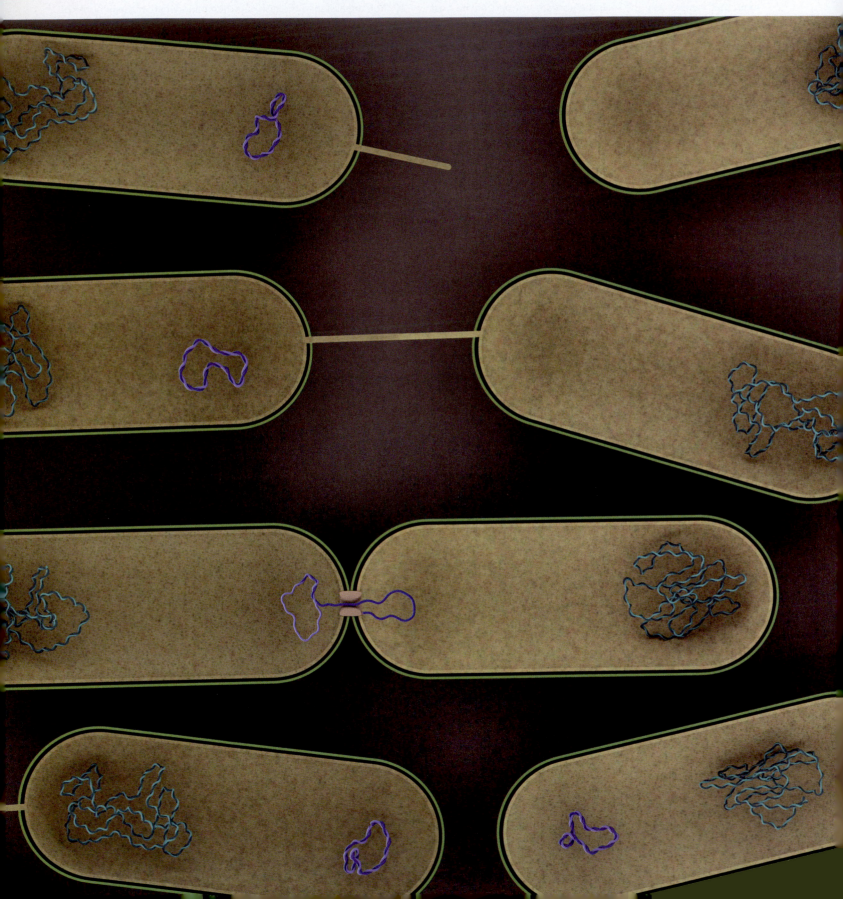

In 1946, a young Columbia University medical student named Joshua Lederberg took a leave of absence from his studies and joined the laboratory of Edward Tatum at Yale University. Spurred by the question of whether genetic mechanisms operated in bacteria as they did in other organisms such as plants and animals, Lederberg had been consumed with the idea of determining whether mating and genetic exchange could occur in bacteria. Is there bacterial sex? At the time, bacteria were not recognized as having genes; the evidence for genes depended on the detection and analysis of mating events resulting in progeny whose properties could be compared to the properties of the parents. In multicellular organisms such as corn, pea, or the fruit fly, the tracking of inherited characteristics involved the simple observation of physical traits, or phenotypes, in the progeny. It was not so easy in bacteria. Different physical characteristics were not easily observed. A real breakthrough came with the idea of isolating mutants with distinctly different metabolic traits that could be observed and then used to differentiate between the progeny and the parental phenotypes.

In the laboratory, Lederberg's organism of choice was *Escherichia coli*, a common bacterial inhabitant of the human gut. He reasoned that if he were to mix two cell types showing different phenotypes and then observe progeny with characteristics intermediate between these two cell types, it would be evidence for a type of bacterial mating. The results of the famous series of experiments that he conducted provided clear evidence for the ability of bacterial cells to mate. Lederberg was awarded a Ph.D. for his work, and went on to an illustrious academic career at the University of Wisconsin, Stanford University, and Rockefeller University where he served as president from 1978 to 1990. Although the phenomenon of bacterial mating initially was presumed to be similar to sexual crossing in single-celled eukaryal organisms such as *Paramecium*, it eventually became clear that the process in bacteria was fundamentally different. Lederberg's uncovering of a genetic mechanism used by bacteria for the transfer of genes displaced the prevailing dogma of the time, that bacteria were primitive, simplistic life-forms that could not be used to investigate the genetic basis of life.

CHAPTER NAVIGATOR

As you study the key topics, make sure you review the following elements:

Studies of bacterial genetics and bacterial genome structures have been of fundamental importance in microbiology.
- Microbes in Focus 9.1: *Escherichia coli*
- Figure 9.1: Nutritional mutants in defined media
- Table 9.1: Some genetic factors carried by plasmids
- Table 9.2: Examples of bacterial genome configurations
- Figure 9.2: Plasmid incompatibility

Bacterial geneticists make extensive use of mutations.
- Figure 9.3: Mutation and phenotype
- Figure 9.4: Phenotypic selection
- Figure 9.5: Phenotypic screening
- Figure 9.6: Process Diagram: Replica plating
- Toolbox 9.1: Isolating nutritional mutants

DNA can be inserted into cloning vectors and amplified.
- Figure 9.14: Process Diagram: Cloning a gene
- Figure 9.18: Cosmids: Plasmid/phage hybrid vectors

DNA is transferred between bacterial cells by distinct mechanisms.
- 3D Animation: Transformation
- 3D Animation: Conjugation
- 3D Animation: Transduction
- 3D Animation: Transposition
- Figure 9.19: Process Diagram: DNA recombination
- Microbes in Focus 9.2: *Bacillus subtilis*
- Figure 9.31: Types of transposable elements
- Mini-Paper: The discovery of transduction

CONNECTIONS for this chapter:

Bacterial genome and plasmid origins of replication (Section 7.2)

Bacterial toxin genes delivered by phage (Section 21.3)

Expression vectors and the production of recombinant proteins (Section 12.2)

Introduction

Historically, the study of bacterial genetics focused on organisms of practical importance. Scientists were interested mainly in bacteria that caused disease or were associated with the human body, and the bulk of bacterial genetics research was carried out on *Escherichia coli* and *Salmonella* sp. Researchers were also interested in bacteria used for the industrial production of compounds, like the antibiotic-producing *Streptomyces*, or bacteria that had special properties amenable to genetic study, like *Bacillus* with its endospore formation. In later years, systems of genetic analysis were developed in other types of bacteria, such as *Rhizobium*, which forms nitrogen-fixing symbiotic relationships with plants, and *Pseudomonas* sp., which can degrade several industrial contaminants like toluene. Much of the fundamental knowledge and understanding in bacterial genetics is based on studies with such model systems.

Today, bacterial genetics is of fundamental importance in the study of microbiology. Cell function is determined by gene function and gene regulation, so basic processes in bacteria are often studied at the level of the gene. In this chapter, we will follow the rich development of bacterial genetics from its birth in the mid-twentieth century to the present days of genomics. As we will see, basic research in bacterial genetics has given rise to molecular biology. By understanding the genetic mechanisms that occur naturally in bacteria, researchers have developed powerful tools for molecular biology and genetic research. We will introduce some of these genetic tools in this chapter. In the chapters that follow, we will see how microbiologists use these tools as they address the many questions of microbiology. To get started, let's consider the following questions:

Why study genetics in bacteria? (9.1)
How are mutations used in bacterial genetics? (9.2)
How is DNA cloned? (9.3)
How do bacterial cells transfer DNA? (9.4)

9.1 Bacteria as subjects of genetic research

Why study genetics in bacteria?

The study of bacterial genetics was a bit late in getting started, compared to the study of genetics in organisms like the fruit fly, corn, and pea. Bacteria, however, present a number of advantages in genetics research. Each cell is a complete organism, and bacterial growth is measured by the duplication of these cells. Bacterial cells usually grow fast, and biomass can increase greatly in a very short time. When cellular differentiation occurs to create specialized cell types such as the endospore, fruiting body, or heterocyst, it is comparatively simple and details of the processes are easy to study.

Bacterial genomes typically are organized in a simple fashion, too. Bacteria generally contain only one complete copy of their genome. The genetic makeup of each chromosome is unique, so each gene that it contains is usually the only copy of that gene. This arrangement is ideal for genetic studies because a mutation in a gene will often produce a new readily observable trait, or phenotype of the cell. In contrast, most eukaryal organisms have more than one copy of the genome. When a mutation occurs in one gene copy, the product of the unmutated gene copy often masks the effect of the mutation in the other copy of the gene. In this section, we will explore more fully the components and organization of bacterial genomes. First, though, we will examine the birth of bacterial genetics.

The birth of bacterial genetics

In the first half of the twentieth century, the science of genetics became well established, but it primarily involved the study of multicellular eukaryal organisms. Bacteria were thought to be quite different life-forms, perhaps too simple to have a system of genetic exchange. While the genetic process of sexual exchange was understood to be an essential part of the life cycle of eukaryal organisms, a similar process had not been demonstrated in bacteria. At the time, genetic experiments with eukaryal organisms involved measuring the inheritance of phenotypes in offspring that arose following a sexual event. In bacteria, no comparable phenotypes clearly demonstrated genetic exchange in these relatively simple organisms.

Returning to our opening story of the discovery of bacterial mating, Lederberg wondered whether sexual crossing, or some similar mechanism providing for genetic exchange, occurred in bacteria. If it did occur, then how could it be demonstrated experimentally? Lederberg knew that strains of a bacterial species possess slight genetic differences that may result in slight phenotypic differences. Transfer of genes between different bacterial strains of the same species, he hypothesized, should result in progeny that share the phenotypic characteristics of both parental strains. Lederberg designed experiments to test this hypothesis using strains of the bacterium *Escherichia coli* (**Microbes in Focus 9.1**).

For his experiments, Lederberg used different types of *E. coli* **nutritional mutants**, strains exhibiting altered metabolic requirements. These strains, also known as auxotrophic mutants, or auxotrophs (see Section 6.1), have mutations that disrupt specific metabolic pathways, making them unable to synthesize amino acids or vitamins normally produced by the parental, or prototrophic strain. The prototrophic strain of *E. coli*, for instance, will grow on a simple defined growth medium

If we culture two different nutritional mutants together and progeny cells result that can thrive on media lacking nutritional supplements, then we can infer that the transfer of genetic material from one mutant to the other has occurred. As an example, let's look at the two nutritional mutants just described. We could culture *met⁻pro⁺* and *met⁺pro⁻* strains of *E. coli* on growth medium lacking both methionine and proline. Any resulting growth (*met⁺ pro⁺*) would be evidence that the two parental strains had exchanged genes for synthesis of these two compounds. Lederberg developed an experiment just like this one. In performing these mating experiments, however, a problem became evident. When control cultures of the individual mutant strains were plated separately on unsupplemented media, *met⁺pro⁺* colonies arose. How could this result occur? The parent strains spontaneously developed mutations (see Section 7.5) that allowed them to recover the ability to synthesize the missing amino acids. The frequency of this change, or **reversion**, was low, typically a frequency of one cell out of every 10^6 to 10^7 cells. If 10^8 cells were plated, then between 10 and 100 colonies per plate would appear. Thus, the reversion frequency was high enough to mask any gene transfer that might occur.

To avoid the problem described above, Lederberg used strains possessing two or three different nutritional mutations. These double and triple mutant strains would be able to grow on unsupplemented growth medium only if all of the mutations reverted. The frequency of two mutations spontaneously reverting in a single bacterial cell would be the product of both mutations reverting independently—an extremely low number. Instead of a reversion frequency of 10^{-6} to 10^{-7} (10–100 colonies per plate for 10^8 cells plated), the reversion frequency in a double mutant strain would be 10^{-12} to 10^{-14} (0.0001–0.000001 colonies per plate), and the reversion frequency in a triple mutant strain would be 10^{-18} to 10^{-21} (virtually no colonies per plate). When Lederberg co-cultured two triple mutant strains and plated them on unsupplemented growth medium, colonies did appear. Based on these results, he concluded that the genes required for the synthesis of the missing compounds were transferred from one cell to another. In other words, bacterial mating or gene exchange had occurred.

that contains appropriate sources of carbon, nitrogen, and phosphorus, but no added vitamins or amino acids (see Section 6.1). Nutritional mutants grow only if the growth medium contains the appropriate amino acids or vitamins. For example, a methionine-negative (*met⁻*) strain of bacteria cannot synthesize the amino acid methionine, and a proline-negative (*pro⁻*) strain cannot synthesize the amino acid proline (**Figure 9.1**). The *met⁻* strain needs methionine but not proline to grow in the medium; this strain can be described as *met⁻pro⁺*. The *pro⁻* strain requires proline but not methionine and can be described as *met⁺pro⁻*.

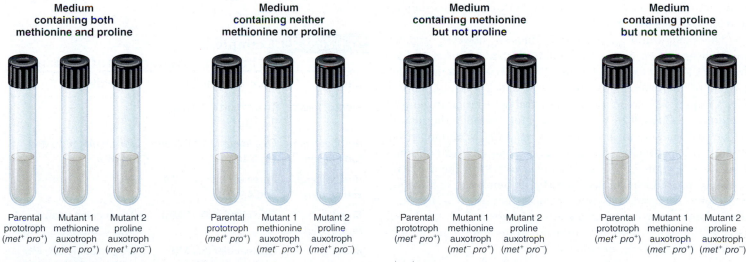

Figure 9.1. Nutritional mutants in defined media The parental prototrophic strain (*met⁺pro⁺*) can grow in defined media without supplements. Mutants that lack the enzymes to produce a required nutrient cannot grow unless that nutrient is supplemented in the growth medium. A methionine auxotroph (*met⁻pro⁺*) can grow in defined medium supplemented with methionine. A proline auxotroph (*met⁺pro⁻*) can grow in defined medium supplemented with proline.

Of course, Lederberg had to rule out the possibility that DNA liberated by dead cells of one of the mutant strains had been taken up by and incorporated into the genome of the other strain. He therefore prepared cell extracts from one strain and added them to the other strain during growth. Very few colonies resulted when plated on unsupplemented medium compared to when the strains were grown together. Uptake of external DNA, he concluded, could not account for the higher frequency of gene transfer.

Lederberg's groundbreaking research is often considered the turning point in the development of bacterial genetics as a scientific discipline. In recognition of his contributions, he shared the 1958 Nobel Prize for Physiology or Medicine with George Beadle and Edward Tatum. In the years following the discovery of bacterial mating, bacterial geneticists used this process to determine the relative positions of many genes on the genetic map of the *E. coli* chromosome. This genetic analysis, coupled with the biochemical and physiological analysis of the mutants, resulted in *E. coli* becoming the best characterized and most intensively studied of all organisms. It paved the way for the genomics revolution.

Organization of bacterial genomes

Compared to the genomes of most eukaryal organisms, bacterial genomes are small. They are also genetically compact. In other words, most of the DNA encodes functional proteins. In contrast, a large amount of DNA typically found in eukaryal organisms does not code for any known products. Early genetic studies suggested that bacteria contain a singular, circular chromosome. Additionally, many bacteria carry smaller circular DNA molecules called **plasmids** that replicate independently of the chromosome (see Figure 2.6). Generally, genes carried on the chromosome are required for the basic metabolic processes of the organism. Plasmids, conversely, generally are not required for cell viability, but instead contain genes for accessory functions, such as antibiotic resistance and degradation of toxic substances **(Table 9.1)**. Plasmids also may contain genes that allow the bacterium to infect a host cell and cause disease or form a symbiosis. The **genome** is composed of all of the hereditary material of a bacterial cell, including both the chromosome(s) and any existing plasmid(s).

The configuration of bacterial genomes, however, is much more diverse than originally thought. Examination of the genomes of an ever-expanding range of bacteria by methods such as genetic mapping, DNA sequencing, and gel electrophoresis (see Chapter 10) has revealed considerable variation in genome architecture **(Table 9.2)**. We now know, for example, that not all bacterial chromosomes are circular. *Streptomyces*, a common group of antibiotic-producing soil bacteria, possesses a linear chromosome (see Section 7.2), as does *Borrelia burgdorferi*, the causative agent of human Lyme disease (see Microbes in Focus 18.5). Moreover, some bacterial cells contain more than one chromosome. The plant tumor-inducing bacterium *Agrobacterium tumefaciens* (see Section 12.5) contains a circular chromosome, a linear chromosome, and two large circular plasmids.

In bacteria, each chromosome and plasmid can be classified as a **replicon**, a DNA molecule that replicates from a single origin of replication. At this site, the enzyme DNA polymerase initiates DNA synthesis. In cells containing plasmids, each cell

TABLE 9.1 Some genetic factors carried by plasmids

Plasmid	Trait	Host
pSym	Nitrogen-fixing nodule formation on legume plant roots	*Rhizobium*
pTi	Tumor formation on plants	*Agrobacterium*
pTol	Toluene degradation	*Pseudomonas putida*
pR773	Arsenic resistance	*Escherichia coli*
pWR100	Entry into host cells	*Shigella flexneri*

has a controlled number of plasmid molecules, known as the **copy number**. Different plasmids have characteristically different copy numbers, ranging from one to several hundred copies per cell. Because DNA polymerase synthesizes DNA at a constant rate, initiation of DNA replication controls the plasmid copy number. Mechanisms to control initiation may involve antisense RNA molecules or inhibitory proteins. The accumulation of a gene product that mirrors plasmid concentration in the cell can inhibit initiation of replication of that plasmid.

CONNECTIONS Bacterial origins of replication, including *E. coli* chromosomal *oriC*, consist of several repeats of a common DNA motif that serve as binding sites for DnaA, a DNA binding protein that begins melting of the double helix to initiate replication (**Section 7.2**). Other repeat sequences and AT-rich sequences may also be found in plasmid origins of replication.

Much of our understanding of plasmid replication control arose out of the investigation of a phenomenon known as "plasmid incompatibility." Two different plasmids are incompatible if they cannot exist stably within a population of cells. Plasmid incompatibility occurs when two different plasmids use the same control mechanisms for the initiation of replication, including similar origins of replication. In this case, the controlling machinery is interchangeable, and replication of the plasmid

TABLE 9.2 Examples of bacterial genome configurations

Bacteria	Chromosome(s)	Plasmid(s)
Vibrio cholerae	Two circular (3.0 and 1.1 Mb)	
Agrobacterium tumefaciens	One linear (2.1 Mb), one circular (2.8 Mb)	Two circular (0.45 and 0.21 Mb)
Burkholderia xenovorans	Three circular (1.5, 3.4, and 4.9 Mb)	
Borrelia burgdorferi	One circular (0.91 Mb)	Nine circular, twelve linear (0.005–0.053 Mb)
Bacillus anthracis	One circular (5.2 Mb)	Two circular (0.095 Mb, and 0.18 Mb)
Streptomyces coelicolor	One linear (8.7 Mb)	One linear (0.36 Mb) and one circular (0.03 Mb)

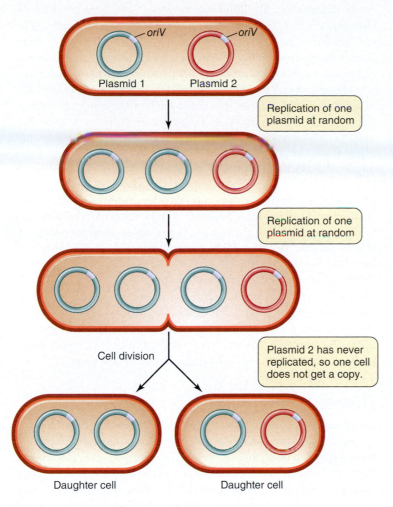

Figure 9.2. Plasmid incompatibility Plasmids that share similar control mechanisms and origins of replication are not distinguished, resulting in eventual loss of one of the plasmids from the population as the plasmids are chosen at random for replication. Plasmids 1 and 2 have a copy number of 4 at cell division. If random initiation results in unequal numbers of the two plasmids when the copy number reaches 4, then one daughter cell does not receive one of the plasmids at cell division. Descendants of the daughter cell on the left will lack plasmid 2.

Labels in figure: *oriV*, *oriV*, Plasmid 1, Plasmid 2, Replication of one plasmid at random, Replication of one plasmid at random, Cell division, Plasmid 2 has never replicated, so one cell does not get a copy, Daughter cell, Daughter cell

these two plasmids and the plasmids are chosen at random for replication, in some cases only plasmid 1 will replicate. When the cell divides there will be four copies in total: three copies of plasmid 1 and only one copy of plasmid 2. If each daughter cell receives two of the four plasmids, then one cell will receive no copies of plasmid 2. This random process will eventually result in the loss of one of the plasmids from the population.

Plasmid incompatibility represents an automatic consequence of this random selection process, rather than a specific competitive mechanism that prevents replication of one of the plasmids. Plasmids that cannot co-exist are assigned to the same incompatibility group. For two or more plasmids to exist in a cell, they must be from different incompatibility groups. Plasmids in different incompatibility groups use different mechanisms to control initiation of replication, so no interference occurs between them. Studying plasmids from different incompatibility groups has provided insights into how plasmids regulate their replication in the cell.

Bacteriophage DNA represents another source of DNA sometimes found in bacteria. The genetic material of viruses sometimes can exist for extended periods of time within the host cell, either as a circular DNA molecule similar to a plasmid, or as DNA integrated into the bacterial chromosome. Both bacteriophages and plasmids play influential roles in the evolution of bacterial chromosomes by providing mechanisms for moving segments of the bacterial genome from one bacterial cell to another. We will examine some of these mechanisms in Section 9.4.

CONNECTIONS Some human pathogens, like *Clostridium botulinum*, possess toxin-encoding genes that have been delivered by phages. In **Section 21.3**, we will examine further the roles of phages, plasmids, and other vectors of DNA movement in pathogen evolution.

9.1 Fact Check

1. How do the genome arrangements of bacterial and eukaryal cells differ?
2. What are bacterial nutritional mutants, or auxotrophs? Give an example.
3. A bacterial genome typically is composed of what types of molecules?
4. What is copy number, and how is it controlled in the bacterial cell?
5. Describe what is meant by plasmid incompatibility.

---

will be initiated at random. The two different plasmids will be treated by the replication control as a single plasmid type, resulting in the cell "losing count" of how many copies of each of the two plasmids exist within the cell. As shown in **Figure 9.2** for example, plasmid 1 and plasmid 2 are incompatible plasmids that have very similar origins of replication and are not distinguished from each other. These plasmids normally have a plasmid copy number of 4 at cell division, but since their origins are not distinguished, the total copy number of both plasmids combined would be 4. If a cell starts with a single copy of each of

9.2 Mutations, mutants, and strains

How are mutations used in bacterial genetics?

In their studies, bacterial geneticists use large numbers of genetically distinct strains of bacterial species that can be distinguished from other strains of the same species. Each strain can be traced back to a **wild-type strain**, which usually possesses

the typical or representative characteristics of the species. In some cases, the wild-type strain may be the original isolate from nature. In other cases, the wild-type strain may be the common laboratory strain from which all of the mutants in a particular study have been derived. For example, *Escherichia coli* K12, isolated from human feces in 1922, is the wild-type strain from which most strains of *E. coli* used in molecular biology have been derived. The most commonly studied wild-type strain of the symbiotic nitrogen-fixing bacterium *Sinorhizobium meliloti* is an antibiotic-resistant derivative of an isolate from an alfalfa root nodule (see Microbes in Focus 13.1). These strains and their mutant derivatives have been cataloged and archived in deep freezers or in a freeze-dried state in individual research laboratories and in central strain collections. Advances in genetics research require access to and sharing of defined strains by scientists from around the world.

In this section, we will first review basic concepts of mutants and mutations. We will then explore how researchers detect and isolate specific mutants. Next, we will examine how mutations lead to evolution in the laboratory. Finally, we will look briefly at how mutations arise.

Mutants and mutations

The words "mutant" and "mutation" may appear similar, but the terms are not interchangeable. "Mutant" refers to a cell or strain possessing a mutation, or a change in its DNA sequence relative to the comparable sequence in the wild-type strain. Change that occurs within the DNA of a mutated gene results in a different **allele** or form of the gene, and in some cases disrupts or changes the function of the gene (see Section 7.5). Mutations may result in loss of gene function, the regaining of function of a previously mutated gene, or modification of gene function, often resulting in a phenotypic change **(Figure 9.3)**. While mutations arise spontaneously due to errors in DNA replication that are not repaired successfully, bacterial geneticists take advantage of the known mutagenic properties of UV light and certain DNA-damaging chemicals to increase the rate of mutation so that fewer colonies must be screened to identify mutations with a desired phenotype.

Researchers use precise genetic terminology to describe strains of bacteria used in the laboratory. Genes are given a three-letter, lowercase, italicized name. When more than one gene is involved in the same biochemical process, then the different genes are distinguished from each other by adding an uppercase, italicized letter

to the designation. We differentiate genes from gene products by changing the italicization and capitalization. For example, the gene *hisC* produces the enzyme HisC. Some gene products have more specific names to describe their biochemical function. HisC is also called by its enzyme name, histidinol phosphate aminotransferase.

The **genotype** describes the alleles within an organism. Because it would be impractical to list each of the alleles in a given organism (*E. coli* has about 4,300 genes, *Streptomyces coelicolor* has about 7,800 genes), we generally describe only the alleles that differ from the alleles of the wild-type parental strain. For example, the alleles of a nutritional mutant strain that contains a disrupted *hisC* would be identical to the alleles of the wild-type organism except that it does not contain a functional *hisC* gene. A superscripted minus sign ($hisC^-$) indicates that the function of the gene is disrupted. So, what is the observable phenotype of this organism? Because HisC is involved in the biosynthesis of histidine, and its function is disrupted, the observable phenotype will be a lack of the ability to make histidine. This cell will only grow if histidine is provided, and so will not grow on defined media without supplemental histidine. We can easily differentiate such auxotrophs from the prototrophs, and then characterize them **(Toolbox 9.1)**. Much of our understanding of biochemical pathways has come from studying various auxotrophs.

Detection of mutants

Auxotrophs represent just one example of bacterial mutants useful for investigating biological processes. Mutants exhibiting other phenotypes can be equally as useful. For example, some pathogenic bacteria produce an extracellular polysaccharide capsule that makes the colonies appear mucoid or smooth. Some bacteria, such as *Serratia marcescens*, produce pigments that result in red-colored colonies. Some bacteria are motile due to the action of flagella. Some antibiotic-resistant mutants have mutations that alter the cellular target of the antibiotic. Mutants exhibiting changes in all of these phenotypes can be isolated and studied.

To isolate phenotypic mutants, researchers must be able to quickly and easily detect cells displaying the desired phenotype. Phenotypic selection represents one method of detecting potentially

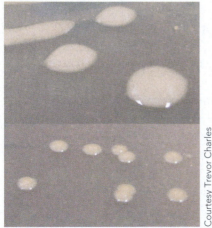

A. Exopolysaccharide mutant phenotype

B. Auxotrophic mutant phenotype

C. Carbon utilization mutant phenotype

Courtesy Trevor Charles

Figure 9.3. Mutation and phenotype Some mutations cause easily detectable changes in phenotype, as demonstrated by these *Sinorhizobium meliloti* mutants. **A.** A mutant that overproduces exopolysaccharide is far more mucoid than the parental strain. **B.** An auxotrophic (*above*) mutant is unable to grow on defined growth medium. **C.** A mutant that is unable to hydrolyze a sugar is white on media containing a chromogenic substrate, compared to the blue-green colonies of the parental strain.

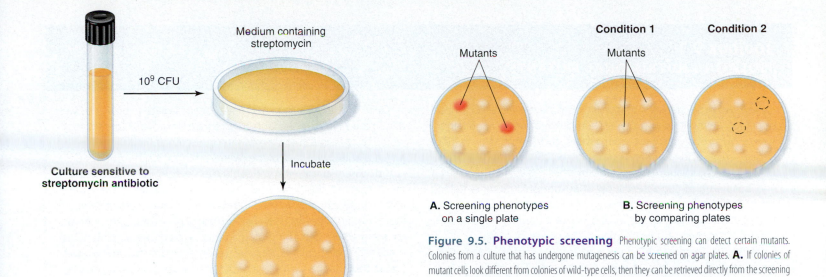

A. Screening phenotypes on a single plate

B. Screening phenotypes by comparing plates

Figure 9.5. Phenotypic screening Phenotypic screening can detect certain mutants. Colonies from a culture that has undergone mutagenesis can be screened on agar plates. **A.** If colonies of mutant cells look different from colonies of wild-type cells, then they can be retrieved directly from the screening plate. **B.** Mutant colonies may be inoculated onto two different types of plates, such as Condition 1, containing complex medium, and Condition 2, containing defined medium with no supplements. Some mutants may be unable to form colonies under Condition 2, but can be retrieved from the Condition 1 screening plate.

Figure 9.4. Phenotypic selection Very rare events can be detected using selection methods. In this example, 10^9 colony-forming units (CFU) from culture of a streptomycin-sensitive strain are plated on agar medium containing streptomycin. Most of the cells cannot grow, but if a cell contains a rare mutation conferring streptomycin resistance, then it will form a colony.

rare mutants **(Figure 9.4)**. This process involves plating bacteria on a growth medium that allows only strains with a particular combination of phenotypic characteristics to grow and form a colony. To isolate an antibiotic-resistant mutant from a culture, for example, cells can be plated on media containing the antibiotic. Only the antibiotic-resistant mutants will survive and form colonies.

The process of screening can also be used to detect rare mutants **(Figure 9.5)**. Screening does not prevent the growth of cells with the wild-type phenotype, but instead all cells form colonies under at least one of the tested conditions. Colonies with the desired phenotype, such as a different colony color, must then be identified by the investigator. Screening is usually less efficient than selection because large numbers of colonies must be screened to find the rare phenotype.

Microbial geneticists frequently employ **replica plating** to screen colonies for phenotypes that cannot be identified by phenotypic selection. In replica plating, colonies are lifted from one plate using a piece of sterile velvet cloth and then deposited onto a fresh plate **(Figure 9.6)**.

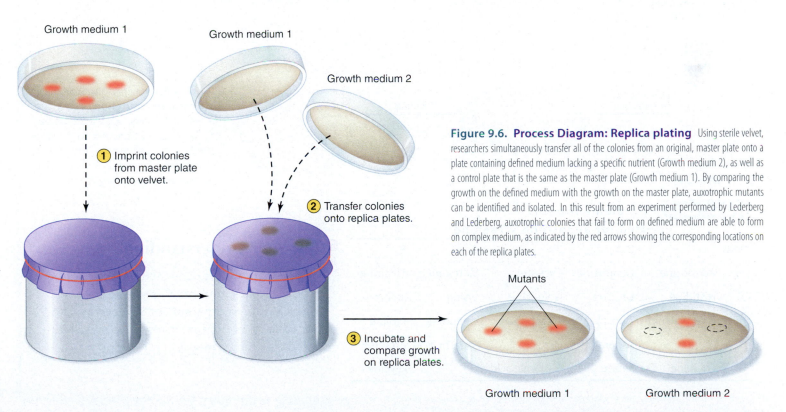

Figure 9.6. Process Diagram: Replica plating Using sterile velvet, researchers simultaneously transfer all of the colonies from an original, master plate onto a plate containing defined medium lacking a specific nutrient (Growth medium 2), as well as a control plate that is the same as the master plate (Growth medium 1). By comparing the growth on the defined medium with the growth on the master plate, auxotrophic mutants can be identified and isolated. In this result from an experiment performed by Lederberg and Lederberg, auxotrophic colonies that fail to form on defined medium are able to form on complex medium, as indicated by the red arrows showing the corresponding locations on each of the replica plates.

Toolbox 9.1
ISOLATING NUTRITIONAL MUTANTS

As we saw in Section 6.1, many microbes have fairly simple nutritional requirements and grow well on defined media containing a source of carbon, nitrogen, phosphorus, and simple salts, without added amino acids or vitamins. Various biochemical pathways enable the microorganisms to synthesize these necessary molecules. The isolation and characterization of auxotrophic mutants helped microbiologists understand these biochemical pathways. These mutants can be used to map the location of genes on the chromosome, and to isolate the wild-type version of the genes for study.

How do investigators isolate auxotrophic mutants? Mutations arise spontaneously, but at a rate far too low for a researcher to reliably isolate mutants with specific phenotypes. Fortunately, we can increase the mutation rate by treating the cells with DNA-damaging chemicals like ethane methanesulfonate (EMS) or nitrosoguanidine (NTG) or with DNA-damaging radiation like ultraviolet (UV) light (Figure B9.1). Auxotrophic mutants can be identified from a population of mutagenized cells. Because auxotrophs cannot grow without a specific nutrient, this identification presents a challenge. Replica plating (see Figure 9.6) overcomes this problem.

Once nutritional mutants have been isolated, how do we determine which supplements they require for growth? One strategy would be to test each of the mutants on a series of media individually supplemented with each of the twenty amino acids, five nucleotides, and six essential vitamins. We could then determine which supplement restored the growth of the mutant. This approach would involve testing numerous media and would be quite cumbersome. A more common approach involves the use of pools of supplements as an intermediate step. The strategy uses a series of overlapping pools such that most supplements are represented in only two of the pools (Table B9.1). In this example, a tryptophan auxotroph would only be able to grow on pools 3 and 8. An aspartic acid auxotroph would only be able to grow on pools 4 and 9. Finally, these phenotypes could be confirmed by testing for growth on media with the single supplement. Through these mutagenesis and screening experiments, large collections of auxotrophic mutants have been added to genetic stock collections where they form a permanent resource for research in genetics and biochemistry.

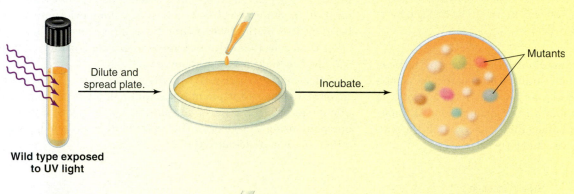

Wild type exposed to UV light

Dilute and spread plate. → Incubate. → Mutants

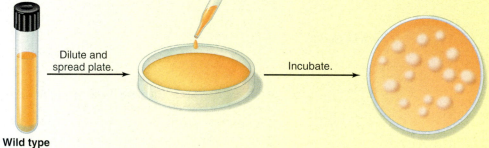

Wild type

Dilute and spread plate. → Incubate.

Figure B9.1. Using mutagenesis to increase mutation rates Cells can be exposed to various mutagens, like ultraviolet (UV) light, to increase the mutation rate. Following mutagenesis, the culture is diluted appropriately so that spread plating gives rise to single colonies. If there is no mutagenesis, mutants are very rare.

TABLE B9.1 Supplement pools used for auxotroph screening

Pool	1	2	3	4	5
6	Adenosine	Guanosine	Cysteine	Methionine	Thiamine
7	Histidine	Leucine	Isoleucine	Lysine	Valine
8	Phenylalanine	Tyrosine	Tryptophan	Threonine	Proline
9	Glutamine	Asparagine	Uracil	Aspartic acid	Arginine
10	Thymine	Serine	Glutamic acid		Glycine

● **Test Your Understanding** ··········

Imagine that you have generated what you believe to be a leucine auxotroph. What media would you use to confirm that this mutant is indeed auxotrophic for leucine? Explain the results you would expect on the chosen media.

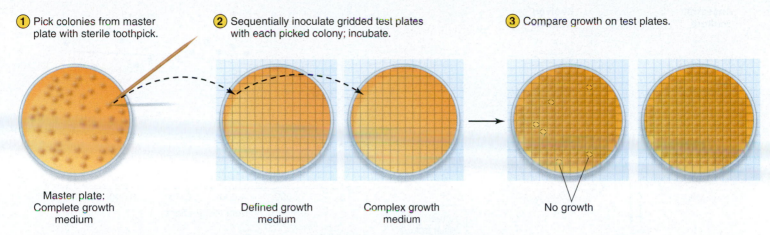

① Pick colonies from master plate with sterile toothpick.

② Sequentially inoculate gridded test plates with each picked colony; incubate.

③ Compare growth on test plates.

Master plate:
Complete growth
medium

Defined growth
medium

Complex growth
medium

No growth

Figure 9.7. Process Diagram: Replica plating using "patching" Sterile plates are placed agar side down on a numbered paper grid and sequentially inoculated with toothpicks, each of which has stabbed a single colony growing on a master plate. This master plate contains colonies plated from a mutagenized culture growing on complex medium. The colonies are transferred to a set of defined growth medium and complex growth medium plates. Those colonies that do not grow on the defined medium plates, but do grow on the complex medium plates, are potential auxotrophic mutants.

Alternatively, some scientists prefer a process called "patching" in which sterile toothpicks rather than velvets are used to transfer individual colonies to the different types of test plates. Numbered grid templates for the plates are used to keep track of where individual colonies are "patched" **(Figure 9.7)**. This technique allows more precision than the velvet method and more reproducible results. With a little practice and dedication, researchers can screen several thousand colonies using patching in a single day of work. Indeed, toothpicks and grid templates may be among the most important tools in the bacterial genetics laboratory! More recently, robotic systems have been developed that allow much higher rates of screening, on the order of tens of thousands of colonies per day.

To better understand the usefulness of replica plating, let's investigate the isolation of auxotrophic mutants that cannot synthesize arginine. Following treatment of cells with a mutagen, colonies are first grown on plates containing arginine to ensure the mutants will grow. These colonies can then be replica-plated onto medium not supplemented with arginine. The two plates can then be compared. Any colonies that appeared on the first plate, but not on the second, would be arginine auxotrophs. The selection and screening protocols used for isolating bacterial mutants provided the first crucial insights into adaptation and evolution. We'll investigate this topic next.

Evolution in a test tube

People often perceive mutations as negative changes in the genetic material that have detrimental outcomes, such as genetic diseases. While mutations can result in diseases, they can also be beneficial. Indeed, they ensure the vitality of all life on Earth. The genetic material in all organisms, including microorganisms, is constantly changing or mutating. These changes provide the raw material upon which environmental pressures act, continually selecting for those organisms with the most successful life strategies. Consider bacteria. Each bacterial cell is a complete organism, and bacterial colonies are composed of millions of genetically identical organisms, or **clones**, each of which originated from a single cell. Mutations arising within a single cell in a population of bacteria will be passed on to its progeny. If the mutation confers an advantage to those cells, then the progeny will be able to out-compete cells that lack the mutation, and their representation within the population will increase.

Biology is rooted in the process of evolution by natural selection, for which we have much evidence, including the fossil record and comparative DNA sequence analysis. Although this evidence is compelling, wouldn't it be satisfying to observe evolution in progress? Scientists can, in fact, demonstrate the evolutionary consequences of mutations experimentally by using a microbial system. On February 15, 1988, Richard Lenski established 12 parallel cultures of *E. coli* in defined growth media with glucose as the carbon source. These cultures were subcultured once per day into fresh growth media. Every 75 days, aliquots were removed and frozen for later analysis. After 10,000 generations over about 1,500 days, cell size and survivability of the evolved populations were compared with these same properties in the ancestral cells, which had been prevented from evolving by keeping them in a frozen, non-growing state. Both cell size and survivability increased in the evolved cultures **(Figure 9.8)**.

Some of these fitness gains resulted from mutations in genes involved in the cellular stress response. Presumably, these mutations allowed the cells to survive the stresses of growing under the culture conditions used in the laboratory. Similar studies have demonstrated that fitness gains are usually specific to given environmental conditions. In other words, a population that has evolved under certain culture conditions will exhibit enhanced fitness under those conditions, but not necessarily under different conditions. These straightforward but carefully conducted experiments showed that it is possible to demonstrate evolutionary processes as a result of mutations in a microbial experimental system.

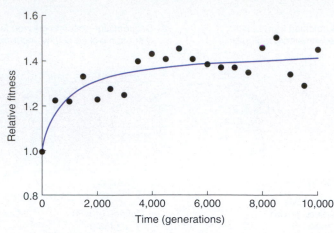

B. Relative fitness increases over time.

Figure 9.8. Experimental evolution of *E. coli* When subcultured repeatedly in growth medium, the relative fitness of *E. coli* cells increases. **A.** Ancestral and evolved cultures are grown separately, then mixed together in a 1:1 ratio. To determine the actual initial ratios, the mixed culture is diluted and plated for single colonies. A neutral marker in one of the strains allows the two strains to be easily differentiated by colony color. After the mixed culture is incubated, it is again diluted and plated for single colonies. The relative fitness is determined from the relative growth rates of the strains. **B.** Relative fitness in a population of *E. coli* during 10,000 generations of experimental evolution measured against the ancestral strain.

S. F. Elena and Richard E. Lenski. 2003. *Nature Reviews Genetics* 4:457. Reprinted by permission from Macmillan Publishers, Ltd.

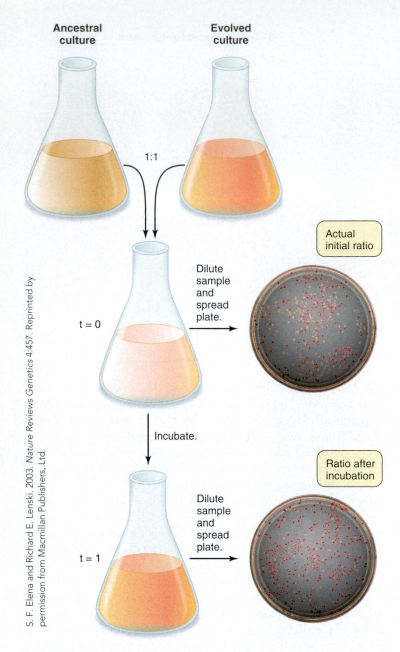

A. Ratio of red to white colonies is a measure of relative fitness.

Mutations and selection

Let's now consider how mutations arise in the first place. Do mutations occur in response to selective environmental conditions experienced by the cell, or do they occur spontaneously even in the absence of selective conditions? Scientists had observed that if a culture of *E. coli* grows in the presence of an antibiotic such as streptomycin, a few resistant variants appear. When these cells are then grown for a time in the absence of streptomycin, these variants retained their streptomycin resistance phenotype. Was the interaction between the antibiotic and the bacterial cell required to cause the mutation producing streptomycin resistance? A key experiment to answer this question was performed in the early 1950s by Esther Lederberg, who was married to Joshua Lederberg at the time.

Esther Lederberg demonstrated that streptomycin-resistant mutations arise spontaneously in the absence of selective pressure from the antibiotic streptomycin. She first spread a culture of *E. coli* on an agar plate that did not contain streptomycin **(Figure 9.9)**. Then she used replica plating to transfer these bacteria to an agar plate containing streptomycin. After overnight incubation, only a few colonies were visible on the streptomycin plate.

Lederberg then removed bacterial cells from precisely the region of the antibiotic-*free* plate that corresponded to the regions in which the colonies were growing on the antibiotic-*containing* plate. Because these cell samples were removed from a lawn, they contained a mixture of cells, but Esther Lederberg suspected somewhere in this mix were mutant cells that were resistant to streptomycin. She added these cells to medium containing streptomycin. She also removed bacterial cells from the antibiotic-free plate from a region distant from the areas corresponding to the streptomycin-resistant colonies. She added these cells to medium containing streptomycin. After incubating these samples, she observed that the cells from the region of the antibiotic-free plate corresponding to the streptomycin-resistant colony grew in the presence of streptomycin, while the cells from a distant region of the plate did not. From this result, she concluded that mutations can occur in the absence of selective pressure.

While Esther Lederberg's mutation experiments were convincing, Salvador Luria and Max Delbrück used another approach to investigate whether mutations originate randomly and spontaneously without regard to selective pressure. They

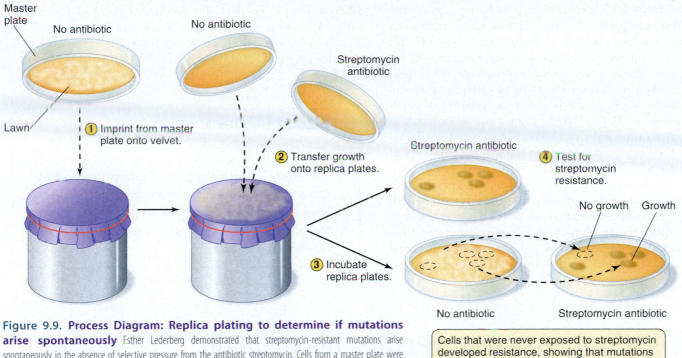

Figure 9.9. Process Diagram: Replica plating to determine if mutations arise spontaneously Esther Lederberg demonstrated that streptomycin-resistant mutations arise spontaneously in the absence of selective pressure from the antibiotic streptomycin. Cells from a master plate were replica-plated on solid medium containing streptomycin and also on another plate containing no antibiotic. Some streptomycin-resistant colonies grew on the streptomycin plate. Cells from the antibiotic-lacking plate corresponding to the location of the resistant colonies grew on plates containing streptomycin, indicating that the mutations existed before exposure to the selective agent.

Cells that were never exposed to streptomycin developed resistance, showing that mutations can arise in the absence of the selective agent.

investigated the generation of resistance in *E. coli* to infection with the bacteriophage T1 (**Figure 9.10**). They reasoned that each random spontaneous mutation imparting phage resistance in a culture would be passed on to the progeny of the cells in which the original mutation occurred. If the mutation occurred early in the growth of the culture, then more progeny containing the mutation would be found in the culture than if the mutation occurred later in the culture.

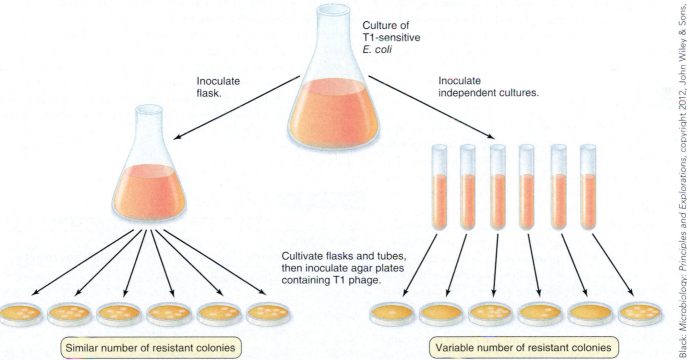

Figure 9.10. The fluctuation test To determine if mutations occur randomly in the absence of selective pressure, Luria and Delbrück grew phage T1-sensitive *E. coli* in a large volume and plated aliquots on agar medium containing phage T1. They also grew many small cultures of T1-sensitive *E. coli* and plated aliquots from each culture on agar containing T1 phage. While most aliquots from the single, large culture contained similar numbers of phage-resistant bacteria, the number of resistant bacteria differed dramatically in the various small cultures.

Black: *Microbiology: Principles and Explorations*, copyright 2012, John Wiley & Sons, Inc. This material is reproduced with permission of John Wiley & Sons, Inc.

Luria and Delbrück's experiment, which they called the fluctuation test, compared the frequency of phage resistance arising in independent cultures of *E. coli*. They removed aliquots from several individual cultures and applied these *E. coli* aliquots to agar plates along with bacteriophage. They predicted that random mutations occurring at different times in different cultures would result in different numbers of resistant colonies on the plates related to how early the mutations occurred in the original culture. When they conducted this experiment, they observed a large variance in the numbers of phage-resistant colonies on the plates from the individual cultures. As a control, a single large culture of cells was grown from initially non-resistant cells, and aliquots from this single culture were applied to several agar plates and subjected to phage challenge. Because the progeny of any phage-resistant mutants would be distributed throughout this culture, they hypothesized that similar numbers of resistant colonies would appear on the

different plates. Their results confirmed this hypothesis. Like Esther Lederberg, Luria and Delbrück had demonstrated that mutations occur spontaneously and randomly in the absence of selective pressure.

9.2 Fact Check

1. What is a wild-type strain?
2. Describe how mutant genes and proteins are named.
3. Explain the processes of selection and screening.
4. What is replica plating, and how was it used to show that antibiotic-resistant mutants arise in the absence of selection pressure?

9.3 Restriction enzymes, vectors, and cloning

How is DNA cloned?

As we saw in Section 7.1, Avery, MacLeod, and McCarty experimentally showed DNA to be the heritable material in 1944. By the 1960s, researchers routinely extracted DNA from cells and viruses. Soon after, researchers demonstrated that DNA ligase, an enzyme extracted from *Escherichia coli* infected with T4 phage, could join together DNA fragments. Then, in 1973, investigators used restriction enzymes, first isolated in 1970, to construct the first recombinant plasmid. DNA cloning was born. In this section, we will explore this remarkable achievement. Specifically, we will discuss restriction enzymes and vectors.

Restriction enzymes

To create recombinant DNA molecules, researchers must be able to cleave DNA and stitch together pieces of DNA very precisely. The discovery of **restriction enzymes**, bacterial proteins that cleave DNA at specific sequences, allowed microbiologists to fulfill both of these needs. Their *in vitro* use led to the development of molecular biology. Here, we will briefly examine how these enzymes function. We will also investigate their role in molecular biology.

Each restriction enzyme recognizes a specific, short DNA sequence, usually of 4, 6, or 8 base pairs, called a **restriction site**. Because they cut at these precise positions in DNA, restriction enzymes can be thought of as "molecular scissors." At these sites, the enzymes make double-stranded cuts, typically generating fragments with single-stranded overhangs called "sticky" or cohesive ends **(Table 9.3)**. These "sticky" ends are valuable in making **recombinant DNA**, or DNA generated in the laboratory that does not exist naturally. Note that the name of a restriction enzyme reflects its source. The first letter reflects the genus, the second and third letters reflect the species, and the fourth letter reflects the strain. The Roman numeral refers to its order of isolation from that strain.

Notice in Table 9.3 that the two strands of DNA at restriction sites have the same sequence of bases if both are read in a 5′ to 3′ direction. In other words, the sequences are identical if one strand is read from left to right and the other strand is read from right to left; they are palindromes. Two fragments of DNA that have been cut with the same restriction enzyme, *Eco*RI, for example, will have single-stranded ends that can anneal with one another **(Figure 9.11)**. The enzyme **DNA ligase** is then added to re-form the phosphodiester bonds between adjacent 5′ phosphate groups and 3′ hydroxyl ends. Blunt-ended fragments, produced by the restriction enzyme *Sma*I for example, can join to any other DNA fragment with blunt ends. Because the ends do not anneal by base-pairing, the sequence of these ends does not matter; any blunt-ended fragment will do.

TABLE 9.3 Properties of selected restriction enzymes

Restriction enzyme	Source	Restriction site	Cleavage result	
*Eco*RI	*Escherichia coli*	5′-GAATC-3′ 3′-CTTAG-5′	5′-G 3′-CTTAA	AATC-3′ G-5′
*Bam*HI	*Bacillus amylo-liquefaciens*	5′-GGATCC-3′ 3′-CCTAGG-5′	5′-G 3′-CCTAG	GATCC-3′ G-5′
*Hind*III	*Haemophilus influenzae*	5′-AAGCTT-3′ 3′-TTCGAA-5′	5′-A 3′-TTCGA	AGCTT-3′ A-5′
*Sma*I	*Serratia marcescens*	5′-CCCGGG-3′ 3′-GGGCCC-5′	5′-CCC 3′-GGG	GGG-3′ CCC-5′

Figure 9.11. Annealing and ligation of sticky ends Sticky ends created by a restriction enzyme are complementary. When fragments generated by the same restriction enzyme are mixed, the sticky ends can anneal. DNA ligase then covalently links the fragments, creating an intact, double-stranded DNA molecule.

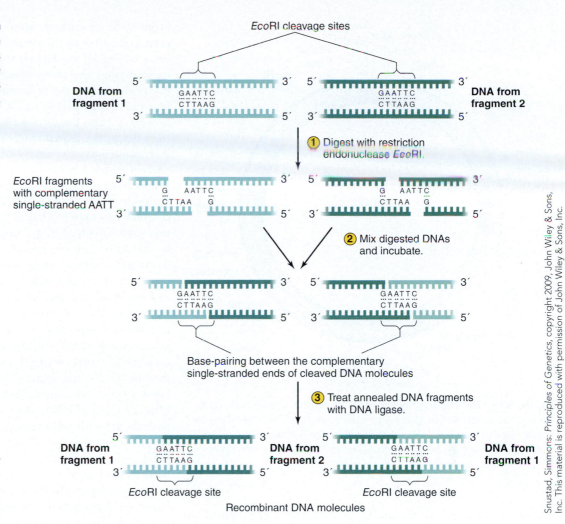

EcoRI cleavage sites

DNA from fragment 1

5´ GAATTC 3´
CTTAAG
3´ 5´

DNA from fragment 2

5´ GAATTC 3´
CTTAAG
3´ 5´

① Digest with restriction endonuclease *EcoRI*.

EcoRI fragments with complementary single-stranded AATT

5´ G AATTC 3´
CTTAA G
3´ 5´

5´ G AATTC 3´
CTTAA G
3´ 5´

② Mix digested DNAs and incubate.

5´ GAATTC 3´
CTTAAG
3´ 5´

5´ GAATTC 3´
CTTAAG
3´ 5´

Base-pairing between the complementary single-stranded ends of cleaved DNA molecules

③ Treat annealed DNA fragments with DNA ligase.

DNA from fragment 1

5´ GAATTC 3´
CTTAAG
3´ 5´

EcoRI cleavage site

DNA from fragment 2

5´ GAATTC 3´
CTTAAG
3´ 5´

EcoRI cleavage site

DNA from fragment 1

Recombinant DNA molecules

Snustad, Simmons: Principles of Genetics, copyright 2009, John Wiley & Sons, Inc. This material is reproduced with permission of John Wiley & Sons, Inc.

Figure 9.12. The first plasmid-cloning experiment Using digestion with the restriction enzyme *EcoRI* followed by treatment with the enzyme DNA ligase, Cohen and colleagues combined fragments of pSC101 and pSC102, constructing a new plasmid, pSC105. They then inserted this recombinant DNA into *E. coli*.

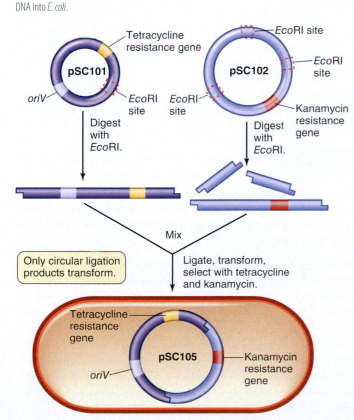

Cloning vectors

As we have just seen, restriction enzymes allow researchers to cleave DNA at specific sites and then stitch together the resulting DNA fragments, thereby creating recombinant DNA molecules. To replicate these recombinant molecules, a process referred to as **molecular cloning**, or **DNA cloning**, researchers generally insert the exogenous DNA into a **cloning vector**, a DNA molecule that can be genetically manipulated and replicates within cells. While many different types of cloning vectors have been developed, plasmids, phages, and cosmids have been used most extensively for DNA cloning. Here, we will examine more thoroughly the usefulness of these entities as cloning vectors.

Plasmid Cloning Vectors

Stanley Cohen and colleagues at Stanford University ushered in the age of molecular biology when they performed one of the first DNA cloning experiments using restriction enzymes and plasmids **(Figure 9.12)**. They began by cleaving a tetracycline-resistant plasmid, pSC101 (named using the initials of the scientist), with the restriction enzyme *EcoRI*. Because this plasmid contained a single *EcoRI* restriction site (5´-GAATTC-3´), the plasmid became linear following cleavage. Another plasmid, pSC102,

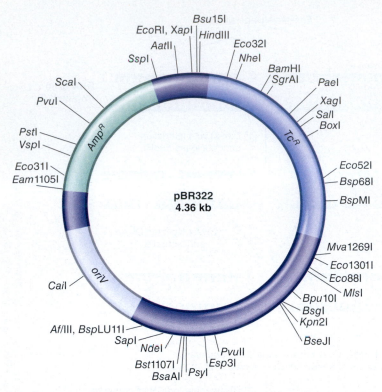

Figure 9.13. The cloning vector pBR322 This plasmid contains several restriction sites for cloning, and ampicillin (*Amp*^{R}) and tetracycline (*Tc*^{R}) resistance genes. The *oriV* site indicates the origin of replication. Unique restriction sites exist within the antibiotic resistance genes. Many plasmid vectors used in molecular biology for routine cloning and for production of recombinant proteins are derived from this plasmid.

conferred resistance to the antibiotic kanamycin and yielded 3 different fragments when digested with *Eco*RI, indicating that it contained 3 *Eco*RI restriction sites. Researchers combined the restriction fragments of these two plasmids, treated the resulting mixture with DNA ligase, and then introduced the DNA into *E. coli* cells. Cells exhibiting resistance to both tetracycline and kanamycin, they reasoned, received a recombinant plasmid that contained the tetracycline resistance gene from pSC101 and the kanamycin resistance gene from pSC102.

To further investigate this result, the researchers then isolated from these doubly resistant cells the DNA of a new plasmid, designated pSC105. When digested with *Eco*RI, this plasmid produced two fragments. These fragments corresponded to the linearized pSC101, encoding tetracycline resistance, and one of the pSC102 fragments, encoding kanamycin resistance. In other words, the pSC105 plasmid was a recombinant molecule consisting of both pSC101 and pSC102 DNA. The age of molecular biology was born.

One of the first widely used cloning vectors was pBR322 **(Figure 9.13)**. This relatively small plasmid (less than 4,400 base pairs) has unique restriction sites that can be used for inserting fragments of DNA following cleavage with the corresponding restriction enzymes. It also has two different antibiotic resistance genes, one for ampicillin resistance and one for tetracycline resistance. Both of these antibiotic resistance genes contain restriction sites within them.

Let's use this plasmid to illustrate a typical cloning experiment. Suppose we want to clone, or create many copies of the α-amylase gene from a starch-degrading *Bacillus* sp.

From previous work, we know that this gene is flanked by *Bam*HI restriction sites in the *Bacillus* genome, but does not contain an internal *Bam*HI restriction site. In pBR322, a single *Bam*HI site exists within the tetracycline resistance gene **(Figure 9.14)**. To insert the α-amylase gene into pBR322, we first can cleave the plasmid with *Bam*HI so that it is cut within the tetracycline resistance gene and becomes a single linear DNA molecule. We can then digest the genomic DNA of the *Bacillus* sp. with this same restriction enzyme, releasing thousands of DNA fragments, each of which contains *Bam*HI sticky ends. These *Bacillus* DNA fragments are combined with the linearized vector DNA, and the mixture is treated with the enzyme DNA ligase to covalently combine pieces of DNA that had joined together by complementary base-pairing. A large number of possible molecules can form by this joining process. Some molecules will be linear, others will be circular, and some will contain multiple pieces of DNA joined together. Many molecules will be made up of the vector alone, re-formed after its two ends have ligated together.

Out of all these molecules that may form, how can we select the one type that interests us: recombinant pBR322 that now contains the α-amylase gene? We can start by adding the mixture of molecules to *E. coli* cells and plating these cells on medium that contains ampicillin. To grow on this medium, cells must express the ampicillin resistance gene present on the pBR322 plasmid DNA. Moreover, the added DNA must be able to replicate within these cells. Only circular plasmid molecules that contain an origin of replication will replicate; fragments of DNA and circular molecules lacking an origin of replication will be degraded and/or lost from the cells. By simply plating our cells on medium containing ampicillin, we can identify cells that have received a functioning plasmid (see Figure 9.14). However, not all circular molecules formed will contain DNA fragments from the *Bacillus* sp. How can we distinguish between recircularized and recombinant pBR322 molecules? Recircularized pBR322 molecules will contain an intact tetracycline resistance gene. In recombinant molecules, this gene will be disrupted and non-functional. Ampicillin-resistant cells could be screened for tetracycline sensitivity. Cells that exhibit resistance to ampicillin, but sensitivity to tetracycline, contain a recombinant pBR322 plasmid in which a DNA fragment has been inserted into the *Bam*HI site.

CONNECTIONS When the genome of an organism is cleaved with a restriction enzyme, all resulting fragments can be ligated into copies of a plasmid cleaved with the same restriction enzyme. A collection of recombinant plasmids, then, could contain all segments of the foreign genome. The creation and use of such genomic libraries will be discussed in **Section 10.2**.

We now have one last challenge. We must identify the recombinant plasmid that contains our gene of interest—the α-amylase gene. To achieve this goal, we can use a functional screen. We can screen each of the ampicillin-resistant, tetracycline-sensitive colonies for the ability to hydrolyze starch, the function of α-amylase. A simple assay reveals this activity (see Figure 9.14). Although *E. coli* itself possesses a gene encoding α-amylase, it has very low levels of enzyme activity. Cells that exhibit a high level of α-amylase activity contain a plasmid with our gene of interest.

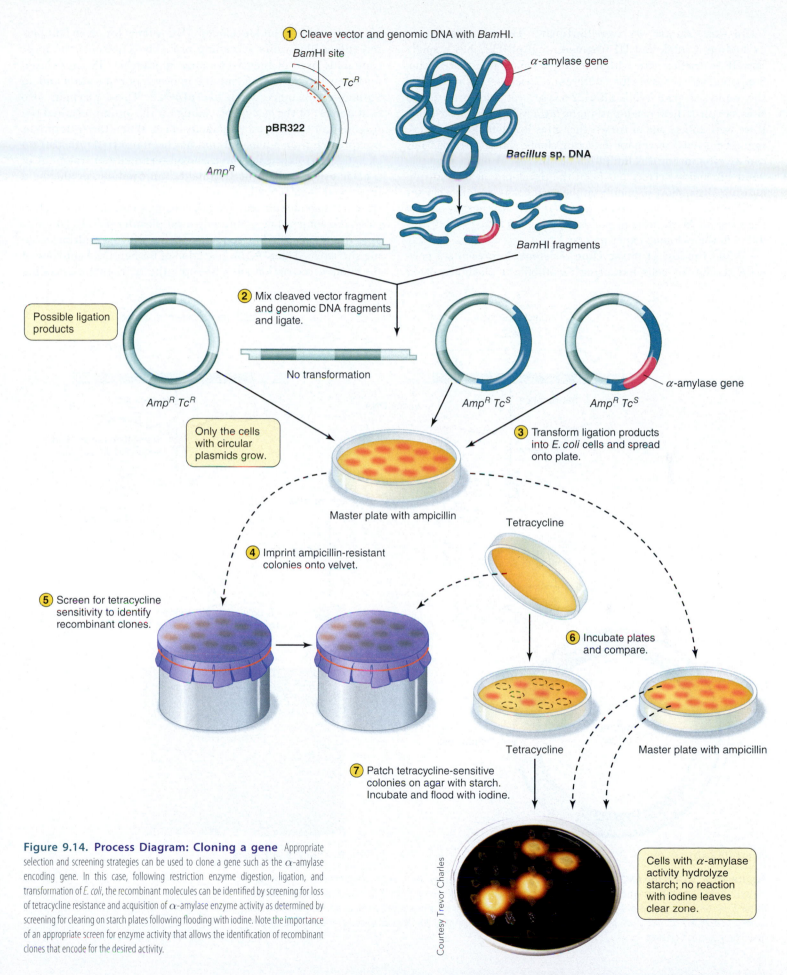

① Cleave vector and genomic DNA with *Bam*HI.

*Bam*HI site

Tc^R

pBR322

Amp^R

α-amylase gene

***Bacillus* sp. DNA**

*Bam*HI fragments

② Mix cleaved vector fragment and genomic DNA fragments and ligate.

Possible ligation products

No transformation

Amp^R *Tc*^{R}

Amp^R *Tc*^{S}

Amp^R *Tc*^{S}

α-amylase gene

Only the cells with circular plasmids grow.

③ Transform ligation products into *E. coli* cells and spread onto plate.

Master plate with ampicillin

Tetracycline

④ Imprint ampicillin-resistant colonies onto velvet.

⑤ Screen for tetracycline sensitivity to identify recombinant clones.

⑥ Incubate plates and compare.

Tetracycline

Master plate with ampicillin

⑦ Patch tetracycline-sensitive colonies on agar with starch. Incubate and flood with iodine.

Cells with α-amylase activity hydrolyze starch; no reaction with iodine leaves clear zone.

Courtesy Trevor Charles

Figure 9.14. Process Diagram: Cloning a gene Appropriate selection and screening strategies can be used to clone a gene such as the α-amylase encoding gene. In this case, following restriction enzyme digestion, ligation, and transformation of *E. coli*, the recombinant molecules can be identified by screening for loss of tetracycline resistance and acquisition of α-amylase enzyme activity as determined by screening for clearing on starch plates following flooding with iodine. Note the importance of an appropriate screen for enzyme activity that allows the identification of recombinant clones that encode for the desired activity.

In this example, we were fortunate. The α-amylase gene existed on a single *Bam*HI fragment, and pBR322 has a single *Bam*HI restriction site within its tetracycline resistance gene. Suppose restriction sites for a different restriction enzyme flanked the α-amylase gene. While pBR322 has several unique restriction sites, not all of them reside within the tetracycline resistance gene. If we were to use one of these other sites for cloning, then there would be no easy screen for detecting clones with insertions. How can we circumvent this limitation of pBR322? A major improvement in cloning vectors has been the development of **multiple cloning sites**, short segments on the vector that contain a cluster of different restriction enzyme sites that each appear only once in the plasmid. With these plasmids, researchers have greater flexibility in their cloning experiments.

While the loss of tetracycline resistance represents a powerful screen for cells harboring recombinant plasmids, other plasmids have been developed that allow for even simpler screening. Blue-white screening probably represents the most commonly used of these screening approaches. Plasmids used for blue-white screening contain an origin of replication and an antibiotic resistance gene, like pBR322. These plasmids also contain part of the *lacZ* gene, coding for the amino-terminal (α) fragment of the enzyme β-galactosidase. When the α fragment is expressed from a plasmid that has been inserted into a strain of *E. coli* that expresses the carboxyl-terminal (ω) fragment of β-galactosidase, then the fragments can combine, producing a functional β-galactosidase enzyme. This enzyme breaks down the chromogenic substrate X-gal, causing the formation of blue-colored colonies. The commonly used plasmid pUC18 contains an α fragment-encoding sequence within which a multiple cloning site exists **(Figure 9.15)**. Insertion of foreign DNA into one of the unique restriction sites disrupts the α fragment-encoding

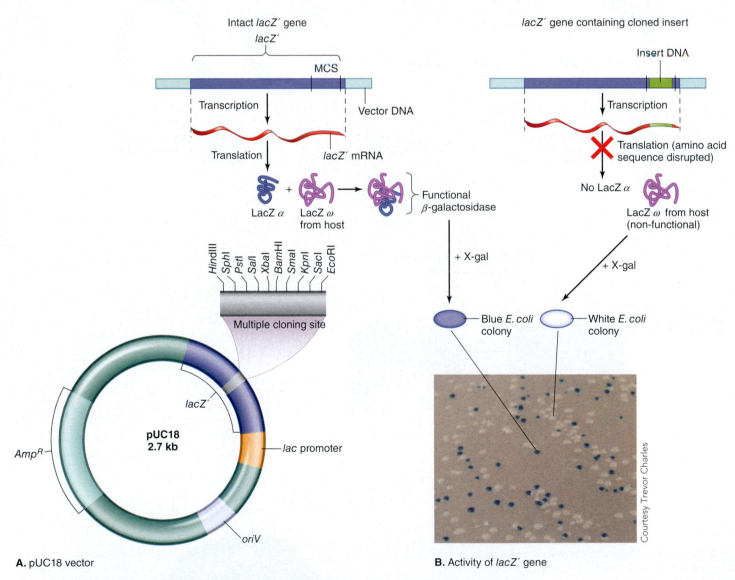

A. pUC18 vector

B. Activity of *lacZ′* gene

Courtesy Trevor Charles

Figure 9.15. Using the *lacZ* gene to improve cloning efficiency **A.** The plasmid pUC18 contains a multiple cloning site within the α fragment-encoding part of the *lacZ* gene (*lacZ′*). Insertion of foreign DNA within this site disrupts *lacZ* expression. **B.** Following ligation, *E. coli* cells that contain recombinant molecules are detected by screening for absence of β-galactosidase activity on agar media containing the chromogenic substrate X-gal. Plasmids that have inserts will give rise to white colonies on X-gal plates, while plasmids that do not have inserts will give rise to blue colonies on X-gal plates.

sequence, thus abolishing β-galactosidase activity. Transformants with inserts therefore appear white when grown in the presence of X-gal, whereas transformants without inserts appear blue. Researchers can easily identify and isolate a white colony growing in the presence of many blue colonies.

Today, a great number of plasmid vectors for a variety of host cells exist. Many of these vectors are commercially available; others are shared within the scientific community. Regardless of their unique properties, modern cloning plasmids typically have the following features:

- an *origin of replication* that functions in the host cell(s)

- a *selectable marker* that imparts a phenotype, usually antibiotic resistance, on cells carrying vector DNA

- a *multiple cloning site* that facilitates cloning and allows screening of cells containing cloned DNA

- a *small size* that maximizes transfer into host cells

- a *high copy number* so many copies of the cloned DNA exist in a small number of cells

Most of the plasmid vectors used in molecular biology are derived from the *E. coli* ColE1 plasmid, which contains the *oriV* origin of replication. Because this origin functions only in *E. coli* and closely related bacteria, these plasmids have a restricted host range. To broaden the usefulness of plasmid vectors, researchers have developed **shuttle vectors**, vectors that can replicate in a more diverse range of hosts. Typically, these vectors contain more than one origin of replication **(Figure 9.16)**, allowing them to replicate in more than one type of host. They also often contain two different selectable markers. In a typical experiment, *E. coli* usually is used for the cloning and amplification steps to construct the recombinant plasmid so one of these origins of replication is able to function in *E. coli*, while the other functions in the organism under study. Shuttle vectors have been used for many disparate purposes, including the *in vivo* assembly of synthesized overlapping oligonucleotide fragments in yeast. Because yeast cells can recombine DNA segments with high efficiency, researchers can insert into yeast cells multiple overlapping DNA fragments and a linearized shuttle vector. Within the transformed yeast cells, the DNA pieces undergo recombination, forming a larger, continuous DNA fragment, which becomes ligated into the shuttle vector. Following growth of the cells, this plasmid can be isolated and transferred to *E. coli*, where it can be further manipulated. The process allows researchers to produce very large segments of DNA with minimal *in vitro* steps.

Phage Vectors

While researchers use plasmid vectors quite frequently, they also use phage vectors. These vectors take advantage of the natural ability of viruses to infect and efficiently deliver their genomes into cells. DNA fragments of interest and phage DNA fragments are mixed and ligated. The resulting recombinant DNA can be packaged into phage particles *in vitro*, using purified phage head and tail assemblies that are commercially available. The resulting phages can be added to *E. coli*, and then plated on a solid medium to form plaques (see Section 5.3).

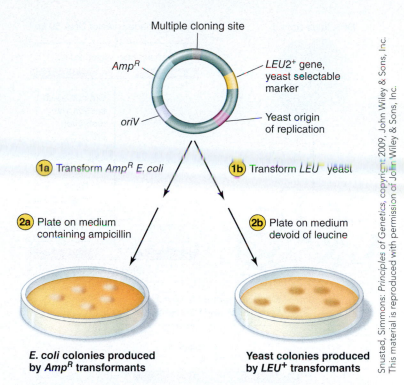

Snustad, Simmons: *Principles of Genetics*, copyright 2009, John Wiley & Sons, Inc. This material is reproduced with permission of John Wiley & Sons, Inc.

Figure 9.16. Shuttle vectors for cloning in two different cell types
Shuttle vectors typically include two origins of replication, one that functions in *E. coli*, and one that functions in the organism being studied (yeast in this schematic).

The best examples of phage vectors are based on phage lambda (λ), a temperate phage that can integrate into the *E. coli* chromosome (see Section 5.3). When integrated, progeny phages are not produced and, as a result, a cloned fragment would not be amplified. To avoid this obvious shortcoming, researchers have removed from the phage λ genome the genes for integration and excision, genes unnecessary for viral replication **(Figure 9.17)**. Removal of this middle 20 kb of sequence from phage λ leaves left and right "arms" of linear DNA into which about 20 kb of foreign DNA fragments can be ligated.

To be packaged correctly, the recombinant DNA must contain the phage *cos* site. As DNA enters the capsid head, a viral endonuclease present at the entry to the capsid head cleaves the DNA at the *cos* site to ensure that about 50 kb of DNA enters each viral particle. Cleavage at the *cos* site produces a linear DNA molecule with a single-stranded overhang at each end that can anneal and recircularize the DNA once inside the cell. A *cos* site exists at the end of each arm of the phage DNA, supplying recombinant DNA fragments with the means to be packaged and propagated appropriately.

Cosmid Vectors

As we have just seen, the main advantage of phages is their ability to efficiently deliver cloned DNA fragments into host cells. Because packaging of 50 kb of DNA into preassembled capsid heads and tails only requires the presence of *cos* sites, the rest

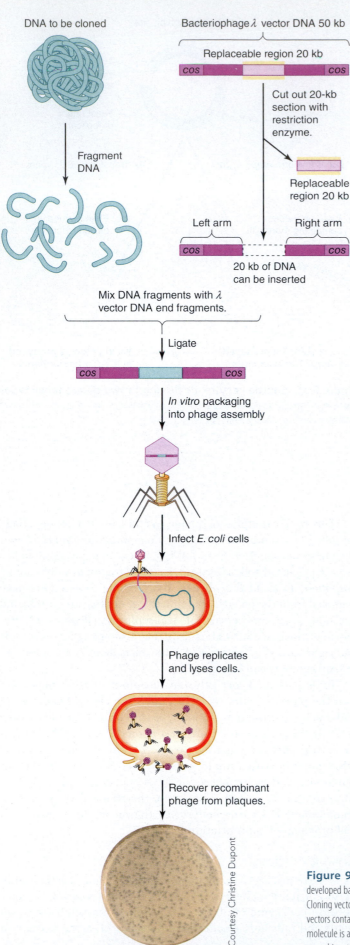

DNA to be cloned

Bacteriophage λ vector DNA 50 kb

Replaceable region 20 kb

cos   cos

Cut out 20-kb
section with
restriction
enzyme.

Fragment
DNA

Replaceable
region 20 kb

Left arm   Right arm

cos   cos

20 kb of DNA
can be inserted

Mix DNA fragments with λ
vector DNA end fragments.

Ligate

cos   cos

In vitro packaging
into phage assembly

Infect E. coli cells

Phage replicates
and lyses cells.

Recover recombinant
phage from plaques.

Courtesy Christine Dupont

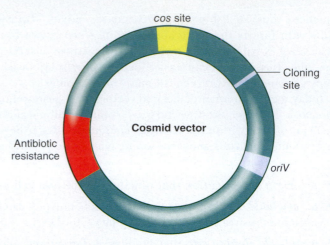

Figure 9.18. Cosmids: Plasmid/phage hybrid vectors Plasmid vectors can be designed to contain the *cos* recognition site, which allows *in vitro* packaging using λ extracts that contain head and tail proteins. These are called <u>cos</u>mid vectors. Just as with the λ phage vectors, only molecules of about 50 kb can be packaged. This allows the selection for clones that contain large DNA inserts, especially if the cosmid vector DNA itself is substantially smaller than 50 kb.

of the phage sequence can be omitted. Hybrid plasmid/phage vectors called **cosmids** have been designed (**Figure 9.18**). The only remnant of the phage genome left in these cloning vectors is the lambda (λ) *cos* packaging recognition sites. The rest of the vector DNA typically consists of a multiple cloning site, an origin of replication, and an antibiotic selection marker. Foreign DNA fragments of approximately 35–45 kb can be inserted into cosmids, depending on the vector size, so that the total size of each clone is about 50 kb. After ligation, recombinant molecules flanked with *cos* are incubated with λ phage head and tail protein mixtures, which will result in the packaging of DNA molecules. The introduction of these packaged clones into an *E. coli* host occurs with very high efficiency. Once in the host cell, the DNA is maintained as a plasmid.

CONNECTIONS One of the reasons to clone a gene in a vector is to produce large amounts of a recombinant protein by expressing that gene in *E. coli* or another host. In **Section 12.2**, we will discuss vectors specially designed for this purpose.

9.3 Fact Check

1. Describe the activity of restriction enzymes.
2. What purpose does a selectable marker play in a plasmid vector?
3. Explain blue-white screening.
4. What is a shuttle vector?
5. What advantages do phage vectors have?
6. What are cosmids?

Figure 9.17. Phage lambda (λ) as a cloning vector A large variety of cloning vectors have been developed based on λ bacteriophage. Less than half of the 50 kb λ genome is required for λ to function as a lytic bacteriophage. Cloning vectors can be developed that exclude part of the genome, thereby providing room for the insertion of foreign DNA. These vectors contain *cos* sites, which are required for packing into phage particles. The λ genome is only packaged if the size of the DNA molecule is about 50 kb, and this size constraint is used to select for clones that contain a DNA insert. Following infection of *E. coli*, recombinant phage clones are recovered from plaques.

9.4 Recombination and DNA transfer

How do bacterial cells transfer DNA?

As we saw in the opening story of this chapter, Joshua Lederberg demonstrated that genetic exchange could occur in bacteria. Subsequent analyses of bacterial genomes have made it abundantly clear that segments of DNA can move between bacteria. Foreign DNA can become fixed in a new host's genome and confer new traits. Such movement of DNA between microbes is called lateral or **horizontal gene transfer** (see Section 10.3), differentiating it from vertical gene transfer, the inheritance of a gene from a direct ancestor. Many of the genetic techniques used by microbiologists are based on naturally occurring mechanisms of horizontal gene transfer in bacteria: transformation, conjugation, transduction, and transposition. Regardless of how DNA gets into the cell, some form of recombination commonly occurs to incorporate the foreign DNA into the bacterial genome. In this section, we will examine these four DNA transfer mechanisms. We'll begin, though, by investigating recombination.

Recombination

Once DNA enters a cell, what becomes of it? To remain in the cell, it either must replicate on its own, or it must join with the recipient cell's DNA. In other words, it must have access to an origin of replication or it will be degraded. Chromosomes and plasmids can replicate on their own, but DNA fragments must be incorporated into the recipient cell's DNA by the process of recombination. This involves breaking and rejoining different DNA strands and may have evolved to repair occasional breaks in DNA that occur during replication. Incorporation of foreign DNA into the bacterial genome appears to be a spin-off of this repair mechanism. Here we examine two forms of recombination: homologous and non-homologous. We will then examine how microbiologists use recombination.

Homologous Recombination

Homologous recombination, also called crossover, occurs when two segments of DNA with identical or very similar sequences pair up and exchange or replace some portion of their DNA. This process is best understood for *Escherichia coli*. A single recombination event, called a single crossover, occurs when double-strand breaks exist in DNA, such as for fragments of foreign DNA that enter the cell **(Figure 9.19)**. To incorporate this DNA, the enzyme RecBCD binds to one blunt end in the first DNA molecule and migrates along the strand, unwinding one of the strands of the double-stranded DNA molecule. Unwinding continues until RecBCD encounters a Chi site (crossover hotspot instigator), where the enzyme nicks the strand, producing an internal free 3′-ended strand. Chi sites of *E. coli* have the sequence 5′-GCTGGTGG-3′ and exist randomly in DNA. Further unzipping of the cleaved molecule produces a long single-stranded end that can pair with complementary DNA. The RecA protein then loads onto this 3′ end. The attached RecA scans the sequence of the second molecule for regions of homology and aligns the single strand by invading and displacing the homologous strand. The displaced strand of the second molecule is

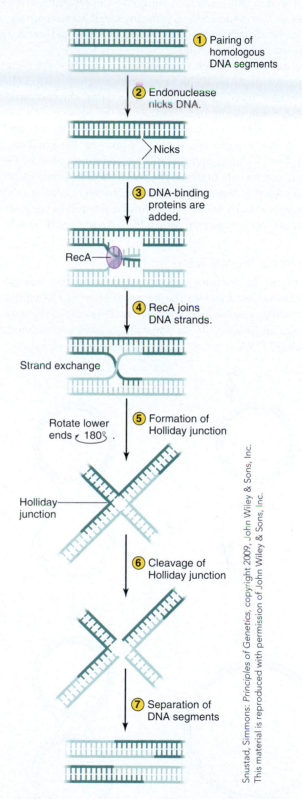

① Pairing of homologous DNA segments

② Endonuclease nicks DNA.

Nicks

③ DNA-binding proteins are added.

RecA

④ RecA joins DNA strands.

Strand exchange

Rotate lower ends 180°.

⑤ Formation of Holliday junction

Holliday junction

⑥ Cleavage of Holliday junction

⑦ Separation of DNA segments

Snustad, Simmons: *Principles of Genetics*, copyright 2009, John Wiley & Sons, Inc. This material is reproduced with permission of John Wiley & Sons, Inc.

Figure 9.19. Process Diagram: DNA recombination Homologous recombination is initiated by the formation of single-stranded breaks. The RecA protein binds to the resulting single-stranded DNA and searches for regions of homology where the DNA can be aligned, displacing one of the strands at that region. A reciprocal event occurs on the other strand. This event produces the Holliday junction, where the two DNA molecules are joined. Cleavage of the junction results in separation of the DNA molecules in a manner that results in recombined molecules.

then cut at the 5′ end. This second free single strand can then anneal to the gap left in the first DNA fragment, producing the Holliday junction, named for the scientist Robin Holliday who first proposed this model. To complete the recombination event, two nicks are made in both non-displaced strands at the Holliday junction, and the joining strands are ligated, freeing the two double-stranded molecules.

In bacterial genetics, the most important types of recombination involve single or double crossovers. Single crossovers between two circular molecules, such as a circular chromosome and a plasmid, result in the incorporation of the plasmid DNA into the chromosome. Double crossovers between one circular and a circular or linear molecule, such as a circular chromosome and a plasmid or DNA fragment, results in the integration of the DNA of the first molecule between the two sites of recombination in the second molecule **(Figure 9.20)**. A single crossover between a circular chromosome and a linear fragment would result in the generation of a single linear molecule, which likely would not be stable.

Non-homologous Recombination

In contrast to homologous recombination, **non-homologous recombination** involves the recombination of DNA pieces that exhibit little or no sequence similarity. This type of recombination occurs in all forms of life and usually involves certain viruses and specialized DNA sequences called "transposable elements." As we

discussed in Section 8.3, temperate bacteriophages can undergo lytic replication, in which new copies of the virus are produced within an infected host cell, or lysogenic replication, in which the phage genome becomes integrated into the host chromosome. This integration occurs via recombination at specific but non-homologous DNA sequences, a type of non-homologous recombination called **site-specific recombination**. Enzymes produced by the viruses cleave host DNA and facilitate insertion of the viral DNA.

CONNECTIONS For temperate bacteriophages like phage λ, the switch between lytic and lysogenic replication involves a tightly controlled molecular signaling pathway for integration into the host genome by site-specific recombination **(Figure 8.21)**. Stressors to the host cell, such as UV light or chemical mutagens, often trigger the conversion.

Uses of Recombination

Naturally occurring recombination occurs, as we have noted, quite frequently and certainly contributes to evolutionary change. Researchers also use this process to intentionally alter the genetic content of cells. The ability to make defined mutations, and to replace segments of the genome with segments that have been altered, is crucial to many genetic manipulations. Researchers also use recombination to disrupt genes, a type of mutation called

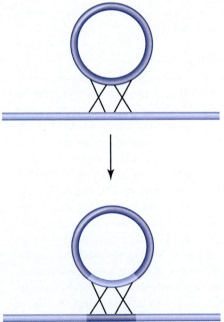

Figure 9.20. Single and double crossovers Single and double crossovers result in different outcomes. **A.** A single crossover between two circular molecules results in a circular molecule. **B.** A double crossover between two circular molecules results in reciprocal exchange of the region between the two crossover events. **C.** A double crossover between a circular and linear molecule results in reciprocal exchange of the region between the two crossover events.

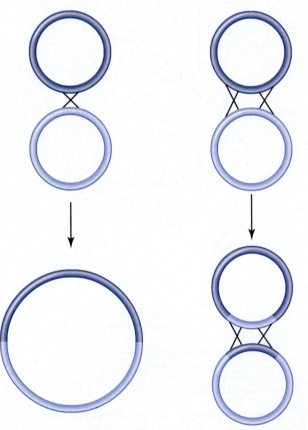

A. Single crossover

B. Double crossover between two circular molecules

C. Double crossover between circular and linear molecules

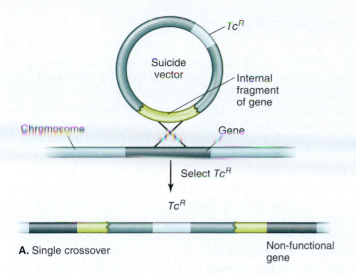

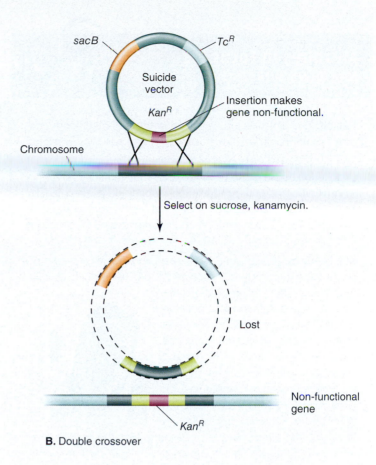

Figure 9.21. Generation of knockout mutants Recombination can be used to generate knockout mutants, where the gene of interest has been rendered non-functional. Suicide vectors, lacking *oriV* that is functional in the strain background, are used. **A.** A single crossover between a clone that contains an internal portion of a gene and the corresponding gene on the chromosome will result in two incomplete copies of the gene separated by the plasmid vector. In this example, vector-carried tetracycline resistance (Tc^R) is used to select for the recombination product. **B.** A double crossover between a clone that contains an insertion within the cloned gene and the corresponding gene on the chromosome will result in a single copy of the gene on the chromosome, but the gene will be interrupted by the insertion. Selection for the double-crossover event is done using the antibiotic resistance that disrupts the gene, in this case kanamycin resistance (Kan^R), and the sucrose sensitivity gene *sacB* carried by the suicide vector.

a **knockout mutation**. The resulting interruption of the gene renders it non-functional. By disrupting a gene in this fashion, researchers often learn important information about the normal function of the gene. In this section, we will explore this particular use of recombination more fully.

In a single crossover knockout mutation, the cloned DNA must be an internal fragment of the gene targeted for disruption. Additionally, this fragment must be cloned into a **suicide plasmid**, a specially designed plasmid vector that, once introduced into the cell, cannot replicate in that host. Often, these plasmids contain an antibiotic resistance selectable marker and an origin of replication that functions in the cloning host but not the target cell.

If we want to knock out a gene in *Bacillus subtilis*, for example, we might use a plasmid that contains an origin of replication that functions only in *Escherichia coli*. The cloning steps could be completed in *E. coli*. Upon transfer of the plasmid to *B. subtilis*, only cells in which a single recombination event has led to the formation of a **cointegrate**, in which the plasmid and chromosome are joined, would remain antibiotic-resistant; unintegrated copies of the plasmid would not be maintained. The single crossover event leads to an antibiotic-resistant cell in which the gene of interest has been disrupted **(Figure 9.21)**.

Gene knockouts using double-crossover events result in more stable insertions than knockouts produced by single crossovers but are more difficult to make. Usually, double-crossover knockouts require the insertion of a selectable marker gene into the target gene to be disrupted. As shown in Figure 9.21B, for example, the plasmid contains a kanamycin resistance gene inserted within the gene of interest, making this gene non-functional. When this suicide construct is introduced into the target organism, a double-crossover event results in the replacement of the functional gene within the target cell with the disrupted gene from the plasmid.

Plasmids used for these types of experiments often also contain a second, different antibiotic resistance gene, and a counter-selectable marker such as *sacB*, which confers sensitivity to sucrose due to its production of an enzyme called "levansucrase," which hydrolyzes sucrose. This produces units of fructose that are polymerized by cells into large levan polymers. Accumulation of levan is lethal to many bacteria. In cells where a double crossover occurs, the second antibiotic resistance gene and the *sacB* gene are lost. As shown in Figure 9.21B, recombinant cells would be resistant to sucrose, sensitive to tetracycline, and resistant to kanamycin. Recombinant cells first can be identified by selecting for kanamycin resistance, followed by selection for sucrose resistance. Tetracycline-sensitive, kanamycin-resistant colonies can then be identified. These cells have undergone a double-recombination event, resulting in the knockout of the gene of interest.

Microbes in Focus 9.2
BACILLUS SUBTILIS

Habitat: First isolated and described in the mid-nineteenth century. Common, non-pathogenic inhabitant of soil. Closely related to the human pathogen *Bacillus anthracis*.

Description: Gram-positive bacterium approximately 3 μm in length in the Firmicutes phylum. Rod-shaped cells possess flagella, grow aerobically, and can form heat- and desiccation-resistant endospores. The 4.2-Mbp genome sequence of strain 168, encoding 4,100 genes, was reported in 1997.

Key Features: *Bacillus subtilis* has been adopted as a model organism for the study of fundamental processes such as DNA replication and cellular differentiation. Its ease of genetic manipulation has facilitated detailed study of this organism and the developmental processes associated with nutrient starvation. These processes include the development of competence for DNA transformation, the secretion of enzymes such as proteases and amylases, the production of antibiotics, and the formation of endospores. Many of its enzymes, like subtilisin used in laundry detergents, have been produced commercially, making *B. subtilis* a very important industrial microbe.

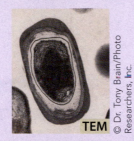

TEM © Dr. Tony Brain/Photo Researchers, Inc.

Transformation

Transformation refers to the introduction of extracellular DNA directly into an organism. This process does not require cell-to-cell contact for DNA uptake. Extracellular DNA released from dead cells exists naturally in the environment. Many bacteria, including species of *Streptococcus*, *Haemophilus*, *Neisseria*, and *Bacillus* (**Microbes in Focus 9.2**) are naturally **competent**, meaning they can take up free DNA from their surroundings. These bacteria have specialized receptor and transport systems for the uptake of DNA, often involving degradation of one of the strands so that only a single strand enters the cell (**Figure 9.22**). This is an active process, involving binding of DNA to receptor proteins on the cell wall, transfer of one strand through a channel formed by specialized pore proteins in the cytoplasmic membrane, and degradation of the strand that is not transferred. One of the driving forces for natural transformation may be that DNA released from other organisms can be used as a nutrient, but in some cases, this DNA is integrated into the chromosome instead of being metabolized.

CONNECTIONS To effectively take up free DNA fragments, competent cells express a series of polypeptides that bind to the external DNA and facilitate its passage across the plasma membrane. Quorum sensing, a process we'll examine in **Section 11.5**, controls the regulation of the genes encoding these polypeptides.

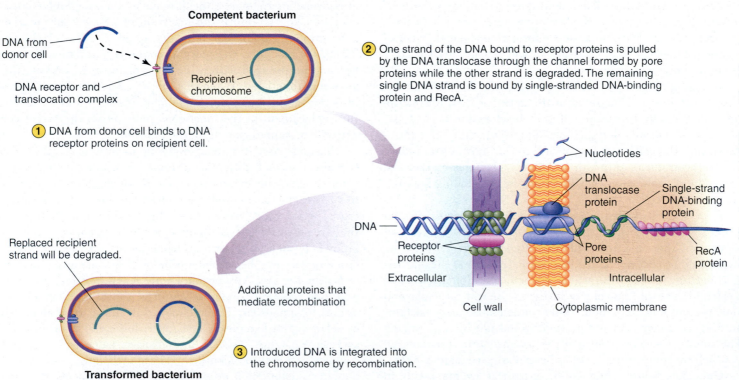

Figure 9.22. Transformation of competent bacterial cells In naturally competent cells, like *Bacillus subtilis*, external DNA first binds to a DNA-binding receptor on the cell surface. This DNA is then transported across the plasma membrane, and one strand becomes degraded. The remaining strand becomes integrated into the host chromosome via recombination.

Some bacteria, such as *E. coli*, do not exhibit natural competence; they lack the specialized machinery for DNA uptake. However, cells of *E. coli* can be made competent by treatment with a solution containing cations like calcium. This treatment probably makes the membrane temporarily more permeable to large molecules, like DNA. Sometimes, researchers use an electric pulse to increase the efficiency of transformation, a process called **electroporation**. In this technique, researchers apply an electric current to cells, which transiently generates holes in cellular membranes through which the DNA can enter. A major advantage of the electroporation technique is that it can be applied to a much broader range of bacteria.

The process of transformation has played an important role in the history of modern microbiology. Indeed, transformation made possible the very discovery of DNA as the heritable material. Recall the studies in the 1920s by Frederick Griffith. He investigated the ability of the pneumococcus bacterium *Streptococcus pneumoniae* to cause disease (see Figure 7.1). He provided evidence that a then unknown substance he called the "transforming principle," isolated from a heat-killed virulent bacterial preparation, could transform bacteria from non-virulent to virulent. We now know that this phenotypic change resulted from the transfer of DNA from one cell type to the other by transformation. Transformation has also been vitally important to the development of molecular biology. The transformation of bacteria and yeast with plasmids now occurs commonly in research and teaching laboratories. Researchers continue to use transformation in new and innovative ways, increasing our understanding of biology with the help of this basic bacterial process.

while other strains, called "recipient or F⁻ strains," could receive DNA. Hayes mated different F⁺ and F⁻ strains with multiple antibiotic resistance and auxotrophic alleles. Examination of the progeny made it clear that the genetic material moved in a particular direction. Moreover, the F⁺ property was infectious; F⁻ infertile cells frequently converted to F⁺ cells following a conjugation event.

We now know that conjugation requires the fertility factor F, a very large (94,500-bp) circular dsDNA plasmid, called the **F plasmid**. When transferred by conjugation from an F⁺ cell, the F plasmid converts an F⁻ cell lacking the plasmid to an F⁺ cell. The F plasmid contains all the genes needed for conjugation and plasmid maintenance in the cell (**Figure 9.23**). The more than 30 different *tra* (for transfer) genes encode proteins needed for the production of conjugational structures that link the cells together and transport the DNA. Indeed, loss-of-function mutations in the *tra* genes result in loss of plasmid transfer. Some of the *E. coli* strains used by Joshua Lederberg carried the F plasmid. If this had not been the case, then he might never have discovered bacterial conjugation.

In 1950, Bernard Davis used immersed filters that prevented cellular contact between donor and recipient cultures to show that conjugation requires physical contact. While many researchers have studied the conjugation apparatus, we still do not know the precise structures involved. We do know that contact is mediated by special **sex pili** that are encoded by some of the *tra* genes. After contact between donor and recipient cells occurs, a complex of *tra*-encoded proteins called the "mating bridge" forms. In the donor cell (F⁺), an endonuclease associated with the mating bridge makes

Conjugation

While transformation involves the uptake of free exogenous DNA by competent cells, **conjugation** involves the specialized transfer of DNA from one cell to another via direct cell-to-cell contact. Unlike transformation, the process of conjugation transfers relatively large segments of DNA between cells. Additionally, we now know that conjugation can facilitate the transfer of DNA between diverse organisms, even between organisms in different domains (see Figure 1.21). In some cases, conjugation facilitates the transfer of virulence factors between cells, resulting in the conversion of non-pathogenic organisms into deadly human pathogens **(Perspective 9.1)**. To better understand this important method of DNA transfer, we will first discuss how conjugation occurs. Then, we will investigate some specific uses of conjugation in the laboratory.

Mechanism of Conjugation

While Joshua Lederberg discovered conjugation, he really only studied its end results. In subsequent experiments, William Hayes demonstrated that gene transfer occurs in a single direction. Experiments in *E. coli* showed that certain strains, called "donor or F⁺ strains" (F for "fertility"), could donate DNA

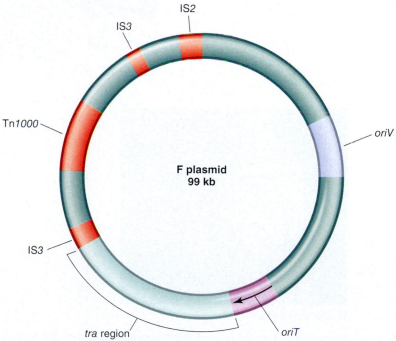

Figure 9.23. F plasmid Nearly 100 kb and self-transmissible, the F plasmid contains two different types of IS elements, IS2 and IS3, and the transposon, Tn*1000*. It also contains an origin of transfer (*oriT*) and *tra* genes, which allow it to conjugate. In addition to conjugating on its own, the F plasmid can also cause the transfer of chromosomal DNA.

We probably all are somewhat familiar with *Bacillus anthracis*. Identified by Robert Koch in the 1870s as the cause of anthrax, this bacterium remains intensively studied. When inhaled by people, spores of *B. anthracis* germinate and can cause a highly lethal form of anthrax **(Figure B9.2)**. As we saw in the opening story of Chapter 6, terrorists used these spores as a weapon in the United States in 2001. In contrast, *Bacillus cereus* is a common soil microbe that occasionally causes food poisoning in humans, and *Bacillus thuringiensis* causes disease in insects.

What do these three *Bacillus* species have in common? So much, in fact, that a better question might be, "What makes them different?" The *B. cereus*, *B. anthracis*, and *B. thuringiensis* genomes and ribosomal RNA sequences are so similar that one could argue that they really should be considered the same species except for the dramatic differences in pathogenic behavior! The pathogenic differences appear to depend on genes present on plasmids, the kind that can be transferred via conjugation.

In *B. anthracis*, genes for the anthrax neurotoxin and production of the protective capsule exist on two large plasmids, pXO1

and pXO2. *B. thuringiensis* contains plasmids that express genes that produce a crystalline toxin, which allows the bacterium to penetrate the insect gut and induce spore germination. Different versions of these toxin-encoding genes produce toxins with specificity for different groups of insects. *B. thuringiensis* strains can contain multiple plasmids and produce different combinations of toxins.

The virulence plasmids of *B. anthracis* and *B. thuringiensis* are transferable by conjugation. Therefore, we can envision the creation of virulent "*B. anthracis*" or "*B. thuringiensis*" strains from non-virulent *B. cereus* cells by the acquisition of a plasmid expressing virulence-related traits. These bacteria all exist in soil, a crowded habitat for microbes, where conjugation may happen frequently. Because the *B. cereus* genome, even without the virulence plasmids of *B. anthracis* and *B. thuringiensis*, contains a few genes that can contribute to pathogenicity, perhaps it is already an opportunistic pathogen of insects. Plasmid acquisition via conjugation just enhances and targets its virulence toward distinct hosts.

A. *Bacillus anthracis*

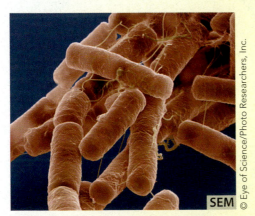

B. *Bacillus cereus*

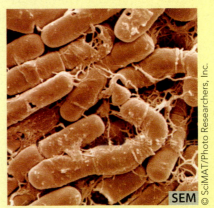

C. *Bacillus thuringiensis*

Figure B9.2. Closely related *Bacillus* organisms Three organisms in the *Bacillus* genus are very closely related, yet interact with humans in very different ways. **A.** *B. anthracis* causes anthrax, a deadly human disease. **B.** *B. cereus* is a common soil inhabitant that occasionally causes food poisoning in humans. **C.** *B. thuringiensis* can infect insects but not humans.

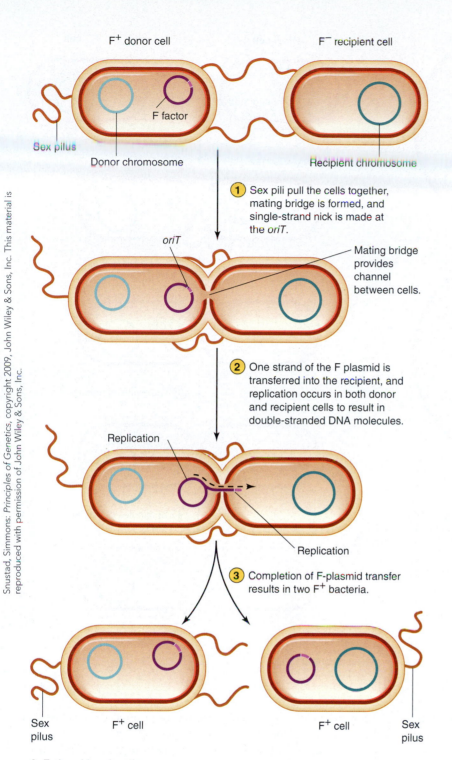

F⁺ donor cell

F⁻ recipient cell

F factor

Sex pilus

Donor chromosome

Recipient chromosome

① Sex pili pull the cells together, mating bridge is formed, and single-strand nick is made at the *oriT*.

oriT

Mating bridge provides channel between cells.

② One strand of the F plasmid is transferred into the recipient, and replication occurs in both donor and recipient cells to result in double-stranded DNA molecules.

Replication

Replication

③ Completion of F-plasmid transfer results in two F⁺ bacteria.

Sex pilus

F⁺ cell

F⁺ cell

Sex pilus

A. F-plasmid conjugation process

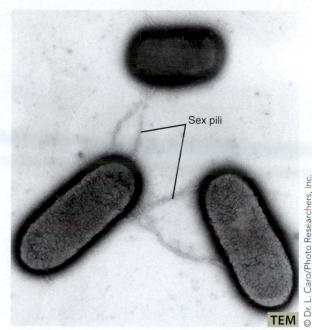

B. Conjugation in *E. coli*

Sex pili

TEM

© Dr. L. Caro/Photo Researchers, Inc.

Figure 9.24. Process Diagram: Formation of mating pair and transfer of DNA Conjugation occurs between cells that form mating pairs, a donor (F⁺) and recipient (F⁻). **A.** F-plasmid-encoded pili help the cells to form intimate physical contact and facilitate transfer of DNA, which occurs through a mating bridge. **B.** *E. coli* cells undergoing conjugation.

synthesizes the complementary strand, making the plasmid double stranded. A similar process occurs in the donor cell, replacing the DNA strand that has been transferred. Following transfer, the end of the plasmid in both the donor and the recipient cell releases from the mating bridge and attaches to the other end of the plasmid, regenerating the circular form of the plasmid in both cells.

As described in this process, conjugation transfers only the F plasmid without involvement of chromosomal genes. So, how did the chromosomal genes that Lederberg was tracking get into recipient cells? Occasionally the F plasmid integrates into the bacterial chromosome by a single-crossover event. This event is reversible, so the F plasmid again can become extrachromosomal. Sometimes, though, the F plasmid remains integrated when conjugation begins. When the F plasmid initiates transfer in this integrated state, attached chromosomal genes can be transferred to the recipient cell. Typically, this type of conjugation does not result in the transfer of a complete extrachromosomal F plasmid to the recipient cell. Rather, transferred chromosomal genes integrate into the recipient chromosome by homologous

a single-stranded nick at *oriT* (origin of transfer) of the F plasmid **(Figure 9.24)**. One end of the nicked DNA strand attaches to a protein of the mating bridge, and the remainder of the strand gradually passes to the cytoplasm of the recipient cell (F⁻) through the mating bridge. If the mating bridge remains intact long enough, an entire single-stranded copy of the plasmid is transferred into the recipient cell. Once the single-stranded DNA enters the cell, the recipient cell's DNA polymerase

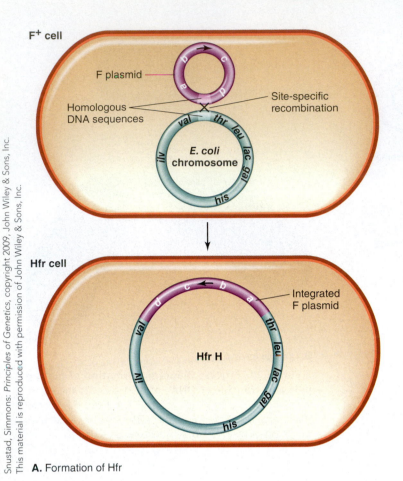

F⁺ cell

F plasmid

Homologous
DNA sequences

Site-specific
recombination

E. coli
chromosome

Hfr cell

Integrated
F plasmid

Hfr H

A. Formation of Hfr

Figure 9.25. Conjugation mediated by integrated F plasmid produces high frequency of recombinants (Hfr). An F plasmid may become integrated into the host chromosome resulting in an Hfr strain. **A.** Recombination between the F plasmid and the chromosome can occur between homologous DNA sequences, such as transposons, located on both the F plasmid and the chromosome. The resulting Hfr is a cointegrate of the F plasmid and the chromosome, which is able to transfer chromosomal DNA because of the F plasmid *oriT*. By convention, the *oriT* arrow points away from the first DNA that is transferred. This particular Hfr is called Hfr H, and *thr* will be one of the first chromosomal genes that it transfers. **B.** The Hfr strain can transfer part of the host chromosome into recipient cells, where it can recombine with the recipient chromosome. Longer matings result in more of the chromosome being transferred, and donor genes near the site of F insertion are transferred at a higher rate than genes that are farther from the site of F insertion.

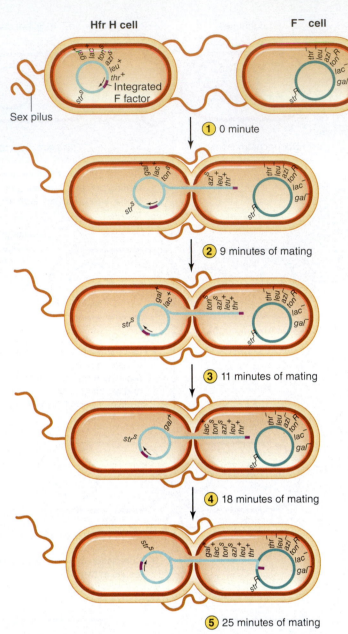

Hfr H cell   **F⁻ cell**

Sex pilus   Integrated
F factor

① 0 minute

② 9 minutes of mating

③ 11 minutes of mating

④ 18 minutes of mating

⑤ 25 minutes of mating

B. Gene transfer by Hfr

recombination, resulting in a **transconjugant**, a cell that has incorporated DNA from another cell via conjugation. This process produced the prototrophs seen by Lederberg in his crosses between auxotrophic mutants.

On rare occasions, donor strains transfer some chromosomal genes with much higher frequency than other genes. These donors, called **Hfr strains**, for <u>h</u>igh <u>f</u>requency of <u>r</u>ecombination **(Figure 9.25)**, form when the entire F plasmid incorporates into the chromosome. Unlike normal F⁺ strains, Hfr strains do not convert F⁻ recipients to F⁺ because only part of the F plasmid transfers. When transfer begins, *oriT* is nicked and only half of the F plasmid is transferred. The rest of the F plasmid remains at the other end of the chromosome. For all the genes of the F plasmid to be transferred, the entire chromosome would need to move into the recipient cell. This extensive transfer rarely happens because mating pairs often separate before the entire

chromosome can be transferred. Chromosomal genes closer to the site of insertion transfer at a higher frequency than genes farther away from the insertion site, resulting in a gradient of transfer **(Figure 9.26)**.

The gradient of transfer can be demonstrated experimentally by purposeful interruption of mating during conjugation. Such interruption experiments using crosses with Hfr strains with F inserted at different locations were used to map the position of the genes on the *E. coli* chromosome, long before DNA sequencing was available. The results of such experiments demonstrated that during conjugation, the chromosomal DNA transfers at a constant rate, and it takes 100 minutes to transfer the entire chromosome. The genetic maps thus were marked off in minutes **(Figure 9.27)**. Because the *E. coli* K12 genome contains a 4.6-Mbp chromosome, each minute corresponds to approximately 46 kbp of DNA.

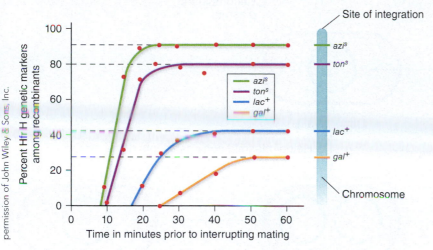

Figure 9.26. Time and sequence of gene transfer in conjugation between Hfr and F⁻ The time required for the transfer of a given genetic marker from donor to recipient increases with distance from the site of integration of the F plasmid in the Hfr strain. This relationship can be used to determine locations of genes on the chromosome. In this example, the distance from the location of the site of the inserted F plasmid in the Hfr H strain can be deduced by the time of transfer.

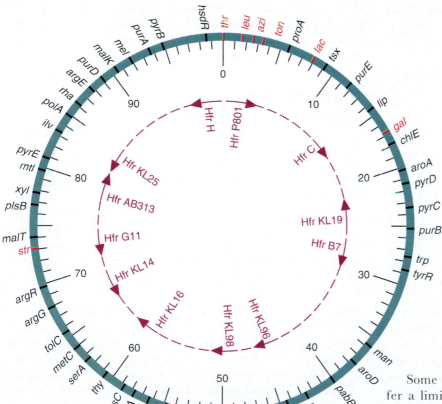

Figure 9.27. The circular genetic map of *E. coli* (distance in minutes) Researchers derived the genetic map of *E. coli* by using Hfr-mediated conjugation. The locations and direction of transfer of commonly used Hfr insertions is shown in the inside circle. The total distance of 100 minutes reflects the gradient of transfer during conjugation. Genetic loci are shown on the outside, with those in red having been used in previous figures.

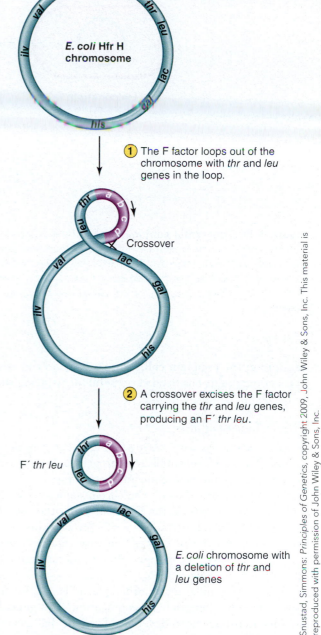

Figure 9.28. F′ plasmids F′ plasmids are formed when recombination occurs between chromosomal DNA on either side of the F insertion, resulting in looping out of a substantial segment of the chromosome that also contains the F plasmid. The products of this event are an F plasmid that contains a segment of chromosomal DNA, and a corresponding deletion in the chromosome.

Some donor cells, called **F-prime (F′) strains**, only transfer a limited number of markers in conjugation, but they do so at extremely high frequency. These F′ strains result from rare events in which an integrated F plasmid excises out of the chromosome. During excision, sections of chromosomal DNA are removed with the plasmid. The resulting plasmid/chromosome hybrids can contain large segments of chromosomal DNA **(Figure 9.28)**. When these F′ strains mate with F⁻ strains, they transfer this modified F plasmid into the recipient. The transferred chromosomal DNA from the donor remains part of the F

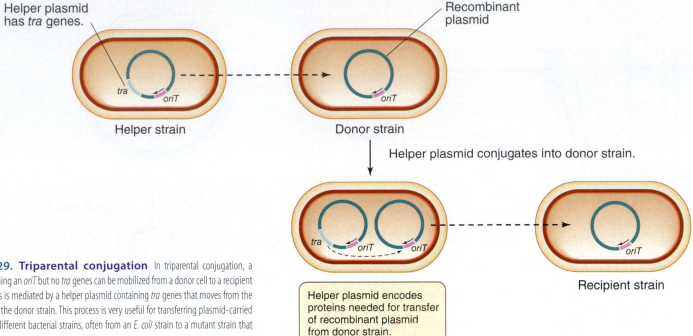

Figure 9.29. Triparental conjugation In triparental conjugation, a plasmid containing an *oriT* but no *tra* genes can be mobilized from a donor cell to a recipient cell. This process is mediated by a helper plasmid containing *tra* genes that moves from the helper strain to the donor strain. This process is very useful for transferring plasmid-carried DNA between different bacterial strains, often from an *E. coli* strain to a mutant strain that is being studied.

plasmid in the recipient cell. As we will see in the next section, this arrangement has been very useful for studying the genetics of bacteria.

Uses of Conjugation

F′ plasmids were very useful in the early genetic studies of *E. coli*. Using F′ plasmids, researchers can construct strains that carry two different gene alleles or copies of a chromosomal region. One copy exists on the chromosome, while the second, different copy exists on the F′ plasmid. In other words, the cell is a partial diploid, or **merozygote**—it contains two copies of some of its DNA. Using this system, investigators can characterize genes involved in common cellular processes and learn more about their expression.

Like F plasmids, many other plasmids can be transferred from one cell to another by conjugation. The genetic analysis systems that have been developed for many bacteria have often included the use of broad host range or shuttle-vector recombinant plasmids in which genes of interest are cloned. Once the recombinant plasmid has been constructed, conjugation or transformation can be used to transfer it from *E. coli* to the strain of interest. For some bacteria, conjugation is much more efficient than transformation.

How is conjugation used to transfer recombinant plasmids? Conjugation requires that the plasmid contains both *oriT* and *tra*, but the inclusion of these elements in the plasmid often makes it too large for easy manipulation using recombinant DNA techniques. For this reason, the method of triparental conjugation **(Figure 9.29)** has been devised. In this process, the *tra* genes are provided by a helper plasmid in a mobilizer strain, while the donor strain contains the recombinant plasmid to be transferred to the recipient strain. When cultures of the three strains are mixed, the helper plasmid transfers into the donor strain, where

the *tra* genes are expressed, causing the recombinant plasmid to be transferred to the recipient strain via conjugation. The recipient strain containing the recombinant plasmid can then be selected by using media on which none of the three original strains can grow.

Transposition

A third method of genetic transfer in bacteria is **transposition**, the movement of DNA via mobile genetic elements. These mobile genetic elements, or **transposable elements**, mediate genome rearrangements by moving within and between genomes. First detected in corn by Barbara McClintock, transposable elements have generated controversy. To many earlier researchers, they appeared to contradict the principles of Mendelian genetics stating that genes occupy specific points on a genetic map and are passed to progeny in a predictable fashion. Today, though, researchers universally accept the existence of transposable elements. Indeed, genome-sequencing data show that transposable elements occur in virtually all genomes (see Section 10.3), and, like the other methods of DNA transfer that we have examined, greatly affect evolutionary change. In this section, we will examine simple transposable elements, referred to as insertion sequences, or IS elements, which encode only the proteins necessary for transposition, and more complex **transposons**, transposable elements that contain other genes such as antibiotic resistance genes.

Mechanisms of Transposition

Transposable elements facilitate much of the genome rearrangement and horizontal gene transfer that occurs in bacteria. These elements move from one part of the genome to another, using non-homologous recombination. The transposition

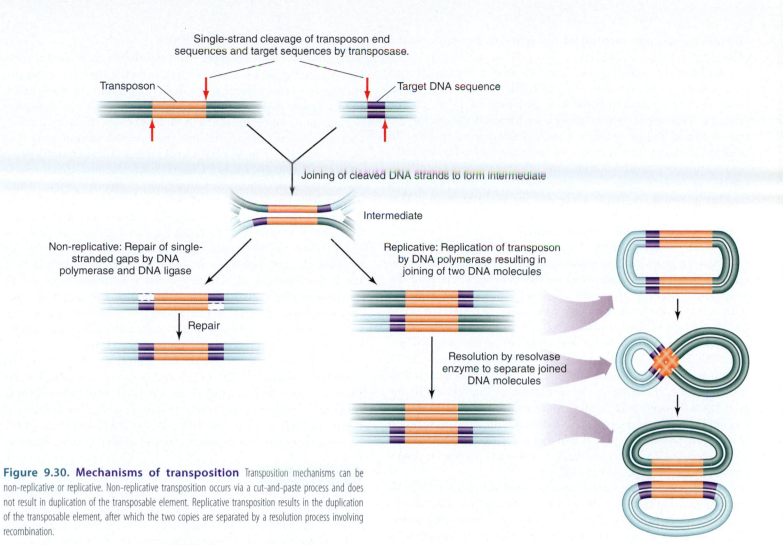

Figure 9.30. Mechanisms of transposition Transposition mechanisms can be non-replicative or replicative. Non-replicative transposition occurs via a cut-and-paste process and does not result in duplication of the transposable element. Replicative transposition results in the duplication of the transposable element, after which the two copies are separated by a resolution process involving recombination.

Within the figure:

Single-strand cleavage of transposon end sequences and target sequences by transposase.

Transposon — Target DNA sequence

Joining of cleaved DNA strands to form intermediate

Intermediate

Non-replicative: Repair of single-stranded gaps by DNA polymerase and DNA ligase

Repair

Replicative: Replication of transposon by DNA polymerase resulting in joining of two DNA molecules

Resolution by resolvase enzyme to separate joined DNA molecules

---

process requires only the **transposase** enzyme and terminal inverted repeat sequences on the transposon that are recognized by the transposase enzyme. Non-replicative, or cut-and-paste, transposable elements excise from one location and insert into another location. Replicative transposable elements require a replication step and leave a copy behind at the original site **(Figure 9.30)**. In both cases, the transposable element may move to a new location within its original DNA molecule or to a different DNA molecule.

The simplest transposable elements generally consist of the transposase gene flanked by inverted repeats, short sequences of DNA that exist in an inverted orientation relative to each other **(Figure 9.31)**. Referred to as **insertion sequences**, or **IS elements**, these transposable elements generally are less than 2,500 bp in length. Different inverted repeat sequences, usually between 10 and 40 bp in length, exist in different IS elements. Transposons are more complex transposable elements, including not only the inverted repeats and transposase gene, but

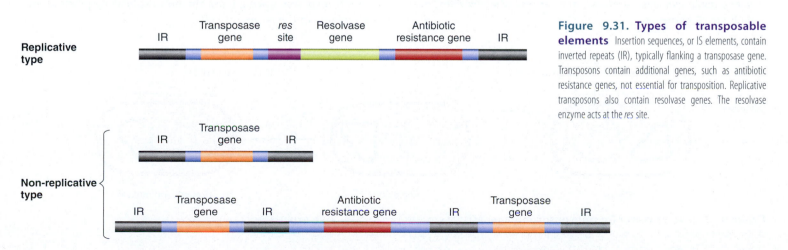

Replicative type

IR | Transposase gene | *res* site | Resolvase gene | Antibiotic resistance gene | IR

Non-replicative type

IR | Transposase gene | IR

IR | Transposase gene | IR | Antibiotic resistance gene | IR | Transposase gene | IR

Figure 9.31. Types of transposable elements Insertion sequences, or IS elements, contain inverted repeats (IR), typically flanking a transposase gene. Transposons contain additional genes, such as antibiotic resistance genes, not essential for transposition. Replicative transposons also contain resolvase genes. The resolvase enzyme acts at the *res* site.

also other genes not essential for the actual transposition of the element.

Antibiotic resistance genes have often been associated with transposons. In many cases, the antibiotic resistance genes on plasmids exist within transposons that happen to be on the plasmid. This arrangement provides additional means for the transfer of antibiotic resistance by horizontal gene transfer. In addition to being transferred with the plasmid by, for instance, conjugation or transformation, the transposon-associated antibiotic resistance genes can also move from one plasmid to another plasmid, or from a plasmid to a chromosome. This movement increases the distribution of antibiotic resistance genes throughout bacterial communities.

Typically, transposition does not exhibit any target site specificity; transposable elements can insert anywhere in the genome. Each transposition event results in a short repeat of the host DNA sequence at the site of insertion. Current evidence indicates that the transposase enzyme makes a double-stranded cut at the end of the terminal inverted repeat sequence of the transposon, while it makes a staggered cut at the site of insertion. The DNA molecules are then joined by transposase. DNA polymerase and DNA ligase fill in the gaps and covalently link the free ends, resulting in the formation of the repeats in the host DNA. In replicative transposons, replication continues, resulting in joining of the two DNA molecules and also duplication of the transposon. The joined molecules are then separated into two molecules by an enzyme called **resolvase**, acting at a specific DNA sequence called the *res* site.

When a transposon resides on a plasmid, it can leave the host cell, via plasmid conjugation, and move to another cell, even to another type of bacteria. Through this process, multidrug antibiotic resistance can spread rapidly (see Section 24.3). Transposons also play a role in evolution by promoting rearrangement of genomes, and by mediating the movement of host DNA sequences to new locations. Some transposons exist in multiple copies in a single genome, and can promote homologous recombination, resulting in deletions, inversions, and translocations.

CONNECTIONS Retrotransposons are transposable elements that move from one location to another via an RNA intermediate, employing the activity of reverse transcriptase. The duplication of these transposable elements resembles the process whereby retroviruses replicate (**Section 8.3**).

Uses of Transposable Elements

Transposable elements have been developed as powerful tools for bacterial genetics. These elements do not have insertion site specificity, so they insert randomly in the genome. In many cases, their transposition is regulated such that only a single insertion per genome can occur. If these elements insert within a gene, then they almost always disrupt the function of that gene. For this reason, they can be very useful for generating mutations.

In 2009, scientists at the University of Alabama, Birmingham used large-scale transposon mutagenesis to determine which genes within the *Mycoplasma pulmonis* genome are essential for life. To conduct this study, they examined 1,700 mutants of *M. pulmonis* that contained insertions of Tn4001T, a well-characterized transposon. Because this transposon inserts randomly within the genome, the researchers reasoned that every gene should be a target for insertion. Moreover, because transposon insertion almost invariably leads to the inactivation of the targeted gene, they hypothesized that cells containing insertions within essential genes would not be viable. In other words, only cells containing insertions within non-essential genes could replicate.

When they characterized the mutants, they identified Tn4001T insertions in 321 of the 782 *M. pulmonis* genes. The remaining genes, they concluded, are essential. Disruption of these genes apparently prevents replication of the cells. The function of these essential genes is now being determined, and these genes are being compared to essential genes identified in other species.

CONNECTIONS Many microbiologists are interested in defining the essential genes in bacterial genomes. In the **Chapter 12 Mini-Paper**, we will see how researchers used information about essential genes to construct a synthetic genome. Some people have described this accomplishment as the creation of synthetic life.

The isolation of transposon insertion mutants usually involves introducing the antibiotic resistance transposon on a suicide vector into the recipient host cell (**Figure 9.32**). Recall that a suicide vector cannot replicate in the recipient cell, and, as a result, has a very short life span within the culture. Thus, antibiotic resistance will arise in the recipient cell only by insertion of the transposon into the host chromosome. Transfer of the plasmid into a target bacterial cell, followed by application of the antibiotic, produces colonies that have arisen as a result of the transposon moving from the non-replicating plasmid into the genome of the recipient cell.

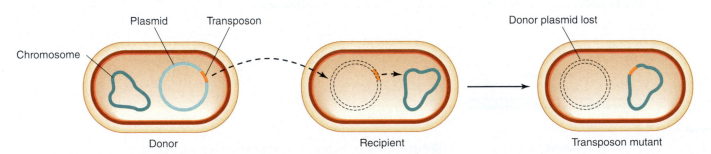

Figure 9.32. Suicide vector for transposon mutagenesis Transposon-carrying plasmid that is unable to replicate in the recipient strain can be used for transposon mutagenesis. Selection for the antibiotic resistance of the transposon will select for the transposition event from the plasmid to the chromosome, allowing the donor plasmid to be lost from the cell.

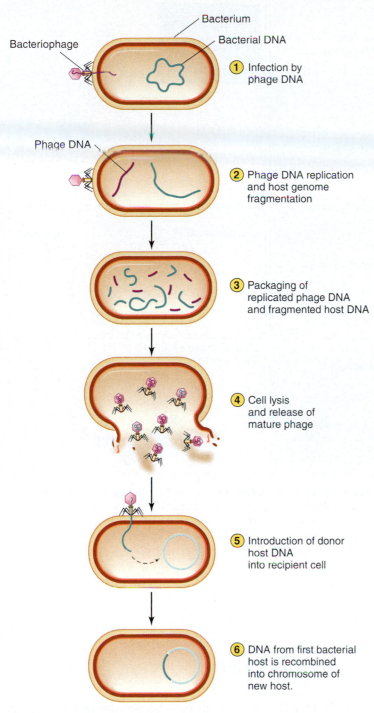

1 Infection by phage DNA

Bacteriophage · Bacterium · Bacterial DNA

2 Phage DNA replication and host genome fragmentation

Phage DNA

3 Packaging of replicated phage DNA and fragmented host DNA

4 Cell lysis and release of mature phage

5 Introduction of donor host DNA into recipient cell

6 DNA from first bacterial host is recombined into chromosome of new host.

Figure 9.33. Process Diagram: Formation of transducing particles
Transducing particles form when bacterial genomic DNA becomes packaged into phage heads. The resulting particles can inject the bacterial DNA into a suitable recipient cell and can be used for strain construction and for genetic-mapping experiments.

Transduction

As we conclude our survey of DNA transfer in bacteria, let's examine **transduction**, the transfer of bacterial DNA from one cell to another by a bacteriophage. Again, we will begin by examining how this transfer of genetic material occurs. Then, we will look at how this process can be used in the laboratory.

Mechanism of Transduction

As we learned in Chapter 5, bacteriophages infect bacteria and then use the cellular machinery of the bacterial cells to copy themselves. Occasionally, in the frenzy to assemble new phage particles, segments of genomic DNA from the infected bacterial cell become packaged into the capsid. This "error" produces transducing particles, phages that carry bacterial DNA (**Figure 9.33**). Transducing particles are usually defective for infection; they often do not contain all the phage genes needed to form new, infectious viral particles. These transducing particles, though, can attach to recipient bacterial cells and inject their DNA into these cells. If this DNA undergoes homologous recombination with the genome of the recipient strain, then a segment of the recipient genome becomes replaced by a segment of the donor strain. A **transductant**, or bacterial cell that contains DNA obtained via transduction, results. Transduction is a very powerful and convenient tool for transferring genetic markers between strains and has also played an important role in the construction of genetic maps. Not all bacteriophages, however, can carry out efficient transduction, and bacteriophages in general have a very restricted host range.

Norton Zinder, a graduate student of Joshua Lederberg, first discovered transduction in *Salmonella typhimurium* (now called *Salmonella enterica* serovar Typhimurium, designated *S. Typhimurium*). He used techniques similar to the ones Lederberg had used to detect genetic recombination in *E. coli*. He observed that prototrophic recombinants resulted from mixing two auxotrophic strains even when a filter separated these strains. The responsible filtrable agent, then called FA was later found to be the *Salmonella* bacteriophage P22 (**Mini-Paper**).

Uses of Transduction

Transduction is one of the most useful genetic tools for those bacteria that have transducing phages. Transducing lysates, cleared cultures that result from phage infection, can be produced easily in the laboratory by infecting a bacterial culture with a small amount of an existing lysate. During a short incubation, the phages will infect the cells and lyse them. A dilution of the resulting lysate, which contains a mixture of virulent phage particles and transducing particles, is then added to a culture of the intended recipient bacterial strain. Both types of particles adsorb to the recipient cells, and inject their DNA into the cells. The cells that have taken up the bacterial DNA-containing transducing particles will have a chance to incorporate that DNA into their genome by double-crossover recombination.

Historically, transduction has been used to map genetic markers by examining co-transduction frequencies. Co-transduction frequency refers to the frequency with which an unselectable marker gene is transduced along with a given selectable marker gene. This frequency depends on the distance between marker genes; close genetic loci are more likely to be transduced together than are genetic loci that are farther apart. By examining the rates of co-transduction between various genetic markers, geneticists could develop genetic maps of bacteria.

Today, transduction remains an effective experimental tool. In a paper published in 2010, for instance, researchers from South Africa and the United States hypothesized that transduction of *Lactobacillus* strains could be used to combat

Observations: *Lactobacillus gasseri* ADH is a normal resident of the vaginal mucosa. CCL5 exhibits anti-HIV properties.

Hypothesis: Transduction could be used to genetically alter *L. gasseri* ADH, enabling these bacteria to express CCL5.

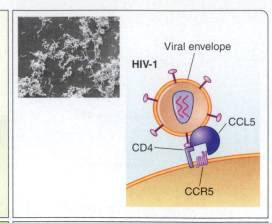

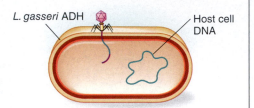

Experiment: Use transducing phages to introduce CCL5 into *L. gasseri* ADH.

Results: Transduced bacteria produced CCL5.

Conclusion: Transduction represents an effective way of introducing foreign genes into *L. gasseri* ADH.

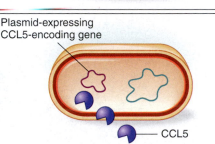

Photo: From I. Jankovic et al. 2003. J Bacteriol 185:3288–3296. Reproduced with permission from American Society for Microbiology.

Figure 9.34. Using transduction to genetically alter *Lactobacillus gasseri* ADH To investigate the possibility of developing a bacterial strain that contributes to HIV prevention, researchers used transduction to introduce genes for the chemokine CCL5 into *Lactobacillus gasseri* ADH. Because this chemokine can block entry of HIV into human cells and because *L. gasseri* ADH is a natural inhabitant of the vaginal mucosa, these altered bacteria may represent a novel HIV prevention strategy.

HIV/AIDS. Let's look more closely at their experiment. The researchers noted that *Lactobacillus* species normally reside within the vagina of healthy women. If these resident bacteria could be genetically altered to express anti-HIV compounds, the investigators postulated, then women in whom these bacteria existed might possess some level of protection from HIV. In other words, the resident bacteria would be altered to become a long-lasting microbicide.

To engineer bacteria in this way, the investigators used transduction. They constructed transducing phages that contained the genes for CCL3 and CCL5, two human chemokines shown to exhibit some anti-HIV properties. They then used these phages to transduce *Lactobacillus gasseri* ADH, a bacterial species commonly found on the vaginal mucosa of women. Following transduction, these bacteria produced the chemokines (**Figure 9.34**). While these studies were done *in vitro*, they do demonstrate that transduction can be used to genetically alter *L. gasseri* ADH. Much more research needs to be done before this strategy can be used clinically. As the authors noted, though, their approach may lead to a novel means of preventing HIV infection.

CONNECTIONS In humans, the chemokines CCL3 and CCL5 interact with the chemokine receptor CCR5. CCR5 also serves as a co-receptor for HIV (**Section 5.1**). Increased expression of CCL3 and CCL5 inhibits HIV infection by binding to CCR5 and blocking the attachment of the virus.

9.4 Fact Check

1. Distinguish between horizontal gene transfer and vertical gene transfer.

2. Describe the process of homologous recombination.

3. Distinguish between single- and double-crossover events. How are they used for genetic manipulation in bacteria?

4. What is DNA transformation, and how are competent cells involved in the process?

5. Create a table explaining F^+, F^-, Hfr, and F′.

6. What are transposons?

7. What is transduction? Explain the role of bacteriophage in the process.

Mini-Paper: A Focus on the Research
THE DISCOVERY OF TRANSDUCTION

N. D. Zinder and J. Lederberg. 1952. Genetic exchange in *Salmonella*. J Bacteriol 64:679–699.

Context

In 1948, 19-year-old Norton Zinder embarked on his Ph.D. studies at the University of Wisconsin. He was the first graduate student of the 23-year-old Professor Joshua Lederberg. At this time, scientists were still digesting the implications of Avery's careful demonstration of DNA as the heritable material and Lederberg's description of conjugation in bacteria. Because *Salmonella* was closely related to *Escherichia*, but was a much more dangerous pathogen, Lederberg wanted to investigate conjugation in *Salmonella*. Medical microbiologists had isolated and characterized thousands of different types of *Salmonella*, and these strains were a resource of tremendous potential value. As the topic of a Ph.D. dissertation, the project appeared ideal.

The experiments from Norton Zinder's Ph.D. dissertation were published in the *Journal of Bacteriology*. The paper begins with the following introductory statement:

Genetic investigations with many different bacteria have revealed parallelisms and some contrasts with the biology of higher forms. The successful application of selective enrichment techniques to the study of gene recombination in *Escherichia coli* (Tatum and Lederberg, 1947; Lederberg et al., 1951) suggested that a similar approach should be applied to other bacteria. This paper presents the results of such experiments with *Salmonella typhimurium* and other *Salmonella* serotypes. The mechanism of genetic exchange found in these experiments differs from sexual recombination in *E. coli* in many respects so as to warrant a new descriptive term, transduction.

Experiment

To investigate genetic transfer in *Salmonella*, Zinder first generated a series of double auxotrophic mutants of *Salmonella* Typhimurium. These mutants were then mixed in different combinations, and he screened the progeny for prototrophs. Initially, no prototrophs appeared, but Zinder did not give up. After almost two years of careful work, he found a combination of strains that yielded prototrophs. One strain, a derivative of strain LT-22, required tyrosine and tryptophan. The other strain, a derivative of LT-2, required methionine and histidine. When Zinder combined these strains, he obtained prototrophic recombinants in frequencies similar to the frequencies observed by Lederberg in his *E. coli* mating experiments. Important differences in the results, however, did exist. First, Zinder found no evidence of linkage or gradient of transfer. Second, the recombinant strains always contained the LT-22 genetic background.

Furthermore, Zinder showed that, in contrast to conjugation, genetic transfer in *Salmonella* did not require cell-to-cell contact. To explain this finding, Zinder and Lederberg concluded that the transfer of genetic material involved some sort of "filtrable agent" (FA). Additional studies revealed that FA was only produced when the LT-2 and LT-22 derivatives were co-cultivated. The researchers postulated that FA produced by the LT-2 derivative was induced by the LT-22 derivative. Moreover, they showed that FA was not naked DNA; treatment with a nuclease

The effect of FA from LT-7 and its derivatives upon LT-7 derivatives

CELLS/*FA*	LT-7	SW-184	SW-188	SW-191	BOILED *FA*
SW-184	203	26	247	253	31*
SW-188	62	76	0	68	0
SW-191	198	210	236	18*	10*
LA-22 (control)	230	242	202	275	0

* Presumably spontaneous reverse mutations.
Figures are transductions from auxotrophy to prototrophy per plate.

From N. D. Zinder et al. 1952. J Bacteriol 64:679–699. Reproduced with permission from American Society for Microbiology.

Figure B9.3. Conversion of gal⁻ mutants to gal⁺ by filterable agents Several gal⁻ mutants were analyzed. The filterable agent from LT-7 converted these mutants to a gal⁺ phenotype. A filterable agent derived from one mutant could convert other mutants, but not itself, to a gal⁺ phenotype.

did not abolish its genetic transfer abilities. How can we explain these results? Although Zinder and Lederberg did not explicitly state this conclusion in their paper, it is now clear that FA is a bacteriophage.

Zinder and Lederberg observed that active FA could be produced during co-cultivation of LT-22 derivatives and other S. Typhimurium isolates, including LT-7. FA was also induced in several strains by treatment of the cells with penicillin or crystal violet. Crosses carried out using FA from strain LT-7 and a number of auxotrophic derivatives demonstrated the inability to transduce prototrophic markers from the corresponding auxotroph.

Crosses between independently isolated galactose utilization mutants **(Figure B9.3)** revealed that, similar to what was observed for the auxotrophs, they could not be transduced by their own FA. In some cases, they could not be transduced by FA from other galactose utilization mutants. This result was taken to be evidence for the mutations being in the same genes.

Impact

Norton Zinder received his Ph.D. in 1952 and immediately took a position at Rockefeller University in New York City, where he spent his entire career conducting research in bacterial and bacteriophage genetics. Transduction developed into one of the most useful methods for fine structure genetic mapping, and also for strain construction. Enthusiastic genetic analysis of strain LT-2 resulted in it becoming, after *E. coli* K12, one of the best-characterized strains of bacteria. An extensive collection of mutant strain derivatives has been maintained at the Salmonella Genetic Stock Centre at the University of Calgary. These strains are distributed freely for research purposes.

● Questions for Discussion ·················

1. The original crosses involved the transfer of two markers (*tyr*⁺, *trp*⁺) from the LT-2 derivative to the double auxotroph LT-22 derivative. How could this be, if the transduction process only transfers single markers?

2. Explain why a prototrophic marker cannot be transduced from a strain that is auxotrophic for that marker.

Image in Action

WileyPLUS Shown are *Escherichia coli* cells undergoing conjugation. The donor cell at the top forms a pilus that can attach to a recipient cell and bring the cells together (second pair of cells from the top). One strand of the F plasmid moves from the donor cell to the recipient cell (third pair of cells from the top). DNA synthesis then occurs in each cell to give rise to a double-stranded DNA plasmid (bottom pair of cells).

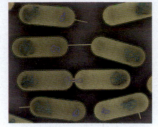

1. If this image were to zoom in on the F plasmid itself, identify what genes would likely be shown and explain their roles in the conjugation process.

2. Bisphosphonates are compounds that have been found to inhibit the enzymes that make the single-stranded nick at *oriT* of the F plasmid. Explain how this image would be different if bisphosphonates were present. Why might these compounds be considered as a means of battling the rise in antibiotic resistance?

The Rest of the Story ●●●●●●

Earlier, we explored Joshua Lederberg's Nobel Prize-winning research on bacterial mating, or conjugation. Fundamental genetic research, like these early conjugation experiments, have led to the development of applications that have contributed to various aspects of human well-being. The study of conjugation by Lederberg resulted in the widespread use of *E. coli*, which eventually became the workhorse of molecular biology. The study of the F plasmid and λ bacteriophage that existed in strain K12 contributed to the understanding of fundamental genetic mechanisms such as conjugation and transduction. They also led to the development of tools for gene cloning, gene expression, and protein purification.

Applications of bacterial genetic mechanisms have not been limited to molecular biology. More recent applications include bacterial conjugation-based technologies aimed at combating antibiotic-resistant pathogens. For example, genetic information carried on a plasmid can be transferred via conjugation and expressed in a recipient pathogenic bacterium. The transferred plasmids are engineered to carry instructions for destruction of the recipient cells. The approach has proven successful in treating drug-resistant *Acinetobacter baumanii* wound infections in mice, and work continues on perfecting this conjugation-based approach.

Summary

Section 9.1: Why study genetics in bacteria?

Because of their genomic structure, bacteria are ideal candidates for genetic studies.

- Researchers used **nutritional mutants**, also known as auxotrophic mutants or auxotrophs, in some of the earliest genetic studies of bacteria.
- To avoid problems associated with **reversion**, researchers used double and triple mutants, thereby decreasing the probability that spontaneous revertants would arise.

The bacterial **genome** typically consists of a single, circular chromosome and **plasmids**.

- The chromosome and each plasmid is a **replicon** and the plasmid **copy number** within the cell is closely regulated.
- Plasmid incompatibility refers to plasmids that utilize the same copy number control mechanism and, generally, cannot co-exist stably within a cell.

Section 9.2: How are mutations used in bacterial genetics?

By studying mutants, microbiologists can learn much about the normal workings of bacterial cells.

- By comparing **wild-type strains** and mutant strains of bacteria, scientists can identify **alleles** of genes.
- The **genotype** describes the alleles of an organism.
- Mutants often can be distinguished from wild-type strains by phenotypic selection and phenotypic screening.
- **Replica plating** can be used to isolate mutants with certain phenotypic properties. **Clones** of these mutants then can be analyzed.
- Bacterial strains are described using specific genotypic notation, which ensures that their characteristics are known and understood.
- Several lines of experimental evidence have indicated that mutations arise spontaneously, in the absence of selective pressure, and provide the phenotypic variety on which natural selection acts.

Section 9.3: How is DNA cloned?

With recombinant DNA techniques, specific fragments of DNA can be amplified and modified, a process referred to as DNA cloning or, simply, cloning.

- **Restriction enzymes** cleave DNA at specific **restriction sites**. **DNA ligase** can be used to link cleaved DNA fragments, creating **recombinant DNA** molecules.
- **Molecular cloning**, or **DNA cloning** generally requires a **cloning vector**. Commonly used cloning vectors include plasmids, phages, and cosmids.
- Plasmid cloning vectors often contain a **multiple cloning site**, a short section of DNA that contains cleavage sites for several restriction enzymes.
- While many plasmid vectors have a narrow host range, **shuttle vectors** can be used in a wider range of organisms.
- Because phages normally transfer genetic material into host cells, phage vectors facilitate efficient cloning and allow the production of high quantities of cloned DNA.
- **Cosmid** vectors, which contain plasmid DNA linked to phage *cos* sites, can be packaged by phage λ, allowing more efficient cloning.

Section 9.4: How do bacterial cells transfer DNA?

Bacteria naturally exchange DNA, a process referred to as horizontal gene transfer, and scientists have exploited these mechanisms of exchange to develop recombinant DNA techniques.

- **Horizontal gene transfer**, in which DNA moves between bacterial cells, contributes to genome changes and rearrangements.
- Transferred DNA molecules can join together via **homologous** or **non-homologous recombination**. Non-homologous recombination may occur via **site-specific recombination**.
- Recombination can be used for many purposes, including the inactivation of a specific gene **cointegrate** formed by a **suicide plasmid**. Such an alteration is referred to as a **knockout mutation**.

- **Transformation** refers to the uptake of DNA by bacterial cells from the surrounding environment. In the laboratory, **competent** cells can be transformed in various ways, including **electroporation**.
- **Conjugation** refers to a specialized method of DNA transfer that requires cell-to-cell contact. The **F plasmid** contains genes encoding proteins needed to form the **sex pilus**, which forms a bridge between two cells.
- **Transconjugants**, **Hfr strains**, and **F-prime (F′) strains** and the formation of **merozygotes** have been used extensively in gene-mapping studies.
- **Transposition** refers to the movement of DNA by **transposable elements** like **insertion sequences**, or **IS elements** and **transposons**. Transposition typically requires two enzymes: **transposase** and **resolvase**.
- **Transduction** refers to the transfer of DNA by bacteriophages, leading to the formation of a **transductant**.

● Application Questions

1. A strain of *E. coli* is labeled as HB101 *recA⁻*. Based on this designation, what do you know about the wild-type strain and the phenotype of this strain?

2. You want to determine whether an *E. coli* strain is prototrophic or auxotrophic for histidine synthesis. What agar plates and techniques would you need to make your determination?

3. Scientists observe that antibiotic-resistant bacterial colonies appear when bacteria are grown on agar supplemented with an antibiotic. Does the antibiotic cause the antibiotic resistance mutations to develop? Explain your answer.

4. You are trying to screen a plate of bacterial colonies for resistance to the antibiotic ampicillin. What methods can you use to carry out this screen? Explain each approach to screening.

5. Imagine that a strain of *E. coli* produces a pink pigment. You streak an agar plate with the bacteria, expose it to UV light for one minute and incubate your plate. Following incubation, you observe some colonies that lack the pink pigment. Explain what most likely occurred at the molecular level to produce this result.

6. To extend your findings, you streak an agar plate with the pink pigment-producing bacteria, expose it to UV light for two minutes and incubate the plate overnight. When you examine the plate the next morning, no colonies are present. How would you explain this result?

7. Why are genes that are nearest the *oriT* of the integrated F in an Hfr strain more likely to be transferred to the recipient cell than genes that are farther from the *oriT*?

8. The evolution of antibiotic-resistant bacteria has presented a major medical problem. A recent development involves administering a drug that prevents conjugation by interfering with the *tra* genes. Explain how this approach might reduce the number of resistant bacteria in a population.

9. In the laboratory, you purify an *E. coli* strain and attempt to cleave its chromosomal DNA with the restriction enzyme that it produces, *Eco*RI. Electrophoretic analysis shows that the chromosome was not digested at all. Explain what occurred at the molecular level to produce these results.

10. In the laboratory, a group of scientists attempt to generate mutant strains of bacteria that they then can screen for effects. Many tools can be used to generate mutations.
    a. Describe a chemical mutagen you suggest they use and explain its effects.
    b. Explain how transposons could be used to generate the mutants.
    c. How would they screen and select for mutants using the approach described in part b?

Suggested Reading

Avery, O. T., C. M. MacLeod and M. McCarty. 1944. Studies on the chemical nature of the substance inducing transformation of pneumococcal types. J Exp Med 79:137–158.

Brock, T. D. The Emergence of Bacterial Genetics. 1990. Cold Spring Laboratory Press, Cold Spring Harbor, New York.

Cairns, J., J. Overbaugh and S. Miller. 1988. The origin of mutants. Nature 335:142–145.

Damelin, L. H., D. Mavri-Damelin, T. R. Klaenhammer and C. T. Tiemessen. 2010. Plasmid transduction using bacteriophage Φadh for expression of CC chemokines by *Lactobacillus gasseri* ADH. Appl Environ Microbiol 76:3878–3885.

Elena, E. and R. E. Lenski. 2003. Evolution experiments with microorganisms: The dynamics and genetic bases of adaptation. Nat Rev Genet 4:457–469.

French, C. T., P. Lao, A. E. Loraine, B. T. Matthews, H. Yu and K. Dybvig. 2008. Large-scale transposon mutagenesis of *Mycoplasma pulmonis*. Mol Microbiol 69:67–76.

Gibson, D. G. 2009. Synthesis of DNA fragments in yeast by one-step assembly of overlapping oligonucleotides. Nucl Acids Res 37:6984–6990.

Hayes, F. 2003. Transposon-based strategies for microbial functional genomics and proteomics. Annu Rev Genet 37:3–29.

Lederberg, J. and E. M. Lederberg. 1952. Replica plating and indirect selection of bacterial mutants. J Bacteriol 63:399–406.

Lenski, R. E. and M. Travisano. 1994. Dynamics of adaptation and diversification: A 10,000-generation experiment with bacterial populations. Proc Natl Acad Sci USA 91:6808–6814.

Redfield, R. J. 2001. Do bacteria have sex? Nature Reviews Genetics 2:634–639.

Zinder, N. D. and J. Lederberg. 1952. Genetic exchange in *Salmonella*. J Bacteriol 64:679–699.

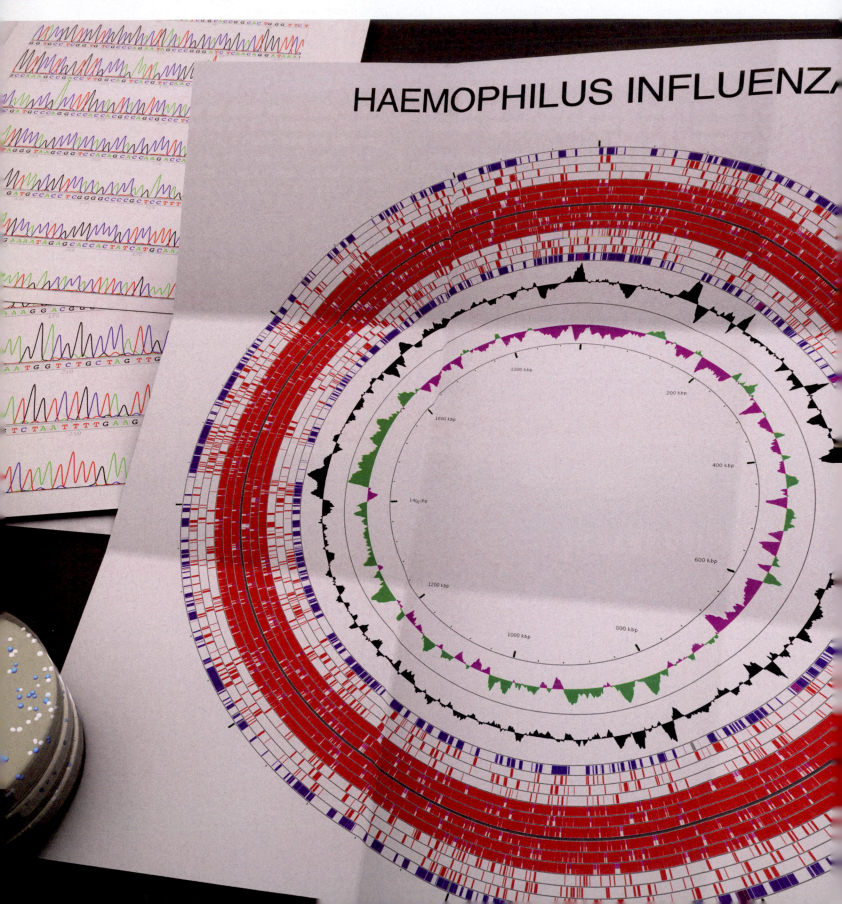

HAEMOPHILUS INFLUENZA

In the summer of 1995, Craig Venter, Hamilton Smith, Claire Fraser, and their team of scientists at The Institute for Genomic Research (TIGR) shook the microbiology and genetics research communities when they announced that they had decoded the complete genome sequence of a bacterium, *Haemophilus influenzae*. Using an innovative strategy called "shotgun sequencing" that took advantage of the speed and efficiency of the newly developed automated DNA-sequencing instruments and the power of modern computers, they completed their task in a fraction of the time it might have taken using a more traditional approach. They determined the genome sequence after less than a year of work, and finished years ahead of other efforts to sequence the *Escherichia coli* and *Bacillus subtilis* genomes, projects that had begun several years earlier. They completed the sequencing so quickly, in fact, that shortly before they published the sequence they received word that their grant proposal to do this work had been rejected. Reviewers doubted that the strategy would be successful. Fortunately, Venter and colleagues had received venture capital funding to initiate the sequencing effort. Following the success of this project, shotgun sequencing became the standard method used to sequence microbial genomes.

The work of Venter and colleagues represents an important milestone for bacterial genetics. In just over 50 years, bacterial geneticists have gone from the identification of DNA as the heritable material and the demonstration that bacteria really did have genes, to the detailed description of each of those genes in a single bacterium. Indeed, the development of shotgun sequencing has contributed to the sequencing of genomes of complex eukaryal organisms, including humans, and many microorganisms. As of February 2012, the complete sequences of nearly 3,000 bacteria and archaeons had been reported. We truly live in the Age of Genomics.

CHAPTER NAVIGATOR

As you study the key topics, make sure you review the following elements:

Sequencing techniques and computing power have made whole genome sequencing possible.

- 3D Animation: Measuring gene expression using pyrosequencing
- Figure 10.1: Sanger, or dideoxy, sequencing
- Perspective 10.1: Rate of DNA sequencing
- Figure 10.6: Process Diagram: Whole genome shotgun sequencing
- Toolbox 10.1: Genome databases

Genomic analysis also requires the study of transcripts and proteins within a cell.

- 3D Animation: Genomic analysis using microarrays
- Figure 10.7: Process Diagram: Generating a genomic library
- Figure 10.8: Process Diagram: Generating a cDNA library
- Microbes in Focus 10.2: *Yersinia pestis*
- Figure 10.10: *Yersinia pestis* infection cycle and analysis by microarray
- Figure 10.13: Proteomic analysis of response to salt concentration

By comparing the genomes of different organisms, we can learn more about their properties and evolutionary history.

- Perspective 10.2: The minimal genome
- Figure 10.15: Paralogs and orthologs
- Mini-Paper: Genome sequence of a killer bug
- Figure 10.17: Horizontal gene transfer

With DNA sequencing, we can learn much about organisms that cannot be cultivated.

- Figure 10.18: Metagenomic analysis

CONNECTIONS for this chapter:

The discovery of reverse transcriptase (**Mini-Paper, Chapter 8**)

Microarrays and virus identification (**Perspective 8.1**)

Exchange of antibiotic resistance genes (**Section 24.3**)

Pathogenicity islands and human bacterial pathogens (**Section 21.3**)

Introduction

The genomics revolution has had an important impact on microbiology. The genome sequences allow microbiologists to link directly the genetic characteristics of individual organisms with their physiological properties and their ecological niches. Microbiologists can make hypotheses based on genome sequence information, and then test these hypotheses experimentally. Comparison of genomes from closely related organisms can reveal the genetic basis of key physiological differences between those organisms. Genome comparisons also provide profound insight into the evolutionary relationships among all living organisms.

In Chapter 9, we introduced a number of genetic tools, such as conjugation, transduction, and transposition, which microbiologists use in their research. We considered the fundamental genetic concepts upon which these tools are based, and the mechanisms behind the major genetic processes that occur in bacteria. In the current chapter, we will explore studies of complete genome sequences. Historically, most of the research in

bacterial genetics has taken place in a limited number and range of organisms, the so-called model organisms. This approach has resulted in detailed knowledge and powerful research tools. It has also established the fundamentals of the genetics approach. As we explore the unique characteristics that contribute to the genetic, biochemical, and physiological diversity of microbial life, it will be useful to make comparisons to the genetic processes that have been outlined for the model organisms. Let's now take a more detailed look at the impact of genomics on microbiology and consider the following questions:

How are genome sequences determined? (10.1)
How is gene expression measured using genomics tools? (10.2)
What information can be obtained from comparing genomes? (10.3)
Can we apply genomics to the study of uncultivated microbes? (10.4)

10.1 Genome sequencing

How are genome sequences determined?

The establishment of molecular biology as a science was cemented with the advent of recombinant DNA techniques and the development of methods to determine DNA sequences. Initially, the sequencing of single genes required massive and costly efforts. Only specialized laboratories could conduct these experiments. Subsequent technological improvements increased the speed and efficiency of DNA sequencing, while at the same time lowering the cost, until it became possible to determine the complete genome sequence of an organism. Improvements in DNA-sequencing methods continue, and what was once specialized now has become almost routine. This dramatic change in DNA sequencing has drastically changed the nature of microbiology research.

In this section, we will explore **genomics**, the determination and study of complete genome sequences, focusing on some important techniques. First, we will investigate various DNA-sequencing methods, including the process of shotgun sequencing that we mentioned in the opening story. Then, we will explore the computational tools that have been developed to allow us to analyze the massive amounts of sequence data that we can generate. As we will see, the combination of innovations in DNA-sequencing technologies and advances in computing power has resulted in an explosion of available DNA sequence data.

DNA-sequencing methods

As we saw in Section 9.3, relatively simple molecular biology techniques can be used to clone specific fragments of DNA in

plasmid vectors. Often, though, researchers also need to determine the sequences of these DNA pieces. In the late 1970s, researchers developed two innovative means of achieving this task. Walter Gilbert and his student Allan Maxam at Harvard University developed a chemical degradation method. Around the same time, Fred Sanger and colleagues at Cambridge University developed an enzymatic method using DNA polymerase. Today, most researchers still use some variation of the Sanger method, but this is rapidly changing with advances in sequencing technology. In this section, we will first explore Sanger sequencing. We will then investigate newer sequencing technologies and approaches, which have led to the high-throughput, automated methods used today.

Sanger Sequencing

Sanger, or **dideoxy**, **sequencing** takes advantage of the primer-specific DNA synthesis activity of DNA polymerase (see Section 7.1). Briefly, the procedure requires three steps: cloning of the fragment to be sequenced, DNA synthesis, and gel electrophoresis **(Figure 10.1)**. Researchers first isolate the plasmid containing the cloned DNA and denature it to provide a single-stranded template. To this template, they add a radiolabeled short oligonucleotide primer, DNA polymerase, and the four deoxyribonucleotide triphosphates, collectively referred to as "dNTPs," where N refers to the bases G, C, T, and A (deoxyguanosine triphosphate, dGTP; deoxycytidine triphosphate, dCTP; deoxythymidine triphosphate, dTTP; deoxyadenosine triphosphate, dATP). The primer, designed to be

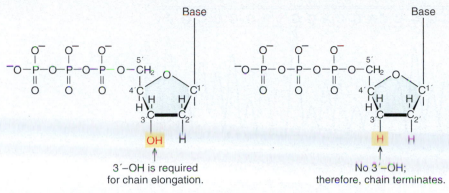

Base   Base

3′—OH is required
for chain elongation.

No 3′—OH;
therefore, chain terminates.

A. Deoxyribonucleoside triphosphate and dideoxyribonucleoside triphosphate

Figure 10.1. Sanger, or dideoxy, sequencing A. The Sanger-sequencing method relies on the use of special dideoxyribonucleotide bases that can be inserted into a growing DNA strand by DNA polymerase, but block the insertion of further nucleotides. **B.** The Sanger-sequencing method is based on the synthesis of a DNA strand, using a template strand to guide the insertions of bases. The reaction contains a certain proportion of dideoxynucleotides, which when inserted cannot support the further elongation of the synthesized strand. Because the reaction contains radioactive nucleotides (^{32}P-dCTP, highlighted in yellow), the products of the reaction can be detected by exposure to X-ray film after electrophoresis. For many years, this technique was carried out on thin polyacrylamide gels that were able to resolve fragments that differed in size by a single nucleotide. The fragments were then tabulated by eye and converted into DNA sequence.

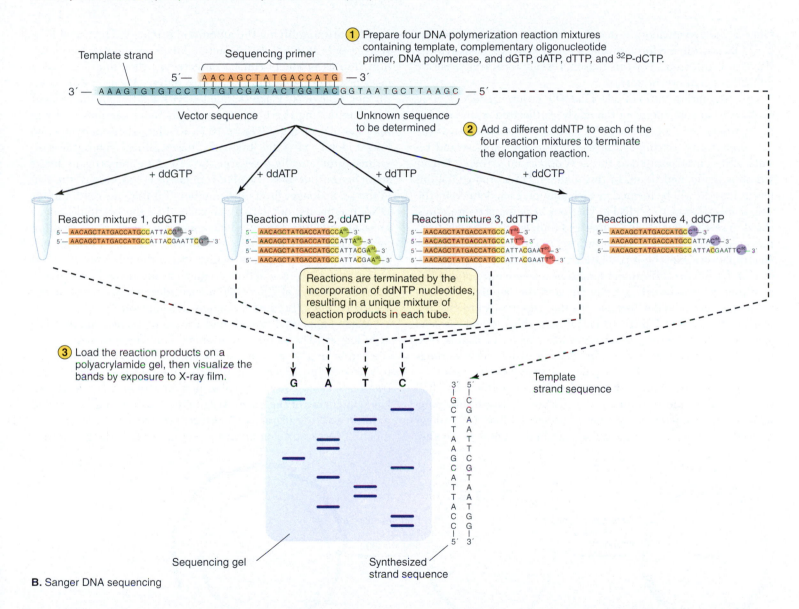

B. Sanger DNA sequencing

complementary to the vector sequence flanking one end of the cloned DNA, anneals to the single-stranded template, providing a base-paired free 3′—OH to which DNA polymerase can add additional nucleotides. As DNA polymerase continues to add new nucleotides, it makes a piece of DNA complementary to the single-stranded template.

As we just have described the Sanger DNA-sequencing method, a strand of DNA complementary to our DNA of interest

can be produced *in vitro*. How, though, can we determine the actual sequence of this newly synthesized strand of DNA? First, the mixture described above is divided into four separate aliquots. To each aliquot, researchers also add one more key reagent: *di*deoxyribonucleotide triphosphates (ddNTPs). One aliquot receives ddATP, one receives ddTTP, one receives ddCTP, and one receives ddGTP. DNA polymerase can add a dideoxynucleotide to the growing DNA chain just as it adds a deoxynucleotide.

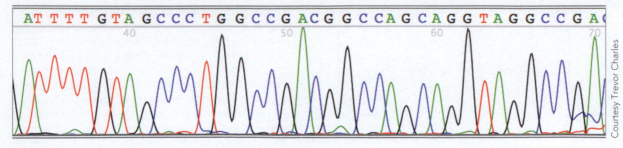

AT TTT GTA GCCC TG G CCG ACG G C CAG C AGG TA GG CC GA(

Figure 10.2. Trace from an automated DNA sequencer Automated methods have improved the efficiency of DNA sequencing and removed the requirement for the researcher to read the gel directly. In this method, fluorescent dye tags are used in place of radioactivity, enabling the automated detection of DNA fragments during electrophoresis. In the resulting output, a different color represents each of the four bases: black for G, green for A, red for T, blue for C. The trace of the peaks can be interpreted by a computer program.

Once a dideoxynucleotide becomes incorporated, however, no 3′—OH is available for addition of the next dNTP. DNA synthesis, as a result, stops. For this reason, researchers often describe ddNTPs as chain terminators.

An appropriate ratio of ddNTP:dNTP ensures a ddNTP is incorporated at some point in the newly synthesized strand during the synthesis reaction. If the ddNTP becomes incorporated at the first opportunity during synthesis, then a very short strand results. If it is incorporated at the next opportunity, then a longer strand is made, and so on. In this manner, DNA products of different lengths will be formed in each aliquot, each defined by the site of insertion of the ddNTP. The four reaction sets are denatured to release the newly synthesized radiolabeled strands from the unlabeled template strands. The mixture is separated by electrophoresis on a high-resolution polyacrylamide gel that can discern fragments differing in length by a single nucleotide. The resulting gel is exposed to X-ray film and the radiolabeled fragments produce a ladder-like pattern that can be read to give the DNA sequence (see Figure 10.1). Initially, this method resolved just over 100 nucleotides of sequence in a single run, that is, on a single gel. Subsequent improvements in electrophoresis methods have increased this number to greater than 400 nucleotides.

Since its initial development, the Sanger method has had several major improvements. The use of thermostable polymerases, the same ones used for PCR (see Toolbox 1.1), allows multiple rounds of synthesis from a single template strand,

significantly amplifying the amount of products. The use of fluorescent dyes instead of radioactive labels eliminates the X-ray film step as products can be detected as they migrate past a laser near the end of the gel. To streamline this procedure even further, each ddNTP can be labeled with a different fluorescent dye, eliminating the need to carry out four reactions and run four samples on a gel (**Figure 10.2**). Reading the sequence is no longer done manually. Instead, "base-calling" computer programs automatically interpret the raw data and provide direct sequence output. Because electrophoresis is continuous, the size of the gel no longer limits the number of fragments read. Larger DNA fragments, not clearly separated in the original Sanger method, undergo longer electrophoretic separation, yielding more sequence data. These methods are called **automated-** or **cycle-sequencing methods**, and although they are still based on Sanger's use of dideoxynucleotides, it is now routine to obtain sequences of 700–1,000 nucleotides in a matter of hours, for the cost of a fraction of a penny per nucleotide.

What happens if the cloned fragment to be sequenced is even longer? The sequence obtained from the first sequencing run can be used to design a second primer that, when subjected to a second run of sequencing, will anneal farther along in the fragment. Consecutive runs of sequencing and primer design allow sequencing of the entire fragment. This technique is called **primer walking (Figure 10.3)**. Of course, there is a limit to how much DNA can be cloned into a vector. Most cloning plasmids

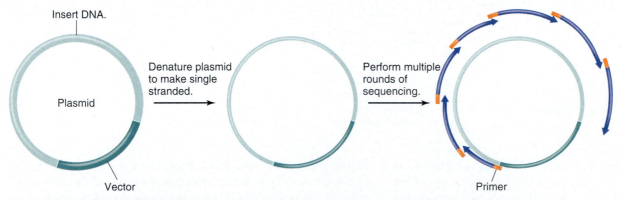

Figure 10.3. Primer walking Sanger sequencing requires binding of a sequencing primer to initiate DNA polymerase activity. As demonstrated in Figure 10.1, the ends of cloned fragments are usually sequenced using sequencing primers complementary to sequences within the vector (in blue). A cloned fragment (insert DNA) of about 3 kb is sequenced using the primer walking method, with each sequencing reaction giving between 500 and 1,000 bp of sequence. The dark blue arrows indicate the sequence that is generated, and the short gold lines at the base of each arrow correspond to the sequencing primers. Starting from the first sequences obtained from the end of the insert using a primer to the vector cloning site sequence, each subsequent sequencing reaction, with the same template, is carried out using primers designed based on the preceding sequence.

Perspective 10.1
RATE OF DNA SEQUENCING

The core principles of dideoxy DNA sequencing have remained the basis for DNA sequencing well into the genomics age. The rate of DNA sequencing, though, certainly has not stayed the same! The base pairs of DNA sequence deposited into public databases has increased exponentially since the mid-1990s **(Figure B10.1)**.

Today, we can sequence genomes more quickly and more cheaply than ever before. The doubling time for completion of bacterial genomes is 20 months. In other words, about every 20 months, the number of sequenced bacterial genomes increases by a factor of 2 **(Figure B10.2)**. This exponential increase in throughput has been accompanied by a reduction in cost, and this cost reduction actually has accelerated in recent years **(Figure B10.3)**.

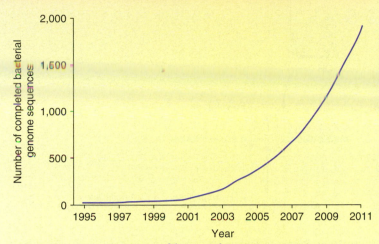

Figure B10.2. Sequencing of bacterial genomes Since the late 1990s, the number of sequenced bacterial genomes has been increasing exponentially, doubling about every 20 months.

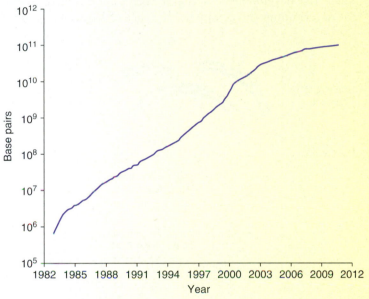

Courtesy of NCBI'S GenBank, October 2007

Figure B10.1. Growth in DNA sequencing The amount of sequenced DNA continues to increase rapidly. The number of base pairs deposited in the GenBank database each year is increasing exponentially.

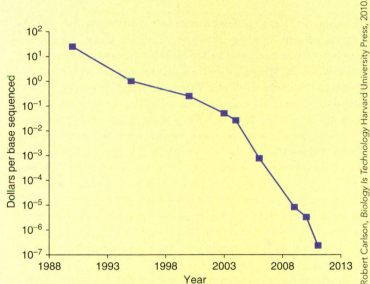

Figure B10.3. Cost of DNA sequencing The cost of DNA sequencing has decreased dramatically over the years. Moreover, the rate of decrease has increased dramatically since the early 2000s. In 1990, an investment of $10 would have yielded a fraction of a base of sequence, while in 2012, it could have funded hundreds of millions of bases of sequence, depending on the sequencing technology used.

Robert Carlson, *Biology Is Technology* Harvard University Press, 2010.

can hold around 10 kb of cloned DNA, while cosmids (see Section 9.5) can hold about 45 kb of cloned DNA. Even one of the smallest bacterial genomes to be sequenced, that of the insect symbiont *Candidatus* Carsonella ruddii (see Section 17.5), consists of 169 kb—too big to be cloned in one piece. To sequence a complete genome, multiple, overlapping clones must be sequenced sequentially and analyzed. The process involved is very iterative. The story, though, does not end here. Researchers continue to develop new improvements to the process.

Next Generation Sequencing

During the early days of genomics, few, if any, major breakthroughs in DNA sequencing technology occurred. Around 2004, though, a major cost reduction occurred **(Perspective 10.1)**. What caused this abrupt change? New sequencing technologies became available to the scientific community around this time—post-Sanger, **next generation sequencing methods** that were not reliant on gel electrophoresis. Originally developed by Mostafa Ronaghi at the Royal Institute of Technology in

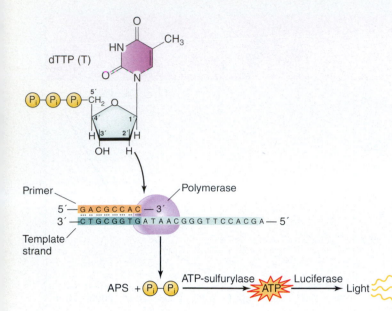

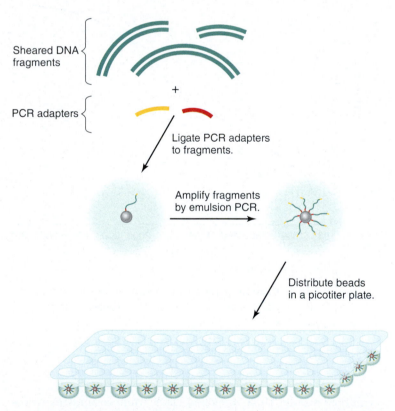

Figure 10.4. Pyrosequencing DNA polymerase adds dTTP to the growing DNA strand. ATP-sulfurylase reacts the pyrophosphate released upon incorporation with adenosine phosphosulphate (APS) to generate ATP. This ATP, in turn, interacts with luciferase, releasing a burst of light. The automated detection of this light by a camera indicates that the nucleotide was added to the growing DNA strand.

Stockholm in 1996, **pyrosequencing** represents one of the most significant next generation methods. Like Sanger sequencing, pyrosequencing relies on DNA polymerase-mediated synthesis of the complementary strand of a single-stranded template. It differs from the Sanger method, however, in that instead of using chain-terminating ddNTPs, the four dNTPs are added one at a time. If the dNTP is complementary to the template base, then it becomes incorporated into the growing piece of DNA **(Figure 10.4)**. This incorporation releases pyrophosphate (PPi), which can be converted by ATP-sulfurylase to ATP. This ATP drives a luciferase reaction, producing light, which can be detected with a charge-coupled device (CCD) camera. This light production gives rise to the name "pyrosequencing." After washing, the reaction is repeated, except with a different dNTP base. Light will only be produced if the dNTP is complementary to the next template base.

An adaptation of pyrosequencing—**454 DNA sequencing**— dramatically increases the throughput. The name "454" has no special meaning; it comes from the code name of the project that resulted in its development. In 454 sequencing, unlike the Sanger method, the DNA fragments do not have to be cloned into vectors. The DNA is then fragmented into short pieces, and ligated with adapters that facilitate the trapping of the fragments on separate beads **(Figure 10.5)**. Each bead, containing a single, short DNA fragment, can be separated into its own aqueous droplet of PCR reaction mixture within an oil emulsion, and PCR is carried out. The result is millions of DNA molecules on each bead. Hundreds of thousands of the beads are distributed into the wells of a flow cell picotiter plate, a slide that has almost 2 million 75-picoliter-volume wells. Each well contains a single bead and its bound DNA. The DNA-sequencing reagents are added to each well, and the sequencing reactions are carried out by sequential addition of each of the four dNTPs by horizontal flow over the wells. A CCD camera captures the light. While

short lengths of individual reads initially limited the usefulness of 454 DNA sequencing, subsequent improvements have led to read lengths greater than 500 bp, making it competitive with the Sanger method. Now, over 500 million bases of sequence data can be generated from a single machine in a 10-hour run.

Since the development of 454 DNA sequencing, other high-throughput DNA sequencing methods have been developed. In Illumina sequencing, also known as "sequencing by synthesis," DNA polymerase adds fluorescently labeled nucleotides to amplified, immobilized fragments of DNA. Following imaging, the fluorescent label is cleaved from the terminal nucleotide and another fluorescently labeled nucleotide can be added. In Ion Torrent or ion-semiconductor sequencing, a sensor detects hydrogen ions that are released during the synthesis of DNA. Biologists, chemists, engineers, and mathematicians are working together to develop other innovative approaches to sequencing DNA. Continual reductions in cost and increases in throughput will certainly have major impacts on microbial genomics and microbiology in general.

Shotgun Sequencing

The ability to generate high-throughput sequence data enabled the development of the **shotgun-sequencing** method for determining the complete sequence of bacterial genomes.

Figure 10.5. 454 DNA Sequencing This high-throughput pyrosequencing method starts with shearing of the DNA into fragments. Short adapter sequences are then ligated to the end of each fragment. One of these adapter ends is attached to agarose beads, and the fragment is then amplified by PCR within water droplets, a process called "emulsion PCR." Individual beads with the amplification products are distributed in a flow cell, where repeated pyrosequencing reactions are carried out as in Figure 10.4, alternating between each of the four dNTP bases.

This process involves shearing the DNA into short pieces and then sequencing those fragments from their ends (**Figure 10.6**). After determining the sequences of individual fragments, researchers use sophisticated computer programs to identify regions of sequence overlap. With this information, individual sequences can be linked together. Because the sequences are derived from a random distribution of fragments, the amount of sequence data that must be obtained to ensure complete coverage of the genome is on the order of ten times the genome size.

Microbes in Focus 10.1
MYCOPLASMA GENITALIUM

Habitat: Originally isolated in 1980 from a human urethra. Commonly resides in epithelial cells of human genital and respiratory tracts.

Description: Member of the Mollicutes class of the Firmicutes phylum. Has one of the smallest genomes of any known free-living bacterium, with 582,970 bp encoding 521 genes. Second genome to be completely sequenced, after *Haemophilus influenzae*.

Key Features: Like other *Mycoplasma*, *M. genitalium* lacks a cell wall, and thus is resistant to many antibiotics that target the synthesis of the cell wall. The cells are typically very small, usually less than 1 μm in diameter, and irregularly shaped. Its genetic code differs slightly from the standard genetic code, in that TGA codes for tryptophan, rather than a stop codon. Because of its small genome size, it has been studied intensively as part of the Minimal Genome Project, and its genome is the first genome of a cellular organism to be completely synthesized.

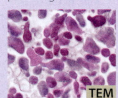
TEM
© Science Photo Library/ Photo Researchers, Inc.

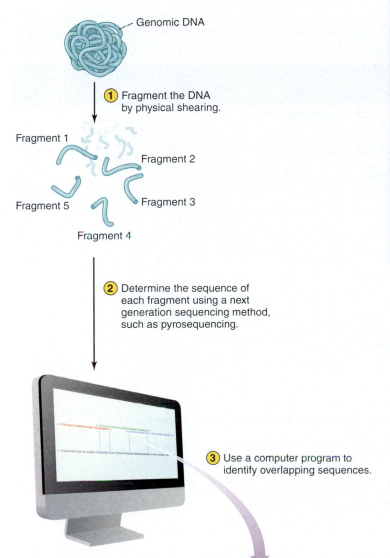

Genomic DNA

1 Fragment the DNA by physical shearing.

Fragment 1
Fragment 2
Fragment 5
Fragment 3
Fragment 4

2 Determine the sequence of each fragment using a next generation sequencing method, such as pyrosequencing.

3 Use a computer program to identify overlapping sequences.

Soon after the development of these automated DNA-sequencing methods, several research centers established high-throughput genome-sequencing facilities that could generate sequence data at an astounding pace. Initially, microbiologists began sequencing the genomes of human pathogens, because of their fundamental effects on human health. Now, we are determining the genome sequences of a broader range of microbes more representative of the diversity and ecology of the microbial world. These genome sequences have reinforced our prediction of the variation in genome size and complexity among the microbes. For example, a 15-fold size difference exists between the genomes of *Mycoplasma genitalium* (**Microbes in Focus 10.1**), a parasitic bacterium associated with urethritis in humans

3' GCTCGCCGTAGCCGAACCGCTAGCGGGCCTGGGACCGCTTCCGGCCACGG 5'
5' ATGACCAAGACTGCGGTGATAACGGGTTCCACGAGCGGCATC 3'   5' ACCCTGGCGAAGGCCGGTGCCAATATCGTCCTGAACGGCTTCGGTGCGCC 3'

4 Assemble sequence from the overlaps.

5' ATGACCAAGACTGCGGTGATAACGGGTTCCACGAGCGGCATCGGATTGGCGATCGCCCGGACCCTGGCGAAGGCCGGTGCCAATATCGTCCTGAACGGCTTCGGTGCGCC 3'

Figure 10.6. Process Diagram: Whole genome shotgun sequencing Whole genome shotgun sequencing is the strategy of choice for determining the sequence of bacterial genomes. It involves the direct sequencing of random clones or DNA fragments. The development of this method was made possible by the improvement of DNA-sequencing methods, and the availability of computational power to find overlaps between the obtained sequences.

To be most useful to scientists, DNA sequence data should be freely accessible over the Internet. Indeed, three major DNA sequence databases do exist: GenBank, DNA Data Bank of Japan, and the EMBL Nucleotide Sequence Database. Moreover, scientific journals require that DNA sequence data be deposited in one of these public databases before publication in the journal. The databases then share the deposited data, making it readily available to researchers throughout the world.

While an important function of these databases is to allow distribution of the sequence data, they also serve as portals to a broad range of bioinformatics tools. Using these tools, scientists can query the data in several ways. They can compare sequences at the level of nucleotide or protein, determine whether a gene sequence is similar to any sequences in the database, identify motifs or patterns within DNA and protein sequences, and link to relevant literature in journal databases. The most widely used search tool is BLAST, the Basic Local Alignment Search Tool (**Figure B10.4**).

Using the BLAST programs, researchers can search nucleotide and protein sequence databases with either nucleotide or protein

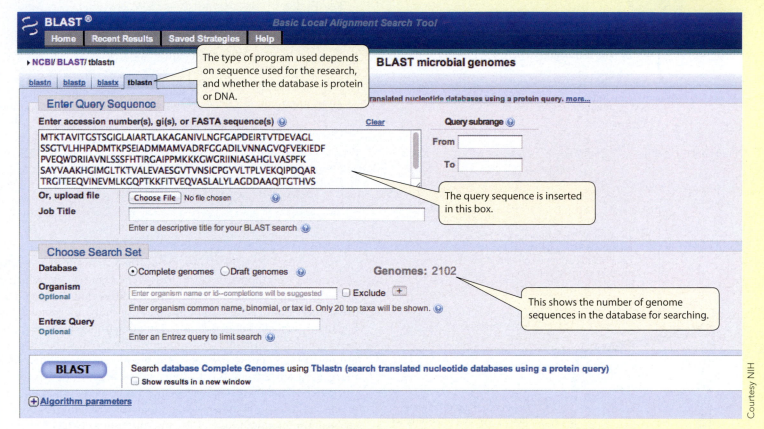

Figure B10.4. BLAST Maintained by the National Center for Biotechnology Information (NCBI), the Basic Local Alignment Search Tool (BLAST) allows researchers to enter a query sequence and search for similarities within the existing database. In this example, an amino acid sequence is compared to genomic sequences using the TBLASTN program, which translates the database DNA sequences in all reading frames before comparison with the amino acid query sequence.

(582,970 bp) and *Bradyrhizobium japonicum*, a soil bacterium that forms symbiosis with the soybean plant (9,105,828 bp). Even within a species, there can be considerable variation in size. *Escherichia coli* genomes may differ in size by as much as 30 percent!

Bioinformatics

As we have seen in the preceding section, automated DNA-sequencing methods allow researchers to generate DNA sequence data at a rate unimaginable just 20 years earlier. Now comes the really difficult part: analyzing all of this information.

Bioinformatics, a burgeoning area within the information technology field, allows us to process large quantities of biological data, and has developed in tandem with advances in DNA-sequencing. Many computational tools have been developed, and these tools have contributed to the intensity of bioinformatics research. In a process called **annotation**, for instance, researchers use programs employing specific algorithms to predict the beginning and end of protein-encoding sequences, or **open reading frames (ORFs)**. Other programs compare the sequences of predicted protein products with the sequences of known proteins available in databases. Based on observed

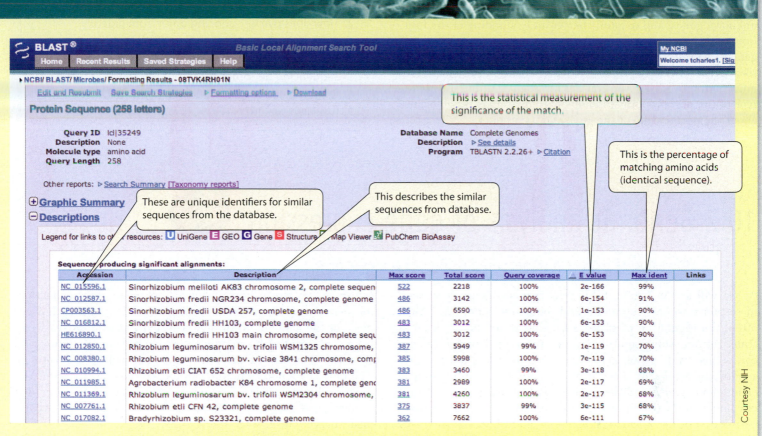

Figure B10.5. Results of a BLAST search After comparing the query sequence to sequences within the database, BLAST provides the researcher with a list of previously reported sequences that best match the query sequence. In this example, the query sequence matches the sequence of *Sinorhizobium meliloti*.

sequences. An input sequence gets compared to previously deposited sequences in the database. A list of the most likely matches is then generated **(Figure B10.5)**.

While the resources available at sites like the National Center for Biotechnology Information certainly are impressive, a number of third party online supplementary databases and bioinformatics tools also exist. Over 1,000 of these databases and applications have been catalogued by the journal *Nucleic Acids Research*. As we have seen throughout this section, an Internet connection has become essential for conducting research in microbiology!

● **Test Your Understanding** ·················

You have sequenced a gene and entered the nucleotide sequence into BLAST. You identify a closely related gene that has been well defined with a very high similarity to your gene sequence. How can you use this information for the next step in your research?

similarities, functions for newly discovered proteins may be suggested.

Of course, analysis of sequence data requires access to both the data and the analysis programs. As described in **Toolbox 10.1**, several repositories of DNA, RNA, and protein sequence data and suites of analysis programs exist and can be accessed via the Internet. Interestingly, these sequence repositories actually were initiated before the advent of high-throughput sequencing and the widespread availability of the Internet.

Often, sequence analyses can allow researchers to predict the class of protein that a newly discovered gene will produce.

Because of sequence similarities, for example, researchers may speculate that a newly identified gene encodes a transcriptional factor, a dehydrogenase enzyme, or a transport protein. Identifying the biological molecules with which this protein interacts, however, cannot be predicted as reliably. Indeed, in many cases, sequence analysis cannot even assign a gene to a class. Even for a bacterium as well studied as *E. coli*, many of the genes uncovered by genome sequencing encode gene products whose functions remain unknown. Indeed, as many as a third of the predicted genes in a newly sequenced bacterial genome may be of unknown function. Perhaps more surprisingly, about a quarter of the genes may be unique—present in no other genome that has been sequenced.

What do these "new" genes do? The goal of **functional genomics** is to determine the biological functions of the unknown genes, and to gain a better understanding of the functions of all gene products. There are a number of functional genomics projects underway that involve large-scale mutagenesis activities. In these projects, researchers attempt to introduce mutations into each of the genes. They then examine the phenotype of the mutants. Any phenotypic changes may provide clues about the normal role of the gene and its product.

> ### 10.1 Fact Check
> ....
>
> 1. Describe the Sanger method of DNA sequencing.
> 2. How does pyrosequencing differ from dideoxy sequencing?
> 3. What is shotgun sequencing, and how is it used in sequencing genomes?
> 4. What is bioinformatics, and in what ways can it be used in research?

10.2 Genomic analysis of gene expression

How is gene expression measured using genomics tools?

As we will see in Chapter 11, cells regulate gene expression in response to environmental and metabolic conditions. Most often, the regulation of gene expression occurs at the level of transcription; some genes may be transcribed under certain conditions, while other genes may be transcribed under other conditions. Similarly, some proteins may be present within a cell under certain conditions, while other proteins may be present under other conditions. The production of diptheria toxin by isolates of *Corynebacterium diptheriae* and Shiga-like toxin by strains of *Escherichia coli*, for instance, increases when available iron concentrations are low. Other *E. coli* proteins, including iron-binding siderophores, are also produced only when environmental iron is lacking.

> **CONNECTIONS** Why would gene expression be regulated by environmental iron concentrations? As we will see in **Section 19.2**, iron is a limiting nutrient in many environments, including the human body. Pathogens, as a result, have evolved various strategies to acquire it.

To more completely understand the properties of an organism, then, it may be useful to examine the transcripts and proteins present in a cell under specific environmental conditions. In this section, we will investigate genomics tools that allow us to monitor both types of molecules. We will also see what this information tells us about cellular functions. Before we tackle these issues, however, we'll look at another important tool—the creation of genetic libraries.

> **CONNECTIONS** In **Chapter 11**, we will investigate several well-studied gene regulation systems, and delve into their mechanisms. The ability to study these systems using genomics tools provides a much clearer picture of the many regulatory pathways of the cell.

Genomic libraries

For many genetic analyses, researchers require a **genomic library**, a collection of cloned DNA fragments that, in total,

comprise the entire genome of an organism. This collection, for instance, can be used for whole genome sequencing. If every piece of the genome of an organism exists within at least one clone and the DNA fragments within each clone are sequenced, then the sequence of the entire genome will be determined. The only challenge remaining would be to correctly order the individual pieces of DNA sequence. As we saw in Section 10.1, sophisticated software programs allow researchers to accomplish this very task.

While the concept of a genomic library may seem fairly straightforward, the details, at first glance, may appear more problematic. How can we obtain clones of every single piece of an entire genome? To address this question, let's revisit the basics of DNA cloning. In Section 9.3, we examined the methodology required to clone a specific gene from a bacterium. The process begins with the digestion of the host genome with a restriction enzyme. The resulting DNA fragments are then mixed with a plasmid that has been digested with the same restriction enzyme. As we noted, some plasmids may religate, without incorporating any foreign DNA, some plasmids may incorporate a piece of DNA containing the gene of interest, and other plasmids may incorporate other fragments of DNA from the genome (see Figure 9.14). These plasmids are then introduced into a new host, such as *E. coli*, and we screen the *E. coli* for the foreign gene of interest.

Because all fragments of the digested genome can ligate with the digested plasmid, all pieces of the digested genome should exist within at least one recombinant host cell **(Figure 10.7)**. A large collection of recombinant cells, then, should contain a complete genomic library of the organism of interest. The exact number of clones needed for a complete genomic library depends, not surprisingly, on the size of the genome and the average size of each individual cloned fragment. We can estimate the necessary genomic library size with the following equation:

$$N = \frac{\ln(1 - P)}{\ln(1 - f)}$$

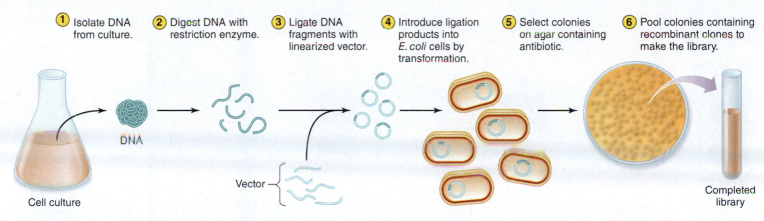

Figure 10.7. Process Diagram: Generating a genomic library To create a genomic library of bacteria, the genome of interest and a vector plasmid are digested with a restriction enzyme. The resulting fragments are mixed and ligated. The ligation products are introduced into suitable host cells, usually *E. coli*. Cells containing a plasmid are selected. Using a mathematical formula, a researcher can estimate the number of clones needed to ensure that every fragment of the original chromosome probably exists in at least one of the clones.

In this equation, N is the number of clones required, P is the desired probability of generating a complete library, and f is the average size of each cloned fragment divided by the total size of the genome. To see how this equation can be used, let's look at an example. Suppose we wanted to be 99 percent confident that we generated a complete genomic library for a bacterium with a genome of approximately 4.5 Mb, roughly the size of the *E. coli* genome. Further, let's assume that we want individual fragments to be about 10,000 bp long. The value of P, then, would be 0.99. The value of f would be 10,000/4,500,000, or about 0.000444. Using these values, $N = 2,000$. In other words, we would need about 2,000 individual clones to be 99 percent certain that the entire genome were contained within our library.

Genomic libraries certainly do not represent the only type of clone libraries useful to microbiologists. In many cases, a **complementary DNA library**, or **cDNA library**, is quite useful. Unlike a genomic library, a cDNA library represents a collection of all the mRNA species present within a cell under certain conditions **(Figure 10.8)**. To generate a cDNA library, researchers first isolate mRNA from the cell of interest. This mRNA is then converted into complementary DNA, or cDNA, using the enzyme reverse transcriptase. These DNA fragments are then ligated into an appropriate cloning vector and introduced into an appropriate host cell. Again, we can predict that a sufficiently large collection of recombinant host cells will contain the cDNA versions of all the mRNA molecules present in the original cell. As we will see in the next section, cDNA libraries can be quite useful for examining gene expression in cells.

CONNECTIONS Discovered independently by David Baltimore and Howard Temin **(Mini-Paper, Chapter 8)**, reverse transcriptase represents an essential enzyme in the normal replication of retroviruses, like HIV. This viral enzyme is also the target of effective antiretroviral drugs, like AZT.

Transcriptomes

While whole genome sequencing tells us much about the properties of and evolutionary relationships between organisms, understanding the expression patterns of genes allows us to better understand how organisms function under different conditions. Transcription patterns, in other words, can reveal much about an organism. Here, we will see how researchers can analyze the **transcriptome**, or set of transcripts encoded

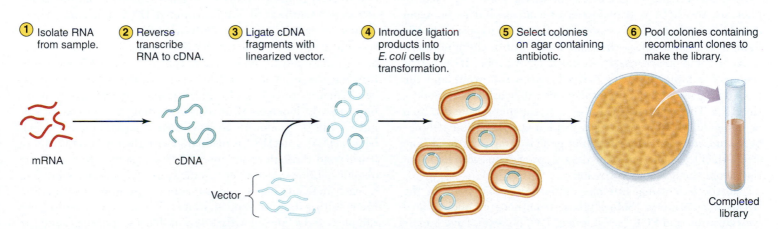

Figure 10.8. Process Diagram: Generating a cDNA library To generate a cDNA library from a eukaryal organism, one first isolates mRNA from cells of interest. The RNA is then used to generate cDNA via reverse transcriptase. This cDNA is ligated into a plasmid and the products are introduced into suitable host cells. Cells containing a plasmid are selected.

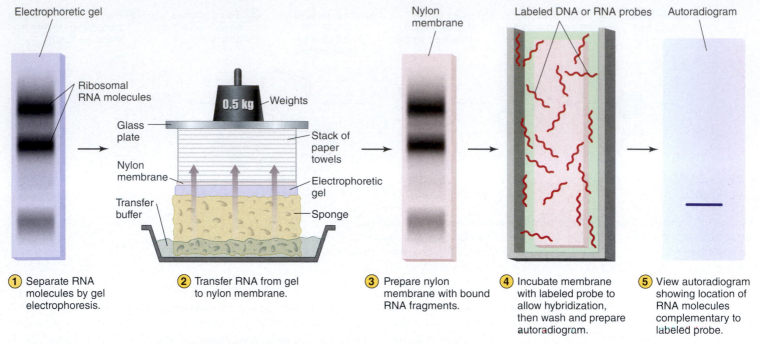

Electrophoretic gel

Ribosomal RNA molecules

0.5 kg — Weights

Glass plate

Stack of paper towels

Nylon membrane

Electrophoretic gel

Transfer buffer

Sponge

Nylon membrane

Labeled DNA or RNA probes

Autoradiogram

① Separate RNA molecules by gel electrophoresis.

② Transfer RNA from gel to nylon membrane.

③ Prepare nylon membrane with bound RNA fragments.

④ Incubate membrane with labeled probe to allow hybridization, then wash and prepare autoradiogram.

⑤ View autoradiogram showing location of RNA molecules complementary to labeled probe.

Figure 10.9. Process Diagram: Northern blotting RNA is isolated from cells of interest. Electrophoresis is used to separate the molecules by size. The major bands are ribosomal RNA, the most abundant RNA in the cell, while the smear represents the collection of other RNA, mostly mRNA. The RNA molecules are then transferred to a nitrocellulose or nylon membrane and probed with a labeled oligonucleotide probe that is complementary to the mRNA of the gene whose expression is being examined.

by each of the genes within a genome. How can we analyze the transcriptome of a cell? One answer to this question relies on a fundamental technique in the molecular biology repertoire—nucleic acid hybridization.

Northern Blotting

In 1975, E. M. Southern described a method, now known as the **Southern blot** in honor of the inventor, for detecting specific DNA sequences by hybridization using labeled probes of complementary sequence. A derivation of this technique, termed the **Northern blot**, can be used to measure the transcriptional expression of individual genes **(Figure 10.9)**. In this procedure, researchers first isolate RNA from cells and then separate these molecules by agarose gel electrophoresis. Following electrophoresis, the RNA is transferred to a nitrocellulose or nylon membrane, to which it is covalently linked by exposure to UV light. The membrane is then incubated with a chemiluminescent, or radiolabeled single-stranded DNA probe. If sufficient sequence complementarity exists between the probe and the RNA on the membrane, then the probe will hybridize, or bind to, the RNA. The location of the signal on the membrane can then be detected by autoradiography or photography. The existence of a signal tells us that a particular gene is being expressed. The strength of the signal, to some degree, provides us with information about the level of expression.

Northern blotting still has important applications in the research laboratory. Solution-based methods of nucleic acid hybridization and PCR (see Toolbox 1.1), however, have largely supplanted this technique for high-throughput analyses. As we will see in Chapter 11, bacteria have evolved complex mechanisms of gene regulation to ensure that cells express specific genes at specific times in response to changing environmental conditions. With Northern blotting, tracking the expression patterns of many genes becomes quite cumbersome. Other techniques allow us to analyze more easily the expression patterns of many genes at once.

Microarrays

Advances in photolithography, a technique related to computer chip production, and the availability of complete genome sequences have resulted in the development of the **microarray**, a method used to examine the transcriptional activity of all genes in the genome simultaneously. We can think of microarrays as essentially the reverse of Northern blot hybridization. The process starts with the microarray—a glass slide that contains short, artificially synthesized single-stranded DNA molecules called "oligonucleotides," or short, PCR-amplified gene segments representative of each of the transcripts produced from the genome of a given organism. Researchers convert total mRNA isolated from an organism into cDNA, incorporating a fluorescent label. When more copies of a given transcript exist, then more fluorescently labeled cDNA corresponding to that particular mRNA will be produced. To detect the relative amounts of labeled cDNAs, the cDNA is incubated with the microarray, where it will bind through complementary base-pairing to one of the oligonucleotides. A scanner can measure the amount of binding to each individual spot and a computer program can analyze the resulting data. In some systems, cDNAs from two different samples can be labeled with different fluorescent labels, and the hybridizations to a single microarray are done simultaneously. By merging the signals, the relative abundance of each transcript can be determined by the color output.

To more fully explore the usefulness of microarray analysis, let's consider *Yersinia pestis*, the bacterium associated with the bubonic plague **(Microbes in Focus 10.2)**. This pathogenic bacterium normally infects a variety of rodents, including rats, squirrels, and prairie dogs. The bacterium also replicates within fleas, and flea bites serve as the primary mode of transmission between rodent hosts. The microbe can be transmitted to humans, either via fleas or through contact with an infected rodent **(Figure 10.10)**. Humans, we know, have a body temperature of 37°C. The internal temperature of a flea, conversely, is much lower. *Y. pestis*, then, must be able to survive at two quite different temperatures.

We could hypothesize that the temperature shift *Y. pestis* encounters during its transmission from a flea to a warm-blooded host results in a change in gene expression. To test this hypothesis, researchers in China isolated RNA from *Y. pestis* cells growing at 26°C and 37°C. Using DNA microarray analysis, they identified over 400 genes, including genes encoding virulence factors and proteins involved in metabolism, that were expressed differently at the two different temperatures. Expression of some

Microbes in Focus 10.2
YERSINIA PESTIS

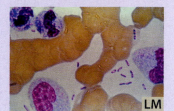

Habitat: Replicates in a wide range of rodent species and can be transmitted between hosts by fleas. Can be transmitted to humans, either by flea bites or by contact with an infected rodent.

Description: Gram-negative, rod-shaped bacterium in the *Enterobacteriaceae* family. One of three members of the *Yersinia* genus (along with *Yersinia pseudotuberculosis* and *Yersinia enterocolitica*) that can cause disease in humans.

Key Features: During the 1300s, the plague, or Black Death, swept through Europe, killing roughly one-third of the population. While the plague is not common today, it certainly has not gone away. In the United States, according to the Centers for Disease Control and Prevention (CDC), roughly 10–15 cases of the plague occur annually, mostly in the rural areas of the Southwest. Annually, 1,000–3,000 cases occur in countries throughout the world.

© Science Source/Photo Researchers, Inc.

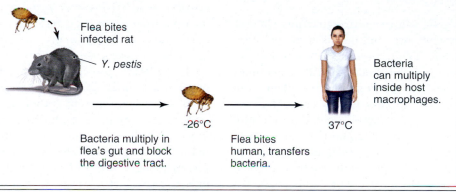

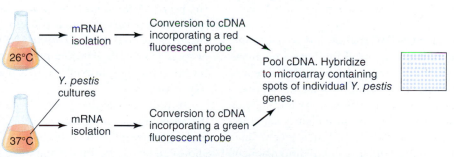

Observation: *Yersinia pestis* causes different diseases in humans and fleas. Humans have a higher body temperature than fleas.

Flea bites infected rat

Y. pestis

Bacteria multiply in flea's gut and block the digestive tract.

~26°C

Flea bites human, transfers bacteria.

37°C

Bacteria can multiply inside host macrophages.

Hypothesis: Specific genes involved in the transition from one host to the other are induced at different temperatures.

Experiment: Perform microarray analysis to compare mRNA expression in cultures grown at 26°C and 37°C.

26°C

37°C

Y. pestis cultures

mRNA isolation

Conversion to cDNA incorporating a red fluorescent probe

mRNA isolation

Conversion to cDNA incorporating a green fluorescent probe

Pool cDNA. Hybridize to microarray containing spots of individual *Y. pestis* genes.

Results: Many genes were expressed under one temperature, but not the other.

Conclusion: The identified genes might be involved in the different disease processes in humans and in fleas.

Yellow color indicates expression under both conditions.

Figure 10.10. *Yersinia pestis* infection cycle and analysis by microarray The hosts that are infected by *Y. pestis* differ in internal temperature, with the human body temperature being much higher that the internal temperature of a flea. To identify possible factors that might be involved in the transition from one host to the other, microarray analysis was performed on *Y. pestis* cultures of different temperatures. The identified genes are candidates for future research.

genes increased at the higher temperature, while the expression of other genes decreased at the higher temperature. These genes, the authors note, may be important determinants of *Y. pestis* pathogenicity.

RNA-seq Technology

RNA sequencing, or RNA-seq, represents the most recent advancement in studying transcriptomes. In this approach, researchers use high-throughput next generation sequencing technologies to sequence RNA molecules isolated from cells being studied. Briefly, RNA is isolated from cells of interest and converted to cDNA. Sequencing linkers, short segments of DNA of known sequence, are attached to the cDNA fragments and then sequenced. Bioinformatics tools can be used to align small fragments and compare the sequences to known RNA sequences. In this fashion, a detailed analysis of the RNA within a cell can be achieved.

In 2008, Michael Snyder and colleagues at Yale University used RNA-seq to examine the transcriptome of a model eukaryal organism—*Saccharomyces cerevisiae*. Since then, this technique has been used to examine the transcriptomes of other organisms. While still in its infancy, RNA-seq will probably continue to develop in tandem with advances in bioinformatics and sequencing technologies.

Proteomes

Not only may bacterial cells contain different transcripts at different times, but they may also contain different proteins, according to the cells' needs. The **proteome** refers to the collection of proteins present in a cell under specific conditions. Differences in protein types and levels may reflect changes in gene expression and/or protein stability. Analysis of proteins by gel electrophoresis, however, can be complicated; different proteins have different overall charges, based on their amino acid composition. Therefore, both the mass and charge of a protein would affect its migration through a gel. In the early 1970s, biologists developed a method for the electrophoretic separation of protein molecules based solely on their molecular mass—SDS-PAGE (see Figure B8.1).

As we saw in Section 8.1, SDS-PAGE allows us to overcome the inherent difficulty of separating proteins based on mass alone. However, many of the bands resolved in SDS-PAGE represent multiple polypeptides that have similar molecular masses. Because of this issue, scientists desired improved methods of resolution, where it would be possible to more clearly separate the thousands of proteins present in the cell at a given time. Perhaps, some researchers reasoned, a second dimension of separation could be employed, in which the proteins would be separated based on another property. What other property could be used to separate proteins?

Remember that SDS-PAGE mitigates the effects of protein charge so that proteins can be separated based on mass alone. In the absence of SDS, though, each protein has a characteristic pI, or **isoelectric point**, that is a function of its amino acid composition. The isoelectric point is the pH at which the protein contains no charge. In a pH gradient to which an electrical current is applied, a protein will tend to migrate to the pH

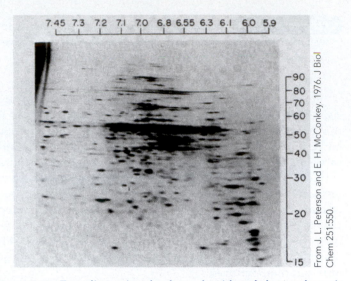

Figure. 10.11. Two-dimensional polyacrylamide gel electrophoresis
The protein 2D gel electrophoresis method separates polypeptides in two dimensions. The first dimension is based on isoelectric point (pI), while the second dimension is based on molecular mass. The patterns of spots represent the proteins present in the samples. By comparing the difference in patterns resulting from different strains or culture conditions, inferences can be made regarding the involvement of specific proteins in different processes.

From J. L. Peterson and E. H. McConkey. 1976. J Biol Chem 251:550.

equivalent of its pI. Using this property of proteins, researchers developed **2D-PAGE**, a two-dimensional separation scheme incorporating separation based on both isoelectric point and mass. By separating proteins in this fashion, proteins with similar molecular weights can be differentiated.

In 2D-PAGE, proteins are first subjected to isoelectric focusing. The sample is applied to a pH gradient on a polyacrylamide strip. An electric current is then applied to the strip. The proteins migrate along the gradient to the point on the strip where there is no charge on the protein, in other words, to where the pH of the strip equals the pI of the protein. The strip is then placed along the top of an SDS polyacrylamide gel, and the proteins are electrophoresed. This second dimension separates the proteins based on their molecular mass. Following these sequential separations, each protein exists as a spot within the two-dimensional matrix (**Figure 10.11**). Comparing the patterns of spots from cells grown under different conditions, or with different genotypes, provides a measure of the differences in gene expression.

Initially, researchers visually identified pattern differences. Today, computer programs can compare gel scans, match spots, and indicate pattern differences. The identities of individual spots can be determined using mass spectrometry methods coupled to knowledge of the genome sequence (**Figure 10.12**). With mass spectrometry, one can even determine the amino acid sequences of portions of the polypeptides. This method can be a very powerful approach to the identification of genes involved in particular processes, since the products of these genes are often only produced, or are produced at higher levels, under the conditions in which they function or when they are needed by the cell.

Let's now examine how 2D-PAGE can be used to address a biological question. Perhaps we are interested in the response

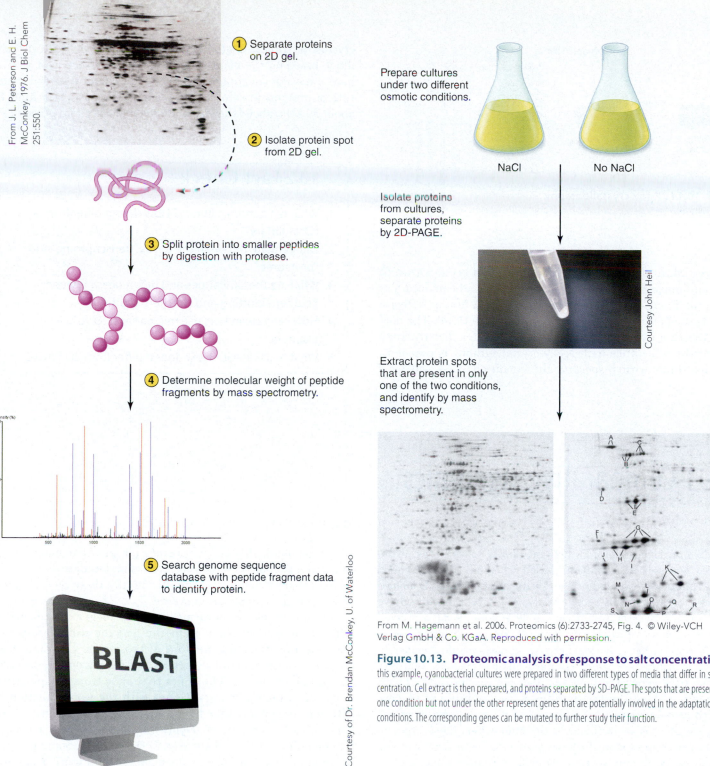

① Separate proteins on 2D gel.

From J. L. Peterson and E. H. McConkey. 1976. J Biol Chem 251:550.

② Isolate protein spot from 2D gel.

③ Split protein into smaller peptides by digestion with protease.

④ Determine molecular weight of peptide fragments by mass spectrometry.

Intensity (%)

⑤ Search genome sequence database with peptide fragment data to identify protein.

BLAST

Courtesy of Dr. Brendan McConkey, U. of Waterloo

Figure 10.12. Process Diagram: Mass spectrometry to identify proteins The identity of individual spots in a 2-D protein gel can be determined using mass spectrometry. This is made possible for organisms with sequenced genomes by comparing the amino acid sequence with the genome sequence database. In this method, spots are first extracted from the gel and then digested into fragments using proteases. The fragments are analyzed by mass spectrometry to determine the amino acid sequence, which can then be compared with the sequence database to identify the corresponding protein.

Prepare cultures under two different osmotic conditions.

NaCl   No NaCl

Isolate proteins from cultures, separate proteins by 2D-PAGE.

Courtesy John Heil

Extract protein spots that are present in only one of the two conditions, and identify by mass spectrometry.

From M. Hagemann et al. 2006. Proteomics (6):2733-2745, Fig. 4. © Wiley-VCH Verlag GmbH & Co. KGaA. Reproduced with permission.

Figure 10.13. Proteomic analysis of response to salt concentration In this example, cyanobacterial cultures were prepared in two different types of media that differ in salt concentration. Cell extract is then prepared, and proteins separated by SD-PAGE. The spots that are present under one condition but not under the other represent genes that are potentially involved in the adaptation to the conditions. The corresponding genes can be mutated to further study their function.

of an organism to changes in salt concentrations. To address this question, we could compare the protein patterns from cells growing under different osmotic conditions **(Figure 10.13)**. This type of analysis often leads to genetic investigations in which the corresponding genes can be mutated, and the phenotypes of the resulting mutants can be investigated to see if their function is actually relevant to the conditions being studied.

While 2D-PAGE allows researchers to identify differences in protein production, other powerful tools allow researchers to determine protein structures. Because the three-dimensional structure of a protein ultimately determines its function, knowing the structure of a protein can be very valuable in the functional analysis of that protein. **X-ray crystallography** and **nuclear magnetic resonance (NMR)** spectroscopy represent the two main methods for determining protein structure at the level of individual atoms. Each of these techniques has advantages and disadvantages. X-ray crystallography involves

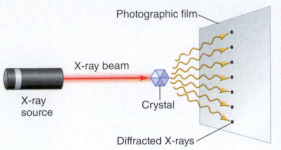

Karp: Cell and Molecular Biology: Concepts and Experiments, copyright 2010, John Wiley & Sons, Inc. This material is reproduced with permission of John Wiley & Sons, Inc.

Figure 10.14. X-ray crystallography This diagram shows how the diffraction of an X-ray beam by a crystal can be detected. The pattern is used to infer the molecular structure of the molecule.

the crystallization of proteins (the growing of crystals), which are then subjected to X-ray diffraction. An X-ray beam interacts with protein crystal, resulting in scattering. A film or detector records the pattern of diffracted X-rays **(Figure 10.14)**. The data from the resulting patterns can be used to infer the structure. One of the difficulties with X-ray crystallography is that most proteins differ with respect to the conditions that will yield

crystals of sufficient quality for X-ray analysis. NMR, a technique based on the measurement of distances between atomic nuclei, has the advantage of being able to determine the structure of proteins in solution. NMR, however, has a size limit of about 30 kDa, about the size of an average bacterial protein.

10.2 Fact Check

1. What is a genomic library? How does it differ from a cDNA library?
2. Differentiate between the terms "transcriptome" and "proteome."
3. What are the similarities and differences between Southern blotting and Northern blotting?
4. How can researchers use microarrays? Give an example.
5. What is SDS-PAGE? How does it differ from 2D-PAGE?

10.3 Comparative genomics

What information can be obtained from comparing genomes?

We saw in previous chapters how the ability to carry out phylogenetic analyses of organisms based on small subunit rRNA gene sequencing revealed previously unrecognized relationships between organisms. In recent years, these relationships have been further investigated at the whole genome level, a process referred to as **comparative genomics**. By comparing genomes of different species, it has become apparent that evolutionary relationships are more complex than previously realized. Comparative genomics, however, does not just provide us with information about the relationships between different species. We can also compare the sequences of different strains of the same species. As we will see in the **Mini-Paper**, these types of studies can help us identify genes associated with virulence and pathogenicity. In this section, we will introduce some aspects of this important area of study. We will briefly examine some means by which genetic differences arise. We will also discuss some mechanisms by which bacteria can share genetic material. In **Perspective 10.2**, we will investigate an interesting theoretical question: What is the minimal genome necessary for life? Finally, in the Mini-Paper, we will see how comparative genomics can provide meaningful insight into microbial pathogenesis.

Genetic variability

Let's consider how new genetic capabilities arise in a genome. As we noted earlier, genetic variability ultimately arises from mutations—changes in the DNA sequence. When we examine

whole genomes, though, it becomes obvious that many genes in a given genome belong to related gene families that we can assume share a common origin. Potentially, these related genes arose from gene duplication events.

A number of mechanisms can lead to gene duplication. After such a duplication event the genome will contain two copies of the affected gene. One of those copies becomes free to evolve novel functions, leaving the other copy to perform the original function. The genes that arise from a duplication event are called **paralogs**. Genes with the same function in different organisms that have evolved from the same ancestor are called **orthologs (Figure 10.15)**. Families of paralogs can be quite large, as illustrated in **Figure 10.16**. Some paralog families, such as the ABC transporters (see Figure 2.12) and short-chain dehydrogenase families, can have dozens of members in a single genome. While the gene products of individual family members probably will carry out similar functions, the specific substrates of the catalyzed biochemical reactions usually will differ.

The evolutionary relationships of orthologs mirror the evolutionary history of their respective genomes. We may assume that genes in two different genomes that encode proteins with high levels of sequence identity are orthologs, but the true test of orthologs is the demonstration that they carry out the same functions. For example, the enzymes malate dehydrogenase and lactate dehydrogenase are members of a large family of related enzymes that carry out NAD- or NADP-dependent reactions.

A fundamental question in biology concerns the core requirements for life. How many genes does a cell require to survive and replicate? In other words, what is the minimal genome for a cell? We can address this question with two different and complementary approaches—the bioinformatics approach and the experimental approach. In the bioinformatics approach, using comparative genomics, orthologs shared in each genome can be considered to be candidates essential genes. Because they exist in all cells, they may be required for cell viability. Expected in this class of genes would be genes encoding proteins necessary for processes such as transcription, translation, replication, membrane transport, and energy generation.

The first comparative genomics study involved a comparison of the genomes of the first two available bacterial genomes, *Haemophilus influenzae* and *Mycoplasma genitalium* (see Microbes in Focus 10.1). This study revealed 256 shared orthologs between these two genomes. In a later study, however, researchers identified only 63 shared orthologs common to 100 different genomes. Most of these genes encoded proteins involved in transcription and translation. Investigators concluded that comparative genomics probably could not provide a reliable estimate of the minimal genome after all. Most likely, several genes that carry out the same essential functions exist in different organisms. These genes, though, are not sufficiently similar in sequence to be recognized as orthologs.

To further identify essential genes, researchers have utilized various experimental methods. Gene inactivation experiments invariably have revealed a much larger number of genes that can be considered essential. Most likely, each type of organism will have a unique minimal gene set, related to its evolutionary history and environmental niche. In several ongoing genome reduction projects, genetic methods are being used to remove segments of the genome sequentially, with the goal of eventually obtaining the smallest genome that can support the life of a free-living cell. Another strategy to define the minimal genome involves using synthesized DNA. We will learn more about this interesting endeavor in Chapter 12.

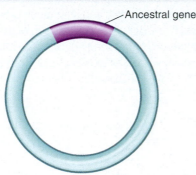

Ancestral gene

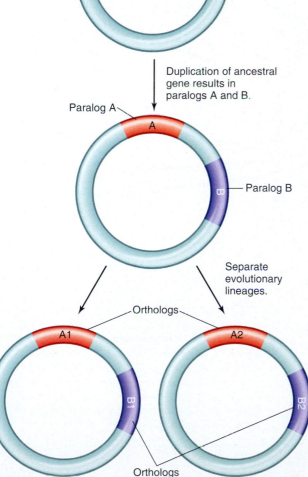

Duplication of ancestral gene results in paralogs A and B.

Paralog A

A

Paralog B

B

Separate evolutionary lineages.

Orthologs

A1

A2

B1

B2

Orthologs

Figure 10.15. Paralogs and orthologs The duplication of genes is important in the evolution of genomes. Two genes that arise via duplication within the same genome, such as A and B in this example (e.g., malate dehydrogenase and lactate dehydrogenase) are called "paralogs," while genes that evolve in separate lineages, such as A1 and A2 in this example (e.g., malate dehydrogenase in two different genomes) are called "orthologs."

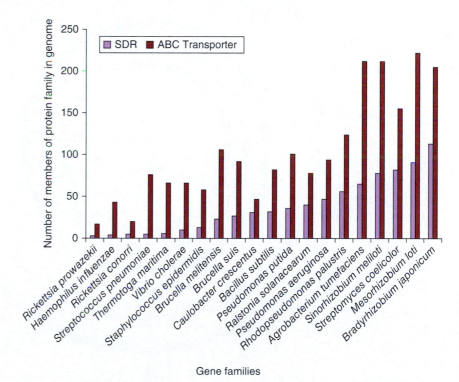

Gene families

Figure 10.16. Gene families Genome sequence analysis results in the identification of paralogous gene families. This figure compares the numbers of members of two specific paralogous gene families, ABC (ATP-binding cassette) transporters and short-chain dehydrogenases (SDR) in a selection of bacterial genomes. The determination of the functions of individual paralogs is an important goal of functional genomics.

N. T. Perna et al. 2001. Genome sequence of enterohaemorrhagic *Escherichia coli* O157:H7. Nature 409:529–533.

Context

We all have an intimate association with the common intestinal bacterium *Escherichia coli*. It colonizes the human gut shortly after birth. Most *E. coli* strains cause no harm but exist as constituents of the gut microbial community. Several strains of *E. coli*, however, do cause diseases in humans. Pathogenic strains of *E. coli* cause most urinary tract infections. Even more serious intestinal diseases result from infections with other pathogenic strains. For example, the strain O157:H7 can cause a potentially lethal bloody diarrhea in humans. This strain commonly is found in the GI tract of animals such as cattle, although it does not cause disease symptoms in these animals. Not surprisingly, it is also commonly found in manure from these animals. There have

been well-publicized outbreaks related to undercooked ground beef, where the intestinal microbes find their way into the meat during processing. Outbreaks also have been linked to vegetables that become contaminated through the application of manure in the fields, and improperly maintained drinking water systems contaminated by farm runoff. Surprisingly, this disease was not widely recognized until the early 1980s. In fact, the first association with human disease was an outbreak in 1982 involving contaminated hamburgers.

Following ingestion by a human, the bacterial cells bind to intestinal epithelial cells, mediated by the bacterial protein intimin. Here, the bacteria produce Shiga toxins, toxins also produced by related bacteria that cause dysentery. These toxins cause serious damage to the lining of the intestine, resulting in severe abdominal pain and bloody diarrhea. Kidney failure can develop.

What causes *E. coli* O157:H7 to be so virulent, while other strains of *E. coli* exhibit limited or no pathogenicity? Several genetic factors probably contribute to the virulence and pathogenicity of O157:H7. Few virulence factors other than the Shiga toxin and the intimin protein, though, have been identified. With the aim of gaining a more thorough understanding of how *E. coli* O157:H7 causes disease, Fred Blattner and colleagues at the University of Wisconsin used a comparative genomics approach. They sequenced the O157:H7 genome and compared it to the sequence of the non-pathogenic *E. coli* strain K12. A comparison of these genomes, they reasoned, could help them identify other genes associated with *E. coli* O157:H7 pathogenesis.

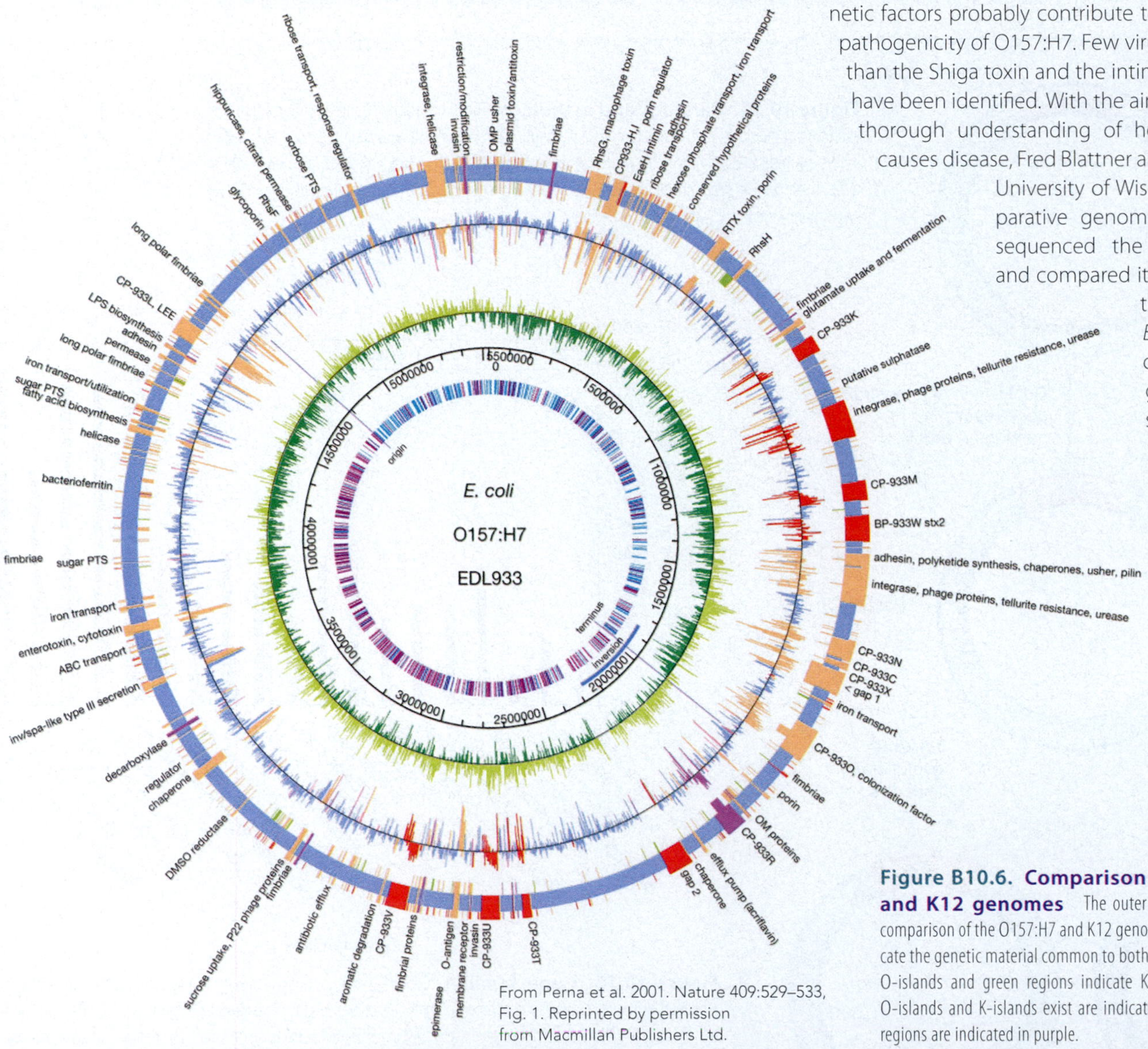

Figure B10.6. Comparison of *E. coli* O157:H7 and K12 genomes The outer ring of this image shows a comparison of the O157:H7 and K12 genomes. Light blue regions indicate the genetic material common to both strains. Red regions indicate O-islands and green regions indicate K-islands. Areas where both O-islands and K-islands exist are indicated in tan and hypervariable regions are indicated in purple.

From Perna et al. 2001. Nature 409:529–533, Fig. 1. Reprinted by permission from Macmillan Publishers Ltd.

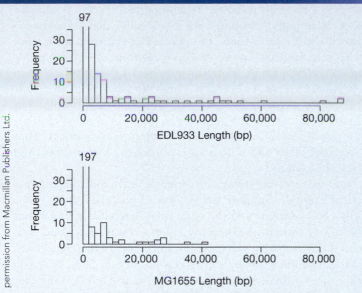

From Perna et al. 2001. Nature 409:529–533, Fig. 3. Reprinted by permission from Macmillan Publishers Ltd.

Figure B10.7. Size distribution of O- and K-islands These histograms show the frequency of islands based on size in O157:H7 (labeled EDL933) and K12 (labeled MG1655). More medium- and long-length islands exist in O157:H7.

Experiment

An *E. coli* isolate from ground beef associated with the 1982 outbreak was the chosen strain for the genome sequence study. This strain contains two plasmids, and the sequence of one of these plasmids had been determined previously (Burland et al. 1998. Nucl Acids Res 26:4196–4204). Blattner and colleagues determined the sequence of the chromosome by whole genome shotgun sequencing. By comparing this sequence to the non-pathogenic K12 sequence, the researchers quickly recognized that the O157:H7 genome was about 15 percent larger than the K12 genome.

Perhaps more importantly, they determined that large segments of the genome existed in one strain, but not the other. The two strains, which are thought to have diverged from a last common ancestor 4.5 million years ago, share a set of genes, or common backbone, consisting of about 4.1 Mb of DNA. These common genes have DNA sequence similarity of at least 98 percent, and 25 percent of the gene pairs encode identical amino acid sequences. The authors called the unique segments found only in O157:H7 "O-islands," and called the segments found in K12 "K-islands." They identified 177 O-islands, consisting of 1.34 Mb of DNA, and 234 K-islands, consisting of 0.53 Mb of DNA **(Figure B10.6)**.

The O- and K-islands vary considerably in size. While some islands are quite short, others are over 80 kb in length. Moreover, the size distribution of these islands differs between the two strains. While most of the islands are short, more medium or long islands exist in the pathogenic strain **(Figure B10.7)**.

The researchers hypothesized that genes associated with pathogenicity would reside within the O-islands and, as a result, analyzed these genes. Not surprisingly, many of the predicted proteins encoded by O-island genes exhibit similarity to known virulence-associated proteins in other pathogenic *E. coli* strains. These genes include the gene for intimin. Other genes of interest appear to encode virulence factors

such as RTX toxin, urease enzyme, a Type III secretion system, fimbrial synthesis, iron uptake and utilization, and numerous toxins. Genes predicted to be involved in the catabolism of diverse substrates and genes that confer antibiotic resistance also exist in these O-islands. All of these genes, one could argue, might be important for the ability of the bacteria to survive and compete in soil and water habitats, and in association with animals.

Impact

Knowing the genome sequence of *E. coli* O157:H7 has provided many opportunities for research that will lead to greater understanding of how this organism causes disease. Of direct interest are the several genes encoding potential virulence factors that have been identified. To confirm that these genes are related to pathogenicity, it is essential that appropriate experiments be performed. These experiments would include the creation of mutations in each of these genes, followed by testing for effects on the ability to cause disease. Of course, these types of experiments would have to be conducted in suitable animal models that mimic the disease as it occurs in humans. The identification of these genetic factors will provide useful targets for the development of novel treatments for the disease.

Also very important will be the experimental identification of genes critical for O157:H7 cells to survive in soil and water environments, and in animal hosts. The knowledge of the functions of such genes will help to develop methods and strategies to control the organism, and possibly to develop vaccines to reduce its incidence in the GI tract of cattle. A thorough understanding of the physiology of the organism that is essential to such studies will be tremendously facilitated by the knowledge of the complete genome sequence.

This study shows how comparative genomics can be used in the prediction of genes that might be important for virulence. By comparing a pathogenic genome with a non-pathogenic genome, candidate virulence genes can be readily identified. It will be fascinating to see which of these genes are confirmed to be important for causing disease, how this knowledge helps in our understanding of the disease process, and ultimately how this information helps us control this disease.

● Questions for Discussion ·····················

1. This study established that while most of the genes were shared by both genomes, and had a very high level of similarity, certain genes were present in one strain and not in the other. These genes were found in clusters called "islands." Discuss some of the genetic mechanisms that might have led to the presence of islands. Where might these islands have originated?

2. Discuss the application of comparative genomics to identify genes that might be important for pathogenicity.

While lactate dehydrogenases catalyze the conversion of lactate to pyruvate, malate dehydrogenases catalyze the conversion of malate to oxaloacetate. Typically, each genome encodes several members of this family of enzymes that have resulted from gene duplication events of ancestral genes, followed by evolution of novel substrate specificities. The enzymes that have different substrate specificities and are encoded in the same genome are paralogs. So malate dehydrogenase (substrate malic acid) and lactate dehydrogenase (substrate lactic acid) encoded in the same genome would be paralogs, while malate dehydrogenase encoded in two separate genomes would be considered orthologs.

Horizontal gene transfer

One of the revelations of comparative genomics is the sharing of genetic material by microbes. By comparing the phylogenetic relationships of different genes in different organisms, we can surmise that a fair amount of horizontal gene transfer occurs among all forms of life. In thinking about how this transfer of genetic material might occur, we can revisit our discussions of gene transfer mechanisms such as conjugation, transformation, transduction, and transposition from Section 9.5. Because of these gene transfer events, genomes as they currently exist can be considered mosaics. Extant genomes have arisen from evolutionary changes and horizontal gene transfers.

CONNECTIONS Most microbes exist in complex communities, as we will discuss in **Chapter 15**. Within these communities, ample opportunities for horizontal gene transfer between species exist. In **Section 24.3**, we'll learn that horizontal gene transfer is a major mechanism for transfer of antibiotic resistance genes from one microbial cell to another.

For decades, microbiologists have known that bacteria can exchange genetic material via horizontal gene transfer. Until the genomics era, however, many scientists downplayed its significance as an evolutionary mechanism in nature. It is easy to understand this skepticism. Genes from different microbe types often have distinct promoters and regulatory features, making it seem doubtful that transferred genes would function very well, particularly if they moved into a distantly related organism. Even if the transfer of DNA occurred, and the DNA segment became fixed in the recipient's genome, such events would have little practical consequence if the recipient could not express the new trait. Acquisition of new DNA might even be deleterious to the organism. The incoming DNA segment could disrupt existing genes or encode a gene that interferes with cellular metabolism. We now know, though, that horizontal gene transfer *does* occur in nature, it *does* have practical consequences, and it *does* play a significant role in microbial evolution.

Several types of evidence demonstrate the process of horizontal gene transfer. Plasmids that can transfer DNA from one cell to another represent the most obvious vehicle for horizontal gene transfer, but plasmids do not represent the whole story. Genome sequencing has revealed that the genomes of every type of microbe contain foreign genes. What is the basis for this conclusion? One indication that a gene or genomic region may have been transferred via horizontal gene transfer is if the base-pair composition of the gene or region differs significantly from the rest of the chromosome. Genomes vary in the frequency of G:C base pairs, compared with A:T pairs. Researchers express this value as **% G+C**, the percent of the total base pairs in the genome that are G:C. The *Escherichia coli* genome, for example, is 50% G+C, while the genome of the soil bacterium *Streptomyces coelicolor* is 72% G+C, and the genome of the yeast *Saccharomyces cerevisiae* is 38% G+C. G+C content usually does not vary much across the genome, so genes or regions exhibiting a significant difference in G+C content could indicate that the variant DNA had a distinct evolutionary history from the rest of the genome.

To illustrate this point more clearly, let's consider a hypothetical example. Suppose the genome of bacterial strain A has a 50% G+C content, and the genome of bacterial strain B has a 70% G+C content. Sequencing studies reveal that gene X in strain B has a sequence that is virtually identical to gene X from strain A, including having a G+C content of only 50%. Based on G+C content, then, gene X in strain B more closely resembles the genome of strain A. To explain this observation, we can hypothesize that strain B acquired gene X from strain A. More sophisticated methods for detecting horizontal gene transfer based on nucleotide composition have been developed (e.g., patterns of nucleotide pairs, and codon usage patterns), but the basic idea of comparing DNA sequence characteristics remains the same.

By comparing the genome sequences of related microbes, including gene order and location within the genome, it often becomes apparent that large segments of DNA exist in one genome but not in another, otherwise closely related genome (Mini-Paper). This observation implies that the introduction or removal of stretches of DNA, presumably by the standard gene transfer mechanisms of transformation, conjugation, and transduction, with the assistance of transposable elements, has occurred. Introduced DNA segments greater than about 10 kb up to about 200 kb are called **genomic islands (Figure 10.17)**, and often are associated with tRNA genes, transposable elements, plasmids, or bacteriophages.

CONNECTIONS Genomic islands frequently contain genes related to pathogenesis, in which case investigators refer to them as pathogenicity islands (PAIs). In **Section 21.3**, we will discuss several examples of PAIs found in human bacterial pathogens.

10.3 Fact Check

1. What are paralogs and orthologs?
2. What is % G+C?
3. How can % G+C be used to analyze the possibility of horizontal gene transfer events?

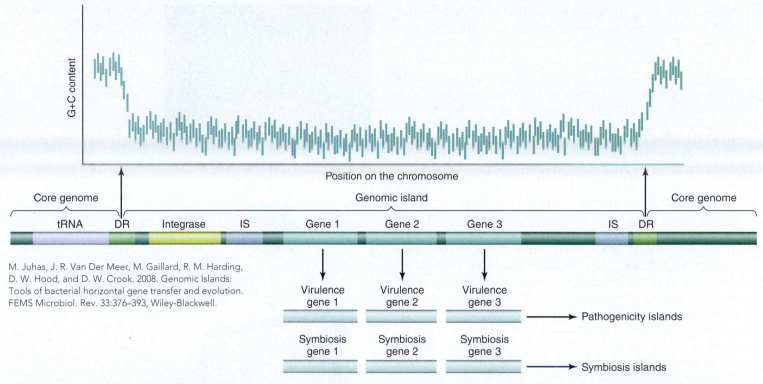

Figure 10.17. Horizontal gene transfer Genomic islands are detected by comparison of genome sequences and analysis of nucleotide composition. They often contain G+C content that is different from the majority of the genome. They also often integrate at tRNA genes, and contain plasmid or bacteriophage DNA, as well as insertion sequences (IS) and direct repeat sequences (DR). The pathogenicity islands contain virulence genes, while symbiosis islands contain genes for symbiosis.

M. Juhas, J. R. Van Der Meer, M. Gaillard, R. M. Harding, D. W. Hood, and D. W. Crook. 2008. Genomic Islands: Tools of bacterial horizontal gene transfer and evolution. FEMS Microbiol. Rev. 33:376–393, Wiley-Blackwell.

10.4 Metagenomics

Can we apply genomics to the study of uncultivated microbes?

As we have seen in Chapter 9, several approaches to analyzing and modifying the DNA of bacteria have been developed. Before such genetic analysis can be conducted, however, the organism, one might think, must be cultivated. Only a small fraction of the microorganisms within a given microbial community, though, can be cultured on microbiological growth media (see Section 6.4). The rest of the organisms cannot be cultivated. Presumably, these uncultivated organisms would be interesting to study, but how can we approach the genetic analysis of these microbes if they cannot be grown in the laboratory?

In almost every introductory microbiology course, instructors stress the importance of working with pure cultures. If cultures become contaminated, then valid conclusions cannot be drawn from any experiments, no matter how good the experimental technique. So, how can we experimentally examine a collection of microbes that cannot be cultivated in a pure form? Fortunately, genomics technology has provided a solution—**metagenomics**—a process in which DNA is extracted directly from microbial communities and sequenced or cloned into vectors to make libraries. Metagenomics is quickly affecting all areas of microbiology research.

CONNECTIONS Consider the juxtaposition of pure culture microbiology, as discussed in **Chapter 6**, and the natural existence of microbes in complex ecosystems, which we will cover in **Chapter 15**. Advances in sequencing technology enable DNA from uncultivated organisms to be examined in detail, thereby providing the means to bring these two worlds together.

In the 1990s, Ed DeLong and coworkers at Monterey Bay Aquarium Research Institute began an investigation that would have far-reaching impact. Studying marine microbial communities, they isolated genomic DNA directly from marine samples, inserting large fragments of this DNA into vectors, and then determined the sequence of these genome fragments. Analysis of these sequences resulted in the discovery of a number of novel genes, including genes encoding a novel light-harvesting protein, proteorhodopsin. In some cases, the phylogenetic position of the original organism that donated the cloned genome fragment could be identified because the fragment contained a small subunit, or SSU, rRNA gene. We have seen in Chapter 1 that the widely used tree of life is based on comparison of these SSU rRNA gene sequences.

While DeLong's group developed culture-independent methods for analyzing marine microbial communities, other scientists studied microbial communities using different molecular techniques. As we are well aware, microbial communities on Earth occupy an almost limitless diversity of habitats. In many cases, researchers wanted to survey various communities and catalog the diversity of microbes present. The most common approach was to extract genomic DNA directly from samples taken from these habitats, use PCR to amplify the SSU rRNA genes, clone the amplified genes, and determine the DNA sequences. The DNA sequences were then compared to DNA sequence databases. In many cases, the sequenced SSU rRNA gene matched an existing sequence in the database.

Occasionally, though, the newly sequenced SSU rRNA gene did not closely match any known SSU gene. In some cases, the newly sequenced SSU rRNA genes differed significantly from any known SSU rRNA genes. In these instances, researchers hypothesized that the SSU rRNA genes had been amplified from organisms representing branches of the tree of life that had never before been described. Indeed, such cloned gene sequences represent the only information we have about several branches in the tree of life.

Using these molecular data, the tree of life continues to develop, incorporating information about uncultivated microbes. However, even as this type of research helps us to better understand the diversity of microbes, it has its limitations. Often, researchers cannot make firm predictions about the characteristics of an organism based solely on the available sequence information. Also, the sequence data alone may not provide adequate information about how organisms interact with each other and the environment.

To more fully understand the physiology of diverse microbial communities, a new approach has arisen. This approach involves constructing clone libraries from community DNA, and then analyzing that DNA. In some cases, the clones are sequenced directly to determine the types of genes present in the genomes within the community. With the huge diversity of organisms represented in many microbial communities, you can imagine what a Herculean task it would be to determine the complete DNA sequence of these libraries. In many cases, clones are isolated based on properties of interest, and then those clones are characterized and sequenced. This approach often involves screening the clones for particular enzyme activities followed by characterization of the clones. In this way, many new genes have been isolated and characterized from organisms that have never been cultured. This strategy, pioneered by Jo Handelsman and Robert Goodman at the University of Wisconsin, has transformed areas of microbiology. Gene libraries constructed from uncultivated microbial communities found in soil, freshwater and marine environments, wastewater treatment systems, acid mine drainage, and mammalian intestines have yielded new genes encoding novel enzymes and antibiotics, many of which have applications in biotechnology.

As the cost of DNA sequencing continues to drop, and the capacity rises, it has become ever more popular to skip the screening step, and just to sequence randomly chosen clones of the metagenomic library (**Figure 10.18**). This approach has even resulted in the determination of genome sequences of the most

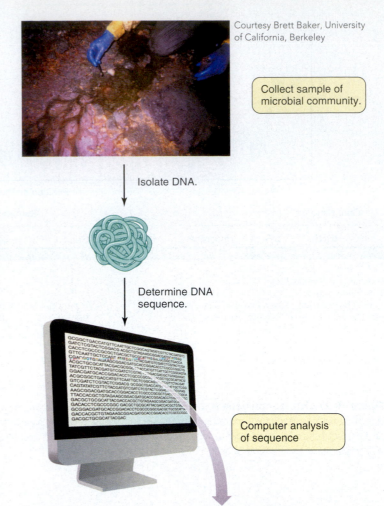

Courtesy Brett Baker, University of California, Berkeley

Collect sample of microbial community.

Isolate DNA.

Determine DNA sequence.

Computer analysis of sequence

GCGGCTGACCATGTTCAATTGCTCGGCAGTATATCGTTCTACGATGTCGAT CTCGTACTCGGACGACGCTGTAGAAGCGGACGATGCACCGGACACCTCG CCCGCGCTGACGCTGCGCATTACGAC

Figure 10.18. Metagenomic analysis Analysis of the genetics of complex microbial communities has been made possible by the development of metagenomic techniques. In sequence-based metagenomics, DNA that has been extracted directly from the community is sequenced directly and that sequence is then analyzed.

abundant organisms in the sample, without ever cultivating them. The large amounts of data generated by this type of work can often be overwhelming. For example, a series of publications of metagenomic sequence data from marine environments has provided sequence data equivalent to several times the size of the human genome. This data set suggests the existence of thousands of entirely new protein families, as well as novel branches of the tree of life. Most likely, we have only scratched the surface of the knowledge we can gain from metagenomic analysis. We will take a more in-depth look at the use of metagenomics methods in Chapter 15.

10.4 Fact Check

1. What is meant by the term "metagenomics"?
2. How can DNA isolated from a community be analyzed?
3. Describe how a metagenomics approach might affect biotechnology.

Image in Action

WileyPLUS The laboratory bench of a researcher involved in sequencing the *Haemophilus influenzae* genome is shown. Using the LB agar plates on the left, the investigator can identify bacterial colonies that have been successfully transformed with plasmids containing fragments of genomic DNA to be sequenced. The printout in the upper left illustrates the results of an automated sequencing

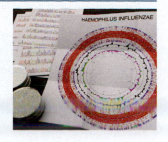

reaction. The color of each peak indicates the identity of the base in the sequence. The resulting genome map on the right displays open reading frames (red lines) and regions predicted to encode proteins (blue lines), as well as indicating whether the region is GC rich or not (purple/green).

1. Identify and describe the reagents that the researcher utilized to generate the sequencing results shown in the upper left. How was this process conducted?

2. Now that this stage of the researcher's work has been completed, what would be the next likely step for utilizing these results?

The Rest of the Story

The successful determination of the genome sequence of *Haemophilus influenzae* was an impressive feat and a springboard into the world of genomics research. Looking back on this event, it is clear that it was a landmark on the way to transitioning microbiology into a digital science. Dr. Craig Venter and his research teams have continued to be pioneers in the digital science of genomics research. The J. Craig Venter Institute (JCVI) was formed in October 2006 as a large multidisciplinary genomics-focused organization. JCVI research groups use their genomics tools to explore human genomics medicine, infectious disease, plant, microbial, and environmental genomics, and synthetic biology.

The Institute's Global Ocean Sampling Expedition project has focused on sampling, sequencing, and analyzing the DNA of the microorganisms (many previously unknown) living in the world's oceans. Their work on this project has paved the way for similar analyses of microbes that inhabit the human body, the human microbiome. JCVI has played a leading role in the Human Microbiome Project. In May 2010, the JCVI reported on their successful construction of the first self-replicating, synthetic bacterial cell (see Mini-Paper in Chapter 12). This synthetic bacterial cell is called *Mycoplasma mycoides* JCVI-syn1.0, and it began as a digitized genome. Indeed, sequencing technology and genomics have come a long way, as this creation has proven that genomes can be designed using computers, chemically made in the laboratory, and transplanted into a recipient cell that then is controlled by a synthetic genome!

Summary

Section 10.1: How are genome sequences determined?

Innovations in DNA-sequencing techniques and information technology have resulted in a rapid increase in DNA sequence data.

- The field of **genomics** has been spurred by the development of recombinant DNA protocols, improved DNA-sequencing techniques, and the creation of tools to analyze large data sets.
- **Sanger**, or **dideoxy**, **sequencing**, and **primer walking** allow researchers to sequence large pieces of DNA.
- **Shotgun sequencing** allows researchers to sequence complete genomes more quickly.
- With **automated-** or **cycle-sequencing methods**, a large amount of sequence data can be generated.
- **Next generation sequencing methods**, especially **pyrosequencing** and the related **454 DNA sequencing**, have further increased the speed with which DNA can be sequenced.
- **Bioinformatics** greatly facilitates the analysis of genome sequences and the prediction of gene function.

- The **annotation** of genome sequences can help researchers identify **open reading frames (ORFs)**.
- **Functional genomics** studies aim to determine the biological functions of unknown genes.

Section 10.2: How is gene expression measured using genomics tools?

Various techniques allow researchers to analyze the transcripts and proteins within cells.

- The generation of **genomic libraries** and **complementary DNA libraries**, or **cDNA libraries**, facilitates whole genome studies.
- The **transcriptome** of a cell can be analyzed by means of the **Northern blot**, a technique similar to the **Southern blot**.
- **Microarrays** can be used to analyze the levels of expression of each gene in the genome at a given time.
- **Proteome** study involves the use of **2D-PAGE**, which separates polypeptides based on both **isoelectric point** and size.

- The individual polypeptides can be identified by mass spectrometry, and this information can be used to determine differences in gene expression patterns.
- **X-ray crystallography** and **nuclear magnetic resonance (NMR)** can be used to determine the three-dimensional structure of proteins.

Section 10.3: What information can be obtained from comparing genomes?

Genome comparisons can reveal much about the evolutionary history of organisms.

- **Comparative genomics** yields new insight into evolutionary processes.
- **Paralogs** arise from gene duplication in a single genome, while **orthologs** encode the same functions but in different genomes.
- Horizontal gene transfer can be detected by comparison of sequence features such as **% G+C**.

- One intriguing outcome of comparative genomics has been the identification of **genomic islands**, stretches of DNA transferred from one species to another.

Section 10.4: Can we apply genomics to the study of uncultivated microbes?

Advancements in DNA sequencing allow us to analyze microbes that can't be cultivated.

- **Metagenomics** involves the construction of gene libraries from DNA extracted directly from the complex microbial communities.
- Metagenomic libraries can be analyzed by screening for specific gene function and by large-scale DNA sequencing.
- Large-scale metagenomic-sequencing projects have resulted in the detection of novel protein families and microbial types.

● Application Questions

1. Explain why automated sequencing results in longer reads than non-automated dideoxy-sequencing methods.
2. In some ways, the dideoxynucleotides used in Sanger sequencing are structurally similar to nucleoside analogs used as antiviral treatments. Explain why.
3. Explain how you could use microarrays to study a strain of pathogenic bacteria before and after it enters the host's lungs.
4. A laboratory technician sets up a sequencing reaction, using the enzymatic method (Sanger method). You see her adding several reagents to her reaction tubes. What is she adding, and what will she do next after adding all the necessary reagents?
5. Many complete genomes have been sequenced using a shotgun-sequencing approach. Outline how a researcher can use this approach to sequence a newly discovered soil bacterium.
6. You are asked to determine whether a newly discovered strain of a bacterial pathogen possesses a gene common to many pathogens. Your boss tells you to use the Southern blot to make your determination. Explain what you will need to do and how this process will work.
7. Describe how you could use 2D-PAGE to analyze expression of genes induced at low pH.
8. Eating a hamburger that is cooked rare could be deadly, while eating a steak that is cooked rare is perfectly safe. Discuss what developments would be necessary to make the consumption of a rare hamburger a safe experience.
9. Two bacterial strains are being studied to determine if horizontal gene transfer of an antibiotic resistance gene (*bla* gene) has occurred. What would you conclude based on the results below? Explain your answer.

% G+C Analysis of strain X and strain Y

DNA sampled	% G+C
Strain X	57%
Strain X *bla* gene	48%
Strain Y chromosome	48%
Strain Y *bla* gene	48%

10. A microbial geneticist isolates a new mutant of *E. coli* and wishes to map and sequence the mutation accurately.
    a. What genomics techniques could she use?
    b. Using what you have learned in Chapters 9 and 10, compare the approach that she might have taken in 1990 with the approach that she could take in 2012. Explain your answer.
    c. What are the pros and cons of the approaches described above?

Suggested Reading

Carbone, A. 2006. Computational prediction of genomic functional cores specific to different microbes. J Mol Evol 63:733–746.

Ekins, R., and F. W. Chu. 1999. Microarrays, their origins and applications. Trends Biotech 17:217–218.

Fleischmann, R. D., M. D. Adams, O. White, R. A. Clayton, E. F. Kirkness, A. R. Kerlavage, C. J. Built, J. F. Tomb, B. A. Dougherty, J. M. Merrick, K. McKenney, G. Sutton, W. FitzHugh, C. Fields, J. D. Gocayne, J. Scott, R. Shirley, L. Liu, A. Glodeck, J. M. Kelley, J. F. Weidman, C. A. Phillips, T. Spriggs, E. Hedblom, M. D. Cotton, T. R. Utterback, M. C. Hanna, D. T. Nguyen, D. M. Saudek, R. C. Brandon, L. D. Fine, J. L. Fritchman, J. L. Fuhrmann, N. S. M. Geoghagen, C. L. Gnehm, L. A. McDonald, K. V. Small, C. M. Fraser, H. O. Smith, and J. C. Venter. 1995. Whole genome random sequencing and assembly of *Haemophilus influenzae* RD. Science 269:496–512.

Galperin, M. Y., and G. R. Cochrane. 2009. Nucleic Acids Research annual Database Issue and the NAR online Molecular Biology Database Collection in 2009. Nucl Acids Res 37:D1–D4.

Koonin, E. V. 2003. Comparative genomics, minimal gene-sets and the last universal common ancestor. Nat Rev Microbiol 1:127–136.

Koonin, E. V., and Y. I. Wolf. 2008. Genomics of bacteria and archaea: The emerging dynamic view of the prokaryotic world. Nucl Acids Res 36:6688–6719.

MacLean, D., J. D. G. Jones, and D. J. Studholme. 2009. Application of "next-generation" sequencing technologies to microbial genetics. Nat Rev Microbiol 7:287–296.

Ochman, H., and L. M. Davalos. 2006. The nature and dynamics of bacterial genomes. Science 311:1730–1733.

Perna, N. T., G. Plunkett III, V. Burland, B. Mau, J. D. Glasner, D. J. Rose, G. F. Mayhew, P. S. Evans, J. Gregor, H. A. Kirkpatrick, G. Pósfai, J. Hackett, S. Klink, A. Boutin, Y. Shao, L. Miller, E. J. Grotbeck, N. W. Davis, A. Lim, E. T. Dimalanta, K. D. Potamousis, J. Apodaca, T. S. Anantharaman, J. Lin, G. Yen, D. C. Schwartz, R. A. Welch, and F. R. Blattner. 2001. Genome sequence of enterohaemorrhagic *Escherichia coli* O157:H7. Nature 409:529–533.

Ronaghi, M. 2001. Pyrosequencing sheds light on DNA sequencing. Genome Res 11:3–11.

Rondon, M. R., P. R. August, A. D. Bettermann, S. F. Brady, T. H. Grossman, M. R. Liles, K. A. Loiacono, B. A. Lynch, I. A. MacNeil, C. Minor, C. L. Tiong, M. Gilman, M. S. Osburne, J. Clardy, J. Handelsman, and R. M. Goodman. 2000. Cloning the soil metagenome: A strategy for accessing the genetic and functional diversity of uncultured microorganisms. Appl Environ Microbiol 66:2541–2547.

Schena, M., D. Shalon, R. W. Davis, and P. O. Brown. 1995. Quantitative monitoring of gene expression patterns with a complementary DNA microarray. Science 270:467.

Southern, E. M. 1975. Detection of specific sequences among DNA fragments separated by gel electrophoresis. J Mol Biol 98:503–517.

11 Regulation of Gene Expression

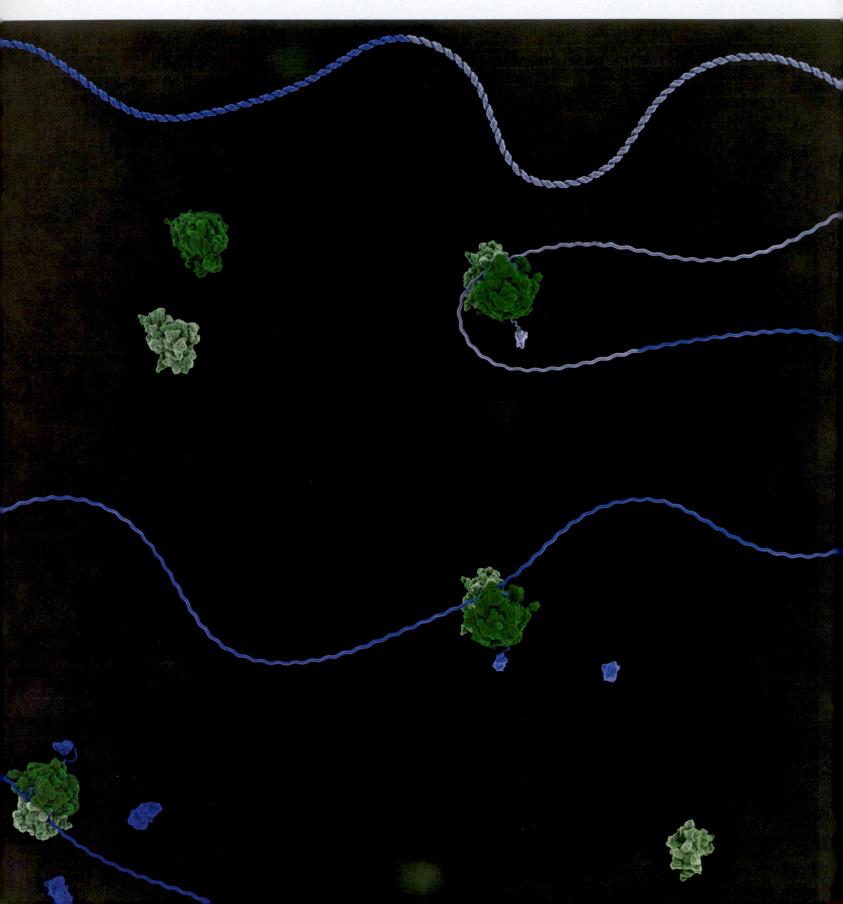

The second half of the twentieth century witnessed perhaps the greatest series of developments in the history of the biological sciences. In earlier chapters, we learned how researchers determined that DNA was the genetic material, discovered genetic mechanisms in bacteria, and determined the structure of DNA. They also formulated the idea of gene regulation using this knowledge. Indeed, the 1965 Nobel Prize for Physiology or Medicine was awarded to François Jacob, André Lwoff, and Jacques Monod in recognition of their discoveries about genetic regulation in bacteria and bacteriophages.

These French researchers, working at the Institut Pasteur in Paris, proposed the "operon concept," which suggests a mechanism by which regulatory genes can direct cell metabolism by altering rates of transcription of structural genes. The discovery of gene regulation in *Escherichia coli* had far-reaching impact on the understanding of the living world, as foreshadowed by the famous quote from Jacques Monod, "What is true for *E. coli* is true for the elephant."

Through a series of elegant genetic experiments, these scientists showed for the first time that genes can be turned on and off. By studying the control of metabolism of lactose, or milk sugar, and also the control of infection of *E. coli* by the lambda bacteriophage, they deduced the fundamental nature of gene regulation. Since these early discoveries, scientists have studied numerous gene regulatory systems. The diversity of these systems is remarkable, although probably not unexpected, given the immense variety of microbes and their habitats.

CHAPTER NAVIGATOR

As you study the key topics, make sure you review the following elements:

Enzymes can be regulated in several ways.
- Figure 11.1: Control of gene expression

Regulatory proteins control transcription.
- 3D Animation: Transcriptional regulation in the *lac* operon
- Figure 11.3: The *lac* operon
- Table 11.1: Key components of operons
- Figure 11.6: Repressible expression: The *trp* operon
- Figure 11.8: Control of *lac* operon expression
- Mini-Paper: Tuning promoters for use in synthetic biology

Multiple genes can be coordinately regulated.
- Perspective 11.1: The use of lactose analogs in gene expression studies
- Figure 11.15: The SOS response control and the LexA repressor

Messenger RNA levels can be regulated after transcription.
- 3D Animation: RNA interference and attenuation
- Toolbox 11.1: Using RNA molecules to decrease gene expression
- Figure 11.18: Riboswitches

Bacteria can "communicate" with their neighbors.
- 3D Animation: Quorum sensing
- Microbes in Focus 11.1: *Vibrio fischeri*
- Figure 11.21: Quorum sensing and luciferase expression

Two-component systems are used widely by bacteria.
- Table 11.4: Examples of two-component regulatory systems
- Figure 11.22: Process Diagram: Using an environmental signal to alter gene expression
- Figure 11.23: A two-component regulatory system and expression of virulence genes

By regulating gene expression, bacteria can regulate their behavior.
- Figure 11.26: Process Diagram: Chemotaxis

CONNECTIONS for this chapter:

Allosteric regulation of enzymes (Section 13.2)
Chaperonins in hyperthermophilic archaeons (Section 4.3)
Chloramphenicol mode of action (Section 24.2)
Genetic engineering of plants with *Agrobacterium* (Section 12.5)

Introduction

As we have seen in previous chapters, microbes live in diverse, ever-changing habitats. Survival depends upon their ability to utilize the available resources and to respond to changes in environmental conditions. Consider the waterborne human pathogen *Vibrio cholerae*. It can thrive in aquatic environments, but can also infect humans. Or consider the symbiotic nitrogen-fixing *Rhizobium* bacterium. It can live in the soil, but can also live symbiotically in compatible plants, where it fixes nitrogen. How do microbes respond to these dramatic environmental shifts?

Often, shifts in the activity or availability of enzymes coincide with these shifts in environmental conditions. Changes in enzymatic activity typically result from changes in the structures of the enzymes. While this type of regulation is quite effective and can occur quite quickly, it does have a significant downside. The cell has used energy and resources in the transcription and translation of genes coding for enzymes that may not be needed.

How could a cell avoid this apparent waste of energy and resources? A more energy efficient way to control metabolic pathways might be to regulate the transcription and/or translation of genes. In other words, the cell could regulate the amount of a protein present by controlling gene expression, rather than by regulating the activity of the protein. As we will see throughout this chapter, control of gene expression plays a central role in governing responses to environmental changes.

In this chapter, we will present a variety of well-researched examples of how microorganisms control gene expression. Some control mechanisms affect one gene at a time, while other control mechanisms affect groups of genes. At the end of the chapter, we will examine three relatively complex bacterial behaviors—quorum sensing, virulence, and chemotaxis—and uncover the roles of both gene expression and protein modification in these behaviors. As we progress through this chapter, we will consider the following questions:

How are gene expression and enzyme activity controlled? (11.1)
How do regulatory proteins control transcription? (11.2)
How does a single environmental stimulus control multiple operons? (11.3)
How can mRNA be controlled? (11.4)
How do bacteria communicate with their neighbors? (11.5)
How do environmental conditions affect gene expression? (11.6)
How can bacterial cells regulate their behavior? (11.7)

11.1 Differential gene expression

How are gene expression and enzyme activity controlled?

In Chapter 7, we learned how the processes of transcription and translation result in the production of gene products with specific functions. As we consider the activities of a cell, though, it should be obvious that not all gene products are required at all times. Cells may need some gene products, such as the enzymes involved in glycolysis, transcription, and translation, all the time. After all, these gene products are needed for metabolism and the production of proteins, essential processes of a cell. Therefore, the genes encoding these products are always expressed, a state referred to as **constitutive expression**. A cell may need other gene products only under certain conditions. For instance, a cell needs enzymes to metabolize a particular substrate only when that substrate is available. Similarly, it needs to produce an essential molecule only when sufficient amounts of that molecule are not present in the environment. Expression of genes encoding these products, then, may be turned on or off as needed, a state referred to as **inducible expression**.

How can genes be turned on or off? The answer is quite interesting, and quite complex **(Figure 11.1)**. The binding of RNA polymerase to the promoter can control the initiation of transcription. Enzymatic degradation of mRNA may be regulated, thus influencing the amount of transcript in the cell. Translation of mRNA may be inhibited. Even after translation, individual proteins may be activated or inhibited. Let's begin our exploration of this topic by examining how the activity of enzymes can be regulated.

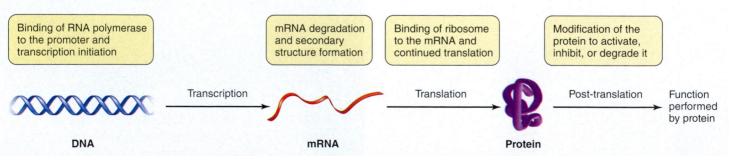

Figure 11.1. Control of gene expression Gene expression can be controlled at several levels, ranging from control of transcription to control of translation to activation or inhibition of the final gene product.

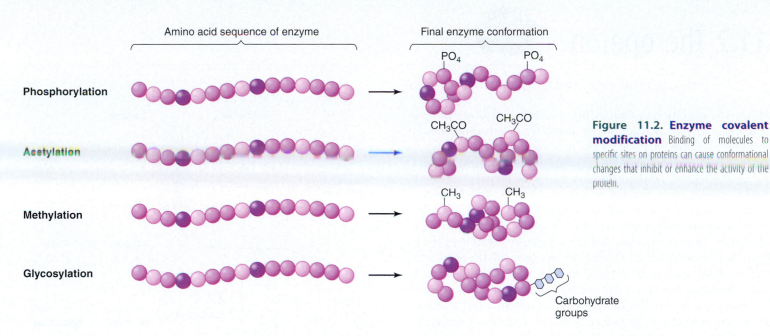

Phosphorylation

Acetylation

Methylation

Glycosylation

Amino acid sequence of enzyme

Final enzyme conformation

PO₄   PO₄

CH₃CO   CH₃CO

CH₃   CH₃

Carbohydrate groups

Figure 11.2. Enzyme covalent modification Binding of molecules to specific sites on proteins can cause conformational changes that inhibit or enhance the activity of the protein.

Activity of enzymes

Organisms can regulate the activity of enzymes by altering their structure and therefore their binding properties. Enzymes already present in the cell can be turned off rapidly if they suddenly are not needed. In a pathway of many steps, the final product often inhibits the activity of the enzyme at the first step of the product synthesis pathway. The inhibitory product may bind to the enzyme, changing its conformation so that the substrate is unable to bind at the active site. This type of regulation is called **allosteric inhibition**, and shuts down the pathway (see Figure 13.9). In other cases, enzyme activity can be regulated by **covalent modification**, such as phosphorylation or methylation of the enzyme **(Figure 11.2)**. Bacteria use both of these forms of modulation to control chemotaxis, a topic we will investigate in Section 11.7.

CONNECTIONS In biochemistry, the binding of a molecule to an enzyme resulting in a change of shape and therefore change in activity of the enzyme is referred to as "allosteric regulation." In **Section 13.2**, we will investigate allosteric regulation and its role in ATP production.

Still, in these cases the cell has already used energy and resources to produce these enzymes that now are not needed. Even though enzymes can be quickly deactivated by allosteric or covalent modifications, a more efficient way to control metabolic pathways would be to regulate transcription of the genes that code for these enzymes, thus saving energy and resources.

Production of enzymes

Indeed, the initiation of transcription can be regulated in several ways. Often this control involves regulatory molecules and DNA-binding proteins that affect the ability of RNA polymerase to bind to the DNA. By modulating the ability of RNA polymerase to bind to DNA, these molecules modulate the amount of transcription that occurs. When transcription is prevented, no mRNA is produced and no protein is made. The amount of transcript can also be affected by factors that influence mRNA stability. Normally, however, mRNA degrades within minutes of its production. The window for controlling this mechanism, therefore, is narrow.

RNA polymerase binds to DNA at the start of a gene, interacting with the promoter. For a given gene, RNA polymerase will bind most tightly to the optimal promoter sequence (consensus sequence), resulting in a very high basal level of gene expression in the absence of any regulatory proteins. RNA polymerase will bind less well to weak promoters, promoters with sequences that differ from the consensus sequence, resulting in a low basal level of activity. Activator proteins can increase the binding of RNA polymerase to weak promoters, while repressor proteins can obstruct the binding of RNA polymerase to promoters.

CONNECTIONS In bacteria, the component of the RNA polymerase that actually binds to the promoter is the sigma factor. **Toolbox 7.1** describes how DNA-binding proteins like the sigma factor can be identified using a gel shift assay.

As we will see later in this chapter, other regulatory processes occur after the initiation of translation. In some cases, the mRNA molecule is produced, but translation is inhibited. These mechanisms may be the least common regulatory processes; they are neither the most efficient nor the quickest. Nevertheless, mechanisms of translation inhibition involving small RNA molecules are under intense study. We will explore these intriguing molecules in Section 11.4.

11.1 Fact Check

1. Describe the difference between constitutive and inducible genes. What types of genes might be constitutive, and what types of genes might be inducible?
2. Microbial cells can regulate the amount of protein produced. At what level is this control most efficient? At what level is it quickest?
3. Explain the various ways enzyme activity and/or metabolic pathways can be regulated.

How do regulatory proteins control transcription?

While the initiation of transcription clearly can affect the production of a single polypeptide, it may often be advantageous for a cell to coordinately regulate the production of several polypeptides. For instance, if the cell requires polypeptides A, B, and C for a single process, then the production of polypeptides A, B, and C should occur (or not occur) at the same time. Operons facilitate this type of coordinated regulation. An **operon** is a transcriptional unit consisting of (1) a series of structural genes that code for polypeptides, and (2) regulatory elements that affect their transcription. Often, the products of the structural genes of an operon participate in a single biochemical pathway. Simultaneous transcription, therefore, facilitates efficient operation of that pathway. Operons commonly exist in bacteria where mRNA can be polycistronic, or capable of coding for more than one polypeptide. They generally are not present in eukarya, where each mRNA transcript typically codes for only one polypeptide.

To investigate how operons function, let's explore the *lac* operon of *Escherichia coli*, the study of which contributed greatly to the development of the operon concept (**Figure 11.3**). The *lac* operon consists of three structural genes, *lacZ*, *lacY*, and *lacA*, and an **operator**, a DNA sequence to which regulatory proteins can bind. The product of the *lacY* gene, permease, facilitates the uptake of lactose by the cell, while the product of the *lacZ* gene, β-galactosidase, facilitates its enzymatic breakdown (**Figure 11.4**). These genes are encoded on a single transcript, along with the *lacA* gene, which encodes β-galactoside transacetylase. The role of this enzyme in lactose metabolism remains unclear.

How can the cell regulate the expression of these structural genes? A common promoter controls the initiation of

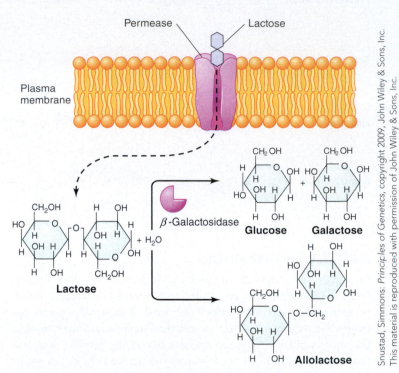

Figure 11.4. Action of β-galactosidase Lactose can be cleaved into glucose and galactose that can be further metabolized by the cell, or it can be converted to allolactose, which can turn on the *lac* operon.

Snustad, Simmons: *Principles of Genetics*, copyright 2009, John Wiley & Sons, Inc. This material is reproduced with permission of John Wiley & Sons, Inc.

Figure 11.3. The *lac* operon The *lac* operon includes three structural genes (*lacZ*, *lacY*, and *lacA*) along with a common promoter and an operator. A regulatory gene, *lacI*, located upstream of the operon, codes for a protein, the LacI repressor, that controls the operator.

TABLE 11.1 Key components of operons

Feature	Definition
Promoter	Site on the DNA bound by the RNA polymerase; directs the initiation of transcription
Activator	Protein that binds to a site on the DNA; assists binding of the RNA polymerase to the promoter, resulting in increased transcription initiation
Activator binding site	Site on the DNA bound by the activator
Repressor	Protein that binds to the operator site on the DNA, inhibiting transcription
Operator	Site on the DNA bound by the repressor
Effector	Small molecule that binds to activator or repressor proteins, modifying their gene regulation activity
Inducer	Effector that increases transcription by enabling an activator or disabling a repressor
Corepressor	Effector that decreases transcription by enabling a repressor

transcription in an operon. Regulatory proteins that bind the operator, along with associated effector molecules, modulate the ability of RNA polymerase to bind to the promoter and initiate transcription (**Table 11.1**). The genes encoding these regulatory proteins may be located adjacent to the target genes on the chromosomal DNA, or they can be located elsewhere on the chromosome. As we will see in this section, these regulatory proteins can inhibit operon transcription, a process referred to as "negative control," or facilitate operon transcription, a process referred to as "positive control." Additionally, as we will see in Section 11.3, some regulatory proteins have a number of target genes distributed throughout the genome.

Negative control of transcription

In general, higher rates of transcription initiation result in higher levels of mRNA transcript and higher levels of protein production. **Negative control** refers to regulatory mechanisms involving a **repressor** that blocks transcription. The repressor binds to the operator as a dimer and inhibits the binding of RNA polymerase to the promoter. **Effectors**, molecules that are often intermediates of the related metabolic pathways, interact with repressor proteins and modulate their ability to bind to the operator. In these systems, the promoter is usually strong in the absence of the repressor protein. Binding of the repressor to the operator blocks the path of RNA polymerase, thus inhibiting transcription. Effector molecules can be either **inducers** that inhibit binding of the repressor protein to the operator or **corepressors** that enhance binding to the operator.

The *lac* operon of *E. coli* represents a good example of an inducible catabolic pathway involving negative control of transcription (**Figure 11.5**). In the absence of lactose, a repressor binds to the operator as a dimer, blocking the promoter site. When lactose is present in the environment, however, allolactose, a metabolite of lactose (see Figure 11.4), acts as an effector, binding to the repressor and preventing it from binding to the operator. Allolactose, then, can be classified as an inducer; it counteracts repression and allows transcription of the structural genes that enable the use of lactose. Negative control shuts down the catabolic pathway in the absence of the substrate, lactose.

The anabolic pathway for the production of tryptophan in *E. coli* presents a slightly different example of a repressible

Figure 11.5. Inducible expression: The *lac* operon Inducible operons can be repressed in different ways. **A.** When the repressor binds to the operator, it blocks transcription by the RNA polymerase. **B.** When an inducer (allolactose) binds to the repressor (LacI) and prevents the repressor from binding to the operator, transcription proceeds.

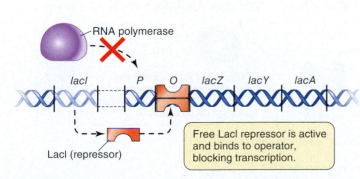

Free LacI repressor is active and binds to operator, blocking transcription.

A. Absence of inducer: Repression

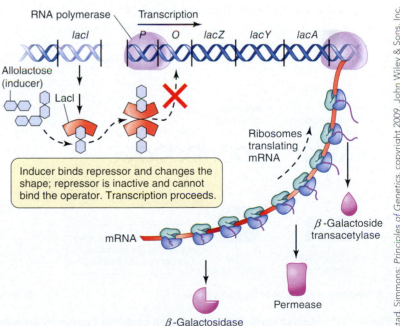

Inducer binds repressor and changes the shape; repressor is inactive and cannot bind the operator. Transcription proceeds.

B. Presence of inducer: Induction

Snustad, Simmons: *Principles of Genetics*, copyright 2009, John Wiley & Sons, Inc. This material is reproduced with permission of John Wiley & Sons, Inc.

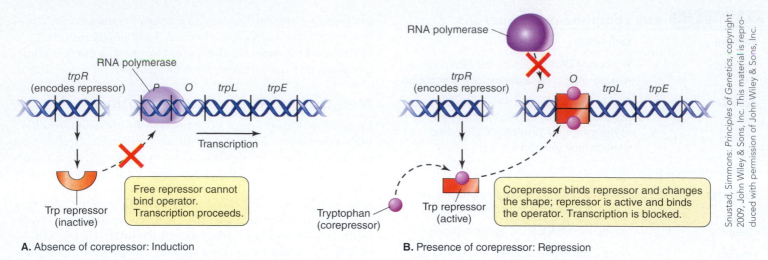

A. Absence of corepressor: Induction

Free repressor cannot bind operator. Transcription proceeds.

Trp repressor (inactive)

B. Presence of corepressor: Repression

Corepressor binds repressor and changes the shape; repressor is active and binds the operator. Transcription is blocked.

Tryptophan (corepressor)

Trp repressor (active)

Figure 11.6. Repressible expression: The *trp* operon The *trp* operon contains six structural genes; not all are shown. **A.** Transcription proceeds in the absence of the corepressor tryptophan, when the repressor cannot bind to the operator. **B.** When tryptophan binds to the repressor, it can bind to the operator and prevent transcription.

pathway involving negative control of transcription (**Figure 11.6**). In the absence of tryptophan, the repressor does not bind to the operator of the tryptophan (*trp*) operon and the cell produces tryptophan. If tryptophan is present in the environment, however, then the tryptophan acts as a corepressor. It binds to the repressor, allowing it to bind to the operator and block transcription. In this example, negative control shuts down the anabolic pathway in the presence of the essential molecule, tryptophan.

Positive control of transcription

In contrast to negative control, operons can also be regulated by **positive control**, in which regulatory molecules lead to increased transcription. For these operons, an **activator molecule** increases the affinity of the RNA polymerase to promoters that otherwise do not bind them very strongly. Effector molecules can alter the conformation of the activator protein, allowing it to bind to an **activator binding site** on the DNA. In this case,

the pathway may be shut down by an inhibitor that removes the activator from the activator binding site on the DNA. While an inducer normally enables binding of the activator protein and increases transcription of the DNA, removal of the inducer shuts down the pathway.

As we saw in the previous section, the *lac* operon exhibits negative control. The *lac* operon, however, also exhibits positive control. *E. coli* cells grow better using glucose than using lactose. Quite simply, cells expend less energy during glucose metabolism. It enters glycolysis directly. Conversely, lactose must first be converted to galactose and glucose. As a consequence, the presence of glucose overrides the signal to activate the *lac* operon. Let's examine how this positive control works.

For transcription of the *lac* operon to occur, an **activator protein (cAMP receptor protein**, or **CRP)** must bind to the activator binding site to enhance binding of RNA polymerase to the promoter and initiation of transcription. Furthermore, CRP requires a coactivator (cyclic AMP, or cAMP) before it can bind to the activator binding site. In the presence of glucose,

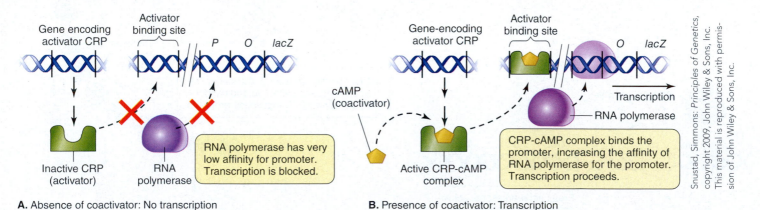

A. Absence of coactivator: No transcription

RNA polymerase has very low affinity for promoter. Transcription is blocked.

Inactive CRP (activator)

RNA polymerase

B. Presence of coactivator: Transcription

cAMP (coactivator)

Active CRP-cAMP complex

CRP-cAMP complex binds the promoter, increasing the affinity of RNA polymerase for the promoter. Transcription proceeds.

RNA polymerase

Figure 11.7. Positive control of transcription of the *lac* operon by activators An activator protein (CRP) must bind to the activator binding site for transcription to proceed. **A.** In the absence of coactivator (cAMP), CRP does not bind the activator binding site, the RNA polymerase does not bind well to the promoter, and there is no activation of transcription. **B.** Under low glucose conditions, the amount of cAMP will increase. When cAMP binds CRP, it is able to bind the activator binding site, increasing the affinity of RNA polymerase for the promoter, resulting in transcription.

Glucose concentration	Lactose concentration	cAMP concentration	Operon status	Level of *lacZ*, *lacY*, and *lacA* transcription	Lactose metabolized?
Low	High	High		High	Yes
Low	Low	High		Low	No
High	Low	Low		Low	No
High	High	Low		Low	No

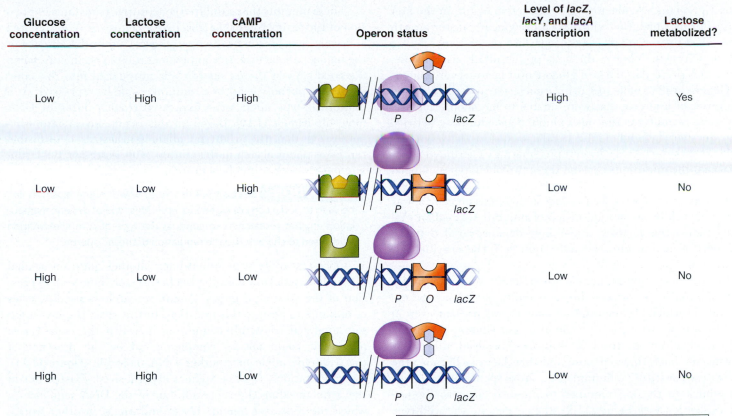

Figure 11.8. Control of *lac* operon expression The binding of LacI repressor by allolactose and binding of CRP activator by cAMP regulate expression of the *lac* operon so that maximal expression occurs in the presence of lactose and absence of glucose.

- RNA polymerase
- LacI (repressor)
- Allolactose (inducer)
- CRP (activator)
- cAMP (coactivator)

levels of cAMP are low. As a result, CRP does not bind to the activator binding site with great affinity. RNA polymerase, in turn, cannot bind to the *lac* operon promoter, and transcription of the *lac* operon genes does not occur **(Figure 11.7)**. Conversely, when glucose levels are low, cAMP levels increase, CRP facilitates binding of RNA polymerase to the promoter, and the cell can make the enzymes needed to use lactose (see Figure 11.7B). Of course, the enzymes are still not produced in the absence of lactose because allolactose would not be present to remove the repressor from the operator site **(Figure 11.8)**.

Investigating the *lac* operon

In the preceding two sections, we gained a better understanding of how the *lac* operon is regulated. In this section, we will address a different question about this operon. How do we know that the operon is regulated in these ways? The nature of the *lac* operon has been revealed through the work of many scientists whose experimental analyses have defined the study of genetics in bacteria.

Let's begin by investigating the role of glucose on the *lac* operon. While wild-type *E. coli* cells can use lactose to support growth, glucose represents the preferred carbon source for these cells. When growth medium contains both glucose and lactose, the cells will exhaust the supply of glucose before they start using lactose, resulting in a **diauxic growth curve (Figure 11.9)**. Between the two growth phases, a lag occurs during which *lacZ*

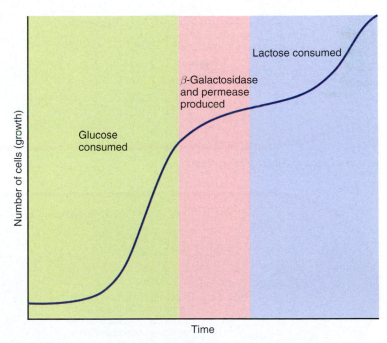

Figure 11.9. Diauxic growth curve of *E. coli* When growing in medium containing glucose and lactose, *E. coli* preferentially uses glucose. When glucose becomes depleted, β-galactosidase and permease must be produced before the cells can begin to use lactose. This requirement results in a lag in growth. When they are using glucose, the organisms do not produce these enzymes.

transcription and β-galactosidase production begin. In the first phase, during which the cells metabolize glucose, faster growth occurs than the second phase, during which the cells metabolize lactose. Given the choice, the cells preferentially use glucose. How is this preferential use of glucose over lactose coordinated?

Genetic studies of lactose utilization began with the isolation of a series of mutants that were unable to metabolize lactose. One class of mutants did not exhibit β-galactosidase activity, while another class did exhibit β-galactosidase activity, but still could not grow on lactose. Interrupted mating experiments indicated that the mutations causing these two phenotypes mapped close to each other (*lacZ* and *lacY* in Figure 11.3). The first group of mutants, in which β-galactosidase activity was abolished, was called "Z," while the second group of mutants was called "Y." Using radioactive lactose, researchers demonstrated that the transport of lactose into the cell required Y, the enzyme now known as permease.

In subsequent experiments, researchers also used a mutagenic approach to address the inducibility of the *lac* operon. Mutants in which β-galactosidase activity was present even in the absence of lactose were isolated by enrichment through a cycling subculture protocol. This process involved alternately growing the bacteria in glucose and then lactose. Because the few lactose constitutive mutants that arose spontaneously were not subject to the lag required to produce enzymes for lactose utilization (see Figure 11.9), they overcame the wild-type strains under these growth conditions and came to dominate the population.

These mutants that exhibited constitutive (constant) expression of the *lac* operon in the absence of lactose were designated i^- because expression of the operon occurred in the absence of the inducer allolactose. In complementation experiments, when a second copy of the *lac* operon was introduced into the same strain, the phenotype of the i^- mutants could be repressed, even if the wild-type allele were expressed from a different DNA molecule (**Figure 11.10**). Based on this result, investigators hypothesized that the wild-type allele synthesizes a diffusible repressor that is absent in the mutant. This repressor, LacI, subsequently was shown to be a protein.

CONNECTIONS **Section 9.4** describes how F-prime plasmids can be used to add a copy of a gene to a cell. This type of genetic manipulation enables researchers to conduct the type of complementation study used to characterize the regulation of the *lac* operon.

Once researchers identified LacI, another question needed to be addressed. How does the LacI repressor shut down expression of the structural genes? Again, researchers used a series of mutants to answer this question. In this case, the investigators analyzed constitutive mutants in which the constitutive phenotype could not be complemented by the presence of the wild-type allele on another DNA molecule (**Figure 11.11**). These mutations, the researchers hypothesized, existed not in the gene encoding the repressor, but in the DNA sequence to which the repressor bound. These mutations, in other words, defined the operator. Because the repressor binds specifically to the *lac* operator, changes in the operator sequence would

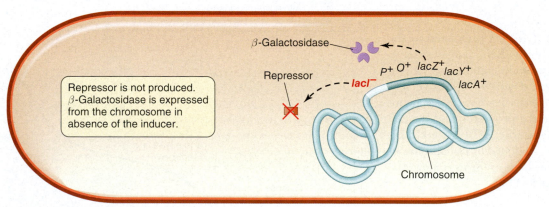

Repressor is not produced. β-Galactosidase is expressed from the chromosome in absence of the inducer.

A. No *lacI* mutant: Constitutive expression

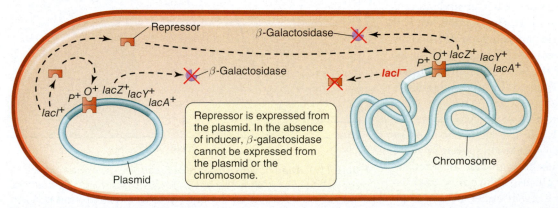

Repressor is expressed from the plasmid. In the absence of inducer, β-galactosidase cannot be expressed from the plasmid or the chromosome.

B. Addition of *lacI*⁺: Repression restored

Figure 11.10. Constitutive LacI repressor mutants of the *lac* operon By analyzing constitutive mutant strains, researchers identified mutations of LacI repressor. **A.** In the absence of an inducer, the cell can synthesize β-galactosidase. **B.** The wild-type *lacI* gene carried on a plasmid can interact and repress (complement) a *lacI* mutant (i^-) even though it resides on a separate piece of DNA, indicating that the LacI repressor is diffusible.

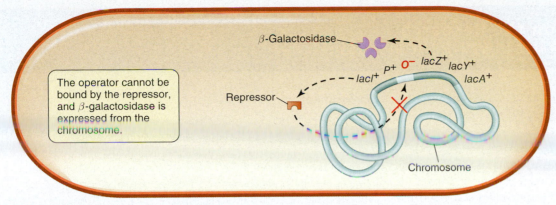

The operator cannot be bound by the repressor, and β-galactosidase is expressed from the chromosome.

β-Galactosidase

Repressor

P^+ O^- $lacZ^+$ $lacY^+$
$lacI^+$ $lacA^+$

Chromosome

A. Operator mutant: Constitutive expression

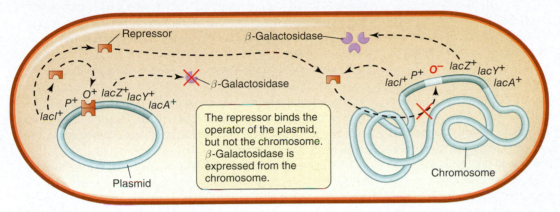

Repressor

β-Galactosidase

β-Galactosidase

P^+ O^+ $lacZ^+$ $lacY^+$
$lacI^+$ $lacA^+$

The repressor binds the operator of the plasmid, but not the chromosome. β-Galactosidase is expressed from the chromosome.

$lacI^+$ P^+ O^- $lacZ^+$ $lacY^+$
$lacA^+$

Chromosome

Plasmid

B. Addition of $lacI^+$: Repression not restored

Figure 11.11. Constitutive operator mutants of the *lac* operon

By analyzing constitutive mutant strains, researchers identified mutations of the *lac* operator. **A.** In the absence of an inducer, the cell can synthesize β-galactosidase. **B.** The addition of a second copy of the *lac* operon to these constitutive mutants was not able to restore repression, thus distinguishing this class of mutants from the LacI repressor mutants. In these new mutants, the operator is unable to bind LacI repressor.

prevent repressor binding and result in constant expression of *lacZ* and *lacY*. During their studies of the *lac* operon, researchers also identified a set of mutants that never expressed *lacZ* or *lacY*. These mutations defined the promoter. Changes in the DNA sequence of the promoter region can prevent binding of RNA polymerase. In the absence of RNA polymerase binding, transcription, of course, does not occur.

These early studies thus laid the foundation for our current understanding of gene structure and gene regulation. The advances since these initial experiments have been remarkable. As we will see in the rest of this chapter, we now have an even better understanding of gene expression. Perhaps more importantly, we now have the ability to intentionally piece together existing genetic elements, allowing us to modify gene regulation systems in creative and useful ways **(Mini-Paper)**.

11.2 Fact Check

1. Describe the process of negative control of gene expression. Explain the role that repressor proteins play in this process.

2. Describe the process of positive control of gene expression. Explain the role that activator proteins play in this process.

3. What role does lactose play in *lac* operon regulation? What role does glucose play in this regulation?

4. What role does the LacI repressor play in *lac* operon regulation?

11.3 Global gene regulation

How does a single environmental stimulus control multiple operons?

As we saw in the previous section, *Escherichia coli* (in the presence of lactose and absence of glucose) begins expressing the structural genes of the *lac* operon. Other environmental changes, though, may require altering the expression of more than just a few genes. When a cell is stressed, for example, it produces a number of heat shock proteins that help to stabilize other proteins in the cell. The heat shock proteins allow the cell to survive. Additionally, the optimal cellular response to some changes in the environment may require the increased expression of some genes and the reduced expression of other genes. The collections of genes regulated in this global manner are called **regulons**.

R. S. Cox III, M. G. Surette, and M. B. Elowitz. 2007. Programming gene expression with combinatorial promoters. Mol Sys Biol 3:145.

Context

The early part of the twenty-first century represents another golden age for microbiology, as molecular biology matures and allows the detailed study of microbial cells and communities at the genomic level. What will the future hold? Today, much excitement surrounds synthetic biology, a developing field in which engineers and biologists are working together to design and build simpler, more predictable, and more reliable biological systems. Investigators hope that this approach can be used to develop useful biotechnology applications.

A central requirement of synthetic biology involves the optimization of gene expression within the system. To accomplish this goal, researchers must have a good understanding of how promoters operate, and have at their disposal a well-characterized collection of promoters to use for the design of novel gene regulatory circuits. To better understand the functions of promoters, one could take the DNA-binding sites of existing promoters, recombine them in different ways, and then examine the regulatory characteristics of the new constructs. Robert Sidney Cox, a graduate student working with Michael Elowitz at the California Institute of Technology (Caltech) and in collaboration with Michael Surette of the University of Calgary, utilized such a combinatorial approach. Their work was published in *Molecular Systems Biology*. Its relevance and potential impact are described next.

Experiment

Many genes respond to more than one transcription factor. These transcription factors can be either repressors or activators. To better understand the mechanism of such gene expression, and to identify rules for the operation of these systems, the investigators designed an experiment in which parts of different naturally occurring promoters could be combined artificially.

They began with 16 naturally occurring promoters of the sigma-70 family with well-defined transcription factor binding sites. These promoters respond to signals through either the AraC or LuxR activators, or the LacI or TetR repressors. To create new, potentially better promoters, the investigators first divided each of these promoter sequences into a 45-bp distal region, corresponding to the region immediately upstream of the −35 box, a 25-bp core region, encompassing the region between the −35 and −10 boxes, and a 30-bp proximal region downstream of the −10 box. They then developed a novel strategy to create new hybrid promoters containing random combinations of these parts.

To engineer new promoters from these parts, the researchers first synthesized complementary oligonucleotides for each of these regions such that after annealing, 5′ overhangs would exist. When these pieces were mixed together, they reasoned, any distal region, core region, and proximal region could interact, forming a new promoter **(Figure B11.1)**. In theory, 16^3 or 4,096 possible combinations could result. After ligation, the constructs were cloned into a luciferase reporter plasmid, which provided a means to easily measure gene expression. A total of 217 of the 288 analyzed constructs were

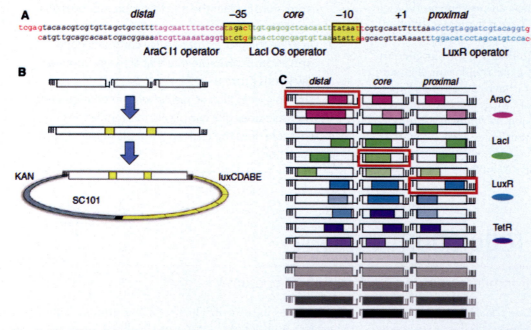

Figure B11.1. Construction of novel promoters By randomly combining parts of previously characterized promoters, the researchers created new promoters. **A.** Existing promoters were divided into distal, core, and proximal regions, which were then combined to make hybrid promoters, such as the one shown here. **B.** The hybrid promoters were inserted into a plasmid containing the luciferase reporter gene. **C.** Any combination of these three regions, from any of the original promoters, could result.

From R. S. Cox III, M. G. Surette, and M. B. Elowitz. 2007. Mol Sys Biol 3:145, Fig. 1. Reproduced with permission of M. Elowitz.

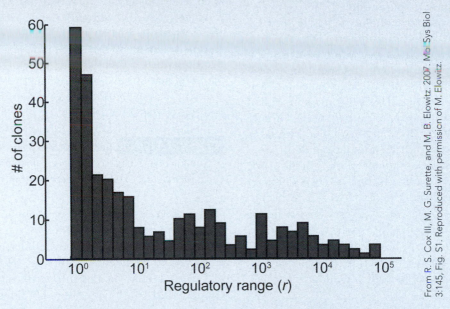

From R. S. Cox III, M. G. Surette, and M. B. Elowitz. 2007. Mol Sys Biol 3:145, Fig. S1. Reproduced with permission of M. Elowitz.

Figure B11.2. Activity of novel promoter constructs Luciferase activity of novel promoter constructs was determined. Most engineered plasmids were functional. Half of the constructs led to a 10-fold increase in the induced compared to uninduced activity, in production of luciferase.

confirmed by DNA sequencing to be unique, and were subjected to further study.

Following the construction of these novel promoters, the investigators then asked the obvious question: Do these synthesized promoter constructs work? Perhaps surprisingly, most of them were functional. Detectable expression of luciferase was seen in 83 percent of the constructs. Half of the constructs responded to one of the four regulating signals with a regulatory range, defined as the ratio of induced to uninduced activity, of 10 or more **(Figure B11.2)**. Further analysis resulted in the development of useful rules that can be used in the design of completely synthetic promoters with desired properties. The authors demonstrated that it is possible to remove promoter parts from their original context without losing function.

Impact

The work reported in this paper is pioneering, and it will be interesting to see in the next few years how these advanced methods of DNA manipulation are adopted and adapted in the continuing development of synthetic biology. It soon may become possible to

design regulatory systems for specific purposes and applications. Probably more important, though, is the insight that this type of work provides into the function of naturally occurring promoters, and the detailed mechanisms that have evolved in nature to carry out complex regulatory processes. It eventually may be possible to develop models to predict the behavior of naturally occurring promoters.

Questions for Discussion

1. The paper describes combining parts of different promoters to create novel regulatory units. Can you suggest another level at which such a combinatorial approach might be useful?

2. Discuss the possible evolutionary consequences of the generation of novel promoters by natural combinatorial processes.

CONNECTIONS **Section 7.4** explains the role of chaperonins, a type of heat shock protein, that aid in the correct folding of proteins. These proteins may be particularly important in hyperthermophilic archaeons, as discussed in **Section 4.3**.

We have already seen an example of global control in our study of the *lac* operon. The presence of glucose in the environment lowers the level of cAMP that can bind to CRP and shuts down the production of enzymes that would allow utilization of lactose. The effects of glucose, though, are more extensive. Glucose also shuts down operons that produce enzymes utilizing a number of other nutrients. This phenomenon, called **catabolite repression**, ensures that *E. coli* preferentially utilizes glucose. In *E. coli*, the CRP-cAMP complex regulates well over 100 operons, thus constituting a major regulon.

In this section, we will explore regulons more fully. To start, we will examine how bacterial cells respond to DNA damage, focusing on the experimental evidence for a DNA repair mechanism. As part of this discussion, we will also examine the use of the *lac* operon and lactose analogs in expression studies. Finally, we will discover how sigma factor/promoter interactions can affect gene regulation. Throughout this section, we will see how cells can quickly and effectively effect global changes in their gene expression patterns.

Experimental evidence for the existence of regulons

To understand regulons more fully, let's consider the response of the cell to DNA damage. A number of environmental stresses, such as ultraviolet (UV) radiation and various chemicals, can damage DNA. If a cell experiences severe DNA damage, then the cell will die. Not surprisingly, then, cells possess mechanisms to repair damaged DNA. One type of DNA repair response, the **SOS response**, functions as a final option for rescuing cells whose genomes have been seriously damaged. As might be expected, this mechanism does not become operational until the cell senses the DNA damage.

CONNECTIONS **Section 7.5** explains how various mutagens can damage DNA. For reasons that remain unclear, the bacterium *Deinococcus radiodurans* exhibits extreme resistance to damage from radiation.

Jean-Jacques Weigle, a Swiss physicist who became a phage biologist, performed early studies of the cellular response to DNA damage while working at the California Institute of Technology (Caltech) in Pasadena, California. He investigated the effect of UV treatment on the infectivity of lambda phage particles **(Figure 11.12)**. After exposure to UV light, the phage particles had

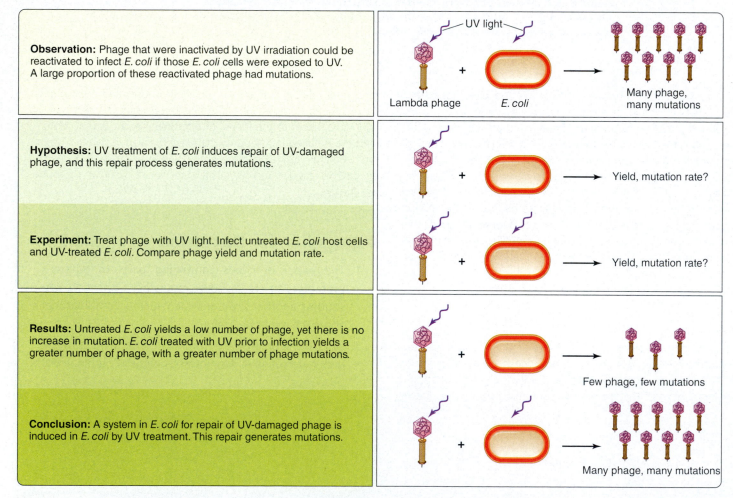

Observation: Phage that were inactivated by UV irradiation could be reactivated to infect *E. coli* if those *E. coli* cells were exposed to UV. A large proportion of these reactivated phage had mutations.

Hypothesis: UV treatment of *E. coli* induces repair of UV-damaged phage, and this repair process generates mutations.

Experiment: Treat phage with UV light. Infect untreated *E. coli* host cells and UV-treated *E. coli*. Compare phage yield and mutation rate.

Results: Untreated *E. coli* yields a low number of phage, yet there is no increase in mutation. *E. coli* treated with UV prior to infection yields a greater number of phage, with a greater number of phage mutations.

Conclusion: A system in *E. coli* for repair of UV-damaged phage is induced in *E. coli* by UV treatment. This repair generates mutations.

UV light

Lambda phage   *E. coli*   Many phage, many mutations

Yield, mutation rate?

Yield, mutation rate?

Few phage, few mutations

Many phage, many mutations

Figure 11.12. Infectivity of lambda phage particles after UV light treatment Exposure of the bacteriophage to UV light causes DNA damage and reduces their ability to infect *E. coli*. If the *E. coli* cells are also exposed to UV light prior to infection, then the rate of phage infection remains high. The phage also exhibit an increased rate of mutations.

a reduced rate of infection of *E. coli*, and no increase in mutation rate, or mutagenesis. Presumably, the UV light exposure damaged the phage DNA, making them unable to replicate. If the *E. coli* host cells, however, were also treated with UV prior to infection, then the rate of phage infection was high, as was the rate of mutations observed in the phage. This correlation between infectivity and mutagenesis of the phage suggested that UV treatment of the host cells enhances their ability to repair the UV-induced damage of the phage DNA, thus allowing the phage to replicate. However, the repair mechanism is error-prone, resulting in an increased number of mutations in the phage.

In another set of experiments, researchers investigated the effects of the antibiotic chloramphenicol on DNA repair. This antibiotic specifically blocks protein synthesis in bacteria. Investigators observed that chloramphenicol also inhibits DNA repair. From this observation, they concluded that DNA repair requires protein synthesis. They then asked an obvious follow-up question. What proteins need to be produced for DNA repair to occur?

To identify which genes must be expressed, the researchers used a genetic approach to identify genes that are members of the SOS regulon responsible for the production of these proteins. First, a *lacZ* promoter probe transposon was used to mutate *E. coli*. The promoter probe transposon contains a **reporter gene**, a promoterless *lacZ* gene that is not expressed unless it inserts within an actively transcribed gene, as well as an antibiotic resistance gene used for selection. For most genes, the level of expression does not result in readily measurable phenotypic changes, but use of a reporter gene with an easily detectable phenotype provides a good proxy measurement. As we have discussed earlier in this chapter, the *lacZ* gene encodes the enzyme β-galactosidase, the activity of which is very easy to detect and measure using colorimetric methods (**Perspective 11.1**). In **Figure 11.13** the introduction of the promoter probe transposon resulted in some colonies that were blue on the chromogenic substrate X-gal, because the insertions were in genes that were expressed, while some were white on X-gal, carrying insertions in genes that were not expressed.

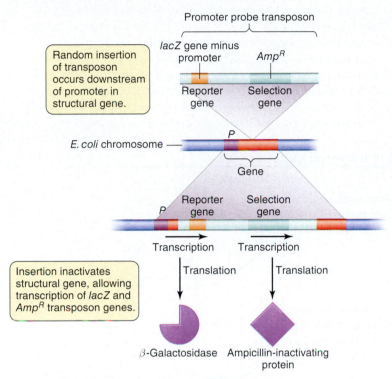

A. Creation of *E. coli* mutants using promoter probe

CONNECTIONS **Section 24.2** outlines how the antibiotic chloramphenicol binds to the 23S rRNA of the 50S ribosomal subunit, thus interfering with peptide bond formation during translation. This mode of action makes it a very useful reagent in experiments designed to determine if certain biological processes require protein synthesis.

Investigators first screened these promoter probe-induced mutant colonies to identify those colonies that exhibited *lacZ*-encoded β-galactosidase activity in the presence of the

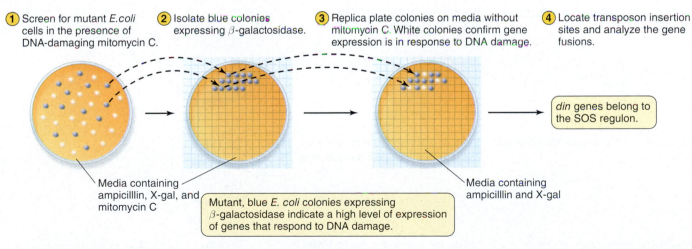

B. Screening process to identify genes involved in the SOS regulon

Figure 11.13. Process Diagram: Monitoring the induction of gene expression in response to DNA damage By using *lacZ* gene fusions, researchers can monitor the expression of genes in the SOS regulon. **A.** A *lacZ* promoter probe transposon is used to mutate *E. coli*, facilitated by selection for antibiotic resistance. Insertion of the transposon within expressed genes results in β-galactosidase activity, yielding blue colonies on X-gal. **B.** Fusions to *din* genes are strongly induced immediately after the introduction of the DNA-damaging agent mitomycin C (MC). They are identified through a two-step screening process to determine which promoter probe fusions are only expressed in the presence of mitomycin C, and not expressed in the absence of mitomycin C.

Not only was the *lac* operon one of the first operons to be studied and characterized, but it has also been developed as a powerful tool for investigating the function and regulation of other genes. Many experiments that led to our understanding of the nature of genes and gene regulation were performed on the *lac* operon, and the use of synthetic lactose analogs as substitute substrates was instrumental to these studies. The usefulness of the *lac* operon as a genetic tool has been enhanced considerably by the availability of these synthetic lactose analogs, and as a result, components of the *lac* operon have found widespread use as integral parts of cloning vectors, expression vectors, and reporter gene fusions. Let's investigate the use of these analogs further.

In addition to its ability to cleave the glucose-galactose disaccharide lactose, the *lacZ*-encoded enzyme β-galactosidase can break down several synthetic analogs of lactose that contain β-linkages between galactose and chromogenic, lumigenic, or fluorogenic molecules in place of glucose. Cleavage of these molecules causes color change, luminescence, or fluorescence respectively. **Table B11.1** lists several of these analogs and their properties. ONPG (*ortho*-nitrophenyl-β-D-galactoside) is a particularly useful molecule; cleavage of this colorless compound results in the release of ONP (*ortho*-nitrophenol), which is yellow in color. The intensity of the color in a reaction can be measured using a spectrophotometer. Greater enzyme activity results in a more intense color. ONPG thus forms the basis for a very sensitive assay for β-galactosidase activity. Because β-galactosidase activity is directly related to the level of expression of the *lac* operon, this analog has become a very convenient method to measure gene expression, especially using plasmid and transposon gene fusion tools.

TABLE B11.1 Artificial lactose analogs used in bacterial genetics

Analog	Characteristics	Structure
Ortho-nitrophenol-β-D-galactoside (ONPG)	Chromogenic substrate, yellow in color; used for quantitative measurement of β-galactosidase activity	
5-Bromo-4-chloro-3-indolyl-β-D-galactoside (X-gal)	Chromogenic substrate conferring blue color, non-inducing; commonly used for detection of β-galactosidase activity in bacterial colonies	
4-Methylumbelliferyl-β-D-galactoside (MUG)	Fluorogenic substrate, cleavage releases fluorescence; used for enzyme assays	
Isopropyl β-D-1-thiogalactopyranoside (IPTG)	Inducer, not metabolized; used for induction of *lac* operon expression	
Phenyl-β-D-galactoside (Pgal)	Substrate, non-inducing; *lacY* not required for uptake. Used for the isolation of constitutively expressing *lacZ* mutants.	

X-gal (5-bromo-4-chloro-3-indolyl-β-D-galactoside) represents another widely used lactose analog. Upon cleavage, this analog releases an indole dye, which leads to the formation of a blue precipitate. On solid media containing X-gal, colonies expressing β-galactosidase appear blue. A number of screens are based on the ability to easily differentiate blue and white colonies. Blue colonies have *lacZ*-encoded β-galactosidase activity, while white colonies do not. We have seen in Section 10.1 that this type of screen is particularly useful for the detection of recombinant plasmids in which DNA fragments have been successfully cloned, and it is widely used in cloning vectors.

Less widely used than ONPG or X-gal, 4-methylumbelliferyl-β-D-galactoside (MUG) represents another lactose analog. This compound provides researchers with a fluorogenic measurement of β-galactosidase activity. Upon its cleavage, MUG releases 7-hydroxy-4-methylcoumarin, which absorbs light at 360 nm and then emits light at 449 nm.

The design and implementation of effective, often clever, genetic screens showcases the creative abilities of the bacterial geneticist. Because such screens rely on the phenotypes associated with mutations, they are limited to cases with selectable or easily detectable phenotypes. To overcome this limitation, gene fusion technology often is used, in which reporter genes that do not contain upstream regulatory regions are fused to the genes whose expression is being analyzed. Connecting the expression of a gene of interest to a readily detected reporter gene such as *lacZ* makes it possible to monitor gene expression using the phenotype of the reporter gene. This procedure facilitates screening for factors that influence expression of the gene of interest. Once researchers identify interesting relationships by the X-gal screening, they can then use ONPG to more accurately quantify the gene expression profile.

Some other lactose analogs, such as IPTG (isopropyl β-D-1-thiogalactopyranoside), can induce the expression of the *lac* operon without being subject to degradation by the β-galactosidase enzyme. This feature of IPTG has been important in studies of the regulation of the *lac* operon. As a molecular biology tool, IPTG induction is commonly used in expression vectors to produce large amounts of recombinant proteins. Such expression vectors typically contain the *lac* promoter and ribosome-binding site alongside which the ORF of the gene to be expressed is cloned; activation of the promoter results in expression of the gene. In the presence of *lacI*, the gene will be repressed, and the addition of IPTG will relieve this repression. Variation of the amount and timing of IPTG addition allows careful control of the desired level of gene expression to produce the product. In contrast to IPTG, phenyl-β-D-galactoside (Pgal) can be metabolized by *E. coli*, but does not induce the *lac* operon.

It truly is a testament to the quality of the research undertaken by the early microbial geneticists who unraveled the workings of the *lac* operon that this set of genes continues to find widespread use throughout the life sciences. While artificial analogs of other substrates have been used in microbial genetics, *lac* is still the model. After well over a half century, this icon has found a permanent place in the science of microbiology.

DNA-damaging agent, mitomycin C. These were blue on X-gal. A second screen of these positive mutants was performed in the absence of mitomycin C to determine which mutants were white on X-gal, and thus dependent on the mitomycin C for β-galactosidase activity (see Figure 11.13B). These mutants contained promoter probe transposon insertions in several different regions of the chromosome, as determined by gene-mapping experiments. Mitomycin C, the researchers concluded, activated the expression of these genes that were interrupted by the promoter probe transposon. These genes, called "*din* genes" for *d*amage *in*ducible, thus defined the potential SOS regulon.

After identifying genes of the SOS regulon, researchers then addressed another question. How does a cell sense DNA damage? *In vitro* studies using purified proteins showed that bacteria do not actually detect the agent that causes the DNA damage. In fact, any number of different agents can cause damage. Rather, the cells detect a molecule that results from DNA damage: single-stranded DNA. The SOS repair system uses alternative DNA polymerases that can fill in missing DNA strands but lacks the ability to proofread. This mechanism repairs the structure of the chromosome, but results in a high rate of mutation.

If the SOS genes all respond to the same stimulus, then what common regulatory system results in their expression? Using the inducible mutants identified in the previous experiment, another round of screening was done following mutagenesis with another transposon, this time searching for mutations in which SOS induction was altered (**Figure 11.14**). Two types of mutations

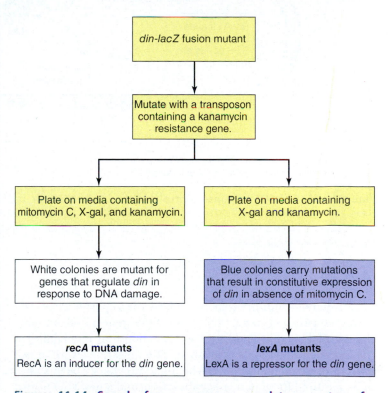

Figure 11.14. Search for a common regulatory system for expression of SOS genes The genes responsible for the control of the SOS response were identified by mutagenesis of a strain containing a *lacZ* fusion to a DNA damage inducible (*din*) gene, followed by screening on X-gal in the presence and absence of mitomycin C. Two types of mutants were sought; ones that exhibited expression in the absence of mitomycin C (blue on X-gal without mitomycin C), and ones that did not express in the presence of mitomycin C (white on X-gal in presence of mitomycin C). This work resulted in the identification of *lexA* (blue colonies in absence of mitomycin C) and *recA* (white colonies in presence of mitomycin C) as the key genes responsible for the SOS response.

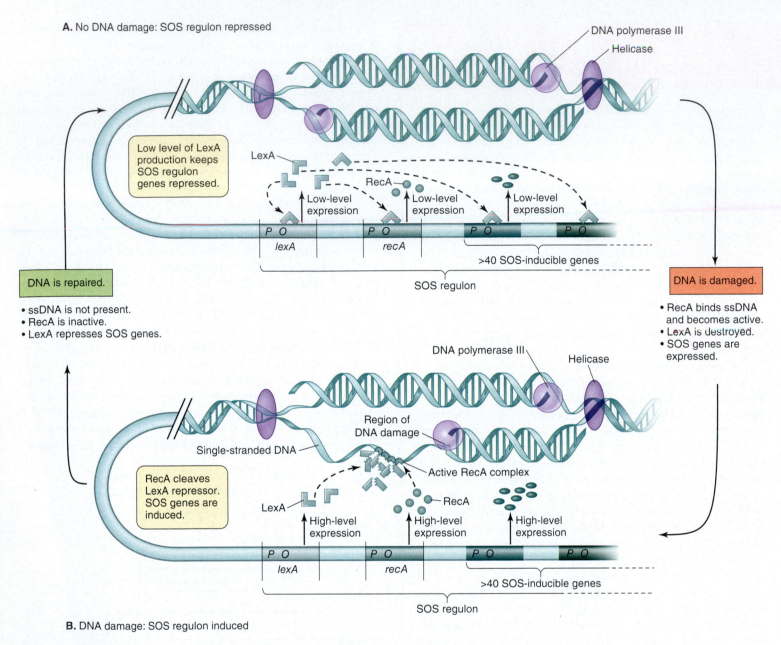

A. No DNA damage: SOS regulon repressed

DNA polymerase III

Helicase

Low level of LexA production keeps SOS regulon genes repressed.

LexA

RecA

Low-level expression

Low-level expression

Low-level expression

P *O*

lexA

P *O*

recA

P *O*

P *O*

>40 SOS-inducible genes

SOS regulon

DNA is repaired.

- ssDNA is not present.
- RecA is inactive.
- LexA represses SOS genes.

DNA is damaged.

- RecA binds ssDNA and becomes active.
- LexA is destroyed.
- SOS genes are expressed.

DNA polymerase III

Helicase

Region of DNA damage

Single-stranded DNA

Active RecA complex

RecA cleaves LexA repressor. SOS genes are induced.

LexA

RecA

High-level expression

High-level expression

High-level expression

P *O*

lexA

P *O*

recA

P *O*

P *O*

>40 SOS-inducible genes

SOS regulon

B. DNA damage: SOS regulon induced

Figure 11.15. The SOS response control and the LexA repressor Expression of the members of the SOS regulon is controlled by LexA. **A.** The LexA repressor inhibits the expression of genes under SOS control. **B.** When DNA damage occurs, the RecA protein interacts with single-stranded DNA produced by damage to DNA, causing it to mediate cleavage of the LexA repressor. This action results in relief of the SOS genes from repression, activating their expression.

were isolated: mutants that turned on the reporter gene, signaling constitutive induction of the SOS genes, and mutants that turned off the reporter gene, signaling that inducibility of the SOS genes had been abolished. The mutations were in two protein-encoding genes, *recA*, which had previously been shown to be required for genetic recombination, and a newly identified gene, *lexA*. Further experimental analysis revealed the protein LexA to be a repressor that binds to operator sequences of members of the SOS regulon. RecA controls the activity of LexA by inducing a proteolytic cleavage of LexA in response to DNA damage. The cleaved LexA protein can no longer bind the operators of the SOS genes, resulting in induction of gene expression (**Figure 11.15**).

CONNECTIONS Certain antibiotics exert their antibacterial activity by damaging DNA, thereby inducing the SOS response. Many prophages have evolved the ability to detect this induction of the SOS response in their host cells. In **Perspective 21.4**, we'll see why this adaptation is advantageous for these temperate phages, and why treating some bacterial infections with DNA-damaging antibiotics may do more harm than good.

DNA-damaging events induce the genes in the SOS regulon. Members of this regulon also possess an operator sequence to which LexA can bind, and induction of the genes must be LexA-RecA dependent. Various experiments indicate that the SOS regulon of *E. coli* consists of at least 40 genes distributed

TABLE 11.2 Members of the *E. coli* SOS regulon

Gene	Function
umuDC	Error-prone DNA polymerase V
dinB	Error-prone DNA polymerase IV
polB	DNA polymerase II, repair of DNA interstrand crosslinks
uvrA	Nucleotide excision repair
uvrB	Nucleotide excision repair
uvrD	DNA helicase, involved in DNA repair
sbmC	DNA gyrase inhibitor
ruvAB	Homologous recombination, Holliday junction resolution
lexA	DNA-binding transcriptional repressor of SOS regulon genes
recA	Recombination and regulation of SOS response, binds single-stranded DNA
dinI	DNA damage inducible protein, regulation of RecA protease activity
recN	DNA recombinational repair protein
Ssb	Single-stranded DNA-binding protein
dinG	ATP-dependent DNA helicase
sulA	SOS cell division inhibitor
ftsK	Cell division protein
pcsA	Phosphatidylcholine synthase

Based on Fernández de Henestrosa et al. 2000. Identification of additional genes belonging to the LexA regulon in *Escherichia coli*. Mol Microbiol 35:1560–1572.

throughout the genome. As we might expect, many of these genes encode polypeptides that function in the repair of DNA damage **(Table 11.2)**.

Alternative sigma factors

While regulons effectively allow cells to control the expression of a large number of genes, alternative sigma factors provide another means of global gene regulation. As we saw in Section 7.3, initiation of transcription in bacteria requires the sigma factor, a polypeptide that allows RNA polymerase to identify the promoter. The sigma factor responsible for recognizing most promoters in *E. coli* is sigma-70, named according to its molecular weight, 70 kDa. Most bacterial genomes encode several alternative sigma factors in the sigma-70 (or RpoD) family that recognize *different* promoters. Increased synthesis of those sigma factors can activate entirely different sets of genes. *Escherichia coli*, for instance, has six sigma-70 members. *Streptomyces coelicolor*, an antibiotic-producing bacterium with a complex developmental pathway, has over 60 alternative sigma-70 family members.

Some bacteria also have one or two alternative sigma factors in the sigma-54 (or RpoN) family. Most of the approximately 30 *E. coli* promoters recognized by sigma-54 family members

deal with some aspect of nitrogen utilization. Members of the sigma-70 and sigma-54 families differ in the mode by which they promote transcription initiation. Members of the sigma-54 family require an activating protein and ATP to initiate transcription; members of the sigma-70 family do not. Sigma-54 family members also recognize promoter regions at different distances from the initiation of transcription while sigma-70 family members do not.

CONNECTIONS Section 7.3 describes the regions of the promoter recognized by sigma factors to initiate transcription. While sigma-70 recognizes sequences at regions 10 and 35 upstream from the first transcribed base, sigma-54 recognizes sequences at positions −12 and −24.

Generally, alternative sigma factors are active only under specific conditions and are not required for cell viability. For example, sigma-32, the heat shock sigma factor also called RpoH, recognizes about 30 different promoters in the *E. coli* genome. RpoH normally is degraded very quickly after it is synthesized. When the temperature rises, however, other proteins in the cell begin to unfold and compete with RpoH for proteases that will degrade them. As a result, the concentration of RpoH increases. As RpoH accumulates, it initiates transcription of a wide variety of genes that code for proteins that can help the cell to survive temperature increases **(Table 11.3)**.

TABLE 11.3 Sigma-32 regulated genes in *E. coli*

Gene	Product and function
ibpB	Small heat shock protein, chaperone
ibpA	Small heat shock protein, chaperone
hslJ	Heat shock protein
htpX	Integral membrane heat shock protein
clpB	ATP-dependent protease, protein disaggregation chaperone
hslU	Heat inducible ATP-dependent protease component involved in degradation of misfolded proteins
hslV	Heat inducible ATP-dependent protease component involved in degradation of misfolded proteins
clpX	Component of ATP-dependent protease and chaperone
gapA	Glyceraldehyde-3-phosphate dehydrogenase A
ftsJ	23S rRNA methyltransferase, involved in cell division and growth
htpG	Heat shock chaperone
groEL	Binds and regulates RpoH
groES	Binds and regulates RpoH
grpE	Heat shock protein
dnaK	DNA synthesis
dnaJ	DNA synthesis
Lon	DNA-binding, ATP-dependent protease

Based on Zhao et al. 2005. J Biol Chem 280:17758–17768.

Another sigma factor that controls expression of a large number of genes is sigma-38 or RpoS, the product of the *rpoS* gene. This sigma factor is responsible for differential gene expression in *E. coli* and related bacteria in response to general stress conditions. Such conditions include high cell density, nutrient starvation, low temperature, high osmolarity, and oxidative stress. Dozens of target genes respond to RpoS, including genes involved in protein processing, general stress adaptation, membrane stability and transport, and metabolism. Expression of these genes depends on the amount of sigma-38 protein present in the cell. The products of these genes result in entry of the cell into the stationary phase as cells prepare for survival over extended periods of time.

CONNECTIONS Section 6.2 describes conditions that might cause cells to enter the stationary phase. One response to these conditions is the stringent response, which shifts use of cellular resources from growth to restoration of amino acid pools that will support baseline protein synthesis.

Just as sigma factors can control entire regulons, and alternative sigma factors can allow the cell to activate different regulons, sigma factors themselves are also regulated. For example, proteins called "anti-sigma factors" can bind to the sigma factor and prevent RNA polymerase from binding to its promoters, thus turning off the genes it controls. These processes may be useful in controlling the timing of global gene expression. When the anti-sigma factors are eliminated, then the entire regulon can be quickly and coordinately activated.

11.3 Fact Check

1. What is a regulon?
2. What is catabolite repression? How is this process involved in the actions of the *lac* operon?
3. Describe the SOS response.
4. What is the effect of a mutation in *lexA* on the SOS response? What is the effect of a mutation in *recA*?
5. Explain the roles of the two different sigma factors known as RpoH and RpoS.

11.4 Post-initiation control of gene expression

How can mRNA be controlled?

Sigma factors and regulatory proteins, like repressors and activators, typically control the initiation of transcription. This type of regulation saves resources and energy used to produce mRNA coding for proteins that the cell does not need. Still, protein synthesis may also be controlled at other points. Transcription may be inhibited after initiation, for example. Enzymatic degradation of mRNA may be regulated, thus influencing the amount of transcript in the cell. Translation of mRNA may be inhibited. Even after translation, the activity of individual proteins may be inhibited, as we saw in Section 11.1. In this section, we will consider some mechanisms of gene regulation that occur after the initiation of transcription. Specifically, we will explore the roles of regulatory RNAs and mRNA secondary structure.

Regulatory RNAs

The genomes of archaeons, bacteria, and eukarya contain regions that code for RNA that is not translated into proteins. These regions include genes for rRNA and tRNA molecules, along with a more recently identified class of RNAs with regulatory functions. These small non-coding RNAs of about 50–400 nucleotides are called **sRNA**. Identified by such names as microRNA, interfering RNA, and antisense RNA, sRNA operates by a variety of methods, but typically affects gene expression very quickly by interacting with existing mRNA. Genes can be expressed without taking the time to transcribe genes into new mRNA. Similarly, expression can be stopped without waiting for existing mRNA to be degraded.

To see how regulatory RNAs function, let's examine the regulation of a specific sigma factor: sigma-38, or RpoS. In *Escherichia coli*, this sigma factor activates genes associated with stress. Control at the level of translation of the *rpoS* mRNA has a major influence on the levels of RpoS molecules present in the cell. Molecules of **antisense RNA**, single-stranded RNA that can interact with specific mRNA molecules through complementary base-pairing, direct this control. In the absence of antisense RNA molecules, the secondary structure of a region called "the leader" at the 5′ end of the *rpoS* transcript inhibits translation. Certain antisense RNA molecules that interact with the leader, however, accumulate when the cell experiences various stresses **(Figure 11.16)**. The 85-nucleotide sRNA DsrA, for instance, accumulates to high levels in the cells in response to low temperature, while the 105-nucleotide sRNA RprA accumulates to high levels in the cell in response to cell surface stresses. In conjunction with the RNA-binding protein Hfq, both of these antisense RNAs bind to the *rpoS* transcript, disrupt the inhibitory secondary structure, and allow translation to proceed. Levels of sigma-38 increase and activate the regulon. The newly produced gene products, in turn, help the cell thrive under these conditions. In this case, the regulatory RNA enhances translation, but in other cases it can inhibit translation. RNA that is complementary to mRNA of a given gene can be used to decrease gene expression **(Toolbox 11.1)**. This technique has become an important tool in molecular biology.

Not too many years ago, scientists referred to DNA that did not code for proteins as "junk DNA." Sure, some DNA coded for important RNA molecules, like rRNA and tRNA, instead of proteins, but researchers generally did not assign a major function to other regions of DNA. Today, we recognize the importance of this "junk." As we have seen in this section, bacteria frequently use small RNAs to regulate the expression of genes. Moreover, researchers have developed ways of utilizing small RNAs to experimentally regulate gene expression.

In 1998, Andrew Fire and Craig Mello demonstrated that small pieces of RNA could downregulate the expression of genes, a process they referred to as "RNA interference." For this intriguing discovery, they were awarded the 2006 Nobel Prize for Physiology or Medicine. Today, scientists throughout the world are using this discovery to investigate basic aspects of cell biology and to develop treatments for human diseases.

The discovery of RNA interference by Fire and Mello has led to the development of gene knockdown techniques. In these experiments, scientists typically engineer cells to produce a short piece of RNA complementary to the mRNA of the gene being studied. When expressed in the cell, this small RNA interacts with the target mRNA, leading to a decrease in protein production. A knockdown in expression occurs. For researchers interested in basic biology, this technique allows them to better understand the role of the target gene and protein.

This technique may also be clinically relevant. For any case in which the overexpression of a gene results in disease, knockdown technology may provide a mechanism whereby the overexpression is corrected. The first human trials of RNA interference to treat a disease began in 2004 when investigators explored the effectiveness of RNA interference as a treatment for macular degeneration, a debilitating eye disease that results in a loss of vision. Today, researchers are exploring the use of RNA interference as a treatment for a number of human diseases, including Huntington's disease and certain cancers.

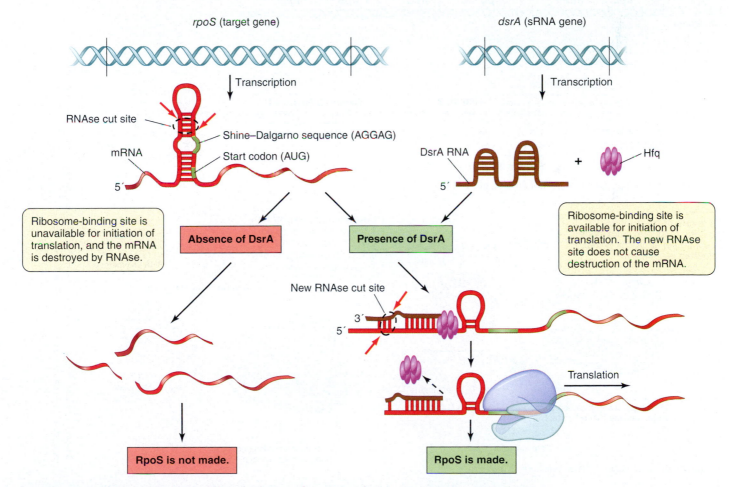

Figure 11.16. sRNA molecules and the regulation of *rpoS* translation The *rpoS* transcript contains a 5′ leader sequence that forms a hairpin stem-loop structure, obscuring the translation start site (shown in green). The Hfq protein promotes binding of sRNA molecules, thus releasing the translation start site so that it can be accessed.

Control of transcription by mRNA secondary structure

In Section 11.2, we saw that a repressor protein blocks transcription of the *trp* operon when tryptophan is present in the environment (see Figure 11.6B). In the early 1970s, scientists observed evidence of another form of regulation of this operon. Deletion mutations in the upstream regulatory region of the *trp* operon resulted in several-fold increases in the level of transcription regardless of the presence or absence of the repressor protein. Further studies revealed this process to be an elaborate type of regulatory mechanism that involves interactions between transcription and translation. This mechanism is called **attenuation**.

Regulation by attenuation occurs after the initiation of transcription, but before transcription of the operon has been completed. Attenuation of the tryptophan biosynthetic operon involves a translated leader sequence on the mRNA. The leader is encoded by *trpL*, found following the operator but before the five structural genes required for synthesis of tryptophan. In bacteria and archaeons, transcription and translation are coupled, so the 5′ end of a transcript containing the leader sequence can be translated before synthesis of that transcript is complete. The leader sequence contains two tryptophan codons. If tryptophan is abundant, then it will be incorporated into the leader peptide chain. If tryptophan is scarce, then the tryptophan aminoacyl-tRNA will be rare, and it will be difficult for the leader peptide to be completely synthesized.

How does this feature of *trpL* affect production of the tryptophan biosynthetic polypeptides? To address this question, we first should note that the leader sequence contains several regions of internal sequence complementarity. Base-pairing among these regions in the leader sequence mRNA results in the formation of double-stranded stem-loop structures affecting the secondary structure of the RNA. If these regions are numbered 1, 2, 3, and 4 from the upstream end, then region 1 contains the tryptophan codons. Region 3 can form a stem-loop with region 2 or with region 4, but not with both. After region 4, a uracil-rich region called "the attenuator" exists that can interact with a 3–4 stem-loop to terminate transcription before reaching the structural genes **(Figure 11.17)**.

If tryptophan is plentiful, then the leader sequence will be fully translated. When the ribosome reaches the stop codon for *trpL*, it blocks formation of a stem-loop between region 2 and region 3. As a result, region 3 binds to region 4, terminating

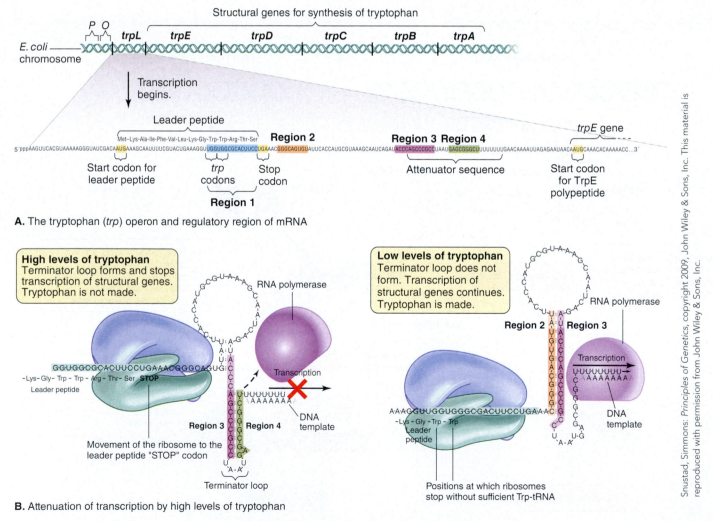

A. The tryptophan (*trp*) operon and regulatory region of mRNA

B. Attenuation of transcription by high levels of tryptophan

Figure 11.17. Control of the *trp* operon by attenuation Regulation by attenuation occurs after transcription initiation. **A.** The *trpL* gene is the first gene in the operon, encodes a leader peptide, and contains an attenuator sequence. **B.** Under high levels of tryptophan, region 3 binds to region 4 to stop transcription. Under low levels of tryptophan, region 3 binds to region 2, allowing transcription to proceed.

Snustad, Simmons: *Principles of Genetics*, copyright 2009, John Wiley & Sons, Inc. This material is reproduced with permission from John Wiley & Sons, Inc.

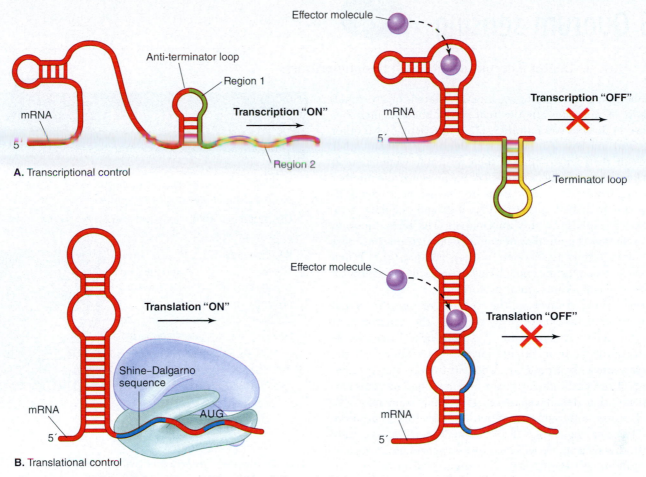

Figure 11.18. Riboswitches Messenger RNA can bind small effector molecules that regulate gene expression. **A.** Some riboswitches affect further transcription. **B.** Some riboswitches regulate translation by inhibiting ribosome binding.

Within the figure:
- Effector molecule
- Anti-terminator loop
- Region 1
- mRNA
- 5′
- **Transcription "ON"**
- Region 2
- **A.** Transcriptional control
- **Transcription "OFF"**
- mRNA
- 5′
- Terminator loop
- **Translation "ON"**
- Shine–Dalgarno sequence
- mRNA
- 5′
- AUG
- **B.** Translational control
- Effector molecule
- **Translation "OFF"**
- mRNA
- 5′

transcription after the leader sequence. Hence, tryptophan production stops.

If tryptophan is scarce, then the ribosome pauses during translation of region 1, waiting for an aminoacyl-tRNA carrying tryptophan. In that position, it blocks region 1 from interacting with region 2. Region 3 then binds with region 2, forming a stem-loop. Under these conditions, region 3 is no longer free to form a stem-loop with region 4, so the attenuator is not activated. Transcription proceeds past the stop codon for *trpL* and transcribes the structural genes for tryptophan synthesis.

The *trp* operon represents an example of the regulation of transcription through the influence of stalled ribosomes on RNA secondary structure. The *trp* operon, though, certainly does not represent the only operon regulated in this fashion. The transcription of a number of other amino acid biosynthesis operons is also influenced by the translation of leader sequences rich in the corresponding amino acid. Computational tools have been developed to scan genome sequences for regions that may encode RNA secondary structures. Based on this type of analysis, it is becoming apparent that attenuation-like mechanisms add considerably to the overall complexity of gene regulation.

CONNECTIONS Regulation by attenuation may represent a direct link to a past world of RNA-based life. **Perspective 1.1** presents the concept of ribozymes, RNA molecules that act as catalysts. These molecules provide more evidence that early life may have relied on RNA functions.

The action of antisense RNA and the formation of stem-loops within mRNA show that segments of RNA can bind to one another through complementary base-pairing. We can also extend our definition of effectors to include regulatory molecules that bind RNA. Some mRNA molecules can bind effectors, like the activators and repressors that we examined earlier. These systems of mRNA and effector molecules collectively have been termed **riboswitches** (**Figure 11.18**). Changes in secondary structure of the upstream 5′ end of the mRNA transcript caused by binding or release of an effector are a widespread means of regulating gene expression. Binding of the effector to mRNA may inhibit further elongation of the transcript as we saw in attenuation, or it may block mRNA binding to the ribosome, affecting translation.

11.4 Fact Check

1. What is sRNA, and what is its role in cells? Give an example.
2. Explain the mechanism of attenuation and its role in regulation of gene expression.
3. Describe the role of stem-loop structure in the regulation of the *trp* operon.

How do bacteria communicate with their neighbors?

In the laboratory, we usually work with isolated, often clonal, populations of microbes. In the real world, though, microbes usually live in diverse communities, often interacting with other microbes and macroscopic organisms. Information about other organisms in their environment, then, probably would benefit them. In fact, many microbes can detect the presence and number of other organisms in a population by the use of a chemical-signaling system called **quorum sensing**, and regulate gene expression in response to this information. The term "quorum sensing" refers to the number of members of a group that must be present in order to conduct business (a quorum). When microbes achieve a quorum, enough cells are present to accomplish a task that one cell alone cannot perform.

Quorum sensing involves the release of specific signal molecules, known as **autoinducers**, into the environment by bacterial cells. The amount of autoinducer increases as the density of the bacterial population increases. Therefore, the bacteria can assess their population density by detecting the concentration of the autoinducer in the surrounding environment. As the population density changes, gene expression may also change. Because different populations employ different autoinducers for this signaling, the communication can be very specific. In this section, we will examine one particular quorum-sensing system, and then briefly investigate the prevalence of quorum-sensing systems in the microbial world.

Lux, a prototypical quorum-sensing system

The Lux system in *Vibrio fischeri* **(Microbes in Focus 11.1)** represents one of the first quorum-sensing systems to be described.

Microbes in Focus 11.1
VIBRIO FISCHERI

Habitat: Found in oceans throughout the world, most often in symbiosis with other marine organisms

Description: Gram-negative rod-shaped bacterium, approximately 2 μm in length

Key Features: *Vibrio fischeri* is well known for its bioluminescent capabilities. Through a quorum-sensing mechanism, cells of *V. fischeri* can produce light. Presumably, this light production aids its host by camouflaging the host from predators during moonlit nights.

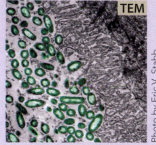

Photo by Eric V. Stabb

The bacterium *V. fischeri* has two distinct lifestyles. It can exist in a free-living state in the marine environment, or as a symbiont of the Hawaiian bobtail squid, *Euprymna scolopes*, which likely uses it for camouflage **(Figure 11.19)**. Only within the light organ of the squid do *V. fischeri* cells emit light, in an energy-intensive set of reactions involving the enzyme luciferase. How does the *V. fischeri* regulate the production of luciferase? Investigations of this question led to the discovery of bacterial quorum sensing.

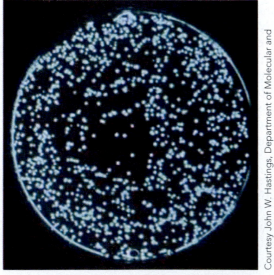

Courtesy John W. Hastings, Department of Molecular and Cellular Biology, Harvard University

A. Luminescent *V. fischeri* colonies

Image taken by C. Frazee; provided by Margarent McFall-Ngai; University of Wisconsin, Madison.

B. Hawaiian bobtail squid

Figure 11.19. Bioluminescence Some bacteria can use chemical energy to produce light. **A.** *Vibrio fischeri* colonies on an agar plate photographed in the dark demonstrate luminescence from the bacteria. **B.** The squid contains a light organ that luminesces after it has been colonized by bacteria.

Luminescence can be induced in free-living *V. fischeri* cells. However, this production of light only occurs when the cultures have been grown to high cell densities, roughly mimicking the population density within the squid light organ. Growth of a colony on solid agar fulfills this requirement. The cells produce a specific molecule, an **N-acyl-homoserine lactone (AHL)**, that acts as an autoinducer based on its ability to induce luminescence (**Figure 11.20**). The LuxI protein, sometimes called autoinducer synthase, catalyzes synthesis of this molecule. Cells living at low cell density also produce the autoinducer, but the molecule diffuses away from the cell as soon as it is produced and never reaches sufficiently high intracellular concentration to induce luminescence. Addition of sufficient amounts of autoinducer to low cell-density cultures results in luminescence.

How do cells detect AHL? A transcriptional activator protein called the "LuxR regulator" interacts with the autoinducer

Length and chemical modification to side chain generates a large family of AHLs.

Figure 11.20. N-acyl-homoserine lactone (AHL) Versions of this molecule act as autoinducers in a number of bioluminescent organisms.

molecule once the molecule reaches a critical threshold concentration and enters the cell. This complex then binds to the *lux* regulatory site on the DNA, called the "*lux* box," situated upstream of the *lux* operon, resulting in activation of the expression of the *lux* structural genes (**Figure 11.21**). These genes

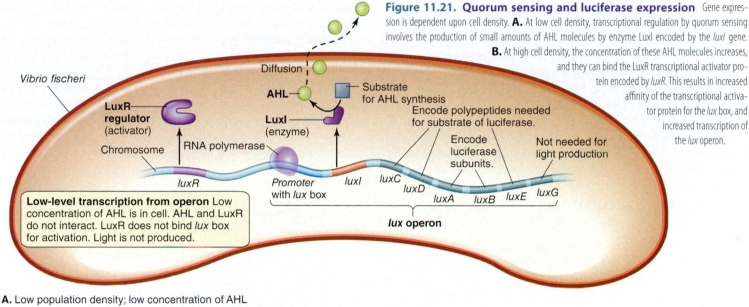

Figure 11.21. Quorum sensing and luciferase expression Gene expression is dependent upon cell density. **A.** At low cell density, transcriptional regulation by quorum sensing involves the production of small amounts of AHL molecules by enzyme LuxI encoded by the *luxI* gene. **B.** At high cell density, the concentration of these AHL molecules increases, and they can bind the LuxR transcriptional activator protein encoded by *luxR*. This results in increased affinity of the transcriptional activator protein for the *lux* box, and increased transcription of the *lux* operon.

Low-level transcription from operon Low concentration of AHL is in cell. AHL and LuxR do not interact. LuxR does not bind *lux* box for activation. Light is not produced.

A. Low population density; low concentration of AHL

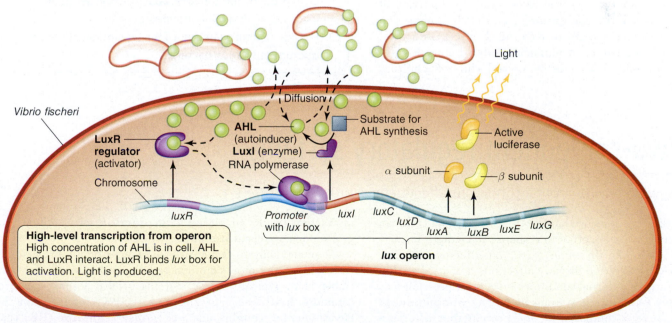

High-level transcription from operon High concentration of AHL is in cell. AHL and LuxR interact. LuxR binds *lux* box for activation. Light is produced.

B. High population density; high concentration of AHL

include *luxA* and *luxB*, which encode the luciferase enzyme that catalyzes the production of light. The operon also includes the *luxI* gene, which encodes LuxI, the enzyme that catalyzes the synthesis of AHL. The system is thus reinforced under high cell density, resulting in amplified synthesis of the autoinducer, which then results in yet higher gene expression and consequently increased light production.

AHL molecules seem to be used as signaling molecules most frequently in Gram-negative bacteria. Gram-positive bacteria, in contrast, utilize small peptides for quorum sensing. Like the AHL molecules we just examined, these peptides are transported out of the cell, into the intercellular milieu. There, they can bind to specific cell surface receptors, triggering a cell-signaling pathway that affects gene expression.

Now that we have a general idea of how this system works, let's ask another question. How does this quorum-sensing mechanism benefit the organisms involved? Within the light organ, the squid provides the bacteria with nutrients. The bacteria require considerable energy and reducing power for the luciferase reaction, and for production of the aldehyde substrate for luciferase. In return, the squid's survivability and therefore reproductive success increases because the light produced by *V. fischeri* provides camouflage from predators that cannot detect the illuminated squid from below against a background of the filtered light above. The ultimate advantage to the bacterium remains unclear. The symbiosis may allow the bacteria to grow to very high density followed by dispersal of large numbers of cells back into the water. This question is under investigation by several research groups around the world.

Widespread occurrence of quorum sensing

Quorum-sensing systems have been discovered in a broad range of microbes and have been found to control a number of different processes. These processes include motility, conjugation, biofilm formation, secondary metabolite production, and pathogenesis. Many of these events involve the colonization or infection of multicellular organisms. For example, one *Vibrio cholerae* cell cannot produce enough toxin to have much effect on a host. In the presence of a sufficient number of other *V. cholerae* cells, however, toxin production might be a worthwhile expenditure

of energy and resources. A larger dose of toxin could enable the pathogen to avoid destruction by the host's immune system.

CONNECTIONS **Section 21.1** describes how successful infection often requires the coordinated production of virulence factors in order to overcome host immune systems. Genes for the production of such virulence factors are expressed in response to environmental conditions consistent with the presence of the pathogen within a suitably susceptible host.

Some organisms contain multiple quorum-sensing systems, which use different signal molecules that have different sets of target genes. While these systems most commonly regulate gene expression within genetically identical populations, these systems may also affect gene expression in other species. In the formation of biofilms, for example, different types of microorganisms might produce similar autoinducers that stimulate one another. Molecules synthesized by one type of organism could also disrupt a quorum-sensing pathway in another type of organism. These interactions could play an important role in competition between populations, or in host resistance against infection by pathogens.

CONNECTIONS As we will see in **Section 24.2**, resistance to today's antimicrobial drugs represents a very serious problem. Quorum-sensing systems involved in some host/pathogen interactions provide the potential for the development of novel antimicrobial medications based on disrupting the quorum-sensing systems of infecting pathogens.

11.5 Fact Check

1. What is quorum sensing, and how are autoinducer molecules involved?
2. In *Vibrio fischeri*, describe how the autoinducer, the LuxR protein, and the *luxA*, *luxB*, and *luxI* genes are involved in producing luminescence.
3. Provide several examples of the advantages/uses of quorum sensing in bacteria.

11.6 Two-component regulatory systems

How do environmental conditions affect gene expression?

In the preceding sections, we have seen how bacterial cells can regulate gene expression, often in response to changes in their external environment. In some cases, an effector molecule binds to DNA or RNA and directly affects its activity. In other cases, the effects of external conditions on gene expression may be less obvious. Often, these systems involve a **two-component regulatory system** composed of one protein that acts as a sensor and another protein that regulates transcription. Their

interaction results in **signal transduction**, or a cellular response to an external stimulus.

Genome analyses reveal two-component regulatory systems to be among the most common regulatory systems in bacteria **(Table 11.4)**. Typically, these systems involve a sensor kinase, commonly **histidine protein kinase (HPK)**, for detecting the environmental stimulus, and a **response regulator (RR)** that regulates transcription **(Figure 11.22)**. Kinases are enzymes that

TABLE 11.4 Examples of two-component regulatory systems

Sensor	Regulator	Function	Organism
VirA	VirG	Tumor formation	*Agrobacterium tumefaciens*
DctB	DctD	Dicarboxylic acid transport	*Sinorhizobium meliloti*
PhoR	PhoB	Phosphate detection	Several
FixL	FixJ	Oxygen detection	*Sinorhizobium meliloti*
EnvZ	OmpR	Osmolarity	*Escherichia coli*
DegS	DegU	Degradative enzymes	*Bacillus subtilis*
ArcB	ArcA	Aerobic respiration control	*Escherichia coli*
RcsC	RcsB	Expolysaccharide synthesis	*Erwinia amylovora*
PilS	PilR	Pilin expression	*Pseudomonas aeruginosa*
PhoQ	PhoP	Virulence	*Salmonella*

phosphorylate other molecules, usually by adding phosphate from ATP. In many cases, the target genes whose expression is regulated by these systems have not been determined, and in most instances the environmental stimulus is also unknown.

A prime example of a two-component regulatory system involves the virulence of the plant tumor-inducing bacterium *Agrobacterium tumefaciens*. This remarkable bacterium functions as a natural genetic engineer, transferring genes from a plasmid, called the **Ti plasmid** (from tumor inducing), into the nucleus of a plant cell. Once these genes have been transferred to the nucleus, they are expressed, and the resulting gene products cause the formation of a tumor by altering the balance of hormones in the plant. *Agrobacterium*-induced plant tumors were once studied as a possible model for human cancer, but it soon became apparent that the mechanisms behind the two types of proliferative growth differed significantly.

Ground-breaking discoveries made in the laboratories of Eugene Nester, Mary-Dell Chilton, and Jeff Schell in the 1970s revealed that the plant tumor formed due to the transfer of a segment of DNA, called "T-DNA" (for transferred DNA), from the Ti plasmid in *Agrobacterium* into the plant cell nucleus. This transfer led to the expression of the T-DNA genes encoding enzymes that direct the production of plant hormones. This system has been extensively developed for the genetic engineering of plants. The system can be manipulated to deliver any gene into the nucleus of a susceptible plant. We now know that the DNA is transferred between the bacterial and plant cells in a manner very similar to bacterial conjugation.

CONNECTIONS In **Section 12.5**, we discuss how scientists have used *Agrobacterium* to introduce genetic material into plants. This technique has been used for improving varieties of crop plants and studying fundamental biological mechanisms.

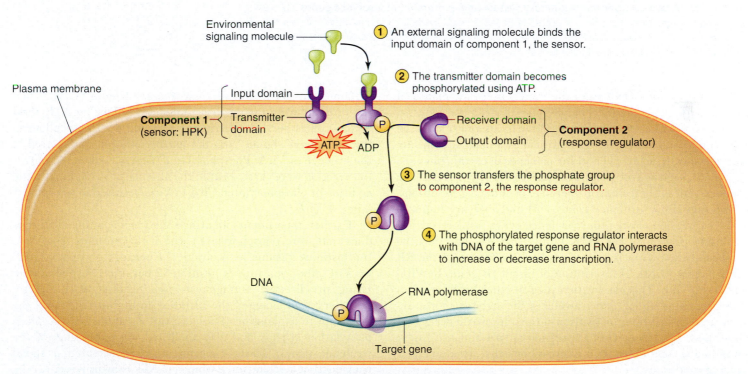

Figure 11.22. Process Diagram: Using an environmental signal to alter gene expression Two-component regulatory systems are composed of two proteins: a sensor histidine protein kinase (HPK), and a response regulator (RR). Upon sensing the input signal, the sensor becomes phosphorylated, and then phosphorylates the response regulator. Phosphorylation of the response regulator results in a change in its output domain, which may activate or repress transcription of specific target genes.

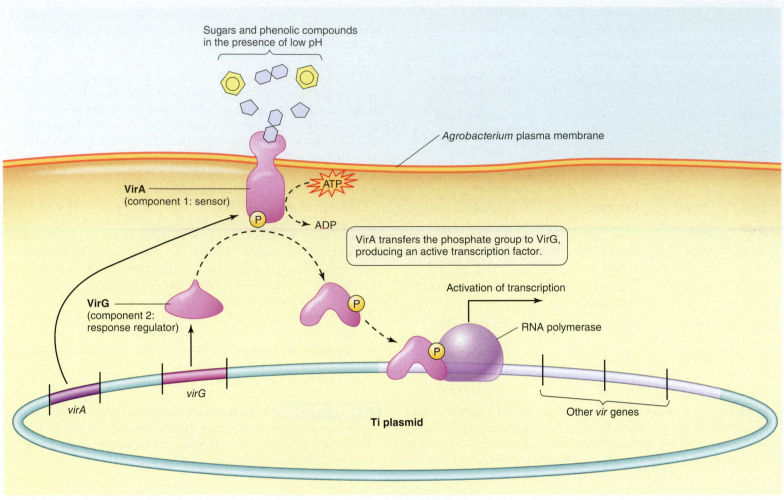

Figure 11.23. A two-component regulatory system and expression of virulence genes The regulation of the *Agrobacterium* signal transduction pathway is mediated by the VirA/VirG signal transduction system. On detection of signal molecules such as sugars and phenolic compounds from the injured plant, as well as acidic pH, the VirA protein is phosphorylated. The phosphorylated VirA in turn phosphorylates VirG, which then activates transcription of the *vir* genes.

The transfer of T-DNA from the bacterium to the plant cell nucleus requires the products of a number of virulence genes on the Ti plasmid. These genes, called *vir* genes, are only expressed under conditions similar to those present at a plant wound site. More specifically, acidic pH and the presence of certain phenolic compounds, such as acetosyringone, lead to the expression of the virulence genes. These are considered to be inducing conditions.

Two regulatory virulence genes, *virA* and *virG*, are required for the expression of the other virulence genes. VirA is a transmembrane protein of the HPK family that extends into the periplasm, while VirG is a transcriptional activator of the RR family **(Figure 11.23)**. Transcription of target *vir* genes by VirG depends on VirA. This observation suggests that VirA perceives an extracellular signal, and transduces the signal to VirG, resulting in transcriptional activation of the other virulence genes. While there is a low level of expression of *virA* and *virG* under non-inducing conditions, their expression is also elevated under inducing conditions.

To better understand the interactions between these proteins, Shougouang Jin, a graduate student working in the laboratory of Eugene Nester at the University of Washington in the late 1980s, conducted *in vitro* experiments using VirA and VirG proteins purified from *Escherichia coli* **(Figure 11.24)**. When the proteins were mixed with radioactively labeled ATP, VirG became phosphorylated, an event he could detect by following the radioactive signal. VirG did not become phosphorylated in the absence of VirA. Phosphorylated VirA was detected only if the reaction were carried out in the absence of VirG. Finally, the addition of phosphorylated VirA to VirG resulted in phosphorylation of VirG.

Further experiments identified the specific amino acid residues that were phosphorylated within the proteins. For VirA, the histidine residue at position 474 became phosphorylated; for VirG, the aspartate at position 52 became phosphorylated. Using site-directed mutagenesis (see Section 12.2), the researchers then changed the histidine to a glutamine, and the aspartate to asparagine. These mutations abolished the phosphorylation *in vitro*, and also abolished the expression of the virulence genes. These experiments demonstrated a signal transduction mechanism in which (1) ATP is hydrolyzed, (2) the phosphate is transferred to the histidine residue, and (3) the phosphate is transferred to the aspartate residue. This same histidine-aspartate phosphotransfer mechanism occurs in many two-component regulatory systems.

Observation: When protein VirA and VirG are mixed with radiolabeled ATP ([^{32}P]ATP), VirG becomes phosphorylated.	VirA + VirG + [^{32}P]ATP $\longrightarrow$ VirG-^{32}P	**Figure 11.24. Sequence of phosphorylation in the VirA/VirG signal pathway** *In vitro* phosphorylation experiments were performed using purified VirA and VirG proteins and radioactively labeled ATP. VirA could be phosphorylated directly by ATP, but VirG could not. VirG was phosphorylated only in the presence of either VirA and ATP, or previously phosphorylated VirA (VirA-P), indicating that phosphotransfer occurred between VirA-P and VirG.
Hypothesis: Phosphotransfer occurs between VirA and VirG.		
Experiment: Add paired combinations of ATP, unphosphorylated and phosphorylated VirA and VirG and observe the transfer of radioactive phosphate(^{32}P)	? + ? $\longrightarrow$ VirG-^{32}P	
Results: VirA-P is only produced in the presence of ATP. VirG-P is only produced in the presence of VirA-P.	VirG + [^{32}P]ATP $\longrightarrow$ VirG VirA + [^{32}P]ATP $\longrightarrow$ VirG-^{32}P	
Conclusion: VirA is first phosphorylated using ATP. The phosphate is then transferred from VirA to VirG.	[^{32}P]ATP + VirG $\longrightarrow$ VirG-^{32}P	

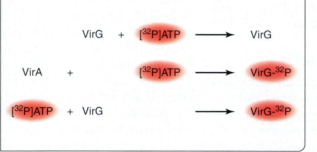

The HPK protein can be subdivided into a variable N-terminal sensor module, and a conserved C-terminal transmitter module, while the RR protein can be subdivided into a conserved N-terminal receiver module, and a variable C-terminal output module. Different two-component regulatory systems differ in the environmental stimulus that causes the phosphorylation of the sensor, and the target of the response regulator. Despite the shared signal transduction mechanism, the correct operation of these systems is dependent on a sufficiently high level of specificity. Phosphotransfer, in other words, occurs between certain pairs of HPKs and RRs, but not between other pairs. The basis of this phosphotransfer specificity remains a subject of much investigation.

11.6 Fact Check

1. Describe the two types of protein components involved in a two-component regulatory system. What are their roles?
2. What is the role of the VirA and VirG proteins in *Agrobacterium* cells?
3. Explain how phosphotransfer mechanisms play a role in VirA and VirG function.

11.7 Chemotaxis

How can bacterial cells regulate their behavior?

One of the most thrilling experiences for the beginning microbiologist involves the use of phase contrast microscopy to observe live bacterial cells. As some cells dart across the field of view, other cells move at a leisurely pace, or not at all. Why are some bacterial cells motile, while others are not? Motility must confer a selective advantage, otherwise it would not be worth the effort to expend the required energy. Chemotaxis, introduced in Sections 2.5 and 4.2, involves the movement of motile bacteria toward favorable chemicals, such as preferred nutrients, or away from detrimental chemicals, such as toxins or poisons. A two-component regulatory system regulates chemotaxis in

bacteria. Unlike most of the regulatory systems that we have explored in this chapter, though, this system does not involve transcriptional or translational regulation. Regulation of chemotaxis involves protein switches.

The mechanism of chemotaxis

Once again, the regulation of chemotaxis is best understood in *Escherichia coli*. Recall from Section 2.5 that *E. coli* cells have several peritrichous flagella that form a bundle during counterclockwise (CCW) rotation, propelling the cells forward. During clockwise

(CW) rotation, the flagella separate and rotate independently. In the absence of attractant or repellent, the cell alternates between CCW and CW rotation of the flagella. This switch occurs on average about once per second. CCW rotation results in smooth swimming or a run. CW rotation causes the cell to tumble end over end resulting in a change of direction. A combination of runs and tumbles results in a random walk (see Figure 2.26).

In the presence of attractant or repellent, the ratio of CCW rotation to CW rotation changes. If the attractant concentration is increasing or the repellent concentration is decreasing, then CCW rotation is favored, resulting in smooth swimming (a run). If the attractant concentration is decreasing or the repellent concentration is increasing, then CW rotation is favored, leading to tumbling and change of direction. A chemical gradient can influence the duration of runs, resulting in net movement up or down the concentration gradient.

How do bacteria detect changes in chemical gradients? To understand the molecular mechanism of chemotaxis, researchers again began with the analysis of mutants. In this case, they analyzed mutants in which chemotaxis was disrupted.

Study of chemotaxis using mutants

The genes involved in chemotaxis were initially identified through mutant analysis. Chemotaxis mutants can be isolated using a capillary tube filled with semisolid agar containing a higher concentration of attractant than the surrounding environment. A normal cell that can perform chemotaxis will migrate into the tube (Figure 11.25). A mutant that cannot change the direction of flagellar rotation in response to an attractant concentration will not be able to move up the tube. Such mutants typically exhibit a bias towards either CCW or CW rotation. Mutations in the *cheA*, *cheY*, *cheW*, and *cheR* genes result in CCW rotation, while mutations in the *cheB* and *cheZ* genes cause CW rotation.

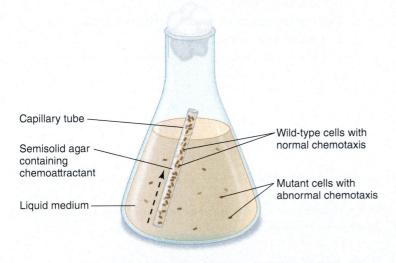

Figure 11.25. Isolating chemotactic mutants A capillary tube filled with semisolid agar and a higher concentration of attractant chemical than found in the surrounding medium is used to screen for mutant bacteria that are deficient in chemotactic ability. In this example, the wild-type strain is attracted to the compound, and it tends to move up the tube to higher concentrations of chemical, while the chemotactic mutant does not exhibit this behavior.

Labels on figure:
- Capillary tube
- Semisolid agar containing chemoattractant
- Liquid medium
- Wild-type cells with normal chemotaxis
- Mutant cells with abnormal chemotaxis

TABLE 11.5 Functions of chemotaxis proteins

Methyl-accepting chemotaxis proteins (MCPs) → sensory proteins or transducers; can be methylated by CheR

CheR → methylates MCPs

CheA → sensor kinase

CheY → response regulator: controls direction of flagellar rotation

CheB → response regulator, demethylates MCP

CheZ → dephosphorylates CheY-P

CheW → involved in transduction of signal from MCP to CheA

The corresponding Che proteins are involved in a series of signal transduction events that regulate chemotaxis. Their functions are summarized in (Table 11.5). The central proteins CheA and CheY are conserved in all chemotactic bacteria. CheA functions as the sensor kinase in the system and is phosphorylated to CheA-P at a specific histidine residue in response to the presence or absence of a particular attractant or repellent. CheA-P then phosphorylates CheY, the response regulator, at a specific aspartate residue to CheY-P, which interacts with the flagellar motor to determine the direction of rotation (Figure 11.26). The phosphorylation circuit is a histidine-aspartate phosphorelay, typical of two-component regulatory systems.

The study of chemotaxis is an excellent example of how combining the approaches of genetics, molecular biology, and biochemistry can lead to a deeper understanding of mechanisms than would be possible using any single approach alone. The generation of mutants resulted in the discovery of the chemotaxis genes. Once these genes were cloned and sequenced, scientists realized that the chemotaxis systems shared much in common with the two-component regulatory systems, even though regulation was not carried out at the level of transcription. Using site-directed mutagenesis, scientists identified key functional motifs in the chemotaxis proteins that were involved in the protein switches. *In vitro* biochemical analysis with purified chemotaxis proteins unraveled the phosphorylation and methylation circuits that are described below.

Transmembrane sensory proteins, called **methyl-accepting chemotaxis proteins (MCPs)**, sense the presence of attractant or repellent molecules. While mutations in the *che* genes affect chemotaxis in response to all chemical signals, mutations that affect the synthesis of MCPs abolish chemotaxis only toward a specific chemical signal. The MCPs interact with the cytoplasmic protein CheW, and the resulting complex modulates the level of autophosphorylation of CheA. Attractants decrease the level of phosphorylation, while repellents increase the level of phosphorylation (see Figure 11.26). The phosphorylated CheA (CheA-P) transfers the phosphoryl group to the protein CheY, resulting in CheY-P. CheY-P causes clockwise rotation and tumbling until CheZ removes the phosphoryl group.

Another level of regulation of chemotaxis, called "adaptation," involves methylation of the MCP proteins, hence their name. Adaptation determines the system's sensitivity to attractants and repellents and prevents it from becoming saturated, which would prevent it from responding to changes in chemical concentration. The protein CheR can methylate the MCPs, decreasing the ability of the MCP to respond to attractant (Figure 11.27). Another protein, CheB, is phosphorylated by CheA-P. The resulting CheB-P molecules demethylate the MCPs, increasing their sensitivity to attractant. As an example, the Tsr MCP can respond to a few serine molecules when fully demethylated, but when highly methylated, a response requires thousands of serine molecules. Thus, in the presence of a continually high level of attractant, resulting in lower levels of CheA-P, CheY-P, and CheB-P, the extent of methylation will increase because of a lack of CheB-P mediated demethylation. Highly methylated MCPs only respond to very high

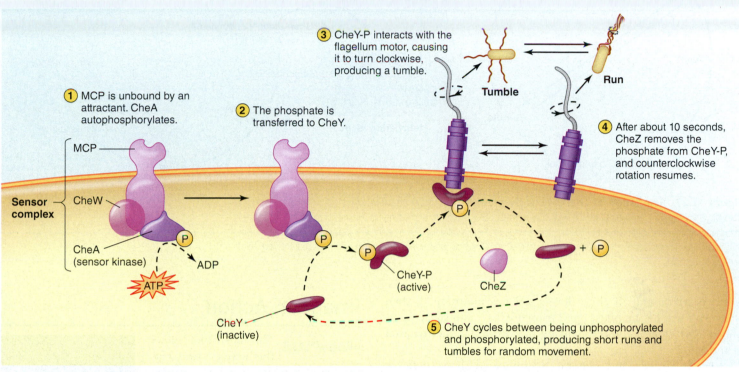

1. MCP is unbound by an attractant. CheA autophosphorylates.

2. The phosphate is transferred to CheY.

3. CheY-P interacts with the flagellum motor, causing it to turn clockwise, producing a tumble.

4. After about 10 seconds, CheZ removes the phosphate from CheY-P, and counterclockwise rotation resumes.

5. CheY cycles between being unphosphorylated and phosphorylated, producing short runs and tumbles for random movement.

MCP
Sensor complex
CheW
CheA (sensor kinase)
ATP
ADP
CheY (inactive)
CheY-P (active)
CheZ
Tumble
Run

A. Chemotaxis in the absence of attractant

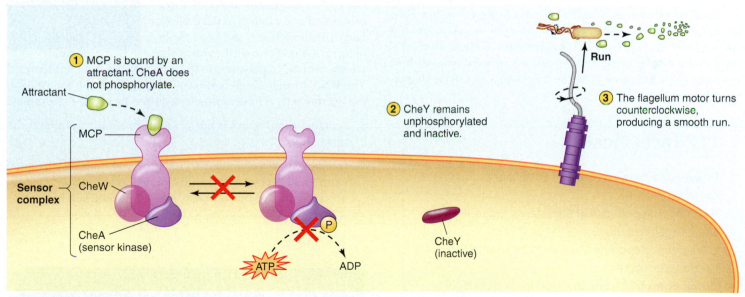

1. MCP is bound by an attractant. CheA does not phosphorylate.

2. CheY remains unphosphorylated and inactive.

3. The flagellum motor turns counterclockwise, producing a smooth run.

Attractant
MCP
Sensor complex
CheW
CheA (sensor kinase)
ATP
ADP
CheY (inactive)
Run

B. Chemotaxis in the presence of attractant

Figure 11.26. Process Diagram: Chemotaxis Phosphorylation patterns can affect chemotactic behavior. **A.** In the absence of attractant, CheW promotes phosphorylation of CheA, which in turn phosphorylates CheY. The resulting CheY-P interacts with the flagellum, signaling it to rotate clockwise (CW), causing a tumble (*left*). Removal of the phosphate from CheY-P by CheZ disrupts interaction with the flagella, resulting in counterclockwise rotation, and the cell runs. **B.** In the presence of attractant, phosphorylation of CheA is inhibited, so CheY is not phosphorylated either. The absence of CheY-P results in counterclockwise rotation (CCW) of the flagella, causing a longer run (*right*).

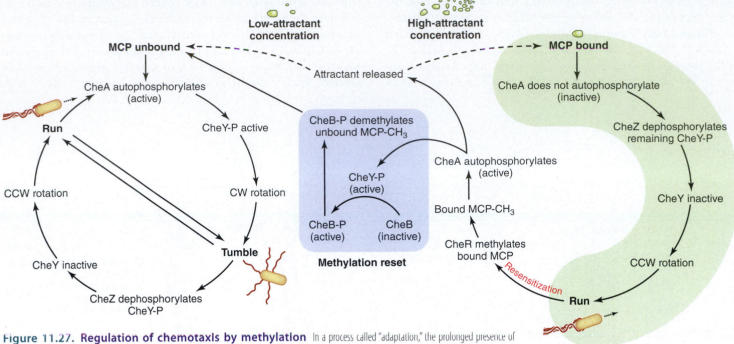

Figure 11.27. Regulation of chemotaxis by methylation In a process called "adaptation," the prolonged presence of attractant results in increased methylation of MCP to MCP-CH$_3$, caused by CheR. This methylation reduces the sensitivity of the MCP to the attractant. Resetting allows it to exit from the run as the signaling proteins are again phosphorylated (*right*). If the concentration of attractant decreases, increased phosphorylation of CheY results in increased phosphorylation of CheB. The CheB-P demethylates MCP-CH$_3$ (*center*), so that it is once again available to be bound by attractant.

levels of attractant. If these levels are not maintained, then the result will be the phosphorylation of CheA and subsequent phosphorylation of CheB, resulting in eventual demethylation of the MCP.

Recent studies and genome sequence analyses indicate that the basic mechanism of chemotaxis studied in *E. coli* is conserved across bacteria and archaeons, as similar genes are found in many organisms. It is becoming clear, however, that *E. coli* possesses one of the simpler systems. As scientists address, through experiments, the diversity in chemotaxis systems revealed by genome sequences, we will understand how the remarkable behavior of these organisms has been shaped through evolution.

11.7 Fact Check

1. Differentiate between CW and CCW flagella rotation in bacteria.
2. How does the presence of attractants and repellents affect flagella rotation?
3. Explain the roles of the CheA and CheY protein in chemotactic bacteria.
4. How is phosphorylation involved in the activity of CheA and CheY?
5. What is the role of methyl-accepting chemotaxis proteins (MCPs)?
6. How do MCPs work with the Che proteins to regulate chemotaxis?

Image in Action

WileyPLUS This illustration shows the active transcription and translation of an operon that codes for three distinct polypeptides. The *lac* operon encodes three proteins: LacZ (shown in dark blue), LacY (medium blue), and LacA (light blue). DNA is shown at the top of the screen, with RNA polymerase (purple) moving from left to right. The RNA strand, which is shown twisting from the top right to the bottom left, is undergoing translation by several ribosomes.

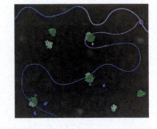

1. Imagine that lactose is not present. In what ways would the image differ?

2. Imagine that you wanted to monitor and quantify the gene expression represented in this image. What lactose analogs are available for this purpose, and how would you use them?

The Rest of the Story

In this chapter, we have seen several well-known examples of gene regulatory systems, many of them in the model organism *Escherichia coli*. These systems operate at several different levels and ensure that cells respond appropriately to environmental and nutritional conditions. Microbial metabolism, for example, is a well-regulated and complex process. A microbe's metabolism is tightly controlled to ensure that enzymes and proteins are only made when they are needed and in the amount they are needed, as the available nutrients and surrounding environmental conditions change. From an evolutionary point of view, an overproduction would be a waste of energy and could impact an organism's ability to compete and survive. Genetic regulation ensures that such wastes and inefficiencies are minimized.

The science of metabolic engineering involves manipulating regulatory pathways to override the natural genetic regulatory and efficiency controls so that microbes overproduce desired products. Deregulation of a particular biosynthetic pathway can be accomplished by overexpressing positive regulators or by inactivating repressors. A classic, early example involved optimizing production of the drug penicillin. Researchers carried out random mutagenesis on the organism *Penicillium chrysogenum* using radiation and chemical agents in the laboratory. They then screened for "deregulated" mutants with the ability to overproduce penicillin. Some of these mutations affected genes related to penicillin biosynthesis; mutants were screened for higher output than parental cells.

As the knowledge of gene regulation and biosynthetic pathways has grown, more rational approaches are also being used to engineer a variety of commercially desirable compounds including lysine and carotenoids as food supplements. These rational approaches involve specific, site-directed mutation and manipulation of the biosynthesis genes.

Indeed, Monod's maxim, "What is true for *E. coli* is true for the elephant," remains accurate today. Early genetic regulation studies in *E. coli* now apply beyond the realms of the "elephant" and impact nearly all areas of science. What has been accomplished in the biosciences, medicine, and even the field of metabolic engineering arguably would not have been possible without having studied this unassuming bacterium, *E. coli*.

Summary

Section 11.1: How are gene expression and enzyme activity controlled?
Many mechanisms for controlling enzymatic activity have been discovered.

- Under a given set of conditions, a cell only requires a small subset of all possible gene products.
- **Constitutive expression** refers to genes that always are expressed, while **inducible expression** refers to genes that are expressed only under certain conditions.
- Gene expression can be regulated at several different levels, from transcription initiation to protein activity.
- Protein activity can be regulated by **allosteric inhibition** or **covalent modification**.

Section 11.2: How do regulatory proteins control transcription?
The initiation of transcription can be controlled in various ways.

- The **operon** is the unit of transcription, whose expression is under the control of discrete regulatory elements including promoters and **operators**.
- In **negative control**, a **repressor** blocks transcription. **Effectors** can bind to repressors and modulate their activity. These effectors can function as **inducers** or **corepressors**.
- In **positive control**, an **activator molecule** may bind to an **activator binding site**, leading to increased transcription. In the *lac* operon, **activator protein (cAMP receptor protein**, or **CRP)** serves as an activator.
- In the presence of glucose and lactose, *Escherichia coli* exhibits a **diauxic growth curve**.

Section 11.3: How does a single environmental stimulus control multiple operons?
Larger sets of genes can also be coordinately regulated.

- **Regulons** are coordinately regulated genes that respond to the same regulatory systems.

- **Catabolite repression** and the **SOS response** represent two well-studied regulon systems.
- Members of the SOS regulon initially were identified using *lacZ* fusions as **reporter genes**.
- Lactose analogs have also been instrumental in studies of regulons and gene expression.
- Alternative sigma factors bind to specialized promoters, resulting in transcription initiation.

Section 11.4: How can mRNA be controlled?
Small, non-translated RNA molecules also regulate gene expression.

- Small non-coding RNAs, or **sRNA**, can control gene expression at transcription or translation.
- **Antisense RNA** regulates translation by complementary base-pairing with mRNA.
- **Attenuation** involves changes in the secondary structure of mRNA.
- **Riboswitches** are a system of mRNA and effector molecules that can increase or decrease the level of gene expression.

Section 11.5: How do bacteria communicate with their neighbors?
Bacteria can regulate gene expression based on their density.

- A chemical signaling system called **quorum sensing** allows bacteria to communicate population status through the detection of chemical signals.
- Studies on the luminescence system in the squid symbiont *Vibrio fischeri* provided much of the initial knowledge about quorum-sensing systems.
- The signal for the quorum-sensing system is the **autoinducer**. In *V. fischeri*, **N-acyl-homoserine lactone (AHL)** acts as the autoinducer.
- Quorum sensing may be important in the infection of multicellular organisms, the formation of biofilms, and other processes that involve cooperation among bacterial cells.

Section 11.6: How do environmental conditions affect gene expression?

Bacteria can also regulate gene expression in response to environmental signals.

- The activity of **two-component regulatory systems** usually results in **signal transduction**, facilitating cellular response to external stimulus.

- Two-component regulatory systems, composed of sensor **histidine protein kinases (HPKs)** and **response regulators (RRs)**, are among the most abundant regulatory systems.

- In *Agrobacterium tumefaciens*, environmental signals result in the transfer of the **Ti plasmid** from the bacterium to the plant cell nucleus.

Section 11.7: How can bacterial cells regulate their behavior?

Complex bacterial behaviors can be modulated by shifts in protein activity.

- Chemotaxis is controlled at the level of protein activity, rather than by changes in gene expression.

- Chemotactic bacteria sense chemical gradients by detecting changes in concentration over time.

- Key genetic studies that contributed to the understanding of the mechanism of chemotaxis involved the isolation of mutations in the *che* genes, which remove the chemotactic ability of the cell but do not affect motility.

- Chemicals are detected by specific **methyl-accepting chemotaxis proteins (MCPs)**, which ultimately affect the direction of flagellar rotation and thus whether the cell runs or tumbles.

● Application Questions

1. You are working as an intern for a researcher who studies "housekeeping" genes in bacteria, particularly those genes that encode the enzymes required for cell maintenance and energy production. Explain the type of gene expression that most likely occurs in these genes.

2. Certain bacteria have the ability to produce capsules that allow them to hide from a host's defenses. Before bacteria were introduced into a host mouse in the laboratory, the capsule genes were not expressed. Within minutes of being introduced into the mouse's bloodstream, however, the genes were actively expressed. Explain the type of gene expression that most likely is involved in this response.

3. You are performing an experiment in which you monitor expression levels of the enzyme β-galactosidase in *Escherichia coli*. At the beginning of the experiment, the bacteria have large amounts of lactose, but at the end of the experiment, no lactose remains. Describe what happens to the β-galactosidase levels during the course of the experiment.

4. Compounds called furanones show promise as antibacterial drugs in humans. The compounds decrease the virulence or harmfulness of some bacteria by inhibiting their quorum-sensing ability. This inhibition of quorum sensing makes it easier for the bacteria to be cleared from the body. Based on your knowledge of quorum sensing, how could furanones make bacteria less harmful?

5. While watching a nature documentary with a friend, you learn of a Hawaiian squid species with the ability to emit light because a bacterial species that undergoes bioluminescence colonizes it. These bacteria can also be found free-floating, but they only luminesce when in the light organ of the squid. Explain the behavior of the bacteria to your friend.

6. Imagine that you discover a new two-component regulatory system. You know that certain aspects of these systems tend to be conserved among different bacterial species. What would you expect to see in your system that also has been described in other systems? What features might be unique to your system?

7. Many bacteria are resistant to the antibiotic ciprofloxacin (Cipro). At the molecular level, ciprofloxacin inhibits DNA gyrase, leading to dsDNA breaks and the activation of the SOS response. Explain how SOS induction could cause drug resistance. What role would the LexA and RecA proteins play in this response?

8. In your study of *E. coli* chemotaxis, you add increasing levels of an experimental repellent. If the repellent is truly effective, what will happen to the Che proteins? Will the addition of the repellent result in running or tumbling of the *E. coli*? How will that action affect its direction of movement?

9. You observe a mutant that is deficient in the CheZ protein, carrying a mutant form of the *cheZ* gene. What specific effect will this mutation have on chemotaxis regulation, and how will this mutation affect the cell's chemotactic response? Explain.

10. You read about a protein that remains attached to an operator site most of the time. When a certain amino acid is present, the protein changes shape and detaches from the operator. What type of gene regulatory system does this behavior describe? Explain.

11. You are assigned the task of describing the chemotactic response of some new *che* gene mutants in the presence of an attractant chemical. You elect to use the capillary tube method described in Section 11.7.
   a. Describe your experimental procedure using a capillary tube and attractant.
   b. What result would be observed in a bacterial cell with normal chemotaxis? What would you conclude?
   c. Imagine that one of the mutant bacteria contains a mutation in the *cheY* gene. What result will you observe in this mutant? What would you conclude?

Suggested Reading

Alon, U., M. G. Surette, N. Barkai, and S. Leibler. 1999. Robustness in bacterial chemotaxis. Nature 397:168–171.

Bassler, B. L., and R. Losick. 2007. Bacterially speaking. Cell 125:237–246.

Chilton, M.-D. 2001. Agrobacterium. A memoir. Plant Physiol 125:9–14.

Gruber, T. M., and C. A. Gross. 2003. Multiple sigma subunits and the partitioning of bacterial transcription space. Annu Rev Microbiol 56:441–466.

Henkin, T. M., and C. Yanofsky. 2002. Regulation by transcription attenuation in bacteria: How RNA provides instructions for transcription termination/antitermination decisions. BioEssays 24:700–707.

Merino, E., and C. Yanofsky. 2005. Transcription attenuation: A highly conserved regulatory strategy used by bacteria. Trends Genet 21:260–264.

Parkinson, J. S., and S. E. Houts. 1982. Isolation and behavior of *Escherichia coli* deletion mutants lacking chemotaxis functions. J Bacteriol 151:106–113.

Wadhams, G. H., and J. P. Armitage. 2004. Making sense of it all: Bacterial chemotaxis. Nature Rev Mol Cell Biol 5:1024–1037.

Walker, G. C. 1984. Mutagenesis and inducible responses to deoxyribonucleic acid damage in *Escherichia coli*. Microbiol Rev 48:60–93.

Waters, L. S., and G. Storz. 2009. Regulatory RNAs in bacteria. Cell 136:615–628.

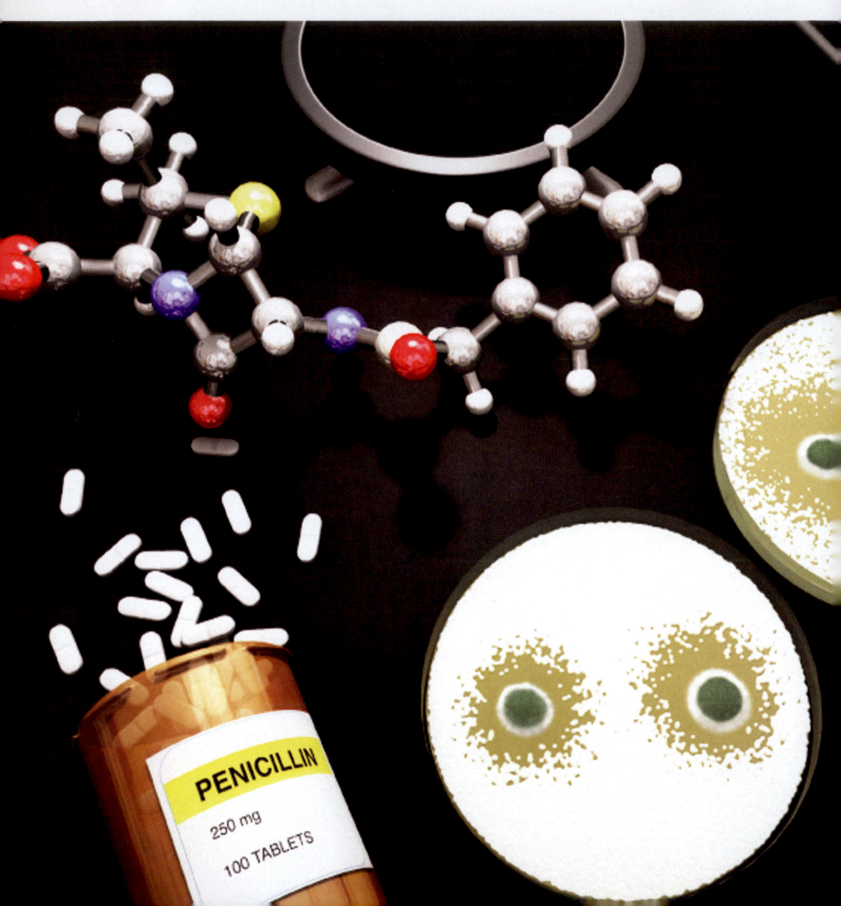

For most of humankind's existence, infectious diseases have been the major cause of mortality. During World War I, many soldiers died not from their battle wounds directly, but from the ensuing infections. Little could be done to treat these patients. Their recovery depended on the ability of their immune systems to clear the infection. While the link between microbes and disease had been made several years earlier and better public sanitation improved the situation somewhat, infectious diseases remained a serious and largely untreatable problem. In September 1928, a chance observation in the laboratory changed all that.

Alexander Fleming, a Scottish bacteriologist working at St. Mary's Hospital in London, noticed that a fungus was inhibiting the growth of *Staphylococcus* on some culture plates. Many scientists would have thought nothing of this observation. Fleming, however, decided to further explore this serendipitous finding. He identified the mold as a *Penicillium*, a filamentous fungus. Culture filtrates of *Penicillium*, which he termed "penicillin," could inhibit the growth of many Gram-positive bacterial pathogens as well as the common sexually transmitted Gram-negative bacterium *Neisseria gonorrhoeae*.

Fleming also demonstrated, using animal tests, that penicillin was not toxic. Despite this finding, the clinical potential of penicillin was not realized until the late 1930s. At that time, a research group at Oxford led by Howard Florey and Ernst Boris Chain scaled up production so that sufficient amounts of the penicillin culture filtrate were available for human clinical trials. The results were so successful that the Allied forces in World War II employed the antibiotic, reducing mortality due to infected wounds on the battlefield. This success spurred massive investment in the development of large-scale production methods, resulting in drastic reductions in cost of production by over 100-fold. In recognition of the discovery and development of penicillin as the first clinically useful antibiotic, Fleming, Florey, and Chain shared the 1945 Nobel Prize for Physiology or Medicine.

The discovery of penicillin and its development as an antibiotic to treat infections clearly was a major breakthrough in the medical field. The impact of this breakthrough, though, is not limited to medicine. Work on penicillin led to the design and development of large vessels that enable large-scale growth of microbial cultures under defined conditions. This advance contributed to an explosion of biotechnology products that have enriched human lives and contributed to the improvement of living conditions worldwide.

CHAPTER NAVIGATOR

As you study the key topics, make sure you review the following elements:

Microbes used for biotechnology may be obtained from existing collections or isolated from nature.

- Figure 12.2: Microbial culture collections
- Perspective 12.1: Bioprospecting: Who owns the microbes?
- Figure 12.4: Types of bioreactors

To improve their usefulness, microbes can be genetically modified in numerous ways.

- 3D Animation: Using molecular biology tools to improve microbial strains
- Microbes in Focus 12.2: *Penicillium chrysogenum*: The mold that started the antibiotic revolution
- Toolbox 12.1: Site-directed mutagenesis
- Figure 12.8: Process Diagram: Directed enzyme evolution
- Figure 12.13: Process Diagram: Design and construction of a synthetic organism
- Mini-Paper: Making a synthetic genome

Red biotech involves the use of biotechnology in the pharmaceutical sector.

- Figure 12.15: Production of recombinant insulin

White biotech involves the use of biotechnology in the industrial sector.

- Figure 12.18: Commercial ethanol production using different feedstocks
- Perspective 12.3: Biofuels: Biodiesel and algae
- Figure 12.22: Structures of bacterial polyhydroxyalkanoates (PHAs)

Green biotech involves the use of biotechnology in the agricultural sector.

- 3D Animation: *Agrobacterium* in agricultural biotechnology
- Toolbox 12.3: Plant transformation using bacteria
- Figure 12.33: Process Diagram: *Bacillus thuringiensis* crystals and mode of action

CONNECTIONS for this chapter:

Using metagenomics to find potential genes for biotechnology applications in unculturable microbes (Section 10.4)

Use of GFP fusion proteins (Section 3.1)

Effects of endotoxin on humans (Section 19.3)

Interactions between plants and soil microbes (Section 17.2)

Introduction

Quite simply, **biotechnology** refers to the use of biological processes or organisms for the production of goods such as antibiotics, or services such as food preservation, wastewater treatment (see Section 16.5), and bioremediation of polluted soils (see Section 15.4). For centuries, humans have relied on biotechnology to support their existence. The rise of agriculture and the domestication of animals were accompanied, perhaps less conspicuously, by the use of microbes for the production or preservation of foods. Beer, wine, bread, pickles, sauerkraut, and fermented milk products like cheese and yogurt all were produced using naturally occurring microbes. Indeed, the yeast *Saccharomyces cerevisiae* has been used for centuries in the production of wine and beer. Not until the mid-nineteenth century, though, did Louis Pasteur demonstrate that alcoholic fermentation was not purely a chemical process but was carried out by living yeast cells. These yeasts are found on the skin of grapes and result in a natural fermentation after the grapes are crushed. While such natural fermentation is still carried out in some traditional wineries, fermentations started by inoculation with known yeasts are now the standard as it results in a more rapid process and consistent product.

Today, as we will see in this chapter, biotechnology affects virtually all aspects of our lives. This phenomenal growth in biotechnology has been possible because of our basic understanding of microbial genetics. As we saw in previous chapters, we now know much about gene regulation, transcription, and translation. Understanding these basic processes has led to the development of powerful molecular biology tools. These tools, in turn, have made possible the widespread use of biotechnology. Basic research into the genetic machinery of simple microbes, in other words, has revolutionized our world.

Current applications of biotechnology can be grouped broadly into **red biotechnology** related to medical applications, **white biotechnology** related to industrial applications, and **green biotechnology** related to the agricultural sector **(Figure 12.1)**. The application of biotechnology to the environment is discussed in the context of bioremediation in Chapter 15. Microbes participate in all sectors of biotechnology, and as these represent the major activities of current economies, microbial biotechnology is integral to the workings of modern societies. In this chapter, we will consider each of the types of biotechnology and some of the ongoing research in microbiology that is leading to new products and applications. These include replacements for many of the products that are currently derived from fossil fuels **(Table 12.1)**.

TABLE 12.1 Examples of microbial products

Microbial product	Examples
Food additives	Amino acids, organic acids, fatty acids, vitamins
Solvents	Acetone, butanol, ethanol
Enzymes	Proteases, amylases, cellulose, lipases
Biofuels	Ethanol, methane, hydrogen
Agrochemicals	Feed additives, biopesticides
Whole cells	Baker's yeast, crop inoculants
Fine chemicals	Antibiotics, chemotherapeutic agents, nucleic acids, proteins, enzymes, biochemicals, optically pure chiral molecules

Courtesy Trevor Charles

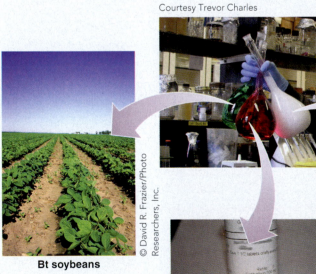

© David R. Frazier/Photo Researchers, Inc.

Bt soybeans

Antibiotics

Courtesy Christine Dupont

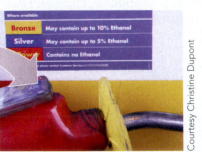

Courtesy Christine Dupont

Ethanol-blended fuel

Figure 12.1. Red, white, and green biotechnology The applications of biotechnology are very diverse, including the agricultural (green, represented by soybeans, an important crop), medical and pharmaceutical (red, represented by antibiotic capsules), and industrial (white, represented by ethanol fuel) sectors.

As we explore biotechnology, we will examine how our understanding of microbial genetics has improved our ability to use microbes for the production of desired products. We will also examine some of the strategies used for the large-scale production of these products, the tools that are employed for improving product efficiency and yield, and the prospects for the development of new products. Along the way, we will address the following questions:

12.1 Microbes for biotechnology

What microbes and genetic materials are available for biotechnology applications?

Today, our agricultural system depends on animals and plants that have been developed slowly over thousands of years from naturally occurring organisms. The corn grown by farmers throughout the United States, for instance, bears little resemblance to maize, its native ancestor. Over centuries, farmers selected variants of maize that produced larger ears of corn, grew better in local environments, or were resistant to diseases. Similarly, the microbes used in biotechnology applications today are modified versions of naturally occurring organisms. These microbes, taken from ecosystems in which they are optimally adapted, most likely are not optimally adapted for biotechnology processes. To be useful in biotechnology processes, microbes must, at the very least, replicate well under standard laboratory conditions. Additionally, naturally occurring strains may not produce high levels of a desired product.

To combat these problems, microbiologists can employ simple selection strategies. Genetic variants with improved characteristics can be isolated, starting from the original strain. These improved strains may replicate more quickly or produce greater amounts of desired product. Sometimes, though, genetic selection alone is not sufficient. Recombinant DNA technology, coupled with a better understanding of basic microbial genetics, biochemistry, and metabolism, can facilitate the construction of more useful microbial strains. Because microbes have short generation times and can be handled easily, the selection of microbes with superior qualities can be achieved within years, rather than decades, as is the case for livestock and crop plants. With powerful recombinant DNA tools at our disposal, we can generate microbes with novel properties in the laboratory.

In this section, we will look at how researchers obtain microbes for use in biotechnology processes. We will then investigate the large-scale use of these microbes.

CONNECTIONS You can review basic recombinant DNA technology and the roles of restriction enzymes and plasmids in **Figure 1.22**.

Sources of microbes

While many potentially useful microbes certainly exist, we must know about them before we can begin to use them. To facilitate the widespread availability of microbial strains, microbiologists throughout the world deposit microbes that have been isolated and characterized in publicly available archives called **culture collections** (**Figure 12.2**). These collections, containing freeze-dried, frozen, or otherwise preserved living samples of the microbial cultures, allow scientists from around the world to obtain these organisms at minimal cost. As a result, experimental findings reported in the scientific literature can be confirmed by independent research, and those findings can be extended through additional studies. Moreover, scientists do not have to always isolate microbial strains directly from the environment. They can instead screen strains from culture collections for properties and products of interest. Such an open source model, to which all scientists have access, and to which they are expected to contribute biological material, is critical for the growth of microbial biotechnology, as it is for all microbial science.

While it is convenient to obtain microbes that have been previously characterized, in many cases well-suited organisms with the appropriate physiological characteristics for a given biotechnology application may not be available. For example, suppose that a researcher wanted to produce methane at very low temperatures. Initially, he or she could see if any psychrophilic methanogens had been isolated and were available in culture collections. Suppose, though, that no such organisms had been described. The

Figure 12.2. Microbial culture collections Microbiology cultures are archived in international culture collections where they may be obtained for study or use in commercial applications. Depending on the organism, they are usually stored in a freeze-dried or frozen state, and must be revived after shipment to the end user.

Courtesy Fisher BioServices

researcher could then search anoxic, cold environments for novel organisms that survive at low temperatures and produce methane. Similarly, one might search in a rotting log, or in the gut of a termite or cow to find an organism that can produce cellulase, an enzyme that degrades cellulose. To find an organism that produces a new antibiotic, one might search in soils where many organisms, particularly species of *Streptomyces* bacteria, are known to produce antimicrobial compounds that help them compete in these rich ecosystems (**Microbes in Focus 12.1**; see Table 24.1). Termed **bioprospecting**, such searches for novel organisms, biological materials, or biological processes in nature has fueled innovations in biotechnology (**Perspective 12.1**).

Using a variety of search strategies, microbiologists can isolate novel microbes from diverse environments and then screen them for specific activities, like the production of methane. While this process may sound quite simple, the success rate of bioprospecting remains somewhat limited, given that only a small fraction of microbes can be cultivated (see Section 6.4). The uncultivated majority of microbial life is a rich, untapped resource of biological material for biotechnology applications that is now being accessed using metagenomics approaches.

CONNECTIONS Discussion of metagenomics in **Section 10.4** considers the application of genome-sequencing technology to determine DNA sequences present in complex microbial communities. By combining DNA sequencing with phenotypic selection and screening of libraries, genes of interest for biotechnology applications are accessible from unculturable microbial communities.

Industrial culture of microbes

In industrial microbiology, **fermentation** refers to any industrial process involving the culture of microorganisms, either aerobic or anaerobic, for the production of desired substances. As we'll see

Microbes in Focus 12.1
STREPTOMYCES: A GOLD MINE OF PRODUCTS

Habitat: Soil, compost

Description: Gram-positive, obligate aerobes producing a fungal-like branching mycelium with hyphae of 0.5–1 μm in width and spores called "conidia"; member of the family Actinomycetes.

Key Features: Like fungi, *Streptomyces* have a complex life cycle. Similar to fungi, they also have a linear chromosome rather than a circular one as in most other bacteria. Metabolically versatile, they degrade and metabolize a wide variety of substrates, producing a host of useful products including anticancer agents and immunosuppressant drugs used to prevent the rejection of transplanted organs and to treat autoimmune diseases. Most antibiotics used today originate from *Streptomyces*. The genome of *Streptomyces coelicolor* was sequenced in 2002, and is one of the largest bacterial genomes ever decoded, at 8.7 Mbp.

Courtesy David Capstick and Marie Elliot / McMaster University

in Section 13.3, metabolic or biochemical fermentation refers to specific catabolic reactions that produce ATP without oxygen. For example, in the production of fermented beverages such as wine and beer, yeast produces ethanol only through biochemical fermentation reactions. Industrial use of the term "fermentation" probably arose from such historical use of biochemical fermentations in biotechnology. Many modern industrial "fermentations"

Perspective 12.1
BIOPROSPECTING: WHO OWNS THE MICROBES?

One could imagine a bioprospecting researcher to be the Indiana Jones of microbiology. Looking for enzymes that can function at low temperatures? Then hunt for microbes in Antarctica. Need an enzyme that remains active under high temperature and high sulfur conditions? Scale an active volcano in Hawaii and collect microbes that live just inches from the molten lava. Just like Indiana Jones scoured the globe searching for valuable antiquities, bioprospectors may scour the globe searching for valuable microbes.

Aficionados of the Indiana Jones movies may remember that Dr. Jones always insisted that the relics he found should be displayed in a museum for everyone to see, and not used for any one person's personal gain. This same basic issue surrounds bioprospecting. Should potentially valuable microbes be made available to everyone, or should they become the property of the person who discovers them?

Bioprospecting can have enormous economic benefits for a company. The discovery of a microbe that produces a potent anticancer compound, for instance, could increase the revenue of a pharmaceutical company manyfold. The question remains: Does the company own that microbe? Many countries say "No." Since the 1993 United Nations Convention on Biological Diversity statement that nations have sovereignty over their biochemical resources, countries increasingly have implemented benefit-sharing laws. Generally, these laws state that money earned by a private company from biological material obtained within the country must be shared with the government.

Probably the best-known and most successful benefit-sharing agreement exists in Costa Rica. In this country, companies must provide the government with a share of the profits derived from biological materials obtained within the country. However, implementation and enforcement of such agreements can be problematic. Also, what about the academic bioprospector who, much like Indiana Jones, searches for novel microorganisms just because they are neat? For him or her, the issue of microbial ownership remains murky.

are carried out in the presence of oxygen, and biochemical fermentation may not even be occurring. For example, large-scale culture of *Escherichia coli* for recombinant protein production is almost always done in the presence of oxygen. *E. coli* will use cellular respiration rather than fermentation under these conditions. Yet in industry, this large-scale culture is still referred to as fermentation. One of the most important challenges in the scale-up from laboratory to industrial culture is the transfer of oxygen and maintenance of adequate mixing at those very large volumes.

CONNECTIONS Fermentation in metabolism is defined as the transfer of electrons from a reduced electron carrier such as NADH to an organic molecule such as pyruvate as shown in **Figure 13.17**. The process of fermentation does not require oxygen, but it is not restricted to anaerobic microorganisms, nor does it require strict anaerobic conditions.

Industrial fermentations take place in large-culture vessels called **bioreactors** or fermenters that are designed to maximize cell density and product yield **(Figure 12.3)**. Bioreactors are specially designed so that environmental conditions, including nutrients, oxygen, pH, and temperature can be controlled precisely. Much of the optimization of production usually involves manipulation of the fermentation conditions. The most commonly used bioreactors are open systems to which nutrients are continually added. The two basic designs are a fed-batch reactor and a chemostat that was introduced in Section 6.2.

Generally, bioreactors are designed to support maximum production of the desired product. A **fed-batch reactor** (Figure 12.4) supports very high cell densities by providing the culture with a growth-limiting nutrient, such as a carbon source, over time. This process controls the growth rate and can often prevent the production of non-desired side products. In continuous bioreactors, like the **chemostat** shown in Figure 12.4, an equivalent amount of culture is removed as new medium is added. This continuous addition of fresh medium results in a very precisely controlled constant growth rate of the organisms and the maintenance of a physiological steady state. The choice of a bioreactor is very much dependent on the type of microbe employed and the nature of the desired end product. In practice, the fed-batch reactor is much more dependable and reproducible, so it is used more often than the chemostat in industry.

CONNECTIONS In **Section 6.3**, we looked at many of the critical factors affecting the growth of microorganisms including oxygen availability, nutrient types and concentration, pH, temperature, and osmotic pressure. These variables can be tightly regulated and monitored in a bioreactor.

Figure 12.3. Bioreactor This bioreactor at the Iogen Corporation, a biotechnology company located in Ottawa, Canada, has a capacity of 200,000 L.

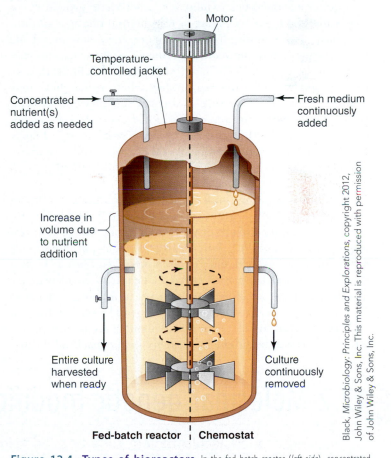

Figure 12.4. Types of bioreactors In the fed-batch reactor (*left side*), concentrated nutrient (feed) is added in a controlled manner until maximum concentration of cells is reached, and then the biomass is harvested. In the chemostat (*right side*), some amount of biomass is continually removed (effluent) as the same amount of nutrient solution is added (feed). The addition of nutrient solution to the chemostat can continue indefinitely.

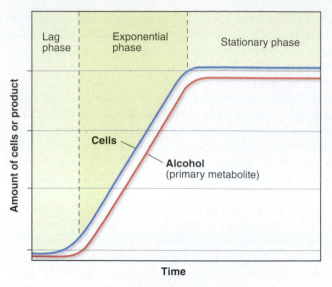

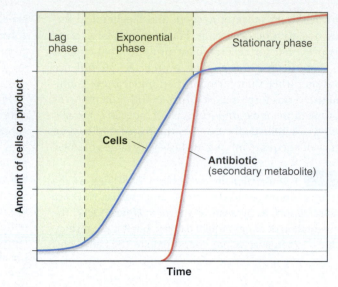

A. Primary metabolite produced during exponential phase

B. Secondary metabolite produced during stationary phase

Figure 12.5. Primary and secondary metabolite production A. The production of primary metabolites, such as ethanol in a yeast anaerobic culture, mirrors the increase in biomass. **B.** The production of secondary metabolites, such as antibiotics, is usually induced in stationary phase.

By controlling the conditions within a bioreactor, scientists can also control the growth of microbes within the vessel. Within a closed system, in which no new nutrients are added or cells removed, microbes exhibit a characteristic growth curve consisting of a lag phase, a log or exponential phase, a stationary phase, and a death phase (see Figure 6.18). The metabolic pathways used, and metabolites produced by the cells, can vary significantly depending on the growth phase. To obtain the optimal production of a desired product, then, researchers may need to maintain the culture in a particular phase of growth. For example, the main source of commercial ethanol is fermentation of sugars by the yeast *Saccharomyces cerevisiae* (see Figure 13.18). To maximize production of ethanol, yeast cells are maintained in exponential phase with low oxygen and with glucose as a carbon and energy source.

For yeast growing under anaerobic conditions, ethanol represents a **primary metabolite**, a product of metabolic processes required for growth of yeast under anaerobic conditions **(Figure 12.5)**. **Secondary metabolites**, such as antibiotics, are not required for growth and are often produced during the stationary phase of the growth curve. Concentrations of primary metabolites may be difficult to increase in a culture. Because they are intrinsically linked to energy-production pathways, high concentrations of primary metabolites frequently interfere with growth by producing toxicity or feedback inhibition.

Even recombinant *S. cerevisiae* can tolerate only about 15 percent ethanol before the culture ceases to grow. On the other hand, secondary metabolites can often be overproduced without affecting the growth of the culture.

CONNECTIONS Feedback inhibition **(Section 13.3)** occurs when the product of an enzyme or series of enzyme reactions accumulates in the cell and inhibits the action of an enzyme. Typically, binding of the product causes a conformational change in the enzyme that diminishes its ability to bind or act on the substrate, inhibiting production.

12.1 Fact Check

1. How are culture collections used in biotechnology?
2. What is bioprospecting?
3. Explain how bioreactors are used for biotechnology fermentation processes.
4. Distinguish between fed-batch bioreactors and chemostats.
5. Distinguish between a primary and a secondary metabolite.

12.2 Molecular genetic modification

How can molecular biology tools be used to improve microbial strains?

Whether researchers obtain microbes from a culture collection or isolate them directly from the environment, the usefulness of these naturally occurring microbes is often limited. As we noted earlier, the naturally occurring strains may not grow efficiently in the laboratory or may not produce large amounts of a desirable product. How can these inefficiencies be overcome? Production may be improved by altering the genetics of the strain. It may be possible to manipulate the expression of genes for certain enzymes in a pathway, resulting in the enhanced production of desired metabolites. Alternatively, strains might be altered to downregulate competing pathways or pathways that lead to the production of unwanted substances. In some cases, a strain could be altered so that it grows on a lower cost carbon nutrient. All of these types of genetic alterations in a microbe could lead to the production of more material, more cheaply.

In this section, we will examine several means of improving microbes used in biotechnology. We will start by investigating two methods of mutating microbial strains—random mutagenesis and site-directed mutagenesis. We will then examine methods for producing recombinant proteins. Finally, we will explore an area that seemed like science fiction only a decade ago—the synthesis of designer organisms.

Mutagenesis

How can we generate beneficial genetic alterations in naturally occurring microbes? We can rely on **random mutagenesis**. As we saw in Section 9.3, the exposure of cells to X-rays, UV light, or DNA-damaging chemicals such as nitrosoguanidine or ethyl methanesulfonate can cause mutations in the DNA of a cell. The resulting mutants can be screened for desired phenotypes, like the increased production of a certain enzyme **(Figure 12.6)**. A major

drawback of this approach is that the resulting mutations are undefined with little information about their nature, and mutations that have no effect, or even detrimental effects, can also occur. Additionally, it is often difficult to effectively screen the resulting mutants for the desired phenotype. Despite these drawbacks, this type of approach is still employed today. Although large numbers of mutants must be screened, mutagenesis approaches can be very successful in developing more desirable strains.

The history of penicillin production provides a wonderful example of the effectiveness of random mutagenesis. The strain of *Penicillium* isolated by Alexander Fleming, *Penicillium notatum*, produced small amounts of penicillin. Through a type of bioprospecting, researchers identified another naturally occurring strain, *Penicillium chrysogenum*, which produced higher levels of the antibiotic **(Microbes in Focus 12.2)**. This strain was then subjected to X-ray and UV light mutagenesis. Variants that produced even higher amounts of the drug were isolated. Eventually, researchers identified a mutant strain of *P. chrysogenum* that produced much higher amounts of penicillin than Fleming's original isolate of *P. notatum*.

While random mutagenesis can be used quite effectively to improve microbial strains, **site-directed mutagenesis (Toolbox 12.1)**, a technique in which researchers can make specific mutations at specific sites within a DNA molecule, allows researchers to alter microbes in a much more rational way. Using this technique, researchers can introduce precise genetic changes into specific microbes. By altering a specific nucleotide, researchers can change a specific amino acid within a protein.

Let's now ask a more practical question. What types of mutations would researchers want to make? Often, scientists want to improve the activity of an enzyme. Improved functionality may

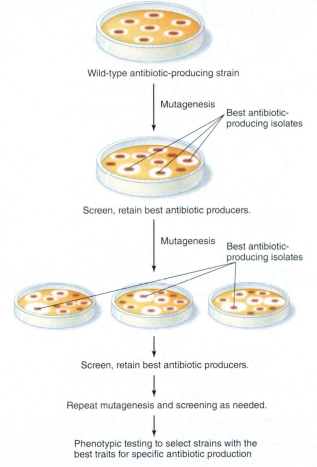

Wild-type antibiotic-producing strain

↓ Mutagenesis

Best antibiotic-producing isolates

Screen, retain best antibiotic producers.

↓ Mutagenesis

Best antibiotic-producing isolates

Screen, retain best antibiotic producers.

↓

Repeat mutagenesis and screening as needed.

↓

Phenotypic testing to select strains with the best traits for specific antibiotic production

Figure 12.6. Strain improvement by random mutagenesis A random mutagenesis strain-improvement scheme typically involves multiple rounds of mutagenesis combined with screening for superior phenotypic characteristics. The best performers after each round of screening are retained for additional rounds of mutagenesis and screening. These are then subjected to additional phenotypic analysis to identify the strains with the best traits for production of the desired product.

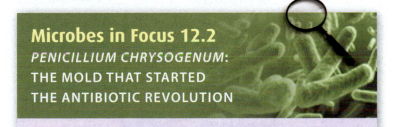

Microbes in Focus 12.2
PENICILLIUM CHRYSOGENUM:
THE MOLD THAT STARTED
THE ANTIBIOTIC REVOLUTION

Habitat: Wide distribution in nature, preferring moist conditions. Commonly found in soil, and in foods as a contaminant. Also found in large industrial bioreactors.

Description: Filamentous fungus forming long, narrow hyphae up to 5 μm wide. Bears conidia in specialized hyphae called "conidiophores." Hyphae provide a fuzzy texture, usually white, blue, or green in color. The 32-Mbp genome sequence was reported in 2008.

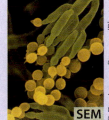

SEM

© Dr. Jeremy Burgess/Photo Researchers, Inc.

Key Features: This fungus is best known for its production of the antibiotic penicillin, the first mass-produced antibiotic. It also produces several other antimicrobial compounds as secondary metabolites. It is generally not pathogenic to humans, although the conidia sometimes act as allergens.

Pioneered by the Canadian biochemist Michael Smith in the late 1970s, site-directed mutagenesis has become a powerful molecular biology tool. In recognition of the significance of this development, Smith shared the 1993 Nobel Prize for Chemistry with Kary Mullis, the inventor of PCR (see Toolbox 1.1). Similar to PCR, site-directed mutagenesis involves the use of a short oligonucleotide primer. In this case, though, the primer used for the initiation of replication contains an altered, or mutated, DNA sequence. After extension with DNA polymerase, the resulting double-stranded molecule contains this mutation. This fragment of double-stranded DNA can then be introduced into *Escherichia coli* by transformation, where it replicates. This series of steps results in the stable incorporation of the desired genetic change. Designed oligonucleotides can also be used in PCR to introduce the desired alteration in a DNA sequence, a technique referred to as PCR site-directed mutagenesis **(Figure B12.1)**.

Once the desired sequence changes in genes of interest have been made, then the altered genes can be introduced into the production organism. Popular hosts include the bacteria *E. coli* and *Bacillus subtilis*, the yeasts *Saccharomyces cerevisiae* and *Pichia pastoris*, Chinese hamster ovary (CHO) cells, and insect cells using baculovirus vectors. In some cases, the genes can simply be introduced on a plasmid expression vector that can replicate in the host organism. In other cases, it is desirable to replace the wild-type copy of the gene with the altered gene. This gene replacement requires recombination events between the plasmid vector that contains the altered gene and the corresponding genome region. A number of systems have also been developed based on virus integrase systems that allow site-specific integration within the genome of the host cell.

● **Test Your Understanding**

What is the function of the oligonucleotide primer used in site-directed mutagenesis?

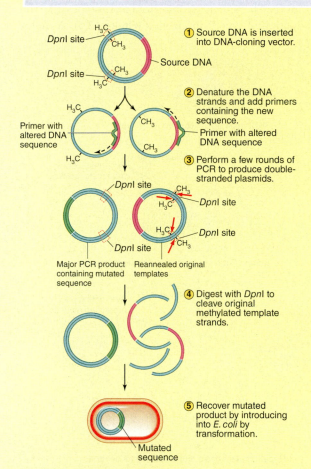

A. Oligonucleotide site-directed mutagenesis

B. PCR site-directed mutagenesis

Figure B12.1. Process Diagram: Site-directed mutagenesis
A. The original oligonucleotide-mediated site-directed mutagenesis procedure required the cloning of the DNA to be mutated into a vector that produces single-stranded DNA molecules. A complementary oligonucleotide containing the desired sequence change is allowed to anneal, and then DNA synthesis is carried out by DNA polymerase. Once the complementary strand is completed, double-stranded plasmids carrying the desired mutation can be recovered after transformation of *E. coli*. **B.** In PCR site-directed mutagenesis, researchers design complementary primers with the desired mutation, and conduct PCR with these primers. The product is subjected to digestion with the *Dpn*I restriction enzyme, which cleaves its recognition site (GATC) only when it is methylated. Thus, it will not cleave the DNA that was synthesized in the PCR reaction, only the original template DNA that has been replicated in the *E. coli* cell. This will enrich for the desired mutated product. The products are introduced into a host cell by transformation, and the presence of the correct mutation can easily be confirmed by DNA sequence analysis.

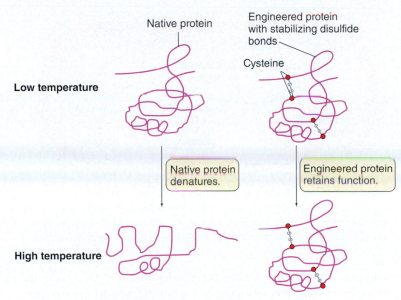

Figure 12.7. Increasing protein stability Site-directed mutagenesis can be used to add disulfide bonds to proteins to increase stability. When the structure of a protein is known, codons in the gene can be purposefully changed to new codons that impart a beneficial characteristic in the protein. Changing two sets of amino acids that are near each other in the folded peptide chain to cysteine amino acids allows internal disulfide bonds to form.

arise from the addition of cysteine amino acids in key positions within the polypeptide to facilitate disulfide bond formation in the final product. Disulfide bonds stabilize proteins, making them less susceptible to denaturing conditions of extreme pH, as might be experienced in drugs taken orally, or high heat, as would be experienced by enzymes produced for washing detergents **(Figure 12.7)**. For example, site-directed mutagenesis has been used to add three disulfide bonds in T4 lysozyme, an enzyme originally derived from phage but now produced in recombinant *E. coli* cells. This enzyme, an effective antimicrobial agent, is used widely as a food preservative. It is also used in infant formula and as sprays on meats and vegetables destined for the grocery store. The increased stability of engineered T4 lysozyme allows it to maintain its enzymatic function longer under various conditions.

Not surprisingly, the successful alteration of a protein via site-directed mutagenesis usually requires a fairly detailed knowledge of the three-dimensional structure of the protein. Knowing the structure allows researchers to target codons for amino acids in binding or catalytic sites. Additionally, the amino acids targeted for substitution must not be critical for the normal functioning of the protein.

In many instances, not enough structural information about a protein of interest is known to use site-directed mutagenesis. In these cases, combined methods have been developed to carry out random mutagenesis on a targeted gene. This approach, termed **directed enzyme evolution**, applies rounds of random mutation and selection to achieve stepwise desired changes in a gene of interest **(Figure 12.8A)**. This approach has been used to

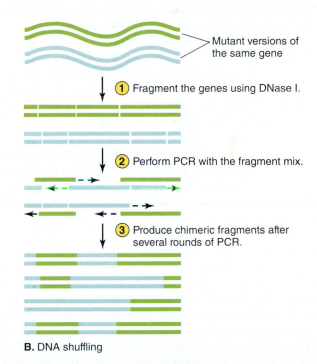

B. DNA shuffling

Figure 12.8. Process Diagram: Directed enzyme evolution A. Starting with a gene encoding the enzyme of interest, mutagenesis is carried out to generate a library of variant genes. These libraries can then be screened for improved gene function, with the best-performing genes being subjected to mutagenesis again, and the screening process repeated. The process is similar to that described in Figure 12.7, but here it is limited to a single gene rather than the entire genome. **B.** DNA shuffling allows the *in vitro* recombination of multiple mutants of the same gene, dramatically extending the range of mutant genes that can be recovered. In this process, a collection of mutant genes is cleaved randomly by DNase I. The remaining fragments are mixed together and subjected to PCR amplification, which results in recombination of the mutated gene segments.

A. Directed evolution

improve strains by generating more active enzymes, and imparting new or broadened ranges of enzyme activities for particular substrates.

As an example, let's consider a dehydrogenase enzyme that catalyzes the oxidation of a substrate such as glycerol in the following reaction:

$$\text{Glycerol} + \text{NAD}^+ \xrightarrow{\text{glycerol dehydrogenase}} \text{dihydroxyacetone} + \text{NADH} + \text{H}^+$$

The product of this reaction, dihydroxyacetone, is the main ingredient in sunless tanning products. From a biotechnology perspective, obtaining a variant of glycerol dehydrogenase that more effectively catalyzes this reaction could be quite profitable. A form of directed enzyme evolution could be used to produce such a variant.

Initially, a collection of glycerol dehydrogenase variants can be constructed using a technique called **error-prone PCR**. In this process, the native glycerol dehydrogenase gene is PCR amplified under conditions that result in a very high error rate. The PCR products, in other words, quite frequently contain mutations. These amplified fragments are then cloned into a plasmid vector, and the plasmids are introduced into a host strain that does not produce the glycerol dehydrogenase enzyme. The resulting colonies, each containing a glycerol dehydrogenase gene variant, are then screened for enhanced glycerol dehydrogenase activity. Researchers select the best-performing clones and repeat the process. In many ways, this type of artificial evolution resembles the selective breeding used in animals and plants; only clones producing improved enzymes are allowed to reproduce. Like natural evolution, the improvements typically occur in small steps, making it a time-consuming process.

To improve the efficiency of directed enzyme evolution, **DNA shuffling** can be used. In this approach, groups of mutants generated from directed enzyme evolution, or groups of related genes from different organisms, serve as the starting material (see Figure 12.8B). The investigators randomly fragment these gene variants, usually using the enzyme DNase I, which cleaves DNA. The fragments are mixed together, denatured, and reassembled into full-length genes by PCR. The result is chimeric genes, products of recombination between the input fragments. This process provides a way to combine multiple beneficial variations into single clones and can result in dramatically improved phenotypes.

In microbial biotechnology, production is ultimately limited by the genetic and physiological characteristics of the microbial strains that are used. The process can often be improved by making changes to the metabolic characteristics of the strain. This can be accomplished by introducing genes or altering their expression. For example, production costs might be lowered if genes are introduced that allow organisms to use a cheap and readily available source of carbon. Knowledge of metabolic pathways may suggest ways to manipulate the expression of genes for enzymes in the pathway, enhancing production of the desired metabolites. Alternative strategies might repress competing pathways or pathways that lead to the production of unwanted substances. For example, high cell density of *E. coli* is often hard to achieve during fermentation because of the accumulation of waste products, in particular acetate, which inhibits cell growth

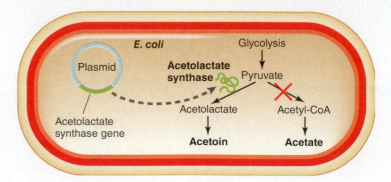

Figure 12.9. Manipulating a pathway Modifying the metabolic pathway reduces production of the inhibitory waste product acetate. Introduction of a plasmid encoding the gene for acetolactate synthase into *E. coli* results in conversion of pyruvate into acetolactate, which is then converted into acetoin. Acetoin is less inhibitory to growth than acetate, allowing cells to reach higher densities in culture.

and protein production. One potential solution to this problem is to divert the precursor pyruvate from the acetate production pathway. When the gene is added for acetolactate synthase, pyruvate is instead converted to acetolactate, which is then converted to the less inhibitory end product, acetoin **(Figure 12.9)**.

Biosynthetic pathways can also be manipulated. The production of the penicillin-like antibiotic cephalosporin, for instance, increased significantly when researchers inserted into the fungus *Acremonium chrysogenum* additional copies of *cefEF* and *cefG*, genes involved in the biosynthesis of cephalosporin. With increased copies of these genes in its genome, the recombinant *A. chrysogenum* strain produced more cephalosporin than the non-modified strain. While this type of approach can be very successful, it requires a certain level of specialization in molecular biology and a knowledge of the genetics and physiology of the organism being used.

CONNECTIONS Recombinant DNA techniques like restriction enzyme digests, PCR, and ligations have become established tools for molecular biology and biotechnology. These methods are discussed in **Section 9.4**.

Production of recombinant proteins

As we have seen, various mutagenesis techniques can be used to alter DNA within a cell. Other molecular biology tools can be used to introduce foreign DNA into a cell (see Section 9.4). While vectors used for this purpose allow for the amplification of cloned DNA in cells, they may not be designed to express foreign genes. **Expression vectors** are specifically designed to produce recombinant proteins. In *Escherichia coli*, expression vectors can produce recombinant protein to levels of 20 percent or more of the total cell protein. In contrast, even the most abundant native protein in *E. coli* constitutes only about 2 percent of the total cell protein. These advantages make expression vectors a first choice for commercial production of recombinant proteins.

First designed for use in *E. coli*, expression vectors now exist for a number of different cell types, including yeast, insect, and mammalian cells. These expression vectors are used for production of therapeutic human proteins such as insulin, human growth hormone, antiviral interferons, and interleukin-2. In humans, specialized cells typically produce such proteins, and usually under specific conditions and in very small quantities. Their large-scale, commercial production could not occur without the use of expression vectors.

Expression of a eukaryal protein in a bacterial host cell like *E. coli* will occur only if the foreign gene contains the correct bacterial elements for transcription and translation. The eukaryal gene, in other words, must be re-formatted to look like a bacterial gene. To better understand this issue, let's first think about transcription. RNA polymerase binds to a promoter to initiate transcription. Eukaryal promoters, however, differ significantly from bacterial promoters (see Section 7.3). The RNA polymerase of *E. coli* will not recognize a eukaryal promoter. Instead, the expression vector must contain an *E. coli* promoter (**Figure 12.10**). Transcription termination also differs between bacteria and eukarya. *E. coli* expression vectors must also contain a bacterial transcription terminator (see Figure 7.17). If an expression vector contains a bacterial promoter and transcription terminator, then the cloned sequence between these elements can be transcribed to produce an mRNA transcript. Moreover, it should be possible to regulate the promoter so that production of the

protein can be controlled. Transcription regulation can be accomplished by adding an appropriate operator sequence to the vector that will respond to inducer or repressor proteins.

CONNECTIONS Understanding how expression vectors work requires a solid understanding of the gene structure of bacteria and eukarya. In **Sections 7.3** and **7.4**, we examined the functions and placement of various transcriptional and translational elements. In **Section 11.2**, we explored the regulation of these processes.

For expression of a eukaryal gene in a bacterial host, the gene itself must also be re-formatted. Normally, eukaryal genes contain introns (see Figure 7.23). Eukaryal cells remove these non-coding regions during RNA processing in the cell nucleus before translation of the mRNA. For correct expression of a eukaryal gene in bacteria, then, the inserted DNA must be free of introns. How can researchers remove these introns? The intron-free sequence could be artificially synthesized. Alternatively, an mRNA transcript from the eukaryal cell can be used as a template for the reverse transcriptase enzyme (see Sections 8.3 and 10.2) to produce a DNA strand that lacks the introns. This **complementary DNA (cDNA)** can then be inserted into the expression vector.

Let's now consider translation. The ribosomes of the *E. coli* recognize the Shine–Dalgarno sequence on the mRNA and initiate translation at the downstream AUG start codon. Will eukaryal genes have this element? No. The Shine–Dalgarno sequence must be contained within the vector, or added to the cloned DNA sequence at the appropriate place. A eukaryal coding sequence, including the stop codon, can then be cloned so that the translation start codon, ATG, exists approximately 6–10 bases downstream of the Shine–Dalgarno sequence. With the addition of these various elements, the *E. coli* host will recognize the foreign DNA and make the protein.

As we noted in Section 7.3, the sequences of bacterial gene promoter elements vary, altering the affinity of the sigma factor for the promoter. Promoter sequences that interact strongly with sigma factors undergo high levels of transcription. Expression vectors typically contain promoter sequences designed to drive a high level of transcription of the cloned sequence. Bacterial genes also show variation in their ribosome binding sites. These variations affect the level of translation. Bacterial expression vectors therefore contain ribosome binding sites with preferred Shine–Dalgarno sequences and a perfectly placed ATG codon 6–10 bases downstream to maximize initiation of translation.

To facilitate cloning from a variety of sources and make a truly universal expression vector, a multiple cloning site containing an array of several unique restriction sites (see Section 9.4), is often placed directly after the ATG start codon in the vector. As long as the coding sequence of interest is cloned in the correct reading frame for translation, then *E. coli* will produce the recombinant protein. A universal expression vector can often express *any* given coding sequence—eukaryal, bacterial, or archaeal.

While the expression of recombinant proteins may seem fairly simple, several caveats exist. In eukarya, for instance, proteins often undergo glycosylation, the addition of complex sugar residues to polypeptides (see Section 7.4). Bacterial hosts, though, often are not capable of carrying out the specific type

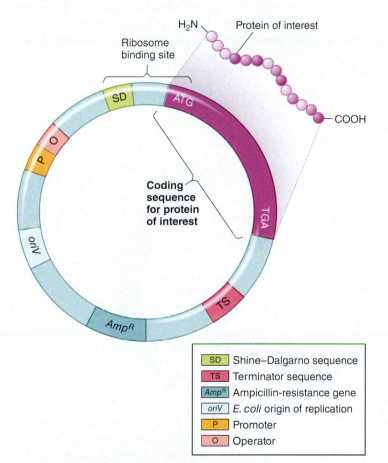

SD	Shine–Dalgarno sequence
TS	Terminator sequence
AmpR	Ampicillin-resistance gene
oriV	*E. coli* origin of replication
P	Promoter
O	Operator

Figure 12.10. Expression vector Expression vectors have customized promoters to drive a high level of transcription. An operator, such as O^{lac} of the *lac* operon, regulates the level of transcription. Expression vectors also contain an optimized ribosome binding site (RBS) including a Shine–Dalgarno (SD) sequence and a start codon ATG. The coding sequence of the protein of interest is inserted 6–10 bases downstream of the SD to maximize initiation of translation. The coding sequence for the protein follows, including the start codon ATG and stop codon TAA, TGA, or TAG. Transcriptional terminator sequences end transcription. Expression plasmids also contain a selectable marker gene, shown here as an ampicillin resistance gene, and an origin of replication for the expression host, shown here as *oriV*. Most cloned protein coding sequences can be transcribed and translated, including those from eukaryal genes.

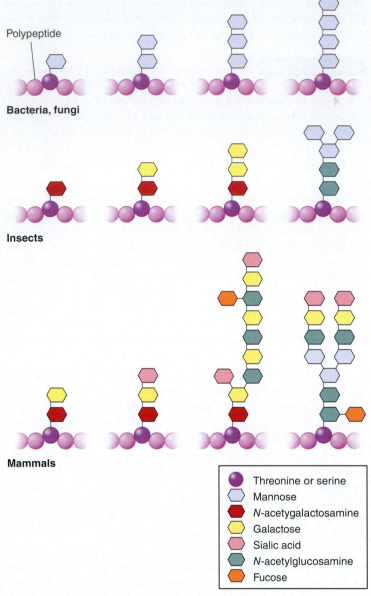

Polypeptide

Bacteria, fungi

Insects

Mammals

●	Threonine or serine
⬡	Mannose
⬢	*N*-acetygalactosamine
⬡	Galactose
⬡	Sialic acid
⬡	*N*-acetylglucosamine
⬡	Fucose

Figure 12.11. Glycosylation These post-translational modifications vary between groups of organisms. The addition of specific sugars to particular amino acids is often required for proper protein folding and function. Shown here are common glycosylations added to threonine or serine amino acids in proteins from bacteria and fungi, insects, and mammals. The choice of host expression system will depend on the requirement for glycosylation.

such as Chinese hamster ovary (CHO) cells, and cultured human cells, such as HeLa cells. These host systems are more expensive than bacterial systems and often do not produce as much recombinant protein; a trade-off exists between function and amount.

Vectors designed for expression in eukaryal organisms follow the same general principles as vectors used in bacterial species. They must contain a eukaryal promoter and a transcription terminator that also signals the addition of a polyA tail to the transcript. Eukaryal mRNAs, as we noted previously, do not contain a Shine–Dalgarno binding site. Instead, eukaryal ribosomes bind the 5′ cap structure of the mRNA and scan the mRNA until they reach an AUG codon, where translation begins. A Shine–Dalgarno sequence, as a result, is not needed in a eukaryal expression vector.

While expression vectors can be used to produce individual proteins, expression vectors can also be used to produce **fusion proteins**, proteins that contain domains of two or more proteins (**Figure 12.12**). Sometimes called "tagged" proteins, fusion proteins typically contain a protein of interest fused to a portion of another protein that adds a beneficial characteristic to the recombinant protein. For example, an **affinity tag**, a peptide that facilitates purification, can be added to a protein of interest to simplify the selective enrichment of the recombinant protein from the other proteins produced by the cell (**Toolbox 12.2**).

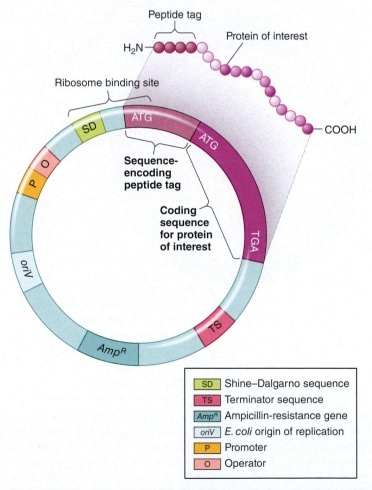

SD	Shine–Dalgarno sequence
TS	Terminator sequence
Amp^R	Ampicillin-resistance gene
oriV	*E. coli* origin of replication
P	Promoter
O	Operator

Figure 12.12. Expression of "tagged" proteins Expression vectors designed for producing fusion proteins have a built-in coding sequence for a peptide tag that is fused to the coding sequence of the cloned protein of interest. Following translation, the peptide tag will be part of the recombinant protein. Tags may be used for purification of the target protein.

of glycosylation that many eukaryal proteins require for correct folding and function. Many mammalian proteins, for instance, are glycosylated with sialic acid, whereas bacteria and fungi tend to glycosylate proteins with mannose units (**Figure 12.11**). If a eukaryal protein requires specific glycosylations to function correctly, then a eukaryal expression system, rather than a bacterial expression system, may be required.

Disulfide bond formation may also be problematic in *E. coli*. Without correctly formed disulfide bonds, proteins may not fold correctly and, as a result, may be inactive (see Section 7.4). Again, if extensive disulfide bond formation is necessary, then a eukaryal host may be needed. Recombinant protein production in insect cells uses modified insect baculoviruses as vectors. Similarly, viruses that infect mammals can be modified for use as vectors to allow production of recombinant proteins in cell lines,

Expression vectors can produce large amounts of a desired protein for research and commercial purposes. Following expression of the recombinant protein, the task of separating it from all the other proteins in the cell begins. With tagged proteins, an efficient method for purification involves affinity chromatography. In this process, researchers employ receptors, like antibodies or metal ions, which specifically bind the tag fused to the protein of interest. Using chemical techniques, these receptors are attached to small beads, which are then added to a column. Cells expressing the tagged proteins of interest are lysed, and the cellular lysate is applied to the column. Proteins containing the tag will bind to the receptors. Unwanted proteins will not bind and can be washed through the column (**Figure B12.2**). With this relatively simple approach, recombinant proteins can often be purified 100-fold, with greater than 90 percent recovery (**Figure B12.3**).

Following purification, the tag must often be removed from therapeutic recombinant proteins before they receive approval by regulatory agencies for use. This step provides a safeguard against unforeseen reactions the tag may cause in the body. To accomplish this tag removal, vectors can be designed to code for a short amino acid sequence between the tag and protein of interest that is recognized by specific proteases that cleave this sequence following purification. Thus, the tag can be removed from the final product.

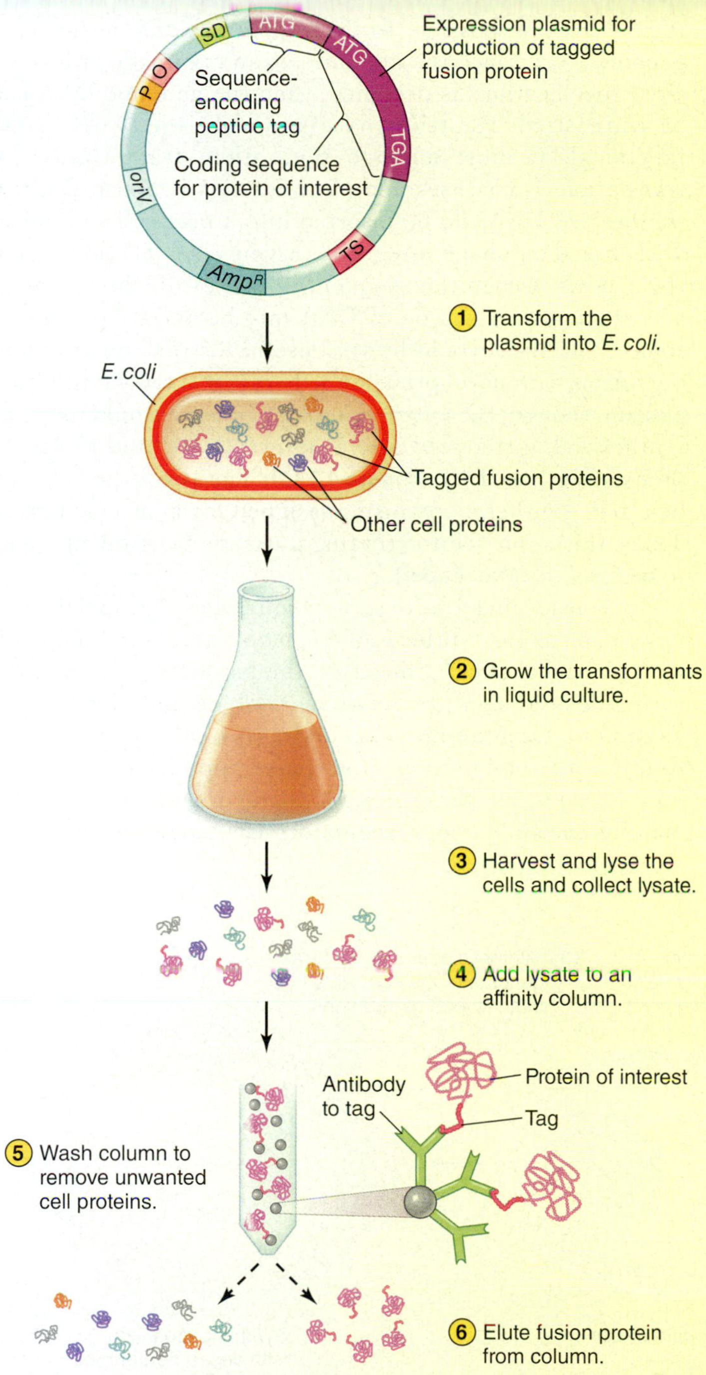

Expression plasmid for production of tagged fusion protein

SD ATG ATG — Sequence-encoding peptide tag

Coding sequence for protein of interest

P O oriV AmpR TS TGA

① Transform the plasmid into *E. coli*.

E. coli

Tagged fusion proteins

Other cell proteins

② Grow the transformants in liquid culture.

③ Harvest and lyse the cells and collect lysate.

④ Add lysate to an affinity column.

Antibody to tag — Protein of interest — Tag

⑤ Wash column to remove unwanted cell proteins.

⑥ Elute fusion protein from column.

Figure B12.2. Process Diagram: Separation of tagged fusion proteins using affinity chromatography Cells containing the recombinant fusion protein are lysed, and the resulting mixture of proteins is passed through a column containing an antibody to the tag. Tagged proteins will remain bound in the column while other proteins pass through.

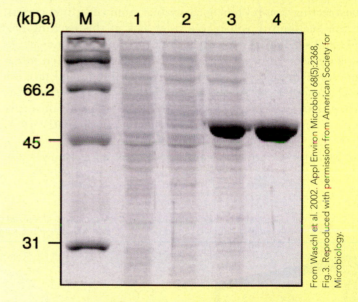

From Waschl et al. 2002. Appl Environ Microbiol 68(5):2368, Fig.3. Reproduced with permission from American Society for Microbiology.

Figure B12.3. Polyacrylamide gel electrophoresis shows the purified protein. The successful purification of a tagged protein can be visualized by SDS-PAGE, followed by staining with Coomassie Brilliant Blue dye. Note that the purified protein is present as a single band, while the original cell lysate from which the tagged protein was purified contains multiple bands representing *E. coli* proteins.

● **Test Your Understanding** ....

How might a protease remove a tag from a therapeutic protein before it is used by humans?

TABLE 12.2 Common affinity tags used to purify fusion proteins

Affinity tag	Ligand used for purification
Protein A	Antibody
Histidine (His) amino acids	Ni^{2+}
Maltose-binding protein (MBP)	Maltose
Glutathion S-transferase (GST)	Glutathion
Strep-tag	Streptavidin

Many affinity tags have been developed. Expression vectors and purification systems for producing and isolating tagged recombinant proteins are available commercially **(Table 12.2)**.

CONNECTIONS Not all fusion proteins are designed to aid in purification. As we saw in **Section 3.1**, GFP fusion proteins are often used to monitor the location of a protein within a living cell.

Designer organisms

While traditional molecular biology techniques involve modifying existing organisms, **synthetic biology** involves constructing a novel biological system from constituent parts. In this section, we will take a brief look at this exciting new field and see how it may affect the future of biotechnology. As we have seen in other sections, PCR and site-directed mutagenesis techniques require the ability to design and produce single-stranded DNA oligonucleotides of specific sequence. When DNA synthesis technology was first developed, university departments often ran their own oligonucleotide synthesizers, shared by several research groups. As DNA synthesis technology has matured, most DNA synthesis is now done commercially.

Oligonucleotides can be ordered online. The companies synthesize the oligonucleotides almost immediately, and often ship them by courier within a day. Some companies now specialize in the production of longer, double-stranded DNA molecules. The dsDNA of a typical 1-kb gene can be synthesized for less than $500. Even larger DNA molecules in the size range of 10 kb can also be constructed, and by connecting these larger molecules, the complete syntheses of first viral and then bacterial genomes have been achieved.

We have entered an age in which novel genomes with desired specifications can be designed and then synthesized. It may become possible to do computer-aided design at the genome level. Once the genome for an organism with the desired specifications is designed, then the genomic DNA could be synthesized. The tricky part is to make the synthetic DNA functional. It must undergo replication, transcription, and translation. For viruses and plasmids this is not difficult; the synthesized DNA can be inserted into a host cell to replicate. Tools based on phage integrase enzymes (see Section 8.3) and their DNA recognition sequences are available for precise and efficient integration of DNA into bacterial host chromosomes, and these could be very useful for the construction of organisms with novel properties. To create a truly synthetic organism, though, the entire microbial genome would need to be synthesized, introduced into a cell, replicate, and replace the preexisting host DNA **(Figure 12.13)**. The resulting cell would be a true synthetic organism, designed for a specific purpose. Today, this approach to creating a designer organism appears to be feasible **(Mini-Paper)**.

In considering the design of microbes for specific functions, such as the synthesis of a valuable secondary metabolite with pharmaceutical properties, the production of an enzyme with a desired catalytic activity, or the degradation of an environmental contaminant, genetic elements must be joined in different combinations and under appropriate regulation. We can consider these genetic elements, whether they are enzyme-encoding genes, regulatory DNA sequences, or genes

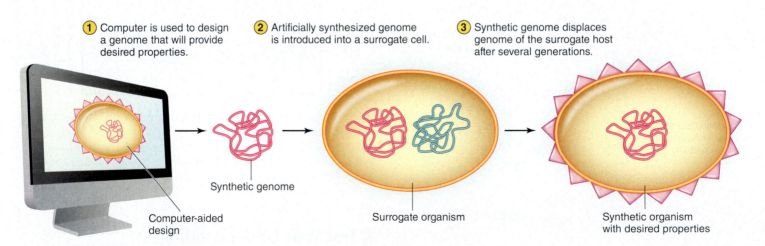

① Computer is used to design a genome that will provide desired properties.

② Artificially synthesized genome is introduced into a surrogate cell.

③ Synthetic genome displaces genome of the surrogate host after several generations.

Synthetic genome

Computer-aided design

Surrogate organism

Synthetic organism with desired properties

Figure 12.13. Process Diagram: Design and construction of a synthetic organism The synthesis of complete microbial genomes is technically possible, and has been achieved, but the challenge is to reliably generate viable organisms after introduction of the synthesized genome into the surrogate host. Once these problems have been solved, the design and production of synthetic microbes will likely find a multitude of applications in biotechnology.

encoding regulatory proteins, as biological parts. The goal is to use these parts in a reproducible manner so that they can be combined using computer-aided design software, and the corresponding synthesized DNA will have predictable behavior once it is introduced into a cell. Proponents of this vision of synthetic biology are actively working on developing the standards in an open source environment where they share experimental protocols and resources, and provide tools for online collaboration.

The concept of biological parts has its roots in engineering disciplines, where standardized parts may not necessarily be identical, but do adhere to certain specifications. Using these parts, engineers can design objects that will perform predictably without necessarily understanding the details of how each of the individual parts functions. Biology, consisting of collections of natural entities shaped by evolutionary processes, does not adhere to any such standards. As a result, standard molecular genetics research involving the manipulation of genes with tools such as restriction enzymes and PCR, can be a very tedious process that often proceeds through trial and error. It is very difficult, if not impossible, to design a biological entity, carry out the construction based on that design, and have the resulting genetic construct behave as initially predicted.

To address this biological reality, researchers have created the Registry of Standard Biological Parts. It contains descriptions of genetic parts such as promoters and gene regulatory elements, ribosome binding sites and protein open reading frames, and devices that are combinations of parts for protein production, reporter genes, and cell signaling. Each part has been inserted into a vector and is flanked by *Eco*RI and *Xba*I restriction sites on one side, and *Spe*I and *Pst*I restriction sites on the other side **(Figure 12.14)**, in an arrangement called a BioBrick™. The 6-base *Xba*I and *Spe*I restriction sites share the same core 4-base sequence, but they differ in the flanking nucleotides on either side. Cleavage yields compatible sticky ends, and combination of *Xba*I and *Spe*I sites by ligation results in loss of both restriction enzyme recognition sequences.

While the BioBrick concept is very useful for physical sharing of biological parts, as the cost of DNA synthesis drops, it will become possible to design devices and construct them without needing physical access to the DNA. Eventually, we may be able to use computer software programs to design biological systems based on standard biological parts, have the DNA for those systems synthesized, and then expect the resulting organisms to perform exactly as predicted. Currently, however, we do not understand fully how even simplified biological organisms operate. As our knowledge continues to increase, and we combine this knowledge with engineering expertise, synthetic biology surely will have a great impact on biotechnology applications **(Perspective 12.2)**.

Now that we have seen how microbes can be modified, let's ask an important question. How can we use these genetically modified organisms? In the next sections of this chapter, we will look at three categories of biotechnology—red biotech, white biotech, and green biotech. As we will see, all three areas employ similar techniques, but have different foci. Within each category, we will emphasize a few specific examples.

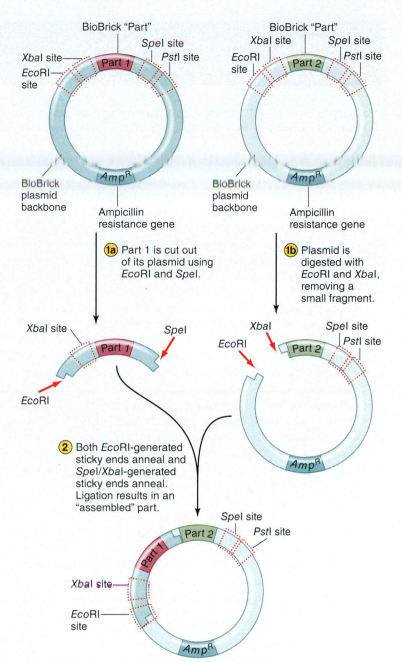

Figure 12.14. Process Diagram: Standard assembly using BioBricks
The BioBrick plasmid backbone contains restriction sites that allow removal or insertion of the part from the plasmid backbone. Parts available for assembly include such things as promoters, protein coding sequences, ribosome binding sites, and transcription terminators.

12.2 Fact Check

1. Explain the benefits and drawbacks of using random mutagenesis in biotechnology.
2. What are some of the challenges in the production of recombinant eukaryal proteins in bacterial cells?
3. What is synthetic biology?

Mini-Paper: A Focus on Research
MAKING A SYNTHETIC GENOME

D. G. Gibson et al. 2010. Creation of a bacterial cell controlled by a chemically synthesized genome. Science 329:52–56.

Context

Advances in DNA sequencing and DNA synthesis have made these methods inexpensive and available to most scientists. We now know the complete genome sequences of thousands of microbes, and it is possible to design genes right down to the individual base pair. Genes can be altered at will so that they operate optimally within a given organism and strain. There is an upper limit to the length of double-stranded DNA that can be produced reliably, however, so that only about a dozen or so synthesized genes can be prepared at a time on one double-stranded DNA fragment. What if it were possible to design the complete genome of a microbe, have it synthesized from scratch, and then regenerate a live organism using that DNA? This could allow complete control over each of the details of the designed microbe. Once gene functions are better understood, the biotechnology applications will be vast.

A series of efforts in gene synthesis started with Gobind Khorana's construction of the 72-bp gene for alanine tRNA in 1970. Synthesis of the genes for human insulin in 1978, the 7.7-kb genome of the poliovirus, and the 5.4-kb genome of the ΦX174 bacteriophage in 2003 further extended the lengths of synthetic DNA. Eventually, the construction of a complete bacterial genome could be contemplated. In preparation for this, scientists have tried to determine the minimum genome size required for a bacterial cell to live and replicate (see Perspective 10.2). The genomes of species of *Mycoplasma* are the smallest known bacterial genomes, ranging from 400 to 500 Mbp. This was within the range of a complete synthesis effort. A strategy for piecing together the synthesized DNA fragments into longer pieces just needed to be developed. A strategy involving homologous recombination after transformation of yeast was described in Toolbox 9.2. Now, we describe the use of this strategy to successfully construct a complete bacterial chromosome, and have that chromosome replicate after introduction into an existing cell, displacing the preexisting chromosome.

Experiment

This work started with a design for a synthetic genome, designated JCVI-syn1.0. The design was based on the known genome sequence of *Mycoplasma mycoides* subspecies *capri*. Added to the natural sequence were four watermarks, or introduced sequences. One of these watermarks contained the cipher that is required for decoding the other three watermarks. After successful deciphering, those watermarks indicated names of scientists who had contributed to the project, as well as quotations from literature. Note that these watermark sequences were not functional sequences intended to be translated to gene products, but instead were introduced as markers or tags for the synthetic genome to confirm the identity of the cells obtained at the end of the assembly. Including these watermarks, there were 25 sequence differences between the synthesized genome and the original genome.

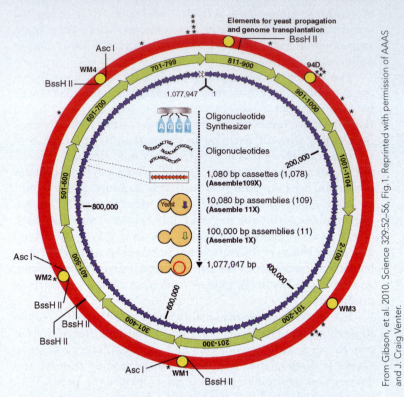

From Gibson, et al. 2010. Science 329:52–56, Fig.1. Reprinted with permission of AAAS and J. Craig Venter.

Figure B12.4. Assembly of a synthetic genome The flow chart shows the progression from synthesis of overlapping single-stranded oligonucleotides to assembly into 1-kb double-stranded cassettes, which are then recombined in yeast to make 10-kb assemblies (blue arrows) and then 100-kbp assemblies (green arrows), which are finally recombined to make the 1-Mb genome. The locations of the four watermarks (WM1–WM4) and relevant restriction enzyme sites are indicated in the outer ring. Asterisks show the locations of other differences in sequence from the wild-type sequence.

As indicated in **Figure B12.4**, the genome was assembled in several steps starting from the synthesized oligonucleotides, with yeast being used for the *in vivo* recombination steps. The correct assembly had to be extensively checked by PCR to ensure that the correct fragments were present in the correct order. To enable replication in yeast, all fragments were constructed in a vector that contained a yeast origin of DNA replication called an "autonomously replicating sequence," a yeast centromere, a URA3 yeast selection marker, as well as a tetracycline-resistance marker and the *lacZ* gene encoding the β-galactosidase enzyme needed for selection and expression in the final bacterial cell host.

Once the complete synthesized genome had been constructed, the next step was to introduce it into another strain of *Mycoplasma* by transformation, and hopefully have it replicate and take over control of the cell from the chromosome that was already present. The strain that was chosen to receive the synthetic genome was a restriction endonuclease gene mutant of *Mycoplasma capricolum*. To ensure that the chromosome extracted from the yeast cells remained intact, the cells were broken open only when they were embedded in an agarose plug to reduce the chance of chromosome breakage by shear forces, and the DNA carefully purified from the agarose. This DNA was then

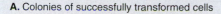

A. Colonies of successfully transformed cells

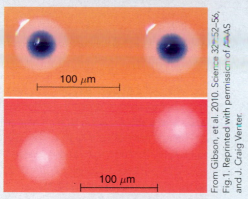

From Gibson, et al. 2010. Science 329:52–56, Fig.1. Reprinted with permission of AAAS and J. Craig Venter.

B. Colonies of untransformed cells

Figure B12.5. Identification of transformants A. Colonies of successful tetracycline-resistant transformant cells appear blue when grown on an X-gal-containing medium. **B.** Colonies of cells that have not taken up the exogenous DNA appear white.

added to competent *M. capricolum* recipient cells, and after a suitable period of incubation, the mixture was placed under selection on agar media containing tetracycline and X-gal. Blue colonies, indicating the presence of the *lacZ* marker of the introduced DNA, were visible within a few days, as shown in **Figure B12.5**.

Analysis by restriction enzyme digest and PCR were consistent with the inheritance of the complete synthesized chromosome by the blue transformants. The complete genome sequence of one clone, named JCVI-syn1.0, was determined using the Sanger shotgun method (see Figure 10.2). Interestingly, there were 10 minor sequence differences between the designed sequence and the sequence from the transformant. Most of these were single-nucleotide changes, but one was an insertion of the *E. coli* transposable element IS1, which was likely obtained during the 10-kb assembly that included an intermediate passage through *E. coli*. Comparison of the proteomes by 2D-PAGE, as shown in **Figure B12.6**, revealed no discernable differences between the patterns of protein spots between JCVI-syn1.0 and the wild-type strain.

Impact

The announcement of this feat attracted considerable global media attention. Much of the discussion revolved around the implications of synthetic life. There was even talk about the possibility of designer humans! While this study was certainly an impressive technical accomplishment, it needs to be put into perspective since it is still a very long way to the design and synthesis of complex multicellular organisms. Humans have been synthesizing genes for decades, and individual mutations can be placed into many organisms at will. Genomes can be manipulated with high accuracy by recombination methods. Complete genome synthesis simply demonstrates that this type of work can be done at the genome level. While the *Mycoplasma* genome is relatively small compared to the genomes of most other microbes, it is expected that the technology will be scaled up for use in most microbes, especially those most relevant to biotechnology applications.

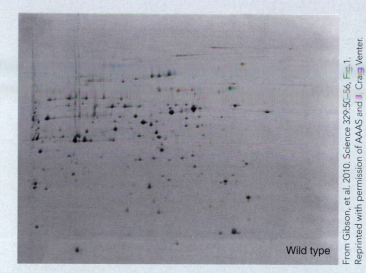

Wild type

From Gibson, et al. 2010. Science 329:52–56, Fig.1. Reprinted with permission of AAAS and J. Craig Venter.

JCVI-syn1.0

From Gibson, et al. 2010. Science 329:52–56, Fig.1. Reprinted with permission of AAAS and J. Craig Venter.

Figure B12.6. Proteomes of transformant and wild-type cells Comparison of the proteomes of the synthetic genome transformant and the wild-type strain. There are no readily detectable differences.

● Questions for Discussion ....

1. Why was it important to use a restriction endonuclease mutant as the recipient?

2. What is the function of the *lacZ* gene in the transformation experiments?

3. Why do you think these experiments were carried out in *Mycoplasma* rather than an organism that is more relevant to biotechnology?

Perspective 12.2

THE INTERNATIONAL GENETICALLY ENGINEERED MACHINE (IGEM) COMPETITION, STANDARD BIOLOGICAL PARTS, AND SYNTHETIC BIOLOGY

Each fall, on a weekend in November, teams of undergraduate students from around the world descend on Cambridge, Massachusetts, for the iGEM World Championship Jamboree. These students represent a number of academic disciplines, from molecular biology to mathematics, computer science, chemical engineering, and electrical engineering. They visit Massachusetts Institute of Technology (MIT) to present the research projects that they have been working on at their home campuses, to share ideas, and to contribute to the growth of synthetic biology. Their aim is straightforward—simplify complex biological organisms so that they can be truly engineered. This synthetic biology movement is a very open culture in which all developed material is made available for sharing, and genetic material that is produced by the teams is contributed to a central repository.

The iGEM Jamboree has grown from a January course at MIT to a small group of five teams in the summer of 2004 to 185 teams representing over 30 countries in 2012 **(Figure B12.7)**. Only the top teams compete at the World Championship Jamboree, following regional competitions in Asia, Europe, Latin America, North America East, and North America West. There is now even a high school division with 40 teams competing in 2012. Through the various research projects, the participants aim to demonstrate that diverse constructs can be designed and built within living cells **(Table B12.1)**.

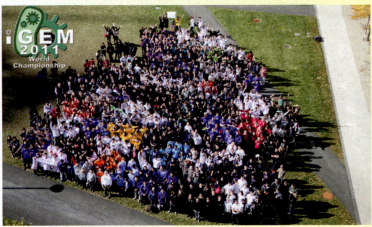

Courtesy iGEM and David Appleyard

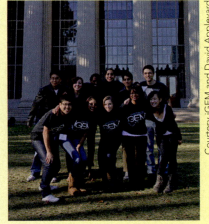

Courtesy iGEM and David Appleyard

Figure B12.7. Some of the participants at the iGEM 2011 World Championship Jamboree

TABLE B12.1 The iGEM competition

Year	Number of teams	Grand prize winning team	Winning project title and description
2004	5	No grand prize	
2005	13	No grand prize	
2006	32	Slovenia	*Engineered Human Cells*: Human embryonic kidney cells were transfected with constructs designed to induce artificial tolerance in the TLR-signaling pathway, thus reducing the chance of sepsis following bacterial infection
2007	54	Peking	*Towards Self-differentiated Bacterial Assembly Line*: Development of devices that allow genetically identical cells to differentiate so that they perform complementary tasks
2008	84	Slovenia	*Immunobricks*: Production of a designer vaccine against *Helicobacter pylori*
2009	112	Cambridge	*E. Chromi*: Design and construction of novel biosensor systems that respond to the concentration of inducer with different color output
2010	130	Slovenia	*DNA Coding Beyond Triplets*: Development of a system for DNA-guided assembly of functional proteins
2011	160	University of Washington	*Make It or Break It*: Diesel production and gluten destruction, the synthetic biology way

12.3 Red biotechnology

The inexpensive production of antibiotics has revolutionized public health worldwide. As we saw at the start of this chapter, the first antibiotic to be produced in mass quantities and at a reasonable price was penicillin. In less than 10 years during the 1940s, pharmaceutical companies developed processes to manufacture penicillin on a large scale. By the late 1940s, the cost of penicillin production was approximately $5.00 per kg. Production improvements continued, and by the 1970s, the cost of the drug had dropped to an astonishing $0.03 per kg. Red biotech, though, does not only involve the production of antibiotics. During this introduction to red biotech, we will look at two specific applications—secondary metabolites and human protein therapeutics.

First, a number of other secondary metabolites from microbes, besides antibiotics, have been discovered to have therapeutic effects. However, microbes are not the only original source of useful secondary metabolites. The antimalarial compound artemisinin is traditionally extracted from the plant *Artemisia annua*, but as we'll see at the end of the chapter, it can now be produced by recombinant microbes. Second, as we have seen, recombinant DNA methods have been used to provide reliable sources of synthetically produced human proteins, mostly hormones such as insulin. Let's first consider the secondary metabolites, and we'll come back to the human proteins later in this section.

Secondary metabolites as therapeutics

Penicillin, of course, is not the only antibiotic. In the early 1940s, two other antibiotics, actinomycin and streptomycin, were discovered in culture filtrates of actinomycetes, saprophytic filamentous soil bacteria, sometimes mistaken for fungi because of their filamentous growth and complex life cycle. *Streptomyces* (see Microbes in Focus 12.1), the most widely used actinomycete, produce many antibiotics as secondary metabolites, compounds produced by an organism but not directly needed for the survival of the organism. In fact, members of this genus produce most of the known antibiotics, including antibacterial and antifungal agents. *Streptomyces* also produce many other medically important compounds, including anticancer agents and immunosuppressants.

CONNECTIONS Antibiotics are compounds derived from microorganisms that inhibit the growth of other microorganisms, although the term is sometimes used to encompass similarly acting synthetic compounds. Many antibiotics are discussed in **Section 24.2**.

Not all drugs based on secondary metabolites are antimicrobial in nature. One of the most profitable classes of drugs on the market today is a secondary metabolite that inhibits the activity of hydroxymethylglutaryl (HMG)-CoA reductase, a key enzyme in the cholesterol synthesis pathway. These HMG-CoA reductase inhibitors, called "statins," were introduced in 1987 as a treatment to lower cholesterol linked to cardiovascular disease. They act by binding to the HMG-CoA reductase binding site, thus blocking interaction with the substrate.

Akira Endo, working for the Japanese pharmaceutical company Sankyo, became interested in the link between cholesterol and heart disease. Many scientists recognized that the ability to specifically inhibit HMG-CoA reductase would be the key to reducing cholesterol production. Endo and his team decided to screen several thousand fungal cultures for their ability to inhibit the enzyme extracted from rat livers. After two years of testing, they found that an extract from a *Penicillium citrinum* strain contained a potent inhibitor of HMG-CoA reductase. They had identified the first statin. Similar compounds subsequently were found in a number of other fungi, and even in some *Streptomyces* strains of bacteria. Clinical trials showed that these compounds not only reduced cholesterol levels, but also reduced the risk of heart attack. In the United States, roughly one-quarter of all people over the age of 45 take these drugs.

Human proteins as therapeutics

While secondary metabolites constitute a large portion of red biotech products, various human proteins also constitute a significant part of the pharmaceutical industry. Examples of these therapeutics include Type I interferons with antitumor and antiviral properties (see Section 19.4), blood coagulation factor XIIIa used for the treatment of hemophilia, and epidermal growth factor that aids in treating burns and organ damage. Some of these human factors may be expressed in bacterial cells. Other factors are expressed in eukaryal cells, as we discussed in Section 12.2.

One of the earliest successes of recombinant DNA technology involved the production of a human hormone. Hormones are signaling molecules that travel through the bloodstream and regulate the activity of target tissues or organs. Insulin, a protein hormone produced by the pancreas, regulates blood sugar. In people with type 1 diabetes, the destruction of pancreatic cells leads to the decreased production of insulin and a corresponding increase in blood sugar levels. Treatment typically involves the administration of exogenous insulin, usually via repeated injections or an insulin pump that provides a more constant supply of the hormone.

Prior to the advent of recombinant DNA technology, the insulin used to treat people with type 1 diabetes was extracted from pigs or cattle. In 1978, shortly after the first demonstrations of DNA cloning using restriction enzymes in 1973 (see Section 9.4), researchers at Genentech announced the cloning and expression of human insulin in *Escherichia coli*

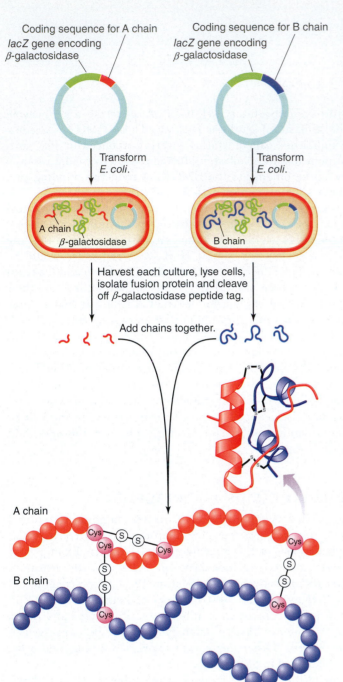

Figure 12.15. Production of recombinant insulin Insulin was the first recombinant human protein to be successfully produced and marketed. The coding sequences for each of the two chains were cloned to make translational fusions with the *lacZ* gene encoding β-galactosidase. Following expression in *E. coli*, the polypeptides were purified and separated, the β-galactosidase tag was removed, and the resulting peptides were combined to make active insulin.

Labels in figure:
Coding sequence for A chain
lacZ gene encoding β-galactosidase
Coding sequence for B chain
lacZ gene encoding β-galactosidase
Transform *E. coli.*
A chain
β-galactosidase
B chain
Harvest each culture, lyse cells, isolate fusion protein and cleave off β-galactosidase peptide tag.
Add chains together.
A chain
B chain
Cys

(Figure 12.15). This achievement marked a milestone in molecular biotechnology. The recombinant human insulin was safer and more plentiful than the pig- and cattle-derived alternatives. Today, various insulin analogs, versions of human insulin experimentally modified to improve their characteristics, are available for people with type 1 diabetes.

CONNECTIONS Today, most of the world's recombinant human insulin is produced in yeast. Production in yeast reduces the costly purification procedures needed to rid the *E. coli* product of residual endotoxin (lipopolysaccharide, or LPS), which can produce severe inflammatory side effects when injected into patients. See **Sections 19.3** and **21.1** for the effects of endotoxin on the body and **Toolbox 19.2** to learn about the assay for endotoxin in products destined for human use.

12.3 Fact Check

1. Describe how the pharmaceutical industry can use microbes to produce drugs that are effective at guarding against some of these disease-causing microbes.
2. How were microbes involved in the development of statin drugs for treating high cholesterol levels?
3. Identify some of the human proteins produced by microbes using recombinant DNA technology.
4. Why would recombinant human insulin be preferred over insulin from animals?

12.4 White biotechnology

What role do microbes play in industrial biotechnology?

White biotech encompasses the many and diverse applications of biotechnology for industrial purposes, from the production of household cleaners to cosmetics to fuel for your car. Currently, the production of many of these products requires the use of fossil fuels. Our fossil fuel reserves, though, are limited and the burning of fossil fuels has environmental impacts. The increased use of microbes in these processes may alleviate some of these obvious problems.

Over the past century, society has transitioned from an agriculture-based economy to a fossil fuel-based economy. The use of coal, oil, and natural gas as fuels or in the production of goods has expanded rapidly, and now forms the basis of modern economies. The oil and gas reserves, however, will eventually be depleted. As demand increases and reserves fall, extraction will become more difficult and the cost of extraction will continue to rise. Perhaps more importantly, burning of fossil fuels increases the atmospheric concentration of gases like CO_2 and N_2O. The resulting greenhouse effect is increasing the temperature of Earth's surface, resulting in human-driven climate change. Of course, billions of years before humans walked Earth, microorganisms radically altered the atmosphere when photosynthetic bacteria began releasing oxygen. That change, however, occurred over millions of years; human industrial activities have added enough greenhouse gases to the atmosphere to affect the global climate in less than two centuries.

Can we alter our society's "energy metabolism" to reduce our dependence on declining oil reserves and reduce the impact of fossil fuel wastes? Scientists hope microbes can help. Bacteria, archaeons, yeast, and algae are being tapped to produce biofuels that can power transportation or generate electricity, and be used in the production of various chemicals. Perhaps microbes can contribute to an energy economy that is more environmentally friendly than the current fossil fuel-based system. In this section, we will first explore the biorefinery concept. Then, we will examine how microbes can be used to produce fuels like ethanol and butanol. We will also investigate how microbes can be used to produce bioplastics. To end this section, we will explore two other areas of white biotech—the production of industrial enzymes and the production of vitamins and amino acids.

The biorefinery concept

Today, oil refineries convert crude petroleum into a number of more usable products such as gasoline, kerosene, wax, and asphalt. Many of these refined products, in turn are used in the production of other materials, like plastics. Our supplies of petroleum, as we noted, are dwindling. Furthermore, the refinery process leads to the production of various pollutants.

A **biorefinery (Figure 12.16)** converts biomass, living or recently living biological material, into a number of products, including chemicals, energy, and materials. Unlike crude petroleum, biological materials as a **feedstock**, or raw starting material, are virtually limitless, providing us with a renewable source of starting materials. Moreover, the harvesting and processing of biomass generally have fewer associated environmental concerns. The biomass used in biorefineries can include crop plants specifically grown for this purpose, as well as agricultural and forestry waste that otherwise would be discarded and eventually decompose.

Crop plants that are specifically grown as biomass feedstock to support the biorefinery industry include the perennial grasses miscanthus and switchgrass and fast-growing trees like poplar. Within the biorefinery, the cellulose/hemicellulose fraction of plant biomass provides the starting material for the microbiological production of fuels such as ethanol and butanol, basic biochemicals such as succinic acid and acetone, and other biopolymers. Cellulose, a complex polymer of β-1,4-linked D-glucose units, provides structure to the plant cell wall and is the most abundant organic molecule on Earth. This fiber is very difficult to break down, but some microbes produce collections of glycoside hydrolase enzymes that can degrade cellulose to its component sugars. Because microbes used for fermentation in the biorefinery usually cannot break cellulose, it must be degraded to its component sugars in advance. This degradation can be achieved through physical processes using heat and alkaline treatment or by glycoside hydrolase enzymes.

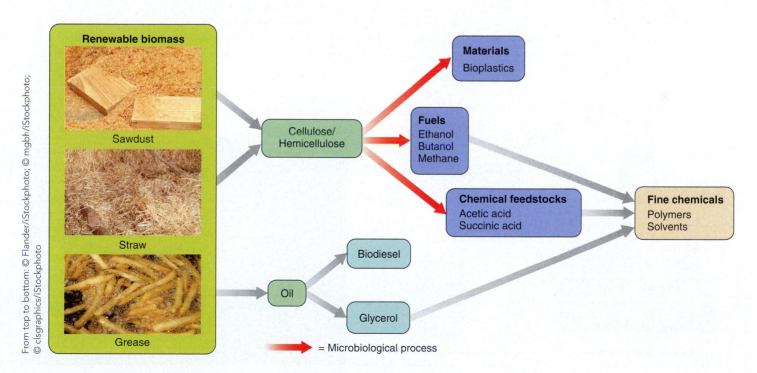

Figure 12.16. The biorefinery The biorefinery concept is modeled on the petroleum refinery; biomass is processed into a number of usable products. In the biorefinery, crude biomass feedstock would be converted to biomaterials, bioenergy, and biochemicals.

CONNECTIONS In **Section 15.4** we will see how natural decomposition is an essential aspect of soil microbiology, and in **Chapter 17** we will examine the symbiotic communities of microbes that can degrade cellulose in animal guts.

Can the products of biorefineries be viable substitutes for fossil fuel-derived products? Perhaps. The efficiencies inherent in the biorefinery concept should contribute to the competitiveness of biofuels and bioproducts, leading to reduced reliance on fossil fuels. Next, we will explore the production of biofuels, biorefinery products that may reduce our dependence on fossil fuels.

Biofuels

Perhaps the largest current pursuit involving biotechnology is the production of renewable biofuels, fuels produced via biomass conversion. These materials could be the key to independence from fossil fuels. While biofuels currently are a hot topic, it might be surprising to know that they are not really new. At the beginning of the twentieth century, as the automobile came into prominence, biofuels actually were the preferred transportation fuels. The original diesel engines ran on biodiesel produced from vegetable oil, and the Ford Model T automobile ran on ethanol produced by corn fermentation. Petroleum, however, proved to be a more plentiful and less expensive source of fuel. Gasoline and diesel soon replaced the biofuels. The resulting development of the petrochemical industry eventually formed the basis of much of the industrial expansion throughout the world. As the world's petroleum reserves become increasingly depleted, the interest in ethanol, biodiesel, and other biofuels as transportation fuels has increased again.

Ethanol as a Biofuel

Ethanol, a 2-carbon organic alcohol, has many uses in the industrial sector, and has garnered much attention in recent years as a biofuel to replace or supplement gasoline as a transportation fuel. As we have seen earlier, strains of yeast like *Saccharomyces cerevisiae* generate ethanol during the fermentation of sugars during anaerobic growth (see Section 13.3). Humans have exploited this process for thousands of years for the production of alcoholic beverages. While our ancestors relied on naturally occurring yeast strains, current commercial production of alcohol is carried out by specific yeast strains that have been chosen for their robustness and their efficient production characteristics. Even industrial yeast strains, however, can only tolerate a maximum of 15 percent ethanol. Increased concentrations of ethanol require distillation, in which manufacturers heat the culture to evaporate the ethanol, which then condenses around cooling coils in the still and is collected **(Figure 12.17)**.

Because internal combustion engines can burn ethanol with little or no modification, this molecule may be a reasonable alternative to gasoline. The cost of commercial ethanol production depends in large part on the cost of the biomass feedstock that is used **(Figure 12.18)**. In Brazil, sugarcane provides an inexpensive source of sucrose, composed of one glucose molecule and one fructose molecule, that supports a large commercial ethanol industry. Most of the motor vehicles sold in Brazil burn ethanol. In North America, much of the ethanol mixed with gasoline is produced from cornstarch that has been broken down to glucose by the addition of amylase enzymes, another microbially produced commercial product (see high fructose corn syrup later in this section). The production of ethanol from these starting materials, though, may have unintended consequences. An increased demand for sugarcane and cornstarch may lead to increased prices for foods containing these products. Additionally, large amounts of fertilizer are used to produce crops such as corn, and the environmental costs of this fertilizer usage may outweigh the benefits of ethanol as a fuel.

To avoid these downsides, waste biomass could be used as feedstock. Most waste biomass from agriculture and forestry operations is composed of a mixture of polymeric cellulose,

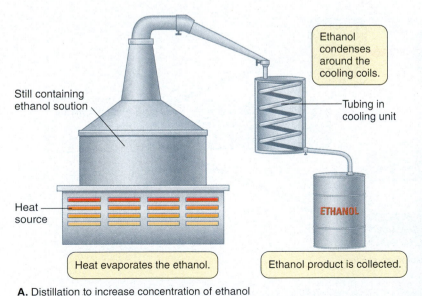

A. Distillation to increase concentration of ethanol

© Pat Corkery/Dept. of Energy/National Renewable Energy Laboratory

B. Industrial ethanol still

Figure 12.17. Ethanol still A. In an ethanol still, heat is used to evaporate the ethanol from the fermented solution, which is then condensed around cooling coils and collected. **B.** Large stills are used for the distillation of industrial ethanol following fermentation.

From top to bottom: (top two photos) Courtesy Christine Dupont; narvikk/iStockphoto; © Ray M. Carson/© AP/Wide World Photos

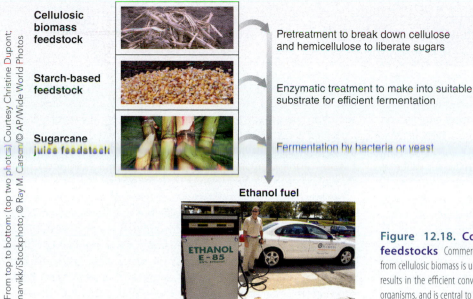

Cellulosic biomass feedstock

Pretreatment to break down cellulose and hemicellulose to liberate sugars

Starch-based feedstock

Enzymatic treatment to make into suitable substrate for efficient fermentation

Sugarcane juice feedstock

Fermentation by bacteria or yeast

Ethanol fuel

Figure 12.18. Commercial ethanol production using different feedstocks Commercial ethanol is commonly produced from sugar and starch; ethanol production from cellulosic biomass is under development, with some demonstration plants in operation. Pretreatment results in the efficient conversion of the cellulose to a form that can be fermented by alcohol-producing organisms, and is central to the successful production of ethanol from biomass.

hemicellulose, and lignin **(Figure 12.19)** collectively called "lignocellulose." Due to its function as a structural material for the plant cell, this material has evolved to be very resistant to microbial degradation. While it cannot be metabolized directly by yeast, cellulase enzymes, acid treatment, or physical methods such as steam explosion can be used to liberate sugars from the lignocellulose polymer. These sugars can then be fermented by yeast to produce ethanol. The growth of the cellulosic ethanol technology has been rapid **(Perspective 12.3)**, and some cellulosic ethanol biorefineries are already operating in North America.

Butanol and Acetone

While the development of ethanol as a broadly used transportation fuel is promising, ethanol may not be the best possible biofuel. It only contains about 70 percent of the energy as gasoline, and is more corrosive than gasoline, making long-term storage difficult.

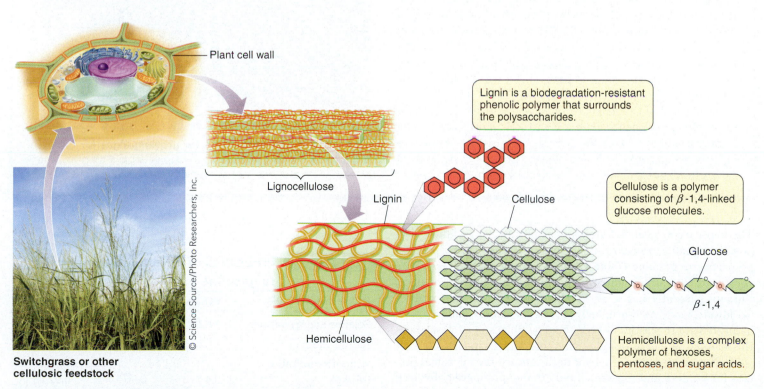

Plant cell wall

Lignocellulose

Lignin is a biodegradation-resistant phenolic polymer that surrounds the polysaccharides.

Lignin

Cellulose

Cellulose is a polymer consisting of β-1,4-linked glucose molecules.

Glucose

β-1,4

Hemicellulose

Hemicellulose is a complex polymer of hexoses, pentoses, and sugar acids.

© Science Source/Photo Researchers, Inc.

Switchgrass or other cellulosic feedstock

Figure 12.19. Lignocellulose biomass Lignocellulose is usually subjected to pretreatment by enzymatic, chemical, or physical methods to release the sugars. These sugars can then be used as feedstock for bioproduction.

Even if ethanol consumption grows dramatically, for the near future it will share the transportation market with diesel fuels. Diesel engines predominate in heavy vehicles, and thanks to more efficient engines and cleaner-burning fuels, are regaining popularity in automobiles. Renewable forms of diesel—biodiesel— are currently made from plant oils. Soybean, palm, and rapeseed (canola) oils lead the market in different parts of the world. The use of plant-derived oils for biodiesel, though, is probably unsustainable using current practices. None of these crops can supply enough oil to significantly reduce global diesel consumption without displacing cropland currently devoted to food crops, or inducing farmers to put more virgin land under cultivation. This is already happening with oil palm in Southeast Asia, where tropical rain forests are being cut for palm plantations and local biodiversity is threatened.

Microbes could offer a more sustainable approach to biodiesel production. Many microorganisms store excess carbon as lipids. Diesel fuel, like gasoline, is simply a mixture of hydrocarbons. The hydrocarbons in petroleum-derived diesel contain 12 carbons on average, whereas plant oils primarily consist of triglycerides with esterified fatty acids of 16 or 18 carbons. Do any microorganisms produce lipids similar to the lipids found in these plants? Yes. The so-called microalgae, photosynthetic unicellular eukarya, can accumulate lipids to tremendously high levels under certain conditions. Some species of microalgae have been shown to accumulate up to 80 percent of their dry weight as lipids. To these cells, lipids are simply a storage depot for photosynthetically fixed carbon. Laboratory-scale experiments have shown that these lipids can be extracted easily from the algae and converted to a fuel mix that is as acceptable for diesel engines as plant-based biodiesel. The potential for algal biodiesel is vast—like plants, they require sunlight and atmospheric CO_2, but the lipid yield per unit of land area is up to 15-fold higher for algae (about 100,000 liters of oil per hectare) than for oil palm, the most productive oil crop (about 6,000 liters of oil per hectare). Canola and soybean only yield about 25 percent and <10 percent the amount of oil per hectare as palm, so they are far inferior to the potential algal oil yield.

So why don't we see vast ponds filled with algae making biofuels? Much research still needs to be done to scale up algal growth for oil production, and make the process economically competitive with petroleum-based fuels and ethanol. During the oil crises of the 1970s, this field flourished. The field languished as oil prices retreated in the next few decades, only to surge again recently as concerns over oil prices and the climate effects of fossil fuel consumption have grown. Techniques for genetically engineering algal strains useful for oil production are still limited, so there must be more investment in studying the basic genetics and molecular biology of these organisms if we hope to improve oil production through biotechnology.

Researchers continue to study many different microalgal species, looking for higher oil yields under different nutritional and environmental regimes. Wastewater is being examined as a possible nutrient source, and CO_2-rich exhaust gases from power plants might turbocharge carbon fixation to boost oil production. Engineers are designing large-scale cultivation systems that maximize light exposure and nutrient availability. Open ponds would be the least expensive growth venue, but would likely experience significant problems with contamination, grazing by protozoa, viral infections, and water evaporation. Closed systems (Figure B12.8) bring those problems under control, but are more expensive and are challenging to illuminate with natural lighting. Plenty of work needs to be done before microalgal biodiesel comes to a pump near you, but the problems seem manageable and the potential benefits may be huge.

Courtesy Solix Biosystems, Inc.

Figure B12.8. A closed system for growing algae for biofuel production Bioreactor systems such as this increase the amount of culture that can be penetrated by light and also help reduce contamination by other microbes.

The four-carbon alcohol, butanol, may be a better and more economically viable transportation biofuel. It has similar physical properties to gasoline (Table 12.3) and can be used directly as a substitute for gasoline in current internal combustion engines. Butanol currently has a $7–8 billion per year annual market worldwide, being used primarily for latex, enamels, and lacquers, as an additive to plastics to keep them flexible, and as a solvent in the manufacture of antibiotics, vitamins, and hormones.

In the early years of the twentieth century, the microbial process for butanol and the associated ethanol production involved fermentation by *Clostridium acetobutylicum*, so named for these metabolic products. Certain isolates of this anaerobic bacterium

TABLE 12.3 Properties of butanol in comparison to gasoline and ethanol

Properties of fuels	Butanol	Gasoline	Ethanol
Relative energy density	91.25	100	61.25
Air-fuel ratio	11.2	14.6	9
Heat of vaporization (MJ/kg)	0.43	0.36	0.92
Motor octane number	78	81–89	102

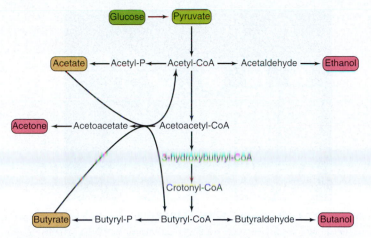

Figure 12.20. Production of acetone, butanol, and ethanol by *Clostridium acetobutylicum* The initial fermentation phase produces butyrate and acetate (brown boxes) that can be used by *Clostridium acetobutylicum* in metabolic pathways that lead to the formation of acetone, butanol, and ethanol (pink boxes).

Figure 12.21. Non-biodegradable plastics washed up on a beach The increasing use of plastic and its resistance to degradation has led to a growing problem with plastics waste. The increased production and use of biodegradable plastics would go a long way toward solving this problem.

naturally produce large amounts of these substances, and with improved fermentation methods, production rapidly expanded in the 1920s. Butanol became a valuable solvent for the production of lacquer used to prepare the finish of automobile bodies. Acetone became important in making cordite, the propellant used in military ammunition during World War I and World War II. The production method was first developed by Chaim Weizman, the future first president of the State of Israel, working in England around 1914. The method used strains of the *C. acetobutylicum* and a cooked corn mash or molasses feedstock. Fermentation occurred in large bioreactors with capacities of 200,000 to 1,000,000 liters. CO_2 was bubbled through the culture to ensure that O_2 was excluded. Fermentation was typically biphasic, with the first phase being acidogenic, forming acetate, butyrate, hydrogen, and CO_2. The cells then use these acids to produce butanol, acetone, and ethanol **(Figure 12.20)**. After 40 to 60 hours, a typical yield was 12–20 g/L of solvent, in a ratio of 6:3:1 butanol/acetone/ethanol. The solvent was removed by distillation, and the remaining microbial dried solids were used as high nutrient animal feed. While this process was robust and relatively efficient, production declined during the 1950s and 1960s as it was out-competed by solvent production from the petrochemical industry. As the cost of fossil fuels rises, however, renewed interest in the biological production of butanol has spurred new research to further improve the process.

Bioplastics

The twentieth century witnessed a tremendous increase in the use of plastics, versatile polymers with a wide range of physical characteristics. While these synthetic materials have many disparate uses, they share one important characteristic—their resistance to biodegradation. As a result, they accumulate in the environment, often with detrimental effects on wildlife and ecosystems **(Figure 12.21)**. To combat this problem, several municipal governments worldwide have banned the use of plastic shopping bags or have begun charging people for their use. In Washington, DC, for instance, the government now charges customers five cents for every disposable bag they use. During

the first year of this regulation, the number of disposable plastic bags used dropped an estimated 80 percent.

Imagine if naturally biodegradable plastics could replace these non-degradable plastics. Microbes may help us obtain this ideal. In the mid-1920s, researchers discovered that the bacterium *Bacillus megaterium* produces **polyhydroxybutyrate (PHB)**, a natural polyester. We now know various microbes can produce an assortment of polyesters, collectively known as **polyhydroxyalkanoates (PHAs)**, that are similar in many ways to the synthetic polyester used today in everything from clothes to plastic bottles to sleeping bags. PHB is the simplest and most common of these polymers formed by microbes. It subsequently has been shown that bacteria store these hydrocarbon polymers under conditions of carbon abundance **(Figure 12.22)**. The cell can use them as sources of carbon and energy under environmental conditions where they would otherwise starve. The short chain length (scl) PHAs are made up of carbon monomers with 3 to 5 carbon atoms, while the carbon monomers of medium chain length (mcl) polymers range from 6 to 16 carbons. Interest in the commercial promise of these bioplastics has led to intensive investigation of the biochemistry and genetics of PHA synthesis.

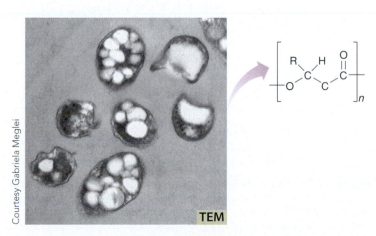

Figure 12.22. Structures of bacterial polyhydroxyalkanoates (PHAs) PHA deposits are accumulated within bacterial cells as carbon storage polymers. Diverse structures of the PHAs result in a broad range of physical properties that are similar to those of existing petroplastics.

Industrial PHA production is well established, with several companies involved in large-scale fermentation. Since PHA is accumulated under conditions of carbon abundance, high levels of PHA can be achieved by supplying limiting amounts of nutrients such as phosphorus or nitrogen, while providing ample amounts of carbon. This can be done in either batch or continuous culture, and is an example of using physiological manipulation to improve production of the desired end product.

CONNECTIONS We encountered PHB granules in **Section 2.2**; they are an example of an inclusion body in the cytoplasm and can form over 50 percent of a cell's dry weight.

One of the factors making fossil fuel-derived plastics so ubiquitous is the large variety of physical properties associated with these materials. Fortunately, such variety also exists in PHAs. While PHB is relatively hard and brittle, alterations in the length of the polymer molecule or the nature of the monomeric constituents can introduce tremendous variability in the physical characteristics. For example, longer chain length PHAs exhibit more elasticity and have lower melting temperatures. Copolymers that contain more than one type of monomer exhibit even more variation in their physical properties. Indeed, almost every major fossil fuel-derived plastic probably could be replaced by a corresponding PHA bioplastic.

The key enzyme for the synthesis of PHA molecules is PHA synthase. This enzyme incorporates monomers into the growing polymer. Genes encoding PHA synthase enzymes have been characterized in a number of bacteria, including cyanobacteria, and some halophilic archaea. For PHB, metabolism of sugars forms 2-carbon acetyl-CoA molecules. When excess acetyl-CoA molecules are formed, two are combined to make the four-carbon acetoacetyl-CoA (**Figure 12.23**) by a ketothiolase enzyme. The acetoacetyl-CoA is then reduced to form the monomer 3-hydroxybutyryl-CoA by an acetoacetyl-CoA reductase enzyme. The monomer is polymerized by PHA synthase into intracellular PHA granules. Bacterial

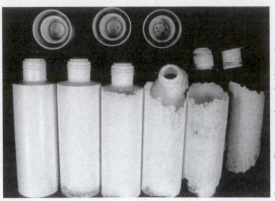

Figure 12.24. Biodegradability of PHA plastic Extracellular PHA depolymerase is active on polymer that has been released from dead cells in the environment. Production of this enzyme by environmental microbes is responsible for the biodegradable nature of the PHA bioplastics.

mutants in any of the three genes encoding these enzymes cannot synthesize the polymer. Introduction of these three genes into some bacteria that naturally do not produce PHA granules, such as *Escherichia coli*, results in PHA production.

Another aspect of the PHA cycle deals with degradation. Materials constructed of PHA are naturally biodegradable (**Figure 12.24**). Remember, cells normally use PHA as a hedge against future starvation conditions that the cell might experience. For the cell to benefit from the accumulated deposits, it must be able to degrade the polymer and catabolize the resulting breakdown product. The key degradative enzyme is PHA depolymerase, which degrades PHA to D-hydroxyalkanoates. These are further oxidized to acetyl-CoA, which can enter the energy-generating TCA cycle (see Figure 13.21). In contrast, fossil fuel-derived plastics are not subject to this type of biodegradation.

The main goals of PHA bioplastics research have been (1) to lower the costs of production to make it competitive with fossil fuel-derived plastics, and (2) to produce plastics with properties appropriate for specific uses. Several companies around the world are now developing such products. More specialized products, such as medical devices like sutures, vascular stents, and scaffolds for tissue engineering have an inherently high commercial value, making their cost of production less significant (**Figure 12.25**). Other materials, like shopping bags, must be

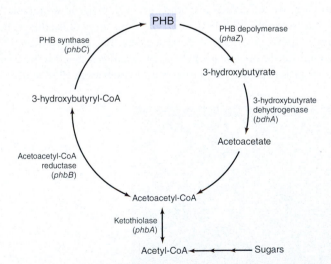

Figure 12.23. The poly-3-hydroxybutyrate (PHB) synthesis cycle Knowledge of the genetics of PHA biosynthesis is important for the production of polymers with desired properties. This figure shows the pathway for synthesis of the most common PHA, poly-3-hydroxybutyrate (PHB). The key enzymes ketothiolase and acetoacetyl-CoA reductase are responsible for building the CoA-activated form of the monomer; synthase carries out the polymerization reaction. The substrate specificity of the PHB (or PHA) synthase enzyme influences the final form of the polymer. The polymer within the cell can be broken down and used as a source of carbon and energy.

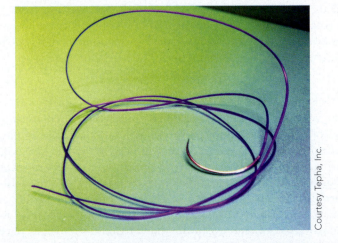

Figure 12.25. Surgical suture made from poly-4-hydroxybutyrate High-value specialty polymers such as this surgical suture are more likely to be cost competitive with petroleum-based plastics, as the raw material cost has relatively little impact on final cost of production.

produced as cheaply as possible. The banning of plastic shopping bags in a number of jurisdictions around the world should promote the adoption of biodegradable PHA bioplastic bags, even if they may be slightly more expensive to produce.

The sun may be a source of energy to produce bioplastics. Photosynthesis in photoautotrophic organisms harnesses the sun's energy that is then used to reduce atmospheric CO_2 to sugars (see Section 13.6). Some photosynthetic bacteria, most notably the cyanobacteria, naturally accumulate large quantities of PHA deposits during photoautotrophic growth. Not surprisingly, researchers have expressed considerable interest in harnessing these organisms for the commercial production of bioplastics. Terrestrial plants might also be able to produce PHA bioplastics efficiently. Being eukaryal organisms, they are unable to naturally produce PHA granules, but bacterial PHA synthesis pathway genes have been successfully introduced into and expressed in a number of different types of plants. These **transgenic plants** (see Section 12.5) containing DNA from another type of organism, successfully produce bioplastics. The amount of production has been relatively low in most cases, so the process is not yet commercially viable, but the results of ongoing research show promise. It will be especially valuable to target PHA production to specific parts of the plant such as the seed, leaf, or stem. It might even be possible to produce the PHA as a coproduct of a food crop. As the cost of fossil fuel plastics increases due to the inevitable dwindling of oil supplies, and the cost of PHA production by microbial fermentation on waste biomass and by photosynthesis in transgenic plants decreases, we may indeed find that the use of environmentally friendly biodegradable plastics becomes widespread.

Figure 12.26. Common consumer products based on industrially produced enzymes HFCS is found in soft drinks and many other foods. A number of different enzymes are used in household detergents to enhance the stain removal properties.

Industrial enzymes

The production of many commercial products, including foods, detergents, textiles, and paper, requires the use of enzymes. Many of these enzymes, often referred to as **biocatalysts**, originate from microbes. In the production process, these microbial enzymes have several advantages over organic chemistry alternatives. The biocatalysts quite frequently exhibit high specificity and high efficiency. Additionally, they are biodegradable. Some examples of these enzymes are lipases, proteases, glycosidases, hydroxylases, nitrilases, acylases, and amidases.

How exactly do these enzymes affect our day-to-day lives? Well, the next time you quench your thirst with a soft drink **(Figure 12.26)**, reflect on the fact that it contains a sugary mix of glucose and fructose called high fructose corn syrup (HFCS). HFCS is produced by the action of amylase enzymes on cornstarch to make corn syrup, followed by treatment with glucose isomerase to adjust the ratio of glucose to fructose. The amylase enzymes are purified from cultures of *Bacillus* sp., while the glucose isomerase is purified from cultures of *Streptomyces* sp. HFCS is much cheaper than sugar, and about 10 million megatonnes of high fructose corn syrup is produced each year in the United States.

When you do your laundry, note that the detergent likely includes a potent mix of lipases, amylases, proteases, glycosidases, and oxidases that work together to remove dirt and stains at lower wash temperatures. In fact, the detergent industry represents the largest single market for microbial enzymes, with other key markets being baking, beverage, and dairy. Because of these applications, the enzymes themselves are valuable products, and considerable effort is expended in finding new enzymes with improved properties such as higher substrate turnover rates and desired substrate and product specificity. Sometimes, companies find better enzymes through bioprospecting. Other times, researchers at these companies try to improve existing enzymes through directed and random mutagenesis approaches like the ones described in Section 12.2.

Vitamins and amino acids

As we conclude this section, let's explore another group of white biotech products—vitamins and amino acids. Today, manufacturers often fortify processed foods by adding nutritive compounds like vitamins to them during production. The reasons for this fortification may be several-fold. The nutritive compounds may be lost from the raw food during processing, or the processed food may be unbalanced in its nutritive value. This strategy of adding nutritive compounds is gaining popularity with the introduction of functional foods that provide specific health benefits beyond their nutritional value. Many breads, pastas, and cereals, for instance, have folic acid (vitamin B_9) added to them to help prevent neural tube defects in the developing fetus. Many individuals also attempt to balance their

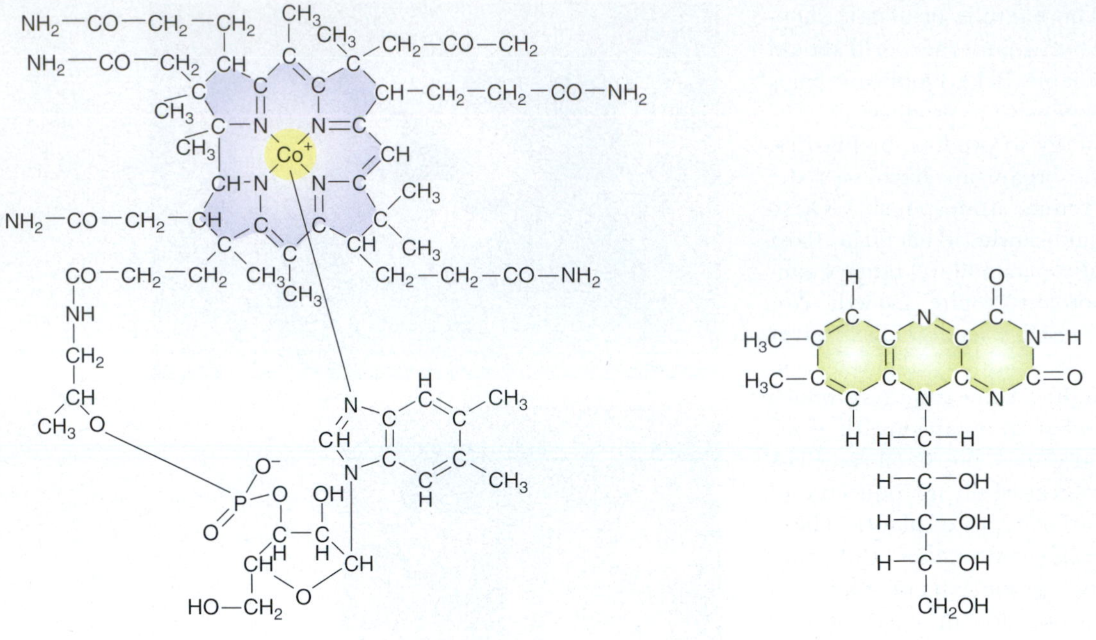

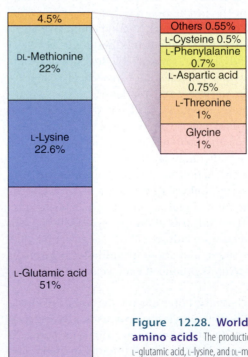

A. Vitamin B₁₂

B. Vitamin B₂

Figure 12.27. Vitamin B₁₂ and vitamin B₂ A. Vitamin B₁₂ (cyanocobalamin) and **B.** vitamin B₂ (riboflavin) are two examples of vitamins produced from microbial cultures for supplementing the human diet.

nutritive intake by supplementing their diet with daily vitamin doses. Most vitamins used for these purposes are synthesized chemically, but several are synthesized as secondary metabolites using microbial culture **(Figure 12.27)**.

To explore the role of biotechnology in vitamin production, let's first look at vitamin B₁₂. Plants do not synthesize or require this compound. Animals, conversely, need vitamin B₁₂, also known as cyanocobalamin. Normally, we obtain a sufficient supply of this vitamin through our food intake and the activities of our gut microbiota. A deficiency of B₁₂ in humans, though, can lead to an illness called pernicious anemia. Usually, a deficiency results from malabsorption in the gut as occurs in intestinal disorders like Crohn's disease and inflammatory bowel syndrome. People with these disorders often take supplemental B₁₂, and manufacturers also add this vitamin to certain fortified foods and multivitamin pills. The chemical synthesis of vitamin B₁₂, however, is extremely complicated. Instead, almost all of the commercially produced vitamin B₁₂, some 10 tonnes per year, is produced by *Pseudomonas* or *Propionibacterium* bacterial strains.

Unlike vitamin B₁₂, vitamin B₂, or riboflavin, can be chemically synthesized relatively easily. Indeed, most commercial B₂ production occurs via chemical synthesis. However, it can also be synthesized in a microbial process that can be economically competitive with the chemical synthesis. Vitamin manufacturers typically maintain the capacity to do both chemical and microbial synthesis. Cultures of two fungi, *Ashbya gossypii* and *Eremothecium ashbyii*, can yield as much as 20 g/L of B₂ over a 7-day growth period. In both instances, natural fungal isolates have been selected over the years to develop industrially useful overproducing strains.

In addition to vitamins, amino acids can also be harvested from microbes. In fact, the industrial production of a wide variety of amino acids amounts to millions of tonnes per year **(Figure 12.28)**. These products have many uses in the food,

animal feed, and nutritional supplement industries, and in the production of synthetic chemicals. L-glutamic acid, for instance, is used as a flavor enhancer, while DL-methionine and L-lysine are used as animal feed supplements. Microbial synthesis contributes to much of this production. An advantage of microbial production of amino acids is stereospecificity (see Figure 2.16). Amino acids produced by microbes are always L-isomers that can be used by the human body, rather than D-isomers, which cannot be used. For example, the common sugar substitute

Figure 12.28. Worldwide production of amino acids The production of amino acids is dominated by L-glutamic acid, L-lysine, and DL-methionine, which are used mostly as additives for food and animal feed.

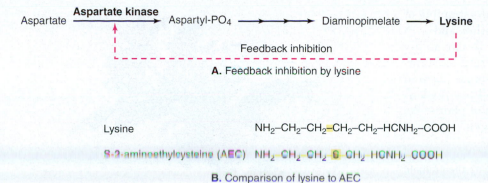

A. Feedback inhibition by lysine

Lysine $NH_2-CH_2-CH_2-CH_2-CH_2-HCNH_2-COOH$

S-2-aminoethylcysteine (AEC) $NH_2-CH_2-CH_2-S-CH_2-HCNH_2-COOH$

B. Comparison of lysine to AEC

Figure 12.29. Biosynthesis of lysine **A.** The production of lysine is enhanced in mutant cells that possess an aspartate kinase enzyme that is no longer subject to feedback inhibition by lysine. **B.** S-2-aminoethylcysteine (AEC) is an antimetabolite lysine analog that will bind to aspartate kinase and shut down the pathway. Mutants of *Corynebacterium glutamicum* that overproduce lysine are obtained by screening for AEC-resistant cells. AEC-resistant mutants overproduce lysine because feedback inhibition is abolished.

aspartame consists of chemically combined L-aspartic acid and L-phenylalanine, produced by *Bacillus flavum* and *Clostridium glutamicum*, respectively.

Most of the microbial strains used in industrial processes for the production of metabolic products have been genetically manipulated, and are in essence metabolically crippled. They cannot survive and compete in the natural environment because their normal metabolic processes have been biased toward producing one or a few metabolites. However, growing as pure cultures within the highly controlled bioreactor environment, they produce relatively large amounts of the desired product. As an example, consider L-lysine synthesis by overproducing strains of *Corynebacterium glutamicum*. Lysine is one of a family of amino acids synthesized from the precursor aspartate (see Figure 13.47). In the cell, aspartate kinase is a key regulatory enzyme in the pathway to lysine, and in wild type *C. glutamicum* aspartate kinase is subject to feedback inhibition as shown in **(Figure 12.29)**. Regulatory mechanisms such as this are important in controlling the relative amounts of amino acids present in the cytoplasmic amino acid pool. Should lysine begin to accumulate in the cytoplasm, its further synthesis would be prevented. Maintenance of cellular balance is not useful in commercial amino acid production, however. Accordingly, a good lysine-producing strain is one in which the regulatory allosteric site of aspartate kinase is non-functional, but the catalytic active site remains functional (see Figure 13.9). Under these conditions, lysine will be made continuously as long as the substrate for its synthesis, aspartate, is present in the culture medium.

Directed enzyme evolution using error-prone PCR can be used to produce such a strain (see Section 12.2).

To select mutants capable of such higher production, an **antimetabolite** of lysine is used. An antimetabolite is a compound that closely resembles the structure of a natural compound. In other words, it is an analog, and it interferes with physiological reactions (see Figure 12.29). The regulatory site of aspartate kinase can recognize and bind to the lysine analog S-2-aminoethylcysteine (AEC) as if it were the true lysine, shutting down activity of aspartate kinase. The strategy used to isolate overproducers entails inoculation of lysine-free, AEC-containing growth medium with large numbers of *C. glutamicum* cells. Wild-type cells are unable to grow under these conditions because they cannot produce lysine in the presence of AEC, and since lysine is not in the medium, they starve. However, a cell that has an altered aspartate kinase regulatory site will be unaffected by AEC, and will be capable of synthesizing lysine and so will grow in the medium. These mutant cells will synthesize lysine continuously because their aspartate kinase enzyme is no longer subject to feedback inhibition. Several other modifications have also been made to lysine overproducers to maximize the amount of cellular carbon flux devoted to lysine synthesis.

12.4 Fact Check

1. What is a biorefinery, and how can it be used?
2. Describe how waste biomass can be used as feedstock.
3. Explain why butanol may be a better biofuel than ethanol.
4. How was *Clostridium acetylbutylicum* originally used to produce acetone, butanol, and ethanol?
5. What are PHAs, and why is there interest in them?
6. Describe a role for photosynthesis in bioplastic production.
7. What are the advantages of using microbial enzymes for commercial applications?
8. Explain how AEC serves as an antimetabolite.

12.5 Green biotechnology

What role do microbes play in agricultural biotechnology?

So far, we have examined the importance of biotechnology in industrial and pharmaceutical settings. Biotechnology is also critically important to modern agriculture. Green biotech refers to the use of biotechnology in this setting. Today, modern agriculture involves the use of large amounts of pesticides, herbicides, and synthetic fertilizers. In addition to being economically costly, this type of intensive agriculture poses human health concerns and environmental repercussions. Green biotech can benefit plant agriculture in several significant ways. The soils that support the growth of plants are complex microbial ecosystems (see Section 15.4). An understanding of the interactions of these microbes with each other and with the subterranean parts of the plant is crucial to our ability to improve plant productivity. Additionally, bacterial systems can be used to introduce genes into plants that impart desired properties to the plants. In this section, we will first look at the use of bacteria to alter the genetic makeup of plants. We will then examine the usefulness of these transgenic traits.

CONNECTIONS In **Section 17.2**, we will investigate the intimate associations that many plants have formed with soil microbes that live as endosymbionts inside the plant. Some of these bacteria can fix nitrogen, and the cultivation of several important crops takes advantage of this bacterial capability. Because of these bacteria, the application of nitrogen fertilizer is not required for optimal crop yield. The agricultural inoculant industry provides cultures of endosymbiotic nitrogen-fixing root nodule bacteria, as well as free-living bacteria and fungi with plant growth-promoting properties such as phosphate solubilization.

Figure 12.30. Crown gall Tumors tend to form at the crown of the plant, where the roots meet the stem, but they can also form elsewhere, even below ground on the root. These tumors result from infection by *Agrobacterium tumefaciens*.

© Nigel Cattlin/Alamy Limited

Agrobacterium—Nature's genetic engineer

Crop plants have been bred for centuries, resulting in the modern forms of plant-based foods that are consumed throughout the world. The application of biotechnology to plant breeding broadens and extends the traits that can be introduced or changed. These traits include taste, yield, nutritional content, pest and pathogen resistance, and shelf life. The introduction of traits depends on the ability to introduce DNA into the plant genome, and then have it expressed. The different methods used to achieve this are all called "transformation," but should not be confused with the term "transformation" in bacterial genetics, which refers to the uptake of exogenous DNA by cells. One of the most efficient ways to introduce DNA into plants is transformation by *Agrobacterium*.

At the beginning of the twentieth century, the researchers Erwin Smith and C. O. Townsend demonstrated that a bacterium they called *Bacterium tumefaciens* caused crown gall disease **(Figure 12.30)**, which produces tumors in many dicot plants. Scientists later changed the name of this plant pathogen to *Agrobacterium tumefaciens* (see Microbes in Focus 11.2). Active research over the years demonstrated that the bacterium was required for the initiation of the tumor. The bacterial cells, though, could be killed after tumor initiation without interfering with tumor growth. In other words, the tumors continued to grow in the absence of bacteria.

Researchers began to understand the basis of crown gall tumor formation in the 1970s and 1980s. First, they showed that tumorigenic bacteria contain a large plasmid and removal of this plasmid from the bacteria prevented tumor formation. It was concluded that the plasmid was necessary for tumor formation, and it was named pTi, for "tumor-inducing plasmid." The plasmid could move between strains of *A. tumefaciens* by conjugation, and this conjugation was induced by opines, amino acid-like compounds produced by *Agrobacterium*. The real breakthrough in understanding the tumor formation process was made when researchers showed that part of pTi, called "transfer DNA" or "T-DNA," was transferred from the bacteria

to the plant cells and integrated into the plant's genome. Eventually, it became clear that the expression of genes on T-DNA in the plant nucleus resulted in the production of the opines, as well as plant hormones (also referred to as phytohormones) like auxins and cytokinins. These materials result in plant cell proliferation. The bacteria, in essence, are natural genetic engineers. They provide the plant with the genetic ability to produce a tumor, within which the bacteria then reside. Moreover, the tumor cells produce large amounts of opines that the bacteria can use as nutrients.

The mechanism of tumor formation by *A. tumefaciens* is fascinating. It represents one of the first demonstrated examples of trans-kingdom genetic transfer. Perhaps more importantly, though, it turns out that *Agrobacterium* can be used for the genetic engineering of plants. Experiments quickly showed that the opine and phytohormone biosynthetic genes within the T-DNA could be replaced by any other genes, and the T-DNA could still be transferred to the plant nucleus, carrying with it these foreign genes. Following this observation, researchers immediately realized that novel traits could be introduced into plants through T-DNA-mediated DNA transfer. These investigators quickly developed methods for *Agrobacterium*-mediated plant transformation. From the transformed cells, transgenic plants, genetically engineered to contain foreign DNA, can be generated. Such transgenic crops are now widely grown in North America. We will examine the technique of *Agrobacterium* transformation as a biotechnology tool in **(Toolbox 12.3)**.

While *Agrobacterium*-based transformation systems efficiently and effectively introduce foreign DNA into the cells of some plants, the cells of other plants cannot be readily transformed. Unfortunately, plants in this group include agriculturally important crops such as wheat, barley, and rice. These monocots are not naturally susceptible to crown gall. Diligent research, often focused on immature embryos, has resulted in the optimization of *Agrobacterium* transformation for some of these plants. Other methods of introduction of DNA into plant cells have also been developed. These methods include protoplast transformation, in which the cell wall is removed prior to introduction of DNA, and particle gun biolistics transformation, in which metal fragments coated with DNA are fired into the cells. Once inside the cell, the DNA can be taken up by the nucleus and then integrated into the genomic DNA.

Applications of transgenic plants

Once it had been demonstrated that transgenic plants could be generated, researchers then considered which traits to introduce into the major crop plants such as corn, soybean, cotton, canola, and rice. Two traits targeted for engineering into crop plants were resistance to nontoxic herbicides and resistance to insect pests. In both cases scientists turned to microbes to find useful genes for these traits. Here, we will briefly examine both of these examples.

Herbicide Resistance

Weeds are the bane of all farmers, big and small. For the home gardener, weeds can be physically removed by plucking them one at a time. That approach certainly will not work for farmers who oversee crops covering hundreds or thousands of hectares! To make the large-scale farmers' job a bit easier, the U.S. company Monsanto produces a broad-spectrum herbicide marketed as Roundup. The active ingredient in this herbicide, glyphosate, is a derivative of glycine (*N*-phosphonomethyl-glycine; **Figure 12.31**). Plant tissues efficiently absorb glyphosate, which then inhibits the activity of the plant 5-enolpyruvylshikimate-3-phosphate (EPSP) synthase, an enzyme in the aromatic amino acid synthesis pathway. Inhibition of this enzyme blocks the production of the aromatic amino acids phenylalanine, tryptophan, and tyrosine. No longer able to produce these amino acids, plants treated with Roundup die. Humans and other mammals do not possess this biosynthetic

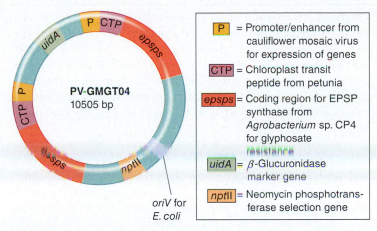

Figure 12.32. Plasmid map The plasmid construct used to introduce the *Agrobacterium* sp. CP4 EPSP synthase gene into soybean plants contains two copies of EPSP synthase gene fused to petunia chloroplast transit peptide (CTP) and under the control of plant virus promoters. The *uidA* gene from *Escherichia coli* provides a marker for the detection of successful transformation events.

pathway and, as a result, must acquire these essential amino acids from their diet. The lack of this aromatic amino acid synthesis pathway also means that glyphosate is not toxic to mammals, a very desirable trait for any pesticide.

Unfortunately, glyphosate is toxic to most crop plants, which meant that farmers could spray it on their fields to kill weeds only before planting crops. Scientists tried for many years to isolate glyphosate-resistant plants by selection so they could use glyphosate after planting too. These efforts met with little success. Then microbes came into the picture. Several soil microbes exhibit resistance to glyphosate and actually metabolize it, rapidly removing it from the soil. Scientists at Monsanto isolated the EPSP synthase gene from one of these bacterial isolates, *Agrobacterium* sp. CP4, and confirmed the resistance for the expressed enzyme to glyphosate. Next, they attempted to introduce this gene into plants. Because plant EPSP enzymes localize to chloroplasts, the researchers fused the *Agrobacterium* version of the EPSP gene to DNA encoding the chloroplast transit portion of the petunia EPSP synthase gene (**Figure 12.32**). When expressed in plant cells, the resulting fusion protein, they reasoned, would be transported to the chloroplast. The investigators then introduced the gene fusion construct into soybean cell particle gun biolistics transformation and successfully generated transformed plants. Screening of these transgenic plants identified lines exhibiting glyphosate resistance. In other words, these plants exhibited resistance to Roundup. Since the development of these first "Roundup Ready®" plants, glyphosate resistance has been incorporated into several major crop plants including corn, cotton, and canola. Farmers planting these herbicide-resistant transgenic plants can now use Roundup to control weeds throughout the growing season.

Insect Resistance

Like weeds, insect pests are also the bane of farmers. To prevent extensive crop loss due to these pests, farmers typically spray various insecticides on their fields. While these insecticides reduce insect damage to the crops, thereby increasing crop yields, this increased yield comes at a cost. Many of the

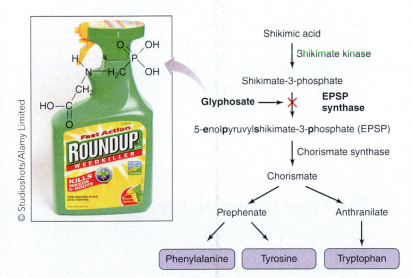

Figure 12.31. Glyphosate Glyphosate is a specific inhibitor of the EPSP synthase enzyme, which catalyzes a key step in the synthesis of aromatic amino acids. It is a very effective herbicide. Humans and other mammals do not contain the aromatic amino acid synthesis pathway, as they do not produce these amino acids. This greatly reduces the likelihood of human toxicity.

chemical insecticides used in agriculture are potent neuro-toxins and may have detrimental health effects on humans, livestock, and wildlife. Additionally, many of these insecti-cides affect non-pest species of insect, too. For these reasons, numerous groups sought alternative strategies for the man-agement of insect pests. In particular, the study of eco-friendly insect pathogens increased in intensity during the mid-twen-tieth century.

One bacterial species, *Bacillus thuringiensis* (**Microbes in Focus 12.3**), has received considerable interest. This spore-forming soil bacterium produces intracellular protein crystals called "δ-endotoxins," or "Cry proteins," associated with its spores (**Figure 12.33**). These Cry proteins, often referred to as **Bt toxin**, have highly specific insecticidal activity against lepidopteran (moths and butterflies), dipteran (flies and mosquitoes), or coleopteran (beetles) larvae. The full-length Cry proteins are not active, but upon ingestion by the larvae, the protein crystals dissolve in the alkaline conditions of the insect midgut, where they are cleaved by proteases. The resulting active polypeptides bind to specific receptors on the gut epithelial cells, producing pores in the cell membrane, thus disrupting the osmotic bal-ance and killing the insect. In a natural setting, killing of insect larvae by *B. thuringiensis* likely provides a source of nutrients for vegetative growth of the bacteria. When it was found that some *B. thuringiensis* strains contained multiple Cry proteins of dif-fering specificities, scientists could target specific agricultural

Microbes in Focus 12.3
BACILLUS THURINGIENSIS: THE NATURAL, SAFE INSECTICIDE

Habitat: First isolated and described in Japan in 1902; soil bacteria are pathogenic to insects but not mammals. Often found in insect guts and on plants.

Description: This rod-shaped member of the Firmicutes phylum has a typical Gram-positive cell structure, averaging about 1 μm in width by 5 μm in length. Heat and desiccation-resistant en-dospores are formed, as well as spore-associated crystal proteins called μ-endotoxins, which are insecticidal. The genome size rang-es between 5.3 and 6.2 megabases, depending on the strain.

Key Features: Commonly known simply as Bt, the major distin-guishing feature of this microbe is the production of μ-endotoxins. Upon ingestion by an insect, the high pH of the insect midgut activates the toxin. The toxin then binds to the in-testinal epithelial cells, forming pores that cause cell lysis. Each toxin affects a defined range of insects. Bt com-monly is used as an insecticide spray,  SEM © SciMAT/Photo Researchers, Inc.
and the toxin genes have been expressed in transgenic plants, resulting in resistance of the plants to insect pests.

Toolbox 12.3
PLANT TRANSFORMATION USING BACTERIA

Known for its efficiency, *Agrobacterium*-mediated transformation involves replacing the normal genes within the T-DNA with genes to be introduced into a plant. This can be done using a binary vec-tor system. One *Agrobacterium* strain houses a modified pTi with *vir* genes that encode the DNA transfer ability but from which the T-DNA region has been deleted (**Figure B12.9**). A separate plas-mid contains the 25-bp borders of the T-DNA between which the DNA to be transferred is cloned. This plasmid can be easily maintained in *Escherichia coli* until the desired gene is transferred to a plant. At that time, the plasmid containing the gene is intro-duced into the *Agrobacterium* strain containing the vector with the *vir* genes. Transfer and expression of the modified T-DNA does not cause tumor formation because the tumor formation genes have been deleted. The modified T-DNA does include genetic markers such as genes encoding antibiotic resistance and β-glucuronidase enzyme activity. Transformed cells can easily be identified using these markers.

The process of *Agrobacterium*-mediated transformation starts with the preparation of leaf discs using a standard office single-hole puncher (**Figure B12.10**). The discs are then immersed in a suspen-sion of *Agrobacterium* cells that have been pretreated with an ap-propriate phenolic compound such as acetosyringone to ensure induction of the *vir* genes. This step is done under partial vacuum, which helps the *Agrobacterium* cells enter the plant tissue in a process called "infiltration." The leaf discs are then placed on solid plant growth regeneration medium supplemented with phytohormones to promote plant growth and antibiotics to inhibit the growth of *Agrobacterium* and non-transformed plant cells. Cell masses called "calli" that grow out of the disc and are stained blue by X-glucuronide due to β-glucuronidase activity are candidate transformants. PCR and Southern hybridization methods can confirm that the introduced DNA is integrated into the plant genome. Because plant cells are totipotent, meaning that com-plete organisms made up of different cell types can develop from individual mature plant cells, fully functioning plants can be generated from the calli. These plants can then be analyzed to confirm the traits encoded by the introduced genes and bred with other varieties to combine the introduced genes with different genetic backgrounds.

● **Test Your Understanding** ....

Imagine you forgot to add antibiotics to the leaf discs' growth regeneration medium. Explain why your *Agrobacterium*-mediated plant transformation process would not be successful.

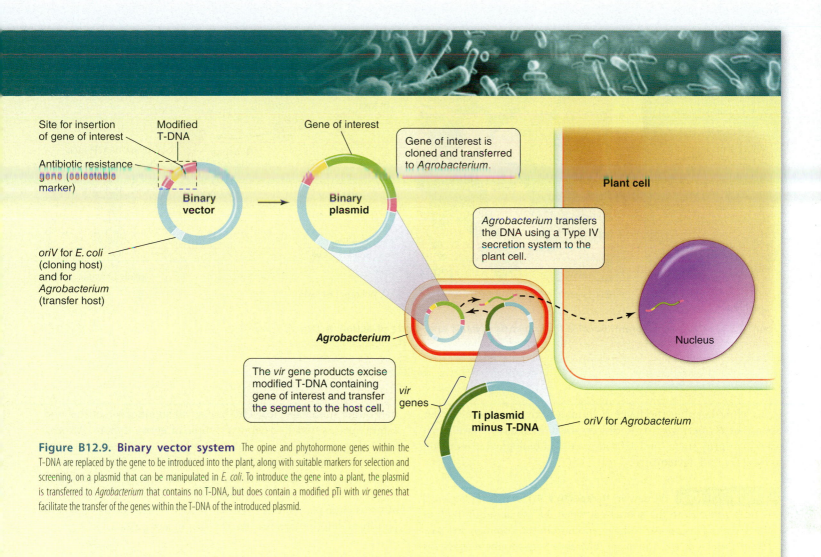

Site for insertion of gene of interest

Modified T-DNA

Antibiotic resistance gene (selectable marker)

Binary vector

oriV for *E. coli* (cloning host) and for *Agrobacterium* (transfer host)

Gene of interest

Binary plasmid

> Gene of interest is cloned and transferred to *Agrobacterium*.

Plant cell

> *Agrobacterium* transfers the DNA using a Type IV secretion system to the plant cell.

Agrobacterium

Nucleus

> The *vir* gene products excise modified T-DNA containing gene of interest and transfer the segment to the host cell.

vir genes

Ti plasmid minus T-DNA

oriV for *Agrobacterium*

Figure B12.9. Binary vector system The opine and phytohormone genes within the T-DNA are replaced by the gene to be introduced into the plant, along with suitable markers for selection and screening, on a plasmid that can be manipulated in *E. coli*. To introduce the gene into a plant, the plasmid is transferred to *Agrobacterium* that contains no T-DNA, but does contain a modified pTi with *vir* genes that facilitate the transfer of the genes within the T-DNA of the introduced plasmid.

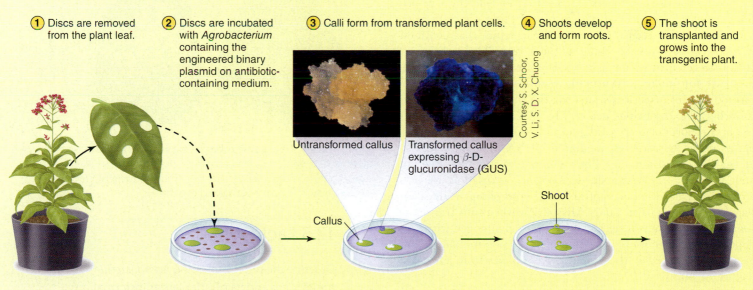

1 Discs are removed from the plant leaf.

2 Discs are incubated with *Agrobacterium* containing the engineered binary plasmid on antibiotic-containing medium.

3 Calli form from transformed plant cells.

4 Shoots develop and form roots.

5 The shoot is transplanted and grows into the transgenic plant.

Untransformed callus

Transformed callus expressing β-D-glucuronidase (GUS)

Courtesy S. Schoor, V. Li, S. D. X. Chuong

Callus

Shoot

Figure B12.10. Process Diagram: *Agrobacterium*-mediated transformation After their formation, calli are transferred to media that promote the growth of shoots and roots.

A. Packaging for commercially prepared Bt crystals

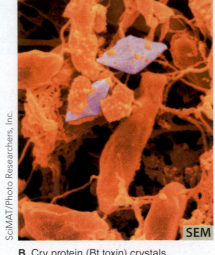

B. Cry protein (Bt toxin) crystals

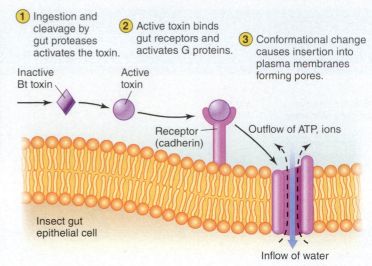

① Ingestion and cleavage by gut proteases activates the toxin.

② Active toxin binds gut receptors and activates G proteins.

③ Conformational change causes insertion into plasma membranes forming pores.

C. Action of Bt toxin

Figure 12.33. Process Diagram: *Bacillus thuringiensis* **crystals and mode of action A.** Bt crystals are commercially produced and widely marketed for insecticide use. **B.** This electron micrograph shows intracellular protein crystals that are produced during sporulation and have insecticidal activity that is only activated after ingestion by the larvae. **C.** After the toxin is activated, it binds to cadherin receptors in the membranes of gut epithelial cells, where it forms a pore that increases permeability to water, cations, and ATP, resulting in cell lysis.

pests while keeping the level of human toxicity extremely low. Spores of *B. thuringiensis* or the crystals themselves have been used in a spray since the 1920s as a natural biological insecticide in forestry and agriculture.

CONNECTIONS We saw in **Perspective 9.1** that *B. thuringiensis* is closely related to the animal pathogen *Bacillus anthracis*. Both of these pathogens appear to have arisen from the common soil bacterium *Bacillus cereus*. The major difference between these species is the presence or types of virulence plasmids, which are transferable by conjugation.

Soon after discovering the genes that encode the Cry proteins, researchers successfully expressed these genes in other bacteria like *Pseudomonas fluorescens*, a bacterium that commonly interacts with plants. Some of these recombinant bacterial strains were cultured for use as sprays for plant protection. As methods for making transgenic plants were developed, the introduction of insect resistance via the *B. thuringiensis* Cry proteins became an obvious priority. In 1987, scientists published the first reports of successful generation of Bt toxin transgenic plants using *Agrobacterium*-mediated T-DNA transfer. Scientists at the biotech companies Plant Genetic Systems in Belgium and Agracetus in Wisconsin worked with transgenic tobacco, a commonly used experimental model crop plant, while scientists at Monsanto in St. Louis, Missouri, worked with transgenic tomato. When the transgenic plants were challenged with lepidopteran larvae, not only was there a reduction in damage, but also larvae that fed on these plants were killed within a few days **(Figure 12.34)**.

Following these demonstrations, scientists soon introduced Bt toxin genes into major crops such as corn and cotton, and these insect-resistant transgenic crops gained popularity very quickly. As good as it sounds, there is caution that these transgenic plants may have some safety issues. Bt toxins can affect closely related

non-target insect species. The biological significance of this non-target species toxicity remains a topic of active debate. Also, we know that insects tend to evolve resistance to conventional insecticides, and not surprisingly, resistance to Bt toxin in transgenic cotton has been reported for some insect species in India. Most likely, more examples of resistance will follow. The generation of transgenic plants that express Bt toxin certainly has not completely ended the farmers' battle with insect pests.

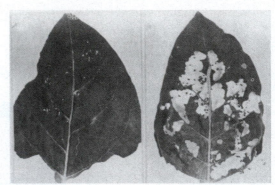

Figure 12.34. Transgenic tobacco expressing Bt toxin The expression of Bt toxin in transgenic plants such as transgenic tobacco (*left*) showing resistance to tobacco hornworm larvae, compared to the non-transgenic leaf (*right*) can be very effective at preventing major crop losses due to insect predation while avoiding the use of chemical insecticides.

12.5 Fact Check

1. Describe the process of using *Agrobacterium tumefaciens* and its pTi for plant transformations.
2. Explain the process that was used to generate "Roundup Ready" plants.
3. What is Bt toxin, and how is it used in agriculture?

Image in Action

The petri dishes on the lab bench have a thick white lawn of *Staphylococcus* bacteria growing on them, as well as large colonies of *Penicillium chrysogenum* (green with white edges). *Penicillium* produces an antibiotic that inhibits the growth of *Staphylococcus*, resulting in a clear halo

that surrounds the colonies. The structure of penicillin, the antibiotic derived from *Penicillium*, is shown in the ball-and-stick model above, and a prescription bottle filled with penicillin tablets is shown in the lower left.

1. Explain why the microbe in this image would naturally produce a secondary metabolite that inhibited bacterial growth, and why this understanding leads bioprospectors to continue searching for antimicrobial compounds in nature.

2. Imagine you are asked to use mutagenesis techniques to improve the penicillin production in this mold. Identify and explain some mutagenesis options that are available to you for this task.

The Rest of the Story

After the commercialization of penicillin, the antibiotic industry blossomed. Several classes of antibiotics were discovered, mostly produced by actinomycetes of the genus *Streptomyces*. Currently, though, the discovery of new antibiotics has slowed. Given the increasing incidence of pathogenic bacteria that exhibit resistance to all existing antibiotics, the need for new classes of antibiotics has become critical.

Malaria is one of the most serious human diseases in the world. Transmitted by the bites of infected mosquitoes, it is caused by the *Plasmodium falciparum* parasite (see Section 23.2). The parasite multiplies in the liver and then infects red blood cells. Symptoms, including fever, headache, and vomiting, appear 10–15 days after infection, and if not treated, can become life threatening by disrupting blood supply to vital organs. Half of the world's population is at risk for malaria. The disease causes 1–3 million deaths annually, and accounts for 20 percent of all childhood deaths in Africa. The main therapeutic in use is chloroquine, which interferes with the parasite's ability to detoxify the heme by-product of its hemoglobin breakdown. While this treatment is quite effective, resistant strains of the parasite pump the chloroquine back out of the cell.

In searching for an alternative therapy, researchers observed that extracts from the sweet wormwood plant *Artemisia annua*, as used in traditional Chinese medicine, effectively cleared the parasite. Extraction of the active ingredient, artemisinin, however, is very labor intensive and not very reproducible. In a heroic effort, using the principles of synthetic biology, Jay Keasling and coworkers at the University of California at Berkeley and Amyris Biotechnologies re-engineered the pathway for artemisinin synthesis in bacteria, using mostly bacterial and yeast genes. For one of the key last steps, they cloned the key cytochrome P450 encoding gene from the sweet wormwood plant and introduced it into their production strain to complete the pathway. The resulting recombinant bacteria efficiently produced artemisinic acid, from which artemisinin can be obtained by a one-step chemical conversion. Using biotechnology, researchers generated a bacterium that could produce a biological compound difficult to obtain in nature. Using biotechnology, researchers may have found a cheap, easy way to produce a life-saving drug.

Summary

Section 12.1: What microbes and genetic materials are available for biotechnology applications?

Microbes that are used for biotechnology applications are all derived from isolates that originate in natural environments. Most of them have been improved by mutagenesis, screening for enhanced properties, and genetic engineering.

- Beginning with agriculture, over thousands of years humans have taken organisms from the wild and used them for production of food and materials.
- Microbes originating from natural environments are the basis of **biotechnology**, but they are not optimized for biotechnology processes; molecular biotechnology involves genetic engineering of organisms for the production of commercial products.

- **Red biotechnology** is related to medical applications, **white biotechnology** is related to industrial applications, and **green biotechnology** is related to the agricultural sector.
- **Culture collections** provide sources of different types of microbes that can be used in biotechnology.
- **Bioprospecting** is often necessary to obtain microbes with characteristics that are useful or critical for specific biotechnology applications.
- Because most organisms have not been cultivated, the use of cultivation-independent methods such as metagenomics can provide rich sources of genetic material that can be useful in biotechnology applications.

- **Fermentation** in biotechnology refers to the conversion of biomass from one form to another by microbial cell cultures.
- **Bioreactors** are production systems in which cell growth is closely regulated through control of culture conditions.
- **Fed-batch reactors** add a growth-limiting nutrient over time, controlling the growth rate and resulting in very high cell densities.
- **Chemostats** are continuous bioreactors in which an equivalent amount of culture is removed as new medium is added, resulting in very precise control of growth rate and maintenance of physiological steady state.
- **Primary metabolites** are produced during exponential growth; **secondary metabolites** are produced in stationary phase.

Section 12.2: How can molecular biology tools be used to improve microbial strains?

Random mutagenesis followed by phenotypic selection or screening can be very effective at yielding strains that are better producers, but since detrimental mutations can also accumulate, this often results in strains that have reduced vigor. The use of molecular biology allows the genetic changes to be targeted to a particular gene or collection of genes, reducing the chance of creating unwanted genetic changes.

- **Site-directed mutagenesis** allows the construction of precise sequence changes in a gene or in an organism's genome.
- **Directed enzyme evolution**, using methods such as **error-prone PCR** and **DNA shuffling**, is used to improve strains by limiting mutations to a particular gene of interest.
- **Expression vectors** have been developed for several different types of cells, ranging from bacterial to mammalian cells, and have been developed as popular host organisms to be used as the basis of production systems.
- Expression of a eukaryal gene in a bacterial host often uses **complementary DNA (cDNA)**, produced from an mRNA transcript using reverse transcriptase, inserted into an expression vector.
- **Fusion proteins** contain domains of more than one protein, often including an **affinity tag** that helps in protein purification.
- **Synthetic biology** is an emerging discipline in which the goal is to design and build novel organisms to defined specifications to carry out desired tasks.

Section 12.3: What role do microbes play in pharmaceutical biotechnology?

The pharmaceutical industry is heavily reliant on microbes as production organisms. Although some of the products are derivatives of natural products, others are based on human proteins that can be produced in microbes using genetic engineering methods.

- The two major uses of microbes in the pharmaceutical industry are as the producers of secondary metabolites with therapeutic properties, and as hosts for the production of recombinant human proteins such as hormones.
- Antibiotics that are naturally produced by bacteria and fungi have been major contributors to improved public health.
- Improvements of strains and fermentation techniques have resulted in drastic reductions in production costs for antimicrobial drugs.
- Statins, produced by several fungi, are an important class of drugs that inhibit cholesterol synthesis.

- *Streptomycetes*, which are filamentous bacteria, are the major producers of antibiotics.
- Several human proteins have been produced using recombinant methods, and are used as therapeutic drugs.

Section 12.4: What role do microbes play in industrial biotechnology?

Most industrial biotechnology is based on microbial conversion of low cost biomass to products that have higher value. Biotechnology is likely to become more important as society transitions to an economy that is based more on biomass than on fossil fuels.

- In **biorefineries**, a broad range of chemicals and **feedstocks** can be produced through fermentation of biomass feedstock.
- A key to successful fermentation of biomass feedstock is the ability to break down cellulose, which is very resistant to microbial degradation, but it can be achieved with the help of cellulose-degrading enzymes.
- Production of ethanol from lignocellulosic biomass rather than from cornstarch would result in greater environmental and energy benefits, but this requires the conversion of lignocellulose to fermentable substrate.
- Butanol and acetone fermentation was well established in the early part of the twentieth century, but this production was supplanted by petrochemical sources, although there is renewed interest as fossil fuel costs rise.
- Bioplastics based on **polyhydroxyalkanoates (PHAs)** such as **polyhydroxybutyrate (PHB)**, natural storage compounds within bacterial cells, can be used as biodegradable replacements for fossil fuel-derived plastics.
- **Transgenic plants** have been engineered to produce bioplastics.
- Enzymes, as **biocatalysts**, are used in many industrial processes due to their high specificity and low energy requirements.
- **Antimetabolites**, which interfere with physiological reactions due to their similar structure to natural compounds, can often be used to select for mutants that overproduce desired products.

Section 12.5: What role do microbes play in agricultural biotechnology?

The use of biotechnology to improve crop plants has resulted in the introduction of traits such as yield increase, insect resistance, and resistance to nontoxic herbicides that can then be used for enhanced weed control without harming the crop plant.

- One of the most effective methods to introduce genes into plants is through the natural genetic engineer, the bacterium *Agrobacterium tumefaciens*.
- Other methods of plant transformation, such as protoplast transformation and biolistics transformation, have also been developed and are widely used.
- Crop plant resistance to the herbicide glyphosate, based on introduction of the bacterial EPSP synthase gene, allows the use of glyphosate throughout the growing season to control weeds.
- **Bt toxin** has been used for several years as a spray insecticide, and genes for Bt toxin are now used extensively for the production of crop plants that are resistant to insect damage.
- Hundreds of different crop plant species have been produced by transformation methods, and this has revolutionized plant biology research and the ability to produce improved breeds of crop plants.

● Application Questions ....

1. The Nature Conservancy reports that 70 percent of the plants identified by the U.S. National Cancer Institute as useful in the treatment of cancer are found only in rain forests. Explain how a bioprospector might have played a role in determining this statistic. Next, explain how biotechnology would play a role after the bioprospecting is carried out.

2. Imagine you are taking part in an online discussion about the safety of rBGH (recombinant bovine growth hormone). One person comments that she's heard recombinant proteins are bad but she doesn't even know what they are. Explain what a recombinant protein is and why her blanket statement isn't completely true by describing an example of a recombinant protein used in medicine.

3. Bioplastics are biodegradable, while fossil fuel-derived plastics are resistant to degradation by microbes. Explain what is thought to account for this difference.

4. A novice researcher is doing research on ethanol production using a strain of yeast. She gives her yeast strain the right conditions, and they begin producing ethanol, but ethanol production always stops at a certain point and no more is made. Explain to the researcher why this is happening. How could she use biotechnology methods to get around this?

5. You are working for a new biotechnology company as a lab technician. Because the company is competing with others, their work is proprietary and kept secret. As a technician, you are given individual tasks but aren't always sure what they will be used for. You are currently working on cloning the gene for PHA polymerase and expressing it in a new strain of *Escherichia coli*. What is this enzyme used for, and what does this tell you about the goals of this start-up company?

6. High fructose corn syrup (HFCS) is a widely used sweetener produced from cornstarch. Your lab partner has heard of this chemical but doesn't understand what it is or where it comes from. Explain to her how it is produced, including enzymes used in its production, what reactions those enzymes catalyze, and how microbes are involved in the process.

7. Explain the action of an antimetabolite. Then explain how one might be used as an antimicrobial treatment.

8. Imagine you are asked to use an antimetabolite in a biotechnology lab that is attempting to overproduce a specific vitamin by altering the synthesis pathway. What role might the antimetabolite play in this situation?

9. While there are those who oppose the use of transgenic crop plants, there are also arguments that they have the potential to reduce the amount of synthetic chemicals used in plant agriculture. Describe an example to illustrate this point.

10. A research team is attempting to develop an edible vaccine. Their overall strategy is to get certain microbial proteins expressed in a tomato plant (specifically, the tomato fruit itself). Outline and explain how *Agrobacterium tumefaciens* could be used to introduce these microbial proteins into the tomato plant.

11. The enzyme glutaminase may be a possible cancer treatment, as it has some ability to inhibit tumor growth. Outline and describe steps you would use to carry out directed enzyme evolution to produce an improved cancer-fighting glutaminase enzyme. Be specific in describing your mutagenesis and screening approaches.

12. A microbiologist has isolated a *Streptomyces* strain with a novel antimicrobial activity. You have identified a unique enzyme in this strain responsible for the activity.
    a. You believe the activity could be increased to improve the antimicrobial. Outline several biotechnology approaches you could use to increase the efficiency of this enzyme's action.
    b. Explain how you would go about producing this enzyme using biotechnology methods.

Suggested Reading

Beloqui, A., P. D. de Maria, P. N. Golyshin, and M. Ferrer. 2008. Recent trends in industrial microbiology. Curr Opin Microbiol 11:240–248.

Demain, A. L., and J. L. Adrio. 2008. Contributions of microorganisms to industrial biology. Mol Biotechnol 38:41–55.

Fortman, J. L., S. Chhabra, A. Mukhopadhyay, H. Chou, T. S. Lee, E. Steen, and J. D. Keasling. 2008. Biofuel alternatives to ethanol: Pumping the microbial well. Trends Biotechnol 26:375–381.

Funke, T., H. Han, M. L. Healy-Fried, M. Fischer, and E. Schönbrum. 2006. Molecular basis for the herbicide resistance of Roundup Ready crops. Proc Natl Acad Sci USA 103:13010–13015.

Gibson, D. G., J. I., Glass, C. Lartigue, V. N, Noskov, R. Chuang, M. A. Algire, G. A. Benders, M. G. Montague, L. Ma, M. M. Moodie, C. Merryman, S. Vashee, R. Krishnakumar, N. Assad-Garcia, C. Andrews-Pfannkoch, E. A. Denisova, L. Young, Z. Qi, T. H. Segall-Shapiro, C. H. Calvey, P. P. Parmar, C. A. Hutchison, III, H. O. Smith, and J. C. Venter, 2010. Creation of a bacterial cell controlled by a chemically synthesized genome. Science 329:52–56.

Goeddel, D. V., D. G. Kleid, F. Bolivar, H. L. Heyneker, D. G. Yansura, R. Crea, T. Hirose, A. Kraszewski, K. Itakura, and A. D. Riggs. 1979. Expression in *Escherichia coli* of chemically synthesized genes for human insulin. Proc Natl Acad Sci USA 76:106–110.

Keasling, J. D. 2008. Synthetic biology for synthetic chemistry. ACS Chem Biol 3:64–76.

Lee, S. Y., J. H. Park, S. H. Jang, L. K. Nielsen, J. Kim, and K. S. Jung. 2008. Fermentative butanol production by Clostridia. Biotechnol Bioeng 101:209–228.

Tracewell, C. A., and F. H. Arnold. 2009. Directed enzyme evolution: Climbing fitness peaks one amino acid at a time. Curr Opin Chem Biol 13:3–9.

Verlinden, R. A., D. J. Hill, M. A. Kenward, C. D. Williams, and I. Radecka. 2007. Bacterial synthesis of biodegradable polyhydroxyalkanoates. J Appl Microbiol 102:1437–1449.

13 Metabolism

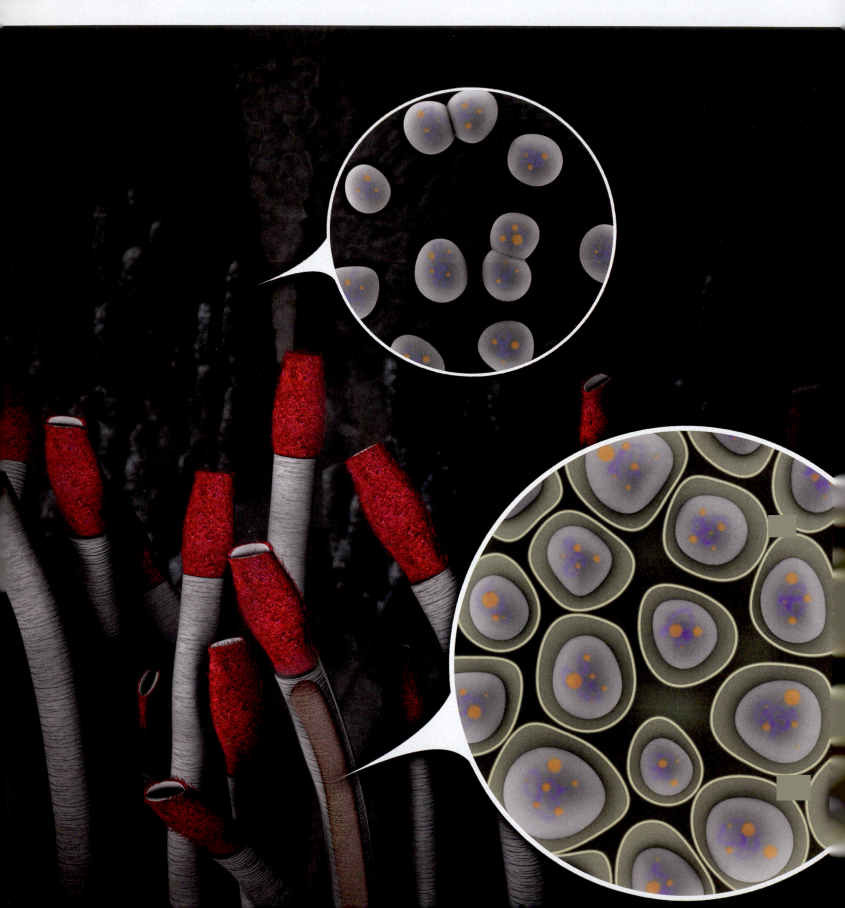

In the depths of the ocean, several kilometers below the surface, there are places where volcanic gases spew from Earth's crust into the dark, cold water. These hydrothermal vents release hydrogen sulfide and methane, deadly toxins for most forms of life. When deep sea submersibles first discovered and explored such sites in the late 1970s, to the initial astonishment of the scientific community, rich communities of invertebrates, fish, and microorganisms were revealed, thriving in a harsh environment largely disconnected from the biosphere far above. The most striking residents of this ecosystem are giant tube worms *Riftia pachyptila*, averaging 2 meters in length and estimated to live 250 years.

Closer analysis of the tube worms, using samples returned by submersibles such as the *Alvin*, found that they have no mouth and no digestive tract! Unlike their smaller, shallow-water relatives, these giants are totally dependent on chemolithoautotrophic bacteria living inside their bodies for their nutrition. Chemolithoautotrophs are microorganisms that can fix inorganic carbon, CO_2 in this case, to make organic molecules using chemical energy obtained through the oxidation of inorganic reduced compounds like H_2S. The chemolithoautotrophs harbored by the tube worms support this unique ecosystem in complete darkness.

This remarkable relationship between the tube worms and the bacteria begins with tube worm larvae, which do have digestive tracts and feed on bacteria in the water surrounding the vents. As the larvae develop and settle, they become infected with a single species of free-living bacteria from the surrounding area. The bacteria move through the tube worm's skin into deeper tissue. The infected tissue organizes into a specialized organ called the trophosome. This organ virtually fills the central core of the tube worm's body. Once the bacteria are established, the worm's gut and mouth degenerate as they commit to an intimate partnership with the bacteria. The worms depend completely on the bacteria to supply compounds for the biosynthesis of all their sugars, fatty acids, and amino acids, as well as oxygen for cellular respiration. How do the bacteria produce compounds to feed their host in the absence of light for photosynthesis? The answer is not completely known, since the tube worms do not survive in the laboratory, but molecular biologists have started to analyze the genome of both the tube worm and the bacteria to uncover the genes involved in metabolism.

Many intriguing questions surround the tube worm and bacteria relationship. For example, the tube worm captures H_2S, a compound that is toxic to most animal cells, from the surrounding waters and delivers it to the bacterial partner for oxidation. Shouldn't this activity kill the tube worm? Even more curious is the fact that the concentration of H_2S in the waters of the ecosystem isn't sufficient to support the large number of bacteria found in the tube worms. How then do these bacteria maintain an adequate supply of H_2S? Are the worms somehow not only able to tolerate but *concentrate* this toxic compound for the bacteria?

CHAPTER NAVIGATOR

As you study the key topics, make sure you review the following elements:

ATP transfers energy from catabolism to reactions needed for biosynthesis.

- Figure 13.5: Metabolism: Energy transfer between catabolism and anabolism by ATP
- Figure 13.6: ATP-coupled endergonic reaction

Microbes have many ways of obtaining energy and electrons to generate ATP.

- Figure 13.11: ATP production from glucose
- Table 13.1: Major metabolic groups of microorganisms

Glucose is a common source of energy and carbon for many cells.

- Figure 13.13: Process Diagram: The Embden–Meyerhof–Parnas (EMP) pathway of glycolysis
- Figure 13.16: Recycling of NADH by fermentation
- Figure 13.21: Process Diagram: The tricarboxylic acid (TCA) cycle

Electron transport systems produce energy to move protons across the membrane to be used for ATP production or work.

- 3D Animation: Electrons generated by glycolysis and the TCA cycle
- Figure 13.24: The aerobic electron transport system of mitochondria
- Figure 13.27: Generation of the proton motive force
- Animation: Cell respiration

Microbes can metabolize a variety of organic carbon sources.

Phototrophs convert light energy into chemical energy.

- Figure 13.34: Process Diagram: General structure and function of a photosystem
- Figure 13.38: Overview of phototrophy
- Figure 13.39: Process Diagram: The Calvin cycle
- Mini-Paper: Genome sequence of a deep sea symbiont

All organisms need nitrogen and sulfur compounds to build biomolecules.

Macromolecules are built from subunits synthesized from common precursors.

CONNECTIONS for this chapter:

Evolution of photosynthesis in eukaryal cells (Section 1.2)
Nitrogen-fixing bacteria and the survival of many plant species (Sections 14.3 and 17.2)
Antimicrobial drugs that target nucleic acid synthesis (Section 24.2)

Introduction

The diversity of microbial metabolic pathways is truly astonishing. Indeed, the ability of microorganisms to inhabit extraordinarily diverse habitats on Earth depends on their ability to use a wide variety of energy and nutrient sources. The reactions any organism can carry out are ultimately limited by the set of enzymes that are encoded in the genome. Many, perhaps most, of the enzymes that *could* be synthesized from the set encoded in any microbial genome are *not* produced in appreciable amounts at any given time because they are useful, and therefore expressed, only in a limited range of environmental conditions.

Macromolecule synthesis consumes most of a cell's metabolic resources during active growth. However, the metabolic diversity of microbes truly manifests itself in the variety of ways that cells acquire nutrients, extract energy, and generate the precursors for macromolecular synthesis. The number of metabolic pathways collectively available to microorganisms means they can utilize a vast array of molecules unavailable to most multicellular organisms. This allows microbes to operate in environments that most multicellular organisms cannot tolerate, such as anaerobic environments.

All organisms must utilize energy sources available in their environment in order to produce adenosine triphosphate (ATP), which is needed for all mechanical and biosynthetic processes in the cell. The source of energy needed to synthesize ATP in animals is limited to preformed organic matter, such as carbohydrates and proteins, obtained by consumption. In animals, these molecules are also the sole source of carbon. Microbes are not so limited. Collectively, they have a multitude of metabolic pathways that allow them to use inorganic as well as organic sources of energy. Like plants, many can use the energy of light and CO_2 to generate chemical energy and carbon compounds. A large number of bacteria obtain all their energy needs from inorganic molecules found in their environment, such as elemental sulfur, H_2S, NO_3^-, and Fe^{3+}. This independence from preformed organic molecules means many microbes can exploit habitats that other forms of life cannot.

A single chapter can't possibly provide full coverage of microbial metabolism. Instead, we'll try to provide a useful overview, present metabolic principles that microbes abide by, and showcase some of the most important metabolic processes and pathways of the microbial world. In this chapter, we'll address the following questions concerning microbial metabolism:

What are the basic principles of catabolism and energy that control metabolism? (13.1)

How do cells make ATP? (13.2)

How do microbes utilize organic carbon? (13.3)

What becomes of the electrons generated by glycolysis and the TCA cycle? (13.4)

How do microbes get nutrition from compounds other than glucose? (13.5)

How do microbes capture light energy for ATP synthesis and carbon fixation? (13.6)

How do microbes use nitrogen and sulfur? (13.7)

How do microbes synthesize new cellular components? (13.8)

13.1 Energy, enzymes, and ATP

What are the basic principles of catabolism and energy that control metabolism?

Metabolism encompasses all the biochemical reactions occurring in a living cell. Most aspects of metabolism can be categorized as **catabolism** or **anabolism**. Catabolism refers to the breakdown and oxidation of larger molecules, yielding energy needed for anabolism, the biosynthesis of macromolecular cell components from smaller molecular units. In cells, nearly all these metabolic functions rely on enzymes to facilitate the multitude of chemical reactions necessary for metabolism. An enzyme is a molecule or complex that catalyzes the conversion of specific reactants called **substrates** into products. As a catalyst, an enzyme may participate in the reaction, but once the reaction is completed it is returned to its prior state. Like all catalysts, an enzyme is not consumed in a reaction and can be used repeatedly.

Thousands of biochemical reactions have been identified that occur in microbial cells. Of course, not all of these reactions occur in every microbe. The energy and nutrient sources that different bacteria use depend on the specific enzymes that they can produce. An enzyme increases the rate of a specific

 Metabolism ANIMATION

chemical reaction. Without the enhancement of biochemical reaction rates provided by enzymes, life could not exist. Let's explore why this is.

Energy

Enzymes must obey the **laws of thermodynamics** that govern all energy conversions. Enzymes can only catalyze reactions that are thermodynamically possible. The first law of thermodynamics states that energy is always conserved; it cannot be created or destroyed. Energy can, however, be converted from one form into another. The second law of thermodynamics states that in all energy exchanges, where no energy enters or leaves the system, the potential energy of the resulting state will always be less than that of the initial state. This means energy will spontaneously flow from a more ordered state to a less ordered state. This is referred to as entropy, and is a measure of disorder. Large molecules will naturally break down into smaller molecules, and in the process of energy transfer some energy will dissipate as heat. The flow of energy maintains life. Cells must take in energy to maintain their organized state by building more ordered molecules (for example, nucleic acids and proteins) from less ordered molecules (such as simple sugars and ammonia). Entropy takes over when cells cease to take in energy and die.

As governed by the laws of thermodynamics, in order for the reaction A + B → C + D to proceed, the products of the reaction (C and D) must have a lower "free energy" content than the reactants (A and B). When this is true, the conversion of A and B to C and D releases energy. The amount of free energy released in a chemical reaction is represented by the **Gibbs free energy, G**, with the International System of Units given in joules or kilojoules per mole ($kJ \cdot mol^{-1}$). Calories or kilocalories per mole can also be used. The change in free energy in a reaction is represented as ΔG, where $\Delta G = G_{products} - G_{reactants}$. Usually ΔG of a reaction is measured under standard conditions of temperature, concentration, pressure, and pH, in order to make comparisons between reactions. This is indicated as $\Delta G^{o\prime}$. An energy-yielding chemical reaction is termed **exergonic**, and has a negative $\Delta G^{o\prime}$ value. An energy-absorbing reaction is termed **endergonic** and has a positive $\Delta G^{o\prime}$ value. The value of $\Delta G^{o\prime}$ must be negative in order for a reaction to proceed spontaneously.

Spontaneity implies nothing about the rate at which a reaction will occur. Reaction rate is determined by a property called **activation energy (E_A)**. We can think of E_A as the energy that must be invested before a reaction can occur. (For example, to bring the chemical bonds in the substrates to a transition state where they can be broken, E_A must be invested before bonds can be re-formed in a new way to generate the product.) The greater E_A is, the more slowly a reaction will proceed. For most biochemical reactions to occur in the absence of an enzyme, the E_A needed for reactants to move to a transition state to form the final products is greater than the kinetic energy of the reactant molecules as they randomly collide. Even if the reaction has a very negative resulting $\Delta G^{o\prime}$ value, the reaction will almost never occur spontaneously. This is where enzymes come in.

Enzymes and activation energy

Enzymes can only catalyze reactions that are thermodynamically favorable, meaning that the free energy (G) of the products is less

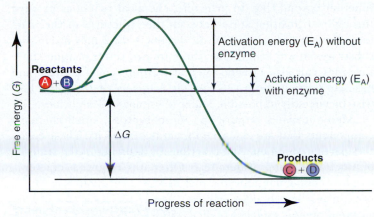

Figure 13.1. Enzymes lower activation energy. In an enzyme-catalyzed reaction the activation energy is significantly reduced compared to the energy required to form the products in the absence of the enzyme. Reducing the activation energy favors the production of products from substrates, thus enhancing the reaction rate.

than that of the substrates. No matter how favorable the reaction may be, getting to the products requires a transient increase in energy—the activation energy—in order to break and rearrange bonds. The secret to the effect of enzymes on chemical reactions is that they bind the substrates or reactants in specific ways to lower the activation energy barrier (**Figure 13.1**). Conceptually, one can think of activation energy as a dam separating water in two reservoirs. Reservoir 1 represents the reactants and reservoir 2 the products. The difference in height between the two reservoirs illustrates ΔG. When reservoir 2 is lower than reservoir 1, ΔG is negative, and water should flow spontaneously from 1 to 2—reactants are converted to products. What about the dam? It represents the activation energy that must be supplied before water can move from reservoir 1 to reservoir 2. When the dam is high, water is less likely to move. By lowering the dam (that is, the required activation energy), an enzyme increases the rate at which products are formed.

Enzymes have many ways of enhancing reaction rates through their **active site**, the physical location where the substrates bind to the enzyme (**Figure 13.2**). One mechanism for increasing reaction rate is to physically position substrates properly for the desired reaction to occur, but enzymes generally

① The enzyme binds the substrates at binding sites within the active site.

② Binding brings the substrates into close proximity with one another to form the enzyme-substrate complex.

③ Bond rearrangement forms a transition-state complex, resembling both substrates and products.

④ The transition-state complex decomposes, releasing final products, and the enzyme returns to its initial state.

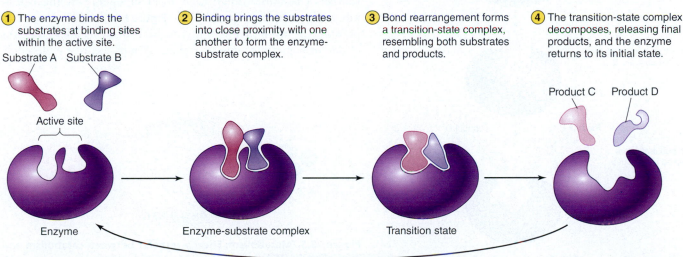

Figure 13.2. Process Diagram: Enzyme-catalyzed reaction The production of products from substrates happens in a series of steps.

function beyond this. In principle, chemical reactions proceed through a transition state between the structures of the reactants, and the structures of the products. Enzymes often bind substrates in ways that move them toward the transition state. Enzymes can also provide chemical groups such as acids and bases to participate in reaction mechanisms that would otherwise be virtually impossible under physiological conditions.

Many enzymes require specific **cofactors**, small chemical components that are essential for catalytic activity. These cofactors participate in enzyme reactions by assisting in the transfer of functional groups **(Figure 13.3)**. Often, the required cofactor is a tightly bound inorganic ion, such as Mg^{2+}, Fe^{2+}, Zn^{2+}, or Mn^{2+}, but can also be a small loosely associated organic molecule called a **coenzyme**. Many coenzymes are made from precursor molecules, such as vitamins. Vitamins are broadly defined as organic compounds whose functions are essential for cell growth, but cannot be synthesized in sufficient quantities by the organism and must be either consumed in the diet or produced by microorganisms harbored by the host **(Perspective 13.1)**. A vitamin isn't an enzyme or a nutrient as it is not metabolized for energy, nor does it provide raw materials for biosynthesis. The availability of cofactors is one way enzyme activity can be regulated.

We can surmise that enzymes increase the rate at which reactants possessing a high potential free energy content become products with a lower potential free energy content. In

@ Functions of enzymes ANIMATION

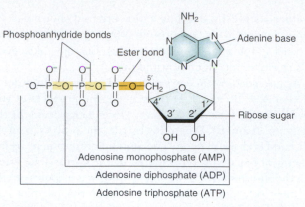

Figure 13.4. Structure of adenosine phosphate molecules ATP, ADP, and AMP all consist of a ribose sugar and an adenine base covalently bound to one (AMP), two (ADP), or three (ATP) phosphate groups.

this process, energy is released, resulting in a negative $\Delta G^{o\prime}$ value. However, many anabolic biosynthetic chemical reactions that cells carry out, such as those involved in synthesizing DNA and proteins, require energy. They produce positive $\Delta G^{o\prime}$ values and are endergonic. How then do endergonic reactions in cells occur? In our dam analogy, this would be equivalent to reservoir 2, containing the product, being higher than reservoir 1, containing the reactants, and water still moving from 1 to 2! This can't happen directly, but it can happen if two reactions are connected or coupled together by an enzyme so that one reaction releases energy to drive another. This is analogous to someone carrying water from reservoir 1 up to 2, expending their own energy in the process. To carry out anabolic processes, cells must use enzymes that couple endergonic reactions with exergonic reactions. When the energy released from the exergonic reaction is greater than that needed to complete the endergonic reaction, the endergonic reaction will be driven to completion.

ATP—Energy currency of the cell

We can see that cells need some internal source of energy to drive these biosynthetic coupled reactions. What is this source? The most widely used form of energy in the cell is **adenosine triphosphate (ATP) (Figure 13.4)**. ATP contains two phosphoanhydride bonds that can be broken to allow transfer of one or sometimes two phosphate groups to drive endergonic

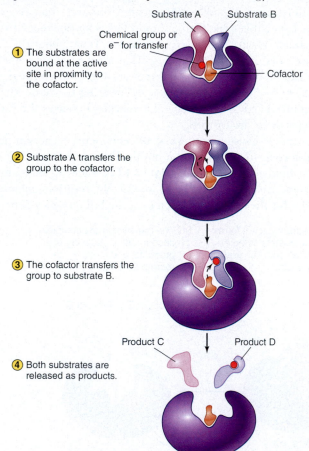

Figure 13.3. Process Diagram: The function of cofactors The cofactor is associated with the enzyme. Substrate A has a chemical group or electron to be transferred by the cofactor to substrate B.

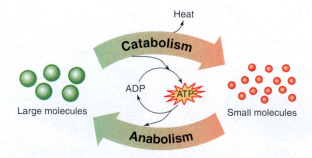

Figure 13.5. Metabolism: Energy transfer between catabolism and anabolism by ATP Catabolism is the breakdown of larger substrate molecules to generate energy and smaller metabolic products. Some energy is released as heat. Energy from catabolism can be used to synthesize ATP, which can be used for anabolism, the biosynthesis of molecules useful to the cell.

Growing up, your parents might have impressed on you the importance of vitamins in your diet, either from eating the "right" foods or maybe just taking Flintstones chewable vitamins. Do microbes need vitamins like people do? If so, why? In addition to their protein units, many enzymes require coenzymes, and several important coenzymes are derived from vitamins.

Different organisms require different vitamins in their diet, the most common of which are listed in **Table B13.1**. Humans need 13 distinct dietary vitamins, but most microorganisms need far fewer, and many have no dietary vitamin requirements, being fully capable of synthesizing all the coenzymes they need from simpler nutrients. In fact, our intestinal bacteria usually produce and excrete enough

extra biotin and vitamin K2 (menaquinone) that we don't need these compounds directly in our diet. At the other extreme, some microbes such as lactic acid bacteria (see Section 13.3) need nearly as many vitamins as humans and are therefore limited to habitats rich in complex organic nutrients.

In microbiology, vitamins are often referred to as "growth factors." When planning media, the need for growth factor supplements depends on the particular species and strain being used. The use of complex media components such as yeast extract often eliminates the need for supplemental vitamins, since the extract contains enough growth factors to support most microbes. When needed, vitamins are typically added to media at concentrations of 0.1 mg/L or less.

TABLE B13.1 Common coenzymes produced from vitamins and their functions

Vitamin	Coenzyme	Function
Biotin (B_7)	Biocytin	Used in carboxylase enzymes involved in fatty acid synthesis and glucose production
Cobalamine (B_{12})	5'-deoxyadenosylcobalamin (coenzyme B_{12})	Used in synthesis of folic acid, required for reactions involving methyl group transfer, including nucleotide and methionine synthesis
Folic acid (B_9)	Tetrahydrofolate	Used in reactions involving methyl group transfer, such as during purine and thymine synthesis. Paraminobenzoic acid (PABA), a precursor of folic acid, can also be used as a growth factor.
Nicotinic acid (niacin, B_3)	Nicotinamide adenine	NAD and NADP are major electron carriers in cells.
Pantothenate (B_5)	Coenzyme A	Critical in carbon and energy metabolism, and in lipid synthesis and catabolism
Pyridoxine (B_6)	Pyridoxal phosphate	Used for enzymes involved in transamination reactions during amino acid biosynthesis, and certain deamination and decarboxylation reactions
Riboflavin (B_2)	Flavin adenine dinucleotide (FAD) and flavin mononucleotide (FMN)	Used for enzymes catalyzing numerous redox reactions
Thiamine (B_1)	Thiamine pyrophosphate	Used by enzymes that catalyze dehydrogenation and decarboxylation, including pyruvate dehydrogenase

reactions. The phosphate group of AMP is bound by an ester linkage to the sugar and cannot be easily transferred.

All cells produce enzymes that allow them to convert energy sources they receive from their environment, such as glucose, in order to produce ATP for biosynthesis and to carry out mechanical work such as transport of molecules across membranes, or movement of flagella. One can think of ATP as the common chemical currency of the cell. In metabolism, ATP transfers energy from exergonic reactions, resulting from catabolism, to endergonic reactions needed for anabolism **(Figure 13.5)**. Anabolism uses products of catabolism, or precursor molecules acquired from the environment for the assembly of cellular

constituents that ultimately lead to the production of new cells. This is analogous to individuals (cells) earning money (ATP) to purchase needed goods (DNA, RNA, and proteins). In some instances, functionally equivalent molecules are used to transfer energy as in the example of guanosine triphosphate (GTP).

The potential free energy content of ATP is much higher than the potential free energy content of its hydrolysis products adenosine diphosphate (ADP; see Figure 13.5) and inorganic phosphate (P_i). This is evident from the resulting large, negative $\Delta G^{o'}$ value when ATP is hydrolyzed:

$$\text{ATP} + \text{H}_2\text{O} \rightarrow \text{ADP} + P_i \qquad \Delta G^{o'} = -30.5 \text{ kJ/mol}$$

Figure 13.6. ATP-coupled endergonic reaction The conversion of glucose to glucose 6-phosphate is endergonic, but is favorable when coupled to the highly exergonic conversion of ATP to ADP. The reaction results in the transfer of a phosphate group from ATP to glucose.

Enzymes can use this energy to drive otherwise unfavorable endergonic reactions, such as the biosynthesis of macromolecules from smaller precursors. As long as there is more energy released in the conversion of ATP to ADP and P_i than is necessary to achieve the endergonic reaction, an overall negative ΔG will result, which drives the reaction strongly forward toward the products. We can look at the conversion of glucose to glucose 6-phosphate, the first step in the catabolism of glucose. The reaction can be written as

Glucose $+ P_i \rightarrow$ glucose 6-phosphate $+ H_2O$
$$\Delta G^{o'} = +13.8 \text{ kJ/mol}$$

As can be seen from the positive $\Delta G^{o'}$ value, this reaction is not thermodynamically favorable, but if it is carried out together, that is, *coupled*, with hydrolysis of ATP ($\Delta G^{o'} = -30.5$ kJ/mol), the overall reaction results in $\Delta G^{o'} = -16.7$ kJ/mol. The reaction will now favor the formation of glucose 6-phosphate. We can write the combined reaction as

Glucose $+$ ATP $\rightarrow$ glucose 6-phosphate $+$ ADP
$$\Delta G^{o'} = -16.7 \text{ kJ/mol}$$

Coupled reactions are commonly illustrated as in **Figure 13.6**, which more clearly shows how the two reactions interact. It is important to note that in the enzyme-coupled reaction, the two reactions do not occur separately. If ATP were hydrolyzed to ADP, all the energy would immediately be released as heat, and could not drive a chemical reaction. The enzyme that couples the reaction facilitates the stepwise transfer of the phosphate group from ATP to the substrate glucose, raising its free energy potential, and releasing the remaining free energy as heat. The products ADP and glucose 6-phosphate are then displaced and released from the enzyme.

13.1 Fact Check

1. Distinguish between the terms metabolism, catabolism, and anabolism.
2. Enzymes must obey the laws of thermodynamics. Describe the laws.
3. Differentiate between exergonic and endergonic reactions and identify their relative $\Delta G^{o'}$ values.
4. Describe how enzymes enhance reaction rates.
5. What are cofactors and coenzymes and what are their roles in enzymatic reactions?
6. What is the role of ATP in a cell?

13.2 Central processes in ATP synthesis

How do cells make ATP?

ATP is the primary energy currency used to power virtually every activity of cells from bacteria to human. Although we have been discussing ATP as having a major role in driving endergonic chemical reactions, the potential free energy in ATP is also used to do mechanical work. How then is ATP produced by cells? Cells must carry out exergonic reactions that release even greater amounts of energy to drive the synthesis of ATP from ADP and P_i, shown below:

ADP $+ P_i \rightarrow$ ATP $+ H_2O$ $\Delta G^{o'} = +30.5$ kJ/mol

As can be seen from the large positive $\Delta G^{o'}$ value, the formation of ATP is a strongly endergonic reaction. Where does the energy to synthesize ATP come from? The simplest way to produce ATP is by harnessing the potential free energy of a high-energy phosphate group from a phosphorylated intermediate molecule. This is called **substrate-level phosphorylation**.

Substrate-level phosphorylation

Substrate-level phosphorylation is an enzymatically coupled reaction that produces ATP by the transfer of a phosphate group from a reactive intermediate generated during catabolism to ADP. This reaction can occur because the transfer of the phosphate group from a molecule with high potential free energy (the intermediate) creates one of lower potential free energy (ATP). In cells, substrate-level phosphorylation occurs in chemoorganotrophs in the cytoplasm under both aerobic and anaerobic conditions, and begins with the breakdown of an organic molecule. Although chemoorganotrophs can use a wide variety of molecules to generate energy, one that is used by many is glucose. We will use the catabolism of glucose to illustrate how ATP can be produced by substrate-level phosphorylation.

CONNECTIONS Recall that we were introduced to the different metabolic types of microbes, including chemoorganotrophs in **Section 6.1**. Chemoorganotrophs use preformed chemicals as an energy source, and organic molecules as a source of electrons. We will revisit these types in some detail later in this section.

A commonly used pathway that begins glucose catabolism for ATP production is **glycolysis (Figure 13.7)**. Glycolysis involves the catabolism of glucose to pyruvic acid (pyruvate) and can produce a total of two ATP molecules from each glucose molecule. The details of glycolysis will be examined in Section 13.3. One of the intermediate molecules generated during glycolysis is 1,3-bisphosphoglycerate. The production of 1,3-bisphosphoglycerate is one point where ATP is generated by substrate-level phosphorylation because 1,3-bisphosphoglycerate has an even greater potential free energy than ATP. From the reaction below, we can see that the transfer of the phosphate group from 1,3-bisphosphoglycerate to ADP is thermodynamically favorable in a coupled reaction:

$$\text{1,3-bisphosphoglycerate} + H_2O \rightarrow \text{3-phosphoglycerate} + P_i$$
$$\Delta G^{o\prime} = -52 \text{ kJ/mol}$$

$$ADP + P_i \rightarrow ATP + H_2O \qquad \Delta G^{o\prime} = +30.5 \text{ kJ/mol}$$

Overall reaction:

$$\text{1,3-bisphosphoglycerate} + ADP \rightarrow \text{3-phosphoglycerate} + ATP$$
$$\Delta G^{o\prime} = -21.5 \text{ kJ/mol}$$

In the above reaction, the catabolism of 1,3-bisphosphoglycerate releases an inorganic phosphate (P_i) and produces 3-phosphoglycerate. The energy released, indicated by the negative $\Delta G^{o\prime}$ value, is used to produce a covalent bond between P_i and ADP to produce ATP. It costs energy to synthesize ATP, but as can be seen from the overall negative $\Delta G^{o\prime}$ value of this reaction, more energy is supplied than is needed to produce ATP, ensuring the reaction is driven forward to completion to produce ATP. The excess energy is released as heat. The molecule 1,3-bisphosphoglycerate is a very high energy molecule, meaning it has a high potential free energy, enough to drive the endergonic reaction of making ATP, itself a high energy molecule. Another intermediate molecule in glycolysis that can be used to generate ATP by substrate-level phosphorylation is phosphoenolpyruvate.

The enzyme that catalyzes the coupled reaction to produce ATP using phosphoenolpyruvate in the cell is pyruvate kinase. We can look in more detail at how this enzyme carries out the coupled reaction between phosphoenolpyruvate and ADP to produce ATP through substrate-level phosphorylation. Kinases are a group of enzymes that transfer phosphate groups from high-energy donors to substrates of lower potential free energy. Specific kinases are used extensively during biosynthesis to transfer phosphate groups from ATP in coupled reactions.

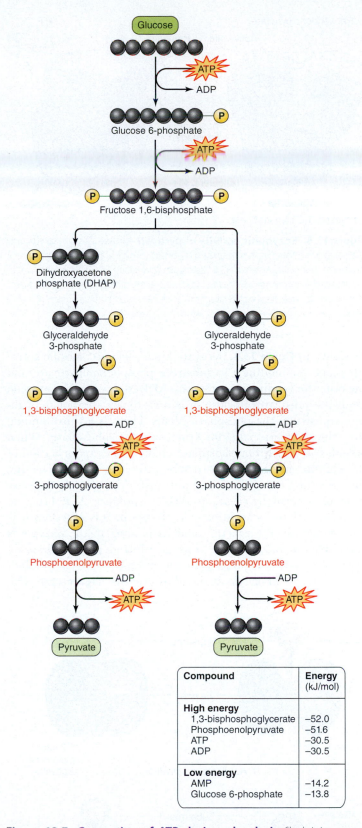

Compound	Energy (kJ/mol)
High energy	
1,3-bisphosphoglycerate	−52.0
Phosphoenolpyruvate	−51.6
ATP	−30.5
ADP	−30.5
Low energy	
AMP	−14.2
Glucose 6-phosphate	−13.8

Figure 13.7. Generation of ATP during glycolysis Glycolysis is a common first step in the pathway of glucose catabolism, producing two molecules of pyruvate from one molecule of glucose. Phosphoenolpyruvate and 1,3-bisphosphoglycerate are intermediate molecules produced during glycolysis. Both molecules produce very high free energy upon hydrolysis to release a phosphate group (see table insert). This energy can drive the formation of ATP from ADP in a coupled reaction.

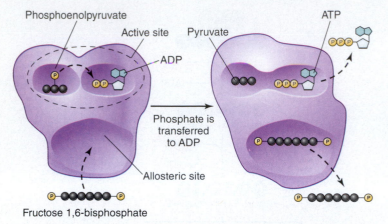

Fructose 1,6-bisphosphate

Figure 13.8. Enzymatic activity of pyruvate kinase The active site of pyruvate kinase binds the substrate phosphoenolpyruvate at one binding site and ADP at an adjacent binding site. A third site, the allosteric site, binds fructose 1,6-bisphosphate. When fructose 1,6-bisphosphate is bound, the conformation of the enzyme is modified and the two substrates are brought together, lowering the activation energy required to transfer the phosphate group from phosphoenolpyruvate to ADP to produce ATP. Fructose 1,6-bisphosphate is not changed in the reaction and its binding is reversible.

As shown in **Figure 13.8**, pyruvate kinase has two specific active sites, one for binding the substrate phosphoenolpyruvate and the other for binding the substrate ADP. Notice how the enzyme binds the substrates in close proximity to one another to facilitate transfer of the phosphate group. There is a third binding site, the allosteric site, for fructose 1,6-bisphosphate. When bound, fructose 1,6-bisphosphate changes the conformation of the enzyme, bringing the two substrates even closer, lowering the activation energy required to transfer the phosphate group from phosphoenolpyruvate to ADP to produce ATP. Fructose 1,6-bisphosphate is an effector molecule in this reaction; it is not changed in the reaction and its binding to the enzyme is reversible. In biochemistry, this is referred to as **allosteric regulation**—the regulation of an enzyme's activity by binding an effector molecule at the allosteric site **(Figure 13.9)**. The

term "allosteric" is derived from the Greek word *allos*, meaning "other" and *stereos*, meaning "solid," interpreted as "other molecule" since it is bound at a site distant from that of the active site. Effector molecules, such as fructose 1,6-bisphosphate, that increase an enzyme's activity are known as allosteric activators. Allosteric inhibitors can decrease activity. One such example is inhibition by ATP of the enzyme phosphofructokinase, discussed in Section 13.3 (see Fermentation). Regulation of enzyme activity is critical for microorganisms as they compete for survival and growth.

At this point, it will be useful to have a brief overview of the process of ATP generation from the complete catabolism of glucose. In the next sections we will discuss the details and variations on this central theme. Catabolism of glucose begins with glycolysis, producing two ATP molecules and two pyruvate molecules from each molecule of glucose. Each pyruvate proceeds through the **tricarboxylic acid (TCA) cycle**, a series of reactions that further dismantles the carbon intermediates into six CO_2 molecules (see Section 13.3). Substrate-level phosphorylation in the TCA cycle produces two more ATP (one for each pyruvate), for a total of four ATP molecules. This is a minor contribution compared to the total number of ATP molecules, up to 38, that can be produced from the complete catabolism of glucose, as shown by the equation below:

$$C_6H_{12}O_6 + 6\,O_2 + 38\,ADP + 38\,P_i$$
$$\rightarrow 6\,CO_2 + 6\,H_2O + \text{38 ATP}$$

Where do the other 34 ATP come from?

The chemiosmotic model

The complete catabolism of glucose into carbon dioxide and water generates a very large amount of free energy:

$$C_6H_{12}O_6 + 6\,O_2 \rightarrow 6\,CO_2 + 6\,H_2O$$
$$\Delta G^{o\prime} = -2{,}840\ \text{kJ/mol}$$

Investigators originally postulated that the large number of ATP molecules produced from glucose in cells occurred at membrane sites through substrate-level phosphorylation. However, for this to occur, there had to be another high-energy phosphorylated intermediate molecule generated, other than phosphoenolpyruvate and 1,3-bisphosphoglycerate, which were intermediates of glycolysis only. None was ever found. In 1961, Peter Mitchell proposed an alternative hypothesis for the production of ATP. It stated that ATP generation occurred at membrane sites where electrons were transferred along a chain of molecules, and this was coupled to the translocation of protons in a process called chemiosmosis, which could be used to produce ATP from ADP and P_i in the cell. Today, we know many details of this **chemiosmotic process**. Unlike substrate-level phosphorylation, this method of ATP generation is a complex process, rather

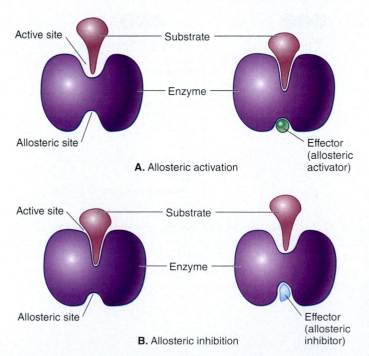

A. Allosteric activation

B. Allosteric inhibition

Figure 13.9. General scheme of enzyme allosteric regulation A. Allosteric activation requires binding of an effector molecule (the allosteric activator) to the allosteric site. Binding causes a conformational change in the enzyme that results in increased ability to bind or act on the substrate. **B.** Allosteric inhibition requires binding of an allosteric inhibitor to the allosteric site. The resulting conformational change diminishes the enzyme's ability to bind or act on the substrate.

than a simple reaction. It involves synthesis of ATP from free ADP and P_i using energy and electrons to create a proton gradient across a membrane. The flow of protons back across the membrane releases energy that is harnessed to produce ATP. As we'll see later in phototrophs, this method of ATP generation requires light energy and is termed **photophosphorylation** (see Section 13.6). In chemotrophs, the energy is derived from the oxidation of chemical substrates and is termed **oxidative phosphorylation**.

The components for the generation of ATP via the chemiosmotic process varies in different organisms, but is accomplished through coordination of three interconnected steps or processes, which we will examine later. For now, a preview of these processes used in oxidative phosphorylation is as follows:

1. Electrons are transferred to an electron carrier molecule, such as the coenzyme **nicotinamide adenine dinucleotide (NAD)** (Figure 13.10).

2. NAD transports electrons to a chain of electron-transporting molecules in the membrane, called the **electron transport system**. The electron transport system transfers electrons in a series of steps, yielding energy to move protons (hydrogen ions, H^+) across the membrane.

3. The protons flow back across the membrane at sites containing the enzyme **ATP synthase**. The kinetic energy of the flowing protons is converted by ATP synthase into chemical energy and ATP is synthesized from free ADP and P_i.

Figure 13.11 gives an overview of the production pathway for 38 ATP from the complete catabolism of one molecule of glucose.

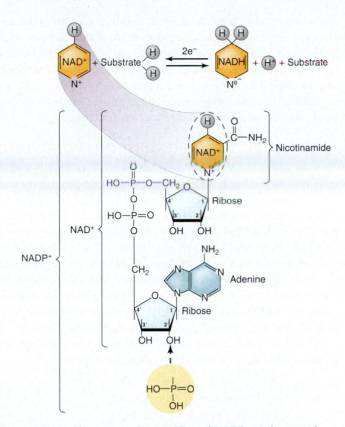

Figure 13.10. Electron carriers NAD and NADP NAD^+ and $NADP^+$ can accept two electrons and a proton on the nicotinamide ring (circled). Using a substrate as an electron source, NAD^+ accepts two electrons and a proton, producing NADH and a proton (H^+). NADH can donate electrons to an electron acceptor to convert back to NAD^+.

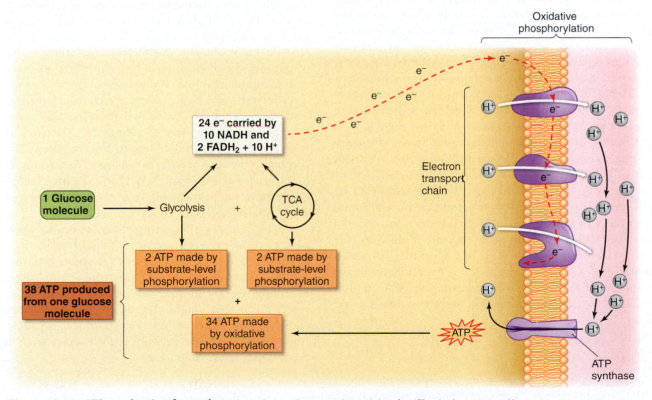

Figure 13.11. ATP production from glucose As glucose undergoes complete catabolism, four ATP molecules are generated by substrate-level phosphorylation. Oxidative phosphorylation produces 34 ATP using electrons removed from the catabolic intermediates of glucose. The electrons are transported by electron carriers NADH and $FADH_2$ to the membrane electron transport system. The energy released by the transfer of electrons is used to pump protons across the membrane. ATP synthase uses the inward flow of H^+ ions to synthesize ATP.

In organotrophs, electrons for transfer are made available when organic nutrients, such as glucose, are dismantled in small steps through metabolic pathways that carry out oxidation and reduction reactions. Electrons can also be generated using light absorption in phototrophs, the subject of Section 13.6. To understand how energy is yielded through transfer of electrons, we need to examine oxidation and reduction reactions.

Reduction and oxidation (redox) reactions

Reduction and oxidation reactions involve the transfer of electrons from one molecule to another, and are commonly referred to as **redox reactions**. Electron transfer reactions are a fundamental mechanism by which energy can be liberated or captured in cells. As we have just seen, this energy can be used to generate ATP. By definition, an electron donor—the molecule that *loses* electrons—is said to be *oxidized*. An electron acceptor—the molecule that *gains* electrons—is *reduced*. A simple mnemonic device to help remember the difference between oxidation and reduction is "LEO the lion says GER", representing Loss of Electrons is Oxidation, Gain of Electrons is Reduction. **Oxidation** of a molecule results in loss of an electron and **reduction** of a molecule results in a gain of an electron. Examples:

$$2\,H^+ + 2\,e^- \rightarrow H_2$$
$$\text{oxidized} \qquad \text{reduced}$$

$$NAD^+ + 2\,H^+ + 2\,e^- \rightarrow NADH + H^+$$
$$\text{oxidized} \qquad\qquad \text{reduced}$$

Nicotinamide adenine (NAD) is an important electron carrier molecule in the cytoplasm of cells (see Figure 13.10). Reduction of NAD^+ to NADH is important for many catabolic reactions, not just oxidative phosphorylation. A closely related molecule, NADPH, is typically used as an electron donor for biosynthetic (anabolic) reactions. Having two distinct electron carriers may help the cell to organize reducing power, in order to allocate it to catabolism or anabolism in a physiologically optimal manner.

The tendency of a molecule to acquire electrons is its **reduction potential** or **redox potential, E**, which is measured in units of volts (V). Molecules with more negative redox potentials are therefore better electron donors than those with more positive redox potentials. The acceptor and donor pair in a redox reaction is called a "redox couple." Redox potentials for many biologically relevant redox couples are shown in **Figure 13.12**. For purposes of comparison, redox potentials are measured under standardized conditions; the standard reduction potential (E_o') is based on the oxidized and reduced forms of the molecule both being present at a concentration of 1 M, at pH 13.0, and a temperature of 25°C. These couples are conventionally shown with the number of electrons transferred and the oxidized form of the molecule, or acceptor, on the left, and the reduced donor on the right.

$$\textbf{NAD}^+ + \textbf{H}^+ + 2\,e^- \rightarrow \textbf{NADH}$$
$$\text{acceptor/oxidized} \qquad\qquad \text{donor/reduced}$$

The redox couples in Figure 13.12 are organized into a tower with more negative E_o' values on top and more positive E_o' values below. The more positive a redox potential is, the greater its affinity for electrons is, and it will have a higher tendency to become reduced. Another way to think of this is that

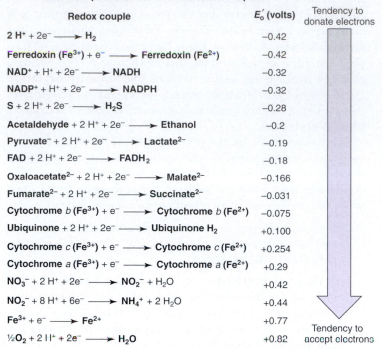

Reduction potentials for common redox couples

Redox couple		E_o' (volts)
$2\,H^+ + 2e^- \longrightarrow H_2$		−0.42
Ferredoxin $(Fe^{3+}) + e^- \longrightarrow$ Ferredoxin (Fe^{2+})		−0.42
$NAD^+ + H^+ + 2e^- \longrightarrow$ NADH		−0.32
$NADP^+ + H^+ + 2e^- \longrightarrow$ NADPH		−0.32
$S + 2\,H^+ + 2e^- \longrightarrow H_2S$		−0.28
Acetaldehyde $+ 2\,H^+ + 2e^- \longrightarrow$ Ethanol		−0.2
Pyruvate$^- + 2\,H^+ + 2e^- \longrightarrow$ Lactate^{2-}		−0.19
FAD $+ 2\,H^+ + 2e^- \longrightarrow$ FADH$_2$		−0.18
Oxaloacetate$^{2-} + 2\,H^+ + 2e^- \longrightarrow$ Malate^{2-}		−0.166
Fumarate$^{2-} + 2\,H^+ + 2e^- \longrightarrow$ Succinate^{2-}		−0.031
Cytochrome b $(Fe^{3+}) + e^- \longrightarrow$ Cytochrome b (Fe^{2+})		−0.075
Ubiquinone $+ 2\,H^+ + 2e^- \longrightarrow$ Ubiquinone H$_2$		+0.100
Cytochrome c $(Fe^{3+}) + e^- \longrightarrow$ Cytochrome c (Fe^{2+})		+0.254
Cytochrome a $(Fe^{3+}) + e^- \longrightarrow$ Cytochrome a (Fe^{2+})		+0.29
$NO_3^- + 2\,H^+ + 2e^- \longrightarrow NO_2^- + H_2O$		+0.42
$NO_2^- + 8\,H^+ + 6e^- \longrightarrow NH_4^+ + 2\,H_2O$		+0.44
$Fe^{3+} + e^- \longrightarrow Fe^{2+}$		+0.77
$\frac{1}{2}O_2 + 2\,H^+ + 2e^- \longrightarrow H_2O$		+0.82

Tendency to donate electrons → Tendency to accept electrons

Figure 13.12. Reduction potentials for redox couples Redox couples, indicated in bold, are arranged in a tower with those near the top having large negative reduction potentials and those near the bottom having large positive reduction potentials. Electrons will flow from donors near the top to acceptors near the bottom.

a positive redox potential means that the oxidized molecule "wants" to accept electrons, as the reaction arrow pointing to the right indicates. A negative E_o' value for a couple indicates that the equilibrium is in the other direction; under standard conditions, the reduced form gives up electrons in an energetically favorable reaction. Reduction potential determines which way electrons will flow as they are transferred from one molecule to another. Electrons will move toward molecules with a higher redox potential. In doing this, electrons are moving from a higher potential energy to a lower potential energy, according to the laws of thermodynamics. As a result, free energy is released. For example, NADH will donate electrons to oxygen to form water, releasing free energy. Chemotrophs can use this available energy to synthesize ATP.

An electron transfer in the cell always involves *two* redox couples, because electrons are never found free in the cytoplasm. The reduced form of the couple with the more negative reduction potential donates one or more electrons to the oxidized form of the couple with the more positive E_o'. The energy released by the reaction is proportional to the difference in reduction potential ($\Delta E_o'$) between the couples. For transfer of a pair of electrons from NADH to O_2, the relevant redox couples are NADH to O_2, the relevant redox couples are

$$\textbf{NAD}^+ + \textbf{H}^+ + 2\,e^- \rightarrow \textbf{NADH}$$
$$E_o' = -0.32\,\text{V}$$
$$\tfrac{1}{2}\,\textbf{O}_2 + 2\,\textbf{H}^+ + 2\,e^- \rightarrow \textbf{H}_2\textbf{O}$$
$$E_o' = +0.82\,\text{V}$$

The first reaction has a much more negative reduction potential than the second, so the reduced form of the first couple (NADH) will donate a pair of electrons to the oxidized form of the second couple (O_2). The overall reaction will be

$$NADH + H^+ + \tfrac{1}{2} O_2 \rightarrow NAD^+ + H_2O \qquad \Delta E_o' = +1.14 \text{ V}$$

The large *positive* $\Delta E_o'$ indicates the redox reaction is energetically favorable and a great deal of energy is released. You will recall from earlier in this section that a *negative* $\Delta G^{o'}$ also indicated the reaction is favorable. The mathematical relationship between $\Delta E_o'$ and $\Delta G^{o'}$ is:

$$\Delta G^{o'} = -nF(\Delta E_o')$$

where n = number of electrons, F = Faraday's constant (96.5 kJ/mole·volt), and $\Delta E_o'$ = the difference in reduction potential between the two couples. A *positive* $\Delta E_o'$ translates into a *negative* $\Delta G^{o'}$ reflecting the release of energy.

The large $\Delta E_o'$ for the transfer of electrons from NADH to O_2 is the basis for energy generation for ATP synthesis during aerobic respiration, the subject of Section 13.4, and accounts for the large number (38) of ATP molecules that can be generated from the complete oxidation of a molecule of glucose. However, the transfer of electrons from NADH to O_2 proceeds through several intermediates, with incremental $\Delta E_o'$ values for each transfer. These intermediate electron transfers occur within the membrane electron transport system and are coupled to proton movement across a membrane that can be used to drive ATP synthesis.

Metabolic groups of microbes

Microbes can be categorized based on the source of energy and the source of electrons they use for redox reactions to produce ATP. As we saw in Section 6.1, **chemotrophs** acquire energy through oxidation of preformed chemicals that they obtain from their environment. Conversely, **phototrophs** acquire energy by capturing photons of light to raise electrons to higher energy levels that can drive otherwise unattainable redox reactions (see Figure 6.1). **Organotrophs** remove electrons from organic molecules, such as glucose, whereas **lithotrophs** remove electrons from inorganic reduced molecules, such as H_2S, H_2, or elemental sulfur (S^0).

The source of carbon an organism uses is another major consideration when classifying metabolic types. **Heterotrophs** obtain carbon from organic molecules, such as sugars, obtained from their environment. **Autotrophs** use inorganic carbon compounds, usually CO_2. As shown in **Table 13.1**, combining information on energy and electron sources with the source of carbon used for biosynthesis gives an overall metabolic profile of a microorganism. In many cases, chemoorganotrophic heterotrophs use the same molecule as the carbon, energy, and electron source. Sometimes the shortened term "chemoorgano-heterotroph" or just "heterotroph" is often used to describe this combined use of a molecule. Glucose is such an example. As we'll see in the following sections, its consumption through glycolysis and the tricarboxylic acid cycle provides essential precursors for macromolecular synthesis *and* powers ATP synthesis. In contrast, chemolithotrophic autotrophs oxidize inorganic compounds, such as ferrous iron (Fe^{2+}), elemental sulfur, hydrogen gas (H_2), hydrogen sulfide (H_2S), ammonium (NH_4^+), and nitrite (NO_2^-) to yield energy and obtain electrons. Their principal carbon source is CO_2. At the end of the chapter, we'll see that the bacterium associated with the giant tube worm described in the chapter opening story is an example of a chemolithotrophic autotroph. It uses H_2S from its environment as an electron donor to generate energy to produce ATP in order to synthesize organic molecules from CO_2.

13.2 Fact Check

1. Compare and contrast ATP generation by substrate-level phosphorylation, oxidative phosphorylation, and photophosphorylation.
2. Distinguish differences of oxidation and reduction.
3. Describe the mathematical relationship between $\Delta E_o'$ and $\Delta G^{o'}$.
4. Explain how glycolysis, the TCA cycle, and the electron transport system are involved in ATP generation in a cell.
5. Describe chemotrophs, phototrophs, organotrophs, lithotrophs, heterotrophs, and autotrophs.

TABLE 13.1 Major metabolic groups of microorganisms

Metabolism	Energy source	Electron source	Carbon source	Microorganisms
Chemoorganoheterotrophy	Chemicals	Organic e⁻ donor (same as carbon source)	Organic carbon	Most non-photosynthetic microbes
Chemolithoheterotrophy	Chemicals	Inorganic e⁻ donor	Organic carbon	Some sulfur-oxidizing bacteria
Chemolithoautotrophy	Chemicals	Inorganic e⁻ donor	CO_2	Some sulfur-oxidizing bacteria, iron-oxidizing bacteria, methanogens, nitrifying bacteria
Photolithoautotrophy	Light	Inorganic e⁻ donor	CO_2	Cyanobacteria, purple and green sulfur bacteria
Photoorganoheterotrophy	Light	Organic e⁻ donor	Organic carbon	Purple non-sulfur and green non-sulfur bacteria

13.3 Carbon utilization in microorganisms

Carbon is the most abundant element in biological macromolecules and acquiring it is critical for cell growth and replication. The general process by which cells import a molecule and incorporate it into cellular constituents is called **assimilation**. As mentioned in the previous chapter, most of Earth's biosphere is built on carbon originally assimilated by autotrophic organisms. The organic molecules they produce, through mechanisms that will be covered in Section 13.6, are consumed and assimilated by heterotrophs (see Table 13.1). Heterotrophic microbes have adapted to transport and incorporate carbon from many kinds of organic molecules—carbohydrates, amino acids, lipids, organic acids, alcohols, and more. Our overview of carbon utilization will focus on what happens when glucose—a sugar assimilated by nearly all cells—is used as a source of carbon and energy.

Glycolysis

Glucose is one of the most abundant organic molecules in the biosphere. It is the building block for cellulose, the main polymer in plant cell walls, and modified forms of glucose are major contributors to bacterial and fungal cell walls. In Section 13.1, we introduced glycolysis as the catabolism of glucose to pyruvate. There are several ways this can be done, and different chemoorganoheterotrophs use different glycolytic (sugar-splitting) pathways for breaking down glucose to obtain energy and carbon compounds for biosynthesis.

The Embden–Meyerhof–Parnas (EMP) Pathway

The most common glycolytic pathway is the **Embden–Meyerhof–Parnas (EMP) pathway**, named for the German biochemists Gustav Embden and Otto Meyerhof, who worked out this pathway in the early twentieth century. The name of Russian biochemist Jakub Parnas is sometimes also added. This pathway, shown in **Figure 13.13**, generates pyruvate, ATP, and NADH. The overall reaction is

$$\text{Glucose} + 2\,\text{ADP} + 2\,P_i + 2\,\text{NAD}^+ \rightarrow 2\,\text{pyruvate} + 2\,\text{ATP} + 2\,\text{NADH} + 2\,\text{H}^+$$

The EMP pathway can be divided into two phases. In Phase I (Steps 1–5 in Figure 13.13), ATP is used to phosphorylate the six-carbon backbone of glucose to form glucose 6-phosphate. Another ATP molecule is invested after isomerization to fructose 6-phosphate, to produce a high-energy molecule of fructose 1,6-bisphosphate, which is cleaved into two three-carbon compounds (Step 4). Dihydroxyacetone phosphate is converted into another molecule of glyceraldehyde 3-phosphate by triose phosphate isomerase (Step 5), resulting in two identical three-carbon units that proceed through Phase II (Steps 6–10). Catabolism

of the two glyceraldehyde 3-phosphate molecules will produce even more potential free energy; four ATP molecules are synthesized by substrate-level phosphorylation. The investment of two ATP molecules for the initial phosphorylation of glucose results in a net profit of two ATP for the cell. Before any energy earnings can be made, energy must be invested.

Phase II generates energy in the form of electrons as well as ATP, and begins with the first oxidation (Step 6). Electrons are removed in a coupled reaction that produces NADH and adds another phosphate to each 3-carbon glyceraldehyde 3-phosphate to generate 1,3-bisphosphoglycerate. Notice that this phosphate doesn't come from ATP. Instead, the electron removal is coupled with formation of a phosphate bond from a free inorganic phosphate.

In reaction 7, a phosphate group is transferred from 1,3-bisphosphoglycerate to ADP, yielding ATP. You should recognize this as a substrate-level phosphorylation reaction (Section 13.1), since a phosphate is being transferred to ADP from a high-energy molecule (1,3-bisphosphoglycerate). As there are two molecules of 1,3-bisphosphoglycerate from each glucose molecule, the cell has now recovered its initial investment of ATP during the preparatory phase. The key accomplishment is to capture the energy from the redox reaction in Step 6 as an ATP molecule in Step 7.

Steps 8 and 9 reconfigure and dehydrate the three-carbon molecule to increase the potential energy in the covalent bond attaching the phosphate group of phosphoenolpyruvate. A second substrate-level phosphorylation generates another ATP in Step 10, along with pyruvate. This final step, carried out on each of the three-carbon units produced from a glucose molecule, gives the EMP pathway a net yield of two molecules of ATP per molecule of glucose. It also yields two NADH molecules per glucose molecule.

The EMP pathway is found in bacteria, archaeons, and eukaryal microbes. It is thought to have originated very early in the history of life, prior to the divergence of the three domains. Later, we will see that the pyruvate resulting from glycolysis can be further oxidized in other pathways, like the TCA cycle, to provide more energy. The EMP pathway is not used just for energy metabolism. Several intermediates, including glucose 6-phosphate, fructose 6-phosphate, glyceraldehyde 3-phosphate, phosphoenolpyruvate, and pyruvate, serve as precursors for synthesis of other key molecules, such as amino acids (covered in Section 13.8). Because of this, the EMP pathway is of central importance as a replenishing **anaplerotic reaction** for the production of intermediates for other metabolic pathways.

The Entner–Doudoroff Pathway

The **Entner–Doudoroff pathway** is less commonly used for glucose catabolism but is often used for catabolism of other carbohydrates containing aldehyde groups, such as gluconate, which cannot be readily processed by the EMP pathway. The

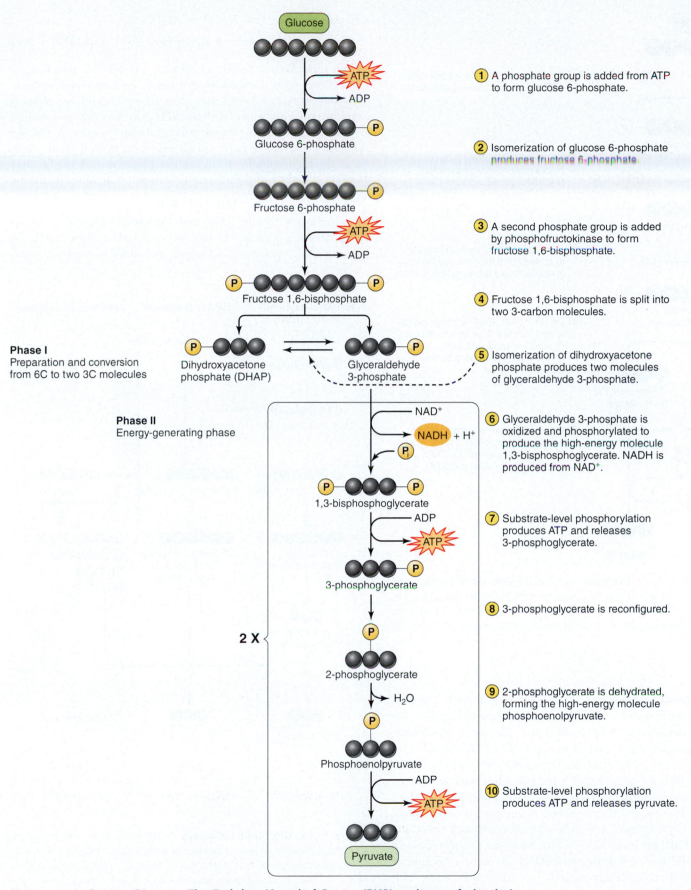

Figure 13.13. Process Diagram: The Embden–Meyerhof–Parnas (EMP) pathway of glycolysis

Glucose is converted to two molecules of pyruvate. Glucose is prepared for dismantling in Phase I, which consumes two molecules of ATP. Phase II generates four molecules of ATP by substrate-level phosphorylation. This results in a net gain of two molecules of ATP from each glucose molecule.

Within the figure:

Glucose

① A phosphate group is added from ATP to form glucose 6-phosphate.

Glucose 6-phosphate

② Isomerization of glucose 6-phosphate produces fructose 6-phosphate.

Fructose 6-phosphate

③ A second phosphate group is added by phosphofructokinase to form fructose 1,6-bisphosphate.

Fructose 1,6-bisphosphate

④ Fructose 1,6-bisphosphate is split into two 3-carbon molecules.

Phase I
Preparation and conversion from 6C to two 3C molecules

Dihydroxyacetone phosphate (DHAP)

Glyceraldehyde 3-phosphate

⑤ Isomerization of dihydroxyacetone phosphate produces two molecules of glyceraldehyde 3-phosphate.

Phase II
Energy-generating phase

NAD^+ → $NADH + H^+$ + P_i

⑥ Glyceraldehyde 3-phosphate is oxidized and phosphorylated to produce the high-energy molecule 1,3-bisphosphoglycerate. NADH is produced from NAD^+.

1,3-bisphosphoglycerate

ADP → ATP

⑦ Substrate-level phosphorylation produces ATP and releases 3-phosphoglycerate.

3-phosphoglycerate

⑧ 3-phosphoglycerate is reconfigured.

2 X

2-phosphoglycerate

H_2O

⑨ 2-phosphoglycerate is dehydrated, forming the high-energy molecule phosphoenolpyruvate.

Phosphoenolpyruvate

ADP → ATP

⑩ Substrate-level phosphorylation produces ATP and releases pyruvate.

Pyruvate

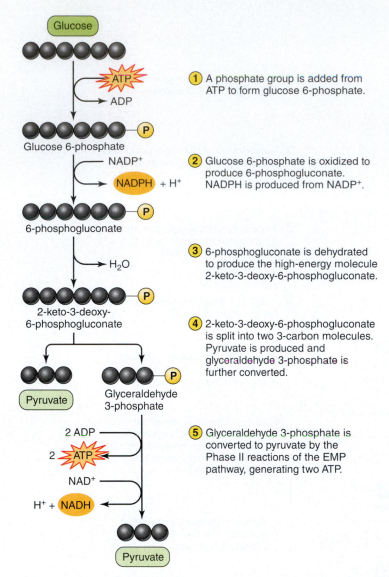

1. A phosphate group is added from ATP to form glucose 6-phosphate.

2. Glucose 6-phosphate is oxidized to produce 6-phosphogluconate. NADPH is produced from NADP$^+$.

3. 6-phosphogluconate is dehydrated to produce the high-energy molecule 2-keto-3-deoxy-6-phosphogluconate.

4. 2-keto-3-deoxy-6-phosphogluconate is split into two 3-carbon molecules. Pyruvate is produced and glyceraldehyde 3-phosphate is further converted.

5. Glyceraldehyde 3-phosphate is converted to pyruvate by the Phase II reactions of the EMP pathway, generating two ATP.

Figure 13.14. Process Diagram: The Entner–Doudoroff pathway of glycolysis Steps 1 through 4 convert glucose to pyruvate and glyceraldehyde 3-phosphate, with the consumption of one ATP. Glyceraldehyde 3-phosphate enters the Phase II reactions of the EMP pathway to produce a second molecule of pyruvate and two ATP. The pathway results in one molecule of ATP from each glucose molecule.

pathway, shown in **Figure 13.14**, begins with Step 1, like the EMP pathway, with the ATP-driven conversion of glucose to glucose 6-phosphate. Step 2 oxidizes glucose 6-phosphate to 6-phosphogluconate and generates NADPH. Recall that NADPH is a common electron donor for biosynthetic reactions. At Step 3, dehydration of 6-phosphogluconate generates 2-keto-3-deoxy-6-phosphogluconate. This molecule is then split (Step 4) into two three-carbon units, pyruvate and glyceraldehyde-3-phosphate, which can proceed through reactions 6–9 of the EMP pathway to generate a second pyruvate, along with two ATP molecules and an NADH. Because only one of the three-carbon units generated by reaction 4 yields ATP directly, the ATP yield of the Entner–Doudoroff pathway, summarized as follows, is less than the EMP pathway.

$$\text{Glucose} + \text{NADP}^+ + \text{NAD}^+ + \text{ADP} + \text{P}_i$$
$$\rightarrow 2 \text{ pyruvate} + \text{NADPH} + 2 \text{ H}^+ + \text{NADH} + \text{ATP}$$

Bacteria known to use the Entner–Doudoroff pathway include *Pseudomonas*, *Caulobacter*, *Rhizobium*, and *Azotobacter* species. Why this strategy evolved isn't clear. These species are obligate aerobes that gain most of their ATP from glucose metabolism by conveying NADH to the electron transport system, so the lost ATP molecule has a fairly minor effect on overall yield. Some facultative anaerobic bacteria, such as *Zymomonas*, use the Entner–Doudoroff pathway for fermentation under anaerobic conditions.

The Pentose Phosphate Pathway

An alternative to the EMP and Entner–Doudoroff pathways for processing glucose is the **pentose phosphate pathway** (**Figure 13.15**), so named because it produces five-carbon sugar

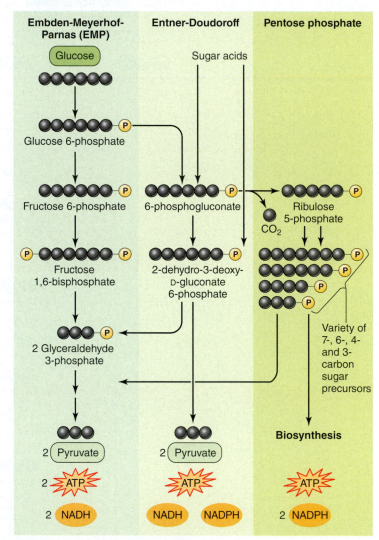

Figure 13.15. Glycolytic pathways The EMP pathway is most useful for generating ATP and NADH. The pentose phosphate pathway produces NADPH, a source of reducing power often used for biosynthetic reactions, as well as a number of seven- to three-carbon compounds that can be used as precursors for biosynthesis. The intermediate product, glucose 6-phosphate from the EMP pathway, can be routed to the Entner–Doudoroff pathway to provide NADPH. The pathways are interconnected and can be used to meet metabolic demands.

intermediates. This is technically not glycolysis, as it does not form pyruvate, but we will include it here as an important glycolytic pathway for processing glucose. The pentose phosphate pathway works with the EMP pathway, rather than replacing it. Its primary use is in generating the electron donor NADPH and essential precursors for biosynthesis. As such, it is an important anaplerotic reaction. For example, a microbe operating with glucose as its sole carbon source must convert some of it into the five-carbon ribose molecule to make the ribonucleotides for RNA synthesis and deoxyribonucleotides for DNA synthesis. Erythrose 4-phosphate is also an essential precursor for synthesis of the aromatic amino acids tryptophan, tyrosine, and phenylalanine, as well as other aromatic metabolites. Most microbes execute some version of this pathway.

The pentose phosphate pathway starts with the intermediate product glucose 6-phosphate, generated by Step 1 of the EMP pathway (see Figure 13.13), and so begins with the same investment of ATP to phosphorylate glucose. Glucose 6-phosphate is oxidized to 6-phosphogluconate and generates NADPH in the process. A second oxidation step generates NADPH, and simultaneously decarboxylates the six-carbon molecule to yield the pentose ribulose 5-phosphate. The ribulose backbone is converted to two different pentoses, ribose and xylulose. These are rearranged to produce three-carbon (glyceraldehyde) and seven-carbon (sedoheptulose) units, which in turn recombine to give six-carbon (fructose) and four-carbon (erythrose) units. Erythrose phosphate can be combined with xylulose phosphate to generate fructose 6-phosphate and glyceraldehyde 3-phosphate. The latter molecule can enter the EMP pathway to generate pyruvate, ATP, and NADH. The result, shown in the equation below, is the conversion of three molecules of glucose to two molecules of fructose 6-phosphate and one glyceraldehyde 3-phosphate, with the production of six molecules of NADPH.

$$3 \text{ glucose 6-phosphate} + 6 \text{ NADP}^+ + 3 \text{ H}_2\text{O}$$
$$\rightarrow 2 \text{ fructose 6-phosphate} + \text{glyceraldehyde 3-phosphate}$$
$$+ 3 \text{ CO}_2 + 6 \text{ NADPH} + 6 \text{ H}^+$$

The pathway is of significant use as it generates NADPH, a common electron donor for biosynthetic reactions. The fate of the electrons carried by NADH or other reduced electron carriers generated from glycolytic pathways depends on the genetic capacity of the organism and the conditions it experiences. As discussed in Section 13.2, the electrons can be used to generate ATP by oxidative phosphorylation, but other options, such as fermentation, are also used by some microbes. At this point, we'll look at a summary of these options, and examine them in more detail in later sections.

Fermentation

There is a finite pool of NAD$^+$/NADH in a cell. As glycolysis proceeds, NADH would accumulate and the supply of NAD$^+$ would be exhausted if NADH weren't converted by oxidation back to NAD$^+$. What would happen to glycolysis if there were no NAD$^+$ in the cell? Enzymes that require NAD$^+$ for coupled reactions would

cease to function. As a result, glycolysis would stop, and cells may die. Recycling of NADH to NAD$^+$ must happen for the cell to continue to consume carbohydrates, so there must be an acceptor to take electrons from NADH. Chemoorganoheterotrophs do this conversion by either of the following:

1. **fermentation**, where electrons are passed directly to an organic terminal electron acceptor, and the TCA cycle and the electron transport system is not utilized.

2. cellular **respiration**, where the electrons are passed through an electron transport system and on to an inorganic or sometimes an organic terminal electron acceptor.

Microbes can live as long as they have a nutrient source with more energy than the terminal electron acceptor they are capable of using. They can use a multitude of metabolic pathways to achieve this. Fermentation involves fewer processes than respiration, and we will cover it first. Fermentation can be defined as the transfer of electrons from NADH, or some other reduced electron carrier, to an organic molecule, commonly pyruvate (**Figure 13.16**). As oxygen is not used during fermentation, it is an anaerobic process, but many organisms can carry out this process in the presence of oxygen. In other words, fermentation is not restricted to obligate anaerobic organisms. For example, *Escherichia coli* and members of the genus *Streptococcus* can carry out fermentation in the presence of oxygen. The process of fermentation begins with glycolysis and the resulting fermentation products, such as lactic acid (lactate) or ethanol, become waste to be eliminated. The purpose of fermentation is to replenish NAD$^+$ stocks so that glycolysis can continue to produce a small, but constant stream of ATP from glycolysis.

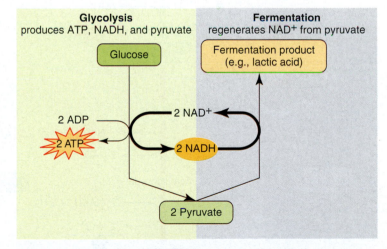

Figure 13.16. Recycling of NADH by fermentation For each molecule of glucose catabolized by the EMP glycolysis pathway, two ATP and two NADH are generated from two ADP and two NAD$^+$. Fermentation recycles NADH by reducing pyruvate, or a similar organic acid, which accepts electrons from NADH, replenishing NAD$^+$ for continued glycolysis and releasing a reduced waste product, such as lactic acid.

Lactic Acid Fermentation

Lactic acid fermentation is a very common process in fermenting bacteria and protists. Lactic acid is produced from pyruvate in a very simple process requiring only lactate dehydrogenase **(Figure 13.17)**. Lactic acid is the reduced product of the transfer of an electron from NADH and is transported from the cell as the waste product lactate. In solution, lactic acid loses a proton to become the negatively charged counterpart lactate.

CONNECTIONS Lactate may be a waste product for some fermenting bacteria, but as described in **Section 17.4**, can be a nutrient for other microbes living in the gut of herbivorous animals, such as cows.

The lactic acid fermentation pathway is the dominant route of glucose catabolism in a large group of Gram-positive bacteria known as **lactic acid bacteria (LAB)**, which includes genera such as *Streptococcus*, *Lactobacillus*, and *Bifidobacterium*. The streptococci include several important pathogens of humans (see Microbes in Focus 17.2). In contrast, various lactobacilli and bifidobacteria are commensals of the human intestinal and genitourinary tracts that are known to have beneficial functions (see Section 17.4). Lactic acid bacteria are used in the food industry for the fermentation of milk to produce yogurt and sour cream (see Section 16.3).

CONNECTIONS The production of lactic acid in foods inhibits other microbes that would otherwise cause food spoilage in carbon-rich foods. See **Section 16.3**.

Lactic acid fermentation can occur using one of two pathways from glucose, either homolactic fermentation or heterolactic fermentation. Homolactic fermenting LAB use the EMP pathway to make pyruvate and then reduce it directly to lactate (see Figure 13.17, brown pathway). The overall reaction is:

$$C_6H_{12}O_6 + 2\,ADP + 2\,P_i \rightarrow 2\,C_3H_6O_3 + 2\,ATP$$

Recall that the balance of $NAD^+/NADH$ is restored and does not appear in the final equation.

Some LAB, such as *Leuconostoc*, undertake a more complex process of heterolactic fermentation that generates both lactic acid and ethanol **(Figure 13.18)**. Initial reactions of this pathway are identical to the pentose phosphate pathway (see Figure 13.15). Note, that a total of only one ATP is produced by heterolactic fermentation. The pathway illustrated here is an example of the diverse strategies bacteria have evolved for anaerobic sugar metabolism. Mixtures of fermentation products from the same microbe are quite common.

Perhaps the most fascinating aspect about the metabolism of LAB is that although they lack respiration pathways and are restricted to fermentation, an anaerobic process, they are able to carry out fermentation under aerobic conditions. They are classified as aerotolerant anaerobes. Given the energetic advantages of respiratory metabolism, it is perplexing on the surface that they would forgo this option completely. Let's consider this issue from another perspective. In light of their extensive nutritional requirements, LAB are adapted and limited to habitats rich in organic material, such as soil, decaying vegetation,

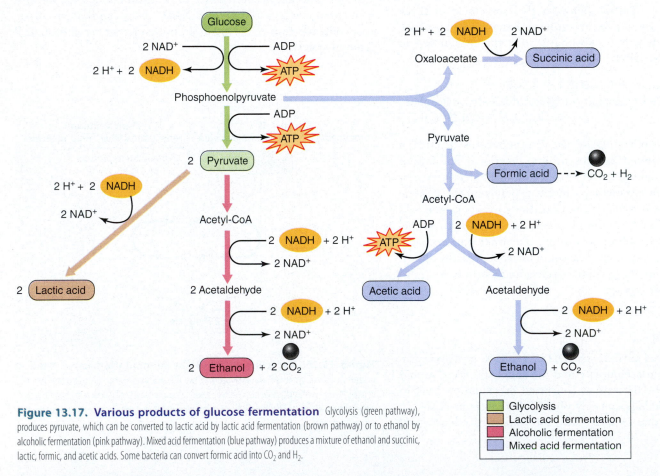

Figure 13.17. Various products of glucose fermentation Glycolysis (green pathway), produces pyruvate, which can be converted to lactic acid by lactic acid fermentation (brown pathway) or to ethanol by alcoholic fermentation (pink pathway). Mixed acid fermentation (blue pathway) produces a mixture of ethanol and succinic, lactic, formic, and acetic acids. Some bacteria can convert formic acid into CO_2 and H_2.

and intestinal tracts. They have fairly small cells and small genomes, and they turn out prodigious amounts of lactic acid as a metabolic by-product. Because they are acid-tolerant, this characteristic alters the environment to favor their growth and inhibits the growth of non-acid-tolerant competitors. Dental cavities are a common outcome of this activity, described in detail in Section 17.4. Their restriction to fermentative metabolism, streamlined genomes, and small cells may enable LAB to reproduce quickly and compete effectively with other microbes in carbohydrate-rich habitats with limited oxygen.

CONNECTIONS Recall from **Section 6.3** that non-respiring aerotolerant anaerobes, such as LAB, must possess defenses against toxic oxygen species, such as superoxide anions, that are generated in an oxygenated environment. Without protection, these oxygen species can damage enzymes and nucleic acids. Obligate anaerobes lack adequate defenses to such damage.

Alcoholic Fermentation

Some microbes produce ethanol and CO_2 as major fermentation products. This is known as alcoholic fermentation. Many types of yeast and some bacteria use this process (see Figure 13.17). Pyruvate, a three-carbon molecule, is first split to CO_2 and acetaldehyde by pyruvate decarboxylase. Acetaldehyde is then reduced to ethanol by alcohol dehydrogenase to accomplish the desired oxidation of NADH to NAD^+. Ethanol is a small, weakly polar molecule that rapidly diffuses across the plasma membrane to leave the cell as a waste product. The overall reaction beginning with glucose is

$$C_6H_{12}O_6 + 2\,ADP + 2\,P_i \rightarrow 2\,C_2H_5OH + 2\,CO_2 + 2\,ATP$$

Alcoholic fermentation has been used by humans for the production of beer and wine for several thousand years, long before people knew that microbes such as yeast were responsible for this activity. In the brewing of beer, starch from grains is broken down to glucose to be used as a substrate for fermentation. During the wine-making process, sugars such as fructose and glucose in the juice of grapes or other fruits are fermented. The yeasts that carry out fermentation are naturally present in fruit, although specific strains of yeast are typically used in the modern wine and beer industries. These yeast are capable of aerobic respiration, and will use fermentation under anaerobic conditions. They are facultative anaerobes (see Section 6.3). Beer and wine fermentation must be carried out in sealed containers to exclude air. The yeast initially consume the dissolved oxygen, then switch to fermentative metabolism and start producing alcohol. Growth and metabolism naturally stop when the fermentable carbon sources are exhausted or when the ethanol concentration becomes so toxic (12–15 percent range) that metabolic processes stop (see Chapter 6).

Mixed Acid Fermentation

Other fermentation pathways have been identified in various bacteria and fungi that yield a variety of organic end products from pyruvate or a similar organic acid. Many bacteria belonging to the family Enterobacteriaceae use the mixed acid fermentation pathway (see Figure 13.17). Under anaerobic conditions,

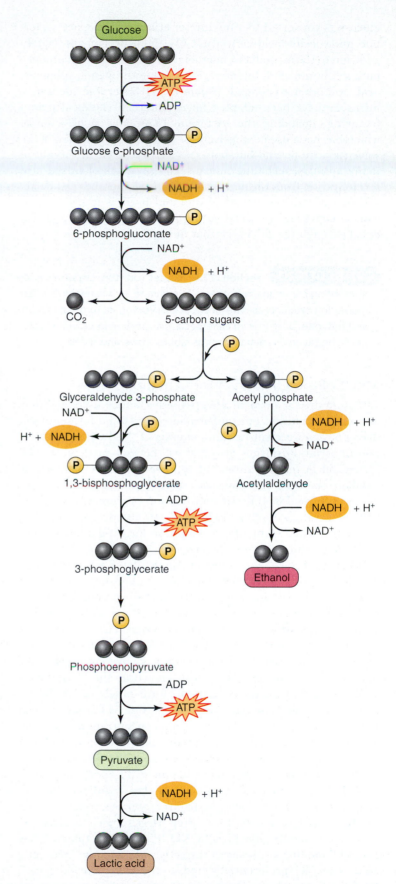

Figure 13.18. Heterolactic fermentation Five-carbon compounds from the pentose phosphate pathway are converted into multiple products. Shown here are lactic acid and ethanol.

glucose is converted to a mixture of ethanol, lactic, acetic, formic, and succinic acids as well as CO_2, and hydrogen gas (H_2). In addition to these products, microbial fermentations can produce butanol, butanediol, butyric acid, acetone, propanol, propionic acid, and other compounds **(Figure 13.19)**. Several fermentative microbes have been adapted for commercial chemical manufacturing, including the bacterium *Clostridium acetylbutylicum,* which has been used extensively for industrial production of the solvents acetone and butanol (see Section 12.4). In the gut of many animals, including humans, the products of mixed acid fermentation from the activity of intestinal microbes can be absorbed through the intestinal wall and used as nutrients that can be converted to acetyl coenzyme A (acetyl-CoA) and oxidized through the TCA cycle, our next subject.

CONNECTIONS **Section 17.4** focuses on the importance of fermenting communal microbes in human and animal health. The fermentation products acetate (acetic acid), propionate (propionic acid), and butyrate (butyric acid) are especially important in providing calories in herbivorous animals, such as rabbits, cows, and sheep.

Metabolic Regulation of Fermentation

Although diverse fermentation pathways can be used by different microorganisms, one unifying outcome is that fermentation does not generate substantial amounts of ATP. Substrate-level phosphorylation during glycolysis is the *only* process of ATP formation in fermentation, and glucose cannot be completely catabolized. Only two molecules of ATP per molecule of glucose are formed by this reaction, about 5 percent of the energy yielded through aerobic respiration. To compensate for this, organisms using fermentation consume large amounts of substrate to sustain their metabolic needs.

Fermentation is much less energetically efficient than respiration, but it is a common mode of metabolism in facultative anaerobes that routinely experience anaerobic conditions. Yeast, such as *Saccharomyces cerevisiae*, can produce energy using either aerobic respiration or fermentation. Louis Pasteur observed that aerating a culture of yeast led to a reduction in the amount of glucose consumed. Under low oxygen conditions, much more glucose was consumed. The effect, known as the "Pasteur effect," is due to an increase in the rate of glycolysis during fermentation because only two molecules of ATP are produced per molecule of glucose. When oxygen concentrations increase, aerobic respiration takes place producing 38 moles of ATP per mole of glucose. Thus, in aerobically respiring cells, a high rate of glycolysis is not needed to provide the cell's energy needs.

What is the molecular mechanism that regulates the rate of glycolysis? The enzyme phosphofructokinase, a key enzyme in glycolysis (see Figure 13.13, Step 3) is subject to allosteric regulation (see Section 13.2) by ATP. Phosphofructokinase has two ATP binding sites, one in the active site and one in an allosteric site. ATP preferentially binds to the active site along with fructose 6-phosphate. During low levels of ATP, experienced during fermentation, ATP binds only to the active site, allowing production of fructose 1,6-bisphosphate and the continuation of glycolysis for fermentation. High levels of ATP experienced during aerobic respiration enables ATP to also bind the allosteric

site. In so doing, ATP becomes an allosteric inhibitor, causing a conformational change in the enzyme, shutting down its ability to convert fructose 6-phosphate. This reduces the amount of fructose 1,6-bisphosphate available for glycolysis and explains the decreased rate of metabolism of glucose observed during respiration. Regulation of this key enzyme in glycolysis allows the cell to balance ATP production with its energy needs. This is an example of **feedback inhibition** of a metabolic pathway **(Figure 13.20)**, where the product of an enzyme(s) in a pathway inhibits the reaction of an earlier enzyme through allosteric inhibition (see Figure 13.9). In this case, the accumulation of ATP produced later in the respiratory pathway inhibits glycolysis, a much earlier step in the energy-generating pathway. Some metabolic pathways use **precursor activation** when a substrate in an enzyme pathway increases the activity of one or more enzymes further on in the series.

@ **End product inhibition**
ANIMATION

Microorganisms may have over a thousand biochemical reactions occurring in the cell at any given time. If these reactions are not carefully controlled, chaos and waste would ensue. Generally, microbes regulate metabolism in order to:

- avoid making enzymes for which substrates aren't available
- ensure that enzymes are only active when their products are needed, so that substrates aren't wasted and excessive products are not generated

In general, enzyme activities are adjusted so that the product concentration is near optimal for growth in the current habitat. Metabolic regulation is accomplished in the following ways:

- *Regulation of gene expression*. RNA and protein synthesis consume cellular resources. If an enzymatic activity is not useful in a particular environment, it makes sense to reduce or eliminate gene expression (see Chapter 11).
- *Control of enzyme activity*. For enzymes that have already been synthesized, activity can be increased or decreased by allosteric regulation. Enzyme activity can also be affected by regulating the stability of the protein.

As a general principle, metabolic pathways are regulated at the first "committed" step, where an intermediate is produced that has no other function. This is logical in terms of energy efficiency. By controlling the first step of the pathway, the cell can restrict unnecessary metabolite flux into the pathway. The signal for regulation varies, depending on the process being regulated as we saw in Chapter 11. For catabolic pathways dedicated to the breakdown of a particular molecule, it may make little sense to express the enzymes of the pathway in the absence of that molecule.

Respiration

In microbes using fermentation, the electrons carried by NADH are essentially waste that must be eliminated to recycle NAD^+. In microbes capable of respiration, these electrons can be used to fuel more ATP production using oxidative phosphorylation. Respiration is the process by which electrons generated from the oxidation of an energy substrate are transferred through

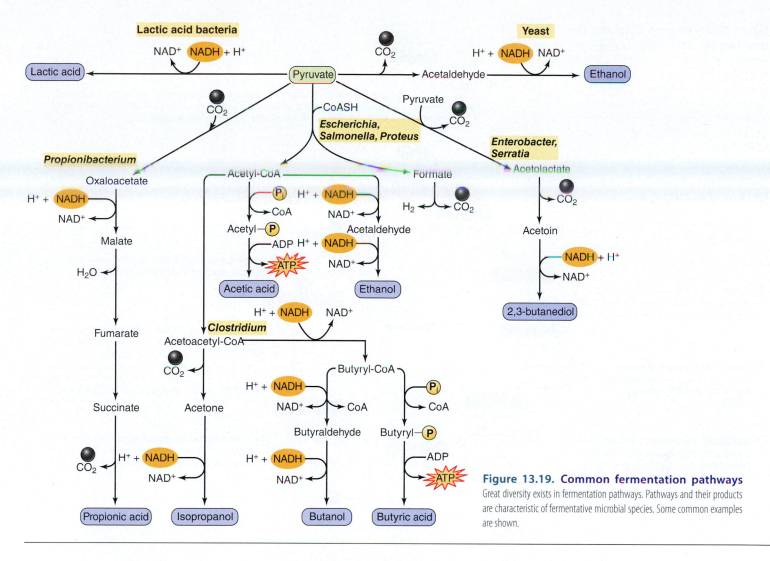

Figure 13.19. Common fermentation pathways
Great diversity exists in fermentation pathways. Pathways and their products are characteristic of fermentative microbial species. Some common examples are shown.

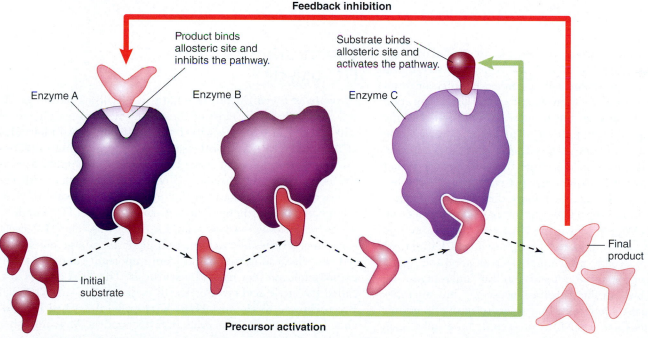

Feedback inhibition

Product binds allosteric site and inhibits the pathway.

Substrate binds allosteric site and activates the pathway.

Enzyme A

Enzyme B

Enzyme C

Initial substrate

Final product

Precursor activation

Figure 13.20. Feedback inhibition and precursor activation Feedback inhibition occurs when the product of a series of enzyme reactions accumulates in the cell and inhibits the action of an enzyme earlier in the pathway. Further production of the product is halted. Precursor activation occurs when there is an abundance of an early substrate that can specifically increase the action of an enzyme further along the pathway. Continued use of the substrate ultimately reduces its concentration in the cell. Both mechanisms rely on allosteric regulation of enzymes to control their activity.

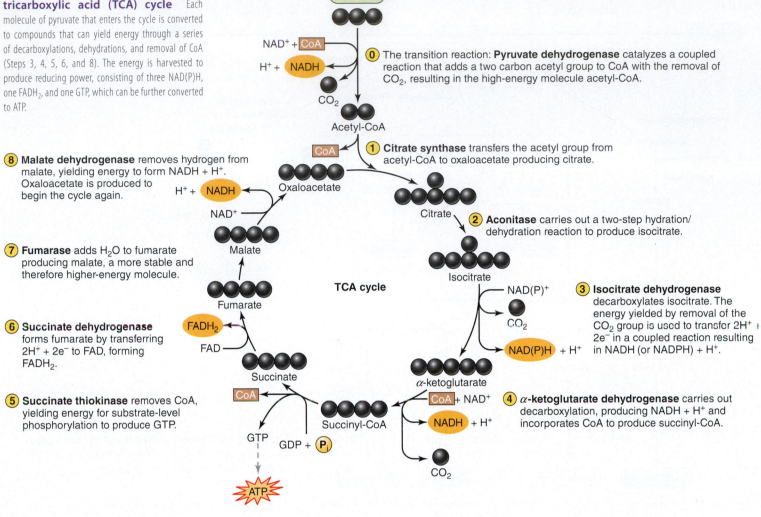

Figure 13.21. Process Diagram: The tricarboxylic acid (TCA) cycle Each molecule of pyruvate that enters the cycle is converted to compounds that can yield energy through a series of decarboxylations, dehydrations, and removal of CoA (Steps 3, 4, 5, 6, and 8). The energy is harvested to produce reducing power, consisting of three NAD(P)H, one FADH$_2$, and one GTP, which can be further converted to ATP.

0 The transition reaction: **Pyruvate dehydrogenase** catalyzes a coupled reaction that adds a two carbon acetyl group to CoA with the removal of CO$_2$, resulting in the high-energy molecule acetyl-CoA.

1 Citrate synthase transfers the acetyl group from acetyl-CoA to oxaloacetate producing citrate.

2 Aconitase carries out a two-step hydration/dehydration reaction to produce isocitrate.

3 Isocitrate dehydrogenase decarboxylates isocitrate. The energy yielded by removal of the CO$_2$ group is used to transfer 2H$^+$ + 2e$^-$ in a coupled reaction resulting in NADH (or NADPH) + H$^+$.

4 α-ketoglutarate dehydrogenase carries out decarboxylation, producing NADH + H$^+$ and incorporates CoA to produce succinyl-CoA.

5 Succinate thiokinase removes CoA, yielding energy for substrate-level phosphorylation to produce GTP.

6 Succinate dehydrogenase forms fumarate by transferring 2H$^+$ + 2e$^-$ to FAD, forming FADH$_2$.

7 Fumarase adds H$_2$O to fumarate producing malate, a more stable and therefore higher-energy molecule.

8 Malate dehydrogenase removes hydrogen from malate, yielding energy to form NADH + H$^+$. Oxaloacetate is produced to begin the cycle again.

the electron transport system to yield energy. Cellular respiration can be either aerobic, that is, O$_2$ is used as a terminal electron acceptor, or anaerobic, where some other organic or inorganic molecule is used. We'll investigate anaerobic respiration options later in Section 13.4, but for now let's continue looking down the energy-generating pathway of an aerobically respiring chemoorganotroph. This entire respiration process is depicted in Figure 13.11. Respiration is a large three-step process that begins first with glycolysis to generate pyruvate, which is further oxidized to CO$_2$ in the TCA cycle. Finally, the electrons generated during glycolysis and the TCA cycle are passed to an electron carrier, such as NAD, and down the electron transport system to generate more ATP by oxidative phosphorylation. Ultimately, the electrons are passed to a terminal electron acceptor. Again, the electron acceptor used depends on the microorganism; those that carry out aerobic respiration use O$_2$, but microbes that carry out anaerobic respiration use alternative inorganic molecules, such as CO$_2$ and NO$_3^-$, or organic molecules, such as fumarate. In the next section (Section 13.4), we will see that the biggest advantage of respiration is that pyruvate is further metabolized to generate significant amounts of ATP.

The tricarboxylic acid (TCA) cycle

Pyruvate is a central intermediate in glycolysis pathways. At the end of glycolysis, most of the chemical energy of glucose is still trapped in pyruvate with only a maximum two molecules of ATP and two NADH released so far. Fermenting microbes lose out on this potential energy, as pyruvate or a similar derivative is used as an electron acceptor, not to be further oxidized. However, many microbes can further oxidize pyruvate to yield a large amount of energy. The tricarboxylic acid (TCA) cycle is a central pathway in respiration for doing this. The TCA cycle consists of a series of reactions that generate a large number of reduced electron carriers, and is an important source of small carbon molecules for use in biosynthesis. The TCA cycle is also called the citric acid cycle or Krebs cycle, after the 1953 Nobel Prize winning biochemist Hans Krebs who described the biochemical reactions of the cycle in mitochondria. Most microbes capable of respiration use some version of the TCA cycle, but we will examine only the most commonly used pathway here **(Figure 13.21)**.

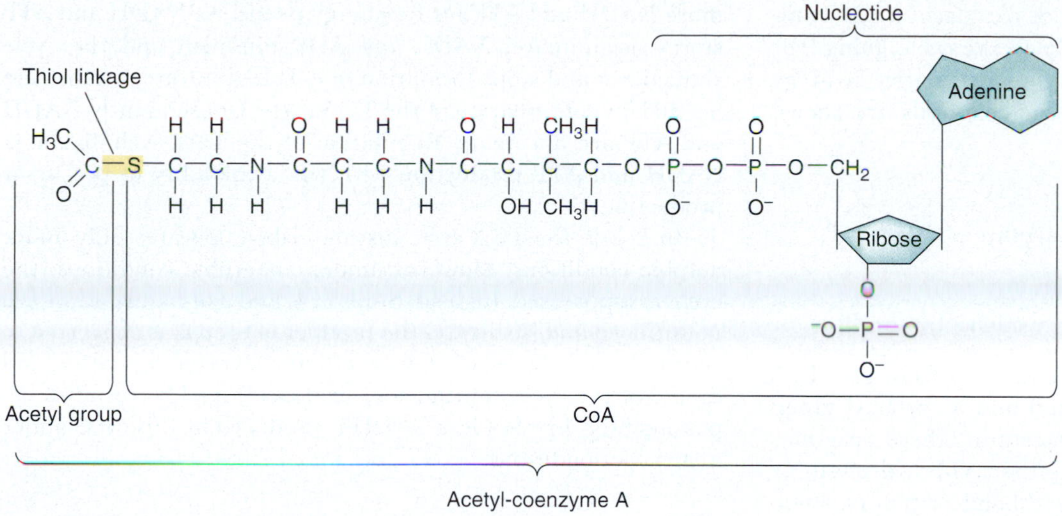

Figure 13.22. Acetyl-coenzyme A (acetyl-CoA) The acetyl group is easily transferred to other molecules by the enzyme-catalyzed hydrolysis of the thiol bond.

Prior to entering the cycle, pyruvate is first oxidized to produce acetyl-CoA. This reaction is sometimes referred to as the transition reaction. Pyruvate dehydrogenase (see Figure 13.21, Step 0) catalyzes a complex, multistep reaction in which pyruvate is both oxidized and decarboxylated, meaning that a molecule of CO_2 is lost. The overall reaction is deceptively simple:

$$\text{Pyruvate} + NAD^+ \rightarrow \text{acetyl-CoA} + NADH + H^+$$

Not shown in this equation are the *five* separate enzymatic steps that occur in the course of this reaction. Pyruvate dehydrogenase contains multiple active sites to accommodate the mechanistic complexity of the reaction. In fact, this is the largest enzyme, other than the ribosome, in most microbial cells. In addition to NAD^+, pyruvate dehydrogenase depends on coenzymes such as flavin adenine dinucleotide (FAD), an electron carrier related to NAD, and lipoamide that are not in the final equation. Lipoamide was the main target of organic arsenic-containing drugs such as Salvarsan, an early antimicrobial drug used to treat syphilis caused by the bacterium *Treponema pallidum* (see Section 1.1). Coenzyme A (CoA), a complex cofactor based, like NAD and FAD, on adenine, is also required. Coenzyme A becomes covalently linked to the two-carbon acetyl group remaining from pyruvate. The acetyl group is attached via a thiol linkage to CoA **(Figure 13.22)**, making acetyl-CoA a high-energy molecule, ready to transfer its acetyl group to oxaloacetate.

The TCA cycle as shown in Figure 13.21 includes four redox reactions (Steps 3, 4, 6, and 8), two decarboxylations (Steps 3 and 4), and a substrate-level phosphorylation to generate GTP (Step 5), which can be converted to ATP. The initial step joins a two-carbon acetyl unit to oxaloacetate to generate citrate, a six-carbon molecule. The overall reaction for the complete catabolism of pyruvate is

$$\text{Pyruvate} + 4\,NAD + FAD + ADP + P_i$$
$$\rightarrow 3\,CO_2 + 4\,NADH + FADH + GTP\,(ATP)$$

The TCA cycle completes the oxidation of glucose, in the form of two pyruvate molecules, to six CO_2. Note that for the complete oxidation of one molecule of glucose two turns of the TCA cycle, one for each pryruvate, are required. The majority of the oxidative reactions applied to the original six-carbon skeleton of glucose occur during the TCA cycle and generate large amounts of reduced electron carriers (NADH and $FADH_2$). This is a prime function of the TCA cycle since transfer of these electrons to the electron transport system fuels ATP production. We'll examine this in more detail in Section 13.4.

Anaplerotic Reactions and the TCA Cycle

The TCA cycle is critical for generating macromolecular building blocks **Table 13.2**.

This use of TCA cycle intermediates raises an interesting question: How can a "cyclic" pathway be sustainably operated when intermediates are being continuously diverted for other purposes? Every molecule of α-ketoglutarate, succinyl-CoA, or oxaloacetate that gets consumed by a reaction *outside* the

TABLE 13.2 TCA cycle intermediates used for anabolism

Cycle intermediate	Reaction step in cycle	Anabolic use	Products
α-ketoglutarate	3	Amino acids	Glutamate, proline, arginine
Succinyl-CoA	4	Porphyrins	Photosynthetic pigments
Oxaloacetate	8	Amino acids, nucleotides	Aspartate, asparagine, threonine, lysine, isoleucine, methionine, cytidine, thymidine, uridine

TCA cycle means one less molecule of oxaloacetate is available to combine with acetyl-CoA to keep the cycle going. For this to be sustainable, a cell must have an alternative strategy to generate oxaloacetate. Two anabolic reactions are shown below:

Pyruvate carboxylase reaction:

$$\text{Pyruvate} + \text{ATP} + CO_2 \rightarrow \text{oxaloacetate} + \text{ADP} + P_i$$

Phosphoenolpyruvate carboxylase reaction:

$$\text{Phosphoenolpyruvate} + CO_2 \rightarrow \text{oxaloacetate} + P_i$$

Both reactions take a three-carbon compound from glycolysis (pyruvate or phosphoenolpyruvate) and add a carboxyl group derived from CO_2 to produce oxaloacetate. These reactions require energy, which is provided by either ATP hydrolysis, or breakage of the high-energy phosphate bond of phosphoenolpyruvate. Microbes generally use one or the other of these reactions, but not both, to maintain a sufficient minimum pool of oxaloacetate to keep the TCA cycle flowing as intermediates are diverted for anabolism.

Regulation of the TCA Cycle

The TCA cycle is regulated through several enzymes. Isocitrate dehydrogenase is one of these (see Figure 13.21, Step 3). It links the TCA cycle with the electron transport system (see Section 13.4) by being the first point of NADH production from the TCA cycle. Isocitrate dehydrogenase is regulated allosterically through both feedback inhibition and precursor activation (see Figure 13.20). Feedback inhibition occurs by the allosteric inhibitors ATP and NADH. Precursor activation is carried out by the allosteric activators ADP and NAD^+. When the concentration of NADH, and correspondingly ATP, in the cell is low, the precursors NAD^+ and ADP are high and the TCA cycle operates to feed electrons into the electron transport system to produce more NADH and ATP for its energy needs. As NADH and ATP stores accumulate, NAD^+ and ADP diminish and the cycle shuts down and stops formation of α-ketoglutarate, a substrate needed for continuation of the TCA cycle. Consequently, NADH and ATP are not made. Regulation of the TCA cycle balances NADH and ATP production with the availability of precursor production.

In *E. coli*, the TCA cycle just described operates fully under aerobic conditions. However, under anaerobic conditions, the cycle is suppressed. Instead, Steps 3, 4, and 5 of Figure 13.21 are not utilized and isocitrate, the product of Step 2, is converted to succinate and glyoxylate. Glyoxylate is converted to malate, and Step 8 of the cycle continues to oxaloacetate. The result of bypassing Step 4 is that less NADH is produced by this cycle under anaerobic conditions.

13.3 Fact Check

1. Compare and contrast these three glycolytic pathways: Embden–Meyerhof–Parnas (EMP) pathway, Entner–Doudoroff pathway, and pentose phosphate pathway.
2. Differentiate between fermentation and cellular respiration.
3. Describe an overview of lactic acid fermentation, and provide examples of lactic acid fermenting bacteria.
4. Provide an overview of the process of alcoholic fermentation, including examples of microbes that carry out this process.
5. Describe how the following are involved in the TCA cycle: pyruvate, CO_2, NADH, $FADH_2$, and ATP.

13.4 Respiration and the electron transport system

What becomes of the electrons generated by glycolysis and the TCA cycle?

In the previous section, we learned how oxidation of organic compounds such as glucose can generate ATP through substrate-level phosphorylation during glycolysis and the TCA cycle. In addition, both the TCA cycle and, to a much lesser extent, glycolysis generate a considerable amount of reduced electron carriers, NADH and $FADH_2$. The electrons carried by these molecules still have a great deal of potential energy, as indicated by their negative E_o' values (see Figure 13.12), and can be used to reduce other molecules to generate further energy. As you will recall, cellular respiration is the process by which electron carriers, such as NADH, generated from the oxidation of an energy substrate, transfer electrons through the electron transport system to yield energy. Respiration allows the recycling of NAD (NADH to NAD^+) to be used again and the generation of ATP.

In fermentation, ATP and NADH are generated through glycolytic pathways. NADH is converted back to NAD^+ by donating an electron to an organic end product, such as pyruvate. Fermentation does not produce additional ATP because the TCA cycle and electron transport system are not used. Therefore, respiration is more energetically beneficial. Cellular respiration can be either anaerobic, where O_2 is not used as a terminal electron acceptor, or aerobic, where O_2 is used as the terminal electron acceptor. Aerobic respiration produces a maximum of 38 molecules of ATP for each molecule of glucose versus a maximum of two molecules of ATP from fermentation. **Figure 13.23** shows the possible catabolic fates of glucose and the pathways to fermentation and aerobic and anaerobic respiration. Some microbes can use all three options for generating ATP. Central to the process of cellular respiration is the electron transport system, our next topic.

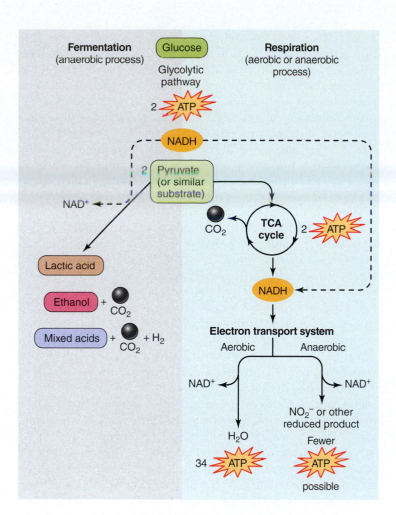

Figure 13.23. Possible catabolic fates of glucose for generating energy Glucose is catabolized in a glycolytic pathway under aerobic or anaerobic conditions to form pyruvate or similar molecules and a maximum of two ATP. Microbes using fermentation (*left*) to reduce glycolytic products form organic waste molecules, CO_2 and H_2. The TCA cycle is not utilized. Microbes using cellular respiration (*right*) use the TCA cycle to catabolize pyruvate in a series of oxidation reactions resulting in CO_2 and NADH. Electrons carried by NADH enter the electron transport system and are used to generate a proton motive force, and ultimately ATP. Under aerobic conditions (in the presence of O_2), glucose can be completely catabolized to carbon dioxide and produce 38 ATP. Some microbes are capable of using anaerobic respiration when O_2 is not available, which generates fewer ATP. The number varies with the electron acceptor used.

The electron transport system

The ultimate purpose of respiration is to facilitate the formation of ATP through oxidative phosphorylation by using the energy carried by the electron carriers generated during glycolysis and the TCA cycle. The transfer of electrons from NADH to a terminal electron acceptor, such as O_2, produces a very large difference in reduction potential (~1.1 volts). The use of an incremental series of redox reactions in the electron transport system ensures that the energy released by electron transfers is captured with maximal efficiency. This energy is used to drive the energetically unfavorable movement of protons across the membrane, to be used for ATP production or to do mechanical work. To accomplish this, the electron transport system is used. Electrons are progressively passed down a chain of electron acceptors with increasing redox potentials, much like water flowing down sets of rapids. As the electrons move from one acceptor to the next, protons are pumped out of the membrane at multiple sites along the chain (**Figure 13.24**). This creates a strong potential difference across the membrane, which forms the **proton motive force**. The proton motive force drives a flow of H^+ through ATP synthase to produce ATP.

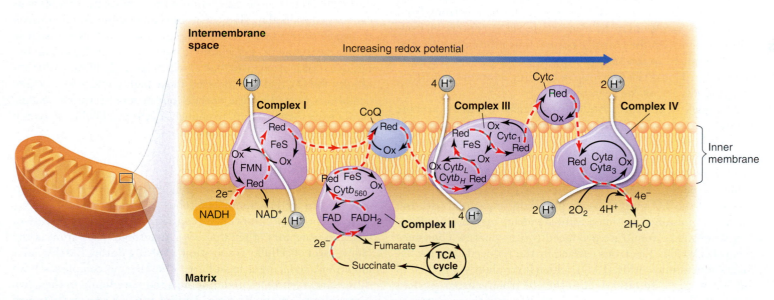

Figure 13.24. The aerobic electron transport system of mitochondria Electrons are passed from NADH and $FADH_2$ to a series of membrane-associated molecules and complexes that are arranged in close proximity to one another in order of increasing redox potential to move the electrons along. At points along the chain, protons are transferred across the mitochondrial inner membrane, creating a proton motive force that can be used to synthesize ATP. The red dashed line shows the flow of electrons ending with the reduction of O_2 forming H_2O.

Respiratory electron transport systems are composed of integral membrane proteins and lipids and their associated cofactors. To recap, a cofactor is a molecule that binds to an enzyme, either stably or transiently, and is needed for the activity of the enzyme. The cofactors of enzymes involved in the electron transport system participate directly in electron transfer reactions. The major groups of electron carriers are:

- flavoproteins (which also carry hydrogen)
- quinones (which also carry hydrogen)
- iron-sulfur proteins
- cytochromes

In bacteria and archaeons, the electron transport system is located in the plasma membrane. In eukaryal microbes, it is located in the inner mitochondrial membrane. There are differences in the composition of the electron carriers between archaeons, bacteria, and eukaryal microbes, but the function is fundamentally the same in all. There are also variations in the aerobic and anaerobic respiration electron transport systems, but again, they all function to yield energy from the transport of electrons.

Aerobic Respiration

Aerobic respiration specifically uses O_2 as a terminal electron acceptor and can only be carried out under aerobic conditions. Molecular oxygen is a strong oxidant and excellent terminal electron acceptor (see Figure 13.12), allowing the release of a great deal of energy as electrons move from NADH to O_2. We can examine some of the details of the electron transport system by following the example of aerobic respiration as it occurs in mitochondria, a well-studied system. The system is organized into four membrane complexes (see Figure 13.24). The pair of electrons carried through the mitochondrial matrix by NADH is transferred to Complex I, also known as NADH dehydrogenase or NADH:Coenzyme Q oxidoreductase. NADH dehydrogenase contains a bound flavoprotein cofactor called flavin mononucleotide (FMN) that accepts the electrons from NADH and passes them to iron-sulfide (FeS) centers of iron-sulfur proteins. As the transfer occurs, four protons are ejected from Complex I to the opposite side of the mitochondrial inner membrane. This is the first step in building a proton concentration gradient across the membrane.

Complex II, the succinate dehydrogenase complex, accepts electrons carried by $FADH_2$, which was generated from the TCA cycle. Complex II is associated with succinate dehydrogenase, which you may recall from the TCA cycle (see Figure 13.21, Step 6) and is the site of a redox reaction generating the reduced cofactor $FADH_2$. Succinate dehydrogenase is the only enzyme of the TCA cycle that is physically associated with the plasma membrane, or inner mitochondrial membrane of eukaryal cells. The fumarate produced by this enzyme continues in the TCA cycle. Electrons carried by the reduced cofactor flavin adenine dinucleotide (FAD), which is covalently attached to the enzyme, are fed directly into the electron transport system by transferring them to coenzyme Q. Because these electrons don't go through Complex I, they miss the proton-pumping event it carries out. Thus, the yield of ATP from electrons entering the chain from Complex II only is reduced.

Electrons from Complex I and/or Complex II are passed to the carrier coenzyme Q (CoQ). CoQ, also known as ubiquinone, belongs to a class of lipids called quinones. Quinones are mobile in membranes and carry hydrogen and electrons from Complexes I and II to Complex III of the electron transport system. After reduction, ubiquinone becomes ubiquinol, which then passes the electrons to Complex III, the ubiquinol:cytochrome c oxidoreductase, sometimes called the cytochrome bc_1 complex. This is the second proton-pumping step. Cytochromes are electron-transporting proteins that contain heme groups, complex organic ring structures with an iron (Fe) atom bound in the center. The iron atom alternates in its oxidation state between Fe(III) and Fe(II) as it accepts and transfers an electron. Three types of cytochrome proteins are involved in aerobic respiratory electron transport, cytochromes a, b, and c. Each of these contains a slightly different type of heme group—heme a, b, or c, and there are multiple variants of each cytochrome type, indicated by a subscript number following the letter. Electrons passed to Complex III move through two b-type heme groups, an iron-sulfur center, and a c-type cytochrome, before being transferred to a separate cytochrome c associated with the external surface of the membrane. From cytochrome c, electrons are transferred to Complex IV, also called cytochrome c oxidase, which catalyzes oxidation and reduction reactions using O_2 as the terminal electron acceptor to produce two H_2O molecules using electrons and H^+ (see Figure 13.24). Complex IV contributes to proton motive force generation by catalyzing the transfer of two protons from the cytoplasm across the membrane. Note that the number of protons transferred at each point is not precisely known. Approximately eight to ten protons are pumped for each NADH oxidation along with transfer of four electrons through the chain.

The mitochondrial respiratory electron transport system is functionally similar to that of many aerobically respiring bacteria and archaeons under normal atmospheric levels of oxygen (approximately 21 percent). However, the proteins and carriers of the electron transport system used by various microorganisms can vary considerably. Many microbes can fine-tune their electron transport system by switching proteins and cofactors in response to the environment. For example, the bacterium *Escherichia coli* has two terminal oxidases, one of which has a significantly higher affinity for O_2, and is expressed most highly under low-oxygen conditions. Some bacteria encode four or more terminal oxidases, allowing them to adapt to changes in the availability of oxygen and other terminal electron acceptors.

Anaerobic Respiration

If the concentration of O_2 is sufficiently low to severely limit aerobic respiration, many microbes can use alternative electron acceptors for **anaerobic respiration**, defined as the use of electron acceptors other than O_2. Anaerobic respiration is the oldest form of cellular respiration, evolving during Earth's early anaerobic, reducing atmosphere. The enzyme complexes have different cytochromes and quinones than are found in the aerobic transport system. Enzyme complexes involved in reduction of the terminal electron acceptor are called reductases rather than oxidases because oxygen is not the terminal acceptor. One of the most common alternative acceptors is nitrate (NO_3^-), which is reduced to nitrite (NO_2^-), as shown in **Figure 13.25**. Some bacteria stop at nitrite while others employ additional enzymes to further reduce nitrite to dinitrogen (N_2) in a process called **denitrification**. Denitrification is harmful to soil productivity as it converts nitrate that can be used by plants into N_2 that diffuses into the atmosphere and is lost to the plant roots. Nitrogen-cycling reactions of microbes is covered in Section 14.3. Under anaerobic conditions, some bacteria donate excess electrons to unusual electron acceptors. As **Perspective 13.2** describes, for some bacteria this can result in conducting an electrical charge to surrounding metal surfaces.

Other potential respiratory electron acceptors include sulfate (SO_4^{2-}), elemental sulfur (S^0), ferric iron (Fe^{3+}), and organic molecules such as fumarate. Fumarate can be generated internally by the TCA cycle. These molecules all have less positive reduction potentials than O_2, but significantly more positive reduction potentials than NADH, so energy is still released by letting electrons flow to them. Recall that the energy released by a redox reaction is proportional to the difference in reduction potentials between the donor and acceptor **(Figure 13.26)**. Compared to aerobic respiration from glucose ($\Delta G^{o\prime} = -28842$ kJ), less energy is released during anaerobic respiration. For example, the $\Delta G^{o\prime}$ values for anaerobic respiration using electron acceptors NO_3^-, SO_4^{2-}, and S^0 are -1796 kJ, -453 kJ, and -333 kJ, respectively. This means fewer protons are pumped per electron, and less ATP is generated.

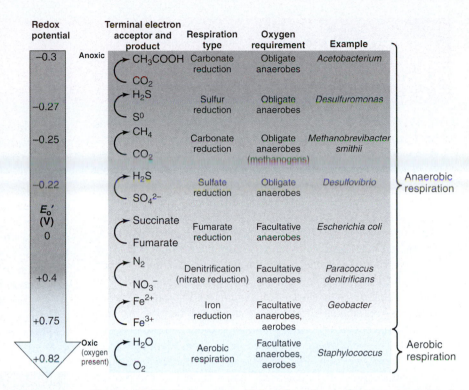

Figure 13.26. Redox potentials of various electron acceptors/donors Molecular oxygen, used only in aerobic respiration, has the highest redox potential and is capable of generating much more energy than the alternative acceptors used in anaerobic respiration.

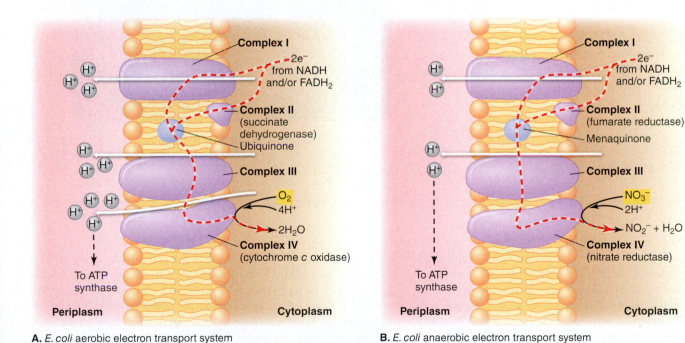

A. *E. coli* aerobic electron transport system

B. *E. coli* anaerobic electron transport system

Figure 13.25. Comparison of aerobic and anaerobic respiration in *E. coli* A. Aerobic respiration uses an electron transport system that transfers electrons to O_2, the terminal electron acceptor. **B.** Anaerobic respiration uses an electron transport system composed of similar but not identical molecules. The terminal electron acceptor is not O_2, and the cytochrome *c* oxidase complex (Complex IV) is replaced by a reductase enzyme. Shown here is nitrate reductase, which transfers an electron to nitrate to produce nitrite. Compared to aerobic respiration, fewer protons are pumped per electron, resulting in less ATP generation.

In 1911, British microbiologist M. C. Potter reported a fascinating application for *E. coli*. Observing that *"the disintegration of organic compounds by microorganisms is accompanied by the liberation of electrical energy,"* Potter managed to generate a small electrical current from an *E. coli* culture, and in doing so created the first microbial fuel cell. As Potter realized, oxidation of organic compounds such as glucose initiates a flow of electrons. Depending on the microbe, there are many possible destinations for these electrons, including being transferred to O_2 during respiration. Construction of a microbial fuel cell simply required redirecting the electrons to an external anode, from which they could flow through an electrical circuit. Potter's culture generated miniscule electrical currents with no practical utility. However, contemporary microbiologists are revisiting the microbial fuel cell concept in the hope of generating a useful, real-world device.

Bacteria have been identified that are far more adept than *E. coli* at transferring electrons from organic electron donors to insoluble external acceptors and generating usable current in the process. Such bacteria, which include members of the Gram-negative genera *Shewanella* and *Geobacter*, are called "electricigens." In anaerobic environments, these microbes look for a place to dump electrons obtained from oxidation of organic compounds. When soluble electron acceptors aren't available, insoluble metals such as Fe(III) and Mn(IV), which have relatively positive reduction potentials, can provide this service. Note that this is distinct from lithotrophy, where metal ions serve as electron *donors*, rather than *acceptors*. The structural challenge electricigens face is to transfer electrons from redox centers in the plasma membrane to insoluble metal oxides beyond the cell wall and cell surface.

In 2005, Derek Lovley's lab at the University of Massachusetts–Amherst discovered that one of the best-studied electricigens, *Geobacter sulfurreducens*, produces pilus-like structures that serve as *nanowires* to conduct electricity from the bacterium to metal surfaces **(Figure B13.1)**.

Other labs soon demonstrated that *Shewanella oneidensis* also depends on conductive pilus-like appendages for electron transfer to insoluble ferric oxide. Precisely how these structures connect with the electron transport system and serve as efficient electrical conductors isn't known yet, but the goal seems clear—conductive nanowires expand the physical reach of the cell for electron transfer reactions and eliminate the need for direct cell surface contact with an acceptor. There is still much to learn about this phenomenon. With respect to applications, results so far have been modest. The highest current generated was roughly 50 watts for a cubic meter of fuel cell volume, about enough to power a light bulb. However, scientists and engineers are intrigued by the potential to develop improved microbial fuel cells that generate electricity while consuming organic wastes. Perhaps such systems could be used for bioremediation of organic pollutants in anaerobic sediments or groundwater. Even if large scale applications don't become a reality, microbial fuel cells could be useful in more limited contexts, such as powering devices in remote, harsh environments where solar cells aren't appropriate.

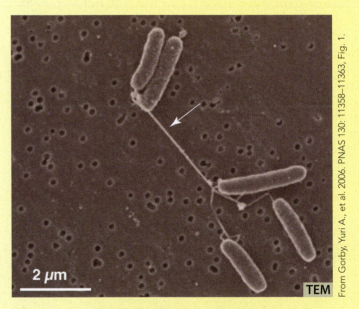

From Gorby, Yuri A., et al. 2006. PNAS 130: 11358–11363, Fig. 1.

Figure B13.1. Current-conducting pili nanowires Pilus-like structures (arrow) of *Shewanella oneidensis* conduct electrons from the cytoplasm to iron oxide solids present in the culture medium.

Microbes capable of both aerobic and anaerobic respiration prefer to use O_2 when it is available. Expression of terminal reductases for anaerobic respiration is generally repressed under aerobic conditions. Only under anaerobic conditions is expression of an alternative terminal reductase induced by the presence of its substrate. We can look at *E. coli* for an example. *E. coli* can alter the composition of its electron transport system to suit the conditions it experiences. Under anaerobic conditions, it expresses reductases and represses expression of oxidases. Fumarate reductase and nitrate reductase are alternatives that can be used to allow *E. coli* flexibility depending on what terminal electron acceptor is available. If you compare the redox potentials of fumarate and nitrate in Figure 13.12, you will see that nitrate is a preferred electron acceptor for anaerobic respiration as its reduction yields more energy than fumarate. If both compounds are present, nitrate reductase will be preferentially expressed. *E. coli* ensures this by employing a nitrate detection system that, when nitrate is detected, induces expression of transcription factors that upregulate transcription of nitrate reductase and actively represses genetic expression of fumarate reductase. Finally, if a suitable terminal electron acceptor for anaerobic respiration is not available or is of low concentration, then fermentation is used to generate small amounts of ATP to sustain growth until conditions change. To facilitate this, pyruvate dehydrogenase, the enzyme that converts pyruvate to acetyl-CoA, is repressed under anaerobic conditions. Instead, pyruvate will be converted to the mixed acid fermentation products acetate, lactate, and ethanol through the combined action of other enzymes.

We must distinguish between transfer of electrons to acceptors such as nitrate and sulfate for energy metabolism and reduction of these molecules so that they can supply the nitrogen and sulfur needs of the cell. When nitrate and sulfate are used as respiratory electron acceptors, the process is known as dissimilative reduction. Assimilative reduction occurs when these molecules are reduced so they can be incorporated into biomolecules such as amino acids. Dissimilative electron transfer reactions are carried out by membrane-bound reductases linked to the electron transport system and generate large amounts of reduced waste products, such as N_2 or H_2S, that are eliminated from the cell. Assimilative electron transfer reactions occur in the cytoplasm using soluble electron donors such as NADPH or $FADH_2$, and produce only as much product as is needed by the cell. We will return to assimilative processes in Section 13.7.

The proton motive force

As the electron transport system moves electrons from one acceptor to the next, protons are pumped out of the membrane at sites along the chain (see Figure 13.25). This creates a potential difference across the membrane, which forms the proton motive force. The proton motive force drives a stream of H^+ through ATP synthase to produce ATP, but can also be used to do mechanical work for the cell. How are these activities achieved?

First, let's look at what the proton motive force actually is. During respiratory electron transport, there is a net movement of H^+ across a membrane (Figure 13.27). The topology of this movement differs among microbes. In Archaea and Gram-positive bacteria, protons move from the cytoplasm out of the cell. In Gram-negative bacteria, the movement is from cytoplasm to periplasm. In the mitochondria of eukaryal cells, protons start in the mitochondrial

matrix, and move to the intermembrane space, between the inner and outer mitochondrial membranes (see Figure 13.25). A proton motive force is also generated during photosynthesis, as we will see in Section 13.6.

Transmembrane proton pumping results in both an H^+ concentration difference *and* a charge difference between the two sides of the membrane. The "outside" to which protons have been pumped has an acidic pH, that is, a higher H^+ concentration relative to the inside, and a more positive charge. Together, these factors yield the proton potential (Δp), a quantitative expression of the force pushing protons to re-enter the cell or mitochondrial interior. Protons outside of the membrane can't simply diffuse back into the cell, as their charge forbids them from crossing to the hydrophobic interior of the membrane.

One of the most important uses of the transmembrane proton gradient is to drive ATP synthesis through the membrane ATP synthase, the focus of the next section. In bacteria, proton motive force has other applications as well. Rotation of the flagellar filament to propel cells is driven by H^+ influx through the flagellar motor in the membrane. Proton motive force also energizes many transmembrane transport systems. For example, in *E. coli* proton symporters—channels that mechanistically couple the inward movement of H^+ to simultaneous import of another molecule—are used to transport several amino acids, lactose, xylose, arabinose, galactose, citrate, and sucrose as well as other nutrients (see Figure 2.11).

@ **Cell respiration** ANIMATION

The H^+ gradient is functionally replaced in some microbes by a sodium gradient, or sodium motive force (SMF), which is likewise used for ATP synthesis, flagellar motor rotation, and nutrient transport. SMF-dependent processes and transporters for the most part utilize proteins that are evolutionarily homologous to proton motive force-dependent systems. Several *Vibrio* species, including the pathogen *Vibrio cholerae* (see Microbes in Focus 21.7), rely extensively on SMF, at least in marine environments. There's clearly more to learn on this subject, which is made even more intriguing by a theory that the use of sodium gradients for energy metabolism actually preceded the use of H^+ gradients, and that the change in ion specificity occurred many times independently during the evolution of various microbial lineages.

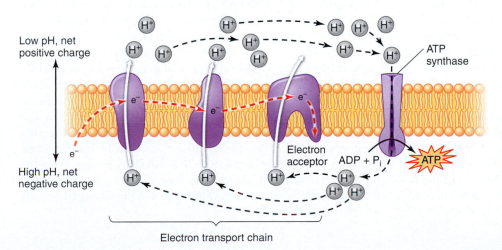

Low pH, net positive charge

High pH, net negative charge

ATP synthase

Electron acceptor ADP + P_i ATP

Electron transport chain

Figure 13.27. Generation of the proton motive force The electron transport system pumps protons across the membrane. This creates a pH and charge differential due to a net increase in protons in the mitochondrial intermembrane space compared to the matrix. In bacteria and archaeons, the proton increase occurs outside the plasma membrane compared to the cytoplasm. The differentials across the membrane create a motive force that compels protons to flow back across the membrane. The flow of protons is channeled through the ATP synthase complex to produce ATP from ADP and P_i.

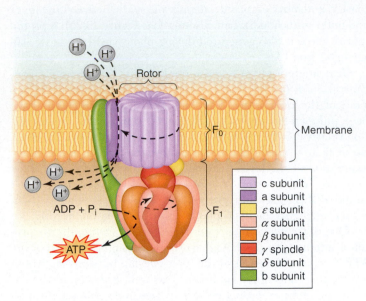

	c subunit
	a subunit
	ε subunit
	α subunit
	β subunit
	γ spindle
	δ subunit
	b subunit

Figure 13.28. ATP synthase function The flow of three protons through F_0 causes rotation of the γ spindle and conformational changes in the β subunits. Each rotation causes a change in a β subunit from a weak to a strong affinity for the substrates ADP and P_i, resulting in the production of ATP. After ATP is released, the conformation cycles back to a weak affinity and the process repeats for the next β subunit upon passage of the next three protons.

ATP synthase

The potential energy of proton motive force is captured by allowing protons to flow back down their charge and concentration gradient through ATP synthase **(Figure 13.28)**. The ATP synthase enzymes are remarkably similar in structure and function in all organisms, having been conserved through evolution. ATP synthase functions as a rotary motor. The F_0 portion is embedded in the membrane and is composed of an "a" subunit and a rotor composed of "c" subunits. The F_0 portion provides the path through which protons cross the hydrophobic interior of the membrane. The F_0 portion is held to the cytoplasmic F_1 portion via the b and δ subunits. The F_1 portion consists of pairs of α and β subunits arranged around a central γ spindle. The ε subunit is also part of F_1. A proton first contacts the a subunit, then moves on to contact a c subunit. This causes the c subunit to rotate, and the proton is ejected to the cytoplasm. Rotation of a c subunit turns the entire rotor assembly including the large γ spindle of the F_1 component. The movement of approximately three protons turns the rotor and the spindle 120°, which is enough to produce one ATP molecule. The actual mechanism of ATP production involves changes in conformation of the F_1 β subunits with rotation. With one 120° turn of the spindle, the affinity of one of the three β subunits for the substrates ADP and P_i increases significantly and ATP is made. The β subunit then returns to its low-affinity conformation. Movement of the next three protons acts in turn on the next β subunit.

The ATP synthase system can run in reverse to pump H^+ out of the cell (via F_0) at the expense of ATP. If ATP is present in the cytoplasm and there is no proton motive force pushing inward, proton pumping to the outside can happen spontaneously, simply because the hydrolysis of ATP becomes energetically favorable. Some bacteria intentionally use F_1F_0 as a proton pump to generate proton motive force to drive the cytoplasmic accumulation of ADP and P_i and to maintain ionic balance.

Eukaryal cells often use related systems called V-type ATPases as dedicated ATP-driven proton pumps. V-type ATPases are important for generating acidic conditions in organelles such as lysosomes. Recall from Chapter 3 that lysosomes are important for breaking down engulfed food particles and "old" organelles. The various hydrolase enzymes responsible for digesting these materials, including proteases and nucleases, only function under acidic conditions, in order to protect the more neutral cytoplasm should they escape.

CONNECTIONS Lysosomes are also important for degrading microbes that have been taken in by phagocytic cells such as macrophages, which we'll encounter in **Chapter 19**.

Chemolithotrophy

So far we have focused on the energetics of chemoorganotrophs, which use organic molecules as an electron source and generate ATP either by fermentation or respiration. There are two other ways to generate electrons and ATP, and neither uses an organic molecule. Photolithotrophy will be covered in Section 13.6, and we will briefly cover chemolithotrophy here. Chemolithotrophs, like the giant tube worm bacterial symbiont, include archaeons and bacteria that can use reduced inorganic sources from their environment for electrons and energy. These sources include metals such as Fe^{2+}, non-metals such as S^0, and gases such as H_2, NH_3, and H_2S. Since inorganic compounds cannot supply these organisms with carbon compounds, most chemolitotrophs are autotrophs and synthesize carbon compounds from CO_2. Consequently, respiration in chemolithoautotrophs does not use glycolysis and the TCA cycle to generate reducing power and ATP. However, some chemolithotrophs can also use heterotrophy, utilizing glycolytic breakdown of glucose and the TCA cycle when suitable carbon substrates are present—a metabolism sometimes called "mixotrophy."

Chemolithotrophs directly couple ATP synthesis by oxidative phosphorylation to oxidation of the inorganic source. Like chemoorganotrophs, they pass electrons down an electron transport system to generate ATP via the proton motive force (see Figure 13.27). This type of respiration can be aerobic, using O_2 as a terminal electron acceptor if it is present in sufficient quantities, or it can be anaerobic, using other terminal electron acceptors. The use of O_2 is energetically preferred as it gives the highest free energy yield and thus will generate the most ATP.

The respiratory electron transport system of chemolithotrophs contains specific membrane-bound enzymes to facilitate the removal of electrons directly from the inorganic source. For example, microbes that oxidize hydrogen gas have a large hydrogenase enzyme in addition to a collection of cytochromes. Consequently, electron carriers like NADH are not required for ATP generation in chemolithotrophs. The major respiration differences between chemolithotrophy and chemoorganotrophy are shown in **Figure 13.29**.

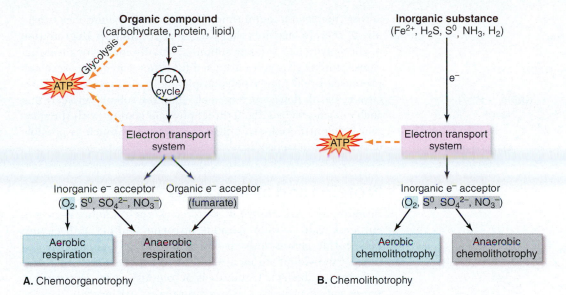

A. Chemoorganotrophy

B. Chemolithotrophy

Figure 13.29. Comparison of respiration in chemoorganotrophs and chemolithotrophs **A.** In chemoorganotrophs, respiration involves the removal of electrons through oxidation of an organic compound, such as glucose via glycolysis and the TCA cycle. ATP is generated at each of these steps by substrate-level phosphorylation. The electrons are carried by electron carrier molecules to the electron transport system to generate further ATP by oxidative phosphorylation. The final electron acceptor is O_2 in aerobic respiration. Anaerobic respiration uses an alternative inorganic or organic electron acceptor. **B.** Bacteria and archaeons that use chemolithotrophy remove electrons from a reduced inorganic substance (see examples) and pass the electrons directly to an electron transport system. Production of ATP occurs only through oxidative phosphorylation. Most chemolithotrophs can use O_2 as a terminal electron acceptor for aerobic chemolithotrophy. In anaerobic chemolithotrophy, some other oxidized inorganic molecule is used (see examples).

13.4 Fact Check

1. What are two main results or benefits of respiration?
2. Describe the flow of electrons in the electron transport system of organisms using cellular aerobic respiration.
3. What is anaerobic respiration, and what can serve as terminal electron acceptors in this process?
4. Differentiate between dissimilative reduction and assimilative reduction.
5. What is the proton motive force, and what is it used for in a cell?
6. Describe the role of ATP synthase.
7. Describe the differences in ATP generation between chemolithotrophs and chemoorganotrophs.

13.5 Metabolism of non-glucose carbon sources

How do microbes get nutrition from compounds other than glucose?

So far we have focused our discussions on metabolism in the context of glucose catabolism. Through this, we have covered ways of generating ATP through glycolysis, fermentation, and respiration, including the TCA cycle and the electron transport system. Although glucose is used by many microorganisms as an energy, carbon, and electron source, and is very abundant in the biosphere, particularly as a component of cellulose, there are many other carbohydrates. How are these compounds metabolized?

Sugars other than glucose

Rather than evolve completely new metabolic pathways for metabolizing non-glucose sugars, most organisms simply add a few enzymatic steps to convert alternative sugars into glucose or a glycolytic intermediate, at which point the product can enter an established glycolytic pathway. For example, fructose, a component of sucrose and a common hexose (a six-carbon sugar) found in fruits, is phosphorylated by the enzyme fructokinase

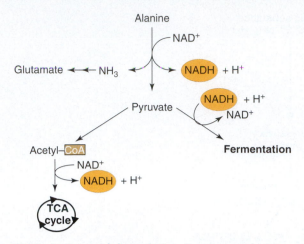

Figure 13.30. Amino acid degradation The amino group of an amino acid, such as alanine (shown here), is removed to form glutamate. Pyruvate is produced, which can enter into further reactions, such as fermentation reactions or the TCA cycle.

to produce fructose 6-phosphate, an intermediate in the EMP pathway of glycolysis (see Figure 13.13).

Another sugar commonly encountered by bacteria that live in mammalian intestines is the disaccharide lactose, the major sugar in milk. In Section 11.2 we learned about the *Escherichia coli lac* operon, which contains genes encoding enzymes essential for transport and metabolism of lactose. Disaccharides such as lactose are first split by hydrolysis or phosphorolysis into monosaccharides, which are handled separately. Lactose is split by β-galactosidase into galactose and glucose. Glucose immediately enters glycolysis, whereas galactose is phosphorylated, then converted through additional reactions to glucose 6-phosphate. Other common disaccharides include sucrose, maltose, trehalose, and cellobiose.

Pentose sugars are common in many habitats and are readily consumed by many microbes. In plant cell walls, xylans and arabinogalactans composed of the pentoses D-xylose and L-arabinose together make up hemicellulose, which is nearly as abundant as cellulose. As with glucose metabolism, there are diverse approaches to pentose metabolism. In *E. coli* and many other bacteria, xylose is isomerized to xylulose, which is then phosphorylated to xylulose 5-phosphate, an intermediate in the pentose phosphate pathway. Arabinose undergoes a similar process, being isomerized to ribulose then phosphorylated to ribulose 5-phosphate, another intermediate in the pentose phosphate pathway. Some bacteria, including *Caulobacter* and *Pseudomonas* species, take an alternative approach, metabolizing xylose and arabinose through a modified version of the Entner–Doudoroff pathway. Yeast take yet another approach, reducing xylose to xylitol using the enzyme xylose reductase. Xylitol is then converted to xylulose and phosphorylated.

Polysaccharides

Polysaccharides are abundant in the biosphere. They are used as structural components of cell walls in many organisms, as structural components of multicellular organisms, such as insect exoskeletons, and as carbon storage depots. Polysaccharides are typically enormous molecules, too big to be transported intact

across the plasma membrane. The solution for microbes is usually to secrete enzymes that carry out preliminary degradation of the polymer to smaller subunits, usually mono- or disaccharides. Similar strategies are used for other polymers—secreted proteases break down polyphosphoenolpyruvatetides, and nucleases break down nucleic acids. The microbial cell secreting such enzymes is usually in direct physical contact with the polymer being attacked, attempting to capture as much as possible of the breakdown product.

Proteins and amino acids

Proteins are rich nutrient sources, as their amino acid constituents can provide carbon, nitrogen, sulfur, and energy. Polypeptide chains must be broken down into small polypeptides or individual amino acids to be transported into microbial cells for catabolism. Catalysis is accomplished by secreted enzymes called proteases. Not all proteases are used to acquire nutrients; some pathogens secrete proteases to defend themselves from attack by host proteins, such as antibodies. The first step in the breakdown of most amino acids inside the cytoplasm is deamination in which the amino group is removed by transfer to glutamate (transamination) **(Figure 13.30)**. The organic acid, such as pyruvate, that remains after deamination is often utilized in the TCA cycle, although this depends on the amino acid. Not surprisingly, synthesis of many amino acids works in the opposite direction with TCA cycle intermediates as the starting point, to which the amino group is added as the final step. We'll return to this in Section 13.8.

The profile of carbon sources a microorganism is capable of utilizing can assist in its identification. Rapid bacterial identification systems, described in **Toolbox 13.1** allow simultaneous detection of a microbe's metabolic functions using a variety of media contained on a single strip or tube. These tests require very small quantities of media, reagents, and culture and are simple to use and reliable.

Lipids

Lipids are highly reduced carbon molecules that are rich in potential energy, and are actively catabolized by many microbes. Phospholipids and triglycerides are initially attacked by enzymes called lipases that separate fatty acid chains of various lengths from the glycerol or modified phosphoglycerol backbone. The backbone unit is then metabolized independently for carbon and energy, but the big energetic payoff is from the fatty acids as they can donate many electrons to fuel oxidative phosphorylation. The primary pathway for fatty acid degradation is called the **β-oxidation pathway**—so called because oxidation and cleavage between carbon 2 and 3 (the β carbon) occurs **(Figure 13.31)**. Each fatty acid chain of varying length is first linked to coenzyme A (CoA), forming an acyl-CoA molecule that then enters the pathway. Each acyl-CoA is progressively degraded into two-carbon acetyl-CoA units with each pass through the pathway. Reduced electron carriers are generated in the pathway and the acetyl-CoA products enter the TCA cycle for production of more reduced electron carriers and ATP. All that generated reducing power can be harvested

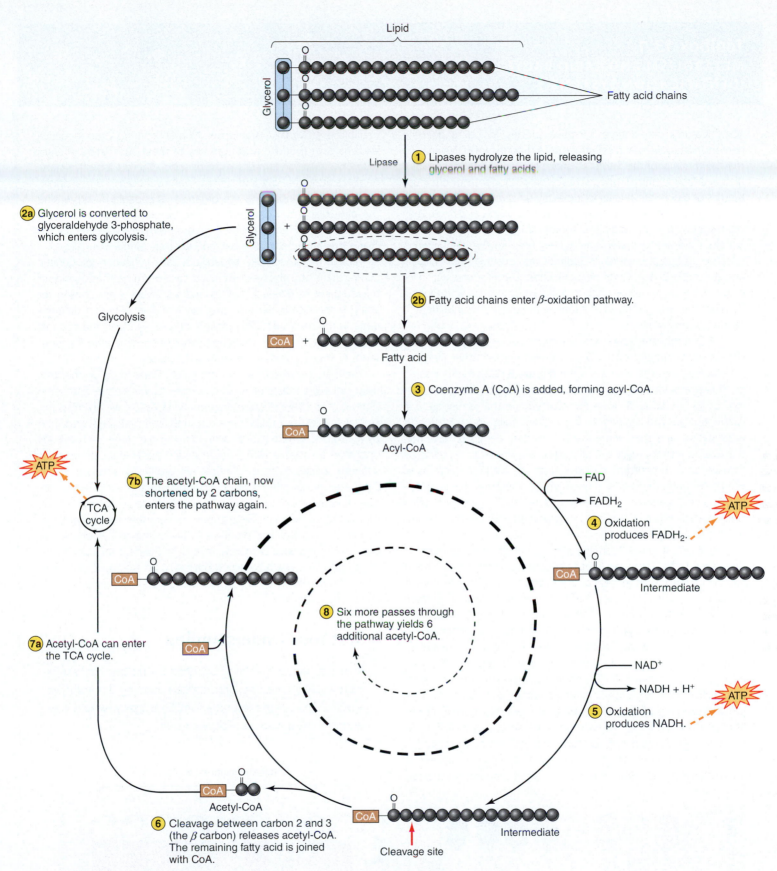

Figure 13.31. Process Diagram: Lipid degradation and β-oxidation of fatty acids Lipids are first hydrolyzed by lipases. The resulting glycerol backbone can enter glycolysis and the TCA cycle to ultimately generate ATP. The released fatty acid chains are joined to coenzyme A (CoA), forming acyl-CoA molecules that enter the β-oxidation pathway. Each pass of an acyl-CoA molecule through the pathway removes two carbons from the chain, producing the two-carbon molecule acetyl-CoA. The dashed arrow indicates successive passes of the remaining acyl-CoA through the pathway. The released acetyl-CoA units can enter the TCA cycle to produce ATP. With each pass, NADH and FADH$_2$ are also produced, which can be used for oxidative phosphorylation to produce more ATP.

Many features differentiate microbial species, including genome content, metabolism, and cell structure. Although rapid DNA analysis may be the way of the future, at present when clinical microbiologists need to identify a suspected bacterial pathogen, they often use a series of biochemical tests to indirectly uncover specific genes or sets of genes that are expressed under controlled conditions. They often start with a Gram stain (see Toolbox 2.1), then turn to metabolic traits. Commercial systems are widely used for the identification of some groups of bacteria, including Gram-positive staphylococci and streptococci, and Gram-negative Enterobacteriaceae, which includes *E. coli* and relatives such as *Salmonella*, *Shigella*, and *Proteus*. We'll use the latter group as an example of how these rapid identification systems work.

The Enterobacteriaceae are common causes of gastrointestinal disorders, urinary tract infections, and certain other illnesses. Let's look at how the API20E system **(Figure B13.2)**, manufactured by BioMerieux Vitek, works to identify members of this group. Before using this kit, a clinician must determine that an isolate is a Gram-negative rod, by performing a Gram stain and microscopic examination, and then verify that it is oxidase negative. Note that this assay doesn't actually test all oxidase enzymes—it is specifically focused on a cytochrome c oxidase that can transfer electrons to the dye tetramethyl-*p*-phenylenediamine. The Enterobacteriaceae largely lack this enzyme complex, which is expressed by most other Gram-negative rods that might be associated with the human gastrointestinal tract.

The API20E strip has 20 small wells, each of which contains dehydrated media for assaying a specific biochemical activity. The strip is inoculated with a suspension of the bacteria. The first eleven wells examine a variety of enzymes. The first well assays beta-galactosidase activity using the synthetic lactose analog ONPG, which releases bright yellow o-phenol when cleaved. Enzymes involved in amino acid catabolism are examined in assays for arginine dihydrolase (ADH), lysine decarboxylase (LDC), ornithine decarboxylase (ODC), tryptophan deaminase (TDA), and indole production from tryptophan (IND). Other wells assess citrate utilization (CIT), urease activity (URE), hydrogen sulfide production (H$_2$S), the production of acetoin from pyruvate (VP), and the ability to break down gelatin (GEL). The last nine wells contain a pH indicator that assesses acid production from fermentation of specific sugars and sugar-alcohols (shown in order: glucose, mannitol, inositol, sorbitol, rhamnose, saccharose,

mellibiose, amygdalin, arabinose). Non-fermenting Gram-negative bacteria can't produce acid from any of these substrates, whereas Enterobacteriaceae ferment most of them. The pH indicator starts out blue at neutral pH, and shifts to yellow if the pH drops significantly below 6, indicating the organism is capable of fermenting the particular substrate. Identities can be obtained by converting the profile of results of the unknown organism to a seven-digit code that is then matched with a bacterial species using a manual.

Variation in metabolic capabilities is a reflection of genetic composition and thus can be used to determine the identity of microorganisms. These enzymes and pathways were chosen for analysis specifically because they are variable between different genera and species. Activity profiles for these reactions, and optional tests such as for nitrate reduction, have been generated for thousands of known strains to generate a database.

There are drawbacks to this approach. Depending on the tests used, not every isolate of a particular species will yield precisely the same profile. There is extensive genetic variation within most microbial species, and the various enzymes and pathways examined by the API20E system vary as well. For example, there are hundreds of profiles associated with *E. coli*. It's even possible for strains from different species to have identical API profiles, which is why the database readout often indicates multiple possible species, and provides an estimate of the likelihood of each based on previous data. If such ambiguity is a problem, there are many alternative identification methods. Fatty acid profiles can be examined by commercial laboratories and rapid DNA sequence-based genetic tests are on the horizon. But for now, streamlined kits based on metabolic traits offer a fast and reliable approach to the identification of many clinically important bacteria.

● Test Your Understanding ·····················

Your lab partner has used an API20E strip but her results were inconclusive. She later realized that she did not have a pure culture of the test organism. Specifically explain how and why she got an inconclusive result.

Sugar and sugar-alcohols

Courtesy BioMarieux

Figure B13.2. API20E biochemical results for a *Corynebacterium* isolate Each well contains a medium designed to detect a metabolic function. After inoculation and incubation, the test can be read and the isolate identified. Results for fermentation of nine different sugars are shown.

in the electron transport chain for ATP production. For example, in mitochondria, β-oxidation of a 16-carbon fatty acid can yield 130 molecules of ATP. For each carbon atom, 48 ATP are yielded compared to 38 ATP for glucose. Of course, the catabolic products and reduced electron carriers can also be used for biosynthesis of other carbon compounds, such as proteins.

Catabolism of fats and proteins
ANIMATION

Microbes require lipids for membranes. Even when lipids or fatty acids are available for uptake from the environment, lipid synthesis and modification still occurs to tailor a cell's lipid content to its needs. Lipid metabolism must be tightly controlled to ensure that lipids can be used as carbon and energy sources when appropriate, but prevent inappropriate degradation of a cell's own membrane constituents or lipid precursors. Most microbes limit the expression of enzymes for lipid catabolism to times when exogenous lipids are available. Another way that catabolic and anabolic fatty acid metabolism are separated is by linking fatty acids to distinct carriers for synthesis or degradation. Fatty acid chains undergoing degradation are linked to coenzyme A, whereas chains being synthesized are linked to acyl carrier protein (ACP). ACP-linked chains aren't subject to attack by the β-oxidation pathway.

13.5 Fact Check

1. Describe how sugars other than glucose are metabolized.
2. How are large polymers, including polysaccharides, proteins, lipids, and nucleic acids, utilized as nutrients by microbial cells?
3. How are amino acids typically broken down for use in the TCA cycle?
4. What is the β-oxidation pathway, and how does it yield energy?

13.6 Phototrophy and photosynthesis

How do microbes capture light energy for ATP synthesis and carbon fixation?

Previous sections have focused on energy generation by catabolism of chemical substrates such as glucose by chemoorganotrophs, or inorganic molecules by chemolithotrophs. Oxidation of these substrates is the source of energy and electrons for generation of ATP. Phototrophs use light energy, or photons, instead of preformed chemicals as an energy source to produce ATP (see Figure 6.1 and Table 13.1). As you will recall, in phototrophy ATP is generated by photophosphorylation using captured light energy, an electron transport system, and the chemiosmotic process in conjunction with ATP synthase. **Photosynthesis** is a two-step process used by photoautotrophs. The first step of photosynthesis is photophosphorylation. The next step uses the ATP produced by photophosphorylation to fix CO_2 to make carbohydrates. It is important to distinguish between photophosphorylation and photosynthesis as some photoorganoheterotrophs can carry out photophosphorylation to make ATP, but do not use it for carbon fixation, instead obtaining carbon from organic sources (see Table 13.1).

Chlorophyll-based photosynthesis is a bacterial invention that has found its way into eukaryal algae and plants thanks to endosymbiotic cyanobacteria that evolved into chloroplasts. The biomass generated by photosynthesis forms the foundation of the food chains that sustain nearly all life on Earth. We'll treat modern cyanobacterial and chloroplast photosynthesis together in the section on oxygenic (oxygen-producing) photosynthesis, since they share mechanisms and machinery. Other bacteria carry out anaerobic, or anoxygenic photosynthesis. Oxygenic

(oxygen-generating) photosynthesis is important both historically and on contemporary Earth. As discussed in Chapter 1, rising levels of oxygen in Earth's atmosphere began about 3.5 billion years ago thanks to oxygenic photosynthesis, and radically altered the evolutionary trajectory of life on the planet. We now depend on oxygenic photosynthetic organisms for oxygen *and* food. In this section, we'll see how light energy is absorbed and transformed into chemical energy that can be used to drive the synthesis of carbon compounds by photosynthetic microorganisms.

CONNECTIONS Recall from **Section 1.2** that phototrophic bacteria brought solar power to eukaryal cells through endosymbiotic events. Evolution resulted in algae and plants.

Phototrophy is found in five distinct lineages within the domain Bacteria. Four of these phyla are Gram-negative: Cyanobacteria, Proteobacteria (α-proteobacteria, or the purple non-sulfur bacteria, and γ-proteobacteria, or the purple sulfur bacteria), Chlorobi (green sulfur bacteria), and Chloroflexi (green non-sulfur bacteria). The only Gram-positive group is the genus *Heliobacterium*, belonging to phylum Firmicutes. Of the bacteria, only cyanobacteria carry out oxygenic photosynthesis. Other bacteria do not generate O_2 from photosynthesis. We'll discuss this in more detail in the section on anoxygenic photosynthesis.

Photophosphorylation: ATP synthesis and the "light" reactions

Photophosphorylation converts light energy to chemical energy that can be used by the cell. The reactions are often referred to as the **light reactions**, as they require the presence of light. The light-absorbing machinery is always located in a membrane, since one of the primary functions of the light reactions of photosynthesis is to generate a transmembrane proton concentration gradient. The photosynthetic membrane is folded inward to expand the surface area for light absorption, forming disk-shaped or layered structures called **thylakoids** (Figure 13.32). In chloroplasts, the photosynthetic membrane is continuous with the inner membrane (see Figure 3.4). The space outside the thylakoid is called the stroma in chloroplasts. The inner chamber of a thylakoid is called the lumen. Bacteria do not contain chloroplasts. Instead, the thylakoids are formed by extensive inward folding of the plasma membrane. In Gram-negative bacteria, the lumen is an extension of the periplasm.

Photopigments

Photophosphorylation depends on light-absorbing photopigments to capture photons. The primary pigment used by cyanobacteria and chloroplasts is **chlorophyll** (Figure 13.33), which consists of a hydrocarbon component linked to a chromophore that absorbs photons.

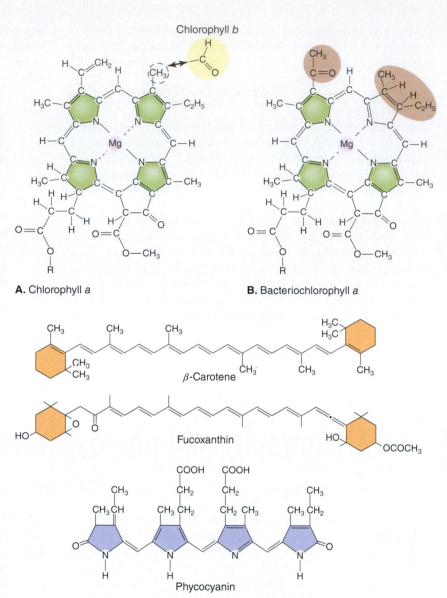

A. Chlorophyll *a*

B. Bacteriochlorophyll *a*

β-Carotene

Fucoxanthin

Phycocyanin

C. Accessory pigments

Figure 13.33. Photopigments The light-absorbing chromophore structure of chlorophylls contains four pyrrole rings (green) and a central magnesium ion. **A.** Chlorophyll *a* contains a methyl group (dashed circle). Chlorophyll *b* has an ester group substitution (yellow). **B.** Bacteriochlorophyll *a* has two group substitutions (brown). **C.** Accessory pigments β-carotene and fucoxanthin (a carotenoid) and phycocyanin (a phycobiliprotein).

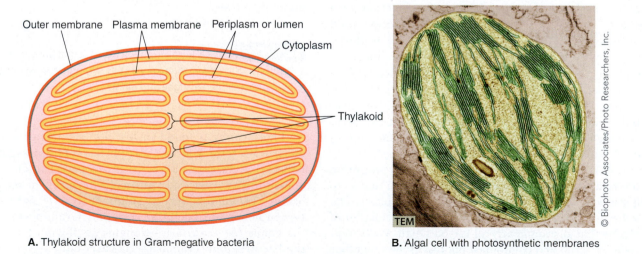

A. Thylakoid structure in Gram-negative bacteria

B. Algal cell with photosynthetic membranes

Figure 13.32. Thylakoids A. In bacteria, the stacked thylakoids are composed of membranes that are continuous with the plasma membrane. The lumen is an extension of the periplasm in Gram-negative bacteria. **B.** Color-enhanced chloroplast of the alga *Nitella* showing interconnected thylakoids appearing as stacked membranes.

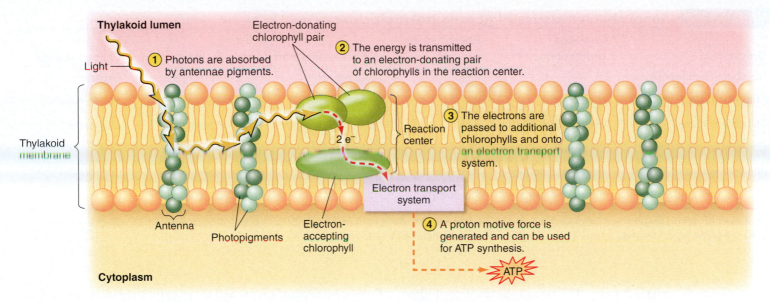

Figure 13.34. Process Diagram: General structure and function of a photosystem A photosystem is composed of many antennae containing photopigments that surround a reaction center. Photons are absorbed by the antenna photopigments and the energy is transferred from pigment to pigment to the reaction center containing a pair of electron-donating chlorophyll molecules. The electrons are passed to a chlorophyll electron acceptor and onto an associated electron transport system for ATP production.

The chromophore is based on a flat, porphyrin ring structure very similar to the heme component of cytochromes, with the notable exception that chlorophyll coordinates a central magnesium (Mg^{2+}) ion instead of iron. Most cyanobacteria produce only chlorophyll a, but the chloroplasts of most eukaryal algae and plants produce both chlorophyll a and a slightly different version called chlorophyll b. The slight chemical difference between these chromophores alters the wavelengths of light they absorb best. Chlorophyll a absorbs maximally in two peaks centered at 430 nm and 662 nm. Chlorophyll b is shifted to peaks centered at 453 and 642 nm. The presence of both pigments in algae and plants expands the light absorption spectrum of the reaction center. Neither pigment absorbs green or yellow light strongly. Another way of looking at this is that the green/yellow region of the spectrum is transmitted and reflected most effectively. As a result, cultures of actively photosynthesizing cyanobacteria appear green, similar to plant leaves containing chloroplasts packed with chlorophyll.

Other versions of chlorophyll, called **bacteriochlorophylls**, are produced by anaerobic photosynthetic bacteria (see Figure 13.33B). Compared to cyanobacterial chlorophylls, bacteriochlorophylls tend to absorb most strongly in more extreme regions of the visible spectrum, below 400 nm, approaching the ultraviolet region, or greater than 750 nm, the infrared region. These microbes tend to inhabit aquatic sediments below where oxygenic phototrophs are found. Their photopigments have evolved to function with light wavelengths not absorbed effectively by cyanobacteria and chloroplasts.

Chlorophylls aren't the only light-absorbing pigments used by photosynthetic microbes. A variety of accessory pigments help absorb light over a broader range of wavelengths than are captured by the primary chlorophylls, particularly in the green-yellow-orange region from 500–620 nm (see Figure 13.33C). Carotenoids exhibiting absorption maxima around 500 nm

are found in cyanobacteria and photosynthetic eukaryal protists. Cyanobacteria and various eukaryal algae also produce phycobiliproteins containing pigments such as phycoerythrin and phycocyanin. Phycoerythrin absorbs most strongly in the greenish-yellow range that escapes chlorophylls, while phycocyanin absorption is shifted more to orange/red light (620–640 nm). These accessory pigments are regulated by the wavelength of light to which the organism is exposed, to maximize the efficiency of light capture and minimize the production of photopigments that can't operate effectively with the available light spectrum. Microbes that rely on bacteriochlorophyll and accessory pigments such as carotenoids, phycoerythrin, and phycocyanin usually appear red, brown, or purple, depending on the mix of photopigments they produce at any given time.

Photosystems

The light reactions depend on an organized structure called the **photosystem** that absorbs energy from photons of light and transfers the generated electrons to an associated electron transport system **(Figure 13.34)**. A photosystem consists of numerous antennae containing photon-absorbing pigments and a reaction center containing electron-donating chlorophylls.

An antenna is composed of an array of chlorophylls or bacteriochlorophylls and accessory photopigments. When an antenna pigment absorbs a photon, the energy excites electrons, but antenna pigments don't directly transfer these electrons. Instead, the excitation energy is shared with other antenna molecules, and eventually the energy makes its way to a reaction center. The reaction center is surrounded by a number of antennae that funnel the energy from the photons to a special pair of chlorophyll molecules. The received excitation energy causes the redox potential of chlorophyll to become very negative, causing it to donate an electron to a nearby chlorophyll

TABLE 13.3 Some characteristics of photoautotrophic bacteria

Group	Common group name	Photosystem present	Common electron donor	O_2 production	Carbon source
Heliobacterium (genus)		PS I	Organic	Anoxygenic	Organic
Chlorobi (phylum)	Green sulfur	PS I	H_2S, S^0, H_2, $S_2O_3^{2-}$	Anoxygenic	CO_2, organic
Chloroflexi (class)	Green non-sulfur	PS II	$H_2S^{\#}$, H_2^a, organic[a]	Anoxygenic	CO_2, organic
Chromatiaceae (family)	Purple sulfur	PS II	H_2S^a, H_2^a, S^{0a}	Anoxygenic	CO_2
Rhodospirillaceae (family)	Purple non-sulfur	PS II	H_2^a, organic[a]	Anoxygenic	CO_2, organic
Cyanobacteria (phylum)	"Blue-green algae"[b]	PS I and PS II	H_2O	Oxygenic	CO_2

[a] Not electron donors for photosynthesis reactions, but are used as a source of electrons to make NADH or NADPH through the electron transport system.

[b] The use of the word "algae" is a former, incorrect reference to these photosynthetic bacteria.

molecule in the reaction center. The fate of the electron differs depending on the design of the photosystem. Ultimately, the electron is passed to an electron transport system and a proton motive force is produced across the photosynthetic membrane, which can be used to synthesize ATP.

There are two basic types of photosystems: photosystem I (PS I) and photosystem II (PS II). Chloroplasts and cyanobacteria have both PS I and PS II systems. Other bacteria have one or the other **(Table 13.3)**. Photosystems I and II have similarities in component parts, indicating they have evolved from a common ancestor. To illustrate how these photosystems function, we'll start with PS I from the anaerobic green sulfur bacterium *Chlorobium tepidum*, which carries out anoxygenic photosynthesis **(Microbes in Focus 13.1)**.

Anoxygenic Photosynthesis

Bacteria that can carry out **anoxygenic photosynthesis** that does not produce O_2, evolved much earlier than those using oxygenic photosynthesis, corresponding with the initial anoxic atmosphere of Earth. The photosynthetic ability in the earliest branching bacterial phylum Firmicutes, specifically members of the genus *Heliobacterium*, indicates that the earliest organisms to evolve were likely photosynthetic. Photosystem I is the sole photosystem in anaerobic photosynthetic heliobacteria and green sulfur bacteria and is responsible for anoxygenic photosynthesis (see Table 13.3). The bacteriochlorophyll *c* associated with PS I antennae in *Chlorobium* best absorbs long wavelength (infrared) photons, with an absorption maximum around 750 nm. As shown in **Figure 13.35**, two photons are absorbed to excite two electrons in the reaction center, called P840, containing the electron-donating bacteriochlorophyll *a* pair. The electrons are passed from P840 to a different bacteriochlorophyll *a* electron acceptor, and finally to the iron-sulfur protein ferredoxin. Here, the enzyme ferredoxin reductase uses electrons from ferredoxin to reduce either NAD^+ to NADH, or $NADP^+$ to NADPH. Thus, light energy is converted to reducing power. The reduced carrier (NADH or NADPH) can then be used for other metabolic needs including, for *Chlorobium*, carbon fixation

for synthesizing of sugars and other carbon compounds. Carbon fixation occurs in the "dark" reactions described at the end of this section.

To keep PS I operating, the electrons separated from the bacteriochlorophylls of P840 must be replaced from an external source. Thus, the electron transport pathway in PS I

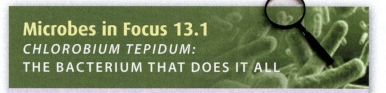

Microbes in Focus 13.1
CHLOROBIUM TEPIDUM: THE BACTERIUM THAT DOES IT ALL

Habitat: Hot springs and thermal mud pools and other warm bodies of water that contain sufficient hydrogen sulfide; often forms dense mats

Description: Gram-negative, thermophilic photolithoautotrophic bacterium with small ovoid-shaped cells 5–8 μm in length that grow in long chains, belonging to the green sulfur bacteria group

Key Features: *Chlorobium tepidum* carries out carbon (CO_2) fixation via a reductive TCA cycle, and anoxygenic photosynthesis using sulfides, such as H_2S, as electron donors. Elemental sulfur is a waste product deposited in granules outside the cell. It also fixes atmospheric nitrogen. *Chlorobium tepidum* possesses a unique combination of bacteriochlorophylls and carotenoid pigments that allow it to carry out photosynthesis in low light conditions, avoiding exposure to UV radiation. The characteristics of *C. tepidum* are thought to be like those possessed by the first phototrophs when Earth was much hotter, the atmosphere was anoxic, and organic compounds were scarce.

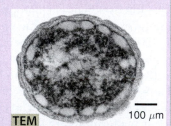

TEM   100 μm

From Niels-Ulrik Frigaard, Ginny D. Voight, Donald A. Bryant. 2002. J Bact 184:3368–3376. Reproduced with permission from American Society for Microbiology

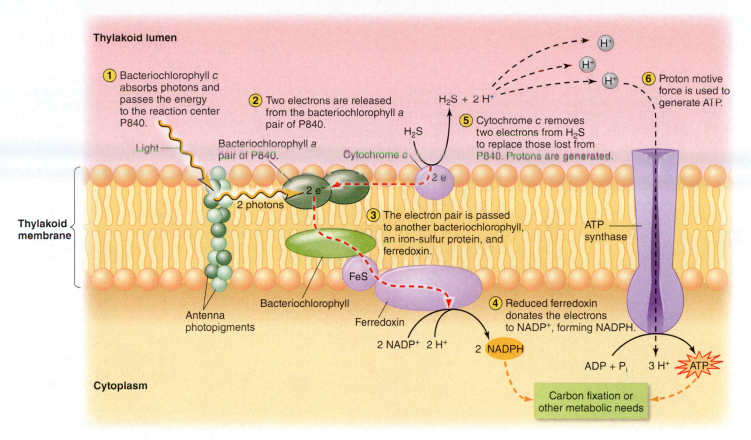

Figure 13.35. Process Diagram: Function of photosystem I of *Chlorobium* Absorption of two photons excites two electrons that are released from the P840 reaction center. The electrons pass through a series of carriers to produce two NADPH. This provides reducing power to carry out metabolic needs, such as carbon fixation. The electrons lost from P840 are replaced using an external source of electrons. Shown here is the oxidation of H_2S to elemental sulfur (S^0) and two protons, which are expelled outside the membrane, creating a proton motive force that can be used to drive ATP synthesis.

is non-cyclic. Green sulfur bacteria such as the photolithoautotroph *C. tepidum* oxidize electron sources such as hydrogen sulfide (H_2S), elemental sulfur (S^0), and hydrogen (H_2). The bacteriochlorophyll of the PS I reaction center is exposed to the periplasmic side of the photosynthetic membrane, but the electrons are passed to carriers on the cytoplasmic side. As replacement electrons are obtained from H_2S, for example, S^0 is produced and protons are released into the periplasm, increasing its acidity and net positive charge. Meanwhile, on the cytoplasmic side, reduction of $NADP^+$ is accompanied by incorporation of a proton. Thus, H^+ ions are liberated to the periplasm and consumed from the cytoplasm, creating a concentration differential across the membrane. As in other bacteria, the resulting proton motive force is used to support ATP synthesis and other proton motive force-dependent processes, such as nutrient transport.

Photosystem II is used for anoxygenic phototrophy in Proteobacteria such as purple non-sulfur *Rhodospirillum* and *Rhodopseudomonas*. The molecular structure of the *Rhodospirillum* PS II reaction center has been intensively studied. In the PS II reaction center called P870, a bacteriochlorophyll *a* pair absorbs photons, and the excited electrons are transferred

first to bacteriopheopytin, then to a quinone carrier (**Figure 13.36**). As the quinone is reduced to quinol, two protons are incorporated from the cytoplasm. These protons are shed to the lumen as the electrons are passed to cytochrome *bc*, generating a proton motive force. The electron pair is transferred next to a soluble cytochrome *c*-type carrier protein in the lumen, then back to the PS II reaction center. Because the electron transport pathway is cyclic, no external electron donor is needed. This is a major difference from the non-cyclic electron transport in PS I, where replacement electrons came from an external source. Another outcome of cyclic electron transport is that reduced electron carriers NADH or NADPH are not produced. The sole function of PS II is to serve as a light-driven proton pump. However, all cells need reducing power to carry out metabolic reactions, such as CO_2 fixation, so microbes employing PS II must generate NADH or NADPH by other mechanisms. The mechanisms vary, but one way is to reverse the electron flow in the photosynthetic electron transport chain, at the expense of ATP production, by removing electrons from iron-sulfur proteins. The electrons are replaced using other inorganic or organic electron donors, depending on the organism.

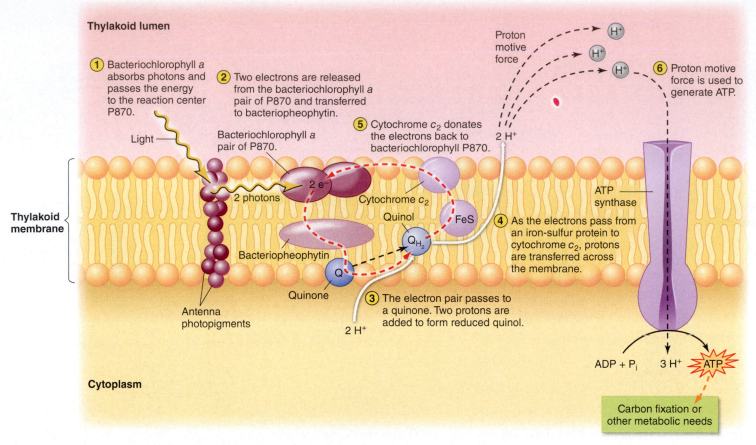

Figure 13.36. Process Diagram: Photosystem II of *Rhodospirillum* Absorption of two photons excites two electrons that are released from the P870 reaction center. The electrons are passed to bacteriopheophytin, then to quinone, reducing it to quinol with the removal of two protons from the cytoplasm. As the electrons pass from an iron-sulfur protein, the protons are transferred across the membrane. The electrons lost from P870 are replaced internally through oxidation of a cytochrome *c* molecule in the transport system. The transferred protons create a proton motive force that can be used to drive ATP synthesis.

Inside figure:

Thylakoid lumen

1. Bacteriochlorophyll *a* absorbs photons and passes the energy to the reaction center P870.

2. Two electrons are released from the bacteriochlorophyll *a* pair of P870 and transferred to bacteriopheophytin.

5. Cytochrome c_2 donates the electrons back to bacteriochlorophyll P870.

6. Proton motive force is used to generate ATP.

Proton motive force

Light

Bacteriochlorophyll *a* pair of P870.

2 H^+

Thylakoid membrane

2 e^-

2 photons

Cytochrome c_2

Quinol

FeS

ATP synthase

4. As the electrons pass from an iron-sulfur protein to cytochrome c_2, protons are transferred across the membrane.

Bacteriopheophytin

QH_2

Q

Quinone

Antenna photopigments

3. The electron pair passes to a quinone. Two protons are added to form reduced quinol.

2 H^+

Cytoplasm

$ADP + P_i$ 3 H^+ ATP

Carbon fixation or other metabolic needs

Oxygenic Photosynthesis

Cyanobacteria and chloroplasts combine PS I with PS II in **oxygenic photosynthesis**, a non-cyclic process that obtains electrons from water and produces both reducing power and proton motive force. On the electron-acquisition side of the process, the generation of one dioxygen (O_2) molecule requires the splitting of two water molecules:

$$2 H_2O \rightarrow 4 H^+ + 4 e^- + O_2$$

The structure and function of PS II in oxygenic photosynthesis differs from *Rhodospirillum* in that, instead of the excited electrons ultimately flowing back to the PS II in a cyclic pathway, they are passed to PS I **(Figure 13.37)**. For this to work, PS II must have an exogenous source of electrons to replace the electrons it sends to PS I; H_2O is used as an electron donor and O_2 is released as a product. It is hard to overstate the practical significance of being able to use water as an electron donor—it covers most of the surface of Earth, at high concentrations, and is thus far more widely available than alternative electron donors.

The combined pathway makes several contributions to proton motive force. Protons are released into the lumen of the chloroplast or cyanobacterial thylakoid when PS II splits the electron donor H_2O. The excited electrons of PS II pass from an electron-accepting chlorophyll to a quinone carrier, which incorporates two H^+ from the cytoplasmic, or stromal side as it is converted to a quinol. The quinol donates electrons to reduce the complex containing cytochrome *b*, an iron-sulfur center, and cytochrome *f*. This transfer is accompanied by translocation of two protons to outside the membrane. The cytochrome *bf* complex transfers electrons to a plastocyanin carrier, which uses them to replenish the reaction center of PS I. At that point, the electrons can be excited again by photon absorption, and travel through essentially the same pathway discussed earlier for *Chlorobium* PS I. The electrons end up incorporated into NADPH, along with two protons released to the cytoplasm or stroma. Bacteria other than cyanobacteria do not generate O_2 from photosynthesis, either because the electrons used in their photosystem follow a cyclic pathway back to the original reaction center as in PS II, or because the electron donor for PS I non-cyclic pathways is a molecule other than water. An overview of phototrophic energy capture pathways is shown in **Figure 13.38**.

As far as we know, no archaeons use light energy to support autotrophic growth. However, some halophilic (salt-loving) archaeons employ a non-chlorophyll pigment called bacteriorhodopsin that functions as a light-powered proton pump to

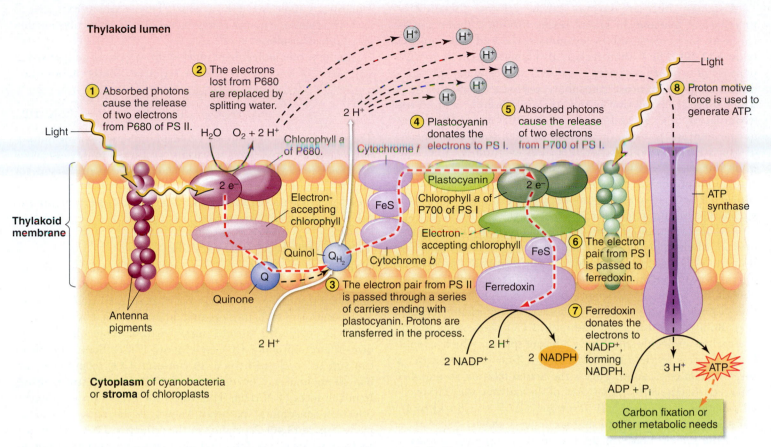

Figure 13.37. Process Diagram: Oxygenic photosynthesis in cyanobacteria and chloroplasts Both PS I and PS II are present, and electrons are passed from PS II to PS I. Water is used as a donor for electrons resulting in the production of O_2. Protons are released to the lumen at two points in the system, efficiently generating a proton motive force used for ATP synthesis. The presence of PS I means reducing power is simultaneously generated, and can be used for carbon fixation or other metabolic processes.

create a proton-motive force to produce ATP and simultaneously move Na^+ out of the cell (see Microbes in Focus 4.3). The light-powered transporter ships sodium out of the cell in exchange for protons and K^+ coming in, to maintain electric potential and osmotic balance. These archaeons use the energy of light as a survival mode, rather than a central metabolic function, but that makes it no less interesting.

The Calvin cycle and carbon fixation: The "dark" reactions

The ability to carry out **carbon fixation**, the utilization of carbon dioxide from the atmosphere as the sole carbon source for growth, gave microbes a tremendous opportunity to exploit new habitats. It is no surprise that photoautotrophs, both microbes and plants (endowed with this ability by cyanobacterial endosymbionts-turned-chloroplasts) dominate the illuminated biosphere. In this section, we'll look at how photosynthetic organisms use the ATP and reducing power produced by the light reactions of photosynthesis to turn CO_2 into useful carbohydrates. The reactions that convert CO_2 into organic carbon compounds are collectively referred to as the **dark reactions**, as they do not require the presence of light, nor do they require darkness to operate.

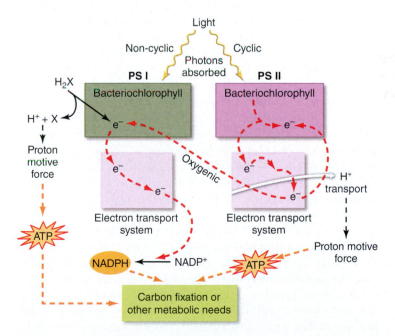

Figure 13.38. Overview of phototrophy Anoxygenic photosynthesis is carried out by PS I or PS II using bacteriochlorophyll. PS I produces reduced electron carriers and a proton motive force that can be used to generate ATP. Electrons are replaced through oxidation of an electron source, represented by H_2X. PS II does not generate reduced electron carriers as electrons are cycled within the system. A proton motive force is still produced. Oxygenic photosynthesis combines PS I and PS II and uses chlorophyll. Water is used to supply electrons to the system.

Although there are several known pathways for autotrophic carbon fixation, in global terms the most important is the **Calvin cycle** used by cyanobacteria and chloroplasts **(Figure 13.39)**. The Calvin cycle is also known as the reductive pentose phosphate pathway, because it shares many intermediates with the pentose phosphate pathway (Section 13.3). The essential distinction is that the Calvin cycle is used for assimilation of CO_2, and consumes NADPH and ATP, whereas the pentose phosphate pathway is used for dissimilation of glucose and produces NADPH and, eventually, ATP. The main purpose of the Calvin cycle is making the three-carbon compound glyceraldehyde 3-phosphate. This product is then converted into a variety of macromolecules, such as hexose sugars and amino acids.

The key reaction of the Calvin cycle is carried out by the enzyme ribulose bisphosphate carboxylase, often referred to as Rubisco. This enzyme is probably the most abundant protein on Earth. In cyanobacteria and plant leaves, it is certainly the most abundant protein. It has been estimated that four *billion* tons of this protein are synthesized by photosynthetic organisms each year. Rubisco catalyzes the addition of CO_2 and water to the five-carbon molecule ribulose 1,5-bisphosphate, which is split in the process to yield two molecules of 3-phosphoglycerate, a three-carbon compound. This is the actual fixation step and is referred to as the *carboxylation phase* of the Calvin cycle. The overall reaction is energetically favorable ($\Delta G^{o\prime} = -35$ kJ/mol), but it is a complex reaction that occurs very slowly.

In the second phase of the Calvin cycle, the *reduction phase*, ATP and NADPH generated by the light reactions are consumed. A molecule of 3-phosphoglycerate is phosphorylated to generate 1,3-bisphosphoglycerate, which is then reduced to glyceraldehyde 3-phosphate. These reactions are essentially the reverse of reactions 6 and 7 of the EMP glycolytic pathway (see Figure 13.13). NADPH serves as the reductant in the Calvin cycle, whereas NAD^+ is the oxidant in the EMP pathway, illustrating the partitioning of these electron carriers between catabolic and anabolic processes.

After glyceraldehyde 3-phosphate generation, the Calvin cycle gets more complicated, just as the pentose phosphate pathway does. Rather than going through the reactions in detail, suffice it to say that roughly five out of every six glyceraldehyde 3-phosphate molecules generated continue in the cycle. Only one glyceraldehyde 3-phosphate is available to leave the cycle for other biosynthetic reactions. The processing of three ribulose 1,5-bisphosphate by the cycle is required in order to regenerate the Rubisco substrate ribulose 1,5-bisphosphate. If this is not remade in the cycle, the Calvin cycle cannot actually function as a cycle. Counting carbons will help show this requirement. We'll start with the entry of one ribulose 1,5-bisphosphate. One five-carbon molecule plus one CO_2 yields six carbons. The three-carbon product (glyceraldehyde 3-phosphate) exiting the cycle leaves only three carbons—not enough to re-make the five-carbon substrate ribulose 1,5-bisphosphate. Similarly, two, five-carbon molecules plus two CO_2 gives twelve carbons. Removing a three-carbon product leaves only nine carbons—not enough to re-make two five-carbon molecules (10 carbons). Three ribulose 1,5-bisphosphate molecules (15 carbons) join with three CO_2 to produce a total of six three-carbon glyceraldehyde

3-phosphate molecules (18 carbons). Three carbons exit, and 15 carbons remain—enough to regenerate three molecules of ribulose 1,5-bisphosphate.

To make one molecule of glucose, or any other hexose sugar, six "turns" of the Calvin cycle are required. Total reactions are

$$6 \ CO_2 + 18 \ ATP + 12 \ NADPH + 12 \ H^+ + 12 \ H_2O$$
$$\rightarrow C_6H_{12}O_6 + 18 \ ADP + 18 \ P_i + 12 \ NADP^+$$

The regeneration of ribulose 1,5-bisphosphate involves numerous reactions, producing a variety of three- to seven-carbon intermediates. Many of these molecules serve as precursors for biosynthesis, including erythrose 4-phosphate used for aromatic amino acid synthesis, and ribose 5-phosphate used for nucleotide synthesis (see Section 13.8).

The reductive TCA cycle

While the Calvin cycle is used by chloroplasts and cyanobacteria, some autotrophic microbes, including both phototrophs and non-phototrophs, use pathways other than the Calvin cycle for incorporation of CO_2 into organic molecules. We have seen that the TCA cycle takes various organic carbon molecules and oxidizes them to produce CO_2 (see Figure 13.21). Some organisms can run the cycle in reverse to consume CO_2 to produce carbon compounds at the expense of ATP and electrons, usually in the form of NADH. This turns the cycle into a reductive (electron-consuming) pathway, known as the **reductive TCA cycle (Figure 13.40)**. A product of the reductive TCA cycle is acetyl-CoA, which can be fed into lipid synthesis, or used for glucose production, called "gluconeogenesis." Intermediates of the reductive cycle can also be used for biosynthesis, just like the TCA cycle operating in the oxidative direction. Some autotrophs, such as *Sulfolobus* (see Microbes in Focus 7.2), a non-photosynthetic archaeon, and photosynthetic green sulfur bacteria, such as *Chlorobium* (see Microbes in Focus 13.1), use simple inorganic compounds present in their environments, such as hydrogen or sulfates, as electron donors to run the reductive TCA cycle. As you will recall, *Chlorobium tepidum* uses photosystem I for anoxygenic photosynthesis. The ATP and NADPH that result are used to drive carbon fixation via the reductive TCA cycle.

The reductive TCA cycle goes in the opposite direction of the TCA cycle, beginning with oxaloacetate. The entry of CO_2 is shown in Figure 13.40, as are the steps where an investment of ATP and electrons, carried by reduced molecules such as NADH, are necessary to force the cycle in reverse. Most of the enzymes used in the reductive TCA cycle are the same enzymes used in the normal TCA cycle. How can this be? Reactions that have small $\Delta G^{o\prime}$ values can be driven in one direction or the other simply by availability of the reactants. To make this clearer, consider the following reaction:

$$X \xleftrightarrow{\text{enzyme}} Y$$

If this reaction has a very small $\Delta G^{o\prime}$, it is reversible for practical purposes. Consider that under "normal" physiological conditions, this reaction occurs in the context of a metabolic pathway, where X and Y are produced sequentially. X is produced first in high concentrations by a metabolic pathway and then converted to Y, which is rapidly consumed by another

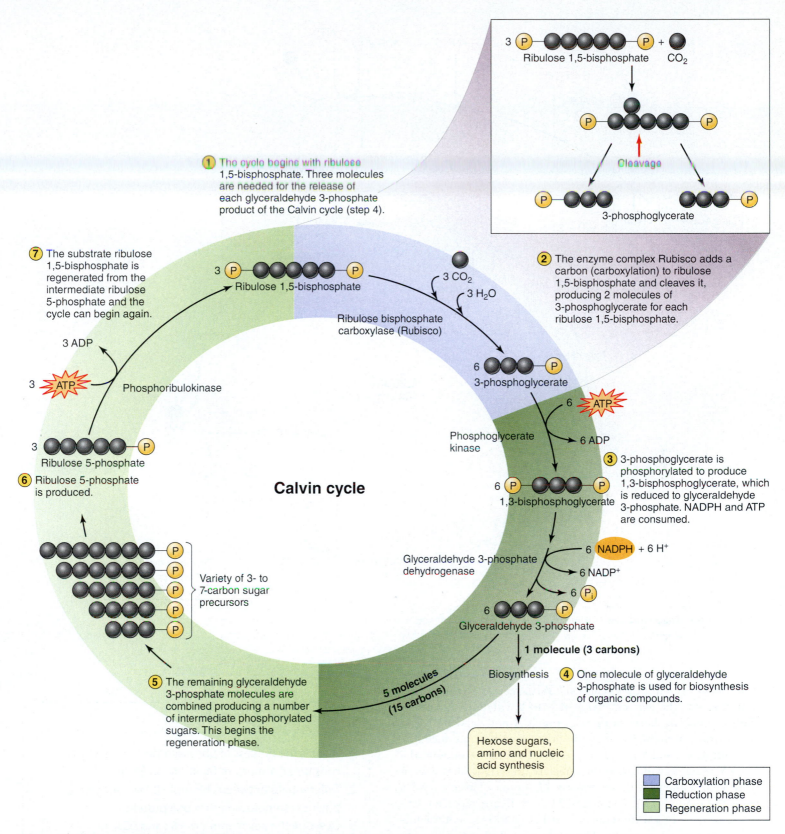

Figure 13.39 Process Diagram: The Calvin cycle Many phototrophs use the Calvin cycle to produce organic carbon compounds from CO_2. The carboxylation phase is catalyzed by Rubisco and incorporates CO_2 into a five-carbon ribulose 1,5-bisphosphate molecule that is cleaved into two three-carbon molecules. A detail (*upper right*) of the reactions carried out by Rubisco on ribulose 1,5-bisphosphate is shown. The reduction phase uses ATP and NADPH to produce the product glyceraldehyde 3-phosphate. One molecule of glyceraldehyde 3-phosphate is available for other biosynthetic processes, including production of glucose. To continue the cycle, ribulose 5-phosphate is regenerated through a complex series of reactions that use five out of every six glyceraldehyde 3-phosphates made, and produces various intermediate sugars.

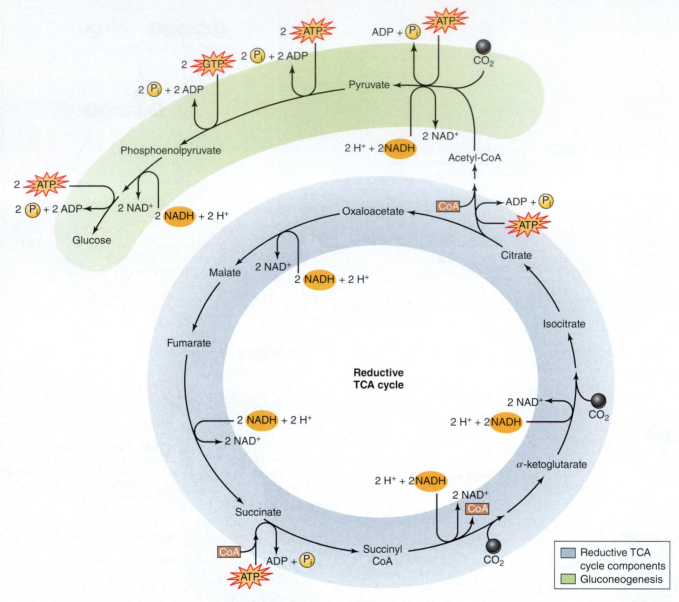

Figure 13.40. The reductive TCA cycle The cycle (blue) begins with the reduction of the four-carbon compound oxaloacetate, by using electrons donated from a reduced compound, such as NADH. A product is the two-carbon compound acetyl-CoA, which enters into further biosynthetic reactions (green), depending on the metabolic capabilities of the organism. Shown here are some of the steps in gluconeogenesis, the production of glucose.

reaction or even multiple reactions. With respect to this particular reaction, the net conversion of X to Y will dominate under these conditions, because there is much X and very little Y. If circumstances change so that the other reactions producing X are shut down, and Y instead is produced by other reactions, the X ↔ Y reaction will run in the opposite direction. That is basically what happens with the reverse TCA cycle. Three enzymes from the oxidative version of the TCA cycle are replaced with new enzymes: fumarate reductase, α-ketoglutarate oxidoreductase, and ATP citrate lyase. The replacement enzymes execute the energy-demanding reactions of the reductive version of the TCA cycle, so that it *incorporates* two molecules of CO_2 to form a new molecule of acetyl-CoA each time around. The dependence of the giant tube worms on the ability of the chemoautotrophic bacteria to use a reductive TCA cycle to supply them with carbon compounds is supported by metagenomic analysis, described in this chapter's **Mini-Paper**.

13.6 Fact Check

1. What is the process of photosynthesis, and why is it considered important for all life on Earth?
2. Differentiate between chlorophyll, bacteriochlorophyll, carotenoid, and phycobiliprotein.
3. Describe the key features of PS I and PS II in anoxygenic phototrophy.
4. Compare and contrast oxygenic photosynthesis and anoxygenic photosynthesis.
5. Provide an overview of the Calvin cycle.
6. What is the reductive TCA cycle, and how is it used by microbes?

13.7 Nitrogen and sulfur metabolism

How do microbes use nitrogen and sulfur?

Organic carbon sources needed by heterotrophs or CO_2 needed by autotrophs are essential for building organic molecules (such as polysaccharides, lipids, proteins, and nucleic acids) needed by all cells. In addition, all organisms need access to nitrogen and sulfur to build critical biomolecules. Nitrogen is an essential part of the chemical structure of amino acids and nucleotides, and sulfur is required for the amino acids cysteine and methionine, and for other components such as the iron-sulfur centers used in many electron transfer reactions. Assimilation of nitrate (NO_3^-), dinitrogen (N_2), sulfate (SO_4^-), or elemental sulfur (S^0) requires an investment of energy and electrons to reduce these compounds to the oxidation state at which they will be incorporated into biomolecules. We'll start our discussion of nitrogen metabolism with nitrogen fixation, and move to other methods of assimilation, before touching briefly on sulfur metabolism.

Nitrogen fixation

The largest source of nitrogen in the biosphere is atmospheric nitrogen gas, N_2. The problem for living organisms is that N_2 is extremely stable. The triple covalent bonds holding dinitrogen together must be broken in order to incorporate nitrogen atoms into biological molecules, but reducing this molecule takes a major investment of energy. "Diazotrophic microorganisms" (nitrogen fixers) are rewarded for their efforts with a significant competitive advantage in environments where there is little biologically available nitrogen (ammonia, nitrate, nitrite, or organic molecules such as amino acids). Because of its energetic cost, the enzymatic machinery for nitrogen fixation is highly regulated and is only expressed when alternative forms of nitrogen are exhausted or unavailable.

The conversion of N_2 into compounds for cell use is termed **nitrogen fixation**. Nitrogen fixation is purely a microbial process, carried out only by bacteria and archaea. Many phototrophs can fix nitrogen, including cyanobacteria and anoxygenic phototrophs such as Heliobacteria, *Rhodospirillum*, and *Chlorobium* species (see Microbes in Focus 13.1). Non-phototrophic diazotrophs include bacteria with agricultural significance, such as species that form symbiotic partnerships with plant roots. These bacteria fix nitrogen to supply their own and their host plant's nitrogen needs. In return, they receive organic compounds from their host to supply their carbon and energy requirements. Given that gene sets for nitrogen fixation have been shared between many bacterial species through horizontal gene transfer, it's a bit mystifying that no eukaryal organism seems to have acquired this ability.

CONNECTIONS Symbiotic nitrogen-fixing bacteria are crucial to the survival of many plant species, including agriculturally important soybeans, peas, and alfalfa. The bacteria invade the cells of the plant roots where they remain safely encased to carry out nitrogen fixation. This partnership is further discussed in **Sections 14.3** and **17.2**.

The key enzyme for nitrogen fixation is **nitrogenase**, which catalyzes the reaction:

$$N_2 + 8\,H^+ + 8\,e^- \rightarrow 2\,NH_3 + H_2$$

This reaction is accompanied by hydrolysis of at least 16 ATP molecules. The stoichiometry of the reaction is unusual, since only six electrons and six protons are actually incorporated into the two ammonia molecules produced. Why H_2 is generated is not clear, but it is thought that N_2 may displace formed H_2 in the enzyme active site. The hydrolysis of ATP reflects the investment of energy necessary to reduce the highly stable N_2 molecule.

Nitrogenase is a large multi-subunit enzyme whose core consists of two types of subunits: dinitrogenase reductase, which provides dinitrogenase with reducing power, and dinitrogenase, which actually reduces N_2. The active sites of both subunits contain metal-bound organic cofactors involved in electron transfer. The dinitrogenase cofactor contains iron and molybdenum, and is referred to as "FeMo-co" **(Figure 13.41)**. Complete reduction

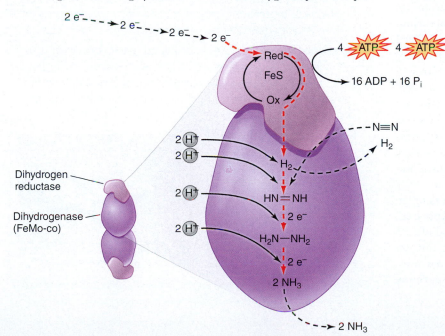

Figure 13.41. Nitrogenase The dihydrogenase complex is composed of two dinitrogenase reductase subunits (pale purple) and two dihydrogenase subunits (dark purple). Cofactors within the subunits facilitate the transport of electrons. Hydrogen (H_2) is first formed by two protons and two electrons. Nitrogen (N_2) is thought to enter the active site upon exit of the H_2 from the enzyme. Several steps sequentially break the triple bond of N_2. A succession of six more electrons and six protons enters the enzyme. Two electrons, two protons, and four ATP are needed to catalyze each of the three reduction steps to form ammonia.

Figure 13.42. Nitrogen-fixing heterocysts in *Anabaena* filaments
The large brownish heterocysts shown here (5–6.5 μm diameter), are developed in the absence of ammonia or nitrate for N_2 fixation. The thickened cell walls of the heterocysts limit exposure to O_2 to protect the oxygen-sensitive nitrogenase enzyme. The heterocyst cells do not have a functioning PS II system, ensuring O_2 is not generated in the cell.

of N_2 to ammonia is a multistep process. First, a pair of electrons combines with two protons to form H_2. Then three pairs of electrons are transferred from the electron transport chain in sequential steps through the dinitrogenase reductase subunit, which gets them from donor proteins such as flavodoxin or ferredoxin, depending on the microbe. Dinitrogenase reductase must be activated by ATP binding, in order for the active site to reach a conformation with a sufficiently negative reduction potential to transfer electrons to the dinitrogenase FeMo-co cofactor. For *each electron* transferred, two ATP molecules are hydrolyzed.

The FeMo-co cofactor, located near the surface of the enzyme, is extremely sensitive to oxidation, which makes it very difficult to operate in an aerobic environment. This is a conundrum for diazotrophic microbes that live in aerobic habitats and need aerobic respiration, or that produce O_2 during photosynthesis. Such microbes must take special precautions to protect nitrogenase when they need to fix nitrogen. One adaptation has been studied extensively in the filamentous cyanobacterium *Anabaena*. Nitrogen fixation in *Anabaena* occurs in specialized cells called heterocysts (**Figure 13.42**). In the presence of ammonia or nitrate, chains of *Anabaena* have no heterocysts. Nitrogen limitation induces the formation of heterocysts, which are recognizable by microscopy in filaments because they thicken and expand their cell walls to limit inward O_2 diffusion. Heterocysts are spaced roughly every 10 to 15 cells, and provide fixed nitrogen to the surrounding cells. Heterocysts cannot divide—they sacrifice their reproductive potential to provide an essential nutrient to their compatriots in the filament.

Heterocysts induce a new program of gene expression in which genes for nitrogenase production are highly expressed, and many genes for photosynthetic activities are turned off. PS II is degraded so that O_2 won't be generated. PS I is not degraded, but instead shifts to a cyclic electron transport pathway, like that used by purple bacteria with PS II. This cyclic pathway

pumps protons to generate proton motive force for ATP synthesis. Heterocysts also increase expression of glycolytic enzymes to metabolize carbohydrates such as sucrose from adjacent cells. In return, the heterocysts share nitrogen in the form of amino acids. Sugars and amino acids are readily transported between cells. This is a fascinating example of cooperation among microbial cells, and considerable research has gone into understanding the elaborate regulatory system that allows *Anabaena* cells to communicate and decide which cell will differentiate into a heterocyst.

Other aerobic diazotrophs take different approaches to protecting FeMo-co. *Azotobacter* cells increase respiratory metabolism so that O_2 is consumed rapidly and also add a protective redox-active protein to the nitrogenase complex that counteracts oxidants such as O_2. Symbiotic nitrogen-fixing rhizobia, a group of bacteria detailed in Section 17.2, are partially protected by the plant tissue surrounding them, which lowers O_2 concentration through respiration, and by physically slowing diffusion. **Microbes in Focus 13.2** presents a member of the Rhizobiaceae family. These bacteria, though, need oxygen for respiration. To accommodate their bacterial partners, leguminous plants such as soybeans that host rhizobia in root nodules produce an oxygen carrier in root tissue called leghemoglobin.

CONNECTIONS Leghemoglobin, described in **Section 17.2**, is structurally similar to mammalian hemoglobin and sequesters O_2 in the microbial cell, dramatically lowering the free O_2 concentration and protecting nitrogenase. The rhizobia respiratory cytochrome oxidase apparently has a high enough affinity for O_2 that it is can operate under these conditions.

Microbes in Focus 13.2
SINORHIZOBIUM MELILOTI

Habitat: Free-living *Sinorhizobium meliloti* are found in soil but form a symbiotic partnership with alfalfa (*Medicago trunculata*), in which the bacterium forms specialized cells called bacteroids in small nodules that emerge from the plant roots (shown here).

Description: Gram-negative, rod-shaped bacterium from the γ-subdivision of the Proteobacteria with aerobic metabolism, but in the bacteroid state tolerates a near-anaerobic environment to carry out nitrogen fixation. In soil, rods are 1–2 μm long, but enlarge to 5–10 μm once inside the plant host.

Key Features: Molecular communication between the plant and bacteria ensures specificity of symbiosis. Bacteroids fix nitrogen and provide ammonia to the plant root tissue, which in turn provides carbohydrates and oxygen to the bacteria.

© Inga Spence/Photo Researchers, Inc.

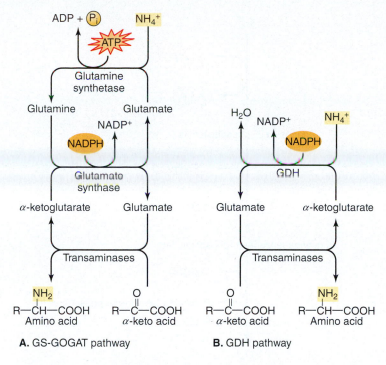

A. GS-GOGAT pathway

B. GDH pathway

Figure 13.43. Ammonia assimilation A. The GS-GOGAT pathway uses dissolved ammonium to incorporate nitrogen into amino acids using a sequence of reactions catalyzed by glutamine synthetase and glutamate synthase. This pathway consumes ATP. **B.** The GDH pathway features glutamine dehydrogenase (GDH), which does not require ATP to produce glutamate from ammonium. Both pathways produce glutamate, which is a substrate for specific transaminases that transfer the amino group onto α-keto acid amino acid precursors, forming all common 20 amino acids. Note R represents the various side chains of specific amino acids.

Ammonia assimilation

Ammonia (NH_3) and its protonated form, ammonium ion (NH_4^+), are the only forms of inorganic nitrogen incorporated directly into biological molecules. Other forms of nitrogen are first converted to ammonia before they are used for biosynthesis. When free ammonia/ammonium is available in the medium, it is transported into the cytoplasm. Uncharged NH_3 moves relatively freely across membranes. However, in water at neutral pH, equilibrium favors the protonated NH_4^+ form, which is *not* membrane permeable and must enter the cell through a transport system. Once in the cytoplasm, ammonium is incorporated into glutamate or glutamine. There are two common microbial pathways for incorporation of ammonium:

- the GS-GOGAT pathway
- the GDH pathway

The GS-GOGAT pathway uses a combination of two enzymes: glutamine synthetase (GS) and glutamate synthase **(Figure 13.43A)**. GO-GAT is an abbreviation for an alternative name for glutamate synthase: glutamine:oxoglutarate aminotransferase. The GS reaction is driven by hydrolysis of an ATP molecule, and GOGAT uses NADPH as an electron donor. The GDH pathway uses the enzyme glutamate dehydrogenase,

for which it is named, to add ammonium to α-ketoglutarate (see Figure 13.43B). This reaction requires NADPH as an electron donor, but does not consume ATP.

Both of these pathways for ammonium incorporation are found in many microbes. Why? It is not known for sure. In general, glutamate dehydrogenase is preferentially used at high NH_4^+ concentrations, and GS-GOGAT is used during low ammonium availability. GS has a higher affinity for ammonium ($K_m < 0.2$ mM for *Escherichia coli* GS) than glutamate dehydrogenase ($K_m \approx 3$ mM for the *E. coli* enzyme), so GS is more active at sub-millimolar NH_4^+ concentrations. In many microbes, these differences are reflected in the regulation of the enzymes, with GDH expression repressed and GS-GOGAT induced in low nitrogen medium, and the converse when ammonium is more abundant. However, other factors affect these enzyme activities as well. For example, experiments have shown that when *E. coli* is stressed for carbon and energy, GDH takes on a more important role in ammonia assimilation, even if ammonium concentrations are quite low. This makes sense in light of the fact that GS directly requires ATP to function, whereas GDH does not.

The amino group added to glutamate by GDH or GS-GO-GAT can be donated to other molecules by enzymes called "transaminases." There are many transaminases in microorganisms, such as those used for synthesis of amino acids by amination of α-keto acid precursors, as we'll see in Section 13.8. The GDH reaction is reversible, so the enzyme can function for catabolic deamination of glutamate, generating NADPH and α-ketoglutarate. In mammalian cells, GDH is largely found in mitochondria, where it operates primarily as a deaminase, and is regulated by energy metabolism. Mutations that cause defective regulation of GDH appear to result in excessive insulin secretion, through a complex cascade of physiological changes emanating from mitochondria of pancreatic cells.

Utilization of nitrate and nitrite

Nitrate (NO_3^-) can be used as a terminal electron acceptor for anaerobic respiration (see Figure 13.25). This process is called *dissimilative* nitrate reduction, since the products aren't incorporated into cellular material. More common is the process of *assimilative* nitrate reduction, in which NO_3^- is reduced to NH_3 **(Figure 13.44)**. In the cytoplasm, ammonia becomes ammonium and is then incorporated into biomass by routes illustrated in

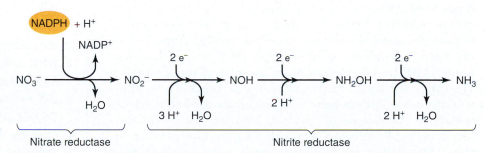

Figure 13.44. Assimilative nitrate reduction When ammonium is not available for biosynthetic needs, NO_3^- is reduced to ammonia in a series of reactions that begins with nitrate reductase and is completed by nitrite reductase.

Figure 13.43. Assimilative nitrate reduction occurs in the cytoplasm and is not linked to energy metabolism. Cytoplasmic nitrate reductases usually use NADH or NADPH as electron donors, although there are other electron donors. The expression of assimilatory nitrate reductase is generally dependent on the presence of nitrate in the medium, and on the nitrogen needs of the cell. Ammonia/ammonium is preferred as a nitrogen source, since it needs no further reduction; if ammonia/ammonium levels are adequate to supply the biosynthetic needs of the cell, little or no nitrate reductase is expressed even if nitrate is present in high levels.

Nitrate reduction to nitrite is the first step in making ammonia/ammonium for assimilation. Production of nitrite (NO_2^-) sets the stage for nitrite reductase to go into action. Nitrite reductases used for nitrogen assimilation catalyze sequential transfer of three pairs of electrons, using reduced ferredoxin as electron donors. Each transfer is accompanied by uptake of two protons. The end result is ammonia that is incorporated into glutamate as described above. Again, it is important to distinguish assimilative and dissimilative processes. Denitrifying bacteria also generate nitrite from nitrate, but the dissimilatory nitrite reductase instead generates nitric oxide (NO), which is further reduced to nitrous oxide (N_2O), and ultimately to dinitrogen (N_2), none of which are incorporated into biomass.

Sulfur metabolism

As with nitrogen, biological molecules contain sulfur in a reduced oxidation state. The most common inorganic form of sulfur in most environments is sulfate ion (SO_4^-), which is highly oxidized and must be extensively reduced before it can be incorporated into cysteine. There are exceptions to this, of course. Some habitats are rich in elemental sulfur, or even hydrogen sulfide, in which case microbes undertake different assimilatory strategies. We will only discuss sulfate reduction here. As with nitrogen, assimilative sulfate reduction, which occurs in the cytoplasm and is used to incorporate sulfur into cellular components, is distinguished from dissimilative sulfate reduction, which is a form of respiration coupled to energy metabolism. Assimilative sulfate reduction is very common in microbes; dissimilative sulfate reduction is rare.

Sulfate (SO_4^{2-}) is a stable molecule to which it is difficult to add electrons. To make it a better electron acceptor, sulfate is covalently attached to ATP by the enzyme ATP sulfurylase, generating adenosine 5′-phosphosulfate (APS), as shown in **Figure 13.45**. APS is used for both dissimilative and assimilative sulfate reduction. In the assimilative system, APS is converted to phosphoadenosine 5′-phosphosulfate (PAPS). From PAPS, the sulfate group is reduced to sulfite (SO_3^{2-}) with electrons from NADPH, and released from the rest of the molecule. The sulfite is reduced further to hydrogen sulfide (H_2S), a potentially toxic intermediate that is quickly used to displace acetate in acetylserine to generate cysteine. The dissimilatory process also proceeds through a sulfite intermediate to H_2S, which is excreted as a waste product from the cell. Sulfate reducers that use the dissimilatory pathway for energy

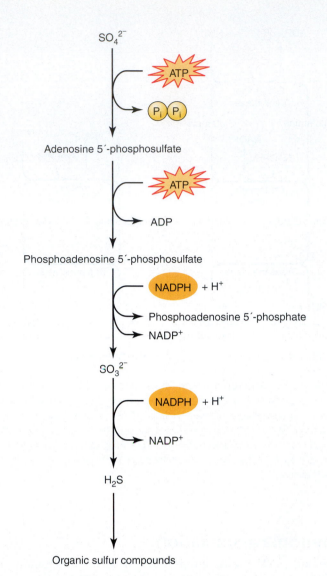

Figure 13.45. Sulfate assimilation Sulfate is a poor electron acceptor and must be converted to adenosine 5′-phosphosulfate in an ATP-consuming reaction. A second phosphate is donated by another ATP, forming phosphoadenosine 5′-phosphosulfate. Reduction to H_2S then occurs through a series of reactions. Hydrogen sulfide is then incorporated into sulfur-containing molecules, such as the amino acid cysteine.

metabolism can supply other microbes with reduced sulfur, and often live in close proximity to anaerobic phototrophs that consume H_2S as an electron donor for photosynthesis.

13.7 Fact Check

1. Why are nitrogen and sulfur necessary for cells?
2. What is nitrogen fixation, and what role does nitrogenase play in the process?
3. Differentiate between these two microbial pathways: GS-GOGAT pathway and GDH pathway.
4. Describe the process of assimilative sulfate reductions.

13.8 Biosynthesis of cellular components

How do microbes synthesize new cellular components?

Most of the dry weight of a microbial cell is composed of macromolecules: proteins and nucleic acids, plus some lipid and cell wall material of varying amounts and composition, depending on the organism. These macromolecules are built from just a few subunits: twenty amino acids, four ribonucleotides, four deoxyribonucleotides, and some fatty acids and carbohydrates for the lipids and cell walls. A central theme of this section is that these macromolecular subunits are themselves synthesized from a common pool of precursors. The precursors in this pool are drawn from the EMP pathway (see Figure 13.13), the TCA cycle (see Figure 13.21), and the pentose phosphate pathway (see Figure 13.15). In this section, we'll provide an overview of the pathways for synthesis of amino acids, nucleotides, lipids, and peptidoglycan, and try to show how these anabolic pathways are linked. We will not go into much detail on the biochemistry or regulation of these pathways, but students interested in exploring this fascinating field in more depth are encouraged to investigate microbial physiology and biochemistry references.

Amino acid synthesis

The biosynthetic pathways that build amino acids in microorganisms start with one or more of the common precursor pool (Figure 13.46). We'll start with the glutamate family. Glutamate

has broad significance for amino acid synthesis, since it is the amino group donor for several transaminases used in the synthesis of other amino acids. In Section 13.7 (see Figure 13.43), we mentioned two reactions for synthesizing glutamate, which serve as the entry point for ammonia into organic molecules. GS-GOGAT and GDH pathways generate glutamate by addition of amino groups to α-ketoglutarate, a product of the TCA cycle. Glutamine is produced by glutamine synthetase, the first enzyme of the GS-GOGAT system. Glutamate also gives rise to the cyclic amino acid proline through a three-step pathway. Finally, arginine is the product of a long, nine-step and energy-intensive pathway starting with glutamate. The arginine pathway is also important because one of its intermediates, carbamoylphosphate, is a precursor for the synthesis of pyrimidine bases for nucleotide synthesis. Some enzymes involved in arginine synthesis are regulated by the demand for pyrimidines as well as arginine, to ensure an appropriate distribution of carbamoyl phosphate.

The aspartate family of amino acids begins synthesis with oxaloacetate, another intermediate of the TCA cycle (see Table 13.2). Oxaloacetate receives an amino group from glutamate to produce aspartate, which also serves as a precursor for synthesis of five more amino acids. Pyruvate, the terminal product of glycolysis, is

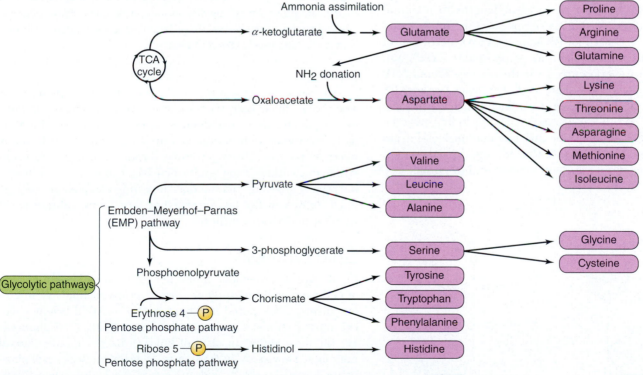

Figure 13.46. Anabolic precursors and pathways to amino acid synthesis The top two synthesis pathways use products of the TCA cycle as precursors for amino acid synthesis. The glutamate family uses ammonia assimilation to produce glutamate from α-ketoglutarate. Glutamate is a precursor for the synthesis of proline, arginine, and glutamine. Glutamate also donates an amino group to oxaloacetate to produce aspartate, which is a precursor for several other amino acids. Glycolytic pathways provide precursors for the other amino acids.

the starting point for making three relatively small amino acids, alanine, valine, and leucine. The molecule 3-phosphoglycerate, an intermediate of the EMP pathway, initiates production of serine. Serine gives rise to glycine and cysteine, a sulfur-containing amino acid produced by the action of O-acetylserine sulfhydrylase, noted earlier in sulfur assimilation. Cysteine then serves as the sulfhydryl donor for methionine synthesis, another sulfur-containing amino acid. Cysteine also serves as the sulfur donor during synthesis of sulfur-containing cofactors such as biotin and lipoic acid (see Table 13.1).

The last major family of amino acids is the aromatic amino acids: tryptophan, tyrosine, and phenylalanine. Aromatic amino acid synthesis is a long, complex process that starts with phosphoenolpyruvate (from the EMP pathway) and erythrose-4-phosphate, a four-carbon molecule derived from the pentose phosphate pathway. These precursors are combined to generate a seven-carbon molecule that proceeds through several additional steps to generate chorismate. From chorismate, three separate pathways branch out to generate the individual aromatic amino acids. The overall process for making these three amino acids is energy intensive, and is subject to multilayered regulation. The common intermediate, chorismate, is also the starting point for synthesis for folic acid and quinones. Lastly, synthesis of the amino acid histidine uses the precursor ribose 5-phosphate, a product of the pentose phosphate pathway.

Nucleotide synthesis

Nucleotides are the precursors for RNA and DNA synthesis. Given the importance of gene expression and chromosome replication, microbial cells must go to great lengths to maintain adequate nucleotide pools. This is complicated by the fact that nucleotides are not just used for RNA and DNA synthesis. One of the pyrimidine nucleotides, adenosine triphosphate (ATP), also serves as the primary energy currency in the cell (see Section 3.1). The other, guanosine triphosphate (GTP), provides energy for polypeptide synthesis. Derivatives of uridine participate in some biosynthetic processes as carrier molecules, including peptidoglycan synthesis. Cyclic adenosine and guanine nucleotides are used in cellular signaling processes. Considering these demands, maintaining and regulating nucleotide pools is quite complex.

Nucleotides, first mentioned in Chapter 7 (see Figure 7.4A), have three components—a nitrogenous base, a sugar (ribose in RNA, deoxyribose in DNA), and a phosphate group (**Figure 13.47**).

Cells must either synthesize or acquire five bases: two purines (adenine and guanine), and three pyrimidines (cytosine, thymine, and uracil; see Figures 7.4 and 7.15). Purine and pyrimidine nucleotides are built in distinct ways. Pyrimidine bases are constructed as a cyclic precursor first, then linked to ribose 5-phosphate. In contrast, purines are built up from precursors already linked to ribose 5-phosphate (**Figure 13.48**).

All nucleotides start as *ribo*nucleotides. The *deoxy*ribose sugar is generated by reduction of the 2′ carbon of the ribose, which replaces the 2′—OH group of the ribose component with a hydrogen atom (see Figure 7.15). This reaction is carried out on all four ribonucleotide triphosphates, despite the fact that deoxyribouridine triphosphate (dUTP) is not intended to be used in DNA synthesis. In the cell, dUTP is quickly converted to the monophosphate dUMP, which can't be added to a DNA chain by DNA polymerases. The enzyme thymidylate synthetase then methylates the uracil ring to generate the thymidine, in the form of deoxythymidine monophosphate (dTMP). This is then phosphorylated twice to give dTTP, the appropriate precursor for DNA synthesis. Inherent in the pathway to adenosine, guanine, and thymidine synthesis is the donation of carbons from tetrahydrofolic acid, a derivative of folic acid.

CONNECTIONS Sulfonamides and trimethoprim are antimicrobial drugs that act on two separate reactions in the pathway to tetrahydrofolic acid synthesis. They are used together to synergistically block DNA and RNA synthesis by diminishing pools of thymidine, guanine, and adenine. Antimicrobial drugs targeting tetrahydrofolic acid synthesis are discussed in **Section 24.2**.

Microbes do not always need to synthesize nucleotides *de novo*. Intact nucleotide triphosphates are rare outside of cells, but ribose and free nitrogenous bases are more common, at least in habitats rich in organic nutrients. Consequently, many microbes possess salvage pathways through which they can construct nucleotides from preformed bases and ribose. When a microbe is grown on rich media containing components such as yeast extract in the laboratory, they largely use salvage pathways for nucleotide synthesis.

Lipid synthesis

Lipids are hydrophobic molecules that are the foundation of cellular membranes. There are a variety of lipids in microorganisms, as we noted in Chapters 2 to 4. We'll briefly cover the most common classes of membrane lipids. Phospholipids are the most abundant class of lipids in bacteria and eukaryal microbes. They are built from fatty acids linked to glycerol phosphate derivatives. Fatty acids are essentially hydrocarbon chains with a carboxyl group at one end. Within this general description, there is diversity in terms of chain length (16–18 carbons being most common in bacterial membranes), whether double bonds are present (an *unsaturated* chain), and whether branches are present.

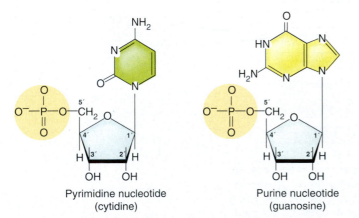

Figure 13.47. Nucleotide structure Nucleotides are composed of a nitrogenous base covalently bound to the 1′ carbon of a ribose sugar. The 5′ carbon is bound to a phosphate group.

Pyrimidine nucleotide (cytidine)

Purine nucleotide (guanosine)

Figure 13.48. Process Diagram: Nucleotide synthesis **A.** The ring structure of the nitrogenous base of pyrimidine nucleotides is built first, beginning with a condensation between the amino acid aspartic acid and carbamoyl phosphate. Ribose 5-phosphate is added last. This produces a uridine monophosphate that is the precursor for DNA (thymidine and cytosine) and RNA (uracil) synthesis. Tetrahydrofolate is required for synthesis of thymidine. The antibacterial drugs trimethoprim and sulfonamides inhibit its production, reducing DNA synthesis and cell growth. **B.** The synthesis of purine nucleotides begins with ribose 5-phosphate and the purine ring structure is built up from this in several complex steps to form inosinic acid, which can then be modified to produce the two final purine nucleotides. Again, tetrahydrofolate is required for synthesis. Sulfonamide drugs prevent the formation of the precursor folic acid.

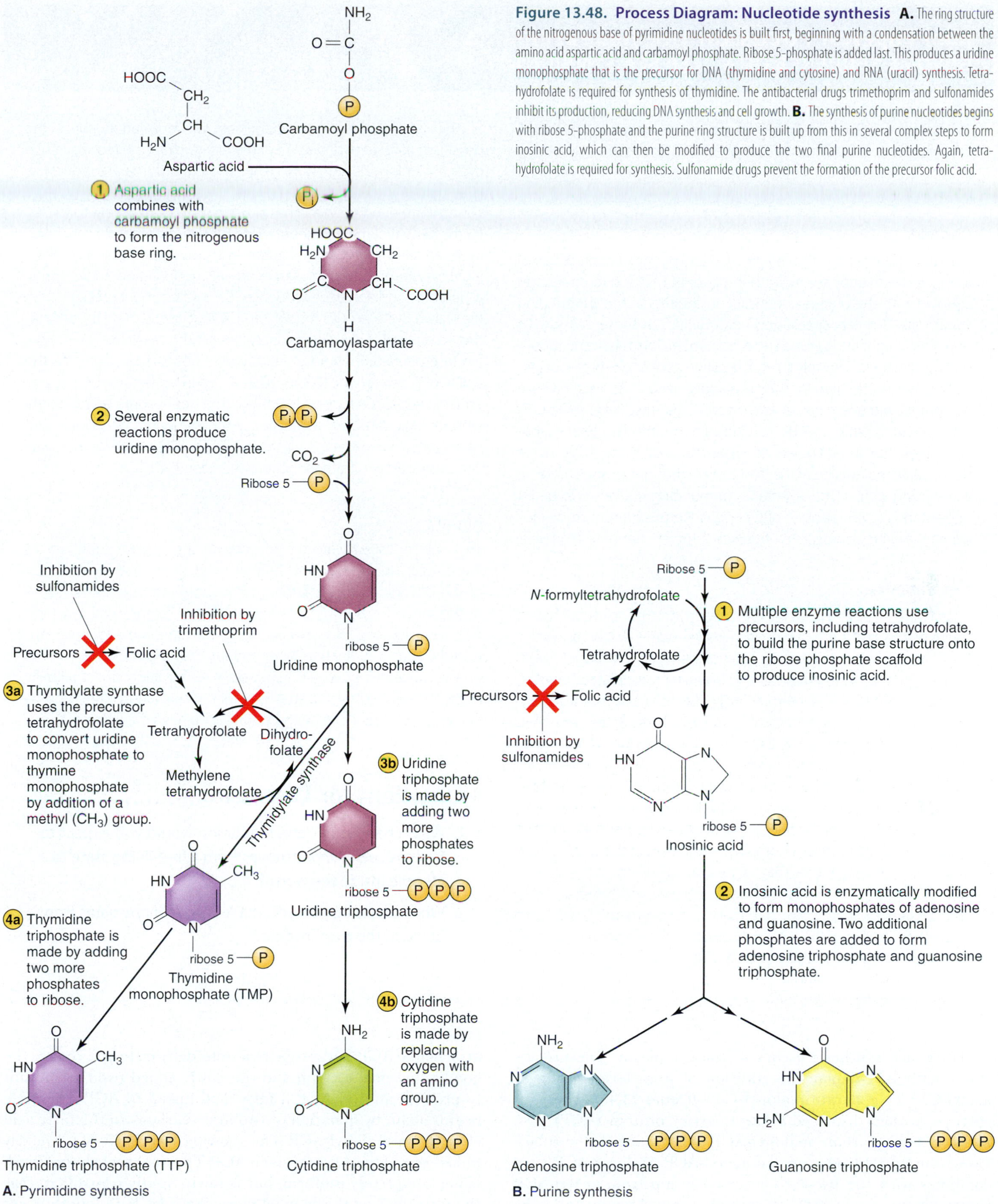

NH₂

O=C

O

(P)

Carbamoyl phosphate

HOOC

CH₂

CH

H₂N — COOH

Aspartic acid

(1) Aspartic acid combines with carbamoyl phosphate to form the nitrogenous base ring.

(Pᵢ)

HOOC

H₂N — CH₂

C — CH — COOH

O= N

H

Carbamoylaspartate

(2) Several enzymatic reactions produce uridine monophosphate.

(Pᵢ)(Pᵢ)

CO₂

Ribose 5 — (P)

O

HN

O — N

ribose 5 — (P)

Uridine monophosphate

Inhibition by sulfonamides

Inhibition by trimethoprim

Precursors ✗ Folic acid

(3a) Thymidylate synthase uses the precursor tetrahydrofolate to convert uridine monophosphate to thymine monophosphate by addition of a methyl (CH₃) group.

Tetrahydrofolate   Dihydro-folate

Methylene tetrahydrofolate

Thymidylate synthase

O

HN — CH₃

O — N

ribose 5 — (P)

Thymidine monophosphate (TMP)

(4a) Thymidine triphosphate is made by adding two more phosphates to ribose.

O

HN — CH₃

O — N

ribose 5 — (P)(P)(P)

Thymidine triphosphate (TTP)

A. Pyrimidine synthesis

(3b) Uridine triphosphate is made by adding two more phosphates to ribose.

O

HN

O — N

ribose 5 — (P)(P)(P)

Uridine triphosphate

(4b) Cytidine triphosphate is made by replacing oxygen with an amino group.

NH₂

N

O — N

ribose 5 — (P)(P)(P)

Cytidine triphosphate

B. Purine synthesis

Ribose 5 — (P)

N-formyltetrahydrofolate

Tetrahydrofolate

(1) Multiple enzyme reactions use precursors, including tetrahydrofolate, to build the purine base structure onto the ribose phosphate scaffold to produce inosinic acid.

Precursors ✗ Folic acid

Inhibition by sulfonamides

O

HN — N

N — N

ribose 5 — (P)

Inosinic acid

(2) Inosinic acid is enzymatically modified to form monophosphates of adenosine and guanosine. Two additional phosphates are added to form adenosine triphosphate and guanosine triphosphate.

NH₂

N — N

N — N

ribose 5 — (P)(P)(P)

Adenosine triphosphate

O

HN — N

H₂N — N — N

ribose 5 — (P)(P)(P)

Guanosine triphosphate

13.8 Biosynthesis of Cellular Components **449**

J. C. Robidart, S. R. Bench, R. A. Feldman, A. Novoradovsk, S. B. Podell, T. Gaasterland, E. E. Allen, and H. Felbeck. 2008. Metabolic versatility of the *Riftia pachyptila* endosymbiont revealed through metagenomics. Environ Microbiol 10:727–737.

Context

The giant tube worm *Riftia pachyptila* is a specially adapted animal associated with deep sea hydrothermal vents. The high temperatures coupled with the unique chemical constituents of the surrounding water make for very specialized communities in these ecosystems deprived of light and carbon substrates. Tube worms have no mouth or digestive tract. They rely solely on sulfide oxidizing γ-proteobacteria that live inside them for their metabolic needs. The bacteria have no contact with the external environment and thus rely on the host for providing certain growth substrates and removing waste products. These bacteria, known as *Endoriftia persephone*, have never been cultivated outside of the host but are thought to also survive in a free-living state in the sediments surrounding the vents. To better understand the metabolism of *E. persephone*, this present study took a metagenomic approach to obtain its complete genome sequence.

Experiments

Tube worm samples were collected near hydrothermal vents from a depth of 2,550 m at the East Pacific Rise, latitude 9°N. The East Pacific Rise is a ridge that formed at the junction between the Pacific tectonic plate and the North American and Rivera tectonic plates. Microbes were collected from the internal body tube, the trophosome, of two worms. The DNA was extracted, cleaved with restriction enzymes, and cloned into vectors. The DNA of individual clones was then sequenced using the Sanger method, yielding 14x coverage of the estimated 3.2 Mb genome. The resulting sequence was assembled into 2,472 contiguous or non-overlapping fragments, which were then analyzed. One of the first questions was whether the collective sequence represents the genome of a single genome, or if it includes DNA from other organisms that might also be present in the trophosome. Only a single 16S rRNA gene sequence was detected, and very little sequence variation was detected in the overlapping sequence traces that comprised the contiguous *Endoriftia persephone* appeared to reside in the trophosome as a pure culture.

The sequence was next examined for information about the physiology of *E. persephone*. All of the enzymes for the reductive TCA cycle, including the signature genes coding for fumarate reductase, α-ketogluterate oxidoreductase, and ATP citrate lyase, were present. This strengthened the hypothesis that *E. persephone* is able to fix CO_2 for the benefit of the tube worm host. Two key enzymes for the Calvin cycle, an alternative carbon-fixation pathway, were missing, indicating this system was not in use. For organisms operating under largely anaerobic conditions, the reductive TCA cycle is more efficient than the Calvin cycle, so the loss of the Calvin cycle is likely an adaptation to anaerobic conditions. Enzymes for the complete TCA cycle, fructose degradation and glycolysis were identified, supporting the ability of *E. persephone* to survive as a free-living heterotroph. It was proposed that CO_2 fixation is active in the trophosome of the worm, while organic carbon is assimilated in the free-living state. Finally, the genes for sulfide oxidation using a reverse sulfate reduction pathway and dissimilatory nitrate reduction were identified.

Impact

The inability to cultivate the microsymbiont of giant tube worms found near deep sea hydrothermal vents limits the investigation of the physiological relationship between these organisms. It also restricts investigation of the transition between the symbiotic and free-living life of *E. persephone*. While the genome sequence obtained through cultivation-independent means provides a window on the metabolic potential, its greatest contribution might actually be that it helps to develop strategies for the cultivation of this unique bacterium so that it can be thoroughly studied in the laboratory.

● Questions for Discussion ....

1. What changes in gene regulation would you expect to find as *E. persephone* moves from a free-living state to a lifestyle inside the worm?

2. What advantage do you think *E. persephone* gains from being in the trophosome?

Fatty acid synthesis occurs in the cytoplasm. The process starts with ATP-dependent addition of a carboxyl group to acetyl-CoA to generate malonyl-CoA **(Figure 13.49)**. The CoA cofactor is then displaced by acyl carrier protein (ACP), creating a three-carbon malonyl-ACP. ACP can also displace CoA from acetyl-CoA to make acetyl-ACP. Malonyl-ACP then combines with the two-carbon acetyl group from acetyl-ACP, removing the newly added terminal carboxyl group from the malonyl unit in the process. The result is a four-carbon unit attached to ACP. Successive reactions dehydrate and reduce the bond between the chain and the newly added two-carbon unit to produce an acyl-ACP, a fatty acid linked to ACP. The cycle begins again by donation of two more carbons to the fatty acid chain from malonyl-ACP. The growing fatty acid is covalently linked throughout synthesis to ACP. The final acyl chain length is not completely uniform, but is fairly tightly controlled. *How* the fatty acid synthetic machinery controls chain length is not well understood.

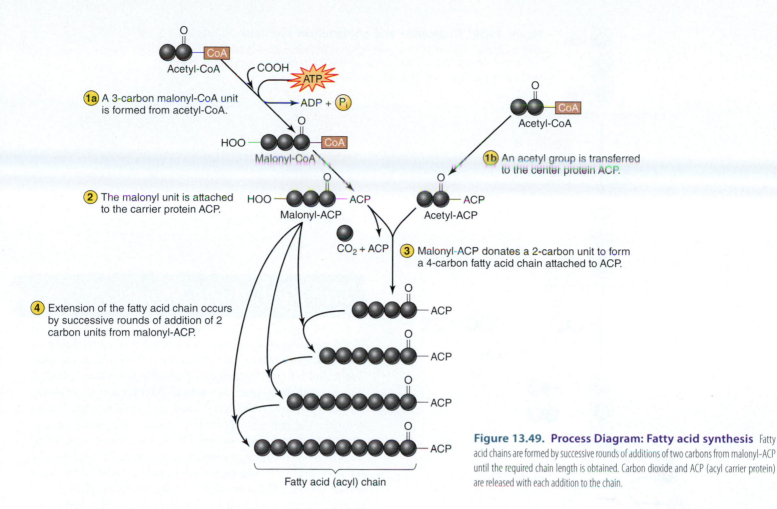

1a A 3-carbon malonyl-CoA unit is formed from acetyl-CoA.

1b An acetyl group is transferred to the center protein ACP.

2 The malonyl unit is attached to the carrier protein ACP.

3 Malonyl-ACP donates a 2-carbon unit to form a 4-carbon fatty acid chain attached to ACP.

4 Extension of the fatty acid chain occurs by successive rounds of addition of 2 carbon units from malonyl-ACP.

Fatty acid (acyl) chain

Figure 13.49. Process Diagram: Fatty acid synthesis Fatty acid chains are formed by successive rounds of additions of two carbons from malonyl-ACP until the required chain length is obtained. Carbon dioxide and ACP (acyl carrier protein) are released with each addition to the chain.

Fatty acid chains are eventually transferred by membrane-bound enzymes to a glycerol 3-phosphate backbone to form lipids for carbon and energy storage (triglycerides) or for cell wall synthesis (phospholipids) **(Figure 13.50)**. The glycerol 3-phosphate comes from NADH-dependent reduction of dihydroxyacetone phosphate (DHAP), an intermediate from glycolysis. Two fatty acids are added to each glycerol 3-phosphate molecule to generate phosphatidic acid, which is then a substrate for enzymes that modify the polar end of the molecule. The enzyme CDP diglyceride synthase activates the phosphate group by addition of cytosine triphosphate (CTP) to generate CDP-diacylglycerol. The CDP moiety can then be replaced with other chemical units such as serine to generate phosphatidylserine, which in turn can be decarboxylated to make phosphatidylethanolamine. These are two of the most common phospholipids in *Escherichia coli* and many other bacteria.

13.8 Fact Check

1. Cellular macromolecules are synthesized from a pool of precursors in the cell. Describe the sources of these precursors.
2. What nitrogenous bases must cells be able to synthesize and why?
3. Describe how fatty acids are synthesized.

Image in Action

WileyPLUS Giant tube worms are found living near hydrothermal vents on the sea floor. The trophosome of one of the tube worms is shown (cylindrical mass in the tube), and a zoom in shows the bacterial endosymbionts with small orange vesicles containing elemental sulfur, each encased in their own host-derived membrane. A close-up of the vents in the background shows free-living forms of the bacteria on the surface of the vent.

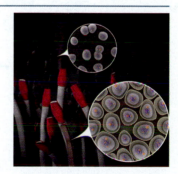

1. Imagine that you are attempting to culture these bacteria in the laboratory. Why would you need to provide H_2S? Explain how H_2S would be used by the bacteria.

2. If this image were to have a zoom in to highlight the metabolic processes of the bacteria, it would show them using the reductive TCA cycle to supply the worms with carbon compounds. Explain this process and how it could be depicted.

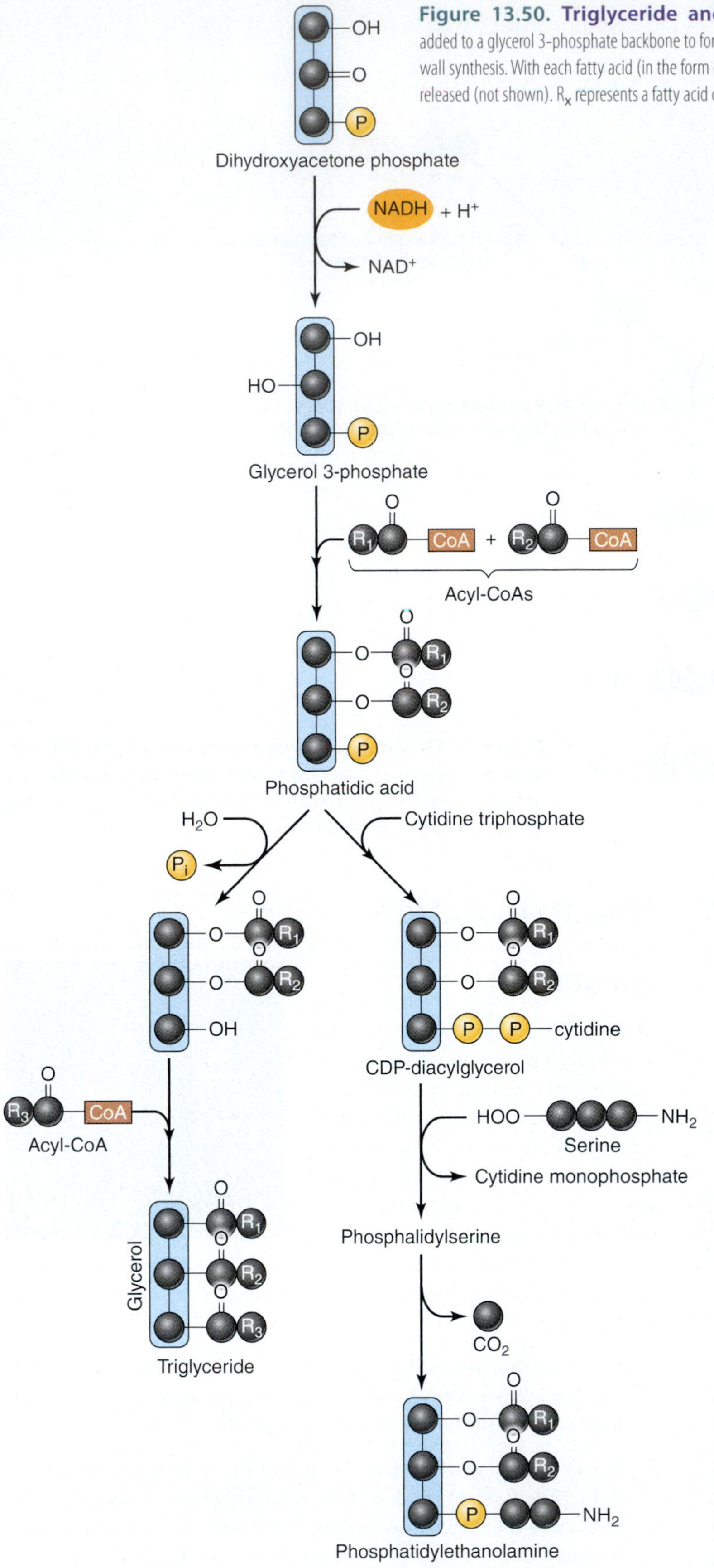

Figure 13.50. Triglyceride and phospholipid synthesis Fatty acids are added to a glycerol 3-phosphate backbone to form triglycerides for storage or to form phospholipids for cell wall synthesis. With each fatty acid (in the form of an acyl-CoA), added to the glycerol backbone, CoA-SH is released (not shown). R_x represents a fatty acid carbon chain of variable length.

Dihydroxyacetone phosphate

Glycerol 3-phosphate

Acyl-CoAs

Phosphatidic acid

CDP-diacylglycerol

Acyl-CoA

Serine

Cytidine monophosphate

Triglyceride

Phosphalidylserine

Glycerol

CO_2

Phosphatidylethanolamine

The Rest of the Story

In the opening story of this chapter, we were introduced to a very unusual relationship between a deep sea giant tube worm and its tiny inhabitant, *Endoriftia persephone*, a bacterial chemolithotrophic autotroph. Metagenomic analysis, as described in the chapter Mini-Paper, confirms the bacterium has the genetic capability to obtain electrons from the oxidation of hydrogen sulfide (H_2S), and transfer these to the electron transport system, generating ATP. It is thought the ATP and reducing power generated is used by the bacterium to fix CO_2, a product of tube worm respiration, to produce organic carbon compounds using the reductive TCA cycle. As you will recall, whereas the TCA cycle, found in the majority of life-forms, uses oxidation to break down organic carbon compounds to produce ATP and CO_2, the reductive TCA cycle functions in the opposite direction, using ATP and CO_2 to produce carbon compounds and O_2, both of which are used by the worm.

The tube worm has evolved several unusual characteristics in order to cultivate its indispensable bacterial inhabitants. Its digestive system converts to a growth chamber (the trophosome) for the microbe, and the tube worm produces a unique form of hemoglobin containing zinc instead of iron that can bind both O_2 and H_2S. The hemoglobin is also extracellular, so bound H_2S cannot enter cells. This is an important adaptation because H_2S is toxic to animals, inhibiting the electron transport system in mitochondria. The massive hemoglobin-filled feathery red plumes of the adult tube worm are used like gills to efficiently collect and concentrate H_2S from the surrounding waters and deliver it to the bacteria inside, as well as delivering O_2 to the tube worm cells. It is incredible to realize that bacterial oxidation of H_2S is the source of energy that provides for all of the nutritional needs of the amazing giant tube worm. Other large citizens that exist in the complete darkness of the tube worm forests include eel-like eelpout fish and pure white vent crabs. They, like the tube worms, depend on the metabolic activities of the microbes that harvest chemical energy provided by the released products of the vent, converting it to energy stored as organic molecules.

Summary

Section 13.1: What are the basic principles of catabolism and energy that control metabolism?

Metabolism includes **catabolism** of larger molecules to obtain energy and obtain smaller precursor molecules for **anabolism** of macromolecular cell components.

- Enzymes catalyze nearly all biochemical reactions in cells, when the **active site** binds to specific **substrates** and converts them into products.

- For a chemical reaction to occur spontaneously, the **laws of thermodynamics** dictate that the products of the reaction must have a lower free energy content than the reactants. The free energy released in a chemical reaction is the **Gibbs free energy (G)**. The change in free energy during a reaction measured under standard conditions is $\Delta G^{o'}$.

- An energy-yielding chemical reaction is **exergonic** and has a negative $\Delta G^{o'}$ value.

- Energy-absorbing **endergonic** reactions will have positive $\Delta G^{o'}$ values.

- An endergonic reaction in a cell must be coupled with an exergonic reaction that supplies enough energy to drive the coupled reaction to completion.

- The rate at which a reaction will occur is determined by its **activation energy (E_A)**.

- Enzymes increase the rate of a reaction by lowering its activation energy (E_A).

- **Cofactors**, such as organic **coenzymes**, assist in transferring functional groups from a substrate to a product.

- The most widely used form of chemical energy in the cell is **adenosine triphosphate (ATP)**.

Section 13.2: How do cells make ATP?

ATP can be synthesized by three processes in cells. These are substrate-level phosphorylation, oxidative phosphorylation, or photophosphorylation.

- **Substrate-level phosphorylation** occurs by transfer of a phosphate group from a reactive intermediate generated during catabolism to ADP. This occurs during **glycolysis** and the **tricarboxylic acid (TCA) cycle**.

- **Oxidative phosphorylation** uses the **chemiosmotic process** and requires the transfer of electrons through the **electron transport system**, which generates a proton gradient. The energy from the flow of protons can be used to synthesize ATP from ADP and P_i using **ATP synthase**. Oxidative phosphorylation is used by chemotrophs.

- **Photophosphorylation** also produces ATP using a proton gradient and an electron transport system. It requires light energy and is used by phototrophs.

- Many enzymes are controlled through **allosteric regulation**, involving molecules that induce a conformational change in the enzyme.

- **Redox reactions** involve the transfer of electrons from one molecule to another.

- **Oxidation** of a molecule results in loss of an electron and **reduction** of a molecule results in a gain of an electron.

- The tendency of a molecule to acquire electrons is its **reduction potential** or **redox potential (E)**.

- Energy is released when electrons flow from donors with more negative redox potentials to acceptors with more positive redox potentials.

- Electron carrier molecules, such as **nicotinamide adenine dinucleotide (NAD)**, are a source of reducing power used by cells to drive macromolecular biosynthetic reactions and provide energy for ATP synthesis by oxidative phosphorylation.

- The TCA cycle is a major producer of reduced electron carriers in chemoorganotrophs.

- Microbes can be grouped based on how they obtain energy (**chemotrophs** or **phototrophs**), electrons (**organotrophs** or **lithotrophs**), and carbon (**heterotrophs** or **autotrophs**).

Section 13.3: How do microbes utilize organic carbon?

Heterotrophs use many kinds of organic molecules for **assimilation** of carbon. Glycolytic pathways that catabolize glucose are important **anaplerotic reactions** that generate precursors for other metabolic pathways, including fermentation and respiration.

- Different glycolytic pathways can be used by different organisms. The pathways are interconnected and can be used depending on the demands of the cell. These are the **Embden–Meyerhof–Parnas (EMP) pathway**, the **Entner–Doudoroff pathway**, and the **pentose phosphate pathway**.

- Many organisms, such as **lactic acid bacteria (LAB)**, are capable of using fermentation under low oxygen conditions.

- **Fermentation** uses pyruvate, or a similar organic derivative, to accept electrons from NADH generated by glycolysis. Fermentation regenerates NAD^+, allowing glycolysis to continue to produce ATP.

- Fermentation produces only two moles of ATP for one mole of glucose.

- To keep up with energy needs, glycolysis occurs at a much faster rate during fermentation than during respiration. Allosteric regulation of phosphofructokinase by ATP controls the rate of glycolysis.

- **Respiration** is an alternative to fermentation that is used by chemoorganotrophs for regenerating NAD^+, which allows glycolysis to continue. Electrons of NADH are passed to an electron transport system and on to an inorganic acceptor.

- The TCA cycle is used during respiration to further catabolize and oxidize pyruvate into into CO_2. The TCA cycle yields a large amount of energy and generates carbon intermediates needed for biosynthetic reactions.

- Microbes avoid making enzymes when substrates aren't available or products are not needed or are in excess. Metabolic pathways are commonly regulated through **feedback inhibition** or **precursor activation**.

Section 13.4: What becomes of the electrons generated by glycolysis and the TCA cycle?

Respiration captures the energy in organic molecules by transferring electrons through the electron transport system. Two formats of respiration are **aerobic respiration**, which uses O_2 as a terminal electron acceptor, and **anaerobic respiration**, which uses some other electron acceptor.

- Nitrate (NO_3^-) is a common terminal electron acceptor used in anaerobic respiration. **Denitrification** occurs when microbes remove nitrate from soils and reduce it to gaseous nitrogen (N_2).

- In chemoorganotrophs, the aerobic electron transport system, in conjunction with glycolysis and the TCA cycle, can generate 38 ATP from each glucose molecule.
- Anaerobic respiration produces less ATP because available terminal electron acceptors have less positive redox potential than oxygen, therefore less energy is available to create a **proton motive force**.
- The proton motive force allows protons to flow down the charge gradient back across the membrane through ATP synthase, which couples this proton flow to the production of ATP by oxidative phosphorylation.
- Chemolithotrophs get energy and electrons by oxidizing inorganic molecules, which are directly passed down the electron transport system to generate a proton motive force for ATP production.

Section 13.5: How do microbes get nutrition from compounds other than glucose?

Monosaccharides other than glucose, as well as polysaccharides, proteins, lipids, and nucleic acids that must first be broken down into smaller units that can be transported into cells, can be converted into substrates suitable for processing in glycolytic pathways.

- Inside the cell, deamination of amino acids produces pyruvate as a product, which can enter the TCA cycle or fermentation reactions.
- Fatty acids are processed through the *β-oxidation pathway* and can be used to generate significant amounts of ATP through production of NADH and $FADH_2$, as well as acetyl-CoA that can enter the TCA cycle.

Section 13.6: How do microbes capture light energy for ATP synthesis and carbon fixation?

Phototrophs convert light energy to chemical energy by photophosphorylation.

- **Photosynthesis** combines phototrophy and carbon fixation for production of carbon compounds.
- The **light reactions** capture light energy and convert it to excited electrons that are passed to an electron acceptor. A proton motive force is generated that can be used to synthesize ATP.
- The site of light capture is the **photosystem** located in specialized **thylakoids**.
- **Chlorophyll** and **bacteriochlorophyll** are the major photopigments that capture photons in the photosystem. Accessory pigments are also used.
- The reaction center of the photosystem holds a special pair of chlorophyll or bacteriochlorophyll molecules, which absorb the transferred energy and donate excited electrons.
- Photosystem I (PS I) requires an external electron donor to replace the electron removed from the reaction center. Reduced electron carriers, NADH or NADPH, and a proton motive force for ATP synthesis are produced.
- In photosystem II (PS II), electrons are replaced internally by the photosystem. PS II generates a proton motive force, but no NADH or NADPH.

- **Oxygenic photosynthesis** uses both PS I and PS II and is carried out in chloroplasts and cyanobacteria. Water is used as an electron donor and O_2 is generated. All other bacteria use **anoxygenic photosynthesis** and possess either PS I or PS II. Oxygen is not produced.
- The reducing power and ATP that is generated during photosynthesis can be used by photoautotrophs to carry out the **dark reactions** for **carbon fixation**.
- The **Calvin cycle** used by plants and cyanobacteria is a dark reaction that consumes NADPH and ATP for production of glyceraldehyde 3-phosphate, which is used for the synthesis of sugars.
- The **reductive TCA cycle** is an alternative reaction used by some autotrophs and for carbon fixation. It is a reversed version of the TCA cycle that consumes NADH (or NADPH) and ATP.

Section 13.7: How do microbes use nitrogen and sulfur?

Nitrogen is required for amino acids and nucleotide synthesis. Sulfur is required for the amino acids cysteine and methionine, and the iron-sulfur centers used in proteins of electron transfer systems.

- **Nitrogen fixation** converts N_2 into compounds, such as ammonia, and is carried out by the **nitrogenase** enzyme, produced by some species of bacteria and archaea. The nitrogenase enzyme must be protected from exposure to oxygen, which inactivates it.
- The triple bond of dinitrogen gas is extremely stable, and an enormous amount of ATP must be consumed to separate the atoms.
- When ammonium is not available, nitrate is first reduced to ammonia, which consumes ATP. Ammonia, or its charged ion ammonium, is incorporated into glutamate or glutamine and from here, into amino acids.
- The sulfate ion (SO_4^-) is a common inorganic source of sulfur and must be reduced before it can be used in biosynthesis.

Section 13.8: How do microbes synthesize new cellular components?

Proteins, nucleic acids, lipids, and various cell wall materials are built from a small number of common subunits that can be synthesized from precursors in the cell.

- Amino acids are built from intermediates of the TCA cycle and glycolysis.
- ATP is the primary energy currency in the cell and must be replaced when it is incorporated into nucleic acid.
- The ring structure of pyrimidine nucleotides are built first using the amino acid aspartic acid, which is then linked to ribulose 5-phosphate of the pentose phosphate pathway. Purine nucleotides begin with ribulose 5-phosphate and the ring structure is added.
- Fatty acid synthesis begins with actyl-CoA, produced from pyruvate, to which carboxyl groups are added. Fatty acids are then added to glycerol 3-phosphate, an intermediate of glycolysis, to make triglycerides for storage or phospholipids for cell wall synthesis.

Application Questions ···

1. When preparing culture media in the lab, it is important that components such as Mg^{2+}, Fe^{2+}, and Mn^{2+} are available. Explain why these components are necessary in order for the microbes to grow on the media.

2. In a biochemistry lab, a researcher is closely examining a microbial chemical reaction. During his experiment, he determines that the value of $\Delta G^{\circ\prime}$ is approximately +38 kJ/mol. Explain what this value tells him about the reaction being examined.

3. You are examining an organism that is in anoxic conditions with no available electron acceptors for NADH. Describe how this organism will go about catabolizing glucose under these conditions.

4. Imagine a microbial cell that normally carries out aerobic respiration has encountered a low O_2 environment and may die if the O_2 levels do not increase. Describe the key role that O_2 plays for this microbe in the aerobic respiration process and what will occur differently without oxygen that could lead to death.

5. Some microbes, such as *Escherichia coli*, have flexible metabolic systems that allow them to generate energy under aerobic and anaerobic conditions. If *E. coli* is growing in the presence of O_2, nitrate, and fumarate, explain its metabolic processes, including which of these it will use first and why.

6. Soil ecologists are very interested in the process of denitrification. Explain how denitrification is linked to anaerobic respiration, and why it is an important consideration in soils.

7. Fluoride and triclosan are compounds that inhibit the oral pathogen *Streptococcus mutans*, which causes cavities (dental caries). These compounds have been shown to inhibit pyruvate kinase.

Describe the direct metabolic effect of inhibiting this particular enzyme, and how this could then lead to a reduction in cavities.

8. Sometimes cells are left without oxygen and cannot carry out aerobic respiration to produce large amounts of ATP. These cells may then utilize fermentation that produces much smaller quantities of ATP but also allows for the very important recycling of NADH to NAD^+. Explain why this recycling is so important to the cell.

9. You are watching a movie in which a meteor is headed toward Earth and is predicted to cause a dust cloud that will block the sun for months. The main character declares that this could destroy life on Earth! Explain the impact this dust cloud would have on the light and dark reactions of photosynthesis. Specifically, what would still occur and what would no longer occur without the sunlight? Could this effect destroy life on Earth?

10. The enzyme Rubisco has been described in this chapter as the most abundant protein on Earth. Imagine a toxin entered the atmosphere that inhibited this enzyme. Describe the specific effects this toxin would have on photosynthetic cells.

11. A newly discovered microbe has been studied and is being called a chemolithotrophic autotroph.
    a. Dissect this term to explain what it means about the microbe's metabolism.
    b. Imagine that this microbe cannot be cultured. How could you use metagenomic analysis to determine these characteristics of the microbe?
    c. Postulate what genes may be found during a metagenomic analysis to characterize the microbe in this way. Explain your answer.

Suggested Reading

Bryant, D. A., and N.-U. Frigaard. 2006. Prokaryotic photosynthesis and phototrophy illuminated. Trends Microbiol 14:488–496.

Gorby, Y. A., S. Yanina, J. S. McLean, K. M. Rosso, D. Moyles, A. Dohnalkova, T. J. Beveridge, I. S. Chang, B. H. Kim, K. S. Kim, D. E. Culley, S. B. Reed, M. F. Romine, D. A. Saffarini, E. A. Hill, D. Shi, D. A. Elias, D. W. Kennedy, G. Pinchuk, K. Watanabe, S. Ishii, B. Logan, K. H. Nealson, and J. K. Fredrickson. 2006. Electrically conductive bacterial nanowires produced by *Shewanella oneidensis* strain MR-1 and other microorganisms. Proc Natl Acad Sci USA 103:11358–11363.

Lovley, D. R. 2008. Extracellular electron transfer: Wires, capacitors, iron lungs, and more. Geobiology 6:225–231.

Markert, S., C. Arndt, H. Felbeck, D. Becher, S. M. Sievert, M. Hügler, D. Albrecht., J. Robidart, S. Bench, R. A. Feldman, M. Hecker, and T. Schweder. 2007. Physiological proteomics of the uncultured endosymbiont of *Riftia pachyptila*. Science 315:247–250.

Mulkidjanian, A., P. Dibrov, and M. Y. Galperin. 2008. The past and present of sodium energetics: May the sodium-motive force be with you. Biochim Biophys Acta. 1777:985–992.

Walter, J. M., D. Greenfield, C. Bustamante, and J. Liphardt. 2007. Light-powering *Escherichia coli* with proteorhodopsin. Proc Natl Acad Sci USA 104:2408–2412.

CHAPTER 14

Biogeochemical Cycles

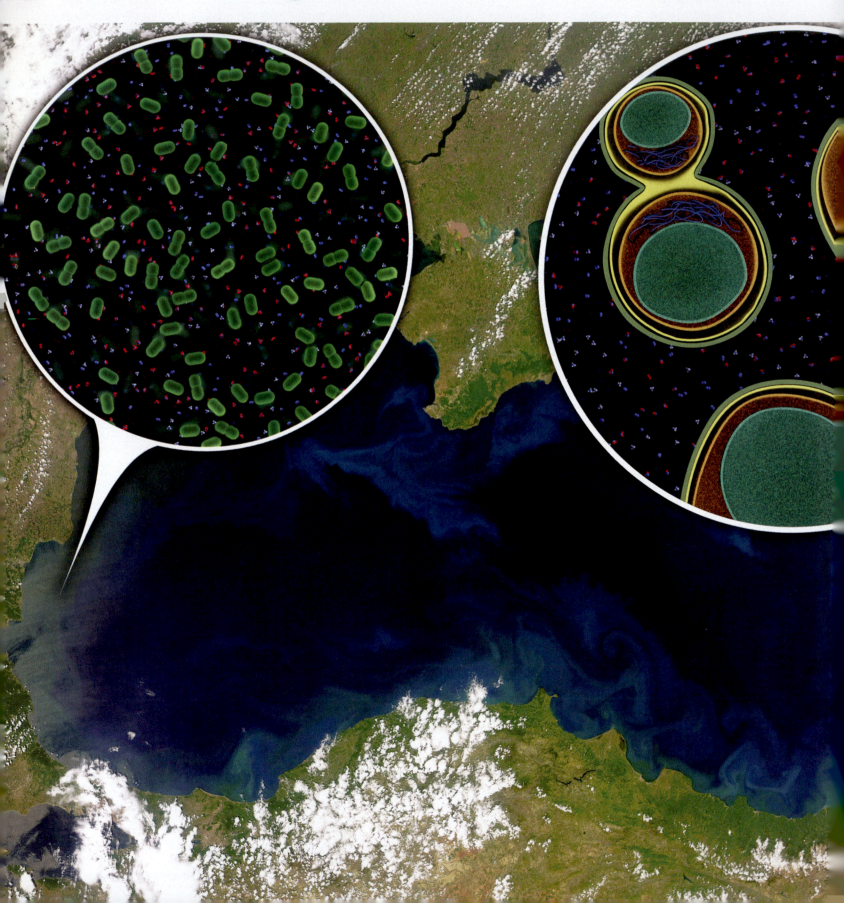

Toward the end of the nineteenth century, the Ukrainian microbiologist Sergei Winogradsky showed that certain aerobic bacteria could oxidize ammonium to nitrite. This conversion proved to be a key step in the nitrogen cycle, and appeared to occur only in the presence of O_2. Microbiologists quickly learned that microbes play fundamental roles in the transformation and recycling of matter, and the study of these transformations formed the core of environmental microbiology research over the years. For about a hundred years, microbiologists assumed that the oxidation of ammonium only occurred under aerobic conditions.

In the 1970s, however, that assumption changed. The Austrian chemist Engelbert Broda predicted, based on thermodynamic grounds, that ammonium oxidation in the absence of O_2 should be possible, even though microbes able to perform this task had not been isolated. Additionally, oceanographers observed that the amount of ammonium predicted to be present in anoxic basins like the Black Sea in Europe greatly exceeded the amount actually measured, suggesting that the ammonium was being removed. Despite these hints, it was not until 1995 that researchers unequivocally documented a type of ammonium oxidation that occurs under anaerobic conditions. Interestingly, this first demonstration occurred not in a natural ecosystem but in an anaerobic reactor designed to treat wastewater.

Microbiologists have explored the details of this activity since its discovery. We now recognize that this so-called anammox reaction, for anaerobic ammonium oxidation, probably contributes significantly to the global cycling of nitrogen. Indeed, about half of the loss of nitrogen from the oceans by the formation of N_2 gas may be due to the anammox reaction. The microbes responsible for this reaction have yet to be isolated in pure culture. Given our inability to cultivate the majority of the microbes on Earth, many other processes driving our planet's biogeochemical cycles may still await discovery.

CHAPTER NAVIGATOR

As you study the key topics, make sure you review the following elements:

Nutrients that fuel the activities of microbial cells are cycled through the biosphere.

- Figure 14.1: The major reactions involved in biogeochemical cycles
- Figure 14.2: Elemental composition of a bacterial cell
- Figure 14.3: Disruption of the carbon and nitrogen cycles by human activities
- Toolbox 14.1: Using microarrays to examine microbial communities: The GeoChip

Microbes play key roles in carbon cycling.

- Figure 14.6: The carbon cycle
- Perspective 14.1: CO_2 as a greenhouse gas and its influence on climate change
- Microbes in Focus 14.1: *Methanococcus maripaludis*: An archaeal methane producer
- Microbes in Focus 14.2: *Methylosinus trichosporium*: A methane eater
- Figure 14.11: New Zealand hot spring

Microbes play central roles in nitrogen cycling.

- Figure 14.12: The nitrogen cycle
- Microbes in Focus 14.3: *Nitrobacter winogradskyi*
- Mini-Paper: The first isolation and cultivation of a marine archaeon

Biogeochemical cycles are interconnected.

- Perspective 14.2: The microbiology of environmentally toxic acid mine drainage

CONNECTIONS for this chapter:

Deep sea hydrothermal vents and the oxidation of H_2S (Section 15.5)

Methane-producing archaeon *Methanobrevibacter smithii* in the human GI tract (Section 4.3)

Nitrification and denitrification during wastewater treatment (Section 16.5)

Introduction

Throughout the previous chapters, we have discussed what individual microbes do as they conduct their lives. Their activities include the uptake of nutrients, metabolic processes, and the synthesis of structural materials. In this chapter, we will consider the consequences of microbial activities on a *larger scale*. As we will discover, the metabolic functions of these tiny organisms can have a global impact, ultimately determining the physical properties of this planet.

All living organisms consist of chemical elements bound together to form biological molecules. Normal metabolic processes result in the production and degradation of biomass. During these processes, the atoms of various elements continually cycle through the biosphere, the part of Earth that supports life. While all organisms participate in this cycling, microbes play crucial roles in moving atoms between living organisms and the physical environment. The resulting linked sets of cycles are termed the biogeochemical cycles (**Figure 14.1**). Without the microbial contributions to these essential cycles, all other life on Earth would not exist; the elements would not be recycled to forms that can be used as nutrients by other cells.

How do microbes contribute to these cycles? In most instances, they do so simply by conducting their normal metabolic activities. Their diverse physiological capabilities, and the rapid rates of their enzymatic reactions, drive these cycles. While microbes may not be directly visible, their activities can move substantial quantities of material, and certain microbial groups perform important, unique reactions that no other living organisms can.

In this chapter, we will begin by considering the use of nutrients by cells and how we can measure this use. We will then focus on the major cycles of carbon (C) and nitrogen (N). To conclude, we will briefly examine the interconnectedness of various cycles. As we investigate these topics, we will address the following questions:

How are the nutrients used by cells cycled in the biosphere? (14.1)
How do microbes influence the cycling of carbon? (14.2)
How do microbes influence the cycling of nitrogen? (14.3)
How are various biogeochemical cycles connected? (14.4)

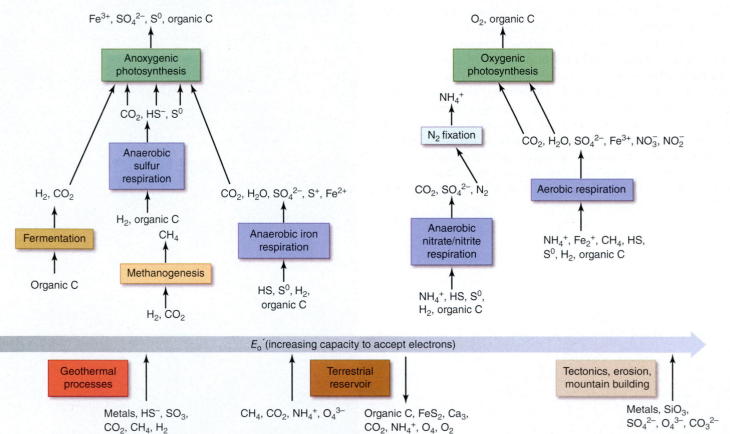

Figure 14.1. The major reactions involved in biogeochemical cycles Microbial metabolism drives biogeochemical cycles. These important microbial processes include oxygenic photosynthesis, anoxygenic photosynthesis, nitrogen fixation, fermentation, methanogenesis, and respiration involving iron, sulfur, oxygen, and nitrogen. The electron acceptors are arranged according to capacity to accept electrons, increasing from left to right. The cycles are interconnected due to the metabolic linkages.

From *Science*, vol 4, issue 9, April 29, 2010. Reprinted with permission from AAAS.

14.1 Nutrient cycling

How are the nutrients used by cells cycled in the biosphere?

We often discuss **biogeochemical cycles**, the movement of elements between various locations and forms, separately. We should start, though, by recognizing that these cycles are interconnected. They do not occur in isolation. For instance, consider how aerobic respiration results in the release of CO_2 from organic compounds (like glucose, $C_6H_{12}O_6$), while also reducing oxygen (O_2) to water (H_2O). Or consider how the production of new organic material during photosynthesis by algae or cyanobacteria involves the formation of O_2 from water. These two complementary reactions of respiration and photosynthesis involve cycling of carbon, oxygen, and hydrogen. Most elements have a cycle that may involve various biological transformations and chemical reactions, as well as changes in physical state between gas, liquid, and solid. Some cycles, like the ones for carbon and nitrogen, are complex, while other cycles, like the one for iron, are relatively uncomplicated. In this section, we will review some basic information about nutrient cycling. We will also look at how investigators can study these processes.

Cycling of elements

As we begin to explore nutrient cycling, we should remember that all living organisms require approximately the same suite of elements. All biomass consists of these essential or **biogenic elements** in roughly similar proportions. Consider a pure culture of *Escherichia coli* that has been recovered by centrifugation, washed, and then dried, resulting in a concentrated pellet of biomass. An elemental assay of this cellular material would reveal approximately half of the cellular mass to consist of carbon, a fifth oxygen, a sixth nitrogen, and less than a tenth hydrogen **(Figure 14.2)**, with lesser amounts of phosphorus, sulfur,

potassium, calcium, iron, magnesium, and chlorine. Other elements, including sodium, manganese, cobalt, copper, zinc, nickel, and molybdenum would be found in trace amounts comprising less than 0.3 percent of the cell dry mass combined. As we saw in Chapter 6, organisms require carbon, nitrogen, phosphorus, and sulfur in relatively large amounts as components of macromolecules, and cell growth is ultimately limited by lack of sufficient quantities of one or more of these elements. Potassium, sodium, magnesium, calcium, iron, and chloride ions are the most abundant ions in the cell, involved in functions such as control of osmotic balance and enzyme activity.

Cycling of these elements often involves changes in the oxidation state of the elements. Living organisms contribute to these changes by conducting redox reactions. These reactions involve the transfer of electrons between chemical elements, resulting in changes to the oxidation state of the atoms. To better understand this concept, let's examine the oxidation states of carbon **(Table 14.1)**. During photosynthesis, carbon in CO_2 has an oxidation state of $+4$, but carbon in the photosynthesis product—glucose—has an oxidation state of 0. These changes in oxidation state often alter the physical characteristics of the element, and may affect whether that form of the element is metabolically available to cells.

To understand the cycling of an element, we also need to determine the sizes of the major reservoirs, or pools, of the element on the planet. On a global scale, Earth can be divided into the terrestrial reservoir, the aquatic reservoir, and the atmospheric reservoir. The terrestrial reservoir, as we will see later in this section, represents an important pool for carbon, which exists both in the soil and in plants and animals. Various elements, though, often transition between these reservoirs. The terrestrial reservoir, for instance, acts as a significant source of inorganic carbon in the atmosphere because the decomposition of dead organic matter by soil microbes releases CO_2 to the atmosphere. This movement between reservoirs can be expressed as **flux**, the amount flowing through a unit area or volume per unit of time.

The natural transfers of elements between reservoirs undoubtedly have shifted over Earth's history. We know that approximately 2.7 billion years ago the first phototrophs producing O_2 arose, and their activities over the ensuing 300 million years

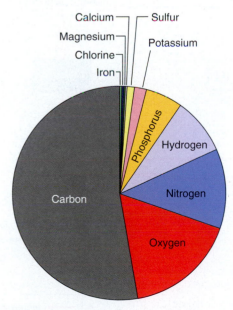

Figure 14.2. Elemental composition of a bacterial cell About half of the mass of a typical bacterial cell consists of carbon, followed in order by oxygen, nitrogen, and hydrogen as the most abundant elements.

TABLE 14.1	Oxidation states for carbon
Oxidation state	**Example compound**
-4	CH_4 (methane)
-2	CH_3OH (methanol), $[-CH_2]_n$ (hydrocarbons, e.g., C_8H_{18}, octane)
0	$[CH_2O]$ (carbohydrates, e.g., $C_6H_{12}O_6$, glucose)
$+2$	$HCOOH$ (formic acid)
$+4$	CO_2 (carbon dioxide), HCO_3^- (bicarbonate ion)

led to a change in Earth's atmosphere from its original reducing condition to the current oxidizing state (see Section 1.2). The movement of oxygen over that time was decidedly unidirectional. The liberation of O_2 by these phototrophs greatly exceeded the removal of O_2 from the atmosphere. Now, however, oxygen and other elements tend to cycle.

Evidence indicates that these cycles have maintained a dynamic equilibrium over the long term. With the rise of industrialization, though, human activities have perturbed this equilibrium, most evidently in the carbon and nitrogen cycles **(Figure 14.3)**. In general, smaller, rapidly cycling reservoirs are most susceptible to perturbation. With respect to carbon, fossil fuel combustion and changes in land use, such as deforestation and conversion of natural grasslands to agriculture or urban use, have had a marked influence on the atmospheric CO_2 level. The high energy requirements for the manufacture of fertilizers and cement, and the production of lime by heating limestone, resulting in the release of CO_2, are examples of industrial activities that also contribute to increasing levels of atmospheric CO_2. The responses of the overall carbon cycle to these anthropogenic perturbations are the subjects of intense, ongoing research.

To truly understand the role of microbes in biogeochemical cycling, we need to consider the biochemical activities that occur in a cell. In the individual cell, these biochemical activities determine growth and survival. Natural selection refines the genetic control of these biochemical pathways, and ultimately determines the process of evolution. Intricate sets of linkages and dependencies thus interconnect genetics, biochemistry, and the environment. The application of molecular techniques to the study of microbes in biogeochemical cycling has increased our knowledge greatly. For example, microarray analysis can be used to determine what types of metabolic genes exist within a given microbial community **(Toolbox 14.1)**.

Molecular techniques like the GeoChip can provide valuable insight into the metabolic potential of a community. Many other techniques can also be used by researchers interested in biogeochemistry. In the next two sections, we will explore more closely how microbial activity affects the cycling of two important elements—carbon and nitrogen. Before we examine these cycles, though, let's address a more fundamental question. How can scientists study the cycling of elements?

Monitoring chemical cycling

To understand the cycling of elements, we need to accurately and reliably measure the concentrations of these elements and their conversion from one form to another. Many techniques

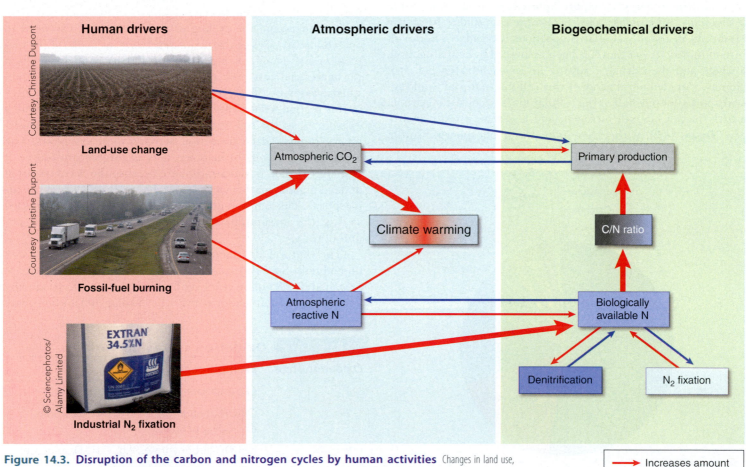

Figure 14.3. Disruption of the carbon and nitrogen cycles by human activities Changes in land use, burning of fossil fuels, and industrial N_2 fixation result in perturbations in the carbon and nitrogen cycles. The most noticeable effects are in increases in atmospheric CO_2 and biologically available N.

Microarrays have been used extensively to analyze the transcriptional activity of genes. As we saw in Perspective 8.1, this technique can even be used to identify viruses. Microarray technology can also be used to study the composition of microbial communities. With the vast amount of sequence information available, and the capacity of microarray chips, researchers can design probes to functional genes that represent the possible metabolic processes that are being carried out in a given environment. The GeoChip, designed and produced by Jizhong Zhou and colleagues at the University of Oklahoma, exemplifies this type of approach.

GeoChip 3.0, first described in 2010, contains 28,000 50-nucleotide probes, representing genes involved in carbon, nitrogen, phosphorus, and sulfur cycles, as well as energy metabolism, resistance to antibiotics and metals, and degradation of organic contaminants **(Figure B14.1)**. The probes span the sequence diversity of known genes for these processes, from Bacteria, Archaea, and Fungi.

To use the GeoChip, researchers first isolate DNA from an environmental sample, and label it by a random primer method. In this process, DNA polymerase synthesizes single-stranded fragments using the isolated DNA as a template, incorporating fluorescently tagged nucleotides to label the newly produced DNA. In the hybridization step, the resulting single-stranded labeled DNA is then incubated with the microarray slide to allow binding to the probe, after which the unbound labeled DNA is washed off. The microarrays are scanned in the microarray scanner to determine which spots were bound, and how strongly. The data are analyzed, presenting the investigators with a picture of which genes are present within the microbial community. This information, by extension, provides the investigators with clues about the metabolic potential of the community.

The great advantage of this technique is that it provides information about not just the phylogenetic diversity, but also the genes that actually determine the metabolic characteristics of the organisms present in the community, especially with regard to the biogeochemical cycles. It is also very rapid, much faster than metagenomic DNA sequence analysis. As functional genomics provides more knowledge about the functions associated with gene sequences, future designs of the GeoChip and similar microarrays will contain that much more information.

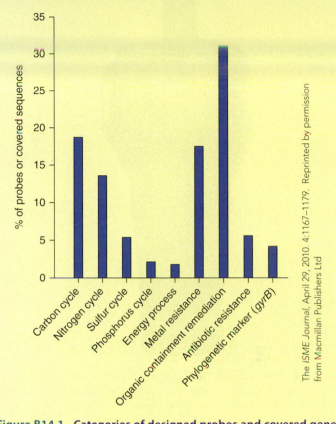

The *ISME Journal*, April 29, 2010, 4:1167–1179. Reprinted by permission from Macmillan Publishers Ltd

Figure B14.1. Categories of designed probes and covered gene sequences Genes for probe design were chosen to ensure good representation of gene function.

Test Your Understanding

You use the GeoChip 3.0 to analyze the microbial community structures and possible metabolic characteristics of two different soils. The results indicate that 95 percent of the genes you identify are found in both soils. What does this result suggest to you about the soils?

have been developed to obtain these data. Let's look specifically at the measurement of carbon dioxide. As we will see in Section 14.2, numerous research groups have reported that the level of CO_2 in the atmosphere is increasing at an unprecedented rate. Moreover, we know that CO_2 absorbs infrared energy radiating off of Earth, reflects it back, and, in the process, contributes to the warming of our planet. To fully understand this process, researchers must first be able to accurately measure

CO_2 concentrations. They must also be able to determine the cycling of elements.

Detecting Atmospheric Gases

How do scientists determine the concentration of CO_2 in the atmosphere? The infrared absorption of CO_2 provides the key. After collecting air in a sealed cylinder, researchers expose the air to infrared radiation. Within the device, an infrared detector

Courtesy of TSI

Figure 14.4. Detection of CO₂ To determine CO_2 levels in the atmosphere, samples are collected and measured using an infrared detector. Infrared light is strongly absorbed by CO_2. Increased CO_2 levels lead to reduced amounts of infrared light reaching the detector.

Courtesy Lawrence Berkeley National Laboratory, Roy Kaltschmidt photographer

Figure 14.5. Mesocosms These systems allow researchers to easily manipulate conditions within a natural environment.

monitors the amount of input infrared radiation that passes through the sample. Increased amounts of CO_2 lead to increased infrared radiation absorption, and decreased detection of the infrared radiation **(Figure 14.4)**. By calibrating the monitoring device with known concentrations of CO_2, researchers can convert detection levels to a more meaningful number—parts per million (ppm). A CO_2 reading of 350 ppm, for instance, indicates that CO_2 represents 350 of every 1 million gas molecules within a sample.

While data clearly show an increase in atmospheric CO_2 levels over the past 50 years, many scientists have argued that CO_2 levels have been rising significantly since the beginning of the Industrial Revolution. These claims raise another important question. How can we determine historical levels of atmospheric gases? Many of these data come from analyses of gas bubbles trapped in ice cores obtained from Antarctica. Deeper ice samples formed longer ago and the air bubbles trapped within these ice samples represent pockets of "old" atmosphere. The concentrations of gases within these samples can then be determined.

Similar types of analyses can be done under more controlled conditions. By creating **mesocosms**, small, contained enclosures designed to replicate certain ecosystem characteristics,

researchers can measure changes in gas concentrations over time **(Figure 14.5)**. They can determine, for instance, how the introduction of more nutrients into a system affects the amount of a gas present in the atmosphere of the mesocosm.

Radioisotope Tracers

Mesocosms can also be useful in studies involving **radioisotope tracers**. In these studies, investigators monitor the fate of specific molecules by labeling the molecules with detectable radioisotopes. Let's say, for instance, that researchers were interested in the biodegradation of toluene, a solvent used as a paint thinner. Chemists could synthesize toluene (C_7H_8) containing radioactive carbon (^{14}C). After adding this radioisotope tracer to an appropriately constructed mesocosm, scientists could then monitor the atmosphere within the mesocosm for the appearance of $^{14}CO_2$, as opposed to the more common $^{12}CO_2$. One could then confidently predict $^{14}CO_2$ would arise only from the oxidation of the radiolabeled toluene.

14.1 Fact Check

1. What four key elements make up most of the cellular mass of *E. coli*, and what additional elements are also found in the cell?
2. What three reservoirs is Earth divided into when considering biogeochemical cycling?
3. What are mesocosms, and how can they aid in the biogeochemical cycling studies?
4. Explain the usefulness of radioisotope tracers.

14.2 Cycling driven by carbon metabolism

How do microbes influence the cycling of carbon?

Life on Earth is carbon-based. From DNA to sugars to proteins, carbon represents the backbone of virtually all biological molecules. However, despite the importance of carbon to life, few of us probably contemplate the biogeochemical cycling of large amounts of carbon within the terrestrial, aquatic, and atmospheric reservoirs. The main stores of carbon include biomass, soil organic matter (SOM), atmospheric CO_2, fossil fuels, sedimentary rock deposits, calcium carbonate-containing shells, and dissolved CO_2 in the oceans. Today, there is keen interest in potential reservoirs where CO_2 might be stored, and how that carbon may cycle, due to the recognition that CO_2 serves as a **greenhouse gas**, a gas, that when present in the atmosphere, causes thermal absorption **(Perspective 14.1)**.

Carbon cycling can be considered in terms of long-term and short-term cycling. Short-term carbon cycling tends to be most interesting to microbiologists, and will be the main focus of this section. Before exploring the short-term cycling of carbon, though, we will briefly examine the global carbon cycle. We will then examine several processes affecting short-term carbon cycling, including the related processes of photosynthesis, respiration, and decomposition. We will also explore the microbial processes of CH_4 formation and degradation that drive the contribution of CH_4 to the carbon cycle.

The global carbon cycle

To better understand the long-term, global cycling of carbon, let's examine the various carbon reservoirs and the flux of carbon between these reservoirs **(Figure 14.6)**. Atmospheric CO_2 exchanges relatively freely and with high turnover rate with the planet's living organisms, the soil, and the upper reaches of the oceans. In contrast, carbon in carbonate-based sedimentary rocks, marine sediments, and fossil fuels cycles very slowly, and constitutes by far the largest carbon reservoirs on Earth. An exception to this generalization, of course, involves the recovery and combustion of fossil fuels by humans, which sends carbon from a rather static pool into the atmosphere at an unnaturally rapid rate.

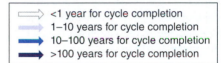

- ⟹ <1 year for cycle completion
- ⟹ 1–10 years for cycle completion
- ⟹ 10–100 years for cycle completion
- ⟹ >100 years for cycle completion

Figure 14.6. The carbon cycle Carbon transfers between reservoirs of different sizes. The different sizes of the carbon reservoirs are reflected in the time it takes for cycle completion, represented by the color of the arrows. Arrow thickness indicates the relative amount of carbon transferred over a given time. Terrestrial vegetation, for instance, is estimated to contain about 600 gigatonnes of carbon (GtC) and about 120 GtC/yr from the atmosphere is taken up into new plant biomass during photosynthesis. About 60 GtC/yr is returned to the atmosphere due to decomposition of plants as a result of microbial respiratory and fermentative activities, as well as plant and animal respiration. The carbon that is not returned to the atmosphere is accumulated in the soil as organic matter, which accounts for about 1,600 GtC. Another 60 GtC/yr is returned to the atmosphere through decomposition of this soil organic matter. Although approximately the same total amounts of these pools are returned to the atmosphere each year, the rate of the exchange of carbon from the plant reservoir is much faster than the rate of exchange from the soil reservoir. This is due to the difference in sizes of the plant and soil carbon pools.

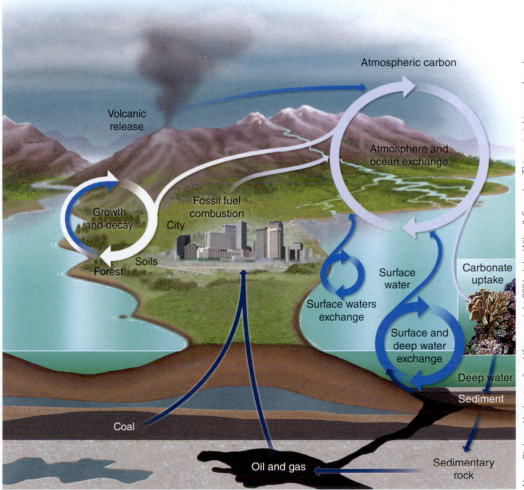

Perspective 14.1
CO₂ AS A GREENHOUSE GAS AND ITS INFLUENCE ON CLIMATE CHANGE

Since the Industrial Revolution began in the nineteenth century, human activities have increased the atmospheric CO_2 concentration by about 30 percent. **Figure B14.2** illustrates this consistent upward trend in atmospheric CO_2 level over the past half century. Fuel combustion accounts for about three-quarters of the increase, and land use changes, such as clearing of land for agricultural activities that disrupts the carbon reservoir of the natural vegetation, accounts for the other quarter.

Why should we be concerned about this increase in atmospheric CO_2? CO_2 is a greenhouse gas. CO_2 and other greenhouse gases strongly absorb radiation in the infrared range, although they are transparent to visible light. When visible sunlight strikes Earth, a portion irradiates back toward space as longer-wavelength infrared radiation **(Figure B14.3)**. Gases like CO_2 act as a sort of thermal blanket, trapping this infrared radiation and thereby influencing the planetary climate. Accumulating atmospheric CO_2 therefore most likely contributes to climate change, and has shaped the climate on Earth throughout its history.

Given the documented perturbation in atmospheric CO_2 content linked to industrialization, we need an accurate accounting of global carbon, and we need to understand what regulates carbon transfer between various compartments. This is necessary so that strategies can be devised to deal with a shifting carbon cycle. Although we know quite a lot about how carbon cycling operates today and how it has worked in the past, we still do not have enough quantitative data to construct an accurate global carbon budget. Less carbon exists in the atmosphere than can be accounted for. When using current information to model the global carbon cycle, some carbon appears to be missing. How can we explain this inconsistency?

Several possible scenarios could allow us to account for the apparent missing carbon. Perhaps, some of the data are inaccurate or some carbon reservoir remains uncounted. Scientists at the Woods Hole Oceanographic Institution have used data from the 1990s to summarize the global carbon cycle as follows:

Atmospheric increase = emissions from fossil fuels
+ net emissions from changes in land use
− oceanic uptake − missing carbon reservoir

$$3.2 \,(\pm 0.2)\, GtC = 6.3 \,(\pm 0.4)\, GtC + 2.2 \,(\pm 0.8)\, GtC$$
$$- 2.4 \,(\pm 0.7)\, GtC - 2.9 \,(\pm 1.1)\, GtC$$

When using estimated values for each of the first four categories, the cycle remains unbalanced unless the *missing* category of about 2.9 GtC is added to the right side of the equation. The unidentified reservoir is required to balance estimates of global releases with global accumulations of carbon over about 150 years **(Figure B14.4)**. Researchers are reasonably confident in their fossil fuel combustion and land use-related data and the atmospheric accumulation data, but there are considerable uncertainties in the understanding of microbe-controlled global photosynthesis and respiration rates, and carbon storage. Could some unrecognized type of microbial carbon metabolism be responsible for the unaccounted carbon in the models? The answer to this question will only come once we have much better knowledge of the microbial control of the biogeochemical cycles.

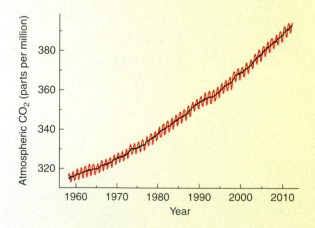

Figure B14.2. Atmospheric CO₂ at Mauna Loa Observatory in Hawaii The black line represents a steady rise in CO_2 since 1958 when measurements began. The annual up-and-down pattern occurs because primary producers take up relatively more CO_2 during the growing season in the forested Northern Hemisphere. During winter in the Northern Hemisphere, primary producers in the relatively smaller land masses in the Southern Hemisphere take up less CO_2, and the atmospheric level rises.

Courtesy of the National Oceanic & Atmospheric Administration

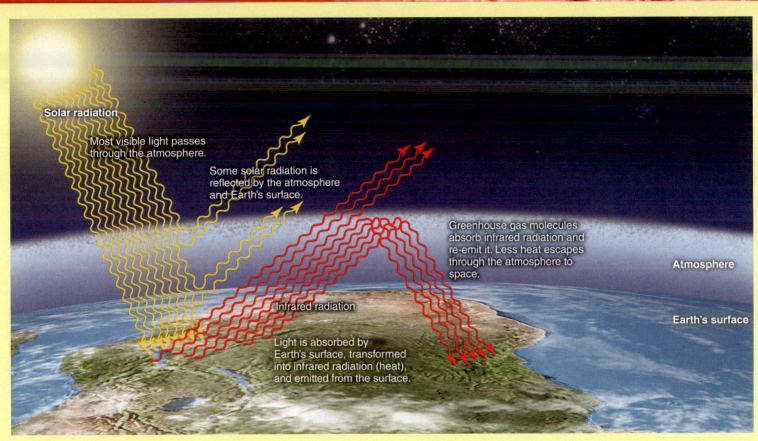

Figure B14.3. The greenhouse effect The absorption of infrared radiation by greenhouse gases prevents some of it from escaping to space, contributing to the warming of Earth's atmosphere and surface.

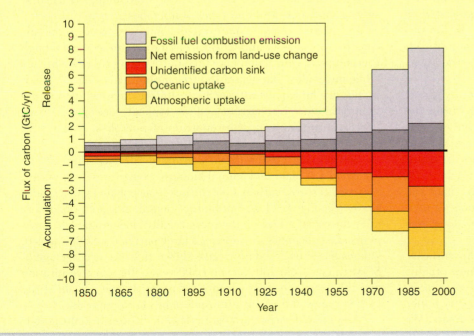

Figure B14.4. The global carbon budget Estimated global releases and global accumulations over the past 150 years balance only if an unidentified reservoir is included. The reservoir is most likely due to microbial photosynthesis and respiration, which are not completely understood.

Republished with permission of Annual Reviews, Inc., from *Annual Review of Earth and Planetary Sciences*, May 2007. R. A. Houghton, "Balancing the Global Carbon Budget," 35:313–347. Permission conveyed through Copyright Clearance Center, Inc.

Deep ocean water links to the upper waters and then to the other reservoirs, but the exchange of deep ocean carbon with the more rapidly cycling surface water carbon takes place over hundreds of years. Atmospheric CO_2 dissolves into marine waters. Here, it can be incorporated into corals, clams, protozoa, and other organisms in the form of calcium carbonate ($CaCO_3$). When these organisms die, the carbonate structures sink, contributing to the eventual formation of sedimentary carbonate rock such as limestone. Over millions of years, this rock may be drawn into Earth's crust and undergo metamorphosis into calcium silicate rock. This reaction releases CO_2, which can find its way back to the atmosphere via volcanic release. Photosynthesis by land plants also captures atmospheric CO_2. In this case, the inorganic carbon becomes incorporated into new plant biomass (see Figure 14.6).

As we can see from this introduction to global carbon cycling, carbon moves between various reservoirs via a series of interconnected processes. Big picture representations such as this one are useful when focusing on geological or environmental perspectives. We can also focus on the relatively short-term cycling of carbon associated with various microbial activities. Elemental processing by microbes has had a profound effect on Earth over geological time spans. As we will see in the next section, though, many interesting biological questions involve the relatively rapid transfers of carbon between compartments.

Photosynthesis, respiration, and decomposition

A brief overview of the global carbon cycle illustrates the balance between incorporation of carbon from the atmosphere into new plant biomass during photosynthesis, and the release of carbon into the atmosphere through the decomposition of this biomass. The metabolic activities of organisms are paramount in this short-term carbon cycling. The bulk of the action is composed of the complementary activities of photosynthesis and respiration, activities detailed in Chapter 13. Photoautotrophic organisms use light-derived energy to fix inorganic carbon into organic matter, the process of photosynthesis. In the terrestrial compartment, plants dominate this conversion of CO_2 to [CH_2O], although microorganisms such as terrestrial algae, cyanobacteria, and lichens also contribute.

In aquatic environments, microbes dominate photosynthetic activity. Phytoplankton are the primary producers within the water column. Only in near-shore habitats where the water is shallow enough to enable them to root can macroscopic aquatic plants thrive. Within these aquatic environments, oxygenic phototrophs such as plants, eukaryal algae, and cyanobacteria dominate photosynthesis. Anoxygenic phototrophic bacteria, conversely, predominate in specialized habitats such as non-mixing lakes or other places where light is available along with H_2S or S as electron donors.

In addition to primary production via photosynthesis, chemolithotrophy contributes to the formation of new organic matter by fixing CO_2 at the expense of inorganic energy sources. This form of carbon fixation, however, represents a very minor component of primary production in most environments. We should note, though, that some specialized communities on Earth thrive in the absence of sunlight by relying on primary production by chemolithotrophic microbes.

CONNECTIONS In **Section 15.5**, we will explore ecosystems, such as deep sea hydrothermal vents, that are driven by non-solar forms of energy. In these ecosystems, the oxidation of inorganic molecules such as H_2S, drives the generation of energy.

As we discussed in Chapter 6, respiration involves the release of CO_2 and water from the oxidative metabolism of organic compounds, resulting in energy generation. When the total production of organic material and community respiration are in complete balance, no net gain or loss of organic matter occurs in the community. When the organic matter produced via photosynthesis exceeds the amount of organic carbon converted back to CO_2, however, then positive **net community productivity** occurs. In other words, organic matter accumulates. In contrast, when the amount of organic carbon converted back to CO_2 exceeds the amount of organic material produced, then negative net community productivity occurs. This situation could occur if an external source of organic matter enters a community. For example, leaves falling into a stream could supply extra organic carbon to heterotrophic stream microbes.

An important component of the photosynthesis-respiration interaction shown in Figure 14.6 is the decomposition of organic material by chemoorganotrophic microbial activity. Like respiration by plants and animals, microbial respiration has the ultimate effect of converting organic carbon back to CO_2. This CO_2 represents the major natural source of CO_2 in the atmosphere and again becomes available to autotrophic organisms—the primary producers. In an idealized food web, decomposition of organic material by microbes occurs via aerobic respiration, with O_2 serving as the terminal electron acceptor. In the absence of O_2, anaerobic respiration may occur if suitable alternative electron acceptors such as sulfate, nitrate, or ferric iron are available. Fermentation (see Section 13.3), however, often occurs more frequently than anaerobic respiration in O_2-limited environments. Fermentative activities generate partially oxidized organic end products like acids and alcohols as well as CO_2. These compounds, in turn, may serve as nutrients for other types of microbes.

Eventually, organic material decomposes to its inorganic constituents of H_2O and CO_2 through decay processes, though this complete conversion may take quite a long time. Relatively labile molecules, like sugars and free amino acids, degrade readily, but insoluble polymers such as cellulose, hemicelluloses, and lignin are more recalcitrant and are typically susceptible to attack by a relatively small proportion of microbes. As a result, they degrade more slowly.

Soil organic matter (SOM), primarily derived through the breakdown of plant and animal biomass, constitutes a significant reservoir of stored carbon. Like the insoluble polymers, it degrades slowly (see Figure 14.6). Soil humus, organic material remaining after microbial attack on the more readily biodegradable components of dead biomass, comprises a large part of the SOM. Indeed, studies based on ^{14}C-dating have shown that the average age of SOM ranges from 150 to over 1,500 years. Together, respiration and the decay processes account for most, though not all, of the carbon removed from the atmosphere by photosynthesis.

As we noted earlier, atmospheric CO_2 diffuses into the aquatic reservoirs on Earth, where it can remain as dissolved CO_2, or be converted into bicarbonate (HCO_3^-) or carbonate (CO_3^{2-}). This source of inorganic carbon is fixed into organic material by

TABLE 14.2 Substrates used by methanogens

Acetate	CH_3COO^-
Carbon dioxide	CO_2
Dimethylamine	$(CH_3)_2NH$
Dimethylsulfide	$(CH_3)_2S$
Formate	$HCOO^-$
Methanol	CH_3OH
Methylamine	CH_3NH
Methylmercaptan	CH_3SH
Pyruvate	CH_3COCOO^-
Trimethylamine	$(CH_3)_3N$

Microbes in Focus 14.1
METHANOCOCCUS MARIPALUDIS: AN ARCHAEAL METHANE PRODUCER

Habitat: Commonly found in marine environments, often in salt marshes, where it produces methane. Also fixes N_2.

Description: Coccus-shaped archaeon with a cluster of flagella, known for its rapid growth. The 1.66-Mbp genome sequence, encoding 1,722 protein-encoding genes, was reported in 2004.

Key Features: This microbe can grow autotrophically on H_2, CO_2, and N_2. It is a popular experimental organism, with a well-developed genetic system, including methods for introduction of DNA by transformation, and construction of mutants.

TEM

Courtesy Shin-Ichi Aizawa and Ken F. Jarrell

the oceans' phytoplankton community (algae, cyanobacteria), and subsequently this organic carbon is subject to respiration and eventual decay by heterotrophic microbes. The marine biota, however, constitutes a much smaller carbon pool than terrestrial biota (see Figure 14.6). As a result, activities of the terrestrial microbes have a relatively greater influence on biological carbon cycling than aquatic microbes. The huge volumes of the world's oceans, however, mean that the aquatic compartments store vast amounts of dissolved inorganic carbon.

Methanogenesis

As we saw in the preceding section, CO_2 represents a major end product of organic matter breakdown. Another gaseous end product, however, can result from the biodegradation of organic carbon in anoxic environments—methane (CH_4). In a type of metabolism called **methanogenesis**, several archaeons, referred to as **methanogens**, produce CH_4 as an end product. The methanogen *Methanococcus maripaludis* is profiled in **Microbes in Focus 14.1**. The actions of these organisms are important to global carbon cycling. The gaseous CH_4 they produce in environments devoid of O_2 may diffuse into adjacent oxic habitats where it can be metabolized by other microbes. This movement of CH_4, then, prevents carbon from becoming irreversibly trapped in anoxic environments.

CONNECTIONS In some superficial respects, archaeons "look" like bacteria. As we saw in **Chapter 4**, though, archaeons differ significantly from bacteria and eukarya. For this reason, Archaea comprise one of the three domains of life.

The reactions below compare the conversion of 6-carbon sugars to end products under aerobic (reaction 1) and methanogenic conditions (reaction 2):

$$C_6H_{12}O_6 + 6\ O_2 \rightarrow 6\ CO_2 + 6\ H_2O;\ \Delta G^{o\prime} = -2{,}870\ kJ/mol\ (1)$$
$$C_6H_{12}O_6 \rightarrow 3\ CO_2 + 3\ CH_4;\ \Delta G^{o\prime} = -390\ kJ/mol\ \quad (2)$$

In both reactions, the Gibbs free energy values ($\Delta G^{o\prime}$) are negative, indicating energy-releasing reactions. The reaction under methanogenic conditions, however, releases only about 15 percent as much energy as does the aerobic biodegradation. This difference demonstrates that considerable energy remains stored within the CH_4 molecule. On a related note, it also means that relatively little biomass can be produced per molecule of CH_4 evolved. Generally, biomass yields under anaerobic conditions are small compared to the yields under aerobic conditions; many metabolic end products of anaerobes remain incompletely oxidized.

Now that we have some understanding of how methanogenesis occurs, let's ask another question. Where does this reaction typically occur? Methanogenesis generally occurs in habitats that contain plenty of organic matter but little or no oxygen. Specific locations include freshwater sediments, swamps and bogs, waterlogged soils including rice paddies, and in the anoxic components of sewage treatment systems such as anaerobic sludge digestors. While methanogenesis represents the end of the microbial food web in such O_2-restricted habitats, the methanogens rely on other organisms to process complex organic materials and provide them with substrates. Methanogens use only a restricted range of substrates (**Table 14.2**). Consequently, methanogenesis involves a **consortium**, or interacting community, of interdependent microorganisms. These organisms form **syntrophic partnerships** in which the metabolic activities of the organisms are mutually dependent since only in combination are they thermodynamically favorable. Probably, much of the biodegradation in aerobic environments is also community driven, but unlike anaerobes, many aerobic microbes possess metabolic pathways enabling them to completely mineralize carbon-containing compounds to CO_2 and H_2O.

The decomposition of organic compounds in a methanogenic environment occurs in three stages (**Figure 14.7**). First, organisms hydrolyze polymers such as cellulose and proteins. The resulting monomeric sugars and amino acids are then fermented to fatty acids, organic acids, and alcohols. This stage, sometimes termed "primary fermentation," is carried out by primary fermenter organisms. In the second stage, the fatty acids, organic acids, and alcohols are used as substrates for secondary fermentation, which may result in the formation of hydrogen, acetate, formate, and other related compounds. These reactions are carried out by secondary fermenters, or syntrophs. In the third stage, methanogens metabolize these products, producing CH_4 and CO_2. **Hydrogenotrophic methanogens** produce CH_4 from CO_2 and H_2, whereas **acetotrophic methanogens** produce

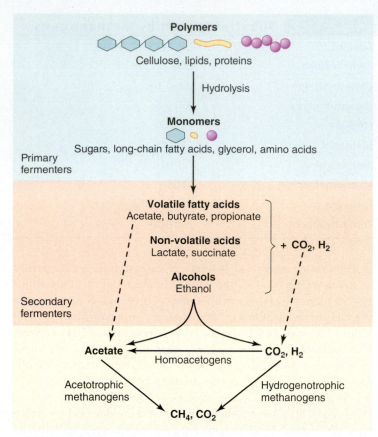

Figure 14.7. Decomposition of organic compounds in anoxic methanogenic environments This process requires different types of microbes working together in syntrophic partnerships. Primary fermenters produce fatty acids, organic acids, and alcohols that syntrophic bacteria further degrade to acetate, H_2, and CO_2, the substrates of methanogenesis.

CH_4 from acetate. The outcome of the overall, three-stage process is the conversion of polymeric carbon-containing organic materials to the simple end products CO_2, CH_4, and H_2O. The degradation of proteins, phospholipids, and nucleic acids leads to the generation of these products, as well as NH_3, HPO_4^{3-}, and SO_4^{2-}.

As we can see from this description and Figure 14.7, the microbial activities of the later stages depend on the occurrence of the previous stages. Each stage in the upper portion of Figure 14.7 also relies on the occurrence of the stage below. In particular, a tight syntrophic association exists between H_2-producing microbes and methanogens, involving a process known as "interspecies hydrogen transfer." In this type of relationship, H_2 produced by one organism is used by a hydrogenotrophic methanogen **(Figure 14.8)**. The H_2-producing microbes carry out oxidation reactions such as the following:

$$CH_3CH_2CH_2COO^- + 2\,H_2O \rightarrow 2\,CH_3COO^- + H^+ + 2\,H_2;$$
(butyrate) $\qquad\qquad\qquad\quad \Delta G^{o\prime} = +48.1\;kJ \quad (3)$

$$CH_3CH_2COO^- + 3\,H_2O + CH_3COO^- + HCO_3^- + H^+ + 3\,H_2;$$
(propionate) $\qquad\qquad\qquad \Delta G^{o\prime} = +76.1\;kJ \quad (4)$

$$CH_3CH_2OH + H_2O \rightarrow CH_3COO^- + H^+ + 2\,H_2;$$
(ethanol) $\qquad\qquad\qquad\quad \Delta G^{o\prime} = +9.6\;kJ \quad (5)$

Each of reactions (3) through (5), though, is thermodynamically unfavorable under standard conditions (solutes 1 M, gases 1 atm, 25°C) at pH 7.0, as indicated by the positive Gibbs free

energy value ($\Delta G^{o\prime}$). How, then, can these organisms gain energy by conducting such reactions? They do so by living in a close, mutualistic relationship with another microbe that utilizes H_2.

Let's examine the benefits of this partnership more closely. The H_2-utilizing microbes consume H_2 as soon as it is produced by the H_2-producing microbes, thereby keeping the partial pressure of H_2 in their immediate environment very low ($\sim10^{-4}$ atm or less). The low H_2 partial pressure pulls the above reactions toward H_2 production. In bioenergetic terms, the actual ΔG of the H_2-producing microbe's reaction under the environmental conditions in which the microorganisms are living is negative, or energy-yielding, and not positive. Reaction (1), for instance, would have an actual ΔG of about -17.6 kJ in a typical anoxic freshwater ecosystem with ~1 mM butyrate, ~0.6 atm CH_4, and $\sim10^{-4}$ atm H_2 present. In the community depicted by Figure 14.7, the methanogens and the homoacetogens play the role of H_2 utilizers. Without these H_2-utilizing groups at the end of the anaerobic microbial food chain, the earlier reactions could not occur. Thus, all the organisms in the process are interdependent.

Methanogenesis also occurs in the gastrointestinal tract of animals. Production of CH_4 by acetotrophic methanogens, however, is less important in this environment. In fact, these microbes grow too slowly to maintain themselves in the gut to any extent. Instead, acetate and the other incompletely degraded organic molecules, such as fatty acids, are absorbed by the animal, thus serving as energy sources. In Chapter 17, we will investigate the importance of these processes in ruminant animals.

CONNECTIONS Methanogenesis plays an important role in our gastrointestinal tract. As we saw in **Section 4.3**, the archaeon *Methanobrevibacter smithii* represents a major inhabitant of the human GI tract. This microbe produces CH_4, probably using H_2 and formate as substrates.

Methanotrophy

As we have seen in the previous section, methanogens can convert organic material to CH_4. Let's now ask a closely related and quite important question. What happens to this CH_4? Approximately half of the organic carbon degraded in the absence of O_2 ends up as CH_4, and organisms that rely on aerobic metabolism require the eventual return of this carbon to the aerobic side. The fate of CH_4 is also important to consider because of its high infrared absorbance. CH_4 is about 30 times more potent than CO_2 as a greenhouse gas.

Only a small fraction of biologically produced CH_4, though, ever reaches the atmosphere. Most of the CH_4 produced by

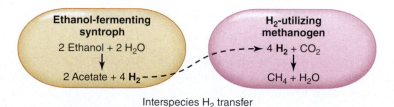

Figure 14.8. An example of a syntrophic partnership In a syntrophic partnership based on interspecies H_2 transfer, the consumption of H_2 by the methanogen keeps the concentration of H_2 low. This makes the oxidization of ethanol thermodynamically favorable. The physiological processes of the two organisms are interdependent.

methanogens is efficiently used by **methanotrophs**, bacteria that can metabolize CH_4. These microbes are a subgroup of the **methylotrophs**, members of the Proteobacteria that can oxidize organic compounds like CH_4, formate, and methanol that do not contain C—C bonds. Some methylotrophs cannot use CH_4, and thus are not methanotrophs, and some methylotrophs are also able to metabolize compounds that do contain C—C bonds.

Basically, the methanotrophs metabolize CH_4, converting the carbon in this molecule into forms that can be utilized by other organisms. The methanotrophs, then, allow the carbon present in CH_4 to be cycled more completely. In this section, we will examine two specific forms of methanotrophy—aerobic and anaerobic.

Aerobic Methanotrophy

Aerobic methanotrophs oxidize CH_4 to CO_2 as follows:

$$CH_4 \rightarrow CH_3OH \rightarrow CH_2O \rightarrow HCOO^- \rightarrow CO_2 \quad (6)$$

(methane $\rightarrow$ methanol $\rightarrow$ formaldehyde $\rightarrow$ formate $\rightarrow$ carbon dioxide)

These organisms do not fit neatly into the usual heterotrophic/autotrophic and organotrophic/lithotrophic classifications of carbon and energy metabolism. They have unique metabolic capabilities all their own. Carbon is assimilated from formaldehyde and CO_2 via either of two different pathways that contain certain enzymes unique to aerobic methanotrophs **(Figure 14.9)**. The serine pathway produces 2-carbon units as acetyl-CoA from formaldehyde plus CO_2 in a reaction sequence that requires 2 ATP and reducing power. The ribulose monophosphate pathway produces 3-carbon units (glyceraldehyde 3-phosphate) from 3 formaldehyde molecules, using 1 ATP per glyceraldehyde 3-phosphate molecule synthesized. The produced 2- or 3-carbon units feed into central metabolic pathways, enabling anabolic metabolism and the production of new biomass. Reducing power in the form of NADH reduces CO_2 to the oxidation level of cell biomass in the serine pathway. Formaldehyde, conversely, is already at the same oxidation

^a Sometimes considered Type X; *Methylococcus capsulatus* (Bath) is a well-studied model, and possesses Calvin cycle enzymes (incl RubisCO) as well.
^b Facultative; lack well-developed intracytoplasmic membrane system and pMMO; also relatively acidophilic with pH optimum of 5. Although Alphaproteobacteria, not closely related to *Methylocystis or Methylosinus*.

TABLE 14.3 Some characteristics of aerobic methanotrophs

Characteristic	Type I	Type II
Phylogenetic group	Gammaproteobacteria	Alphaproteobacteria
Internal membranes	Bundles of disc-shaped vesicles throughout cell	Paired membranes along cell periphery
C assimilation pathway	Ribulose monophosphate	Serine
TCA cycle	Incomplete (lack α-ketoglutarate dehydrogenase)	Complete
Resting stage	Cyst-like body, if present	Exospore
N₂ fixation	Typically cannot (*Methylococcus*^a does fix N₂)	Yes
Example genera (cell morphology)	*Methylomonas* (rod), *Methylomicrobium* (rod), *Methylobacter* (coccobacillus), *Methylococcus* (coccus)	*Methylocystis* (rod), *Methylosinus* (rod or vibrio), *Methylocella*^b (rod)

level as cell biomass. That, and its lesser ATP requirement, makes the ribulose monophosphate pathway the more efficient of these two routes for incorporating carbon.

The majority of known aerobic methanotrophs can be separated into two groups, designated Type I and Type II methanotrophs, based on physiological and structural characteristics of the cells, including carbon assimilation pathway used, the arrangement of internal membrane structures, and phylogeny **(Table 14.3)**. The Type I methanotrophs using the ribulose monophosphate pathway are Gammaproteobacteria.

Figure 14.9. Pathways for assimilation of carbon by aerobic methanotrophs Either of two pathways can be used to assimilate the formaldehyde and CO₂ generated by CH₄ oxidation. The serine pathway uses ATP and reducing power generated from the TCA cycle to produce the 2-carbon molecule acetyl-CoA, which is assimilated to biomass. The ribulose monophosphate pathway produces the 3-carbon molecule glyceraldehyde 3-phosphate from three formaldehyde molecules.

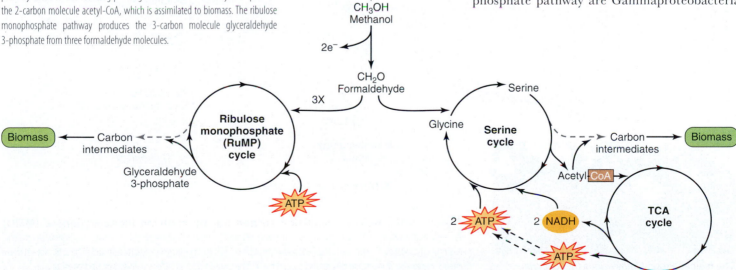

The Type II methanotrophs with the serine pathway, conversely, are Alphaproteobacteria. A Type II methanotroph, the well-studied *Methylosinus trichosporium*, is profiled in **Microbes in Focus 14.2**.

The ability of aerobic methanotrophs to oxidize methane depends on methane monooxygenase (MMO), the enzyme responsible for converting CH_4 to CH_3OH (methanol). Accordingly, non-methanotrophic methylotrophs lack the MMO enzyme. MMO is a very interesting enzyme, renowned for its ability to act on a broad range of substrates, including aromatic molecules and halogenated hydrocarbons. In a reaction requiring a reductant **(Figure 14.10)**, MMO adds an O atom derived from O_2 to CH_4 producing CH_3OH plus H_2O. In the later oxidative steps of the pathway, electrons are gained from oxidizing CH_2O to CO_2, and enter the electron transport chain, generating a proton motive force that permits ATP synthesis. The requirement for O_2 as a reactant in the MMO reaction is the reason these bacteria are obligate aerobes, and of course O_2 is the terminal electron acceptor in their energy metabolism as well. Methanotrophic cells possess extensive internal membrane structures, which are readily visualized by electron microscopy (see Figure 14.10B) and contain the site of MMO activity, the particulate MMO, or pMMO. Additionally, some methanotrophs produce soluble MMO, or sMMO, that is not membrane associated.

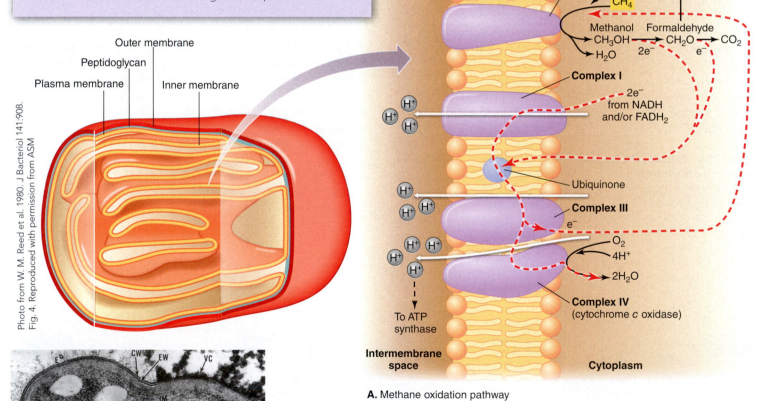

Outer membrane
Peptidoglycan
Plasma membrane
Inner membrane

Photo from W. M. Reed et al. 1980. J Bacteriol 141:908. Fig. 4. Reproduced with permission from ASM

A. Methane oxidation pathway

Carbon assimilation
MMO O_2
CH_4
Methanol   Formaldehyde
CH_3OH → CH_2O → CO_2
H_2O $2e^-$ e^-
Complex I
$2e^-$ from NADH and/or $FADH_2$
H^+
Ubiquinone
Complex III
e^-
O_2
$4H^+$
$2H_2O$
Complex IV (cytochrome *c* oxidase)
To ATP synthase
Intermembrane space
Cytoplasm

B. The methanotroph *Methylosinus trichosporium*

TEM

Figure 14.10. Methane oxidation carried out by methane monooxygenase (MMO) **A.** The reaction catalyzed by MMO, oxidizing CH_4 to methanol, requires O_2. This direct involvement of O_2 accounts for the obligately aerobic nature of CH_4 oxidation. Oxygen is also required for the generation of ATP via the electron transport chain and ATPase. **B.** The extensive internal membranes of methanotroph cells that support the MMO activity are clearly visible by transmission electron microscopy.

Methanotrophs exist in widely distributed areas. They have been isolated from muds, swamps, peat bogs, sediments, fresh and saline waters, rice paddies, sewage sludge, and soils. Not surprisingly, populations of methanotrophs exist in locales where CH_4 accumulates as a result of methanogenesis or from a geologic or human source. Examples of such sites include flooded soils, wetlands, soil overlays on landfills, and near hydrocarbon seeps or natural gas pipeline leaks. CH_4 oxidation by the methanotrophs reduces the amount of CH_4 that these sources ultimately release to the atmosphere. Often, the methanotrophs can be isolated simply by inoculating some mineral salts medium with a soil or water sample, and then adjusting the headspace of the closed flask to contain ~80 percent CH_4. Evidence also suggests that readily culturable aerobic methanotrophs may not be representative of the complete spectrum of natural methanotroph populations. Methanotroph distribution can be surveyed by probing for the *pmoA* gene using molecular techniques. This gene is recognized as a robust marker for this group of methanotrophs, encoding a conserved 24 kDa subunit of the membrane-associated pMMO.

Methanotrophs can optimize substrate availability by positioning themselves along chemical gradients at the interface between anoxic and oxic environments. At this location, they can obtain CH_4 from anaerobic methanogenic activity and O_2 from the atmosphere or aerated water. Methanotrophic populations in lakes with methane-emitting sediments tend to occupy a relatively narrow band of water where the CH_4 rising up through the water column meets O_2 diffusing downward from the lake surface.

Locations of natural hydrocarbon release represent another habitat favored by methanotrophs. Cold gas seeps occur in certain near-shore locales in the Atlantic and Pacific oceans, and deep sea hydrothermal vents situated in the vicinity of the mid-oceanic ridge can emit geologically derived methane. Some methanotrophs live independently within these habitats, while other species live in symbiotic relationships with various invertebrates, including mussels, sponges, and tube worms. Type I methanotrophs have been identified near cold gas seeps in the Gulf of Mexico where they live as intracellular endosymbionts within mussel gill tissue, taking up dissolved CH_4 and O_2 from seawater passing over the gills. The use of radiolabeled $^{14}CH_4$ to track the flow of carbon has demonstrated that methane-derived carbon is incorporated into mussel tissue. These symbioses therefore appear to be similar in concept to the S-oxidizing bacteria tube worm symbioses studied in the deep sea hydrothermal vent environment. In fact, several research groups have described the co-occurrence of methanotrophic and S-oxidizing bacteria as symbionts within the same deep sea mussel.

CONNECTIONS The relationship between giant tube worms and their bacterial symbionts is described in the opening story and **The Rest of the Story** in **Chapter 13**.

Until recently, methanotrophs were characterized as unique groupings of obligately aerobic, obligate users of reduced C_1

Figure 14.11. New Zealand hot spring Methane-rich geothermal springs such as this are prime locales for searching for novel methanotrophs.

substrates such as CH_4 that belonged to either the Alpha or Gamma subclasses of the Proteobacteria. However, recent research has revealed the existence of novel types of methanotrophic activity, and so we must revise our thinking. In 2007, the existence of acidophilic methanotrophic bacteria were independently reported by the research groups of Arjan Pol and Peter Dunfield. These bacteria were isolated from a solfatara volcano mudpot near Naples, Italy, and a geothermal, methane-rich area in Hell's Gate, New Zealand, and belong to the poorly characterized phylum Verrucomicrobia, not the Proteobacteria **(Figure 14.11)**. They can grow on CH_4 at pH of less than 1, far below the pH optimum of other known methanotrophs. Intriguingly, very similar environmental 16S rRNA gene sequences have been obtained from thermophilic habitats in Yellowstone Park, suggesting that these newly discovered methanotrophs may be widespread in extreme habitats.

Anaerobic Methanotrophy

While aerobic methanotrophy affects the processing of some methane, CH_4 can also be processed in the absence of oxygen. Over the past quarter century or so, microbiologists have realized that **anaerobic oxidation of methane (AOM)** occurs and is of great significance to the global carbon cycle. AOM particularly influences the fate of CH_4 evolved in near-shore marine sediments. In fact, nearly all of the CH_4 produced globally every year in anoxic marine sediments undergoes anaerobic oxidation. As a result, relatively little of this CH_4 reaches the atmosphere. The microbes responsible for AOM have yet to be isolated, but culture-independent molecular and biogeochemical tools have enabled us to learn much about how they accomplish AOM. Clearly, the oxidative reaction sequence catalyzed by the aerobic methanotrophs and involving MMO cannot be at work here; no O_2 is available. Another means of oxidizing the CH_4 molecule must be involved.

The organisms capable of AOM include archaeal **anaerobic methane oxidizers (ANME)**. Three phylogenetically distinct groups of these organisms have been recognized—ANME-1,

TABLE 14.4 Potential reactions performed by archaeons and bacteria involved in anaerobic oxidation of methane

Proposed reaction mechanisms for the CH₄-consuming archaeons

$$CH_4 + 2\,H_2O \rightarrow CO_2 + 4\,\textbf{H}_2 \tag{a}$$

$$CH_4 + 4\,HCO_3^- + 2\,H^+ \rightarrow CO_2 + 4\,\textbf{HCOOH (formate)} + 2\,OH^- \tag{b}$$

$$CH_4 + CO_2 \rightarrow \textbf{CH}_3\textbf{COOH (acetate)} \tag{c}$$

$$2\,CH_4 + 2\,H_2O \rightarrow \textbf{CH}_3\textbf{COOH} + 4\,\textbf{H}_2 \tag{d}$$

Associated reactions catalyzed by SRB (syntrophic partners)

$$SO_4^{2-} + 4\,\textbf{HCOOH} \rightarrow S^{2-} + 4\,CO_2 + 4\,H_2O \tag{e}$$

$$SO_4^{2-} + 4\,\textbf{H}_2 \rightarrow S^{2-} + 4\,H_2O \tag{f}$$

$$SO_4^{2-} + \textbf{CH}_3\textbf{COOH} \rightarrow 2\,HCO_3^- + H_2S \tag{g}$$

Source: Valentine, D. L. 2002. Biogeochemistry and microbial ecology of methane oxidation in anoxic environments: A review. Anton van Leeuw 81:271–282.

ANME-2, and ANME-3. Organisms in all three groups carry out reverse methanogenesis. Rather than producing CH_4, these archaeons consume CH_4 by essentially running the methanogenic reactions backwards. Exactly how they do so remains to be determined. Sequencing of SSU RNA genes has shown the ANME-2 archaeons to be closely related to the *Methanosarcinales*, a group of largely methylotrophic methanogens that includes all known species able to produce CH_4 using acetate as substrate.

In addition, the *mcrA* gene, which encodes a subunit of the methanogen-specific enzyme methyl coenzyme M reductase, has been identified in ANME-1 and ANME-2 enrichments. The ANME archaeons occur in close association with sulfate-reducing bacteria (SRB) like *Desulfobacterium autotrophicum*. Together, the microbes accomplish AOM in anoxic zones where areas of sulfate reduction and methanogenesis intersect. In these areas, CH_4 released from methanogenesis encounters available SO_4^{2-}. Contributions from the partners in this syntrophic association lead to the net chemical reaction associated with AOM:

$$CH_4 + SO_4^{2-} \rightarrow HCO_3^- + HS^- + H_2O \tag{7}$$

How the microbial association accomplishes this conversion has not been firmly established. Proposed mechanisms assume that the archaeal oxidation of CH_4 is coupled to the generation of reduced intermediates (H_2, formate, acetate) that are then oxidized by syntrophic SRB **(Table 14.4)**. The carbon of the CH_4 molecule eventually cycles back to the form of CO_2, which can be fixed into organic compounds by phototrophs, thus closing the carbon cycle.

The reactions in Table 14.4 indicate the potential syntrophic interspecies **hydrogen**, **formate**, or **acetate** transfer. Reactions (a) and (d) are thought to be the most viable mechanisms for CH_4 consumption. Reactions (e) through (g) are known metabolic reactions found in sulfate-reducing bacteria.

14.2 Fact Check

1. What is the global carbon cycle?
2. Distinguish between positive net community productivity and negative net community productivity.
3. What is methanogenesis, and in what environments does it occur?
4. How can methanotrophy occur with oxygen or without oxygen?

14.3 Cycling driven by nitrogen metabolism

How do microbes influence the cycling of nitrogen?

Before humans began to exert such influence on Earth, microbes almost entirely controlled the planet's nitrogen cycle. Microbial processes remain key to processing the nitrogen vital to all living organisms. Today, however, human actions related to fossil fuel utilization, profoundly affect the nitrogen cycle. As shown in **Figure 14.12**, Earth's atmosphere comprises the largest reservoir of nitrogen, mostly in the form of highly inert dinitrogen gas, N_2, which makes up about 79 percent of atmospheric gas. Living biomass accounts for about 1-millionth of the nitrogen found in the atmosphere, mostly in terrestrial plant material,

but also in land animals and marine plants and animals. The soil reservoir is of roughly the same magnitude as the biomass reservoir.

Within these various locations, major forms of nitrogen include organic nitrogen within biomass, and inorganic nitrogen such as NH_4^+ (ammonium), N_2 (dinitrogen), NO_3^- (nitrate), NO_2^- (nitrite), and N_2O (nitrous oxide). As we saw with carbon in the preceding section, these molecules contain nitrogen in different oxidation states **(Table 14.5)**. These molecules, additionally, may or may not be metabolically available to cells.

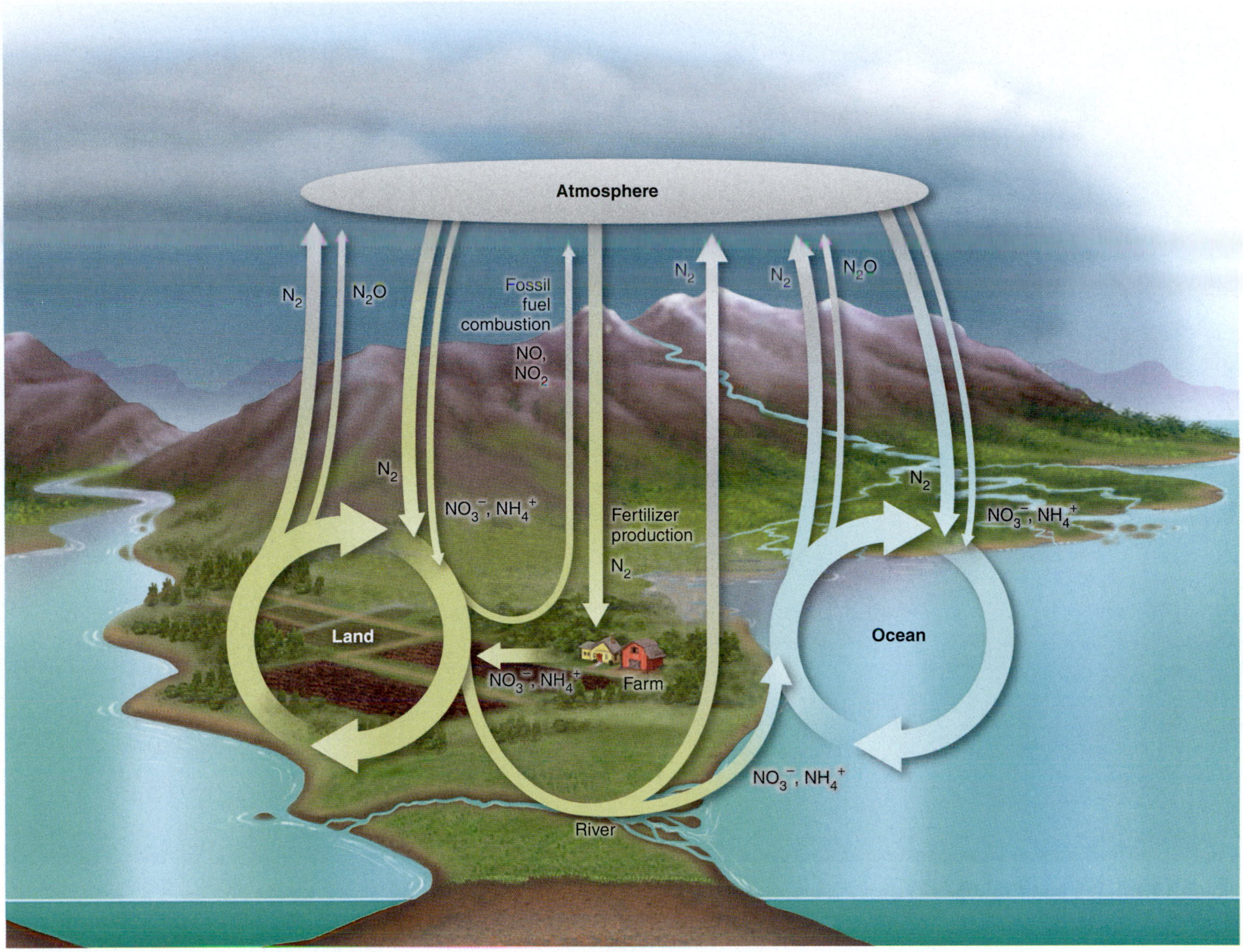

Figure 14.12. The nitrogen cycle Only a very small amount of the N on Earth is biologically available. This diagram demonstrates how the N cycles on land and in the oceans are connected by rivers and atmospheric deposition. Anthropogenic input is mostly from burning of fossil fuels and industrial N_2 fixation for fertilizer production.

TABLE 14.5	Oxidation states of nitrogen
Oxidation state	**Typical compounds**
+5	HNO_3, NO_3^-
+4	NO_2, N_2O_4
+3	HNO_2, NO_2^-, NO^+, NCl_3
+2	NO
+1	N_2O, $H_2N_2O_2$, $N_2O_2^{2-}$, $NHCl_2$
0	N_2
$-\frac{1}{3}$	HN_3, N_3^-
−1	NH_3OH^+, NH_2OH, NH_2Cl
−2	$N_2H_5^+$, N_2H_4
−3	NH_4^+, NH_3

The nitrogen cycle consists of several key steps (**Figure 14.13**), each carried out by specific organisms with characteristic metabolic qualities. In **nitrogen fixation**, atmospheric N_2 is reduced to NH_4^+. Some of the NH_4^+ becomes incorporated into biomass, while some of it is transformed in a process called **nitrification** into nitrites (NO_2^-) and then nitrates (NO_3^-). Nitrates can be directly assimilated by plants and a number of microbes, thus becoming incorporated into biomass, or they can be converted back to N_2 in a process called **denitrification**. In this section, we will look more closely at the processes of nitrogen fixation, nitrification, and denitrification. We will also examine the effects of human activities on nitrogen cycling.

Nitrogen fixation

As we noted previously, N_2 comprises nearly 80 percent of the atmosphere. Yet the vast majority of living organisms cannot use N_2 directly in their biochemical reactions. Rather, they depend

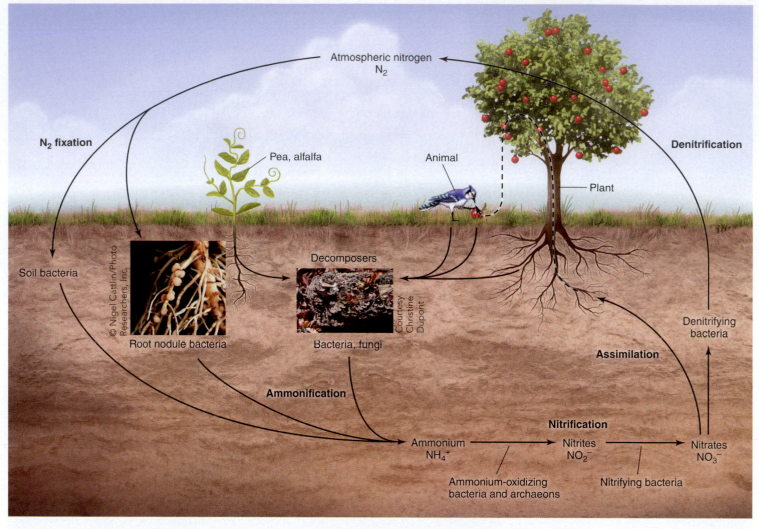

Figure 14.13. The biotic nitrogen cycle The major steps in the nitrogen cycle controlled by living organisms are depicted, including N_2 fixation, ammonification, nitrification, and denitrification. Each of these steps is carried out by organisms with characteristic metabolic properties.

on a select group of microbes that can convert N_2 into a more biologically useful form, through the actions of an enzyme complex called the **nitrogenase complex**. Currently, all known nitrogen-fixing organisms are either bacteria or archaeons, and the nitrogen-fixation genes are highly conserved (see Section 13.7). Despite the conservation of the nitrogen-fixation system itself, the nitrogen fixers exhibit incredible diversity, both physiologically and phylogenetically. They may be free-living or they may live in symbiotic relationships with plants, most notably legumes (see Section 17.2). Some of them are aerobes, while others are anaerobes. They may be phototrophs or chemotrophs, either organotrophic or chemolithotrophic. They may be terrestrial or aquatic. In all cases, though, O_2 irreversibly inactivates the nitrogenase enzymes, and the N_2-fixing microbes have evolved a number of strategies to deal with this limitation. As we discussed in Chapter 13, these strategies include the limitation of nitrogen fixation to anaerobic conditions, the formation of specialized cells such as heterocysts in cyanobacteria that physically separate oxygenic photosynthesis from the nitrogen-fixation reaction, and the formation of plant root nodules that protect the bacteria from the harmful effects of O_2.

CONNECTIONS In **Section 17.2**, we will take a look at the use of symbiotic N_2-fixation bacteria with legume crop plants as a major part of the agricultural inoculant industry.

Nitrification and denitrification

Following the conversion of N_2 to NH_4^+ by nitrogen fixers, nitrification can occur. In this two-step process, NH_4^+ (ammonium) is oxidized to NO_2^- (nitrite), which is then oxidized to NO_3^- (nitrate).

Ammonium oxidation: $NH_4 + 2\,H_2O \rightarrow NO_2^- + 8\,H^+ + 6\,e^-$ (8)

Nitrite oxidation: $NO_2^- + 2\,H_2O \rightarrow NO_3^- + 2\,H^+ + 2\,e^-$ (9)

Both steps are chemoautotrophic processes, carried out by different sets of microbes. Ammonium oxidizers, such as members of the Betaproteobacterial *Nitrosomonas* and Gammaproteobacterial *Nitrosococcus* genera, convert ammonium to nitrite (8). Two enzymes are involved in this process. First, ammonia monooxygenase, AMO, a 3-subunit enzyme evolutionarily related to pMMO, catalyzes the oxidation of ammonium to hydroxylamine, NH_2OH. Hydroxylamine is then oxidized to NO_2^- by the

Microbes in Focus 14.3
NITROBACTER WINOGRADSKYI

Habitat: Originally isolated by the great pioneering microbial ecologist Sergei Winogradsky in 1892. Widely distributed in the environment due to its diverse metabolic capabilities.

Description: Alphaproteobacterial microbe. Short, rod-shaped cells. Divides by asymmetric polar budding. Contains intracellular PHB granules and microcompartments called carboxysomes where CO_2 fixation takes place.

Key Features: *N. winogradskyi* can grow chemolithoautotrophically, obtaining energy for CO_2 reduction by oxidizing nitrite to nitrate, using the enzyme nitrite reductase located on the extensive intracellular membranes. It can also grow as a chemoorganoheterotroph, and anaerobically use nitrate as terminal electron acceptor. Researchers are interested in this organism because of its key role in the nitrogen cycle, and its potential for removing harmful nitrogen compounds from aquatic habitats and the generation of the potent greenhouse gases, nitric oxide, NO, and nitrous oxide, N_2O. The 3.4-Mbp genome encodes 3,143 predicted proteins.

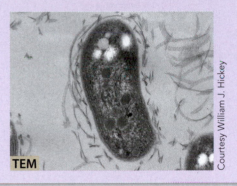

TEM

Courtesy William J. Hickey

heme-containing enzyme hydroxylamine oxidoreductase. As we will see in the **Mini-Paper**, some archaeons may also serve as ammonium oxidizers. The role of these organisms in nitrogen cycling remains to be determined.

Nitrifiers, such as the Alphaproteobacterial *Nitrobacter winogradskyi* (**Microbes in Focus 14.3**), carry out nitrite oxidation, the second step of nitrification. The key enzyme in this process is nitrite oxidoreductase, NOR, located on the inner face of the cytoplasmic membrane. Nitrifiers can be found in many different aerobic environments that contain ammonium. These environments include agricultural fields to which ammonium fertilizer has been applied, wastewater treatment systems, and aquariums. Nitrifiers are particularly important in wastewater treatment systems for nitrogen removal. In these systems, the produced nitrate is processed by denitrifiers, resulting in N_2 that returns to the atmosphere.

The process of denitrification completes the nitrogen cycle, but causes nitrogen to be essentially lost from the habitat. The ability to carry out denitrification is widespread within the Bacteria, and, as we saw in Chapter 13, is usually a heterotrophic

process in which an organic substrate, under anoxic conditions, is oxidized with nitrate as a terminal electron acceptor. Through a series of reduction steps, carried out by reductase enzymes, N_2 is eventually formed:

$$NO_3^- \rightarrow NO_2^- \rightarrow NO \rightarrow N_2O \rightarrow N_2 \qquad (10)$$

The key enzymes in this process are the corresponding reductase enzymes for each step, that is, nitrate reductase, nitrite reductase, nitric oxide reductase, and nitrous oxide reductase.

While denitrification that removes nitrogen from wastewaters and surface waters is considered beneficial since it will reduce the accumulation of oxygen-depleting microbial biomass in these waters, high denitrification activities in agricultural fields can reduce the effectiveness of nitrogen fertilizer applications. Denitrification can be minimized by ensuring soil aeration, thus preventing anoxic conditions. Another approach to limiting denitrification is to use nitrapyrin [2-chloro-6-(trichlormethyl) pyridine], an inhibitor of ammonium oxidation. Application of this chemical to soils can prevent the production of substrate for denitrification, resulting in more of the applied nitrogen remaining in the soil and available to the plants.

CONNECTIONS In **Section 16.5**, we will examine the use of nitrification and denitrification for the removal of nitrogen during wastewater treatment.

Human impact on the nitrogen cycle

Has human activity affected the natural balance of the nitrogen cycle? Just as with the carbon cycle, very large changes in the nitrogen cycle have occurred in a very short time, primarily due to human activities such as burning of fossil fuels and the use of synthetic fertilizers. One of the major advances in modern agriculture that has enabled food production to keep pace with human population growth has been the development of the **Haber–Bosch industrial process** for the production of synthetic nitrogen fertilizer (**Figure 14.14**). This process relies on a series of chemical reactions that extract H_2 from CH_4 so that it can be combined with N_2 under high temperature and pressure to form NH_3.

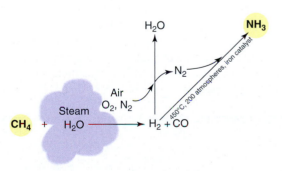

Figure 14.14. Industrial synthetic nitrogen fertilizer production The production of synthetic nitrogen fertilizer was a major stimulus for advances in agricultural productivity in the past half-century. The Haber–Bosch process uses CH_4 to generate hydrogen, which is then reacted with N_2 and an iron catalyst at high temperature under high pressure, yielding ammonia.

M. Könneke, A. E. Bernhard, J. R. de la Torre, C. B. Walker, J. B. Waterbury, and D. A. Stahl. 2005. Isolation of an autotrophic ammonia-oxidizing marine archaeon. Nature 437:543–546.

Context

In the opening story to Chapter 15, we will revisit the remarkable discovery, in the early 1990s, of the marine Crenarchaeota. Despite the fact that these organisms are now recognized as abundant members of the microbiota of cold ocean waters, accounting for up to 40 percent of the microbial community in deep ocean waters, and are widely distributed throughout the diverse habitats on Earth, the non-thermophilic Crenarchaeota had never before been isolated in pure culture. Through SSU sequence surveys, microbiologists have estimated that these organisms make up approximately 20 percent of the microbiota of Earth's oceans. Now that we know that these organisms exist in the oceans, numerous questions arise. What are the physiological characteristics of these organisms? How do they affect the biogeochemical cycles? Are they autotrophic or heterotrophic?

Because SSU sequences of Crenarchaeota can commonly be found in nitrifying environments such as estuary sediments and aquariums, researchers hypothesized that they are involved in nitrification. Metagenomic studies revealed the presence of sequences for *amoA* genes encoding one subunit of the key nitrification enzyme ammonia monooxygenase in uncultivated Crenarchaeota DNA. Could the marine Crenarchaeota be driving nitrification in the oceans by carrying out the first step, ammonia oxidation? To address this question, and to properly study the physiology of these organisms, researchers needed pure cultures for examination and experimentation. In this paper, members of David Stahl's group at the University of Washington in Seattle describe a successful strategy to cultivate marine Crenarchaeota.

Experiment

Starting with the premise that the organisms were capable of ammonia oxidation, investigators developed enrichment growth media by taking Seattle Aquarium water, filtering it to remove existing cells, and supplementing it with 1 mM NH_4Cl. This medium was then inoculated with gravel from a tropical marine tank in which Crenarchaeotal sequences had previously been detected. The resulting cultures were incubated in the dark, and subcultures were repeatedly made into fresh growth media. After six months, examination of SSU sequences indicated that the resulting enrichment

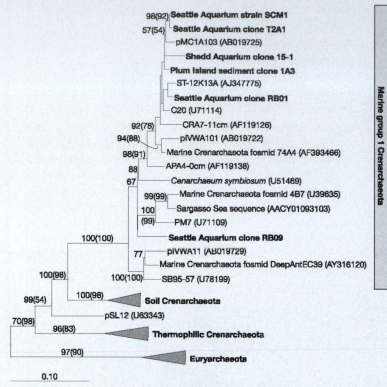

From M. Könneke et al. 2005. Nature 437:543–546, Fig. 1. Reprinted by permission from Macmillan Publishers, Ltd.

Figure B14.5. Phylogenetic analysis of strain SCM1 The sequence comparison demonstrated that SCM1 is a member of the marine group I Crenarchaeota.

cultures contained approximately 90 percent Crenarchaeota, and they were able to oxidize ammonia to nitrite. These cultures were then used to inoculate defined medium that contained bicarbonate and ammonia as sole carbon and energy sources. Through a series of dilutions, a pure culture was obtained, and was designated SCM1. The purity of this culture was supported by the lack of detectable bacterial SSU sequences, and the inability to promote heterotrophic growth by adding yeast extract and peptone. PCR amplification of the SSU sequences followed by sequence analysis confirmed the clonal nature of the population. Comparison to other known Crenarchaeota sequences **(Figure B14.5)** indicated high similarity with other marine Crenarchaeota, less similarity with the soil Crenarchaeota, and even less with thermophilic Crenarchaeota.

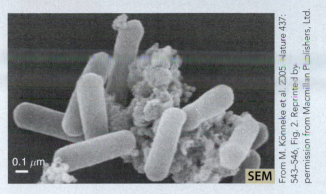

Figure B14.6. Visualization of SCM1 cells This SEM image of a cluster of cells shows them to be straight rods without visible flagella.

From M. Könneke et al. 2005. Nature 437: 543–546, Fig. 2. Reprinted by permission from Macmillan Publishers, Ltd.

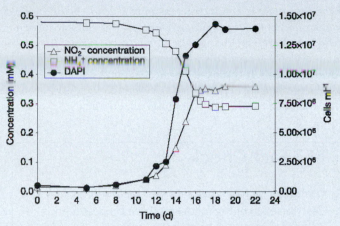

From M. Könneke et al. 2005. Nature 437:543–546, Fig. 3. Reprinted by permission from Macmillan Publishers, Ltd.

Figure B14.7. Conversion of ammonia to nitrite by SCM1 The culture was grown with ammonium chloride as sole energy source and bicarbonate as sole carbon source. Growth was followed by microscopic detection of DAPI-stained cells. Consumption of ammonia was balanced by nitrate production.

Electron microscopic examination indicated that the cells were uniformly straight rods, 0.17–0.22 μm in width, 0.5–0.9 μm in length **(Figure B14.6)**. The maximum density obtained by the cultures was 1.4×10^7 cells per mL, with a minimum generation time of 21 h, and the addition of organic compounds apparently inhibited growth. The growth of the cultures was accompanied by reduction in NH_4^+ concentration and increase in NO_2^- concentration, indicating conversion of NH_4^+ to NO_2^- **(Figure B14.7)**. PCR was used to amplify AMO-encoding genes, and subsequent sequence analysis confirmed that the genes were very similar to other Crenarchaeotal sequences, and of low similarity to the bacterial AMO-encoding genes.

Impact

Strain SCM1 is the first known chemolithoautotrophic nitrifier in the domain Archaea, and the first mesophilic Crenarchaeotal isolate. Prior to this work, known nitrifying organisms belonged to the Betaproteobacteria and Gammaproteobacteria. The significance of the NH_4^+-dependent chemolithoautotrophic growth is that a large amount of primary production in the oceans is carried out by ammonia oxidation in oligotrophic environments without sunlight or organic sources of energy, and that ammonia oxidation maintains the concentration of ammonia at low levels. These organisms are therefore key players in the cycling of both carbon and nitrogen. The ability to work with pure cultures opens the door to detailed genetic, genomic, biochemical and physiological studies of these environmentally important organisms.

● Questions for Discussion ·················

1. After reading this paper, what do you think are the main characteristics of the marine Crenarchaeota that have contributed to their difficulty in cultivation?

2. The isolation of SCM1 in pure culture confirms that at least some of the non-thermophilic Crenarchaeota are chemolithoautotrophic organisms. Do you think that they are all chemolithoautotrophs, or might some of them be other metabolic types?

3. Since SCM1 was isolated from an aquarium rather than from a natural environment, are you concerned that it might not be representative of the abundant oceanic Crenarchaeota? How did the authors of the paper address this concern?

Incredibly, this fertilizer production now accounts for a major portion of all biologically available nitrogen (**Figure 14.15**). Burning of fossil fuels also affects the nitrogen cycle; it results in formation of large amounts of N_2O, or nitrous oxide, a potent greenhouse gas. As we will see in Chapter 15, where we discuss the aquatic dead zones that are becoming increasingly widespread on a global basis, the input of large amounts of biologically available nitrogen into the biosphere can change the nutrient balance with drastic consequences for ecosystems whose productivity we depend upon to provide our food. It is the biochemical activities of the microbes that keep the biogeochemical cycles running efficiently, even in the face of massive influence from human activities.

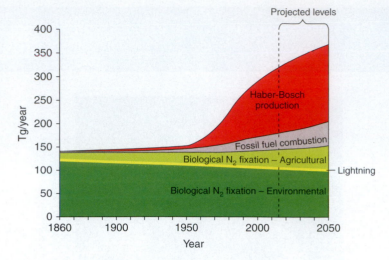

Nielsen. R. W. 2005. *Can we feed the world?*

Figure 14.15. Impact of synthetic nitrogen production Comparison of the sources of fixed nitrogen indicates that production of ammonia by the Haber–Bosch process is responsible for by far the largest input of biologically available nitrogen worldwide. It is increasing at a tremendous rate, challenging the total amounts of nitrogen fixation by all biological processes. The production of the greenhouse gas N_2O through the burning of fossil fuels is also increasing rapidly.

14.3 Fact Check

1. In what major forms can nitrogen be found on Earth?
2. Describe each of the following three processes and explain how they are part of the nitrogen cycle: nitrogen fixation, nitrification, denitrification.
3. Explain the steps of the ammonium oxidation process.
4. What is a denitrifier, and what is its role?
5. What is the Haber–Bosch process, and how is it used?

14.4 The interconnectedness of cycles

How are various biogeochemical cycles connected?

Although the previous two sections have dealt primarily with carbon and nitrogen cycling, you probably have noticed that the cycling of these elements does not occur in isolation. The presence or absence of molecular oxygen (O_2), for instance, has a marked influence on the fate of carbon atoms. Indeed, as we discussed carbon cycling, we also considered aspects of the oxygen cycle. The two cycles are intimately connected. In this section, we will explore the interconnectedness of cycles more fully. First, we will look at the oxygen cycle. Then we will examine the cycling of sulfur and phosphorus. As we examine the cycling of these elements, we will focus most on the interrelatedness of these cycles.

Cycling of oxygen

Actively cycled reservoirs of oxygen include atmospheric and dissolved O_2, which is produced via photosynthesis, and is the terminal electron acceptor used in aerobic respiration. These reservoirs also include CO_2 produced via respiration and decomposition of organic matter, and fixed by autotrophs. H_2O, an end product of many metabolic activities, including respiration, some methanogenic reactions, and aerobic and anaerobic CH_4 oxidation, represents another major source of oxygen. Living and dead biomass constitute yet another reservoir of actively cycled oxygen atoms.

Oxygen is even involved in anaerobic respiration. Sulfate-reducing bacteria, for instance, use SO_4^{2-}, which contains oxygen, as the terminal electron acceptor in their electron transport chain. This process releases oxygen as CO_2, as well as the rotten-egg-smelling hydrogen sulfide (H_2S) gas, which dissociates in water to the weak acid HS^-. Sulfate minerals are plentiful on Earth, so the oxygen in sulfate cycles through the biosphere relatively slowly. The utilization of NO_3^- as a terminal electron acceptor results in the production of nitrite (NO_2^-). Compared to sulfate, nitrate is a small and rapidly cycled oxygen reservoir. Connections can be made between the various cycles: carbon and oxygen, carbon and nitrogen, oxygen and nitrogen, and sulfur and oxygen, to name just a few. Sometimes, human activity also plays a role in the interactions of these cycles as seen in **Perspective 14.2**. These interconnections should be expected. They occur in a shared biosphere, and involve organisms with extensive histories of coevolution.

Cycling of sulfur and phosphorus

While carbon, nitrogen, and oxygen represent three abundant elements in biomass, sulfur and phosphorus also represent crucial elements in all cells. Not surprisingly, some microbes actively acquire these elements from inorganic sources, while other microbes aid in the return of these elements from organic

A natural phenomenon that occurs worldwide is the oxidation of the abundant mineral pyrite, FeS_2, yielding sulfuric acid. This process occurs following exposure of pyrite to oxygen and water. As you can imagine, this reaction causes a drastic reduction in the pH of the water, which becomes stained orange as a result of the iron hydroxide that forms. What can live in these acidic aqueous habitats? Microbes can, of course! They actually contribute to the acidification process. In the absence of sources of fixed carbon, acidophilic autotrophic iron-oxidizing microbes such as the bacteria *Thiobacillus* and *Leptospirillum*, and the archaeon *Ferroplasma*, use Fe^{2+} as the electron donor in their quest to obtain energy. The resulting Fe^{3+} contributes to the further oxidation of pyrite, releasing even greater amounts of sulfuric acid. Because these microbes are acid loving, they thrive in this environment, often forming thick biofilms.

Pyrite oxidation and the resulting acidification have been exacerbated by mining activities that expose previously buried pyrite to water and oxygen. Referred to as "acid mine drainage," or "AMD," the acidic water flowing from these mines also results in the leaching of heavy metals such as zinc, copper, cadmium, and arsenic from the surrounding rocks. The combined toxicity of the acidification and increased heavy metal concentrations can be severe. Contamination of adjacent waterways can decimate non-microbial life, and also poses threats to sources of drinking water. AMD can remain long after the mines are abandoned, a nasty legacy of human civilization's extraction of metals from the ground.

The microbiology of AMD from one abandoned mine, the Iron Mountain Mine near Redding, California, has been studied intensively. Iron Mountain was active from the 1860s until the early 1960s, being mined for iron, silver, gold, copper, and zinc. A half century after the mine closed, AMD still threatens the sources of drinking water for the surrounding human populations. The pH of some samples has been recorded to be an astounding −3.6! In the dark depths of the mine, pink biofilms growing in green pools of highly acidic water were studied in one of the first applications of microbial community sequencing **(Figure B14.8)**. This study was carried out by Gene Tyson, a graduate student in Jillian Banfield's lab at the University of California, Berkeley, and involved a number of collaborators.

By applying the random shotgun method (see Chapter 10) to samples collected in 2000, Tyson and colleagues demonstrated that the community was relatively simple, consisting mostly of five types of known iron-oxidizing microbes, *Leptospirillum ferriphilum* (Group II), *Leptospirillum* Group III, *Sulfobacillus*, *Ferroplasma* Type I, *Ferroplasma* Type II, and traces of eukaryal sequences. The dominant microbe was *Leptospirillum ferriphilum* (Group II). They even regenerated near complete genome sequences for two of the organisms, *Leptospirillum ferriphilum* (Group II) and *Ferroplasma* Type II, and mapped out their metabolic processes based on those sequences. It was confirmed that both organisms had the genetic capacity for carbon fixation, but not for nitrogen fixation.

Nitrogen fixation is essential to AMD biofilm communities; no source of fixed nitrogen exists in these acidic habitats. Neither *Leptospirillum ferriphilum* (Group II) nor *Ferroplasma* Type II have genes for N_2 fixation, but N_2-fixation genes were found in sequences associated with *Leptospirillum* Group III. This observation led to further studies in which *Leptospirillum* Group III was successfully isolated in pure culture based on knowledge of its likely metabolic characteristics. It is likely to be an essential member of these biofilms, providing the other components of the biofilm with essential nitrogen nutrients.

AMD sites are of serious environmental concern. Understanding of the microbial-controlled biogeochemistry will be essential to cleanup of these toxic sites, and also prevention of their formation in the first place. They are also fascinating examples of the connections between the biogeochemical cycles.

Courtesy Brett Baker, University of California, Berkeley

Figure B14.8. Collection of biofilm samples Metagenomic studies found five types of iron-oxidizing microbes at Iron Mountain Mine, and one group that could fix nitrogen.

stores to the environment. As we saw for carbon, nitrogen, and oxygen, microbes play an active role in the cycling of sulfur and phosphorus. As we have also seen while examining the oxygen cycle, these cycles are interconnected in many ways.

Unlike carbon, nitrogen, and oxygen, very little phosphorus or sulfur exists in the atmosphere. Instead, the major reserves of these elements exist in rock or are dissolved in the oceans. Weathering of rocks can release phosphate ions (PO_4^{3-}, HPO_4^{2-}, $H_2PO_4^{1-}$, or H_3PO_4, depending on the pH of the environment) from phosphorus-containing minerals like apatite. In the absence of biological activity, these phosphate ions will be transported to the oceans, eventually settle to the ocean floor, and become part of newly formed rock, thereby completing the cycle. This entire process—from mineralized phosphorus to soluble phosphate to mineralized phosphorus—takes millions of years.

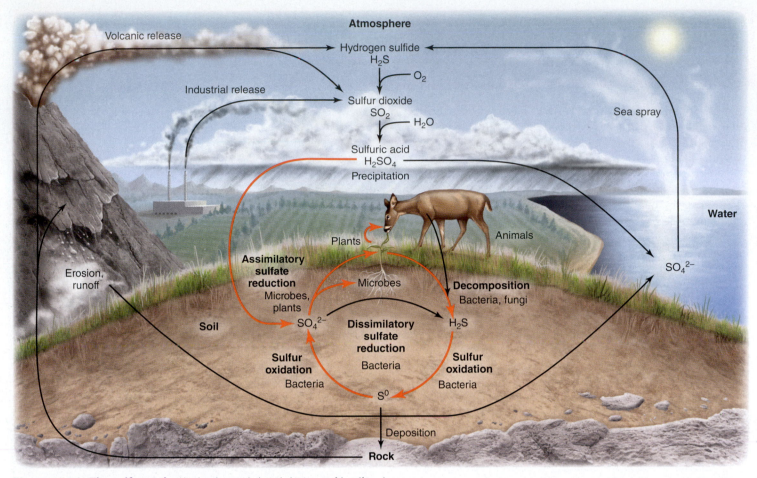

Figure 14.16. The sulfur cycle Microbes play central roles in the biotic parts of the sulfur cycle.

Along the journey, phosphorus can cycle between living organisms over and over again. As we have seen in earlier chapters, P is an essential element for cells, existing in DNA, RNA, ATP, and a host of other molecules. How do cells acquire this essential element? Phosphate ions released from rocks into the soil and water can be taken up by plants. Once incorporated into organic molecules, phosphorus can be transferred between organisms by processes of ingestion and decay.

The normal concentration of usable phosphate in the soil, however, is typically quite low. To counteract this paucity of usable phosphate, farmers often apply large amounts of phosphate-containing fertilizers to their fields, thereby increasing the growth of their crops. Evidence suggests, though, that bacteria may be able to aid in this process. Several genera of bacteria, including *Pseudomonas*, *Bacillus*, and *Rhizobium*, have the ability to solubilize inorganic phosphate compounds. Moreover, these same bacteria have been shown to improve the growth of plants. These bacteria, then, may represent a type of natural fertilizer. Their use could decrease our need to produce usable phosphate from natural stores.

In some ways, the sulfur cycle resembles the phosphorus cycle. Very little sulfur exists in the atmosphere. In the absence of biological cycling, sulfur leaves rocks via weathering and travels to the ocean, where it exists primarily as SO_4^{2-}. Volcanoes and other geothermal features release other forms of sulfur, most

notably sulfur dioxide (SO_2) and hydrogen sulfide (H_2S) into the atmosphere **(Figure 14.16)**. These compounds quickly react with oxygen and water and return to Earth through precipitation.

Plants and microbes can assimilate some of these sulfur-containing products. Additionally, as we have seen in this chapter, microbes can utilize molecules like H_2S directly. Anoxygenic photosynthetic bacteria, for instance, use this molecule as a source of electrons, releasing elemental sulfur (S^0). The geochemical cycling of sulfur interacts with the biochemical cycling of sulfur. Moreover, the biogeochemical cycling of sulfur is interconnected with the biogeochemical cycling of other elements essential to life on Earth.

14.4 Fact Check

1. How is the cycling of oxygen connected to the cycling of other elements?
2. What is acid mine drainage?
3. In what ways do the cycling of S and P differ from the cycling of C, O, and N?
4. Can we really examine the cycling of one element?

Image in Action

In the background is a NASA satellite image of the Black Sea that reveals large, swirling algal blooms. On the right, a zoom-in window shows *Candidatus* Kuenenia stuttgartiensis showing its unusual cell organization containing a typical cell wall (green) and cyto-

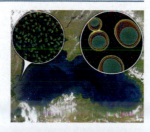

plasmic membrane (yellow), as well as an additional intracytoplasmic membrane (dark orange), which contains the genetic material (blue), and a membrane-bounded compartment called the anammoxosome (teal), where the anammox reaction occurs. Behind the cells, molecules of ammonium, nitrite, and dihydrogen are shown. On the left, cells of *Synechococcus*, a type of nitrogen-fixing cyanobacteria, are illustrated.

1. If in the laboratory you grew the nitrogen-fixing *Synechococcus* shown in this image, what products would you expect it to generate in your culture? Explain.

2. Imagine you are able to more closely examine the environmental conditions and the metabolism of the *Candidatus* Kuenenia stuttgartiensis shown in this image. What conditions would you expect to be present (and need to provide *in vitro*) and, as you begin your examination, what chemical compounds would you hypothesize to find in the anammoxosome?

The Rest of the Story

Interestingly to this day, no microbiologist has successfully obtained pure cultures of anammox microbes in the laboratory. This difficulty is partly because they grow extremely slowly with doubling times of about 11 days under ideal laboratory conditions, and two to three weeks or longer in their natural habitats. However, we have learned a lot about these bacteria through the use of specialized enrichment culture procedures and molecular approaches to probe their metabolic capabilities. Researchers have even determined the complete genome sequence of one bacterium, *Candidatus* Kuenenia stuttgartiensis, within this category. This genome sequence was reconstructed from the metagenomic sequence data of the community of an anoxic laboratory bioreactor where the anammox reaction was taking place, and in which this organism comprised approximately 75 percent of the cells.

These microbes have proven to be very unusual, belonging to a bacterial group called the Planctomycetes, although the anammox isolates are not closely related to other known Planctomycetes. A decidedly unique bacteria, Planctomycetes do not have a peptidoglycan-containing wall, dividing by budding rather than binary fission. Their cytoplasm contains different membrane-enclosed compartments, rather like the membrane-bounded organelles we associate with the Eukarya. These compartments seem to be devoted to different cellular functions. In the case of the anammox bacteria, the anaerobic NH_4^+ oxidation takes place in a compartment called the anammoxosome. An important intermediate of the oxidation reaction is hydrazine, N_2H_4, an extremely toxic component of rocket fuel and a molecule not previously known to be a biological product. Another peculiarity of the anammox bacteria is their possession of ladderane lipids, which are cyclobutane-containing molecules not found in other organisms. The ladderane lipids consist of a mix of ester- and ether-linked lipids. In other words, these lipids possess bacterial and archaeal characteristics. Researchers who work with the bacteria speculate that this lipid characteristic may reflect the early divergence of these species in the bacterial lineage. Ladderane lipids are difficult to produce synthetically but may have potential technical applications. Perhaps one day anammox bacteria will contribute to biotechnology.

Summary

Section 14.1: How are the nutrients used by cells cycled in the biosphere?

As microbial cells use nutrients, the associated metabolism causes the transformation of molecules. The **biogeochemical cycles** are interconnected due to the interdependence of chemical reactions and transformations during metabolism.

- All biomass is composed of essential or **biogenic elements** in similar proportions.

- Earth can be divided into compartments that hold major reservoirs for each element.

- The movement of elements between reservoirs, or **flux**, is an important part of cycling, and the cycles have achieved equilibrium over the long term.

- Human activities threaten to perturb the equilibria that have been established.

- Redox reactions carried out as part of cellular metabolism directly support several of the biogeochemical cycles.

- **Mesocosms**, which replicate ecosystems under controlled conditions, can be used for experimental study of ecosystems.

- The fate of specific molecules in an ecosystem can be monitored using **radioisotope tracers**.

Section 14.2: How do microbes influence the cycling of carbon?

The carbon-based nature of life is reflected in the direct influence of the abundant microbial biomass on the carbon cycle.

- The global carbon cycle maintains a balance between organic and inorganic carbon reservoirs through the complementary metabolic activities of photosynthesis and respiration and decomposition.

- Human activity has disrupted these cycles, causing an increase in the atmospheric levels of **greenhouse gases**.

- The metabolic activities of photosynthesis, respiration, and decomposition contribute to the short-term cycling of carbon. Organic matter can accumulate if CO_2 fixation exceeds respiration, resulting in a positive **net community productivity**. Fermentation results in the production of partially oxidized organic end products that can serve as nutrients for other organisms.

- CH_4 represents a major biodegradation end product under anoxic conditions.

- **Methanogens** produce CH_4 through a type of metabolism referred to as **methanogenesis**.

- **Hydrogenotrophic methanogens** produce CH_4 from CO_2 and H_2. **Acetotrophic methanogens** produce CH_4 from acetate (CH_3COO^-).

- CH_4 production from biomass often involves a **consortium** of metabolically interdependent microbes that form **syntrophic partnerships**.

- In some anoxic environments, inorganic electron acceptors other than CO_2 may be used by microbes.

- CH_4 can be efficiently oxidized by **methanotrophs**, a subgroup of the **methylotrophs**, under aerobic conditions.

- **Anaerobic oxidation of methane (AOM)** is now recognized as an important part of the global carbon cycle, even though the responsible archaeal organisms—**anaerobic methane oxidizers (ANME)**—are only known by DNA sequence and have not yet been cultured.

- Other elemental cycles linked to the carbon cycle include oxygen, sulfur, and nitrogen.

Section 14.3: How do microbes influence the cycling of nitrogen?

Each of the key steps in the nitrogen cycle involves transformations mediated by or influenced by microbial metabolism.

- The atmosphere, made up of almost 80 percent N as dinitrogen gas, is by far the largest reservoir of nitrogen.

- **Nitrogen fixation**, carried out only by certain bacteria and archaea, is essential for converting atmospheric dinitrogen to forms in which it can be incorporated into biomass.

- The nitrogen fixation process is O_2 sensitive due to irreversible activation of the **nitrogenase complex** by O_2.

- **Nitrification**, the conversion of ammonium to nitrite and then nitrate, is carried out by the sequential activities of ammonium oxidizers and **nitrifiers**.

- **Denitrification** involves the conversion of nitrate to dinitrogen gas through a series of reduction steps carried out by reductase enzymes during anaerobic respiration.

- Synthetic nitrogen fertilizer, often produced via the **Haber–Bosch industrial process**, now represents a major source of biologically available nitrogen on Earth, and while microbes are able to keep the nitrogen cycle running, the changes in nutrient balance can sometimes have adverse affects on ecosystems.

Section 14.4: How are various biogeochemical cycles connected?

By briefly examining the cycling of oxygen, phosphorus, and sulfur, we can see that all the biogeochemical cycles are interconnected.

- Major reservoirs of oxygen include O_2, CO_2, and H_2O. All of these molecules are involved in photosynthesis and aerobic respiration.

- Acid mine drainage leads to the oxidation of FeS_2, yielding sulfuric acid.

- Sulfur and phosphorus cycle mainly through terrestrial and marine reservoirs.

- Microbes aid in making sulfur and phosphorus available to living organisms.

● Application Questions

1. Imagine that you present information on the carbon cycle to a class of eighth graders. You need to explain how CO_2 in the atmosphere is connected to CO_2 in the oceans and the role this connection plays on Earth.

2. When studying and calculating global carbon cycling, scientists find a "missing sink" that doesn't allow their equations to balance. Some researchers theorize that uncultivated or uncharacterized microbes might explain this phenomenon. Explain what this means.

3. What role does plant and animal respiration play as a "natural" way of providing CO_2 to the atmosphere? Describe an "unnatural" way of adding CO_2 to the atmosphere.

4. A biogeochemist graduate student wants to study methanogens in more detail. Based on the conditions needed to support methanogenesis, what habitats would be best for collecting samples? Explain why.

5. Your new research advisor studies interspecies hydrogen transfer in a biofilm community. Describe what she is studying and what is occurring in the biofilm.

6. A high school student hopes to do a research project to survey a site for the presence of methanogens using a molecular tool. As a mentor for the student, explain what gene or genes he could use for this molecular survey.

7. Why do some farmers inoculate their soil with nitrogen-fixing bacteria when planting their crops?

8. At a tour of your local wastewater treatment facility, you learn that they use different microbes at various stages of the treatment process. You hear that at one stage they use nitrifiers. Why might they be using nitrifiers, and what would the nitrifiers do in the treatment process?

9. You learn in a soil ecology class that a chemical is often added in agriculture to inhibit ammonium oxidation. What does this chemical do, and why might it be useful for the soil and the crops?

10. You recently have read about the cultivation of crenarchaeotal nitrifiers from the Seattle aquarium, and you suspect that the nitrifiers in your tropical fish tank are archaeons, rather than bacteria.
    a. What would you need to look for to determine if the microbes were nitrifiers and were indeed archaeons?
    b. What lab techniques might you use to make these determinations?

Suggested Reading

Dunfield, P. F., A. Yuryev, P. Senin, A. V. Smirnova, M. B. Stott, S. Hou, B. Ly, J. H. Saw, Z. Zhou, Y. Ren, J. Wang, B. W. Mountain, M. A. Crowe, T. M. Weatherby, P. L. Bodelier, W. Liesack, L. Feng, L. Wang, and M. Alam. 2007. Methane oxidation by an extremely acidophilic bacterium of the phylum Verrucomicrobia. Nature 450:879–883.

Falkowski, P. G., T. Fenchel, and E. F. DeLong. 2008. The microbial engines that drive Earth's biogeochemical cycles. Science 320:1034–1039.

Francis, C. A., J. M. Beman, and M. M. M. Kuypers. 2007. New processes and players in the nitrogen cycle: The microbial ecology of anaerobic and archaeal ammonia oxidation. ISME J 1:19–27.

Gruber, N., and J. N. Galloway. 2008. An Earth-system perspective of the global nitrogen cycle. Nature 451:293–296.

He, Z., Y. Deng, J. D. Van Nostrand, Q. Tu, M. Xu, C. L. Hemme, X. Li, L. Wu, T. J. Gentry, Y. Yin, J. Liebich, T. C. Hazen, and J. Zhou. 2010. GeoChip 3.0 as a high-throughput tool for analyzing microbial community composition, structure, and functional activity. ISME J 4:1167–1179.

Lloyd, K. G., L. Lapham, and A. Teske. 2006. An anaerobic methane-oxidizing community of ANME-1b Archaea in hypersaline Gulf of Mexico sediments. Appl Environ Microbiol 72:7218–7230.

McInerney, M. J., C. G. Struchtemeyer, J. Sieber, H. Mouttaki, A. J. M. Stams, B. Schink, L. Rohlin, and R. P. Gunsalus. 2008. Physiology, ecology, phylogeny, and genomics of microorganisms capable of syntrophic metabolism. Ann NY Acad Sci 1125:58–72.

Pol, A., K. Heijmans, H. R. Harhangi, D. Tedesco, M. S. Jetten, and H. J. M. Op den Camp. 2007. Methanotrophy below pH 1 by a new Verrucomicrobia species. Nature 450: 874–878.

Theisen, A. R., M. H. Ali, S. Radajewski, M. G. Dumont, P. F. Dunfield, I. R. McDonald, S. N. Dedysh, C. B. Miquez, and J. C. Murrell. 2005. Regulation of methane oxidation in the facultative methanotroph *Methylocella silvestris* BL2. Mol Microbiol 58:682–692.

15 Microbial Ecosystems

In the early 1990s, Ed DeLong, a microbiologist working at Woods Hole Oceanographic Institution in Massachusetts, made a striking discovery. He detected archaeons in oxygenated coastal waters off the coast of North America. As we noted in Chapter 4, archaeons historically had been considered to be "extremophiles." Indeed, until DeLong's discovery, almost all archaeons found in the ocean were either thermo- or hyperthermophiles living near deep sea hydrothermal vents, or methanogens living in anaerobic sediments, habitats quite different from the oxygenated surface waters. Based on DeLong's work, however, microbiologists began to reassess their views of archaeons. Perhaps they weren't relegated to environments that we consider extreme. While this discovery of archaeons in coastal waters may have been unexpected, the way in which DeLong and colleagues identified these organisms makes an even more interesting story.

You may recall that we first encountered Ed DeLong in Section 10.4 during our discussion of metagenomics. He and his colleagues showed that they could analyze marine microbial communities in a culture-independent fashion. By isolating and then sequencing genomic DNA obtained from the environment, the researchers began to understand what types of bacteria were present. DeLong decided to use these same techniques to look at the composition of microbial communities in coastal marine surface waters. Using a series of PCR primers designed to specifically amplify archaeal SSU rRNA genes, he detected two very different types of archaeal DNA in several surface seawater samples. Phylogenetic analysis showed that some sequences belonged in the phylum Euryarchaeota, while other samples belonged in the phylum Crenarchaeota. Moreover, these archaeal species appeared to account for a substantial number of the microbes in both Atlantic and Pacific coastal marine surface waters. In some cases, archaeal sequences accounted for over 2 percent of the total SSU rRNA genes found.

The discovery of archaeons in this environment raised many questions. What roles did they play in the ecosystem as they lived alongside bacterial and eukaryal organisms? Were they also found deeper in the ocean and away from the coasts? Did archaeal species exist in different habitats, such as soils, or were they limited to the marine environment? These are only a few of the questions that resulted from this relatively simple application of molecular methods to microbial ecology. What else is in store? Probably much more than anyone would have predicted.

Introduction

Earth is very much a microbial planet, and its microbial residents have been evolving for about 4 billion years. In the preceding chapter, we learned how microbial metabolism controls the development and maintenance of the biosphere. In this chapter, we will consider the interactions of microbes with their surroundings, a scientific discipline known as "microbial ecology." Microbes naturally interact with other organisms, both directly and indirectly. In some cases, their own metabolic activities depend on the metabolic activities of their neighbors. In other cases, they compete with their neighbors for common resources. In the process of making a living, microbes often greatly modify their environments. In the study of microbial ecology, links can be made between genetics, physiology, and ecological roles, and how the evolution of microbial capabilities and characteristics results from genetic selection within the environment. Whether we consider a wastewater treatment system (see Section 16.5), the human gut (see Section 17.4), an acid mine drainage system, or fertile agricultural soil, the interactions between individual members of the microbial community are considerable.

While the various environments on Earth that support life obviously display great range and diversity, we can group them within the major geographic divisions of marine, terrestrial, and subsurface. In this chapter, first we will examine general ideas about microbial communities. We will then look at techniques used by microbiologists to study these communities. Finally, we will investigate several specific types of environments. As we investigate these aspects of microbial ecology, we will focus on the following questions:

What factors affect microbial communities? (15.1)
How can microbial communities be studied? (15.2)
What is the nature of life in water? (15.3)
What is the nature of life on land? (15.4)
Can life exist in the absence of photosynthesis-driven primary production? (15.5)

15.1 Microbes in the environment

What factors affect microbial communities?

In Chapters 6 and 13, we examined the concept of microbial growth and explored the incredible diversity of metabolic ways in which microbes make a living. Most of the studies that have led to our understanding of microbes were conducted on cells growing in pure culture, often at mid-log phase. In fact, pure culture studies have been the foundation of the science of microbiology. Microbes, however, do not exist in nature under ideal laboratory conditions. Instead, they live in complex communities where they exchange nutrients, wastes, and genetic material with their neighbors in their continual quest to grow and divide. For example, let's look at *Escherichia coli*. You may be working with a specific strain of *E. coli* in the laboratory right now, growing it in LB broth at 37°C. Do you think the Petri dish in your laboratory is the natural habitat of *E. coli*? Of course not. We know that this bacterium normally resides, among other places, in our gut along with numerous other microbes. In this location, it is subject to the variations of our diet, our overall health, and many other external factors.

To fully understand *E. coli*, then, we should investigate how it survives in a more natural environment. Moreover, we should investigate how it interacts with other species. As we will see throughout this chapter, these goals can be quite difficult to achieve. Despite our focus on cultivation-based microbiology, for example, it has become widely accepted that culturable microbes represent a small subset of the organisms in most microbial communities (Table 15.1). We will address this conundrum in Section 15.2. For now, let's begin our study of microbial ecology with a brief exploration of ecosystems.

Ecosystems

Life on Earth represents a collection of **ecosystems**, a term first used in the 1930s to describe the interactions and exchange of materials between organisms and their surrounding environment. An ecosystem thus consists of a **community** of organisms living in a specified area, interacting with each other and the environment. An ecosystem includes both biotic factors, such as other organisms, and abiotic factors, such as minerals, gases, and water. The interactions typically are quite dynamic. As environmental conditions change, organisms react to the changes. Changing physiological activities of the organisms in turn influence the environmental conditions. We can consider several

TABLE 15.1   Culturable microbiota

Habitat	Culturability (%)[a]
Seawater	0.001–0.1
Freshwater	0.25
Unpolluted estuarine waters	0.1–3
Activated sludge	1–15
Sediments	0.25
Soil	0.3

[a]Culturable bacteria are measured as colony-forming units (CFU).
Source: Amann, R. I., W. Ludwig, and K. -H. Schleifer. 1995. Microbiol Rev 59:143–169.

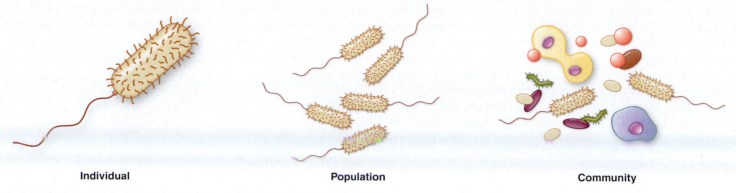

| Individual | Population | Community |

Figure 15.1. Different levels within the microbial community Microbial cells can be considered at several different levels, including the level of the individual cell, populations of similar cells, and communities of diverse cells.

different levels within the microbial community, ranging from individual cells through populations of similar cells, and finally the complete community that typically consists of a collection of organisms (Figure 15.1).

How can we categorize the various members of an ecosystem? We can focus on the roles different organisms within an ecosystem play within the context of energy flow. Familiar **primary producers**, like plants and photosynthetic microbes, capture energy through photosynthesis. This captured energy drives the incorporation of inorganic carbon, usually from CO_2, into organic molecules. As we will see in Section 15.5, however, non-photosynthetic primary producers also exist, typically in habitats devoid of sunlight. In contrast to producers, heterotrophic organisms function as **consumers**, ingesting and utilizing the stored photosynthetic energy. **Decomposers** break down dead organic matter and recycle the components back into the environment.

In many cases, it may be useful to further separate organisms into functional **guilds**, groups of organisms that carry out similar processes. All anoxygenic phototrophs, for instance, harvest light energy under anaerobic conditions without producing O_2, while thermophilic methanogens all produce methane under high-temperature conditions. Although members of a guild play similar roles in a given ecosystem, they need not be genetically related. Often, though, a degree of genetic similarity to other organisms provides a rough guide to the likely characteristics of an organism.

Within a community, each individual organism must obtain nutrients to support the energy and material needs of the cell. In most ecosystems that have been studied, energy for metabolism and organic carbon for cell growth originates from sunlight and CO_2 through photosynthesis carried out by the producers. The harvested energy and carbon is then channeled to the heterotrophic consumers. Some ecosystems, such as those on the ocean floor or in the deep subsurface below the seafloor or soil surface, do not have direct access to light energy. Instead, they rely on other sources of energy and organic carbon. For example, the giant tube worms introduced in Chapter 13 rely on chemolithotrophic bacteria that oxidize H_2S for energy. This type of process is often termed **chemosynthesis**, as shown here:

$$CO_2 + O_2 + 4\,H_2S \rightarrow CH_2O + 4\,S + 3\,H_2O$$

Other organisms obtain energy in different ways. To fully understand the intricate workings of an ecosystem we not only need to know what guilds are present, we must also understand how individual species obtain nutrients. To facilitate this goal, we need to more fully understand the physiology of the ecosystem.

Ecosystem physiology

To fully understand the workings of an ecosystem, we need to know more than simply which producers, consumers, and decomposers exist within the ecosystem. We also need to know how each organism interacts with the environment. A fundamental concept of ecology is that of the **niche**, which describes the specific functional role of an organism within an ecosystem. An organism's niche includes its physiological interactions with its physical habitat and with other co-existing organisms. For a particular organism, the type, quality, and quantity of sustaining resources defines the niche. The evolutionary pressures of selection and competition result in the specialization of organisms for their prime niches, and ensure that all niches are occupied. The definition of niche and its relationship to ecosystem and communities is clarified further by the requirement that *no two populations of organisms can occupy the same niche*.

In characterizing microbial habitats, we need to think of microenvironments. For example, gradients of O_2, pH, light, or nutrient may exist over very short distances, and these gradients may have significant effects on different microbes. Metabolic activities of the inhabitants can also have profound effects on local physical and chemical conditions. Conditions also change with time. While microbes might be specialized for a prime niche, their ultimate survival and fitness depends on their ability to adapt to changes in their habitat and to exploit the available resources better than competing organisms. Through gene regulation, microbes can modify their metabolism according to the external conditions.

CONNECTIONS In **Chapter 11**, we learned about several mechanisms by which microbial cells modulate their gene expression in response to changes in environmental conditions. The resulting metabolic changes are critical in determining the cell's ability to survive when competing for resources.

While an astounding number of niches exist on the planet, and innumerable combinations of environmental and physiological factors affecting microbial growth and success exist within every niche, certain basic environmental components or conditions need to be considered when investigating any given niche. Within ecosystems, factors affecting the primary production of organic matter by photoautotrophs and chemolithoautotrophs (see Section 13.2), the decomposition of organic matter by chemoorganotrophs (see Section 14.2), and the associated reactions that drive the biogeochemical cycling of elements that were described in Chapter 14 are probably the most important. When studying microbes in nature, we typically focus more on which metabolic types are present rather than the taxonomic identities of the organisms. Nevertheless, there is often some correlation between the taxonomic identity of an organism and its metabolic type. All species within the *Desulfovibrio* genus, for instance, are sulfate-reducing bacteria, while all bacteria in the *Azotobacter* genus are aerobic N_2 fixers. In many cases, the types of organisms that might be present in a given habitat can be predicted, such as the likely presence of methanogens in anaerobic freshwater sediments or the intestinal tracts of animals (see Section 17.4).

CONNECTIONS Recall from **Chapter 14** that the major nutrients required by microbes are carbon, nitrogen, phosphorus, and sulfur compounds, as these are required for generation of energy and for growth. Metallic elements, such as Fe, Mg, Mn, K, Ca, Na, and Cl are also important.

To understand how certain microbes succeed in different environments, it is often best to consider the elements found in a particular niche and examine how microbes obtain these different nutrients. To put it very simply, microbes generate metabolic energy by oxidizing (removing electrons from) certain compounds, and then reducing (passing electrons to) other compounds. The oxidized compound can be thought of as an electron donor, and the reduced compound

as an electron acceptor, as was discussed in Chapter 13. In this way, the microbial cell works like a living catalyst facilitating the transfer of electrons between different compounds, while at the same time harvesting the energy released by these reactions.

Biofilms

As we noted in Section 2.5, microbes tend to accumulate and grow on solid surfaces as biofilms, structured communities held together by extracellular polysaccharide (see Figure 2.35). That "goo" inside the shower drain, the plaque on the surface of your teeth, and the slime on rocks in a stream all represent biofilms. Nutrients often bind to these solid surfaces, supporting higher cell concentrations than typically exist in the bulk fluid. Moreover, the members of the biofilm interact with and support each other. The formation of these biofilms can be initiated within minutes of placing an object within a microbe-containing fluid. Mature biofilms can be very complex microbial communities, in which the majority of the cells have good access to nutrients and protection from environmental stresses **(Figure 15.2)**. Because some of these stresses are from human activity aimed at reducing microbial numbers, such as the application of antimicrobial agents, the formation of biofilms is often of practical concern.

CONNECTIONS In **Section 16.5**, we will see that the proper formation of biofilms is important in the treatment of sewage and the purification of drinking water.

During natural biofilm formation, the first microbes to adhere to a surface are often appendaged bacteria, such as *Caulobacter* or *Hyphomicrobium*, specialists in attaching to surfaces **(Figure 15.3)**. As these primary colonizers divide and form microcolonies, other microbes, the secondary colonizers, join the growing biofilm. Together, they secrete exopolysaccharide or EPS (also referred to as extracellular polymeric substances).

50 μm

LM

Courtesy Christine Dupont

Figure 15.2. Biofilms Biofilms are microbial communities that are often found on solid surfaces within aquatic environments, such as on this rock in a stream. As shown in the light micrograph, the biofilm is made up of a community of microbes. Members of the biofilm benefit by improved nutrient acquisition and avoidance of antimicrobials. Biofilms are usually held together by extracellular polysaccharide produced by their inhabitants.

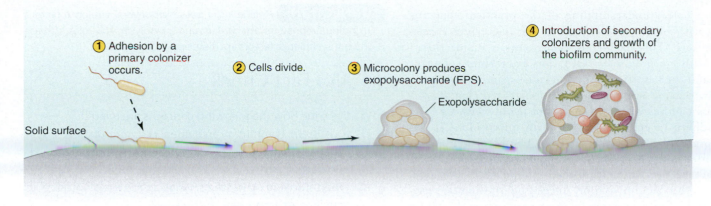

① Adhesion by a primary colonizer occurs.

② Cells divide.

③ Microcolony produces exopolysaccharide (EPS).

④ Introduction of secondary colonizers and growth of the biofilm community.

Exopolysaccharide

Solid surface

Figure 15.3. Process Diagram: Formation of biofilm Biofilms typically form through a stepwise process. First, the surface is populated by adhesion of the primary colonizers. Growth results in the formation of microcolonies that produce exopolysaccharide (EPS). Secondary colonizers are introduced as the biofilm matures. Throughout this development, intercellular communication occurs through molecular signaling, such as quorum sensing.

The exopolysaccharide forms a gel-like matrix around the microcolonies. Most likely, this matrix aids in the formation of water-filled channels that, in turn, aid in the transport of nutrients and wastes throughout the biofilm.

Indeed, several researchers have shown through experiments the importance of EPS production in biofilm development. Roberto Kolter and colleagues at Harvard University examined the importance of exopolysaccharide production in biofilm formation by *Escherichia coli* K12. This bacterium produces the exopolysaccharide colanic acid during biofilm formation. To investigate the importance of this substance, the researchers examined *E. coli* K12 mutants defective in the production of colanic acid. When grown in the presence of a solid surface, these bacteria attached to the surface normally but did not form a complex, three-dimensional biofilm like the wild-type strain. From these results, the scientists concluded that, under their experimental conditions, *E. coli* requires the exopolysaccharide for biofilm development **(Figure 15.4)**.

It's worth noting that in *E. coli*, colanic acid production appears to occur specifically during biofilm formation. In other words, a product essential for biofilm formation is only produced when biofilm formation may occur. How might this form of regulation occur? It appears that cells within a growing biofilm community sometimes communicate by quorum sensing (see Section 11.5). Chemical signaling allows the cells to assess their population density and induce specific gene expression profiles in response to cell density changes.

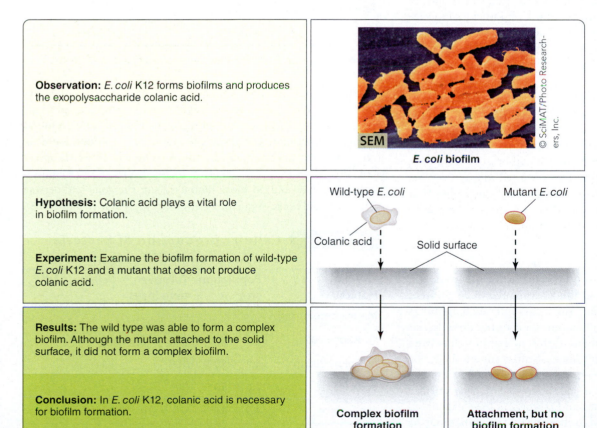

Observation: *E. coli* K12 forms biofilms and produces the exopolysaccharide colanic acid.

SEM

© SciMAT/Photo Researchers, Inc.

E. coli biofilm

Hypothesis: Colanic acid plays a vital role in biofilm formation.

Wild-type *E. coli*

Mutant *E. coli*

Colanic acid

Solid surface

Experiment: Examine the biofilm formation of wild-type *E. coli* K12 and a mutant that does not produce colanic acid.

Results: The wild type was able to form a complex biofilm. Although the mutant attached to the solid surface, it did not form a complex biofilm.

Conclusion: In *E. coli* K12, colanic acid is necessary for biofilm formation.

Complex biofilm formation

Attachment, but no biofilm formation

Figure 15.4. Role of colanic acid in *E. coli* biofilm formation To test the role of colanic acid production on biofilm formation by *E. coli* K12, researchers isolated a mutant that did not produce this exopolysaccharide. When exposed to a solid surface, cells of this strain attached to the surface but did not form a complex biofilm. In contrast, wild-type cells that did produce colanic acid attached to the surface and formed a biofilm.

To get a better understanding of the practical concerns associated with biofilms, let's look at the bacterium *Pseudomonas aeruginosa*. This bacterium represents a major threat to people with cystic fibrosis (CF). The bacterium colonizes the lungs and the resulting infection causes significant tissue damage. Increasing evidence has demonstrated that *P. aeruginosa* forms biofilms within the lungs of people with CF. The bacteria adhere to epithelial cells, form microcolonies, and secrete alginate, an exopolysaccharide. The resulting biofilm probably affords the bacteria several advantages. First, it offers the bacteria protection from the immune system. Second, cells within the biofilm exhibit increased antibiotic resistance. Thus, biofilm formation by *P. aeruginosa* directly affects its virulence.

CONNECTIONS *P. aeruginosa* possesses several virulence factors, including fimbriae for attachment and a cytolysin. The roles of these and other virulence factors are discussed in **Section 21.1**.

15.1 Fact Check

1. Describe how the following concepts are related: ecosystem, community, individual cells, population.
2. Differentiate between primary producers, consumers, and decomposers.
3. When studying ecosystems, what basic physiological activities are important to consider?
4. What is a biofilm, and where can biofilms be found?

15.2 Microbial community structure

How can microbial communities be studied?

Researchers studying microbial ecology generally pursue two broad objectives. First, they want to understand the biodiversity, or variety of microbial life, in nature. Second, they want to determine the activities of the organisms in nature and understand the effects of those activities on ecosystems. Until recently, however, we did not even have a clear view of what microbes existed in most communities. Cultivation-dependent approaches only detect those organisms that easily grow under laboratory conditions. These organisms may not be very active in the environment at the time of their isolation; they are just better suited to the culture conditions. Moreover, as we saw in Section 15.1, individual species do not live in isolation in nature. Even if we can grow a particular microbe in the laboratory, its activities there may not reflect its activities in nature. Uncertainties about exactly which organisms are active in an ecosystem and the nature of their activities, have severely limited our understanding of microbial ecosystems.

In this section, we will explore techniques used by microbial ecologists to determine what microbes exist in ecosystems. We will focus primarily on techniques used to analyze the microbes that cannot be cultivated easily in the laboratory. First, we will discuss enrichment cultures. Then, we will discuss several culture-independent techniques used to identify microbes that cannot be cultivated. As we will see, these techniques help us to better understand complex microbial communities.

Enrichment cultures

We know that microbes play essential roles in the environment, but the true extent of microbial life on this planet is far from obvious. How can we determine the total microbial numbers and biomass on Earth? Each cell cannot be counted directly, but the microbial life residing in many different habitats has been sampled and quantified by a number of different methods. By extrapolating direct measurements of microbial cells from the representative major habitats of aquatic, soil, and subsurface environments, William Whitman and coworkers have calculated that the number of bacteria and archaeons on the planet may be an astounding 4–6×10^{30} cells, representing 3.5–5.5×10^{17} g

TABLE 15.2 Cell numbers and biomass in different environments on Earth

Environment	Number of prokaryotic cells $\times 10^{28}$	Grams of carbon $\times 10^{15}$
Aquatic habitats	12	2.2
Oceanic subsurface	355	303
Soil	26	26
Terrestrial subsurface	25–250	22–215
Total	415–640	353–546

Source: W. Whitman et al. 1998. Proc Natl Acad Sci USA 95:6578–6583.

of carbon, about the same amount of carbon that exists in plants (**Table 15.2**). While this figure might seem surprising, perhaps more surprising is the amount of nitrogen and phosphorus found in Earth's microbes. For these life-essential elements, bacteria and archaeons may contain 10 times as much as exists in all plant life. Moreover, the collective deep subsurface microbial inhabitants likely contain more biomass than the soils and oceans combined, and these organisms grow extremely slowly (**Table 15.3**).

TABLE 15.3 Cellular production of Bacteria and Archaea in different habitats

Habitat	Population size	Turnover time (days)	Cells/yr $\times 10^{29}$
Marine heterotrophs			
Above 200 m	3.6×10^{28}	16	8.2
Below 200 m	8.2×10^{28}	300	1.1
Soil	2.6×10^{29}	900	1.0
Subsurface	4.9×10^{30}	5.5×10^5	0.03

Source: W. Whitman et al. 1998. Proc Natl Acad Sci USA 95:6578–6583.

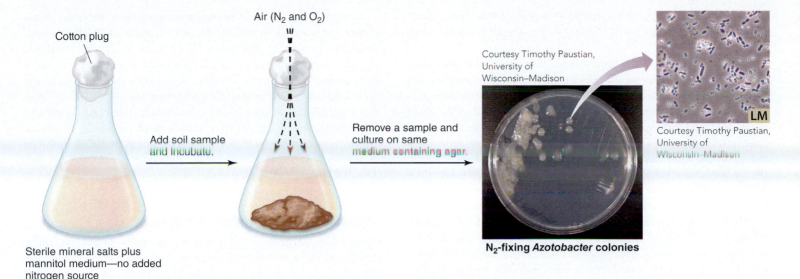

Cotton plug

Air (N₂ and O₂)

Add soil sample and incubate.

Remove a sample and culture on same medium containing agar.

Sterile mineral salts plus mannitol medium—no added nitrogen source

Courtesy Timothy Paustian, University of Wisconsin–Madison

Courtesy Timothy Paustian, University of Wisconsin–Madison

LM

N₂-fixing *Azotobacter* colonies

Figure 15.5. Enrichment culture for nitrogen-fixing bacteria To isolate nitrogen-fixing bacteria, soil can be added to liquid medium that lacks any biologically available nitrogen sources like NH_4^+ or NO_3^-. During incubation in the presence of air, nitrogen-fixing bacteria will become dominant in the culture. An aliquot of this enriched culture can then be plated on solid medium that lacks biologically available nitrogen. Again, only nitrogen fixers will grow.

How can we isolate slow growing or rare organisms? Enrichment culture, the use of culture medium, and conditions designed to favor the growth of particular microbes (see Section 6.4) can be very effective at isolating organisms with particular physiological properties. For example, *Azotobacter* and other aerobic N₂-fixing bacteria can be isolated from soils even if they are not very active in those soils at the time of isolation. Let's see how. *Azotobacter* can convert N₂ into biologically available molecules. Unlike most microbes, then, it can grow in the absence of biologically available nitrogen-containing organic material. If we simply inoculate a nitrogen-free liquid medium with soil, then only organisms present in the soil that can assimilate nitrogen from the atmosphere will grow. Nitrogen fixers like *Azotobacter* will increase in number or become enriched. We can then plate a small amount of this enriched broth on an agar plate. Again, if the solid medium does not contain any biologically available nitrogen, then only nitrogen fixers like *Azotobacter* will grow. Individual colonies growing on the plate can be isolated, grown in pure culture, and analyzed **(Figure 15.5)**.

While researchers can use enrichment cultures to isolate microbes with specific metabolic properties, enrichment cultures are also commonly used to introduce younger students to microbial diversity. Named after the Russian biologist Sergei Winogradsky, the Winogradsky column allows students of microbiology to watch enrichment occur. Making a Winogradsky column is quite simple. Mud from a lake or pond is added to a cylindrical container, like a two-liter soda bottle or a graduated cylinder. Carbon and sulfur are also added. These elements can come from a variety of sources, including shredded newspaper, grass clippings, or straw (carbon) and eggs (sulfur). Water from the pond is added to partially fill the container. It is then sealed and placed in a sunny spot. Over time, microorganisms initially present in relatively low numbers, will proliferate in different sections of the tube.

Near the top of the tube, in the oxygenated water, cyanobacteria may proliferate. Oxygen levels decrease, though, as the distance from the surface increases. In these oxygen-poor

regions, green sulfur bacteria and purple sulfur bacteria may proliferate. Deeper down, in the anoxic regions at the bottom of the column, anaerobic sulfur-reducing bacteria thrive. Each of these types of bacteria display a distinct color within the tube, providing students with a visual cue that enrichment of different species has occurred **(Figure 15.6)**.

Cultivation-independent techniques

While enrichment cultures allow us to selectively isolate microbes with specific physiological properties, this process does not help us identify the non-culturable inhabitants of an ecosystem. How can we identify organisms that we cannot grow? In recent years, the application of powerful molecular tools has helped us address this question. Most of these methods have utilized SSU rRNA gene sequences as a sort of barcode to determine which types of organisms exist in a given sample (see Toolbox 1.1). These techniques have been especially influential in microbial ecology, and have made it possible to

Courtesy Trevor Charles and Josh Neufeld

Figure 15.6. Enrichment in a Winogradsky column Sediment from a pond or lake, a carbon source, a sulfur source, and water are added to a cylindrical glass container. Over time, different types of bacteria proliferate in different regions of the cylinder. As seen here, different zones within the cylinder have different colors, demonstrating the enrichment of different species with different metabolic capabilities.

detect organisms that we currently cannot grow. With these techniques, the majority of life no longer remains hidden from view (**Figure 15.7**).

Culture-independent studies have reinforced the idea that the majority of the microbial cells in a given environment are not easily culturable, and in many cases they carry out metabolic processes completely overlooked in earlier ecological studies. With molecular techniques, some of the microbes that had been isolated from environmental samples by enrichment methods have been shown to be very minor components of the community. Indeed, the predominant culturable isolates from seawater often cannot be detected easily using molecular methods, indicating that they exist in very small numbers in their natural environment.

Here, we will investigate several molecular techniques used to analyze microbial communities. Typically, these techniques rely on the analysis of SSU rRNA genes. The extensive secondary structure of rRNA molecules, with hairpins and loops (see Figure 15.7), results in some highly conserved regions and some quite variable regions. The dissimilarity in the variable regions allows determination of the phylogenetic relationships among organisms, while the conserved regions enable the design of oligonucleotide primers with different levels of specificity to the SSU rRNA genes. Within SSU rRNA gene sequences can be found stretches of sequence specific to different taxonomic levels. For example, universal primers (see Figure 15.7) can amplify SSU rRNA genes from all organisms. In contrast, bacterial-specific primers only amplify SSU rRNA genes from bacteria.

Direct Sequencing

Today, the most effective technique for microbial community analysis involves direct sequencing of DNA isolated from a community. A typical analysis would start with collection of the sample from the environment followed by transport and storage of the sample under refrigeration to minimize changes to the microbial community. In the laboratory, DNA can be extracted from the sample and used as a template for PCR using SSU rRNA gene primers (see Toolbox 1.1). The resulting amplified DNA molecules then are cloned into plasmid vectors (see Section 9.4), and individual clones are sequenced (see Section 10.1). With some of the newer sequencing methods, the vector-cloning step is not necessary.

Comparison of the DNA sequences to the collection of sequences in publicly available databases, such as the Ribosomal Database Project at Michigan State University, or GenBank, provides a determination of the identity of each clone, or at least indicates the most similar sequences in the databases. Such analysis provides detailed information about the taxonomy of organisms present in the community. The results can also allow estimates of the relative numbers of each type of organism in the community. If, for instance, 50 out of 100 sequenced clones from a given environment correspond to *Escherichia coli*, then one could predict that about 50 percent of the organisms in that environment are *E. coli*.

Denaturing Gradient Gel Electrophoresis

Historically, large-scale sequencing projects could be quite expensive. As a result, researchers developed less costly indirect methods of comparing molecules. **Denaturing gradient gel electrophoresis (DGGE)** represents one common indirect method. This technique can be used to separate small DNA fragments (200–700 base pairs) that have been isolated from, for example, an environmental community. Restriction endonuclease fragments or amplified PCR products are commonly used. The separation is based on the fact that different alleles, or versions of a gene from different organisms, will have different DNA sequences. The SSU rRNA genes of bacteria, for example, contain blocks of sequence that are unique to particular species. No matter what gene sequence is chosen for analysis, sequence differences of even one base pair influence how tightly the

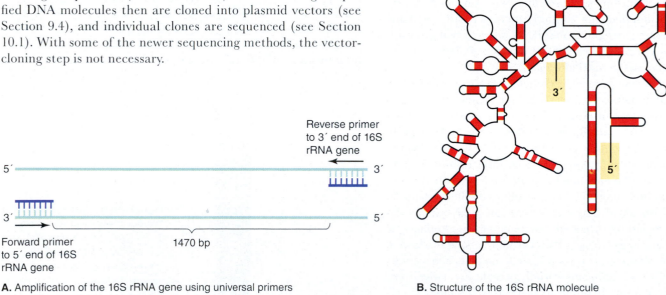

A. Amplification of the 16S rRNA gene using universal primers

Reverse primer to 3′ end of 16S rRNA gene

5′   3′

3′   5′

Forward primer to 5′ end of 16S rRNA gene   1470 bp

B. Structure of the 16S rRNA molecule

3′

5′

Karp: *Cell and Molecular Biology, Concepts and Experiments*, copyright 2010, John Wiley & Sons, Inc. This material is reproduced with permission of John Wiley & Sons, Inc.

Figure 15.7. SSU rRNA molecule A. Universal primers can be used to amplify the whole gene by PCR. **B.** The secondary structure, involving intramolecular base-pairing shown in red, results in different degrees of sequence conservation within the SSU rRNA gene. The design of oligonucleotide primers for PCR amplification and probes for DNA and RNA hybridization studies can provide different levels of specificity.

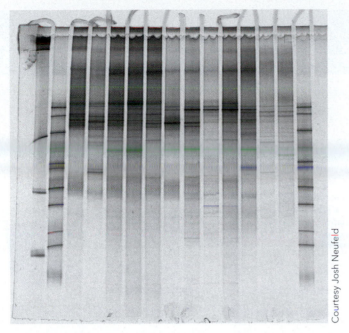

Courtesy Josh Neufeld

Figure 15.8. The DGGE method DNA fragments that differ by as little as a single nucleotide can be resolved. This method can be used to compare microbial communities by SSU rRNA sequence.

double-stranded DNA molecule is held together by hydrogen bonding.

Unlike more traditional gel electrophoresis, DGGE employs a gradient of denaturing conditions, usually achieved by a combination of temperature and chemical denaturants such as urea and formamide, in a polyacrylamide gel matrix. As the DNA fragments migrate down the gel during electrophoresis, they encounter a gradient of increasing denaturing conditions. As the double-stranded DNA begins to separate into two single strands, it deforms, and the altered shape prevents the fragment from migrating further in the gel. The denaturing conditions at which this denaturation happens depend on the DNA sequence. The resulting banding patterns **(Figure 15.8)** can then be analyzed for diversity, indicated by numerous bands in a single lane, or for community comparison between different samples, indicated by the presence of band variation between lanes. Other information, such as the relative abundance of the various organisms in a sample, can also be deduced.

Depending on the DNA samples used for DGGE, individual bands can be purified directly from the gel, cloned, and sequenced to determine the taxonomic identity (e.g., SSU rRNA genes), or physiologic capabilities (e.g., a gene required for photosynthesis) of the organisms. The technique can uncover the presence of a multitude of microorganisms from a single sample, many of which may be unculturable and not previously identified. While the DGGE method is quick and inexpensive, it is being replaced by direct sequence-based methods that provide more information.

Terminal Restriction Fragment Length Polymorphism

Terminal restriction fragment length polymorphism (TRFLP) represents another indirect method of comparing DNA. As for DGGE, this method is also being replaced by direct sequence-based methods. In this technique, researchers profile microbial communities by examining the position of a restriction site closest to a labeled end of an amplified gene, usually the SSU rRNA gene. Initially, PCR is performed on the target gene using one or both primers labeled with a fluorescent molecule on their 5′ ends. Then, the mixture of PCR-amplified variants of the gene is digested with a restriction enzyme. The resulting fragment mixture is separated by electrophoresis. Because only the terminal fragments contain the labeled primer, only they will fluoresce and their sizes can be determined by a fluorescence detector, as is used for reading DNA sequence fragments in cycle sequencing (see Section 10.1). The result is an intensity graph presenting the sizes of the fragments on the x-axis and their fluorescence intensity on the y-axis **(Figure 15.9)**. Each DNA fragment is represented as a peak, the size of which represents the quantity of that type of organism present. TRFLP therefore can give a profile of the diversity and relative numbers of different microbes present in the sample.

While the SSU rRNA gene has become the standard for microbial community analysis, either via direct sequencing, DGGE, or TRFLP, researchers may also examine genes that code for proteins that benefit the organism in a particular ecosystem. For example, *pmoA* genes encoding methane mono-oxygenase could be amplified from methanotroph-containing ecosystems, or *nif* genes encoding the subunits of the nitro-genase enzyme could be amplified from ecosystems in which N_2 fixation is taking place. The conserved sequences of these genes can also be used directly as probes to microscopically detect their expression or presence in cells by using fluorescently labeled oligonucleotide probes in a technique called **fluorescent *in situ* hybridization** (**FISH**; see Figure 6.2). Multiple, differently labeled probes can be used for simultaneous detection of cells with different genetic characteristics. This powerful technique can be used to examine the interactions of organisms at the cellular level. Fluorescence can also be used to screen cells by a method called **flow cytometry (Toolbox 15.1)**.

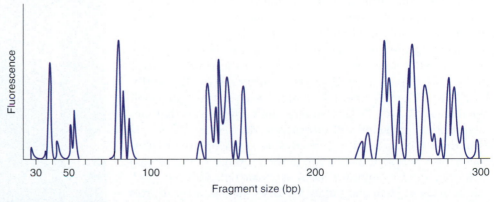

Figure 15.9. TRFLP data Different terminal fragment sizes are detected. The size of each peak represents the relative quantity of specific organisms in the sample.

Cells can be fluorescently labeled in a number of ways. Fluorescent dyes coupled to antibodies that recognize specific molecular structures on the outside of the cell can be used. The production of fluorescent proteins, such as green fluorescent protein (GFP), also results in fluorescence. The use of promoterless fluorescence protein genes such as the GFP gene makes it possible to couple the gene expression to promoters of interest so that the cells will fluoresce only when those genes are expressed. The technique of flow cytometry was developed to detect cells by their fluorescence. As shown in **Figure B15.1**, labeled cells in a liquid pass by a laser beam. Detectors in a fluorescence-activated cell sorter (FACS) measure the fluorescence and scatter of light. On this basis, it is possible to determine whether the cells are fluorescently labeled, the nature of the fluorescence, and even the volume of the cell. These determinations happen very quickly, with thousands of cells passing the laser per second. Once past the laser, the flow can be adjusted so that each droplet carries a single cell. Moreover, each cell-containing droplet can be sorted out to a container, based on the detected properties. Cells that fluoresce one color, indicating a certain property, can go in one container, while cells that fluoresce another color, indicating another property, can go in another container. The method is nondestructive; viable cells can be recovered from the different sorted batches. The live cells in these sorted batches can then be used in additional experiments.

● Test Your Understanding ....

Suppose that you want to determine the presence of *Nitrobacter winogradskyi*, a Gram-negative alphaproteobacterium capable of nitrogen fixation, in a soil sample. How could you use this technique to find cells that might be your organism, and what types of experiments would you perform next to confirm its identity?

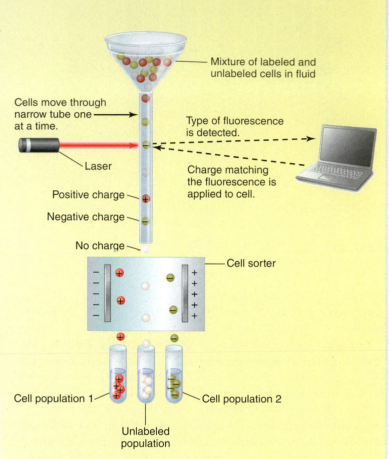

Figure B15.1. Flow cytometry In the fluorescence-activated cell sorter (FACS), cells can be sorted according to their fluorescence. In this example, the differentially labeled cells are separated into two different populations.

Stable Isotope Method

The inability to cultivate many members of microbial communities poses difficulties in determining the metabolic roles of different organisms. We can address this issue by examining the fate of radiolabeled molecules. Organisms capable of metabolizing a given compound may be determined by using stable isotopes. Two complementary variations of the **stable isotope probe (SIP)** method use DNA (DNA-SIP) or RNA (RNA-SIP). In DNA-SIP, environmental samples are exposed to small amounts of ^{13}C- or ^{15}N-labeled molecules. Organisms that metabolize the compound will incorporate the heavier ^{13}C or ^{15}N from these molecules into their DNA, thus labeling their DNA. Following a period of incubation in the presence of the labeled molecule, total community DNA is extracted. Labeled DNA will be heavier than the unlabeled DNA and, as a result, can be separated from unlabeled DNA by cesium chloride gradient ultracentrifugation (**Figure 15.10**). This heavy DNA can then be

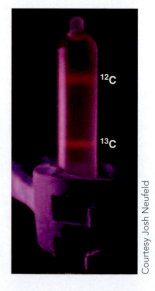

Courtesy Josh Neufeld

Figure 15.10. DNA stable isotope probing (DNA-SIP) The spiking of microbial habitats with key nutrients labeled with heavy nitrogen or carbon molecules results in those molecules being incorporated into the DNA of the organisms that take up and metabolize those molecules. The labeled DNA can be separated from non-labeled DNA, and through sequence analysis of that DNA, it is possible to determine the identity of the metabolically active members of the community. Note the separate bands of ^{13}C- and ^{12}C-labeled DNA that have been resolved by ultracentrifugation on a CsCl gradient.

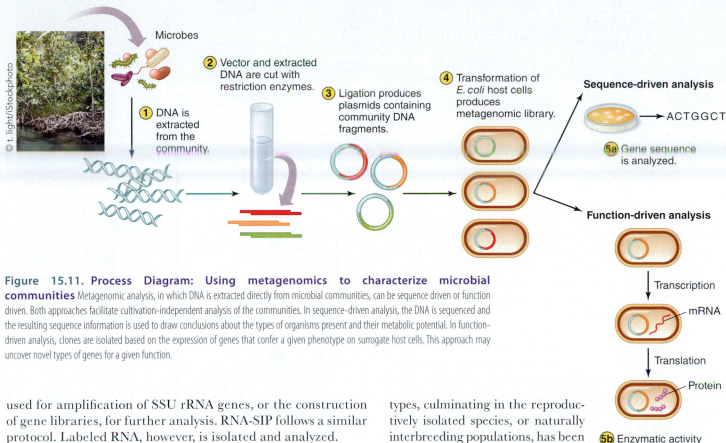

Figure 15.11. Process Diagram: Using metagenomics to characterize microbial communities Metagenomic analysis, in which DNA is extracted directly from microbial communities, can be sequence driven or function driven. Both approaches facilitate cultivation-independent analysis of the communities. In sequence-driven analysis, the DNA is sequenced and the resulting sequence information is used to draw conclusions about the types of organisms present and their metabolic potential. In function-driven analysis, clones are isolated based on the expression of genes that confer a given phenotype on surrogate host cells. This approach may uncover novel types of genes for a given function.

Labels in figure: Microbes; ② Vector and extracted DNA are cut with restriction enzymes. ① DNA is extracted from the community. ③ Ligation produces plasmids containing community DNA fragments. ④ Transformation of *E. coli* host cells produces metagenomic library. Sequence-driven analysis — ACTGGCT; ⑤a Gene sequence is analyzed. Function-driven analysis; Transcription — mRNA; Translation — Protein; ⑤b Enzymatic activity is analyzed.

used for amplification of SSU rRNA genes, or the construction of gene libraries, for further analysis. RNA-SIP follows a similar protocol. Labeled RNA, however, is isolated and analyzed.

In addition to all of these tools, metagenomics has also become a core tool for microbial ecologists **(Mini-Paper)**. As discussed in Chapter 10, metagenomics involves the construction of gene libraries from DNA extracted directly from the environment. Such libraries represent the total set of genomic information within the community. Genes can be isolated based on their function by screening for activity in surrogate hosts **(Figure 15.11)**. An advantage of this functional approach is that it may result in the isolation of genes encoding proteins with the given activity that might be missed by sequence comparison studies alone. Metagenomics also involves the use of high-throughput DNA sequencing methods such as pyrosequencing (see Section 10.1) in direct analysis of the community DNA.

Microbial diversity

As we have seen in this section, molecular techniques allow us to "see" many more microbial species than do culture techniques alone. So, what have we learned from these molecular studies? Quite simply, we share this planet with a lot of microbes! Four billion years of microbial evolution on Earth, with microbes actively contributing to the development of the planet's biosphere, has resulted in the astonishing breadth of genetic and metabolic diversity within today's microbial world. The extent of this diversity, however, has not been well recognized. Cultivation-independent techniques, especially those techniques involving direct sequencing, are telling us more about this diversity. Categorizing this new diversity represents the next problem.

Outside of the microbial world, the standard measure of diversity is the number of species. Division of organisms into different

types, culminating in the reproductively isolated species, or naturally interbreeding populations, has been central to the formulation and understanding of evolutionary processes. Microbiologists also use the species designation, but, because of the absence of true sexual reproduction, a bacterial species clearly is not equivalent to a plant or animal species.

Because the species concept may be difficult to apply to bacteria and archaeons, the term **operational taxonomic unit (OTU)** has been proposed. By definition, an operational taxonomic unit is any group of organisms that shares at least 97 percent SSU rRNA gene sequence identity. This measurement has been used as the microbial equivalent of a species. We should note, though, that if this same measure were applied to the class Mammalia, then it would include all members of the class, from rats to cats to chimpanzees!

In 2002, Thomas Curtis and colleagues famously reported their calculations showing that more species exist in a tonne of soil than in all the oceans combined, using the OTU as the definition of a microbial species. Most of these species had not been cultured. Only a small bit of DNA sequence data, usually just a portion of the 16S rRNA gene, was known. Should we classify these organisms as species with so little information? Should these bacteria even be given names? **Perspective 15.1** gives some insight into the naming systems used for uncultured and uncharacterized microbes.

As more data become available through direct DNA sequence analysis of different microbial communities, the relative abundance of the different types of organisms within these communities will become clearer. Probably more relevant to the operation of microbial ecosystems will be the ability to determine the exact composition of microhabitats, and the changes in diversity that occur over time and with changes in the microenvironment.

T. Uchiyama, T. Abe, T. Ikemura, and K. Watanabe. 2004. Substrate-induced gene-expression screening of environmental metagenome libraries for isolation of catabolic genes. Nature Biotech 23:88–93.

Context

The development of metagenomics has enabled investigators to better understand microbial communities. Studies usually follow one of two approaches: function-based screening or sequence-based screening. In the function-based approach, DNA clones are inserted into a surrogate host. These hosts are then screened for a specific phenotype such as a desired enzyme activity. In the sequence-based approach, DNA fragments are sequenced on a random basis, and the resulting sequence data are analyzed using bioinformatics tools for genes that might be of interest.

While both the function- and sequence-based approaches have yielded valuable information about microbial communities and provided novel genes that are of potential value for biotechnology applications, each of these strategies has its limitations. The function-based approach is limited by the inability of many genes to be expressed in specific surrogate hosts, and if phenotypic selection is not possible, the screening can be very laborious. While the sequence-based approach is not limited by the requirement for expression in a surrogate host, it is hampered by the fact that there is arguably more value in the identification of novel genes that are more distantly related to known genes, and such genes will not be identified by sequence similarity methods. Is there a way to enrich the metagenomic libraries for clones that are more likely to be involved in specific processes? One approach was described in 2005 by Kazuya Watanabe and colleagues, based on the likelihood that genes encoding catabolic pathway enzymes would be controlled by and located in close proximity to genes encoding transcriptional regulators that respond to the presence of the substrates of those catabolic pathways. This strategy was called "substrate-induced gene-expression screening," and abbreviated as "SIGEX."

Experiment

The implementation of the SIGEX strategy required the design and construction of a promoter probe plasmid carrying a reporter gene that could be fused to metagenomic DNA fragments. The commonly used green fluorescent protein (GFP) gene was chosen as the reporter gene, and cloned into the pUC18 plasmid to make the SIGEX cloning vector p18GFP. In this vector, the expression of the GFP gene was driven by the *tac* promoter of pUC18, resulting in detectable cell fluorescence. Individual cells could then be isolated based on their fluorescence by flow cytometry (**Figure B15.2**; see Toolbox 15.1). The cloning of metagenomic fragments into the *Bam*HI site between the *tac* promoter and the GFP gene disrupted the *tac*-driven expression of GFP, and instead put it under the potential control of regulatory genes present on the inserted DNA fragment.

To test the system, the researchers extracted DNA from groundwater below a crude-oil storage facility, and partially digested it with

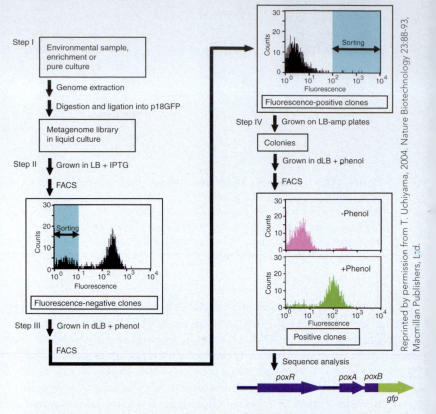

Figure B15.2. A type of flow cytometry was used to separate cells based on their expression of GFP. The SIGEX method involves the construction of metagenomic libraries by cloning DNA fragments into a vector that generates a fusion to a *gfp* gene, resulting in production of GFP in individual clones being controlled by the regulatory elements contained on those DNA fragments. First, FACS is used to screen out the clones that do not show GFP expression, and this subpopulation is then exposed to the desired inducing substrate. The clones that produce GFP in the presence of the inducing substrate contain genes whose expression is controlled by that substrate.

Reprinted by permission from T. Uchiyama, 2004. Nature Biotechnology 23:88–93, Macmillan Publishers, Ltd.

the *Sau*3A restriction enzyme. *Sau*3A enzyme has a 4-base recognition site, so it cuts more frequently than a restriction enzyme with a 6-base recognition site such as *Bam*HI. The sticky ends that result from *Sau*3A digestion are the same as the sticky ends resulting from *Bam*HI digestion, so the *Sau*3A fragments could be ligated into the *Bam*HI site of p18GFP. The recombinant plasmids were then transformed into *Escherichia coli* to produce a library that could be screened.

The goal was to identify metagenomic clones from the resulting library that contained genes expressed only in the presence of benzoate. Benzoate is a key intermediate in aromatic hydrocarbon catabolic pathways and is known to be an inducer of such pathways. Flow cytometry was used to first identify and then discard cells that fluoresced in the absence of benzoate; in these cells, expression of the GFP reporter gene was not dependent on the presence of benzoate. The remaining population of cells was then exposed to benzoate. The cells that fluoresced in the presence of benzoate were retrieved from the population by flow cytometry and spread on agar plates to obtain colonies. These colonies contained DNA fragments that contained benzoate-responsive regulatory genes, as well as the target genes, in an operon arrangement with the GFP reporter **(Figure B15.3)**. DNA sequence analysis of the insert DNA confirmed the operon structure of the genes, and demonstrated that most clones contained at least one transcriptional regulatory gene. Determination of taxonomic affiliation based on the DNA sequence characteristics indicated that most fragments were from members of the *Proteobacteria*, and one was determined to be from *Actinobacteria*.

The best match to known sequences was determined for each of the genes. Some of the genes were homologous to known aromatic degradative genes, while some of the clones contained genes predicted to be for catabolism of non-aromatic compounds, or for transporter proteins. One of the putative aromatic degradative genes, BZO71-8, was predicted to encode a cytochrome P450 enzyme. This gene was subcloned into an expression plasmid to obtain high-level expression, and the resulting enzyme activity was active against a specific compound, 4-hydroxy-benzoate, converting it to another compound protocatechuate.

Impact

This novel method of gene identification based on their expression characteristics is potentially very useful for the enrichment of genes that might not be isolated by the function-based or sequence-based approaches alone. As for other expression-based metagenomics strategies, the genes might not be expressed in the chosen host, so it would be desirable to carry out SIGEX in different hosts. A possible limitation to the method is that transcriptional regulator genes might not be located near the genes that they regulate. The overall success of metagenomics will depend on the development and use of novel screening methods, and each will have its unique strengths and limitations.

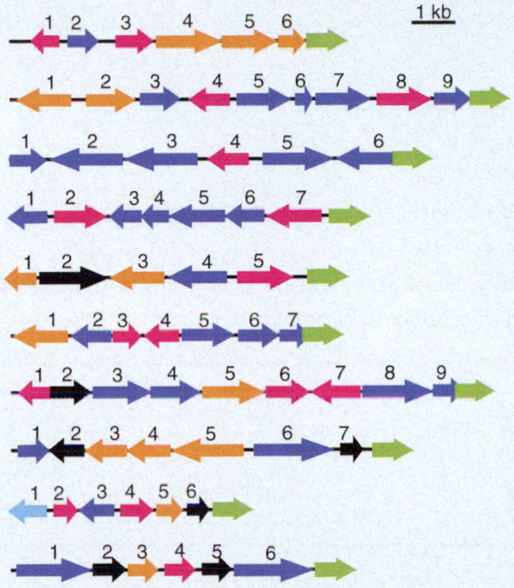

Name	Operon structure	Phylogenetic affiliation (%)			Induction efficiency (fold)
		1st	2nd	3rd	
BZO23		BP(96)	GP(4)		40
BZO26		GP(100)			302
BZO32		AC(96)	BP(4)		6
BZO47		GP(100)			152
BZO62		GP(66)	BP(23)		13
BZO70		AP(91)	GP(7)	BP(2)	7
BZO71		GP(87)	AP(9)	BP(4)	160
BZO135		DP(81)	AP(14)	BP(5)	11
NAP1		GP(97)	AP(1)	AC(1)	12
NAP3		BP(94)	BP(6)		53

Figure B15.3. Genes isolated by SIGEX Using SIGEX to screen for genes that respond to the presence of benzoate resulted in the discovery of a number of novel aromatic degradative operons. These operons were characterized by DNA sequence analysis.

Reprinted by permission from T. Uchiyama. 2004. Nature Biotechnology 23:88–93, Macmillan Publishers, Ltd.

● Questions for Discussion

1. The SIGEX method can be considered a selection for genes that are expressed in response to the presence of a specific substrate. How does this compare to the DNA-SIP method of enrichment?

2. The expression of some catabolic genes is not regulated by the compounds they are involved in degrading. Would such genes be isolated by SIGEX?

3. Some of the cloned genes did not appear to be involved in degradation of aromatic compounds. Discuss reasons why these were isolated by SIGEX.

Classification of living organisms is based on phylogeny—their evolutionary relationships, ideally based on their genetic relatedness or *phylogenetics*. For microorganisms, SSU rRNA genes can be used to assign evolutionary relatedness (see Chapter 1 Mini-Paper), but assigning a formal taxonomic name—a genus, a species, sometimes a subspecies—to a new bacterium ideally requires that the bacterium be isolated, cultured, and thoroughly described in the International Journal of Systematic and Evolutionary Microbiology, the official journal of record for new bacterial taxa. A culture of the bacterium should then be deposited in a culture collection for others to study. One of the largest culture collections of microbes is the American Type Culture Collection (ATCC), which maintains and distributes cultures of bacteria as well as protozoa, fungi, algae, viruses, plants, animal and human cells, and even genomic DNA, DNA probes, and antibodies.

Newly discovered uncultured bacteria that lack some of these characterizing details are not assigned a formal name. Yet there is often a practical need for some sort of designation so that scientists can compare results and build information regarding new isolates. A system for naming such putative new taxa has been recommended in the International Journal of Systematic and Evolutionary Microbiology. The designation "*Candidatus*," meaning "candidate," is used to identify uncultured bacteria that are characterized only partially. It does not include putative species identified only by a single sequence submission. For *Candidatus* designation, information, such as structure, metabolism, reproduction, and the environment in which the organism is naturally located, must accompany genetic data. The names usually are written with the category *Candidatus* in italics and the unique, proposed genus not italicized. If a species name is proposed, then this name is added and not italicized. An example of a proposed species name is *Candidatus* Carsonella ruddii (see Section 17.5).

So what do we call microbes that have been identified only by a stretch of nucleic acid sequence? These putative species commonly are referred to as "ribotypes" if the sequence is based on an RNA complement. Sometimes the term "phylotype" is used to designate a group of ribotypes that show close phylogenetic relationships. These terms are used in the recognition that limited sequence data identifies these microbes as distinct from one another, but little else is known about them.

15.2 Fact Check

1. Describe how SSU rRNA gene sequences are commonly used as a phylogenetic tool.

2. What is an operational taxonomic unit (OTU), and what gave rise to the term?

3. Explain how FISH can be used to characterize a microbial community.

4. Explain how the stable isotope probe method is used in studying microbial metabolism.

15.3 Aquatic ecosystems

What is the nature of life in water?

Earth is very much an aquatic planet, with over 70 percent of its surface covered by water. Most of that water, not surprisingly, exists in the oceans. Freshwater environments, though, obviously exist too, and provide quite different microbial ecosystems. In this section, we will begin to explore microbial life in aquatic habitats. We will begin by examining marine environments. Then, we will investigate freshwater environments. Along the way, we will look at how microbes survive in these aquatic habitats and how they interact with the environment.

Marine ecosystems

Earth's oceans are home to some amazing creatures. Blue whales, great white sharks, and the elusive giant squid all capture our imagination in myriad ways. Yet the great majority of life in our oceans consists of microbes. Bacteria, archaeons, and eukaryal microbes collectively constitute over 98 percent of the total ocean biomass. While the oceans certainly are not uniform, ocean water typically contains about 35 grams per liter of salt with about two-thirds of the salt consisting of Na^+ and Cl^- ions, and comparatively minor amounts of Mg^{2+}, Ca^{2+}, K^+, SO_4^{2-}, and HCO_3^-. In general, conditions are oxic, so aerobic respiration can occur, but levels of available nitrogen, phosphorus, and iron are low. As a result, microbes are often in a state of **oligotrophy**, or using nutrients at very low concentrations. Increasing the nutrient concentration in marine waters increases microbial growth, which quickly can lead to anoxic conditions, the ecological impacts of which can be devastating (**Perspective 15.2**).

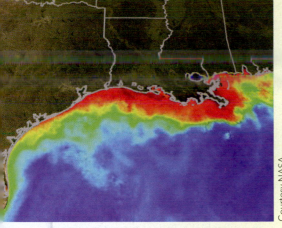

Courtesy NASA

A. Mississippi River Dead Zone (red)

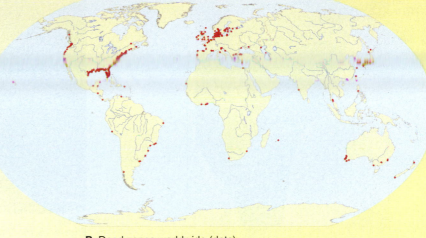

Based on Diaz and Rosenberg, 2008. Modified by Rabalais et al., 2010. Courtesy Robert J. Diaz

B. Dead zones worldwide (dots)

Figure B15.4. Dead zones A. The Mississippi River Dead Zone in the Gulf of Mexico is a coastal region that is heavily impacted by chemical fertilizer runoff. **B.** Dead zones around the world tend to be found in coastal regions at the outfall of rivers flowing through agricultural regions that are heavily fertilized.

Productive fisheries, many of them in coastal regions, have been important contributors to the sustenance of human populations for thousands of years. Increasingly, however, massive disappearances of fish and other sea life in coastal regions have been reported, especially in areas where rivers flow into the ocean. A prominent example of this problem is the Gulf of Mexico at the mouth of the Mississippi River **(Figure B15.4)**. These areas have been called "dead zones." Over 400 of these dead zones have been documented worldwide, and the numbers are increasing rapidly. Marine animal life cannot survive because of hypoxia, or low O_2 levels.

What causes the appearance of these dead zones? Within the oceans, the scarcity of key nutrients like nitrogen and phosphorus generally limits photosynthesis. As the Mississippi River flows through the intensively farmed and fertilized land of the American Midwest, though, runoff from the land deposits nitrogen and phosphorus into the river. As the river water enters the Gulf of Mexico, it brings with it a high dose of nutrients. Nitrogen and phosphorus are deposited in the Gulf of Mexico, converting the oligotrophic conditions to a nutrient-rich eutrophic state. Primary producers proliferate. The resulting massive input of energy and organic carbon into the marine ecosystem supports the increased growth of aerobic heterotrophic microbes. The aerobic respiration of these heterotrophs consumes O_2, causing hypoxia **(Figure B15.5)**.

If the soils of the Midwestern United States have been farmed for hundreds of years, then why are dead zones only now becoming such a problem? The answer can be linked directly to the increased use of synthetic and mined inorganic fertilizers that easily leach from the soil, instead of organic fertilizers like manure and compost and the bacterial symbiotic N_2 fixation in root nodules of legume crop plants (see Section 17.2). This increased use of synthetic fertilizers has become necessary to support intensive cultivation of crops like corn and wheat.

Of course, the dead zones are far from dead. They contain active communities of microbes. We know relatively little, though, about how the species in these communities cause hypoxia and how they then respond to the resulting low O_2 conditions. How might they adapt to life in the absence of O_2? How do their activities affect the cycling of marine nutrients and greenhouse gases such as CO_2? Do these hypoxic events have long-term effects on the microbial communities? We now have some tools needed to address these questions. The answers almost certainly will be extremely interesting.

© J. Baylor
Roberts/NG Image
Collection

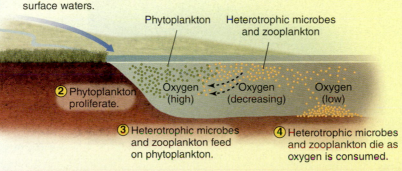

① Runoff from farmland adds nitrogen and phosphorus to surface waters.

Phytoplankton

Heterotrophic microbes and zooplankton

② Phytoplankton proliferate.

Oxygen (high)

Oxygen (decreasing)

Oxygen (low)

③ Heterotrophic microbes and zooplankton feed on phytoplankton.

④ Heterotrophic microbes and zooplankton die as oxygen is consumed.

Figure B15.5. Process Diagram: Dead zone formation There is a direct connection between dead zones and agricultural activities. Runoff from farmland increases the amounts of nitrogen and phosphorus in surface waters that empty into marine waters. The resulting increased eutrophication causes phytoplankton blooms that support increased growth of aerobic heterotrophic microbial populations. Consumption of phytoplankton cells by bacteria and zooplankton can severely reduce the level of oxygen in the water.

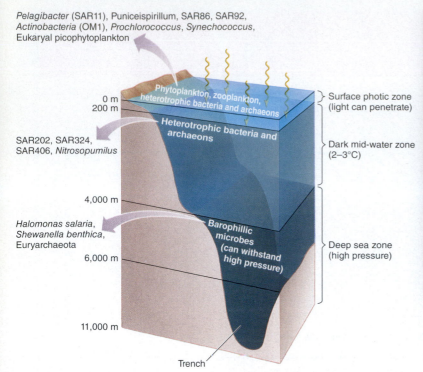

Pelagibacter (SAR11), Puniceispirillum, SAR86, SAR92,
Actinobacteria (OM1), Prochlorococcus, Synechococcus,
Eukaryal picophytoplankton

SAR202, SAR324,
SAR406, Nitrosopumilus

Halomonas salaria,
Shewanella benthica,
Euryarchaeota

Phytoplankton, zooplankton, heterotrophic bacteria and archaeons

Heterotrophic bacteria and archaeons

Barophilic microbes (can withstand high pressure)

0 m
200 m

4,000 m

6,000 m

11,000 m

Trench

Surface photic zone (light can penetrate)

Dark mid-water zone (2–3°C)

Deep sea zone (high pressure)

Berg, Hager, Hassenzahl: *Visualizing Environmental Science*, copyright 2011, John Wiley & Sons, Inc. This material is reproduced with permission of John Wiley & Sons, Inc.

Figure 15.12. Ocean zones The oceans can be conveniently divided into zones based on depth. These zones are characterized by amount of light, temperature, and pressure. Sampling and DNA sequence analysis of oceanic waters has revealed the major groups of organisms found in different zones. Most of these microbes have not been cultivated, so their physiology is not known and their roles are not well understood.

As we noted earlier, the oceans hardly are uniform. We can divide the ocean (**Figure 15.12**) into the surface zone, which includes the top few hundred meters where light can penetrate, the dark mid-water zone, where the temperature remains a constant 2–3°C, and the deep sea zone where extremely high pressures exist. SSU rRNA sequencing studies have revealed that the distribution of microbes changes as a function of depth (**Figure 15.13**).

Courtesy Jed Fuhrman, University of Southern California, Los Angeles

Figure 15.14. Viral particles and microbial cells in a seawater sample
The number of viral particles found in ocean waters is at least an order of magnitude higher than the number of cells. This figure shows the visualization of bacterial and/or archaeal cells and viral particles using fluorescence microscopy. The larger dots are the cells, while the smaller dots are the viral particles.

Within the ocean, numbers of microorganisms decline with depth as distance from the site of primary production increases. However, the proportion of archaeal cells increases with depth. Indeed, for much of the ocean volume, the biomass consists of more archaeons than bacteria, with upwards of 10^{29} microbial cells in all of the oceans combined. As large as this number seems, it has been estimated that the numbers of virus particles are an order of magnitude greater (**Figure 15.14**). We'll investigate their role at the end of this section. For now, let's briefly investigate the three major zones of the ocean.

Figure 15.13. Comparison of the relative abundance of bacteria and archaeons in ocean water DNA sequence analysis of samples collected from ocean depths has suggested that the proportion of archaeons increases with depth, and that there might actually be more archaeal cells than bacterial cells in the world's oceans.

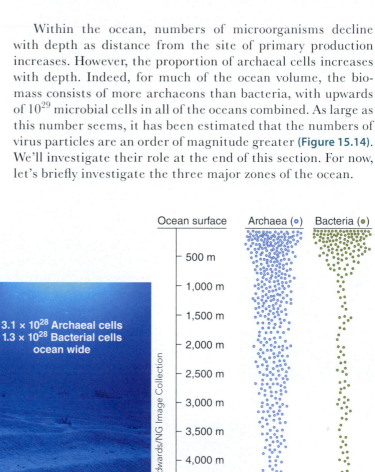

3.1×10^{28} Archaeal cells
1.3×10^{28} Bacterial cells
ocean wide

Ocean surface   Archaea (•)   Bacteria (•)

500 m
1,000 m
1,500 m
2,000 m
2,500 m
3,000 m
3,500 m
4,000 m
4,500 m
5,000 m

© Jason Edwards/NG Image Collection

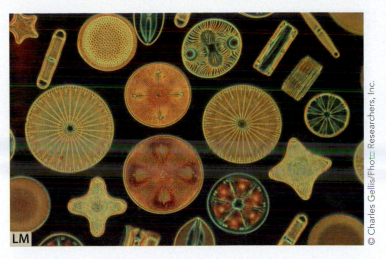
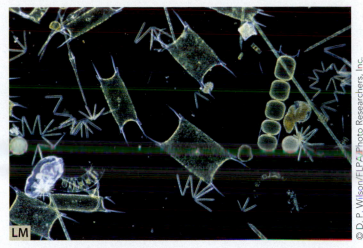

Figure 15.15. Micrographs of different diatom species A collection of diatom cells, demonstrating the uniquely patterned cell walls composed of silica and pectin.

Surface Zone

When you gaze out at crystal blue waters on a tropical beach, probably the last thing that comes to mind is that each teaspoon of water is teeming with hundreds of thousands of microbial cells. Despite the oligotrophic nature of the oceans, the typical cell counts in coastal marine environments can reach 10^8 cells per mL, and can be even higher in marine sediment. The oceans are perhaps the most intensively studied, and yet surprisingly not very well understood, ecosystems on the planet.

Photosynthetic microbial inhabitants of the surface waters of the oceans support all other life in the ocean, from other microbes to blue whales. Of course, the primary producers in the oceans are not higher plants like they are on land, but **phytoplankton**, photosynthetic microbes like the cyanobacteria *Prochlorococcus* and *Synechococcus*, diatoms like *Thalassiosira* **(Figure 15.15)**, and dinoflagellates like *Ceratium*. During photosynthesis, these organisms contribute to the oxygenation of the water. Many of the cyanobacteria can also fix N_2, increasing the nitrogen available in the system. Analysis of SSU rRNA genes indicates that a diverse collection of other, presumably heterotrophic, bacteria and archaeons accompany these phototrophs. Because most of these organisms have not been cultivated, their roles in the ecosystem remain unclear. Non-photosynthetic **zooplankton**, many of them protists and small crustaceans, graze on the primary producers and the other microbes in the surface zone, thus distributing the carbon and energy throughout the ecosystem.

Dark Mid-water Zone

Despite the seemingly inhospitable environmental conditions of the dark mid-water zone where photosynthesis is not possible, a wide variety of microbial life exists here, being fed by the organic matter produced in the upper surface waters. Organic carbon, released by viral lysis of the phytoplankton becomes available to support the growth of heterotrophic bacteria and archaeons. This released carbon tends to fall, often attached to small particles, and supports the heterotrophic microbes in the dark mid-water zone as well. It has been estimated that about 1 percent of the primary production reaches the seafloor, where anoxic conditions contribute to its slow breakdown and accumulation. Over hundreds of millions of years, this raining detritus has resulted in the formation of extremely rich organic sediment on the seafloor that contains very high cell numbers. Many of the oil and gas resources that our economies rely upon originate from such sediments.

Deep Sea Zone

The average depth of the seafloor is about 3,500 meters, but trenches can reach much greater depths. Indeed, the deepest point known, the Mariana Trench in the western Pacific, reaches to 11,000 meters below the sea surface. The atmospheric pressure there is about 1,000 times that at sea level. Microbes that can reside at these depths and withstand extremely high atmospheric pressures are called **piezophiles**, or barophiles. Some of them are obligate piezophiles that must be kept under pressure in order to grow at all.

Life under extreme pressure presents myriad problems. For example, high pressure causes the plasma membrane fluidity to decrease. Enzymes also tend to work less well at high pressures. To combat these problems, piezophiles often have high levels of polyunsaturated fatty acids (PUFA) in their plasma membranes. These PUFAs probably improve the fluidity of the membranes. Other adaptations of these microbes are still being investigated.

Viruses of the oceans

Probably 10 times more viruses than bacteria and archaeons combined, 10^{30} total virions, exist in the oceans. Yet we know relatively little about them or their roles in this vast ecosystem. Our level of knowledge is changing, however. In August 2011, researchers at the Bermuda Institute of Ocean Science, the University of South Florida, and the University of California, Santa Barbara, reported the results of a 10-year-long study of viruses in the ocean. During this period, they regularly monitored the viruses present at various depths within the Sargasso Sea. Although they sampled only the top 300 m of the ocean, their results demonstrate very definite temporal and spatial variations in the types of viruses present. Perhaps not surprisingly, viruses of cyanobacteria appear to dominate in the surface zone.

What role do these viruses play? Viruses of cyanobacteria replicate in, and typically lyse, cyanobacteria. When these cells lyse, fragments of the plasma membrane, DNA, proteins, and assorted nutrients are released into the environment. Other microorganisms can then acquire these organic molecules. The viruses, in other words, supply the ecosystem with nutrients. We still know very little about how these viruses interact with the environment. It is becoming clear, though, that viruses play an integral role in the normal functioning of the complicated marine ecosystem. Viruses probably play an equally important role in the normal functioning of other ecosystems, too.

Cultivation of oligotrophic microbes from seawater

The science of microbiology was founded on the ability to isolate and maintain pure cultures from single cells. The introduction of solid growth media that supports the growth of discrete colonies was central to pure culture. As discussed in Chapter 6, the development of microbial ecology was based on the application of enrichment culture techniques involving the design of conditions aimed at isolating cells that have particular physiological characteristics and metabolic abilities. However, many microorganisms are not readily isolated in pure culture, and there are a number of reasons why this is so. Some grow too slowly to form colonies in a reasonable amount of time without being overwhelmed by the faster

TABLE 15.4 Elemental composition of seawater and marine broth 2216

Element	Seawater (grams per liter)	Marine broth 2216 (grams per liter)
Chlorine	19	19.8
Sodium	10.5	8.8
Magnesium	1.35	2.2
Sulfur	0.885	0.73
Calcium	0.4	0.65
Potassium	0.38	0.32
Bromine	0.065	0.054
Carbon	0.028[a]	4.78
Boron	0.0046	0.0047
Silicon	0.003	0.00085
Fluorine	0.0013	0.00109
Nitrogen	0.0005[a]	0.72
Phosphorus	0.0007[a]	0.045
Iron	0.00001	0.0226

[a]Contains particulate and dissolved components. Note that biologically available organic carbon in seawater is usually less that 12 μg per liter.

Courtesy Cheryl Chow, University of Southern California Marine and Environmental Biology

Figure 15.16. Process Diagram: Dilution to extinction method This method involves first using a microscope to count the cells, then diluting the sample in small volumes of autoclaved seawater so that there is on average less than one cell per starting culture. Cultures that show growth are subcultured to large volumes of autoclaved seawater and incubated. If cultures grow, they can be harvested by centrifugation and analyzed further.

① Seawater sample is collected and direct microscopic counts are made.

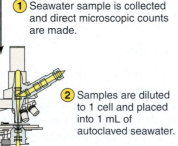

② Samples are diluted to 1 cell and placed into 1 mL of autoclaved seawater.

③ Tubes showing growth are subcultured in a larger volume.

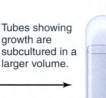

④ Cells are harvested by centrifugation for analysis.

Courtesy Trevor Charles

growing microbes in the community. Some rely on the presence of multicellular organisms or other microbes with which they form obligate intimate associations (see Chapter 17). Others, like marine microbes, are optimized for low nutrient, oligotrophic conditions.

The study of microbial communities in marine environments provides a particularly good example of the imprecise view provided by the use of cultivation and enrichment-based approaches. Standard commercially available marine isolation media contain elemental profiles similar to seawater, except that carbon, nitrogen, phosphorus, and iron are present at levels several orders of magnitude greater than the levels found in seawater (Table 15.4). While these media are very good for isolating some marine organisms such as *Vibrio fischeri* and *Vibrio harveyi*, they do not replicate the growth conditions of the natural habitat. Seawater generally contains extremely low levels of dissolved organic carbon, often less than 1 mg of carbon per liter. Moreover, microbial cells cannot assimilate much of this carbon easily. A fundamental question, therefore, needs to be addressed: Do the oceans harbor microbes specialized for growth under very low available carbon concentrations?

Even before the widespread adoption of nucleic acid-based techniques, microbiologists realized that only a small fraction of the cells detectable by direct microscopic observation of ocean water formed colonies on marine agar. Environmental DNA sequence studies subsequently confirmed that the dominant organisms in these communities are not the organisms easily isolated on standard marine isolation media. Instead, these communities consist mostly of microbes that cannot be cultivated.

Genetic information obtained from metagenomic studies may shed light on the metabolic characteristics of the members of microbial communities and their contributions to biogeochemical cycles and ecosystem dynamics. Physiological studies, however, require the availability of pure cultures. Efforts to obtain cultures of marine oligotrophs are hindered by the fact that low levels of nutrients do not yield large amounts of biomass, and certainly do not support the formation of visible colonies on solid growth media.

So, how do we isolate and grow the predominant but cultivation-resistant cells? We should start by recognizing that these cells might be obligately oligotrophic and unable to withstand high concentrations of carbon. We should also separate them from cells that may be less abundant but also less sensitive to higher levels of carbon. In the presence of higher carbon levels, these cells would quickly outgrow the obligate oligotrophs. The standard method of separating the cells for cultivation, developed in the laboratory of Don Button at the University of Alaska in the early 1990s, involves **dilution to extinction**. Seawater samples are diluted so that very few cells remain in each aliquot. Inoculations are then made from these dilutions into autoclaved seawater, followed by incubation. Pure cultures of obligate oligotrophs have been isolated using this approach (**Figure 15.16**). Titers are usually limited to about 10^6 cells per mL, which corresponds to the total cell concentrations in a typical natural seawater community.

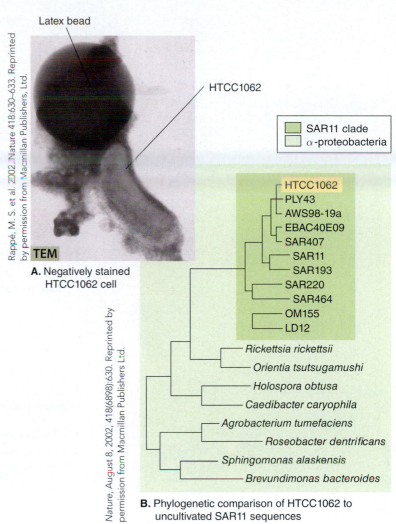

A. Negatively stained HTCC1062 cell

B. Phylogenetic comparison of HTCC1062 to uncultivated SAR11 sequences

Figure 15.17. Successful cultivation of members of SAR11 The dilution to extinction method was used to cultivate these previously uncultivated but abundant marine bacteria. The phylogenetic tree compares the SSU rRNA sequence of one of the cultures, strain HTCC1062, with SAR11 SSU sequences that had previously been determined from seawater samples without cultivation.

In 2002, Stephen Giovannoni and co-workers at Oregon State University used the dilution to extinction method to isolate the widespread SAR11 type of microorganism from the Sargasso Sea in pure culture (**Figure 15.17**). This microbe type has been estimated to make up 25–50 percent of the cells in seawater microbial communities. Prior to growing it in the laboratory, researchers knew little about the role of SAR11 in the ocean. The availability of pure cultures has allowed microbiologists to determine its genome sequence, study its growth, biochemistry, and physiology, investigate its role in biogeochemical cycles, and learn more about its connections with other organisms.

Ocean community genomics

The extremely low culturability of marine microbes (see Table 15.1) has led to substantial sequence-based metagenomics studies of marine ecosystems (see Figure 15.14). In these experiments, researchers collect samples of seawater and recover the biomass by filtration. DNA is extracted from the cells and then prepared for sequencing. The resulting DNA sequence data can then be

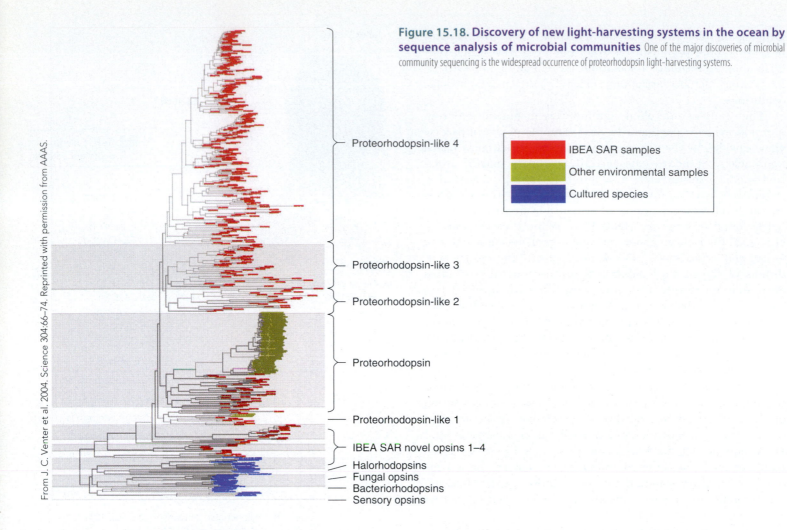

Figure 15.18. Discovery of new light-harvesting systems in the ocean by sequence analysis of microbial communities One of the major discoveries of microbial community sequencing is the widespread occurrence of proteorhodopsin light-harvesting systems.

Proteorhodopsin-like 4

Proteorhodopsin-like 3

Proteorhodopsin-like 2

Proteorhodopsin

Proteorhodopsin-like 1

IBEA SAR novel opsins 1–4

Halorhodopsins

Fungal opsins

Bacteriorhodopsins

Sensory opsins

🟥	IBEA SAR samples
🟩	Other environmental samples
🟦	Cultured species

compared to DNA sequence databases. Craig Venter and other investigators have used this approach to catalog the genomic information in marine surface waters. Their studies have revealed new organisms and new proteins and suggested new types of metabolic activities. One surprising finding was the apparently widespread occurrence of proteorhodopsin in marine bacteria **(Figure 15.18)**. Proteorhodopsin commonly functions as a light-driven proton pump to create a proton motive force used to drive ATP synthesis (see Section 13.6). It may also have sensory functions in some bacteria. Perhaps this energy-capturing process contributes significantly to ocean carbon and energy levels.

Even with all the new sequence data, scientists have only begun to understand the complexity of ocean ecosystems. The functions of newly determined sequences often cannot be inferred because they are not similar enough to genes in characterized organisms. While the widespread use of DNA-sequencing studies has greatly furthered our knowledge, cultivation of oligotrophic members of seawater communities remains necessary and continues to present a unique challenge.

Freshwater ecosystems

Freshwater ecosystems, including lakes, rivers, and streams, differ in many ways from their marine counterparts. In general, freshwater environments contain much higher nutrient levels

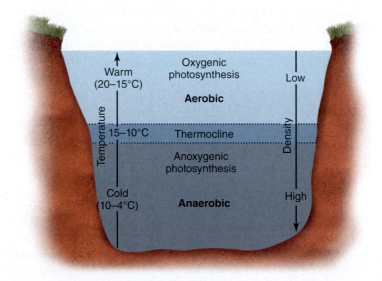

Figure 15.19. Structure of a stratified lake Deep lakes can form discrete layers. Typically, aerobic upper zones are separated from anaerobic lower zones by the thermocline. The surface waters support oxygenic photosynthesis; below that is a region of aerobic heterotrophic activity. At the thermocline, anoxygenic photosynthesis by H_2S-oxidizing organisms occurs. The deeper regions support anaerobic heterotrophic activity and NO_3^- reduction. Anaerobic decomposition takes place in the sediment.

than marine environments. They normally receive higher levels of organic matter due to runoff from land rather than relying on biomass production by the photosynthetic microbes.

In deep lakes, stratification into separate aerobic and anaerobic zones can occur **(Figure 15.19)**. This stratification results from the temperature-dependent density of water. Typically, an upper layer of warmer, low-density water is separated from the lower, colder, high-density water by a thin layer called the **thermocline**, a region of rapid temperature and density change. Aerobic conditions exist in the upper layer, while anaerobic conditions exist in the lower layer. Primary production through oxygenic photosynthesis occurs in the uppermost zone and supports aerobic heterotrophic activity directly below it. Just below the thermocline, the anaerobic conditions can allow anoxygenic photosynthetic bacteria to contribute to primary production. The deepest layers contain considerable CH_4 oxidation and NO_3^- reduction. The sediments support significant methanogenesis and SO_4^{2-} reduction.

Many deep tropical lakes and relatively fewer temperate lakes have permanent stratification with no mixing of the layers. In colder regions, lake stratification occurs seasonally, becoming established in the spring and lasting throughout the summer. During the winter, the top aerobic layer cools. The water becomes densest when the temperature drops to 4°C and sinks to the bottom. The more anaerobic water that was on the bottom then rises to the top where it may cool to temperatures below 4°C. This new stratification lasts until the surface water warms enough for the "spring turnover." The ensuing mixing of the layers results in the redistribution of nutrients from the lower layers, and can result in natural algal and cyanobacterial blooms in the summer **(Figure 15.20)** as nutritional limitations to growth suddenly are relieved. Similar to the marine dead zones, these freshwater blooms can lead to anoxia and fish kills. However, the organisms in blooms are not all bad; *Synechocystis* sp. **(Microbes in Focus 15.1)**, a common component of cyanobacterial blooms, has been intensively studied as a model organism for photosynthesis and the photosynthetic production of biofuels and bioproducts. The relatively easy genetic manipulation of this microbe lends itself very well to this type of application.

Microbes in Focus 15.1
SYNECHOCYSTIS SP. PCC6803—FROM UNPLEASANT CYANOBACTERIAL BLOOMS TO SCIENCE AND BIOTECH WORKHORSE

Habitat: Isolated from a Berkeley, California, freshwater lake in 1968; has become one of the best studied of the cyanobacteria and a model system for oxygenic photosynthesis.

Description: Unicellular, with spherical cells ranging in size from 2 to 7 μm. The genome sequence, determined in 1996, consists of a 3.57-Mb chromosome and seven plasmids ranging in size from 120 kb to 2.3 kb.

Photo provided by Trevor Charles. Courtesy Bithi Eshaque

Key Features: The organism can carry out oxygenic photosynthesis, and can grow either autotrophically or heterotrophically. In the dark, it can also fix nitrogen. It is easily transformable by DNA, can act as a conjugation recipient, and can be mutated by transposons or through homologous recombination methods. Researchers are studying this organism to gain insight into the fundamentals of oxygenic photosynthesis, as well as for the development of systems for the light-driven production of biofuels and bioproducts. *Synechocystis* sp. PCC6803 was the first photosynthetic organism to have its genome completely sequenced.

Figure 15.20. Algal bloom Overgrowth of algae occurring along the Susquehanna River in Pennsylvania.

© Michael P. Gadomski/Photo Researchers, Inc.

15.3 Fact Check

1. What is oligotrophy?
2. Describe the conditions found in the three main zones of the ocean.
3. What are phytoplankton, and what is their role in ocean waters?
4. Explain the role of zooplankton and viruses in surface ocean waters.
5. Describe the layers and associated microbial life found in freshwater ecosystems.

15.4 Terrestrial ecosystems

What is the nature of life on land?

The various terrestrial ecosystems on Earth have been divided into several **biome** types, categories based on vegetation characteristics **(Figure 15.21)**. Temperature and precipitation distinguish these biomes. Latitude, elevation, and geological features, in turn, influence these factors. In these biomes, photoautotrophic plants play central roles as the dominant primary producers, providing nutrients to the soil via microbial decomposition and photosynthesis-derived root exudates. Because the health of soils is a key determinant in the functioning of terrestrial ecosystems, we will begin this section by examining soils. Then we will investigate the rhizophere, look at microbial communities in terrestrial ecosystems, and examine how successful bioremediation relies on our knowledge of these communities.

Soils

Soil, a relatively thin band over the terrestrial surface of Earth, contains significant microbial diversity. Soil microbial communities are also the most active microbial communities, with the highest rate of cell turnover (see Table 15.3). Soils form through the microbial decomposition of plant and animal matter. This material then combines with abiotic material, such as minerals from weathering of rocks. Plants also contribute to the soil by excreting nutrients from their roots that encourage the growth of diverse microbial communities, especially communities of ammonifiers and denitrifiers (see Chapter 14). Many of these microbes directly benefit the plant by transforming various chemical compounds to forms that can be better assimilated by the plant. This interplay between plants and their microbial partners is central to the development of soils.

Due to the importance of soils in agriculture and forestry, soil scientists have developed extensive systems for categorizing soil types. Soils, much like oceans, are divided into distinct zones. A typical temperate zone soil can be divided into layers, or **horizons** **(Figure 15.22)**, based on certain characteristics. The deepest layers, close to the bedrock and composed primarily of inorganic material, are called the C horizon. Directly above this layer is the B horizon, sometimes called "subsoil," that contains some organic matter. The A horizon, representing the uppermost surface layers, is known as "topsoil." The organic matter on the surface of the soil composes the O horizon. The topsoil

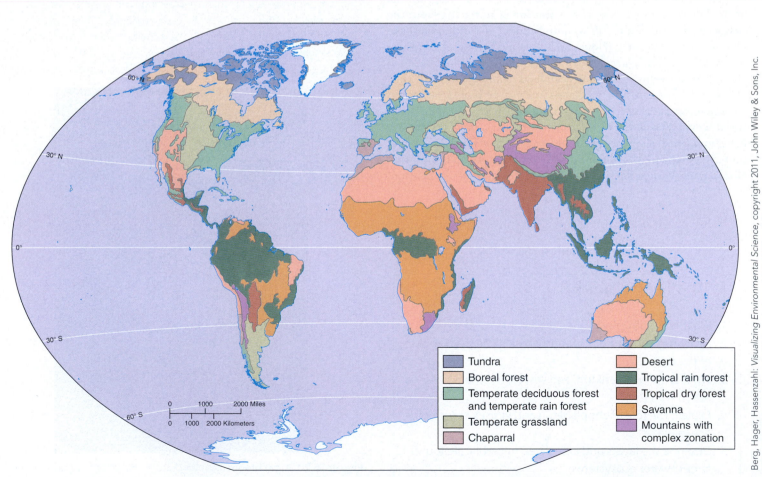

Legend:
- Tundra
- Boreal forest
- Temperate deciduous forest and temperate rain forest
- Temperate grassland
- Chaparral
- Desert
- Tropical rain forest
- Tropical dry forest
- Savanna
- Mountains with complex zonation

Figure 15.21. Global map of terrestrial biomes Terrestrial ecosystems are divided into biome types based on vegetation characteristics.

TABLE 15.5 FAO soil classification

FAO classification	General characteristics
Acrisols	Highly weathered low base saturation (<50%), clay horizon
Andosols	Derived from volcanic ejecta, usually low bulk density
Arenosols	Sandy soils, poorly developed, no diagnostic horizons
Cambisols	Weakly developed soils with a cambic or umbric (light to dark brown) horizon
Chemozems	Deep, nearly black mellic horizon
Ferralsols	Highly weathered tropical soils
Fluvisols	Young soils developed on alluvial deposits
Gleysols	Very wet soils, but not organic
Greyzems	Deep, dark grey mollic horizon
Histosols	Organic soils
Kastanozems	Deep, dark brown surface horizon, high calcium content
Lithosols	Shallow (<10 cm) over hard rock
Luvisols	Argillic horizon
Nitosols	Argillic horizon with even distribution of translocated clay
Phaeozems	Deep, dark grey mollic horizon
Planosols	Flat, poorly drained soils
Podzols	Translocation of humus downward
Podzoluvisols	Transition between podzols and luvisols
Rankers	Steep, shallow soils
Regosols	Thin, weakly developed soils
Rendzinas	Deep, dark surface horizon with high lime
Solonchaks	Saline desert soils
Solunetz	Sodic desert soils
Vertisols	Contains >30% swelling clays
Xerosols	High calcium desert soils
Yermosols	Desert soils

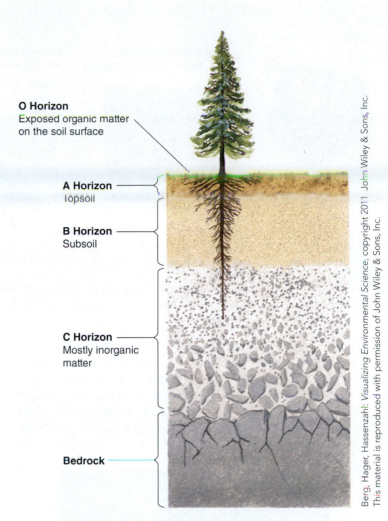

O Horizon
Exposed organic matter on the soil surface

A Horizon
Topsoil

B Horizon
Subsoil

C Horizon
Mostly inorganic matter

Bedrock

Berg, Hager, Hassenzahl: *Visualizing Environmental Science*, copyright 2011 John Wiley & Sons, Inc. This material is reproduced with permission of John Wiley & Sons, Inc.

Figure 15.22. Soil horizons Soils are divided into layers based on depth. The amount of organic matter is highest in the upper levels, and decreases with increasing depth.

has more organic material than the lower layers due to its proximity to plant roots and decomposing biomass. Accordingly, microbial activity and numbers, like that in oceans, are highest near the surface.

In addition to the soil horizons, soils are classified according to a soil taxonomy. The United Nations Food and Agriculture Organization soil taxonomy system describes 26 different soil classes based on physical characteristics (Table 15.5). Soil texture represents the most important physical characteristic and is defined by the proportion of particle sizes that make up the mineral components of the soil. Sand particles range from 0.5 to 2 mm in diameter, silt from 0.002 to 0.5 mm, and clay less than 0.002 mm. Twelve textural classes can be defined based on the combination of particle sizes (Figure 15.23).

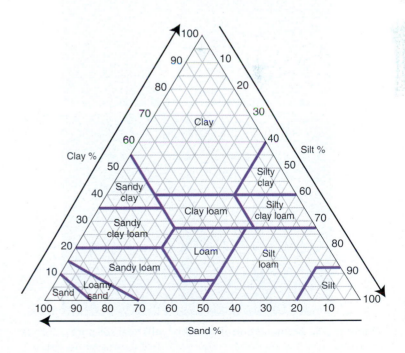

Figure 15.23. The twelve textural classes of soils The classes of soils are categorized based on the relative amounts of clay (< 0.002 mm particle size), silt (0.002–0.05 mm particle size), and sand (0.05–2 mm particle size).

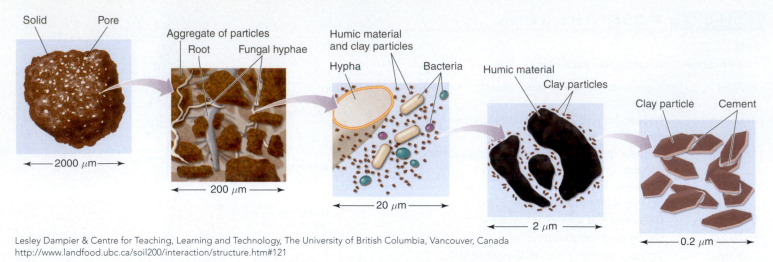

Solid   Pore

Aggregate of particles
Root   Fungal hyphae

Humic material
and clay particles

Hypha   Bacteria

Humic material
Clay particles

Clay particle   Cement

← 2000 μm →

← 200 μm →

← 20 μm →

← 2 μm →

← 0.2 μm →

Lesley Dampier & Centre for Teaching, Learning and Technology, The University of British Columbia, Vancouver, Canada
http://www.landfood.ubc.ca/soil200/interaction/structure.htm#121

Figure 15.24. Model of soil aggregate formation mediated by microbial cellular material In a typical soil from lower magnification (*left*) to higher magnification (*right*), it is apparent that microbial cells and microbial debris hold the clumps together and influence the overall structure.

The aggregation of soil, that is, its tendency to form clumps, determines its structure with the main binding agents being substances derived from microbial cells, as well as from **humic material** that results from incomplete breakdown of plant biomass **(Figure 15.24)**. Microbes, although found at very high concentrations, typically on the order of 10^9 cells/g, are not evenly distributed in the soil. They tend to accumulate within or near soil pores, containing air and water but little solid material where nutrients might be more readily available **(Figure 15.25)**.

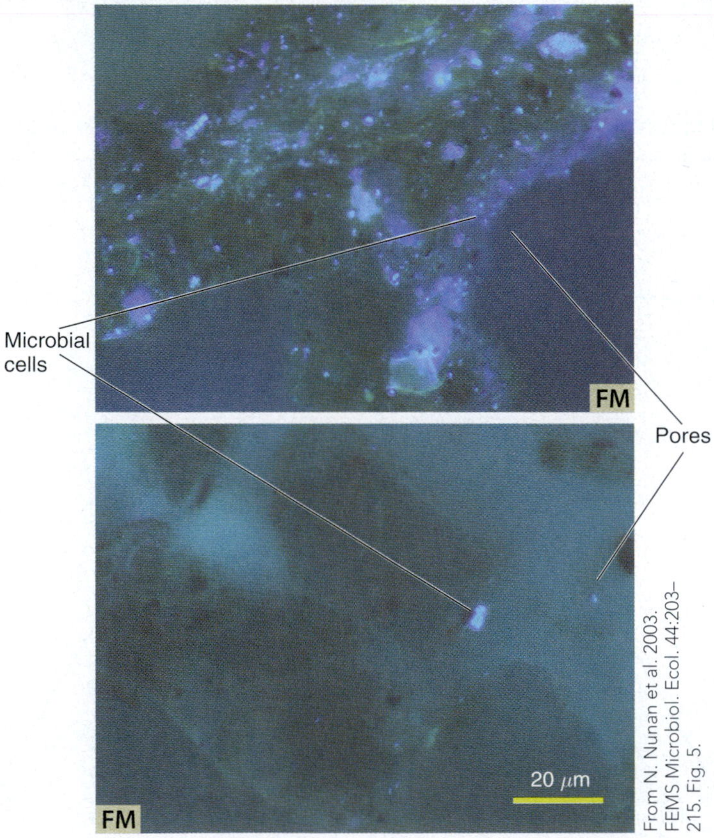

The organic compounds found in soils mostly result from the decomposition of plant matter. Soluble components, such as sugars and amino acids, can be assimilated and metabolized by various organisms. Extracellular enzymes must break down high molecular weight lignocellulose molecules consisting of cellulose, hemicellulose, and lignin. The resulting breakdown products can then be assimilated, resulting in formation of energy and new organic matter within the microbial cell. Cellulase and hemicellulase can degrade, respectively, cellulose and hemicellulose. Lignin, a complex macromolecule composed of covalently linked phenolic-alcohol molecules, is quite resistant to degradation by all but a few types of microbes. As a result of this resistance, the relative proportion of lignin increases during decomposition as the soil forms.

Microbial cells

FM

Pores

From N. Nunan et al. 2003.
FEMS Microbiol. Ecol. 44:203–215. Fig. 5.

20 μm

FM

Figure 15.25. Accumulation of microbial cells near soil pores Microscopic examination of soils indicates that, rather than being evenly distributed throughout the soil, microbial cells tend to accumulate near soil pores.

Rhizosphere

Besides contributing their biomass to the microbial decomposers, plants influence soil microbial communities in other important ways. The **rhizosphere**, that part of the soil immediately surrounding plant roots, represents an interesting and quite important ecosystem. In comparison to the non-root-associated bulk soil, the rhizosphere contains large amounts of organic carbon due to the considerable excretion of plant root exudate. These exudates contain a mixture of sugars, sugar alcohols, and organic acids. Additionally, root cap cells that slough off the growing plant roots contribute organic carbon to the rhizosphere. As a result of this extra organic carbon, the rhizosphere typically contains higher numbers of microbes than bulk soil **(Table 15.6)**. This rhizosphere effect provides evidence for the direct influence of plant roots on the soil microbial community.

Bacterial group	Rhizosphere soil (number/g)	Bulk soil (number/g)	Relative abundance of microbes
Ammonifiers	5×10^8	4×10^6	125:1
Gas-producing anaerobes	3.9×10^5	3×10^4	13:1
Anaerobes	1.2×10^7	6×10^6	2:1
Denitrifiers	1.26×10^8	1×10^5	1,260:1
Aerobic cellulose degraders	7×10^5	1×10^5	7:1
Anaerobic cellulose degraders	9×10^3	3×10^3	3:1
Endospore formers	9.3×10^5	5.75×10^5	1:1

TABLE 15.6 The influence of rhizosphere on the microbial community

A large proportion of the carbon fixed as a result of photosynthetic activity in plants moves out of the roots into the rhizosphere as root exudate. These rich nutrients support the activity and proliferation of large and diverse microbial communities. Because microbes vary in their abilities to utilize different carbon sources, the mixture of carbon compounds largely determines the microbial community structure in a given rhizosphere. Obviously, the microbes that proliferate in this region receive a clear nutritional benefit. Let's look at the rhizosphere from a different perspective. Do the plants benefit by providing these nutrients to the microbes?

Complex rhizosphere microbial communities can influence plant growth both positively and negatively. While some microbes are pathogenic for plants, the majority are not. Many microbes have beneficial effects on the growth of plants. Moreover, an extensive and sophisticated two-way communication exists between the partners. Some of the mechanisms used by soil bacteria to promote plant growth include:

- production of phytohormones or compounds that interact with phytohormones, which directly stimulate plant growth
- production of antibiotics that prevent the growth of pathogenic microbes
- fixation of atmospheric nitrogen (N_2) to a form that the plants can use
- solubilization of phosphate so that it can be taken up by the plant roots

CONNECTIONS In **Section 17.2**, we will consider several examples of symbiotic microbes specialized for growth in the rhizosphere, including organisms that promote the growth of plants by converting atmospheric nitrogen into compounds plants can use directly or producing antibiotics to inhibit plant pathogens.

Microbiological communities in soils

Soils are fascinatingly complex and heterogeneous, in both the abiotic and the biotic sense. The immense microbial diversity within soils and the distribution of these microbes

reflects this complexity. The use of high-throughput DNA sequence analysis **(Figure 15.26)** has allowed the microbiota of different soils to be compared. We now know that soils that appear to be very similar in the physical sense can sometimes share very few, if any, of the same microbes. This realization, made possible only through the use of high-resolution microbial community DNA sequence analysis, challenges the fundamental ideas about the distribution of microbes within and between ecosystems.

Think about the last time you took a walk in the woods and came upon an old stump or fallen tree. Microbes cannot easily degrade some of this plant biomass, like the lignocellulose that

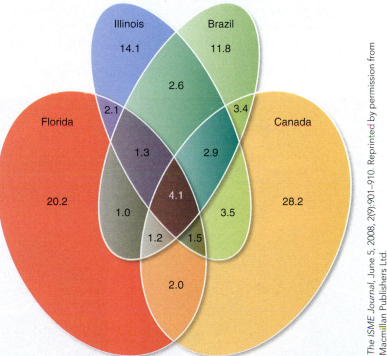

The *ISME Journal*, June 5, 2008, 2(9):901–910. Reprinted by permission from Macmillan Publishers Ltd.

Figure 15.26. Overlap of microbial OTUs from different soils It is apparent from this Venn diagram that very few microbes are common to all soils. The data for this comparison are SSU rRNA sequences found in different soils. Note that the numbers add up to 100 and represent the overall OTU distribution.

we introduced earlier. Lignocellulose eventually does break down. Much of it, though, does not completely **mineralize**, or break down to CO_2 and water. Such incomplete mineralization contributes to soil organic matter. While most microbes cannot degrade the complex lignocellulose, fungi such as the brown rot and white rot basidiomycete fungi specialize in lignocellulose degradation. The fruiting bodies of these fungi are often visible on rotting tree stumps **(Figure 15.27)**. While the white rot fungi can break down lignin and cellulose, the brown rot fungi can only break down cellulose. In both types of organisms, cellulase and hemicellulase enzymes convert cellulose and hemicellulose polymers to their component sugars. The white rot fungi also produce and secrete lignin-degrading enzymes such as peroxidases and oxidases. These organisms, in fact, are the only microbes that can depolymerize lignin. They therefore play an essential role in the decomposition of plant matter.

The actinomycete bacteria are well known for their abundance in soils and the special roles they play in soil ecosystems. In Chapter 12, we looked at the prominence of these bacteria in the pharmaceutical industry and their role in the production of antibiotics and other types of drugs. The production by these organisms of so many secondary metabolites that happen to be of pharmaceutical interest must have some link to their natural habitat. Recall that secondary metabolites are small organic molecules produced by cells but not essential for the normal growth of the organism. The actinomycetes, such as members of the genus *Streptomyces*, produce and secrete a number of enzymes like hydrolases that break down a number of different compounds, proteases that degrade protein, chitinases that break down chitin, cellulases that break down cellulose, and amylases that break down starch. Together, these enzymes effectively break down much of the plant organic matter, providing the actinomycetes with ample nutrients. The production of antibiotic secondary metabolites by actinomycetes inhibits the growth of potential competitors for these nutrients.

The incomplete breakdown of plant biomass, especially lignin, results in the accumulation in soil of recalcitrant organic molecules. These molecules, humic substances, contribute to the important aggregation properties of soils, resulting in the formation of clumps and a complex spatial heterogeneity. The diversity of microenvironments provides opportunities for many different types of specialist microbes, resulting in high diversity. Typically, microbial growth is limited in bulk soils, away from the influence of the plant root rhizosphere, as the organisms are in a state of famine in which all available nutrients are quickly used. Organisms that can survive for long periods under these conditions, including microbes like the *Firmicute Bacillus* sp. that form heat- and desiccation-resistant endospores, will be selected for. Input of nutrients from plant exudates and decomposing plant matter supports the proliferation of those organisms that can take advantage of these nutrients with high growth rates.

Bioremediation

As we have seen in the previous section, different soil conditions support different microbial communities. In every location, microbes that adapt to the conditions, especially the available nutrients, proliferate. While we have focused in this section on naturally occurring nutrients, suppose that human-made materials proliferate in the soil. Microbiologists are exploring how to use microbes to clean up this contamination, a process referred to as **bioremediation (Figure 15.28)**.

Brown rot basidiomycetes

White rot basidiomycetes

© Michael P. Gadomski/Photo Researchers, Inc.

© Robert & Jean Pollack/Photo Researchers, Inc.

Figure 15.27. White rot and brown rot basidiomycetes Fruiting bodies of white rot and brown rot fungi growing on dead tree trunks.

A. Petroleum-contaminated site before bioremediation

B. Contaminated site following treatment with hydrocarbon-degrading microbes

Figure 15.28. Bioremediation Microbes can be used to help clean up sites contaminated with harmful contaminants, such as petroleum hydrocarbons. **A.** Petroleum contamination caused barren conditions and poor growth of native plants. **B.** Following bioremediation treatment including application of hydrocarbon-degrading bacteria, this site is safe for cultivation of food crops.

Human activities generate considerable waste, much of it ending up in the soil. Diverse communities of soil microbes degrade some of this waste quite easily. As we saw in Chapter 14, microbes recycle this waste through the biogeochemical cycles. Since industrialization and the development of chemical industries, however, more and more human-generated waste cannot be degraded easily by typical soil inhabitants. Tens of thousands of synthetic organic chemicals are in commercial use. These **xenobiotics**, chemicals not naturally found in or produced by organisms, are often toxic, carcinogenic, mutagenic, teratogenic, or otherwise detrimental to human and animal health. Moreover, because these chemicals do not occur as part of natural processes, microbes have not evolved the ability to degrade them efficiently.

Major types of xenobiotics of concern include the petroleum-derived polycyclic aromatic hydrocarbons (PAHs), chlorinated solvents such as polychlorinated biphenyls (PCBs) and trichloroethylene (TCE), and nitroaromatics such as 2,4,6-trinitrotoluene (TNT) used as explosives (**Figure 15.29**). Generally, laws regulate the disposal of such chemical wastes. Nevertheless, many sites throughout the world are extensively contaminated. Soil microbes, as we have seen, can degrade most organic compounds. Over time, the natural microbial communities probably will degrade even the most resistant of these pollutants. Time, however, may not be available. The land may be slated for development or the contaminants may be causing health problems in nearby human populations. So, how can we clean up these areas? One way is for the contaminated soil to be physically removed and transferred to another location where it can degrade naturally over time. Alternatively, it can be incinerated, which can release toxins to the environment. Both of these options are expensive and problematic. The third option of bioremediation may be a less expensive and less problematic alternative.

What are the strategies used for bioremediation? As we noted earlier, we can simply let nature take its course. Such natural biodegradation, however, is slow and often limited by the lack of available O_2; the best degrading enzymes typically require oxygen. **Biostimulation** involves providing missing O_2 and limiting nutrients such as nitrogen and phosphorus to increase the activity of the resident microbes. Sometimes, degradation pathways for a given xenobiotic have not yet evolved, but natural substrates can be added to the environment to induce degradation pathways with broad substrate range. For example, the

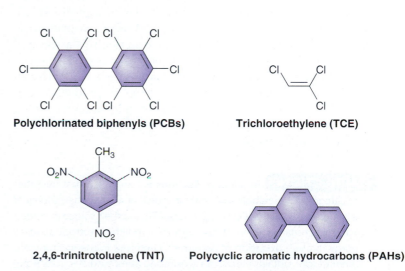

Polychlorinated biphenyls (PCBs)

Trichloroethylene (TCE)

2,4,6-trinitrotoluene (TNT)

Polycyclic aromatic hydrocarbons (PAHs)

Figure 15.29. Important targets of bioremediation This figure shows the structures of some of the major contaminants that are subject to bioremediation efforts.

methane or toluene pathways can stimulate trichloroethylene (TCE) degradation. In an example of a **cometabolism** process, methane can be added to a TCE-contaminated area. Naturally occurring microbes that utilize methane as a carbon source are stimulated, and these microbes, fortuitously, produce metabolic enzymes that degrade TCE. In **bioaugmentation**, bacteria known to degrade the contaminant can be added to a contaminated environment. Many scientists apply molecular methods to study these processes with the hope of better understanding them so that bioremediation can be used more consistently. A constant challenge will be keeping ahead of the continual development of new xenobiotics.

15.4 Fact Check

1. How do soils form?
2. Distinguish between A, B, and C soil horizons.
3. What physical characteristics differentiate the soil classes?
4. Differentiate between the rhizosphere and bulk soil.
5. Explain how lignocellulose is degraded in the environment.
6. What are humic substances, and why are they important in soil ecology?
7. Why are microbes generally not able to easily degrade xenobiotics?

15.5 Deep subsurface and geothermal ecosystems

Can life exist in the absence of photosynthesis-driven primary production?

We tend to think of the sun as the source of all energy supporting life on Earth. Indeed, photosynthesis accounts for a significant amount of the energy harvesting that occurs on our planet. We also know that a significant amount of biomass exists in Earth's deep subsurface. What supports life in these ecosystems well below the reach of sunlight? Photosynthetic-derived organic carbon, O_2 and NO_3^- can be carried down into some subsurface habitats by groundwater flow and sedimentation, thereby supporting heterotrophic metabolism. However, many deep subsurface systems have minimal input from these photosynthetic-derived nutrients. Instead, these ecosystems rely on bacterial and archaeal chemolithoautotrophic primary producers that harvest geothermal energy. In this section, we will explore a few of these unusual environments. First, we will investigate deep subsurface microbiology. Then, we will explore the microbial communities found around deep sea hydrothermal vents. Finally, we will look at the communities associated with terrestrial hot springs. Throughout our investigation of these ecosystems, we will see how communities can thrive in the absence of photosynthesis-driven primary producers. Understanding these systems is critical to understanding life on Earth. Studies of these ecosystems also raise central questions about the evolution of life.

CONNECTIONS Recall from **Sections 6.1** and **13.2**, that microbes are categorized metabolically on their source of energy, their source of electrons for redox reactions to produce ATP, and their source of carbon for forming organic material for growth. A chemolithoautotroph acquires energy through oxidation of preformed chemicals in the environment (chemotroph), obtains electrons by oxidizing (removing electrons from) inorganic reduced **molecules**, such as H_2S, or Fe^{2+} (lithotroph), and gets its carbon from **inorganic carbon compounds**, like CO_2.

Deep subsurface microbiology

The field of geomicrobiology combines geology and microbiology by investigating the role of microbes in geological and geochemical processes. The field began in the mid-1970s with the discovery of deep sea vent communities and expanded in the mid-1980s to include a range of extreme habitats including deep aquifers, hot springs, hypersaline environments, oil wells, and even subterranean rock. Microbiological surveys performed during oil and gold drilling operations made it clear that the deep subsurface, rather than being a sterile environment, often contained substantial biological material **(Figure 15.30)**. We now realize that the

SEM

Figure 15.30. Microbial biofilm from a gold mine Microbial communities thriving on rocks extracted from deep underground are not uncommon.

Courtesy Maggy Lengke and Gordon Southam

biosphere extends to great depths, in some cases several kilometers below the surface of Earth. At these depths, negligible input from surface nutrients derived from photosynthesis occurs. Additionally, many of these environments are very hot as they approach or breach Earth's mantle. In one study, hyperthermophiles recovered from hot oil wells lived at temperatures up to 115°C. In a study published in 2003, researchers reported the identification of an iron-reducing archaeon isolated from a vent in the Juan de Fuca Ridge of the Pacific Ocean that grew well at 100°C and could continue to replicate at 121°C, the temperature of an autoclave. Named "Strain 121" in honor of this remarkable ability, it could survive for an impressive two hours at 130°C but did not replicate at this higher temperature.

CONNECTIONS Extremophilic achaeons exhibit numerous adaptations to prevent membrane disruption and the denaturation of proteins and nucleic acids. Some of these adaptations include the presence of tetraether lipids or lipid monolayers in their membranes and abundant α-helical regions, salt bridges, and side chain interactions in their proteins to maintain three-dimensional structure in the face of high temperature. In **Section 4.3**, we learned more about these heat-proofing characteristics of hyperthermophiles.

Over the past decade, samples taken from depths as great as 3 km below Earth's surface in various sites around the world have uncovered novel and diverse species of bacteria and archaeons through SSU rRNA gene sequence analysis. The microbes living here depend on chemolithoautotrophy for primary production of organic compounds, which respiring microbes then can utilize. Because of the chemical processes these microbes carry out, they influence the composition of these habitats by altering water chemistry and the mineral and physical properties of rock strata. Compared to respiring organisms, chemolithotrophs

generate less energy, pumping fewer protons per electron, resulting in less ATP generation. This characteristic, along with the scant nutrients available, explains the slow growth rate observed for these microbial communities.

To quantify various metabolic processes and growth in core samples **(Figure 15.31)**, researchers often use radiolabeled nutrients. For example, researchers can study basic metabolic processes by measuring the rates of incorporation of radiolabeled acetate into lipids or radioactive thymidine into DNA. Although we still know relatively little about the communities that make up subsurface systems, the basis of life probably begins with chemical and physical interactions between rock and water that release soluble and insoluble minerals as well as H_2 that can fuel microbial energy metabolism. Metals in the rock, such as iron, react with water to form H_2 gas:

$$FeO + H_2O \rightarrow H_2 + FeO_{3/2}$$

While this reaction is completely abiotic, it results in the liberation of hydrogen gas that can be used by archaeal methanogens to produce methane gas:

$$H_2 + CO_2 \rightarrow CH_4 + H_2O$$

The detection of DNA from methanogens in subsurface samples supports this idea. Some chemolithotrophs can oxidize metal ions such as Fe^{2+} and sulfur compounds such as H_2S or S^0 to fuel their energy needs, producing Fe^{3+} and SO_4^{2-} respectively, which other microbes can use as terminal electron acceptors for their own energy pathways. The output of metallic metabolism is the release of metal by-products. While life in the depths exists at a slow pace, the deposition of these metals by biofilm microbes concentrates them in ores. The majority of the world's iron, uranium, and gold may have been formed this way over millions of years. Similarly, commercial sulfur deposits result from bacterial

Taking a sample of core sediment

© AFP PHOTO/William WEST/NewsCom

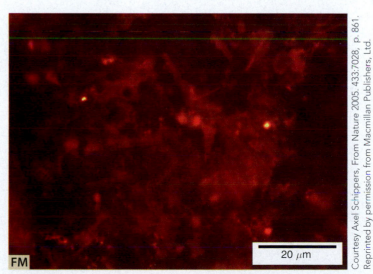

FM

20 μm

Microbial communities from deep below the ocean floor

Courtesy Axel Schippers, From Nature 2005. 433:7028, p. 861. Reprinted by permission from Macmillan Publishers, Ltd.

Figure 15.31. Core from marine subsurface Sediment cores from deep below the ocean floor have been found to harbor active microbial communities.

sulfate reduction, as is the production of magnetite (Fe_3O_4) and pyrite (FeS_2) by reduction of Fe^{3+} by sulfate-reducing bacteria. The activity of the microbes in these deep subsystems means we can view them as important catalysts in rock weathering, a process previously thought to be a largely abiotic process of erosion by water and pressure. The metabolic activities of subsurface microbes also means they contribute to the formation of fossils as they can replace organic materials with minerals. Another important facet of their activity is the technological application of lithotrophic microbes to extract metals from mine wastes.

In some parts of the world, mineral-oxidizing bacteria may be responsible for the release of heavy metals into deep groundwater reservoirs, the source of many people's drinking water. The conversion of arsenate AsO_4^{3-} to the more soluble and therefore more mobile arsenite AsO_3^{3-} has tainted the waters in Bangladesh that run through arsenic-containing rock. Millions of people rely on these aquifer systems.

Hydrothermal vents

Recall from our Chapter 13 opening story that, in the late 1970s, scientists made a remarkable discovery in the deep waters off the Galapagos Islands. In the darkness of the ocean floor, 2.5 kilometers below the surface, a massive collection of never before seen animals existed around volcanic hot vents. These animals included giant tube worms as well as giant clams and mussels. They all thrived in an ecosystem devoid of light and photosynthetically derived nutrients. The existence of such a huge ecosystem without any contribution from the sun surprised most biologists. What energy source drove this ecosystem, and what organisms served as the primary producers? To answer these questions, first we need to look at the geology and chemistry of the vent regions.

Most hydrothermal vents exist on mid-ocean ridges where volcanic activity and plate movement cause the often violent release of minerals and gases from the interaction of molten rock with seawater (**Figure 15.32**). The released compounds include CO_2, CH_4, and sulfides like H_2S. Additionally, large amounts of metals including iron, manganese, lithium, and cesium exist near these vents. Because photons are not available as an energy source in these locations, you may have deduced correctly that one or more of these minerals must serve as the primary energy and electron sources to drive carbon fixation, leading to the production of organic carbon compounds needed to support this ecosystem. Bacteria are the primary producers here and they use these compounds, most importantly H_2S, as major electron donors, performing chemosynthesis (**Figure 15.33**). Sulfide oxidizers such as the betaproteobacteria *Thiobacillus* sp. and the gammaproteobacteria *Beggiatoa* sp. are examples of free-living, biofilm-forming primary producers. When new vents are formed, ecosystems quickly develop. The biofilms, which can grow into mats several inches thick, first attract small grazing crustaceans such as amphipods and copepods, then larger snails, shrimp, crabs, anemones, clams, mussels, tube worms, fish, octopi, and jellyfish.

Many invertebrate-associated sulfide oxidizers also exist in these environments. The bacterial symbiont of the giant tube worm also relies on H_2S as a source of energy and electrons. Through its CO_2 fixation, this bacterium supports the growth of

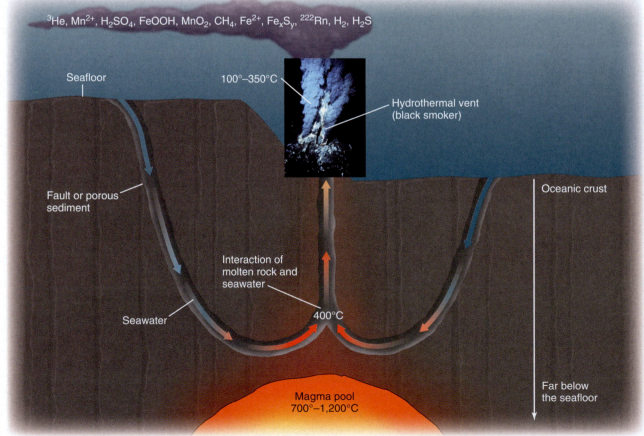

^{3}He, Mn^{2+}, H_2SO_4, FeOOH, MnO_2, CH_4, Fe^{2+}, Fe_xS_y, ^{222}Rn, H_2, H_2S

Seafloor

100°–350°C

Hydrothermal vent (black smoker)

Fault or porous sediment

Oceanic crust

Interaction of molten rock and seawater

Seawater

400°C

Far below the seafloor

Magma pool 700°–1,200°C

Figure 15.32. Deep sea hydrothermal vent chemistry The vent fluids contain rich mixtures of chemicals. Note that seawater is drawn beneath the seafloor and is superheated as it encounters molten rock. As the fluids are cooled, the chemicals precipitate.

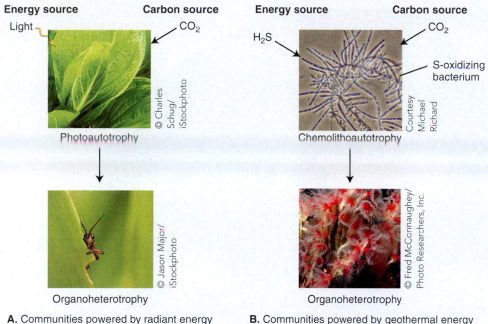

Energy source   **Carbon source**
Light CO_2

Photoautotrophy

© Charles Schug/iStockphoto

Organoheterotrophy

© Jason Major/iStockphoto

A. Communities powered by radiant energy

Energy source   **Carbon source**
H_2S CO_2

S-oxidizing bacterium

Chemolithoautotrophy

Courtesy Michael Richard

Organoheterotrophy

© Fred McConnaughey/Photo Researchers, Inc.

B. Communities powered by geothermal energy

the giant tube worms. While the H_2S oxidizers, both free-living and animal associated, are the main primary producers, a variety of free-living thiosulfate-, methane-, hydrogen-, iron-, and manganese-oxidizing bacteria also have been identified (**Table 15.7**). With all the organic material resulting from this primary production, many chemolithoheterotrophs also live here. They oxidize the abundant inorganic compounds of the vent for energy and electrons, but they derive their carbon from organic sources supplied by the primary producers and the various animal species of the ecosystem.

Figure 15.33. Chemosynthesis primary production at hydrothermal vents The generation of energy by S-oxidizing bacteria at deep sea hydrothermal vents can be contrasted with the generation of energy in light-driven habitats. **A.** Photoautotrophs use light to generate energy by photosynthesis. **B.** In deep sea hydrothermal vents, the carbon and energy produced by the S-oxidizing bacteria are used to support the growth of animals such as tube worms.

TABLE 15.7   **Metabolic reactions in hydrothermal vent ecosystems**

Metabolism	Reaction	Examples in vent environments
Anaerobic		
Methanogenesis	$4 H_2 + CO_2 \rightarrow CH_4 + 2 H_2O$ $CH_3CO_2^- + H_2O \rightarrow CH_4 + H$ $CO_3^- + 4 HCOO^- + H^+ \rightarrow 3 HCO_3^- + CH_4$	*Methanococcus* spp. common in magma-hosted vents; Methanosarcinales at Lost City
S^0 reduction	$S^0 + H_2 \rightarrow H_2S$	Lithotrophic and heterotrophic; hyperthermophilic archaeons
Anaerobic CH_4 oxidation	$CH_4 + SO_4^{2-} \rightarrow HS^- + HCO_3^- + H_2O$	*Methanosarcina* spp. and epsilonproteobacteria at mud volcanoes and methane seeps
Sulfate reduction	$SO_4^{2-} + H^+ + 4 H_2 \rightarrow HS^- + 4 H_2O$	Deltaproteobacteria
Fe reduction	$8 Fe^{3+} + CH_3CO_2^- + 4 H_2O \rightarrow 2 HCO_3^- + 8 Fe^{2+} + 9 H^+$	Epsilonproteobacteria, thermophilic bacteria, and hyperthermophilic Crenarchaeota
Fermentation	$C_6H_{12}O_6 \rightarrow 2 C_2H_6O + 2 CO_2$	Many genera of bacteria and archaeons
Aerobic		
Sulfide oxidation[a]	$HS^- + 2 O_2 \rightarrow SO_4^{2-} + H^+$	Many genera of bacteria; common vent animal symbionts
CH_4 oxidation	$CH_4 + 2 O_2 \rightarrow HCO_3^- + H^+ + H_2O$	Common in hydrothermal systems; vent animal symbionts
H_2 oxidation	$H_2 + 0.5 O_2 \rightarrow H_2O$	Common in hydrothermal systems; vent animal symbionts
Fe oxidation	$Fe^{2+} + 0.5 O_2 + H^+ \rightarrow Fe^{3+} + 0.5 H_2O$	Common in low-temperature vent fluids; rock-hosted microbial mats
Mn oxidation	$Mn^{2+} + 0.5 O_2 + H_2O \rightarrow MnO_2 + 2 H^+$	Common in low-temperature vent fluids; rock-hosted microbial mats; hydrothermal plumes
Respiration	$C_6H_{12}O_6 + 6 O_2 \rightarrow 6 CO_2 + 6 H_2O$	Many genera of bacteria

[a]Some epsilonproteobacteria from subseafloor hydrothermal vents, including newly erupted vents, can oxidize H_2S to S^0.

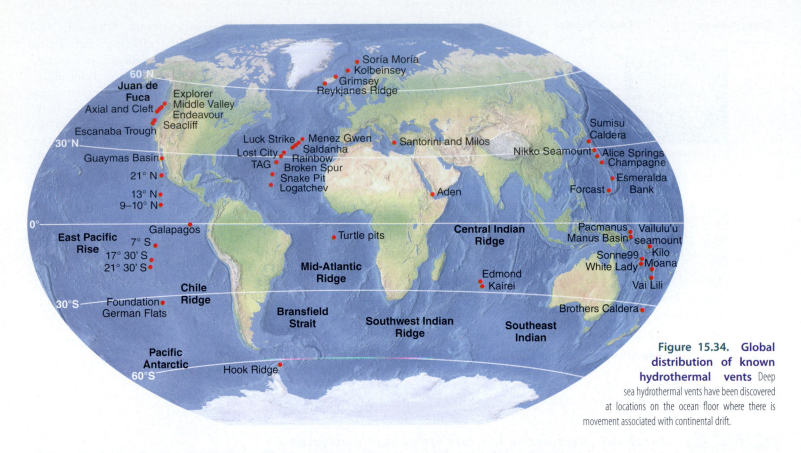

Figure 15.34. Global distribution of known hydrothermal vents Deep sea hydrothermal vents have been discovered at locations on the ocean floor where there is movement associated with continental drift.

A variety of vents have been discovered worldwide (Figure 15.34) and all of them support dense communities of microbes. The vent fluids are highly reduced and acidic, of pH 2.8–5. Although the fluids spewing out of the vent are usually very hot, ranging from 60°C up to as high as 464°C, the water temperature at the ocean floor is about 2°C, and the rapid cooling results in the precipitation of some of the dissolved chemicals causing a smoke-like effect, and the formation of chimney-like vents (see Figure 15.32). Each vent system is unique, with different temperatures and combinations of elements dissolved in the vent fluid. "Black smokers," like the Galapagos vents, are very hot and emit high concentrations of sulfides, which hit the cold seawater and immediately precipitate, producing dark plumes (see Figure 4.1). "White smokers" are at lower temperatures, and although they spew a similar chemical brew, the barium, calcium, and silicon precipitate out at these lower temperatures, producing white plumes and pale chimney stacks. The quick cooling results in a temperature gradient that supports organisms of varying temperature optima. The vents thus provide several unique niches for specialized organisms. Microbial growth chambers can be deployed in the flow around the vents to study the colonizing populations of bacteria generated by the microbial mats. After three to eight days, the chambers can be recovered and the DNA extracted for analysis.

Sequence analyses of SSU rRNA genes reveal various free-living and invertebrate-associated H_2S oxidizers. Some studies have shown that gammaproteobacteria represent the dominant primary producers, while other studies have shown that the dominant primary producers are H_2S-oxidizing members of the epsilonproteobacteria. Most studies show that the vent biofilm mats, while of substantial biomass, are usually composed of only a few OTUs, although different mats can be composed of different types at different locations around the vent. Since the primary producers may vary from site to site, it follows that the animals they support also vary. While various species of tube worms are the dominant invertebrates of the Pacific thermal vents, a small white shrimp is the dominant species in Atlantic vents, and it too harbors symbiotic bacteria. Whereas tube worms house their symbionts internally, the shrimp appear to farm the bacteria on their surface. They even battle for the best positions near the vent in order to maximize the growth of the bacteria, which they harvest as their main source of food. They and other invertebrates also graze the free-living epsilonproteobacteria that cover the surface of the vents. Such symbiotic relationships are a fundamental necessity for animals to exist in these dark and isolated environments.

Before leaving this topic, we should note that not all vents are hot. Cold seeps, first discovered in the Gulf of Mexico in 1983, release various sulfides like H_2S, methane, and other hydrocarbons, but these compounds seep slowly from Earth's crust at about the same temperature as the surrounding seawater. These ecosystems are much slower growing and longer lived because the geological formations that form at these low temperatures are more stable than those of the volatile hydrothermal vents. The methane and seawater react slowly to form carbonate rock instead of vents. Cold seeps have a similar community makeup as the hydrothermal vents. *Beggiatoa* mats form and tube worms exist, and like their Pacific vent cousins, they also harbor dense populations of chemolithoautotrophic bacterial symbionts. Thus, the biology and ecology of deep sea seep and vent habitats are intricately linked to the geology and chemistry of the surrounding ocean.

Terrestrial hot springs

Geothermal springs are the terrestrial equivalent of the deep sea hydrothermal vents. Unlike their marine counterparts, however, at temperatures below mid 70°C and at neutral pH, they can have significant primary production from cyanobacterial oxygenic photoautotrophs and green nonsulfur anoxygenic phototrophs. These organisms are key components of thick microbial biofilm mats with photosynthetic organisms in the upper layers (**Figure 15.35**). The photosynthetic cells provide organic carbon and energy for the underlying layers, supporting diverse communities that include mostly aerobic chemoorganotrophs in the lower layers.

In hotter, acidic springs often associated with volcanic activity, the conditions are not hospitable to photosynthetic organisms. These types of environments, typically rich in inorganic energy sources such as H_2, H_2S, elemental sulfur, and $Fe(II)$, can be dominated by hyperthermophilic archaeons such as the crenarchaeote *Sulfolobus*. These chemolithoautotrophs can oxidize elemental sulfur to H_2SO_4 (sulfuric acid), and their metabolism thus contributes directly to the acidity of the water. This low pH means that only acidophilic microbes are able to survive in these environments, and the community diversity is quite low compared to other, less challenging habitats.

15.5 Fact Check

1. What environmental conditions are present in hydrothermal vents; how are these vents thought to form?
2. How does the community around a cold seep differ from that around a hydrothermal vent?
3. Describe geothermal springs as an ecosystem, including the physical conditions and microbes present.

A. Octopus Spring, Yellowstone National Park

Courtesy David M. Ward

Figure 15.35. Microbial mats at geothermal springs Geothermal springs often harbor interesting microbial communities. **A.** The Octopus Spring, an alkaline spring at Yellowstone National Park. **B.** Cyanobacterial mat from Octopus Spring. **C.** Dominant cyanobacteria from mat.

Image in Action

WileyPLUS A cross-sectional view of the ocean shows that microbial life can be found at all depths. On the sunny ocean surface, photosynthetic microbes exist, while heterotrophic microbes thrive in the dark, colder lower waters. Deep sea vents provide yet another habitat for many other forms of microbial life. DNA sequencing and analysis allow scientists to study these microbial communities even when individual microbes cannot be grown in culture.

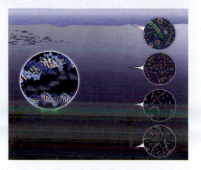

1. Imagine you are asked to provide additional detail to this image. Where would you need to identify the presence (and relative amounts) of the following types of microbes: piezophiles, archaeons, bacterial cells, and viruses? Explain your answer.

2. Imagine you have been asked to examine the impact of the 2010 Gulf of Mexico oil spill on microbial biodiversity. Outline your design for using a non-cultivation method to conduct this research.

3. What results might you expect from your study in Question 2 if the disaster had a major impact on the total number of organisms in surface waters? Explain.

B. Section of microbial mat

Courtesy David M. Ward

C. Cyanobacteria of the mat

Courtesy David M. Ward

The Rest of the Story ●●●●●○ ○

The detection of large numbers of non-thermophilic archaeons in marine samples in the early 1990s was unexpected but not an isolated incident. As additional mesophilic and psychrophilic environmental samples have been examined in the ensuing years, many more non-thermophilic archaeal organisms have been found. They appear to be widespread in soils, oceans, sediments, deep subsurface, freshwater lakes, and as symbionts of sponges and other animals. While they are widespread, their cultivation has been extremely difficult. Given their numbers, they must be making substantial contributions to ecosystem function, but to truly understand their function in the ecosystem we must determine how to study these microbes.

The answer seems to lie in the advancement of non-cultivation-based sequencing technologies as described in Chapter 10. In that chapter's Rest of the Story section, the ongoing research at The J. Craig Venter Institute (JCVI) was highlighted. The institute's Global Ocean Sampling Expedition, for example, has involved research teams circumnavigating the globe and collecting samples since 2003. Microbiologists on the research vessel *Sorcerer II* collect 400-liter water samples every 200 miles while also documenting the environmental conditions at each site. The water collected is passed through smaller and smaller filters to capture and separate different size organisms. These samples are frozen and sent back to land-based research labs for metagenomic analysis. Such a systematic and molecular approach is the new frontier in determining the true biodiversity of microbial ecosystems and a first step in understanding the functions of the organisms found there.

Summary

Section 15.1: What factors affect microbial communities?

The physiology of individual microbes within microbial communities helps to define their roles in those communities.

- Although the foundation of microbiology is the study of microbes in pure culture, in the environment microbes exist in complex communities.

- The cultured microbes do not accurately represent microbial communities.

- An **ecosystem** is a **community** of organisms, including **primary producers**, **consumers**, and **decomposers**, interacting with each other and their environment.

- Members of communities can be grouped into functional groups called **guilds**.

- Microbial ecology involves the study of biodiversity of microbes and their activities within ecosystems, including specialized communities such as the ones that rely on **chemosynthesis**.

- Microbes, being small, exist within microenvironments in which conditions change over short distances.

- The success of microbes and their ability to occupy particular **niches** depends on their ability to obtain different nutrients to produce energy and biomass.

- To access nutrients and protection from antimicrobials in aquatic habitats, biofilm communities form on solid surfaces, held together by exopolysaccharide and engaging in

intercellular communication through mechanisms such as quorum sensing.

Section 15.2: How can microbial communities be studied?

Many tools are now available for the dissection of microbial communities in ecosystems.

- The SSU rRNA gene has become the standard marker for cultivation-independent detection of the presence of microbes in the environment, and many molecular tools such as **denaturing gradient gel electrophoresis (DGGE)** and **terminal restriction fragment length polymorphism (TRFLP)**, **fluorescence *in situ* hybridization (FISH)**, and **flow cytometry** have been developed based on the detection of SSU and other sequences.

- It has been estimated that the total combined biomass on Earth from Bacteria and Archaea is approximately equivalent to the biomass represented by plants.

- **Stable isotope probing (SIP)** is a method that can provide information about the metabolic activities of different organisms in the environment.

- The **operational taxonomic unit (OTU)** concept, based on SSU sequence identity, is becoming the standard for measuring microbial diversity. Use of this tool has made it possible to compare the levels of diversity of different environments.

Section 15.3: What is the nature of life in water?

Most of the life in aquatic ecosystems is microbial in form and living in a state of **oligotrophy**.

- Primary production in marine ecosystems predominantly takes place through photosynthesis by **phytoplankton** in surface waters.

- Carbon and energy are distributed through the ecosystem through grazing of the phytoplankton by **zooplankton** and viral-mediated lysis.

- Approximately 1 percent of the primary production reaches the seafloor where anoxic, cold conditions render it fairly resistant to biodegradation, resulting in its accumulation over hundreds of millions of years to form rich sediment on the sea floor.

- The atmospheric pressure is extremely high at great depths, where **piezophiles**, microbes that are able to survive such high pressures, are found.

- Characterization of marine communities using metagenomic methods has revealed the existence of new organisms with novel proteins and metabolic activities, and some of these organisms have been successfully cultivated using **dilution to extinction** methods.

- Freshwater ecosystems that are part of terrestrial environments typically are less oligotrophic than marine environments.

- Lakes typically stratify into upper, warmer aerobic zones, and lower, colder anaerobic zones, separated by a layer called the **thermocline**.

Section 15.4: What is the nature of life on land?

Microbes are important components of terrestrial environments with especially rich and diverse soil microbial communities.

- Terrestrial ecosystems can be divided into several **biome** types, and collectively they represent the most diverse habitats on Earth.

- The dominant primary producers in terrestrial environments are plants.

- Soils typically harbor very active microbial communities that, through their decomposition of plant and animal matter, direct the formation of new soil.

- Soils have been categorized into specific types based on physical characteristics, and they are typically divided into horizontal layers called **horizons**.

- The **rhizosphere** directly surrounding plant roots is very rich in carbon due to the excretion of plant root exudates. This results in much higher numbers of microbial cells compared to what is found in the bulk soil.

- The brown rot and white rot fungi are specialized in lignocellulose degradation, while the actinomycetes are able to break down a number of other forms of plant biomass through the production of extracellular enzymes.

- Recalcitrant organic molecules, such as **humic material**, do not completely **mineralize**, accumulating in soils and contributing to the aggregation properties of the soils.

- **Bioremediation** involves the use of microbes to clean up environmental contamination.

- Degradation of **xenobiotics**, chemicals that are not normally found in nature, often does not occur easily.

- Commonly used bioremediation strategies include **biostimulation**, **cometabolism**, and **bioaugmentation**.

Section 15.5: Can life exist in the absence of photosynthesis-driven primary production?

Much of Earth's biomass is not subject to sunlight, and yet it is metabolically active. Primary production can occur in the absence of sunlight.

- It has been estimated that there is more total biomass in subsurface habitats than there is above ground.

- Microbial life has been detected at depths of several kilometers where it is likely that chemolithoautotrophic forms of metabolism are responsible for primary production.

- Chemolithoautotrophic metabolism also supports life at deep sea hydrothermal vents in the ocean floor and in terrestrial hot springs.

Application Questions

1. A group of microbial ecologists is researching the biodiversity of a local habitat. According to this description, explain what this means and discuss what might be the next research steps following the initial biodiversity studies.

2. Oxidation and reduction are considered key chemical reactions in microbes. Explain how these processes play a particularly important role in microbial generation of energy and give an example.

3. You are attempting to create an anti-biofilm chemical. Identify a component of a biofilm you might consider targeting and describe the specific effect it would have on the microbes and the biofilm overall.

4. A research team is writing a grant and proposing the use of SSU rRNA genes as a phylogenetic tool in the study of several newly discovered marine microbes. How would they justify that this is the best tool to use for their purposes?

5. Outline and explain how a researcher could go about using fluorescent in situ hybridization (FISH) to examine the expression of a nitrogen fixation gene in several previously unknown soil microbes.

6. You are on a research vessel and the sampling data show the area to be an oxic, oligotrophic marine environment. In simple terms, explain what this statement tells you about the characteristics of the environment and the metabolic activities of the microbes living there.

7. Explain how the metabolic activities of microbes in surface ocean waters are connected to those in dark mid-water zones. Give an example to illustrate your explanation.

8. Imagine that you have taken the same sample of seawater and used standard marine isolation medium and the dilution to extinction method to examine the microbial contents of the sample. How would your results be different using these two methods? Be specific and explain the reasons for the differences.

9. Imagine a new agricultural pesticide is now being shown to negatively impact the metabolic activities of white rot fungi in the soil and is actually decreasing the white rot fungal numbers. Soil ecologists are therefore arguing that this pesticide should not be used. What would be their argument, and what would they specifically say about the importance of these fungi in soils?

10. As a new employee of a company called Actinocorp, you are learning at your orientation that the company focuses on recovering and studying novel marine actinomycetes. Why would a company focus on recovering new actinomycetes, and from this brief overview what would you determine the company is trying to do? Explain.

11. A conservation research team is surveying the microbial populations in a deep freshwater lake in New York that consistently has a thermocline throughout the summer. Describe the types of microbes they might expect to find based on the water conditions that will most likely be present in each location (layer) in the lake.

12. A microbiologist working for an oil-drilling company is attempting to obtain samples deep within the drilling site. A camera view shows what appears to be a microbial mat at the current sample site. Explain what metabolic characteristics would be expected in microbes at this site. Once a site sample is obtained, how could he go about characterizing these bacteria back in the laboratory, that is, what tools could be used and how?

13. With the bacteria described in Question 11, imagine that it is your job in the lab to conduct an SSU rRNA analysis to determine phylogeny.
    a. Would you elect to use universal primers or bacterial primers? Explain your reasoning.
    b. Detail each of the steps you would use in your SSU rRNA analysis. Explain the purpose of each step you outline.
    c. Based on the location of the sample, what results might you expect in your phylogenetic analysis?

Suggested Reading

Cardenas, E., and J. M. Tiedje. 2008. New tools for discovering and characterizing microbial diversity. Curr Opin Biotechnol 19:544–549.

Curtis, T. P., W. T. Sloan, and J. W. Scannell. 2002. Estimating prokaryotic diversity and its limits. Proc Natl Acad Sci USA 99:10494–10499.

DeLong, E. F. 1992. Archaea in coastal marine environments. Proc Natl Acad Sci USA 89:5685–5689.

Diaz, R. J., and R. Rosenberg. 2008. Spreading dead zones and consequences for marine ecosystems. Science 321:926–929.

Fry, J. C., R. J. Parkes, B. A. Cragg, A. J. Weightman, and G. Webster. 2008. Prokaryotic biodiversity and activity in the deep subseafloor biosphere. FEMS Microbiol Ecol 66:181–196.

Fuhrman, J. A., K. McCallum, and A. A. Davis. 1992. Novel major archaebacterial group from marine plankton. Nature 356:148–149.

Jorgensen, B. B., and A. Boetius. 2007. Feast and famine—Microbial life in the deep-sea bed. Nat Rev Microbiol 5:770–781.

Neufeld, J. D., J. Vohra, M. G. Dumont, T. Lueders, M. Manefield, M. W. Friedrich, and J. C. Murrell. 2007. DNA stable-isotope probing. Nat Protoc 2:860–866.

Teske, A. P. 2005. The deep subsurface biosphere is alive and well. Trends Microbiol 13:402–404.

Venter, J. C., K. Remington, J. F. Heidelberg, A. L. Halpern, D. Rusch, J. A. Eisen, D. Wu, I. Paulsen, K. E. Nelson, W. Nelson, D. E. Fouts, S. Levy, A. H. Knap, M. W. Lomas, K. Nealson, O. White, J. Peterson, J. Hoffman, R. Parsons, H. Baden-Tillson, C. Pfannkoch, Y. H. Rogers, and H. O. Smith. 2004. Environmental genome shotgun sequencing of the Sargasso Sea. Science 304:66–74.

Ward, D. M., M. J. Ferris, S. C. Nold, and M. M. Bateson. 1998. A natural view of microbial biodiversity within hot spring cyanobacterial mat communities. Microbiol Mol Biol Rev 62:1353–1370.

Whitman, W. B., D. C. Coleman, and W. J. Wiebe. 1998. Prokaryotes: The unseen majority. Proc Natl Acad Sci USA 95:6578–6583.

16 The Microbiology of Food and Water

Sandwiches of packaged luncheon meats are a lunchtime staple for many people. Sitting in their plastic-sealed containers on the supermarket shelf, these meats appear anything but dangerous. If the meat has been contaminated during packaging, however, the microbes that survive and thrive can sicken or kill those who eat it. Such was the case in Canada in August 2008 when meats contaminated by slicing machinery at a Maple Leaf Foods facility in Toronto, Ontario, were distributed to supermarkets nationwide. Fifty-seven people fell ill with listeriosis after eating meat that had been contaminated by a bacterium called *Listeria monocytogenes*. Of these, 22 people died, almost 40 percent of those who became ill. This high percentage of deaths is not unusual since *L. monocytogenes* is one of the more dangerous sources of foodborne disease. In the United States, about 1,600 people get seriously ill with listeriosis annually and 260 of these die.

Listeria monocytogenes has a short history as a pathogen, being first identified as the cause of systemic illness in a 1981 outbreak in Halifax, Nova Scotia. For most of us, the risk posed by *L. monocytogenes* is very low. For the very young, the elderly, pregnant women, and people whose immune system is not working properly, the situation can be disastrously different. Pregnant women may suffer spontaneous abortion. Others can suffer from vomiting, nausea, cramps, diarrhea, severe headache, constipation, and fever. The bacteria can access the bloodstream to reach the brain and spinal cord, eyes, and lungs.

The widespread pattern of the 2008 Canadian outbreak is an unfortunate consequence of our modern method of food preparation. Often food prepared at one or several large facilities is shipped to distant locations. When the system works, it provides an economical method of getting food to many people. When something goes wrong, however, the consequences can be national or even international in scope.

CHAPTER NAVIGATOR

As you study the key topics, make sure you review the following elements:

Foods are a source of nutrients and must be protected or altered to inhibit spoilage.
- Table 16.1: Factors affecting microbial growth in foods
- Figure 16.2: Water activity (a_w)
- Table 16.2: Typical spoilage patterns of selected foods
- Figure 16.3: Alteration of milk as a microbial growth medium

Food preservation involves chemical or physical alteration of food and/or modification of food packaging.
- Figure 16.7: Flash-heating pasteurization
- Figure 16.12: The hurdle technology concept

Fermentation of food can preserve and impart flavors.
- Microbes in Focus 16.1: *Lactococcus lactis*
- Figure 16.15: Process Diagram: Steps in cheese making
- Figure 16.18: Process Diagram: Production of soy sauce

Illnesses can occur when pathogens or microbial toxins are present in food or water.
- Table 16.6: Top five annual foodborne-disease estimates for the United States (2011)
- Figure 16.22: Cholera—a waterborne disease
- Figure 16.23: Gas production from microbial growth

Wastewater and drinking water treatments reduce or remove microbes and chemical contaminants.
- Figure 16.25: Classes of contaminants in wastewater
- Figure 16.26: Process Diagram: General schematic of a wastewater treatment plant
- Toolbox 16.1: Measuring biochemical oxygen demand (BOD)
- Microbes in Focus 16.2: *Cryptosporidium parvum*
- Figure 16.30: Process Diagram: General schematic of drinking water purification
- Toolbox 16.2: Most probable number (MPN) method

CONNECTIONS for this chapter:

The microbial growth curve and food spoilage (Section 6.1)

Use of UV light for killing microbes in food and water (Section 7.5)

The metabolism of lactic acid bacteria and fermentation of foods (Section 13.3)

Pathogens and their products in waterborne illnesses (Chapter 21)

Biofilm-forming microbes and wastewater treatment (Sections 2.5 and 15.1)

Introduction

Everyone has had the experience of trying to identify a long-forgotten food recovered from the depths of the pantry or refrigerator. Microbial activity can yield unpleasant surprises. Yet, should we be surprised? Since our foods provide us with nutritional resources, we might logically conclude that they may be exploited by microbes for the very same reason. When the result of that exploitation is undesirable, we say that our food is spoiled. The spoilage of foods by microorganisms is one component of the field known as food microbiology. Food preservation refers to a group of methods developed to prevent food spoilage and minimize the chance that foods will become contaminated with disease-causing microbes. Food safety is an important aspect of the food industry that deals with minimizing the incidence of foodborne diseases caused by microorganisms. Disease may result from mishandling foods during processing or preparation, either in a commercial setting or in the home.

Not all microbial activity in food is undesirable, however. Humans have used microbial fermentation for food production to preserve relatively unstable food materials since ancient times. For instance, when converted to cheese, milk remains consumable much longer than it does in its more perishable liquid form. Indeed, fermentation of milk to produce cheese has been described by author Clifton Fadiman as "milk's leap toward immortality."

Today, even though we do not need to preserve foods by fermentation when a good refrigerator or freezer will do the job, the practice persists. The taste, texture, and aroma of fermented food products contribute a pleasing variety to our diet. Many enjoy fermented fruits in the form of wine or fermented grains in the form of beer.

In addition to nutrients from food, living organisms require water, and it too must be free of pathogens and toxic compounds. Like carbon and nitrogen, water is cycled between living and non-living reservoirs. Water removed from natural waters for human use is eventually returned to its source. To protect our drinking water supply and to maintain the natural ecology of water systems, water used for everyday living as well as for agriculture and industry must be treated before it is released back into the environment. As we will see in this chapter, microbes are one of the main tools in removing chemical and biological pollutants from wastewater.

This chapter will specifically focus on several questions:

How can microorganisms spoil food? (16.1)
What techniques are used to minimize food spoilage? (16.2)
How are microorganisms used in food production? (16.3)
How can microorganisms in our food or water cause illness? (16.4)
How can we keep our water supplies safe? (16.5)

16.1 Food spoilage

How can microorganisms spoil food?

Many microorganisms will grow in our food if provided with the opportunity (**Figure 16.1**). Such growth does not necessarily lead to unsafe food, but most consumers consider such food to be unpalatable or unattractive and refer to it as spoiled. **Perishable foods** can easily support the growth of microorganisms. These include many fresh foods such as green salad ingredients or fresh

© Craig Veltri/iStockphoto

A. Fungal rot

© Queen's Printer for Ontario, 2009. Reproduced with permission.

B. Bacterial "soft rot"

Figure 16.1. Spoilage of fruit by microbes A. Fungi produce spores that are easily dispersed and can survive in an inert form for relatively long periods. When they land on a nutrient source, such as food, the spores germinate and the fungus proliferates. **B.** Bacterial soft rot produces discolored and water-distended lesions that develop and expand rapidly. Other microbes may invade and a soft, slimy mass of lysed cells and bacteria fills the inside of the fruit.

meat. **Semi-perishable foods** such as nuts or potatoes are not likely to spoil as quickly. Foods that cannot support microbial growth, often because they lack available water, are called **non-perishable foods**, and they may remain edible for long periods of time, even years. Examples of these foods include flour, sugar, and dried beans.

The likelihood that a food will spoil is related to its qualities as a microbial growth medium, which in turn depends on the characteristics of the food and its surroundings **(Table 16.1)**. **Intrinsic factors** relate to characteristics of the food itself such as its nutrient content or pH. These also include structures such as rinds or shells that protect the food from microbial invasion. We simulate that protection when we wrap our foods in plastic to protect it from spoilage. **Extrinsic factors** relate to the conditions under which a food is stored. For example, cold temperatures in a refrigerator will slow the growth of human pathogens that grow best in warmer temperatures of the human body.

Moist foods are more perishable than dry foods. Water content alone, however, is not the key factor in spoilage. Rather, one must consider the amount of water that is available and accessible to microorganisms. This is called the **water activity (a_w)**. Most fresh foods have an a_w above 0.99, which is sufficient to support the growth of most microbes. Most spoilage bacteria require an a_w above about 0.91 **(Figure 16.2)**. Foods that are dry or possess a high solute content, usually of sugar or salt, have reduced a_w values. Fungal spoilage organisms, such as molds, can grow at lower a_w of approximately 0.8. Foods with lower a_w values are

TABLE 16.1	Factors affecting microbial growth in foods
Intrinsic factors	**Extrinsic factors**
Water availability	Temperature
Osmolarity	Humidity
Nutrient content	Presence and concentration of gases
pH and buffering capacity	
Antimicrobial constituents	
Biological structures such as rinds or shells	

generally quite stable and have long shelf lives. Though spoilage bacteria require an a_w above about 0.91, one notable salt-tolerant bacterium is *Staphylococcus aureus*, which can grow at an a_w around 0.86. This can be significant for foods such as ham with a relatively high salt content that reduces available water, because some strains of *S. aureus* secrete microbial **enterotoxins** that act on the intestine leading to foodborne illness. This is why your picnic ham should not remain out of the cooler for too long.

CONNECTIONS Enterotoxins act on the gut (*entero* refers to intestine), and those produced by *S. aureus* can induce severe vomiting. Enterotoxigenic strains of *S. aureus* discussed in **Section 21.1** are a leading cause of foodborne illness. They are most commonly found in foods that have been improperly cooked or insufficiently refrigerated.

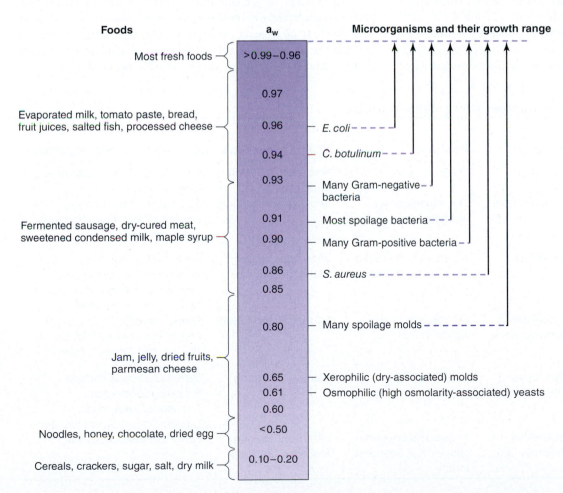

Figure 16.2. Water activity (a_w) Foods with a high a_w value, approaching 1.0, have readily available water and can support microbial growth. Foods with a low a_w value, approaching 0, contain very little available water; these foods are resistant to microbial growth.

The nutrient content of a food affects its likelihood to spoil, and also determines the types of microbes that might contribute to the spoilage (Table 16.2). Fruits and vegetables are rich in carbohydrates, but low in protein. As a consequence, their spoilage patterns reflect metabolic pathways that microbes use to degrade carbohydrates resulting in accumulation of acids and gas from fermentation of sugars (see Section 13.3). Foods that contain the structural polysaccharide pectin, which helps hold the cells of plant tissue together, may be attacked by microbes capable of pectinolysis (see Table 16.2). The result is a release of methanol or uronic acid, characteristic of pectin breakdown. The fruit or vegetable undergoes "soft rot" and loses its structure, becoming soft and mushy with a distinctly unpleasant taste (see Figure 16.1).

Fruits are typically more acidic than vegetables. Because of this, mold spoilage is more common on fruit. The slower-growing fungi are generally more tolerant of lower pH and can grow on fruits that are too acidic to support bacterial growth. Faster-growing bacteria out-compete molds at near neutral pH, as found in many vegetables. Lemons, for example, spoil extremely slowly, but will eventually be covered by mold if abandoned in the back of the refrigerator.

A variety of microorganisms can degrade the carbohydrate, protein, or fat components of milk. Souring caused by *Lactobacillus* sp. and *Streptococcus* sp. can cause formation of curds and a sour taste. Some of these reactions are desirable and produce cheese. Contamination of milk by *Clostridium* sp. or intestinal bacteria like *Escherichia coli* produces gas. Growth of *Alcaligenes* sp. or *Klebsiella* sp. causes the milk to become slimy. This variety of reactions reflects the nutrient-rich nature of milk.

Potential spoilers must have access to a food item. Microbes that spoil fruits and vegetables are often found in the soil; such foods are unlikely to be spoiled by intestinal microorganisms.

However, fresh produce can sometimes be contaminated with incorrectly treated manure that has been used as fertilizer in the field, or untreated rinse water containing intestinal pathogens from sewage. It is not unusual to find intestinal inhabitants on poultry or fresh meats. Their source is typically the carcass from which the tissue has been derived or other animals nearby in the feedlot, farm, or slaughterhouse. Workers and airborne dust within food processing plants are additional sources of microbes that can contaminate and spoil foods.

Food spoilage typically becomes evident when the contaminating microbe(s) reach late exponential growth. In the late exponential phase, the metabolic activities of the increased numbers of microbes will consume significant amounts of the food-related nutrients. Accumulation of metabolic end products, along with microbial biomass, result in detectable food defects that include foul odors, sliminess, or visible tissue destruction. Exponential growth may occur early due to the plentiful supply of substrates, but damage to the food will not be evident because the microbial population is relatively small. In general, biologically induced spoilage is not detectable at microbial densities below one million (10^6) per milliliter (mL) or gram (g) of product. Above that, some sensory evidence of deterioration may be evident, and at $>10^9$/mL or g definite structural change is apparent.

CONNECTIONS The microbial growth curve described in **Section 6.1** is useful for characterizing the behavior of a microorganism under defined conditions.

The concept of exponential growth is important to food preservation, affecting both the quality and safety of food products. A newly purchased carton of pasteurized milk can remain at room temperature for several hours with little adverse effect. If the initial population is about 10^3/mL and doubles every 30

TABLE 16.2 Typical spoilage patterns of selected foods

Food	Microbial substrates	Spoilage reactions	Products	Effects of spoilage	Common spoilage organisms
Fruits, vegetables	Pectin	Hydrolysis	Methanol, uronic acids	Soft rot, loss of fruit structure	Bacteria: *Erwinia, Pseudomonas, Corynebacterium*
	Sugars	Fermentations	CO_2, organic acids, alcohols	Souring, acidification	Fungi: *Aspergillus, Botrytis, Rhizopus, Penicillium, Alternaria,* various yeasts
Fresh meat, poultry, seafood	Proteins	Proteolysis, deamination	Amino acids, peptides, amines, H_2S, ammonia, indole	Bitterness, souring, bad odor, sliminess	Bacteria: *Acinetobacter, Aeromonas, Campylobacter, Escherichia, Listeria, Micrococcus, Pseudomonas, Salmonella*
	Carbohydrates	Hydrolysis, fermentations	CO_2, organic acids, alcohols	Souring, acidification	Fungi: *Candida, Cladosporium, Mucor, Penicillium, Rhizopus, Rhodotorula, Sporotrichium*
Milk	Lactose	Hydrolysis, fermentations	Lactic acid, CO_2	Souring, clumping	Bacteria: lactic acid bacteria (including *Streptococcus, Leuconostoc, Lactococcus, Lactobacillus*), *Pseudomonas, Proteus*
	Proteins (casein)	Proteolysis, deamination	Amino acids, peptides, amines, H_2S, ammonia, indole	Bitterness, souring, foul odor, sliminess	Fungi: *Candida, Geotrichum*

minutes at countertop temperature, the population might reach around 10^4/mL before the carton is discovered. This microbial density is not great enough to effect visible damage. However, growth at room temperature would increase the number of microorganisms in the milk that was finally re-refrigerated. This could damage the milk quality and lead to quicker spoilage if the milk is left on the counter again.

The lag phase at the beginning of the growth cycle when organism numbers remain fairly constant, and the exponential phase during which the number of living organisms doubles in a given time are most critical to determining microbial spoilage (see Figure 6.9). Often a product is already noticeably spoiled before growth stabilizes in the stationary phase. By extending the lag phase or reducing the rate of microbial growth a product's shelf life can be extended. For example, microbial growth can be minimized during food handling through the use of chlorinated water to wash soil from harvested plants, the early and efficient removal of damaged or diseased products from the collected harvest, or use of scrupulously clean utensils during milking. All of these will ensure that the microbial load on a product is small. The lag phase is extended and more time will pass before the microbes reach exponential phase where the number of microbes is high enough to cause noticeable product deterioration. Certain preservation methods such as heat processing can also reduce and even eliminate the microbial load on foods. Other preservation practices aim to produce conditions that are not optimal for microbial growth by manipulating either intrinsic factors such as pH or extrinsic parameters such as food storage temperature. This increases microbial generation time, extending the lag phase and delaying entry into exponential phase. **Figure 16.3** illustrates how various types of processing and use of refrigeration for preservation can alter the characteristics of raw milk to produce dairy products that are considerably less perishable.

16.1 Fact Check

1. List the intrinsic and extrinsic factors that affect microbial growth in food.
2. What is water activity (a_w), and how is it related to food spoilage?
3. Explain how the microbial growth phases can be manipulated to reduce food spoilage.

Raw milk has a neutral pH, contains the nutrients lactose, fat, protein, minerals, and naturally harbors microbes such as lactic acid bacteria.

Figure 16.3. Alteration of milk as a microbial growth medium Raw milk is suitable for the growth of natural microbiota as well as contaminating microbes. Processing of milk into various products aims to change the physical and chemical form of milk to make it less favorable for growth of spoilage microbes.

© Tanuki Photography/iStockphoto

© craftvision/iStockphoto

© Viktor Lugovskoy/iStockphoto

© Juanmoino/iStockphoto

© Rubén Hidalgo/iStockphoto

Refrigerated, pasteurized milk begins with a reduced microbial load from heat processing, and storage at cold temperature increases generation time.

Salted butter has a low aw, a solid structure to reduce microbial mobility, and low pH.

Cheese has a low aw (salt, little water), and its solid structure reduces microbial mobility.

Yogurt has a pH ≤ 4.4, which is inhibitory to many contaminating microbes.

16.2 Food preservation

What techniques are used to minimize food spoilage?

We have seen that reducing the growth of contaminating microbes in food can be achieved by altering the food product to make it less optimal for growth of these spoilage microbes. Food preservation techniques prevent, or at least discourage, microbial growth and maintain food safety by destroying or restricting the growth of foodborne pathogens. Some techniques of food preservation such as drying meat and fruit in the sun have been practiced for most of human history. This ancient method has found new life in the form of the sun-dried tomatoes that adorn your pizza or salad. Another old preservation technique is to add high concentrations of salt or sugars to foods. Heating food and refrigeration are other preservation techniques that extend the length of time a product can be stored before becoming unsuitable for consumption.

Reduction of the water activity (a_w) of food

Microorganisms need available water to flourish, so removing water from food can prevent spoilage, sometimes indefinitely. Oven-drying can replace sun-drying for large-scale commercial production **(Figure 16.4)**. Drying is sometimes used in conjunction with wood smoking to impart flavor to certain products. Chemicals in the smoke, including phenolics, also inhibit microbial growth and contribute to preservation.

In **freeze-drying** or **lyophilization**, the food is first frozen and then dried under a vacuum, with subsequent storage in watertight packaging. The process is expensive but efficient and can be useful for special purposes. Freeze-dried foods are compact and lightweight, making them convenient for transport.

Campers, for example, often carry freeze-dried food in a backpack and add water on site to rehydrate it before consumption.

Drying reduces a_w by removing water. An alternative for decreasing the amount of available water in foods is to bind water molecules by the addition of solute, such as sugar or salt, making water unavailable to microorganisms. For thousands of years humans have practiced food preservation by the addition of large amounts of sugar or salt (NaCl) to food. Neither option is viewed as particularly healthy today, but this strategy is still seen in foods like jams, jellies, and salted meat **(Figure 16.5)**. Other compounds that may be added to some processed foods to tie up water include sugar alcohols such as glycerol, sorbitol, or xylitol. Unfortunately, ingestion of large amounts of products like sorbitol, as found in some sugar-free foods, can also lead to diarrhea as it draws water from the cells lining the intestine.

Control of temperature

In industrialized countries, surely the most prominent food preservation method is cold temperature, or **refrigeration**. Like sugaring or salting, this practice has deep roots; the ancient Romans reportedly used ice and snow to pack perishable seafood. Modern household refrigerators are usually set at about 4°C, whereas commercial chillers may operate at 1 to 2°C. However, cold storage is not foolproof. Some bacteria can tolerate refrigerator temperatures; such *psychrotolerant* microbes can grow and spoil food even while the food is refrigerated. An example is *Listeria monocytogenes*, introduced in the opening story of this chapter. Fortunately, most

© Merih Unal Ozmen/iStockphoto

A. Sun-dried tomatoes

© Paul Prescott/Alamy Limited

B. Heat-dried smoked salmon

Courtesy Christine Dupont

C. Freeze-dried (lyophilized) soup

Figure 16.4. Drying reduces water activity (a_w). A. Fruits can be dried using low heat in ovens, or by the sun. **B.** Smoking used with heat-drying adds additional chemicals inhibitory to microbes. **C.** Freeze-dried foods are sealed in watertight evacuated packs. Water can be added, as in the right half of the bowl, before heating.

A. Ham has high salt concentration.

B. Jelly has high sucrose concentration.

Figure 16.5. Adding solutes reduces water activity (a$_w$). High concentrations of solutes bind water molecules, making water unavailable for microbial use. **A.** Many meats are preserved by soaking in NaCl (brine) or covering in dried salt. **B.** Jams and jellies contain high concentrations of sucrose.

human pathogens do not grow well at refrigerator temperatures, but prefer the warmer temperatures of a living body.

Extended storage at freezing temperatures is possible for some foods, although not all. Ice crystal formation during the freeze-thaw process can damage food tissue, producing a mushy thawed product that bears little resemblance or taste to the original fresh material. Apples and citrus fruits, for example, undergo this transformation after freezing. Although such a soft texture is tolerated by infants, the majority of us prefer more robust textures in our food. A typical home freezer operates at about $-20°C$, below the temperatures that permit most microbial growth. In addition, a$_w$ is reduced by the conversion of liquid water to ice crystals. It is important to monitor how long frozen food has been stored, because the cold, dry atmosphere of the freezer will lead to physical defects such as freezer burn. Moreover, some slow microbial growth may be possible in small pockets of fluid that do not freeze because of their high solute content. Frozen foods will remain unspoiled for months, but not indefinitely.

High temperature processes also have a role in food preservation. Most obviously, thorough cooking of a food will kill most or all of its microbial community, although once cooled the food item is subject to recontamination. **Canning** is a high heat process that involves heating a food to $100°C$ or above for an extended period of time, usually 20 to 100 minutes and under pressure of 10 to 15 psi. Traditional canning processes often use a temperature of $121°C$, the temperature of a laboratory autoclave that is used to sterilize equipment and solutions, also known as "retort" temperature in the canning industry. Traditional canning, sometimes called *appertization*, has its roots in a process devised by Frenchman Nicolas Appert, in the time of Napoleon. From about 1795 to 1810, Appert developed a method in which food sealed inside a corked glass jar was heated in boiling water for several hours. Treated products were stable indefinitely, an advantage exploited by Napoleon's armies as their supply routes extended across great distances of European territory. Sealed glass jars became available by the 1880s, and home canning of foods became a common domestic chore **(Figure 16.6)**.

Although improvements were made to commercial appertization processes the commercial canning industry nearly foundered in the early part of twentieth century because of problems with the disease **botulism**. Botulism is a paralytic and deadly illness caused by botulinum toxin produced by *Clostridium botulinum*—normally an innocuous bacterial inhabitant of soils and aquatic sediments (see Microbes in Focus 21.2). *Clostridium botulinum* is a strictly anaerobic, endospore-forming chemoorganotroph, and so the low oxygen and high nutrient content of canned foods provides ideal conditions for germination of spores and growth of the organism. Botulinum toxin is the most toxic natural substance known. Early commercial canning procedures sometimes did not entirely eliminate *C. botulinum* endospores from foods. As a result, the industry was plagued with periodic outbreaks of botulism, resulting in severe illness and regular fatalities among those who consumed the contaminated food. Intense consumer reaction to botulism outbreaks led to research by commercial canners in the 1920s to establish protocols designed to vanquish endospore survival in canned products. The concept of decimal reduction time, or D value, arose from this research. Recall from Section 6.5, the D value is the time required to kill 90 percent of the target organisms under specific conditions. Knowing the D value for a target organism, like endospores of *C. botulinum*, under set conditions, such as temperature or pH, allows treatment of that product to render it safe for consumption.

Figure 16.6. Home canning During World War I, Europe could not supply enough food for Allied forces, so the U.S. National War Garden commission was established to encourage Americans to use idle land for growing fruits and vegetables to send to troops. Households were educated on how to preserve food through the distribution of detailed manuals and local canning lessons.

CONNECTIONS Endospores of *C. botulinum* may germinate under
anaerobic conditions in cans or sealed jars and produce botulinum
neurotoxin. The toxin blocks release of acetylcholine at the neuromuscular
junction causing flaccid paralysis, which can lead to death because the
muscles needed for breathing, including the diaphragm, no longer
function. See **Section 21.1** for details.

Today, commercial canners aim for "commercial sterility,"
but not necessarily absolute sterility as a lab microbiologist would
require of bacteriological media. Canned goods are, however, generally free of danger from pathogenic microorganisms. Either no
viable organisms can be detected by the usual culturing methods,
or the very low number is inconsequential because other intrinsic
or extrinsic factors are not conducive to growth. Botulism is now
an extremely rare foodborne illness in industrialized countries,
and is almost never linked to commercially canned products.
Improper home canning of foods, however, still claims lives. Botulinum toxin is inactivated by heat so home-canned foods should
be thoroughly heated before eating. Acidic canned foods are safer
as they inhibit the growth of *C. botulinum*.

Heat can adversely affect the flavor or texture of some food
products making them unsuitable for high heat processing.
Heat processing, or **pasteurization**, was developed by Louis
Pasteur to avoid such problems. Pasteur's original goal was to
develop a process to prevent wine from souring without causing irreparable damage to the wine from heat. He found that
mild heating could prevent souring without greatly affecting the
flavor of the wine. His success was primarily due to killing the
majority of alcohol-metabolizing acetic acid bacteria and other spoilage microbes in the wine. Today, pasteurization uses a
variety of techniques to destroy microorganisms without cooking the food product. Pasteurization processes include heating,
irradiation, or applying high pressure. You may know that milk
is typically pasteurized, but so are other products such as fruit
juices, dried fruit, beer, honey, and liquid eggs. The specific time
and temperature used in pasteurization depends on the item
being treated, but the typical goal is to kill 99 to 99.9 percent of
pathogens and spoilage bacteria and fungi.

The original process of milk pasteurization involved heating
milk in batches to a temperature of at least 62.8°C and holding
it there for 30 minutes. This is termed the **low temperature
hold (LTH)** method (**Table 16.3**). Modern dairies often use **flash
heating**, which forces a small volume of milk between metal

TABLE 16.3   Pasteurization processes for milk

Process	Temperature and time employed
Low temperature hold (LTH)	63°C (145°F) for 30 min
Flash heating	
High-temperature, short-time (HTST)	72°C (161°F) for 15 sec
Ultra-high temperature (UHT)	138–150°C (280–304°F) 2 sec or less

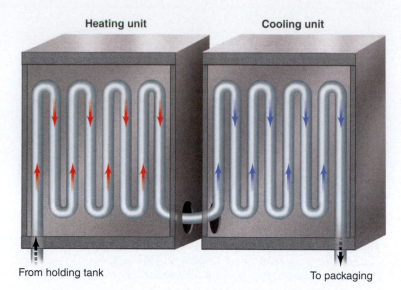

Heating unit   Cooling unit

From holding tank   To packaging

Figure 16.7. Flash-heating pasteurization Flash heating allows continuous flow-
through. In order to kill contaminating microbes, a small volume of raw liquid flows for a specified but very
short period of time through metal plates or tubing (*left*) heated to a very high temperature. Liquid then
passes through a cooling unit (*right*) to quickly lower the temperature to prevent excessive heat denaturation
of nutrients. The product is then packaged.

plates or pipes heated to very high temperatures (**Figure 16.7**).
Small volumes reach the required temperatures much more
quickly and evenly than large volumes. Exposure for only a very
short time to high temperatures will kill the required number of
microbes. Another advantage of flash heating is that nutrients,
such as vitamins, are not exposed to long periods of heat that may
break the vitamins down (as occurs in the LTH method).

A common flash heating method uses the *high-temperature,
short-time (HTST)* process, where milk flows for at least
15 seconds at a temperature of 72°C. This is the minimum
temperature and time required for fluid milk sold for
consumption in the United States; however, commercial
processors often use variations of this protocol. *Ultra-high
temperature (UHT)* processing (138 to 150°C for a few seconds or
less) is commonly used for milk products and juices, and when
combined with aseptic packaging, such products can be safely
stored at room temperature for several weeks (usually 90 days).
Coffee creamers, juice boxes, beer, and heat-and-serve soups are
examples of products that may be processed this way.

An early target of milk pasteurization was *Mycobacterium bovis*,
the agent of bovine tuberculosis (TB), which can also cause
TB in humans (see Section 21.2). Today, *M. bovis* is controlled
by strict veterinary practices, but outbreaks still occur. Other
pathogens potentially carried in raw milk include species of
Brucella, *Salmonella*, *Campylobacter*, *Coxiella burnetti*, and pathogenic *Escherichia coli* strains. All of these are non-spore-forming
bacteria that are killed by pasteurization.

Increase in acidity of food

Microbial spoilage occurs most readily in a pH ranging from
6 to 8, where many microbes prefer to grow. Food is preserved
from spoilage if the pH can be made more acidic or more alkaline without impairing the food's appeal. Humans ingest very
few alkaline foods due to their bitter taste; egg white is one.

We don't seem to mind acidity, however, and perceive it as tartness. Too much acid in a food, however, can be perceived as sour. In a process known as **pickling**, the pH of a food can be reduced by storing or marinating it in a chemical such as dilute acetic acid (vinegar) or by allowing lactic or acetic acid production to occur naturally through microbial fermentation. Pickling achieves a pH of less than 4.6, a condition that deters the growth of most microorganisms **(Figure 16.8)**. During pickling of vegetables or meats, spices or sugar may be added to fuel microbial metabolism and impart specific flavors. Foods have been preserved through pickling for much of human history.

Addition of chemical preservatives

Acetic acid is one of several weak organic acids used as food preservatives. In 1908, sodium benzoate was the first chemical food preservative permitted by the U.S. Food and Drug Administration (FDA). Propionates and sorbates have also been used in some types of food for decades. All of these acids have *"generally recognized as safe" (GRAS)* status awarded by the FDA, permitting their widespread use. The preservative effect of these weak acids is restricted to relatively acidic foods, because the acid molecule must be in a protonated form to cross the plasma membrane and disrupt microbial activity. Sodium benzoate, for example, is used in products such as fruit juices, pickles, ketchup, salad dressings, and soft drinks. Significantly lowering pH inhibits growth of even yeasts and molds that can generally tolerate relatively low pH.

Another chemical that is widely used in cured meat products is sodium nitrite ($NaNO_2$), which has multiple effects. Nitrite inhibits some spoilage microorganisms as well as *C. botulinum*, and other gas-producing clostridia. In cured meats sodium nitrite also stabilizes the red color of the product and contributes to flavor development. Nitrite seems to exert its inhibitory effect by interfering with cellular respiration by targeting non-heme iron-sulfur proteins, such as ferredoxin, in susceptible microorganisms (see Section 13.4). Unfortunately, nitrite reacts with secondary amines to form nitrosamines, some of which are known carcinogens. Nitrosamines have been detected at low levels in cured meat products. As a result, nitrite levels are controlled, and often isoascorbate is added concurrently to prevent nitrosamine formation.

A more recent approach to food preservation is biocontrol through the use of microbial products called **bacteriocins**, produced by some bacteria that have been used in food fermentations for millennia. Bacteriocins are small proteins that do not affect the organisms that produce them, but exert a negative effect on other closely related bacteria. This microbial version of chemical warfare provides a means for one type of microbe to

Pickled cucumbers

Pickled eggs

Pickled beets

Pickled herring

Pickled (corned) beef

Figure 16.8. Common pickled foods Pickling involves the production or addition of lactic or acetic acid in food to reduce the pH of food to less than 4.6, which is inhibitory to microbial growth.

out-compete another for available resources. Nisin and pediosin are the two bacteriocins most investigated. Nisin has GRAS status for certain uses and is commercially available. It is used in products such as milk, cheeses and other dairy products, canned goods, and baby foods. It inhibits *C. botulinum* and can also affect other Gram-positive spoilage bacteria. Pediocin synthesized by *Pediococcus* species is effective against *Listeria monocytogenes*. An alternative to adding the bacteriocin is to add bacteriocin-producing bacteria strains to the food, an approach that fits nicely with the production of fermented food products.

Irradiation of food

For decades exposing food to radiation, termed **irradiation**, has been recognized as an effective way to reduce the number of microorganisms on the surface of or, to a limited extent, within fresh food products. Most consumers do not understand how radiation works and may associate radiation of food with radioactive food, thinking this technology is dangerous to them. World War II, atomic bomb testing, and nuclear accidents like Chernobyl and Fukushima have likely left consumers wanting little to do with radiation of any description. In reality, food irradiation is extremely safe and effective and is widely used. Radioactive material is not added to food; the food is simply exposed to the radiant energy of an external radiation source.

Radiation harmful to microbes includes ionizing radiation and non-ionizing radiation **(Figure 16.9)**. **Ionizing radiation** refers to radiation of short wavelength with sufficient energy to remove electrons from molecules in a medium, producing ions. In cells, ionizing radiation produces oxidative damage and toxic free radical generation, all of which directly or indirectly damage nucleic acid. Ionizing radiation in the form of X-rays or gamma rays can be used to pasteurize or even sterilize foods, particularly in regions where refrigeration is not widely available. Sources of ionizing radiation include gamma rays from Cobalt-60 or Cesium-137, X-rays, or electron beams. Cobalt-60 sources used for medicine are sometimes "recycled" for use in the food industry once they have become too depleted for medical use. (Cobalt-60 has a half-life of about 5 years.)

Ultraviolet (UV) radiation has a wavelength range of 10 to 380 nm. UV radiation in the wavelength range of 240 to 280 nm is the most lethal form of **non-ionizing radiation**. It causes the formation of damaging thymine dimers in deoxyribonucleic acid (DNA), which prevents transcription and DNA replication. UV radiation does not penetrate materials very well, and so its use is limited in the food industry to surface sterilization only. UV lights may be installed in rooms where cheeses or sausages are stored for aging or ripening to reduce the microbial load in the air. UV light may also be used to reduce the viable microbes on surfaces such as warehouse shelving or in equipment such as plastic milk bags used for aseptic filling of liquid foods. Technological advances over the last quarter century have led to more powerful UV lamps, with better penetrating power. Consequently, UV is being increasingly used in water disinfection protocols (see Section 16.5).

CONNECTIONS Exposure to natural UV light causes thymine dimers, and as we learned in **Section 7.5**, cells can repair this damage when it happens on a small scale. Extensive formation of thymine dimers, as would occur through exposure to UV disinfecting lights, causes DNA damage that exceeds the cell's repair capabilities, resulting in cell death.

Irradiation is of considerable benefit in the handling of raw meats, especially hamburger and poultry, to eliminate foodborne pathogens such as *Escherichia coli* O157:H7, *Salmonella*, and *Campylobacter*. These have been the cause of several large and infamous outbreaks of illness usually related to undercooking of raw meat (see The Rest of the Story, Chapter 21). Proper handling and thorough cooking eliminates the risk, but there is no guarantee this will occur in restaurants or even at home. Raw products devoid of these microbes are impossible to obtain with certainty at the meat processor level, because they are frequent inhabitants of the animal gut and can contaminate carcasses during butchering and processing. Radiation treatments to eliminate potential pathogens have been permitted in the United States since 1990 for poultry and 1997 for raw and frozen ground beef. With 48 million cases of

Figure 16.9. The electromagnetic spectrum Ionizing radiation on the short-wavelength end of the spectrum includes some ultraviolet (UV) light (less than 124 nm), X-rays and gamma rays. The only form of non-ionizing radiation used for food irradiation is longer wave (240 to 280 nm) UV light.

Snustad, Simmons: *Principles of Genetics*, copyright 2009, John Wiley & Sons, Inc. This material is reproduced with permission of John Wiley & Sons, Inc.

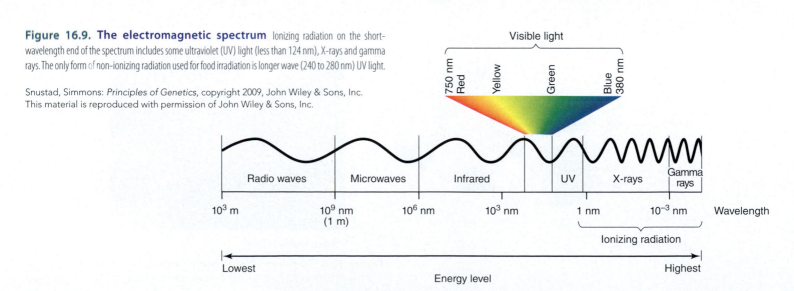

Figure 16.10. The Radura The Radura is the international symbol for food products irradiated with ionizing radiation. It resembles a plant in a circle, and is usually green. The U.S. Food and Drug Administration version is shown here. Along with the Radura, irradiated foods must have written labeling such as "treated with radiation" or "treated by irradiation."

Figure 16.11. Modified atmosphere packaging (MAP) for fresh produce The pepper on the left was stored for three weeks at room temperature with no packaging. The pepper on the rigth was refrigerated for three weeks in its original MAP package. This form of MAP is utilized extensively for produce shipped long distances.

foodborne illness causing 3,000 deaths annually in the United States (2011), the Centers for Disease Control and Prevention (CDC) supports the increased use of food irradiation. Food irradiated with ionizing radiation must be labeled as such, using the Radura, the international symbol for radiation, and appropriate wording **(Figure 16.10)**.

Use of Modified Atmosphere Packaging (MAP)

Many foods are rapidly spoiled by the activities of aerobic microorganisms. One way to slow such activity is to remove the oxygen (O_2) required by the growing microbes. This strategy is employed in **modified atmosphere packaging (MAP)**, or vacuum packaging. MAP is frequently coupled with refrigerated storage. In vacuum-packed foods, microbial respiration and respiratory activity of fresh vegetables and fruits leads to O_2 depletion and CO_2 accumulation within the sealed package. One example of MAP, used for products such as cured luncheon meats or wieners, uses relatively impermeable plastic pouches, which are then evacuated and sealed. As a result, growth of aerobic spoilage bacteria and fungi is discouraged, and growth of relatively innocuous fermentative microbes such as lactic acid bacteria (LAB) is favored. When significant microbial growth does eventually occur, usually following the stamped expiration date, the food item may become slimy, sour, and off-smelling so that it is unlikely to be eaten. If the food is eaten, such microbes are unlikely to cause illness.

CONNECTIONS In **Section 13.3**, we were introduced to LAB, a loose grouping of Gram-positive bacteria that produce lactic acid as a main product of their metabolism, which is always fermentative. See **Figure 13.19** for a review of the metabolic pathways for microbial production of lactic acid.

In other MAP processes, the atmosphere surrounding the food is altered and maintained by gassing the package or container with an appropriate gas mixture before sealing. For fresh fruits and vegetables, aerobic microorganisms are largely responsible for spoilage. MAP packaging for such produce typically uses high levels of CO_2 (greater than 10 percent), low levels of O_2 (less than 10 percent), with N_2 making up the balance. Coupled with refrigeration, this significantly reduces spoilage **(Figure 16.11)**. High-O_2 MAP is used for packaging raw red meats. A high O_2 level oxidizes hemoglobin and aids in maintaining the color of the meat, while reducing anaerobic growth from contaminating organisms such as *Clostridium*.

Hurdle technology

You may wonder if a sealed pouch made of relatively impermeable plastic, filled with food and depleted in O_2 might prove a suitable habitat for *Clostridium botulinum*, a strict anaerobe. Indeed, if MAP technology is used with foods where the presence of *C. botulinum* is possible, then additional preservative methods must also be applied. This is an example of **hurdle technology** that employs multiple constraints or *hurdles*, each of which the microorganism in question must overcome in order to proliferate **(Figure 16.12)**. The application of combined multiple hurdles works synergistically to control the growth of pathogens and

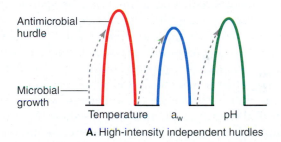

A. High-intensity independent hurdles

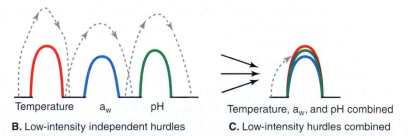

B. Low-intensity independent hurdles

C. Low-intensity hurdles combined

Figure 16.12. The hurdle technology concept Each constraint applied is a hurdle the microorganism must overcome in order to proliferate. **A.** When used independently, each antimicrobial hurdle must be of sufficient intensity to inhibit microbial growth, represented by the dashed line. Thorough cooking, salting or dehydration, or pickling can each prevent microbial growth, but significantly alters the food. **B.** Reducing the intensity of the hurdle is desirable to maintain food freshness, but does not sufficiently inhibit growth when each is used independently. **C.** When hurdles are used together, they sufficiently weaken microbial activity enough to allow lowering of the intensity of the preservation treatment and thereby maintain food freshness.

spoilage microorganisms in a food. To control *C. botulinum*, for example, any one of the following hurdles may be employed as each is inhibitory to *C. botulinum*:

- pH ≤ 4.6 (pickling or marinating in acidic solutions)
- a_w < 0.96 (salt-curing; see Figure 16.2)
- high NO_2 or O_2 gas concentration (MAP packaging)
- temperature < 4.4°C (freezing, refrigeration)

These individual hurdles are relatively intense and any of them can significantly alter the food. When these hurdles are applied together, however, their intensities can be much milder. This is the basis of hurdle technology that maintains food freshness over longer periods of time, avoids or reduces food processing, and reduces chemical additives—all of which modern consumers demand. Understanding the physiology of various spoilage microbes and the knowing pathogens associated with certain foods has allowed such advances in food technology.

Consider some of the products in your own kitchen. Can you determine how they are preserved, and whether multiple hurdles are in use? Do you have honey in the cupboard? It is likely pasteurized. What other feature does it possess that enables you to leave it sitting on a shelf at room temperature for a long time without spoiling?

16.2 Fact Check

1. How can a_w be reduced in food products, and why will this prevent spoilage?
2. Describe the various ways temperature is used to control microbial growth in food.
3. Explain how the chemicals acetic acid, sodium benzoate, and sodium nitrite function as food preservatives.
4. How can biocontrol methods such as the use of bacteriocins be used in food preservation?
5. Explain how different radiation methods are used to control or destroy microbes in food products.
6. What is modified atmosphere packaging (MAP) and how is it used in the food industry? Give an example.
7. Describe hurdle technology and how it can be used in preventing botulism.

16.3 Food fermentation

How are microorganisms used in food production?

Although microorganisms in food are often associated with spoilage, microorganisms are also used purposefully to create food and food products with unique qualities. As specific microbes metabolize specific foods, their metabolic by-products can impart desirable flavors, aromas, and textures that are not found in natural foods. Examples of this microbial bounty include breads, wines, cheeses, pickled vegetables, olives, and dry sausages, as well as many other popular foods. Indeed, one could survive a lifetime eating only foods that are the products of microbial activity.

Historical writings and other records show that, millennia ago, human societies relied on microbial fermentation to preserve or stabilize foods, including milk as cheese and yogurt, grains as bread and beer, and fruits as wine. These ancient practitioners had no concept of "microorganisms" or "fermentative metabolism," yet they did recognize that if certain steps were followed, a predictable sequence of events occurred in a raw food material, converting it to a desired end product. That end product might be valuable because it had superior spoilage resistance compared to the original material, or because it introduced a pleasing sensation on the palate, or, in the case of alcoholic beverages, because it evoked an interesting change of mental state. **Table 16.4** lists a variety of fermented foods.

The concept of a **starter culture** or inoculum was established in these early societies. Starter cultures are preparations of microorganisms that are added to food to aid in the production of fermented products. For example, taking leftover dough from a previous bread-making session and adding it to a new dough preparation ensures that bread will rise. In today's

Microbes in Focus 16.1
LACTOCOCCUS LACTIS

Habitat: Naturally found on green vegetation and in raw milk

Description: Gram-positive chain-forming coccus *(top)*, 0.5–1.5 µm diameter; an aerotolerant anaerobe with strictly fermentative metabolism

Key Features: Two subspecies (*L. lactis* subsp. *lactis* and subsp. *cremoris*) are widely used in making buttermilk, yogurt, and cheese. *Lactococcus lactis* grows well between 20 to 30°C, and hydrolyzes lactose and casein, producing lactic acid to a concentration of 1 percent, reducing pH to approximately 4.5 in suitable conditions. A wide variety of Gram-positive spoilage and foodborne pathogens such as *Staphylococcus* and *Clostridium*. The presence of nisin in foods containing *L. lactis* subsp. *lactis* helps prevent the growth of spoilage organisms. Nisin is believed to destabilize the plasma membrane. Nisin is produced commercially from *L. lactis* subsp. *lactis* cultures for addition to a wide variety of foods.

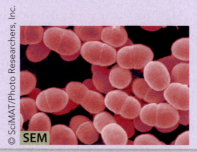

© SciMAT/Photo Researchers, Inc.

SEM

TABLE 16.4

TABLE 16.4 Fermented foods and beverages and associated microorganisms

Food	Starting material	Fermenting microorganisms
Beer	Barley, hops	*Saccharomyces cerevisiae, Saccharomyces uvarum*
Cheeses	Milk	Lactic acid bacteria and molds
Chocolate	Cacao beans	*Candida rugosa*, lactic acid bacteria, *Acetobacter, Gluconobacter*
Coffee	Coffee beans	*Erwinia dissolvens, Saccharomyces* spp.
Kimchi	Cabbage and other vegetables	Lactic acid bacteria
Miso	Soybeans	*Aspergillus oryzae*
Green olives	Unripened olives	*Leuconostoc mesenteroides, Lactobacillus plantarum*
Sake	Rice	*Aspergillus oryzae*
Sauerkraut	Cabbage	*Leuconostoc mesenteroides, Leuconostoc plantarum, Leuconostoc brevis*
Summer sausage	Meat	*Pediococcus* spp. and other lactic acid bacteria
Soy sauce	Soybeans	*Aspergillus oryzae* or *Aspergillus soyae, Lactobacillus delbrueckii*
Surströmming (pickled herring)	Herring (fish)	*Haloanaerobium* bacteria
Tempeh	Soybeans	*Rhizopus oligosporans, Rhizopus oryzae*
Yogurt	Milk	Lactic acid bacteria
Wine	Grapes	*Saccharomyces cerevisiae, Candida* spp.

industrialized age we have numerous other food preservation strategies from which to choose, yet fermented foods continue to be very popular worldwide because their diverse sensory properties add variety to the diet.

Fermentation of milk products

An amazing array of microbes, and variants on metabolic pathways, contribute to food fermentations, and the exact biochemistry that is taking place in each of the thousands of fermented products found globally is still not completely known. However, we are aware of some of the main metabolic pathways and groups of microbes involved. One group particularly prevalent in food fermentation are the lactic acid bacteria (LAB). Certain LAB, such as *Lactococcus lactis* (**Microbes in Focus 16.1**), are capable of fermenting the milk sugar lactose, a disaccharide of glucose and galactose, and are used in producing fermented dairy products such as cheese, buttermilk, koumiss, kefir, sour cream, and yogurt.

In commercial production of yogurt, milk is first centrifuged to remove some of the fat (**Figure 16.13**). Milk solids are increased by evaporation concentration or the addition of powdered milk solids to enhance the rich texture of the final product. The preparation is pasteurized and a defined starter culture is added. *Lactococcus thermophilus* and *Lactobacillus delbrueckii* subspecies (subsp.) *bulgaricus* are used because they produce volatile aromatic compounds that produce the distinctive aroma and flavor of yogurt. Lactic acid production during fermentation coagulates milk proteins (mostly casein) to thicken the milk. *Lactococcus thermophilus* and *Lactobacillus*

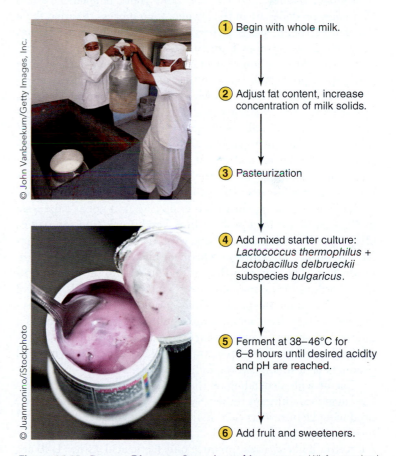

© John Vanbeekum/Getty Images, Inc.

© Juanmonino/iStockphoto

1. Begin with whole milk.

2. Adjust fat content, increase concentration of milk solids.

3. Pasteurization

4. Add mixed starter culture: *Lactococcus thermophilus* + *Lactobacillus delbrueckii* subspecies *bulgaricus*.

5. Ferment at 38–46°C for 6–8 hours until desired acidity and pH are reached.

6. Add fruit and sweeteners.

Figure 16.13. Process Diagram: Steps in making yogurt Milk fats are reduced by centrifugation and milk solids are increased. Pasteurization minimizes populations of microbes present in the whole milk and a defined starter culture of LAB is added to initiate fermentation. Lactic acid production coagulates milk proteins to thicken the milk. Fermentation proceeds until 0.9 percent acidity and pH 4.4 are reached.

delbrueckii subsp. *bulgaricus* are specifically used in yogurt production because they produce volatile aromatic compounds, particularly acetaldehyde, providing the distinctive aroma and flavor of yogurt. Fermentation proceeds at controlled temperatures until at least 0.9 percent acidity and pH 4.4 are reached (minimum standards for yogurt manufacture in the United States).

Other LAB are involved in producing fermented vegetables and meats. *Leuconostoc mesenteroides* is active in sauerkraut and *Lactobacillus brevis* in fermented meats. *Pediococcus acidilacti* is often a component of starter cultures used to produce dry or semi-dry fermented sausages and salamis. Sugars are added to these meat mixtures to provide a carbohydrate source for fermentation. Molds such as *Penicillium* often grow naturally on the surfaces of these meat products, producing other distinctive flavors. The principal function of LAB in food fermentation is to produce acid, which is responsible for the preservation of food. Of course, these bacteria also impart tangy flavors and pleasing aromas to fermented food products.

Historically, the production of fermented products relied on the activities of microbes found in the raw materials, and today this is still sometimes true, especially with ethnic foods produced in relatively small quantities for local use. However, a reliance on the natural microbiota can cause problems. If only a few cells are present, very weak fermentation may result, and if normally less abundant microbes happen to dominate a particular batch of starting material, the entire process may go awry. Who wants cheese tasting strongly of acetone, butanol, and ethanol? This could happen if certain *Clostridium* predominate in the fermentation process. To avoid these imbalances, commercial-scale fermentations usually employ specific, defined starter cultures. Large numbers of microbes of a desired strain are added to ensure operation of the desired fermentative pathways at predictable rates, so that consistent product characteristics result. Starters are often mixed cultures. In dairy fermentations, one or more primary strains may act as a dominant acid producer. Acid production in milk results in coagulation and stabilization of milk proteins and is responsible for preservation. Other secondary strains of microbes are added to produce metabolites that contribute flavor or other qualities to the food. These secondary strains usually account for only 10 to 20 percent of the starter culture.

This use of starter cultures is expensive, but not as expensive as the cost of a ruined product line. Food processing companies often purchase frozen starters with desirable characteristics from another company whose primary business is starter culture production. The starter is then added directly to the raw materials according to the manufacturer's instructions. This protocol in itself has made the production of fermented foods a more predictable enterprise, though some would argue that it produces a monotonously consistent product. The desire by some for batch-to-batch variety will help maintain the distinctive fermented products on a small, cottage industry scale.

Lactobacilli, depending on the species, may metabolize different sugars via different routes. Some LAB are homofermentative, producing lactic acid as their major (>85 percent) end product of fermentation (see Figure 13.17). For example, LAB in the genera *Lactococcus* and *Pediococcus* produce only lactic acid. Other LAB are heterofermentative, producing smaller amounts (~50 percent) of lactic acid in addition to various other fermentation end products (see Figure 13.18). Milk contains 0.15 to 0.2 percent citric acid,

and metabolism of this substrate may be a key contributor to flavor and aroma, producing CO_2, lactic and acetic acids, and diacetyl, the component responsible for buttery flavor. A particular *Propionibacterium* strain is used as a secondary strain in the manufacture of Swiss cheese because of its ability to produce propionic acid, acetic acid, and CO_2 from lactic acid (**Figure 16.14**). The acids contribute

Primary acid-producing strains

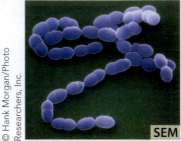

© Hank Morgan/Photo Researchers, Inc.

Streptococcus thermophilus

Secondary flavor-producing strain

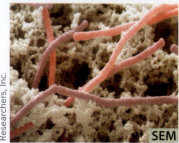

© Andrew Syred/Photo Researchers, Inc.

Propionibacterium freundenreichii spp. *shermanii*

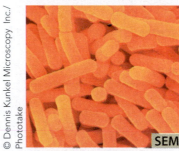

© Dennis Kunkel Microscopy Inc./ Phototake

Lactobacillus helveticus

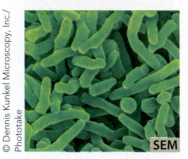

© Dennis Kunkel Microscopy, Inc./ Phototake

L. delbrueckii spp. *bulgaricus*

© Robyn Mackenzie/iStockphoto

Swiss cheese

Figure 16.14. Examples of starter cultures for cheese making Starter cultures consist of high numbers of primary lactic acid-producing strains of bacteria (LAB) whose principal role is to lower the pH during fermentation. Primary strains can also play a secondary role in flavor by release of metabolites. Secondary bacteria and/or fungi are present in lower proportions in starter cultures. Their primary role is to provide additional distinctive flavors and aromas.

to the taste we associate with this type of cheese, while the gas produced is responsible for the holes in the finished product.

Fungi sometimes contribute to dairy fermentations, too. *Penicillium* molds are secondary strains required for Camembert and blue cheeses. The proteolytic and lipolytic activities of the molds contribute greatly to the sensory qualities of these products. Mold is allowed to penetrate the blue cheese, whereas Camembert and Brie-type cheeses are surface-ripened. **Figure 16.15** shows the steps used in cheese making, which begins with pasteurized milk to rid the product of undesirable microbes. The milk is homogenized to evenly distribute the fat. Starter culture is added to induce acid production. Acid production coagulates the milk, forming solid *curds* made up mostly of the milk protein *casein* and its hydrolysis products, but also containing trapped liquid *whey*. Whey contains soluble nutrients, such as proteins, sugars, and minerals. Milk can also be coagulated using citric acid. Addition of the protease *rennin*, otherwise known as chymosin, facilitates coagulation. Rennin digests casein, producing a precipitated protein product. Rennin was initially harvested from calf stomachs but is now produced by recombinant DNA technology; equivalents are produced from various fungi or plants. Whey is kept with

the curd to make loose unripened (unaged) cheeses, such as ricotta or cottage cheese. Processing of these cheeses stops here.

For other cheeses, the curd is cut into small 1-cm cubes to aid in whey release. Heating removes more fluid and causes shrinkage, then the curd is salted in brine. For hard cheeses such as Cheddar, the curd is pressed to express whey and bind the curds together. For cheeses with a creamy texture, such as blue cheese, the curd is placed into molds to drain whey by gravity. The microbial cultures are not destroyed by these processes, and their activities are allowed to continue during storage or aging, to produce ripened cheeses. For blue cheese, the cheese mass is pierced with steel needles to form holes through which air penetrates to allow *Penicillium roquefortii* mold to grow, producing the blue "veins" found in this type of cheese. Some cheeses are unripened and do not go through an aging process.

Koumiss is traditionally prepared from mare's milk, which is fermented by a mixture of lactobacilli and a lactose-fermenting yeast, producing both lactic acid (about 1 percent) and as much as 2 percent ethanol in the product; yeast-derived CO_2 provides effervescence to the product. Ethanol production by yeast is inhibitory to the growth of many microbes and is responsible for the preservation of such alcoholic beverages.

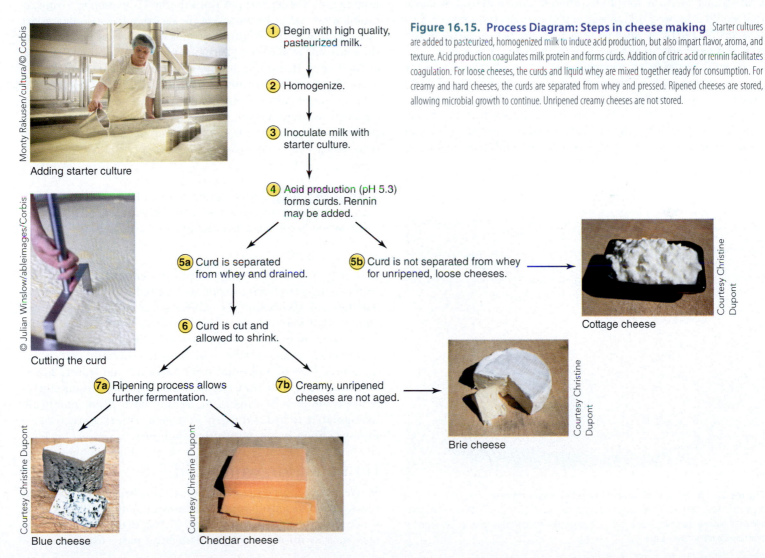

Monty Rakusen/cultura/© Corbis

Adding starter culture

© Julian Winslow/ableimages/Corbis

Cutting the curd

Figure 16.15. Process Diagram: Steps in cheese making Starter cultures are added to pasteurized, homogenized milk to induce acid production, but also impart flavor, aroma, and texture. Acid production coagulates milk protein and forms curds. Addition of citric acid or rennin facilitates coagulation. For loose cheeses, the curds and liquid whey are mixed together ready for consumption. For creamy and hard cheeses, the curds are separated from whey and pressed. Ripened cheeses are stored, allowing microbial growth to continue. Unripened creamy cheeses are not stored.

1 Begin with high quality, pasteurized milk.

2 Homogenize.

3 Inoculate milk with starter culture.

4 Acid production (pH 5.3) forms curds. Rennin may be added.

5a Curd is separated from whey and drained.

5b Curd is not separated from whey for unripened, loose cheeses.

6 Curd is cut and allowed to shrink.

7a Ripening process allows further fermentation.

7b Creamy, unripened cheeses are not aged.

Courtesy Christine Dupont

Cottage cheese

Courtesy Christine Dupont

Brie cheese

Courtesy Christine Dupont

Blue cheese

Courtesy Christine Dupont

Cheddar cheese

Product defects may occur when the wrong microbes dominate the fermentation. With starting materials rich in carbohydrates, a bacterial lactic acid fermentation may ensue when a yeast ethanolic fermentation was desired, or vice versa. In other words, LAB often spoil products like wine, and an undesired alcoholic taste and frothiness in lactic acid-based fermented products is frequently due to contaminating yeasts attacking the available substrates. Heterofermentative bacteria such as *Clostridium* and *Leuconsostoc* species, can be problematic because they tend to cause *blown* or *gassy* cheese caused by excessive CO_2 and/or H_2 accumulation during cheese making or after cheese has aged for several weeks. These bacteria can ferment lactose and the end products of desired fermentations, such as lactate and citrate. Occasionally, one may observe a package of late blown cheese in the grocery store (**Figure 16.16**). The use of only high quality pasteurized milk, good starter cultures, and excellent sanitation protocols throughout the cheese-making process, help reduce the chances of gas defects.

An infection by lytic bacteriophages can cause a fermentation to halt due to destruction of the population of fermenting bacteria. Obviously, cheese manufacturers derive no profit from huge vessels of milk and lysed bacteria unable to convert lactose into lactic acid. One cause of late-blown cheese is contamination with LAB bacteriophage that significantly reduce populations of desired LAB strains, allowing for proliferation of contaminating bacteria that would otherwise not present a problem. Bacteriophage infection has plagued the dairy product fermentation industry in the past, but today scientists know how to alleviate the problem.

Several methods are used to avoid phage infection of starter cultures. Good manufacturing practices, including aseptic technique, especially during starter culture production, are primary. Second is the use of media that are not conducive to phage infection. LAB bacteriophages need calcium ions (Ca^{2+}) to adsorb to their hosts; accordingly low-calcium phage-inhibitory media are

Starter	Plasmid/transposon-linked traits
Lactococcus lactis	Lactose hydrolysis and transport
	Proteinase activity
	Citrate hydrolysis and transport
	Bacteriocin genes
	Resistance to specific bacteriophages
	Resistance to several antibiotics
	Metabolism of several carbohydrates (including galactose)
Pediococcus	Sucrose hydrolysis
	Bacteriocin genes
Lactobacillus species	Lactose hydrolysis
	Maltose hydrolysis
	Bacteriocin genes
	Mucin production

TABLE 16.5 Examples of traits found on mobile genetic elements in starter culture bacteria

(Modified from Ray & Bhunia, 2008, after Montville & Matthews, 2005)

used to grow the bacteria. A third method is to employ strain rotation, which takes advantage of the fact that bacteriophages are often strain-specific. Multiple strains of bacteria that can fulfill the same role in starter culture function are rotated into and out of the mix so that an accumulation of phages that can infect a particular bacterial strain does not occur. Microbiologists who develop starter cultures continually create more phage-resistant strains through mutation, to add to the pool of rotating strains. In other words, microbiologists strive to keep their starter cultures at least one step ahead of the destructive viruses.

CONNECTIONS Bacteriophages are found in large numbers wherever their bacterial hosts exist. See **Sections 5.1** and **8.2** to review how bacteriophages infect and lyse bacterial cells.

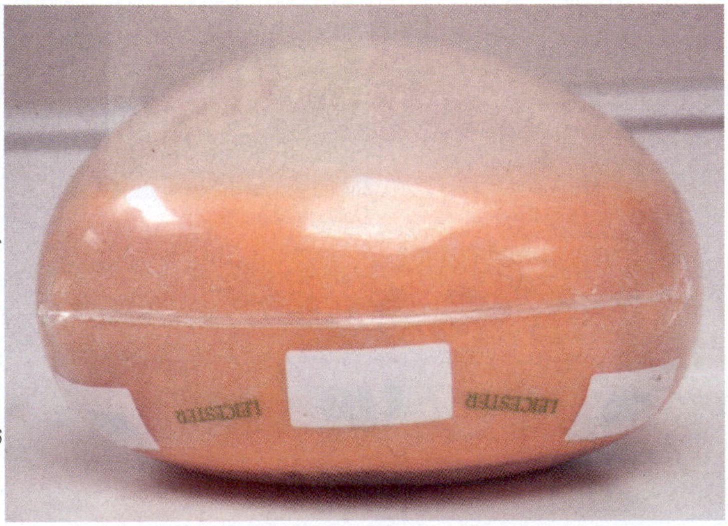

Starter culture inoculum failure is another factor that may lead to a defective food product, or no product at all. Before the genetics of LAB became the subject of research, it was not recognized that many of the critical features of fermenter strains were coded on mobile genetic elements like plasmids, and not on the chromosome (**Table 16.5**). This is especially true in the lactococci. Loss of a plasmid from a strain could lead to loss of a critical feature of that bacterium as a fermentation starter. Some commercially important strains have been genetically manipulated, to insert the genes for such critical characteristics into the chromosome, preventing their loss.

Fermentation with mold

In Western countries, food products resulting from LAB and yeast fermentations dominate. However, mold fermentation of solid substrates is used extensively in Asia to produce food products such as miso, a fermented soybean paste used as a base

Figure 16.16. Wheel of late-blown Leicester cheese Note the accumulation of gas (CO_2 in this case) within the package, indicating undesired fermentation was still active after packaging. The causative species was shown to be a salt-tolerant strain of *Leuconostoc mesenteroides*, present as a contaminant in the starter culture.

Figure 16.17. Miso and tempeh **A.** Miso and **B.** tempeh are fermented soybean products, originating from Japan and Indonesia, respectively.

A. Miso

B. Tempeh

and nucleic acids in the mash to produce a liquid rich in soluble nutrients. This now enables associated halophilic LAB and osmophilic yeasts to utilize the released nutrients to carry out lactic acid and ethanol fermentation. Fermentation continues for

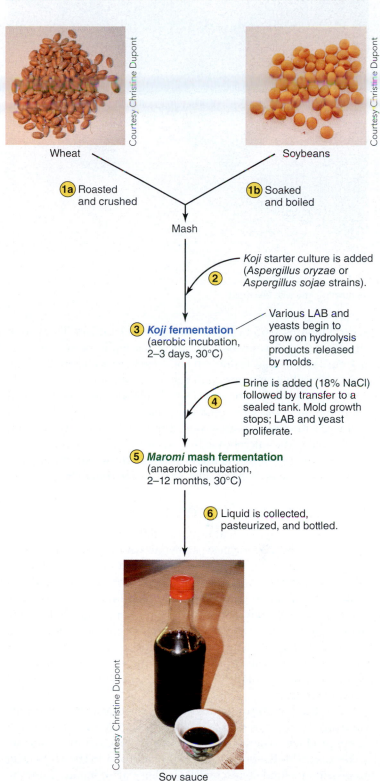

Wheat

Soybeans

1a Roasted and crushed

1b Soaked and boiled

Mash

2 *Koji* starter culture is added (*Aspergillus oryzae* or *Aspergillus sojae* strains).

3 *Koji* **fermentation** (aerobic incubation, 2–3 days, 30°C)

Various LAB and yeasts begin to grow on hydrolysis products released by molds.

4 Brine is added (18% NaCl) followed by transfer to a sealed tank. Mold growth stops; LAB and yeast proliferate.

5 *Maromi* **mash fermentation** (anaerobic incubation, 2–12 months, 30°C)

6 Liquid is collected, pasteurized, and bottled.

Soy sauce

Figure 16.18. Process Diagram: Production of soy sauce Prepared wheat and soybeans are inoculated with *koji*—a mixture of cultivated mold strains. The *koji* fermentation step encourages growth of the aerobic mold throughout the substrate to maximize extracellular enzyme production. The resulting breakdown products from starches and proteins are utilized by LAB and yeast in the next anaerobic *moromi* mash fermentation stage. The end product contains ethanol, low pH (lactic acid), and salt, all of which are inhibitory to growth of microbial contaminants.

for soups and sauces **(Figure 16.17)**. Other examples are tempeh and soy sauce. Tempeh originated in Indonesia and is prepared from soybeans, although other legumes, cereals, and agricultural by-products may also serve as starting materials.

Traditionally, soy sauce was produced in wooden tanks by a static fermentation process that took 1 to 2 years. Today, computer-controlled fermentations using large enamel or stainless-steel tanks are the norm, but the process is still lengthy and requires two stages **(Figure 16.18)**. The first stage is called a *koji* fermentation. *Koji* refers to various cultivated molds used in making fermentation products, such as soy sauce, miso, and sake or rice wine. The seed *koji* mold contains a mixture of strains pre-grown on a similar substrate to produce enough fungal mass to mix into the soybean-wheat mix. For soy sauce, a mixture of soaked, steamed soybeans and roasted cracked wheat (in a ratio of about 1:1) is inoculated with spores of the aerobic mold *Aspergillus oryzae*, or *Aspergillus sojae*, and spread in layers about 5 cm deep. Over 2 to 3 days, the mold grows throughout the solid substrate, producing the extracellular proteases and amylases it requires to hydrolyze the macromolecular material in the substrate. The distinctive flavor of soy sauce comes from the amino acids that are released from the hydrolyzed soybean proteins.

The second stage in the production of soy sauce, known as the *moromi* fermentation or mash stage, involves addition of salt water (brine) and depletion of available oxygen by sealing the mash in air-tight vats to prevent further mold growth. The mold-derived enzymes continue to hydrolyze proteins, polysaccharides,

Inoculating rice with *koji* mold

Cultivating *koji* mold

Figure 16.19. Cultivating *koji* for sake brewing A *Toji*, or brewing artisan, is cultivating *koji* in a traditional manner. In this case steamed rice is inoculated with *Aspergillus oryzae* spores (*left*). The inoculated rice is then placed into wooden containers to cultivate the *koji* mold (*right*). Individual manufacturers have their own proprietary versions of *koji*, some of which are centuries old.

several months at 35 to 40°C in the sealed containers. Thereafter, the *moromi* mash is pressed to collect the liquid phase, which is pasteurized and may be filtered and aged in casks before bottling. The end product contains up to 2 percent ethanol, lactic acid (pH 4.8), and salt (18 percent), all of which inhibit growth of microbial contaminants. Modern processes may use specific bacterial and yeast inocula to better control product characteristics, but, traditionally, the microbes active in secondary fermentation are derived from the substrate mixture. A similar two-stage process is required for the brewing of sake, or rice wine **(Figure 16.19)**.

Vinegar manufacture

The origin of vinegar is almost certainly rooted in the souring of wine. In the distant past, someone who was initially disappointed to discover their wine cache had spoiled ultimately recognized that the resultant liquid was not harmful and was not objectionable in taste, at least in small quantities. Further investigation revealed that the liquid was a good flavoring agent, paving the way for the current popularity of vinegar as a condiment. Vinegar is also used in the food industry as a pH control agent, as a preservative, and as a flavoring agent. Ketchup, sauces, salad dressings, pickles, and mayonnaise all contain vinegar.

Vinegar's main component is dilute acetic acid. Yet, simply diluting concentrated (glacial) acetic acid does not yield the same product. True vinegar is produced through microbial activity that adds flavor components in addition to the acetic acid to provide its final characteristics. These arise in part from the alcohol source used. The starting material is some type of alcoholic solution metabolized by bacteria belonging to a group known as the *acetic acid bacteria*, to produce acetic acid. Diluted pure ethanol is used for distilled or white vinegar; wines (from grapes in Europe, rice in Asia), fruit ciders, and malt liquors are other starting materials. Normally, a vinegar manufacturer

produces these alcoholic starting materials in the same general manner used in the beverage alcohol business, except that the aroma and other aesthetic aspects that are so important to alcoholic beverages such as wines are of no consequence in vinegar because, in the second stage of the process, the alcoholic solution is purposely spoiled as it is converted to an acetic acid solution.

The acetic acid bacteria are strict aerobes, and the oxidation of ethanol is one route that they use to obtain energy and carbon:

$$CH_3CH_2OH \rightarrow CH_3CHO \rightarrow CH_3COOH$$
$$\text{ethanol} \qquad \text{acetaldehyde} \qquad \text{acetic acid}$$

Acetic acid bacteria can also metabolize sugars and are tolerant of pH values below 5, as might be expected of microbes that produce organic acids as the end products of their metabolism. All known acetic bacteria are α-Proteobacteria and belong to one of two genera, *Acetobacter* and *Gluconobacter*. Species of *Acetobacter* can further oxidize acetic acid to CO_2 and water (H_2O), and are called complete oxidizers. *Gluconobacter* strains are used to make vinegar because they are incomplete oxidizers, and accumulate acetic acid.

The earliest and simplest method of vinegar manufacture, known as the Orleans process, is a surface culture technique where the acetic acid bacteria grow as a biofilm, or adherent community, on the surface of the ethanolic solution, gradually converting the alcohol to acid. Once a suitable level of acidity is reached, about two-thirds of the total volume is drawn from the bottom of the vessel to avoid disturbing the biofilm bacteria, and is replaced with fresh ethanolic starting material. The operation is a semi-continuous process, albeit not a very efficient one, requiring about 2 weeks for sufficient acetification to occur. It is still used, though only at small scale. The high-tech method of making vinegar uses a submerged culture technology with a bioreactor called the Frings Acetator, a far more efficient semi-continuous operation that produces product in 24 to 48

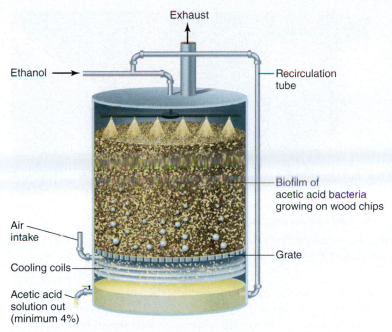

Exhaust

Ethanol →

Recirculation tube

Biofilm of acetic acid bacteria growing on wood chips

Air intake

Cooling coils

Grate

Acetic acid solution out (minimum 4%)

Figure 16.20. Trickle or Quick method of vinegar production The ethanol-containing liquid is introduced onto a bed of inert support material such as wood twigs or shavings. As the fluid passes through the bed by gravity, the adherent biofilm of acetic acid bacteria metabolize the ethanol to acetic acid. Air is constantly diffusing up through the bed, to ensure the bacteria have sufficient oxygen. The acidic liquid drips through the grate at the bottom, and is recirculated as necessary to obtain a 4 percent minimum acetic acid concentration. The process requires approximately 4 to 5 days.

hours under highly controlled conditions. **Figure 16.20** depicts the Quick or Trickle method of vinegar production, a "medium-tech" approach that employs acetic acid bacterial biofilms growing on an inert support material such as beech wood chips or birch twigs.

16.3 Fact Check

1. How are lactic acid bacteria (LAB) used in food production? Provide examples.
2. What are starter cultures, and how are they used to prevent undesirable results in microbial food products?
3. Describe examples and causes of the defects that may occur in products such as cheese due to the presence of undesirable microbes.
4. Describe the steps that can be taken to prevent destruction of LAB by phages.
5. What is the role of mold (*koji*) in soy sauce production?
6. Explain why the moromi mash stage requires removal of oxygen and addition of salt.
7. Explain how bacteria are used to produce vinegar.

16.4 Foodborne and waterborne illness

How can microorganisms in our food or water cause illness?

The provision of food that is safe to eat is a major focus of the food industry, and a primary goal of food handling, processing, and preservation technologies. Inevitably, however, foodborne illnesses do occur and, while the occurrence of a lethal illness such as botulism is extremely rare, foodborne illness unfortunately is not. The true number of foodborne illnesses that occur on a yearly basis is unknown, because many cases are never reported. As stated in Section 16.2, the CDC estimated 48 million cases of foodborne illness led to 3,000 deaths in 2011. **Table 16.6** shows the top five most common pathogens associated with foodborne illness. Foodborne illness may range from a nuisance to death. The severity of disease depends on the microbial species or strain, host-related characteristics including genetics and state of health, and other factors such as the type of food involved.

CONNECTIONS **Chapter 21** describes how several pathogens associated with foodborne illness cause disease.

Outbreaks of foodborne or waterborne illness are known as *common source outbreaks* as the incidence pattern typically reflects the fact that many people are exposed to the causative agent and become ill at about the same time from the same source (see Figure 18.22). The common vehicle of transmission that spreads

the pathogen could be a contaminated food item ingested by a large number of people at an event such as a party or picnic. After an incubation period, a significant proportion of people who ingested the food will become ill within a fairly short window of time. Consumption from a shared contaminated water source

TABLE 16.6 Top five annual foodborne-disease estimates for the United States (2011)

Bacteria	Number of reported illnesses/year	Foods
Salmonella spp.	1,027,561	Poultry, eggs
Clostridium perfringens (enterotoxin)	965,958	Reheated meats
Campylobacter spp.	845,024	Poultry, dairy
Staphylococcus aureus (enterotoxin)	241,148	Gravies, meats, puddings
Viruses		
Norovirus	5,461,731	Many foods

A. *Salmonella*-contaminated eggs

B. Hand washing reduces illness.

Figure 16.21. Salmonellosis is a common foodborne disease. A. *Salmonella* spp. are commonly found in the intestinal tracts of chickens and can contaminate eggs and meat. **B.** Thorough hand washing with hot, soapy water is recommended after handling these raw foods.

has the same effect. A tragic example occurred in the summer of 2000, in the town of Walkerton, Ontario, where contamination of one of the town's wells by *E. coli* O157:H7 sickened hundreds of people and killed seven (see Chapter 21 opening story). This bacterium more commonly causes foodborne outbreaks, usually from contaminated ground beef. Once the food has been consumed, or the defective water supply is discovered and replaced by a safe source, the number of cases diminishes rapidly.

A particular pathogen might tend to be transmitted more often through food or more often through water, but in many cases transmission by either route may occur. For example, typhoid fever, a serious disease caused by *Salmonella* Typhi (see Microbes in Focus 18.2), is classically considered a waterborne disease, but it can be spread through contaminated food. On the other hand, salmonellosis, a milder salmonella-caused illness, is most frequently foodborne, although outbreaks of salmonellosis resulting from contaminated water certainly have been reported **(Figure 16.21)**.

CONNECTIONS Common source outbreaks show an epidemiological pattern of cases that is distinctive from those caused by pathogens transmitted from person-to-person. See **Section 18.3** for further discussion.

Although there are exceptions, agents of foodborne illness most commonly cause sudden or acute symptoms that reflect

disruption of the gastrointestinal tract. The stomach and/or intestine may become inflamed. Gastritis, gastroenteritis, and enterocolitis are terms used to describe inflammation of the stomach, intestine, and colon, respectively. Symptoms such as nausea, vomiting, and diarrhea are common results of this trauma.

Enterotoxins, mentioned in Section 16.1, are secreted microbial toxins that cause or contribute to the intestinal symptoms in affected individuals. This frequently occurs because enterotoxin action stimulates water and electrolyte release from the cells lining the gut, resulting in diarrhea and dehydration. *Oral rehydration therapy (ORT)* that increases fluid intake by mouth as opposed to intravenously may be an effective means of combating these effects. ORT can be particularly important in infants and small children, who do not have a large volume of body fluids, and in individuals suffering from extended bouts of severe diarrheal illness, as occurs with the waterborne disease cholera **(Figure 16.22)**.

CONNECTIONS *Vibrio cholerae*, the bacterial agent of cholera, attaches to intestinal cells following ingestion of contaminated water. Once attached, it secretes a potent enterotoxin that causes ion imbalances in intestinal cells, causing significant water loss across cell membranes, resulting in massive diarrhea and dehydration. ORT replaces lost fluids and restores ion balance. See **Chapter 21** for details.

A. *Vibrio cholerae*-contaminated water

B. Oral rehydration therapy (ORT)

Figure 16.22. Cholera—a waterborne disease A. *Vibrio cholerae*-contaminated water following the 2010 earthquake in Haiti.
B. Haitian girl receiving ORT, a solution of sodium salts in water to replace that lost from severe diarrhea.

Colloquially, foodborne illness is often referred to as "food poisoning," a reflection of the fact that the first food illnesses linked to microbes truly were poisonings. Staphylococcal food illness was recognized in 1914, and definitively linked to an enterotoxin produced by *S. aureus* in 1930. In reality, food poisoning may be an intoxication or an infection.

A **foodborne intoxication** refers to illness caused by ingestion of microbial exotoxins as is the case with *S. aureus* food poisoning. Toxins produced by microbes may be present in food even though the organisms that produced them are no longer present when the contaminated food is eaten. Generally in foodborne intoxication, the microbes themselves may be harmless, and ingestion of the cells alone may be safe; it is the toxin that they produce that causes illness. A classic example of food intoxication is illness that follows consumption of ham sandwiches or potato salad that is left out too long and fosters growth of *S. aureus* that produces a potent enterotoxin (see Figures 21.6 and 21.7). Toxins are not necessarily destroyed unless food is thoroughly cooked or reheated properly. Since many foods that come in contact with *S. aureus* are only warmed or served cold, it is important to follow good hygiene practices such as hand washing when preparing foods. Illness from *S. aureus* enterotoxin is usually short-lived, but a more lethal example of food intoxication occurs after consuming foods in which *C. botulinum* has produced the toxin that results in deadly botulism. The growth of *C. botulinum* in cans produces gas, which is evidenced by bulging of the can (**Figure 16.23**). Because the toxin will be present in the food, symptoms may appear relatively quickly after ingestion, sometimes within thirty minutes and usually within six hours.

A **foodborne infection** requires consumption of the organisms themselves. When these living pathogens are ingested, they actively multiply within the body, typically in the small or large intestine. These organisms may also produce a toxin that initiates symptoms when the level becomes high enough. Symptoms of food infections typically do not appear immediately as the organisms will enter a lag phase of growth before showing the exponential increase causing illness. Foodborne infections may appear the day after ingestion of the offending food. Bacteria that commonly cause foodborne infections include *Salmonella*,

© Michael P. Gadomski/Photo Researchers, Inc.

Figure 16.23. Gas production from microbial growth Bulging cans result from accumulated gas and should be suspected of containing *C. botulinum* toxin.

Shigella, *E. coli*, and *Campylobacter*. Thorough cooking of food will usually prevent foodborne infections. It is particularly important to cook ground meat well so that bacteria that originated on the surface of the meat, but are now distributed throughout, will be killed. Ineffective cooking of hamburger patties has led to deadly outbreaks of *E. coli* O157:H7.

16.4 Fact Check

1. Describe the incidence pattern typically observed for a food or waterborne outbreak.
2. How do foodborne or waterborne pathogens cause diarrhea?
3. Differentiate between foodborne infection and foodborne intoxication. Give some examples of pathogens associated with each.

16.5 Microbiological aspects of water quality

How can we keep our water supplies safe?

Although we may be exposed to pathogens in the food we eat, far more often we face the risk of widespread exposure to pathogenic microbes in the water we drink. Moreover, foodborne pathogens may be transmitted to new human hosts through water contamination, since water is involved in food irrigation, and processing and preparation. People who live in areas of the world where waste is treated before it is emptied into natural waters and where drinking water is monitored for safety generally have more than an adequate supply of

drinkable, or **potable**, water. Potable water is free from intestinal bacteria and pathogens. In many developing areas of the world, however, potable water is not always available, and waterborne diseases still kill thousands of people every year. For example, in 2008 more than 60,000 cases of cholera were reported in the African country of Zimbabwe, and almost 3,000 people lost their lives from drinking unsafe water containing the pathogen *Vibrio cholerae* (see Microbes in Focus 21.7). The 2010 outbreak in Haiti resulted in over 20,000 cases and more than 900 deaths.

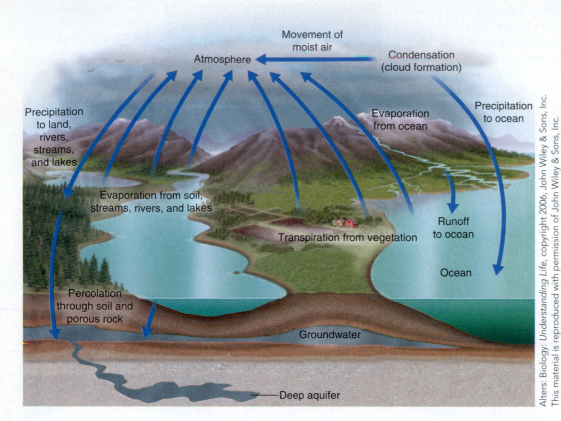

Figure 16.24. The water cycle Recycling of Earth's freshwater supply occurs through the water (hydrologic) cycle. Aquifers and surface water are sources of drinking water. Excessive removal from these sources can prevent their renewal.

About 70 percent of the surface of our planet is covered by water. Of that, about 97 percent is saltwater and 3 percent is the freshwater that sustains us. Water cycles as it evaporates from oceans, rivers, and lakes. Water also evaporates from plants—a process called transpiration. Water vapor condenses in the atmosphere to form clouds and falls as precipitation (**Figure 16.24**). When precipitation, usually rain or snow, falls on land, water may flow back into oceans, rivers, lakes, or streams, carrying organic materials with it, or it may pool in permeable underground geological formations known as "aquifers." Aquifers are replenished from water percolating slowly through porous soil and rock down from the surface of the earth, but they can be depleted if the amount of water withdrawn exceeds that recharging the aquifer. For example, the Ogalla aquifer that underlies the agricultural heartland of the midwestern United States is being depleted, mostly for irrigation, at a rate predicted to exhaust it within a few decades. Freshwater found in lakes, rivers, streams, and aquifers is useful and accessible to human populations, but about 70 percent of Earth's freshwater is locked away in glaciers, permanent snow, ground ice, and permafrost.

As humans explored and spread across this once sparsely populated planet, new settlements were often established near a body of water. This practice not only provided a ready mode of transport and an escape route from enemies, but it also ensured an adequate supply of potable water. The flow of water also conveniently washed away the various waste materials accumulated by occupants of the settlement. When waste levels were small, surface waters could diminish them naturally over time through microbial activity, oxidation, or photochemical breakdown from exposure to ultraviolet light from the sun. Today, many large

cities are still located near a water source. The time when water could dilute away the wastes of those residing in the settlement on the shore, however, are gone. The natural regenerative powers of lakes and rivers cannot deal with wastes produced by the masses of humans living in modern urban areas.

In heavily populated regions, raw water from a source such as a river, may be used for some purpose and then released back into the source, only to be taken up and reused for another purpose a short while later, or a short distance downstream. This pattern may be repeated many times as the water moves down the length of the river. In essence, the water cycle becomes shortened. There is less and less time and opportunity for the water to be cleansed by the biodegradative activities of its natural microbial inhabitants, for oxygenation, and for natural chemical and photochemical reactions to occur. The reality along a river is that communities downstream inherit and rely on a raw water source comprised partly of water that was released by their upstream neighbors. In some regions, so much water is removed from rivers for industrial use or diverted for agricultural use, especially irrigation, that the river may be reduced to a mere trickle, or end miles upstream of its natural mouth. For example, in theory the Colorado River discharges into the Gulf of California but in reality its flow often fails to reach that far.

No water body on the planet is guaranteed to be pristine, regardless of how remote from civilization it may seem. Chemical contaminants may disperse through the air as well as through surface and subsurface water flows. Wild animals may deposit feces and urine containing pathogenic microbes directly into a water body, or onto the land where subsequent runoff into a nearby stream potentially carries these pathogens. We can never confidently and safely quench our thirst by drinking directly from a lake or stream. Instead, we must protect our raw water sources, treat our wastewater, and purify and monitor our drinking water to ensure its potability and our health.

Protection of our water sources employs strategies to prevent contamination by industrial chemicals or agricultural runoff through severely restricting development in areas where water drains into rivers and lakes. New York City enacted these strategies to protect its water supply in the nineteenth century. The Croton, Catskill, and Delaware reservoir systems in upper New York State provide exceptionally high quality raw water to New York City residents from the largest unfiltered surface water supply in the world.

Wastewater, or **sewage**, is water carrying soluble or solid wastes originating from human activities, including farming and manufacturing. Household or *domestic wastewater*, consists of toilet waste, bathing water, cooking wastes, laundry wastes, and other waste products originating from everyday living. In addition, surface runoff, or *storm water*, from urban streets is often treated before it is allowed to enter water reservoirs. In urban centers, wastewater treatment plants help to minimize the amounts of contaminating material from urban centers entering the source water. In some locations, water goes through another purification process before being made available for drinking. In rural settings, water is treated on a much smaller scale. The discussion below focuses primarily on the treatment of urban domestic wastewater. However, the same general processes may be used in the on-site treatment of industrial wastes. As we shall see, as in the natural regeneration of freshwater, the treatment of wastewater relies heavily on the talents of microbes.

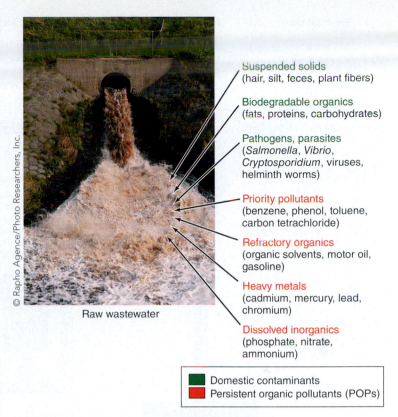

Raw wastewater

Suspended solids
(hair, silt, feces, plant fibers)

Biodegradable organics
(fats, proteins, carbohydrates)

Pathogens, parasites
(*Salmonella*, *Vibrio*, *Cryptosporidium*, viruses, helminth worms)

Priority pollutants
(benzene, phenol, toluene, carbon tetrachloride)

Refractory organics
(organic solvents, motor oil, gasoline)

Heavy metals
(cadmium, mercury, lead, chromium)

Dissolved inorganics
(phosphate, nitrate, ammonium)

◼ Domestic contaminants
◼ Persistent organic pollutants (POPs)

Figure 16.25. Classes of contaminants in wastewater The composition of wastewater varies greatly, depending on the sources of the waste entering the treatment facility. Typical domestic contaminants are handled relatively well by conventional treatment technologies. Persistent organic pollutants (POPs) are not as effectively degraded or neutralized. POPs are generally industry-associated contaminants, but individuals are also responsible for releasing some of these into wastewater by, for example, improperly disposing of drugs, cleaning solutions, solvents, and other chemicals down the sink or toilet.

Principles of wastewater treatment

Wastewater contains a literal stew of chemicals and microorganisms, and for health and aesthetic reasons, it must be treated before it is released into environmental water sources. Wastewater from the petrochemical industry, plastics manufacturers, metallurgy and electronics industries, or pharmaceutical manufacturers may, in addition, be laden with toxic substances such as cyanides, heavy metals, or organic solvents. Industrial plants frequently maintain their own treatment or pre-treatment systems, designed and optimized to deal with the characteristics of their particular wastes. This is often needed because industry-derived toxic substances may be lethal to the microbes of the municipal treatment system, which generally treats mostly domestic wastewater. A large-scale dairy operation would not produce toxic chemicals, but the high organic load of its outflow could exceed the capacity of a local sewage treatment facility, again necessitating pre-treatment on site prior to release into the municipal system.

A wastewater treatment plant is designed to achieve several goals before the outflow, or **effluent**, of cleaned wastewater exits the plant and is discharged into the receiving water such as a lake or stream. These goals include:

- reduction in the total organic content (TOC) of the wastewater
- removal or inactivation of any harmful microbes in the wastewater

- reduction in the level of inorganic nutrients (nitrogen- and phosphorus-containing compounds; primarily ammonium, nitrate, and phosphates)

Plants treating wastes receive a variety of contaminants **(Figure 16.25)**. Some of these chemicals, such as synthetic pesticides and pharmaceutical drugs excreted in urine are difficult to break down, and are referred to by the acronym **POP**, which stands for **persistent organic pollutant**. If present in wastewater, certain of these organics may be found in effluents exiting the plant. They may also be detected in natural waters. The concentrations are typically very low, and the significance of POPs in natural waters is uncertain. Evidence is accumulating, however, that even vanishingly small quantities of some POPs may affect reproduction and development of fish and mammals as they tend to accumulate in tissues.

The treatment of large-scale wastewater is a multistep process typically encompassing two or even three levels of treatment at the municipal facility. Ideally, municipal treatment facilities handle only wastewater from designated sanitary sewers—those receiving domestic sewage. In many urban areas, however, the storm sewer systems that handle surface runoff are not totally separate from the sanitary sewer system. As a result, after storms the treatment plant may be inundated with large volumes of water originating from combined sewer lines (partly domestic wastewater and partly storm water). If the incoming

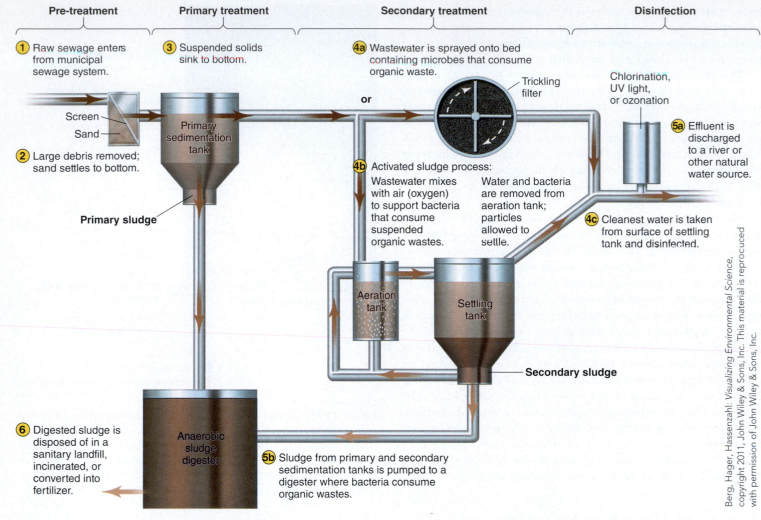

Berg, Hager, Hassenzahl: *Visualizing Environmental Science*, copyright 2011, John Wiley & Sons, Inc. This material is reprocuced with permission of John Wiley & Sons, Inc.

Pre-treatment

① Raw sewage enters from municipal sewage system.

Screen

Sand

② Large debris removed; sand settles to bottom.

Primary treatment

③ Suspended solids sink to bottom.

Primary sedimentation tank

Primary sludge

Secondary treatment

④a Wastewater is sprayed onto bed containing microbes that consume organic waste.

or

Trickling filter

④b Activated sludge process: Wastewater mixes with air (oxygen) to support bacteria that consume suspended organic wastes.

Water and bacteria are removed from aeration tank; particles allowed to settle.

④c Cleanest water is taken from surface of settling tank and disinfected.

Aeration tank

Settling tank

Secondary sludge

Disinfection

Chlorination, UV light, or ozonation

⑤a Effluent is discharged to a river or other natural water source.

⑥ Digested sludge is disposed of in a sanitary landfill, incinerated, or converted into fertilizer.

Anaerobic sludge digester

⑤b Sludge from primary and secondary sedimentation tanks is pumped to a digester where bacteria consume organic wastes.

Figure 16.26. Process Diagram: General schematic of a wastewater treatment plant In an effective treatment plant, wastewater moves through a pre-treatment screening, followed by physical primary treatment, and biological and physical secondary treatment. Effluent from these treatments is subjected to final disinfection before being released to a natural water system. Treated, collected solids (sludge) can be added to soils.

volume overwhelms the capacity of the plant, partially treated or untreated sewage may be released directly into the receiving water. Cities with such combined sewers are working toward replacing them with separate sanitary and storm sewer lines, but this is a slow and expensive undertaking. **Figure 16.26** shows an overview of the flow of wastewater through a plant operating a pre-treatment step followed by primary and secondary treatments and final disinfection of the effluent.

Pre-treatment of Wastewater

The pre-treatment step of wastewater treatment sends incoming material through a series of grates and coarse metal screens to ensure that large objects such as branches, stones, or pieces of metal that might damage or clog plant equipment are removed from the incoming wastewater. Screening of the incoming wastewater also removes much of the gravel and grit that may be present, particularly if the plant is treating waste from combined sewers. Solid debris is collected by periodically scraping the grates and screens and is frequently deposited in landfills.

Primary Treatment of Wastewater

Primary treatment of wastewater involves the physical removal of particulate matter by sedimentation and oils by skimming. Pre-treated wastewater flows into large tanks, where it is typically retained for several hours to separate these materials. Oil and grease and any material less dense than the water float to the surface and are removed by skimmers. Suspended materials that are denser than the water fall to the bottom of the tank and are removed by a raking apparatus that scrapes the material from the tank bottom into a hopper forming the **primary sludge**, which is collected and then pumped to the **anaerobic sludge digester**. The digester processes this waste under anaerobic conditions and will be described later. About 40 to 60 percent of the suspended solids are removed during sedimentation, along with the microbes that adhere to the particles. However, not all microbes are removed, nor are dissolved organic and inorganic compounds. The effluent derived from the primary treatment step is further subjected to secondary treatment.

Secondary Treatment of Wastewater

Secondary treatment of wastewater follows primary treatment and makes use of microbial activities to degrade the organic content of the sewage, further improving the quality of the wastewater. The two main strategies for secondary treatment include the *trickling filter* and the *activated sludge* process. A particular plant will likely employ either trickling filters or an activated sludge system, but not both. Trickling filters and activated sludge units contain communities of microorganisms that would be found in natural waters, but under the conditions provided in the treatment plant, these populations are far denser and break down waste much faster.

Trickling Filter The advantage of a **trickling filter** is that it is simple and relatively passive in operation, making it inexpensive to run **(Figure 16.27)**. Trickling filters are used for municipal and on-site wastewater treatment, such as for farms and industrial plants. It consists of a bed of solid material such as crushed rock or special molded plastic units that are contained in a cement tank. Effluent from primary treatment is pumped to a rotary spray boom that sprays the effluent across the filter bed surface. The fluid percolates through the bed by gravity, and is collected beneath. As the fluid trickles downward, it serves as a source of substrates for a complex microbial biofilm community growing on the bed matrix. This community includes a variety of bacteria, fungi, and protozoa that feed on components of the effluent, as well as more complex organisms such as rotifers, nematodes, and worms that graze on the inhabitants of the biofilm. Grazing activity helps to keep the biofilm in a state of constant renewal, which in turn helps to maintain efficient removal of potential organic and inorganic growth substrates in the effluent trickling through the filter.

> **CONNECTIONS** Biofilms can be complex communities of various adherent microbes, in which the majority of the cells are metabolically active. See **Figure 2.33** and **Sections 2.5** and **15.1** for some examples of biofilm communities.

The top and bottom of the filter bed are exposed to the air, and air may also be piped into the interior of large beds, to ensure that oxygen is available to the organisms making up the biofilm. Recall from Chapter 13 that aerobic metabolism is more efficient than anaerobic metabolism, so biofilm microbes will be more effective in removing materials from the effluent. Regions in the interior of most biofilms may still be anaerobic. Substances are removed from the effluent as they are assimilated by the biofilm. Organic compounds are mineralized by the biofilm. That is, microbial utilization of the organic portion of matter leaves behind the inorganic mineral components. **Mineralization** is the end result of decomposition of organic matter. Mineralized matter is adsorbed by the slimy biofilm matrix composed of proteins, nucleic acid, and **exopolysaccharide**, a high-molecular-weight polymer composed of various sugars that is secreted by microbes into the surrounding medium. The adsorption of organic and inorganic matter by exopolysaccharide facilitates their metabolism by

Figure 16.27. Trickling filter A. Overview of a trickling filter system. Wastewater is sprayed evenly over a filtration bed of rock, shown in **B**, about 2 m deep, or plastic matrices 20–30 m deep. The nutrients in the percolating wastewater support a microbial biofilm on the bed matrix, as shown in **C**. Biofilm formed by microbes that have colonized the extensive inner surface area of a small 2.5 cm diameter plastic matrix (manufactured by Headworks BIO, Inc.) forming the bed of a treatment system. Microbial metabolism removes pollutants from the effluent.

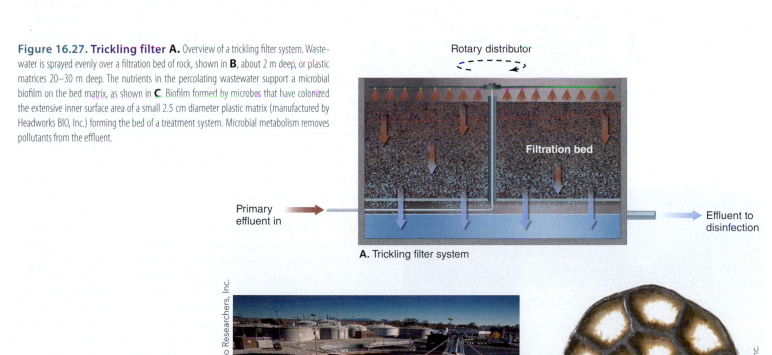

Rotary distributor

Filtration bed

Primary effluent in

Effluent to disinfection

A. Trickling filter system

B. Rock bed trickling filter

© Jonathan A. Meyers/Photo Researchers, Inc.

C. Wastewater biofilm on bed matrix

Courtesy Headworks BIO, Inc.

the biofilm microbes and removes these pollutants from the effluent. The effluent from the trickling filter then moves to the disinfection stage.

One of the drawbacks of trickling filters is seen when the incoming effluent is quite high in organic compounds that can lead to excessive growth of the biofilm. The exopolysaccharide-containing biofilm (see Figure 16.27C) can plug the filter bed, leading to accumulation of water on the surface. The filter then must be dismantled, cleaned, and reassembled. Trickling filters are also sensitive to the surrounding air temperature, and are not very effective in the cold. This limits their use to regions with mild winters.

Activated Sludge Activated sludge systems are used for municipal wastewater treatment plants. An activated sludge process relies on the formation of **flocs**—clumps of biomass consisting of adsorbed material and biofilm microorganisms. **Activated sludge** is comprised of accumulated flocs, the formation of which is critical to efficient function of the activated sludge unit. The word "activated" refers to the effect that the perfusion of oxygen throughout the otherwise thick and oxygen-free mixture has on the resident microorganisms in the sludge. An activated sludge unit is composed of two parts; an *aeration tank*, which mechanically mixes the sludge and purges it with air or sometimes pure O_2, and a *settling tank* or *clarifier* that collects sludge (see Figure 16.26). Primary treatment effluent enters the aeration tank and is retained there for 5 to 10 hours **(Figure 16.28)**. The effluent is then moved into the settling tank where the flocs settle as sludge. Some of the activated sludge is used to reseed the aeration tank. Excess sludge is moved to the anaerobic digester. Some mineralization of readily biodegradable organic matter does occur, but, as in the trickling filter, it is assimilation of materials by microorganisms that drives the purification process.

Floc formation occurs because of the presence of exopolysaccharide-forming microbes, such as *Zoogloea*. As in the trickling filter, these organisms are important to biofilm formation. The exopolysaccharide matrix they form acts as glue to form the floc into a firm, dense particle. As these slime-formers grow, they tend to entrap other microbes as well, including pathogens. The flocs serve as substrate to which other microbes including bacteria, fungi, and protozoa can attach and grow. It is essential that proper floc formation occurs to support the large number of microbes responsible for degradation of biological material, removal of fine solids and heavy metals, and to promote settling of floc. Well-formed flocs range in size from <1 μm (one-millionth of a meter) to more than 1,000 μm (a millimeter). The floc microbes, along with free-floating (planktonic) microbes within the roiling liquid of the aeration tank actively take up the substrates available in the effluent and grow. The microbial community within an activated sludge system is similar to that found in a trickling filter, although fungal populations are smaller.

After being held for a suitable time in the aeration tank, the floc-laden **secondary sludge** is pumped to the settling tank, where a rapid settling of floc is desirable to separate the sludge from clarified effluent. The clarified effluent is next sent to the disinfection process to kill remaining microbes. Part of the settled floc is used as a starter culture and is recycled to inoculate the next volume of incoming effluent in the aeration tank. Once in the tank, it becomes part of the activated sludge, some of which is sent to the anaerobic sludge digester for degradation. Sludge recycling is quite important to maintain the very high levels of microbial activity in the aeration tank. Similarly, flocs that settle are important to settling tank function. If flocs do not settle well, clumps of solid material and microbes will remain in the fluid effluent and cannot be properly disinfected. Moreover, the material contains bound pollutants, such as heavy metals. This problem, referred to as *sludge bulking*, is described in **Perspective 16.1**.

The effluent exiting the activated sludge unit is much reduced in its total organic content (TOC) because much of the organic content of the water that entered the aeration tank is now retained in the settled floc that forms the secondary sludge. The reduction in organic carbon in the effluent is reflected in a reduction in the **biochemical oxygen demand (BOD)** or biological oxygen demand. BOD is the amount of dissolved oxygen required by microorganisms to decompose the organic matter in sewage or water. As you might expect, the BOD of raw sewage entering the treatment plant is relatively high. If treatment is effective, the BOD of the effluent should reflect the lower concentration of organic compounds and be relatively low. Incoming effluent from a slaughterhouse or a dairy could exert significant pressure on a treatment plant designed to handle only domestic sewage **(Table 16.7)**.

Since most microorganisms use carbon compounds for their growth and survival, water that is loaded with carbon-containing pollutants is apt to support the growth of microbes, including ones that can make us ill. Although BOD measures only the amount of carbon that can be metabolized by microorganisms rather than the total organic content of the sample, it is the most common measure of organic pollutants in water. Much research has determined the correlation between BOD and the nature of the water source. In general, a natural water source such as a lake, river, or stream has a low BOD indicating that the organic content of the water, and hence the number of bacteria in the water, are relatively low. A high BOD indicates the water is polluted with organic material and supports high numbers of bacteria. These bacteria may not be a health concern, but, more often than not, high numbers of bacteria are a sign that the water has been recently contaminated with feces, and perhaps by

© GI Photostock/Photo Researchers, Inc.

Figure 16.28. Aeration tank of activated sludge system Effluent enters the aeration tank where a constant stream of air oxygenates and mixes the wastewater to promote the aerobic growth of floc-forming microbes, responsible for the removal of organic and inorganic matter.

A fundamental goal of sewage treatment involves activated sludge that is formed when air is percolated through the mixture of sewage and sludge in the secondary treatment tank. In the presence of oxygen, the microbes in sludge typically can metabolize nutrients faster than they can in the near or total absence of oxygen. The action of the microbes in the activated sludge drives sewage treatment. But the process does not always run smoothly. For example, when the level of organics in the wastewater is high and the oxygen level is low, such as in effluent from a brewery, conditions are optimum for growth of filamentous bacteria, which can enter the treatment system in the incoming wastewater. Large strands of filamentous bacteria, such as *Sphaerotilus* or *Thiothrix*, provide a large surface area for the sludge to collect in loose associations instead of tight, dense particles. This sludge may not settle, but instead result in sludge bulking **(Figure B16.1)**. Sludge bulking can stall the treatment process.

Sludge bulking is the bane of sewage plant operators worldwide, principally because, even though the problem has been known since the 1920s, science has not yet provided effective ways to avoid it. Solving the problem in a wastewater treatment system, with its huge volume and presence of numerous compounds, is far more complicated than solving a bacterial infection in the body, where the use of an antibiotic may be all that is needed. Moreover, despite the example provided above, the root of sludge bulking is by no means clear. Studies have linked both low dissolved oxygen and high dissolved oxygen, and both low and high ratios of nutrient-to-microorganisms to sludge bulking.

Since no certain solution is known, avoidance requires the experience and ingenuity of the plant operator. Good background data on the environment of the treatment tanks can help to detect changes. Manipulating the conditions of the treatment system to return the altered parameters to their former values may be needed. The list of parameters is long, so the solution to sludge bulking may not be easy.

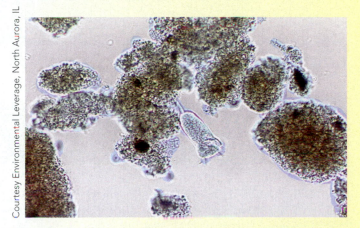
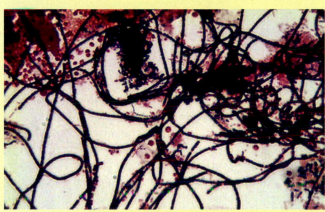

Courtesy Environmental Leverage, North Aurora, IL

Figure B16.1. Flocs Well-formed floc particles (*left*) are compact and will settle. Poorly formed floc (*right*) is loose and of low density and will tend to remain in suspension.

TABLE 16.7 BOD₅[a] reduction in a wastewater treatment plant

Waste type	BOD₅ (mg/L)
Biologically treated effluent	10–<20
Vegetable processing	200–500
Raw domestic sewage	200–600
Dairy wastewater	900
Slaughterhouse wastes	1,000–4,000
Cattle shed effluent	20,000
Pigsty effluent	25,000

[a]BOD₅ refers to measurement of this demand over a 5-day period at 20°C in the dark.

feces-dwelling pathogenic bacteria. BOD is also useful in monitoring how well a sewage treatment plant is operating. The BOD should drop as water progresses through the treatment steps.

BOD relates directly to the dissolved oxygen (DO) in the water. The DO is used by bacteria when organic nutrients are present in the water; the combination of oxygen and carbon-containing nutrients enables the bacterial population to grow and divide. As microorganisms break down the organic nutrients, the oxygen content of the water drops. Clearly, the levels of oxygen and carbon in the water are linked. High levels of carbon-containing compounds support large populations of microorganisms that use oxygen to metabolize them, and thereby decrease the level of oxygen in the water. When oxygen levels get too low, the water will not support aquatic species like invertebrates and fish. See **Toolbox 16.1** for a description of how to measure BOD.

In addition to a decrease in the total organic content, a significant reduction in the number of intestinal pathogens in the effluent occurs during secondary treatment. Microbes that are adapted to conditions in the treatment plant will out-compete intestinal pathogens for resources in this environment. Thus, the pathogens will grow poorly, if at all. Predation is also a factor. Although all microbial populations will face similar grazing pressure, the smaller populations of intestinal pathogens will be decimated more quickly than other microbes. Finally, many of the surviving intestinal pathogens will settle with the floc. Properly formed activated sludge can remove about 90 percent of enteric bacteria, 90–99 percent of enteric viruses, and about 90 percent of protozoan parasites such as *Giardia* and *Cryptosporidium parvum* (**Microbes in Focus 16.2**) that may be present in wastewater. However, considering only percent reduction can be misleading. Removing 90–99 percent of a microbial population whose density is 10^6 per 100 milliliters still leaves behind an appreciable number of organisms.

The conventional activated sludge process may be modified to increase the removal of nitrogen (N) and/or phosphorus (P) from the clarified effluent. These elements may ultimately reduce the water quality in natural waters where they are often a limiting nutrient for natural plant and microbial populations. If they are released with effluent from a treatment plant, they may result in overgrowth of populations leading to **eutrophication** of the lake or stream. Eutrophication results from increasing the availability of limiting nutrients, such as nitrogen and phosphate compounds, in an aquatic ecosystem, leading to rapid growth in autotroph populations, particularly algae. Microbial decomposition of the increased biomass creates anoxic conditions, resulting in severe reduction of water quality, negatively impacting aquatic animal populations. The strategy for nitrogen removal is to encourage

nitrification followed by *denitrification*. Several protocols have been developed to achieve this, but the overall idea is to replace the single aeration tank with several tanks to permit sequential aerobic (for nitrification) and anaerobic (for denitrification) activities, achieving the following overall reaction sequence:

Nitrification **Denitrification**

[ammonia → nitrite → nitrate] followed by [nitrate → nitrite → NO → N_2O → N_2]

CONNECTIONS While denitrification is a desired process in wastewater treatment, it is not desirable in agricultural soils as it removes nitrogenous nutrients, such as nitrate and ammonia often supplied by fertilizers, needed for plant growth. See **Section 14.3** for a description of the denitrification process and how it contributes to the global cycling of nitrogen.

Phosphorus removal can be conducted biologically by using an anoxic/oxic process in place of conventional aeration. Here, a short-term anoxic stage is used prior to the aerobic phase to encourage inorganic phosphate release in the absence of O_2 as a result of polyphosphate hydrolysis from volutin. Volutin is a granular inorganic phosphorus-containing storage substance found in the cytoplasm of various bacterial and fungal cells. During the subsequent period in the aeration tank, soluble phosphorus is taken up by vigorously growing microbes, and so is removed from the effluent when floc particles settle in the settling tank. Other physicochemical methods can enhance removal of nitrogen and phosphorus. For instance, raising the pH of ammonium (NH_4^+)-containing effluent will encourage formation of gaseous ammonia (NH_3), which can then be stripped from the water. At an appropriate point in the treatment process, chemicals such as alum ($KAl(SO_4)_2 \cdot 12H_2O$), calcium sulfate ($CaSO_4$), or ferrous sulfate ($FeSO_4$) can be added to the wastewater, leading to Al/Ca/Fe phosphate formation. These phosphates are insoluble and precipitate out of solution, thereby removing phosphorus from the fluid phase. The **Mini-Paper** describes research aimed at identifying and understanding the microbes and conditions important in phosphate removal.

Following the activated sludge process or treatment by a trickling filter, the fluid phase or effluent is typically disinfected and released from the treatment plant. In the next few paragraphs, however, we will follow the trail of the collected primary and secondary sludge that moves into the anaerobic sludge digester.

Anaerobic Sludge Digester Completion of secondary treatment of wastewater involves the anaerobic sludge digester—essentially a large anoxic bioreactor operating in a semi-continuous mode—that allows final anaerobic digestion of remaining organic wastes **(Figure 16.29)**. Sludge is typically held from 2 to 4 weeks in the digester, which can be designed to operate in a mesophilic mode, functioning optimally at 35–37°C, or thermophilic mode, functioning optimally at about 50°C. The latter has the advantage of more rapid reaction rates and a greater degree of pathogen kill. However, this advantage is offset by the disadvantage of the energy required to maintain high reactor temperatures. Particularly in temperate regions, the reactor may be partly underground to take advantage of the insulating properties of the surrounding soil. Over the course of the incubation period, organic materials in the sludge are degraded by a large and diverse anaerobic microbial community.

Figure 16.29. Anaerobic sludge digester Anaerobic microbial digestion of organic material in sludge takes place under anoxic and heated conditions. The majority of the digester is underground to help maintain temperature. The roof is suspended so that it may be vented in response to accumulated CO_2 and CH_4 (methane) within the reactor. Methane can be collected as a fuel source.

In the anaerobic sludge digester, organic material is broken down to organic acids, alcohols, carbon dioxide (CO_2), and hydrogen gas (H_2) by primary fermenters. Microbes cooperatively process acids and alcohols to acetate and H_2. Homoacetogens and methanogens constitute the final groups in the microbial food web. You may recall the discussion of similar anaerobic biodegradation of organic substrates in Chapter 14 (see Figure 14.6). The biogases methane (CH_4) and CO_2 are products of an anaerobic sludge digester. CH_4 is frequently captured and burned to provide energy to the plant and surrounding community. Processed anaerobic sludge is the other main product, consisting of indigestible materials and microbial biomass. Anaerobes produce far less biomass per unit of substrate metabolized than do aerobes because their metabolism is not as efficient. This is one reason why anaerobic digestion is usually preferable to aerobic processing of this material. A treatment plant, however, must still handle a considerable volume of sludge. Disposal of this leftover material is a significant problem in waste treatment. Aerobic sludge digestion is sometimes practiced, particularly with industrial wastes, because it is more rapid than anaerobic digestion. However, it requires an O_2 supply and does not yield a supply of combustible biogas.

A considerable number of pathogenic microbes may be killed during the digestion process, but their populations are only reduced, not eliminated. Moreover, if wastewater entering the plant contains recalcitrant organic compounds or considerable amounts of metals, these too may be concentrated in the sludge. Thus, sludge from the anaerobic digester can potentially be considered hazardous material. This is one reason why it is undesirable for industrial wastes to enter plants treating mostly domestic wastewater.

Digested sludge is typically dehydrated before meeting one of several fates. It may be dried and then incinerated, with an associated energy recovery. Or it may be added to a landfill. Sludge added to the environment in this way is first stabilized by the addition of materials that transform it so that the hazardous constituents, such as heavy metals, are immobilized or converted to a less toxic form. The sludge can be treated with lime to

H. G. Martín, N. Ivanova, V. Kunin, F. Warnecke, K. W. Barry, A. C. McHardy, C. Yeates, S. He, A. A. Salamov, E. Szeto, E. Dalin, N. H. Putnam, H. J. Shapiro, J. L. Pangilinan, I. Rigoutsos, N. C. Kyrpides, L. L. Blackall, K. D. McMahon, and P. Hugenholtz. 2006. Metagenomic analysis of two enhanced biological phosphorus removal (EBPR) sludge communities. Nat Biotechnol 24:1263–1269.

Context

A major concern in communities around the world is the large amount of environmentally damaging phosphorus that is found in municipal and industrial wastewater in the form of inorganic phosphate. If this phosphorus is not removed during the water treatment process, it finds its way into the effluent, causing eutrophication in waterways and environmental degradation. A process called "enhanced biological phosphorus removal (EBPR)" has been widely adopted in wastewater treatment plants over the past 30 years to reduce the amount of phosphorus in the wastewater. The EBPR process involves naturally occurring bacteria in activated sludge that under aerobic conditions accumulate phosphorus as intracellular polyphosphate. The polyphosphate-containing cells are then allowed to settle as floc, resulting in greatly reduced phosphorus levels in the effluent water.

While the EBPR process is used worldwide, its performance is not always reliable. Incredibly, the assumption that bacteria of the genus *Acinetobacter* (that were easily cultured from wastewater treatment systems) are primarily responsible for EBPR has now been shown to be incorrect. Instead, the EBPR-active organisms turn out to be a newly discovered but dominant uncultured member of the β-Proteobacteria that has been termed *Accumulibacter phosphatis*. Even though this organism had not yet been isolated in pure culture, metagenomic sequencing methods might allow reconstruction of complete genomes of these bacteria. With this information in hand, as well as the knowledge of other organisms present in the EBPR community, it would be possible to examine the metabolic relationships that develop during the EBPR process. The study presented in this paper was a metagenomic analysis of two EBPR sludges set up in laboratory-scale reactors. The results from this lab-scale study provide data that can be meaningfully extended to the real-world scale of the treatment plant.

Experiments

Two lab-scale batch reactors were seeded with water obtained from treatment plants in Wisconsin and Australia and operated under typical EBPR cycling conditions. DNA was isolated from sludge samples and subjected to shotgun sequencing (see Section 10.1), followed by analysis of the sequence reads to find overlaps, and to identify 16S rRNA genes. *Accumulibacter phosphatis* was clearly the dominant organism in the EBPR community, and the near-complete assembly of its genome from the sequence data showed it to be approximately 5.6×10^6 bp in size. This genome sequence made it possible to reconstruct the metabolic pathways of *A. phosphatis*. Examination of the proposed pathways raised several questions about the details of their operation; these will have to be addressed through a combination of metabolic modeling and experimentation. Perhaps surprisingly, *A. phosphatis* was found to have genes for N_2 fixation, CO_2 fixation, and high affinity phosphate transporters that would be very useful under phosphorus-limited conditions. These functions are clearly not required in the EBPR systems, but are perhaps important for survival in alternative, nutrient-poor habitats such as freshwater.

Impact

This metagenomic analysis opens many avenues for the study of EBPR, ultimately leading to a practical understanding and improvement of the process. The availability of the *A. phosphatis* genome sequence will allow examination of the gene expression changes during EBPR cycling, which could result in sequence-based diagnostic tools for following the process in wastewater treatment plants. Such diagnostic tools would be especially useful when based on a thorough understanding of the metabolic processes that are developed through metabolic modeling and testing of hypotheses through experimentation. This study is a terrific example of the use of the sequence-based metagenomic approach to provide valuable information about complex microbial communities.

● Questions for Discussion ....

1. This research was carried out using Sanger sequencing. How do you think the outcome might have differed if some of the more recent higher throughput sequencing methods had been used?

2. What is the rationale of using samples from more than one treatment plant?

3. The sequences of many other microbes were found in the EBPR sludge. What might be some of their functions within the EBPR system?

---

raise the pH and stabilize the material, or it can be allowed to naturally decompose to stabilize it. If this material meets certain standards imposed by the relevant jurisdictional agency, it can be termed a **biosolid**. Stabilized biosolids may be used as soil conditioners in horticulture and agriculture, supplying nutrients to the soil and improving its texture and water-holding capacity. These materials have also been used in the remediation of severely disturbed landscapes such as abandoned mine sites.

The nutrients found in biosolids include N, P, K, and trace elements such as Ca, Cu, Fe, Mg, Mn, S, and Zn, all of which are needed by crop plants. It is critical that biosolids applied to soils do not add hazardous materials, including pathogenic microbes to the environment. Class A biosolids are considered safe, as they do not have detectable levels of *Salmonella*, enteric viruses, and viable helminth (worm) eggs. This type of biosolid can be applied freely and with minimal restrictions in small amounts to garden soils or in bulk amounts to larger croplands. The use of Class B biosolids is more restricted because these are of lesser quality due to their low but still detectable numbers of pathogens. Other regulated qualities of biosolids include the metal content and the degree of vector attraction. The latter refers to minimizing access of flies, mosquitoes, fleas, rodents, and birds to the biosolids because these vectors could transmit pathogens to humans. The United States recycles to the land about 50 percent of all biosolids under the regulation and quality standards of the U.S. Environmental Protection Agency (EPA).

Effluent Disinfection

Disinfection is the last step in wastewater treatment. Disinfection of water relies on the addition of chemicals to inactivate infectious agents that may be present. The aim of disinfection is to kill bacteria and viruses that normally reside in the intestinal tract, and which can be present in feces-contaminated water, but were not eliminated in previous stages of treatment. A simple method of disinfection is **chlorination**. The water is dosed with a hypochlorite solution, either $Ca(OCl)_2$ or $NaOCl$ (bleach). Adding hypochlorite to water forms hypochlorous acid (HOCl):

$$NaOCl + H_2O \rightarrow HOCl + NaOH^-$$

Hypochlorous acid is a strong oxidant that will oxidize the organic matter in the water, including microbes. It will also oxidize any reduced inorganic compounds that may be present, such as NH_4^+ or reduced Fe. Hypochlorous acid may also be produced by percolating chlorine (Cl_2) gas through the water where it will react with H_2O to form HOCl.

A disadvantage of chlorine use is formation of chlorinated disinfection by-products, such as trihalomethanes and haloacetic acids that result from oxidation of organic matter by reactive chlorine ions. These may harm the environment or act as weak carcinogens. Accordingly, alternative disinfection protocols for waters containing significant levels of dissolved organic carbon are becoming more common. **Ozonation** uses ozone (O_3) to disinfect water, and is one alternative that produces fewer by-products than chlorination. Ozone is an inherently unstable molecule that generates a free oxygen radical, which is a powerful oxidizing broad-spectrum disinfectant. It is fairly costly to run an ozone generator, but one advantage over chlorine use is that the ozone can be generated as needed. Chlorine gas is highly poisonous but has to be stored on-site, creating a potential hazard in the event of an accidental release. An ozone generator produces O_3 by passing air over a high potential electrode.

Ultraviolet (UV) radiation is another method of disinfection. The development of more powerful UV lamps that can send the short wave UV rays deeper into the volume being treated has led to increased use of this strategy. The ability of UV to cause damage to important biomolecules such as nucleic acids and proteins is well known, but the poor penetrating power of low energy UV lamps was a hurdle that had to be overcome before this strategy could be successful. Many disinfection procedures use combinations of chlorination or ozonation with UV radiation.

Tertiary Treatment of Wastewater

Tertiary treatment of wastewater, which is rarely used, employs additional biological or physicochemical protocols to effluent water prior to disinfection. Modern technologies exist that will permit the production of very clean wastewater treatment effluents, but they are quite costly and not widely implemented. The techniques typically involve some types of physicochemical processes, including filtration technologies, activated carbon adsorption, treatments such as reverse osmosis, or additional disinfection or coagulation protocols. The targets are the same as in the primary and secondary treatments, but tertiary treatment seeks to further reduce waste levels. Reduction of one or more waste products may be targeted: the organic content, turbidity, N or P content, metal concentration, or pathogen levels.

Tertiary treatment may be implemented to protect a receiving water body with a very low volume, or one of particularly high value for environmental or economic reasons. Important wildlife habitats or water-related tourist areas are examples. In water-impoverished regions treated effluent may be reused for irrigation of crops, or for drinking water. It may be piped directly to the drinking water purification plant, or applied to the land surface and allowed to infiltrate into the soil. Infiltrated water can be stored underground in depleted aquifers and then drawn upon when needed.

Drinking water purification

Ensuring a supply of clean drinking water includes selecting the best available source, for example, a river, lake, or aquifer, and protecting that source from contamination. Although microbial numbers can be reduced quite significantly in a well-operated water treatment plant using conventional wastewater treatment technologies, microbes still remain in the effluent. These microbes are introduced into the receiving water when the effluent is discharged from the plant. Drinking water sources may also be contaminated by wild or domestic animals that discharge potential pathogens into a river, lake, or stream. For these reasons, water is often purified before it is made available to citizens for drinking.

Purification of drinking water varies significantly depending on the quality of the water source. Compared to surface waters, aquifer water is usually relatively free from sediment, plant, animal, and microbial material, and requires fewer steps to clarify it. Most purification procedures incorporate many of the same steps used in wastewater treatment, with similar but more stringent results (**Figure 16.30**).

Raw water, especially from surface water sources, may first be screened to remove large particulate matter, such as sticks. It then enters a flocculation basin where a chemical such as aluminum sulfate (alum) is added. Alum forms gel-like floc precipitates that trap suspended particles, including microbes and other organic material. The floc-containing water is then moved to a sedimentation tank, where the flocs are allowed to settle as sludge. The clarified water is then filtered.

The most common type of filter is a sand filter, which uses physical and biological activity to purify water. Water moves vertically down through a deep layer of sand, which often has a layer

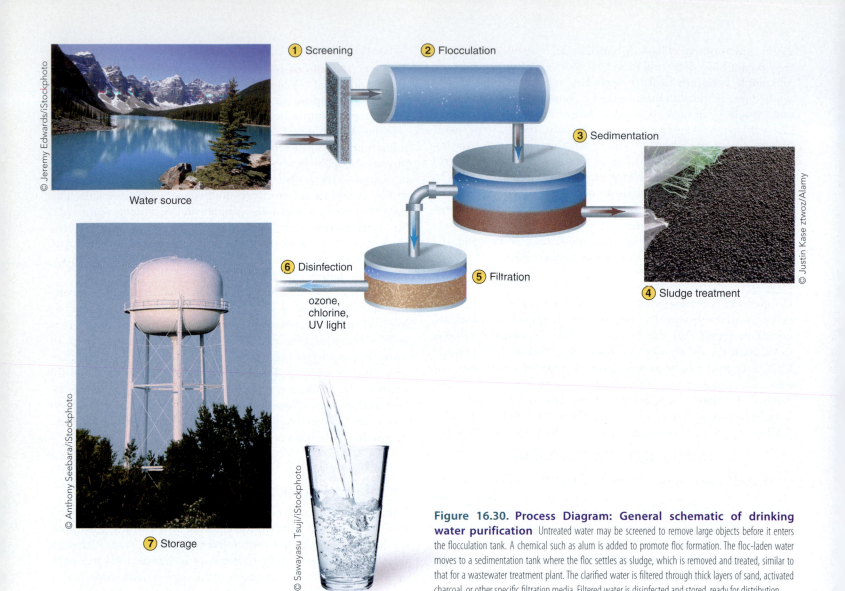

① Screening ② Flocculation ③ Sedimentation ④ Sludge treatment ⑤ Filtration ⑥ Disinfection — ozone, chlorine, UV light ⑦ Storage

Water source

© Jeremy Edwards/iStockphoto

© Anthony Seebara/iStockphoto

© Sawayasu Tsuji/iStockphoto

© Justin Kase ztwoz/Alamy

Figure 16.30. Process Diagram: General schematic of drinking water purification Untreated water may be screened to remove large objects before it enters the flocculation tank. A chemical such as alum is added to promote floc formation. The floc-laden water moves to a sedimentation tank where the floc settles as sludge, which is removed and treated, similar to that for a wastewater treatment plant. The clarified water is filtered through thick layers of sand, activated charcoal, or other specific filtration media. Filtered water is disinfected and stored, ready for distribution.

of activated carbon on the surface to remove dissolved chemical contaminants. The high surface area of the sand supports biofilm formation that adsorbs incoming microbes, including viruses, and a variety of organic and inorganic materials that may contribute to foul taste and odor. The sand also presents a physical barrier to the passage of particulate matter including spores, cysts, and microbial cells.

Membrane filters are often employed instead of or in combination with sand filtration. Specially constructed membrane filters can remove particles larger than 0.2 μm. This may be the only sure way to remove *Cryptosporidium* oocysts that are resistant to chlorination and ultraviolet radiation (see Microbes in Focus 16.2).

The final step in water purification is disinfection by chlorination or ozonation, often combined with UV irradiation. The advantage of chlorination is its residual effect—a small amount of chlorine remains in treated water and can protect it from recontamination during storage and as it flows through

distribution channels. Ozonation and UV light have no residual effect, but have the advantage that they do not tend to produce by-products, as can occur with chlorination.

Before water is released for drinking, it is tested to ensure its purity. Public health drinking water standards state that drinking water should be entirely free from pathogenic microorganisms. However, routine examination of water for the presence of all human pathogens would be a time-consuming, difficult, and expensive task. Microbiologists do not test for every potential pathogen that might remain in the water but instead use a presumptive test for **indicator organisms**. These are microbes whose presence in a water sample indicates the presence of fecal matter, and thereby, potential intestinal pathogens. Since many pathogens associated with drinking water are of fecal origin, having been introduced by sewage, bacteriological tests are designed to detect the indicator organisms called **fecal coliforms**—a group of bacteria found in the intestinal tract of humans and other animals.

Coliforms are Gram-negative, non-spore-forming, facultative anaerobic bacilli belonging to family *Enterobacteriaceae*. The most numerous fecal coliform bacterium in intestines is *E. coli*, although it only constitutes about 1 percent of microbes present in the human intestine.

Most fecal coliform bacteria are not pathogenic or are of low virulence, but as their presence in water indicates the presence of fecal matter, then pathogens, such as *Vibrio cholerae*, *E. coli* O157:H7, *Salmonella* Typhi, and *Cryptosporidium*, may also be present. A widely used test to estimate coliform numbers in water is the **most probable number (MPN)** method, as described in **Toolbox 16.2**.

16.5 Fact Check

1. Distinguish between domestic wastewater and storm water.
2. What are the goals of wastewater treatment?
3. Outline and explain each of the steps of wastewater treatment.
4. What is the significance of exopolysaccharide as it relates to floc formation and wastewater treatment?
5. What is biochemical oxygen demand (BOD) and how is it involved in sewage treatment?
6. Outline the steps in drinking water purification.
7. How are the terms "indicator organism" and "fecal coliform" related to the process of water treatment?

Image in Action

WileyPLUS The sandwich and water at this picnic all contain several species of microbes. On the slice of ham, *Listeria monocytogenes* (maroon), *Salmonella* Typhi (tan), *Escherichia coli* (green), *Staphylococcus aureus* (purple), and *Leuconostoc mesenteroides* (light blue) are present. The water is contaminated with bacteria including *Escherichia coli* (green) and *Vibrio cholerae* (teal), and protozoa, including cysts of *Giardia lamblia* (yellow) and oocytes of *Cryptosporidium parvum* (blue). Additionally, viruses such as rotovirus (pink) can also be carried in drinking water.

1. Imagine a child is spending time playing at the park on this sunny, warm day before finally coming to the table to enjoy this picnic lunch his parent has set out for him. During that extended period of time, the *S. aureus* bacteria on the ham have reproduced significantly. Describe the key features of the foodborne illness that may result due to this organism.

2. What preservation methods should have been employed during the manufacture of the ham to prevent the growth of pathogens and spoilage organisms? Provide an example for each.

3. What disinfection methods could have been used to prevent contamination of the water?

The Rest of the Story

Listeria monocytogenes is found widely in soils, vegetation, water, sewage, milk, and feces of animals. Usually, but not always, common sanitary procedures used in meat-processing plants are effective in killing the organism. How *L. monocytogenes* comes to contaminate meats is not precisely known, but since it is killed by heating during the meat processing stage, contamination likely occurs afterwards in the packaging and slicing process. *L. monocytogenes* does not grow in foods having a water activity (a_w) value of less than 0.92 (see Figure 16.2) and/or a pH less than 4.39. Moist, ready-to-eat meats, like cold cuts and wieners therefore present the most risk. Unlike most pathogens, cool temperatures do not prevent the growth of *L. monocytogenes*. Lunchmeats are not commonly cooked in the home before being consumed and wieners are frequently not sufficiently heated before being consumed.

The listeriosis outbreak that occurred in Canada in the summer of 2008 cost Maple Leaf Foods $27 million, paid to the relatives of those who died after eating the contaminated meats.

Moreover, consumer confidence was shaken in this once stalwart Canadian food company. The outbreak spurred a report to the Canadian government that found both Maple Leaf Foods and the Canadian Food Inspection Agency at fault for lapses in inspection policy and cost-cutting measures that enabled *Listeria* problems to persist. The damage suffered spurred this and other companies to implement more stringent equipment inspection and cleaning policies, established in the aftermath of the 2008 outbreak.

Testing and surveillance for *L. monocytogenes* includes monitoring ready-to-eat food products, as well as packaging and slicing equipment, floors, drains, and storage areas. For many years, surveillance was difficult and time-consuming, with contamination being discovered many days later because of the time it takes for the growth of cultures. Current methods are reliable and more rapid, making use of immunodetection and DNA-based technologies. These *Listeria* surveillance techniques can provide results in 2 or 3 days rather than a week. Because *Listeria* is ubiquitous, eradicating the organism is probably not possible, but a commitment to maintaining constant surveillance of the entire food production process, utilizing modern, rapid detection systems can prevent a repeat of the Maple Leaf Foods outbreak in the future.

The most probable number (MPN) method is commonly used to generate a statistical estimate of microbial numbers in water samples, as well as in samples of foods, soils, milk, and juice. Such samples can contain heterogeneous populations of microbes, therefore exact cell numbers of a particular organism, like *E. coli*, can be impossible to determine. The MPN method estimates the size of a microbial population, based on biochemical attributes of an indicator organism, such as coliform bacteria. It does not directly quantify the exact number of microbes in a sample or confirm fecal coliforms are indeed present. It is considered a *presumptive* test, useful for screening purposes only.

The sample to be tested is added to growth medium. An enrichment broth such as lauryl tryptose broth is used to encourage the growth of fecal coliforms. *Escherichia coli*, along with *Enterobacter*, *Proteus*, *Citrobacter*, and *Klebsiella*, produce acid and gas from lactose in the broth. Gas production indicating the growth of such coliforms can be visualized in an inverted gas-collection tube called a "Durham tube" **(Figure B16.2)**. In theory, if at least one coliform bacterium was present in a tube of broth, gas should be in that tube.

There are many variations on the MPN method, and the volumes and number of replicate tubes tested will vary depending on the type of sample. The pattern of positive and negative test results for gas production is used to derive a statistical estimate of the population.

To better understand the MPN method, let's examine how drinking water is tested. As shown in **Figure B16.3**, 15 tubes containing 10 mL of broth are inoculated with the sample being tested — five tubes are inoculated with 10 mL of sample, five tubes are inoculated with 1 mL of sample, and five tubes are inoculated with 0.1 mL of sample. Note that the broth in the first set of replicate tubes is prepared at double (2x) concentration to accommodate a 10-mL sample. This will result in a 1x concentration for optimal growth of coliforms. The rest of the tubes contain a 1x concentration of broth. The inoculated tubes are incubated for 48 hr at 35 °C and are noted for the presence or absence of gas production. The tubes exhibiting gas production for each set of replicate tubes are counted to generate a score. In our example, a number series of **4 3 1** is produced. Using the MPN reference table, this score gives a value of 33 coliforms per 100 mL of water. The values in the table are generated by probability estimates. Most countries require residential water supplies to contain zero coliforms/

Gas

Figure B16.2. Test for gas production Prior to sterilization, a small inverted Durham tube is placed inside a larger tube containing the growth media. Following sterilization, a sample is added to the tube and incubated. Any gas produced will collect in the tube and can be easily visualized.

Courtesy Christine Dupont

100 mL. The water source in our example is therefore not suitable for drinking. For testing wastewater, a higher dilution series would be used as the number of coliforms would be expected to be greater.

The MPN test gives an estimation of numbers of possible fecal coliforms. However, not all coliforms are of fecal origin. *Enterobacter aerogenes* is a common soil bacterium, so its presence in a water sample does not necessarily indicate fecal contamination. For this reason, samples that are positive for gas production are further tested to confirm the presence of fecal coliforms. If confirmed, the species is then identified.

- A *confirmed* test is used to confirm fecal coliform bacteria are present from the MPN test. A confirmed test is only done if the MPN test is positive. All tubes showing gas formation are selected, and a loopful from each is inoculated to a medium more selective for fecal coliform detection, such as brilliant green lactose bile broth, or EMB agar. If the tube shows gas production or growth

Summary

Section 16.1: How can microorganisms spoil food?

Food spoilage is related to microbial growth and will vary depending on the number of microbes present, with spoilage occurring more rapidly at higher microbial densities.

Food spoilage depends on the type of food, type of microorganism, chemical/physical parameters of the food, and storage conditions:

- **Perishable foods** such as some fresh fruit are rapidly spoiled.

- **Semi-perishable foods**, such as nuts and potatoes, degrade more slowly.
- **Non-perishable foods** such as flour, sugar, and dried beans rarely spoil.

Intrinsic factors, such as nutrient content, **water activity (a$_w$)**, pH, and oxidation-reduction potential, are characteristics of the food. **Extrinsic factors**, such as storage temperature and presence of gases, are characteristics of the environment of the food.

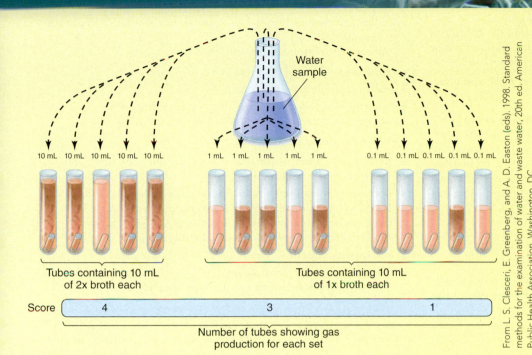

Most probable number reference table (MPN/100 mL) for five replicate tubes of sample sizes 10 mL, 1 mL, and 0.1 mL

Original sample volume			
10 mL	1 mL	0.1 mL	MPN/100 mL
4	2	0	22
4	2	1	26
4	3	0	27
4	3	1	33
4	4	0	34
5	0	0	23
5	0	1	31
5	0	2	43
5	1	0	33
5	1	1	46
5	1	2	63
5	2	0	49
5	2	1	70
5	2	2	94
5	3	0	79
5	3	1	110
5	3	2	140
5	3	3	180
5	4	0	130

Water sample

10 mL 10 mL 10 mL 10 mL 10 mL   1 mL 1 mL 1 mL 1 mL 1 mL   0.1 mL 0.1 mL 0.1 mL 0.1 mL 0.1 mL

Tubes containing 10 mL of 2x broth each

Tubes containing 10 mL of 1x broth each

Score   4   3   1

Number of tubes showing gas production for each set

From L. S. Clesceri, E. Greenberg, and A. D. Easton (eds). 1998. Standard methods for the examination of water and waste water, 20th ed. American Public Health Association, Washington, DC.

Figure B16.3 Procedure for the most probable number method A series of sample volumes are inoculated in suitable growth media. Each sample volume is repeated five times to produce five replicate tubes. Following incubation, the tubes are recorded for gas production to generate a score, which is applied to a reference table to generate an estimate of the numbers of bacteria.

occurs on EMB agar, the confirmed test is considered positive. Other selective media can be used.

- A *completed* test identifies the particular species present from the confirmed test. It uses more specific biochemical tests, such as rapid bacterial identification systems (e.g., API20E; see Toolbox 13.1) .

Other methods can be employed to estimate coliform numbers. The membrane filtration technique filters a water sample through a membrane. The membrane is then transferred to an appropriate growth medium for quantification and identification. A more recent method for coliform testing is an enzymatic assay for β-galactosidase, since all bacteria belonging to the coliform group possess this enzyme.

● **Test Your Understanding** ·····················

Imagine you are working in a public water testing lab and your coworker is analyzing the daily MPN test results. She walks in with a concerned look and asks you to confirm her results because she believes she has a positive presumptive test. What does this mean and, if she is correct, what should she do next?

Proper storage of food can inhibit production of microbial **enterotoxins** that can cause foodborne illness.

Section 16.2: What techniques are used to minimize food spoilage?

Reducing a_w is an important means of food preservation, since available water is important for microbial growth. This can be done by:

- heat-drying, **freeze-drying (lyophilization)**
- adding high concentrations of solutes, such as sugar or salt

Temperature control is important since microbes vary in their tolerance for temperature.

- **Refrigeration** severely inhibits the growth of many microbes, but not all (e.g., *L. monocytogenes*).
- Freezing can be efficient, although freeze-thaw cycles can damage food cells due to ice crystal formation.
- Temperature can be useful in killing microbial cells and spores and include; **canning**, or appertization, and **pasteurization**.
- **Flash-heating pasteurization** is more efficient than **low temperature hold (LTH)** methods.

Chemical preservatives can inhibit the growth of spoilage microorganisms. Examples include sodium benzoate, proprionate, sorbates, sulfur dioxide, nitrite, and acids. A pH ≤ 4.6 is inhibitory to spoilage microbes. **Botulism** is prevented by applying high temperature and low pH to kill endospores of *C. botulinum*. Natural microbial preservatives include **bacteriocins** (e.g., nisin), lactic acid, and acetic acid. **Pickling** uses acetic acid to preserve food by reducing the pH to less than 4.6. **Irradiation** of food serves to disrupt genetic material and/or cause oxidative damage to contaminating microbes and includes:

- **non-ionizing radiation** in the form of **ultraviolet (UV) radiation**
- **ionizing radiation** (gamma rays and X-rays)

Irradiation can be used for pasteurization. **Modified atmosphere packaging (MAP)** uses plastic films to maintain an altered atmosphere inside the food package. High CO_2 levels inhibit the growth of acrobes and high O_2 reduces growth of anaerobes. **Hurdle technology** uses multiple individual constraints, or hurdles, to control microbial growth. This avoids the severity needed to inhibit growth using just one constraint.

Section 16.3: How are microorganisms used in food production?

Microbial fermentation of food produces end products, such as lactic acid, that inhibit the growth of spoilage microbes. Metabolites can also add pleasing flavors and aromas to foods.

Industrial fermentations often use **starter cultures** or inoculums containing certain microorganisms, such as lactic acid bacteria (LAB) to maintain batch-to-batch consistency. The primary purpose of acid production by microbes is to retard the growth of unwanted microorganisms while promoting the growth of desired microbes. Vinegar manufacture is an ancient use of microorganisms, which relies on species of acetic acid bacteria that oxidize the ethanol in the starting liquid to acetic acid.

Section 16.4: How can microorganisms in our food or water cause illness?

Foodborne illness can be caused by **foodborne intoxication** due to microbial release of toxins.

An example is botulism or production of enterotoxins in the food. A microbial **foodborne infection** occurs when growth of the ingested pathogen in the body occurs. The same principles and causes of foodborne illness also apply to contamination of drinking water. Outbreaks of foodborne or waterborne illness are caused by a common source of transmission, such as a supply of hamburger or a contaminated water supply.

Section 16.5: How can we keep our water supplies safe?

Contamination of water can be biological or chemical.

Potable water is usually only attainable following water treatment. **Wastewater**, or **sewage** from domestic and industrial sources needs to be treated before being released to the natural environment to reduce microbial and chemical loads to acceptable levels in order to protect water supplies.

Wastewater treatment involves:

- reducing total organic carbon (TOC)
- reducing numbers of microorganisms
- reducing inorganic compounds that contain nitrogen and phosphorus to prevent **eutrophication** of natural waters

Wastewater treatment includes:

- **primary treatment**—physical removal of sediments and grease that form **primary sludge**
- **secondary treatment**—relies on formation of complex microbial biofilm communities capable of **mineralization** of organic compounds and binding pollutants to improve water quality. Uses a **trickling filter** unit or an **activated sludge** unit
- **tertiary treatment** is sometimes used (e.g., filtration)
- disinfection using **chlorination**, UV light, or **ozonation** before final release of **effluent** to a natural water supply

An activated sludge process relies on **flocs**: clumps of biomass consisting of microbes, **exopolysaccharide**, and absorbed material. **Secondary sludge** consists of settled flocs and is further treated along with primary sludge in an **anaerobic sludge digester**, which completes the digestion of organic material by anaerobic microbial communities. Protozoa, principally *Cryptosporidium* and *Giardia*, produce oocysts that are resistant to chlorination and UV light. When properly done, wastewater treatment reduces **biochemical oxygen demand (BOD)** by reducing organic matter that acts as substrate for microbes. **Persistent organic pollutants (POPs)** are synthetic chemicals that are resistant to degradation by microbes and can bioaccumulate in tissues when released in effluent. A by-product of waste treatment is treated **biosolid**, derived from digested sludge. With approval based on stringent criteria, it can be used as a nutrient for crops. Water to be released for drinking is tested for the presence of **fecal coliforms**, a group of **indicator organisms** that suggests intestinal pathogens may be present. The **most probable number (MPN)** method is commonly used to test for fecal coliform bacteria.

Application Questions

1. *Listeria monocytogenes* is a psychrotolerant bacterium. Given this fact, explain why food manufacturers focus their *Listeria* testing efforts on ready-to-eat foods.

2. As both support microbial populations, explain the differences between spoiled milk and yogurt. Be specific.

3. Imagine you are training food handlers at the local restaurant. Identify two or three key points you should teach them regarding safe food handling. Explain the microbiology behind each.

4. Dairy farmers disinfect their milking equipment each day to keep the initial microbial load in the fresh milk very low. Describe how this will help minimize spoilage of the milk for the consumer.

5. Those training to be wine or cheese makers must be educated on the proper sanitation of their equipment. Explain why this is important to them, and describe some specific effects if their sanitation efforts are not thorough.

6. You just bought some fresh sliced ham for your daily lunch sandwich. Imagine you accidently left the unpackaged ham out of the refrigerator overnight. Explain the type and source of illness that may likely arise due to eating ham that has not been refrigerated. Contrast that with the type and source of illness that may likely arise due to eating poorly cooked hamburger.

7. Imagine you are talking with your friend who is against the use of chemical preservatives in foods because of perceived ill health effects from consuming these chemicals. You explain to your friend that although there may be valid reasons for these concerns, there are also valid reasons for adding preservatives to foods. Provide two or three examples of preservatives and explain to your friend how those chemicals act to prevent spoilage or illness.

8. During a tour of a waste treatment plant, the guide explains that aerobic digestion is preferred in the activated sludge process and anaerobic digestion is preferred for processing of sludge in the anaerobic digester. Explain the reasoning.

9. Last summer, an outbreak of *Cryptosporidium parvum* occurred at a water park. The water was recycled and reused in the park, but park officials thought the water was safe because it was chlorinated. Explain how the outbreak could have occurred even though there was chlorination, and what could have been done instead to prevent the outbreak.

10. Early settlers of the American West did not have refrigerators in their covered wagons to keep food fresh along the trail. Many times they would salt their meat to keep it from spoiling. Explain how this would keep their meat safe and prevent microbial growth.

11. Your friend read online that unpasteurized milk is healthier to drink because it contains more vitamins. She wonders if this information can be trusted and doesn't want to risk her own safety until she knows for sure. She asks your advice. Describe to her the reasoning behind the practice of pasteurizing milk and then explain how the different pasteurization processes for milk can address her concern about nutrients and vitamins.

12. Your task is to develop an inexpensive but effective mechanism for treating drinking water in the developing country of Zimbabwe where cholera caused by contaminated water recently killed thousands of people.
    a. What challenges do you face in trying to establish a safe water supply for a village in Zimbabwe?
    b. What would be some of your basic goals for treating their drinking water?
    c. Given the challenges and your overall goals, outline and explain the minimum procedures you would try to establish for a water treatment facility.

Suggested Reading

Benidickson, J. The Culture of Flushing: A Social and Legal History of Sewage. 2007. UBC Press, Vancouver.

Demain, A. L., and S. Sanchez. 2007. Microbial synthesis of primary metabolite: Current advances and future prospects. *In* E. M. T. El-Mansi, C. F. A. Bryce, A. L. Demain, A. R. Allman, (eds), Fermentation microbiology and biotechnology. CRC Press, Boca Raton, FL, pp. 99–130.

Hartel, R. H., and A. Hartel. Food Bites: The Science of the Foods We Eat. 2008. Springer, New York.

Maier, R. M., I. L. Pepper, and C. P. Gerba. 2009. Environmental Microbiology, 2nd ed. Academic Press, Oxford, UK.

Montville, T. J., and K. R. Matthews. 2008. Food Microbiology: An Introduction, 2nd ed. ASM Press, Washington, DC.

Murooka, Y., and M. Yamshita. 2008. Traditional healthful fermented products of Japan. J Ind Microbiol Biotechnol 35:791–798.

Ray, B., and A. Bhunia. 2008. Fundamental Food Microbiology, 4th ed., CRC Press, Boca Raton, FL.

Sigee, D. C. 2005. Freshwater Microbiology. John Wiley & Sons, Chichester, England.

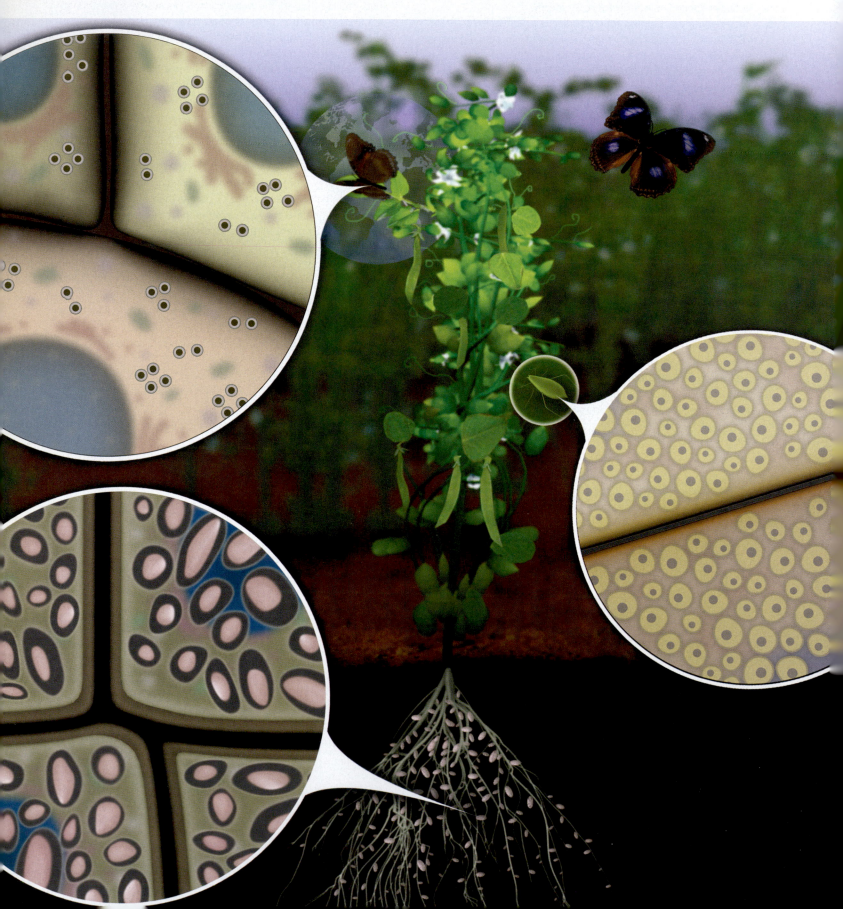

An infection that makes females? In the early 1920s, researchers studying butterflies found a curious trait in some populations of the blue moon butterfly *Hypolimnas bolina* in the Fiji islands. Many populations were composed almost exclusively of females, like the legendary Greek Amazons. Males were rare. At first, this did not seem so unusual. It was known that females of many insect species, such as aphids, commonly produce only daughters by parthenogenesis, an asexual mode of reproduction not requiring males. This was not what was happening in the butterflies, however. In 1924, a New Zealand researcher named Hubert Simmonds showed that in these populations mating with males *did* occur, and genetically male embryos *were* produced but died before hatching. Males died in astounding proportions, resulting in females vastly outnumbering males. Through meticulous breeding experiments over many generations, Simmonds and others reported that the "female-only" trait was passed directly from mother to daughter and was not due to parthenogenesis. What could be causing this?

A few interested scientists tried to determine the mode of the male-killing. They had an idea that population domination by females was due to an infection that affected male embryos but not females. However, without modern molecular technologies and using only simple culture methods, they were unable to isolate any infectious agent. Around the same time in 1924, scientists M. Hertig and S. B. Wolbach were studying the mosquito *Culex pipiens*. They observed a tiny microorganism living within the tissues of the mosquito, which they described and formally named *Wolbachia pipientis*. About 50 years later, in 1971, scientists found that in matings between *Wolbachia*-infected male mosquitoes and uninfected females, all the embryos mysteriously died before hatching. They hoped that unlocking the secrets of this killing mechanism, which appeared to be linked to *Wolbachia* infection, might lead to a control for mosquito populations. Suddenly, interest in this previously obscure bacterium began to increase.

Wolbachia has since been found to infect an astounding number of insect species, including the blue moon butterfly. Sequence analyses estimate that *Wolbachia* is present in up to 76 percent of insect species—but it may be higher. *Wolbachia* is present in a number of other arthropod groups, including crustaceans (shrimp and lobsters), spiders, and centipedes. It currently holds the honor of being the most widely spread infection on Earth!

With the help of modern molecular tools it has become apparent that *Wolbachia* infection is actually inherited through maternal lines directly from infected females. In some species, all females are infected. To maintain successful inheritance, *Wolbachia* alters the reproductive capabilities of its hosts to favor females and, in some cases, can even convert genetically male embryos to females!

CHAPTER NAVIGATOR

As you study the key topics, make sure you review the following elements:

Some microbial symbionts are damaging to the host, while others are beneficial. Microbial symbionts of plants use photosynthetic products from their plant partners.

- Figure 17.8: Process Diagram: Nodule formation by infecting rhizobia
- Figure 17.11: Exchange of carbon and nitrogen compounds between plant host cells and bacteroids

Lichens are microbial partnerships composed of a fungus and algal or cyanobacterial cells. Vertebrates possess beneficial microbial symbionts that colonize outer and inner body surfaces.

- Toolbox 17.1: Germfree and gnotobiotic animals
- Microbes in Focus 17.2: *Streptococcus mutans*, a sweet tooth, and cavities
- Perspective 17.1: Probiotics—Do they work?
- Table 17.2: Major groups and metabolic activities of intestinal microorganisms of humans
- Mini-Paper: Do microbes cause obesity?
- Figure 17.26: Process Diagram: Overview of the major contributions of rumen symbionts in cellulose conversion

Insects often possess nutritional microbial symbionts that colonize the gut or are housed intracellularly in the host.

- Table 17.5: Insects and their symbionts
- Figure 17.32: Process Diagram: Overview of the major metabolic activities of symbiotic microbes of lower termites
- Perspective 17.4: Death of coral reefs

CONNECTIONS for this chapter:

The surfaces of animals and microbial consortia (Section 6.4)
Beneficial intracellular symbionts and pathogens (Chapter 21)
Colonizing symbionts and disease (Sections 19.3 and 21.2)
Microbial symbionts of the digestive tract host energy needs (Sections 13.3 and 13.4)
Legume plants, protein and N_2-fixing bacteria (Section 13.7)

Introduction

Media and medical attention given to disease-causing microbes often results in a negative bias toward the relationships between microbes and humans. In reality, the majority of host-microbe interactions are non-harmful, and even beneficial. The relationships between these microbes and their hosts are extraordinary in terms of their interdependence and complexity. Yet compared to many disease-causing microbes, we know little about the fascinating life of these beneficial microbes. Many of these microbes have been associated with their host species over thousands, even millions, of years in a dynamic interaction where both parties have formed an intimate, and in many cases, required relationship, where one species cannot live without the other.

Just how dependent are animals and plants on their microbial partners? Judged against microbes, animals and plants are metabolically challenged in terms of the range of food sources they can collectively utilize. If you took away the microorganisms that animals harbor and which contribute to their nutrition, there would be little hope of survival for the current diversity of animal species. They simply would not be able to exploit the range of food sources that they do, much of which can only be broken down by the microbes they harbor. For example, the vast majority of animal species are herbivores and are directly dependent on the consumption of plant material, primarily cellulose. However, few of these animals can digest cellulose. They do not possess the genetic information to make the cellulase enzymes that can hydrolyze the bonds linking the glucose units of cellulose. A few invertebrates, like termites and crayfish, produce some inherent cellulase, but even they require a significant contribution from their microbial inhabitants to hydrolyze enough cellulose to survive. Without the microbes that do this for them, these animals could not obtain their carbon and energy requirements and with the vast numbers of herbivores this planet sustains would not exist **(Figure 17.1)**. Plants also depend upon microbes for their success. Even though they use photosynthesis to meet their energy needs and fix CO_2 to meet their carbon requirements, most depend on the microbes in the surrounding soil, and sometimes in the plant tissues themselves, to provide other nutrients (such as nitrogen and phosphorus compounds) that are required for growth and reproduction.

Although many close relationships exist between free-living macroscopic species, the intimate associations we will examine in this chapter focus on the host-microbe kind. Many involve microbes that are not free-living but are found exclusively associated with a host that they live in. Incredibly, every type of multicellular organism is likely to harbor at least one microbe whose existence is limited to only that host. One may think of a plant or animal as an independently functioning unit, but that is rarely ever the case—most do not exist as single species at all but are inextricably united with microbes on and in them. Most plants and animals are an ensemble of species.

This chapter will explore a spectrum of symbiotic relationships involving microbes and their plant and animal hosts by addressing the following questions:

What is a symbiont? (17.1)
What roles do microbial symbionts play in plants? (17.2)
Can symbiotic microbes produce new types of organisms? (17.3)
What are the roles of symbionts in vertebrates? (17.4)
What are the roles of symbionts in invertebrates? (17.5)

Figure 17.1. Catabolism of cellulose supports herbivores Cellulose is the most abundant carbohydrate on the planet; however, herbivores cannot digest it themselves. They depend on the production of cellulase from their symbiotic microbes.

© Accent Alaska.com/Alamy Limited

A. Caribou

© Bob Carter/Alamy Limited

B. Rabbits

© Donald Swartz/iStockphoto

C. Geese

17.1 Types of microbe–host interactions

What is a symbiont?

A **symbiont** is an organism that has entered into a *long-standing and intimate* relationship, termed **symbiosis**, with an organism of another species. There are many types of symbiotic relationships. **Mutualism** occurs when there is benefit to both species. **Commensalism** occurs when one species benefits while the other is unaffected. **Parasitism** occurs when one species benefits while the other is harmed. Although these are clear definitions, symbiotic relationships do not always neatly fall into these three categories; many symbiotic relationships can show aspects that range on a spectrum from mutualism to parasitism **(Figure 17.2)**. The degree of damage or benefit is not always apparent, as it is frequently relative to the circumstances used for study. For example, lysogenic viruses (see Section 5.3) like bacteriophage lambda, are symbionts that move between a parasitic (lytic phase) and commensal (integrated prophage) relationship with their host. Some lysogenic phages, like the one that infects pathogenic strains of *Corynebacterium diphtheriae*, may even be mutualists as they provide toxins or other invasion factors for their host bacteria, making them more successful pathogens (see Section 21.2). Some "commensals" and even some "mutualists" can damage their hosts under certain conditions. For example, many animal species, including humans, harbor benign strains of *E. coli* in the gut. The relationship can be considered mutualistic; a cooperative "back-scratching" in which the animal provides a protected environment and nutrients, and *E. coli* provides vitamin K_2 in return. However, vitamin K_2 can be acquired from food, so *E. coli* is not a required source. Thus, the relationship can be more commensal than mutualistic. Under the right circumstances, even benign strains of *E. coli* can cause severe disease if they gain access to body tissues, through intestinal injury, for example. Here the relationship becomes parasitic. Indeed, many mutualistic relationships appear to have developed from original parasitic relationships. In this chapter, we will use the term "parasitic" to infer that the symbiotic microbe is obviously or predominantly damaging to its host. Use of the terms "commensal" or "mutualist" will infer that it is not obviously or predominantly parasitic under normal circumstances.

Ectosymbionts (*ecto* meaning "outer") live on the surface of their host. Examples include the normal bacteria and yeast

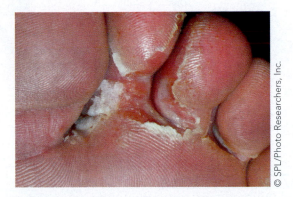

Figure 17.3. Fungal parasitic ectosymbionts Athlete's foot is caused by several species of parasitic fungi associated with humans.

that exist on human skin, but also include parasitic fungal species, such as those that cause athlete's foot and skin infections **(Figure 17.3)**.

Parasitic ticks, lice, and fleas **(Figure 17.4)** are examples of insect ectosymbionts that live in association with many animal species. The term "ectoparasite" better describes the harmful nature of these ectosymbionts. **Endosymbionts** (*endo* meaning "within" or "inside") live within the body tissues of a host, sometimes even within the host cells themselves. We will normally use the term "endosymbiont" for internal microbes that have a mutualistic or commensal relationship with their host. Parasitic endosymbionts, such as the malarial parasites of genus *Plasmodium* (see Section 23.2), are described as endoparasites.

Within symbiotic relationships, more than one species of symbiont may be present. It is not uncommon for animals to harbor a **consortium** of microbes—a stable community in which the growth or survival of members is interdependent (see Section 6.4). The digestive tracts of many animal species harbor such consortia of symbiotic microbes that, in addition to providing nutritional benefits, are necessary for maintaining homeostasis in their host. Disruptions in normal symbiotic microbial populations may result in overpopulation of one particular type of microbe, presenting serious implications for the health of the host.

> **CONNECTIONS** As we learned in **Section 6.4**, microbial consortia are often present in sediments. In this chapter we will see that consortia are also commonly found associated with animals, especially in the digestive tracts of vertebrates.

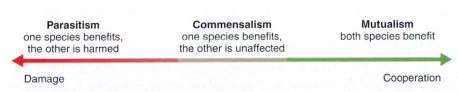

Parasitism	Commensalism	Mutualism
one species benefits, the other is harmed	one species benefits, the other is unaffected	both species benefit

Damage   Cooperation

Figure 17.2. Symbiosis: A continuum between damaging and benefiting the host
Many symbiotic relationships exhibit a spectrum of characteristics, showing aspects of parasitism, commensalism, and mutualism. Organisms may be considered to be mutualists under some conditions, but commensals or even parasites under other conditions. In relationships involving many symbiotic partners it is especially difficult to determine where each symbiont may lie on the continuum.

Deer tick

Cat flea

Human body louse

Figure 17.4. Multicellular ectoparasites These externally located symbionts have a long-standing intimate but damaging symbiotic relationship with their hosts. The parasites feed on blood from the host, which is damaging in itself, but their bites also cause inflammation and infections.

17.1 Fact Check

1. Differentiate between the types of symbiotic relationships: mutualism, commensalism, and parasitism. Describe an example of each.

2. Differentiate between ectosymbionts and endosymbionts.

3. What is meant by the term "consortium of microbes"?

17.2 Symbionts of plants

What roles do microbial symbionts play in plants?

We have defined a symbiont as an organism that has evolved a long-standing and intimate relationship with another species. Not all relationships are symbiotic as many microbes can form transient and loose associations with other species, including plants. Soil bacteria can obtain numerous advantages by living near, on, or in living plants. As we've seen in Section 15.4, saprophytic soil microbes, which convert dead plant material to organic nutrients, form loose but important relationships with plants. The microbes take advantage of dead plant material, and nearby living plants use the resulting organic nutrients for growth. As a result, the soils that surround plants are some of the most complex communities known. Recall from Chapter 15 that the diversity of microbes in a ton of soil (from 2×10^6 to 4×10^6 species) likely exceeds that of all the oceans on Earth combined (2×10^6 species).

As we know, the vast majority of plants carry out photosynthesis, using energy from the sun to convert CO_2 into useful carbon compounds (see Section 13.6). Plant-associated microbes that can directly use these products of photosynthesis have an advantage over those that cannot. These microbes exhibit a variety of mechanisms to get and use photosynthetic products from their plant partners. Some are pathogens, forming parasitic relationships with their host plant, producing damage to obtain plant products. Others form mutualistic relationships in which they provide growth-promoting compounds for the plant in return for the carbon-rich nutrients. Mutualistic plant endosymbionts that aid in the uptake or manufacture of nutrients *inside* the plant are some of the most successful plant symbionts **(Table 17.1)**.

As we saw in Section 15.4, the rhizosphere, the part of the soil immediately surrounding plant roots, is the richest area of the soil in terms of available nutrition. This is in contrast to the non-plant associated soil, in which carbon nutrients are generally more scarce. The released root material, the root exudate, is typically rich in carbon-containing amino acids and other organic acids that support the activity and proliferation of large and diverse microbial communities in the rhizosphere. However, microbes are not limited to life on the outside of plants. Some actually live within the tissues of plants, and they do so without causing apparent disease. These organisms are called **endophytes**, a specific term that refers to plant endosymbionts. Modern molecular tools confirm that a huge number of different types of bacteria and fungi live within the tissues of various plant species. The endophytic microorganisms are typically discovered by sterilizing the surface of the plant, carefully cutting through the plant tissue to avoiding contamination with environmental microbes, and placing the tissue on microbial growth media. Endophytic microbes that have been isolated in this manner represent a hidden world ripe for scientific investigation. Relatively few endophytic relationships have been investigated, but the best-characterized systems are mutualistic. One group has been intensively studied for many years—those that carry out symbiotic N_2 fixation.

Symbiotic N_2 fixation

Recall that in the nitrogen cycle (see Section 14.3) a critical step is the conversion of atmospheric N_2 to NH_3. This is an energy-intensive process called "nitrogen fixation" (shown below), which is carried out by the **nitrogenase** enzyme (see Section 13.7).

$$N_2 + 8\,e^- + 8\,H^+ + (12 - 24\,ATP) \xrightarrow{\text{nitrogenase}} 2\,NH_3 + H_2 + (12 - 24\,ADP + 12 - 24\,P_i)$$

TABLE 17.1 Some microbial endosymbionts of plants

Symbiont	*Frankia* filamentous bacteria	Mycorrhizal fungi	*Nostoc* filamentous cyanobacteria	Rhizobia (species of *Rhizobium*, *Bradyrhizobium*, *Sinorhizobium*, *Mesorhizobium*)
Plant	Actinorhizal plants	Most plant families	*Gunnera*, herbaceous flowering plants, often with very large leaves	Members of the Leguminosae family (peas, beans, clover, alfalfa)
	Alder tree	Tropical plants	Gunnera plant	Sweet peas
Description of relationship	Infection of the root. Host provides fixed carbon to support N_2 fixation by the endosymbiont.	Infection of the root. Host provides fixed carbon to the fungus. Fungus aids in uptake of nutrients such as phosphorus from the soil.	Intracellular infection of glands at the base of the leaves. Host provides fixed carbon to support N_2 fixation by the endosymbiont.	Infection of the root. Host provides fixed carbon to support N_2 fixation by the endosymbiont.

The nitrogenase enzyme, found only in bacteria and archaeons, is irreversibly inactivated by O_2. Nitrogen-fixing organisms can only carry out this process under anaerobic conditions, or they require some way to prevent the detrimental effects of oxygen on the nitrogen fixation pathway. Endosymbiotic nitrogen-fixing bacteria, called **bacteroids**, use their relationship with the plant to solve the problems of both high energy requirements and oxygen toxicity by inducing the plant root to produce an encasing structure called the **root nodule**, in which the plant provides a continual supply of nutrients to the bacteria and shields them from high concentrations of O_2 **(Figure 17.5)**.

A. Spherical nodules

B. Lobate nodules

Figure 17.5. Root nodules and symbiotic N_2-fixing bacteria Nodule shape is dependent on the plant species. **A.** Spherical nodules 2–5 mm in diameter. **B.** Lobate nodules ranging in size from three to several mm in length. **C.** Membrane-bound intracellular bacteroids. **D.** Root nodule section filled with orange-stained bacteroids inside infected cells. The presence of the nitrogen-fixing bacteria allows plants to grow even in the absence of soil nitrogen.

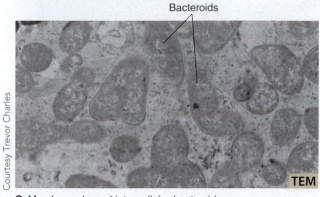

Bacteroids

TEM

C. Membrane-bound intracellular bacteroids

LM

D. Bacteroid-filled

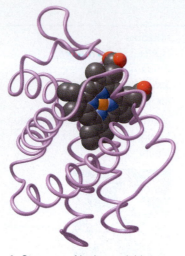

A. Structure of leghemoglobin

Courtesy Hauke Hennecke, Institute of Microbiology, ETH Zurich

LM

B. Root nodule filled with leghemoglobin

Figure 17.6. Leghemoglobin Leghemoglobin is produced by the plant host and keeps O_2 from binding with the bacterial nitrogenase enzyme, which would destroy it. **A.** Structure of leghemoglobin molecule showing the heme structure (grey and blue atoms) with central Fe atom (orange), and bound oxygen atoms (red). **B.** Sliced root nodule showing red coloration due to leghemoglobin.

Inside the root nodule, oxygen, even at low concentration, must be prevented from reaching the nitrogenase enzyme. This is managed through the use of a plant-derived hemoprotein called **leghemoglobin (Figure 17.6A)**. The leghemoglobin tightly binds O_2 (similar to the way that blood hemoglobin binds O_2) and keeps it from reacting with the critical nitrogenase enzyme. However, the bacterial symbionts must have oxygen for the TCA cycle and electron transport system, which they use to generate energy from carbon nutrients that are provided by the plant. In this metabolic process, leghemoglobin serves a dual purpose: delivering a steady supply of O_2 to the membrane-associated electron transport system while keeping it away from the critical nitrogenase enzyme. Leghemoglobin gives the interior of the nodule its characteristic red color (Figure 17.6B).

Legume plants of the family Leguminosae, such as beans, peas, alfalfa, clover, and peanuts, are notable for their ability to form root nodules harboring bacteria collectively known as **rhizobia**. The majority of rhizobia are from the Rhizobiales order of the α-proteobacteria, and some from the β-proteobacteria. Much effort has been devoted to the study of legume nodules and their bacterial partners, due to the agricultural importance of crop and animal forage plants such as soybeans, peas, and alfalfa. Throughout recorded human history, cultivation of legume plants has been used to enrich agricultural soils, and crop rotation involving legumes is a long-standing practice for maintaining soil fertility. Not until the late nineteenth century, however, was it shown that root nodules were responsible for soil fertility, and that bacteria within the nodule convert atmospheric nitrogen to a biologically available form. Fixed nitrogen provides a nutritional advantage to the plants on marginal soils where available nitrogen sources are limited. The isolation and cultivation of the bacteria from the root nodules quickly developed into an agricultural inoculant industry.

Before planting, roots of seedlings are dipped in the bacterial inoculant suspension, or seeds are coated with a dry preparation containing the bacteria. Upon growth, the bacteria invade the root to form nodules. **Figure 17.7** shows a dramatic demonstration of the effect of inoculating legume crops with root nodule bacteria. Considering that the production of nitrogen fertilizer requires large amounts of increasingly costly fossil fuels and the environmentally detrimental consequences of nitrogen fertilizer runoff from fields to aquatic environments (see Perspective 15.1), the agricultural use of symbiotic nitrogen-fixing bacteria is very appealing.

Different types of rhizobia have characteristic host ranges, meaning they only form nodules on particular types of legumes. For instance, *Sinorhizobium meliloti* (see Microbes in Focus 13.2) forms nodules on alfalfa, but not on beans, peas, or soybeans, whereas *Bradyrhizobium japonicum* forms nodules on soybeans, but not on beans, peas, or alfalfa. Later in this section, we will examine some of the reasons for this host specificity.

Development of Legume Root Nodules

Root nodule formation is an exquisitely controlled developmental process, with well-defined stages. It begins when a free-living rhizobial cell in the soil migrates towards the root hairs of a

Courtesy Brian Driscoll, McGill University

Figure 17.7. Effect of inoculation of *Sinorhizobium meliloti* on alfalfa Both plants were grown in nitrogen-deficient soil. Plant grown in the absence of *Sinorhizobium meliloti* (*left*). Plant grown in the presence of *Sinorhizobium meliloti* (*right*). Enhanced growth is a result of the ability of *S. meliloti* to fix nitrogen and convert it to nitrogenous compounds for plant use.

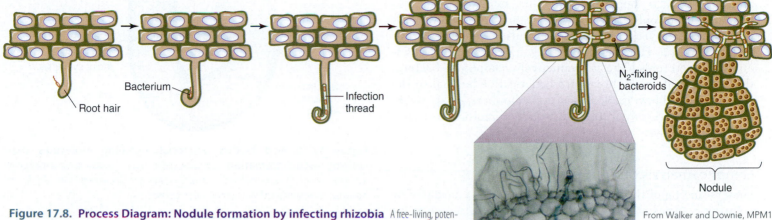

① **Bacterium attaches** to the root hair.

② **Root hair curls around bacterium** forming shepherd's crook.

③ **Infection thread forms and bacteria replicate.**

④ **Infection thread extends into root cortex as root cells divide.**

⑤ **Infection thread branches; bacteria enter cells, lose their cell wall, and become N₂-fixing bacteroids.**

⑥ **Nodule consisting of plant cells is stimulated to grow by infecting bacteroids.**

Root hair

Bacterium

Infection thread

N₂-fixing bacteroids

Nodule

LM

Figure 17.8. Process Diagram: Nodule formation by infecting rhizobia A free-living, potential symbiotic bacterium is attracted to the root hair. Successful attachment induces a series of changes in the root and the bacterium, producing a plant wall-bound infection thread from which the replicating bacteria enter the root cells to form an enlarged root nodule. These infecting bacteria remain encased in a host-derived membrane, change their cell walls, and further differentiate into non-replicating bacteroids, capable of fixing N₂.

From Walker and Downie, MPM1 13:754–762, Fig. 2. Image provided courtesy of J. Allan Downie. Reproduced with permission of The American Phytopatholical Society

suitable plant host. The first visible outcome of this association is curling of the root hair around the cell to form a structure called a "shepherd's crook" **(Figure 17.8)**. Interestingly, if culture supernatant from a compatible bacterial symbiont grown *in vivo* is applied to a legume root, the root hairs will start to curl. However, to form the shepherd's crook, attachment of the bacterial symbionts themselves is required.

Next, the bacterium must infect the plant tissue. "Infect" may seem like a strange term to use as it is usually associated with disease, but symbiont cell entry has many parallels to infection by intracellular pathogens. Many use type III or type IV secretion systems and endocytosis to gain access to host cells (see Figure 2.23). These systems appear to have been selectively maintained in the symbiont from what was once an ancestral pathogen.

CONNECTIONS Type III or type IV secretion systems are used by many Gram-negative pathogens, such as *Bordetella pertussis*, *Salmonella*, and pathogenic *E. coli* strains, to deliver toxins to host cells. See **Chapter 21**.

The infection process begins with an invagination of the plant cell wall that forms a narrow tube called the **infection thread (Figure 17.9)**. As the infection thread lengthens, the bacteria divide so that the infection thread becomes filled with them. As the plant root cells continue to divide, the nodule structure starts to form, and the infection thread extends toward the center of the developing nodule and begins to branch. Near the extending tip of the infection thread, bacteria are taken up by the host cells and are surrounded by a host plasma membrane derived from the infection thread. Once inside the cell, the bacteria differentiate, gain the ability to fix N₂, and are now termed bacteroids. While the bacteroids are within the plant host cell, they remain separated from the cytoplasm by the surrounding plant cell membrane. During the differentiation process, the bacteroids become larger, lose their flagella, modify their cell wall, and sometimes change from rod shaped to multilobed. The fully differentiated N₂-fixing bacteroids are unable to divide, becoming "terminally differentiated."

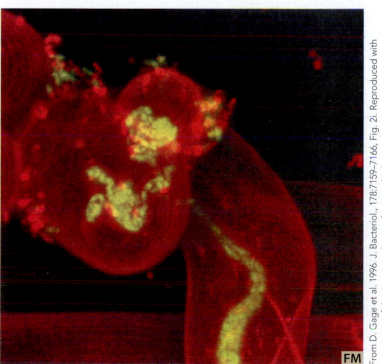

FM

Figure 17.9. Infection thread Green rhizobia cells can be seen filling the infection thread in a root hair.

From D. Gage et al. 1996. J. Bacteriol, 178:7159–7166, Fig. 2i. Reproduced with permission of the American Society for Microbiology.

The bacteroids can fix N_2 because they are provided with protection from the nitrogenase-toxic effects of O_2 and are given the necessary metabolic resources from the plant. It is clear that the bacteroids perform nitrogen fixation for the plant, but what advantage is provided for bacteria? The ability to fix N_2 is coupled with inability to divide further, so while the bacteroids are viable, they cannot proliferate. This would appear to be a tremendous evolutionary disadvantage for the bacteria. However, this is only a dead end for the individual bacteroids. The reproductively competent bacteria that fill the infection thread are genetically identical to the bacteroids, and having increased their numbers greatly here, release more progeny to the soil to infect another host once the nodule decays.

Genetic Analysis of Nodulation

Understanding of the nodulation processes was elucidated by creating a series of mutant rhizobia that were screened for alterations in symbiotic phenotype, such as the inability to induce root nodules, or the formation of root nodules that lacked nitrogen-fixing ability. This involved painstakingly inoculating single mutant colonies one by one onto legume seedlings that had been germinated from surface-sterilized seeds, raising those seedlings without an added nitrogen source, and observing over time for the production of nodules and nitrogen-fixing ability. The mutant bacterial strains that showed a symbiotic phenotype different than the wild type were then examined to determine which genes were interrupted. These experiments resulted in the identification of several *nod* genes that are required for nodulation, *nif* genes that are directly involved in the synthesis of the nitrogenase enzyme, and *fix* genes that are required for synthesis of other products needed for nitrogen fixation.

The discovery of the *nod* genes was a breakthrough in the understanding of the symbiotic process. Recall that one of the first changes to be detected in the formation of a legume root nodule is curling of the root hairs. This does not require the presence of the bacterial cells, as it can be induced by exposure to culture supernatant from a potential bacterial symbiont. This provided strong evidence for the production of a diffusible compound by the bacteria. Also, culture supernatant from the *nod* mutant strains did not induce root hair curling. Interpretation of these experimental results suggested that the *nod* genes were involved in the production of a lipochitin oligosaccharide signaling compound that controlled the nodule formation process—the Nod factor. Many different Nod factors are produced and are structurally unique to different species of rhizobia. Different Nod factors can be recognized by specific types of host plants, and this contributes to the plant host range on which the rhizobia can form nodules.

While the Nod factors signal from the bacteria to the plant, signaling also takes place in the other direction, from plant to bacteria. Determination of the type of plant on which the rhizobia can form nodules requires specific plant-secreted compounds called flavonoids that are involved in *nod* gene expression in the bacteria. Specific plant flavonoid compounds are detected by the

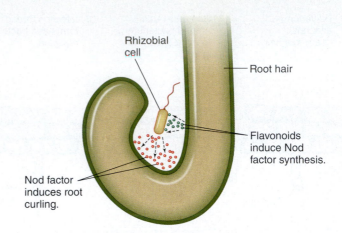

Figure 17.10. Nod factors: Bacterial signaling molecules that control nodule formation Flavonoid secretion by the plant host induces expression of *nod* genes in rhizobia in a species-specific manner. Nod factor is in turn secreted by the bacterium to induce root curling in the plant host, beginning nodule formation.

bacterial NodD transcriptional activator protein, which then induces *nod* gene transcription, resulting in production of the Nod factor, which induces root curling (**Figure 17.10**). For example, the alfalfa flavonoid luteolin is recognized by *Sinorhizobium meliloti* NodD, while genistein from soybeans is recognized by *Bradyrhizobium japonicum* NodD.

Nodule Metabolism

The nitrogen fixation process requires much energy in the form of ATP (see Section 13.7). The rhizobia receive nutrients from the plant and metabolize them to generate the required ATP. What carbon compound is metabolized by the bacteroids as they fix nitrogen? In the plant, sucrose, a product of carbon fixation, is transported from the leaves, the site of photosynthesis, to the roots. You might expect that the bacteroids metabolize sucrose directly, but this doesn't seem to be the case. First, genetic experiments demonstrate that rhizobia mutants deficient in sucrose, glucose, and fructose metabolism can still fix nitrogen within nodules. It is therefore not likely that immediate products of photosynthesis are directly fueling nitrogen fixation.

Following these experiments, researchers working with clover and pea rhizobia showed that mutants that cannot incorporate C_4-dicarboxylic acids, such as succinate, fumarate, and malate, form nodules that cannot fix N_2. It thus appears that the carbon nutrient that is provided to the symbiotic bacteroid cells is in the form of C_4-dicarboxylic acids, which are used in the TCA cycle (see Figure 13.21). **Figure 17.11** provides an overview of the metabolic relationship between the legume plant and the bacterial symbiont. In return for C_4-dicarboxylic acids provided by the plant, the bacteroids provide nitrogen in a form that the plant can use directly for its metabolism. Depending on the host/symbiont species pairing, this supply is in the form of ammonia and/or amino acids.

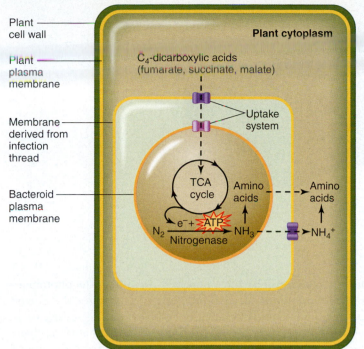

Plant cell wall

Plant plasma membrane

Plant cytoplasm

C_4-dicarboxylic acids (fumarate, succinate, malate)

Membrane derived from infection thread

Uptake system

Bacteroid plasma membrane

TCA cycle

Amino acids → Amino acids

N_2 — e⁻ + ATP → NH_3 --→ NH_4^+
Nitrogenase

Figure 17.11. Exchange of carbon and nitrogen compounds between plant host cells and bacteroids C_4-dicarboxylic acids, such as malate, succinate, and fumarate, are manufactured by plant cells and are taken up from the plant cytoplasm by bacteroids to enter into the TCA cycle. The electrons and ATP that are generated are used by nitrogenase to convert nitrogen gas to ammonia (nitrogen fixation). Ammonia is used by the bacteroid to manufacture amino acids. Excess ammonia diffuses across the membrane into the plant cytosol, where it is reduced to ammonium which is then used to manufacture amino acids for protein synthesis in the plant cell.

17.2 Fact Check

1. Explain the role some endophytes play in plants.
2. Why do nitrogen-fixing bacteria induce formation of root nodules?
3. What role does the protein leghemoglobin play inside the nodule?
4. Outline and describe the stages of root nodule formation. Include an explanation of the term "bacteroid" in your outline.
5. What genes are known to be involved in the nodulation process?
6. In the metabolic relationship between plants and bacteroids, what does each provide for the other?

17.3 Lichens

Can symbiotic microbes produce new types of organisms?

So far, we have examined symbiotic microbes that form associations with multicellular organisms, like plants. Microbial species can also form symbiotic relationships with other microbes. Lichens are such a case. Take a walk in the forest, and you will certainly notice the mossy and paper-like growths on the bark of trees and on rocks **(Figure 17.12)**. These are not plants, but lichens, species of which are also the favorite meal of reindeer and caribou in the barren arctic regions where plant growth is sparse. While **lichens** have been classified as individual species, each one is actually a composite organism, composed of two or three different life-forms. One of these life-forms is fungal, and is called the **mycobiont** (*myco* refers to fungi). The other one (sometimes two) has photosynthetic

capabilities, and is called the **photobiont**. The photobiont is usually a green alga, but in about 10 percent of the cases is cyanobacterial. Cyanobacteria and eukaryotic algae are typically photosynthetic aquatic organisms with limited cellular differentiation. When they associate with fungi in the lichen, the result is a life-form that is quite plant-like, and in fact plays the role of the plant in many ecosystems where plants cannot survive. Despite the widespread occurrence of lichens, and their importance in ecosystem functioning, relatively little research has been carried out on the basic mechanisms that govern the complex interactions between the fungal mycobiont and algal and/or cyanobacterial photobiont partners.

© Gregory G. Dimijian/Photo Researchers, Inc.

A. Lichens on a branch

© Thomas Porett/Photo Researchers, Inc.

B. Lichens on a rock

© Articphoto/Alamy Limited

C. Hanging lichen

Figure 17.12. Lichens: Symbiosis between fungi and algae or cyanobacteria Lichens are composite organisms, composed of a fungus and one or two types of photosynthetic algae or cyanobacteria. **A.** Several lichens differing in form and color grow on a branch. **B.** Many types of lichens can grow on bare rock. **C.** Some lichens like this *Usaea repena* can look like hanging moss.

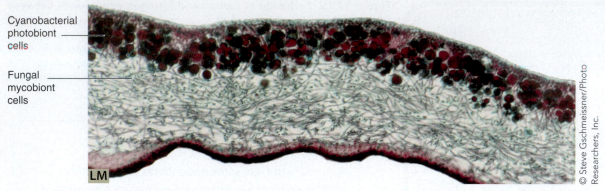

Cyanobacterial photobiont cells

Fungal mycobiont cells

LM

© Steve Gschmeissner/Photo Researchers, Inc.

Figure 17.13. Cells of the different lichen partners form layers. A section of a lichen thallus showing the distinct layer of dark red cyanobacterial photobionts and fungal mycobiont cells beneath.

Structure of lichens

The main, leafy part of the lichen body is called the "thallus," and the different forms taken are given descriptive names, such as crustose (flat), foliose (leafy), or fruticose (branched). Some lichens also have root-like structures that do not contain photobionts but serve to adhere the lichen to solid surfaces such as trees, rock, or soil. The fungal mycobiont forms the major structure of the thallus and encases one or two photobiont partners. While the partner cells are intimately associated, this is always an extracellular interaction where the photobiont remains outside of the fungal cells. Often, the photobionts are arranged in a distinct zone or layer exposed to sunlight within the lichen where their main function is to produce carbon nutrient for the mycobiont through photosynthesis **(Figure 17.13)**.

Benefits to the lichen partners

The partnership formed by the lichen enables each of the symbionts to grow in nutrient-poor and inhospitable habitats, such as on tree bark, soils, or wind-swept barren rock (see Figure 17.12B). In fact, lichens are the dominant form of photosynthetic organism in polar regions. Their ability to survive in these conditions so much better than the individual fungal or algal cells is an intriguing mystery that deserves investigation.

The lichen relationship is rooted in metabolic cooperation. Instead of growing within the food source, as other fungi typically do, lichen mycobionts have an internal source of nutrient, courtesy of their photosynthetic partner. Carbon compounds produced by the photobiont through photosynthesis, often ribitol or glucose, are provided to the mycobiont. In some cases, where the photobiont is cyanobacterial, N_2 fixation can take place, resulting in a source of available nitrogen as well. But what do the photobionts gain from interaction? Organic acids excreted from the fungal partner release inorganic nutrients from the substrate on which the lichen resides, and these are used by the photobiont for metabolism. Encased in the lichen structure, the photobiont is protected from wind and rain that would otherwise wash it from surfaces. The lichen structure can also provide protection from high light intensity, especially critical during dessication. In the absence of sufficient water for an electron donor, for example, photosystem II (see Section 13.6) of

the photobiont can become damaged. To protect the photobiont from desiccation-induced photosynthetic damage, the fungal partner undergoes physiologic changes during desiccation, resulting in an increase in light-absorbing pigments to shade the photobiont from light. Without protection from the fungal partner, survival of the photobiont on its own would be difficult in the extreme conditions in which lichens live.

Lichen development and reproduction

While lichens are historically classified as individual species, the mycobiont and photobiont partners are not necessarily completely dependent upon each other. They can often exist independently as well. When on its own, the structure of the mycobiont is that of a typical fungus, without the formation of the leaf-like thallus. Just like other fungi, they can often reproduce sexually. However, when in symbiosis, lichens reproduce only asexually by **soredia**, which are reproductive structures made up of a few algal cells packaged by fungal hyphae **(Figure 17.14)**.

When brought together, free-living mycobionts and photobionts can also form a lichen. Lichens are typically very slow-growing, with individual lichens sometimes persisting for thousands of years. When lichens dry out, their metabolic processes are brought to a standstill, but they can quickly resume metabolic activity upon rehydration. Amazingly, considering the importance of lichens in key ecosystems, we still do not have a good understanding of the mechanisms that control the formation of the symbiosis or promote the formation of the thallus structure after the association of the partners.

17.3 Fact Check

1. What life forms compose lichens?
2. Define the terms "mycobiont" and "photobiont." How are they related to the term "lichen"?
3. Explain how each member of the lichen benefits from the relationship.

A. Soredia on thallus

B. Soredia

SEM

Figure 17.14. Soredia: Lichen reproductive structures A. Lichen thallus showing cup-shaped structures containing reproductive soredia appearing as brown powder. **B.** Soredia composed of green algal cells contained in a mesh of fungal hyphae beginning to colonize a green moss.

17.4 Symbionts of vertebrates

What are the roles of symbionts in vertebrates?

Vertebrates do not appear to harbor endosymbionts. However, most possess symbionts that colonize outer or inner body surfaces, such as the skin and intestinal tract. Of all the vertebrate body surfaces available for colonization, the digestive tract is the richest in microbes, both in species diversity and numbers. Why is this? In general, the gut provides a relatively protected and constant environment for microbes, and of course it is a good place to access nutrients. What the gut microbes do for the animal is not precisely defined. In general, experiments using germfree and gnotobiotic animals have demonstrated that symbiotic microbes provide nutritional, immunological, and developmental benefits. **Germfree** animals are born and raised under completely sterile conditions. They have no symbionts or any other microbial inhabitants. **Gnotobiotic** animals are initially germfree, but are introduced to a few select or known (*gnotos* is Greek for "known") species for controlled study. **Toolbox 17.1** briefly describes how germfree and gnotobiotic animals are produced. Comparisons of conventionally raised

---

Toolbox 17.1
GERMFREE AND GNOTOBIOTIC ANIMALS

Controlled comparative studies using germfree and gnotobiotic animals and their normally colonized counterparts have helped shed some light on the importance of symbiotic microbes. In the case of mammals, usually mice, rats, or pigs, the young are surgically removed aseptically to prevent the contamination that invariably happens from the birthing process. Just passing through the birth canal inoculates the newborn with vaginal and fecal microbiota. The animals are then maintained in isolation chambers on sterile bedding, fed sterile semi-synthetic food, and given sterile water. Even the air they breathe is sterile. Animal keepers must enforce this sterility by wearing gowns, gloves, and masks.

Gnotobiotic animals are produced from germfree animals by purposeful inoculation of just one or a few known species of microbes. Inoculation can be done by supplying the microbe(s) in food or water, or swabbing them onto the skin. The animals are then maintained under sterile conditions to study the effects of the inoculated microbes. Protocols for germfree fish, birds, and reptiles have also been developed. Sterile eggs are fertilized *in vitro* and raised in sterile conditions.

● Test Your Understanding ....

Maintaining germfree animals leaves little room for error. Imagine you are training someone who will be working in your lab to raise germfree mice for research on intestinal flora. What aseptic techniques would you highlight in your training, and how would you explain why they are important?

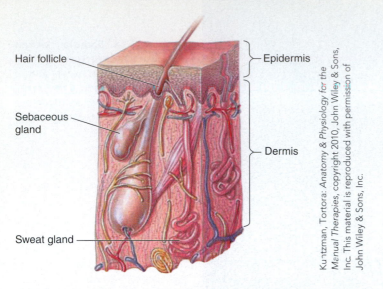

Hair follicle

Sebaceous gland

Sweat gland

Epidermis

Dermis

Kuntzman, Tortora: Anatomy & Physiology for the Menual Therapies, copyright 2010, John Wiley & Sons, Inc. This material is reproduced with permission of John Wiley & Sons, Inc.

Figure 17.15. Cross section of skin Schematic showing the different layers, glands, and a hair follicle. Oil-secreting sebaceous glands and hair follicles form deep pores in the dermis of the skin, providing an environment for the growth of anaerobes.

animals with these animals can give insights into the roles of symbiotic bacteria.

Trying to decipher the roles of the individual symbionts in the gut, like in any complex ecosystem, is a monumental task. Controlled studies designed to determine the contributions from each species in such a diverse population are often difficult, if not impossible. Studying symbionts in isolation, such as in gnotobiotic animals, often does not fully explain their role within the dynamic structure of the gut ecosystem. Symbionts of the digestive tract can be even more difficult to study because most species have not been cultured, and have not even been characterized for most vertebrate species, including humans. As we learned in Chapter 6, characterization and understanding of the microbial symbionts have been enhanced by new molecular methods, such as small subunit (SSU) rRNA gene sequencing (see Chapter 1 Mini-Paper and Toolbox 1.1) and FISH analyses (see Toolbox 6.5). These techniques allow a glimpse into the true nature of these inhabitants—something that culture methods alone cannot do. Much of the work in this area has focused on the digestive tracts of cattle, because of their economic importance, and humans. We'll first examine human symbionts.

Symbionts of humans

The human **microbiome**, the total of all the microbes living on and in the body, is enormous in terms of sheer number and species diversity. Currently, a U.S. National Institutes of Health initiative launched in 2007 called the Human Microbiome Project has a goal to define all the species inhabiting the gut, the skin, and the female urogenital tract, as well as the nasal and pharyngeal cavities. It is an immense undertaking because of the huge number of species that are postulated to inhabit humans. The size of their collective genome dwarfs the human genome. On a smaller scale, a European Union initiative is examining human intestinal microbiota. Both these projects will enlighten the relationship between human health and the microbial populations we carry.

Even though only a small number of people have been studied so far, it is apparent that there is no consistently definable microbial population in terms of species within and among humans. We each seem to have our own unique species composition of microbiota, which can change with age and diet. While it is doubtful that there will ever be a defined species profile that represents the typical human, it may be possible to broadly define microbial populations in terms of prevalent phyla, and maybe common genera and species. A useful beginning to determining the relationship between health and our microbial symbionts is to look at the common microbial groups that inhabit the body.

Skin

To us, skin may appear to be a uniform surface, but to microbes it is as varied as the earth's surface. Sebaceous glands and hair follicles form deep canyons where anaerobes can find a home (**Figure 17.15**). Areas dense in sebaceous glands are like seas of oil. Simply swabbing the skin, a common sampling method employed in the majority of skin microbiota studies, may not accurately reflect the microbes present in the depths of pores and hair follicles. As you may rightly guess, the microbiota of skin varies locally, depending on oiliness, sweatiness, hairiness, and environmental exposure. Differences in numbers and variety of microbes exist between the toes (moist and warm) where microbes abound, to the inner forearm (dry and cool) where densities are significantly lower.

Although attempts to identify microorganisms of human skin are inconsistent in sampling methods and sites, some generalities seem to be emerging. As shown in **Figure 17.16**, skin is generally dominated by members of the Gram-positive phyla Actinobacteria and Firmicutes followed by the Gram-negative phyla Proteobacteria and Bacteroidetes. Not surprisingly, the dominant genera found depend on the sampling site. For oily areas such as the face, *Propionibacterium acnes* (as its name suggests, can cause acne) dominates, while on the arm, the genus *Pseudomonas* dominates.

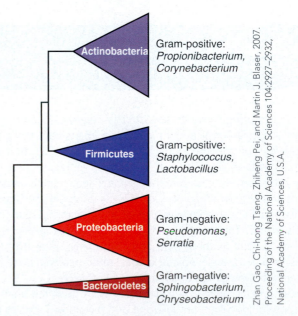

Actinobacteria — Gram-positive: *Propionibacterium, Corynebacterium*

Firmicutes — Gram-positive: *Staphylococcus, Lactobacillus*

Proteobacteria — Gram-negative: *Pseudomonas, Serratia*

Bacteroidetes — Gram-negative: *Sphingobacterium, Chryseobacterium*

Zhan Gao, Chi-hong Tseng, Zhiheng Pei, and Martin J. Blaser, 2007. Proceeding of the National Academy of Sciences 104:2927–2932, National Academy of Sciences, U.S.A.

Figure 17.16. Major bacterial phyla found on skin Relative proportions are shown, based on 16S rRNA gene analyses. While the dominance of particular genera depends on the site sampled, the proportions of phyla tend to remain unchanged in healthy tissue.

Few studies have comprehensively examined single-celled eukarya of the skin. Members of the yeast genus *Malassezia* (Microbes in Focus 17.1) appear to be the most common fungal inhabitants of skin. Examination of 18S rRNA genes show *Malassezia* is not present on the skin of all persons, and is only a minor skin component in healthy persons, as well as many other mammalian species. Some *Malassezia* species are a major cause of dandruff in humans, and other skin conditions in animals.

CONNECTIONS Like the yeast *Malasezzia*, potential pathogens such as bacterial species *Staphylococcus aureus*, *Streptococcus pyogenes*, and *Pseudomonas aeruginosa* frequently can be isolated from skin. See **Chapter 21** for more details on these pathogens.

Although there is some information regarding the general microbial composition of human skin, little is known about the significance of the various populations. It is well established that skin microbiota participate in the development of proper immune responses and the prevention of disease from pathogens, many of which can be found as minor inhabitants on skin. We'll examine the general relationship between symbionts and immune function and disease prevention later in this section, when we cover the digestive tract.

Vagina

As the vagina is in contact with skin and is situated near the anus, it is not uncommon to find bacteria present from both surfaces. For example, *Staphylococcus epidermidis*, a common skin bacterium and yeast species of genus *Candida* are frequently found in the vagina. Fecal bacteria such as *Enterococcus faecalis* and *E. coli* are also commonly, but not consistently, found here. Many of these microbes are likely to be transient, rather than resident, as they do not appear to replicate here to appreciable numbers under ordinary circumstances. Lactobacilli originating from the intestinal tract are the exception and appear to truly colonize the vagina of healthy women as symbionts. Most women studied are colonized with one or two species of *Lactobacillus*, most commonly *L. acidophilus*.

Many studies examining colonization of the vagina and vulva have focused on *Candida*, as some species, such as *C. albicans*, cause vulvo-vaginal yeast infections, estimated to affect 75 percent of women over 40 years of age. *Candida* species are also commonly associated with diaper rash in infants. Favorable conditions for growth of *Candida* include a moist environment and a near neutral pH. The composition of vaginal microbiota is affected by sexual practices, diabetes, antibiotic treatment, age, and hormone levels that affect pH. It is well known that in menopausal women of approximate age 50, the pH of the vagina is nearer to neutral, favoring the growth of *Candida*. Consequently, yeast infections caused by *Candida* species rise sharply in this age group. By contrast, women of child-bearing age tend to have acidic vaginal pH, favoring the growth of lactobacilli that in turn maintain this low pH by releasing acidic metabolic products, such as lactic acid. Maintenance of this homeostasis means yeast infections are relatively uncommon in this age group.

Oral Cavity

The human oral cavity is a heavily colonized area of the body, second only to the colon, or large intestine. Like the colon it provides a warm, moist, nutrient-rich environment for growth and its varied landscape (teeth, tongue, mucosal surfaces) provides a variety of surfaces for diverse populations of microbes. Over 700 bacterial species have been detected in the mouth by 16S rRNA gene analysis, over half of which have not been cultivated. A few yeasts, such as *Candida* species, can also be found here. As on skin, population profiles vary by location. For example, the tongue is dominated by lactobacilli, the exposed tooth surface is dominated by a diversity of Gram-positive streptococci, commonly referred to as "viridans streptococci." Viridans streptococci include species such as *Streptococcus sanguinis* and *Streptococcus mutans*, the agent of dental caries or cavities (Microbes in Focus 17.2). Although most of the species found in the oral cavity of healthy humans are considered to be commensals or even mutualists, many have pathogenic potential.

Most of the research done on the oral cavity microbiota has focused on pathogens, particularly those involved in dental diseases. Virtually all of the oral microbiota are adherant to a surface in some way. Many readily adhere to other cells, which leads to biofilm formation (see Section 2.5). Most of us are familiar with the biofilm **plaque** (see Figure 2.33), the adherent layer of bacteria and their organic compounds that forms on the tooth surface. If left unattended by regular brushing and flossing, it can have serious consequences.

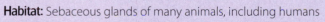

Microbes in Focus 17.1
MALASSEZIA

Habitat: Sebaceous glands of many animals, including humans

Description: Aerobic eukaryal yeast, 5–7 μm in diameter, producing spherical conidia and filamentous hyphae

Key Features: *Malassezia* produces lipases that break down the oils on the skin into free fatty acids, some of which it can consume. Excessive growth of *Malassezia* causes a buildup of fatty acids on skin, producing irritation, hair loss, and flaking skin. Different species have been associated with dandruff, psoriasis, and other skin conditions in humans and animals.

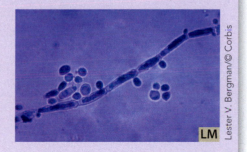

LM

Lester V. Bergman/© Corbis

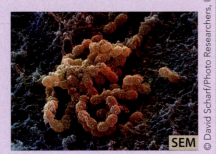
Plaque formation and cavities are two negative outcomes of the activities of our oral symbiotic bacteria. Early colonizers of the tooth surface are dominated by viridans streptococci (**Figure 17.17**). They adhere via pili to the pellicle, an amorphous coating on the tooth surface composed of non-cellular materials, largely proteins from saliva and components shed and secreted by bacteria. Below the gum, or gingivus, anaerobic species of *Fusobacterium* in turn adhere to colonizing streptococci. Their filamentous form provides a large surface area for further adherence of other colonizers. The adherent cells become embedded within a strong, self-produced matrix called "extracellular polymeric substance," including proteins, polysaccharides, and DNA. Plaque populations and the associated matrix will continue to increase if not kept in check by brushing three or four times daily. Unchecked growth results in an increase in biofilm thickness, reduction in oxygen concentration, and a change of the bacterial profile from aerobic Gram-positive species to Gram-negative anaerobes, in particular *Fusobacterium*.

Plaque buildup is a major cause of periodontal disease that can result in a loss of gum and bone tissue, infection and eventual tooth loss (**Figure 17.18**). *Gingivitis*, or inflammation of gingiva, is usually the first sign of trouble. As plaque below the gingivus builds up, bacterial components, such as enzymes and cell wall material, cause inflammation and damage to the epithelial cells surrounding the tooth. This extends the gap between the tooth and tissue, allowing further invasion of plaque bacteria. Some of the anaerobic plaque bacteria, such as *Porphyromonas gingivalis*, produce proteases that further damage epithelial cells, encouraging the invasion of bacteria into tissue and causing further inflammation.

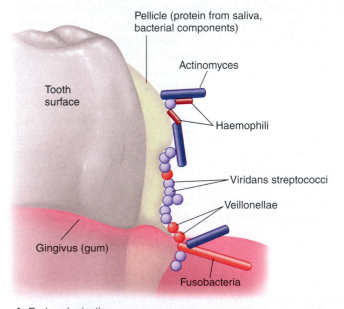

A. Early colonization

Pellicle (protein from saliva, bacterial components)

Actinomyces

Tooth surface

Haemophili

Viridans streptococci

Veillonellae

Gingivus (gum)

Fusobacteria

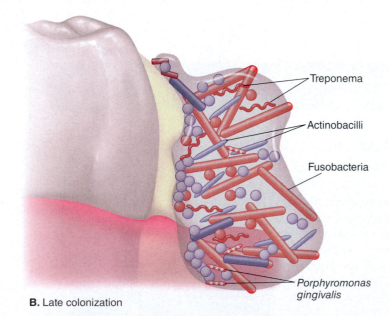

B. Late colonization

Treponema

Actinobacilli

Fusobacteria

Porphyromonas gingivalis

Figure 17.17. Formation of a plaque biofilm Gram-positive bacteria are shown in shades of blue and Gram-negative bacteria are shown in shades of red. **A.** Early colonizers are composed largely of Gram-positive viridans streptococci, which adhere to the non-cellular pellicle of the tooth above and below the gum (gingivus). Other colonizers include smaller numbers of actinomycetes, *Haemophilus* species, and veillonellae. Fusobacteria are anaerobes found in small numbers on the tooth surface below the gingivus. **B.** If left undisturbed, plaque populations continue to increase, becoming dominated by Gram-negative anaerobic groups, in particular fusobacteria, which provide an extended surface for attachment allowing other bacteria to thicken the plaque layer and extend it below the gum line. Invasion of plaque below the gum line leads to inflammation of the gingivus and eventual loss of bone supporting the teeth.

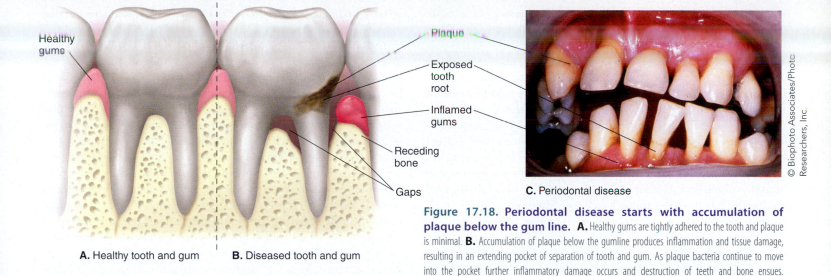

Healthy gums

Plaque

Exposed tooth root

Inflamed gums

Receding bone

Gaps

A. Healthy tooth and gum   **B.** Diseased tooth and gum

C. Periodontal disease

© Biophoto Associates/Photo Researchers, Inc.

Figure 17.18. Periodontal disease starts with accumulation of plaque below the gum line. **A.** Healthy gums are tightly adhered to the tooth and plaque is minimal. **B.** Accumulation of plaque below the gumline produces inflammation and tissue damage, resulting in an extending pocket of separation of tooth and gum. As plaque bacteria continue to move into the pocket further inflammatory damage occurs and destruction of teeth and bone ensues. **C.** Periodontal disease showing inflamed and receded gums and plaque buildup.

CONNECTIONS Inflammation is a common tissue response to bacterial components in which defensive molecules and cells are brought in to clear infection. When clearance cannot be achieved, chronic inflammation occurs, leading to further damage. See **Sections 19.3** and **21.2** for details.

Digestive Tract

It is estimated that GenBank, the world's largest database of genetic sequences, contains at least 800 distinct, and mostly uncharacterized, 16S rRNA gene sequences from the human intestinal tract. Each sequence represents a bacterial species, or operational taxonomic unit (OTU; see Section 15.2). Some researchers believe the true number of gastrointestinal bacterial species may be in excess of 2,500. Viruses, especially phages, are also likely to be a large part of this community. In one study, over

1,200 different viral genomes were identified in human feces. Of course, some of these may just be "passing through" rather than associated with the microbes colonizing this environment. The finding of plant-specific viruses in feces indicates the transient nature of some of these viruses, but resident phages are certain to exist as they are known to be major contributors to genetic diversity in resident bacterial populations by facilitating exchanges of genetic information (see Sections 9.5 and 21.4).

The digestive tract is a continuous tube that goes from mouth to anus. Compared to the digestive tract in many other species (like ruminants), the human digestive tract is quite simple. The major organs consist of the stomach, the small intestine (consisting of the duodenum, jejunum, and ileum), and **colon** or large intestine **(Figure 17.19)**. The stomach is a single

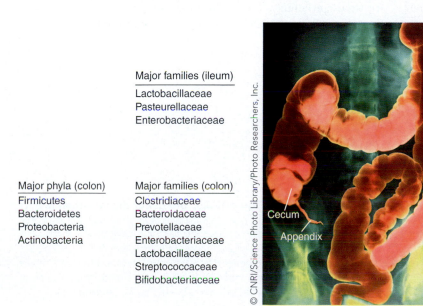

Major families (ileum)

Lactobacillaceae
Pasteurellaceae
Enterobacteriaceae

Major phyla (colon)

Firmicutes
Bacteroidetes
Proteobacteria
Actinobacteria

Major families (colon)

Clostridiaceae
Bacteroidaceae
Prevotellaceae
Enterobacteriaceae
Lactobacillaceae
Streptococcaceae
Bifidobacteriaceae

Cecum

Appendix

© CNRI/Science Photo Library/Photo Researchers, Inc.

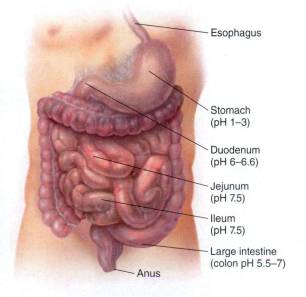

Esophagus

Stomach (pH 1–3)

Duodenum (pH 6–6.6)

Jejunum (pH 7.5)

Ileum (pH 7.5)

Large intestine (colon pH 5.5–7)

Anus

Figure 17.19. Human digestive tract structure and major bacterial groups
The acidic pH of the stomach prevents colonization by most microbes. The less acidic ileum and colon contain significant numbers of bacteria; the major microbial phyla and most prevalent families found here are shown on the left.

Tortora, Derrickson: *Principles of Anatomy and Physiology*, 13e, copyright 2012, John Wiley & Sons, Inc. This material is reproduced with permisson of John Wiley & Sons, Inc.

large chamber, where food is hydrolyzed by hydrochloric acid. The pH of the stomach is approximately 2 and few microbes can withstand this environment. For many intestinal pathogens like *Salmonella*, *Vibrio*, and most *E. coli* strains, ingestion of ten thousand or more bacteria is needed to cause disease. Some pathogens can withstand or even neutralize the acid to increase their survival in the stomach. For these, ingestion of only a few cells can cause disease. For example, as few as ten *Shigella dysenteriae*, an acid-resistant relative of *E. coli*, can survive transit through the stomach to colonize the intestine. *Helicobacter pylori*, a parasitic symbiont commonly infecting humans worldwide, can actually colonize the stomach. It protects itself by neutralizing acid during its transit to the stomach wall (see Microbes in Focus 18.1). Researcher Dr. Barry Marshall made medical history when he drank a concentrated culture of *H. pylori* to prove it caused gastric and duodenal ulcers, and indeed he developed ulcers (see Perspective 6.1). He then treated himself with antibiotics and the infection and the ulcers disappeared—a rather risky demonstration of cause and effect (see Koch's postulates, Section 18.4).

As we move from the stomach through to the colon, the digestive system becomes increasingly less acidic and more anaerobic. This environmental change is accompanied by increasing numbers and species of microorganisms. Few studies have been done on isolated areas of the small intestine, and it appears from these that the duodenum and jejunum have only scant colonization. The ileum shows progressively more numbers of bacteria and is dominated by families Lactobacilliaceae, Pasteurellaceae, and Enterobacteriacea, although the proportions of these vary with individuals. Members of the genera *Lactobacillus* and *Bifidobacterium* have received much attention as they are common components of yogurt and have been linked to health benefits—although, as **Perspective 17.1** points out, evidence for this is controversial.

At the junction of the ileum and colon is the **cecum**, a dead-end pocket that extends a short distance from the junction. Humans have a small pouch projecting from the cecum called the *appendix* (see Figure 17.19). Like the colon and cecum, the appendix contains anaerobic bacteria. Persons who have had their appendix removed do not report a negative impact, so it appears

to have little importance in humans. In many animal species, like rabbits, mice, and beavers, the cecum is greatly enlarged and has critical digestive functions, which will be described later.

The human colon contains one of the highest densities of microorganisms known for any microbial habitat—up to 10^{12} cells per mL of colon wash or gram of fecal matter. That gives an estimate of 10^{14} bacterial cells in the colon, with an estimated total weight of 1 kg. It's hard to imagine just how huge this population is. In comparison, the human body is made of approximately 10^{13} cells (see Chapter 1, The Rest of the Story). This means you possess approximately ten times more bacterial cells in your colon than the total of all your body cells! As we learned in Chapter 6, the colon is an example of a continuous culture system as a portion of the microbes that inhabit it are continuously expelled in feces and are replaced by growth of the remaining colonic inhabitants.

Humans, like other omnivores and carnivores, use **colonic fermentation**. Material that is not absorbed in the small intestine is passed to the colon where the microbes take over. Most of the undigested substrates for microbial breakdown are dietary fiber consisting of complex carbohydrates, such as cellulose, hemi-cellulose, and pectins found in plant structural materials that are not broken down by host enzymes. Some of these can be degraded and fermented by symbiotic microbes via mixed acid fermentation (see Section 13.2) producing "volatile fatty acids," predominantly acetic acid, propionic acid, and butyric acid, in order of predominance. In solution, these acids readily lose a proton to produce the ions acetate, propionate, and butyrate, which are absorbed by the colon and transported to the liver where they are used as energy sources. It is estimated that between 5 to 10 percent of our caloric uptake may be due to absorbance of these microbial fermentation by-products in the colon.

Colonizing microorganisms of the intestinal tract are not generally found attached to the cells lining the intestine surface. Attachment appears to be a feature largely reserved for pathogens (see Section 21.2). Instead, symbiotic microorganisms form a biofilm on the periphery of the mucus layer that coats the intestinal epithelium. The biofilm extends into the lumen of the intestine **(Figure 17.20)**, where most of the population resides.

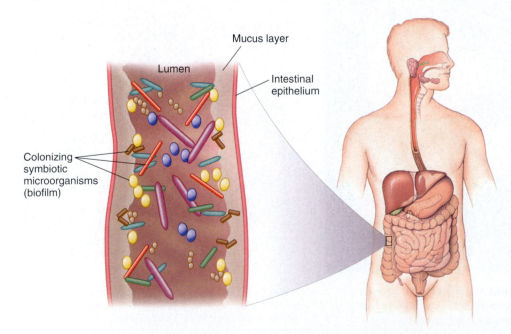

Mucus layer

Lumen

Intestinal epithelium

Colonizing symbiotic microorganisms (biofilm)

Figure 17.20. Biofilm of the intestinal mucus
Colonizing symbiotic microorganisms are not attached to the epithelial cells lining the lumen of the intestine. Instead, they form a biofilm on the outer edge of the mucus layer that extends into the lumen of the intestine. Colonization of the epithelium would result in infection and subsequent inflammation and damage. The ability to colonize and/or invade the intestinal epithelium is a feature of specialized intestinal pathogens.

Allen, Harper: *Laboratory Manual for Anatomy and Physiology*, copyright 2009, John Wiley & Sons, Inc. This material is reproduced with permission of John Wiley & Sons, Inc.

Who would believe that people would actually choose to eat or drink concentrated bacteria? This would seem to be, at the least, distasteful to most, but clever marketing has changed attitudes. Descriptions like "friendly," "good," and "beneficial" bacteria have helped remove previous misconceptions that all bacteria are bad. Even better are descriptions of "live active culture," and of course "probiotics" as this sounds less like bacteria. The truth is, for the greater part of over a century, humans have been unknowingly eating bacterial cultures. It's the dirty little secret of fermented products like yogurt and sour cream. Bacteria give them their flavor and texture (see Chapter 16).

Probiotics are defined by the Food and Agriculture Organization (FAO) of the United Nations as "live microorganisms administered in adequate amounts which confer a beneficial health effect on the host." Prebiotics are non-living dietary supplements that are preferentially metabolized by those bacteria thought to promote good health, and include dietary fiber, which humans can't digest, but many bacteria can. Synbiotics are a marriage of the two, containing probiotics and the prebiotics that supposedly feed them. But do these actually help? The answer is yes, no, and maybe.

The effects of probiotics have been studied for many diseases and conditions, including Crohn's disease, diverticulitis, constipation, diarrhea, irritable bowel syndrome, allergies, cancer, urogenital infections, and obesity. However, the effects have varied considerably between studies, few have been repeated, varying conditions are used, different age groups of people are used, they are not consistent in the strains or numbers of microorganisms used, and sometimes not enough people are included to make results statistically significant. However, probiotics seem to provide fairly consistent benefit in humans suffering from symptoms of lactose intolerance, diarrhea precipitated by antibiotic use, and childhood diarrhea caused by infectious disease. Because of the inability to compare results, these studies have been unable to definitively establish the effectiveness of probiotics for any condition. They do, however, collectively show that probiotic therapy may be a promising avenue for treatment and possibly for prevention.

If the effect of probiotics is real, how might they act? Currently, some evidence suggests that probiotic microbes may interfere with the colonization of pathogens in the gut and/or enhance the immune system, but the mechanisms of these activities have not been established. Interestingly, none of these studies have shown that the probiotic bacteria actually colonize the gut. They are present in the feces, but this may be due to passive transfer following ingestion. Supporting this, several studies show that when the probiotic consumption stops, the bacteria disappear from feces, indicating colonization does not occur.

Lack of colonization may not be critical in providing a benefit. Perhaps any probiotic effect may be due to ingestion of bacterial products, including fermentation metabolites, rather than ingestion of live bacteria. If numbers of live bacteria are important, then what is this number? Many companies advertise the numbers and species of live microorganisms in their products, but this information varies greatly. Members of the bacterial genera *Lactobacillus* and *Bifidobacterium* are most commonly used in these products, but there is significant variation in the species and strains used. This is important because different strains may show extensive differences in physical sensitivities and metabolic activities. Many of the probiotic bacteria in grocery store yogurts (**Figure B17.1**) are destroyed in the acidic environment of the stomach and by proteases and lipid-destroying bile salts farther on in the digestive tract. In addition, yogurts and prepared capsules of live cultures commonly show drastic drops in the number of live bacteria from the time of manufacture (the number quoted on the product label) to the expiration date. This is not surprising for *Bifidobacterium*, as it is an obligate anaerobe and will be destroyed upon contact with oxygen. So one must wonder if biologically relevant numbers of these bacteria manage to stay alive to even reach the colon? Given there are approximately 10^{14} bacteria in the colon, it is hard to believe the addition of a few thousand or even a few hundred thousand can make a difference. Future studies are needed to address these issues. In the meantime, be wary of the supposed health benefits some companies advertise. They have yet to prove that there actually are health benefits gained from consuming these products.

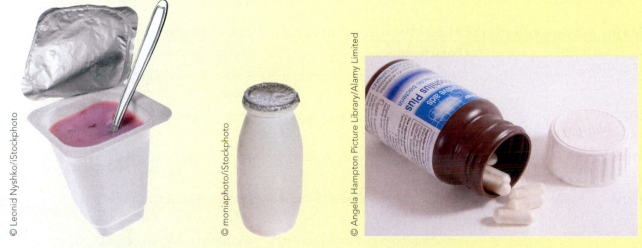

Figure B17.1. Consumer products containing probiotic cultures *Left to right:* Yogurt, drinks, and capsules.

© Leonid Nyshko/iStockphoto

© moniaphoto/iStockphoto

© Angela Hampton Picture Library/Alamy Limited

This is probably why feces contains such a large number of microbial cells. On average, feces contains as much microbial weight as undigested foodstuffs. In healthy persons, the deeper areas of the mucus layer in the colon contain very few microorganisms.

CONNECTIONS Coating of the body's internal passageways with mucus is an important defense against invasion from microbes. This is discussed in **Section 19.2.**

The single most problematic aspect of collecting samples representing specific areas of the intestinal tract is that surgical procedures are required. Only a few studies have been conducted from samples obtained directly from areas of the intestine. In these cases, samples were obtained from patients who required intestinal surgery for other reasons. For some animal species, like cows, a surgical plug, called a "cannula," is inserted directly through the skin and through the digestive tract wall to form a conduit for collecting samples directly from a specific area of the digestive tract (**Figure 17.21**). In humans and most other vertebrates, the most convenient and therefore most frequently used way to collect intestinal microbiota is from feces. Most of the microorganisms represented in feces are of course from the colon. Fecal microbiota may not, however, be completely representative of the colon. Feces probably contains some organisms carried from the small intestine and is likely predominated by lumen species, with species in mucus less represented. Although these studies may not be ideal and are subject to some interpretation, they give some insight into the diversity and variation of the human microbial inhabitants of the gut.

Analyses of rRNA genes show the gut microbiota is dominated by bacteria. Only three or four species of archaeons, all methanogens, have been found along with a small number of eukaryotic species, mostly yeast. Little disagreement is reported among studies on the major groups and their proportions, but the specific genera and their numbers vary widely. As shown in Figure 17.19, in young healthy adults the most abundant bacteria found in the colon, and in feces, are anaerobes belonging to phylum Firmicutes, from class Clostridia, followed by phylum Bacteroidetes, predominated by members of the family Bacteroidaceae and Prevotellaceae. Following this are facultative anaerobes of phylum Proteobacteria, of which family Enterobacteriacea dominate,

Figure 17.21. Cow with surgically installed permanent cannula
A cannula allows samples to be directly collected from the intestinal tract.

including *E. coli* and relatives, and phylum Actinobacteria, including members of the genus *Bifidobacterium*. Lactobacillaceae and Streptococcaceae are other members of phylum Firmicutes but show much lower numbers.

What do symbionts do in the human digestive tract? Obviously a great deal of metabolism is occurring (**Table 17.2**). The majority of gut microbes are metabolically diverse, utilizing and producing a wide variety of molecules, making it difficult to decipher the exact relationship or dependence. Metabolites produced from some microbes are utilized by others. For example, members of the genus *Clostridium* produce lactate as an end product of fermentation, and this can be used by Enterobacteriaceae. You may recall from the Chapter 4 Mini-Paper that *Methanobrevibacter smithii*, the only dominant archaeon found in the human intestine, uses formate, H_2, and CO_2 (all end products of fermentations from bacteria) to produce methane. The methanogens increase fermentation efficiency by removing the inhibiting end product H_2.

Currently, we know very little about the dynamics of the consortium of intestinal microorganisms, how it is controlled, and how it relates to intestinal health. Modern molecular techniques are just beginning to provide some clues. Differences in the microbial populations of healthy persons versus patients with ulcerative colitis, inflammatory bowel disease, Crohn's disease, and colon cancer have been noted. It remains to be established if these differences in microbial populations are the result or the cause of perturbations in the microbial populations. The Mini-Paper presents one surprising way in which the balance of intestinal bacteria may affect our health.

TABLE 17.2 Major groups and metabolic activities of intestinal microorganisms of humans

Major group	Characteristics	Metabolic activities
Phylum Firmicutes Class Clostridia	Gram-positive anaerobic spore-forming bacilli	Fermentation of starch, glucose to butyrate and acetate
Phylum Bacteroidetes Family Bacteroidaceae, Family Prevotellaceae	Gram-positive non-spore-forming rods or cocci	Fermentation of plant-derived carbohydrates
Phylum Proteobacteria Family Enterobacteriaceae	Gram-negative non-spore-forming bacilli	Fermentation of various sugars
Phylum Actinobacteria	Gram-positive non-spore-forming rods or cocci	Fermentation of starch to lactate
Domain Archaea *Methanobrevibacter smithii*	Only archaeon consistently present in the gut	Methane production from CO_2 and H_2

R. E. Ley, P. J. Turnbaugh, S. Klein, and J. I. Gordon. 2006. Human gut microbes associated with obesity. Nature 444:1022–1023.

Context

Obesity rates in humans are on the rise. Individual genetic makeup is part of the scenario, but the recent increase in obesity rates cannot be attributed to this alone since genetic changes in human populations would not happen in the relatively short time that we are seeing this increase. Most experts agree that modern lifestyles foster inadequate activity and consumption of calorie-rich diets, and these are major contributors to the increased rates of obesity. Now it seems there is another facet to putting on weight—microbes may directly contribute to obesity. To address this and other questions, microbiologists started applying the DNA sequence-based techniques of microbial ecology to analyze the microbes that inhabit the gastrointestinal tract of humans and model animals such as mice.

Studies in mice had shown that adding normal intestinal microbiota to adult germfree animals results in an increase in body fat, without an increase in dietary intake. Based on these results, researchers hypothesized that microbial metabolism in the gut may influence the amount of energy that is extracted from food. To address this, they compared populations of intestinal microbes of obese and lean animals fed the same diet. When the 16S rRNA sequence data of the two groups were compared, they found striking differences. The intestinal microbiota of obese animals had a much lower population of members of the bacterial phylum Bacteroidetes, a higher proportion of the bacterial phylum Firmicutes, and increased numbers of methanogenic archaeons. Lean animals had just the opposite. The researchers also found increased concentrations of major microbial fermentation products in the obese animals, and these could be readily used as nutrients by the host. Supporting this, feces from the obese animals contained fewer remaining calories compared to feces from lean animals.

Examination of gene fragments that were mass-sequenced from lean and obese animals showed the obese intestinal microbiome was enriched for genes coding for enzymes that could break down otherwise indigestible food material, such as complex starches and dietary fiber. When germfree animals were transplanted with microbiota from either obese or lean individuals, they found after two weeks those with "obese" microbiota had significantly more body fat than those that received "lean" microbiota, even though all received the same amount of food. This strongly suggests "obese" microbial populations contribute to weight gain. These are interesting findings, but do humans follow similar patterns?

Experiment

Researchers examined the gut microbial community profiles of obese humans by 16S rRNA sequence analysis of fecal samples. The results showed that, incredibly, about 70 percent of the sequences were unique to each person at the operational taxonomic unit (OTU) level. However, there was a clearly increased ratio of Firmicutes to Bacteroidetes in obese compared to lean individuals. Placing twelve obese people on a calorie-reduced diet over the course of a year

produced a reduction in Firmicutes populations and an increase in Bacteroidetes, approaching a ratio more like that of the lean people **(Figure B17.2)**. Even so, the individual communities maintained many of their individual characteristics over time, so they could be recognized as coming from the individual human. It was not possible to make a correlation of weight loss to specific microbes, but there was certainly a correlation with the ratio of Firmicutes to Bacteroidetes. The obvious question: Is the composition of the microbial population a cause or an effect of obesity?

Impact

While the composition of microbes in the human intestinal tract appears to be a contributing factor to obesity, it has not been resolved whether it *causes* obesity. Undoubtedly, this will remain an active area of research for some time as many questions are still unanswered. How does changing diet cause shifts in the symbiont population? Are certain microbial profiles more beneficial or detrimental than others in regard to human health? If so, how can these profiles be manipulated and maintained to support health? Could exchanging the microbial population of obese people with that of lean people through a fecal transplant, or a microbe-laced cocktail, help with weight reduction or even cure intestinal or other diseases?

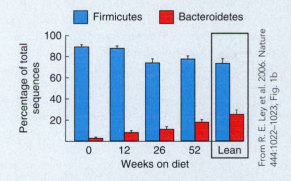

Figure B17.2. Effect of dieting on the relative abundance of Firmicutes and Bacteroidetes The profile of the two major bacterial intestinal phyla in obese dieting individuals becomes more like that of lean controls (boxed) over time.

● Questions for Discussion ···················

1. The use of antibiotics to clear infections can have drastic effects on microbial communities. Antibiotics are commonly fed to livestock because their use has been associated with increased weight gain. Is there a possible link between human obesity rates and the use of antibiotics?

2. If the Firmicutes are associated with obesity, what types of metabolic characteristics can you think of that might be responsible for this?

Besides nutritional benefits, what other roles do the symbionts of our intestine provide? Not surprisingly, germfree mice must ingest almost twice as many calories to maintain their body weight, and frequently need vitamin K_2 and vitamin B supplements, compared to normal mice. However, this is not just due to a reduction in colonic fermentation products by microbes. Compared to their normal counterparts, germfree mice show marked impairment of nutrient absorption through the intestinal wall. Other functional and morphological abnormalities of germfree mice include a much thinner intestinal wall and reduced mucus secretion. In addition, repair of intestinal injury is markedly impeded. Reconstitution of germfree mice with symbiotic bacteria obtained from conventionally raised mice results in profound improvement of digestive function and morphology.

Intestinal microorganisms are also important for preventing intestinal infections. With such a high density of colonizers in the colon, competition for nutrients and space is intense—most newcomers cannot easily become established. This is known as **competitive exclusion** and is based on the principle that different species requiring the same resources cannot coexist. Many intestinal pathogens, like those of genera *Salmonella*, *Shigella*, *Vibrio*, and pathogenic strains of *E. coli*, colonize the less crowded small intestine by attaching to epithelial cells. Symbiotic intestinal microbiota also produce a variety of antimicrobial compounds such as bacteriocins, which are small peptides that have toxic activity on related microorganisms (see Section 16.2). For example, some strains of *E. coli* can produce bacteriocins against competing strains of *E. coli*, including pathogenic strains, reducing their ability to colonize.

You may be surprised to know that microbial symbionts are also important for the correct functioning of the immune response. This is another way colonizing microbes help protect against infection and disease. Germfree animals are more prone to infections, immune-mediated diseases, and allergies—all evidence of a dysfunctional immune response. Transplantation of these animals with microbiota from their normal counterparts improves immune function. Such studies have led to the "hygiene hypothesis," which presumes that the lack of exposure to a wide spectrum of microbes including those causing common childhood infections results in an increased risk of allergies and autoimmune diseases, both of which are increasing in incidence, particularly in the developed world.

Why would intestinal microbes be needed for proper immune function? An overview of what many scientists believe happens between the immune system and resident microorganisms may help to answer this. As mentioned earlier, symbiotic microbes of the intestine tend to colonize only the edge of the mucus layer and lumen. The basal layer of mucus is much too viscous for them to effectively penetrate. However, small numbers of these inhabitants and their components do routinely get through the mucus layer to contact the epithelium. Many of these intestinal inhabitants are closely related to pathogens. For example, there are commensal strains of *E. coli* and pathogenic strains of *E. coli*. Commensal species and strains do not have the genetic capacity to aggressively invade and damage intact intestinal tissues or to overcome normal host immune defenses, as can intestinal pathogens (see Chapter 21). Commensal organisms frequently possess surface molecules that are similar to those of related pathogens, and these can be recognized by the immune system. When these molecules come in contact with immune cells, they can stimulate an immune response that can also protect against related pathogenic strains. This important routine contact stimulates immune responses on a continuous basis, keeping it "primed" and ready for rapid defense. When a pathogenic microorganism does invade, immune defenses can quickly mobilize for attack. In this way, the rather benign normal microbiota provide practice for building immune defenses. Chapters 19 and 20 will describe the complex mechanisms of immune stimulation.

Contact with symbionts is also essential for training the immune system to recognize and tolerate these normal microbial inhabitants. Many allergies and autoimmune diseases have been linked to an imbalance or deficit in normal microbiota, preventing the immune system from distinguishing normal ever-present molecules from potentially harmful ones. Subsequently inappropriate and often aggressive immune responses occur to an otherwise insignificant stimulus. This overzealous response to a threat that does not really exist damages the host.

Symbionts of herbivore digestive systems

Herbivorous vertebrates, like cattle, zebras, geese, and rabbits, depend on the consumption of leafy plants for their energy needs. However, like humans, they do not possess the genes that produce the cellulase enzymes necessary to break down cellulose, the major material of plants. Cellulose, like starch, is composed of long chains of glucose molecules. However, starch consists of glucose units joined by α-1,4-glycosidic linkages that are easily hydrolyzed by the enzyme amylase produced by many animals, while cellulose is composed of β-1,4-glycosidic linkages that requires a different enzyme, cellulase, for hydrolysis **(Figure 17.22)**. Without cellulase, the glucose in cellulose cannot be released for use by the animal. This is not of concern for omnivores (like humans) or carnivores, because our nutrition does not rely on the conversion of cellulose into usable material. Herbivores, however, rely almost completely on this conversion, and they depend on their symbiotic microorganisms to do this. Consequently, their digestive systems have evolved to cultivate symbiotic microorganisms on a grand scale. Functionally, herbivores are fermentation vats with legs. Their bodies are dominated by a digestive tract containing a large fermentation chamber needed to accommodate their symbiotic microbes and the large amounts of forage they must ingest to thrive **(Figure 17.23)**.

The major fermentation chamber of omnivores and carnivores—the colon—is relatively small in comparison. While colonic fermentation works well for diets that are primarily digestible by the host alone, it is not sufficient for herbivores. To extract nutrients, primarily volatile fatty acids, microbial fermentation is relied upon. The two basic templates for accommodating fermentation in herbivores are **cecal fermentation** and **rumen fermentation**. Recall, the cecum is a dead-end pouch at the junction of the ileum and colon (see Figure 17.19

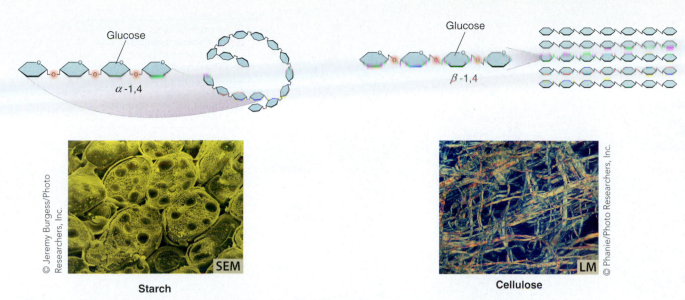

Glucose

α-1,4

Glucose

β-1,4

Starch
SEM
© Jeremy Burgess/Photo Researchers, Inc.

Cellulose
LM
© Phanie/Photo Researchers, Inc.

Figure 17.22. Structure of cellulose and starch Both cellulose and starch are carbohydrates made by plants, composed of units of glucose. Starch is an energy storage polymer with glucose units joined by α-1,4-glycosidic linkages, which are easily hydrolyzed by the enzyme amylase produced by animals. Cellulose is an extracellular structural polymer with glucose units linked by β-1,4-glycosidic linkages. The tight packing of linear, not helical, molecules contributes to their strength but impedes their enzymatic digestion. With a few exceptions, animals do not possess the enzyme cellulase to hydrolyze this bond.

and part A of **Figure 17.24**). In cecal fermentation systems, the cecum is the major site of fermentation (see Figure 17.24B). The **rumen** (see Figure 17.24C) is a major digestive organ found in a group of herbivores, called **ruminants**. In rumen fermentation, the primary site of fermentation is the rumen. Some examples of vertebrates and the different fermentation systems they use are shown in **Table 17.3**.

Cecal Fermentation

A cecal fermentation system is really just an elaboration of the colonic fermentation system (see Figure 17.24). To accommodate cecal fermentation, the cecum is greatly enlarged to become a fermentation vessel for extracting nutrition from plant material. Both the cecal and colonic fermentation systems

are referred to as "hindgut" fermentation systems. Hindgut fermentation relies on host digestive processes in the stomach and small intestine first, and microbial fermentation processes last. The fermentation products of the cecum and colon that are not utilized by the microbes themselves are then available to be absorbed by the host, along with microbial products such as vitamins. The rest is expelled in feces. Approximately 20 to 30 percent of the caloric requirement of various cecal fermenters may be provided by absorption of volatile fatty acids, predominantly acetate, propionate and butyrate, released through microbial fermentation processes.

Herbivorous cecal fermenters need to consume large amounts of food to survive because relatively little of the total mass consumed is converted into usable products for the host. Some cecal fermenters, like rabbits and beavers, compensate somewhat for this inefficiency by practicing *coprophagia*, or eating of feces. You may not think of feces as a valuable food source, but feces from hindgut producers is a concentrated package of

Figure 17.23. Idealized cow showing a body dominated by a digestive system "The Durham Ox" by John Boultbee, 1802. Desirable features of livestock were commonly exaggerated in eighteenth- and nineteenth-century paintings. For cattle, a large capacity for digestion was a characteristic of good breeding.

Historical Picture Archive/© Corbis

TABLE 17.3 Fermentation systems used by vertebrates

Colonic fermentation	Cecal fermentation	Rumen fermentation
Humans	Rabbits	Deer
Dogs	Koalas	Antelope
Bears	Guinea pigs	Cattle
Cats	Mice	Buffalo
Carnivorous fish, such as sharks	Herbivorous fish, such as angel fish	Camels
Carnivorous birds, such as eagles	Herbivorous birds, such as geese	Sheep

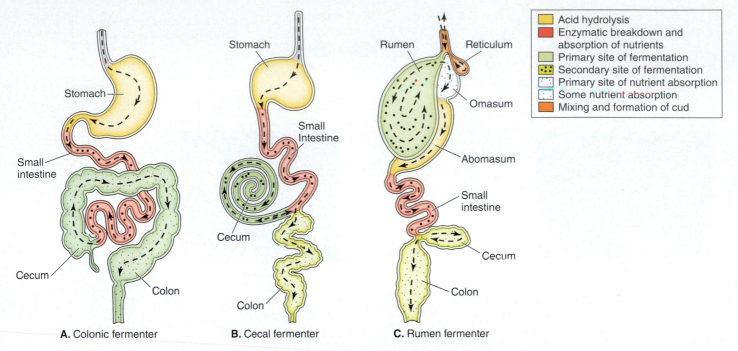

Acid hydrolysis

Enzymatic breakdown and absorption of nutrients

Primary site of fermentation

Secondary site of fermentation

Primary site of nutrient absorption

Some nutrient absorption

Mixing and formation of cud

A. Colonic fermenter

B. Cecal fermenter

C. Rumen fermenter

Figure 17.24. Vertebrate digestive systems for digestion of different diets Different digestive systems are desig-
nated for colonic fermentation, cecal fermentation, and rumen fermentation. **A.** Colonic fermenters carry out most microbial fermentation in the colon. Food
enters the stomach and undergoes hydrolysis by acid, then is passed to the small intestine where it is digested by host enzymes and absorbed. Undigested
material is passed to the colon where it is fermented by microbes. In colonic fermenters the cecum is rudimentary. **B.** Cecal fermenters carry out most fermentation
in the cecum. Digested and undigested food is passed to the cecum where it undergoes extensive fermentation. Any remaining undigested material is passed to
the colon where further fermentation may take place. **C.** Rumen fermenters (ruminants) carry out most fermentation in the rumen. Before reaching the rumen,
large pieces of food sink to the reticulum where they are mixed into cud, regurgitated and chewed, then swallowed to enter the rumen. In the rumen, the pulver-
ized food fragments are fermented. Fermented, liquified material and microbes flow to the omasum and on to the abomasum, an acid stomach. Here, microbes
from the rumen are digested along with fermented material. The rest of the ruminant digestive process is similar to that of cecal fermenters. Note the cecum is
shorter and is not a major fermentation chamber in ruminants.

potential nutrients, being full of dead and live microorganisms
and undigested and partially digested food particles. As unpalatable as it seems, coprophagia serves three important purposes
for these animals.

- Microbes in feces are a source of nutrients, especially nitrogenous compounds like amino acids, which may be
lacking in a plant-based diet.
- Ingestion by young herbivores of their mother's or herd
mates' feces quickly provides a mature population of intestinal symbiotic microbiota, allowing them to utilize a diet
of plants.

A. Rabbit "first" feces

B. Rabbit "second" feces

Figure 17.25. Nutrient extraction by coprophagia **A.** Rabbit feces passed
through the digestive system the first time is soft and moist and contains incompletely fermented materials,
including microbes, which are source of nitrogen compounds. This feces is re-ingested for complete fermentation and further nutrient and water absorption. **B.** Feces that is passed the second time is hard and dry
and contains little usable material.

- Feces passed the first time is rich in partially digested food
and water, so consuming it a second time provides another
round of fermentation and extra opportunity for water and
nutrient extraction. This "second" feces is expelled as a
hard, dry pellet (**Figure 17.25**).

Rumen Fermentation

Rumen fermentation provides a more efficient way for herbivores
to extract nutrients from plant material. Up to 70 percent of the
caloric requirements of ruminants is provided by absorption of
volatile fatty acids, major products of microbial fermentation.
Unlike cecal and colonic fermenters, which use hindgut fermentation, ruminants use "foregut" fermentation, fermenting their
food first using microorganisms. Any undigested, unabsorbed material is then passed through the stomach, the small intestine,
cecum, and colon, which function in the conventional manner.
This arrangement allows ruminants to benefit nutritionally from
their symbionts in two ways:

1. Ruminants receive fermentation and other products produced by the microbes, such as vitamins.
2. They digest rumen microbes as a proportion of them pass
into the stomach and intestine along with the fermented
slurry. Hindgut fermenters cannot digest their own intestinal microbes and so miss this potential source of nutrition.

Ruminants have a number of other interesting digestive processes to increase fermentation efficiency. Ruminants do not chew

their food as it is ingested. They tear it off and swallow, which allows them to feed quickly and move on. The large, heavy food first sinks to an open chamber called the "reticulum," and the lighter, smaller material goes to the rumen, where major fermentation processes occur (see Figure 17.24C). Particulate material in the rumen forms a heavy slurry that eventually gravitates to the reticulum to mix with the larger food material there forming "cud." Cud from the reticulum is regurgitated then chewed into a paste and swallowed again (you may think this is another pretty unpalatable way to eat). Reducing the plant material to small pieces increases the surface area available for microbial attack, and thus maximizes fermentation product yields. Consequently, ruminants spend significant amounts of time (up to 40 percent of their day) chewing their cud, getting it just right.

Well-fermented, liquified material and microbes flow to the omasum. Here, water and electrolytes are absorbed. It is important to note that rumen material that is not thoroughly broken down cannot flow into the omasum, and remains in the rumen where it is further fermented. From the omasum, product moves into the abomasum, which is a true stomach. The abomasum secretes acid and also lysozyme that breaks down bacterial cell walls (see Figure 2.18), facilitating harvest of the nutrition contained in the bacteria. The rest of the digestive process is a similar story to that of cecal fermenters. In the small intestine, enzymes degrade substrates, including material of microbial cells, and further nutrient absorption occurs. A smaller amount of fermentation happens again in the cecum and colon, but contributes less than 15 percent of the total digestion. Overall, it's a very efficient system for capturing the energy locked in plant material, with up to 80 percent of total plant carbohydrates being converted into usable energy. Approximately 40 percent of the energy comes from cellulose alone. Compare that to humans eating a vegetarian diet who extract virtually no energy from cellulose.

What are the symbionts of the ruminant digestive tract? One of the best-studied ruminant systems is that of the cow, although studies in sheep, goats, and deer reveal similar community profiles. Rumen microbial communities, in order of proportion, are: bacteria (predominately Firmicutes and Bacteroidetes), protozoa (mostly ciliates), fungi (mostly yeasts), and archaeons (methanogens). Except for some yeasts, all are anaerobes. Yeasts scavenge O_2 that enters the rumen and maintain the anaerobic conditions needed for

efficient fermentation. Although Firmicutes and Bacteroidetes are the predominant phyla in the rumen, as they are in the human colon, the species profile is not the same. Rumen bacteria are predominated by cellulolytic species and others that degrade structural carbohydrates of plants. Although some of these species have been found in some humans, they are a very small minority of the bacterial population. Some species, such as *E. coli*, are ubiquitous, existing in the digestive tracts of all vertebrates. Methanogenic archaeons, including *Methanobrevibacter smithii* (see Microbes in Focus 4.4), are also found in the rumen. However, *M. smithii*, the dominant archaeon in humans, is not a dominant archaeon in the ruminants. The different microbial communities reflect, by way of their metabolic capabilities, the vastly different diets of these two host species. Thus, humans and cows have different gastrointestinal ecologies.

What do symbionts do in the ruminant digestive tract? The biochemical reactions that occur in the rumen are complex and involve a wide variety of microorganisms. As we will see later in this section, the community profile can change with diet. Modern farming focuses on rapid growth, which means domestic ruminants are often fed starch-rich grains and protein-rich legumes. The natural diet of ruminants, however, is cellulose-based grasses and leaves. **Figure 17.26** shows an overview of cellulose conversion in the ruminant.

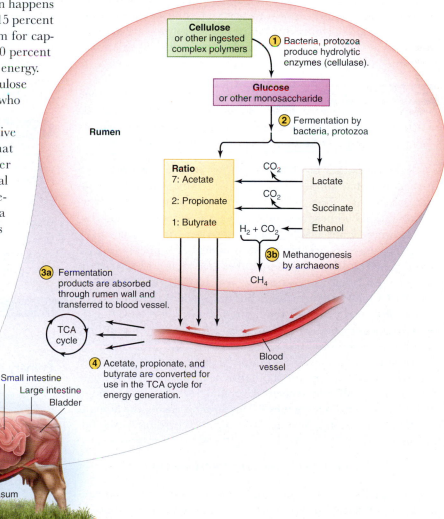

Figure 17.26. Process Diagram: Overview of the major contributions of rumen symbionts in cellulose conversion Cellulose, or similar polymers, are degraded into glucose by rumen microbes. Glucose is taken up by microbes and fermented. The end products of some microbes are converted by others. For example, lactate is converted to acetate and CO_2; succinate is converted to propionate and CO_2; ethanol is converted to H_2 and CO_2. The major waste products include the volatile fatty acids: acetate, propionate, and butyrate. These are absorbed through the rumen wall, diffuse into the bloodstream, and are delivered to host cells where they enter the TCA cycle as an energy source for generating ATP. The gases CO_2 and H_2, also fermentation waste products, are converted to CH_4 by methanogenic archaeons, and expelled by the host.

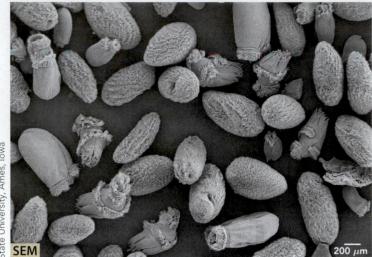

SEM   200 μm

Figure 17.27. Predatory protozoa of the rumen Micrograph of protozoa from a cow rumen. Many protozoa feed on the bacterial inhabitants of the rumen, maintaining homeostasis.

A variety of bacteria and protozoa hydrolyze cellulose to glucose, and then ferment this to organic acids. The major organic acids that are the end-products of rumen fermentation are the volatile fatty acids acetate, propionate, and butyrate (see Figure 13.19). These are absorbed directly through the rumen into the bloodstream to become the primary energy source for the ruminant. Bacteria appear to be responsible for the majority of fermentation in ruminants. Many of the protozoa are actually predators of the bacteria, maintaining the ecological balance in the rumen (**Figure 17.27**).

CONNECTIONS In eukaryal cells, acetate, propionate, and butyrate are converted to acetyl-CoA and oxidized through the TCA cycle in mitochondria, ultimately releasing ATP for the animal's metabolic needs. See **Section 13.3** for a review of the TCA cycle.

Numerous other biochemical pathways form a complex web of reactions that support the ecology of rumen microbial populations. Many rumen microbes produce other compounds as fermentation end products such as lactate, succinate, and ethanol. However, appreciable amounts of these are not normally found in rumen contents. These products are quickly further metabolized by other rumen microbes (see Figure 17.27). Methanogens can reduce excess acetic acid to CH_4 (methane) and convert CO_2 and H_2 into CH_4. You may recall from our investigation of fermentation in the human colon, the removal of H_2 by methanogens helps fermentation efficiency by keeping this rate-limiting end product in low concentration. This results in the production of more acetate, propionate, and butyrate, and therefore more energy for the herbivore. Some of the gut bacteria can anaerobically convert CO_2 and H_2 into acetic acid, a process known as "acetogenesis," further increasing the amount available for uptake. Acetogenesis occurs according to the following reaction:

$$2\,CO_2 + 4\,H_2 \longrightarrow CH_3COOH + 2\,H_2O$$

An equally valuable contribution of the rumen microorganisms is their ability to convert the nitrogen compounds urea and ammonia into amino acids (see Figure 13.43). Dietary sources of urea and ammonia come from protein and other nitrogenous components of digested plant material, but are also produced endogenously by the ruminant as metabolic waste products from the breakdown of amino acids (**Figure 17.28**). Urea and ammonia diffuse into the bloodstream and are transported into the rumen where the microbes recycle them into amino acids (see Figure 13.46). Cellulolytic bacteria are extremely efficient at this nitrogen conversion. It is estimated that 50 to 80 percent of microbial protein in the rumen is produced using urea and ammonia. Ruminants benefit from this microbial activity in two major ways:

1. Supplying cellulolytic microbes with these nitrogen compounds facilitates their growth, increasing the digestion of cellulose and enhancing fermentation. This in turn, increases the harvest of fermentation products by the host.
2. Digesting the rumen microbes provides a source of protein for the ruminant. In turn, this allows ruminants to exist on poor quality food such as grass, which is low in protein and amino acids.

Most of the amino acids harvested by the host from digestion of rumen microbes are incorporated into muscle protein, an important consideration for the commercial production of food animals. Recapturing amino acids from microbes gives ruminants an advantage when feeding on protein-poor diets in comparison to cecal and colonic fermenters, who pass most of their microbes in an undigested form in feces. Consequently, cecal and colonic fermenters must either eat more or be fed higher quality and more expensive food. Pound for pound, it is more expensive to feed a pig (colonic fermenter) or a horse (cecal fermenter) than a cow or sheep (both ruminants), all other conditions being equal.

CONNECTIONS Recall from **Section 17.2**, legumes like clover and alfalfa have mutualistic relationships with bacteria that can fix N_2 and convert it into amino acids for plant use. Therefore, these plants are highly nutritious, being rich in protein and amino acids.

Non-cellulose carbohydrates, such as starch and pectin (composed of units of galacturonic acid), are also present in plants and these are also fermented by various microbes. Not surprisingly, changing the diet of ruminants can alter the specific proportions of microbes in the rumen. **Table 17.4** shows

TABLE 17.4 Diet and changes in microbial populations in the rumen of cows

Primary diet	Primary catalytic process	Dominating bacteria
Grass	Cellulose degradation to glucose	*Fibrobacter succinogenes* *Prevotella ruminicola* *Ruminococcus albus*
Alfalfa/clover	Pectin degradation to galacturonic acid	*Lachnospira multiparus*
Corn/barley	Starch degradation to glucose	*Ruminobacter amylophilus* *Succinomonas amylolytica* *Streptococcus bovis*

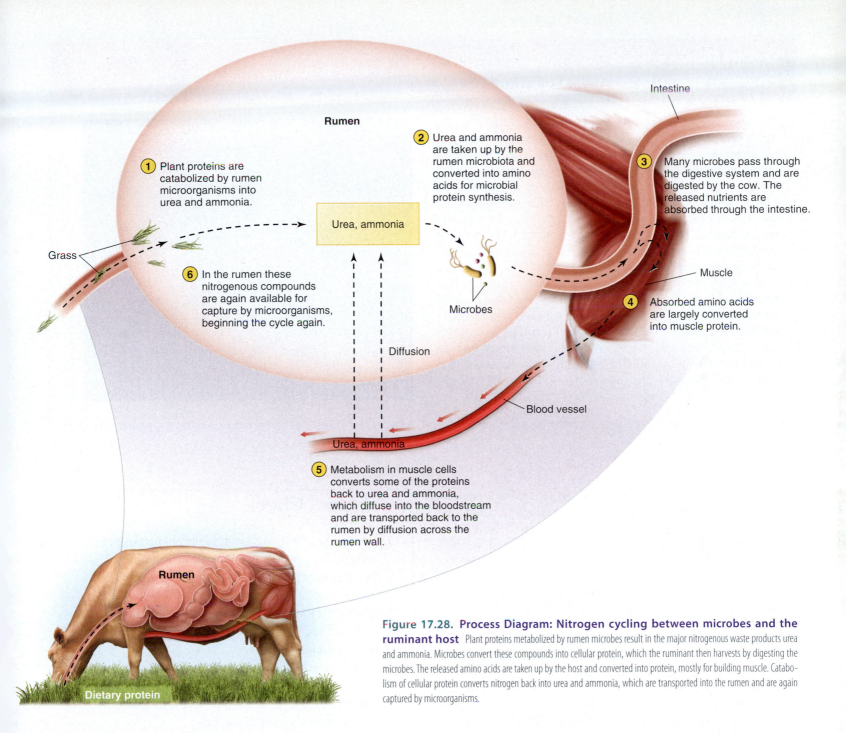

Rumen

① Plant proteins are catabolized by rumen microorganisms into urea and ammonia.

② Urea and ammonia are taken up by the rumen microbiota and converted into amino acids for microbial protein synthesis.

Intestine

③ Many microbes pass through the digestive system and are digested by the cow. The released nutrients are absorbed through the intestine.

Grass

⑥ In the rumen these nitrogenous compounds are again available for capture by microorganisms, beginning the cycle again.

Urea, ammonia

Microbes

Muscle

④ Absorbed amino acids are largely converted into muscle protein.

Diffusion

Blood vessel

Urea, ammonia

⑤ Metabolism in muscle cells converts some of the proteins back to urea and ammonia, which diffuse into the bloodstream and are transported back to the rumen by diffusion across the rumen wall.

Rumen

Dietary protein

Figure 17.28. Process Diagram: Nitrogen cycling between microbes and the ruminant host Plant proteins metabolized by rumen microbes result in the major nitrogenous waste products urea and ammonia. Microbes convert these compounds into cellular protein, which the ruminant then harvests by digesting the microbes. The released amino acids are taken up by the host and converted into protein, mostly for building muscle. Catabolism of cellular protein converts nitrogen back into urea and ammonia, which are transported into the rumen and are again captured by microorganisms.

some bacteria known to dominate the rumen contents of cows fed with different diets.

Sometimes there are serious consequences to these diet-related changes in the microbial community. A rapid switch to starch-rich grains from grass can cause a drastic drop of the rumen pH, a condition known as "acidosis." This is due to the rapid fermentation of easily digestible carbohydrates leading to a flood of organic acids. This has two negative impacts:

1. Increased acid is absorbed across the rumen wall into the bloodstream, exceeding the buffering capacity of blood and other body fluids, which can lead to death.

2. Released acids exceed the absorption capacity of the rumen wall, and the rumen begins to acidify below its normal pH of 6 to 7. This encourages the growth of acid-tolerant

lactic acid bacteria such as *Streptococcus bovis*. Under these conditions, *S. bovis* grows quickly and produces large quantities of lactic acid, further acidifying the rumen. When the pH drops to 5.5, a large proportion of rumen microbes are killed. Rumen food cannot be further processed and does not move out of the rumen. The animal stops feeding and can weaken and die.

Avoiding acidosis can be accomplished by a gradual change to a grain-rich diet. This allows the rumen populations to reach a new and stable equilibrium that maintains normal levels of organic acids.

Microbial fermentation produces gaseous waste products and ruminants can generate startling quantities, consisting mostly of carbon dioxide (45 to 70 percent) and methane

In reading the previous section, you may have recognized that the major gases produced by cows—carbon dioxide and methane—are greenhouse gases. When present in the atmosphere, these gasses trap radiant energy, preventing its dissipation to space. Thus, cows pose a concern for climate change as methane is a potent greenhouse gas. Cow populations have increased greatly over the past century as a result of human demand for meat and dairy products. Currently, there are an estimated 1.3 billion cattle on the planet—a source of approximately 15 percent of the total methane released into the atmosphere. When you consider the biochemistry involved in methane production in cows, it is their microbes that present the problem, although the two can hardly be separated. Researchers have been trying to find ways to reduce methane production in cows. They have found that high quality feeds, like alfalfa pasture, results in significantly less methane than cattle kept on grass pasture. There is a cost in this, however. High protein diets are more costly to supply and increase the danger of bloat.

Courtesy Christine Duport

(20 to 30 percent). For cows, gas production is estimated to be between 200 and 400 liters *per day*. This gas must be expelled regularly through "eruction," which is similar to belching (another interesting mannerism of ruminants imparted by their microbial inhabitants). With such large quantities of gas being generated, anything that interferes with the expulsion of gas can be deadly to the ruminant. A condition called "bloat" occurs when gas cannot be expelled and the rumen quickly expands to an extreme size, restricting internal organs, resulting in respiratory collapse and death. In most cases, bloat is caused by improper diet and is usually brought on by a sudden switch from cellulose-based grass to fresh clover or alfalfa, foods rich in protein. It is thought that large quantities of protein that are rapidly released by microbial degradation act as foaming agents that mix with gas to create froth in the rumen slurry. This effectively traps the gas instead of allowing it to rise freely for release by eruction. In older times, it was common practice to relieve the pressure caused from bloat by stabbing the animal in the rumen. Surprisingly, many lived through this! Nowadays, anti-foaming agents are commonly administered. **Perspective 17.2** describes another consequence of rumen gas production.

All animals studied so far have microbial symbionts. In most cases, the host and the symbionts are inextricably connected. From a molecular and ecological perspective, we need to view animals, as well as many plants, as a conglomerate of species. Adopting this holistic view is certainly a step forward in understanding the checks and balances that affect host health and how disruptions in the host/microbe ecology contribute to disease.

17.4 Fact Check

1. How are germfree and gnotobiotic animals used to study symbiotic bacteria? Provide an example.
2. What types of microbes are found as symbionts on human skin?
3. Explain the environmental conditions that affect the numbers and types of symbionts presenting in the human vagina.
4. What connects *Streptococcus mutans*, sugar, and cavities?
5. Explain how the pH of the digestive tract affects the number and diversity of microbes.
6. What types of microbes are symbionts in the human digestive tract? Explain some of their roles.
7. What is the hygiene hypothesis? Explain how the relationship between the immune system and the microbial symbionts of the digestive system may support this hypothesis.
8. Explain the advantage of hindgut fermentation versus foregut fermentation for herbivores.
9. What are the major end products of fermentation in the digestive system of animals? What is their importance?
10. Explain why ruminants can exist on poorer quality food than cecal fermenters.

17.5 Symbionts of invertebrates

What are the roles of symbionts in invertebrates?

Symbiotic relationships are of widespread occurrence in invertebrates, and many invertebrates carry endosymbionts. The dependency between microbe and host varies from species to species, from a general but not required benefit, to obligately required for life and reproduction. The housing for these symbionts ranges from colonization in the gut lumen, to the formation of specialized cells and even organs whose exclusive function is cultivating endosymbionts in the host. You may recall that this was the case for deep sea giant tube worms, introduced in Chapter 13. For many of these invertebrates, the symbioses are inseparable, obligate relationships that have developed through evolution.

Commonly, the symbionts of invertebrates, like those of vertebrates, are not culturable. Therefore, much of what is known about their existence is often based largely on small subunit (SSU) rRNA gene sequencing. Such molecular technologies have given new insights into the world of microbes by allowing us to glimpse some of their genes, and through this, some of their biochemical activities. Estimates based on such molecular analyses show that 15 to 20 percent of all insect species have nutritional bacterial symbionts and often harbor multiple symbionts. Fewer species carry yeast or protozoa (Table 17.5). Common to almost all of these insects is an incomplete diet that lacks essential nutritional components such as amino acids, sterols, or vitamins. As you might guess, the symbiotic microbes produce these required or supplementary nutrients, which are needed for proper development and reproduction. For example, aphids extract plant phloem or xylem fluid (sap), which is rich in sugars but low in amino acids and other nitrogenous compounds required by the aphids. Tsetse flies feed exclusively on vertebrate blood, which lacks several vitamins needed for fly fertility. Termites eat cellulose from wood and other dead plant material, which they cannot efficiently digest themselves. They rely on protozoa in the gut to digest this material and convert it to carbon and nitrogen compounds that they can then use. In most of these partnerships the symbiont and the host cannot survive without each other. It is an obligate mutualistic relationship. These are some of the most complex and fascinating relationships known to exist between microbes and multicellular organisms. Comprehending the specific ways symbionts contribute to the health of these insect pests and those that spread disease also sets a foundation for the development of more creative and more effective control strategies.

How do scientists determine what nutrients are supplied by the symbionts? Most studies use insects that have been rendered symbiont-free, or aposymbiotic, by treatment with antibiotics. They are then maintained on chemically defined diets that mimic their natural diet, but have additional components that support insect growth and reproduction. Researchers then compare these diets to similarly defined diets required by symbiotic aphids. Nutrients needed by the aposymbiotic insects but not needed by the symbiotic insects are assumed to be supplied by the symbionts. Such dietary comparisons have shown aphids cannot manufacture any of the essential amino acids they need. Presumably, these are made by their symbionts because plant sap lacks amino acids. Interestingly, the performance on artificial diets of aposymbiotic aphids compared to symbiotic insects on similar diets is usually suboptimal, indicating the symbiotic benefit may be more than just nutritional.

In most insect studies, there is often little direct evidence to prove that the symbionts make these essential nutrients because most symbionts cannot be cultured *in vitro* for confirmatory metabolic studies. Genome sequence, when available, can be useful in confirming that the symbiont is at least capable of manufacturing the nutrient. Whether or not the nutrient is made available to insect tissues is another question. In the pea aphid, which is the best-studied insect symbiosis, this has been confirmed for the sulfur (S)-containing essential amino acid methionine. Supplying symbiotic aphids with ^{35}S allows tracking of molecules into which S is incorporated, and has shown the endosymbiont *Buchnera aphidicola* incorporates it into methionine, and releases this to aphid tissues (Figure 17.29).

A. Aphids feed on plant sap.

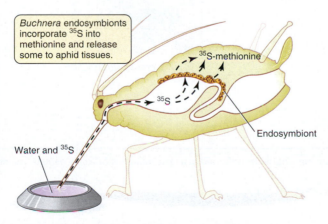

Buchnera endosymbionts incorporate ^{35}S into methionine and release some to aphid tissues.

^{35}S-methionine

^{35}S

Water and ^{35}S

Endosymbiont

B. ^{35}S is converted to methionine by endosymbionts.

Figure 17.29. Evidence for methionine production by the aphid endosymbiont *Buchnera* **A.** Aphids feed on plant sap, which is lacking in amino acids. Aphids cannot synthesize any of their essential amino acids, such as methionine. **B.** Supplying aphids with the radioactive element ^{35}S shows ^{35}S appears first in the amino acid methionine only within the bacterial endosymbionts. Later, ^{35}S-methionine is found in the surrounding aphid tissues, indicating that it is released from the bacteria to the aphid cells.

TABLE 17.5 Insects and their symbionts

Host	Endosymbiont	Host diet	Host	Endosymbiont	Host diet
Aphids Pea aphid *Acyrthosiphon pisum*	*Buchnera* species *Arsenophonus* species	Sap	**Termites (lower)** Eastern subterranean termite *Reticulitermes flavipes*	*Trichomonas* spp.[b] *Trichonympha* spp.[b] *Trichomitopsis* spp.[b] *Methanobrevibacter* spp.[b]	Wood
Carpenter ants *Camponotus* spp.[b]	*Candidatus* Blochmannia spp.[a]	Detritis, plant nectar	**Ticks** American wood tick *Dermacentor variabilis*	*Francisella* spp.[b]	Blood
Kissing bugs Kissing bug *Triatoma rubida*	*Rhodococcus rhodnii*	Blood	**Tsetse fly** *Glossina brevipalpis*	*Wigglesworthia glossinidia* *Sodalis glossinidius*	Blood
Lice (Human head and body lice) Human head louse *Pediculus humanus*	*Candidatus* Riesia pediculicola[a]	Blood	**Weevils** Boll weevil *Anthonomous grandis*	*Candidatus* Nardonella[a]	Flower buds
Plant lice (psyllids) Tomato psyllid *Paratrioza cockerelli*	*Candidatus* Carsonella[a] *Arsenophonus* species	Sap	**Wood cockroaches** *Cryptocercus punctulatus*	*Blattabacterium cuenoti*	Wood

[a] The designation *Candidatus* indicates this bacterium has not been confirmed to a taxa and is not officially named. The name(s) following *Candidatus* are the proposed names. See Perspective 15.1.

[b] The designation "spp." indicates a group of species within the genus.

Further studies show aphids harboring *Buchnera* produce methionine while aposymbiotic aphids do not. Together, these results provide strong evidence that methionine is made by *B. aphidicola* and passed to the insect. Tryptophan, leucine, and the vitamin riboflavin are other compounds provided by *B. aphidicola*. Up to 50 percent of these compounds are released to the host.

Studies that compare structure, biochemistry, and genetic information of related modern insect species and their symbionts can be used to trace the evolutionary origins of the symbiosis events. Based on these comparisons, scientists group endosymbionts into two major types. **Primary endosymbionts** have evolved along with their hosts over millennia, while **secondary endosymbionts** are more recently acquired. As we will see, the nature of the relationship between a host and its primary endosymbionts is quite different from the relationship between a host and its secondary endosymbionts.

Aphids

Aphids are one of the best-studied insect/microbe symbiotic systems, and we will use them as an example to understand endosymbionts.

Primary Endosymbionts

Primary endosymbionts show evidence of **co-speciation** with their hosts, whereby an ancestral host/microbe association subsequently diversified into a number of different but related host-symbiont species. This is known to have occurred for aphids and their endosymbionts. Almost all species of modern day aphids are associated with primary endosymbionts belonging to the genus *Buchnera*. The few aphid species that don't have a *Buchnera* species instead have a yeast primary endosymbiont. The aphid-*Buchnera* symbioses that we see today appear to have arisen from a single event approximately 200 million years ago. Development of symbiosis and diversification of both host and bacteria is thought to have evolved by the following scenario.

A free-living bacterium infected an aphid-like insect and both managed to survive and propagate. The bacterial species was perhaps somewhat more successful inside the host than outside. Those bacteria that were less damaging and more beneficial to the host, by providing extra nutrients, were more likely to be passed along from host to host. Hosts that were able to better maintain the bacteria in their cells were also more successful. The association increased the fitness of both species. To secure propagation of the intracellular microbe, the host cells became specialized to cultivate the bacteria. Compared to the outside world, the intracellular environment was relatively nutrient-rich and protected. Any bacterial genes used to make molecules the bacteria could readily get from the host became redundant. Mutational events that inactivated these genes were of little consequence. Over time, the bacteria lost many of the genes needed for independent living, becoming more and more dependent on the host. The host became dependent on the bacterium for essential nutrients, allowing it to feed on plants that it could not otherwise exploit as a food source. This evolution did not happen along a single path. Many different successful adaptations occurred at the genetic level in both the host and the symbiont, and these branched off to become unique pairings of species of aphids and species of *Buchnera*, which are all phylogenetically related to one another. This co-speciation is common to the development of all primary endosymbionts of insects.

All primary endosymbionts of insects have a number of common characteristics that are evidence of evolutionary events that occur during co-speciation.

- They have highly reduced genomes.
- They are found exclusively in specialized host cells, called **bacteriocytes**.
- They are maternally transmitted through generations from mother to offspring.
- They are required by the host for survival and/or fertility.

We can examine some of these characteristics in more detail.

All insect primary endosymbionts that have been sequenced have highly reduced genomes. For example, the average genome size for *Buchnera* species is 660,000 base pairs, about seven times smaller than its relative *E. coli* K12 at 4.6 million base pairs. *Buchnera* lacks the genes required for DNA repair and recombination, and genes for making the cell wall constituent LPS (see Chapter 2). The genes for biosyntheses of amino acids essential for the aphid hosts are present, but most others are missing. Only some of the genes for the TCA cycle are present. *Buchnera* takes up TCA cycle intermediates from the host instead of generating them by the TCA cycle, which commits it to being an obligate intracellular endosymbiont.

Interestingly, gene loss and dependence on the host for biosynthetic compounds is also a common feature of obligate intracellular bacterial pathogens. Examples of such parasitic endosymbionts include *Rickettsia*, the causative agent of arthropod-borne diseases such as typhus and spotted fevers; *Chlamydia trachomatis* and *Mycoplasma genitalium*, both sexually transmitted pathogens that can cause pelvic inflammatory disease and sterility; *Mycobacterium leprae*, the causative agent of leprosy; *Treponema pallidum*, causing syphilis; and *Borrelia burgdorferi*, the agent of Lyme disease (see Microbes in Focus 18.5). All of these organisms have highly reduced genomes and depend on host cells to provide them with certain molecules they can no longer manufacture themselves.

The bacterium *Candidatus* Carsonella ruddii, the primary endosymbiont of a psyllid insect (see Table 17.5), has the smallest genome of any cell known to date (**Microbes in Focus 17.3**). Astoundingly, DNA sequence analysis has shown it has lost the ability to carry out independent DNA replication, transcription, and translation. These are three of the most fundamental requirements for life. How can this microbe possibly exist? Researchers postulate that this microorganism likely relies on using the missing gene products from the host cell, in the same way that mitochondria rely on the products encoded in the cell chromosome for carrying out processes they similarly lack (see Section 1.2). If this is true, it means *Candidatus* Carsonella ruddii may be a transition form in the road to a new cellular organelle. It is not yet known what this endosymbiont does for the host cell.

CONNECTIONS Recall from **Perspective 15.1** that *Candidatus* designation is used for proposing species names for uncultured microorganisms for which DNA sequence information exists along with some information regarding the microbe's structure, metabolism, reproduction, and the environment in which the organism is naturally found.

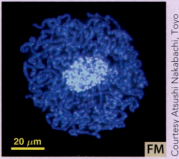
Primary endosymbionts are not found in a free-living form, being present only inside host bacteriocytes. Bacteriocytes appear to have no other function but to house the endosymbiont (**Figure 17.30**). In some aphid species, *Buchnera* in bacteriocytes constitute up to 10 percent of the aphid's mass. In aphids, the formation of bacteriocytes is genetically preprogrammed, and they are formed even in aposymbiotically maintained lines of aphids, indicating the profound influence this ancient relationship has had on shaping the genome, and consequently the development of the host.

If primary endosymbionts are not found free-living in the environment, how do insects acquire them? This cannot be left to chance occurrence since survival of the insect species requires possession of the primary endosymbionts. In most cases, the insects inherit them from their mother, a process termed **maternal transmission**. For aphids, which can give birth to live offspring, *Buchnera* migrate from bacteriocytes and infect the embryo as it develops inside the ovary (**Figure 17.31**). Other modes of maternal transmission involve the deposition of the microbial symbionts or infected bacteriocytes directly on or in eggs, causing the hatchlings to become infected as they develop. In comparison, males do not efficiently transmit the symbionts and so become a dead end for symbiont propagation. As primary endosymbionts are required for their hosts, they are found in both males and females. This is the case for aphids, tsetse flies, and ticks. Maternal transmission also occurs for some secondary endosymbionts.

Secondary Endosymbionts

Many insect species harbor one or sometimes two primary endosymbionts, plus one or several other secondary endosymbionts. It is possible that some secondary endosymbionts may eventually become primary endosymbionts. Secondary endosymbionts show much more diversity in terms of their association with the host.

- They are more recently acquired and do not show evidence of co-speciation with their host.
- They can sometimes be found in the environment as free-living forms.
- They are not restricted to bacteriocytes and can sometimes be found in a variety of cells.
- They may live extracellularly in the insect gut or the body cavity (hemocoel).
- They are not always present or essential.
- They are not always maternally transmitted.

Some secondary endosymbionts that rely on maternal transmission are found only in females, as they are not always required by their hosts. You may recall from the chapter opening story that *Wolbachia*, a secondary endosymbiont of many insect species, is capable of skewing sexual ratios to favor females in order to perpetuate its own transmission. Depending on the host species, *Wolbachia* selects for females by one of the following:

- "Male-killing," resulting in death of infected male embryos (as is the case for the blue moon butterfly)
- Cytoplasmic incompatibility, where infected males are not killed but are unable to produce viable embryos when mated with uninfected females or even females infected with a different strain of *Wolbachia* (as is the case for the mosquito *Culex pipiens*)
- Parthenogenesis (no males required), where unfertilized eggs develop into embryos (as happens with some mite and wasp species)
- Feminization, which, depending on the species, converts genetic males to fertile females or sterile pseudo-females (as happens in the Asian corn borer moth)

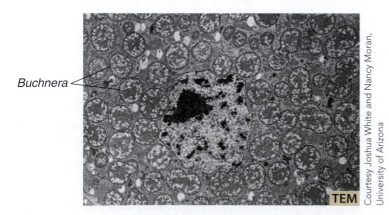

Buchnera

TEM

Courtesy Joshua White and Nancy Moran, University of Arizona

Figure 17.30. Specialized aphid bacteriocyte cells harboring *Buchnera* endosymbionts A section of a bacteriocyte cell showing *Buchnera* cells, each approximately 35 µm in diameter, packing the cytoplasm.

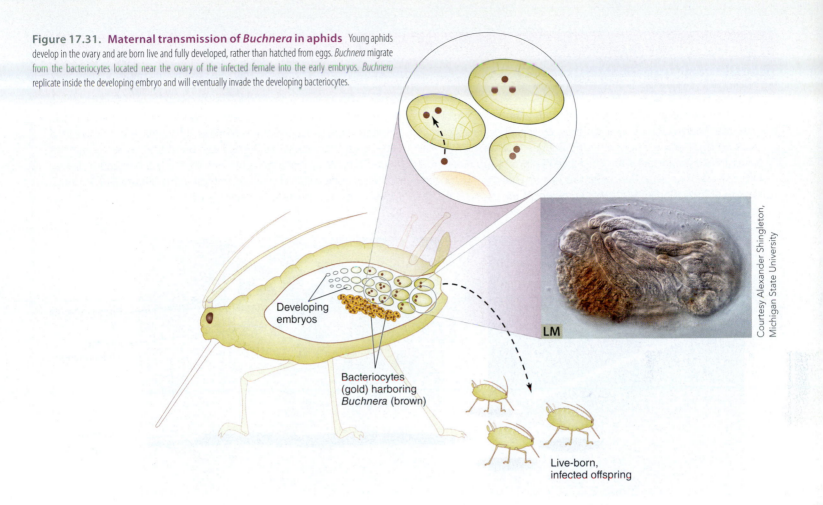

Figure 17.31. Maternal transmission of *Buchnera* in aphids Young aphids develop in the ovary and are born live and fully developed, rather than hatched from eggs. *Buchnera* migrate from the bacteriocytes located near the ovary of the infected female into the early embryos. *Buchnera* replicate inside the developing embryo and will eventually invade the developing bacteriocytes.

Developing embryos

Bacteriocytes (gold) harboring *Buchnera* (brown)

LM

Courtesy Alexander Shingleton, Michigan State University

Live-born, infected offspring

How *Wolbachia* actually carries out any of the reproductive manipulations is not yet known.

For some endosymbionts, being a mutualist or a parasite may be a matter of degrees. Where the *Wolbachia* symbiosis falls on the symbiosis continuum is a matter of speculation. In *Drosophila*, *Wolbachia* infection appears to diminish the fly's life span by 50 percent, suggesting it is parasitic. On the other hand, *Wolbachia* kills males, and reducing male populations may give a competitive advantage to the females for food and habitat, in turn increasing the success of *Wolbachia* and *Drosophila* as a species in a mutualistic manner. The European tick *Ixodes ricinus* is an ectoparasitic symbiont of several animal species and is an important vector species of human Lyme disease, caused by *Borrelia* species in western Europe. The tick harbors a bacterial endosymbiont, of proposed name *Candidatus* Midichloria mitochondrii, which lives not only inside the tick, but *inside* the mitochondria themselves (**Perspective 17.3**). The endosymbionts enter cells of the ovary and appear to take over many of the mitochondria in a parasitic manner, using the membrane-bound organelle as a nutrient source and brood chamber within which to multiply. Despite this takeover, they are not obviously damaging to ticks and may be commensals. *Candidatus* Midichloria mitochondrii are present in all female ticks and to ensure their transmission, they invade the eggs in the ovary and establish themselves in mitochondria of the developing ovarian tissue of new female offspring.

Termites

Termites fall into two evolutionary groups: the "higher" termites primarily feed on plant materials found in soil or on fungus, while the "lower" termites feed primarily on wood. Both groups of termites have a metabolic dependence on the microbes they harbor in their digestive systems. Despite being first observed in 1856, little is known about these microbes, largely because their populations are hugely diverse and the majority are unculturable. The symbionts found in lower termite guts consists of a consortium of protozoa, bacteria, and archaeons that live freely in the lumen of the termite gut, but most surprising is the finding that some bacteria reside *inside* many of the symbiotic protozoa—endosymbionts of endosymbionts! Higher termites do not possess protozoa or archaeons, but only bacteria. In this section, we'll look specifically at the lower termites.

Lower termite species are major insect pests, most notorious for eating houses. They selectively eat dead plant material, and lumber makes an attractive meal. As such, their diet consists primarily of cellulose, the nutrient-rich contents of the cells being long gone. An exclusive diet of cellulose lacks nitrogen, which is needed for the biosynthesis of compounds, such as amino acids, vitamins, and nucleotides. How then do termites survive on such a poor diet?

Over 400 different species of protozoa have been documented from various lower termite species, and are estimated to make up an incredible 30 to 50 percent of the weight of the termite (see Figure 3.24).

Even George Lucas, the creator of *Star Wars*, was impressed with the endosymbiotic relationships of microbes. His fictional midichlorians—those endosymbiotic entities that exist in all life-forms and communicate with "The Force"—played a prominent role in the *Star Wars* universe. Jedi knights possessed the highest concentrations of midichlorian symbionts in their cells. Mr. Lucas based the concept of midichlorians on the endosymbiotic theory that gave rise to mitochondria and chloroplasts of eukaryal cells (see Section 3.3). To honor the concept of midichlorians, a bacterial endosymbiont of the tick species *Ixodes ricinus* has been given the proposed name of *Candidatus* Midichloria mitochondrii. It's the first bacterium discovered to reside *within* animal cell mitochondria **(Figure B17.3)**, which was once an endosymbiont itself.

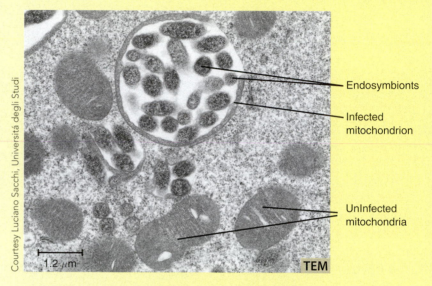

Figure B17.3. *Candidatus* **Midichloria mitochondrii** Endosymbionts can be seen within mitochondria. Intact mitochondria are also present in the host cell.

Symbiotic relationships between termites, their protozoa, and bacteria are quite different from those found in herbivores. The major fermenting microbes in the termite gut are protozoa, not bacteria as in herbivores. Wood particles are first chewed into small pieces by the termite. Termites are one of a very few animals capable of producing their own cellulase. Cellulase is secreted by the termite salivary glands and attaches to the ingested cellulose particles, and these move through the midgut and into the anaerobic hindgut where most of the microbial action occurs. **Figure 17.32** shows a summary of the major contributions of symbionts thought to occur in the termite.

Termites need nitrogen compounds for biosynthesis, and these are not found in their cellulose diet. Nitrogen-fixing bacteria can convert N_2 into nitrogenous compounds, such as ammonia for termite use. These bacteria utilize the acetate released from protozoan fermentation as an energy source. Nitrogen compounds are also contributed by fermentative bacteria that break down the termite waste product uric acid and convert it to ammonia, which can be released to the termite. While released ammonia can be directly assimilated by the termite, the symbionts can convert the ammonia into other nitrogenous compounds, such as amino acids for their own growth. To access this microbial nitrogen, termites routinely regurgitate and share their hindgut contents, upon request from their fellow termites, who then digest it in their foregut and midgut.

The hydrogen and CO_2 that are waste products of the protozoa are captured by methanogenic archaeons. They use hydrogen as an electron donor to convert CO_2 into methane gas. Recall that methanogens increase fermentation efficiency by removing the inhibiting end product H_2. This results in the production of more acetate, and therefore more energy for the termite. As in the rumen, some termite gut bacteria carry out anaerobic acetogenesis by converting CO_2 and H_2 into acetate. This further increases the amount of acetate available for uptake. Interestingly, the mix of methane, acetate, and other metabolic products of the termite digestive system can be detected by sniffer dogs. Well-trained dogs have a very high rate of success in locating termite infestations.

CONNECTIONS Acetate is a common end product of mixed-acid fermentation reactions, and can be further utilized by organisms capable of carrying out the TCA cycle and aerobic respiration. See **Sections 13.3** and **13.4** for a review of these metabolic pathways.

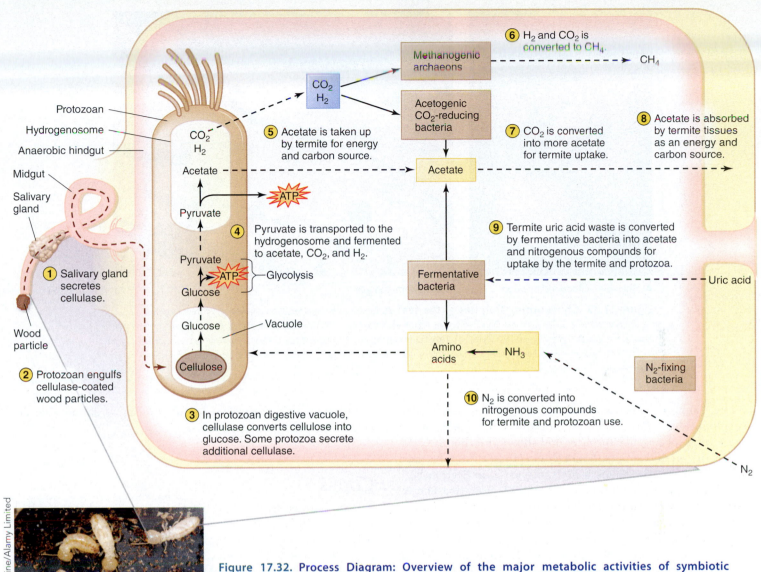

Figure 17.32. Process Diagram: Overview of the major metabolic activities of symbiotic microbes of lower termites Termite salivary glands produce cellulase that coats chewed cellulose particles. The particles move into the anaerobic hindgut (size exaggerated) and are engulfed by protozoa. This begins the protozoan conversion of cellulose to the products acetate, CO_2, and H_2. Acetate is an energy and carbon source for the termite. H_2 and CO_2 are converted by bacterial symbionts to more acetate, and to CH_4 by archaeons. Nitrogen compounds are synthesized by bacteria for termite and protozoan use.

The role of the intracellular bacterial inhabitants of the protozoa remains a matter of speculation. They do not appear to contribute to the cellulolytic activity in their protozoan hosts. Protozoa ridded of their bacterial endosymbionts by antibiotic treatment can still degrade cellulose. The intracellular location of the bacteria is also not a result of phagocytosis or infection. Analysis of 16S rRNA genes show these endosymbiotic bacteria are related to each other, but are not related to bacteria residing free in the gut. Their presence in the protozoa is likely therefore symbiotic.

Shipworms

Shipworms are termites of the sea. There are about 65 different marine species, and many are known for eating the timber hulls of ships and wood pilings of wharves. Shipworms are actually bivalve mollusks, related to clams. Their soft bodies look like worms, but they have a shell that they use to scrape shavings of wood for food, making burrowing holes that soon riddle the wood, much like that seen in houses with termites (**Figure 17.33**). Like termites, their diet consists of cellulose. Whereas termites rely on a multitude of microbes in their digestive tracts for their nutritional needs, shipworms have only a few species of bacteria, and surprisingly, they are not in the digestive tract. Instead, the bacteria are housed in the gills inside bacteriocytes that form a specialized gland-like organ from which the endosymbionts secrete cellulases. The duct of the gland runs into the digestive tract, where the cellulose is degraded into glucose for use by the shipworm. Shipworms also need nitrogenous compounds, and as you have likely guessed they get these from N_2-fixing endosymbionts. Again, this is similar to termites. However, in shipworms, a single microorganism, *Teredinibacter turnerae*, can secrete cellulase and fix nitrogen—a metabolic feat rarely found in nature.

Waste pellets

Siphon

© E. Eugenia Patten/California Academy of Sciences

A. Shipworm imbedded in wood

© Gary G. Gibson/Photo Researchers, Inc.

B. Shipworm tunnels

Figure 17.33. Shipworms: "Termites of the sea" **A.** *Bankia setacea* shipworms on a submerged timber. The main body of the shipworm mollusk is hidden inside the timber. Their siphons, through which they take water in and expel waste material (the pinkish pellets), are all that can be seen. **B.** Damage is caused by consumption of wood particles as the shipworms tunnel through the wood. Shipworms harbor bacteria in a specialized organ from which the endosymbionts secrete cellulases. One of these endosymbionts, *Teredinibacter turnerae*, can fix nitrogen as well as digest cellulose.

The shipworm endosymbiont *T. turnerae* has some commercial interest. First, cellulases produced by symbionts of wood-eating species, like termites and shipworms, may be valuable for the production of ethanol from cellulose crop wastes, a renewable alternative to fossil fuels, discussed in Section 12.3. Second, the rare combination of cellulose degradation and N_2 fixation in a single, culturable bacterium makes it a candidate for producing protein from waste cellulose without the need for nitrogen fertilizers. It may therefore be useful for producing a cheap protein-rich animal feed supplement. Third, *T. turnerae* produces a very stable protease that remains active at high temperatures (50°C for at least 60 minutes), tolerates a wide range of pH (4 to 12), is functional in high salt concentrations (up to 3M NaCl), and is stimulated by oxidizing agents, such as hydrogen peroxide. These characteristics would make this protease a useful addition to laundry detergents and other cleaning agents.

Corals

All corals possess photosynthetic algal endosymbionts, called **zooxanthellae**, which are mutualistic partners of the coral polyps. The majority of zooxanthellae are dinoflagellates of a single genus *Symbiodunium*. In the coral, they are contained within specialized intracellular vacuoles, called "symbiosomes" found in the cells surrounding the gut **(Figure 17.34)**. In many cases, the zooxanthellae make up the majority of the cellular mass of the coral, indicating it has a major role in coral biology. What is this role?

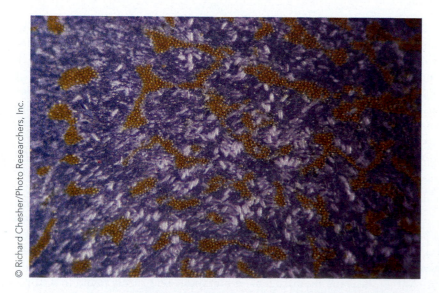

© Richard Chesher/Photo Researchers, Inc.

Figure 17.34. Corals: A symbiosis between photosynthetic algae and coral polyps Golden brown photosynthetic algal symbionts called "zooxanthellae" are contained within cells of the coral polyp.

Zooxanthellae are the pigment producers of corals, with different species giving distinctive colors to various types of coral. Death or expulsion of the zooxanthellae results in loss of color, termed "bleaching" **(Figure B17.4)**. Prolonged bleaching results in death of the coral, leaving only the intricate skeleton. Corals can regain their symbionts from the environment and recover, as long as bleaching has not been prolonged.

The exact physiological events that mediate bleaching are not precisely known. Laboratory studies indicate changes in pH, UV radiation, salinity, and temperature can cause the coral to expel and often digest their zooxanthellae, but this has not been definitively demonstrated in nature. Pathogen outbreaks have sometimes been the cause of bleaching events. While these factors fluctuate with natural events, such as hurricanes and deep water upwellings, most scientists agree that the rate and extent of recent bleaching events is not completely normal.

Prior to the 1980s, coral bleaching was a rare and isolated event. Since then, massive bleaching mortalities of corals and their reef communities have expanded to many areas. Most analyses point to climate change, largely due to human activities, as the main cause of these bleaching events. Climate change is considered the primary threat to the world's remaining reefs. In 1998, the increased ocean temperatures brought by el Niño caused significant bleaching of 60 percent of the world's reefs. Changes in the average reef water temperatures of just 1 or 2°C for several consecutive weeks is not tolerated, and the corals die.

© Michael Patrick O'Neill/Photo Researchers, Inc.

© Mark Conlin/Alamy Limited

Figure B17.4. Effects of coral bleaching The coral on the *left* has intact endosymbiotic zooxanthellae, and consequently the reef community is diverse with numerous animals. The coral on the *right* has undergone bleaching—the result of the death of the zooxanthellae. The coral dies, resulting in loss of productivity and leaving fewer animals of less diversity.

The lifeblood of the coral is the photosynthetic activity of the zooxanthellae, and this is the foundation for the spectacular diversity of vertebrate and invertebrate species found in coral reefs. The zooxanthellae supply the coral polyps with carbon compounds, primarily in the form of glycerol, through the fixation of CO_2 (see Section 13.6). Up to 95 percent of the glycerol is transferred to the coral tissues. Without the zooxanthellae, the coral will starve because most species cannot acquire enough carbon from their captured diet to meet their needs **(Perspective 17.4)**. The coral, in turn, supplies the zooxanthellae with concentrated CO_2 from cellular respiration, which is needed to maximize the photosynthetic biosynthesis of carbon compounds by the zooxanthellae. The coral, of course, also supplies the algae with a protected environment—the carbonate skeleton produced by most coral species is a formidable fortress, protecting the zooxanthellae from currents and predators. The mutualistic resource coupling between the coral and zooxanthellae supports the multitude of fish, invertebrates, and even other algal species that feed off the coral, and these in turn are prey for other inhabitants of the reef.

17.5 Fact Check

1. Differentiate between the terms "primary" and "secondary" endosymbionts.
2. Outline and explain how the aphid-*Buchnera* symbiotic relationship is believed to have developed.
3. Why do endosymbionts evolve to have highly reduced genomes?
4. What is maternal transmission? Explain why this process is necessary for primary endosymbionts.
5. Describe the role of symbiotic protozoa in lower termites. What metabolic products are provided by the protozoa? Explain.
6. Describe the location and role of symbiotic bacteria in shipworms.
7. What are zooxanthellae, and what role do they play as symbionts?

Image in Action

WileyPLUS This illustration shows three different instances of microbial symbiosis. The pea plant acts as a host to nitrogen-fixing bacteria known as rhizobia, which infect the plant root and eventually form elongated nodules. The pea plant is also host to a pea aphid (*right*). *Buchnera aphidicola* are bacterial endosymbionts typically found in specialized cells known as bacteriocytes in aphids. A male and female Blue Moon butterfly are shown at the top with the ovary of the female butterfly infected with the bacteria known as *Wolbachia*.

1. If you were to genetically analyze the mature root nodules in this image, what bacterial genes related to nodule formation would be expressed? What genes are expressed to initiate nodule formation? Explain.

2. While examining the genome of *Buchnera aphidicola*, the aphid symbionts in this image, you determine that several genes for production of certain amino acids and many needed for the TCA cycle are missing. What does this genetic finding tell you about the symbiotic relationship between the aphid and *Buchnera*?

3. Like many endosymbionts of insects, *Wolbachia* relies on maternal transmission. What is meant by this term, and do you think this is a successful infection strategy? Defend your answer.

The Rest of the Story

The *Wolbachia* story has yet another interesting twist. Filarial diseases in humans, such as River Blindness (onchocerciasis), and elephantiasis (lymphatic filariasis), and heartworm disease in dogs, are caused by species of filarial nematodes that are transmitted by biting flies and mosquitoes. River Blindness alone is estimated to affect over 18 million people in Africa, the Arabian peninsula, and Latin America. In 1994, the Filarial Genome Project was established, funded by the World Heath Organization. Researchers hoped that sequencing the genomes of these nematodes would give new insights into disease mechanisms and lead to new controls and therapies to combat these devastating diseases. To their surprise, some of the cloned cDNAs isolated from nematode cells contained gene sequences identical to those of *Wolbachia*.

Wolbachia symbionts have now been found in almost every species of filarial nematode where they appear to be obligate mutualists. As in arthropods, *Wolbachia* is maternally transmitted through the reproductive tract of female nematodes. Since it is found in 100 percent of worms in affected species, *Wolbachia* is a potential target for filariasis control. Even more amazing is that recent studies strongly indicate that much of the damage from filarial diseases is not caused by the worm but by *Wolbachia*. Components of the bacterium are released by dead worms and initiate damaging inflammatory and immune responses (see Section 19.4). Since the worms require *Wolbachia*, treating patients with antibacterial drugs that can be taken up by the worm could be an effective treatment for these diseases. Indeed, recent clinical trials using antibiotics have shown great success in preventing disease and stopping reproduction of the worms in the host. Researchers are hopeful that antibiotic treatment of patients will also prevent transmission of filarial nematodes from humans to mosquitoes, breaking the cycle.

Summary

Section 17.1: What is a symbiont?

A **symbiont** is an organism that has, through evolution, developed an intimate and long-standing association with another species. The relationship is referred to as **symbiosis**. There are three general types of symbiosis:

- **Parasitism**: where one species does obvious harm to the other
- **Commensalism**: where one species benefits and the other is not obviously harmed or benefited
- **Mutualism**: where both species benefit

   Many symbioses do not fall neatly into these three categories at all times.

   Many symbioses involve multiple species of interdependent symbionts existing together as a **consortium**.

   Symbionts can be described according to their placement on or in the host:

- **Ectosymbionts** live on the surface of their host.
- **Endosymbionts** live within the tissues, or even within the cells of their host, and have a mutualistic or commensal relationship with the host.

Section 17.2: What roles do microbial symbionts play in plants?

Most plants have symbiotic relationships with soil microorganisms.

   Endophytes are plant endosymbionts. **Rhizobia** are endophytes that can infect plant root cells to become N_2-fixing endosymbionts known as **bacteroids**. The formation of a **root nodule** requires cooperation between the rhizobium and the plant.

- The expression of bacterial *nod* genes needed to initiate nodule formation occurs only in the presence of a compatible host plant.
- Invasion of the plant root begins with formation of an **infection thread**.
- The root nodule provides a protected and low-oxygen environment required by the **nitrogenase** enzyme of bacteroids.
- Production of **leghemoglobin** by the root cells further protects the nitrogenase enzyme.
- Bacteroids in the root nodule are dependent on the plant cell for carbon compounds, which are used in the TCA cycle to generate ATP needed to drive nitrogen fixation.

- In exchange, the bacteroids provide the nitrogen compounds ammonia and/or amino acids to the plant.

Section 17.3: Can symbiotic microbes produce new types of organisms?

Lichens are a symbiotic partnership between a **mycobiont** and a **photobiont**.

- Carbon compounds are supplied to the mycobiont by the photobiont through photosynthesis.
- The photobiont receives protection from the environment within the structure provided by the mycobiont.
- In cyanobacterial lichen partnerships, nitrogen compounds are also supplied to the mycobiont through nitrogen fixation.

Lichens reproduce asexually by **soredia**.

Section 17.4: What are the roles of symbionts in vertebrates?

Vertebrates do not harbor endosymbionts, but they commonly cultivate large numbers of ectosymbionts on their body surfaces, especially in their digestive tracts.

Ribosomal RNA gene sequencing projects have given insights into the nature of the uncultivated masses of symbiont populations.

Germfree and **gnotobiotic** animal models are useful in studying the effects of the absence of symbionts, or effects of particular symbionts, respectively.

Symbionts are intrinsic to the good health of vertebrates. They have roles in:

- Exclusion of pathogens by **competitive exclusion** and production of bacteriocins
- Normal development of the intestinal mucosa
- Normal development of immune responses
- Priming of the immune system
- Digestion of foodstuffs, synthesis of vitamins and essential amino acids, and nutrient conversion (synthesis of amino acids from urea and ammonia)

The human **microbiome** is not completely defined, but is enormous in terms of shear number and species diversity.

Moist, warm locations generally harbor the densest and most diverse microbial populations.

- The **colon** is the most heavily colonized site of the human body. It is one of the largest and most diverse microbial populations known.
- The oral cavity is the second most populated microbial habitat in humans.
- The skin is dominated by the bacterial phyla Actinobacteria and Proteobacteria, but the profile of microbiota varies locally.
- The vagina and vulva are largely colonized by species of *Lactobacillus*, which maintain an acid environment. Small numbers of transients from skin and the colon can commonly be found.

Plaque is a biofilm of bacteria that forms on the tooth surface. An overgrowth of plaque leads to anaerobic conditions and a change in the population from predominantly Gram-positive streptococci to Gram-negative anaerobic bacteria. This leads to periodontal disease.

Streptococcus mutans causes dental carries as it is a potent producer of lactic acid. Sucrose increases the ability of *S. mutans* to colonize the tooth.

The symbiotic populations found within the lumen of the intestine form a biofilm on the outer mucus layer. They generally lack the ability to colonize the epithelium lining the digestive tract—this is an ability possessed by pathogens.

- The major products of microbial fermentation are butyrate, propionate, and acetate. These are absorbed and used as energy sources by the body.
- Methanogenic archaeons convert CO_2 and H_2 to methane. This increases the efficiency of fermentation by keeping H_2 concentrations low.

Human intestinal microbiota consist primarily of anaerobic bacteria from phyla Firmicutes and Bacteroidetes, although individuals have their own unique species profile.

Humans use **colonic fermentation** to extract some of the nutrients from otherwise undigested foodstuffs.

Herbivores depend on their intestinal symbionts to hydrolyze cellulose to glucose.

- **Cecal fermentation**, like colonic fermentation, relies first on the host digestive enzymes to break down and absorb food compounds. Microbial breakdown and fermentation of any remaining compounds happens afterwards, in the **cecum**.
- Some cecal fermenters, like rabbits, practice coprophagia to increase their nutrient uptake.
- **Rumen fermentation** relies first on food digestion and fermentation by the microbes in the **rumen** of **ruminants**. Fermentation products are then absorbed through the rumen. Any remaining compounds, including rumen microbes, are then digested in a manner similar to that of cecal fermenters.
- An important function of the rumen microbiota is the conversion of urea and ammonia into amino acids for microbial protein.
- Rumen fermentation is advantageous because the rumen microbes can be digested. This allows ruminants to compete more successfully than cecal fermenters on poorer quality food, low in nitrogen compounds.

Section 17.5: What are the roles of symbionts in invertebrates?

Microbial endosymbionts are of widespread occurrence in insects, particularly those whose diet lacks required nutritional components.

Insects raised in a sterile environment or having been cleared of their symbionts by antibiotic treatment fail to thrive and eventually die of nutrient starvation.

Primary endosymbionts are intracellular endosymbionts that:

- have highly reduced genomes, a characteristic shared by many obligate intracellular parasites
- are found in specialized host cells called **bacteriocytes**
- are passed through the generations by **maternal transmission**
- are required by the host for survival and/or fertility
- show evidence of **co-speciation** whereby both host and symbiont have evolved into new species together

Insects with primary endosymbionts sometimes also harbor **secondary endosymbionts** that:

- are recently acquired and do not show evidence of co-speciation with their host
- can sometimes be found free-living in the environment
- are not always found in bacteriocytes and can sometimes be found in other cells or be found extracellularly in the host body cavity
- are not always maternally transmitted
- are not always present or essential

Lower termites harbor large numbers of symbiotic anaerobic protozoa, bacteria, and archaeons in their hindgut.

- Protozoa assist in the degradation of cellulose, converting it to CO_2, H_2, and acetate.
- Acetate is used as an energy source by the termite and is used by nitrogen-fixing bacteria that supply ammonia to the termite for its nitrogen needs.
- Other bacteria convert the nitrogenous compound uric acid into amino acids, vitamins, and nucleotides.
- Methanogenic archaeons keep H_2 concentrations low, increasing the efficiency of fermentation.

- Some bacteria convert CO_2 into acetate, increasing the amount available for uptake.

Shipworms possess an obligate bacterial endosymbiont that secretes cellulase, releasing free glucose for host uptake. The symbiont also fixes N_2 to produce nitrogenous compounds for the host.

Zooxanthellae are photosynthetic obligate intracellular algal endosymbionts of coral polyps that supply the coral with carbon compounds, primarily glycerol, through the fixation of CO_2.

● Application Questions

1. Legumes are often planted by farmers to enrich soil. Imagine you are teaching agriculture students the microbiological rationale behind this practice. Explain how this works to enrich the soil as opposed to using nitrogen-rich fertilizers.

2. Describe how mutagenesis was used to determine the role of *nod* genes in nodulation.

3. In the United States, about 75 percent of liquid soaps are labeled as "antibacterial soaps." Most of these contain triclosan or triclocarbon, which inhibit bacterial and fungal enzymes involved in fatty acid synthesis. While these soaps do not appear to greatly reduce the spread of infections, what might they do to normal skin microbiota? Is this a concern?

4. Some patients are prescribed proton pump inhibitors for controlling gastric acid levels. Explain why long-term use of acid inhibitors can be associated with stomach and intestinal infections.

5. Coral bleaching is considered an indicator of coral reef stress. Explain the events that lead to coral bleaching and the role of the endosymbionts in this process.

6. Some endosymbionts are known for possessing highly reduced genomes. What types of genes are often missing? Explain how these organisms survive in their absence.

7. We learned that archaeons are found in the digestive tracts of many animal species. What is their role in maintaining the ecology of the digestive system? What would happen if the archaeons were removed?

8. Imagine you are advising a farmer who is raising grass-fed beef. He would like to change to corn feed. Advise the farmer of the microbiology implications of this change.

9. Although herbivore ruminants may consume protein-poor diets, they still obtain necessary amino acids. Explain the role of microbial symbionts in this process.

10. An Associated Press report in 2008 revealed the presence of many different antibiotics in waterways. Although the report focused on the human health implications, identify and explain some of the effects these drugs may have on other vertebrates and invertebrates (based on your understanding of microbial symbionts from this chapter).

11. In this chapter, you learned how the role of bacterial symbionts is often studied by killing them with an antibiotic such as tetracycline and observing the effects in the host. You immediately thought of your friend who believes that antibiotics are miracle drugs and constantly asks for antibiotics from her doctor even when she is told they will not help. Explain to her what will happen to her bacterial symbionts when she takes such drugs and the health benefits she may lose.

12. A new species of aphid has been discovered and your lab project focuses on examining the microbial symbionts present in this species.
    a. Initially you attempt standard culturing methods but they do not yield much data. Why might this be?
    b. How might you go about pursuing your project goals without using culture methods? What lab methods are available to you and what method would you choose?
    c. Describe how the method you chose above will actually work to reach your project goal. Be specific.

13. Your assigned lab project is to determine the precise contributions of an individual bacterial species to the overall ecology in the mouse digestive tract.
    a. Explain why this might be a difficult task to undertake.
    b. Based on your reading in this chapter, what approach can you take in the lab to accomplish this goal?

Suggested Reading

Denison, R. F. 2000. Legume sanctions and the evolution of symbiotic cooperation by rhizobia. Am Nat 156:567–576.

Dimijian, G. G. 2000. Evolving together: The biology of symbiosis, part 1. Proc (Bayl Univ Med Cent) 13:215–226.

Dimijian, G. G. 2000. Evolving together: The biology of symbiosis, part 2. Proc (Bayl Univ Med Cent) 13:381–390.

Dubilier, N., C. Bergin, and C. Lott. 2008. Symbiotic diversity in marine animals: The art of harnessing chemosynthesis. Nat Rev Microbiol 6:725–740.

Farnsworth, E. R. 2008. The evidence to support health claims for probiotics. J Nutr 138:1250S–1254S.

Rajilić-Stojanović, M., H. Smidt, and W. M. de Vos. 2007. Diversity of the human gastrointestinal tract microbiota revisited. Environ Microbiol 9:2125–2136.

Ruby, E. G. 2008. Symbiotic conversations are revealed under genetic interrogation. Nat Rev Microbiol 6:752–762.

Simon, L. 2008. The role of *Streptococcus mutans* and oral ecology in the formation of dental caries. Lethbridge Undergraduate Research Journal 2 ISSN 1718-8482.

Tamames, J., R. Gil, A. Latorre, J. Pereto, F. S. Silva, and A. Moya. 2007. The frontier between cell and organelle: Genome analysis of *Candidatus* Carsonella ruddii. BMC Evol Biol 7:181.

Werren, J. H., L. Baldo, and M. E. Clark. 2008. *Wolbachia*: Master manipulators of invertebrate biology. Nat Rev Microbiol 6:741–751.

18 Introduction to Infectious Diseases

Polio Cases in the United States

POLIOMYELITIS VACCINE
aqueous preparation
9 cc.

MALARIA

Artemisinin
100 mg

50 capsules

POLIO

BCG VACCINE
For Intracutaneous Injection Only

Rv
300 mg
100 TABLETS

The summer of 1952 was especially unsettling for many Americans. The Cold War was escalating, the United States was embroiled in the Korean War, and another enemy lurked at home—poliovirus. Outbreaks of poliovirus infection had occurred in the United States since the late 1800s. By the middle of the twentieth century, these outbreaks became more common and affected more people. In 1952 alone, over 21,000 cases of paralytic polio occurred. No treatment existed then, or now.

Most often, poliovirus infection results in no overt disease or a mild disease characterized by fever, headache, and nausea. In about 1 in 200 infected people, however, the viral infection leads to paralysis and, in some cases, death. During the first half of the twentieth century, survivors of paralytic poliovirus often required the use of crutches or wheelchairs. In many people, the poliovirus-associated paralysis resulted in severe breathing problems. These individuals were placed in iron lungs, metal cylinders that generated negative pressure to artificially draw air into the person's lungs. In the 1940s and early 1950s, these iron lungs filled entire hospital wards.

While no treatment existed, people did understand that poliovirus was easily transmitted and that transmission often occurred through contaminated water or direct contact. In many communities, swimming pools were closed during July and August of 1952 to prevent the spread of the virus. In some cases, parents did not allow their children to play with the neighbors or attend summer camp. Fear of this infectious disease was rampant.

Fear, however, was not the only response to this infectious disease. Scientists and physicians responded, and their work eventually led to the development of effective vaccines. Perhaps more interestingly, the poliovirus epidemics spurred other responses. Franklin D. Roosevelt, for instance, founded the National Foundation for Infantile Paralysis, the fund-raising organization now known as the March of Dimes. Through this group, millions of Americans contributed money to combat the disease, helping to fund important research. Additionally, individuals affected by paralytic poliovirus, individuals often reliant on crutches or wheelchairs, led efforts to decrease discrimination against people with disabilities. They fought, for instance, to increase access to public buildings. These efforts eventually resulted in the passage of the Americans with Disabilities Act in 1990. The poliovirus epidemics of the early twentieth century, then, have had long-lasting effects on areas as diverse as the public funding of research and the basic civil rights of all Americans.

CHAPTER NAVIGATOR

As you study the key topics, make sure you review the following elements:

Certain microbes cause infectious diseases.
- Table 18.1: Signs and symptoms of specific infectious diseases
- Microbes in Focus 18.1: Measles: A highly contagious disease
- Toolbox 18.1: Measuring the virulence of pathogens
- Figure 18.3: Genetics of a pathogen and its virulence

While mechanisms of pathogenesis differ, some common features exist.
- Table 18.3: Virulence actions of various pathogens
- Figure 18.6: Evading host defenses through antigenic variation
- Table 18.4: Selected strategies for evading host defenses
- Perspective 18.1: Host defenses in bacteria

Microbial pathogens can be transmitted in various ways.
- Table 18.5: Selected zoonotic diseases
- Figure 18.9: Prevalence and incidence: Epidemiological measures of infectious diseases in a population
- Microbes in Focus 18.3: Monkeypox: An emerging disease
- Figure 18.16: Epidemic curve and temporal pattern of an outbreak

We can determine how specific microbes cause disease.
- Figure 18.21: Genetic variations among individual hosts and susceptibility to pathogens
- Perspective 18.2: The armadillo: An ideal animal model?
- Figure 18.22: Acquisition of virulence factor genes

Pathogens evolve via various mechanisms.
- Mini-Paper: Epidemiology of an infectious disease
- Figure 18.26: Emerging diseases and increased exposure to reservoir species
- Table 18.8: Selected U. S. infectious disease surveillance programs
- Figure 18.27: Morbidity and Mortality Weekly Report (MMWR)

CONNECTIONS for this chapter:

Host genetic differences and viral pathogenesis (Section 8.1)

Horizontal gene transfer and pathogenicity islands (Section 21.4)

Operons and the regulation of gene expression (Section 11.2)

Life cycle of *Plasmodium falciparum* (Section 23.2)

Introduction

The human story has been closely linked with infectious diseases. Bubonic plague. Smallpox. Influenza. These infectious diseases and many others—diseases all caused by microbes—have shaped our history in profound ways.

In the 1340s, the bubonic plague, a disease caused by the bacterium *Yersinia pestis*, ravaged human populations throughout the world. Historians estimate that 75 million people worldwide may have died from this disease, also known as the Black Death, in just a few years. The population of Europe plummeted, decreasing by approximately one-third. The impact of the disease extended beyond the deaths of millions. With the extreme loss of life in Europe, cheap land became available, ultimately changing the feudal system that was predominant before the plague.

When Europeans first visited the Americas in the fifteenth and sixteenth centuries, they brought with them more than just muskets, axes, and clothes. They also brought with them a host of disease-causing microbes. Because the people of the New World had never been exposed to these infectious agents, they were especially susceptible to them. Historians now speculate that 95 percent of the Native American populations may have died from these infectious diseases. The ravages of smallpox, perhaps more than the guns of the Spaniards, may have led to the end of the Inca Empire, considered by some historians to have been one of the greatest civilizations ever.

In the span of less than two years, from 1918 to 1919, an H1N1 strain of influenza virus, known as the Spanish flu, traversed the globe, killing an estimated 50 million people **(Figure 18.1)**. During this time, more people died from influenza than had been killed in World War I, which, at the time, had been the most destructive war in the history of humankind. In the United States alone, an estimated 675,000 people died from the flu.

For the most part, though, we made great strides against infectious diseases throughout the twentieth century. Because of advances in health care and sanitation, people in developed countries experienced fewer infectious diseases. We also developed effective means of diagnosing and preventing infections and/or treating the associated diseases. With the development and mass production of penicillin, deaths in the United States due to bacterial infections dropped dramatically. The poliovirus vaccine, approved for use just a few years after the frightening summer of 1952 mentioned in this chapter's opening story, ended a major viral scourge. As deaths due to infectious diseases decreased, the average life expectancy in developed countries increased significantly (see Figure 18.1). The changes were so remarkable that, in 1968, the Surgeon General of the United States, William H. Stewart, announced, "The war against infectious diseases has been won."

Of course, we now realize that his statement of victory was premature; we clearly have not vanquished infectious diseases **(Figure 18.2)**. According to the World Health Organization (WHO), approximately 26 percent of all deaths worldwide in 2002 could be attributed to infectious disease. Nearly 2 million people worldwide die from tuberculosis every year. A similar number of people die each year from malaria. Roughly 14,000 people every day become infected with HIV.

As we have seen with SARS and, more recently, the H1N1 strain of influenza virus, the ease of travel today means that an infectious agent can spread around the globe in a matter of days, if not hours. As our population increases, our exposure to new environments and new animals increases, leading to more frequent transfers of infectious agents from animals to humans. Because of global climate change, the ranges of certain

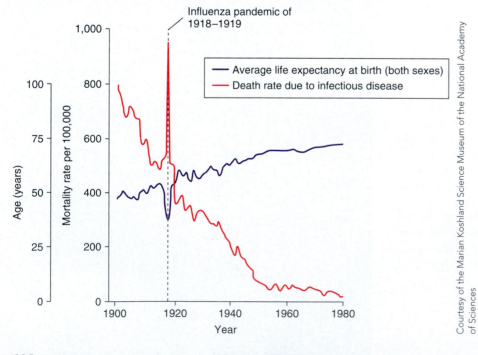

Figure 18.1. **Mortality associated with the 1918–1919 influenza pandemic** Throughout the twentieth century, improvements in public health, sanitation, and medical treatments caused a steady decrease in the rate of death due to infectious diseases in the United States. In association with this decrease, the average life expectancy rose significantly. The dramatic increase in infectious disease deaths in 1918 and 1919 resulted from the 1918 influenza pandemic, which resulted in the deaths of between 500,000 and 675,000 Americans.

Courtesy of the Marian Koshland Science Museum of the National Academy of Sciences

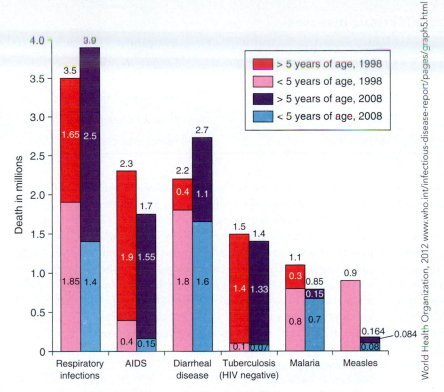

Figure 18.2. **The continuing toll of infectious diseases** Despite the progress made in our abilities to prevent and treat infectious diseases, the infectious disease burden remains high, especially in developing countries. As shown here, deaths from some infectious diseases, like measles, decreased significantly from 1998 to 2008. Deaths from other infectious diseases, like respiratory diseases and tuberculosis, remained fairly constant over this time period.

World Health Organization, 2012 www.who.int/infectious-disease-report/pages/graph5.html

Phototrophic organisms, as we learned in Chapter 13, obtain the energy needed to synthesize organic molecules from sunlight and, as a result, need to reside in environments where sunlight is accessible. Some microbes, as we saw in Chapter 17, exist in partnerships with other organisms. In these partnerships, the interaction between the organisms allows the microbe to obtain needed resources and, again, dictates its preferred environment.

Microbes that cause disease typically obtain resources through a partnership, of sorts, with another organism, the host organism. This host organism, then, can be thought of as their environment. In many respects, an infectious disease merely represents the consequence of a relationship between a host and a parasite that lives at the expense of the host. Unlike most of the microbes that we discussed in Chapter 17, these microbes damage the host in their quest for resources.

Let's look at this idea of infectious disease using poliovirus as an example. As we noted in the opening story of this chapter, poliovirus infection can lead to paralysis. How? To replicate, the virus needs many basic resources, including nucleotides, amino acids, and the machinery needed for the production of polypeptides. It obtains these resources by entering certain human cells—its preferred environment. When replicating within a neuron, the virus uses an extraordinary amount of the cell's resources and energy, eventually leading to the destruction of that cell, which cannot be replaced. When enough neurons die, paralysis results. The disease, in other words, occurs because of the microbe's quest for resources.

disease vectors, like the mosquitoes that spread malaria, West Nile virus, and yellow fever, may change. How potential changes in mosquito distribution would affect disease spread remains unclear. Finally, and perhaps most discomforting, our use of the very drugs lauded by Surgeon General Stewart and others, has led to the evolution of drug-resistant strains of many bacteria. Today, because of the evolution in bacteria of antibiotic resistance, we may be less able to combat diseases caused by *Staphylococcus aureus* than we were 30 years ago. No, unfortunately, the battle against infectious diseases has not been won.

As we have seen throughout our exploration of microbiology, the microbial world exhibits a dizzying array of diversity. Bacteria, archaeons, viruses, and eukaryal microbes come in a multitude of shapes and sizes. All of these microbes, though, share at least one fundamental feature—they need basic resources in order to replicate. This need for resources generally determines the environment in which a given microbe exists.

In subsequent chapters, we will focus on how hosts protect themselves from infectious diseases, how various types of microbes cause disease, and what strategies we have developed to protect ourselves from these disease-causing microbes. In this chapter, we will focus on more general topics related to infectious diseases. Specifically, we will address these questions:

What is an infectious disease? (18.1)

How do microbes cause disease? (18.2)

How are infectious diseases transmitted? (18.3)

How can we determine the cause of an infectious disease? (18.4)

How do new infectious diseases arise? (18.5)

18.1 Pathogenic microbes

What is an infectious disease?

As we mentioned in the introduction, some microbes obtain nutrients from a host, and in the process, damage the host. **Disease**, a disturbance in the normal functioning of an organism, ensues. A disease caused by a microbe that can be transmitted from host to host is an **infectious disease**. This definition includes diseases transmitted from species to species. **Zoonotic diseases** are infectious diseases of animals that can cause disease when transmitted to humans. For example, rabies

TABLE 18.1 Signs and symptoms of specific infectious diseases

Disease	Cause	Signs	Symptoms
Acquired immunodeficiency syndrome (AIDS)	Human immunodeficiency virus	Opportunistic infections, decreased CD4⁺ T cell count	Fatigue, muscle aches
Malaria	Plasmodium falciparum	Fever, anemia	Muscle pain, headaches
Flu	Influenza virus	Fever	Muscle aches, malaise
Salmonellosis	Salmonella Typhimurium	Diarrhea, fever	Abdominal cramps

is a zoonotic viral disease that is occasionally transmitted to humans through the bite (saliva carries the virus) of an infected animal, such as a raccoon or a bat.

Most infectious diseases can be transmitted from person to person. Influenza virus, for instance, may be transmitted from one person to another via contaminated respiratory droplets. HIV can be transmitted via the direct transfer of blood or semen from one person to another. Other diseases caused by microbes typically are not transmitted. Often, these diseases result from the ingestion of a toxin formed by a microbe. Some types of food poisoning, for instance, may result from the ingestion of toxins produced by bacteria.

Microbes that cause diseases are referred to as **pathogens**; the term **pathogenesis** refers to the process by which a pathogen causes disease. **Infection** refers to the replication of a pathogen in or on a host. We generally refer to subjective disease states—muscle aches, for instance—as **symptoms** of a disease. Symptoms are difficult to quantify. More objectively defined disease states that can be readily observed or measured—like a rash or fever—are referred to as **signs** of a disease. From a medical perspective, specific disturbances typically characterize specific diseases. A sore throat and muscle aches are symptoms that characterize influenza. Vomiting and diarrhea may be signs of food poisoning **(Table 18.1)**.

Not all pathogens have the same intrinsic ability to cause disease. **Primary pathogens** tend to produce disease readily in healthy hosts. **Opportunistic pathogens**, conversely, generally only cause disease when they become displaced to an unusual site within the host or the host possesses a weakened immune system **(Table 18.2)**. Among humans, individuals who are immunocompromised, or have a weakened immune system, due to HIV disease represent the single largest group of people most susceptible to opportunistic pathogens. Individuals who have received an organ transplant or are receiving chemotherapy also may be immunocompromised.

CONNECTIONS HIV weakens the immune system by infecting and ultimately destroying CD4⁺ cells—a key component of the human immune system. As we saw in **Section 8.1**, the virus infects these cells by binding to a receptor and a co-receptor. The drug maraviroc slows viral replication by interfering with the interaction between HIV and one of the co-receptors.

The distinction between primary and opportunistic pathogens seems relatively clear, but the relationship between infection and disease is quite complex. Poliovirus, the pathogen that we introduced in the opening story, can be

considered a primary pathogen; it causes disease in otherwise healthy, immunocompetent individuals. Every single person exposed to poliovirus, though, does not become infected and, as we noted, paralytic poliomyelitis only occurs in a fraction of the people who become infected. The rate of infection depends on the proportion of hosts that are susceptible to the pathogen. Factors influencing a host's susceptibility to a pathogen may include, for instance, prior exposure to the microbe through disease or vaccination. Genetic factors also may affect a host's susceptibility to a pathogen. Hosts who are not susceptible to a pathogen may be exposed to the pathogen, but do not become infected.

CONNECTIONS In **Section 8.1**, we examined the susceptibility of two mouse strains to mouse hepatitis virus (MHV). Because of genetic differences between the mouse strains, this virus can bind to and infect cells of one mouse strain but not the other.

The rate of disease among infected individuals also depends on the **case-to-infection (CI) ratio**, or the proportion of infected people who develop the disease. The CI ratio of different microbes can differ dramatically. Measles virus **(Microbes in Focus 18.1)**,

Microbes in Focus 18.1
MEASLES: A HIGHLY CONTAGIOUS DISEASE

Habitat: Humans; infects many cell types including respiratory epithelial cells and leukocytes

Description: Enveloped ssRNA virus in the *Paramyxoviridae* family. Approximately 150–300 nm in diameter, with helical nucleocapsid symmetry

Key Features: Measles virus can be transmitted by respiratory secretions, and aerosolized viral particles can remain infectious for several hours. Measles infection is characterized by a prominent rash and spots on the inner cheek surface. The disease can lead to severe complications that may be fatal.

TEM

© AMI Images/Science Photo Library/Photo Researchers, Inc.

TABLE 18.2 Examples of primary and opportunistic bacterial pathogens

Category	Bacterium	Disease
Primary pathogen	*Bacillus anthracis*	Anthrax
	Bordetella pertussis	Whooping cough
	Borellia burgdorferi	Lyme disease
	Corynebacterium diphtheriae (lysogenized strains)	Diphtheria
	Escherichia coli O157:H7	Hemorrhagic colitis, kidney disease
	Helicobacter pylori	Gastritis and ulcers
	Mycobacterium tuberculosis	Tuberculosis
	Rickettsia typhii	Typhus
	Salmonella Typhi[a]	Typhoid fever
	Salmonella Typhimurium[a]	Salmonellosis
	Treponema pallidum	Syphilis
	Vibrio cholerae	Cholera
	Yersinia pestis	Plague
Opportunistic pathogen	*Clostridium botulinum*	Botulism
	Clostridium difficile	Pseudomembranous colitis
	Clostridium perfringens	Gas gangrene
	Clostridium tetani	Tetanus
	Enterobacter aerogenes	Urinary, respiratory infections
	Escherichia coli uropathogenic strains	Urinary tract infections
	Haemophilus influenzae	Meningitis, pneumonia
	Klebsiella pneumoniae	Pneumonia
	Legionella pneumocystis	Pneumonia, septicemia
	Listeria monocytogenes	Meningitis, septicemia
	Mycoplasma pneumonia	Atypical pneumonia
	Neisseria meningitidis	Meningitis
	Pseudomonas aeruginosa	Pneumonia, skin infections
	Serratia marcescens	Urinary tract infections, endocarditis
	Staphylococcus aureus (various strains)	Skin infections, toxic shock syndrome, pneumonia, endocarditis, vascular infections
	Streptococcus agalactiae	Newborn meningitis
	Streptococcus mutans	Tooth decay, endocarditis
	Streptococcus pneumoniae	Meningitis, pneumonia
	Streptococcus pyogenes	Pharyngitis, scarlet fever, rheumatic fever, necrotizing fasciitis, toxic shock syndrome

[a]*Salmonella* Typhi and *Salmonella* Typhimurium are approved short forms for *Salmonella enterica* subspecies *enterica* serovar Typhi and *Salmonella enterica* subspecies *enterica* serovar Typhimurium, respectively. They were previously designated as species *S. typhi* and *S. typhimurium*. The new naming reflects recognition that both of these organisms belong to a single species and subspecies.

for instance, has a CI ratio of approximately 0.95 in previously unexposed people. In other words, about 95 percent of all people who become infected with measles virus will develop the disease. Poliovirus, conversely, has a very low CI ratio. Less than 1 percent of people infected with poliovirus will develop paralytic poliomyelitis. Many factors, such as the immune status of the individuals, affect the CI ratio.

By determining the case-to-infection ratio of a pathogen, researchers can begin to quantify the **virulence** of that pathogen. Virulence is a measurement of the severity of disease that a

pathogen causes. Without a doubt, the 1918 strain of influenza virus, often referred to as the "Spanish flu," caused severe disease. While modern influenza infections may be severe and cause a significant number of deaths annually, they usually do not cause widespread deaths in the majority of the population. The virulence of the 1918 influenza virus strain, then, was much greater than the virulence of most other strains of influenza virus. We should note that comparing the virulence of one pathogenic species to the virulence of another species is problematic. One cannot, for instance, determine if the diarrheal disease caused by *Salmonella* is more or less severe than the respiratory disease caused by influenza virus. Researchers, however, can compare the virulence of different strains or isolates of a specific pathogen (**Toolbox 18.1**). These types of studies provide quantitative data about the ability of pathogens to cause disease in a host organism.

Genetic differences between two strains of a pathogen may account for differences in virulence. For example, the bacterium *Yersinia pestis*, the causative agent of plague, produces a series of nucleoside monophosphate kinases (NMPKs), enzymes involved in the synthesis of nucleotides. One of these enzymes, adenylate kinase (AK), is involved specifically in the biosynthesis of AMP and dAMP, precursors of the nucleotide adenosine. To investigate the importance of this enzyme to *Y. pestis* virulence, researchers at the Pasteur Institute mutated certain bases within the *Y. pestis* AK gene. A specific mutation, which resulted in a change in amino acid 87 from a proline to a serine, caused the adenylate kinase enzyme to be more sensitive to elevated temperatures and proteolytic digestion than the wild-type enzyme. To determine the effects of this mutation on the ability of the bacterium to cause disease in an animal host, the investigators determined the number of colony-forming units (CFU) required to kill mice. Following intravenous injection with fewer than 50 CFU of wild-type *Y. pestis*, all mice died. When the researchers inoculated mice with the mutant *Y. pestis*, no mice died even when they were

How can the virulence of a particular strain of a pathogenic microbe be determined? One could simply infect a susceptible animal and record the severity of disease that ensues. It should be obvious, though, that this approach poses problems. Most notably, it is very subjective. Conditions that one person considers severe may be considered mild by someone else. To avoid this reliance on judgment, one could measure specific disease states. A body temperature between 99 and $101°F$ (37.2 and $38.3°C$), for instance, might constitute mild disease, while a temperature over $101°F$ ($38.3°C$) might constitute severe disease. While this approach may provide a more objective measurement of disease, it also is more time-consuming as the temperature of each infected animal must be measured.

To more easily provide accurate and objective measures of virulence, researchers may determine the LD_{50}, or lethal dose 50. This measurement shows the amount of a pathogen or toxin that kills an average of 50 percent of the test subjects. To determine the LD_{50} of a bacterial toxin, for instance, groups of test animals would be inoculated with different concentrations of the toxin. At very low doses of toxin, few or none of the animals would die. At very high doses, 100 percent of the animals would die. By analyzing the data, researchers can determine the amount of toxin predicted to kill, on average, 50 percent of the animals (see Figure 5.25). Alternatively, researchers can use this type of assay to determine the ID_{50}, or infectious dose 50, of a microbe. With this test, researchers can determine the amount of microbes needed to cause an infection in 50 percent of the inoculated hosts.

Why use 50 percent rather than the 100 percent to characterize the lethal dose? With the scatter of values for each measurement, it is laborious and often unclear exactly what represents a given dose value giving 100 percent response. However, it is possible to identify asymptotically the value to which high doses are forming a plateau. This value is taken as the 100 percent response, and the 50 percent value can be calculated that is both accurate and consistent for defined conditions.

Both LD_{50} and ID_{50} measurements convey information about the virulence of a microbe or the toxicity of a toxin. Tetanus toxin, a potent neurotoxin produced by the bacterium *Clostridium tetani*, has an LD_{50} of approximately 0.001 $\mu g/kg$. Cholera toxin, produced by the bacterium *Vibrio cholerae*, conversely, has an LD_{50} of approximately 250 $\mu g/kg$. Tetanus toxin, in other words, is roughly 250,000 times more potent than cholera toxin.

Because LD_{50} and ID_{50} measurements require the use of many animals, researchers have been using these tests much less frequently. Instead, researchers studying the virulence of viruses may determine a somewhat related value, the $TCID_{50}$, or tissue culture infectious dose 50 (see Section 5.3). By using this assay, researchers can determine the amount of virus that leads to a productive infection in cells 50 percent of the time.

Test Your Understanding ....

Researchers examined the LD_{50} of three different strains of *Staphylococcus epidermidis* that caused sepsis in hospital patients. Using a mouse LD_{50} assay, they found the LD_{50} of strain A to be 2.54×10^9, strain B to be 2.35×10^9, and strain C to be 4.71×10^9. What do each of these results mean, and what conclusions can you draw about each of the three strains? Explain.

---

inoculated with 1.5×10^4 CFU of the bacteria **(Figure 18.3)**. This mutation resulted in the extreme **attenuation**, or decrease in virulence, of this pathogen. In this particular case, the virulence of the microbe decreased so markedly that it no longer caused disease with any noticeable frequency. It became avirulent.

When an attenuated strain of a pathogen infects a host, the microbe still replicates. The host, as a result, still mounts an immune response to the infectious agent. This immune response to a weakened form of the pathogen allows the host to better combat future infections with more virulent forms of the pathogen. Attenuated strains of pathogens, in other words, often are effective **vaccines**, agents that can stimulate the immune system's response to a particular microbe and provide some immunity to that microbe. This ability to measure virulence and identify less virulent strains of pathogens, in fact, has been instrumental in the development of vaccines.

The oral poliovirus vaccine, also known as the Sabin vaccine, represents a classic example of an attenuated vaccine. To create this strain, Dr. Albert Sabin allowed wild-type strains of poliovirus to replicate repeatedly in a monkey epithelial cell culture line (see Toolbox 5.1 for a review of cell culture techniques). Sabin then measured the virulence of the virus and observed that it had become greatly attenuated. When used as a vaccine, this attenuated form of poliovirus rarely causes disease, but provides the recipient with a strong, long-lasting protection against wild-type poliovirus.

Finally, some people may become infected with a pathogenic microbe but never exhibit any overt signs or symptoms of disease. In these cases, the person is said to be asymptomatic, even though the microbe may still be replicating within these hosts. The host, as a result, often acts as a **carrier** that can transmit the infectious agent to others. Probably the most famous carrier of a pathogenic microbe was Mary Mallon, also known as Typhoid Mary. Ms. Mallon was a carrier of *Salmonella* Typhi **(Microbes in Focus 18.2)**, the bacterium that causes typhoid fever **(Figure 18.4)**. During the early part of the twentieth century, at least 53 people developed typhoid fever after coming in contact with Mary Mallon,

Observation: The *Yersinia pestis* adenylate kinase (AK) enzyme is involved in the biosynthesis of adenosine.

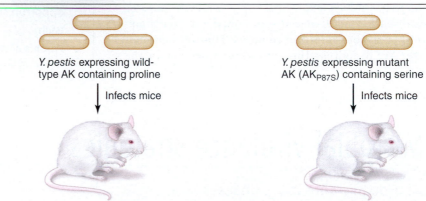

$$\text{ATP} + \text{AMP} \xrightarrow{\text{AK}} \text{ADP} + \text{ADP}$$

Hypothesis: Thermosensitive mutants of AK will lead to decreased growth at the non-permissive temperature 37°C.

Y. pestis expressing wild-type AK containing proline → Infects mice

Y. pestis expressing mutant AK (AK$_{P87S}$) containing serine → Infects mice

Experiment: Mutate codon 87 of AK gene, changing the amino acid from proline to serine. Compare pathogenesis of wild-type and mutant strains.

Results: AK$_{P87S}$ mutant exhibited greatly decreased pathogenesis.

Conclusion: Adenylate kinase contributes to the pathogenesis of *Y. pestis*.

Y. pestis strain	CFU injected (iv)	% Lethality
Wild type	0	0
	3.6	20
	36	100
AK$_{P87S}$	0	0
	150	0
	15,000	0

Figure 18.3. Genetics of a pathogen and its virulence To investigate the role of adenylate kinase in the replication of *Yersinia pestis*, researchers constructed a mutant strain of *Y. pestis* in which amino acid 87 of adenylate kinase was changed from a proline to a serine (AK$_{P87S}$). When injected into mice, less than 50 colony-forming units (CFU) of the wild-type strain resulted in the deaths of all the mice. The mutant strain, in contrast, was avirulent; even when injected with 1.5×10^4 CFU of bacteria, none of the mice died.

Microbes in Focus 18.2
SALMONELLA TYPHI: THE CAUSE OF TYPHOID FEVER

Habitat: Infects humans. No other known hosts. Transmitted by fecal contamination of food and water

Description: Gram-negative, motile bacterium, approximately 2 μm in length, within the family *Enterobacteriaceae*. Commonly referred to as *Salmonella* Typhi, a subspecies of *Salmonella enterica*.

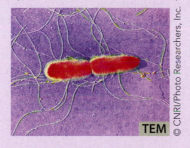

Key Features: Typhoid fever affects roughly 20 million people annually. Increasingly, antibiotic-resistant isolates of this pathogen are being identified. A recent study, for instance, showed that 95 percent of *Salmonella* Typhi isolates from Vietnam exhibited resistance to the antibiotic nalidixic acid.

Figure 18.4. Typhoid Mary: Forcibly quarantined to prevent spreading typhoid Mary Mallon of New York City was determined to be a carrier of *Salmonella* Typhi, the causative agent of typhoid fever. Because she did not voluntarily cease work as a cook, in order to prevent the spread of the bacterium to others, she was quarantined twice. From 1915 until her death in 1938, she was confined to an island in the East River. Upon her death, an autopsy determined that she was still infectious.

even though she did not show any signs or symptoms of the disease. Because she refused to adequately prevent transmission of the pathogen from herself to others, the New York City Health Department twice placed her in quarantine. The second time, she remained in quarantine until her death over 20 years later.

In this section, we have examined some common features of infectious diseases and reviewed some important terminology. However, an obvious question remains: How do microbes cause disease? We'll address this important question in the next section.

18.2 Microbial virulence strategies

How do microbes cause disease?

As we will see in subsequent chapters, bacteria, viruses, and eukaryal microbes often cause disease via dramatically different mechanisms. Indeed, even among members of any one of these groups, the mechanisms of pathogenesis may differ greatly. Despite this great variation, we still can identify some commonalities. Most notably, pathogens:

- Gain entry to the host
- Attach to and invade specific cells and/or tissues within the host
- Evade host defenses
- Obtain nutrients from the host
- Exit the host

Frequently, cellular damage to the host occurs as a result of these processes. Moreover, pathogenic microbes typically produce products that facilitate several of these events (Table 18.3).

Often referred to as "virulence factors" (see Section 21.1), these products enhance the pathogen's ability to cause disease and may distinguish a pathogenic microbe from a closely related non-pathogenic relative. In Section 23.1, we will examine *Ophiostoma novo-ulmi*, the fungus that causes Dutch elm disease. This pathogen produces xylanase, an enzyme that allows that fungus to degrade plant cell walls, thereby allowing the pathogen to obtain nutrients. In Section 21.1, we will examine *Neisseria gonorrhoeae*, the causative agent of gonorrhea. This pathogenic bacterium produces IgA protease, an enzyme that destroys host IgA antibodies, thereby helping the pathogen evade the host defenses.

TABLE 18.3 Virulence actions of various pathogens

Virulence action	Viruses	Bacteria	Eukaryal microbes
Attachment and invasion	HIV: Viral gp120 protein attaches to CD4 molecule found on certain immune system cells.	*Escherichia coli*: Enteropathogenic and enterohemorrhagic strains produce intimin, an adhesion molecule.	*Giardia lamblia*: Ventral disk allows adherence to epithelial cells of small intestine.
Evade host defenses	Herpes simplex virus: Following acute replication, the viral genome exists as an episome in the host cell nucleus, undergoing limited transcription and translation.	*Neisseria gonorrhoeae*: The organism produces IgA protease, an enzyme that destroys the host IgA antibodies.	*Trypanosoma brucei*: The organism undergoes antigenic variation to periodically change its surface proteins.
Obtain nutrients from the host	Human papillomavirus: Viral E6 and E7 proteins inhibit cellular p53 and Rb tumor suppressor proteins, thereby driving the cell to replicate and produce molecular building blocks needed by the virus.	*Mycobacterium tuberculosis*: Mycobactins, iron-binding proteins, allow the pathogen to acquire iron from the host.	*Plasmodium falciparum*: Nucleoside transporter proteins allow the pathogen to obtain purines from the host.
Cause cellular damage	Poliovirus: Replication induces apoptosis of host cells.	*Clostridium botulinum*: The organism produces botulinum toxin, a powerful neurotoxin that results in flaccid paralysis.	*Cochliobolus carbonum*: The organism produces HC-toxin that inhibits host cell histone deacetylases.

As we begin our examination of pathogenesis, we will discuss events that occur within the host, saving a discussion of entry and exit for the next section. We will focus in this section on the molecular bases of invasion, evasion of host defenses, and cellular damage. Drawing from the viral, bacterial, and eukaryal worlds, we will look at a few examples of how pathogens achieve these goals. The following chapters will provide a more thorough understanding of these topics.

Attachment, invasion, and replication

For many pathogens, attachment to and invasion of specific cells and/or tissues precedes pathogen replication and the onset of disease. Bacteria use various features, including fimbriae, pili involved specifically in attachment, and lipoteichoic acid, a cell wall component of Gram-positive bacteria, for attachment. Viruses generally attach to cells via a very specific interaction between a viral attachment protein and a cellular receptor (see Figures 8.1 and 8.3). This attachment often triggers entry of the virus or its genome into the cell, a necessary event for all viruses. Some pathogens attach in a less specific manner. Spores of the rice blast fungus, for instance, adhere to most hydrophobic surfaces, including cells. For other fungal pathogens of plants, a more specific interaction between glycoproteins on the spore surface and molecules on the plant cell surface occur. Following attachment, some pathogenic fungi of plants develop appressoria, specialized invasion structures that can penetrate between host cells, allowing the fungus to penetrate the plant cuticle, a waxy protective covering present in most plants (see Figure 23.11).

The ability to attach to, invade, and replicate within the host often determines the **host range** of a pathogen—the group of organisms that the pathogen can infect (see Figure 8.1). Molecular changes in the pathogen can alter which cells a pathogen can attach to or invade. Such changes, in turn, can alter the host range of a pathogen. To explore this concept, let's examine the emergence of canine parvovirus, a small, single-stranded DNA virus that causes severe vomiting and diarrhea in puppies. Canine parvovirus, or CPV, was first recognized around 1978 and quickly spread throughout the world. Sequence analysis studies have shown that CPV is very closely related to feline panleukopenia virus (FPLV), a parvovirus that infects cats. Moreover, researchers have demonstrated that FPLV can replicate in feline, but not canine cells, whereas CPV can replicate in both feline and canine cells (**Figure 18.5**). This phenotypic difference is determined by

Figure 18.5. The emergence of canine parvovirus Feline panleukopenia virus (FPLV) can replicate in feline, but not canine, cells. Modern canine parvoviruses (CPV-2a and CPV-2b) can replicate in both feline and canine cells. Based on DNA and protein sequence analyses, a mutation in an ancestral strain of virus changed the capsid protein permitting it to replicate in canine cells, thereby altering and enlarging its host range to include dogs. Two hypotheses exist to explain the apparent evolution of feline-like parvoviruses to canine parvoviruses.

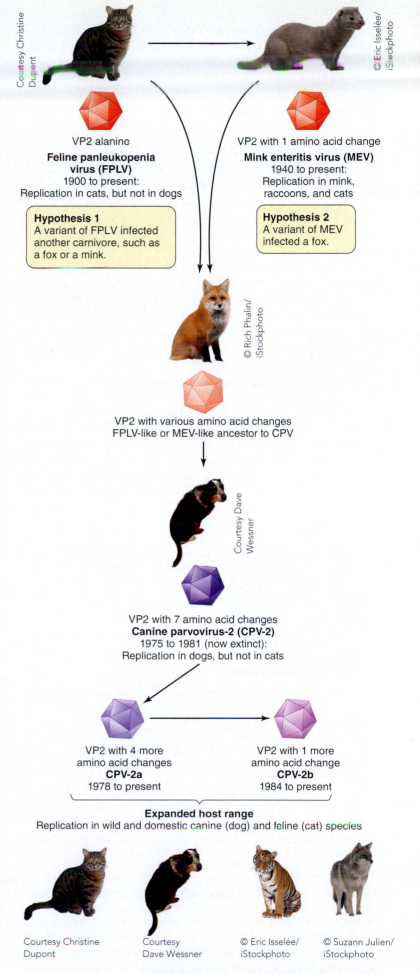

VP2 alanine
Feline panleukopenia virus (FPLV)
1900 to present:
Replication in cats, but not in dogs

Hypothesis 1
A variant of FPLV infected another carnivore, such as a fox or a mink.

VP2 with 1 amino acid change
Mink enteritis virus (MEV)
1940 to present:
Replication in mink, raccoons, and cats

Hypothesis 2
A variant of MEV infected a fox.

VP2 with various amino acid changes
FPLV-like or MEV-like ancestor to CPV

VP2 with 7 amino acid changes
Canine parvovirus-2 (CPV-2)
1975 to 1981 (now extinct):
Replication in dogs, but not in cats

VP2 with 4 more amino acid changes
CPV-2a
1978 to present

VP2 with 1 more amino acid change
CPV-2b
1984 to present

Expanded host range
Replication in wild and domestic canine (dog) and feline (cat) species

Courtesy Christine Dupont

Courtesy Dave Wessner

© Eric Isselée/ iStockphoto

© Suzann Julien/ iStockphoto

specific mutations within the VP2 capsid protein of parvoviruses like FPLV and CPV. It appears then, that a few specific mutations within FPLV altered the host range of this virus. Once FPLV became capable of replicating within canine cells, its host range expanded and the virus quickly spread throughout this new host population.

Evading host defenses

Pathogens not only must attach to and/or invade appropriate cells and tissues, but also they must evade host defenses. Pathogens evade host immunity in various ways. The protozoan *Trypanosoma brucei*, which causes sleeping sickness, evades the host immune system by altering the composition of its surface antigens—proteins on the surface of the pathogen to which the host immune system responds. Termed **antigenic variation**, this process allows the pathogen to avoid being eliminated by the host **(Figure 18.6)**. When the host immune system begins eliminating pathogens displaying one particular surface antigen, the surface antigens on the pathogens change, thereby providing the pathogen with a mechanism for evading the immune response.

To further understand this process of antigenic variation, let's examine *Trypanosoma brucei* in more detail. During an infection, the surface of this extracellular pathogen is coated with over 1×10^7 molecules of variable surface glycoprotein, or VSG. The *T. brucei* genome contains genes for over a thousand different VSG molecules. Only one of these molecules, however, is expressed at a given time. This thick coating shields other cell surface molecules from the host immune response. A strong immune response to the VSG molecules themselves, though, is generated. To evade this response, *T. brucei* cells can switch their VSG molecules. The expression pattern periodically changes. When a new VSG is expressed, the host's existing immune response becomes much less effective. For a period of time, the pathogen can replicate relatively unencumbered by the host. Eventually, the host's immune system responds to this new VSG and the numbers of *T. brucei* cells within the host again begin to decline until a new VSG is expressed.

CONNECTIONS The bacterial pathogen *Salmonella* Typhimurium alters one of its major surface antigens—the flagellin subunit of its flagellum—by turning off the expression of one flagellin gene and turning on the expression of another flagellin gene. This switch occurs through the inversion of the regulatory region of an operon. A detailed description of operons can be found in **Section 11.2**.

Herpes simplex viruses avoid detection by the host immune system in a very different way. Following a period of replication in epithelial cells, some virus particles enter sensory neurons. The viral capsid travels to the cell nucleus and establishes **latency**; the viral genome remains present within these cells but transcription becomes very limited. Because few, if any, viral proteins are produced, the host immune response to the virus also becomes very limited. Periodically, though, **reactivation** occurs **(Figure 18.7)**. This process often occurs when the host experiences some type of cellular stress. The virus travels down the neuron, leaves the cell, and infects an epithelial cell. As viral replication then resumes, cellular damage also resumes, resulting in an outbreak of cold sores or genital lesions.

Antigenic variation and latency represent two means of evading a particular host defense—the immune system. Microbial pathogens have evolved many other means of evading various host defenses **(Table 18.4)**. As we saw in Perspective 7.1, for example, the Vif protein allows HIV to evade the retrovirus-directed defense mechanism of human cells. Even bacterial cells have defense mechanisms targeting bacteriophages **(Perspective 18.1)**.

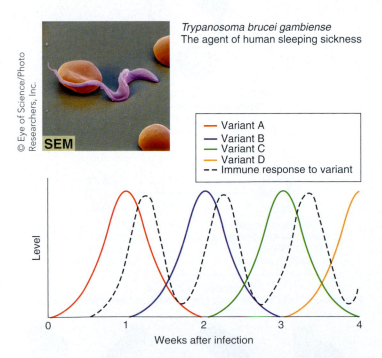

Trypanosoma brucei gambiense
The agent of human sleeping sickness

© Eye of Science/Photo Researchers, Inc.

SEM

— Variant A
— Variant B
— Variant C
— Variant D
- - - Immune response to variant

Level

Weeks after infection

Figure 18.6. Evading host defenses through antigenic variation
During a trypanosome infection, the immune system mounts a response, primarily targeting the major surface antigen. During the infection, though, the trypanosomes alter their surface antigen (indicated by the color changes between variants A, B, C, and D). When the surface antigen shifts, the host immune system is less effective and must play catch-up, giving the pathogen a chance to resume replicating.

TABLE 18.4 Selected strategies for evading host defenses

Evasion strategy	Example	Function
Antigenic variation	Influenza virus *Trypanosoma brucei*	Avoid antibody-mediated immunity of host
Latency	Herpes simplex virus Epstein Barr virus (EBV)	Avoid recognition by host immune system
Capsule formation	*Streptococcus pneumoniae*	Inhibit phagocytosis
Biofilm formation	*Pseudomonas aeruginosa*	Prevent access to bacterial cells
Replication within macrophages	*Listeria monocytogenes*	Facilitates cell-to-cell spread

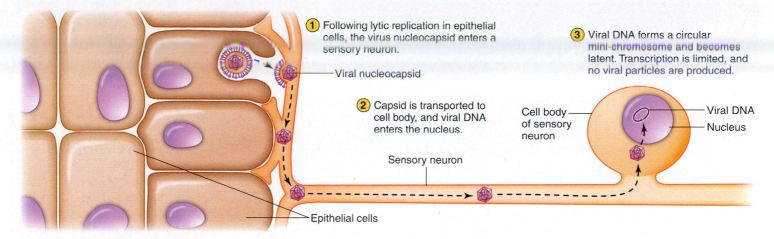

① Following lytic replication in epithelial cells, the virus nucleocapsid enters a sensory neuron.

Viral nucleocapsid

② Capsid is transported to cell body, and viral DNA enters the nucleus.

③ Viral DNA forms a circular mini-chromosome and becomes latent. Transcription is limited, and no viral particles are produced.

Cell body of sensory neuron

Viral DNA

Nucleus

Sensory neuron

Epithelial cells

A. Establishment of latency

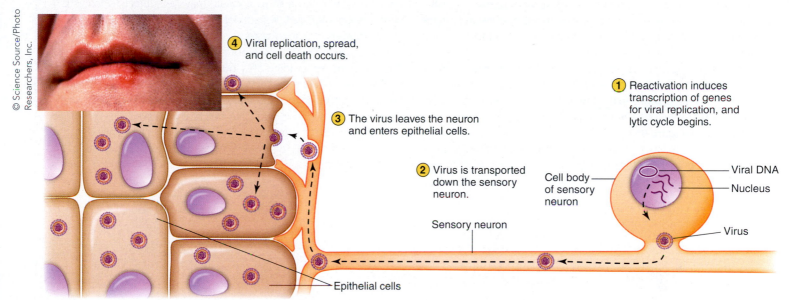

© Science Source/Photo Researchers, Inc.

④ Viral replication, spread, and cell death occurs.

③ The virus leaves the neuron and enters epithelial cells.

② Virus is transported down the sensory neuron.

① Reactivation induces transcription of genes for viral replication, and lytic cycle begins.

Cell body of sensory neuron

Viral DNA

Nucleus

Sensory neuron

Virus

Epithelial cells

B. Reactivation and lytic replication

Figure 18.7. Process Diagram: Herpes simplex virus latency and reactivation A. Following an initial period of replication and acute disease, the virus enters a sensory neuron. It travels to the cell nucleus and the viral genome exists in a latent state. **B.** In response to some stress encountered by the host, the virus can undergo reactivation. The virus leaves the sensory neuron, infects an epithelial cell and resumes replication. Transcription and translation of viral genes lead to the production of more virus particles and a recurrent outbreak of cold sores or genital lesions.

Pathogenesis

Not surprisingly, different microbial pathogens cause disease in different ways. Moreover, most pathogens possess several properties that, in concert, lead to pathogenesis. Often, pathogenic bacteria cause disease through the production of **toxins**, substances that damage the host. Some bacteria, like *Clostridium botulinum*, produce exotoxins, toxins made inside the cell that are released or transported outside the cell. These toxins can have various effects on their target cells (see Table 21.5 and Table 21.7). The lipopolysaccharide component of the cell wall of Gram-negative bacteria (also referred to as "endotoxin") and the lipoteichoic acid component of Gram-positive bacteria also can act as toxins. We will explore the specific actions of various bacterial toxins more thoroughly in Section 21.1.

Viruses generally do not produce toxins. Rather, diseases caused by viruses often result from the intracellular replication of these microbial pathogens. Let's examine this concept in more detail. As we noted in the opening story of this chapter, poliovirus can cause paralysis. The disease occurs because infected motor neurons, the neurons that interact with muscle cells, die. One could assume that infected motor neurons die because the replicating virus lyses the host cell. It appears, though, that there is a different cause of cell death. Poliovirus-infected cells undergo **apoptosis**, or programmed cell death. Infection activates a specific cellular process that results in the destruction of the infected cell. Presumably, destruction of infected cells in this manner reduces the spread of the virus to other cells. In this case, however, disease results from the host's attempts to limit the infection.

In our discussions about infectious diseases, it is important to remember that not only humans, or vertebrates, for that matter, are subject to infections. Insects can be infected by various microbial pathogens, like the protozoal pathogen *Plasmodium falciparum* that infects mosquitoes. Additionally, probably all microbes can be infected by other microbes. In fact, as we noted in Chapter 5, viruses infect every known form of life. Because of these pathogenic threats, all forms of life possess defenses against infections.

Bacteria possess several means of defending themselves against bacteriophages, viruses that can infect bacteria. For instance, bacteria may prevent or impede bacteriophage attachment. The bacterial cells may lack an appropriate receptor or possess an external structure, like a slime layer or capsule, that blocks bacteriophage attachment. Additionally, bacteria possess a novel intracellular host defense system—restriction enzymes.

As we saw in Section 9.3, restriction enzymes have played a major role in the development of molecular biology techniques. Restriction enzymes, though, do not exist simply to benefit our research endeavors. Rather, they represent an effective host defense for bacteria against viruses **(Figure B18.1)**.

Because restriction enzymes cleave double-stranded DNA at specific locations, they destroy any DNA molecules that contain these particular recognition sites. When the DNA of a bacteriophage enters a bacterial cell, host restriction enzymes cleave it and the infection is blocked. While this strategy seems straightforward, an obvious question arises. How does the bacterial cell protect its own DNA from degradation by the restriction enzymes? The answer may seem familiar. As we learned in Section 7.5, DNA repair mechanisms often depend on the methylation of DNA. Similarly, bacteria protect their DNA from degradation by adding methyl groups ($-CH_3$) to specific bases within their DNA. The restriction enzymes do not cleave this methylated DNA, thereby preventing them from cleaving the host DNA. Invading bacteriophage DNA, conversely, may not be methylated and, as a result, may be subject to restriction enzyme cleavage and degradation. This process is referred to as the "restriction-modification system."

Of course, the story does not end here. Bacteriophages possess mechanisms for evading this host defense system. Certain bacteriophages encode proteins that can inhibit host cell restriction enzymes. Other phages stimulate the host cell's methylases in an effort to methylate their own DNA before it is cleaved. The genomes of still other phages contain few or no sequences recognized by the host cell restriction enzymes. All of these evolutionary adaptations allow the phages to subvert bacterial restriction-modification systems.

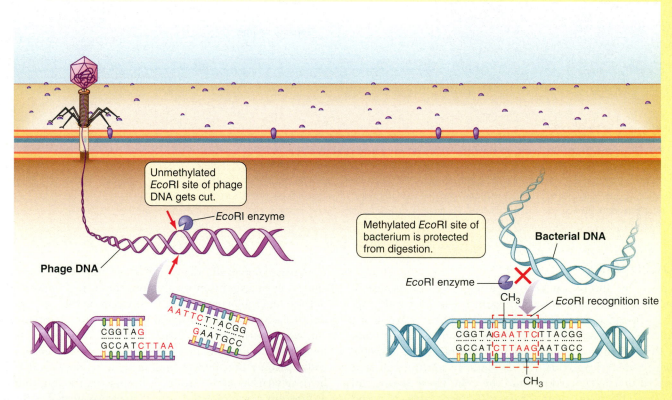

Figure B18.1. Process Diagram: Restriction enzymes—bacterial defenses against viruses Following their attachment to a host cell, bacteriophages inject their genome into the cell. To prevent infection, host cell restriction enzymes can cleave the invading phage genome, thereby preventing the generation of new phage particles and protecting the cell. To protect their own DNA from cleavage by endogenous restriction enzymes, bacteria modify their DNA, usually by adding a methyl group ($-CH_3$) to bases within the restriction enzyme recognition sites. The restriction enzymes do not cleave this modified DNA. Invading bacteriophage DNA may not possess these DNA modifications and, as a result, is subject to cleavage.

Other microbial pathogens also cause disease by inducing apoptosis in their host. As we will see in **Section 21.1**, some bacterial pathogens produce cytolysins, toxins that damage the plasma membrane. For example, α-toxin, a cytolysin produced by some strains of *Staphylococcus aureus*, creates pores in the plasma membrane of host cells, which then leads to apoptosis.

In this section, we have seen how example pathogens—bacterial, viral, and eukaryal—invade hosts, evade the host defenses, and cause cellular damage. We also must consider how pathogens enter and exit hosts and spread within a population. We will investigate these issues in the next section.

18.2 Fact Check

1. Describe some microbial features that facilitate attachment or entry.
2. Explain how attachment and host range of a pathogen are related. Provide an example.
3. Identify and describe microbial strategies for evading host defenses.
4. How can apoptosis induce disease?

18.3 The transmission of infectious diseases

How are infectious diseases transmitted?

For a pathogen, simply replicating within a host is not sufficient for continued survival. An individual host organism eventually will die. The microbe, then, must be transmitted successfully between susceptible hosts. Some pathogens, like HIV, are transmitted directly from one host to another. Other microbes, like influenza virus, must remain stable in the outside environment during transmission, remaining capable of initiating an infection when they encounter new susceptible hosts. Still other pathogens, like *Vibrio cholerae*, may replicate in a specific host, causing disease, and also replicate in the external environment. As we will discover in this section, the transmission of pathogens from one individual to another depends on where the pathogen replicates, how stable it is in the environment, and how it exits its host. Transmission between individuals, in turn, affects how pathogens spread within populations.

Routes of transmission

How a pathogen moves from one host to another represents an interesting and important aspect of studying infectious diseases. **Transmission**, the spread of an infectious agent from one host to another or from its source, or **reservoir**, to a host, can occur in many ways. Most human infectious diseases spread via contact. **Direct contact transmission** refers to spread that occurs via physical contact between an infected person and a susceptible person. **Indirect contact transmission** refers to spread that occurs from one person to another via another object, often a contaminated inanimate object or an insect. Other, more specific categories of transmission also can be described.

Generally, routes of pathogen transmission are related to how the pathogens enter and exit their hosts. The site of replication in the host, in turn, often determines entry and exit. Pathogens that can replicate within the intestinal tract, like the bacterium *Vibrio cholerae* and the virus rotavirus, are eliminated through the feces. The pathogens then may contaminate water supplies, be transferred to hands, or be transferred to **fomites**,

inanimate objects like doorknobs. Another susceptible host may ingest the pathogens, either by drinking contaminated water or transferring the pathogens from a fomite to his or her mouth. This type of transmission, in which a pathogen is excreted in the feces of one individual and then ingested by another individual, is referred to as **fecal-oral transmission**. Pathogens that replicate in the respiratory tract usually exhibit **respiratory transmission**, also referred to as "aerosol transmission." In these instances, infectious pathogens leave an infected host when the host coughs or sneezes. The pathogens also may leave a host when he or she laughs, burps, or simply talks. The pathogens then can be inhaled directly by a second host or become deposited on a fomite and transferred to the nose or mouth of a second host.

Other pathogens may undergo **vector-borne transmission**. In these cases, another species, often an insect, transfers the pathogen between hosts. For some pathogens, such as *Ophiostoma novo-ulmi*, the fungus that causes Dutch elm disease, the insect vector merely transfers the pathogen from host to host. For other pathogens, such as *Plasmodium falciparum*, the protozoan that causes malaria, the pathogen replicates and develops within mosquitoes, its vector. When an infected mosquito then bites a susceptible human, the pathogen enters the new host and continues its replication cycle.

Plasmodium falciparum undergoes a complex life cycle that requires replication on both mosquitoes and humans. The details of this life cycle, and how the replication of *P. falciparum* leads to disease, will be addressed in **Section 23.2**.

Sexual transmission refers to transmission that occurs during vaginal, anal, or oral sex. Infectious diseases that are spread sexually are referred to as **sexually transmitted infections**, or **STIs**. Some sexually transmitted pathogens reside on the surface of the penis or within the vaginal or anal

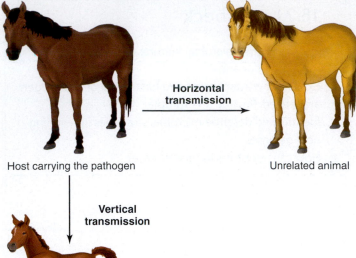

Host carrying the pathogen

Horizontal transmission

Unrelated animal

Vertical transmission

Offspring

Figure 18.8. Horizontal and vertical transmission of pathogens Generally, pathogens are transmitted from one organism to another in a horizontal fashion. For some pathogens, however, vertical transmission from a parent, usually the mother, to the offspring can occur. The pathogen usually is transmitted *in utero*, during birth, or via breast milk.

canals and are transmitted through direct contact during sex. Other sexually transmitted pathogens, like HIV, are present in semen and vaginal fluids and can be transferred from these fluids through small abrasions in the vagina, anus, or penis to the bloodstream of another person.

HIV also can undergo **vertical transmission**, passing from parent to child. Most commonly, vertical transmission of pathogens occurs from mother to child *in utero*, during birth, or shortly after birth. Because HIV can exist in the breast milk of an HIV⁺ woman, breastfeeding can lead to the vertical transmission of this pathogen. In the absence of preventive measures, approximately one-third of infants born to an HIV⁺ woman will acquire the virus. **Horizontal transmission**, in contrast, refers to transmission of a pathogen between members of a species other than parent to offspring (**Figure 18.8**).

Finally, humans may become infected by pathogens through **zoonotic transfers**. In these cases, the pathogen is transferred from its natural host, or **reservoir host**, to a human (**Table 18.5**). Often, humans serve as **dead-end hosts**, or incidental hosts, for zoonotic transfers; the pathogen may replicate within an infected person and cause severe disease, but is not efficiently transferred from one person to another. Human-to-human spread of the pathogen, in other words, does not occur or occurs inefficiently.

The H5N1, or avian, influenza virus presents a good case with which to illustrate this concept. Between 2003 and 2011, approximately 580 confirmed human cases of H5N1 influenza were reported. Of these infections, 340 resulted in death, making the virus quite virulent in humans. In almost all of the cases, however, the virus was transmitted to humans from poultry. Very few cases of human-to-human transmission have been reported. Researchers and public health officials remain very concerned

about this virus. The virus could mutate in such a way that transmission between humans becomes more likely. Indeed, in the fall of 2011, researchers reported that they had identified specific mutations within the H5N1 influenza virus genome that makes it much more transmissible between ferrets. These same mutations could make it more transmissible between humans (see Perspective 22.4).

Epidemiology

To more fully understand how pathogens are transmitted, we need to examine not just the interactions between a pathogen and a single individual. Rather, we need to examine the interactions between pathogens and populations. **Epidemiology** is the study of patterns of disease in populations. Epidemiologists investigate all diseases—cancers, chemical poisoning, obesity, genetic diseases, and, of course, infectious diseases. Epidemiologists concern themselves with determining the source of diseases, identifying disease risk factors, designing and evaluating infection control policies, and predicting the future spread of disease. They also investigate the **morbidity rate**, or rate of disease, and **mortality rate**, or rate of death, associated with diseases. Statistics and mathematical modeling of disease patterns form a large part of epidemiology, and provide valuable predictions of disease spread and identification of the major parameters involved in the transmission and control of disease. The ultimate goal is quite simple: the more we know about how infectious diseases spread, the better equipped we are to prevent them.

Organizations, such as the United States Centers for Disease Control and Prevention (CDC), rely on communication with local health officials to collect and report relevant data from populations to assist them in identifying the frequency of disease and death, and in evaluating treatment and prevention practices. As you can imagine, information cannot be gathered from every single member of a large population because of practical and monetary restrictions. Frequently, only a subset of the population can be observed. For example, to get information on HIV infection rates in South Africa, researchers from various organizations commonly monitor women by visiting maternity wards. Researchers then use these data to predict levels of infection in the wider population.

TABLE 18.5	Selected zoonotic diseases	
Disease	**Pathogen**	**Animal host(s)**
Rabies	Rabies virus	Many mammals (raccoons, bats, skunks, foxes)
Hantavirus pulmonary syndrome	Hantavirus	Rodents
Ebola hemorrhagic fever	Ebola virus	Unknown (bats?)
Anthrax	*Bacillus anthracis*	Cattle, sheep, goats
Tularemia	*Francisella tularensis*	Rodents, rabbits
Psittacosis	*Chlamydia psittaci*	Many bird species

Epidemiologists report the status of an infectious disease in a population using specific quantitative and qualitative terms. To understand how a pathogen may affect a population, we must first understand these terms.

Case

In the context of infectious disease, a case must be specifically defined in order to unambiguously diagnose and count affected individuals. A case may be defined as an individual who exhibits the disease. Alternatively, a case may be defined as an individual who is infected, but may or may not be showing signs or symptoms of disease. These asymptomatic individuals may be identified through diagnostic testing, such as PCR, culture, or ELISA (see Toolbox 20.1).

Among people, asymptomatic and subclinical infections usually do not come to the attention of health care workers and therefore frequently do not get recorded as cases. In farmed animals, however, routine diagnostic testing is carried out to determine the infection status of animals for certain notifiable agents such as foot-and-mouth disease virus, which causes foot-and-mouth disease, and *Mycobacterium bovis*, which causes tuberculosis in cattle. With these tests, researchers can detect subclinical infections—infections that have not yet progressed to a noticeable disease state—in the animals.

Incidence and Prevalence

Incidence refers to the number of new cases appearing in a population during a specific time period. Incidence may be reported as a proportion or a percentage. For example, in Africa, the incidence of tuberculosis during 2006 was 363 per 100,000 people (or roughly 0.36 percent) according to the World Health Organization (WHO). For 1990, the incidence was 162 per 100,000 people.

Prevalence refers to the total number of cases in a population at a particular point in time or during a particular time period. Again, prevalence often is reported as a proportion or rate. For example, in Africa the prevalence of tuberculosis during 2005 was 547 per 100,000 people according to the WHO. In 1990, the prevalence was 333 per 100,000. The prevalence of disease is higher than the incidence because it includes both new and preexisting cases **(Figure 18.9)**.

Incidence and prevalence reporting represent the most common epidemiological measures of infectious disease in populations. These measurements not only tell us the level of infectious disease in a population, but also allow us to make comparisons within and between populations over time. Incidence rate is one of the best indicators of disease risk in a susceptible population, and also tells us how fast a disease is spreading in a population. For the year 2007, for example, China and India had the largest overall number of new tuberculosis cases. Does that mean China and India are hotspots for tuberculosis? Not necessarily. The number of new cases does not reflect the total population size. To compare the risk of tuberculosis between populations, we need to compare the incidence rates—how many new cases occurred within a given population size.

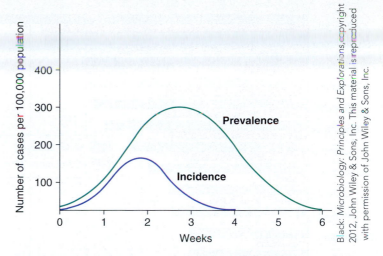

Figure 18.9. Prevalence and incidence: Epidemiological measures of infectious diseases in a population Prevalence is an estimate of how common or widespread a disease is within a population over a defined period of time. Incidence gives information about the risk of contracting the disease within a population over a defined period of time. Prevalence will always be higher than incidence because it includes all existing cases, not just new cases.

China and India are the two most populated countries in the world, and even a low incidence rate corresponds to very large numbers of new cases. While China and India have the overall largest number of new tuberculosis cases (incidence), they did not have the highest incidence rate (number of new cases per 100,000 people, for instance). The incidence of tuberculosis is much higher in several African countries and Cambodia, indicating a higher density of disease and therefore more risk of infection from the disease in these countries.

To better understand how an infectious disease is spreading in a population, we need to compare the incidence and incidence rate over time. **Table 18.6** shows the incidence and incidence rate of tuberculosis in China, South Africa, and the Netherlands for the years 1990 and 2010. In 1990, over 15 times more people in China than in South Africa became infected with tuberculosis. The incidence rate, however, was only about 2 times higher in South Africa. From these numbers, we could speculate that the risk of acquiring tuberculosis was fairly similar in both countries. The situation changed drastically by 2010. China's incidence rate dropped slightly, and South Africa's skyrocketed. In 2010, the incidence rate of tuberculosis was 13 times higher in South Africa than in China. During this same period, the incidence and incidence rate of tuberculosis remained very low in the Netherlands. What caused the dramatic increase in the tuberculosis incidence rate in South Africa? Between 1990 and

TABLE 18.6	Comparison of incidence and incidence rates for tuberculosis					
Country	**Population[a]**		**Incidence**		**Incidence rate[b]**	
	1990	**2010**	**1990**	**2010**	**1990**	**2010**
South Africa	37	50	110,000	490,000	300	980
China	1,145	1,341	1,700,000	1,000,000	150	75
Netherlands	15	17	1,500	1,200	10	7

[a] In millions.
[b] Per 100,000.

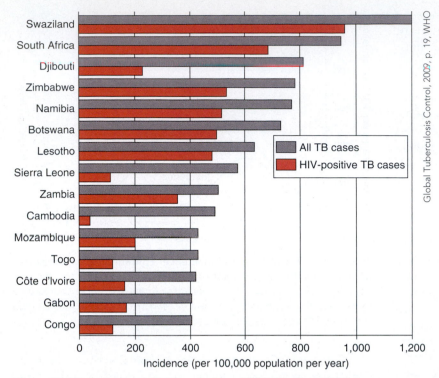

Global Tuberculosis Control, 2009, p. 19, WHO

Legend:
- All TB cases
- HIV-positive TB cases

Incidence (per 100,000 population per year)

Figure 18.10. Incidence of tuberculosis and HIV infection, 2006 Many of the countries with the highest incidence rates of tuberculosis (TB) have a large proportion of tuberculosis cases associated with HIV.

in other words, are more susceptible to tuberculosis. In 2006, 44 percent of all new tuberculosis cases in South Africa were accompanied by HIV infection (**Figure 18.10**). In China, which experienced a lower incidence rate of HIV, only 0.3 percent of people with tuberculosis also were HIV$^+$.

Incidence also can be used to identify **emerging diseases** that previously are unknown diseases, or known diseases with a recent significant increase in incidence. The pathogens associated with these diseases likewise are referred to as emerging pathogens. Examples of pathogens that cause emerging diseases include HIV, Ebola virus, *Helicobacter pylori*, *Escherichia coli* O157:H7, SARS-CoV, *Borrelia burgdorferi*, *Clostridium difficile*, and monkeypox virus (**Figure 18.11**), a virus closely related to the smallpox virus that can infect humans (**Microbes in Focus 18.3**). By analyzing incidence patterns, we also can uncover re-emerging diseases/pathogens—those known diseases that have recently increased in incidence after a previous significant decline. Examples of re-emerging diseases include tuberculosis, necrotizing

2010, the number of people with HIV/AIDS increased dramatically in South Africa. The incidence of tuberculosis showed a positive correlation with HIV infection rates. This correlation is explained by the destruction of immune cells by HIV. Specifically, HIV leads to the destruction of T helper (T$_H$) cells, which are central to the adaptive immune response (see Chapter 20). A diminished T$_H$ cell population means immunity to tuberculosis is severely compromised. People with HIV/AIDS,

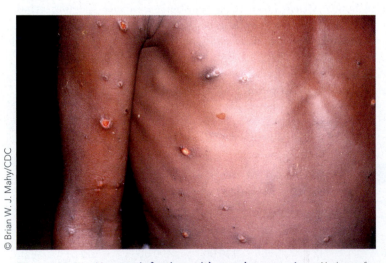

Figure 18.11. Human infection with monkeypox virus Monkeypox first was reported in humans in 1970. In 2003, an outbreak occurred in the United States, where the people most likely contracted the virus from pet prairie dogs.

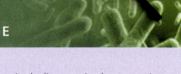

Microbes in Focus 18.3
MONKEYPOX: AN EMERGING DISEASE

Habitat: Infects a variety of hosts including squirrels, rats, mice, rabbits, monkeys, humans (first reported in 1970). Originates from central and western Africa.

Description: Large, enveloped virus, about 200 by 250 nm in size, belonging to family *Poxviridae*, and closely related to smallpox (variola) virus

Key Features: Monkeypox virus produces a disease in humans similar to smallpox, but milder. It usually is transmitted to humans from bites of infected animals, and is communicable between humans. In 2003, several people in the United States contracted monkeypox from pet prairie dogs (small ground squirrels), which had contracted the disease when exposed to a pet store-bound shipment of small rodents originating from Ghana. Some of the animals were carrying the virus. Monkeypox appears to be an emerging zoonotic disease in Africa. The range of monkeypox virus recently has increased, and the incidence of monkeypox in humans has escalated.

TEM

© Hazel Appleton, Health Protection Centre for Infections/ Photo Researchers, Inc.

© Brian W. J. Mahy/CDC

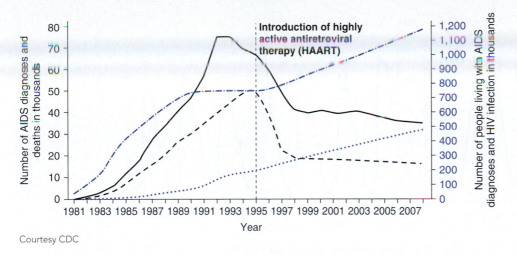

Courtesy CDC

Figure 18.12. Comparison of prevalence, incidence, and mortality of HIV/AIDS Data are shown for persons >13 years of age. AIDS incidence and deaths increased rapidly through the 1980s, and began to decline around 1995, when effective antiretroviral therapies became more readily available. The decline in mortality results in an increase in HIV and AIDS prevalence as fewer people with AIDS are dying.

diseases caused by *Streptococcus pyogenes*, and invasive diseases caused by *Staphylococcus aureus*.

Incidence may tell us how fast a disease is spreading within a population, but prevalence gives a practical indication of the disease burden in a population. It describes how many individuals are living with the disease at a given time. Prevalence information is particularly useful for health care systems that need to plan for the care of patients. Prevalence rates are also useful for monitoring the effectiveness of infection control and therapy programs. Let's look at how prevalence rates can be used to monitor AIDS.

AIDS was first recognized in the United States in 1981. The time from HIV infection to death from AIDS in adults, without treatment, varies greatly between individuals, with a median time of 10 years. Throughout the 1980s and early 1990s, incidence and prevalence of AIDS increased greatly, because more people became infected with HIV and developed AIDS than were dying from AIDS. Around 1995 when highly active antiretroviral therapy (HAART) became available, mortality decreased, but prevalence continued to increase greatly. Why did this happen? The effective HIV/AIDS therapies lengthen the duration of AIDS and delay mortality. As a result, more people were living with AIDS, pushing prevalence higher. An increase in prevalence in this case means treatment strategies are working. The AIDS incidence curve is confounded because in 1993, the clinical diagnosis for AIDS was changed, and as a result, a greater number of cases were diagnosed. This change in the case definition explains the apparent peak and subsequent drop in incidence in 1993 **(Figure 18.12)**.

Patterns of infectious disease

When do health officials become concerned about the incidence of an infectious disease? For some diseases, like extremely drug-resistant tuberculosis, a single case can be alarming because it cannot be effectively treated and is contagious. For other diseases, like measles, it takes an unusually large number of cases to raise the alarm. In this section we will define terms used by epidemiologists to describe the presence of infectious disease in populations.

Endemic Disease

An **endemic disease** or agent is one that is habitually present in a population. For example, rabies is endemic to North American foxes, raccoons, skunks, and bats. For many endemic diseases, cyclical patterns of increased and decreased incidence occur. This periodicity is usually a function of (1) seasonal changes or (2) variations in population immune status. For example, in the United States and Canada, the incidence of human encephalitis from West Nile virus infection peaks in late summer and fall following the height of local mosquito (the vector) populations **(Figure 18.13)**. Immune status also can cause fluctuations in

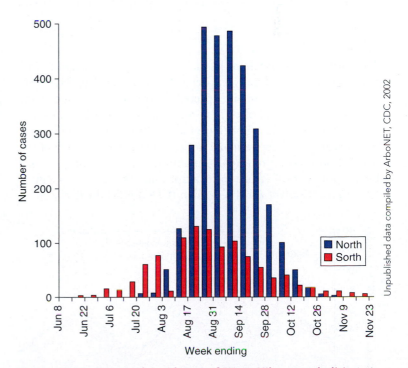

Unpublished data compiled by ArboNET, CDC, 2002

Figure 18.13. Seasonal incidence of West Nile encephalitis In the northern United States, incidence of disease rises sharply in August, and declines through the fall months, closely paralleling vector populations of mosquitoes. In southern states, cases appear over a wider range of months because there is less variation in mosquito populations.

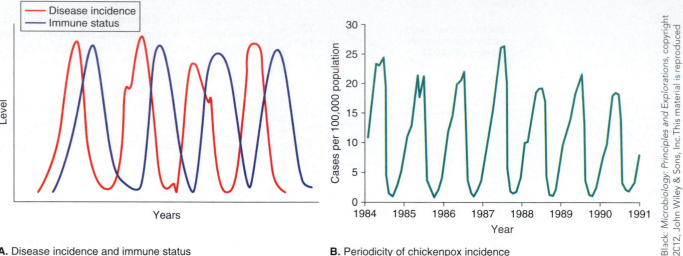

A. Disease incidence and immune status

B. Periodicity of chickenpox incidence

Figure 18.14. Periodicity of disease incidence When a pathogen enters a susceptible, non-immune population, there will be an initial high incidence of disease. Survivors of infection will be immune and incidence will drop to a low level. **A.** The incidence of disease may exhibit a cyclical pattern. As young are born into the population, the proportion of immune individuals decreases and the incidence increases. This periodicity commonly is seen for many endemic diseases. **B.** Varicella-zoster virus, the cause of chickenpox, is endemic to the United States and shows cyclical occurrence in incidence.

incidence. In a non-immune population, a previously high incidence of an infectious disease results in a population composed largely of immune survivors. Any susceptible individuals in this population will be widely dispersed so the agent is no longer frequently transmitted. Consequently, incidence drops to a low level **(Figure 18.14)**. As non-immune offspring join the population, the population shifts again to a high level of susceptible individuals, and another peak in incidence occurs. Such periodicity is found in several endemic diseases, such as chickenpox, influenza, pertussis, and measles.

Epidemic

An **epidemic** occurs when the incidence of a disease significantly rises above that normally expected for a population. The term **outbreak** refers to an unexpected cluster of cases appearing within a short period of time in a localized population. There is no absolute dividing line between an outbreak and an epidemic. In general terms, an epidemic involves a larger proportion of the population than an outbreak. Sometimes epidemics are defined by a certain cut-off value. For example, in 1990, the U.S. Centers for Disease Control and Prevention (CDC) officially declared an

influenza epidemic in the United States when the mortality rate from influenza rose above the expected 6.7 percent set in previous years. Endemic diseases can reach epidemic levels if enough susceptible individuals are present **(Figure 18.15)**.

Pandemic

A **pandemic** is an epidemic concurrently affecting populations globally, usually on more than one continent. Some pandemics have covered the world, such as the H1N1 Spanish flu of 1918 **(Table 18.7)**. In June of 2009, the World Health Organization officially classified H1N1 influenza disease, commonly dubbed "swine flu" as a pandemic. They made this designation after

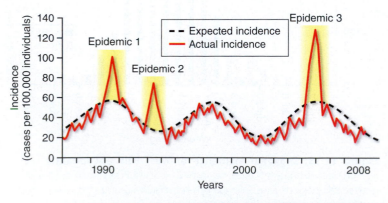

Figure 18.15. Endemic diseases may become epidemic The dashed line shows the expected cyclical incidence, resulting from historical patterns of endemic disease. Epidemics, shown as peaks 1, 2, and 3, arise when the incidence of endemic disease rises above that expected.

TABLE 18.7   Notable pandemics of humans

Disease	Causative agent	Years
Plague	*Yersinia pestis*	541–747
		1346–1451
		1665–1666
		1855–1859
Smallpox	Variola virus	1493–1591
		1665–1666
		1870–1875
Flu	Influenza virus H1N1	1918–1920
	Influenza virus H2N2	1957–1958
	Influenza virus H3N2	1968–1969
	Influenza virus H1N1	2009–2010
Cholera	*Vibrio cholerae*	1816–1826
		1829–1851
		1961–current
Typhus	*Rickettsia prowazekii*	1528–1566
AIDS	Human immunodeficiency virus (HIV)	1981–current

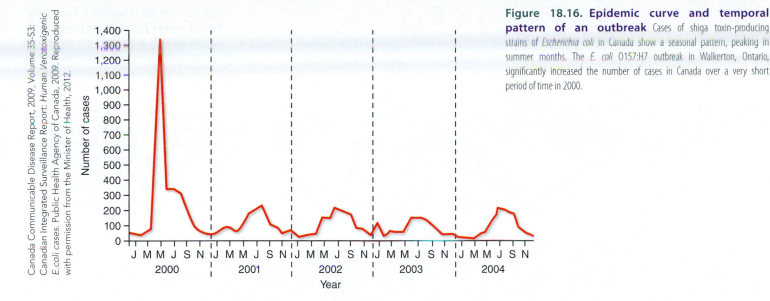

Canada Communicable Disease Report, 2009, Volume 35-S3: Canadian Integrated Surveillance Report: Human Verotoxigenic E.coli cases. Public Health Agency of Canada, 2009. Reproduced with permission from the Minister of Health, 2012.

Figure 18.16. Epidemic curve and temporal pattern of an outbreak Cases of shiga toxin-producing strains of *Escherichia coli* in Canada show a seasonal pattern, peaking in summer months. The *E. coli* O157:H7 outbreak in Walkerton, Ontario, significantly increased the number of cases in Canada over a very short period of time in 2000.

observing conclusive evidence of sustained transmission of the virus on at least two continents.

Types of epidemics

Cases of infectious diseases within populations do not occur randomly; they occur in patterns. These patterns describe how an outbreak or epidemic moves temporally (with time) and spatially (with location) in populations. The temporal pattern of an outbreak or epidemic is presented in the form of an epidemic curve (**Figure 18.16**). The frequency of cases is plotted against an appropriate time interval. The shape of the curve can help to reveal the nature of the epidemic. Spatial patterns can be tracked on a map to help visualize the pattern of cases.

Common-Source Epidemics

A **common-source epidemic** or outbreak occurs when there is a single source of infection to which the population is exposed

(**Mini-Paper**). For example, outbreaks of *E. coli* O157:H7 frequently can be traced to contaminated food or sometimes water (read about the Walkerton outbreak in the opening story of Chapter 21). Similarly, *Cryptosporidium* outbreaks often can be traced to contaminated freshwater supplies. Because a large number of susceptible individuals are exposed to the source of infection in a short period of time, common-source epidemics typically show a sharp increase in disease incidence. A slower but marked decline then occurs; fewer individuals in the exposed population remain susceptible (**Figure 18.17**). The time over which cases are reported tends to reveal the disease **incubation period**, the period between pathogen entry and the appearance of illness. Cases in common-source epidemics tend to remain spatially isolated near the local source of infection, which facilitates its identification (**Figure 18.18**). Successful location of the source usually allows the outbreak to be rapidly controlled. For example, removal of contaminated food or chlorinating the drinking water can stop outbreaks of *E. coli*.

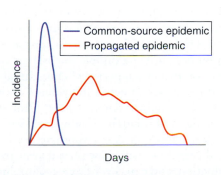

Figure 18.17. Propagated and common-source epidemics Common-source epidemics (blue line) commonly display a single peak in incidence because the majority of cases will have encountered the source of infection within a short period of time. The time over which cases appear corresponds to the incubation period. In propagated epidemics (red line), each case becomes a source, causing several other cases, which results in numerous cases over several incubation periods and multiple small peaks.

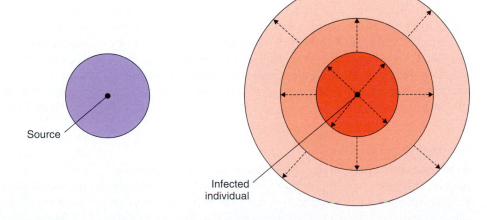

A. Common-source epidemic   **B.** Propagated epidemic

Figure 18.18. Movement of common-source and propagated epidemics A. In a common-source epidemic, cases will tend to cluster near the source of infection, which remains relatively static. **B.** In a propagated epidemic, the source is moved with each new infection. Cases first will appear near the source, usually an infected individual, but will begin to radiate away from this source as transmission moves the infection from individual to individual.

Black: Microbiology: Principles and Explorations, copyright 2012, John Wiley & Sons, Inc.
This material is reproduced with permission of John Wiley & Sons, Inc.

Figure 18.19. Cholera and the Broad Street pump: Mapping cases of cholera to a source In London in 1854, John Snow linked cholera with contaminated drinking water by carefully plotting known deaths on a map. He noticed cases were tightly clustered around the Broad Street pump. His systematic investigation confirmed all cases had obtained their water from this particular pump, which under microscopic examination was shown to contain organic matter. Snow suspected the contamination came from a sewer that was located above and just a few feet away from the shallow well that supplied the Broad Street pump. For his innovative thinking and actions, John Snow is considered to be the father of epidemiology.

⊗ Pumps
∴ Deaths from Cholera

The cholera outbreak of 1854 in the Soho area of London represents one of the most famous examples of a common-source epidemic. John Snow, widely regarded as the first epidemiologist, showed that cholera resulted from contaminated drinking water. By tracking the location of cholera cases on a map **(Figure 18.19)**, Snow linked the cases to a single source of contaminated water obtained from the Broad Street pump. People living near another neighborhood pump, from which they obtained their water, did not become ill. Snow convinced the Board of Guardians to remove the handle from the pump to prevent access to the contaminated water supply. Although new cases of the disease probably already were declining, Snow's observation often receives credit for ending the outbreak.

Propagated Epidemics

A **propagated epidemic** or outbreak results from infection passing from one host to another, either directly or indirectly. Propagated epidemics usually stem from the introduction of an infected individual into a susceptible population. Diseases easily transmitted between individuals produce propagated epidemics. Examples of such propagated epidemics include measles, influenza, chickenpox, and tuberculosis. Propagated epidemics also occur for some vector-transmitted diseases, such as yellow fever and malaria. Both agents are delivered by mosquitoes and replicate very efficiently in humans, resulting in a pattern of spread similar to diseases in which the pathogen is transmitted directly from person to person.

Propagated epidemics typically show a gradual increase in incidence as the infection is spread to more and more individuals, eventually reaching a peak. At this point, a significant proportion of the population is infected, has recovered and is immune, or has died. Fewer members of the population remain susceptible to infection. As opposed to common-source epidemics, cases in a propagated epidemic will continue to appear over more than one incubation period. For example, measles, a contagious viral disease, has an incubation period of approximately ten days. During this time, newly infected individuals do not show clinical signs of disease, and are not yet cases. After this time, they become cases and a new batch of individuals will be infected, resulting in a resurgence of cases approximately every ten days. Propagated epidemics will show spread away from the initial site of introduction, and can cover large geographical areas (see Figure 18.18).

Many infectious diseases do not show classic common-source or propagated epidemic curves. For example, cholera usually begins as a common-source epidemic, but in crowded, unhygienic conditions, it spreads from person to person, resulting in a mixed epidemic of propagated and common-source transmission. Investigating the epidemiology of an infectious disease, therefore, is not a simple task. Determining what causes an infectious disease and how the microbe is transmitted, however, is quite important. Effective treatment and prevention strategies depend on accurate information about these topics. In the following analysis of a specific investigation, we will begin to see how these issues can be addressed.

18.3 Fact Check

1. Describe the concepts of fecal-oral transmission, respiratory transmission, vector-borne transmission, and sexual transmission, and provide examples of each.
2. Differentiate between vertical and horizontal transmission.
3. How are the following related: zoonotic transfer, reservoir host, and dead-end host?
4. Distinguish between morbidity and mortality.
5. Describe how the terms "case," "incidence," and "prevalence" are used in epidemiology.
6. What are emerging and re-emerging pathogens? Give examples.
7. Differentiate between outbreak, epidemic, and pandemic.

18.4 Proving cause and effect in microbial infections

How can we determine the cause of an infectious disease?

The study of epidemiology, as we just saw, allows us to determine how infectious diseases spread through a population. From these studies, we may be able to determine the mode of transmission, infectivity, and natural vector of a pathogen. Such studies do not, however, allow us to determine which specific microbe causes a specific disease. This piece of information is absolutely necessary as prevention and treatment options depend on our ability to conclusively determine which microbe causes a given infectious disease. In this section, we will investigate this important topic. Additionally, we will look at a molecular corollary of this topic. More specifically, we will begin to address how we can determine which genes within a pathogenic microbe contribute to its pathogenicity.

Koch's postulates

Today, we know that microbes cause various diseases and may take for granted the evidence supporting this fundamental knowledge. A nineteenth-century microbiologist, Robert Koch, addressed the causality of infectious diseases. A physician, Koch investigated the link between bacteria and disease. He identified *Bacillus anthracis*, the causative agent of anthrax, *Mycobacterium tuberculosis*, and *Vibrio cholerae*, and demonstrated that bacteria could be grown in the laboratory in pure culture, separated from any other bacterial species. In other words, a single bacterial species could be grown in isolation, free from any other bacterial species. Koch's most famous contribution to microbiology and medicine, however, was his development of **Koch's postulates**, a guideline for demonstrating that a specific microbe causes a specific disease. These postulates directly led to the development of the germ theory of disease, and still represent important guides to determining the causes of infectious diseases.

According to Koch, to clearly prove that a microbe causes a disease, researchers must be able to:

1. Identify the suspected microbe in every person with the disease, but not those without the disease
2. Isolate the suspected microbe and grow it in pure culture
3. Show that experimental inoculation of a healthy host with the suspected microbe causes the same disease
4. Recover the suspected microbe from the experimentally inoculated host

To some extent, you may have seen Koch's postulates in action when visiting your physician. Strep throat is caused by *Streptococcus pyogenes* and is characterized by a severe sore throat, enlarged tonsils, and a fever. Other bacteria and viruses, though, also cause diseases with similar signs and symptoms. To determine if a person actually has strep throat, a physician often will conduct one of two tests. First, a throat swab may be used for a rapid strep test. In this test, the throat swab is analyzed for the presence of specific molecular determinants found in Group A streptococci, the group to which *S. pyogenes* belongs. The results of this test are available within minutes. Second, the throat swab may be streaked on agar plates. Following a period of incubation, the plates can be observed for the presence of *S. pyogenes* colonies. With these tests, then, the physician is applying Koch's first postulate—the causative microbe must be identified in every person with the disease.

CONNECTIONS Today, clinicians have at their disposal relatively rapid tests for numerous human pathogens. Often, these tests rely on enzyme-linked immunosorbent assays (ELISAs). In **Toolbox 20.2**, we will investigate these assays more thoroughly.

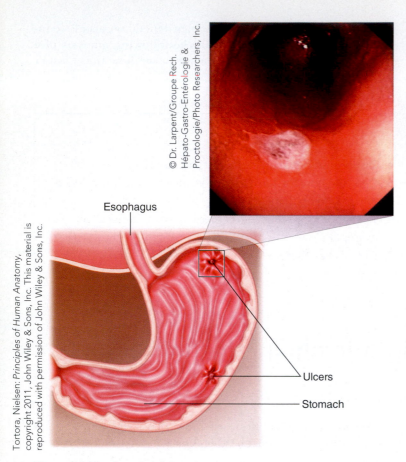

Esophagus

Ulcers

Stomach

Figure 18.20. Gastric ulcers Physicians used to assume that the ulcers formed from too much acid. Now, most physicians agree that the bacterium *Helicobacter pylori* plays a significant role in the development of ulcers.

Microbes in Focus 18.4
LIFE IN THE STOMACH: *HELICOBACTER PYLORI*

Habitat: Stomach and small intestine of several types of animals, including humans, cheetahs, dogs, and cats

Description: Slow-growing, spiral-shaped, Gram-negative bacterium about 3 μm long with flagella

Key Features: Roughly 50 percent of all humans are infected with *H. pylori*, although infection rates are higher in developing countries and lower in developed countries. The bacterium produces urease, an enzyme that converts urea to bicarbonate and ammonia. These molecules can neutralize acid, thereby creating an acid-free zone around the bacillus, which protects it during transit from the lumen of the stomach into the more neutral layers of mucus, where it begins its colonization of the stomach wall. *H. pylori* is equipped with several highly active flagella to help it drill through the thick mucus layer lining the stomach.

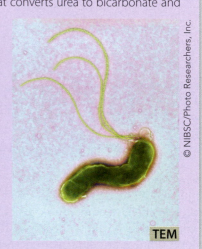

TEM

Our current understanding of ulcers presents a fascinating example of the importance and modern relevance of Koch's postulates. Until quite recently, physicians assumed that gastric ulcers, lesions in the mucosal lining of the stomach, resulted from too much acid (**Figure 18.20**). Additionally, some physicians also thought that the overproduction of acid resulted from stress. Treatment typically involved attempts to decrease stress and neutralize the excess acid. Usually, antacids were recommended. For many years, though, microbiologists had noticed curved bacteria in tissue samples taken from the stomachs of people with ulcers. The significance of these bacteria, however, was not adequately explored. No one isolated or characterized these organisms.

In the 1980s, two researchers at the University of Western Australia, Drs. Robin Warren and Barry Marshall, changed that. Warren observed that he *always* saw these curved bacteria associated with gastric tissue that showed signs of inflammation, or *gastritis*. Furthermore, the number of bacteria seemed to correlate with the severity of the inflammation. To explore the possible role of these bacteria, Warren and Marshall isolated the bacterium and grew it in the laboratory. Koch's first two postulates had been fulfilled. The suspected microbe, now known as *Helicobacter pylori* (**Microbes in Focus 18.4**), could be identified in everyone with the disease and the microbe could be grown in a pure culture.

To address Koch's third and fourth postulates (inoculation of a healthy host with the microbe causes the disease, and the

microbe then can be isolated from the inoculated host), the researchers took a very bold approach, an approach that almost seems more likely to occur in a Hollywood movie than in a real microbiology laboratory. They grew a pure liquid culture of *H. pylori*. Then Marshall and another volunteer drank it. Both of them developed an inflammation of the stomach lining, or gastritis. Furthermore, *H. pylori* could be isolated from biopsies of their stomachs. This experiment showed a strong link between *H. pylori* infection and gastritis. More thorough studies involving large numbers of people showed a link between infection and both gastritis and ulcers.

In this particular case, researchers used Koch's postulates to do more than just identify the microbe responsible for a disease. Warren and Marshall also changed dramatically the standard treatment regimen for ulcers. As we stated earlier, ulcers had been treated most commonly with antacids. After all, if excess acid causes ulcers, then decreasing the amount of acid should be the best treatment. Once physicians realized that a bacterium causes ulcers, though, antibiotics became the preferred treatment regimen. Today, people with ulcers typically are treated with a series of antibiotics. Warren and Marshall, in fact, showed that 80 percent of people with ulcers could be cured if *H. pylori* were eradicated. Because of these discoveries, Warren and Marshall received the Nobel Prize in Physiology or Medicine in 2005.

Despite the obvious value of Koch's postulates, there are some legitimate criticisms of them. Koch himself admitted the

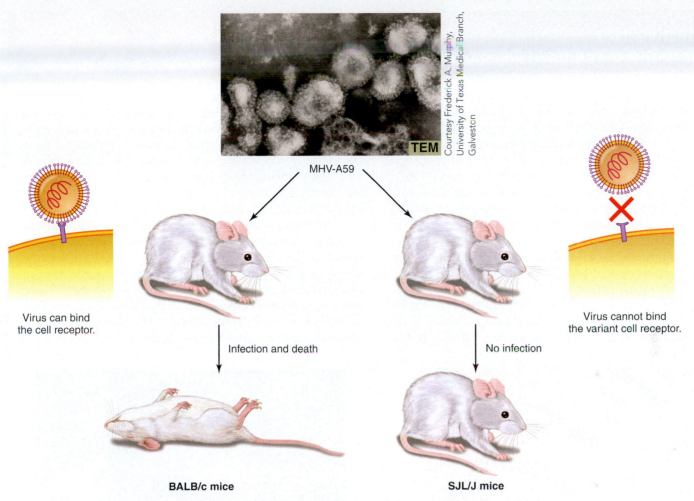

Courtesy Frederick A. Murphy, University of Texas Medical Branch, Galveston

MHV-A59

Virus can bind the cell receptor.

Virus cannot bind the variant cell receptor.

Infection and death

No infection

BALB/c mice

SJL/J mice

Figure 18.21. Genetic variations among individual hosts and susceptibility to pathogens Mouse hepatitis virus strain A59 (MHV-A59) readily infects the BALB/c strain of mice. Conversely, another common laboratory strain of mice, SJL/J, is largely resistant to infection by MHV-A59. This dramatic difference in susceptibility results from a relatively simple genetic difference between the two mouse strains. SJL/J mice produce a variant of the MHV-A59 receptor protein to which the virus cannot bind.

potential limitations of his third postulate—if a healthy host is inoculated with the suspected microbe, then the host will develop the disease. On its surface, this postulate states that every host experimentally inoculated with a suspected pathogen should develop the disease. Koch observed that not all individuals respond equally to pathogens. Some individuals may exhibit resistance to a pathogen. To put it another way, not every individual inoculated with a potential pathogen will develop the associated disease. Genetic variations among the host organisms influence their susceptibility to various microbes. We encountered an excellent example of this concept in Chapter 5 when we discussed mouse hepatitis virus. The A59 strain of mouse hepatitis virus (MHV-A59) readily infects BALB/c mice. Conversely, SJL/J mice are largely resistant to this virus **(Figure 18.21)**. Why? The SJL/J mice possess an allele of the MHV receptor gene that codes for a protein not readily recognized by the virus.

The third postulate also necessitates that a suitable experimental host for the disease exists. If we are studying a dangerous human pathogen for which no suitable treatment exists, then obviously we cannot experimentally inoculate humans to demonstrate that the potential pathogen causes the disease (the example of Warren and Marshall notwithstanding!). Rather, we need to identify a suitable animal **(Perspective 18.2)** that

(1) can be infected by the microbe, and (2) consistently develops a disease as a result of the infection. We should note, though, that the disease state in an animal model does not always perfectly mirror the disease state in humans. Some people also may argue that we technically cannot fulfill the second postulate—the suspected microbe should be grown in pure culture—with obligate intracellular pathogens. These microbes only can replicate in the presence of susceptible host cells. One could argue, then, that obligate intracellular parasites never can be grown in pure culture. This argument, though, seems to be an issue of semantics rather than a legitimate criticism of Koch's postulates.

Molecular Koch's postulates

Despite being developed over a century ago, Koch's postulates continue to influence our ability to demonstrate the causal relationship between microbes and diseases. Today, however, advances in molecular biology allow us to examine the influence of specific microbial genes and gene products on pathogenesis. In recognition of this change, Stanley Falkow, a noted microbiologist at Stanford University, proposed that we could use molecular tools to identify specific molecules produced by pathogens that enhance their ability to cause disease. Furthermore,

Studies with animals have provided us with invaluable information about human diseases. As we just mentioned in the previous section, animals can be instrumental in proving Koch's postulates. We also depend on animals for basic studies on the transmission, prevention, and treatment of infectious diseases. Indeed, much of what we know about human pathogens depends on studies conducted in various model organisms. Quite often, these studies are conducted in mice. In fact, if there were a biomedical hall of fame, then mice probably would be among the inaugural inductees. Armadillos, on the other hand, probably are not high on many people's lists of important model organisms. Much of what we know about leprosy, or Hansen disease, however, we owe to the nine-banded armadillo.

The causative agent of leprosy, *Mycobacterium leprae*, has proven to be especially difficult for researchers to study. We know that the bacterium, discovered in 1873 by Gerhard Armauer Hansen, is rod shaped, and is an obligate intracellular pathogen **(Figure B18.2)**.

In other words, it only replicates within cells. Genetic studies have shown that *M. leprae* lacks a number of genes needed for, among other activities, metabolism and DNA replication. Presumably, these genes have been lost as *M. leprae* has evolved from a free-living organism to an intracellular parasite. This dependence on a host cell also makes it difficult to study *M. leprae*. Researchers, in fact, have not been able to develop any kind of medium or identify any cell line that supports *M. leprae* replication.

Luckily, the bacterium does replicate in armadillos **(Figure B18.3)**. Researchers can inject bacteria into the footpad of an armadillo. Over a period of time, the bacteria replicate and can be isolated from the animal. Additionally, *M. leprae* causes a disease in the armadillos, so the effectiveness of potential drugs can be tested in them. Finally, there is some evidence that wild armadillos are infected by *M. leprae*. Furthermore, a study published in April 2011 provides evidence that transmission of *M. leprae* from armadillos to humans may occur.

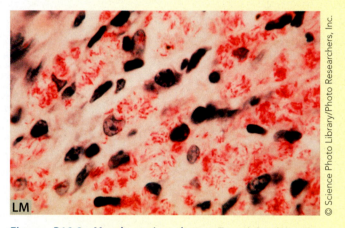

Figure B18.2. *Mycobacterium leprae* This rod-shaped bacterium is extremely slow growing, with a doubling time of approximately 13 days. Although the bacterium was discovered over 125 years ago, researchers still have not developed a means of cultivating this microbe in the laboratory on prepared media and under controlled conditions.

Figure B18.3. Using armadillos to grow the leprosy bacterium The bacterium that causes leprosy, *Mycobacterium leprae*, still cannot be cultivated in the laboratory. To circumvent this problem, researchers have used a rather unusual research animal—the armadillo. *M. leprae* can be injected into the footpad of armadillos. It replicates in, and later can be isolated from, this host organism.

he proposed a list of **molecular Koch's postulates**, modeled after the original postulates of Koch. According to Falkow:

- The virulence factor should be present in the pathogen.
- Experimental inactivation of the virulence factor gene should lead to a decrease in virulence.
- Experimental reversion of this inactivating change should result in a restoration of virulence.
- The virulence factor gene should be expressed during an infection.
- Immunity to the pathogen must provide protection.

Falkow postulated, in other words, that we should be able to demonstrate the contribution of specific genes, and their

products, to the pathogenesis process. Through molecular techniques, we should be able to inactivate or remove suspected virulence factor genes from a pathogenic organism. In the absence of essential virulence factors, the microbe should exhibit decreased virulence. A microbe, in other words, is not inherently pathogenic. Specific molecular attributes contribute to the pathogenic phenotype.

These virulence factors may affect some of the processes that we discussed earlier and include:

- adhesion factors
- invasion factors
- toxins
- secretion factors

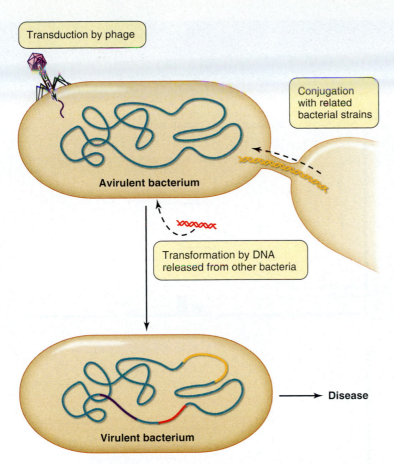

Transduction by phage

Conjugation with related bacterial strains

Avirulent bacterium

Transformation by DNA released from other bacteria

Disease

Virulent bacterium

Figure 18.22. Acquisition of virulence factor genes An avirulent bacterium may obtain genes for virulence factors, such as adhesion factors, invasion factors, or toxins, through a variety of mechanisms, including conjugation, transduction, and transformation. After acquiring these genes, the previously avirulent bacterium may become virulent and produce disease.

allow the microbe to better exit an infected host. Exiting a host, of course, is a necessary prerequisite to infecting another host and allowing the microbes to continue replicating. By obtaining the genes for some or all of these factors, the bacterium becomes, in a manner of speaking, more aggressive. It becomes more virulent.

The genes for these virulence factors can be transferred from one cell to another via several different mechanisms, including conjugation, transduction, and transformation (**Figure 18.22**). Interestingly, the chromosomes of many pathogenic strains of bacteria contain **pathogenicity islands**, or **PAIs**, regions of the chromosome that contain multiple virulence factor genes. These pathogenicity islands contain various genetic elements that have been obtained from various sources. Evolutionary pressure, it appears, has selected for the linkage of these individual virulence factor genes into a cohesive genetic structure. We will explore the origins and evolution of pathogenicity islands more thoroughly in Chapter 21. In the next section of this chapter, we will look more generally at the evolution of pathogens.

CONNECTIONS We now know that horizontal, or lateral, gene transfer commonly occurs among various organisms and can affect greatly the genomes of these organisms. Experimental evidence demonstrating this process was discussed in **Section 10.3**.

18.4 Fact Check

1. What are the key postulates put forth by Robert Koch?
2. How were Koch's postulates specifically used in determining that *H. pylori* was the cause of ulcers?
3. What are some of the limitations of Koch's postulates?
4. Describe Falkow's molecular Koch's postulates.

In many cases, these virulence factors allow the microbe to better obtain nutrients. Invasion factors, for instance, allow microbes to enter cells, thereby providing them with access to needed resources. In other cases, these virulence factors may

18.5 The evolution of pathogens

How do new infectious diseases arise?

In the preceding sections, we have explored (1) the concept of an infectious disease, (2) how microbes cause diseases, (3) what factors affect the spread of infectious diseases within a population, and (4) how we can determine that a specific microbe causes a specific disease. To complete our introduction to infectious diseases, another major puzzle needs to be addressed. We need to investigate how new infectious diseases arise.

Some infectious diseases, like toxic shock syndrome, Legionnaire's disease, Ebola, Lyme disease, SARS, and AIDS are classified as emerging or re-emerging. As we will see in this section, several events can lead to the emergence of a new disease or re-emergence of a previously recognized disease. In

this section, we will focus on two particular scenarios. First, a host population may come in contact with a pathogen it has not encountered previously, much as we saw with the golden lion tamarins. Second, a microbe can become more virulent. We'll end this section by briefly examining what we can do to remain vigilant of these emerging infectious diseases.

Encountering a new population

A new or emerging disease may arise when an existing pathogen encounters a new population, perhaps through a zoonotic transfer. Following such a transfer, the microbe is in a new

R. J. Montali, C. A. Scanga, D. Pernikoff, D. R. Wessner, R. Ward, and K. V. Holmes. 1993. A common-source outbreak of callitrichid hepatitis in captive tamarins and marmosets. J Inf Dis 167(4):946–950.

Context

As we mentioned in the opening story of Chapter 5, outbreaks of an often fatal hepatitis affecting golden lion tamarins (GLTs) had been reported by veterinarians at several zoos in the United States and the United Kingdom during the early 1980s. Because these primates are endangered, the zoo officials were especially eager to determine the cause of this disease.

Initial studies showed that a virus caused the disease. Researchers demonstrated this point most conclusively by inoculating common marmosets, non-endangered relatives of GLTs, with a bacteria-free liver homogenate obtained from a GLT that had died from the disease. The marmosets developed a similar disease, leading the researchers to conclude that a virus caused the disease. Subsequently, the researchers determined that the virus was an arenavirus (see Section 5.4 for information about different groups of viruses). This determination was made through a series of immunological and genetic studies. A major question, however, still remained. How were the GLTs becoming infected? In other words, the virologists and veterinarians wanted to know the source of the virus and how it was transmitted.

A subsequent outbreak of this disease at the Fort Worth Zoo in Texas allowed researchers to address this very important question. The answer allowed zoos to better protect their GLTs from future outbreaks.

Experiments

In April of 1991, an outbreak of hepatitis among marmosets and tamarins, referred to as callitrichid hepatitis (CH), occurred at the Fort Worth Zoo. Four of seven GLTs became sick; three of them died. Five of seven pygmy marmosets (PMs) also became sick. All five of them died. The animal handlers observed that all of the animals initially showed signs of disease about two weeks after having been fed neonatal mice (**Figure B18.4**). The outbreak, then, appeared to be a common-source outbreak, and the source of the infection appeared to be the neonatal mice.

The mice appeared to be the source of the infection in the GLTs and PMs, but the researchers next wanted to demonstrate that the mice contained the infectious agent—a strain of lymphocytic choriomeningitis virus (LCMV). To accomplish this goal, the researchers used a solid-phase immunoassay. Briefly, liver homogenates from the mice, affected primates, and uninfected primates were spotted onto a nylon membrane. The membrane then was probed with serum from an animal that had been experimentally inoculated with the virus (CHV), LCMV antibodies (LCMV), serum from an uninfected GLT (Tamarin control), and serum from an uninfected guinea pig (G. pig control). As seen in **Figure B18.5**, CHV and LCMV antibodies reacted

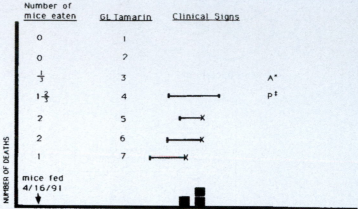

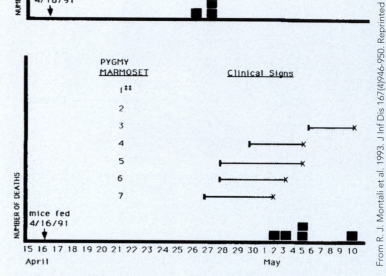

From R. J. Montali et al. 1993. J Inf Dis 167(4)946–950. Reprinted by permission of Oxford University Press.

Figure B18.4. Time course of disease among tamarins and marmosets at Fort Worth Zoo Mice were fed to the animals on April 16, 1991. Animals began showing clinical signs of disease about two weeks later. Blocks represent fatalities.

strongly with all of the affected GLTs and marmosets, the fetus of one of the tamarins, and the feed mice. These antibodies, though, did not react with liver homogenates from an uninfected marmoset or an uninfected mouse. Additionally, the control sera reacted only weakly with all of the samples. Based on these results, the researchers concluded that LCMV antigens were present in all of the affected animals and the potential vector—the feed mice.

Impact

These findings had profound implications for zoos. If the virus were being transmitted to GLTs from mice used as food supplements, then the outbreaks should end if mice were no longer fed to the primates. This information was conveyed to zoos throughout the country and, not surprisingly, outbreaks of callitrichid hepatitis stopped.

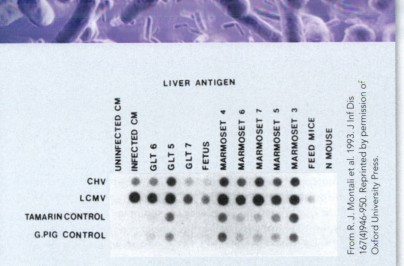

LIVER ANTIGEN

Figure B18.5. Solid-phase immunoassay detection of viral antigens Liver homogenates were spotted onto a nylon membrane and probed with serum from an animal previously infected with CHV, LCMV-specific serum, tamarin control serum, and guinea pig control serum. Liver samples from all of the affected animals reacted strongly with CHV-specific and LCMV-specific serum.

From R. J. Montali et al. 1993. J Inf Dis 167(4)946-950. Reprinted by permission of Oxford University Press.

The findings of this paper, though, also raised additional questions. Could wild mice harbor this virus? Could these wild mice, which are quite common in zoos, transmit the virus to GLTs? Could infected mice transmit the virus to humans? Could the virus be transmitted from infected GLTs to zoo personnel or visitors to the zoo?

Outbreaks of callitrichid hepatitis have occurred in several zoos that did not use mice as a food source. Presumably, these outbreaks occurred when primates came in contact with infected wild mice. Wild mice, therefore, may be a source of infections. In none of the outbreaks, however, has there been any evidence of primate-to-primate transmission. The potential role of this virus in human diseases still needs to be determined.

● Questions for Discussion ··················

1. The aborted fetus of an infected GLT contained viral antigens, as shown in Figure B18.3 What does this finding tell you about the transmission of the virus?

2. The onset of disease, as shown in Figure B18.2, is typical of a common-source outbreak. What would this graph look like if the virus were propagated?

environment. It still needs to obtain resources in order to replicate. In some cases, the microbe may cause only limited disease. In other cases, it may cause significant disease. During the hepatitis outbreak among golden lion tamarins that we discussed in the preceding Mini-Paper, researchers were concerned about the possible zoonotic transfer of LCMV from primates to humans. They did not know how the virus would interact with a human host. Apparently, the concerns were unwarranted; even veterinary pathologists who worked very closely with the dying animals did not become infected.

To explore more fully the emergence of new human infectious diseases, we will briefly examine the emergence of HIV/AIDS and Lyme disease. In the case of HIV, as we will see, a virus was transferred from non-human primates to humans. In the case of Lyme disease, a change in human activities caused humans to come in contact with a pathogen more frequently.

HIV/AIDS

HIV/AIDS represents a horrific example in which a zoonotic transfer has led to a worldwide emerging disease. Through extensive sequence analysis studies, researchers have concluded that HIV has evolved from **simian immunodeficiency virus (SIV)**, a virus found in various non-human primates throughout Africa. Possibly, the virus was transferred from a non-human primate to a human during a hunting trip. Initially, the virus may have replicated poorly within infected humans, or been transmitted poorly from human to human. Over time, though, the virus mutated to become the human-adapted HIV. Again, based on sequence comparisons of SIV and HIV strains, researchers estimate that the initial transfer of the virus to humans may have occurred around 1930.

The zoonotic transfer of SIV to humans, and eventual evolution of HIV, probably occurred more than once. We know that two main types of HIV, HIV-1 and HIV-2, currently infect humans. HIV-1 is, by far, the more prevalent of these viruses, and is the virus most typically referred to when we discuss AIDS. Conversely, HIV-2 tends to be found primarily in western Africa, and is associated with a less severe form of disease. Four genetically discernible groups of HIV-1, the M, N, O, and P groups, have been identified, and group M viruses can be further subdivided into nine discernible subtypes, or **clades** (A–D, F–H, J, and K). Additionally, recombinant forms of these clades, referred to as "circulating recombinant forms (CRFs)" have been identified **(Figure 18.23)**. While the distribution of these clades

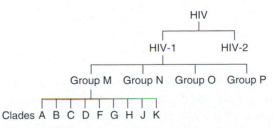

Figure 18.23. Genetically distinct groups of HIV with different geographic distributions Two distinct viruses, HIV-1 and HIV-2, have been identified. HIV-1 can be further subdivided into four groups: M, N, O, and P. Ninety percent of HIV infections are due to HIV-1 group M. Nine distinct subgroups, or clades, of group M have been identified. Recombinant forms of these clades (CRFs) also have been identified. AIDS typically is associated with viruses within the M group of HIV-1.

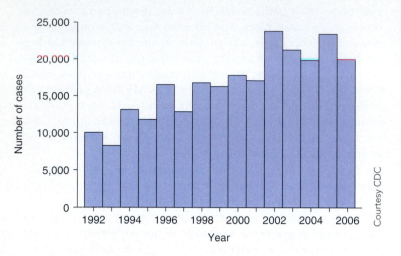

A. Incidence of Lyme disease in the United States

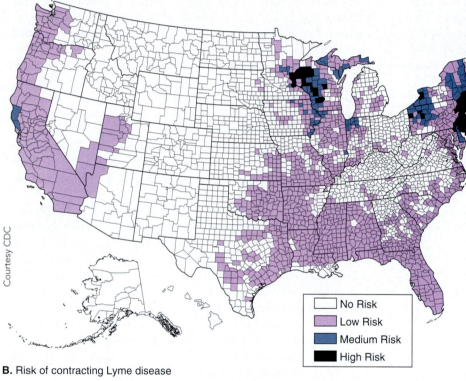

No Risk
Low Risk
Medium Risk
High Risk

B. Risk of contracting Lyme disease

Figure 18.24. Increasing incidence of Lyme disease in the United States The increase in Lyme disease has resulted primarily from the expansion of human habitation. As housing developments have spread into previously wooded areas, contact with Lyme disease vectors has become more common. **A.** The annual number of reported cases of Lyme disease increased dramatically between 1982 and 2006. **B.** The risk of contracting Lyme disease varies from state to state.

is worldwide, certain clades are more common in certain geographic areas. Clade B HIV-1, for instance, predominates in the United States, while clade C predominates in southern Africa.

The identification of these HIV variants raises an interesting and important question: What is the origin of HIV? Genetic studies show that HIV-2 most closely resembles SIV isolated from sooty mangabeys (Old World monkeys), SIV_{smm}, while HIV-1 group M isolates most closely resemble SIV isolated

from chimpanzees (SIV_{cpz}), indicating that these two human viruses arose from separate zoonotic transfers. Further, many researchers also hypothesize that the other three HIV-1 groups (N, O, and P) arose from three other transfers of SIV to humans. While interesting in its own right, this hypothesis raises another very compelling question: Could a new form of HIV arise through yet another zoonotic transfer?

Lyme Disease

The emergence of Lyme disease in the United States presents a very different situation. This disease, caused by the bacterium *Borrelia burgdorferi* (**Microbes in Focus 18.5**), first caught the widespread attention of the medical community and American population in 1975. At that time, physicians reported a cluster of cases of an arthritis-like disease in Old Lyme, Connecticut. Today, there are roughly 18,000 reported cases of Lyme disease throughout the United States every year (**Figure 18.24**).

What led to the emergence of this disease? To address that question, we need to know a little more about the infectious agent—*B. burgdorferi*. In the northeastern United States, this bacterium normally resides within deer and mice, its natural reservoirs. It spreads from one individual to another via the blacklegged tick, *Ixodes scapularis*. The tick ingests blood containing the bacterium from one host and then transfers the bacterium to the next host that it bites. If the tick bites a human, then it can transmit the bacterium to the human, and disease occurs in this new host (**Figure 18.25**). Humans, though, are incidental hosts; the pathogen is not transmitted from person to person.

In all likelihood, *B. burgdorferi* has been infecting deer, mice, and other animals for many, many years. Transmission of the bacterium from ticks to humans also has been occurring for many, many years. Indeed, a historical review of medical literature reveals several cases of what was probably Lyme disease throughout the 1900s. What changed over the last 30 years? The emergence of Lyme disease can be directly tied to increased human development. As our population expands, suburban developments are expanding into previously wooded areas. Contact between people and deer, mice, and other potential reservoirs of *B. burgdorferi* are increasing (**Figure 18.26**). This disease has emerged, not because the infectious agent has changed in any way, but because our exposure to it has changed.

Microbes becoming more virulent

While some emerging diseases, like HIV/AIDS and Lyme disease, arise primarily when an existing microbe comes in contact with a new host, other emerging diseases arise when a microbe becomes more virulent. The microbe, in other words, changes. The microbe can undergo various types of changes. A microbe,

Before feeding

After blood meal

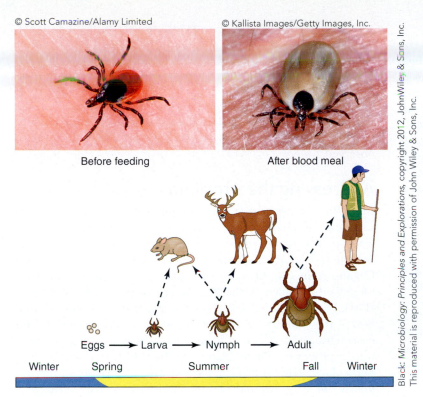

Eggs → Larva → Nymph → Adult

Winter   Spring   Summer   Fall   Winter

A. Life cycle of the deer tick; humans are incidental hosts

B. Bull's-eye rash typical of Lyme disease

Figure 18.25. Lyme disease Caused by a bacterium transmitted by ticks, Lyme disease can be a serious human disease. **A.** The pathogen associated with Lyme disease, the bacterium *Borrelia burgdorferi*, can be spread from ticks to mice, deer, or humans. **B.** In humans, the infection commonly causes a bull's-eye rash around the site of the tick bite.

Microbes in Focus 18.5
BORRELIA BURGDORFERI AND LYME DISEASE

Habitat: Ticks, deer, mice, humans, and other mammals

Description: Spirochete, Gram-negative bacterium; may be over 20 μm long, but less than 1.0 μm wide

Key Features: Lyme disease is the most common tick-borne illness in America. Unlike most other bacterial pathogens, *B. burgdorferi* does not need iron. Typically, the relative absence of free iron within humans presents a major limitation for bacterial growth. Many bacterial pathogens, as a result, have evolved specific mechanisms to acquire this essential element from the host. *B. burgdorferi* has addressed this problem in another way. It does not contain iron-associated proteins normally found in other organisms and can grow in the absence of this metal.

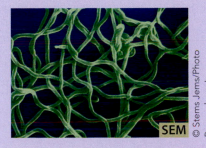

SEM

for instance, may acquire a gene that codes for a virulence factor. Alternatively, a microbe that we can control through the use of drugs may develop a means of evading our control measures. The virulence of the microbe, per se, does not change, but it becomes more of a threat because of its ability to avoid control. To illustrate both of these scenarios, we will look at two specific examples, pathogenic *Escherichia coli* and methicillin-resistant *Staphylococcus aureus*.

Figure 18.26. Emerging diseases and increased exposure to reservoir species As human populations and their habitats (suburban living spaces) have expanded, our contact with other species has increased. The increased incidence of Lyme disease is related to our increased contact with deer, mice, and other potential reservoirs of *Borrelia burgdorferi*.

Pathogenic *Escherichia coli*

Escherichia coli normally inhabits our intestinal tract. Usually, this bacterium is non-pathogenic, causing us no harm. In fact, it benefits us in many ways, helping us digest food and providing us with some important nutrients and vitamins. We know, though, that *E. coli* also is associated with severe intestinal diseases, or enteritis. How can a single type of microbe be both a common commensal species and a major pathogenic threat? The pathogenic strains of *E. coli* have obtained various virulence factors. As a result, these strains of *E. coli* no longer reside within our intestinal tract as harmless commensals. Rather, they cause disease.

Let's look at *E. coli* strain O157:H7 as an example. This strain of *E. coli* can cause a potentially life-threatening diarrhea. To a large extent, the increased virulence of this strain results from its production of shiga-like toxins, toxins derived from *Shigella* species. The toxin genes probably were transferred from *Shigella* to *E. coli* via a transducing bacteriophage. The shiga toxins are secreted by the bacteria, bind to a specific receptor on the host cell, and are internalized. Within the host cell, the toxin interacts with host cell ribosomes, halting translation. Destruction of the host cells ensues. Let's ask another question, though. Of what benefit to the bacteria are these toxins?

In a paper published in 2006, microbiologist Dr. Alison O'Brien and colleagues provided some insight into this important evolutionary question. They demonstrated that a mutant strain of *E. coli* O157:H7 in which the shiga toxin gene had been inactivated adhered less well to epithelial cells *in vitro* than a strain with a functional shiga toxin gene. Furthermore, the mutant strain was less able to colonize mice than the strain expressing shiga toxin. The toxin, in other words, helps the bacteria remain in a hospitable environment.

CONNECTIONS Along with diphtheria toxin, cholera toxin, and tetanus toxin, to name just a few, Shiga toxin is classified as an A-B toxin. While various A-B toxins affect host cells in different ways, these exotoxins share a common structure, containing an enzymatically active A subunit and a membrane-binding B subunit **(Section 21.1)**.

Methicillin-Resistant *Staphylococcus aureus*

Like *Escherichia coli*, *Staphylococcus aureus* is a very common microbe. Quite frequently, it resides as a commensal on the skin or in the nasal passages. It can cause superficial skin infections, such as abscesses, when the protective layer of the skin is breached. As we saw with *E. coli*, certain strains of *S. aureus* have obtained various virulence factors, and these strains can cause a variety of human diseases. For example, strains of *Staphylococcus aureus* expressing toxic shock syndrome toxin (TSST) can cause the rare but potentially fatal disease known as toxic shock syndrome, while strains expressing enterotoxins often cause food poisoning. An even bigger medical concern has been the development of **methicillin-resistant Staphylococcus aureus (MRSA)**. This strain of *S. aureus* is resistant to a wide

range of antibiotics and, as a result, is very difficult to treat. The first reports of MRSA came out in 1961, just two years after methicillin was first used to treat *S. aureus*. Since the 1990s, the number of MRSA infections has exploded. Treatment with standard antibiotics no longer works. Rather, physicians must treat MRSA infections with less common antibiotics, like vancomycin, to which most strains of MRSA are still sensitive. Unfortunately, vancomycin-resistant *S. aureus* strains, or VRSA, have been observed.

Addressing the problem

In the face of this threat of emerging diseases, what can we do to protect ourselves? Surveillance may be our most effective weapon. If we become aware of an epidemic quickly, then we may be able to curtail its spread. Internationally, the World Health Organization (WHO) organizes many surveillance programs. Organizations such as the Pan American Health Organization (PAHO) coordinate some regional programs. In the United States, the Centers for Disease Control and Prevention (CDC) coordinate or participate in several infectious disease- monitoring programs **(Table 18.8)**. The Border Infectious Disease Surveillance (BIDS) Project, for instance, represents a program jointly run by the United States and Mexico. As part of this program, facilities on both sides of the U.S. and Mexican border routinely track and report several different infectious diseases in an attempt to better understand the spread of diseases across the border. The prevalence of antibiotic-resistant strains of various bacteria is documented by the National Antimicrobial

TABLE 18.8 Selected U.S. infectious disease surveillance programs

Name	Agencies	Focus
National Antimicrobial Resistance Monitoring System	CDC FDA Department of Agriculture State and local health departments	Monitor antibiotic resistance of *Salmonella*, *Escherichia coli* O157:H7, and *Campylobacter*
National West Nile Virus Surveillance System	48 states and 4 cities	West Nile virus in humans, wild birds, and sentinel chicken flocks
United States Influenza Sentinel Physicians Surveillance Network	Physicians throughout the country	Flu-like illnesses
Waterborne-Disease Outbreak Surveillance System	CDC U.S. Environmental Protection Agency Council of State and Territorial Epidemiologists	Water-borne disease outbreaks in drinking water and recreational water

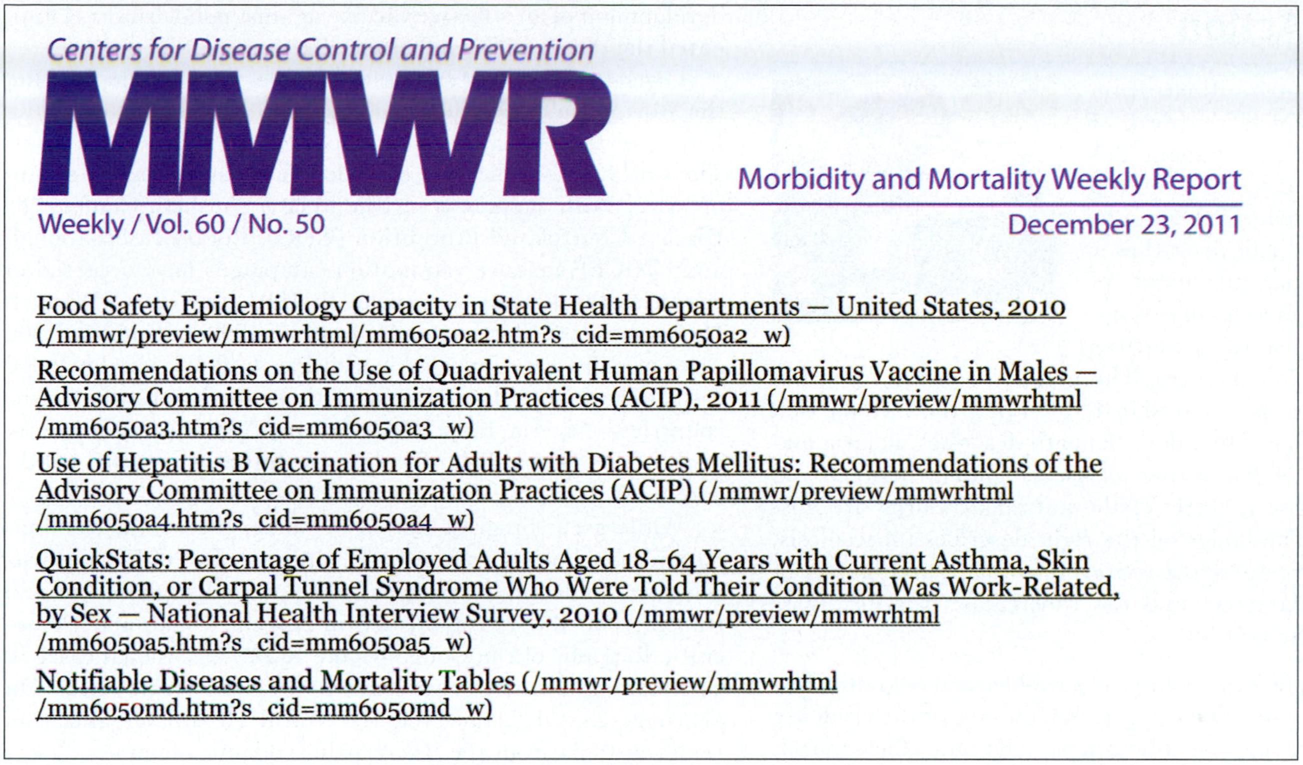

Centers for Disease Control and Prevention

MMWR

Morbidity and Mortality Weekly Report

Weekly / Vol. 60 / No. 50   December 23, 2011

Food Safety Epidemiology Capacity in State Health Departments — United States, 2010 (/mmwr/preview/mmwrhtml/mm6050a2.htm?s_cid=mm6050a2_w)

Recommendations on the Use of Quadrivalent Human Papillomavirus Vaccine in Males — Advisory Committee on Immunization Practices (ACIP), 2011 (/mmwr/preview/mmwrhtml /mm6050a3.htm?s_cid=mm6050a3_w)

Use of Hepatitis B Vaccination for Adults with Diabetes Mellitus: Recommendations of the Advisory Committee on Immunization Practices (ACIP) (/mmwr/preview/mmwrhtml /mm6050a4.htm?s_cid=mm6050a4_w)

QuickStats: Percentage of Employed Adults Aged 18–64 Years with Current Asthma, Skin Condition, or Carpal Tunnel Syndrome Who Were Told Their Condition Was Work-Related, by Sex — National Health Interview Survey, 2010 (/mmwr/preview/mmwrhtml /mm6050a5.htm?s_cid=mm6050a5_w)

Notifiable Diseases and Mortality Tables (/mmwr/preview/mmwrhtml /mm6050md.htm?s_cid=mm6050md_w)

Courtesy CDC

Figure 18.27. Morbidity and Mortality Weekly Report (MMWR) Published weekly by the Centers for Disease Control and Prevention (CDC), the MMWR provides health professionals with up-to-date information about various infectious diseases. It also includes information about other diseases and health-related issues.

Resistance Monitoring System (NARMS), a program that includes members of, among others, the CDC, the Food and Drug Administration (FDA), and the U.S. Department of Agriculture (USDA).

In addition to these specific programs, the United States also monitors the incidence of a series of infectious diseases through the National Notifiable Diseases Surveillance System. Originally designed as a means of monitoring the occurrence of smallpox, plague, cholera, and yellow fever among American officials serving in other countries, the program now involves the reporting of a lengthy list of infectious diseases (Table 18.9) in all 50 states, the District of Columbia, and Puerto Rico. This information is published regularly in Morbidity and Mortality Weekly Report (MMWR), a weekly publication of the CDC (Figure 18.27). Additionally, the CDC presents an annual account of this information. The World Organization for Animal Health (OIE) monitors and reports on notifiable diseases in animals.

Through these surveillance programs, we can quickly become aware of an outbreak of an infectious disease. With today's communication capabilities, information about the disease can be distributed rapidly to physicians and local or regional government officials. Hopefully, a rapid response can lead to an equally rapid curtailment of the outbreak.

TABLE 18.9 Selected notifiable infectious diseases, United States (2012)

Anthrax	HIV infection	Plague
Botulism	Lyme disease	Salmonellosis
Cholera	Malaria	Severe acute respiratory syndrome (SARS)
Diphtheria	Measles	Syphilis
Gonorrhea	Mumps	Tetanus

18.5 Fact Check

1. Describe the conditions that may cause an emerging or re-emerging disease to arise, and provide examples.
2. Describe how *E. coli* O157:H7 is thought to have emerged and persisted.
3. Identify some of the infectious disease surveillance systems in place and describe their objectives.

Image in Action

WileyPLUS This illustration depicts the desk of  a medical researcher studying different microbial diseases. The top folder describes polio and contains an image of the poliovirus (lower right picture) and an image of motor neurons undergoing apoptosis during an active poliovirus infection (lower left picture). The graph shows the rapid fall of new polio cases per year shortly following the introduction of the polio vaccine. The folder beneath describes malaria and holds an image of *Plasmodium falciparum* gliding between red blood cells. Nearby, a bottle of the antimalaria drug artemisinin is present. The folder on the right describes tuberculosis, including an image of *Mycobacterium tuberculosis*, and bottles of the antibiotic rifampicin and the tuberculosis vaccine, BCG (Bacillus Calmette-Guérin).

1. Imagine you open the poliomyelitis folder and begin reading about the disease and pathogen. You encounter the virulence section and see that the organism has a CI ratio of less than 1. Explain the meaning of this information and how this number is related to the virulence of the poliovirus.

2. As you examine the tuberculosis folder, you encounter an epidemiology report on the incidence and prevalence of TB in New York City. What do these two terms mean, and how may the epidemiologists use this information to understand and study the disease in New York City?

development of an effective vaccine against poliovirus has eliminated the threat of this infectious disease.

Given the unbridled success of the U.S. vaccination program, the World Health Organization (WHO) in 1988 announced the launch of the Global Polio Eradication Initiative. Their goal? The worldwide eradication of polio. The results have been impressive. With support from the WHO, the U.S. Centers for Disease Control and Prevention (CDC), Rotary International, and UNICEF, massive vaccination campaigns have occurred in countries throughout the world. In 1994, the entire Western Hemisphere was declared polio-free. By 1999, only about 7,000 cases of polio were reported worldwide. In 2006, roughly 2,000 cases were reported, and most of them occurred in only four countries—Nigeria, India, Pakistan, and Afghanistan. The lofty goals of the Global Polio Eradication Initiative seemed within reach.

While great progress has been made, the complete elimination of polio has proven to be difficult. In 2011, poliovirus remained endemic in Nigeria, Pakistan, and Afghanistan and transmission was re-established in Angola, Chad, and Democratic Republic of the Congo **(Figure 18.28)**. Occasional cases in surrounding countries also continue to occur **(Table 18.10)**. On February 25, 2012, however, the World Health Organization removed India from the list of polio endemic countries. Not a single case of naturally occurring polio had been documented in over a year. In Angola, as shown in Table 18.10, five cases occurred in 2011, the last of which was reported in July 2011. While Angola was not considered to be an endemic nation, transmission of the virus had been occurring. As of September 2012, however, no new cases had been reported in over a year. Health officials certainly welcome this good news, but remain vigilant.

The Rest of the Story

The poliovirus scare of the early 1950s in the United States quickly dissipated, thanks to the development of two effective vaccines. In 1955, an inactivated poliovirus vaccine, developed by Dr. Jonas Salk, was approved for use. Other researchers, including Drs. Albert Sabin and Hilary Koprowski, developed attenuated vaccines, which were approved for use in the early 1960s. With the development these vaccines, the number of poliomyelitis cases in the United States dropped dramatically. In 1952, 58,000 cases were reported. In 1957, following a nationwide immunization program, there were only about 5,600 cases. In 1964, 121 cases were reported. The last case of wild poliomyelitis in the United States occurred in 1979. Since then, only sporadic cases, caused by the vaccine itself or imported to the United States from other countries, have been reported. The

TABLE 18.10 **Polio cases worldwide: 2011**

Country	Number of cases	Country	Number of cases
Afghanistan	80	CAR	4
Pakistan	198	Cameroon	1
Nigeria	60	China	21
India	1	Guinea	3
Chad	132	Kenya	1
DR Congo	93	Cote d'Ivoire	36
Angola	5	Mali	7
Niger	5	Gabon	1
		Total cases	648

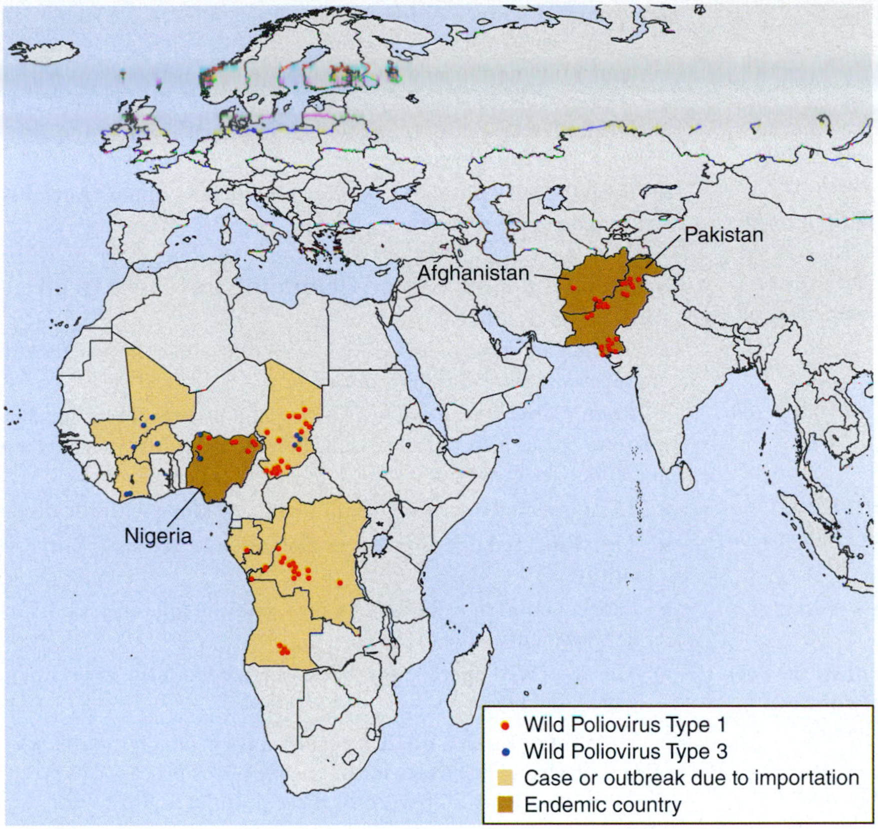

From R. B. Aylward and T. Yamada. "The Polio Endgame," 2011. N Engl J Med 364:2273-2275. Reprinted with permission from Massachusetts Medical Society.

Pakistan

Afghanistan

Nigeria

- ● Wild Poliovirus Type 1
- ● Wild Poliovirus Type 3
- ▢ Case or outbreak due to importation
- ▢ Endemic country

Figure 18.28. Cases of poliovirus today Since the development of the poliovirus vaccines, the number of poliovirus infections has decreased dramatically. Many parts of the world, in fact, are considered to be polio-free. Despite the efforts of many governmental and non-governmental agencies, however, this pathogen has not been completely eliminated. Infections continue to occur in several countries, most notably, Pakistan, Nigeria, and Afghanistan.

Summary

Section 18.1: What is an infectious disease?

A **disease** reflects disturbances in the normal functioning of an organism.

- **Infectious diseases** are caused by microbes that can be transmitted from host to host.
- **Zoonotic diseases** can be transmitted from animals to humans.
- Specific **symptoms** and **signs** generally are associated with specific diseases.

Microbes that are frequently associated with disease production are **pathogens**. Not all pathogens have the same ability to cause disease. The process by which pathogens cause disease is **pathogenesis**. **Infection** refers to the replication of a pathogen in or on its host.

- **Primary pathogens** have an intrinsic ability to produce disease in a proportion of uncompromised hosts.
- **Opportunistic pathogens** generally include environmental or commensal organisms that can cause disease when the host is somehow compromised by immune deficits, or preexisting damage, such as a wound, or when the pathogen becomes displaced from its normal residence.

The **virulence** of a pathogen is a measure of the severity of disease that it causes.

- More severe disease generally results in a larger **case-to-infection (CI) ratio**, or the proportion of infected people who develop the disease after exposure to the pathogen.
- Strains of a pathogen that have experienced **attenuation** exhibit decreased virulence. If the pathogen can no longer cause disease, it is avirulent. Attenuated strains of a pathogen may be effective **vaccines** that stimulate the immune system to respond to that pathogen.

Individuals may act as **carriers** who can transmit the infectious agent to others even though they themselves may be asymptomatic.

Section 18.2: How do microbes cause disease?

To cause an infection, most pathogens must attach to and/or invade specific host cells or tissues. Following attachment and invasion, the pathogen must replicate.

- Attachment may occur through specific protein:protein interactions.

- Attachment may result from more general interactions.
- Attachment often determines the **host range** of a pathogen.

Pathogens must evade host defenses.

- Some pathogens, like *Neisseria gonorrhoeae* and *Trypanosoma brucei*, evade the host immune response by continually altering their surface molecules, a process known as **antigenic variation**.
- Herpes viruses avoid host immune responses by establishing **latency**. However, periodic **reactivation** may occur.
- Even bacteria possess host defenses to protect themselves from pathogens. Restriction enzymes probably evolved to help bacteria protect themselves from bacteriophages.

Pathogens cause disease in a variety of ways. Some bacteria produce **toxins** that lead to death of target cells. Some virus-infected cells experience programmed cell death, or **apoptosis**.

Section 18.3: How are infectious diseases transmitted?

In addition to causing disease within an individual host, pathogens also are transmitted between hosts.

- **Transmission** may occur between hosts or between a source, or **reservoir** and a host.
- **Direct contact transmission** requires physical contact between an infected person and a susceptible host. **Indirect contact transmission** may occur through an inanimate object or **fomite**.
- Pathogens that replicate within the intestinal tract often are transmitted via a **fecal-oral transmission** route. Pathogens are excreted in the feces of the host and subsequently ingested by another host.
- Other pathogens exhibit **respiratory transmission**. They leave an infected host through coughs and sneezes.
- In **vector-borne transmission**, another species, quite often an insect, transmits the pathogen.
- Still other pathogens undergo **sexual transmission**. This results in **sexually transmitted infections**, or **STIs**.
- **Vertical transmission** occurs from parent to child. **Horizontal transmission** occurs between other members of a species.
- **Zoonotic transfer** involves transmission of a pathogen from its natural animal host, or **reservoir host**, to a human that serves as a **dead-end host** for that pathogen.

Epidemiology is the study of patterns of disease in populations. Epidemiologists study the rate of disease in a population, or **morbidity rate**, as well as the rate of deaths due to diseases, or **mortality rate**.

- Two important measures of diseases in population are **incidence**, which is the number of new cases in a population within a specific period of time, and **prevalence**, which is the total number of cases at a particular point in time.
- By monitoring infectious disease rates within populations, epidemiologists can identify **emerging diseases** and re-emerging diseases.
- Diseases habitually present in a population are referred to as **endemic diseases**. An **epidemic** occurs when the incidence of a disease increases significantly above the normally expected levels, while **pandemic** refers to an epidemic that spans a large global region. An **outbreak** refers to a cluster of diseases appearing within a short period of time in a localized population.

- **Common-source epidemics** occur when there is a single source of infection, such as a contaminated water supply, to which a large number of people are exposed in a relatively short period of time. The time over which cases are reported is related to the **incubation period**, the period between pathogen entry and appearance of illness.
- **Propagated epidemics** occur when the infection is transmitted from one host to another.

Section 18.4: How can we determine the cause of an infectious disease?

Prevention and treatment options necessitate that the specific cause of an infectious disease is determined.

Koch's postulates can be used to demonstrate that a specific microbe causes a specific disease. These postulates state that cause and effect are proven if:

- The suspected microbe is identified in everyone with the disease.
- The suspected microbe can be isolated and grown in pure culture.
- Experimental inoculation of a host with the microbe results in development of the disease.
- The suspected microbe can be re-isolated from the experimentally inoculated host.

Molecular Koch's postulates list ways in which the role(s) of specific virulence factors in the pathogenesis process can be demonstrated. Most importantly, these postulates state that:

- Experimental inactivation of the virulence factor should lead to a decrease in virulence.
- Experimental reversion of the virulence factor inactivation should result in a restoration of virulence.

Virulence factors may include adhesion factors, invasion factors, toxins, and secretion factors. Many pathogenic strains of bacteria contain multiple virulence factor genes on regions of the chromosome called **pathogenicity islands**, or **PAIs**.

Section 18.5: How do new infectious diseases arise?

Emerging or re-emerging diseases may occur when a pathogen encounters a new population.

- A new population may be encountered through a zoonotic transfer, as we have seen with HIV. This human virus arose from **simian immunodeficiency virus (SIV)**, a virus found in non-human primates in Africa. HIV-1 subsequently evolved within humans into several **clades**, or genetically distinct subtypes.
- A new population may be encountered when the range of the microbe or the range of the host change, as we have seen with Lyme disease.

Emerging diseases may occur when a pathogen becomes more virulent.

- The virulence of a microbe may increase when it acquires specific virulence factors, a toxin gene, for instance.
- The virulence of a microbe also may increase when the microbe develops resistance to available treatments. This has led to the development of **methicillin-resistant *Staphylococcus aureus* (MRSA)**.

Extensive surveillance networks are employed to alter physicians, state, and local authorities to the emergence or re-emergence of infectious diseases.

● Application Questions ....

1. Consider whether each of the following bacteria would be a primary pathogen or an opportunistic pathogen. Give reasons for your answers.
   a. Enterotoxigenic *Staphylococcus aureus*
   b. *Escherichia coli*
   c. *Vibrio cholera*
   d. *Clostridium botulinum*

2. Explain the statement that disease is an unintended consequence of obtaining nutrients.

3. Explain how a microbe might result in a disease that is not infectious. Give an example.

4. What factors can affect the virulence of a pathogenic microbe?

5. In simple terms, explain why the zoonotic transfer of a pathogen may cause severe disease within the recipient host. Why is the recipient often a dead-end host?

6. Are emerging diseases *new* diseases? Defend your answer.

7. Emerging and re-emerging diseases often are seen in areas afflicted by war. Why?

8. Several virulence factor genes often are grouped together in the pathogen genome, forming a pathogenicity island. Explain how this genetic unit might form. What are the evolutionary benefits of linking together several virulence factor genes?

Suggested Reading

Falkow, S. 1988. Molecular Koch's postulates applied to microbial pathogenicity. Rev Infect Dis 10 (Suppl 2):S274–S276.

Garrett, L. The coming plague: Newly emerging diseases in a world out of balance. 1994. Farrar, Straus and Giroux, New York.

Henig, R. M. A dancing matrix: Voyages along the viral frontier. 1993. Alfred A. Knopf, Inc., New York.

Kolata, G. Flu. 1999. Farrar, Straus and Giroux, New York.

Munier-Lehmann, H., V. Chenal-Francisque, M. Ionescu, P. Christova, J. Foulon, E. Carniel, and O. Barzu. 2003. Relationship between bacterial virulence and nucleotide metabolism: A mutation in the adenylate kinase gene renders *Yersinia pestis* avirulent. Biochem J 373:515–522.

Robinson, C. M., J. F. Sinclair, M. J. Smith, and A. D. O'Brien. 2006. Shiga toxin of Enterohemorrhagic *Escherichia coli* type O157:H7 promotes intestinal colonization. Proc Natl Acad Sci USA 103:9667–9672.

Innate Host Defenses Against Microbial Invasion

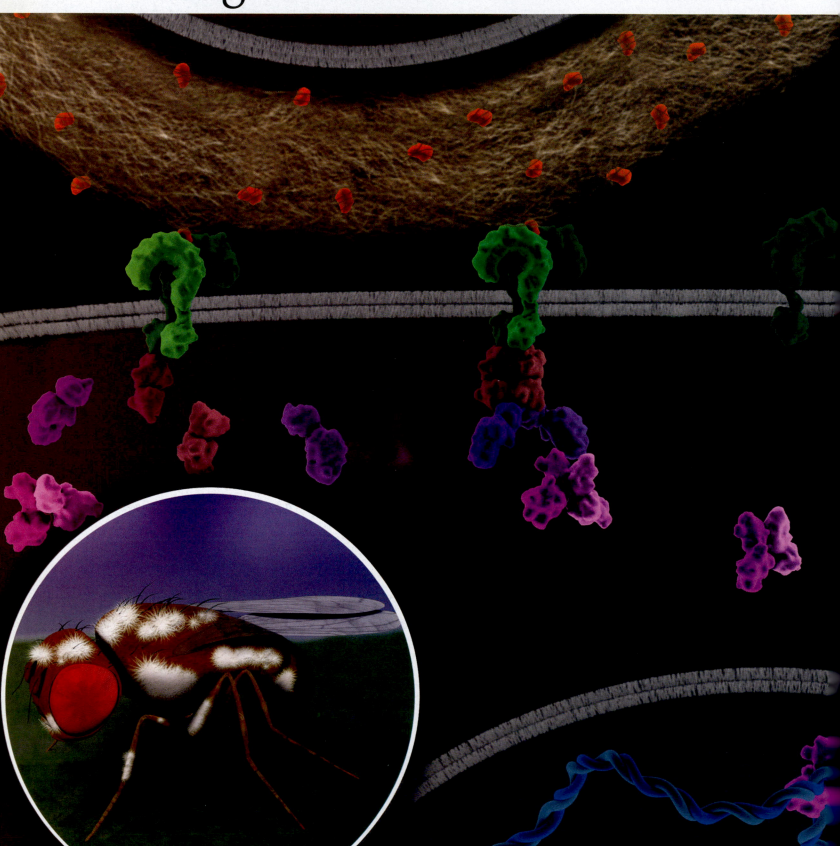

In 1991, a group of researchers was studying differentiation in the developing embryo of *Drosophila melanogaster*, the common fruit fly. They discovered a cellular molecule that was involved in the determination of the dorsal and ventral orientation in the embryo. Although they did not know it at the time, this molecule, which they named Toll (after the German slang word for "fantastic"), had another astounding function; it was part of an evolutionarily ancient system capable of recognizing a range of microbial components and initiating immune defenses in many multicellular species. In *Drosophila*, Toll turned out to be important in adult flies for resistance to fungal infections and some bacterial infections.

Since 1996 a panel of proteins structurally similar to Toll was found in vertebrates, and these were named Toll-like receptors (TLRs). Like the protein found in *Drosophila*, they too were transmembrane proteins shown to have important immune functions. The first characterized mammalian TLR was TLR4. It was found to recognize the bacterial endotoxin LPS. Portions of this detection and signaling system have also been found in plants and nematodes. In all multicellular species examined, these proteins appear to be at the root of pathogen sensing. As you will see, in mammals TLRs detect pathogens by taking advantage of the biochemical signatures that are found in microbes, but not in host cells. In mammals, only a very small number of different kinds of these sensors exist, but they are found in almost all cells. They function as an early warning system but also carry out the monumental task of activating appropriate innate and adaptive immune defenses. After decades of hunting for the molecular mechanism of how cells first detect a microbial invasion, it appeared that the missing puzzle pieces of how pathogens are first detected were found.

CHAPTER NAVIGATOR

As you study the key topics, make sure you review the following elements:

The cornerstone of the functioning of the immune system is the recognition of self versus non-self components.

The body has many innate physical and chemical barriers to prevent infection.
- Figure 19.2: Chemical and physical defenses of skin
- Table 19.1: Physical and chemical innate defenses associated with mucosal membranes
- Perspective 19.1: Messy mucus

Inflammation can limit infection by bringing immune system components to the damaged area.
- Figure 19.4: Process Diagram: The local inflammatory process

Multicellular organisms are equipped with molecules that can detect common biochemical features of microbes.
- Figure 19.5: Model of human Toll-like receptors (TLRs), their binding targets, and action
- Figure 19.6: Opsonization of a bacterial cell by mannose-binding lectin
- Figure 19.7: Process Diagram: The classical complement activation pathway and the formation of the membrane-attack complex (MAC)
- Figure 19.11: Type I interferons

Phagocytes and natural killer cells target extracellular microbes and infected host cells.
- Figure 19.16: Process Diagram: Phagosome formation and intracellular killing
- Animation: Endocytosis and exocytosis
- Figure 19.22: Activities of NK cells
- Microbes in Focus 19.1: Covert operations of human cytomegalovirus

Innate immune defenses of invertebrates function similarly to those of vertebrates.
- Mini-Paper: Mammalian cells can recognize bacterial DNA

CONNECTIONS for this chapter:
Role of normal microbiota in defense from pathogens (Section 17.4)
MHC I molecules and cytotoxic T cells (Section 20.3)
Use of adjuvants in vaccines (Sections 20.4 and 24.5)

Introduction

We live in a world teeming with microbes. Every second of every day, we breathe, touch, and ingest them. We are immersed in microbes, and microbes are opportunists. If our bodies provide a suitable environment for their growth, they will exploit this opportunity to prosper and spread, often at our own expense. We begin Chapter 19 with an introduction to the subject of immunology, followed by an overview of the physical and chemical barriers that help to prevent infections. The main focus will be on the innate immune system, which includes the cells and molecules that first recognize and respond to microbial components. Chapter 20 will focus on adaptive immunity and will provide an underpinning of knowledge needed to understand the host/pathogen interactions discussed in Chapters 21, 22, and 23.

In all multicellular organisms, the first line of immune defense against microbial invasion is conducted by innate mechanisms that are pre-programmed from birth and through evolution to sense and even identify the nature of microbial invaders. Innate immune defense mechanisms appear in plants, invertebrates, and vertebrates. Innate immune defenses are of primitive origin, appearing long before adaptive immunity of vertebrates. As we will see, being of ancient roots does not mean innate immunity is a rudimentary or simple system. During our exploration of innate defenses, we will answer the following questions:

Are there different types of immunity? (19.1)
What are the built-in barriers to infection? (19.2)
What is an early response of the body to infection? (19.3)
How do we sense or detect the presence of pathogens? (19.4)
What cells first defend us against infectious agents? (19.5)
What kinds of immune defenses do invertebrates have? (19.6)

19.1 Immunity

Are there different types of immunity?

An organism's defense mechanisms are composed of various cell types and molecules that are responsible for clearing the body of foreign invaders and abnormal cells, such as cancer cells. The **immune system**, composed of cells, molecules, and associated tissues involved in identifying foreign agents, fighting infection, and ridding the body of abnormal cells, is one of the most complex systems known in terms of its interactions with and effects on other cells in the body. **Immunology** is the study of the components and processes of the immune system. The cornerstone of the functioning of the immune system is the recognition of *self* versus *non-self* components.

The outcome of an infection is often the result of a tug-of-war between the microbe, which is trying to reproduce, and the host, which is trying to clear the body of the invader and prevent disease. A healthy host has many effective strategies for passively and actively preventing, clearing, or reducing most infections. Because of this, many **pathogens**, microbes that cause disease, have developed strategies designed for evading, destroying, or overwhelming host immune defenses. Therefore, it is of utmost importance to gain an understanding of the normal operation of the immune system if we are to comprehend how pathogens cause disease.

The term **immune** can be broadly defined as the state of being resistant to infection and subsequent disease. An immune individual may be *passively* resistant by nature. For example, recall from Chapter 5, the first requirement in animal viral infections is attachment of a virus to the correct receptors on the cell surface. Dogs are immune to infection with human immunodeficiency virus (HIV) because, as a species, they lack the correct cell surface receptors. From an immunological perspective, being immune means one can *actively* mount a defensive **immune response** to an invading microbe and prevent it from causing disease. An immune response defends the body against foreign agents, such as bacteria, fungi, parasitic worms, protozoa, viruses, toxins, and cancerous cells. New information on immune defense is constantly generated by research, and it shapes our approach in combating infectious as well as non-infectious diseases, such as cancer and asthma. These insights have given rise to new advances in the fields of vaccine design and genetic improvement of livestock and crops, as the example in **Figure 19.1** demonstrates.

From Milan Osusky et al. 2000. Nature Biotechnology 18:1162–1166. Reprinted by permission from Macmillan Publishers, Ltd. Photos courtesy Santosh Misra.

A. Diseased potato seedling   B. Diseased-resistant seedling

Figure 19.1. Engineering disease-resistant transgenic potatoes Researchers produced disease-resistant potatoes by integrating a synthetic gene that codes for a small antimicrobial peptide. The structure of the peptide was based on two potent antimicrobial peptides found naturally in some insects. **A.** The control potato with no genetic modification succumbs to infection from the fungus *Fusarium solani* on the lower stem, which has caused it to fall over. **B.** The potato possessing the new antimicrobial peptide gene shows resistance to the pathogen as evidenced by its erect growth. In the future, crops like these may combat the rising use of pesticides and improve crop yields.

Although most of us may commonly think of a successful immune response as leading to destruction of the microbial agent, this is not always the case, even in a healthy individual. The term "sterile immunity" describes a response that results in clearance by destruction of the agent. For example, when a person becomes infected with measles virus (see Microbes in Focus 18.1), the body mounts an immune response and, assuming the person survives the disease, the virus is then completely cleared from the body. In other instances a person may be infected, but the infection is kept in check by the immune system so that disease does not occur. This can be described as "non-sterile immunity." An example of non-sterile immunity is the case for infection with the bacterium *Mycobacterium tuberculosis*, the causative agent of tuberculosis. There is a high level of natural immunity to this microbe in humans. Approximately 90 percent of people who become exposed to *M. tuberculosis* (about one-third of the world's population is infected!) mount an immune response that is capable of preventing active disease. However, some individuals can continue to harbor non-replicating *M. tuberculosis* in lung tissue for many, many years. Thus, the immune response that is mounted results in non-sterile immunity that protects from disease but does not necessarily clear the bacterium. In Chapter 21, we will look more closely at the pathogenesis of *M. tuberculosis*. Sterile immunity is always a desirable goal for vaccine development for infectious diseases because such vaccines prevent the carrier states that serve as reservoirs of infection for unvaccinated or immune-compromised individuals (see Section 18.1). Regardless of the final result of an infection, an immune response will occur. The outcome depends on how effective the immune response is in protecting from disease.

Innate and adaptive immunity

In vertebrates, immunity can occur at two levels: **innate (non-specific) immunity** and **adaptive (specific) immunity**. As the term implies, innate immunity involves immune mechanisms that are present at birth and do not require previous exposure to the microbe. Thus, the innate immune response is a first line of immune defense, being mobilized within minutes of detection of a foreign invader. At the genetic level, innate immunity is hard-wired, involving germline genes that encode various proteins that can distinguish self from non-self components. Detection of foreign material is non-specific, accomplished by recognition of common, broad-based foreign features. In contrast, adaptive immunity is acquired only *after* exposure to a particular microbial agent or other foreign substance. It involves rearrangements of genes within specific cells of the adaptive immune system. These cells are called T cells and B cells, which we will learn about in Chapter 20. These rearrangements allow the manufacture of selected immune proteins that are tailored to bind to specific components of the particular infectious agent. An amazing result of an adaptive immune response is immunological memory. If the invader is encountered again, those tailor-made molecules can be quickly produced to prevent infection. In humans, immunity to the chickenpox virus is acquired this way. As we will see, innate and adaptive immune mechanisms are not mutually exclusive in vertebrates. They work together in response to infection to provide a one-two punch to destroy the invader and to mobilize immune forces quickly the next time the invader is encountered.

CONNECTIONS Most people know that you can get chickenpox once but you usually do not get it again. Immunity to the virus has been acquired through the adaptive immune system, which prevents successful reinfection. However, sterile immunity to the chickenpox virus is not always achieved. The virus finds its way into peripheral nerves and can remain dormant for many years. If the individual's immune system becomes compromised, as often occurs in the elderly, the virus can use this opportunity to reactivate infection and cause a different disease called shingles. This disease is discussed in **Section 22.1**.

19.1 Fact Check

1. What is meant by the term "immune"?
2. Define sterile and non-sterile immunity.
3. Describe the key features of innate and adaptive immunity.

19.2 Barriers to infection

What are the built-in barriers to infection?

To cause an infection, a microbe must first establish itself in the body. This is not always an easy task as the body has many built-in barriers to microbial invasion. If a microbe manages to overcome these barriers, host immune responses will begin. If contact with pathogenic agents takes place at an inner or outer body surface, these surfaces have a number of physical and chemical barriers that help prevent infection by pathogenic microorganisms. It is important to keep in mind that these barriers, although innately produced, are not part of the immune system. They do not identify or actively fight infection, but their significance in protection from infection is not to be underestimated.

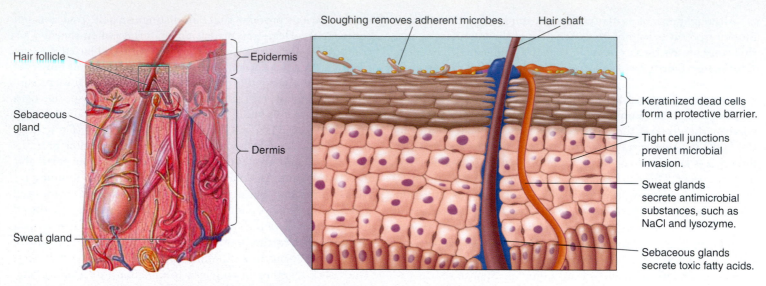

Hair follicle

Sebaceous gland

Sweat gland

Epidermis

Dermis

Sloughing removes adherent microbes.

Hair shaft

Keratinized dead cells form a protective barrier.

Tight cell junctions prevent microbial invasion.

Sweat glands secrete antimicrobial substances, such as NaCl and lysozyme.

Sebaceous glands secrete toxic fatty acids.

Figure 19.2. Chemical and physical defenses of skin The outer epidermis consists of interwoven dead, flattened, keratinized cells that are constantly sloughed. Underlying cells are connected by tight cell junctions. Sweat glands secrete antimicrobial substances, and sebaceous glands release oils containing fatty acids on the surface of the skin.

Skin

Intact human skin, and indeed the skin of most animals, is not an inviting environment for colonization by most microorganisms **(Figure 19.2)**. The skin of most mammals is generally cool, dry, and acidic (pH 5.0). It is covered in an armor of keratinized cells called the "stratum corneum." Above this outer layer is a layer of oil (sebum) secreted by sebaceous glands. The oils are composed of fatty acids that are inhibitory to many microorganisms. In addition, sweat glands, when present, deposit other antimicrobial substances, such as NaCl and lysozyme, an enzyme that degrades the peptidoglycan of bacterial cell walls (see Section 2.3). Underneath this outer layer are layers of epithelial cells with tight cell junctions that hold adjacent cells firmly together, effectively sealing any gaps to prevent microbes from penetrating. The constant sloughing of the outer epithelial cells, called "desquamation," helps to rid the skin of pathogenic microorganisms. Constant cell division in the basal layer of the epidermis replaces the sloughed cells. The entire epidermis is replaced in about 48 days. Few microorganisms can actually breach the barriers presented by skin, and usually some form of trauma, such as a cut, is required for entry.

Skin is also home to large numbers of symbiotic yeast and bacteria that form the normal microbiota of skin. These are well adapted to living in this cool, dry, acidic environment and colonize much of the available surface. In this respect, skin is really a miniature ecosystem, with all niches being filled by resident skin microbiota. The resident microbiota compete with each other for space and nutrients, making it difficult for new microorganisms, including pathogens, to establish themselves in this habitat. Therefore, we should view the microbes living on our skin as integral to our good health.

CONNECTIONS Competitive exclusion is an important function of skin microbiota. The normal microbiota of the body is important, not only for preventing pathogens from causing disease, but it is also key to normal immune system functioning. See **Section 17.4** for a review.

Mucosal membranes

The inner surfaces of the body, such as the respiratory tract, digestive tract, urogenital tract, and conjunctiva of the eye socket, are warm and moist and are not shielded by a casing of keratinized cells. These are much more inviting living conditions for microbes. To help protect these surfaces from infection, various particle removal systems are present that incorporate diverse intrinsic physical, mechanical, and chemical barriers **(Figure 19.3)**. These surfaces, sometimes called "mucosal surfaces" or "mucous membranes," are lined with **mucosal membranes** composed of tightly packed epithelial cells with tight junctions. Mucosal membranes are kept constantly moist by fluids, such as saliva, tears, and/or **mucus**. Mucus is a viscous, adherent mixture of hydrated glycoproteins called "mucins," secreted by specialized goblet cells found scattered in mucosal membranes of the upper respiratory tract, the stomach, intestine, and genital tract. Mucus coats the membrane surface, forming a gelatinous layer to reduce contact of microbes with the epithelium. Its thick, sticky texture traps microbes, reducing their ability to move. Constant removal of mucus controls microbial colonization of these surfaces.

Special ciliated epithelial cells line the respiratory tract and vagina and rhythmically beat their cilia to mechanically move mucus and entrapped particles out of these passages. This mechanism is referred to as **mucociliary clearance**, and serves a very important function as discussed in **Perspective 19.1**. The intestine is not lined with ciliary epithelium. Instead, it uses the mechanical motion of **peristalsis**, wave-like contractions of smooth muscle to keep intestinal contents, including mucus, moving along. This action reduces the ability of ingested microbes to attach to the mucosal surface. The flushing action of tears, urine, and saliva also helps to dislodge and remove mucus and microbes from respective mucosal surfaces.

Mucus may also contain antimicrobial substances, such as lysozyme, defensins, and lactoferrin, which are secreted by epithelial

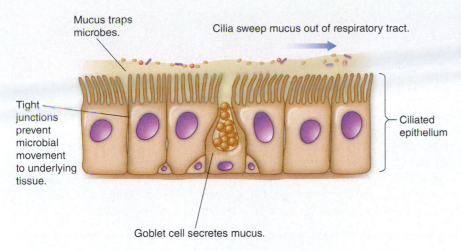

Mucus traps microbes.

Cilia sweep mucus out of respiratory tract.

Tight junctions prevent microbial movement to underlying tissue.

Ciliated epithelium

Goblet cell secretes mucus.

A. Ciliated mucosal surface of upper respiratory tract

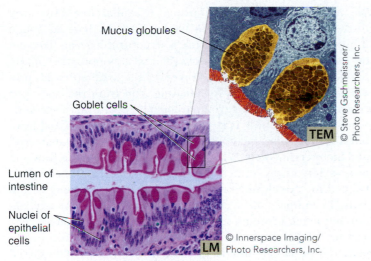

Mucus globules

Goblet cells

TEM

© Steve Gschmeissner/ Photo Researchers, Inc.

Lumen of intestine

Nuclei of epithelial cells

LM

© Innerspace Imaging/ Photo Researchers, Inc.

B. Mucosal surface of small intestine

Figure 19.3. Defenses of mucosal membranes lining inner body surfaces A. The mucosal membrane of the human upper respiratory tract consists of a thin outer layer of epithelial cells with tight junctions. Goblet cells, not associated with all mucosal membranes, secrete mucus that traps particulate matter, including microbes, and prevents their contact with the surface. Cilia of the upper respiratory tract continuously remove mucus and trapped particles, further preventing colonization of the surface. **B.** Goblet cells of the small intestine can be seen interspersed among epithelial cells lining the lumen. The color-enhanced micrograph insert shows globules of mucus packing a goblet cell.

cells. Defensins are one family of proteins belonging to a larger group of **antimicrobial peptides**. Antimicrobial peptides have been identified in plants, invertebrates (see Section 19.4), and vertebrates. They are low molecular weight proteins secreted by various cell types that appear to disrupt microbial membranes, resulting in gaps or pores. **Lactoferrin** is an iron-sequestering protein secreted by various cells, including epithelial cells of the respiratory tract. Withholding elemental iron from microorganisms inhibits their growth, as bacteria require it for metabolism. We'll discuss the role of iron later on in this section. Examples of physical and chemical defenses of specific mucosal membranes are listed in **Table 19.1**.

TABLE 19.1 Physical and chemical innate defenses associated with mucosal membranes

Site	Mucosal surface	Defense
Eye	Conjunctiva	Filtering by eyelashes
		Lysozyme
		NaCl
		Flushing by tears
Digestive tract	Mouth	Lysozyme
		Antimicrobial peptides
		Desquamation
		Mucus
	Stomach	Mucus
		pH 2–3
	Small intestine	Peristalsis
		Digestive enzymes
		Bile salts
		Desquamation
		Mucus
		Antimicrobial peptides
	Colon	Large numbers of resident microbiota
		Mucus
		Peristalsis
Respiratory tract	Nasal passage	Filtering by nostril hairs
		Mucus
	Trachea and bronchi	Ciliated epithelium
		Mucus
	Lungs	Lactoferrin
		Lysozyme
		Antimicrobial peptides
Urogenital tract	Urethra	Flushing by urine
	Vagina	Antimicrobial peptides
		pH 5
		Mucus
		Ciliated epithelium

The importance of mucociliary clearance is evident in patients with cystic fibrosis, a genetic disease. An aspect of the disease is poorly hydrated mucus, which is abnormally thick and sticky. Mucus, even the normal kind, is very good at trapping microbes that enter the respiratory tract. In cystic fibrosis patients, the extreme viscosity of the mucus overwhelms the ability of the cilia to transport it out of the upper respiratory tract, and so it becomes a sink for inhaled microbes and descends to the lower areas of the lungs. Consequently, cystic fibrosis patients suffer almost continual respiratory infections. Lung infections with *Pseudomonas aeruginosa* are common and are notoriously difficult to treat as the bacteria can establish an impervious layer called a biofilm (see Section 2.5), which is resistant to penetration with antimicrobial drugs. Such infections are a major cause of death in cystic fibrosis patients. Their inability to clear mucus from the upper respiratory tract becomes a detrimental condition and illustrates the importance of mucociliary clearance as a barrier to infection.

Recall from Section 17.4 that skin and mucosal surfaces, especially the intestine, are colonized with large numbers of microbes that form the normal microbiota. Competitive exclusion by these populations serves to deter colonization by new microbes, including pathogens, and keeps any one population from dominating this environment. Many species inhabiting the intestine produce essential vitamins needed for good health in humans and some animal species. As we saw in herbivores (see Section 17.4), these microorganisms are essential for breaking down fibrous substances, such as cellulose, which cannot otherwise be utilized by the host.

Disturbances in normal microbiota, such as can occur with oral antibiotic use, may reduce some microbial species and increase others. This may result in short-tem digestive disturbances, such as diarrhea. Other surfaces such as the lung and the uterus are not normally colonized. The effective barriers in the upper respiratory tract and the vagina prevent most microbes from entering these regions. Transient microbes that do enter are removed by mucociliary clearance or are destroyed by various antimicrobial substances (see Table 19.1).

Natural barriers of the body prevent most microbes from establishing infection. However, if a breach in one of the protected surfaces occurs, then immune responses begin. An injury, such as a cut or a burn, allows entry of opportunistic pathogens that could not otherwise gain access to tissues. As we will see in Chapters 21 to 23, some pathogenic microbes have evolved strategies to colonize or invade intact barriers.

Iron: The limiting element

We mentioned earlier that limiting the availability of free iron in the body has an important bacteriostatic effect. Although limiting iron does not constitute a structural barrier, like skin or mucosal surfaces, it is nonetheless a noteworthy innate barrier to infection. Indeed, as we will see in Chapter 21, producing virulence factors for stealing iron from the body is a common feature of bacterial pathogens. Most iron in the body is not in free form. It is kept sequestered by specialized iron-binding molecules, including hemoglobin, ferritin, transferrin, and lactoferrin (Table 19.2). Iron is important for the function of a variety of proteins in bacteria, as it is for all cells. It is needed for cytochromes that catalyze redox reactions within cells. In animals, iron is an integral component of oxygen-binding heme proteins such as hemoglobin of red blood cells and myoglobin of muscle cells. In humans, most iron in the body (50 to 80 percent) is associated with hemoglobin of red blood cells. Lactoferrin binds iron particularly strongly, a feature that is an advantage in fluid secretions, including milk, to prevent iron from being made available to microbes that might otherwise abound in these rich fluids. There is a wealth of convincing evidence that iron deficiency actually protects against many infectious diseases such as malaria, plague, and tuberculosis. Equally, there is substantial evidence that iron supplementation can predispose one to bacterial infection and exacerbate some diseases, such as tuberculosis (see Chapter 21). Free iron can also be dangerous. It is a potent catalyst for the formation of toxic free radicals. This is another reason for cells to keep iron in a chemically bound form.

TABLE 19.2 Iron-binding molecules of the body

Iron-binding protein	Location	Function
Hemoglobin	Red blood cells	Binds oxygen
Myoglobin	Muscle and heart cells	Binds oxygen
Transferrin	Plasma	Transport of iron around the body
Lactoferrin	Milk, tears, saliva	Antimicrobial activity in secretions by binding iron
Ferritin	Cells, small amounts present in plasma	Storage of iron

19.2 Fact Check

1. Explain how skin and mucosal membranes protect the body from microbes.
2. What antimicrobial roles do lysozyme, defensins, and lactoferrin play in the body?
3. How does the body limit iron availability, and how does this protect against microbes?

19.3 The inflammatory response

What is an early response of the body to infection?

What happens if a microbe manages to get past the barriers and other passive defenses of the body? We can first look at a relatively common occurrence and one with which we are all familiar—an infected cut. Although a few microbes may directly invade intact skin, most access the underlying tissues only through breaks. Initially, once microbes enter tissue, several simultaneous events

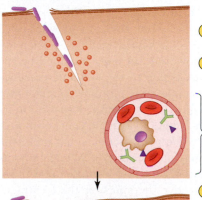

1. Bacteria enter the tissue through a cut.

2. Vasodilator molecules are released such as histamine, prostaglandins, and serotonin.

   Blood vessel

3. Bacteria multiply and invade tissue.

4. Concentrations of vasodilators increase.

5. Vessel dilates, allowing more blood into the area. An increase in local temperature and redness occurs.

6. Vessel walls become more permeable. Fluid moves into the tissue, causing swelling. Extravasation of cells and molecules occurs.

7. Antibody and complement factors bind the bacteria.

8. Phagocytes engulf and destroy the bacteria. Cytokines are released, attracting more leukocytes.

9. Adhesion molecules are expressed on vessel walls to facilitate extravasation of immune cells.

10. Clotting is initiated to restrict access of bacteria to circulatory system.

A. The inflammatory process

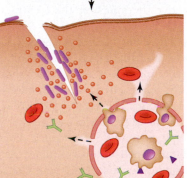

Pus

Legend:
- — Bacteria
- Phagocyte
- Y Antibody
- ▲ Complement factor
- ● Red blood cell
- · Vasodilator
- · Cytokine
- ‖ Adhesion molecules

B. Neutrophil accumulation due to inflammation

occur. Tissue injury caused by the external trauma results in release of various cell-signaling molecules, called "effector" molecules, which have direct effects on surrounding cells. Examples produced from local cells include histamine and prostaglandins that are important effectors in initiating **inflammation**. Inflammation is a progression of physiologic events that activate cells and the release of molecules involved in defense and repair to the damaged or infected site. We can identify the clinical signs of inflammation that accompany an infected cut—swelling, redness, and heat. Inflammation itself is not an immune response. Rather, it is a frequent precursor that facilitates immune responses. Immune responses can occur without inflammation. Indeed, not all pathogens will evoke an inflammatory response. Examples include HIV, rabies virus, and *Vibrio cholerae* (see Section 8.2).

Sometimes the invading microbe can initiate release of effector molecules from the surrounding cells by producing substances such as lipopolysaccharide (LPS, an endotoxin), or various exotoxins. Release of histamine from tissue mast cells (see Section 19.5) also activates the plasma-derived fibrin and kinin clotting pathways that restrict infection and tissue damage by forming a mesh-like barrier to wall off damaged blood vessels and associated tissue to prevent the spread of infection. Activation of these pathways triggers release of more **proinflammatory molecules** that stimulate inflammation and attract cells with immune functions to the site. An important family of proinflammatory molecules produced by cells during an infection are **cytokines**. Cytokines are secreted, soluble, low molecular weight glycoproteins that are produced by cells involved in immune reactions. Examples of cytokines include interleukins, tumor necrosis factor-alpha (TNF-α), and interferons (IFNs). Cytokines are the chief signaling molecules used by cells to communicate information during an immune response. They initiate cell-signaling pathways that result in the expression of certain cellular genes controlling a variety of immune functions. Some cytokines recruit **leukocytes**, commonly called "white blood cells," to come to the damaged site for cleanup. Leukocytes, such as macrophages, neutrophils, and T cells and B cells (described later) have immune functions. Release of cytokines from the site of inflammation/infection forms a chemical gradient that guides the migration of nearby leukocytes to the site. Imagine the scent of a cooling apple pie left on a windowsill that attracts a distant bear out of the forest. This cytokine gradient functions in a similar way.

The initiation of the inflammatory response is not fully understood but involves three major processes: (1) vasodilation, (2) increased vessel permeability, and (3) **extravasation**, the movement of cells and molecules from blood vessels into tissues **(Figure 19.4)**. **Vasodilation** is an increase in the diameter of the blood vessels. Vasodilation at the site of infection brings

Figure 19.4. Process Diagram: The local inflammatory process
A. Tissue damage and exposure to microbes begins inflammation and involves release of vasodilator effector molecules, an increase in vessel permeability, and migration or extravasation of defensive cells and molecules from the circulatory system. Phagocytes, such as neutrophils, can engulf and destroy invading bacteria. Clotting factors form a mesh-like barrier to wall off the infected/injured area from the rest of the body to limit the spread of infection. **B.** Neutrophils are the most numerous leukocytes in blood and can accumulate as pus.

more blood, and with it, more defense cells and molecules to the area. Increased blood volume that results produces clinical signs of redness and increased temperature in the local area. Prostaglandins and histamine are potent vasodilators, released from mast cells upon encountering certain stimulatory molecules associated with damage, such as hydrogen peroxide and activated complement components. Cells release serotonin following sensory nerve stimulation.

Increased vessel permeability of the dilated vessel walls is another characteristic of the inflammatory response. It is a result of the loosening of the cell junctions that line the vessels, creating gaps that allow fluid and cells from the blood to move into the tissues where they can combat the infection. Extravasation is facilitated by dilation and increased permeability of the vessel walls. Fluid carries plasma components such as antibodies and complement factors, discussed later, into the affected area to aid in the destruction of microbes. Excess fluid in inflamed tissues appears as swelling or edema of the local area.

As the inflammatory response progresses, concentrations of inflammatory effector molecules increase in the local area. This induces cells of the vessel walls to produce surface molecules that bind passing leukocytes from the blood, allowing them to adhere and pass through the vessel walls, actively facilitating extravasation. As more blood enters the area, the adherent cells, mostly phagocytes such as neutrophils and macrophages, accumulate in the affected tissue. **Phagocytes** are specialized leukocytes that engulf and destroy microbes. Once phagocytes have entered the tissue, they begin to clean up foreign particles and damaged cells. Active phagocytes release chemoattractants, such as the cytokine interleukin 8 (IL-8), that attract more cells with immune function to the area. Phagocytes, especially neutrophils, can increase to large numbers as they are recruited to the site, often seen as pus in infected cuts.

Some local inflammatory responses are large enough to induce further responses from the body. For example, the cytokines IL-1, IL-6, and tumor necrosis factor-alpha (TNF-α) can signal cells of the liver to produce other secreted molecules, such as C-reactive protein (discussed later) that take part in fighting the early phase of infection/damage. These three cytokines work on the hypothalamus to produce **fever**, a prolonged increase in the core body temperature of endothermic animals. This temperature increase can restrict the growth of many pathogens, allowing the immune system time to clear them. Fever benefits the immune system directly by increasing production and activity of leukocytes as well as speeding up tissue repair. Too high a fever can have detrimental effects due to denaturation of critical enzymes needed for cellular metabolism.

Inflammation is a process that works in favor of the body, at least when it is contained locally to the infected site. However, if the infecting microbe escapes the initial infection site and distributes to other sites (including the bloodstream) in large enough numbers, a **systemic inflammatory response** can occur. This takes place when the chemical signals that cause inflammation are released on a grand scale producing widespread inflammation in the body. This type of inflammation often leads to physiological shock, characterized by a drastic drop in blood pressure as large quantities of fluid leak from the circulatory system. In turn, vital organs, such as the brain, heart, kidney, and liver, no longer receive full supplies of blood, resulting in organ damage and, in many cases, death. **Septic shock** is a systemic inflammatory response induced by the systemic or widespread presence of bacteria in the body. Septic shock is a common outcome to untreated infections that are not contained by the immune system. **Toxic shock** is due to the systemic presence of microbial exotoxins that stimulate the immune system. Once septic shock or toxic shock begins, it is very difficult to prevent its progression. High death rates result (30 to 50 percent) even with intensive care. Identifying the microbial products that evoke the inflammatory response and mapping the body's chemical signals that amplify the inflammatory response are intense areas of research toward developing effective therapies for the treatment and prevention of septic and toxic shock.

Two potent proinflammatory cytokines are IL-1 and TNF-α. Injection of large amounts of either of these cytokines in animal models has shown they can induce clinical signs of septic shock in the absence of infection. LPS, commonly known as endotoxin, produced by Gram-negative bacteria such as *Escherichia coli* and *Pseudomonas*, and the so-called "superantigens," produced by some strains of *Streptococcus pyogenes* and *Staphylococcus aureus* (see Chapter 21), stimulate immune cells to overproduce these and other proinflammatory cytokines to very high levels. Consequently, serious infection with these organisms often leads to a systemic inflammatory response, causing septic or toxic shock.

19.3 Fact Check

1. What is inflammation, and what are the results of the inflammatory response?
2. Describe the role of cytokines in response to infection, and provide examples.
3. Define systemic inflammatory response, septic shock, and toxic shock, and explain how these terms are related.

19.4 The molecules of the innate system

How do we sense or detect the presence of pathogens?

Inflammation is a common early response to damage and infection that can facilitate immune responses by transporting a variety of immune cells and secreted defense molecules to the site of infection. Now we will examine molecules of innate

immunity to see how they actually detect the presence of pathogens and signal this information to initiate immune responses. As innate immunity is the first line of immune defense, it must act quickly and be able to readily recognize foreign substances in the body without having encountered them before. Multicellular organisms have evolved relatively simple but very effective means to accomplish this, as we will see next.

Pathogen-associated molecular patterns

Human beings recognize foreign objects and organisms visually. We assign "foreignness" to things that are notably very different from other things we encounter on a frequent basis. For example, a moose grazing on the roadside in Australia would be recognized as being foreign to Australians even if they didn't know it was a moose. The innate immune system is also based on the recognition of foreign matter. The body does not perceive foreign matter visually but does so biochemically. Microbes contain a variety of unusual molecules that are easily recognized by the host as being foreign. These **pathogen-associated molecular patterns (PAMPs)** are broad-based molecular signatures or motifs commonly found in many pathogens but not normally found in the host. Examples include the lipid A component of Gram-negative endotoxin (LPS), teichoic and lipoteichoic acid from Gram-positive cell walls, and mannose of bacterial and yeast cell walls. These and other molecules are recognized by **pattern recognition receptors (PRRs)** produced by host cells. PRRs are encoded by the host and can be secreted or membrane-bound proteins. Most are present all the time and act as sensors for microbes in the host. Although only a small variety of PRRs are produced by the host, they recognize a diverse array of microbes including bacteria, fungi, viruses, and protozoa. This makes them *nonspecific* in terms of identification of

a particular microbe but gives them the broad spectrum necessary for surveillance of infection. In evolutionary terms, innately produced PRRs appear to be ancient. They exist in various forms in invertebrates, vertebrates, and plants and are a fundamental aspect of all foreign-organism recognition systems.

Toll-Like Receptors

Important PRRs involved in the initiation of immune responses in invertebrates and vertebrates are the **Toll-like receptors (TLRs**, see the chapter opening story). Toll receptors were originally described in *Drosophila melanogaster* embryos. Since then, genomic sequencing of various organisms from plants to humans has uncovered many similar genes, all coding for Toll-like proteins with important immune functions. TLRs are transmembrane proteins with exterior domains that recognize different pathogen-associated molecular patterns (PAMPs), and cytoplasmic domains that activate distinct components of signaling pathways. Each type of receptor has a specific binding target or **ligand** to which it binds. Over ten different TLRs have been reported in humans and mice but not all have had their ligands identified. TLRs have evolved as an early warning system for the detection of pathogens.

In vertebrates, TLRs are expressed in diverse cell types such as epithelial cells of the gut, lung, and skin, cells of the brain, and the front-line cells of the immune system, such as neutrophils, monocytes, macrophages, and dendritic cells. The amount and profile of TLRs expressed is characteristic of the cell type and even the stage of development of the cell. As shown in **Figure 19.5**, individual

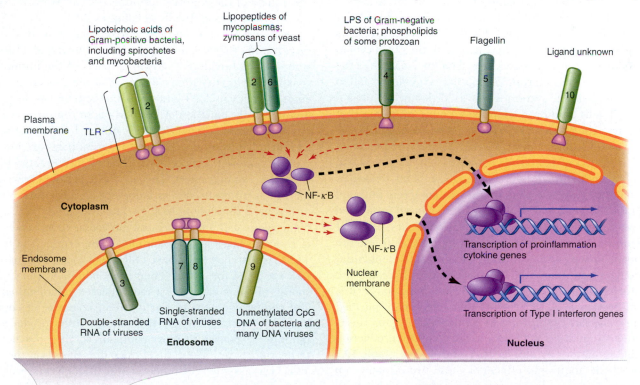

Figure 19.5. Model of human Toll-like receptors (TLRs), their binding targets, and action TLRs 1 through 10 are found on many cell types, including immune cells. The PAMPs that bind TLR10 have not yet been identified. TLRs on the surface of cells recognize many surface PAMPs of microbes. TLRs exposed in endosome bind released nucleic acids of ingested and degraded microbes. Upon binding of PAMPs to TLRs, distinct signaling pathways are activated to recruit transcription factors, such as NF-κB. The transcription factors enter the nucleus and initiate transcription cytokine genes whose products stimulate and direct immune responses. The two major sets of genes activated are Type I interferon genes or genes encoding proinflammatory cytokines, such as IL-1, IL-12, and TNF-α.

TLRs recognize and bind to specific PAMPs. For example, TLR4 binds LPS of Gram-negative bacteria and TLR5 binds fimbrin, the structural protein of bacterial fimbria (see Section 2.5). TLR9 recognizes *unmethylated* pairs of adjacent cytidine and guanosine nucleotides. These are referred to as **CpG dinucleotides**, commonly found in bacterial and some viral DNA. The discovery that host cells could recognize biochemical differences between microbial and host DNA was astounding, and is described in the Mini-Paper of this chapter. Many of the PAMPs that bind TLRs have been identified through *in vitro* assays. As such, it is unclear if all of these recognitions are relevant *in vivo*.

TLRs not only recognize a panel of PAMPs but also signal the general nature of the pathogen to the immune system. In so doing, binding of a TLR to its particular PAMP target initiates cellular signaling pathways that allow the immune system to more effectively deal with the invader. A transcription factor called "nuclear factor-kappa B" (NF-κB) is recruited through this pathway, along with other customized sets of transcription factors, to enter the nucleus. These induce transcription of a customized set of genes whose products protect the host against the category of pathogen identified. There appear to be two general categories of cytokines induced: one group called **Type I interferons (IFNs)** that have numerous antiviral activities, and proinflammatory cytokines. We'll learn about Type I IFNs near the end of this section. To see how proinflammatory cytokines facilitate clearance of the pathogen, let's look at what happens when a phagocyte contacts a Gram-negative bacterium. TLR4 binding of LPS turns on expression of genes encoding the cytokines IL-1, IL-12, and TNF-α. Recall that IL-1 and TNF-α induce beneficial fever in the host. TNF-α also activates genes that increase phagocytic abilities and stimulate the phagocyte to divide to produce an army of phagocytes. TNF-α also induces the phagocytes to secrete more proinflammatory cytokines to call in other cells of the immune system. IL-12 secretion by the phagocyte activates T cells to initiate adaptive immune responses that selectively target the particular Gram-negative pathogen.

You may recall from Chapter 17 that filarial diseases associated with infection of nematode worms is largely due not to the worms but to the worm bacterial endosymbiont *Wolbachia*. The inflammatory response that produces damage is initiated through TLR4 recognition of *Wolbachia* LPS released from dead worms. In river blindness, this TLR4 activation occurs in cells of the eye, resulting in chronic infiltration of immune cells into the cornea, causing opacity and eventually blindness. In elephantiasis, treatment with antinematodal drugs releases *Wolbachia* into the circulatory system and severe systemic inflammation occurs. Similarly, in Section 19.3, we learned that injection of purified LPS can cause the same symptoms as septic shock. Again, this is through direct stimulation of TLR4 and the resultant production of the proinflammatory cytokines IL-1 and TNF-α. **Table 19.3** shows some of the additional effects of these two cytokines on innate functions and the adaptive immune system, which we will explore in more detail in Chapter 20.

Another example of how TLR binding begins immune responses is through TLR3 detection of double-stranded RNA, a common intermediate produced by viral replication. Double-stranded RNA in the cell signals the presence of a virus and turns on expression of Type I IFNs in the cell. These IFNs have

TABLE 19.3 Effects of cytokines TNF-α and IL-1 produced by macrophages

Effect	TNF-α	IL-1
Production of fever	+	+
Proliferation of phagocytic macrophages	+	−
Production of molecules for extravasation of leukocytes	+	+
Increases vascular permeability	+	+
Stimulates production of C-reactive protein (binds microbes)	+	+
T-cell activation (for adaptive immunity)	+	+
Induction of IL-8 and IL-6 (stimulates adaptive immunity)	+	+

antiviral activities that help cells resist viral infection, and will be discussed at the end of this section.

TLR signaling is also crucial for the activation of adaptive immunity and appears to be a major process linking the innate and adaptive systems. For example, an innate function of macrophages is to engulf and destroy microbes. Activation of TLR-signaling pathways induces macrophages to become more efficient killers, but also stimulates production of the surface molecules needed for their interaction with T cells, the main controllers of the adaptive immune system. We will explore the specific functions of T cells in Chapter 20.

You may wonder how scientists determined the binding targets and other immune functions of individual TLRs. Understanding the function of individual parts of the immune system is akin to trying to assign roles to all of the parts of a modern jet engine when you don't know how the system was designed. One way to tell the function of the parts is to remove specific pieces one at a time and then observe the effect. To confirm the function and provide insight into how the different parts are interrelated, you can remove the piece that you think it interacts with and see if you get a similar effect. This approach is called "reverse genetics" and is widely used in immunology and disease research, just as it is used throughout biology. The availability of complete genome sequences and associated molecular tools makes this type of research feasible. For example, to determine the function of TLR2, researchers produced TLR2 knockout mice engineered to lack the gene for TLR2. Then they examined the immune response of these mice to various injected microbial compounds and compared them to the immune response of control non-knockout mice. Although the mutant mice produced normal immune responses to most compounds, they did not respond to peptidoglycan. The control mice did. Based on these results, TLR2 was assigned the function of peptidoglycan recognition.

Identification of the molecules that bind TLRs and the resulting type of immune response that is stimulated by this binding has been instrumental in the design of new vaccine formulations. One of the components often included in vaccine preparations is an **adjuvant**. An adjuvant enhances the immune response to the microbial components in the vaccine formulation by stimulating the innate immune system. For example, the adjuvant Ribi.259 has been shown to specifically activate TLR4 in a manner similar to bacterial endotoxin (LPS). Unlike LPS, it does not evoke inflammation and is safe for use in veterinary applications. Synthetic unmethylated CpG dinucleotides that mimic bacterial DNA exhibit adjuvant activity and can stimulate innate cellular responses by binding to TLR9. The **Mini-Paper** describes the initial characterization of immunogenic DNA sequences. CpG dinucleotides are the basis for the newly developed DNA vaccines, many of which have shown a high level of protection from experimental infections in animal models.

CONNECTIONS Adjuvants greatly enhance the effectiveness of vaccines in a variety of other ways described in **Section 20.4**. Modern vaccine formulations routinely use adjuvants to boost immune stimulation. See also **Section 24.5** for the history of vaccination and details of vaccines, including DNA vaccines.

Other research on TLR-signaling pathways promises new therapies for human chronic inflammatory diseases, such as inflammatory bowel syndrome and Crohn's disease. By interfering with binding of the molecules identified in induction of inflammation to their TLRs, or disruption of the signaling pathways, chronic inflammation may be reduced or eliminated. Prevention of LPS-induced septic shock by similar blocking techniques also holds great promise.

Mannose-Binding Lectin and C-Reactive Protein

TLRs are just one type of pattern recognition receptor (PRR). Some PRRs are secreted instead of being membrane-bound. **Table 19.4** lists an example of a secreted PRR: **mannose-binding lectin (MBL)**. **Lectins** are proteins that bind carbohydrates. The carbohydrate mannose is found in the cell walls of a variety of microorganisms such as the yeast *Candida albicans* and bacterial *Salmonella* species. Molecules composed of repeating units of mannose are not a constituent of higher eukaryotic cell surfaces; they are recognized by the immune system as a pathogen-associated molecular pattern (PAMP). When mannose is encountered, MBL binds to it at the microbial surface (**Figure 19.6**). This coating of the microbial surface is called **opsonization**, and it leads to two important events: activation of complement (described shortly), leading to leakage and death of the microbial cell, and **phagocytosis** by macrophages. Phagocytosis involves

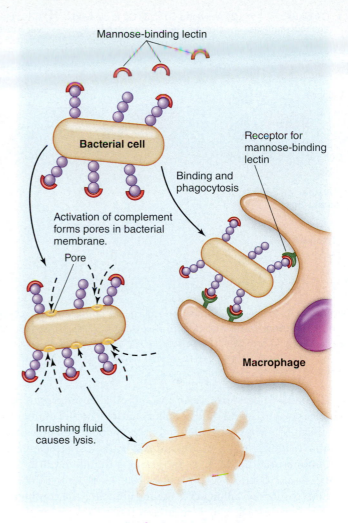

Figure 19.6. Opsonization of a bacterial cell by mannose-binding lectin Coating of a microbial cell with mannose-binding lectin can facilitate phagocytosis by macrophages, which possess receptors for the bound lectin. Opsonization with mannose-binding lectin also activates complement, leading to destruction of the microbe by lysis.

the formation of large cytoplasmic extensions that surround the particle and then fuse to engulf the microbial cell. Phagocytosis is facilitated by opsonization as macrophages possess receptors for bound mannose-binding lectin that allow them to adhere to the opsonized microbial cell. The processes of opsonization and phagocytosis are discussed in more detail later (Section 19.5).

The importance of secreted PRRs in immunity is evident in patients or animal models where mutations exist in genes coding for secreted PRRs. MBL deficiency is considered to be the most common cause of non-HIV human immunodeficiency. As it is an innately produced protein, MBL is thought to be particularly important for protection in very young children whose adaptive immune system is not yet fully developed. Recent investigations have indeed shown that MBL deficiency is correlated with increased susceptibility and frequency of respiratory infections in children 5 to 18 months of age, the age of weaning from mother's antibody-containing milk. Clinical evidence also indicates that adults with low serum levels of MBL suffer from more frequent and severe bacterial and fungal infections than do individuals with normal MBL levels. Consequently, trials using recombinant and serum-derived MBL as therapeutic treatments for individuals who suffer from these frequent infections are underway.

TABLE 19.4 **Some secreted pattern recognition receptors (PRRs) of vertebrates**

Receptor	Molecule bound
Mannose-binding lectin (MBL)	Mannose of bacterial and fungal cell walls
C-reactive protein	Phosphatidylcholine/phosphocholine of bacterial and fungal membranes
LPS-binding protein	Lipopolysaccharide (endotoxin) of Gram-negative cell walls
Complement factor C3b	Variety of cell-wall components, including lipoteichoic acid of Gram-positive cell walls, and LPS of Gram-negative bacteria

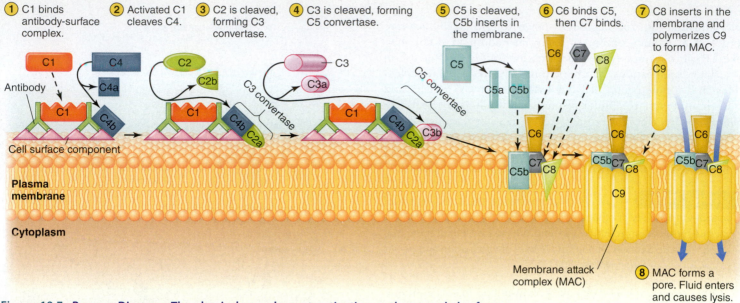

Figure 19.7. Process Diagram: The classical complement activation pathway and the formation of the membrane-attack complex (MAC) The classical pathway is initiated by antibody binding to the microbial surface. Complement factor C1 binds the terminus of bound antibody and forms an active complex that initiates a series of cleavages, beginning with C4 to form MAC. MAC forms an open pore in the membranes of bacteria (shown here), or of enveloped viruses, permitting the free diffusion of water into the microbe. The result is disruption of the membrane and lysis.

C-reactive protein is another secreted lectin present in serum. It recognizes and binds to phosphatidylcholine of microbial membranes and phosphocholine, a component in the capsules of some bacteria and some fungal cell walls. C-reactive protein is manufactured by the liver in response to inflammation. Like MBL, bound C-reactive protein opsonizes the microbial cell, leading to uptake by phagocytosis, and activates the complement cascade (see below), which results in lysis of the microbial cell. As C-reactive protein concentrations increase with inflammation, this protein is commonly used as a non-specific biomarker of inflammation. A C-reactive protein test is most often done to monitor certain inflammatory diseases such as rheumatoid arthritis. More recently, C-reactive protein is used as an indicator of a person's risk for coronary disease and hypertension, both of which are characterized by inflammation.

Complement

Complement is a collective term that refers to a group of more than 30 serum proteins involved in binding microbial surfaces and lysing membranes. Complement components exist as inactive precursor proteins in the blood. Enzymatic cleavage and conformational changes of the precursors occur through a cascade, whereby one activated complement factor acts on another precursor in a sequential manner. An end result of the complement cascade is formation of the pore-forming **membrane-attack complex (MAC)**, which leads to microbial death by lysis (see Figure 19.6).

The complement cascade is an innate mechanism, designed to occur early in infection to kill microbes before they get a chance to proliferate. In the presence of microbes, it can be activated by several simultaneous mechanisms, which helps to ensure a wide variety of pathogens will trigger the cascade and lysis by MAC will occur. As we will see, the complement system can be activated by harnessing both innate and adaptive immune responses, and so occurs whether the host can or cannot specifically recognize the particular infecting microbe. As well as directing lytic activity, complement factors also allow for more efficient uptake of microbes by phagocytes, and induce inflammatory and immune responses. Activation of the complement cascade is accomplished through any one of three different activation pathways: the **classical pathway**, the **alternative pathway**, and the **lectin pathway**.

The Classical Pathway

The classical pathway, so called because it was the first complement activation pathway described, leads to the use of all of the complement factors, named complement factor 1 (C1) through complement factor 9 (C9) **(Figure 19.7)**. The other two activation pathways funnel into this same cascade at various points, so examination of the classical pathway provides a good overview of how this system works. Initiation of the classical pathway involves the binding of C1 to microbial membranes.

C1 C1 is a complex composed of several peptides. C1 binds to the surface of some pathogens by directly recognizing certain surface components, such as lipoteichoic acid of Gram-positive cell walls and LPS of Gram-negative walls. As such, it acts as a PRR, but it can also bind surfaces that have been coated with C-reactive protein or **antibody**. Antibodies are secreted glycoproteins that bind to microbes and their components and are produced by B cells of the adaptive immune system. Antibodies and B cells will be described in more detail in Chapter 20. A common diagnostic test that takes advantage of the ability of C1 to recognize any antibody bound to a microbial surface is called the complement fixation test (CFT), described in **Toolbox 19.1**. This test is used to indirectly detect the presence of antibodies specifically produced as a result of infection with a pathogen. The ability of the adaptive immune response to initiate the innate complement cascade is just one example of the synergy between these two systems. Once sufficiently bound at a microbial surface, C1 undergoes autocatalysis, cleaving itself to produce an active protease that cleaves the next complement component in the cascade, C4.

C4 Cleavage of C4 by activated C1 results in the products C4a, which is released, and C4b. The product C4b can bind to the surface of the target cell where it remains associated with C1.

C2 In its bound and active form C4b binds C2, bringing it into position for active C1 to cleave into C2a and C2b. The smaller C2b fragment is released while C2a remains surface-bound to form the complex C4b2a, also known as the "C3 convertase complex."

C3 The C3 convertase complex acts upon C3, converting it into C3a and C3b. As the C3 convertase complex has enzymatic activity, it can cleave *many* molecules of C3 upon contact. Some of the produced C3b diffuses away and the rest remains associated with C4b2a at the cell surface. The resultant complex formed is C4b2a3b, also known as the "C5 convertase complex."

C5 The enzymatic C5 convertase complex cleaves molecules of C5 into C5a and C5b. The larger C5b fragment associates with the next several components to form the membrane-attack complex (MAC).

C6, C7, C8, C9, and MAC MAC is formed by the sequential interaction of C5b, C6, C7, C8, and C9 with the microbial membrane. The complex C5bC6C7 forms an anchor to the membrane via insertion of the hydrophobic site of C7 into the lipid bilayer. C8 next associates with this complex, similarly inserting into the membrane and now able to initiate polymerization of up to 15 C9 molecules in the membrane. This large complex of C9 polypeptides forms pores in microbial membranes **(Figure 19.8)**. In Gram-negative bacteria, disruption of the outer membrane allows lysozyme present in serum to destroy the peptidoglycan layer. Water can enter the resulting protoplast and the cell lyses (see Section 2.4). A summary of the complement components of the classical pathway is shown in **Table 19.5**.

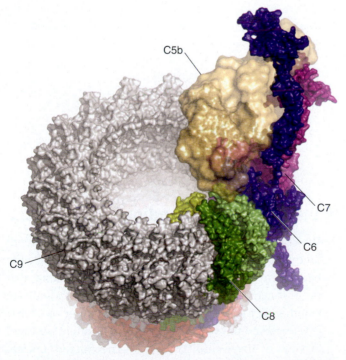

Figure 19.8. Membrane attack complex (MAC) Molecular model of the assembled complex, showing the large central pore that leads to membrane disruption and lysis. From "Structure of Complement C6 (in blue) in the context of a hypothetical model of the entire Membrane Attack Complex," © 2012 The American Society for Biochemistry and Molecular Biology. Courtesy Alex Aleshin and Robert Liddington.

TABLE 19.5 Activation and functions of complement components

Component	Result/Function
C1	Binds at microbial surface; self-activates to form C1 protease to cleave C4
C4a[a]	Degranulation of mast cells and basophils
C4b	Binds to microbial surface with C1 protease, then binds C2 to form a complex
C4b + C2 + C1 protease complex	C1 protease cleaves C4b-bound C2
C2a	Binds to C4b
C2b	Increases vascular permeability
C4b2a (C3 convertase)	Cleaves C3
C3a	Leukocyte chemoattractant; degranulation of mast cells and basophils
C3b	Binds to C4b2a, opsonin
C4b2a3b (C5 convertase)	Cleaves C5
C5a	Degranulation of mast cells and basophils; phagocyte chemoattractant; stimulates antibacterial activities of phagocytes
C5b	Associates with membranes; binds C6 to begin MAC formation; neutrophil chemoattractant
Bound C6	Binds C7
Bound C7	Inserts in membrane; binds C8
Bound C8	Inserts in membrane; polymerizes C9
C9	In presence of C8, several C9 units polymerize to form a pore in the microbial membrane; completes formation of MAC
MBL-surface complex (C1-like activity)	Cleaves C4
Factors B, D	Combines with C3b to form C3bBb (C3-like convertase)
C3bBb (C3-like convertase)	Cleaves C3 to form C3bBbC3b (C5-like convertase)
C3bBbC3b (C5-like convertase)	Cleaves C5 to begin MAC formation

red = components unique to classical pathway
blue = components unique to alternative pathway
purple = components shared by classical and lectin pathways only
green = components unique to lectin pathway
black = components shared by all pathways

[a] Note that the various complement factors were named according to their order of discovery, so their numerical designations do not necessarily relate to their position in the cascade.

Most Gram-positive bacteria appear somewhat resistant to lysis by complement due to their thick peptidoglycan layer, which retards access of MAC components to the underlying plasma membrane. Gram-positive bacteria can still effectively initiate the complement cascade to produce accessory complement molecules with important immune functions, as seen later.

Direct destruction of enveloped viruses via complement-fixing antibody and MAC formation is an important antiviral defense. Lysis of virus-infected cells can also occur because prior to exit from the host cell, many enveloped viruses first insert viral transmembrane proteins into the host cell membrane in preparation for envelope formation (see Section 8.5). When these cells are recognized by complement-fixing antibody, they can be lysed by MAC formation, preventing the formation of infectious viral particles.

The Lectin Pathway

We were earlier introduced to mannose-binding lectin (MBL). It binds mannose-containing pathogen-associated molecular patterns (PAMPs) found on the surface of many microbes. Once bound, MBL further complexes with other serum proteins to impart a C1-like activity on the complex, which then activates the complement cascade (see Table 19.5). The mannose-binding pathway is not activated by antibodies, and therefore its initiation can occur in the absence of an adaptive immune response. **Figure 19.9** shows an overview of each of the pathways.

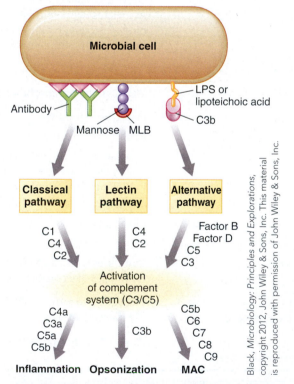

Black, Microbiology: Principles and Explorations, copyright 2012, John Wiley & Sons, Inc. This material is reproduced with permission of John Wiley & Sons, Inc.

Figure 19.9. Complement activation pathways and actions of released complement factors The pathways differ in mode of initiation but share several common steps leading to the formation of MAC in microbial membranes. Initiation of the classical pathway involves binding of C1, often in the presence of antibodies, to the microbial surface. The lectin pathway begins with serum mannose-binding lectin (MBL) attaching to mannose units on the surface of the microbe. In the alternative pathway, initiation begins when spontaneously produced C3b binds certain microbial components, such as LPS. As well as forming the MAC, several of the complement products have important activities in inflammation, and opsonization.

The complement fixation test (CFT) is an important diagnostic serological test used to determine the presence of antibodies specific to a particular microorganism from a sample of serum. This in turn *indirectly* indicates infection or recent past infection. The test takes advantage of the ability of C1 to react with almost any antibody-bound surface. So, while the antibody is specific in its recognition of the particular pathogen, complement binding to antibody is non-specific in terms of pathogen recognition.

By adding a preparation of a specific microbial component, or the microbe itself, antibodies are detected because antibody-bound complexes "fix" or bind a specific amount of complement added in the test well. **Figure B19.1A** shows the steps in a complement fixation test. To detect whether complement has been fixed, "sensitized" sheep red blood cells are used as indicators for lysis. These cells are prepared by coating them with specific antibody that binds sheep red blood cells (anti-sheep red blood cell antibody) to make them sensitive to lysis if complement is present. If specific antibodies to the microbe in question are present in the test serum, then complement is fixed and is no longer available to take part in lysis of the antibody-coated red blood cells. These cells will settle out of the suspension and form a "button" in the well, giving a positive test that indicates specific antibodies *are* present for the microorganism being tested. If specific antibodies are *not* present, then complement is not fixed and is available to lyse the indicator red blood cells. Consequently, no button forms in the well, resulting in a negative test. A complement fixation test requires two days to complete and requires highly standardized and specific microbial preparations. Despite these limitations, it is the gold standard of diagnosis for several infectious agents (see Figure B19.1B).

● Test Your Understanding ·················

The complement fixation test was used to test a patient for the zoonotic disease known as Q fever, caused by the bacterium *Coxiella burnetii*. The patient was told that the test came back positive but that it did not indicate whether he had a current or past Q fever infection. What was observed to call the results for this patient positive? In view of these results, why are the doctors unclear whether the infection is current or occurred in the past?

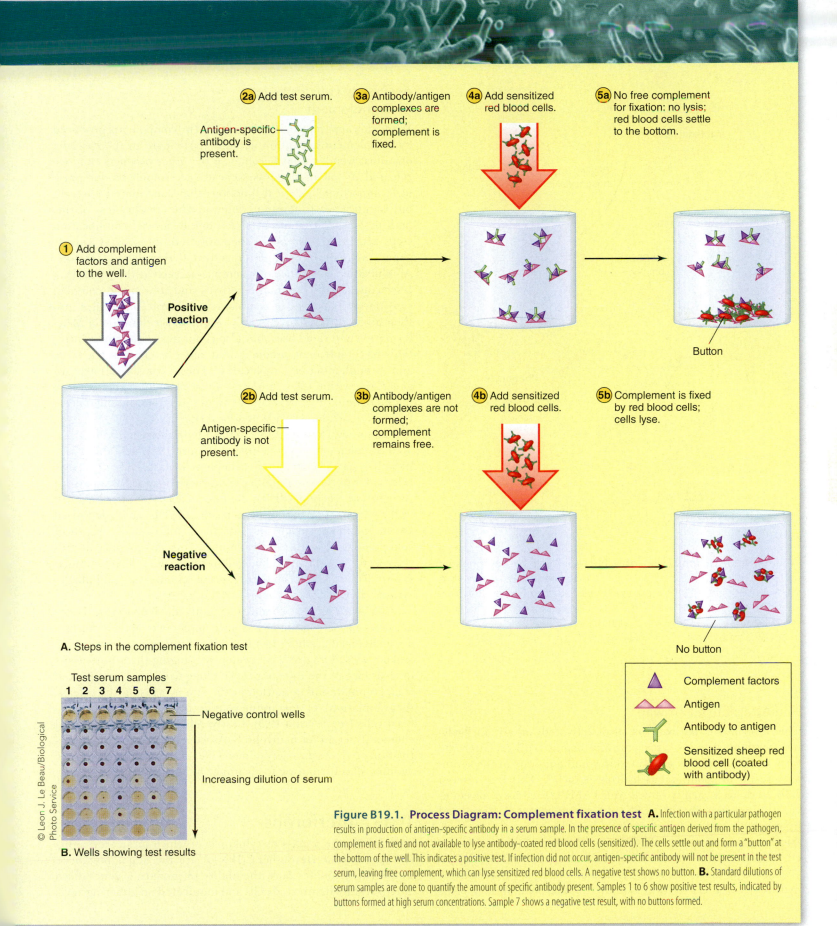

2a Add test serum.

Antigen-specific antibody is present.

3a Antibody/antigen complexes are formed; complement is fixed.

4a Add sensitized red blood cells.

5a No free complement for fixation: no lysis; red blood cells settle to the bottom.

1 Add complement factors and antigen to the well.

Positive reaction

Button

2b Add test serum.

Antigen-specific antibody is not present.

3b Antibody/antigen complexes are not formed; complement remains free.

4b Add sensitized red blood cells.

5b Complement is fixed by red blood cells; cells lyse.

Negative reaction

A. Steps in the complement fixation test

No button

Test serum samples

1 2 3 4 5 6 7

Negative control wells

Increasing dilution of serum

© Leon J. Le Beau/Biological Photo Service

B. Wells showing test results

▲ Complement factors

△ Antigen

⅄ Antibody to antigen

Sensitized sheep red blood cell (coated with antibody)

Figure B19.1. Process Diagram: Complement fixation test **A.** Infection with a particular pathogen results in production of antigen-specific antibody in a serum sample. In the presence of specific antigen derived from the pathogen, complement is fixed and not available to lyse antibody-coated red blood cells (sensitized). The cells settle out and form a "button" at the bottom of the well. This indicates a positive test. If infection did not occur, antigen-specific antibody will not be present in the test serum, leaving free complement, which can lyse sensitized red blood cells. A negative test shows no button. **B.** Standard dilutions of serum samples are done to quantify the amount of specific antibody present. Samples 1 to 6 show positive test results, indicated by buttons formed at high serum concentrations. Sample 7 shows a negative test result, with no buttons formed.

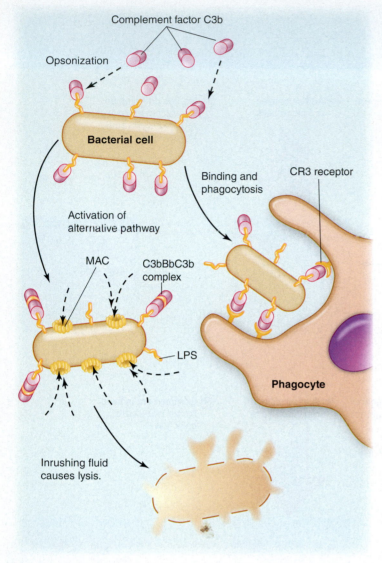

Figure 19.10. Opsonization of a bacterial cell by complement factor C3b Coating of a microbial cell with C3b molecules facilitates phagocytosis by neutrophils and macrophages. These phagocytes possess CR3 receptors (CR = complement receptor) on their surface that specifically bind to the C3b molecules coating the bacterial cell. Opsonization with C3b also activates the complement cascade, leading to the membrane-attack complex (MAC) and destruction of the microbe by lysis.

The Alternative Pathway

The alternative pathway is also activated in an antibody-independent manner. This pathway begins with an alternate form of C3b, which is generated in small amounts in the serum as a result of spontaneous hydrolysis of C3. This free altered C3b binds to a variety of cell wall PAMPs, such as zymosan of yeast, lipopolysaccharide (LPS) of Gram-negative bacteria, and lipoteichoic acid of Gram-positive bacteria. Bound C3b binds another serum protein called "factor B." Bound factor B is acted upon by serum "factor D" to form the cell surface complex C3bBb. This forms a C3 convertase-like complex that actively hydrolyzes free C3 to form C3b at the microbial surface. This

C3b then associates with C3bBb to form C3bBbC3b, a complex analogous in function to C5 convertase, which generates the MAC. The unique components of the alternative pathway are shown in Table 19.5.

Accessory Functions of Complement

While it is apparent that only some of the components produced by complement activation actually participate in the cascade, one must wonder what becomes of the other products that are released. As you might guess, some of these soluble products also have important biological functions that contribute indirectly, but significantly, to immune defense. The effectiveness of this built-in conservation of purpose is due to several amplification steps within the system, so once initiated, the process quickly generates many thousands of activated complement factors. Amplification occurs because complexes, such as the C3 and C5 convertases, have enzymatic activity, and so a single complex can convert hundreds of complement substrate molecules into their active forms. This not only results in acceleration of MAC formation and rapid lysis, but also generates critical concentrations of complement products that serve as chemoattractants and proinflammatory molecules. These concentrations are generated within a few minutes of complement activation in a manner similar to an avalanche, where just a few falling objects trigger a much larger slide at a local site.

Let us now examine some of the other functions of the generated complement components (see Figure 19.9 and Table 19.5). As well as participating in the complement cascade, some C3b is released into the surrounding area where it serves as a pattern recognition receptor (PRR), binding a variety of microbial cell wall components. Millions of molecules of C3b can be deposited on the microbial surface in a matter of minutes as a result of complement activation. This resulting opsonization facilitates phagocytosis of the microbial cell by phagocytes of the immune system **(Figure 19.10)**. The processes of opsonization and phagocytosis will be discussed in more detail later.

C3a serves as a chemoattractant for leukocytes. Large amounts of C3a, produced from complement activation, create a chemical gradient from the site of infection toward which leukocytes will migrate, conveniently bringing them in to assist in clearance of the infection. C3a, along with released C4a and C5a, stimulate inflammation at the site by causing the release (degranulation) of histamine-containing vesicles from tissue mast cells (see Section 19.5). Lastly, released C5b is a powerful chemoattractant for neutrophils. As we can see, complement activation produces a variety of molecules with important immune actions, all contributing to clearance of infection.

Type I interferons

Type I IFNs are a class of cytokines, of which interferon-alpha (IFN-α) and interferon-beta (IFN-β) are the major members. These two cytokines have the ability to produce a generalized antiviral state in infected cells and uninfected neighboring cells, which protects them from viral infection (**Figure 19.11**). More than 300 genes show increased levels of expression when cells are exposed to Type I IFN. Some of the outcomes of this gene induction are:

- **Inhibition of ribosome binding:** As we saw in Chapters 5 and 8, all viruses rely on host cell ribosomes for translation of their proteins, so progeny virus particles are not produced if ribosomes do not bind mRNA efficiently.

- **Changes in membrane components:** Cell surface changes discourage viral attachment and entry into cells.

- **Degradation of RNA:** This impedes cellular replication, limiting the availability of available host cells and restricts metabolic functions of infected cells making them unsuitable hosts. As you can imagine, this can kill the infected cell as well, but is of overall benefit for the host.

- **Proliferation of cytotoxic natural killer cells** (see Section 19.5): These cells can destroy infected host cells, preventing the manufacture and spread of virus.

IFN-α is produced mostly by leukocytes, and IFN-β is produced by numerous cell types, including leukocytes. A third type of IFN (classified as Type II IFN) is interferon-gamma (IFN-γ). It is an important cytokine produced by T cells of the adaptive immune system. It has very different activity from the Type I IFNs, as we will see in Chapter 20.

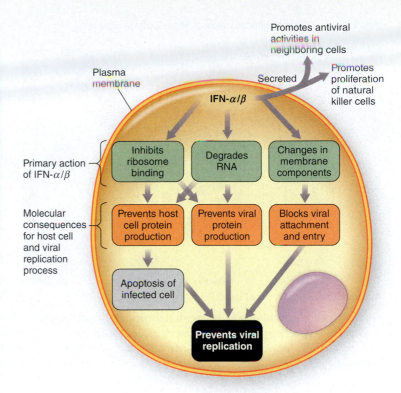

Figure 19.11. Type I interferons Type I interferons IFN-α and IFN-β activate a variety of defenses against viruses in infected and neighboring cells. Secretion of Type I IFNs promotes production of natural killer cells, which are able to destroy virus-infected cells.

19.4 Fact Check

1. Explain how pathogen-associated molecular patterns (PAMPs) and pattern recognition receptors (PRRs) interact in the body.
2. What are the functions of Toll-like receptors (TLRs) and how do they link innate and adaptive immune responses?
3. Describe the immunological role of the PRRs mannose-binding lectin and C-reactive protein.
4. What is complement, and how does it protect against microbial pathogens?
5. Compare and contrast the following complement-activation pathways: classical pathway, alternative pathway, and lectin pathway.
6. Describe the effect of Type I interferons (IFNs) on host cells.

19.5 The cells of innate immunity

What cells first defend us against infectious agents?

The molecules involved in innate immunity have the ability to recognize pathogen-associated molecular patterns (PAMPs), and work in synergy with each other and with cells of the immune system. We have already seen that binding of C-reactive protein and MBL to their target PAMPs initiates the complement cascade and results in opsonization, which facilitates phagocytosis by neutrophils and macrophages of the immune system. Cell-associated TLR recognition of PAMPs results in the expression of genes that are needed for cellular immune responses, such as the production of Type I IFNs that provide immunity to

viral infections. Chemoattractants produced through the complement cascade, like C3a, attract neutrophils to the site of infection, and the inflammatory mediators C4a and C5a stimulate inflammation. All work together in the defense against pathogens. In this section, we will focus on the functions of the cells involved in innate immunity.

Cellular defense is initially carried out by phagocytes and natural killer (NK) cells. These cells can respond quickly to an invasion, before the cells of the adaptive immune system can act. These cell types have distinct roles in immune function and

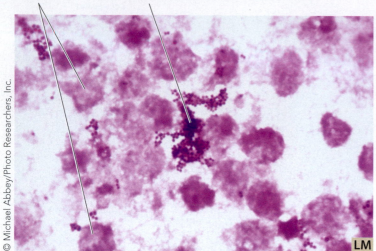

Host cells   *Staphylococcus aureus* grows outside host cells.

A. Extracellular growth of pathogen

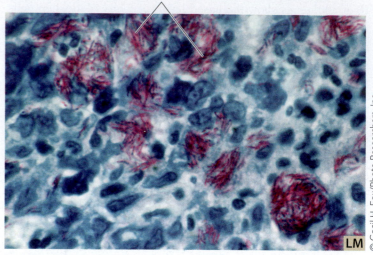

Mycobacterium intracellularae grows inside host cells.

B. Intracellular growth of pathogen

Figure 19.12. Extracellular or intracellular growth of pathogens **A.** Extracellular growth: *Staphylococcus aureus* cells, stained purple, can be seen in clusters outside host cells, and are susceptible to phagocytosis and killing by phagocytes. Bacterial cells can be seen in clusters outside host cells, and are susceptible to phagocytosis and killing by phagocytes of the immune system. **B.** Intracellular growth: *Mycobacterium intracellularae*, shown stained red, filling host cells. Intracellular bacteria are not accessible for uptake and killing by phagocytes.

together they are able to respond to the presence of extracellular microbes and infected host cells (intracellular microbes). Invading microbes have two basic strategies for existence and replication once inside the body. They can be extracellular, that is, existing outside the cell, or they can be intracellular, existing inside cells **(Figure 19.12)**. To be effective, innate immune defenses need to combat both. Some microorganisms such as *Salmonella* species can adopt both lifestyles (see Figure 21.21), making them especially successful at eluding the immune system. Of course, the success with which microbes establish themselves in the environment of the body depends on how well they are equipped with the physiologic and structural tools for a pathogenic existence. For example, pathogenic strains of *Streptococcus pneumoniae* have an outer capsule that prevents phagocytosis by cells of the immune system, allowing them to proliferate extracellularly (see Section 21.1). Recall in Chapter 7, Griffith described the phenotype of pathogenic, encapsulated strains as smooth or S strain, and they readily killed mice. Griffith's R strain, without the capsule, were readily phagocytosed and destroyed and did not cause disease. Another example is the intracellular pathogen *Mycobacterium tuberculosis*. This pathogen produces various neutralizing compounds that reduce intracellular killing by phagocytic cells, allowing it to exist *inside* macrophages. Such strategies will be discussed in Section 21.1.

Phagocytes

A "phagocyte" or "phagocytic cell" is a functionally descriptive term used to describe wandering cells of the immune system that engulf large extracellular particles, such as bacterial cells and dead host cells that they come across in their travels. Neutrophils, along with monocytes and macrophages are phagocytes. Once activated by Toll-like receptor (TLR) binding and cytokine signaling, these phagocytes become adept killing machines.

Neutrophils

Neutrophils are present in large numbers in the blood of vertebrates and are the most abundant cell of the immune system found in normal human blood. Their large numbers allow them to efficiently patrol the body to find invading microbes. Consequently, they are often the first to arrive *en masse* at the site of a bacterial infection. Neutrophils **(Figure 19.13)** are characterized by a multilobed ("poly") nucleus of different shapes ("morph"), giving them an alternative name "polymorphonuclear leukocytes" (PMNs). They also contain numerous,

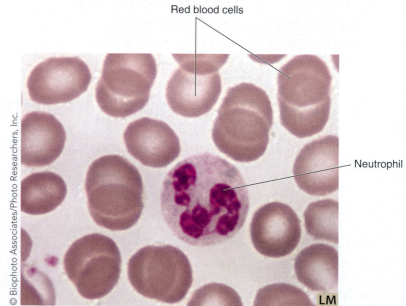

Red blood cells

Neutrophil

Figure 19.13. Neutrophil The darkly stained multilobed nucleus, from which these cells derive their alternative name of polymorphonuclear leukocyte (PMN), is clearly visible. Neutrophils are the most plentiful immune cell in human blood.

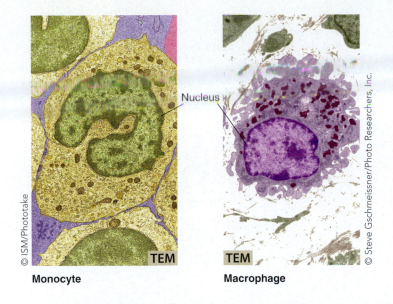

Nucleus

TEM   TEM

Monocyte   **Macrophage**

© ISM/Phototake

© Steve Gschmeissner/Photo Researchers, Inc.

Figure 19.14. Monocytes differentiate into macrophages. Monocytes (*left*) circulate in the blood before migrating to the tissues where they differentiate into macrophages (*right*). The large nucleus can be seen in both cells. The macrophage is larger than the monocyte, has more surface extensions and organelles, including lysosomes, and is more active in phagocytosis and intracellular killing.

distinctive membrane-bound, chemical-containing vesicles called **granules** in their cytoplasm. Granules are easily seen when specifically stained. Neutrophils get their name from the fact that their granules do not readily stain with the acidic dye eosin or the basic dye methylene blue. This is due to the *neutral* nature of the substances contained in their granules, such as lactoferrin and lysozyme. Granules containing the enzyme lysozyme are specifically referred to as **lysosomes**.

Monocytes and Macrophages

Monocytes and **macrophages** are collectively referred to as "mononuclear cells" because of their large, single, round or horseshoe-shaped nucleus (**Figure 19.14**). Mononuclear cells are produced in the bone marrow and are released to the blood as monocytes. Monocytes circulate in the blood for one to two days, after which time they move into tissues and further differentiate into mature macrophages (see Appendix G: Origin of Blood Cells). Macrophages are "beefed up" versions of monocytes. They are larger, contain more organelles, especially lysosomes, have numerous projections, and are more efficient in the processes of phagocytosis and intracellular killing.

Macrophages can be found wandering throughout the body or as permanent residents in certain tissues. Some of these resident macrophages have specialized functions and are given names specific to their place of residency. Examples include alveolar macrophages of the lungs, mesengial cells of the kidney, Kupffer cells of the liver, and histiocytes of connective tissue.

Microbial Capture by Phagocytes For a phagocytic cell to kill an extracellular microbe, it must first identify it as being foreign and then capture and internalize it to begin its destruction. Although some extracellular microbes can be directly bound by specific receptors present on the phagocyte surface, most are first opsonized. As previously mentioned, opsonization is the covering of the microbial surface with molecules called **opsonins** that bind foreign substances. Examples of opsonins are C-reactive protein, antibodies, MBL (see Figure 19.6), and complement

component C3b (see Figure 19.10). Some of these are recognized by receptors on the surface of various phagocytes. C3b opsonin is recognized by complement receptors CR1, CR2, and CR3, found on the surface of neutrophils and macrophages. The exposed portion of opsonizing antibody, the F_C portion, is recognized by a specific receptor, the F_C receptor (F_CR) of the macrophage (see Section 20.6, Figure 20.34 for details). Opsonization by lectins, such as MBL and C-reactive protein, activates the complement cascade, and the resulting C3b molecules further coat the target for opsonization. These and other common methods of opsonization used by vertebrates are shown in **Figure 19.15**. Therefore, opsonization allows the phagocyte to indirectly bind to the microorganism for effective phagocytosis. The effect is analogous to the addition of the small bumps on the surface of a basketball that allow your fingertips to get a better grip on its otherwise slippery surface.

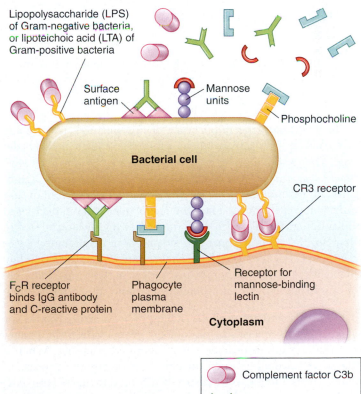

Figure 19.15. Opsonization of bacterial cells Phagocytes possess a variety of specific receptors on their surface for binding to various defense opsonins that coat microbial cells.

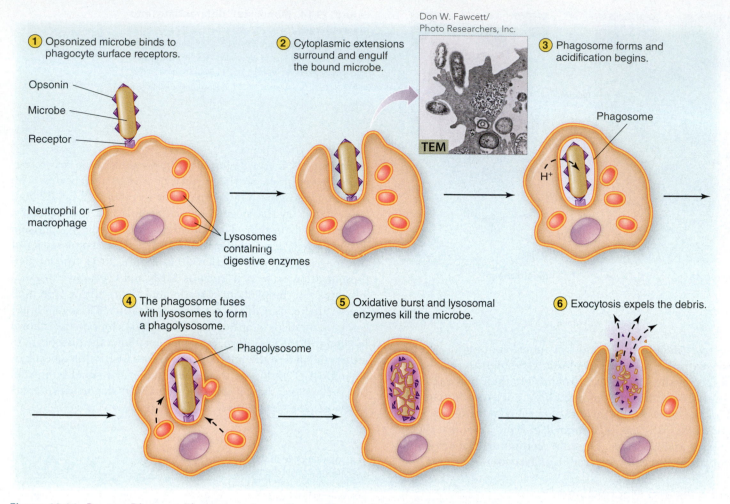

① Opsonized microbe binds to phagocyte surface receptors.

Opsonin
Microbe
Receptor
Neutrophil or macrophage
Lysosomes containing digestive enzymes

② Cytoplasmic extensions surround and engulf the bound microbe.

Don W. Fawcett/
Photo Researchers, Inc.

TEM

③ Phagosome forms and acidification begins.

Phagosome
H⁺

④ The phagosome fuses with lysosomes to form a phagolysosome.

Phagolysosome

⑤ Oxidative burst and lysosomal enzymes kill the microbe.

⑥ Exocytosis expels the debris.

Figure 19.16. Process Diagram: Phagosome formation and intracellular killing The foreign object, the bacterial cell in this case, is often first opsonized. A phagosome forms and enters the endocytic pathway. Degradative enzymes in combination with toxic compounds from the respiratory burst kill and degrade the microbe.

To investigate the importance of opsonization, researchers have used strains of mice deficient for C3 from which C3b is derived. For example, experimental infection of these mice with *Pseudomonas aeruginosa* resulted in the accumulation of significantly higher numbers of bacteria in the blood compared to that of normal control mice. These deficient mice showed decreased ability of neutrophils to bind and ingest *P. aeruginosa*, compared to their normal counterparts. Not surprisingly, documented cases of human C3-deficient patients show recurrent and severe infections with a variety of organisms, indicating an important role for opsonization by C3b for phagocytosis and subsequent killing.

Phagocytosis and Intracellular Killing The starting point for phagocytosis is a series of physiologic events within the phagocyte that results in destruction of engulfed microbes. When a phagocyte encounters such an object, it begins the process of phagocytosis. Phagocytosis is specifically used to ingest large particulate matter such as bacterial cells. Phagocytosis is a specialized form of **endocytosis**, a generalized process for the uptake of external materials by cell membrane extension or invagination and fusion to form a vesicle inside the cytoplasm of the cell. This vesicle is an **endosome**. Endocytosis of matter is carried out by most cells of the body and can be accomplished by other mechanisms, detailed in Chapter 20 and Figure 20.12. The endosome formed

by phagocytosis is called a **phagosome**. Phagocytes secrete lactoferrin into the phagosome to keep iron away from bacteria they may have engulfed (see Section 21.5). Destruction of the particle begins within the phagosome. The resulting phagosome enters into the **endocytic pathway** where a sequence of fusions with other cytoplasmic vesicles occurs to digest the endocytosed material. After a series of biochemical changes, the digested material is expelled by membrane fusion with the plasma membrane. This process is referred to as "exocytosis" **(Figure 19.16)**.

@ **Endocytosis and exocytosis** ANIMATION

Within the endocytic pathway, phagocytes employ several strategies to kill phagocytosed microorganisms:

- degradation by enzymes
- inactivation by acidification
- lysis by formation of pores in bacterial membranes
- destruction by toxic oxygen compounds

First, the phagosome begins to acidify to pH 4.0–5.0. At this pH, most microbes are inactivated but not necessarily destroyed. To effectively kill microbes, the phagocyte engages two more strategies. The phagosome fuses with lysosomes to form

TABLE 19.6	Antimicrobial compounds of lysosomes and their effects
Molecule	**Killing action**
Lysozyme	Hydrolysis of peptidoglycan of the bacterial cell wall
Proteases	Degradation of proteins
Defensins	Form pores in bacterial membranes

TABLE 19.7	Toxic oxygen products of the respiratory burst
Chemical formula	**Common name**
H_2O_2	Hydrogen peroxide
HOCl	Hypochlorite
OH·	Hydroxyl radical
O_2^-	Superoxide anion
NO	Nitric oxide

a **phagolysosome**. As mentioned previously, lysosomes contain lysozyme and a variety of other antimicrobial agents including proteases and antimicrobial peptides, such as defensins (**Table 19.6**). Recall that defensins are antimicrobial peptides that are thought to form pores in microbial membranes to cause leakage and lysis. These antimicrobial substances empty into the phagosome upon fusion with lysosomes, and the low pH activates these compounds to begin digestion of the microbe. The phagolysosome can be thought of as the "stomach" of the cell, because the end result is destruction and degradation of the ingested material.

In addition to acidification, phagocytes carry out a unique process called the **respiratory (oxidative) burst**, named because of its large requirement for cellular oxygen. The respiratory burst produces many highly toxic oxygen compounds inside the phagolysosome, some of which are listed in **Table 19.7**.

The combined action of these toxic oxygen compounds is usually sufficient to quickly kill and degrade most engulfed microbes. In many cases, the infection is rapidly cleared by neutrophils without the need for adaptive immune responses. If the infection persists, the concentrations of cytokines, such as IL-8 and IL-1, from activated phagocytes and damaged cells increases. This causes migration of more neutrophils to the site and begins the recruitment of monocytes from the blood for assistance. In the tissues, the monocytes mature into macrophages. Once stimulated by the activities of binding and phagocytosis, the macrophage exhibits enhanced phagocytosis, increased respiratory burst, and production of antimicrobial compounds such as defensins and lysozymes. In this manner, macrophages participate in the innate immune response much as neutrophils do. However, in addition to destroying microorganisms, macrophages also directly interact with cells of the adaptive immune system.

The end result is the degradation of the ingested microbe into small molecular components. For full activation (that is, maximally enhanced phagocytosis and killing), neutrophils and macrophages require the help of cytokines, such as tumor necrosis factor-beta (TNF-β) and interferon-gamma (IFN-γ), produced from T cells of the adaptive system. Again, this illustrates the interdependence of the innate and adaptive immune systems. It is hard to imagine any microbe capable of surviving such an arsenal of killing mechanisms, but some pathogens have devised ways of preventing intracellular killing.

CONNECTIONS Many pathogenic bacteria, such as *Salmonella* Typhimurium, *Mycobacterium tuberculosis*, and *Mycobacterium leprae*, actually live inside the phagosome of macrophages. They avoid destruction by producing microbial products that interfere with the fusion of the phagosome with lysosomes and other endocytic vesicles. See **Figure 21.21** and **Section 21.2** (Pathogen Study 2: *Mycobacterium tuberculosis*) for details.

Eosinophils, basophils, and mast cells

Eosinophils, **basophils**, and **mast cells** generally act against extracellular parasitic worms, many of which are far too large for phagocytosis (**Figure 19.17**). Eosinophils are found in tissues and contain numerous granules that stain red with the dye eosin. Basophils are also found in the blood and carry granules that stain with basic dyes. The differential staining of the granules of these two cell types indicates they are of different chemical properties. Mast cells are found widely in tissues.

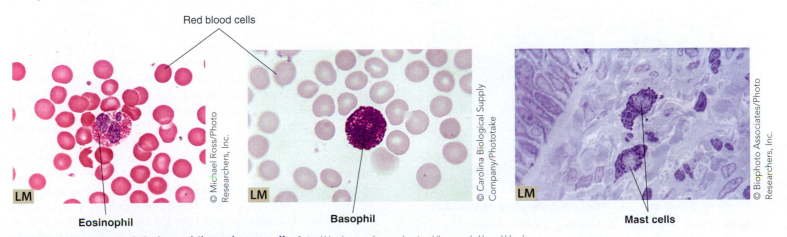

Red blood cells

LM — Eosinophil — © Michael Ross/Photo Researchers, Inc.

LM — Basophil — © Carolina Biological Supply Company/Phototake

LM — Mast cells — © Biophoto Associates/Photo Researchers, Inc.

Figure 19.17. Eosinophils, basophils, and mast cells Stained blood smear of a central eosinophil surrounded by red blood cells (*left*). The many granules in the cytoplasm are stained with the dye eosin and appear red. The multilobed nucleus is blue. Stained blood smear of a central basophil surrounded by red blood cells (*middle*). The many basic-staining granules (blue) pack the cytoplasm and obscure the nucleus. Stained smear of mast cells in tissue (*right*). The numerous granules in the cytoplasm are stained dark blue. Nuclei are stained pale blue.

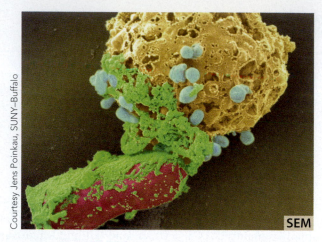

Figure 19.18. Eosinophils degranulating to attack a parasitic fungus
The eosinophil, shown in yellow, has released toxic protein (called "major basic protein"), onto the surface of a germinating fungal spore, shown in red. Concentrated toxic protein is shown in blue as it leaves the eosinophil, changing to green as it disperses. The numerous dark indents on the surface of the eosinophil are due to the fusion of granules with the plasma membrane during degranulation.

Eosinophils, basophils, and mast cells are **cytotoxic cells**. In other words, they kill their target (the parasite in this case) by releasing their toxic granular contents directly onto the target cell in a process called **degranulation**. To do this, mast cells and basophils usually require the presence of IgE antibodies produced by the adaptive immune system (see Section 20.6). IgE is specifically made in response to allergens and parasites to recruit the killing action of basophils and mast cells. These cells can bind free, unbound IgE using surface receptors for the IgE molecule. Eosinophils have receptors for IgG antibody and will bind to IgG-coated parasites. This causes degranulation in a manner similar to mast cells and basophils (**Figure 19.18**). Despite being able to interact with specific antibodies, these cell functions are strictly innate as they do not recognize the specific infectious agent but rather only recognize the antibodies produced because of the infection. This is yet another example of how the innate and adaptive arms of the immune system work together.

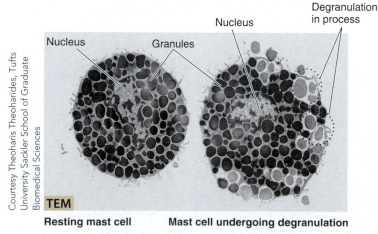

Resting mast cell   Mast cell undergoing degranulation

Figure 19.19. Mast cell undergoing degranulation Intact granules are electron-dense and appear as dark spheres filling the cytoplasm. A resting mast cell is shown on the left. A mast cell degranulating is shown on the right. Fusion of granule and plasma membranes results in the release of granular components. Degranulation results in release of the electron-dense contents; these vesicles appear paler.

When the IgE that is bound to mast cells or basophils contacts its specific target—the parasite or allergen—the cell is triggered into releasing its granular components. Basophils and mast cells are also stimulated to degranulate in the presence of the complement components C3a and C5a even before the adaptive immune response has had a chance to produce IgE (**Figure 19.19**). Many of the released chemical components are not only cytotoxic, but are also potent stimulators of inflammation. An example is histamine, introduced in Section 19.3. Induction of a strong inflammatory response at the site of infection irritates the parasites, making it difficult for them to become established in the host. Other released chemicals have the ability to cause muscle contractions and excess mucus secretion, both of which help to dislodge the parasite from host tissues. Mast cells, eosinophils, and basophils are also involved in allergy responses in vertebrates.

Natural killer (NK) cells

So far we have examined the immune response to extracellular microbes, but how does innate immunity deal with intracellular pathogens? Obviously, there is no easy way to get at a microbe once it is inside a host cell. It would be analogous to you trying to retrieve your keys locked inside your car without actually entering the vehicle. In an emergency you may have to damage your car to get them out. This is what **natural killer (NK) cells** do. Their targets are the infected cells themselves. NK cells are cytotoxic cells; they are *not* phagocytic and they cannot internalize their target. Rather, they attach to infected cells and release their toxic granular contents, resulting in death of the infected host cell and maybe the microbe along with it. Any released intact microbes can then be effectively killed by phagocytes, such as neutrophils and macrophages. The rather recent discovery of NK cells is described in **Perspective 19.2**.

Two of the granular components of the NK cell's cytotoxic arsenal are **perforin** and **granzymes**. Perforin is a peptide that is thought to produce pores in the target cell membranes. In this model, it is proposed that perforin forms channels in the target cell that assist in the uptake of the released granular components, such as granzymes. Perforin is found in many cells with cytotoxic activities, including cytotoxic T cells of the adaptive immune system (see Section 20.3). Granzymes induce **apoptosis**, or programmed cell death, in target host cells. Apoptosis is a highly ordered cascade of chemical signaling events that results in cellular suicide (**Figure 19.20**). It is characterized by fragmentation of DNA and the cell itself into small membrane-bound apoptotic bodies to be disposed of by phagocytic cells. Apoptosis is a tightly *controlled* process resulting in a "quiet death" without the release of proinflammatory cytokines and further damage to cells. **Necrosis** is a different type of cell death that occurs when cells are injured by structural damage or loss of oxygen supply due to trauma or infection. It results in cell rupture and the release of toxic cellular contents and cytokines to the surrounding tissues producing further damage and inflammation. It is the equivalent of a cellular train wreck, resulting in the collateral damage of many healthy neighboring cells.

Natural killer (NK) cells were first described in 1976. Their discovery came unexpectedly, as many scientific discoveries do, from the observation that mice with preexisting tumors could kill the tumor cells when they were cultured then injected (Figure B19.2). Thus, mice sensitized to tumor cells could kill them. To the surprise of researchers, they found the tumor-free control mice could also efficiently kill the tumor cells. It appeared that this killing function was an innate response, as the mice did not require previous exposure to the tumor cells. Further investigations of this antitumor activity uncovered a population of large lymphocytes containing many granules that demonstrated cytotoxicity toward a variety of tumors. The natural ability to kill abnormal cells without the requirement for previous exposure gave these cells their name—natural killer cells. Other researchers were later able to show NK cells could destroy cells infected with certain viruses, bacteria, fungi, and protozoa.

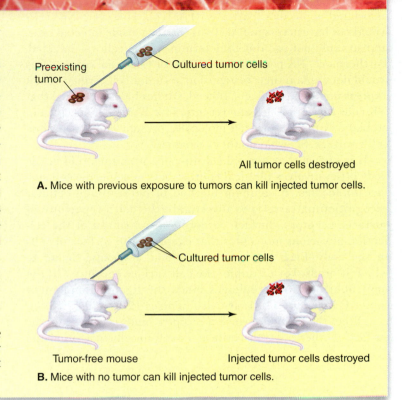

A. Mice with previous exposure to tumors can kill injected tumor cells.

B. Mice with no tumor can kill injected tumor cells.

Figure B19.2. Discovery of NK cells **A.** Mice with preexisting tumors were able to kill injected tumor cells. **B.** Normal mice not previously exposed to tumors were injected with tumor cells and were also able to kill the tumor cells. This indicated that mice have a natural ability to detect and destroy abnormal cells that is not dependent on previous exposure to the abnormal cells.

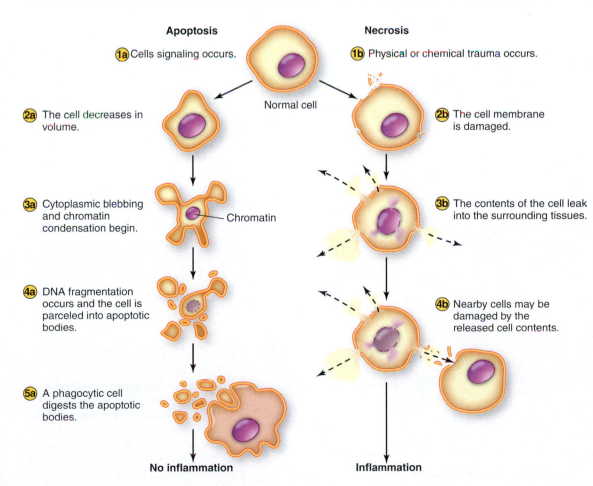

Apoptosis

1a Cells signaling occurs.

2a The cell decreases in volume.

3a Cytoplasmic blebbing and chromatin condensation begin. — Chromatin

4a DNA fragmentation occurs and the cell is parceled into apoptotic bodies.

5a A phagocytic cell digests the apoptotic bodies.

No inflammation

Normal cell

Necrosis

1b Physical or chemical trauma occurs.

2b The cell membrane is damaged.

3b The contents of the cell leak into the surrounding tissues.

4b Nearby cells may be damaged by the released cell contents.

Inflammation

Figure 19.20. Process Diagram: Apoptosis and necrosis Apoptosis (*left*) is programmed cell death and results from the induction of specific chemical signaling pathways. Apoptosis results in the systematic deconstruction of the cell into small, membrane-bound apoptotic bodies, which are then taken up by phagocytes and digested. As cellular contents are not released to the surrounding tissues, apoptosis does not induce inflammation or cause damage to surrounding cells. Necrosis (*right*) occurs when the cell has sustained significant physical or chemical damage, such as happens during infection. Uncontrolled death of the cell results in the release of cellular contents, many of which can damage other cells and induce inflammation.

In order for NK cells to be effective in fighting infection and limiting damage to the host, they must distinguish normal cells from infected cells. One way to do this might be by using pattern recognition receptors (PRRs) that recognize pathogen-associated molecular patterns (PAMPs). However, NK cells would not have access to PAMPs as they are inside infected cells, and so could not use this to determine if the cell was infected. Furthermore, NK possession of PRRs for microbial components would not explain their recognition of tumor cells, which do not express microbial products. Indeed, studies have confirmed that NK cells appear to lack any receptors for PAMPs or specific microbial or tumor-expressed molecules present on an abnormal cell's surface. So, what do they recognize?

It is known that normal cell-to-cell communication in the body often involves physical contact between specific cell surface molecules. A common method of determining cellular recognition patterns is to investigate the cell's repertoire of surface receptors. Surface receptors are a subset of the collection of various molecules that decorate the surface of cells in the body. Each type of receptor has a specific ligand to which it binds. For example, the cell surface receptor CR3 of neutrophils and macrophages binds the ligand complement factor C3b (**Figure 19.21**; see also Figures 19.10 and 19.16).

Often, the ligand to which a receptor binds is located on the surface of other cell types, critical for cell-to-cell recognition and communication. The number and kind of surface molecules on a cell are specific to the cell type and even the developmental stage of the cell. As such, these surface molecules are used as markers to identify and classify cells. Characterization of cell surface molecules can give important information as to what the cell is capable of binding to and so gives clues as to its function.

Earlier studies of the types of tumor cells that are killed by NK cells revealed these tumor cells had little or no expression of a surface marker called **class I major histocompatibility (MHC I)** molecule. MHC I is expressed on the surface of all normal nucleated vertebrate cells. Individuals have their own unique form of MHC I, and this allows the immune system to detect non-self cells such as transplanted cells and cells such as cancerous cells that may have abnormal MHC I. From studies on tumors, it appeared that the *lack* of MHC I on the tumor cells triggered degranulation of NK cells. Other studies using human MHC I-deficient cell lines showed that NK cells readily killed these. However, when genes for expression of MHC I were added back to the cells, they were no longer targets for NK killing. This led to a hypothesis for NK targeting whereby NK cells are able to bind to MHC I molecules on the surface of normal cells. In this event, an inhibitory signal is sent and killing is not initiated. Yet, when an NK cell comes in contact with a cell without MHC I molecules, killing *is* initiated. Of course for this to be true, NK cells must possess a receptor for MHC I. This was indeed confirmed as receptor CD94.

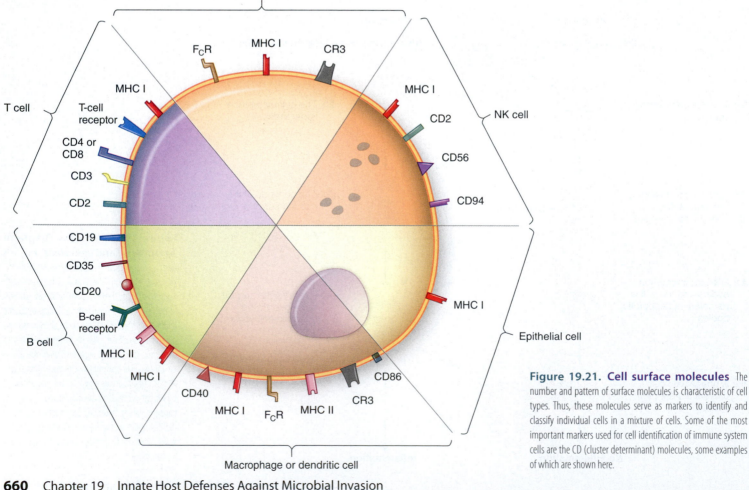

Figure 19.21. Cell surface molecules The number and pattern of surface molecules is characteristic of cell types. Thus, these molecules serve as markers to identify and classify individual cells in a mixture of cells. Some of the most important markers used for cell identification of immune system cells are the CD (cluster determinant) molecules, some examples of which are shown here.

As the default setting in NK cells is to undergo degranulation in the absence of an inhibitory signal following contact with another cell, then they must also have an activating receptor that registers contact with another cell before degranulation proceeds. NK cells recognize several different cell surface ligands to signal they are in cell contact (**Figure 19.22**).

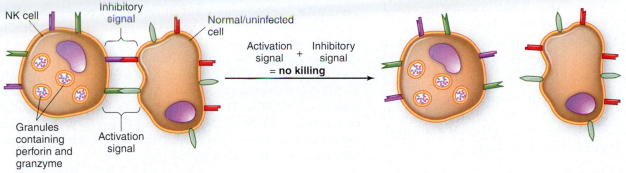

A. Normal MHC I expression on cell surface

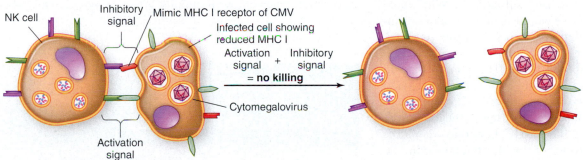

B. Reduced MHC I expression on cell surface

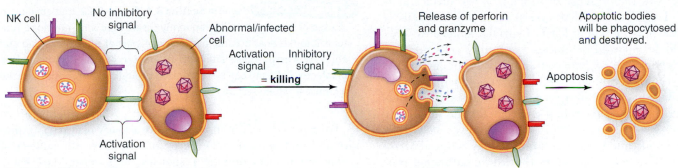

C. Cytomegalovirus mimicry of MHC I

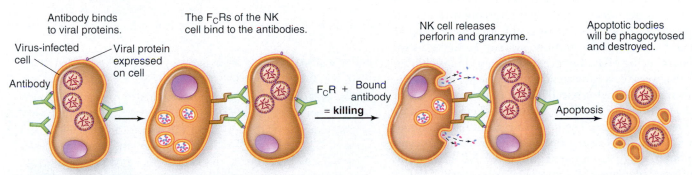

D. Antibody-dependent cell-mediated cytotoxicity (ADCC)

Figure 19.22. Activities of NK cells A. NK cells recognize normal cells through binding of the NK cell CD94 surface molecule to MHC I molecules on the surface of other body cells. This produces an inhibitory signal that counteracts the bound NK cell activation signal, preventing degranulation and killing of the cell. **B.** Infected or abnormal cells typically show reduced MHC I expression, consequently inhibitory MHC I binding does not happen. The activation signal triggers degranulation, resulting in apoptosis of the infected/abnormal cell. **C.** Molecular mimicry is carried out by human cytomegalovirus (CMV). Infection with the virus causes reduced MHC I expression on the surface of the host cell. Within the infected cell, CMV produces a peptide that mimics MHC I that binds CD94 receptors on the NK cell. This produces an inhibitory signal in the NK cell, allowing the infected cell to escape detection. **D.** Antibody-dependent cell-mediated cytotoxicity (ADCC) by NK cells requires the binding of antibody to an infected cell. ADCC by NK cells can kill virus-infected cells.

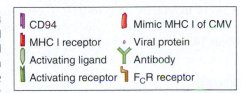

- CD94
- MHC I receptor
- Activating ligand
- Activating receptor
- Mimic MHC I of CMV
- Viral protein
- Antibody
- F_CR receptor

From accumulated evidence, the current model, diagrammed in parts A and B of Figure 19.22, is as follows:

- When an NK cell contacts another cell, binding between an activating receptor on the NK cell and an activating ligand on the surface of the contacted cell occurs.
- When an NK cell simultaneously also binds to MHC I of the contacted cell, inhibitory signals are produced.
- This *prevents* initiation of killing.
- When no MHC I is encountered upon cell contact, an inhibitory signal is not produced and killing is initiated; degranulation and the release of perforin and granzymes occurs to cause apoptosis of the bound cell.

NK cells can therefore discriminate normal cells from abnormal cells by the presence or absence of the surface molecule MHC I. One of the characteristics of infected cells is often the loss or reduction of MHC I surface molecules. This has been widely reported in studies examining cells infected with various viruses and intracellular bacterial pathogens. Controlled studies of cellular MHC I expression in phagocytes show that phagocytosis of inert foreign substances, such as latex beads or individual microbial components, do not result in the repression of MHC I expression. However, cells that harbor pathogenic microbes commonly display reduced MHC I expression. Detailed studies show that these pathogens produce molecules that actively interfere with host cell MHC I production and/or transport to the cell surface. Why would pathogens do this? After all, this would seem to be a disadvantage as it makes their infected host cell vulnerable to attack by NK cells.

We can hypothesize that reducing MHC I expression in a host cell is somehow advantageous to the pathogen—and indeed this is the case. Normally, an infected cell couples peptide components released from the infecting pathogen during growth to MHC I molecules. The MHC I:peptide complexes are then presented on the surface of the infected host cell. The presentation of MHC I:peptide signals cells of the adaptive immune system, specifically cytotoxic T cells (see Section 20.3), to destroy the presenting infected cell. By interfering with normal MHC I expression, pathogens are thus able to avoid killing of their host cell by the adaptive immune system. However, as we have learned, reducing MHC I expression signals innate killing by NK cells. So, we have a functional synergy between NK cells of the innate system and cytotoxic T cells of the adaptive system that provides for the destruction of any infected cells missed by the other. The pathogen can't win, or so it would seem. Some viruses have, through natural selection and evolution, developed ingenious ways to circumvent killing by NK cells while at the same time reducing the expression of host cell MHC I molecules to avoid detection by cytotoxic T cells (see Figure 19.22C). An example is human cytomegalovirus, presented in **Microbes in Focus 19.1.**

CONNECTIONS MHC I molecules are involved in adaptive immune responses and are used to specifically tag infected cells for destruction by cytotoxic T cells. **Section 20.3** describes this activity.

What about infected cells that still display normal MHC I markers? It would seem they would escape NK cell detection. However, NK cells use a second mechanism that works in conjunction with the adaptive immune system to target these MHC I-presenting infected cells. This process is called **antibody-dependent cell-mediated cytotoxicity (ADCC).** NK cells, like macrophages, possess F_C receptors on their surface that can bind to antibody (see Figure 19.15). Binding of NK cell F_C receptors to an antibody-coated infected cell leads to NK cell degranulation causing apoptosis of the target cell (see Figure 19.22D). The combined mechanisms of ADCC and recognition of abnormal MHC I allows NK cells to detect and kill a broad spectrum of infected cells without having to specifically recognize the pathogen.

CONNECTIONS Recall in **Section 8.4, Figure 8.22**, that enveloped viruses of eukaryotic cells frequently insert their viral envelope proteins into the host plasma membrane as part of their packaging and egress processes. These viral proteins are thus exposed on the surface of infected cells and can be recognized by antibodies produced by the adaptive immune system.

We have covered many important aspects of innate immunity and immunity in general. Before we leave the subject of vertebrate innate immune responses, let's consolidate some of the information we have learned and begin to link it to our next chapter on adaptive immunity. We are exposed to microbes

Microbes in Focus 19.1
COVERT OPERATIONS OF HUMAN CYTOMEGALOVIRUS

Habitat: Human cytomegalovirus infects 60 to 90 percent of adult humans, but disease is rare. Various tissues can be infected, and the virus can remain latent for many years, not causing active infection.

Description: Enveloped, icosahedral, double-stranded DNA virus of 150-nm diameter belonging to family *Herpesviridae*.

Key Features: Human cytomegalovirus (CMV) produces a mimic MHC I peptide. This copycat peptide is expressed on the surface of infected cells and binds to the NK CD94 receptor (recall this binds to MHC I) and "fools" the NK cell by making the infected cell appear to have normal MHC I distribution. As a result, cytotoxic killing of infected host cells is reduced and the virus can continue to commandeer cells, making it a highly successful pathogen, but thankfully, one of low virulence in healthy adults. CMV infection in the developing fetus can cause death or neurological damage in 5 to 10 percent of infants of infected mothers.

© CNRI/Photo Researchers, Inc.

TEM

every day and therefore it is inevitable that some of these microbes will gain access to our body tissues, mostly by accidental breaches in our protective coverings—skin and mucosal membranes. When this happens, innate immune responses are the first line of defense to prevent spread of the infection and the ensuing damage caused. The degree to which innate immune responses occur depends on a number of parameters, including how many and what type of microbes are present. In many cases, innate immune defenses are sufficiently effective in clearing a microbial invasion so that adaptive immune responses do not occur to any large extent.

Microbes that produce molecules such as LPS that efficiently stimulate inflammation, produce a sizable inflammatory response. The result is that very large numbers of immune cells can be brought into the affected area. The large concentrations of cytokines that are subsequently released by cells with immune activities attract and activate cells of the adaptive immune system (Chapter 20). The adaptive immune system is capable of producing a wide variety of molecules that are tailor-made to components of the particular infectious agent. Thus, molecules that are outside the repertoire of the innate system can be targeted, leading to a greater ability to destroy these microbes. In this way, the adaptive immune system works together with the body's innate responses to clear infections. Of course, pathogenic microbes have found ways to circumvent the immune system and can go on to cause disease. We will introduce some of these in Chapters 21, 22, and 23. We've largely been referring to innate immunity in vertebrates, so before we leave this chapter, we'll look briefly at the immune defenses of invertebrates.

19.5 Fact Check

1. What is the role of phagocytes in innate immunity?
2. How are phagocytes activated?
3. Distinguish between the different types of phagocytes.
4. Describe the steps involved in the process of phagocytosis.
5. Describe the respiratory burst and its effect on microbes.
6. What are the roles of eosinophils, basophils, and mast cells?
7. What are NK cells, and what cell types do they target?
8. What is MHC I, and how does it influence NK cells?
9. What is ADCC?

19.6 Invertebrate defenses

What kinds of immune defenses do invertebrates have?

Invertebrates are far more numerous than vertebrates in terms of species, diversity, and populations, yet little is known regarding the spectrum of defense mechanisms used by these creatures. Defense mechanisms in invertebrates appear to be strictly innate, involving both cells and molecules. Most of what is known about invertebrate immune functions has been derived from just a few model organisms, such as *Drosophila melanogaster*, and *Limulus polyphemus*, the American horseshoe crab, shown in **Figure 19.23**. The horseshoe crab is a large aquatic invertebrate related to scorpions, native to the Atlantic Ocean from Maine to the Mexican Yucatan. As different to us as crabs and flies may seem, they show many parallels to vertebrates, possessing intricate signaling processes and numerous functional homologies. Valuable genome sequencing projects have also uncovered many molecular homologies with genes involved in the innate immune responses of vertebrates. For many of these genes, and others that have been shown to show increased expression during the infection process in various invertebrates, functions have not yet been assigned.

Horseshoe crab   Hemocyte

LM

Figure 19.23. The hemocyte: A phagocytic cell of invertebrates Much of the research on hemocyte function and invertebrate immune defenses has been performed on *Limulus polyphemus*, the horseshoe crab (*left*). The large size of horseshoe crabs makes it convenient to collect hemolymph (invertebrate blood) and return the crab to the water. A hemocyte (*right*) is found in hemolymph and body tissues where it recognizes, engulfs, and destroys foreign cells using lysozyme, cytotoxic peptides, and toxic forms of oxygen.

© intek1/iStockphoto

Courtesy J. Adam Frederick, Maryland Sea Grant Extension Program

A. M. Kreig, A.-K. Yi, S. Matsin, T. J. Waldschmidt, G. A. Bishop, R. Teasdale, G. G. Koretzky, and D. M. Klinman. 1995. CpG motifs in bacterial DNA trigger direct B-cell activation. Nature 374:546–549.

Context

The 1998 finding that the first mammalian Toll-like receptors (TLRs) appeared to initiate immune responses in cells by recognition of specific bacterial components began a number of intensive searches for other TLRs and their binding targets. It had been known for some time that bacterial DNA could induce immune responses in mammals. The first accounts of this phenomenon began in 1984 with the recognition of apparent antitumor activity of DNA extracted from *Mycobacterium bovis* BCG, widely used in humans as a live vaccine against tuberculosis. Subsequent studies showed DNA extracted from eukaryotic cells did not evoke immune responses; it was not immunogenic. The immunogenicity of DNA came as a surprise to researchers. After all, how could a molecule as ubiquitous as DNA evoke immune responses? Yet, there definitely appeared to be some way for cells to distinguish bacterial DNA from eukaryotic DNA. These initial reports prompted researchers to investigate the biochemical differences between bacterial DNA and DNA from a variety of eukaryal organisms including plants.

As early as 1943, scientific literature reported an interesting finding: extensive methylation existed in mammalian DNA. We now know that eukaryotes commonly methylate their DNA in order to package it into the nucleus and as a mechanism for transcriptional silencing. This became an early clue into a possible discriminatory feature between bacterial DNA and that of eukaryotes. In the early 1990s, researchers began to test a variety of synthetic oligonucleotides composed of particular base sequences for their ability to evoke immune responses. Not all unmethylated sequences tested were found to be immunogenic. The common denominator deduced from immunogenic oligonucleotides appeared to be adjacent C and G nucleotides, referred to as "CpG dinucleotides" ("p" refers to the phosphodiester bond between the nucleotides).

These early findings prompted Arthur Kreig and Dennis Klinman and their colleagues to determine the basis for the immunogenic activity of bacterial DNA. In 1995, they published their results in the scientific journal *Nature*. They write:

In considering explanations for the mitogenicity (activation of lymphocytes) of bacterial DNA, we noted that CpG dinucleotides are generally present at the expected frequency of 1 per 16 dinucleotides in microbial DNA, but they are only about one-quarter as prevalent in vertebrates. In addition to this "CpG suppression," the cytosines in CpG dinucleotides are highly methylated in vertebrates, but not microorganisms.

Could prokaryotic unmethylated CpG motifs be a pathogen-associated molecular pattern recognized by cells of vertebrates?

Experiments

Kreig and Klinman's group conducted a series of experiments designed to progressively define the immunogenic activity of DNA. In their first experiment, they established that unmethylated *Escherichia coli* DNA, but not methylated *E. coli* DNA, was immunogenic. They extracted *E. coli* genomic DNA and chemically methylated an aliquot *in vitro*. The extent of methylation was determined by exposing the DNA to *Hpa*II, a restriction enzyme that can only cleave unmethylated restriction sites. As a control, the DNA was also subjected to *Msp*I, which can cut both methylated and unmethylated restriction sites, to show that DNA could be cleaved. The DNA was completely protected from digestion with *Hpa*II but not *Msp*I, establishing the *E. coli* DNA was fully methylated.

Methylated and unmethylated *E. coli* DNAs were then added separately to cultured B cells of the immune system and the cells were examined for immune stimulation. A hallmark of B-cell immune stimulation is rapid cell division, and this was easily assessed by measuring the incorporation of radiolabeled thymidine (^{3}H-thymidine) in newly replicated B-cell DNA. The striking results, shown in **Figure B19.3**, showed only that unmethylated DNA (open boxes) stimulated the B cells in a dose-dependent manner. Methylated DNA (filled boxes) did not stimulate at any of the concentrations used.

The researchers then established that unmethylated CpG motifs were responsible for the observed B-cell activity. They began by constructing more than 300 different synthetic oligonucleotides, some containing one or more CpG dinucleotides with and without methylation. Some of these are shown in **(Table B19.1)**. These were individually tested for B-cell stimulation by comparing levels of gene expression, measured by ^{3}H-uridine incorporation into newly synthesized RNA and antibody production (IgM) from the B cells.

The researchers found oligonucleotides lost their ability to stimulate B cells when the CpG dinucleotide was specifically eliminated. For example, compare oligonucleotide 2 to 2a, 3D to 3Dc, and 3M to 3Ma in Table B19.1. They also found loss of immune stimulation if the cytosine of the CpG pair was methylated, as would be expected in eukaryotic DNA. For example, compare oligonucleotide 3D to 3Dd, and 3M to 3Mb. They further determined if the position of the dinucleotide pair and the surrounding sequences had any influence on immune stimulation. The best immunostimulatory effects were found in oligonucleotides with centrally located unmethylated CpG dinucleotides surrounded by two upstream purines and two downstream pyrimidines. For example, compare oligonucleotide 1, having no methylation and strong immune activity, to 1c. Methylation of the last CpG did not significantly

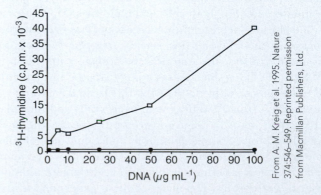

Figure B19.3. B cell immune stimulation as measured by the incorporation of radiolabeled thymidine (^{3}H-thymidine) Unmethylated DNA (open boxes) stimulated B cells while methylated DNA (filled boxes) did not.

TABLE B19.1 B-cell stimulation by CpG oligonucleotides

Oligonucleotide	Sequence	Stimulation Index	
		^{3}H-Uridine	IgM
1	GCTAGA<u>CG</u>TTAG<u>CG</u>T	6.1 ± 0.8	17.9 ± 3.6
1a	· · · · · · · T · · · · <u>CG</u> ·	1.2 ± 0.2	1.7 ± 0.5
1b	· · · · · · · Z · · · · <u>CG</u> ·	1.2 ± 0.1	1.8 ± 0.0
1c	· · · · · · <u>CG</u> · · · · Z · ·	10.3 ± 4.4	9.5 ± 1.8
1d	· · AT · · <u>CG</u> · · GAGC ·	13.0 ± 2.3	18.3 ± 7.5
2	TCAA<u>CG</u>TT	6.1 ± 1.4	19.2 ± 5.2
2a	· · · · GC · ·	1.1 ± 0.2	1.5 ± 1.1
2b	· · · G<u>CG</u>C ·	4.5 ± 0.82	9.6 ± 3.4
2c	· · · T<u>CG</u>A ·	2.7 ± 1.0	ND
2d	· · T<u>TCG</u>AA	1.3 ± 0.2	ND
2e	T · · · GC · ·	1.3 ± 0.2	1.1 ± 0.5
3D	GAGAA<u>CG</u>CTGGACCTTCCAT	4.9 ± 0.5	19.9 ± 3.6
3Da	· · · · · <u>CG</u> · · <u>CG</u> · · · · · · · · ·	6.6 ± 1.5	33.9 ± 6.8
3Db	· · · · · <u>CG</u> · · <u>CG</u> · · · · <u>CG</u> · ·	10.1 ± 2.8	25.4 ± 0.8
3Dc	· · · C · A · · · · · · · · · · · ·	1.0 ± 0.1	1.2 ± 0.5
3Dd	· · · · · · Z · · · · · · · · · · ·	1.2 ± 0.2	1.0 ± 0.4
3De	· · · · · · <u>CG</u> · · · · · · Z · · · · · ·	4.4 ± 0.2	18.8 ± 4.4
3Df	· · · · · · <u>CGA</u> · · · · · · · · · · ·	1.6 ± 0.1	7.7 ± 0.4
3Dg	· · · · · <u>CG</u> · · CC · G · ACTG · ·	6.1 ± 1.5	18.6 ± 1.5
3M	TCCATGT<u>CG</u>GTCCTGATGCT	4.1 ± 0.2	23.2 ± 4.9
3Ma	· · · · · · CT · · · · · · · · · · ·	0.9 ± 0.1	1.8 ± 0.5
3Mb	· · · · · · · Z · · · · · · · · · · ·	1.3 ± 0.3	1.5 ± 0.6
3Mc	· · · · · · <u>CG</u> · · Z · · · · · · ·	5.4 ± 1.5	8.5 ± 2.6
3Md	· · · · · A<u>CG</u>T · · · · · · · ·	17.2 ± 9.4	ND
3Me	· · · · · · · <u>CG</u> · · · · · · C · · A ·	3.6 ± 0.2	14.2 ± 5.2
LPS (30 mg/μL)		7.8 ± 2.5	4.8 ± 1.0

CpG dinucleotides are underlined.

Dots represent identical nucleotides with those oligonucleotide sequences shown for each group.

Z = methylated cytosine

ND = not done

affect immune activity of the oligonucleotide. In contrast, methylation of the central CpG in 1b greatly reduced its immune activity. Comparison of oligonucleotides 2 and 2c, and 3M and 3Md confirm the CpG dinucleotide is most immunogenic when flanked upstream by purines and downstream by pyrimidines. Bacterial DNA appeared to be specifically recognized as foreign by virtue of its unmethylated CpG motifs, a feature not found in eukaryotic cells.

Impact

The importance of the immune stimulatory effects of CpG motifs has not been lost on researchers. Development of a whole new generation of DNA-based therapeutics and vaccines is rapidly progressing. In 2000, TLR9 was identified and confirmed to initiate immune responses to CpG dinucleotides. Identification of the cells expressing TLR9 has allowed researchers to better understand the immune responses to bacterial DNA. CpG motifs can directly activate macrophages, B cells, and dendritic cells, which in turn produce cytokines to activate other cells of the immune system. It has even been demonstrated that administration of DNA containing CpG dinucleotides can redirect inappropriate host immune responses, such as occurs with autoimmune diseases. When used in conjunction with components derived from the infectious agent, CpG motifs demonstrate strong adjuvant activity, stimulating the immune response against the target pathogen components. Addition of CpG-containing DNA to some of our conventional vaccines has been shown to boost their effectiveness in animal models. These exciting findings have led to the design of plasmid-based DNA vaccines with built-in CpG motifs that carry genes for the expression of specific microbial components. The continuous production of microbial product from the plasmid within cells enhances immune responses and requires delivery of very little vaccine.

Extensive studies using various DNA vaccines have been carried out in animal models, establishing the effectiveness, safety, and general utility of CpG-based vaccines. Several experimental DNA vaccines have now advanced to human clinical trials including those against West Nile virus, HIV, and Ebola virus with many more on the horizon. Once again, DNA has turned out to be a surprising molecule.

● Questions for Discussion ···················

1. The following oligonucleotides have been designed for incorporation into a DNA vaccine. Which one would you expect to be the most immunogenic? Defend your answer by contrasting it to the others.

   5′ CTAGGA^{m}CGTCAATGA
   5′ CTAGGACGTCAATGA
   5′ CTCGGACCTCAATGA
   5′ CTAGTCCGAGAATGA
   5′ CT^{m}CGGA^{m}CCTCAATGA
   5′ ^{m}CTCCGACCT^{m}CAATGA

2. Adenoviruses are large DNA viruses that replicate their DNA in the host cell nucleus. Would you expect this viral DNA to be immunogenic? What about DNA from poxviruses that replicate in the cytoplasm?

3. A safety concern that is often raised regarding DNA vaccines is the potential for recombination with host DNA. What other concerns would you have regarding DNA vaccines?

Interestingly, investigations in sponges and copepod crustaceans have revealed a memory response, which is the hallmark of vertebrate adaptive immune responses. The nature of this invertebrate memory response is not yet known. Cells homologous to T and B cells, involved in the adaptive response in vertebrates, have not been identified in invertebrates. Similarly, no genes have been found that show any homology to those used in adaptive response, such as genes encoding antibodies or MHC molecules. Despite the lack of adaptive immunity, innate defense mechanisms in invertebrates are surprisingly varied, adaptable, and efficient in combating infection.

In vertebrates we have seen that the processes of inflammation, complement activation, production of antimicrobial peptides, opsonization, killing by phagocytes and cytotoxic NK cells are important in innate immunity. We have also seen that the initial detection of the presence and nature of a pathogen in vertebrates is done through TLR signaling. A brief tour of some key features of invertebrate defenses will illustrate some of the functional parallels with vertebrate innate defenses. What is striking is how little information has been gathered about these invertebrate systems compared to what is known about those of vertebrates. Perhaps this is not surprising as the vast amount of research in immunology and infectious disease continues to focus on humans.

Invertebrate molecules of defense

Just as we have seen for vertebrates, specific molecules are produced by the cells of invertebrates that act in synergy with cells to stimulate immune responses. We'll begin our comparison with inflammation, which you'll recall is a common physiologic response in vertebrates to damage and infection.

Inflammation-Like Responses

Invertebrates do not possess blood vessels, so there is no close parallel to all of the inflammatory responses. However, there are a number of cytokines released by cells upon damage and infection that have the ability to attract phagocytic cells called **hemocytes** (see Figure 19.23) to the infected area. These hemocytes are similar in function to cytokine interleuken 8 (IL-8) of vertebrates. There is also a clotting system similar in function to our own fibrin clotting system. Melanin encapsulation and the coagulogen system in *Limulus* are examples of clotting systems that will be discussed later.

Complement-Like Molecules

A major purpose of complement activation is lysis of microorganisms. However, invertebrates have not been found to produce complement. Investigation of genetic databases of some invertebrates has uncovered some C1-like genes and some C6-like genes, but a role for these has not been deduced. Invertebrates synthesize a large number of antimicrobial peptides that lyse microbial cells in a manner similar to complement. As described at the end of this chapter (see The Rest of the Story) some of these antimicrobial peptides are induced by activation of Toll receptors. Unlike the TLRs of vertebrates, Toll receptors of invertebrates do not function as PRRs. Instead, they are activated *after* detection of a microbe.

Antimicrobial Peptides

A large number of antimicrobial peptides have been identified in insects, crustaceans, and nematodes. Experimental infection of *Drosophila* with various microorganisms has shown that the innate system can recognize and discriminately respond to various groups of microbes with a selection of antimicrobial peptides produced in response to such infections. The known targets of some of the antimicrobial peptides of *Drosophila* are listed in **Table 19.8**. Unlike most vertebrate antimicrobial peptides, these peptides are induced and are not normally present at all times. For example, fungal infections result in strong increase in expression of the antifungal peptide drosomycin. In infections with Gram-negative bacteria, the anti-Gram-negative peptides diptericin and drosocin are induced.

Opsonization

A number of secreted molecules, including lectins and peptidoglycan-recognition proteins, have been found in various invertebrate species that can opsonize microorganisms for phagocytosis. Genetic sequencing in several invertebrate species has found up to nine genes encoding MBL-like lectins. How these proteins are used in these species is not yet known.

Invertebrate cellular immune responses

Unlike vertebrates, invertebrates do not have a wide variety of specialized immune cells. Again, comparatively little is known about the roles of these cells, and few species have been intensively studied.

Cytotoxic Cells

Earthworms have cytotoxic cells called "coelomocytes" that may be similar to NK cells, at least in function. However, little is known about these cells or how invertebrates in general combat intracellular pathogens.

| TABLE 19.8 | *Drosophila* antimicrobial peptides and their targets |

Peptide	Target
Attacin	Gram-negative bacteria
Diptericin	Gram-negative bacteria
Drosocin	Gram-negative bacteria
Defensin	Gram-positive bacteria
Metchinikowin	Fungi
Drosomycin	Fungi

Phagocytes

The invertebrate main line of cellular immune defense is a phagocytic cell called the "hemocyte" (see Figure 19.23). Hemocytes form a collection of wandering cell sets that migrate from hemolymph, the invertebrate equivalent of blood, into the tissues. Hemocytes can recognize and bind foreign particles and engulf them. Like vertebrate phagocytes, these cells also produce a repertoire of toxic oxygen intermediates, antimicrobial peptides, and enzymes, such as lysozyme that can destroy many microbes.

Cellular defense has been most intensely investigated in the horseshoe crab, *Limulus polyphemus* (see Figure 19.23). Its relatively large size makes it a convenient source of hemolymph for study. When hemocytes are stimulated by contact with a bacteria, or even just LPS, they undergo dramatic physiological and morphological changes. They assume an irregular, amoeba-like shape with numerous cytoplasmic projections. This phenotypic change gives these activated cells the name **amoebocytes**. The various granules produced in amoebocytes contain numerous clotting factors and antimicrobial peptides.

In addition to their phagocytic functions, amoebocytes have two very important innate functions that differentiate them from phagocytes of vertebrates. First, upon the recognition of a foreign substance, microorganisms included, a chemical signaling pathway is induced that results in the secretion of **melanin**. As well as having direct toxic effects, melanin polymerizes to coat or encapsulate microorganisms.

This is a primary defense against microbial invasion in invertebrates. The second non-phagocytic function of amoebocytes involves the formation of a clot to wall off the invader and prevent its spread. This sequestering of the microbes also facilitates destruction by antimicrobial peptides released from the amoebocyte. One of the PAMPs that triggers clotting is LPS. This characteristic is used in the *Limulus* amoebocyte lysate assay, as described in **Toolbox 19.2**. The assay ensures substances such as injectable drugs, fluids, and surgical tubing are free of endotoxins.

The importance of innate mechanisms is evidenced throughout the animal kingdom and can even be found in plants. We observe many parallels in innate defense between vertebrates and invertebrates. The fact that invertebrates are one of the most successful groups of organisms on Earth speaks to the effectiveness of innate mechanisms.

19.6 Fact Check

1. What organisms have been used the most to study immunity in invertebrates?
2. What role do antimicrobial peptides, opsonization hemocytes, melanin, and amoebocytes play in the invertebrate immune response?

Toolbox 19.2
THE *LIMULUS* AMOEBOCYTE ASSAY FOR LPS

Amoebocytes are very sensitive to bacterial LPS. When an amoebocyte encounters a Gram-negative organism, LPS is detected by surface receptors. This causes migration of the amoebocyte granules to the plasma membrane. Degranulation occurs in a manner similar to NK or mast cells. Clotting factors are released from the granules and become activated upon contact with LPS. The end result of the clotting cascade is conversion of the protein coagulogen into a polymerized, insoluble gel designed to trap microorganisms, prevent their spread, and reduce their access to nutrients. The ability of coagulogen to form clots is the basis of a simple *in vitro* commercial assay to detect residual amounts of bacterial endotoxin on medical equipment or fluids used *in vivo*. In this simple assay, the sample from the equipment or solution is mixed with lysate containing clotting factors and coagulogen prepared from horseshoe crab amoebocytes. If the mixture coagulates after 45 minutes, then the test is positive for endotoxin, indicating the product being tested is contaminated.

● Test Your Understanding

You are the owner of a medical supply company and when training new lab technicians you emphasize the importance of running the *Limulus* amoebocyte lysate assay on your company's products. Explain what would happen in a patient if a lab mistake resulted in a positive *Limulus* amoebocyte lysate assay outcome being falsely interpreted as negative.

Image in Action

WileyPLUS Toll receptors are found in *Drosophila* and help provide protection from fungal infections, such as the one shown in the fly in this image. Numerous related Toll-like receptors are found in vertebrates. Toll-like receptors are important pattern recognition receptors involved in the innate immune response and signal the adaptive immune response to begin. In this image of a mouse cell, Toll-like receptors (green proteins) have bound to zymosan (orange protein) on a yeast cell. Binding sets off a signaling cascade of cellular proteins (shades of purple), ultimately resulting in the activation of the transcription factor NF-κB (pink dimer). NF-κB is shown binding to DNA in the nucleus at the bottom, where it is initiating transcription of the genes involved in immune defense against fungal pathogens.

1. What TLRs are expected to bind to zymosan? What is the outcome of this binding?

2. What is another pathogen-associated molecular pattern that would be present in yeast? What innate pattern recognition receptors would bind to it?

3. What kind of innate immune cells would you expect to be activated during an extracellular fungal infection in mice?

4. Explain how you might design an experiment to investigate the role of the different TLRs in responding to fungal infections in mice.

The Rest of the Story

Since their recent discovery, Toll-like receptors (TLRs) of vertebrates have been on center stage in immune research. Not only can these molecules sense the presence of pathogens, they can also determine the nature of the microbe and chemically communicate this to the host. Based on this information, the host is then able to utilize the best defense mechanisms to deal with the particular pathogen. Research into the functioning of TLRs has revealed their roles in mammalian immune dysfunctions, such as autoimmune diseases and allergies, as well as new insights for the design of improved vaccines and therapies for infectious diseases.

What about invertebrates? How do Toll receptors function in immunity? Investigation of the originally discovered *Drosophila* Toll receptor has revealed that its role in the immune response is somewhat different from the TLRs in vertebrates. It is not directly involved in the initial binding of PAMPs, but rather is activated at the *end* of a signaling cascade that is presumably initiated by other largely unidentified cell surface receptors. The result of Toll activation is the transcription of various antimicrobial peptides against fungi and Gram-positive organisms. Nine Toll receptors have been identified in *Drosophila*, but not all have yet been investigated.

Toll-independent-signaling pathways for immune response also exist in *Drosophila*. This was discovered by studying a strain of flies that was shown to be highly susceptible to experimental infections with Gram-negative bacteria, but not with Gram-positive bacteria. It was found that these flies had a recessive homozygous mutation in a gene named *imd* for immune deficiency gene. Activation of this *imd* pathway results in transcription of the anti-Gram-negative antimicrobial peptides diptericin, drosocin, and attacin. So it appears that the PRRs of invertebrates, like those of vertebrates, are able to effectively communicate to the cell the nature of the pathogen and with this information, signal production of the most appropriate antimicrobial products.

Summary

Section 19.1: Are there different types of immunity?

Immunology is the study of the components and processes of the immune system. An individual who is protected from infection is said to be **immune**. Immunity is provided by intrinsic barriers of the body and specialized molecules and cells of the **immune system**.

- The immune system distinguishes foreign substances from normal host components.
- Vertebrates have two kinds of immunity: **innate (non-specific) immunity** and **adaptive (specific) immunity**. Both immune defense mechanisms work together to prevent or minimize disease caused by **pathogens**.

- An **immune response** actively defends the body against foreign agents and cancerous cells.
- Innate immune defenses are found in all multicellular organisms and provide a first line of detection and defense against microbial invasion.
- Innate immune defenses exist in all multicellular eukaryotes and can recognize biochemical differences between microbes and host cells.
- The innate system detects microbes as being foreign and belonging to a certain group of microbes but cannot discern the precise identity of the pathogen.

The adaptive immune system is found only in vertebrates. It is designed to work with the innate immune system to provide an increased level of protection against infection.

- It requires previous exposure to the pathogen.
- It is specific in its recognition of the pathogen.
- It responds to pathogens by producing molecules that bind to specific pathogens.

Section 19.2: What are the built-in barriers to infection?

Skin and **mucosal membranes** are body surfaces exposed to microbes.

- Skin is constructed of epithelial cells with tight cell junctions that prevent passage of most microbes.
- The outer layer of cells on skin surfaces is constantly sloughed to help prevent pathogens from establishing infection.

  Skin prevents colonization by most pathogens because:

- it is cool and acidic
- it produces antimicrobial toxic fatty acids, salts, and lysozyme
- it is heavily colonized by resident microbiota

Mucosal membranes line the inner surfaces of the body, such as the respiratory and intestinal tracts. These surfaces are warm and moist—conditions attractive to microbes. To keep these surfaces free of pathogens, mucosal membranes:

- secrete substances with antimicrobial actions, such as **lactoferrin** and **antimicrobial peptides**
- produce **mucus**
- remove contaminated mucus by **peristalsis** (intestine), flushing by fluid (conjunctiva, urinary, and intestinal tracts), and **mucociliary clearance** (urogenital and respiratory tracts)

Section 19.3: What is an early response of the body to infection?

Inflammation is an important early physiologic response to microbial invasion and damage. It is triggered by release of **proinflammatory molecules,** such as histamine and **cytokines**, from local cells.

- Three immediate consequences of inflammation are **vasodilation**, **extravasation**, and an increase in vessel permeability. These effects facilitate immune responses by bringing in cells, such as **phagocytes**, and molecules with immune functions to the damaged site where microbes have likely entered.
- Release of cytokines, primarily by **leukocytes**, is the primary mode of communication used to signal information regarding an infection. Cytokines can produce **fever**, enhance inflammation, and stimulate further immune responses.
- Excessive inflammation can be damaging to the host, and can produce a **systemic inflammatory response**, leading to **septic shock** or **toxic shock**.

Section 19.4: How do we sense or detect the presence of pathogens?

The molecules involved in innate immunity have the ability to recognize **pathogen-associated molecular patterns (PAMPs)**. These are broad-based molecular motifs associated with microbes but not normally found in the host.

- Molecules that recognize and bind PAMPs are called **pattern recognition receptors (PRRs)**. They can be secreted or membrane-bound.
- An important group of membrane-bound PRRs are **Toll-like receptors (TLRs)**. TLRs are molecular sensors that can detect microbial presence, identify the nature of the pathogen, and signal appropriate immune responses, including induction of adaptive immune responses.
- There are a number of TLRs, each of which can bind to a particular **ligand** of microbial origin.
- TLR9 detects unmethylated **CpG dinucleotides** in DNA of bacterial and viral origin.
- **Adjuvants** used in vaccines often contain molecules recognized by TLR.
- The result of TLR binding is the expression of a number of genes that are needed for cellular immune responses.

Two important secreted PRRs are the **lectins**, **C-reactive protein** and **mannose-binding lectin (MBL)**. Binding of C-reactive protein and MBL to microbes results in:

- **opsonization**, which facilitates **phagocytosis**
- activation of **complement**

Inactive complement components in serum are enzymatically cleaved and activated upon detection of microbes. The end result is:

- formation of the **membrane-attack complex (MAC)** on microbial surfaces, which leads to lysis of microbial membranes
- enhancement of inflammation
- opsonization of microbes
- attraction of leukocytes

Complement can be activated through the **classical pathway**, often utilizing **antibody**, the **lectin pathway**, or the **alternative pathway**.

Type I interferons (IFNs) are cytokines that are produced by cells when a viral infection has been detected. They confer a generalized antiviral state in nearby infected and uninfected cells.

Section 19.5: What cells first defend us against infectious agents?

Phagocytes (**neutrophils**, **monocytes**, and **macrophages**) are the first line of defensive cells involved in clearance of extracellular pathogens.

- **Phagocytosis**, a specialized type of **endocytosis**, is used to engulf large particles, such as bacteria.
- Phagocytosis is facilitated by opsonization. **Opsonins**, such as antibody or C3b, coat the microbe.
- Phagocytosis forms an **endosome** called the **phagosome** that enters the degradative **endocytic pathway**.
- The phagosome fuses with **lysosomes** to form a **phagolysosome** where killing and destruction of microorganisms is accomplished by the production and action of acidic pH, lysozyme, proteases, antimicrobial peptides and toxic oxygen species formed by the **respiratory (oxidative) burst**.

Eosinophils, **basophils**, and **mast cells** are **cytotoxic cells** containing numerous granules that provide immunity against large extracellular parasites, such as worms. When activated, the combined effect is designed to irritate, dislodge, and damage the parasite. Largely through IgE binding, they release:

- cytotoxic chemicals
- stimulators of inflammation (e.g., histamine)
- molecules that stimulate muscle contractions and mucus secretion

Natural killer (NK) cells are the first line of defensive cells involved in clearance of intracellular infections.

- NK cells are cytotoxic cells that possess **granules** containing **perforin** and **granzymes**.
- Release of the contents of the granules by **degranulation** causes **apoptosis** of the contacted abnormal/infected cell. Any surviving microorganism is then ingested and destroyed by phagocytes.
- Unlike **necrosis**, apoptosis does not cause damage to nearby cells or invoke inflammation.
- Abnormal cells are recognized by NK cells through the lack of **class I major histocompatibility (MHC I)** molecules on their surface.

- Cells infected with enveloped viruses can be also be targeted for destruction by the **antibody-dependent cell-mediated cytotoxicity (ADCC)** action of NK cells, which uses F_C receptors (F_CR) to bind antibody-coated cells.

Section 19.6: What kinds of immune defenses do invertebrates have?

Immune defense in invertebrates is strictly innate. Immunity is provided by:

- opsonization by PRRs, such as lectins and peptidoglycan-binding proteins
- phagocytosis and intracellular killing by **hemocytes**
- microbial encapsulation by **melanin**
- clotting by coagulogen produced by activated **amoebocytes**
- lysis by induced antimicrobial peptides
- cytotoxic cells produced in some species

Toll receptors, at least in *Drosophila*, are used only after the pathogen has already been detected to conduct information regarding the nature of the pathogen to cells involved in immune responses. They are not PRRs as TLRs are in vertebrates.

Application Questions

1. Smokers have been shown to have cilia damage and reduced cilia beat frequency. Explain how this may result in increased vulnerability to lung infections in smokers.

2. People who are taking oral antibiotics often experience intestinal upsets, such as mild cramping and diarrhea. Using your knowledge of the role of normal microbiota, explain why this may occur.

3. Postoperative patients are told to look for warning signs of infection at their surgical incision site such as redness, swelling, and heat. What activities are involved in causing these clinical signs and why might these indicate an infection?

4. Based on your understanding of the role of TLR4, describe how a TLR4 knockout mouse's immune system would react differently to LPS injection than the immune system of a normal mouse.

5. A young researcher is attempting to identify an adjuvant that can be safely added to a new vaccine. The researcher is considering LPS. Provide an analysis of whether LPS would serve as a desirable vaccine adjuvant.

6. Your mother had a blood test to check her levels of C-reactive protein and the results show her levels are above the normal average. What would this result reveal, and how would you explain this to her?

7. Both hepatitis A and hepatitis B can be treated by administering interferon-alpha (IFN-α) and antiviral drugs. Explain to a hepatitis patient what the interferon is doing for them as part of their treatment.

8. An experimental cancer treatment involves removal of MHC I on tumor cells and is being tested in a mouse model. What effects would you expect this treatment to have on the NK cells in the mouse model? Explain.

9. Imagine a patient's innate immune system suffers from an inability to produce mannose-binding lectin. What effects would be observed and why?

10. Some strains of the opportunistic pathogen *Streptococcus pyogenes* cause necrotizing fasciitis, a serious and sometimes fatal skin infection that will be examined in Chapter 21. Common anti-inflammatory medications, such as ibuprofen, have been implicated in the precipitation of this disease. Explain this link.

11. Imagine you are in a lab studying the specific effects of cytokines. How might you go about designing an experiment to study the chemoattractant activity of IL-8? Explain what results you would expect.

12. You are asked to design a hypothetical pathogenic bacterium that is able to effectively overcome or avoid three innate immune defenses or detection mechanisms.
    a. What three innate defenses will your pathogen be able to overcome?
    b. What do those defenses normally do?
    c. How will your pathogen overcome each of those defenses?

Suggested Reading

Akira, S., S. Uematsu, and O. Takeda. 2006. Pathogen recognition and innate immunity. Cell 124:783–801.

Hemmi, H., O. Takeuchi, T. Kawai, T. Kaisho, S. Sato, H. Sanjo, M. Matsumoto, K. Hoshino, H. Wagner, K. Takeda, and S. Akira. 2000. A Toll-like receptor recognizes bacterial DNA. Nature 408:740–745.

Rock, F. L., G. Hardiman, J. C. Timans, R. A. Kastelein, and J. F. Bazan. 1998. A family of human receptors structurally related to *Drosophila* Toll. Proc Natl Acad Sci USA 95:588–593.

Takeuchi, O., O. Hoshino, T. Kawai, H. Sanjo, H. Takada, T. Ogawa, K. Takeda, and S. Akira. 1999. Differential roles of TLR2 and TLR4 in recognition of Gram-negative and Gram-positive cell wall components. Immunity 11:443–451.

von Müller, L., A. Klemm, M. Weiss, M. Schneider, H. Suger-Wiedeck, N. Durmus, W. Hampl, and T. Mertens. 2006. Active cytomegalovirus infection in patients with septic shock. Emerg Infect Dis 12:1517–1522.

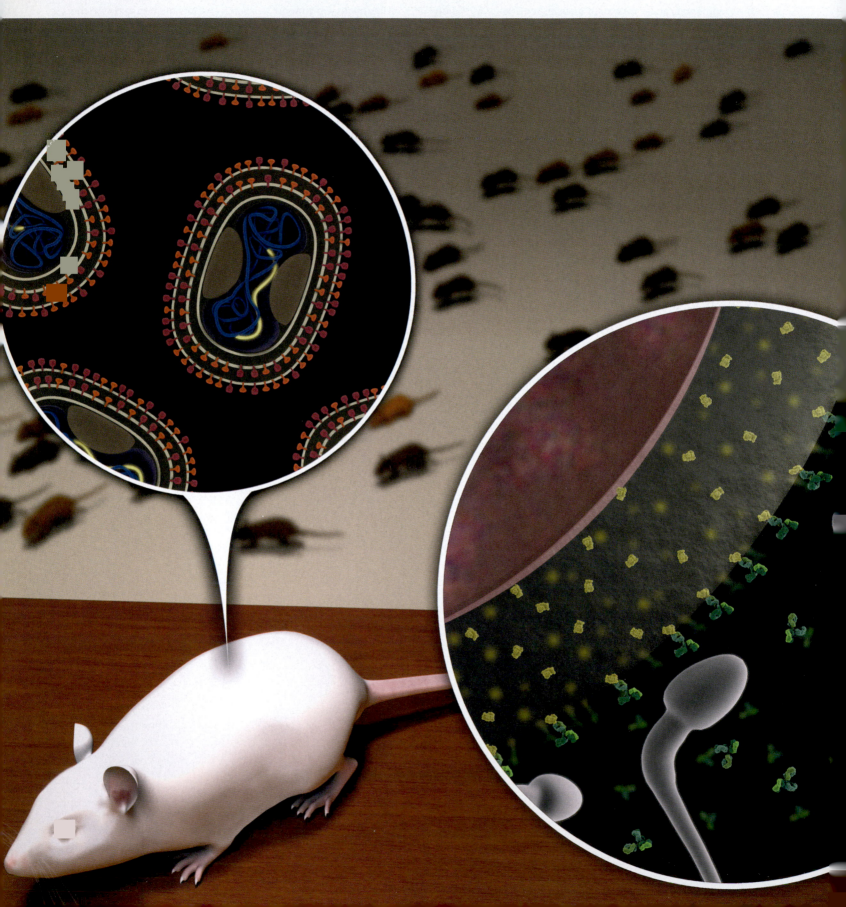

In 1998, a research project in Australia became the focus of scientific and political communities. The scientists were working on a method of contraception to help control plagues of rodents in grain barns in Australia. Their research focused on developing a method to use the mouse's own immune system to produce infertility—a concept known as immunocontraception. To do this, they re-engineered ectromelia virus, a relatively benign mouse virus that causes the disease mousepox. Into the genome of the virus, they inserted the gene for a mouse egg surface protein called mZP3, a receptor for sperm. The hope was that the virus would now produce the sperm receptor mZP3 protein and infected female mice would produce antibody against it. Antibody would block the sperm receptor, rendering the female mice infertile. To achieve high rates of infertility, researchers needed to produce high levels of antibody against mZP3. To do this, they used their knowledge of adaptive immune responses and supplied the virus with a second gene that codes for a small protein called interleukin-4 (IL-4). IL-4 is known to boost the production of antibody in mammals—a response desired by the researchers to maximize the contraceptive effect.

However, things didn't go as planned. When the engineered virus was experimentally used to infect mice, it unexpectedly killed them. The researchers had turned a relatively harmless virus into a lethal pathogen. Even more surprising, the engineered virus killed a large proportion of mice that had been previously exposed to normal ectromelia mouse virus. These mice should have been resistant to the engineered virus.

This case immediately raised red flags. Should this research be published and released to the public arena where it could be used for the intentional engineering of organisms with increased lethality? Parallels were quickly drawn to apocalyptic science fiction movies such as *Resident Evil* and *I Am Legend*, where engineered viruses are released upon the human race with disastrous results. Concerns were further fueled because ectromelia virus is related to variola virus, the human smallpox virus. This virus is already naturally virulent and is on the short list of potential agents for biological warfare. In the wrong hands, could this information be used to make the smallpox virus even more deadly? As you read this chapter, you will gain insight into the workings of the adaptive immune response and decide for yourself whether or not these are legitimate concerns. We will return to our story "of mice and men" near the end of the chapter.

CHAPTER NAVIGATOR

As you study the key topics, make sure you review the following elements:

Adaptive immunity is based on the production of specialized immune receptors that bind foreign antigen.
- Figure 20.1: Immune receptors of T cells and B cells
- Figure 20.5: Primary and memory immune responses

T cells are the central control cells in adaptive immunity.
- Figure 20.6: Activation and clonal expansion of a naïve T cell
- Figure 20.9: General process of T-cell activation by an antigen-presenting cell

Peptides must be processed inside cells and presented on MHC molecules to activate T cells.
- Figure 20.12: Process Diagram: The endocytic pathway for presentation of exogenous peptides on MHC II
- Figure 20.16: Process Diagram: The endogenous pathway for presentation of cytoplasmic peptides on MHC I

Specialized antigen-presenting cells present peptides to activate T cells.
- Table 20.2: Activities of antigen-presenting cells

The two main arms of the adaptive immune response are cell-mediated immunity and humoral immunity.
- 3D Animation: What is the cell-mediated immune response?
- 3D Animation: How does humoral immunity fight infection?
- Figure 20.24: Development of a naïve $CD4^+$ T cell into T_H1 or T_H2 cells
- Mini-Paper: Attempting to engineer a virus to improve immunocontraception

Activation of B cells produces antibody.
- Figure 20.25: Summary of T-cell and B-cell activation
- Figure 20.27. Process Diagram: T_H2-dependent production of antibody against non-protein antigen
- Microbes in Focus 20.1: Slippery *Haemophilus influenzae*
- Figure 20.28: Activation of a B cell by a T-independent antigen
- Perspective 20.2: Vaccines against T-independent antigens
- Toolbox 20.2: Enzyme-linked immunosorbent assay (ELISA)

CONNECTIONS for this chapter:

Cytokines and the direction of immune responses (Section 19.3)

Vaccines and immunological memory (Section 24.5)

Toll-like receptors and the development of immune responses (Section 19.4)

Capsules and immune evasion by pathogens (Section 21.1)

Introduction

Despite the best efforts of our innate defenses, a myriad of microbes are still capable of invading our bodies. Mostly, these are pathogens that have evolved strategies to elude innate immune defenses. This is where the adaptive immune system of the jawed or higher vertebrates steps in. This arm of immunity is designed to specifically recognize almost any foreign molecule, including those possessed by pathogens. It is a very effective, highly coordinated system, capable of producing a spectacular diversity of immune recognition molecules far beyond that of innate immunity. Best of all, this system can genetically "remember" these molecules and, in many cases, provide lifelong protection, making vaccination a feasible strategy for control of many infectious diseases (see Chapter 24).

Successful pathogens must evolve specific strategies to overcome, outlast, or circumvent adaptive immune responses. Many of these mechanisms will be covered in Chapters 21, 22, and 23. We need to understand the workings of the adaptive immune system first, and we will structure our exploration of adaptive immunity around the following questions:

What is the advantage of adaptive immunity? (20.1)
What are the central cells in adaptive immunity? (20.2)
How are adaptive immune responses initiated? (20.3)
What cells can launch adaptive immune responses? (20.4)
Are there different responses to extracellular and intracellular pathogens? (20.5)
How are antibodies produced (20.6)?

20.1 Features of adaptive immunity

What is the advantage of adaptive immunity?

In Chapter 19, we defined immunological immunity as the defensive response to microbial invasion that occurs at a molecular and cellular level. We further defined innate or non-specific immunity as involving immune mechanisms that are already present at birth and do not require previous exposure to the microbe. It is based on molecules that are genetically hardwired in each cell to distinguish self from non-self components. As you will recall, many of these molecules are pattern recognition receptors (PPRs) that recognize pathogen-associated molecular patterns (PAMPs), which are broad-based molecular motifs commonly found in microorganisms. Innate systems cannot recognize the precise identity of the microbe, only its general nature and that it is foreign. Adaptive or specific immunity can precisely recognize and "remember" a particular type of invading microbe. This immune response is found only in vertebrates, and involves the production of specialized immune receptor molecules through dynamic genetic rearrangements in response to an infecting agent. Although this may seem like an incredible feat, we will see that the effectiveness of the adaptive immune response is based on relatively simple principles.

Immune receptors and antigen

The innate and adaptive immune responses are functionally intertwined and thus the separation of the two categories is really based on the genes that are employed by the two systems. We can make this definition even more precise: The adaptive immune response is immunity based on the production of specialized **immune receptors** that bind **antigen**, the foreign material. Immune receptors include **T-cell receptors (TCRs)** and **immunoglubulin (Ig) molecules**. TCRs are on the surface of T cells. Immunoglobulins are produced by B cells and include surface **B-cell receptors**

(BCRs) and the secreted version of BCRs, **antibody (Figure 20.1).** Immune receptors arise from random genetic rearrangements in the genes that code for the peptides making up these immune receptors (see Section 20.6). T cells and B cells are leukocytes that are central to adaptive immunity. This chapter will present the process by which these cells and their immune receptors can recognize foreign antigen and generate adaptive immune responses.

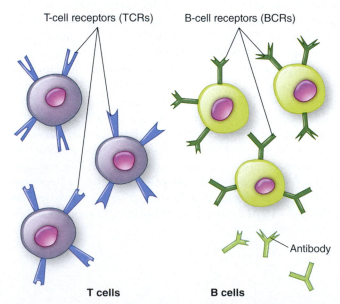

Figure 20.1. Immune receptors of T cells and B cells Membrane-bound T-cell receptors (TCRs) of T cells and B-cell receptors (BCRs) of B cells recognize and bind to foreign antigens. Antibodies, produced by specialized B cells, are a secreted form of BCRs, and similarly bind to antigens. Individual T cells and B cells produce only one antigen-binding type (shape) of immune receptor that binds to a particular antigen.

CONNECTIONS Recall from **Section 19.4** that antibodies are secreted glycoproteins that bind to microbes and their components and are produced by B cells of the adaptive immune system.

We need to clarify some related terms: "antigen," "ligand," "epitope," and "immunogen." Antigen is any component that can be specifically bound by a specialized immune receptor, namely a TCR or an immunoglobulin. In Chapter 19, the general term "ligand" was used to describe molecules that bind to receptors. Antigens are ligands that are bound by immune receptors. Innate immunity does not involve the production of immune receptors (TCRs or immunoglobulins), and therefore targets bound by molecules and cells of the innate immune system are correctly defined as ligands, but not as antigens. The distinction between these terms is important. For example, you would not use the term "people" when you are only referring to adult males. "Antigen" is a much more precise term than "ligand." The specific interaction between antigen and immune receptors is the cornerstone of adaptive immunity.

The term "antigen" is often incorrectly used to broadly describe any component that induces an immune response. The correct term for a substance that induces an immune response is "immunogen." It is a general term just as "ligand" is a general term. Immunogens are immunogenic—they induce innate or adaptive immune responses, or both.

Individual T cells and B cells of the adaptive immune system produce only one antigen-binding species of immune receptor. The unique shape of that receptor makes it specific to a single **epitope** of the antigen (**Figure 20.2**). An epitope is a single site on a larger antigen molecule that is recognized by an immune receptor. It is the smallest unit that can be bound. Different T and B cells can produce different immune receptors that recognize different epitopes.

The production of immune receptors by the adaptive immune system involves rearrangements of genes that code for the TCR in T cells and immunoglobulin molecules in B cells (see Section 20.6). These rearrangements are limited to T cells and B cells and do not occur in germline cells, that is, eggs and sperm. Consequently, adaptive immunity is not passed along to offspring. Each infant must build his or her own T- and B-cell profile as he or she moves through life with its many inevitable exposures to a multitude of infectious agents.

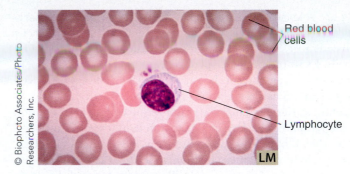

Figure 20.3. Lymphocyte Stained cells from blood showing a central lymphocyte surrounded by red blood cells.

Lymphocytes and lymphoid tissues

T cells, along with B cells and natural killer (NK) cells (see Section 19.5), are classified as **lymphocytes**, small, round leukocytes that bear specialized receptors that bind other cells to direct immune responses (**Figure 20.3**). Lymphocytes are generated from stem cells in the bone marrow and mature in the primary lymphoid organs—the bone marrow and thymus (see Appendix G). T cells mature in the thymus, for which they are named, the "T" standing for thymus (**Figure 20.4**). Mature T cells with unique

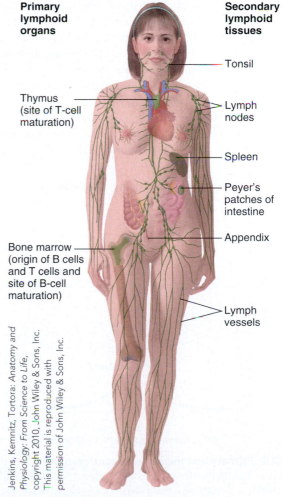

Figure 20.4. Lymphoid tissues of the human body Lymphocytes originate in bone marrow and mature in the primary lymphoid organs where they produce their unique immune receptors. B cells mature in bone marrow while T cells mature in the thymus. Both cell types circulate through the lymph vessels and secondary lymphoid tissues as well as the blood vessels (not shown).

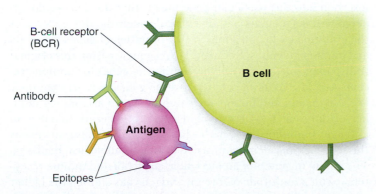

Figure 20.2. Epitopes bound by specific immune receptors Different epitopes of an antigen can be recognized by different immune receptors. Shown here is binding of a B-cell receptor (BCR) and antibodies to different epitopes.

TCRs are released into the circulatory system. B cells similarly mature in the bone marrow. From the circulatory system, T cells and B cells access secondary lymphoid tissues distributed around the body. Examples are the spleen, lymph nodes, tonsils, appendix, and Peyer's patches of the small intestine. These specialized tissues house dense populations of recirculating T and B cells and detain antigen to allow exposure required for the activation of these lymphocytes. Organ surfaces in contact with microbes, such as the skin, pharynx, and intestine have collections of associated lymphoid tissues, including peripheral lymph nodes, secondary lymphoid organs, and diffuse collections of dendritic cells and macrophages. Examples include SALT (skin-associated lymphoid tissue), MALT (mucosal-associated lymphoid tissue), and GALT (gut-associated lymphoid tissue).

Primary and memory immune responses

When an infectious agent is encountered by cells of the adaptive immune system for the first time, only a few cells can recognize the antigens carried by the pathogen. Again, this is because only a few cells will carry immune receptors capable of recognizing the particular epitopes of the pathogen molecules. These cells respond by greatly increasing their activity and numbers over a few weeks in a process called a **primary immune response**. During this response, vast numbers of cells specific to the agent are generated through **clonal expansion**. Clonal expansion refers to the many rounds of cell division that follow activation of a T cell or B cell. The end result is a tailored population of immune cells all capable of recognizing the *same* foreign molecule. Unlike many innate immune responses that are already present and can act immediately, a primary immune response requires time to generate these specific cell populations, producing a peak in activity approximately two weeks following first exposure to the antigen **(Figure 20.5)**. Some of the cells generated by

clonal expansion survive for months or years following the exposure. These **memory cells** confer an immunological **memory** or **anamnestic response** ("anamnestic" meaning "recall to memory" in Greek) originating from the stimulation of memory cells and characterized by a faster and more vigorous adaptive immune response upon a repeated exposure to antigen. When the same agent is encountered a second time, memory cells give rise to a memory, sometimes called a "secondary immune response" in a larger and more accelerated fashion. Of course, many pathogens are encountered more frequently than this, and the terms "memory" or "secondary immune response" are often used interchangeably to refer to all subsequent memory responses to that antigen. Each time the antigen is encountered, the memory response gains in strength. The details of immunological memory will be covered in Section 20.2. Often, the memory gives rise to such a fast and strong memory response that the pathogen is stopped in its tracks upon another encounter, even years later. In many cases, we are unaware that we have even been exposed again. This is a principal advantage of the adaptive immune system.

Evolution of adaptive immunity

Let's consider how adaptive immunity appears to have arisen in vertebrates. All aspects of adaptive immunity are rooted in the formation of immune receptors—TCRs and immunoglobulins. As you will recall from Section 19.1, adaptive immunity is not present in invertebrates. Adaptive immunity is exclusive to vertebrates, but not all vertebrates. In the evolutionary tree of life, it appears abruptly in jawed or cartilaginous fish. Jawless fish, like lamprey, have no adaptive immunity. They lack the genes used to make immune receptors. As a consequence, jawless fish and invertebrates cannot produce antibody or TCRs, and they lack the associated T and B cells. Of course, they do not have immunological memory.

Most researchers who study the genetic evolution of adaptive immunity agree that a transposable element, probably a retrotransposon (see Section 5.2), was responsible for delivering the ability to perform gene rearrangements that characterizes modern immune receptors. The transposable element is thought to have invaded a germline cell in an ancestor of the cartilaginous fish, and randomly inserted itself into an immunoglobulin-like precursor gene. Transposable elements, as you may recall from Section 9.4, are famous for their ability to move genes around, cause genetic rearrangements, and even duplicate genes. The insertion into this particular gene meant that the gene could undergo rearrangements and diversify through the descendents of the evolutionary line, leading to the formation of genes for TCRs and immunoglobulins. Remnant DNA sequence from this original transposable element is still present in the genes for immune receptors, and confers the ability of these genes to rearrange their DNA segments in different ways to produce an astounding number of different receptor combinations to generate matches to antigens. Immunologist Charles Janeway of Yale University likens this transposable element invasion to an immunological Big Bang. The ability to generate an incredible collection of immune receptors to suit a multitude of foreign antigens has allowed the higher vertebrates who possess adaptive immunity to recognize and defend themselves against a larger spectrum of potential antigens compared to their innate-restricted lower kin and invertebrates.

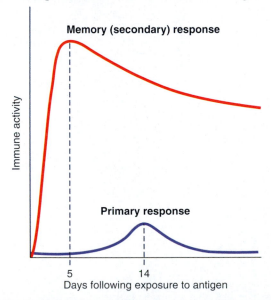

Figure 20.5. Primary and memory immune responses A primary immune response (blue line) occurs following a first-time exposure to an antigen. It produces a relatively low level of immune activity that peaks at 10 to17 days and is of short duration. From this primary response, long-lived memory cells are generated. When memory cells are subsequently exposed to that same antigen, a memory response (red line) occurs. The memory response is characterized by a high level of immune activity that peaks quickly at two to seven days, and lasts for a much longer time. By the second day, the activity has already surpassed the peak level of a primary response.

20.1 Fact Check

1. What is adaptive immunity and how is it different from innate immunity?
2. How are the terms "antigen," "ligand," "epitope," and "immunogen" involved in an immune response?
3. Differentiate between primary immune response and memory immune response.
4. Describe how the adaptive immune response is thought to have evolved in vertebrates.

20.2 T cells

What are the central cells in adaptive immunity?

Initiation of adaptive immunity is a complex process requiring antigen presentation, cell signaling, and production of stimulatory molecules. To understand the adaptive immune response, we will need to take a closer look at the cells involved and their functions. A central theme in the initiation of an adaptive immune response is the activation of **T cells** or **T lymphocytes**. T cells are the orchestrators of adaptive immune responses.

Cells involved in adaptive immune responses, including T cells, need to be activated before they can participate in adaptive immune responses. Cells that have never encountered antigen are referred to as "naïve cells." This means they have not been activated through antigen binding. When a T cell is activated it undergoes clonal expansion and differentiation, resulting in a population of long-lived memory T cells and short-lived effector T cells (**Figure 20.6**). Memory cells are responsible for the previously mentioned immunological memory that generates the much faster and more aggressive memory immune response.

Effector cells are fully armed cells that directly carry out immune functions. In the case of T cells, the effector activity varies with different T cells and will be described later in this section.

T-cell activation

Activation of a T cell first requires binding of the T-cell receptor (TCR) with an antigen in the form of a peptide. T cells, however, cannot bind free peptide; it must be "presented" to them by other cells known as **antigen-presenting cells (APCs)**. APCs are macrophages, dendritic cells, and B cells. Recall from Section 19.5 that macrophages are phagocytic leukocytes of the innate immune system, but because they can also present antigen, they can activate the adaptive immune system. **Dendritic cells** are wandering leukocytes that specialize in sampling smaller extracellular material to similarly present it to the adaptive immune system. **B cells** or **B lymphocytes** are circulating lymphocytes

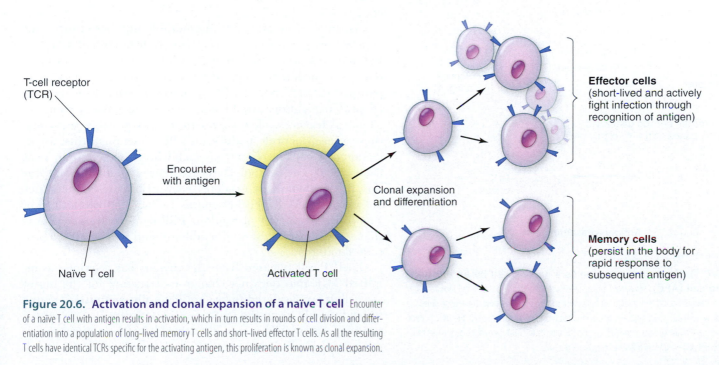

Figure 20.6. Activation and clonal expansion of a naïve T cell Encounter of a naïve T cell with antigen results in activation, which in turn results in rounds of cell division and differentiation into a population of long-lived memory T cells and short-lived effector T cells. As all the resulting T cells have identical TCRs specific for the activating antigen, this proliferation is known as clonal expansion.

Labels in figure: T-cell receptor (TCR); Naïve T cell; Encounter with antigen; Activated T cell; Clonal expansion and differentiation; Effector cells (short-lived and actively fight infection through recognition of antigen); Memory cells (persist in the body for rapid response to subsequent antigen)

that give rise to antibody, but they also specifically bind foreign materials for presentation to T cells. Dendritic cells and B cells will be covered in more detail later in the chapter. APCs activate T cells by specifically promoting T-cell binding directly to the peptide antigen they present on their surface. The different roles each of these cell types plays in stimulation of immune responses will be discussed in Section 20.4.

The peptide antigen presented is bound to **major histo-compatibility (MHC) molecules** expressed on the surface of APCs **(Figure 20.7)**. MHC molecules are proteins that are generated intracellularly and bind peptides produced from the breakdown of microbial proteins inside cells (see Section 20.3). Once an MHC molecule binds to a peptide, the MHC:peptide complex is transported to the cell surface where it can present the peptide to nearby T cells.

Why can only specific APCs activate T cells? Imagine what would happen if all cells of the body could induce adaptive immune responses by presenting antigen to T cells. It could easily cause immune overreaction. Consider what would happen if a large group of people alerted the police to a crime: police might think the criminals were numerous and overreact. Instead, a few accurate reports by reliable witnesses could direct the police into taking appropriate action. Specialized APCs ensure that activation of T cells occurs in a controlled manner and only when there is sufficient need. Only these cells possess the right kinds of surface molecules to fully activate T cells.

If microbial invaders are present in large enough numbers, or induce significant inflammation (see Section 19.3), they will be contacted by APCs, and through them, cells of the adaptive immune system will be signaled. In another scenario, when a single microbe enters a tissue, it may be destroyed by innate immune mechanisms such as complement (see Section 19.3). It

would not survive and proliferate, and therefore would not come into contact with APCs. Adaptive immune responses would not be evoked or required.

CONNECTIONS Inflammation is a progression of physiologic events that efficiently brings molecules such as complement factors and cells of defense, such as APCs, to the damaged or infected site. See **Section 19.3** for details.

A class of MHC molecules known as MHC I was introduced in Chapter 19 (see Section 19.5). Recall that MHC I molecules are made by all nucleated cells of the body and allow any type of infected cell to present peptides to T cells of the immune system. Presentation of peptide antigen by these cells results in an immune response by a T cell only if T cells have been previously activated by binding the MHC:peptide complex presented by an APC. Although many cell types can present antigen to T cells, only APCs possess all the characteristics required for *activation* of T cells. These include:

- expression of high numbers of MHC molecules on their surface
- proficiency in the processing and presentation of peptides to T cells
- display of the surface stimulatory molecules required for T-cell activation

Binding of the T cell with antigen occurs between the TCR and a compatible peptide presented by an MHC molecule on the APC. Binding is non-covalent, based on matching shape and charge between the peptide and the TCR. Since the TCR binds antigen, it is easier to think of it as the T-cell antigen receptor. Binding between the TCR and the MHC:peptide complex depends on the specificity of the TCR for the peptide being presented. The correct "fit" provided by chemical and conformational configuration must be present for binding to occur.

As you can imagine, an innumerable number of different peptides are generated when cells degrade microbial proteins. In innate immunity, foreign molecules are recognized by their possession of intact common pathogen-associated molecular patterns (PAMPs; Section 19.4). However, none of the degraded protein peptides would be expected to possess common chemical features. How then can T cells possibly recognize these if each peptide is different? To accomplish this, a collection of T cells is generated, each with its own unique TCR **(Figure 20.8)**. The custom TCRs are generated during maturation in the thymus by random rearrangements of the genes encoding the TCR peptide chains in each individual T cell. The collective variety of random TCR conformations ensures that some of the TCRs will be able to bind specifically to presented antigen.

Each T cell released from the thymus has a unique, committed TCR that does not change its structure for the life of the cell. It is considered to be "antigenically committed"—able to bind to only a specific epitope of an antigen. If a presented peptide from an APC can bind to the TCR, then activation can proceed. If binding does not occur, then initiation of a T-cell

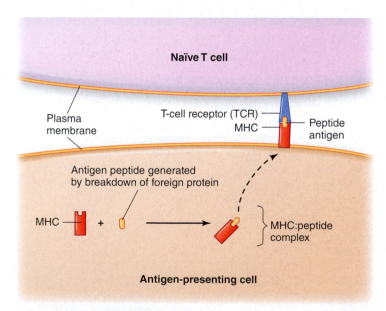

Figure 20.7. Antigen presentation to a naïve T cell by an antigen-presenting cell (APC) For naïve T-cell activation, the membrane-bound T-cell receptor (TCR) on the surface of mature T cells binds foreign peptides presented by MHC molecules. The foreign protein is processed inside an APC into short peptides, which are then bound by MHC molecules. The MHC:peptide complex is transported to the surface of the APC. Antigen binding by the T cell is governed by the specificity of the TCR and the peptide bound by the MHC molecule of the APC.

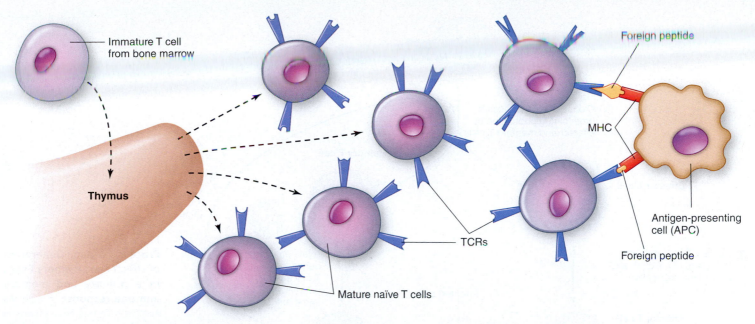

Figure 20.8. Maturation of T cells in the thymus Each unique TCR of the naïve T-cell population is generated during maturation in the thymus. Thus, each mature T cell has TCRs that are capable of recognizing a particular antigen; the cells are said to be antigenically committed. The enormous variety of TCR conformations ensures that some of the TCR conformations will specifically bind to a particular peptide presented by APCs.

response will depend on the arrival of another T cell with the correct receptor conformation. This may seem like a lottery but the number of different TCRs generated in the body is vast—estimated to be 2.5×10^7—so there is a strong possibility that some of the available TCRs will be able to bind to the presented peptide and begin a response.

Specific binding between the TCR and the presented peptide is only one requirement for successful activation of the T cell. Activation also requires the T-cell surface markers CD4 or CD8. These co-receptors bind to the MHC portion of the MHC:peptide complex of the APC **(Figure 20.9)**. Important T-cell subsets carrying these co-receptors are $CD4^+$ and $CD8^+$ T cells. **$CD4^+$ T cells** carry the CD4 surface marker and **$CD8^+$ T cells** carry the CD8 surface marker. The effector functions of these two subsets are very different from one another and will be described further in Section 20.3. When all criteria for activation have been met, several T-cell genes are expressed, including those for certain cytokines. An important cytokine produced as a result of successful activation of $CD4^+$ T cells is **interleukin-2 (IL-2)**. IL-2 functions as a growth factor for T cells, inducing resting cells to enter into cell division. Activation also results in the expression of the IL-2 receptor (IL-2R) on the T-cell surface, which binds IL-2. Clonal expansion of the T cell produces populations of memory cells and effector cells (see Figure 20.6). Because of the antigen commitment of the original activated T cell, the resulting memory and effector cells all carry identical antigen-specific TCRs; they are all genetically identical clones of one another. Each is therefore capable of recognizing the same epitope of an antigen. We can now see that the seemingly incredible feat of producing specialized molecules designed and manufactured in response to an infecting agent is really based on the relatively simple principle of antigen commitment and clonal expansion.

Cell-mediated immunity ANIMATION

The specificity of the TCR is therefore the cornerstone of the customized adaptive immune response that follows.

CONNECTIONS Cytokines are small, secreted glycoproteins that are produced by various cells during an immune reaction (see **Section 19.3**). They are the major communication molecules used to direct immune responses. Examples include interleukins, interferons, and tumor necrosis factor-alpha (TNF-α).

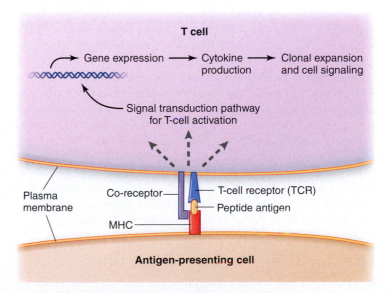

Figure 20.9. General process of T-cell activation by an antigen-presenting cell The TCR must bind with peptide antigen, and the CD4 or CD8 co-receptor must bind with the MHC molecule presenting the antigen. Activation results in expression of cytokines and other molecules used for cell proliferation and signaling to other cells. Some of the products of the expressed genes such as IL-2, induce clonal expansion of the activated T cell to produce a population of effector cells and memory cells, all carrying the identical TCR.

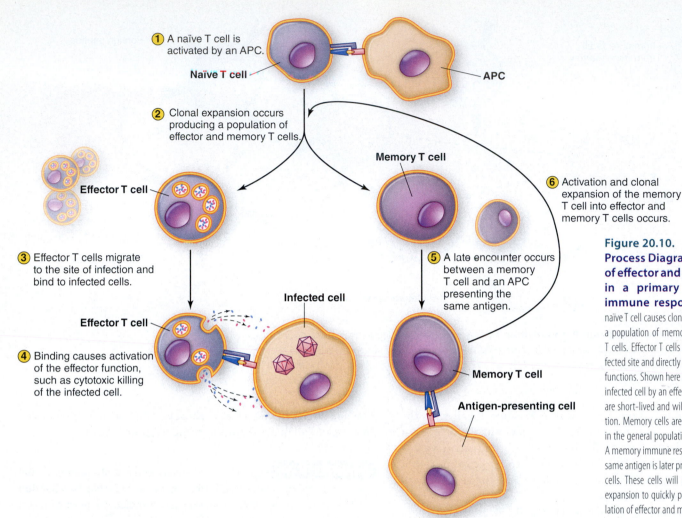

① A naïve T cell is activated by an APC.

Naïve T cell

APC

② Clonal expansion occurs producing a population of effector and memory T cells.

Effector T cell

Memory T cell

⑥ Activation and clonal expansion of the memory T cell into effector and memory T cells occurs.

③ Effector T cells migrate to the site of infection and bind to infected cells.

⑤ A late encounter occurs between a memory T cell and an APC presenting the same antigen.

Infected cell

Effector T cell

④ Binding causes activation of the effector function, such as cytotoxic killing of the infected cell.

Memory T cell

Antigen-presenting cell

Figure 20.10.
Process Diagram: Generation of effector and memory T cells in a primary and memory immune response Activation of a naïve T cell causes clonal expansion producing a population of memory T cells and effector T cells. Effector T cells will migrate to the infected site and directly carry out their immune functions. Shown here is cytotoxic killing of an infected cell by an effector T cell. Effector cells are short-lived and will not remain in circulation. Memory cells are long-lived and remain in the general population of circulating T cells. A memory immune response begins when the same antigen is later presented to the memory cells. These cells will rapidly undergo clonal expansion to quickly produce a second population of effector and memory T cells.

Memory cells and the memory response

The establishment of immunological memory is an important and usually desired result of the activation of adaptive immune responses. It is dependent on the production of memory cells as a result of clonal expansion. The production of long-lived memory T cells and memory B cells (see Section 20.4) often results from strong stimulation of adaptive immune responses. The presence of these cells produces a faster and more vigorous adaptive immune response when the same antigen is encountered again. Memory cells generally have fewer requirements for activation than do their naïve counterparts. They are primed for action so that when they encounter the antigen again, they can rapidly undergo clonal expansion and differentiate into effector cells **(Figure 20.10)**. This clonal expansion contributes to a faster and larger adaptive response and in many cases prevents the pathogen from establishing a repeated infection.

Effector T cells and their functions

Effector cells are action cells. They are short-lived cells armed with direct immune functions. For T cells, the effector activity depends on whether they carry the CD4 or CD8 co-receptor on their surface. As we saw earlier, these co-receptors are needed for activation of T cells by interacting with the MHC molecules

of APCs. The effector function of activated $CD4^+$ T cells is primarily the secretion of large concentrations of cytokines. The profile of cytokines secreted by active effector $CD4^+$ T cells enhances and directs the immune functions of various other cells, including $CD8^+$ T cells. For this reason, effector $CD4^+$ T cells are commonly referred to as **T helper cells (T_H cells)**. Cytokine help from effector $CD4^+$ T cells is an absolute requirement for full function of adaptive immune responses.

The effector function of activated $CD8^+$ T cells is primarily cytotoxic killing of infected cells. For this reason, effector $CD8^+$ T cells are referred to as **cytotoxic T cells (T_C cells)**. Effector $CD8^+$ T cells can directly kill infected cells that they contact by releasing granzymes and perforin that cause apoptosis. This is similar to the action of natural killer (NK) cells (see Section 19.5). T_C killing is described in more detail in Section 20.3.

20.2 Fact Check

1. Explain the process of T-cell activation, including how co-receptors, TCRs, antigen-presenting cells, and MHC molecules are involved.
2. Distinguish between $CD8^+$ and $CD4^+$ effector T cells.

20.3 Antigen processing

Let's continue to further trace some of the specific events that lead to a T-cell immune response. As we know, the initiation of T-cell activation requires the presentation of antigen by an APC. The pathway used to process antigen within the APC generally depends on the source of the antigen, that is, whether it is from an extracellular or an intracellular microbe. These factors will impact the immune response that is generated.

Exogenous antigens and the endocytic pathway— Activation of CD4$^+$ T cells by MHC II:peptide

Exogenous antigens, or extracellular antigens are antigens that have entered the cell from the extracelluar environment. As we saw in Chapter 19 (see Section 19.5), the process of uptake of exogenous particles is called "endocytosis." Endocytosis occurs in three ways **(Figure 20.11)**:

- Phagocytosis, a form of endocytosis, is used to take in a large object such as a bacterial cell. Of the APCs, only macrophages and certain types of dendritic cells carry out phagocytosis.

- Receptor-mediated endocytosis is carried out by many cell types, and occurs when uptake is initiated through the binding of particles such as viruses or bacterial components to specific cell surface receptors (see Section 8.2).

- Pinocytosis is used by all cells to take in macromolecules such as nutrients. This can be thought of as "drinking" by the cell.

Recall from Section 19.5 (see Figure 19.12), the vesicles formed through phagocytosis (phagosomes) and receptor-mediated endocytosis (endosomes) enter the endocytic pathway for digestion of their contents. Although many microbial components are degraded in the endocytic pathway, proteins are of primary importance for triggering adaptive immune responses. Only they can activate T cells. The proteins are digested in the endosome into small peptides 13 to 18 amino acids in length.

The peptides are then bound by or "loaded" onto specialized **class II major histocompatibility (MHC II) molecules**, supplied by vesicles of the Golgi apparatus participating in

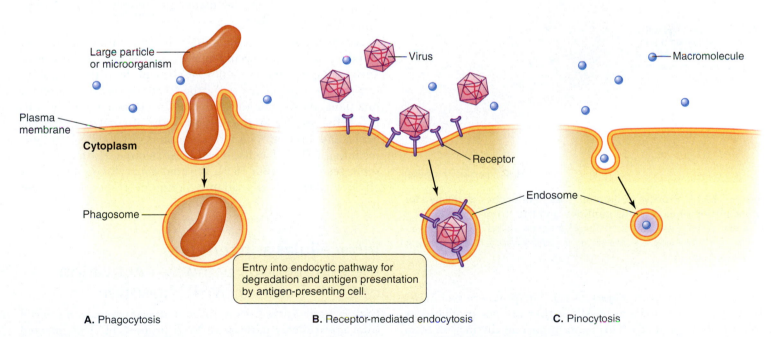

A. Phagocytosis   **B.** Receptor-mediated endocytosis   **C.** Pinocytosis

Figure 20.11. Endocytosis
A. Phagocytosis is used to engulf large particles such as bacterial cells. The plasma membrane surrounds the particle to produce an endocytic vesicle called a "phagosome."
B. Receptor-mediated endocytosis occurs when smaller particles such as bacterial components or virus particles bind to specific cell surface receptors. This initiates invagination in the plasma membrane leading to formation of an endosome.
C. Pinocytosis uses tiny invaginations of the plasma membrane to bring small amounts of extracellular fluid-containing nutrients and other small macromolecules into the cell. Small endosome vesicles are formed.

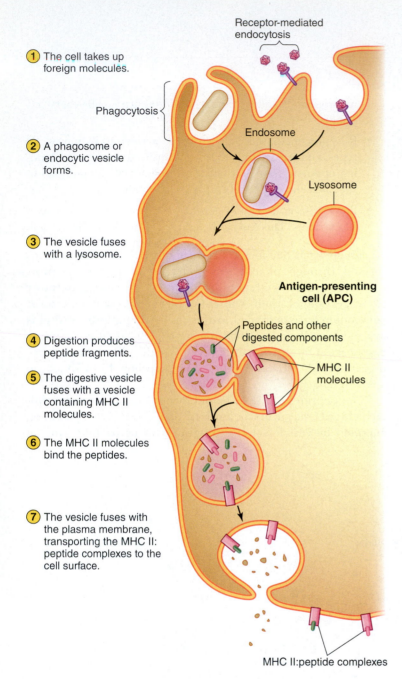

① The cell takes up foreign molecules.

Receptor-mediated endocytosis

Phagocytosis {

Endosome

② A phagosome or endocytic vesicle forms.

Lysosome

③ The vesicle fuses with a lysosome.

Antigen-presenting cell (APC)

④ Digestion produces peptide fragments.

Peptides and other digested components

⑤ The digestive vesicle fuses with a vesicle containing MHC II molecules.

MHC II molecules

⑥ The MHC II molecules bind the peptides.

⑦ The vesicle fuses with the plasma membrane, transporting the MHC II: peptide complexes to the cell surface.

MHC II:peptide complexes

Figure 20.12. Process Diagram: The endocytic pathway for presentation of exogenous peptides on MHC II Exogenous antigens such as extracellular microbes are taken up by the cell. The vesicle formed enters a series of processes resulting in digestion of the microbial components and presentation of the peptide fragments on MHC II molecules at the surface of the APC.

the endocytic pathway **(Figure 20.12)**. The production of MHC II molecules is restricted to APCs; they are not produced by other cells of the body. This is an important distinction from MHC I molecules, which you will recall are produced by most cells of the body.

Peptide-loaded MHC II molecules are brought to the surface of the APC to await binding with a compatible T-cell receptor (TCR) of a CD4$^+$ T cell. Only CD4$^+$ T cells can bind peptides associated with MHC II. Once binding between the TCR and the presented peptide has taken place, the CD4 co-receptor

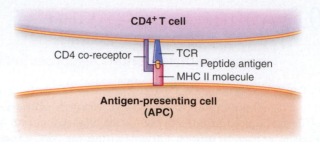

CD4$^+$ T cell

CD4 co-receptor —— TCR

Peptide antigen

MHC II molecule

Antigen-presenting cell (APC)

Figure 20.13. CD4$^+$ T-cell interaction with an antigen-presenting cell Activation of a CD4$^+$ T cell requires specific binding of MHC II-presented peptide with the TCR. Simultaneous binding of the T-cell CD4 co-receptor with the MHC II molecule of the APC must also occur. The requirement of CD4-MHC II binding of CD4$^+$ T cells restricts these T cells to interaction with peptide presented by the endocytic pathway.

binds to the MHC II molecule of the APC **(Figure 20.13)**. Recognition of the MHC II molecule by CD4 is a pivotal step in activating a bound CD4$^+$ T cell. Without this binding, activation of the CD4$^+$ T cell will not proceed even though the TCR has bound the peptide antigen presented by the MHC II molecule. CD4$^+$ T cells are therefore MHC II-restricted.

Production of IL-2 and IL-2R by activated CD4$^+$ T cells causes autostimulatory clonal expansion and differentiation into a population of effector CD4$^+$ T cells and memory CD4$^+$ T cells **(Figure 20.14)**. These cells carry all necessary co-stimulatory surface molecules needed for further action when they encounter the same antigen again on other APCs. We

@ Cell-mediated immunity: Helper T cells ANIMATION

have already seen that memory cells will initiate a memory response when this occurs upon a subsequent exposure to antigen. Upon contact with MHC II:peptide at the site of infection, the armed CD4$^+$ effector T cells deploy their effector functions. As we saw in Section 20.2, effector CD4$^+$ T cells (T$_H$ cells) secrete large concentrations of cytokines that enhance and direct the immune functions of various other cells including CD8$^+$ T cells, B cells (see Section 20.6), and macrophages (see Section 19.5). Without this cytokine help from effector CD4$^+$ T cells, adaptive immune responses will not progress.

Intracellular antigens and the endogenous pathway—Activation of CD8$^+$ T cells by MHC I:peptide

Endogenous antigens are those that originate *within* the cell from intracellular pathogens. Viral proteins of virus-infected cells are commonly processed by the endogenous pathway as these proteins are produced by the cell's own translational machinery in the cytoplasm. When other intracellular microorganisms infect a cell, their molecular components enter the cytoplasm as they replicate or die.

Endogenous proteins are not processed inside endosomes, but in the cytoplasm by a specialized complex of proteolytic

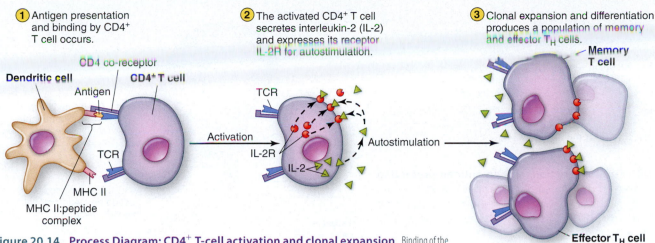

① Antigen presentation and binding by CD4⁺ T cell occurs.

② The activated CD4⁺ T cell secretes interleukin-2 (IL-2) and expresses its receptor IL-2R for autostimulation.

③ Clonal expansion and differentiation produces a population of memory and effector TH cells.

CD4 co-receptor
Dendritic cell
Antigen
CD4⁺ T cell
TCR
Memory T cell
Activation
IL-2R
Autostimulation
IL-2
MHC II
MHC II:peptide complex
Effector TH cell

Figure 20.14. Process Diagram: CD4⁺ T-cell activation and clonal expansion Binding of the antigen-specific TCR to peptide presented by MHC II molecules on a macrophage or dendritic cell begins a signaling process that results in expression of the T-cell growth factor IL-2 and its surface receptor IL-2R by the CD4⁺ T cell. Binding of IL-2 on the CD4⁺ T cell begins autostimulation, resulting in rounds of cell division (called clonal expansion), and the production of a population of antigen-specific effector TH cells and memory cells.

enzymes, the **proteasome (Figure 20.15)**. Proteasomes degrade cellular proteins as well, such as those that are misfolded or no longer needed. The cell tags these proteins with another small protein called "ubiquitin," which targets them for proteasome degradation. This is crucial for cellular housekeeping in the same way that we must dispose of leftover consumable items in our own homes. Proteasomes degrade proteins into smaller peptides, six to nine amino acids in length. The peptides are then transported from the cytoplasm to the endoplasmic reticulum where they associate with MHC I molecules produced by the cell (see Section 19.5). From here, the MHC I:peptide complex is presented on the cell surface (**Figure 20.16**).

Figure 20.15. Degradation of proteins by the proteasome
Cytoplasmic protein is designated for degradation by association with ubiquitins, small proteins that attach to targeted proteins. Ubiquitin-associated protein is transported to the unfolding subunit of the proteasome. Here, the protein is linearized for insertion into the central channel of the proteolytic subunit. Several specific proteases contained within this subunit break various peptide bonds of the chain, producing peptides six to nine amino acids in length. The peptides will bind to MHC I molecules and be transported to the cell surface.

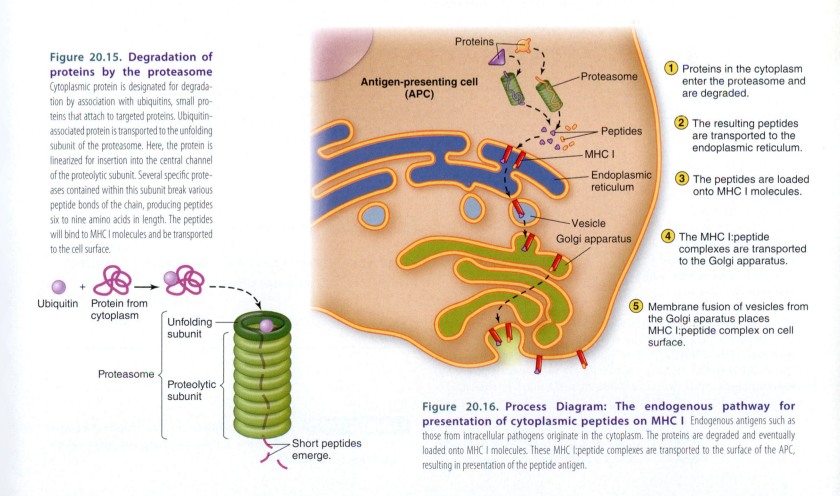

Ubiquitin + Protein from cytoplasm

Unfolding subunit
Proteasome
Proteolytic subunit
Short peptides emerge.

Proteins
Antigen-presenting cell (APC)
Proteasome
Peptides
MHC I
Endoplasmic reticulum
Vesicle
Golgi apparatus

① Proteins in the cytoplasm enter the proteasome and are degraded.

② The resulting peptides are transported to the endoplasmic reticulum.

③ The peptides are loaded onto MHC I molecules.

④ The MHC I:peptide complexes are transported to the Golgi apparatus.

⑤ Membrane fusion of vesicles from the Golgi aparatus places MHC I:peptide complex on cell surface.

Figure 20.16. Process Diagram: The endogenous pathway for presentation of cytoplasmic peptides on MHC I Endogenous antigens such as those from intracellular pathogens originate in the cytoplasm. The proteins are degraded and eventually loaded onto MHC I molecules. These MHC I:peptide complexes are transported to the surface of the APC, resulting in presentation of the peptide antigen.

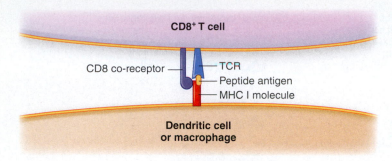

Figure 20.17. CD8⁺ T-cell interaction with an antigen-presenting cell Activation requires specific binding of MHC I-presented peptide with the TCR. Simultaneous binding of the T-cell CD8 co-receptor with the MHC I molecule of a dendritic cell or macrophage must also occur. The requirement of CD8-MHC I binding of CD8⁺ T cells restricts these T cells to interaction with antigen largely presented by the endogenous pathway.

Recall that nearly all somatic cells, including cells with immune functions, produce MHC I molecules and can present endogenous antigen on their surface. However, only dendritic cells and macrophages (both are APCs), are proficient at producing enough stimulatory molecules, including MHC I:peptide complexes, to activate naïve CD8⁺ T cells. Because CD8⁺ T cells only recognize peptide antigens presented in association with MHC I, they are "MHC I-restricted." This restriction is due to the direct binding of the CD8 co-receptor molecule with the MHC I molecule of the MHC I:peptide complex (**Figure 20.17**).

Successful interaction between the TCR of the CD8⁺ T cell and MHC I:peptide on a dendritic cell or macrophage is the first step in initiation of clonal expansion of the CD8⁺ T cell into a population of memory CD8⁺ T cells and effector CD8⁺ T cells. Recall that effector CD8⁺ T cells have cytotoxic activity (T_C cells). They migrate to sites of infection and can directly kill any infected cell they find that presents the *same* peptide in association with MHC I on its surface. Presentation of MHC I:peptide by infected cells therefore targets them for destruction, like waving a red flag in front of a bull. Cells presenting MHC I:peptide are conventionally referred to as "target cells." Killing of a target cell is accomplished by binding of the TCR of the T_C to the target cell's MHC I:peptide complex (**Figure 20.18**). This is followed by the release of toxic molecules contained in the granules of the T_C, including perforin and granzymes, which results in apoptosis of the target cell (see Section 19.5).

Killing by T_C cells is a very important response to infections involving intracellular pathogens such as viruses. Destroying this infected cell restricts the replication of the pathogen as it no longer has a host cell. Any viable pathogens released from the dead cell can then be taken up by local neutrophils and macrophages for destruction. **Table 20.1** summarizes the characteristics of effector T-cell types.

@ Cell-mediated immunity: Cytotoxic T cells
ANIMATION

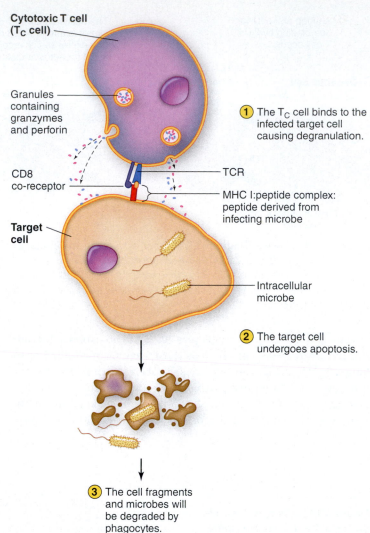

Figure 20.18. Process Diagram: Killing of an infected cell by a cytotoxic T cell A cytotoxic T cell (T_C cell) binds to its target cell in an antigen-specific and MHC I-restricted manner. This binding signals the release of perforin and granzymes at its bound surface, resulting in apoptosis of the target cell.

TABLE 20.1 **Characteristics of effector T cells**

T-cell type	Effector function	Co-receptor	MHC class restriction
T helper cell (T_H)	Cytokine secretion	CD4⁺	MHC II
Cytotoxic T cell (T_C)	Cytotoxic killing	CD8⁺	MHC I

CONNECTIONS Perforin is thought to produce pores in the target cell membrane that allow the uptake of granzymes that induce apoptosis. See **Section 19.5**.

The clonal expansion of naïve CD8$^+$ T cells, like that of naïve CD4$^+$ T cells, requires the cytokine IL-2. A major source of this cytokine is from activated CD4$^+$ T$_H$ cells. The requirement for IL-2 for T$_C$-cell production has been demonstrated using mice engineered to lack the IL-2 gene (IL-2 knockout mice). These mice are unable to efficiently produce T$_C$ cells. While there appears to be more than one way to activate CD8$^+$ T cells, the general scheme of activation, shown in **Figure 20.19**, is as follows:

- Binding of a CD8$^+$ T cell TCR with MHC I:peptide presented by a dendritic cell or a macrophage, leads to expression of the receptor IL-2R on the CD8$^+$ T-cell surface.

- In the presence of a sufficient number of IL-2-secreting T$_H$ cells, IL-2 is bound by the CD8$^+$ T cell.

- This induces clonal expansion of the activated CD8$^+$ T cell into a population of effector T$_C$ cells and memory CD8$^+$ T cells, all carrying the identical antigen-specific TCR.

- When T$_C$ cells subsequently contact other infected cells displaying the same MHC I:peptide complex, they can bind by their TCR in an MHC I-restricted manner.

- This binding signals the T$_C$ cell to release its toxic components, resulting in apoptosis of the target cell.

As dendritic cells and macrophages can express MHC I:peptide on their surface, logically then, they can be targets for T$_C$ cells as well as activators for naïve CD8$^+$ T cells. This may serve a self-policing function. As the numbers of T$_C$ cells accumulate, clearance of infected cells takes place, so the need for further activation of CD8$^+$ T cells by an APC is no longer required. In fact, excessive activation of these cytotoxic cells can be detrimental to the host (see Section 20.5). As the number of infected cells decreases, the APCs become targets for

destruction, further cytotoxic stimulation does not occur, and the response diminishes.

Antigen cross-presentation—Activation of both CD4$^+$ and CD8$^+$ T cells

In some cases, exogenous antigens have been shown to stimulate CD8$^+$ T cells. At first, this would seem to violate everything we have just covered. So far we have learned that macrophages and dendritic cells endocytose foreign substances that they find in their environment. They next degrade the proteins of these larger substances into small peptides. The peptides are loaded onto MHC II molecules and transported to the surface of the cell (see Figure 20.12). From here, they await recognition by a TCR on a CD4$^+$ T cell (recall that CD4$^+$ T cells are MHC II-restricted). If this happens, the CD4$^+$ T cell becomes activated. However, activation of CD8$^+$ T cells requires antigen to be presented by MHC I molecules. This happens when foreign protein from within the cell cytoplasm is degraded and presented through the endogenous pathway (see Figure 20.16). It would follow then that activation of CD8$^+$ T cells only occurs when a macrophage or a dendritic cell is infected by an intracellular pathogen. In many cases, however, pathogens infect cells other than APCs such as epithelial cells or muscle cells. How then is an adaptive immune response mounted to such an intracellular pathogen?

Part of the answer lies in the innate ability of macrophages and dendritic cells to endocytose dead or dying infected host cells or their parts. One might expect these proteins to be processed through the exogenous pathway and presented as MHC II:peptide. As these infected cells contain parts of the pathogen, or even the pathogen itself, proteins from the pathogen will also get processed and presented by MHC II. This can activate CD4$^+$ T cells. So far in this scenario, CD8$^+$ T cells have not been activated and no T$_C$ cells have been generated. The intracellular pathogen is still on the loose.

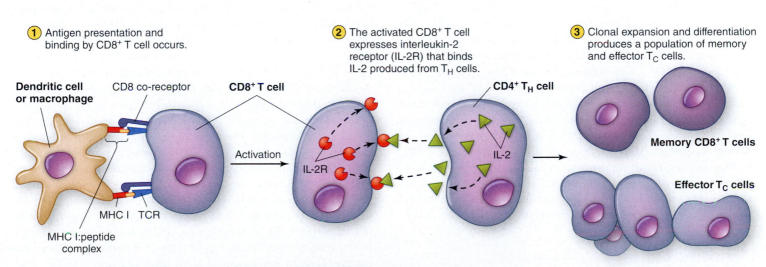

① Antigen presentation and binding by CD8$^+$ T cell occurs.

② The activated CD8$^+$ T cell expresses interleukin-2 receptor (IL-2R) that binds IL-2 produced from T$_H$ cells.

③ Clonal expansion and differentiation produces a population of memory and effector T$_C$ cells.

Dendritic cell or macrophage — **CD8 co-receptor** — **CD8$^+$ T cell**

CD4$^+$ T$_H$ cell

Activation

IL-2R

IL-2

Memory CD8$^+$ T cells

Effector T$_C$ cells

MHC I   TCR

MHC I:peptide complex

Figure 20.19. Process Diagram: CD8$^+$ T-cell activation and clonal expansion Binding of the antigen-specific TCR to peptide presented by MHC I molecules on a macrophage or dendritic cell begins activation resulting in expression of IL-2R by the CD8$^+$ T cell. Assistance is generally required from activated T$_H$ cells, which provide the cell growth factor IL-2. The result is a population of effector T$_C$ cells with cytotoxic capabilities, and a population of memory CD8$^+$ T cells.

As we know, activation of CD8⁺ T cells requires antigen to be presented by MHC I. There must then be a mechanism to load exogenously acquired antigen onto MHC I molecules in APCs. This is accomplished through a process known as antigen **cross-presentation** that shuttles proteins from the endocytic pathway into the endogenous pathway. It is not yet known exactly how the shunt functions, and there appears to be more than one pathway involved. The importance of cross-presentation is that it allows both CD4⁺ and CD8⁺ T-cell activation for an effective adaptive response to intracellular pathogens. Now a population of antigen-specific T_C cells can be generated to seek out any host cell that is infected. Antigen cross-presentation is especially important for generating a protective immune response to viruses that do not infect APCs.

20.3 Fact Check

1. What are exogenous antigens and how are they processed?
2. Explain how an APC interacts with a CD4⁺ T cell. What is the result on the T cell?
3. Describe the effects of IL-2 production from CD4⁺ T cells.
4. What are endogenous antigens and how are they processed?
5. Explain how an APC interacts with a CD8⁺ T cell. What is the result on the T cell?
6. Explain how a CD8⁺ T cell recognizes and kills infected target cells.
7. What is the purpose of antigen cross-presentation?

20.4 Antigen-presenting cells

What cells can launch adaptive immune responses?

As we have seen, most cells in the body are capable of presenting endogenous antigens on MHC I molecules. However, only APCs can launch adaptive responses. Recall that APCs are dendritic cells, macrophages, and B cells. Only these cells express MHC II molecules and large numbers of MHC I molecules and the required accessory binding molecules to interact with T cells to begin adaptive responses. As shown in **Table 20.2**, these different APC types have different capabilities for T-cell activation.

Dendritic cells can initiate primary immune responses by activating naïve T cells, and memory responses by activating memory T cells. Dendritic cell interaction with circulating effector CD4⁺ T cells induces these cells to secrete cytokines (T_H functions) for further immune functions. Interaction of any cell presenting MHC I:peptide with circulating effector CD8⁺ T cells will cause the CD8⁺ T to deploy cytotoxic killing (T_C functions) of the cell. Macrophages and B cells are generally not adept at activating naïve T cells and consequently are not proficient at initiating primary immune responses. Macrophages are, however, potent activators of memory CD4⁺ T cells, and thus can initiate memory responses. Macrophages can also induce memory CD8⁺ T-cell responses, resulting in clonal expansion into effector CD8⁺ cells for cytotoxic killing of target cells. B cells, as we will see in this section, are limited to interactions with effector CD4⁺ T cells. We will now examine the different APCs in more detail.

Dendritic cells

Dendritic cells are a collection of multifunctional cells whose predominant role is to present peptides to the adaptive immune system for activation. Their activities are strictly innate. They take in substances from their environment in a non-specific manner and present them to T cells in the form of peptide fragments associated with MHC II molecules on their cell surface. Dendritic cells can bind ligands for phagocytosis through a variety of pattern recognition receptors (PRRs) that are found on their surface. They can also take in extracellular macromolecules from their environment by engulfing extracellular fluid in a process called "pinocytosis" (see Figure 20.11). They are not armed to be efficient killers of microbes compared to the "professional" phagocytes: macrophages and neutrophils. They lack the ability to generate the respiratory (oxidative) burst, but they do possess lysosomes with a supply of hydrolytic enzymes that degrade proteins and present the resulting peptides. Like other cells of the body, if a dendritic cell becomes infected, it can also present peptides on MHC I molecules.

Dendritic cells derive their name from the many long cytoplasmic extensions possessed by mature cells that resemble the dendrites of nerve cells (**Figure 20.20**). The majority of dendritic cells are derived from monocytes (see Appendix G). Dendritic cells are produced in the bone marrow and mature in diverse locations. Eventually, the cells move into various tissues and take up residency. Here, they further differentiate into a number of distinct subsets, characterized by variation in cell surface markers and cytokine secretion. The subsets are given names such as Langerhans cells of the skin, interstitial dendritic cells of mucosal membranes and internal organs, and interdigitating dendritic cells of the thymus and spleen. These strategic placements provide ample opportunity for dendritic cells to encounter foreign substances as they enter the body.

Dendritic cells are particularly competent in their abilities to express MHC molecules, present antigen, and produce various co-stimulatory molecules needed to stimulate T cells. Studies have conclusively shown that dendritic cells are the most efficient at activating naïve CD4⁺ T cells. Recall that naïve cells are

TABLE 20.2 **Activities of antigen-presenting cells**

Antigen-presenting cell	Activated cell types	MHC restriction	Outcome
Dendritic cell	Naïve CD4$^+$ T cells	MHC II	Primary response; generation of memory and effector CD4$^+$ T cells
	Naïve CD8$^+$ T cells	MHC I	Primary response; generation of memory and effector CD8$^+$ T cells
	Memory CD4$^+$ T cells	MHC II	Memory response; generation of effector CD4$^+$ T cells
	Memory CD8$^+$ T cells	MHC I	Memory response; generation of effector CD8$^+$ T cells
	Effector CD4$^+$ T cells	MHC II	T$_H$ cytokine secretion
	Effector CD8$^+$ T cells	MHC I	T$_C$ cytotoxic killing
Macrophage	Naïve CD8$^+$ T cells[a]	MHC I	Primary response; generation of memory and effector CD8$^+$ T cells
	Effector CD4$^+$ T cells	MHC II	T$_H$ cytokine secretion
	Effector CD8$^+$ T cells	MHC I	T$_C$ cytotoxic killing
	Memory CD4$^+$ T cells	MHC II	Memory response; generation of effector CD4$^+$ T cells
	Memory CD8$^+$ T cells	MHC I	Memory response; generation of effector CD8$^+$ T cells
B cell	Effector CD4$^+$ T cells	MHC II	T$_H$2 cytokine secretion and B-cell proliferation

[a]Only in certain instances.

resting cells that have not yet come into contact with antigen. The ability of dendritic cells to initiate this appears to be due to the stringent requirements of naïve T cells for several accessory molecules found only on the surface of dendritic cells. We earlier discussed the need for controlling initiation of adaptive immune responses. If the right signals are not produced, initiation of T-cell responses will not be initiated. Macrophages and B cells have been demonstrated *in vitro* to activate naïve CD4$^+$ T cells, but

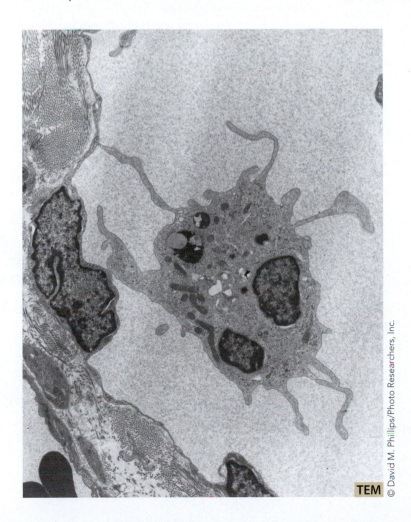

© David M. Phillips/Photo Researchers, Inc.

TEM

Figure 20.20. Dendritic cell Long cytoplasmic extensions are characteristic of dendritic cells and are used for capturing foreign substances.

only when present in very large numbers. In contrast, activation can be accomplished with very few dendritic cells.

The general process of dendritic cell activation and presentation to CD4$^+$ T cells is shown in **Figure 20.21**. In their place of residence, dendritic cells are thought to be at rest. Once a dendritic cell encounters and ingests a particle, it becomes activated through Toll-like receptor (TLR) signaling (see Section 19.4). Dendritic cells typically express a large number of many different TLRs. Activation results in three effects:

- a dramatic increase in the ability to present foreign peptides
- increased expression of a wide variety of cytokines and stimulatory cell surface molecules such as MHC II, for interaction with T cells
- migration to the regional lymph nodes where T cells are concentrated

In the lymph nodes, activated dendritic cells present antigen to T cells and initiate adaptive immune responses.

Macrophages

As we have already learned, macrophages are cells important in both innate and adaptive immunity. Their innate functions include phagocytosis and intracellular killing (see Section 19.5), but as professional APCs, they also have the ability to initiate memory immune responses through presentation of processed peptides to memory CD4$^+$ T and CD8$^+$ T cells. Along with dendritic cells, macrophages also play a key role in determining the course of CD4$^+$ T-cell differentiation into subsets that support either cell-mediated immune functions or B-cell humoral functions. This topic is discussed in Section 20.6.

B cells

B cells are lymphocytes that derive their name from early studies in birds. Removal of a small organ unique to birds called the "bursa of Fabricus," or "bursa," resulted in an inability to produce antibody. The antibody-producing cells were named B cells for bursa-derived cells. The B cells of mammals are produced and mature in the bone marrow (see Appendix G). Therefore, bone marrow is a primary lymph organ for B cells (see Section 20.1). In a manner similar to that of T cells, B cells migrate from here to circulate in secondary lymphoid tissues such as the spleen and lymph nodes where they await encounters with antigen. Because B cells commonly reside in secondary lymphoid tissues, they frequently bind unprocessed antigens carried to these tissues on the surface of dendritic cells. They can also bind free antigen, an especially important function of B cells in the spleen, which screens large volumes of blood for antigen.

B cells have two very important roles in adaptive immunity:

- they produce antibody
- they present MHC II:peptide to CD4$^+$ T cells

In this section, we will limit our examination of B cells to their antigen-presenting functions. Antibody production will be examined in Section 20.6. B cells express unique B-cell receptors (BCRs) on their surface, much as T cells express TCRs. BCRs and antibody belong to a large family of proteins known as immunoglobulins (Ig), all of which can bind antigen in a specific manner. Antibody is a secreted version of the BCR. Like TCRs, the BCRs found on mature B cells are randomly generated. B cells have two mechanisms of generating diversity in their immunoglobulin chains:

1. random rearrangements of the genes that encode the immunoglobulin polypeptide chains

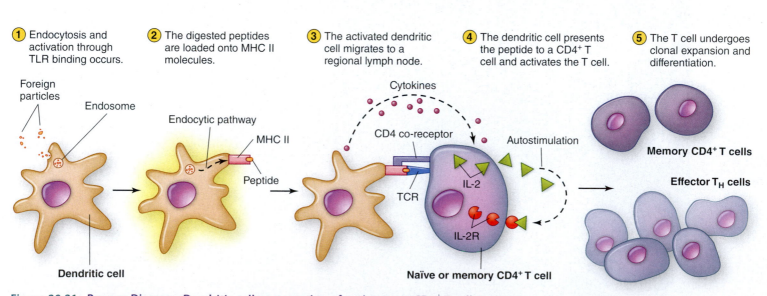

① Endocytosis and activation through TLR binding occurs.

② The digested peptides are loaded onto MHC II molecules.

③ The activated dendritic cell migrates to a regional lymph node.

④ The dendritic cell presents the peptide to a CD4$^+$ T cell and activates the T cell.

⑤ The T cell undergoes clonal expansion and differentiation.

Figure 20.21. Process Diagram: Dendritic cell presentation of antigen to a CD4$^+$ T cell Dendritic cell activation by exogenous antigen begins with binding and endocytosis of the particle. Toll-like receptors (TLRs) are activated, causing increased MHC II and cytokine production. Digestion occurs through the endocytic pathway and peptide is loaded onto MHC II. The dendritic cell begins its migration from tissue to a regional lymph node where it will contact T cells. Successful binding of presented peptide with a CD4$^+$ T cell results in T-cell activation, clonal expansion, and differentiation into CD4$^+$ memory cells and effector T$_H$ cells.

2. point mutations and deletions that change the sequence of the immunoglobulin antigen-binding domain

Thus, a wide variety of BCRs is constantly generated in the bone marrow, and mature B cells are antigenically committed. The large variety of receptors at hand helps ensure that when a foreign substance is present, like a microbe or toxin, some of the B-cell receptors will be able to bind it.

BCRs can bind antigens such as microbial toxins or cell wall components directly from the surrounding environment. In this, B cells are unique. They do not require antigens to be presented in an MHC format as T cells do. The general process of B-cell antigen presentation is shown in **Figure 20.22**. Binding of free antigen to a BCR begins the process of endocytosis. More specifically, this process is receptor-mediated endocytosis as the BCR is a specific receptor for the antigen (see Figure 20.11). Once internalized, the antigen is broken down in the endocytic pathway and the peptides are presented by the B cell on MHC II molecules, just as we have seen with other APCs (see Figure 20.12). Unlike other APCs, B cells do not activate naïve T cells. Instead, they activate T$_H$ cells which have been produced through activation of a naïve or memory CD4$^+$ T cell by a dendritc cell or macrophage. When a B cell presenting MHC II:peptide binds a T$_H$ cell, it stimulates the T cell to become a cytokine-secreting effector T$_H$ cell. The cytokines secreted by the T$_H$ cell (see Section 20.5) in turn stimulate the B cell to undergo clonal expansion and differentiation into antibody-secreting **plasma cells** and memory B cells. This proliferation and differentiation is absolutely dependent on interaction with T$_H$ cells as it requires cytokines supplied by the T$_H$ cell. In this way, the T cell "helps" the B cell.

Characteristics of antigens presented by antigen-presenting cells

Generally speaking, to evoke a primary adaptive immune response, it is crucial that antigen be taken up, processed, and presented by dendritic cells. Certain characteristics of microbial components can dictate the efficiency with which these activities proceed. Microbial components that are very small and soluble such as small peptides, are not typically good substrates for cell uptake and thus are poor stimulators of adaptive immune responses. Components that are resistant to the degradative effects of the endocytic pathway or the proteasome are also poor stimulators as they cannot be broken down into the required peptide lengths for loading onto MHC molecules. These factors need to be considered in designing a vaccine.

Many vaccines are composed of specific molecules purified from the pathogens for which they are designed. However, the immunogenicity of these molecules is often reduced when they are isolated away from the more complex structure of the pathogen. Frequently, administration of these purified components does not produce memory cells in sufficient numbers to provide the long-term immunological memory that is required of effective vaccines. Small molecules such as peptides, can be made more immunogenic—a term we have used earlier to describe the relative ability to stimulate an immune response—by mixing them with adjuvants. Recall from Section 19.4, adjuvants are commonly added to vaccine formulations to enhance the immune response to an antigen. Adjuvants are composed of inorganic salts or organic compounds such as oils or microbial components. When adjuvants are mixed with the target antigen, a significant number of long-lived memory cells are generated. This means we

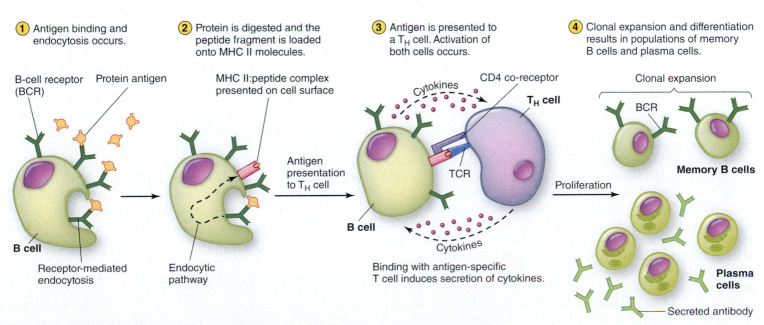

① Antigen binding and endocytosis occurs.

② Protein is digested and the peptide fragment is loaded onto MHC II molecules.

③ Antigen is presented to a T$_H$ cell. Activation of both cells occurs.

④ Clonal expansion and differentiation results in populations of memory B cells and plasma cells.

B-cell receptor (BCR)   Protein antigen

MHC II:peptide complex presented on cell surface

Cytokines   CD4 co-receptor   T$_H$ cell

Clonal expansion

BCR

B cell

Antigen presentation to T$_H$ cell

TCR

B cell

Memory B cells

Receptor-mediated endocytosis

Endocytic pathway

Binding with antigen-specific T cell induces secretion of cytokines.

Cytokines

Proliferation

Plasma cells

Secreted antibody

Figure 20.22. Process Diagram: B-cell antigen presentation to a T$_H$ cell Binding of the B-cell receptor (BCR) with antigen begins the process of receptor-mediated endocytosis. Digestion of the antigen occurs through the endocytic pathway, and the resulting peptides are loaded onto MHC II molecules and presented to T$_H$ cells. Binding of the B-cell MHC II:peptide with a T$_H$ cell induces secretion of cytokines causing the B cell to undergo clonal expansion and differentiation into antibody-secreting plasma cells and memory B cells.

require fewer injections of vaccine "boosters" to provide us with long-term immunity. Adjuvants act in a number of ways:

- Precipitation of a substance to make it a physically larger particle appears to facilitate uptake by dendritic cells and macrophages. An example of this type of adjuvant is aluminum potassium sulfate, commonly referred to as alum. Alum is the most common adjuvant universally licensed for use in humans, but some organic compounds such as squalene are used in some countries.

- Slow release of antigen appears to give the immune system more opportunity to interact with it. Longer antigen exposure is generally required for production of immunological memory. Mixing antigen with mineral oil (incomplete Freund's adjuvant), commonly used in animals, results in release of antigen over time and produces the immunological memory required for an effective vaccine.

- Binding Toll-like receptors (TLRs; Section 19.4) of dendritic cells and macrophages promotes uptake of antigen and provides for more efficient MHC presentation. TLR stimulation also enhances the production of proinflammatory cytokines to stimulate inflammation.

In addition to mineral oil, Freund's complete adjuvant contains mycobacterial cell wall components that can directly activate TLR2 (see Section 20.5). This potent adjuvant is not commonly used as its potency often results in severe and chronic inflammation at the injection site. DNA vaccines owe their adjuvant effects to the DNA molecule itself through TLR9 signaling, (Mini-Paper, Chapter 19). The DNA is of bacterial origin and is recognized as foreign by the innate immune system. In most cases, the particular pathogen gene that has been inserted into the DNA vaccine for expression in host cells simply provides an antigen to direct the specificity of the adaptive immune response.

CONNECTIONS Vaccines are preparations of microbes or their components used to immunize against microbial diseases by conferring immunological memory through the production of memory cells. For more details on vaccines and formulations, see **Section 24.5**.

> ## 20.4 Fact Check
>
> 1. What are the key features of dendritic cells, and how do they function?
> 2. Describe the roles of B cells in adaptive immunity.
> 3. How do BCRs play a role in antigen presentation?
> 4. What are adjuvants, and what are they capable of doing in the body?

20.5 Humoral and cell-mediated immune responses

Are there different responses to extracellular and intracellular pathogens?

We have so far surmised that cytokine secretion by T_H cells is required to help in the activation of both B cells for antibody production *and* T_C cells for cytotoxic killing. Recall from Section 19.4 that opsonization of extracellular microbes by antibody activates the classical complement pathway and results in lysis of the pathogen. Opsonization by antibody also facilitates phagocytosis of microbes by neutrophils and antigen-presenting macrophages. There is often little use for antibody production in the elimination of intracellular pathogens such as viruses or *Mycobacterium tuberculosis*. Antibodies cannot enter host cells unless they attach to the pathogen while it is extracellular and are brought into the cell. Cytotoxic killing of infected host cells by T_C cells is often a requirement for clearance of intracellular bacterial and viral infections. As we have learned, destroying infected host cells restricts the replication of the pathogen, and any released pathogens from the dead host cells can then be taken up by local neutrophils and macrophages for destruction. Activating T_C cells would not help in clearing extracellular microorganisms such as *Streptococcus pyogenes*, as T_C cells can only recognize and kill infected host cells. It is therefore wasteful to have both these cell types activated at the same time.

Examinations of the relative levels of antibody and the cell populations that are produced in immune responses to specific pathogens are, in fact, known to be generally polarized to either a **humoral immune response** for enhanced antibody production or a **cell-mediated immune response** involving little antibody, killing of infected cells by T_C cells, and increased intracellular killing by macrophages. What decides which type of response is produced? That clearance of extracellular pathogens requires antibody production while clearance of intracellular pathogens requires a cell-mediated response may be obvious, but how does the immune system "decide" this?

T_H1 and T_H2 T-cell subsets

Studies in mice have shown activated T_H cells secrete distinct patterns of cytokines that are correlated with either a humoral or a cell-mediated immune response. In mice, the different profiles have been extensively investigated and have led to the functional classification (no differential cell markers have yet been identified) of T_H cells as being of either subset **T_H1 cells**, correlating with a cell-mediated response, or subset **T_H2 cells**, correlating with a humoral response. The pattern of cytokine secretion from mouse T_H1 cells is dominated by the proinflammatory cytokines IL-2, tumor necrosis factor-alpha (TNF-α), and interferon-gamma (IFN-γ). That of T_H2 cells is predominantly IL-4, IL-5, IL-6, IL-10, and IL-13. The pattern of cytokine production differs in other species such as cattle and humans **(Table 20.3)**. However, most experts agree these species do show *overall* differential cytokine patterns related to the type of immune response. The designation of T_H1 and T_H2 has therefore become the paradigm for describing activated T_H-cell functions. The influences of some of the T_H1 and T_H2 cytokines on the immune response are summarized in **Figure 20.23**.

TABLE 20.3 Cytokines produced by human T_H1 and T_H2 subsets

Cytokine	T_H1	T_H2
IL-2	+	−
IFN-γ	+	−
TNF-α	+	−
IL-3	+	+
IL-4	−	+
IL-5	−	+
IL-6 [a]	+	++
IL-10 [a]	+	++
IL-13 [a]	+	++

[a] T_H1 cells of mice do not produce IL-6, IL-10, or IL-13.

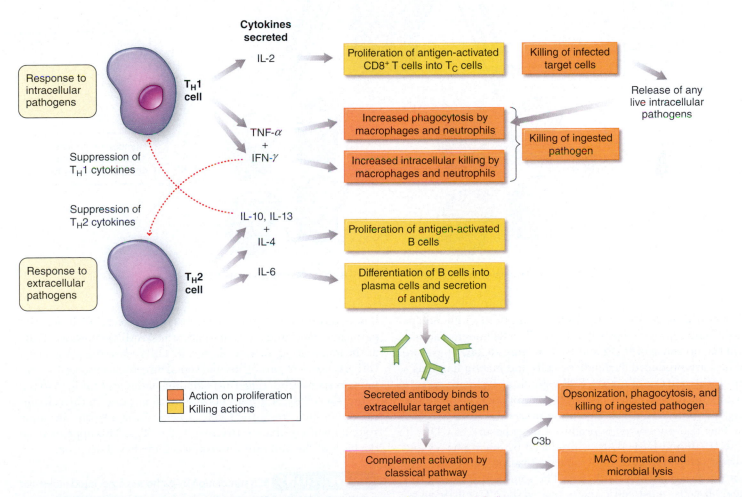

Figure 20.23. Effects of some T_H1 and T_H2 cytokines In humans, the production of IL-10 and IL-13 by T_H2 cells actively suppresses the production of T_H1 cytokines and promotes a humoral immune response to target extracellular pathogens. The production of IFN-γ from T_H1 cells suppresses production of T_H2 cytokines and promotes a cell-mediated immune response to target intracellular pathogens. Other cytokines induce expression of specific genes associated with various humoral or cell-mediated immune responses as depicted.

Figure 20.24. Development of a naïve CD4$^+$ T cell into T$_H$1 or T$_H$2 cells Activation of a naïve CD4$^+$ T cell begins with interaction with an antigen-presenting dendritic cell (*center*). **A.** Clonal expansion and differentiation of CD4$^+$ T cells into T$_H$1 cells is decided by the relative concentration of IL-12 and IL-18 produced by neighboring activated macrophages and dendritic cells. T$_H$1 cells secrete the cytokines IFN-γ, IL-2, and TNF-α involved in cell-mediated responses. **B.** Clonal expansion and differentiation into T$_H$2 cells occurs in the presence of IL-4, produced by activated mast cells and B cells, and IL-10, produced by nearby activated macrophages and dendritic cells. T$_H$2 cells secrete cytokines that serve to enhance humoral responses.

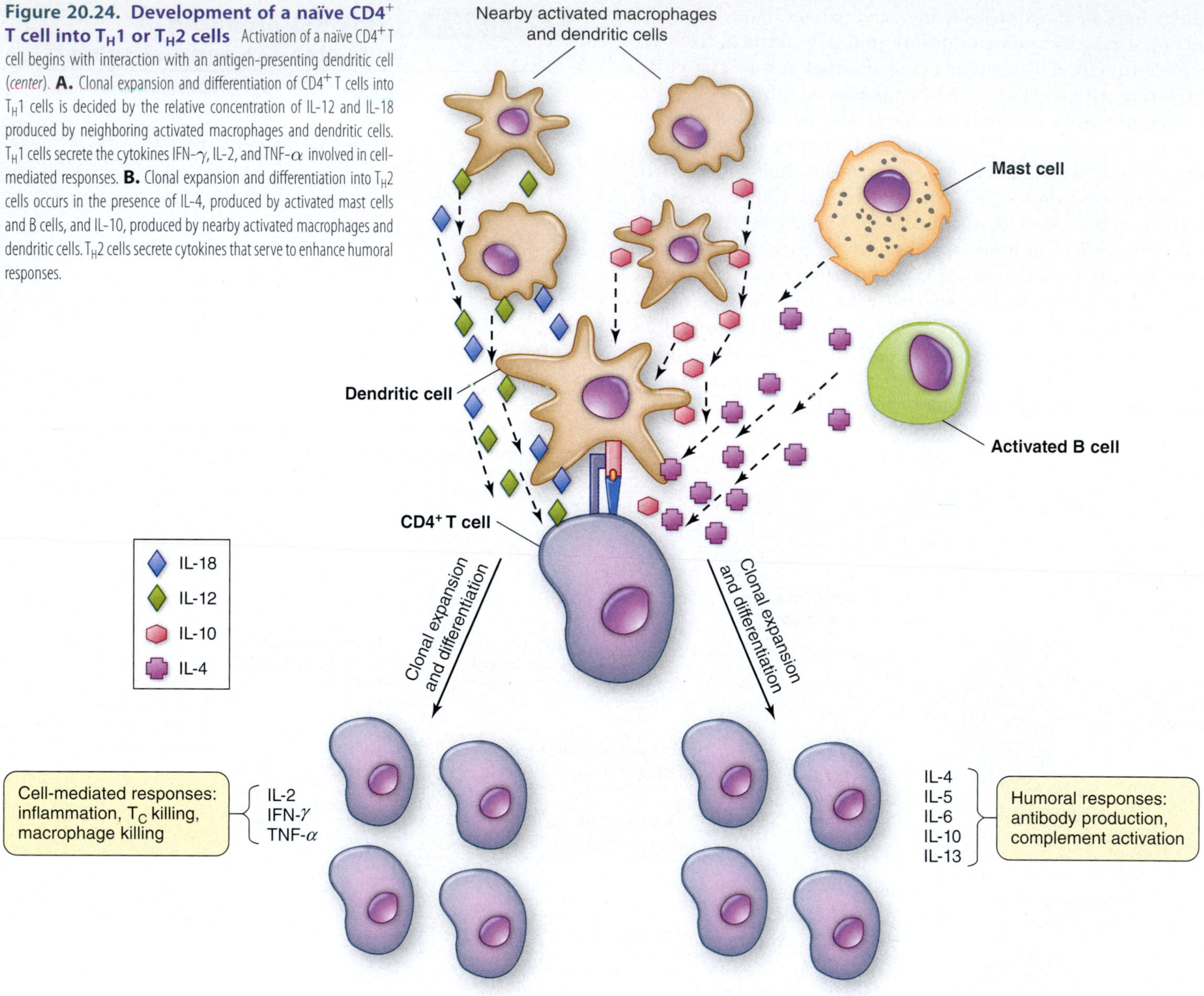

Nearby activated macrophages and dendritic cells

Mast cell

Dendritic cell

Activated B cell

CD4$^+$ T cell

IL-18

IL-12

IL-10

IL-4

Clonal expansion and differentiation

Clonal expansion and differentiation

Cell-mediated responses: inflammation, T$_C$ killing, macrophage killing

IL-2
IFN-γ
TNF-α

IL-4
IL-5
IL-6
IL-10
IL-13

Humoral responses: antibody production, complement activation

A. CD4$^+$ T-cell differentiation to T$_H$1 cells

B. CD4$^+$ T-cell differentiation to T$_H$2 cells

Generally, *in vitro* studies have shown that the T$_H$1 phenotype in most mammals develops when CD4$^+$ T cells undergo activation in the presence of IL-12 and IL-18 **(Figure 20.24)**. These two cytokines are produced by dendritic cells and macrophages as a result of the pattern of innate TLR signaling. T$_H$1 cells secrete the cytokine IFN-γ, which enhances macrophage intracellular killing functions and MHC II:peptide presentation to CD4$^+$ T cells. The T$_H$2 phenotype is produced by exposure of CD4$^+$ T cells to IL-10 and IL-4. IL-10 is produced by macrophages and dendritic cells that have undergone activation through a different pattern of TLR stimulation. The source of IL-4 remains unclear; B cells, mast cells, and activated T$_H$2 T cells have all been implicated. It is also unclear if memory CD4$^+$ T cells are irreversibly committed to either the T$_H$1 or T$_H$2 subset upon subsequent antigen challenge. In any case, the innate system has a strong influence on the direction of the adaptive response.

It is logical that cells of the innate system, which first contact foreign substances, should inform the adaptive system on how to destroy those substances. As we have seen in Chapter 19, pathogen-associated molecular patterns (PAMPs) stimulate TLRs at the innate level. As you will recall, TLRs sense the presence of various classes of pathogens and signal directions for adaptive immune responses by triggering signaling pathways that lead to the expression of select groups of cytokine genes that in turn chemically communicate to the CD4$^+$ T cells of the adaptive immune system to differentiate into either T$_H$1 or T$_H$2. This appears to be the basis of the humoral versus the cell-mediated decision.

CONNECTIONS TLR stimulation by pathogen-associated molecular patterns (PAMPs) is central in the development of immune responses. See **Section 19.4**.

Once a T$_H$1 or T$_H$2 profile has been established it appears to be self-perpetuating, ensuring other CD4$^+$ T cells undergoing activation are similarly polarized. This is accomplished by the mutually antagonistic effect of some of the cytokines that are involved in polarization. For example, *in vitro* studies have shown IL-4 and IL-10 from T_H2 cells inhibit the production of IL-12 from activated macrophages. In turn, the loss of IL-12 slows production of the cytokines TNF-α, IL-2, and IFN-γ in T_H1 cells. Conversely, IFN-γ from stimulated T_H1 cells inhibits the T_H2 cytokine IL-4 (see Figure 20.23). Thus, the perpetuation of the profile of the cytokine environment guarantees the specific immune response is maintained until the clearance of the pathogen. This "locking in" of a humoral or a cell-mediated immune response may sometimes have negative repercussions, as illustrated in **Perspective 20.1**. The **Mini-Paper** describes the potent ability of IL-4 to maintain suppression of the protective cell-mediated immune response in mice infected with mousepox virus.

Immune-mediated damage

While the immune response has protective functions against infectious agents, it can also cause damage. For example, *Neisseria meningitidis*, a cause of meningitis, produces severe inflammation of the meninges, the lining that surrounds the brain and spinal cord. The influx of fluid and cells creates excessive pressure that damages the nerves, leading to paralysis and often death. Similarly, the inflammatory damage that occurs with *Neisseria gonorrhoeae* infection can cause sterility (see Section 21.1). Some strains of *Staphylococcus aureus* and *Streptococcus pyogenes* produce toxins called superantigens, which activate large numbers of CD4$^+$ T cells by bypassing the need for specific peptide recognition. The proliferation of activated CD4$^+$ T cells results in excessive production of potent proinflammatory cytokines such as interleukin-1 (IL-1), interferon-gamma (IFN-γ), and tumor necrosis factor-alpha (TNF-α). This produces a systemic inflammatory response (see Section 21.1). Unfortunately, the robust inflammatory response to these infections affords little protection against these pathogens.

Another type of immune damage occurs as a result of the generation of cross-reactive antibodies. These are antibodies produced against a specific microbial component that can also bind to host components with similar epitopes. Infection with certain strains of *S. pyogenes* can lead to rheumatic heart disease after the infection is cleared due to the continued binding of the cross-reactive antibody to heart valves. This is

thought to lead to complement fixation and lysis of the tissue, and more inflammatory damage (see Section 21.2). Infection with *Mycobacterium tuberculosis* often causes a chronic cell-mediated response because of the continued release of TNF-α produced by infected macrophages that are unsuccessfully attempting to destroy the bacilli. The continued inflammatory recruitment of immune cells to the focus of infection results in a pathological lesion, which damages the lung tissue (see Section 21.2).

Viruses can also cause immune-mediated damage to host tissues. Hepatitis B virus itself does not appear to result in any significant cell damage, but liver disease results from the cytotoxic T-cell destruction of infected hepatocytes. Infection with this virus can also cause kidney damage produced by the accumulation of antigen:antibody complexes in the kidney, resulting in capillary blockages and complement activation (see Section 22.2).

In the future, the redirection of an inflammatory cell-mediated immune response to a humoral response may be therapeutic for treating septic shock and chronic inflammatory diseases such as Crohn's disease and ulcerative colitis of the intestine. Similarly, redirection from a T_H2 to a T_H1 response in antibody-mediated autoimmune diseases such as lupus and type 1 diabetes mellitus, or allergic diseases such as asthma has the potential to offer valuable treatments for individuals suffering from these ailments. Stimulation of specific sets of TLRs with their ligands has shown promising results in shifting T_H1/T_H2 immune responses.

20.5 Fact Check

1. Compare and contrast humoral responses and cell-mediated responses.
2. Distinguish between T_H1 and T_H2 cells.
3. Describe how the following would work in conjunction to respond to a pathogen: PAMPs, TLRs, cytokines, and CD4$^+$ T cells.

R. J. Jackson, A. J. Ramsay, C. D. Christensen, S. Beaton, D. F. Hall, and I. A. Ramshaw. 2001. Expression of mouse interleukin-4 by a recombinant ectromelia virus suppresses cytolytic lymphocyte responses and overcomes genetic resistance to mousepox. J Virol 75:1205–1210.

Context

As we have seen in the opening story of this chapter, mouse plagues in Australia severely impact agriculture, causing extensive losses to grain crops and grain stores. There has been much interest in developing methods to control these rodents, and one idea has been to use the mouses's own immune system to render them infertile. This process, called "immunocontraception," has been used to control populations of other animals in the wild such as deer. One of the most popular cellular targets used in immunocontraception is the zona pellucida protein, which forms a membrane surrounding the oocyte and is involved in binding spermatozoa. This protein has been developed as a drug that causes antibodies to be produced against the animal's own oocytes. Wild animals can be remotely injected using darts. The study that is reported in this paper describes experiments performed by Australian scientists that were designed to increase the immunogenicity of a mousepox virus that had been engineered to express the mouse zona pellucida protein. How was this study carried out?

Experiment

The mousepox virus was engineered with a copy of the mouse cytokine interleukin-4 (IL-4), which is known to boost the production of antibody, and this is the response that was desired by the researchers to maximize the contraceptive effect. Let's take a look at how this recombinant virus was engineered. The mousepox virus

has a large genome of 208 kb, and because it is a virus, the genome is packaged into virus particles in the cell and is not found as an extrachromosomal replicon in cells, like plasmids would be. Once extracted from the viral particles, the naked genome cannot be placed back into cells to begin viral replication. Instead, insertion of DNA sequences into the viral genome are first constructed by *in vitro* molecular methods and introduced into these viral genomes by homologous recombination events within the eukaryotic host cell, resulting in recombinant viruses.

Mammalian cells growing in culture are able to produce the essential nucleotide thymidine monophosphate using two alternative pathways: one involving thymidylate synthetase enzyme and the other involving thymidine kinase (TK) enzyme. TK has been developed as a selectable marker, and is commonly used in the construction of recombinant viruses. This gene is naturally present in poxviruses and herpesviruses, and although it is required for reproduction of the virus in animals, it is dispensable for growth in mammalian cell culture. The host TK enzyme is also not required for mammalian cell growth, as long as the thymidine monophosphate pathway is still functional. Conveniently, the thymidylate synthetase pathway, but not the TK pathway, can be inhibited by drugs such as methotrexate or aminopterin, and this provides a method to select for infection of TK− cells by TK+ viruses. TK− viruses can still replicate in TK− cells in the absence of the inhibitor drugs.

The recombinant virus construction for this study started with a mousepox vector, ECTV-602, which was TK− due to an insertion of the *Escherichia coli lacZ* gene within the TK gene. Mouse TK− cells were infected with this virus, and the resulting infected cells were subsequently transfected with a previously constructed plasmid pFB-TK-IL4, which contained the IL-4 gene expressed under the control of a viral

20.6 B cells and the production of antibody

How are antibodies produced?

In Section 20.4 we learned that binding of specific antigen to the BCR is the first step in initiation of B-cell activation. Antigen binding induces the B cell to express MHC II molecules and cytokine receptors on its surface. A striking ability of B cells is that they can recognize protein antigens as well as non-protein antigens such as polysaccharides and lipids. They can also bind *free* antigen; it does not have to be presented to them on MHC molecules. Recall that T cells do not generally have this ability—they are MHC-restricted and so require antigen to be in the form of MHC-bound peptides for

recognition. Non-protein components, although they may be taken in and degraded by APCs, are not loaded onto MHC molecules for presentation.

B-cell response to protein antigens

As we saw in Figure 20.23, BCR-bound antigen is internalized in the B cell by receptor-mediated endocytosis. The next critical step in B-cell activation is antigen-specific binding between the B-cell-presented MHC II:peptide and the TCR of

TABLE B20.1	Virus titers in spleens of C57BL/6 mice seven days after infection with recombinant control ECTV or ECTV expressing IL-4		
Virus	**Titer (PFU/g) in spleen**		
	10^2 PFU	10^3 PFU	10^4 PFU
Control ECTV [ECTV-602(TK+)]	3.0×10^4	2.8×10^4	1.5×10^2
ECTV-IL4(TK+)	2.5×10^5	3.5×10^5	6.6×10^7

promoter as well as a TK+ marker from the herpesvirus HSV. Recombination between the ECTV-602 vector and the plasmid resulted in a recombinant virus designated ECTV-IL4(TK+) that expressed IL-4 and had a TK+ phenotype, allowing selection for growth in TK– cells in the presence of aminopterin.

When this engineered virus was tested on mice by footpad inoculation, it unexpectedly killed them. As shown in **Table B20.1**, the numbers of virions in the spleens of infected animals were much higher after infection with the IL-4 expressing virus ECTV-IL4(TK+) compared to the control virus ECTV-602(TK+) that did not express IL-4. Even more surprising, the IL-4 expressing virus killed a large proportion of mice that were immunized with the normal mouse virus and should have been resistant to infection. As we have learned, IL-4 drives the antibody response at the expense of the other arm of the immune system known as the cell-mediated response. This cellular response is involved in the production of cytotoxic T cells that are necessary for killing infected cells to contain the spread of the virus. Without an effective cellular response, the virus was not controlled and high mortality resulted.

Impact

The results from this study were very interesting from a scientific point of view since they directly demonstrated the suppression of cell-mediated immune responses by IL-4 as well as inhibition of memory T-cell development. However, this study also showed that a single gene could convert a relatively harmless virus to a deadly virus that even killed animals that had been vaccinated. In other words, the pathogenicity of the virus was dramatically increased. Some have called for caution with respect to the development of immune-regulating vaccines. Could immunization with certain vaccines lock in a pattern of immune response when exposed to other microorganisms? Further research in these areas continues to be very valuable to our understanding of the immune response and disease. This study is surely not the last one to yield surprising and unexpected results. Make sure to read The Rest of the Story to find out more about the controversy that followed the publication of this study.

● Questions for Discussion

1. Why was it necessary to carry out the recombinant virus construction using a bacterial plasmid?

2. What aspects of immunology and infectious disease do you think this line of research will help us to understand more completely?

a T_H cell. As we learned in the previous section, bound T_H2 cells are much more effective at stimulating B cells than are T_H1 cells. Binding to a B cell begins a cytokine dialogue that stimulates the T_H2 cell to secrete cytokines such as IL-4 and IL-6, necessary for B-cell clonal expansion and differentiation into antibody-secreting plasma cells (see Figure 20.22). This proliferation and differentiation is absolutely dependent on interaction with T_H2 cells as it requires cytokines supplied by the T_H2 cell. In presenting antigen to the T_H2 cell, the B cell uses the T_H2 cell for its own proliferation and differentiation for the ultimate purpose of producing antibodies. The interaction between the antigen-specific T_H2 cell and the B cell can be seen as a request for permission to proceed with antibody production. True to the theme of the adaptive immune response, this two-way activation produces a tightly regulated antigen-specific response to invasion. A summary of T-cell and B-cell activation by antigen is given in **Figure 20.25**.

Antibody is a secreted form of the immunoglobulin BCR and therefore has the same epitope specificity. Memory B cells are also produced as a result of activation of B cells with T_H2 cells. They give rise to a memory response when the same antigen is encountered again. **Figure 20.26** shows a humoral primary and memory response. The second encounter with the same antigen results in higher levels of antibody production within a shorter time and of greater duration than the first encounter. Subsequent exposures to antigen often elicit responses that are even faster and of greater magnitude.

The specificity of the immune response means that during a memory response to a particular antigen, exposure to a second, new antigen will generate a primary immune response. The two

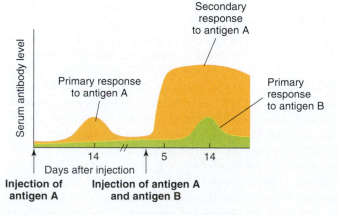

Figure 20.25. Summary of T-cell and B-cell activation CD4⁺ T cells are first activated by binding with an antigen-presenting dendritic cell. The CD4⁺ T cell gives rise to a population of T_H cells. T_H1 cells secrete cytokines that stimulate the development of CD8⁺ T cells into T_C cells for a cell-mediated immune response. T_H2 cells secrete cytokines that stimulate the development of B cells into plasma cells for a humoral response. Memory cells are also produced through activation.

CD4⁺ T-cell activation

Antigen presentation by dendritic cell

Naïve CD4⁺ T cell

MHC II:peptide

CD4⁺ T helper cell (T_H)

Memory CD4⁺ T cell

Antigen presentation by dendritic cell or macrophage

Cell-mediated immune response

Humoral immune response

T_H1 cell

T_H2 cell

Naïve B cell

Antigen presentation by dendritic cell to T_H1 cell

CD8⁺ T-cell activation

IL-2
IFN-γ
TNF-α

IL-4
IL-5

B-cell activation

Antigen uptake

Antigen presentation by B cell to T_H2 cell

Antigen uptake

Naïve CD8⁺ T cell

MHC I:peptide

Antigen presentation by dendritic cell or macrophage to memory CD8⁺ T cell

T_H2 cell

MHC II:peptide

CD8⁺ cytotoxic T cell (T_C)

Memory CD8⁺ T cell

Plasma cell

Memory B cell

IgM, IgG, IgA, IgE antibody

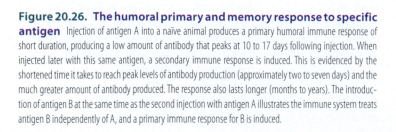

Figure 20.26. The humoral primary and memory response to specific antigen Injection of antigen A into a naïve animal produces a primary humoral immune response of short duration, producing a low amount of antibody that peaks at 10 to 17 days following injection. When injected later with this same antigen, a secondary immune response is induced. This is evidenced by the shortened time it takes to reach peak levels of antibody production (approximately two to seven days) and the much greater amount of antibody produced. The response also lasts longer (months to years). The introduction of antigen B at the same time as the second injection with antigen A illustrates the immune system treats antigen B independently of A, and a primary immune response for B is induced.

Secondary response to antigen A

Primary response to antigen A

Primary response to antigen B

Serum antibody level

14

5

14

Days after injection

Injection of antigen A

Injection of antigen A and antigen B

different antigens are treated as independent events. This means that one may mount a primary immune response at the same time that memory responses are mounted to other antigens.

B-cell responses to non-protein antigens

We have learned that B cells have the unique ability to recognize non-protein antigens such as polysaccharide components of bacteria. Recognition and production of antibody to these non-protein antigens is critical for immunity against many encapsulated pathogens such as *Haemophilus influenzae* (**Microbes in Focus 20.1**) and *Streptococcus pneumoniae* (see Microbes in Focus 7.1), both of which cause meningitis and pneumonia. We have also learned that B cells present peptide antigens, not carbohydrate antigens, to T_H2 cells for antibody production. How then do B cells make antibody to substances such as carbohydrates?

The BCR itself is not restricted to binding peptides. It can bind to non-protein antigens too, as long as the BCR-antigen

conformation is compatible. **Figure 20.27** shows the process of antibody production to a carbohydrate antigen. When a BCR binds a non-protein antigen, receptor-mediated endocytosis begins. In the process, other molecules—including proteins attached to the carbohydrate or in close proximity—will also be endocytosed. This is a pivotal step for the production of antibody against non-protein components because the peptides can be presented on MHC II molecules to T_H2 cells. Following this, T_H2 cells are stimulated to produce the cytokines needed for clonal expansion and differentiation of the B cells into plasma cells and memory B cells. The role of the protein is to activate the T_H2 cell so that the B cell can proliferate. Because of clonal expansion, the plasma cells produced will secrete antibody specific to the non-protein antigen—the same antigen that was first bound by the BCR of the original B cell. So B cells can produce antibody to *any* type of antigen. They are not restricted to binding just proteins.

Mice or humans without a thymus cannot produce functional T cells but can sometimes still produce antibody. At first, this

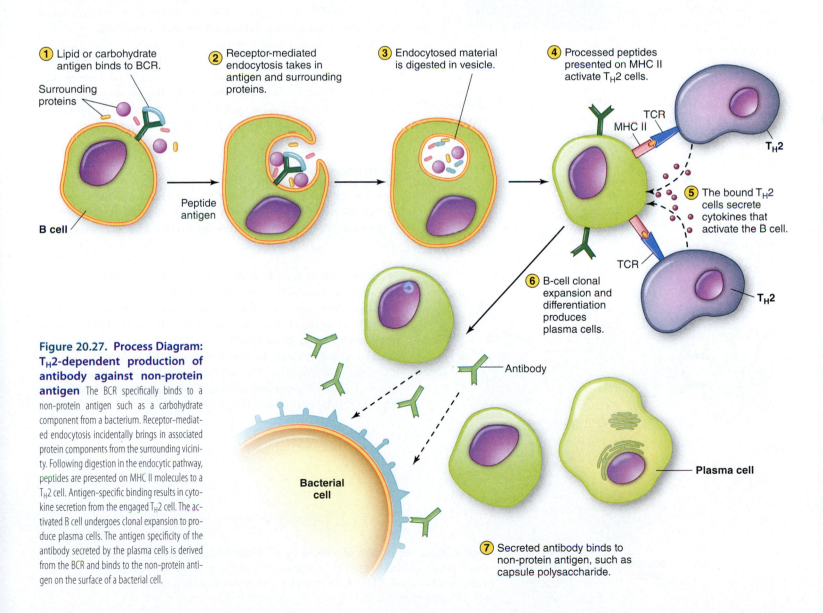

Figure 20.27. Process Diagram: T_H2-dependent production of antibody against non-protein antigen The BCR specifically binds to a non-protein antigen such as a carbohydrate component from a bacterium. Receptor-mediated endocytosis incidentally brings in associated protein components from the surrounding vicinity. Following digestion in the endocytic pathway, peptides are presented on MHC II molecules to a T_H2 cell. Antigen-specific binding results in cytokine secretion from the engaged T_H2 cell. The activated B cell undergoes clonal expansion to produce plasma cells. The antigen specificity of the antibody secreted by the plasma cells is derived from the BCR and binds to the non-protein antigen on the surface of a bacterial cell.

1 Lipid or carbohydrate antigen binds to BCR.

Surrounding proteins

B cell

Peptide antigen

2 Receptor-mediated endocytosis takes in antigen and surrounding proteins.

3 Endocytosed material is digested in vesicle.

4 Processed peptides presented on MHC II activate T_H2 cells.

TCR

MHC II

T_H2

5 The bound T_H2 cells secrete cytokines that activate the B cell.

TCR

T_H2

6 B-cell clonal expansion and differentiation produces plasma cells.

Antibody

Bacterial cell

Plasma cell

7 Secreted antibody binds to non-protein antigen, such as capsule polysaccharide.

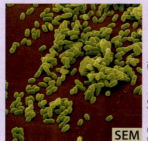
highly repetitive structures such as flagellin protein and polysaccharides commonly found in bacterial capsules have the ability to crosslink several BCRs on a single cell at once **(Figure 20.28)**. When a critical number (10–21) of BCRs become linked, the B cell is stimulated to undergo proliferation into antibody-secreting plasma cells without interacting with T_H2 cells. Antigens that can stimulate antibody production without engaging T_H2 cells are called **T-independent antigens**, for thymus-independent antigens.

As we have seen in Chapter 2, many capsules are made of carbohydrates. Capsules commonly have the property of preventing phagocytosis, and pathogens possessing capsules are not readily cleared by phagocytes. A critical protective response against such pathogens is production of opsonizing antibodies against the capsule. These coat the capsule to facilitate ingestion by phagocytes. However, waiting for significant antibody levels to be produced by conventional T_H2 interaction can take many days in a primary immune response (see Figure 20.26). Pathogens can quickly reach lethal numbers during this immunological interval. The ability of B cells to begin producing antibody immediately upon contacting these antigens can reduce this time and is often critical for protection from encapsulated pathogens. As described in **Perspective 20.2**, some pathogens that cause meningitis in children can evade clearing by the adaptive immune system by coating themselves with T-independent antigens. To develop effective vaccines against these pathogens, the capsular polysaccharides are chemically bound, or conjugated, to a peptide antigen to form **conjugate vaccines**. Section 24.5 describes some commonly used conjugate vaccines.

seems to pose a dilemma. If antibody production requires T_H2 cells and athymic individuals cannot make mature T cells, how does antibody production occur? Certain types of antigens with

---

CONNECTIONS Possession of a capsule is a common feature of extracellular pathogens such as *Streptococcus pneumoniae* and *Haemophilus influenzae*. Capsules help pathogens to escape immune detection and avoid phagocytosis. See **Section 21.1** for details.

Activation of B cells with T-independent antigens is generally weaker than activation by T_H2 cells. As a result, memory B cells are not formed and fewer antibodies are produced. Also, the type of antibody that is first produced, called IgM (immunoglobulin M), is of relatively low affinity and does not bind antigen strongly. Large amounts of high-affinity antibody, called IgG (immunoglobulin G), and memory B-cell production requires T_H2-dependent interaction. Examination of the different types of antibody produced by B cells will be our next topic.

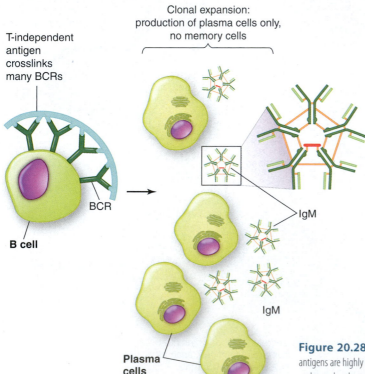

T-independent antigen crosslinks many BCRs

Clonal expansion: production of plasma cells only, no memory cells

BCR

B cell

IgM

IgM

Plasma cells

Figure 20.28. Activation of a B cell by a T-independent antigen T-independent antigens are highly repetitive in structure, allowing crosslinking of many BCRs on a B cell. Linking activates the B cell to undergo clonal expansion and differentiation into antibody-secreting plasma cells without the help of T_H2 cells. Low affinity pentameric IgM antibodies are made and memory B cells are not formed.

Pathogenic strains of *Streptococcus pneumoniae*, *Haemophilus influenzae*, and *Neisseria meningitidis* all possess capsules and are major culprits in meningitis. There are approximately 90 different strains of *S. pneumoniae*, six major strains of *H. influenzae*, and six major strains of *N. meningitidis*, based on differences in the polysaccharides that make up the capsules. The capsule interferes with phagocytosis by preventing C3b opsonization of the bacterial cells (see Sections 19.4 and 19.5). Protection against these strains is totally dependent on host production of antibody specific to the capsular polysaccharide. Capsular polysaccharides with their repetitive structure are T-independent antigens. The T_H2-independent response is of major importance in disease protection from these pathogens. Those who can mount this response have significantly less risk of developing serious disease and permanent disabilities such as deafness, which often results if they recover. The group at greatest risk is young children. Why? It seems that young children under five years of age do not have the full complement of all the subpopulations of B cells found in mature immune systems. Before this age, B cells do not respond well to T-independent antigens. However, young children can mount conventional T_H2-dependent antibody responses.

Vaccinating infants using conjugate vaccines, capsular polysaccharides that are conjugated to foreign protein molecules such as inactivated diphtheria toxin (see Section 21.1), essentially converts the polysaccharides from T-independent antigens to T-dependent antigens. The peptide addition is now displayed by the B cell on MHC II molecules, resulting in antibody production to the capsular polysaccharide in a T_H2-dependent manner. This provides protection for young children by producing immunological memory to the capsular polysaccharide. Immunized children can quickly mount a protective antibody response when they encounter *S. pneumoniae*, *H. influenzae*, or *N. meningitidis*. Currently, the conjugated vaccines used in children for protection against *S. pnuemoniae* include only seven of the nearly 90 different polysaccharide types (see Figure 24.18). For *H. influenzae*, the vaccine contains a single species of conjugated type B capsular polysaccharide. These are the most prevalent polysaccharides found in pathogenic strains of these bacteria. The conjugated vaccines are routinely given a few weeks after birth; their widespread use has significantly reduced the incidence of serious disease in very young children. The most recent *N. meningitidis* vaccine, currently used in children over 11 years of age, contains a mixture of four conjugated capsular polysaccharides and is expected to soon be licensed for use in infants.

Antibody production by plasma cells

Plasma cells are antibody factories, possessing large amounts of cytoplasm to accommodate the high levels of translation and secretion needed to turn out great numbers of antibody molecules, estimated at up to 2,100 molecules per second per cell (Figure 20.29). Plasma cells become terminally differentiated and cannot undergo cell division; all are produced from the proliferation and differentiation of activated naïve or memory B cells. Plasma cells undergo apoptosis after about two days, but the antibodies they secrete are often quite stable and can remain in circulation for many weeks.

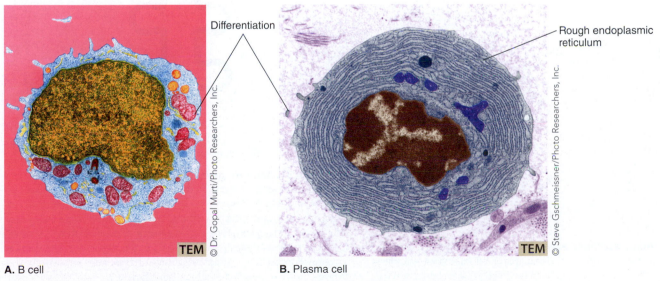

A. B cell

B. Plasma cell

Figure 20.29. B cell and plasma cell A. The B cell shows a small amount of cytoplasm surrounding the large, dark nucleus. **B.** The plasma cell shows elaborate rough endoplasmic reticulum and extensive cytoplasm to accommodate large amounts of antibody production.

All the antibodies produced from a single plasma cell are identical to each other. They can bind to only a particular epitope on an antigen. Recall that antibody is a secreted form of the immunoglobulin BCR, so it has the same specificity for epitope as the original BCR. Again, this specificity is genetically predetermined from the original maturation events involving gene rearrangements in individual B cells. Different epitopes are bound by different antibodies produced from different plasma cell populations **(Figure 20.30)**.

The ability of individual antibody molecules to bind to a specific epitope makes antibody a useful tool for isolating and detecting a specific antigen from a complex mixture. To produce a single species of antibody in large amounts, a single antibody-secreting plasma cell could be isolated and cloned in the lab to provide a ready source of specific, homogeneous antibody termed **monoclonal antibody (mAb)**. This may sound simple enough but a complication arises because antibody-producing B cells are short-lived, dying in seven to ten days, and so cannot be sustained in culture. In 1975, Cesar Milstein and Georges Kohler, working at the University of Cambridge, figured out how to make hybrid B cells called "hybridomas," which could be cultured to provide mAb. For this milestone discovery, they won the Nobel Prize for Medicine in 1984. **Toolbox 20.1** describes the method for producing hybridomas and mAb, and discusses some of the applications of mAb technology.

Immunoglobulin structure and diversity

A molecule of IgG antibody is made up of four polypeptide chains: two identical "light" or short chains and two identical "heavy" or large chains **(Figure 20.31)**. The chains each possess several internal disulfide bonds, and disulfide bonds also exist between the chains, holding them together. The numerous disulfide bonds stabilize the molecule, making it resistant to denaturation and proteolysis. Antibody can therefore remain functional for long periods of time in the body (the half-life for IgG is approximately 20 days in serum).

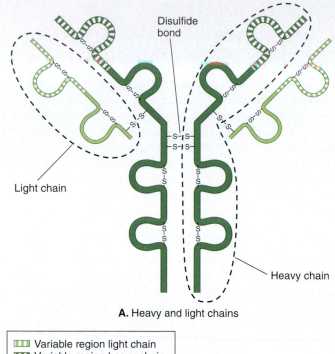

A. Heavy and light chains

- Variable region light chain
- Variable region heavy chain
- Constant region light chain
- Constant region heavy chain

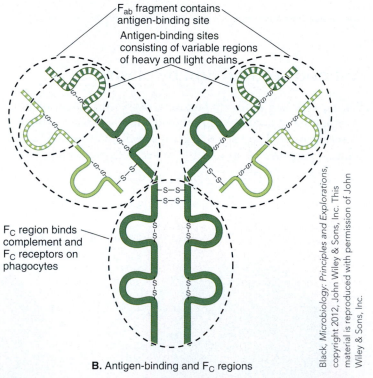

F_{ab} fragment contains antigen-binding site

Antigen-binding sites consisting of variable regions of heavy and light chains

F_C region binds complement and F_C receptors on phagocytes

B. Antigen-binding and F_C regions

Black, Microbiology: Principles and Explorations, copyright 2012, John Wiley & Sons, Inc. This material is reproduced with permission of John Wiley & Sons, Inc.

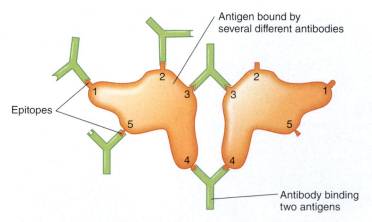

Figure 20.30. Many epitopes exist on a single antigen. Epitopes 1 to 5 are regions of the antigen that are actually bound by immunoglobulin (antibody is shown here). Different epitopes are bound by different antibodies as dictated by the specificity of the antigen-binding site. A single antigen can be bound by five different antibodies recognizing five different epitopes, and a single antibody can bind two antigens possessing the same epitope.

Figure 20.31. Immunoglobulin G (IgG) **A.** The IgG molecule is made up of two identical light chains and two identical heavy chains, held together by disulfide bonds. The heavy and light chains also have internal disulfide bonds. Each light chain consists of a variable (V) region and a constant (C) region. Each heavy chain consists of a V region and three C regions. **B.** The antigen-binding site consists of the V regions of a heavy and light chain. IgG fragmented by treatment with the protease papain yields three fragments; two of these fragments are called the F_{ab} portions, forming the branches of the Y, and are associated with antigen-binding. F_{ab} fragments are composed of one C region and one V region of a heavy and a light chain held together by disulfide bonds. Each antibody has two F_{ab} portions, and can simultaneously bind two identical epitopes. The third fragment is the F_C portion of the antibody. It possesses biological functions such as complement binding and interacting with F_C receptors of phagocytes.

Each IgG molecule is roughly Y-shaped, consisting of two F_{ab} (**f**ragment, **a**ntigen-**b**inding) portions that contain an antigen-binding site each, allowing for a single antibody to simultaneously bind two identical epitopes of one or two antigens. The actual site involved in antigen binding consists of the variable (V) regions of a heavy and a light chain. Sequence variation in just this small area of the genes encoding the heavy and light chains takes place in B cells to produce a variety of antibody specificities. This results in different amino acids in the variable regions that determine the specificity of epitope binding. The remainder of the light and heavy chains consists of constant (C) regions, so named because the amino acid sequence in these regions is always the same among IgG molecules. The base of the Y comprises the F_C (**f**ragment **c**rystallizable) site. This site also has important biological properties, as we will see.

Generation of Immunoglobulin Diversity

The diversity of antibodies that can be generated in a human is estimated to be between 10^6 to 10^8 different antibodies. Given that there are only about 15,000 genes in the human genome, how can such diversity be produced? Generating this astounding repertoire is done through a series of recombination events involving just a few gene regions. Let's examine these regions in humans. Like most mammals, humans have two different copies of the light-chain coding region, or locus, in DNA. As shown in **Figure 20.32**, one copy exists on chromosome 2 and is called the kappa (κ) locus. The other is on chromosome 22 and is called the lambda (λ) locus. Each mature B cell expresses either κ or λ light chains, producing a ratio of approximately 60 to 40, respectively, in humans. Each light-chain locus contains numerous V (variable) gene regions, about 40 for each κ and λ region, and multiple J (joining) segments for joining V and constant (C) regions together by recombination. There is one C region for the κ locus and four C regions for the λ locus. The single heavy-chain locus is on chromosome 14. The heavy-chain locus has five C regions, each of which determines the class of antibody produced, that is, IgA, IgD, IgE, IgG, or IgM. The heavy locus also contains about 65 V regions, 6 J regions, and 65 D (diversity) regions. The V, J, and D regions all form the variable region of the heavy-chain polypeptide. The D regions are not found in light chains.

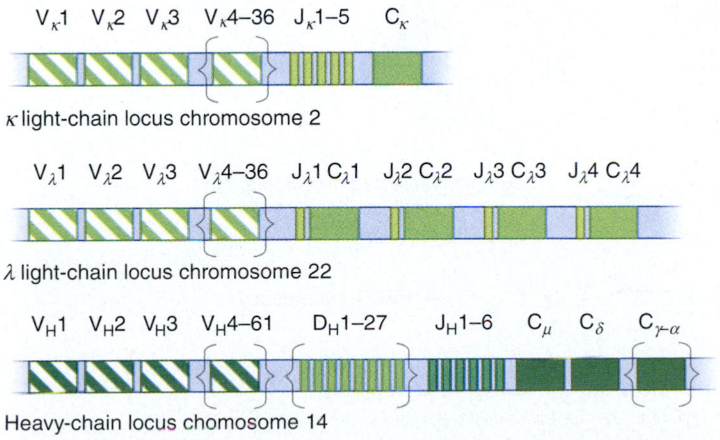

$V_\kappa1$ $V_\kappa2$ $V_\kappa3$ $V_\kappa4$–36 $J_\kappa1$–5 C_κ
κ light-chain locus chromosome 2
$V_\lambda1$ $V_\lambda2$ $V_\lambda3$ $V_\lambda4$–36 $J_\lambda1$ $C_\lambda1$ $J_\lambda2$ $C_\lambda2$ $J_\lambda3$ $C_\lambda3$ $J_\lambda4$ $C_\lambda4$
λ light-chain locus chromosome 22
V_H1 V_H2 V_H3 V_H4–61 D_H1–27 J_H1–6 C_μ C_δ $C_{\gamma\text{-}\alpha}$
Heavy-chain locus chromosome 14

Figure 20.32. Arrangement of immunoglobulin light and heavy chain loci in germline DNA The two light-chain loci are located on different chromosomes and vary from each other in V, J, and C regions. The heavy-chain locus is located on a third chromosome and has an additional region, the D (diversity) region, that forms part of the V region of the heavy chain. Heavy-chain C regions determine the class of antibody that will be produced; IgM uses C_μ, IgD uses C_δ, IgG uses C_γ, IgE uses C_ε, and IgA uses C_α. These loci undergo recombination only in maturing B cells to produce a diverse population of immunoglobulins.

To understand how the recombination events produce different immunoglobulin outcomes, we can briefly examine the assembly of just one chain in a developing B cell: the κ light chain. In the assembly of the variable region, one randomly selected V gene segment is joined to one of the five J (joining) segments adjacent to the light-chain constant (C) gene segment. Several successive recombination events produce a single continuous DNA region for expression of the κ light chain **(Figure 20.33)**.

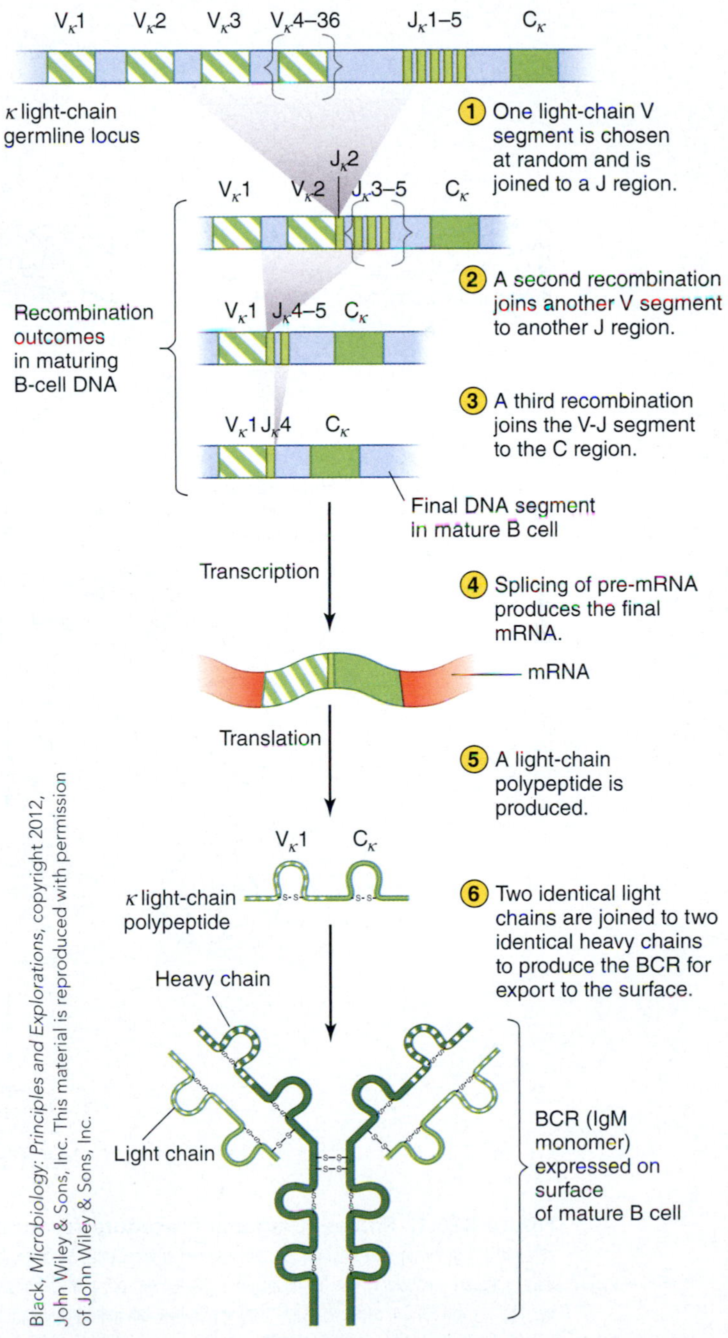

$V_\kappa1$ $V_\kappa2$ $V_\kappa3$ $V_\kappa4$–36 $J_\kappa1$–5 C_κ
κ light-chain germline locus
1 One light-chain V segment is chosen at random and is joined to a J region.
$J_\kappa2$
$V_\kappa1$ $V_\kappa2$ $J_\kappa3$–5 C_κ
Recombination outcomes in maturing B-cell DNA
2 A second recombination joins another V segment to another J region.
$V_\kappa1$ $J_\kappa4$–5 C_κ
3 A third recombination joins the V-J segment to the C region.
$V_\kappa1$ $J_\kappa4$ C_κ
Final DNA segment in mature B cell
Transcription
4 Splicing of pre-mRNA produces the final mRNA.
mRNA
Translation
5 A light-chain polypeptide is produced.
$V_\kappa1$ C_κ
κ light-chain polypeptide
6 Two identical light chains are joined to two identical heavy chains to produce the BCR for export to the surface.
Heavy chain
Light chain
BCR (IgM monomer) expressed on surface of mature B cell

Black, *Microbiology: Principles and Explorations*, copyright 2012, John Wiley & Sons, Inc. This material is reproduced with permission of John Wiley & Sons, Inc.

Figure 20.33. Process Diagram: Immunoglobulin κ light-chain formation Recombination events join one of several light-chain V gene segments to one of several J regions that are located adjacent to the κ chain C gene. Lambda light chains and heavy chains are built in a similar manner from different DNA locations, similarly encoding a variety of gene segments. Together, the light-chain V region and the heavy-chain V region determine the specificity of antigen binding.

The general procedure for production of monoclonal antibody is outlined in **Figure B20.1**. First, a mouse is inoculated with antigen against which the mAb is to be directed. Next, the mouse is tested for production of antibody to the antigen. This can be done with a simple blood test using an indirect ELISA test, described in Toolbox 20.2. The mouse is next euthanized and the spleen (a source of B cells) is removed and B cells are collected. Antibody-producing B cells are short-lived and cannot be cultured indefinitely. To prevent death and

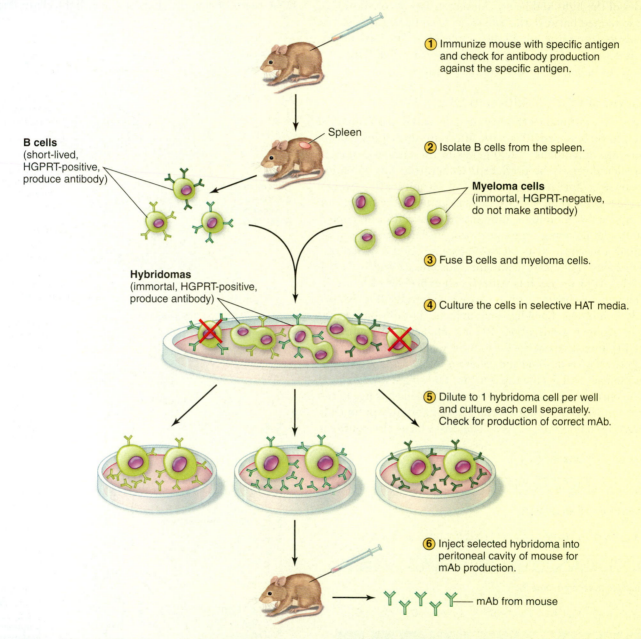

① Immunize mouse with specific antigen and check for antibody production against the specific antigen.

Spleen

B cells
(short-lived, HGPRT-positive, produce antibody)

② Isolate B cells from the spleen.

Myeloma cells
(immortal, HGPRT-negative, do not make antibody)

③ Fuse B cells and myeloma cells.

Hybridomas
(immortal, HGPRT-positive, produce antibody)

④ Culture the cells in selective HAT media.

⑤ Dilute to 1 hybridoma cell per well and culture each cell separately. Check for production of correct mAb.

⑥ Inject selected hybridoma into peritoneal cavity of mouse for mAb production.

mAb from mouse

Figure B20.1. Process Diagram: Procedure for monoclonal antibody (mAb) production Mice are inoculated with the antigen to which the mAb is to be formed. B cells, which produce antibody but are short-lived, are harvested and fused to myeloma cells to impart immortality. The hybridomas are selected on special medium containing hypoxanthine, aminopterin, and thymidine (HAT). Only the hybridomas will survive as they possess the enzyme hypoxanthine–guanine phosphoribosyltransferase (HGPRT) from the B cells and are long-lived. Hybridoma clones producing mAbs that bind the antigen are isolated. Selected hybridomas are individually seeded in the peritoneal cavity of another mouse where they will grow into a population of identical cells producing a single type of mAb.

allow the B cells to continue replicating to produce antibody, the B cells are fused to myeloma cells. These are transformed B cells that do not produce their own antibody but are immortal—they have undergone genetic mutations that allow them to continue to replicate indefinitely. Fusion of the two cell types produces a hybridoma that has the antibody-secreting characteristics of the original B cell and the ability to continue to replicate, conferred by the myeloma cell. Union of the cells is accomplished by adding polyethylene glycol (PEG), which facilitates membrane fusion.

Of course, the mix of B cells, myeloma cells, and hybridomas means there must be a way to select only the hybridoma cells. Elimination of the myeloma cells is accomplished by using a line of myeloma cells that are defective for the enzyme hypoxanthine-guanine phosphoribosyltransferase (HGPRT). When grown in the presence of culture media containing the compounds hypoxanthine, aminopterin, and thymidine (known as HAT medium), cells lacking the HGPRT gene cannot survive and thus die upon incubation. As shown in **Figure B20.2**, aminopterin blocks synthesis of GTP and TTP needed for DNA synthesis. Thymidine allows for TTP production. Cells producing HGPRT can use hypoxanthine to make GTP.

B cells produce HGPRT but are short-lived and will die. Myeloma cells lack the HGPRT enzyme and cannot use hypoxanthine, and thus cannot synthesize GTP and will die in HAT medium. The hybridomas contain a functional HGPRT gene from the B cell genome and are conferred with the ability to replicate indefinitely from the myeloma genome, and so they will be the only cell type to survive prolonged incubation in HAT media.

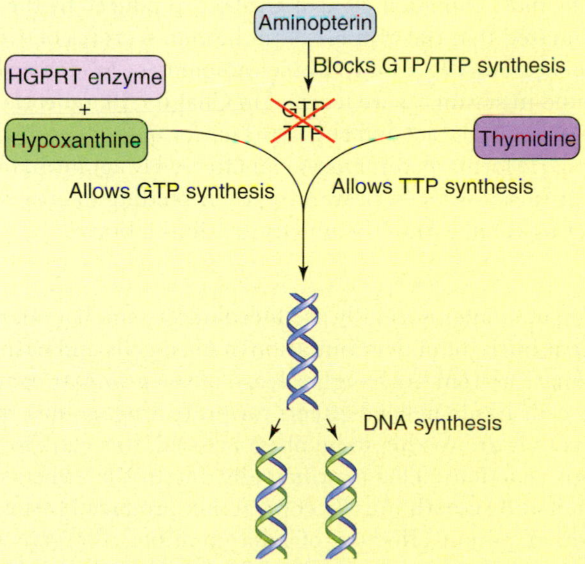

A collection of different hybridomas will result after selection and must be screened to see which are producing an antibody to the original antigen. The hybridomas collected from the HAT media are diluted to an appropriate concentration such that each well into which they will be placed contains a single hybridoma cell. The diluted cells are transferred to individual wells of a multi-well culture plate and are allowed to grow. The culture supernatant is collected and tested for the presence of antibody to the antigen. Again, this can be done using an ELISA test (see Toolbox 20.2). As all the cells in a single well are clones, they produce only one species of mAb capable of recognizing a single epitope. The collection of mAbs are tested for specificity and binding strength to the target antigen before large-scale culture.

Hybridomas have been notoriously difficult to culture in large amounts *in vitro*, and consequently less mAb results from cell culture, making it more cost-effective to use a live mouse or a rabbit as a culture vessel. The chosen hybridoma line is injected into the peritoneal cavity of the mouse and grows as a tumor, pumping mAb into the peritoneal fluid, which can be harvested after killing the mouse. Improved techniques for *in vitro* culture of mAb are soon to replace this method.

Monoclonal antibodies are useful because they ensure maximum specificity for the target antigen, which eliminates or reduces binding with related antigens that may be present. Monoclonal antibody technology is being used in medicine for therapy against chronic inflammatory diseases such as intestinal Crohn's disease. In this case, delivery of a mAb against the proinflammatory cytokine tumor necrosis factor-alpha (TNF-α; see Chapter 19) can greatly improve the patient's condition. For research and commercial purposes, mAbs are frequently used for detection and purification of proteins and virtually any other substance to which an antibody can be made. As Toolbox 20.2 shows (see direct ELISA), the specificity of mAbs makes them invaluable reagents for testing samples for the presence of enzymes, hormones, infectious agents such as HIV, and toxins such as pesticides.

Figure B20.2. Selection for hybridomas In the presence of medium containing hypoxanthine, aminopterin, and thymidine (HAT medium), only hybridomas are able to replicate. They possess the gene encoding HGPRT from the fused B cell, which allows them to synthesize GTP for DNA synthesis. They also possess the ability to replicate from the fused myeloma cell.

● **Test Your Understanding** ···················

Imagine the HAT medium being used in your lab was mistakenly prepared without hypoxanthine. What effects would this have during the preparation of your hybridomas? Explain.

Following transcription, the pre-mRNA is spliced and the exons of the V, J, and C regions are joined to form a continuous coding region for translation into the final κ light chain polypeptide. Lambda light chains and heavy chains undergo recombination events in a similar manner. The final immunoglobulin is constructed of two identical heavy chains and two identical light chains held together by disulfide bonds. The specificity of antigen binding is determined by the combination of the light chain V region and the heavy-chain V region of the immunoglobulin. Gene rearrangement occurs in a similar manner for generating diversity in TCRs of T cells. Mature B cells leave the bone marrow each with their unique BCRs. Presentation of peptide antigens and binding to a T_H cell leads to B-cell activation and further changes in the immunoglobulin molecule. These include antibody class-switching and affinity maturation.

Antibody Class-Switching

B cells, unlike T cells, do not proliferate in a perfectly clonal manner. The first class of antibody to be secreted in a primary immune response is immunoglobulin M (IgM) **(Table 20.4)**. As the antibody that is produced is a reflection of the membrane-bound BCR, it follows that the BCR of naïve B cells is therefore a membrane-bound version of IgM monomer. As B-cell clonal expansion and differentiation progresses, a new type or class of antibody, such as IgG, IgA, or IgE, is synthesized. This is referred to as **class-switching**. As shown in Table 20.4, each antibody class has different characteristics and somewhat different functions so the humoral response can be tailored to the specific type

of pathogen. The class of antibody produced is determined by the structure of the heavy-chain constant region and appears to be influenced by the cytokine environment of the B cell at the time of activation. A new antibody class is generated by joining the existing variable region with a new heavy-chain constant region.

IgM

Secreted IgM is a very large molecule composed of five monomer units joined together by disulfide bonds and a small polypeptide. This multimeric structure provides exceptional capacity to simultaneously bind several identical epitopes at once, resulting in maximum agglutination of particles to help prevent dissemination of the pathogen. IgM also fixes complement very efficiently. These characteristics make it well suited for collecting and lysing pathogens present in an infected area. Low amounts of IgM in the serum can bind and fix complement more effectively than similar amounts of IgG. This is especially important early in infection, when serum antibody levels are low.

IgG

IgG is the major antibody present in serum. It is a monomer and because of its relatively small size, it readily enters tissues and diffuses quickly to move out to contact antigen. Its small size also allows it to cross the placenta to provide protection against infections in newborns. As the antibody is not of the newborn's own making, this immunity is referred to as passive immunity.

IgA

The majority of IgA is produced as a dimer, made of two units held together by a small polypeptide chain. Dimerized IgA is secreted from mucosal surfaces and into body fluids such as tears, saliva, and milk. IgA in milk provides passive immunity for the newborn. It is the most abundant class of antibody produced by the body. It is estimated that the average adult human secretes between 5 to 15 g of secretory IgA a day! Lesser amounts of monomer IgA can be found in serum. As we learned in Chapter 19, mucosal surfaces are a main contact and entry surface for pathogens. Because of its dimeric form, secretory IgA can effectively agglutinate pathogens at these surfaces, preventing their binding to host cells and collecting them within the mucus for elimination.

IgE

IgE is a monomer normally produced in very small amounts. It is a potent initiator of degranulation of mast cells and basophils of the innate system. Although release of the granular contents of these cells has beneficial effects for protecting against parasitic worms that are too big to be phagocytosed, this can also induce allergic reactions. This is principally due to the effects of vasodilation and smooth muscle contraction produced by histamine and other potent effectors of inflammation. Allergens such as pollen stimulate the production of IgE rather than IgG from B cells. The cause of this is not precisely known but appears to involve a defect in cell regulation of antibody production. Just like parasites, allergens crosslink the IgE molecules on the surface of basophils and mast cells, signaling the cell to degranulate.

IgD

IgD is a monomer that is co-expressed along with IgM on the surface of mature B cells prior to class-switching. It is not normally secreted and its functions have not been characterized.

TABLE 20.4 **Classes of secreted antibody and their main functions**

Class	Structure	Features	Primary functions
IgA	Dimer	Secretory antibody of mucosal membranes, tears, milk, saliva	Agglutination of antigen; binding to phagocyte F_C receptors for phagocytosis
IgE	Monomer	Binds surface receptors of mast cells and basophils	Crosslinking with antigen induces degranulation in mast cells, eosinophils, basophils; involved in parasitic infections and allergic reactions
IgG	Monomer	Major antibody in serum; crosses placenta	Complement activation; opsonization for phagocytosis; toxin neutralization; ADCC killing by NK cells
IgM	Pentamer	First antibody made in primary immune response; present in serum	Complement activation; agglutination of antigen for phagocytosis

Black, *Microbiology: Principles and Explorations,* copyright 2012, John Wiley & Sons, Inc. This material is reproduced with permission of John Wiley & Sons, Inc.

Affinity maturation

Another amazing consequence of the activation of B cells by T_H2 cells is **affinity maturation** that progressively increases the binding strength of the immunoglobulin for its epitope. As B-cell proliferation and class-switching occurs, random mutations in the variable region of the genes encoding the heavy and light chains take place. This limited mutation, referred to as "somatic hypermutation," generates a population of B cells that produce BCRs that differ in just one or a few amino acids in the variable portion of the chain. The result is a collection of B cells that have slightly different affinities for the same epitope. The specificity is maintained but the binding strength changes. Those cells with very high-affinity BCRs bind epitope more tightly when it is encountered. This produces a stronger and more prolonged B-cell stimulation, resulting in more expansion of this particular population. Consequently, they out-compete B cells with weaker-affinity BCRs. Affinity maturation results in a population of B cells that have "learned" to better recognize the pathogen. The antibody that is produced during a humoral response therefore progressively increases in binding strength; it adapts to the antigen. This selection is akin to Darwin's process of natural selection, resulting in adaptation and increased fitness in species. The memory B cells produced also increase in fitness, expressing BCRs with very high affinity. Upon second encounter with the same antigen, this stronger binding of the BCRs stimulates more B-cell proliferation, producing more memory B cells, more plasma cells, and therefore more antibody. The antibodies that result bind much stronger than those that are first produced in a primary response. Even at low concentrations, they are very effective in binding antigen. They come out of the gate running, which contributes to the overall efficiency of the humoral response in clearing infections.

Detection of specific antibody, commonly from a blood sample, is a valuable diagnostic indicator of recent exposure to microbial pathogens. Quantitation of antibody levels and detection of the different classes of antibody can give further information as to the immune status of an individual. For example, detection of IgM indicates a very recent infection, resulting from a primary exposure. A simple test that detects antibody in a serum sample is the indirect **enzyme-linked immunosorbent assay (ELISA)**. Another use of antibody is for the direct ELISA. A direct ELISA takes advantage of the ability of antibody to specifically bind to its target antigen and can be used for the identification of almost any substance such as an illegal drug or an environmental contaminant to which an antibody can be made. These two ELISA tests are described in **Toolbox 20.2**.

Protection by antibodies

How does antibody binding protect the host from infection? Protection is achieved in several ways, depending on the type of antibody. Binding of an antigen by antibody can result in:

- blocking of the binding of pathogens or action of toxins on host cells
- complement fixation and lysis
- opsonization
- agglutination (clumping of antigen)
- activation of eosinophils, basophils, and mast cells
- ADCC killing of infected cells by NK cells (see Figure 19.22D)

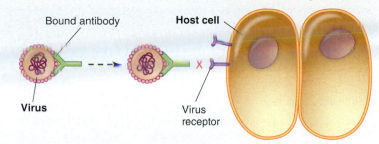

A. Prevention of binding to receptors

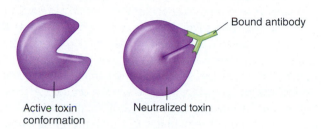

B. Prevention of conformational changes

Figure 20.34. Antibody neutralization of pathogens and toxins
A. Binding of antibody can physically block agents (virus shown here) from binding to receptors on host cells. **B.** Antibody binding to foreign agents such as viruses and toxins can neutralize their activity by preventing them from undergoing conformational changes needed for their activities.

Many pathogens and toxins require specific receptors to attach to host cells for colonization or uptake. Binding of antibody to their surface can physically block their ability to bind to these receptors and can also constrain the ability of toxins and viruses to undergo necessary conformational changes needed for their activity, thus neutralizing them **(Figure 20.34)**. This is similar to attaching a padlock to a door to prevent it from opening.

Bound antibodies can also activate the complement cascade by the classical pathway, resulting in disruption of the viral or bacterial membrane by formation of the membrane-attack complex (MAC) as shown in **Figure 20.35**.

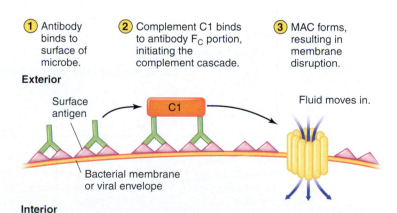

① Antibody binds to surface of microbe.

② Complement C1 binds to antibody F_C portion, initiating the complement cascade.

③ MAC forms, resulting in membrane disruption.

Figure 20.35. Process Diagram: Antibody initiation of complement and lysis Binding of antibody to the surface of a bacterial cell permits serum complement factor C1 to bind to the F_C portion of the antibody. Binding begins the complement cascade, resulting in formation of the membrane-attack complex (MAC) and lysis of the bacterium or disruption of the viral envelope. Cell lysis kills the bacterium, while damage to the viral envelope prevents viral attachment.

Toolbox 20.2
ENZYME-LINKED IMMUNOSORBENT ASSAY (ELISA)

Indirect ELISA Test

Indirect ELISA tests are used to detect infectious agents indirectly by directly detecting specific antibody that is produced as a result of the infection. Almost any body fluid where antibody would be expected can be used as a sample for testing. Depending on the infectious agent, urine, saliva, or blood serum can be tested.

A schematic of the indirect ELISA test is shown in **Figure B20.3**. This test uses antigen prepared from the infectious agent (often the entire disrupted microbe) to coat the surface of a test well. Into this coated well, the serum sample is added and the antibody it may contain is allowed to attach to the antigen on the surface of the well. Any unattached antibody not specific to the test agent is then removed by washing with buffer. To detect the specific bound or primary antibody, a secondary antibody is used. This antibody is conjugated to an enzyme for detection and is added and allowed to attach to the primary antibody. Any unbound secondary antibody

is then washed away. As the secondary antibody is directed against the primary antibody, it must be made in an animal species different from the species being tested. For example, Figure B20.3 shows the detection of human IgG, the primary antibody, from a serum sample. The secondary antibody is goat anti-human IgG antibody produced by injecting pooled human IgG into a goat. As human IgG is a foreign protein to a goat, it will produce antibodies against it. Goat serum will then contain antibodies directed against *any* human IgG molecule. The goat antibodies are collected and an enzyme is covalently bound to them. Production of conjugated secondary antibody is usually done commercially and purchased ready to use for ELISA tests. Detection of the goat secondary antibody in the well is then done with addition of substrate to detect the bound enzyme. Conversion of the substrate by the attached enzyme results in a colored product that can be readily measured by spectroscopy.

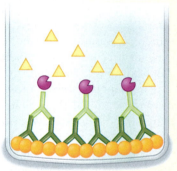
Positive test

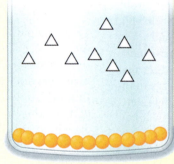
Negative test

© Hank Morgan/Photo Researchers, Inc.

Yellow wells are positive for serum antibody to the antigen.

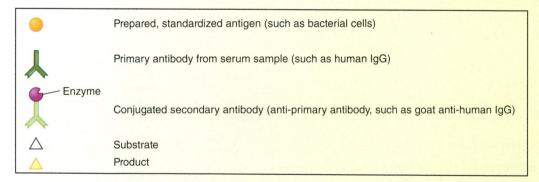

⬤ Prepared, standardized antigen (such as bacterial cells)

Y Primary antibody from serum sample (such as human IgG)

Enzyme — Conjugated secondary antibody (anti-primary antibody, such as goat anti-human IgG)

△ Substrate

△ Product

Figure B20.3. Indirect ELISA Antigen is first coated on the surface of the well. The test serum that may contain antibody (the primary antibody) to the antigen is next added. After washing, a secondary antibody conjugated to an enzyme is added, then washed again. If antibody against the antigen is present in the sample, it will be bound by the secondary antibody, and a colored product results after addition of substrate. Shown is the schematic for a positive test (*left*), a negative test (*center*), and an actual ELISA test showing positive and negative wells (*right*).

Direct ELISA Test

A direct ELISA, or sandwich ELISA, as it is sometimes called, is used for the detection and/or quantitation of almost any substance to which an antibody can be made. The sample is usually of industrial or toxicological importance. For example, in beer brewing, a direct ELISA is often used to check concentrations of microbial by-products and enzymes to monitor quality. Specific herbicides in water samples or illegal drugs in serum or urine samples are other examples applicable to direct ELISA testing. Direct ELISAs are not commonly used to test for infectious agents, as these agents are usually concentrated in tissues and not found free in high enough numbers in fluid samples to be detected this way.

The direct ELISA test begins with coating the surface of a test well with antibody prepared against the target substance (Figure B20.4). Often, the antibody used in this first step is a monoclonal antibody (see Toolbox 20.1). It is specially selected to bind to only one particular epitope of the antigen. This ensures maximum specificity for the target in order to eliminate or reduce binding with related substances that may be in the sample. Into this coated well, the sample suspected of containing the target substance is added and incubated to allow the target to attach to the antibody. Any unattached substances are then removed by washing with buffer. To detect the bound target, another layer of antibody that is also specific to the target is then added. This can be a second monoclonal antibody directed to a different epitope of the same antigen, but often serum collected from an animal immunized with the target substance is used. This antiserum contains polyclonal antibody, a collection of antibodies directed against different epitopes of the target antigen. To detect the sandwiched target substance, a secondary antibody conjugated to an enzyme is added and allowed to attach to the primary antibody. Any unbound secondary antibody is then washed away. Substrate to the conjugated enzyme of the secondary antibody is then added and the colored product is measured by spectroscopy.

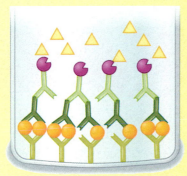
Positive test

Negative test

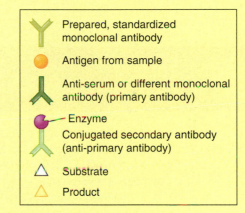

- Prepared, standardized monoclonal antibody
- Antigen from sample
- Anti-serum or different monoclonal antibody (primary antibody)
- Enzyme / Conjugated secondary antibody (anti-primary antibody)
- Substrate
- Product

Figure B20.4. Direct ELISA An antigen-specific monoclonal antibody is first coated on the surface of the well. The test sample that may contain target antigen is next added. After washing, standardized antiserum or another monoclonal antibody is added (primary antibody), then washed. Secondary antibody conjugated to an enzyme is added and washed. If antigen is present in the sample, it will be recognized by the primary antibody, and this in turn is bound by the secondary antibody. A colored product results after addition of substrate. The schematic shows a positive and a negative test.

● Test Your Understanding ....

You believe that a newly discovered plant from South America contains a medically important protein that could be used in cancer treatment. How would you use a direct ELISA to demonstrate that the protein is present in the plant extracts? Be sure to explain the specific "tools" you need to do this.

Complement is an important innate first line of defense against many pathogens, and antibody binding to pathogen surfaces is an important activator of the complement cascade. See **Section 19.4** for details.

Coating of antibody on a bacterial cell surface (opsonization) facilitates phagocytosis through binding of the exposed F_C portion of the bound antibody with the F_C receptor (F_CR) on the phagocyte **(Figure 20.36)**. Some pathogens produce their own surface molecules that mimic the action of F_CRs **(Perspective 20.3)**. These antibody receptors render antibody ineffective by binding them upside down by their F_C portions and thus preventing opsonization, phagocytosis, and complement fixation. Agglutination or clumping of antigen is accomplished by the ability of a single antibody to bind to more than one antigen **(Figure 20.37)**. Agglutination produces a much larger particle, which not only facilitates uptake by phagocytes and dendritic cells, but also provides stronger immune stimulation through TLR signaling.

Antibody-mediated immunity ANIMATION

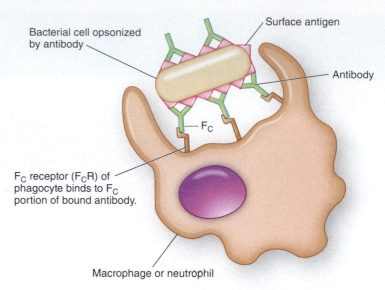

Figure 20.36. Antibody opsonization and phagocytosis Coating of antibody on a bacterial cell surface (opsonization) facilitates phagocytosis through binding of the F_C portion of the antibody molecule with the F_C receptor (F_CR) of the phagocyte.

Perspective 20.3
TURNING ANTIBODY UPSIDE DOWN

The extracellular pathogens *Staphylococcus aureus* and *Streptococcus pyogenes* produce surface molecules—protein A and M protein, respectively—that mimic the action of F_CRs. Both of these pathogens are associated with a number of diseases, many of which involve abscess formation. Avoiding phagocytosis and lysis by complement gives them time to establish an abscess, which helps to wall them off from the defenses of the host, including the diffusion of antibody into the abscess. Protein A and M protein are mounted onto the surface of the pathogens, where they bind the F_C portion of antibody,

rendering it ineffective **(Figure B20.5)**. This prevents opsonization, phagocytosis, and complement fixation.

Antibody such as IgG possesses two antigen-binding sites at the ends of the branched structure that bind to pathogens. When bound this way, the F_C portion of the antibody remains free to bind to F_C receptors (F_CR) of phagocytes, and to bind complement for lysis. *S. aureus* coats its surface with protein A, which binds to the F_C region of antibody, holding the antibody in reverse orientation. The F_C region is no longer available for complement fixation or to bind with phagocytes.

Figure B20.5. Action of *Staphylococcus aureus* protein A Protein A binds antibody in reverse orientation, blocking access to the F_CR and preventing it from facilitating phagocytosis and complement fixation.

F_C binds to phagocyte F_CR, which enhances phagocytosis and fixes complement.

Protein A binds F_CR, which inhibits phagocytosis and complement fixation.

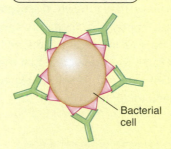

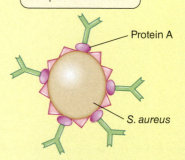

Antigen-binding sites

F_C portion

IgG antibody

Bacterial cell

Normal binding of IgG to bacterial cell

Protein A

S. aureus

S. aureus protein A binding to IgG

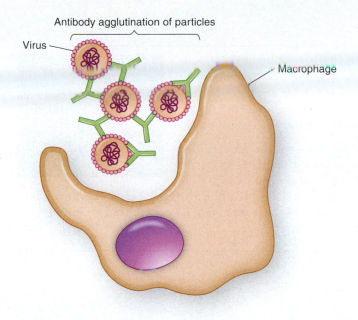

Antibody agglutination of particles

Virus

Macrophage

Figure 20.37. Agglutination by antibody and phagocytosis Antibody binding that links together several antigens (virus shown here) creates a physically larger target for more efficient phagocytosis.

Antibody, specifically of the IgE class, helps to prevent parasitic worm infections and is also involved in allergic reactions. When produced, IgE attaches to receptors for the F_C portion of the IgE molecule on the surface of basophils and mast cells (see Section 19.5). When several of these IgE antibodies are bound, the cell degranulates, releasing chemical irritants such as histamine and toxic proteins (**Figure 20.38**). These can kill the parasite, or at least encourage it to move to another host. Eosinophils have F_C receptors for IgG. Attachment of tissue eosinophils to IgG-coated parasites causes degranulation in a manner similar to mast cells and basophils.

20.6 Fact Check

1. What are T-independent antigens, and how do they stimulate antibody production?
2. What are monoclonal antibodies?
3. Describe the general structure of IgG antibody.
4. Distinguish between the structures and functions of the different classes of antibodies.
5. How are class-switching and affinity maturation involved in primary and memory responses?
6. Explain the various protective effects of antigen-antibody binding.

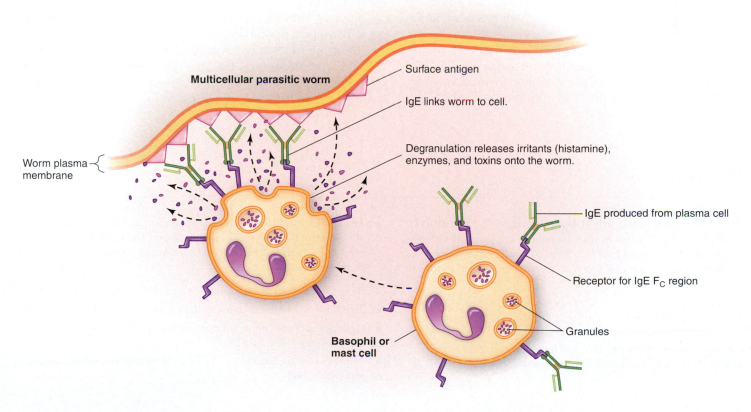

Multicellular parasitic worm

Surface antigen

IgE links worm to cell.

Degranulation releases irritants (histamine), enzymes, and toxins onto the worm.

Worm plasma membrane

IgE produced from plasma cell

Receptor for IgE F_C region

Granules

Basophil or mast cell

Figure 20.38. Immunoglobulin E (IgE) action against a parasitic worm Antibody of the IgE class coats the surface of basophils and mast cells. When these antibody-coated cells contact antigen several antibodies on the surface are crosslinked, signaling degranulation of the cell. Degranulation releases chemical irritants such as histamine and toxic proteins onto the parasite.

Image in Action

WileyPLUS A white lab mouse has been inoculated with an engineered mousepox virus containing the mouse egg-sperm receptor *mZp3* gene (yellow) and IL-4 gene (aqua). The inset on the right shows how this immunocontraception method was designed to work. Following the inoculation of the female mouse with the virus, antibodies (green) produced against the mZp3 sperm receptor attach to this receptor on the surface of the egg. Sperm are blocked from binding and entering the egg, preventing fertilization.

1. Imagine you were asked to develop an animation to accompany this image and provide additional detail. The

animation would feature how the antibodies involved in the immunocontraception process were produced in the mouse following the viral inoculation.

2. The virus in the image also contains the IL-4 gene to boost anti-mZp3 antibody production. Describe the normal role of IL-4 in the mouse immune response and how it contributed to lethality in the inoculated mice.

The Rest of the Story

In hindsight the lethal effect of the IL-4 engineered ectromelia virus shouldn't have been so surprising. We've read that IL-4 drives production of antibody and suppresses the cell-mediated response. What kind of adaptive immune response do you think is needed to control viral replication? You are correct if you answered cell-mediated immunity. This cellular response is involved in the production of the cytotoxic T cells (T_C cells) that are necessary for killing infected cells to contain the spread of virus. Without an effective cellular response, the ectromelia virus was not controlled and high mortality in the mice resulted.

Anyone (microbiology students included) who understands basic immunology could have predicted lethality would occur. In fact, the researchers themselves did not overlook this possibility. The mousepox lethality experiments were quickly and

sharply scrutinized because of the fear of bioweapons development. Some public groups began calling for general censorship of scientific research to prevent the dissemination of "sensitive" scientific information to the public domain, including to other scientists. This left many scientists outraged. It has since been deemed that the research did not violate the United Nations Biological and Toxin Weapons Convention, which forbids "the development, production, stockpiling, or acquirement of biological agents or toxins that have no justification for peaceful or defensive purposes." In fact, prior to the publication of the engineered mousepox virus, several papers had been published by other researchers describing the ability of IL-4 to enhance lethality of viruses. As you can see, it is difficult or impossible to draw a definitive line in the sand as to whether scientific information poses more of a security risk or more of a benefit. In the future, this new knowledge may be used to help design better vaccines and immune-regulating therapies for allergies and autoimmune diseases.

Summary

Section 20.1: What is the advantage of adaptive immunity?

Innate immunity involves genes that are genetically hard-wired, being identically coded in each cell, while adaptive (specific) immunity relies on rearrangements within genes encoding the **immune receptors** of T cells and B cells.

- Immune receptors are the **T-cell receptors (TCRs)** of T cells and **immunoglobulin (Ig) molecules** of B cells, that include the **B-cell receptors (BCRs)**, and **antibody**.
- Only immune receptors can bind **antigen**. The smallest part of an antigen that can be recognized by an immune receptor is called an **epitope**.

- The specificity of immune receptors for antigen is the cornerstone of customized adaptive immune response. The variety of immune receptors generated ensures that almost any foreign invader will be recognized.

B cells and T cells are **lymphocytes** produced in the bone marrow.

- T cells mature in the thymus and B cells mature in the bone marrow.
- Genetic rearrangements produce unique TCRs and BCRs that are antigenically committed.
- Lymphocytes are housed in secondary lymphoid tissues distributed around the body.

Exposure to a new infectious agent produces a **primary immune response** involving **clonal expansion** of activated cells.

- The immunity produced can take several days to peak and produces long-lived **memory cells** that convey immunological memory.
- Exposure to the same agent a subsequent time results in a **memory** or **anamnestic response**, sometimes called a secondary immune response, which will be faster and more potent.

Adaptive immunity is unique to vertebrates, with the exception of the jawless fish.

Section 20.2: What are the central cells in adaptive immunity?

Adaptive immune responses begin with the activation of **T cells** or **T lymphocytes**.

- Only **antigen-presenting cells (APCs)** can activate T cells. These include macrophages, **dendritic cells**, and **B cells**.
- T-cell activation requires correct binding of the TCR with specific peptide presented on **major histocompatibility (MHC) molecules** of APCs.
- CD4 or CD8 co-receptors of **CD4$^+$ T cells** and **CD8$^+$ T cells**, respectively, must also correctly interact with the MHC molecules of the APC to activate the T cell.
- Activation of a T cell results in production of **interleukin-2 (IL-2)**, which induces clonal expansion to produce a population of genetically identical memory T cells and **effector cells**, all possessing the same antigen-specific TCR.
- Effector T cells include **T helper cells (T$_H$ cells)** and **cytotoxic T cells (T$_C$ cells)**. Each has different immune activities. They migrate to the site of infection and exert their particular effector activity when they encounter the same antigen that was used to stimulate them originally.

Section 20.3: How are adaptive immune responses initiated?

Exogenous antigens are derived from extracellular microbes that are ingested and degraded in the endocytic pathway by APCs.

- Peptides derived from exogenous proteins are presented to CD4$^+$ T cells by **class II major histocompatibility (MHC II) molecules** on the surface of the APCs.
- CD4$^+$ T cells are MHC II-restricted due to required binding of the CD4 co-receptor with MHC II.
- Activated CD4$^+$ T cells undergo clonal expansion to give rise to memory CD4$^+$ T cells and effector T helper cells (T$_H$ cells).
- T$_H$ cells secrete large amounts of cytokines that serve to direct further immune responses.

Endogenous antigens are present in the cytoplasm and are derived from intracellular pathogens.

- The **proteasome** degrades endogenous proteins into small peptides. These are presented to CD8$^+$ T cells by MHC I molecules.
- CD8$^+$ T cells are MHC I-restricted due to required binding of the CD8 co-receptor with MHC I.
- Activated CD8$^+$ T cells undergo clonal expansion to give rise to memory CD8$^+$ T cells and cytotoxic T cells (T$_C$ cells).
- T$_C$ cells recognize MHC I:peptide presented on the surface of infected cells. They then release perforin and granzyme to cause apoptosis of the infected cell.

- **Cross-presentation** of antigen allows exogenous antigen, derived from phagocytosis of dead or dying cells carrying intracellular pathogens, to be presented on MHC I molecules for the activation of T$_C$ cells that will kill other infected cells.

Section 20.4: What cells can launch adaptive immune responses?

APCs—dendritic cells, macrophages, and B cells—can express a high number of MHC:peptide complexes to activate T cells.

- The most important APC in primary immune responses is the dendritic cell, which possesses all the co-stimulatory signals necessary to efficiently stimulate naïve T cells.
- Macrophages and dendritic cells can stimulate memory T cells to produce the memory response.
- B cells possess antigen-specific B-cell receptors (BCRs) on their surface that bind free antigen, including non-protein substances.
- BCRs are immune receptors that arise from random genetic rearrangements in the genes that code for the peptides making up the BCR. Each B cell carries a unique BCR that can bind to a specific antigen. Like T cells, they are antigenically committed.
- Binding of antigen to the BCR begins endocytosis and processing of protein antigens for MHC II presentation to T$_H$ cells.
- B-cell interaction with T$_H$ cells results in T$_H$-cell cytokine production that in turn causes B-cell clonal expansion and differentiation into antibody-secreting **plasma cells** and memory B cells.
- Antibodies are immune receptors. They are secreted versions of the membrane-bound BCR.

The addition of adjuvant to components, especially small soluble peptides, significantly increases their immunogenicity.

Section 20.5: Are there different responses to extracellular and intracellular pathogens?

T$_H$ cells are the main cytokine-secreting cells of the adaptive immune system. The profile of cytokines they secrete categorizes these cells into the functional subsets of **T$_H$2 cells** or **T$_H$1 cells**.

- Humoral responses are generally evoked by extracellular pathogens.
- The antibodies that are produced are needed for opsonization and complement fixation.
- Cell-mediated immune responses are generally directed against intracellular pathogens.
- T$_C$ cells are needed to kill infected cells, and macrophages are needed to engulf and destroy the resulting cell fragments that may carry the pathogen.

A **humoral immune response** is governed by T$_H$2 cells. It is characterized by:

- B-cell clonal expansion and differentiation into plasma cells
- large amounts of antibody production from plasma cells
- suppression of cell-mediated immune responses

A **cell-mediated immune response** is governed by T$_H$1 cells. It is characterized by:

- production of T$_C$ cells
- increased macrophage functions
- suppression of humoral immune responses

Section 20.6: How are antibodies produced?

BCR binding of an antigen occurs by recognition of an eptitope. Binding of antigen by BCRs initiates the following:

- endocytosis of the antigen by the B cell
- degradation in the endocytic pathway
- presentation of peptides on MHC II molecules
- binding of the MHC II:peptide complexes to compatible T_H2 cells
- stimulation of the T_H2 cells to secrete cytokines
- B-cell clonal expansion and differentiation into memory cells and plasma cells
- production of antibodies with the same specificity as the original BCRs

Antibody can be made to many different substances including polysaccharides, lipids, and proteins. Antibody production to non-protein substances occurs as follows:

- The BCR binds the substance and endocytosis begins
- Any other substance that is associated with the bound substance will also be taken in. Some of these will be proteins.
- The substances are degraded. The protein will be digested into peptides and loaded onto MHC II molecules
- Compatible T_H2 cells are bound
- Clonal expansion of the B cell results in plasma cell and antibody production
- The antibody that is produced is of the same specificity as the original BCR and will bind to the non-protein substance

In a primary humoral response, the first immunoglobulin class produced is IgM because the BCRs of naïve B cells are single units of membrane-bound IgM.

- As the humoral response progresses, B cells undergo **class-switching** and **affinity maturation**.
- Class-switching results in immunoglobulin of classes IgG, IgA, or IgE.
- Affinity maturation increases the affinity of the BCR for the epitope, resulting in a larger stimulation signal for B-cell proliferation, a larger population of plasma cells and memory cells, and more antibody and of higher affinity.
- A humoral memory response begins with production of IgG, IgA, or IgE.

B cells can produce antibody to **T-independent antigens**; highly repetitive molecules such as capsule polysaccharides that crosslink many BCRs.

- T-independent B-cell activation does not involve T_H2 cells.
- It does not induce class-switching or affinity maturation.
- It does not produce memory B cells.
- It produces only IgM.
- Capsule polysaccharides of pathogens are chemically bound to a peptide antigen to form **conjugate vaccines**, allowing interaction with T_H2 cells to produce memory B cells to the T-independent antigen.

The specificity of antibodies is the basis for **enzyme-linked immunosorbent assay (ELISA)** diagnostic tests. **Monoclonal antibodies (mAbs)** are commonly used in the direct ELISA.

● Application Questions ···

1. An organ transplant recipient is being given Tacrolimus, an important immunosuppressant drug that reduces IL-2 production. Describe the specific effect this would have in the body, and why this effect would be necessary in a transplant patient.

2. A knockout mouse has been engineered to lack the IL-12 gene. Explain the ways in which this mouse's immune system would behave differently than one with an intact IL-12 gene.

3. In order to determine the effects of various T_H1 cytokines on T_H2 cells, a researcher is culturing a subset of T_H2 cells *in vitro*. She adds IFN-γ to the culture flask. What effect might this have on T_H2 cells? Explain why this effect might have evolved in these types of cells.

4. *Mycobacterium tuberculosis* is an intracellular bacterial pathogen that multiplies inside the phagosome of macrophages. What type of adaptive immune response would be important in clearing this pathogen? Explain your answer.

5. *Streptococcus pneumoniae* is an encapsulated extracellular bacterial pathogen that can cause pneumonia and meningitis. What type of adaptive immune response would be important in clearing this pathogen? Explain your answer.

6. Imagine you are trying to design a new vaccine. How could you generate memory B cells from a bacterial polysaccharide? Describe the immune mechanisms you must target to do this.

7. Imagine someone else in your lab is trying to design a different vaccine. However, his challenge is to generate memory B cells from a small bacterial peptide. How might he do this? What immune mechanisms must he target in his approach compared to what you suggested in Question 6?

8. Babies born with DiGeorge syndrome often lack thymus function. Given this diagnosis, what type of antigens would their immune systems respond to and in what ways would they be immunodeficient? Explain your answer.

9. IgA deficiency is a common immunodeficiency in humans. Describe the typical role of IgA antibodies. Explain what effects this deficiency may have on a person's immunity.

10. HIV is initially diagnosed using an indirect ELISA. Using a diagram and a written description, explain how an indirect ELISA for HIV testing would be carried out. Be sure to describe the specific components used in testing, including their source.

11. Crohn's disease involves significant intestinal inflammation that appears to be caused by an immune response shift that causes excessive amounts of proinflammatory cytokines such as tumor necrosis factor-alpha (TNF-α). What subset of T cells most likely plays a major role in this disease? Explain.

12. If you needed to design a treatment for Crohn's disease to inhibit the excess TNF-α, how could you use your knowledge of cytokines from this chapter to shift the profile of the cytokine environment in these patients? Explain the rationale and the specifics of your approach.

Suggested Reading

Baumgarth, N. 2000. A two-phase model of B-cell activation. Immunol Rev 176:171–180.

Crane, I. J., and J. V. Forrester. 2005. Th1 and Th2 lymphocytes in autoimmune disease. Crit Rev Immunol 25:75–102.

Deyriendt, B., B. G. De Geest, and E. Cox. 2011. Designing oral vaccines targeting intestinal dendritic cells. Expert Opin Drug Deliv 4:467–483.

Huang, L., M. C. Khuls, and L. C. Eisenlohr. 2011. Hydrophobicity as a driver of MHC class I antigen processing. EMBO J 30:1634–1644.

Lesinski, G. B., and M. A. Westerink. 2001. Vaccines against polysaccharide antigens. Curr Drug Targets Infect Disord 1:325–334.

Lindblad, E. B. 2004. Aluminium compounds for use in vaccines. Immunol Cell Biol 82:497–505.

Mowen, K.A., and L. H. Glimcher. 2004. Signaling pathways in Th2 development. Immunol Rev 212:213–218.

Partidos, C. D., A. S. Beignon, J. P. Briand, and S. Muller. 2004. Modulation of immune responses with transcutaneously deliverable adjuvants. Vaccine 21:2385–2390.

Pulendran, B. 2004. Modulating TH1/TH2 responses with microbes, dendritic cells, and pathogen recognition receptors. Immunome Res 29:187–196.

Zubler, R. H. 2001. Naïve and memory B cells in T cell-dependent and T-independent responses. Springer Semin Immunopath 23:405–419.

Bacterial Pathogenesis

In May 2000, a torrential downpour soaked the small town of Walkerton, Ontario, Canada. A few days later, many of the townspeople began to complain of bloody diarrhea, vomiting, cramps, and fever. By the end of two weeks, hundreds were sick. Seven people died. The cause of the illness was shown to be *Escherichia coli* serotype O157:H7. Most *E. coli* strains are harmless commensals of the large intestine of healthy individuals, but not this strain. This strain of *E. coli* acquired genetic elements that allowed it to colonize and invade the intestinal lining where it produced deadly toxins called Shiga toxins—a recent addition for *E. coli*. As their name suggests, these toxins are closely related to toxins produced by the pathogen *Shigella dysenteriae*, which also causes severe intestinal disease.

Clinical signs of infection with *E. coli* O157:H7 in humans include bloody diarrhea, severe cramping, and dehydration. Even with treatment, between 2 and 7 percent will go on to develop further life-threatening complications approximately one week after onset of gastrointestinal symptoms. These include widespread destruction of red blood cells, diabetes, renal failure, seizures, stroke, and coma.

Cattle are a major reservoir for *E. coli* O157:H7. Interestingly, infection in cattle is subclinical—they are not affected by the Shiga toxins and do not show any signs of disease. In Walkerton, the downpour washed the pathogen from manure on a nearby cattle farm into the town's shallow well. Tests of water sampled just before the illnesses began revealed *E. coli* contamination, but the town's two water system managers failed to notify the public health authorities or the townspeople. During the investigations that followed, the two men admitted to faking microbiological tests and chlorination records. They were formally charged and sentenced.

This was not the first or last time O157:H7 has tragically reared its ugly head. Another highly publicized outbreak took place in the United States in 1993 and involved undercooked hamburgers from a supplier for fast-food restaurants in four western states. Over 600 people fell ill, and four died. This case made clear the necessity for thoroughly cooking ground beef. Since then, other outbreaks have been traced to a variety of sources including salami, apple juice, lettuce, onions, and potatoes. How is it that *E. coli*, a normally harmless resident of the intestine, suddenly became a deadly pathogen?

CHAPTER NAVIGATOR

As you study the key topics, make sure you review the following elements:

Virulence factors are used to attach to host cells, avoid immune clearance, and damage host cells to obtain nutrients.

- Table 21.1: Some major virulence factors of selected pathogens
- Mini-Paper: *E. coli* injects its own receptor
- Toolbox 21.1: Serotyping
- Figure 21.8: Action of some A–B toxins
- Perspective 21.1: The good, the bad, and the ugly side of botulinum toxin
- Table 21.7: Source, action, targets, and diseases produced by cytolysins
- Perspective 21.3: Iron, vampires, fashion, and the white plague

Pathogens employ a variety of mechanisms to establish infection.

- Table 21.11: Virulence factors of *Streptococcus pyogenes*
- Toolbox 21.2: The tuberculin test for tuberculosis

Virulence factors are commonly established through horizontal gene transfer.

- Figure 21.32: General structure of a pathogenicity island (PAI)
- Microbes in Focus 21.7: *Vibrio cholerae*
- Table 21.12: Genes introduced to bacterial cells by temperate phages
- Perspective 21.4: Antibiotics trigger toxins?

CONNECTION for this chapter:

Systemic microbial components can cause deadly inflammation (**Sections 19.3** and **19.4**).

CD4+ T cells are the central lymphocytes involved in the adaptive immune response (**Sections 20.2, 20.3,** and **20.5**).

Role of capsules in avoidance of host immune responses (**Perspective 20.2**)

Temperate or lysogenic phages can transfer virulence factors (**Sections 5.3** and **8.3**).

Introduction

In this chapter, we will examine bacterial pathogens. The diseases produced in various species of host organisms by bacterial pathogens are as diverse as the pathogens themselves and we will not attempt to cover all of them. Rather, we will cover common aspects of bacterial **pathogenesis**—the processes used by pathogens to produce disease. In doing this, some common bacterial pathogens and the diseases they cause will be used to illustrate some common aspects of pathogenesis.

Like any organism, pathogenic microorganisms have, through evolution, developed strategies to secure nutrients from their surroundings. Bacterial pathogens can survive in a living host by obtaining nutrients through *damaging tissue*. This distinguishes them from non-pathogenic bacteria that may also be living on or within the body, but do not damage tissue to obtain nutrients. Pathogens are not evil entities that set out to purposefully kill their host, although that may sometimes be the outcome; the host is simply a nutrient source and a place to live, and pathogens have adapted to exploit this habitat. It is really no different than the way humans exploit the resources of the planet. This also results in damage, but we don't do this for the purpose of destroying the environment, although harm may be an outcome.

Bacterial pathogens can produce a spectrum of diseases that range from damage due to production of a single toxin, to diseases produced largely because of a robust but relatively ineffective host immune response. In many cases, the damage is caused by both the pathogen and the host response to the pathogen. Most bacterial pathogens follow a common scheme that leads to the establishment of infection; the pathogen must gain access and often attach to host tissue, avoid host defenses, and get nutrients needed to multiply. These requirements are some of the key aspects of bacterial pathogenesis that we will investigate. As we shall see, many of the products produced by pathogens that allow them to carry out these processes are encoded on specific genetic elements that can be transferred between strains and even other species. This genetic mobility is a principal feature in pathogen evolution.

Our investigation of bacterial pathogenesis will focus on answering the following questions:

What are common features of bacterial pathogens? (21.1)
How do bacterial pathogens cause disease? (21.2)
How do some bacteria become pathogens? (21.3)

21.1 Bacterial virulence factors

What are common features of bacterial pathogens?

We previously came across the term **virulence factor**. A virulence factor is a product made by a pathogen that enhances its ability to cause disease (see Section 18.4). As you might imagine, the range of virulence factors that pathogenic bacteria produce is vast **(Table 21.1)**, but surprisingly, many of these virulence factors act in common ways. Any bacterial pathogen must perform a few basic tasks to establish itself in the host. First, it must gain access to tissues. Then, it must evade, or overcome host defenses. Finally, it has to get nutrients, usually by damaging cells and/or stealing nutrients away from the host. For example, the sexually transmitted primary pathogen *Neisseria gonorrhoeae* first attaches to cervical or urethral epithelial cells using fimbriae, then invades the underlying tissues. Here, production of the cell wall component endotoxin evokes an intense inflammatory response in the host (see Section 19.3). The inflammation damages tissue, such as the fallopian tubes, and produces most of the symptoms of gonorrhea. *Neisseria gonorrhoeae* also produces various enzymes that directly damage cells **(Figure 21.1)**. Tissue damaged by inflammation and enzymes facilitates invasion and provides nutrients. To successfully do all this, the bacterium must evade host immune responses. One of its key strategies is to constantly change its surface antigens, particularly the pilin proteins of fimbriae, which renders antibodies ineffective. In addition, the bacterium produces an IgA protease for degrading

host IgA antibodies. Fimbriae, endotoxin, and IgA protease are major virulence factors of this pathogen.

Attachment factors

Attachment to host tissues is a critical step for most pathogens. Without attachment most bacteria would be routinely removed by friction, fluids, or even be displaced by other microbes. Many pathogens possess mechanisms that allow them to attach to host cells directly. Non-pathogenic bacteria tend to colonize extracellular surfaces of the body, such as mucus or the dead outer layers of skin. You may recall from Section 17.4 that non-pathogenic commensal strains of gut bacteria colonize the lumen of the intestine and do not normally attach directly to intestinal cells. Pathogenic strains, such as *Escherichia coli* O157:H7 and *Helicobacter pylori* (see Microbes in Focus 17.3) have specific mechanisms to attach directly to cells of the intestine.

Attachment factors that are commonly used by bacteria include fimbriae, fibronectin-binding proteins, and various membrane-associated molecules, such as lipoteichoic acid of Gram-positive bacteria (see Section 2.4) and outer membrane proteins of Gram-negative bacteria. In addition to these common molecules for adherence, some bacteria produce specialized proteins for attachment. Molecules produced by bacteria

TABLE 21.1   Some major virulence factors of selected pathogens

Organism	Disease	Virulence factor	Action
Bordetella pertussis	Whooping cough	Fimbriae Pertussis toxin Invasive adenylate cyclase	Attachment Disrupts cell ion balance Disrupts cell ion balance
Escherichia coli O157:H7	Hemorrhagic colitis and kidney failure	Intimin Tir Type III secretion system Shiga toxins	Attachment Receptor for attachment Injects Tir for attachment Stops translation in host cells
Helicobacter pylori	Gastritis, ulcers	Urease Vacuolating cytotoxin A (VacA) Flagella Cytotoxin-associated antigen (CagA) CagA type IV secretion system	Neutralizes gastric acid Host cell death, inflammation Transport through mucus Disrupts host cell cytoskeleton Injects CagA
Neisseria gonorrhoeae	Gonorrhea	Fimbriae IgA protease LOS (a form of endotoxin)	Attachment and immune evasion Destruction of IgA antibody Evokes inflammatory damage
Streptococcus pneumoniae	Pneumonia, meningitis	Capsule Pneumolysin Autolysin	Anti-phagocytic Forms pores in host cells Lysis of bacterial cell to release peptido-glycan (produces inflammation)
Streptococcus pyogenes	Various skin, throat, and systemic infections	Capsule M protein Hyaluronidase Streptokinase	Anti-phagocytic Prevents binding by antibody Degrades connective tissue Degrades fibrin clots

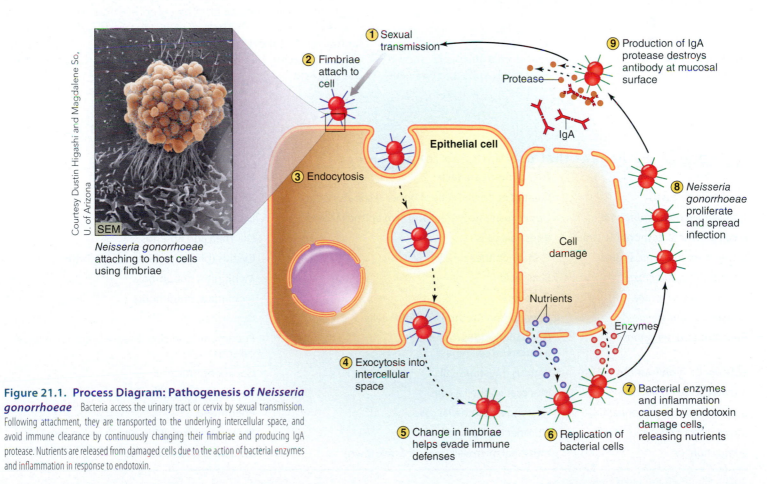

Courtesy Dustin Higashi and Magdalene So, U. of Arizona

SEM

Neisseria gonorrhoeae attaching to host cells using fimbriae

① Sexual transmission

② Fimbriae attach to cell

③ Endocytosis

Epithelial cell

Protease

IgA

⑨ Production of IgA protease destroys antibody at mucosal surface

⑧ *Neisseria gonorrhoeae* proliferate and spread infection

Cell damage

Nutrients

Enzymes

④ Exocytosis into intercellular space

⑤ Change in fimbriae helps evade immune defenses

⑥ Replication of bacterial cells

⑦ Bacterial enzymes and inflammation caused by endotoxin damage cells, releasing nutrients

Figure 21.1. Process Diagram: Pathogenesis of *Neisseria gonorrhoeae* Bacteria access the urinary tract or cervix by sexual transmission. Following attachment, they are transported to the underlying intercellular space, and avoid immune clearance by continuously changing their fimbriae and producing IgA protease. Nutrients are released from damaged cells due to the action of bacterial enzymes and inflammation in response to endotoxin.

TABLE 21.2 Pathogens that bind host fibronectin

	Pathogen	Disease
Gram-positive	*Corynebacterium diphtheriae*	Diphtheria
	Clostridium difficile	Colitis
	Clostridium perfringens	Gas gangrene
	Mycobacterium tuberculosis	Tuberculosis
	Staphylococcus aureus	Various skin and invasive diseases
	Streptococcus pyogenes	Various invasive and toxic diseases
Gram-negative	*Escherichia coli*	Urinary and gastro-intestinal diseases
	Treponema pallidum	Syphilis
	Borrelia burgdorferi	Lyme disease
	Porphyromonas gingivalis	Gum disease

that allow them to bind to and colonize host tissues are generally referred to as **adhesins**. Extracellular bacterial pathogens can adhere directly to cells or to the host *extracellular matrix*, a mesh-like support material composed of fibrous molecules, such as collagen and fibronectin proteins that anchor cells to the matrix, and gel-like substances that fill intercellular spaces, such as hyaluronic acid and sulfate proteoglycans.

Fibronectin-binding Proteins

Fibronectin is a large glycoprotein found circulating in plasma and as fibers of extracellular matrix material where it anchors cells to support surfaces, such as basement membranes and collagen. The ability to bind host fibronectin is important in establishing colonization for many bacterial species, but is probably best studied for *Streptococcus pyogenes* and *Staphylococcus aureus*. *Staphylococcus aureus* produces several **fibronectin-binding proteins** that are predominant adherence factors for skin, nasal mucosa, and endothelial cells lining blood vessels. Once *S. aureus* gains access to the bloodstream, usually from a skin infection, fibronectin-binding proteins allow it to attach to and invade the vessel walls, moving into surrounding tissues from there. This activity is responsible for the ability of *S. aureus* to colonize heart valves, producing endocarditis. The pathological importance of fibronectin-binding proteins has been demonstrated by expressing the *S. aureus* fibronectin-binding protein on the surface of *Lactococcus lactis*, a normally non-invasive, non-pathogenic Gram-positive organism. Recombinant *L. lactis* expressing *S. aureus* fibronectin-binding protein can readily infect heart valves in animal models. Other pathogens that use fibronectin-binding proteins as part of their pathogenesis are listed in **Table 21.2**.

Fimbriae

Many bacteria adhere to host surfaces using **fimbriae**, pili that are primarily used for attachment (**Table 21.3**). They adhere to host tissues and non-biological surfaces, including medical implants. The end of each fimbria strand has a specific adhesive tip composed of a tip protein used for attachment (**Figure 21.2**). Fimbriae are assembled sequentially, beginning with the tip protein to which pilin subunits are added from the bottom up. Finding fimbrial binding targets is often done *in vitro* by selective addition of various oligosaccharides to cultures of bacteria. If subsequent attachment to target host cells is blocked, this indicates the oligosaccharide has been bound, and is a binding target for the adhesin.

TABLE 21.3 Pathogens that use fimbriae for adherence

Bacterium	Attachment surface	Disease
Bordetella pertussis	Ciliated epithelial cells of upper respiratory tract	Pertussis (whooping cough)
Corynebacterium diphtheriae	Upper respiratory tract and skin epithelial cells	Diphtheria
E. coli intestinal pathogenic strains	Intestinal cells	Diarrhea and enteritis
E. coli uropathogenic strains	Urinary epithelial cells	Cystitis (bladder infection), urethritis
Haemophilus influenzae	Sialic acid	Meningitis, pneumonia
Klebsiella pneumoniae	Collagen	Meningitis, pneumonia
Neisseria gonorrhoeae	Cervical and urethral epithelial cells	Gonorrhea
Pseudomonas aeruginosa	G_{M1} gangliosides found on a variety of cell types	Pneumonia and skin disease in compromised persons
Salmonella Typhimurium	Various intestinal cells	Gastroenteritis
Streptococcus agalactiae	Mucosal epithelial cells	Neonatal meningitis
Streptococcus pneumoniae	Pharyngeal epithelial cells	Pneumonia
Streptococcus pyogenes	Fibronectin, collagen	Pharyngitis, tonsillitis, invasive skin diseases
Vibrio cholerae	Intestinal microvilli (receptor unknown)	Cholera (severe diarrhea)

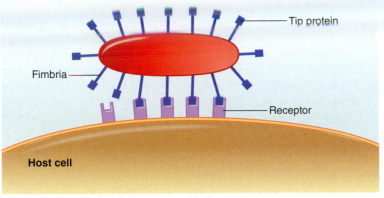

A. Attachment by fimbriae

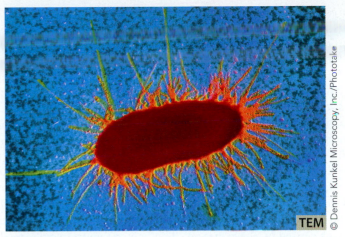

B. Fimbriae of *E. coli*

Figure 21.2. Attachment by fimbriae **A.** Fimbriae attach to receptors on cell surfaces using an adhesive tip protein. **B.** Fimbriae of enterotoxin-producing *E. coli* appear orange in this color-enhanced micrograph.

Why are fimbriae commonly used for bacterial adhesion? A major barrier to bacterial attachment is electrical charge. Host cell and bacterial cell surfaces are covered with molecules that confer an overall net negative charge, causing electrostatic repulsion between the two. Fimbriae may assist in attachment by spanning the distance of the repulsion, overcoming the need for the bacterium to get close to the cell. Fimbriae are long enough to make this relatively distant contact.

Most Gram-negative bacteria and some Gram-positive bacteria produce fimbriae if provided with the correct conditions. Commonly, several different types of fimbriae can be produced by a single bacterium, and the type of fimbriae expressed is frequently modified in response to environmental conditions. A well-studied example of this occurs in *Neisseria gonorrhoeae* (**Microbes in Focus 21.1**). As noted in the introduction, fimbriae are a major virulence factor of this sexually transmitted bacterium, being needed for adherence to epithelial cells of the urethra and cervix. Once infection has been established, expression of fimbriae can be turned on and off. This is known as "phase variation." Expressed fimbriae can also undergo amino acid changes via genetic recombination. This is known as "antigenic variation." The appearance and disappearance of fimbriae and the constant change in protein sequence presents a moving target to the adaptive immune system. By the time specific antibody is made to one type of fimbriae, another is produced, or the fimbriae disappear altogether for a time, rendering present anti-fimbrial antibody ineffective.

Uropathogenic *E. coli* strains, which cause the majority of urinary tract infections in women, also exhibit phase variation. These strains typically express numerous types of fimbriae, with some sets being expressed under some conditions, and other sets expressed under other conditions. Certain fimbria sets are needed to establish infection, and others are needed to maintain infection during the ensuing changes in the urinary tract. Regulation of fimbria expression patterns is accomplished by interactions between different fimbria genes, and between genes and signaling molecules in the environment.

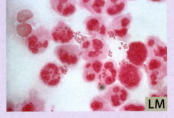
Special Adherence Proteins Tir and Intimin: Home Renovations by *E. coli*

Intestinal enteropathogenic and enterohemorrhagic strains of *E. coli* share a particularly interesting way of anchoring themselves to the microvilli of enterocytes, the epithelial cells lining the intestinal mucosal surface. Attachment of these strains happens on two levels. One level is by attachment using various fimbriae that bind to cell surface glycoproteins, and a second level is mediated by interaction with a specific *E. coli* surface protein named intimin and its specific receptor, Tir, on enterocytes.

Enteropathogenic *E. coli* strains are a significant cause of non-bloody diarrhea in young children in developing countries. Enterohemorrhagic strains, such as O157:H7, cause disease worldwide ranging from non-bloody diarrhea to severe hemorrhagic colitis often complicated with development of hemolytic uremic syndrome (HUS). This syndrome is caused by Shiga toxins, sometimes called "virotoxins" to distinguish them from those produced by *S. dysenteriae*. These enteropathogenic and enterohemorrhagic serotypes use type III secretion systems (see Section 2.4) to inject proteins that disrupt host cell signaling and induce cytoskeletal changes that produce the formation of unusual pedestal-like structures under the attached *E. coli* cells **(Figure 21.3)**. It is thought that remodeling of intestinal cells affords *E. coli* space away from competing microbes and projects them farther into the lumen of the intestine, allowing better access to nutrients. The lesions that are related to formation of pedestals are called "attachment and effacing (AE)" lesions. AE lesions seriously impact the functions of the gut, resulting in tissue degeneration and loss of nutrient absorption that are characteristic of the diseases caused by these strains. This may be advantageous to the bacteria by making more nutrients available for them.

The receptor, Tir, which is needed for pedestal formation, was first identified on the surface of infected cells and therefore was naturally thought to be host cell-encoded. Interestingly, Tir was only found on cells with *E. coli* attached and was never found on uninfected control cells. It appeared that the attachment of *E. coli* somehow induced the host cell to express Tir. Researchers began to investigate the nature of this unusual interaction. To their surprise, they found that Tir was not a host cell product at all, but instead was injected by *E. coli* (see Figure 21.3A and **Mini-Paper**, p. 722). *E. coli* was making its own host cell receptor!

Bacterial toxins

A toxin is a poisonous substance that is produced by an organism. Bacterial pathogens produce a wide variety of toxins. If the toxin is made inside the bacteria and released or transported

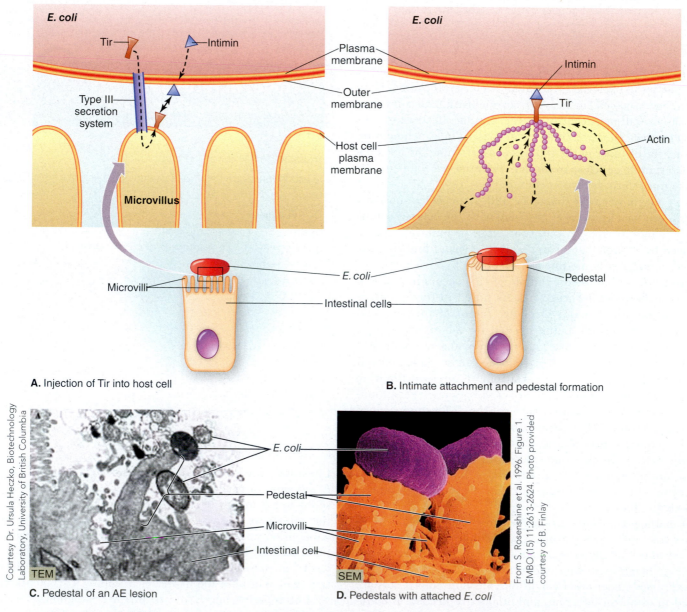

A. Injection of Tir into host cell

B. Intimate attachment and pedestal formation

Courtesy Dr. Ursula Heczko, Biotechnology Laboratory, University of British Columbia

C. Pedestal of an AE lesion

From S. Rosenshine et al. 1996. Figure 1. EMBO (15) 11:2613-2624. Photo provided courtesy of B. Finlay

D. Pedestals with attached *E. coli*

Figure 21.3. Formation of attachment and effacing (AE) lesion by enteropathogenic and enterohemorrhagic *E. coli* **A.** Contact with the host microvillus induces *E. coli* expression of the surface adhesin protein intimin and the receptor protein Tir. Tir is translocated into the host cell using a type III secretion system. **B.** Intimin binds Tir and Tir initiates polymerization of actin in the host cell, deforming the microvilli and forming a large pedestal-like structure characteristic of AE lesions. **C.** Pedestal formation on an AE lesion on rabbit intestinal cells. **D.** Pedestal formation by *E. coli*, shown purple in this color-enhanced micrograph on cultured human HeLa cells.

outside the cell, it is called an *exotoxin*. If the toxin is part of the cell wall structure, it is called an *endotoxin*, but this is somewhat inconsistent because the term "endotoxin" is historically reserved only for Gram-negative bacteria. Lipoteichoic acid (LTA), which is a Gram-positive cell wall-associated toxin, is not referred to as endotoxin; it is just called lipoteichoic acid. These toxins will be investigated in the next section. Toxins that act directly on host cells are called **cytotoxins**. Cell wall toxins are not cytotoxins because they do not exert their toxic activity directly on cells. Rather, they induce inflammation in the host.

Many pathogens cause disease strictly by virtue of the toxins they produce. Diseases due to toxin production are seen with pathogens such as *Clostridium botulinum*, *Clostridium tetani*, *Corynebacterium diphtheriae*, and *Bacillus anthracis*, the causative agents of botulism, tetanus, diphtheria, and anthrax, respectively. Without the toxins, these bacteria are non-pathogenic or are of very low pathogenicity. Other pathogens produce toxins that exacerbate (make worse) disease. Some cytotoxins, commonly called "leukocidins," kill host immune cells (leukocytes) facilitating the replication and spread of the pathogen. Other cytotoxins, commonly called "hemolysins," destroy red blood cells, releasing nutrients such as iron, needed for metabolism and growth of the pathogen.

Endotoxin and Lipoteichoic Acid

As early as the 1890s, scientists recognized a common heat-stable extract that was responsible for many of the toxic effects of systemic diseases caused by Gram-negative bacteria. They called this toxin "endotoxin," thinking it originated from inside bacteria and was released upon lysis. *Endotoxin*, the most common form of which is **lipopolysaccharide (LPS)**, is actually a component of the outer membrane of Gram-negative bacteria (see Figure 2.23). During an infection, it is released through bacterial replication and death. Released endotoxin is highly toxic to mammals, especially when it is present systemically. It is innately recognized as a pathogen-associated molecular pattern (PAMP) by the host. Recognition of endotoxin by cellular Toll-like receptors (TLRs) triggers the release of various cytokines that induce inflammation. Large-scale release of endotoxin during severe infections can trigger a systemic inflammatory response, producing septic shock and death. This is a common outcome when Gram-negative pathogens gain access to the circulatory system.

CONNECTION Inflammation is a common, early response to infection and is induced when microbial components are recognized by host TLRs. While localized inflammation is often beneficial, systemic inflammation can be deadly. For a review of septic shock, PAMPs, and TLRs, see **Sections 19.3** and **19.4**.

Severe inflammation in some tissues can cause organ damage. For example, *Neisseria meningitidis* is a cause of meningitis or inflammation of the meninges, the lining that surrounds the brain and spinal cord. When *N. meningitidis* gains access to the central nervous system from the bloodstream, it colonizes the meninges. Inflammation is produced primarily by the release of endotoxin. As cells and fluids move into the closed environment of the central nervous system, they rapidly increase the pressure on sensitive neurons, which become permanently damaged. If damage is extensive, death will occur. Inflammatory damage due to endotoxin release also occurs for its close relative, *N. gonorrhoeae* (see Microbes in Focus 21.1). When this pathogen reaches the fallopian tubes, released endotoxin evokes inflammation, which damages cells and leads to sterility. Unfortunately, this robust inflammatory response affords little protection against these pathogens.

Toxicity of the various forms of endotoxin produced by different Gram-negative pathogens varies with species and strain and also the location in the body. LPS is composed of three chemically distinguishable sections:

- a basal lipid A section containing unusual fatty acids anchored to the outer membrane

- a middle or core polysaccharide section of approximately 5 sugars with side chains

- an outer O-antigen section of repetitive polysaccharides, approximately 40 units in length, joined by O-linkages

The structure of LPS is presented in Figure 2.23. Each section is associated with specific biological properties, shown in **Figure 21.4**. Analysis of purified endotoxins from various species has shown that the lipid A component is responsible for the toxic properties. The core polysaccharide is usually conserved between genera or species. The O-antigen is a major antigenic target in the body, and antibodies are produced against it. Because the

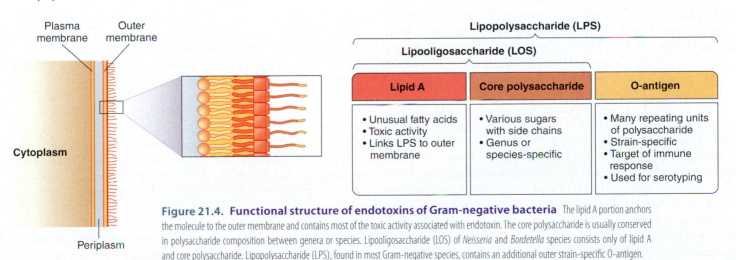

Figure 21.4. Functional structure of endotoxins of Gram-negative bacteria The lipid A portion anchors the molecule to the outer membrane and contains most of the toxic activity associated with endotoxin. The core polysaccharide is usually conserved in polysaccharide composition between genera or species. Lipooligosaccharide (LOS) of *Neisseria* and *Bordetella* species consists only of lipid A and core polysaccharide. Lipopolysaccharide (LPS), found in most Gram-negative species, contains an additional outer strain-specific O-antigen.

B. Kenny, R. DeVinney, M. Stein, D. J. Reinscheid, E. A. Frey, and B. B. Finlay. 1997. Enteropathogenic *E. coli* (EPEC) transfers its receptor for intimate adherence into mammalian cells. Cell 91:511–517.

Context

Enteropathogenic and enterohemorrhagic strains of *E. coli* deliver several virulence factors by their type III secretion systems. These are involved in the development of degenerative AE lesions, which are characteristic of these strains. The virulence factors needed for AE lesion development, including all the genes for production of the type III secretion system, are encoded adjacent to one another on a block of sequence known as the locus of enterocyte effacement (LEE). The LEE locus is a pathogenicity island (see Section 21.3). Included in the LEE locus is the gene encoding the bacterial adhesin *intimin*, needed for strong, "intimate" attachment to host cells. Mutants defective for intimin do not form pedestals or fully developed AE lesions.

Researchers knew that the initiation of this disease process occurs when intimin binds to a membrane receptor protein that was originally named Hp90 and later Tir. However, there were some strange aspects of this receptor. Hp90 was only present in cells with attached *E. coli* and was absent in uninfected cells. The research team, led by Brett Finlay at the University of British Columbia in Canada, set out to investigate the nature of the receptor Hp90. What they found was unprecedented.

Experiment

The first clues that something strange was occurring between host cells and *E. coli* came from earlier investigations done by Finlay's group. They produced polyclonal antibody to an enteropathogenic (EPEC) strain of *E. coli* by injecting the bacteria into a rabbit and later collecting the serum. They used this antibody mix on EPEC-infected and uninfected HeLa cells, a human epithelial cell line. Unexpectedly, the anti-EPEC antibody recognized a protein from infected cells that was the same size (90 kDa) as Hp90. Why would antibody raised to bacterial antigens bind to a host cell protein? At first, they thought this was due to antibody cross-reaction, where antibody made to an EPEC protein was recognizing a similar region on Hp90. However, they did not find the cross-reactive band in uninfected HeLa cells, or those infected with EPEC mutants lacking the type III secretion apparatus.

To investigate further, they made antibody that could recognize a specific region of Hp90. Hp90 has a phosphorylated tyrosine residue and antibody (anti-phosphotyrosine) was made to this. They then labeled EPEC proteins with the isotope ^{35}S, by growing cells in the presence of ^{35}S-methionine. When they infected HeLa cells with labeled *E. coli*, they found the anti-phosphotyrosine antibody bound to a labeled EPEC protein that was identical in size to Hp90. These results made the researchers think that maybe Hp90 was of bacterial origin and was delivered via type III secretion from EPEC.

Assuming Hp90 was actually of bacterial origin and that it was secreted, they then tried to locate it in EPEC culture supernatants. They could not find a secreted 90 kDa protein, but did find one that was slightly smaller in size (78 kDa) that had not been previously identified. They managed to find this protein by culturing EPEC under specific conditions to maximize secretion of proteins. They next purified the 78 kDa protein and made antibody to it. Using this anti-78 kDa antibody they located a 90 kDa EPEC cytoplasmic protein. Anti-phosphotyrosine antibody also recognized this EPEC 90 kDa protein, but it did not recognize the 78 kDa protein, indicating the 90 kDa protein was a phosphorylated cytoplasmic version of the 78 kDa secreted protein. The anti-78 kDa antibody and anti-phosphotyrosine antibody recognized Hp90 in the membranes of infected cells, indicating Hsp90 and the 90 kDa EPEC protein were one and the same **(Table B21.1)**. Hp90 was not a mammalian protein after all, and was renamed Tir for translocated intimin receptor.

Next, the researchers set out to find the gene for Tir. They designed a DNA probe based on the N-terminal amino acid sequence of the 78 kDa unphosphorylated version of Tir. The gene *tir* was

---

polysaccharide structure of the O-antigen is frequently unique to a particular strain, it can be used for identification by **serotyping (Toolbox 21.1**, p. 724). Serotyping is a diagnostic method that uses antibodies made to these substances for identifying closely related strains that differ in their surface antigens. A strain identified by serotyping is called a **serotype**.

LPS is found in enteric Gram-negative bacteria, such as *E. coli* and *Salmonella*. However, some Gram-negative genera, such as *Neisseria* and *Bordetella*, do not have the outer O-antigen component. These shortened non-LPS endotoxins—called **lipooligosaccharides (LOS)**—maintain toxic activity equivalent to that of LPS as they still contain the lipid A component (see Figure 21.4). Endotoxins have several biological activities, listed in **Table 21.4**. Many of the effects produced by low concentrations of endotoxin are important for stimulation of immune responses in the host that in turn can prevent disease.

One of endotoxin's most important properties is its ability to evoke inflammation. As we saw in Section 19.3, inflammation in a local area recruits defense cells and molecules needed to combat the pathogen, but systemic production of endotoxin can turn this protective response deadly.

Lipoteichoic acid (LTA) is a cell wall molecule found anchored to the plasma membrane in most Gram-positive pathogens. Similar to endotoxin, it is recognized by TLRs and can induce inflammation. If significant numbers of bacteria are present systemically, the release of lipoteichoic acid can, like endotoxin, induce septic shock. In this regard, it is the functional equivalent of Gram-negative endotoxin.

Exotoxins

Exotoxins are soluble proteins that are transported or released outside the bacterial cell. Many are actively secreted to the

	Membrane fraction		Insoluble fraction[a]	
	EPEC-infected HeLa cells	Uninfected HeLa cells	EPEC-infected HeLa cells	Uninfected HeLa cells
Anti-phosphotyrosine antibody	+	–	+	–
Anti-78 kDa antibody	+	–	++	–

[a] Insoluble fraction contains HeLa cell cytoskeleton proteins and EPEC proteins from attached bacteria of infected cells.

+ 90 kDa band observed

+ 78 and 90 kDa bands observed

– no bands observed

found to reside in the chromosome within the LEE locus along with all the other genes responsible for pedestal formation in infected cells, including the type III secretion system.

To confirm that Tir was of bacterial origin, they constructed a *tir* mutant by deleting a segment of the coding region. When cells were infected with the EPEC *tir* mutant, Tir could not be detected in infected cells, whereas it could be located in cells infected with wild-type EPEC. Furthermore, no pedestals were produced with the *tir* mutants, indicating Tir was required for this process. To confirm that Tir was injected by the LEE type III secretion system, they mutated one of the genes needed for assembly of the secretion system. Tir was not found in cells infected with this EPEC mutant, confirming it was injected into host cells.

Impact

This is the first time a bacterial receptor has been reported to be translocated into the host cell. With so few bacterial receptors having been characterized, it is not known if this may be a more widespread phenomenon. So far, the only other pathogen for which

injection of a receptor has been reported is *Citrobacter rodentium*, an intestinal pathogen of mice. *Citrobacter rodentium* is closely related to *E. coli*, and also contains the LEE locus. It makes similar AE lesions, and as you might have guessed, also translocates its Tir receptor into host cells.

● Questions for Discussion · · · · · · · · · · · · · · · ·

1. Propose reasons why *E. coli* mutants defective for intimin do not form pedestals or fully developed AE lesions.

2. Hemolytic uremic syndrome (HUS) is produced by Shiga toxins, which are not delivered by type III secretion. Instead, they are released by bacterial lysis. Do you think that ingestion of an intimin mutant strain of O157:H7 might still cause HUS? Give reasons why or why not.

Biological effects of endotoxin and lipoteichoic acid

TABLE 21.4

Effect	Protective activity
Fever	Inhibition of pathogen replication Increase in immune cell activities
Complement activation	Lysis by MAC formation Induction of inflammation
Inflammation	Transport of immune cells and molecules to site of infection
B cell proliferation	Antibody production
IFN-γ expression from T cells	Activation of macrophages and NK cells
Stimulation of clotting cascade	Prevention of pathogen spread

bacterium's external environment, while others are passively released upon bacterial lysis. Exotoxins have diverse origins, structures, and activities. They are commonly grouped by their structure (such as A-B toxins), for their molecular activity (such as superantigens), or for their cellular activity (such as cytotoxins).

Some of the most potent bacterial toxins are those that enter host cells and act selectively on cellular pathways. One of the most powerful toxins known is botulinum toxin or botulin (**Microbes in Focus 21.2**, p. 725). As little as 1 ng kg^{-1} is lethal to humans. This is an amount equivalent to about one billionth (0.000000001) of a grain of salt. Two kilograms, about the weight of three bricks, is estimated to be enough to kill every human on the planet. Other potent bacterial toxins include tetanus toxin or tetanospasmin of *Clostridium tetani*, and diphtheria toxin of *Corynebacterium diphtheriae*. These bacterial toxins have common structural features and are known as A-B toxins.

Serotyping uses antibodies to differentiate closely related strains or serotypes of a species, as they are often associated with particular diseases. It is frequently used to identify pathogens, link cases in a disease outbreak, and trace the origin and movements of a strain in populations. Serotyping is done on a glass slide by mixing a sample of the bacterial culture with specific antiserum—serum containing specific antibodies obtained by injecting a rabbit with a particular surface antigen or microbe carrying that antigen. If the unknown strain possesses the particular antigen, then antibody from the antiserum will bind to it and crosslink the bacterial cells, producing a visible agglutination reaction (**Figure B21.1**).

Serotyping can be done for just about any species of bacteria or any other microbe for which there is strain variation in surface antigens and to which antibody can be made. Serotyping for Gram-negative bacteria is commonly based on the LPS O-antigen, the capsule antigen (if one is present), and the flagella antigen. For *E. coli* and *Salmonella* species, this corresponds to the designation of O:K:H, respectively, for naming different serotypes. For example, *E. coli* O157:H7 identifies the O-antigen as reacting with antibody specific for type 157, and the flagella it possesses reacts with antibody raised to type 7 flagella protein. At least 160 different O variations exist for *E. coli* and more that 60 exist for *Salmonella*. The combinations of K and O antigens are unique enough that they can be used to identify specific strains of bacteria. In many cases, there is a relationship between the particular serotype designation and disease that is caused. For example, *E. coli* O157:H7 is an enterohemorrhagic strain that causes severe intestinal disease, while O15:K52:H1 is a uropathogenic strain, causing urinary tract infections (notice the K indicates this strain is also identified by its specific capsule components). Serotyping is really an indirect way

of identifying genetic constitution, as the surface antigens that the antibodies recognize are a reflection of the different genes possessed by these strains.

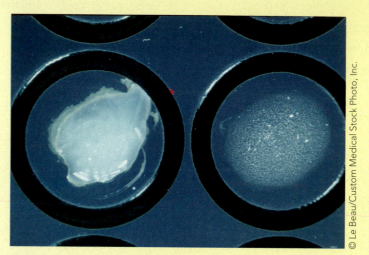

Figure B21.1. Serotyping A positive agglutination reaction is shown on the *right*, and a negative agglutination reaction is shown on the *left*.

© Le Beau/Custom Medical Stock Photo, Inc.

● **Test Your Understanding** ....

In 2010, an outbreak caused by *E. coli*-contaminated lettuce caused a nationwide recall in the U.S. The CDC reported the cause to be a Shiga toxin-producing *Escherichia coli* O145 strain. What does the O145 designation mean; how would serotyping be used to characterize this strain?

A-B Toxins A-B toxins are exotoxins that derive their name from their structure. They are generally composed of two structural parts, an enzymatically active A peptide and a membrane-binding B portion. A-B toxins are produced as inactive precursor proteins that must be cleaved by specific proteases to produce the A and B subunits that make up the active toxin. The location of toxin cleavage depends on the toxin. Cleavage can be done by proteases secreted by the host cell that act on the released toxin precursor. Alternatively, cleavage by proteases can occur inside the bacterium and the subunits remain held together by a disulfide bond. Some A-B toxins are composed of a single A and B subunit whereas others have multiple B subunits associated with a single A subunit peptide. Many variations exist, as shown in **Figure 21.5**.

The assembled A-B toxin must cross the plasma membrane of the host target cell to carry out its activity. The host cell plasma membrane is, however, an effective barrier to the passage of most large molecules. Those that do get through rely on specific transport mechanisms of the cell. A-B toxins take advantage of endocytosis for entry. Receptor-mediated endocytosis is induced by binding of the B portion of the toxin to specific cell-surface receptors. These receptors are usually cell-specific and therefore dictate the tissue

preference and species specificity of the toxin. For example, *E. coli* O157:H7 produces Shiga toxins, which are A-B toxins nearly identical to the Shiga toxins produced by *Shigella dysenteriae*. These *E. coli*

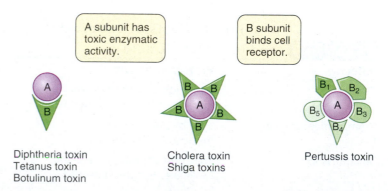

Figure 21.5. Structural variation of some A-B exotoxins Diphtheria toxin from *Corynebacterium diphtheriae*, tetanus toxin from *Clostridium tetani*, and botulinum toxin from *Clostridium botulinum* are composed of single A and B subunits. Cholera toxin from *Vibrio cholerae*, Shiga toxins from *Shigella dysenteriae*, and enterohemorrhagic *E. coli* O157:H7 are composed of a single A subunit with five identical B subunits. Pertussis toxin from *Bordetella pertussis* is composed of a single A subunit with five non-identical B subunits.

Microbes in Focus 21.2
CLOSTRIDIUM BOTULINUM

Habitat: Aquatic sediments and soils

Description: Anaerobic, Gram-positive bacillus, approximately 2–5 μm in length, forming terminal endospores

Key Features: Incidental ingestion of botulinum toxin or endospores (shown here as pale ovals within the bacilli) can cause deadly paralysis. Various strains produce biochemically distinct A-B type neurotoxins (types A to G) that affect various vertebrate species. The toxins are produced under anaerobic conditions in metabolically active vegetative cells. Near-identical non-toxigenic counterparts are taxonomically classed as different species, although the only major genetically distinguishing feature is the lack of the toxin. The genes encoding the toxins are acquired by horizontal gene transfer.

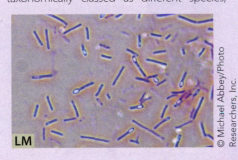
LM
© Michael Abbey/Photo Researchers, Inc.

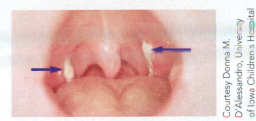

Courtesy Donna M. D'Alessandro, University of Iowa Children's Hospital

Figure 21.6. Diphtheria pseudomembrane Infection with toxigenic strains of *C. diphtheriae* frequently produces characteristic fibrin-encased, adherent lesions in the pharynx.

It exerts its lethal effects by preventing protein translation in these cells. The degree of damage incurred by individuals is related to the number and distribution of the receptor.

Figure 21.7 illustrates the translocation of diphtheria toxin. Inside the endosome, acidification causes conformational changes in the diphtheria toxin, allowing an exposed hydrophobic portion

Shiga toxins cause HUS. The Shiga toxin receptor, called globotriaosylceramide (Gb3), is found on various cell types including epithelial cells of the kidney and endothelial cells lining the blood vessels and organs. Once taken up, Shiga toxins disrupt protein synthesis by cleaving RNA of the ribosome. This can directly kill cells, but also indirectly causes activation of blood clotting in affected platelets. The clots that form obstruct the small capillaries in organs such as the kidney. As red blood cells attempt to flow through the narrowed passages, they rupture. Renal failure occurs when blood can no longer access kidney tissues. By this same pathology, other organs with rich microvasculature, such as the brain, can also become damaged temporarily or permanently.

The Shiga toxin receptor is enriched in the kidneys of humans but is lacking in cattle, deer, sheep, and pigs. These species serve as reservoirs of infection as they can carry these O157:H7 strains without becoming ill themselves. The amount and distribution of Shiga toxin receptors in humans varies and may help to explain why only 10 to 15 percent of persons infected with *E. coli* O157:H7 develop the complications of HUS.

Once the A-B toxin is internalized inside a membrane-bound endosome, the active A subunit must escape to the cytoplasm to exert its toxic activity within the cell. Different toxins have different ways of doing this. One of the best understood pathways is that of diphtheria toxin produced by *Corynebacterium diphtheriae*, the causative agent of diphtheria (**Microbes in Focus 21.3**, p. 727). In the pharynx, local cell death due to the toxin results in a distinctive fibrous, walled-off lesion called a pseudomembrane (**Figure 21.6**). The secreted toxin diffuses and is absorbed into the circulatory system where it is distributed to other tissues possessing toxin receptors, such as the heart, kidneys, liver, and spleen.

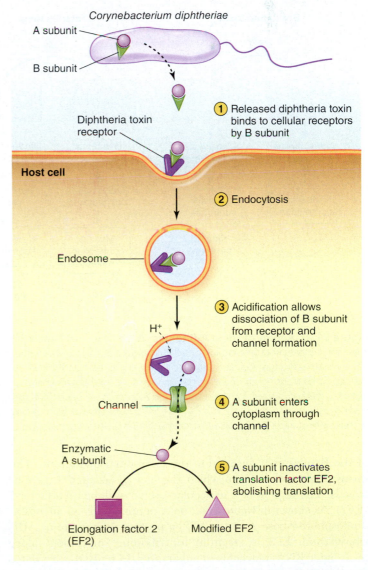

Figure 21.7. Process Diagram: Entry and action of diphtheria toxin Diphtheria toxin is taken up by host cells possessing the diphtheria toxin receptor. Acidification in the endosome causes a conformational change in B subunit, and a channel is formed to allow the A subunit to reach its molecular target EF2 in the host cytoplasm.

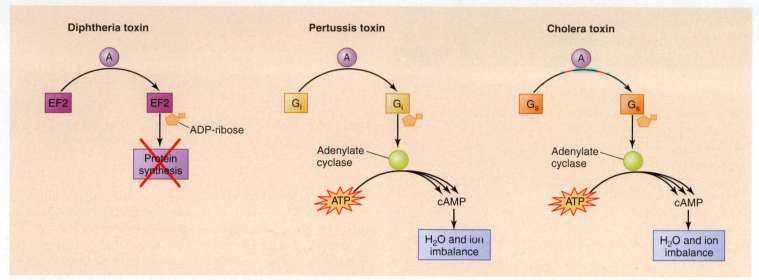

A. ADP-ribosylation of target

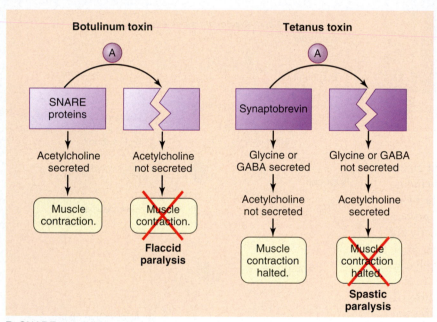

B. SNARE protein cleavage

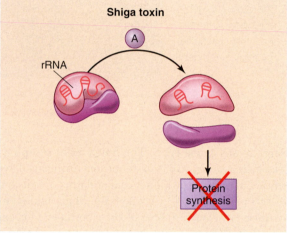

C. Ribosomal RNA cleavage

Figure 21.8. Action of some A-B toxins A. Diphtheria toxin *(left)* adds an ADP-ribose unit to translation factor EF2 halting protein synthesis, which damages cells. Pertussis toxin *(middle)* adds an ADP-ribose to the signaling protein G_i, locking it into an activated form, keeping host cell enzyme adenylate cyclase active, and converting ATP into cyclic AMP (cAMP). Excess cAMP disrupts the regulation of water and ion flow in cells, severely impairing cell functions. Cholera toxin *(right)* adds an ADP-ribose to G_s protein, an enzyme with function similar to G_i, resulting in increased cAMP levels in intestinal cells. **B.** Botulinum toxin cleaves SNARE proteins but in stimulatory nerve terminals, preventing muscular contraction. Tetanus toxin cleaves the SNARE protein synaptobrevin in inhibitory nerve terminals, causing muscular contraction to be continuous. **C.** Shiga toxins cleave ribosomal RNA (rRNA), preventing protein synthesis.

of the B subunit to insert into the endosomal membrane. This forms a channel through which the now dissociated A peptide travels. The A peptide's molecular target is elongation factor 2 (EF2). As shown in **Figure 21.8**, the A peptide adds an adenosine diphosphate-ribose group to EF2, a process referred to as ADP-ribosylation. EF2 is essential for ribosome movement along messenger RNA molecules in eukaryotes. In its ADP-ribosylated form, EF2 cannot function and protein production is halted. As the A subunit is an enzyme, a single molecule can catalyze the ADP-ribosylation of many EF2 proteins, resulting in inhibition of protein production and death of the cell. Many other A-B toxins

carry out ADP-ribosylation of their target molecules **(Table 21.5)**. Others, as in Figure 21.8, directly cleave proteins or ribosomal RNA molecules.

Two examples of A-B toxins that cleave host proteins are botulinum toxin and tetanus toxin, causing botulism and tetanus, respectively. Both are neurotoxins that are closely related in structure and function and act on members of a group of proteins, called SNARE proteins, which are needed for the release of neurotransmitters. Neurotransmitter molecules are contained in vesicles within neurons. To release neurotransmitters, the vesicles must fuse with the plasma membrane of the neuron.

TABLE 21.5 A-B toxins, source, activity, target cells, and disease

Toxin	Bacterium	Activity	Target cells (receptor-restricted)	Disease
Diphtheria toxin	*Corynebacterium diphtheriae*	ADP-ribosylation of EF2; inhibition of protein synthesis	Upper respiratory tract, heart, nerves, kidney	Diphtheria
Botulinum toxin	*Clostridium botulinum*	Cleaves SNARE proteins needed for release of the neurotransmitter acetylcholine	Motor neurons	Botulism
Cholera toxin	*Vibrio cholerae*	ADP-ribosylation of G_s protein; increases cAMP levels; cellular ion disruption	Cells of the intestinal tract where the bacterium colonizes	Cholera
Invasive adenylate cyclase	*Bordetella pertussis*	Increases cAMP levels; cellular ion disruption	Ciliated cells of the respiratory tract where the bacterium colonizes	Whooping cough (pertussis)
LT (heat labile)	Enterotoxigenic *E. coli* strains	ADP-ribosylation of G_s protein; increases cAMP levels; cellular ion disruption	Cells of the intestinal tract where the bacterium colonizes	Diarrhea
Pertussis toxin	*Bordetella pertussis*	ADP-ribosylation of G_i protein; increases cAMP levels; cellular ion disruption	Ciliated cells of the respiratory tract where the bacterium colonizes	Whooping cough (pertussis)
Shiga toxins	*Shigella dysenteriae* and enterohemorrhagic *E. coli*	Cleaves host ribosomal RNA; inhibition of cell protein synthesis	Kidney, liver, neurons	Hemolytic uremic syndrome (HUS)
Tetanus toxin	*Clostridium tetani*	Cleaves SNARE protein synaptobrevin needed for release of neuroinhibitors	Inhibitory neurons	Tetanus (lockjaw)

Microbes in Focus 21.3
CORYNEBACTERIUM DIPHTHERIAE

Habitat: Exclusively found in humans and is carried in the pharynx of some adults, less commonly on the skin.

Description: Gram-positive, non-motile, irregular non-endospore-forming rod 2–6 μm in length. Bacilli are often seen positioned at right angles to each other, giving them the appearance of lettering similar to cuneiform or Chinese script.

Key Features: Some strains of *C. diphtheriae* are toxigenic, producing the disease diphtheria. The A-B toxin that is responsible for the disease is actually encoded by a lysogenic phage that delivers the toxin gene to *C. diphtheriae*. The diphtheria toxin is secreted by *C. diphtheriae* cells colonizing the pharynx, nose, or sometimes skin. Before immunization became widespread in the 1940s, the death rate from diphtheria was approximately 10 percent, mostly in in young children.

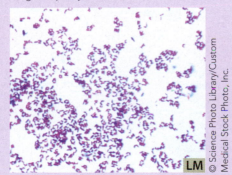

© Science Photo Library/Custom Medical Stock Photo, Inc.

LM

The SNARE proteins synaptobrevin, SNAP25, and syntaxin facilitate this membrane fusion. Synaptobrevin is found inserted in the vesicle membrane, and syntaxin is found inserted in the neuron plasma membrane. SNAP25 binds to both forming a complex that allows the membranes to fuse. Cleaving any of the SNARE proteins (Table 21.6) prevents the release of neurotransmitters that mediate muscle movement. Although these toxins act on the same cellular process, they produce clinically different diseases. Botulinum toxin prevents muscle contractions, resulting in "flaccid" paralysis, while tetanus toxin produces continuous muscle contraction, resulting in "spastic" paralysis. Several types of botulinum toxins are produced by different strains of *C. botulinum*, with certain types preferentially affecting different animal species and acting on a different set of SNARE proteins. Although botulinum toxin and tetanus toxin both bind to stimulatory nerve terminals of motor neurons, which release the stimulatory neurotransmitter acetylcholine, they use

TABLE 21.6 Specific SNARE protein targets of botulinum and tetanus toxins

SNARE target	Toxin
Synaptobrevin	Tetanus toxin, botulinum toxin type B, D, F, G
SNAP-25	Botulinum toxin type A, C, E
Syntaxin	Botulinum toxin type C

different receptors and are transported to different target neurons. Botulinum toxin acts within the motor neuron, while tetanus toxin is transported from the motor neuron to the central nervous system to inhibitory nerve terminals that release the inhibitory neurotransmitters glycine and gamma aminobutyric acid (GABA).

Once the A peptides of the toxins are released to the cytoplasm, they cleave their SNARE protein target, preventing the fusion of neurotransmitter-filled vesicles with the neuron plasma membrane. For botulinum toxin, this prevents the acetylcholine from being released at the neuromuscular junction, and muscle contraction does not occur **(Figure 21.9)**.

For tetanus toxin, once the A subunit enters inhibitory neurons, it prevents the release of glycine or gamma-amino butyruic acid (GABA). Glycine and GABA act on motor neurons to block acetylcholine release, ending contraction. Thus, tetanus toxin allows acetylcholine to be continually released from stimulated neurons, causing the associated muscles to persistently contract **(Figure 21.10)**. Anyone who has suffered a severe leg-cramp might be able to imagine how painful this would be. Any small stimulus can trigger a massive contraction so powerful that tendons and even bones can snap **(Figure 21.11)**.

Tetanus is also known as "lockjaw" because the large masseter muscle is often the first to be noticeably affected. In both botulism and tetanus, death is usually due to respiratory failure as the muscles involved in breathing no longer function normally. Both toxins can act on a variety of animal species. Botulinum toxins are responsible for periodic massive die-offs of fish and aquatic birds, as described in **Perspective 21.1**, p.730.

Humans are probably inadvertent victims of botulism. Endopores of *C. botulinum*, like those of *C. tetani*, sometimes contaminate deep wounds and can germinate under anaerobic conditions to produce the toxin. More commonly, however, botulism is a result of the ingestion of contaminated food (see Section 16.2). In adults, ingestion of the pre-formed toxin causes disease whereas ingestion of *C. botulinum* itself, or its endospores, does not. The human intestinal microbiota provides a sufficiently competitive environment to prevent germination and growth. However, infants lack protective adult microbiota. Consequently, germination of endospores occurs in the anaerobic environment of the infant colon, and the toxin is produced. Infant botulism is the commonest form of the disease and is suspected to be one possible cause of sudden infant death syndrome.

A-B toxins are often targeted for use as vaccines because they are principally responsible for the diseases that result from exposure to or infection by the bacteria that produce the toxins. For example, vaccines for diphtheria, tetanus, and pertussis contain denatured A-B toxins called **toxoids**.

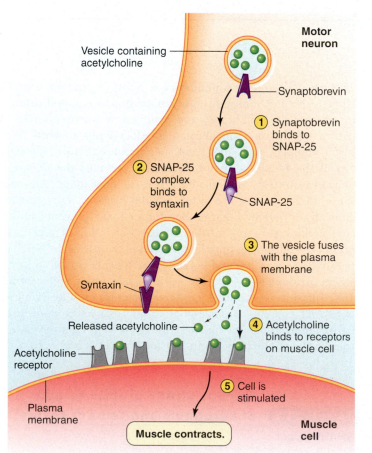

A. Normal function of motor neuron

B. Action of botulinum toxin type A on motor neuron

Figure 21.9. Process Diagram: Action of botulinum toxin **A.** Normal muscular contraction involves release of the neurotransmitter acetylcholine from vesicles in terminals of motor neurons. **B.** Botulinum toxin type A cleaves the SNARE protein SNAP-25 needed for release of acetylcholine. Consequently, muscular contraction does not occur.

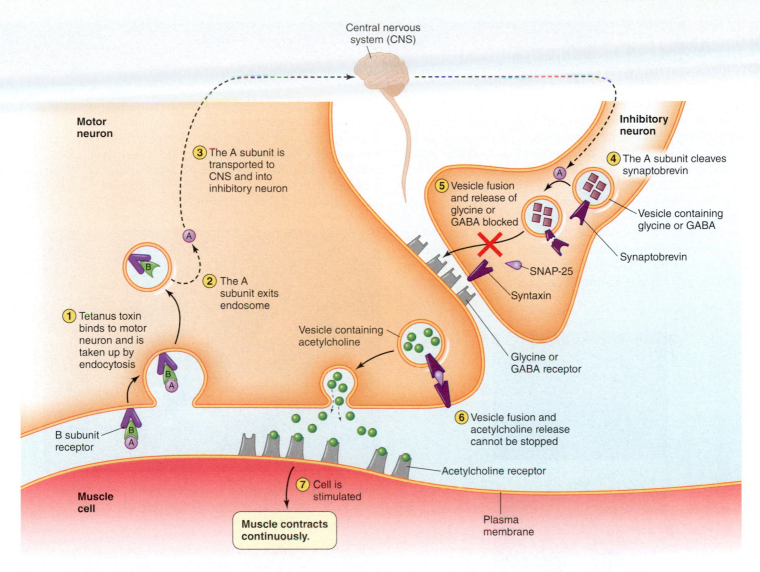

Figure 21.10. Process Diagram: Action of tetanus toxin Once stimulated, normal muscular contractions can be halted by the release of the neuroinhibitory neurotransmitters glycine or gamma-amino butryric acid (GABA) from associated inhibitory neurons. Tetanus toxin is taken up by motor neurons and is transported to the central nervous system, then to inhibitory nerve terminals that release inhibitory neurotransmitters. The A subunit cleaves the SNARE protein synaptobrevin in inhibitory neurons needed for release of glycine and GABA. Consequently, muscular contraction is continuous.

Charles Bell (1774–1842), Oposthotunus. Reproduced by kind permission of The Royal College of Surgeons of Edinburgh

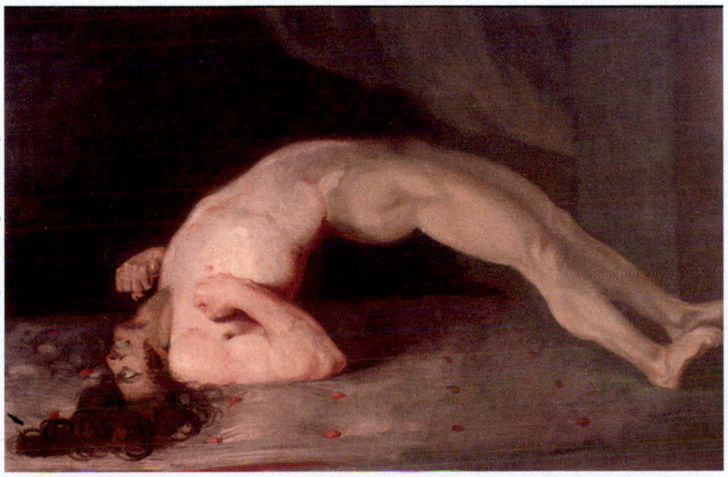

Figure 21.11. Effect of tetanus toxin Muscular contractions can be powerful enough to damage the spine and sever tendons. In this early nineteenth-century painting, Scottish anatomist Sir Charles Bell portrays a soldier with tetanus.

Toxoids lack the structure required for toxic activity but retain the ability to generate protective antibody against the active toxins. Denaturing the toxin either prevents binding and translocation of the B subunit into the cell and/or destroys enzymatic activity of the A subunit. Diphtheria and tetanus vaccines (the D and T in the DTaP vaccine blend) are two of the safest and most effective vaccines produced today. They are highly immunogenic, producing a long-lasting protective humoral response, and are composed of a single protein rather than a mixture. This factor reduces the risk of side effects. Pertussis vaccine was developed against *Bordetella pertussis*, the agent of whooping cough. The vaccine is an acellular pertussis (the aP in DTaP) formulation composed of denatured pertussis toxin in combination with two or more purified proteins used for adherence to ciliated epithelial cells of the upper respiratory tract. These proteins are filamentous hemagglutinin, pertactin, or fimbriae. Why are these added? Remember that production of protective

Botulinum toxin isn't all bad. Since the 1980s, it has been used to treat severe muscle spasms (dystonia), tension headaches, and various other neuromuscular conditions. It is perhaps now more popularly known as Botox®, used for the cosmetic diminishing of "frown" wrinkles caused by contraction of small muscles of the face **(Figure B21.2)**. Local injection of small amounts of the toxin blocks neuroactivation of muscle contraction in the area for several months. Botox® is composed of type A botulinum toxin, as are some of its commercial equivalents. Other cosmetically used formulations are composed of type B toxin. Recent studies have shown that equivalent cosmetic application of type A botulinum toxin in rats is transported to the brain. It remains to be seen if there are any long-term adverse effects in humans who use Botox®.

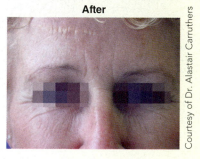

Before

After

Courtesy of Dr. Alastair Carruthers

Figure B21.2. Botox® cosmetic treatment for wrinkles Botox® is injected to relax facial muscles associated with "frown" wrinkles.

Type A botulinum toxin is responsible for most human deaths related to botulism. However, major outbreaks caused by type E botulinum toxin have occurred in fish and aquatic birds. The fish likely feed on endospores or toxin-contaminated insects or decaying plant material found in sediments. Botulinum toxin-affected fish make easy prey but contain the deadly botulism cargo. Fish-eating birds such as red-breasted mergansers **(Figure B21.3)**, horned grebes, loons, and seagulls are often victims of type E fish kills. In the fall of 1999, several thousand such migrating aquatic birds died of botulism in Lakes Erie, Huron, and Ontario. Their bodies littered the shorelines.

Courtesy Ian K. Barker, Canadian Cooperative Wildlife Health Centre

Figure B21.3. Red-breasted merganser victims of botulism A botulism outbreak caused by type E botulinum toxin on Lake Erie in 1999 killed several thousand aquatic birds.

In 1997, a botulism outbreak caused by type C1 botulinum toxin occurred in a shallow lake in Saskatchewan, Canada. An estimated 1 million ducks died there during their spring return migration. The toxin and *C. botulinum* endospores accumulated in decomposing material along marshy shorelines and lakes and were taken up by insect larvae and snails. These were then ingested by dabbling ducks, such as mallards and redheads. Domestic animals such as horses and cattle that have access to shorelines may also contract the disease this way. A classic sign of avian botulism is a flaccid neck, and the disease is often referred to as "limberneck" **(Figure B21.4)**. The inability of the fowl to hold their heads above water results in drowning.

The role of *C. botulinum* and its non-toxigenic counterparts appears to be that of an environmental decomposer. The deadly toxin

antibody to the toxin will prevent disease but will otherwise do nothing to prevent infection by the pathogen. These proteins are added to reduce the ability of the pathogen to colonize the respiratory tract, thus preventing its spread in populations.

CONNECTION The immunization schedule for diphtheria and tetanus using the DTaP combined vaccine as well as other currently recommended vaccines is presented in **Figure 24.19**. Vaccines are discussed in more length in **Section 24.5**.

Cytolysins Cytotoxins that act on plasma membranes are called **cytolysins**. They are commonly, and rather arbitrarily, named according to their origin, for example streptolysin is from *Streptococcus pyogenes*. They may also be named according to their order of discovery (as in α-toxin), or the cells

Figure B21.4. "Limberneck" Flaccid paralysis of the neck is characteristic of waterfowl botulism.

presumably ensures production of a carcass that will supply both the anaerobic conditions and nutrients needed for the proliferation of *C. botulinum* (**Figure B21.5**).

Botulinum toxin is considered to be one of the top three biological weapon threats, behind anthrax and smallpox. Terrorists have already attempted to use the toxin as a bioweapon. In 1990, the Aum Shinrikyo cult planned to disperse aerosolized toxin at multiple sites in Japan. They built a *C. botulinum* culture facility and even stables for horses from which they planned to produce antisera (antibody) to the toxin. Thankfully, they did not properly culture the bacterium and the toxin was not produced. However, the Aum group turned its attention to other agents. In 1995, they released sarin, a chemical nerve gas, in the Tokyo subway. Twelve people died, and many more were severely injured.

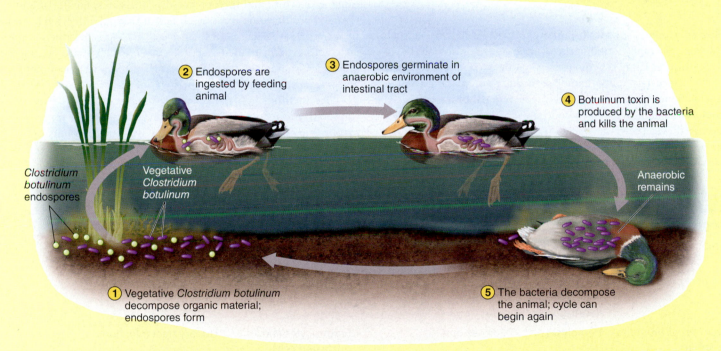

② Endospores are ingested by feeding animal

③ Endospores germinate in anaerobic environment of intestinal tract

④ Botulinum toxin is produced by the bacteria and kills the animal

Clostridium botulinum endospores

Vegetative *Clostridium botulinum*

Anaerobic remains

① Vegetative *Clostridium botulinum* decompose organic material; endospores form

⑤ The bacteria decompose the animal; cycle can begin again

Figure B21.5. Process Diagram: Ecological role of botulinum toxin Vegetative *C. botulinum* cells decompose organic plant and animal matter in anaerobic aquatic soils. Endospores form and animals, such as ducks feeding in shallow waters, incidentally ingest the endospores and/or the toxin. Endospore germination occurs in the anaerobic environment of the lower intestine and toxin production occurs from vegetative cells. The toxin kills the host and the body becomes a new habitat for decomposition, beginning the cycle again.

they can destroy (for example, hemolysins lyse red blood cells). The ability of many cytolysins to lyse a variety of eukaryotic cell types is readily visualized by culture on agar containing red blood cells, commonly called blood agar. Usually, sheep red blood cells are used for this media. Lysis of blood cells, called **hemolysis**, produces a zone of clearing. Hence, many cytolysins have been named hemolysins because of this reaction.

Hemolysis patterns are characteristic of some species and are often used as a diagnostic tool by clinical bacteriologists. This is particularly the case for identification of *Streptococcus* species, which can produce three patterns of hemolysis on blood agar. Lysis that results in a zone of complete, colorless clearing is called "β hemolysis" (**Figure 21.12**). Beta hemolysis is produced by pathogenic strains of *Streptococcus pyogenes*. Production of a zone with a dark greenish tinge is called "α hemolysis." Alpha hemolysis is produced by

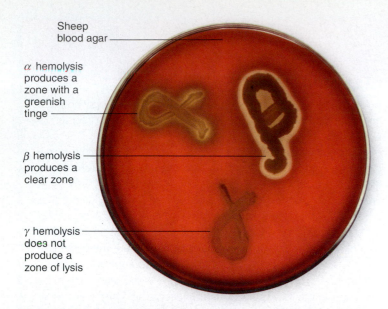

Sheep blood agar

α hemolysis produces a zone with a greenish tinge

β hemolysis produces a clear zone

γ hemolysis does not produce a zone of lysis

Figure 21.12. Hemolysis patterns of *Streptococcus* species *Streptococcus pyogenes* produces β hemolysis, a zone of colorless, complete clearing. Most other streptococci produce α hemolysis, a zone with a distinct greenish tinge. The absence of hemolysis, produced by *Enterococcus faecalis*, is called γ hemolysis. (Courtesy Christine Dupont)

Streptococcus pneumoniae as well as many species of *Streptococcus*, such as *S. mutans*, normally found in the oral cavity (see Microbes in Focus 17.2). Interestingly, α hemolysis is not due to complete cell lysis. The greenish cast seen in α hemolysis is due to peroxide-mediated chemical changes that cause discoloration of partially decomposed hemoglobin. The absence of hemolysis is called "γ hemolysis"; a misnomer if ever there was one. Gamma hemolysis is characteristic of the bacterium *Enterococcus faecalis*, a close relative of the genus *Streptococcus*, and can be used to help distinguish it from pathogenic streptococci. The molecular activities of cytolysins fall into two general categories: pore-forming toxins and membrane-degrading toxins (lecithinases). **Table 21.7** lists some examples of membrane-damaging cytolysins and their activities.

Pore-forming cytolysins are produced as monomers that polymerize in the membrane to form a circular pore. They anchor to molecules that are specifically found concentrated in eukaryotic membranes and not in bacterial membranes. Alpha-toxin from *S. aureus* is secreted as monomers that bind to unspecified lipid receptors. The monomers oligomerize, joining together and refolding to form a complex, making a transmembrane segment that directly inserts into the lipid bilayer **(Figure 21.13)**. Perfringolysin of *Clostridium perfringens* **(Microbes in Focus 21.4)** binds to membrane cholesterol, which is found concentrated in rafts within the plasma membrane. In both cases, pore formation leads to ion fluxes that disrupt membrane function and, with extensive pore formation, can lead to cell lysis. *Clostridium perfringens* perfringolysin contributes to the tissue destruction in gas gangrene, a disease characterized by muscle necrosis, fever, pain, swelling, and intense gas production resulting from vigorous carbohydrate fermentation under anaerobic conditions. Gaseous pockets form within tissues, causing pressure and further necrosis of tissues, facilitating invasion and proliferation in tissue. If left untreated, infection leads to shock and death **(Figure 21.14)**. *Streptococcus pyogenes* produces a similarly acting toxin named streptolysin O, which shares extensive amino acid homology with perfringolysin (60 percent homology), and *pneumolysin* (48 percent homology) of *Streptococcus pneumoniae*.

Pore-forming cytolysins can damage cells in a variety of ways. At high concentrations, they form many large holes in plasma membranes, causing cell lysis. The release of cellular contents, especially the vesicle contents of phagocytes, causes damage and necrosis to neighboring cells and induces

TABLE 21.7 Source, action, targets, and diseases produced by cytolysins

Pathogen	Toxin	Activity	Target cell	Disease
Clostridium difficile	Toxin A Toxin B	Form pores	Colon cells	Colitis and diarrhea
Clostridium perfringens	α-toxin[a] (lecithinase) Perfringolysin	Degrades membrane lipids Forms pores	Many cell types	Gas gangrene
Escherichia coli uropathogenic strains	Hemolysin A	Forms pores	Erythrocytes and nucleated cells	Urinary tract infections
Helicobacter pylori	VacA	Forms pores	Gastric mucosa	Gastritis and ulcers
Listeria monocytogenes	Listeriolysin Lecithinase	Forms pores in endosome Degrades membrane lipids	Macrophages Many cell types	Meningitis and septicemia
Pseudomonas aeruginosa	*Pseudomonas aeruginosa* cytotoxin	Forms pores	Many cell types	Skin infections and pneumonia
Staphylococcus aureus	α-toxin[a] PV leukocidin	Forms pores Forms pores	Many cell types Phagocytes	Skin infections Necrotizing skin and lung infections
Streptococcus pneumoniae	Pneumolysin	Forms pores	Lung and blood vessel cells	Pneumonia
Streptococcus pyogenes	Streptolysin O	Forms pores	Variety of cells	Pharyngitis/tonsillitis, rheumatic fever, rheumatic heart disease

[a] Although both of these toxins are named α-toxin, they are unrelated in structure, function, and origin. To avoid confusion, it is necessary to state the microbial origin of the toxin.

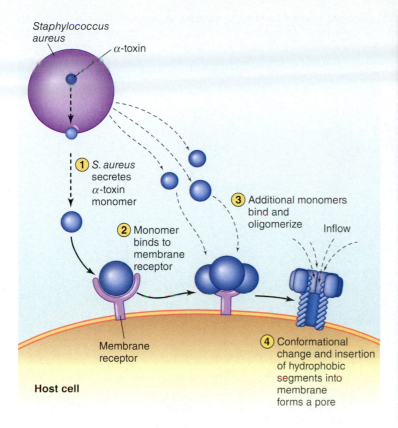

Figure 21.13. **Process Diagram: Pore formation by cytolysins**
Staphylococcus aureus α-toxin, like many pore-forming cytolysins, binds to host cells as monomers that oligomerize causing a conformational change in the peptide units that insert to form a pore in the membrane.

The diagram shows:

Staphylococcus aureus

α-toxin

① *S. aureus* secretes α-toxin monomer

② Monomer binds to membrane receptor

③ Additional monomers bind and oligomerize

Inflow

Membrane receptor

Host cell

④ Conformational change and insertion of hydrophobic segments into membrane forms a pore

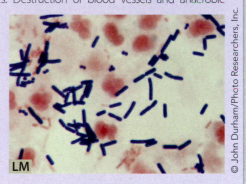
inflammation. Direct lysis may not be the only mechanism of cell death. Low concentrations of pore-forming toxins often produce fewer and smaller pores that can cause ion fluxes that admit calcium ions, which signal cellular production of vasoreactive substances that induce inflammation. The severe inflammation in the lung that characterizes pneumonia produced by *Streptococcus pneumoniae* is due to production of pneumolysin, which acts in this manner.

Sustained calcium influx produced by membrane damage induces cell death by apoptosis. Damaging cells allows pathogens to invade tissues. An example is *Staphylococcus aureus* (**Figure 21.15**). Most strains produce α-toxin, sometimes called α-hemolysin, which is a major virulence factor of this pathogen. Mutant strains lacking α-toxin are less pathogenic. In serious infections with *S. aureus*, cell damage by α-toxin generates inflammation that can lead to septic shock, especially if the toxin is present in the circulatory system. Many pathogens use cytolysins to damage tissues, allowing them to infiltrate, spread, and obtain nutrients.

Some cytolysins can efficiently form pores only under acidic conditions, as is found in the endosome. An example is listeriolysin, produced by the food-borne pathogen *Listeria monocytogenes*. Following ingestion, *L. monocytogenes* is taken up by intestinal macrophages. Production of listeriolysin by phagocytosed bacteria allows disruption of the endosome membrane, preventing intracellular killing and allowing escape of the bacterium into the cytoplasm of the macrophage, where it can replicate (see Figure 21.15B). Other cytolysins, such as VacA of *Helicobacter pylori*, and Panton-Valentine (PV) leukocidin of *S. aureus*, can enter through the pores they make in the plasma membrane and damage the membranes of mitochondria, inducing apoptosis of the cell (see Figure 21.15C). PV leukocidin is produced as two different polypeptides that combine to form pores in susceptible cells. The toxin primarily attacks phagocytes and epithelial cells, causing extensive damage by apoptosis and necrosis, resulting in necrotizing skin and lung infections. Approximately 2 to 5 percent of

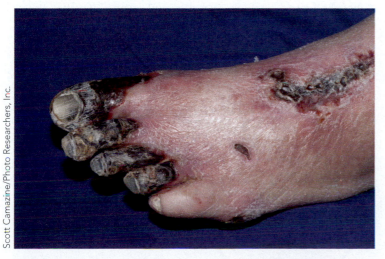

Scott Camazine/Photo Researchers, Inc.

Figure 21.14. Gas gangrene Tissue necrosis, as evidenced by blackening of the toes and swelling on the top of the foot, results from perfringolysin and gas production by *C. perfringens*.

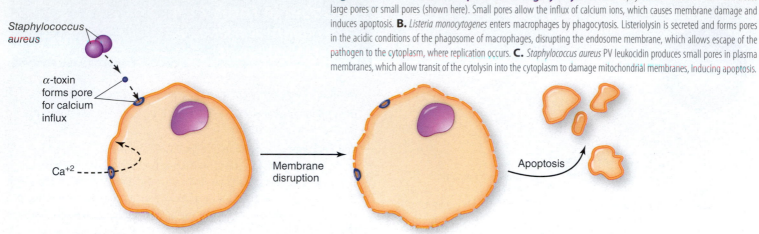

Figure 21.15. Actions of pore-forming cytolysins **A.** *Staphylococcus aureus* α-toxin can create large pores or small pores (shown here). Small pores allow the influx of calcium ions, which causes membrane damage and induces apoptosis. **B.** *Listeria monocytogenes* enters macrophages by phagocytosis. Listeriolysin is secreted and forms pores in the acidic conditions of the phagosome of macrophages, disrupting the endosome membrane, which allows escape of the pathogen to the cytoplasm, where replication occurs. **C.** *Staphylococcus aureus* PV leukocidin produces small pores in plasma membranes, which allow transit of the cytolysin into the cytoplasm to damage mitochondrial membranes, inducing apoptosis.

A. Action of *Staphylococcus aureus* α-toxin

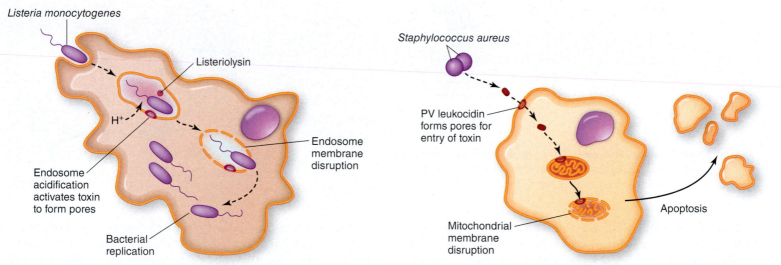

B. Action of *Listeria monocytogenes* listeriolysin

C. Action of *Staphylococcus aureus* PV leukocidin

S. aureus strains produce the PV leukocidin, which is encoded by a lysogenic phage (see Section 21.3). An estimated 90 percent of severe *S. aureus* necrotizing skin and lung infections are due to these lysogenized PV leukocidin-producing strains. Fortunately, most people can mount protective immune responses to these strains of *S. aureus* and prevent spread of infection. Immunocompromised persons, however, are placed at serious risk when infected with these strains. More recently, PV leukocidin-producing strains have acquired resistance to multiple antibiotics, including methicillin, making them especially difficult to treat and control spread in hospitals and the community.

CONNECTION Methicillin-resistant *S. aureus* (MRSA) is a serious pathogen because it is often not diagnosed until conventional antibiotic treatment fails. For more information on methicillin resistance, see **Section 24.3**.

In some cases, it appears that pore formation has an auxiliary function for translocation of other bacterial molecules into the cell. The best understood case is streptolysin O, produced by *Streptococcus pyogenes*, which translocates the bacterial enzyme NAD-glycohydrolase into the host cell. Here, the enzyme disrupts various cell signaling pathways, contributing to cell death. Pore

formation by Gram-positive pathogens may thus be the functional equivalent of the type III secretion system used by many Gram-negative pathogens to deliver their virulence factors to host cells (see type III secretion systems in this section; also Figure 2.25).

Another group of cytolysins are the **lecithinases**, also called "phospholipases," which degrade eukaryotic plasma membranes. Lecithins are a group of various phospholipids, such as phosphatidylcholine, found in eukaryotic membranes. Lecithins are enriched in egg yolk, and egg yolk-containing agar provides a simple test for identification of bacteria capable of producing lecithinases **(Figure 21.16)**. Lecithinases cause disruption

Figure 21.16. Detection of *Clostridium perfringens* lecithinase (α-toxin) The right side of the egg-yolk agar plate shows a zone of precipitated insoluble diglycerides around *C. perfringens* colonies and is the result of hydrolysis of phospholipids (lecithins) by α-toxin. Antitoxin antibody has been added to the left side of the plate prior to inoculation as a control.

© CDC/Dr. Stuart E. Starr

of membranes, damaging cells and stimulating the production of various proinflammatory cytokines. *Listeria monocytogenes* produces a lecithinase homologous to that of *Bacillus cereus* and *C. perfringens* α-toxin (note that this is one of those cases where a toxin has been named rather arbitrarily; it is not at all related to *S. aureus* α-toxin).

Superantigens Superantigens are exotoxins with the ability to activate CD4+ T cells. As we learned in Section 20.2, activated CD4+ T cells are potent producers of *cytokines*. Cytokines are secreted cell signaling molecules that stimulate immune responses and inflammation. Normally, CD4+ T cells are activated through the binding of their T cell receptors (TCRs) with specific peptide presented on MHC II molecules of antigen-presenting cells, such as macrophages, dendritic cells, or B cells. Activation

also requires secondary interaction between the MHC II molecule of the antigen-presenting cell and the CD4 of the bound T cell (Figure 21.17; see also Figure 20.14). The proportion of T cells expected to have TCRs that can recognize a particular peptide is very low, estimated to be less than 0.001 percent of the body's T cells. Thus, normal T cell activation results in limited T cell proliferation and localized cytokine production needed for appropriate stimulation and direction of immune response at the site of infection.

CONNECTION CD4+ T cells are the central lymphocytes involved in the adaptive immune response. For a review of normal activation of CD4+ T cells, see **Sections 20.2, 20.3**, and **20.5**.

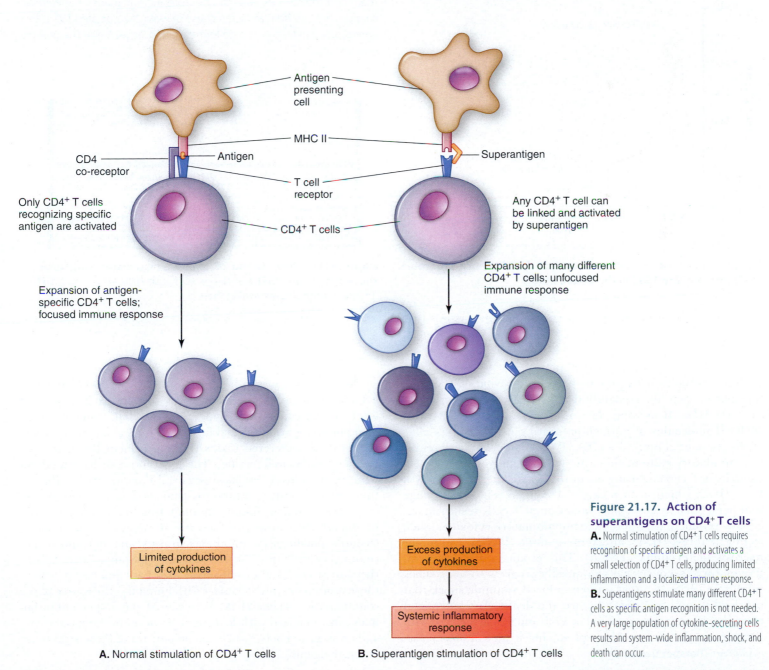

Figure 21.17. Action of superantigens on CD4+ T cells
A. Normal stimulation of CD4+ T cells requires recognition of specific antigen and activates a small selection of CD4+ T cells, producing limited inflammation and a localized immune response.
B. Superantigens stimulate many different CD4+ T cells as specific antigen recognition is not needed. A very large population of cytokine-secreting cells results and system-wide inflammation, shock, and death can occur.

Labels within figure:
Antigen presenting cell
MHC II
CD4 co-receptor
Antigen
T cell receptor
CD4+ T cells
Superantigen

Only CD4+ T cells recognizing specific antigen are activated
Any CD4+ T cell can be linked and activated by superantigen

Expansion of antigen-specific CD4+ T cells; focused immune response
Expansion of many different CD4+ T cells; unfocused immune response

Limited production of cytokines
Excess production of cytokines
Systemic inflammatory response

A. Normal stimulation of CD4+ T cells
B. Superantigen stimulation of CD4+ T cells

Toxic shock syndrome (TSS) was first recognized in 1978. It is a rare but potentially fatal disease that results from bacterial production of superantigen toxins known as TSS-toxins (TSST). In 1980, there were several hundred cases of TSS in the United States. The vast majority of these were associated with *Staphylococcus aureus* and the use of a brand of high-absorbency tampons called Rely® that were introduced to the market at this time **(Figure B21.6)**.

Synthetic polyacrylate fibers used in the Rely® brand tampons were shown to increase the risk of TSS. These materials are no longer used, and Rely tampons were withdrawn from sale by 1981. The incidence of menstrual-related TSS fell dramatically. Tampon manufacturers are now required to provide product labeling and recommended usage information to help women avoid the disease **(Figure B21.7)**.

How do these toxigenic strains end up in tampons? *Staphylococcus aureus* routinely colonizes skin and nasal mucosa. It is thought that tampon-associated TSS occurs when TSST-producing *S. aureus* strains happen to colonize the vagina, probably having been transferred from the skin. The nutrient-rich environment provided by the tampon and the neutral conditions of the vagina during menstruation support the growth of *S. aureus*, allowing it to multiply and produce the toxin, which enters the bloodstream. High-absorbency tampons tend to be changed less frequently, therefore increasing the risk of TSS. Today, approximately 50 percent of TSS cases are still associated with tampon use. Other cases of TSS are associated with wound infections in men, women, and children. *Staphylococcus aureus* is not the only pathogen that can produce TSS. Superantigen-producing strains of *Streptococcus pyogenes* are frequently the cause of wound-associated TSS. Like *S. aureus*, *S. pyogenes* can also be found on skin, as well as in the pharynx (see Section 21.2).

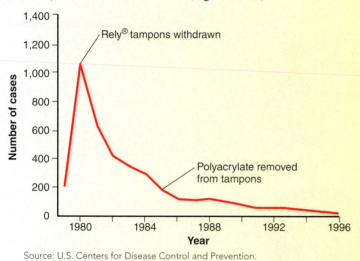

Source: U.S. Centers for Disease Control and Prevention.

Figure B21.6. Cases of tampon-associated toxic shock syndrome (TSS) from 1979 to 1998 in the United States

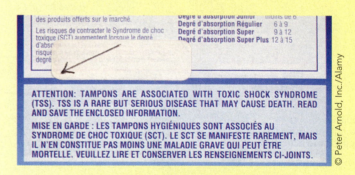

Figure B21.7. Rely® brand of tampons associated with toxic shock syndrome (TSS) Rely® tampons were voluntarily withdrawn from sale in 1981 after hundreds of cases of TSS were linked to their use.

Superantigens activate CD4+ T cells by avoiding the need for specific peptide recognition as well as the requirement for CD4:MHC II binding. They directly crosslink TCRs with MHC II molecules of antigen-presenting cells, such as macrophages (see Figure 21.17B). When superantigens enter the circulatory system, they can activate up to 20 percent of the body's T cells, causing them to secrete the T cell growth factor IL-2, which in turn causes T cells to replicate to huge numbers. Such large-scale stimulation of T cells results in excessive production of potent proinflammatory cytokines, such as interleukin-1 (IL-1), interferon-gamma (IFN-γ), and tumor necrosis factor-alpha (TNF-α). This produces a systemic inflammatory response that can quickly progress to toxic shock syndrome (TSS), a superantigen-induced systemic shock that can lead to multiple organ failure and death. An epidemic of toxic shock syndrome occurred in 1980 and was associated with tampon use and superantigen-producing strains of *Staphylococcus aureus* **(Perspective 21.2)**.

Not all superantigen-producing bacteria are associated with TSS; different superantigens vary greatly in their potency and level of expression. The superantigens that cause common food poisoning are produced by strains of *S. aureus* that encode **enterotoxins**, secreted toxins that act principally on the gut ("entero" refers to intestine). These enterotoxins have weak superantigen activity, but have powerful emetic activity, that is, they induce vomiting. Enterotoxigenic strains of *Staphylococcus aureus* are commonly found contaminating foods that have been improperly cooked or insufficiently refrigerated. Mayonnaise, custards, puddings, cream fillings, gravies, and lunch meats are common culprits in staphylococcal food poisoning, especially as they are often found in self-serve salad bars and buffets where there is often extensive contact with humans—the source of the contamination **(Figure 21.18)**. Between 20 and 50 percent of humans are colonized with *Staphylococcus aureus* in the nasal cavity and on skin, and between 20 and 50 percent of these strains are enterotoxigenic.

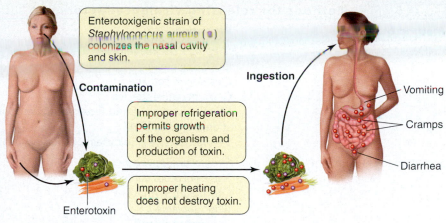

Enterotoxigenic strain of *Staphylococcus aureus* (●) colonizes the nasal cavity and skin.

Contamination

Improper refrigeration permits growth of the organism and production of toxin.

Improper heating does not destroy toxin.

Enterotoxin

Ingestion

Vomiting

Cramps

Diarrhea

A. *Staphylococcus aureus* enterotoxin can be found in foods

B. Enterotoxigenic *Staphylococcus aureus* is transferred from humans to food and causes illness

Figure 21.18. ***Staphylococcus aureus* enterotoxin commonly causes food poisoning A.** Many of the rich sauces and dressings present in self-serve buffets are ideal for growth of enterotoxigenic strains of *S. aureus*, particularly if the food is not adequately refrigerated or cooked. **B.** *Staphylococcus aureus* commonly colonizes the nasal cavity and skin and can be transferred from here to foods. Improper refrigeration allows replication of *S. aureus*. If the strain is enterotoxigenic, then the toxin will be produced. Underheating the food may kill *S. aureus*, but may not destroy the toxin. Ingestion of the food results in absorption of the toxin, which causes vomiting, diarrhea, and cramps.

Enterotoxins are not easily denatured by heat but *S. aureus* can be destroyed by heating food thoroughly (60°C, 140°F or above), and proper refrigeration (7.2°C, 45°F or below) will inhibit its growth, preventing production of the toxin. The amount of enterotoxin ingested is rarely enough to cause toxic shock and symptoms usually disappear within 48 hours.

CONNECTION *Staphylococcus aureus* is salt tolerant and can survive in salt-preserved foods, like lunch meats and soups. See **Section 16.1** for a review of factors affecting microbial growth in foods.

Co-infection with Gram-positive superantigen producers, such as *S. aureus* or *S. pyogenes*, and Gram-negative bacteria are of particular concern as endotoxin may also be released to the bloodstream. As previously mentioned, endotoxin alone can precipitate septic shock. As both endotoxin and superantigens act to induce inflammation, their dual presence produces a much more potent toxic effect than either toxin alone **(Figure 21.19)**.

You may wonder what advantage superantigen production affords the pathogen. Superantigen production provokes an unfocused immune response that does not efficiently target the pathogen, allowing it to continue to reproduce and thus successfully spread. In addition, stimulated T cells are short-lived, and such extensive proliferation means a large proportion of CD4$^+$ T cells undergo apoptosis. The result is temporary immunosuppression, which may also afford advantages to the pathogen.

Type III and type IV secretion systems

While some toxins and other virulence factors are released by lysis of the bacterial cell, many, like *S. aureus* α-toxin, are purposefully secreted to the environment, and still others are actually injected directly into the host cell **(Table 21.8)**.

Type III and type IV secretion systems are major virulence factors that are often hallmarks of Gram-negative pathogens (see Table 21.1). **Type III secretion systems** are injection assemblies that deliver virulence factors, such as toxins, directly into the target cell by forming a tunnel that passes through both the plasma membrane and outer membrane of the pathogen and through the host cell plasma membrane. Type III secretion systems are the more common of the two injection systems and are often found encoded together with the gene products they will inject. Later in the chapter (Section 21.3), we will see that these blocks of sequence are found on pathogenicity islands, which we were introduced to in Section 10.3. It is not unusual for some strains of pathogens to possess more than one kind of type III secretion system. For example, *Salmonella* Typhimurium, a foodborne pathogen that causes enteric infection (salmonellosis) carries two different type III secretion systems. One is used

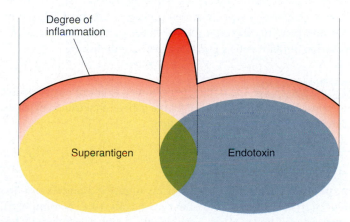

Degree of inflammation

Superantigen

Endotoxin

Figure 21.19. Synergistic effect of endotoxin and superantigens Superantigens and endotoxin can each induce the production of cytokines causing inflammation. When present together, they cause excessive cytokines release, increasing inflammation and enhancing the potential for shock and death.

TABLE 21.8 Methods for releasing virulence factors

Method of release	Pathogen	Virulence factor
Cell lysis	*Streptococcus pneumoniae*	Pneumolysin
	Clostridium tetani	Tetanus toxin
	Clostridium botulinum	Botulinum toxin
	Gram-negative species	Lipopolysaccharide
	Shigella dysenteriae and *E. coli* enterohemorrhagic strains, e.g., O157:H7	Shiga toxins
	Several invasive *E. coli* strains, including O157:H7	Enterohemolysin
Secretion to environment	*Streptococcus pyogenes*	Streptolysin O; superantigens
	Staphylococcus aureus	α-toxin; superantigens
	Clostridium perfringens	α-toxin, perfringolysin
	Vibrio cholerae	Cholera toxin
	Bordetella pertussis	Pertussis toxin
	Helicobacter pylori	VacA
	Listeria monocytogenes	Listeriolysin
Injection into host cell	*E. coli* enterohemorrhagic and enteropathogenic strains	Tir
	Salmonella Typhimurium	SipA, SopE; various proteins preventing phagolysosome fusion
	Pseudomonas aeruginosa	Exoenzyme S, exoenzyme U
	Helicobacter pylori	CagA
	Agrobacterium tumefaciens	Ti plasmid, VirE2, VirD2

for injecting the proteins SipA and SopE, involved in remodeling of the intestinal cells to which it attaches to induce uptake of the cells by intestinal enterocytes (**Figure 21.20**). The other is used inside the phagosome for injecting proteins needed to prevent phagosome fusion with lysosomes. The bacilli survive and replicate in the phagosome of macrophages and enterocytes, where they can avoid host immune defenses.

Equivalent in function are **type IV secretion systems**, which have evolved from conjugation systems and can transport a variety of substances, including toxins directly across both bacterial membranes. Some type IV secretion systems deposit these into the immediate environment. For example, *Bordetella pertussis* secretes pertussis toxin at the host cell surface this way. Others have type IV systems that, like type III systems, can traverse the host plasma membrane. For example, *Helicobacter pylori* uses a type IV secretion system to inject CagA toxin into gastric mucosal cells. The plant pathogen *Agrobacterium tumefaciens* (see Microbes in Focus 11.2) uses a type IV secretion system to inject Ti plasmid DNA and protein virulence. Both factors are needed for transformation of plant cells to produce crown gall disease.

CONNECTION The natural transforming ability of *Agrobacterium tumefaciens* has been harnessed for the genetic engineering of plants. See **Section 12.5** for more information.

Capsules

A **capsule** is an extracellular, loosely organized matrix of polymers that surrounds some bacterial cells (see Section 2.5). Capsules are most often composed of polysaccharides, but sometimes can be polypeptides or a mixture of both. Their role in pathogenicity is primarily to protect the bacterium from phagocytosis and lysis from complement, both of which are important early immune defenses (see Sections 19.4 and 19.5). Many pathogens owe their success to possession of a capsule (**Table 21.9**, p. 740), and it serves as their primary virulence factor. Mutants of these strains that do not produce capsules are non-pathogenic as they are easily phagocytosed and destroyed by the host. You may recall Griffith's demonstration of this with *Streptococcus pneumoniae* from Section 7.1.

Capsules can interfere with phagocytosis and complement activation in several ways. Host opsonization molecules, such as complement C3b and antibody against cell wall antigens, can diffuse through the capsule and bind to the cell surface. However, their opsonization role is rendered ineffective as they remain buried in the capsule and cannot be contacted by phagocyte receptors needed for efficient phagocytosis and killing (**Figure 21.21**; see Figures 19.16 and 20.36). Anti-capsule antibody provides protection from encapsulated pathogens because opsonization of the capsule allows effective uptake and killing by phagocytes.

Successful avoidance of phagocytosis by extracellular pathogens means that few bacterial components enter the endocytic pathway of macrophages and therefore few are presented to T cells. Without strong T cell stimulation, B cell clonal expansion does not occur and consequently, less antibody directed against the pathogen will be made. Without sufficient antibody to bind to the cell, complement factor C1 of the classical pathway is not efficiently fixed and formation of MAC (membrane attack complex) does not occur (see Figures 19.7 and 19.9). In this way, the capsule also protects the pathogen from lysis by complement.

The alternative pathway of complement activation does not depend on antibody binding, so presumably MAC formation can still occur to lyse the bacterial cells. While this appears to be the case with many capsule-forming strains of bacterial pathogens,

Figure 21.20. Process Diagram: Pathogenesis of *Salmonella* Typhimurium

Attachment to enterocytes of the intestinal mucosa is by fimbriae. Two different type III secretion systems are used for injection of virulence factors that induce the formation of pseudopodia for uptake of the bacilli and prevent phagosome fusion with lysosomes. Death of intestinal cells and macrophages is due to apoptosis, which releases *S.* Typhimurium and spreads infection.

① *Salmonella* Typhimurium attach by fimbriae

Enterocyte (intestinal epithelial cell)

② Type III secretion system injects SipA and SopE, inducing phagocytosis

Type III secretion system

③ A second type III secretion system injects proteins that prevent phagosome fusion with lysosomes

Type III secretion system

④ Bacteria replicate in phagosome

⑤a *Salmonella* exit by exocytosis

⑤b Enterocyte undergoes apoptosis, releasing *Salmonella* to infect more cells

Apoptotic body

⑥ Uptake by macrophage

Macrophage

⑦ Bacteria replicate in macrophage phagosome

⑧ Macrophage undergoes apoptosis, releasing *Salmonella* to infect more cells

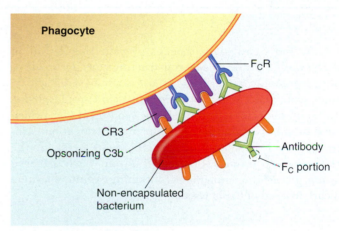

A. Non-encapsulated bacterium

F_CR

CR3

Opsonizing C3b

Antibody

F_C portion

Non-encapsulated bacterium

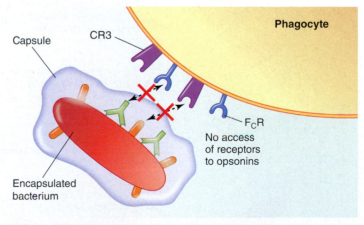

B. Encapsulated bacterium

Capsule

CR3

Phagocyte

F_CR

No access of receptors to opsonins

Encapsulated bacterium

Figure 21.21. Capsules inhibit opsonization and phagocytosis
A. Non-encapsulated bacteria are opsonized with C3b and are readily phagocytosed by binding with C3b receptors (CR3) of phagocytes. Similarly, antibody-coated bacteria are bound by phagocyte receptors (F_CR) that bind the exposed antibody F_C portions to facilitate phagocytosis. **B.** In encapsulated bacteria, any C3b or antibody that moves through the capsule to attach to the cell surface will not be available for phagocyte binding.

TABLE 21.9 Bacterial pathogens possessing capsules as a major virulence factor

Pathogen	Capsule composition	Disease
Bacillus anthracis	Poly-D-glutamic acid	Anthrax
Campylobacter jejuni	Various polysaccharides	Invasive intestinal disease
Enterococcus faecalis	Various polysaccharides	Urinary tract infections, various hospital-acquired invasive diseases
Escherichia coli K1	Sialic acid	Neonatal meningitis
Haemophilus influenzae	Six different capsule types; one is polyribitol phosphate	Pneumonia, meningitis
Klebsiella pneumoniae	>77 different capsule polysaccharide types	Pneumonia, meningitis
Neisseria meningitidis	Sialic acid	Meningitis
Pseudomonas aeruginosa	Alginate (biofilm formation)	Pneumonia
Salmonella Typhi	*N*-acetylglucosamine uronic acid	Typhoid fever
Staphylococcus aureus	Various amino sugars	Invasive skin diseases, pneumonia
Streptococcus agalactiae	Sialic acid	Neonatal meningitis
Streptococcus pneumoniae	>90 different capsule oligosaccharide types	Meningitis, pneumonia
Streptococcus pyogenes	Hyaluronic acid	Variety of invasive and toxic infections

some of the most virulent pathogens prevent this activation pathway as well. They do this by constructing capsules made of molecules commonly found in their host, such as sialic acid or hyaluronic acid. Sialic acid and hyaluronic acid capsules, being made of "self" molecules, discourage the binding of C3b, which is the first step in the alternative complement activation pathway. Prevention of the deposition of complement factors on the bacterial surface is termed **serum resistance**. Such bacterial cells, when placed in complement-containing serum, continue to replicate instead of being lysed.

Pathogens that construct their capsules of host "self" molecules are afforded a further degree of protection. Immune responses, including antibody production and complement activation, are not normally elicited by the host's own components, making the pathogen a wolf in sheep's clothing. Thus, possession of a molecular "self" capsule prevents production of opsonizing capsule antibody, which would otherwise allow efficient phagocytosis. The various ways capsules protect pathogens from host immune responses are summarized in **Figure 21.22**.

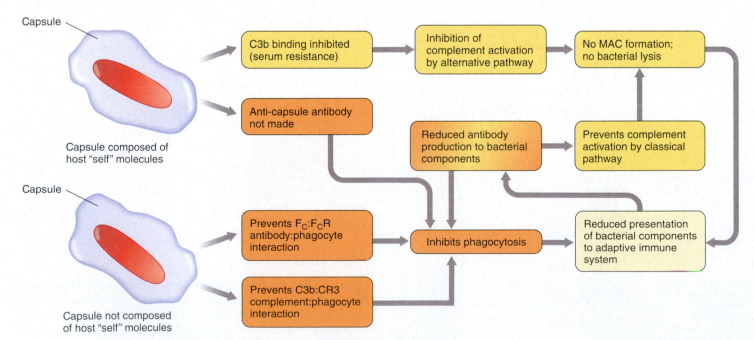

Figure 21.22. Protective mechanisms of capsules Capsules commonly inhibit processes involved in complement activation (shown in yellow) and phagocytosis (shown in orange). Consequently, activation of the adaptive response is inhibited and clearance is reduced.

Avoiding some or all of the host's early defenses, such as complement activation and phagocytosis, buys an opportunity for the pathogen to replicate to numbers that can outpace host immune defenses. By the time a protective adaptive immune response is mounted to the capsule, the pathogen is often already well established. As we learned in Section 20.6, T-independent immune responses to capsule components are essential in bridging the time between first infection and the production of protective anti-capsule antibodies. This response allows rapid production of anti-capsule antibodies before a routine T cell-dependent primary adaptive response can be made. The memory response is similarly important in protecting the host from these pathogens. The faster and larger memory immune response shortens the time to production of protective antibody (see Figure 20.6).

CONNECTION The major virulence factor of *Streptococcus pneumoniae* and *Haemophilus influenzae*, and *Neisseria meningitidis* is their capsule, which allows these pathogens to avoid provoking a T cell immune response. See **Perspective 20.2** for details.

Iron-binding proteins

In Section 19.2, we mentioned that iron is a limiting nutrient for pathogens in the body since it is kept tightly bound by host iron-binding proteins. How then do bacterial pathogens obtain iron? Pathogens have evolved a variety of general strategies to get the iron they need **(Figure 21.23)**.

1. They produce their own iron-binding molecules, called **siderophores** that can scavenge iron from the host by competing with host iron-binding proteins, such as lactoferrin. Some siderophores bind iron very strongly and can steal it away from host iron-binding proteins.

2. They produce proteins that can bind host iron-binding proteins, and then transport the iron into the cell for themselves, letting the host do the iron binding for them.

3. They produce metabolites that lower the pH at the site of infection, which reduces the ability of host iron-binding proteins to bind iron, making more available.

4. They produce cytolysins that lyse host cells to release their iron stores. Producing hemolysins to lyse red blood cells, for example, is an effective way to obtain iron as well as other nutrients.

Comparative studies have shown that while many bacterial species have iron-binding proteins for binding free iron, pathogens commonly have specialized proteins for the task of wrestling iron away from the host. These iron-binding proteins are virulence factors as they are not produced by non-pathogenic species, and gene knockouts are impeded in their ability to establish disease.

A well-studied mechanism of iron acquisition is that of *Mycobacterium tuberculosis*. Inside the phagosome, the bacterium produces citric and salicylic acid, which cause the release of iron in the phagosomes. The iron then binds to siderophores called "mycobactins," for transport into the bacterium. Mycobactins can also directly steal iron from host lactoferrin, present in the phagolysosome. As described in **Perspective 21.3**, supplementing iron in tuberculosis patients can actually facilitate disease.

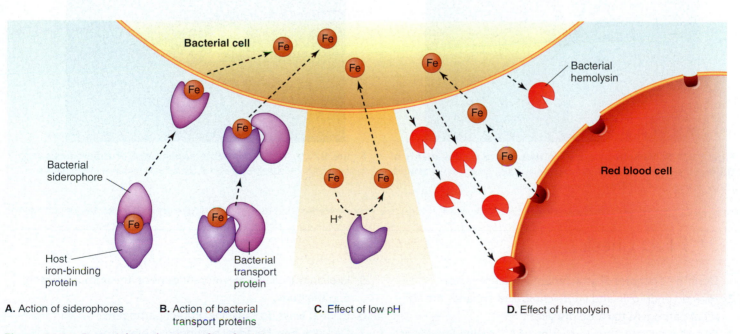

A. Action of siderophores   **B.** Action of bacterial transport proteins   **C.** Effect of low pH   **D.** Effect of hemolysin

Figure 21.23. Bacterial mechanisms for obtaining iron A. Siderophores compete with host iron-binding proteins for iron. **B.** Bacterial transport proteins bind host iron-binding proteins and transport the iron complex to the cell. **C.** Bacterial metabolic products lower the pH, inhibiting the iron-binding ability of host proteins, making it available for bacterial use. **D.** Hemolysins destroy host cells, such as red blood cells, to release intracellular iron.

Tuberculosis was once considered to be a disease of the nineteenth and mid-twentieth centuries, before the advent of the antibiotic streptomycin. Back then (in fact, up until around the 1960s) tuberculosis patients were sent to treatment facilities called sanatoriums where they received fresh air, nursing care, and rest. Those who could afford it often ate iron-rich meat. This habit served to essentially feed iron to *Mycobacterium tuberculosis*, causing these patients to often fare worse than their poorer fellow patients who remained on iron-poor diets. As early as 1872, frequent relapses were documented in patients given a diet rich in iron while recovering from clinical tuberculosis.

Tuberculosis was commonly called "the white plague" due to the pallor that patients frequently developed. "Consumption" was another popular name, as those suffering from tuberculosis appeared to be being consumed, or wasting away —a biological effect of the overproduction of the cytokine tumor necrosis factor alpha (TNF-α).

Physical wasting, weakness, pale skin, and coughing up blood were thought to be the hallmarks of vampires, and many believed tuberculosis patients were actually being visited and slowly consumed by these creatures of myth **(Figure B21.8)**.

Tuberculosis was a disease that did not discriminate between rich or poor, famous or nameless. From 1930 to 1950, the look of tuberculosis actually became fashionable as it reached the most elite of society. Pale, thin, and physically weak (therefore, vulnerable) became the new standards of beauty. Many popular film stars of the time adopted the appearance **(Figure B21.9)**.

Today, tuberculosis is back and it has a new look. It is a disease associated with poverty, drug use, overcrowding, and HIV. Many strains are resistant to multiple drugs commonly used to treat tuberculosis, including streptomycin, and these strains are spreading fast. The white plague has returned. Today, approximately one-third of the human population is infected with *M. tuberculosis*.

Bettman/© Corbis

Figure B21.8. Vampires were once thought to be the agents of tuberculosis The clinical signs of tuberculosis, such as physical wasting, pale skin, and bloody sputum, were considered to be evidence of vampire visitations.

© Clarence Sinclair Bull/John Kobal Foundation/Getty Images, Inc.

Figure B21.9. The fashionable appearance of tuberculosis Many movie stars of the mid-1900s affected the look of tuberculosis, as shown here by Marlene Dietrich.

21.1 Fact Check

1. Describe examples of attachment virulence factors.
2. How do endotoxin and lipoteichoic acid enhance the ability of pathogens to cause disease?
3. Describe the structure and action of A-B toxins. Provide an example.
4. What are the two main modes of action of cytolysins? Provide examples.
5. Explain the effects of superantigens on the immune system and the body.
6. In what ways do capsules protect pathogens?
7. Describe how bacterial pathogens obtain iron.

21.2 Survival in the host: Strategies and consequences

How do bacterial pathogens cause disease?

We have learned that pathogens must carry out three basic processes in order to establish themselves in the host. They must (1) gain access to host tissues, (2) overcome host defenses, and (3) get nutrients. We have examined several virulence factors, more or less in isolation, that help pathogens do this. Let's now look at two comprehensive examples that illustrate how bacterial pathogens can cause disease. The first pathogen we will examine is *Streptococcus pyogenes*, an extracellular opportunistic pathogen that many of us harbor unwittingly. The second, *Mycobacterium tuberculosis*, is an intracellular primary pathogen. Both of these pathogens cause significant human disease worldwide, but use very different means to do so.

Pathogen study 1: *Streptococcus pyogenes*

Streptococcus pyogenes (**Microbes in Focus 21.5**) is a particularly resourceful opportunistic pathogen, thanks to the remarkable variety of virulence factors it can produce, many of which have multiple functions. Think of it as a microscopic Swiss army knife of virulence factors. *Streptococcus pyogenes* is commonly found colonizing the pharynx in 10 to 15 percent of people, can also be found on the skin, and sometimes in the vagina. Most of those colonized have no clinical signs of *S. pyogenes* disease, but the pathogen can be said to be biding its time, waiting for an opportunity. Stated another way, it's often a disease waiting to happen. *Streptococcus pyogenes* is one of the most common bacterial pathogens isolated from infections, and is probably best known as a dominant cause of pharyngitis and tonsillitis in children.

The variety of diseases this bacterium causes (**Table 21.10**) is attributable to (1) the large degree of genetic variation that exists in different strains, providing not only numerous combinations of virulence factors, but also variation in expression of these factors, and (2) various host conditions that allow these strains to produce disease. Predisposing conditions to *S. pyogenes* infection generally involve wounds, weakened immune functions, and poor circulation. Matching the many streptococcal diseases with particular strains or virulence factors has proven to be elusive. For example, cases of streptococcal necrotizing fasciitis, commonly called "the flesh-eating disease" (**Figure 21.24**) often tend to occur in clusters, which would suggest association with particular strains, but there is little evidence for this globally, and molecular analysis of the isolates causing necrotizing fasciitis so far has not led to any consensus on the virulence factors that are primarily responsible for this disease. Examination of the host conditions leading to necrotizing fasciitis similarly shows great diversity, but can be linked to immune status. These include preexisting diabetes mellitus, HIV infection, genetic immune deficiencies, and immunodeficiency produced by medications. Even common anti-inflammatory medications, such as ibuprofen, have

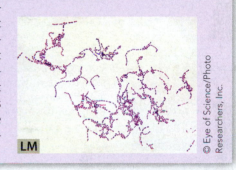
been implicated in the precipitation of this disease. A singular commonality seems to be that it often begins at an otherwise innocuous wound site. Treatment with antibiotics may be ineffective as the death of blood vessels makes penetration into the tissues difficult. The anaerobic conditions in the damaged tissue prevents migration of phagocytes and impedes intracellular killing mechanisms. Often surgical removal of dead tissue, including amputation, is the only effective treatment.

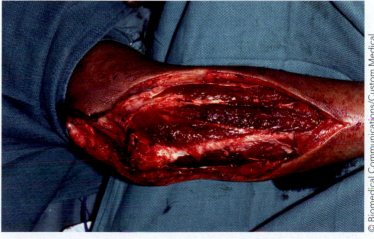

Figure 21.24. Necrotizing fasciitis Necrotizing fasciitis is an invasive skin disease that results in severe, rapid tissue necrosis.

© Biomedical Communications/Custom Medical Stock Photo, Inc.

TABLE 21.10 Diseases produced by *Streptococcus pyogenes*

Disease	Description
Cellulitis © Dr. P. Marazzi/Photo Researchers, Inc.	Inflammation of deep tissues
Erysipelas © ISM/Phototake	Delineated infection of the dermis. Intense inflammation, pain. Infection can spread to produce cellulitis.
Glomerulonephritis 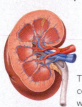 Tortora, Derrickson: *Principles of Anatomy and Physiology*, 13e, copyright 2012. John Wiley & Sons, Inc. This material is reproduced with permission of John Wiley & Sons, Inc.	Inflammatory sequela following *S. pyogenes* infection. Pathogenesis is thought to involve autoimmune processes that damage glomeruli of kidneys. The condition is not limited to *S. pyogenes* infection.
Impetigo 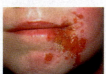 © Phototake	Contagious, spreading infection of the skin, frequently found in association with *S. aureus* in children. Begins with small wound.
Necrotizing fasciitis  © Dr. M. A. Ansary/Photo Researchers, Inc.	Necrotic destruction of tissues, spreading along fascia layer covering the muscles
Puerperal fever © Sara Wight/Getty Images, Inc.	Historic postpartum infection of women resulting from poor hygienic practices of doctors moving between mothers in hospital wards. In 1848, Semmelweis, a Hungarian doctor, was the first to recognize the disease as an infection. He required strict handwashing among maternity ward physicians, drastically reducing rates of infection.
Pharyngitis/tonsillitis © Scott Camazine/Phototake	Contagious infection of the throat/tonsils characterized by intense inflammation and pus-filled nodules. Usually self-limiting.

Disease	Description
Rheumatic heart disease and acute rheumatic fever 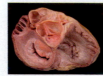 © Mark Nielsen	Inflammatory sequelae following pharyngeal infection with SPE-producing strains. Ongoing fever, pain in joints, and fatigue. Can affect skin, joints, central nervous system, and heart. Pathogenesis is thought to be due to autoimmune processes brought on by streptococcal antigens, such as M5 (see Figure 21.25). Permanent damage to heart valves is known as rheumatic heart disease.
Scarlet fever 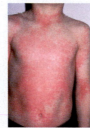 © Biophoto Associates/Photo Researchers, Inc.	Diffuse red skin rash; fever produced by toxigenic strains producing erythrogenic toxins SPE A, B, C.
Septicemia  Tortora, Derrickson: *Principles of Anatomy and Physiology*, 13e, copyright 2012. John Wiley & Sons, Inc. This material is reproduced with permission of John Wiley & Sons, Inc.	Infection that has gained access to the bloodstream; can cause septic shock. The condition is not limited to *S. pyogenes*.
Toxic shock syndrome (TSS)  © Phototake	Wound-associated infection with toxigenic strains producing superantigens. Similar disease is caused by *S. aureus*.

Some virulence factors of *S. pyogenes* are associated with particular diseases. For example, strains possessing genes encoding superantigens (*S. pyogenes* possesses many of these) can cause wound-related TSS (see Perspective 21.2). Some of these superantigens are named SPE toxins, for streptococcal pyrogenic erythrogenic toxin ("pyro," meaning fire, refers to fever, and "erythro," meaning red, refers to rash). They are frequently associated with invasive infections, including scarlet fever. Several variations of SPE toxin genes exist, most being carried by lysogenic phages, but only types A, B, and C are associated with scarlet fever (see Table 21.10).

Access to Host Tissues

The primary residence of *Streptococcus pyogenes* is the pharynx, where it exists either transiently or permanently with the same or different strains present over time. *Streptococcus pyogenes* is probably seeded from the pharynx to skin and occasionally the vagina. Major attachment factors of *S. pyogenes* include fimbriae and a variety of other surface components (most of which have not been well characterized) that allow attachment to host fibronectin and collagen. M protein, a component of fimbriae, allows aggregation of cells and contributes to the formation of microcolonies on the skin and pharynx.

Multiple factors, having as much to do with strain variation as host conditions, likely keep the pathogen at bay. Normal microbiota probably have a significant impact on colonization. If changes in microbiota populations occur, *S. pyogenes* may expand its population, or change expression of genes, initiating inflammation at the site. This development can lead to damage and initiation of infection. Physical damage to skin or mucosa allows *S. pyogenes* to directly access tissue, and is often how skin infections begin.

A recent finding has shown that *S. pyogenes* can induce its own uptake in epithelial cells of the mucosa, thought to be a significant reservoir of the pathogen. Here, it escapes microbial competition as well as host immune defenses—particularly antibody—and even antibiotics. Recall that *Salmonella* species also induce their own uptake in epithelial cells to avoid immune defenses, although the mechanism used to induce uptake is different (see Figure 21.20).

Disease Consequences

Pharyngitis and tonsillitis are usually painful but self-limiting infections characterized by inflammation and accumulation of neutrophils that form pus-filled nodules on the tonsils or throat (see Table 21.10). Protective IgA and IgG antibodies against M protein are eventually produced, and the infection is cleared. If you've ever had streptococcal pharyngitis ("strep" throat), it usually takes about ten days to start feeling better without antibiotics. This coincides with peak antibody production from a primary adaptive immune response (see Figure 20.25). However, *Streptococcus pyogenes* may still remain in small numbers, either inside epithelial cells or extracellularly colonizing the site. There are over 150 distinct types of M proteins, and antibody protection against one type does not necessarily afford protection against other M types, making reinfection with *S. pyogenes* of a different M type a common occurrence. Many of us know this well, having suffered multiple bouts of streptococcal pharyngitis.

Occasionally, pharyngitis or tonsillitis can become serious. This can happen in two ways. First, systemic infection can occur when inflammation is severe enough to allow *S. pyogenes* entry into local blood vessels, and from here, into the circulatory system. This can lead to septic shock. Second, post-infection **sequelae** occur. Sequelae (from "sequel," meaning an after-the-fact consequence) happen *after* the *S. pyogenes* infection are cleared and are thought to be immune-mediated. The most common sequelae associated with *S. pyogenes* are glomerulonephritis, acute rheumatic fever, and its more serious chronic manifestation, rheumatic heart disease (see Table 21.10). How *S. pyogenes* precipitates these diseases is poorly understood although various theories exist. Damage to heart tissue is thought to be due to antibody cross-reaction arising from regions of amino acid homologies between some types of M protein and membrane proteins of myocardium and myosin protein **(Figure 21.25)**. It is also proposed that SPE toxins may directly damage heart tissue, initiating inflammation in the heart. Immune complexes involving bundles of *S. pyogenes* antigens, antibody, and complement are also thought to accumulate in these tissues and evoke inflammatory damage. This is thought to contribute to glomerulonephritis, an inflammatory disease of the kidney filtering units, the glomeruli.

Skin infections are another common consequence of *S. pyogenes* infection. Cuts and scratches on skin allow seeding into tissue. This can lead to *impetigo*, a superficial disease characterized by oozing lesions that can quickly spread across skin (see Table 21.10). Production of enzymes contributes to spread and cell damage that releases nutrients. Impetigo is also frequently caused by *S. aureus* and sometimes by both species together.

Sometimes infection goes deeper, especially if bruising and compromised blood circulation is present. This allows *S. pyogenes* to move through the damaged tissue faster, creating inflammation and further destruction of tissue and vessels. Production of streptokinase and hyaluronidase contributes to spread in tissues. Streptokinase breaks down fibrin clots, which would otherwise help to wall off the spread of infection. Hyaluronidase degrades hyaluronic acid, a common gel-like substance that fills intercellular spaces. Erysipelas, a superficial inflammation of the dermis, and cellulitis, a more extensive inflammation involving the connective tissue, are often associated with bruising, obesity, diabetes, and aging. In some cases, large amounts of tissue are destroyed, producing necrotizing diseases. Strains

Streptococcus pyogenes M5 protein

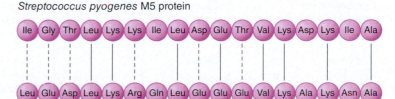

Heart muscle myosin

Figure 21.25. Comparison of streptococcal M5 protein and myosin amino acid sequences M5 and myosin proteins each contain a segment with extensive amino acid homology that can be seen when the sequences are aligned. Functionally similar amino acids are indicated with dashed lines. Identical amino acids are shown with solid lines. Dissimilar amino acids have no lines between them. Antibodies generated to the bacterial M5 protein may cross-react with similar sequences in myosin of heart muscle, resulting in autoimmune damage to the heart.

producing potent superantigens can induce shock even before getting to this stage, as the superantigens are secreted and can readily gain access to the circulatory system.

Necrosis that occurs on the outer layers of skin is termed "necrotizing cellulitis." If necrosis involves all the layers of skin, down to the muscle, it is termed "necrotizing fasciitis" because it spreads along the fascia lining the surface of muscles. Production of pore-forming streptolysin O and streptolysin S results in lysis of a variety of cells, including cells of the immune system. This action contributes to spread and necrosis. The amount expressed of both these toxins varies greatly with strains, and their contribution to disease is not easy to determine in isolation. They are just another facet of the total virulence that is elaborated by particular strains in certain conditions. **Table 21.11** summarizes some of the known virulence factors of *S. pyogenes*.

How do these necrotizing infections occur? There's no concise answer, other than it involves a significant level of invasive tissue destruction. Destruction of blood vessels in the area diminishes oxygen levels and provides great conditions for growth of *S. pyogenes*, a facultative anaerobe. Anaerobic conditions precipitated by *S. pyogenes* frequently allow other opportunistic pathogens, such as the anaerobe *Clostridium perfringens*, to establish concurrent infection. Reduced blood supply means fewer immune cells can access the infected site and loss of oxygen diminishes the ability of phagocytes to mount the oxidative burst—an oxygen-demanding process. Tissue destruction also restricts antibiotics from reaching the site and, like gas gangrene (see Microbes in Focus 21.4), often surgical removal of dead, anaerobic tissue is required to stop the spread.

CONNECTION Recall from **Section 19.5**, the oxidative or respiratory burst in phagocytes produces many highly toxic oxygen compounds inside the phagolysosome, which destroys endocytosed microbes.

Overcoming Host Defenses

Clearance of extracellular pathogens largely involves opsonization, which leads to phagocytosis and subsequent killing in the phagolysosome. Complement, an innate immunity defense product (see Section 19.4), and antibody, an adaptive immunity defense product (see Section 20.6), are major opsonins. As we have seen, for encapsulated pathogens, the capsule is the major target for efficient opsonization as it provides the outermost contact. However, the capsule of *S. pyogenes* is not opsonized by either of these methods because it is made of the "self" molecule hyaluronic acid **(Figure 21.26)**. Capsules that mimic host molecules are serum resistant as C3b does not bind to them, and of course antibody is not made to components that are also part of the host. Consequently, complement lysis and phagocytosis is hindered, reducing both intracellular killing and antigen presentation by macrophages and/or dendritic cells to CD4$^+$ T cells of the immune system (see Sections 20.2 and 20.5). This in turn reduces the adaptive response, and subsequent antibody production against other antigens of *S. pyogenes*, including M protein, is diminished. Naturally, some bacterial cells do get phagocytosed eventually, and protective antibody is made. Deterring phagocytosis allows a window of opportunity for *S. pyogenes* to expand its population.

TABLE 21.11 Virulence factors of *Streptococcus pyogenes*

Activity	Virulence factor	Function
Adherence	Fimbriae	Attachment to epithelial cells
	M protein	Contributes to the establishment of microcolonies on the skin and pharynx
Anti-phagocytosis	Hyaluronic acid capsule	"Self" capsule prevents antibody formation; capsule prevents contact between antibody bound to wall components and F$_C$R of phagocytes
	M protein	Binds antibody in inverted orientation, discourages opsonization; some forms implicated in rheumatic heart disease
Serum resistance	Hyaluronic acid capsule	"Self" capsule discourages complement fixation
	M protein	Binds serum factor H, which discourages complement activation
Complement degradation	C5a peptidase	Degrades complement factor C5a (Table 19.5), preventing migration of phagocytes
	SIC	Prevents and may disassemble MAC
Enzymes	DNase	May contribute to liquefaction and spread through pus; hydrolyzes connective tissue; may contribute to spread
	Hyaluronidase	Degrades hyaluronic acid in host tissues; may contribute to spread
	IdeS	Degrades all IgG antibody
	Streptokinase	Converts serum plasminogen to plasmin, which dissolves fibrin clots, contributing to spread
Toxins	Streptolysin O	Pore-forming oxygen-labile cytolysin acting on a variety of cells; contributes to spread and necrosis; death of leukocytes
	Streptolysin S	Similar to streptolysin O but is oxygen-stable
	SPE toxins	Family of superantigen proteins inducing fever and inflammation

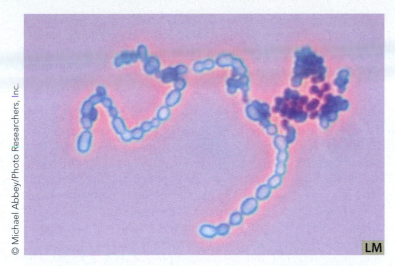

Figure 21.26. Capsule and fimbriae of *Streptococcus pyogenes* The surrounding hyaluronic acid capsule on this color-enhanced image is pink and the fimbriae layer is white (1,600x).

The capsule is not the only defense against phagocytosis. M protein is a major virulence factor of *S. pyogenes*. It functions in a similar manner to *S. aureus* protein A by binding host antibodies upside down by the F_C portion (see Perspective 20.3), reducing opsonization by antibodies. M protein also binds serum factor H, the presence of which discourages complement binding, and contributes further to serum resistance. Avoidance of clearance by phagocytes is further afforded by production of a C5a peptidase, which diminishes the migration of phagocytes into the area. *Streptococcus pyogenes* also produces a unique protein called SIC (streptococcal inhibitor of complement), which can prevent the late stages of MAC formation. In addition, *S. pyogenes* produces numerous enzymes that degrade IgG and IgA (see Table 21.11). Recall *Neisseria gonorrhoeae* also possesses an IgA protease, as does *Haemophilus influenzae*.

Although *S. pyogenes* is capable of producing numerous virulence factors, they are not all of equal importance in pathogenesis. Studies using cultured cells or *in vivo* studies show that knocking out any single virulence gene often does not greatly impact the pathogenicity of *S. pyogenes*. This indicates that it is likely the collection of virulence factors are significant, and different strains appear to mix and match these. The fact that most people who become infected with *S. pyogenes* will suffer only self-limited, mild disease is a testament to the immune defenses of the host in light of such an onslaught of bacterial tactics to undermine these defenses.

Pathogen study 2:
Mycobacterium tuberculosis

Mycobacterium tuberculosis is the causative agent of tuberculosis, a disease characterized by destruction of lung tissue, where it usually manifests, but can also occur in other parts of the body, including bone, lymph nodes, and brain. Relatives of *M. tuberculosis* include *M. bovis*, which causes tuberculosis more commonly in animals, and *M. leprae*, the causative agent of leprosy, or Hansen's disease, which is in many ways similar to tuberculosis but manifests in the skin and associated nerves. All three of these closely related microorganisms use similar tactics to produce disease.

All members of the genus *Mycobacterium* have unusual cell walls that contain a waxy layer of mycolic acids (from which their genus name has been derived) that resist penetration by most dyes, and cannot be Gram-stained. Instead, these cells are stained with carbol fuchsin, a red dye that can permeate the mycolic acid-rich cell wall and is not decolorized by acid-alcohol rinse (hence the name "acid-fast"). The cells stain red, as can be seen in **Microbes in Focus 21.6**.

Access to Host Tissues

Mycobacterium tuberculosis is spread by inhalation of aerosols from individuals with active tuberculosis. The bacterium replicates inside the phagosome of lung macrophages. This intracellular existence affords protection from complement and antibody. We've seen this sort of strategy used by *Salmonella*, *Neisseria*, and even for *S. pyogenes*, which does this on a more limited basis. Like these, *M. tuberculosis* also facilitates its own uptake into target cells. *Mycobacterium tuberculosis* does this by encouraging opsonization with host complement component C3b and mannose-binding lectin, which in turn bind to CR3 and mannose receptors, respectively, on the macrophage (see Figure 19.16).

Overcoming Host Defenses

As we saw in Chapter 19, the fate of bacteria in the phagosome is usually death by way of acidification, the oxidative burst, and fusion with lysosomes. *Mycobacterium tuberculosis*, like many intracellular pathogens, can obstruct many of these activities in imprecisely defined ways. Phagocytosed *M. tuberculosis* cells interfere with two very important immune functions: (1) they prevent fusion between phagosomes and lysosomes and (2) reduce MHC II presentation of antigen to CD4$^+$ T cells. These

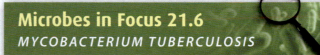

Microbes in Focus 21.6
MYCOBACTERIUM TUBERCULOSIS

Habitat: Humans are the natural reservoir for *M. tuberculosis*. *M. bovis* is found in various mammals and is very closely related, occasionally causing tuberculosis in humans, usually from drinking unpasteurized milk.

Description: Gram-positive bacillus 2–6 μm in length. The bacilli are acid-fast, shown here as red in a lung tissue sample of a tuberculosis patient. Host cells are blue. It is very slow growing, with a doubling time of 24 hours in culture and several days *in vivo*.

Key Features: *Mycobacterium tuberculosis* is an intracellular pathogen and is the causative agent of tuberculosis. It is spread through close contact with aerosol droplets from individuals with clinical disease.

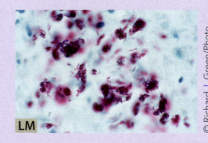

two pivotal events can otherwise protect against *M. tuberculosis* infection and indeed against *Mycobacterium* species in general.

Studies show that the protective immune response to mycobacterial infections is a cell-mediated response, involving (1) activation of $T_H 1$ CD4$^+$ T cells and (2) upregulation of macrophage killing functions. As we learned in Section 20.5, these two immune activities are closely tied to one another. Macrophages are important antigen-presenting cells. Engulfed pathogens are normally broken down in the endocytic pathway and the resulting peptides are loaded onto MHC II molecules for presentation to CD4$^+$ T cells. For mycobacterial infections, production of the cytokines IL-12 and TNF-α by the macrophage signals activated CD4$^+$ T cells to secrete IFN-γ. This cytokine directly stimulates macrophages to increase phagocytosis, oxidative burst activities, and MHC II antigen presentation (**Figure 21.27**). Individuals who can efficiently mount this cell-mediated immune response often achieve protection from clinical tuberculosis disease. It is not clear if *M. tuberculosis* is completely destroyed in this scenario, or if it is just kept inactive by this response. As you might imagine, there is substantial variation in the immune response between individuals.

By preventing phagosome fusion with lysosomes, *M. tuberculosis* avoids the otherwise acidic and degradative environment of the phagolysosome. Arresting further endocytic processing prevents MHC II:peptide presentation so CD4$^+$ T cells are not activated and IFN-γ is not produced. This affords *M. tuberculosis*, a very slow-growing pathogen, the opportunity to grow in the phagosome. However, *M. tuberculosis* must overcome other biocidal effects of the phagosome. Pathogenic mycobacteria produce lipoarabinomannan (LAM), a complex cell wall component that forms a substantial outer layer around the bacterium. LAM downregulates the oxidative burst and neutralizes many of the toxic oxygen species produced by this process (see Table 19.6). LAM is one of the few virulence factors identified for *M. tuberculosis*. Other virulence factors include superoxide dismutase and catalase that also protect it from oxidative damage, contributing to its survival in the phagosome.

Disease Consequences

The damage caused by infection with *Mycobacterium tuberculosis* is a direct result of the host immune response as it attempts to isolate the infected cells and prevent spread of the bacilli. A cell-mediated chronic inflammatory response ensues because of the cytokine TNF-α produced by activated, infected macrophages that are unable to successfully destroy the bacilli. This recruits immune cells to the area and results in production of a **granuloma**, a fibrous encasement composed of fibrin and lymphocytes containing infected and dead macrophages (**Figure 21.28**). The granuloma is the prime pathological feature of tuberculosis, and in advanced cases the dense, fibrous structure of these lesions can often be detected on an X-ray. Granulomas calcify with age but may still contain live but non-contagious bacilli. Granulomas produce damage in alveoli, the primary gas-exchange surfaces of the lung, resulting in reduced gas exchange.

Although granuloma formation is damaging, it can keep *M. tuberculosis* at bay for many years, primarily by enforcing a siege that keeps *M. tuberculosis* in short supply of nutrients and oxygen. If immune conditions change in the host and the cell-mediated response weakens, the granulomas break down. Breakdown is characterized by caseous necrosis ("caseous" meaning cheese-like), a description for the partial liquefaction of necrotic host tissues inside the granuloma due to the release of the contents of the dead macrophages within (**Figure 21.29**). This necrotic "soup" provides nutrients for the bacilli,

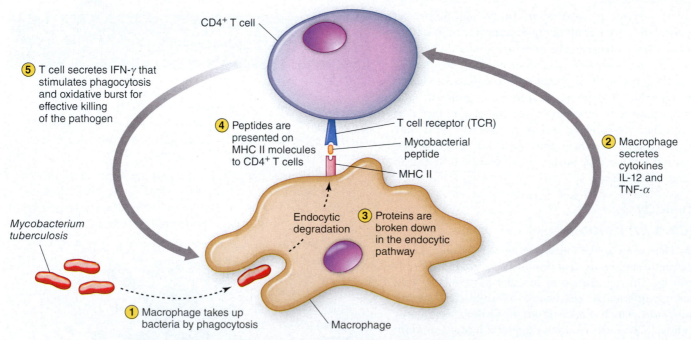

Figure 21.27. Process Diagram: The protective response to *Mycobacterium tuberculosis*

Mycobacterium tuberculosis can live inside macrophages. The protective immune response requires activation of $T_H 1$ CD4$^+$ T cells by macrophages, which in turn stimulate upregulation of macrophage killing functions to destroy the pathogen. Production of the cytokines IL-12, TNF-α, and IFN-γ are crucial to this response.

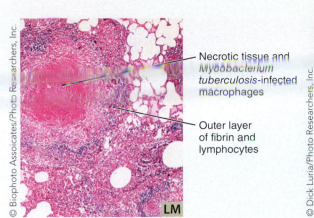

Necrotic tissue and *Mycobacterium tuberculosis*-infected macrophages

Outer layer of fibrin and lymphocytes

© Biophoto Associates/Photo Researchers, Inc.

LM

A. Granuloma lesion

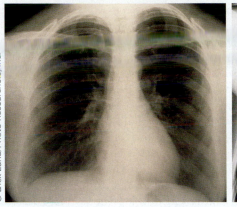

© Dick Luria/Photo Researchers, Inc.

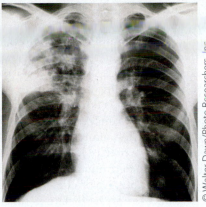

© Walter Dawn/Photo Researchers, Inc.

B. X-rays of normal and diseased lungs

Figure 21.28. Granuloma produced by *Mycobacterium tuberculosis* A. The center of the lesion contains infected macrophages surrounded by a layer of fibrin and active lymphocytes. The granuloma is produced as a result of a chronic inflammatory response due to the inability of macrophages to kill *M. tuberculosis*. **B.** Calcified granulomas can be detected on X-ray. The normal chest X-ray (*left*) shows clear lungs. The heart can be seen as the dense bulge to the right. The other X-ray (*right*) shows a patient with pulmonary tuberculosis. Dense granulomas appear as diffuse white patches.

ending the siege, and allowing replication to ensue. The necrotic contents are a concoction of proteolytic enzymes, acids, and toxic oxygen species that cause substantial damage to the integrity of the granuloma and surrounding tissues. As the damaging material leaks out, it destroys lung tissue, producing cavities in the lungs. This is accompanied by full clinical signs of tuberculosis: night sweats and severe coughing with bloody sputum carrying infectious *M. tuberculosis*. This is the most contagious stage of tuberculosis. **Figure 21.30** shows an overview of the disease process for tuberculosis.

Despite the high prevalence of *M. tuberculosis* infection worldwide, there appears to be a substantial level of natural resistance in most individuals to the pathogen. It is estimated that approximately 90 percent of those exposed to *M. tuberculosis* do not develop clinical disease. Close contact and multiple exposures

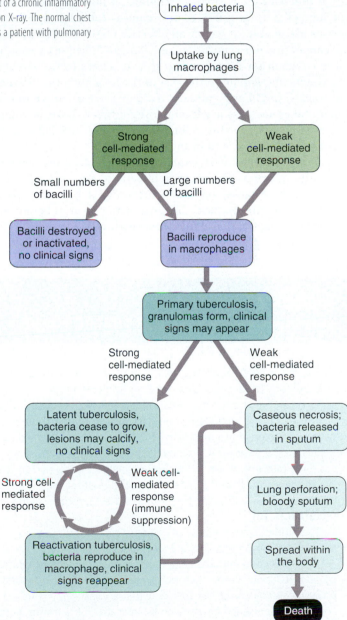

Figure 21.30. *Mycobacterium tuberculosis* pathogenesis in pulmonary tuberculosis The outcome of infection with *M. tuberculosis* is related to the host immune function. Individuals who are able to mount a strong cell-mediated immune response can often contain progression of disease, while those unable to mount this response cannot.

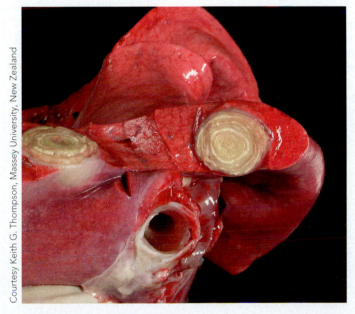

Courtesy Keith G. Thompson, Massey University, New Zealand

Figure 21.29. Caseous necrosis The area of caseous necrosis in the infected lung, shown cut in two, displays a pale, cheese-like consistency.

are usually needed to establish infection. Most who are exposed to low levels of bacilli for the first time seem to either clear the bacterium or maintain it in a non-replicative state, the result of an effective cell-mediated immune response. Of those individuals exposed to high levels of *M. tuberculosis*, or who have weaker immunity, the bacilli multiply in the macrophages and granulomas form. This is known as "primary tuberculosis." Most individuals with primary tuberculosis will successfully hold the microbe at bay for many years without progressing to further disease. This period is known as "latent tuberculosis" and is evidenced by a continuously positive tuberculin skin test (Toolbox 21.2) but absence of clinical signs of tuberculosis. Five to ten percent of those with latent tuberculosis will develop the clinical disease at a later time. This is known as "reactivation tuberculosis." Some people can cycle between latent and reactivation tuberculosis. A small percentage of infected individuals will immediately progress to full clinical tuberculosis. These persons are unable to mount an effective CD4+ T cell response and cannot prevent replication and damage. In these cases, it is not uncommon for *M. tuberculosis* to gain access to the circulatory system and spread to the rest of the lung and other tissues and organs, producing "miliary" (widely seeded) tuberculosis. Miliary tuberculosis is a common form of tuberculosis in young children who lack the mature immune responses needed to keep *M. tuberculosis* sequestered in the lung.

The proportion of each category of tuberculosis can change drastically in areas where other infections, such as HIV, are found. In these areas, primary tuberculosis becomes more prevalent, and the majority of those with latent tuberculosis will develop reactivation tuberculosis with HIV infection. HIV

Humoral response
Non-protective

Low IL-4, IL-6, IL-10, antibody

High IL-4, IL-6, IL-10, antibody

High IFN-γ, activated CD4+ T cells and macrophages

Low IFN-γ, few activated CD4+ T cells and macrophages

Protective
Cell-mediated response

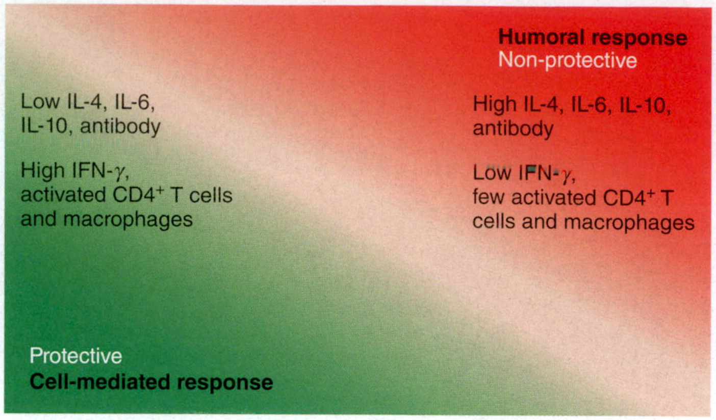

Figure 21.31. Comparison of a cell-mediated response to a humoral response in *M. tuberculosis* infection A strong cell-mediated response is protective, limiting the growth and spread of *M. tuberculosis*. As this response changes to a predominant humoral response, protection wanes. This is accompanied by changes in cell profiles and levels of key cytokines and antibody directed to *M. tuberculosis*.

selectively destroys CD4+ T cells, so any hope of defense against *M. tuberculosis* is removed with HIV infection.

Reactivation tuberculosis and clinical disease is accompanied by a drop in IFN-γ levels, waning of the cell-mediated response, and an increase in the humoral response cytokines IL-10 and IL-6, along with an increase in antibody levels for *M. tuberculosis* (Figure 21.31). This spells a poor prognosis against the patient. Antibody has no protective effect against intracellular pathogens and any extracellular bacilli that are opsonized by antibody are easily phagocytosed by macrophages—an advantage for *M. tuberculosis*.

Toolbox 21.2
THE TUBERCULIN TEST FOR TUBERCULOSIS

Tuberculin, or PPD (purified protein derivative), is a solubilized mix of cultured *M. tuberculosis* proteins that is injected just under the skin, usually on the anterior side (underside) of the forearm. The test measures the immune response to these antigens. Those who have been exposed to *M. tuberculosis* will exhibit a cell-mediated hypersensitivity reaction, an inflammatory response consisting of rapid recruitment of activated immune cells to the area. The test is considered positive if a raised, hardened welt (called an "induration") appears within 72 hours and is 10–15 mm in diameter (Figure B21.10).

The tuberculin skin test is not perfect; it can miss those who may be immunocompromised or are recently infected as they may not yet have activated immune cells. False positive tests can be caused by cross-reaction of tuberculin with similar antigens from other non-tuberculosis mycobacterial species. Often a chest X-ray is used to confirm the presence of granulomas, indicative of active or latent tuberculosis.

Figure B21.10. Positive tuberculin skin test A hard, raised nodule larger than 10 mm is considered positive for tuberculosis.

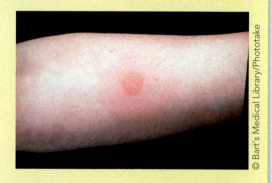

© Bart's Medical Library/Phototake

● **Test Your Understanding** ....

Your friend had a tuberculin test as required by his employer. His test came back positive. However, a chest X-ray showed no granulomas, and he is not sick. What does his positive tuberculin result mean, and why might his employer ask him to take antibiotics?

21.3 Evolution of bacterial pathogens

How do some bacteria become pathogens?

In the previous section, we learned that pathogens have distinctive features known as virulence factors, which are needed to cause disease. Virulence factors allow them to attach, invade, and damage host tissue providing nutrients. The presence of genes encoding virulence factors is the major distinguishing characteristic between pathogens and their non-pathogenic relatives. In this section we will look at ways bacteria acquire virulence factors, and how pathogens may arise from originally non-pathogenic prototypes.

The traditional view of evolution holds that random mutations in genomes, those chance mistakes made by DNA polymerase during replication and sequence changes caused by mutagenic agents, are acted upon by natural selection. By this, under certain environmental conditions, mutations that have a neutral effect or impart a selective advantage to the organism are kept and passed along to subsequent generations. Those that impart a disadvantage are lost as their possessors are out-competed by those lacking the mutation. Over time, selection forces act on such random gene mutations, molding benign organisms into exploitive pathogens. As you can imagine, this is a painstakingly slow process, even in rapidly multiplying populations. Is this the only evolutionary process that accounts for the rise of pathogens?

Horizontal gene transfer: Evolution by quantum leaps

One of the first clues that another significant process contributed to pathogen evolution was the acquisition of antibiotic resistance. In the 1950s, soon after the widespread use of penicillin, it became apparent that the rapid development of resistance to the same antibiotics in both related and unrelated bacterial species could not be solely explained by random mutational events. Scientists soon showed that antibiotic resistance genes were acquired by transfer of plasmids carrying resistance genes. In 1969, 12,500 people died in Guatemala during an epidemic of *Shigella dysenteriae* when treatment with antibiotics unexpectedly failed. This bacterial strain harbored a plasmid carrying resistance to four antibiotics! Although resistance to antibiotics does not make a pathogen, the realization that whole genes could suddenly be acquired led scientists to reconsider the evolutionary processes used by microbes.

Horizontal gene transfer, the transfer of DNA by conjugation, transposable elements, plasmids, or lysogenic phages, is common in bacteria (see Section 9.5). As we have already seen, some pathogens vary from their non-pathogenic or less pathogenic counterparts only because they possess one or a few virulence factors that may be acquired by horizontal gene transfer. In the case of *C. diphtheriae*, one gene transfer turns this microbe from harmless to deadly. The genomics era has since provided much insight into just how prevalent horizontal gene transfer is in bacteria and how profound its impact has been on pathogen evolution.

CONNECTION Comparative genomics of *Bacillus anthracis* and its much less pathogenic relative *B. cereus* exemplify the significance of horizontal gene transfer. The major difference between these two species is just a few genes present in *B. anthracis* that have been delivered by a plasmid (see **Perspective 9.1**).

Pathogenicity Islands: Treasure Troves for Pathogens

From early sequencing projects, it became apparent that large segments of DNA were often present in pathogenic strains but not in their non-pathogenic counterparts. Analysis of these segments has revealed some amazing characteristics. The genes contributing to virulence are often found clustered within these segments. The segments themselves have unusual features not found in the rest of the genome; they are conspicuous islands in an otherwise more uniform genomic sea. These blocks of sequence were named **pathogenicity islands (PAIs)**, and they commonly encode genes for:

- adhesion to host cells
- invasion of host tissue
- toxins for damaging host cells
- type III or type IV secretion systems for injection of virulence factors
- proteins that facilitate metabolic processes, such as iron-binding proteins

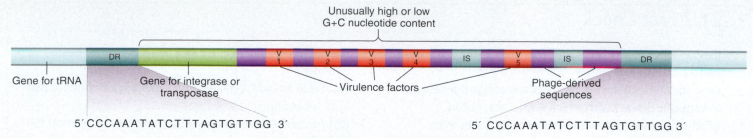

5´ CCCAAATATCTTTAGTGTTGG 3´   5´ CCCAAATATCTTTAGTGTTGG 3´

Figure 21.32. General structure of a pathogenicity island (PAI) PAIs are frequently found next to a tRNA gene and are bounded by direct repeats (DR) sequences. Mobile genetic elements, such as phage-derived sequences and insertion sequences (IS) from other bacteria, commonly exist and are thought to carry the various genes coding for virulence factors (V1–5). These sequences are frequently of higher or lower guanosine and cytosine content than the rest of the genome. A gene encoding an integrase or transposase enzyme is present. The enzyme can insert and excise the PAI element or its parts by recognition of the repeat regions within the IS or DR sequences.

An example of a fully sequenced PAI is the locus of enterocyte effacement (LEE) found in some pathogenic strains of *E. coli*, such as O157:H7. The genes encoded on the LEE PAI are directly responsible for the AE intestinal lesions produced by enteropathogenic and enterohemorrhagic strains (see Figure 21.3). LEE encodes genes for intimin production and a type III secretion system for the injection of Tir. Other gene products of the locus are of phage origin and have not yet been characterized.

In addition to the possession of genes related to virulence, **Figure 21.32** shows some other common features of PAIs:

- unusually high or low cytosine and guanosine nucleotide content, as compared to that of the general chromosome (averaging 40 to 60 percent GC)
- flanking direct repeat regions (used by transposable elements for insertion)
- multiple mobile genetic elements (integrated phage genomes or insertion sequences)
- genes that encode enzymes used for movement and integration of DNA segments (integrases or transposases from phages and transposable elements, respectively)
- adjacent tRNA genes (a common site used by phages for DNA integration)

The common presence of these unusual genetic features provides clues as to how these islands of sequence are acquired. They are delivered by mobile genetic elements: integrative phages, plasmids, and transposable elements, such as insertion sequences. Comparative genomic studies have revealed that the elements that compose PAIs, including virulence genes, appear to originate from multiple foreign sources, such as other bacterial species, phages, and even eukaryotes. As we will see in this section, phages in particular have played a prominent role in the acquisition of virulence factors and the evolution of pathogens.

Many of the genetic elements found in PAIs appear to be derived from gene transfer "pools" that are available in ecological niches, such as the gut, skin, and the nasopharynx. The sheer number and concentration of sequences involved in virulence traits makes the habitat of the body a treasure chest of virulence options unlikely to be found outside the body. These transfer pools contain the genetic building blocks that, through time, have undergone rearrangements and further mutations to become the PAIs associated with particular pathogens found in

Microbes in Focus 21.7
VIBRIO CHOLERAE

Habitat: Widely found in aquatic environments, also in many invertebrates. Some strains periodically cause disease in humans.

Description: Gram-negative bacillus 1–3 μm in length, often having a curved rod shape; motile by means of single polar flagellum. Only strains possessing the lysogenic phage CTX, which encodes cholera toxin, cause significant disease in humans.

Key Features: Cholera toxin is responsible for most of the disease symptoms of cholera. The toxin-induced ion imbalance produces water loss from intestinal cells that results in massive diarrhea that can lead to death. Estimates of cholera deaths are likely greatly underestimated at over 200,000 annually. The *V. cholerae* receptors for the CTX phage are pili, the genes for which are encoded on a PAI acquired from another phage. Interestingly, the main subunit of the pili is also a coat protein for the phage that encodes it! Recent sequence analysis of the deadliest epidemic strains has shown significant genomic diversity, including the existence of two coexisting circular chromosomes.

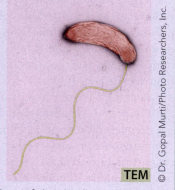

© Dr. Gopal Murti/Photo Researchers, Inc.

TEM

these regions. For example, pathogenic strains of *Vibrio cholerae*, causing severe epidemic diarrhea, possess prophage-encoded cholera toxin (**Microbes in Focus 21.7**). Cholera toxin is closely related to the *E. coli* heat-labile toxin (LT) encoded on a plasmid-borne PAI found in enterotoxigenic strains of *E. coli*, the common cause of "traveler's diarrhea." Another example of transference of communal sequences is found in the *cag* PAI of the pathogen *Helicobacter pylori*, which causes gastric and duodenal ulcers. Sequence comparisons show high similarity between several mobile genetic elements found on the *cag* PAI and those found in other gastrointestinal pathogens, including *Clostridium perfringens*, *E. coli*, and *Salmonella* Typhimurium. Also included in this cornucopia of *cag* PAI sequences are homologous genes that code for various virulence factors found in *Bordetella pertussis*, *Salmonella* Typhimurium and *Agrobacterium tumefaciens* (a pathogen of plants).

It seems pathogen evolution favors the construction of new virulence factors from existing parts rather than solely relying on the much slower process of random mutation. In many cases, it is not known where the sequences for the parts originated, having changed with time over the course of pathogen evolution. The multitudinous insertions of the various mobile genetic elements within pathogen genomes provides significant genetic plasticity. Their presence facilitates recombination events, resulting in genetic rearrangements and chromosomal segment loss, both of which are commonly found in the genomes of pathogenic bacteria.

Phages: Delivery Vans for Virulence Factors

Bacteriophages have recently been realized to be major players in the evolution of many bacterial pathogens. This is not surprising as phages infect every known and probably unknown bacterial species on the planet. Not only do phages participate in natural selection of their bacterial hosts, but they facilitate gene transfer between bacteria by transduction (see Section 9.5). As we saw earlier, certain types of DNA phages, such as the *E. coli* phage lambda, can integrate their genome into the bacterial chromosome as part of their life cycle. They are termed "temperate" or "lysogenic" phages. Temperate phages are now known to have special significance for many bacterial pathogens.

CONNECTION Temperate or lysogenic phages are capable of integrating their genome into the bacterial chromosome, or sometimes into a plasmid. Virus progeny are not produced from the integrated prophage, and few viral genes are expressed. The prophage excises under certain, usually unfavorable, conditions to begin the production of progeny virions. See **Sections 5.3** and **8.3** for a review.

The relationship among phages, their bacterial hosts, and disease has until recently been overlooked in terms of the ecology and evolution of bacterial pathogens. In the 1950s, it was realized that diphtheria was caused only by strains of *C. diphtheriae* that carried a certain phage. In the 1980s, it was discovered that the genes for the two versions of Shiga toxin found in enterohemorrhagic *E. coli* strains, such as O157:H7, were transferred there by lambda-like phages from *Shigella dysenteriae*, the causative agent of epidemic dysentery. In 1996, the deadliest strains of *Vibrio cholerae* were found to harbor a particular toxin-producing phage (see Microbes in Focus 21.7). While this was a surprising discovery, even more surprising is just how prevalent phage-supplied virulence factors are (**Table 21.12**). Ever-increasing amounts of genetic information have given

TABLE 21.12 Genes introduced to bacterial cells by temperate phages

Encoded gene product	Bacterial host	Impact
Botulinum toxin types C1, D	*Clostridium botulinum*	Flaccid paralysis
Cholera toxin	*Vibrio cholerae*	Cholera (massive watery diarrhea)
Diphtheria toxin	*Corynebacterium diphtheriae*	Diphtheria: localized (throat) cell death or systemic damages (heart, kidney)
Enterohemolysin	*Escherichia coli*	Invasion, intestinal cell death (severe diarrhea)
Enterotoxins (superantigens)	*Staphylococcus aureus*	Food poisoning
Exfoliatin toxin	*S. aureus*	Scalded skin syndrome
Hyaluronidase	*Streptococcus pyogenes*	Contributes to tissue invasion and spread
O-antigen acetylase	*Shigella flexneri*	Shigellosis (dysentery)
O-antigen synthesis enzymes of LPS	*Salmonella enterica*	Alters O-antigens to avoid immune clearance, contributes to enteritis
Outer membrane proteins	*E. coli*	Serum resistance
Pilus	*Vibrio cholerae*	Allows attachment to host cells
Pseudomonas aeruginosa cytotoxin	*Pseudomonas aeruginosa*	Pneumonia and skin infections
PV leukocidin	*S. aureus*	Necrotizing infections
Shiga toxins	*Shigella dysenteriae*	Hemolytic uremic syndrome (kidney damage)
Shiga toxins (verotoxins)	*E. coli* O157:H7	Hemolytic uremic syndrome (kidney damage)
Streptococcal pyrogenic toxin (SPE A, C)	*S. pyogenes*	Scarlet fever
Streptococcal TSSTs (superantigens)	*Streptococcus pyogenes*	Toxic shock
TSST (superantigen)	*S. aureus*	Toxic shock syndrome

Shiga toxins, produced by some strains of *Shigella* and enterohemorrhagic strains of *E. coli*, are encoded on temperate lambda-like phage genomes. Production of the toxin occurs only after the phages excise their DNA from the chromosome. When the cell is later lysed to release the progeny phage, the toxin is also released. Recall that Shiga toxins attach to certain lipid receptors enriched in human tissues, such as kidney tissue. From here they are taken up and cause cell death resulting in hemolytic uremic syndrome (HUS).

In 2000, a study of the effectiveness of quinolone antibiotic therapy for persons infected with *E. coli* O157:H7 indicated that the risk of developing HUS was greater when antibiotics were given. How can this be? Quinolones are a class of broad-spectrum antibiotics commonly used to treat severe diarrhea. They act on bacterial DNA gyrase or topoisomerase IV, thereby interfering with bacterial DNA replication, resulting in DNA damage. Significant DNA damage induces a ubiquitous bacterial DNA repair system called the "SOS regulon" (see Section 11.3). The SOS response allows bacteria to survive such assaults to their DNA. As you can imagine, damage to the bacterial host cell is a threat to the integrated prophage: If the cell dies, so ends the phage's existence. Many types of prophages are chemically attuned to induction of the SOS response and will excise when it occurs. This begins transcription of phage genes for the replication of progeny virions that then lyse the cell, abandoning ship. Treatment with quinolone antibiotics induces the SOS response, and the lambda-like Shiga toxin-encoding prophage is excised. Transcription of phage genes begins and Shiga toxin is produced. Thus, quinolone antibiotic treatment greatly increases the risk of HUS.

A similar case occurs for *Streptococcus canis*, a normally commensal opportunistic pathogen that colonizes the skin of dogs. Many strains harbor a prophage that encodes a protein capable of non-specific activation of T cells. It appears that the recent use of DNA-damaging antibiotics has precipitated severe cases of skin disease and toxic shock syndrome in dogs infected with these strains of *S. canis*.

Researchers are concerned that the indiscriminate use of antibiotics may be helping to spread virulence genes by encouraging excision and movement of phages. Indeed, quinolones have widespread veterinary use in production animals, such as cattle, pigs, and poultry. Reports of Shiga toxin-producing strains of several enteric bacterial species has alarmed many scientists. Since 1995, Shiga toxins have been reported in *Citrobacter freundii*, *Enterobacter cloacae*, *Aeromonas*, and *Shigella sonnei*, where they were previously not found.

---

microbiologists new insights into the role of phages in the acquisition of virulence factors. It has also become obvious that toxin production can often be triggered by excision of the toxin-encoding prophage, and that pathogen evolution may be enhanced by therapeutic agents that stimulate excision, allowing movement of toxin genes and other virulence factors. **Perspective 21.4** describes how antibiotics can actually make some diseases worse by triggering prophage toxin production.

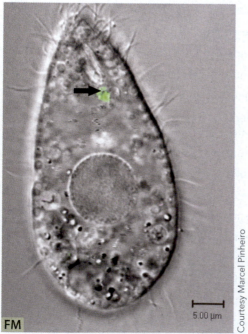

Figure 21.33. Protozoan ingesting a microbe Many protozoa are predators of microbial cells and viruses. The T4 phage particles (fluorescent green) are ingested by *Tetrahymena*. Shiga toxin-producing strains of *E. coli* survive when co-cultured with *Tetrahymena*.

Courtesy Marcel Pinheiro

Do phages cause disease? Why would bacteriophages carry virulence genes that cause disease in eukaryotic organisms? Phages can only infect and lyse prokaryotes, so why cause damage to eukaryotic cells? One school of thought regarding phage-derived toxins is that production of the toxin by the phage is directed against some other organism commonly found in the environment. Plants or animals or humans may be inadvertent victims of these toxins.

This theory appears to have some merit. Shiga toxins bind to a lipid receptor found on certain eukaryotic cells. As mentioned earlier, the receptor is enriched in kidney and liver tissue in humans. But the same receptor is also found on a predatory protozoan called *Tetrahymena* that is commonly found in aquatic habitats, including sewage treatment facilities and composting animal wastes, where enteric microbes like *E. coli* abound. *Tetrahymena* feed on bacteria and viruses, breaking them down and harvesting their chemical constituents **(Figure 21.33)**. An *E. coli* strain harboring the Shiga toxin gene can survive when co-cultured with *Tetrahymena*. These cells appear to be able to kill the predators. Conversely, an *E. coli* strain lacking the toxin gene fared poorly when co-cultured with *Tetrahymena*. Blocking the binding of the Shiga toxin to *Tetrahymena* prevents killing. In this case, it would appear that the advantage to the phage is to kill the protozoa that would otherwise consume phage particles and their bacterial host cells. Maybe disease caused by Shiga toxin-producing bacteria really is an inadvertent result of infection with these phage-harboring strains.

Another interesting phage-host relationship involves diphtheria toxin. Within the *Corynebacterium diphtheriae* phage β lies an operon that encodes the toxin gene and within the

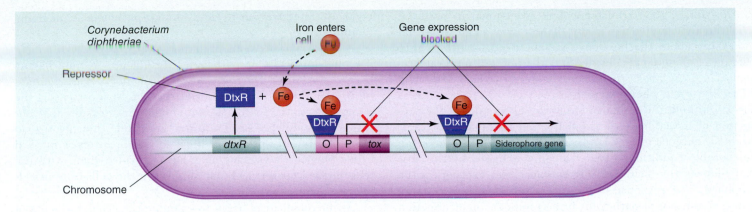

A. Toxin production in high iron concentration

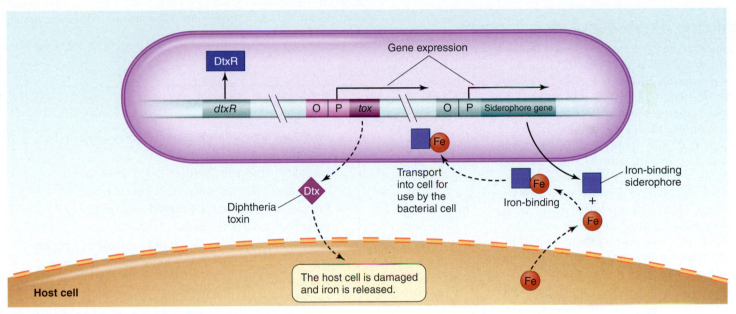

B. Toxin production in low iron concentration

Figure 21.34. Effect of iron on expression of diphtheria toxin gene *Corynebacterium diphtheriae* bacterial genetic elements are shown in blue. Prophage genetic elements are shown in purple. The *dtxR* gene produces a repressor protein DtxR. **A.** DtxR binds free iron, which produces a shape change in DtxR, allowing it to bind the bacterial operators. This prevents binding of RNA polymerase to respective promoters, blocking expression of the prophage-encoded diphtheria toxin and the bacterial iron-binding siderophore. **B.** In the absence of free iron, DtxR cannot bind the operators and diphtheria toxin (Dtx) and the siderophore are produced as a consequence. The toxin destroys host cells, releasing iron that the siderophore can bind and transport into *C. diphtheriae*.

bacterium lies another operon encoding an iron-binding siderophore. Expression of both the toxin and the siderophore is negatively controlled by iron **(Figure 21.34)**. High concentrations of iron repress transcription of these genes. As available iron concentrations are always low in the body, the toxin and siderophore are expressed in lysogenized *C. diphtheriae*. The toxin kills body cells, releasing small amounts of free iron that are in turn bound and transported by the siderophores for the bacterium's own use. Presumably it is to the phage's benefit to supply the bacterium with iron in order to perpetuate its own replication—a phage's way of bacterial farming. The level of codependence between the phage and the bacterium is further evident in that expression of the phage operon is dependent on expression of a bacterial gene, which produces a repressor protein called DtxR (diphtheria toxin repressor). Scientists are just beginning to realize the interdependence between

production of disease, the bacterial pathogen, and the phage that infects it. It seems some bacterial pathogens have a mutualistic relationship with their phages. They need their phages as much as the phages needs them.

Horizontal gene transfer by phages is no doubt an important phenomenon in the acquisition of virulence factors, genetic diversity, and evolution in many pathogens. Studies of the genomes of various bacterial species show that a high prevalence of prophage and phage-derived elements exist in bacterial chromosomes or in the plasmids they carry. For example, genetic analysis of isolates of *E. coli* O157:H7 (see Chapter 10 Mini-Paper) has revealed that the genome contains between 15 and 18 *different* prophages. This adds nearly 1 million extra base-pairs to its genome, for a total of 5.4 million base-pairs, compared to its non-pathogenic counterpart *E. coli* K12 of 4.6 million base-pairs.

Horizontal gene transfer events are known to occur at much higher frequencies *in vivo* than *in vitro*, suggesting that something about the host environment itself facilitates this. One piece of supportive evidence is the observation that prophages in *E. coli* and *Salmonella* are induced to excise and begin a lytic life cycle when exposed to reactive oxygen species, as one would find in the phagolysosome. This results in the production of infectious virus particles that can transport their virulence cargoes to new host cells. Perhaps this is one of the unfavorable conditions exacted upon pathogens by the host that induces the bacterial SOS response and excision of the prophage. The response of the host itself therefore contributes to evolution of pathogens by accommodating acquisition of new virulence factors.

Protozoa: Prep school for pathogens

It appears that the intended targets for several animal and human bacterial pathogens are actually protozoa. The best-known example of this is *Legionella pneumophila*, the causative agent of Legionnaires' disease **(Microbes in Focus 21.8)**. The disease was first recognized after an outbreak of severe pneumonia among those attending a convention of the American Legion in Philadelphia in 1976. Over 200 people staying in a hotel became ill, and 34 died. The outbreak was linked to the hotel air-conditioning system, from which the bacterium was isolated. That was not the most surprising finding, however. The bacterium was found to reside inside single-celled amoebae, also commonly found in human-made water systems. The bacterium and its apparent host turned out to be common environmental inhabitants. Studies have shown that the *L. pneumophila* genes that are required for uptake and growth inside their amoebae hosts are also required for growth inside lung macrophages. *Legionella pneumophila*

received its education to be a human pathogen from living in amoebae.

Associations between predatory protozoa, such as amoebae, and intracellular pathogens are not uncommon, having been reported for *Shigella*, *Listeria*, *Chlamydia*, *Francisella tularensis*, and various mycobacterial species. In most cases, the pathogens appear to be endosymbionts; they are able to replicate and establish a stable interaction inside their protozoan hosts. Their intra-protozoal lifestyle, which involves survival inside phagocytic vesicles, may have provided them with the tools needed for intracellular survival in the cells of multicellular eukaryotes.

As microbiologists begin to examine the ecological roles of pathogens inside and outside the environment of the body, it is becoming apparent that the evolution of the host/pathogen relationship is much more multidimensional than previously thought.

21.3 Fact Check

1. How does horizontal gene transfer play a role in the evolution of bacterial pathogens?
2. What are pathogenicity islands, and what types of genes are contained within these sequences?
3. Explain how phages have played a role in the evolution of bacterial pathogens.
4. Some bacterial pathogens have a mutualistic relationship with phages. Explain what this statement means, and provide an example to illustrate this concept.
5. Explain the role of protozoa in the evolution of some human pathogens.

Microbes in Focus 21.8
LEGIONELLA PNEUMOPHILA:
THE ACCIDENTAL PATHOGEN

Habitat: Found in association with various species of free-living amoebae in aquatic habitats. In humans, it can replicate in lung macrophages, but is not contagious.

Description: Gram-negative non-endospore-forming bacillus 1–3 μm in length

Key Features: *Legionella pneumophila* is the causative agent of Legionnaires' disease, a severe respiratory infection of compromised persons, and "Pontiac fever," a milder, flu-like systemic disease. It is often found in building water systems, such as air-conditioning systems, and hot tubs. The contaminated aerosols are inhaled.

TEM

© CAMR/A. Barry Dowsett/Photo Researchers, Inc.

Image in Action

WileyPLUS Enteropathogenic and enterohemorrhagic strains of *Escherichia coli*, attach themselves to the intestine by injecting their own receptor into host cells. An example is *E. coli* O157:H7. This strain is associated with cattle and caused a deadly outbreak in Walkerton, Ontario, Canada. The enterohemorrhagic *E. coli* in this image are attaching to intestinal microvilli and inducing pedestal formation. This disrupts intestinal structure and function.

1. Identify and describe the potential benefits for the bacteria in the image that are inducing pedestal formation.
2. What proteins are being represented by the green and the pink molecules in the image, and what are their roles in infection and disease?

Eschrichia coli O157:H7 was first isolated in 1975 but was not conclusively connected to enteric disease until 1982. At that time, it was determined to be the causative agent of hemorrhagic colitis in four patients in the United States who had all eaten hamburgers at fast-food chain restaurants. Many of the Walkerton, Ontario, residents who fell ill in 2000 are still suffering the effects of *E. coli* O157:H7 as a result of hemolytic uremic syndrome (HUS) produced by the Shiga toxins. Their chronic ailments include kidney disease, ongoing diarrhea, arthritis, hypertension, and dysfunctions of the liver, pancreas, heart, and brain.

These outbreaks are some of the worst public health disasters because of the combination of the severity of the disease, its potential for fatality, the frequency of severe complications, and the lack of specific therapeutic interventions. Public health efforts include education in food preparation and the development of rapid tests for the pathogen in cattle, humans, and food and water samples. Research is also focusing on eliminating the organism in cattle by vaccination. One experimental approach involved vaccination of cattle with *E. coli* O157:H7-specific proteins. The results were encouraging, showing significantly reduced numbers of these bacteria shed in feces and also fewer bacteria present in the feedlot. A 2009 study utilized a different vaccine formulation that resulted in a 92 percent decrease in *E. coli* O157:H7 colonization, compared to non-vaccinated cattle. The publication of the complete genome sequences of two outbreak isolates of *E. coli* O157:H7 may help identify additional virulence genes by comparing these genomes to that of the non-pathogenic strain *E. coli* K-12. Such comparisons may lead to the development of better methods of detection and understanding of the evolution of this recently emerged pathogen. Studying the evolution of *E. coli* O157:H7 will allow for a more complete knowledge of the disease process and, hopefully, better treatment options that may spare future patients the long-term consequences seen in patients from Walkerton.

Summary

Section 21.1: What are common features of bacterial pathogens?

Virulence factors are products made by pathogens that enhance their ability to cause disease. They are usually essential for disease **pathogenesis**. Pathogens use virulence factors to aid in their:

- access and spread in tissues
- evasion/disruption of host defenses
- ability to obtain nutrients from the body

Virulence factors include:

- molecules for attachment
- toxins
- capsules
- type III or type IV secretion systems
- iron-binding proteins

Cytotoxins encompass a variety of toxins that act on host cells.

Endotoxins are wall-associated molecules produced by Gram-negative pathogens. Endotoxin is released through growth or death of bacterial cells. The lipid A portion of endotoxin evokes inflammation. The O-antigen is often strain-specific and can be used for identifying **serotypes** using **serotyping** techniques.

Lipopolysaccharide (LPS) is the most common form of endotoxin, but some Gram-negative pathogens produce **lipooligosaccharide (LOS)** that lacks the O-antigen component.

Lipoteichoic acid (LTA) is the Gram-positive functional equivalent of endotoxin as it also evokes inflammatory responses in the host.

Exotoxins are toxins that are produced inside the cell and are secreted or released upon lysis. They include:

- A-B toxins
- cytolysins
- superantigens

A-B toxins act intracellulary to modify specific molecular targets in host cells. They consist of a B subunit that binds to a host receptor, and an enzymatic A subunit that is taken up by the cell. **Toxoids** are denatured toxins that can be used in vaccines.

Cytolysins act on cell membranes. They produce damage by:

- forming pores, often detected by **hemolysis** on blood agar media
- degrading lecithin of membranes (**lecithinase**)

Superantigens activate T cells in a non-specific manner by linking TCRs with MHC II molecules of antigen-presenting cells. Some superantigens act systemically to produce toxic shock syndrome. Others are **enterotoxins**, acting on the gut.

Type III secretion systems are a trademark of Gram-negative pathogens that are used for targeted delivery of virulence factors directly into the host cell. Some Gram-negative pathogens possess **type IV secretion systems** for similar purposes.

Gram-positive pathogens use pore-forming cytolysins to transfer other virulence factors into host cells.

Capsules are major virulence factors of many pathogens and protect pathogens from immune defenses by:

- blocking contact between cell-bound antibodies and the F_cR of phagocytes
- blocking contact between complement C3b and the CR3 of phagocytes.
- reducing entry in the endocytic pathway, thus lessening antigen presentation to the adaptive immune response
- mimicking host "self" molecules, which do not evoke antibody production, and which provide **serum resistance** to complement

Attachment to host cells or tissues is a critical event for colonization/infection. Attachment factors or **adhesins** that are commonly used include:

- **fimbriae**
- **fibronectin-binding proteins**

Some pathogens have specialized adhesins. Enteropathogenic and enterohemorrhagic strains of *E. coli* produce the adhesin intimin and the receptor Tir.

Iron is a limiting factor to bacterial metabolism and growth in the body. In the host, it is sequestered by molecules, such as lactoferrin, transferrin, and is locked inside cells.

Pathogens acquire iron from the host by destroying cells, such as red blood cells, and producing **siderophores**, bacterial iron-transporting proteins that can steal iron away form the host iron-binding proteins.

Section 21.2: How do bacterial pathogens cause disease?

Streptococcus pyogenes is an opportunistic pathogen that produces a variety of diseases. It possesses a number of virulence factors that are similarly found in other extracellular pathogens. These include:

- cytolysins (streptolysins O and S) to destroy host cells to get nutrients and invade tissue
- superantigens (TSSTs and SPE toxins) that help to prevent a focused adaptive immune response
- a capsule composed of "self" molecules (hyaluronic acid). Antibodies are not made to the "self" capsule and complement is not fixed. The capsule helps to block antibody and C3b that bind wall-associated molecules and prevents interaction with receptors of phagocytes.
- fimbriae and fibronectin-binding proteins for attachment to host cells
- enzymes that degrade host immune molecules (C5a peptidase and proteases to destroy antibodies)
- a protein (M protein) that binds antibody in an inverted orientation
- enzymes (streptokinase and hyaluronidase) that degrade host tissues and contribute to spread and invasion

Removing any one virulence factor often has little impact on the ability of *S. pyogenes* to cause disease. Some strains of *S. pyogenes* cause post-streptococcal **sequelae**, such as glomerlonephritis and rheumatic heart disease. The possession of so many virulence factors contributes to the variety of diseases *S. pyogenes* produces and its success as a human pathogen.

Mycobacterium tuberculosis is a primary, intracellular pathogen that causes tuberculosis. *M. tuberculosis* shares many features characteristic of intracellular pathogens. These include:

- facilitation of its own uptake in host cells (using C3b and mannose-binding lectin)
- interference with phagosome and lysosome fusion (by unspecified mechanisms)
- obstruction of antigen presentation to CD4$^+$ T cells (LAM downregulates MHC II peptide presentation)

Some virulence factors produced by *M. tuberculosis* include LAM, catalase, and superoxide dismutase. These help to downregulate and neutralize the toxic oxygen species produced by the oxidative burst.

- The pathogenesis of *M. tuberculosis* is largely determined by the host immune response.
- A strong cell-mediated immune response is protective against this intracellular pathogen and can eliminate it or keep it inactive for life. A humoral response is not protective against intracellular bacterial pathogens.
- The hallmark of tuberculosis is **granuloma** formation, which can keep *M. tuberculosis* from spreading, but causes damage to host tissue.
- If the cell-mediated response changes, growth occurs and the granuloma disintegrates. This is characterized by caseous necrosis, which destroys the integrity of the lung and results in distribution of *M. tuberculosis* to other tissues and other people.

Section 21.3: How do some bacteria become pathogens?

Evolution of pathogens involves:

- horizontal gene transfer by conjugation, transposable elements, lysogenic phages, and plasmids that can very suddenly introduce new virulence factors.
- random mutational events, including recombination, that result in the modification of genes over time.

Genes that contribute to virulence are frequently found grouped together in **pathogenicity islands (PAIs)**. PAIs are usually absent in non-pathogenic or less pathogenic strains or related species.

PAIs contain sequences from foreign sources that are delivered by mobile genetic elements (phages, transposable elements, plasmids). The organization and characteristics of the sequences within PAIs set them apart from the rest of the genome.

Many of the genes found in PAIs are shared among bacteria living in similar niches in the body.

Lysogenic phages are major players in the development of many bacterial pathogens.

DNA-damaging compounds, including some antibiotics that activate the SOS response, can encourage excision of prophages, which can result in toxin production, disease, and movement of phages.

There is increasing evidence that phages provide toxins to pathogens in order to increase the successful replication of the pathogen.

Some pathogens have gained the capacity for intracellular survival in multicellular hosts, from living in single-celled protozoa, such as amoebae, in the environment. Hosts, such as humans, may be inadvertent victims.

Application Questions ....

1. Actions of some virulence factors, such as hemolysins, cause host tissue damage. If it is accepted that a pathogen is not necessarily trying to destroy the host, then what is considered the reason or selective advantage for having this type of effect on the host? Explain.

2. A new strain of *E. coli* with increased levels of Shiga toxins is found to be the cause of an outbreak. Describe the effects these microbes would have on an infected patient. How could this strain be serotyped?

3. A strain of *Staphylococcus aureus* isolated from a skin infection is resistant to penicillin because it possesses a gene that encodes the enzyme penicillinase that destroys penicillin. Would you consider penicillinase to be a virulence factor? Explain your answer. How might you use an *S. aureus* knockout mutant to prove this?

4. Parents rush to the emergency room with their sick baby. The doctor's diagnosis is infant botulism after he runs tests and learns the baby has ingested honey. The honey, he explains, may contain *C. botulinum* endospores. The parents are shaken and confused because they also ate the same honey, but they are fine. As the doctor, how would you explain to the parents why their baby is ill and they are not?

5. The actor Christopher Reeve (best known for playing Superman) became a paraplegic after a bad fall from his horse during an equestrian competition. He died years later in 2004 of complications due to an infected bedsore. The culprit was a methicillin-resistant strain of *S. aureus*. The actual cause of death was heart failure. Explain the link between his skin infection with *S. aureus* and his resulting heart failure.

6. During infection, there is often a "battle for iron" between the host and the infecting pathogen. Explain how the human body restricts the amount of free iron available to a pathogen. Then explain how some bacteria still manage to get iron from the body.

7. In the lab, a researcher generated a knockout strain of *S. aureus* lacking the alpha-toxin gene. Describe the effects this mutagenesis would most likely have on these bacteria.

8. A young child is at the pediatrician's office to receive his DTaP vaccination. His parents read the information sheet provided and become concerned when they see that the vaccine contains bacterial toxins. As the child's doctor, explain more fully what the vaccine actually contains and why their child isn't going to receive dangerous toxins.

9. Production of antibody has not been found to be protective against *M. tuberculosis* infection. Explain why this would be.

10. You encounter what appears to be a newly emerged strain of bacteria. As you examine its DNA sequence, you determine it has genes for a type III secretion system. What would this sequence analysis tell you about the pathogenesis of these bacteria?

11. Enterotoxigenic strains of *E. coli* can cause diarrhea in travelers to regions where these strains exist as common gut inhabitants in local populations. Yet they do not usually cause disease in local populations. Suggest why these strains are a problem for travelers, but not for the locals.

12. If you could design your vision of the ideal pathogenic bacterium that was able to gain access to a host, avoid immune destruction, and obtain nutrients, what virulence factor genes would you insert in its genome? Explain your reasoning for each.

Suggested Reading

Arnon, S. S., R. Schechter, T. V. Inglesby, D. A. Henderson, J. G. Bartlett, M. S. Ascher, E. Eitzen, A. D. Fine, J. Hauer, M. Layton, S. Lillibridge, M. T. Osterholm, T. O'Toole, G. Parker, T. M. Perl, P. K. Russell, D. L. Swerdlow, and K. Tonat. 2001. Botulinum toxin as a biological weapon. Medical and public health management. JAMA 285:1059–1070.

Brüssow, H., C. Canchaya, and W-D. Hardt. 2004. Phages and the evolution of bacterial pathogens: From genomic rearrangements to lysogenic conversion. Microbiol Mol Biol Rev 68:560–602.

Casadevall, A., and L. A. Pirofski. 1999. Host-pathogen interactions: Redefining the basic concepts of virulence and pathogenicity. Infect Immun 67:3703–3713.

Clark-Curtiss, J. E., and S. E. Haydel. 2003. Molecular genetics of *Mycobacterium tuberculosis* pathogenesis. Annu Rev Microbiol 57:517–549.

Doherty, C. P. 2007. Host-pathogen interactions: The role of iron. J Nutr 137:1341–1344.

Falnes, P. O., and K. Sandvig. 2000. Penetration of protein toxins into cells. Curr Opin Cell Biol 12:407–413.

Gal-Mor, O., and B. B. Finlay. 2006. Pathogenicity islands: A molecular toolbox for bacterial virulence. Cell Microbiol 8:1707–1719.

Raskin, D. M., R. Seshadri, S. U. Pukatzki, and J. J. Mekalanos. 2006. Bacterial genomics and pathogen evolution. Cell 124:703–714.

Schmidt, H., and M. Hensel. 2004. Pathogenicity islands in bacterial pathogenesis. Clin Microbiol Rev 17:14–56.

Travis, J. 2003. All the world's a phage: Viruses that eat bacteria abound—and surprise. Science News 164:26–27.

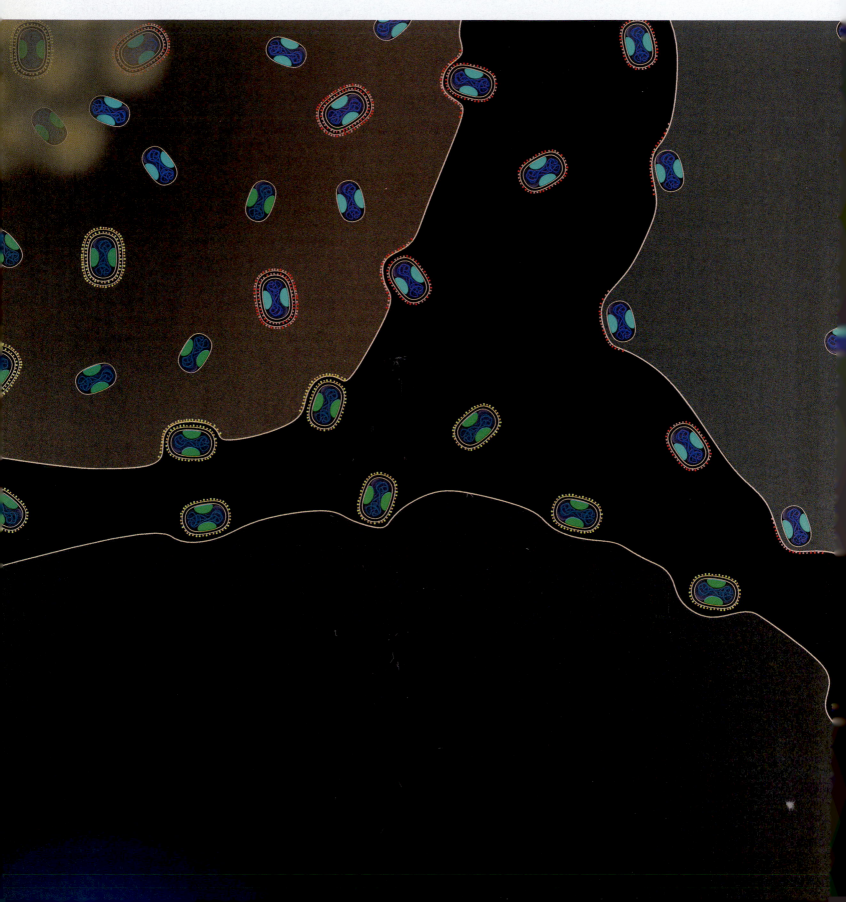

We probably have all heard about "introduced species." The term refers to species that are released into a non-native environment. Often, these introduced species become invasive, spreading extensively in the new habitat and causing economic, ecological, and/or human health problems. Sometimes, the introduction of these species occurs accidentally. In other cases, the species are introduced intentionally, without much forethought into their ultimate role in the new location. The introduction of rabbits into Australia falls into the latter category.

It is thought that Europeans introduced rabbits into Australia several times in the 1700s and 1800s. The current problem with rabbits, though, probably results from the 1859 release of a few rabbits by residents of western Australia. They hoped to generate a population of wild rabbits for hunting. Within a decade, however, the rabbits had spread and multiplied so extensively that hunters could not possibly control their numbers. In the following years, the rabbits caused extensive crop loss and probably led to the extinction of several native plant species. The Australian government tried to contain the rabbits in several ways, including building a 2,000-mile-long fence in 1907, the so-called "rabbit-proof fence." None of their approaches worked.

Then, in 1950, the government decided to employ a type of biological warfare. They intentionally released myxoma virus, a member of the *Poxviridae* family. This virus caused a relatively mild disease in its natural host, the cottontail rabbits of South America, but was highly virulent in the European rabbit that was plaguing Australia. Initially, the strategy seemed to be a success. Over 99 percent of infected rabbits died within two weeks of being infected, and the total rabbit population in Australia plummeted from 600 million to 100 million in just a couple of years. To understand more thoroughly how this virus affected the rabbit populations, we need to explore the general properties of viral pathogenesis. Also, we will begin to understand why this control measure turned out to be a very temporary solution.

CHAPTER NAVIGATOR

As you study the key topics, make sure you review the following elements:

Viruses can cause many types of infections and can be transmitted from one host to another in a variety of ways.

- Microbes in Focus 22.1: Human rhinoviruses and the common cold
- Figure 22.4: Diseases caused by human herpesviruses
- Table 22.2: Common horizontal transmission strategies
- Perspective 22.1: Vertical transmission of HIV

The replication strategies of viruses lead to cellular damage and disease.

- 3D Animation: Interactions with the host: Strategies and consequences
- Figure 22.11: Inhibition of host cell translation by poliovirus
- Figure 22.12: Process Diagram: Inhibition of host cell mRNA translation by bunyaviruses and orthomyxoviruses
- Microbes in Focus 22.5: Tomato spotted wilt virus: A major agricultural pest
- Perspective 22.2: Viral induction of apoptosis
- Figure 22.14: Formation of inclusion bodies in cells infected by viruses
- Figure 22.15: Effects of rhinovirus infection on levels of the inflammatory cytokines IL-6 and IL-8

Some viral infections lead to the development of cancer.

- Table 22.3: Mammalian viruses associated with transformation or tumorigenesis
- Figure 22.18: Gardasil: The first vaccine against cancer
- Figure 22.24: Activation of a proto-oncogene by a *cis*-acting retrovirus
- Mini-Paper: Viruses that cause cancer by affecting cellular proliferation
- Figure 22.25: Acquisition of cellular proto-oncogenes by transducing retroviruses

Several different genetic events can influence the virulence of a virus.

- 3D Animation: Evolution of viral pathogens
- Figure 22.29: Reassortment of viral gene segments
- Perspective 22.4: Ethical concerns about avian flu research

CONNECTIONS for this chapter:

Control of the SOS regulon in *E. coli* (Section 11.3)

Poliovirus capsid disassembly and replication (Section 8.2)

Proinflammatory molecules and the immune response (Section 19.3)

DNA repair mechanisms (Section 7.5)

Introduction

As we noted in the preceding chapter, bacteria, like all other organisms, have evolved various strategies to secure nutrients from their surroundings. Bacterial pathogens live in or on another organism and generally secure nutrients by damaging host tissues, thereby causing some form of disease. Non-pathogenic bacteria also may live in or on another organism, but they acquire nutrients without damaging the tissues of their host. Viruses are somewhat different. As we discovered in Chapter 5, all viruses are obligate intracellular parasites. In other words, all viruses must replicate within a host cell and, to some degree, all viruses harm their host cell in the process.

While all viruses harm their host cells in some way, the processes by which these viruses cause disease vary greatly. Some viral infections are acute, or relatively short-lived. Other viruses cause chronic or persistent infections in which the virus may never be completely eliminated from the host. Still other viruses may cause latent infections, in which viral replication ceases after an initial acute infection, only to resume replicating later, perhaps years after the initial period of replication. In other cases, the infection does not result in any overt clinical symptoms in the host.

Despite the different types of infections that viruses can cause, all viruses, like their bacterial pathogen counterparts, must solve certain key problems. They must get into a permissive host. They must acquire the resources needed for replication. They must evade host defenses. Finally, they must spread to a new host. As viruses replicate in a host, they can cause many types of diseases. These diseases may result from tissue damage caused directly by the replicating virus, or they may result from the immune response of the host to the replicating virus. In some cases, viruses cause their host cells to become transformed, or immortalized, resulting in the formation of warts or tumors.

In this chapter, we will investigate these various aspects of viral pathogenesis. We will focus not on specific viral diseases, but, rather, on general paradigms. A few selected viruses and their associated diseases will be examined in depth to illustrate our main points. While we will focus primarily on mammalian viruses, we also will examine the pathogenesis of plant viruses and bacteriophages, viruses that infect bacteria. We will conclude by investigating the evolution of viral pathogenesis. As we explore these topics, we will address the following questions:

How do viruses cause disease? (22.1)
How do viruses interact with host cells? (22.2)
How do some viruses cause cancer? (22.3)
How do some viruses become highly virulent? (22.4)

22.1 Recurring themes in viral pathogenesis

How do viruses cause disease?

As we mentioned previously, all viruses must replicate within host cells. This viral replication leads to some damage of the host. The type and extent of damage, though, can differ dramatically. So, what exactly do we mean by viral pathogenesis and disease? To answer this question, we need to define a few terms. In Chapter 18, we defined infection as the replication of a parasitic microbe on or in a host. Virologists often define infection differently. To the virologist, **infection** refers to the entry of a virus into a cell. Clearly, we often are exposed to, but not infected by, viruses. To understand this distinction, simply think about the number of viruses—viruses that naturally infect other animals, plants, or bacteria—present in the ocean. When you go swimming, you will be exposed to many of these viruses. Very few, though, if any, will infect you.

If a virus does enter a cell, one of two outcomes can occur. In a **productive infection**, new infectious viral particles will be produced. In contrast, in an **abortive infection**, few if any new infectious viral particles will be produced **(Figure 22.1)**. Generally, infectious disease occurs during a productive infection. As we will see later in this chapter, though, a specific disease state—cancer—may result from an abortive infection. Finally, as with other pathogens, virulence refers to the relative ability of a virus to cause disease.

In this section, we will look at different types of productive infections: acute, latent, and persistent. We also will examine how viruses spread from one organism to another, focusing on the horizontal, vertical, zoonotic, and mechanical transfer of infectious agents. In a later section, we will explore the link between viruses and cancer.

Types of infections

As we noted in previous chapters, all viruses are parasites. They replicate within host cells, using host cell materials. Viral replication, then, damages the infected cell. As we saw in Chapters 19 and 20, the host immune system is designed to eliminate these pathogens, thereby preventing or minimizing the amount of damage done by the virus. The overall outcome of a viral infection, in other words, depends on two interrelated activities: the replication of the virus and the immune response of the host. The interplay between these two activities determines the type of infection that occurs.

In this section, we will explore three different types of infections: acute, latent, and persistent. In each case, we will see how the interplay between viral replication and the host immune response affects the process. In each case, we also will consider how the virus benefits from the particular infection strategy.

Acute Infections

When people typically think about viral infections, they probably think about **acute infections**—infections that have a short duration. Indeed, acute infections most likely represent the majority of viral infections that we experience. We become infected

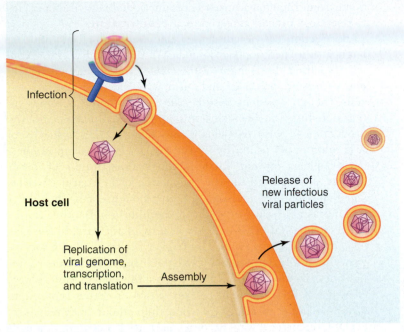

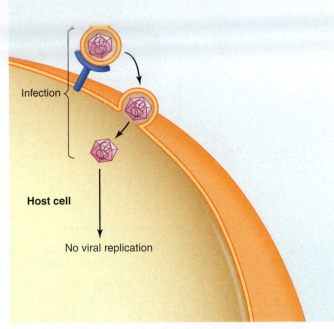

A. Productive infection: Permissive cell

B. Abortive infection: Non-permissive cell

Figure 22.1. Productive versus abortive infections "Infection" refers to the entry of a viral genome into a cell. **A.** If new viral particles are produced, then the infection is productive and the infected cell is referred to as permissive. **B.** If no new viral particles are produced, then the infection is abortive and the infected cell is referred to as non-permissive.

and show signs of disease. After a short duration, we clear the viral infection or, in some cases, the viral infection results in death. Some researchers have described this process as a state of disequilibrium; the virus replicates quickly, producing many new viral particles, while the immune system, after a short time delay, produces a strong response to the infection. The severity of disease may be affected both by the replicative ability of the virus and the strength of the immune response.

As an example of an acute infection, let's examine a typical rhinovirus infection. Rhinoviruses **(Microbes in Focus 22.1)** are the major causes of the so-called common cold. Viral replication occurs in cells of the upper respiratory tract, and the typical signs of disease include coughing, sneezing, and increased nasal secretions. These signs, though, are quickly resolved. Experimental studies, in which volunteers were intentionally infected with rhinovirus, showed that the amount of virus present in nasal washes, as determined by a tissue culture infectious dose 50, or $TCID_{50}$ assay (see Section 5.3), peaked 2–3 days after inoculation **(Figure 22.2)**. The amount of virus, and the clinical signs of disease, decreased significantly by 10 days after inoculation.

Microbes in Focus 22.1
HUMAN RHINOVIRUSES AND THE COMMON COLD

Habitat: Humans

Description: Non-enveloped, icosahedral virus in the *Picornaviridae* family. About 30 nm in diameter with a single-stranded positive sense RNA genome of approximately 8,000 nucleotides.

Key Features: Roughly 100 serotypes of human rhinoviruses have been characterized. These viruses probably cause between a third and a half of all cases of the common cold. While the common cold generally is not a major health threat, the Centers for Disease Control and Prevention (CDC) estimates that children in the United States miss 22 million days of school annually due to colds.

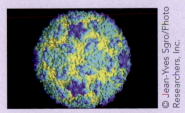

© Jean-Yves Sgro/Photo Researchers, Inc.

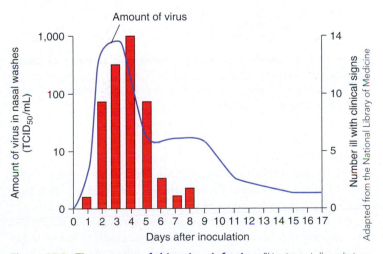

Figure 22.2. Time course of rhinovirus infection Rhinovirus typically results in an acute infection. In this experiment, 17 volunteers were inoculated with rhinovirus. The amount of virus present in nasal secretions, as determined by a tissue culture infectious dose 50 (or $TCID_{50}$) assay, peaked 2–3 days after infection and decreased markedly within 1–2 weeks. Clinical signs of a cold (red bars) showed a similar time course.

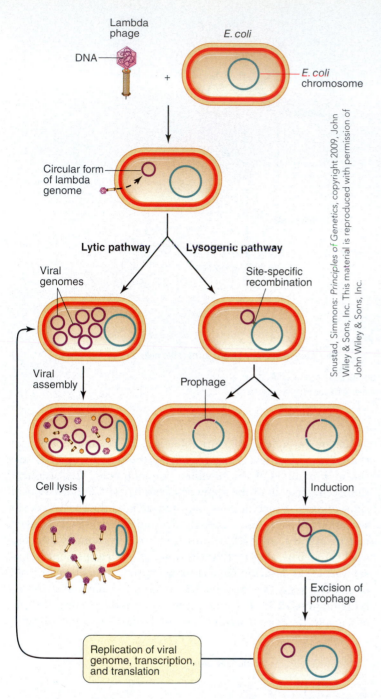

Lambda phage

DNA

E. coli

E. coli chromosome

+

Circular form of lambda genome

Lytic pathway   **Lysogenic pathway**

Viral genomes

Site-specific recombination

Viral assembly

Prophage

Cell lysis

Induction

Excision of prophage

Replication of viral genome, transcription, and translation

Snustad, Simmons: Principles of Genetics, copyright 2009, John Wiley & Sons, Inc. This material is reproduced with permission of John Wiley & Sons, Inc.

Figure 22.3. Lysogenic versus lytic replication strategy of phage lambda (λ) After infection, the phage can undergo lytic replication or enter a lysogenic state as a prophage. Various events, including exposure of the cells to ultraviolet light or antibiotics, can trigger the excision of the prophage and a new cycle of lytic replication.

Latent Infections

Not all viruses exhibit the hit-and-run replicative strategy indicative of acute infections. Other viruses have evolved such that they remain present, and capable of replicating, in the host for extended periods of time. Some of these viruses produce **latent infections**, in which the viral genome remains present in infected cells, but viral replication occurs only sporadically. These viruses typically undergo a period of acute replication immediately after infecting a host. Subsequently, the production of new viral particles stops, but the viral genome remains in some cells.

Some event then may cause the production of new viral particles to resume, a process referred to as **reactivation**.

In Section 8.3, we discussed one well-studied example of latency—the bacteriophage lambda (λ). When λ infects *Escherichia coli*, its natural host, many of the cells become acutely infected; new viruses are produced and, ultimately, the cell lyses. In some cells, though, lysogeny results; the viral genome becomes integrated into the host chromosome as a prophage, and the production of viral particles ceases **(Figure 22.3)**. Lysogeny of λ is maintained primarily by the cI or λ repressor protein, a bacteriophage protein that inhibits transcription of other viral genes required for the production of new viral particles (see Figure 8.21). The lysogenic state, though, can be thought of as a metastable state, or a delicate equilibrium. In other words, the lysogenic state easily can convert to a lytic state. In the case of λ, many different events can trigger the reversal of lysogeny. Experimentally, exposure of lysogenic bacteria to ultraviolet light or certain antibiotics may trigger the switch. Molecularly, cleavage of the cI protein relaxes the cI-mediated repression. Often, cleavage of cI occurs when the cellular SOS response, a DNA repair pathway, becomes activated. Lysogeny, then, is maintained as long as the host cell remains healthy. When the host cell becomes stressed, as indicated by the activation of the SOS pathway, the prophage is excised from the host chromosome and the virus switches to the lytic replication pathway.

CONNECTIONS In *E. coli*, the SOS response involves the coordinated expression of a large number of genes. As we saw in **Section 11.3**, this regulon can be triggered by the detection of single-stranded DNA within the cell.

Probably the best-studied mammalian viruses that exhibit latency are the herpesviruses. Members of the *Herpesviridae* family **(Table 22.1)** cause a variety of diseases in humans **(Figure 22.4)**. Human herpesviruses 1 and 2, also known as herpes simplex virus types 1 and 2 (HSV-1 and HSV-2), primarily cause cold sores

TABLE 22.1 **Human herpesviruses**

Designation	Common name	Disease
HHV-1	Herpes simplex virus 1 (HSV-1)	Cold sores, genital ulcers
HHV-2	Herpes simplex virus 2 (HSV-2)	Genital ulcers, cold sores
HHV-3	Varicella zoster virus (VZV)	Chickenpox, shingles
HHV-4	Epstein–Barr virus (EBV)	Mononucleosis, Burkitt lymphoma
HHV-5	Cytomegalovirus (CMV)	Various, mainly immuno-compromised hosts
HHV-6	N/A	Roseola
HHV-7	N/A	Not known
HHV-8	Kaposi sarcoma herpesvirus (KSHV)	Kaposi sarcoma

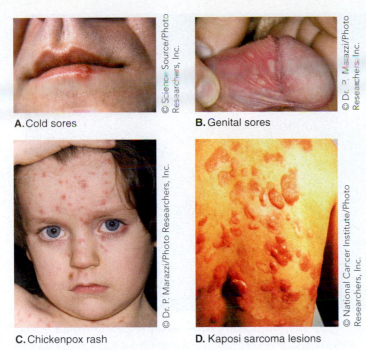

A. Cold sores

B. Genital sores

C. Chickenpox rash

D. Kaposi sarcoma lesions

Figure 22.4. Diseases caused by human herpesviruses The human herpesviruses cause a variety of disease, including: **A.** Cold sores caused by HHV-1 (HSV-1). **B.** Genital sores caused by HHV-2 (HSV-2). **C.** Chickenpox caused by HHV-3 (varicella zoster virus). **D.** Kaposi sarcoma caused by HHV-8.

and genital sores, respectively **(Microbes in Focus 22.2)**. Human herpesvirus 3, also known as varicella zoster virus (VZV), causes chickenpox and shingles. The recently characterized human herpesvirus type 8 (HHV-8) causes Kaposi sarcoma (KS), a defining cancer of people with HIV/AIDS. While the diseases caused by these viruses differ, their replication strategies are similar. All of these viruses display the classic latency pattern—a period of acute infection, followed by latency, and periodic reactivation.

How is this latency and subsequent reactivation achieved? Let's address this question by examining HSV-1. Typically, the

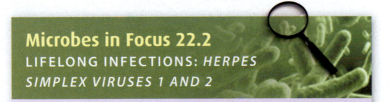

Microbes in Focus 22.2
LIFELONG INFECTIONS: *HERPES SIMPLEX VIRUSES 1 AND 2*

Habitat: Human

Description: Large, enveloped virus with a double-stranded DNA genome.

Key Features: HSV-1, or HHV-1, often is referred to as oral herpes, and usually is spread by kissing. HSV-2, or HHV-2, often is referred to as genital herpes, and usually is spread sexually. Both types can cause oral and genital infections. Several drugs are available to reduce viral shedding and decrease outbreak severity. Generally, though, the infection is lifelong.

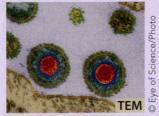

virus spreads via direct, intimate contact between an infected person and a susceptible person. Primary replication occurs in mucosal cells at the site of infection. Often, this primary infection is asymptomatic. In other cases, blisters, or cold sores form on the lips or inside the mouth or nose. Newly produced viral particles then enter local neurons and move to the nuclei of these cells (see Figure 18.8). Within the nucleus, the viral DNA exists as an **episome**, a circular piece of DNA separate from the host chromosomes. At this stage, expression of most viral genes stops and no new virus particles are made. During latency, one class of genes is transcribed, giving rise to the **latency-associated transcripts (LATs)**. It remains unclear exactly how the LATs affect latency, although some researchers speculate that they may repress the expression of other viral genes. At some point, reactivation occurs and viral replication resumes. The expression of the LAT genes decreases and expression of other viral genes increases. The virus particles travel back down the neuron to the periphery, causing characteristic cold sores. This switch away from latency may be triggered by many different stress events, including exposure to UV light, studying for exams, and, in women, menstruation. As we saw with bacteriophage λ, then, reactivation in some way seems tied to the health of the host.

Persistent Infections

In contrast to latent infections, **persistent infections** (also referred to as chronic infections) result in the continuous production of new viral particles. The host does not clear the virus in a reasonable amount of time. Of course, "reasonable" is a subjective term. For most classic persistent infections in humans, the production of viral particles continues for years after the initial infection. This continuous replication, not surprisingly, requires a balance between the host and the virus; the virus must replicate in the presence of the host's innate and adaptive immune responses.

Several different mechanisms for the maintenance of a persistent infection have been postulated. In some cases, the virus may dampen the immune response, thereby limiting the host's ability to clear the virus. In other cases, the virus may repress programmed cell death, thereby keeping infected cells alive. In still other cases, persistence may result from infection by a viral mutant that exhibits altered replication. Recent studies with mammalian reoviruses have shown that the establishment of a persistent infection in cell culture may involve genetic changes in either the virus or the host cell, thereby emphasizing the interplay between the two members of this relationship. Mutations in the virus that affect its ability to undergo proteolytic disassembly within a host cell have been associated with persistence. Similarly, mutations in the host cell affecting cathepsin B, a cellular protease involved in reovirus disassembly, also have been associated with persistent reovirus infections. Thus, perturbations of the normal reovirus disassembly pathway, whether caused by a mutation in the virus or the cell, may lead to a persistent infection.

CONNECTIONS For many viruses, disassembly of the viral capsid must occur prior to replication. For poliovirus, specific conformational changes in the capsid occur after binding of the virus to its receptor. These conformational changes facilitate the exit of the viral genome from the virion **(Section 8.2)**.

TABLE 22.2	Common horizontal transmission strategies	
Transmission route	**Examples**
Respiratory	Rhinovirus, influenza virus
Fecal-oral	Poliovirus, hepatitis A virus
Sexual	HIV, human papillomavirus

Types of transmission

An essential part of any viral disease is transmission from one host to another. Because viruses are obligate intracellular pathogens, they must be able to effectively leave one infected host and enter another. This process of transmission presents several problems for the virus. First, the virus particles must be expelled from the initial host. Second, the virus particles must remain infectious until they encounter a new host. Third, they must gain access to the appropriate cells within the new host. We addressed this issue briefly in Chapter 5. Most human viruses, we noted, undergo horizontal transmission from one person to another. Some viruses, though, can be transmitted vertically from mother to fetus or infant. Other viruses can spread from non-human hosts to humans. Finally, still other viruses spread via a less direct way, requiring the mechanical transfer from one susceptible host to another. In this section, we briefly will examine these mechanisms of transfer.

Horizontal Transmission

Several different strategies of **horizontal transmission**, or transmission other than from parents to their offspring, exist **(Table 22.2)**. In each case, the processes of exit and entry are closely linked and depend on the types of cells that the virus infects. Rhinovirus, the most frequent cause of the common cold, represents a typical example of a virus transmitted horizontally via the respiratory route. Replication occurs in cells of the upper respiratory tract. The disease signs associated with a rhinovirus infection—coughing and sneezing—allow newly produced viral particles to exit the infected host **(Figure 22.5)**. New hosts then become infected by directly inhaling aerosolized droplets containing infectious virions or by transferring virions deposited on hands or inanimate surfaces to the nose. When the virus particles come in contact with permissive cells in the new host, the infection cycle begins again.

Vertical Transmission

For a few viruses, **vertical transmission**, or transmission from the parent (typically the mother) to fetus or newborn may occur. This type of transmission can occur in one of several ways. Virus in the blood may be transferred from the mother to the developing fetus through the placenta or during birth. This type of transmission occurs with several important viral pathogens, including rubella (German measles), hepatitis B virus, hepatitis C virus, and HIV. Some viruses, most notably HIV, also can be transmitted from mother to infant via breast milk. Vertical transmission of HIV, regrettably, remains a major issue **(Perspective 22.1)**.

Figure 22.5. Transmission of viruses via the respiratory route Rhinovirus replicates in cells of the upper respiratory tract. Sneezing, a common sign of rhinovirus infection, allows newly formed viral particles to exit the host and enter a new host.

Some retroviruses, like mouse mammary tumor virus (MMTV) also exhibit a different form of vertical transmission—transmission via the germ cells. As we noted in Chapters 5 and 8, retroviruses integrate a DNA copy of their RNA genome—the provirus—into the chromosomal DNA of an infected cell. If a retrovirus infects a spermatocyte or oocyte, the cells that give rise to sperm and eggs, respectively, then the proviral DNA will integrate into the DNA of the gametes and is referred to as an **endogenous retrovirus**. If a sperm or egg containing proviral DNA becomes part of a zygote, then every cell within the developing embryo will contain the endogenous retrovirus **(Figure 22.6)**.

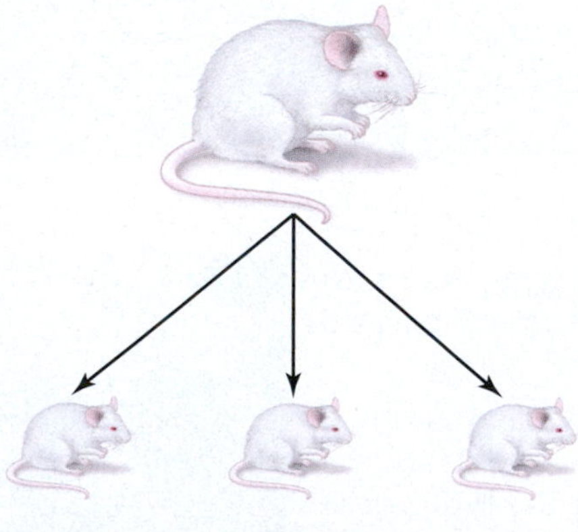

Germ cells of BALB/c mouse
contain MMTV provirus.

Offspring contain MMTV provirus in all cells.

Figure 22.6. Vertical transmission of endogenous retroviruses If an integrated retrovirus is present in the germ cells, then the virus is transmitted to offspring. Every cell within the offspring will contain a copy of the retrovirus. BALB/c mice contain an endogenous copy of mouse mammary tumor virus (MMTV). Expression of viral genes and production of virus is stimulated by hormones in the mammary gland. In some mice, MMTV is transmitted by breast milk in a vertical fashion.

The HIV/AIDS pandemic has wrought untold suffering on populations throughout the world. Sadly, children have not escaped the horrors. According to statistics compiled by the World Health Organization (WHO), roughly 2.5 million children were living with HIV at the end of 2009 and 365,000 children became newly infected during that year. According to the WHO, many of these children probably became infected through mother-to-child transmission (MTCT), or transmission of the virus from an infected woman to her child. MTCT can occur through several mechanisms. The virus may pass through the placenta, resulting in *in utero* infection. More commonly, the virus is transmitted from the mother to child during the birthing process or after delivery, through breastfeeding **(Figure B22.1)**.

In the United States, the rate of MTCT has decreased dramatically over the past 15 years. In 1994, according to the CDC, 25 percent of

children in the United States born to HIV+ women became infected. By 2007, that number dropped to 2 percent **(Figure B22.2)**. Three simple interventions resulted in this dramatic change. First, pregnant women who are HIV+ receive antiretroviral drugs prior to delivery. Second, C-sections, rather than vaginal deliveries, may be used. Third, HIV+ mothers are encouraged to feed their infants formula and avoid breastfeeding. As a result of these three interventions, MTCT in the United States has decreased significantly. Unfortunately, the infection of children remains a major problem in other parts of the world.

To a large extent, transmission of HIV from mother to child can be prevented. Several international programs, including the President's Emergency Plan For AIDS Relief (PEPFAR) and the Global Fund to Fight AIDS, Tuberculosis, and Malaria, have identified the prevention of MTCT as a major goal. Hopefully, we soon will see the success evident in the United States in other countries throughout the world.

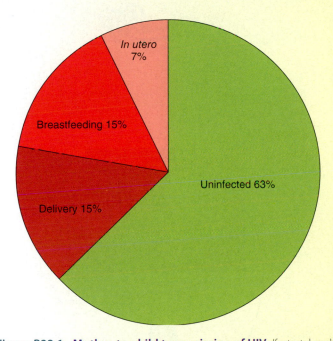

Figure B22.1. Mother-to-child transmission of HIV If untreated, roughly a third of HIV+ pregnant women will transmit the virus to the infant. Transmission generally occurs during delivery or through breastfeeding. More rarely, the virus may be transmitted *in utero*.

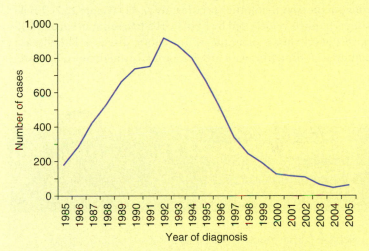

Figure B22.2. Reduction in mother-to-child transmission (MTCT) of HIV in the United States According to the Centers for Disease Control and Prevention, between 1992 and 2005, the number of children in the United States with AIDS from MTCT of HIV dropped approximately 95 percent. This dramatic decrease resulted from increased testing and the use of various preventive strategies. In August of 1994, the United States Public Health Service recommended use of the antiviral drug zidovudine to prevent MTCT. Bottle-feeding infants and C-section deliveries can further reduce the incidence of MTCT.

The inheritance of endogenous retroviruses does not represent a major means of retrovirus transmission. Genetic studies have clearly demonstrated, though, that many retroviruses have become integrated into our chromosomes over our evolutionary history. Through sequence analyses, researchers estimate that as many as 98,000 human endogenous retroviruses (HERVs) or HERV fragments exist within our genome, comprising about 8 percent of our DNA. Most of these HERVs contain major deletions or gene rearrangements and are not replication competent. Rather, they are leftover remnants of proviral DNA that integrated into our ancestral DNA millions of years ago. So,

what, if any, function does this DNA serve? Possibly, it is simply junk. Alternatively, these viral sequences may be a source of new DNA sequences that have shaped, and continue to shape, our evolutionary history.

Zoonotic Transmission

In some cases, human diseases occur when a viral pathogen is transferred from a non-human animal to a human, a process referred to as **zoonosis**. The natural reservoirs for zoonotic viruses typically are vertebrates. Usually, humans are dead-end hosts. While the virus may cause disease in an infected person, transmission

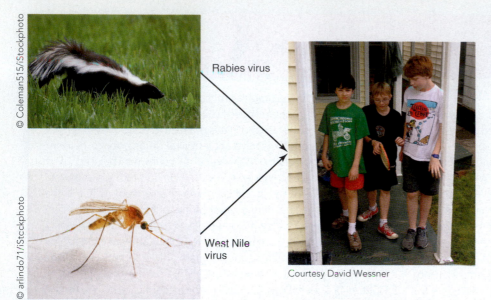

Rabies virus

West Nile virus

Courtesy David Wessner

Figure 22.7. Zoonotic transmission Transmission of a pathogen from a non-human animal to a human may result in a zoonotic disease. In many cases, the pathogen is not readily transmitted from human to human. Humans, therefore, can be considered a dead-end host.

from one person to another occurs only rarely **(Figure 22.7)**. Ebola hemorrhagic fever, caused by Ebola virus **(Microbes in Focus 22.3)**, represents a classic example of a zoonotic disease. The disease in humans is dramatic, with an extraordinarily high mortality rate.

Spread among humans through contaminated blood can occur, but not with great efficiency. Most likely, the virus is spread from another animal to humans. Increasing evidence suggests that the reservoir for this virus may be bats. Infectious Ebola virus has been isolated from various fruit bats and from bat excrement. The virus could be spread directly from bats to humans or the spread could occur through an intermediate organism **(Figure 22.8)**. Finally, we should note that zoonotic viruses may evolve into more typical human viruses. HIV is a perfect example. HIV evolved from a strain of SIV, simian immunodeficiency virus, that infected a human (see Figure 18.30). Over time, this simian virus became an efficient human virus, very capable of replicating in human cells and being transmitted from one human to another.

Microbes in Focus 22.3
EBOLA: A ZOONOTIC DISEASE OF HUMANS

Habitat: Causes zoonotic disease in humans. Natural reservoir unknown, but may be species of bats.

Description: Member of the *Filoviridae* family, approximately 80 nm in diameter and 800 nm in length. Enveloped, with a pleomorphic shape and a linear genome of negative sense single-stranded RNA.

Key Features: The first recorded outbreak of Ebola hemorrhagic fever occurred in 1976 in Zaire (now the Democratic Republic of the Congo). The disease is characterized by fever, headache, diarrhea, and vomiting. In many cases, extensive hemorrhaging, or bleeding, also occurs. Although person-to-person transmission is inefficient, the fatality rate is extremely high. Because of its extreme virulence, several popular books and movies about Ebola have been produced.

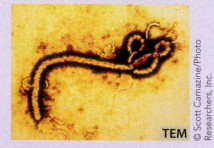

TEM

© Scott Camazine/Photo Researchers, Inc.

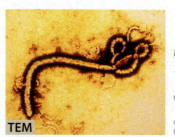

TEM

© Scott Camazine/Photo Researchers, Inc.

A. Ebola virus

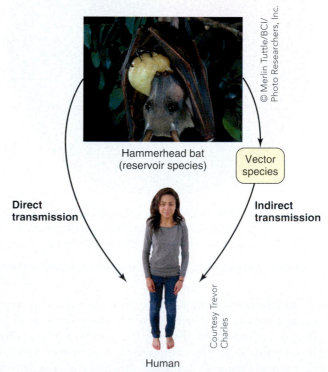

Hammerhead bat (reservoir species)

© Merlin Tuttle/BCI/Photo Researchers, Inc.

Vector species

Direct transmission

Indirect transmission

Human

Courtesy Trevor Charles

B. Possible transmission of Ebola virus

Figure 22.8. Spread of Ebola virus to humans While the natural reservoir of Ebola virus has not been determined conclusively, the virus may be transmitted to humans by bats. **A.** Electron micrograph of Ebola virus. **B.** Recent studies have identified the virus in several species of bats, including *Hypsignathus monstrosus*, the hammerhead bat. The virus could be transmitted directly from bats to humans or indirectly via another species.

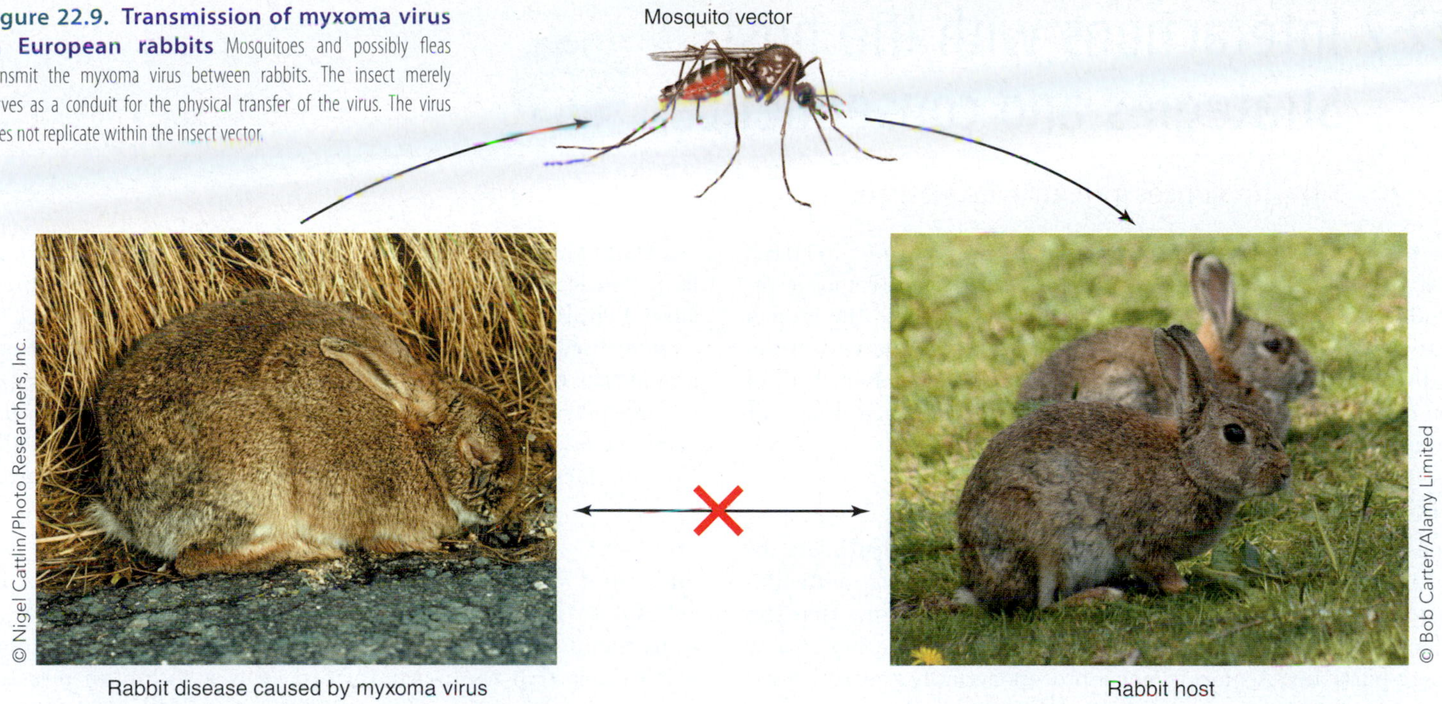

Figure 22.9. Transmission of myxoma virus in European rabbits Mosquitoes and possibly fleas transmit the myxoma virus between rabbits. The insect merely serves as a conduit for the physical transfer of the virus. The virus does not replicate within the insect vector.

Mosquito vector

Rabbit disease caused by myxoma virus

Rabbit host

Mechanical Transmission

Before we end our discussion of transmission strategies, let's briefly mention one last method of transmission: **mechanical transmission**. In this type of transmission, the virus does not move directly from one susceptible host to another. Rather, some vector facilitates the transfer. The myxoma virus that we discussed in the beginning of this chapter presents an excellent example. The virus spreads from rabbit to rabbit via mosquitoes or, possibly, fleas. When a mosquito bites an infected animal, the mouth of the biting insect may become contaminated with viral particles. When this insect subsequently bites another animal, the infectious virions can be transferred **(Figure 22.9)**. It should be noted that the myxoma virus does not replicate in the mosquito or flea. Other viruses transmitted by insects, like yellow fever virus, do replicate within the insect vector **(Figure 22.10)**.

CONNECTIONS As we discussed in **Section 5.1**, many plant viruses are transmitted by mechanical means, either employing insect vectors or farm machinery to aid in their spread. Physical disruption of the plant cell cuticle allows for the entry of the viral particles into the cytoplasm of a new cell.

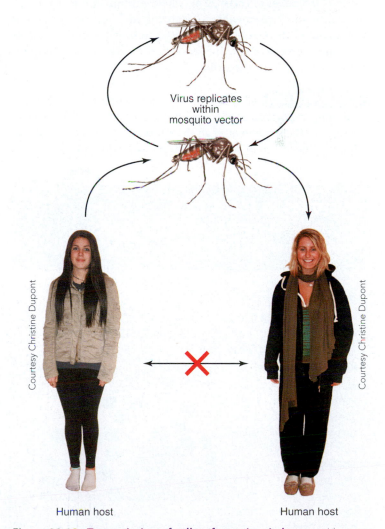

Virus replicates within mosquito vector

Human host

Human host

Figure 22.10. Transmission of yellow fever virus in humans Like myxoma virus, yellow fever virus is transmitted by mosquitoes. Unlike myxoma virus, yellow fever virus replicates within the insect vector. Transmission is the same in monkeys, the natural mammalian host for the virus.

22.1 Fact Check

1. Compare and contrast acute and latent viral infections.
2. Compare and contrast horizontal and vertical transmission.
3. What is a dead-end host?

22.2 Interactions with the host: Strategies and consequences

In the preceding section, we examined the types of infections that viruses can cause and how viruses can spread from one individual to another. In this section, we will examine the effects that viruses have on infected cells and how the infection results in disease. Perhaps surprisingly, this topic is much less clear than you might imagine. As we will see in this section, viruses do not always cause disease simply by killing infected cells.

Viral-induced cellular destruction

One obvious outcome of a viral infection is cell death. As we saw in Chapter 5, lytic bacteriophages very clearly cause the destruction of infected cells. It might be easy to imagine that the infected cell simply bursts because of the accumulation of new phage particles. A more precise process actually controls lysis. Phage lambda (λ) encodes at least two proteins that coordinate cell lysis (Microbes in Focus 22.4). The S protein renders the cell membrane more permeable, permitting the movement of the R protein out of the cytoplasm. The R protein specifically weakens the peptidoglycan layer. In concert, then, these two proteins facilitate the lysis of the cell.

CONNECTIONS All viruses possess mechanisms for entering and, following replication, exiting appropriate host cells. For bacteriophages, these actions require an ability to interact with peptidoglycan. The structure of this molecule is discussed in **Section 2.2**.

Many animal viruses also cause the destruction of cells. In fact, researchers often use cytopathic effect (CPE) as a measure of viral infections of cells in culture (see Figure 5.17). The exact causes of viral-induced cytopathology, both in cell culture and *in vivo*, remain unclear. Furthermore, the role that cytopathology plays in disease also is unclear. However, it is clear that viral infections often lead to cellular destruction. Generally, this cellular destruction occurs via one of two main pathways.

Necrosis

Some viral infections lead to **necrosis**, or death of the infected cell. Quite simply, the cell bursts. Given our previous discussions about viral replication in Chapters 5 and 8, we easily could hypothesize that the newly formed viruses within an infected cell simply cause the cell to swell, eventually leading to the rupturing of the cell. The process, in reality, is not that simple. For numerous viruses, it appears that necrosis actually results from some toxic outcome of the infection. The virus, in order to obtain necessary materials within the host cell, subverts the normal host cell machinery. The host cell, as a result, becomes less viable and dies. Viruses have evolved several mechanisms of subverting different host cell processes. We will explore a few specific mechanisms.

Poliovirus cytopathology occurs primarily because the virus effectively inhibits the translation of host cell mRNAs. The viral protease 2A cleaves the host eIF-4G polypeptide. This host polypeptide, in concert with other host elongation initiation factors (eIFs), forms the caps binding complex that recognizes the 5′ cap present on mRNA molecules and aids in the initiation of translation. Once the poliovirus protease cleaves eIF-4G, translation of host mRNA decreases dramatically **(Figure 22.11)**. The poliovirus RNA, though, can be recognized by host cell ribosomes in a cap-independent manner. Thus, the virus subverts the translational machinery, allowing for the rapid production of viral proteins. Most likely, this poliovirus-induced inhibition of cellular protein production leads to the death of the cell.

Bunyaviruses, like the tomato spotted wilt virus (TSWV) **(Microbes in Focus 22.5)**, do not directly shut down the translation of host cell mRNAs. Rather, these viruses cause the degradation of host cell mRNA molecules. Animal and plant viruses in this family cleave the 5′ capped end of host cell mRNA molecules and then use this short piece of the host cell mRNA as a primer for the synthesis of viral mRNA, a process referred to as "cap-snatching" **(Figure 22.12)**. As a result, the virus produces mRNA that will be recognized by the host cell's ribosomes. Additionally, the host cell mRNA, now devoid of a 5′ cap, will not be translated and, ultimately, will be degraded. For the virus, then, additional resources are available for it to use during its replication. For the cell, the destruction of necessary mRNA molecules leads to necrosis.

Microbes in Focus 22.4
PHAGE LAMBDA (λ): A MOLECULAR BIOLOGY WORKHORSE

Habitat: *Escherichia coli*

Description: Temperate bacteriophage with a head 50 nm in diameter and a 150-nm tail. Double-stranded DNA genome of approximately 48,000 base pairs.

Key Features: Discovered first in 1950, phage lambda (λ) has been studied and used extensively by molecular biologists. Since its discovery, it has been the model for studies of temperate phage biology and specialized transduction. It also is an important tool, being used extensively for cloning and the generation of genomic libraries.

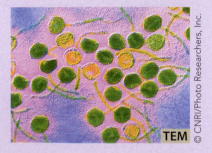

TEM

© CNRI/Photo Researchers, Inc.

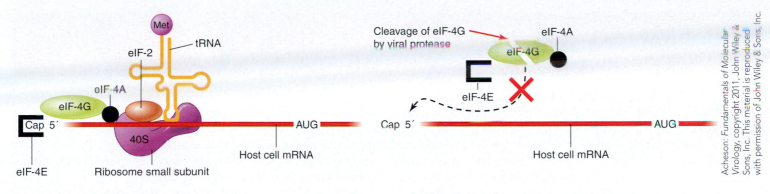

Acheson: *Fundamentals of Molecular Virology*, copyright 2011, John Wiley & Sons, Inc. This material is reproduced with permission of John Wiley & Sons, Inc.

A. Normal initiation of translation

B. Cleavage of eIF-4G by picornaviruses: No translation

Figure 22.11. Inhibition of host cell translation by poliovirus A. Normal eukaryal initiation of translation. **B.** Several picornaviruses, including poliovirus and foot-and-mouth disease virus, inhibit host cell translation by cleaving eIF-4G, an essential translation initiation factor. Cleavage of this elongation factor results in a greatly decreased level of host cell protein synthesis.

Poxviruses, like the myxoma virus we discussed in the opening story, exhibit several specific means of obtaining needed materials. Like the viruses just mentioned, poxviruses also inhibit the translation of host cell mRNA. An essential viral protein binds to the 40S ribosomal subunit and, in the process, interferes with the initiation of translation. Additionally, another poxvirus protein degrades single-stranded DNA, thereby interfering with host cell DNA replication. Both of these viral activities make more needed materials (amino acids and deoxynucleotides) available for viral replication.

Apoptosis

While it may seem logical that viruses normally kill cells by simple necrosis, it turns out that cell death often occurs via a well-defined pathway. **Apoptosis**, or programmed cell death, involves a highly coordinated series of events in a cell that eventually leads to the destruction of that cell. A number of different signaling molecules regulate the induction of these events. The process itself is characterized most notably by three main morphological changes (see Figure 19.20). First, the nucleus becomes fragmented. Second, the cell membrane develops irregular buds, or blebs. Eventually, these blebs break apart and

Acheson: *Fundamentals of Molecular Virology*, copyright 2011, John Wiley & Sons, Inc. This material is reproduced with permission of John Wiley & Sons, Inc.

Viral RNA polymerase complex

Cleavage

Host mRNA

Cap 5′ — A ———————— 3′

① The 5′ end of host mRNA is bound by the viral RNA polymerase complex.

Viral polymerase

Cap 5′ — A 3′
3′ UC ————— UUUUUU 5′

Viral RNA genome

② Adenosine (A) of released host cap pairs with U of viral RNA.

Cap 5′ — AG
3′ UC ————— 3′ ————— UUUUUU 5′

③ A viral protein of the complex extends the 3′ end of the snatched cap using the viral genome as template.

Viral mRNA

Cap 5′ — AG ———————— AAAAAAAAAA 3′

④ A capped viral mRNA is ready for translation.

Figure 22.12. Process Diagram: Inhibition of host cell mRNA translation by bunyaviruses and orthomyxoviruses In a process referred to as "cap-snatching," viral proteins cleave 10–20 5′ nucleotides from the host mRNA. This sequence is then used to prime synthesis of viral RNA, providing the virus with stable, 5′-capped mRNA. For bunyaviruses, this activity occurs in the cytoplasm. For orthomyxoviruses, such as influenza virus, it occurs in the nucleus.

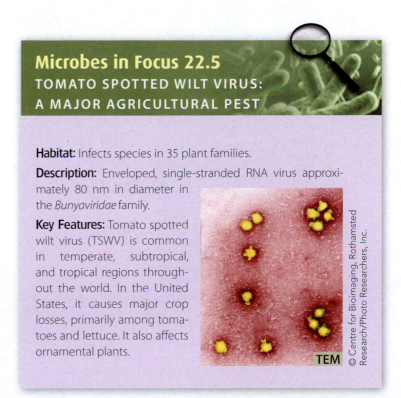

Microbes in Focus 22.5
TOMATO SPOTTED WILT VIRUS: A MAJOR AGRICULTURAL PEST

Habitat: Infects species in 35 plant families.

Description: Enveloped, single-stranded RNA virus approximately 80 nm in diameter in the *Bunyaviridae* family.

Key Features: Tomato spotted wilt virus (TSWV) is common in temperate, subtropical, and tropical regions throughout the world. In the United States, it causes major crop losses, primarily among tomatoes and lettuce. It also affects ornamental plants.

TEM

© Centre for Bioimaging, Rothamsted Research/Photo Researchers, Inc.

Perspective 22.2
VIRAL INDUCTION OF APOPTOSIS

It may seem logical that apoptosis would benefit an infected host. If infected cells are destroyed early in the infectious process, then fewer new viral particles will be produced. It also may be logical that viruses would benefit from preventing apoptosis. If infected host cells are kept alive, then more new viral particles can be produced. Let's look at another scenario. Could apoptosis be beneficial to a virus?

For ascoviruses, the answer to this question appears to be "yes." These large, enveloped viruses, in the *Ascoviridae* family, contain a double-stranded DNA genome of 120,000 to 180,000 base pairs, exhibit a complex shape, and infect various moth and butterfly species. Ascoviruses appear not to be transmitted directly from host to host. Rather, parasitic wasps transmit the viruses. Female wasps normally deposit eggs in the larvae of appropriate moths or butterflies, an activity referred to as "oviposition." When the eggs hatch, the wasp larvae then feed on the host larvae. If a wasp deposits eggs in an ascovirus-infected host, then the wasp's ovipositor may become contaminated with virus. When the wasp then deposits eggs into another larvae, the virus can be transmitted.

In infected larval cells, a viral protein induces apoptosis, resulting in nuclear membrane fragmentation and DNA cleavage. Additionally, the cell breaks down, forming subcellular vesicles. These vesicles contain viral particles and adhere to the wasp ovipositor, thereby facilitating the spread of the virus. For these viruses, then, induction of apoptosis represents an essential part of their replication cycle.

the cell disintegrates into a group of membrane-bound vesicles, or apoptotic bodies. Third, the DNA becomes fragmented into a characteristic "ladder." Endonucleases selectively cleave the DNA between nucleosomes, resulting in fragments of DNA that are multiples of 180 base pairs in length (180, 320, etc.). These DNA fragments can be separated by gel electrophoresis and are easily visualized **(Figure 22.13)**. This process represents a normal part of development, allowing an organism to specifically destroy unneeded cells. Additionally, cells that become damaged may be targeted for apoptotic destruction. Many environmental triggers, such as ionizing radiation, also activate the apoptosis pathway.

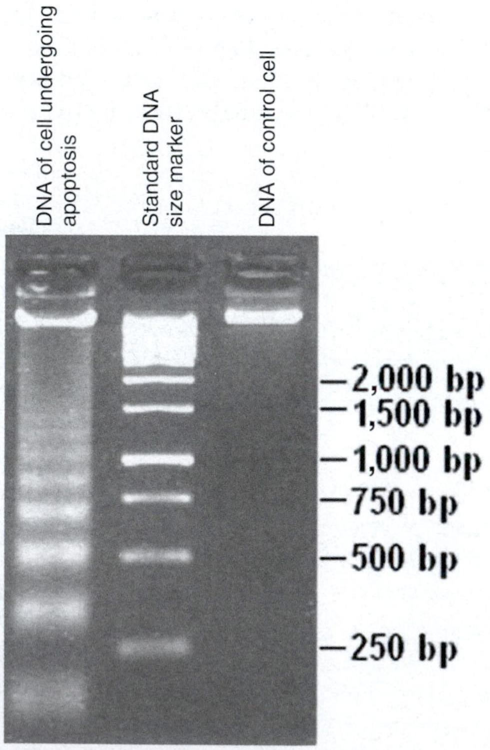

Image from Shinryuu, http://en.wikipedia.org/wiki/File:Apoptotic_DNA_Laddering.png

Figure 22.13. DNA laddering during apoptosis During apoptosis, the DNA becomes fragmented. Here, DNA from cells undergoing apoptosis (*left lane*) and DNA from cells not undergoing apoptosis (*right lane*) were added to an agarose gel and separated based on size by electrophoresis. The DNA then was stained with ethidium bromide. The DNA from apoptotic cells has been cleaved into fragments that are multiples of roughly 180 base pairs in length. The center lane contains a molecular weight marker for size determination. Sizes, in base pairs, of these fragments are indicated on the right.

The death of virally infected cells often occurs via apoptosis. As we have seen, viruses bind to cellular receptors, replicate their genomes, and produce viral proteins. All of these events, through various mechanisms, could trigger the apoptotic pathway. For the host, this induction of apoptosis may be useful in limiting the severity of an infection. Even if the induced cell death results in some localized tissue damage, destruction of an infected cell may prevent the production of more virus particles by that cell and, ultimately, limit the spread of the virus within the host. For the virus, conversely, inhibiting apoptosis may be desirable. If the host cannot destroy infected cells early in the infection, then the virus can continue its replication cycle, producing more virions. Indeed, some herpesviruses and adenoviruses produce proteins that mimic antiapoptotic proteins normally produced by their host cells. These viruses, then, inhibit the host's attempts to destroy infected cells. Finally, and perhaps surprisingly, some viruses actually induce apoptosis **(Perspective 22.2)**.

Syncytia and Inclusion Bodies

While necrosis and apoptosis may be the most obvious signs of viral-induced cellular damage, some viruses cause cellular damage through other means. Several viruses, including HIV, measles virus, and respiratory syncytial virus, form syncytia. Infected cells fuse with adjoining infected or uninfected cells, resulting in the formation of a multinucleated giant cell (see Figure 5.15). Typically, cells in the syncytium ultimately die. Other viruses, such as rabies virus, adenovirus, and reovirus, may result in the formation of nuclear or cytoplasmic **inclusion bodies**, microscopically visible structures within the cell made

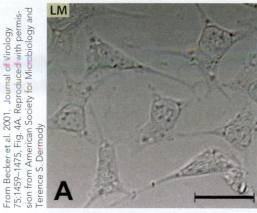

From Becker et al. 2001. *Journal of Virology* 75:1459–1475, Fig. 4A. Reproduced with permission from American Society for Microbiology and Terence S. Dermody

A. Reovirus infection of mouse fibroblasts showing inclusion body formation

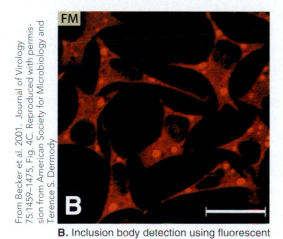

From Becker et al. 2001. *Journal of Virology* 75:1459–1475, Fig. 4C. Reproduced with permission from American Society for Microbiology and Terence S. Dermody

B. Inclusion body detection using fluorescent antibody specific to reovirus proteins

Figure 22.14. Formation of inclusion bodies in cells infected by viruses Some viruses lead to disruptions in cellular morphology and, as a result, cellular function. **A.** Inclusion bodies can be seen as dark cytoplasmic spots using differential interference contrast microscopy. **B.** Cells stained with fluorescent-tagged antibodies against reovirus proteins clearly reveal inclusion bodies in the cytoplasm of the same cells. Scale bar: 25 μm.

of viral or cellular components **(Figure 22.14)**. Again, these structures often impair the normal functioning of the cell and lead to cell death.

Immune responses and disease

Viral infections, as we just saw, can lead directly to cell death. In many cases, though, the clinical signs of disease associated with a viral infection result more from the host's immune response than from the virus itself. As we discussed in Chapters 19 and 20, the innate and adaptive arms of the immune system can target infected cells for destruction. This activity, in some respects, is a double-edged sword. On one hand, destruction of infected cells limits the production of new viral particles and limits the spread of the virus. On the other hand, destruction of cells could cause damage to the tissue. The immune system, then, plays a delicate balancing act, balancing the amount of cells it destroys to limit the infection versus the amount of damage this cellular destruction causes to the body. As we will see in this section,

common clinical signs of viral infections often result from actions of the immune system. We also will investigate a couple instances in which the immune system may go too far.

We all know what it's like to get a common cold. Sneezing becomes a way of life—at least for a few days. Rhinoviruses, members of the *Picornaviridae* family, are the most frequent causes of colds. The viruses typically infect cells of the upper respiratory tract. Perhaps surprisingly, the viral infection itself does not seem to result in much, if any, cytopathology. So what causes us to sneeze? Studies have shown that people with colds have elevated levels of several proinflammatory immune system molecules, including interleukins 6 and 8 (IL-6 and IL-8). Most likely, this inflammatory immune response leads to a buildup of fluids in the respiratory tract. This increased fluid triggers sneezing **(Figure 22.15)**.

CONNECTIONS In humans, invading microbes often trigger the release of proinflammatory molecules like tumor necrosis factor alpha and interferons. These cytokines facilitate communication between cells of the immune system **(Section 19.3)**.

The immune response also represents the major cause of disease associated with hepatitis B virus. Again, the virus itself does not appear to result in any significant cytopathology. In most cases, HBV infections are effectively controlled by the immune system, with no obvious disease occurring. Hepatitis, or liver disease, results from the cytotoxic T lymphocyte-mediated destruction of infected hepatocytes (see Chapter 20 to review the action of CTLs). In some cases, HBV infection also leads to kidney damage. In these cases, disease results not from the CTL response, but rather from the antibody response. HBV antigen:antibody complexes form and become deposited in the

Antimicrobial Agents and Chemotherapy, May 1, 2000, vol. 44, reproduced/amended with permission from American Society for Microbiology.

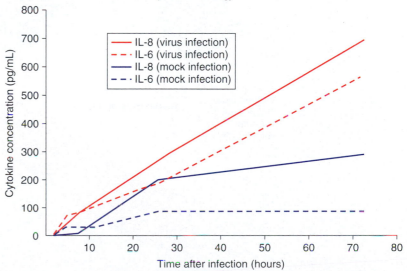

Figure 22.15. Effects of rhinovirus infection on levels of the inflammatory cytokines IL-6 and IL-8 In this experiment, cells were inoculated with rhinovirus (red lines) or mock infected (blue lines). At specified times postinoculation, the concentrations of IL-6 and IL-8 in the cell culture supernatant were determined. Rhinovirus infection leads to a marked increase in both of these cytokines.

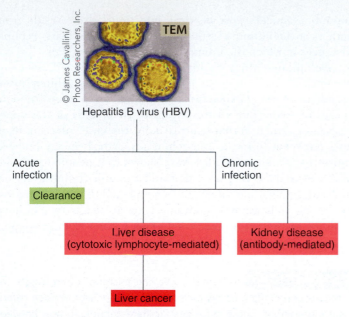

Hepatitis B virus (HBV)

Acute infection → Clearance

Chronic infection
- Liver disease (cytotoxic lymphocyte-mediated) → Liver cancer
- Kidney disease (antibody-mediated)

Figure 22.16. Potential outcomes of hepatitis B virus infection In some cases, the immune system clears the virus. In other cases, the infection becomes chronic. During a chronic infection, CTL-mediated attacks of infected hepatocytes can cause liver damage. Additionally, antigen:antibody complexes may cause kidney damage. Chronic infections of the liver may lead to hepatocellular carcinoma—liver cancer.

kidney, resulting in tissue damage. The clinical outcomes of HBV infection, thus, can be varied, depending to a large extent on the immune response (**Figure 22.16**).

Finally, viruses may indirectly cause disease through a process known as **autoimmunity**. Autoimmune diseases occur when the immune system incorrectly targets normal cellular components. The body, in effect, attacks itself. Several diseases, including multiple sclerosis (MS) and rheumatoid arthritis, are autoimmune diseases. In MS, for instance, the immune system targets the myelin sheath surrounding nerve cells, resulting in impaired neurological functioning. So, how can viruses be involved in this process? Researchers have postulated that autoimmune diseases may be triggered by viral infections through a process known as **molecular mimicry**. Basically, viral antigens that resemble host antigens activate immune cells that then cross-react with the similar host antigens (**Figure 22.17**). In other words, the immune response directed against the virus also attacks the host. Molecular mimicry has been proposed as a cause of several human autoimmune diseases. Conclusive proof, though, is lacking.

22.2 Fact Check

1. How do bunyaviruses inhibit host cell translation?
2. For the host cell, what are the advantages and disadvantages of viral-induced apoptosis?
3. In what ways can the immune response to a viral infection lead to disease?

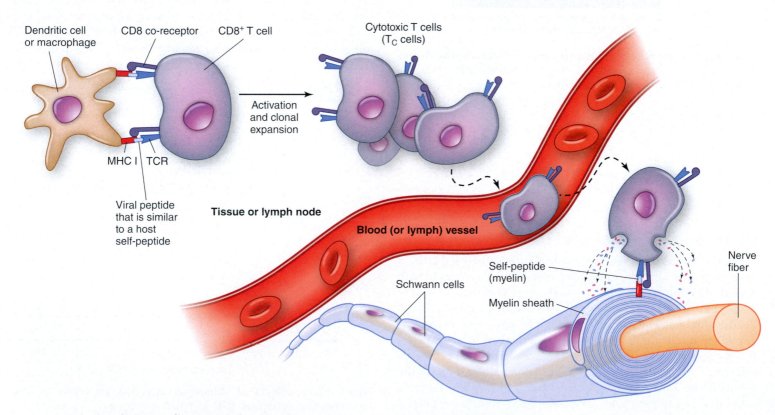

Figure 22.17. Autoimmune diseases and molecular mimicry Possibly, viral infections trigger some forms of autoimmunity. If viral peptides are structurally similar to host cell proteins, then reactive T cells may cross-react with self-peptides, leading to destruction of host tissue. In this example, the viral peptide resembles myelin that is produced by Schwann cells surrounding nerve cells, resulting in symptoms of multiple sclerosis.

22.3 Viral infections and cancer

How do some viruses cause cancer?

To many people, the association between viruses and cancer may seem odd. Viruses, many people might think, cause the flu or the common cold; viruses do not cause cancer. As we discussed in Chapter 8, though, Peyton Rous identified a tumor-causing retrovirus in chickens, Rous sarcoma virus, in 1911. Since then, many viruses have been associated with cancer. In fact, today, viruses in many different virus families have been shown to cause the transformation of cells in culture and/or tumorigenesis in humans or animals (**Table 22.3**).

Transformation refers to changes within a cell that cause the cell to become cancer-like. These changes often include:

- change in morphology
- immortalization
- loss of contact inhibition
- loss of anchorage dependence
- increased sugar transport
- reduced growth factor requirements
- chromosomal aberrations

Tumorigenesis refers to the formation of a tumor. In this section, we will examine mechanisms by which different viruses cause tumor formation. We will focus primarily on two families of DNA viruses—the *Papillomaviridae* and *Adenoviridae*—and one family of RNA viruses—the *Retroviridae*.

DNA tumor-causing viruses

Members of virtually every mammalian DNA-containing virus family have been shown to cause tumor formation in animals or transformation of cells in culture. As we have seen previously, viruses in these families are quite diverse, differing significantly

in size, shape, and replication strategies. Despite these differences, though, the mechanisms by which they transform cells are surprisingly similar. In most cases, the genomes of these viruses contain genes that can be considered **oncogenes**—genes capable of transforming animal cells. As we will see in this section, these viral oncogenes are essential for viral replication.

Papillomaviruses

In June of 2006, the FDA approved for use in girls and young women Gardasil (**Figure 22.18**), a vaccine against certain types,

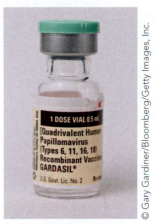

A. Gardasil vaccine   **B.** Genital warts

Figure 22.18. Gardasil: The first vaccine against cancer Human papillomaviruses (HPVs) cause most cases of cervical cancer. **A.** Approved for use in girls and women by the FDA in 2006, Gardasil provides protection against four types of HPV, including types 16 and 18, which cause about 70 percent of all cervical cancers. **B.** Gardasil also provides protection against HPV strains 6 and 11, which cause about 90 percent of all genital warts.

TABLE 22.3	Mammalian viruses associated with transformation or tumorigenesis		
Virus family	**Type**	**Disease**	**Notes**
Papillomaviridae	Human papillomavirus types 16 and 18	Cervical, anal, penile cancer	HPVs may cause up to 10% of all human cancers
Polyomaviridae	SV40	Sarcomas in rodents	Can transform some human cell lines
Adenoviridae	Adenovirus	Transformation of cells in culture	Both E1A and E1B proteins are needed for transformation
Herpesviridae	HHV-8	Kaposi sarcoma	Transforms endothelial cells; defining cancer of people with HIV/AIDS
Herpesviridae	Epstein–Barr virus (HHV-4)	Burkitt lymphoma	Also causes infectious mononucleosis
Hepadnaviridae	Hepatitis B virus	Hepatocellular carcinoma	An effective vaccine is available
Flaviviridae	Hepatitis C virus	Hepatocellular carcinoma	60–85% of people infected with HCV develop a chronic infection
Retroviridae	Rous sarcoma virus	Sarcomas in chickens	Led to the identification of proto-oncogenes

or strains, of human papillomavirus (HPV). In October of 2009, another HPV vaccine, Cervarix, also received FDA approval for use in girls and young women. Both vaccines provide protection against HPV strains that cause **cervical carcinoma**, malignant cancer of the cervix. These strains of HPV also cause anal, penile, vaginal, and certain oral and pharyngeal cancers. These vaccines are the first vaccines against cancer, and we'll return to a discussion of them shortly. First, let's examine the mechanism by which HPV causes cancer.

Viruses in the *Papillomaviridae* family are non-enveloped, icosahedral viruses with a capsid diameter of roughly 60 nm. They contain a circular double-stranded DNA genome of approximately 8,000 nucleotides. Members of the family infect a wide range of animals, and over 100 types of human papillomaviruses (HPVs) have been identified. About 30 HPV types are transmitted sexually.

The association between papillomaviruses and abnormal cellular growth is long-standing. The name, in fact, derives from the observation in the early part of the twentieth century that an infectious agent could transmit warts, or **papillomas**. Peyton Rous—yes, the same Peyton Rous who identified the oncogenic retrovirus Rous sarcoma virus (see Section 8.3), showed in 1935 that papillomaviruses could cause skin cancer in rabbits. More recently, extensive evidence has linked several types of HPV to cervical cancer. According to data presented by the National Cancer Institute, approximately 12,000 women were diagnosed with cervical cancer in the United States in 2012, and 4,200

women died from it. About 250,000 women worldwide die from cervical cancer annually. HPV now is recognized as the major cause of this cancer.

To understand how these viruses cause cancer, we need to think about their basic replication cycle **(Figure 22.19)**. Papillomaviruses typically infect keratinocytes—cells that form the outer layer of the skin and certain mucosal surfaces. Quite often, these keratinocytes are **quiescent cells**, or not actively dividing. Within quiescent cells, many macromolecular building blocks exist in very limited amounts. Only small amounts of deoxyribonucleotides, for instance, may be present. If the cell is not actively replicating, then there is no need to manufacture the building blocks of DNA. For the virus, though, this lack of deoxyribonucleotides presents a problem. The virus cannot replicate its genome without a supply of these molecules. To circumvent this problem, the virus employs a remarkable strategy: it stimulates the cell to enter the S phase of the cell cycle **(Figure 22.20)**. The cell then prepares itself for DNA replication, making the deoxyribonucleotides needed by the virus.

More specifically, papillomavirus proteins interfere with the normal functioning of two cellular proteins: retinoblastoma or Rb, and p53. Both Rb and p53 function as **tumor suppressors**, normal cellular proteins that inhibit cellular replication. Rb binds to specific transcription factors, ultimately blocking the synthesis of cellular factors needed for entry into the S phase of the cell cycle. Phosphorylated p53

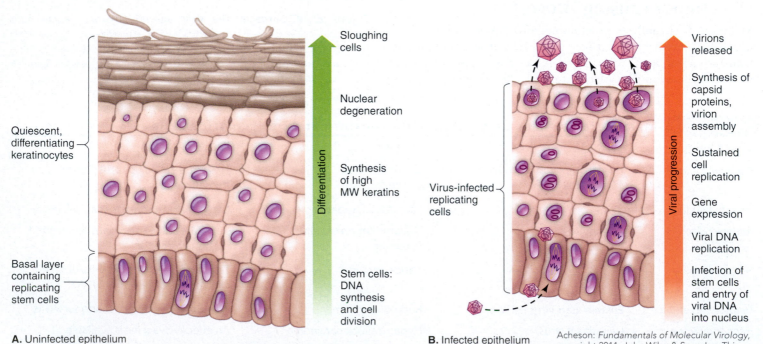

A. Uninfected epithelium

B. Infected epithelium

Acheson: *Fundamentals of Molecular Virology*, copyright 2011, John Wiley & Sons, Inc. This material is reproduced with permission of John Wiley & Sons, Inc.

Figure 22.19. Papillomavirus infection of differentiating epithelial cells **A.** Epithelial cells produced by the basal layer normally differentiate and stop replicating as they approach the surface. **B.** Papillomaviruses infect basal layer cells. The viral DNA is replicated and distributed to daughter cells. Gene products of the viral DNAs interfere with cell cycle control and the daughter cells continue to replicate rather than differentiating. Virions are released from cells of the surface of the epithelium.

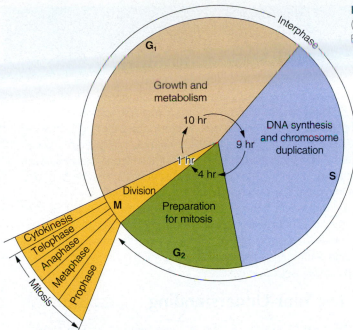

Figure 22.20. The cell cycle Dividing cells progress through the cell cycle, undergoing a growth phase (G₁), DNA replication (S), a second growth phase (G₂), and mitosis (M). Quiescent cells typically are arrested in G₁. Because they are not actively dividing, deoxyribonucleotides, the building blocks of DNA, usually are in short supply.

Snustad, Simmons: *Principles of Genetics*, copyright 2009, John Wiley & Sons, Inc. This material is reproduced with permission of John Wiley & Sons, Inc.

promotes the synthesis of p21, which, in turn, prevents cell replication. Phosphorylated p53 also induces apoptosis in damaged cells (**Figure 22.21**).

In the presence of HPV E6 and E7 proteins, Rb and p53 do not inhibit cellular replication (**Figure 22.22**). The E7 gene product binds to and inactivates Rb. The E6 gene product binds to p53, targeting it for destruction, thereby preventing the transcription of p21. As we will see in **Toolbox 22.1**, researchers can detect these molecular interactions using a form of **immunoprecipitation (IP)**, a technique in which an antibody is used to precipitate an antigen. The interactions between the HPV proteins and Rb and p53 cause the infected cell to exit the quiescent state and begin replicating. To prepare for replication,

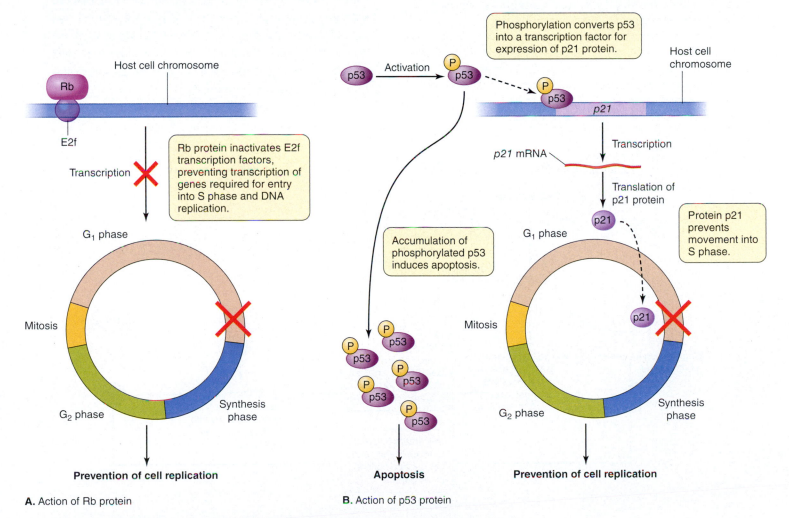

A. Action of Rb protein

B. Action of p53 protein

Figure 22.21. Normal action of tumor suppressor proteins Rb and p53 **A.** Rb blocks entry into S phase by binding to, and inhibiting, a family of transcription factors called "E2fs." To enter S phase, Rb is phosphorylated, allowing E2fs to facilitate the transcription of genes required for entry into the cell cycle. **B.** Upon DNA damage and cell stress, p53 is activated by phosphorylation. Activated p53 is a transcription factor needed for expression p21. Protein p21 prevents movement into S phase. Phosphorylated p53 also induces apoptosis.

Researchers, using techniques like the Western blot, can utilize antibody:antigen interactions to identify proteins of interest (see Toolbox 8.1). These same molecular interactions also can be used to isolate a specific protein from a mixture and to identify proteins that bind to each other. Using immunoprecipitation (IP) techniques, researchers determined that the human papillomavirus (HPV) E6 and E7 proteins interact with, respectively, the p53 and Rb human tumor suppressor proteins.

To better understand this process, let's first examine the basics of immunoprecipitation. With a relatively simple form of IP, researchers can isolate a specific protein from a mixture of proteins, such as a cell lysate. First, antibodies to the protein of interest are chemically attached to small beads. These beads then are incubated with the mixture of proteins. Because of the specificity of antibody:antigen binding, only proteins in the mixture that interact with the antibodies will attach to the immobilized antibodies. By centrifuging the mixture, investigators can precipitate the antibody/antigen/bead complex. All other proteins will remain in the supernatant. The protein of interest, in other words, effectively has been separated from all other proteins in the original milieu.

To identify protein:protein binding interactions, co-immunoprecipitation (co-IP) can be used. In this technique, often referred to as a pull-down assay, investigators begin by attaching antibodies specific to one member of a suspected protein:protein binding pair to beads. The beads, again, are incubated with a mixture of proteins, often from a cell lysate. As just discussed, the protein of interest will bind to the antibody and can be immunoprecipitated. Other proteins that bind to this protein also may be precipitated. Binding partners of the antigen, in other words, will be co-immunoprecipitated. The pelleted beads can be physically or chemically treated to disrupt all weak protein-binding interactions. The released proteins, the ones bound to the immunoprecipitated protein, then can be identified.

Using this technique, researchers showed that p53-specific antibodies pulled down HPV E6 proteins. From this observation, we can hypothesize that the HPV E6 protein interacts with p53. This initial finding contributed to our current understanding of the how human papillomaviruses cause cancer.

● Test Your Understanding ....

While HPV strains 16 and 18 have been shown to cause cancer, other strains of HPV have not been associated with tumorigenesis. To explain this observation, one could hypothesize that the E6 and E7 proteins of these non-tumorigenic HPV strains do not interact with p53 and Rb. Explain how you could use an immunoprecipitation assay to test this hypothesis.

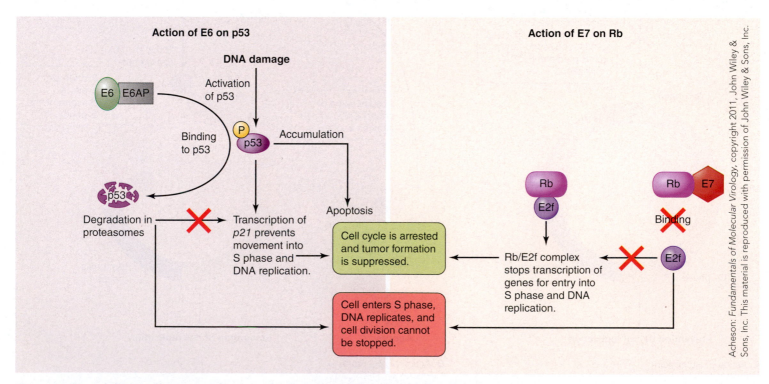

Figure 22.22. Inhibition of tumor suppressors by HPV E6 and E7 proteins Strains of HPV can lead to the development of cancer by inhibiting tumor suppressors. The HPV E6 protein (*left*), assisted by the cellular protein E6AP, binds to the cellular tumor suppressor protein p53 and targets it for destruction, which, in turn, prevents the transcription of p21. The HPV E7 protein (*right*) binds to another cellular tumor suppressor protein, Rb, rendering it inactive. Normally, both p53 and Rb inhibit cell growth. Once these proteins are inactivated by HPV, cellular replication proceeds. When the viral genome integrates, virus particles are not formed and the host cells are not killed, allowing transformation.

SV40, or simian virus 40, is a member of the *Polyomaviridae* family and naturally infects certain non-human primates. The polyomaviruses closely resemble the papillomaviruses. Until recently, in fact, these viruses frequently were classified together in a single family. Much like we saw for papillomaviruses and adenoviruses, polyomaviruses infect quiescent cells and cause the infected cells to proceed to the S phase, or DNA replication phase, of the cell cycle. As a result of this shift to the S phase, the cell produces more deoxynucleotides, which the virus then uses for the replication of its genome. For SV40, a single viral protein appears to be responsible for this shift. The SV40 large T antigen binds to and inactivates both Rb and p53, the same two cellular tumor suppressors affected by papillomaviruses and adenoviruses. While SV40 does not cause tumors in macaques, it is tumorigenic in rodents and can infect and transform some human cells in culture.

Between 1955 and 1963, both the Sabin (oral) and Salk (injectable) poliovirus vaccines, produced in SV40-infected monkey kidney cells, became contaminated with SV40. Nearly 100 million people in the United States, as a result, became exposed to SV40. Because SV40 causes tumor formation in rodents, there has been much debate about its potential to cause cancer in humans.

Recent studies have shown that SV40 DNA sequences can be detected in cells from several types of human tumors, including ependymomas (tumors of the ependyma, a central nervous system tissue) and osteosarcomas (tumors of the bone). Despite the presence of SV40 DNA in these tumors, though, epidemiological studies have not shown an increased prevalence of these tumors among people who received the SV40-contaminated poliovirus vaccine. Based on these results, most scientists contend that there is no relationship between the accidental exposure to SV40 and the development of cancer. Some scientists and advocacy groups, however, disagree.

---

the cell produces deoxyribonucleotides, which the virus then uses to replicate its genome. The virus, in other words, alters the cell state in order to obtain necessary molecules. In many cases, viral replication then occurs, leading to the production of new viral particles and, most likely, death of the infected cell. In some cases, though, a productive infection does not occur. Viral replication is aborted and portions of the viral genome integrate in the host genome. If intact E6 and E7 genes become integrated, then the E6 and E7 proteins continue to inhibit the actions of Rb and p53. The cell continues to divide. A tumor develops **(Mini-Paper)**.

As we mentioned in the beginning of this section, vaccines against HPV now exist. Gardasil, manufactured by the pharmaceutical company Merck and Co., provides protection against four types of HPV: HPV 6, 11, 16, and 18. Cervarix, manufactured by GlaxoSmithKline, provides protection against strains 16 and 18. HPV strains 16 and 18 cause approximately 70 percent of all cervical cancer cases and HPV 6 and 11 cause about 90 percent of all genital warts cases. According to an American Cancer Society news release about Gardasil, this HPV vaccine represents "one of the most important advances in women's health in recent years." In late 2009, the FDA approved the use of Gardasil in boys and men between the ages of 9 and 26. The vaccine has been shown to prevent genital warts in males. It also may reduce the incidence in males of anal and penile cancer, two cancers associated with HPV. More importantly, though, the vaccination of males may reduce the sexual transmission of the virus from males to unvaccinated females.

Adenoviruses

Like the papillomaviruses, the adenoviruses are icosahedral, non-enveloped viruses with a double-stranded DNA genome. Unlike the papillomavirus genome, though, the adenovirus genome is linear and between 25,000 and 45,000 base pairs in length. These viruses infect a wide range of animals. In humans, adenovirus infections can have a variety of effects, ranging from cold-like symptoms to diarrhea, depending on the infecting strain.

Adenoviruses generally infect quiescent epithelial cells. These viruses, then, encounter the same problem faced by papillomaviruses: necessary resources, most notably deoxynucleotides needed for DNA replication, are in short supply. Like the HPV and the polyomavirus SV40 **(Perspective 22.3)**, the adenoviruses circumvent this problem by causing infected cells to enter the S phase of the cell cycle. The adenoviruses also cause the cells to resume replicating by inactivating the same two tumor suppressors, Rb and p53. The adenovirus E1A protein, like the HPV E7 protein, inactivates the cellular Rb, while the adenovirus E1B protein, like the HPV E6 protein, binds to the cellular p53 protein.

No human cancers have been associated with adenovirus infections. Most likely, infection of a human cell with a human adenovirus leads to a productive infection and, eventually, cell death. When the viral E1A and E1B genes are transfected into rodent cells in culture, however, the cells become transformed. When these transformed cells then are injected into animals, tumors form. Interestingly, complete transformation and tumor formation only occur when the cells receive both E1A and E1B. When E1A alone is introduced into baby rat kidney (BRK) cells, an abortive transformation occurs; the cells temporarily show signs of undergoing transformation, but then die. These results indicate that both the Rb and p53 proteins need to be inactivated for cancer to arise **(Figure 22.23)**.

RNA tumor-causing viruses

Unlike the DNA-containing viruses, very few viruses with RNA genomes appear to cause cancer. In fact, the only non-retrovirus RNA-containing virus associated with the development of

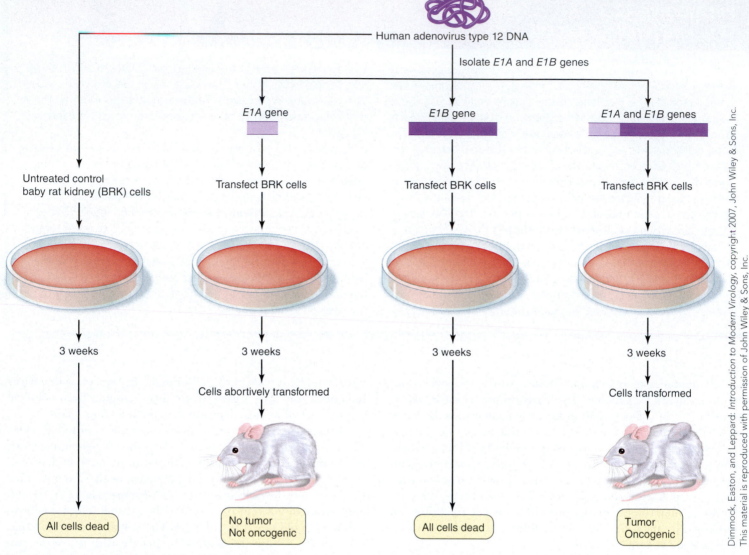

Figure 22.23. Transformation of cells by adenovirus The adenovirus type 12 E1A and E1B proteins are needed to transform cells. When E1A alone is added to baby rat kidney (BRK) cells, an abortive transformation occurs. When both E1A and E1B are added to BRK cells, the cells become fully transformed and cause tumor formation when injected into animals.

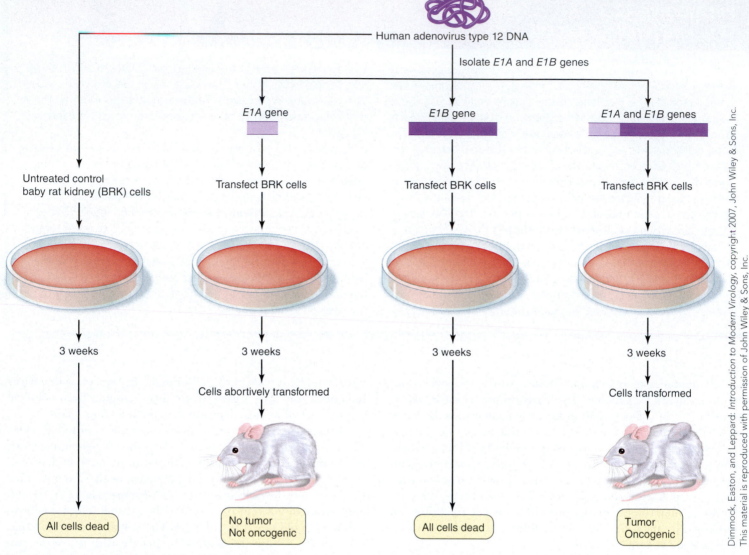

cancer is hepatitis C virus, a member of the *Flaviviridae* family. While infection with hepatitis C virus certainly increases one's risk of developing liver cancer, it seems unlikely that the virus itself transforms the cells. Rather, cancer probably results as a by-product of the chronic infection. Because of the long-term damage caused to the liver, and the liver's constant attempts to replace damaged cells, the genetic material of some hepatocytes eventually becomes compromised and cancer results.

Retroviruses also do not cause cancer directly, at least not in the way that DNA-containing viruses do. As we will see in this section, retroviruses do not contain essential oncogenes. Rather, tumorigenesis is an unintended outcome of the retrovirus replication strategy. Because retroviruses integrate their proviral DNA into the host chromosomes (see Figure 8.19), they can alter cellular **proto-oncogenes**—genes that code for proteins involved in regulating the cell cycle. When the expression or sequence of these genes is altered, the resulting proteins may

function incorrectly, thereby altering the replicative behavior of the cell. The proto-oncogenes, in other words, may become oncogenes. This proto-oncogene to oncogene conversion can occur via at least two different ways. First, the proviral DNA may integrate near a proto-oncogene and alter the expression of that proto-oncogene, causing the infected cell to become transformed. Second, the retrovirus genome may inadvertently acquire a cellular oncogene and cause transformation by then introducing this oncogene into another cell. Studies of both these ***cis*-acting retroviruses**, retroviruses that activate a cellular proto-oncogene after integrating, and **transducing retroviruses**, retroviruses that have acquired a cellular gene and, as a result, can transfer it to a new cell, have been instrumental in our understanding of the normal cell cycle and have led to the identification of numerous proto-oncogenes. Thus, while retroviruses themselves may not be tumorigenic, the study of these tumor-causing retroviruses has greatly increased our knowledge of cancer and cell biology.

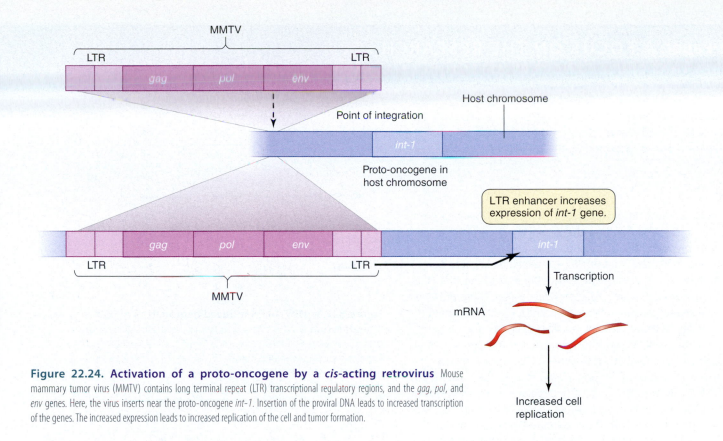

Figure 22.24. Activation of a proto-oncogene by a *cis*-acting retrovirus Mouse mammary tumor virus (MMTV) contains long terminal repeat (LTR) transcriptional regulatory regions, and the *gag*, *pol*, and *env* genes. Here, the virus inserts near the proto-oncogene *int-1*. Insertion of the proviral DNA leads to increased transcription of the genes. The increased expression leads to increased replication of the cell and tumor formation.

Cis-acting Retroviruses

As we saw in Section 8.3, the retroviral replication cycle involves conversion of the viral RNA genome into DNA and then integration of this proviral DNA into the host chromosome. This integration event disrupts the host chromosome and can be thought of as a mutational event. The effect of this mutational event depends, of course, on the location of the integration. If the proviral DNA integrates near a cellular proto-oncogene, then the insertion could alter the expression of the proto-oncogene. Altered expression of a gene involved in regulating the cell cycle could lead to aberrant cellular replication, or tumor formation. Because, in this scenario, the retrovirus affects only an adjoining piece of host DNA, it is referred to as "*cis*-acting."

Probably the best-studied *cis*-acting retrovirus is the mouse mammary tumor virus (MMTV). This virus infects mice and can be spread from a mother to her pups via milk. In infected animals, it causes tumors of the breast tissue. The MMTV genome, though, does not contain an oncogene. Rather, MMTV causes tumor formation by integrating near a cellular proto-oncogene. The virus preferentially integrates near one of three cellular genes, *int-1* (also referred to as *wnt-1*), *int-2*, and *int-3*, and increases their expression **(Figure 22.24)**. While these three genes, and their products, are unrelated, all three genes are thought to be proto-oncogenes. Increased expression leads to altered cell growth.

The identification of a virus that causes mammary tumors in mice presents a very interesting question. Could some cases of breast cancer in women have a viral etiology? Researchers have actively investigated this possibility. A human equivalent of MMTV has not been identified. In late 2007, though, researchers demonstrated that MMTV could infect and replicate in cultured human breast cells. Additionally, some researchers have reported the identification of MMTV-like sequences in DNA isolated from human breast cancer samples. These intriguing results present us with additional research questions. Can MMTV infect primary human breast cells? Could MMTV be transmitted from mice to humans? Does a human mammary tumor virus exist? Answers to these questions may provide important insight into tumorigenesis and the basic biology of viruses.

Transducing Retroviruses

Unlike the *cis*-acting retroviruses, the transducing retroviruses contain an oncogene obtained from a cell. When they infect another cell, then, they transfer this genetic material into the new cell, a process known as "transduction." Studies of Rous sarcoma virus (RSV) and other transducing retroviruses have led to our current understanding of the genetic basis of cancer and our basic understanding of the cell cycle. It should be noted that transducing retroviruses are very rare in nature and, as a result, these viruses are not major causes of cancer. Their contributions to the fields of cancer and cell biology, however, are immeasurable. Many of the proto-oncogenes identified thus far, like the *src* proto-oncogene ultimately discovered in RSV, have been identified through studies of transducing retroviruses.

To obtain an oncogene, a retrovirus first must integrate near a cellular proto-oncogene. During the production of new viral particles, transcription of the proviral DNA occurs, resulting in the production of new viral genomes. If this process occurs incorrectly, then an RNA copy of the adjacent cellular proto-oncogene may become incorporated into the newly formed viral RNA. In some cases, the cellular sequences become added onto an otherwise normal viral genome.

M. Scheffner, B. A. Werness, J. M. Hulbregtse, A. J. Levine, and P. M. Howley. 1990. The E6 oncoprotein encoded by human papillomavirus types 16 and 18 promotes the degradation of p53. Cell 63:1129–1136.

Context

HPV infections cause approximately half of all human cancers caused by viruses. Chief among these cancers is cervical cancer. Understandably, once the association between HPV and cervical cancer was demonstrated in the 1980s, it became critical to determine the reason for this association. Investigators found that HPV E6 and E7 proteins are expressed in cervical cancers. These proteins also can cause the transformation of cell lines. The transformation-inducing ability of these proteins qualifies them as oncoproteins. Because p53 is a well-known tumor suppressor that has a central role in controlling cell growth, researchers investigated the interaction between E6 and p53. Through *in vitro* studies, they confirmed that E6 from cancer-promoting HPV associates with p53. In the paper that we are looking at here, researchers investigated the outcome of this association. They examined whether the E6:p53 interaction had any effect on p53 levels or activity. This understanding could be fundamental to developing therapeutics to control or prevent HPV-associated tumors.

Experiment

This study focused on HPV-16 and HPV-18, two strains of HPV associated with cancer and whose E6 protein binds p53, in contrast to HPV-6 and HPV-11, relatively benign strains of HPV whose E6 does not bind p53. E6 and p53 proteins were expressed from an expression vector in an *in vitro* transcription system, followed by translation of the resulting transcript using a rabbit reticulocyte extract. Such transcription and translation systems are available commercially, and allow production of protein that can be radioactively labeled during translation. In this study, radioactive [L-^{35}S]cysteine was added in order to produce labeled protein. This labeling allowed easy detection of the protein on SDS-PAGE.

To investigate the interaction between the two proteins, the scientists simply incubated them together for 3 hours, and then examined the mixture by SDS-PAGE. The levels of the p53 protein decreased after incubation with E6 from HPV-18 (**Figure B22.3**). The incubation of p53 with the E6 protein from HPV-11, conversely, did not cause a reduction in p53 levels. A time course experiment (**Figure B22.4**) showed that the HPV-18 E6-dependent degradation of p53 was continuous with time.

Because protein degradation usually requires ATP as a source of energy, the investigators next examined the role of ATP in E6-induced p53 degradation. Because the *in vitro* translation system used to produce the proteins contained ATP, it was not possible to exclude ATP from the reaction mixtures. Instead, the researchers added AMP, the product of ATP hydrolysis, or ATPγS, an analog of

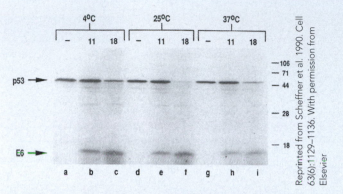

Figure B22.3. HPV-18 E6-induced degradation of p53 Radiolabeled p53 protein was incubated at various temperatures for 3 hours in the absence of any other proteins (-) or with radiolabeled E6 protein from HPV-11 (11) or HPV-18 (18). Following the incubation, the proteins were separated by SDS-PAGE and then detected. When incubated with E6 from HPV-18, but not HPV-11, the amount of p53 decreased.

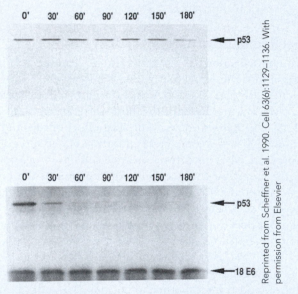

Figure B22.4. Time-dependent degradation of p53 by HPV-18 E6 Radiolabeled p53 was incubated at 25°C for various amounts of time either in the absence (*top*) or presence (*bottom*) of HPV-18 E6. Following the incubations, the proteins were separated by SDS-PAGE and detected. In the presence of HPV-18 E6, p53 degraded in a time-dependent fashion.

ATP that cannot be hydrolyzed, to the reaction mixtures. The presence of both of these molecules, they reasoned, would inhibit the hydrolysis of the ATP present in the reaction. As shown in **Figure B22.5**, the intensity of the p53 bands does not decrease when AMP or ATPγS are added, consistent with the requirement of ATP hydrolysis for p53 degradation.

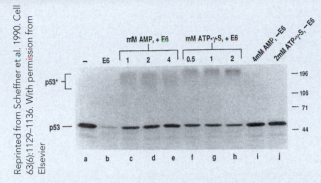

Reprinted from Scheffner et al. 1990. Cell 63(6):1129–1136. With permission from Elsevier

Figure B22.5. Role of ATP in p53 degradation Radiolabeled p53 was incubated with HPV-18 E6 and increasing amounts of AMP or ATPγS. In the presence of these molecules, p53 degradation was not observed, indicating that ATP hydrolysis is involved in the degradation. Some higher molecular weight versions of p53 (p53*) were observed.

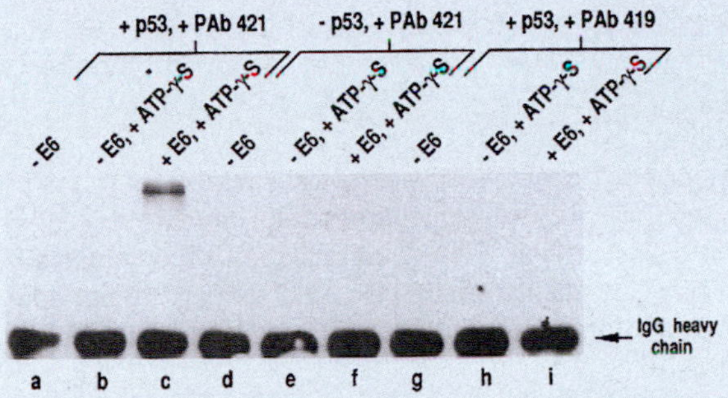

Reprinted from Scheffner et al. 1990. Cell 63(6):1129–1136. With permission from Elsevier

Figure B22.6. HPV-18 E6-associated ubiquitination of p53 The tumor suppressor protein p53 was incubated with or without HPV-18 E6 in the presence of ATPγS. The p53-specific antibody PAb 421 or a control antibody (PAb 419) was used to isolate p53 from the reaction. The isolated p53 then was detected with a ubiquitin-specific antibody. When incubated with HPV-18 E6, p53 became ubiquitinated (lane C).

Interestingly, the presence of AMP and ATPγS also caused some of the p53 to run higher in the gel, indicating that it had a higher molecular weight. This result only occurred in the presence of E6. The researchers hypothesized that the higher molecular weight forms of p53 might be intermediates of p53 degradation. A common proteolytic degradation pathway in eukarya involves the ubiquitination of the protein to be degraded, followed by movement of that protein to the proteasome (see Section 20.3). To determine whether the higher molecular weight forms of p53 are ubiquitinated, a p53 was pulled out of a reaction mixture containing HPV-18 E6 and ATPγS using a p53-specific antibody, run on SDS-PAGE, and then examined in a Western blot (see Toolbox 8.1) using antibodies specific to ubiquitin **(Figure B22.6)**. The results clearly show that the higher molecular weight form of p53 is ubiquitinated, and ubiquitin therefore very likely is involved in the E6-dependent degradation of p53.

Impact

The story of p53 is an interesting one. It was initially thought to be an oncogene itself, since early studies were done with a mutant form that was involved in cell transformation. Subsequent work with the wild-type p53 gene has revealed that it is a major tumor suppressor protein, and its mutation often is associated with different types of cancers. The finding put forward by this study, that E6 protein from cancer-causing HPV (called "high risk" HPV in the paper) interacts with p53, leading to its degradation, is consistent with the lower levels of p53 in HPV-associated tumor cells. It was the first study clearly demonstrating the E6-associated degradation of p53, suggesting a mechanism for the formation of HPV-associated tumors. While HPV vaccines are now available, there still is a need for the development of therapeutics such as antiviral compounds active against existing HPV-induced cancers.

Subsequent work has shown that E6 does not work on its own to bind p53 and cause its ubiquitination. A cellular protein, the E6-associated protein (E6AP), catalyzes the covalent linkage of ubiquitin to p53, but does so only in the presence of E6. E6 apparently directs E6AP to p53. While E6AP was first discovered in this context, it since has been shown to be involved in ubiquitination of a number of different proteins, targeting them for proteolytic degradation. The absence of E6AP activity due to mutation now has been linked to another human disease, Angelman syndrome, a severe neurological genetic disorder. So, the study of virus-induced cancer indirectly has led to some understanding of the basis of another human disease.

● Questions for Discussion · · · · · · · · · · · · · · ·

1. The experiments described in this paper used proteins that were produced in a rabbit reticulocyte system. Another strategy would have been to use proteins produced by overexpression in *Escherichia coli*, and then purified using affinity tags. What differences would you expect to see in experiments using such purified proteins?

2. It has been suggested that a potential target for development of therapeutics against cervical cancer might be E6AP. What would be a major concern for using such an approach?

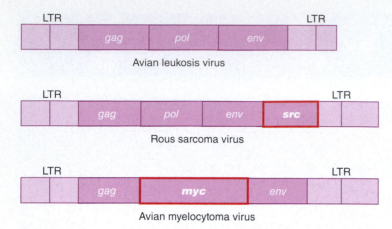

Figure 22.25. **Acquisition of cellular proto-oncogenes by transducing retroviruses** Avian leukosis virus is a normal (non-transducing) retrovirus, containing the long terminal repeat (LTR) transcriptional regulatory regions, and the *gag*, *pol*, and *env* genes. The transducing retrovirus Rous sarcoma virus contains the src proto-oncogene, added to an otherwise normal retrovirus genome. The transducing retrovirus avian myelocytoma virus contains the proto-oncogene *myc*, but is missing the entire *pol* and parts of the *gag* and *env* genes.

In other cases, the cellular sequences replace parts of the viral genome **(Figure 22.25)**. When these altered retroviruses then infect another cell, the newly acquired cellular sequences will be transduced into the infected cell.

While this brief description demonstrates how a retrovirus can acquire a cellular gene, another question remains. How can this process lead to the conversion of a previously normal proto-oncogene into an oncogene? We can imagine two answers to this question. First, the transduced gene simply may be overexpressed. Strong viral promoters may lead to increased transcription of the gene and this increased expression may lead to transformation of the cell. Second, the cellular gene may become altered in some way. Often, the transduced gene is truncated. In other words, it is missing some 5′ or 3′ sequences. These alterations result in the production of altered proteins that may function aberrantly. The *erbB* gene provides us with an excellent example. This proto-oncogene produces epidermal growth factor receptor (EGFR), a transmembrane protein that binds to epidermal growth factor (EGF). In the presence of this ligand, EGFR undergoes a conformational change and autophosphorylation results. This phosphorylation of EFGR, in turn, activates several cell signaling pathways that ultimately lead to cell proliferation. A mutated form of *erbB* carried by avian erythroblastosis virus-H, however, produces an altered form of EGFR that lacks part of the extracellular ligand-binding domain. As a result of this deletion, the protein continuously undergoes autophosphorylation and continuously signals the cell to divide **(Figure 22.26)**.

22.3 Fact Check

1. How do human papillomaviruses cause cancer?
2. How can retroviruses cause cancer?
3. Compare and contrast proto-oncogenes and tumor suppressor genes.

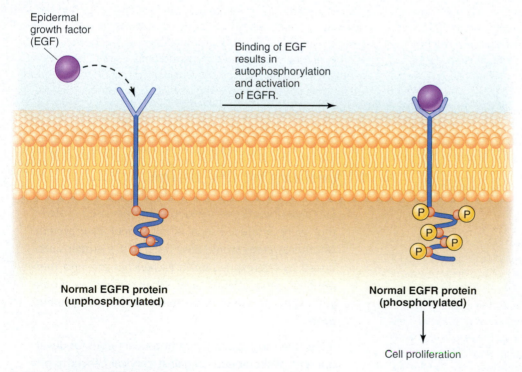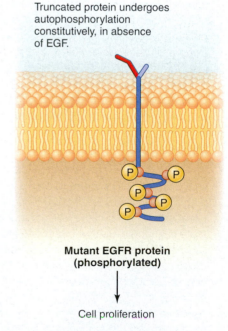

A. Activity of normal EGFR protein

B. Activity of mutant EGFR protein

Figure 22.26. **Effects of mutating a proto-oncogene** Mutations in the cellular *erbB* (*c-erbB*) gene, a proto-oncogene, alter the functioning of the gene's product: epidermal growth factor receptor (EGFR). **A.** When the wild-type protein interacts with its ligand—epidermal growth factor (EGF)—it undergoes autophosphorylation and activates cell-signaling pathways that result in cell proliferation. **B.** The transducing retrovirus avian erythroblastosis virus-H contains a truncated form of this gene that produces an altered form of EGFR. The mutant EGFR lacks most of the ligand-binding domain and, as a result, continuously undergoes autophosphorylation and continuously signals the cell to proliferate.

How do some viruses become highly virulent?

As we noted in the introduction to this chapter, all viruses infect cells and all viruses, to some degree, damage these cells. The degree of damage, though, may differ. In some cases, the amount of damage that a virus causes may be dependent on host factors. The HIV co-receptor CCR5 provides an interesting example. To infect a cell, the gp120 protein on the surface of HIV initially binds to CD4, a signaling molecule found on the surface of certain immune system cells. After this initial binding event, the virus then interacts with a second molecule, a co-receptor. CCR5, another signaling molecule present on the surface of certain immune system cells, serves as the co-receptor for certain strains of HIV. By studying individuals who were exposed to HIV repeatedly, but did not become infected, researchers made an interesting observation. Some of these exposed but not infected individuals were homozygous for a 32-base-pair deletion in the CCR5 gene. In other words, both copies of their CCR5 gene were mutated. The mutant allele, referred to as CCR5-Δ32, results in the production of a mutant protein that is not expressed on the surface of the cells. The virus, as a result, cannot infect the host cells. People possessing this mutation, then, exhibit a high degree of resistance to HIV. Perhaps surprisingly, people with this deletion exhibit no discernible defects in their immune systems.

Population studies have revealed that this mutation is more prevalent in some ethnic groups. Among Caucasians of European ancestry, roughly 1 percent of the population is homozygous for this CCR5 deletion and 10 percent of the population is heterozygous. Conversely, the mutation is very rare among people of African and Asian descent. We also should note that being homozygous for this mutation does not provide a person with absolute protection from HIV. Different strains of HIV utilize different, CCR5-related, co-receptors. Clearly, though, the genetic makeup of a person may affect his or her susceptibility to HIV.

CONNECTIONS As we have seen in **Section 8.1**, researchers have used genetic studies of viral and host cell proteins to identify specific regions of these molecules that are necessary for an infection to occur.

Before leaving our discussion of CCR5-Δ32, let's address another interesting question: Why does this mutation exist? Because the CCR5-Δ32 allele exists in relatively high amounts in certain populations, researchers have speculated that this allele provided protection to some horrific previous infectious disease. If a few people possessing the CCR5-Δ32 allele survived a highly lethal plague, then the allele would have been passed on to their descendants, leading to its high frequency today. Initially, microbiologists and historians speculated that CCR5-Δ32 may have offered some protection against the bubonic plague, which swept through northern Europe during the fourteenth century. Currently, researchers hypothesize that individuals possessing CCR5-Δ32 alleles may have exhibited some protection from smallpox infections.

While host genetics play a part in determining the virulence of viruses, viral genetics, of course, also play a part. As we will see in this section, viral genomes can change via several different mechanisms. To some extent, these mechanisms of change are dependent on the type of genome possessed by a virus and/or the replication strategy utilized by a virus. In the remainder of this section we will examine three different ways in which viral genomes can change: mutations, recombination, and reassortment. We will conclude this section by asking another question: What factors affect the evolution of viruses?

Mutations

As we noted in Chapter 6, mutations—changes in the genome of an organism—are absolutely necessary for evolution. Minor changes in the genome of an individual may alter the phenotype, or characteristics, of that individual. These phenotypic changes, in turn, may alter the individual's fitness. Finally, an individual with greater fitness may be more likely to reproduce or replicate. These basic processes occur in eukarya, bacteria, archaeons, *and viruses*.

Among eukarya, bacteria, and archaeons, numerous mechanisms exist to minimize the mutation rate. While mutations are necessary in the long term, a stable genome certainly is desired in the short term. Among many viruses, especially RNA viruses, these same mechanisms to reduce the mutation rate do not exist. As a result, the mutation rate in many RNA viruses is extraordinarily high.

To understand this high mutation rate, let's begin by examining the replication enzymes used by RNA viruses. As you may remember from our discussion of viral replication in Chapter 8, viruses with an RNA genome cannot rely on the host cell DNA polymerase to replicate their genomes. Rather, they rely on virus-specific RNA-dependent RNA polymerases or, in the case of retroviruses, reverse transcriptases. These enzymes, unlike the DNA polymerases used by eukarya, bacteria, and archaeons, lack a proofreading function. In other words, viral RNA-dependent RNA polymerases and reverse transcriptases cannot remove bases that were incorrectly added to a nucleic acid strand. What's the significance of this? RNA viruses typically have a mutation rate of 10^{-3} to 10^{-5} per base per genome replication. To put it another way, every time one of these viruses replicates its genome, 1 out of every 1,000 to 100,000 bases is altered. *Escherichia coli*, in contrast, has a mutation rate of approximately 10^{-8} per base per genome replication.

CONNECTIONS As we saw in **Section 7.5**, the DNA polymerases used for DNA replication in bacteria, archaeons, and eukarya possess an effective proofreading activity. Other processes, like the bacterial mismatch repair system, further decrease the numbers of mutations that occur.

What are the implications of this high mutation rate? Most importantly, whenever a virus replicates in a cell, the newly formed viruses will be genetically heterogeneous. When one of these viruses replicates, a viral swarm, or quasispecies, forms. Presumably, some of the mutants will be better able to infect other cells. Other mutants may be more environmentally stable and better able to be transmitted from one individual to another. Still other mutants may be completely non-infectious; the mutation(s) will result in particles unable to enter host cells or replicate within these host cells. This high mutation rate also means that viruses are evolving at a fast rate.

This rapid viral evolution has implications for our ability to prevent viral infections. To understand this point, let's examine influenza virus. Various studies have shown that human influenza has a mutation rate of approximately 10^{-5}. As the virus replicates, then, mutations accumulate throughout the influenza genome, including within the genes encoding hemagglutinin (HA) and neuraminidase (NA). These membrane-bound proteins are the major targets for our immune system. The structural changes within these proteins usually are so significant that we can be repeatedly re-infected with influenza; after a few years of genetic changes within the virus, our immune response no longer is effective. This process of mutational changes within the surface proteins is referred to as **antigenic drift**. This process also necessitates the development of a new influenza virus vaccine every year.

Rapid viral evolution also has implications for the treatment of viral infections, particularly for viruses that cause chronic or persistent infections. To understand this point, let's examine HIV. While a rather homogeneous population of HIV particles may infect a person initially, a viral swarm quickly develops within the infected person. The virus actually evolves within the person. This evolution has several important implications. First, the virus may evolve to infect different cells over time. The virions most commonly found in seminal fluid, and transmitted sexually, typically use the cell surface protein CCR5, a chemokine receptor found on macrophages, as a co-receptor to infect cells and are referred to as monocyte (or macrophage) tropic viruses. Over time, though, viral mutants that utilize a related protein, CXCR4, as a co-receptor often evolve **(Figure 22.27)**. These viruses preferentially infect T cells and are referred to as T-cell tropic viruses. Second, the virus may develop resistance to various antiretroviral therapies. As the virus replicates within a person on antiretroviral therapy, mutants that exhibit some resistance to the drugs will possess a selective advantage. Over time, these resistant viruses will become the predominant strains present within the individual. Over time, then, the original drug therapy will begin to fail. Because of this common problem, clinicians typically monitor the viral load in a person infected with HIV (see Perspective 5.1). When they detect a rise in the viral load, they usually prescribe a new set of drugs.

Recombination

While point mutations provide the basic variation needed for evolution, more dramatic changes in viral genomes can occur. Not surprisingly, these larger changes can affect the virulence of these viruses. Recombination, or the process of combining two different pieces of nucleic acid, may occur more frequently among viruses than once thought. Most of us probably are most familiar with recombination within the context of meiosis. Among diploid organisms, homologous chromosomes recombine during meiosis, a process often referred to as crossing over. Among viruses, recombination can occur when a cell simultaneously becomes infected by two genetically distinct strains of the same virus. During replication of these two different viruses within the same cell, a type of recombination event may occur. Progeny viruses that have recombinant

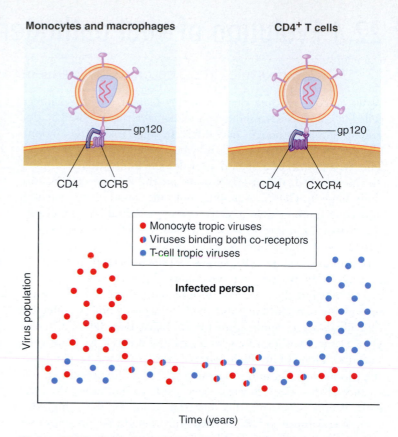

Figure 22.27. Evolution of HIV within an infected person Because of the high mutation rate exhibited by HIV, a person infected with the virus quickly contains a viral swarm, or population of viral mutants. During the course of an infection, the predominant variant of the virus may change from one that preferentially infects monocytes or macrophages (monocyte or macrophage tropic) to one that preferentially infects T cells (T-cell tropic). This switch occurs because of mutations in the viral gp120 protein that determine if the virus can utilize the CCR5 or CXCR4 co-receptor. This switch also correlates with the depletion of CD4$^+$ cells and the onset of AIDS.

genomes then form—part of the genome arises from the genome of one input virus and part of the genome arises from the genome of the other input virus.

Coronaviruses, enveloped viruses with a single-stranded positive sense RNA genome, provide an interesting case study. Studies with mouse hepatitis virus (MHV), a well-studied coronavirus, have shown that recombination occurs at a high rate in cell culture and in animals. If cells or mice are experimentally coinfected with two strains of MHV, as much as 25 percent of the progeny viruses may contain recombinant genomes. This high rate of recombination probably results from the method of genome replication and mRNA production utilized by these viruses. RNA synthesis in coronaviruses appears to be discontinuous; the RNA polymerase begins RNA synthesis from an RNA template and then dissociates from that template RNA strand. The polymerase and nascent RNA molecule then attach to a new location on the same, or another, template strand and RNA synthesis resumes, a process known as template switching **(Figure 22.28)**. Let's examine the implications of this replication mechanism. If the polymerase begins using one template strand and switches to a genetically different template strand, then part of the resulting molecule will be complementary to the first template and part of the molecule will be complementary to the second template. In other words, the newly formed piece of RNA will be recombinant.

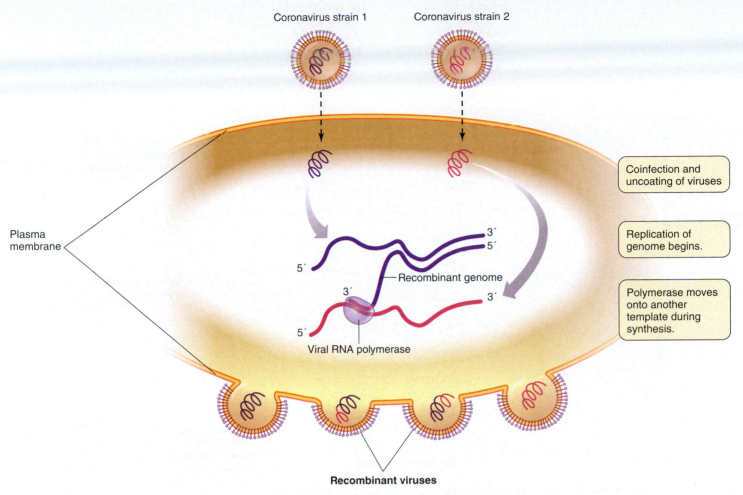

Coronavirus strain 1 Coronavirus strain 2

Coinfection and uncoating of viruses

Replication of genome begins.

Polymerase moves onto another template during synthesis.

Plasma membrane

3′
5′

5′

Recombinant genome

3′

3′

5′

Viral RNA polymerase

Recombinant viruses

Figure 22.28. Recombination in coronaviruses Following coinfection of a cell with two strains of coronavirus, genome replication begins. Occasionally, the viral polymerase, along with its nascent RNA molecule, will dissociate from the template RNA and bind to another template molecule. Genome replication then resumes. The resulting genome is a recombinant, possessing some sequences from one parental genome and some sequences from the other parental genome.

While we now understand how recombination can occur in coronaviruses, let's ask a more fundamental question: Does it ever occur naturally? We noted in the preceding paragraph that recombination occurs when cells or animals are coinfected by two strains of MHV. It may seem unlikely that coinfection would occur in nature. It may seem that MHV recombination is a laboratory curiosity only and not a real-world concern. Recent studies with the coronavirus associated with SARS, however, indicate that recombination among coronaviruses does occur in nature and may affect the evolution and pathogenesis of these viruses.

As we mentioned in Section 5.4, a coronavirus causes severe acute respiratory syndrome (SARS). This virus has been named human SARS-CoV, or Hu-SCoV. Initial studies isolated a genetically identical virus from civet cats, leading to the hypothesis that the virus was transmitted from these cats to humans. Subsequent studies of civet cat populations, though, identified SCoV in very few individual cats, leading to the conclusion that civet cats are not the natural host of this virus. Subsequent studies isolated several SARS-like coronaviruses (SLCoV) from different bat species, indicating that bats may be the natural reservoir for these viruses. While several viruses isolated from horseshoe bats (bat SARS-like coronaviruses, or Bt-SLCoV) closely resemble Hu-SCoV, none appeared to be the direct ancestor of SCoV. Recently, though, researchers have obtained evidence demonstrating that a Bt-SLCoV genome arose via recombination. Part of the genome very closely resembles the genome of Hu-SCoV. These researchers have postulated that the virus they discovered arose via recombination between two coronaviruses that infect bats, one of which may be the direct ancestor of Hu-SCoV. This analysis of viral genome recombination, then, has provided important insight into the natural reservoir of this potentially dangerous pathogen.

Reassortment

Some viruses, as we noted previously, have segmented genomes. Rather than possessing a single genomic molecule, the genomes of these viruses consist of several discrete nucleic acid pieces, or segments. For most viruses with segmented genomes, each genome segment encodes for one or two polypeptides. A different means of generating genetic variation exists for these viruses—**reassortment**, or the combining of gene segments from two or more parental viruses. Like recombination, reassortment only occurs if a cell becomes coinfected with two or more genetically distinct strains of a virus. Unlike recombination, though, individual genome molecules do not undergo recombination, forming a hybrid molecule. Rather, progeny virus particles form that contain some genome segments from one parental virus and some genome segments from another parental

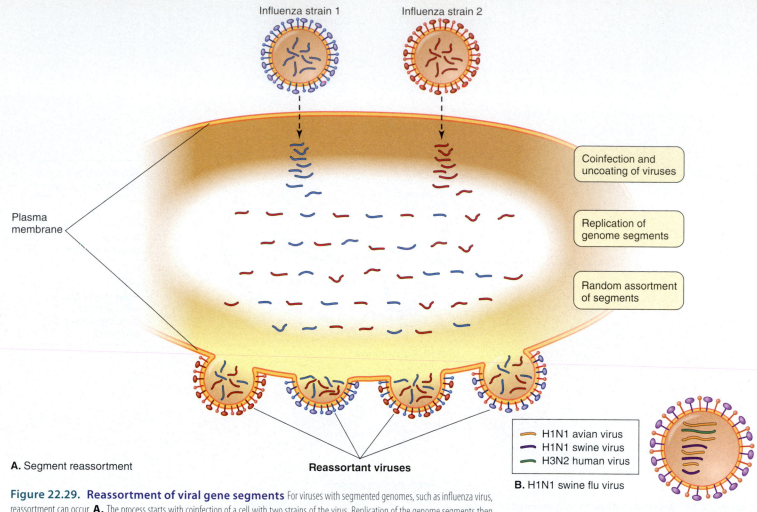

A. Segment reassortment   **Reassortant viruses**

- ⌒ H1N1 avian virus
- ⌒ H1N1 swine virus
- ⌒ H3N2 human virus

B. H1N1 swine flu virus

Figure 22.29. Reassortment of viral gene segments For viruses with segmented genomes, such as influenza virus, reassortment can occur. **A.** The process starts with coinfection of a cell with two strains of the virus. Replication of the genome segments then occurs. When new viral particles assemble, progeny viruses may form that contain some genome segments from one parental virus (indicated in red) and some genome segments from the other parental virus (indicated in blue). **B.** The 2009 H1N1 "swine" flu virus was a triple reassortant virus containing genome segments from avian (orange), pig (purple), and human (green) strains of influenza A virus.

virus **(Figure 22.29)**. If we think of recombination as analogous to crossing over in meiosis, then we can think of reassortment as analogous to independent assortment in meiosis. Just like every gamete receives one copy of each chromosome during meiosis, each infectious progeny viral particle receives one copy of each gene segment. They may contain all the gene segments from one parental virus, all the gene segments from the other parental virus, or some segments from one parental virus and some segments from the other parental segment.

Reassortment can result in viruses that differ significantly from either of the parental strains. As we noted earlier, influenza viruses have a high mutation rate. The accumulation of point mutations can cause significant changes in the surface glycoproteins, resulting in antigenic drift. Influenza viruses also can undergo reassortment. Influenza viruses contain eight genome segments. The HA and NA gene segments encode the hemagglutinin (HA) and neuraminidase (NA) proteins, respectively, the two major antigenic determinants on the surface of influenza particles. Within influenza virus type A, the type of influenza responsible for most human influenza virus infections, sixteen genetically distinct subtypes of the HA gene segment (H1–16) and nine subtypes of the NA gene segment (N1–9) have been identified. If an individual becomes coinfected with viruses of two different subtypes (H1N1

and H2N2, for instance), then reassortment could occur, leading to the formation of a new subtype (H1N2 or H2N1, for instance). Theoretically, 144 different combinations of these two gene segments could exist. The dramatic change in viral antigens caused by reassortment is referred to as **antigenic shift**.

Although 144 influenza A virus subtypes theoretically can exist, only a single subtype of influenza A virus typically causes most of the infections throughout the world during any given flu season. From 1968 until 2009, the H3N2 subtype had been the predominant strain. In March and April of 2009, that situation changed dramatically. Officials in Mexico and the United States reported people infected with a different influenza A subtype: H1N1. Because the antigens of this H1N1 virus differed significantly from the antigens associated with the H3N2 virus, very few people had immunity to the new subtype. Consequently, the virus spread with an amazing rapidity. By June 11, 2009, over 30,000 cases throughout the world had been reported. The World Health Organization declared an influenza pandemic, the first since the H3N2 strain appeared in 1968.

How do these antigenic shifts occur? It's important to realize that influenza A viruses do not only infect humans. These viruses also infect a wide range of animals, including wild birds, domestic poultry, pigs, and horses. Many different influenza A virus

While much of the popular press accounts of influenza virus since 2009 have focused on the recently emerged H1N1 strain, researchers remain concerned about the H5N1 subtype of avian influenza that appears to be widespread in much of Asia. Officials remain concerned about this virus for two main reasons. First, direct bird-to-human transmission has occurred. Second, the virus has been especially virulent in humans. To date, though, human-to-human transmission of this virus occurs very inefficiently. Spread within human populations, therefore, has been limited. If the virus became better suited for human-to-human transmission, though, then a major human pandemic could occur.

Could the H5N1 subtype of avian influenza become better suited to human-to-human transmission? During the latter half of 2011, researchers from two different institutions reported that the answer to this question is "yes." Yoshihiro Kawaoka and colleagues at the University of Wisconsin, Madison, and the University of Tokyo mutated the H5N1 HA gene and then generated a reassortant virus that contained this gene segment and gene segments from the 2009 H1N1 pandemic virus. This reassortant virus could be transmitted easily between ferrets, which are assumed to be a good model for human transmission of influenza virus. They submitted a manuscript detailing their experiments to the journal *Nature*. Ron Fouchier and colleagues at Erasmus Medical Center in the Netherlands also generated a H5N1 variant that

could be transmitted easily between ferrets. They achieved this result by serially passing the virus from ferret to ferret. In the end, they reported, five mutations in the H5N1 virus made it transmissible between ferrets. They submitted a manuscript detailing their experiments to the journal *Science*.

While these results are fascinating, they also are somewhat unsettling. Have these researchers created a super virus that could kill millions of people? Should this research have been conducted? Should the results be made public? Can we prevent a bioterrorist from replicating their experiments? The advancement of science depends on the open sharing of ideas. These experiments, though, have started an interesting discussion about the limits of this openness. Some research may be too dangerous to conduct or report.

In December of 2011, the National Science Advisory Board for Biosecurity (NSABB) in the United States recommended that the papers be published only if details about the methods and results were withheld. Complete and open sharing of this research should not occur. In March of 2012, however, the Board reversed this decision. Noting that the information in the articles might help public health officials track the natural evolution of more dangerous influenza strains, the NSABB recommended that the research should be published. The work by Dr. Kawaoka and colleagues was published in *Nature* in May of 2012 and the work by Dr. Fouchier and colleagues was published in *Science* in June of 2012.

subtypes certainly exist in these various species. Reassortment could occur in any of these species, leading to the formation of additional subtypes. One of these viruses could be transmitted to humans, leading to the introduction of a new subtype into the human population—an antigenic shift. This possibility has fueled the current concerns about avian flu.

Researchers also have exploited the process of reassortment to investigate aspects of viral pathogenesis. Mammalian reoviruses have been used extensively in this capacity. Because reoviruses have segmented genomes, it is possible to coinfect cells with two different strains of reovirus and isolate reassortant progeny viruses—viruses that contain some genome segments from each of the parental viruses. If the two parental viruses exhibit a phenotypic difference, then it would be possible to screen a panel of reassortants and classify them as being like one or the other of the parental viruses. By analyzing the phenotypes of the reassortants and the origins of their genome segments, it may be possible to associate a phenotype with a specific genome segment. Through the use of this approach, researchers have demonstrated that specific reovirus genome segments, and, by extension, specific reovirus proteins affect various aspects of reovirus replication. These basic studies have contributed significantly to our understanding of viral pathogenesis.

Evolution

In this section, we have investigated methods by which genetic variation can occur in viruses. As with all organisms, genetic diversity in viruses originates from point mutations. Additionally, certain families of viruses exhibit recombination or reassortment. Both

of these processes increase the degree of genetic diversity present in viruses. As we conclude this section, let's think about a more esoteric question: What selective forces drive evolution in viruses?

Within an infected organism, the immune response almost undoubtedly represents the major selective pressure. Viral mutants that have altered antigenic determinants will be recognized less well by antibodies directed against the original infecting virus. These mutants, then, may be better able to evade the immune response and infect another cell. If an individual is taking drugs to combat a viral infection, the antiviral therapy also represents a major selective pressure. Viral mutants that exhibit increased resistance to a particular drug will be better able to replicate. This type of pressure is a major problem for viruses that cause chronic or persistent infections, where drug use may be long-term.

We could hypothesize that mutations that increase the rate of production of new viral particles would be beneficial. If a viral mutant arises that can replicate faster than the other viruses, then it should become the predominant virus within an infected individual. Alternatively, we could hypothesize that mutations that limit replication in a host and, as a result, limit the severity of disease, would be beneficial. After all, it's in the virus's best interest for the host to stay alive so that the virus can continue to replicate. Of course, viruses also must be transmitted from one host to another. We could hypothesize that mutations affecting the transmission of a virus between hosts would benefit the virus. It should be clear, then, that many different factors affect the evolution of viruses. Likewise, various mutations could affect the virulence of a virus. Recently, the ethical implications of researching these mutations has been a hot topic (**Perspective 22.4**).

Recent studies with poliovirus have shed some light on this complex issue of virus evolution. Raul Andino, a virologist at the University of California, San Francisco, hypothesized that the high mutation rate exhibited by many RNA viruses may, in fact, be necessary for the overall success of a virus. To test this hypothesis, they analyzed a mutant poliovirus strain possessing a less error-prone polymerase. This poliovirus strain exhibited decreased neurotropism and decreased virulence. The high mutation rate of the wild-type virus, in other words, seems to be necessary for the pathogenicity of the virus. Error-prone polymerases may provide viruses with the genetic plasticity needed to adapt to ever-changing environments.

Some viruses, indicated by yellow surface proteins and green lateral bodies, interact with the cell on the bottom but cannot enter it. Other viruses, indicated by pink surface proteins and aqua lateral bodies, enter the cell on the right but do not replicate. Still other viral mutants could form that infect host cells and replicate very well.

1. Explain how the viral diversity shown in this image could be used to explain why the biological control of the invasive rabbit species ultimately failed.

2. The myxoma virus shown here causes myxomatosis in rabbits. Imagine that you learn it is spread via mechanical transmission. Explain what this means.

22.4 Fact Check

1. What is antigenic drift?
2. How does recombination in coronaviruses occur?
3. What is viral reassortment?
4. Why might a high mutation rate be advantageous to viruses?

Image in Action

WileyPLUS As viruses replicate within infected cells, mutant viral particles form that may exhibit very different properties. The cell in the upper left is infected with myxoma virus. From the viral factory (yellow, cloud-like haze in upper-left corner), intracellular 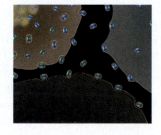 mature virions form, with some wrapped in a double membrane envelope. Among these newly formed virions, genetic differences are indicated by different surface protein and lateral body colors.

The Rest of the Story

Initially, the use of myxoma virus to combat the rabbits in Australia seemed to work. As we noted in the chapter's opening story, the virus was highly virulent in these rabbits, killing a stunning 99 percent of the infected animals. The situation soon changed, however. A year after the myxoma virus was released, its virulence decreased considerably. Studies over the next 30 years showed that fewer and fewer isolates of the virus were virulent in laboratory strains of European rabbits. In 1950, 100 percent of myxoma virus isolated killed 99 percent of infected rabbits. By 1959, less than 2 percent of myxoma virus isolates exhibited this degree of virulence. A less pathogenic form of the virus evolved.

The rabbit population also evolved. In 1950, nearly all wild European rabbits exposed to a virulent strain of myxoma virus died. Today, less than 30 percent of wild rabbits in Australia die when exposed to the same strain of the virus. What do the results of these studies tell us? Most notably, viral pathogenesis is a complex topic, dependent on both the virus and the host.

Summary

Section 22.1: How do viruses cause disease?

The type of infection that develops when a viral genome enters a host cell depends upon two interrelated activities: the replication of the virus and the immune response of the host.

- An **infection** occurs when a viral genome enters a host cell. **Productive infections** produce new infectious viral particles; **abortive infections** do not.

- **Acute infections**, such as a typical rhinovirus infection, have a short duration.

- In **latent infections**, a period of acute infection is followed by a period of latency. **Reactivation** occurs in response to some environmental or biochemical trigger. During latency, the herpesvirus genome exists within the host cell as an **episome**. **Latency-associated transcripts (LATs)** may help maintain this latent state.

- In **persistent infections**, the virus is not eliminated within a reasonable amount of time. Rather, the continuous production of new viral particles occurs.

Viruses are obligate intracellular pathogens; to transmit disease they must be able to effectively leave an infected host and enter another host.

- **Horizontal transmission** refers to the transmission of a virus between individuals; **vertical transmission** refers to the transmission of a virus from a mother to infant and includes the transmission of **endogenous retroviruses**.

- **Zoonosis** refers to a disease that can be transmitted from a non-human host to a human.

- **Mechanical transmission** refers to the facilitated transfer of a virus from one host to another by some vector.

Section 22.2: How do viruses interact with host cells?

Viruses use host cell machinery to replicate and, in the process, damage host cells.

- Viral pathogenesis results from viral-induced cellular destruction. This cellular destruction may occur via **necrosis** or **apoptosis**.

- Cellular damage may occur via viral-induced syncytia formation producing a multinucleated giant cell that eventually dies, or the formation of **inclusion bodies** that impair cell function.

- Viral pathogenesis also may result from the host immune response. **Autoimmunity** in the host may be triggered by **molecular mimicry**. Viral antigens that resemble host antigens activate immune cells that then cross-react with the similar host antigens.

Section 22.3: How do some viruses cause cancer?

Virus replication may lead to aberrant cellular replication, resulting in cancer.

- Many DNA viruses can cause **transformation** of cells in culture, producing changes within a cell that cause it to become cancer-like. Some viruses cause **tumorigenesis** or the formation of a tumor.

- The genomes of transforming viruses typically contain **oncogenes**, or genes that are capable of transforming cells.

- Tumorigenesis may result from an essential viral protein that causes a **quiescent cell** to enter the cell cycle.

- Viral proteins may inhibit p53 and/or Rb, two **tumor suppressors** that normally inhibit cellular replication in the host cells. Certain adenovirus strains inhibit these proteins, causing **cervical carcinomas** and **papillomas**. **Immunoprecipitation (IP)** assays can be used to identify these types of protein interactions.

- Retroviruses also can cause tumorigenesis. *Cis*-acting retroviruses integrate near a cellular **proto-oncogene**, altering that gene's expression. **Transducing retroviruses** acquire a cellular proto-oncogene and then introduce that gene into a new cell.

Section 22.4: How do some viruses become highly virulent?

Mutations, recombination, and reassortment can alter a virus's virulence.

- Random mutations provide the basis for the evolution of all viruses. RNA viruses have an especially high mutation rate, leading to **antigenic drift**.

- Many viruses also undergo recombination. This process may result in progeny viruses containing genomes derived from each parental virus.

- For viruses with segmented genomes, **reassortment** can occur, leading to **antigenic shift** in viruses like influenza virus.

● Application Questions

1. Explain, from a virus's perspective, the potential advantages of acute and persistent infections.
2. Why might stress cause the reactivation of a latent human virus?
3. Explain why apoptosis of infected cells may be beneficial to an organism.
4. We mentioned that cancer may be an unintended outcome of viruses attempting to obtain resources. Explain.
5. Define antigenic drift and antigenic shift. How are these two events similar? How are they different?
6. Explain how bacteriophage lysogeny occurs.
7. What is an endogenous retrovirus?
8. An HIV-infected cell can fuse with adjoining, non-infected cells, forming a syncytium. Explain how this process might occur.
9. Some researchers are concerned that the current H1N1 human influenza virus and the H5N1 avian influenza virus may give rise to a more deadly human virus. How could this occur?
10. For a virus, which is more advantageous: evolving to become highly virulent or evolving to become less virulent? Defend your answer. Provide examples.
11. For a respiratory virus like rhinovirus, transmission often occurs through the air—viral particles exit an infected individual when the host coughs or sneezes and then are inhaled by another susceptible host. The viral particles, then, must remain infectious during this exposure to the outside environment. Devise an experiment that would allow you to measure infectivity of a virus after exposure to various environmental insults.

Suggested Reading

Ebert, D. H., S. A. Kopecky-Bromberg, and T. S. Dermody. 2004. Cathepsin B is inhibited in mutant cells selected during persistent reovirus infection. J Biol Chem 279:3837–3851.

Lau, S. K. P., P. C. Y. Woo, K. S. M. Li, Y. Huang, H.-W. Tsoi, B. H. L. Wong, S. S. Y. Wong, S.-Y. Leung, K.-H. Chan, and K.-Y. Yuen. 2005. Severe acute respiratory syndrome coronavirus-like virus in Chinese horseshoe bats. Proc Natl Acad Sci USA 102:14040–14045.

Li, W., Z. Shi, M. Yu, W. Ren, C. Smith, J. H. Epstein, H. Wang, G. Crameri, Z. Hu, H. Zhang, J. Zhang, J. McEachern, H. Field, P. Daszak, B. T. Eaton, S. Zhang, and L.-F. Wang. 2005. Bats are natural reservoirs of SARS-like coronaviruses. Science 310:676–679.

Liu, R., W. A. Paxton, S. Choe, D. Ceradini, S. R. Martin, R. Horuk, M. E. MacDonald, H. Stuhlmann, R. A. Koup, and N. R. Landau. 1996. Homozygous defect in HIV-1 coreceptor accounts for resistance of some multiply exposed individuals to HIV-1 infection. Cell 86:367–377.

Rous, P. 1910. A transmissible avian neoplasm (sarcoma of the common fowl). J Exp Med 12:696–705.

23

Eukaryal Microbe Pathogenesis

In North America, majestic elm trees are—or used to be—one of the most important urban landscape shade trees. They are perfectly suited to this role. Elms display a large and thick canopy, attain heights of almost 100 feet (30 m), grow in a variety of climates and soil conditions, and are relatively fast growing with lifespans of some 300 years. Today, though, elm trees are uncommon in North America. A fungus has killed most of the trees.

The causative agent of Dutch elm disease was first identified in 1921 in the Netherlands, hence the name of the disease. The fungus originally was designated *Ophiostoma ulmi*. It is proposed to be native to Asia because Asian species of elm are somewhat resistant to the fungus, suggesting that they have co-existed with it for a long period of time. European and American elm species, conversely, are highly susceptible to the disease; a mature elm tree can die in as little as two weeks after infection. For this reason, scientists believe the fungus is a recent introduction to these continents. Genetic and physiological analyses of the fungus involved in the most recent European outbreak (mid-1960s) have shown it to be very different from previous strains, and so it is described as a new species: *Ophiostoma novo-ulmi*. In European and American elm trees, this new species is much more virulent than its predecessors.

In North America, Dutch elm disease was first reported in Ohio in 1930. Its origins have been traced to a shipment of infected logs from Europe. Within two years, the disease had reached the Atlantic seaboard. Spread from tree to tree by the elm bark beetle, the fungus reached the Pacific Coast by 1970. Today, it is estimated that over 70 percent of American elm trees have died of the disease. In many areas, once plentiful elms no longer can be found.

Interestingly, this ongoing outbreak of Dutch elm disease may not be the elm's first encounter with this disease. By analyzing levels of fossilized pollen found in peat deposits, researchers have concluded that elm populations in Europe decreased dramatically about 6,000 years ago. Possibly, this decline was caused by climate change or the activities of humans. Alternatively, it may have been caused by a pathogen. Over the past few decades, at least two pieces of evidence have emerged that support this last hypothesis. First, the existence of elm bark beetles from this period has been documented. Second, researchers have identified pieces of elm wood from this period that exhibit elm bark beetle markings. Perhaps these beetles transferred from tree to tree a fungus similar to *O. novo-ulmi*, just as they do today. Perhaps a historical analysis of elm trees and elm bark beetles can shed light on our current elm tree decline.

CHAPTER NAVIGATOR

As you study the key topics, make sure you review the following elements:

Like viral and bacterial pathogens, pathogenic eukaryal microbes must gain access to a host, evade the host defenses, and obtain nutrients.

- Table 23.1: General patterns exhibited by pathogenic viruses, bacteria, and eukaryal microbes
- Mini-Paper: An experimental system for the genomic study of Dutch elm disease
- Microbes in Focus 23.1: An AIDS-defining opportunistic infection: *Toxoplasma gondii*
- Figure 23.9: Fungal invasion of plants
- Figure 23.14: Role of toxin production by dinoflagellates

The eukaryal microbe *Plasmodium falciparum* causes malaria, one of the most devastating infectious diseases of humans.

- Figure 23.15: Global presence of malaria
- Figure 23.16: Process Diagram: *Plasmodium falciparum* life cycle
- Figure 23.17: Merozoite surface proteins (MSPs) and *Plasmodium falciparum* invasion of erythrocytes

Macroscopic eukarya, especially helminths, cause significant human disease.

- Table 23.5: Representative parasitic worms of humans
- Microbes in Focus 23.5: *Ascaris lumbricoides*: A roundworm parasite of humans

Pathogenic eukaryal microbes, like their bacterial and viral counterparts, exhibit specific properties associated with virulence.

- Perspective 23.3: Chytrid fungus: An emerging fungal pathogen
- Figure 23.22: Sickle-cell disease and misshapen red blood cells
- Figure 23.23: Geographic distribution of malaria and sickle-cell disease

CONNECTIONS for this chapter:

Endospore formation in bacteria (Section 2.4)

Fermentation and lactic acid production (Section 13.3)

Mechanism of action of botulinum toxin (Section 21.1)

Viral reassortment and changes in virulence (Section 22.4)

Introduction

As we noted in Chapter 3, eukaryal microbes share a number of fundamental properties. They all possess a membrane-bound nucleus and other membrane-bound organelles. They all exhibit a complex replication strategy, alternating between diploid and haploid genetic stages. They also undergo mitosis and meiosis. Within the eukaryal microbes, though, a great deal of diversity also exists. Fungi, algae, and slime molds, to name just three categories of eukaryal microbes, all differ from each other dramatically. So, what can we predict about eukaryal pathogens? Do eukaryal pathogens exhibit a great deal of diversity? Do all eukaryal pathogens share any attributes? Do eukaryal pathogens share any attributes with bacterial or viral pathogens?

We should begin our investigation of these microorganisms by reminding ourselves that disease really can be thought of as an unintended outcome of replication. Eukaryal microbes, as we discussed for bacteria and viruses, do not want to cause disease (of course, microbes actually do not *want* to do anything!). Rather, microbes need to obtain nutrients in order to replicate. Many eukaryal microbes obtain nutrients without interacting with a host. Algae, for instance, obtain most of their required nutrients via photosynthesis. A host organism is not required and, as a result, these algae are not pathogens. Many fungi are saprophytes, obtaining nutrients from dead and decaying organic matter. Again, the fungi do not interact with a living host and, consequently, are not pathogens. Other eukaryal microbes may interact with a host in a symbiotic, but not parasitic, fashion. As we discussed in Section 17.3, for instance, lichens represent a symbiotic relationship between fungi and certain algal or cyanobacterial species. Other eukaryal microbes, most notably various protozoa and yeast, reside within the digestive tract of cows and other ruminants, aiding in their digestive processes. In these cases, again, the eukaryal organisms are not pathogens.

As we saw with bacteria and viruses, though, some eukaryal microbes do cause harm to a host organism and, as a result, are classified as pathogens. In fact, a single eukaryal microbe, *Plasmodium falciparum*, is one of the deadliest of all human pathogens. It causes 300 to 500 million cases of malaria, resulting in 1 to 2 million deaths, annually. In this chapter, we will investigate *P. falciparum* and other eukaryal pathogens. In the process, we will note the basic similarities between these pathogenic organisms and the pathogenic bacteria and viruses that we examined in the two preceding chapters. While we will not provide a comprehensive listing of eukaryal pathogens, we will provide information about a few key organisms that illustrate our main points. To organize our discussion, we will focus on these fundamental questions:

How do eukaryal microbes cause disease? (23.1)
How does *Plasmodium falciparum* cause malaria? (23.2)
What do we know about macroscopic eukaryal pathogens? (23.3)
How do some eukaryal microbes become pathogens? (23.4)

23.1 Mechanisms of eukaryal microbe pathogenesis

How do eukaryal microbes cause disease?

Perhaps it should not be surprising that we see general patterns in how bacteria, viruses, and eukaryal microbes cause disease. As we noted in Chapter 21, pathogens must (1) gain access to tissues within a susceptible host, (2) evade or overcome host defenses, and (3) get nutrients, usually by damaging cells and/or stealing them away from the host (Table 23.1). Like pathogenic bacteria and viruses, pathogenic eukaryal microbes must be transmitted from host to host. Many of these pathogens possess specific mechanisms for getting into a host, and certain eukaryal pathogens also possess specific adherence factors that allow them to attach to or enter specific cell types. Once eukaryal pathogens are attached to or are within a host cell, they need to evade host defense mechanisms. We can classify some eukaryal microbes, for instance, as opportunistic pathogens. These microorganisms only cause disease when the immune system of the host becomes compromised. Other eukaryal microbes more actively evade host defense mechanisms by continually altering their antigens, molecules on their cell surface that elicit a host immune response. Like bacteria and viruses, eukaryal pathogens also obtain nutrients from their hosts in very specific ways. Some, as we will see in this chapter, produce specific enzymes that digest structural components of the host cell. Others simply penetrate the host cell through brute force. Finally, eukaryal pathogens often produce toxins, factors that damage the host. The role of toxins produced by eukaryal pathogens, though, is less well understood than the role of their bacterial counterparts.

Transmission, entry, and adhesion

To initiate an infection, a pathogen must be transmitted between appropriate hosts, gain entry into these hosts, and then, in many cases, attach to specific cells within that host. As we saw in Chapters 21 and 22, bacteria and viruses enter hosts through a variety of mechanisms. Specific protein:protein interactions between an attachment protein on the pathogen and a cellular receptor generally facilitate attachment to specific cells within the host. Pathogenic eukaryal microbes utilize these same basic methods.

Ophiostoma novo-ulmi, the causative agent of Dutch elm disease that we introduced in the opening story of this chapter,

TABLE 23.1   General patterns exhibited by pathogenic viruses, bacteria, and eukaryal microbes

Virulence action	Viruses	Bacteria	Eukaryal microbes
Transmission, entry, adhesion	HIV: Viral gp120 protein attaches to CD4 molecule found on certain immune system cells.	*Escherichia coli*: Enteropathogenic and enterohemorrhagic strains produce intimin, an adhesion molecule.	*Giardia lamblia*: Ventral disk allows adherence to epithelial cells of small intestine.
Evade host defenses	Herpes simplex virus: Following acute replication, the viral genome exists as an episome in the host cell nucleus, undergoing limited transcription and translation.	*Neisseria gonorrhoeae*: This bacterium produces IgA protease, an enzyme that destroys the host IgA antibodies.	*Trypanosoma brucei*: This microbe undergoes antigenic variation to periodically change its surface proteins.
Obtain nutrients from the host	Human papillomavirus: Viral E6 and E7 proteins inhibit cellular p53 and Rb tumor suppressor proteins, thereby driving the cell to replicate, and produce molecular building blocks needed by the virus.	*Mycobacterium tuberculosis*: Mycobactins, iron-binding proteins, allow the pathogen to acquire iron from the host.	*Plasmodium falciparum*: Nucleoside transporter proteins allow the pathogen to obtain purines from the host.
Produce toxins		*Clostridium botulinum*: This bacterium produces botulinum toxin, a powerful neurotoxin that results in flaccid paralysis.	*Cochliobolus carbonum*: This microbe produces HC-toxin that inhibits host cell histone deacetylases.

relies on the help of an insect to be transmitted from tree to tree. The fungus has an interesting relationship with certain bark-burrowing beetle species of which the European elm bark beetle is the most common. Typically, the fungus spreads throughout the tree in xylem tissue, clogging the water-conducting vessels as it grows, eventually killing the tree. The beetles burrow into the bark of the dead branches to lay their eggs. The egg-filled brood chambers then provide open spaces for the fungus to grow and form spores, while the beetles hatch and develop. Adult beetles emerge from the tree covered in fungal **spores**, environmentally stable reproductive cells, and fly off to feed on new, young bark. The spores gain entry to a new tree in this way, using the beetles to transmit them. As a result, the fungus spreads rapidly **(Figure 23.1)**.

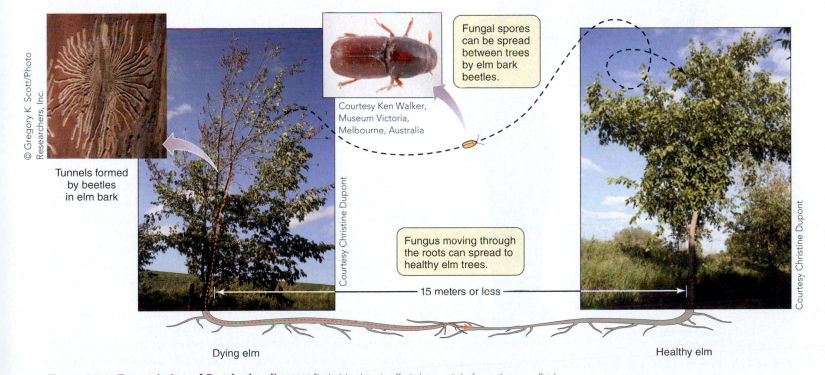

Figure 23.1. Transmission of Dutch elm disease Elm bark beetles quite effectively transmit the fungus that causes Dutch elm disease. The beetles burrow in the bark of diseased trees, forming tunnels, and lay their eggs. Sticky fungal spores become deposited in these tunnels. When beetles hatch and emerge from the tree, they carry with them spores that can infect healthy trees. If trees are close together, the fungus can move between trees via the roots.

In contrast to *Ophiostoma novo-ulmi*, the protozoan *Plasmodium falciparum* exhibits a **complex life cycle**; it needs to infect two or more species to complete its life cycle. This pathogen, the major cause of malaria, spends part of its life cycle in mosquitoes and part of its life cycle in humans. Female *Anopheles* mosquitoes serve as the **definitive host**, the host in which the pathogen replicates sexually. When taking a blood meal, these mosquitoes then can transfer the parasites to a human, the **intermediate host**. Within the intermediate host, the parasite replicates asexually and differentiates **(Figure 23.2)**. We will explore the details of this complex life cycle later in this chapter.

Unlike *Plasmodium falciparum*, the eukaryal microbe *Giardia lamblia*, a protozoan we first encountered in Section 3.2, exhibits a **simple life cycle**; it completes its life cycle entirely within a single species. This parasite, which causes diarrhea, typically is transmitted through the ingestion of water or contaminated foods that contain giardial **cysts**, environmentally stable, non-motile, infective forms of the organism encased within a protective cell wall. Mature cysts contain four nuclei. Following ingestion, excystation occurs, releasing two binucleated **trophozoites**, the active feeding form of spore-forming protozoa, from each cyst. The trophozoites replicate within the small intestine. As the trophozoites progress through the small intestine and enter the large intestine, they undergo encystation, forming cysts, which then are eliminated in the host's feces **(Figure 23.3)**.

Once pathogens enter a host, they often attach to and, in some cases, enter specific cells within the host. This interaction with the host cells, in many cases, plays a very direct role in disease. The trophozoites of *Giardia*, for example, attach to the epithelial cells of the small intestine. In some cases, the trophozoites virtually cover the lining of the small intestine **(Figure 23.4)**. The attachment, it appears, is mediated by a specific giardial structure—the ventral disk. Several studies have demonstrated that disruption of this structure decreases the ability of the trophozoites to adhere to cells. More importantly, the adherence of *Giardia* to the intestinal epithelia probably contributes to its pathogenesis. While researchers do not fully understand how *G. lamblia* causes diarrhea, many investigators postulate that the attachment of the trophozoites to the lining of the small intestine disrupts the nutrient adsorption that normally occurs there. Because of this disruption, diarrhea results. We can see from this example that a better understanding of *Giardia* attachment may lead to the development of new treatments for this pathogen. If we can block the attachment of giardial cells, then we may prevent the signs of giardial disease.

Not surprisingly, host organisms have evolved various ways of preventing, or attempting to prevent, the entry and adherence of eukaryal microbes. A more detailed understanding of these natural processes may allow us to develop more effective prevention and treatment strategies. To more fully understand the interactions between hosts and pathogens, however, a good *in vitro* experimental system often is necessary; research on whole organisms often is not feasible. To better understand how elm trees naturally respond to fungal infections, researchers have begun investigating the responses of elm tree cell cultures to *Ophiostoma novo-ulmi* **(Mini-Paper)**. These studies eventually may lead to the development of resistant elm varieties and the return of this majestic tree.

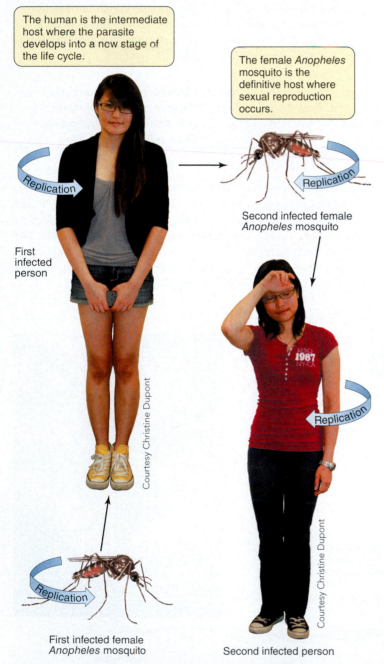

The human is the intermediate host where the parasite develops into a new stage of the life cycle.

The female *Anopheles* mosquito is the definitive host where sexual reproduction occurs.

Replication

Replication

Second infected female *Anopheles* mosquito

First infected person

Courtesy Christine Dupont

Replication

First infected female *Anopheles* mosquito

Replication

Courtesy Christine Dupont

Second infected person

Figure 23.2. Complex life cycle of *Plasmodium falciparum* The protozoan that causes malaria, *Plasmodium falciparum*, requires two hosts to replicate—mosquitoes and humans. When an infected female *Anopheles* mosquito, the definitive host, bites a human, the protozoa enter this new host. Within the human, the intermediate host, the parasites differentiate and replicate asexually. *P. falciparum* released into the bloodstream of the infected human can be ingested by another feeding mosquito, where the pathogen then completes its life cycle and can be transmitted to a second person.

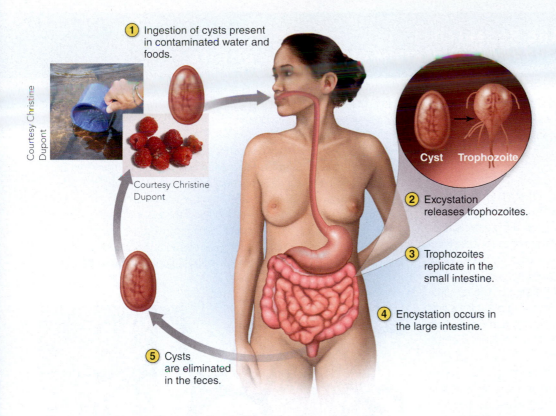

① Ingestion of cysts present in contaminated water and foods.

Courtesy Christine Dupont

Courtesy Christine Dupont

Cyst **Trophozoite**

② Excystation releases trophozoites.

③ Trophozoites replicate in the small intestine.

④ Encystation occurs in the large intestine.

⑤ Cysts are eliminated in the feces.

Figure 23.3. Process Diagram: Simple life cycle of *Giardia lamblia* Humans may ingest giardial cysts by drinking contaminated water or eating foods covered with the cysts. Within the body, excystation occurs, releasing trophozoites. This active feeding form of the organism replicates within the small intestine in the host. In the large intestine, encystation occurs and the resulting cysts are eliminated in the feces.

Evading host defenses

For a pathogen, simply entering a host and attaching to appropriate cells within the host is not enough. To cause disease, a pathogen also must evade various host defenses. Bacterial and viral pathogens achieve this goal in many diverse ways. Eukaryal microbes are no different. Some eukaryal microbes

© AMI Images/Photo Researchers, Inc.

SEM

Figure 23.4. Attachment of *Giardia* to the lining of the small intestine During an infection, large numbers of *Giardia* attach to the lining of the small intestine. Some researchers hypothesize that this attachment interferes with normal adsorption of nutrients in the host, leading to disease.

become pathogenic only when the host's immune system is suppressed. Others cause disease when the ecology of the host changes in some fundamental way. Still other pathogens actively deceive the host immune system. To investigate these mechanisms of circumventing host defenses, we will focus on three representative eukaryal pathogens of humans—the fungi *Pneumocystis jirovecii* and *Candida albicans*, and the protozoan *Trypanosoma brucei*—examining how each of these organisms survives within a host.

The Immunocompromised Host

As we discussed in Section 18.1, some microbes can be classified as opportunistic pathogens; they generally do not cause disease in a healthy host, but can cause disease if the ecology of the host is altered in some way. More specifically, they cause disease when conditions within the host are altered such that replication of the microbe can increase. A decrease in the effectiveness of the host's immune system represents a classic change in the host that facilitates the increased replication of a pathogen.

In Chapters 19 and 20, we investigated the intricacies of the immune system. As we noted, all eukarya possess a complex and highly effective means of controlling microbial infections. Under certain circumstances, however, this barrier to infectious diseases can be compromised. In humans, the immune system may be intentionally muted. When people receive an organ transplant, for instance, they typically receive **immunosuppressive drugs**, drugs that inhibit or decrease the activity of certain aspects of the immune system. By intentionally impairing the immune system in this way, physicians can decrease the likelihood that the organ recipient will reject the donated organ. Not surprisingly, this type of treatment also makes the patient more susceptible to opportunistic infections. Microbes that normally would be controlled by the body's immune system now can replicate more easily. In other words, the patient becomes immunocompromised. A similar type of induced immunosuppression also often occurs during treatment of autoimmune diseases. As we noted in Section 22.2, autoimmune diseases, such as rheumatoid arthritis and multiple sclerosis, occur when the immune system incorrectly targets for destruction normal host cells or proteins. To treat these diseases, physicians may prescribe immunosuppressive drugs. These drugs will lessen the severity of the autoimmune disease, but also will dampen the immune response against foreign bodies.

While physicians may prescribe drugs specifically designed to suppress the immune system, immunosuppression also may be caused by a microbial infection. As we discussed in Section 8.2, HIV replicates in, and ultimately destroys CD4$^+$ cells, a critical component of the human immune system. By destroying this key component of the human immune system, HIV leads to the development of acquired immunodeficiency syndrome, or AIDS.

M. Aoun, D. Rioux, M. Simard, and L. Bernier. 2009. Fungal colonization and host defense reactions in *Ulmus americana* callus cultures inoculated with *Ophiostoma novo-ulmi*. Phytopathology 99:642–650.

Context

The catastrophic losses of elm trees due to Dutch elm disease, caused by the pathogenic fungus *Ophiostoma novo-ulmi*, have spurred efforts to develop disease-resistant elm varieties. These efforts have resulted in the identification and planting of a number of partially resistant varieties of the American elm, *Ulmus americana*. Traditionally, studies of this host-pathogen interaction have been conducted on young saplings, but this approach limits the extent of the research that can be conducted. The controlled manipulation of saplings and the controlled inoculation of these saplings with *O. novo-ulmi* can be difficult. *In vitro* studies provide researchers with the ability to control experimental conditions much more precisely.

Luckily, cells of the American elm, like the cells of many other plants, can be grown in the laboratory. The replication of these cells in culture requires externally supplied plant hormones that promote the growth and division of the plant cells without causing them to differentiate into different cell types. When these cultures grow on solid media, researchers call them "callus cultures." Because the environmental conditions of callus cultures can be controlled very precisely, callus cultures provide researchers with the opportunity to do thorough studies of Dutch elm disease. The availability of the completed genome sequence of *O. novo-ulmi*, whose determination currently is being determined, will allow the application of functional genomics to address key questions about the pathogen-host interaction. In this paper, the researchers demonstrate that elm callus cultures can become infected by *O. novo-ulmi*. The investigators also examined the cellular responses to fungal colonization of elm callus cultures. They show that these cultures appear to respond to the fungus in a manner similar to cells of intact elm saplings.

Experiments

To begin these experiments, the researchers first had to establish the callus cultures. The cultures were initiated from tissue cuttings from either buds of mature trees or young leaves that emerged immediately after seed germination. The original plants were not resistant to infection by *O. novo-ulmi*. The formation of callus cultures from these cuttings on solid media was promoted by the addition to the plant growth media of naphthaleneacetic acid and benzyl adenine, which act as plant hormones. In the presence of these hormones, buds gave rise to hard callus cultures, while young leaves gave rise to

friable, or less compact, callus cultures. Once a callus culture formed, the researchers could propagate it by inoculating fresh media with small pieces of the callus.

Initially, the investigators determined whether or not the fungus could infect these *in vitro* cultures. To infect the callus cultures with *O. novo-ulmi*, the investigators simply added a small amount of fungal culture to a piece of filter paper, and then laid it on the elm cells. While the fungal cells grew very poorly on the plant growth media in the absence of the elm cells, and did not form spore-bearing conidia, they grew well on the callus cultures and also formed conidia. Microscopic study of the infected callus cultures **(Figure B23.1)**

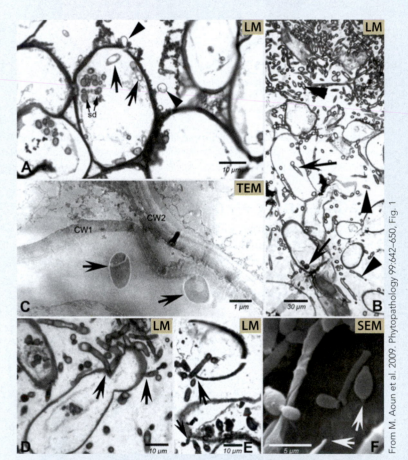

From M. Aoun et al. 2009. Phytopathology 99:642–650, Fig. 1

Figure B23.1. Infection of callus cultures by *Ophiostoma novo-ulmi*
Following inoculation of callus cultures with *Ophiostoma novo-ulmi*, fungal cells, indicated by the arrow heads in panels A, B, and C, were observed between and within the plant cells. Additionally, fungal cells appeared to penetrate the cell walls of elm callus cells, as noted by the arrows in panels D, E, and F. sd: starch deposits. CW1 and CW2: callus cell walls.

showed the presence of fungal cells both between cells and within plant cells, indicating that the fungus initiated a genuine infection.

To analyze the effect of this infection on the callus cells, the researchers investigated the response of the callus cultures to the fungal infection. Previous researchers had shown that elm saplings and leaves respond in very defined ways to fungal infections. Most notably, cells in elm trees infected with *O. novo-ulmi* produce various compounds, including phenolic compounds and tannins. Additionally, infected trees produce increased amounts of suberin, a hydrophobic waxy substance, and lignin, a complex chemical substance that contributes to the strength of wood. All of these compounds function as defense mechanisms for the trees. These same compounds accumulated in the callus cells following infection

(Figure B23.2). The callus cultures appear to respond to infection in a similar manner as the plant.

In addition to producing compounds typically associated with infections, cells of the callus cultures also exhibited increased expression of the PAL gene. This gene encodes the enzyme phenylalanine ammonia lyase, a central enzyme involved in the pathway leading to the formation of several different phenolic compounds. Elevated PAL expression is commonly associated with host defense response. The levels of PAL transcript in infected callus cultures were evaluated using the RT-PCR technique as described in Chapter 5. Increased expression clearly was associated with infection, as it would be in whole plants.

Impact

In anticipation of the completion of the *O. novo-ulmi* genome sequence, researchers hope to conduct detailed studies on the pathogenicity of Dutch elm disease, using functional genomic methods like transcriptomics and proteomics. These studies would be easier to conduct in callus cultures, rather than in mature plants or seedlings. Most obviously, many more replicates, under many different conditions, could be done. With this in mind, investigators first needed to demonstrate that infection studies could in fact be done in callus culture. As Aoun and colleagues have shown, the fungus clearly infects the callus cells, and the infection results in many of the expected responses, including upregulation of PAL, a key infection response-related gene. Based on this study, it should be possible to use the genomics tools along with microscopy to study the molecular mechanisms of infection. Particularly important studies will involve comparing the response to infection of callus cells from pathogen-susceptible and pathogen-resistant lines. These types of studies could result not only in the development of improved resistant lines of American elm, but also in a greater understanding of how fungal plant pathogens operate.

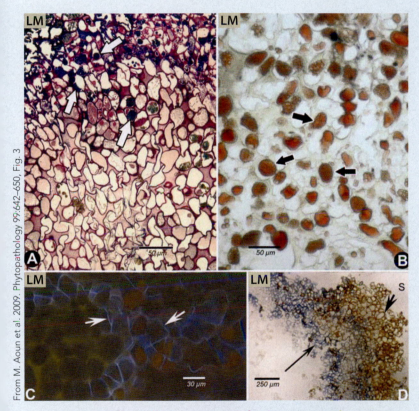

From M. Aoun et al. 2009. Phytopathology 99:642–650, Fig. 3

Figure B23.2. Callus cell responses to infection After being infected with *O. novo-ulmi*, cells in the callus cultures responded by producing various substances produced by infected trees. **A.** Blue stain (indicated by arrows) indicates deposits of phenolic compounds. **B.** Deposits of tannins, like catechin, in cells are indicated by arrows. **C.** Suberin (purple, indicated by arrows) accumulates around the cell walls. **D.** In a section of a callus culture, lignified cells (indicated by short arrow) appear along the surface of the callus, while suberin (indicated by the long arrow) appears in cells in the internal part of the callus.

● Questions for Discussion ....

1. The callus cultures that were used in this study were derived from plants that were not resistant to *O. novo-ulmi* infection. Would you expect a callus derived from resistant plants to respond to inoculation with the fungus differently than these callus cultures do?

2. Why do you think it is necessary to provide plant hormones in order for a callus to form?

TABLE 23.2 Fungal and protozoal opportunistic infections associated with AIDS

	Organism	Disease
Fungi	*Candida albicans*	Thrush
	Histoplasma capsulatum	Pneumonia
	Pneumocystis jirovecii	Pneumonia
Protozoa	*Toxoplasma gondii*	Encephalitis
	Cryptosporidium muris	Gastroenteritis
	Isospora belli	Gastroenteritis

Note: A number of viruses and bacteria also result in opportunistic infections in people with AIDS. The Centers for Disease Control and Prevention (CDC) maintains a list of all AIDS-defining opportunistic infections.

A number of fungal and protozoal opportunistic infections can occur as a result of the decreased number of CD4$^+$ cells present **(Table 23.2)**.

Opportunistic infections have become a defining characteristic of HIV disease. In fact, physicians noted the first cases of AIDS, reported in June of 1981, primarily because the patients had one or more opportunistic infections. Most notably, *Pneumocystis jirovecii* (previously referred to as *Pneumocystis carinii*) and *Toxoplasma gondii* **(Microbes in Focus 23.1)**, two eukaryal microbes, often infected the patients.

Pneumocystis jirovecii, a fungus, appears to reside frequently in the lungs of humans. Studies have shown that *P. jirovecii*-specific antibodies can be found in most children, indicating typical exposure to this microbe at a young age. A more recent study demonstrated the presence of *P. jirovecii* in 20 percent of healthy

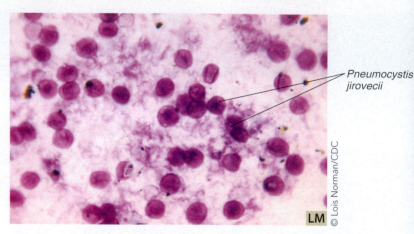

Pneumocystis jirovecii

LM © Lois Norman/CDC

Figure 23.5. *Pneumocystis jirovecii* When *Pneumocystis jirovecii* replicates within the lung, it fills the air spaces resulting in impaired breathing. Shown here are *P. jirovecii* cells isolated from lung tissue.

immunocompetent adults. This suggests that a portion of the population may be asymptomatic carriers of this microorganism. Based on these data, researchers hypothesize that airborne transmission of *P. jirovecii* from person to person probably occurs.

While *P. jirovecii* may be a relatively common inhabitant of our lungs, *P. jirovecii*-associated disease is not common. In immuno-compromised people, however, replication of the fungus is poorly regulated. As it replicates, it fills the air spaces within the lungs, resulting in *Pneumocystis* pneumonia, or PCP **(Figure 23.5)**. Patients develop difficulty breathing and, in the absence of treatment, the disease typically is fatal. Despite advances in treatment and prevention, *P. jirovecii* remains a leading opportunistic infection among people with HIV disease **(Perspective 23.1)**.

Altered Microbiota of the Host

The fungus *Candida albicans* is another opportunistic pathogen commonly associated with HIV disease. Like *P. jirovecii*, *C. albicans* replicates more extensively in people with a weakened immune system. Because the fungus often resides within the mouth and throat, a condition known as oral thrush develops when the fungus grows unchecked in these locations. White lesions appear on the tongue and cheeks **(Figure 23.6)**. In severe

Microbes in Focus 23.1
AN AIDS-DEFINING OPPORTUNISTIC INFECTION: *TOXOPLASMA GONDII*

Habitat: Has been found in many species of mammals and birds, although cats are considered the primary host. Oocysts can develop only in members of the family Felidae. Roughly 20 percent of humans may be asymptomatic carriers of *T. gondii*.

Description: Protozoan with infectious oocysts approximately 10 μm in diameter and tachyzoites, which replicate asexually and cause disease, measuring 6 μm × 2 μm.

Key Features: In people with HIV disease, *T. gondii* can spread to the brain, causing cerebral toxoplasmosis. Toxoplasmosis, in turn, can lead to inflammation of the brain, which can lead to vision problems, severe headaches, and in extreme cases coma and death. Some reports also have linked *T. gondii* infections to schizophrenia.

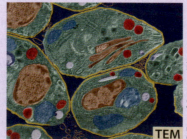

TEM © Eye of Science/Photo Researchers, Inc.

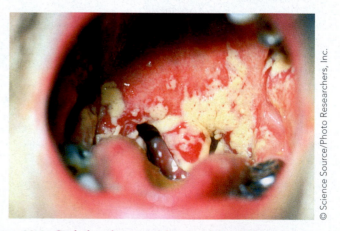

© Science Source/Photo Researchers, Inc.

Figure 23.6. Oral thrush caused by *Candida albicans* The opportunistic pathogen *Candida albicans* is a frequent resident of the human mouth and throat. Normally, the host controls its growth. In immunocompromised hosts, however, growth increases, resulting in thick white lesions, such as those seen here.

Perspective 23.1

PNEUMOCYSTIS JIROVECII OR CARINII: THE EVOLVING FIELD OF TAXONOMY

The name *Pneumocystis carinii* has permeated the infectious disease literature since 1981, when scientific reports of HIV/AIDS were first published. Indeed, the very first report on what we now know as HIV/AIDS was titled, "*Pneumocystis* Pneumonia–Los Angeles," and focused on five men with "*Pneumocystis carinii* pneumonia," or PCP **(Figure B23.3)**. At the 2001 International Workshop on Opportunistic Protists, though, investigators proposed that the causative agent of this pneumonia be renamed. Today, we refer to it as *Pneumocystis jirovecii*. The name pays homage to the Czech microbiologist Otto Jirovec, the person credited with first documenting *Pneumocystis* infections in humans.

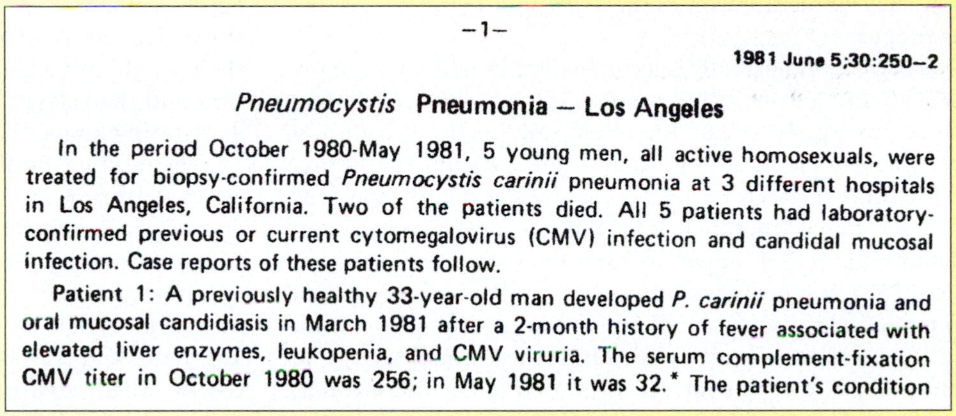

One could argue that this name change matters little. After all, as Shakespeare once wrote, "What's in a name? That which we call a rose by any other name would smell as sweet." While a name may not matter, the taxonomy, or classification, of an organism is very important. Since the 1980s, our understanding of the taxonomy of the causative agent of PCP has changed. Prior to 1988, researchers were unsure if this organism was a fungus or a protozoan. The confusion stemmed from several atypical features of the organism. Most notably, it

- lacks some features common to fungi
- structurally resembles some protozoa
- exhibits resistance to some standard antifungal drugs
- exhibits sensitivity to some standard antiprotozoal drugs

SSU rRNA sequence analysis, however, demonstrates quite clearly that *Pneumocystis* species are fungi **(Figure B23.4)**.

—1—

1981 June 5;30:250—2

Pneumocystis Pneumonia — Los Angeles

In the period October 1980-May 1981, 5 young men, all active homosexuals, were treated for biopsy-confirmed *Pneumocystis carinii* pneumonia at 3 different hospitals in Los Angeles, California. Two of the patients died. All 5 patients had laboratory-confirmed previous or current cytomegalovirus (CMV) infection and candidal mucosal infection. Case reports of these patients follow.

Patient 1: A previously healthy 33-year-old man developed *P. carinii* pneumonia and oral mucosal candidiasis in March 1981 after a 2-month history of fever associated with elevated liver enzymes, leukopenia, and CMV viruria. The serum complement-fixation CMV titer in October 1980 was 256; in May 1981 it was 32.* The patient's condition

Figure B23.3. *Pneumocystis carinii* pneumonia: The beginning of the AIDS pandemic The first scientific report related to HIV/AIDS was published in June of 1981 in *Morbidity and Mortality Weekly Report (MMWR)*, a publication of the Centers for Disease Control and Prevention. The article profiled five young men who had *Pneumocystis carinii* (now *jirovecii*) pneumonia, a disease normally seen only in immunocompromised people.

More recently, additional DNA sequence analyses have shown that *Pneumocystis* organisms infecting humans are genetically distinct from *Pneumocystis* organisms isolated from other mammalian hosts. Researchers have interpreted this information as evidence that a unique species of *Pneumocystis* infects humans. Because a unique species infects humans, researchers proposed the new name for the *Pneumocystis* variant found in humans—*P. jirovecii*. The initials PCP, however, have become so entrenched in the biological literature that researchers continue to use them. These initials now refer simply to *Pneumocystis* pneumonia.

The story of *Pneumocystis jirovecii* highlights the changing field of biological taxonomy. With newer and more sophisticated molecular tools at our disposal, the taxonomic classification of an organism no longer is based solely on its physical attributes. Increasingly, molecular information helps us better understand the classification of organisms.

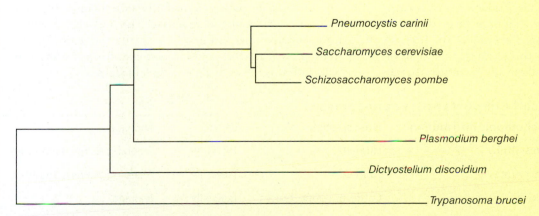

Figure B23.4. The causative agent of PCP—a fungus Until 1988, there was disagreement among microbiologists about the classification of *Pneumocystis carinii* (now *jirovecii*). SSU rRNA sequence analysis showed that *P. carinii* is closely related to *Saccharomyces cerevisiae* and *Schizosaccharomyces pombe*, which are fungi, and more distantly related to protozoa such as *Plasmodium berghei*, *Dictyostelium discoidium*, and *Trypanosoma brucei*. This evidence confirmed that *P. jirovecii* is a fungus.

From J. C. Edman, J. A. Kovacs, H. Masur, et al. 1988. Ribosomal RNA sequence shows *Pneumocystis carinii* to be a member of the fungi. Nature 334:519–522. Reprinted with permission from Macmillan Publishers Ltd.

cases, the fungal growth may clog the esophagus, leading to difficulty swallowing.

This microbe also causes disease quite frequently in immunocompetent people. In these cases, the host ecology again is altered, but the immune system remains intact. Rather, the normal microbial inhabitants of the host change and, as a result, the environment becomes more hospitable for the growth of this fungus.

Candida albicans normally is a commensal inhabitant of many people, benefiting from its interactions with us without harming us in the process. It can be found not only in the mouth and throat, but also in the rectum and vagina. Within the vagina of healthy women, bacteria in the *Lactobacillus* genus represent some of the major microbial inhabitants. These bacteria convert lactose and other sugars to lactic acid via a fermentation reaction. As a result of this lactic acid production, the vagina normally is slightly acidic, with a pH of ~4.5. The lactobacilli, not surprisingly, grow quite well under these conditions. The growth of *C. albicans*, in contrast, is limited under these conditions. The presence of lactobacilli, in other words, limits the growth of this fungus.

CONNECTIONS Why would fermentation result in an acidic environment? Lactic acid bacteria (LAB), which include genera such as *Streptococcus*, *Lactobacillus*, and *Bifidobacterium*, use the enzyme lactate dehydrogenase to convert pyruvate to lactic acid (**Section 13.3**).

Given this information, we can reason that the elimination of lactobacilli from the vagina would result in an increase in the vaginal pH. The increased pH, in turn, would make the environment more hospitable for *C. albicans*. When the population of *C. albicans* in the vagina increases, disease ensues. Itching or burning of the vagina results from irritation of the vaginal lining caused by the yeast. So, what would lead to the elimination of lactobacilli in the vagina? Often, the lactobacillus population decreases when a woman takes antibiotics. The antibiotics may be taken to eliminate another species of bacteria, but the lactobacilli also are eliminated. As a result of their population decline, *C. albicans* flourishes.

Antigenic Variation in the Pathogen

While *Pneumocystis jirovecii* generally does not cause disease unless the host's immune system fails, other organisms cause disease even in the presence of a strong immune response. These organisms have evolved a mechanism for evading the immune response—**antigenic variation**. In this strategy, infectious organisms periodically change selected surface proteins, thereby changing the antigens seen by the host's immune system. The result of this periodic antigen switching is that the host immune response must continually adjust to a new target. For the pathogen, the benefit is immense; it can continue to replicate within the host without being eliminated by a robust immune attack.

Trypanosomes present a remarkable example of this strategy (see Figure 18.6). These organisms are characterized by a single flagellum and a **kinetoplast**, a large mass of DNA within their single mitochondrion. Species within the genus *Trypanosoma* and the related genus *Leishmania* cause some of the most devastating human diseases (**Table 23.3**).

The most medically relevant trypanosomes have complex life cycles that involve replication in both an insect and a mammal. *Trypanosoma brucei* spp. (**Microbes in Focus 23.2**) present a perfect example. When a tsetse fly ingests these protozoa after biting an infected mammal, the parasites differentiate and replicate within the midgut of the tsetse fly. The parasite then differentiates more and travels from the midgut to the salivary gland of the fly, where additional replication occurs. When an infected fly bites a mammal, the trypanosomes can be transmitted to the bloodstream of the mammal, where they differentiate, replicate, and can be transferred throughout the body. The life cycle can begin again when another tsetse fly feeds on an infected mammal (**Figure 23.7**). Subspecies of *T. brucei* cause African sleeping sickness in humans and nagana in cattle. In humans, the disease is characterized by recurring waves of high fever, along with headaches and joint pain. During the late stages of the disease, the parasites affect the central nervous system, resulting in extreme lethargy. A coma and, eventually, death ensue. Not surprisingly, the geographic distribution of the tsetse fly mirrors the distribution of these trypanosomal diseases. It also is worth noting that nagana has effectively prevented the raising of cattle within large areas of the African continent. Within Africa, cattle generally are raised outside of areas in which the tsetse fly resides (**Figure 23.8**).

The recurring bouts of fever associated with sleeping sickness correlate with a cyclic variation in **parasitemia**, or the number of parasites present in the blood of an infected person. The fever is a manifestation of the immune system's response to the parasites. The cyclic decrease in parasitemia results from the destruction of parasites present in the blood by the host immune system. So, let's ask a basic question: If the immune system recognizes the parasites and mounts a strong response, then why do the numbers of parasites in the blood rebound? Why aren't the parasites completely eliminated? The answer, it appears, is that the parasite can continually change its antigenic properties, thereby presenting the immune system with a moving target. *T. brucei* hides, as the saying goes, in plain sight.

TABLE 23.3   Selected medically important trypanosomes

Parasite	Geographic distribution	Host species	Vector	Disease
Trypanosoma brucei gambiense	West Africa	Humans	Tsetse fly	Sleeping sickness (chronic)
Trypanosoma brucei rhodesiense	East Africa	Humans	Tsetse fly	Sleeping sickness (acute)
Trypanosoma brucei brucei	Africa	Cattle, antelope, horses	Tsetse fly	Nagana
Trypanosoma cruzi	Central and South America	Humans	Kissing bug	Chagas disease
Leishmania major	Africa, Middle East	Humans	Sandfly	Dermal leishmaniasis

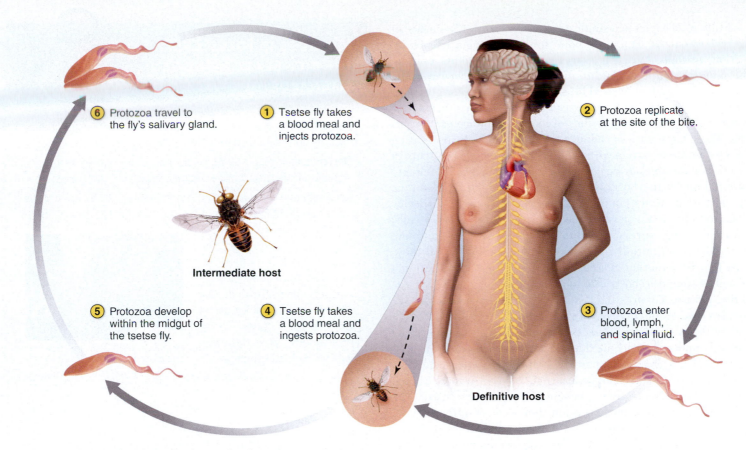

⑥ Protozoa travel to the fly's salivary gland.

① Tsetse fly takes a blood meal and injects protozoa.

② Protozoa replicate at the site of the bite.

Intermediate host

⑤ Protozoa develop within the midgut of the tsetse fly.

④ Tsetse fly takes a blood meal and ingests protozoa.

③ Protozoa enter blood, lymph, and spinal fluid.

Definitive host

Figure 23.7. Process Diagram: Complex life cycle of *Trypanosoma brucei* ssp Like *Plasmodium falciparum*, this protozoal pathogen completes part of its life cycle in an insect host and part of its life cycle in a vertebrate host. The pathogen differentiates and replicates within the midgut of the tsetse fly, and then travels to the insect's salivary gland. Further differentiation and replication occur at this site. When the fly feeds, the parasite is transmitted to a mammal host. Within the mammal, the parasite differentiates and replicates. It also undergoes antigenic variation to evade the host immune system. When another fly bites an infected mammal, the cycle begins again.

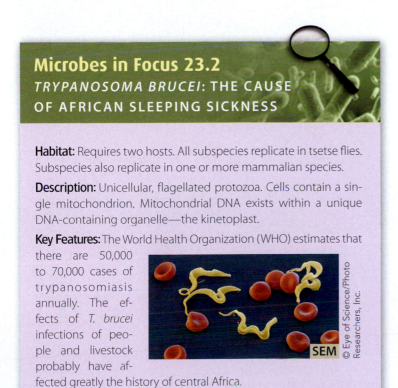

Microbes in Focus 23.2
TRYPANOSOMA BRUCEI: THE CAUSE OF AFRICAN SLEEPING SICKNESS

Habitat: Requires two hosts. All subspecies replicate in tsetse flies. Subspecies also replicate in one or more mammalian species.

Description: Unicellular, flagellated protozoa. Cells contain a single mitochondrion. Mitochondrial DNA exists within a unique DNA-containing organelle—the kinetoplast.

Key Features: The World Health Organization (WHO) estimates that there are 50,000 to 70,000 cases of trypanosomiasis annually. The effects of *T. brucei* infections of people and livestock probably have affected greatly the history of central Africa.

© Eye of Science/Photo Researchers, Inc.

SEM

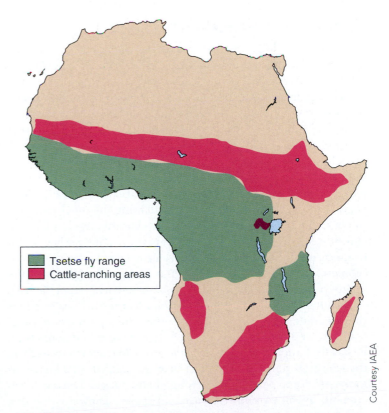

Tsetse fly range
Cattle-ranching areas

Courtesy IAEA

Figure 23.8. Geographic distribution of the tsetse fly in Africa
Because *Trypanosoma brucei* must replicate in the tsetse fly, the distribution of trypanosomiasis in Africa mirrors the distribution of the fly. *T. brucei* also causes disease in cattle. Areas where the tsetse fly are present, shown here in green, are inhospitable for the raising of cattle. Cattle typically are raised only in areas outside of the tsetse fly range (dark pink areas).

Millions of copies of a single glycoprotein cover the surface of every *T. brucei* cell. At a given point in time during a *T. brucei* infection, virtually every cell expresses, and is coated by, copies of the same glycoprotein. At different points in time, however, the cells express, and are coated by, significantly different glycoprotein molecules. During the course of an infection, in other words, *T. brucei* changes the appearance of its glycoprotein coat repeatedly. The effects of these changes are remarkable. By presenting the host immune system with a **variable surface glycoprotein (VSG)**, or variable surface antigen, *T. brucei* can evade complete elimination by the host. During periods of high parasitemia, the host immune system mounts a strong response to a specific VSG and eliminates most of the parasites. Some *T. brucei* cells, however, begin expressing a different VSG. The host immune system does not initially recognize this new antigen, and these cells flourish. Another increase in parasitemia occurs. After a period of time, the host immune system begins effectively targeting parasites expressing the new VSG, causing a decrease in parasitemia. Some *T. brucei* begin expressing a different VSG, and so the pattern continues (see Figure 18.6). Eventually, the body's lymphocytes become depleted and the host loses its ability to combat *T. brucei*. The parasites replicate unchecked.

To achieve this antigenic variation, *T. brucei* devotes a large amount of its DNA to genes encoding VSG variants. In fact, the *T. brucei* genome appears to contain thousands of different VSG genes. Only one of these genes, however, is expressed at any given time. How cells regulate this expression and periodically switch which gene is expressed is an area of intense research.

Obtaining nutrients from a host

After a pathogen has entered a host, it must obtain the nutrients necessary for replication. In some cases, as we saw earlier with the fungus *Candida albicans*, nutrients may become available because of a change in the host. With *C. albicans*, remember, a decrease in the bacterial population of the vagina, and a resulting increase in the vaginal pH, allow the fungus to replicate more extensively.

Other pathogenic fungi secrete various enzymes that break down host cell barriers and allow access to nutrients. Many **phytopathogenic fungi**, or fungi that cause diseases in plants, secrete, for example, cellulases, pectinases, and various proteases. These enzymes degrade the plant cell wall, allowing the fungi to invade the host cells. *Ophiostoma novo-ulmi*, the causative agent of Dutch elm disease, produces several such enzymes. This pathogen, for instance, produces **xylanase**, an enzyme needed to degrade hemicellulose, a major component of the plant cell wall. Once the fungus breaches the cell wall, it can obtain nutrients from the host.

While many phytopathogenic fungi secrete various enzymes that degrade the plant cell wall, other phytopathogenic fungi use more of a brute force approach to obtaining nutrients from host cells. Fungi such as *Magnaporthe grisea* (**Microbes in Focus 23.3**), the causative agent of rice blast, use specialized cells known as **appressoria** to penetrate the tough plant cuticle. During development, hypha extend from asexual spores, known as conidia, and develop appressoria. A penetration peg then develops from this structure. Through extreme turgor pressure, outward pressure exerted against the fungal cell wall caused by water within the cell, these extensions simply penetrate the cuticle, thereby allowing the fungus to invade the plant (**Figure 23.9**).

Microbes in Focus 23.3
THE RICE BLAST PATHOGEN: *MAGNAPORTHE GRISEA*

Habitat: Infects rice, barley, wheat, rye, and other related plants, causing significant crop loss every year.

Description: Fungus that forms hyphae, or long, branching cells. Appressoria develop which infect tissues of the host plants.

Key Features: First seen in the United States in 1996, the fungus has been identified in countries throughout the world. There is some evidence that both the United States and the Soviet Union explored the use of *M. grisea* as a biological weapon during the Cold War. Potentially, it could be utilized by terrorist groups today. This fungus could devastate worldwide food supplies.

Courtesy Donald Groth, Rice Research Station, Louisiana State University

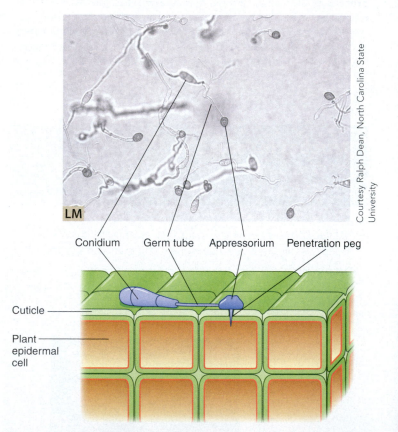

Courtesy Ralph Dean, North Carolina State University

Conidium   Germ tube   Appressorium   Penetration peg

Cuticle

Plant epidermal cell

Figure 23.9. Fungal invasion of plants For some phytopathogenic fungi, asexual spores, or conidia, stick to the plant surface. A germ tube, or hypha, then grows, and develops an appressorium. This structure develops a penetration peg that can penetrate the waxy cuticle of the plant, allowing the fungus to invade.

Specific **mitogens**, chemicals that promote cell division, and a mitogen-activated protein (MAP) kinase pathway control the development of appressoria within *M. grisea*. Researchers have demonstrated that strains of *M. grisea* with mutations in the *pmk1* gene, which encodes a MAP kinase, do not produce appressoria and do not invade rice plants. Further research into this pathway may lead to the development of an effective treatment for this pathogen.

Other pathogens possess mechanisms for obtaining specific resources from the host. Species within the genus *Plasmodium*, for instance, can synthesize pyrimidine rings, the nitrogenous structures in the bases thymine and cytidine, but lack the ability to synthesize purine rings, the nitrogenous structures in the bases adenine and guanine. All organisms, of course, need these chemicals for the synthesis of DNA and RNA. Rather than make their own purines, *Plasmodium* spp. obtain them from the host. The parasites possess a series of purine transporter proteins. The *Plasmodium falciparum* transmembrane nucleoside transporter protein, PfNT1, has been well characterized and appears to transport hypoxanthine, a molecule similar to adenosine, from the red blood cell cytoplasm into the cytoplasm of the parasite. Once hypoxanthine enters the parasite, then various enzymes convert it to nucleic acids **(Figure 23.10)**. Genetic studies suggest that several additional nucleoside transporter proteins, PfNT2, PfNT3, and PfNT4 also exist within this pathogen and may be involved in the transport of other nucleosides and nucleoside-like molecules into the parasite.

Toxins of eukaryal microbes

As we saw in Chapter 21, a number of bacterial pathogens produce toxins. Often, these toxins facilitate the transmission of the microbe. Pathogenic strains of the bacterium *Vibrio cholerae*, for instance, express cholera toxin, a potent enterotoxin that causes the watery diarrhea associated with cholera. Most likely, the effects of the toxin allow the bacteria to exit an infected host and, eventually, infect another susceptible individual. Many eukaryal microbes also express toxins. The effects of these toxins, and their benefits to the microbe, appear to be quite different from the effects and benefits of bacterial toxins.

Several phytopathogenic fungi produce virulence toxins. *Cochliobolus carbonum* **(Microbes in Focus 23.4)**, a fungus that affects maize, produces HC-toxin. Studies have demonstrated that this toxin is necessary for the virulence of the fungus. HC-toxin functions as an inhibitor of **histone deacetylases (HDACs)**, cellular enzymes that remove acetyl groups ($CH_3C\!=\!O$) from histones and, in the process, affect gene expression **(Figure 23.11)**.

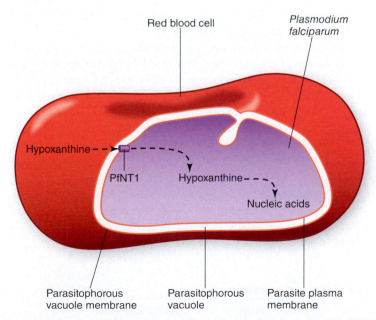

Figure 23.10. Uptake of purine precursors by *Plasmodium falciparum* Within the red blood cell, hypoxanthine crosses the parasitophorous vacuole membrane and then enters the *P. falciparum* cytoplasm through a specific transport protein—PfNT1—embedded in the parasite plasma membrane. Related transport proteins may transport other nucleoside-like molecules into the parasite. Within *P. falciparum*, parasite enzymes convert these molecules into nucleic acids.

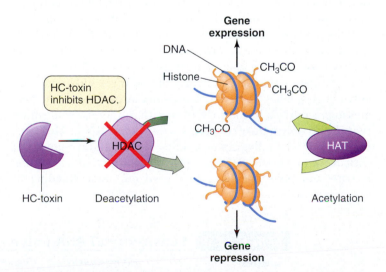

Figure 23.11. Inhibition of histone deacetylases in maize by HC-toxin HC-toxin, produced by the pathogenic fungus *Cochliobolus carbonum*, inhibits host histone deacetylases (HDACs), which remove acetyl groups ($CH_3C\!=\!O$) from histones. Histone acetylases (HATs), conversely, acetylate histones. Generally, HDACs lead to gene repression. The exact mechanism of HC-toxin's toxicity is not completely understood.

A. *Amanita phalloides*

B. α-Amanitin

Figure 23.12. RNA polymerase II inhibitor α-amanitin This potent toxin has been used extensively by cell biologists to elucidate the roles of various RNA polymerases. **A.** Several mushrooms in the genus *Amanitin*, including *Amanita phalloides*, produce this toxin. **B.** Chemical structure of α-amanitin.

It is thought that HC-toxin interferes with the expression of genes for host defense molecules, thereby allowing the fungus to colonize the host.

One of the more well-characterized toxins produced by eukaryal microbes is **α-amanitin**, a toxin produced by mushrooms in the genus *Amanita* (**Figure 23.12**). Unlike other fungi we have discussed, these mushrooms do not fit our previous definition of pathogen. They are not infectious. Rather, they can produce disease through **intoxication**, the ingestion of a substance that leads to a disease. The toxin α-amanitin has an LD_{50} (the dose that is lethal 50 percent of the time) in mammals of approximately 0.1 mg/kg (see Toolbox 18.1 for a review of LD_{50} values). Within a day of ingesting α-amanitin, a person often will develop gastrointestinal problems. More severe effects may occur in the liver and kidneys. The toxin causes the destruction of cells in both organs, eventually leading to liver and renal failure. Death results in about 15 percent of the people who ingest the toxin.

Despite these rather horrible effects, the mechanism of action of α-amanitin actually is quite interesting, and researchers have exploited the toxin quite effectively. Alpha-amanitin binds to RNA polymerase II and inhibits the enzyme. RNA polymerases I and III, conversely, are relatively resistant to the effects of this toxin. During the 1970s, researchers used α-amanitin to investigate the respective roles of the three eukaryal RNA polymerases. They observed that α-amanitin reduces the production of mRNA, but not rRNA or tRNA, thereby confirming that RNA polymerase II forms messenger RNA. More recently, X-ray crystallographic studies of α-amanitin complexed with RNA polymerase II have provided insight into how this enzyme functions and have furthered our understanding of transcription.

CONNECTIONS The roles of RNA polymerases I, II, and III are discussed in **Section 7.3**. While bacteria and archaeons each contain only a single type of RNA polymerase, the archaeal RNA polymerase more closely resembles the eukaryal RNA polymerase II than its bacterial counterpart.

Before leaving this topic, let's ask a somewhat obvious question. If α-amanitin blocks the function of RNA polymerase II and all eukarya possess RNA polymerase II, then how can the *Amanita* mushrooms survive their own toxin? The answer to this question is quite fascinating. The RNA polymerase II enzymes of various organisms, including insects, plants, and mammals, are much more sensitive to the toxin than are the RNA polymerase II enzymes isolated from fungi (**Table 23.4**). The fungi, it appears, are resistant to their own toxin; much more of the toxin is needed to reduce by 50 percent the enzymatic activity of their polymerase. Other fungi, like the mushrooms in the *Psilocybe* genus, also produce a toxin to which they are resistant (**Perspective 23.2**). Most likely, these toxins serve to ward off predators.

Saxitoxin represents another well-characterized toxin produced by eukaryal microbes. Produced by **dinoflagellates**, a group of algae that possess two flagella, this neurotoxin functions as a sodium channel blocker, preventing the normal flow of sodium ions across the cell membrane. Ingestion of sufficient amounts of saxitoxin results in flaccid paralysis, a condition in which the muscles are limp and cannot contract. In extreme cases, respiratory failure results.

CONNECTIONS Botulinum toxin, an AB exotoxin from the bacterium *Clostridium botulinum*, also causes flaccid paralysis. This potent neurotoxin cleaves SNARE proteins in host cells needed for the release of the neurotransmitter acetylcholine (**Section 21.1**).

TABLE 23.4	**Sensitivity of RNA polymerase II to α-amanitin**			
Species	*Homo sapiens* (human)	*Caenorhabditis elegans* (worm)	*Saccharomyces cerevisiae* (yeast)	*Agaricus bisporus* (mushroom)
50% inhibition level[a]	10–25 ng/mL	7 ng/mL	1,000 ng/mL	6,800 ng/mL

[a]Amount of α-amanitin required to inhibit RNA polymerase II activity by 50%.

We probably all have heard about magic mushrooms. Like the mushrooms we examined earlier in this section, these mushrooms are not infectious, but they do produce a substance that can result in intoxication. Consisting mainly of species in the *Psilocybe* genus, these mushrooms contain psilocybin or psilocyn, hallucinogenic compounds **(Figure B23.5)**.

Nathan Griffith/© Corbis

A. *Psilocybe cyanescens*

B. Psilocybin

Figure B23.5. Magic mushrooms Certain mushrooms produce psilocybin, a hallucinogenic compound. **A.** *Psilocybe cyanescens* is a potent psychedelic mushroom found throughout the Pacific Northwest in the United States and in Europe. **B.** Chemical structure of psilocybin. Once ingested, psilocybin is converted to the metabolically active psilocyn (or psilocin) through dephosphorylation.

Native to Mexico and Central America, these mushrooms have been part of certain Native American rituals for thousands of years. Psychedelic mushrooms have been part of the U.S. counterculture since the 1950s. In 1957, R. Gordon and Valentina Wasson authored an article in *Life* magazine describing their experiences with psychedelic mushrooms in Mexico, thereby introducing them to mainstream America for the first time.

According to people who have ingested these mushrooms, they cause heightened senses of sight, smell, sound, and touch. Many people also claim to have spiritual experiences after ingesting them. Others report dizziness, nausea, anxiety, and paranoia. Because of their effects, the possession of these mushrooms is banned in most countries. In the United States, it is illegal to possess, manufacture, or sell mushrooms containing psilocybin or psilocyn without a license. They are classified as Schedule 1 narcotics, like heroin and LSD.

Within the body, psilocybin and psilocyn function as mimics of serotonin, a neurotransmitter that serves to relay messages within the brain. Serotonin affects many bodily functions, including sexual desire, appetite, and sleep. Additionally, serotonin affects our mood. Low levels of serotonin, in fact, often are associated with clinical depression. Many antidepressants, as a result, function by increasing the intercellular serotonin concentrations. While the exact mechanism by which psilocybin and psilocyn produce their hallucinogenic properties is not known, it most likely acts through the natural serotonin pathway.

As we saw with other toxins produced by mushrooms, psilocybin probably exists within the mushrooms to ward off predation. Ingestion of the compound by predators has deleterious effects. For the mushroom, production of the toxin serves as an effective defense mechanism.

Human exposure to saxitoxin generally occurs indirectly. In many marine environments, periodic population explosions, or blooms, of dinoflagellates, cyanobacteria, and algae occur. When the species produce toxins, these blooms often are referred to as **harmful algal blooms (HABs)**, or erroneously, red tides. The increased amount of toxin produced by the larger population can lead to massive fish kills **(Figure 23.13)**. Marine mammals and birds also may be affected. Many filter feeders, especially mollusks and clams, accumulate the toxin within their tissues. If these shellfish are harvested and eaten by humans, then a large dose of toxin can be ingested, resulting in paralytic shellfish poisoning.

As we have seen, both α-amanitin and saxitoxin can severely affect the health of humans if ingested. Of what benefit, though, are these toxins to the organisms that produce them? Unlike the toxins produced by bacteria, which often aid in the transmission of the microbes or aid in the bacteria's acquisition of nutrients, the toxins produced by most eukaryal microbes probably function as defense mechanisms. Toxin production, most likely, functions as a deterrent to predation.

Courtesy NOAA

Figure 23.13. Fish kills caused by harmful algal blooms As populations of dinoflagellates, cyanobacteria, or algae occur, the concentrations of toxins produced by these organisms also increase. In extreme cases, these harmful algal blooms can lead to massive fish kills, such as this one in the Chautauqua National Wildlife Refuge in the United States.

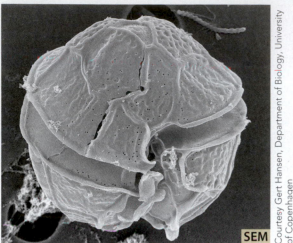

A. *Alexandrium minutum*: A dinoflagellate

Courtesy Gert Hansen, Department of Biology, University of Copenhagen

B. *Acartia tonsa*: A copepod

© Nigel Cattlin/Photo Researchers, Inc.

C. Effect of copepods on toxin production by dinoflagellates

Copyright © 2006, The Royal Society

Figure 23.14. Role of toxin production by dinoflagellates
Dinoflagellates may increase toxin production to protect themselves from predatory copepods. **A.** The dinoflagellate *Alexandrium minutum* can produce paralytic shellfish toxin (PST). **B.** *Acartia tonsa*, a copepod, feeds on dinoflagellates. **C.** *A. minutum* produced very little PST in the absence of copepods. The amount of PST produced by *A. minutum* increased significantly when the dinoflagellates were exposed to starving or grazing copepods.

To experimentally investigate this hypothesis, researchers at Goteborg University in Sweden examined the production of paralytic shellfish toxin (PST) by *Alexandrium minutum*, a dinoflagellate, in the presence and absence of *Acartia tonsa*, a copepod that feeds on dinoflagellates. They observed that *A. minutum* produced approximately 2.5 times as much toxin when exposed to the copepod. Based on these results, the researchers concluded that the dinoflagellates could detect the predator and increase their toxin production in response to this potential threat **(Figure 23.14)**.

23.1 Fact Check

1. Compare and contrast the role of insects in the transmission of *Ophiostoma novo-ulmi* and *Plasmodium falciparum*.
2. How do *Trypanosoma brucei* spp. avoid the immune response of the host?
3. How do some phytopathogenic fungi obtain nutrients from host cells?
4. Should we consider poisonous mushroom pathogens?

23.2 Pathogen study: *Plasmodium falciparum*

How does *Plasmodium falciparum* cause malaria?

In the preceding section, we examined general mechanisms by which eukaryal microbes cause disease. Additionally, we discovered that these mechanisms often mirror, at least in a cursory fashion, the mechanisms utilized by viral and bacterial pathogens. In this section, we will examine a single pathogenic eukaryal microbe—*Plasmodium falciparum*—in more detail. As part of this case study, we will examine how this pathogen replicates, how it causes disease within a host, and how our

knowledge of this microbe may lead to the development of new drugs to combat it.

Replication of *Plasmodium falciparum*

Malaria represents one of the most devastating infectious diseases of humans. Between 300 and 500 million people become infected every year, and over 1 million people worldwide die of this disease annually, most of them in sub-Saharan Africa.

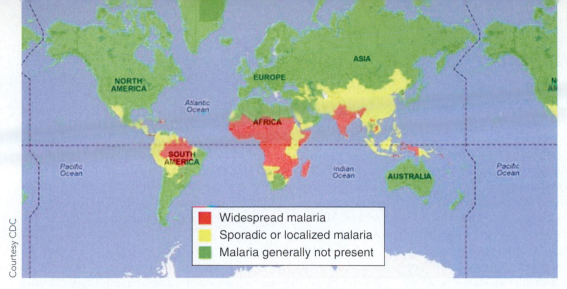

Courtesy CDC

Color	Meaning
Red	Widespread malaria
Yellow	Sporadic or localized malaria
Green	Malaria generally not present

Figure 23.15. Global presence of malaria Despite a worldwide malaria eradication program organized by the WHO, malaria remains endemic in countries throughout the world. As shown in this map from the United States Centers for Disease Control and Prevention (CDC), malaria remains common in many countries (red) and sporadically present in many other countries (yellow).

Several related eukaryal microbes—*Plasmodium falciparum*, *Plasmodium vivax*, *Plasmodium ovale*, and *Plasmodium malariae*—cause malaria. While *P. falciparum* and *P. vivax* cause most infections, *P. falciparum* is the most virulent malarial parasite.

Malaria probably has affected humans for millennia and, historically, has been present in tropical and subtropical areas throughout the world. Until the middle of the twentieth century, malaria remained a significant problem in the southeastern part of the United States. Through the use of a multifaceted eradication program that included spraying insecticides and draining swamps, the U.S. government eradicated the disease from the United States by 1951. A subsequent worldwide malaria eradication program, organized by the World Health Organization

(WHO), has been less successful. Malaria remains endemic throughout sub-Saharan Africa, India, parts of Southeast Asia, and parts of South and Central America (**Figure 23.15**).

As we mentioned earlier, *Plasmodium falciparum* causes the most severe cases of malaria. The parasite's life cycle requires two hosts, mosquitoes and humans. Within the female *Anopheles* mosquito gut, haploid gamete precursors, termed **gametocytes (gamonts)**, are obtained when the mosquito bites an infected person, and the gametocytes mature within the mosquito into gametes. These gametes fuse, producing a zygote, or **ookinete**, which travels to the gut wall and forms an **oocyst**, a cyst-like structure within which the ookinete undergoes meiosis. Eventually, the oocyst ruptures, releasing haploid **sporozoites**, the infective stage of the pathogen. The sporozoites travel to the salivary gland of the mosquito and can be transmitted to a human when the mosquito feeds.

Within the human, the sporozoites initially travel to the liver. Here, they undergo mitosis and are released as **merozoites**, diploid forms of the pathogen capable of infecting red blood cells, also known as **erythrocytes**. The merozoites replicate in erythrocytes, eventually lysing the cells, causing the release of more merozoites, which can infect additional red blood cells. Some of these merozoites, however, do not invade other erythrocytes. Rather, they form gametocytes, which can be ingested by a feeding mosquito to continue the process (**Figure 23.16**).

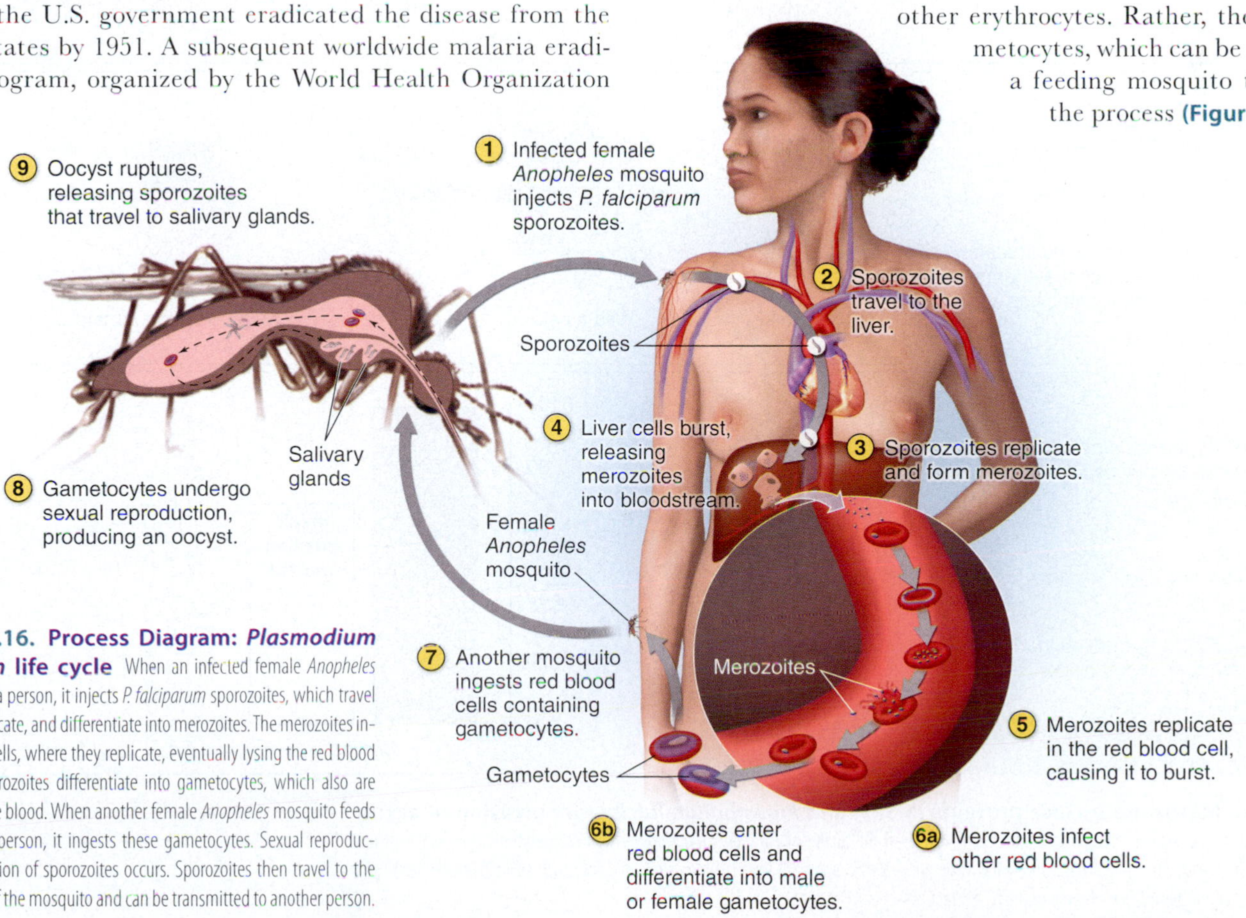

9 Oocyst ruptures, releasing sporozoites that travel to salivary glands.

1 Infected female *Anopheles* mosquito injects *P. falciparum* sporozoites.

2 Sporozoites travel to the liver.

Sporozoites

Salivary glands

8 Gametocytes undergo sexual reproduction, producing an oocyst.

4 Liver cells burst, releasing merozoites into bloodstream.

3 Sporozoites replicate and form merozoites.

Female *Anopheles* mosquito

Merozoites

Figure 23.16. Process Diagram: *Plasmodium falciparum* life cycle When an infected female *Anopheles* mosquito bites a person, it injects *P. falciparum* sporozoites, which travel to the liver, replicate, and differentiate into merozoites. The merozoites infect red blood cells, where they replicate, eventually lysing the red blood cells. Some merozoites differentiate into gametocytes, which also are released into the blood. When another female *Anopheles* mosquito feeds on an infected person, it ingests these gametocytes. Sexual reproduction and formation of sporozoites occurs. Sporozoites then travel to the salivary gland of the mosquito and can be transmitted to another person.

7 Another mosquito ingests red blood cells containing gametocytes.

Gametocytes

5 Merozoites replicate in the red blood cell, causing it to burst.

6b Merozoites enter red blood cells and differentiate into male or female gametocytes.

6a Merozoites infect other red blood cells.

Effects of *Plasmodium falciparum* on humans

Most of the clinical manifestations of malaria in humans occur during this erythrocytic stage, where the merozoites replicate asexually. When these infected erythrocytes lyse, the release of merozoites and various toxic materials present within the infected human triggers a strong immune response. The periodic bouts of fever often associated with malaria result from periodic waves of erythrocyte lysis and merozoite release. Eventually, the repeated destruction of erythrocytes also leads to **anemia**, a decrease in the normal number of erythrocytes. As the number of erythrocytes declines, organs no longer receive enough oxygen.

As we have mentioned in this and other chapters, all microbial pathogens—bacterial, viral, and eukaryal—need to enter appropriate hosts and obtain nutrients. *Plasmodium falciparum* is no different. From its life cycle, it is quite clear how it enters its appropriate hosts. Within each host, the parasite must rely on specific adhesion factors to attach to and invade specific cells. Researchers have begun to understand the mechanism of *Plasmodium* attachment to red blood cells. The surfaces of merozoites,

the blood-borne form of the parasite, contain a series of merozoite surface proteins (MSPs), the most abundant of which are MSP1, MSP6, and MSP7. In *in vitro* studies, researchers have shown that antibodies to MSP7 significantly reduce the ability of merozoites to invade erythrocytes, suggesting that MSP7 is at least partially needed for merozoite invasion. To further these studies, microbiologists examined this issue from a genetic perspective. Using molecular techniques, they disrupted the *msp7* gene in a strain of *P. falciparum*. This strain, they subsequently demonstrated, invaded erythrocytes less well than the wild-type strain. MSP7, it appears, facilitates invasion of *P. falciparum* to erythrocytes **(Figure 23.17)**.

Following attachment, *P. falciparum* enters the erythrocyte and subsequently resides within a specialized vacuole referred to as the parasitophorous vacuole membrane. Within this vacuole, it ingests and then digests host hemoglobin. Hemoglobin (Hb) represents the major protein within red blood cells. Responsible for carrying oxygen, Hb constitutes up to 95 percent of the total protein within a red blood cell. To access this material, *P. falciparum* takes up host cell cytoplasm through **cytostomes**, specialized double-membrane invaginations of

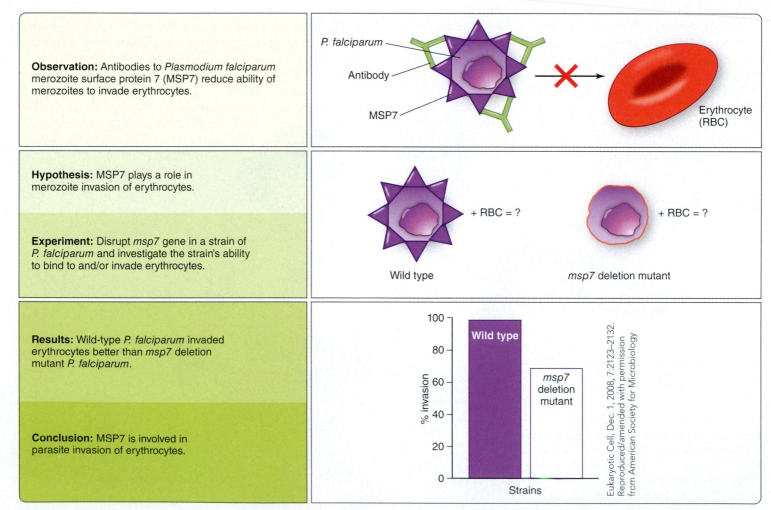

Observation: Antibodies to *Plasmodium falciparum* merozoite surface protein 7 (MSP7) reduce ability of merozoites to invade erythrocytes.

Hypothesis: MSP7 plays a role in merozoite invasion of erythrocytes.

Experiment: Disrupt *msp7* gene in a strain of *P. falciparum* and investigate the strain's ability to bind to and/or invade erythrocytes.

Results: Wild-type *P. falciparum* invaded erythrocytes better than *msp7* deletion mutant *P. falciparum*.

Conclusion: MSP7 is involved in parasite invasion of erythrocytes.

Figure 23.17. Merozoite surface proteins (MSPs) and *Plasmodium falciparum* invasion of erythrocytes Researchers observed that antibodies to MSP7 decreased the ability of *Plasmodium falciparum* to invade erythrocytes. Based on this observation, they hypothesized that MSP7 was involved in invasion. To test this hypothesis, they created a strain of *P. falciparum* that did not express MSP7. This strain invaded erythrocytes less well than the wild-type strain.

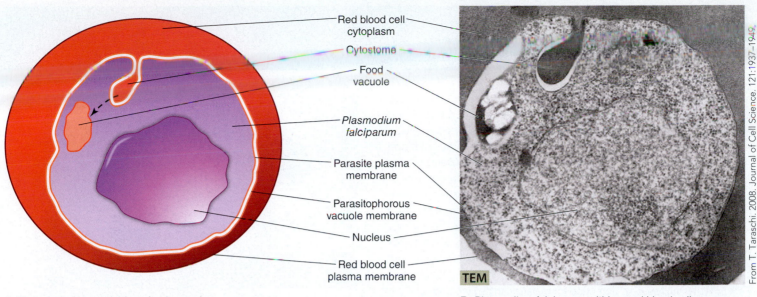

A. Transport of hemoglobin to food vacuole

B. *Plasmodium falciparum* within a red blood cell

From T. Taraschi. 2008. Journal of Cell Science. 121:1937–1949. Fig. 7. Reprinted with permission of The Company of Biologists.

Figure 23.18. Acquisition of hemoglobin by *Plasmodium falciparum* The malaria parasite acquires nutrients by ingesting hemoglobin from the host cell. *Plasmodium falciparum* takes up red blood cell cytoplasm and hemoglobin through cytostomes, double-membraned invaginations of the parasitophorous vacuole membrane and parasite plasma membrane. Individual vesicles fuse into a larger food vacuole. Proteases within the vacuole digest the hemoglobin. Toxic heme then is polymerized into hemozoin.

the vacuole and *P. falciparum* membrane, typically referred to as the parasite plasma membrane **(Figure 23.18)**. The material then is transported to a food vacuole. Within this vacuole, a series of proteases digest the hemoglobin first into peptides and then into amino acids. The amino acids leave the vacuole and enter the parasite cytoplasm, where they can be used during translation.

The digestion of hemoglobin results in the release of free **heme**, a cytotoxic component of Hb. In mammalian cells, several pathways exist to detoxify this material. *Plasmodium* possesses its own mechanism of detoxifying heme. Through a poorly understood process, the parasite polymerizes heme monomers, forming a less toxic, insoluble polymer known as **hemozoin**, or malaria pigment. Thus, not only does the parasite possess a mechanism for obtaining nutrients, but also it possesses a mechanism for ridding itself of a toxic by-product.

Strategies for prevention and treatment of *Plasmodium falciparum* infections

Before we leave this discussion of *Plasmodium falciparum* and malaria, let's ask one more important question. How can we prevent human infections? Our detailed understanding of the *Plasmodium falciparum* life cycle may help us answer this question. By understanding how a pathogen enters a host, binds to cells within the host, and obtains nutrients, we can better protect ourselves from the pathogen.

Preventing interactions between humans and mosquitoes, quite obviously, prevents the spread of malaria. We can utilize various methods to block this interaction. Indeed, the use of insecticides and the elimination of mosquito breeding grounds led to the virtual eradication of malaria from the United States in the mid-twentieth century. Prior to the early 1900s, malaria

was quite common in the southeastern United States. Increased urbanization and the use of screens in homes decreased contact between people and mosquitoes. Additionally, the federal government initiated programs to rid areas of stagnant water—ideal mosquito breeding grounds—and spray insecticides throughout areas with high levels of mosquitoes, primarily in the South. Today's Centers for Disease Control and Prevention (CDC), in fact, began as the Communicable Disease Center, an agency devoted largely to the control of malaria in the United States. The results were remarkable. According to the CDC, 15,000 cases of malaria were reported in 1947. That number dropped to 2,000 reported cases in 1950, and the elimination of malaria from the United States by 1951. Now, some cases of malaria still are reported in the United States every year. Most of these cases, however, occur in immigrants or Americans who have traveled to malaria endemic regions of the world.

Despite the great success of antimalaria campaigns in the United States, the global eradication of malaria has been difficult. In the 1950s, insecticides were used extensively to eliminate the mosquito vector. Resource limitations, political instabilities, and the evolution of mosquitoes resistant to the insecticides all interfered with these efforts. More recently, many groups have advocated the distribution of insecticide-treated mosquito nets to prevent the transmission of *P. falciparum*. The World Health Organization has recommended that long-lasting insecticidal sleeping nets should be distributed freely (or at a low cost) and used by all people in areas where malaria is prevalent. Critics, however, argue that nets will not be used by all people and will not end transmission of the pathogen.

In addition to preventing interactions between mosquitoes and humans, we also can develop better therapies for malaria. Our understanding of *Plasmodium* replication may lead to the development of effective antimalarial drugs. **Chloroquine**,

The current "gold standard" for malaria diagnosis involves the microscopic examination of finger-prick blood smears. A thick blood smear is done first to examine as many red blood cells as possible on a slide. Typically, a health care worker spreads three drops of blood over a large area of a slide, producing a layer approximately ten cells thick. The smear is allowed to air-dry, stained with Romanowsky or Giemsa stain, and then examined for intracellular malarial parasites. Clinicians view approximately 200 visual fields before concluding that a sample is negative. If a sample appears positive, then a similarly stained thin blood smear consisting of a single layer of cells is examined to determine the number and species of *Plasmodium* parasites **(Figure B23.6)**.

While microscopic detection for malaria can be very accurate, the microscopist must be skilled in staining and identifying the developmental forms of the malaria parasite. Skill levels may vary greatly among technicians. Moreover, microscopy is time-consuming and requires a precision microscope capable of 1,000X magnification. Microscopic diagnosis, therefore, is restricted to designated facilities. Rapid tests, commonly known as dipstick tests, have been developed as an alternative to microscopy-based tests. Originally developed for personal use by travelers to malaria-endemic areas, these rapid tests can be performed within minutes anywhere and do not require special equipment or training to perform or interpret.

Rapid dipstick tests employ enzyme-linked immunosorbent assay (ELISA) technology (see Toolbox 20.2), and are best known for their use in home pregnancy tests. Dipstick technology has been adapted for the detection of a variety of antigens, including the products of infectious agents, and specific antibodies produced as a result of infection with certain pathogens. Malaria rapid tests use monoclonal

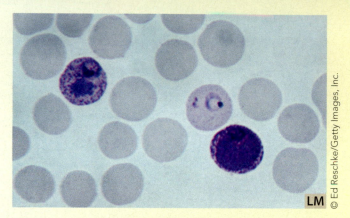

Figure B23.6. Thin blood smear showing malarial parasite
Plasmodium vivax in red blood cells are stained purple.

© Ed Reschke/Getty Images, Inc.

antibodies (see Toolbox 20.1) for the detection of *Plasmodium* antigens from a blood sample. Antigen targets include the secreted histidine-rich protein II of *P. falciparum* and the lactate dehydrogenase enzyme of *Plasmodium* species. Monoclonal antibodies developed to lactate dehydrogenase can distinguish amino acid sequence differences in the enzyme of *P. falciparum* from that other less virulent species of *Plasmodium*. A positive test is indicated by the appearance of a colored band on the supporting strip **(Figure B23.7)**.

Overall, rapid tests are not as sensitive as microscopy. They have a higher rate of false negative results; some individuals actually infected with *P. falciparum* nonetheless receive a negative test result. Because of this occasional inaccuracy, the rapid tests are not

TABLE B23.1	Comparison of tests used for malaria diagnosis		
Test	**Limit of detection (approximate parasite density/μL blood)**	**Ability to discriminate species**	**Accessibility**
Thick blood smear	50	Fair	Limited
Thin blood smear	>100	Good	Limited
Dipstick test	>100	Limited to certain species	Good
PCR	5	Good	Poor

Source: Adapted from Public Health Agency of Canada (2009).

one of the most widely used antimalarial drugs, blocks the formation of hemozoin within *Plasmodium* cells, thereby leading to the accumulation of toxic heme. The increasing prevalence of chloroquine-resistant strains of *Plasmodium*, however, necessitates the need for additional antimalarial drugs. In the future, a better understanding of the role of merozoite surface proteins in

Plasmodium attachment to erythrocytes may lead to the development of a drug that interferes with this part of the parasite life cycle. Additionally, better tests for *P. falciparum* infection are needed so that these drugs can be supplied to infected individuals in a timely fashion **(Toolbox 23.1)**. Basic research today in both of these important areas may lead to the end of this scourge tomorrow.

A. Strip layout before testing

Positive control band
Anti-mouse Ab
fixed to strip

Enzyme substrate
fixed to strip

Test band
Mouse mAb2
fixed to strip

Sample application site
Mouse mAb1 conjugated
to enzyme,
unfixed to strip

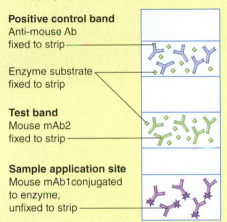

Figure B23.7. Rapid dipstick test for *Plasmodium falciparum* infections Using ELISA technology, researchers have developed a rapid test for *Plasmodium falciparum*. **A.** The rapid test strip contains two mouse monoclonal antibodies (mAb1 and mAb2) that bind a specific *P. falciparum* protein. The antibody mAb1 is covalently bound, or conjugated, to an enzyme. The strip also contains an anti-mouse polyclonal antibody that will bind mAb1 and mAb2. **B.** When a drop of blood containing *P. falciparum* is added to the sample application site, *P. falciparum* protein is bound by some mAb1, which travels up the strip via capillary action. When mAb2 binds the mAb1:protein complex, a colored band results. Unbound mAb1 continues up the strip, interacts with the anti-mouse Ab, and a second colored band results. **C.** A single band indicates a negative result; all mAb1 flows past the test band to the control band.

B. Positive test for *P. falciparum*

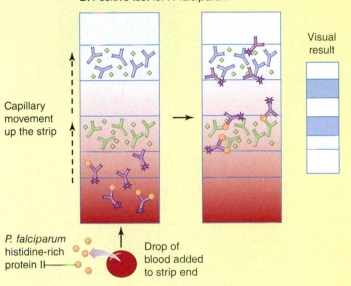

Capillary
movement
up the strip

P. falciparum
histidine-rich
protein II

Drop of
blood added
to strip end

Visual
result

C. Negative test for *P. falciparum*

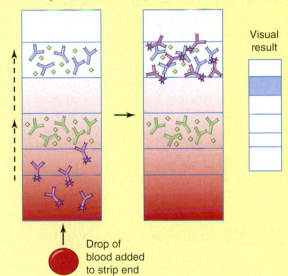

Capillary
movement
up the strip

Drop of
blood added
to strip end

Visual
result

considered acceptable replacements to microscopic examination for malaria diagnosis. Testing by the World Health Organization in 2009, however, showed that in patients with relatively high loads of malarial parasites, the sensitivity of some commercially available rapid tests rivals that of microscopy. Alternatively, PCR tests for amplification of genus and species-specific *Plasmodium* genes have been developed. While extremely sensitive, PCR is very expensive and not routinely used for diagnosis. **Table B23.1** shows a comparison of various tests used.

● Test Your Understanding ....

1. In malaria-endemic areas, why would testing all individuals for the presence of antibodies to *P. falciparum* not be particularly informative?
2. In the rapid test described in Figure B23.7, why is it important to have the positive control band placed last in the sequence?

23.2 Fact Check

1. How does *Plasmodium falciparum* enter red blood cells?
2. How does *Plasmodium falciparum* acquire nutrients as it replicates in red blood cells?
3. How does the antimalarial drug chloroquine affect the replication of *P. falciparum*?

23.3 Macroscopic eukaryal pathogens

What do we know about macroscopic eukaryal pathogens?

Parasitic **helminths**, or worms, technically should not be part of our discussion. They are, after all, macroscopic. These pathogens, however, generally are transmitted via their eggs or larvae, which are microscopic. As we shall see during this brief overview, these macroscopic parasites also share many of the attributes exhibited by their microscopic counterparts. These macroscopic pathogens need to achieve certain goals: entering a host, attaching to specific cells, and obtaining nutrients. In the process of meeting these goals, they cause disease. For these reasons, we will round out our survey of eukaryal pathogens by investigating this interesting group of disease-causing organisms. Specifically, we will examine two human diseases caused by parasitic worms: ascariasis and schistosomiasis.

Ascariasis

Throughout the world, parasitic worms cause a number of human diseases **(Table 23.5)**. The roundworm *Ascaris lumbricoides* **(Microbes in Focus 23.5)** and other species in the *Ascaris* genus infect over 1.5 billion people worldwide. The worms typically mature within the small intestine, adversely affecting the body's ability to adsorb nutrients. Often, the infections result in no acute signs of disease. Rather, **ascariasis**, or infection with an *Ascaris* species, may result in slower growth and decreased weight gain in children. In more severe cases, the worms may lead to a complete blockage of the small intestine.

When living in the small intestine, mature female worms may grow up to 12 inches in length, while the male worms are smaller. The females release eggs that are eliminated from the

Microbes in Focus 23.5

***ASCARIS LUMBRICOIDES*: A ROUNDWORM PARASITE OF HUMANS**

Habitat: Humans. Found in tropical and subtropical regions throughout the world, primarily in areas of poor sanitation.

Description: Largest roundworm parasite of human intestines, with females measuring over 12 inches in length. Transmission occurs when eggs are excreted by an infected person and then later ingested by another human.

Key Features: Mature females may release over 200,000 eggs per day. Humans and *A. lumbricoides* have had a long history—fossilized *Ascaris* eggs have been isolated from caves in France inhabited by humans 30,000 years ago.

© Eye of Science/Photo Researchers, Inc.

host in the feces. If another person ingests these eggs, then transmission of the worms occurs. Within the host, the eggs hatch in the small intestine. Immature worms penetrate the blood vessels of the small intestine, travel to the lungs and throat, are swallowed, and then return to the small intestine, where they mature and mate **(Figure 23.19)**. The adult worms apparently do not possess any specialized means of attaching to the lining of the small intestine. Within this location, however, they can obtain food sources that pass through the human digestive tract.

Often, individuals with minor ascariasis infections exhibit no symptoms or minor gastrointestinal symptoms. People may first become aware of their infection when they observe a worm in their feces. Clinical diagnosis can be made by a microscopic analysis of one's feces. Treatment typically involves taking an antiparasitic drug for a few days.

Schistosomiasis

At least three species of *Schistosoma*—*Schistosoma mansoni*, *Schistosoma haematobium*, and *Schistosoma japonicum*—infect humans, causing **schistosomiasis** in over 200 million people worldwide. Infections occur throughout the world, from Africa to the Caribbean to China . In children, this disease can result in malnutrition and developmental problems. Long-term infections can result in damage to multiple organs, including the liver, intestines, and bladder.

The complex life cycle of *Schistosoma* species differs significantly from that of *Ascaris* species. Most notably, these organisms spend part of their life cycle in freshwater snails, the

TABLE 23.5 Representative parasitic worms of humans

Worm	Location(s)	Disease/Symptoms	Prevalence (people infected)
Roundworm	Tropical and subtropical regions	Ascariasis, intestine blockage	1.5 billion
Whipworm	Tropical and subtropical regions	Diarrhea, anemia	800 million
Hookworm	Africa, Latin America, Southeast Asia, China	Abdominal pain, anemia	740 million
Pinworm	Temperate regions throughout the world	Itching	200 million
Schistosomes	Africa, China, Caribbean	Schistosomiasis, organ damage	200 million
Lymphatic filariae	Tropical and subtropical regions	Elephantiasis	120 million

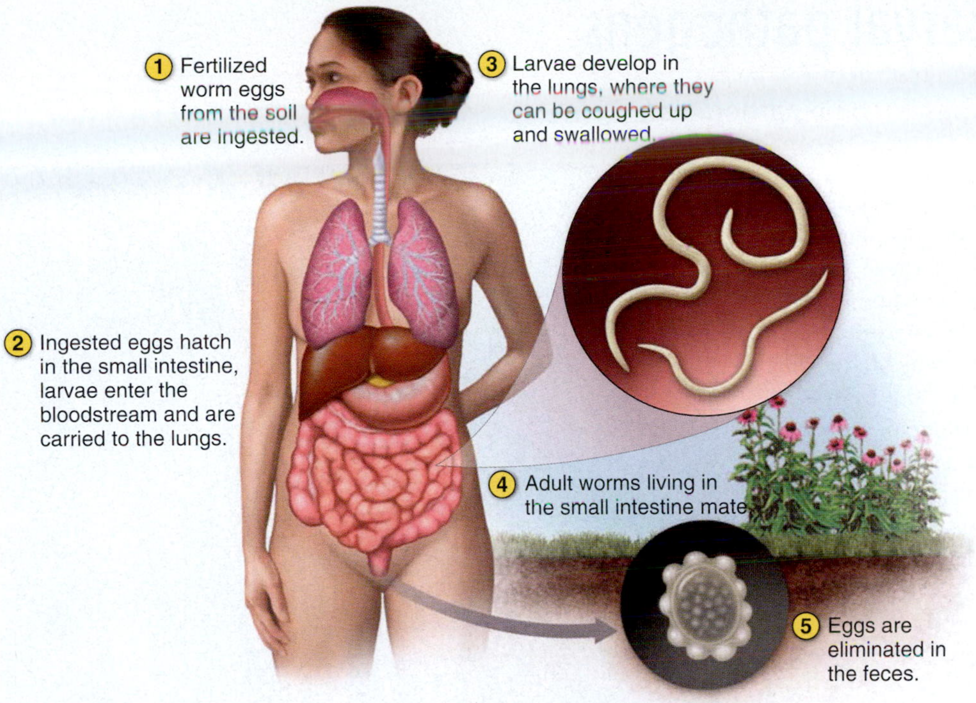

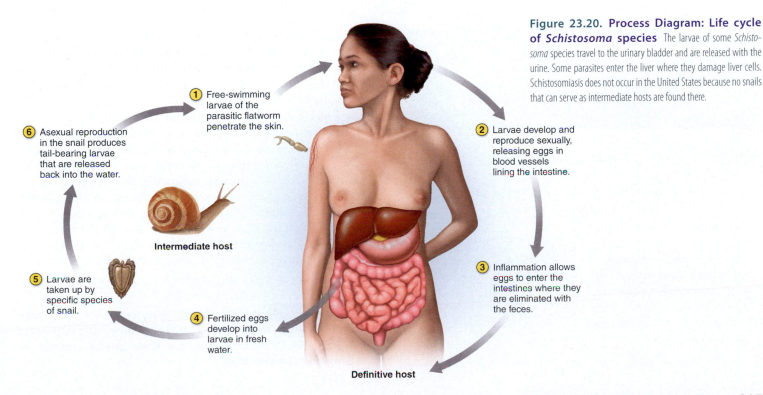

Figure 23.19. Process Diagram: Life cycle of *Ascaris lumbricoides* Female worms may reach up to 12 inches in length. Transmission occurs when eggs released by an infected person are ingested by another person via contaminated hands or food.

① Fertilized worm eggs from the soil are ingested.

② Ingested eggs hatch in the small intestine, larvae enter the bloodstream and are carried to the lungs.

③ Larvae develop in the lungs, where they can be coughed up and swallowed.

④ Adult worms living in the small intestine mate.

⑤ Eggs are eliminated in the feces.

intermediate host. Eggs, excreted in the feces or urine of an infected person, the definitive host, hatch in fresh water and the organisms then enter specific snail hosts. Within the snails, the organisms replicate asexually and differentiate, eventually being released again into the water. This form of the organism can penetrate the skin of people who are swimming or bathing in the contaminated water. Once inside a person, the organisms migrate to the blood system, where they mature and mate. Eggs move toward the intestines or bladder, where they are eliminated **(Figure 23.20).**

23.3 Fact Check

1. List some human diseases caused by worms.
2. Describe the life cycle of *Ascaris lumbricoides*.
3. Describe the life cycle of *Schistosoma* species.

Figure 23.20. Process Diagram: Life cycle of *Schistosoma* **species** The larvae of some *Schistosoma* species travel to the urinary bladder and are released with the urine. Some parasites enter the liver where they damage liver cells. Schistosomiasis does not occur in the United States because no snails that can serve as intermediate hosts are found there.

① Free-swimming larvae of the parasitic flatworm penetrate the skin.

② Larvae develop and reproduce sexually, releasing eggs in blood vessels lining the intestine.

③ Inflammation allows eggs to enter the intestines where they are eliminated with the feces.

④ Fertilized eggs develop into larvae in fresh water.

⑤ Larvae are taken up by specific species of snail.

⑥ Asexual reproduction in the snail produces tail-bearing larvae that are released back into the water.

Intermediate host

Definitive host

23.4 Evolution of eukaryal pathogens

How do some eukaryal microbes become pathogens?

As we have seen throughout this chapter, pathogenic eukaryal microbes possess specific mechanisms to infect their hosts. For instance, specific genes encoding adhesion factors, enzymes, or toxins ultimately determine the virulence of these species. Among bacteria and viruses, these virulence genes can be transferred between species through various types of horizontal gene transfer. The acquisition of a virulence factor gene, then, may lead to the emergence of a new pathogen. This same process, it appears, can lead to the emergence of pathogenic eukaryal microbes. The story, however, does not end there. The host organisms also may undergo genetic changes in response to virulent pathogens. Pathogens, in other words, may affect the evolution of their hosts. In this section, we will examine both of these events.

Acquisition of virulence genes by eukaryal microbes

In 2006, researchers presented strong evidence that horizontal gene transfer led to the increased virulence of *Pyrenophora tritici-repentis* (**Microbes in Focus 23.6**), a fungus that, prior to 1941, exhibited only moderate virulence. In 1941, however, a new severe disease of wheat—tan (or yellow) spot—was described, and *P. tritici-repentis* was determined to be the causative agent (**Figure 23.21**).

How did this fungus quickly evolve into a virulent pathogen? Apparently, the change occurred when *P. tritici-repentis* acquired the gene encoding ToxA, a toxin. Researchers have shown that

© Nigel Cattlin/Alamy Limited

Figure 23.21. Tan (or yellow) spot of wheat Caused by the fungus *Pyrenophora tritici-repentis*, tan (or yellow) spot is characterized by yellow spots on the leaves of wheat. The disease results in significant crop loss.

strains of *P. tritici-repentis* that lack the *ToxA* gene are avirulent. They become virulent if the toxin gene is inserted into their genomes, clearly demonstrating the central role of this toxin in fungal virulence. Another question, though, still remains. How did strains of *P. tritici-repentis* acquire the *ToxA* gene in 1941? *Staganospora nodorum*, another fungal pathogen of wheat, also contains the *ToxA* gene, but the gene has not been identified in other fungal species. Furthermore, *ToxA* genes in *S. nodorum* isolates exhibit more genetic diversity than do *ToxA* genes in various *P. tritici-repentis* isolates, suggesting that the gene has entered the latter species more recently. Based on these pieces of evidence, researchers hypothesize that *ToxA* was acquired by *P. tritici-repentis* from *S. nodorum* sometime around 1940, probably through horizontal gene transfer. The virulence of other emerging fungal pathogens remains less clear (**Perspective 23.3**).

CONNECTIONS Most, if not all, living organisms and viruses possess means of sharing genetic material. Viruses with segmented genomes, like influenza viruses, can share genetic material via reassortment (**Section 22.4**).

Microbes in Focus 23.6
PYRENOPHORA TRITICI-REPENTIS: THE CAUSE OF TAN SPOT IN WHEAT

Habitat: Mainly infects wheat, but also may infect rye and other related grasses.

Description: Fungus that causes brown or yellow spots to form on wheat leaves as a result of necrosis, or localized cell death.

Key Features: A form of the fungus overwinters on wheat stems of other organic matter. Spores are released in the spring and summer. These spores then infect new plants. During a severe outbreak, crop losses may approach 50 percent.

LM

Courtesy Stephen Wegulo, University of Nebraska–Lincoln

Since the 1980s, scientists have reported major declines in amphibian populations throughout the world. The declines among frogs have been most evident in Australia and the Western Hemisphere. Today, researchers estimate that over 30 percent of amphibian species are threatened. The cause of these massive declines has been a matter of lively debate. Many factors, including habitat destruction, pollution, and global climate change, have been postulated as causes. Some investigators, for instance, argue that amphibians are extremely sensitive to variations in moisture and temperature. Slight changes in the climate, therefore, may affect them greatly. Today, most researchers agree that an emerging microbial pathogen, *Batrachochytrium dendrobatidis* (*Bd*), a fungus in the chytrid family, represents the greatest threat to frogs.

First isolated and characterized in 1998, *B. dendrobatidis* now has been identified in frog populations throughout the world. Retrospective studies indicate that it has existed in the United States and Australia since the 1970s. It probably originated, though, much earlier in Africa. The fungus has been identified in museum specimens of *Xenopus laevis*, the African clawed frog, from the 1930s. Since the 1930s, *X. laevis* has been exported to pet stores and research centers throughout the world. Most likely, this trade in the African clawed frog has led to the current global distribution of the pathogen **(Figure B23.8)**.

The pathogenesis of *B. dendrobatidis* remains somewhat unclear. The chytrid fungus spreads through the water and causes lesions on the skin of infected frogs. Most likely, these lesions adversely affect fluid balance and respiration, leading to death. What remains unclear, though, is why the amphibian declines seem to have begun relatively recently. We can formulate at least two reasonable hypotheses. First, the fungus may have obtained a new virulence gene. The fungus, in other words, has evolved. Second, the frogs may be more sensitive to the pathogenic effects of the fungus because of recent increases in pollution or recent changes in the global climate. The fungus itself has not changed, but its host has become more sensitive to its effects. More research is needed to answer this important question.

While the frog declines have been dramatic and, apparently, unrelenting, there are some glimmers of hope. Various species of bacteria produce antifungal compounds. To determine if any bacteria produce compounds toxic to *B. dendrobatidis*, microbiologists have screened bacteria for their ability to limit the growth of the fungus. In 2008, researchers reported that *Janthinobacterium lividum*, a bacterium that normally lives on the skin of certain amphibians, offered some protection to frogs exposed to the chytrid fungus. Perhaps further research into this bacterium will help us stem the loss of frogs.

A. Reported distribution of *Batrachochytrium dendrobatidis*

Figure B23.8. Spread of *Batrachochytrium dendrobatidis* (*Bd*) During the latter half of the twentieth century, *Batrachochytrium dendrobatidis* has spread throughout the world, affecting amphibian populations in many locations. **A.** Today, *B. dendrobatidis*, the cause of chytridiomycosis, has been identified worldwide. Red areas indicate reports of *B. dendrobatidis* associated with amphibian mortalities. Yellow areas indicate reports of *B. dendrobatidis* not associated with mortalities. **B.** *Xenopus laevis*, or the African clawed frog. Retrospective studies have shown evidence of *B. dendrobatidis* in *X. laevis* as early as 1938. Worldwide distribution of the fungus appears to have begun in the 1960s.

B. *Xenopus laevis*

Effects of pathogenic eukaryal microbes on host evolution

Before leaving this topic of evolution, let's investigate a somewhat different question. How have eukaryal pathogens affected the evolution of their hosts? To explore this question, we will focus on two specific examples: the evolution of humans in response to *Plasmodium falciparum*, the causative agent of malaria, and the evolution of fruit flies in response to α-amanitin, the toxin produced by certain fungi.

The human response to *P. falciparum* and malaria represents a classic case of microbial pathogens affecting host evolution. As we noted previously, *P. falciparum* is transmitted from mosquitoes to humans. In humans, the infection ultimately results in the destruction of red blood cells, causing severe anemia. Some people, though, exhibit a partial resistance to the effects of this microbial pathogen. Their resistance to malaria, however, is tied to another human malady—sickle-cell disease.

Sickle-cell disease results from a mutation in the human gene that codes for the β-globin polypeptide, one of the constituents of hemoglobin, the oxygen-carrying protein within red blood cells. In individuals who are homozygous for the mutant allele of this gene, Hb^S_β, the hemoglobin protein is misshapen. It binds oxygen less well and the red blood cells themselves assume a sickled appearance under certain conditions, rather than remaining spherical **(Figure 23.22)**. This sickled shape, in turn, results in the destruction of these cells, causing anemia. Additionally, the misshapen red blood cells tend to become clogged

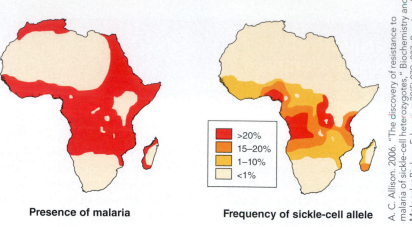

A. C. Allison. 2006. "The discovery of resistance to malaria of sickle-cell heterozygotes," Biochemistry and Molecular Biology Education 30(5):279–287. Reprinted with permission of John Wiley & Sons, Inc.

Presence of malaria

Frequency of sickle-cell allele

>20%
15–20%
1–10%
<1%

Figure 23.23. Geographic distribution of malaria and sickle-cell disease As shown in these maps, the prevalence of the mutant sickle-cell allele largely overlaps with the prevalence of malaria in Africa. Sickle-cell disease is very detrimental, and there should be a strong selective pressure against the mutant allele. Heterozygotes, however, exhibit resistance to malaria. This resistance provides a selective pressure maintaining the mutant β-globin gene within populations where the malarial parasite is prevalent.

in capillaries, resulting in an increased risk of heart problems and stroke.

Because of the severity of this genetic disease, one might assume that the prevalence of the mutant allele within the human population would be quite low. After all, if sickle-cell disease has a significant negative impact on the fitness of people, then it is reasonable to assume that there would be a selective pressure against the allele. Individuals who have sickle-cell disease would be less likely to survive until adulthood and less likely to reproduce. The Hb^S_β allele, then, would be less likely to be passed on to future generations. Yet, the mutant allele and sickle-cell disease are quite prevalent, especially among people of African and Mediterranean descent. Among Americans of African ancestry, for example, approximately one in every 500 individuals is homozygous for the mutant allele and has sickle-cell disease. Additionally, roughly one in twelve African Americans carries the mutant allele. These individuals are heterozygous, possessing one normal, or wild-type copy (Hb^A_β), and one mutant copy (Hb^S_β) of the β-globin gene. These individuals do not have sickle-cell disease. Under typical conditions, their red blood cells function normally. Their red blood cells may sickle, though, under conditions of oxygen stress, such as during exercise at high altitudes. More importantly, these heterozygous individuals have a 50 percent chance of passing the mutant allele on to their offspring. Given the significantly detrimental effects of sickle-cell disease, it is reasonable to wonder why this genetic disease remains so common. It also is reasonable to wonder why it is so much more common among certain populations. Malaria provides the answer to both of these questions. The prevalence of malaria coincides with the prevalence of sickle-cell disease **(Figure 23.23)**. While sickle-cell disease is highly detrimental in all circumstances, sickle-cell trait provides partial protection against malaria. In heterozygous individuals (Hb^A_β/Hb^S_β), *P. falciparum*-infected red blood cells tend to sickle much more readily than uninfected cells. The infected cells, then, are

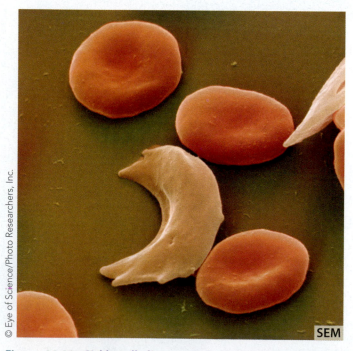

SEM

Figure 23.22. Sickle-cell disease and misshapen red blood cells Wild-type red blood cells exhibit a characteristic biconcave disk shape. Red blood cells from a person with sickle-cell disease have a distorted, sickled shape.

destroyed. As a result of this rapid cellular destruction, the pathogen cannot replicate and spread to other cells. The malarial disease, as a result, is less severe. In other words, Hb^A_β/Hb^S_β heterozygotes possess a selective advantage over Hb^A_β/Hb^A_β and Hb^S_β/Hb^S_β homozygous individuals in areas where malaria is common. Clearly, the evolution of humans has been affected by *P. falciparum*.

CONNECTIONS Throughout our evolutionary history, pathogens certainly have shaped our genome in many ways. Biologists postulate that the CCR5 deletion that provides humans with some protection from HIV infection (**Section 22.4**) may have offered protection to humans from smallpox infections hundreds of years ago.

To investigate another example of how eukaryal pathogens can affect host evolution, let's examine fruit flies. The larvae of several species of fruit flies, most notably *Drosophila falleni* and *Drosophila recens*, exhibit a high degree of resistance to α-amanitin, the fungal toxin we first discussed in Section 23.1. One might assume that this increased resistance to the toxin allows the fruit fly larvae to utilize mushrooms as a food source. Relatively few species of mushrooms, though, produce α-amanitin. Also, many mycophagous, or fungi-feeding, species of fruit flies have been identified that do not exhibit resistance to this toxin. Resistance to the toxin, then, does not correlate in general with mycophagy. So, why would some species of fruit flies evolve a toxin resistance? Perhaps, the ability to feed on fungi that produce α-amanitin provides the fruit flies with a food source that is underutilized. If the larvae of other species cannot feed on these toxic mushrooms, then *D. falleni* and *D. recens* will face less competition.

John Jaenike, a biologist at the University of Rochester, has proposed another explanation for the evolution of α-amanitin resistance in these species. He proposes that increased α-amanitin resistance protects these species of fruit flies from parasitic worms, like those we discussed in Section 23.3. Common parasites of fruit flies, nematodes like *Howardula aoronymphium*, are sensitive to α-amanitin. Fruit flies that breed and feed on mushrooms that produce α-amanitin, Jaenike has shown, have a greatly reduced parasite burden. The evolution of α-amanitin resistance within *D. falleni* and *D. recens* may provide the fruit flies with an indirect resistance to these parasites. The toxin production of the fungi, in other words, has affected the evolution of the invertebrates.

23.4 Fact Check

1. What probably led to the increased virulence of *Pyrenophora tritici-repentis*?
2. Describe the inheritance of sickle-cell disease.
3. Why do the geographic distributions of malaria and sickle-cell disease largely overlap?
4. Provide two possible explanations for the α-amanitin resistance exhibited by some fruit fly species.

Image in Action

WileyPLUS This illustration depicts the spread of Dutch elm disease. The tree on the right is infected, with the infection starting on a lower branch, causing the branch to wither and die. The lower zoom-in window reveals fruiting bodies of *Ophiostoma novo-ulmi* on the tree, with spores being released. The tree on the left is a healthy elm tree being attacked by an elm bark beetle (top zoom-in window), the vector that spreads the disease-causing fungus. The beetle burrows through the bark of the elm, leaving behind spores of *Ophiostoma novo-ulmi*.

1. *Ophiostoma novo-ulmi*, like many other phytopathogenic fungi, produces various enzymes to break down host barriers. Imagine a compound produced by a transgenic elm tree that blocked the fungal enzyme, xylanase. What would be the effects, and why?

2. Imagine that a newly introduced species in the United States is found to be a natural predator of the elm bark beetle. How might this new change in the environment affect the fungal pathogen *Ophiostoma novo-ulmi*?

The Rest of the Story

Today, new cases of Dutch elm disease still occur worldwide. Replanting of elms in areas previously devastated by the disease in past years show the disease abounds and newly planted trees still fall victim. In hopes of repopulating lost elm stands, researchers have tried traditional breeding strategies. Susceptible elm tree species have been crossed with more resistant species. Researchers hope that the hybrid offspring will exhibit the increased resistance to Dutch elm disease of the one parent, while still maintaining all of the desirable traits of the other parental tree. These efforts, unfortunately, have resulted in very limited success.

Scientists also are attacking the problem of Dutch elm disease in a new way. Using some of the molecular biology tools that we discussed in earlier chapters, researchers are attempting to make transgenic elm trees that produce specific antifungal compounds. Genes for these antifungal compounds can be inserted into the genome of an elm tree. As the tree develops, the compounds will be produced, thereby providing the tree with resistance to *Ophiostoma novo-ulmi*. As we learn more about the life cycle and pathogenesis of the fungus, we may identify more genes that produce antifungal compounds. By creating transgenic trees, we may someday again see these majestic trees growing throughout our towns, parks, and forests.

Summary

23.1: How do eukaryal microbes cause disease?

Eukaryal microbes can be transmitted between hosts in various ways.

- For eukaryal pathogens like *Ophiostoma novo-ulmi*, insect vectors passively transport **spores** between hosts.
- Some eukaryal pathogens have a **complex life cycle**, in which they undergo sexual reproduction in the **definitive host** and asexual reproduction and differentiation in the **intermediate host**.
- The eukaryal pathogen *Plasmodium falciparum* must replicate and develop within the mosquito to be transmitted.
- The eukaryal pathogen *Giardia lamblia* undergoes a **simple life cycle**. It is transmitted via the ingestion of **cysts**, which then give rise to **trophozoites**.

Upon entering an appropriate host, eukaryal microbes must evade the host defenses.

- Some opportunistic pathogens, like *Pneumocystis jirovecii*, only cause disease in immunocompromised individuals, such as people with HIV disease or people on **immunosuppressive drugs**.
- Other opportunistic pathogens, like *Candida albicans*, cause disease when the normal microbial inhabitants of the host change.
- Some pathogens, like *Trypanosoma brucei* spp., actively subvert the host defenses through **antigenic variation**. These organisms, characterized by a **kinetoplast** and **variable surface glycoproteins (VSGs)**, exhibit cyclic variations in **parasitemia**.

Eukaryal microbes exhibit various means of obtaining nutrients from their hosts and causing disease.

- Some **phytopathogenic fungi** produce enzymes like **xylanase**, while others use **appressoria** to physically penetrate host cells. In some species, **mitogens** promote appressoria development.
- A number of eukaryal microbes produce toxins. *Cochliobolus carbonum*, a fungus of maize, produces HC-toxin, an inhibitor of **histone deactylases (HDACs)**, which disrupts the expression of host defense genes.
- Various species of mushroom produce toxins such as α-**amanitin**, which can lead to **intoxication**.
- Certain **dinoflagellates** associated with **harmful algal blooms (HABs)** produce saxitoxin.

23.2: How does *Plasmodium falciparum* cause malaria?

Plasmodium falciparum exhibits a complex life cycle, replicating in mosquitoes and humans, ultimately leading to the destruction of erythrocytes in humans.

- Within humans, the intermediate host, **gametocytes (gamonts)** are released from infected **erythrocytes** and ingested by feeding mosquitoes.
- Within the female *Anopheles* mosquito, the definitive host, the gametocytes differentiate into gametes, which fuse, forming a diploid **ookinete**. The ookinete forms an **oocyst** and undergoes meiosis, producing haploid **sporozoites**.

- Sporozoites can be transmitted to a human when an infected mosquito bites.
- In the human, these sporozoites initially infect liver cells, where they replicate, releasing diploid **merozoites**, which then infect erythrocytes.
  In an infected human, *P. falciparum* replication leads to malaria.
- To facilitate attachment to red blood cells, the merozoites express a series of merozoite surface proteins (MSPs).
- Once inside an erythrocyte, the merozoites obtain hemoglobin from the host cell through **cytostomes**. Digestion of hemoglobin by the pathogen releases **heme**, which *P. falciparum* then converts to **hemozoin**.
- Replication of merozoites within erythrocytes leads to the destruction of these cells, resulting in **anemia**.

While malaria remains a huge global problem, methods of preventing malaria and treating it do exist.

- Insecticide-treated mosquito sleeping nets can prevent the transmission of *P. falciparum*.
- The antimalarial drug **chloroquine** blocks the formation of hemozoin.

23.3: What do we know about macroscopic eukaryal pathogens?

Macroscopic eukaryal pathogens, like microbial pathogens, obtain resources from their hosts and, in the process, cause disease.

- Parasitic worms, or **helminths**, cause a number of severe human diseases throughout the world.
- The parasitic worm *Ascaris lumbricoides* is transmitted from human to human via contaminated water and causes **ascariasis**.
- Other parasitic worms, like *Schistosoma* spp., are transmitted via an intermediate host—freshwater snails—and cause **schistosomiasis**.

23.4: How do some eukaryal microbes become pathogens?

Some eukaryal microbes become more virulent by acquiring virulence genes.

- Eukaryal pathogens may acquire virulence genes through horizontal gene transfer.
- *Pyrenophora tritici-repentis* became more pathogenic after acquiring the gene for ToxA.
  Eukaryal pathogens may affect the evolution of their hosts.
- The prevalence of **sickle-cell disease** is due, in large part, to the resistance to malaria conferred by the mutant β-globin allele associated with sickle-cell disease.
- Certain mycophagous species of fruit flies have evolved resistance to α-amanitin.

Application Questions

1. Define "opportunistic pathogen."
2. Lactobacilli are common inhabitants of the vaginas of women. How does the elimination of these bacteria contribute to disease?
3. Would you describe elm bark beetles as hosts of *Ophiostoma novo-ulmi*? Explain.
4. Would you describe mosquitoes as hosts of *Plasmodium falciparum*? Explain.
5. Compare and contrast the probable role of most fungal toxins with the role of most bacterial toxins.
6. Are mushrooms true pathogens? Explain.
7. Certain phytopathogenic fungi penetrate plant cell walls through the use of appressoria. Describe this process.
8. Some phytopathogenic fungi do not form appressoria. How might these fungi penetrate plant cell walls?
9. Provide descriptions of two different mechanisms utilized by *P. falciparum* to obtain nutrients from host cells.
10. Explain two different ways by which a non-pathogenic fungus could become pathogenic.
11. Describe how *P. falciparum* has affected human evolution.
12. To evade the host immune response, *Trypanosoma brucei* varies its surface glycoprotein, thereby presenting the host with a changing antigenic face. Describe how *T. brucei* could regulate the expression of its surface glycoprotein genes. How could you test your explanation?
13. We mentioned that *P. falciparum* merozoites appear to utilize merozoite surface proteins (MSPs) to attach to human erythrocytes. One could postulate that these MSPs interact with a specific protein present on the surface of the erythrocytes. Describe an experiment that could be used to identify this cellular receptor.

Suggested Reading

Edman, J. C., J. A. Kovacs, H. Masur, D. V. Santi, H. J. Elwood, and M. L. Sogin. 1988. Ribosomal RNA sequence shows *Pneumocystis carinii* to be a member of the Fungi. Nature 334:519–522.

El Bissati, K., R. Zufferey, W. H. Witola, N. S. Carter, B. Ullman, and C. Ben Mamoun. 2006. The plasma membrane permease PfNT1 is essential for purine salvage in the human malaria parasite *Plasmodium falciparum*. Proc Natl Acad Sci USA 103:9286–9291.

Friesen, T. L., E. H. Stukenbrock, Z. Liu, S. Meinhardt, H. Ling, J. D. Faris, J. B. Rasmussen, P. S. Solomon, B. A. McDonald, and R. P. Oliver. 2006. Emergence of a new disease as a result of interspecific virulence gene transfer. Nature Genetics 38:953–956.

Jaenike, J. 1985. Parasite pressure and the evolution of amanitin tolerance in drosophila. Evolution 39:1295–1301.

Kadekoppala, M., R. A. O'Donnell, M. Grainger, B. S. Crabb, and A. A. Holder. 2008. Deletion of the *Plasmodium falciparum* merozoite surface protein 7 gene impairs parasite invasion of erythrocytes. Eukaryotic Cell 7:2123–2132.

Tonukari, N. J., J. S. Scott-Craig, and J. D. Walton. 2000. The *Cochliobolus carbonum SNF1* gene is required for cell wall-degrading enzyme expression and virulence on maize. Plant Cell 12:237–248.

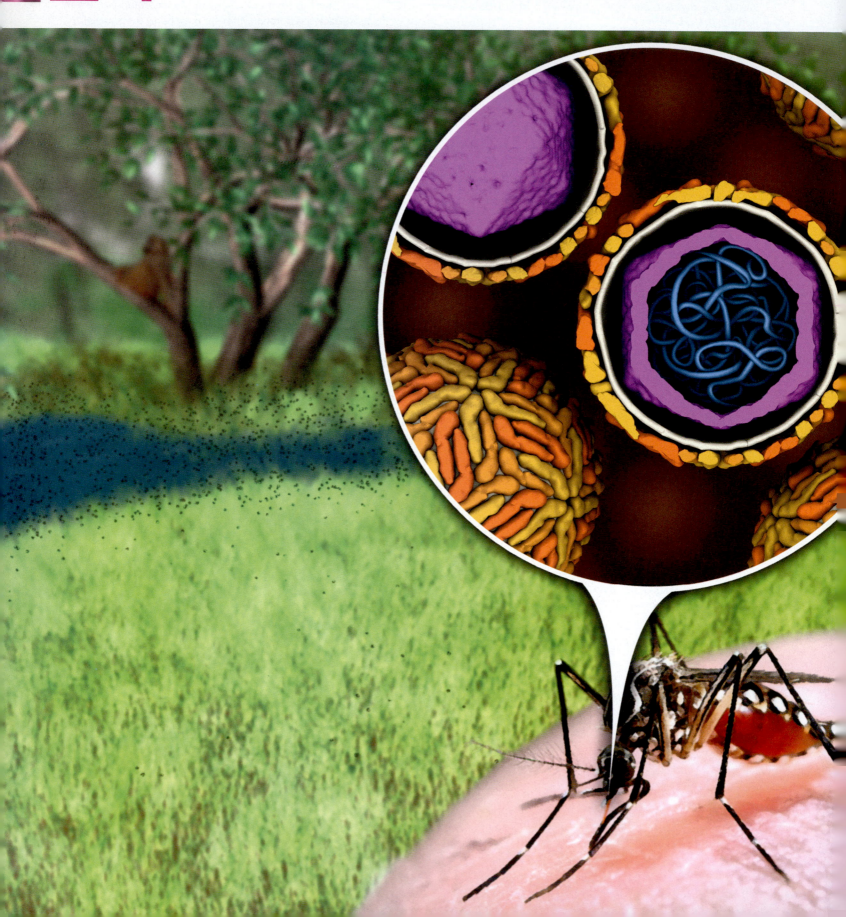

S omeone in your neighborhood has been diagnosed with yellow fever. Should you be concerned about this infectious disease? What would you need to know to decide this? If you were living in eighteenth- or nineteenth-century Europe or the Americas, you would be gravely concerned. The whisper of yellow fever, or "black vomit" or "yellow jack" as it was sometimes called, would strike fear into your heart. If you were smart, you would move hundreds of miles away to a new city, like so many others did, and then move again as the immigrant wave introduced the disease there also.

In 1793, when yellow fever struck Philadelphia, the capital of the United States at that time, the government vacated the city. No one knew what caused the disease or how it was spread, but it would routinely ravage cities when it came. Usually, it arrived with ships carrying infected people—beginning with the slave ships from Africa, the original home of the virus. Vessels flying the yellow jack flag carried suspect yellow fever victims; they were death ships, quarantined and prevented from entering port. Sometimes crew members sick with yellow fever were incorrectly diagnosed with malaria and the ship would be cleared. When such a ship entered the Mississippi River in 1878, it unleashed one of the greatest epidemics in American history, striking New Orleans, Memphis, Charleston, and Savannah. In some cities, 85 percent of those infected died.

Yellow fever is a hemorrhagic viral disease, named because the skin and eyes of infected persons commonly turn a striking bright yellow due to the release of the pigment bilirubin from damaged liver cells. Hemorrhaging of blood vessels causes blood to leak into body cavities, including the stomach, and the partially digested clotted blood turns stomach contents black, hence the name "black vomit."

In 1900, an American commission was launched to scientifically investigate the disease and find a way to control it. It was to be the first comprehensive study to determine the mode of transmission, the causative agent, and the prevention of yellow fever. The Yellow Fever Commission, as the agency was called, was based in American-occupied Cuba, where yellow fever regularly occurred. Appointed to the study were U.S. Army surgeons Drs. Walter Reed, James Carroll, Jesse Lazear, and Aristides Agramonte. A native Cuban, Dr. Carlos Juan Finlay, who also worked with the team, stubbornly held that the disease was transmitted by the common house mosquito, *Aedes aegypti*. This was an unpopular theory in the greater medical community.

Finlay observed the movement of the disease was like that of malaria, known to be transmitted by mosquitoes. Like malaria, yellow fever would attack some households and not others in the same area, and some family members, but not others in the same house. This was not what would be expected for a disease spread by direct person-to-person contact or contaminated water. Finlay's careful observations and refusal to submit to dogmatic thinking set the yellow fever team down a road that led to the discovery of the yellow fever microbial agent, its transmission, and ultimately, control of this dreadful disease.

CHAPTER NAVIGATOR

As you study the key topics, make sure you review the following elements:

In the past, infectious diseases were not effectively treated as they were not recognized as being caused by microbes.

Antimicrobial drugs target specific microbial processes or structures.
- Table 24.1: Antibiotics and their origins
- Figure 24.4: Spectrum differences of antibacterial agents
- Table 24.2: Action, structure, and targets of antibacterial drug groups
- Figure 24.5: Overview of antibacterial drugs and their targets
- Animation: Antivirals

Resistance to antimicrobial drugs is common in microbes.
- Figure 24.10: Antimicrobial drug resistance mechanisms
- Animation: Premature termination of antibiotic treatment
- Toolbox 24.1: Drug susceptibility testing and MIC
- Mini-Paper: Soil microbes possess extensive resistance to antibiotics
- Figure 24.14: Common methods of acquiring resistance genes
- Perspective 24.3: Phage therapy: Biocontrol for infections

Effective public health measures that prevent microbial diseases include immunization, containment of the pathogen, and predicting future epidemics.
- Figure 24.15: Six phases of influenza pandemic alert
- Figure 24.16: Triangle of epidemiology

Improvements in sanitation, vaccines, and antimicrobial drugs have led to the prevention and control of epidemics.
- Figure 24.18: Current immunization schedule
- Figure 24.19: Protection by herd immunity
- Figure 24.21: Outcome of poliovirus infection and immune status
- Microbes in Focus 24.1: Foot-and-mouth disease virus

CONNECTIONS for this chapter:

The life cycle of malaria parasite *Plasmodium falciparum* and the effect of chloroquine (Section 23.2)

The cell wall and resistance to antimicrobial agents (Section 2.4)

Horizontal gene transfer mechanisms and antimicrobial resistance (Section 9.5)

Epidemics, epidemiology, and controlling infectious diseases (Section 18.3)

Exposure to pathogenic microbes from untreated water supplies (Section 16.5)

Introduction

What would you list as the most significant past or present factors involved in the control of infectious diseases? You are correct if you listed improvements in sanitation, hygiene, nutrition, and the development of vaccines and antimicrobial agents. Increased survival rates in humans and domestic animals are closely correlated to these advancements. Of course, underpinning these improvements has been the attainment of economic wealth and education. Today, in developed nations, relatively few people die prematurely of infectious disease **(Figure 24.1)**. This is not the case in developing nations where, according to the World Health Organization (2008), infectious disease is still a major cause of premature death. Sadly, the majority of these deaths occur in young children from common diseases that can be prevented by vaccination or treatment with therapeutic agents. Diarrhea, malaria, measles, and pneumonia kill approximately 4 million children under the age of five each year, accounting for about half of all childhood deaths (see Figure 18.2).

While improved sanitation has significantly reduced infections from foodborne or waterborne agents such as *Vibrio cholerae*, *Escherichia coli*, and rotaviruses, the added combination of vaccines and antimicrobial drugs has had the greatest impact on reducing overall mortality and illness from infectious diseases. These three measures remain the principal weapons in the treatment and control of infectious disease when employed according to strategies based on sound epidemiological foundations. Of course, disease control resulting from such advancements would not have been possible if it weren't for the nineteenth-century discovery of microbes as agents of many diseases.

Antimicrobial drugs, once hailed as the "magic bullet" to end the battle against infectious diseases, have perhaps been one of the biggest disappointments. While undoubtedly advancing the treatment of infectious disease, microbes have not surrendered easily to these drugs, making drug resistance a serious threat in the treatment of many infectious diseases. This underscores the need for disease prevention involving surveillance, planning, and implementation of infection control measures, including strategic vaccination programs. To learn about infectious disease treatment and control, we will focus on answering the following questions:

How did we deal with infectious disease in the past? (24.1)
What kinds of drugs are used to treat infections, and how do they work? (24.2)
How do microbes become drug resistant? (24.3)
How are epidemics predicted and controlled? (24.4)
What are vaccines, and how are they used to control infectious disease? (24.5)

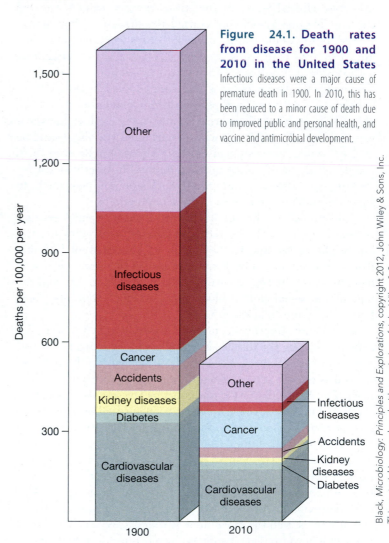

Figure 24.1. Death rates from disease for 1900 and 2010 in the United States Infectious diseases were a major cause of premature death in 1900. In 2010, this has been reduced to a minor cause of death due to improved public and personal health, and vaccine and antimicrobial development.

Black, *Microbiology: Principles and Explorations*, copyright 2012, John Wiley & Sons, Inc. This material is reproduced with permission of John Wiley & Sons, Inc.

24.1 Historical aspects of infectious disease treatment and control

How did we deal with infectious disease in the past?

The epidemic "crowd" diseases such as smallpox, plague, cholera, typhus, and influenza began influencing human populations on a large scale only in recent times. These communicable diseases require suitably large populations to be sustained. Prehistoric hunter-gatherer tribes were small and sparsely distributed. They also did not stay long enough in any one spot to contaminate their water supplies or attract disease-transmitting pests like rats and their fleas. They also did not live with domestic animals, which were the original source of many infectious diseases. Human populations began to accumulate about 10,000 years ago with the advent of crops and animal husbandry. Urbanization and travel associated with trade began in earnest around 3,000 B.C. in Mesopotamia, Egypt, southern China, and Central America. Archeological and recorded evidence of the first large-scale plagues appears with the rise of these civilizations.

For most of mankind's history, religious and superstitious beliefs were the only explanations offered as to why these diseases occurred. Diseases and disasters were heralded as punishment by supernatural beings in response to the immoral, sinful acts of humans. Afflicted persons were often cast out from society, a control measure that was enacted with moral authority as opposed to public safety. Victims of leprosy and other disfiguring diseases were indiscriminately lumped together and banished to live in deplorable camps outside the boundaries of towns. They were considered "unclean" and forbidden to marry or interact with the rest of society. Astonishingly, the last penal leper colony existed in the United States on the Hawaiian island of Molokai and was not transitioned to voluntary residence for treatment until 1969.

The ancient Greeks were the first to separate religious beliefs and disease, although even today superstitions surrounding disease still exist in many cultures. Greek writings document the use of logic and a view of disease as a condition of the physical and not the mystical. Around 400 B.C., Hippocrates, known as the "father of medicine," lectured on disease being a condition of physiological imbalance, and he was on the right track. "Disease" is defined today as a disturbance in the normal functioning of an organism, a condition away from health. What Hippocrates didn't know was the cause of the imbalance or how to reestablish this balance to treat disease. While Hippocrates talked largely of maintaining physiologic balance through diet, others took this to the extreme. In an attempt to reestablish balance within the body and "cure" illness they practiced phlebotomy or venesection, the purposeful lancing of a vein to let "excess" blood drain to cool fever. This may actually have worked temporarily by reducing the level of fever-inducing cytokines such as IL-1 (see Section 19.3). However, it did more overall harm than good as it also diminished immune cells and molecules needed to fight infection. Often, bloodletting reduced blood pressure to the point of unconsciousness. Leeches were similarly used to withdraw blood, and a multitude of imaginative bodily purges were liberally employed. Patients were routinely subjected to such unproven "treatments" at the whim of local health practitioners. These treatments continued well into the nineteenth century until the great microbe hunters of the mid-1800s—Koch, Pasteur, and Lister—provided evidence for the germ theory of disease, which proposed many diseases were caused by microbes.

Quarantines to stop the spread of infectious disease began around the fourteenth century. Ships arriving from areas infected with yellow fever, for example, were required to anchor in a city's harbor for 40 days, or crews were dispatched to nearby islands before being allowed access to port. This was frequently an effective control measure in preventing epidemics. However, epidemics were not limited to humans or even animals. The Great Irish Potato Famine (1845–1851) was due to the airborne water mold *Phytophthora infestans*, which arrived on ships traveling from North America to England. It devastated staple potato crops in Ireland, resulting in the death of 1 million of the 8 million people inhabiting the island. Another 1 million people fled to America, mostly to New York City, to avoid starvation.

One of the most significant weapons in the control of infectious disease has been the development of vaccines. Jenner's development of the smallpox vaccine in 1796 (see Section 24.5) began a rational approach to infectious disease prevention. Robert Koch's development of Koch's postulates (see Section 18.4) were, and still are, instrumental in identifying the infectious agents associated with disease, making it possible to specifically target these agents and apply rational control measures. Lastly, Alexander Fleming's discovery of penicillin in 1928 (see Chapter 12 opening story) started a pharmaceutical revolution that allowed the treatment of a multitude of infectious diseases.

> ### 24.1 Fact Check
>
> 1. Give several reasons why epidemic diseases, such as smallpox, likely did not impact prehistoric human populations.
> 2. Why was bloodletting used to control infection? How could this be harmful?
> 3. What other methods were used to control disease before drugs and vaccines became available?

24.2 Antimicrobial drugs

What kinds of drugs are used to treat infections, and how do they work?

Most of us have been on one type of antimicrobial drug or another at some time in our lives, and many of us would not be alive today if it weren't for these drugs. **Antimicrobial drugs** are compounds used to treat or prevent infections primarily inside the body. They are a subset of antimicrobial agents, which include any substance that inhibits or kills microbes, including viruses. In Chapter 6, we were introduced to antiseptics, those antimicrobial agents that act topically on body surfaces, like skin, and disinfectants, which are used on inanimate surfaces to control the growth of microbes (see Section 6.5). In this chapter, we will examine the mode of action of some of the more common antimicrobial drugs.

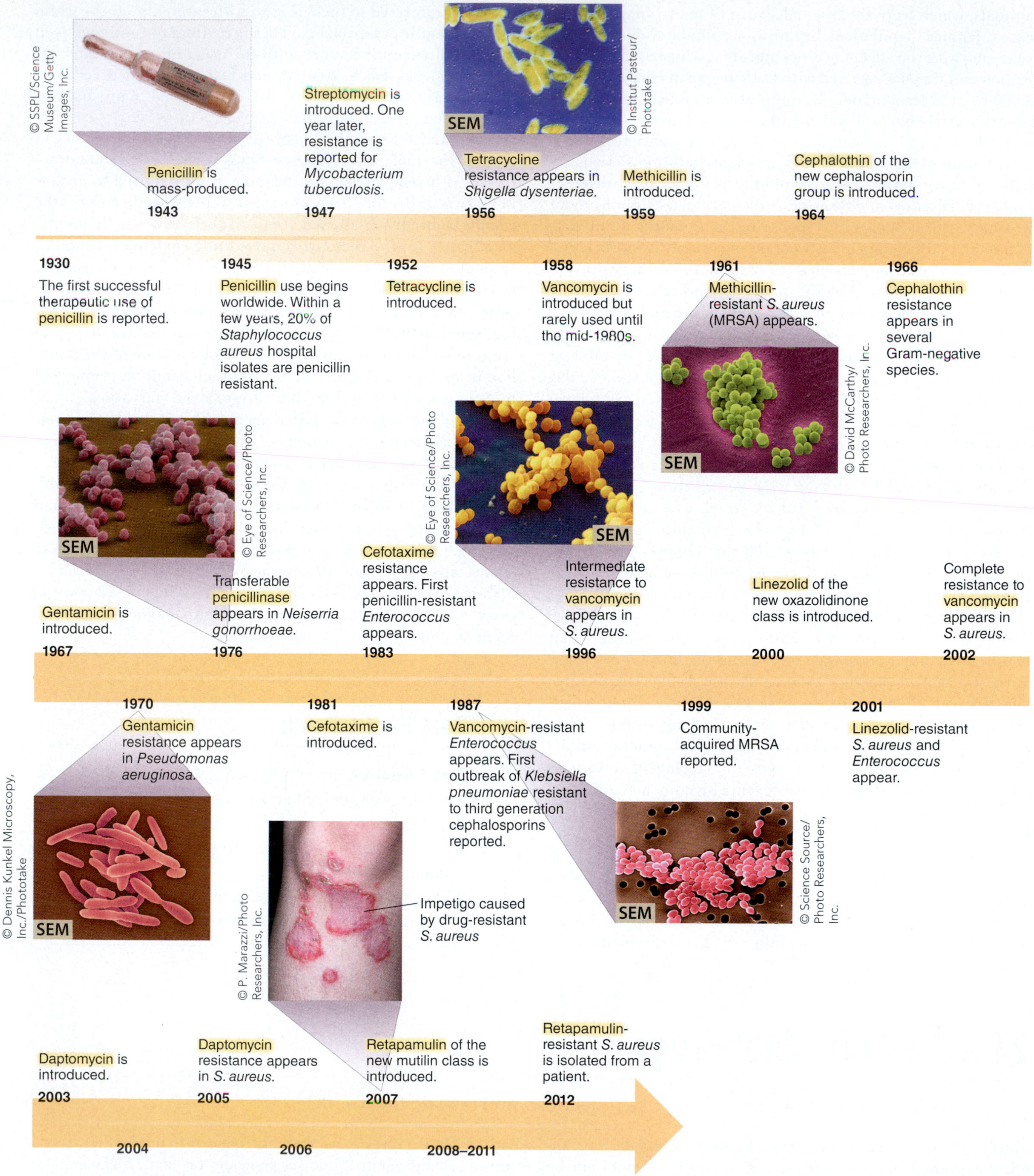

© SSPL/Science Museum/Getty Images, Inc.
1930
The first successful therapeutic use of penicillin is reported.
1943
Penicillin is mass-produced.
1945
Penicillin use begins worldwide. Within a few years, 20% of Staphylococcus aureus hospital isolates are penicillin resistant.
1947
Streptomycin is introduced. One year later, resistance is reported for Mycobacterium tuberculosis.
1952
Tetracycline is introduced.
SEM
© Institut Pasteur/Phototake
1956
Tetracycline resistance appears in Shigella dysenteriae.
1958
Vancomycin is introduced but rarely used until the mid-1980s.
1959
Methicillin is introduced.
1961
Methicillin-resistant S. aureus (MRSA) appears.
SEM
© David McCarthy/Photo Researchers, Inc.
1964
Cephalothin of the new cephalosporin group is introduced.
1966
Cephalothin resistance appears in several Gram-negative species.
SEM
© Eye of Science/Photo Researchers, Inc.
1967
Gentamicin is introduced.
1970
Gentamicin resistance appears in Pseudomonas aeruginosa.
SEM
© Dennis Kunkel Microscopy, Inc./Phototake
1976
Transferable penicillinase appears in Neiserria gonorrhoeae.
1981
Cefotaxime is introduced.
1983
Cefotaxime resistance appears. First penicillin-resistant Enterococcus appears.
SEM
© Eye of Science/Photo Researchers, Inc.
1987
Vancomycin-resistant Enterococcus appears. First outbreak of Klebsiella pneumoniae resistant to third generation cephalosporins reported.
SEM
© Science Source/Photo Researchers, Inc.
1996
Intermediate resistance to vancomycin appears in S. aureus.
1999
Community-acquired MRSA reported.
2000
Linezolid of the new oxazolidinone class is introduced.
2001
Linezolid-resistant S. aureus and Enterococcus appear.
2002
Complete resistance to vancomycin appears in S. aureus.
2003
Daptomycin is introduced.
2004
2005
Daptomycin resistance appears in S. aureus.
2006
Impetigo caused by drug-resistant S. aureus
© P. Marazzi/Photo Researchers, Inc.
2007
Retapamulin of the new mutilin class is introduced.
2008–2011
2012
Retapamulin-resistant S. aureus is isolated from a patient.

Figure 24.2. Antibacterial drug resistance timeline Resistance has developed for every antibacterial drug ever used for therapy, beginning with penicillin in 1945.

All antimicrobial drugs used for treatment possess **selective toxicity**; they are more toxic to an infecting microbe than they are to the host when used at recommended dosages. The reason for this lies in their ability to target specific components of microbes. Drugs that do not exhibit sufficient selective toxicity are not clinically used. Despite exhibiting selective toxicity, some antimicrobial agents still have the potential to cause damage to the host, usually because their mode of action has some activity on host cells as well. When correctly used at therapeutic doses, these toxic side effects are minimized while maintaining useful antimicrobial activity. Many antimicrobial drugs are never approved for use because they are too toxic to the body.

Antibiotics, compounds derived from microorganisms that inhibit the growth of other microorganisms, are an important group of antimicrobial drugs, and are the most widely prescribed therapeutic drugs. The development of antibiotics began with Alexander Fleming in 1928. He recorded his observations of a particular compound from the common bread mold *Penicillium* that could inhibit cultures of *Staphylococcus aureus*. He isolated this compound, *penicillin*, and others began using it to treat infections in the body. Fleming's "wonder drug" saved countless lives during World War II, when it first came into widespread use. However, it was not long before this curative drug became the driving force in the acquisition and maintenance of resistance genes in the very bacteria it previously killed. This pattern has been repeated with every antimicrobial drug since their introduction (**Figure 24.2**).

Although the term "antibiotic" was originally used to describe compounds produced by bacteria or fungi that inhibit the growth of other bacteria or fungi, many naturally produced antibiotics have been modified into semisynthetic compounds by chemical alteration. Many antibiotics are also now produced through cost-effective chemical processes only and are no longer isolated from their associated microorganisms. Synthetic antimicrobial drugs do not originate from microbes. They are designed through technology and have always been man-made. Because of the widespread use of synthetic and semisynthetic therapeutic compounds against microbial cells, the term "antibiotic" is sometimes used to include all antimicrobial drugs that act on the cells of microorganisms, but not viruses or subviral particles. This popular definition reflects the target of the drugs—microbial cells—rather than the origin. In this chapter, we will continue to distinguish antimicrobial agents as antibiotics or their semisynthetic derivatives from synthetic compounds. The distinction serves to illustrate the important contribution of microorganisms in the development of antibacterial and antifungal drugs.

As shown in **Table 24.1**, the bacterial genus *Streptomyces*, common in soils (see Microbes in Focus 12.1), has been particularly valuable for providing antibiotics. Semisynthetic antibiotic derivatives contain modified side chains to improve issues of toxicity, stability, absorption, microbial resistance, or spectrum limitations. For example, as shown in **Figure 24.3**, penicillin G, a naturally occurring antibiotic, is a narrow spectrum drug effective only against Gram-positive bacteria, and is destroyed by acid, so is unsuitable for oral ingestion. Ampicillin, a semisynthetic derivative of penicillin G, is a broad spectrum drug, exhibiting activity against Gram-positive and many Gram-negative bacteria (**Figure 24.4**). It is acid resistant and can be taken orally. Other broad spectrum antibiotics, like methicillin and oxacillin, were developed because they are resistant to attack by penicillin-destroying β-lactamases produced by many strains of *Staphylococcus aureus*, *Enterococcus*, and Enterobacteriaceae.

TABLE 24.1	Antibiotics and their origins		
Source	**Natural antibiotic**	**Semisynthetic derivatives**	
Antibacterial antibiotics			
Penicillium chryosgenum (fungi)	Penicillin G, penicillin V	Methicillin, ampicillin, amoxicillin, carbenicillin, oxacillin	
Cephalosporium species (fungi)	Cephalosporin	Cephalexin, cephradine, cefradoxil, ceftazidime	
Streptomyces griseus	Streptomycin	NA[a]	
Streptomyces aureofaciens	Tetracycline	Doxycycline, oxytetracycline	
Streptomyces venezuelae	Chloramphenicol	NA[a]	
Streptomyces erythreus	Erythromycin	Azithromycin, clarithromycin	
Streptomyces kanamyceticus	Kanamycin	Amikacin, arbekacin	
Streptomyces tenebrarius	Tobramycin	NA[a]	
Streptomyces fradiae	Neomycin	NA[a]	
Streptomyces mediterranei	Rifamycin	Rifampin	
Amycolatopsis orientalis	Vancomycin	Ramoplanin	
Micromonospora species	Gentamicin	NA[a]	
Bacillus licheniformis	Bacitracin	NA[a]	
Bacillus polymyxa	Polymyxins	NA[a]	
Antifungal antibiotics			
Penicillium griseofulvum	Griseofulvin	NA[a]	
Streptomyces nodosus	Polyenes	NA[a]	

[a]NA: not applicable. Not developed or not commonly used.

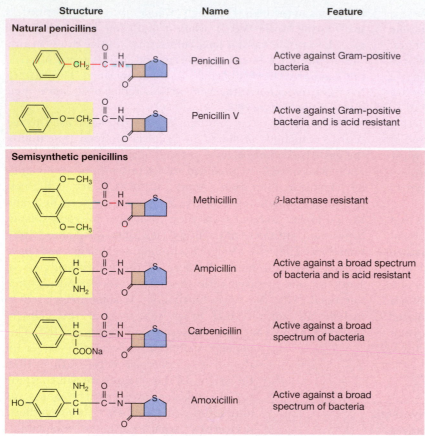

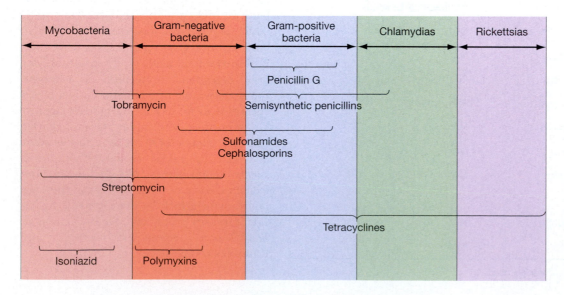

Figure 24.3. Structure of penicillins **A.** Natural penicillins G and V differ only in the structure of the side groups (shown highlighted). Semisynthetic penicillins have modifications of the side groups that can impart resistance to β-lactamases, broaden the spectrum of bacterial groups against which they can be used, and impart acid resistance. Acid resistance permits oral administration. **B.** Penicillins are composed of a 5-member ring structure attached to the β-lactam ring, and a single side group, designated R. Cephalosporins differ by the presence of a 6-member ring that can carry an additional side group, designated R₁.

Black, *Microbiology: Principles and Explorations*, copyright 2012, John Wiley & Sons, Inc. This material is reproduced with permission of John Wiley & Sons, Inc.

Antimicrobial drugs are frequently subdivided into classes based on their structure, which in turn is related to their mode of action. Most antimicrobial drugs act on microbes in common ways, interfering with:

- peptidoglycan synthesis (cell wall synthesis)
- membrane integrity

- DNA synthesis
- transcription
- folic acid synthesis (nucleic acid synthesis)
- ribosome function (protein synthesis)

Table 24.2 gives details of the mode of action and structure of some antibiotics as well as other synthetic antibacterial drugs.

Figure 24.4. Spectrum differences of antibacterial agents Broad spectrum drug groups, such as tetracyclines and semisynthetic penicillins, are effective against diverse groups of unrelated bacteria. Narrow spectrum drugs, such as penicillin G, polymyxins, and isoniazid affect only small groups of related bacteria.

Black, *Microbiology: Principles and Explorations*, copyright 2012, John Wiley & Sons, Inc. This material is reproduced with permission of John Wiley & Sons, Inc.

TABLE 24.2 Action, structure, and targets of antibacterial drug groups

Action	Class/group	Structure	Target	Examples
		Antibacterial drugs		
		Antibiotics		
Inhibition of peptidoglycan synthesis	β-lactams	Penicillin G	Peptidoglycan transpeptidases (PBPs)	Penicillins G and V, methicillin, cephalosporins, monobactams, carbenicillin
	Glycopeptides	Vancomycin	Peptidoglycan peptide subunits	Vancomycin, avoparcin
	Bacitracin (topical use)	Bacitracin	Isoprenyl pyrophosphate	Bacitracin
Disruption of membranes	Polymyxin B (topical use)	Polymyxin B	Membranes	Polymyxin B, polymyxin E
Inhibition of ribosome function	Aminoglycosides	Gentamicin	16S rRNA of 30S ribosome subunit	Gentamicin, neomycin, streptomycin, tobramycin

Black, *Microbiology: Principles and Explorations*, copyright 2012, John Wiley & Sons, Inc. This material is reproduced with permission of John Wiley & Sons, Inc.

(*continues on next page*)

TABLE 24.2 (Continued)

Action	Class/group	Structure	Target	Examples
	Macrolides	Erythromycin	Peptidyl transferase site of 50S ribosome subunit	Erythromycin, spectinomycin, carbomycin
	Tetracyclines	Tetracycline	30S ribosome subunit	Tetracycline, doxycycline, oxytetracycline
	Chloramphenicol	Chloramphenicol	23S rRNA of 50S ribosome subunit	Chloramphenicol
Inhibition of transcription	Rifamycins	Rifampin	β-subunit of bacterial RNA polymerase	Rifampin, rifabutin, rifapentine
		Synthetic drugs		
Inhibition of nucleic acid synthesis	Quinolones	Ciprofloxacin	Gram-negative DNA gyrase or Gram-positive topoisomerase IV	Nalidixic acid, oxolinic acid, fluoroquinolones (e.g., ciprofloxacin)
	Sulfonamides	Sulfanilamide	Dihydropteroate synthetase (folic acid pathway)	Sulfisoxazole, sulfanilamide
	Trimethoprim	Trimethoprim	Dihydrofolate reductase (folic acid pathway)	Trimethoprim

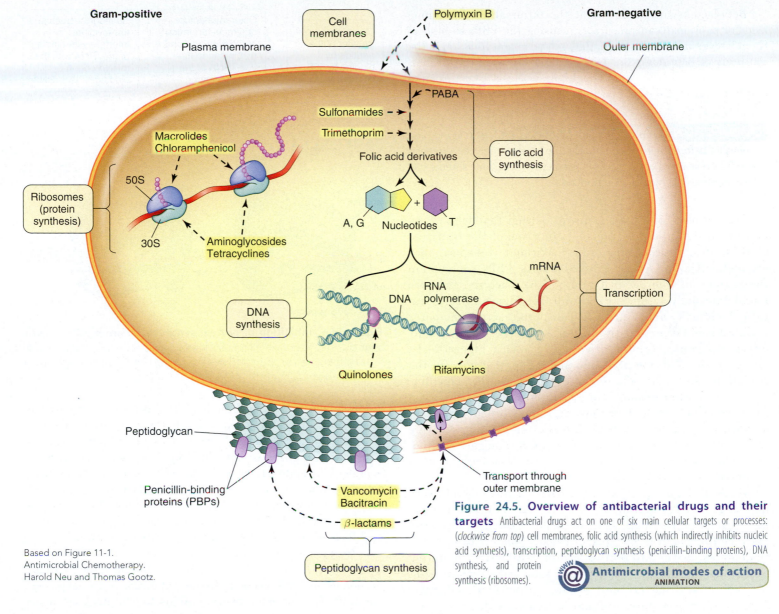

Figure 24.5. Overview of antibacterial drugs and their targets Antibacterial drugs act on one of six main cellular targets or processes: (*clockwise from top*) cell membranes, folic acid synthesis (which indirectly inhibits nucleic acid synthesis), transcription, peptidoglycan synthesis (penicillin-binding proteins), DNA synthesis, and protein synthesis (ribosomes).

Antimicrobial modes of action ANIMATION

Based on Figure 11-1. Antimicrobial Chemotherapy. Harold Neu and Thomas Gootz.

Figure 24.5 shows a schematic overview of the specific targets of the major classes of antibacterial drugs, including antibiotics.

Antibacterial drugs

Antimicrobial drugs can be divided into categories based on the microbial groups they target. These are antibacterial, antifungal, antiprotozoal, and antiviral agents. With few exceptions, antimicrobial drugs do not cross these divisions because they target molecules that are specifically found only in a specific group—a reflection of phylogenetic differences. Antimicrobial drugs are further divided into classes or groups based on their structure, which in turn usually determines their mode of action.

Bacteriocidal drugs are antibacterial drugs that can directly kill bacteria. **Bacteriostatic drugs** prevent replication of bacteria, but do not kill them. These descriptions are based on observations of controlled *in vitro* experiments. A culture treated with a bacteriostatic agent for a short time resumes growth when the agent is removed; the effect is reversible. In a culture treated with a bacteriocidal agent for a short time cells are killed; the

effect is irreversible. Both bacteriostatic and bacteriocidal drugs are effective therapies against infection. Inside the complex environment of the body bacteria are subject to host immune defenses, not just drugs, so any bacterial cells that are unable to grow will usually be destroyed by the immune system within the course of the antimicrobial treatment. We'll examine a few of the most commonly used classes of antibacterial drugs in this section.

Inhibitors of Peptidoglycan Synthesis

Members of the **β-lactam class** of antibacterial drugs share the common structural feature of the β-lactam ring (see Figure 24.3B, also Figure 2.16), required for binding to a group of wall-associated proteins known as **penicillin-binding proteins (PBPs)**. PBPs have a transpeptidase function, required for joining together the subunits that make up peptidoglycan (see Section 2.4). Penicillin G was the first β-lactam antibiotic to be discovered, by Alexander Fleming in 1928, and was the first to come into widespread use, beginning in 1945, as a therapeutic drug to treat infections. Today, penicillins and cephalosporins are among the most commonly prescribed drugs, accounting for almost 50 percent of antibiotics used.

Beta-lactam antibiotics mimic the terminal alanine amino acids of peptidoglycan precursors, causing the PBPs to bind to them irreversibly. In this state, they can no longer function in the synthesis of peptidoglycan. Different β-lactams have different abilities to interfere with peptidoglycan synthesis. This is related to differences in PBPs and the ability of β-lactams to access them.

CONNECTIONS PBPs were named because of their ability to bind penicillin before their transpeptidase role in peptidoglycan synthesis was fully revealed. For a review of the action of β-lactams and the PBP transpeptide enzymes, see **Section 2.4**.

Susceptibility of microorganisms to drugs that act on cytoplasmic or periplasmic targets is a function of the transport of these drugs across membranes. Without transport, they are ineffective. In Gram-negative bacteria, PBPs are found in the periplasm, whereas in Gram-positive bacteria, they are found externally within the cell wall (see Figure 24.5). Natural penicillins G and V are not transported across the outer membrane of Gram-negative bacteria, and therefore are only effective against Gram-positive bacteria (see Figure 24.4). However, the addition of amino and carboxyl groups to the penicillin side chain allows these semisynthetic penicillins to be transported through outer membrane porins to act on the periplasmic PBPs. This imparts a much broader range of activity. For example, carbenicillin is often used for treating infections with *Pseudomonas aeruginosa*, a Gram-negative pathogen, while penicillin G is used for *Streptococcus pneumoniae*, a Gram-positive pathogen.

While PBPs are present in all bacteria that manufacture peptidoglycan, in different species they vary in number (most have several different PBPs), amount, and structure. Similarity between PBPs of different species usually follows phylogenetic relatedness, with different β-lactam antibiotics inhibiting some PBPs more than others. Finally, some bacteria have complex life cycles that prevent β-lactams from being effective. *Chlamydia*, a genus of intracellular pathogens with a two-phase life cycle, are only susceptible to penicillin during differentiation into their infectious, extracellular spore-like elementary body form. The intracellular reticulate body form, which causes damage, does not possess detectable peptidoglycan. It is presumed that *Chlamydia* express PBPs only during this differentiation process to produce the peptidoglycan associated with the inert elementary body. Treatment with penicillin can therefore result in a noninfectious but persistent intracellular state in the host. For this reason, tetracycline is the preferred treatment (see Figure 24.4).

Inhibitors of Ribosome Function

Inhibiting ribosome function interferes with translation and thus protein synthesis in the bacterial cell. Aminoglycosides, such as kanamycin, neomycin, gentamicin, and their derivatives, act on the ribosome by reversibly binding to 16S rRNA, disrupting interaction of mRNA with tRNA, thus interfering with translation. Aminoglycosides are known for their toxicity in humans and animals, particularly to kidneys and the inner ear apparatus. The mechanism of toxicity remains unknown, but is unrelated to protein synthesis in eukaryotes. Lower doses

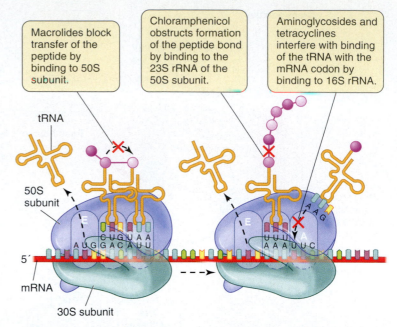

Figure 24.6. Inhibition of protein synthesis by antibiotics Four major groups of antibiotics act on targets of the ribosome to inhibit protein synthesis. These antibiotics are useful in treating bacterial infections because they have a greater affinity for the prokaryotic ribosome than the eukaryotic ribosome.

are less toxic, but not as effective. For this reason, they are frequently used in combination with other antibacterial drugs.

Macrolides act on the prokaryotic ribosome by binding near the 50S peptidyl transferase site, blocking elongation of the growing peptide. Like aminoglycosides, tetracyclines, and chloramphenicol, they inhibit protein synthesis. **Figure 24.6** summarizes the action of many ribosome-inhibiting antibiotics. The macrolide erythromycin is effective against Gram-positive bacteria, but is not efficiently taken up by most Gram-negative bacteria.

Tetracyclines and their derivatives act on the 30S ribosome by binding to the 16S rRNA, blocking interaction of the tRNA anticodon to the mRNA codon, thereby inhibiting bacterial protein synthesis. They have an inherent ability to bind RNA secondary structures, making them somewhat active against mRNA of eukaryotic cells too. Fortunately, tetracyclines are not as efficiently taken up by eukaryotic cells, but can cause toxicity in high doses. Tetracyclines can permanently discolor teeth, especially in children, due to formation of localized calcium/tetracycline complexes **(Figure 24.7)**. Tetracycline is effective against many Gram-negatives, most Gram-positives, both of the developmental forms of *Chlamydia*, and the peptidoglycan-deficient bacteria *Rickettsia* giving it one of the broadest spectrums of any antibacterial drug (see Figure 24.4).

Chloramphenicol binds the 23S rRNA of the 50S subunit (see Figure 7.22), interfering with peptidyl transferase activity of the ribosome, which results in the prevention of peptide bond formation. It has a broad spectrum range, but causes a rare and unpredictable form of aplastic anemia, produced from the destruction of the bone marrow progenitor cells that form erythrocytes and leukocytes. The condition is usually fatal unless a bone marrow transplant, which has long-term complications of its own, is performed. Chloramphenicol is a last-chance drug used only in serious infections when other antibacterial drugs are not effective.

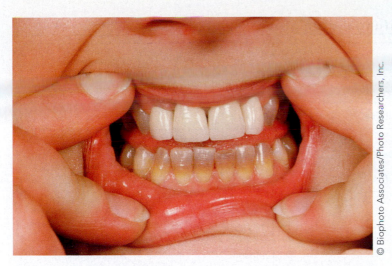

Figure 24.7. Tetracycline staining of teeth Staining is due to the ability of tetracycline to bind calcium. Discoloration is permanent and most pronounced when children under five years of age are given tetracycline. Adults given tetracycline frequently report slight darkening of teeth.

Inhibitors of Nucleic Acid Synthesis

Quinolones and sulfonamides are two of the most commonly used synthetic antibacterial drugs. Quinolones are based on the core structure of naladixic acid, one of the first quinolones to be used. Newer versions are fluorinated to give them better antibacterial activity and less toxicity. Quinolones are broad spectrum drugs that bind to DNA gyrase, particularly of Gram-negative bacteria, and topoisomerase IV of Gram-positive bacteria (see Section 7.2). They can also be used against intracellular pathogens such as *Legionella pneumophila*, *Mycoplasma pneumoniae*, and *Mycobacterium tuberculosis*, because of their ability to enter eukaryotic cells.

Sulfonamides (sulfa drugs) are very small antibacterial drugs that are structural analogs of para-aminobenzoic acid (PABA). PABA is used by many bacteria to make folic acid, a necessary precursor for the synthesis of the nitrogenous bases adenine, guanine, and thymidine, needed for DNA synthesis (**Figure 24.8**; see also Figure 13.48). When sulfonamides competitively bind the enzyme dihydropteroate synthetase that incorporates PABA into the folic acid precursor molecule, enzyme function is inhibited and also a non-functional folic acid derivative results. The cells of animals and humans take up folic acid, also known as vitamin B_9, from their diet and so are unaffected by sulfonamide drugs. Similarly, many single-celled eukaryotes, and some bacteria, that acquire folic acid from their environment are unaffected by these drugs.

Trimethoprim inhibits another enzyme, dihydrofolate reductase, later in the folic acid pathway. Inhibition of this enzyme prevents synthesis of the folic acid derivative tetrahydrofolic acid, preventing synthesis of thymidine (see Figure 13.48). Sulfonamides and trimethoprim are often used together in a drug combination referred to as "co-trimoxazole."

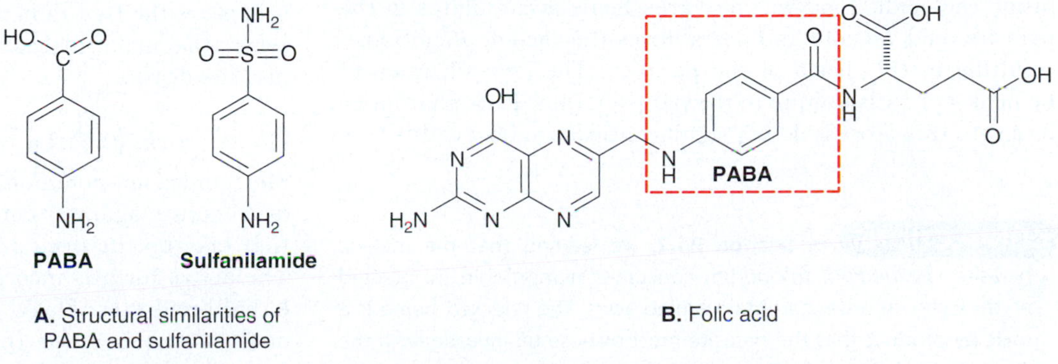

Antifungal drugs

Fungal pathogens are eukaryotes and so their metabolic pathways, DNA, RNA, and protein synthesis machineries are highly similar to those of their host. Treating fungal infections is notoriously difficult because most effective antifungal agents typically show toxicity to host cells as well. Therefore, finding drugs that exhibit sufficient selective toxicity to fungal cells and not host cells is a challenge. Compared to antibacterial drugs, only a modest selection of antifungal agents is available (**Table 24.3**). Additionally, many fungi have the intrinsic ability to modify agents entering their cytoplasm, rendering the drugs less effective.

Disruption of Plasma Membrane

One distinguishing feature of fungi is the presence of ergosterol instead of cholesterol in their plasma membranes. Polyene antibiotics bind ergosterol to form pores in the plasma membrane, causing increased membrane permeability, leading to detrimental ion imbalances. Azoles comprise a major group of antifungal drugs and are also widely used for treating and preventing fungal infections in agricultural crops. They disrupt membrane sterol synthesis, including ergosterol synthesis, leading to membrane damage. Some are used systemically, but most are topical. Azoles, such as fuconazole, are commonly used for *Candida* vaginal and skin infections.

Antiprotozoal drugs

Like fungal infections, combating protozoal infections is complicated by the fact that protozoa share many of the same structural and metabolic features with the cells of their hosts. Consequently, relatively few selectively toxic drugs are available, and the toxicity associated with these commonly causes side effects such as vomiting, diarrhea, and nausea (**Table 24.4**). As we saw in Chapter 23, complicating this is the ability of many protozoa to differentiate into distinct forms through their life cycle, and to migrate to different tissues. As a result, many protozoa present a moving target, and treatment of these can be complicated, requiring a regimen of different drugs.

Inhibition of Heme Detoxification

Quinolines, not to be confused with antibacterial quinolones, are antimalarial agents related to quinine, a compound extracted from the cinchona tree of Peru and Bolivia, but can also be made synthetically. Quinine was commonly used to treat malaria as long ago as

A. Structural similarities of PABA and sulfanilamide

B. Folic acid

Figure 24.8. Sulfonamides and trimethoprim action **A.** Sulfonamides (sulfanilamide shown here) are structural analogs of para-aminobenzoic acid (PABA), which is a precursor for folic acid synthesis. **B.** Folic acid showing PABA component.

TABLE 24.3 Action, structure, and targets of antifungal drug groups

Action	Class/group	Structure	Target	Examples
Antifungal drugs				
Antibiotics				
Inhibition of mitosis	Griseofulvin	Griseofulvin	Tubulin	Griseofulvin
Disruption of plasma membrane	Polyenes	Amphotericin B	Plasma membrane ergosterol	Amphotericin B, nystatin
Synthetic drugs				
	Azoles	Fuconazole	C14-demethylase	Fuconazole, ketoconazole, chlotrimazole, miconazole

the seventeenth century, but it is highly toxic. In the 1940s, it was replaced by related, but less toxic, synthetic drugs such as chloroquine. Resistance to quinine is unusual and it is still used in cases in which the disease-causing agent is resistant to these newer drugs.

Chloroquine diffuses across plasma and lysosomal membranes of host cells and *Plasmodium* parasites, which reside inside erythrocytes during a critical stage in their life cycle (see Section 23.2). Chloroquine binds to hemozoin, a nontoxic derivative of heme formed by the parasite as it consumes hemoglobin. Binding prevents crystallization of hemozoin, and inhibits the ability of the parasite to continue to detoxify heme inside the acidic food vacuole. Free heme accumulates in the parasite food vacuole and destabilizes the vacuole membrane, resulting in the death of the parasite. The crystallization of heme is a process unique to the parasite, thus drugs that interfere with this process do not significantly harm host cells.

CONNECTIONS In **Section 23.2**, we learned that the malaria parasite *Plasmodium falciparum* consumes hemoglobin in infected erythrocytes in order to obtain amino acids. The released heme is a toxic by-product that the parasite must dispose of. Investigating the life cycles and physiology of eukaryal parasites can suggest unique targets for drug development that will minimize host cell damage.

Free Radical Formation

Artemisinin is an antimalarial drug, originally isolated from the plant *Artemisia annua*, but is also synthesized by genetically engineered bacteria (see The Rest of the Story, Chapter 12). Use of artemisinin alone gives rise to artemisinin resistance but when used in combination with other drugs such as mefloquine, remains an effective treatment against chloroquine-resistant strains of *Plasmodium falciparum*. The mode of action of artemisinin is not completely clear, but is active against the erythrocyte stage of the parasite (see Section 23.2). Artemisinin forms reactive free radicals in the presence of ferrous iron (Fe^{2+}), and it appears the free radicals irreversibly inhibit a metabolic enzyme, the malarial calcium-dependent ATPase, resulting in parasite death.

Destruction of DNA

Metronidazole, commonly known as "Flagyl," is a synthetic drug active against some protozoa and anaerobic bacteria that take up the drug and then convert it to its active form. The active form is incorporated into the DNA helix, causing lethal breakages. Flagyl is used to treat intestinal infections of *Giardia*, *Entamoebae* (protozoa), and *Clostridium difficile* (an anaerobic bacteria), as well as *Trichomonas vaginalis*, a sexually transmitted protozoan.

TABLE 24.4 Action, structure, and targets of antiprotozoal drug groups

Action	Class/group	Structure	Target	Examples
Antiprotozoal drugs				
Inhibition of heme detoxification	Quinolines	Chloroquine	Hemozoin (see Section 23.1)	Quinine, chloroquine, primaquine, mefloquine
Free radical formation	Artemisinin	Artemisinin	Malarial calcium-dependent ATPase	Artemisinin, artemether, artesunate
Destruction of DNA	Metronidazole	Metronidazole	DNA	Metronidazole

Antiviral drugs

Finding antiviral drugs that target some critical aspect of the viral life cycle but do not damage host cells is always a challenge because viruses use many of the cell's own systems (Table 24.5). Many broad spectrum antiviral agents are based on **nucleoside analogs** (discussed next), which competitively inhibit DNA synthesis by becoming incorporated into the growing chain during viral replication. These drugs exhibit selective toxicity because they take advantage of the rapid replication rates of viral genomes, resulting in the inhibition of viral replication much more than host cell replication.

TABLE 24.5 Action, structure, and targets of antiviral drug groups

Action	Class/group	Structure	Target	Examples
Antiviral drugs				
Inhibition of DNA synthesis	Nucleoside analogs	Acyclovir (guanine analog)	DNA	Acyclovir, AZT, ribavirin
Inhibition of influenza virus uncoating	Amantadine	Amantadine	Influenza virus M2 protein (an ion channel)	Amantadine, rimantadine

Eukaryotic polymerases also show lower affinities for these analogs than do viral polymerases. There are some tailor-made antiviral drugs that cater to particular viruses or groups of related viruses. The designer drugs amantadine and oseltamivir, commonly known as "Tamiflu," are specifically active against influenza viruses.

CONNECTIONS Amantadine interferes with the viral envelope protein M2, preventing uncoating of the virus, and Tamiflu inhibits viral neuraminidase, an enzyme essential for dissemination of newly budded virions. See **Chapter 8, The Rest of the Story** for more information.

Inhibition of DNA Synthesis

Nucleoside analogs are structurally similar to nucleosides and once phosphorylated in the cell, compete with normal nucleotides for viral DNA polymerases. Acyclovir, a commonly used antiherpes drug, is a dideoxyguanine analog that becomes incorporated into DNA during synthesis (see Figure 8.24). Idoxuridine is a base analog of thymidine. Base analogs produce mutations in subsequent rounds of replication as they cannot be correctly interpreted by polymerase. Azidothymidine (AZT), also called "zidovudine," is a therapeutic drug for HIV infection **(Figure 24.9)**. It is a dideoxythymidine analog that causes DNA chain termination because it lacks the $3'$—OH group needed for nucleotide addition (see Figure 8.23). Incorporation of AZT into DNA during reverse transcription of the viral RNA genome prevents the completion of a DNA copy needed for insertion into the host genome for viral replication (see Section 8.4). AZT is somewhat selective as an antiretroviral drug due to its higher affinity for the reverse transcriptase enzyme than for host cell or other viral DNA polymerases. Resistance to AZT is common in HIV strains due to mutations in the reverse transcriptase gene, a phenomenon that will be examined in the next section.

 Antivirals ANIMATION

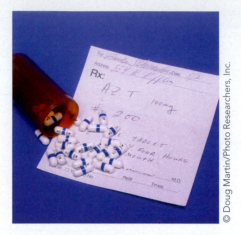

© Doug Martin/Photo Researchers, Inc.

Figure 24.9. Azidothymidine (AZT) AZT prevents nucleotide addition and terminates chain synthesis. Also called "zidovudine," it is prescribed for HIV infection. It is therapeutically effective because it is more active against reverse transcriptase than against host cell DNA polymerases.

24.2 Fact Check

1. What is meant by the term "selective toxicity"?
2. Why is selective toxicity a particular challenge in finding drugs for viruses and eukaryal pathogens?
3. What is the mode of action of β-lactam antibiotics?
4. Which classes of antibiotics are considered ribosome inhibiting, and how do they work?
5. Describe some of the cellular targets and mechanisms of action of antifungal and antiprotozoal drugs.
6. Describe some of the mechanisms of action of antiviral drugs.

24.3 Antimicrobial drug resistance

How do microbes become drug resistant?

Microbes are masters at adapting to adverse conditions because they are experts at genetic change. This genetic adaptability is the reason for their continued success on the planet and their ability to exist in every known habitat ever explored. No other life-form has proven capable of conquering such varied and seemingly adverse conditions (see Section 1.3). Because microbes are adept at protecting themselves from various inhibiting compounds found naturally in their environment, an antimicrobial drug is another toxic substance, a change to their environment that they must overcome. Today, the development of drug resistance by microbes is a serious challenge to health care, not only for the treatment of common infectious diseases, like strep throat

and tuberculosis (see Section 21.3), but for the prevention of infection following medical procedures. The U.S. Centers for Disease Control and Prevention (CDC) report that in the United States alone, about 2 million people contract bacterial infections in hospitals each year. Of these, about 90,000 patients die. The majority of bacterial pathogens (about 70 percent) causing those hospital-acquired infections show resistance to at least one antibacterial drug.

As we learned in Section 9.3, the selection for resistance genes from a population of bacteria by antimicrobial drugs can be demonstrated easily through controlled experiments in the lab. What is the evidence for selection for resistance in clinical settings? The following documented global trends provide

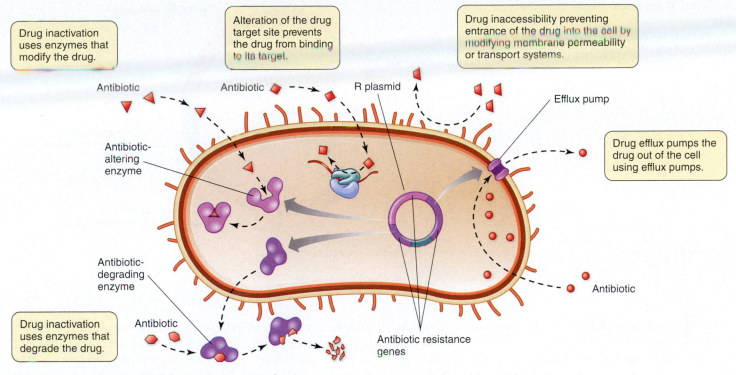

Drug inactivation uses enzymes that modify the drug.

Alteration of the drug target site prevents the drug from binding to its target.

Drug inaccessibility preventing entrance of the drug into the cell by modifying membrane permeability or transport systems.

Antibiotic

Antibiotic

R plasmid

Efflux pump

Antibiotic-altering enzyme

Drug efflux pumps the drug out of the cell using efflux pumps.

Antibiotic-degrading enzyme

Antibiotic

Drug inactivation uses enzymes that degrade the drug.

Antibiotic

Antibiotic resistance genes

Figure 24.10. Antimicrobial drug resistance mechanisms Microbes such as bacteria resist antibacterial drug action by one or more of the following general mechanisms: drug inactivation, alteration of drug target site, drug inaccessibility, or drug efflux. The genes encoding the products for antibiotic resistance are found on R plasmids (shown here) or on the chromosome.

compelling evidence that antimicrobial drug use is linked to development of resistance:

- Changes in antimicrobial drug use are positively correlated with changes in prevalence of resistance.

- Increasing the length of antibiotic treatment in patients is associated with increasing rates of colonization of these patients with resistant organisms.

- Microbial resistance is highest in facilities where antimicrobial drugs are used with the highest frequency.

- Patients with resistant strains have received antibiotics more often than patients with non-resistant strains.

This leaves little doubt that resistance is a direct consequence of antimicrobial drug use. Important questions remain, however. What are the molecular mechanisms of resistance, where do the genes for resistance originate, and how does selection for resistance occur?

Molecular mechanisms of resistance

Resistance mechanisms of bacteria have been the most thoroughly studied because bacterial drug resistance is a widespread phenomenon, associated with almost all bacterial pathogens. As shown in **Figure 24.10**, one or more of the following four mechanisms is generally used to render drugs ineffective:

- drug inactivation by producing enzymes that destroy or structurally modify the drug

- alteration of drug target site, such as modification of its binding target

- drug inaccessibility, achieved by preventing drug entry into the cell

- drug efflux, achieved by pumping the drug out of the cell

Examining just a few examples of drug resistance mechanisms to β-lactams, chloramphenicol, and sulfonamides will provide a model for understanding how resistance to other antimicrobials generally happens.

Resistance to β-Lactams

Resistance to β-lactams is achieved in four major ways:

- production of β-lactamase (drug inactivation)

- alteration of penicillin-binding proteins (PBPs) (alteration of drug target site)

- reduction of membrane permeability to the drug (drug inaccessibility)

- drug removal from the cell (drug efflux)

β-Lactamase Many different types of β-lactamases exist among bacteria, with some more active against certain β-lactam antibiotics than others. All β-lactamases sever the C—N bond of the β-lactam ring (see Figure 2.19), rendering the drugs unable to bind to their PBP targets. Thus, peptidoglycan synthesis is not disrupted. Beta-lactam-mediated resistance in Gram-negative bacteria is most commonly due to possession of **resistance (R) plasmids** carrying genes

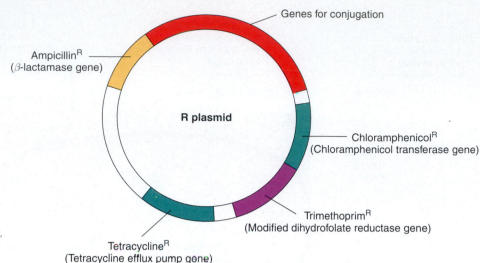

Figure 24.11. R plasmids Resistance (R) plasmids contain multiple genes that code for proteins that impart resistance to a range of antibacterial drugs (noted as superscript R). Some examples of these resistance genes are shown. R plasmids also carry genes needed for their transfer to other cells, such as those used for conjugation.

Genes for conjugation

Ampicillin^R
(β-lactamase gene)

R plasmid

Chloramphenicol^R
(Chloramphenicol transferase gene)

Trimethoprim^R
(Modified dihydrofolate reductase gene)

Tetracycline^R
(Tetracycline efflux pump gene)

for β-lactamases **(Figure 24.11)**. R plasmids encode genes that confer the ability to resist the antimicrobial activity of one or often several drugs and are usually transferred from cell to cell by conjugation. From Section 2.4, you may recall that one way to combat the activity of β-lactamases is to combine β-lactams with clavulanic acid. Clavulanic acid binds to β-lactamases and inhibits their ability to cleave the β-lactam ring, thus rendering β-lactam treatment effective. R plasmids have become common in populations of *Salmonella* Typhi, *Escherichia coli*, *Klebsiella pneumoniae*, and *Proteus vulgaris*.

Altered PBPs The most notorious example of PBP-mediated resistance is that of methicillin-resistant *Staphylococcus aureus*, commonly referred to as *MRSA*. *S. aureus* normally has genes coding for five different PBPs, all highly susceptible to inactivation by β-lactams. However, MRSA strains possess a sixth PBP gene called *mecA* that has been transferred to the chromosome from an unknown source. This new PBP is highly resistant to all β-lactams, yet has sufficient peptidyl transferase activity to provide peptidoglycan synthesis in the absence of activity from the other five.

Reduction of Membrane Permeability Recall that one of the reasons for the development of semisynthetic β-lactams was to improve transport across the outer membrane in Gram-negative bacteria. To counter the ability of these newer drugs to reach the periplasmic PBPs, some Gram-negative bacteria produce altered porin proteins to restrict the transport of these drugs.

CONNECTIONS Recall from **Section 2.4** that porin proteins form channels that allow diffusion of small polar molecules across the membrane into the periplasm, where they are then transported by other proteins into the cytoplasm **(Figure 2.24A)**. In many Gram-negative bacteria the pores are too small for entry of large peptidoglycan-targeting antibiotics like vancomycin.

Drug Removal Microorganisms can use **efflux pumps**, as shown in Figure 24.10, to move a drug directly back out of the cell before it can reach its target. Efflux pumps are protein complexes used to transport compounds directly out of the cell. The presence of the compound inside the cell induces the pump proteins to extrude it through active transport. These systems can transport the drug back across the membrane as quickly as it comes in. Multi-efflux pump systems are promiscuous and can transport a variety of structurally unrelated compounds, including antimicrobial agents, out of the cell. As such, multi-efflux pumps are frequently associated with multidrug resistance. Some clinical strains of *Pseudomonas aeruginosa* possess an operon containing genes for multi-efflux pumps, capable of efficiently transporting not just β-lactams, but a variety of drugs out of the cell.

Resistance to Chloramphenicol

Chloramphenicol resistance, which is common in many bacterial strains, occurs in two major ways:

- production of chloramphenicol acetyltransferase (drug inactivation)
- reduction of membrane permeability to the drug (drug inaccessibility)

A resistance gene coding for chloramphenicol acetyltransferase (CAT) is often found on R plasmids (see Figure 24.11), especially among members of Enterobacteriaceae. The enzyme transfers an acetyl group from acetyl coenzyme A, found in the cell, to chloramphenicol **(Figure 24.12)**. This modified chloramphenicol can no longer bind to its 23S rRNA target. Possession of mutations that increase expression of CAT imparts a high level of resistance to chloramphenicol. Reduced membrane permeability to chloramphenicol is associated with intermediate levels of resistance.

Resistance to Sulfonamides

The major mode of resistance to sulfonamides is through alterations of the enzyme dihydropteroate synthetase (alteration

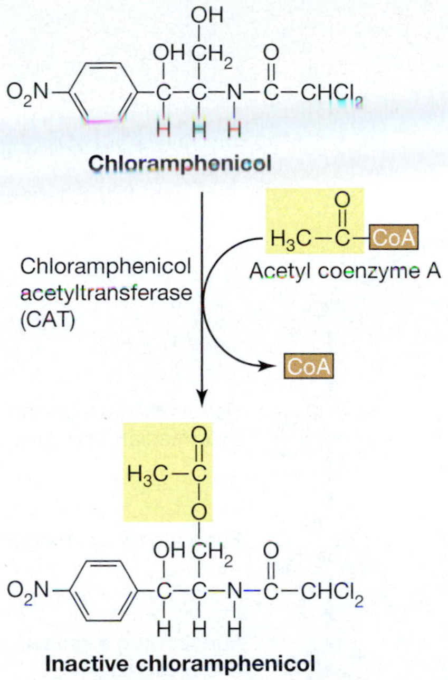

Figure 24.12. Inactivation of chloramphenicol by chloramphenicol acetyltransferase (CAT) Cytoplasmic CAT transfers the acetyl group from acetyl coenzyme A onto chloramphenicol. The presence of the acetyl group prevents chloramphenicol from binding to its ribosome target. Bacterial mutations that increase the expression of CAT produce a high level of resistance to the drug.

of drug target site). Recall from Figure 24.8 that sulfonamides interfere with folic acid synthesis, needed for nucleic acid synthesis, by competing with PABA for the enzyme dihydropteroate synthetase. By producing a modified dihydropteroate synthetase enzyme that has reduced affinity for the drug, folic acid can be synthesized in sufficient amounts. Some bacteria can increase their ability to take up preformed folic acid from their environment, rendering the metabolic pathway for folic acid redundant. Similarly, resistance for the drug trimethoprim, which interferes with the same metabolic pathway, occurs by modification of the enzyme dihydrofolate reductase (see Figure 24.11).

Natural selection and drug resistance

The characteristics of microbes, like all beings, are shaped through natural selection, the evolutionary process first described by Charles Darwin that gave rise to the phrase "survival of the fittest." Fitness is measured by an organism's relative ability to reproduce. Through natural selection, the genotypes of organisms possessing favorable traits—those that allow them to more successfully reproduce under the pressures imposed by their environment—will become more prevalent in successive generations. Natural selection forms the cornerstone of evolutionary processes for all genetic beings, and microbes are no exception. For a microbial population to survive in the face of a selective pressure, like an antimicrobial drug, it must adapt. Of course, microbes cannot be masterminds of this adaptation. No organism is able to self-consciously alter its own genes to suit itself. Nor are the genetic changes *caused* by the presence of the selective pressure. Instead, individual microbes acquire changes in their genetic information by chance through

spontaneous natural processes such as random mutations, recombination, and horizontal gene transfer (see Section 9.5). These chance genetic changes, which happen all the time in microbial populations, *occur in the absence of the selective agent*, the drug in this case.

CONNECTIONS Recall in **Section 9.3** that two milestone experiments were conducted by Ester Lederberg (replica plating using antibiotics) and Luria and Delbrück (the fluctuation test using phage) that demonstrated resistant mutants were present in cultures *before* the addition of the selective agent.

While most random genetic changes will have no impact on drug resistance, a few will. In a large population of replicating microbes it is a biological certainty that a few members will have acquired a genetic change that confers some level of resistance to a given drug, thereby increasing their fitness in the presence of the drug. So, if a few resistant microbes are expected to exist in every population, doesn't that mean some resistant microbes will always remain in a patient after every antimicrobial treatment? Does every patient become a breeding ground for resistant microbes? Of course the answer is no, or antimicrobial drugs would be relatively useless.

Two important aspects must be considered. First, the effectiveness of antimicrobial drugs comes from their ability to work in synergy with the immune system. Antimicrobial drugs don't need to destroy every single infecting microbe to be effective **(Figure 24.13)**. Indeed, bacteriostatic drugs may not achieve killing in the short time many are used. Antimicrobial drugs only need to reduce or halt the spread of the microbial population enough to let immune defenses clear the infection. Recall that it requires 10–14 days to mount a peak adaptive immune response to a new infection, and some microbes can reach high numbers in this time, overwhelming the body (see Figure 20.6 and Figure 21.4). Treatment with antimicrobial drugs can prevent microbes from proliferating to this extent, so when the adaptive response kicks into action, it will not be overcome. Of course, some microbes can effectively evade adaptive immune responses, and will be difficult to treat even with antimicrobial agents. HIV is a prime example of this. Not a single case of absolute clearance of this virus with antimicrobial drug therapy has been recorded.

Second, antimicrobial drugs are used at therapeutic levels that are sufficiently high to discourage selection for resistance while still being safe for use *in vivo*. How does a physician determine which drug, what concentration, and how long it should be used in treating an infection? *In vitro* **susceptibility testing** is done to establish the sensitivity of a particular microorganism to a panel of select antimicrobial drugs and is used to determine which antimicrobial drug will be most successful in treating a bacterial infection *in vivo*. Susceptibility patterns of various bacteria and fungi are examined by both *in vitro* and *in vivo* tests. Prior to drug approval for clinical use, controlled studies are done *in vivo* to determine how useful a particular antimicrobial drug may be. This includes determining how

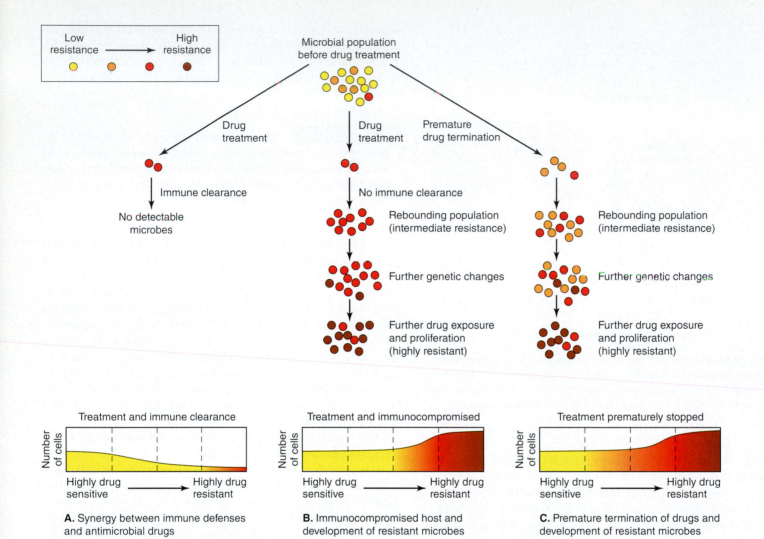

Figure 24.13. Development of resistant microbes In a population of infecting microbes, there may be a variety of resistance levels present. Antimicrobial drug treatment will remove microbes that are susceptible to the drug, leaving microbes with higher resistance levels. **A.** In healthy persons with competent immune systems, the few remaining microbes with intermediate drug resistance will often be cleared. Complete clearance is sometimes not achieved with some microbes such as HIV. **B.** If the host is immunocompromised, drug-resistant microbes not removed by immune defenses may proliferate and acquire other genetic changes that increase resistance. Further exposure to the drug selects for these microbes, giving rise to a highly resistant population. **C.** If the drug treatment is stopped prematurely, even microbes with relatively weak resistance may be left to proliferate and go on to acquire further genetic changes. Further exposure to the drug selects for individual microbes that are more resistant, giving rise to a highly resistant population. The graphs depict the corresponding changes with time in the infecting microbial population.

stable it is inside the body, how quickly it may be metabolized or excreted, and what the effective, therapeutic, nontoxic dose is in patients. This therapeutic dose is related to the **minimal inhibitory concentration (MIC)**. The MIC of a drug is the lowest concentration at which no growth of the microbe being tested occurs *in vitro*, after a standardized period of time. The MIC gives a measure of the drug susceptibility of a microbe.

As described in **Toolbox 24.1**, MIC values for various drugs are determined by *in vitro* testing of hundreds of clinical samples of a particular pathogen. The aim of antimicrobial therapy is to maintain a drug concentration above that of the MIC at the site of infection, as measured in particular tissues or body fluids such as serum, while minimizing toxic side effects. For example, the recommended oral therapeutic dose of penicillin V for treatment of pneumonia caused by *Streptoccocus pneumoniae*

is 43 mg/kg body weight per day. This gives an average serum level of 6.0 μg/mL, which surpasses the established MIC of 0.02 μg/mL for penicillin V-sensitive strains of this pathogen. This higher target concentration is used to minimize selection for resistance.

If the immune system can clear most infections with the help of antimicrobial drugs, then how do drug-resistant microbes arise? The immune system is not always effective in clearing resistant microbes, and some patients do not comply with taking their antimicrobial drugs. These are the two most common ways resistant microbial populations develop. Figure 24.13B depicts events in an immunocompromised patient. Resistant microbes that remain after the drug course is finished can continue to evade immune defenses and proliferate. The descendants of even partially resistant microbes can acquire more genetic changes, some of which may provide an even higher level of resistance. With further drug exposure and selection, the generational line of microbes goes from susceptible to fully drug resistant. A similar situation occurs when patients are noncompliant and stop taking the drug when they feel better, or miss doses (see Figure 24.13C). To treat diseases like malaria and tuberculosis, multiple antimicrobial drugs need to be taken two or three times a day. The average length of treatment for tuberculosis is 6–9 months. It is easy to miss a dose or even an entire day during that period. This increases the risk for selection of drug-resistant microbes.

> **@ Premature termination of antibiotic treatment**
> **ANIMATION**

The origin of drug resistance genes

Where do drug resistance genes originate? Logically, if natural selection is working at the genetic level, then the genes for microbial resistance must have been there in some form before exposure to the drugs. Let's examine where these genes may have originated.

Natural Resistance

Not all microbes become resistant to antimicrobial drugs through the use of these drugs. Indeed, drug resistance is not reserved for pathogens. Many environmental microbes that are not involved in infectious disease also show drug resistance. This natural resistance has arisen through evolution to accommodate selective pressures *unrelated* to drug use.

Intrinsic Drug Resistance Many microbes are resistant to drugs by virtue of their natural structure or metabolism. They do not possess specific resistance genes to target antimicrobial drugs. We have already seen that Gram-negative bacteria are naturally resistant to some antibacterial agents because the drugs cannot pass through the outer membrane. The wall-less bacterial genus *Mycoplasma* is naturally resistant to β-lactams because members do not produce peptidoglycan. The thick layer of hydrophobic mycolic acids that characterizes the genus *Mycobacterium* similarly prevents water-soluble drugs from accessing the peptidoglycan or the plasma membrane, and prevents transport into the cell. In addition, pathogenic mycobacteria produce intracellular infections, so any drug that is active against them needs to enter both the host cell and the bacterium in order to be effective. Only a few drugs are capable of doing this (see Figure 24.4).

As we have already learned, fungi and bacteria commonly possess naturally evolved membrane multi-efflux pumps (see Figure 24.10), which they use to transport various environmental compounds out of the cell. Due to the rather indiscriminate nature of these multi-efflux pumps, antimicrobial drugs can also be transported out of the cell. This intrinsic drug resistance has not evolved from exposure to antimicrobial drugs.

Resistance Genes in the Environment A second way resistance arises in natural settings is through the development of specific antibiotic resistance genes. These genes were present long before the use of antibiotics as drugs. How can this be? Consider that most microorganisms live in complex, highly competitive ecosystems. Antibiotic production in these habitats can secure an advantage by limiting the growth of competitors for resources. Of course, to successfully compete with these antibiotic producers, microbes counter with the development of resistance genes. In fact, some resistant bacteria can even use the antibiotics produced by other microbes as a nutrient source. Recently, several antibiotic-consuming species have been isolated from soil. When grown in the presence of clinically relevant concentrations of antibiotics, these bacteria remained unaffected and indeed were able to grow on media containing the antibiotic as a sole carbon source. Several of these isolates are resistant to multiple antibiotics from different classes. In such natural settings, the pressure of microbial competition leads to antibiotics and antibiotic resistance genes. As we saw in Section 24.1, most antimicrobial drugs used today were developed from such natural antibiotics (see Table 24.1). But what about in clinical settings? What is the source of resistance genes of pathogens? The **Mini-Paper** provides some insight into the possible role of environmental bacteria as a reservoir for the resistance genes found in pathogens.

Resistance Due to Antimicrobial Drug Use

Medically important microbes become drug resistant as a direct consequence of antimicrobial drug use. Even Alexander Fleming noted the appearance of penicillin-resistant *Staphylococcus aureus* in cultures he was working with in the lab. Penicillin-resistant strains of *S. aureus* started appearing in patients receiving the drug shortly after it entered widespread use (see Figure 24.2). Drug resistance genes in clinical strains of microbes come from two sources: preexisting genes and newly acquired genes. The mechanisms of drug resistance are most varied in bacteria, in which they have been most closely investigated. Consequently, we'll focus on these.

Modification of Preexisting Genes Random mutations form the basis of all genetic changes, and occur because of:

- the imperfect ability of polymerases to precisely copy their template, resulting in spontaneous mutations (see Sections 7.5 and 9.3)
- exposure to mutagens, resulting in induced mutations (see Section 7.5)

Although these mutational events happen infrequently (for *E. coli* this is between 10^{-6} and 10^{-7} mutations per gene per generation), the rapid replication rate of most microbes means that mutations can accrue in a population over a relatively short period of time. Recall in Section 9.3, the Luria–Delbrück fluctuation test demonstrated that a random mutation imparting resistance to phage in a single *E. coli* cell gave rise to a population of resistant progeny in a test tube (see Figure 9.10). In such a way, key mutations in a preexisting gene, whose major function is not originally to confer antibiotic resistance, can often bestow resistance. For example, bacteria can develop resistance to β-lactam antibiotics through key nucleotide changes in the genes encoding PBPs (penicillin-binding proteins). This results in amino acid changes that allow continued function of the transpeptidase activity, but prevent binding with β-lactam antibiotics. Similarly,

MIC values can be determined by performing a dilution susceptibility test, as shown in **Figure B24.1**. A dilution series, usually twofold dilutions of a particular antimicrobial drug is inoculated with a standardized culture of the test organism and incubated for a prescribed period of time. The lowest concentration of the drug resulting in no growth is the MIC. The determination of MIC by this method necessitates a separate dilution series for each antimicrobial drug and isolate to be tested. As clinical labs can receive hundreds of samples a day, the tests are commonly formatted for high throughput automated analysis in plates of 96 wells or more that facilitate multiple tests at a time.

MIC determination using a dilution series is not always a practical way to determine drug susceptibility to a panel of antimicrobial drugs. Many labs will not have the automated equipment required. An alternative and more manageable method for determining MIC is done using antibiotic-impregnated strips, each of which contains a gradient of antibiotic concentrations and allows the MIC value to be read directly off the strip. Etest® plastic strips allow a panel of antimicrobial drugs to be tested simultaneously **(Figure B24.2)**.

How does finding the MIC *in vitro* allow one to know if the drug will be effective *in vivo*? Recall that the therapeutic dose needed for effective treatment is not the same as the MIC. The dose is determined by knowing the amount needed to attain a drug concentration higher than the MIC inside the body. A practical way to carry out susceptibility testing and approximating MIC is the Kirby-Bauer disk diffusion test, named after its original developers. The test generates the profile of antimicrobial drugs that a particular isolate is susceptible to, and this in turn determines the drug(s)

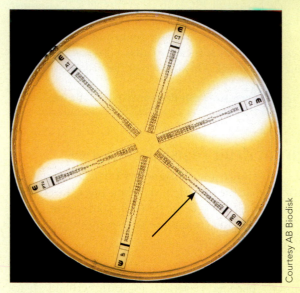

Figure B24.2. Epsilometer test (Etest) Each Etest® strip contains a concentration gradient of an antimicrobial drug. The strips are laid on the inoculated plate with the ends containing the lowest concentration in the center. The MIC is read directly off the strip at the point where the zone of inhibition intersects the strip (see arrow).

Courtesy AB Biodisk

most useful for effective therapy. A panel of small absorbent disks impregnated with known concentrations of antimicrobial drugs is placed aseptically on a Mueller–Hinton agar plate on which a lawn of culture of the microorganism being tested has been seeded at a standardized density **(Figure B24.3)**. The plate is incubated and

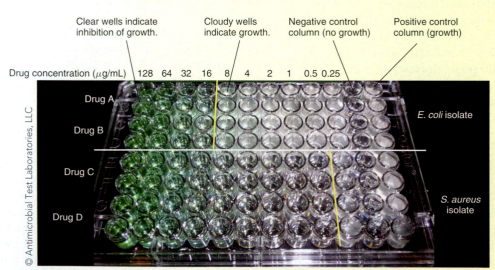

Clear wells indicate inhibition of growth. Cloudy wells indicate growth. Negative control column (no growth) Positive control column (growth)

Drug concentration (μg/mL) | 128 | 64 | 32 | 16 | 8 | 4 | 2 | 1 | 0.5 | 0.25

Drug A
Drug B
E. coli isolate
Drug C
S. aureus isolate
Drug D

© Antimicrobial Test Laboratories, LLC

Figure B24.1. Minimal inhibitory concentration (MIC) determination using dilution susceptibility tests Wells contain ten dilutions, shown along the top of the plate, done in duplicate for four different antimicrobial drugs in a green solution. Two clinical isolates, *Escherichia coli* (*top*) and *Staphylococcus aureus* (*bottom*), were inoculated into the wells and incubated for 18 hours. The concentration of the drug in the wells at which no growth first appears is the MIC. As indicated by the yellow lines, for the *E. coli* isolate the MIC is 16 μg/mL for both drugs A and B. For *S. aureus*, the MIC is 0.5 μg/mL for drugs C and D.

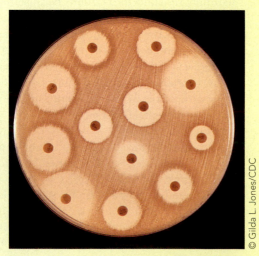

Figure B24.3. Kirby–Bauer disk diffusion test A lawn of the isolated microorganism is plated on Mueller–Hinton agar and disks impregnated with known concentrations of antimicrobial drugs are added. After a suitable incubation time, the diameter of the zone of inhibition for each is measured.

© Gilda L. Jones/CDC

the zones of inhibition, defined as growth-free zones, are measured. The zones are produced because the drug diffuses at a known rate into the agar, resulting in a concentration gradient with the highest concentration near the disk and becoming lower as the distance from the disk increases. For a given drug, the wider the zone, the more susceptible the microbe being tested is. The diameters of the zone of inhibition can therefore be used to determine which drugs the microbe is susceptible to. Each drug diffuses at its own rate so a drug that produces a large zone does not necessarily mean it is more effective at inhibiting growth than a drug showing a smaller zone.

To interpret a Kirby–Bauer test, the zone of inhibition diameter for each antimicrobial drug tested is compared to a table, like that shown in **Table B24.1**, derived from the compilation of hundreds of MIC and zone diameter values for many microbial strains of a particular species. The determination of whether a microbe is resistant, susceptible, or shows intermediate resistance to the various drugs has been determined by correlating these MIC values with the actual therapeutic concentration of the drug found when in the body. For example, if concentrations of ampicillin must reach 9.0 μg/mL in serum to be effective, then any S. aureus isolate with a zone of inhibition diameter less than or equal to 13 mm will have an MIC value greater than 9.0 μg/mL, and therefore will not be sensitive to therapeutic levels of ampicillin.

TABLE B24.1 Zone of inhibition diameter and susceptibility for *Staphylococcus aureus*

Antimicrobial drug	Quantity in disk (μg)	Zone of growth inhibition diameter (mm)		
		Resistant	Intermediate	Susceptible
Ampicillin	10	<12	12–13	>13
Chloramphenicol	30	<13	13–17	>17
Erythromycin	15	<14	14–17	>17
Tetracycline	30	<15	15–18	>18
Streptomycin	10	<12	12–14	>14

● Test Your Understanding ·················

You test a bacterial isolate from a patient using the Kirby–Bauer disk diffusion test. The results of your tests are shown below. Which of the following antimicrobial drugs would you recommend to treat this infection and why?

Antibiotic tested	Diameter of zone of inhibition
Tetracycline	15 mm
Ampicillin	16 mm
Erythromycin	16 mm

exposure to antimicrobial agents frequently selects for mutant genes of natural efflux pump systems that can efficiently move these drugs out, resulting in acquired drug resistance. Each independent random mutation confers a small decrease in affinity to the drug, diminishing the bacteria's susceptibility to the drug, and making the drug less effective. Resistance to the entire penicillin family has occurred in this way for strains of *Staphylococcus aureus* and *Staphylococcus epidermidis*, *Pseudomonas aeruginosa*, *Enterococcus faecalis*, *Enterococcus faecium*, *Haemophilus influenzae*, *Neisseria gonorrhoeae*, and *Neisseria meningitidis*, just to name a few.

Is mutational gene modification leading to resistance unique to bacteria? No. Resistance of the virus HIV to the antiretroviral drug AZT (see Section 24.2) developed through mutations in the gene encoding the viral reverse transcriptase. Recall that AZT is an analog of thymidine and causes DNA chain termination during reverse transcription (see Figure 8.23). Like PBP-mediated resistance, this is not an all-or-nothing event. As HIV accumulates key mutations in the reverse transcriptase gene, its susceptibility to AZT falls markedly. Only four specific mutations at four key sites are needed for complete resistance to AZT. In a virus such as HIV where there is no polymerase proofreading ability, every generation of new viral genomes contains mutations (see Section 22.4). Studies have shown that it takes as little as 10 months for resistance to appear in patients receiving AZT treatment. AZT-resistant HIV strains are now widespread.

Acquisition of New Genes Horizontal gene transfer events involve large-scale sequence changes that are brought about by:

- transposable elements
- transducing phages
- plasmid transfer
- recombination events

CONNECTIONS Recall from **Section 9.5** that microbes, particularly bacteria, can donate, receive, and rearrange DNA to gain new genetic information, including resistance genes, through horizontal gene transfer.

Mobile genetic elements, such as plasmids, transposons, and transducing phages, that facilitate horizontal gene transfer, can efficiently transfer entire resistance genes from one microbial cell to another, almost instantaneously. You'll recall from Chapter 21 that in bacteria, horizontal gene transfer also moves virulence genes around in a rather promiscuous manner, leading to gene transfer in related as well as unrelated species. This happens for resistance genes as well. For example, the transposons Tn*5* and Tn*10*, which encode, respectively, resistance to aminoglycosides and tetracycline, are found in many species of Gram-negative bacteria, particularly the Enterobacteriaceae. *In vitro* experiments show that the tetracycline resistance transposon Tn*5397* can be spontaneously transferred from *Bacillus subtilis*, a common soil bacterium, to an oral *Streptococcus* species. Tn*1546*, which confers vancomycin resistance in *Enterococcus faecium*, can be transferred to

V. M. D'Costa, K. M. McGrann, D. W. Hughes, and G. D. Wright. 2006. Sampling the antibiotic resistome. Science 311:374–377.

Context

The development of antibiotics as therapeutic compounds has had profound beneficial effects on the treatment of infectious disease. Most antibiotics have their origin in soil microbes that produce them to gain a competitive advantage by inhibiting other microbes. Soil microbes also possess antibiotic resistance genes to counter the effects of the antibiotics they encounter. Could these microbes serve as a reservoir for resistance genes that emerge in clinically important pathogens? As a first step in addressing this question, it is necessary to gather some information about the occurrence of antibiotic resistance within environmental communities, and then to relate this to the nature of antibiotic resistance that is found in pathogens or potential pathogens.

The study that is described refers to the collective antibiotic resistance that is found in soil microbes known as the "soil resistome." The authors focused on isolating antibiotic-resistant actinomycetes, since these Gram-positive bacteria are known as the main antibiotic producers. The study was limited to culturable organisms only.

Experiment

With the goal of isolating antibiotic-resistant microbes from diverse soils, soil samples representing urban, agricultural, and forest locations were air dried for several days. They were then suspended in sterile water and dilutions spread on solid *Streptomyces* isolation media containing starch and casein, plus cycloheximide to inhibit fungal growth. This media favors the growth of actinomycetes such as *Streptomyces*. During incubation at room temperature, the plates were monitored for the formation of colonies that had the characteristic morphology of sporulating *Streptomyces*. The resulting strain library had 480 strains, and they were all confirmed by 16S rRNA gene sequence analysis to be members of the *Streptomyces*.

Each of the isolated strains was then screened for the ability to grow in the presence of high levels of each of 21 different antibiotics chosen to be representative of the variety of antibiotics used to treat pathogenic bacteria. These included drugs that have been used for decades and several that have only been available recently. Remember, these strains had been isolated in the absence of these antibiotics. Even so, every strain was found to be resistant to several of the antibiotics, and two strains were resistant to 15 of the 21 antibiotics **(Figure B24.4)**. The 480 strains contained 191 different antibiotic profiles, and the relationships between these profiles are shown in Figure B24.4B, where each black dot represents a profile. A colored line connecting a dot indicates the resistance to the drug. Figure B24.4A shows several antimicrobial drugs were almost universally ineffective against the collection of microbes. These include many *Streptomyces*-produced antibiot-

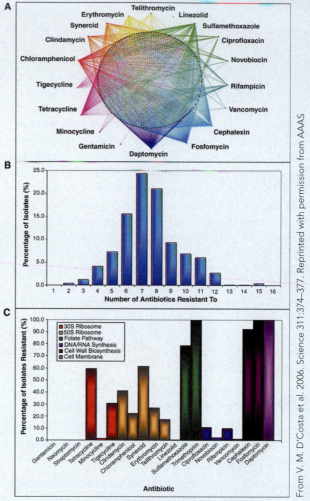

Figure B24.4. Presence of antibiotic resistance of 408 actinomycetes soil strains A. Diagram illustrating density and diversity of resistance to antibacterial drugs. Resistance is indicated by a line connecting a drug and a dot, representing a resistance profile of the isolates tested. **B.** Graph showing percentage of strains resistant to one or more of the 21 drugs tested. **C.** Graph showing percentage of isolates resistant to the drugs tested.

From V. M. D'Costa et al. 2006. Science 311:374–377. Reprinted with permission from AAAS.

ics affecting ribosome function (erythromycin, chloramphenicol, tetracycline, and semisynthetic synercid), and peptidoglycan synthesis (fosfomycin). Widespread resistance was also present for cephalexin, a β-lactam antibiotic produced by a soil fungus, and the ribosome-inhibiting synthetic drug linezolid, which began use only in 2000. Some of the antimicrobial drugs were not effective against any of the 480 isolates (Figure B24.4C), with the most surprising of these being daptomycin, a *Streptomyces*-produced drug permitted for use in 2003, which acts by disruption of the cytoplasmic membrane. Daptomycin is normally thought to be very effective against Gram-positive bacteria, but the soil isolates appeared to readily inactivate it. Resistance to daptomycin had only been observed once before.

Another notable finding was the resistance of five strains to vancomycin, a glycopeptide antibiotic that acts on the peptidoglycan cell wall and is usually reserved as an antibiotic of last resort. Most of the resistant microbes contained the known vancomycin resistance genes whose expression results in altered peptidoglycan, but one strain did not. This strain appears to owe its vancomycin resistance to an undescribed mechanism. The rise of vancomycin resistance in clinical settings is a major problem, and the demonstration of another possible mechanism of vancomycin resistance is worrying.

Impact

This study clearly demonstrated significant and widespread antimicrobial drug resistance in soil bacteria, despite the limited nature of the screening, which only considered cultivated actinomycetes, estimated to compose approximately 5–6 percent of soil bacteria. Intrinsic antibiotic resistance was found for naturally occurring antibiotics, semisynthetic derivatives and completely synthetic drugs. While the high incidence of resistance to naturally occurring antibiotics is perhaps not surprising, resistance to synthetic drugs like linezolid and trimethoprim is unexpected, given the apparent absence of selective pressure. Other studies using functional metagenomics techniques have shown similar patterns of antimicrobial resistance exist in uncultivated microbes. Can these mechanisms of antibiotic resistance find their way into pathogens? This is certainly possible, as antibiotic resistance genes are commonly found on mobile elements such as plasmids. The study may provide a preview of the antibiotic resistance that is likely to appear within human pathogens in the future.

● Questions for Discussion ·····················

1. Why do you think this study focused on the actinomycetes, rather than all culturable bacteria, or another type of bacteria?

2. Are all of these genes likely to function solely for antimicrobial resistance, or might some of them have other functions?

Staphylococcus aureus when the two species are cultured together. The antibiotic resistance plasmid RP1, which was first found in a clinical strain of *Pseudomonas aeruginosa*, is capable of transfer to just about all Gram-negative bacteria by conjugation. Such R plasmids carrying more than one antimicrobial resistance gene (see Figure 24.11) are themselves produced through the insertion of multiple transposable elements.

It is no longer uncommon to find multiple-resistant bacteria—the so-called "superbugs"—that are resistant to virtually all antibacterial drugs licensed for use. For example, strains of *Salmonella* Typhi, *E. coli*, *Klebsiella pneumoniae*, and *Proteus vulgaris* have all been found to carry an R plasmid that is transferable by conjugation that confers resistance to four classes of antibacterial drugs. Maintenance of the entire plasmid is reinforced when exposed to any one of the drugs to which it has resistance. In the same way, resistance cassettes comprising a string of multiple transposable element insertions into the chromosome can also be found. These cassettes can also be mobilized for transport to other cells.

Phages tend to be relatively minor contributors in the transfer of resistance genes between unrelated species because, like most viruses, they tend to be quite host specific. However, transducing phages can sometimes insert their genomes into plasmids within their specific hosts, as do many transposable elements. When this happens, the plasmid need only be transferred through conjugation or transformation to confer resistance in other species.

Recombination events play one of the most important roles in gene transfer and gene rearrangements. We have previously defined recombination as the breaking and rejoining of two different nucleic acid molecules (see Section 9.5). Within cells, recombination can occur between mobile genetic elements and the chromosome, as well as among the mobile genetic elements themselves. All these recombination opportunities mean it is a relatively easy process to move a gene from one cell to another; microbes have incredible genetic dexterity. **Figure 24.14** summarizes some of the ways resistance genes can be transferred between bacterial cells.

With all this gene movement occurring between species, it is not hard to envision how antibiotic resistance genes from environmental sources, such as the soil bacteria mentioned previously, can inadvertently move into microbes carried by humans and animals, and become selected for by antibiotic use. These environmental strains are suspected to be a hidden reservoir for present and future resistance genes. Another source of resistance genes is normal host microbiota. Any antimicrobial drug that is used systemically (taken orally or by injection), acts not only on the target microbe, but on a variety of microbes at various sites, such as the intestinal tract, increasing the possibility of selection for resistance genes in a variety of microbes.

The Rapid Rise of Resistance

On average, resistance to a new antibacterial drug occurs in just four to five years after its introduction (see Figure 24.2). How does resistance appear so quickly? This is due to the combined action of (1) transfer of whole resistance genes by mobile genetic elements, (2) high frequency of random mutations, (3) recombination events, and (4) short generation times. Consider that most medically important microorganisms can double

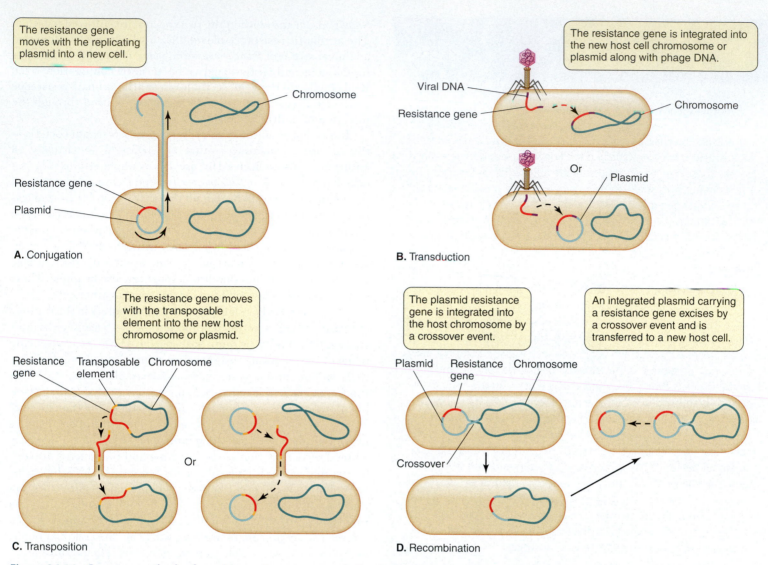

Figure 24.14. Common methods of acquiring resistance genes **A.** Plasmids carrying resistance genes (red) can be transferred to other bacterial cells by conjugation. **B.** Transducing phages can insert resistance genes into chromosomes or plasmids by transduction. **C.** Transposable elements carrying resistance genes on the chromosome or plasmids can be transferred between cells. **D.** Recombination between plasmids and chromosomes can transfer resistance genes within a cell. Once transferred to a plasmid, resistance genes can be easily moved by horizontal gene transfer to other cells.

Labels in figure:

A. Conjugation — The resistance gene moves with the replicating plasmid into a new cell. Chromosome, Resistance gene, Plasmid

B. Transduction — The resistance gene is integrated into the new host cell chromosome or plasmid along with phage DNA. Viral DNA, Resistance gene, Chromosome, Or, Plasmid

C. Transposition — The resistance gene moves with the transposable element into the new host chromosome or plasmid. Resistance gene, Transposable element, Chromosome, Or

D. Recombination — The plasmid resistance gene is integrated into the host chromosome by a crossover event. An integrated plasmid carrying a resistance gene excises by a crossover event and is transferred to a new host cell. Plasmid, Resistance gene, Chromosome, Crossover

their population in less than 24 hours, at least in culture. In comparison, the fastest reproducing multicellular organisms are aphid insects and the nematode *Caenorhabditis elegans*, whose generation time is a lengthy 2.5 days. Relative to humans with a generation time of about 20 years, microbial acquisition of a new characteristic happens in the blink of an eye. With such short generation times, it is no surprise that resistance has been documented for every antimicrobial drug ever used in clinical settings. The more effective and safe an antimicrobial drug is, the more it is prescribed, and the faster resistance spreads, rendering the drug ineffective. It's a Catch-22 that has not gone unnoticed by the pharmaceutical industry. **Perspective 24.1** discusses this dilemma. Rather than continuing to rely solely on the isolation of new compounds by screening products of organisms for antimicrobial activity, the future of antimicrobial drugs includes the rational design of drugs that act on new targets discovered through genomics and structural biology, as will be discussed at the end of this section.

Antimicrobial Resistance in Farmed Animals

The extensive use of antimicrobial drugs over time in human medicine has without doubt led to the rapid rise and spread of resistant microbes. However, agriculture is the second largest consumer of antimicrobial drugs. Approximately 40 percent of all antibiotic use is in farmed animals. What is strikingly different from human applications is how and why these antibiotics are used in animals. Between 40 and 80 percent of the antimicrobial drugs used in farmed animals are not used to treat infections, but to promote the growth of the animals. Massive amounts of data confirm cattle, swine, and poultry fed with subtherapeutic levels of antimicrobial agents show superior weight gain and improved feed efficiencies compared to their antimicrobial-free counterparts. For example, pigs show an increase in weight of 3.3 to 8.8 percent and feed efficiency improves 2.5 to 7 percent, meaning less food needs to be fed to these animals to achieve an equivalent weight gain to their counterparts not receiving antimicrobial drugs. Similar data exist for cattle and poultry. How low levels of antimicrobial drugs

We use more antibiotics now than ever before in history. We use them more liberally, and we often use them incorrectly. For example, antibiotics are commonly prescribed for undiagnosed infections such as sore throats, which commonly are due to viral infections, against which antibiotics are completely ineffective. This abuse has directly increased the prevalence of antimicrobial resistance, making the drugs less useful. Successful new antimicrobial drugs often become widely used, and this leads to their own uselessness—an endless cycle doomed to repeat itself. One must wonder how the pharmaceutical companies are keeping pace. The reality is, they aren't. The number of new antimicrobial drugs approved each year has been decreasing drastically since the 1980s. Although several novel antibiotics have been discovered, few have progressed through clinical trials. Between 1962 and 2000, no new classes of antibiotics were brought to market. In 2000, linezolid, from the class of oxazolidinone synthetic drugs was approved by the FDA, and in 2003 daptomycin from the class of lipopeptide antibiotics followed. The last novel antibacterial drug to make it to market was retapamulin of the mutilin class, introduced in 2007. None of these three classes of antibacterial drugs was actually new, having been discovered in the 1980s. However, for various reasons, usually toxicity issues, they remained undeveloped for human use. They were reexamined largely because of the rise of vancomycin-resistant *S. aureus* and *Enterococcus* strains. Is the antibiotic well drying up? Perhaps, but that's not the only reason for the decline in new drugs. Large, international pharmaceutical companies have lost interest in searching. They simply do not make enough profit in the short time a new drug enters the market to the time resistance appears to warrant the expenditure in research and development. The search for just a single useful new antimicrobial drug requires an estimated $1 billion and takes from between 10 and 25 years to bring to market.

Not surprisingly, pharmaceutical companies have found there is more profit to be made with antidepressants, and drugs for hypertension and cardiac disease. These drugs must be taken for many years, not just a few days, and tend to always remain effective. Antimicrobial drugs have neither advantage. Add to this growing pressures to reduce the use of all antimicrobial drugs, and to keep the most effective ones "on reserve," to be used only when others fail, and the antimicrobial market begins to look even less attractive. Who can blame pharmaceutical companies for moving out of the antibiotic business?

promote growth is not precisely known. Most researchers believe the effects are multifactorial. Studies support evidence for suppression of infectious diseases, such as dysentery, and alteration of normal commensal microbial communities of the gut. It is estimated that as much as 6 percent of the energy in a pig's diet may be harvested by the gut microbiota, instead of by the pig. Reducing microbial populations through antimicrobial use may lead to more nutrient uptake for the animal.

Whatever the mechanisms, subtherapeutic antimicrobial drugs do work to promote growth, and the cost benefits are especially rewarding in large-scale intensive farming operations. Does this application of antimicrobial drugs also contribute to the rise of drug-resistant microbes? The answer in animals is yes, with *Salmonella*, *Campylobacter*, *E. coli*, and *Enterococcus* most commonly associated with resistance. However, it is not clear if the use of growth-promoting antimicrobial drugs has significantly contributed to resistance in microbes important in human health. Data from various countries lack reproducibility and comparability, making it difficult to conclusively implicate the use of antimicrobial agents in farming to resistance in human pathogens. Despite numerous criticisms, it has been established in several cases that drug-resistant bacteria originating from livestock have been transferred to humans, and bacteria that have developed resistance to one antibiotic used in feed have also developed resistance to related antibiotics used in humans. This is known as "cross-resistance" and may have grave consequences in the battle against multidrug-resistant microbes. An example is the glycopeptide antibiotic avoparcin, which is not used in human medicine but is widely used as an additive in poultry feed (see Table 24.2). In 1998, it was noted that isolates of the genus *Enterococcus* from Denmark and Germany were cross-resistant to vancomycin, which is not used in animal feed but is a critical antimicrobial drug used in control of methicillin-resistant *Staphylococcus aureus* (MRSA) strains in humans (see Section 18.5). The transference of this resistance to pathogens like *S. aureus* is a real possibility and will contribute to the dwindling number of options left to combat bacterial pathogens.

For many years, antimicrobial resistance in humans was largely limited to hospital settings; a direct result of the frequent use of antimicrobial drugs there. However, in recent times, resistance has moved into community settings, facilitated by the increasing number of nursing homes, day care centers, and chronic care facilities, use of immunosuppressive drugs, prevalence of HIV infection, and mass transit of people across the globe. In light of this, it can be argued that the contribution of antimicrobial drug use in animals has likely had a minor impact in the rise of resistance in human pathogens. Despite the lack of overwhelming evidence to support the spread of resistance due to farming practices, many countries recognize that antimicrobial agents foster resistance, and microbes frequently do not distinguish between species boundaries. As a result, many countries have banned the routine use of these agents in feed. The European Union banned the routine feeding of all antimicrobial agents to livestock in 2006. Since this, the U.S. Federal Drug and Food Administration (FDA), the U.S. Centers for Disease Control and Prevention (CDC), and the World Health Organization (WHO) have all called for similar bans. The McDonald's Corporation proactively decided in 2003 to purchase chicken only from producers who do not use antimicrobials in feed. This has resulted in the decision of the top ten poultry producers in the United States to voluntarily stop this practice—a significant victory toward the reduction of antimicrobial use.

Even though genetic changes occur at low frequency, the large number of microbes present in, on, and around humans and animals allows a large number of sequence transfer opportunities. These opportunities can reach frightening levels in places like hospitals and long-term-care homes. In the United States alone, these health care-associated infections are estimated to cause 2 million illnesses per year (5–10 percent of hospitalized patients), and cause up to 90,000 deaths per year. As such, the health care facilities represent a major public health problem.

The combination of (1) close contact, (2) infectious microbes, (3) contaminated surfaces, (4) immunocompromised patients, and (5) patients with wounds set the scene for maximizing acquisition of health care-associated infections and genetic transfer events. At a molecular level, mobile genetic elements are moving around just as much as the health care workers themselves. In fact, health care workers are often the vectors for the movement of microbes from patient to patient. Now add to these settings the selective nature of antimicrobial drugs, which are always at high use in these places, and you have a recipe for resistance. Several drug-resistant strains of bacteria have arisen in these settings: vancomycin-resistant strains of enterococci (VRE) and *Staphylococcus aureus*, methicillin-resistant *S. aureus* and *Staphylococcus epidermidis*, cephalosporin-resistant *Klebsiella pneumoniae*, and quinolone-resistant *Clostridium difficile*.

Combating drug resistance

How do we prevent the development of antimicrobial resistance? Stopping the use of antimicrobials is out of the question, but there are several approaches that, if used collectively, may help to reduce the emergence of resistance.

1. *Reduce Use.* Antibacterial drugs in particular are frequently prescribed, indeed demanded by patients for illnesses that are not of bacterial origin. Rapid, cheap, and more accurate diagnostic tests would aid in the decision-making process for prescribing an antimicrobial agent. Educating physicians, patients, and parents about the risk of antimicrobial drug resistance is needed to change these attitudes and practices.

2. *Use Selective Drugs.* Choosing to use antimicrobial drugs with a narrow target range for a specific pathogen rather than a broad spectrum drug, will reduce the selection for resistance genes in non-target species of the normal microbiota. In practice, this is difficult because most physicians cannot wait for identification of the microbe to choose a selective drug. Consequently, broad spectrum drugs are often prescribed when the causative agent of the infection has not been identified. Developing rapid tests for diagnosis will be of benefit.

3. *Use Multidrug Cocktails.* Using a combination of antimicrobial drugs that target different aspects of the microorganism's structure/physiology means that the microbe must develop simultaneous resistance to all the drugs being given in order to survive. This reduces the likelihood of microbial resistance. We have already seen that sulfonamides and trimethoprim are used in combination because they interfere with two different enzymes in the same synthesis pathway for PABA. Multidrug cocktails that target different pathways are used for HIV, tuberculosis, and leprosy.

4. *Use Effective Infection Control.* Hand washing is one of the most effective ways to prevent the spread of contact diseases. Frequent hand washing among care workers, cleaning staff, visitors, and patients is especially important for infection control in health care facilities where immunocompromised patients abound. The density of compromised patients, workers, and visitors increases the risk for infections, many of which are drug resistant **(Perspective 24.2)**. Proper sterilization and quarantine are other important measures.

5. *Develop New Vaccines and Improve Access.* Highly effective vaccines continue to be the cheapest and most effective way to reduce both the spread and impact of disease. Even vaccines that sometimes are not highly protective, such as the yearly formulations of influenza vaccines, still appear to provide enough immunity to reduce severity of disease. This reduces the need to use antimicrobial agents.

6. *Develop Alternatives.* Recombinant DNA technology has recently been used to produce defensins. Defensins are antimicrobial peptides of the innate immune system that have activity against a broad spectrum of Gram-negative and Gram-positive bacteria, fungi, and some enveloped viruses (see Section 19.2). Because they selectively disrupt microbial membranes as opposed to host membranes, they appear to be excellent candidates for developing novel antimicrobial agents. Furthermore, their membrane-disrupting activity is exerted in several ways, which is expected to make it more difficult for the development of resistance by microorganisms. Such recombinant defensins have shown success in clinical and experimental trials when used as therapeutic agents. Another promising alternative to antimicrobial drugs is phage therapy: the administration of specific phage to destroy bacterial pathogens in the body. As described in **Perspective 24.3**, phage therapy is an

Phage can kill bacteria, a fact well known to microbiologists as early as 1915. Why not use them to treat bacterial infections? Indeed, thousands of patients in the early 1900s were treated with phages with apparent success for common bacterial diseases such as dysentery (*Shigella dysenteriae*), cholera (*Vibrio cholerae*), bubonic plague (*Yersinia pestis*), and staphylococcal skin diseases **(Figure B24.5)**. Eastern Europe and Russia were centers for the cultivation of phage stocks to be used by the medical community. In France, L'Oreal, now a famous cosmetics company, marketed several phage remedies, as did Eli Lily in the United States, producing phages in tablet form for easy dispensing and transport.

Phage therapy was pushed aside in the West, as penicillin and other antibacterial agents became mainstream therapies. Despite being widely used for many years, the lack of records and controlled studies has left a checkered history of phage therapy, making it difficult to ascertain just how successful phage therapy was. Recently, the worrisome rise of antibacterial drug resistance has caused researchers to revisit the merits of phage therapy. Phages have several advantages that antimicrobial agents lack:

- They are highly selective for their host, often able to infect only certain strains of a species.
- They remain in the body only as long as the host does.
- They replicate so repeat doses are not needed.
- Most are easily cultivated in the lab.
- Most can be stored for long periods of time and still retain infectivity.
- Lytic phages are absolutely lethal to their host bacterium.

Many studies using phages have been done in animal models and the majority of these have confirmed that phage therapy can be highly effective and very safe. Several issues remain to be addressed before phage therapy becomes a mainstream treatment once again. Some of these concerns include:

- Bacterial phage resistance will likely develop, just as it has for conventional antimicrobial drugs, but could be avoided with multiphage cocktails.
- The species or strain of infecting bacteria will need to be correctly identified in order to use the appropriate phage.
- The phage must maintain its selectivity and not infect non-target microbiota.
- Host production of antibody to the phage may limit its use to topical applications only.

Further study on the efficacy and safety of phages in treating human and animal disease continues, but the production of phages

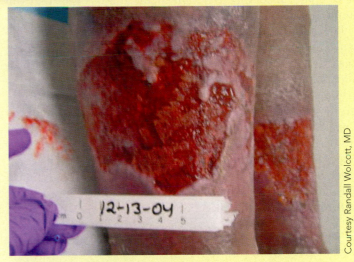

A. Before phage treatment

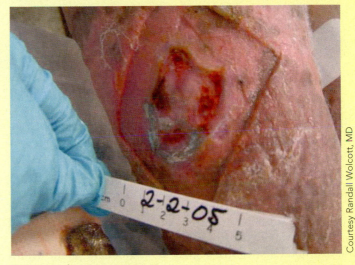

B. During phage treatment

Figure B24.5. Phage therapy A. The ulcerated infected leg shown here did not respond to antibiotic therapy, despite over ten years of treatment. **B.** The leg began to heal with treatment using a phage cocktail containing a mixture of different phage strains against *Staphylococcus aureus*, *Pseudomonas* strains, and *E. coli*. Complete healing was achieved.

for therapy is making a commercial comeback. Biophage of Montreal, Canada, produces several phage therapeutics for *Salmonella* and *E. coli* infections in farmed animals. In 2006, the U.S. Food and Drug Administration approved a spray-on phage cocktail for treating meat for the bacterium *Listeria monocytogenes*, which causes a serious systemic foodborne illness (see Chapter 16 opening story).

old technology that predates penicillin, and has recently been resurrected.

Another approach to find alternatives to the current antimicrobial drugs is to rationally design drugs that will act on new microbial targets. A good place to search for new targets is to start with virulence factors or some other component of microbes needed to interact with host cells. As we learned in Chapter 21, many bacterial pathogens rely on Type III or Type IV secretion systems for injection of components that allow them to colonize and invade host tissues. Many others need fimbriae to attach to host cells. Using knowledge of the structure and biochemistry of common components like these, scientists can use computer modeling to design chemical prototypes that will interact with these structures.

By screening for conserved genes in groups of pathogens, genomics analysis has uncovered many novel potential drug target sites. Most of these genes code for enzymes in critical metabolic pathways or, in viruses, in assembly or genome replication pathways. Genomics and structural studies can also give insight into how easily a microbe may acquire the mutations needed to resist the action of a particular drug. This is an important consideration in drug design as the ability of microbes to modify their genomes to counter drugs is remarkable, to say the least. Although few novel designed drugs have made it to clinical trials, these cutting-edge approaches hold much promise in the future of combating infectious disease.

24.4 Predicting and controlling epidemics

How are epidemics predicted and controlled?

While antimicrobial drugs are invaluable for treating individuals with infectious disease, they are not routinely used as a public health measure for prevention of communicable diseases. As we have just learned, excessive use of antimicrobial drugs leads directly to drug resistance, disrupts normal microbiota balances, and can cause toxicity with prolonged use—not to mention the great expense of such long-term communal use. More practical and effective public health measures are employed to prevent those microbial diseases that present a significant public health risk, that is, communicable diseases capable of causing epidemics. These measures include immunization, the topic of Section 24.5, containment of the pathogen, and importantly, predicting epidemics in order to avert their occurrence.

Epidemics involving known infectious agents are rarely completely unpredictable. The only exceptions are emerging diseases, such as HIV/AIDS and SARS, which have never been previously seen in human populations. However, even these viral pathogens and their associated epidemics did not appear spontaneously (see Sections 18.4 and 22.4). Having studied the viruses, researchers know something about their evolutionary history, how they are transmitted, and who is susceptible. As soon as enough information was gathered in the 1980s regarding HIV, models were developed that predicted HIV would cause significant epidemics in populations practicing unprotected sex with multiple, concurrent partners. Unfortunately, this is exactly what happened in many countries in Africa and Southeast Asia. Health care policies and implementation strategies failed to address the aforementioned host and environmental factors needed to stop the epidemic spread of HIV. Where policies

have been put into place to address these factors, incidence rates of HIV infection are beginning to stabilize or decrease (see Figure 18.12). For all known communicable diseases, a set of predisposing conditions allows the development of an epidemic in a population. Identifying those parameters can help predict, control, and even prevent epidemics. Let's examine some of these parameters.

CONNECTIONS Recall from our introduction to epidemiology in **Section 18.3** that an epidemic occurs when the incidence of cases significantly rises above that normally expected for a population. You may wish to review the terms in that section.

$R_0 = 1$, the epidemic threshold

Epidemiologists have formulated various mathematical equations to model and predict epidemics. Models for propagated epidemics caused by communicable diseases that are spread directly or indirectly between individuals focus on determining the value R_0, **the basic reproduction number**. R_0 is defined as the average number of new infections caused by an infected individual in a population of susceptible individuals. For example, the R_0 value for measles is 12–18, meaning on average, an infected person will spread the infection to between 12 and 18 other people. R_0 values are frequently given as a range, as R_0 values for a particular infectious agent differ from population to population, depending on age, population density, and environmental conditions that may affect the transmission of the agent, such as temperature

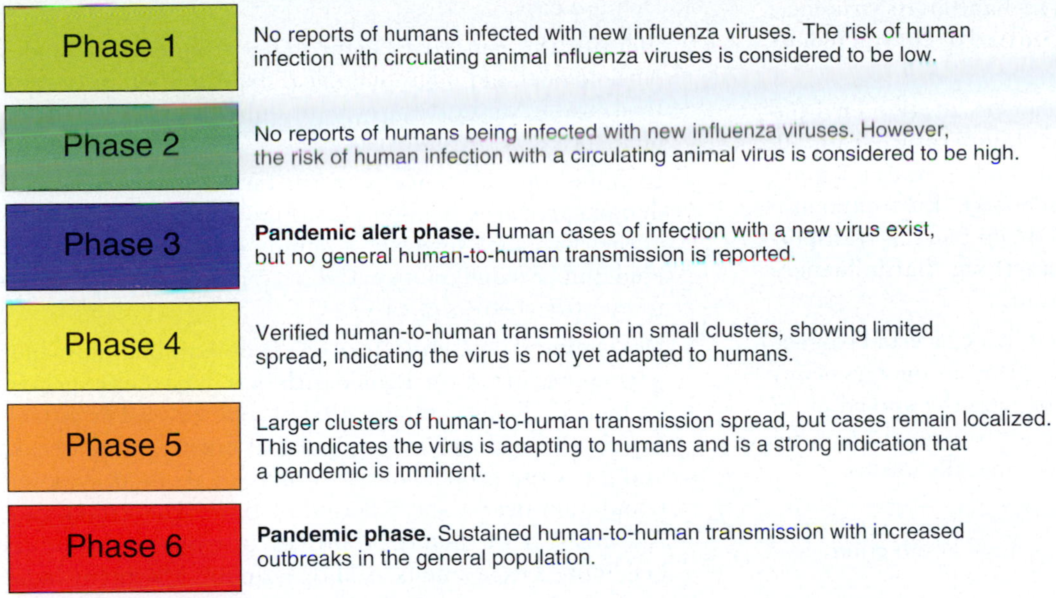

Phase 1	No reports of humans infected with new influenza viruses. The risk of human infection with circulating animal influenza viruses is considered to be low.
Phase 2	No reports of humans being infected with new influenza viruses. However, the risk of human infection with a circulating animal virus is considered to be high.
Phase 3	**Pandemic alert phase.** Human cases of infection with a new virus exist, but no general human-to-human transmission is reported.
Phase 4	Verified human-to-human transmission in small clusters, showing limited spread, indicating the virus is not yet adapted to humans.
Phase 5	Larger clusters of human-to-human transmission spread, but cases remain localized. This indicates the virus is adapting to humans and is a strong indication that a pandemic is imminent.
Phase 6	**Pandemic phase.** Sustained human-to-human transmission with increased outbreaks in the general population.

Figure 24.15. Six phases of influenza pandemic alert The World Health Organization (WHO) uses a six-phased approach for influenza preparedness and response plans. Phases 1 and 2 are the interpandemic period involving surveillance and monitoring of predominantly animal viruses. Phases 3 through 5 indicate a new virus is infecting humans and the level to which the virus can be spread in populations. These are response phases that call for planning and measures to control spread, such as implementation of vaccination, triage recognition in health care settings, and suspension of elective activities such as traveling. Phase 6 declares a pandemic where infections with the new virus have reached the global communities.

and humidity. In general terms, R_0 is a function of (1) density of susceptible contacts, (2) transmissibility of the agent, and (3) the length of time an individual is infectious. Analysis of virtual data and real data from past outbreaks generally confirm that if $R_0 < 1$, the disease will not become an epidemic, and if $R_0 > 1$, it will—assuming no intervention occurs. $R_0 = 1$ then becomes the epidemic threshold, above which an epidemic occurs, and below which it does not. The reason for this can be shown by a simple example. In a population where $R_0 = 18$, one infected individual will give rise to an average of 18 newly infected contacts, who then each infect 18 others, producing 324 more cases (18×18) who then each infect 18 others, for a total of 5,832 new cases (324×18), leading to an exponential increase in the number of cases over time—an epidemic. When $R_0 < 1$, the disease will not become an epidemic because the number of new infections decreases with each round of transmission—the outbreak ceases. Importantly, $R_0 < 1$ does not necessarily mean the disease disappears completely from the population. It can be perpetuated in an endemic manner as long as susceptible individuals continue to be born or immigrate into the population.

In reality, R_0 can rarely be accurately measured, so the answer to our question about whether epidemics can be predicted has to be "not with precision." Still, it is useful to calculate predicted R_0 values. Estimations of R_0 are valuable in assessing the potential risk for an epidemic and the amount of effort that is needed to prevent an epidemic or possibly eliminate an infection from a population. For example, the predicted R_0 value for the H1N1 swine flu was estimated to be approximately 2 as of July 2009. This means on average, each infected individual will infect two others, and up to 80 percent of individuals in an exposed population will become infected. H1N1 swine flu was declared a pandemic by the WHO in June 2009 (see Chapter 8 opening story). The declaration was based on the six defined phases that past influenza pandemics have been observed to follow. Criteria for the six phases are presented in **Figure 24.15**.

As can be seen in Table 24.6, presented in the next section, several communicable diseases have very high estimated R_0 values, indicating they are highly transmissible and therefore likely to produce epidemics in susceptible populations. Understanding

the factors that influence R_0 is therefore critical to the prevention and control of communicable diseases. Frequently, agents with high R_0 values and that cause serious disease are targets for vaccine development.

Principles of epidemic control

How are epidemics controlled or prevented? To answer this question, we need to know the parameters that influence the development of an epidemic. These same parameters will be involved in stopping it.

The Triangle of Epidemiology

A useful conceptual model for controlling infectious disease is the triangle of epidemiology (**Figure 24.16**). This framework illustrates the connection between the three major factors associated with the development of an epidemic. These are:

1. *Host factors*: age, sex, kinship, immigration status, immune status, presence of other diseases, medications. These factors may identify hosts that are most susceptible to the infectious agent.

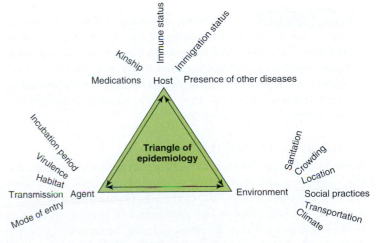

Figure 24.16. Triangle of epidemiology This conceptual model emphasizes the connection between the identified host, agent, and environment in an epidemic. Modifying any one factor or changing a connection, changes the outcome of the disease pattern in a population.

2. *Agent factor*: identity of the agent, its habitat, its virulence, incubation period, how it is transmitted or contracted, its mode of entry into the body.

3. *Environmental factors*: physical aspects of the environment such as changes in climate (temperature, rainfall, drought), sanitation, changes in food or water supply, crowding, location (hospitals, schools). Environmental factors also include cultural and social factors, transportation, and hygienic practices—anything that influences exposure of individuals to the agent.

These three factors will be operating in a concerted balance during an epidemic. Therefore, changing the parameters of any one of the three factors can prevent or stop the spread of an epidemic because it severs a connection in the triangle. Control strategies that target these connections generally involve:

- protection of susceptible populations
- containment or elimination of the source of infection
- disruption of transmission

For example, Lyme disease outbreaks are controlled by restricting access to tick-infected environments. Modifying the environment of the host severs the link between the host and the tick-borne pathogen *Borrelia burgdorferi* (see Microbes in Focus 18.5). Yellow fever can be controlled through vaccination of susceptible populations. Vaccination positively modifies the host immune status, severing the link to the agent without modification of the environment. Modifying the environment by draining standing water and using larvicides contains or eliminates the mosquito vector source of yellow fever virus. This severs the link between the agent and the environment without changing the host factor. Using both approaches concurrently is even more effective. In the case of a cholera outbreak, containment or elimination of the source of *Vibrio cholerae* (see Microbes in Focus 21.7) infection can also be achieved by changing the environment by chlorinating the contaminated water source, or locating a new source of uncontaminated water. Both can break the transmission of the agent. Recall that this is the approach John Snow took when he recommended the removal of the pump handle from the Broad Street pump (see Section 18.3). He prevented access to *Vibrio cholerae* contaminated water and stopped the epidemic.

CONNECTIONS Humans living in developing nations frequently face the risk of exposure to pathogenic microbes in the water they drink. This is especially true for humans living in heavily populated regions where sources of drinking water, such as a river, may be collecting raw sewage. **Section 16.5** describes how wastewater is first treated to protect raw water sources, and how drinking water is purified and monitored to ensure public health.

Investigating an Outbreak

A critical public health measure in the prevention of an epidemic is to have accurate surveillance information systems that report the first possible signs of an outbreak. Identifying a localized outbreak can help contain the pathogen and prevent its spread to epidemic proportions. To investigate an outbreak, a systematic procedure is used to identify the agent and the source of infection. These basic "steps" are presented below, and they are often performed simultaneously during outbreak investigations.

1. Define a case.

2. Identify the causative agent. This frequently includes pathological and microbiological examination of tissues, food, water, and other relevant substrates suspected to be involved in the outbreak.

3. Examine the incidence pattern. First, construct an epidemic curve (see Section 18.3, Figure 18.16). The shape of the curve may suggest a common source or propagated epidemic. Second, plot spatial patterns of cases on a map to identify clusters of cases.

4. Determine the magnitude of the problem. Begin by comparing the current incidence with that of past or expected incidence. Then compute the **attack rate**. The attack rate is an incidence rate calculated during an outbreak, and gives the proportion of ill individuals in an exposed population over a short period of time. The attack rate is a measure of the probability of an exposed individual becoming a case, and is used to estimate how many more individuals are expected to become ill.

$$\text{Attack rate} = \frac{\text{number of new cases during time interval}}{\text{total number of individuals in exposed population}} \times 100$$

5. Examine the data for common host factors, such as similarities in age, gender, immune status, and so on, and common environment factors, such as associations with climate, food, or water supply.

6. From the information collected, develop a working hypothesis, considering the following questions: What is the pattern of the outbreak (common source or propagated; see Section 18.3)? What is the source of the outbreak (water supply, food supply, introduction of an infected individual)? How is the pathogen spread (aerosol, fecal-oral, vector, vehicle)?

7. Test the developed hypothesis by collecting data on all those individuals exposed to the suspected agent or source, and/or perform controlled studies in animal models.

Investigating outbreaks is valuable because it can identify the cause and source of the illness, as well as generate information useful in preventing future outbreaks, improving public health policy, and providing practical training for epidemiologists.

24.4 Fact Check

1. Define R_0, the basic reproduction number. Why is $R_0 = 1$ considered to be the epidemic threshold?

2. R_0 is a function of what factors?

3. What three major factors are associated with the development of an epidemic?

4. What control strategies can be employed to stop the spread of an epidemic?

5. What steps are used to investigate an outbreak?

Long before Edward Jenner's first successful vaccination for smallpox, Asian populations practiced a much riskier method of protection against smallpox called "variolation." Variolation was a purposeful inoculation of live smallpox virus (variola virus) into the skin or into a vein, or by inhalation through the nose. The inoculating material consisted of dried material scraped from smallpox vesicles. When it worked, variolation produced a mild case of smallpox, and the recipient recovered to enjoy a life forever protected from smallpox. When it didn't work, recipients died. Though the risk may seem considerable to us, choices were few prior to the 1800s before Jenner's substitution of smallpox virus for immunization with the much more benign cowpox virus. The odds of dying from smallpox by variolation was 1 in 100; the odds of dying from naturally contracted smallpox was 1 in 3, and back then, smallpox was sure to affect almost every generation.

Variolation was practiced in China as far back as the tenth century, but it took Lady Mary Wortley Montagu, the wife of Britain's ambassador to Turkey in the early 1700s, to bring this practice to the attention of the Western world **(Figure B24.6)**. She reported how Turkish women would hold gatherings (later referred to as "smallpox parties") to inoculate smallpox preparations into the veins of the uninoculated. They selected smallpox material from those who were recovering from the mildest form of smallpox, later called "variola minor." In so doing, the women selected less virulent forms of the virus for variolation. Material from the deadliest and most common form of the disease, variola major, was never used.

Having suffered and survived smallpox herself, Lady Mary had her young son and daughter inoculated. She returned with her family to London where an epidemic of smallpox had begun to take shape. As

Figure B24.6. Portrait of Lady Mary Wortley Montagu Lady Mary was an outspoken advocate of variolation in Europe in the early 1700s. Not portrayed are the disfiguring scars left behind from her bout with smallpox.

© Hulton Archive/Getty Images, Inc.

it reached the households of British royalty, including Lady Mary's own home, several prominent members of the British court, having heard of Lady Mary's successful inoculation of her own children, insisted on having their own children inoculated. News of the success of variolation spread and came to be practiced in other European countries and Russia. Variolation began in Boston in the 1720s. Jenner himself was variolated when he was 8 years old. He developed a mild case of smallpox and was subsequently immune to the disease.

24.5 Immunization and vaccines

What are vaccines, and how are they used to control infectious disease?

One of the most effective tools in preventing microbial infections and epidemics is to protect susceptible populations through mass immunization using vaccines. **Immunization** is the process of generating immunological memory to prevent infection and/or disease through the introduction of an antigen into the body (see Section 20.1). Although often used as a synonymous term to immunization, **vaccination** is the physical act of injecting or inoculating the immunizing substance into the body. **Vaccines** are preparations of microbes or their components that are used in immunization.

History of vaccination

The term "vaccination" was coined by Edward Jenner in 1798, although he was not the first to practice it. **Perspective 24.4** describes the much more ancient vaccination procedure called

"variolation," which was commonly used to immunize against smallpox before Jenner's vaccination method replaced it. Smallpox is a highly contagious, systemic viral disease characterized by the formation of numerous vesicles covering the body, including inside the mouth, throat, and inner eyelids (see Figure 1.1A). Before its eradication in 1980, it exhibited a mortality rate of up to 50 percent in infected individuals.

Jenner's first recorded vaccination was in 1796 of an eight-year-old boy named James Phipps, in England (see Figure 1.31). Jenner produced a scratch on James' arm and into this inoculated the fluid from a pustule of a milkmaid who had cowpox (*vacca* is Latin for "cow"). The cowpox virus produces a mild disease and smallpox-like lesions in humans through infection of scrapes on the skin. The hands of milkmaids often developed cowpox pustules, and the disease was commonly referred to as "milkmaid's disease." Unknown to Jenner, cowpox is caused by

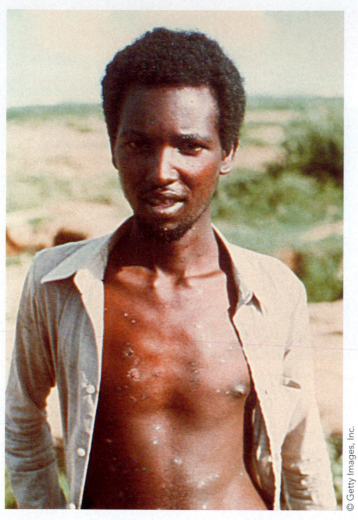

Figure 24.17. The world's last case of naturally occurring smallpox Survivor Ali Maow Maalin, who volunteered to help the WHO locate smallpox cases, contracted smallpox in his native Somalia in 1977. Wild variola virus was declared eradicated by the World Health Assembly in 1980.

a virus (cowpox virus), which is related to variola virus. Jenner was aware that persons who had previously contracted cowpox were frequently immune to the much more serious disease of smallpox. Despite not knowing anything about microbial agents, he correctly reasoned that a purposeful introduction of cowpox material could prevent smallpox. After receiving the vaccination, young James developed mild fever and malaise, and then quickly recovered. Two months later, Jenner inoculated the boy with material from an active smallpox lesion. James did not develop the disease.

Jenner published his results after a few more trials, and by the year 1800, his vaccination method for smallpox had become widespread in Europe and had reached the shores of America. Following a concerted global vaccination effort that began in 1967, the last naturally occurring case of smallpox was reported in 1977 **(Figure 24.17)**. The disease was declared eradicated in 1980—the first and only infectious disease to have successfully been eliminated from the globe. Vaccination for smallpox became unnecessary after this time and was then stopped. Today, a few stocks of wild smallpox virus remain in government labs in the United States and Russia for research purposes.

Interestingly, molecular analysis of the vaccine strain of cowpox virus has shown this virus is not cowpox virus at all, but seems to be a species related to both cowpox virus and variola virus. Vaccinia virus, as it is now known, is postulated to have evolved at some point by recombination between cowpox and variola viruses. It is not hard to imagine recombination happening within persons inoculated with the cowpox virus who were also incubating variola virus. Subsequent mutations resulting in natural attenuation made an effective vaccine against smallpox, but no longer caused the mild illness experienced by those infected with cowpox.

Vaccinia virus can be propagated in cell culture and is still maintained in some research labs for studying the biology of poxviruses. When it was in routine use, vaccinia virus was associated with a relatively high rate of serious complications (1 in 1,000 inoculations as compared to 1 in 10,000 or less for most vaccines in use today), and because it is a live virus, it was spread to unvaccinated individuals. Recently, the perceived risk of a concealed bioweapons store of variola virus has instigated a return to smallpox vaccination in the United States. Not surprisingly, these same undesirable side effects appeared in vaccinia-vaccinated American troops. This includes the spread of sores on parts of the body other than the vaccination site, and can lead to blindness if the eyes are affected. In some cases, the sores become widespread, resulting in significant tissue damage and threat to life. Inflammation of the brain (encephalitis) is another life-threatening side effect of vaccinia vaccination. The CDC Advisory Committee on Immunization Practices currently does not recommend routine use of this vaccine.

Vaccine design

Vaccines confer immunological memory through the production of memory cells—specialized T cells and B cells that are produced following strong stimulation of the adaptive immune system (see Section 20.2). When the same or closely related agent is encountered again, memory cells will recognize the same antigens on the pathogen that were present in the vaccine, and initiate an immune response.

CONNECTIONS You may recall from **Section 20.1** that the memory response occurs upon re-exposure to an antigen, and results in a very rapid and heightened immune response that can prevent a pathogen from causing disease **(Figure 20.6)**.

The ideal vaccine is one that generates a high level of immunological memory to protective antigens—those antigens that can elicit an immune response that protects against infection and/or disease. **Figure 24.18** shows the currently recommended vaccines for routine immunizations in the United States. **Table 24.6** provides descriptions of the vaccines. A good vaccine should also not produce serious side effects when used in large populations. There are a number of different ways vaccines are formulated to expose the immune system to protective antigens. Most vaccines follow the four basic designs outlined below.

Attenuated Vaccines

Attenuated vaccines are composed of a live but weakened version of a microbe that is completely or nearly devoid of pathogenicity.

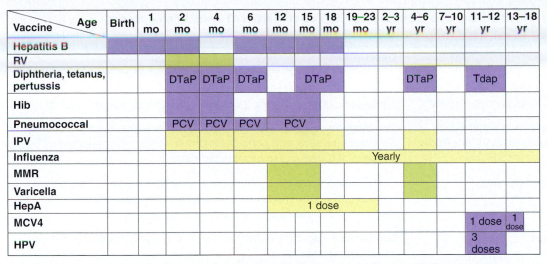

Vaccine \ Age	Birth	1 mo	2 mo	4 mo	6 mo	12 mo	15 mo	18 mo	19–23 mo	2–3 yr	4–6 yr	7–10 yr	11–12 yr	13–18 yr
Hepatitis B														
RV														
Diphtheria, tetanus, pertussis			DTaP	DTaP	DTaP		DTaP				DTaP		Tdap	
Hib														
Pneumococcal			PCV	PCV	PCV	PCV								
IPV														
Influenza						Yearly								
MMR														
Varicella														
HepA						1 dose								
MCV4													1 dose	1 dose
HPV													3 doses	

Subunit vaccine
Attenuated vaccine
Inactivated vaccine

Recommended Immunization Schedule for Persons Aged 0 Through 18 Years
(United States, 2012)

Vaccine \ Age	19–21 yr	22–26 yr	27–49 yr	50–59 yr	60–64 yr	≥ 65 yr
Diphtheria, tetanus, pertussis	One-time dose of Tdap, then Td every 10 years					
Pneumococcal	PPSV 1 or 2 doses					
Meningococcal	At least 1 dose					
Influenza	Yearly					
Varicella	2 doses					
Zoster					1 dose	
MMR	1 or 2 doses			1 dose		
Hepatitis A	2 doses					
Hepatitis B	3 doses					
HPV (female)	If not previously vaccinated, 3 doses					
HPV (male)	3 doses					

Recommended Adult Immunization Schedule (United States, 2012)

Figure 24.18. Current immunization schedule Recommended immunization schedule for all persons specified by the Centers for Disease Control and Prevention. Not included are recommendations for those who have not followed this schedule, or special vaccinations, such as for yellow fever, for travel to some countries, or vaccines for specialized occupations, such as for rabies, or vaccines for persons at high risk, such as immunocompromised persons or those with diabetes.

These vaccines tend to produce high levels of immunity because the attenuated microbe can replicate to some extent in the body, exposing the immune system to its antigens over a period of time, which results in a strong memory response. Consequently, fewer doses or "boosters" are usually needed. Some attenuated vaccines can be shed from the body and can be passed to other individuals. This "contact spread" is both a benefit and a disadvantage for attenuated vaccines; it can help protect others, but in some cases, it can be transmitted to susceptible individuals. For example, live vaccinia virus vaccine can be transmitted from vaccinated persons to newborns, where it causes serious disease. Attenuated microbes may also undergo genetic changes, making them more virulent. This has occurred for the attenuated oral polio vaccine, which we will examine near the end of the chapter.

Inactivated Vaccines

Inactivated, or **killed**, **vaccines** consist of whole cells or viruses that have been physically inactivated by heat, or chemically inactivated by treatment with a denaturing agent such as formalin. The mixture of antigens that are liberated by the inactivation process is then injected into the body. Properly inactivated vaccines cannot replicate in the body, cannot be spread, and of course, cannot regain virulence. Inactivated vaccines typically require several injections to evoke long-term protective immunity because the immune system is exposed to the antigens only for a short period of time. They also tend to produce a higher level of side effects owing to the complex mixture of components that are injected. An example is pertussis vaccine (see *Bordetella pertussis*, Section 21.1). The first generation of pertussis vaccine consisted of killed whole cell preparations. It was an effective vaccine, but because it was a crude preparation of microbial components, convulsions occurred in 0.1 percent of infants given the vaccine, and a smaller percentage of these suffered brain damage. The new generation of pertussis vaccines consists of acellular subunit preparations, described next.

TABLE 24.6 Vaccines recommended by the Centers for Disease Control and Prevention for bacterial and viral diseases

Vaccine	Disease/pathogen target	Description
Vaccines against bacterial diseases		
Hib	*Haemophilus influenzae* type b causing meningitis	Conjugated polysaccharide (polyribitol phosphate) subunit vaccine
DTaP	Against diphtheria (*Corynebacterium diphtheriae*), tetanus (*Clostridium tetani*), and pertussis (*Bordetella pertussis*)	Combination subunit vaccine for children containing inactivated diphtheria, tetanus, and pertussis toxins. Includes other selected antigens of *B. pertussis* to reduce colonization
Tdap	Against diphtheria, tetanus, and pertussis	Combination subunit vaccine for adolescents containing inactivated tetanus and diphtheria toxins. Contains reduced concentrations of inactivated diphtheria and pertussis components
Td	Against diphtheria and tetanus	Combination subunit vaccine for adults containing inactivated tetanus and diphtheria toxins
PCV	*Streptococcus pneumoniae* causing pneumococcal meningitis in children	Conjugated polysaccharide subunit vaccine containing the 7 most common capsule polysaccharides associated with meningitis in children
PPSV	*Streptococcus pneumoniae* causing pneumococcal meningitis and invasive disease in adults	Non-conjugated polysaccharide subunit vaccine containing the 23 most common types of capsule polysaccharides associated with adult disease
MCV4	*Neiserria meningitides* causing meningitis in young adults	Conjugated polysaccharide subunit vaccine containing the 4 most common types of capsule polysaccharides
Vaccines against viral diseases		
RV	Rotovirus causing gastroenteritis in children	Attenuated oral vaccine containing several strains of type A rotovirus.
IPV	Poliovirus causing poliomyelitis	Killed vaccine containing poliovirus types 1, 2, and 3
Influenza	Influenza virus causing seasonal influenza	Polyvalent vaccine containing the dominant circulating strains of influenza A and B types
HepB	Hepatitis B virus causing liver disease	Recombinant subunit vaccine containing the envelope protein HBsAg produced in yeast
MMR	Measels, mumps, and rubella viruses	Killed combination vaccine
Varicella	Herpes varicella-zoster virus causing chickenpox	Live, attenuated vaccine for protection against primary infection
Zoster	Herpes varicella-zoster virus causing shingles	Live, attenuated vaccine for protection against reactivation in adults
HepA	Hepatitis A virus causing liver disease	Killed vaccine
HPV	Human papillomavirus causing cervical, anal, penil, and oropharyngeal cancer	Recombinant subunit vaccine containing capsid proteins of strains 16 and 18, associated with the majority of papillomavirus-induced cancers. The vaccine Gardasil contains capsids from an additional two of the most common strains causing genital warts (11 and 6).

Subunit Vaccines

Subunit, or **acellular**, **vaccines** consist of one or a select number of isolated protective antigens. They are not composed of whole cells or viruses. In general, these are very safe because they are of defined composition rather than a complex mixture. Recall that the currently recommended pertussis vaccines consist of denatured pertussis toxin combined with select purified proteins used for bacterial adherence (see Section 21.1). The *a*cellular, subunit nature of current pertussis vaccines is represented as *a*P in the combined DTaP or Tdap vaccines (see Figure 24.18). Like inactivated vaccines, subunit vaccines typically require several injections (an initial injection followed later by "booster" injections) to evoke long-term immunity. Tetanus and diphtheria toxoids, previously covered in Section 21.2, are examples of subunit vaccines. You can see from Figure 24.18 that the DTaP or Tdap vaccine for diphtheria, tetanus, and pertussis is given frequently to children to maintain adequate levels of protection. Tetanus and diphtheria toxoids are further recommended every 10 years after the age of 18 years.

CONNECTIONS You may recall from **Sections 19.4** and **20.4** that many vaccine formulations use adjuvants to boost their effectiveness. Because small, purified microbial peptides are typically not very immunogenic, they often do not stimulate strong adaptive immune responses. In turn, long-lived memory cells needed for lasting immune memory are not generated in sufficient numbers. Subunit and inactivated formulations therefore frequently include adjuvants.

Conjugate vaccines, introduced in Section 20.5, are modifications of subunit vaccines composed of a polysaccharide antigen bound to an immunogenic protein. Several conjugate vaccines against pathogens causing meningitis are described in Table 24.6. An example is the Hib vaccine, which consists of the *Haemophilus influenzae* type b capsular polysaccharide (polyribitol phosphate) conjugated to denatured tetanus toxin (tetanus toxoid). The capsular polysaccharide is a protective antigen, but like most carbohydrates, it is not immunogenic in young children (see Section 20.6). Conjugating polysaccharides to an immunogenic protein significantly improves their immunogenicity. Similarly, the *Streptococcus pneumoniae* childhood vaccine (PCV) consists of seven different capsular polysaccharides conjugated to diphtheria toxoid (see Perspective 20.2).

CONNECTIONS Recall from **Section 20.6** that capsular polysaccharides are T-independent antigens that do not evoke a B-cell response in young children. Effective immunization against these antigens requires the polysaccharides be bound or conjugated to a foreign protein molecule. This converts the conjugated polysaccharide to a T-dependent antigen, resulting in antibody production and immunological memory.

DNA Vaccines

DNA vaccines (see Chapter 19 Mini-Paper) consist of a cloned gene(s) in a DNA vector, which is then delivered to cells of the body, usually by injection of an engineered virus, or by electroporation of an engineered plasmid using electrical pulses to create pores in cells of muscle or skin. Expression of the cloned gene produces a protective protein antigen that induces an immune response. In this respect, DNA vaccines are similar to recombinant subunit vaccines. However, because the antigen is expressed in cells over a period of time, DNA vaccines evoke a strong memory response, like that of attenuated vaccines. These vaccines do not contain live microbes and are not able to be passed from host to host. DNA vaccines are largely still in experimental stages, but many appear to be very promising. The only currently licensed DNA vaccine is for use in farmed salmon to protect against viral infectious hematopoietic necrosis.

Vaccine efficacy

Are vaccines really effective? Despite the development of many highly effective vaccines, no vaccine is 100 percent protective for every individual. **Vaccine efficacy**, or the effectiveness of a vaccine is an average measure of the degree of protection it affords a population. Most routine childhood vaccines have an efficacy range of 85–90 percent, meaning between 85 and 90 percent of those given the vaccine will be protected from the associated disease. For largely unknown reasons, a certain percentage of individuals will simply fail to establish a high enough level of immunological memory to protect them even when they have been vaccinated. This is known as **vaccine failure**, although "immune failure" is likely the underlying reason. Ali Maow

Maalin, the survivor of the world's last natural case of smallpox, was himself vaccinated before he contracted the disease (see Figure 24.17).

So, even with a vaccine of high efficacy we will expect to find a small number of vaccinated individuals who may contract the disease. To illustrate this, we can look at a hypothetical, but realistic, scenario involving a measles outbreak. In a school of 1,000 students, all but 15 of the students have been vaccinated for measles. The measles vaccine has an efficacy of 90 percent. Assuming in this small, dense population that every student becomes exposed after the introduction of an infectious person (and yes, this is completely realistic; measles has an R_0 value of 12 to 18 and is easily transmitted by aerosols), we would expect all 15 unvaccinated students to contract the disease, as well as those who experience vaccine failure. This last group would comprise 10 percent of the vaccinated student population; that is, about 99 of the 985 vaccinated students. The total number of cases expected is then 99 plus 15: 114 cases. Therefore, 99/114, or about 87 percent of the measles cases in the school are in vaccinated individuals. Unfortunately, when outbreaks of measles do occur, these seemingly high percentages are often quoted by opponents of vaccination as evidence that vaccines are not effective. As you can see, it is a misinformed or maybe purposefully misconstrued interpretation of the data. Just imagine what would happen if no one in the school was vaccinated.

Herd immunity

At the level of populations, mass vaccination reduces or prevents epidemics of communicable diseases, and can even entirely eliminate some infectious diseases from populations. To accomplish this, it is not necessary to vaccinate every individual. Some individuals will be too remote to contact, and others will have underlying health conditions that make vaccination inadvisable. This includes individuals who are immunocompromised, those who have had allergic reactions to previous vaccinations, or those with a history of autoimmune disease. Some people refuse to be vaccinated because of personal or religious choice, and as we have seen, a proportion of vaccinated individuals will experience vaccine failure. Fortunately, a population effect known as "herd immunity" can protect non-immune individuals and stop transmission of a communicable disease.

Herd immunity, sometimes known as "community immunity," occurs when successful immunization of a certain proportion of a population serves to protect the rest of the population from communicable disease. A lack of susceptible individuals in a sufficiently immunized population means that transmission of infection will decline and may cease altogether. Herd immunity effectively establishes a barrier of immune individuals that prevents access of the transmissible agent to susceptible individuals **(Figure 24.19)**. The general aim of public health in guarding against communicable disease is to establish and maintain effective herd immunity through mass immunization.

What number of immunized individuals are needed for herd immunity? The critical proportion of the population required to be immunized to achieve herd immunity is called the

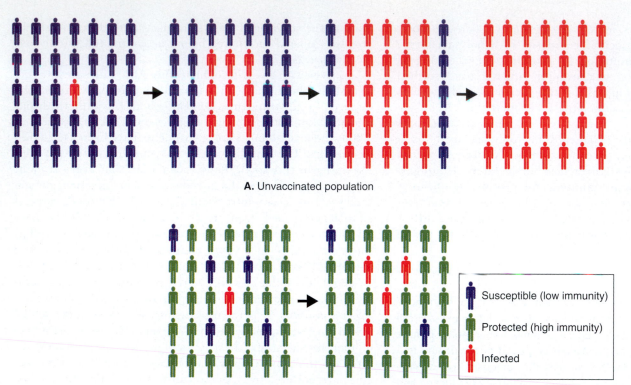

A. Unvaccinated population

B. Vaccinated population

Susceptible (low immunity)

Protected (high immunity)

Infected

Figure 24.19. Protection by herd immunity **A.** In an unvaccinated population, introduction of an infected individual results in infection of all susceptible contacts, who in turn infect other contacts. Soon the entire population becomes infected. **B.** In the vaccinated population, only unvaccinated, susceptible contacts of the infected individual become infected. Infection does not reach all susceptible individuals because they are at a density too low to allow further contact and spread. Consequently, transmission of the pathogen stops.

herd immunity threshold. **Table 24.7** shows herd immunity thresholds for some communicable diseases that are controlled through vaccination. Herd immunity thresholds are frequently derived by calculations based on:

- the susceptibility of the population

| **TABLE 24.7** | R_0 **values and herd immunity thresholds for select communicable diseases** |

Disease	Transmission	Estimated R_0 value	Herd immunity threshold
Measles	Aerosol	12–18	83–94%
Pertussis	Aerosol	12–17	92–94%
Diphtheria	Direct contact	6–7	85%
Smallpox[a]	Direct contact	6–7	83–85%
Polio	Fecal-oral	5–7	80–86%
Rubella	Aerosol	5–7	80–85%
Mumps	Aerosol	4–7	75–86%

[a] Eradicated.

Adapted from: History and Epidemiology of Global Smallpox Eradication. From the training course titled "Smallpox: Disease, Prevention, and Intervention." The CDC and the World Health Organization. Slides 16–17.

- how communicable the disease agent is (estimated by its R_0 value)
- the population density (higher densities must have higher herd immunity)
- vaccine efficacy

The herd immunity threshold allows estimation of the **vaccine coverage rate**, the proportion of individuals in the population that must be immunized to achieve herd immunity. Usually vaccine coverage target rates are set to slightly exceed herd immunity thresholds. For example, the WHO sets vaccine coverage target rates at 95 percent for measles, which has a maximum herd immunity threshold of 94 percent (see Table 24.7). Again, because of vaccine failure, there will always be a small proportion of individuals who fail to be successfully immunized even with a highly effective vaccine.

For some communicable diseases, global eradication can be achieved through vaccination, as has occurred for smallpox. Candidates for eradication must satisfy the following criteria:

- Long-term immunity can be produced through vaccination.
- Only a few strains exist that need to be considered for vaccine design.
- Humans are the only significant reservoir.

Currently, measles, mumps, and rubella (which are conveniently included in the single MMR vaccine), polio, chickenpox, and hepatitis B are on the WHO target list for eradication. In some countries, these diseases have already been eliminated.

For example, recall from Chapter 18 that polio was eliminated in the United States in 1979. The last case of wild poliovirus contracted in the United States occurred in a small community where no one was vaccinated for polio. Other cases of polio that have since been detected have been imported by overseas travelers. In many nations, the cost of vaccination remains a major impediment to the successful eradication of these diseases. In the next sections, we will examine the cases of poliovirus and foot-and-mouth disease virus (FMDV) in order to gain some insight into why communicable diseases are so difficult to eradicate, and what control strategies are used to fight these infectious diseases.

Polio—A near success

The development of poliovirus vaccines is hailed as one of the greatest success stories of medicine. Today, the incidence of disease from poliovirus is but a shadow of its former self, and the virus is on the verge of being eradicated entirely. Yet poliovirus is proving to be more difficult to globally eradicate than was first imagined. Skyrocketing populations, civil wars, and political and social unrest have all contributed to lingering reservoirs of infection.

Polio Disease and Transmission

Poliovirus (see Microbes in Focus 5.2) is transmitted via the fecal-oral route. Humans ingest the virus from contaminated water and food, or through contact with feces-contaminated hands. The virus attaches to a specific immunoglobulin-like receptor known as the poliovirus receptor, which is present on epithelial cells lining the small intestine and pharynx, and also on leukocytes, and motor neurons. Viral replication in epithelial cells of the digestive tract results in shedding of large numbers of infectious virions. In areas where poliovirus is still endemic (see Figure 18.28), it can be found in lakes, rivers, and streams contaminated by raw human sewage.

When a person ingests poliovirus, a number of outcomes are possible. In most cases, dissemination of poliovirus beyond the epithelial lining of the digestive tract does not reach a significant level. These infections are asymptomatic or produce mild fever, nausea, vomiting, sore throat, constipation, cramps, and stiffness. Recovery is usually complete. In some cases however, the virus enters the circulatory system and gains access to the central nervous system. Poliomyelitis then develops as a result of inflammation of the meninges, the layer of tissue surrounding the brain. Most patients recover from this, but in a few, the motor neurons become infected, resulting in nerve damage and paralytic poliomyelitis. Paralytic poliomyelitis results in partial or complete paralysis and is estimated to occur in less than 1 percent of those who develop poliomyelitis. A proportion of paralytic poliomyelitis victims may die due to respiratory failure involving paralysis of the diaphragm. Recall from the opening story of Chapter 18, at the height of the polio epidemic in the 1940s and 1950s, before vaccination, it was common for patients suffering paralysis to be placed in "iron lung" chambers to help them breathe **(Figure 24.20)**.

The Changing Patterns of Polio

You might expect that in an unvaccinated population where high levels of endemic poliovirus exist, the incidence of poliomyelitis, and therefore paralytic poliomyelitis, would be high. However, this is not the case. Poliovirus has existed in human populations for thousands of years, yet the incidence of poliomyelitis has risen to become a public health concern only within the past century. Before modern sanitation facilities and vaccination programs, poliovirus commonly infected very young children. Often, these children did not develop paralytic poliomyelitis because their naturally immune mothers provided them with transplacental IgG and secreted IgA from breast milk (see Section 20.6). When infected with poliovirus during this time of passive immunity,

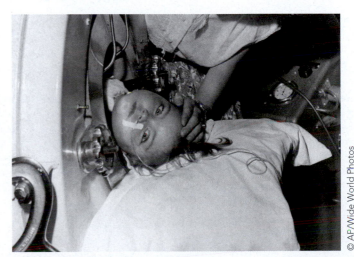

A. Rows of polio patients in iron lung machines

B. Girl receiving breathing assistance

Figure 24.20. Iron lung machine for polio victims **A.** During polio epidemics in the United States (prior to vaccination in 1955), many patients suffering from poliomyelitis were maintained in iron lungs that used changes in pressure to help them breathe. While some patients recovered the use of muscles for breathing, others did not and remained in the machines until their death. **B.** Close-up of a young girl positioned in an iron lung.

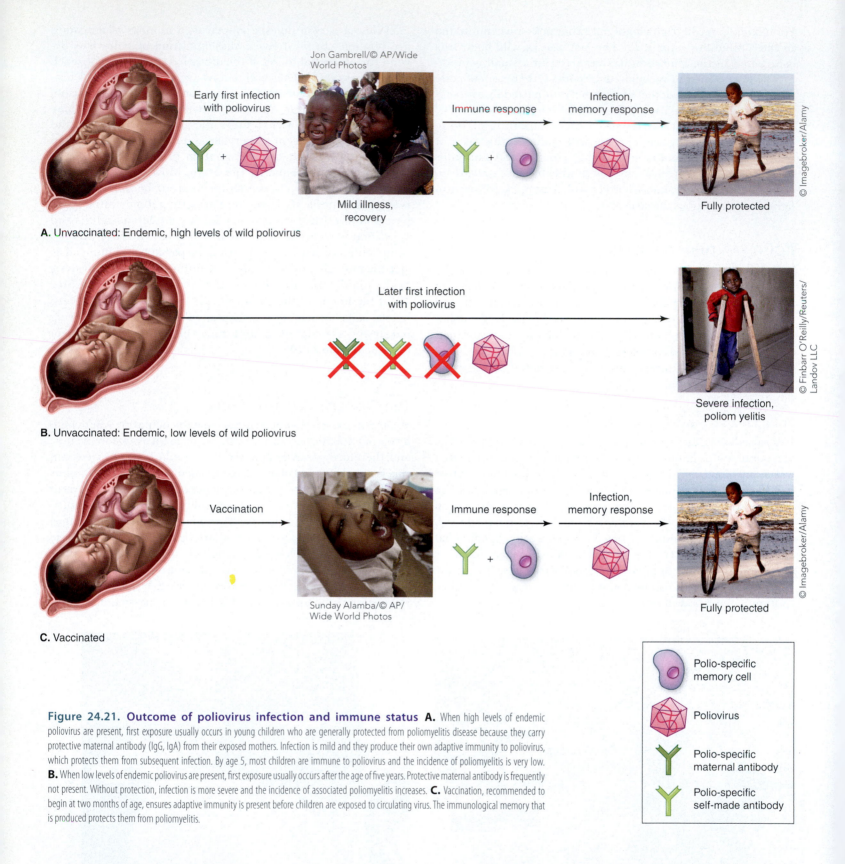

A. Unvaccinated: Endemic, high levels of wild poliovirus

Early first infection with poliovirus

Jon Gambrell/© AP/Wide World Photos

Mild illness, recovery

Immune response

Infection, memory response

Fully protected

© Imagebroker/Alamy

B. Unvaccinated: Endemic, low levels of wild poliovirus

Later first infection with poliovirus

Severe infection, poliom yelitis

© Finbarr O'Reilly/Reuters/Landov LLC

C. Vaccinated

Vaccination

Sunday Alamba/© AP/Wide World Photos

Immune response

Infection, memory response

Fully protected

© Imagebroker/Alamy

Polio-specific memory cell

Poliovirus

Polio-specific maternal antibody

Polio-specific self-made antibody

Figure 24.21. Outcome of poliovirus infection and immune status A. When high levels of endemic poliovirus are present, first exposure usually occurs in young children who are generally protected from poliomyelitis disease because they carry protective maternal antibody (IgG, IgA) from their exposed mothers. Infection is mild and they produce their own adaptive immunity to poliovirus, which protects them from subsequent infection. By age 5, most children are immune to poliovirus and the incidence of poliomyelitis is very low. **B.** When low levels of endemic poliovirus are present, first exposure usually occurs after the age of five years. Protective maternal antibody is frequently not present. Without protection, infection is more severe and the incidence of associated poliomyelitis increases. **C.** Vaccination, recommended to begin at two months of age, ensures adaptive immunity is present before children are exposed to circulating virus. The immunological memory that is produced protects them from poliomyelitis.

they were protected from disease by this antibody and were able to develop their own adaptive immunity to the virus. By the age of five years, most children had acquired adaptive immunity to the virus and were protected from subsequent infections in their later years, lessening the risk of poliomyelitis (**Figure 24.21**).

Under these conditions, the incidence of poliomyelitis among infected individuals is estimated to be 0.05 percent.

As modern waste and water treatment facilities became more commonplace in endemic areas, the levels of poliovirus decreased in the environment and exposure was reduced.

Although no one will claim that providing clean food and water to populations is an overall disadvantage, it had an undesirable impact on the incidence rate of poliomyelitis. In these low level endemic areas, there is a longer time before first exposure to wild poliovirus occurs, typically around five years of age. These older, unexposed children have no adaptive immunity to the virus, nor do they have any maternal antibody left (see Figure 24.21B). Consequently, they acquire more severe infection upon their first exposure. This increases the chance of travel of the virus to the central nervous system, increasing the risk of poliomyelitis and paralysis. From an epidemiological context, the incidence of poliomyelitis in unvaccinated populations in regions where poliovirus is endemic is related to the age of first exposure to poliovirus. This is because the severity of poliovirus infection depends on the immune status of the individual. The advantage of vaccination is that it ensures young children will have adaptive immunity to poliovirus to protect them from subsequent exposure, regardless of the age at which exposure happens.

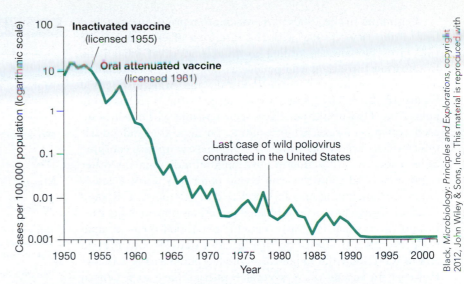

Black, *Microbiology: Principles and Explorations*, copyright 2012, John Wiley & Sons, Inc. This material is reproduced with permission of John Wiley & Sons, Inc.

Figure 24.22. Incidence of poliomyelitis in the United States Introduction of polio vaccines, beginning in 1955, was followed by a drastic drop in the incidence of poliomyelitis. The last case of poliomyelitis contracted from wild poliovirus originating in the United States was in 1979. Sporadic cases appearing since have been vaccine-derived or imported from other countries.

Polio Vaccines

In 1952, American Jonas Salk grew the only three strains of wild, naturally occurring poliovirus in monkey cell culture, and inactivated the virus particles by treatment with formalin. Formalin treatment denatures viral proteins, preventing attachment and entry into cells for infection. Injection of this inactivated vaccine stimulates a protective immune response, preventing disease upon future exposures (see Figure 24.21C). Widespread use of this "Salk" inactivated polio vaccine (IPV) began in 1955, after the vaccine was proven to be effective and safe in small-scale human trials. Almost immediately, the number of new cases of poliomyelitis dramatically fell in vaccinated groups throughout the world. **Figure 24.22** shows the drop in incidence of poliomyelitis in the United States, beginning in 1955.

As successful as the original Salk vaccine was, it had a number of drawbacks. Like most inactivated vaccines, it required several injections to produce sufficient protection. This was expensive and made it difficult for many countries to maintain herd immunity. In addition, injection of the vaccine results in production of serum IgG antibody only. While circulating IgG in blood prevents the virus from reaching the central nervous system, it does not prevent intestinal and pharyngeal cells from being infected by wild virus. So, while individuals vaccinated with inactivated poliovirus do not display disease, they are still able to carry and spread wild virus—a less than desirable outcome toward eradication.

In 1957, another American, Albert Sabin, decided to improve the vaccine. In order to boost immunity and prevent the spread of wild virus, he developed a live, attenuated vaccine derived from the three wild virus strains. The strains he used had spontaneously accumulated various mutations in their RNA genomes over a period of laboratory culture, which caused decreased viral replication, producing less virulent virus. Sabin correctly reasoned that if ingested, such weakened strains would replicate in mucosal cells enough to stimulate a protective immune response, but not enough to cause disease. This vaccine was cost effective for countries who could not afford Salk's inactivated, injectable vaccine.

Perhaps the most significant advantage of the live attenuated virus was that it could prevent replication of wild virulent virus. Replication of wild virus is prevented because the attenuated virus replicates in the gut, stimulating production of protective serum IgG *and* mucosal IgA, which is secreted at the site of infection. The presence of local IgA prevents attachment and infection of wild virus. Individuals vaccinated with the attenuated oral poliovirus vaccine (OPV) are not only protected from disease, but they cannot spread wild virus. Without a reservoir in which to replicate, poliovirus could be completely eradicated. **Table 24.8** compares features of the inactivated and attenuated vaccines.

TABLE 24.8	Comparison of inactivated and attenuated oral polio vaccines
Inactivated polio vaccine (IPV)	**Attenuated, oral polio vaccine (OPV)**
Developed by Jonas Salk	Developed by Albert Sabin
Injected	Oral
Expensive[a] (needles, trained medical personnel, 3–4 boosters)	Inexpensive[a] (no needles, no special training, 2 boosters)
Serum immunity only (IgG)	Mucosal (IgA) and serum immunity (IgG)
Prevents poliomyelitis but not infection from wild virus (carrier)	Prevents poliomyelitis and infection with wild virus (cannot transmit wild virus)
Does not eliminate wild virus	Eliminates wild virus
No risk of vaccine-associated poliomyelitis	Risk of vaccine-associated poliomyelitis
No transmission of vaccine strains	Transmission of vaccine strains

[a]Both formulations require refrigeration.

Beginning in the 1960s mass vaccination programs switched from using IPV to OPV with the intention of eradicating the wild virus globally. According to data from the Global Polio Eradication Initiative, wild poliovirus now remains endemic in only three countries: Nigeria, Pakistan, and Afghanistan (see Figure 18.32). In 2012, India was declared by the WHO as being polio-free. This would not have been possible without attenuated oral vaccine use. Unfortunately, unvaccinated individuals infected with wild virus traveling from these countries continue to be a source of infection in unvaccinated individuals in other regions. Several countries where wild poliovirus was previously eradicated, including Angola, Chad, and the Democratic Republic of the Congo have re-established infections of wild poliovirus. In populations consistently practicing childhood polio vaccination, coverage rates are very high (greater than 90 percent) and risk of infection from importation of wild poliovirus is very low. However, in recent years, several countries have experienced political and cultural shifts that have resulted in resistance to vaccination programs. Nigeria's anti-vaccination movement has arisen from a distrust of Western nations, the sponsors of the polio campaigns. As a result, outbreaks in Nigeria as well as some previously polio-free countries have resulted (see Figure 18.28), an unfortunate illustration of the importance of maintaining herd immunity. Renewed efforts are now being made to target these regions for vaccination. War, poverty, cultural and religious practices, and lack of government-sponsored vaccination programs are other common barriers that lead to epidemics of diseases that are easily controlled by vaccination.

The Polio Vaccine Compromise

As miraculous as attenuated OPV seems, it too has a flaw. Attenuated poliovirus replicates, and any virus that replicates can mutate. Rare cases of poliomyelitis sometimes appeared following vaccination with OPV. Analysis of the virus strains from these cases confirmed they were not wild virus, but that they originated from the vaccine strains. The attenuated viruses regained their lost virulence by accumulating point mutations or by swapping RNA with other related intestinal viruses of the picornavirus family. It is estimated that less than 1 in every 2.5 million first-time doses of OPV results in vaccine-derived paralytic poliomyelitis. Because of this small but real risk, countries where wild poliovirus has been eliminated have recently reverted to using killed, inactivated poliovirus vaccines (IPV). In 2000, the United States banned the routine use of OPV, and several other polio-free countries no longer routinely use OPV. While vaccine-derived poliomyelitis cases are rare indeed, what is perhaps alarming is that these "revertant" forms can be shed and passed to unvaccinated persons. In populations where vaccine coverage rates are low, outbreaks of vaccine-derived poliomyelitis have occurred. Nigeria, a country that experienced a previous lapse in vaccination, saw many vaccine-derived outbreaks of poliomyelitis due to circulation of the vaccine-derived virus in the unvaccinated population **(Table 24.9)**.

The changing epidemiological patterns of polio have demanded a change in the vaccination strategies used. The currently recommended practices for polio vaccination are as follows. First, where wild virus is prevalent, use OPV because the risk of wild virus infection and spread is higher than the risk for vaccine-related poliomyelitis. This will eliminate wild virus. After this, switch to IPV to prevent outbreaks from imported virus from other regions. Otherwise, where wild virus is not prevalent or has been eliminated, use IPV because the risk of becoming a wild virus carrier is very low, and IPV use eliminates the risk of vaccine-related poliomyelitis. Interestingly, examination of vaccine-derived cases of poliomyelitis has revealed that most are from the attenuated type 2 poliovirus (see Table 24.9). There are only three types of wild poliovirus, and standard OPV vaccines contain attenuated versions of all three types. However, wild type 2 poliovirus was declared eradicated worldwide in 1999. Removing type 2 virus from the OPV vaccine may remove the risk of vaccine-related poliomyelitis and prevent subsequent outbreaks of revertant type 2 poliovirus, as was experienced in Nigeria. Indeed, some countries have already switched to an OPV containing only types 1 and 3 poliovirus. Nigeria may need to continue using OPV containing all three types to prevent the spread of circulating type 2 revertant virus. Do we dare to hope that in a few years we will be successful in the endeavor to eradicate poliovirus? If that day comes, polio vaccination, like smallpox vaccination, will disappear like the viruses themselves.

Foot-and-mouth disease: The making of an epidemic

Most of the examples we have been discussing have been human communicable diseases. How are communicable diseases of animals controlled? Let's examine foot-and-mouth disease virus (FMDV), a close relative of poliovirus that has significant impacts on domestic herds **(Microbes in Focus 24.1)**.

In 2001, an outbreak of foot-and-mouth disease (FMD) occurred in the United Kingdom. Media coverage showed

TABLE 24.9 Vaccine-derived poliovirus outbreaks in 2011

Country	Number of cases
Afghanistan	1[a]
Democratic Republic of the Congo	4[a]
Mozambique	2[b]
Niger	1[a]
Nigeria	33[a]
Somalia	7[a]
Yemen	5[a]

[a]Vaccine-derived poliovirus type 2.

[b]Vaccine-derived poliovirus type 1.

Data from Global Polio Eradication Initiative, September 2012.

Microbes in Focus 24.1
FOOT-AND-MOUTH DISEASE VIRUS

Habitat: Infects cloven-hoofed mammals

Description: Small, ssRNA naked virus with icosahedral symmetry belonging to family *Picornaviridae*

Key Features: Foot-and-mouth disease is one of the most highly contagious diseases known. It infects many cell types, particularly epithelial cells of the pharynx, tongue, and clefts of feet where receptors are most prevalent, forming painful vesicles on the mouth, nose, and feet. FMDV rarely infects humans and does not cause serious disease when it does. It causes low mortality in adult animals (3–5 percent), but high morbidity that can result in long-term debilitation of animals, which are usually slaughtered. Mortality is highest in very young animals, up to 90 percent.

© Andrew Syred/ Photo Researchers, Inc.

TEM

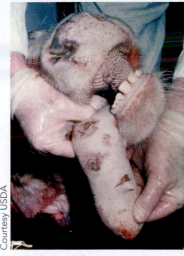
Courtesy USDA

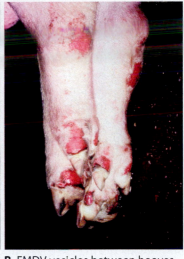
Courtesy USDA

A. FMDV vesicles on tongue

B. FMDV vesicles between hooves

Figure 24.24. Foot-and-mouth disease lesions Foot-and-mouth disease virus is highly contagious, affecting cloven-hoofed mammals and causing mortality in the young and disabling adults. **A.** Painful vesicles form on the mouth and tongue (shown here, cow tongue). **B.** Vesicles and open sores on the feet of a pig.

gruesome scenes of slaughtered cows, sheep, goats, and pigs unceremoniously dumped into pits, covered with lime, and buried or burned in giant funeral pyres in order to stop the spread **(Figure 24.23)**. The epidemic originated from a single pig farm in Britain, the owner of which was later jailed for failure to report this notifiable disease to authorities.

Foot-and-Mouth Disease Transmission

FMDV can spread by every possible means: aerosol, water, contact, soil, bedding, urine, milk, and saliva. Winds can transmit the virus over many kilometers, and with an R_0 value known to be

Barry Batchelor/© AP/Wide World Photos

Figure 24.23. Destruction of carcasses during the 2001 foot-and-mouth disease epidemic In the United Kingdom, several million cattle, sheep, pigs, and goats were slaughtered and disposed of in pits, such as the one shown here, to stop the spread of foot-and-mouth disease virus.

between 36 and 73 it spreads explosively in unprotected herds. FMDV causes high mortality in infected young animals and severely cripples adult animals by producing painful lesions on feet and in the mouth **(Figure 24.24)**, resulting in the long-term disabling of adult animals. The economic losses to agriculture can be enormous and countries fight hard to keep the virus out, not an easy task. When FMDV entered the highly susceptible population of cloven-hoofed animals in the United Kingdom in 2001, it did what any highly contagious pathogen would do—it created a rapidly spreading epidemic.

To stop the epidemic, between 6.5 million and 10 million animals were slaughtered, trade in animals and their products was halted, and tourism and travel of people were restricted for fear of inadvertent transport of the virus. Even federal elections were delayed. Sadly, there are several high efficacy vaccines available to combat FMDV. So why aren't they used?

FMD Vaccination and Control Control of infectious diseases in animals frequently follows policies that are very different from those used for humans. The story of FMDV illustrates this. Animals and their products are commodities, traded on international markets. To protect domestic herds and flocks from imported disease, strict trade policies are established and maintained. Policies for FMDV prevent the export of animals from areas where FMDV is endemic to FMDV-free areas. This is logical, and you can probably appreciate why FMDV-free countries are so vigilant in screening for the presence of this virus, as it may enter from neighboring endemic areas.

Surveillance for FMDV is done by ELISA to detect the presence of antibodies to the virus (see Toolbox 20.2). If the ELISA is positive, the animal is assumed to be infected and is destroyed along with its herd mates. This type of control program is called "test and slaughter," and is used for several notifiable diseases in livestock including avian influenza, bovine tuberculosis caused by *Mycobacterium bovis*, and bluetongue, a serious viral disease of sheep and cattle transmitted by midge insects. Vaccination for FMDV confounds detection because the vaccine produces antibody to the virus making vaccinated and infected animals indistinguishable by ELISA. This results in trade restrictions where vaccination is practiced because authorities cannot guarantee the absence of virus. With this type of economic pressure, it makes sense for countries like the United Kingdom, the United States, and Canada to maintain a FMDV-free, vaccination-free status. For the United Kingdom, maintaining their vaccination-free status cost them an epidemic, loss of their herds, and an estimated $8.5 billion—a cost that many economists consider to have been greater than the economic gains made by not vaccinating. In the wake of the British FMDV disaster, many countries are now reconsidering their vaccination policies and are beginning to vaccinate their herds.

24.5 Fact Check

1. Define the terms "immunization" and "vaccination."
2. Describe the four basic designs of vaccines, and explain the advantages and disadvantages of each.
3. Explain the terms "vaccine efficacy" and "vaccine failure."
4. Define the terms "herd immunity" and "herd immunity threshold," and explain how these terms are related.
5. Describe the differences between the inactivated polio vaccine (IPV) and the attenuated oral polio vaccine (OPV), and briefly explain the controversy surrounding the use of these vaccines.
6. Explain why test and slaughter programs are employed for communicable diseases in farmed animals.

Image in Action

WileyPLUS The mosquito *Aedes aegypti* in this image is transmitting the yellow fever virus from a monkey in the forest background to a human. The envelope of the virus is coated with viral protein. In the cut-through, the membrane is visible (white), and the nucleocapsid protein (purple) encases the single-stranded RNA genome.

1. Imagine that a yellow fever outbreak is occurring in this image. Using parameters of the triangle of epidemiology as a guide, what major factors would you aim to control in this epidemic? Explain and give specific examples that could apply to this outbreak.

2. Imagine a new antiviral drug was developed and, because it was so effective, was highly prescribed during this outbreak. Why would this increased use raise concerns about increased likelihood of resistance to the drug? Explain.

The Rest of the Story

In an attempt to prove yellow fever was transmitted by mosquitoes, Dr. Lazear and Dr. Carroll engaged in self-experimentation with mosquitoes that had fed on the blood of yellow fever victims. Not surprisingly, they both contracted the disease and Lazear died. Unfortunately, their sacrifice did not prove mosquitoes were the agent of transmission—they could have contracted the disease from some other source. Mourning the loss of his good friend, Dr. Reed decided to put the question of transmission by mosquitoes to an end once and for all. He carried out controlled infection studies with volunteer American soldiers and newly arrived Spanish immigrants. (Can you imagine volunteering for this?) Reed placed men in each of two specially built cabins. In the "Infected Clothing Building," mosquitoes were completely absent, but for 20 days, the volunteers wore the dirty clothes and slept in the soiled bedding of those who had died of yellow fever. In the "Infected Mosquito Building," the volunteers were given clean clothes and bedding, but were housed with mosquitoes that had previously fed on yellow fever patients. The men in this building contracted yellow fever, while those in the Infected Clothing Building did not. Reed had successfully proven mosquitoes were the vectors of yellow fever.

Dr. Carroll was able to show that the agent could pass through filters normally used to collect bacteria. By injecting the filtrate prepared from infected mosquitoes into human volunteers, he showed that the agent causing yellow fever was not a bacterium or any other sort of microscopic agent known at the time. We now know the causative agent is yellow fever virus, which belongs to the family *Flaviviridae* (*flavi* means "yellow" in Latin), a group of RNA viruses commonly transmitted by insect vectors.

The repercussions of yellow fever in the New World were, and still are, astounding. During the Spanish-American War (1898) and the American Civil War (1861–1865) more troops died of yellow fever than in combat. In 1803, yellow fever forced the defeat of Napoleon in Haiti after he lost 23,000 soldiers to the disease. This gave him the incentive to sell at a bargain price, an expanse of land to Thomas Jefferson that stretched from Louisiana into Canada—the famous Louisiana Purchase. Imagine how differently things may have turned out in the absence of this tiny microbe.

Let's return to our earlier question posed in the opening story. Should you be worried that your neighbor's case of yellow fever may be the first of many in your area? The answer is no. Thanks to the groundwork of the Yellow Fever Commission, simple mosquito control measures such as screening, insect re-pellants, and removal of standing water has eliminated yellow fever as a public health threat in most countries. Indeed, the last case of urban yellow fever in the United States was in 1942. Your infected neighbor would likely have quickly been hospitalized upon developing symptoms, effectively isolating the virus from native mosquitoes. The development of a highly effective attenuated vaccine in 1938 has further reduced the incidence of yellow fever in many African and South American countries, where the virus is still endemic because monkeys remain a natural reservoir. Vaccination can also protect those who may be traveling to these areas, so they do not bring the yellow fever virus home with them. Unfortunately, it appears your neighbor did not take advantage of this option.

Summary

Section 24.1: How did we deal with infectious disease in the past?

Large-scale epidemics of communicable diseases of humans appeared with the rise of populous civilizations and were spread by trade.

Before antimicrobial drugs and vaccines, control of infectious disease frequently involved banishment, quarantine, bloodletting, and purging.

Section 24.2: What kinds of drugs are used to treat infections, and how do they work?

The majority of **antimicrobial drugs** act on:

- peptidoglycan synthesis (cell wall synthesis)
- membrane integrity
- DNA synthesis
- transcription
- folic acid synthesis (nucleic acid synthesis)
- ribosome function (protein synthesis)

Both **bacteriostatic drugs** and **bacteriocidal drugs** are effective therapies against bacterial infections.

- **Antibiotics** are the most commonly used antimicrobial drugs.
- An important class of antibiotics is the **β-lactam class**, which inhibit peptidoglycan synthesis by binding with **penicillin-binding proteins (PBPs)**.
- Semisynthetic antibiotics contain chemical modifications to increase their effectiveness.

Finding drugs that exhibit sufficient **selective toxicity** for the treatment of fungal, protozoal, and viral diseases is a challenge due to shared features in host cells.

An important class of antiviral drugs are **nucleoside analogs** that interfere with nucleic acid synthesis by viral as opposed to host polymerases.

Section 24.3: How do microbes become drug resistant?

Because of genetic variation, some microbes will always be resistant to the killing effects of an antimicrobial drug. The immune system works in synergy with the drugs to provide clearance.

Susceptibility testing establishes the sensitivity of a particular microorganism to a panel of antimicrobial drugs in order to determine which drug will be most successful in treating an infection.

- The **minimal inhibitory concentration (MIC)** is used to determine if a microorganism will be sensitive or resistant to a particular drug.
- Resistance to antimicrobial drugs is an inevitable consequence of the use of antimicrobial drugs. It is the result of natural selection.

When a drug is not taken as prescribed, or a patient is immunocompromised, the risk for developing drug resistance increases. Resistance genes originate from:

- mutation of naturally occurring precursor genes
- horizontal gene transfer by **mobile genetic elements**

Often, resistance genes are carried on **resistance (R) plasmids**. Not all of the genes that give rise to resistance have developed because of the use of antimicrobial drugs. Microbes may be naturally drug resistant due to:

- intrinsic factors
- genes that have evolved due to the production of antibiotics in natural settings

Resistance can develop quickly because of the fast generation times of most microbes. Resistance to antimicrobial drugs occurs through:

- drug inactivation, by modification of the drug
- alteration of the drug target site
- drug inaccessibility, by preventing entry of the drug
- drug efflux, by pumping the drug out of the cell using **efflux pumps**

Ways to reduce the development of drug resistance include:

- reducing use of the drugs
- using selective drugs
- using drug cocktails
- using effective infection control
- developing new vaccines and increasing access to them
- developing drug alternatives

Section 24.4: How are epidemics predicted and controlled?

Calculation of R_0, **the basic reproduction number**, is used to predict the start or spread of a propagated epidemic. $R_0 = 1$ is the epidemic threshold. When $R_0 < 1$, an epidemic is unlikely to form. When $R_0 > 1$, an epidemic is likely.

Major epidemiological factors influencing infectious disease are:

- host factors
- infectious agent factors
- environmental factors

Changing the parameters of any one of the above factors can prevent and control an epidemic.

Outbreak investigation is a systematic procedure that involves:

- defining a case
- identifying the causative agent
- examining the incidence pattern

- determining the magnitude of the problem by computing the **attack rate**
- looking for common host and environment factors
- formulating a hypothesis
- testing the hypothesis

Section 24.5: What are vaccines, and how are they used to control infectious disease?

Vaccines are designed to expose the adaptive immune system to protective antigens for the purpose of **immunization**. **Vaccination** is carried out to immunize against disease. Vaccine designs include:

- **attenuated vaccines**
- **inactivated**, or **"killed," vaccines**
- **subunit**, or **acellular**, **vaccines**
- **DNA vaccines**

Vaccine failure occurs to some degree for all vaccines because some individuals will fail to be fully immunized after vaccination.

High **vaccine efficacy** does not exist for all diseases, however, where the vaccines do exist, they remain the single most effective way to prevent infectious disease in animals and humans.

Mass immunization against communicable infectious diseases achieves population-wide protection through **herd immunity**.

The **herd immunity threshold** is the critical proportion of the population required to be immunized to achieve herd immunity.

Vaccine coverage rates are usually targeted to exceed the herd immunity threshold in order to compensate for vaccine failure. Eradication through vaccination can be achieved for those infectious agents where:

- long-term immunity can be achieved
- only a few strains exist
- there is a single reservoir host

The disease outcome of infection with poliovirus depends on the immune status of the host.

- The strategic use of the inactivated and attenuated polio vaccines can be used to eradicate wild poliovirus.
- Political and social policies, war, poverty, cultural and religious practices, and lack of government-sponsored vaccination programs can be barriers to attaining vaccination targets.

Testing for the presence of contagious diseases is done to prevent the spread of infectious agents through trade of animals or their products.

In order to maintain their disease-free status for access to markets, many countries do not practice routine vaccination of domestic herds for communicable diseases because the diagnostic tests used cannot distinguish between vaccinated and infected animals.

● Application Questions

1. Imagine you were prescribed a sulfonamide drug to treat a bacterial infection. The doctor explains that the bacterial action of this drug inhibits nucleic acid synthesis. After you leave the office, you start to worry because you realize that your cells also synthesize nucleic acids. Explain why this drug harms the bacteria but isn't going to affect your body cells.

2. You learned that combating fungal and protozoal infections is complicated by the fact that these microbes share many of the same features with the cells of their hosts. Explain how the antifungal drugs known as polyenes and the antiprotozoal drug chloroquine can still show selective toxicity.

3. Your family friend is HIV positive and takes the drug AZT, along with a cocktail of other antiviral medications. You learn that AZT specifically targets retroviruses, but that HIV develops resistance to this drug. Explain the mechanism that this drug uses to combat a retrovirus such as HIV, and how resistance to AZT develops.

4. According to the World Health Organization, drug-resistant strains of *Mycobacterium tuberculosis* are more likely to appear in HIV patients than in tuberculosis patients that are HIV-negative. Provide a hypothesis of why this occurs.

5. You are working in the lab on two different species of bacteria, one of which shows resistance to a particular antibiotic. After mixing and growing the two species together for a time, you notice that they both exhibit resistance when isolated and cultured on media containing the antibiotic. List and explain the mechanisms that may allow a gene for antibiotic resistance originally found on the chromosome in bacteria species A to now appear on the chromosome of bacteria species B.

6. On the news, you hear a CDC scientist explaining that the widespread overuse of antibiotics will do nothing but select for antibiotic-resistant bacteria. You argue with your friend, who is also watching the segment, that the drugs do not *cause* the

resistance. He doesn't believe you. Explain to your friend, then, how excessive use plays a role in resistance.

7. Your child has an ear infection and is given the drug Augmentin, which you read contains amoxicillin and clavulanic acid. Explain why this mixture of drugs helps counter the development of resistance.

8. You are part of a government panel assigned to offer recommendations for reducing antibiotic resistance rates in the United States. One recommendation involves investing research dollars to develop additional rapid diagnostic tests. Explain how the use of such rapid tests could change how antimicrobial drugs are prescribed, and how that change could lead to less resistance.

9. You are part of a research team that has managed to resurrect microbes from the frozen remains of several Ice Age mammals. Out of curiosity, you elect to run a Kirby–Bauer disk diffusion test on one of the bacterial isolates. To your surprise, the Ice Age isolate shows resistance to several antibiotics. Explain why this is possible.

10. In the United States in 1918, the "Spanish flu" resulted in 500,000 deaths, the 1957–1958 "Asian flu" resulted in 70,000 deaths, and the 1968–1969 "Hong Kong flu" produced 34,000 deaths. SARS produced 800 deaths worldwide. All had the same approximate R_0 value. Based on what you learned about the triangle of epidemiology, propose reasons why these diseases produced such different numbers of mortalities.

11. The human papillomavirus (HPV) vaccine is a subunit vaccine that is given in three doses over six months. Imagine that you are a health care provider and that your patient asks you why three doses are necessary to develop immunity to HPV. Use what you've learned in this chapter about subunit vaccines to answer the patient's question.

12. You learned in this chapter that there are varied recommendations for use of the oral and/or inactivated polio vaccines, depending upon the location. Explain the reasoning behind these recommendations in a way that members of the general population could understand and accept.

13. You are assigned the task of determining the minimum inhibitory concentration (MIC) of a newly designed drug that may be effective against a strain of multidrug-resistant *Staphylococcus aureus*.
    a. What does the term "MIC" mean?
    b. Explain how you will set up your *in vitro* testing to determine the MIC.
    c. Imagine you test the following five concentrations of the drug: 0.01, 0.02, 0.04, 0.08, and 0.10 μg/mL. What results would you observe in each of these samples if the MIC was determined to be 0.04 μg/mL?
    d. Explain how you can now use these *in vitro* results from part c to determine a therapeutic dose for an infected patient.

14. You are involved in investigating an outbreak of a new strain of influenza in New York City that is infecting schoolchildren.
    a. Initial estimates indicate that the R_0 value is between 3.0 and 3.5. Exactly what does this tell your team studying the outbreak?
    b. Would these numbers indicate that an epidemic is possible? Explain.
    c. Your job is to now prevent development of this outbreak into an epidemic in the city schools and beyond. List three or four specific control strategies that you would implement.
    d. Explain how each of your strategies above would impact the spread of the infection in the city schools.

Suggested Reading

Anderson, R. M., and R. M. May. 1990. Modern vaccines: Immunization and herd immunity. Lancet 335:641–645.

Baron, C., and B. Coombes. 2007. Targeting bacterial secretion systems: Benefits of disarmament in the microcosm. Infect Disord Drug Targets 7:19–27.

Bennett, P. M. 2008. Plasmid encoded antibiotic resistance: Acquisition and transfer of antibiotic resistance genes in bacteria. Br J Pharmacol 153:S347–S357.

Bull, J., and D. Dykhuizen. 2003. Virus Evolution: Epidemics-in-waiting. Nature 426:609–610.

Fischbach, M. A., and C. T. Walsh. 2009. Antibiotics for emerging pathogens. Science 325:1089–1093.

Guani-Guerra, E., T. Santos-Mendoza, S. O. Lugo-Reyes, and L. M. Terán. 2010. Antimicrobial peptides: General overview and clinical implications in human health and disease. Clin Immunol 135:1–11.

Prescott, J. F. 2008. Antimicrobial use in food and companion animals. Anim Health Res Rev 9:127–133.

Stone, R. 2002. Bacteriophage therapy. Stalin's forgotten cure. Science 298:728–731.

Taubes, G. 2008. The bacteria fight back. Science 321:356–361.

A Reading and Understanding the Primary Literature

Until recently, the majority of our scientific knowledge was spread by word of mouth, by personal letter, or lectures to other scientists. While the first science journals appeared in Europe toward the end of the seventeenth century, mainstream media showed little interest and many discoveries went unnoticed by the general public. The past hundred years have seen a large expansion in journals dedicated to scientific literature, but also an increase in the public's interest in science. As you prepare your lab reports or science research papers, it is important to know what types of information are relevant, appropriate, and available to you. The goal of this guide is to provide a resource to assist you in navigating this wealth of information.

Primary objectives

- What types of journal articles are out there?
- How do I know which type of article I need?
- What types of articles should I use for my research?
- Are some journals better than others?
- How do I find an article of interest?
- How can multimedia tools help me understand the journal articles I've chosen?
- How do I organize my references and avoid plagiarism?
- What is the general format of a primary research article?
- What strategy can I use to read and interpret the data in the primary article I've chosen?

Dr. William Osler, the "Father of Modern Medicine," invented the concept of the Journal Club to help doctors stay current on modern research contributions. After all, there are more than 5,000 new biomedical research articles published every day. As future physicians and scientists, students will find it essential to develop skills for locating, analyzing, and understanding scientific research articles. As you begin your research, you may find that there are several types of journal articles, each with a unique purpose and place in your studies.

Scientific literature is classified as primary, secondary, or tertiary references

Primary (1°) literature consists of all scientific literature presenting novel data or ideas. It may also include a revised analysis of previously published data to advance the theories and hypotheses of the authors. Primary literature is most often published in "peer-reviewed" journals, which means that the articles are evaluated by other scientists prior to publication. This process helps ensure that the data published in the journal has undergone a rigorous evaluation and the data has been deemed accurate, unbiased, and scientifically sound by experts in the field. For example, *The Journal of Virology* is a source of primary journal articles, as seen in Figure B8.3 in the text. The authors were testing a new hypothesis, namely that they could study the MHV receptor by adding MHV to the membrane containing cellular proteins from BALB/c mice and cellular proteins from SJL/J mice. In effect, they provided evidence to support their hypothesis that they could effectively use a virus to identify its own receptor protein. If you wanted to cite their data or this specific approach to identifying a receptor protein, you may choose to cite this article like this:

(*Source:* B. D. Zelus et al. 1998. J. Virol. 72(9):7237–7244.)

Secondary (2°) literature includes summaries of results and ideas previously published in the primary literature, and are written for a scientific audience. These review articles can be a wealth of information if you have recently begun to research a new topic. Some journals are devoted entirely to review articles, while others publish both primary and secondary articles in different sections of the journal. Typically, a review journal can be identified by a name such as "*Trends in…*," "*Advances in…*," or "*Annual Review of….*" Do not be afraid to use the secondary literature in your search. Often, a paper may include a Comment or Letter, which is written by another scientist and provides background and insights into the significance of the primary article. These comments are often a great place to identify questions or concerns about the research that you may want to discuss in your own evaluation of the topic.

Tertiary (3°) literature is generally written for an educated member of the general public. This may include mainstream magazines such as *Discover* or *Scientific American*, science sections in *Time* or *Newsweek*, or science textbooks. While it does offer an evaluation or interpretation of a recent finding, it would not be appropriate to use the tertiary literature in your research in most instances. The main reason for this distinction is that these articles often lack a bibliography, which makes it difficult to distinguish between facts and the author's opinions. Tertiary literature also commonly refers to textbooks or encyclopedias, which are helpful at the earliest stages of your research. However, much of the information in these sources are classified as "common knowledge" and would not necessarily be cited in your research article. Therefore, while you may certainly draw ideas from the tertiary literature, always refer to the primary or secondary article directly if you intend to include these data in your own analysis of a scientific topic. **Table A.1** breaks down the distinctions between the literature.

So, how do you know which types of references are appropriate for your research paper? In general, an introductory science course will pull from the tertiary, secondary, and occasionally, the primary literature. These courses will often use review articles to help clarify a topic from the lecture, but students may not have enough background to understand the primary articles in depth. However, even in your introductory courses, the textbook and encyclopedias are typically not sources you would cite in your references. For advanced science courses, you should plan to rely heavily on the primary and secondary literature. Often, these courses frown upon use of the tertiary literature, particularly if it is used in lieu of the primary articles.

If you are not familiar with a particular research area, it may be difficult to know if the journal is reputable or how rigorous the review process was for publication in that particular journal. What happens if you come across two opposing viewpoints? Is there a way of distinguishing between which journal is better to help you decide which paper is more likely to be correct? This is a common issue, and a difficult question to answer. If you have exhausted other resources such as review articles, Letters, or Comments, there is a simple way to compare different journals within a certain field: *impact factor*. The impact factor is a method for calculating the average number of citations per paper published in that journal during the two preceding years. In effect, it tells you how often a particular journal has published articles that are important enough that they have been cited repeatedly by other researchers. This is not a perfect system, and there are many criticisms of the journal impact factor calculations. It is highly discipline-dependent, and many citations to an article may be made by the author(s) of the original article. Thus, more prolific researchers could potentially overinflate the impact factor for a given journal if they routinely publish in that journal. In addition, journals that publish a large percentage of review articles often have a very high impact factor because they are cited more often.

| TABLE A.1 | Distinguishing between scientific or peer-reviewed journal articles and popular news articles |

	Peer-reviewed articles	Popular/news articles
Authors	Scholars and researchers	Journalists or science writers
Abstract	Included	Rarely included
Bibliography	Included	Rarely included
Format	Standard format, uses language specific to the field of research	No specialized format, contains less technical language
Review process	Peer-reviewed to ensure accuracy	Edited for language/format, not necessarily the content

Locating your article—Beyond Wikipedia

When preparing for a research paper, it's no secret that many students begin their research with a cursory Wikipedia search. That approach is often very helpful if you want to identify key words to help choose a topic or identify appropriate search terms. A quick search of Wikipedia itself reveals that the web site recommends "you should not cite Wikipedia itself, but the source provided; you should certainly look up the source yourself before citing it. If there is no source cited, consider a different method of obtaining this information." To be clear, it is never appropriate to cite Wikipedia or encyclopedias in your research paper.

For research into human diseases and medical topics, there are many web sites that may provide more detailed background information if you are not ready to jump right into a primary journal article. In general, these web sites fall into two categories: government-sponsored agencies and private organizations. The web address will help you distinguish between the two groups. Just look for the .gov or .org suffixes. These web sites can be cited if you are searching for background information about a disease or if you are looking for recent demographic information. Links to a few of the most commonly consulted government agencies are provided in **Table A.2**.

TABLE A.2　Most commonly consulted government agencies

National Institutes of Health	http://health.nih.gov/ http://www.clinicaltrials.gov/	Provides links to MEDLINE®, clinical trials, and the ability to search for specific health topics
PubMed Health	http://www.ncbi.nlm.nih.gov/pubmedhealth/	Based on systematic reviews of clinical trials; provides summaries and full texts of selected recent reviews. Includes general information for consumers and clinicians
Centers for Disease Control and Prevention	http://www.cdc.gov/	Includes information on diseases and epidemiology but also provides links to information about specific topics such as environmental health, emergency preparedness, workplace safety, and travelers' health
World Health Organization	http://www.who.int/en/	WHO is the authority for health within the United Nations system. It is responsible for providing leadership on global health matters and assessing global health trends and up-to-date statistics.
National Cancer Institute	http://www.cancer.gov/	Provides specific information about types of cancer and links to relevant clinical trials

After you have a basic knowledge of your topic and have identified what search terms will help you refine your search, you can now locate a primary article on your chosen topic. There are several peer-reviewed databases that use slightly different approaches to locate primary articles for your scientific research.

Details about each database are summarized in **Table A.3**. You may find different results depending on which database you use, so don't be afraid to try more than one if you are having trouble finding an article that interests you.

TABLE A.3　Several peer-reviewed databases

Name	Dates	Database	Subscription?	Comments
PubMed (National Library of Medicine)	1949–present	4,800 peer-reviewed biomedical journals	Free	Uses MeSH[a] vocabulary to index articles Used extensively in North America
EMBASE® (Elsevier)	1947–1973	7,000 current peer-reviewed journals Biomedical and pharmaceutical literature	Subscription	Finds up to four synonyms for search terms Mostly used by European physicians Lists by publication year or relevance, links to Scopus Biased toward pharmacological usage
SciVerse Scopus (Elsevier)	1996–present	Patent records, theses, conference abstracts, books, systemic reviews. Science, social science, arts and humanities	Subscription	Lists by relevance Tracks citations backwards and forwards Specialized author search
Agricola (AGRI-Cultural OnLine Access) (National Agricultural Library, USDA)	Fifteenth century–present	Extensive collection Catalog and index to the collections of the National Agricultural Library Provides citations (bibliographic references) created by the National Agricultural Library and its cooperators	Free	AGRICOLA is the primary public source for worldwide access to agricultural information. It is a bibliographic database and does not contain the full text. Some citations may include links to full texts available elsewhere from the Internet.
Web of Science (Thomson Reuters)	1984–present	10,000 weekly updates, so it is very current 20–30 journals dropped or changed per month	Subscription	Tracks citations backwards and forwards Specialized author search 77% science; rest social sciences and arts and humanities
Google Scholar (Google)	Current	Scholarly literature, citations/abstracts, peer-reviewed journal articles, dissertations, books, patents, and conference abstracts	Free	Quick and user friendly Not open about how the database works

[a] MeSH is the National Library of Medicine's controlled vocabulary thesaurus.

PubMed is one of the most commonly used resources for locating primary journal articles. It is possible to search by author or topic, and the web site offers a tutorial to assist in refining your search terms to maximize the chances of finding relevant articles and eliminating unwanted ones. For example, if you type in a general search term such as "cancer," PubMed will return over 2.5 million articles. One of the advantages of using PubMed is that the web site offers several helpful hints for narrowing your search and also provides filters for selecting only free full-text articles, review articles, or the most recent articles.

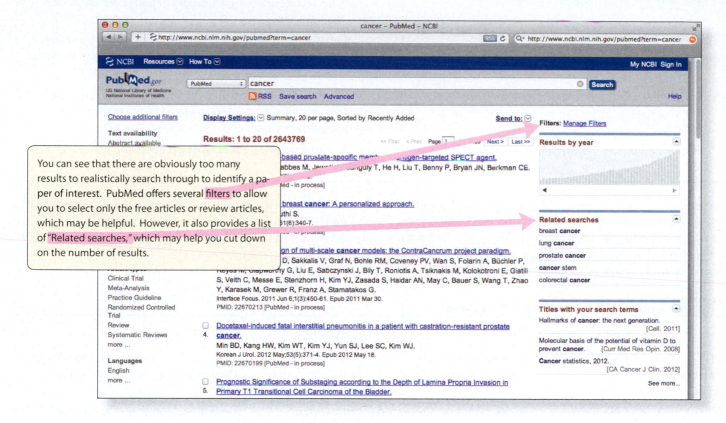

You can see that there are obviously too many results to realistically search through to identify a paper of interest. PubMed offers several filters to allow you to select only the free articles or review articles, which may be helpful. However, it also provides a list of "Related searches," which may help you cut down on the number of results.

You can also set the Limits before your search to help cut down on the irrelevant results. By selecting Limit, you can choose to search for articles based on publication date (most recent), the type of article (review, clinical trial, editorial, etc.), text options (abstract, free full text), and language (published in English or another language). You may also specify articles focusing on a certain species (human vs. animal), gender, or age group.

For example, if you actually want to find an article about recent clinical trials involving pediatric cancer, it may be helpful to set the limits for your search. You can select articles published in the past year that are clinical trials in humans, in the subset of cancer, with free full text, in children ages 0–18.

So, don't be afraid to add limits. This will reduce your results dramatically and save you a lot of time sifting through unhelpful articles.

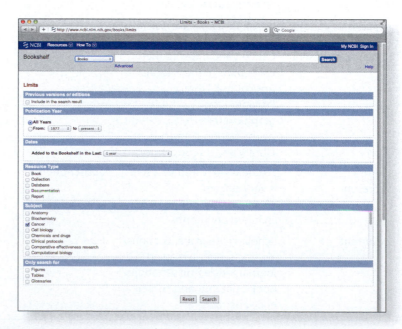

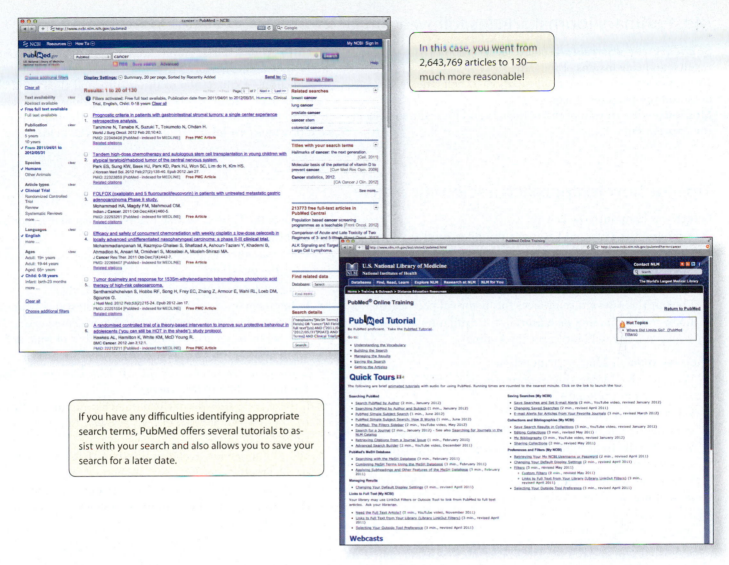

In this case, you went from 2,643,769 articles to 130—much more reasonable!

If you have any difficulties identifying appropriate search terms, PubMed offers several tutorials to assist with your search and also allows you to save your search for a later date.

References or works cited—How to avoid plagiarism and organize your references

After locating your articles, the next step is to begin creating a References or Works Cited section for your paper. It is often much easier to create this section early in the writing process, rather than trying to go back after writing the paper to determine who to cite for each fact. In addition, there are several basic rules for attributions that will help you avoid unintentional plagiarism. First, you do not need to cite information that is common knowledge or part of the basic background you are presenting. For example: Viruses have a lytic or lysogenic cycle. You do, however, need to cite an article if you are using specific information, are referring to a specific theory or idea, or if you use any data, figures, images, and tables from an author's work. You should also use direct quotes *very* sparingly and avoid cutting-and-pasting text. Whenever possible, paraphrase and cite the reference, but make sure it is written in your own words. This will ensure that your paper flows well and is written in your own style and voice and will also help you avoid accidental plagiarism or taking quotes out of context. Here are several helpful hints for how to cite articles within your text:

Most journals use the name-year system. For example:

These results are consistent with previous findings (Drosten et al., 2003) or

These results are consistent with the findings of Drosten et al. (2003).

If there is more than one article in the same year, use letters. For example,

(Drosten et al., 2003a)

If there are two authors, include both names. For example,

(Drosten and Günther, 2003)

If there are several sources, list them chronologically. For example,

(Drosten et al., 2003; Brandenburg et al., 2007; Leiman et al., 2009)

If it is an interview or personal communication, cite the name and date. For example,

(D. Wang, personal communication, 2011 Sept 12)

A final word: Always make sure you have read the relevant section of a paper if you are citing it. This will also allow you to confirm that you are citing the original person or group to make an observation or discovery. The next section will provide some advice about how to approach a paper and read for specific content without being overwhelmed by too many details.

Most primary journal articles follow a similar format, abbreviated as SIMRAD

Summary (or abstract)—concise description of the purpose, results, and conclusions of the paper

Introduction—background information to help the reader understand the purpose of paper

Methods—description of the materials and procedures used in the experiments

Results, And—charts, graphs, and images of the data obtained and experiments performed

Discussion—explanation of the results and their implications to the field of research; corrects misconceptions or previous misinterpretations of the data, and states why the paper is relevant and provides new information relevant to the field.

Tips for how to approach a scientific article

Before you jump into the article:
- Examine the title and keywords to understand the focus of the article.
- Look up definitions of unfamiliar words.
- Read review articles in the field.
- Look at the relevant references for historical perspective.
- Look for Comments or editorials (if available) to see how experts in the field interpret the findings.

Then ask yourself some basic questions:
- Why are they doing this experiment and what is currently known about the topic?
- What is the authors' hypothesis?
- How do the experiments help test the authors' hypothesis?
- What are the basic findings/conclusions? Was the hypothesis correct?
- Why is the authors' data significant? What are the authors' contributions to their field of study?
- What criticisms do you have about the paper? Are experiments missing or unconvincing?

For each of the figures, be able to answer the following questions:
- What question are they asking?
- What was the basic procedure/technique being used to answer the question?
- Are they looking at protein, DNA, RNA, etc.?
- Is the technique appropriate for the questions being asked, and were the experiments properly performed using the correct controls, measurements, and number of repeats?
- What are the results/conclusions?
- Is the data convincing, or does it look like the authors are forcing the data to fit a theory?
- If there are criticisms of the data, what might the authors do to improve the paper and convince you that they are really right?
- Did you notice any trend in the data that was not discussed in the results?
- What questions do you have after reading the paper?
- Do you have any suggestions for follow-up experiments or future directions?

Now for an example taken from your textbook

Mini-Paper: A Focus on the Research
THE DISCOVERY OF REVERSE TRANSCRIPTASE

D. Baltimore. 1970. RNA-dependent DNA polymerase in virions of RNA tumour viruses. Nature 226:1209–1211.

H. Temin and S. Mizutani. 1970. RNA-dependent DNA polymerase in virions of Rous sarcoma virus. Nature 226:1211–1213.

(R-MLV) or Rous sarcoma virus (RSV). Refer to Chapter 5 to review how viruses can be isolated, purified, and quantified. Next, they added these viruses to a standard DNA polymerase assay. The ingredients of this assay included the four deoxyribonucleotide triphosphates (dATP, dTTP, dCTP, and dGTP), the precursors of DNA. Additionally,

In approaching this paper, the first thing you may want to do is to identify key terms and abbreviations to make sure you are familiar with them. By looking them up ahead of time, reading the article will be much easier.

Impact

The identification of this enzyme that converts RNA to DNA, now referred to as reverse transcriptase (RT), had an immediate and profound effect on the fields of virology and, more generally,

Next, you may want to search for related articles that will help provide the background you need to understand this article. You may review the section in your textbook related to the subject, although you would not need to cite that directly in the Works Cited for your paper. To locate relevant articles, you can do a PubMed search.

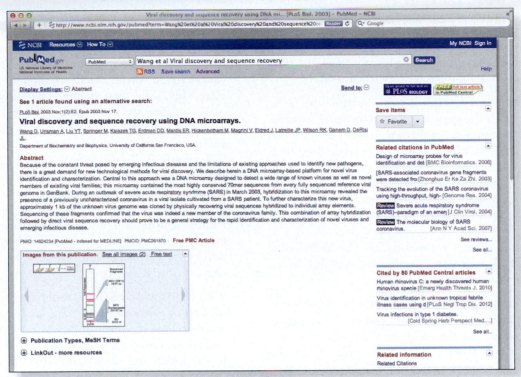

This initial PubMed search will yield several pieces of information. First, the Related Citations includes several review articles published subsequent to this article. These may provide insights into how the technology of DNA microarrays fits into the current research in the field, and also describe its potential significance. Next, PubMed lists 66 citations of this article.

By scanning through this section, you can identify current uses of this technology (as of 2011) and identify potential follow-up experiments by the same research group or another lab. All of this information may be helpful for understanding the article and would be worth assembling early in your writing process.

Context

By the 1960s, several lines of evidence indicated that the replication of RNA tumor viruses, like the tumorigenic virus first described by Peyton Rous in 1911, included a DNA intermediate. First, researchers observed that DNA synthesis inhibitors block replication of these viruses, but only if these inhibitors are added to cells early after their infection. Second, researchers observed that actinomycin D, a drug that inhibits the DNA-dependent RNA polymerase responsible for transcription in mammalian cells, also prevents the formation of new virus particles. Third, several experiments showed that cells infected by RNA tumor viruses contain DNA that hybridizes with, or binds to, viral RNA.

Based on these observations, Baltimore and Temin independently postulated that the replication of RNA tumor viruses proceeds through a DNA intermediate. In other words, the RNA genome is converted into DNA, which later is converted back into RNA genomes. This model would explain the observations previously described. This model also would require the existence of a RNA-dependent DNA polymerase, or an enzyme that could convert RNA into DNA. Such an enzyme had never been observed in any organism. Furthermore, such an enzyme clearly would violate the central dogma of biology: that DNA is converted to RNA, which is converted to protein.

Experiments

To determine if RNA tumor viruses contained RNA-dependent DNA polymerases, the researchers asked two fairly straightforward questions. First, do the viruses produce DNA? Second, what is the template from which this DNA is derived?

To address the first question, both groups started with purified preparations of two RNA tumor viruses: Rauscher mouse leukemia virus

one of the nucleotides, thymidine, was labeled with ^{3}H, a radioactive form of hydrogen. Following an incubation, trichloroacetic acid was added to the reaction and the entire reaction then was filtered through a fine filter. Previous research had shown that DNA was acid-insoluble. In other words, any DNA produced in the reaction would precipitate out of solution when the acid was added and would be retained on the filter. By simply measuring the amount of radioactivity retained on the filter, the researchers could determine if any DNA had been produced.

The reaction did, indeed, result in the formation of an acid-insoluble product. To confirm that this product was DNA, Baltimore then treated the completed reaction with deoxyribonuclease, an enzyme that destroys DNA, ribonuclease, an enzyme that destroys RNA, and micrococcal nuclease, a relatively non-specific nuclease that destroys both RNA and DNA. Deoxyribonuclease and micrococcal nuclease, but not ribonuclease, digested the radioactive product. These two results support the conclusion that purified preparations of R-MLV could produce DNA.

To address the second question, Baltimore and Temin both investigated the sensitivity of the virus to ribonuclease (**Figure B8.6**). In the absence of ribonuclease, the amount of DNA produced increased over time (*line 1*). A similar increase was observed when ribonuclease was added after the virus and deoxynucleotides were allowed to incubate together for several minutes (*line 2*). When ribonuclease was included initially in the reaction mixture, however, the amount of DNA produced decreased markedly (*line 3*). When the virus was pre-incubated with ribonuclease before being added to the polymerase assay reaction, the amount of DNA produced decreased even more dramatically (line 4). These results support the conclusion that RNA is the template for DNA production.

Now, begin answering the basic questions we specified.

- **Why are they doing this experiment and what is currently known about the topic?** You would anticipate finding this information in the introduction section of the paper. The authors describe a previous paper that dealt with the prototype DNA microarray designed for parallel viral detection. Because the paper you are reading builds heavily off of this 2002 work, it will be very helpful to locate this article as well.

Ultimately, the authors' objective is written explicitly: "...to create a microarray with the capability of detecting the widest possible range of both known and unknown viruses." If you were to paraphrase their objective, you might write something like: "Researchers sought to develop a microarray capable of quickly identifying both known and unknown viruses to aid in the global effort to identify the cause of future global health crises." The hypothesis can also be found early in this paper, stated succinctly within the abstract.

To paraphrase, the authors wanted to test the ability of a new and improved DNA microarray technique to rapidly identify a previously uncharacterized virus.

- **How do the experiments help test the authors' hypothesis?** Quite simply, they used the 2003 SARS epidemic as a proof-of-principle, showing that the DNA microarray technology successfully identified a new coronavirus within a culture from a SARS patient. They found this by combining an array hybridization with direct viral sequencing recovery.

- **What are the basic findings/conclusions? Was the hypothesis correct?** Within the Discussion section, the authors state that "In the case of SARS, we were able to ascertain within 24 h that a novel coronavirus was present in the unknown sample, and partial genome sequences of this virus were obtained over the next few days without the need for specific primer design." You may paraphrase this by stating that the authors identified that the unknown virus was a previously unknown coronavirus and quickly identified a partial genomic sequence to aid in identification using this versatile new technique.

- **Why are the authors' data significant? What are the authors' contributions to their field of study?** Again, this information can be found within the Discussion section: "To our knowledge, this is the first demonstration of the feasibility and utility of directly recovering nucleic acid sequences from a hybridized DNA microarray. In light of the continuous threat of emerging infectious diseases, this overall approach will greatly facilitate the rapid identification and characterization of novel viruses." Again, you could restate the significance by saying that this is the first experiment showing that the genomic sequence of an unknown virus can be identified using a hybridized DNA microarray. This is significant, given the need to rapidly identify and characterize novel viral pathogens and prevent the next emerging pandemic before it occurs.

- **What criticisms do you have about the paper? Are experiments missing or unconvincing?** Your criticisms may vary, but it could include concerns about using a retroactive approach and the potential for bias in the results. If the authors knew it was SARS and they found that information within the results, is it possible they were looking at sequences that made it more likely to identify SARS from the data? You may also comment on the fact that the article was very short, which may necessitate some additional research on your part. If you were to present this article for a journal club or research paper, it would definitely be necessary to use additional articles published prior to and subsequent to this article to assure your complete coverage of this topic.

As we noted in Section 1.1, the field of microbiology began with the development of microscopes by Robert Hooke and Anton van Leeuwenhoek. Microscopy has come a long way since van Leeuwenhoek first described microorganisms in vivid detail over three and one-half centuries ago. Let's use this opportunity to explore some of these advances. We'll introduce several important types of microscopy and discuss their uses. As we will see, the microscopy technique used generally depends on the structures being examined and the questions being asked. We'll begin by reviewing some basic principles of microscopy. Then, we will explore more thoroughly two types of microscopy—light, or optical, microscopy and electron microscopy. Throughout this appendix, we will feature images that appear elsewhere in this book. Hopefully, this introduction to microscopy will provide you with a better understanding of how these images were generated and make you think about the advantages and disadvantages of the techniques that were used.

Principles of microscopy

To understand the microbial world revealed by microscopy, we need to be familiar with units of measurement commonly used by scientists. A micrometer (μm), sometimes referred to as a micron, equals one-millionth of a meter, or 10^{-6} m. A nanometer (nm) equals 10^{-9} m. As shown in Figure 2.4, we can see objects larger than about 100 μm with our unaided eyes. Light microscopes allow us to see objects as small as 0.2 μm in diameter, and various forms of electron microscopy allow us to see objects as small as 0.005 μm in diameter.

What factors affect our ability to see objects? The two critical factors are **contrast** and **resolution**. Contrast refers to the difference in light absorbance between two objects or areas. The greater the contrast between an object and its surrounding environment, the easier it is to see that object. Microbiologists can increase the contrast between objects and their surroundings by staining them. Gram staining, which we discuss in Toolbox 2.1, allows us to differentiate Gram-positive and Gram-negative bacteria. The staining procedure also increases the contrast of the bacterial cells. When stained, the light absorbance of the bacteria differs significantly from the light absorbance of the surrounding area **(Figure B.1)**.

Resolution refers to the distance between two objects at which the objects still can be seen as separate **(Figure B.2)**. The closer together two objects are, the higher the resolution necessary to distinguish them will be. Similarly, the smaller an object is, the higher the resolution needed to observe it. The size of most bacterial cells falls below the resolution limit of the human eye. Even if bacterial cells are stained and exhibit significant contrast with their background, they cannot be seen by our unaided eye.

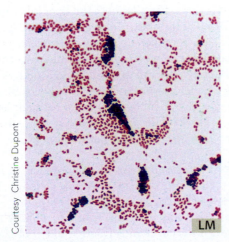

Courtesy Christine Dupont

LM

Figure B.1 Gram-stained bacteria By staining bacteria, we increase the contrast between the bacterial cells and the surrounding area, thereby improving our ability to see them.

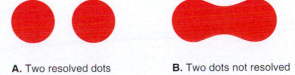

A. Two resolved dots   B. Two dots not resolved

Figure B.2. Resolution The ability to distinguish two objects is referred to as resolution. **A.** Two dots that clearly can be seen as separate are said to be resolved. **B.** If the two dots appear to be fused, then they are not resolved.

To more fully understand resolution, we also need to have some understanding of the properties of light. Light has a number of properties that affect our ability to visualize objects, either with the unaided eye or the light microscope. One of the most important properties of light is its wavelength, or the length of a light ray (Figure B.3). Represented by the Greek letter lambda (λ), wavelength equals the distance between two adjacent crests of a wave. Smaller wavelengths provide greater resolution. In fact, resolution is more important than magnification; the visualization of objects that are magnified but not resolved is not very useful.

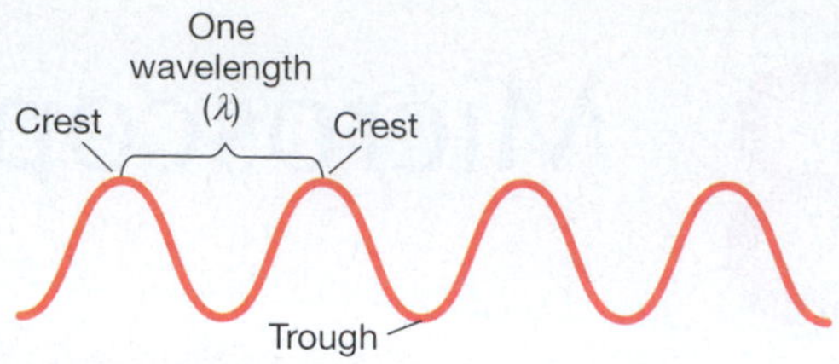

Figure B.3. Wavelength One wavelength is defined as the distance between two crests of a wave. Shorter wavelengths provide greater resolution.

LM Light microscopy

Now that we have explored some of the basics of microscopy, let's look at a specific type of microscopy—light, or optical, microscopy. Robert Hooke and Anton van Leeuwenhoek, microscopists who we first encounter in Chapter 1, developed light microscopes over 350 years ago. With these microscopes, visible light passes through the sample and magnifying lenses. The microscopes crafted by van Leeuwenhoek had a single lens. The quality of these lenses was so high that a magnification of almost 300× was achieved. In other words, the image of the sample was magnified 300 times. This magnification allowed him to view microbial cells. By using more than one lens, even greater levels of magnification can be achieved. A compound light microscope (Figure B.4) basically uses two magnifying lenses, an objective lens and an ocular lens, and the magnification is multiplicative. If the objective lens provides 40× magnification and the ocular lens provides 10× magnification, for instance, then a total magnification of 400× results.

In the simplest form of light microscopy, **bright field microscopy**, light enters the microscope from a source in the base, and often passes through a blue filter, which filters out the long wavelengths of light, leaving the shorter wavelengths and

thus improving the resolution. The light then goes through a condenser, which converges the light beams so that they pass through the specimen. The iris diaphragm controls the amount of light that passes through the specimen and into the objective lens. To see specimens clearly at higher magnifications, greater amounts of light are needed.

The ocular lens usually provides 10× magnification, and objective lenses usually provide from 10× to 100× magnification, resulting in a total magnification of 100× to 1000×. At higher magnifications, however, resolution often decreases. The refractive index, a measure of how light passes through a medium, of the air between the lens and the specimen differs from the refractive index of the lens itself. This difference limits the ability to focus, which determines the resolution. To combat this problem, researchers often add a drop of immersion oil to the specimen when using the 100× objective lens. Immersion oil has a refractive index similar to that of the objective lens. When immersion oil fills the gap between the objective lens and the specimen, the ability to focus is improved. At 1000×, the limit of resolution in a compound microscope is approximately 0.2 μm.

The magnification achieved by a compound microscope would be useless if there were no way to increase the contrast between a bacterial cell and the surrounding medium. Most bacterial cells absorb relatively little light. The refractive index of a bacterial cell also differs only slightly from that of water. As we noted earlier, staining can improve the contrast. Differential stains that react differently with various bacteria can be used. These stains are useful in identification. Other stains react specifically with intracellular structures. For example, the lipophilic stains Nile red and Sudan Black stain lipids. In negative staining, researchers use stains that are not taken up by the cells. These reagents stain the background surrounding the cell, allowing the unstained cell to stand out against this stained background. Staining cells, however, does have drawbacks. Most notably, staining typically requires fixing cells by heat or chemical treatment before adding dye. The fixing process kills and may distort the cells. The image, therefore, may not accurately reflect the image of a live cell.

Phase contrast microscopy Another approach to increase the contrast is to use **phase contrast microscopy** (Figure B.5), which uses a special condenser (that incorporates an annular ring) and objective lens (that incorporates a phase ring) to amplify the slight differences in refractive index of bacterial cells and the surrounding medium to make cells look much darker. This method has the major advantage that the cells remain

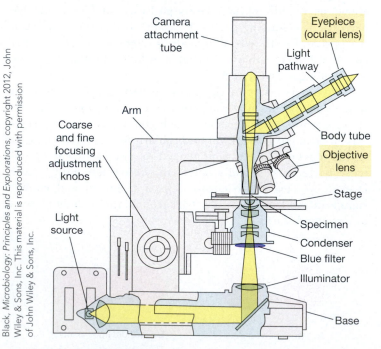

Figure B.4. The compound light microscope In the light microscope, visible light passes through the specimen and magnifying lenses. Often, a filter blocks long wavelength-blue light. The condenser converges the light beams.

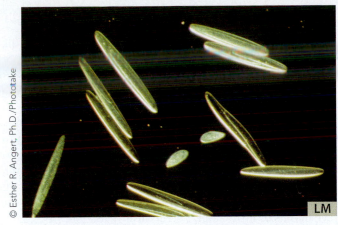

Figure B.5. Phase contrast microscopy By amplifying slight differences in the refractive index of cells and the surrounding medium, phase contrast microscopy allows cells to be visualized without fixation.

alive; fixation or chemical treatment is not required. Active behaviors such as swimming or gliding can be observed, and cell morphology is not subject to fixation artifacts.

Dark field microscopy In dark field microscopy, light does not pass straight through the specimen to the lens, but instead reflects off the specimen at an angle. The approach results in a dark background with bright objects in it (Figure B.6). While this technique usually does not permit one to see intracellular structures, it can greatly increase the contrast when viewing small transparent cells.

FM Fluorescence microscopy In fluorescence microscopy, the energy of the incoming light, usually in the ultraviolet spectrum, causes electrons in special fluorescent molecules in the sample called *fluorophores* to become excited. This excitation results in light being emitted as the electrons lose energy. The emission light typically is of a different wavelength than the incoming excitation light. Optical filters ensure that only the emission light reaches the eye. Some proteins naturally fluoresce, and these fluorophores have become important tools in microscopy. As we discovered in Toolbox 3.1, green fluorescent protein (GFP) and related molecules like yellow fluorescent protein (YFP)

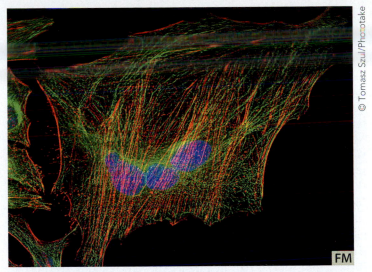

Figure B.7. Fluorescence microscopy By coupling a fluorophore to a protein of interest, the location of this protein within a cell can be visualized. Here, GFP-tubulin fusion protein allows us to see tubulin within these cells.

and red fluorescent protein (RFP) have been used extensively by microbiologists (Figure B.7). These proteins either can be expressed alone in cells or coupled to a normal cellular protein, in which case the behavior and location of the protein of interest can be observed. Molecules also can be tagged with small chemical structures that have fluorescent properties. Fluorescein, for example, emits a yellow-green light when excited by light of a specific wavelength. Antibodies tagged with these fluorescent molecules can bind to their specific antigens, thus allowing the structures containing those antigens to be detected by the fluorescence.

Confocal laser scanning microscopy We all realize that cells are three dimensional. When using conventional light and fluorescence microscopy, however, we cannot focus on multiple planes simultaneously. We cannot visualize specimens, in other words, in 3D. Confocal laser scanning microscopy (CLSM) offers a solution to this problem. In CLSM, a laser focuses on a single plane within the object and collects images within that plane. The laser then moves through the object, focusing sequentially on adjacent planes, collecting two-dimensional images for each plane. Computer programs then can be used to process these 2D images, producing a three-dimensional reconstruction. As we discovered in Perspective 2.1, 3D reconstructions also can be generated by cryoelectron tomography imaging, providing us with a better understanding of how internal structures like magnetosomes are arranged within cells (Figure B.8).

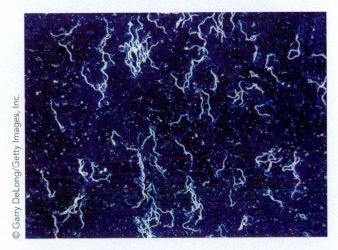

Figure B.6. Dark field microscopy Because light reflects off the specimen at an angle, cells appear bright on a dark background.

Figure B.8. Cryoelectron tomography imaging By collecting and processing 2D images obtained from multiple focal planes, 3D images can be developed. Here, the organization of magnetosomes (yellow) within *Magnetospirillum magneticum* cells can be seen.

Electron microscopy

Because light microscopy cannot resolve objects separated by less than about 0.2 μm, this technique cannot provide much information about subcellular components or viruses. With electron microscopy, better resolution occurs and these subcellular objects become visible. The **electron microscopy (EM)** uses a beam of electrons instead of a beam of light. Because of the very short wavelength of electron beams, the resolution limit of EM is less than 1 nm, a great improvement over light microscopy. Instead of a lens, electromagnets focus the electron beam, and the electrons are detected by their impact with a fluorescent screen, photographic film, or by a charge-coupled device (CCD) camera. Because air molecules interfere with the movement of electrons, the electrons must travel in a vacuum. The first electron microscope was built in 1933 and earned its inventor, Ernst Ruska, the Nobel Prize in Physics in 1986. In this section, we will explore the two most common types of electron microscopy: **transmission electron microscopy (TEM)** and **scanning electron microscopy (SEM)**.

TEM **Transmission electron microscopy** In transmission electron microscopy, or TEM, the object being observed rests between the electron source and the detector. The differential ability of electrons to pass through the object results in an image. For a specimen to be viewed by TEM, it must first be chemically crosslinked to preserve its structure, dehydrated in ethanol or acetone, stained with electron dense stains such as uranyl acetate or osmium tetroxide to enhance contrast, and embedded in a block of resin that then is hardened by heat treatment. The embedded sample is then cut with a glass or diamond knife using an ultramicrotome to produce very thin slices called "sections." These sections must be extremely thin (70–90 nm); otherwise, none of the electrons will be able to pass through to the detector.

TEM can be used for visualizing subcellular structures, like a phospholipid bilayer or viruses **(Figure B.9)**. Even single DNA molecules can be seen with TEM. Additionally, TEM images can be colorized, providing more dramatic images. The technique, however, does have some disadvantages. Most notably, the extensive processing requires that living samples must be killed. The processing also can alter some cell structures, resulting in the appearance of artifacts. Alternative methods of sample preparation can provide a better picture of the true structure. For example, the freeze-fracturing technique, which involves freezing the sample and then fracturing it with a knife, often reveals important surface details of intracellular structures (see Figure B2.7).

SEM **Scanning electron microscopy** In scanning electron microscopy, or SEM, the reflection of a moving electron beam off the surface of a specimen forms a three-dimensional surface view of the specimen. Conceptually, this technique resembles confocal laser scanning microscopy. As with TEM, live samples cannot be visualized by SEM. Specimens must first be fixed and coated with a thin layer of gold or palladium that reflects electrons. Unlike TEM, internal structures are not visible by SEM. Scanning electron microscopy does, however, provide us with a more detailed view of specimen structure **(Figure B.10)**. These images, like TEM images, can be colorized.

In this appendix, we have examined only briefly some of the more commonly used forms of microscopy. In addition to these techniques, microbiologists use other types of microscopy and new developments are continually occurring. The basic principles, though, remain fairly standard. As we have noted here, each microscopy technique has advantages and disadvantages. The type of microscopy used depends on the questions being asked and the information that one wants to obtain. The microscope truly is the microbiologist's best friend.

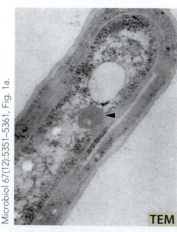

From Gordon C. Cannon et al. 2001. Appl Environ Microbiol 67(12):5351–5361, Fig. 1a.

A. *Synechococcus* bacterium   **B.** T4 bacteriophage

© Department of Microbiology, Biozentrum, University of Basel/Photo Researchers, Inc.

Figure B.9. Transmission electron microscopy (TEM) With transmission electron microscopy, better resolution is possible. **A.** Structures within cells, like the plasma membrane and nucleoid, can be seen more clearly. **B.** Subcellular entities like viruses also can be seen, as in this colorized image.

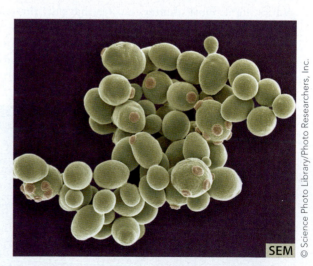

© Science Photo Library/Photo Researchers, Inc.

Figure B.10. Scanning electron microscopy (SEM) Like transmission electron microscopy, SEM provides much better resolution than light microscopy. Additionally, as seen here, SEM provides a three-dimensional view of the specimen, allowing us to see surface details more clearly.

C Classification of Bacteria

Initially, bacterial classification was determined by comparison of cell and colony morphology, growth characteristics, physiology, and biochemistry. The classification of the domain Bacteria is continually expanding as more organisms are isolated from nature and DNA sequence analysis provides information about their evolutionary relatedness. This follows from the initial breakthroughs in molecular phylogenetics made by Carl Woese that offered a much more realistic view of how bacteria are related to each other. In addition to sequence information

from cultured organisms, there is also an increasing amount of data from organisms that have not yet been cultured. Bacteria can be organized into several groups, called "phyla" (singular "phylum"). Shown here are the major phyla of cultured bacteria that are recognized, along with some examples of the better known representative genera of each phylum. Related bacteria often share similar characteristics, although this is not always the case. For this reason, the properties of uncultivated organisms can only be approximated by their membership in a taxonomic group.

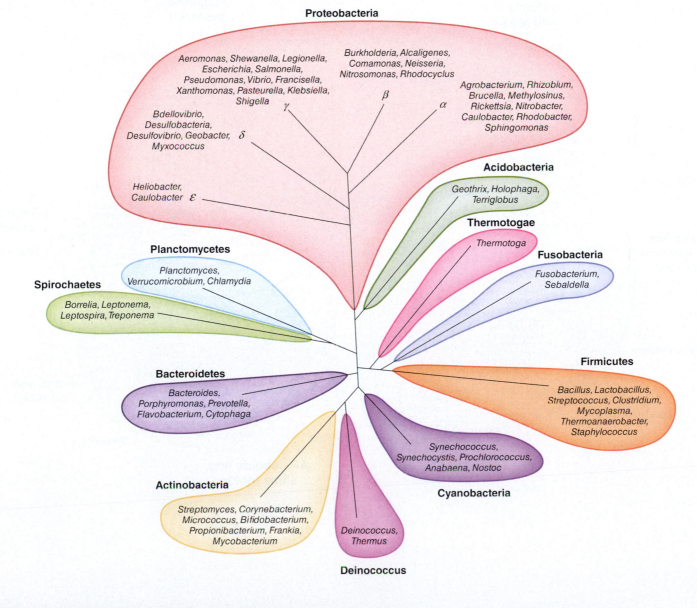

Classification of Eukarya

From the earliest days, the classification of the domain Eukarya was based solely on comparison of shared physical characteristics. This was much more effective for the non-microbial members than for the microbial members. As happened in the domain Bacteria, DNA sequence information has now found its way into eukaryal classification. These sequence data have reinforced how most of the diversity of the eukaryal domain is lodged within its microbial representatives. As can be seen in this tree, the vast majority of organisms in each of the five major eukaryal taxonomic groups is microbial.

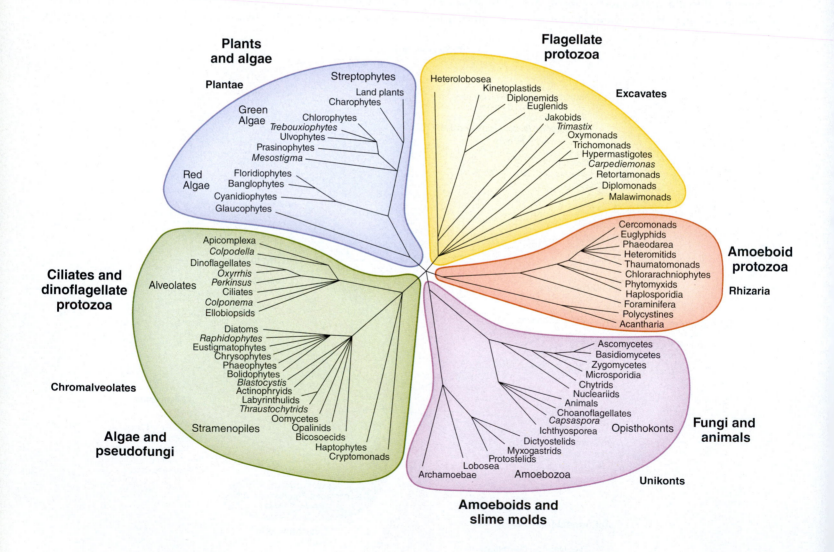

E Classification of Archaea

Classification within the domain Archaea was initially based on shared characteristics, similar to the case with the domain Bacteria. The impact of molecular phylogeny on archaeal classification has been considerable. Given the difficulty of cultivation of so many of the Archaea, the retrieval of environmental sequences has perhaps had more of an influence on the archaeal classification than for the other domains, with many members only known by their DNA sequences. Representative members of the major archaeal phyla are shown here, along with general characteristics.

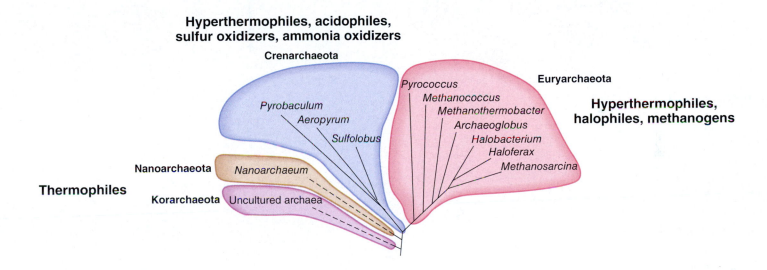

Hyperthermophiles, acidophiles, sulfur oxidizers, ammonia oxidizers

Crenarchaeota

Pyrobaculum
Aeropyrum
Sulfolobus

Pyrococcus
Methanococcus
Methanothermobacter
Archaeoglobus
Halobacterium
Haloferax
Methanosarcina

Euryarchaeota

Hyperthermophiles, halophiles, methanogens

Nanoarchaeota   Nanoarchaeum

Thermophiles

Korarchaeota   Uncultured archaea

Classification of Viruses

DNA animal viruses

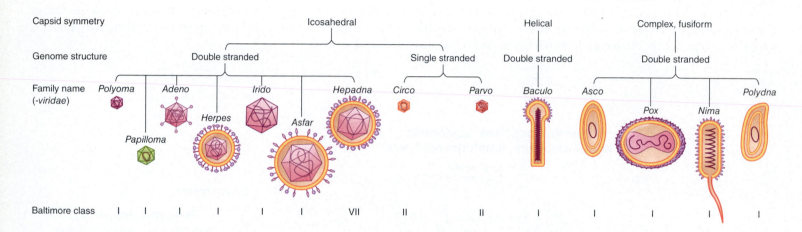

| | | | | | | Icosahedral | | | | | | | | Helical | | Complex, fusiform | | |

Capsid symmetry — Icosahedral — Helical — Complex, fusiform

Genome structure — Double stranded — Single stranded — Double stranded — Double stranded

Family name (-viridae): *Polyoma* *Adeno* *Irido* *Hepadna* *Circo* *Parvo* *Baculo* *Asco* *Pox* *Nima* *Polydna*
Papilloma *Herpes* *Asfar*

Baltimore class: I I I I I I VII II II I I I I I

RNA animal viruses

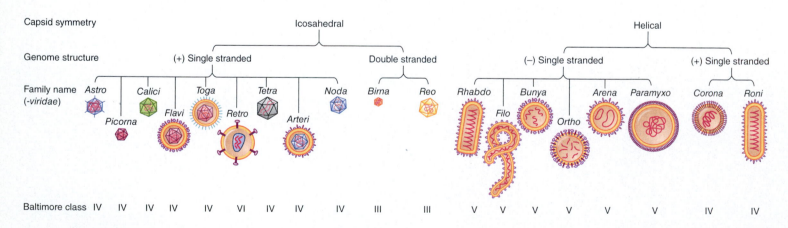

Capsid symmetry — Icosahedral — Helical

Genome structure — (+) Single stranded — Double stranded — (−) Single stranded — (+) Single stranded

Family name (-viridae): *Astro* *Calici* *Toga* *Tetra* *Noda* *Birna* *Reo* *Rhabdo* *Bunya* *Arena* *Paramyxo* *Corona* *Roni*
Picorna *Flavi* *Retro* *Arteri* *Filo* *Ortho*

Baltimore class: IV IV IV IV IV VI IV IV IV III III V V V V V V IV IV

G Origin of Blood Cells

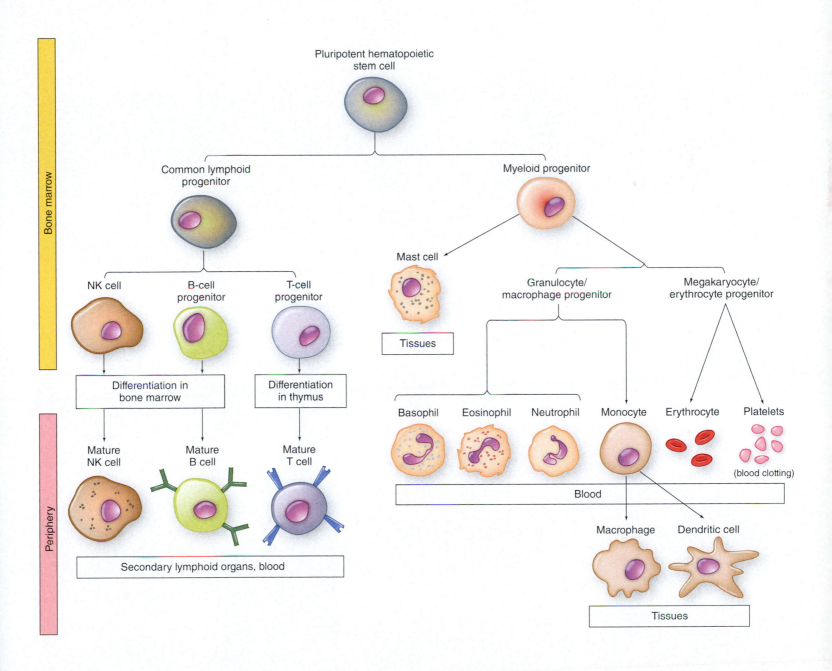

% G+C Measure of the total base pairs in a genome that are G:C

2-deoxyribose Five-carbon sugar found in DNA

2D-PAGE Two-dimensional polyacrylamide electrophoresis in which molecules are separated based on both isoelectric point and mass

3′ poly (A) tail String of adenosine monophosphates found on the 3′ end of a growing RNA molecule

454 DNA sequencing Next generation sequencing method that combines fragmentation of DNA and separation of fragments on separate beads combined with pyrosequencing; *see also* pyrosequencing

5′ cap A modified nucleotide, 7-methylguanosine, found on the 5′ end of a growing RNA molecule

α-amanitin Toxin produced by mushrooms in the genus *Amanita*; inhibits RNA polymerase II

A-B toxin Two-component toxin possessing a receptor-binding B portion and an enzymatic A portion

ABC transporter System of proteins that binds to a substrate, hydrolyzes ATP to provide energy for transport (ATP-binding cassette), and forms a channel for membrane transport

abortive infection Outcome of a viral infection in which few if any new infectious viral particles are produced

absorbance Measure of the light that is absorbed or scattered as it passes through a solution and does not reach the photodetector in a spectrophotometer

acellular vaccine Vaccine composed of one or of a select number of isolated antigens; not composed of whole cells or viruses; *see also* subunit vaccine

acetotrophic methanogen Organism that produces methane (CH_4) from acetate (CH_3COO^-)

acidophile Organism that grows optimally in low pH environments (pH < 5.5)

activated sludge Secondary wastewater treatment using the perfusion of oxygen through accumulated flocs to maintain aerobic digestion by microorganisms

activation energy (E_A) Energy that must be supplied for a reaction to proceed; determines the rate of a chemical reaction

activator binding site DNA sequence that binds an activator molecule, enhancing binding of RNA polymerase to a promoter

activator molecule Molecule that increases the affinity of RNA polymerase to a promoter

active site Physical location where one or more substrates bind to an enzyme

active transport system Membrane proteins that use energy to move solutes against a concentration gradient

acute infection Infection of short duration

adaptive (specific) immunity Immunity found in jawed vertebrates that occurs only after exposure to a foreign substance, usually following innate stimulation; involves antigen-specific lymphocytes (B cells and T cells) that possess unique receptors generated through random genetic changes in the receptor genes; also called "acquired immunity"

adenine (A) Purine nitrogenous base, along with guanine

adenosine triphosphate (ATP) Cellular biochemical source of energy used to drive coupled biosynthetic reactions

adhesin Any microbial surface structure or extracellular substance that is used to bind to host tissues

adjuvant Any substance that enhances an immune response; commonly used in vaccine formulations

aerobe Organism that grows in the presence of oxygen

aerobic respiration Cellular respiration using O_2 as a terminal electron acceptor; can only be carried out under aerobic conditions

aerotolerant anaerobe Organism that does not require oxygen for respiration, but can grow in the presence of oxygen

affinity maturation Progressive increase in the binding strength of antibody for its antigen as the immune response progresses

affinity tag Peptide sequence added to a recombinant protein that will aid in its purification

agar Polysaccharide purified from red algae used as a solidifying agent in microbial growth media

alkalophile Organism that grows optimally in high pH environments (pH > 8.5)

alkyltransferase Enzyme that removes methyl groups from nucleotide bases

allele One of the alternate forms of a gene

allosteric inhibition Inhibition of enzyme action by an effector molecule that binds reversibly to the enzyme allosteric site and is not changed in the reaction

allosteric regulation Regulation of enzyme action by an effector molecule that binds reversibly to the enzyme allosteric site and is not changed in the reaction; regulation can be positive (activation) or negative (inhibition)

alternative pathway Complement cascade activation pathway initiated by the binding of an alternative form of C3b to a variety of cell wall components

amino acid Subunit of a protein molecule

amitochondriate Eukaryal microbe that lacks mitochondria

amoebocyte Invertebrate cell derived from an activated hemocyte, containing antimicrobial peptides and clotting factors to fight infection and reduce tissue damage

anabolism Biosynthesis of macromolecular cell components from smaller molecular units

anaerobic growth Microbial replication in the absence of oxygen

anaerobic methane oxidizer (ANME) Archaeon that oxidizes methane (CH_4) anaerobically in an important part of the global carbon cycle; known only by DNA sequence and has not yet been cultured

anaerobic oxidation of methane (AOM) Important part of the global carbon cycle carried out by anaerobic methane oxidizers (ANME)

anaerobic respiration Cellular respiration using electron acceptors other than O_2; usually carried out under low oxygen concentrations or anaerobic conditions

anaerobic sludge digester Large anoxic bioreactor operating in a semi-continuous mode that allows anaerobic digestion of remaining biowaste sludge following primary and secondary wastewater treatment

anamnestic response Immune response that originates from the stimulation of memory cells and is characterized by a faster and more vigorous adaptive immune response upon repeated exposure to antigen; *see also* memory response

anaplerotic pathway Replenishing pathway or reaction that generates a critical metabolic intermediate in order to sustain another metabolic pathway

anemia Decrease in oxygen-carrying capacity of blood, often due to decrease in number of erythrocytes

annotation Use of computer programs to predict the beginning and end of protein-encoding sequences of DNA

anoxic Lacking oxygen

anoxygenic photosynthesis Bacterial phototrophy that does not produce O_2 and uses either PS I or PS II

antibiotic Compound derived from microorganisms that inhibits the growth of other microorganisms; often encompasses synthetic compounds used for the treatment of bacterial, fungal, and protozoal diseases

antibody Secreted immunoglobulin produced by B cells, specifically plasma cells, of the adaptive immune system

antibody-dependent cell-mediated cytotoxicity (ADCC) Killing mechanism used by NK cells; binding of F_CR receptors on the NK cell to antibody that is bound to the surface of an infected cell causes degranulation and apoptosis of the target cell

anticodon Sequence of three nucleotides on tRNA that binds to complementary codon on the mRNA

antigen Any component that can be specifically bound by an immune receptor, namely a T-cell receptor (TCR) or an immunoglobulin (Ig)

antigen-presenting cell (APC) Cell capable of processing and presenting antigen in association with MHC molecules; includes dendritic cells, macrophages, and B cells

antigenic drift Gradual change in viral antigens caused by random mutation

antigenic shift Dramatic change in viral antigens caused by reassortment

antigenic variation Change in molecules on the surface of a pathogen to which the immune system responds; can allow pathogen to avoid immune detection

antimetabolite Chemical analog that is similar in structure to a natural compound but does not participate in the same chemical reactions

antimicrobial drug Compounds used primarily to treat or prevent microbial infections of the body; includes antibacterial, antifungal, antiprotozoal, and antiviral drugs

antimicrobial peptide A member of a group of low molecular weight proteins secreted by various cell types in plants, invertebrates, and vertebrates that appear to destabilize the membrane structure of microbes, resulting in disruptive gaps or pores

antiparallel Arrangement in double-stranded DNA where the 5′ end of one strand is aligned with the 3′ end of the other strand

antiporter Molecule used in co-transport in which the energy-requiring movement of one molecule is driven by the energetically favorable movement of another molecule in the opposite direction

antisense RNA Single-stranded RNA that can interact with specific mRNA molecules through complementary base-pairing

antiseptic Chemical that can be used on living tissue such as skin to kill infectious microbes

apoptosis Cell death accomplished through a highly ordered and tightly controlled cascade of chemical-signaling events within the cell that results in its destruction without the release of damaging cell constituents or triggering of inflammation

appressoria Specialized cells of fungi that can penetrate the plant cuticle, enhancing potential for infection

aquaporin Protein channel in the plasma membrane that facilitates passage of water

archaeon Member of the domain Archaea

ascariasis Infection with a species in the *Ascaris* genus of roundworms

ascus *pl.* **asci** Structure containing haploid ascospores produced by some fungi

assimilation Process by which cells import a molecule and incorporate it into cellular constituents

ATP synthase Membrane enzyme complex that uses the kinetic energy of flowing protons to synthesize ATP from free ADP and P_i

attack rate Special incidence rate calculated during an outbreak and usually covering a short period of time that gives the proportion of ill individuals in an exposed population; used to estimate how many more individuals are expected to become ill

attenuated vaccine Vaccine composed of a live but weakened, or less virulent, version of a microbe that is completely or nearly devoid of pathogenicity yet retains the ability to generate protective immunity in the host

attenuation Regulatory mechanism that occurs after the initiation of transcription but before transcription of the operon is complete; decrease in virulence

autoclave Instrument used to sterilize materials by application of heat and pressure

autoimmunity Immune reactions directed inappropriately against the body's own cells

autoinducer Signal molecule used in quorum sensing that increases in concentration as the density of the population increases

automated sequencing Technique based on Sanger sequencing that yields sequence of up to 1,000 nucleotides within hours; *see also* cycle sequencing

autotroph Organism that can build complex organic molecules used as nutrients from an inorganic carbon source

auxotroph Organism that requires certain organic precursors from the environment in order to synthesize all necessary cellular constituents

β-lactam Refers to antibiotics containing a β-lactam ring that interferes with synthesis of the bacterial cell wall; includes penicillins and cephalosporins

β-lactamase Enzyme that hydrolyzes the β-lactam ring of some antibiotics rendering them ineffective

β-lactam class Group of antibiotics, all of which possess a β-lactam ring as a common structural feature

β-oxidation Major pathway for the catabolism and oxidation of lipid fatty acid chains into acetyl-CoA, producing significant amounts of NADH and $FADH_2$ for ATP production

B cell (B lymphocyte) Lymphocyte of the adaptive immune system expressing immunoglobulin B-cell receptors (BCRs) on their surface; give rise to antibody-secreting plasma cells

B-cell receptor (BCR) Membrane-bound immunoglobulin expressed on the surface of B cells that can bind antigen in a specific manner

bacillus *pl.* **bacilli** Bacterial cell that is roughly cylindrical; also called a rod

bacteriochlorophyll Primary photopigment used by anaerobic photosynthetic bacteria

bacteriocidal drug Antibacterial medication that directly kills bacteria

bacteriocin Antimicrobial peptide produced by one strain of bacteria and that is harmful to another strain within the same family

bacteriocyte Specialized host cell found only in symbiotic insects; houses endosymbionts

bacteriophage Virus that infects bacteria; literally, "bacteria eater"

bacteriorhodopsin Archaeal membrane protein bound to retinal that functions as a light-driven proton pump to generate ATP

bacteriostatic drug Antibacterial medication that prevents the replication of bacteria but does not directly cause bacterial death

bacteroid Endosymbiotic, terminally differentiated bacteria found in symbiotic plants (frequently in root nodules)

barophile Organism that grows optimally under conditions of high pressure

base substitution Type of mutation in which a single base is changed

basic reproduction number (R_0) Average number of new infections caused by an infected individual in a population of susceptible individuals

basophil Cytotoxic cell of the innate immune system found in tissue and blood and named for the distinct staining of its granules with certain basic dyes; generally acts against extracellular parasitic worms in the presence of IgE or C3a or C5a

batch culture A closed system for microbial growth in which nutrients are not replenished and wastes are not removed

bioaugmentation Introduction of microorganisms that increase the rate of degradation of organic wastes or pollutants

biocatalyst Enzyme that facilitates an industrial chemical reaction, often derived from microbes

biochemical or **biological oxygen demand (BOD)** Measure of the amount of dissolved oxygen required by microorganisms to decompose the organic matter in a sewage or water sample

biofilm Microbial community attached to a surface within a matrix of secreted polymers

biogenic element Elements comprising biomass

biogeochemical cycle Transition of chemicals between organic and inorganic forms that results in cycling through and within parts of the ecosystem

bioinformatics Field that uses computers to analyze large quantities of biological data, such as DNA or protein sequences

biome Major category of ecosystem characterized by vegetation and climate

bioprospecting Searching for novel organisms, biological materials, or biological processes that can be useful in biotechnology

bioreactor Large culture vessel designed to maximize cell density and product yield during industrial fermentation

biorefinery Facility that converts biomass into industrial products or energy

bioremediation Use of microorganisms to degrade environmental pollutants

biosolid Composted or treated sludge that is suitable for use as soil conditioner in horticulture and agriculture

biosphere Regions on Earth that can support life

biostimulation Addition of O_2 or limiting nutrients to increase the metabolic activity of microorganisms that degrade organic wastes or pollutants

biotechnology Use of biological processes or organisms for the production of goods or services

biphytanyl Isoprene polymer in the plasma membrane of some archaeons; 40-carbon hydrocarbon forming a phospholipid monolayer

botulism Foodborne intoxication resulting in flaccid paralysis that occurs when botulinum toxin produced by *Clostridium botulinum* is ingested

bright field microscopy Type of microscopy in which light is passed through a specimen and an image is magnified by lenses before reaching the eye

Bt toxin Intracellular protein crystal formed by sporulating *Bacillus thuringiensis* that acts as an insecticide against moths, butterflies, flies, mosquitoes, and beetles

C-reactive protein Serum pattern recognition receptor (PRR) of the innate immune system produced in early infection

Calvin cycle Common dark reaction pathway used by cyanobacteria and plants for autotrophic carbon fixation

cAMP receptor protein (CRP) Protein that, in a complex with cAMP, binds to the activator binding site in the *lac* operon to enhance binding of RNA polymerase to the promoter

canning High heat process for food preservation where food is heated to 100°C or above for an extended period of time, followed by sealing in a sterilized can or jar

cannula *pl.* **cannulae** Hollow glycoprotein tubes that connect individual cells to form a complex network

capsid Protein structure surrounding the genome of a virus

capsomere Subunit of a virus capsid

capsule Polysaccharide layer surrounding some bacterial cells that may shield pathogens from host defense systems

carbon fixation Conversion of inorganic carbon, usually CO_2, from the environment into organic molecules for growth

carboxysome Intracellular compartment that contains key enzymes involved in the conversion of inorganic carbon into organic matter during photosynthesis

carrier Asymptomatic host that can transmit an infectious agent to others

case-to-infection (CI) ratio Proportion of infected individuals who develop disease

catabolism Breakdown of larger molecules into smaller ones for energy production

catabolite activator protein (cAMP receptor protein, or CRP) Protein that, in a complex with cAMP, binds to the activator binding site in the *lac* operon to enhance binding of RNA polymerase to the promoter

catabolite repression Process of inhibiting operons that use alternate substrates; glucose inhibits pathways producing enzymes that catabolize other nutrients

CD4+ T cell T lymphocyte expressing the surface molecule CD4. *See also* T helper cell

CD8+ T cell T lymphocyte expressing the surface molecule CD8. *See also* cytotoxic T cell

cDNA library Collection of complementary DNA representing all mRNA species present within a eukaryal cell; *see also* complementary DNA library

cecal fermentation Digestive system where the major site of fermentation is the cecum

cecum Dead-end pouch found at the junction of the ileum (distal end of the small intestine) and the colon; vital in many herbivores for microbial-assisted breakdown of partially digested food from the small intestine

cell envelope Structures surrounding the cytoplasm, composed of the plasma membrane, cell wall, and, in Gram-negative cells, an outer membrane

cell Simplest structure capable of carrying out all the processes of life

cell wall Semi-rigid structure at the cell periphery that holds the cell in a certain shape, limits cellular access to chemical compounds, and protects the cell from environmental pressures

cell-mediated immune response Immune response dominated by T_H1 cells, T_C cells, increased activation of macrophages, and inflammation in response to the presence of intracellular pathogens

cellular respiration Metabolic process in which electrons (1) are stripped from a chemical energy source, (2) pass through an electron transport chain to generate a proton gradient that drives the synthesis of ATP, and (3) reduce an electron acceptor

cellulose Polysaccharide in the cell walls of plants and algae with β-1, 4-glycosidic bonds between glucose molecules

cervical carcinoma Cancer of the cervix

chemiosmosis Synthesis of ATP from free ADP and P_i using energy and electrons to create a proton gradient across a membrane; the flow of protons back across the membrane releases energy that is harnessed by ATP synthase to produce ATP

chemoreceptor Receptor that detects chemical signals, such as attractants or repellents

chemostat Continuous culture system in which growth rate and cell density can be manipulated through the nutrient composition of the medium and the flow rate of fresh medium into the culture vessel

chemosynthesis Metabolic process that derives energy from chemicals rather than from light

chemotaxis Process that uses chemical signals from the environment to direct movement

chemotroph Organism that acquires energy through the oxidation of reduced organic or inorganic compounds

chlorination Disinfection process for water treatment that uses the addition of a hypochlorite solution, such as bleach, or percolation of chlorine gas through water to produce hypochlorous acid (HOCl), a strong oxidizing agent

chlorophyll Primary photopigment used by cyanobacteria and chloroplasts

chloroplast Organelle that contains pigments and carries out photosynthesis

chloroquine Widely used antimalarial drug that blocks the formation of hemozoin, leading to the accumulation of toxic heme in the parasite

cilium *pl.* **cilia** Short filamentous appendage used in locomotion that projects from the surface of a eukaryal cell; shorter and usually more numerous than flagella

***cis*-acting retrovirus** Retrovirus that activates a cellular proto-oncogene after integrating into a host chromosome

clade Genetically distinct subtype of a species

class I major histocompatibility (MHC I) molecule Cell surface molecule expressed by many cell types that presents endogenous antigen to CD8+ T cells

class II major histocompatibility (MHC II) molecule Cell surface molecule expressed by antigen-presenting cells that presents exogenous antigen to CD4+ T cells

class-switching Genetic process in activated B cells that results in production of one of the classes of IgA, IgG, or IgE immunoglobulin from the initial production of class IgM

classical pathway Complement cascade activation pathway initiated by C1 binding directly to certain surface components of pathogens, for example, LPS and lipoteichoic acid, or binding to surfaces coated with C-reactive protein or antibody; utilizes complement components C1 through C9

clonal expansion Cell division from a single activated T or B cell resulting in a population of cells with the same specificity for antigen

clone Cell or organism that is genetically identical to another cell or organism

cloning vector DNA molecule to which other DNA fragments can be joined for replication within cells

co-speciation Evolutionary process beginning with an ancestral host/microbe association that subsequently diversified into a number of different but phylogenetically related host–symbiont species

coccus *pl.* **cocci** Bacterial cell that is roughly spherical

codon Sequence of three nucleotides on mRNA that codes for a single amino acid

coenzyme An organic cofactor

coevolution hypothesis Premise that viruses originated prior to or at the same time as primordial cells and continued to coevolve with these hosts

cofactor Small ion or organic molecule essential for enzyme catalytic activity by assisting in the transfer of functional groups

cointegrate Union of plasmid and chromosome as a result of a single recombination event

colon Distal end of the intestinal tract of many vertebrates; frequently harbors large concentrations of symbiotic microorganisms; also called the "large intestine"

colonic fermentation Digestive system where the major site of fermentation is the colon

colony Mound of clonal cells that grows large enough to be seen without a microscope

colony-forming unit (CFU) Viable microorganism that replicates to form a visible colony on a growth medium

cometabolism Metabolic transformation of a molecule that is not used as a carbon or energy source

commensalism Symbiotic relationship in which one partner benefits, while the other is apparently unaffected

common-source epidemic An outbreak that results from exposure to a single source of infection

community All of the organisms that live and interact in an area at the same time

comparative genomics Field of study that determines evolutionary relationships among organisms based upon DNA sequence data

competent Describes a cell that can take up free DNA from its surroundings

competitive exclusion Concept that different species requiring the same resources cannot co-exist; prevents colonization by competing species

complement Group of serum proteins of the innate immune system that can become activated to undergo a cascade of cleavage reactions in the presence of extracellular pathogens; end result of complement activation is formation of the membrane-attack complex (MAC) and generation of complement factors with supplementary immune functions

complementary Refers to nucleotide pairing such that adenine (A) always aligns with thymine (T, in DNA) or uracil (U, in RNA) and cytosine (C) always aligns with guanine (G)

complementary DNA (cDNA) Eukaryal DNA strand produced using mRNA transcript as a template for the reverse transcriptase enzyme; it lacks introns and can be inserted into an expression vector

complementary DNA library Collection of complementary DNA representing all mRNA species present within a eukaryal cell; *see also* cDNA library

complex life cycle Infection of two or more species during the life cycle of a parasite

complex medium Growth medium for which chemical components are not precisely defined in identity or concentration

conjugate vaccine Vaccine composed of a capsular polysaccharide bound to a peptide; used to immunize against pathogens possessing T-independent antigens

conjugation Transfer of DNA from one cell to another via direct cell-to-cell contact; type of horizontal gene transfer

consortium Microbial community in which growth or survival of members is dependent upon interactions within the community

constitutive expression Constant expression of genes that encode products always needed by the cell such as enzymes for glycolysis or transcription

consumer Organism that consumes other organisms as a source of carbon and energy

continuous culture Open growth system to which nutrients are continually added and waste products are continually removed

contrast Difference in light absorbance between two separate objects or areas

copy number Number of plasmids within a cell

corepressor Molecule that enhances binding of a repressor protein to an operator

cosmid Hybrid vector composed of plasmid and phage elements

covalent modification Regulation of enzyme action by addition of a chemical entity such as a phosphate group or a methyl group

CpG dinucleotides Adjacent guanine and cytosine nucleotides in a DNA chain; unmethylated CpG dinucleotides are commonly found in the DNA of bacteria and some viruses, but not in eukaryal cells, and can be innately recognized by human Toll-like receptor 9

cross-presentation Pathway in antigen-presenting cells that shuttles proteins from the endocytic pathway into the endogenous pathway to allow for both CD4$^+$ and CD8$^+$ T-cell activation

cultivation-independent method Process of analyzing microbes in an environment that is not dependent upon standard laboratory plating techniques

culture collection Archive of microbial strains for use in microbiology studies and in biotechnology

cycle sequencing Technique based on Sanger sequencing that yields sequence of up to 1,000 nucleotides within hours; *see also* automated sequencing

cyst Environmentally stable non-motile infective form of an organism encased within a protective cell wall

cytokine Any member of a group of secreted, soluble, low molecular weight glycoproteins used as the primary means of cell-to-cell communication to regulate inflammatory and immune responses; includes interleukins, interferons, and tumor-necrosis factors

cytokinesis Division of the cytoplasm, usually after nuclear division

cytolysin Cytotoxin that causes lysis of host cells by membrane disruption

cytopathic effect (CPE) Visible change in cellular morphology often associated with cell damage or death

cytoplasm Aqueous environment enclosed by the plasma membrane

cytosine (C) Pyrimidine nitrogenous base, along with thymine and uracil

cytoskeleton Network of microfilaments, microtubules, and intermediate filaments within a eukaryal cell that provides structure and directs movement

cytostome Double-membrane structure through which material is taken into a cell

cytotoxic cell Any effector cell of the immune system containing cytotoxic granules, which when released induce apoptosis of target cells; natural killer (NK) cells, mast cells, and cytotoxic CD8$^+$ T (T$_C$) cells are examples

cytotoxic T cell (T$_C$ cell) Effector CD8$^+$ T cell that induces apoptosis in target cells through MHC I:antigen binding followed by the release of perforin and granzyme

cytotoxin Toxin that acts on host cells

dark reactions Chemical reactions that utilize ATP and NADPH produced by phototrophy to convert CO_2 into organic carbon compounds through carbon fixation

dead-end host Individual in which a pathogen may replicate and cause disease but isn't transferred efficiently to another individual; also called an "incidental host"

death phase Period of growth during which the viable cell population declines

decimal reduction time (D value) Time required to kill 90 percent of target organisms under specific conditions

decomposer Organism that consumes dead organisms as a source of carbon and energy, releasing simple organic products

defined medium Growth medium for which the chemical identity and concentration of all constituents is known

definitive host Host in which a parasite reproduces sexually

degranulation Release of the chemical constituents of the granules of cytotoxic cells by vesicle fusion with the plasma membrane

deletion Mutation in which one or more nucleotide bases are removed from a DNA sequence

denaturing gradient gel electrophoresis (DGGE) Technique that allows electrophoretic separation of DNA fragments based upon DNA sequence differences rather than size

dendritic cell Migrating antigen-presenting cell found in most tissues of the body, characterized by extensive cytoplasmic projections

denitrification Anaerobic conversion of nitrate into N_2

diauxic growth curve Representation of two different rates of cell growth in a medium containing two different nutrient sources that are used sequentially

dideoxy sequencing Technique to determine DNA sequence that relies on incorporation of chain-terminating dideoxy nucleotides and includes cloning of the fragment to be sequenced, DNA synthesis, and gel electrophoresis; *see also* Sanger sequencing

differential centrifugation Process of subjecting a medium containing larger and smaller particles to low-speed centrifugation causing the larger particles (often cells and cell debris) to pellet followed by ultracentrifugation causing the smaller particles (often viruses) to pellet

differential medium Growth medium that allows different types of bacteria to be distinguished visually

dilution to extinction Technique that produces a dilute solution from which samples that contain individual cells can be grown to produce a pure culture

dinoflagellate Type of flagellated unicellular organism; many are photosynthetic; some produce saxitoxin that can block sodium channels, resulting in flaccid paralysis

diploid Possessing two sets of chromosomes

direct contact transmission Spread of an infectious agent that occurs via physical contact between an infected person and a susceptible person

directed enzyme evolution Method to improve microbial strains by applying rounds of random mutagenesis followed by selection for desired changes in a particular gene of interest

disease Disturbance in the normal functioning of an organism

disinfectant Chemical applied to non-living objects to kill potentially infectious microorganisms

DNA cloning Replication of recombinant DNA molecules; *see also* molecular cloning

DNA ligase Enzyme that joins a 5′-phosphate end and a 3′-OH end in a DNA strand; *see also* ligase

DNA polymerase Enzyme that catalyzes production of a new DNA strand from an existing DNA strand

DNA shuffling Method in which DNA fragments are mixed together, denatured, and reassembled into full-length chimeric genes that are products of recombination between the input fragments; can dramatically improve phenotypes by combining multiple beneficial variations into single clones

DNA vaccine Vaccine composed of a cloned gene(s) in a DNA vector, such as a plasmid; delivery into the cells of the body results in expression of the immunizing antigen product of the cloned gene

DnaA DNA-binding protein that acts in strand separation required for DNA replication in bacteria

DnaG Synthesizes short segments of RNA needed to prime DNA replication in bacteria; also called "primase"

ecosystem A community and its physical environment

ectosymbiont Symbiont that lives on the exterior surface of its host

effector Molecule that interacts with a repressor molecule and affects its ability to bind to an operator; often an intermediate of the related metabolic pathway

effector cell Armed, fully differentiated immune cell that directly carries out immune functions; examples include T_C, T_H, and plasma cells

effluent Outflow water that is discharged to another treatment tank or released to a water supply

efflux pump Membrane protein complex used to transport various compounds, such as antimicrobial agents, directly out of the microbial cell; found naturally in many bacteria and fungi

electron microscopy (EM) Type of microscopy that uses beams of electrons instead of light to produce images

electron transport system Chain of membrane-associated proteins that transfers electrons in a series of steps, yielding energy to move protons across the membrane

electroporation Application of electric current to cells that generates temporary holes in cellular membranes through which DNA can enter

Embden–Meyerhof–Parnas (EMP) pathway Glycolytic pathway where glucose is converted into two molecules of pyruvate, producing a total of two molecules of ATP from each glucose molecule

emerging disease Disease that was previously unknown or that shows a significant increase in incidence

endemic disease Disease that is normally present in a population

endergonic reaction Energy-absorbing chemical reaction having a positive $\Delta G^{o'}$ value

endocytic pathway The flow of material internalized by endocytosis and degraded through a number of biochemical processes until its expulsion from the cell; involves a series of vesicle fusion events; only in eukaryal cells

endocytosis Generalized process for the uptake of external materials by cell membrane extension or invagination and fusion to form a vesicle inside the cytoplasm of the cell generally known as an endosome

endogenous antigen Peptide antigen that originates from the cytoplasm of a cell and is presented on MHC I molecules

endogenous retrovirus Proviral DNA integrated into the DNA of host gametes and subsequently found in every cell of a developing embryo

endophyte Endosymbiotic microbe that lives inside plant tissue

endoplasmic reticulum (ER) System of membranes in the cytoplasm of a eukaryal cell

endosome General term for a membrane-bound vesicle produced by endocytosis

endospore Metabolically inert structure with increased resistance to harsh environmental conditions formed by some Gram-positive bacteria

endosymbiont Symbiont that lives within the body tissues or cells of a host; usually refers to internal microbes that have a mutualistic or commensal relationship with their host

endosymbiosis Interaction between cells in which one type of cell is engulfed by another, forming a stable relationship

endotoxin Cell-wall component of Gram-negative bacteria; *see also* lipopolysaccharide (LPS) and lipooligosaccharide (LOS)

enrichment medium Growth medium that contains nutrients or other components designed to favor the growth of particular microbes

enterotoxin Microbial secreted toxin that acts on the gut and causes or contributes to the intestinal symptoms in an affected individual

Entner–Doudoroff pathway Glycolytic pathway interacting with the EMP pathway in which glucose is converted to two molecules of pyruvate, producing one molecule of ATP from each glucose molecule

enveloped virus Virion consisting of a capsid surrounding a genome and a cell-derived membrane called the envelope

enzyme Macromolecule that acts as a catalyst to speed up chemical reactions within a cell by lowering the energy of activation

enzyme-linked immunosorbent assay (ELISA) *In vitro* detection system that uses antibodies conjugated to enzymes for the detection of serum antibodies (indirect ELISA) or antigen (direct ELISA)

eosinophil Cytotoxic cell of the innate immune system found in blood and named for the distinct staining of its granules with the dye eosin; generally acts against extracellular parasitic worms in the presence of IgE

epidemic Significant rise in incidence of a disease above that normally expected in a population

epidemiology Study of patterns of disease within populations

episome Circular piece of viral DNA separate from the host chromosome(s) within the nucleus of a host cell

epitope Single antigenic site on a molecule that is recognized and bound by an immune receptor, such as an immunoglobulin (BCR or antibody) or TCR

error-prone PCR Process in which a gene is PCR amplified under conditions that result in a very high error rate

erythrocyte Red blood cell

Eukarya Domain in which members are defined by the presence of a membrane-bound nucleus

eukaryote Organism that is composed of cells that contain a membrane-bound nucleus

eutrophication Increase in the concentration of nutrients, such as nitrogen and phosphate compounds in an aquatic ecosystem, leading to rapid growth in autotroph populations, particularly algae

evolution Inherited change within a population over time

exergonic reaction Energy-yielding chemical reaction having a negative $\Delta G^{o\prime}$ value

exogenous antigen Peptide antigen that enters the endocytic pathway from outside the cell and is presented on MHC II molecules

exon Protein-coding region of a eukaryal gene

exopolysaccharide High-molecular-weight polymer composed of various sugars that is secreted by microbes into the surrounding medium and aids in the formation of a biofilm

exotoxin Microbial toxin that is released into the surroundings or injected into host cells

exponential phase Period of growth during which cells divide to produce two cells at a maximum rate; *see also* log phase

expression vector Agent such as a plasmid that is specifically designed to produce recombinant proteins within a host cell

extravasation Movement of cells and fluid from a blood vessel into the surrounding tissue; occurs during inflammation

extrinsic factor A characteristic of the storage environment of the food, such as temperature and humidity

F plasmid Large, circular dsDNA plasmid required for conjugation; also called the "fertility factor"

F-prime (F′) strain Cells that transfer a limited number of markers at extremely high frequency during conjugation

facultative anaerobe Organism that can use oxygen for respiration when it is present, but also can grow in the absence of oxygen

fecal coliforms Group of coliform bacteria indigenous to the intestinal tract of humans and animals

fecal-oral transmission Spread of an infectious agent that is eliminated in the feces of one individual and then ingested by another individual

fed-batch reactor Production system in which a growth-limiting nutrient is added over time to control growth rate resulting in high cell densities

feedback inhibition Allosteric inhibition of an early enzyme in a metabolic pathway by a product produced later in the series

feedstock Biomass serving as raw starting material in industrial processes or biorefineries

fermentation Anaerobic metabolic pathway where electrons generated from glycolytic processing are passed directly to an organic terminal electron acceptor; an industrial culture process for production of microbial products

fever Sustained increase in average core body temperature in endothermic animals

fibronectin-binding protein Adhesin found on many bacterial pathogens, allowing them to bind and colonize the extracellular fibronectin of skin, mucosal surfaces, and blood vessels

fimbria *pl.* **fimbriae** Refers to cell-surface fibers used for adhesion

flagellin Protein that comprises the flagellum in bacteria and archaeons

flagellum *pl.* **flagella** Long filamentous appendage on the surface of cells used for motility

flash heating Pasteurization method that heats a small volume of liquid to a very high temperature (at least 72°C) for a short period of time (15 s or less); includes high-temperature-short-time (HTST) and ultra-high-temperature (UHT) processes

flocs Clumps of biomass consisting of adsorbed material and microorganisms

flow cytometry Method of detecting cells in suspension by measuring scattering of light as the cells pass by a laser beam

fluid mosaic model Model of a membrane composed of a double layer of phospholipids in which proteins are dispersed

fluorescence *in situ* hybridization (FISH) Technique that detects specific nucleotide sequences using fluorescently labeled oligonucleotide probes

flux Movement of elements between environmental reservoirs

fomite Inanimate object via which pathogens may be transferred to a susceptible host

foodborne infection Disease caused by the ingestion of live pathogens that actively multiply within the body

foodborne intoxication Illness caused by ingestion of food containing microbial exotoxins

frameshift mutation Addition or deletion of one or more nucleotide bases in a DNA sequence resulting in an altered reading frame

freeze-drying Process for preserving food, whereby food is frozen and then dried under a vacuum, allowing removal of water by sublimation; *see also* lyophilization

functional genomics Field of study to determine biological functions of unknown genes determined from DNA sequence data

fusion peptide A short string of hydrophobic amino acids that mediates merging of membranes between virus and host cell

fusion protein Recombinant protein that contains domains of two or more proteins; typically including a protein of interest fused to a polypeptide that adds a beneficial property such as an affinity tag

gametocyte (gamont) Precursor cell of a haploid gamete, or reproductive cell, in a protozoan

gas vesicle Protein structures in cells that can contain gas and provide buoyancy to cells

gel shift assay Electrophoretic technique that separates molecules based on their size; may be used to identify proteins bound to DNA

gene Segment of DNA that is transcribed into single-stranded RNA, along with the associated DNA elements that direct its transcription

gene therapy Treatment of a genetic disorder by introducing a wild-type allele of a mutant gene that codes for an abnormal or missing protein, thereby eliminating the problem

general secretory pathway Proteins in the plasma membrane that move proteins to the outside of the cell

generation time Average time interval between one cell division and the next

genome All of the hereditary material within a cell

genomic island DNA sequence transferred from one species to another; result of horizontal gene transfer

genomic library Collection of cloned DNA fragments that comprise the entire genome of an organism

genomics Determination and study of complete genome sequences

genotype Description of the alleles possessed by an organism

germfree Laboratory animal that is born and maintained in a completely sterile environment free of all microbial contact

Gibbs free energy (G) Potential energy that can be released in a chemical reaction

gliding motility Type of locomotion used by some non-flagellated bacteria to move across a surface

glycolysis Catabolism of glucose to pyruvic acid (pyruvate)

glycosylation Addition of a complex sugar residue to a polypeptide

gnotobiotic Germfree animal that has subsequently been purposefully inoculated with one or a few particular species of microbes

Golgi apparatus Membranous organelle in eukaryal cells in which molecules synthesized within the cell are packaged and distributed

gradient centrifugation Process of subjecting a medium to ultracentrifugation within a tube containing a linear gradient of a salt or sugar solution that will separate particles based upon their relative density within the gradient

Gram stain Technique for staining bacterial cells allowing differentiation into two types of bacteria, Gram-positive bacteria that appear purple and Gram-negative bacteria that appear pink

Gram-negative bacteria Bacterial cells with a thin layer of peptidoglycan in the cell wall surrounded by an outer membrane; lose crystal violet stain in the Gram stain procedure and appear pink

Gram-positive bacteria Bacterial cells with a thick layer of peptidoglycan in the cell wall; retain crystal violet stain in the Gram stain procedure and appear purple

granule Membrane-bound, chemical-containing vesicle found in the cytoplasm of certain cells of the immune system with killing potential

granuloma Pathological lesion resulting from a focus of chronic inflammation and characterized by a nodule-like encasement composed of fibrin and lymphocytes at the periphery and infected and dead macrophages inside

granzyme Peptide produced by cytotoxic cells of the immune system that triggers apoptosis in target cells

green biotechnology Use of advanced biological processes in agricultural applications

greenhouse gas Gas that absorbs infrared radiation in the atmosphere; examples are CO_2 and methane (CH_4)

growth Increase in mass of biological material; an increase in number of microbes

growth curve Pattern observed when cells are grown in a closed system; includes the lag phase, the exponential or log phase, the stationary phase, and the death phase

growth rate Generation time, or average time interval between one cell division and the next

growth yield Number of cells that results from growth over a specific time period

guanine (G) Purine nitrogenous base, along with adenine

guild Group of organisms that carry out similar processes

Haber–Bosch industrial process Process that produces ammonia by a reaction of nitrogen gas with hydrogen gas; enables production of synthetic nitrogen fertilizer

halophile Organism that requires high salt (NaCl) levels for growth

haploid Possessing a single set of chromosomes

harmful algal bloom (HAB) Population explosion of organisms producing toxins that can lead to death of other aquatic organisms

heat shock protein Molecule produced when cells are exposed to stress such as high temperature; often aid in the correct folding of proteins

helical morphology Structure of a virus in which the capsomeres form a helix and the capsid resembles a hollow tube

helicase Aids in unwinding of DNA during DNA replication in bacteria; also called DnaB

helminth Parasitic worm

hemagglutination Clumping of red blood cells

heme Iron-containing pigment of hemoglobin that is cytotoxic in its free form

hemocyte Wandering phagocytic cell of invertebrates that migrates from hemolymph (blood) into tissue

hemolysis Destruction of red blood cells through production of cytolysins called "hemolysins"

hemozoin Insoluble polymer composed of heme monomers produced by the malaria parasite; sometimes called "malaria pigment"

herd immunity Protection from communicable disease of non-immunized, susceptible individuals in a population by immunization of a large proportion of the rest of the population

herd immunity threshold Critical proportion of a population required to be immunized to achieve herd immunity

heterotroph Organism that obtains carbon from organic molecules, such as sugars, obtained from the environment

Hfr strain High frequency of recombination cells formed when an entire F plasmid incorporates into the host chromosome during conjugation

histidine protein kinase (HPK) Common sensor kinase involved in detecting an environmental stimulus in a two-component regulatory system

histone Protein in eukaryal cells and some archaeons that protects and compacts DNA

histone deacetylase (HDAC) Cellular enzyme that affects gene expression by removing acetyl groups ($CH_3C=O$) from histones

homeostasis Active regulation of the internal environment to maintain relative constancy

homologous recombination Process that occurs when two segments of DNA with identical or very similar sequences pair up and exchange or replace some portion of their DNA; also called "crossover"

hopanoid Sterol-like molecule believed to stabilize the plasma membrane of some bacteria

horizon Layer of soil with distinct characteristics

horizontal gene transfer Movement of DNA between microbes; also called "lateral gene transfer"

horizontal transmission Spread of an infectious agent between members of a species that are not parent and offspring

host range The diversity of cells that can be infected by a pathogen or support the replication of a virus or plasmid

humic material Product of incomplete breakdown of plant biomass found in soil

humoral immune response Adaptive immune response dominated by T$_H$2 cells and characterized by production of large amounts of antibody in response to extracellular pathogens or toxins

hurdle technology The use of multiple constraints or *hurdles* to control the growth of pathogens and spoilage microorganisms in foods; targeted microorganism(s) must overcome each hurdle in order to proliferate

hydrogenotrophic methanogen Organism that produces methane (CH$_4$) from CO$_2$ and H$_2$

hyperthermophile Organism that grows optimally at temperatures greater than 80°C

hypha *pl.* **hyphae** Irregularly branching filaments found in most fungi and some bacteria

icosahedral morphology Structure of a virus in which the capsomeres form an icosahedron, or 20-sided polygon, with each capsomere making up a face of the icosahedron

immune The state of being resistant to infection and subsequent disease

immune receptor T-cell receptor (TCR) or B-cell receptor (BCR) capable of binding to antigen

immune response Defensive response mounted by the host involving the immune system

immune system The cells, molecules, and associated tissues and organs involved in identifying foreign agents, fighting infection, and ridding the body of abnormal cells

immunization Process of generating immunological memory to prevent infection and/or disease through the introduction of antigen(s), such as microbes or their components, into the body

immunoglobulin (Ig) molecule Secreted (antibody) or membrane-bound (B-cell receptor; BCR) glycoproteins that bind antigen in a specific manner

immunology The study of the components and processes of the immune system

immunoprecipitation Technique in which an antibody is used to precipitate an antigen

immunosuppressive drug Drug that decreases the activity of one or more immune system responses

inactivated vaccine Vaccine composed of whole cells or viruses that have been physically inactivated by heat or chemical treatment yet retains the ability to generate protective immunity in the host; *see also* killed vaccine

incidence Number of new cases of a disease appearing in a population during a specific time period

inclusion body Microscopically visible granule within the cell; composed of aggregated cellular or viral materials

incubation period Time between entry of a pathogen into a host and appearance of illness

indicator organism Microbe, particularly fecal coliform bacteria, whose presence in a water sample indicates fecal matter is present and, by association, potential intestinal pathogens

indirect contact transmission Spread of an infectious agent that occurs from one person to another via an inanimate object or a vector

inducer Molecule that inhibits binding of a repressor protein to an operator

inducible expression Expression of genes that encode products only when they are needed by the cell

infection Successful colonization and invasion by a pathogen of a host

infection thread Invagination of the plant cell wall that forms a narrow tube produced by a rhizobia-infected root cell; extends into the plant root tissue to deliver the bacteria to root cells

infectious disease Disease caused by a microbe that can be transmitted from host to host

infectious RNA The genome of a positive-sense single-stranded RNA virus that can direct the production of new viral particles after being injected into a host cell

inflammation General term for the tissue accumulation of fluid, proteins, and cells from the blood in response to injury caused by trauma or infection

innate (non-specific) immunity Immunity that occurs as a first response to infection involving cells and proteins that are innately present from birth; found in some form in all multicellular eukaryotes

insertion Mutation in which one or more nucleotide bases are added to a DNA sequence

insertion sequence (IS) element Transposable element consisting of the transposase gene flanked by inverted repeats

integrase An enzyme that mediates the integration of viral DNA into a host cell chromosome

interleukin-2 (IL-2) T-cell growth factor cytokine produced by activated CD4$^+$ T cells that induces resting cells to enter into cell division

intermediate filament Protein polymer 9–11 nm in diameter forming a portion of the cytoskeleton in eukaryal cells

intermediate host Host in which a parasite reproduces asexually and differentiates but does not reproduce sexually

intoxication Ingestion of a toxin that leads to disease

intrinsic factor A characteristic inherent quality of the food, such as water and salt content

intron Non-protein coding region of a eukaryal gene; also called an "intervening sequence"

inversion Change in DNA sequence in which a section of DNA is inverted and reinserted into the genome

ionizing radiation High-energy, penetrating radiation in the form of short UV light (less than 124 nm), X-rays, or gamma rays; commonly used for pasteurization or sterilization of foods

irradiation Exposure to radiation, commonly used to reduce microbial contamination of food and surfaces

isoelectric point The pH at which a protein contains no charge

isoprenoid Hydrocarbon molecule in the archaeal plasma membrane built from 5-carbon isoprene subunits attached to glycerol 1-phosphate (G1P)

killed vaccine Vaccine composed of whole cells or viruses that have been physically inactivated by heat or chemical treatment yet retains the ability to generate protective immunity in the host; *see also* inactivated vaccine

kinetoplast Large mass of DNA found within the mitochondrion of a trypanosome

knockout mutation Recombination that disrupts a gene

Koch's postulates Guideline for demonstrating that a specific microbe causes a specific disease

lactic acid bacteria (LAB) Grouping of Gram-positive bacteria characterized by production of lactic acid as a main product of their fermentative metabolism; includes various members of *Streptococcus*, *Lactobacillus*, *Lactococcus*, and *Bifidobacterium*

lactoferrin Small iron-transporting protein found in body secretions such as milk, tears, and saliva and found intracellularly in the granules of phagocytes

lag phase Initial period of growth during which cells grow slowly as they adjust to new surroundings

lagging strand Strand of replicating DNA that shows discontinuous elongation from the replication fork

latency Infection characterized by a delay or cessation of disease; in viral disease, characterized by limited transcription of the viral genome; results in reduced immune response

latency-associated transcript (LAT) Herpesvirus transcript transcribed during latency

latent infection Infection characterized by a delay or cessation of disease; in viral disease, characterized by limited transcription of the viral genome; results in reduced immune response; *see also* latency

laws of thermodynamics Set of rules that govern all energy conversions

leading strand Strand of replicating DNA that can be extended continuously from the replication fork in a 5′ to 3′ direction

lecithinase Phospholipase cytolysin that disrupts cell membranes by hydrolyzing various phospholipids, such as phosphatidylcholine, found in eukaryal membranes

lectin Protein produced by plants and animals that binds specific, usually foreign, carbohydrates

lectin pathway Complement cascade activation pathway initiated by the complexing of mannose-binding lectin with other serum proteins to produce C1-like activity, which activates the complement cascade

leghemoglobin An O_2-binding protein produced in root nodules by the plant cells, which protects the nitrogenase enzyme of bacteroids from O_2^- inactivation

lethal dose 50 (LD$_{50}$) Amount of virus that kills 50 percent of infected hosts

leukocyte Cell with immune functions; includes lymphocytes, neutrophils, monocytes, and macrophages; also known as "white blood cells"

lichen Symbiotic life-form consisting of a fungal species and one or more species of photosynthetic cyanobacteria and/or algae

ligand Molecule that binds to a cell receptor

ligase Enzyme that joins a 5′-phosphate end and a 3′-OH end in a DNA strand; *see also* DNA ligase

light reactions Reactions that convert light energy to chemical energy

lipid Organic macromolecule that does not dissolve in water; important component of cell membranes

lipid raft Pocket of specific lipids within a membrane that often is associated with specific membrane proteins

lipooligosaccharide (LOS) Non-LPS endotoxin produced by some Gram-negative bacteria, which lacks the outer O-antigen component of LPS but maintains the toxic activity of LPS due to the presence of an intact lipid A component

lipopolysaccharide (LPS) Molecule comprising the outer layer of the outer membrane in Gram-negative bacteria; composed of lipid A, a core polysaccharide, and an O side chain

liposome Vesicle made from cell membrane material

lipoteichoic acid (LTA) Cell wall molecule produced by many Gram-positive bacteria and composed of ribose or glycerol polymers and anchored to the cytoplasmic membrane by a lipid tail

lithotroph Organism that removes electrons from inorganic reduced molecules, such as H_2S, H_2, or elemental sulfur (S^0)

log phase Period of growth during which cells divide to produce two cells at a maximum rate; *see also* exponential phase

lophotrichous Refers to bacterial cells with more than one flagellum at one or both ends

low temperature hold (LTH) Pasteurization method that heats a large volume of liquid to a low temperature (at least 62.8°C) for a long period of time (30 minutes or more)

lymphocyte Type of leukocyte possessing specific receptors for recognizing antigen or MHC I molecules presented on other cells; includes NK cells, B cells, and T cells

lyophilization Process for preserving food, whereby food is frozen and then dried under a vacuum, allowing removal of water by sublimation; *see also* freeze-drying

lysogen Bacterial cell containing a prophage

lysogenic bacteriophage Virus that exists in a latent state within a host bacterial cell as the phage genome is integrated into the bacterial genome; can also replicate lytically

lysosome Intracellular granule containing the enzyme lysozyme

lysostaphin Enzyme that cuts the pentaglycine crossbridge of peptidoglycan in *Staphylococcus aureus*

lysozyme Enzyme that degrades peptidoglycan by hydrolyzing the bond between NAG and NAM

lytic bacteriophage Virulent virus that replicates within a bacterial cell and eventually lyses it

macromolecule Large, complex molecule composed of simpler subunits

macrophage Phagocytic leukocyte found in tissue and derived from a blood monocyte; proficient antigen-presenting cells capable of activating T cells of the adaptive immune system

magnetosome Membrane-enclosed structure that contains magnetite

major histocompatibility (MHC) molecule Cell surface protein that presents self peptides or foreign peptides to T cells of the immune system

malaria Disease characterized by recurring bouts of chills and fever; caused by the protozoan parasite *Plasmodium* transmitted between humans by the female *Anopheles* mosquito

mannose-binding lectin (MBL) Serum pattern recognition receptor (PRR) of the innate immune system that binds mannose polymers of yeast and bacterial cell walls

mast cell Cytotoxic cell of the innate immune system found widely in tissues of the body; potent stimulator of inflammation due to the release of the vasodilator histamine from its granules upon stimulation

maternal transmission Direct inheritance of a genetic trait, infection, or symbiont from mother to offspring

mechanical transmission Spread of an infectious agent from one host to another via a vector that physically carries the infectious agent

meiosis Type of nuclear division in eukaryal cells that results in production of haploid cells from a diploid cell

melanin Pigment synthesized by a variety of organisms; in invertebrates, melanin production is a primary innate defense used to encapsidate foreign agents in the body and has direct toxic effects.

membrane-attack complex (MAC) Pore-forming structure resulting from complement activation, composed of complement factors C5b, C6, C7, C8, and polymerized C9 that forms in microbial membranes to cause lysis

memory cell Long-lived B or T cell that is produced following strong stimulation of a naïve B or T cell

memory response Immune response that originates from the stimulation of memory cells and is characterized by a faster and more vigorous adaptive immune response upon repeated exposure to antigen; *see also* anamnestic response

merozoite Stage of protozoal parasite; in malaria, released from the liver and can invade human red blood cells

merozygote Partially diploid cell produced by conjugation; one allele appears on the chromosome and another appears on an F′ plasmid

mesocosm Contained enclosure used for experimental study of ecosystems under controlled conditions

mesophile Organism that grows optimally at temperatures between 15°C and 40°C

messenger RNA (mRNA) RNA molecule that is translated to produce a polypeptide

metabolism Chemical reactions within a cell that extract nutrients and energy from the environment and transform them into new biological materials

metagenomics Study of genomes isolated from environmental samples using cultivation-independent methods

methanogen Archaeon that produces methane (CH_4)

methanogenesis Anaerobic metabolic process with methane (CH_4) as a product

methanotroph Organism that can metabolize methane (CH_4)

methicillin-resistant *Staphylococcus aureus* (MRSA) Strain of *Staphylococcus aureus* that is resistant to a wide variety of antibacterial medications, including methicillin

methyl-accepting chemotaxis protein (MCP) Transmembrane sensory protein that can sense the presence of attractant or repellent molecules

methylotroph Organism capable of metabolizing 1-carbon compounds such as methane (CH_4) and methanol (CH_3OH)

microRNA (miRNA) Small RNA molecule that regulates the expression of genes

microaerophile Organism that grows optimally at O_2 concentrations lower than are typically present in the atmosphere

microarray Solid support on which single-stranded DNA molecules within a genome are fixed in position; can be used to study the transcriptional activity of all genes in the genome simultaneously

microbe Term that includes both microorganisms and viruses

microbiology The study of microbes

microbiome The total of all the microbes living in a defined environment, such as soil, the human body, or even part of the body

microfilament Polymers of actin about 7 nm in diameter forming a portion of the cytoskeleton in eukaryal cells

micronutrient Chemical such as zinc or copper needed in small amounts for growth

microorganism Microscopic form of life

microtubule Polymers of α- and β-tubulin dimers that assemble into fibers about 24 nm in diameter forming a portion of the cytoskeleton in eukaryal cells

mineralization The end result of decomposition of organic material, which leaves behind inorganic mineral components

minimal inhibitory concentration (MIC) Lowest concentration of a drug at which no growth of the microbe being tested occurs *in*

vitro after a standardized period of time; gives a measure of the drug susceptibility of a microbe

mismatch repair Process that corrects errors in DNA replication by excising several bases around the mismatched base, filling in the resulting gap with the correct bases, and linking together the adjoining fragments

missense mutation Change of a nucleotide base that results in substitution of a different amino acid in a polypeptide

mitochondrion *pl.* **mitochondria** Organelle that contains the electron transport chain and produces the majority of ATP within the cell

mitogen Chemical that promotes cell division

mitosis Type of nuclear division in eukaryal cells that results in production of two genetically identical nuclei

mobile genetic element Genetic vector, such as a plasmid, transposon, or transducing phage, that facilitates horizontal gene transfer from cell to cell

model organism Organism that has been so thoroughly studied that its traits define much of the understanding of other organisms within the group

modified atmosphere packaging (MAP) Packaging technology that substitutes the atmospheric air inside the package with a gas mixture designed to extend the shelf life of a food product

molecular chaperone Protein that helps other proteins to fold or to refold properly

molecular cloning Replication of recombinant DNA molecules; *see also* DNA cloning

molecular Koch's postulates Guideline using molecular tools to demonstrate that a specific microbial product is a virulence factor

molecular mimicry Process in which immune cells react against host antigens that resemble antigens of a pathogen

monocistronic Molecule of mRNA that codes for a single polypeptide

monoclonal antibody Antibody that specifically binds a single unique epitope of an antigen; purposefully produced, and usually of IgG class, for diagnostic or research purposes

monocyte Mononuclear leukocyte of the innate immune system found circulating in the blood and maturing into a macrophage upon migration into the tissue

monotrichous Refers to bacterial cells with a single flagellum

morbidity rate Rate of disease

mortality rate Rate of death associated with disease

most probable number (MPN) Presumptive test used to estimate fecal coliform bacteria numbers in a water sample

mucociliary clearance Movement of mucus and entrapped particles out of body passages by the coordinated rhythmic movement of the cilia of special ciliated epithelial cells

mucosal membrane Any of the body's internal surfaces that are lined with mucus-coated epithelium

mucus Viscous secreted fluid containing mucins produced by and covering mucosal surfaces of the body

multiple cloning site Short DNA segment in a cloning vector that contains a cluster of different restriction sites that each appear only once in the plasmid

mutagen Agent that increases the rate of mutation

mutation Heritable change in the base sequence of the genome

mutualism Symbiotic relationship in which both partners benefit

mycelium *pl.* **mycelia** Three-dimensional network of hyphae found in fungi and some bacteria

mycobiont The fungal partner in a lichen

N-acyl-homoserine lactone (AHL) Autoinducer with the ability to induce luminescence

nanotechnology Development and use of devices that are nanometers in size

natural killer (NK) cell Cytotoxic lymphocyte of the innate immune system that recognizes infected and abnormal cells and kills them by releasing toxic granular contents, causing apoptosis of the target cell

necrosis Cell death that occurs through chemical or physical cell injury or loss of oxygen supply; results in cell rupture and the release of toxic cellular contents and cytokines to the surrounding tissues, producing further cell death and triggering inflammation

negative control Regulation of transcription involving a repressor that blocks transcription

negative-sense RNA or **ssRNA** (−) Viral genomic RNA that is complementary to the viral messenger (positive-sense) RNA

net community productivity Amount of carbon fixed in organic matter minus the amount of carbon lost via respiration

neutrophil Phagocytic cell of the innate immune system containing numerous cytoplasmic granules and characterized by a multilobed nucleus; also known as "polymorphonuclear leukocyte" (PMN)

neutrophile Organism that grows optimally in environments with pH close to neutrality (5.5–8.5)

next generation sequencing method Technique to sequence DNA post-Sanger and not reliant upon gel electrophoresis

niche Specific functional role of an organism within an ecosystem

nicotinamide adenine (NAD) Cytoplasmic electron carrier molecule serving as a coenzyme

nitrification Conversion of ammonia (NH_4^+) into nitrites (NO_2^-) and then into nitrates (NO_3^-)

nitrifier Organism that carries out nitrite (NO_2^-) oxidation in the process of nitrification

nitrogen fixation Conversion of atmospheric N_2 into inorganic compounds, such as ammonia, for cell use; *see also* nitrogenase

nitrogenase Large enzyme complex responsible for the conversion of N_2 to inorganic compounds (nitrogen fixation)

non-enveloped virus Virion consisting only of a capsid surrounding a genome; also called a "naked virus"

non-homologous recombination Process that occurs when two segments of DNA with little or no sequence similarity exchange or replace some portion of their DNA

non-ionizing radiation Radiation of wavelength longer than 240 nm that does not possess sufficient energy to produce ions in solution; non-ionizing UV radiation of wavelength 240–280 nm can damage DNA

non-perishable food Food that does not support microbial growth due to lack of available water

nonsense mutation Change in a nucleotide base that results in production of a stop codon

Northern blot Technique for detecting RNA as a result of gene expression by using DNA or RNA probes

nuclear magnetic resonance (NMR) Technique to determine protein structure based on measurement of distances between atomic nuclei

nucleic acid Polymer of nucleotides composed of a sugar, a phosphate moiety, and a nitrogen-containing base; includes DNA and RNA

nucleocapsid The capsid and genome of a virus

nucleoid Region in bacterial and archaeal cells that contains the chromosomal DNA

nucleoside analog Antiviral structural analog of nucleosides capable of inhibiting DNA or RNA synthesis by incorporation into growing nucleic acid chains by viral polymerases; prevents the addition of further nucleotides, thus interfering with viral nucleic acid replication

nucleus *pl.* **nuclei** Membrane-enclosed organelle that contains the chromosomes in a eukaryal cell

nutritional mutant Organism with altered metabolic requirements

obligate aerobe Organism that requires oxygen for growth

obligate anaerobe Organism that cannot grow in the presence of oxygen

obligate intracellular parasite Infectious agent that can replicate only within a living host cell; cannot replicate independently

Okazaki fragment Newly synthesized pieces of DNA and their RNA primers found on the lagging strand

oligotrophy Use of nutrients present in very low concentrations

oncogene Altered form of a proto-oncogene that can lead to uncontrolled cell growth and the development of tumors

oncolytic virus Virus that infects and kills cancerous cells without harming normal cells

oocyst Structure of protozoal parasites in which meiosis occurs

ookinete Cell formed by fusion of gametes in protozoal parasites like the malaria parasite, forms an oocyst where meiosis occurs

open reading frame (ORF) Sequence of DNA that can be translated to produce a polypeptide

operational taxonomic unit (OTU) Group of organisms that shares at least 97% SSU rRNA gene sequence identity

operator DNA sequence to which regulatory proteins can bind

operon Transcriptional unit consisting of a series of structural genes that code for polypeptides and the regulatory elements that affect their transcription

opportunistic pathogen Microbe that causes disease only when a host has been compromised

opsonin Any component of the immune system capable of binding and coating a microbial surface to enhance its phagocytic uptake; examples include antibody, C3b, C-reactive protein, and mannose-binding lectin

opsonization Coating of a microbial surface with host components, such as complement or antibody, to aid in phagocytosis

optical density Measure from a spectrophotometer that is roughly proportional to population density within a sample

organelle Intracellular compartment that performs a specific function for a cell

organotroph Organism that removes electrons from organic molecules, such as glucose

oriC Origin of replication; specific site on bacterial chromosome where DNA replication begins

ortholog Gene that evolves from a common ancestral gene and performs the same function in different organisms

outbreak Cluster of cases appearing within a short period of time in a localized population

outer membrane Bilayer found outside the peptidoglycan layer in the cell envelope of Gram-negative bacteria; inner layer is primarily phospholipid, outer layer contains lipopolysaccharide (LPS)

oxidation Removal of an electron(s) from a molecule or atom (the electron donor)

oxidative phosphorylation Chemiosmotic process used by chemotrophs to generate ATP using the energy captured from oxidation of a chemical substrate

oxygenic photosynthesis Phototrophy carried out by cyanobacteria and chloroplasts that produces O_2 and combines PS I with PS II

ozonation The use of ozone (O_3), a powerful oxidizing agent, to disinfect water

packaging sequence A motif on the viral genome that interacts with viral proteins to cause packaging into the capsid

pandemic Global epidemic, usually on more than one continent

papilloma Wart; caused by a virus

paralog Gene that arises from a duplication event; paralogous genes may develop different functions

parasitemia Presence of parasites in the blood of an infected host

parasitism Symbiotic relationship in which one partner benefits by damaging the other

pasteurization Any technique using mild heating, irradiation, or high pressure to destroy spoilage or pathogenic microorganisms without cooking the food product

pathogen A microbe that routinely causes disease

pathogen-associated molecular pattern (PAMP) Broad-based molecular signature or motif commonly associated with pathogens but not normally found in the host

pathogenesis Process by which a pathogen causes disease

pathogenicity island (PAI) Region of a chromosome containing multiple virulence factor genes

pattern recognition receptor (PRR) Receptor of the innate immune system capable of recognizing pathogen-associated molecular patterns (PAMPs)

penicillin-binding protein (PBP) Bacterial transpeptidase required for joining together the subunits that make up peptidoglycan of bacterial cell walls

pentose phosphate pathway Glycolytic pathway interacting with the EMP pathway where glucose is converted to a number of 3- to 7-carbon compounds, many of which are used for biosynthesis, and producing NADPH and ATP

peptidoglycan Cross-linked polysaccharide-peptide matrix found in bacterial cell walls

perforin Peptide produced by cytotoxic cells of the immune system that is thought to produce pores in target cell membranes

periplasm Space between the inner membrane and outer membrane in Gram-negative bacteria

perishable food Food that readily supports survival and growth of microorganisms, such as fresh lettuce, strawberries, and meat

peristalsis Sequential contraction of smooth muscles that moves material, such as food, along the digestive tract

peritrichous Refers to bacteria with multiple flagella spread over the surface of the cell

persistant organic pollutant (POP) Chemicals, such as synthetic drugs and pesticides, that are resistant to degradation and can be present in low concentrations in effluent waters; POPs tend to bioaccumulate in animal tissues

persistent infection Viral infection during which new viral particles are continuously produced; also called a "chronic infection"

Petroff–Hausser counting chamber Specially constructed microscope slide with a chamber of defined depth and a grid marking off squares of defined area; used for determining cell concentration

phagocyte Leukocyte, such as a macrophage or neutrophil, capable of engulfing particulate matter, such as bacterial cells

phagocytosis Form of endocytosis involving the cellular engulfment of particulate matter, such as bacterial cells, by the formation of cytoplasmic extensions that surround the particle and fuse to form a membrane-bound endosome vesicle called a "phagosome"

phagolysosome Digestive vesicle formed in the endocytic pathway from the fusion of a phagosome with an intracellular lysosome(s)

phagosome Membrane-bound vesicle produced through phagocytosis and containing engulfed particles

phosphate Inorganic molecule derived from phosphoric acid

phosphorylation Addition of a phosphate group to the —OH group of certain amino acids

photobiont The photosynthetic algal or cyanobacterial partner in a lichen

photolyase Enzyme that repairs thymine dimers by cleaving the covalent bonds linking adjacent thymine bases

photophosphorylation Chemiosmotic process used by phototrophs to generate ATP using light energy

photosynthesis Process by which light energy is captured by organisms to generate chemical energy in the form of ATP to carry out metabolic functions

photosystem Organized structure of photopigments consisting of antennae and a reaction center that absorb energy from photons of light to ultimately generate electrons for production of chemical energy

phototroph Organism that captures light energy, or photons, through the process of photosynthesis to generate chemical energy, such as ATP

phototrophy Acquisition of energy from sunlight

phylogeny Evolutionary history of an organism

phytanyl Isoprene polymer in the plasma membrane of some archaeons; 20-carbon hydrocarbon

phytopathogenic fungus Fungus that causes disease in plants

phytoplankton Drifting photosynthetic microorganisms that are primary producers in the ocean

pickling Acidification of food for storage by adding dilute acetic acid (vinegar) or allowing lactic or acetic acid production to occur naturally through microbial fermentation

piezophile Organism that can withstand high pressure

pilus *pl.* **pili** Proteinaceous fiber that protrudes from the cell surface; often used for attachment

plaque Localized clear area on a lawn of cells caused by cell death; adherent microorganisms and extracellular matrix substances, often produced by both the host and the microbes themselves

plasma cell B-cell-derived, terminally differentiated cell that secretes antibody

plasma membrane Selectively permeable bilayer enclosing the cytoplasm of a cell; composed primarily of phospholipids

plasmid Small, circular extrachromosomal DNA molecule

pleiomorphic Variable in shape

polycistronic Molecule of bacterial mRNA that codes for two or more different polypeptides

polyhydroxyalkanoate (PHA) Natural polyester polymer stored by bacteria when carbon is abundant and used when nutrients are scarce; possible source of biodegradable bioplastic

polyhydroxybutyrate (PHB) Simplest and most common PHA chain produced by microbes

polymerase chain reaction (PCR) Technique that quickly amplifies specific pieces of DNA

polypeptide Macromolecule composed of amino acids joined by peptide bonds; proteins

polysaccharide Macromolecule composed of monosaccharides used in structure or for energy storage

porin Channel in the outer membrane of Gram-negative bacteria that allows movement of essential molecules

positive control Regulation of transcription involving regulatory molecules that increase transcription

positive-sense RNA or **ssRNA (+)** Viral genomic RNA that can be translated by the host cell (messenger RNA)

post-transcriptional modification Events that occur after transcription in the production of functional mRNA from the RNA transcript

post-translational modification Chemical change in a protein that occurs after translation

potable Water suitable for drinking and judged to be free from intestinal bacteria and pathogens

precursor activation Allosteric activation of a late enzyme in a metabolic pathway by a substrate used by an earlier enzyme in the series

prevalence Total number of cases of a disease in a population at a particular time or during a particular time period

primary endosymbiont Obligate endosymbiont that shows evidence of co-speciation with a host and is absolutely required by the host for survival and/or fertility

primary immune response Response of the adaptive system that occurs with first exposure to an antigen, involving activation of naïve T cells and B cells, in the absence of immunological memory

primary metabolite Metabolic product required for growth; produced by microorganisms during the exponential growth phase

primary pathogen Microbe that causes disease in otherwise healthy hosts

primary producer Organism that incorporates simple inorganic substances such as CO_2 into organic molecules, providing potential food sources for other organisms

primary sludge Sludge that is a product of primary treatment of wastewater; consisting of large particulate matter

primary treatment Physical removal of particulate matter by sedimentation to form sludge and oils by skimming

primase Synthesizes short segments of RNA needed to prime DNA replication in bacteria; also called DnaG

primer Short piece of single-stranded RNA attached to single-stranded DNA at the replication fork to which DNA polymerase can add new bases

primer walking Technique in which a DNA sequence obtained from an early sequencing run is used to design a new primer that will anneal farther along the fragment to be sequenced; consecutive runs allow sequencing of the entire DNA segment

prion Infectious agent (proteinaceous infectious particle) composed of protein that can replicate within a host cell and cause transmissible spongiform encephalopathies

productive infection Outcome of a viral infection in which new infectious viral particles are produced

progenote Woese's proposed first living organism containing information stored in genes not yet linked together on chromosomes

progressive hypothesis Premise that viruses originated when genetic material in a cell gained functions that allowed the DNA or RNA to replicate and be transmitted in a semi-autonomous fashion

proinflammatory molecule Any molecule produced by cells, such as certain cytokines, which induces or stimulates the inflammatory response; examples include IL-1, IL-12, TNF-α, and IFN-γ

prokaryote Organism that lacks a membrane-bound nucleus; includes bacteria and archaeons

promoter Region of DNA where RNA polymerase binds to begin transcription

proofreading activity Action of DNA polymerase that identifies and removes an incorrect base that has been incorporated into a growing chain

propagated epidemic Pattern of disease characterized by infection passing from one host to another, either directly or indirectly

prophage Phage DNA integrated into host DNA; latent form of a temperate phage

protease inhibitor Drug that blocks the activity of enzymes that degrade proteins; required for the formation of infectious retroviral particles

proteasome Cytoplasmic complexes in eukaryal cells that degrade cellular proteins into peptides that can be presented on MHC I molecules

proteome Collection of proteins within a cell under specific conditions

proto-oncogene Gene involved in normal regulation of the cell cycle

proton motive force Strong potential difference generated by the displacement of protons to one side of the membrane as they are pumped by an electron transport system; drives a flow of H^+ through ATP synthase to produce ATP

prototroph Organism that can synthesize all necessary cellular constituents from a single organic carbon source and inorganic precursors

provirus Viral DNA that is integrated into the host DNA; latent form of a virus

pseudomurein Polymer similar to peptidoglycan that is found in the cell wall of some archaeons

psychrophile Organism that grows optimally at temperatures less than 15°C

purine Nitrogenous base containing a double-ring structure; includes adenine (A) and guanine (G)

pyrimidine Nitrogenous base containing a single-ring structure; includes cytosine (C), thymine (T), and uracil (U)

pyrosequencing Next generation sequencing method in which incorporation of a pyrophosphate (PP_i) drives a luciferase reaction, producing light that can be detected with a charge-coupled device (CCD) camera; *see also* 454 DNA sequencing

quiescent cell Cell that is not actively dividing

quorum sensing Detection of the presence and number of microbes in the environment that leads to regulation of density-dependent gene expression

radioisotope tracer Radioactive atoms or molecules containing radioactive atoms that can be monitored as they move through a system

random mutagenesis Exposure of genetic material to mutagens such as X-rays, UV light, or chemicals that increase the probability of any mutation

reactivation Resumption of disease after a latent period; can occur due to cellular stress in the host; in viruses, resumption of transcription of a previously latent viral genome

reassortment Packaging of gene segments from two or more parental virus strains into one virus

receptor Membrane protein that binds to a specific molecule in the environment, such as a viral attachment protein

recombinant DNA Linked DNA sequences from two or more organisms forming a DNA molecule that does not otherwise exist in the biological world

red biotechnology Use of advanced biological processes in medical applications

redox reaction Reduction and oxidation reactions involving the transfer of electrons from one molecule to another

reduction Addition of an electron(s) to a molecule or atom (the electron acceptor)

reduction (redox) potential Tendency of a molecule to accept electrons; represented by the electrode potential, E, which is measured in units of volts (V)

reductive TCA cycle Cycle used by many autotrophic bacteria to produce carbon compounds; essentially runs the TCA cycle in reverse to consume CO_2 to produce carbon compounds at the expense of ATP and electrons, usually in the form of NADH

refrigeration Food preservation done at cold temperature, usually 1–4°C

regressive hypothesis Premise that viruses represent a form of "life" that has lost some of its essential features and has become dependent on a host

regulon Collection of genes and operators globally affected by the same regulatory molecule

replica plating Technique to screen colonies for phenotypes that cannot be identified by phenotypic selection; location of mutants on a plate can be determined by location of colonies on a replica plate using selective media or conditions

replication bubble Single-stranded regions where DNA replication proceeds between replication forks

replicon DNA molecule that replicates from a single origin of replication

reporter gene Promoterless gene that is not expressed unless inserted within an actively transcribed gene

repressor Molecule that blocks transcription

reproduction Production of new copies of an organism

reservoir Source of an infectious agent

reservoir host Natural host that supports replication of a pathogen

resistance (R) plasmid Plasmid that confers the ability to resist antimicrobial activity of one or usually several drugs

resolution Ability to distinguish two objects as separate, distinct objects

resolvase Enzyme that separates joined DNA molecules in replicative transposition

respiration Metabolic pathway in which electrons generated from glycolytic processing are passed through an electron transport system and on to an inorganic or sometimes an organic terminal electron acceptor; cellular respiration can be aerobic, using oxygen as a terminal electron acceptor, or anaerobic

respiratory burst (oxidative burst) Generation of highly toxic oxygen compounds, such as nitric oxide and hydrogen peroxide, inside the phagolysosome to kill phagocytosed microbes

respiratory transmission Spread of an infectious agent that replicates in the respiratory tract and leaves the infected host as an aerosol during coughing, sneezing, laughing, or similar activities

response regulator (RR) Protein that regulates transcription in a two-component regulatory system

restriction enzyme Enzyme that cleaves DNA at a specific sequence; may protect host cell from viral infection by cleaving viral DNA

restriction site DNA sequence recognized by a restriction enzyme

retrotransposon Piece of DNA that is converted to RNA that can move to a new location within the genome where it is converted back to DNA by a reverse transcriptase enzyme and reinserted into the genome

reverse DNA gyrase Enzyme in hyperthermophiles that increases the supercoiling of DNA

reverse transcriptase A retroviral enzyme that produces double-stranded DNA using an RNA template

reverse transcriptase PCR (RT-PCR) Process using reverse transcriptase to convert a small amount of RNA into cDNA that can then be amplified using a PCR procedure

reversion Spontaneous recovery of a mutant to perform an action, such as synthesis of an amino acid

rhizobia Refers to various members of the order Rhizobiales, characterized by their ability to form root nodules; the majority of the group are α-proteobacteria, and some are β-proteobacteria

rhizosphere Region of soil immediately surrounding plant roots

ribose Five-carbon sugar found in RNA

ribosomal RNA (rRNA) RNA involved in the structure of ribosomes

ribosome Organelle where protein synthesis, or translation, occurs

riboswitch Sequence of mRNA that binds an effector molecule that regulates transcription of the mRNA molecule

ribozyme RNA molecule that can act as a catalyst within a cell

RNA polymerase Enzyme that transcribes DNA to produce RNA; also called DNA-dependent RNA polymerase

RNA-dependent DNA polymerase A retroviral polymerase that produces double-stranded DNA using a single-stranded RNA template such as reverse transcriptase

RNA-dependent RNA polymerase A viral polymerase that produces RNA from an RNA template

root nodule Encasing structure formed on plant roots that houses endosymbiotic nitrogen-fixing bacteria

rumen Major digestive organ that is the primary site of fermentation in a group of herbivores called "ruminants"

rumen fermentation Digestive system where the major site of fermentation is the rumen

ruminant Herbivorous mammal possessing a digestive organ called the "rumen"

S-layer Crystalline-like layer of protein found on the surface of many bacterial cells

Sanger sequencing Technique to determine DNA sequence that relies on incorporation of chain-terminating dideoxy nucleotides and includes cloning of the fragment to be sequenced, DNA synthesis, and gel electrophoresis; *see also* dideoxy sequencing

satellite RNA Infectious agent of plants composed of an RNA genome and a protein coat that is encoded by a helper virus

satellite virus Infectious agent of plants composed of an RNA genome containing a gene that codes for its protein coat

scanning electron microscopy (SEM) Type of microscopy that produces images as electrons are reflected from the surfaces of specimens coated with metal atoms

schistosomiasis Infection with a species in the *Schistosoma* genus of flukes

secondary endosymbiont Endosymbiont of insects that can sometimes be found free-living and is not always present or essential and is not maternally transmitted

secondary metabolite Metabolic product not required for growth; often produced by microorganisms during the stationary phase of the growth curve

secondary sludge Sludge that is a product of secondary treatment of wastewater, consisting of accumulated floc material

secondary treatment Wastewater treatment procedure following primary treatment that makes use of microbial activities to degrade the organic content of the sewage; examples include use of a trickling filter or activated sludge treatment

secretory pathway System of membranous organelles involved in protein trafficking within a eukaryal cell

selective medium Growth medium designed to allow growth of only certain target organisms

selective toxicity Characteristic of antimicrobial drugs to exhibit a higher level of toxicity against infecting microbes than against host cells

semi-perishable food Food that is semi-dry, making it low in available water, and that does not easily support microbial growth

semiconservative replication DNA replication in which each strand of the original molecule serves as a template for the construction of a new strand; one strand of the original DNA molecule is found in each of the two new DNA molecules

septic shock Systemic inflammatory response due to bacterial infection, usually due to the presence of bacteria in the circulatory system

sequela *pl.* **sequelae** Damage indirectly resulting from an infection in the host but occurring after the infection has been cleared; thought to be immune-mediated

serotype Strain of a microbe identified by serotyping

serotyping Diagnostic method that uses antibodies for identifying and distinguishing closely related microbial strains or serotypes, based on differences in their surface antigens

serum resistance Characteristic of some bacterial pathogens to prevent lysis in the presence of serum containing complement factors

sewage Water carrying soluble or solid wastes originating from human activities, including farming, industry, and domestic wastes; *see also* wastewater

sex pilus *pl.* **pili** Special type of pilus that connects cells during conjugation, the transfer of DNA

sexual transmission Spread of an infectious agent that occurs during vaginal, anal, or oral sex

sexually transmitted infection (STI) Infectious disease that is spread sexually through vaginal, anal, or oral sex

Shine–Dalgarno (SD) sequence Segment of mRNA that binds to the 16S rRNA of the small ribosomal subunit and properly aligns mRNA on the ribosome

shotgun sequencing Technique for sequencing DNA that shears DNA into short fragments that are sequenced, after which computer programs are used to identify regions of sequence overlap

shuttle vector Plasmid that can replicate in more than one host; typically contains more than one origin of replication

sickle-cell disease Systemic human disease resulting from a specific mutation in the gene that codes for the β-globin polypeptide in hemoglobin

siderophore Bacterial-secreted protein that binds iron from the surrounding medium for bacterial use

sigma (σ) factor Portion of RNA polymerase that recognizes the promoter sequence

sign Objective disease state that can be readily observed or measured, such as a rash or fever

signal peptide Short sequence of amino acids at the amino terminal end of a protein that targets it for the general secretory pathway in the plasma membrane

signal transduction Process producing a cellular response to an external stimulus

silent mutation Change in a nucleotide base that does not alter the amino acid sequence of a polypeptide

simian immunodeficiency virus (SIV) Virus found in non-human primates throughout Africa that may have given rise to HIV

simple life cycle Completion of a parasite life cycle within a single host

single-stranded DNA-binding protein (SSB) Molecules that attach to a newly formed single strand of DNA to keep the strands from reannealing

site-directed mutagenesis Method that induces specific mutations within a DNA molecule

site-specific recombination Non-homologous recombination at a particular sequence of DNA

soredia Asexual reproductive structures of lichens consisting of a few algal cells packaged by fungal hyphae

SOS response Type of DNA repair in response to serious damage

Southern blot Technique for detecting specific DNA sequences by hybridization using labeled probes of complementary sequence

species Basic taxonomic group composed of similar strains that differ significantly from other groups of strains

spirillum *pl.* **spirilla** Bacterial cell shaped like a spiral

spore Environmentally stable reproductive cell

sporozoite Haploid infective cell of some protozoan parasites, including the malaria parasite

sRNA Small non-coding RNA molecule that often affects gene expression by interacting with existing mRNA

stable isotope probing (SIP) Technique that exposes organisms to specific nutrient molecules labeled with ^{13}C or ^{15}N, thus labeling their DNA, which can later be extracted and analyzed

stalk Tubular extension of the cell envelope in some Gram-negative bacteria that increases cell surface area and aids in nutrient acquisition

starter culture Microbial inoculum used in the production of cultured dairy products, fruits, or grains

stationary phase Period of growth in which the rate of cell division equals the rate of cell death and population growth stops

stop codon Sequence of three nucleotides on mRNA that signals the end of translation; UAG, UGA, and UAA typically function as stop codons

strain Distinct subtype of a species that differs genetically, and often phenotypically, from other subtypes

streak plate Petri plate in which microbial cells are spread over a solid medium, separating individual cells that eventually produce isolated colonies

substrate Specific reactant converted by an enzyme into a product

substrate-level phosphorylation Enzymatic production of ATP through the transfer of a phosphate group from a reactive intermediate to ADP

subunit vaccine Vaccine composed of one or a select number of isolated antigens; not composed of whole cells or viruses; *see also* acellular vaccine

suicide plasmid Plasmid vector that cannot replicate in a host cell

sulfur globule Intracellular storage form of elemental sulfur

superantigen Toxin capable of evoking a systemic inflammatory response by directly crosslinking the T-cell receptor (TCR) of CD4$^+$ T cells with MHC II molecules of antigen-presenting cells, such as macrophages

supercoiling Highly twisted form of DNA that produces a more compact chromosome

surface array Crystalline-like layer of protein found on the surface of many bacterial cells; *see also* S-layer

susceptibility testing Controlled *in vitro* testing done to establish the sensitivity of a particular microorganism to various antimicrobial agents

symbiont Organism of one species that has entered into a long-standing and intimate relationship with another species

symbiosis Long-standing, intimate relationship between two different species

symport Mechanism of co-transport that couples the movement of one substance across the plasma membrane with movement of another in the same direction

symptom Subjective disease state, such as muscle aches

syncytium *pl.* **syncytia** Multinuclear mass formed by the fusion of cells

synthetic biology Construction of new biological functions and systems from constituent parts

syntrophic partnership Relationship in which the metabolic activities of the organisms are mutually dependent

systemic inflammatory response Inflammatory response involving a large portion of the body, including the circulatory system, that often leads to physiological shock characterized by a drastic drop in blood pressure because of fluid leaking from the circulatory system

T cell (T lymphocyte) Lymphocyte of the adaptive immune system that matures in the thymus and expresses surface T-cell receptors (TCRs)

T helper cell (T$_H$ cell) Effector CD4$^+$ T cell that secretes cytokines involved in directing immune responses. *See also* CD4$^+$ T cell, T$_H$1 cell, and T$_H$2 cell

T-cell receptor (TCR) Membrane-bound immune receptor molecule expressed by T cells that binds specific antigen associated with MHC molecules

T-independent antigen Antigen that can stimulate B cells to proliferate into antibody-secreting plasma cells without requiring B-cell interaction with T cells; possesses highly repetitive structures with the ability to crosslink several BCRs on a B cell

taxon *pl.* **taxa** Named grouping in a hierarchical taxonomic system

teichoic acid Polymer of ribitol phosphate or glycerol phosphate attached to peptidoglycan chains in the cell wall of Gram-positive bacteria

telomerase DNA polymerase that replicates DNA at the ends of chromosomes in eukarya

telomere Non-transcribed region at the ends of each eukaryal chromosome

temperate phage Phage that usually integrates its genome into the genome of the host cell, where it is replicated along with the host DNA

terminal restriction fragment length polymorphism (TRFLP) Technique to profile DNA in a microbial community using position of a restriction site closest to a labeled end of an amplified gene, usually the SSU rRNA gene

tertiary treatment Wastewater treatment protocol sometimes used following secondary treatment, involving additional biological or physicochemical procedures to further remove organic or inorganic material prior to disinfection and release to a natural water source

T$_H$1 cell Cell belonging to a subset of T helper (T$_H$) cells that secretes cytokines, such as IL-2, IFN-γ, and TNF-β, to produce a cell-mediated immune response

T$_H$2 cell Cell belonging to a subset of T$_H$ cells that secretes cytokines, such as IL-4, IL-5, IL-6, IL-10, and IL-13, to produce a humoral immune response

thermocline Layer of water in deep lakes that separates warmer, low-density water from lower, colder, high-density water; characterized by rapid change in temperature and density

thermophile Organism that grows optimally at temperatures greater than 55°C

thermosome Protein complex that functions as a molecular chaperone to refold partially denatured proteins in hyperthermophiles

thylakoid Layered photosynthetic membrane structures

thymine (T) Pyrimidine nitrogenous base, along with adenine and uracil

Ti plasmid Tumor-inducing plasmid in *Agrobacterium tumefaciens* that can transfer genes into a plant cell where they are expressed

tissue culture infectious dose 50 (TCID$_{50}$) Amount of virus that causes cytopathic effects in tissue culture cells 50 percent of the time

titer Concentration of a preparation

Toll-like receptor (TLR) Membrane-associated pattern recognition receptor (PRR) of the innate immune system that recognizes a panel of pathogen-associated molecular patterns (PAMPs) and can induce expression of various cytokines and other products

topoisomerase Enzyme that encourages supercoiling in a chromosome

toxic shock Systemic inflammatory response due to the presence of certain microbial toxins in the circulatory system

toxin Substance of biological origin that damages a host

toxoid Microbial toxin that has been denatured through chemical or physical processes in order to be used in a vaccine

transconjugant Cell that has incorporated DNA from another cell via conjugation

transcription Process through which segments of DNA serve as templates for the production of complementary strands of single-stranded RNA

transcription factor One of a series of proteins required for transcription

transcriptome Set of transcripts encoded by each of the genes within a genome

transducing retrovirus Retrovirus that has acquired a cellular gene and can transfer it to a new cell

transductant Bacterial cell that contains DNA obtained via transduction

transduction Transfer of bacterial DNA from one cell to another by a bacteriophage; a type of horizontal gene transfer

transfer RNA (tRNA) RNA molecule that delivers an amino acid to the ribosome for incorporation into a growing polypeptide during translation

transformation A type of horizontal gene transfer in which a piece of free DNA is taken up by a cell; change within a cell that causes it to become cancer-like

transgenic plant Plant containing DNA that has been transferred from a different type of organism

translation Process of protein synthesis

translocation Change in DNA sequence in which a segment of a chromosome breaks off and reattaches to a different location

transmissible spongiform encephalopathy (TSE) Neurological disease caused by prions

transmission Spread of an infectious agent from one host to another or from its source

transmission electron microscopy (TEM) Type of microscopy that produces images when electrons pass through a specimen

transposable element Mobile genetic element

transposase Enzyme that recognizes inverted repeat sequences on a transposon

transposition Movement of DNA via mobile genetic elements or transposable elements

transposon Transposable element that may also contain other genes such as antibiotic resistance genes

tricarboxylic acid (TCA) cycle Series of metabolic reactions used in the oxidation of carbon compounds, producing CO_2, NADH, and ATP; also called the citric acid cycle or Krebs cycle

trichome Smooth unbranched chain of cells seen in associations of cyanobacteria

trickling filter A form of secondary treatment for wastewater effluent that uses a filter bed of solid material, which supports a microbial biofilm community; effluent is distributed across the bed surface by a rotary spray boom

trophozoite Active feeding form of a spore-forming protozoan

tumor suppressor Cellular protein that inhibits cellular replication

tumorigenesis Formation of a tumor

turbidity Cloudiness of a liquid that can indicate density of suspended particles or cells

two-component regulatory system Regulatory system that includes one protein that acts as a sensor and another protein that regulates transcription

Type I interferon Includes the cytokines interferon-α and -β produced by many cell types upon viral infection to induce an antiviral state in infected and uninfected neighboring cells

Type III secretion system Protein complex found in many Gram-negative pathogens that serves as an injection system to deliver virulence factors, such as toxins, directly into the target cell by forming a tunnel that passes through both bacterial membranes and through the host cell plasma membrane

Type IV secretion system Protein complex found in some Gram-negative pathogens that transports toxins or other components directly across both bacterial membranes into the immediate environment or directly into the host cells

ultraviolet (UV) radiation Light with a wavelength range of 100–400 nm

uncoating Disassembly of viral capsids releasing the viral genome

uracil (U) Pyrimidine base found in RNA that interacts with adenine (A)

vaccination Physical act of injecting or inoculating an immunizing substance into the body

vaccine Preparation containing antigen(s) used to immunize against disease

vaccine coverage rate Proportion of individuals in the population that must be immunized to achieve herd immunity

vaccine efficacy Effectiveness of a vaccine measured by the average degree of protection it affords a population

vaccine failure Development of disease in individuals despite having received vaccination; usually due to the failure of certain individuals to mount an effective immune response to an immunizing agent

variable surface glycoprotein (VSG) Antigens on the surface of a pathogen that may be expressed at one time but not at another; enables pathogen to evade host immune responses

vasodilation An increase of the diameter of the blood vessel, such as during inflammation

vector-borne transmission Spread of an infectious agent from an infected individual to another organism such as an insect that then transmits the infectious agent to a different susceptible host

vertical transmission Spread of an infectious agent from parent to child

vibrio *pl.* **vibrios** Bacterial cell shaped like a curved rod

viral attachment protein Protein expressed by a virus used for attachment to the host cell

virion Complete viral particle

viroid Infectious agent of plants composed of naked RNA

virulence Severity of disease

virulence factor Product made by a pathogen that enhances its ability to cause disease

virulent phage Phage that lyses and kills the host cell when exiting

virus An acellular particle including a DNA or RNA genome surrounded by a protein coat that can replicate only within living host cells

wastewater Water carrying soluble or solid wastes originating from human activities, including farming, industry, and domestic wastes; *see also* sewage

water activity (a$_w$) Vapor pressure of a liquid relative to the vapor pressure of pure water under identical conditions; measure of water availability in an environment

white biotechnology Use of advanced biological processes in industrial applications

wild-type strain Organisms that usually possess the typical or representative characteristics of a species

X-ray crystallography Crystallization of proteins then subjected to X-ray diffraction to determine protein structure at the level of individual atoms

xenobiotic Chemical not naturally found in or produced by organisms; usually not degraded by microorganisms

xylanase Enzyme that degrades hemicellulose, a component of the plant cell wall

Z-ring Structure composed of FtsZ needed for bacterial cell division

zoonotic disease (zoonosis) Infectious disease of animals that can cause disease when transmitted to humans

zoonotic transfer Spread of a pathogen from its natural animal host to a human

zooplankton Drifting non-photosynthetic microorganisms that act as consumers in the ocean

zooxanthellae Endosymbiotic photosynthetic algal of coral polyps

INDEX